D0161421

CLINICAL ANATOMY FOR EMERGENCY MEDICINE

Clinical Anatomy for Emergency Medicine

Richard S. Snell, M.D., Ph.D.
Emeritus Professor
Department of Anatomy
The George Washington University School of Medicine and Health Sciences
Washington, D.C.

Mark S. Smith, M.D.
Professor and Chairman
Department of Emergency Medicine
The George Washington University School of Medicine and Health Sciences
Washington, D.C.

St. Louis Baltimore Boston Chicago London Philadelphia Sydney Toronto

Sponsoring Editor: James F. Shanahan
Associate Managing Editor, Manuscript Services: Deborah Thorp
Production Manager: Nancy C. Baker
Proofroom Manager: Barbara M. Kelly

Mosby-Year Book, Inc.
11830 Westline Industrial Drive
St. Louis, MO 63416

2 3 4 5 6 7 8 9 0 CL/KP 97 96 95 94

Library of Congress Cataloging-in-Publication Data

Snell, Richard S.
Clinical anatomy for emergency medicine / Richard S. Snell, Mark S. Smith.
p. cm.
Includes index.
ISBN 0-8016-6549-3
1. Human anatomy. 2. Emergency medicine. I. Smith, Mark Stephen. II. Title.
[DNLM: 1. Anatomy. 2. Emergency Medicine.
QS 4 S671cb]
QM23.2.S623 1992 92-49973
611'.0024616—dc20 CIP
DNLM/DLC
for Library of Congress

For Maureen and Maxine

ACKNOWLEDGMENTS

The authors thank Little, Brown & Co and Blackwell Scientific Publications for permission to reproduce the following figures from the following sources:

Figures 7–1, 7–4 through 7–9, 7–11 through 7–18, 7–25, 7–31, 7–34, and 7–37 are from Snell RS, Lemp M: *Clinical Anatomy of the Eye*. Cambridge, Mass, Blackwell, 1989.

Figures 2–4, 2–5, 9–1, 9–2, 11–10, 11–13, 11–17, 11–20, 11–21, 11–26, 11–27, 12–2, 13–4, 13–5, 15–7, and 17–1 are from Snell RS: *Gross Anatomy: A Review With Questions and Explanations*. Boston, Little, Brown, 1992.

Figures 2–32, 2–33, 10–31, 10–32, 10–34, and 10–35 are from Snell RS: *Neuroanatomy for Medical Students,* ed 2. Boston, Little, Brown, 1987.

Figures 4–8, 5–9, 5–11, 9–81, 9–83, 11–7, 11–13, 11–22, 11–28, 11–34, 11–42, 11–44, 12–15, 13–2, 13–3, 13–7, 13–9, 13–26, 13–27, 14–5, 14–13, 15–8 through 15–10, 15–14, 15–20, 15–34, 15–37 through 15–39, 16–9, 16–10, 16–12, 17–14, and 17–29 are from Snell RS: *Clinical Anatomy for Medical Students,* ed 4. Boston, Little, Brown, 1992.

PREFACE

This book is written to provide the physician in the emergency department with knowledge of anatomy necessary to practice emergency medicine. It is recognized that this new and important specialty of medicine requires detailed knowledge of anatomy of certain regions of the body, whereas a superficial knowledge of other regions is quite adequate. For example, the anatomy of the upper and lower airways is of paramount importance, while the anatomy of the sole of the foot is of less importance.

Throughout the book, each chapter is constructed in a similar manner to provide easy access to the material. Each chapter is divided into the following categories:

1. *Basic Anatomy.* This section provides useful basic information and scattered throughout are *Clinical Notes* on areas of vital importance. Numerous examples of normal radiographs, CT scans, MRIs, and sonograms are also provided.

2. *Surface Anatomy.* This section provides surface landmarks of important anatomical structures, many of which are located some distance beneath the skin. Photographs of living subjects have been used extensively.

3. *Clinical Anatomy.* This section provides the anatomy of common disease entities found in an emergency situation. It also provides the anatomical knowledge necessary to perform procedures and techniques, and emphasizes the anatomical "pitfalls" commonly encountered. It must be noted that this is not a "how-to" book, but one that ensures that the physician is familiar with the structures that he or she will commonly encounter when treating a patient.

4. *Anatomical-Clinical Problems.* This section provides the emergency physician with many examples of clinical situations in which a knowledge of anatomy is necessary to solve the clinical problem and to institute treatment; solutions to the problems are provided at the end of each chapter.

In addition to the detailed index at the end of the book, each chapter is preceded by a list of contents with page numbers, so that immediate access is possible. To assist in the quick understanding of anatomical facts, the book is profusely illustrated, and the majority of the figures have been kept simple; color has been used where appropriate.

We thank the many colleagues in emergency medicine who have made valuable suggestions regarding the preparation of the manuscript. In particular, a special appreciation goes to Dr. Craig Feied for his extensive and incisive reviews of all the chapters. Other than the authors, he is the only one who read the entire book cover to cover. We thank also Drs. James Scott and Keith Ghezzi for their excellent recommendations on the respiratory and cardiovascular systems chapters.

We are greatly indebted to the late Dr. Alwyn Wyman and Dr. David O. Davis in the Department of Radiology, George Washington University Medical Center, for the loan of radiographs that have been reproduced in different sections of the book. We are also grateful to Dr. Gordon Sze and Dr. Leslie Scoutt of the Department of Radiology at Yale University for examples of CT scans and sonograms. We also thank the Audiovisual Department at George Washington University for their fine photographic work and Laura Farah for handling the complex correspondence between us.

We wish to express our sincere thanks to our artists, Terry Dolan, Virginia Childs, Myra Feldman, and Ira Alan Grunther, for the excellent drawings used in the illustrations.

To the staff of Mosby-Year Book, Inc., in particular our editor, James Shanahan, we wish to express our sincere appreciation for guiding this book through every phase of its production.

Richard S. Snell, M.D., Ph.D.
Mark S. Smith, M.D

CONTENTS

1

The Upper and Lower Airway and Associated Structures

No medical emergency quite produces the drama, urgency, and anxiety of the compromised airway. The emergency physician not only has to make a rapid diagnosis but has to institute almost immediate treatment. All the techniques of airway management, from manual manipulation through endotracheal intubation to cricothyroidotomy, require a detailed knowledge of anatomy.

In addition, the emergency physician commonly has to deal with epistaxis and nasal lacerations. Diseases of the paranasal sinuses and the salivary glands may also have to be considered.

Since the relative anatomy may change as the patient matures, comparisons, when necessary, between anatomic structures in infancy, childhood, and adulthood are presented.

BASIC ANATOMY

The airway extends from the nostrils of the nose and the lips of the mouth to the alveoli of the lungs.

Basic Anatomy of the Nose

The nose consists of the external nose and the nasal cavity, both of which are divided by a septum into right and left halves.

External Nose

The external nose has two elliptical orifices called the *nostrils,* which are separated from each other by the nasal septum (Fig 1–1). In adults each measures about 1.5 to 2 cm anteroposteriorly and 0.5 to 1 cm transversely; they are of course much smaller in children. The lateral margin, the *ala nasi,* is rounded and mobile.

The framework of the external nose is made up above by the nasal bones, the frontal processes of the maxillae, and the nasal part of the frontal bone. Below, the framework is formed of plates of hyaline cartilage, which include the upper and lower lateral nasal cartilages and the septal cartilage (see Fig 1–1).

Blood Supply of the External Nose.—The skin of the ala and the lower part of the septum are supplied by branches from the facial artery. The remainder of the skin of the external nose is supplied by branches of the ophthalmic and the maxillary arteries.

Sensory Nerve Supply of the External Nose.—The chlear and external nasal branches of the ophthalmic nerve and the infraorbital branch of the maxillary nerve.

Clinical Note

In cases of *herpes zoster,* unless the tip of the nose (external nasal branch of the ophthalmic division of the trigeminal nerve) is involved, the cornea (ciliary branches of the nasociliary branch of the ophthalmic division of the trigeminal nerve) cannot be involved.

Nasal Cavity

The nasal cavity extends from the nostrils in front to the *posterior nasal apertures or choanae* behind, where the nose opens into the nasopharynx. Each choana is oval and in the adult measures about 2.5 cm vertically and 1.25 cm transversely.

The nasal cavity is divided into right and left halves by the nasal septum. Each half consists of three regions—the vestibular, the olfactory, and the respiratory.

The *nasal vestibule* is the slightly expanded area of the nasal cavity lying just inside the nostril (Fig 1–2). It is lined with skin and has coarse stiff hairs. The vestibule extends superiorly to a curved ridge called the *limen nasi.* Here the skin of the vestibule becomes continuous with the mucous membrane of the remainder of the nasal cavity.

The *olfactory region* is confined to the superior part of the nasal cavity. The olfactory mucous membrane lines the upper surface of the superior concha and the sphenoethmoidal recess. It also lines a corresponding area on the nasal septum and lines the roof.

The *respiratory region* is the larger region and occupies the remainder of the nasal cavity below the superior concha.

Walls of the Nasal Cavity.—Each half of the nasal cavity has a floor, a roof, a lateral wall, and a medial or septal wall.

The *floor* is formed by the palatine process of the maxilla and the horizontal plate of the palatine bone (see Fig 1–1). It is gently concave transversely and flat anteroposteriorly.

The *roof* is narrow and is formed anteriorly beneath the bridge of the nose by the nasal and frontal bones, in the middle by the cribriform plate of the ethmoid, located beneath the anterior cranial fossa, and posteriorly by the downward sloping body of

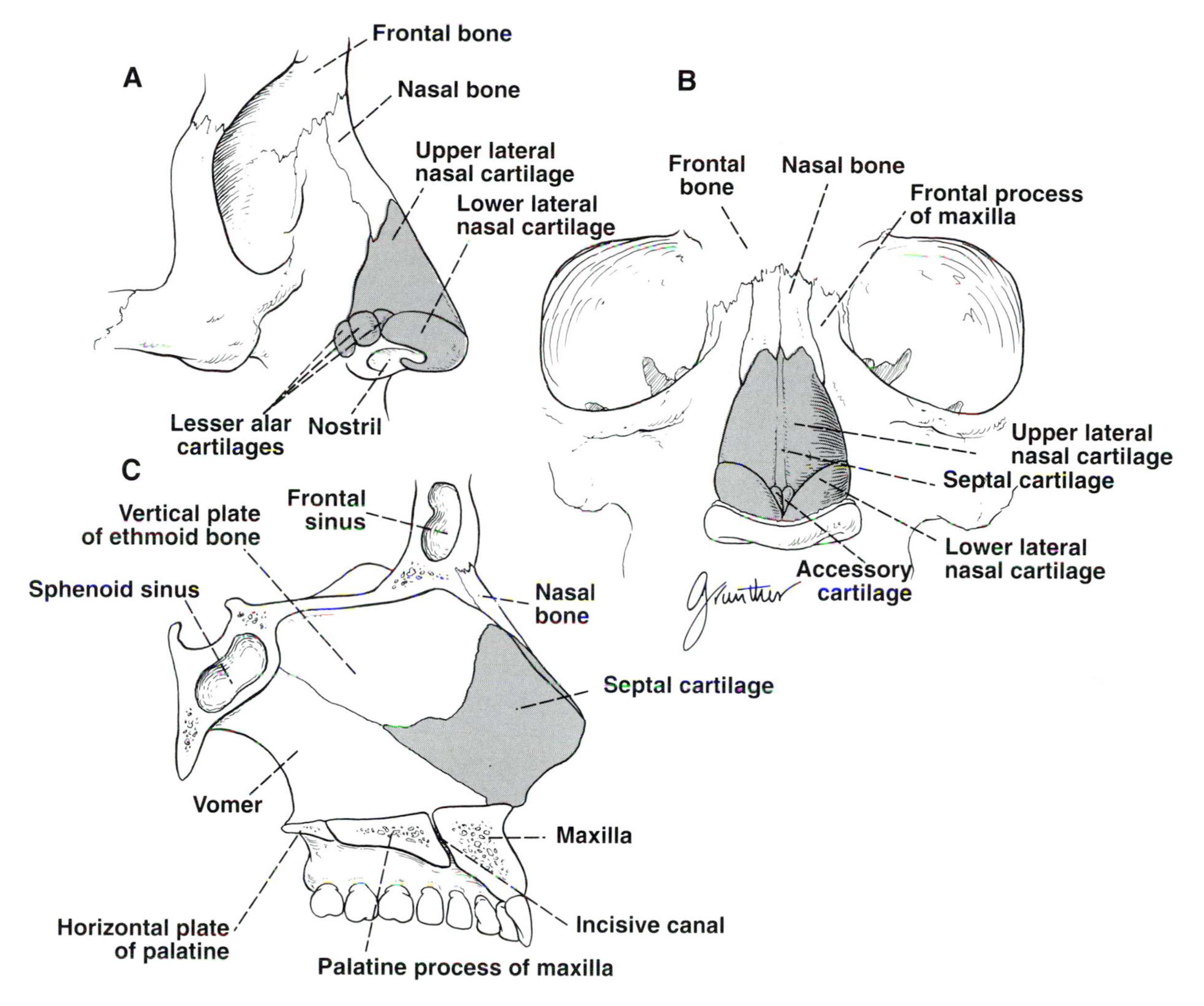

FIG 1–1.
External nose and nasal septum. **A,** lateral view of bony and cartilaginous skeleton of external nose. **B,** anterior view of bony and cartilaginous skeleton of external nose. **C,** bony and cartilaginous skeleton of nasal septum.

the sphenoid (see Fig 1–2). The highest point on the roof is formed by the ethmoid bone.

The *lateral wall* is marked by three horizontal scroll-like ridges of bone called the *superior, middle,* and *inferior nasal conchae* (see Fig 1–2). The area below each concha is referred to as a *meatus.* The *sphenoethmoidal recess* is a small area of the nose that lies above the superior concha and in front of the body of the sphenoid bone. It receives the opening of the *sphenoidal air sinus.*

The *superior meatus* lies below and lateral to the superior concha (see Fig 1–2). It receives the openings of the *posterior ethmoidal sinuses.* The *middle meatus* lies below and lateral to the middle concha and has on its lateral wall a rounded prominence, the *bulla ethmoidalis,* caused by the bulging of the underlying middle ethmoidal air sinuses, which open on its upper border. A curved opening, the *hiatus semilunaris,* lies immediately below the bulla. The anterior end of the hiatus leads into a funnel-shaped channel called the *infundibulum,* which is continuous with the *frontal sinus.* The *maxillary sinus* opens into the middle meatus via the hiatus semilunaris.

The *inferior meatus* lies below and lateral to the inferior concha and receives the opening of the *nasolacrimal duct,* which is guarded by a valve (see Fig 1–2).

Clinical Note

Pupillodilatation.—A vasoconstrictor sprayed into the nasal vestibule can ascend in the nasolacrimal duct to the conjunctival sac where it is absorbed and may produce pupillodilatation.

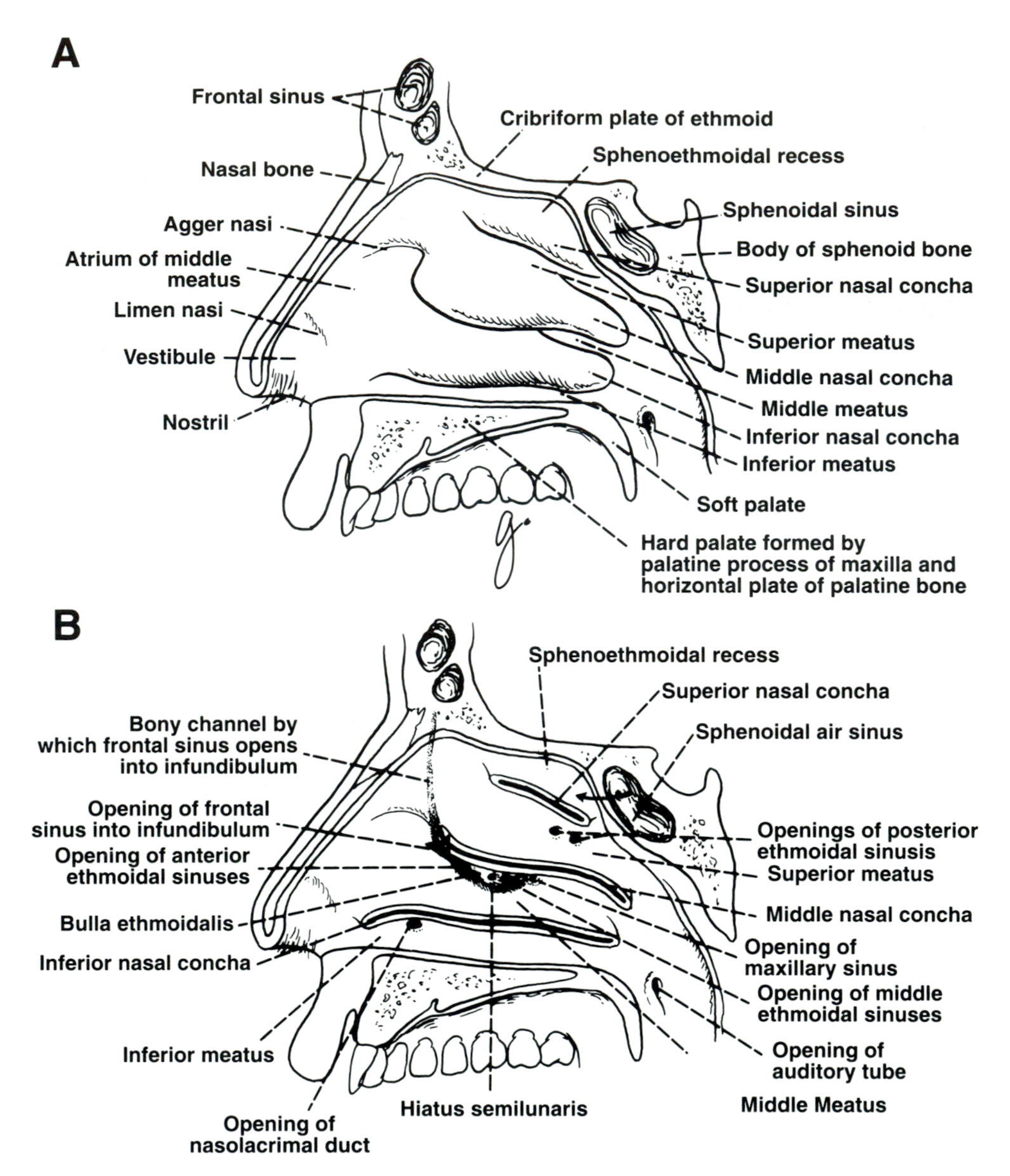

FIG 1–2.
A, lateral wall of the right nasal cavity. **B,** lateral wall of the right nasal cavity; the superior, middle, and inferior conchae have been partially removed to show openings of the paranasal sinuses and the nasolacrimal duct into the meati.

The *medial wall, or nasal septum,* is an osteocartilaginous partition covered by adherent mucous membrane. The upper part is formed by the vertical plate of the ethmoid and the vomer (see Fig 1–1). The anterior part is formed by the septal cartilage. The septum rarely lies in the median plane, thus increasing the size of half of the nasal cavity and decreasing the size of the other. Ridges or spurs of bone often project from the septum.

Clinical Note

Ischemic Necrosis of Nasal Septum.—The blood supply to the septal cartilage can be compromised by bilateral cauterization of the mucous membrane in the treatment of severe epistaxis; ischemic necrosis may also occur in posttraumatic hematoma of the septal cartilage unless the clot is quickly detected and drained.

Mucous Membrane of the Nasal Cavity.—The mucous membrane is closely adherent to the underlying periosteum or perichondrium and is thickest over the conchae and nasal septum and is thinnest in the meatuses. A large plexus of veins in the submucous connective tissue is present in the respiratory region.

Clinical Note

Nasal Congestion.—Congestion of the veins in the submucous connective tissue causes swelling of the mucous membrane, which considerably reduces the size of the nasal cavity and the openings communicating with it.

Nerve Supply of the Nasal Cavity.—The olfactory nerves arise from the special olfactory cells in the olfactory mucous membrane described above. They ascend through the cribriform plate to the olfactory bulbs (Fig 1–3).

Clinical Note

Nasal Polyps.—These are not neoplasms but hypertrophic inflammatory swellings of the mucous membrane of the nose. They may reach several centimeters in diameter and may obstruct the openings of the paranasal sinuses causing chronic sinusitis.

The nerves of ordinary sensation are branches of the nasociliary nerve, a branch of the ophthalmic division (V1) of the trigeminal nerve, and branches from the maxillary division (V2) of the trigeminal

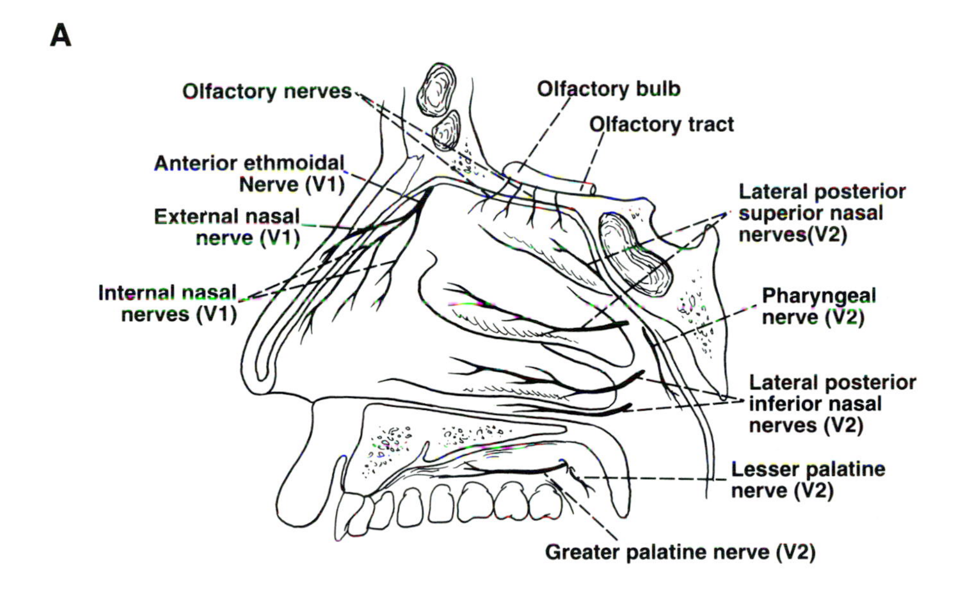

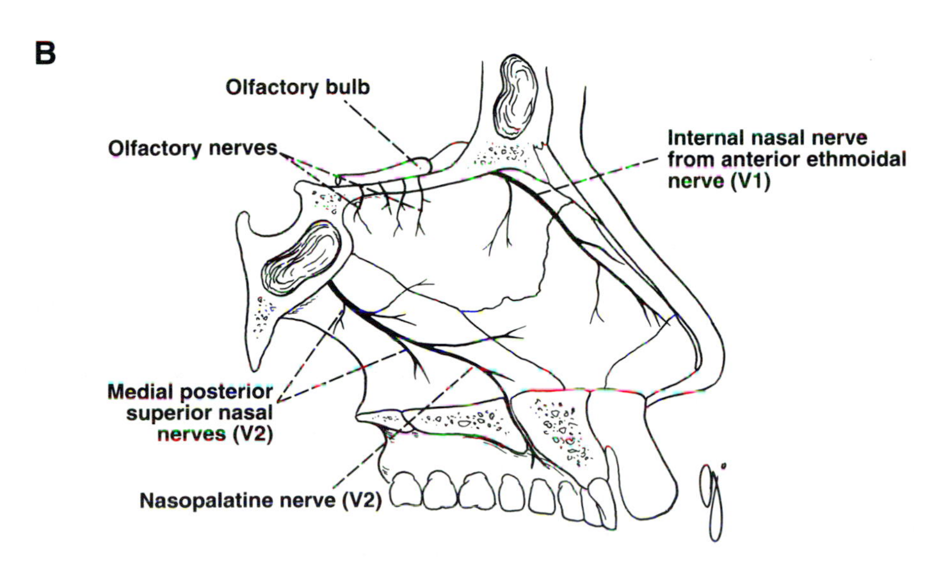

FIG 1–3.
A, lateral wall of nasal cavity showing sensory innervation of mucous membrane. **B,** nasal septum showing sensory innervation of mucous membrane.

nerve. The distribution of these nerves is shown in Figure 1–3.

Clinical Note

The nasal mucous membrane is extremely sensitive and may be the site of reflex responses. *Nasal pack.* Packing a patient's nose bilaterally to control epistaxis may produce intense discomfort. *Nasal "tickle."* The patient's sensitivity to manipulation of the nasal mucosa provides a way of delivering a noxious stimulus—the nasal tickle with a cotton-tipped swab or nasopharyngeal airway. This is an excellent way of assessing the patient's true state of wakefulness and is much less painful than rubbing the sternum or squeezing a nailbed.

Sympathetic postganglionic vasoconstrictor fibers from the superior cervical ganglion supply the nasal arteries and veins. Increased sympathetic activity causes the mucous membrane to shrink, and decreased activity causes the mucous membrane to swell. Parasympathetic postganglionic secretomotor fibers from the pterygopalatine ganglion supply the nasal glands.

Blood Supply to the Nasal Cavity.—The arterial supply to the nasal cavity is derived mainly from branches of the maxillary artery, one of the terminal branches of the external carotid artery. The most important branch is the sphenopalatine artery, which enters the nasal cavity through the pterygopalatine foramen (Fig 1–4). The sphenopalatine artery anastomoses with the septal branch of the superior labial branch of the facial artery in the region of the vestibule.

Clinical Note

The vestibule is a very common site of bleeding from the nose *(Kiesselbach's area).* Note that bleeding from the anterior part of the nose tends to ooze (bleeding from capillary plexus fed by small arteries), whereas bleeding from the posterior part tends to spurt (arterial bleeding fed by large arteries).

The submucous venous plexus is drained by veins that accompany the arteries.

Lymphatic Drainage of the Nasal Cavity.—The lymphatic vessels draining the vestibule end in the submandibular nodes. The remainder of the nasal cavity is drained by vessels that pass to the upper deep cervical nodes.

Basic Anatomy of the Paranasal Sinuses

These are cavities found in the maxilla, frontal, sphenoid, and ethmoid bones (Fig 1–5). They are lined with mucoperiosteum and communicate with the nasal cavity through small openings (Table 1–1).

Basic Anatomy of the Mouth

The Lips

The lips are two fleshy folds that surround the oral orifice (Fig 1–6). They are covered on the outside by skin and are lined on the inside by mucous membrane. The substance of the lips is made up by the orbicularis oris muscle and the muscles that radiate from the lips into the face (Fig 1–7). Also included are the labial blood vessels and nerves, connective tissue, and many small salivary glands. The *philtrum* is the shallow vertical groove seen in the midline on the outer surface of the upper lip. Median folds of mucous membrane—the *labial frenulae*—connect the inner surface of the lips to the gums.

The Mouth Cavity

The mouth extends from the lips to the pharynx. The entrance into the pharynx, the *oropharyngeal isthmus,* is formed on each side by the palatoglossal fold (see Fig 1–6). The *vestibule of the mouth* lies between the lips and the cheeks externally and the gums and the teeth internally. This slit-like space communicates with the exterior through the *oral fissure.* When the jaws are closed, it communicates with the mouth proper behind the third molar tooth on each side. The vestibule is limited above and below by the reflection of the mucous membrane from the lips and cheeks to the gums.

TABLE 1–1.
The Paranasal Sinuses and Their Site of Drainage Into the Nose*

Name of Sinus	Site of Drainage
Maxillary sinus	Middle meatus through hiatus semilunaris
Frontal sinus	Middle meatus via infundibulum
Sphenoid sinuses	Sphenoethmoidal recess
Ethmoidal sinuses	
Anterior group	Infundibulum and into middle meatus
Middle group	Middle meatus on or above bulla ethmoidalis
Posterior group	Superior meatus

*Note that the maxillary and sphenoidal sinuses are present in rudimentary form at birth and enlarge appreciably after the eighth year of life and are fully formed in adolescence.

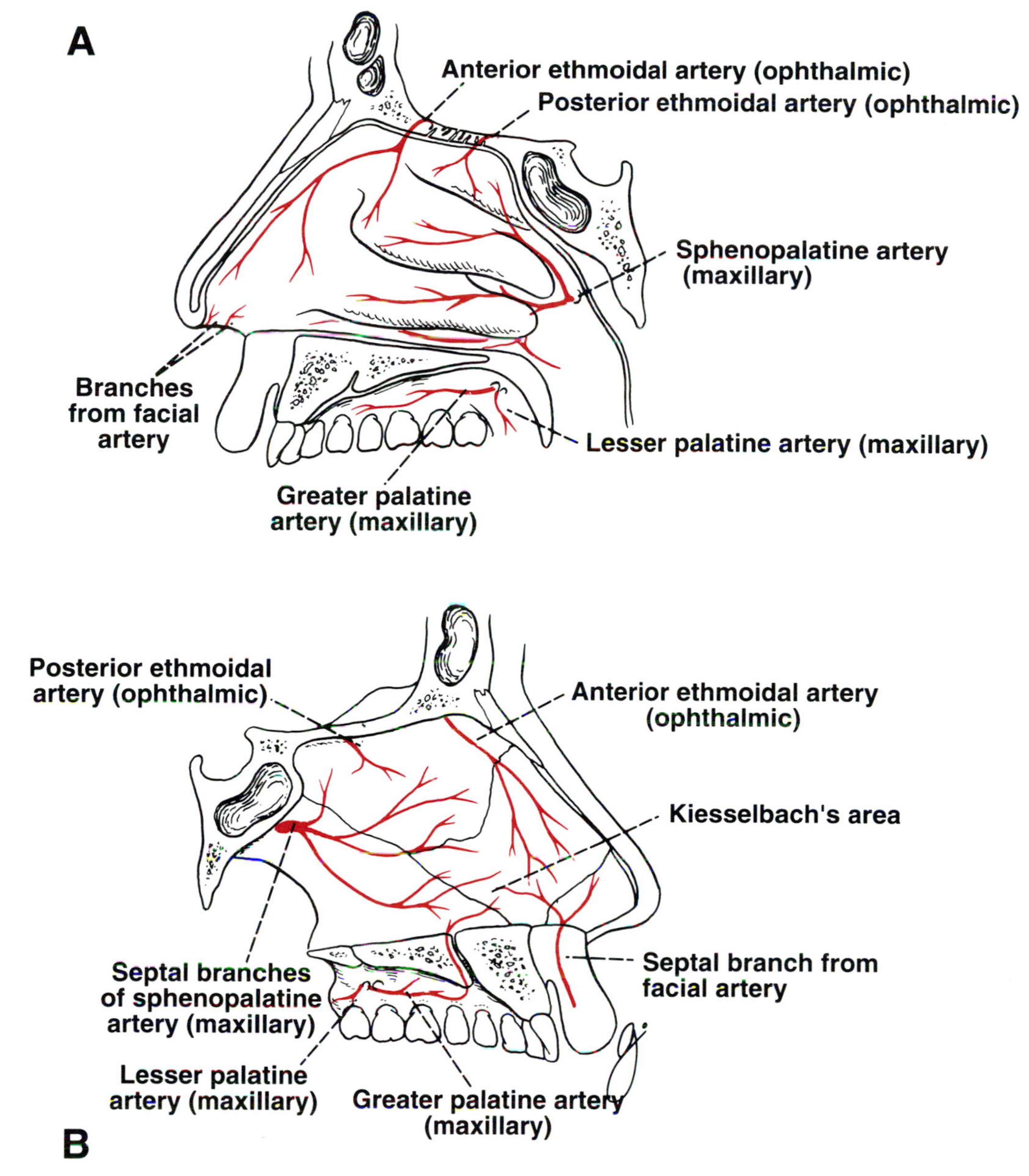

FIG 1–4.
A, lateral wall of nasal cavity showing the arterial supply of the mucous membrane. **B,** nasal septum showing the arterial supply of the mucous membrane.

The lateral wall of the vestibule is formed by the *cheek,* which is made up by the buccinator muscle and is lined with mucous membrane. The tone of the buccinator muscle and that of the muscles of the lips keeps the walls of the vestibule in contact with one another. The *duct of the parotid salivary gland* opens on a small papilla into the vestibule opposite the upper second molar tooth (see Fig 1–10).

The *mouth proper* has a roof formed by the hard palate in front and the soft palate behind. The floor is formed largely by the anterior two thirds of the tongue and by the reflection of the mucous membrane from the sides of the tongue to the gum of the mandible. A fold of mucous membrane called the *frenulum of the tongue* connects the undersurface of the tongue in the midline to the floor of the mouth (see Fig 1–6). Lateral to the frenulum the mucous membrane forms a fringed fold, the *plica fimbriata* (see Fig 1–6).

The *orifice of the submandibular duct* opens onto the floor of the mouth on the summit of a small papilla on either side of the frenulum of the tongue

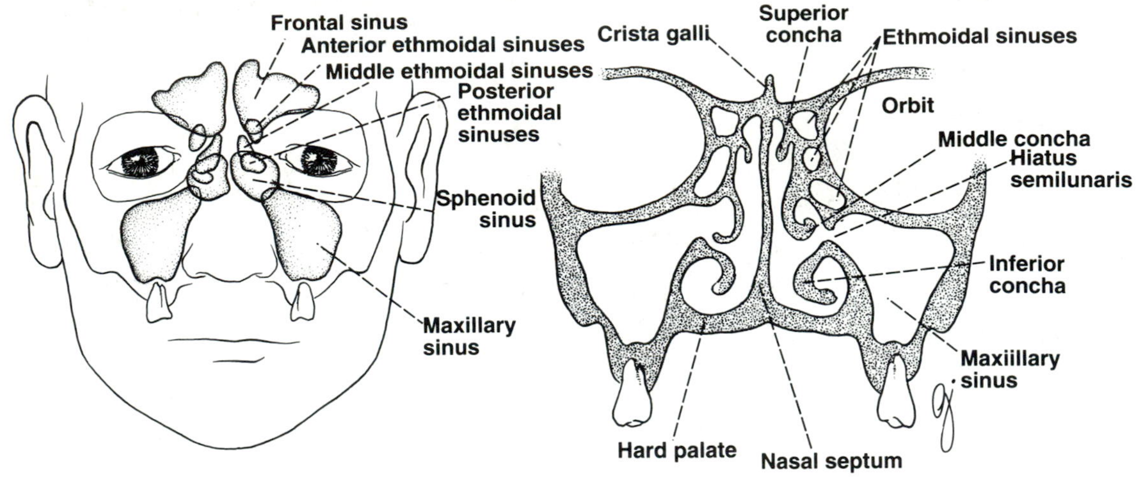

FIG 1–5.
A, the position of the paranasal sinuses in relation to the face. **B,** coronal section through the nasal cavity showing the ethmoidal and the maxillary sinuses.

(see Fig 1–6). The *sublingual gland* projects up into the mouth, producing a low fold of mucous membrane that extends laterally from the submandibular papilla. The numerous ducts of the gland open on the summit of the fold.

Mucous Membrane of the Mouth

In the vestibule the mucous membrane is tethered to the buccinator by elastic fibers in the submucosa that prevent redundant folds of mucous membrane from being bitten between the teeth when the jaws are

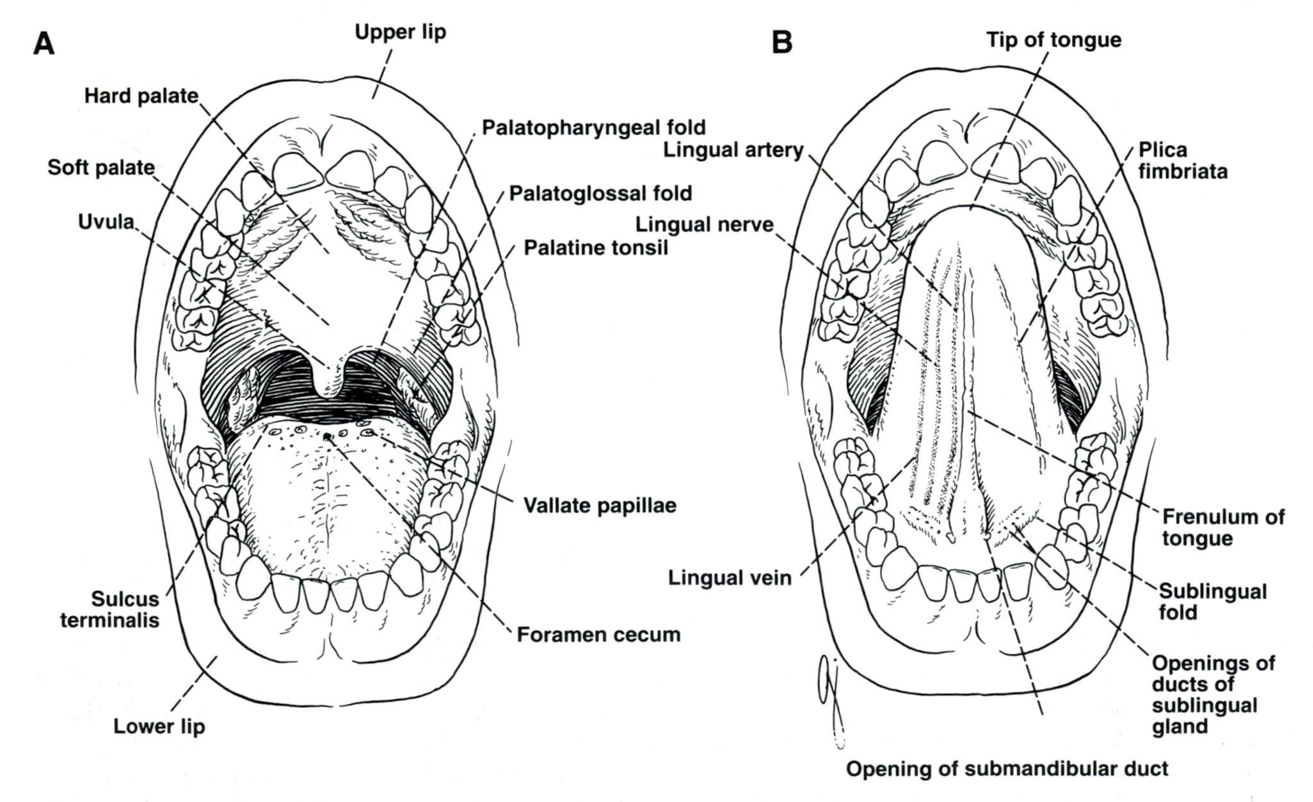

FIG 1–6.
A, mouth open to display cavity of mouth and oral pharynx. **B,** undersurface of tongue.

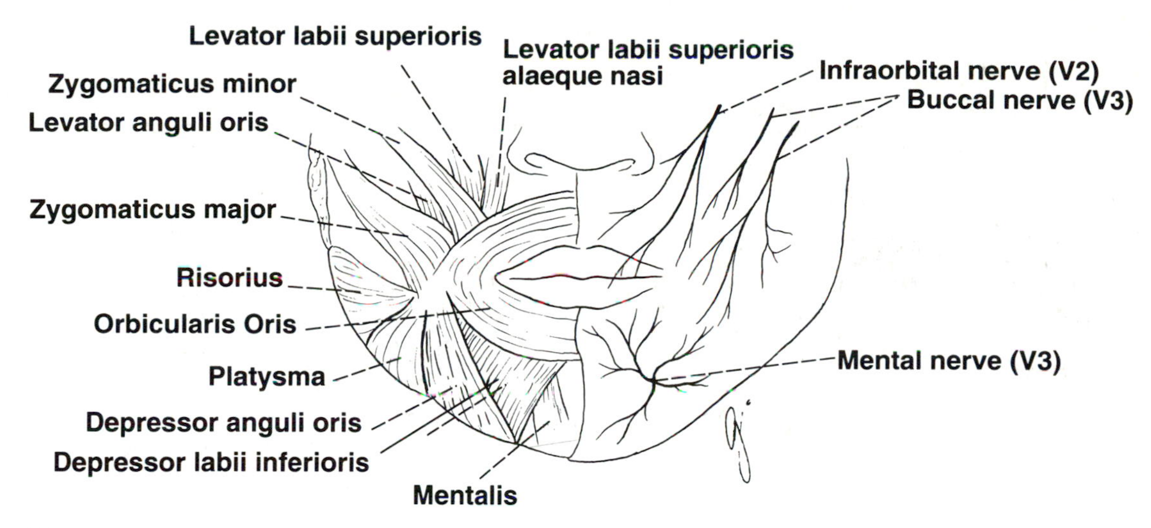

FIG 1–7.
Arrangement of the facial muscles around the lips; the sensory nerve supply of the lips is also shown.

closed. The mucous membrane of the gingiva, or gum, is strongly attached to the alveolar periosteum.

Clinical Note

Ranula.—This is a cystic lesion arising in a distended mucous gland of the mucous membrane. It commonly occurs in the floor of the mouth, and because of its transparent covering it resembles frog skin.

Sensory Nerve Supply of the Mucous Membrane of the Mouth.—The roof is supplied by the greater palatine and nasopalatine nerves (Fig 1–8). The nerve fibers are derived from the maxillary division of the trigeminal nerve.

The floor is supplied by the lingual nerve (common sensation), a branch of the mandibular division of the trigeminal nerve. The taste fibers travel in the chorda tympani nerve, a branch of the facial nerve.

The cheek is supplied by the buccal nerve, a branch of the mandibular division of the trigeminal nerve (the buccinator muscle is innervated by the buccal branch of the facial nerve).

Teeth

There are 20 *deciduous teeth*—four incisors, two canines, and four molars in each jaw. They begin to erupt about 6 months after birth and have all erupted by the end of 2 years. The teeth of the lower jaw usually appear before those of the upper jaw.

There are 32 *permanent teeth,* including four incisors, two canines, four premolars, and six molars in each jaw (Fig 1–9). They begin to erupt at 6 years of age. The last tooth to erupt is the third molar, and this may happen between the ages of 17 and 30. The teeth of the lower jaw usually appear before those of the upper jaw.

Basic Anatomy of the Tongue

The tongue is composed of a mass of striated muscle covered with mucous membrane (see Fig 1–6). The muscles of the tongue are intrinsic and extrinsic. The intrinsic muscles are confined to the tongue and are arranged in longitudinal, transverse, and vertical directions, and they alter the shape of the tongue. The extrinsic muscles of the tongue, which alter the position of the tongue, extend from the tongue to the soft palate and the styloid process of the temporal bone above, and the mandible and the hyoid bone below. The tongue is divided into right and left halves by a median *fibrous septum.* The origin, insertion, nerve supply, and action of the tongue muscles are summarized in Table 1–2.

The mucous membrane of the upper surface of the tongue may be divided into anterior and posterior parts by a V-shaped sulcus, the *sulcus terminalis* (see Fig 1–6). The apex of the sulcus projects backward and is marked by a small pit, the *foramen cecum.* The sulcus serves to divide the tongue into the anterior two thirds, or oral part, and the posterior third, or pharyngeal part. Three types of papillae are present on the upper surface of the anterior two thirds of the tongue—the filiform papillae, the fungiform papillae, and the vallate papillae.

TABLE 1–2.
Muscles of the Tongue

Name of Muscle	Origin	Insertion	Nerve Supply	Action
Intrinsic muscles				
Longitudinal, transverse, vertical	Median septum and submucosa	Mucous membrane	Hypoglossal nerve	Alters shape of tongue
Extrinsic muscles				
Genioglossus	Superior genial spine of mandible	Blends with other muscles of tongue and body of hyoid bone	Hypoglossal nerve	Protrudes apex through mouth
Hyoglossus	Body and greater cornu of hyoid bone	Blends with other muscles of tongue	Hypoglossal nerve	Depresses tongue
Styloglossus	Styloid process of temporal bone	Blends with other muscles of tongue	Hypoglossal nerve	Draws tongue upward and backward
Palatoglossus	Palatine aponeurosis	Side of tongue	Pharyngeal plexus	Pulls root of tongue upward and backward

Clinical Note

Occasionally, when the vallate papillae are very prominent, *fish bones* have been known to lodge behind the papillae on the back of the tongue.

In the midline anteriorly, the undersurface of the tongue is connected to the floor of the mouth by a fold of mucous membrane called the *frenulum of the tongue* (see Fig 1–6).

The mucous membrane covering the posterior third of the tongue is devoid of papillae but has an irregular surface produced by underlying lymphatic nodules, the *lingual tonsil.*

Clinical Note

A *thyroglossal cyst* can appear at any age. It is due to a persistence of a segment of the thyroglossal duct. Such a cyst occurs within the tongue or in the midline of the neck at any point along the thyroglossal tract. The presence of such a cyst should not be forgotten in the differential diagnosis of "lumps in the throat."

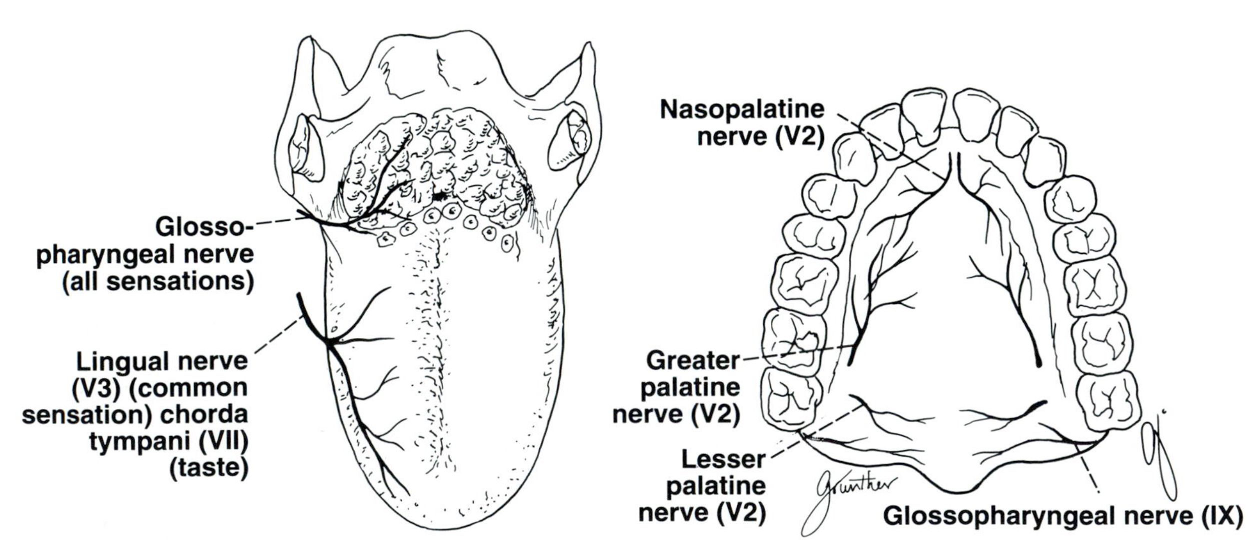

FIG 1–8.
A, sensory nerve supply to the mucous membrane of the tongue. **B,** sensory nerve supply to the mucous membrane of the hard and soft palate; taste fibers run with branches of the maxillary nerve (V2) and join the greater petrosal branch of the facial nerve.

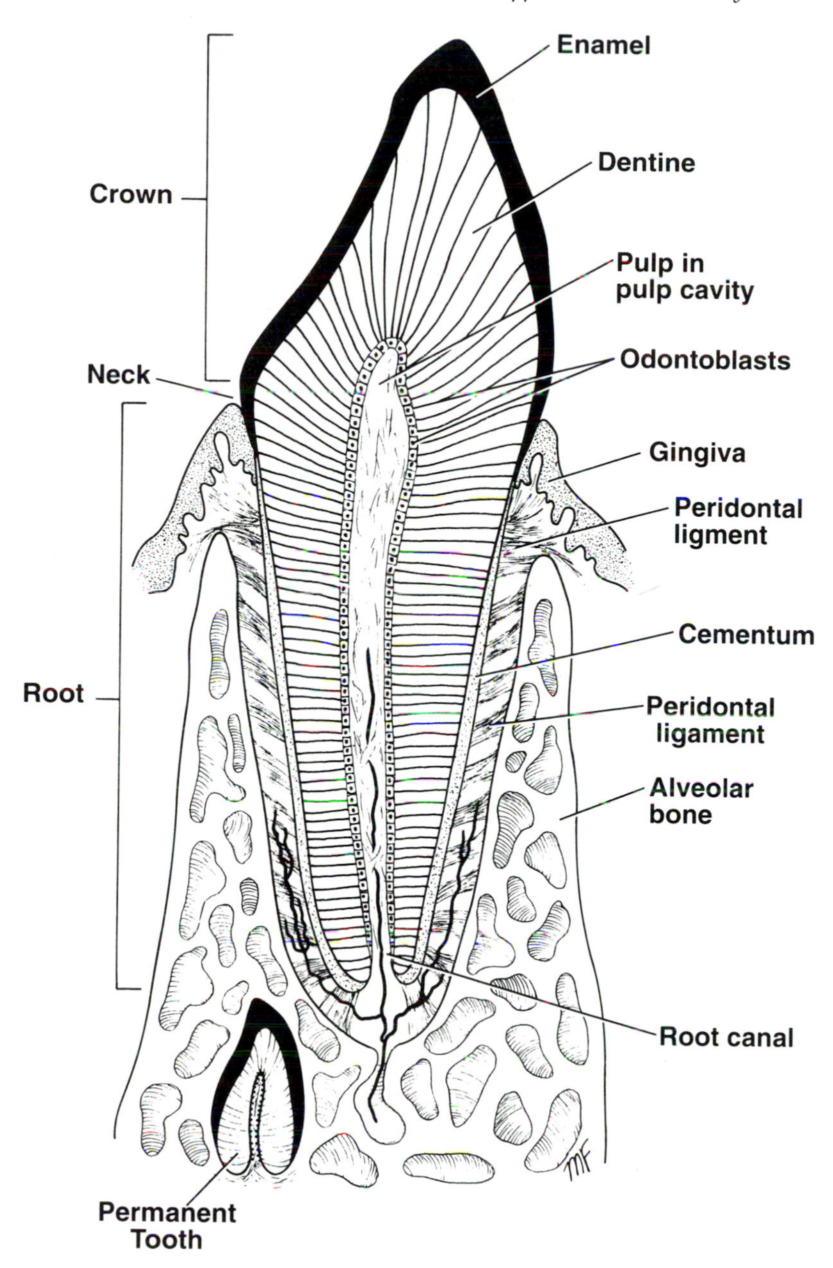

FIG 1–9.
Sagittal section through the lower jaw and gum showing an erupted temporary incisor tooth and a developing permanent tooth.

Blood Supply of the Tongue.—Lingual artery, tonsillar branch of facial artery, ascending pharyngeal artery.

Lymphatic Drainage of the Tongue.—Tip drains into submental lymph nodes; sides of anterior two thirds drain into submandibular and deep cervical nodes.

Sensory Innervation of the Tongue.—Anterior two thirds, lingual nerve (common sensation), chorda tympani (taste). Posterior third, glossopharyngeal for common sensation and taste (see Fig 1–8).

Movements of the Tongue

Protrusion.—Genioglossus muscles on both sides.

Retraction.—Styloglossus and hyoglossus muscles on both sides.

Depression.—Hyoglossus muscles on both sides.

Retraction and Elevation of Posterior Third.—Styloglossus and palatoglossus on both sides.

Shape Changes.—These are produced by the intrinsic muscles.

Basic Anatomy of the Palate

The palate forms the roof of the mouth and the floor of the nasal cavity. It may be divided into two parts—the hard palate in front and the soft palate behind.

The *hard palate* is formed by the palatine processes of the maxillae and the horizontal plates of the palatine bones (see Fig 1–1). The upper and lower surfaces of the hard palate are covered with mucous membrane. The *soft palate* is a mobile fold attached to the posterior border of the hard palate (see Fig 1–6). It is covered on its upper and lower surfaces by mucous membrane and contains an aponeurosis of dense fibrous tissue, muscle fibers, lymphoid tissue, glands, vessels, and nerves. Its free posterior border has in the midline a conical projection called the *uvula.* The soft palate is continuous at the sides with the lateral wall of the pharynx.

Clinical Note

The *uvula* has a core of voluntary muscle, the musculus uvulae, that is attached to the posterior border of the hard palate. Surrounding the muscle is the loose connective tissue of the submucosa that is responsible for the great swelling of this structure secondary to angioedema *(Quincke's uvula).*

From the undersurface of the soft palate a muscular fold extends downwards and forwards on each side and runs to the side of the base of the tongue. It is covered with mucous membrane and is called the *palatoglossal arch or fold* (see Fig 1–6). The muscle contained in the fold is the *palatoglossus muscle.* Another similar arch extends from the undersurface of the soft palate downwards and laterally to join the pharyngeal wall. It is also covered with mucous membrane and is called the *palatopharyngeal arch or fold* (see Fig 1–6). The muscle contained within the arch is the palatopharyngeus muscle. The *palatine tonsils,* which are masses of lymphoid tissue, are situated between the palatoglossal and palatopharyngeal arches (see Fig 1–6).

Clinical Note

Foreign bodies.—Foreign bodies, such as fishbones, may get caught in the palatine tonsils.

The palatoglossal arch on each side serves as a landmark and is the place where the mouth cavity becomes continuous with the oral part of the pharynx. At the posterior border of the soft palate the oral part of the pharynx becomes continuous with the nasal part of the pharynx (see Fig 1–11).

The muscles of the soft palate include the tensor veli palatini, the levator veli palatini, the palatoglossus, the palatopharyngeus, and the musculus uvulae.

Blood Supply of the Palate.—Greater palatine branch of maxillary artery, ascending palatine branch of facial artery, ascending pharyngeal artery.

Lymphatic Drainage of the Palate.—Deep cervical lymph nodes.

Sensory Innervation of the Palate.—Greater and lesser palatine nerves, nasopalatine and glossopharyngeal nerves (see Fig 1–8).

Movements of the Soft Palate

The channel between the oral part of the pharynx and the nasal part of the pharynx can be closed by raising the soft palate. Closure occurs during swallowing to prevent food from entering the nasal part of the pharynx and the back of the nose; closure also occurs during speech.

Basic Anatomy of the Salivary Glands

The three main pairs of salivary glands are called the *parotid, submandibular,* and *sublingual glands,* and, in addition, many small *buccal glands* are scattered throughout the mouth.

Parotid Gland

The parotid gland lies in a deep hollow below the external auditory meatus, behind the ramus of the mandible and in front of the sternocleidomastoid muscle. The *parotid duct* emerges from the anterior border of the gland and passes forward over the lateral surface of the masseter. It enters the vestibule of the mouth upon a small papilla opposite the upper second molar tooth (Fig 1–10).

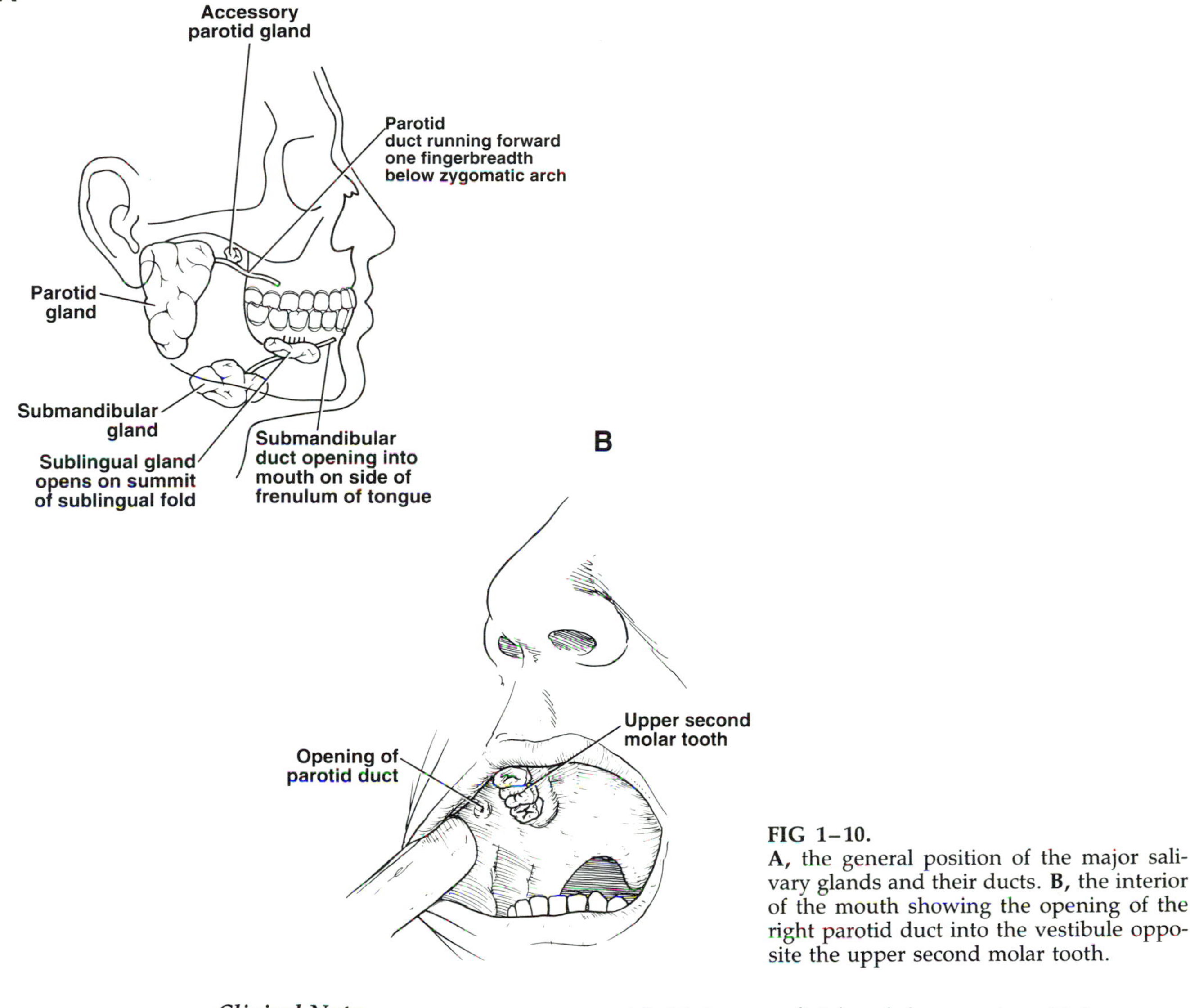

FIG 1–10.
A, the general position of the major salivary glands and their ducts. **B,** the interior of the mouth showing the opening of the right parotid duct into the vestibule opposite the upper second molar tooth.

Clinical Note

The *parotid duct* runs forward from the parotid gland one fingerbreadth below the zygomatic arch (see Fig 1–10). It can be rolled beneath the examining finger at the anterior border of the masseter as it turns medially and opens into the mouth opposite the upper second molar tooth (see Fig 1–10).

The integrity of the parotid duct can be established by wiping the inside of the cheek dry and then pressing on the parotid gland. Look for a drop of viscid saliva to appear on the tip of the papilla in the mouth.

Submandibular Gland

The submandibular gland lies beneath the lower border of the body of the mandible. It can be divided into superficial and deep parts, which are continuous with each other around the posterior border of the mylohyoid muscle. The *submandibular duct* emerges from the anterior end of the deep part of the gland and runs forward beneath the mucous membrane of the mouth. It opens into the mouth on a small papilla, which is situated at the side of the frenulum of the tongue (see Fig 1–10).

Sublingual Gland

The sublingual gland lies beneath the mucous membrane (sublingual fold) of the floor of the mouth close to the frenulum of the tongue. The *sublingual ducts* (8 to 20 in number) open into the mouth on the summit of the sublingual fold, but a few open into the submandibular duct (see Fig 1–10).

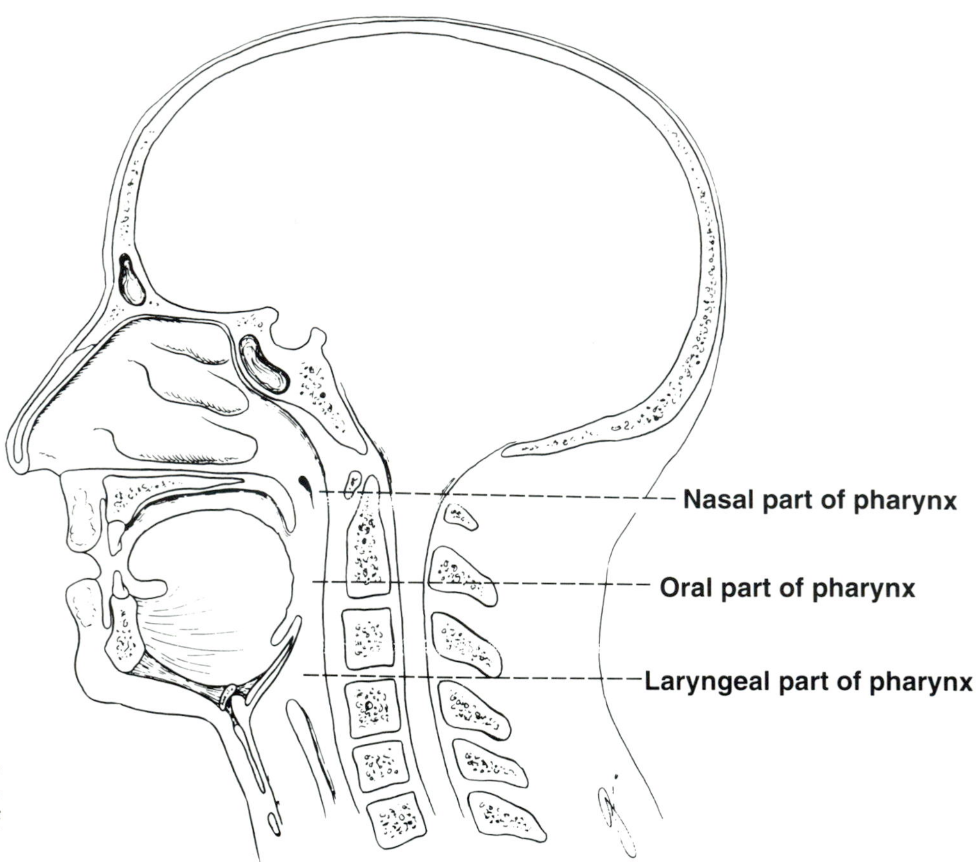

FIG 1–11.
Sagittal section through the nose, mouth, pharynx, and larynx to show the subdivisions of the pharynx.

Basic Anatomy of the Pharynx

The pharynx is situated behind the nasal cavities, the mouth, and the larynx (Fig 1–11) and may be divided into nasal, oral, and laryngeal parts. The pharynx is funnel-shaped, its upper, wider end lying under the skull, and its lower, narrow end becoming continuous with the esophagus opposite the sixth cervical vertebra. The pharynx has a musculomembranous wall, which is deficient anteriorly. Here, it is replaced by the posterior openings into the nose (choanae), the opening into the mouth, and the inlet of the larynx. The mucous membrane is continuous with that of the nasal cavities, the mouth, and the larynx. By means of the auditory tube, the mucous membrane is also continuous with that of the tympanic cavity.

The muscles in the wall of the pharynx consist of the superior, middle, and inferior constrictor muscles (Fig 1–12), whose fibers run in a somewhat circular direction, and the stylopharyngeus and salpingopharyngeus muscles, whose fibers run in a somewhat longitudinal direction.

The three constrictor muscles extend round the pharyngeal wall to be inserted into a fibrous band or raphe that extends from the pharyngeal tubercle on the basilar part of the occipital bone down to the esophagus. The three constrictor muscles overlap each other so that the middle constrictor lies on the outside of the lower part of the superior constrictor and the inferior constrictor lies outside the lower part of the middle constrictor (see Fig 1–12).

The lower part of the inferior constrictor, which arises from the cricoid cartilage, is called the *cricopharyngeus muscle* (see Fig 1–12). The fibers of the cricopharyngeus pass horizontally round the lowest and narrowest part of the pharynx and act as a sphincter. *Killian's dehiscence* is the area on the posterior pharyngeal wall between the upper propulsive part of the inferior constrictor and the lower sphincteric part, the cricopharyngeus (see Fig 1–12).

Clinical Note

Inverted foreign bodies tend to get snared in the region of Killian's dehiscence.

A part of the superior constrictor, which arises from the soft palate and sweeps round the pharyngeal wall close to its upper border, is a band of mus-

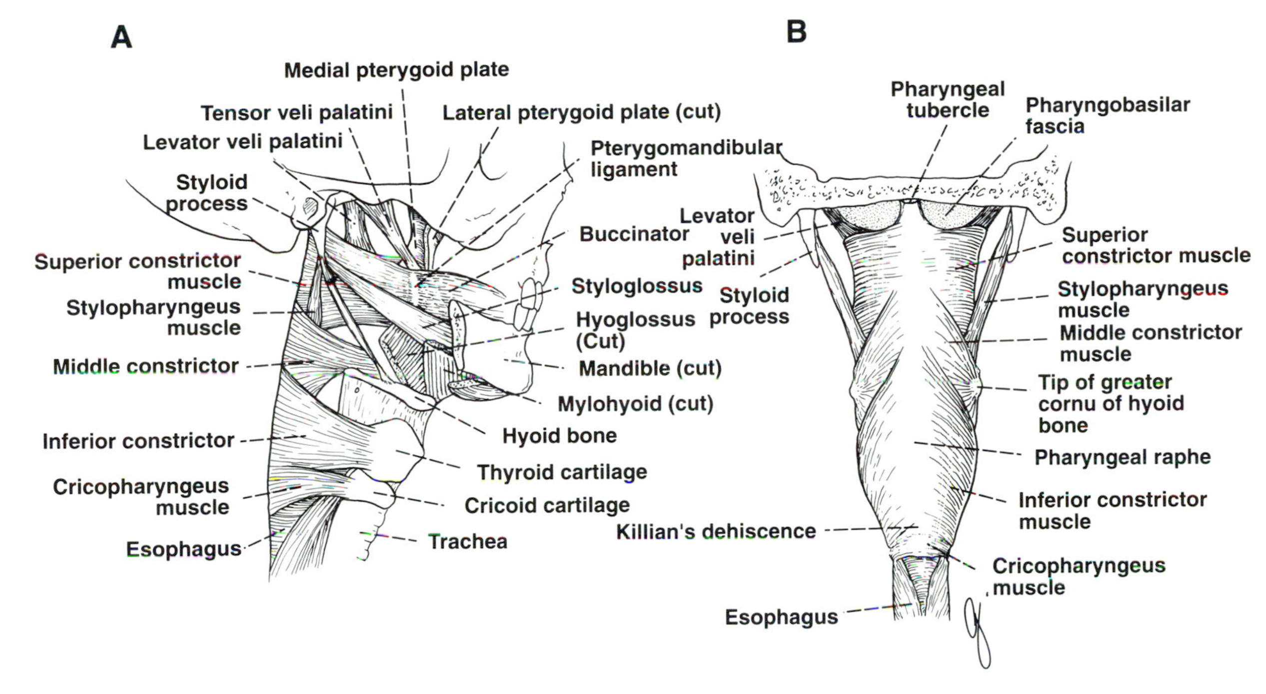

FIG 1–12.
A, pharynx, lateral view, showing the three constrictor muscles. **B,** pharynx, posterior view, showing the three constrictor muscles and the stylopharyngeus.

cle called the *palatopharyngeal sphincter* and produces a rounded ridge called the *ridge of Passavant.*

Interior of Pharynx

The pharynx is divided into three parts—the nasal pharynx, the oral pharynx, and the laryngeal pharynx.

Nasal Pharynx

This lies above the soft palate and behind the nasal cavities (see Fig 1–11). In the submucosa of the roof is a collection of lymphoid tissue called the *pharyngeal tonsil* (Fig 1–13). The *pharyngeal isthmus* is the opening in the floor between the soft palate and the posterior pharyngeal wall. On the lateral wall is the opening of the *auditory tube,* the elevated ridge of which is called the *tubal elevation.* The *pharyngeal recess* is a depression in the pharyngeal wall behind the tubal elevation. The *salpingopharyngeal fold* is a vertical fold of mucous membrane covering the salpingopharyngeus muscle.

Oral Pharynx

This lies behind the oral cavity (see Fig 1–13). The floor is formed by the posterior one third of the tongue and the interval between the tongue and the epiglottis. In the midline is the *median glossoepiglottic fold,* and on each side the *lateral glossoepiglottic fold.* The depression on each side of the median glossoepiglottic fold is called the *vallecula* (see Fig 1–16).

On the lateral wall on each side are the palatoglossal and the palatopharyngeal arches or folds and the *palatine tonsils* between them (Fig 1–14). The *palatoglossal arch* is a fold of mucous membrane covering the palatoglossus muscle. The interval between the two palatoglossal arches is called the *oropharyngeal isthmus* and marks the boundary between the mouth and the pharynx. The *palatopharyngeal arch* is a fold of mucous membrane covering the palatopharyngeus muscle. The *tonsillar sinus,* a recess between the palatoglossal and palatopharyngeal arches, is occupied by the palatine tonsil.

Laryngeal Pharynx

This lies behind the opening into the larynx (see Fig 1–11). The lateral wall is formed by the thyroid cartilage and the thyrohyoid membrane. The *piriform fossa* is a depression in the mucous membrane on each side of the larygeal inlet (Fig 1–15).

Clinical Note

The piriform fossa is a common site for fish bones or other foreign bodies to become lodged.

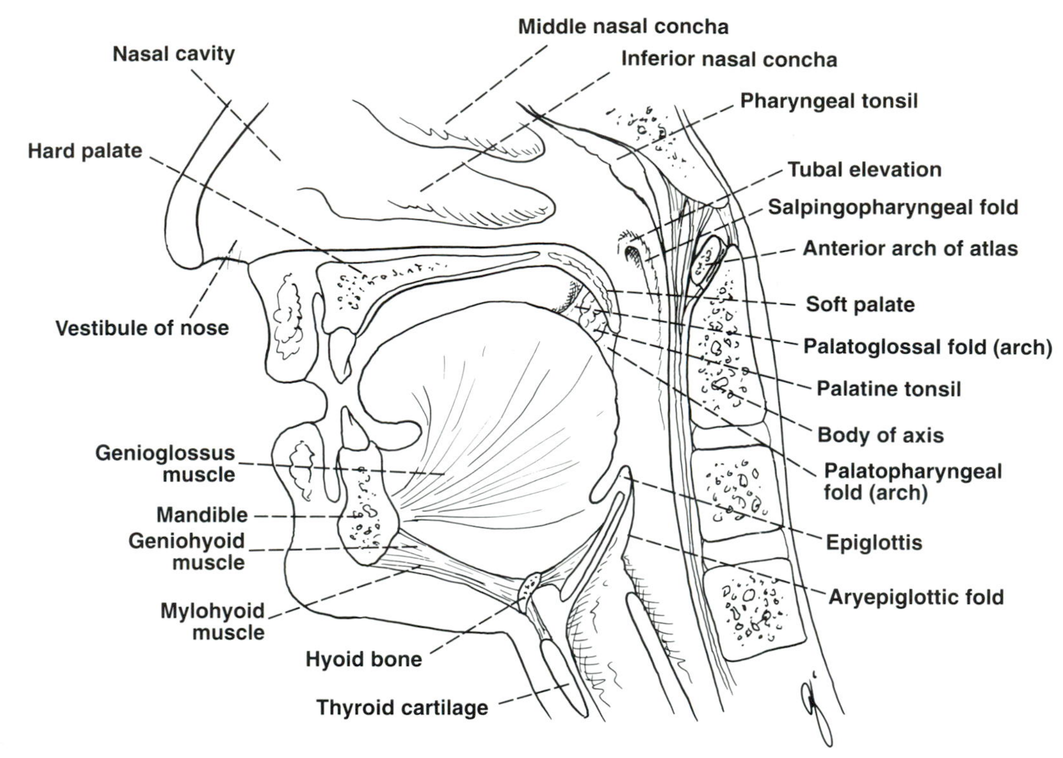

FIG 1–13.
Sagittal section of the head and neck showing the relations of the nasal cavity, mouth, pharynx, and larynx.

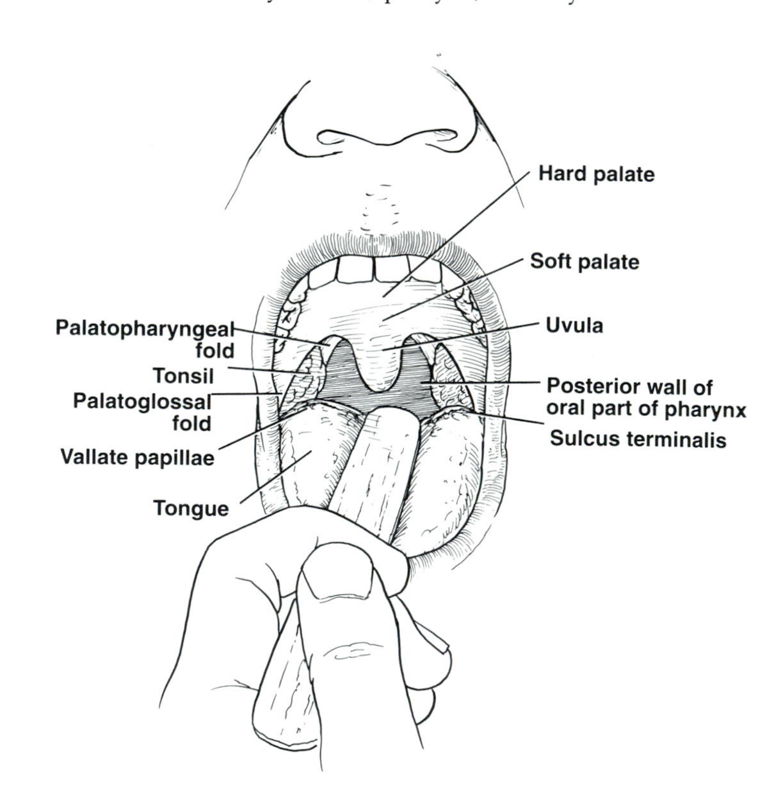

FIG 1–14.
The oral pharynx as seen through the open mouth.

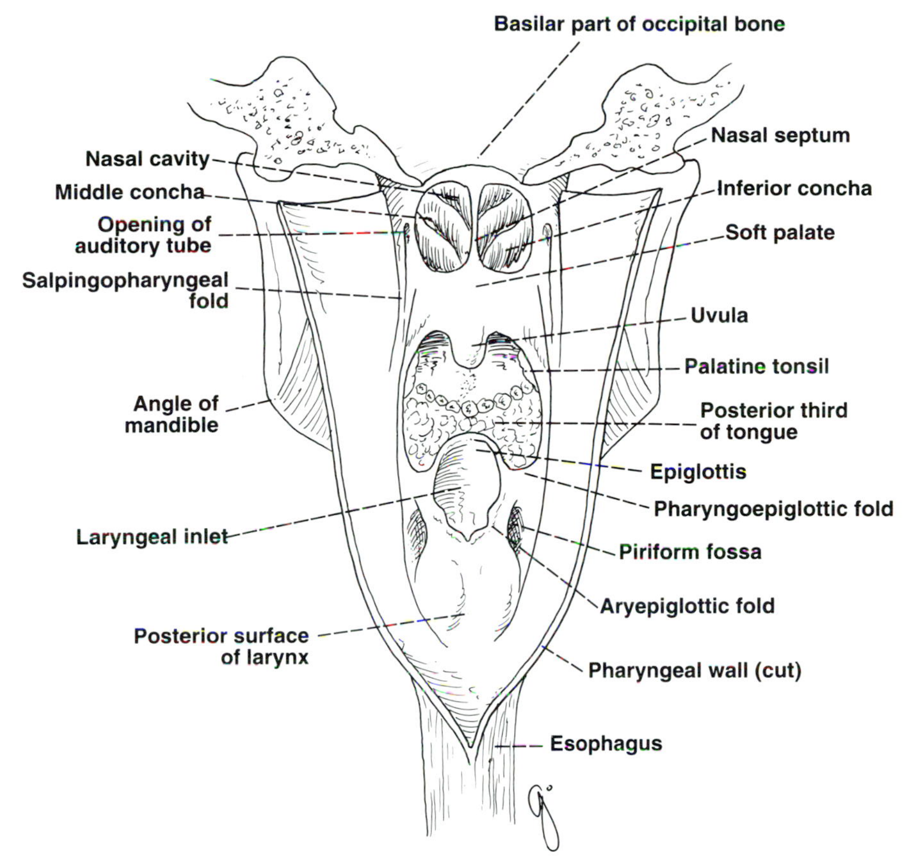

FIG 1–15.
Interior of the pharynx as seen on posterior view. The posterior wall has been cut in the midline, and the right and left halves have been turned laterally.

Sensory Nerve Supply of Mucous Membrane of Pharynx.—The nasal part is mainly supplied by the maxillary nerve (V2), the oral part by the glossopharyngeal nerve. The mucous membrane around the entrance into the larynx is supplied by the internal laryngeal branch of the vagus nerve (see Fig 1–24).

Blood Supply of Pharynx.—Ascending pharyngeal, tonsillar branches of facial arteries, and branches of maxillary and lingual arteries.

Lymphatic Drainage of Pharynx.—Directly into the deep cervical lymph nodes or indirectly via the retropharyngeal or paratracheal nodes into the deep cervical nodes.

Basic Anatomy of the Palatine Tonsils

The palatine tonsils are two masses of lymphoid tissue, each located in the tonsillar sinus on the lateral wall of the oral part of the pharynx between the palatoglossal and palatopharyngeal arches (see Fig 1–14). Each tonsil is covered by mucous membrane, and its free medial surface projects into the pharynx. The surface is pitted by numerous small openings that lead into the *tonsillar crypts*. The upper part of the medial surface of the tonsil has a deep *intratonsillar cleft*. The tonsil is covered on its lateral surface by a layer of fibrous tissue called the *capsule* (Fig 1–16).

The tonsil reaches its maximum size during early childhood, but after puberty it diminishes considerably in size.

Relations of the Palatine Tonsil

Anteriorly.—The palatoglossal arch.

Posteriorly.—The palatopharyngeal arch.

Superiorly.—The soft palate.

Inferiorly.—The posterior third of the tongue.

Medially.—The cavity of the oral part of the pharynx.

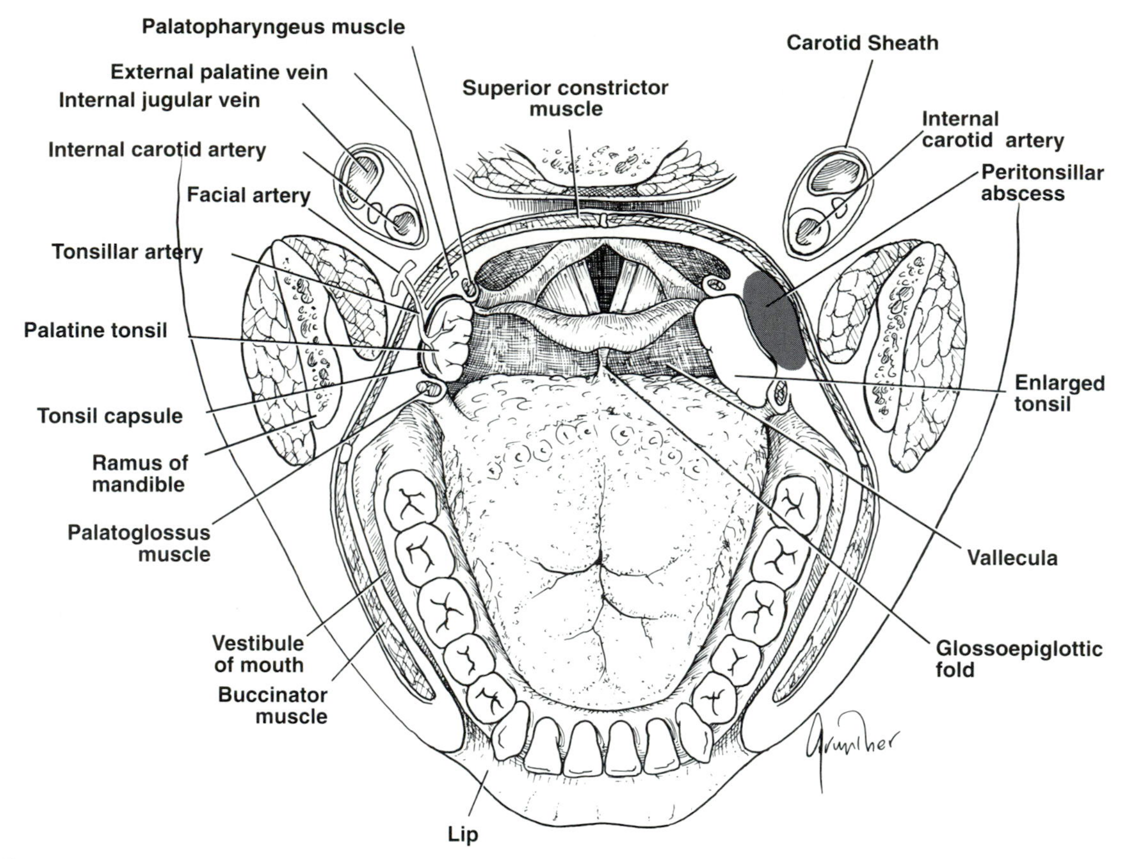

FIG 1–16.
Horizontal section through the mouth and oral pharynx. On the left the normal palatine tonsil and its relationships are shown; on the right the position of a peritonsillar abscess is shown. Note the relationship of the abscess to the superior constrictor muscle and the carotid sheath. The opening into the larynx can also be seen below and behind the tongue.

Laterally.—The capsule is separated from the superior constrictor muscle by loose areolar tissue (see Fig 1–16). The external palatine vein descends from the soft palate in this loose connective tissue to join the pharyngeal venous plexus. Lateral to the superior constrictor muscle lie the styloglossus muscle and the loop of the facial artery. The internal carotid artery lies 1 in. (2.5 cm) behind and lateral to the tonsil.

Lymphatic Drainage of Tonsil.—The tonsil drains into the upper deep cervical lymph nodes, just below and behind the angle of the mandible.

Waldeyer's Ring

The lymphoid tissue that surrounds the opening into the respiratory and digestive systems forms a ring. The lateral part of the ring is formed by the palatine tonsils and tubal tonsils (lymphoid tissue around the opening of the auditory tube in the lateral wall of the nasopharynx). The pharyngeal tonsil in the roof of the nasopharynx forms the upper part, and the lingual tonsil on the posterior third of the tongue forms the lower part.

Anatomy of Swallowing (Deglutition)

Masticated food is formed into a ball or bolus on the dorsum of the tongue and voluntarily pushed upward and backward against the undersurface of the hard palate. This is brought about by the contraction of the styloglossus muscles on both sides, which pull the root of the tongue upward and backward. The palatoglossus muscles then squeeze the bolus backward into the pharynx. From this point onward the process of swallowing becomes an involuntary act.

The nasal part of the pharynx is shut off from the oral part of the pharynx by the elevation of the soft palate, the pulling forward of the posterior wall of the pharynx by the palatopharyngeal sphincter, and the pulling medially of the palatopharyngeal arches. This prevents the passage of food and drink into the nasal cavities.

The larynx and the laryngeal part of the pharynx are pulled upward by the contraction of the stylopharyngeus, salpingopharyngeus, thyrohyoid, and palatopharyngeus muscles. The main part of the larynx is thus elevated to the posterior surface of the epiglottis, and the entrance into the larynx is closed. The laryngeal entrance is made smaller by the approximation of the aryepiglottic folds, and the arytenoid cartilages are pulled forward by the contraction of the aryepiglottic, oblique arytenoid, and thyroarytenoid muscles.

The bolus moves downward over the epiglottis, the closed entrance into the larynx, and reaches the lower part of the pharynx as the result of the successive contraction of the superior, middle, and inferior constrictor muscles. Some of the food slides down the groove on either side of the entrance into the larynx, i.e., down through the *piriform fossae.* Finally, the lower part of the pharyngeal wall (the cricopharyngeus muscle) relaxes and the bolus enters the esophagus.

Basic Anatomy of the Larynx

The larynx is an organ that provides a protective sphincter at the inlet of the air passages and is responsible for voice production. It is situated below the tongue and hyoid bone and between the great blood vessels of the neck and lies at the level of the fourth, fifth, and sixth cervical vertebrae (see Fig 1–30). It opens above into the laryngeal part of the pharynx, and below is continuous with the trachea. The larynx is covered in front by the infrahyoid strap muscles and at the sides by the thyroid gland.

The framework of the larynx is formed of cartilages that are held together by ligaments and membranes, moved by muscles, and lined by mucous membrane.

Cartilages of the Larynx

Thyroid Cartilage.—This is the largest cartilage of the larynx (Figs 1–17 and 1–18). It consists of two laminae of hyaline cartilage that meet in the midline in the prominent V angle (the so-called Adam's apple). The posterior border extends upward into a *superior cornu* and downward into an *inferior cornu.* On the outer surface of each lamina is an *oblique line* for the attachment of the sternothyroid, the thyrohyoid, and the inferior constrictor muscles.

Cricoid Cartilage.—This cartilage is formed of hyaline cartilage and shaped like a signet ring, having a broad plate behind and a shallow arch in front (see Figs 1–17 and 1–18). The cricoid cartilage lies below

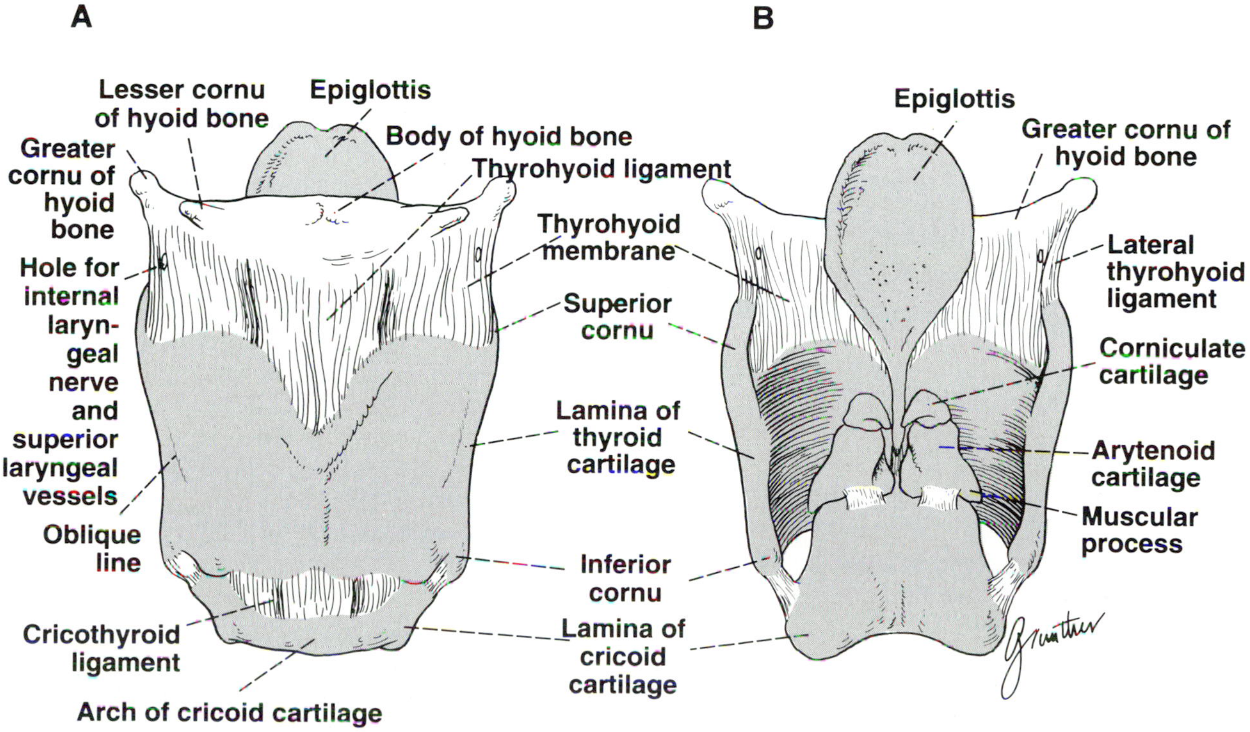

FIG 1–17.
Larynx and its ligaments. **A,** anterior view. **B,** posterior view.

the thyroid cartilage, and on each side of the lateral surface is a facet for articulation with the inferior cornu of the thyroid cartilage. Posteriorly, the lamina has on its upper border on each side a facet for articulation with the arytenoid cartilage. All these joints are synovial.

Clinical Note

The continuous ring structure of the cricoid cartilage is utilized when applying pressure on the cricoid to control regurgitation of stomach contents during the induction of anesthesia in the *Sellick maneuver.*

Arytenoid Cartilages.—The two arytenoid cartilages are small and pyramidal shaped (see Fig 1–17). They are located at the back of the larynx, on the lateral part of the upper border of the lamina of the cricoid cartilage. Each cartilage has an *apex* above that articulates with the small corniculate cartilage, a *base* below that articulates with the lamina of the cricoid cartilage, and a *vocal process* that projects forward and gives attachment to the vocal ligament. A *muscular process* that projects laterally gives attachment to the posterior and lateral cricoarytenoid muscles.

Corniculate Cartilages.—Two small conical-shaped cartilages articulate with the arytenoid cartilages (see Fig 1–17). They give attachment to the aryepiglottic folds.

Cuneiform Cartilages.—These two small rod-shaped cartilages are found in the aryepiglottic folds and serve to strengthen them (Fig 1–19).

Epiglottis.—This leaf-shaped lamina of elastic cartilage lies behind the root of the tongue (see Fig 1–19). Its stalk is attached to the back of the thyroid cartilage. The sides of the epiglottis are attached to the arytenoid cartilages by the aryepiglottic folds of mucous membrane. The upper edge of the epiglottis is free. The covering of mucous membrane passes forward onto the posterior surface of the tongue as the *median glossoepiglottic fold;* the depression on each side of the fold is called the *vallecula* (see Fig 1–16). Laterally the mucous membrane passes onto the wall of the pharynx as the *lateral glossoepiglottic fold.*

Membranes and Ligaments of the Larynx

Thyrohyoid Membrane.—This connects the upper margin of the thyroid cartilage to the upper margin of the posterior surface of the hyoid bone (see Fig 1–17). In the midline it is thickened to form the *median thyrohyoid ligament.* The membrane is pierced on each side by the superior laryngeal vessels and the internal laryngeal nerve, a branch of the superior laryngeal nerve (see Fig 1–17).

Cricotracheal Ligament.—This connects the cricoid cartilage to the first ring of the trachea (see Fig 1–19).

Fibroelastic Membrane.—This lies beneath the mucous membrane lining the larynx. The upper part is thin and called the *quadrangular membrane* (see Fig 1–19). It extends between the epiglottis and the arytenoid cartilages, its thickened inferior margin forming the *vestibular ligament.* The vestibular ligaments form the interior of the *vestibular folds* (see Fig 1–19). The lower part of the fibroelastic membrane is called the *cricothyroid ligament.* The anterior part of the cricothyroid ligament is thick and connects the cricoid cartilage to the lower margin of the thyroid cartilage. The lateral part of the ligament is thin and is attached below to the upper margin of the cricoid cartilage. The superior margin of the ligament, instead of being attached to the lower margin of the thyroid cartilage, ascends on the medial surface of the thyroid cartilage. Its upper margin, composed almost entirely of elastic tissue, forms the important *vocal ligament* on each side. The vocal ligaments form the interior of the *vocal folds (vocal cords)* (see Fig 1–19). The anterior end of each vocal ligament is attached to the thyroid cartilage, and the posterior end is attached to the vocal process of the arytenoid cartilage.

Cavity of the Larynx

The *inlet* of the larynx looks backward and upward into the laryngeal part of the pharynx. The opening is wider in front than behind and is bounded in front by the epiglottis, laterally by the aryepiglottic fold of mucous membrane, and posteriorly by the arytenoid cartilages with the corniculate cartilages. The cuneiform cartilage lies within and strengthens the aryepiglottic fold and produces a small elevation on the upper border. The *piriform fossa* is a recess on either side of the fold and inlet (Fig 1–20). It is bounded medially by the aryepiglottic fold and laterally by the thyroid cartilage and the thyrohyoid membrane.

The cavity of the larynx extends from the inlet to the lower border of the cricoid cartilage, where it is continuous with the cavity of the trachea. It is divided into three regions by the presence of an upper and a lower pair of folds of mucous membrane that project into the cavity from the sides of the larynx.

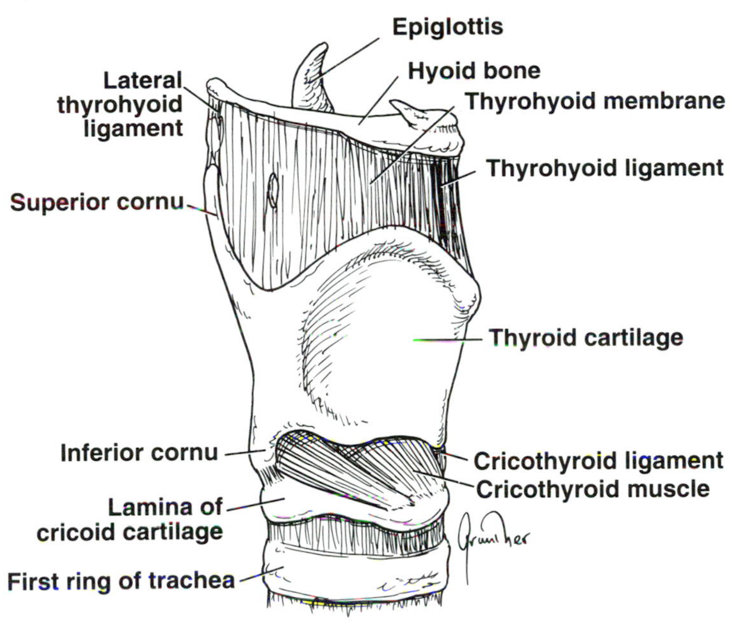

FIG 1–18.
Larynx and its ligaments as seen from the right side.

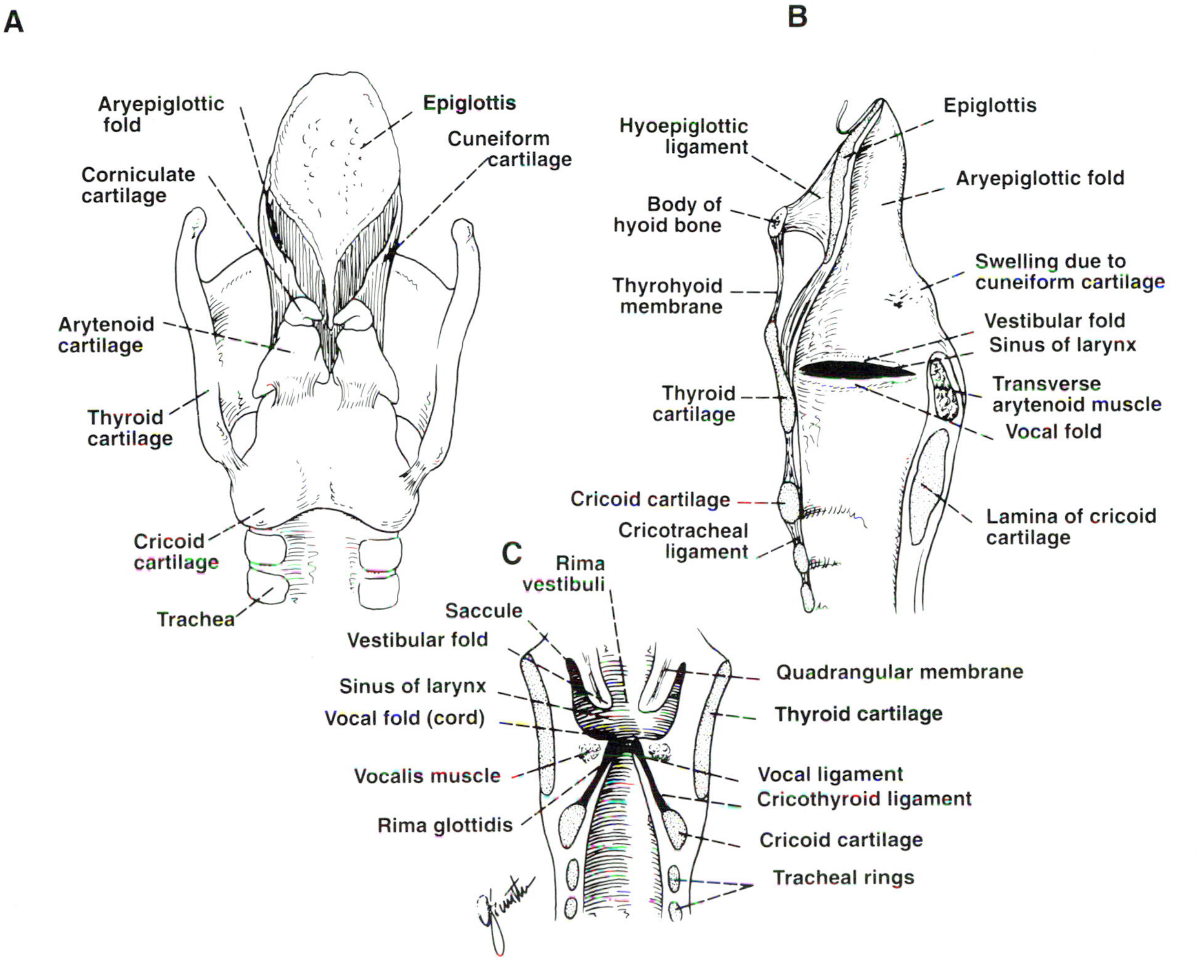

FIG 1–19.
Larynx. **A,** posterior view showing aryepiglottic folds in position. **B,** sagittal section through larynx showing vestibular fold, sinus, and vocal fold. **C,** coronal section through larynx showing vocal ligament and cricothyroid ligament.

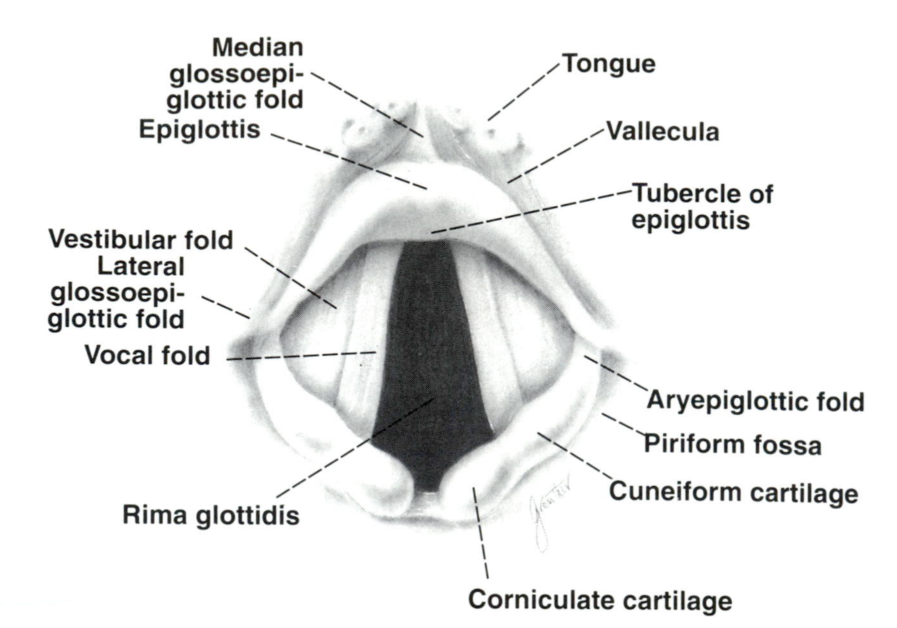

FIG 1–20.
The laryngeal inlet as seen from above.

The upper pair of folds are fixed and called the *vestibular folds* (also called "false cords") and the gap between them is the *rima vestibuli* (see Fig 1–19). The lower pair are mobile and concerned with voice production and are called the *vocal folds;* the gap between them is the *rima glottidis or glottis.* The glottis is bounded in front by the vocal folds and behind by the medial surfaces of the arytenoid cartilages. The glottis is the narrowest part of the larynx and measures about 2.5 cm from front to back in the male adult and less in the female. In children the lower part of the larynx within the cricoid cartilage is the narrowest part.

Clinical Note

In Figure 1–19 note the relationship between the vocal folds and the cricothyroid ligament. It is clear that the folds may be damaged in puncture wounds or in surgical incisions through the ligament to establish an airway.

The three regions of the cavity of the larynx are as follows:

1. The upper region called the *vestibule,* situated between the inlet and the vestibular folds. Its anterior wall is formed by the epiglottis, the lower part of which projects backward as the *tubercle.*
2. The middle region, situated between the vestibular and vocal folds.
3. The lower region, situated between the vocal folds and the lower border of the cricoid cartilage.

The *vestibular fold* on each side is formed of mucous membrane and contains the vestibular ligament, which is the thickened lower edge of the quadrangular membrane. The fold is fixed (i.e., does not move with respiration) and is pink when viewed with a laryngoscope (see Fig 1–20).

The *vocal fold* on each side is formed of mucous membrane and contains the vocal ligament, which is the thickened upper edge of the cricothyroid ligament. The fold is mobile with respiration and is white when viewed with a laryngoscope (see Fig 1–20).

The *sinus of the larynx* is a small recess on each side of the larynx situated between the vestibular and vocal folds. It is lined with mucous membrane (see Fig 1–19).

The *saccule of the larynx* is a diverticulum of mucous membrane that ascends from the sinus of the larynx between the vestibular fold and the thyroid cartilage (see Fig 1–19). The submucosa contains large numbers of mucous glands that open onto the surface of the mucous membrane.

Muscles of the Larynx

The muscles of the larynx may be divided into two groups—extrinsic and intrinsic.

Extrinsic Muscles.—These muscles move the larynx up and down during swallowing. Note that many of these muscles are attached to the hyoid bone, which is attached to the thyroid cartilage by the thyrohyoid membrane. It follows that movements of the hyoid bone are accompanied by movements of the larynx.

The *elevator muscles* of the larynx include the di-

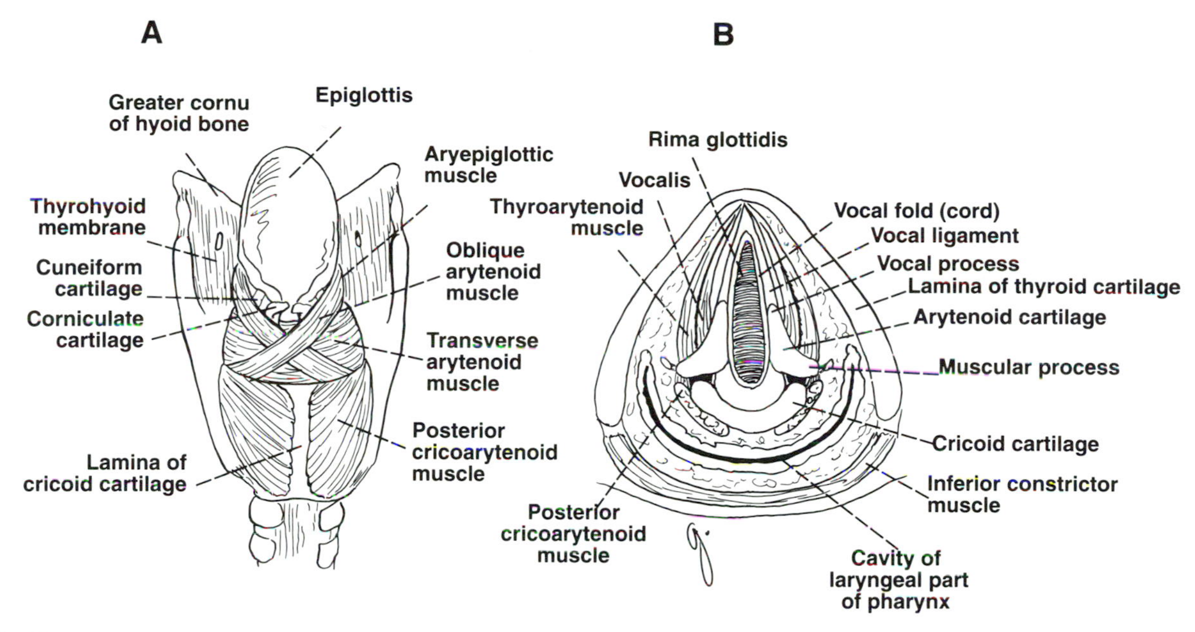

FIG 1–21.
A, intrinsic muscles of the larynx as seen on posterior view. **B,** horizontal section through the larynx at the level of the rima glottidis (glottis).

gastric, the stylohyoid, the mylohyoid, the geniohyoid, the stylopharyngeus, the salpingopharyngeus, and the palatopharyngeus muscles. The *depressor muscles* of the larynx include the sternothyroid, the sternohyoid, and the omohyoid muscles.

Intrinsic Muscles.—These muscles modify the inlet of the larynx and move the vocal folds (Fig 1–21). The details of the origins, insertions, nerve supply, and action are given in Table 1–3.

Movements of the Vocal Folds

The movements of the vocal folds depend on the movements of the arytenoid cartilages, which rotate and slide up and down on the sloping shoulders of the superior border of the cricoid cartilage.

TABLE 1–3.
Intrinsic Muscles of the Larynx

Name of Muscle	Origin	Insertion	Nerve Supply	Action
Muscles controlling the laryngeal inlet				
Oblique arytenoid	Muscular process of arytenoid cartilage	Apex of opposite arytenoid cartilage	Recurrent laryngeal nerve	Narrows the inlet by bringing the aryepiglottic folds together
Thyroepiglottic	Medial surface of thyroid cartilage	Lateral margin of epiglottis and aryepiglottic fold	Recurrent laryngeal nerve	Widens the inlet by pulling aryepiglottic folds apart
Muscles controlling the movements of the vocal folds (cords)				
Cricothyroid	Side of cricoid cartilage	Lower border and inferior cornu of thyroid cartilage	External laryngeal nerve	Tenses vocal cords
Thyroarytenoid (vocalis)	Inner surface of thyroid cartilage	Arytenoid cartilage	Recurrent laryngeal nerve	Relaxes vocal cords
Lateral cricoarytenoid	Upper border of cricoid cartilage	Muscular process of arytenoid cartilage	Recurrent laryngeal nerve	Adducts the vocal cords
Posterior cricoarytenoid	Back of cricoid cartilage	Muscular process of arytenoid cartilage	Recurrent laryngeal nerve	Abducts the vocal cords
Transverse arytenoid	Back and medial surface of arytenoid cartilage	Back and medial surface of opposite arytenoid cartilage	Recurrent laryngeal nerve	Closes posterior part of rima glottidis by approximating arytenoid cartilages

The rima glottidis is opened by the contraction of the posterior cricoarytenoid, which rotates the arytenoid cartilage and abducts the vocal process (Fig 1–22). The elastic tissue in the capsules of the cricoarytenoid joints keeps the arytenoid cartilages apart so that the posterior part of the glottis is open.

The rima glottidis is closed by contraction of the lateral cricoarytenoid, which rotates the arytenoid cartilage and adducts the vocal process (see Fig 1–22). The posterior part of the glottis is narrowed when the arytenoid cartilages are drawn together by contracting the transverse arytenoid muscles.

The vocal folds are stretched by contraction of the cricothyroid muscle (Fig 1–23). The vocal folds are slackened by contraction of the vocalis, a part of the thyroarytenoid muscle (see Fig 1–21).

Movements of the Vocal Folds With Respiration
On quiet inspiration, the vocal folds are abducted and the rima glottidis is triangular in shape with the apex in front (see Fig 1–22). On expiration the vocal folds are adducted, leaving a small gap between them (see Fig 1–22).

On deep inspiration, the vocal folds are maximally abducted, and the triangular shape of the glottis becomes a diamond shape because of the maximal lateral rotation of the arytenoid cartilages (see Fig 1–22).

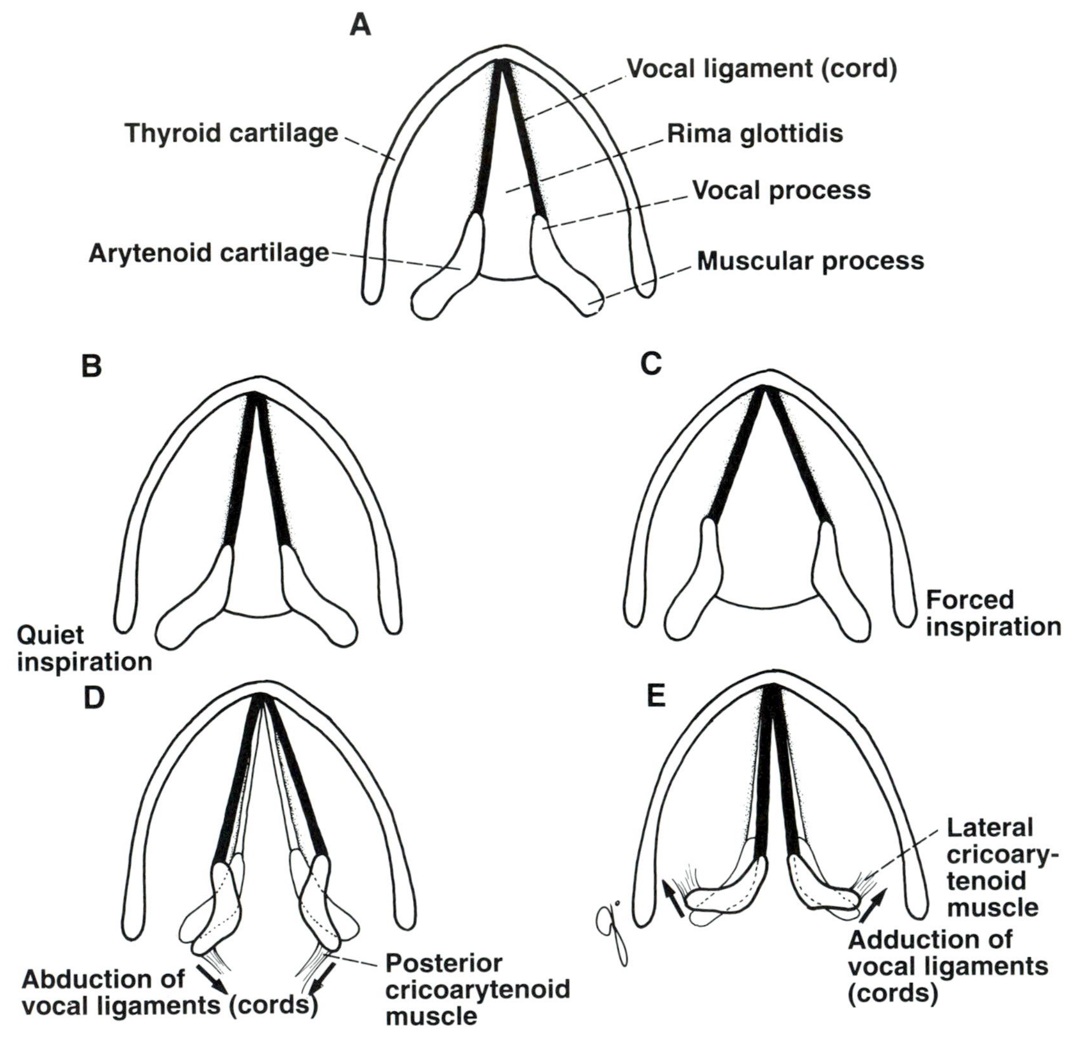

FIG 1–22.
Movements of the vocal ligaments and the arytenoid cartilages. **A,** the attachments of the vocal ligaments. **B,** the position of the vocal ligaments in quiet inspiration; note that the anterior part of the rima glottidis is triangular in shape and the posterior part between the arytenoid cartilages is rectangular in shape. **C,** the position of the vocal ligaments in forced inspiration; the rima glottidis is now diamond shaped. **D,** abduction of the vocal ligaments caused by contraction of the posterior cricoarytenoid muscles. **E,** adduction of the vocal ligaments caused by the contraction of the lateral cricoarytenoid muscles.

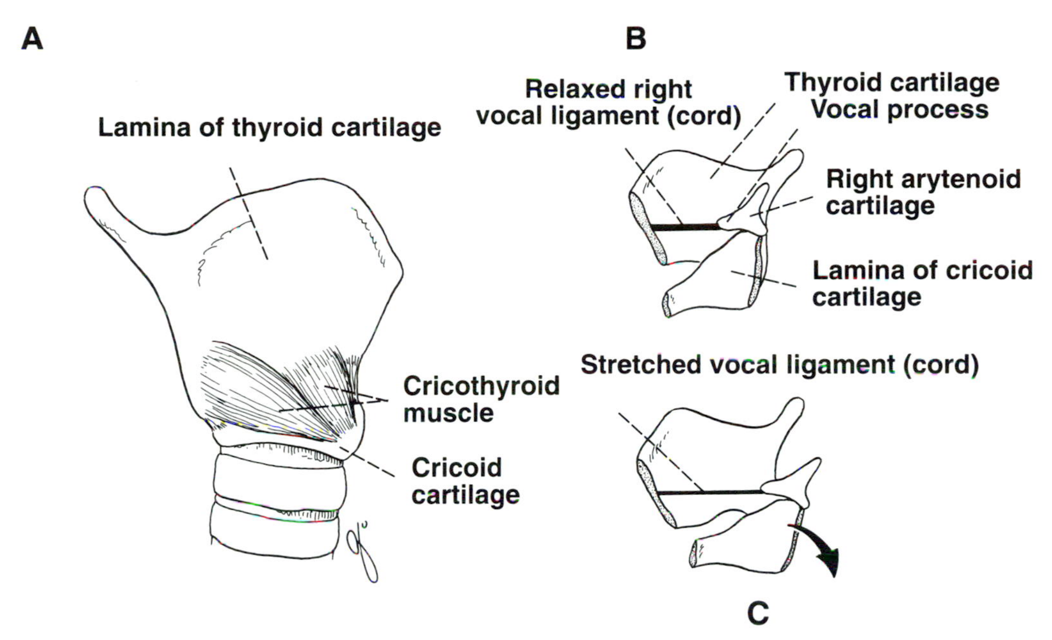

FIG 1–23.
Diagrams showing the attachments and actions of the cricothyroid muscle. **A,** right lateral view of larynx and cricothyroid muscle. **B,** interior view of the larynx showing relaxed right vocal ligament. **C,** interior view of larynx showing right vocal ligament stretched as the result of the cricoid and arytenoid cartilages tilting backward by contraction of the cricothyroid muscles.

Sphincteric Function of the Larynx

There are two sphincters in the larynx; one at the inlet and another at the rima glottidis.

The sphincter at the inlet is used only during swallowing. As the bolus of food is passed backward between the tongue and the hard palate, the larynx is pulled up beneath the back of the tongue. The inlet of the larynx is narrowed by the action of the oblique arytenoid and aryepiglottic muscles. The epiglottis is pulled backward by the tongue and serves as a cap over the laryngeal inlet. The bolus of food, or fluids, then enters the esophagus by passing over the epiglottis or moving down the grooves on either side of the laryngeal inlet, the piriform fossae.

In coughing or sneezing, the rima glottidis serves as a sphincter. After inspiration, the vocal folds are adducted, and the muscles of expiration are made to contract strongly. As a result, the intrathoracic pressure rises, and the vocal folds are suddenly abducted. The sudden release of the compressed air will often dislodge foreign particles or mucus from the respiratory tract and carry the material up into the pharynx, where the material is either swallowed or expectorated.

In the *Valsalva maneuver,* forced expiration takes place against a closed glottis.

In abdominal straining associated with micturition, defecation, and parturition, air is often held temporarily in the respiratory tract by closing the rima glottidis. After deep inspiration the rima glottidis is closed. The muscles of the anterior abdominal wall now contract, and the upward movement of the diaphragm is prevented by the presence of compressed air within the respiratory tract. After a prolonged effort the person often releases some of the air by momentarily opening the rima glottidis, producing a grunting sound.

Voice Production in the Larynx

The intermittent release of expired air between the adducted vocal folds results in their vibration and in the production of sound. The *frequency, or pitch,* of the voice is determined by changes in the length and tension of the vocal ligaments. The *quality* of the voice depends on the resonators above the larynx, namely, the pharynx, mouth, and paranasal sinuses. The quality is controlled by the muscles of the soft palate, tongue, floor of the mouth, cheeks, lips, and jaws. Normal speech depends on the modification of the sound into recognizable consonants and vowels by the use of the tongue, teeth, and lips. Vowel sounds are usually purely oral with the soft palate raised; that is, the air is channeled through the mouth rather than the nose.

Speech involves the intermittent release of expired air between the adducted vocal folds. *Singing* a note requires a more prolonged release of the ex-

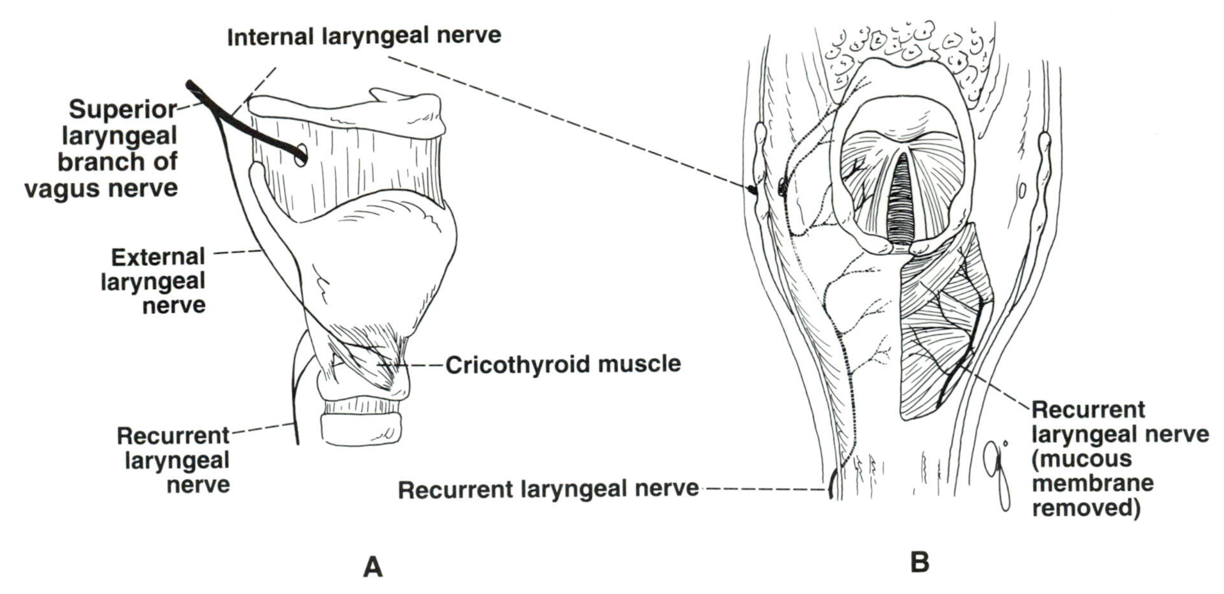

FIG 1–24.
A, lateral view of larynx showing the internal and external laryngeal branches of the superior laryngeal branch of the vagus nerve. **B,** the distribution of the terminal branches of the internal and recurrent laryngeal nerves. The larynx is viewed from above and posteriorly.

pired air between the adducted vocal folds. In *whispering,* the vocal folds are adducted, but the arytenoid cartilages are separated; the vibrations are given to a constant stream of expired air that passes through the posterior part of the rima glottidis.

Mucous Membrane of the Larynx

The mucous membrane of the larynx lines the cavity and is covered with ciliated columnar epithelium. On the vocal folds, however, where the mucous membrane is subject to repeated trauma during pho-

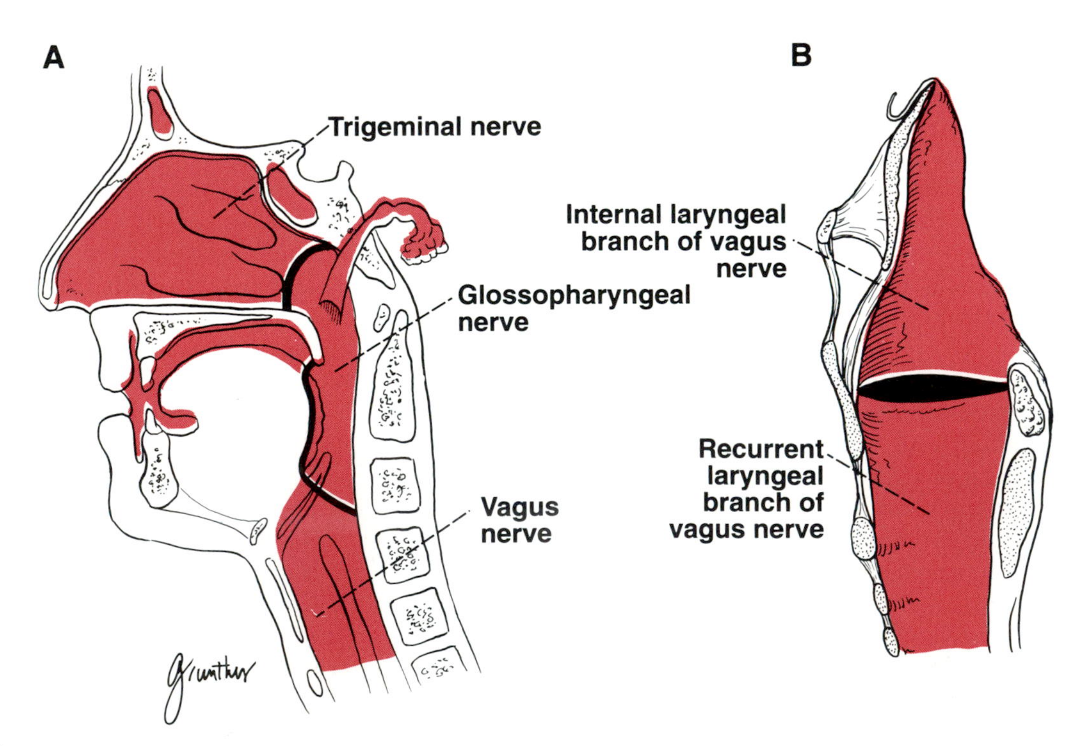

FIG 1–25.
A and **B,** sagittal section of the head and neck showing the sensory innervation of the mucous membrane of the different parts of the pharynx and larynx.

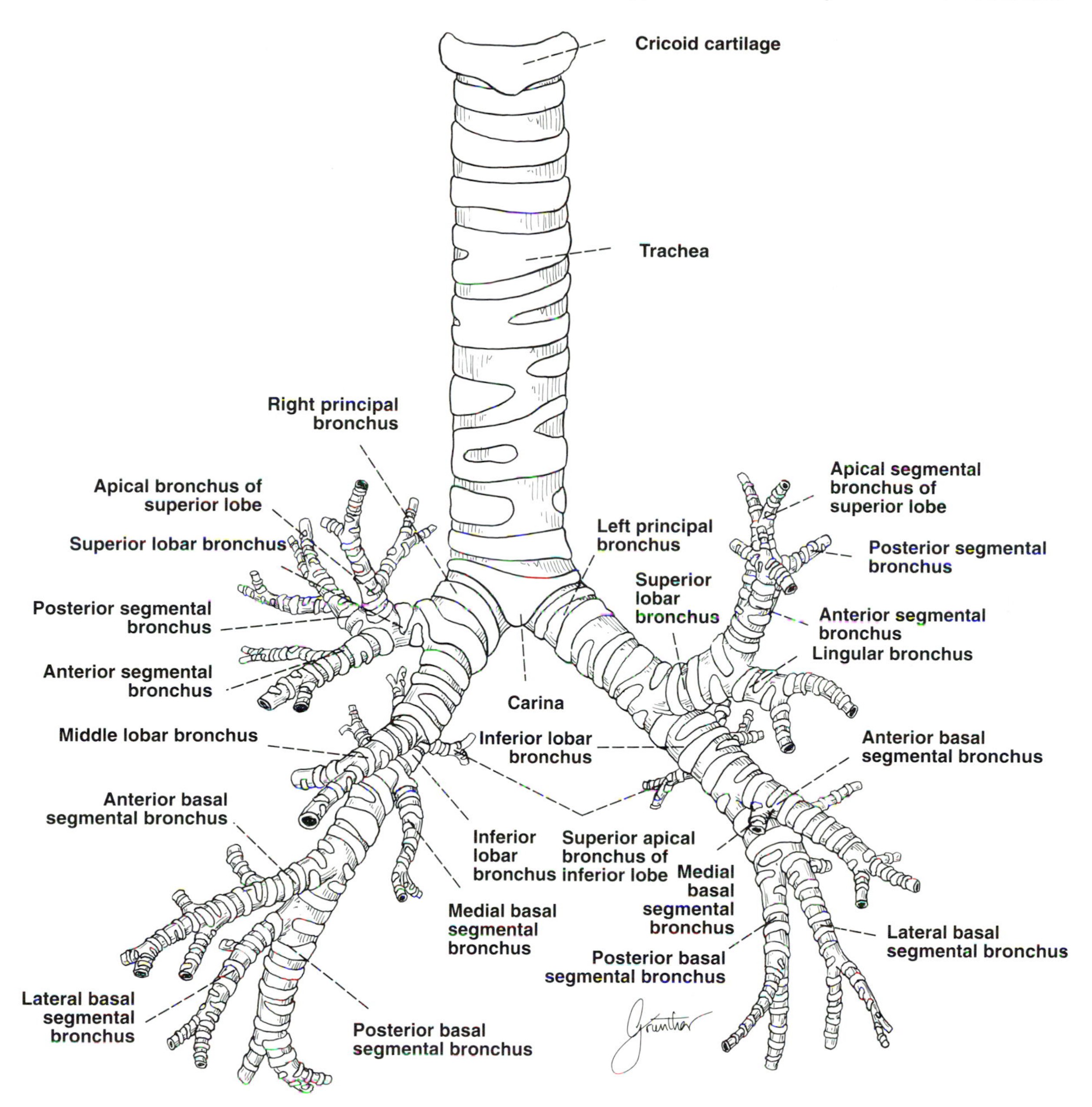

FIG 1–26.
The trachea and the bronchi.

nation, the mucous membrane is covered with stratified squamous epithelium.

Nerve Supply of the Larynx.—The *sensory nerve supply* to the mucous membrane above the vocal folds is the internal laryngeal nerve (Fig 1–24). Below the level of the vocal folds, the mucous membrane is supplied by the recurrent laryngeal nerve, a branch of the vagus (Fig 1–25). The *motor nerve supply* to the intrinsic muscles is the recurrent laryngeal nerve, except for the cricothyroid muscle, which is supplied by the external laryngeal nerve.

Blood Supply and Lymphatic Drainage of the Larynx.—The *arterial supply of the larynx* is from branches of the superior and inferior thyroid arteries. The *lymphatic drainage* is into the deep cervical lymph nodes.

Basic Anatomy of the Trachea

The trachea is a mobile cartilaginous and membranous tube (Fig 1–26). It begins as a continuation of the larynx at the lower border of the cricoid cartilage at the level of the sixth cervical vertebra. It descends

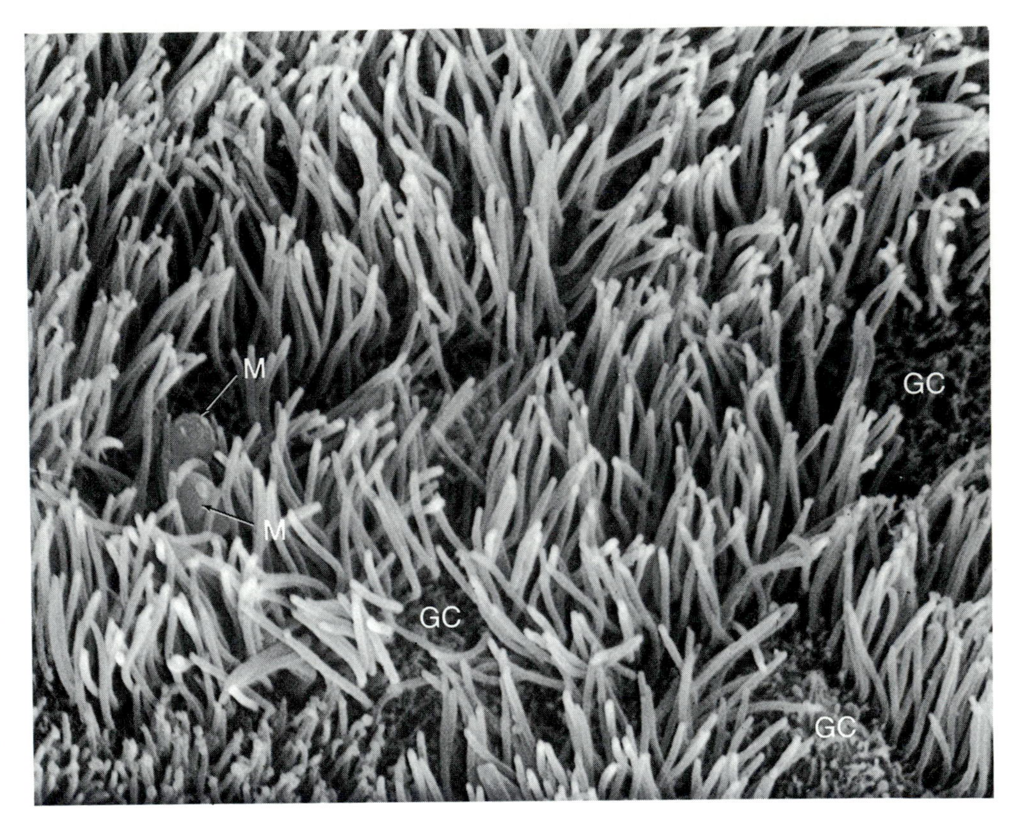

FIG 1–27.
Scanning electron micrograph of surface epithelial cells of the mucous membrane of the trachea showing ciliated columnar cells and goblet cells *(GC)*. *M* = droplet of mucus (× 2,400). Courtesy of Dr. M. Koering.

in the midline of the neck and in the thorax ends at the *carina* by dividing into right and left principal (main) bronchi at the level of the sternal angle (opposite the disc between the fourth and fifth thoracic vertebrae), where it lies slightly to the right of the midline. During expiration the bifurcation rises by about one vertebral level, and during deep inspiration may be lowered as far as the sixth thoracic vertebra, a distance of about 3 cm.

In adults the trachea is about 4½ in. (11.25 cm) long and 1 in. (2.5 cm) in diameter. In infants the trachea is about 1.6 to 2 in. (4 to 5 cm) long and may be as small as 3 mm in diameter. As children grow the diameter in millimeters corresponds approximately to their age in years. The fibroelastic tube is kept patent by the presence of U-shaped cartilaginous bars (rings) of hyaline cartilage embedded in its wall. The posterior free ends of the cartilage are connected by smooth muscle, the *trachealis muscle.*

The mucous membrane of the trachea is lined with pseudostratified ciliated columnar epithelium (Fig 1–27) and contains many goblet cells and tubular mucous glands.

Relations in the Neck

Anteriorly.—Skin, fascia, isthmus of the thyroid gland (in front of the second, third, and fourth rings), inferior thyroid veins, jugular arch, thyroidea ima artery (if present), and the left brachiocephalic vein in children, overlapped by the sternothyroid and sternohyoid muscles.

Posteriorly.—Right and left recurrent laryngeal nerves, esophagus.

Laterally.—Lobes of the thyroid gland and the carotid sheath and contents.

Relations in the Superior Mediastinum of the Thorax

Anteriorly.—Sternum, thymus, left brachiocephalic vein, origins of brachiocephalic and left common carotid arteries, arch of aorta.

Posteriorly.—Esophagus and left recurrent laryngeal nerve.

Right Side.—Azygos vein, right vagus nerve, and the pleura.

Left Side.—Arch of aorta, left common carotid, left subclavian arteries, left vagus nerve, left phrenic nerve, and pleura.

Nerve Supply of the Trachea.—The sensory nerve supply of the upper half of the trachea is from the vagi and the recurrent laryngeal nerves. Sympathetic nerves arise from the sympathetic trunks, and parasympathetic fibers are carried by branches of the vagi.

Blood Supply of the Trachea.—The upper two thirds is supplied by the inferior thyroid arteries and the lower third is supplied by the bronchial arteries.

Lymphatic Drainage of the Trachea.—Into the pretracheal and paratracheal lymph nodes and the deep cervical nodes.

Basic Anatomy of the Bronchi

The trachea bifurcates behind the arch of the aorta into the right and left principal (primary, or main) bronchi (see Fig 1–26). The right bronchus leaves the trachea at an angle 25° from the vertical, and the left bronchus leaves the trachea at an angle 45° from the vertical. In children younger than 3 years, both bronchi arise from the trachea at equal angles.

Clinical Note

Aspiration of foreign bodies and stomach contents.—In adults foreign bodies and stomach contents tend to be aspirated into the right principal bronchus, since this is more in line with the trachea than the left bronchus. In young children, since both bronchi arise from the trachea at equal angles, no predilection for the right bronchus exists.

The bronchi divide dichotomously, giving rise to several million terminal bronchioles that terminate in one or more respiratory bronchioles (see Fig 1–26). Each respiratory bronchiole divides into 2 to 11 alveolar ducts that enter the alveolar sacs. The alveoli arise from the walls of the sacs as diverticula.

Right Principal Bronchus

The right principal bronchus is wider, shorter, and more vertical than the left bronchus and is about 1 in. (2.5 cm) long (see Fig 1–26). The azygos vein arches over its superior border. The *superior lobar bronchus* arises within 2 cm of the commencement and is level with the carina. The right principal bronchus then enters the hilum of the right lung, where it divides into a *middle and an inferior lobar bronchus.*

Left Principal Bronchus

The left principal bronchus is narrower, longer, and more horizontal than the right bronchus and is about 2 in. (5 cm) long (see Fig 1–26). It passes to the left below the arch of the aorta and in front of the esophagus. On entering the hilum of the left lung it divides into a *superior and an inferior lobar bronchus.*

Bronchopulmonary Segments

The bronchopulmonary segments will be discussed with the structure of the lungs in Chapter 2.

Radiographic Appearances of the Airway in the Neck

The radiographic appearances of the larynx and trachea are shown in a lateral radiograph of the neck (see Fig 5–18).

SURFACE ANATOMY

Surface Anatomy of the Airway in the Neck

Preferably the neck should be palpated while the patient is in a supine position, when the muscles overlying the deeper structures are relaxed and the structures become easier to feel.

In the midline anteriorly, the following structures may be palpated from above downward.

The *symphysis menti* may be felt where the two halves of the body of the mandible unite in the midline (Fig 1–28).

The *body of the hyoid bone* lies opposite the body of the third cervical vertebra (Figs 1–28 and 1–29). The hyoid bone is a horseshoe-shaped structure, and the greater cornua can be felt on each side of the neck between the finger and thumb. The hyoid bone moves superiorly when the patient swallows.

The *thyrohyoid membrane* occupies the interval between the hyoid bone and the thyroid cartilage (see Fig 1–29).

The notched upper border of the *thyroid cartilage* lies opposite the fourth cervical vertebra (see Figs 1–28 and 1–29). The anterior border of the thyroid cartilage is more prominent in male adults than in female adults.

The *cricothyroid ligament* fills in the interval between the cricoid cartilage and the thyroid cartilage (see Fig 1–29).

The *cricoid cartilage* lies at the level of the sixth cervical vertebra at the junction of the larynx with

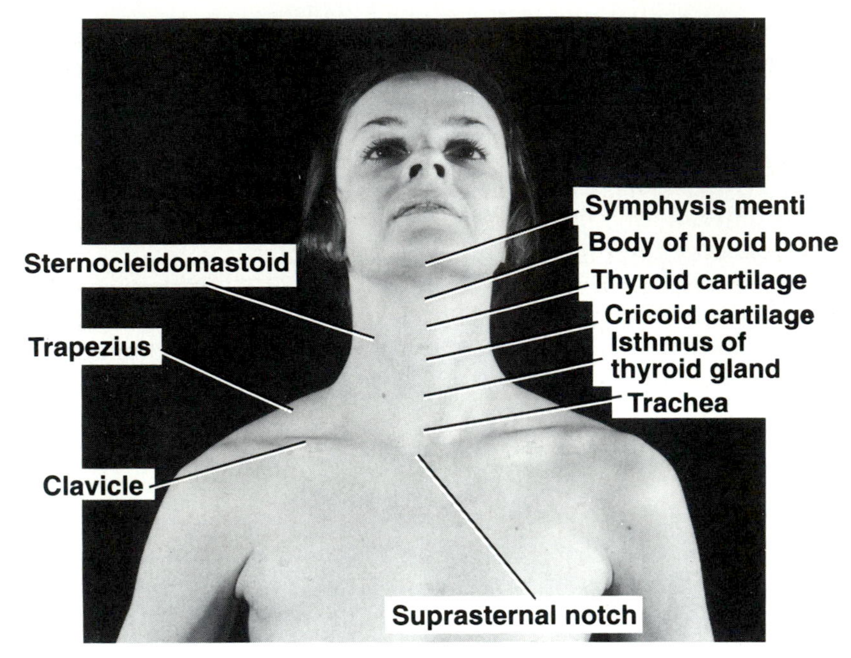

FIG 1–28.
Anterior view of the head and the neck showing important surface landmarks.

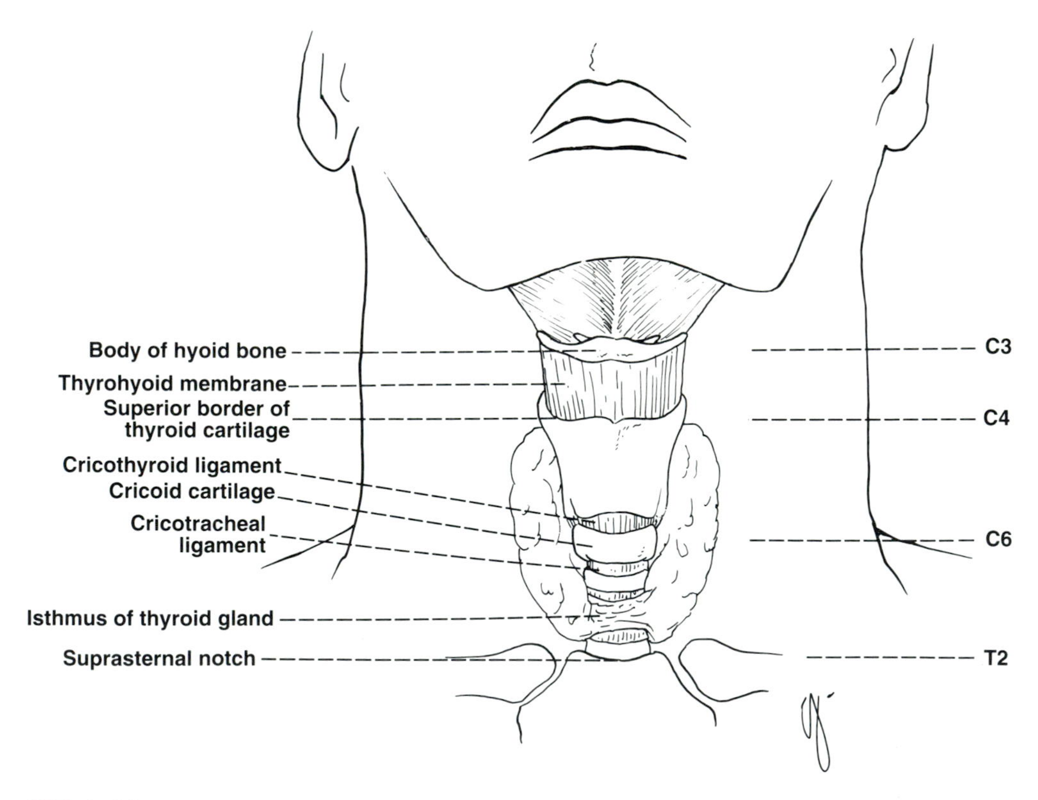

FIG 1–29.
Head and neck of an adult showing the vertebral levels of different parts of the larynx.

the trachea (see Figs 1–28 and 1–29). It is not as prominent as the thyroid cartilage, but it can be identified with gentle palpation when the patient is asked to swallow and the cartilage rises in the neck. In the unresponsive patient, it can be identified as the first cartilaginous ring below the thyroid cartilage.

The *cricotracheal ligament* fills in the interval between the cricoid cartilage and the first ring of the trachea (see Fig 1–29). It can be recognized with gentle palpation.

The *first ring of the trachea* can usually be identified by careful palpation.

The *isthmus of the thyroid gland* can be recognized as a soft structure crossing in front of the second, third, and fourth rings of the trachea (see Figs 1–28 and 1–29).

The *inferior thyroid veins* and the *thyroidea ima artery* (when present), although not palpable, lie in front of the fifth, sixth, and seventh rings of the trachea. The trachea is now receding from the surface as it approaches the root of the neck.

The *brachiocephalic artery*, the *left brachiocephalic vein*, the *thymus gland*, and even the upper margin of the *arch of the aorta* are sometimes present in front of the trachea just above the suprasternal notch in young children.

The *jugular arch* connects the two anterior jugular veins just above the suprasternal notch.

The *suprasternal notch*, which is the upper margin of the manubrium sterni, can be felt between the anterior ends of the clavicles (see Figs 1–28 and 1–29). It lies opposite the lower border of the body of the second thoracic vertebra in the mid–respiratory position.

The *trachea* in the neck lies in the midline, and the examiner should be able to confirm this by inserting the index and middle finger into the grooves on either side of the trachea, between the trachea and the sternocleidomastoid muscles.

In infants many of the important anatomical structures in the neck, referred to above, lie at different vertebral levels from those found in adults (Fig 1–30). Also, in adults the trachea may measure as much as 1 in. (2.5 cm) in diameter, whereas in infants it is much smaller (3 mm).

Surface Anatomy of the Paranasal Sinuses

The maxillary, frontal, and ethmoidal sinuses can be palpated clinically for areas of tenderness. The maxillary sinus can be examined by pressing a finger against the anterior wall of the maxilla below the in-

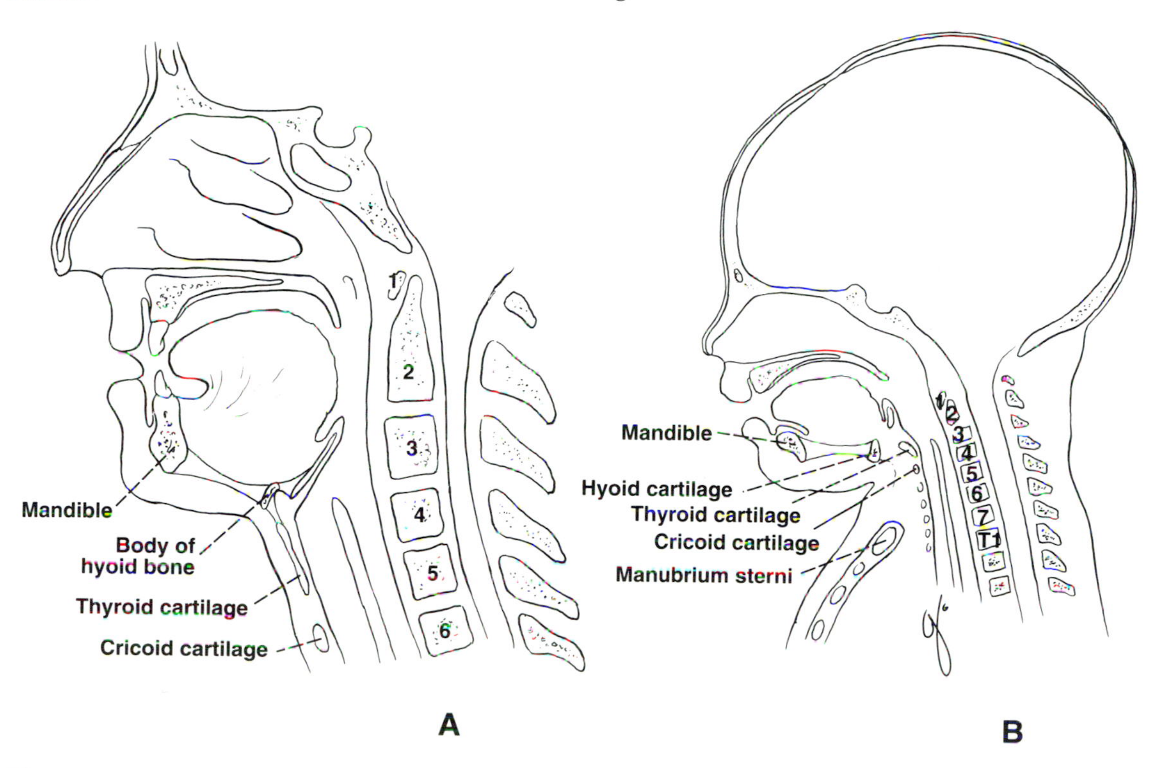

FIG 1–30.
Sagittal sections of the neck of an adult *(A)* and an infant *(B)* shortly after birth. Different vertebral levels in these age groups are shown.

ferior orbital margin; pressure over the infraorbital nerve may reveal increased sensitivity. Because the bony walls are relatively thin, directing a beam of a flashlight either through the roof of the mouth or through the cheek in a darkened room will enable the examiner to determine whether the maxillary sinus is full of inflammatory exudate rather than air.

The frontal sinus can be examined by pressing a finger upward beneath the medial end of the superior orbital margin. Here the relatively thin floor of the frontal sinus is closest to the surface. In a similar manner, the ethmoidal sinuses can be palpated by pressing a finger medially against the medial wall of the orbit.

CLINICAL ANATOMY

Clinical Anatomy of the Nose

Nasal Obstruction

Nasal obstruction can be caused by edema of the mucous membrane secondary to infection, foreign bodies lodged between the conchae (in children), or swelling of the mucous membrane secondary to trauma. Other causes include tumors, polyps, and septal abscesses.

Deflection of the nasal septum is common. It is believed to occur most commonly in males because of trauma in childhood.

The most voluminous part of the nasal cavity is close to the floor, and it is usually possible to pass a well-lubricated tube through the nostril along the inferior meatus into the nasopharynx.

Epistaxis

This common disorder occurs in children, adolescents, and the elderly. Nosebleed is often caused by a spontaneous erosion of one of the superficial mucosal blood vessels near the anterior end of the nasal septum *(Kiesselbach's area)*. The mucous membrane in the anterior part of the nose is particularly susceptible to drying, especially during winter, when the humidity is low because of forced-air heat. Nosebleeds in the posterior part of the nose are commonly seen in the elderly with a history of hypertension and atherosclerotic disease.

The blood supply to the posterior part of the nose is mainly from the sphenopalatine branch of the large maxillary artery. Bleeding from this area tends to spurt, whereas the blood supply to the anterior part of the nose is from branches of smaller arteries (sphenopalatine and facial) and bleeding tends to ooze.

Epistaxis is almost always unilateral in origin, but if the source is posterior, blood can run around the posterior margins of the choanae and emerge through the other nostril.

When an anterior pack is used in an attempt to control bleeding, remember that the vestibule extends predominantly posteriorly and not superiorly.

Beware of *bilateral cauterization* of the nasal septum. It could compromise the blood supply to the perichondrium and cause necrosis of the cartilaginous part of the septum. The blood supply of the septum can also be compromised in posttraumatic septal hematoma unless it is quickly detected and drained.

The great vascularity and delicacy of the nasal mucous membrane dictates that extreme care should be exerted in passing a nasal tube; a severe hemorrhage in this area is difficult to treat. To avoid unnecessary bleeding, a good topical anesthesia with the addition of a vasoconstrictor is helpful; if the nasal cavity is still obstructed, the passageway can be further dilated with progressively larger soft nasal airways.

Anatomy of Procedures for the Nose

Anesthetizing the Nose

The mucous membrane of the nose is best anesthetized with topical anesthesia mixed with oxymetazoline (Afrin) or another vasoconstrictor. The anatomy of the procedure is described in Chapter 19, p. 814.

The skin of the nose may be infiltrated with a local anesthetic with epinephrine as described in Chapter 19, p. 814.

Repair of Nasal Lacerations

Skin Lacerations.— Lacerations confined to the skin are sutured in the usual way. However, remember the following points. Since there is very little excess skin on the nose, this restricts the removal of vascularly compromised tissue to the minimum. Avoid making incisions across depressed areas of skin, such as on the side of the nose or the nasolabial junction, or the lower eyelid, since future scars tend to contract and distort the depression.

Lacerations of the alar nasi are sutured in layers starting with the nasal lining and followed by the skin.

Septal Lacerations.— The normal anatomical alignment is restored and the edges of the nasal cartilage are approximated with sutures if necessary.

Clinical Anatomy of Nasotracheal Intubation

Nasotracheal intubation is used in breathing patients who need to be intubated, in patients who have possible cervical spine injuries, in those with certain airway tumors, and in conditions in which oral intubation would be difficult. Nasotracheal intubation is made easier using topical anesthesia and judicious sedation. It is usually routine, but sometimes the procedure can be very difficult or impossible because of the presence of a deflected nasal septum, bony spurs in the nasal cavity, or a distorted laryngeal anatomy.

The nose, pharynx, and larynx are anesthetized. Sometimes it is desirable to anesthetize the tissues around the entrance into the larynx by blocking the superior laryngeal nerve on both sides. The internal laryngeal branch of the superior laryngeal nerve pierces the thyrohyoid membrane and supplies the mucous membrane of the larynx down as far as the vocal cords (see Fig 1–24). The internal laryngeal nerve is easily blocked as it pierces the thyrohyoid membrane.

With the patient in a supine position and the head slightly extended, the superior cornu of the thyroid cartilage and the greater cornu of the hyoid bone may be palpated on the side of the neck. The anesthetic needle may be inserted midway between these structures to pierce the thyrohyoid membrane. The anesthetic is then injected around the internal laryngeal nerve.

The intubation can be performed in a recumbent or sitting position. The head is best held in the "sniffing" position, since overextension causes the tube to enter the mouth and excessive flexion causes it to enter the esophagus. The tip of the tube is directed toward the less sensitive nasal septum rather than the sensitive chonchae. The tube is passed through the nostril and directed backward along the widest part of the nasal cavity close to the floor.

With the tube passed through the nose, the operator senses that it "gives" as it enters the laryngeal part of the pharynx. Breath sounds should now be easily heard over the open end of the tube. When the breath sounds are maximally audible, the end of the tube is opposite the laryngeal inlet. With one hand the tube is advanced into the larynx, while the other hand feels the larynx for the entrance of the tube within its cavity. The tube is passed through the glottis during inspiration (when the vocal cords are abducted).

To avoid having the tube enter the esophagus, since its entrance lies immediately behind the larynx (see Fig 1–12), it is sometimes helpful to press the larynx posteriorly. Should the tube enter the esophagus the breath sounds cease but passage of the tube is still possible. Should the tube enter the vallecula or piriform fossa (see Fig 1–20), the breath sounds also cease but further passage of the tube is impossible. By appropriate movement of the head and neck, it is possible to alter the position of the advancing tip of the tube so that it enters the larynx and the trachea.

Nasotracheal intubation in infants has certain anatomical differences. The air passages are smaller and the mucous membrane is softer, looser, and more fragile. The lymphatic tissue of the adenoids and tonsils may increase the risk of obstruction. The epiglottis is more vertical in children and the larynx is higher in the neck (see Fig 1–30), lying anterior to the C3 and C4 vertebrae. The lumen within the cricoid cartilage is the narrowest part of a child's larynx (in adults, the glottic opening is the narrowest part of the larynx).

Clinical Anatomy of the Paranasal Sinuses

Sinusitis

Infection of the paranasal sinuses is a common complication of nasal infections. Inflammatory edema of the mucous membrane lining the orifices of the sinuses results in their blockage and predisposes to infection of the stagnant secretions. Frequently maxillary sinusitis is caused by extension from an apical dental abscess.

The maxillary sinus is particularly susceptible to infection because its drainage orifice through the hiatus semilunaris is vulnerably positioned near the roof of the sinus. In other words, the sinus has to fill with fluid before it can effectively drain when the person maintains an upright position.

The frontal sinus drains into the hiatus semilunaris close to the orifices of the ethmoidal and maxillary sinuses on the lateral wall of the nose. Thus, it is not surprising that a patient with frontal sinusitis nearly always has a maxillary sinusitis, or even an ethmoidal sinusitis.

The extreme thinness of the medial wall of the orbit relative to the ethmoidal air cells must be emphasized. Ethmoidal sinusitis is the commonest cause of orbital cellulitis. The infection can easily spread through the paper-thin bone.

Referred Pain from the Paranasal Sinuses

The maxillary sinus is innervated by the infraorbital nerve and the anterior, middle, and posterior superior alveolar nerves. Pain from the sinus is referred to the upper jaw, including the teeth, as well as to the skin of the cheek.

The frontal sinus is innervated by the supraorbital nerves, which also supplies the skin of the forehead and scalp as far back as the vertex. Therefore it is not surprising that patients with frontal sinusitis have pain referred over this area.

Blow-out Fractures of the Maxilla

A severe blow to the eyeball drives it posteriorly, causing a sudden rise in intraorbital pressure. This may result in a blow-out fracture through one of the two weakest points of the orbit—the floor of the orbit, which is the roof of the maxillary sinus, or the medial wall of the orbit, which is formed by the weak ethmoid bone. The blow-out fracture causes herniation of the orbital contents through the defect and may result in entrapment of the extraocular muscle. Fractures involving the floor of the orbit not only cause diplopia but may damage the infraorbital nerve in the roof of the maxillary sinus, resulting in loss of sensation to the cheek and upper lip.

Barotrauma of the Sinuses

If the communication with the nose through the meatus is blocked by swelling of the mucous membrane secondary to infection, a paranasal sinus is subject to barotrauma. This may occur during the descent of an aircraft, when the pressure in the cabin may increase and create a relative vacuum in the paranasal sinuses. Because of the blockage of the meatus, equilibration of pressure through the nose with the outside is impossible. The result is implosion and tearing of the mucoperiosteum, with subsequent hemorrhage.

Clinical Anatomy of the Mouth and Tongue

Lips and Vestibule and Facial Paralysis

Asymmetry of the lips and paralysis of the buccinator with a tendency to accumulate saliva and food in the vestibule indicate a lesion of the facial nerve on that side.

Tongue and Airway Obstruction

In an unconscious patient, there is a tendency for the tongue to fall backward and obstruct the laryngeal opening. This is caused by the loss of tone of the extrinsic muscle and, unless quickly corrected "with a jaw thrust or chin lift maneuver," will lead to all of the sequelae of airway obstruction.

Anatomy of Procedures

Pulling the Tongue Forward in Airway Obstruction

The head should be extended at the atlanto-occipital joint and the neck flexed at the C4 to C7 joints. The extended head stretches the fascia and muscles of the front of the neck and causes a forward and downward movement of the mandible that is correctable by placing a finger below the symphysis menti and pulling the mandible forward and up. Sometimes this is inadequate to relieve the obstruction and should be supplemented by placing the fingers behind the angles of the mandible and exerting forward pressure. This moves the mandible forward, causing displacement of the tongue away from the laryngeal opening, since the mandible is attached to the tongue by the genioglossus muscles.

Oral Endotracheal Intubation

Total visualization of the glottis with a laryngoscope is not necessary for endotracheal intubation. If the epiglottis is visible, the tube is laid on the laryngeal side of the epiglottis and advanced along its surface. Often this procedure alone will allow the tube to go into the trachea. If only the esophagus is visible, and not the vocal cords, the endotracheal tube can be placed "blindly" just anterior to the esophageal opening. Occasionally when the tube is caught at the anterior glottic constriction, the head should be flexed slightly, allowing the pressure of the tongue to displace the endotracheal tube posteriorly and hence move it into the opening of the glottis. Frequently this maneuver has to be supplemented by turning the head slightly to one side or another. The use of styleted endotracheal tubes also may help in this situation. "Trigger tubes" may be used, which allow the tip to be manipulated from above.

When oral endotracheal intubation is impossible in the above situations, nasotracheal intubation may be successful, since the tube approaches the glottis slightly more posteriorly and is directed more towards it.

Oral Endotracheal Intubation and the Incisor Teeth

Interference with endotracheal intubation may be caused by the presence of protruding incisor teeth, often making it necessary to put the endotracheal tube in an extreme lateral position to approach the glottis.

Oral Endotracheal Intubation and the Small Mandible

Patients with receding jaws, secondary to a small mandible, often make intubation difficult, and in some cases the nasal route or a lighted stylet or digital intubation must be used. However, since this anatomic configuration approaches the picture seen in younger children, many times a small straight blade such as a Miller no. 2 or Miller no. 3 can overcome the visual difficulties noted when a curved blade of the Macintosh type is used.

Clinical Anatomy of the Pharynx

Lymphoid Tissue of the Pharynx

Enlargement of the lymphoid tissue of the pharynx may seriously encroach on the airway, especially in children. The pharyngeal tonsil and the palatine tonsils reach their maximum normal size in early childhood. After puberty, together with other lymphoid tissues in the body, they gradually atrophy.

Peritonsillar Abscess

Occasionally an infected tonsil may be complicated by a peritonsillar abscess or quinsy (see Fig 1–16). The infection extends laterally through the tonsillar capsule into the loose connective tissue of the tonsillar bed. The swelling is usually situated anterior and superior to the tonsil just posterior to the palatoglossal arch. The bright red swelling pushes the uvula to the opposite side. The deep cervical lymph nodes, especially the node just below and behind the angle of the jaw, are enlarged and tender. The intense pain causes the patient to resist the opening of the mouth widely during examination of the pharynx.

Anatomy of Complications

Lateral Extension.— Through the superior constrictor muscle to enter the fascial spaces of the neck, to enter the carotid sheath, to erode the internal carotid artery, or cause thrombophlebitis of the internal jugular vein (see Fig 1–16).

Medial Extension.— Rupture into the oral pharynx causing airway obstruction and aspiration of pus.

Anatomy of Drainage of Peritonsillar Abscess

(1) The abscess may be aspirated at the point of maximal swelling or it may be incised with a scalpel. Aspirating first confirms the presence of pus and helps prevent inadvertent damage to the internal carotid artery. Note the closeness of this artery (about 1 in. [2.5 cm]) lateral to a normal tonsil and its location just medial to the superior constrictor muscle. (2) Because of the danger of aspiration of infected material, the patient is sitting and suction is used.

Mechanism of Swallowing

During swallowing in conscious individuals food and fluid cross naturally from the mouth to the esophagus, and movement of air from the nose to the larynx is momentarily stopped. In unconscious individuals, when the reflex mechanisms are not functioning, it is possible for food and fluid to enter the bronchial tree or air to enter the stomach. Moreover, should vomiting occur, the regurgitated gastric contents may be inhaled into the lungs (see below).

Pharyngeal Obstruction of the Upper Airway

This condition frequently occurs in patients during cardiopulmonary arrest or in the decreased level of consciousness that accompanies a major cerebrovascular accident or drug overdose. The obstruction is caused when the atonic tongue falls back and the pharyngeal wall caves in due to loss of tone of the pharyngeal muscles. The obstruction may clear if the patient is placed in the lateral decubitus position, with the neck extended and the jaw pulled forward (which pulls the tongue forward). If the patient must lie in a supine position, an oropharyngeal or nasopharyngeal airway may have to be inserted to counteract the flaccid pharyngeal walls.

Loss of the Gag Reflex

In conscious patients the airway is protected by a number of important reflexes, including the gag reflex, the laryngeal reflex, and the cough reflex. The gag or swallowing reflex occurs in response to stimulation of the pharyngeal mucous membrane, which is innervated by the glossopharyngeal nerve. The laryngeal and cough reflexes (trachea and bronchi) are mediated via the vagus nerve. These protective reflexes are lost in descending order as the patient becomes less and less responsive. In these circumstances the airway may be blocked by aspiration of vomit and gastric and pharyngeal secretions.

Clinical Anatomy of the Larynx

Larynx in Children

In children the neck is shorter and the larynx is more cephalad than in adults (see Fig 1–30). At birth the cricoid cartilage lies at the level of the fourth cervical vertebra, and only at the age of 6 years does it lie opposite the sixth cervical vertebra. The glottis at birth lies opposite the second cervical vertebra.

The epiglottis is U-shaped and less flexible in children, which sometimes makes it difficult to line up the oral, pharyngeal, and tracheal axes when passing a laryngoscope.

The rima glottidis tends to be more anterior in children than in adults. The vocal folds in children have thicker submucosa, so that edema of the folds is more likely to occlude the glottis.

As mentioned previously, the cavity of the larynx is narrowest within the cricoid ring in children, whereas the glottis is the narrowest part of the cavity in adults.

Epiglottitis

An acute inflammatory swelling of the mucous membrane of the epiglottis can compromise the upper airway, which is accentuated by the increased secretions of the mucous glands. The inflammation may spread rapidly in the loosely arranged submucosa down to the vocal cords (supraglottitis). Here the spreading stops because of the mucosa that tightly adheres to the underlying vocal ligaments. The condition is most often seen in children where the very loose and abundant submucosa together with the narrow passageway quickly leads to upper airway obstruction.

Subglottitis

This is an acute inflammatory swelling of the mucous membrane of the larynx below the vocal cords. The inflammation may spread rapidly in the loose submucosa to involve the trachea and large bronchi. It does not spread superiorly to the supraglottic region because of the adherence of the mucous membrane to the vocal ligaments. The airway in children is normally very narrow within the cricoid cartilage. The swollen mucosa together with the increased production of mucus quickly leads to airway obstruction. In newborns 1 mm of circumferential swelling produces about 75% reduction in cross-sectional area of the larynx. In adults the larger size may permit adequate airflow in this condition.

Foreign Bodies in the Airway

The laryngeal and cough reflexes mediated through the vagus nerves are the natural defense mechanisms for expelling foreign bodies from the airway at all ages. If coughing is successfully freeing the obstruction, it should be encouraged to continue. If intervention is necessary, anatomical and physiological age differences dictate treatment.

Anatomical Rationale For Differences in Procedures for Removing Foreign Bodies in Adults and Children.—It is generally agreed that all maneuvers are directed toward the increase in intrathoracic pressure by compressing the intrathoracic gas volume to expel the foreign body from the airway. For children older than 1 year and for adults, the abdominal thrust *(Heimlich maneuver)* should be used. The rapid compression of the abdominal viscera suddenly forces the diaphragm into the thoracic cavity. In infants, the relatively large size of the liver and the delicate structure of the abdominal viscera generally preclude its use. Children younger than 1 year should be placed face down over the rescuer's arm, with the head lower than the trunk, and measured back blows should be delivered between the scapulae. If this fails to open the airway, they should be rolled over, and four rapid sternal compressions should be administered.

It is now accepted that sudden blows to the back in the older age groups, especially in the standing or sitting position, extends the thoracic part of the vertebral column and may displace the foreign body further down the airway, leading to impaction or complete obstruction.

Inspection of the Vocal Cords

The interior of the larynx may be inspected indirectly through a laryngeal mirror passed through the open mouth into the oral pharynx (Fig 1–31). An alternate more satisfactory method is to use either the direct or fiberoptic laryngoscope (see Fig 1–31). The left-to-right and anterior-to-posterior orientation provided by these two views is different and can be confusing to the physician (see Fig 1–31).

The neck is brought forward on a pillow, and the head is fully extended at the atlanto-occipital joints (Fig 1–32). The illuminated instrument can then be introduced into the larynx over the back of the tongue. The valleculae, the piriform fossae, the epiglottis, and the aryepiglottic folds are clearly seen (see Fig 1–20). The two elevations produced by the corniculate and cuneiform cartilages can be recognized. Within the larynx, the vestibular folds and the vocal folds can be seen (see Fig 1–31). The former are fixed, widely separated, and reddish; the latter move with respiration and are white. With quiet breathing, the rima glottidis is triangular, with the apex in front. With deep inspiration, the rima glottidis assumes a diamond shape because of the lateral rotation of the arytenoid cartilages.

If the patient is asked to breathe deeply, the vocal cords become widely abducted, and the inside of the trachea can be seen.

Paralysis of the Vocal Cords

The muscles of the larynx are innervated by the recurrent laryngeal nerves, with the exception of the cricothyroid muscle, which is supplied by the external laryngeal nerve. Both these nerves are vulnera-

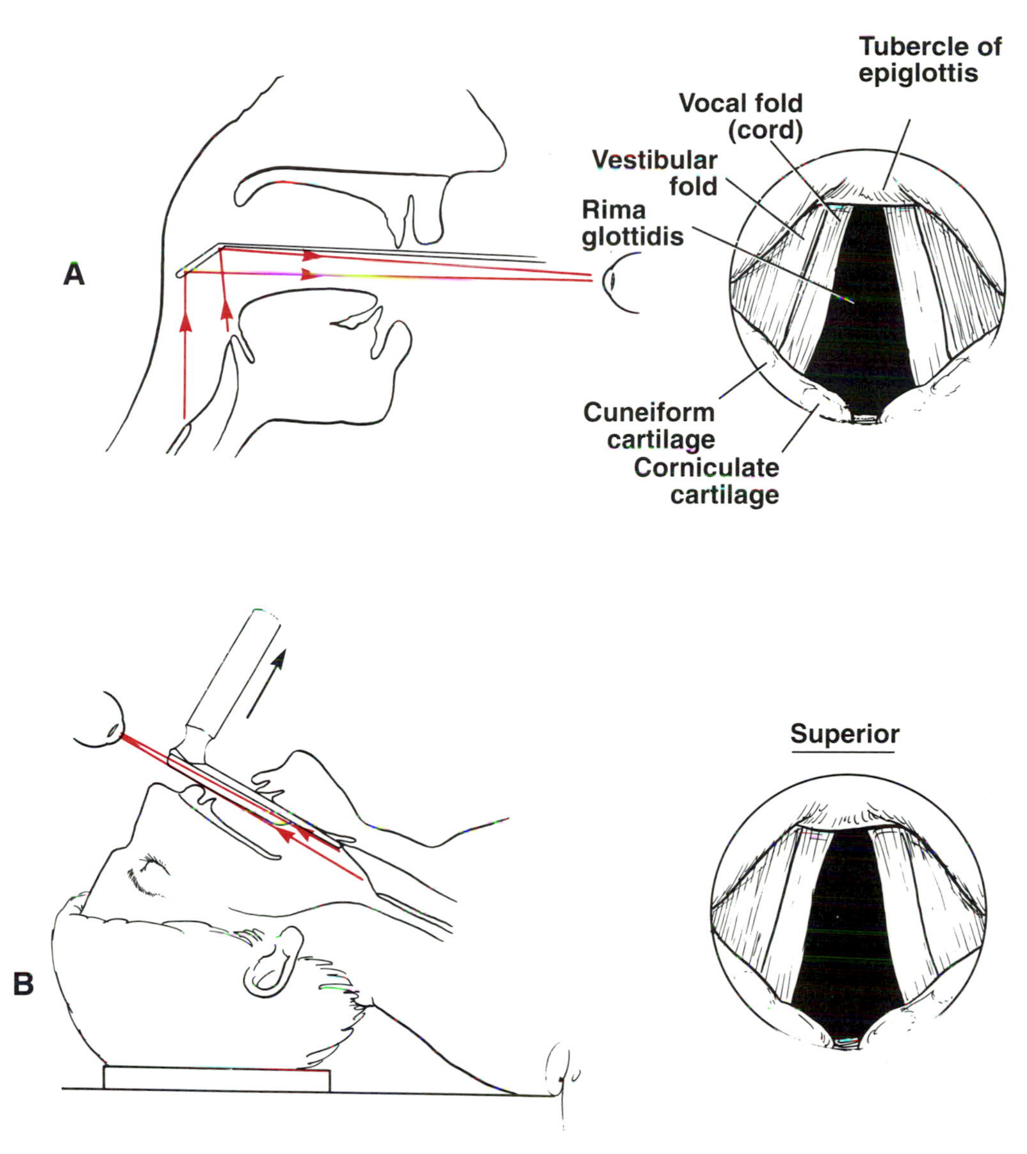

FIG 1–31.
Inspection of the vocal cords indirectly through a laryngeal mirror *(A)* or through a laryngoscope *(B)*. Note the orientation of the laryngeal inlet. A curved laryngeal blade is sometimes more helpful.

ble during operations on the thyroid gland because of the close relationship between them and the arteries of the gland. The external laryngeal nerve may be injured during ligation of the superior thyroid artery, and the recurrent laryngeal nerve may be damaged during ligation of the inferior thyroid artery. The left recurrent laryngeal nerve may be involved in a bronchial or esophageal carcinoma or in secondary metastatic deposits in the mediastinal lymph nodes. The right and left recurrent laryngeal nerves may be damaged by malignant involvement of the deep cervical lymph nodes.

Section of the external laryngeal nerve produces weakness and huskiness of the voice, since the vocal cord cannot be tensed (Fig 1–33). The cricothyroid muscle is paralyzed.

Unilateral complete section of the recurrent laryngeal nerve results in the vocal cord on the affected side assuming the position midway between abduction and adduction (see Fig 1–33). It lies just lateral to the midline. Speech is not greatly affected, since the other vocal cord compensates to some extent and moves toward the affected vocal cord.

Bilateral complete section of the recurrent laryngeal nerve results in both vocal cords assuming the position midway between abduction and adduction (see

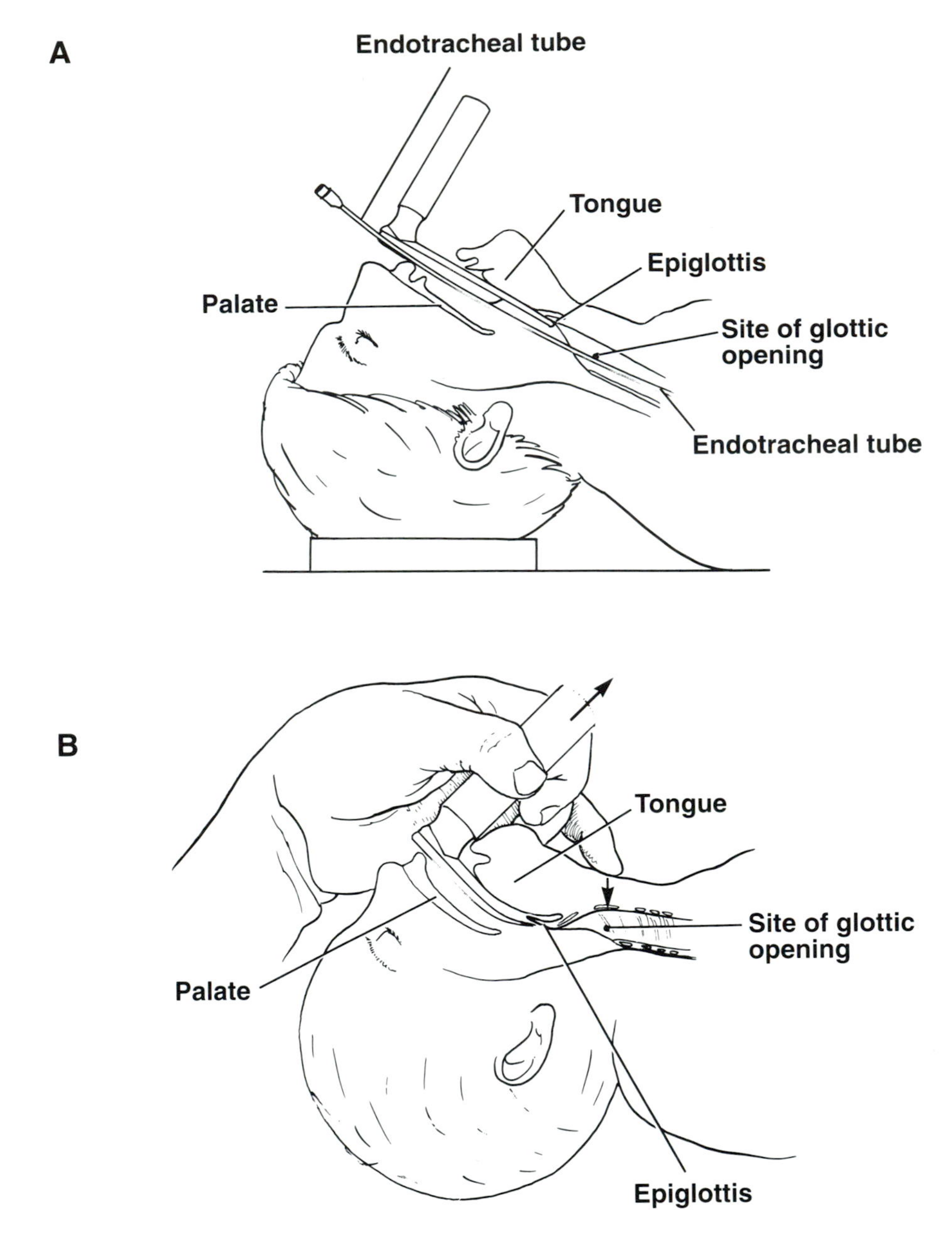

FIG 1–32.
A and **B**, direct laryngoscopy using straight or curved blades. Note that the tip of the blade lifts the epiglottis forward. In children up to the age of 9 years, a straight blade is commonly used. The relatively large size of the tongue, a floppy epiglottis, and a short neck make forward displacement of the tongue more difficult. Occlusion of the esophagus *(B)* by applying backward pressure to the cricoid cartilage may assist in the passage of an endotracheal tube (Sellick maneuver).

Fig 1–33). Breathing is impaired, since the rima glottidis is partially closed and speech is lost.

Unilateral partial section of the recurrent laryngeal nerve results in a greater degree of paralysis of the abductor muscles than the adductor muscles. The affected vocal cord assumes the adducted midline position (see Fig 1–33). This phenomenon has not been explained satisfactorily. It must be assumed that the abductor muscles receive a greater number of nerves than the adductor muscles, and thus partial damage of the recurrent laryngeal nerve results in damage to relatively more nerve fibers to the abductor muscles. Another possibility is that the nerve fibers to the abductor muscles are traveling in a more exposed position in the recurrent laryngeal nerve and are therefore more prone to being damaged.

Bilateral partial section of the recurrent laryngeal

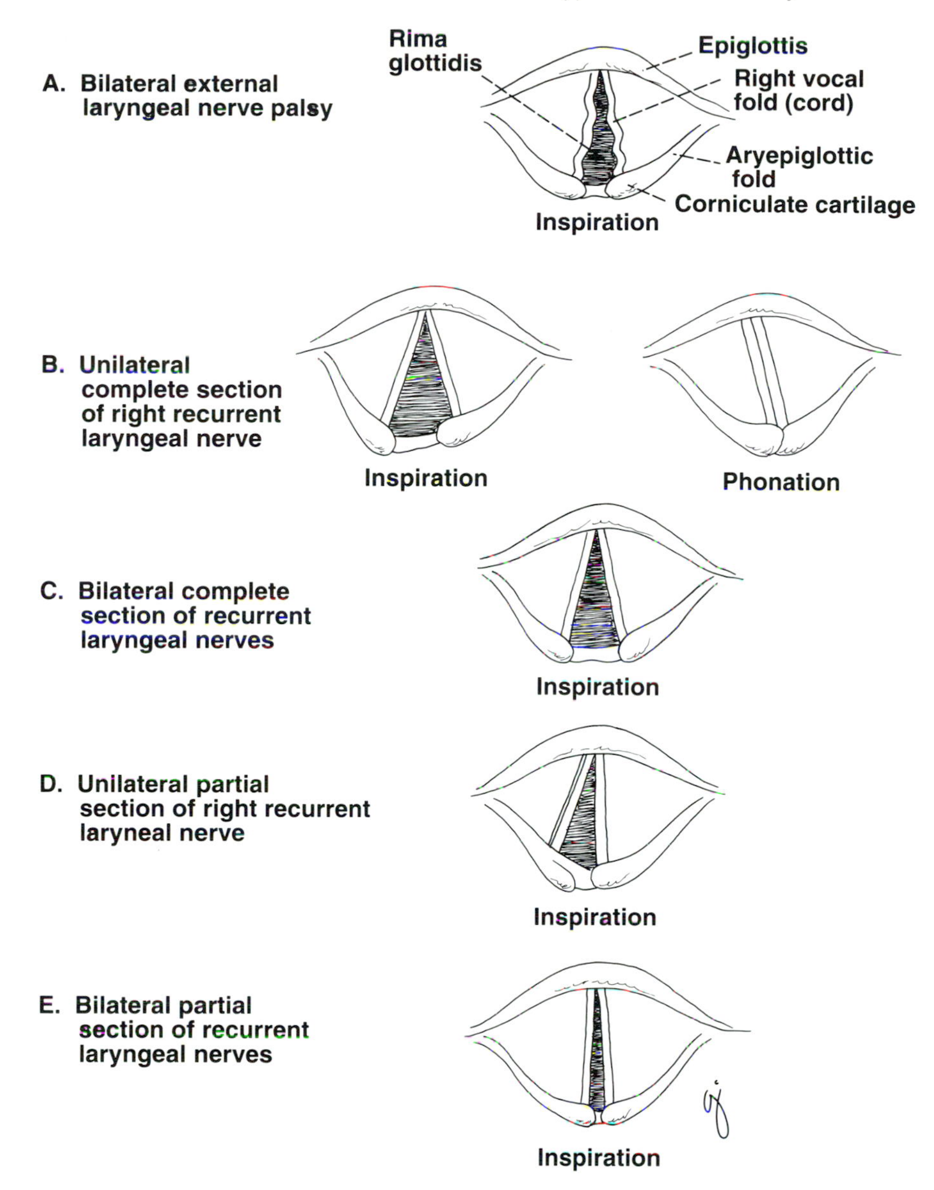

FIG 1–33.
The position of the vocal folds (cords) after damage to the external and recurrent laryngeal nerves.

nerve results in bilateral paralysis of the abductor muscles and the drawing together of the vocal cords (see Fig 1–33). Acute dyspnea and stridor follow, and tracheotomy may be necessary.

Clinical Anatomy of the Trachea

Some Practical Distances

Table 1–4 shows some important distances between the incisor teeth or external nares to anatomic landmarks in the airway in the adult. These approximate figures are helpful in determining the correct placement of an endotracheal tube.

TABLE 1–4.
Important Airway Distances*

Airway	Distance (cm)
Incisor teeth to the vocal cords	15
Incisor teeth to the carina	20
External nares to the carina	30

*Average figures given ± 1 to 2 cm.

Changes in the Tracheal Length With Respiration and Position of the Head and Neck

On deep inspiration the carina may descend by as much as 3 cm. Extension of the head and neck, as when maintaining an airway in an anesthetized patient, may stretch the trachea and increase its length by 25%.

Important Anatomical Axes for Endotracheal Intubation
The upper airway has three axes that have to be brought into alignment if the glottis is to be viewed adequately through a laryngoscope—the axis of the mouth, the axis of the pharynx, and the axis of the trachea (Fig 1–34).

The following procedures are necessary. (1) The head is extended at the atlanto-occipital joints. This brings the axis of the mouth into the correct position. (2) The neck is flexed at cervical vertebrae C4 to C7 by elevating the back of the head off the table, often with the help of a pillow. This brings the axes of the pharynx and the trachea in line with the axis of the mouth.

Reflex Activity Secondary to Endotracheal Intubation
Stimulation of the mucous membrane of the upper airway during the process of intubation may produce cardiovascular changes such as bradycardia and hypertension. These changes are largely mediated through the branches of the vagus nerves.

Anatomy of Cricothyroidotomy

This is performed in the interval between the thyroid cartilage and the cricoid cartilage (Fig 1–35). The trachea may be steadied and positioned more anteriorly by extending the neck over a sandbag.

Two techniques may be used—Vertical incision or Transverse incision. The incision is made through (1) the skin, (2) the superficial fascia (beware of the anterior jugular veins, which lie on either side of the midline close together), (3) the investing layer of deep cervical fascia, (4) the pretracheal fascia (separate the sternohyoid muscles and incise the fascia), and (5) the larynx, which is incised through a horizontal incision through the cricothyroid ligament. *Note* the position of the vocal cords (Fig 1–36). *Avoid* the occasional small branches of the superior thyroid

A

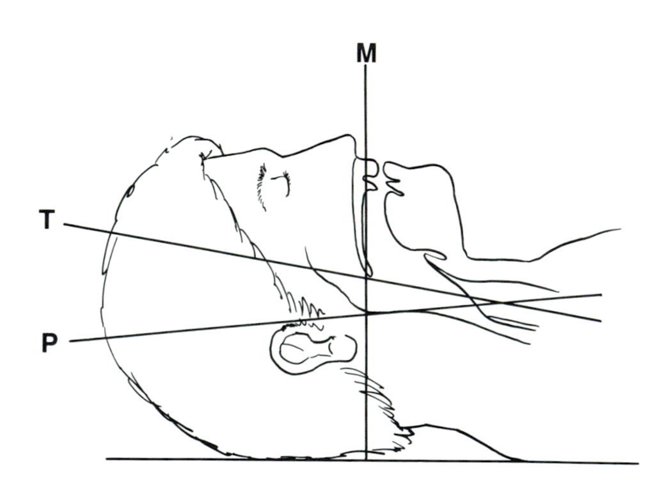

B

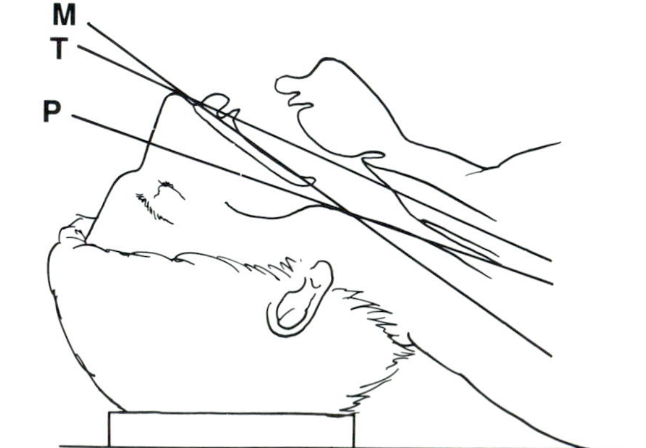

FIG 1–34.
Anatomical axes for endotracheal intubation. **A,** with the head in the neutral position, the axis of the mouth *(M)*, the axis of the trachea *(T)*, and the axis of the pharynx *(P)*, are not aligned with one another. **B,** if the head is extended at the atlanto-occipital joint, the axis of the mouth is correctly placed. If the back of the head is raised off the table with a pillow, thus flexing the cervical vertebral column, the axes of the trachea and pharynx are brought in line with the axis of the mouth.

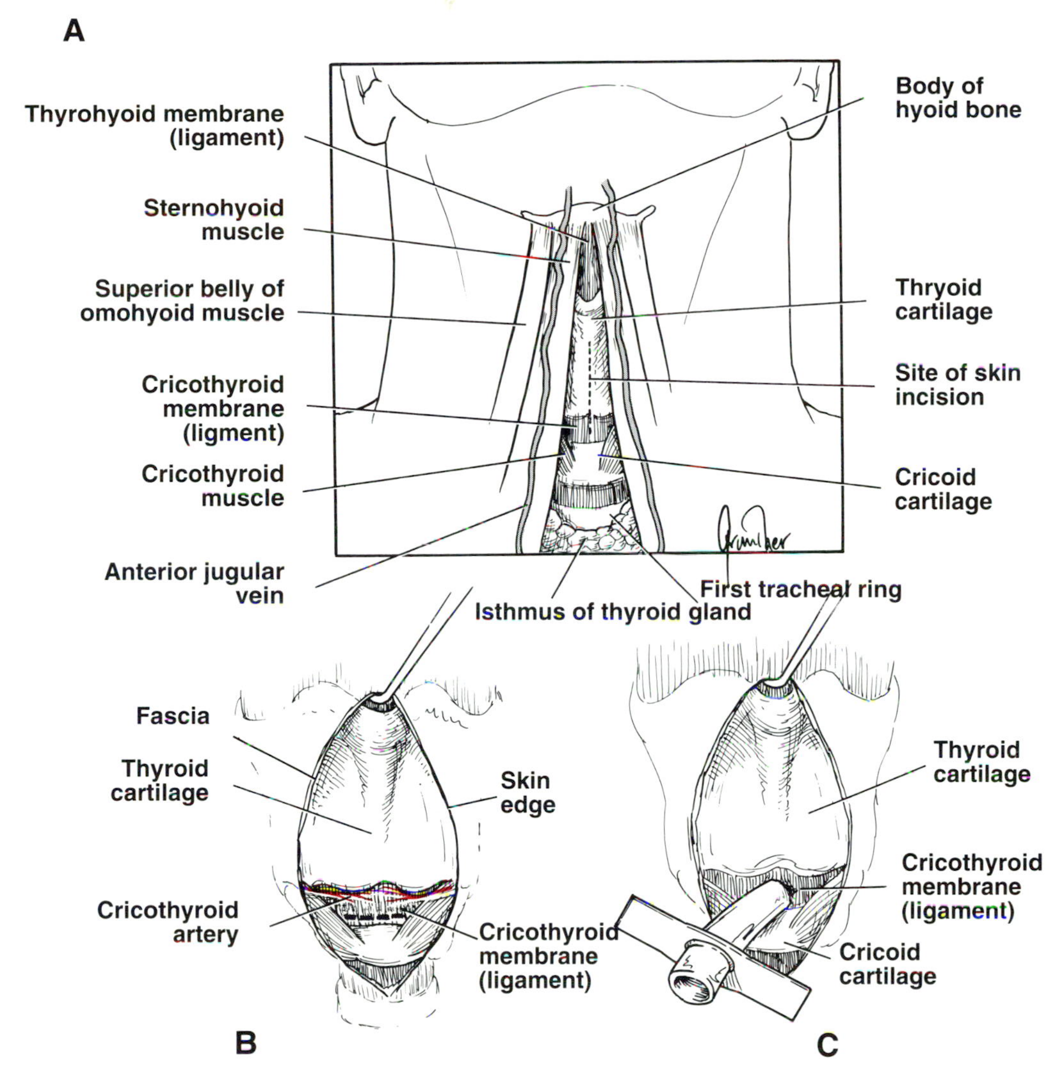

FIG 1–35.
The anatomy of cricothyroidotomy. **A,** a vertical incision is made through the skin and superficial and deep cervical fasciae. **B,** the cricothyroid membrane (ligament) is incised through a horizontal incision close to the upper border of the cricoid cartilage. **C,** insertion of tube is shown.

arteries on the two sides that anastomose with one another across the midline in front of the cricothyroid membrane.

Complications of Cricothyroidotomy

1. *Esophageal perforation.* Since the lower end of the pharynx and the beginning of the esophagus lie directly behind the cricoid cartilage, it is imperative that the scalpel incision through the cricothyroid membrane be not carried too far posteriorly. This is particularly important in young children where the cross diameter of the larynx is so small.

2. *Hemorrhage.* Avoid the small branches of the superior thyroid artery that occasionally cross the front of the cricothyroid membrane.

3. *Vocal cords.* Since the vocal cords are formed from the upper inturned edges of the cricothyroid membrane, care must be exercised to avoid their damage. For details consult Figure 1–36.

4. *Subcutaneous emphysema.* Air may enter the subcutaneous tissues of the neck through the hole in the larynx or through an excessively large hole in the skin.

Anatomy of Tracheostomy

Tracheostomy is rarely performed in the emergency department and is limited to patients with extensive laryngeal fractures and in infants with severe airway obstruction. Because of the presence of major vascular structures, the thyroid gland, nerves, the pleural

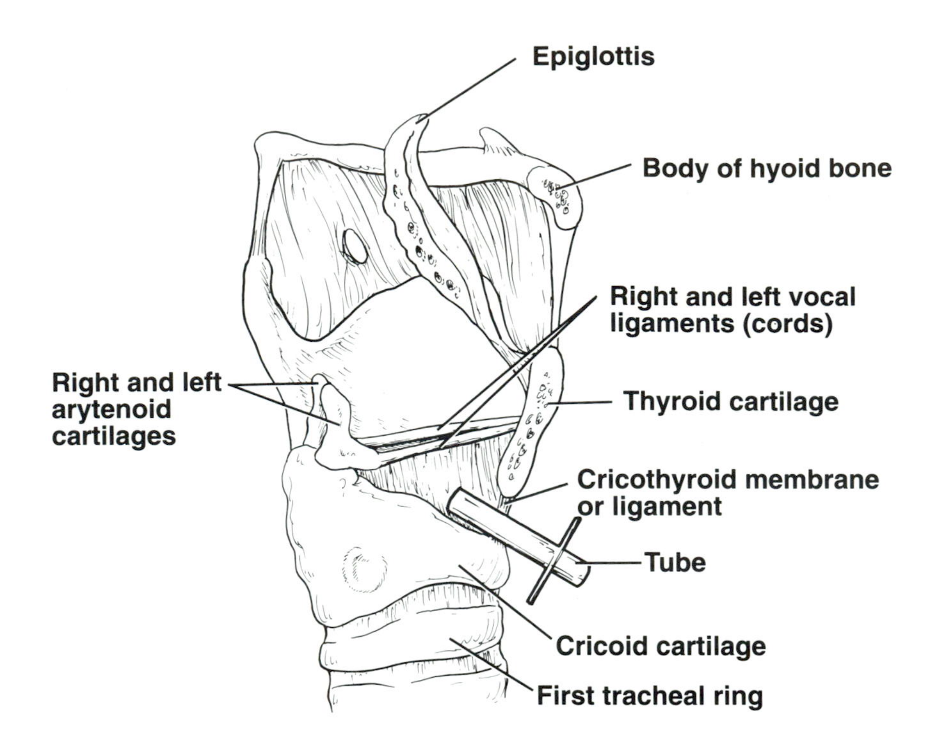

FIG 1–36.
View of the interior of the larynx as seen from the right side (the right lamina of the thyroid cartilage has been removed). Note the closeness of the deep end of the cricothyroidotomy tube to the vocal cords, especially if the tube is directed upward.

cavities, and the esophagus, meticulous attention to anatomical detail should be observed (Fig 1–37).

1. The thyroid and cricoid cartilages are identified and the neck is extended to bring the trachea forward.
2. A vertical midline skin incision is made from the region of the cricothyroid membrane inferiorly toward the suprasternal notch.
3. The incision is carried through the superficial fascia and the decussating fibers of the platysma muscle. The anterior jugular veins in the superficial fascia are avoided by maintaining a midline position.
4. The investing layer of deep cervical fascia is incised.
5. The strap muscles embedded in the pretracheal fascia are split in the midline two fingerbreadths superior to the sternal notch.
6. The tracheal rings are then palpable in the midline or the isthmus of the thyroid gland is visible. If a hook is placed under the lower border of the cricoid cartilage and traction is applied upward, the slack is taken out of the elastic trachea, which stops it from slipping from side to side.
7. A decision is then made as to whether to enter the trachea (1) through the second ring above the isthmus of the thyroid gland; (2) through the third, fourth, or fifth rings by first dividing the very vascular isthmus of the thyroid gland, or (3) through the lower tracheal rings below the thyroid isthmus. At the latter site the trachea is receding from the surface of the neck and the pretracheal fascia contains the inferior thyroid veins and possibly the thyroidea ima artery.
8. The preferred site is through the second ring of the trachea in the midline with the thyroid isthmus retracted inferiorly. A vertical tracheal incision is made, and the tracheostomy tube is inserted.

To avoid late scars that are tethered to deep structures, some authorities recommend carefully placed horizontal skin incisions for cosmetic reasons. In an emergency situation the vertical midline incision is preferred because it permits greater exposure and avoids large lateral blood vessels.

Tracheostomy Complications

The majority of these complications result from the operator not adequately palpating and recognizing the thyroid, cricoid, and tracheal cartilages and not confining the incision strictly to the midline. Failure to steady the elastic-walled trachea with a suitable hook can cause the trachea to move laterally.

1. *Hemorrhage.* Avoid the anterior jugular veins located in the superficial fascia close to the midline

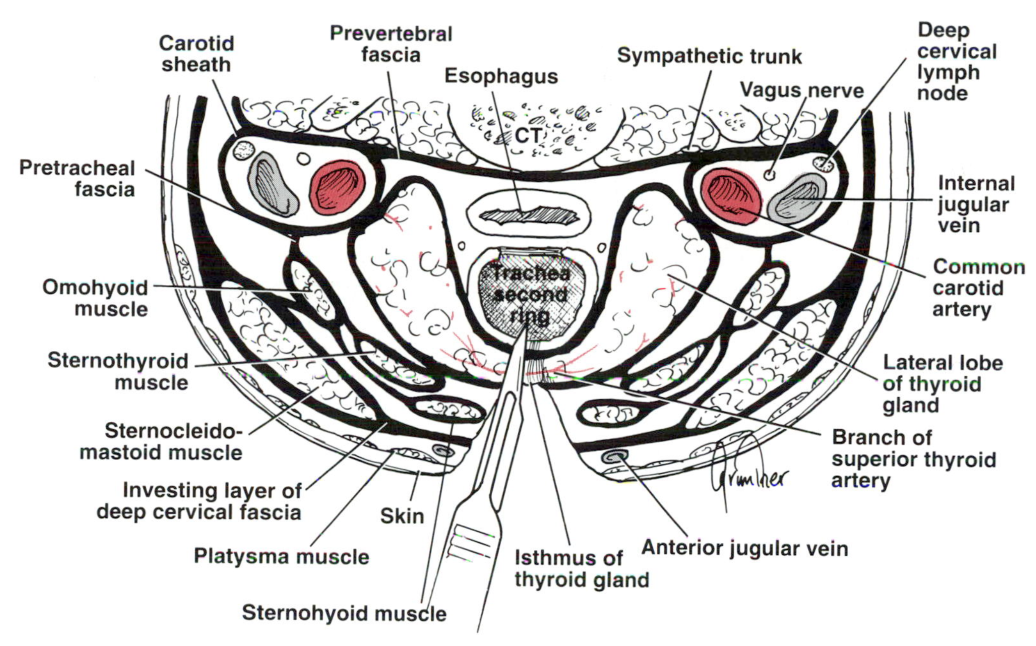

FIG 1–37.
Cross section of the neck at the level of the second tracheal ring. A vertical incision is made through the ring, and the tracheostomy tube is inserted.

of the neck. If the thyroid isthmus is transected, great care must be taken to secure the branches of the superior and inferior thyroid arteries that cross the midline on the isthmus. In patients with severe trauma to the neck with grossly altered anatomy, the carotid arteries have been mistakenly incised for the trachea.

2. *Nerve paralysis.* Damage to the recurrent laryngeal nerves as they ascend lateral to the trachea.

3. *Pneumothorax.* This results from piercing the cervical dome of the pleura and is especially common in children because of the high level of the pleura in the neck.

4. *Esophageal injury.* Damage to the esophagus, which is located immediately posterior to the trachea, occurs most commonly in infants; it follows penetration of the small-diameter trachea by the point of the scalpel blade.

5. *Mediastinal emphysema.* This results from the entrance of air from the trachea or skin opening into the tissue space beneath the pretracheal fascia.

Clinical Anatomy of the Bronchi

Inhalation of Foreign Bodies

Inhalation of foreign bodies into the lower respiratory tract is common, especially in children. Pins, screws, nuts, bolts, peanuts, and parts of chicken bones and toys have all found their way into the bronchi. Parts of teeth may be inhaled while a patient is under anesthesia during a difficult dental extraction. Since the right principal bronchus is the wider and more direct continuation of the trachea (see Fig 1–26), foreign bodies tend to enter the right rather than the left principal bronchus. From there, they usually pass into the middle or lower lobe bronchi. In children younger than 3 years, both bronchi arise from the trachea at equal angles, and therefore no predilection for the right bronchus exists.

Suction Catheters, Endotracheal Tubes, and the Bronchi

Suction catheters and endotracheal tubes are more likely to enter the right more vertical principal bronchus than the obliquely positioned left principal bronchus in adults and older children.

ANATOMICAL-CLINICAL PROBLEMS

1. A 24-year-old man was taken to the emergency department after having been found lying unresponsive in a local park with an empty whisky bottle nearby. He was given oxygen by an open face mask during the 15-minute ride in the ambulance. The paramedic decided to improve the airway by passing a soft nasal tube. On attempting to pass the well-lubricated tube into the patient's nose, the paramedic found it impossible to push it much be-

yond the nasal vestibule on either side. What are the common anatomical causes of obstruction of the nasal airway?

2. A 58-year-old male alcoholic was found unconscious in the street and was brought to the emergency department in an ambulance. Lying supine on the emergency department stretcher, the patient started to snore loudly. What is the most likely anatomical cause for the condition? Explain in detail the rationale for treatment.

3. A 10-year-old girl was brought to the emergency department with a history of fever, malaise, anorexia, and a sore throat. She also had hoarseness, a cough, and rhinitis. On examination there was erythema of the posterior pharyngeal wall, with small ulcers on the palatoglossal folds and soft palate. The tonsils were seen to be red and enlarged, and an obvious white-yellow exudate was seen on the surface of the right tonsil. Examination of the deep cervical lymph nodes showed enlargement and tenderness of the node below and behind the angle of the mandible; the enlargement was greatest on the right side. A diagnosis of viral pharyngitis was made. List the various lymphoid organs found in the nasal and oral parts of the pharynx. What is Waldeyer's ring?

4. The presence of disease of the temporomandibular joints or the cervical part of the vertebral column may make it difficult to establish an adequate airway in an unresponsive patient. Explain why this is so.

5. A 3-year-old child was playing with her toys on the floor when her brother and sister decided to share some peanuts with her. A few minutes later she started to cough and gave a hoarse cry. The cough then became croupy, and aphonia occurred. The mother, hearing the commotion, rushed into the room and quickly realized what had happened. She turned the child upside down and hit her back several times, but with no effect. The child, now in obvious respiratory distress, was rushed to the local emergency department. On examination, she was tachypneic, with suprasternal retractions. She was not coughing, and although she attempted to cry, there was no sound. She would not tolerate being laid down. On the basis of your knowledge of the anatomy of the airway, where do you think the foreign body was lodged? Describe the normal protective reflexes that exist in the airway to prevent the inhalation of a foreign body. What is the anatomical and physiological rationale behind the use of back blows, chest thrusts, and abdominal thrusts (Heimlich procedure) in the management of upper airway obstruction. Which of these procedures is most appropriate for a 3-year-old child?

6. A 7-year-old boy complained of a severe sore throat and dysphagia. Three hours later, his parents noticed that he was having difficulty breathing and had developed inspiratory stridor. He also had great difficulty swallowing and had a tendency to drool. He had a fever of 103°F. On examination in the emergency room, the boy was in extreme respiratory distress with inspiratory stridor. The alae nasi were flared, with inspiratory retraction of the neck above the suprasternal notch and intercostal spaces. Examination of the pharynx showed only mild erythma of the walls, but the tip of the epiglottis, which appeared swollen and cherry red, could just be seen over the back of the tongue. The diagnosis of acute epiglottitis was made. Since the airway of the child was severely compromised, he was taken to the operating room where an anesthesiologist attempted intubation; an ear, nose, and throat physician stood by to perform a "crash" tracheostomy. The anesthesiologist selected what he considered to be the correct tube size and length, and he successfully passed the tube by the swollen epiglottis and through the glottis, but he could not descend it any further. What is the likely anatomical explanation for this?

7. A 16-year-old boy was driving his minibike at high speed along a country lane, when he suddenly saw what he thought was a shortcut through a gap in a hedge. He did not see that the gap was closed by a strand of barbed wire. He struck the wire with his neck and was thrown from the bicycle. On arrival at the emergency department, he had all the signs and symptoms of upper airway obstruction. Using your knowledge of the anatomy of the neck, explain the type of injury that could have occurred in this case. Does the position of the vocal cords at the time of impact influence the type of injury that occurs? What anatomical factors normally protect the upper airway from serious blunt injuries? Does age play a role in the severity of the injury?

8. A 20-year-old drug dealer received a bullet wound in the neck. On arrival at the emergency department, a single small entrance and exit wound was noted on the left side of the neck close to the midline and just below the jaw. He exhibited aphonia, severe inspiratory stridor, and suprasternal and intercostal retractions. Skin crepitation and a hissing sound at the site of the entrance wound indicated subcutaneous emphysema. The anterior part of the neck was extremely swollen and bruised. In anatomical terms, explain the path taken by the escaping air from the damaged larynx. Which large blood vessels

are situated at the side of the larynx and may have been damaged by the bullet?

9. Following an automobile accident, a 36-year-old woman was admitted to the emergency department with suspected obstruction of the upper airway. Contusion to the front of the neck was evident. The larynx appeared to have lost its cartilaginous support. Is it possible that the anatomical structure of the mucosa and submucosa lining the larynx could influence the size of the lumen of the larynx?

10. A 45-year-old man with extensive maxillofacial injuries following an automobile accident was brought to the emergency department. Evaluation of the airway revealed partial obstruction. Despite an obvious fractured mandible, an attempt was made to move the tongue forward from the posterior pharyngeal wall by pushing the angles of the mandible forward. This maneuver failed to move the tongue, and it became necessary to hold the tongue forward directly in order to pull it away from the posterior pharyngeal wall. At times, why is it not possible to pull the tongue forward in the presence of a fractured mandible?

11. A 2-year-old child inhaled part of a toy, which was later found lodged in a principle bronchus on a chest radiograph. In young children, into which bronchus does an inhaled foreign body most commonly lodge? In adults, which bronchus is most likely to receive a dislodged denture if it is inhaled?

12. When a laryngoscope is passed it is important to align the mouth, the oropharynx, and the larynx into one plane. How do you bring the axes of the oropharynx and the larynx in line? How do you bring the axis of the mouth in line with the other axes? Describe the structures in the order that you can view them through a laryngoscope from the base of the tongue down to the trachea.

13. A 17-year-old boy was seen in the emergency department suffering from acute tonsillitis. He complained of throat pain, difficulty with swallowing, and trismus; he also said he had pain over the right side of his head and right ear. On examination the pharyngeal walls and soft palate were red and inflamed. The right tonsil was enlarged and red and there was a bright red swelling above and in front of the tonsil. The uvula was red and edematous and displaced to the left side. Examination of the neck revealed a large tender lymph node below and behind the angle of the jaw on the right side. A diagnosis of right-sided peritonsillar abscess was made. Because the most prominent part of the abscess was yellowish and obviously about to rupture, it was decided to incise and drain the abscess. Using your knowledge of anatomy, name the structures that lie lateral to the abscess. Is there a good anatomical reason for aspirating the abscess before incising it? Explain the cause of the enlarged lymph node. Why did the patient experience trismus and pain over the right side of the head and right ear?

ANSWERS

1. The most common cause for difficulty in passing a nasal tube is a deflected nasal septum. This occurs more commonly in the male, and is thought to be due to previous trauma to the septum during the period of active growth. Nasal spurs and polyps may cause difficulty and swelling of the mucous membrane secondary to infection or chemical irritation, and can also cause blockage. The widest part of the nasal cavity is near the floor.

2. Snoring is caused by a partially blocked airway, and the most likely cause for blockage of the airway in this patient is the tongue falling back to meet the flaccid posterior pharyngeal wall. With the patient unconscious and lying on his back, the tone of the extrinsic muscles of the tongue is lost, and the tongue falls backward simply because of gravity. This may be easily and quickly corrected by pulling the jaw forward. Place two fingers under the symphysis menti and pull forward or press forward the angles of the jaw with the thumbs. Anatomically, as the mandible is pulled forward, the genioglossus muscles, which run from the mandible to the tongue, pull the tongue forward. The patient should also be placed in the left lateral decubitus position on the stretcher, since if he should vomit, he is less likely to inhale the stomach contents.

3. The lymphoid tissue around the openings of the mouth and nasal cavities into the pharynx include (1) the palatine tonsils, (2) the lingual tonsil, (3) the tubal tonsils, and (4) the pharyngeal tonsil. For details of Waldeyer's ring see p. 18.

4. The successful use of a laryngoscope is more complicated if there is difficulty in flexing the cervical part of the vertebral column and extending the atlanto-occipital joints. Also it may prove difficult to visualize the glottis in a patient who has ankylosis of the temporomandibular joint. Alternate methods for intubation, i.e., fiberoptic techniques, a lighted stylet, digital intubation, nasal intubation, and others, should be prepared for.

5. The presence of severe respiratory distress with suprasternal retractions and aphonia indicates the presence of upper airway obstruction, probably

located within the larynx. The airway is protected by a number of important reflexes, including the gag reflex, the laryngeal reflex, and the cough reflex. The gag reflex occurs in response to stimulation of the pharyngeal mucous membrane innervated by the glossopharyngeal nerve. The laryngeal and the cough reflexes are mediated via the vagus nerve. These protective reflexes are lost in descending order as a patient loses consciousness.

All maneuvers that are directed toward freeing an obstruction of the airway by an inhaled foreign body are based on an attempt to increase the intrathoracic pressure by compressing the intrathoracic gas volume, so that the foreign body is expressed from the mouth. The underlying mechanisms involved in the use of back blows, chest thrusts, and abdominal thrusts are discussed on p. 36. It is now generally agreed that the best and safest method to use on a 3-year-old child is the abdominal thrust (see p. 36).

6. The narrowest part of the laryngeal cavity in an adult is the area between the vocal cords. In children younger than 10 years, the narrowest part of the larynx is within the cricoid cartilage i.e., below the vocal cords.

7. The impact of the wire to the front of the neck caused hyperextension of the cervical part of the vertebral column with stretching of the larynx and trachea. This effectively fixed the airway structures in the midline so that they were not deflected laterally at the moment of impact. Under these circumstances the cartilages of the larynx are fractured or crushed. Depending on the speed of the impact, the larynx could be completely avulsed from the trachea. In this situation the tone of the suprahyoid muscles would cause the larynx to be retracted superiorly and the elasticity of the trachea would cause it to retract inferiorly to the root of the neck or behind the sternum.

If the glottis were closed at the time of impact, the raised intraluminal pressure within the upper airway may contribute to the severity of the injury.

The upper airway receives a considerable amount of protection from blows to the front of the neck and chest because of the presence of the mandible and manubrium sterni. With the head and neck in the flexed position, the larynx and trachea are remarkably mobile and often deflected laterally by an anterior blow to the neck.

In children the very flexible nature of the laryngeal and tracheal cartilages and the looseness of the supporting connective tissue reduce the likelihood of severe damage to these structures.

8. The larynx is covered anteriorly by the strap muscles, namely the sternohyoid, sternothyroid, and omohyoid muscles, which are embedded in pretracheal fascia. Anterior to these structures is the thick investing layer of deep cervical fascia that runs across the front of the neck between the sternocleidomastoid muscles. With a penetrating wound that has damaged the integrity of the larynx, air will escape and reach the subcutaneous tissues along the path taken by the bullet across the fascial barriers.

The carotid sheath containing the internal, the external, and the common carotid arteries and the internal jugular vein, lies lateral to the larynx and could easily be damaged by a bullet.

Do not forget that in penetrating wounds in the neck, the possibility exists of damage to the cervical vertebral column. Failure to be aware of this possibility could result in damage to the spinal cord during forcible manipulations of the head, mandible, and neck in the process of securing an adequate airway.

9. Edema fluid and blood from hemorrhaging can easily accumulate in the submucosa of the supraglottic and infraglottic regions of the larynx. The resulting swelling of the mucosa could compromise the airway. It is interesting that swelling in one of these regions is unable to extend to another because the mucous membrane of the vocal cords closely adheres to the underlying vocal ligaments.

10. The root of the tongue is attached anteriorly to the mental spines on the posterior surface of the symphysis menti of the mandible by the right and left genioglossus muscles. If this bony origin were floating because of fractures on both sides of the body of the mandible, pulling the angles of the mandible forward would have no effect on the position of the tongue.

11. Under the age of 3 years, the angulation of the two principal bronchi at the carina is equal on both sides (55°). Both bronchi are therefore equally likely to receive an inhaled foreign body. In adults, however, the right principal bronchus is the most direct continuation of the trachea, and therefore the right principal bronchus is more likely to receive an inhaled foreign body, such as a tooth. (Remember that in children, swallowed foreign bodies often get stuck at the level of the cricopharyngeus or where the aorta or left bronchus crosses the esophagus anteriorly.)

12. The axis of the oropharynx and the larynx are brought into direct line by flexing the cervical part of the vertebral column. The axis of the mouth is brought in line with the oropharynx by extending the atlanto-occipital joints.

The following structures may be viewed: (1) the

base of the tongue; (2) the median glossoepiglottic fold, the two lateral glossoepiglottic folds, and the valleculae on each side of the median fold; (3) the upper edge of the epiglottis, and the opening into the larynx, bounded in front by the epiglottis with its tubercle and laterally by the aryepiglottic folds—the rounded elevations of the cuneiform and corniculate cartilages in the folds can be recognized; (4) the reddish fixed vestibular folds; (5) the whitish mobile vocal cords (folds) with the rima glottidis; and (6) below the glottis the interior of the trachea with the upper two or three rings.

13. The following structures lie lateral to a peritonsillar abscess: (1) superior constrictor muscle, (2) external palatine vein, (3) styloglossus muscle, (4) facial artery, and (5) *internal carotid artery*—1 in. (2.5 cm) behind and lateral to the abscess.

Aspiration of a peritonsillar abscess is always performed by some physicians prior to making a drainage incision to confirm the presence of pus and lessen the chance of hitting the internal carotid artery. This procedure also eliminates the possibility of confusing a pulsatile internal carotid aneurysm with an abscess.

The lymphatic drainage of the tonsil and its immediate surrounding area is into the deep cervical lymph node, which is below and behind the angle of the jaw. This node was enlarged in this patient because of the spread of infection to the node.

Trismus is due to spasm of the muscles of mastication, which are innervated by the mandibular division of the trigeminal nerve. The tonsil and the undersurface of the soft palate receive their sensory innervation from the glossopharyngeal nerve and the lesser palatine nerve, a branch of the maxillary division of the trigeminal nerve. Inflammatory irritation of these tissues resulted in reflex spasm of the muscles of mastication, with the nervous impulses traveling in the nerves involved.

The right-sided head and ear pain was referred along the trigeminal nerve. The brain incorrectly assumed that the pain was originating from the skin over the side of the head (supplied by the mandibular division of the trigeminal nerve), when in fact it originated from the soft palate (supplied by the lesser palatine nerve, a branch of the maxillary division of the trigeminal nerve).

2

The Chest Wall, Chest Cavity, Lungs, and Pleural Cavities

Contained within the chest are the important life-sustaining organs—the lungs, heart, and major blood vessels. Although the chest wall is strong, blunt or penetrating wounds can injure the soft organs contained beneath it. This is especially so in an era in which automobile accidents, stab wounds, and gunshot wounds are commonplace.

The lungs contain the vital interface, the respiratory membrane, that enables gas exchange between inspired air and circulating blood. Anything that interferes with the normal function of this membrane has the potential to develop into a life-threatening situation. A wide spectrum of causes, ranging from inhalation of stomach contents or toxic gases, acute bronchospasm, pulmonary edema, or damage to the lung parenchyma from blunt or penetrating trauma, can precipitate an emergency. The lungs receive the total cardiac output. A sudden impairment of cardiac function may present as a pulmonary emergency, and, conversely, an internal obstruction of blood flow (e.g., pulmonary embolism) may present as a cardiovascular problem.

The normal aeration of the lungs depends on the normal movements of the chest wall and the contraction of the diaphragm; lesions of the central nervous system or local tissue disease involving these structures can seriously impair respiratory function.

Assessment and treatment of chest trauma will depend on the ability of emergency department personnel to relate the pathological mechanisms of the injury to the anatomy of the structures involved. All chest injuries are potentially life threatening.

BASIC ANATOMY

Basic Anatomy of the Chest Wall

The thorax has in its walls a bony, cartilaginous cage formed by the sternum, costal cartilages, ribs, and bodies of the thoracic vertebrae (Fig 2–1). The thoracic cage is cone shaped, having a narrow inlet* superiorly and a broad outlet inferiorly. The cage is flattened in its anteroposterior diameter. It participates in the movements of respiration and protects the underlying thoracic viscera, especially the heart, lungs, and the upper abdominal viscera, namely, the liver, spleen, and stomach.

*Clinicians refer to this opening as the thoracic outlet, since important blood vessels and nerves emerge from the thorax here to enter the neck and upper limb.

Sternum

The sternum lies in the midline of the anterior chest wall and may be divided into the following three parts: (1) manubrium sterni, (2) body of the sternum, and (3) xiphoid process (see Fig 2–1).

The *manubrium* is the upper part of the sternum, and it articulates with the clavicles and the first and upper part of the second costal cartilages on each side. The manubrium has a depression on its superior border called the *suprasternal (jugular) notch.*

The *body of the sternum* articulates above with the manubrium through a fibrocartilaginous joint, the *manubriosternal joint.* Below, it articulates with the xiphoid process at the *xiphisternal joint.* On each side are notches for articulation with the lower part of the second costal cartilage and the third to seventh costal cartilages (see Fig 2–1).

The *xiphoid process,* which is a plate of hyaline cartilage that becomes ossified at its proximal end in adult life, is the lowest part of the sternum. No ribs or costal cartilages are attached to it. However, it does enable attachment for some muscles of the anterior abdominal wall.

Costal Cartilages

Costal cartilages are bars of hyaline cartilage connecting the upper 7 ribs to the lateral edge of the sternum, and the 8th, 9th, and 10th ribs to the cartilage immediately above. The cartilages of the 11th and 12th ribs end in the abdominal musculature (see Fig 2–1).

Ribs

There are 12 pairs of ribs, the majority of which encircle the trunk from the thoracic part of the vertebral column behind to the region of the sternum in front (see Fig 2–1). The upper 7 pairs are attached anteriorly to the sternum by their costal cartilages. The 8th, 9th, and 10th pairs of ribs are attached anteriorly to each other and to the 7th rib by their costal cartilages. The 11th and 12th pairs have no anterior attachment and are referred to as *floating ribs.*

A *typical rib* is a long, twisted flat bone having a rounded, smooth superior border and a sharp, thin inferior border (see Fig 2–1). The anterior part of the inferior border overhangs and forms the *costal groove,* which accommodates the intercostal vessels and nerve.

A rib has a *head, neck, tubercle, angle,* and *shaft.* The *head* has two facets for articulation with the numerically corresponding vertebral body and that of the vertebra immediately above. The *neck* is a constricted portion situated between the head and the

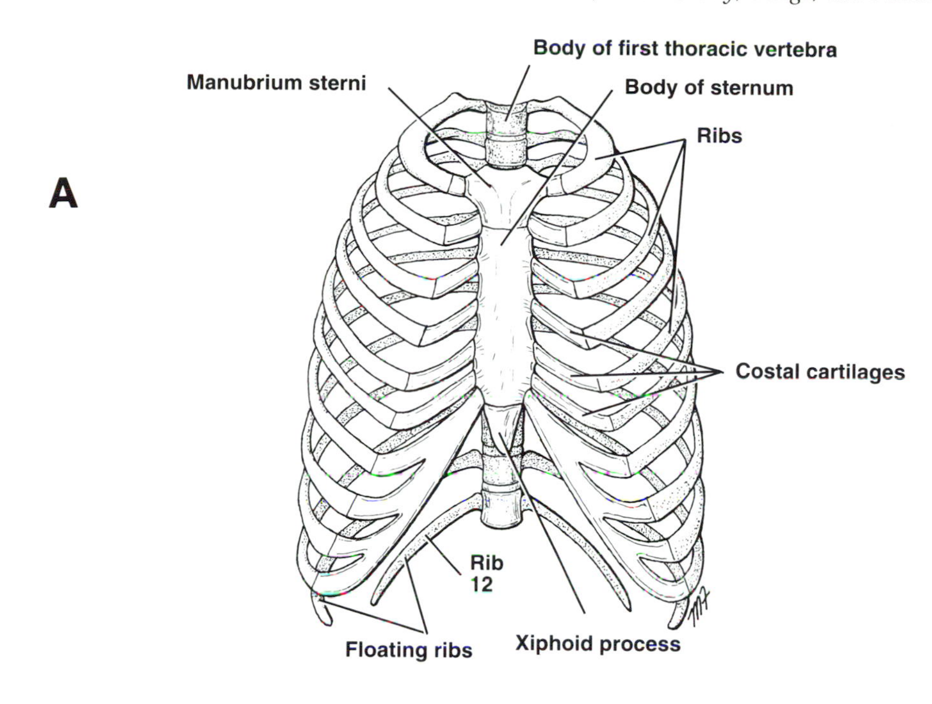

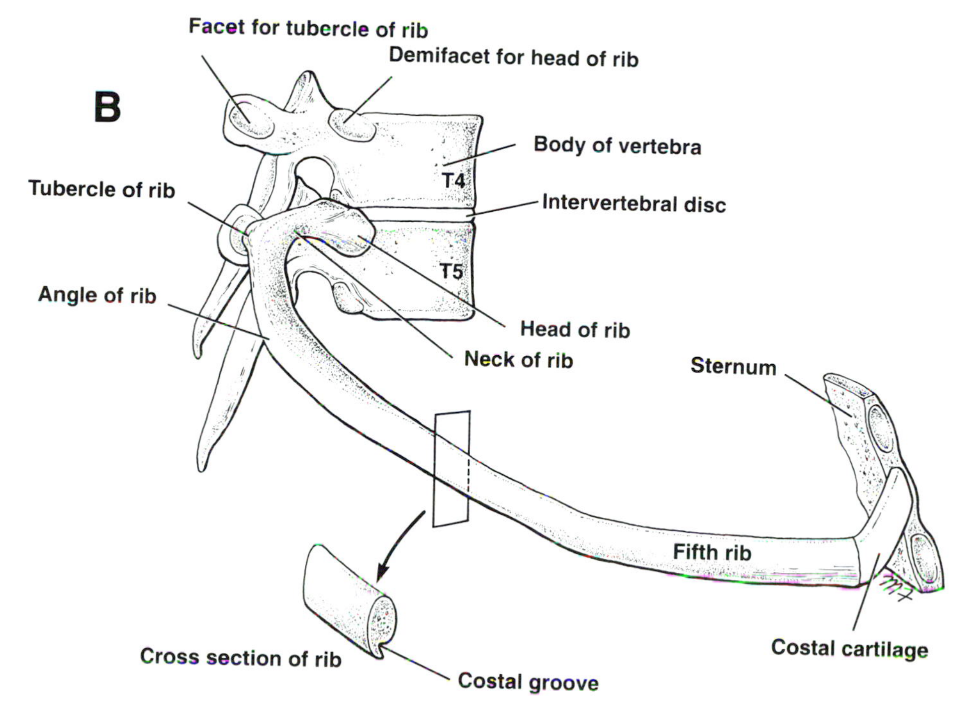

FIG 2–1.
A, anterior view of the bones and cartilages forming the thoracic skeleton. **B,** fifth right rib as it articulates with the vertebral column posteriorly and with the sternum anteriorly. The rib head articulates with the vertebral body of its own number and that of the vertebra immediately above. Note the presence of a costal groove along the inferior border of the rib.

tubercle. The *tubercle* has a facet for articulation with the transverse process of the numerically corresponding vertebra. The *angle* is where the rib bends sharply forward. The *shaft* of each rib is attached to the corresponding costal cartilage.

The *first rib is atypical.* It is important because of its close relationship to the lower nerves of the brachial plexus and the main vessels to the arm, namely, the subclavian artery and vein (see Fig 2–10), all of which cross over the rib (and under the clavicle). This rib is flattened from above downward. The tubercle on the inner border, known as the *scalene tubercle,* receives the insertion of the scalenus anterior muscle.

Joints of the Ribs and Costal Cartilages

Joints of the Heads of the Ribs.—The first rib and the three lowest ribs have a single joint with their corresponding vertebral body. For the second to the ninth ribs, the head articulates by means of a joint on the body of the corresponding vertebra and on that of the vertebra above (see Fig 2–1).

Joints of the Tubercles of the Ribs.—The tubercle articulates with the transverse process of the vertebra to which it corresponds numerically (see Fig 2–1). (It is absent on the 11th and 12th ribs.)

Costochondral Joints.—These are cartilaginous joints, and no movement is possible.

Joints of the Costal Cartilages With the Sternum. The 1st costal cartilages are attached to the manubrium, and no movement is possible (see Fig 2–1). The 2nd costal cartilages articulate with the manubrium and body of the sternum by a mobile synovial joint. The 3rd to 7th costal cartilages articulate with the lateral border of the body of the sternum by synovial joints. (The 6th, 7th, 8th, 9th, and 10th costal cartilages articulate with each other along their borders by small synovial joints. The cartilages of the 11th and 12th ribs are embedded in the abdominal musculature.)

Movements of the Ribs and Costal Cartilages

The first ribs and their costal cartilages are fixed to the manubrium and are immobile. The raising and lowering of the ribs during respiration are accompanied by movements in both the joints of the head and the tubercle, permitting the neck of each rib to rotate around its own axis.

Joints of the Sternum

The *manubriosternal joint* is a cartilaginous joint between the manubrium and body of the sternum. A small amount of angular movement is possible during respiration. The *xiphisternal joint* is a cartilaginous joint between the xiphoid process (cartilage) and the body of the sternum. The xiphoid process usually fuses with the body of the sternum during middle age.

Thoracic Openings

The *thoracic inlet (clinically, the thoracic outlet)* is the opening of the thorax into the root of the neck (see Fig 2–10). This narrow aperture is bounded by the superior border of the manubrium sterni, by the medial borders of the first ribs, and by the body of the first thoracic vertebra. The opening transmits structures that pass between the thorax and the neck (esophagus, trachea, blood vessels, etc) and for the most part lie close to the midline. On either side of these structures lies the apex of the lung, with its visceral and parietal layers of pleura, protected by a deep fascial layer called the *suprapleural membrane* (see Fig 2–11).

The *thoracic outlet (so called by anatomists)* is situated inferiorly. This wide aperture is where the thorax opens into the abdomen (see Fig 2–1). It is bounded by the xiphisternal joint, the costal margin, and by the body of the 12th thoracic vertebra. It is closed by the diaphragm, which is pierced by the structures that pass between the thorax and the abdomen (see Fig 2–5).

Intercostal Spaces

Each space contains three muscles of respiration—the external intercostal muscle, the internal intercostal muscle, and the transversus thoracis muscle (Fig 2–2). Lying internally to this latter muscle is the *endothoracic fascia* and the parietal pleura. The intercostal nerves and blood vessels run between the intermediate and deepest layer of muscles (Fig 2–3). They are arranged in the following order from above downward: intercostal vein, intercostal artery, and intercostal nerve (i.e., VAN).

Intercostal Muscles

External Intercostal Muscle.—This muscle forms the most superficial layer of muscle. Its fibers are directed downward and forward from the inferior border of the rib above to the superior border of the rib below (see Fig 2–2). The muscle extends forward from the rib tubercle behind to the costochondral junction in front, where the muscle is replaced by an aponeurosis, the *anterior intercostal membrane* (Fig 2–4).

Internal Intercostal Muscle.—This muscle forms the intermediate layer of muscle. Its fibers are directed downward and backward from the subcostal groove of the rib above to the upper border of the rib below (see Fig 2–2). The muscle extends backward from the sternum in front to the angles of the ribs behind,

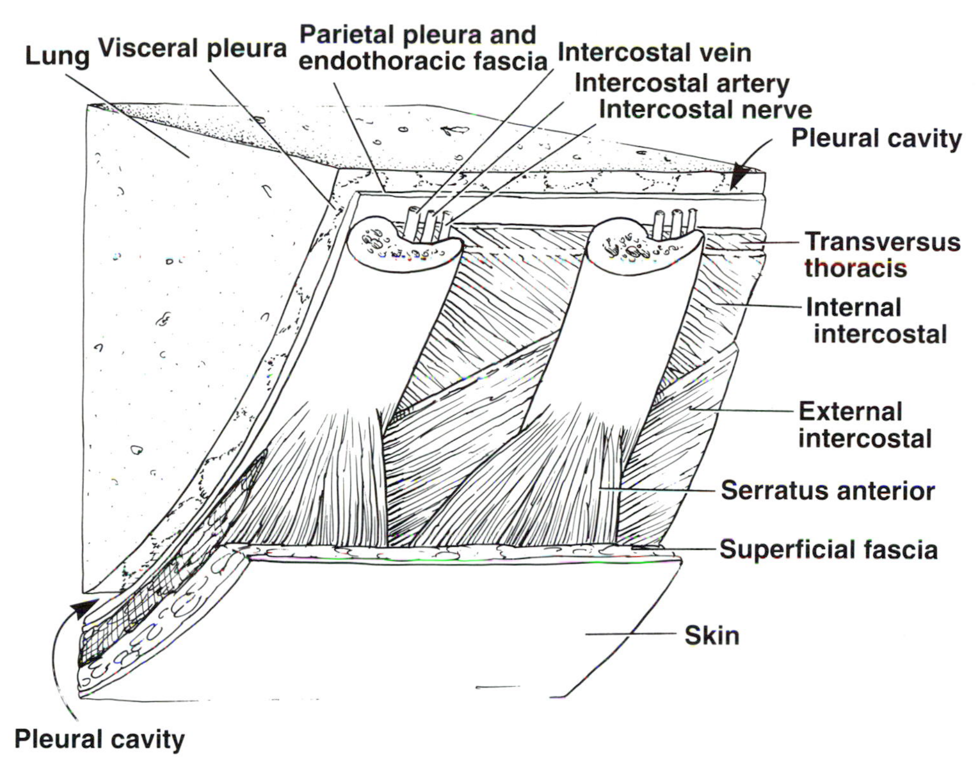

FIG 2–2.
Section through an intercostal space as seen from the right side with the patient in a supine position. Note the protected position of the neurovascular bundle in the subcostal groove and the relative positions of the intercostal vein, the intercostal artery, and the intercostal nerve.

where the muscle is replaced by an aponeurosis, the *posterior intercostal membrane* (see Fig 2–4).

Transversus Thoracis Muscle.—This muscle forms the deepest layer of muscle and corresponds to the transversus abdominis muscle in the anterior abdominal wall. It is an incomplete muscle layer and crosses more than one intercostal space within the ribs. The transversus thoracis muscle is related internally to the endothoracic fascia and the parietal pleura (see Fig 2–2).

Action of Intercostal Muscles.—If the 1st rib is fixed by the neck muscles, the intercostal muscles will raise the 2nd to the 12th ribs toward the 1st rib, as in inspiration. If the 12th rib is fixed by the abdominal muscles, the intercostal muscles will lower the 1st to the 11th ribs toward the 12th rib, as in expiration. The tone of the intercostal muscles during the different phases of respiration serves to strengthen the tissues of the intercostal spaces, thus preventing them from sucking in or blowing out with changes in intrathoracic pressure.

Nerve Supply.—The intercostal muscles are supplied by the corresponding intercostal nerves.

Intercostal Arteries and Veins

Intercostal Arteries.—Each intercostal space contains a large single *posterior intercostal artery* and two small *anterior intercostal arteries.*

The posterior intercostal arteries of the first two spaces are branches from the superior intercostal artery, a branch of the costocervical trunk of the subclavian artery. The posterior intercostal arteries of the lower nine spaces are branches of the thoracic aorta (see Fig 2–3).

The anterior intercostal arteries of the first six spaces are branches of the internal thoracic artery (see Fig 2–4). The anterior intercostal arteries of the lower spaces are branches of the musculophrenic artery, one of the terminal branches of the internal thoracic artery.

Each intercostal artery gives off branches to the muscles, skin, and parietal pleura. In the region of the breast in the female, the branches to the superficial structures are particularly large.

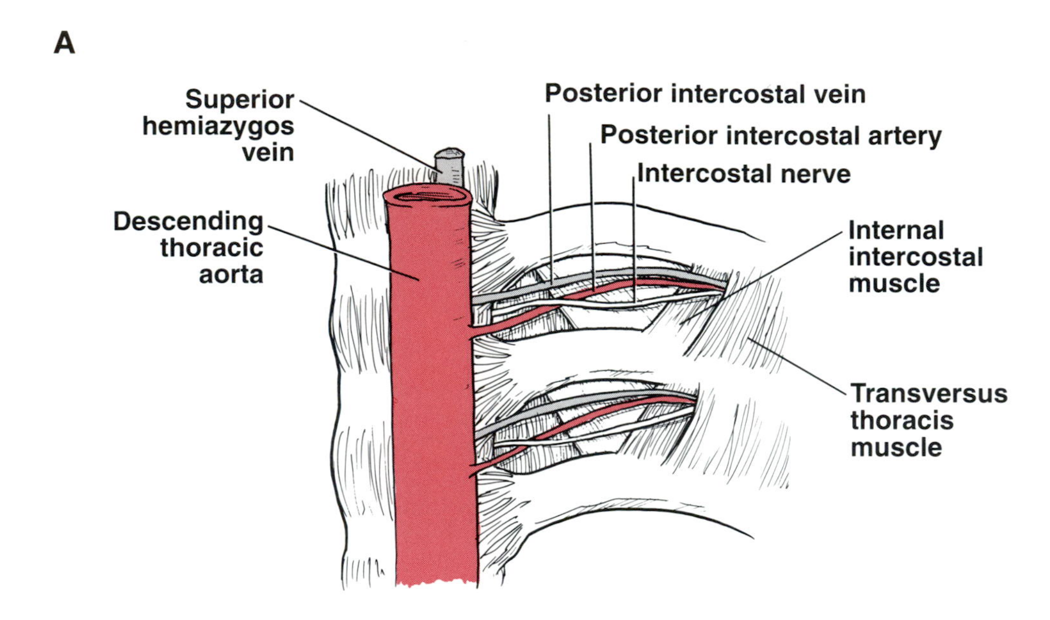

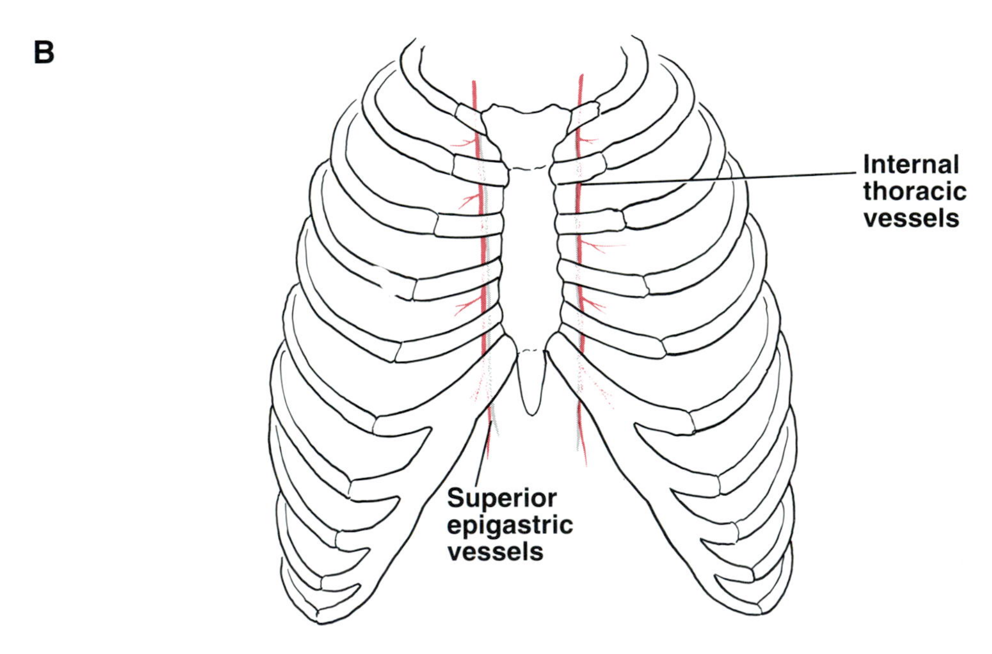

FIG 2–3.
A, internal view of the posterior end of two typical intercostal spaces; the posterior intercostal membrane has been removed for clarity. **B,** anterior view of the chest showing the courses of the internal thoracic vessels. These vessels descend about one fingerbreadth from the lateral margin of the sternum.

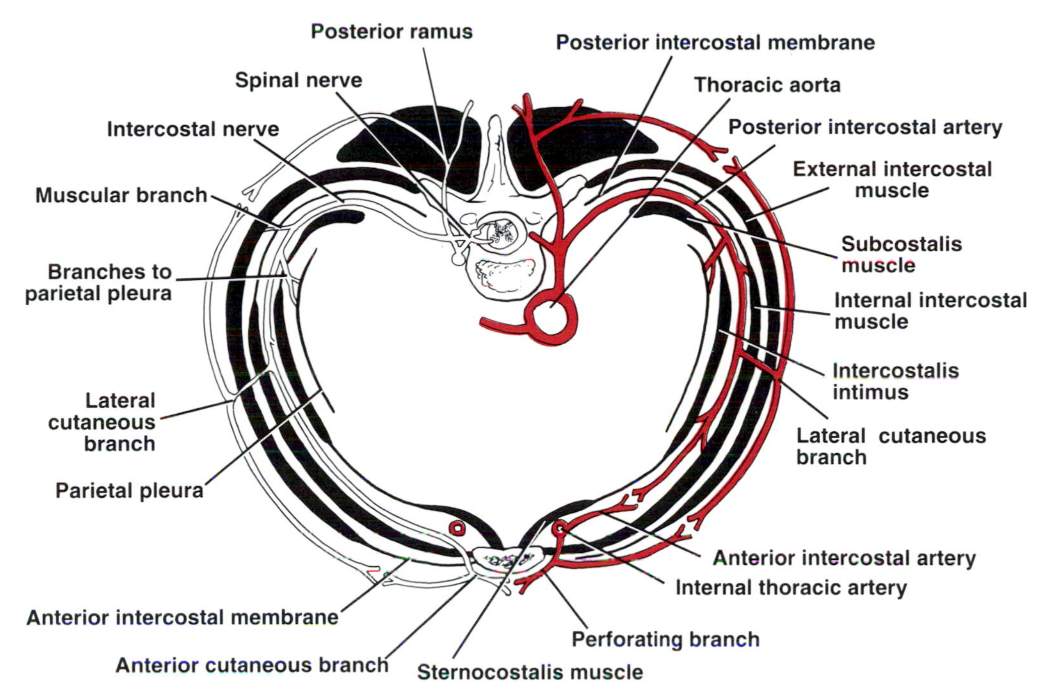

FIG 2–4.
Cross section of the thorax showing the distribution of a typical intercostal nerve on the left and the distribution of a posterior and an anterior intercostal artery on the right.

Intercostal Veins.—The corresponding posterior intercostal veins drain backward into the azygos or hemiazygos veins, and the anterior intercostal veins drain forward into the internal thoracic and musculophrenic veins.

Intercostal Nerves

The intercostal nerves are the anterior rami of the first 11 thoracic spinal nerves. Each intercostal nerve enters an intercostal space (see Fig 2–3) and runs forward inferiorly to the intercostal vessels in the subcostal groove of the corresponding rib, between the transversus thoracis and internal intercostal muscle (see Fig 2–2). The first 6 nerves are distributed within their intercostal spaces. The 7th to the 9th intercostal nerves leave the anterior ends of their intercostal spaces by passing deep to the costal cartilages, to enter the abdominal wall. The 10th and 11th nerves, because the corresponding ribs are floating, pass directly into the abdominal wall.

Branches of the intercostal nerves include the following:

1. *Collateral branch.*—Runs forward below main nerve in lower part of intercostal space.
2. *Lateral cutaneous branch.*—Arises in the midaxillary line and divides into anterior and posterior branches that supply the skin.
3. *Anterior cutaneous branch.*—The terminal portion of the main trunk. It reaches the skin near the midline and divides into a medial and a lateral branch.
4. *Muscular branches.*—Numerous muscular branches originate from the main nerve and its collateral branch.
5. *Pleural branches* to parietal pleura and *peritoneal branches* (7th to 11th intercostal nerves only) to parietal peritoneum.—These are sensory nerves.

The first six intercostal nerves give off numerous branches that supply skin and parietal pleura covering the outer and inner surface of each intercostal space, respectively, and the intercostal muscles of each intercostal space.

In addition, the 7th to 11th intercostal nerves supply skin and parietal peritoneum covering the outer and inner surfaces of the abdominal wall, respectively, and the abdominal muscles, which include the external oblique, internal oblique, transversus abdominis, and rectus abdominis muscles. The first and second intercostal nerves are an exception.

The *first intercostal nerve* is joined to the brachial plexus by a large branch that is equivalent to the lateral cutaneous branch of typical intercostal nerves.

The remainder of the first intercostal nerve is small.

The *second intercostal nerve* is joined to the medial cutaneous nerve of the arm by the *intercostobrachial nerve.* The second intercostal nerve therefore assists in the supply of the skin of the armpit and the upper medial side of the arm. In coronary artery disease, pain is referred along this nerve to the medial side of the arm.

Basic Anatomy of the Diaphragm

The diaphragm is the most important muscle of respiration (Fig 2–5). A dome-shaped muscle, it consists of a peripheral muscular part, which arises from the margins of the lower thoracic outlet, and a centrally placed tendon. The origin of the diaphragm may be divided into the following three parts:

1. A *sternal part,* consisting of small right and left portions arising from the posterior surface of the xiphoid process.
2. A *costal part,* arising from the deep surfaces of the lower six costal cartilages.
3. A *vertebral part,* arising by means of vertical columns or *crura* and from the arcuate ligaments.

The *right crus* arises from the sides of the bodies of the first three lumbar vertebrae and the intervertebral discs; the *left crus* arises from the sides of the bodies of the first two lumbar vertebrae and the intervertebral discs.

Lateral to the crura, the diaphragm arises from the medial and lateral arcuate ligaments. The *medial arcuate ligament* extends from the side of the body of the 2nd lumbar vertebra to the tip of the transverse process of the 1st lumbar vertebra (see Fig 2–5). The *lateral arcuate ligament* extends from the tip of the transverse process of the 1st lumbar vertebra to the lower border of the 12th rib (see Fig 2–5).

The diaphragm is inserted into a *central tendon.* The surface of the tendon is partially fused with the inferior surface of the fibrous pericardium. Some of the muscle fibers of the right crus pass up to the left and surround the esophagus in a slinglike loop. These fibers appear to act as a sphincter and assist in

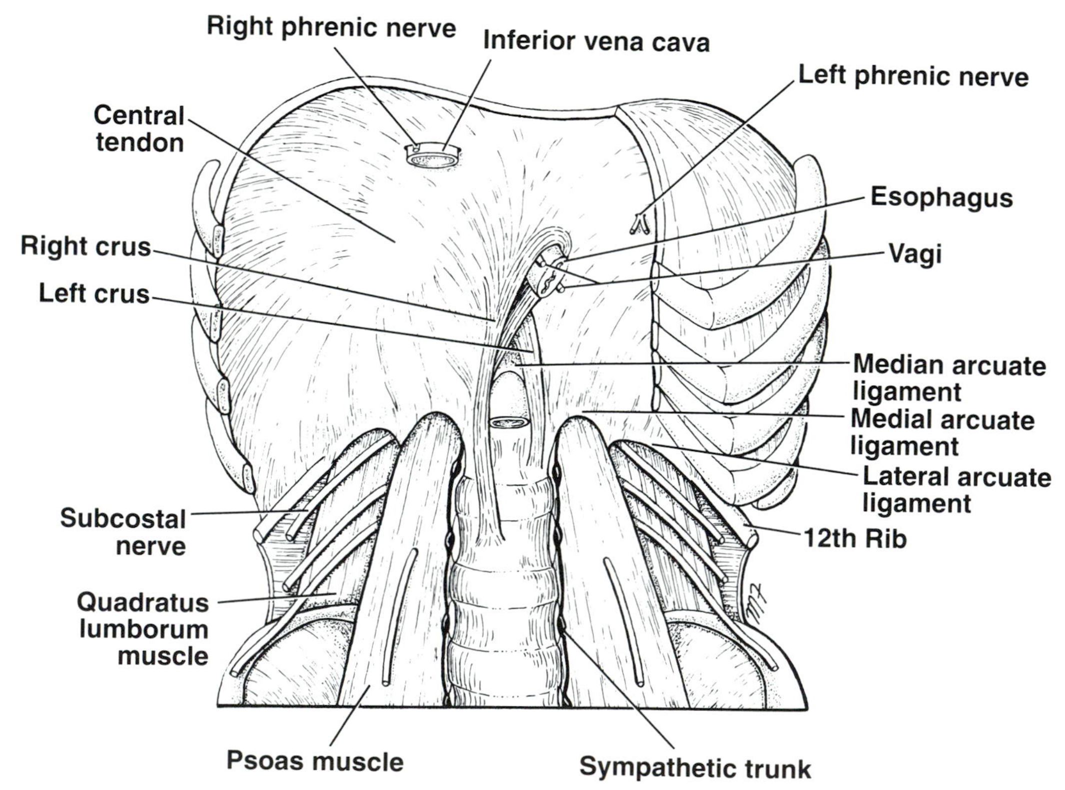

FIG 2–5.
Diaphragm as seen from below. The anterior portion of the right side has been removed. Note the sternal, costal, and vertebral origins of the muscle and the important structures that pass through it.

preventing regurgitation of the stomach contents into the thoracic part of the esophagus.

Shape of the Diaphragm

As seen from the front, the diaphragm curves up into the right and left domes. The right dome reaches as high as the upper border of the fifth rib, and the left dome reaches the lower border of the fifth rib. The central tendon lies at the level of the xiphisternal joint. The domes support the right and left lungs, whereas the central tendon supports the heart.

Nerve Supply of the Diaphragm

The motor nerve supply comes from the phrenic nerves. The sensory nerve supply to the parietal pleura and peritoneum covering the upper and lower surfaces of the central part of the diaphragm is also from the phrenic nerves. However, the sensory nerve supply to the pleura and peritoneum covering the upper and lower surfaces of the peripheral part of the diaphragm is from the lower six intercostal nerves.

Openings in the Diaphragm

The diaphragm has three main openings (see Fig 2–5).

Aortic Opening.—This lies anterior to the body of the 12th thoracic vertebra between the crura. It transmits the aorta, the thoracic duct, and the azygos vein.

Esophageal Opening.—This lies at the level of the 10th thoracic vertebra in a sling of muscle fibers derived from the right crus. It transmits the esophagus, the right and left vagus nerves, the esophageal branches of the left gastric vessels, and the lymphatics from the lower third of the esophagus.

Caval Opening.—This lies at the level of the 8th thoracic vertebra in the central tendon. It transmits the inferior vena cava and the terminal branches of the right phrenic nerve.

Other Openings.—The splanchnic nerves pierce the crura, the sympathetic trunks pass behind the medial arcuate ligaments on each side, and the superior epigastric vessels pass between the sternal and costal origins of the diaphragm on each side.

The radiographic appearances of the diaphragm are shown in Figures 2–23 through 2–25.

Basic Anatomy of the Breasts

The breasts are specialized accessory glands of the skin (Fig 2–6) present in both sexes. In males and immature females they are similar in structure. The *nipples* are small and surrounded by a colored area of skin called the *areola.* The breast tissue consists of little more than a system of ducts embedded in connective tissue that does not extend beyond the margin of the areola.

At puberty in females because of the influence of the ovarian hormones, the mammary glands gradually enlarge and assume their hemispherical shape (see Fig 2–6). The ducts elongate, but the increased size of the glands is mainly from the deposition of the adipose tissue. The base of the breast extends from the second to the sixth rib and from the lateral margin of the sternum to the midaxillary line. The greater part of the gland lies in the superficial fascia. A small part, called the *axillary tail,* extends upward and laterally, pierces the deep fascia at the lower border of the pectoralis major muscle, and enters the axilla. The breasts are separated from the deep fascia covering the underlying muscle by an area of loose areolar tissue known as the *retromammary space* (see Fig 2–6).

In young women the breasts tend to protrude forward from a circular base; in older women they tend to be pendulous. Each breast consists of 15 to 20 *lobes* that radiate from the nipple. There is no capsule. Each lobe is separated from its neighbor by connective tissue septa containing adipose tissue that, in the upper part of the gland, are well developed and extend from the skin to the deep fascia and serve as *suspensory ligaments.* The main lactiferous duct from each lobe opens separately on the summit of the nipple and possesses a dilated *ampulla or lactiferous sinus* just prior to its termination (see Fig 2–6).

The *nipple* is traversed by 15 to 20 lactiferous ducts that open by small orifices on its tip. The areola is an area of pigmented skin that surrounds the base of the nipple. Numerous sebaceous and sweat glands exist in the areola.

Blood Supply of the Breasts

Arteries.—Lateral thoracic and thoracoacromial arteries, branches of the axillary artery; perforating branches of the internal thoracic and intercostal arteries.

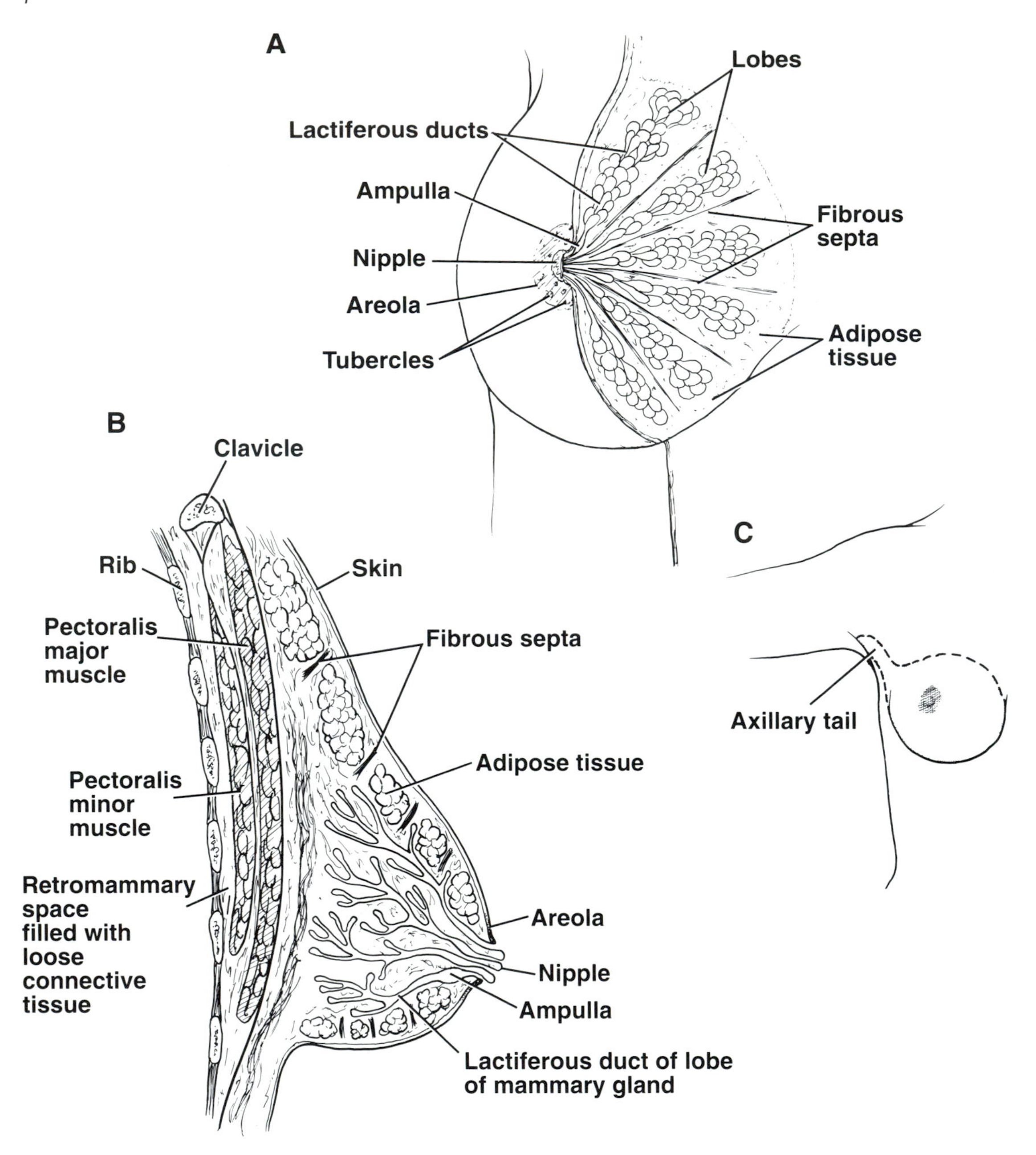

FIG 2–6.
Mature mammary gland in the female. **A,** anterior view with skin partially removed to show internal structure. **B,** sagittal section showing the arrangement of the ducts. **C,** the axillary tail, which pierces the deep fascia and extends up into the axilla.

Veins.—These correspond to the arteries.

Lymphatic Drainage of the Breasts

The lateral part of the gland drains into the anterior axillary or pectoral group of nodes (Fig 2–7) (situated just posterior to the lower border of the pectoralis major muscle). The medial part of the gland drains into the internal thoracic group of nodes (situated within the thoracic cavity along the course of the internal thoracic artery). A few lymph vessels drain posteriorly into the posterior intercostal nodes (situated along the course of the posterior intercostal arteries), whereas some vessels communicate with the lymphatic vessels of the opposite breast and with those of the anterior abdominal wall.

Basic Anatomy of the Muscles Connecting the Upper Limb and the Chest Wall

Pectoralis Major

This thick triangular muscle covers part of the anterior chest wall (Fig 2–8); its lower margin forms the anterior axillary fold.

Origin.—Medial half of clavicle, the sternum, and the upper six costal cartilages.

Insertion.—Its fibers converge and are inserted into the lateral lip of the bicipital groove of the humerus.

Nerve Supply.—Medial and lateral pectoral nerves from the medial and lateral cords of the brachial plexus.

Action.—Adducts and internally rotates the shoulder joint; the clavicular fibers also flex the shoulder.

Pectoralis Minor

This thin triangular muscle lies deep to the pectoralis major muscle (Fig 2–9).

Origin.—From the third, fourth, and fifth ribs.

Insertion.—The fibers converge to be attached to the tip of the coracoid process of the scapula.

Nerve Supply.—Medial pectoral nerve, a branch of the medial cord of the brachial plexus.

Action.—Pulls the shoulder downward and forward; if the shoulder is fixed, it elevates the ribs of origin.

Serratus Anterior

This large, thin muscle covers the lateral chest wall (see Fig 2–9).

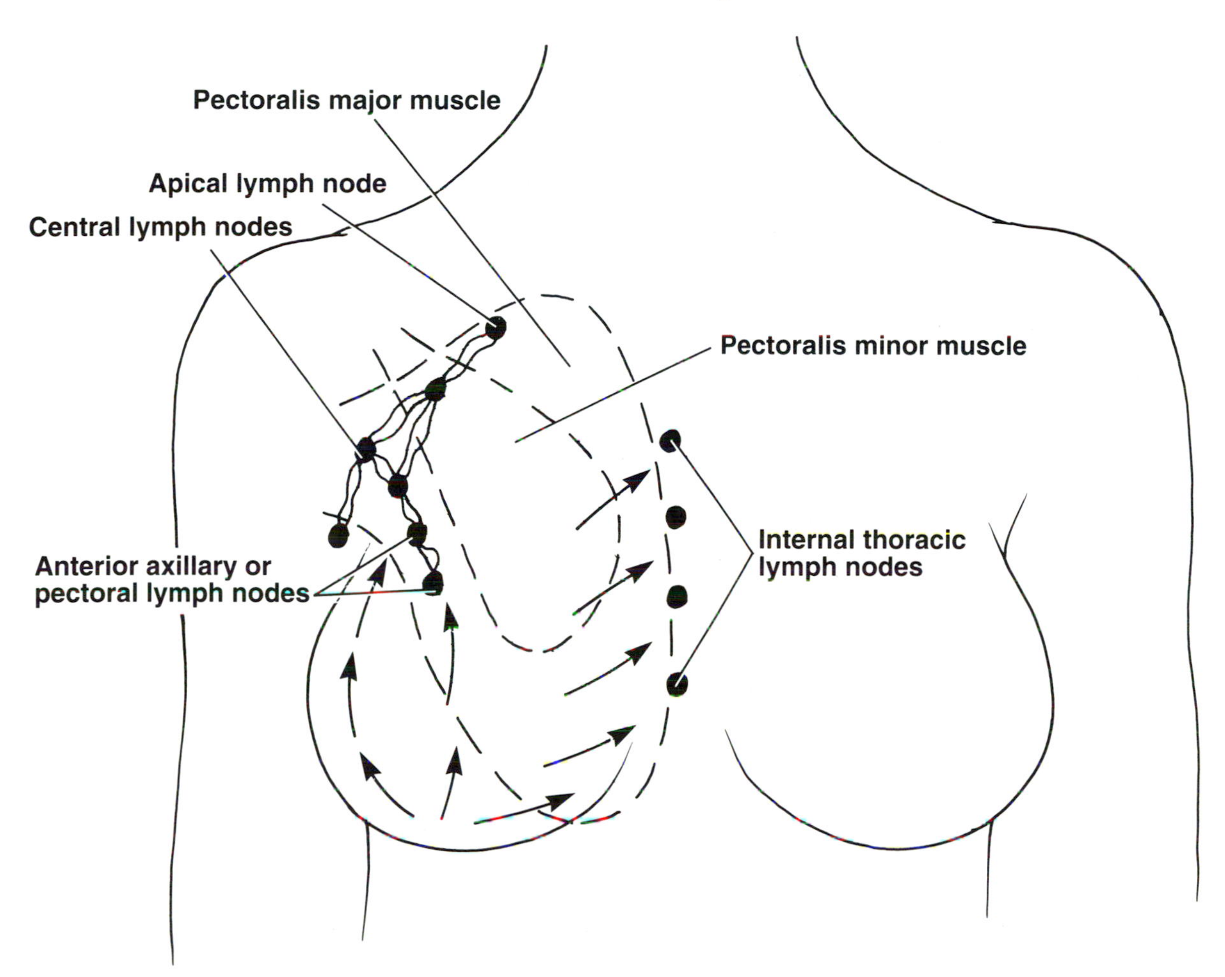

FIG 2–7.
Lymphatic drainage of the mammary gland.

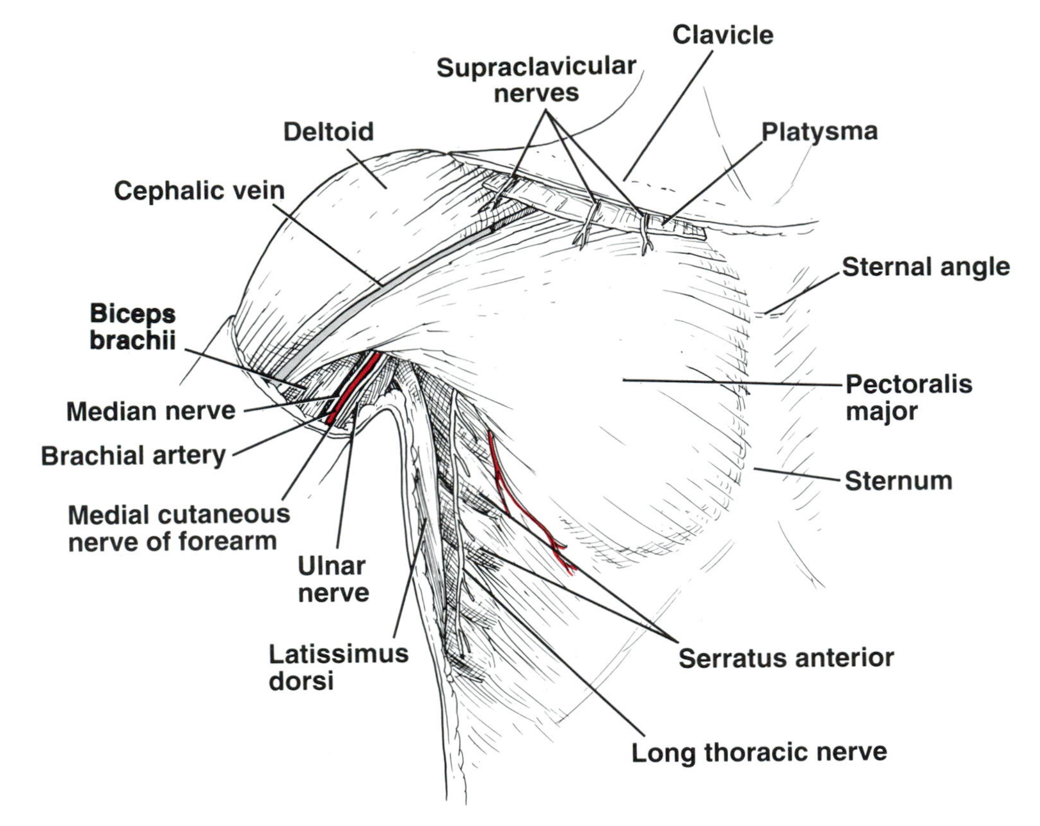

FIG 2–8.
Pectoral region and axilla.

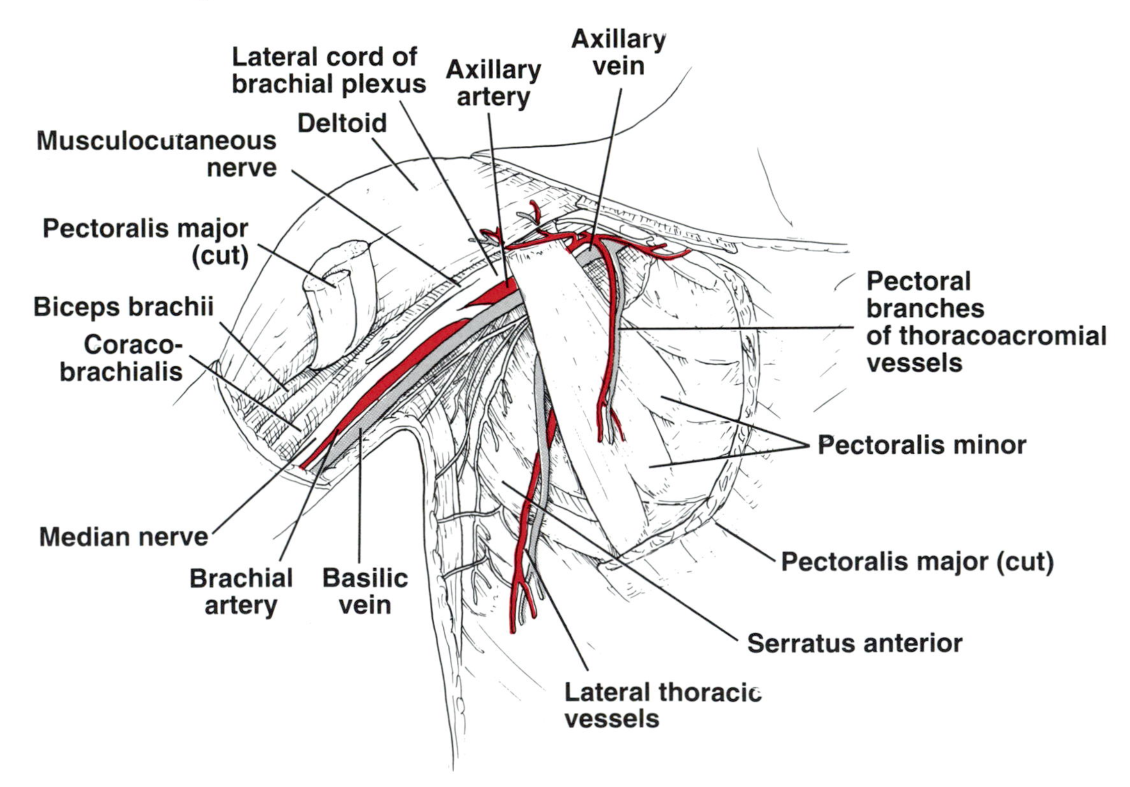

FIG 2–9.
Pectoral region and axilla; pectoralis major muscle has been removed to display the underlying vessels and nerves.

Origin.—From the outer surfaces of the upper eight ribs.

Insertion.—The anterior surface of the medial margin of the scapula, especially near the inferior angle.

Nerve Supply.—The long thoracic nerve from the roots C5, C6, and C7 of the brachial plexus.

Action.—Pulls the scapula forward around the thoracic wall and rotates the scapula.

Basic Anatomy of the Muscles of the Root of the Neck Associated With the First Rib

Scalenus Anterior
A deep muscle on the side of the neck connecting the vertebral column to the first rib. It lies beneath the sternocleidomastoid muscle and runs almost vertically downward (Fig 2–10).

Origin.—Transverse processes of the third, fourth, fifth, and sixth cervical vertebrae.

Insertion.—Inner border of first rib.

Nerve Supply.—Cervical spinal nerves.

Action.—Elevates first rib; laterally flexes and rotates cervical part of vertebral column.

Posterior Relations.—Subclavian artery, brachial plexus, and cervical dome of pleura.

Scalenus Medius
A large muscle connecting the vertebral column to the upper surface of the first rib; it lies posterior to the scalenus anterior (see Fig 2–10).

Origin.—Transverse processes of upper six cervical vertebrae.

Insertion.—Upper surface of the first rib behind the subclavian artery.

Nerve Supply.—Cervical spinal nerves.

Action.—Elevates first rib; laterally flexes and rotates cervical part of vertebral column.

Anterior Relations.—Subclavian artery, brachial plexus, and cervical dome of pleura.

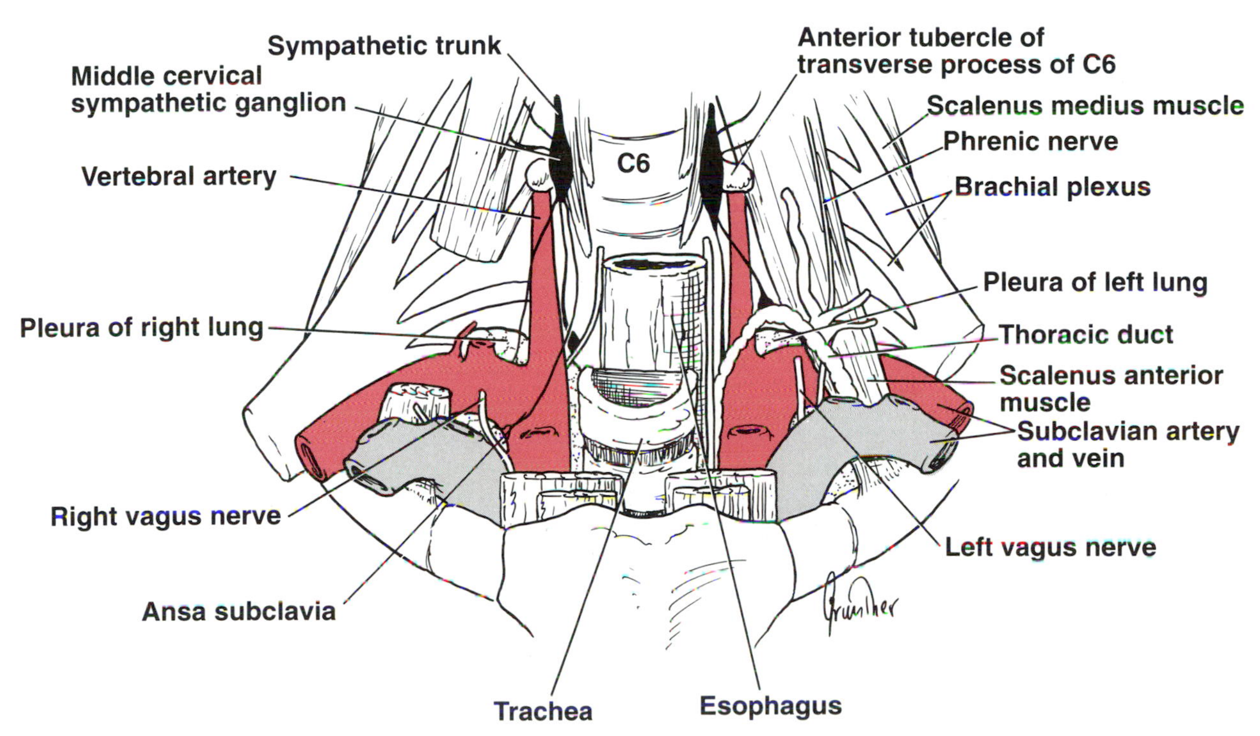

FIG 2–10.
Anterior view of the root of the neck.

Basic Anatomy of the Clavicle and its Relationship to the Thoracic Outlet

The clavicle is a long S-shaped bone that lies horizontally at the root of the neck and articulates with the sternum and first costal cartilage medially, and with the acromion process of the scapula laterally. The clavicle acts as a strut, which holds the upper limb away from the trunk. It transmits forces from the upper limb to the axial skeleton and provides attachment for muscles. The clavicle lies just beneath the skin through its length (Fig 2–11).

The clavicle crosses anterior to the apex of the axilla, and thus it is closely related to the first rib and the underlying brachial plexus and subclavian and axillary vessels (see Fig 2–9).

Basic Anatomy of the Chest Cavity

The chest cavity is bounded by the chest wall, the suprapleural membrane above (closing the thoracic outlet), and the diaphragm below (closing the lower opening of the thorax).

The chest cavity extends upward into the root of the neck about one fingerbreadth above the clavicle on each side (see Fig 2–11). Also, the diaphragm is a very thin muscle and is the only structure (apart from the pleura and peritoneum) that separates the thoracic from the abdominal viscera.

The chest cavity is divided by a midline partition, called the mediastinum, and has laterally placed pleurae and lungs.

Mediastinum

The mediastinum is a movable partition that lies between the pleurae and the lungs (Fig 2–12). It extends superiorly as far as the root of the neck and inferiorly to the diaphragm. It extends anteriorly to the sternum and posteriorly to the thoracic vertebrae.

For descriptive purposes the mediastinum is divided into the *superior mediastinum* and the *inferior mediastinum* by an imaginary plane passing from the sternal angle anteriorly to the lower border of the body of the fourth thoracic vertebra posteriorly (Fig 2–13). The inferior mediastinum is further subdivided into the *middle mediastinum*, which consists of the pericardium and heart; the *anterior mediastinum*, which is the space between the pericardium and the sternum; and the *posterior mediastinum*, which lies between the pericardium and the vertebral column.

Contents of the Superior Mediastinum.—*From anterior to posterior,* include the remains of thymus, bra-

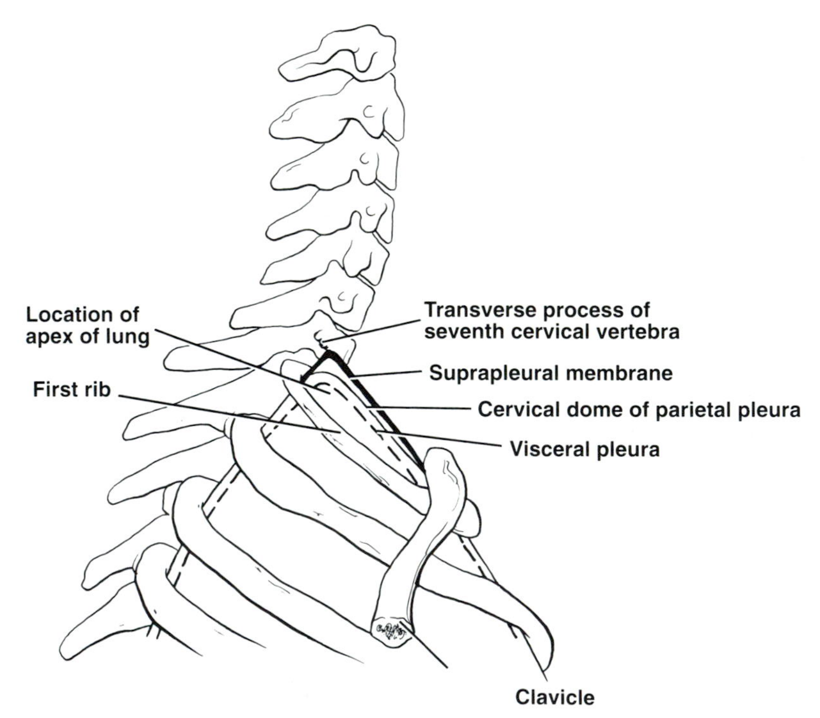

FIG 2–11.
Lateral view of the upper opening of the thoracic cage showing how the apex of the lung projects superiorly into the root of the neck. The apex of the lung is covered with visceral and parietal layers of pleura and protected by the suprapleural membrane, which is a thickening of the endothoracic fascia.

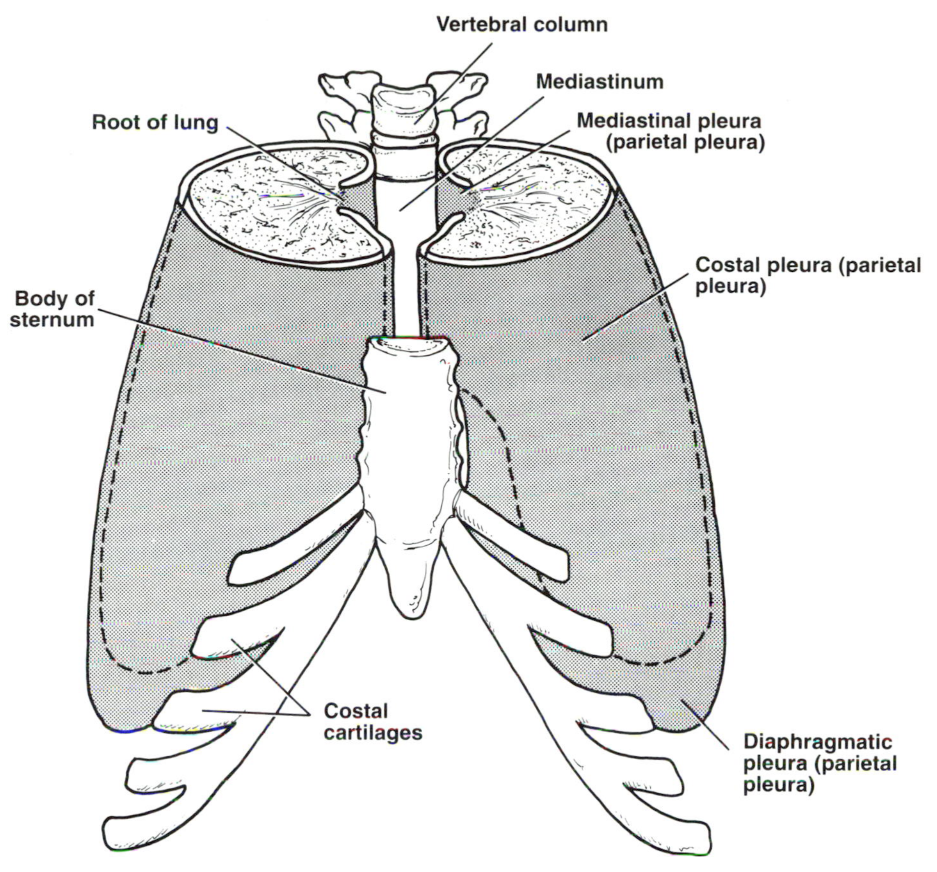

FIG 2–12.
Pleurae from above and in front. Note position of mediastinum and root of each lung.

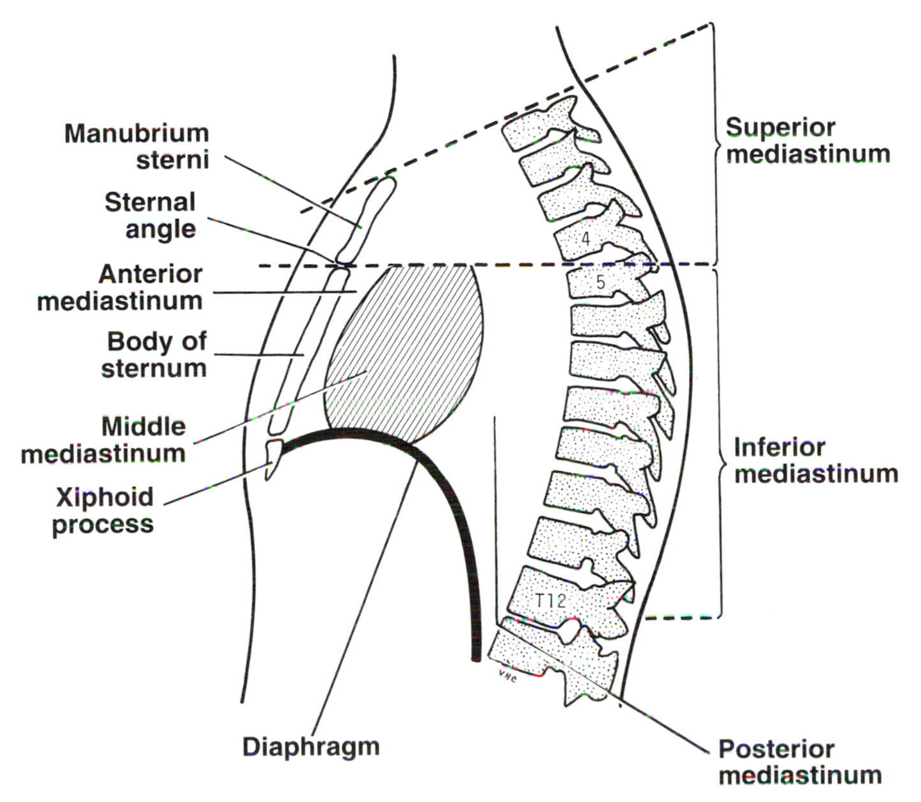

FIG 2–13.
Subdivisions of mediastinum.

chiocephalic veins, upper part of superior vena cava, brachiocephalic artery, left common carotid artery, left subclavian artery, arch of aorta, both phrenic and vagus nerves, left recurrent laryngeal and cardiac nerves, trachea and lymph nodes, esophagus and thoracic duct, and sympathetic trunks.

Contents of the Inferior Mediastinum.—*Contents of the anterior mediastinum* include the sternopericardial ligaments, lymph nodes, and remains of thymus. *Contents of the middle mediastinum* include the pericardium, heart and roots of great blood vessels, phrenic nerves, bifurcation of trachea, and lymph nodes. *Contents of the posterior mediastinum* include the descending thoracic aorta, esophagus, thoracic duct, azygos and hemiazygos veins, vagus nerves, splanchnic nerves, sympathetic trunks, and lymph nodes.

Basic Anatomy of the Pleurae

The pleurae and lungs lie on either side of the mediastinum within the chest cavity (see Fig 2–12). The pleurae are two serous sacs surrounding and covering the lungs. Each pleura has two parts—a *parietal layer,* which lines the thoracic wall and covers the thoracic surface of the diaphragm and the lateral surface of the mediastinum, and a *visceral layer,* which covers the outer surfaces of the lungs and extends into the interlobar fissures.

The parietal layer of each pleura becomes continuous with the visceral layer at the *hilum* of each lung, where they form a cuff that surrounds the structures entering and leaving the lung at the *lung root.* The *pulmonary ligament* is a loose extension of the cuff below the lung root that allows for movement of the pulmonary vessels and large bronchi during respiration (Fig 4–14).

Nerve Supply Of Pleura

Parietal Pleura.—The parietal pleura is sensitive to pain, temperature, touch, and pressure. The costal parietal pleura is innervated by intercostal nerves, the mediastinal pleura is supplied by the phrenic nerve, and the diaphragmatic pleura is supplied over the domes by the phrenic nerve and around the periphery by the lower intercostal nerves (Fig 2–14).

Visceral Pleura.—The visceral pleura covering the lungs receives an autonomic vasomotor supply from the pulmonary plexus and is only sensitive to stretch.

Basic Anatomy of the Pleural Cavity (Pleural Space)

The pleural cavity is a slitlike space that separates the parietal and visceral layers of pleura (see Fig 2–2).

The pleural cavity normally contains a small amount of *pleural fluid* that lubricates the apposing pleural surfaces. The *costodiaphragmatic recess* is the

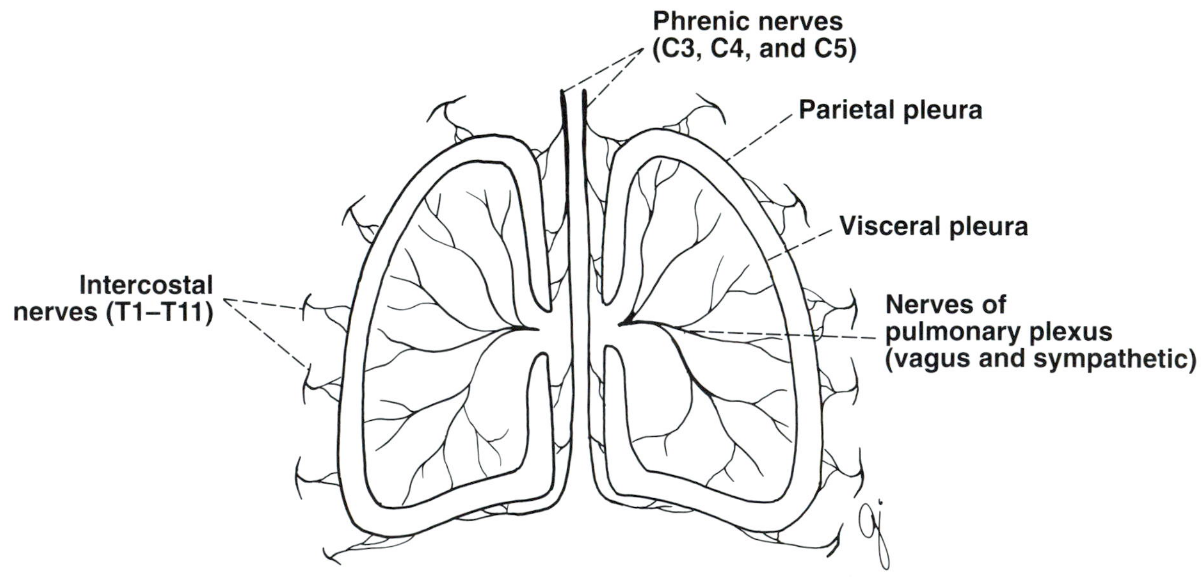

FIG 2–14.
Diagram showing the innervation of the parietal and visceral layers of pleura.

lowest area of the pleural cavity into which the lung expands on inspiration. The *costomediastinal recesses* are situated along the anterior margins of the pleura. They are slitlike spaces between the parietal pleura lining the chest wall and that covering the lateral surface of the mediastinum. During inspiration and expiration the anterior borders of the lungs slide in and out of the recesses.

Pleural Reflections

The boundaries of the pleural cavity can be marked out as lines on the body surface. The lines, which will indicate the limits of the parietal pleura where it lies close to the body surface, are referred to as the *lines of pleural reflection*. The surface markings of the pleural reflections are described on p. 80.

The lower margins of the lungs (and visceral pleura) cross the 6th, 8th, and 10th ribs at the midclavicular lines, the midaxillary lines, and the sides of the vertebral column, respectively. The lower margins of the parietal pleura cross, at the same points respectively, the 8th, 10th, and 12th ribs. The distance between the two borders corresponds to the *costodiaphragmatic recess*.

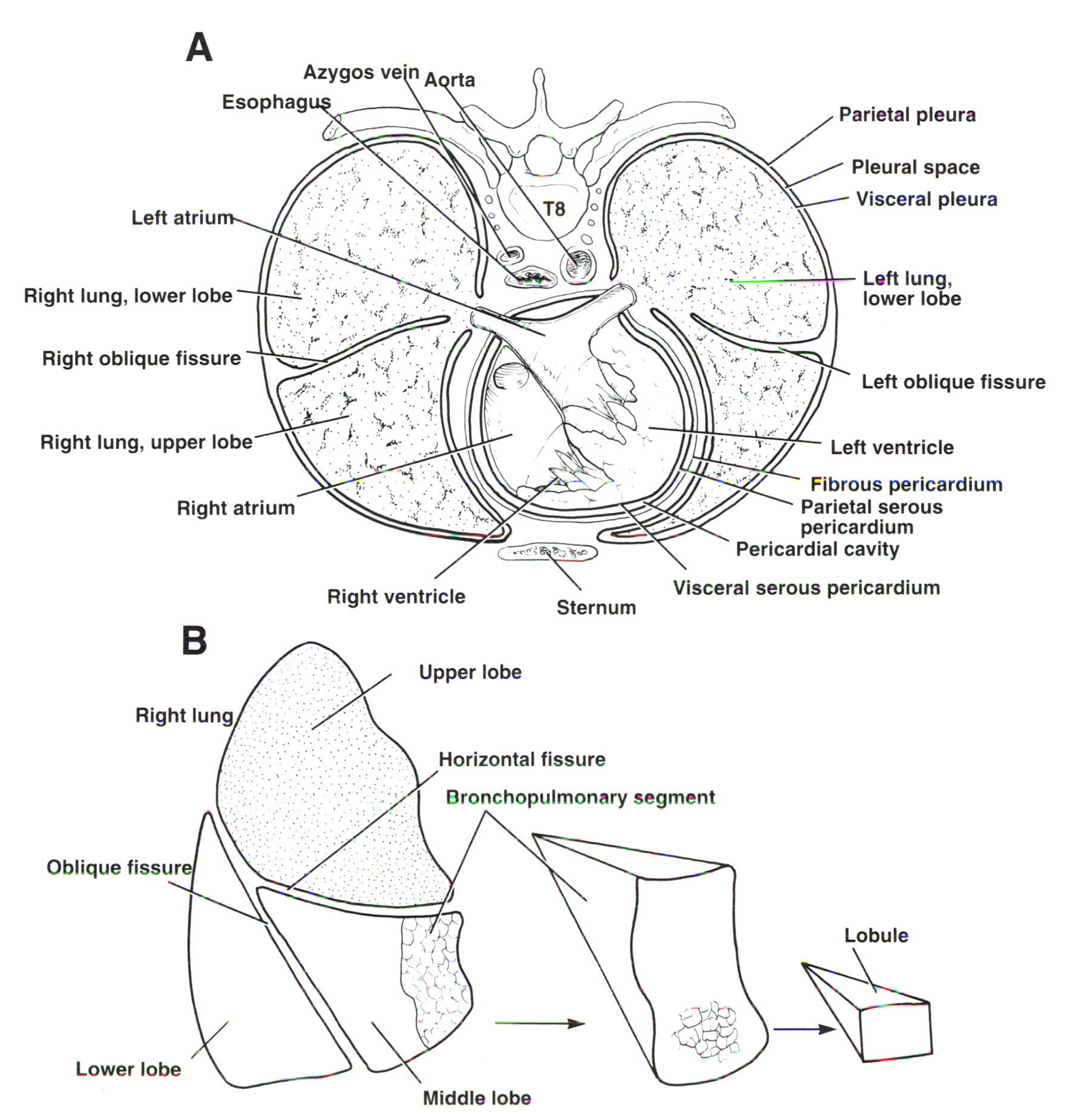

FIG 2–15.
A, cross section of the thorax at the level of the eighth thoracic vertebra. Note the arrangement of the pleura and pleural cavity and the fibrous and serous pericardia. **B**, subdivisions of the right lung passing from the three lobes to a bronchopulmonary segment to a lobule.

Basic Anatomy of the Lungs

The lungs, one on each side of the mediastinum, are separated by the heart, great vessels, and other structures in the mediastinum (Fig 2–15). Each lung has a blunt *apex* that projects upward about 1 in. (2.5 cm) above the clavicle, a concave *base* that overlies the diaphragm, a convex *costal surface* that corresponds to the chest wall, and a concave *mediastinal surface* that is molded to the pericardium and other mediastinal structures (see Fig 2–15). About the middle of this surface is the *hilum,* a depression where the bronchi, pulmonary blood vessels and lymphatic vessels, and nerves enter the lung to form the *root.*

The *anterior border* is thin and overlaps the heart; it is here on the left lung that a notch called the *cardiac notch* is located (Fig 2–16). The *posterior border* is thick and lies beside the vertebral column.

Lobes and Fissures

The right lung is slightly larger than the left lung and is divided by the oblique and horizontal fissures into three lobes—the *upper, middle,* and *lower lobes* (see Fig 2–16). The *oblique fissure* runs from the inferior border upward and backward across the medial and costal surfaces until it cuts the posterior border about 2.5 in. (6.25 cm) below the apex. The *horizontal fissure* runs horizontally across the costal surface at the level of the fourth costal cartilage to meet the oblique fissure in the midaxillary line. The middle lobe is thus a small triangular lobe bounded by the horizontal and oblique fissures (see Fig 2–16).

The left lung is divided by a similar oblique fis-

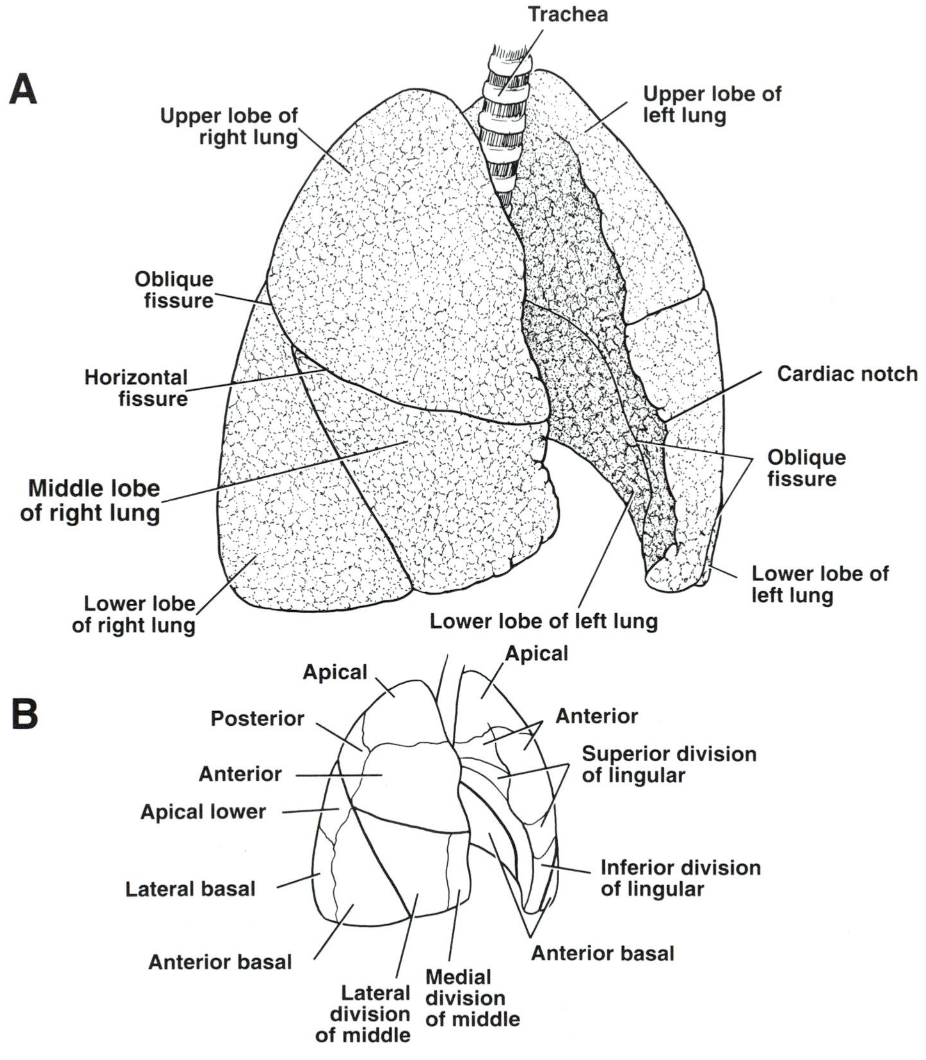

FIG 2–16.
Lungs viewed from the right. **A,** lobes. **B,** bronchopulmonary segments.

sure into two lobes, the *upper* and *lower lobes* (Fig 2–17). There is no horizontal fissure in the left lung.

Bronchopulmonary Segments

Each lobar (secondary) bronchus, which passes to a lobe of the lung, has branches called *segmental* (tertiary) *bronchi* (Fig 2–18). Each segmental bronchus passes to a structurally and functionally independent unit of a lung lobe called a *bronchopulmonary segment*, which is surrounded by connective tissue (Fig 2–19). The segmental bronchus is accompanied by a branch of the pulmonary artery, but the tributaries of the pulmonary veins run in the connective tissue between adjacent bronchopulmonary segments. Each segment has its own lymphatic vessels and autonomic nerve supply.

The main bronchopulmonary segments (see Figs 2–16 and 2–17) are as follows:

Right lung	
Superior lobe	(1) Apical, (2) Posterior, (3) Anterior,
Middle lobe	(4) Lateral, (5) Medial,
Inferior lobe	(6) Superior (apical), (7) Medial basal, (8) Anterior basal, (9) Lateral basal, (10) Posterior basal.
Left lung	
Superior lobe	(1) Apical, (2) Posterior, (3) Anterior (4) Superior lingular, (5) Inferior lingular
Inferior lobe	(6) Superior (apical), (7) Medial basal (8) Anterior basal, (9) Lateral basal, (10) Posterior basal.

Diagrams showing the positions of the bronchopulmonary segments are seen in Figures 2–16 and 2–17.

On entering a bronchopulmonary segment, each segmental bronchus divides repeatedly (see Fig 2–19). As the bronchi become smaller, the U-shaped

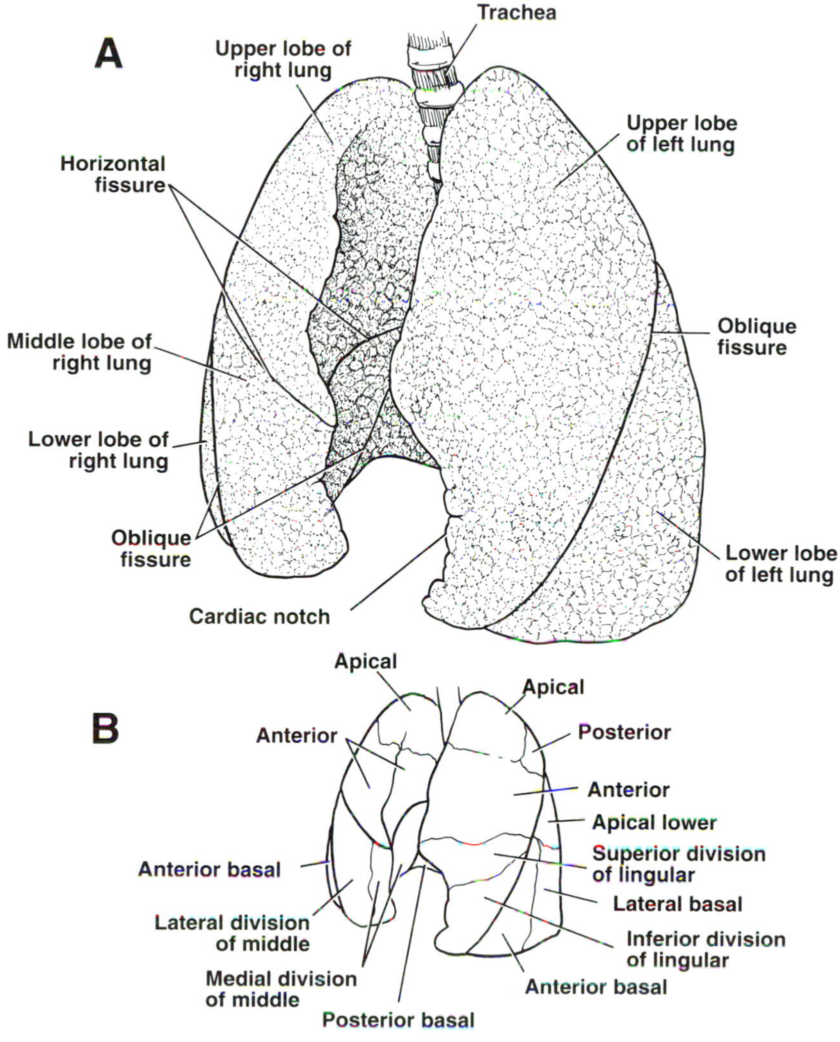

FIG 2–17.
Lungs viewed from the left. **A,** lobes. **B,** bronchopulmonary segments.

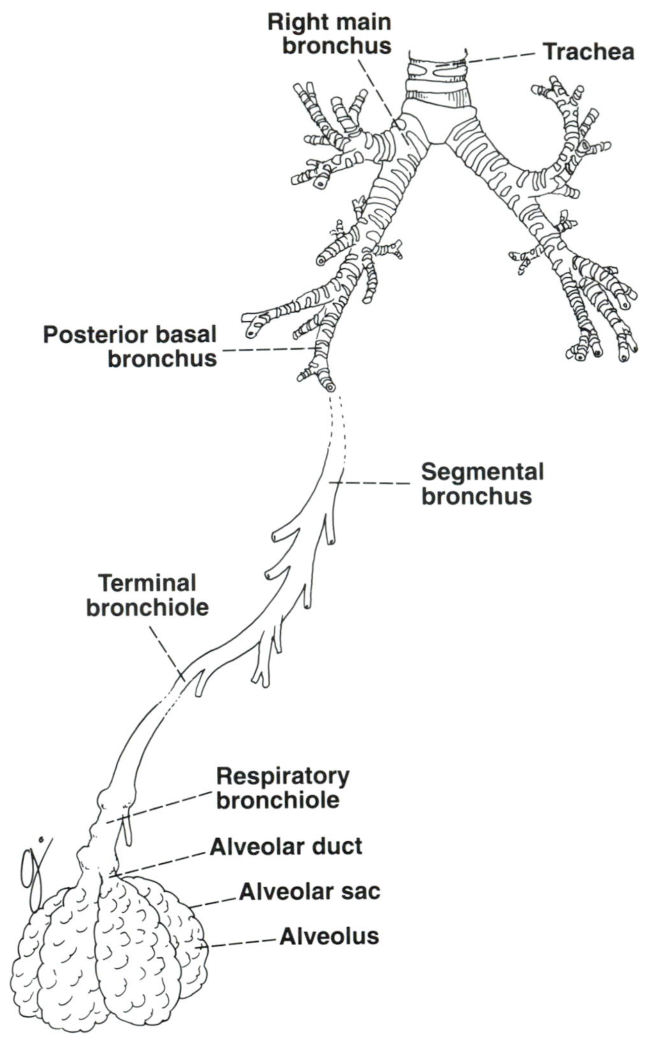

FIG 2–18.
Diagram showing the continuity of the trachea, bronchi, bronchioles, alveolar ducts, alveolar sacs, and alveoli.

bars of hyaline cartilage found in the trachea are gradually replaced by irregular plates of cartilage, which become smaller and fewer in number. The smallest bronchi divide and give rise to *bronchioles,* which are less than 1 mm in diameter (see Fig 2–19). Bronchioles possess no cartilage in their walls and are lined with ciliated columnar cells. The submucosa possesses a complete layer of circularly arranged smooth-muscle fibers.

The bronchioles then divide and give rise to *terminal bronchioles* (see Fig 2–19). The terminal bronchioles divide and give rise to *respiratory bronchioles,* which show delicate, air-containing outpouchings from their walls (see Fig 2–19). Gaseous exchange between blood and air takes place in the walls of these outpouchings, which explains the name respiratory bronchiole. The diameter of a respiratory bronchiole is about 0.5 mm. The respiratory bronchioles end by branching into *alveolar ducts* that lead into tubular passages with numerous thin-walled outpouchings called *alveolar sacs* (Fig 2–20). The alveolar sacs consist of several alveoli opening into a single chamber (see Fig 2–19). Each alveolus is surrounded by a rich network of blood capillaries. Examination of the wall of an alveolus with an electron microscope (Fig 2–21) shows that the alveoli are in fact lined with simple squamous cells (type 1), but an occasional cuboidal cell (type 2) is responsible for secreting *surfactant* into the lumen. The surfactant reduces the surface tension on the lining cells and permits the alveolar walls to separate from one another as air enters during inspiration.

Respiratory Membrane

Electron microscopic study of the alveolar walls (see Fig 2–21) shows that the lumen of the alveolus is separated from the lumen of the blood capillary by the following: (1) surfactant-containing fluid produced by the type 2 alveolar cells, (2) the alveolar squamous epithelium or type 1 cell, (3) epithelial basement membrane, (4) minute tissue space, (5) blood capillary basement membrane, which fuses with the epithelial basement membrane in many places, and (6) capillary squamous endothelium.

The combination of these layers is known as the *respiratory membrane* and averages about 0.5μm in thickness. Gaseous exchange takes place at the respiratory membrane between the air in the alveolar lumen and the blood within the capillary.

The bronchi, the terminal bronchioles, the respiratory bronchioles, the alveolar ducts, the alveolar sacs, the alveoli, the blood and lymphatic vessels, and the nerves are held together by connective tissue to form the lung.

Lung Changes During Respiration

Inspiration.—During inspiration, the root of the lung descends and the level of the bifurcation of the trachea may be lowered by as much as two vertebrae. The bronchi elongate and dilate and the alveolar capillaries dilate, thus assisting the pulmonary circulation. As the diaphragm descends, the costodiaphragmatic recess of the pleural cavity opens up, and the expanding sharp lower edges of the lungs descend to a lower level.

Expiration.—In expiration, the roots of the lungs ascend along with the bifurcation of the trachea. The bronchi shorten and contract. With the upward movement of the diaphragm, increasing areas of the

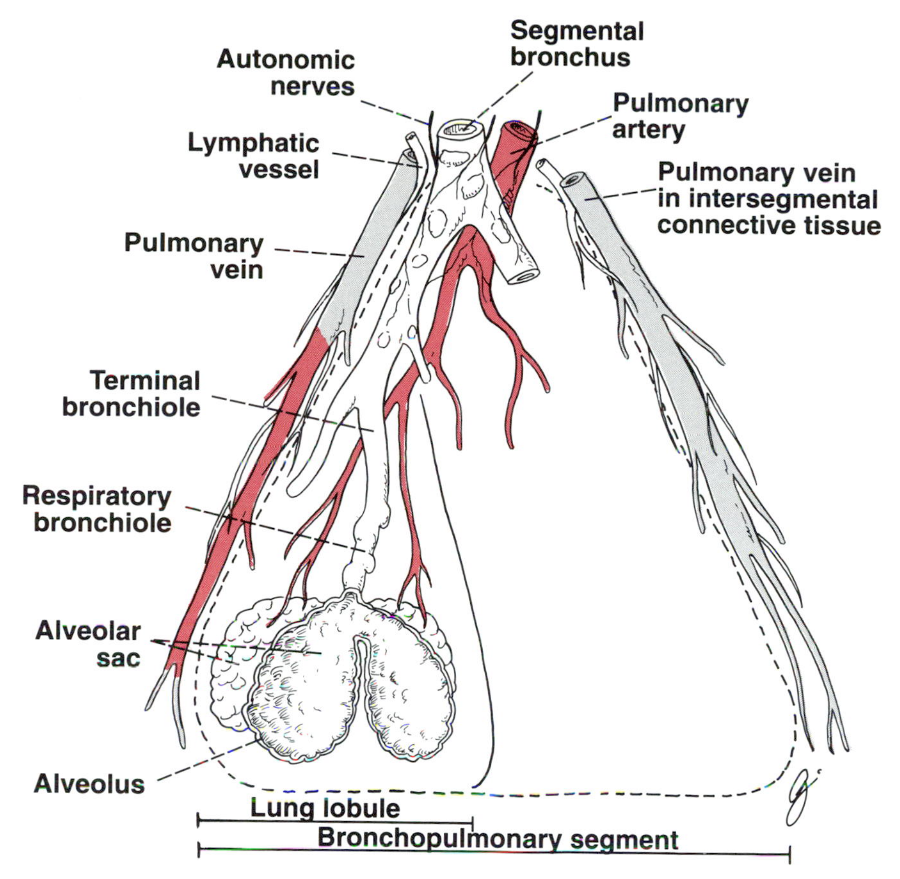

FIG 2–19.
A bronchopulmonary segment and a lung lobule. Note that the pulmonary veins lie within the connective tissue septa that separate adjacent segments.

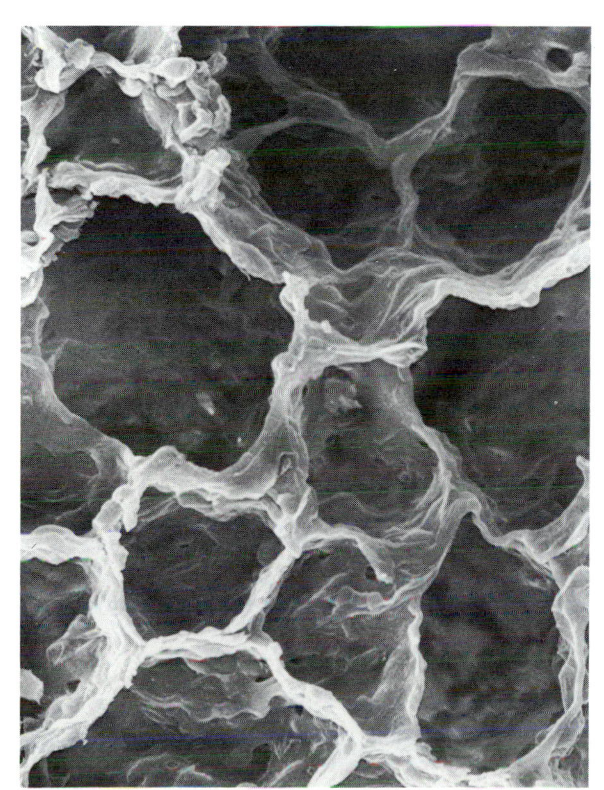

FIG 2–20.
Scanning electron micrograph of the lung showing numerous alveolar sacs. The alveoli are the depressions, or alcoves, along the walls of the alveolar sac (×430). Courtesy of Dr M Koering.

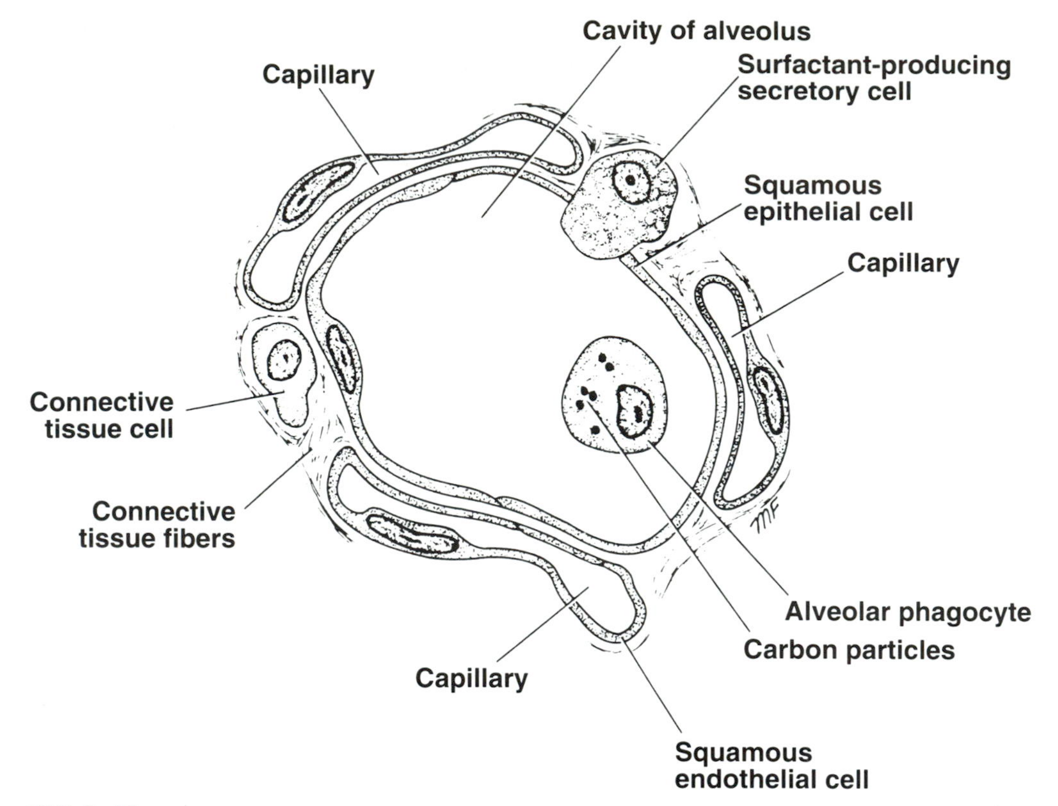

FIG 2–21.
Diagram showing the detailed structure of the respiratory membrane.

diaphragmatic and costal parietal pleura come into apposition, and the costodiaphragmatic recess becomes reduced in size. The lower margins of the lungs shrink and rise to a higher level.

Clinical Note

The changes in the position of the thoracic and upper abdominal viscera and the level of the diaphragm during different phases of respiration relative to the chest wall are of considerable clinical importance. A penetrating wound in the lower part of the chest may or may not damage abdominal viscera, depending on the phase of respiration at the time of injury.

Blood Supply of the Lungs

The bronchi, the connective tissue of the lung, and the visceral pleura receive their blood supply from the bronchial arteries, which are branches of the descending aorta. The bronchial veins are divided into two groups—the deep bronchial veins and the superficial bronchial veins. The *deep bronchial veins* drain the bronchi within the lungs and end in the pulmonary veins. The *superficial bronchial veins* drain the extrapulmonary bronchi, the visceral pleura, and the hilar lymph nodes and drain into the pulmonary veins, and the azygos and hemiazygos veins. The azygos veins lie in the posterior mediastinum and drain their blood into the superior vena cava; they are also connected below to the inferior vena cava.

The alveoli receive deoxygenated blood from the terminal branches of the pulmonary arteries. The tributaries of the pulmonary veins join together, and two pulmonary veins leave each lung root to empty into the left atrium of the heart.

The double blood supply to the lungs from the pulmonary and bronchial arteries is of clinical significance. The bronchial arteries contain oxygenated blood that amounts to about 1% to 2% of the total cardiac output. They are able to supply the lungs adequately and keep the lung tissue viable should the pulmonary artery suddenly be blocked, as in pulmonary embolism. If the pulmonary embolism is small and lodges very peripherally, the bronchial circulation cannot adequately sustain the lungs, and actual pulmonary infarction can result.

Lymph Drainage of Lungs

The lymph vessels originate in superficial and deep plexuses. The superficial plexus lies beneath the visceral pleura, and efferent vessels converge to the *bronchopulmonary nodes* in the hilum. The deep plexus travels along the bronchi and toward the root of the lung. The efferent vessels from the deep plexus drain into *pulmonary nodes,* which are located close to the hilum. All lymph from the pulmonary nodes drains into the *bronchopulmonary nodes* in the hilum. The lymph then drains into the *tracheobronchial nodes* (Fig 2–22) and into the *bronchomediastinal lymph trunks.*

Nerve Supply of the Lungs

Each lung is supplied by the *pulmonary plexus.* The plexus is formed from branches of the sympathetic trunk and receives parasympathetic fibers from the vagus nerve. The sympathetic fibers produce bronchodilation and are vasoconstrictor to the mucosal blood vessels. The parasympathetic fibers produce bronchoconstriction and vasodilation and increase glandular secretion.

Radiographic Appearances of the Lungs and Chest Wall

The more important structures seen in standard posteroanterior and lateral radiographs of the chest are shown in Figures 2–23 and 2–24. An example of an

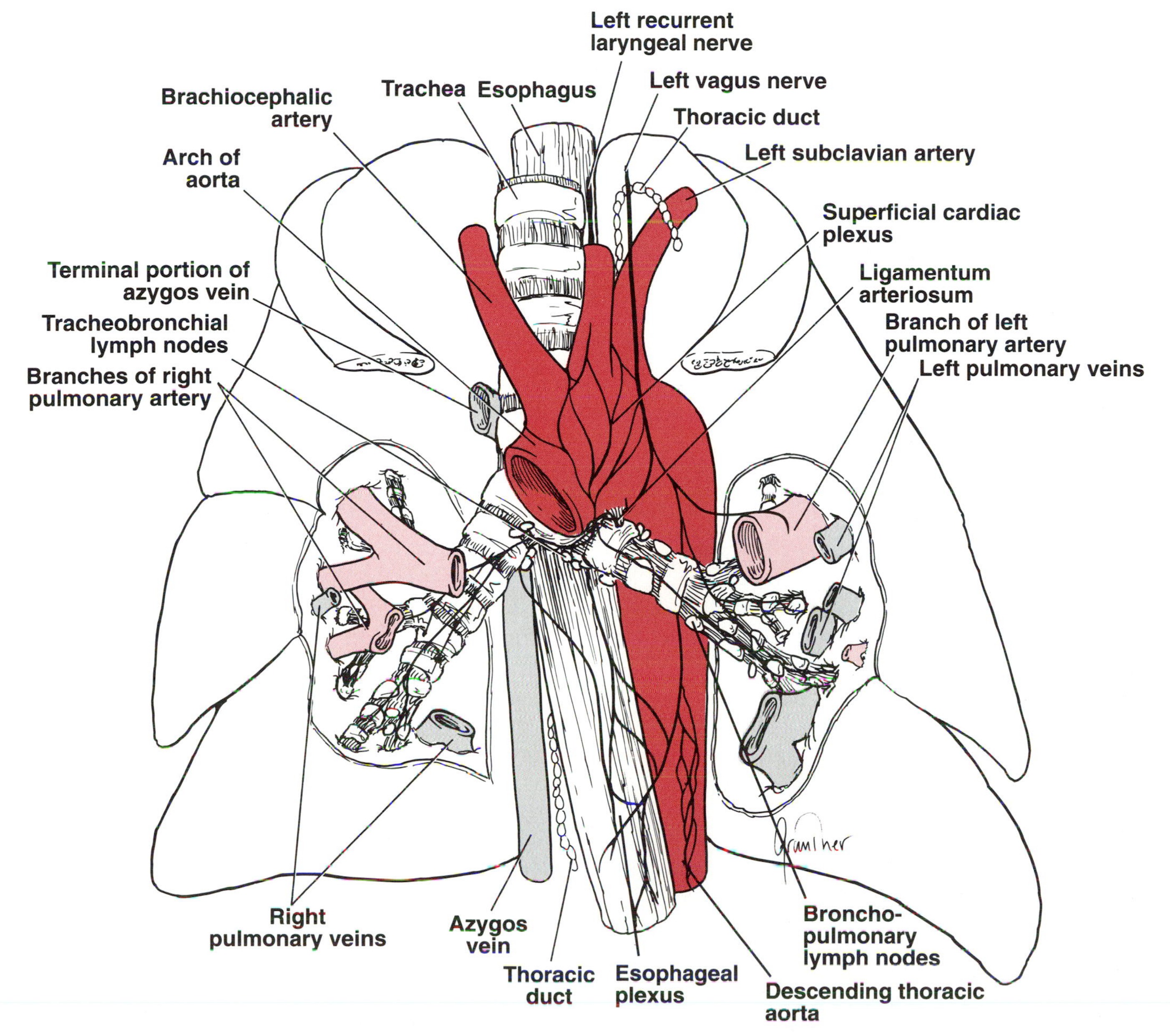

FIG 2–22.
Anterior view of the trachea, main bronchi, and roots of the lungs.

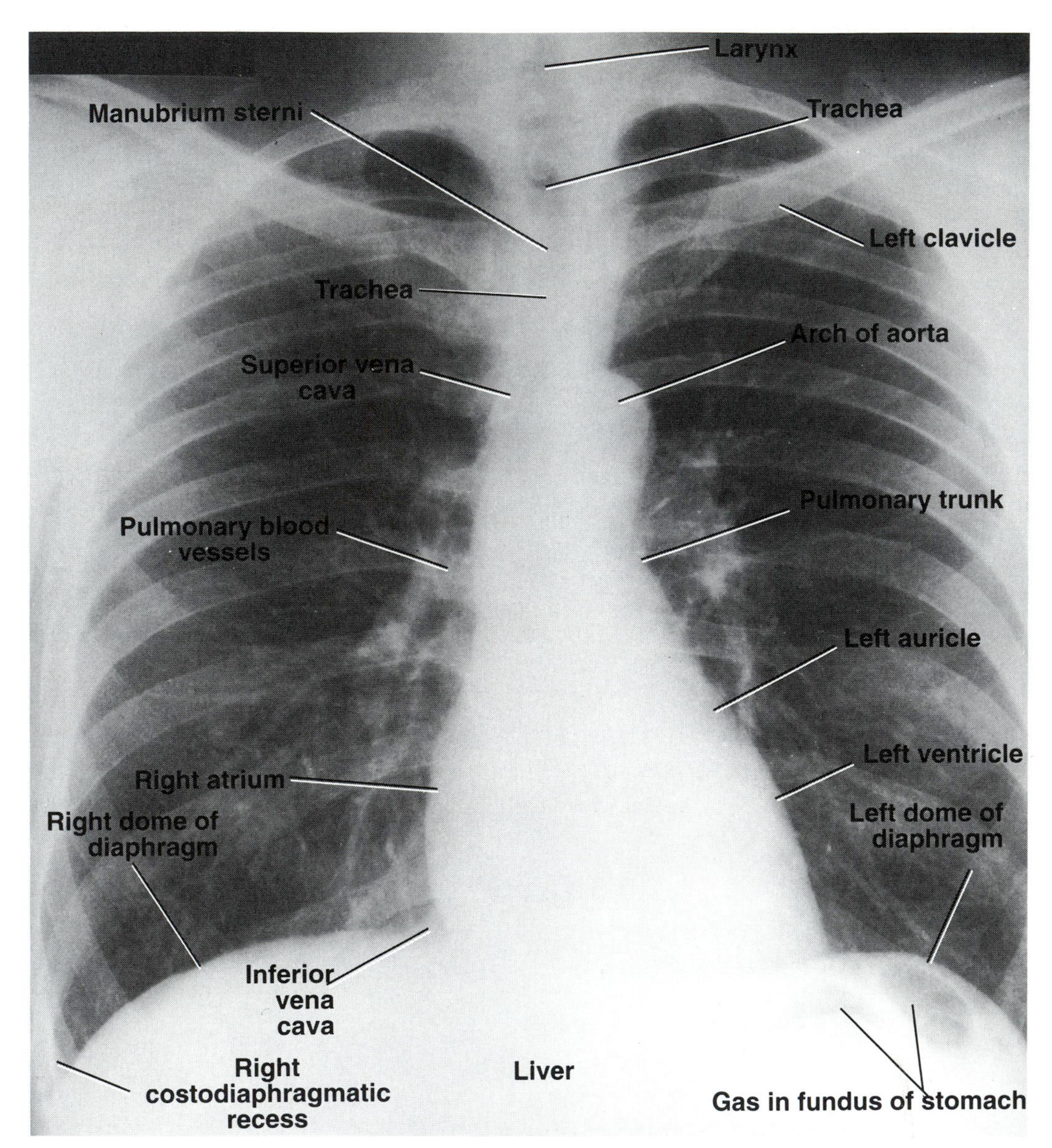

FIG 2–23.
Posterior anterior radiograph of the chest of a normal adult male.

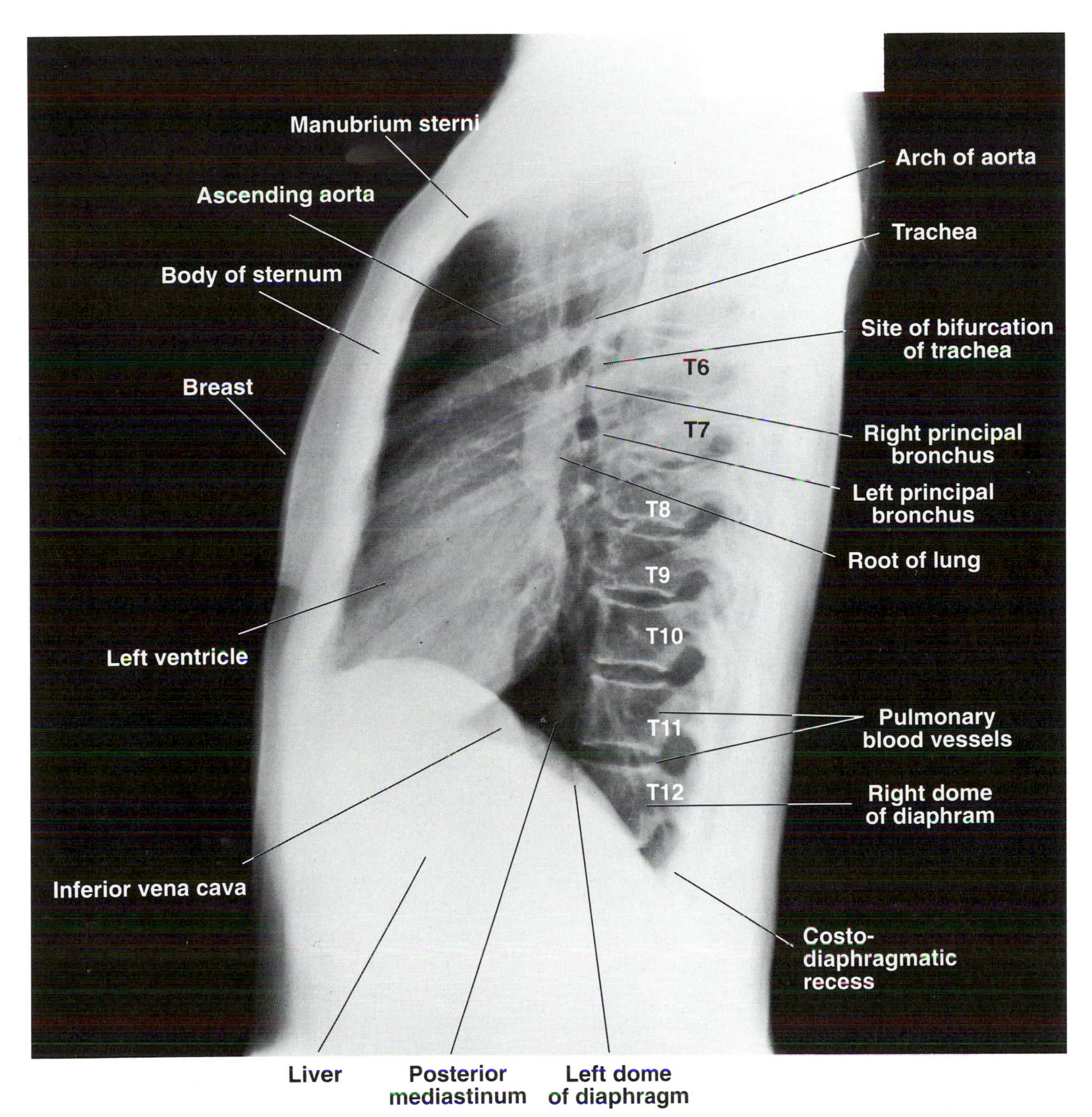

FIG 2–24.
Left lateral radiograph of the chest of a normal adult female.

oblique radiograph obtained with a barium examination is shown in Figures 2–25 and 2–26. Examples of computed tomographic scans of the chest are shown in Figure 2–27.

SURFACE ANATOMY

Surface Anatomy of the Chest Wall

Anterior Surface of the Chest

Suprasternal Notch.—This is the superior margin of the manubrium sterni and is easily palpated between the prominent medial ends of the clavicles in the midline (Fig 2–28). It lies opposite the lower border of the body of the second thoracic vertebra (see Fig 2–13).

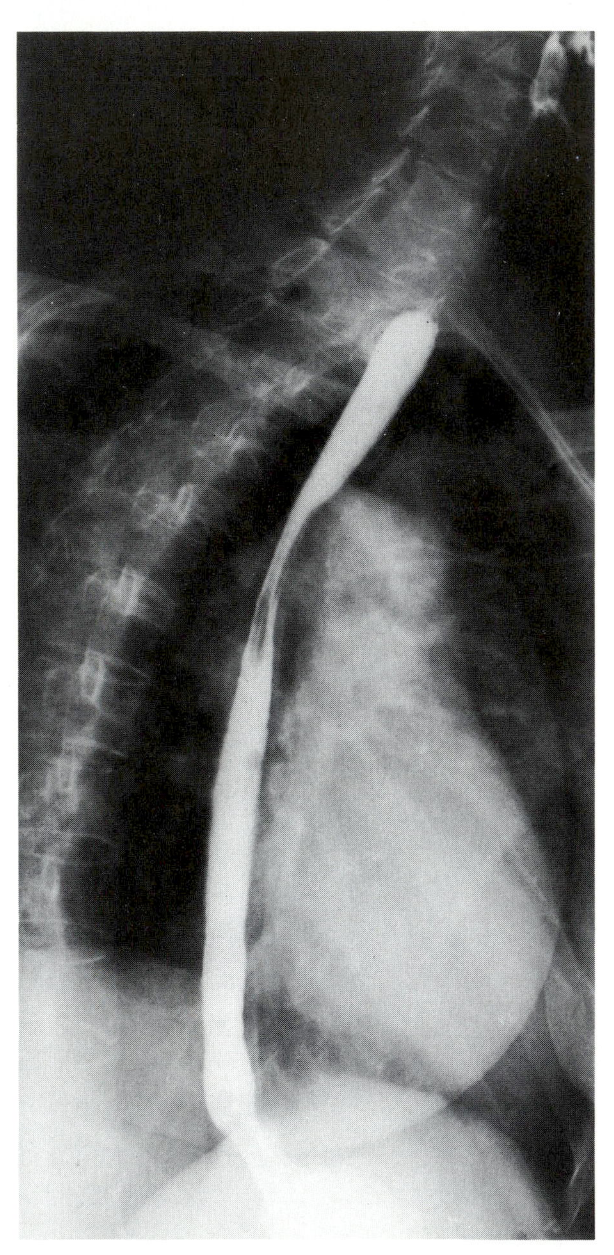

FIG 2–25.
Right oblique radiograph of the chest following a barium swallow.

Sternal Angle (Angle of Louis).—This is the angle between the manubrium and body of the sternum (see Fig 2–28); at this level the *second costal cartilage* joins the lateral margin of the sternum. The sternal angle lies opposite the intervertebral disc between the fourth and fifth thoracic vertebra (see Fig 2–13).

Clinical Note

This is a very important landmark when examining the chest. Its position can easily be felt and is often seen as a transverse ridge. A finger moved to the right or to the left will pass directly onto the second costal cartilage and then the second rib. All other ribs may be counted from this point. Occasionally in a very muscular male, the ribs and intercostal spaces are obscured by large pectoral muscles. In these cases it may be easier to count up from the 12th rib.

Xiphisternal Joint.—This is the joint between the xiphoid process of the sternum and the body of the sternum (see Fig 2–1). It lies opposite the body of the ninth thoracic vertebra (see Fig 2–13).

Subcostal Angle.—This is situated at the inferior end of the sternum, between the sternal attachments of the seventh costal cartilages (see Fig 2–1).

Costal Margin.—This is the lower boundary of the thorax and is formed by the cartilages of the 7th, 8th, 9th, and 10th ribs and the ends of the 11th and 12th cartilages (see Fig 2–28). The lowest part of the costal margin is formed by the 10th rib and lies at the level of the 3rd lumbar vertebra.

Clavicle.—This bone is subcutaneous throughout its entire length and can be easily palpated (see Fig 2–28). It articulates at its lateral extremity with the acromion process of the scapula.

Ribs.—The 1st rib lies deep to the clavicle and cannot be palpated. The lateral surfaces of the remaining ribs can be felt by pressing the fingers upward into the axilla and drawing them downward over the lateral surface of the chest wall. The *12th rib* can be used to identify a particular rib by counting from below. However, in some individuals, the 12th rib is very short and difficult to feel. For this reason an alternative method may be used to identify ribs by

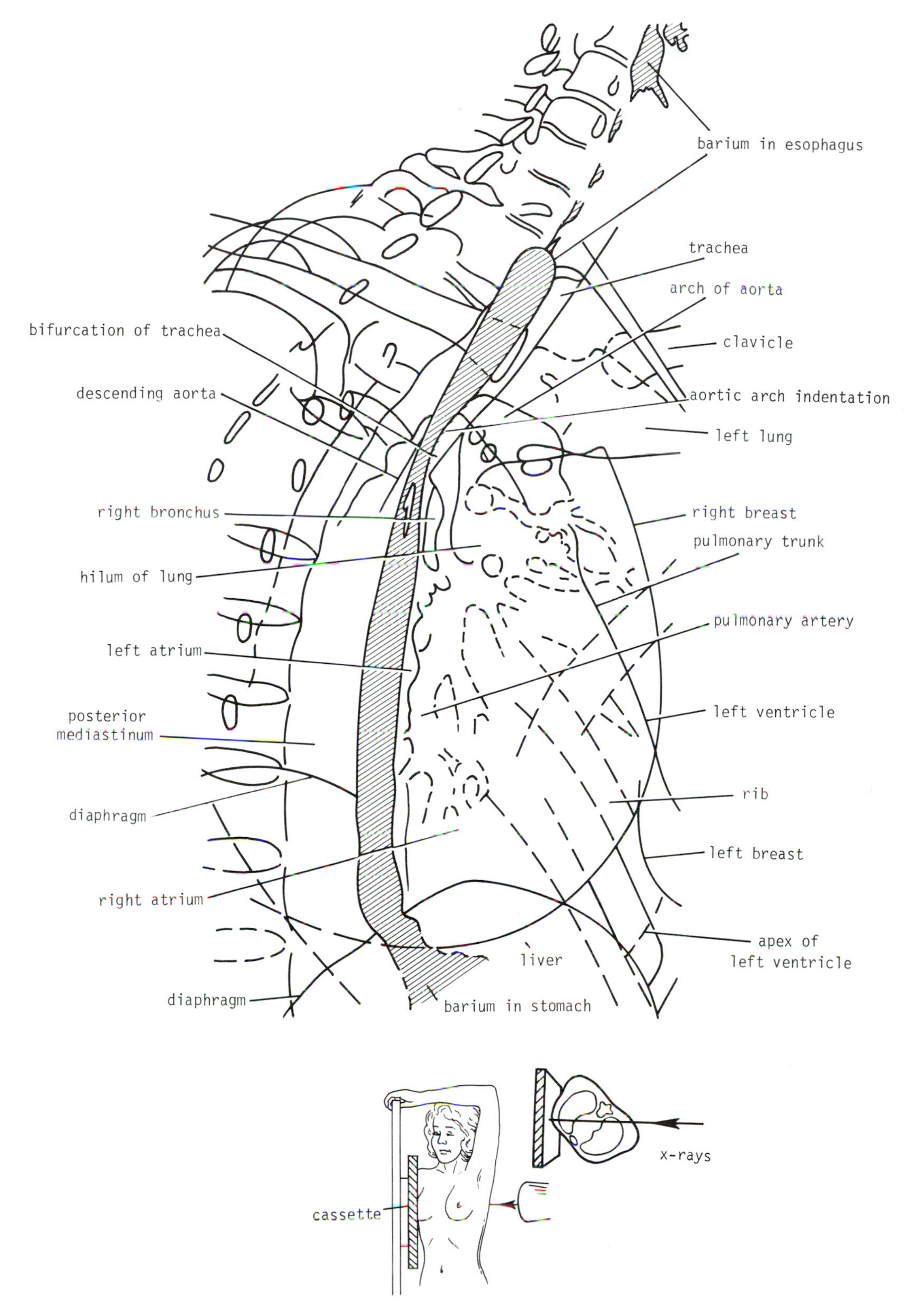

FIG 2–26.
Main features seen in right oblique radiograph in Figure 2–25. Note the position of the patient in relation to the X-ray source and cassette holder.

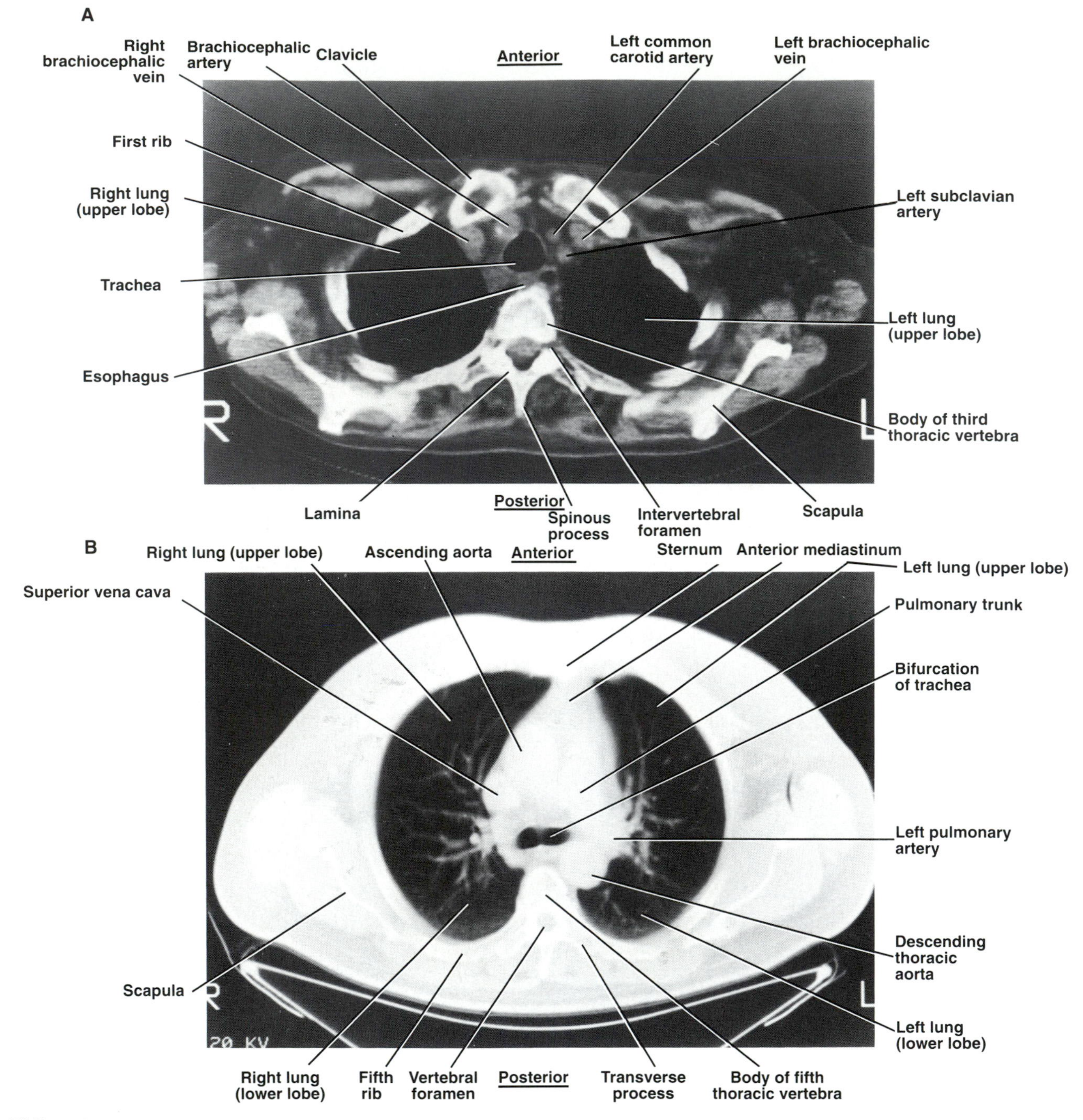

FIG 2–27.
A, computed tomographic (CT) scan of upper part of thorax at the level of the third thoracic vertebra. The section is viewed from below. **B,** CT scan of the middle of the thorax at the level of fifth thoracic vertebra. The section is viewed from below.

first palpating the sternal angle and the 2nd costal cartilage (see above).

Nipple.—In males it usually lies in the fourth intercostal space about 4 in. (10 cm) from the midline. In females its position is not constant.

Apex Beat.—The apex of the heart is formed by the lower portion of the left ventricle. It is normally found in the fifth left intercostal space 3½ in. (9 cm) from the midline. The beat may be accentuated by having the patient lean forward in the sitting position.

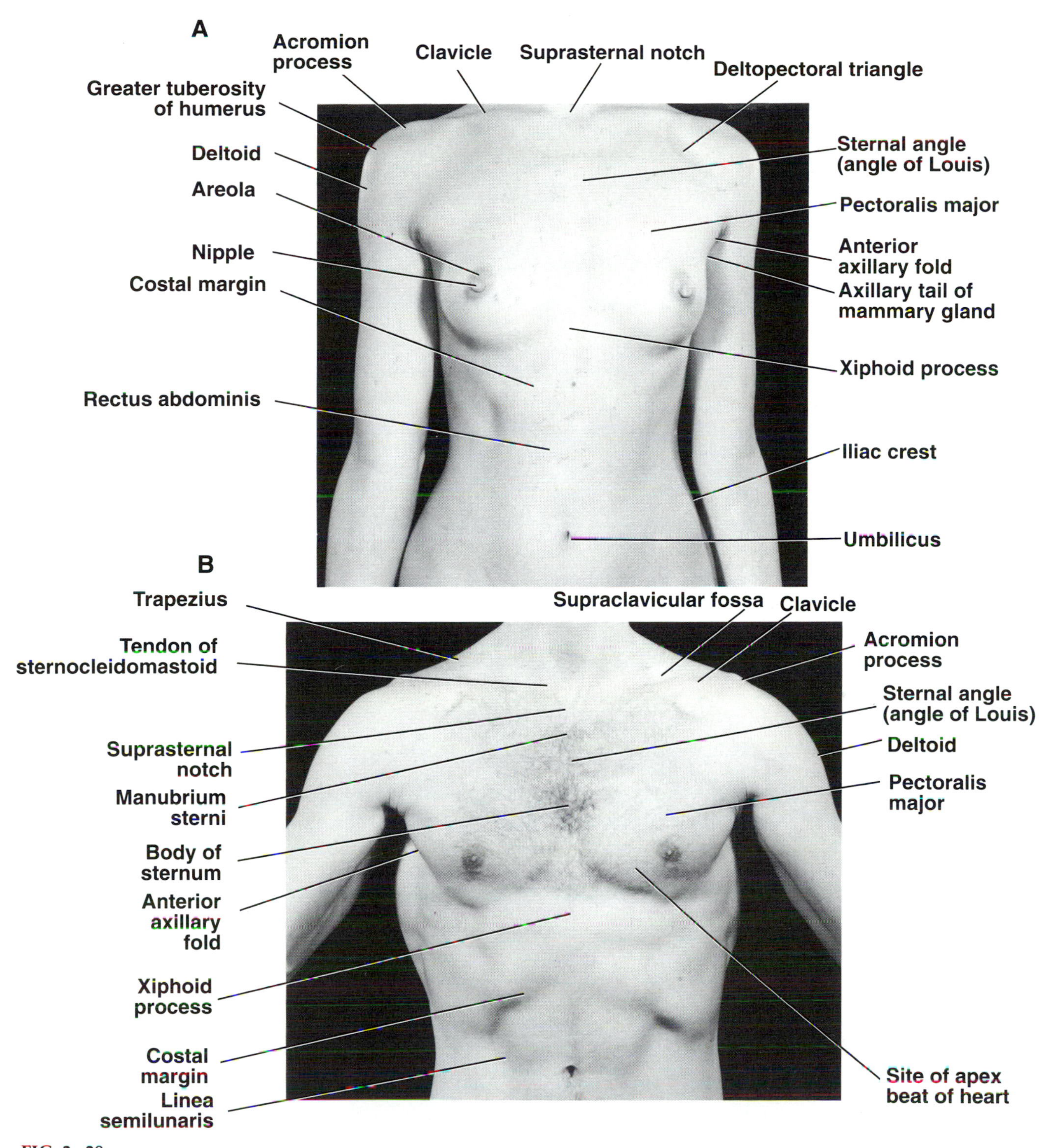

FIG 2–28.
Anterior view of the thorax. **A,** adult female. **B,** adult male.

Axillary Folds.—The *anterior axillary* fold is formed by the lower border of the pectoralis major muscle (see Fig 2–28). This may be made to stand out by asking the patient to press his hand hard against his hip. The *posterior axillary* fold is formed by the tendon of latissimus dorsi as it passes round the lower border of the teres major muscle.

Posterior Surface of Chest

Spinous Processes of Thoracic Vertebrae.—These can be palpated in the midline posteriorly (Fig 2–29). The index finger should be placed on the skin in the midline on the posterior surface of the neck and drawn downward in the nuchal groove. The first spinous process to be felt is that of the seventh cervical vertebra *(vertebra prominens).* Below this level are the overlapping spines of the thoracic vertebrae. Cervical spines 1 through 6 are covered by the large ligament, the ligamentum nuchae. The tip of a spinous process of a thoracic vertebra lies posterior to the body of the next vertebra below.

Scapula.—This bone, which is flat and triangular, is located on the upper part of the posterior surface of the chest. The *superior angle* lies opposite the spine of the second thoracic vertebra. The *spine of the scapula* is subcutaneous, and the root of the spine lies on a level with the spine of the third thoracic vertebra (see Fig 2–29). The *inferior angle* lies on a level with the spine of the seventh thoracic vertebra (see Fig 2–29).

Surface Anatomy of the Breast

In children and men the breast anatomy is rudimentary and the tissue is confined to a small area beneath the pigmented areola. In young female adults it overlaps the second to the sixth ribs and their costal cartilages and extends from the lateral margin of the sternum to the midaxillary line (see Fig 2–28). Its upper lateral edge extends around the lower border of the pectoralis major and enters the axilla. In middle-aged multiparous women the breasts may be large and pendulous and in older women the breasts may be smaller.

Surface Anatomy of the Trachea, Lungs, and Pleura

Trachea

The trachea extends from the lower border of the cricoid cartilage in the neck to the level of the sternal angle in the chest (Fig 2–30). At the root of the neck

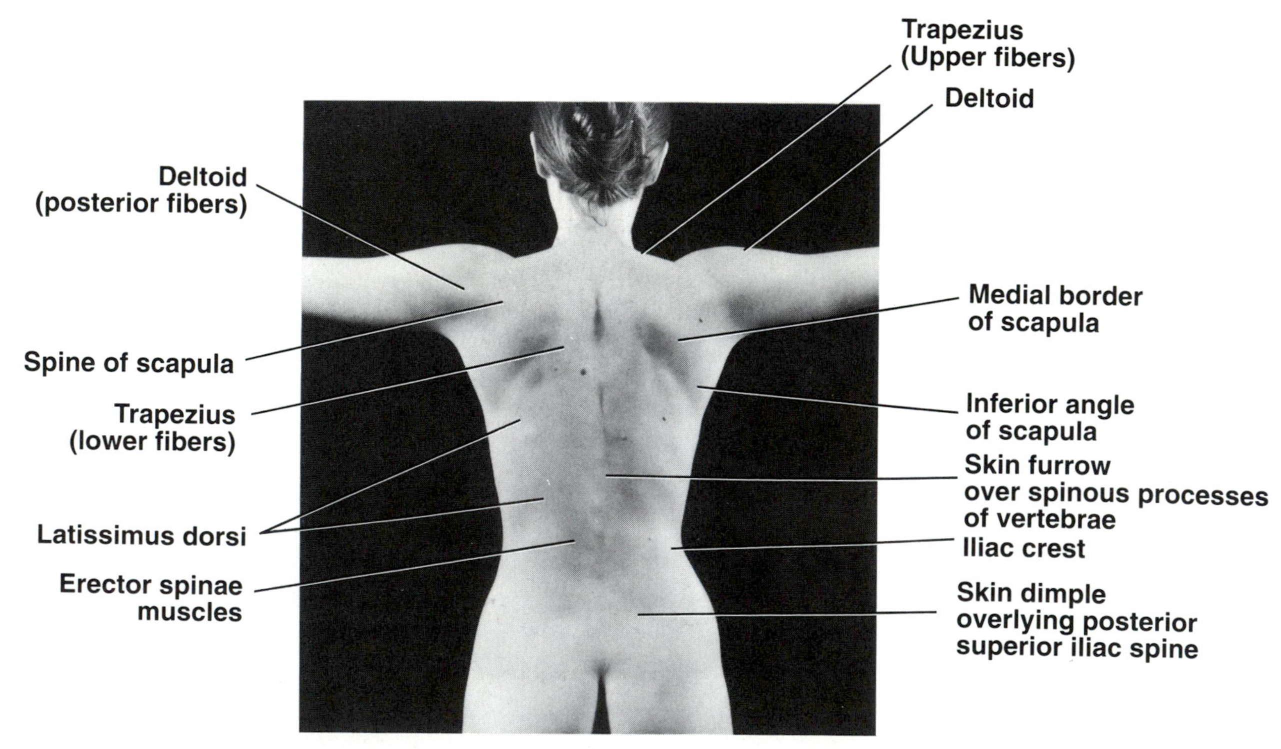

FIG 2–29.
Posterior view of the thorax of an adult female.

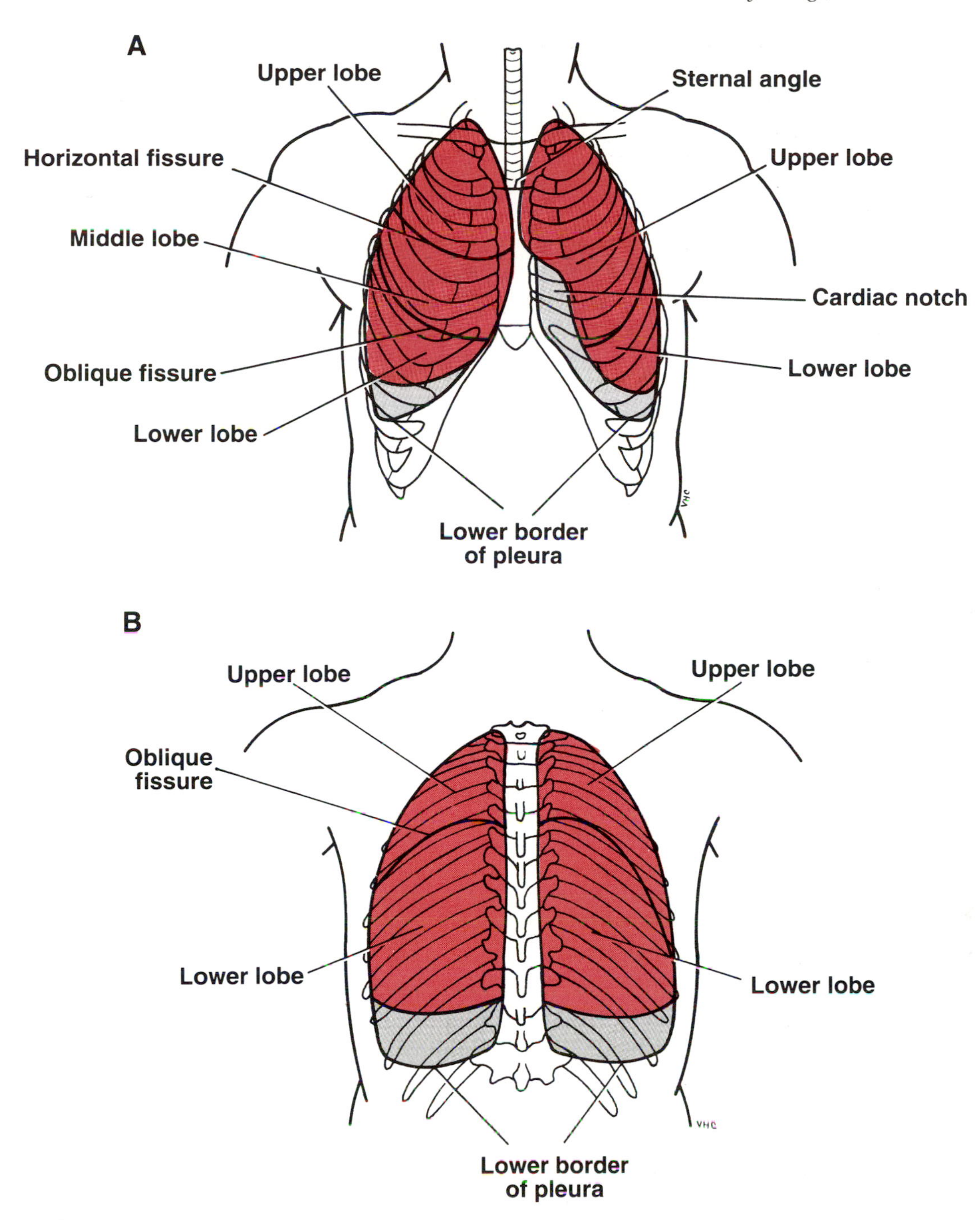

FIG 2–30.
Surface markings of lungs and parietal pleura. **A,** anterior thoracic wall. **B,** posterior thoracic wall.

it may be palpated in the midline in the suprasternal notch.

Lungs

The *apex of the lung* projects up into the neck. It can be mapped out on the anterior surface of the body by drawing a curved line, convex upward, from the sternoclavicular joint to a point 1 in. (2.5 cm) above the junction of the medial and intermediate thirds of the clavicle (see Fig 2–30).

The *anterior border of the right lung* begins behind the sternoclavicular joint and runs downward almost reaching the midline behind the sternal angle. It then continues downward until it reaches the xiphisternal joint (see Fig 2–30). The *anterior border of the left lung* has a similar course, but at the level of the fourth costal cartilage it deviates laterally and extends for a variable distance beyond the lateral margin of the sternum to form the *cardiac notch* (see Fig 2–30). This notch is produced by the heart displacing the lung to the left. The anterior border then turns sharply downward to the level of the xiphisternal joint.

The *lower border of the lung* in midrespiration follows a curving line, which crosses the sixth rib in the midclavicular line and the eighth rib in the midaxillary line, and reaches the tenth rib adjacent to the vertebral column posteriorly (see Fig 2–30). The level of the inferior border of the lung, of course, changes during inspiration and expiration.

The *posterior border of the lung* extends downward from the spinous process of the seventh cervical vertebra to the level of the tenth thoracic vertebra and lies about 1½ in. (4 cm) from the midline (see Fig 2–30).

The *oblique fissure* of the lung can be indicated on the surface by a line drawn from the root of the spine of the scapula obliquely downward, laterally and anteriorly, following the course of the sixth rib to the sixth costochondral junction. In the left lung the upper lobe lies above and anterior to this line; the lower lobe lies below and posterior to it (see Fig 2–30).

In the right lung, the *horizontal fissure* may be represented on the surface by a line drawn horizontally along the fourth costal cartilage to meet the oblique fissure in the midaxillary line (see Fig 2–30). The upper lobe lies above the horizontal fissure and the middle lobe lies below it; below and posterior to the oblique fissure lies the lower lobe.

Pleura

The boundaries of the pleural sac can be marked out on the surface of the chest wall. The lines, which will indicate the limits of the parietal pleura where it lies close to the body surface, are called the *lines of pleural reflexion.*

The *cervical pleura* bulges upward into the neck and has a surface marking identical to that of the apex of the lung. A curved line may be drawn, convex upward, from the sternoclavicular joint to a point 1 in. (2.5 cm) above the junction of the medial and intermediate thirds of the clavicle (see Fig 2–30).

The *anterior border of the right pleura* runs down behind the sternoclavicular joint, almost reaching the midline behind the sternal angle. It then continues downward until it reaches the xiphisternal joint (see Fig 2–30).

The *anterior border of the left pleura* has a similar course, but at the level of the fourth costal cartilage it deviates laterally and extends to the lateral margin of the sternum to form the *cardiac notch.* (Note that the pleural cardiac notch is not as large as the cardiac notch of the lung.) It then turns sharply downward to the xiphisternal joint (see Fig 2–30).

The *lower border of the pleura* on both sides follows a curved line, which crosses the 8th rib in the midclavicular line and the 10th rib in the midaxillary line, and reaches the 12th rib adjacent to the vertebral column, i.e., at the lateral border of the erector spinae muscle (see Fig 2–30). The lower margins of the lungs cross the 6th, 8th, and 10th ribs at the midclavicular lines, the midaxillary lines, and the sides of the vertebral column, respectively, and the lower margin of the pleura cross, at the same points respectively, the 8th, 10th, and 12th ribs. The distance between the two borders corresponds to the *costodiaphragmatic recess* (see p. 64).

Surface Anatomy of the Heart

The heart may be considered to have both an *apex* and *four borders.*

The *apex,* formed by the left ventricle, corresponds to the apex beat and is found in the fifth left intercostal space 3½ in. (9 cm) from the midline (Fig 2–31).

The *superior border,* formed by the roots of the great blood vessels, extends from a point on the second left costal cartilage (level of sternal angle) ½ in. (1.3 cm) from the edge of the sternum to a point on the third right costal cartilage ½ in. (1.3 cm) from the edge of the sternum (see Fig 2–31).

The *right border,* formed by the right atrium, extends from a point on the third right costal cartilage ½ in. (1.3 cm) from the edge of the sternum downward to a point on the sixth right costal cartilage ½ in. (1.3 cm) from the edge of the sternum (see Fig 2–31).

The *left border,* formed by the left ventricle, extends from a point on the second left costal cartilage ½ in. (1.3 cm) from the edge of the sternum to the apex of the heart (see Fig 2–31).

The *inferior border,* formed by the right ventricle and the apical part of the left ventricle, extends from the sixth right costal cartilage ½ in. (1.3 cm) from the sternum to the apex beat (see Fig 2–31).

Surface Anatomy of the Blood Vessels

The *arch of aorta* and the roots of the *brachiocephalic* and *left common carotid arteries* lie behind the manu-

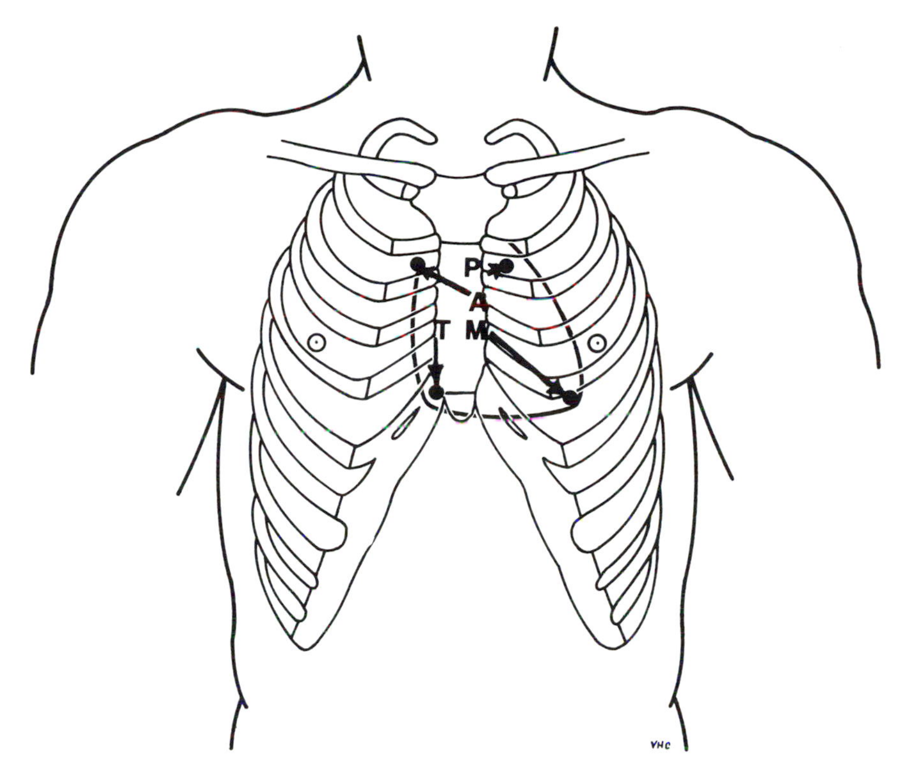

FIG 2–31.
Position of heart valves. *P* = pulmonary valve; *A* = aortic valve; *M* = mitral valve; *T* = tricuspid valve. Arrows indicate positions where valves may be heard with the least interference.

brium sterni (see Fig 2–22). The *superior vena cava* and the terminal parts of the *right* and *left brachiocephalic veins* also lie behind the manubrium sterni. The *internal thoracic vessels* run vertically downward posterior to the costal cartilages, a fingerbreadth lateral to the edge of the sternum (see Fig 2–3), as far as the sixth intercostal space. The *intercostal vessels and nerve* are situated immediately below their corresponding ribs in the following order from above downward—*V*ein, *A*rtery, and *N*erve—VAN.

CLINICAL ANATOMY

Clinical Anatomy of the Chest Wall

Cutaneous Innervation and Dermatomes

The innervation of the skin of the anterior chest wall above the level of the sternal angle is from the *supraclavicular nerves* (C3 and C4). Below this level the anterior and lateral cutaneous branches of the intercostal nerves supply oblique bands of skin (dermatomes) in regular sequence. The skin on the posterior chest wall is supplied by the posterior rami of the thoracic spinal nerves (Figs 2–32 and 2–33).

An intercostal nerve not only supplies areas of skin but also supplies the ribs, costal cartilages, intercostal muscles, and the parietal pleura lining the intercostal space. Furthermore, the 7th to 11th intercostal nerves leave the thoracic wall and enter the anterior abdominal wall, so that they, in addition, supply dermatomes on the anterior abdominal wall, muscles of the anterior abdominal wall, and parietal peritoneum. This latter fact is of great clinical importance, since it means that disease in the thoracic wall may be revealed as pain in a dermatome that extends across the costal margin into the anterior abdominal wall. For example, a pulmonary thromboembolism or a pneumonia with pleurisy involving the costal parietal pleura could give rise to abdominal pain and tenderness and rigidity of the abdominal musculature.

Anatomy of Intercostal Nerve Block

To produce analgesia of the anterior and lateral thoracic and abdominal walls, the intercostal nerve should be blocked before the lateral cutaneous branch arises in the midaxillary line. The details of an intercostal nerve block are fully described in Chapter 19, p. 823.

Herpes Zoster

This represents a reactivation of a latent varicella infection. It often presents as a burning or stabbing pain associated with clusters of vesicles, having a unilateral dermatomal distribution. In more than half the patients the lesions occur on the trunk and commonly involve thoracic or abdominal dermatomes.

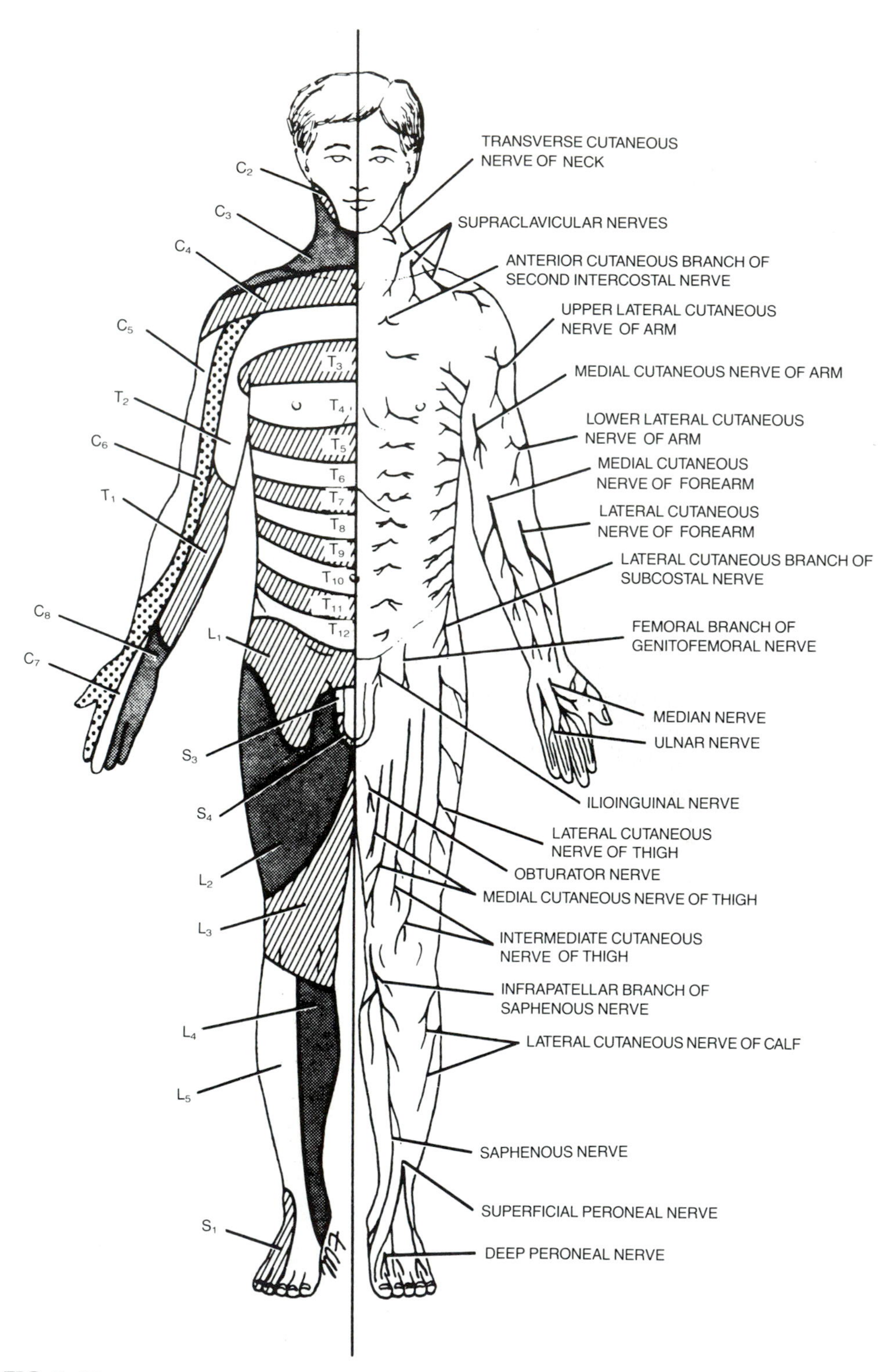

FIG 2–32.
Dermatomes and the distribution of cutaneous nerves on the anterior aspect of the body.

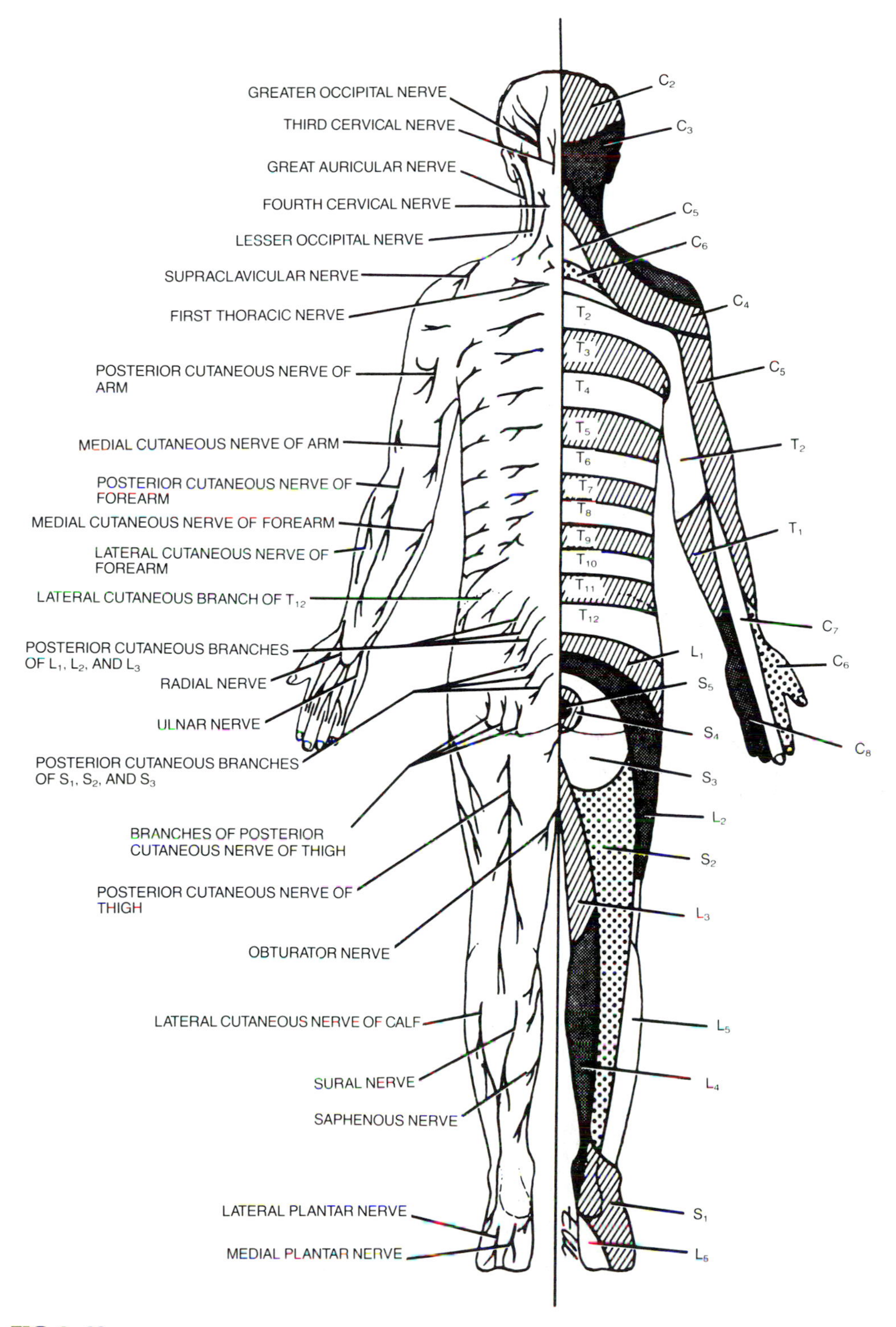

FIG 2–33.
Dermatomes and the distribution of cutaneous nerves on the posterior aspect of the body.

Chest Trauma

Mechanics of Chest Trauma

Chest organ injuries from blunt trauma occur as the result of rapid acceleration or deceleration, by compression, or by a sudden increase in intrathoracic or intra-abdominal pressure. A knife wound piercing the chest wall injures the organs along its path. A bullet wound does not follow a straight path but yaws, tumbles, and may fragment, causing widespread tissue damage. In addition, the kinetic energy generated by a speeding bullet may damage tissue that is distant from the actual path of the bullet. The energy from a bullet can be represented by the formula $F = \frac{1}{2}MV^2$, where F represents the force, M the mass of the bullet, and V the muzzle velocity.

Rib Contusion.—Bruising of a rib, secondary to trauma, is the most common rib injury. In this painful condition a small hemorrhage occurs beneath the periosteum.

Rib Fractures.—Fractures of the ribs are common chest injuries. In children the ribs are highly elastic, and fractures in this age group are therefore rare. Unfortunately, the pliable chest wall in the young can be easily compressed so that the underlying lungs and heart may be injured. With increasing age, the rib cage becomes rigid due to the deposit of calcium in the costal cartilages, and the ribs become brittle. The ribs then tend to break at their weakest part, their angles.

The ribs prone to fracture are those that are exposed or relatively fixed. The most commonly fractured ribs are ribs 5 through 10. The first 4 ribs are protected by the clavicle and pectoral muscles anteriorly, and by the scapula and its associated muscles posteriorly. The 11th and 12th ribs float and move with the force of impact.

Because the rib is sandwiched between the skin externally and the delicate pleura internally, it is not surprising that the jagged ends of a fractured rib may penetrate the lungs and present as a pneumothorax.

Severe localized pain is usually the most important symptom of a fractured rib. The periosteum of each rib is innervated by the intercostal nerves above and below the rib. In order to encourage the patient to breathe adequately, it may be necessary to relieve the pain by performing an intercostal nerve block.

Flail Chest.—In severe crush injuries, a number of ribs may break. If limited to one side, the fractures may occur near the rib angles and also anteriorly near the costochondral junctions. This causes flail chest, in which a section of the chest wall is disconnected to the rest of the thoracic wall. If the fractures occur on either side of the sternum, the sternum may be flail. In either case, the stability of the chest wall is lost, and the flail segment is sucked in during inspiration and driven out during expiration, thus producing paradoxical and ineffective respiratory movements (Fig 2–34).

Fractured Sternum.—The sternum is a resilient structure and is held in position by relatively pliable costal cartilages and bendable ribs. For these reasons fracture of the sternum is not common; however, it does occur in high-speed motor vehicle accidents. Remember that the heart lies posterior to the sternum and may be severely contused by the sternum on impact.

Traumatic Injury to the Back of the Chest.—The posterior wall of the chest in the midline is formed by the vertebral column. In severe posterior chest injuries the possibility of a vertebral fracture with associated injury to the spinal cord should be considered. Remember also the presence of the scapula, which overlies the upper seven ribs. This bone is covered with muscles and is only fractured in cases of severe trauma.

Abdominal Viscera and Chest Trauma.—When the anatomy of the thorax is reviewed, it is important to remember that the upper abdominal organs, namely, the liver, stomach, and spleen, may be injured by trauma to the rib cage; in fact, any injury to the chest below the level of the nipple line may involve abdominal organs as well as chest organs (see Fig 11–31).

Clinical Anatomy of the Diaphragm

Penetrating Injuries of the Diaphragm

These injuries may result from stab or bullet wounds to the chest or abdomen. Any penetrating wound to the chest below the level of the nipples should be suspected of causing damage to the diaphragm until proved otherwise. The arching domes of the diaphragm may reach the level of the fifth rib (the right dome may reach a higher level).

Rupture of the Diaphragm

In severe crushing injuries to the thorax or abdomen, the diaphragm may rupture, usually through

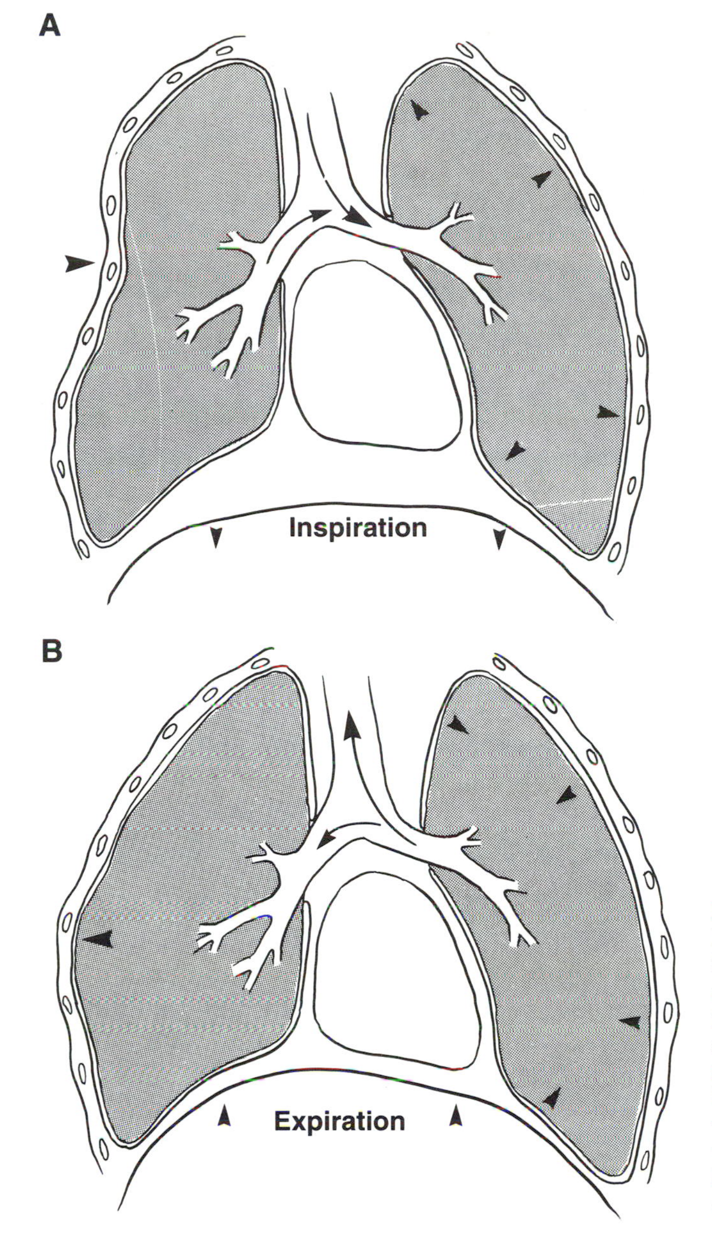

FIG 2–34.
Flail chest is a condition in which a portion of the chest wall is drawn inward during inspiration and bulges outward during expiration; it occurs when several ribs are fractured in two or more places. **A,** on inspiration the fractured ribs are pulled inward as the pressure within the thorax decreases. The inspired air passing down the trachea tends to be drawn into the lung on the unaffected side. **B,** on expiration the fractured ribs are pushed outward as the pressure within the thorax rises. Note that some of the air in the bronchi tends to enter the lung on the affected side as well as passing up the trachea.

the central tendon. Herniation of abdominal viscera into the thorax may occur, especially if the left dome of the diaphragm is the site of the rupture. The rupture of the right dome or the central tendon is usually plugged by the large right lobe of the liver, unless the opening is very great. A ruptured diaphragm, if not repaired, may result in a delayed herniation of abdominal contents.

Clinical Anatomy of the Breasts

Breast Examination

The breasts are first inspected for symmetry, with the patient in a sitting position. Some degree of asymmetry is common and is the result of unequal breast development. Any swelling should be noted. Swelling may be caused by an underlying tumor, cyst, or abscess formation. The nipples should be carefully examined for evidence of retraction. A carcinoma within the breast substance can cause nipple retraction by pulling on the lactiferous ducts.

The breast should be palpated against the underlying chest wall with the patient in a supine position. Then, with the patient sitting up and with both arms raised above her head, dimpling of the skin or retraction of the nipple will be produced if a carcinoma is tethered to the skin or lactiferous ducts.

Supernumerary Nipples
Occasionally supernumerary nipples occur along a line extending from the axilla to the groin; they may or may not be associated with breast tissue. This minor congenital anomaly may result in a mistaken diagnosis of warts or moles.

Retracted Nipple
A long-standing retracted nipple is a congenital deformity due to failure in the complete development of the nipple. A retracted nipple of recent occurrence, however, is usually caused by an underlying carcinoma pulling on the lactiferous duct.

Breast Abscess
Acute infection of the breast may occur during lactation. Pathogenic bacteria gain entrance to the breast tissue through a crack in the nipple. Because the breast is divided into 15 to 20 compartments, each of which contains a lobe of the gland separated by fibrous septa, the infection remains localized, in the beginning, to 1 compartment. Should an abscess occur, it should be drained through a radial incision to avoid the spread of infection into neighboring compartments; a radial incision will also minimize the damage to the radially arranged ducts.

Carcinoma of the Breast and Lymphatic Drainage
The importance of the lymphatic drainage of the breast in relation to the spread of carcinoma from this organ cannot be overemphasized. The medial half of the breast drains into the lymph nodes alongside the internal thoracic artery inside the thorax (see Fig 2–7). The lateral half of the breast drains into the anterior or pectoral group of nodes situated just beneath the lower border of the pectoralis major muscle. Therefore, a carcinoma occurring in the lateral half of the breast will tend to spread to the axillary nodes. Thoracic metastases are difficult or impossible to treat, but the lymph nodes of the axilla may be removed surgically.

Fortunately, about 60% of breast carcinomas occur in the upper lateral quadrant. The lymphatic spread of carcinoma to the opposite breast, to the abdominal cavity, or into lymph nodes in the root of the neck is due to obstruction of the normal lymphatic pathways by malignant cells or destruction of lymphatic vessels by surgery or radiotherapy. The entrance of malignant cells into the blood vessels accounts for the metastases in distant bones.

Clinical Anatomy of the Thoracic Outlet Syndromes

Brachial Plexus
The spinal nerves C5, C6, C7, C8, and T1 leave the vertebral column and enter the neck through the intervertebral foramina. The anterior ramus of each nerve then passes laterally in the interval between the scalenus anterior and the scalenus medius muscles and above the first rib to emerge to form the roots of the brachial plexus (Fig 2–35). The roots then unite to form the trunks of the plexus. The divisions of the trunks reunite to form the cords of the plexus behind the clavicle (see Fig 2–35). As the brachial plexus leaves the scalene muscles, it takes with it a covering of the prevertebral fascia known as the *axillary sheath.* The cords of the brachial plexus, on entering the axilla, become arranged around the axillary artery behind the pectoralis minor muscle.

The roots C8 and T1 unite to form the lower trunk of the brachial plexus, which lies on the upper surface of the first rib behind the subclavian artery (see Fig 2–10). The sensory distribution of C8 and T1 nerves is to the ulnar 1½ fingers and the medial aspect of the forearm. The motor distribution of these nerves is to the small muscles of the hand.

Subclavian Artery
On each side of the thorax, the subclavian artery exits behind the sternoclavicular joint and passes over the upper surface of the first rib in the interval between the scalenus anterior and scalenus medius muscles; here it is related posteriorly to the lower trunk of the brachial plexus (see Figs 2–10 and 2–35). The artery then passes laterally behind the clavicle, and at the outer border of the first rib enters the axilla and becomes the axillary artery.

Subclavian Vein
The subclavian vein is formed at the outer border of the first rib as a continuation of the axillary vein. It runs medially behind the clavicle and in front of the scalenus anterior muscle (see Fig 2–10) and therefore is situated anterior to the subclavian artery, which passes posterior to the scalenus anterior muscle.

Anatomical Sites for Compression of the Brachial Plexus and the Subclavian Vessels
The possible sites for compression of the nerves and the blood vessels are as follows: (1) nerves at the intervertebral foramina by a prolapsed intervertebral disc or by osteophytes in osteoarthritis; (2) nerves

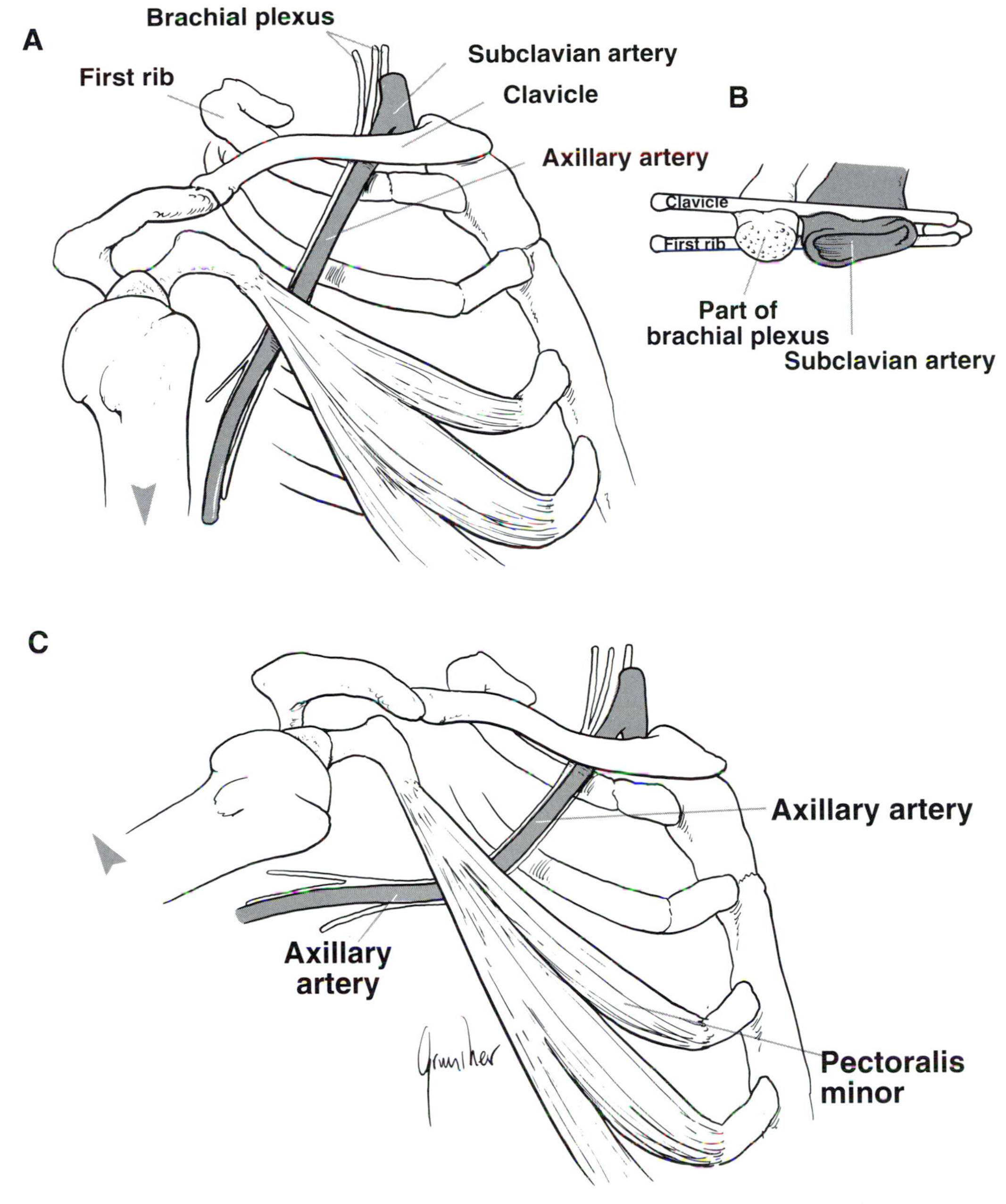

FIG 2–35.
Examples of thoracic outlet syndrome. **A,** the relationship between the brachial plexus, the subclavian and axillary arteries, the clavicle, the first rib, and the pectoralis minor tendon. **B,** how the cords of the brachial plexus and the subclavian artery can be squeezed between the clavicle and the first rib in some individuals. **C,** how the axillary artery and the branches of the brachial plexus might be pressed upon by the pectoralis minor tendon when the arm is abducted at the shoulder joint.

and subclavian artery in the triangular interval between the scalenus anterior and medius and the first rib; (3) subclavian vein between the scalenus anterior and the clavicle; (4) nerves, subclavian artery, and subclavian vein between the first rib and the clavicle (see Fig 2–35); (5) nerves, axillary artery, and axillary vein at the upper border of the pectoralis minor muscle (see Fig 2–35); and (6) nerves, subclavian artery, and subclavian vein by a cervical rib (Fig 2–36). The cervical rib arises from the anterior tubercle of the transverse process of the seventh cervical vertebra. It is a congenital anomaly that occurs in about 0.5% of persons. This additional rib passes forward beneath the lower trunk of the brachial

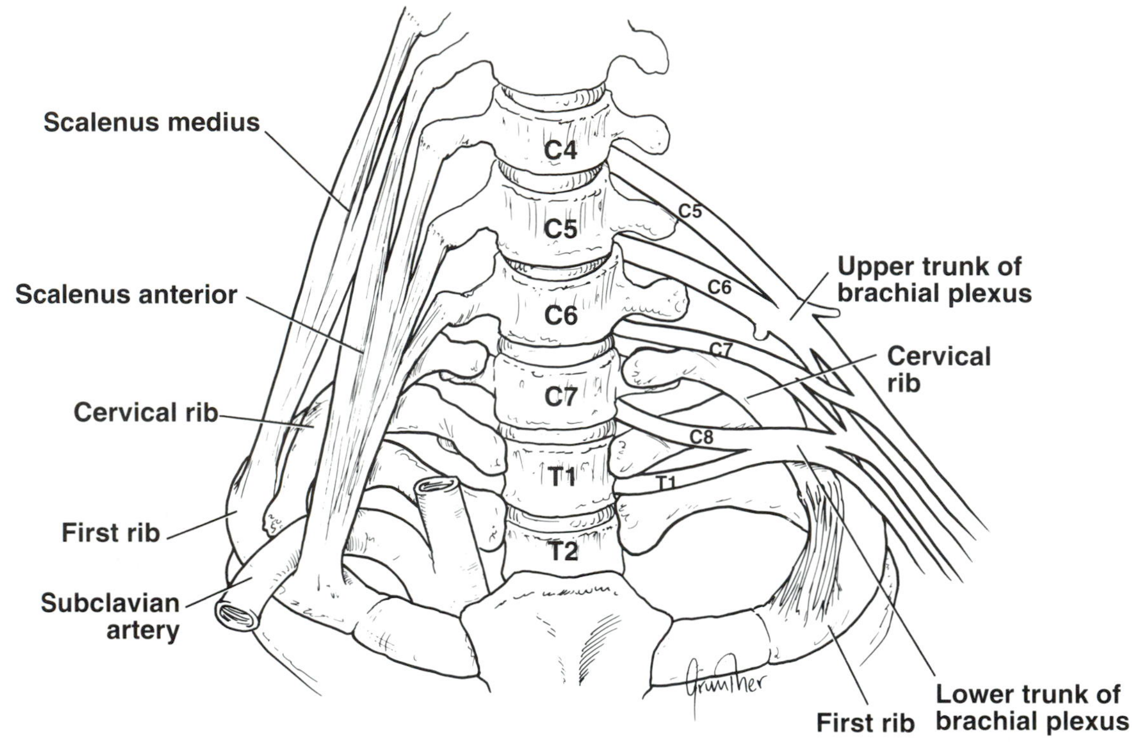

FIG 2–36.
Thoracic outlet as seen from in front. Note the presence of cervical ribs. On the right side, the rib is almost complete and articulates anteriorly with the first rib. On the left side, the rib is rudimentary but continues forward as a fibrous band that is attached to the first rib. The cervical rib may exert pressure on the lower trunk of the brachial plexus and may kink the subclavian artery.

plexus and subclavian vessels exerting pressure from below (see Fig 2–36).

Most often the symptoms are due to pressure on the lower nerve trunk, producing pain down the ulnar side of the forearm and hand, and wasting of the small muscles of the hand. Pressure on the blood vessels compromises the circulation of the upper limb. Pressure on the subclavian-axillary vein produces venous congestion and eventually thrombosis and chronic pain and discomfort.

The Adson Maneuver
This maneuver was commonly used in making the diagnosis of thoracic outlet syndrome; recently the reliability of the test has been questioned. The patient takes a deep breath (raises the first rib), extends the neck (takes up the slack of the brachial plexus and subclavian vessels), and turns his chin to the side being examined (narrows the interval between the scalenae muscles); at the same time the radial pulse is palpated. Disappearance or reduction of the pulse, and possibly coldness and paleness of the hand, would indicate that the subclavian artery is being compressed by the scalene muscles and/or the first (or cervical) rib. In addition to looking for vascular compromise, the physician should also look for replication of the nerve symptoms down the arm.

Clinical Anatomy of the Mediastinum

From the clinical point of view, the mediastinum should be considered a pliable septum that can be easily deviated to the right or left by changes in the intrathoracic pressure (see pneumothorax p. 89). The interior of the septum is a potential space filled with loose connective tissue in which numerous organs and other anatomic structures are embedded. Superiorly, the mediastinal space communicates with the fascial spaces of the neck beneath the deep fascia. Penetrating wounds of the chest involving the esophagus, or infections of the neck, may produce a mediastinitis. In esophageal perforations air escapes into the space and ascends beneath the fascia to the root of the neck, producing *subcutaneous emphysema*.

Clinical Anatomy of the Pleurae and Pleural Cavities

Pleural Fluid

The pleural cavity* normally contains about 5 to 10 mL of clear fluid that lubricates the apposing surfaces of the visceral and parietal pleurae during respiratory movements. The formation of the fluid results from hydrostatic and osmotic pressures. Since the hydrostatic pressures are greater in the capillaries of the parietal pleura than in the capillaries of the visceral pleura (pulmonary circulation), the pleural fluid is normally absorbed into the capillaries of the visceral pleura. Any condition that increases the production of the fluid (e.g., inflammation, malignancy, congestive heart disease) or impairs the drainage of the fluid (e.g., collapsed lung) will result in the abnormal accumulation of fluid, called *pleural effusion.* The presence of 300 mL of fluid in the costodiaphragmatic recess in an adult is sufficient to enable its clinical detection. The clinical signs include decreased lung expansion on the side of the effusion, with decreased breath sounds and dullness on percussion over the effusion.

Pleurisy

Inflammation of the pleura, secondary to inflammation of the lung, results in the pleural surfaces becoming coated with inflammatory exudate, causing the surfaces to be roughened. This roughening produces friction, and a *pleural rub* may be felt or heard with a stethoscope on inspiration and expiration. Often the exudate becomes invaded by fibroblasts, which lay down collagen and bind the visceral pleura to the parietal pleura, forming *pleural adhesions.*

Pneumothorax

In this condition air enters the pleural cavity. Depending on the amount of air that has entered the pleural cavity, the lung will partially collapse under the atmospheric pressure; if only a small amount of air has entered, only minor collapsing occurs and respiration is minimally compromised. If, however, a large amount of air has entered the pleural cavity and if the pressure of the air trapped in the cavity becomes greater than the atmospheric pressure, the lung will completely collapse and the pliable mediastinum will be deflected to the opposite side (Fig 2–37). The resulting signs of pneumothorax are absent or diminished breath sounds over the affected lung and deflection of the trachea to the opposite side.

Open pneumothorax occurs when the air enters the pleural cavity through an opening in the chest wall and may result from stab or bullet wounds (see Fig 2–37). *Sucking pneumothorax* occurs when the hole in the chest wall is larger than the glottis. With each inspiration the negative pressure created is more effective at sucking air in through the chest wound than air entering through the glottis; this produces a sucking sound. The lung cannot be expanded, and respiration is compromised.

Tension pneumothorax occurs when air is sucked into the pleural cavity through a chest wound with each inspiration but does not escape (Fig 2–38). This can occur as the result of clothing and/or the layers of the chest wall combining to form a valve so that air enters on inspiration but cannot exit through the wound. In these circumstances, the air pressure builds up on the wounded side and pushes the mediastinum progressively over to the opposite side. Because of the anatomic thin walls of the great veins (vena cavae) and the atria of the heart, the increase in air pressure within the chest cavity interferes with blood return to the heart; the patient may die because of lack of venous return. The clinical signs are hypertension, hyperresonance to percussion on the affected side, the engorgement of the neck veins, and the evidence of mediastinal deflection. Eventually, hypotension (secondary to lack of venous return) results. The treatment is immediate decompression of the affected side by the insertion of a needle thoracostomy.

Spontaneous pneumothorax is a condition in which air enters the pleural cavity suddenly without an immediately apparent cause. It is usually found that air has entered from a ruptured bulla (see Fig 2–38).

Hemothorax

In this condition blood enters the pleural cavity. It can be caused by stab or bullet wounds to the chest wall, resulting in bleeding from blood vessels in the chest wall, bleeding from vessels in the chest cavity, or bleeding from a lacerated lung.

Root of the Neck Injuries

The cervical dome of the parietal pleura and the apex of the lungs extend up into the neck so that at

*Clinicians are increasingly using the term pleural space rather than the anatomical term pleural cavity. This is probably to avoid the confusion between the pleural cavity (a slitlike space) and the larger chest cavity.

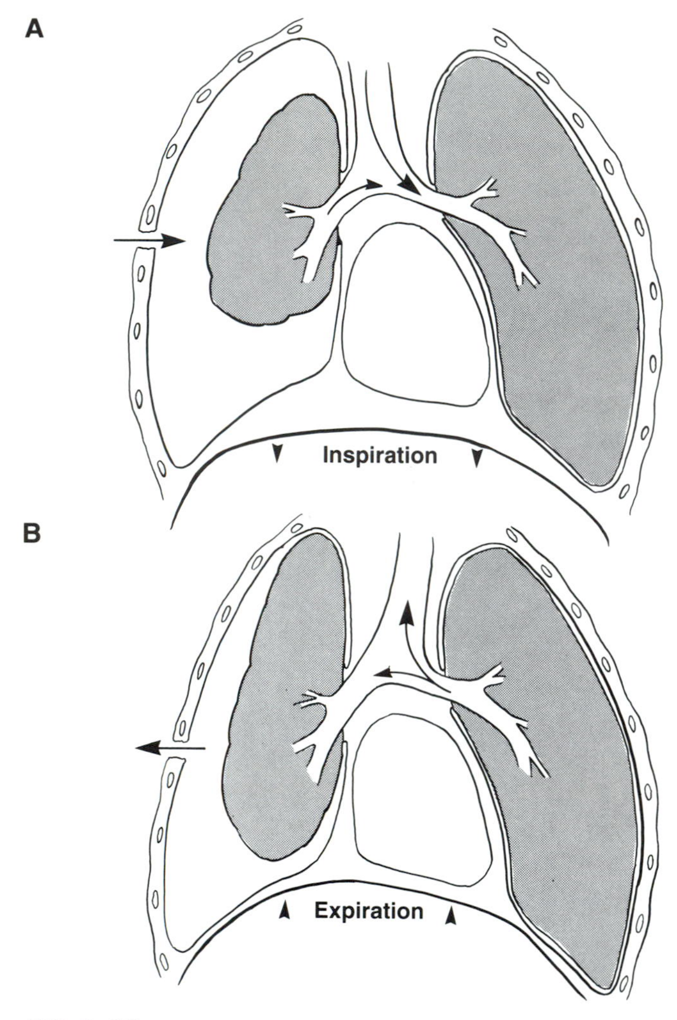

FIG 2–37.
Open pneumothorax without tension. **A,** on inspiration air is drawn in through the wound in the chest wall at atmospheric pressure, and the lung partially or completely collapses (depending on the size of the hole in the chest wall) from its own inherent elasticity; the mediastinum is deflected to the opposite side. **B,** on expiration air passes out of the chest wound as the diaphragm rises and the mediastinum is deflected to the same side. With large chest openings (larger than the cross-sectional area of trachea), air will preferentially use the hole in the chest wall rather than passing up and down the trachea, and respiratory ventilation will cease.

their highest point they lie about 1 in. (2.5 cm) above the clavicles (see Fig 2–30). Consequently they are vulnerable to stab wounds in the root of the neck.

Traumatic Asphyxia

The sudden caving in of the anterior chest wall associated with fractures of the sternum and ribs causes a dramatic rise in intrathoracic pressure. Apart from the immediate evidence of respiratory distress, the anatomy of the venous system plays a significant role in the production of the characteristic vascular signs of traumatic asphyxia. The thinness of the walls of the thoracic veins and the right atrium causes their collapse under the raised intrathoracic pressure, and venous blood is dammed back in the veins of the neck and head. This produces venous congestion, bulging of the eyes, which become injected, and swelling of the lips and tongue, which become cyanotic. The skin of the face, neck, and shoulders becomes purple.

The Anatomy of Cardiopulmonary Resuscitation

Cardiopulmonary resuscitation (CPR), achieved by compression of the chest, was originally believed to succeed because of the compression of the heart between the sternum and the vertebral column. Now it is recognized that the blood flows in CPR because the whole thoracic cage is the pump; the heart functions merely as a conduit for blood. An extrathoracic pressure gradient is created by external chest compressions. The pressure in all chambers and locations within the chest cavity is the same. With compression, blood is forced out of the thoracic cage. The blood preferentially flows out the arterial side of the circulation and back down the venous side because the venous valves in the internal jugular system prevent a useless oscillatory movement. With the release of compression, blood enters the thoracic cage, preferentially down the venous side of the systemic circulation.

Clinical Anatomy of Thoracocentesis

Needle Thoracostomy

This procedure is necessary in patients with tension pneumothorax or with a large hemothorax. The purpose is to remove the air or blood to allow the lung to re-expand.

Anterior Approach.—With the patient in a supine position, the sternal angle is identified, then the second costal cartilage and second rib and the second intercostal space are found in the midclavicular line.

Lateral Approach.—With the patient in a lateral decubitus position, the second intercostal space is identified as above, but the anterior axillary line is used.

The skin is prepared and draped in the usual sterile manner, and a local anesthetic is introduced

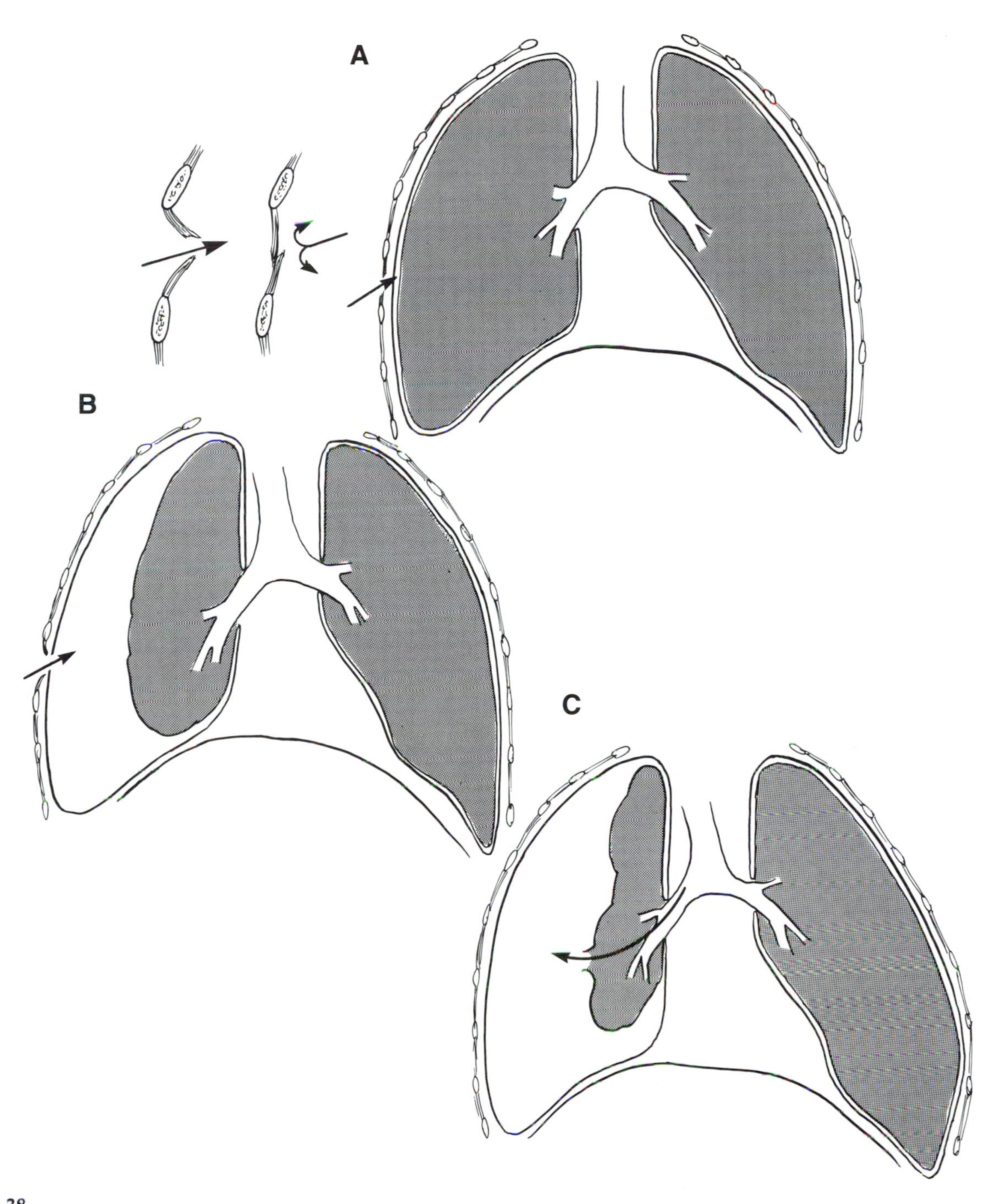

FIG 2–38.
Tension pneumothorax. **A,** following penetration of the chest wall, clothing and/or tissue create a valvelike mechanism that permits air entry into the pleural space during inspiration but prevents exit during expiration. **B,** the lung collapses on the wounded side and the build up of air pressure with each respiration causes severe deflection of the mediastinum to the opposite side. **C,** spontaneous pneumothorax with air entering the pleural space through a ruptured bulla; the lung collapses and the mediastinum is deflected to the opposite side.

along the course of the needle above the upper border of the third rib. The thoracostomy needle will pierce the following structures (Fig 2–39) as it passes through the chest wall.

1. Skin.
2. Subcutaneous tissue (in the anterior approach the pectoral muscles are then penetrated).
3. Serratus anterior muscle.
4. External intercostal muscle.
5. Internal intercostal muscle.
6. Transversus thoracis muscle.
7. Endothoracic fascia.
8. Parietal pleura.

The needle should be kept close to the upper border of the third rib to avoid injuring the intercostal vessels and nerve in the subcostal groove.

Tube Thoracostomy
The preferred insertion site for a tube thoracostomy is the fourth or fifth intercostal space at the anterior axillary line (Fig 2–40). Once the intercostal muscles are separated by blunt dissection, a finger should be introduced through the incision to confirm the entrance into the pleural cavity and to ascertain that the lung is not stuck against the parietal pleura (held in place by an old adhesion). The tube is then introduced and pushed upward and backward toward the apex of the lung for the treatment of pneumothorax, and downward and backward for treatment of hemothorax.

Remember that the neurovascular bundle changes its relationship to the ribs as it passes forward in the intercostal space. In the most posterior part of the space in the region of the posterior axillary line, the bundle lies in the middle of the intercostal space. Later, as the bundle reaches the location of the rib angle, it becomes closely related to the lower border of the rib above and maintains that position as it courses forward.

In most situations it is tempting to introduce a thoracostomy tube or needle through the lower intercostal spaces. This is acceptable provided that the physician remembers that the domes of the diaphragm curve up into the rib cage as far as the fifth rib (higher on the right). Avoid the possibility of damaging the diaphragm and entering the peritoneal cavity and injuring the liver, spleen, or stomach.

Clinical Anatomy of Thoracotomy

In certain situations, such as penetrating chest wounds with uncontrolled intrathoracic hemorrhage, thoracotomy may be a life-saving procedure. The skin is rapidly cleaned with a suitable antiseptic solution and an incision is made over the fourth or fifth intercostal space, extending from the lateral margin of the sternum to the anterior axillary line (Fig 2–41). Whether a right or left incision will be made will depend on the site of the injury. For access to the heart and aorta, the chest should be entered from the left side.

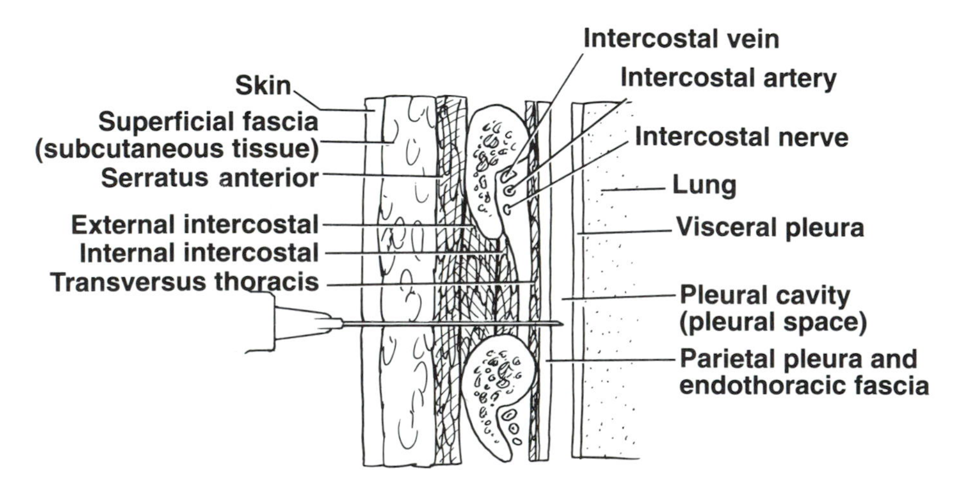

FIG 2–39.
Needle thoracostomy. The various layers of tissue penetrated by the needle before it enters the pleural space are shown. Depending on the site of penetration, the pectoral muscles will be penetrated in addition to the serratus anterior muscle. The needle is kept close to the upper border of the rib to avoid injuring the intercostal vessels and nerve in the subcostal groove.

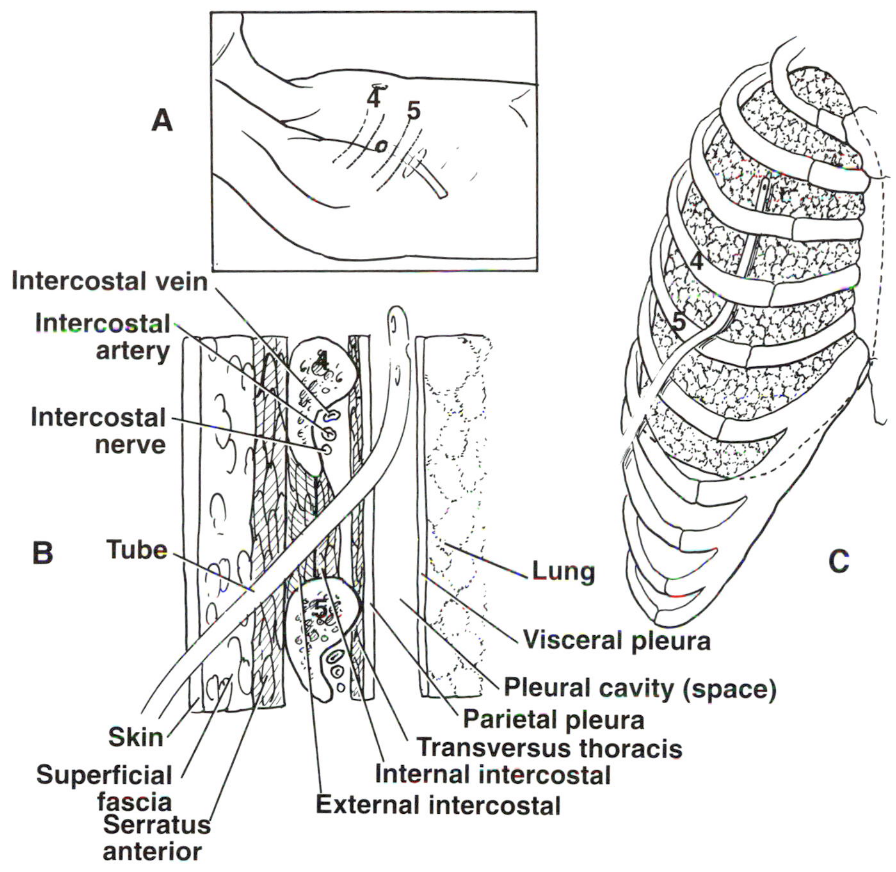

FIG 2–40.
Tube thoracostomy. **A,** the site for insertion of the tube at the anterior axillary line. The skin incision is usually made over the intercostal space one below the space to be pierced. **B,** the various layers of tissue penetrated by the scalpel and later the tube as they pass through the chest wall to enter the pleural space. The incision through the intercostal space is kept close to the upper border of the rib to avoid injuring the intercostal vessels and nerve. **C,** the tube advancing superiorly and posteriorly in the pleural space.

The following tissue layers will be incised (see Fig 2–41).

1. Skin.
2. Superficial fascia.
3. Serratus anterior and the pectoral muscles.
4. External intercostal muscle, anterior intercostal membrane, and covering fascia.
5. Internal intercostal muscle.
6. Transversus thoracis muscle.
7. Endothoracic fascia.
8. Costal parietal pleura.
9. Once you are through the costal parietal pleura, the next structure to be encountered is the lung, which must be pushed up and out of the way in order to visualize the mediastinum.

Avoid the *internal thoracic artery,* which runs vertically downward behind the costal cartilages about a fingerbreadth lateral to the margin of the sternum, and the *intercostal vessels* and *nerve,* which extend forward in the subcostal groove in the upper part of the intercostal space (see Fig 2–40).

Clinical Anatomy of the Lungs

Physical Examination of the Lungs

The upper lobes of the lungs are most easily examined from in front of the chest and the lower lobes from the back because of the position and oblique direction of the oblique fissure. The middle lobe of the right lung is most easily examined from in front because of the position and direction of the horizontal fissure. In the axillae areas of all lobes may be examined (see Fig 2–30).

Lung Apex in the Neck

Remember that the apex of the lung projects up into the neck (1 in., or 2.5 cm, above the clavicle) and

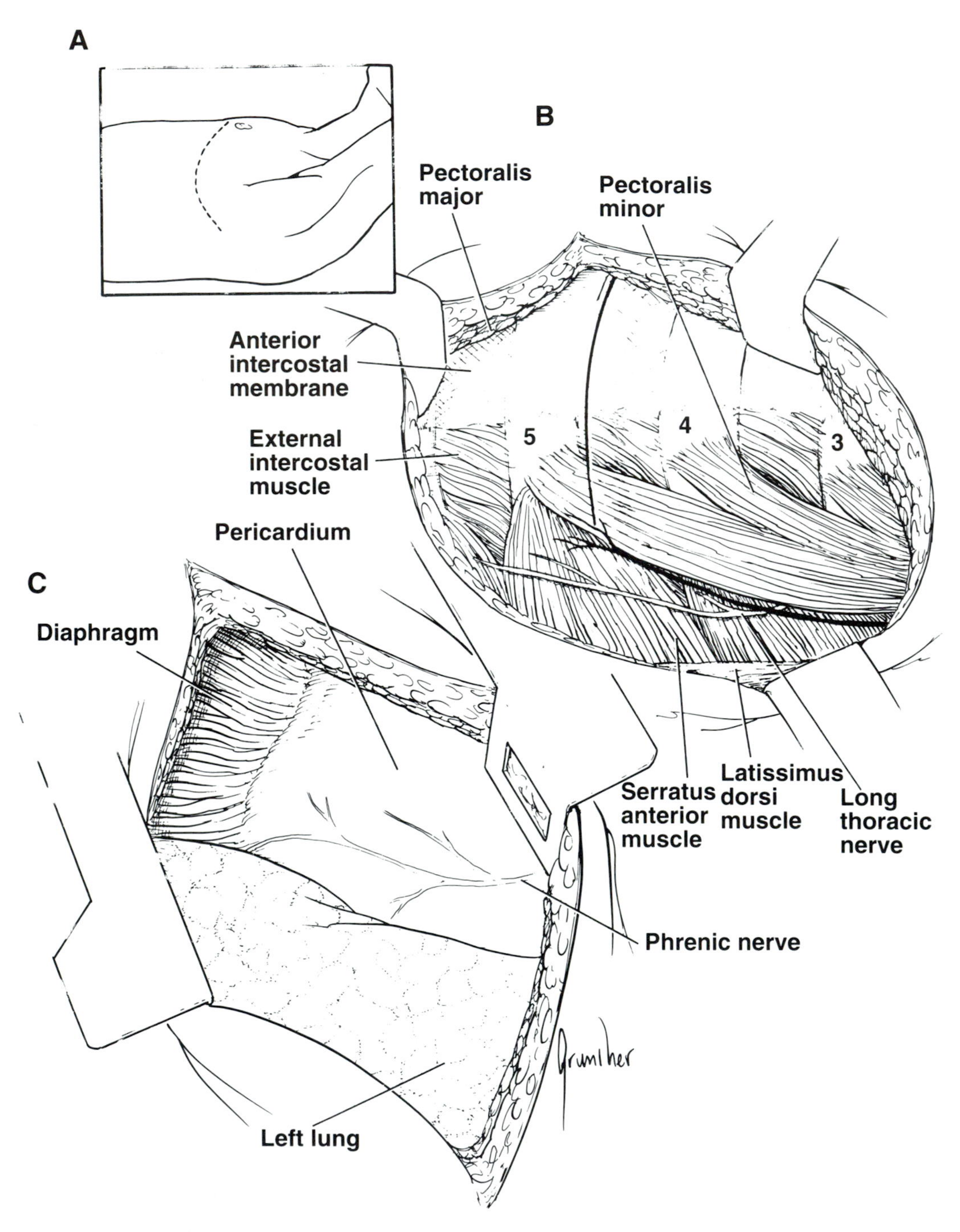

FIG 2–41.
Left thoracotomy. **A**, site of skin incision over fourth or fifth intercostal space. **B**, the exposed ribs and associated muscles. The line of incision through the intercostal space should be placed close to the upper border of the rib to avoid injuring the intercostal vessels and nerve. **C**, the pleural space opened and the left side of the mediastinum exposed. The left phrenic nerve descends over the pericardium beneath the mediastinal parietal pleura. The left lung must be pushed up and out of the way to visualize the posterior mediastinum.

may be injured in stab or bullet wounds in this region.

Fractured Ribs and the Lungs

Although the lungs are well protected by the bony rib cage, a splinter from a fractured rib may nevertheless penetrate the lung and air may escape into the pleural cavity, causing a *pneumothorax* and partial collapse of the lung. Air may also find its way into the connective tissue of the lung and from there travel under the visceral pleura to the lung root. It may then pass into the mediastinum and up into the neck, producing a *subcutaneous emphysema.*

Pain and Lung Disease

Lung tissue and the visceral pleura have no pain-sensitive nerve endings, so that pain in the chest is always the result of conditions affecting the surrounding structures. In pneumonia or tuberculosis, for example, pain may never be experienced.

Once lung disease crosses the visceral pleura and the pleural cavity to involve the parietal pleura, pain becomes a prominent feature. Lobar pneumonia with pleurisy, for example, produces a severe tearing pain, accentuated by deep inspiration or coughing. Since the lower part of the costal parietal pleura receives its sensory innervation from the lower five intercostal nerves, which also innervate the skin of the anterior abdominal wall, pleurisy in this area commonly produces pain that is referred to the abdomen. This sometimes results in a mistaken diagnosis of an acute abdominal process.

Similarly, pleurisy secondary to pneumonia involving the central part of the diaphragmatic pleura, which receives sensory innervation from the phrenic nerve (C3, C4, and C5), may lead to referred pain over the shoulder, since the skin of this region is supplied by the supraclavicular nerves (C3 and C4).

Asthma

One of the problems associated with bronchial asthma is the spasm of the smooth muscle in the wall of the bronchioles. This particularly reduces the diameter of the bronchioles during difficulty in expiring, although inspiration is accomplished normally. If expiratory air trapping occurs chronically, the lungs become distended and the thoracic cage becomes permanently enlarged, forming a so-called *barrel chest.*

Segmental Resection of the Lung

If a localized chronic lesion, such as that associated with tuberculosis or a benign neoplasm, is restricted to a bronchopulmonary segment, it is possible to dissect a particular segment and remove it, leaving the surrounding lung intact. This procedure is called segmental resection.

Pulmonary Contusion

This condition is caused by a sudden rapid compression of the chest wall and underlying lung. It may be produced by blunt trauma or gunshot wounds. Because of the pliability of the chest wall in children, lung contusion is often present in the absence of rib fractures. The localized endothelial damage to the capillaries results in the transudation of fluid into the lung parenchyma, thus compromising lung function. In cases of blunt trauma, the area of damage to the lung will depend on the site of impact and will not be determined by anatomical subdivisions of the lung.

Tracheobronchial Injury

The lung root is the site where the mobile lung is connected by the main bronchi to the relatively fixed lower end of the trachea. It is not surprising, therefore, to find that when a rapid deceleration or shearing force is applied to the lungs, injuries occur to the bronchi. In the majority of patients the tear occurs within 1 in. of the carina.

ANATOMICAL-CLINICAL PROBLEMS

1. Why is it that children with blunt chest injuries have a relatively higher mortality than adults with blunt injuries of the same force? What is the anatomical reason for this difference?

2. A 34-year-old man was involved in a violent quarrel with his wife over another woman. In a fit of rage, the wife picked up a carving knife and lunged forward at her husband, striking his anterior chest wall over the right clavicle. The husband collapsed on the kitchen floor, bleeding profusely from the wound. The distraught wife called an ambulance.

On examination in the emergency department the patient was conscious but in shock. A wound about 1-in. (2.5-cm) wide was found over the right clavicle, from which blood was still oozing. Examination of the chest revealed diminished breath sounds over the right hemithorax. The trachea was deflected slightly to the left. The right upper limb was lying stationary on the table, and active movement of the small muscles of the right hand was absent. In addition the patient was found to be insensitive to pin prick along the medial side of the arm, forearm, hand, and ulnar two fingers (pinky and ring). Using your knowledge of anatomy, explain the widespread symptoms and signs.

3. A 47-year-old woman was seen in the emergency department complaining of pain over the ninth left intercostal space that extended downward and forward across the costal margin to an area on the anterior abdominal wall just above the umbilicus. The pain worsened when the patient took a deep breath. A small amount of abdominal tenderness and guarding was present in the left upper quadrant. Give the anatomical reason why the pain was felt over such a wide area.

4. On examination of the chest, a 52-year-old man was found to have a warm swelling measuring about 2 in. (5 cm) in diameter over the right fourth intercostal space. The swelling was about 3 in. (7.5 cm) from the lateral margin of the sternum. He had a temperature of 99.5° F and complained of malaise. Radiographic examination showed a dense area in the middle lobe of the right lung. Three days later the swelling ruptured, discharging a yellow purulent fluid. Using your knowledge of anatomy, describe how you would accurately localize the swelling on the chest wall by counting the ribs. Name the various structures that the pus would have to traverse in order to reach the skin from the middle lobe of the right lung.

5. An 85-year-old man was seen in the emergency department complaining of burning pain on the right side of his chest. On examination the patient was asked to show the physician with one finger the exact site of the pain. The patient indicated that the pain was in an area that extended posteriorly as far as the posterior axillary line and passed forward over the right fifth intercostal space as far as the midline over the sternum. It was noted that there were several erythematous maculopapules on the skin and a vesicle. Can you comment on these findings? In anatomical terms explain the distribution of the pain and rash.

6. Define what is meant by the term dermatome. What is the dermatomal innervation above the clavicle? In which dermatome is the nipple located?

7. A 32-year-old woman was driving her car across an intersection when another car passing through a red light hit her car broadside at high speed. An ambulance arrived on the scene within a few minutes, and the attendants managed to free the woman from the wreckage. She was found to be deeply cyanotic, and all respiratory movements had ceased. Her heart was still beating, but her pulse was rapid and weak. Given that the woman's spinal cord had been severed at the level of the second cervical segment, explain in anatomical terms why all respiratory movements had ceased.

8. A 51-year-old man was brought to the emergency department after having a head-on collision in his motor vehicle. He had not been wearing his seat belt. The paramedics noted at the site of the crash that the steering wheel was bent. The patient was tachypneic but conscious and alert.

On examination the patient was found to have moderate respiratory distress, but his color was good. His respiratory rate was 30 per minute. Examination of the chest revealed an area of the left anterior chest wall that was bruised and had paradoxical movement. On palpation, the patient complained of acute tenderness over the anterior and posterior parts of the left sixth and seventh ribs, and also there was evidence of bony crepitus over these sites. A clinical diagnosis of flail chest was made. Using your knowledge of anatomy, explain how a flail chest can compromise the mechanics of respiration.

9. A 16-year-old boy was seen in the emergency department with a large boil on the back of his chest near the inferior angle of his right scapula. There was evidence of widespread acne on his chin and on his anterior chest wall and back near the midline. The boy also complained of a tender swelling in his right armpit. Using your knowledge of anatomy, explain the connection between the symptoms and signs.

10. A 31-year-old woman was thrown from a horse. She landed on a fallen tree trunk, striking the lower part of her chest on the left side. When she was seen in the emergency department about 30 minutes later, she was conscious but breathless. The lower left side of her chest showed severe ecchymosis and was extremely tender to touch. Bony crepitus could be detected over the ninth and tenth ribs on the posterior axillary line. In addition, the patient showed tenderness and evidence of rigidity in the left upper quadrant of the anterior abdominal wall. Explain in anatomical terms the signs and symptoms.

11. A 27-year-old man was seen in the emergency department after transportation by ambulance from the scene of a street shoot-out. On examination the patient showed signs of severe hemorrhagic shock. His pulse was rapid, his capillary refill was absent, and his systolic blood pressure was 80 mm Hg. There was a small entrance wound about 1-cm across in the third left intercostal space about 3 cm from the lateral margin of the sternum. There was

no exit wound. The left hemithorax was dull on percussion, and breath sounds were absent on that side of the chest. A chest tube was placed immediately. Because of massive amounts of blood pouring out of the chest tube with no slowing down, it was decided to perform a thoracotomy. On the basis of the physical examination and your knowledge of anatomy, where would you make the incision? Name the layers of tissue that your scalpel will divide in order to enter the pleural space. What structures must you be particularly careful not to divide?

12. A 70-year-old man with chronic obstructive lung disease fell down stairs and fractured the fifth and sixth ribs on his right side. It was decided not to splint the ribs. Despite analgesics, the pain persisted, and it was decided to perform an intercostal nerve block. Why was it decided not to splint the chest wall in this patient? What are the relationships of an intercostal nerve as it lies in the subcostal groove?

13. A 19-year-old man was traveling at high speed in his sports car when he lost control and hit a bridge abutment. He was wearing a waist seat belt (lap belt) only, and he was found unconscious slumped over the steering wheel. On arrival at the emergency department the patient had recovered consciousness and was complaining of pain and tenderness on the right side of his chest. A careful physical examination confirmed the presence of fractures of the fifth and seventh ribs, with diminished breath sounds of the right hemithorax. There was no tracheal shift. Using your knowledge of anatomy, provide three possible explanations for the diminished breath sounds on the right side of his chest.

14. A 24-year-old man was admitted to the hospital with the signs and symptoms of large-bowel obstruction combined with impaired respiration. Apart from a serious automobile accident 5 years previously, which resulted in the fracture of several ribs on his left side, his medical history was normal. Routine chest and abdominal radiographs revealed the presence of gas-filled bowel in the left side of his chest cavity. Can you give a possible explanation for this unusual radiographic finding?

15. A 58-year-old man visited his physician because of a dull, aching pain in the forearm on the right side. The discomfort was made worse by exercising the upper limb, especially in the elevated position. The pain was relieved by rest. He noticed his right hand was sometimes colder than his left, and, when held above his head, his right hand became white. When his right arm was held by his side for any length of time, his hand became blue, especially in cold weather. On examination, his radial pulse was found to be absent on the right side and normal on the left side. The brachial artery pulse was weak on the right side but normal on the left side. The pulsations of the subclavian arteries were normal on both sides of the neck. It was possible to produce the color changes described by the patient by suitable positioning of the right arm. Using your knowledge of anatomy, describe the structure or structures in the neck that could produce these signs and symptoms.

16. A 38-year-old man, who was a carpenter, was seen in the emergency department complaining of a swollen right hand and an aching pain in his right upper limb. He said that his electric saw had broken down and that it had been necessary for him to cut a large amount of wood by hand that morning. On examination, the right upper limb was suffused, and his right hand and arm were swollen. His radial pulses were normal on both sides, with no neurological defects. What is the diagnosis? Using your knowledge of anatomy, explain the possible cause of the problem. Describe the anatomical reasons for performing the Adson maneuver.

17. A 60-year-old man with a known history of bronchitis and emphysema suddenly experienced severe pain in the chest and became breathless. On examination in the emergency department, his left hemithorax was somewhat more hyperresonant to percussion than the right, and breath sounds were decreased. Using your knowledge of anatomy, explain the clinical symptoms and signs.

18. A 36-year-old man had a history of repeated episodes of hemoptysis and recurring attacks of pneumonia in the inferior lobe of the left lung. After performing a bronchoscopy and consulting with a radiologist, the thoracic surgeon made a tentative diagnosis of bronchial adenoma. The lesion was situated in the superior bronchopulmonary segment of the inferior lobe. Since the lesion was localized with no evidence of spread, the surgeon decided to perform a segmental resection. What is a bronchopulmonary segment? Which parts of the lower airway lie within a lung segment? What structures belonging to a lung segment lie in the connective tissue between adjacent segments?

19. A 65-year-old man was seen in the emergency department after a severe fall in the street, resulting in a Colles' fracture to the right wrist. Prior to reducing the fracture, the physician examined the patient's chest and noted an old thoracotomy scar over the right sixth intercostal space. A large hard lump was also found deep to the right sternocleido-

mastoid muscle in the root of the neck. On questioning, the patient said that he had had his right lung removed for cancer 3 years previously. Using your knowledge of anatomy, can you explain the neck swelling. What is the lymphatic drainage of the lung?

20. A third-year medical student, who was rotating through the emergency department, was asked by the physician in charge whether it is possible for a patient to have pneumonia without experiencing chest pains. How would you have answered that question?

21. A 23-year-old woman was being examined in the emergency department because of the sudden onset of respiratory distress. The physician was listening to breath sounds over the right hemithorax and was concerned when no sounds were heard on the front of the chest at the level of the tenth rib in the midclavicular line. Would you be concerned?

22. A 61-year-old woman was brought to the hospital with congestive heart failure. Examination of the chest revealed the presence of a pleural effusion on both sides extending superiorly as far as the fourth rib in both axillae. In areas where there was fluid, the chest was dull on percussion and breath sounds were diminished and distant. Explain the mechanism by which pleural fluid is normally formed and absorbed. What is the function of pleural fluid?

23. A 19-year-old man was admitted to the emergency department following an automobile accident. He was unconscious and in obvious respiratory distress. On examination it was found that the left side of his chest was partially caved in over an area involving the fifth, sixth, and seventh ribs. No breath sounds could be heard over the left hemithorax. A satisfactory airway was established with an endotracheal tube. Because of the respiratory distress and evidence of chest trauma, it was decided to insert a thoracostomy tube. On the basis of your knowledge of anatomy and the physical findings in this patient, where would you insert the tube? List the structures that your scalpel blade and dissecting instruments would penetrate in order to enter the pleural space.

24. A 15-year-old girl was knocked down by a passing car as she was riding her bicycle to school. The ambulance took her to the nearest emergency department, where she was found to be in extreme respiratory distress. She was conscious and crying and complaining of pain on the right side of her chest. The patient had tachypnea and dyspnea, and her lips were slightly cyanotic. Examination of the chest showed some bulging of the right hemithorax. Palpation of the rib cage on the right side showed evidence of a fracture of the sixth rib in the anterior axillary line. On auscultation breath sounds were decreased over the right hemithorax. The trachea was deflected to the left. The patient's condition began to deteriorate. Her respiration became very rapid and shallow and her pulse rapid and thready. The external jugular veins in the neck were distended. No breath sounds could be heard over the right lung. It was decided to perform an immediate needle thoracostomy. What is your diagnosis? Using your knowledge of anatomy, where would you place the thoracostomy? List the structures that are perforated by the needle. What anatomical structures do you have to avoid? Is it necessary in this patient to insert a tube into the pleural space? Should the physician have ordered a chest radiograph before performing a thoracostomy? Explain the cause of the cyanosis and the distended neck veins.

25. Several ambulances were involved in the transport of injured patients to the emergency department in the city hospital following a serious automobile accident on the highway involving many vehicles. While performing an initial assessment of all the patients, the physician in charge noted that one patient, a middle-aged man in respiratory distress, had a bloated look about the face. When the blanket was turned back, it was found that the puffy look extended to the neck. Examination of the patient the following morning disclosed that the puffy look involved the rest of the body, including the legs. What is the probable diagnosis? Explain in anatomical terms the possible connection between the respiratory problem and the bloated frog appearance.

26. A 37-year-old workman was trapped under a building that had suddenly collapsed during the final stages of its erection. Five hours later he was found to be alive under the debris. On examination in the emergency department, he was found to be in respiratory distress and had hemoptysis. Several of the ribs on both sides of the chest were fractured. Palpation of the skin of the neck revealed a peculiar crackling sensation. A diagnosis of subcutaneous emphysema was made. Later, a bronchoscopy was performed and a tear was found in the left main stem bronchus. In anatomical terms explain the etiology of bronchial tears.

ANSWERS

1. In young children the ribs are relatively pliable and the costal cartilages are flexible. This means that a given blunt force applied to the chest is not completely dissipated in the compression of the rib cage. The unexpended force is transmitted to the underlying organs, namely, the heart and lungs, often with serious consequences. With increasing age the ribs become stiff and, in later life, brittle. The costal cartilages lose their flexibility as calcium is deposited within them after middle age.

2. Stab wounds to the chest wall over the clavicle are often deflected by that bone into the root of the neck. There, at the base of the posterior triangle, lie the subclavian artery (the subclavian vein is usually behind and protected by the clavicle), the roots and trunks of the brachial plexus, and the cervical dome of the pleura with the apex of the lung. During surgery in this patient, it was found that the right subclavian artery was damaged by the knife but not completely severed. The lower trunk of the brachial plexus was cut. This latter finding would explain the loss of movement of the small muscles of the right hand and the loss of sensation in C8 and T1 dermatomes. The knife had also pierced the cervical dome of the right pleura, causing a right pneumothorax. This accounted for the diminished breath sounds over the right hemithorax and the slight deflection of the mediastinum to the left as seen by the deflection of the trachea to the left.

3. This patient was diagnosed as having a pneumonia complicated by pleurisy that involved the lower lobe of the left lung. The involvement of the parietal pleura accounted for the pain over the left ninth intercostal space, since it is innervated by that nerve. Because the ninth intercostal nerve also extends into the abdominal wall and supplies the parietal peritoneum and the dermatome just above the umbilicus, this would account for the referred pain to the abdomen. The rigidity of the muscles of the anterior abdominal wall can also be explained since they receive their innervation from the ninth intercostal nerve. The irritation of the parietal pleura lining the ninth intercostal space reflexly increased the tone of the abdominal muscles.

4. One of the most important surface landmarks on the anterior chest wall is the sternal angle. This is the site of the manubriosternal joint and can be easily felt as a low transverse ridge on the front of the sternum. If the examining finger is moved laterally over the ridge, it passes onto the second costal cartilage and then onto the second rib. All ribs and intercostal spaces can be counted from this landmark.

This patient was found to have actinomycosis of the middle lobe of the right lung. An abscess developed at the site of the lesion and spread to the surface through the following structures: (1) visceral pleura, (2) parietal pleura (adherent to the visceral pleura so that the pleural space was not widely involved), (3) endothoracic fascia, (4) transversus thoracis muscle, (5) internal intercostal muscle, (6) external intercostal muscle, (7) serratus anterior muscle, (8) pectoralis major muscle, (9) fascia, and (10) skin.

5. This man was suffering from herpes zoster, which is considered to be reactivation of a prior varicella infection. It is now generally believed that the varicella virus enters the cutaneous sensory nerves and ascends to the neuron cell bodies of the posterior root ganglia where it lies dormant. Later, the virus becomes reactivated and descends along the nerves to the skin, causing dermatomal pain and the eruption of vesicles. In this patient the right fifth intercostal nerve was involved.

6. A dermatome is an area of skin supplied by a single spinal nerve and therefore a single segment of the spinal cord. The skin above the clavicle involves dermatomes supplied by the third and fourth cervical spinal nerves. Branches of these spinal nerves reach the skin in the supraclavicular nerves (C3 and C4). In both sexes the nipple is usually located in the dermatome supplied by the fourth thoracic spinal nerve. Branches of the nerve reach the skin in the fourth intercostal nerve.

7. In this patient, the spinal cord had been severed above the level of the origin of the phrenic nerve (C3, C4, and C5), causing paralysis of the diaphragm. All the intercostal muscles, which are supplied by the intercostal nerves (T1 through T11), were also paralyzed. Without innervation to the intercostal muscles and the diaphragm, all respiration ceases.

8. In flail chest, the stability of the chest wall is lost, and the flail (unattached) segment is sucked in during inspiration and driven out during expiration. This produces paradoxical respiratory movements, and the process of respiration is compromised. The condition is most obvious when the sternum is fractured. It is less noticeable when the ribs are fractured near their angles, since this area is covered by superficial muscles, namely, the serratus anterior, the latissimus dorsi, and the trapezius muscles.

In the past, flail chest was believed to impair respiration by causing the useless movement of air back and forth from the flail-sided lung to the normal lung (pendula movement). Now the following two factors are believed to be responsible: (1) severe contusion of the underlying lung on the flail side of the chest interferes with the oxygenation of the blood within the lung, and (2) the extreme pain caused by the multiple fractures inhibits the respiratory movements. In addition, the lung underlying the flail segment may not be effectively expanded during inspiration. Splinting of the flail segment may improve the mechanics of respiration and may reduce the pain caused by the movements at the fracture sites.

9. The lymphatic drainage of the skin of the anterior and posterior chest walls is into the axillary lymph nodes. The anterior part of the chest drains into the pectoral group of lymph nodes, just beneath the lower border of the pectoralis major muscle. The posterior part of the chest drains into the subscapular (posterior axillary) group of nodes lying on the posterior wall of the axilla in front of the subscapularis muscle. In this boy the infection had spread from the boil into the subscapular nodes, which were swollen and tender.

10. Whenever a fracture of the lower ribs is diagnosed, the possibility of injury to the upper abdominal viscera and damage to the thoracic viscera must always be considered. This patient had fractures of the ninth and tenth ribs near their angles on the left side. Her more serious injury was a ruptured spleen. The blunt trauma to the ribs had also resulted in a tear of the underlying spleen. The presence of blood in the peritoneal cavity had irritated the parietal peritoneum, producing reflex rigidity of the upper abdominal muscles.

11. After skin decontamination, thoracotomy was performed through the fourth left intercostal space. The incision was extended from the sternum to the anterior axillary line. In this patient the chest was not entered through the same intercostal space (third) as the gunshot wound, since it was thought that the fourth space would give a larger, better exposure to the thoracic contents; the source of the hemorrhage was not known prior to the operation. It was later found that the left atrium had been perforated by the bullet.

When the thoracotomy incision was made, the following structures were incised: (1) skin, (2) subcutaneous tissue, (3) pectoral muscles and the serratus anterior muscle, (4) external intercostal muscle and anterior intercostal membrane, (5) internal intercostal muscle, (6) tranversus thoracis muscle, (7) endothoracic fascia, and (8) parietal pleura.

It is important to avoid the internal thoracic artery, which descends vertically behind the costal cartilages about one fingerbreadth from the lateral margin of the sternum. Also remember that the intercostal vessels and nerve—VAN—in that order from above downward, lie in the subcostal groove in the upper part of the intercostal space.

12. Splinting the chest wall in an elderly patient with bronchitis and emphysema is mainly contraindicated, since any form of splinting would further impair respiratory movements and invite the possibility of pneumonia. The neurovascular bundle in the subcostal groove has the following relationships from above downward: (1) intercostal vein, (2) intercostal artery, and (3) intercostal nerve.

13. In this patient the sudden compression of the chest wall against the steering wheel caused traumatic damage to the underlying lung. This resulted in capillary hypertension and transudation in the lung parenchyma. A chest radiograph confirmed the presence of a lung density with ill-defined borders in the area of the middle lobe of the right lung. No hemothorax or pneumothorax was present on chest radiograph, the two other causes of a posttraumatic decrease in breath sounds. The fractures of the right fifth and seventh ribs were confirmed.

14. This patient had fractured the left ninth and tenth ribs as the result of trauma from the steering wheel in the motor vehicle accident 5 years previously. At that time it was determined that there were no apparent injuries to the intrathoracic or abdominal viscera. Unfortunately, the diaphragm was torn at the time of the accident, and in the intervening years the diaphragmatic parietal pleura and peritoneum had slowly stretched up into the chest cavity, forming a hernial sac. At surgery, a small hole was found in the left dome of the diaphragm, and the hernial sac contained a part of the splenic flexure. The hernia was repaired, and the patient recovered uneventfully.

15. This patient was suffering from vascular insufficiency of the right upper limb due to partial constriction of the subclavian artery. An anteroposterior radiograph of the neck revealed the presence of a complete cervical rib on the right side. At surgery the subclavian artery was found to be angulated as it passed over the rib.

16. The diagnosis is thrombosis of the right subclavian vein secondary to repeated trauma to the vein wall caused by the movements of the right clavicle while sawing the wood. The unusual and excessive amount of movement of the clavicle repeatedly compressed the subclavian vein against the anterior

surface of the scalenus anterior at the root of the neck.

In the Adson maneuver, while the radial pulse is palpated, the following procedures are performed: (1) the patient takes a deep breath to raise the first rib, exerting the maximum pressure from below on the lower trunk of the brachial plexus and subclavian vessels; (2) the neck is extended, stretching the brachial plexus and the subclavian vessels; and (3) the chin is rotated to the side being examined, narrowing the interval between the scalenus anterior and the scalenus medius muscles so that the brachial plexus and the subclavian artery are compressed. A diminished or absent radial pulse under these conditions is a positive finding that indicates the subclavian artery is being compressed by neighboring structures at the root of the neck.

17. This man had a left-sided pneumothorax following the rupture of an emphysematous bulla of the left lung. The air entered the left pleural cavity and, since the air pressures in the bronchial tree and pleural cavity were equal, the elasticity of the lung caused it to collapse. The air-filled left pleural cavity was responsible for the hyperresonance to percussion, and since the collapsed lung had no air movement, breath sounds were decreased. The mobile, pliable mediastinum was deflected to the right by the pressure of air in the left pleural cavity.

18. A bronchopulmonary segment is a structurally and functionally independent unit of a lung lobe. It is triangular, with its apex pointing toward the hilum of the lung, and is surrounded by connective tissue. A segmental bronchus, which is a branch of a lobar bronchus, enters each segment; it also has a branch of the pulmonary artery and lymphatic vessels and autonomic nerves. The veins are tributaries of the pulmonary vein and run outside the segment in the connective tissue between adjacent segments.

The parts of the lower airway that lie within a bronchopulmonary segment include the (1) segmental bronchus, (2) terminal bronchiole, (3) respiratory bronchiole, (4) alveolar duct, (5) alveolar sac, and (6) alveolae.

As previously mentioned, the tributaries of the pulmonary vein that lie within the connective tissue between segments are at risk during segmental resection.

19. This patient had a bronchogenic carcinoma of the right lung diagnosed 3 years previously that was treated with a right-sided pneumonectomy. The hard lump in the neck was a metastasis in the deep cervical lymph nodes on the right side.

The lymph vessels of the lung originate in superficial and deep plexuses. The superficial plexus lies beneath the visceral pleura, and the deep plexus travels along the bronchi toward the root of the lung, where they drain into the pulmonary nodes close to the hilum. All lymph from the pulmonary nodes and the superficial plexus drains into the bronchopulmonary nodes and then the tracheobronchial nodes; finally, the lymph drains into the bronchomediastinal trunk. In patients with bronchogenic carcinoma, it is common for a retrograde flow of lymph to move into the neck, where the carcinoma cells come to rest in the lower deep cervical nodes. This process must have occurred prior to the pneumonectomy.

20. The lungs and visceral pleura are innervated by autonomic nerves from the pulmonary plexus. The only sensations originating in the lung are those due to stretch. Pneumonia in its early stages, when it is confined to the lung, is therefore painless. It is only when the disease process crosses the pleural space and involves the parietal pleura that the patient experiences pain. The parietal pleura is innervated by intercostal nerves and the phrenic nerve (see p. 64).

21. The lower border of the lung in the midclavicular line in the midrespiratory position (i.e., midway between full inspiration and full expiration) is at the level of the sixth rib. The parietal pleura in the midclavicular line crosses the eighth rib. This means that the lower edge of the lung on extreme inspiration could only descend in the costodiaphragmatic recess as far as the eighth rib. In this situation, therefore, breath sounds would not be heard at the level of the tenth rib; the stethoscope would be over the liver or stomach.

22. The formation and drainage of pleural fluid and some of the factors that influence this process are fully discussed on p. 89. The normal volume is about 5 to 10 mL. Its function is to lubricate the apposing surfaces of the visceral and parietal layers of the pleura so that they can move easily on one another during respiratory movements.

23. The thoracotomy tube was inserted through the fourth left intercostal space in the anterior axillary line. The following structures would be pierced by the scalpel blade: (1) skin, (2) subcutaneous tissue, (3) serratus anterior muscle, (4) external intercostal muscle, (5) internal intercostal muscle, (6) transversus thoracis muscle, (7) endothoracic fascia, and (8) parietal pleura.

24. This patient had a right-sided tension pneumothorax. The air had escaped from the bronchial

tree following laceration of the right lung by a fragment of the sixth rib. The right lung had collapsed, which would explain the diminished breath sounds, and the right pleural cavity was filled with air, explaining the right-sided hyperresonance. The rising air pressure in the right pleural cavity caused the deflection of the mediastinum to the left, as evidenced by the movement of the trachea and heart to the left. The lacerated lung was acting as a valve, allowing air to enter the pleural cavity but not allowing it to escape. The rising air pressure compressed the thin-walled veins (vena cavae and other large veins) and thin-walled atria, preventing the blood from returning to the heart. This would explain the patient's cyanosis and the engorged neck veins. The rapid, weak pulse resulted from inadequate venous return.

Tension pneumothorax is life threatening, and an immediate needle thoracostomy should be performed. The needle should be inserted in the right second intercostal space in the midclavicular line with the patient 30° up from the supine position. The following structures are perforated by the needle: (1) skin, (2) subcutaneous tissue, (3) pectoral muscles, (4) serratus anterior muscle, (5) external intercostal muscle, (6) internal intercostal muscle, (7) transversus thoracis muscle, (8) endothoracic fascia, and (9) parietal pleura. The needle should be placed close to the upper border of the third rib to avoid damaging the second intercostal vessels and nerve. The needle may be replaced at a later time with a chest tube, as described on p. 92. This is placed in a water seal suction bottle to encourage the lung to reexpand.

Tension pneumothorax is a true emergency; the decision to decompress the chest may be made on clinical grounds; waiting for a radiograph of the chest only delays the start of treatment—a delay that may prove to be fatal.

25. This patient has extensive subcutaneous emphysema secondary to air escaping from the tracheobronchial tree into the surrounding tissues. The condition could be caused by laceration of the lung, or tearing of the lower trachea or main stem bronchi. The air passes beneath the visceral pleura and enters the loose connective tissue of the mediastinum. Then it can ascend to the neck and lie beneath the fascia. From here it may travel extensively over the body from the scalp to the toes.

26. Bronchial tears commonly occur within 1 in. of the carina of the trachea. It can result from a number of causes as follows: (1) shearing forces generated when a sudden deceleration causes the mobile lung to swing forward on its anchored root to the more stable trachea in the mediastinum, (2) airway distension against a closed glottis, and (3) a sudden vertical motion stretching the main stem bronchi.

3

The Heart, Coronary Vessels, and Pericardium

Cardiac emergencies, whether they originate from coronary artery disease, valvular dysfunction, electrical disturbances, myocardiopathies, or problems of the pericardium, present the emergency physician with a diagnostic and therapeutic challenge. The evaluation of chest pain and the management of disturbances in cardiac rhythm and pump function are common problems facing emergency personnel. In children, congenital heart disease may present as a previously undiagnosed condition.

The cardiovascular system is the transport system of the body and conveys nutrients, gases, hormones, and the products of metabolism. The heart is the vital muscular pump in this system and should it function inefficiently, the transport system fails. Should the heart stop beating, within a few minutes the oxygen supply to all tissues of the body is depleted and tissue cells start to die.

The purpose of this chapter is to review the structure of the heart, including its conducting system, its important blood supply, and its surrounding pericardium.

BASIC ANATOMY

Basic Anatomy of the Heart

The heart is somewhat pyramidal in shape and is located in the mediastinum. It lies free within the pericardium, being only connected at its base to the great blood vessels. The heart has three surfaces—the sternocostal surface (anterior), the diaphragmatic surface (inferior), and the base (posterior).

The *sternocostal surface* is formed mainly by the right atrium and the right ventricle, which are separated from each other by the vertical atrioventricular groove (Fig 3–1). The right border is formed by the right atrium, and the left border is formed by the left ventricle and part of the left auricle. The right ventricle is separated from the left ventricle by the anterior interventricular groove.

The *diaphragmatic surface* of the heart is formed mainly by the right and left ventricles separated by the posterior interventricular groove. The inferior surface of the right atrium, into which the inferior vena cava opens, also forms part of this surface.

The *base* of the heart, or posterior surface, is formed mainly by the left atrium, into which the four pulmonary veins open (see Fig 3–1). The right atrium also contributes to a lesser extent to this surface.

The *apex* of the heart, formed by the left ventricle, is directed downward, forward, and to the left (see Fig 3–1). It lies at the level of the fifth left intercostal space 3½ in. (9 cm) from the midline.

Note that the base of the heart is called the base, since the heart is pyramid shaped and the base lies opposite the apex. The heart does not rest on its base, but on its diaphragmatic surface.

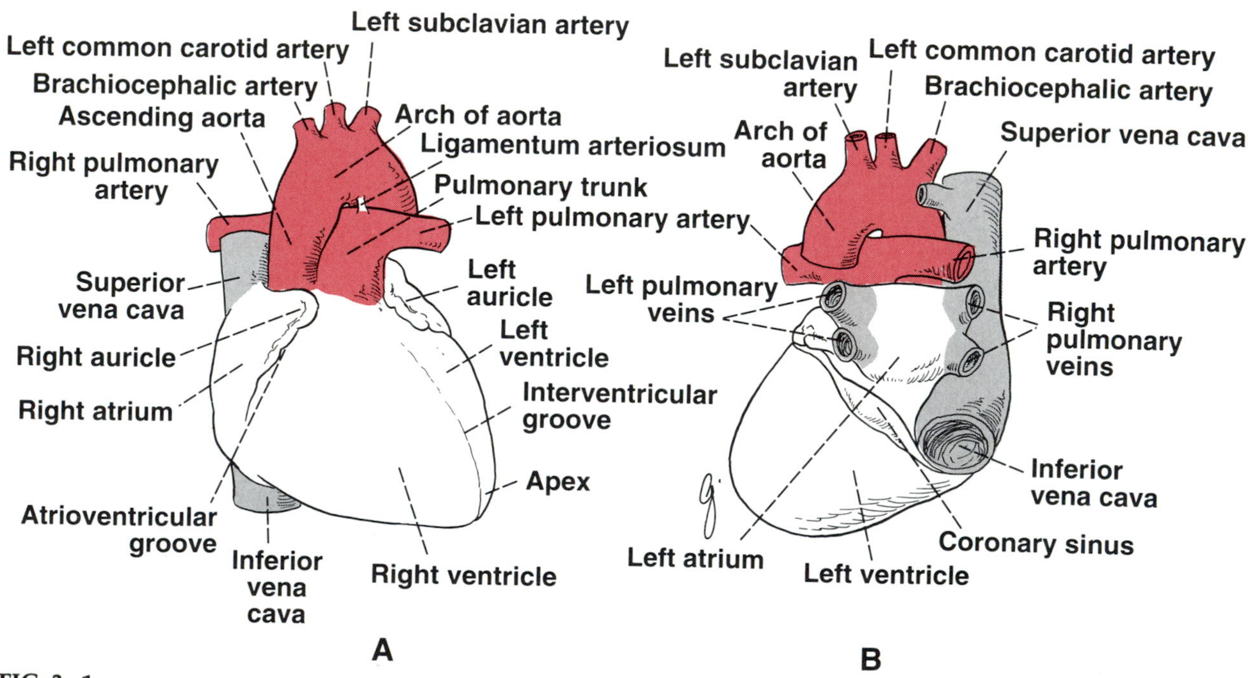

FIG 3–1.
A, anterior surface of the heart and the great blood vessels. **B,** posterior surface or base of heart.

Structure of the Heart

The heart is divided by vertical septa into four chambers—the right and left atria and the right and left ventricles. The right atrium lies anterior to the left atrium, and the right ventricle lies anterior to the left ventricle (see Fig 3–1).

The walls of the heart are composed of cardiac muscle, the *myocardium,* covered externally with serous pericardium, called the *epicardium,* and lined internally with a layer of endothelium, the *endocardium.*

Skeleton of the Heart

The skeleton of the heart consists of fibrous rings that surround the atrioventricular, aortic, and pulmonary orifices (Fig 3–2). These rings are composed of dense fibrous tissue and provide attachment for the cardiac muscle fibers of the atria and ventricles and cusps of the atrioventricular valves. Its function is to limit the diameter of the atrioventricular and arterial orifices and to provide a solid foundation for the valves of the heart during the different phases of the cardiac cycle.

Borders of the Heart

The right border is formed by the right atrium, the left border by the left auricle, and below by the left ventricle (see Fig 3–1). The lower border is formed mainly by the right ventricle, but also by the right atrium and the apex by the left ventricle.

Radiographic Heart Shadow

The normal radiographic anatomy of the heart in posteroanterior and oblique views on radiographs is shown in Figures 3–3 through 3–5.

Chambers of the Heart

Right Atrium.—This consists of a main cavity and a small outpouching, the *auricle,* which together form the atrium (see Fig 3–2). On the outside of the heart at the junction between the right atrium and the right auricle is a vertical groove, the *sulcus terminalis,* which on the inside forms a ridge, the *crista terminalis.* The main part of the atrium, which lies posterior to the ridge, is smooth walled, whereas the interior of the auricle is ridged or trabeculated by bundles of muscle fibers, the *musculi pectinati.*

Openings into the Right Atrium.—The *superior vena cava* (see Fig 3–2) opens into the upper part of the right atrium; it has no valve. The *inferior vena cava,* which is larger than the superior vena cava, opens into the lower part of the right atrium; it is guarded by a rudimentary valve. The *coronary sinus,* which drains most of the blood from the heart wall

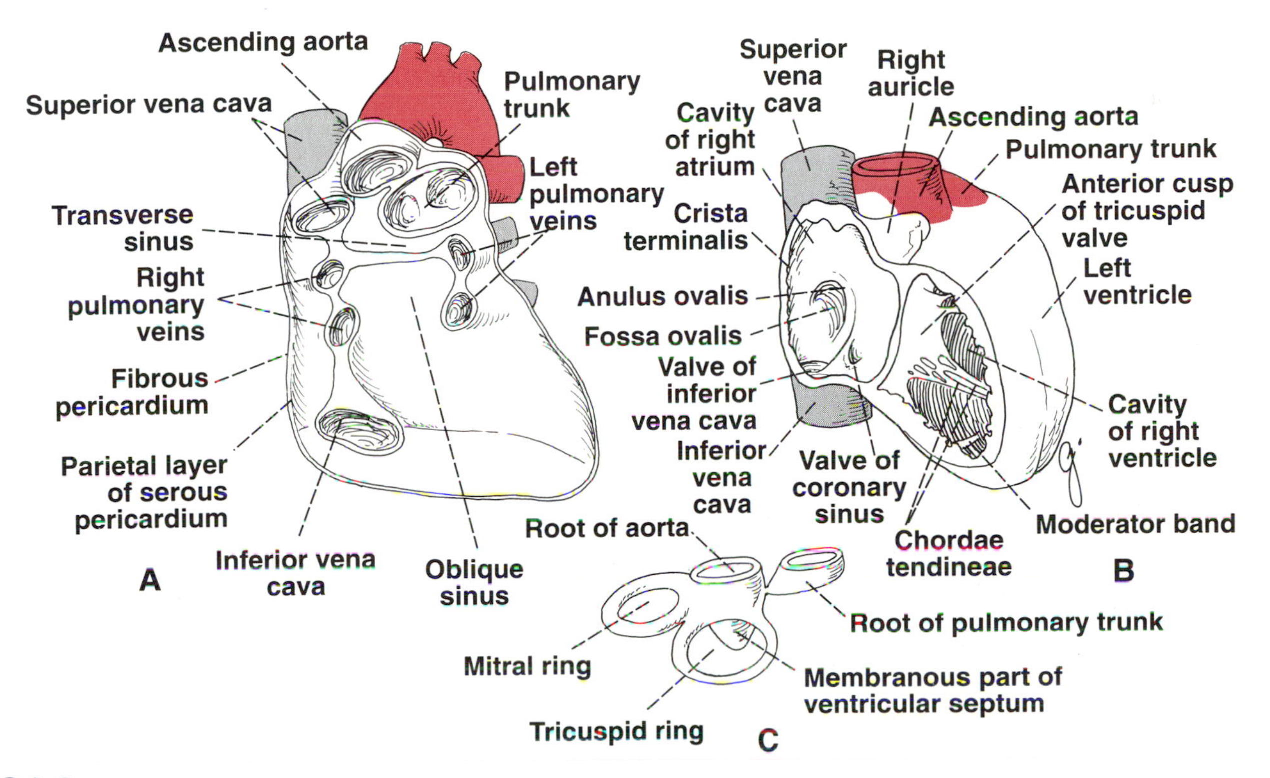

FIG 3–2.
A, interior of the pericardium showing the reflection of the serous pericardium around the great blood vessels. **B,** interior of the right atrium and the right ventricle. **C,** fibrous skeleton of the heart.

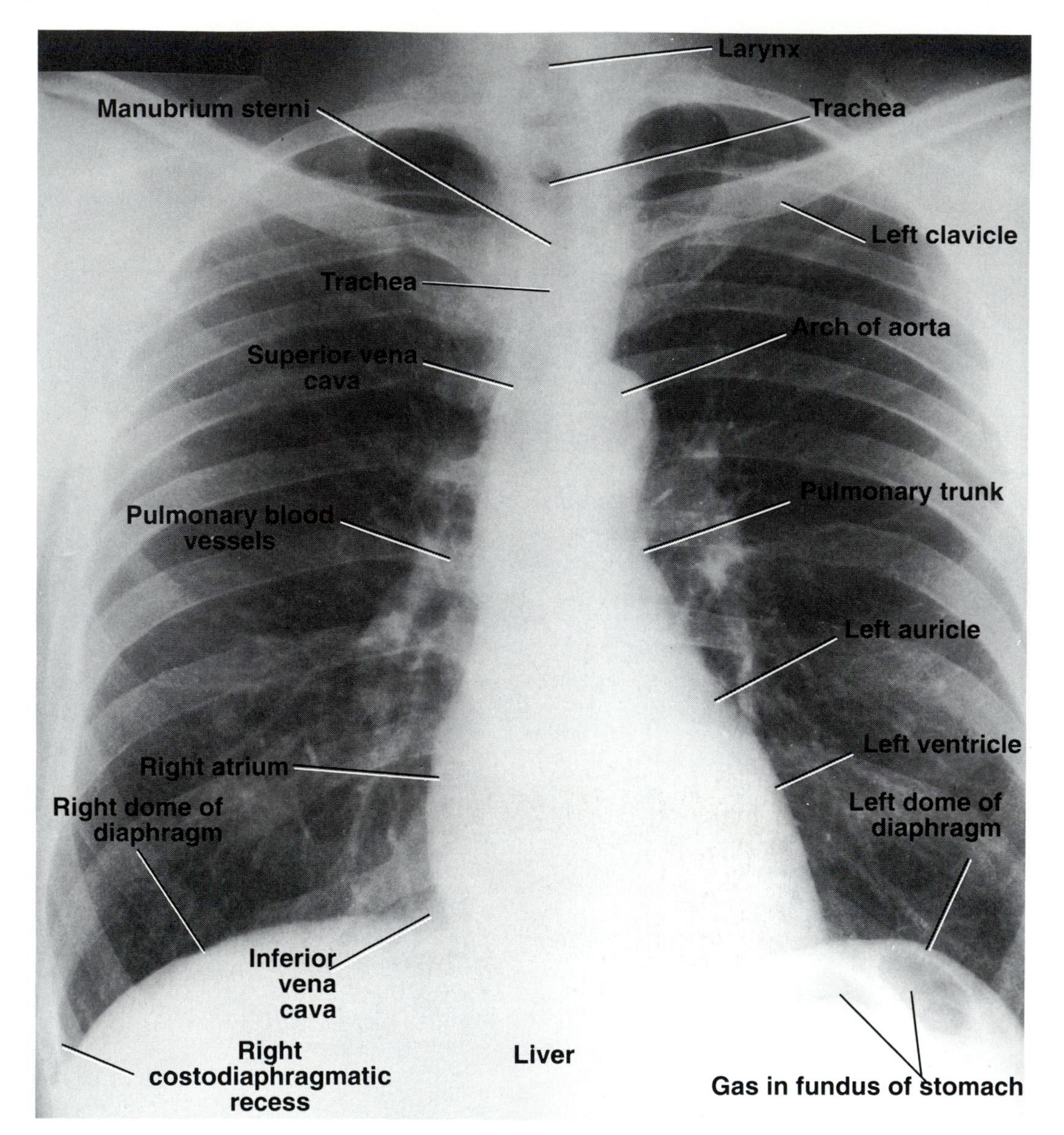

FIG 3–3.
Chest radiograph, posteroanterior view, of a normal adult male.

(see Fig 3–2), opens into the right atrium between the inferior vena cava and the atrioventricular orifice; it is guarded by a rudimentary valve. The *right atrioventricular orifice* lies anterior to the inferior vena caval opening and is guarded by the tricuspid valve (see Fig 3–2). Also, many small orifices of small veins drain the wall of the heart and open directly into the right atrium.

Fetal Remnants.—In addition to the rudimentary valve of the inferior vena cava, there are the *fossa ovalis* and *anulus ovalis.* These latter structures lie on the atrial septum that separates the right atrium from the left atrium. The fossa ovalis is a shallow depression that is the site of the foramen ovale in the fetus (Fig 3–6). The anulus ovalis forms the upper margin of the fossa.

Right Ventricle.—This communicates with the right atrium through the atrioventricular orifice and with the pulmonary trunk through the pulmonary orifice. As the cavity approaches the pulmonary orifice it becomes funnel shaped and is referred to as the *infundibulum.*

The walls of the right ventricle are much thicker than those of the right atrium and show a number of internal projecting ridges formed of muscle bundles. The projecting ridges are known as *trabeculae carneae* and include the following three types.

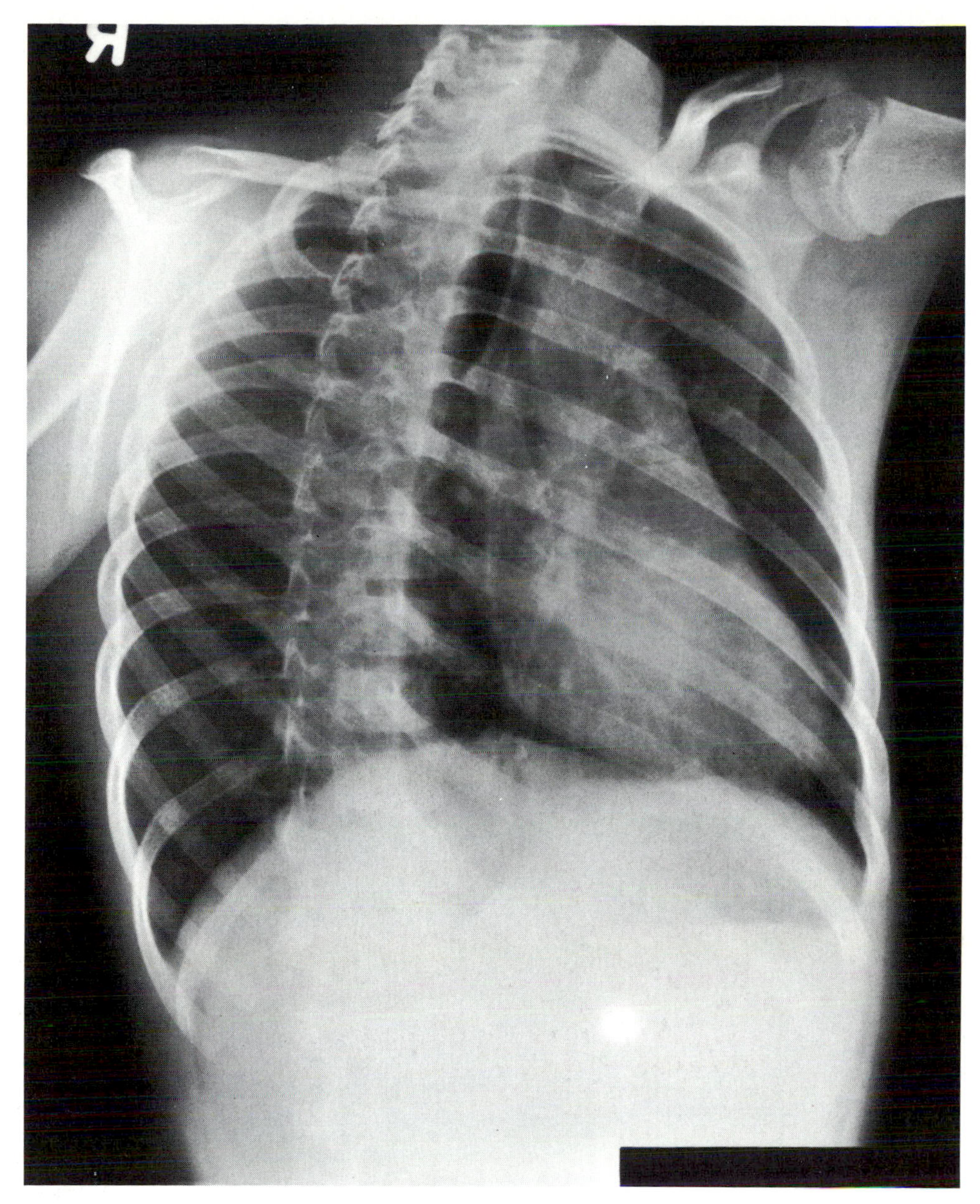

FIG 3–4.
Chest radiograph, right anterior oblique view, of a normal 10-year-old boy.

1. *Papillary muscles* that project inward.—An anterior and a posterior papillary muscle and possibly a smaller variable septal muscle exist. Their bases are attached to the ventricular wall and their apices are connected by fibrous chords, the *chordae tendineae* (Fig 3–7), to the cusps of the tricuspid valve.
2. *Moderator band*.—This crosses the ventricular cavity from the septal wall to the anterior wall (see Fig 3–2), and its ends are attached to the ventricular walls, being free in the middle. It conveys the right branch of the atrioventricular bundle, part of the conducting system of the heart.
3. *Simple prominent ridges*.

The *tricuspid valve* guards the atrioventricular orifice (see Fig 3–7). It consists of three cusps—the anterior, septal, and inferior (posterior) cusps—formed by a fold of endocardium with some delicate fibrous tissue enclosed. The anterior cusp lies anteriorly, the septal cusp lies against the ventricular septum, and the inferior or posterior cusp lies inferiorly.

The bases of the cusps are attached to the fibrous ring of the skeleton of the heart, and the *chordae tendineae* are attached to the free edges and ventricular surfaces of the cusps. The chordae tendineae connect the cusps to the papillary mus-

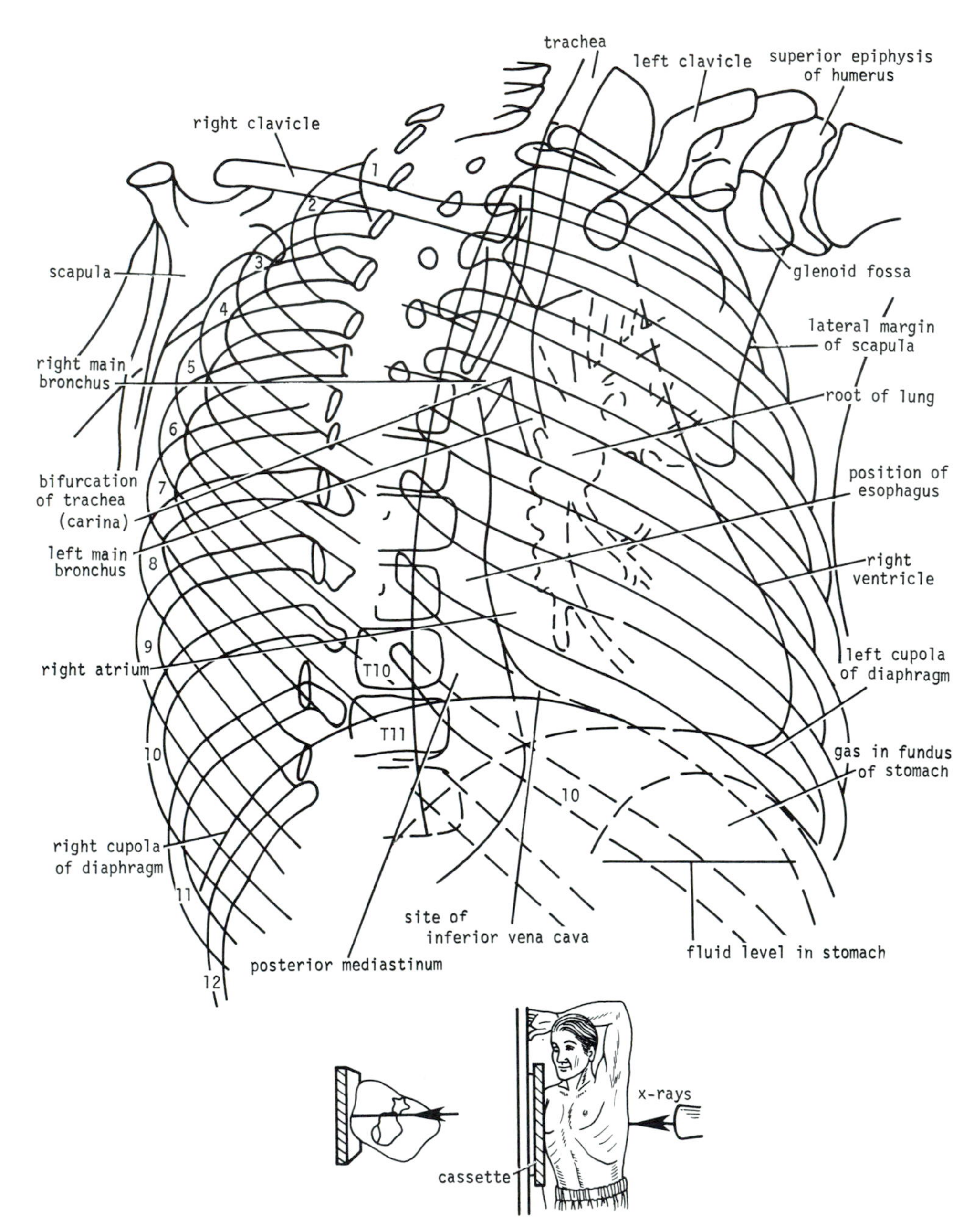

FIG 3–5.
Main features seen in the chest radiograph, right anterior oblique view, in Figure 3–4. Note the position of the patient in relation to x-ray source and cassette holder

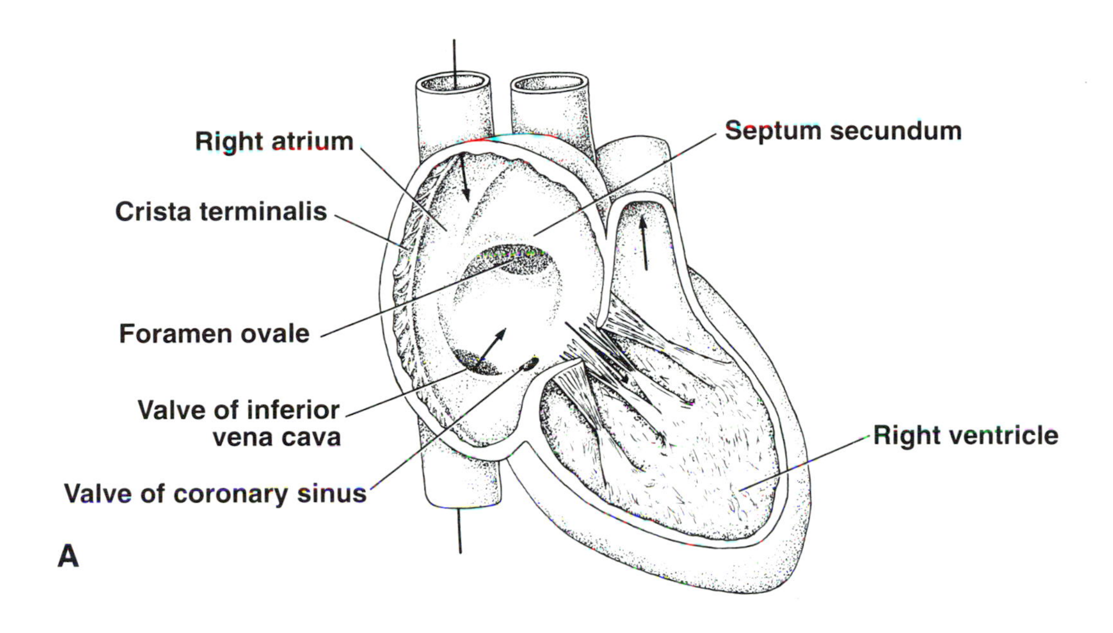

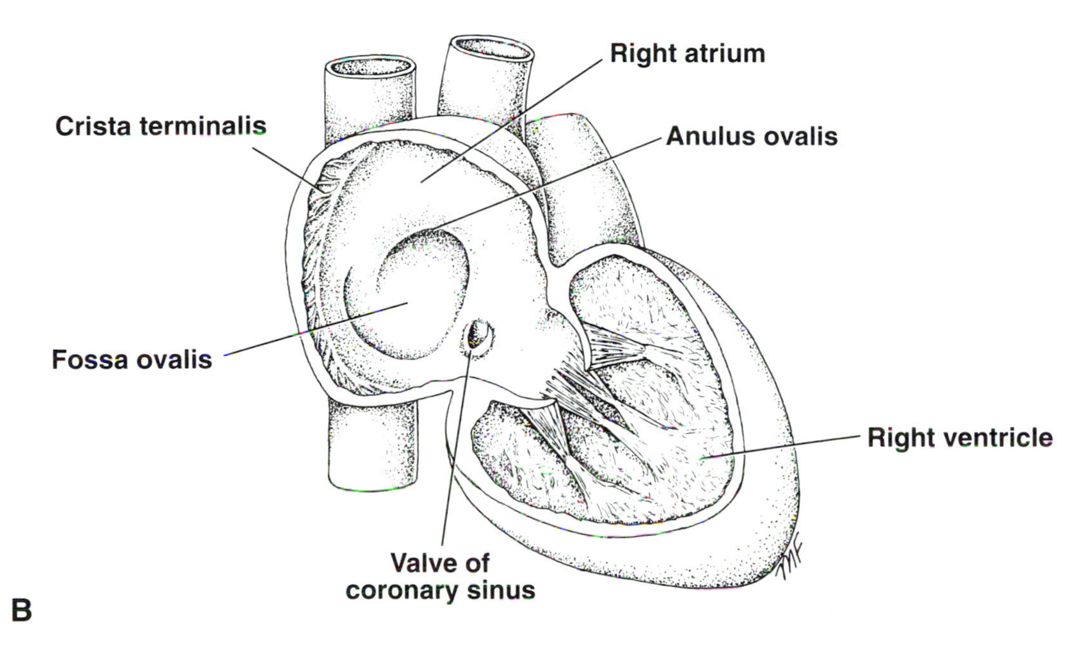

FIG 3–6.
A, the interior of the developing heart showing the formation of the atrial septum; the valve of the inferior vena cava is directing blood to the foramen ovale. **B,** the fully developed heart showing the region of the foramen ovale that closed at birth. Note the anulus ovalis formed from the lower margin of the septum secundum and the fossa ovalis formed of the septum primum.

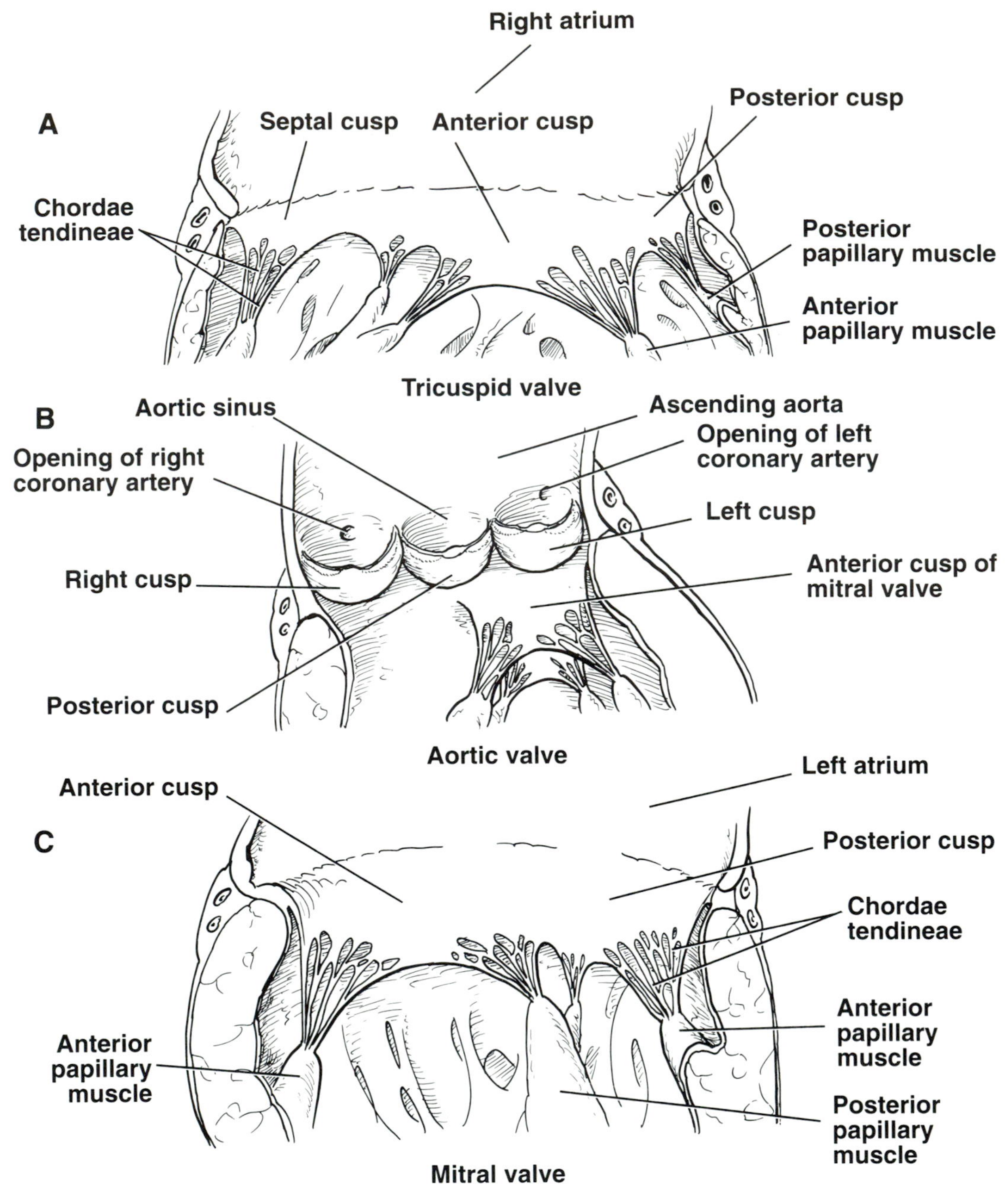

FIG 3–7.
Valves of the heart. **A,** tricuspid valve showing the septal, anterior, and posterior cusps and their chordae tendineae. **B,** aortic valve showing the relationship between the cusps and the openings of the coronary arteries. **C,** mitral valve showing the anterior and posterior cusps and their chordae tendineae.

cles. The anterior papillary muscle is attached by chordae to the adjacent margins of the anterior and inferior cusps. The posterior papillary muscle is attached by chordae to the adjacent margins of the inferior and septal cusps. The third smaller papillary muscle, if present, is attached by chordae to the adjacent margins of the anterior and septal cusps.

The *pulmonary valve* guards the pulmonary orifice (see Fig 3–10) and consists of three semilunar cusps formed by folds of endocardium with some delicate fibrous tissue enclosed. The curved lower margins and sides of each cusp are attached to the arterial wall. The open mouths of the cusps are directed upward into the pulmonary trunk. There are

Clinical Note

Chordae tendineae and papillary muscle rupture.—Rupture of the chordae tendineae by disease, such as acute bacterial endocarditis, may cause sudden valvular insufficiency and cardiac decompensation. Rupture of a papillary muscle, since each has many chordae attached to its apex, is much more serious. Rupture of a papillary muscle may occur in penetrating wounds of the heart.

no chordae or papillary muscles associated with these valve cusps; the attachments of the sides of the cusps to the arterial wall prevent the cusps from prolapsing into the ventricle. At the root of the pulmonary trunk are three dilations called *sinuses,* and one is situated external to each cusp (see aortic valve). The three semilunar cusps are arranged with one posterior (left cusp) and two anterior (anterior and right cusps).*

Left Atrium.—Similar to the right atrium, the left atrium consists of a main cavity and a *left auricle.* The interior of the wall of the left auricle has muscular ridges as in the right auricle.

Openings into the Left Atrium.—The four *pulmonary veins,* two from each lung (see Fig 3–1), open through the posterior wall and have no valves. The *left atrioventricular orifice* is guarded by the mitral valve.

Left Ventricle.—This communicates with the left atrium through the atrioventricular orifice, and with the aorta through the aortic orifice (see Fig 3–7). The walls of the left ventricle are three times thicker than those of the right ventricle (the left intraventricular pressure is six times higher than that inside the right ventricle). There are well-developed trabeculae carneae, two papillary muscles, but no moderator band. The part of the ventricle below the aortic orifice is called the *aortic vestibule.*

The *mitral valve* guards the atrioventricular orifice. It consists of two cusps, one anterior and one posterior, which have a similar structure to those of the tricuspid valve. The attachment of the chordae tendineae to the cusps and the two papillary muscles is similar to the tricuspid valve (see Fig 3–7).

*The cusps of the pulmonary and aortic valves are named according to their position in the fetus before the heart has rotated to the left. This unfortunately causes a great deal of unnecessary confusion.

Clinical Note

Mitral valve prolapse.—In this condition one or both mitral valve cusps balloon up into the left atrium during ventricular systole. The valve cusps are larger than normal and the chorda tendineae may be excessively long. The posterior cusp is always involved; the anterior cusp is involved less frequently.

The *aortic valve* guards the aortic orifice and is similar in structure to the pulmonary valve (see Fig 3–7). One cusp is situated on the anterior wall (right cusp) and two are located on the posterior wall (left and posterior cusps). Behind each cusp the aortic wall bulges to form an *aortic sinus.* The right coronary artery originates from the anterior aortic sinus, and the left coronary artery originates from the left posterior sinus.

Conducting System of the Heart

The conducting system of the heart consists of the sinoatrial node, the atrioventricular node, the atrioventricular bundle, and the Purkinje plexus.

Sinoatrial Node.—The sinoatrial node is located in the wall of the right atrium in the upper part of the sulcus terminalis just to the right of the opening of the superior vena cava (Fig 3–8). The node spontaneously gives origin to rhymical electrical impulses that spread in all directions through the cardiac muscle of the atria and cause the muscle to contract.

The action potential originating in the specialized cardiac muscle fibers of the sinoatrial node spreads to the surrounding muscle fibers and reaches the atrioventricular node in about 0.04 seconds.

Atrioventricular Node.—The atrioventricular node is strategically placed in the lower part of the atrial septum just above the attachment of the septal cusp of the tricuspid valve (see Fig 3–8). From it, the cardiac impulse is conducted to the ventricles by the atrioventricular bundle. The atrioventricular node is stimulated by the excitation wave as it passes through the atrial myocardium.

The speed of conduction of the cardiac impulse through the atrioventricular node (about 0.11 seconds) allows sufficient time for the atria to empty their blood into the ventricles before the ventricles start to contract.

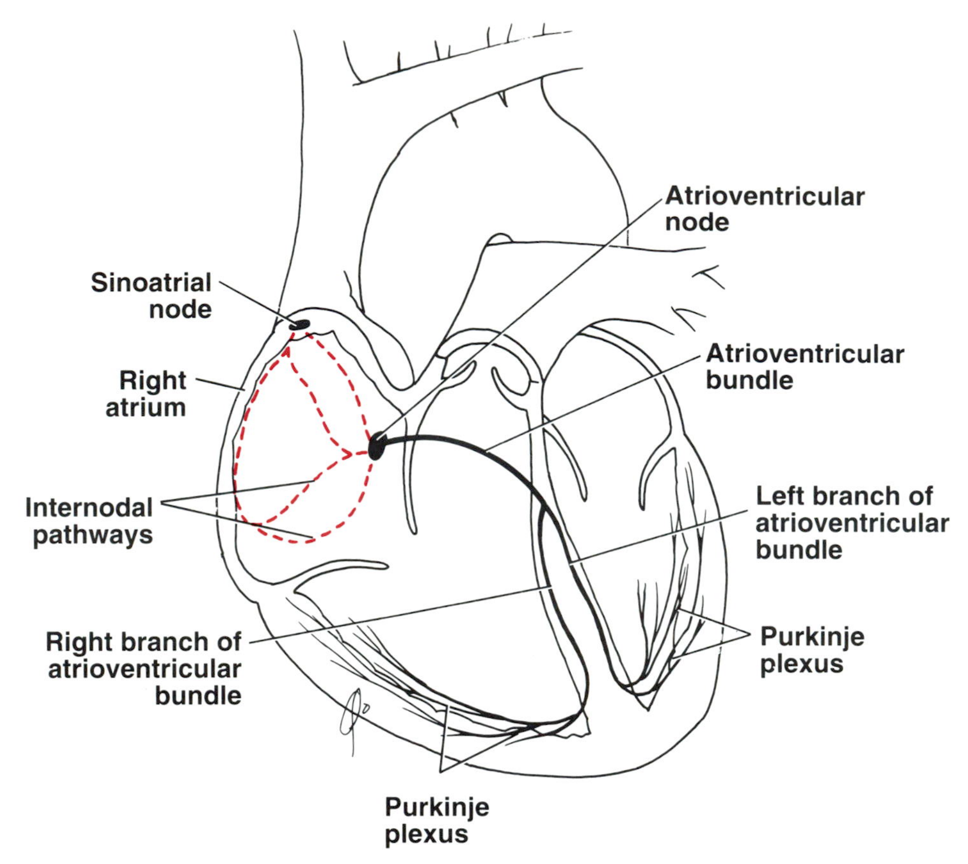

FIG 3–8.
The conducting system of the heart. Note the internodal pathways.

Atrioventricular Bundle.—The atrioventricular bundle (bundle of His) is the only pathway of cardiac muscle that connects the myocardium of the atria and the myocardium of the ventricles; it is thus the only route along which the cardiac impulse can travel from the atria to the ventricles (see Fig 3–8). The bundle descends through the fibrous skeleton of the heart.

The atrioventricular bundle then descends behind the septal cusp of the tricuspid valve to reach the inferior border of the membranous part of the ventricular septum. At the upper border of the muscular part of the septum it divides into two branches, one for each ventricle. The right bundle branch passes down on the right side of the ventricular septum to reach the moderator band, where it crosses to the anterior wall of the right ventricle. Here it becomes continuous with the fibers of the Purkinje plexus (see Fig 3–8).

The left bundle branch pierces the septum and passes down on its left side beneath the endocardium. It usually divides into two branches (anterior and posterior), which eventually become continuous with the fibers of the Purkinje plexus of the left ventricle.

The specialized cardiac muscle fibers, which are fast conducting and make up the atrioventricular bundle, are commonly called the *Purkinje fibers*.

Clinical Note

Accessory atrioventricular bundles are thought to exist. They are believed to be slender and normally have no functional significance. However, in the condition of *accelerated atrioventricular conduction,* the aberrant connection (bundle of Kent) permits one ventricle to be excited early. In this condition the PR interval is shortened and a delta wave appears on the initial part of the QRS complex. Another aberrant connection (bundle of Mahaim) bypasses the atrioventricular node and inserts just distal to the node. In this condition, known as the Lown-Ganong-Levine syndrome, the PR interval is short but the QRS complex is normal.

Purkinje Plexus.—The right and left terminal bundle branches become continuous with small branches that spread around each ventricular chamber and back toward the base of the heart. These fine terminal branches are referred to as the *Purkinje plexus* (see Fig 3–8).

The cardiac impulse travels from the beginning of the atrioventricular bundle to the terminal branches of the Purkinje plexus in about 0.03 seconds. This rapid spread of the cardiac impulse ensures that the entire myocardium of the ventricles is stimulated to contract at almost exactly the same time.

The conducting system of the heart is responsible not only for generating rhythmical cardiac impulses but also for conducting these impulses throughout the myocardium of the heart.

Structure of the Conducting System of the Heart

The sinoatrial and atrioventricular nodes consist of small cardiac muscle fibers embedded in vascular connective tissue. The atrioventricular bundle, its right and left branches, and the subendocardial plexus are composed of cardiac muscle fibers that are usually larger than ordinary cardiac muscle fibers. These specialized fibers are the *Purkinje fibers.* The fibers have relatively few myofibrils; those that are present tend to be concentrated in the periphery of the cytoplasm. The center of the sarcoplasm is pale staining and contains large amounts of glycogen.

The specialized cardiac muscle fibers of the sinoatrial and atrioventricular nodes have a higher rate of intrinsic rhythmical contraction and a slower speed of conduction than do ordinary cardiac muscle fibers. The specialized fibers of the atrioventricular bundle and its terminal branches, however, have a higher speed of conduction.

Internodal Conduction Paths.—Impulses from the sinoatrial node have been shown to travel to the atrioventricular node more rapidly than they could have traveled by passing along the ordinary myocardium. This phenomenon has been explained by the description of special pathways in the atrial wall (see Fig 3–8), having a structure consisting of a mixture of Purkinje fibers and ordinary cardiac muscle cells. The *anterior internodal pathway* leaves the anterior end of the sinoatrial node and passes anterior to the superior vena caval opening. It descends on the atrial septum and ends in the atrioventricular node. The *middle internodal pathway* leaves the posterior end of the sinoatrial node and passes posterior to the superior vena caval opening. It descends on the atrial septum to the atrioventricular node. The *posterior internodal pathway* leaves the posterior part of the sinoatrial node and descends through the crista terminalis and the valve of the inferior vena cava to the atrioventricular node.

Clinical Note

Circus movement.—This abnormal form of conduction allows a wave of excitation to travel continuously in a circle. This ring may occur in the atrioventricular node causing abnormal atrial contractions and paroxysmal nodal tachycardia. If the individual has an accessory atrioventricular bundle the circus movement may pass in one direction through the atrioventricular node and in the opposite direction through the bundle of Kent.

Arterial Supply of the Heart

Coronary Arteries

The right and left coronary arteries arise from the ascending aorta immediately above the aortic valve (Fig 3–9). They and their major branches are distributed over the surface of the heart lying within subepicardial connective tissue. In the majority of individuals (60%) the left artery is larger than the right, in 23% the two arteries are of equal size, and in 17% the right artery is larger than the left.

Right Coronary Artery.—This artery arises from the anterior aortic sinus (Fig 3–10) and runs forward between the pulmonary trunk and the right auricle (see Fig 3–9). It descends almost vertically in the right atrioventricular groove, and at the inferior border of the heart it continues posteriorly along the atrioventricular groove to anastomose with the left coronary artery in the posterior interventricular groove. The following branches from the right coronary artery supply the right atrium and right ventricle and parts of the left atrium and left ventricle and the atrioventricular septum.

Branches:

1. *Right conus artery.* This supplies the anterior surface of the pulmonary conus (infundibulum of right ventricle) and the upper part of the anterior wall of the right ventricle.
2. *Anterior ventricular branches.* Two or three in number, they supply the anterior surface of the right ventricle. The *marginal branch* is the largest and runs

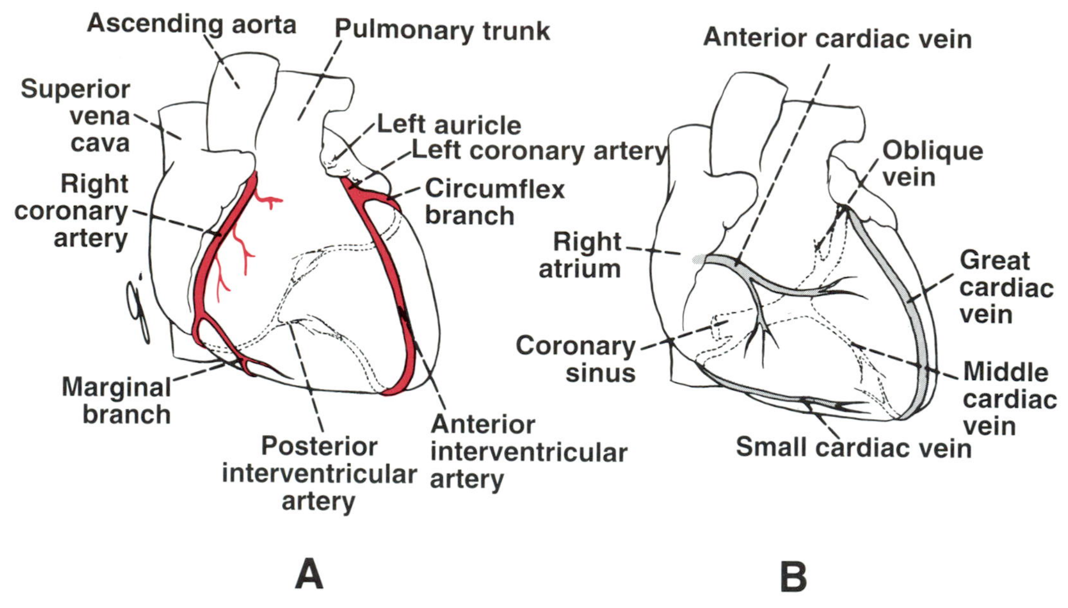

FIG 3–9.
A, the coronary arteries. **B,** the cardiac veins.

along the lower margin of the costal surface to reach the apex.

3. *Posterior ventricular branches.* Usually two in number, they supply the diaphragmatic surface of the right ventricle.

4. *Posterior interventricular (descending) artery.* This artery runs toward the apex in the posterior interventricular groove. It gives off branches to the right ventricle and the left ventricle, including its inferior wall. It supplies branches to the posterior part of the ventricular septum but not the apical part, which receives its supply from the anterior interventricular branch of the left coronary artery. A large septal branch supplies the *atrioventricular node.* In

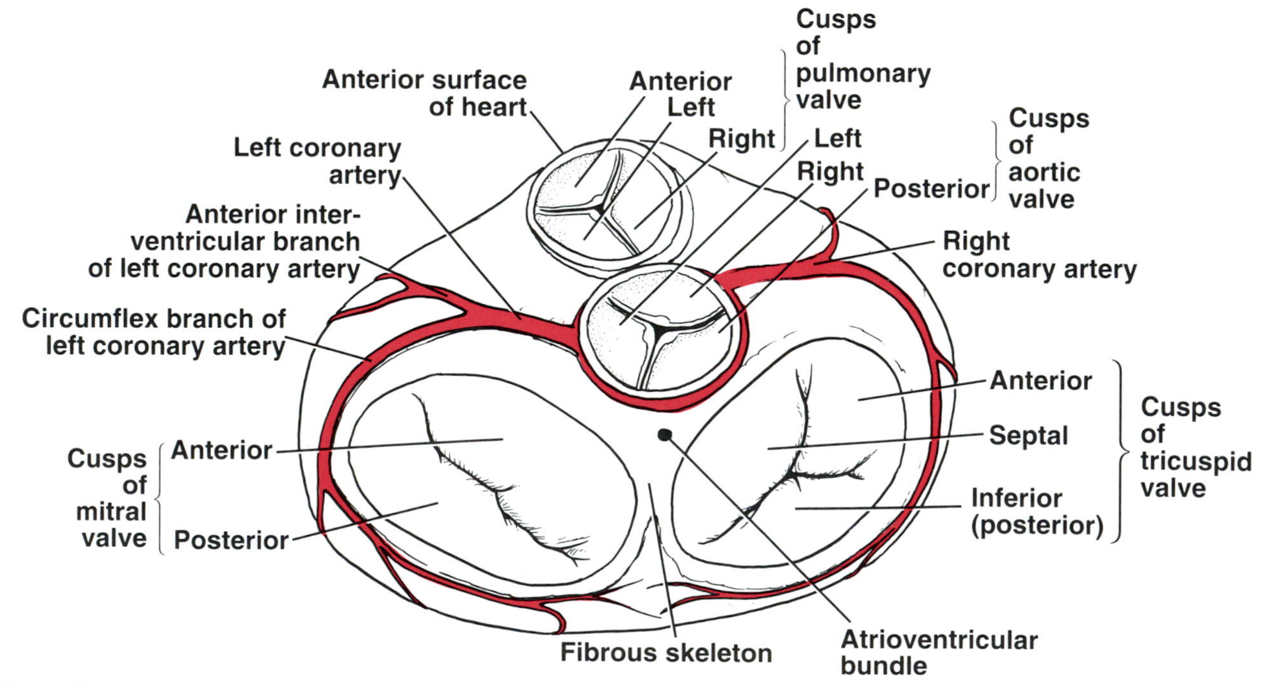

FIG 3–10.
The valves of the heart and the origin of the coronary arteries, superior view. The atria and the great vessels have been removed.

10% of individuals the posterior interventricular artery is replaced by a branch from the left coronary artery.

5. *Atrial branches.* Several branches supply the anterior and lateral surfaces of the right atrium. One branch supplies the posterior surface of both the right and left atria. The *artery of the sinoatrial node* supplies the node and the right and left atria; in 35% of individuals it arises from the left coronary artery.

Left Coronary Artery.—This is usually the larger coronary artery and supplies the major part of the heart, including the greater part of the left atrium, left ventricle, and ventricular septum. It arises from the left posterior aortic sinus (see Fig 3–7) and passes forward between the pulmonary trunk and the left auricle (see Fig 3–9). It then enters the atrioventricular groove and divides into an anterior interventricular branch and a circumflex branch.

Branches:

1. The *anterior interventricular (descending) branch* runs downward in the interventricular groove to the apex of the heart (see Fig 3–9). In the majority of individuals it then passes around the apex of the heart to enter the posterior interventricular groove and anastomoses with the terminal branches of the right coronary artery. In a third of individuals it ends at the apex of the heart. The anterior interventricular branch supplies the right and left ventricles with numerous branches that also supply the anterior part of the ventricular septum. In 30% to 50% of individuals, one of these ventricular branches *(left diagonal artery)* arises directly from the trunk of the left coronary artery. A small *left conus artery* supplies the pulmonary conus.

2. The *circumflex artery* is the same size as the anterior interventricular artery (Fig 3–10). It winds around the left margin of the heart in the atrioventricular groove (see Fig 3–9). A *left marginal artery* is a large branch that supplies the left margin of the left ventricle down to the apex. *Anterior ventricular* and *posterior ventricular branches* supply the left ventricle. *Atrial branches* supply the left atrium.

The *artery to the sinoatrial node* is a branch of the circumflex artery in 35% of individuals. The *artery of the atrioventricular node* is also a branch of the circumflex artery in 10% of patients.

Variations in the Coronary Arteries.—Variations in the blood supply to the heart do occur, and the most common variations affect the blood supply to the diaphragmatic surface of both ventricles. Here the origin, size, and distribution of the posterior interventricular artery are variable (Fig 3–11). In *right dominance* the posterior interventricular artery is a large branch of the right coronary artery. Right dominance is present in the majority of individuals (90%). In *left dominance* the posterior interventricular artery is a branch of the circumflex branch of the left coronary artery (10%).

Coronary Artery Anastomoses.—Anastomoses between the terminal branches of the right and left coronary arteries (collateral circulation) exist, but they are usually *not large enough* to provide an adequate blood supply to the cardiac muscle should one of the large branches become blocked by disease. A sudden block of one of the larger branches of either coronary artery usually leads to myocardial infarction, although sometimes the collateral circulation is enough to sustain the myocardium.

Summary of the Overall Arterial Supply to the Heart in the Majority of Individuals.—The *right coronary artery* supplies all of the right ventricle (except for the small area to the right of the anterior interventricular groove); the variable part of the diaphragmatic surface of the left ventricle; the posterior inferior third of the ventricular septum; the right atrium and part of the left atrium; and the sinoatrial node and the atrioventricular node and bundle. The left bundle branch (LBB) also receives small branches.

The *left coronary artery* supplies most of the left ventricle, a small area of the right ventricle to the right of the interventricular groove, the anterior two thirds of the ventricular septum, most of the left atrium, the sinoatrial node (25%), the right bundle branch (RBB), and the LBB.

Clinical Note

Correlation of myocardial infarction, arterial supply, and the electrocardiogram.—An attempt to correlate the site of a myocardial infarction with the artery occluded and the electrocardiographic findings is shown in Table 3–1.

Arterial Supply to the Conducting System.—The sinoatrial node in the majority of individuals (70%) is supplied by the right coronary artery (the left in 25% and the right and left in 5%). The atrioventricular node and the atrioventricular bundle are supplied by

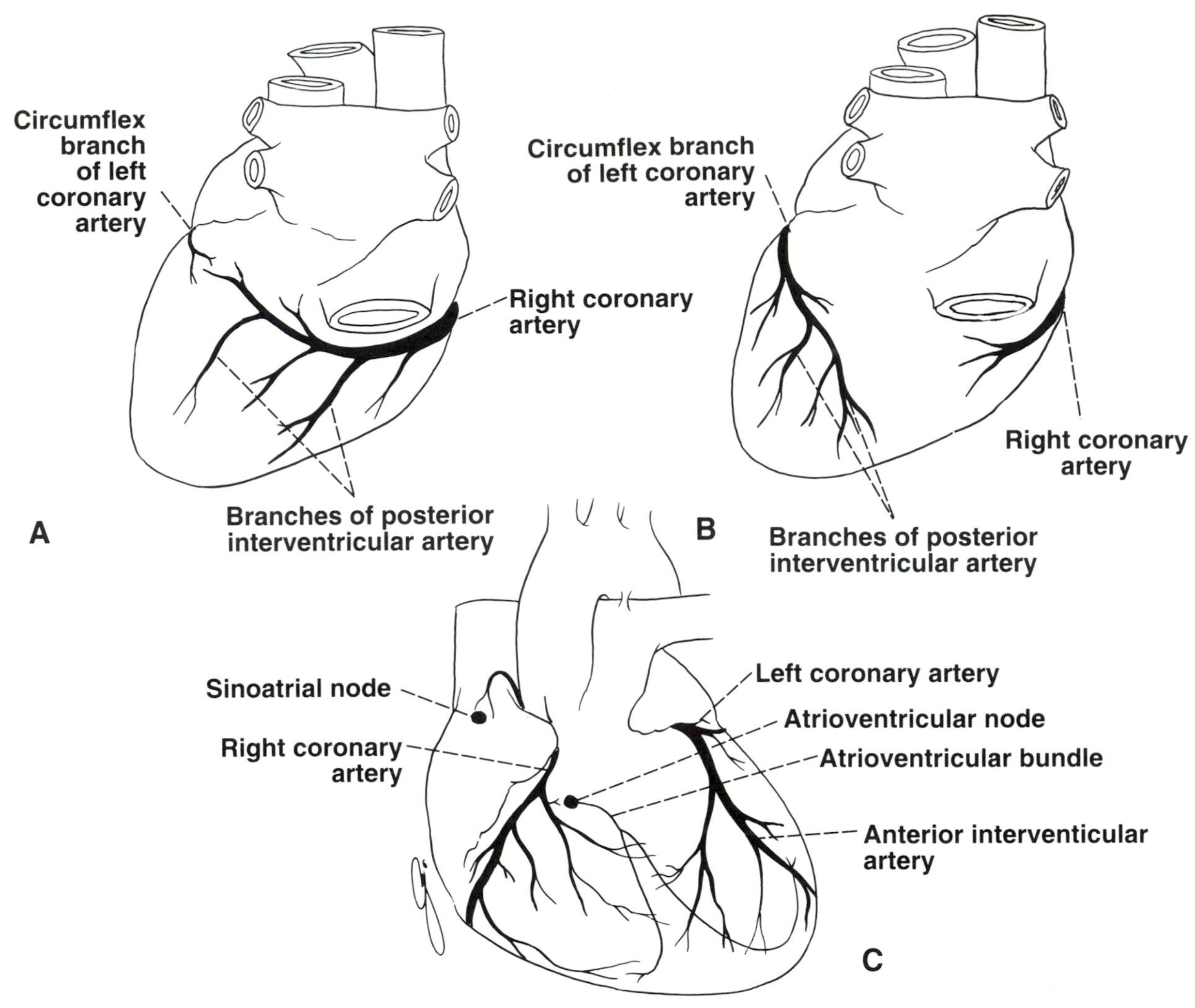

FIG 3–11.
A, posterior view of the heart showing the origin and distribution of the posterior interventricular artery in right dominance. **B,** posterior view of the heart showing the origin and distribution of the posterior interventricular artery in left dominance. **C,** anterior view of the heart showing the relationship of the blood supply to the conducting system.

Clinical Note

Myocardial infarction and papillary muscle rupture.— Rarely, in acute myocardial infarction involving the left ventricle, a papillary muscle may rupture. Rupture of the posteromedial papillary muscle is more common since it is supplied by a single artery, the right coronary artery. The anterolateral papillary muscle is less likely to rupture since it has a dual blood supply from the anterior interventricular and circumflex branches of the left coronary artery.

the right coronary artery. The RBB is supplied by the left coronary artery, and the LBB is supplied by the left coronary artery and small branches from the right coronary artery.

Venous Drainage of the Heart

Most of the venous blood from the heart wall drains into the right atrium through the coronary sinus (see Fig 3–9). The remainder passes directly into the right atrium through the *anterior cardiac vein* and small veins, the *venae cordis minimae.*

Coronary Sinus

This large vein lies in the posterior part of the atrioventricular groove and is a continuation of the great cardiac vein (see Fig 3–9).

Great Cardiac Vein

This ascends from the apex of the heart in the anterior interventricular groove (see Fig 3–9) and then

TABLE 3–1.
Coronary Artery Lesions, Infarct Location, and ECG Signature*

Coronary Artery	Infarct Location	ECG Signature
Proximal LAD	Large anterior wall	ST elevation: I, L, V1–V6
More distal LAD	Anterior-apical	ST elevation: V2–V4
	Inferior wall if "wraparound" LAD	ST elevation: II, III, F
Distal LAD	Anteroseptal	ST elevation: V1–V3
Early obtuse, marginal	High lateral wall	ST elevation: I, L, V4–V6
More distal marginal branch, circumflex	Small lateral wall	ST elevation: I, L, or V4–V6, or no abnormality
Circumflex	Posterolateral	ST elevation: V4–V6; ST depression: V1–V2
Distal RCA	Small inferior wall	ST elevation: II, III, F; ST depression: I, L
Proximal RCA	Large inferior wall and posterior wall	ST elevation: II, III, F
		ST depression: I, L, V1–V3
	Some lateral wall	ST elevation: V5–V6
RCA	Right ventricular	ST elevation: V2R–V4R; some ST elevation: V1, or ST depression: V2, V3
	Usually inferior	ST elevation: II, III, F

*ECG = electrocardiographic; LAD = left anterior descending; RCA = right coronary artery.

enters the atrioventricular groove, curving to the left side and back of the heart. The great cardiac vein becomes continuous with the coronary sinus.

Middle Cardiac Vein

This vein runs from the apex of the heart in the posterior interventricular groove (see Fig 3–9). It drains into the coronary sinus.

Small Cardiac Vein

This vein accompanies the marginal artery along the inferior border of the heart (see Fig 3–9). It drains into the coronary sinus.

Anterior Cardiac Vein

This vein drains the anterior surface of the right atrium and the right ventricle and empties directly into the right atrium (see Fig 3–9).

Coronary Circulation

The coronary blood flow in the normal, resting individual is about 225 mL/min and is continuous throughout the cardiac cycle, although about 75% occurs in diastole because of compression of the small branches of the coronary arteries by the cardiac muscle that takes place during systole. Stimulation of the sympathetic nervous system causes slight vasodilatation of the coronary arteries, whereas parasympathetic stimulation causes slight vasoconstriction. Increased coronary flow mainly results from the increased work of the heart muscle and the local affects of the products of metabolism causing vasodilation.

Nerve Supply of the Heart

Sympathetic Innervation

Sympathetic postganglionic fibers arise from the cervical and upper thoracic portions of the sympathetic trunks (Fig 3–12). Postganglionic fibers reach the heart by way of the *superior, middle,* and *inferior cardiac branches* of the cervical portion of the sympathetic trunk and a number of *cardiac branches* from the thoracic portion of the sympathetic trunk. Nerve fibers pass via the *cardiac plexuses* to terminate on the *sinoatrial* and *atrioventricular nodes,* the cardiac muscle fibers, and on the coronary arteries. Activation of these nerves results in cardiac acceleration, increased force of contraction of the cardiac muscle, and dilation of the coronary arteries. As mentioned previously, coronary dilation is mainly produced in response to local metabolic needs rather than by direct nerve stimulation.

Parasympathetic Innervation

Parasympathetic preganglionic fibers originate in the *dorsal nucleus of the vagus nerve* and descend into the thorax in the vagus nerves (see Fig 3–12). The fibers terminate by synapsing with postganglionic neurons in the *cardiac plexuses.* Postganglionic fibers terminate on the *sinoatrial* and *atrioventricular nodes* and on

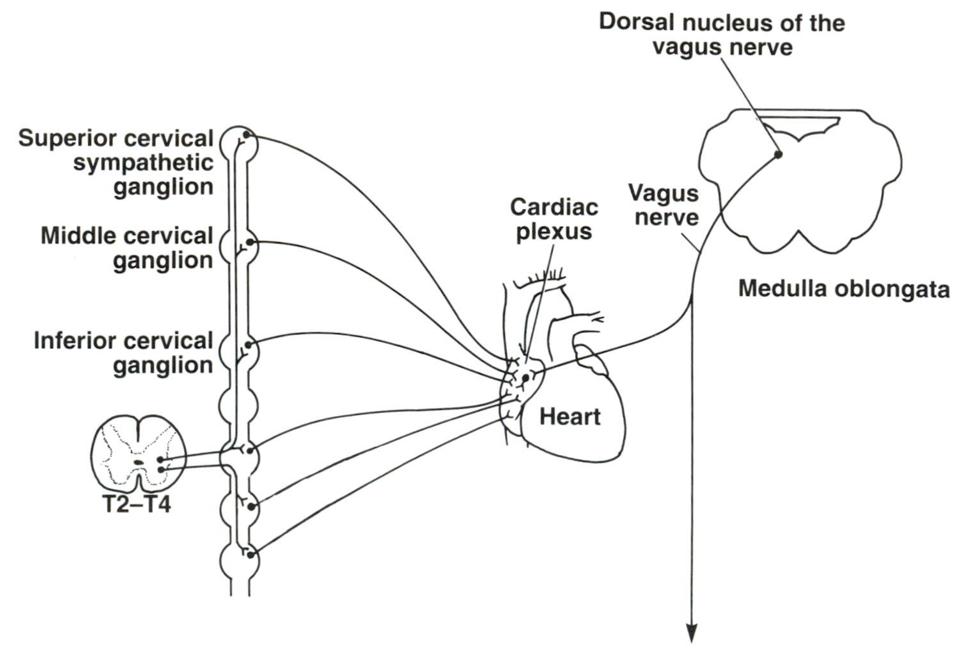

FIG 3–12.
Autonomic innervation of the heart.

the coronary arteries. Activation of these nerves results in a reduction in the rate and force of contraction of the myocardium and constriction of the coronary arteries. Here again the coronary constriction is mainly produced by the reduction in local metabolic needs rather than by neural effects.

Sympathetic Afferent Fibers
Afferent fibers running with the sympathetic nerves carry impulses that normally do not reach consciousness. Should the blood supply to the myocardium become impaired, however, pain impulses reach consciousness via this pathway.

Parasympathetic Afferent Fibers
Afferent fibers running with the vagus nerves take part in cardiovascular reflexes.

Bainbridge Atrial Reflex

Stretch receptors present in the walls of the atria are stimulated by an increase in atrial pressure. The afferent stimuli ascend to the medulla oblongata in the vagus nerves, and the heart rate is increased in response to the diminished activity of the vagi and the increased activity of the sympathetics.

Bezold-Jarisch Reflex

Receptors present in the left ventricular walls are stimulated by certain chemicals, such as nicotine. Afferent impulses ascend to the medulla oblongata in the vagus nerves, and the heart rate is slowed in response to increased vagal activity on the heart. It has been postulated that chemicals released by the degenerating tissue in myocardial infarction may initiate this reflex and contribute to the hypotension in this condition.

Blood Circulation Through the Heart

The heart is a muscular pump. The series of changes that take place within the heart as it fills with blood and empties is called the *cardiac cycle.* The normal heart contracts about 70 to 90 times per minute in the resting adult and about 130 to 150 times per minute in the newborn child.

Blood is continuously returning to the heart, and during ventricular systole, when the atrioventricular valves are closed, the blood is temporarily accommodated in the large veins and atria. Once ventricular diastole occurs, the atrioventricular valves open, and blood passively flows from the atria to the ventricles. About 70% of ventricular filling occurs passively. When the ventricles are nearly full, atrial systole occurs and forces the remainder of the blood in the atria into the ventricles. The sinoatrial node initiates the wave of contraction in the atria, which commences around the openings of the large veins and *milks* the blood toward the ventricles. By this means there is no reflux of blood into the veins.

The cardiac impulse, having reached the atrioventricular node, is conducted to the papillary muscles by the atrioventricular bundle and its branches (see Fig 3–8). The papillary muscles then begin to contract and take up the slack of the chordae tendineae. Meanwhile, the ventricles start contracting and the atrioventricular valves close. The spread of the cardiac impulse along the atrioventricular bundle (see Fig 3–8) and its terminal branches, including the Purkinje fibers, ensures that myocardial contraction occurs at almost the same time throughout the ventricles.

Once the intraventricular blood pressure exceeds that present in the large arteries (aorta and pulmonary trunk), the semilunar valve cusps are pushed aside, and the blood is ejected from the heart. At the conclusion of ventricular systole, blood begins to move back toward the ventricles and immediately fills the pockets of the semilunar valves. The cusps float into apposition and completely close the aortic and pulmonary orifices.

Basic Anatomy of the Pericardium

The pericardium is a sac that encloses the heart and the roots of the great blood vessels. Its function is to restrict excessive movements of the heart as a whole and to serve as a lubricated container in which the different parts of the heart can contract. The pericardium lies within the middle mediastinum (Fig 3–13). It is located posterior to the body of the sternum and the second to the sixth costal cartilages and anterior to the fifth to the eighth thoracic vertebrae.

Fibrous Pericardium

This is the fibrous part of the sac. It is attached anteriorly to the sternum by the *sternopericardial ligaments* and above to the walls of the great blood vessels, namely the aorta, pulmonary trunk, superior vena cava, and pulmonary veins (see Fig 3–13). Below, it is firmly attached to the central tendon of the diaphragm. This fibrous pericardium limits unnecessary movement of the heart.

Serous Pericardium

This lines the fibrous pericardium and coats the heart (see Fig 3–13). It is divided into parietal and visceral layers. The parietal layer lines the fibrous pericardium and is reflected around the roots of the great vessels to become continuous with the visceral layer that closely covers the heart *(epicardium).* The slitlike space between the parietal and visceral layers is called the *pericardial cavity.* The cavity contains a small amount of tissue fluid (about 50 mL), which acts as a lubricant to facilitate movements of the heart.

Transverse Sinus

This is a passage within the pericardial cavity that is located on the posterior surface of the heart. It lies between the reflection of serous pericardium around the aorta and pulmonary trunk and the reflection of serous pericardium around the great veins (see Fig 3–13).

Oblique Sinus

This is a large recess within the pericardial cavity on the posterior surface of the heart. It is formed by the reflection of the serous pericardium around the venae cavae and the four pulmonary veins (see Fig 3–13). The sinuses are formed during the bending of the primitive heart tube during development, and they have no function.

Nerve Supply of the Pericardium

The fibrous pericardium and the parietal layer of the serous pericardium are supplied by the phrenic nerves as they descend in the mediastinum between the pericardium and the mediastinal layer of the parietal pleura (see Fig 3–17). The visceral layer of the serous pericardium is innervated by branches of the sympathetic trunks and the vagus nerves.

Principal Relations of the Pericardium and Heart

Anterior.—Body of the sternum, third to the sixth costal cartilages and the intercostal spaces between them, internal thoracic vessels, anterior borders of the right and left lungs, and the pleural cavities (see Fig 3–13). In young children the thymus lies anterior to the upper part of the pericardium.

Posterior.—Fifth to eighth thoracic vertebrae, esophagus, descending thoracic aorta, main bronchi, and the rounded posterior part of each lung.

Lateral.—Mediastinal parietal pleura, phrenic nerve, and lung and pleural cavities.

Inferior.—Diaphragm, liver, and fundus of the stomach.

SURFACE ANATOMY

Surface Markings of the Heart

Borders

Superior Border.—Line from the second left costal cartilage to the third right costal cartilage (Fig 3–14).

Inferior Border.—Line from the sixth right costal cartilage to the apex.

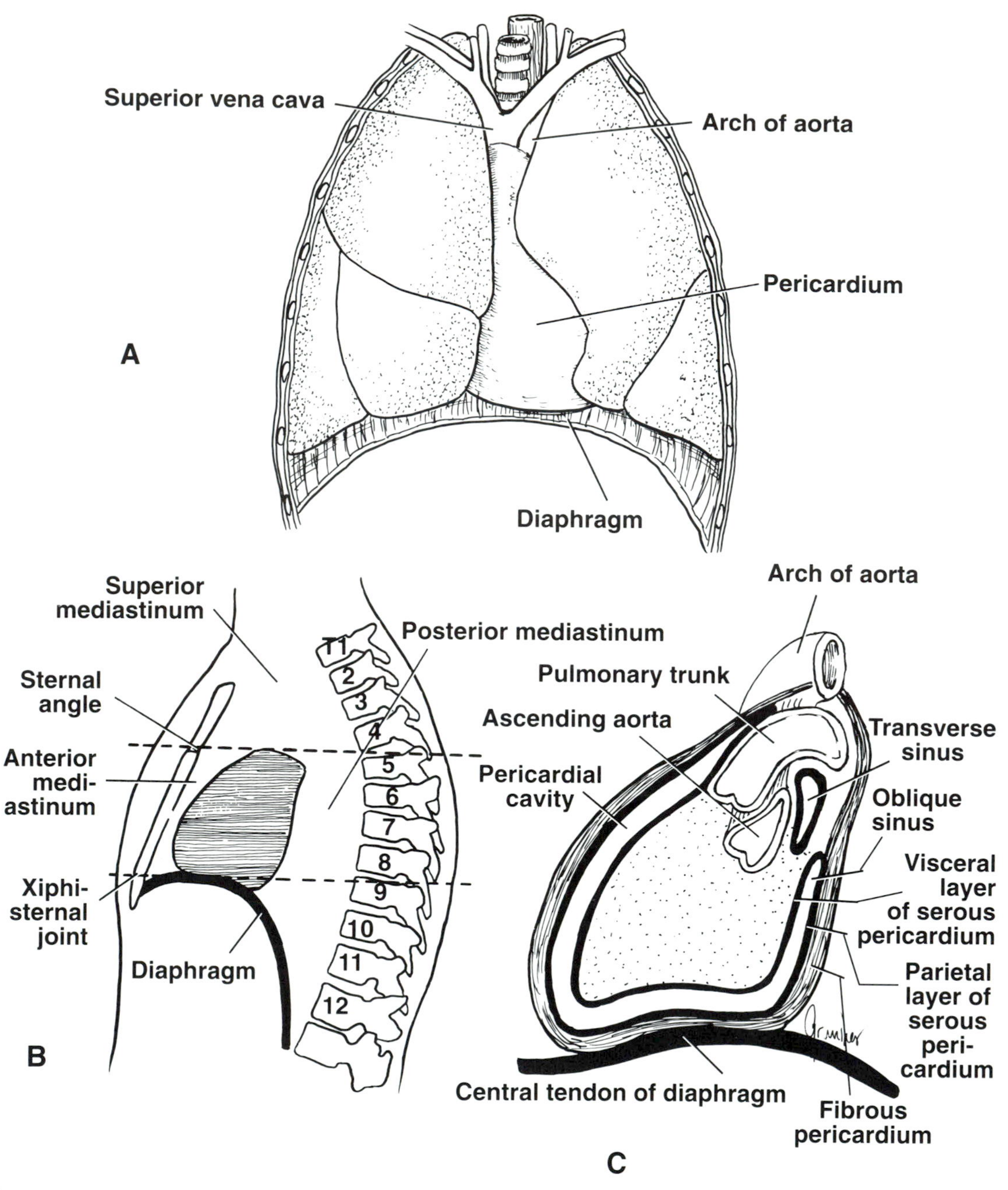

FIG 3–13.
A, anterior view of the thoracic contents. The sternum, costal cartilages, and ribs have been removed. **B,** sagittal section of the mediastinum showing the subdivisions. **C,** sagittal section of the heart and pericardium showing the pericardial cavity with its transverse and oblique sinuses.

Right Border.—Line drawn from a point on the third right costal cartilage 0.5 in. (1.3 cm) from the edge of the sternum downward to a point on the sixth right costal cartilage 0.5 in. (1.3 cm) from the edge of the sternum.

Left Border.—Line drawn from the second left costal cartilage 0.5 in. (1.3 cm) from the edge of the sternum to the apex of the heart.

Apex

This lies at the level of the fifth left intercostal space, 3.5 in. (9 cm) from the midline. In the region of the apex, the apex beat can usually be seen and palpated.

Heart Valves

The heart valves may be auscultated on the chest wall in the following areas (see Fig 3–14).

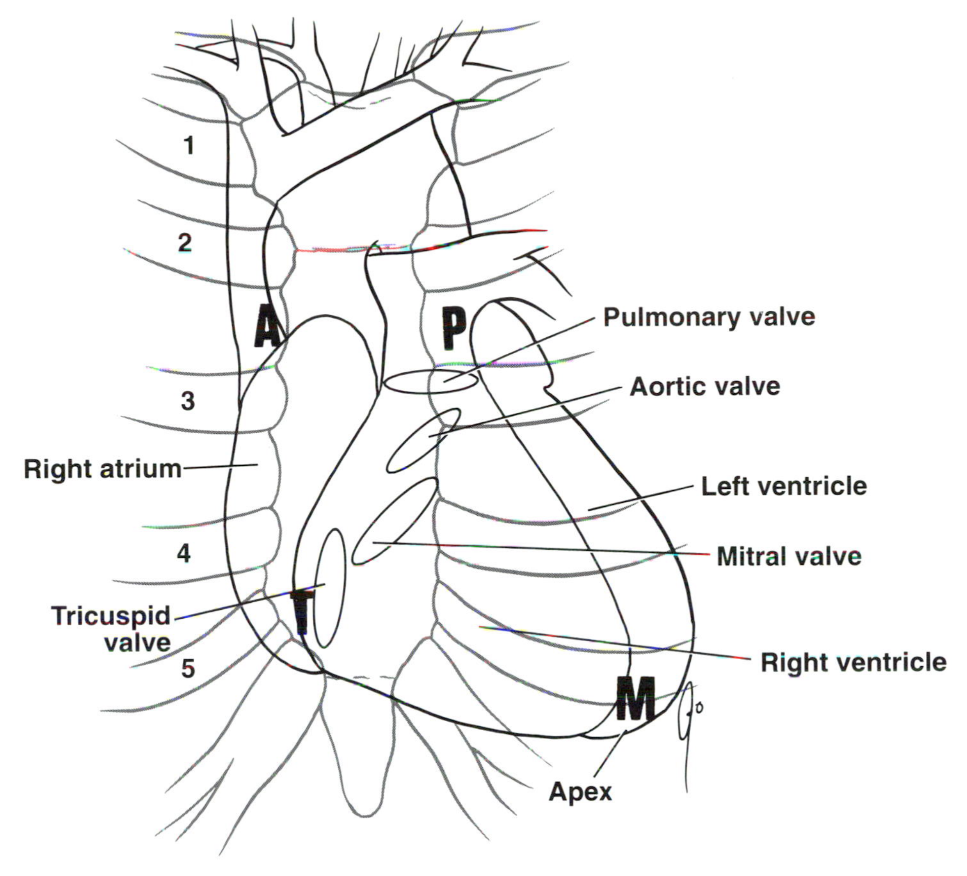

FIG 3–14.
Surface anatomy of the heart and great blood vessels. Note the position of the heart valves relative to the chest wall. The letters indicate positions where valves may be heard with least interference. *A* = aortic valve; *P* = pulmonary valve; *T* = tricuspid valve; *M* = mitral valve.

Tricuspid Valve.—This may be heard over the right half of the lower end of the sternum.

Mitral Valve.—This is most clearly heard over the apex beat of the heart.

Pulmonary Valve.—This is clearly heard over the medial end of the second left intercostal space.

Aortic Valve.—This is heard over the medial end of the second right intercostal space.

Clinical Note

Substitute heart valves.—The surface marking of the actual position of the valves (see Fig 3–14) may be important in the recognition of substitute heart valves on a radiograph.

Tricuspid Valve.—This valve lies behind the right half of the sternum opposite the fourth intercostal space.

Mitral Valve.—This valve lies behind the left half of the sternum opposite the fourth costal cartilage.

Pulmonary Valve.—This valve lies behind the medial end of the third left costal cartilage and the adjoining part of the sternum.

Aortic Valve.—This valve lies behind the left half of the sternum opposite the third intercostal space.

CLINICAL ANATOMY

Clinical Anatomy of the Heart

Important Congenital Anomalies of the Heart

Atrial Septal Defects.—In 25% of hearts, a small opening persists in the atrial septum, but it is usually so minor that it has no clinical significance. Occasionally the opening is much larger and results in oxygenated blood from the left atrium passing over into the right atrium, which consequently overworks the right side of the heart. Wide patency of the sep-

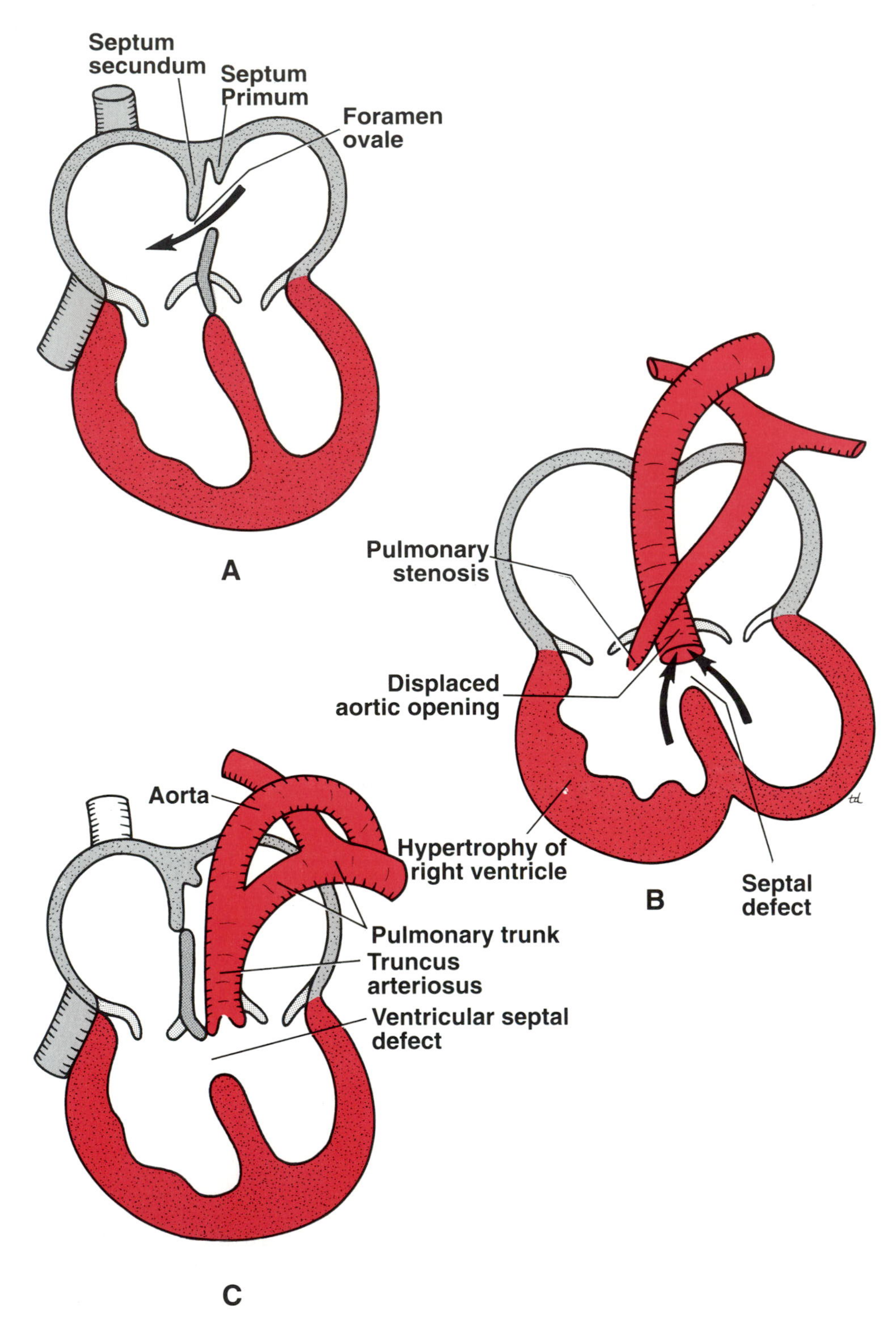

FIG 3–15.
Three important congenital anomalies of the heart. **A,** atrial septal defects. **B,** tetralogy of Fallot. **C,** persistent truncus arteriosus.

tum commonly is associated with other important congenital defects, such as pulmonary stenosis or transposition of the great arterial trunks. In nearly half the cases of wide patency alone, death results from right-sided congestive heart failure.

Atrial septal defects result from a failure of the closure of the foramen ovale (Fig 3–15). Normally after birth, the raised pressure in the left atrium forces the septum primum over to fuse with the septum secundum in such a way that the foramen ovale completely closes. Very rarely, the foramen primum may persist or the entire atrial septum may be absent. Both of these conditions are more serious than patency of the foramen ovale. When there is only one atrium and two ventricles, the mortality is high.

Ventricular Septal Defect.—An isolated ventricular septal defect comprises about 15% to 20% of all congenital heart lesions. They are found almost invariably in the membranous part of the septum. They are oval or round and measure from 1 to 2 cm in diameter. The blood passes through the defect from left to right, causing enlargement of the right ventricle. Cyanosis is rare, and there may be no symptoms. Small septal defects are relatively benign, but large defects can shorten life if surgery is not performed.

Normally, the ventricular septum is formed from two sources. The inferior muscular part grows up from the floor of the primitive ventricle, and the membranous part is formed as the result of fusion of the lower ends of the bulbar ridges and the septum intermedium (endocardial cushions). The complete ventricular septum is formed when the membranous part fuses with the muscular part.

Paradoxical Embolism.—Rarely, in cases of atrial or ventricular septal defects, or even a patent atrial septum, emboli have been known to pass from the left side of the heart to the right side, causing pulmonary embolism.

Tetralogy of Fallot.—This congenital anomaly is responsible for about 9% of all congenital heart disease (see Fig 3–15). The anatomic abnormalities include (1) large ventricular septal defect, (2) stenosis of the pulmonary trunk, which may occur at the infundibulum of the right ventricle or at the pulmonary valve, (3) exit of the aorta immediately above the ventricular septal defect (instead of from the left ventricular cavity only), and, because of the high blood pressure in the right ventricle, (4) severe hypertrophy of the right ventricle. The defects cause congenital cyanosis and considerably limit activity; in severe cases death occurs if untreated. Once the diagnosis has been made, most children can be successfully treated surgically.

Most children find that assuming the squatting position after physical activity relieves their dyspnea. This happens because squatting reduces the venous return by compressing the abdominal veins and increasing the systemic arterial resistance by kinking the femoral and popliteal arteries; both these mechanisms tend to decrease the right-to-left shunt and improve pulmonary circulation.

Mitral Valve Prolapse.—In this condition one or both mitral valve cusps balloon up into the left atrium during ventricular systole. The valve cusps are larger than normal, and the chordae tendineae may be excessively long. The posterior cusp is always involved with the anterior cusp being involved less frequently. The majority of the patients are female (75%), and there may be a familial incidence of the syndrome. The typical symptoms are chest pain and palpitations, and dysrhythmias may occur. Endocarditis may occur following instrumentation. The diagnosis is made by echocardiography.

Bicuspid Aortic Valve.—This is a congenital anomaly occurring in 1% to 2% of the population. At first it causes no functional problems. However, with advancing years and continued wear and tear, the valve cusps become damaged and undergo fibrosing stenosis and calcification. The valve is also prone to the development of infective endocarditis.

Persistent Truncus Arteriosus.—This condition represents about 1% of all congenital heart defects. Only one artery arises from the heart, the pulmonary artery and aorta sharing a common trunk (see Fig 3–15). A large ventricular septal defect is usually present. The child exhibits mild cyanosis and heart failure. Surgical correction of the ventricular septal defect and the establishment of a separate pulmonary and aortic outflow are necessary.

Patent Ductus Arteriosus.—The ductus arteriosus represents the distal portion of the sixth left aortic arch and connects the left pulmonary artery to the descending aorta. During fetal life blood passes through it from the pulmonary artery to the aorta, thus bypassing the lungs. After birth, it normally constricts, later closes, and becomes the *ligamentum arteriosum*.

Failure of the ductus arteriosus to close may occur as an isolated congenital anomaly or may be present in association with congenital heart disease. Persistent patent ductus arteriosus results in high-pressure aortic blood passing into the pulmonary artery, which raises the pressure in the pulmonary circulation (Fig 3–16). A small patent ductus gives rise to no immediate symptoms, and the only sign is the characteristic continuous machinery-like murmur heard at the upper left sternal border. A large patent ductus, on the other hand, gives signs of increased vascularity of the lungs, left ventricular hypertrophy, as well as the murmur. Patent ductus arteriosus is life threatening and should be divided surgically.

Infective Endocarditis.—Endothelial injury caused by blood turbulence secondary to regurgitant and jet streams may be responsible for infective endocarditis. This may be followed by deposits of sterile platelet-fibrin clots on the injured endothelial surface and later the deposit of clumps of blood borne agglutinated organisms. Blood turbulence caused by congenital heart disease or rheumatic mitral stenosis may lead to endocarditis. Endothelial injury caused by the introduction of foreign bodies such as intracardiac catheters, substitute heart valves, or pacemakers may also be factors in initiating endocarditis.

Clinical Anatomy of Chest Pain

The presenting symptom of chest pain is a common and classic problem in emergency medicine. Unfortunately, chest pain is a symptom common to a large number of conditions and may be caused by disease in the thoracic and abdominal walls or in many different thoracic and abdominal viscera. The severity of the pain is often unrelated to the seriousness of the cause. Myocardial pain may mimic esophagitis, musculoskeletal chest wall pain, and other non–life-threatening causes. Unless the emergency physician is astute, a patient may be discharged from the emergency department with a more serious condition than the symptoms indicate. It is not good enough to have a correct diagnosis only 99% of the time with chest pain. An understanding of the anatomy of chest pain will help the physician in the systematic consideration of the differential diagnosis.

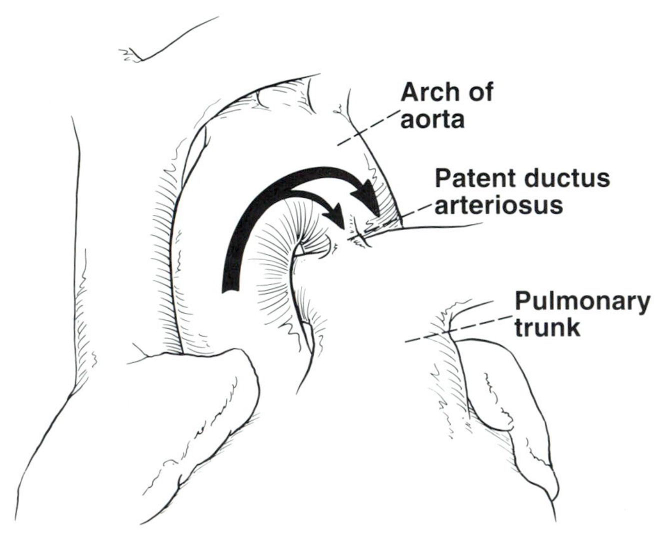

FIG 3–16.
Patent ductus arteriosus.

Somatic Pain

Pain arising from the chest or abdominal walls is intense and discretely localized. Somatic pain arises in sensory nerve endings in these structures and is conducted to the central nervous system by somatic segmental spinal nerves.

Visceral Pain

Visceral pain is diffuse and poorly localized. It is conducted to the central nervous system along afferent autonomic nerves. Most visceral pain fibers ascend to the spinal cord along sympathetic nerves and enter the cord through the posterior nerve roots of segmental spinal nerves. Some pain fibers from the pharynx and upper part of the esophagus and the trachea enter the central nervous system through the parasympathetic nerves via the glossopharyngeal and vagus nerves. The descending colon, the sigmoid colon and rectum, and the bladder reach the sacral spinal cord through the parasympathetic nerves.

Referred Pain

Visceral pain frequently is referred to skin areas that are innervated by the same segment of the spinal cord as is the painful viscus (see Fig 11–45). The explanation for referred pain not known. One theory is that the nerve fibers from the viscus and the dermatome ascend in the central nervous system along a common pathway and the cerebral cortex is incapable of distinguishing between the sites of origin. Another theory is that under normal conditions the viscus does not give rise to painful stimuli, whereas the skin area repeatedly receives noxious stimuli. Because both afferent fibers enter the spinal cord at the same segment, the brain interprets the information as coming from the skin rather than the viscus. Pain arising from the gastrointestinal tract is referred to the midline. This can probably be explained since the tract arises embryologically as a midline structure and receives a bilateral nerve supply.

Thoracic Dermatomes

The dermatomes on the anterior and posterior chest walls are shown in Figures 2–32 and 2–33. A *dermatome* is an area of skin supplied by a single spinal nerve and, therefore, a single segment of the spinal cord. On the trunk, adjacent dermatomes overlap considerably and a given area of skin is innervated by three adjacent spinal nerves.

Each thoracic spinal nerve innervates a large number of structures, including the vertebrae; the ribs and costal cartilages and their joints; the postvertebral, intercostal, and abdominal muscles; and the costal parietal pleura and parietal peritoneum. Any one of these structures could be the source of chest pain.

The skin of the anterior and posterior chest walls, down as far as the sternal angle in front and the spine of the scapula behind, is supplied by the supraclavicular nerves (C3 and C4). The phrenic nerves (C3, C4, and C5) supply the parietal pleura over the central part of the diaphragm and a corresponding area of parietal peritoneum over the lower surface of the diaphragm. Irritation of these areas by disease of neighboring viscera, such as the gallbladder, liver, or stomach could send afferent impulses up to the central nervous system. Because of the phenomenon of referred pain, the patient would presume that the cause of the pain was located over the upper part of the chest wall or shoulder.

Below the level of the sternal angle and the spine of the scapula, the chest skin is innervated by the thoracic segments of the spinal cord. The anterior chest wall is supplied by the intercostal nerves T2 through T6; the 7th intercostal nerve enters the anterior abdominal wall to supply the skin in the xiphoid area. The posterior chest wall is supplied by the posterior primary rami of the thoracic spinal nerves T2 through T11. Remember that the 7th to the 11th intercostal nerves cross the costal margin and innervate the full thickness of the anterior abdominal wall including the parietal peritoneum. This would explain how irritation of the parietal peritoneum caused by disease of abdominal viscera could give rise to pain referred to the chest wall.

Note also the distribution of the T1 and T2 dermatomes; they extend down the medial side of the upper limbs. The second thoracic nerve reaches the skin of the axilla and the medial (ulnar) side of the arm via the *intercostobrachial nerve* (a branch of the second intercostal nerves). T1 reaches the ulnar side of the forearm via the medial cutaneous nerve of the forearm from the brachial plexus.

Clinical Anatomy of Cardiac Pain

Pain originating in the heart as the result of acute myocardial ischemia is assumed to be caused by oxygen deficiency and the accumulation of metabolites, which stimulate the sensory nerve endings in the myocardium. The afferent nerve fibers ascend to the central nervous system through the cardiac branches of the sympathetic trunk and enter the spinal cord through the posterior roots of the upper four thoracic nerves. The nature of the pain varies considerably, from a severe crushing pain to nothing more than a mild discomfort.

The pain is not felt *in* the heart, but is referred to the skin areas supplied by the corresponding spinal nerves. The skin areas supplied by the upper four intercostal nerves and by the intercostobrachial nerve (T2) are therefore affected. The intercostobrachial nerve communicates with the medial cutaneous nerve of the arm and is distributed to skin on the medial side of the upper part of the arm. A certain amount of spread of nervous information must occur within the central nervous system, for the pain is sometimes felt in the neck and the jaw.

Myocardial infarction involving the inferior wall or diaphragmatic surface of the heart often gives rise to discomfort in the epigastrium. One must assume that the afferent pain fibers from the heart ascend in the sympathetic nerves and enter the spinal cord in the posterior roots of the seventh, eighth, and ninth thoracic spinal nerves and give rise to referred pain in the T7, T8 and T9 thoracic dermatomes in the epigastrium.

Since the heart and the thoracic part of the esophagus probably have similar afferent pain pathways, it is not surprising that painful acute esophagitis can mimic the pain of myocardial infarction.

Clinical Anatomy of the Coronary Arteries

The myocardium receives its blood supply through the right and left coronary arteries. Although the coronary arteries have numerous anastomoses at the arteriolar level, they are essentially functional end arteries. A sudden block of one of the large branches of either coronary artery will usually lead to necrosis of the cardiac muscle (myocardial infarction) in that vascular area, and often the patient dies. The great majority of cases of coronary artery blockage are caused by an acute thrombosis on top of a chronic atherosclerotic narrowing of the lumen.

Arteriosclerotic disease of the coronary arteries may present in three ways, depending on the rate of narrowing of the lumina of the arteries: (1) a general degeneration and fibrosis of the myocardium, which occurs over a period of many years and is caused by a gradual narrowing of the coronary arteries; (2) *angina pectoris,* cardiac pain that occurs on exertion and is relieved by rest; in this condition, the coronary arteries are so narrowed that myocardial ischemia occurs on exertion but not at rest; and (3) *myocardial infarction,* in which coronary flow is suddenly reduced or stopped and the cardiac muscle undergoes necrosis. Myocardial infarction is the major cause of death in industrialized nations.

Table 3–1 shows the different coronary arteries that supply the different areas of the myocardium. This information can be helpful when attempting to correlate the site of myocardial infarction, the artery involved, and the electrocardiographic signature.

Failure of the Conducting System of the Heart

The sinoatrial node is the spontaneous source of the cardiac impulse. The atrioventricular node is responsible for picking up the cardiac impulse from the atria. The atrioventricular bundle is the only route by which the cardiac impulse can spread from the atria to the ventricles. Failure of the bundle to conduct the normal impulses will result in alterations in the rhythmic contraction of the ventricles (arrhythmias) or, if there is complete bundle block, complete dissociation between the atrial and ventricular rates of contraction. The common cause of defective conduction through the bundle or its branches is atherosclerosis of the coronary arteries, which results in a diminished blood supply to the conducting system.

Atrioventricular Block

This condition can result from injury to the atrioventricular node or the atrioventricular bundle. The atria continue to beat at their normal rate while the ventricles initially fail to contract. Within a few seconds the ventricles start to beat at their own rate. They have escaped from the influence of the sinoatrial node and then contract in response to the atrioventricular node or an area of the conducting system below the node. Under these circumstances the ventricles contract independently of the atria at a rate that may be as low as 30 beats per minute.

Bundle Branch Block

Right or Left Bundle Branch Block.—In this condition the cardiac impulse passes down the normal branch of the atrioventricular bundle, causing the ventricle on that side to contract, and then passes to other ventricle on the blocked side. Because the left branch of the atrioventricular bundle terminates by dividing into anterior and posterior fascicles, it is possible to have *left anterior hemiblock* and *left posterior hemiblock.* Since the function of the conducting system in the ventricles is to facilitate the almost immediate distribution of the cardiac impulse throughout both ventricles, the various forms of block described above result in the loss of effective ventricular contraction, and the efficiency of the heart as a pump is lost.

The site and type of block can be analyzed with the help of an electrocardiogram. Abnormal left axis deviation in the electrocardiogram indicates left anterior hemiblock, and abnormal right axis deviation indicates left posterior hemiblock. Right bundle branch block produces a QRS greater than 0.12 seconds and a tall R wave in V1. Left bundle branch block produces a QRS greater than 0.12 seconds, a QS in V1, and an upright QRS in V6.

Sinus Bradycardia

This is common in myocardial infarction involving the diaphragmatic wall of the heart. This phenomenon can be explained because the right coronary artery supplies both the sinoatrial node (sinus branch) and the diaphragmatic surface of the heart (posterior interventricular branch).

First- and Second-Degree Heart Block

This may occur with acute inferior myocardial infarction. It can be explained because the inferior or diaphragmatic wall of the heart is supplied by the posterior interventricular branch of the right coronary artery, which also supplies the atrioventricular node and bundle.

Myocardial Infarction and Persistent Hypotension

The persistent hypotension that sometimes occurs in patients with myocardial infarction may be caused by left ventricular pump failure (with a large area of myocardium affected), right ventricular infarction, or volume depletion from vomiting, diaphoresis, and lack of oral intake. It has also been suggested that the Bezold-Jarisch reflex (see p. 118), which causes bradycardia, may be responsible; the reflex is initiated by chemicals liberated from the infarcted area stimulating receptors in the left ventricular wall.

Clinical Anatomy of Cardiac Injuries

Myocardial Contusion

The heart, although protected by the thoracic cage, can be squeezed between the sternum and the vertebral column (see Fig 3–13) when the thorax is subjected to a severe frontal impact. Moreover, if the force is also applied to the anterior abdominal wall, the diaphragm is thrust upwards, impinging on the heart from below. The highly flexible rib cage present in children makes myocardial contusion a common occurrence in this age group. The result of heart muscle damage is precordial pain, similar in nature to a myocardial infarction. Tachycardia often occurs and, if enough cardiac muscle is contused, cardiac output may decrease. Depending on the severity of the injury, there may be arrhythmias and evidence of heart block.

Valve and Septal Injuries

In both blunt and penetrating injuries to the heart, the valve cusps, the papillary muscles, and the chordae tendineae can be damaged. The incidence of valve involvement is in the following order: aortic, mitral, and pulmonary. Acute valvular insufficiency can be diagnosed clinically, but it should be confirmed by cardiac catheterization. In severe cases prompt surgical repair may be necessary.

Penetrating Injuries to the Heart

The anatomy of the heart relative to the front of the thoracic cage determines the common sites of injury. The anterior surface of the heart is formed largely by the right ventricle—the left border formed by the left ventricle and the right border formed by the right atrium. The right ventricle is most commonly injured, followed by the left ventricle and the right atrium. The anterior interventricular (descending) branch of the left coronary artery is the most common artery to be damaged.

Since the pericardium has to be penetrated for the heart to be injured, cardiac tamponade is often present. The hemopericardium quickly presses on the thin-walled atria and large veins and compromises the venous return. The classic triad of (1) distension of the jugular veins of the neck, (2) faint heart sounds (damped down by blood in the pericardial sac), and (3) hypotension may all be present. If there is substantial concomitant blood loss or volume depletion, the jugular veins may not be enlarged. Do not expect to see a greatly enlarged heart shadow on a chest radiograph. Even though the blood in the pericardial sac may be under a high pressure caused by a ventricular leak, the tough fibrous tissue in the wall of the fibrous pericardium prevents its undue distension. Hemopericardium must be relieved surgically.

Clinical Anatomy of the Pericardium

Traumatic Injury

Pericardial Disruption.—This may occur from blunt trauma from a puncture following penetration by a fractured rib and rupture from a sudden increase in pressure. The most common location of rupture is along the lateral margins, especially on the left side. The phrenic nerve may be involved, as it is situated between the mediastinal parietal pleura and the pericardium (Fig 3–17). Since one of the important functions of the pericardium is to anchor the heart, one of the rare complications of pericardial rupture is movement of the heart through the opening, producing a hernia. This could result in torsion of the heart, since it would be free to rotate around its great vessels, with disastrous consequences.

Hemopericardium.—With blunt or penetrating injuries to the heart, blood can escape into the pericardial cavity (see p. 119). The normal volume of pericardial fluid, about 50 mL, is used to lubricate the apposing surfaces of the serous pericardium during heart movements. Once the fluid in the cavity exceeds about 250 mL, the diastolic filling is compromised, a condition known as *cardiac tamponade.*

Clinical Anatomy of Pericardiocentesis

The emergency department therapeutic intervention for cardiac tamponade is pericardiocentesis. The patient is placed in a supine position with the shoulders raised 60° from the horizontal position. The usual skin preparation is now completed. The needle is inserted in the angle between the xiphoid process and the left costal margin. With the positions of the cardiac notches of the lungs and pleura in mind (Fig 3–18), the tip of the needle is directed upward, backward, and to the left in the left xiphocostal angle, i.e., toward the left shoulder. The following structures will be penetrated by the needle.

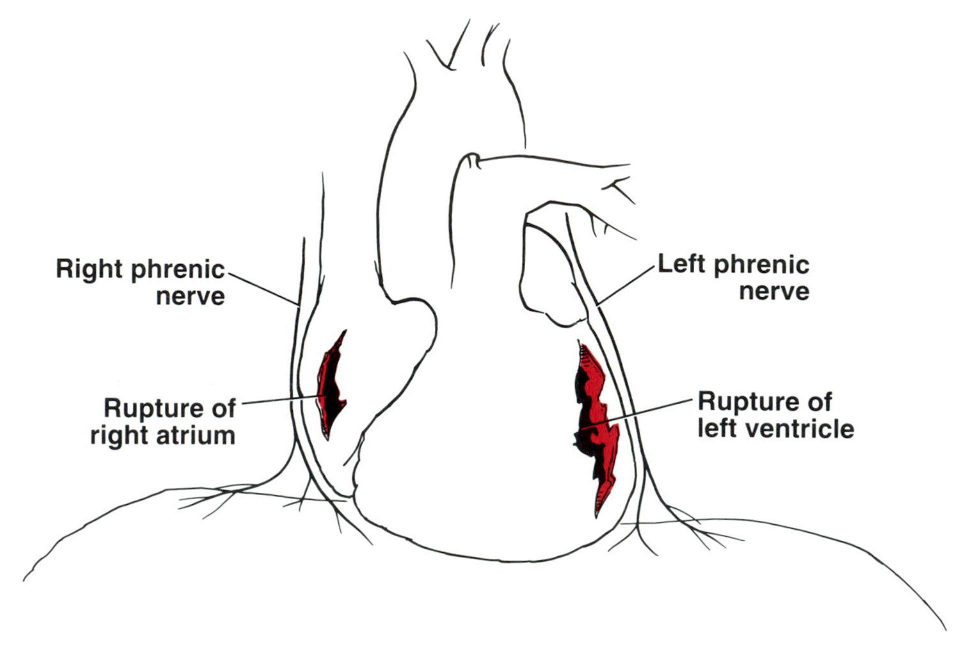

FIG 3–17.
Anterior view of the heart showing common sites for rupture on the anterior surface. Note the positions of the right and left phrenic nerves.

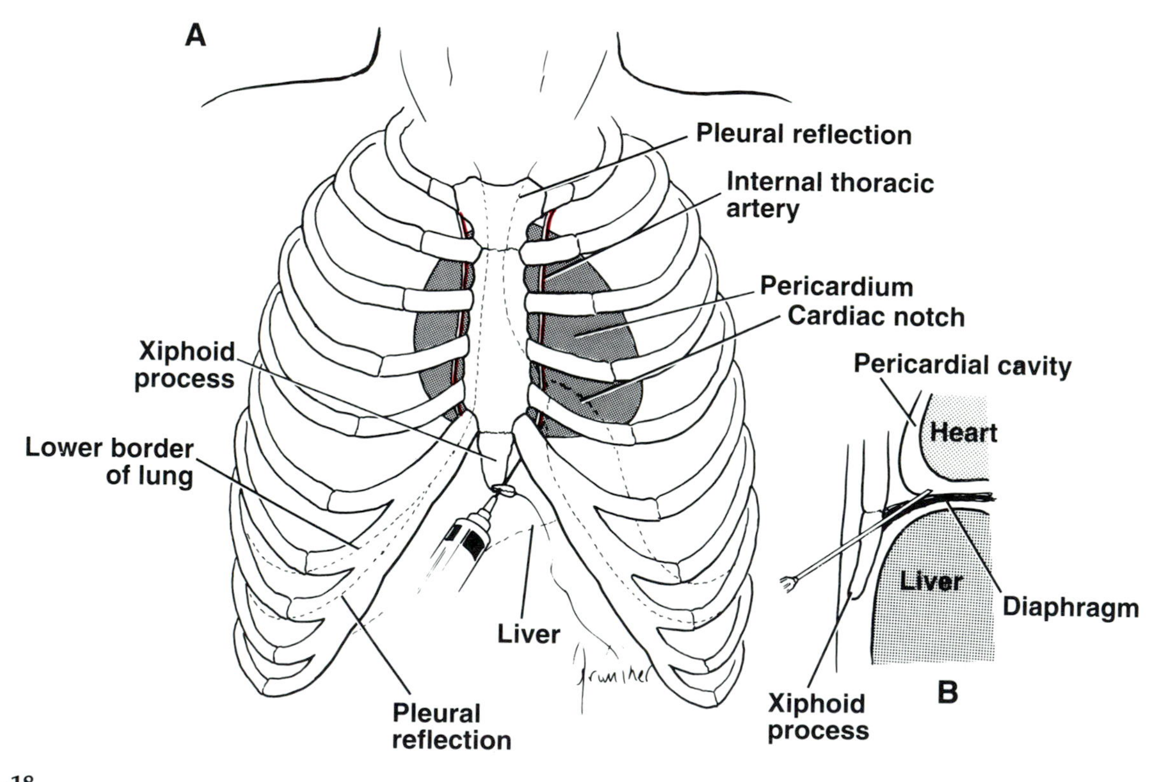

FIG 3–18.
A and **B**, xiphisternal approach for pericardiocentesis. Note that the needle is aimed upward and backward in the direction of the left shoulder.

1. Skin.
2. Subcutaneous tissue.
3. Aponeurosis of the external oblique muscle of the abdomen.
4. Anterior lamina of the aponeurosis of the internal oblique muscle of the abdomen.
5. Rectus abdominis muscle.
6. Posterior lamina of the aponeurosis of the internal oblique and the transversus abdominis muscles.
7. Extraperitoneal fat. Avoid entering the peritoneal cavity.
8. Diaphragm.
9. Pericardium (fibrous pericardium and parietal serous pericardium).

The long large-bore needle is connected to a 20-mL syringe with a three-way stopcock. As the needle is advanced through the various structures, aspirations are performed frequently. Pericardial penetration is indicated by a sudden "give." The procedure, if urgent but not a dire emergency, may be done using portable echocardiographic control.

Anatomical Complications

1. The needle is advanced too far and enters the myocardium of the right ventricle.
2. The needle pierces the anterior descending branch of the left coronary artery.
3. The needle enters the pleural cavity, producing a pneumothorax or a hydropneumothorax.
4. The needle pierces the liver.

Clinical Anatomy of the Pericardial Window

An alternative to pericardiocentesis in the management of cardiac tamponade is the pericardial window. In the treatment of cardiac tamponade caused by hemorrhage into the pericardial cavity, pericardiocentesis fails to draw blood in about 25% of patients. A vertical incision is then made in the midline from the xiphisternal joint downward about 2 in. (5 cm). The following anatomical structures will be incised (Fig 3–19).

1. Skin.
2. Subcutaneous tissue (retract edges).
3. The xiphoid process is then retracted upward to expose the origin of the diaphragm.
4. The diaphragm is then incised to expose the fibrous pericardium. The pericardium will have a blue appearance and will be tight because of the contained blood.
5. The fibrous pericardium and the parietal layer of serous pericardium (fused together as a single layer) are incised, and the blood clots are evacuated.

Anatomical Complications

1. The incision enters the peritoneal cavity because the xiphoid process was not carefully dissected free before being reflected upward (see Fig 3–19).
2. The incision into the diaphragm enters the pleural cavity.
3. The phrenic nerve is injured as it lies on the pericardium. The lateral relationship of the phrenic nerve to the pericardium is important when making pericardial incisions (see Fig 3–17). A vertical incision should be made when possible since a transverse incision is likely to cut the nerve.

Clinical Anatomy of the Transvenous Pacemaker

A transvenous cardiac pacemaker can be inserted in the emergency department (Fig 3–20). The usual routes of venous access are the right internal jugular, the left subclavian, and the right subclavian veins. The right or left femoral veins are often used in conjunction with fluoroscopy.

When the catheter is inserted into the right internal jugular vein (see p. 196), the following veins and heart structures are traversed in order to position the pacemaker: (1) right internal jugular vein, (2) right brachiocephalic vein, (3) superior vena cava, (4) right atrium, (5) tricuspid valve, and (6) right ventricle. The route is direct, with no bends.

When the catheter is inserted into the left subclavian vein (see p. 160), the following veins and heart structures are traversed: (1) left subclavian vein (preferred to the right subclavian vein because the curve of the catheter follows the natural curve of the left vein), (2) left brachiocephalic vein, (3) superior vena cava, (4) right atrium, (5) tricuspid valve, and (6) right ventricle. The route is direct, with no bends.

The following veins and heart structures are traversed when the catheter is inserted into either femoral vein: (1) femoral vein, (2) external iliac vein, (3) common iliac vein, (4) inferior vena cava, (5) right atrium, (6) tricuspid valve, and (7) right ventricle.

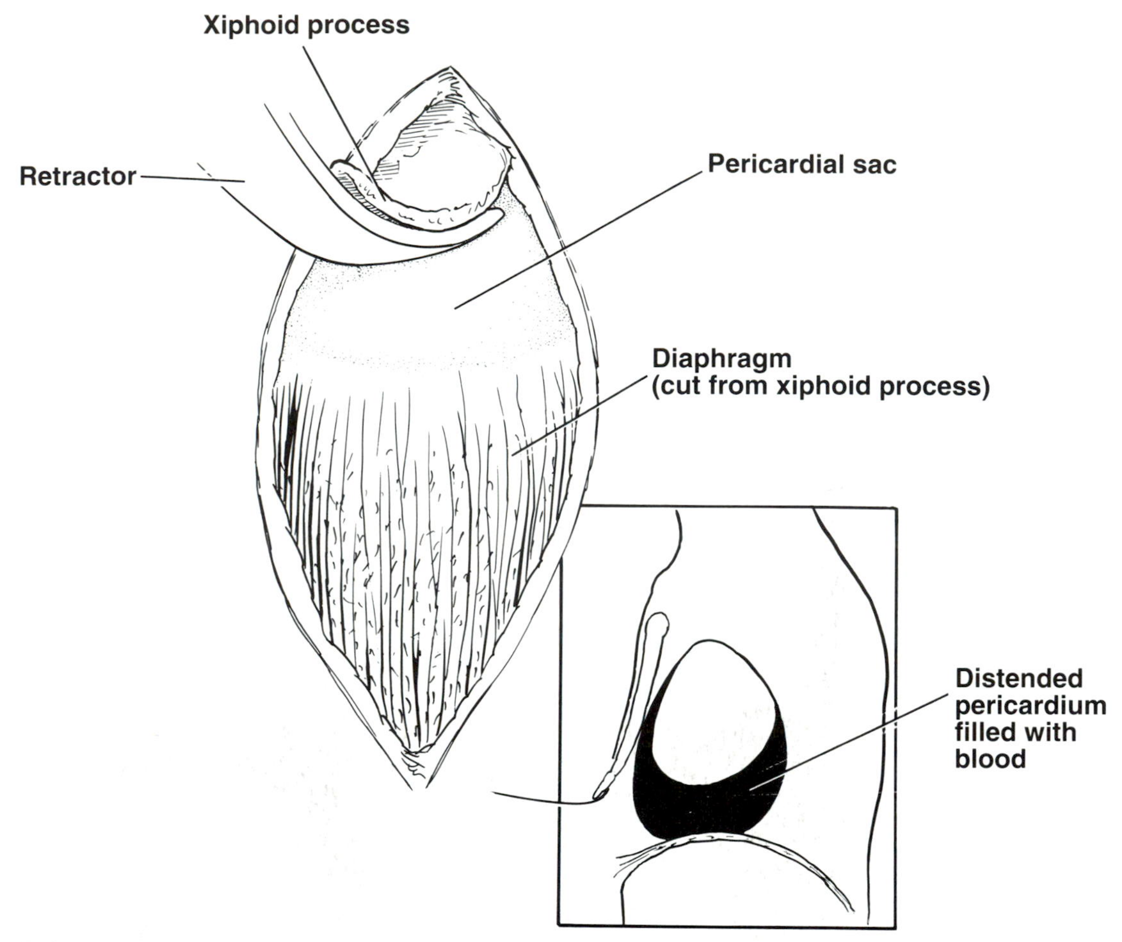

FIG 3-19.
Pericardial window. Note the position of the pericardium and the diaphragm.

In each approach the passage from the right atrium into the right ventricle is aided by inflating the balloon so the catheter floats across the tricuspid valve.

Anatomical Complications

1. The right atrial wall is perforated.
2. The right ventricular wall is perforated.
3. The ventricular septum is perforated.
4. The catheter tip is moved out of the right ventricle back into the right atrium and then down the inferior vena cava into the hepatic veins or the coronary sinus.
5. The tip of the catheter may enter the mouth of the coronary sinus in the right atrium; a chest radiograph, lateral view, would show the wire directed posteriorly and not anteriorly.

ANATOMICAL-CLINICAL PROBLEMS

1. In anatomical terms explain the cause and distribution of chest pain in myocardial ischemia. Explain why pain sometimes can be felt in the epigastrium when the infarct occurs on the inferior or diaphragmatic surface of the heart.

2. A 5-year-old boy was seen in the emergency room following an attack of breathlessness during which he had lost consciousness. The mother, on questioning, said that her child had had several such attacks before and sometimes his skin had become bluish. Recently she had noticed that he breathed more easily when he was playing in a squatting position; he also seemed to sleep more easily with his knees drawn up. On physical examination, the child was found to be thinner and shorter than normal. His lips were cyanotic, and his fingers and toes were clubbed. A systolic murmur was present along the left border of the sternum, and the heart was consid-

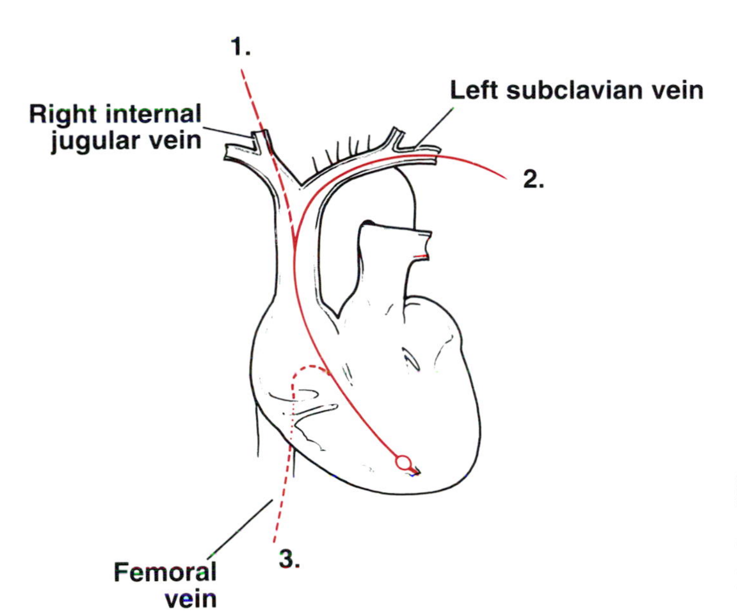

FIG 3–20.
Transvenous pacemaker. The usual routes for insertion are the right internal jugular *(1)*, the left subclavian *(2)*, and the right subclavian veins; the right or left femoral veins *(3)* are also used in conjunction with fluoroscopy.

erably enlarged to the right. During the examination his breathlessness diminished and he looked less anxious. An extensive workup, including angiography, demonstrated tetralogy of Fallot. Using your knowledge of cardiac development, explain this syndrome. Why does the child often assume the squatting position in this condition? What role, if any, can the ductus arteriosus play in determining the degree of cyanosis in a patient with this disease?

3. A 39-year-old woman was seen in the emergency department complaining of a severe localized pain over the left hemithorax. She stated that the pain started quite suddenly, about 1 hour previously, when she was reaching for a book on a high shelf in the public library. On further questioning she disclosed that she had been sailing with her husband the previous day and that they had been caught in a storm. She described the pain as a continuous dull ache that was made worse by taking deep breaths and using the left arm. She indicated that the pain was localized over the sixth left costal cartilage about 2 in. to the left of the sternum, which corresponded to an area of tenderness of the sixth left costochondral junction. What is the possible diagnosis? Explain the nerve supply to the painful area. How would this pain differ from pain originating from a thoracic viscus?

4. An 8-year-old girl was seen in the emergency department because of a nasty cut to her right little finger. Before giving the child a local anesthetic prior to suturing the wound, the physician listened to her chest. To his surprise, he could hear a continuous machinery-like murmur at the upper left border of the sternum. The second heart sound was accentuated, and the radial pulse was bounding. Later, chest radiographs revealed left ventricular hypertrophy and an increased pulmonary vascularity. Doppler echocardiography showed continuous blood flow in the pulmonary artery. A diagnosis of patent ductus arteriosus was made. What is a patent ductus? Explain its presence embryologically. What factor or factors are responsible for its normal closure?

5. Describe the anatomy of the conducting system of the heart. Describe the blood supply of the conducting system. Can coronary artery disease affect the normal rhythm of the heart?

6. A 48-year-old woman was taken from the scene of an automobile accident to the emergency department. She had been driving her car with a seat belt on but without a shoulder strap, and she had hit a utility pole head-on. On examination she was conscious and alert but had sustained severe facial and chest injuries. A careful examination of the thoracic cage revealed the presence of a sternal fracture (middle of the body) and fractures of the third and fourth left ribs near their costochondral junctions. After an extensive workup it was decided that the patient had a ruptured pericardium. Using your knowledge of anatomy, explain why the left phrenic nerve was damaged in this patient. Explain also why the heart and pericardium are subject to such pressures when the thorax and abdomen are exposed to severe blunt trauma.

7. A 61-year-old man was seen in the emergency room complaining of a feeling of pressure within his chest. On questioning he said that he had

had several attacks before and they had always occurred when he was climbing three flights of stairs or digging in the garden. He found that the discomfort disappeared with rest after about 5 minutes. The reason he came to the emergency department was that the chest discomfort had occurred with much less exertion. What is the blood supply to the diaphragmatic surface of the heart? Does this blood supply vary in different individuals? Do the coronary arteries anastomose with one another? What is an anatomical end artery? Does the lumen of a coronary artery change in diameter with exercise?

8. A 33-year-old woman was jogging across the park at 11 p.m. when she was attacked by a gang of youths. After she was brutally mugged and raped; one of the youths decided to stab her in the heart to keep her silent. Later she was found by a policeman and miraculously was seen to be still alive. She was rushed to the emergency department. On examination she was unconscious and in extremely poor shape. A small wound about ½ in. in diameter was present in the left fifth intercostal space about ½ in. from the lateral sternal margin. Her carotid pulse was rapid and weak, her pulse pressure was narrow, and her neck veins were distended. There was no evidence of a left-sided pneumothorax. What is your diagnosis? Assuming that she had a cardiac tamponade, explain the signs and symptoms. Considering the position of the knife wound, explain the absence of a pneumothorax.

9. A 21-year-old man was walking along a city street when suddenly a burst of automatic gunfire came from a passing car. The man, who was an innocent bystander, was struck in the front of the chest by a stray bullet. He was rushed to the emergency department by an ambulance. The position of the entry wound was the fourth left intercostal space about 3 in. from the midline. Using your knowledge of anatomy, explain the following: (1) the likely chamber or chambers of the heart to have been damaged, and (2) which coronary artery was most likely to have been damaged.

10. Name the anatomical structures penetrated by a needle when performing a pericardiocentesis. What anatomical structures are likely to be damaged if care is not taken when performing this procedure?

11. Describe the anatomical reasons for using the left subclavian vein or the right internal jugular vein route when inserting a transvenous pacemaker catheter.

ANSWERS

1. The cause of pain in myocardial infarction is the lack of arterial blood to an area of the myocardium; this is secondary to a blockage of a coronary artery or one of its branches. The pain is produced by oxygen deficiency and the accumulation of metabolites in the field supplied by the blocked artery. This stimulates the sensory nerve endings in the myocardium.

The distribution of pain in myocardial ischemia is directly related to the anatomy of the nerves involved. The afferent nerve fibers ascend to the central nervous system through the cardiac branches of the sympathetic trunk and enter the spinal cord via the posterior roots of the upper four thoracic nerves. The pain is *not felt in the heart* but is referred to skin areas supplied by the corresponding spinal nerves. The dermatomes supplied by the upper four intercostal nerves and by the intercostobrachial nerve (to the medial side of the arm) are therefore affected. Some spread of nervous information may occur in the central nervous system because pain is often felt in the neck.

Myocardial infarction involving the diaphragmatic wall of the heart often causes epigastric discomfort. This could be explained by the fact that the pain fibers from this area of the heart ascend in the sympathetic nerves and enter the spinal cord in the posterior nerve roots of the T7 through T9 spinal nerves; this would give rise to referred pain in the T7, T8, and T9 thoracic dermatomes in the epigastrium.

2. The tetralogy of Fallot consists of the following: (1) a large ventricular septal defect, (2) stenosis of the pulmonary trunk that may occur at the infundibulum of the right ventricle or at the pulmonary valve, (3) an exit of the aorta that lies immediately above the ventricular septal defect and thus communicates with both ventricles, and (4) a right ventricular hypertrophy secondary to the ventricular septal defect and pulmonary stenosis.

Pulmonary stenosis impairs pulmonary circulation so that a right-to-left shunt occurs and the arterial blood is poorly oxygenated. Lack of oxygen causes impaired growth and development, cyanosis, and clubbing of the fingers and toes. Attacks of breathlessness are caused by failure of the pulmonary circulation to provide sufficient oxygenation of the blood on exertion. Maintaining a squatting position during play and sleeping with the knees drawn up increases the peripheral resistance in the systemic circulation, decreases the right-to-left shunt through the ventricular septum, and consequently slightly increases the pulmonary blood flow.

In a child with a slight pulmonary stenosis or a large patent ductus arteriosus, cyanosis is minimal. A large patent ductus allows aortic blood to enter the pulmonary trunk distal to the stenosis and, in this way, enables the blood to enter the pulmonary circulation for oxygenation.

3. The diagnosis was a separation of the sixth left costochondral joint or an acute costochondritis. These injuries may be very painful. The joint is innervated by the sixth intercostal nerve. This case is a good example of somatic pain that may be intense and localized, in contradistinction to visceral pain, which is diffuse and poorly localized.

4. The ductus arteriosus is the remains of the distal portion of the sixth left aortic arch artery and connects the left pulmonary artery to the descending aorta. In fetal life it serves to bypass blood from the pulmonary trunk to the aorta at a time when the lungs are nonfunctional. At birth, the ductus arteriosus normally constricts in response to a rise in arterial oxygen tension. Later, fibrosis occurs and the ductus becomes the ligamentum arteriosum.

5. The conducting system of the heart consists of the sinoatrial node, the atrioventricular node, the atrioventricular bundle and its branches, and the Purkinje plexus. The location, structure, function, and blood supply of the conducting system are fully described on pp. 111, 113, and 115.

6. Severe blunt frontal trauma to the chest wall, especially if there is some shearing force, rarely causes rupture of the pericardium. The site of rupture is often in a coronal plane along the borders, where the phrenic nerve is located between the fibrous pericardium and the mediastinal parietal pleura. The effect of the forces of frontal trauma to the heart and the surrounding pericardium are discussed on p. 127. In the elderly the toughness of the pericardium and lack of elasticity may make rupture more common in this age group.

7. The diagnosis was a classic case of angina pectoris. The sudden change in history, i.e., pain caused by less exertion, should cause the physician concern that the patient now has unstable angina or an actual myocardial infarction.

The diaphragmatic surface of the heart receives its arterial supply from the posterior interventricular (descending) branch of the right coronary artery (right dominance). However, this branch varies in size and origin. In left dominance, which is less common than right dominance (10%), the branch arises from the left coronary artery.

The terminal branches of the coronary arteries do anastomose, but the anostomoses are not large enough to provide an adequate blood supply should one of the larger branches become blocked by disease. For this reason, these terminal branches are called anatomical end arteries. The lumen of a coronary artery adjusts to the metabolic demands of the myocardium, and the products of metabolism, namely, carbon dioxide and lactic acid, bring about the changes locally. In exercise the arteries are largely dilated in response to local metabolites; the sympathetic nerves play only a minor role.

8. This patient did have cardiac tamponade. The knife had struck the upper border of the sixth costal cartilage before piercing the fifth intercostal space. Fortunately much of the force was now expended. The tip of the knife had pierced the pericardium and entered the anterior wall of the right ventricle. Severe hemorrhage then occurred into the pericardial cavity. The blood in the pericardial cavity was under right ventricular pressure and pressed on the exterior of the thin-walled atria and large veins as they traversed the pericardium to enter the heart. This caused a back up of venous blood and was responsible for the congested veins seen in the neck. The poor venous return severely compromised the cardiac output. The presence of the blood around the outside of the heart reduced the loudness of the heart sounds, and they were muffled or distant.

A left-sided pneumothorax did not occur, since the knife barely missed the pleura, passing through the cardiac notch (see Fig 2–30).

9. The greater part of the anterior surface of the heart is formed by the right ventricle. Consequently, this part of the heart is most vulnerable when a bullet enters the frontal surface of the chest. Since the left ventricle is positioned behind the right ventricle, this chamber could also be damaged if the bullet continues on its path posteriorly. The anterior interventricular (descending) branch of the left coronary artery is the most likely artery to be damaged in such a wound.

10. The following structures are pierced by a needle when performing pericardiocentesis in the left xiphocostal angle: (1) skin, (2) subcutaneous tissue, (3) aponeurosis of the external oblique muscle of the abdomen, (4) anterior layer of the aponeurosis of the internal oblique muscle of the abdomen, (5) rectus abdominis muscle, (6) posterior layer of the aponeurosis of the internal oblique and transversus abdominis muscles of the abdomen, (7) extraperitoneal fat, (8) diaphragm, and (9) pericardium (fibrous

pericardium plus the parietal layer of serous pericardium fused together).

If the needle is not pointed upward and toward the left shoulder and is directed more posteriorly, there is a strong possibility that it will enter the peritoneal cavity, where it may damage the liver. Later when the diaphragm is penetrated, the needle may miss the cardiac notch of the parietal pleura and enter the pleural cavity, producing a pneumothorax.

11. The left subclavian vein is a good approach for inserting a transvenous pacemaker. The curve of the vein tends to coincide with the natural tendency for the catheter to curve. This means that the catheter easily follows the anatomical curve of the left subclavian and left brachiocephalic veins into the superior vena cava. On the right side the right subclavian vein joins the right brachiocephalic at a more acute angle, and for this reason the right subclavian vein is less commonly used.

The right internal jugular vein is directly in line with the superior vena cava and is an ideal approach.

4

The Blood Vessels of the Thorax and Upper Extremity

The largest blood vessels in the body are located within the chest cavity—the aorta, the pulmonary arteries, the venae cavae, and the pulmonary veins. Trauma to the chest wall can result in the disruption of these vessels, with consequent hemorrhage, rapid exsanguination, and death. Penetrating injuries to the chest can pierce the vessel walls, and blunt injuries causing sudden body acceleration or deceleration can tear the vessels. Unfortunately, because these vessels are hidden from view within the thorax, the diagnosis of a major blood vessel injury is often delayed, with disastrous consequences to the patient.

The upper part of the thoracic cage provides considerable protection for the aorta and its large branches, but fractures in this area can result in damage to the underlying soft tissues. In particular, sternoclavicular dislocation or fractures of the first and second ribs are often associated with vascular injury.

Arterial injuries in the upper extremity are common. Fortunately the presence of a good collateral circulation ensures a good prognosis. The close proximity of many of the upper limb arteries to veins and nerves may result in multisystem injury, with the subsequent development of arteriovenous fistulas, alterations of sensation, and muscular paralysis. The use of upper limb arteries and veins as the site for vascular access and invasive monitoring can become the subject for iatrogenic injuries.

The purpose of this chapter is to familiarize the emergency physician with the basic anatomy of the blood vessels of the thorax and the upper extremity in order that diagnosis of vascular injury can be made promptly and vascular access can be established quickly and accurately.

BASIC ANATOMY OF THE LARGE ARTERIES OF THE THORAX

Basic Anatomy of the Aorta

The aorta is the main arterial trunk that delivers oxygenated blood from the left ventricle of the heart to the tissues of the body (Fig 4–1). It is divided for purposes of description into the following parts: ascending aorta, arch of aorta, descending thoracic aorta, and abdominal aorta.

Ascending Aorta

The ascending aorta begins at the base of the left ventricle, where it is about 3 cm in diameter. It runs upward and forward and comes to lie behind the right half of the sternum at the level of the sternal angle, where it becomes continuous with the arch of the aorta. The ascending aorta lies within the fibrous pericardium (see Fig 4–1) and is enclosed with the pulmonary trunk in a sheath of serous pericardium. At its root it possesses three bulges, the *sinuses of the aorta,* one behind each aortic valve cusp.

Important Relations (Figs 4–1 and 4–2).—The following relations exist.

Anterior.—Pulmonary trunk, right auricle, edge of right pleura and right lung, remains of thymus, and sternum.

Posterior.—Left atrium, right pulmonary artery, and right principal bronchus.

Right Lateral.—Superior vena cava and right atrium.

Left Lateral.—Left atrium and pulmonary trunk.

Branches.—The *right coronary artery* arises from the anterior aortic sinus, and the *left coronary artery* arises from the left posterior aortic sinus (see p. 113).

Clinical Note

Ascending aorta and the pericardium.—The entire ascending aorta lies within the pericardial sac. Consequently, tear or rupture of this part of the aorta occurs into the pericardial cavity, producing cardiac tamponade.

Arch of Aorta

The arch of the aorta is a continuation of the ascending aorta (see Fig 4–2). It lies in the superior mediastinum behind the manubrium sterni and arches upward, backward, and to the left and in front of the trachea. (Its main direction is backward.) It then passes downward to the left of the trachea, and at the level of the sternal angle becomes continuous with the thoracic aorta.

Important Relations.—The following relations exist.

Anterior and to the Left (Fig 4–3).—Left mediastinal pleura, left phrenic nerve, left vagus nerve, cardiac branches of vagus and sympathetic nerves, left superior intercostal vein, left lung, and pleura.

Posterior and to the Right.—Left recurrent laryngeal nerve, cardiac plexus, trachea, esophagus, and vertebral column.

Superior.—Brachiocephalic, left common carotid, and left subclavian arteries take origin from its convexity.

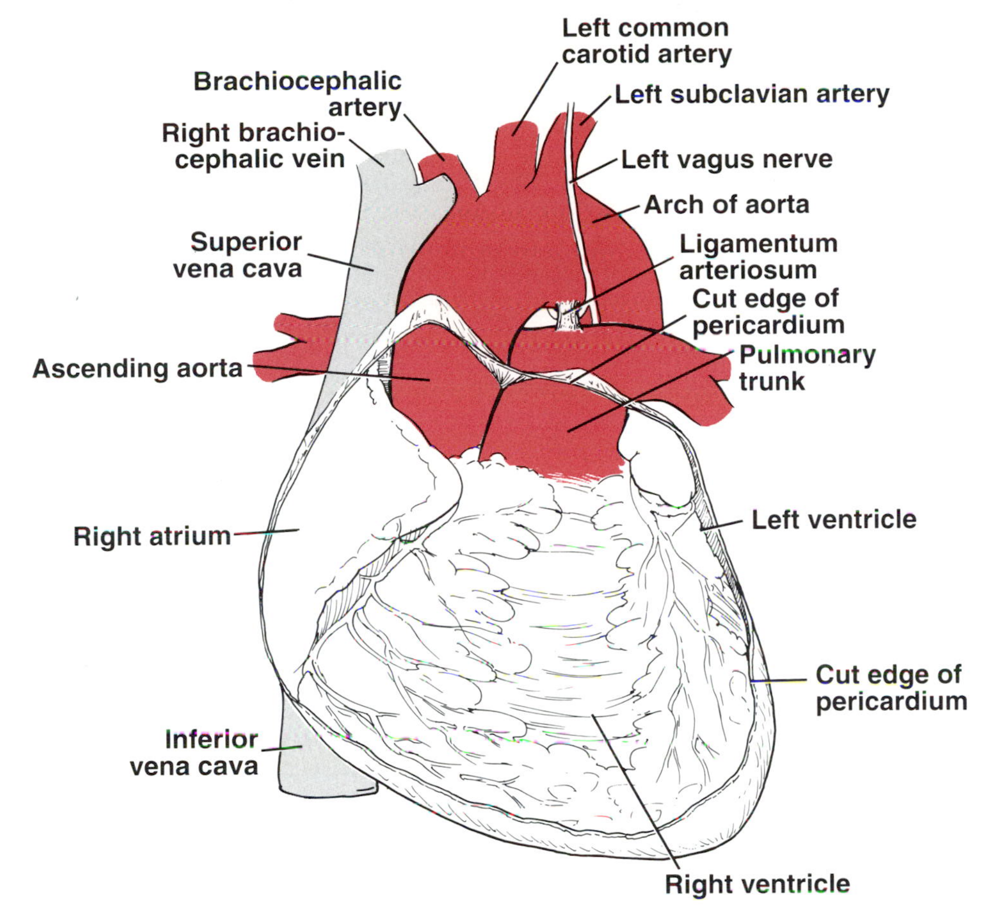

FIG 4–1.
Anterior surface of the heart and the great blood vessels. The pericardial sac has been opened to show the ascending aorta and the pulmonary trunk.

Inferior (see Fig 4–1).—Bifurcation of pulmonary trunk, ligamentum arteriosum, left recurrent laryngeal nerve, and cardiac plexus.

Branches.—The *brachiocephalic artery* arises from the convex surface of the aortic arch and passes upward and to the right of the trachea (see Fig 4–1). It terminates behind the right sternoclavicular joint by dividing into the right subclavian and right common carotid arteries.

The *left common carotid artery* arises from the convex surface of the aortic arch on the left side of the brachiocephalic artery (see Fig 4–1). It runs upward and to the left of the trachea and enters the neck behind the left sternoclavicular joint.

The *left subclavian artery* arises from the aortic arch behind and to the left of the left common carotid artery (see Fig 4–1). It runs upward along the left side of the trachea and the esophagus to enter the root of the neck. It lies in contact with the apex of the left lung, over which it arches.

Descending Thoracic Aorta

The thoracic aorta (see Fig 4–3) lies in the posterior mediastinum and begins as a continuation of the arch of the aorta on the left side of the lower border of the body of the fourth thoracic vertebra (i.e., opposite the sternal angle). It runs downward in the posterior mediastinum, inclining forward and medially to reach the anterior surface of the vertebral column. At the level of the 12th thoracic vertebra it passes behind the diaphragm (through the aortic opening) in the midline and becomes continuous with the abdominal aorta.

Important Relations (see Fig 4–3).—The following relations exist.

Anterior.—Hilum of the left lung, the pericardium, esophagus, and the diaphragm.

Posterior.—Vertebral column and hemiazygos veins.

Right Lateral.—Azygos vein, thoracic duct, right pleura, and right lung.

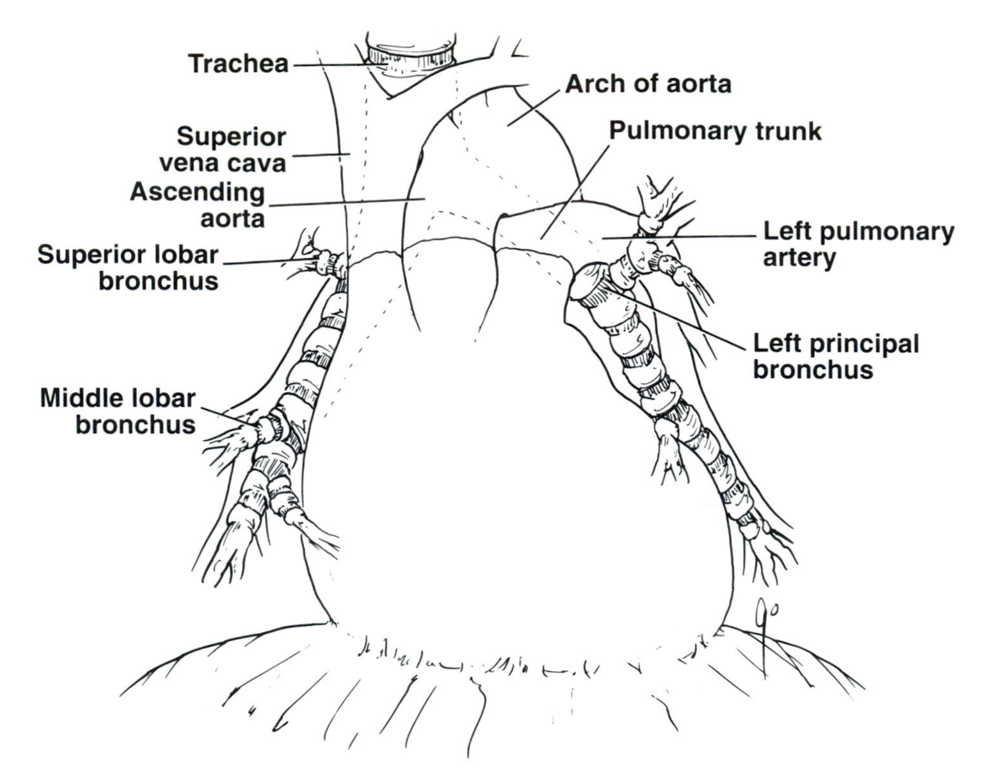

FIG 4–2.
Anterior surface of the heart and the great blood vessels showing their relationship to the bifurcation of the trachea and the main bronchi. The pericardium is intact.

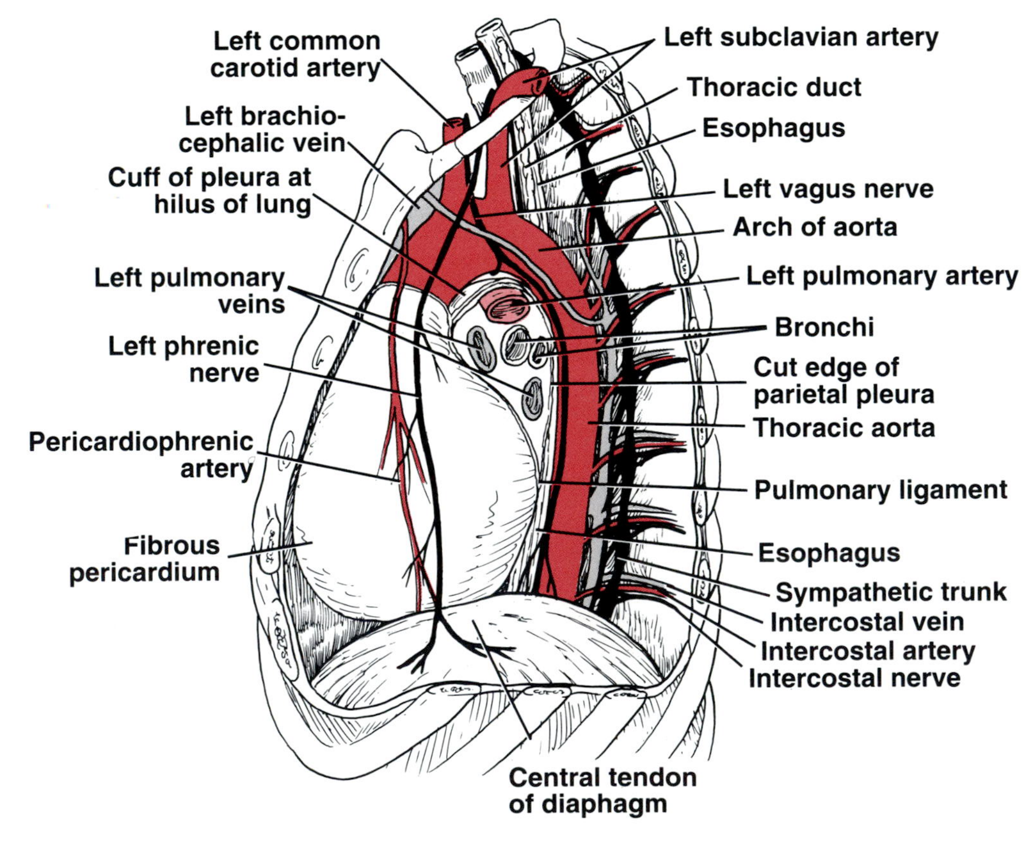

FIG 4–3.
Left side of the mediastinum. Note the position of the hilum of the left lung and the relationship between the esophagus and the descending thoracic aorta.

Left Lateral.—Left pleura and left lung. Note that with respect to the aorta in the posterior mediastinum, the esophagus is right lateral above, anterior lower down, and becomes left anterolateral below. In other words, the aorta and the esophagus cross in the posterior mediastinum.

Clinical Note

This relationship of the aorta and esophagus in the posterior mediastinum is important when distinguishing between these structures during an emergency thoracotomy.

Branches.—Posterior intercostal arteries are given off to the lower nine intercostal spaces on each side (first and second posterior intercostal arteries arise from the costocervical branch of the subclavian artery). The arteries of the right side cross the vertebral column and pass behind the esophagus. On both sides the arteries run behind the sympathetic trunks. On entering the intercostal space, each artery runs forward in the costal groove between the vein above and the corresponding intercostal nerve below (see p. 53). The arteries anastomose with the anterior intercostal arteries from the internal thoracic and musculophrenic arteries. The lowest two intercostal arteries pass forward into the anterior abdominal wall. Each posterior intercostal artery gives off a posterior branch that supplies the muscles and skin of the back, the spinal cord, and the meninges.

Subcostal arteries are given off on each side and run along the lower border of the 12th rib to enter the abdominal wall.

Pericardial, esophageal, and *bronchial arteries* are small branches that are distributed to these organs.

Clinical Note

Descending thoracic aorta and left hemothorax.—The close relationship of the descending thoracic aorta to the left pleura and lung means that rupture of the aorta may occur into the left pleural cavity, producing a massive hemothorax.

Radiographic Appearances of the Arch of the Aorta

The normal radiographic appearances of the arch of the aorta and its branches are shown in an arteriogram in Figures 4–4 and 4–5.

Abdominal Aorta

This is described in Chapter 14, p. 540.

Basic Anatomy of the Pulmonary Trunk

The pulmonary trunk conveys deoxygenated blood from the right ventricle to the lungs (see Fig 4–1). It measures about 3 cm in diameter. The pulmonary trunk leaves the upper part of the right ventricle and runs upward, backward, and to the left, at first in front of the ascending aorta and then to its left. It is about 5 cm (2 in.) long and terminates in the concavity of the aortic arch by dividing into right and left pulmonary arteries. The bifurcation lies anteroinferior and to the left of the tracheal bifurcation (see Fig 4–2).

The pulmonary trunk is contained within the fibrous pericardium and, together with the ascending aorta, it is enclosed in a sheath of serous pericardium.

The *ligamentum arteriosum* is a fibrous band that connects the bifurcation of the pulmonary trunk to the lower surface of the aortic arch (Fig 4–1). It is the remnant of the *ductus arteriosus,* which in the fetus conducts blood from the pulmonary trunk to the aorta, thus bypassing the lungs. Following birth, the ductus closes.

Important Relations.—The following relations exist.

Anterior.—Sternal end of second left intercostal space, left lung and pleura, and pericardium.

Posterior.—Ascending aorta, left coronary artery, and left atrium.

Branches.—The *right pulmonary artery* runs to the right *behind* the ascending aorta and superior vena cava to enter the root of the right lung, where it divides into three primary branches, one for each lobe.

The *left pulmonary artery* runs to the left *in front* of the descending aorta to enter the root of the left lung, where it divides into two primary branches, one for each lobe.

Radiographic Appearance of the Pulmonary Trunk

The normal radiographic appearance of the pulmonary trunk is shown in the arteriogram in Figure 4–6.

Basic Anatomy of the Arteries of the Upper Extremity

Subclavian Artery

This artery is described on p. 177.

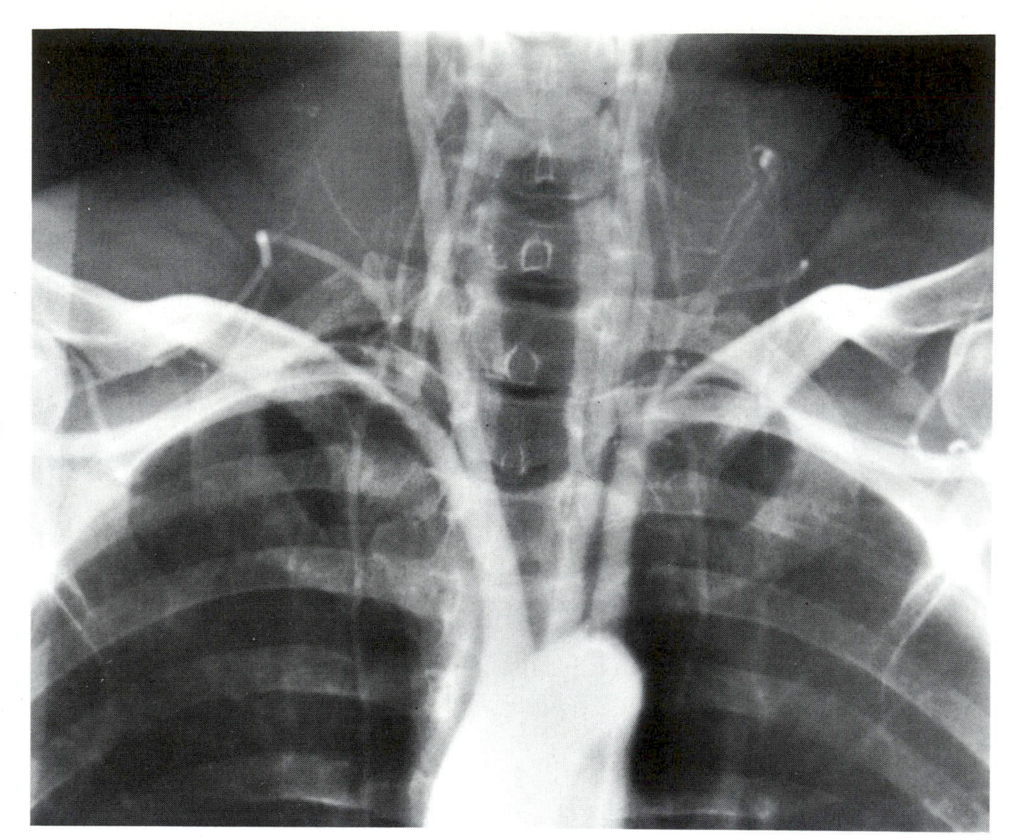

FIG 4–4.
Aortic arch angiogram showing large arteries at the root of the neck.

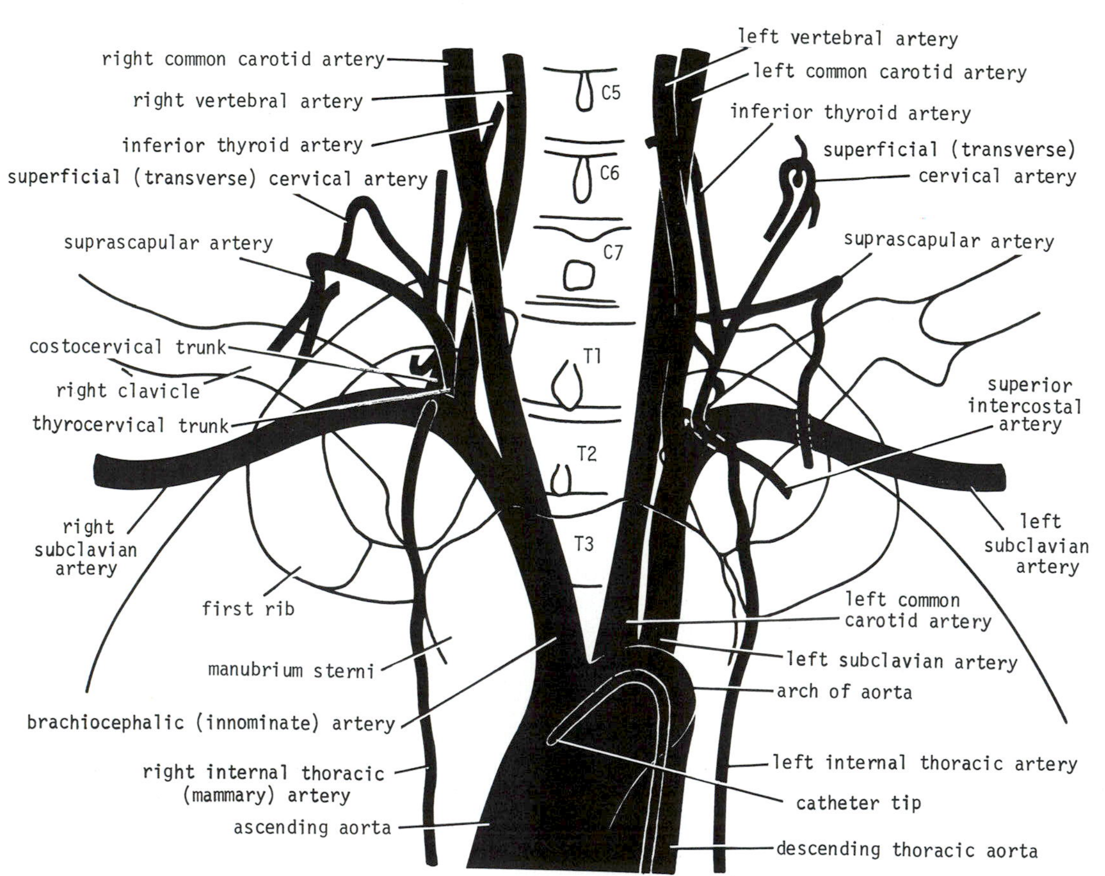

FIG 4–5.
Diagrammatic representation of main features seen in the aortic angiogram in Figure 4–4.

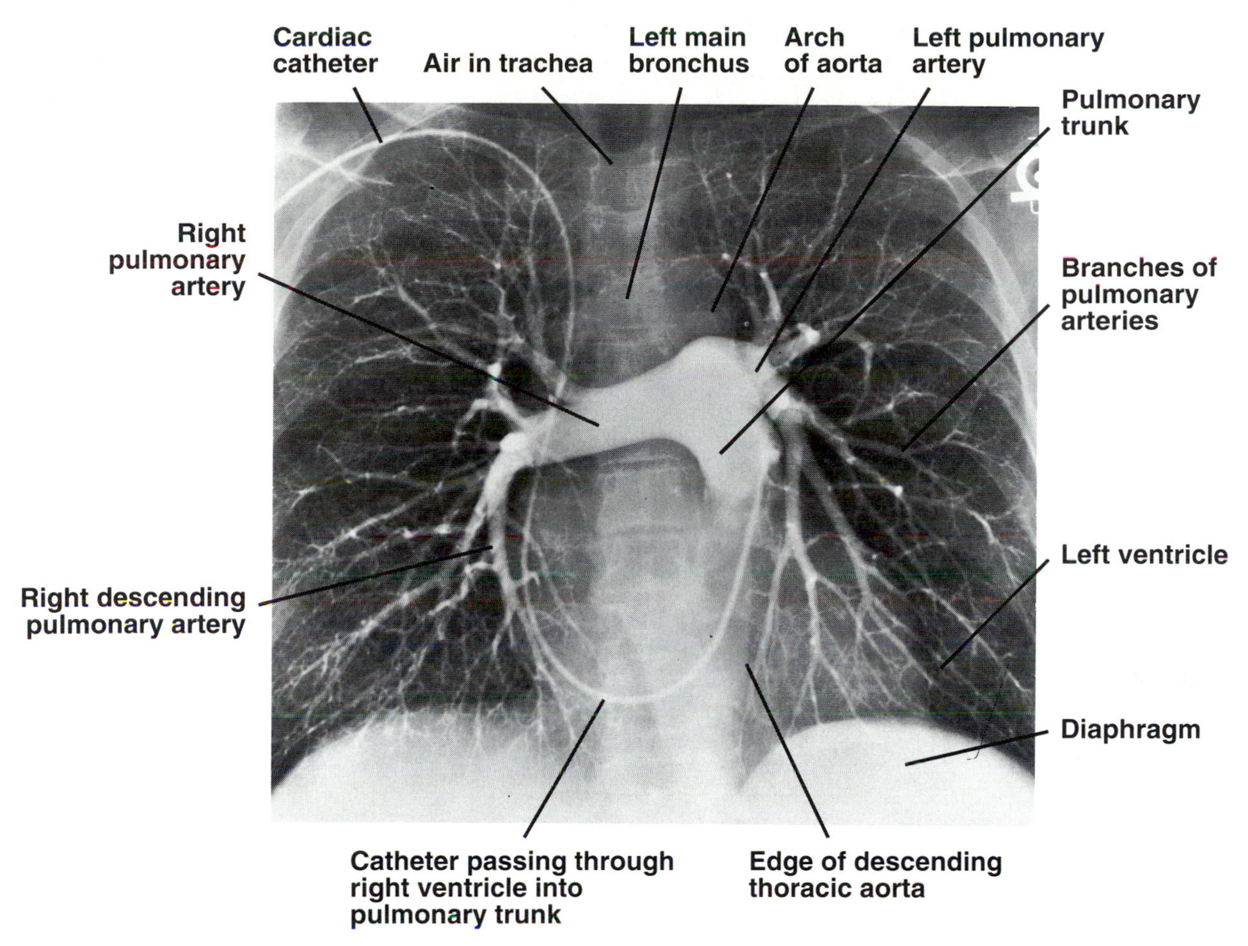

FIG 4–6.
Angiogram of the pulmonary trunk and pulmonary arteries.

Axillary Artery
The axillary artery begins at the lateral border of the first rib as a continuation of the subclavian artery, and at the lower border of the teres major muscle it becomes the brachial artery (see Fig 4–1). Throughout its course, the artery is closely related to the course of the brachial plexus and their branches and is enclosed with them in a connective tissue sheath called the *axillary sheath* (see p. 170).

The pectoralis minor muscle crosses in front of the axillary artery, dividing it into three parts. The first part extends from the lateral border of the first rib to the upper border of the pectoralis minor. The second part is the portion that lies posterior to the pectoralis minor. The third part is the portion from the lower border of pectoralis minor to the lower border of the teres major.

Important Relations of the First Part of the Artery (Fig 4–7).—The following relations exist.

Anterior.—Skin, fasciae, and pectoralis major.

Posterior.—Long thoracic nerve and serratus anterior.

Lateral.—Three cords of the brachial plexus.

Medial.—Axillary vein.

Important Relations of the Second Part of the Artery (see Fig 4–7).—The following relations exist.

Anterior.—Skin, fasciae, and pectoralis major and minor.

Posterior.—Posterior cord of the brachial plexus.

Lateral.—Lateral cord of brachial plexus.

Medial.—Medial cord of brachial plexus and axillary vein.

Important Relations of the Third Part of the Artery (see Fig 4–7).—The following relations exist.

Anterior.—Skin, fasciae, and pectoralis major. The artery is crossed by the medial root of the median nerve.

Posterior.—Axillary and radial nerves.

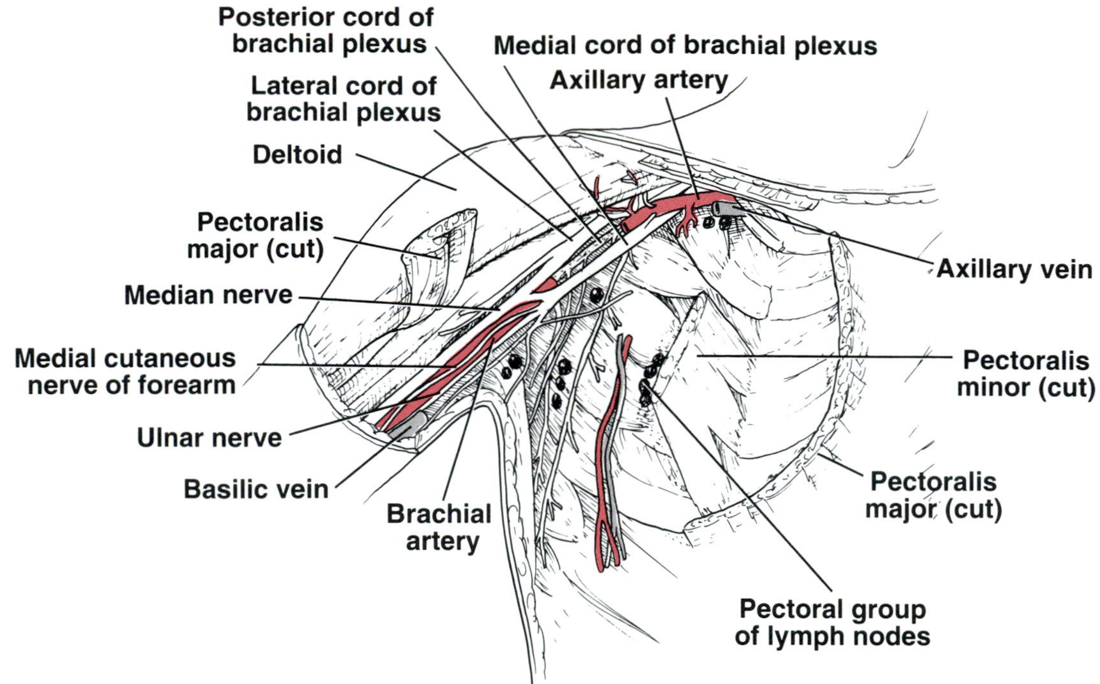

FIG 4–7.
Pectoral region and axilla; pectoralis major and minor muscles have been removed to display the axillary vessels and the brachial plexus. The axillary lymph nodes are also shown.

Lateral.—Lateral root of the median nerve and the musculocutaneous nerve.

Medial.—Ulnar nerve, medial cutaneous nerve of the arm, and the axillary vein.

Branches.—The first part of the axillary artery gives off one branch, the *highest thoracic artery;* the second part gives off two branches, the *thoracoacromial artery* and the *lateral thoracic artery;* and the third part gives off three branches, the *subscapular artery* and the *anterior and posterior circumflex humeral arteries* (Fig 4–8).

Clinical Note

Arterial anastomosis around the shoulder joint.—To compensate for temporary occlusion of the axillary artery during movements of the shoulder joint and to ensure an adequate blood flow to the upper limb (see Fig 4–8), the suprascapular and superficial cervical arteries (branches of the thyrocervical trunk from the first part of the subclavian artery) anastomose with the subscapular and anterior and posterior circumflex humeral arteries (branches of the third part of the axillary artery).

Brachial Artery

The brachial artery begins at the lower border of the teres major muscle as a direct continuation of the axillary artery (Fig 4–9). It ends at the level of the neck of the radius by dividing into the radial and ulnar arteries.

Important Relations (Fig 4–10).—The following relations exist.

Anterior.—The artery is superficial and overlapped from the lateral side by the coracobrachialis and biceps muscles. The median nerve crosses the middle part of the artery, and the bicipital aponeurosis crosses the lower part of the artery.

Posterior.—The triceps, the insertion of the coracobrachialis, and the brachialis.

Lateral.—The median nerve, the coracobrachialis, and the biceps in the upper part of the arm and the tendon of the biceps in the lower part of its course.

Medial.—The ulnar nerve and the basilic vein in the upper part of the arm and the median nerve in the lower part of the arm.

Branches:

Muscular branches to the surrounding muscles.

Nutrient artery to the humerus.

Profunda brachii artery is a large branch that follows the radial nerve into the posterior compartment of the arm in the spiral groove of the humerus (Fig 4–11).

Superior ulnar collateral artery follows the ulnar nerve.

Inferior ulnar collateral artery takes part in the anastomosis around the elbow joint.

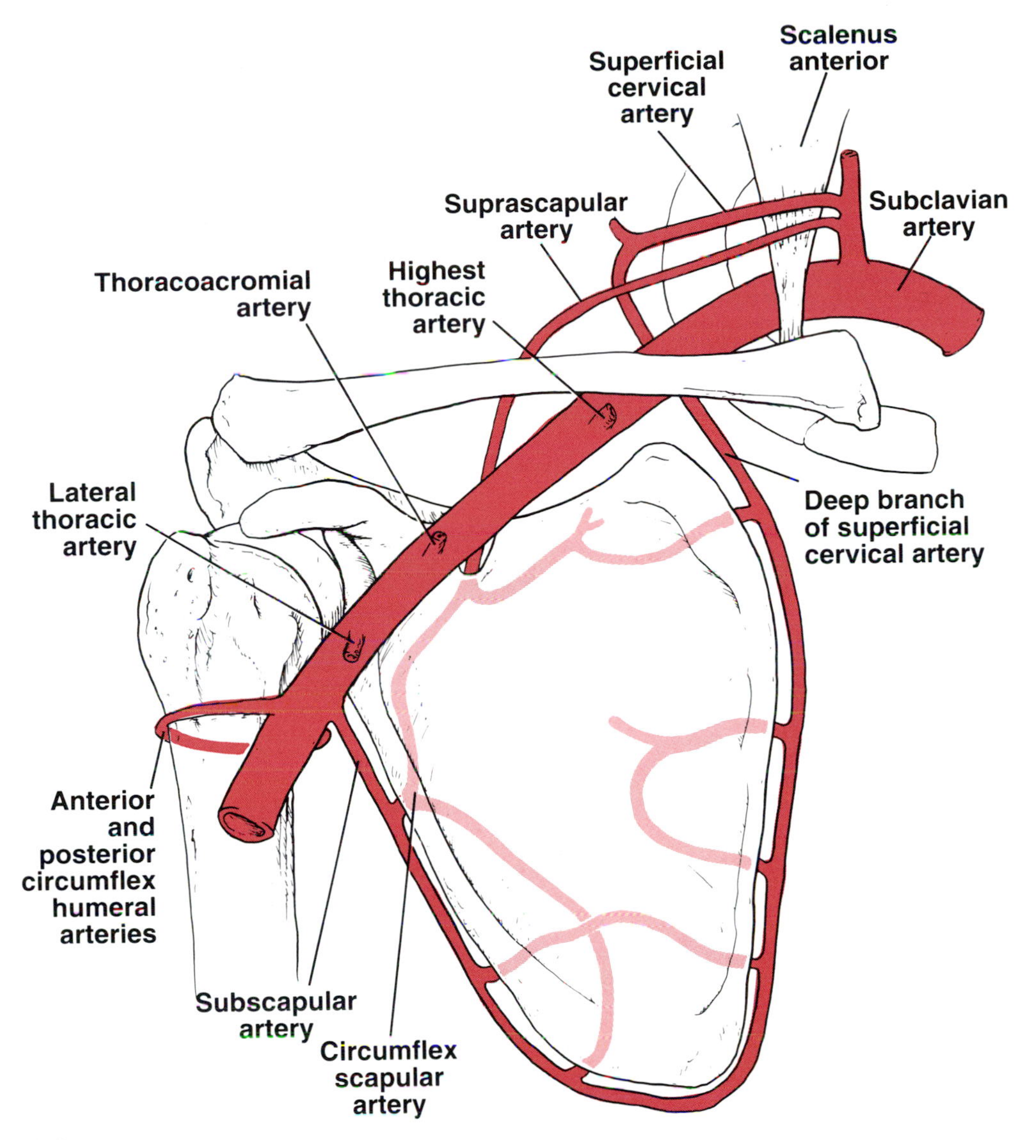

FIG 4–8.
Arteries that take part in anastomosis around the shoulder joint.

Radial and ulnar arteries. Theses are terminal branches (see Fig 4–11).

Clinical Note

Arterial anastomosis around the elbow joint.—To compensate for temporary occlusion of the brachial artery during movements of the elbow joint, the following arteries anastomose with one another. The profunda brachii—the superior and inferior ulnar collateral arteries from the brachial artery anastomose with the radial and ulnar recurrent arteries and the posterior interosseous recurrent artery (branch of common interosseous artery from the ulnar artery) inferiorly (see Fig 4–11).

Radial Artery

The radial artery, one of the terminal branches of the brachial artery, begins in the cubital fossa at the level of the neck of the radius (Fig 4–12). It descends through the forearm lying superficially throughout most of its course. In the region of the wrist it winds backward, around the lateral side of the carpus to the proximal end of the space between the first and second metacarpal bones. Here it passes anteriorly into the palm between the two heads of the first dorsal interosseous muscle (Fig 4–13) and joins the deep branch of the ulnar artery, forming the *deep palmar arch.*

In the lower part of the forearm the radial artery lies on the anterior surface of the radius, covered only by skin and fascia. Here the artery has the ten-

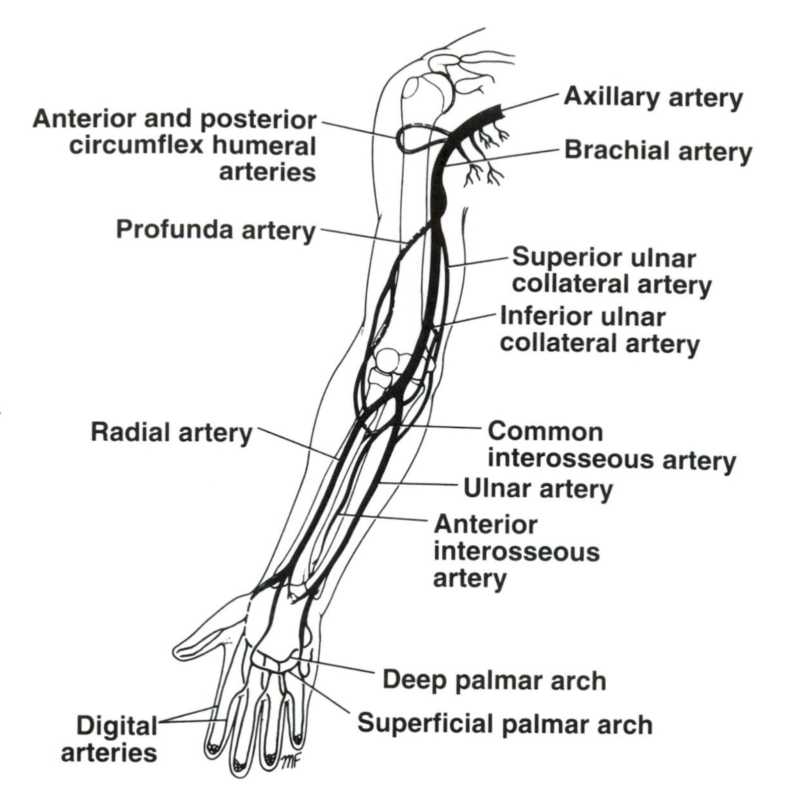

FIG 4–9.
The main arteries of the upper limb.

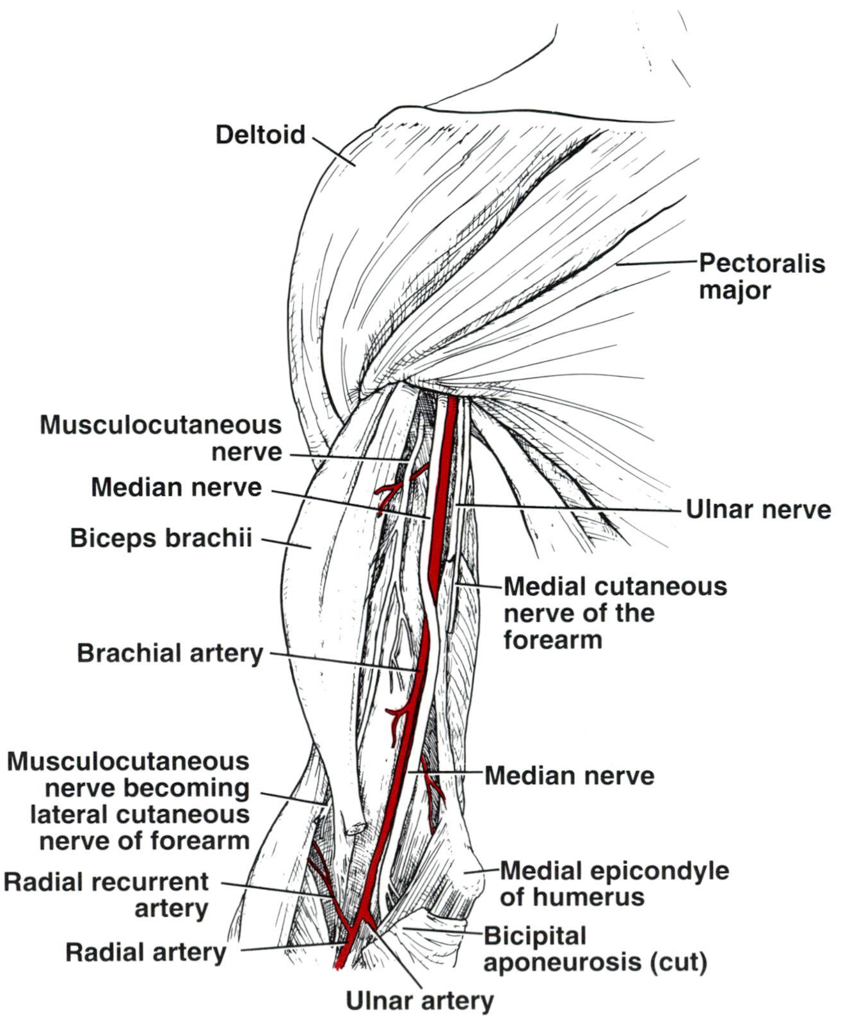

FIG 4–10.
Anterior view of the upper part of the arm. The biceps brachii has been pulled laterally to show the musculocutaneous nerve and the brachial artery.

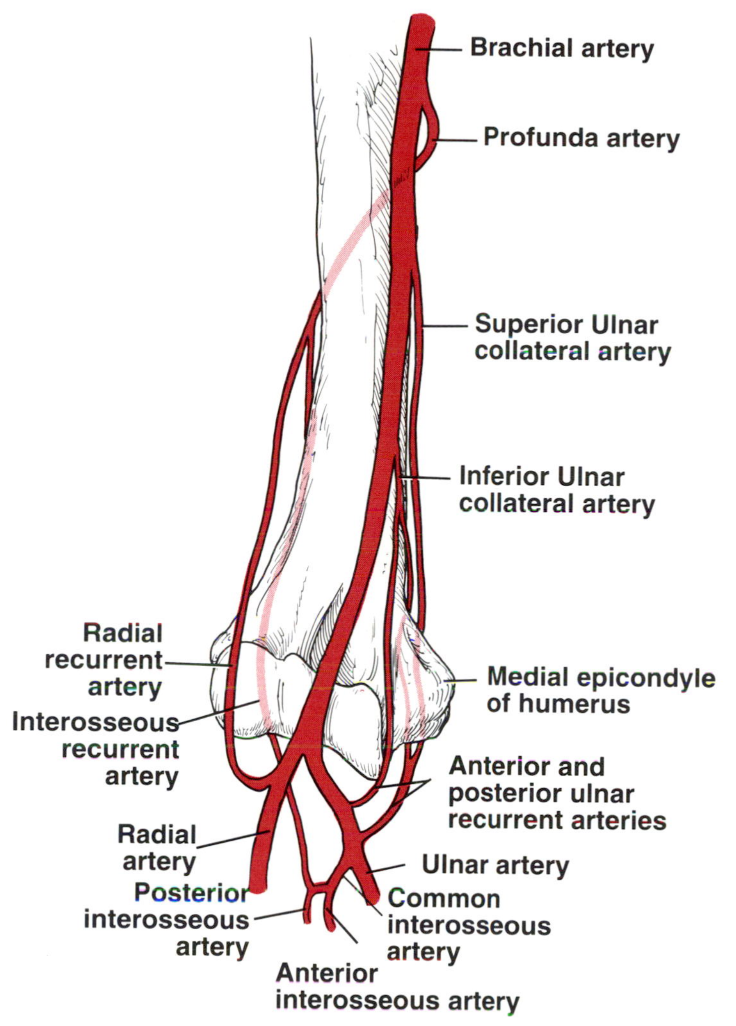

FIG 4–11.
Arteries that take part in anastomosis around the elbow joint.

don of the brachioradialis muscle on its lateral side and the tendon of the flexor carpi radialis on its medial side (see Fig 4–12). The radial pulse is taken at this site. In the middle third of its course the radial nerve lies on the lateral side of the artery.

Branches:

Muscular branches to surrounding muscles.

Recurrent branch that takes part in the arterial anastomosis around the elbow joint (see Fig 4–11).

The *superficial palmar branch* arises just above the wrist. It enters the palm and frequently joins the ulnar artery to form the *superficial palmar arch* (Fig 4–12).

The *first dorsal metacarpal artery* arises just before the radial artery passes between the two heads of the first dorsal interosseous muscle. It supplies the adjacent sides of the thumb and index finger.

The *arteria princeps pollicis* arises from the radial artery in the palm. It divides into two branches that supply the sides of the thumb (Fig 4–13).

The *arteria radialis indicis* arises from the radial artery in the palm. It supplies the lateral side of the index finger (see Fig 4–13).

Variations in the Origin of the Radial Artery.—High origins of the radial artery are common. These arise from the axillary artery or from the upper part of the brachial artery. The aberrant artery is commonly related to the median nerve as it descends to the cubital fossa. Here it may be connected to the brachial artery.

Deep Palmar Arch.—This arch is deeply placed in the palm and extends from the proximal end of the first interosseous space to the base of the fifth metacarpal bone (see Fig 4–13). It is formed by the radial

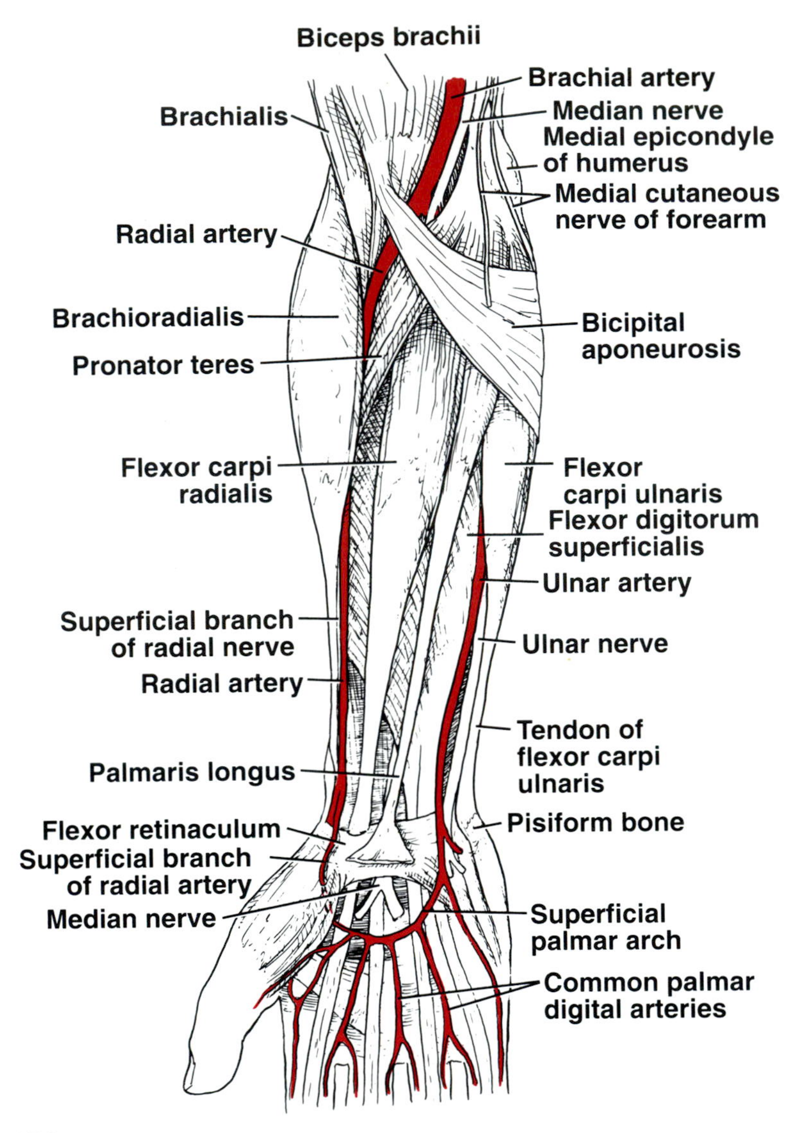

FIG 4–12.
Anterior view of the forearm and hand.

artery and terminates by anastomosing with the deep branch of the ulnar artery.

Branches.—The deep palmar arch gives off the palmar, metacarpal, perforating, and recurrent branches.

Ulnar Artery

The ulnar artery is the larger of the two terminal branches of the brachial artery (see Fig 4–12). It begins in the cubital fossa at the level of the neck of the radius and descends to the wrist and enters the palm *in front of the flexor retinaculum in company with the ulnar nerve*. It ends by forming the superficial palmar arch, often uniting with the superficial palmar branch of the radial artery.

In the upper part of its course, it passes deep to the flexor muscles that arise from the common flexor origin on the medial epicondyle of the humerus. In the lower part of its course, it becomes superficial and lies between the tendons of flexor carpi ulnaris and the tendons of flexor digitorum superficialis (see Fig 4–12). As the artery passes in front of the flexor retinaculum it is covered only by skin and fasciae and lies just lateral to the pisiform bone. It is here that the pulsations of the ulnar artery can be palpated.

Branches:

Muscular branches to the surrounding muscles.

Recurrent branches take part in the arterial anastomosis around the wrist joint.

The *common interosseous artery* arises from the upper part of the ulnar artery and quickly divides into

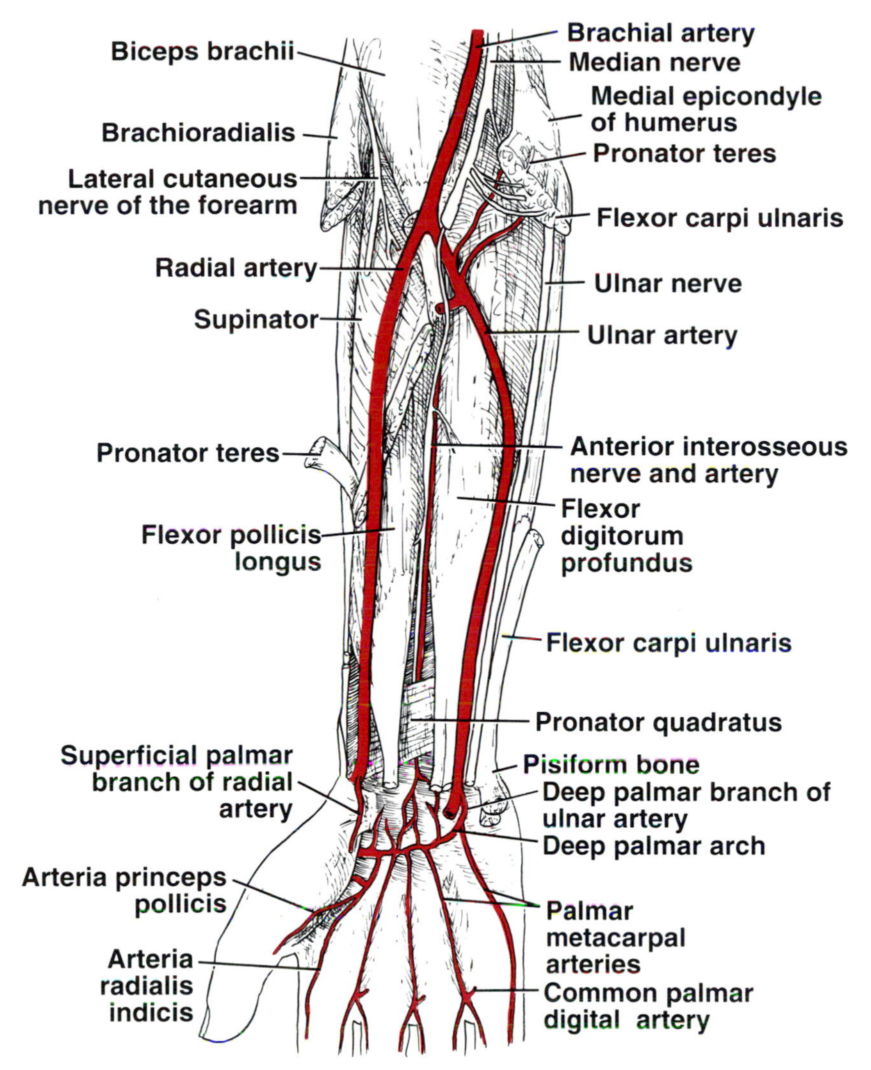

FIG 4–13.
Anterior view of the forearm and hand showing deep structures.

the *anterior and posterior interosseous arteries;* which descend on the anterior and posterior surfaces of the interosseous membrane, respectively (see Fig 4–11).

The *deep palmar branch* arises in front of the flexor retinaculum and joins the radial artery to complete the deep palmar arch (see Fig 4–13).

Superficial Palmar Arch.—This arterial arch lies immediately beneath the palmar aponeurosis and on the long flexor tendons (see Fig 4–12). It is a direct continuation of the ulnar artery. The arch is often completed on the lateral side by the superficial palmar branch of the radial artery. The curve of the arch lies across the palm, level with the distal border of the extended thumb. Four digital arteries arise from the arch and supply the ulnar side of the little finger and the adjacent sides of the little, ring, middle, and index fingers, respectively.

Nerve Supply of the Arteries of the Upper Extremity
The arteries of the upper limb are innervated by sympathetic nerves. The preganglionic fibers originate from cell bodies in the second to eighth thoracic segments of the spinal cord. They ascend in the sympathetic trunk and synapse in the middle cervical, inferior cervical, first thoracic, or stellate ganglia. The postganglionic fibers join the nerves that form the brachial plexus and are distributed to the arteries within the branches of the plexus. For example, the digital arteries of the fingers are supplied by postganglionic sympathetic fibers that run in the digital nerves.

Basic Anatomy of the Large Veins of the Thorax

The Brachiocephalic Veins

The *right brachiocephalic vein* is formed at the root of the neck by the union of the right subclavian and the right internal jugular veins (Fig 4–14). The *left brachiocephalic vein* has a similar origin on the left side of the root of the neck. It passes obliquely downward and to the right behind the manubrium sterni and in front of the large branches of the aortic arch. It joins the right brachiocephalic vein to form the superior vena cava (see Fig 4–14).

Tributaries.—Each brachiocephalic vein has the following tributaries: *vertebral vein, internal thoracic vein, inferior thyroid vein,* and the *first posterior intercostal vein.*

Superior Vena Cava

The superior vena cava collects all the venous blood from the head and the neck and both upper limbs and is formed by the union of the two brachiocephalic veins (see Fig 4–14). It descends vertically to end in the right atrium of the heart. The azygos vein joins the posterior aspect of the superior vena cava just before it pierces the pericardium (see Fig 4–14).

Tributaries.—The right and left brachiocephalic veins.

Azygos Veins

The azygos veins consist of the main azygos vein, the inferior hemiazygos vein, and the superior hemiazygos vein. They drain blood from the posterior parts of the intercostal spaces, the posterior abdominal wall, the pericardium, the diaphragm, the bronchi, and the esophagus (Fig 4–15).

Azygos Vein.—The origin of this vein is variable. It is often formed by the union of the *right ascending lumbar vein* and the *right subcostal vein.* It ascends through the aortic opening in the diaphragm on the right side of the aorta to the level of the fifth thoracic vertebra (see Fig 4–15). Here it arches forward above the root of the right lung to empty into the posterior surface of the superior vena cava (see Fig 4–14).

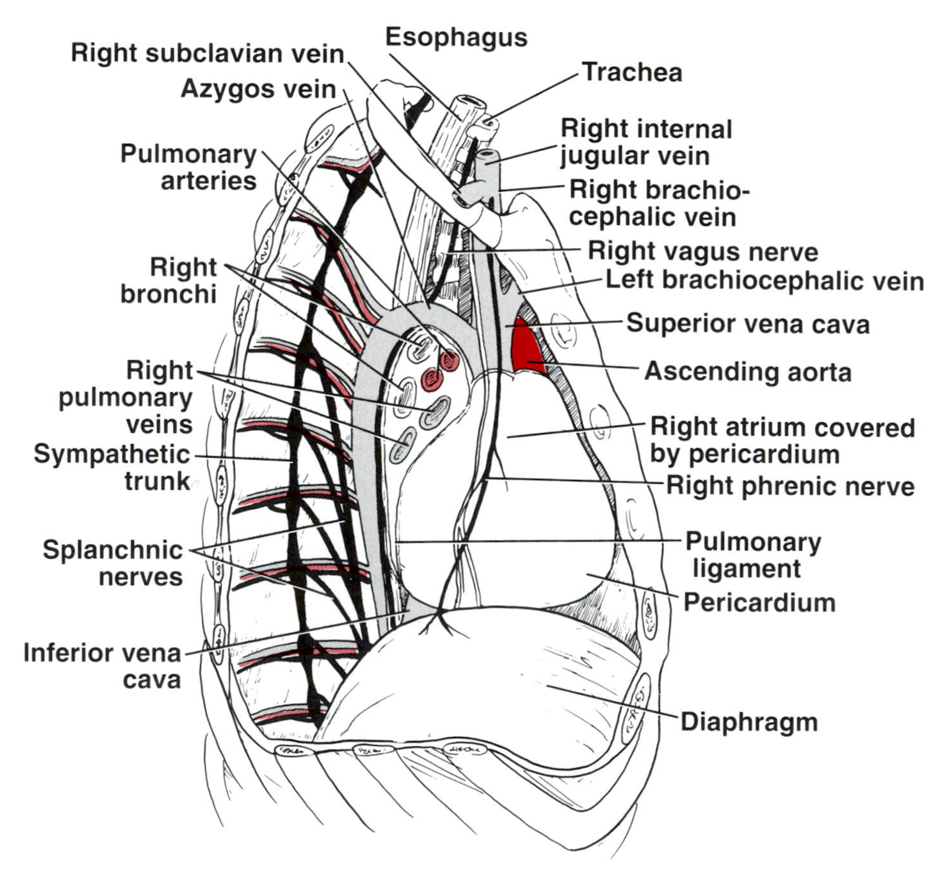

FIG 4–14.
Right side of the mediastinum. Note the position of the hilum of the right lung and the azygos vein draining into the superior vena cava.

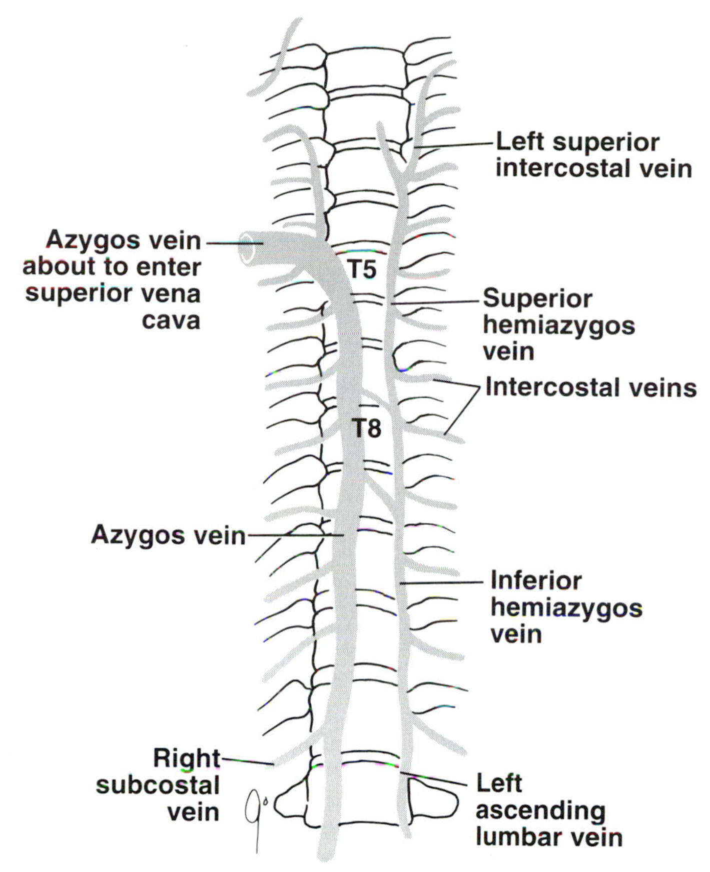

FIG 4–15.
The azygos veins.

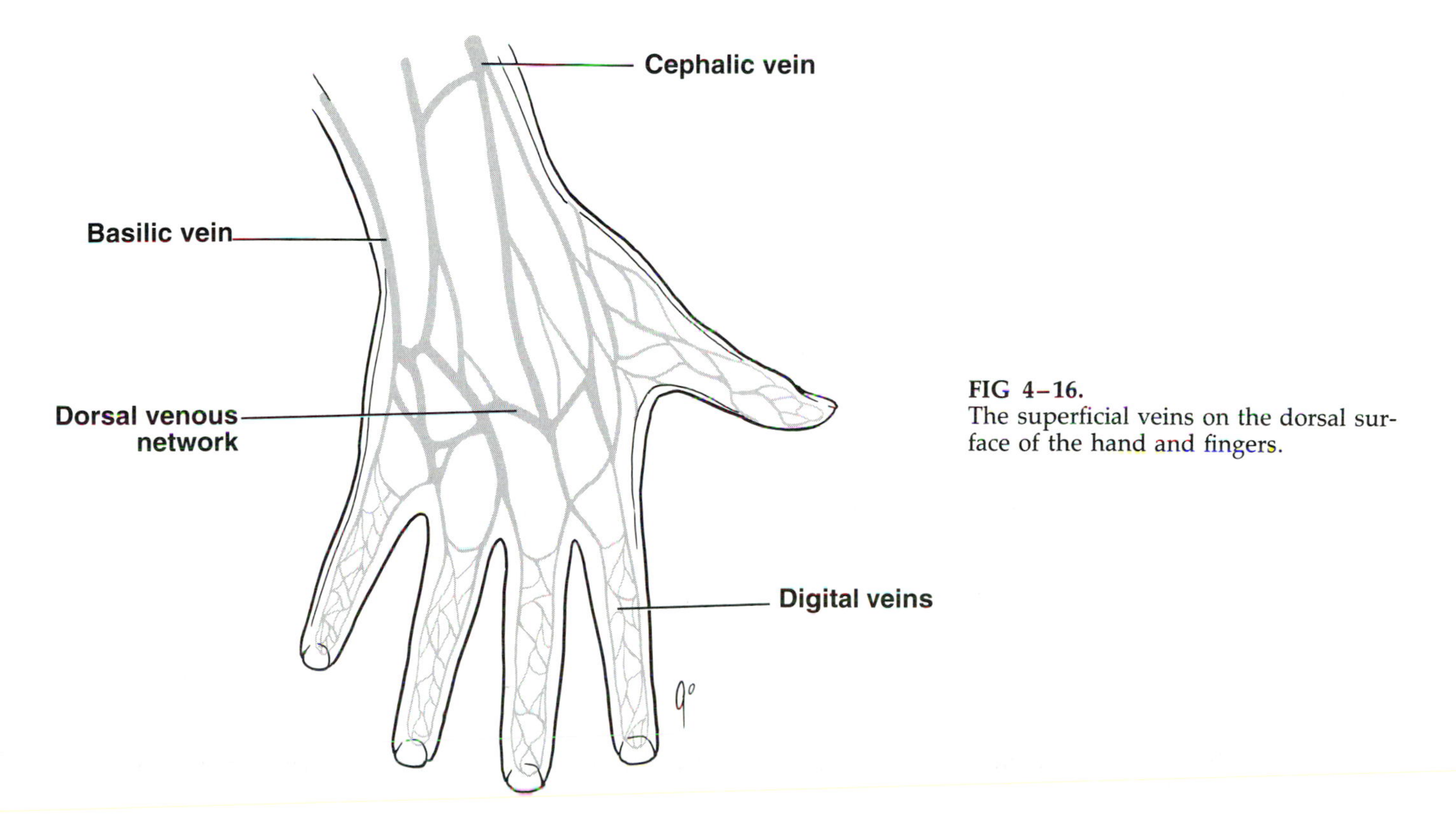

FIG 4–16.
The superficial veins on the dorsal surface of the hand and fingers.

The azygos vein has numerous tributaries that include the *eight lower right intercostal veins,* the *right superior intercostal vein,* the *superior and inferior hemiazygos veins,* and numerous *mediastinal veins.*

Inferior Hemiazygos Vein.—This vein is often formed by the union of the left ascending lumbar vein and the left subcostal vein. It ascends through the left crus of the diaphragm, and at about the level of the eighth thoracic vertebra turns to the right and joins the azygos vein (see Fig 4–15). It receives as tributaries some *lower left intercostal veins* and *mediastinal veins.*

Superior Hemiazygos Vein.—This vein is formed by the union of the fourth to the eighth intercostal veins. It joins the azygos vein at the level of the seventh thoracic vertebra (Fig 4–15).

Inferior Vena Cava

The inferior vena cava is formed in the abdomen and perforates the tendinous part of the diaphragm at the level of the eighth thoracic vertebra (see Fig 4–14). It passes through the pericardium and opens into the lower and back part of the right atrium of the heart. The valve of the inferior vena cava is important in the fetus but rudimentary in the adult (see p. 105).

Pulmonary Veins

There are four pulmonary veins (two on each side). They carry oxygenated blood from the lungs and empty into the left atrium (see Figs 4–3 and 4–14). There are no valves.

Basic Anatomy of the Veins of the Upper Extremity

The veins of the upper limb may be divided into superficial and deep groups. The superficial veins are of great clinical importance and lie in the superficial fascia. The deep veins accompany the main arteries.

Superficial Veins

Dorsal Venous Arch Or Network.—This venous network of superficial veins lies on the dorsum of the hand (Fig 4–16). It receives most of the blood from the hand and the fingers. It is drained on the lateral side by the cephalic vein and on the medial side by the basilic vein.

Cephalic Vein.—This vein arises from the lateral side of the dorsal venous network (Fig 4–17). It ascends around the lateral border of the forearm, just lateral to the styloid process of the radius, to reach the anterior aspect of the forearm. It then ascends into the arm and runs along the lateral border of the biceps to reach the interval between the deltoid and pectoralis major muscles. The cephalic vein ends by piercing the deep fascia to enter the axillary vein.

Basilic Vein.—This vein arises from the medial side of the dorsal venous network and ascends on the posterior surface of the forearm (see Fig 4–17). Just below the elbow it inclines forward to reach the anterior aspect of the forearm. The vein then runs upward medial to the biceps and pierces the deep fascia at about the middle of the arm. The *median cubital vein* links the cephalic and basilic veins in the cubital fossa. The basilic vein joins the venae comitantes, which accompany the brachial artery, at the lower border of the teres major muscle to form the axillary vein.

Median Vein of the Forearm.—This small vein arises in the palm and ascends on the front of the forearm (see Fig 4–17). It drains into the basilic vein, or the median cubital vein, or divides into two branches, one of which joins the basilic vein *(median basilic vein)* and the other joins the cephalic vein *(median cephalic vein).*

Deep Veins

Venae Comitantes.—The deep veins accompany the respective arteries as venae comitantes. The two venae comitantes of the brachial artery join the basilic vein at the lower border of the teres major to form the axillary vein.

Axillary Vein.—This vein is formed at the lower border of the teres major muscle on the posterior wall of the axilla by the union of the venae comitantes of the brachial artery with the basilic vein (see Fig 4–7). It becomes the subclavian vein at the outer border of the first rib. The axillary vein receives tributaries that correspond to the branches of the axillary artery and possesses several valves.

Subclavian Vein.—Although strictly a vein of the root of the neck, it will be considered here. This vein is a continuation of the axillary vein at the outer border of the first rib (see Fig 4–7). It joins the internal jugular vein behind the medial end of the clavicle to form the brachiocephalic vein. The subclavian vein measures about 2 cm in diameter. The subclavian vein is anterior to the subclavian artery and is separated from it by the scalenus anterior muscle. The

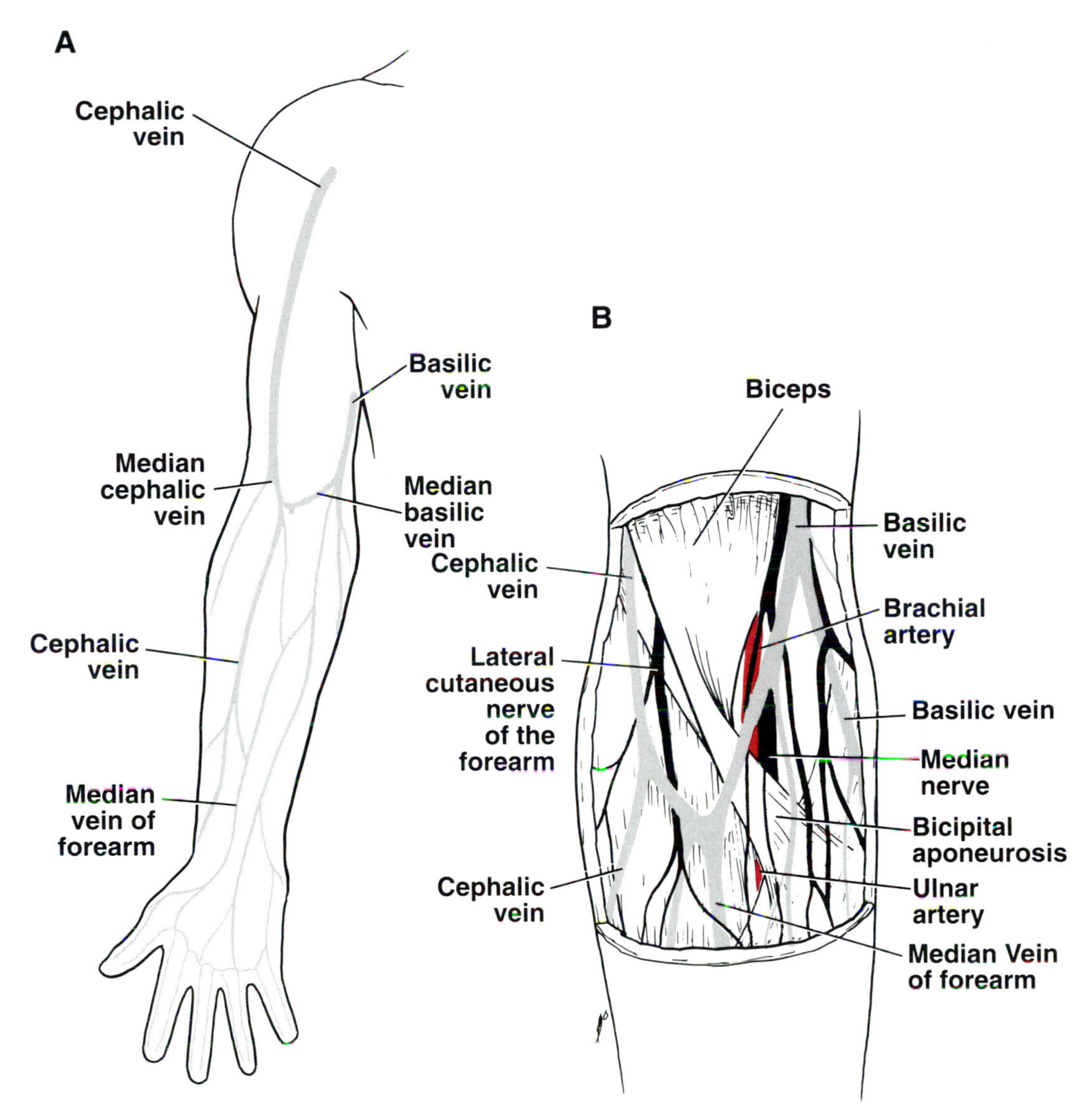

FIG 4–17.
A, the superficial veins of the upper limb. **B,** the superficial veins lying in front of the elbow region. Note the position of the brachial artery and the median nerve as they lie in the cubital fossa.

vein is related above and behind to the roots of the brachial plexus and below and behind to the cervical dome of the pleura. The subclavian vein receives the external jugular vein, and it often receives the *thoracic duct* on the left side, and the *right lymphatic duct* on the right side.

Nerve Supply of the Veins of the Upper Extremity

Like the arteries, the smooth muscle in the wall of the veins is innervated by sympathetic postganglionic nerve fibers that provide vasomotor tone. The origin of these fibers is similar to those of the arteries.

Clinical Note

In extreme hypovolemic shock, excessive venous tone may inhibit venous blood flow and thus delay the introduction of intravenous blood into the vascular system.

SURFACE ANATOMY

Surface Anatomy of the Large Arteries of the Thorax

Aorta

The *ascending aorta* lies behind the right half of the sternum just below the sternal angle (see Fig 3–14).

The *arch of the aorta* and the roots of the *brachiocephalic* and *left common carotid arteries* lie behind the manubrium sterni (see Fig 3–14). The *descending thoracic aorta* begins at the termination of the arch to the left of the midline at the level of the sternal angle. As it descends, the aorta deviates to the midline and passes through the aortic opening in the diaphragm at the level of the 12th thoracic vertebra.

The *pulmonary trunk* bifucates into the *right* and *left pulmonary arteries* to the left of the midline at about the level of the sternal angle (see Fig 3–14).

Surface Anatomy of the Arteries of the Upper Extremity

Subclavian Artery

The subclavian artery, as it crosses over the first rib to become the axillary artery, may be palpated in the root of the posterior triangle of the neck (see Fig 5–5).

Axillary Artery

The *first and second parts of the axillary artery* cannot be palpated since they lie deep to the pectoral muscles. The *third part of the axillary artery* may be felt in the axilla, where it lies in front of the teres major muscle (see Fig 4–7).

Brachial Artery

The brachial artery can be palpated in the arm as it lies on the brachialis muscle and is overlapped from the lateral side by the biceps brachii muscle (see Fig 4–10).

Radial Artery

The radial artery lies superficially in front of the distal end of the radius, between the tendons of the brachioradialis and the flexor carpi radialis muscles (see Fig 4–12). It is here that the radial pulse can be easily felt. If the pulse cannot be detected, try feeling for the radial artery on the other wrist; occasionally a congenitally abnormal radial artery may be difficult to feel. The radial artery may be less easily felt as it crosses the anatomical snuffbox.

Ulnar Artery

The ulnar artery may be palpated as it crosses anterior to the flexor retinaculum along with the ulnar nerve. The artery lies lateral to the pisiform bone, separated from it by the ulnar nerve (see Fig 4–13).

Surface Anatomy of the Large Veins of the Thorax

The *superior vena cava* and the terminal parts of the *right* and *left brachiocephalic veins* lie behind the manubrium sterni.

Surface Anatomy of the Superficial Veins of the Upper Extremity

Cephalic Vein

Wrist Region.—The cephalic vein crosses the anatomical snuffbox (see Fig 4–17) and winds around the lateral side of the forearm to reach the anterior aspect. It is constantly found in the superficial fascia posterior to the styloid process of the radius.

Elbow Region.—The cephalic vein ascends into the arm along the lateral border of the biceps muscle (see Fig 4–17).

Shoulder Region.—The cephalic vein lies in the groove between the deltoid and pectoralis major muscles (see Fig 4–17).

Basilic Vein

Wrist Region.—The basilic vein ascends from the dorsum of the hand by slowly curving around the medial side of the forearm to reach the cubital fossa (see Fig 4–17).

Elbow Region.—The basilic vein ascends from the cubital fossa along the medial border of the biceps muscle; it pierces the deep fascia at about the middle of the arm (see Fig 4–17).

Median Cubital Vein

This vein links the cephalic vein to the basilic vein in the cubital fossa. It is separated from the brachial artery at this site by the bicipital aponeurosis. In about 30% of individuals the median cubital vein is replaced by the median cephalic and median basilic veins (see Fig 4–17).

Median Vein of the Forearm

This vein ascends on the anterior aspect of the forearm along a variable course (see Fig 4–17). It joins either the basilic vein, the median cubital vein, or divides into the median cephalic and median basilic veins.

CLINICAL ANATOMY

Clinical Anatomy of the Large Arteries of the Thorax

Coarctation of the Aorta

Coarctation of the aorta is a congenital narrowing of the aorta just proximal to, opposite, or distal to the site of attachment of the ligamentum arteriosum (Fig 4–18). This condition is believed to result from an unusual quantity of ductus arteriosus muscle tissue being present in the wall of the aorta. When the ductus arteriosus contracts, the ductal muscle in the aortic wall also contracts and the aortic lumen becomes narrowed. Later, when fibrosis takes place, the aortic wall also is involved and permanent narrowing occurs.

Clinically, the cardinal sign of aortic coarctation is absent (extremely rare) or diminished pulses in the femoral arteries of both lower limbs. To compensate for the diminished volume of blood reaching the lower part of the body, an enormous collateral circulation develops, with dilatation of the internal thoracic, subclavian, and posterior intercostal arteries. The condition should be treated surgically.

Blunt Trauma to the Aorta in the Thorax

Deceleration injuries.—The heart is suspended from the aorta much like a weight attached to the end of a curved piece of flexible tubing.

In *horizontal deceleration injuries* the movement of the body is suddenly stopped and the heart continues to move forward within the pericardial sac. The descending thoracic aorta from the point of origin of the left subclavian artery onward is firmly attached to the vertebral column by connective tissue and pleura. As the heart continues to move forward on body impact, the curve of the arch of the aorta is slightly straightened and the vessel shears just distal to the origin of the left subclavian artery. The aorta may be completely *transected* and be accompanied by a massive hemorrhage. If the trauma to the aorta is less, a tear may occur in the tunical intima and media, leaving the more resilient tunica adventitia intact. Blood then dissects between the media and the adventitia, producing a false aneurysm.

In *vertical deceleration injuries,* as in falls from a height, the violent pull of the heart on the end of the aortic arch causes an intimal tear at the root of the ascending aorta. If rupture of the ascending aorta should ensue, the aortic blood would burst into the pericardial sac, causing immediate cardiac tamponade. The ascending aorta together with the pulmonary trunk is surrounded only by a thin sleeve of serous pericardium and by the fibrous pericardium. Once the sleeve of serous pericardium gives way, the blood enters the pericardial cavity, causing cardiac tamponade. Most of these injuries are fatal.

Because of the relationship of the aorta to surrounding structures, a chest radiograph of a patient with aortic rupture may show the following abnormalities:

1. Widening of the mediastinal shadow (capping with deviation of the nasogastric tube)
2. Trachea deviation to the right
3. Right main bronchus displacement inferiorly
4. Aortic knob unclear
5. Fluid in the left pleural cavity.

It must be stressed that a definitive diagnosis is made with a computed tomographic scan or aortogram.

Penetrating Injuries of the Aorta in the Thorax

Penetrating vascular injuries are likely to occur when the entrance and exit wounds are on opposite sides of the chest, indicating that the bullet has crossed the midline of the chest. Penetrating injuries involving blood vessels and not the heart have the highest mortality when they occur in the superior mediastinum; here the aortic arch gives rise to its major branches, and the superior vena cava and brachiocephalic veins are present.

Aortic Dissection

This is the most serious and difficult form of aortic disease to treat (Fig 4–19). Degeneration of the tu-

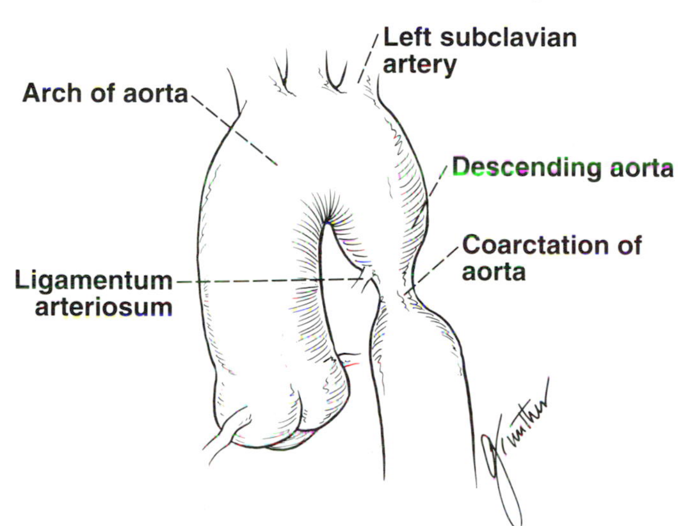

FIG 4–18.
Coarctation of the aorta.

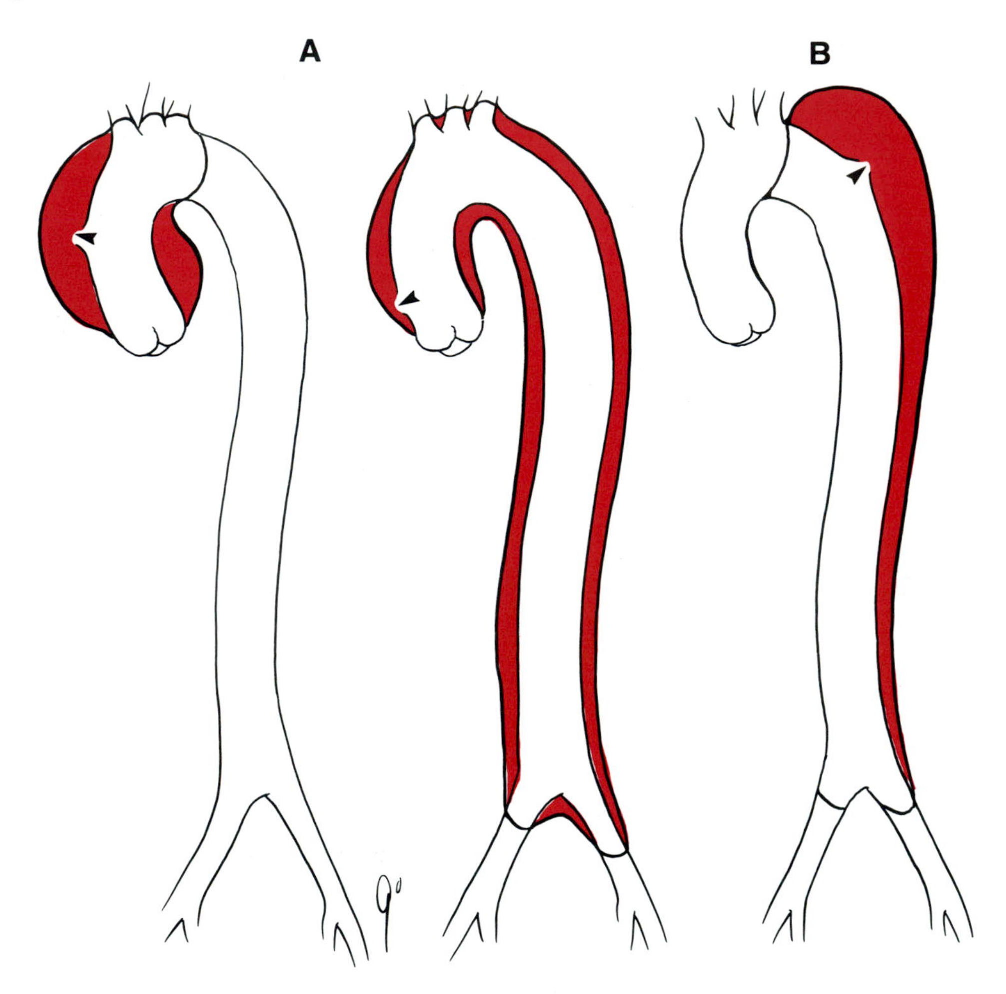

FIG 4–19.
Aortic dissection. Type A (proximal dissection) involves the ascending aorta or ascending and descending aorta. Type B (distal dissection) does not involve the ascending aorta.

nica media is believed to be the basic cause, and the condition is associated with a history of hypertension in the majority of patients; Marfan's disease can also be responsible for the degeneration. Hemorrhage occurs through the tunica intima and extends as an expanding hematoma between the middle and outer thirds of the tunica media. The initial tear may occur anywhere along the thoracic aorta. Type A (proximal dissection) involves the ascending aorta or ascending and descending aorta (see Fig 4–19). Type B (distal dissection) does not involve the ascending aorta and usually begins distal to the left subclavian artery. The clinical signs and symptoms will depend on the type of aneurysm present and the extent of distal propagation.

The sudden onset of excruciating, sharp, tearing pain localized to the front or the back of the chest and back is a typical presenting symptom. It must be assumed that the pain impulses originate in the aortic wall, ascend to the central nervous system along with the sympathetic nerves, and enter the spinal cord through the posterior roots of the segmental spinal nerves. The pain is then referred along the somatic spinal nerves of the same segments. The number of dermatomes involved will depend on the extent of the dissection. If the dissection extends around the aortic arch and involves the brachiocephalic and carotid arteries, the pain may be felt in the neck. If the dissection continues to spread distally, the pain may be felt segmentally in the abdomen, lower back, and legs.

The patient's clinical presentation will depend on the site and extent of the dissection. An aneurysm of the ascending aorta has the highest mortality since it may rupture into the pericardial cavity, producing immediate cardiac tamponade; it may rupture into the mediastinum or pleural cavities; it may extend into the coronary arteries, causing occlusion; or it may extend to the aortic valve, producing acute aortic incompetence. Involvement of the brachiocephalic, left common carotid, or left subclavian arteries at their origin from the aortic arch may give rise to symptoms of cerebral ischemia or upper limb ischemia.

An aneurysm of the descending thoracic aorta can rupture into the pleural cavity on the left side. If the dissection progresses distally to involve the abdominal aorta, occlusion of the mesenteric arteries could result in bowel infarction, and occlusion of the renal arteries could result in renal failure. Ischemic necrosis of the spinal cord resulting in paraplegia could follow blockage of the posterior intercostal arteries arising from the thoracic aorta or the lumbar arteries arising from the abdominal aorta (see blood supply of spinal cord).

Rupture of the aneurysm below the diaphragm may produce catastrophic retroperitoneal hemorrhage.

Clinical Anatomy of the Procedure for Thoracic Aortic Occlusion

Cross-clamping of the aorta just proximal to the diaphragm may be necessary in patients with massive hemorrhage into the peritoneal cavity following penetrating trauma to the abdominal cavity.

Through a left-sided thoracotomy (for details, see p. 92), the descending thoracic aorta is identified by palpation as it lies on the anterior surface of the thoracic vertebrae just above the diaphragm. The aorta descends behind the diaphragm at the level of T12. Because the aortic pulse may be difficult to feel in a severely hypovolemic patient, it may be difficult to distinguish between the aorta and the esophagus at first. The esophagus (Fig 4–20) lies anterior to the

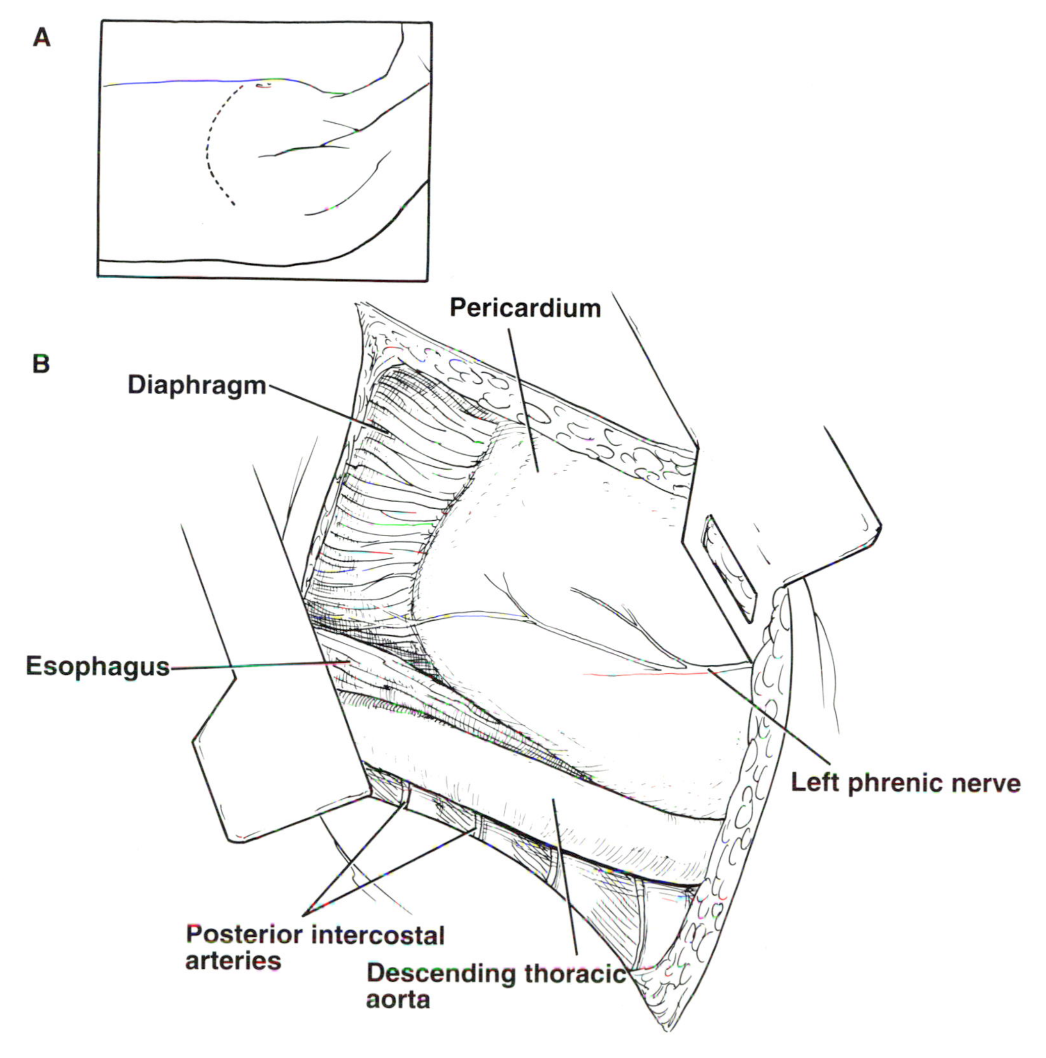

FIG 4–20.
Exposure for aortic cross-clamping. **A,** site for thoracotomy incision. **B,** the chest cavity has been opened and the lung retracted in a superior-medial direction (not shown for clarity). The mediastinal pleura has been removed to show the relationship between the esophagus and the descending thoracic aorta.

aorta in the posterior mediastinum; the esophagus then descends to the left of the aorta before perforating the diaphragm at the level of T10. Because the aorta is covered on its anterolateral surface by the mediastinal pleura, the pleura has to be incised before the vascular clamp is inserted.

Anatomical Complications of Aortic Clamping

The following complications are possible: left ventricular failure (blockage of outflow), bowel ischemia (lack of blood through distal mesenteric arteries), renal failure (lack of blood flow), or spinal cord ischemia (lack of blood through radicular branches of lower intercostal and lumbar arteries).

Clinical Anatomy of the Pulmonary Trunk

Penetrating Injuries to the Pulmonary Trunk

Any missile injury with entry or exit wounds close to the manubrium sterni may damage the pulmonary trunk or any other vessel in the superior mediastinum. Remember that the pulmonary trunk and the ascending aorta together are surrounded by a sheath of serous pericardium within the fibrous pericardium (Figs 4–2 and 4–3), so that a hemorrhage into the pericardial cavity from either of these vessels could result in cardiac tamponade.

Pulmonary Artery (Swan-Ganz) Catheterization

The pulmonary artery catheter is used to assess left and right ventricular function; measure pulmonary artery, pulmonary capillary wedge, and right and left atrial pressures; measure cardiac output; and take samples of right atrial and pulmonary arterial blood.

The catheter is introduced through the right internal jugular vein, right subclavian vein, right basilic vein, or the femoral vein (see pp. 160, 196, and 569). The catheter is advanced into the right atrium, through the tricuspid valve into the right ventricle (Fig 4–21). The balloon is advanced into the pulmonary trunk and then into the left pulmonary artery.

Pulmonary Embolism

In the great majority of patients the pulmonary emboli arise from a thrombosis in the large deep veins of the lower extremity, especially from the femoral veins, and from the internal iliac veins in the pelvis. Thromboses in the veins of the calf muscles and the superficial veins of the lower limb are now believed to be unlikely sites for the origin of pulmonary emboli.

Clinical Anatomy of the Arteries of the Upper Extremity

Allen Test

This test may be used to determine the patency of the ulnar and radial arteries. However, the validity of this test has been questioned in the literature.

With the patient's hands resting in his lap, the radial arteries are compressed against the anterior surface of each radius. The patient then tightly clenches his fists, which closes off the superficial and deep palmar arterial arches. When he opens his hands, the skin of the palms is at first white, and then normally the blood quickly flows into the arches through the ulnar arteries, causing the palms to promptly turn pink. This establishes that the ulnar arteries are patent. The patency of the radial arteries can be established by repeating the test, only with the ulnar arteries compressed where they lie lateral to the pisiform bone.

Penetrating Arterial Injuries in the Upper Extremity

Arterial injuries of the upper limb are common and may occur from guns, knives, automobile accidents, and iatrogenic causes. Basically three types of arterial injury are possible, and the structure of the artery determines the signs and symptoms as well as the type of treatment instituted.

Completely Severed Artery.—In a completely severed artery the circular smooth-muscle fibers of the tunica media contract, immediately slowing the bleeding. In addition, the elastic fibers and longitudinal smooth-muscle fibers of the media contract, causing the ends of the artery to retract. The contraction and retraction of the arterial ends usually slow the blood flow to such an extent that bleeding ceases spontaneously, and a firm blood clot plugs both ends of the severed artery. The loss of distal pulses is immediate.

Partially Severed Artery.—In a partially severed artery the vessel is unable to contract and retract; in fact, any retraction that does occur causes the arterial wound to gape, resulting in serious bleeding. Hemorrhage into the surrounding tissues may produce an enlarging pulsatile hematoma that may slowly expand along fascial planes to reach the surface and cause a severe hemorrhage. Another possibility is that the damaged arterial wall gives way, leaving only the tunica adventitia intact. In these circumstances a *pseudoaneurysm* is formed.

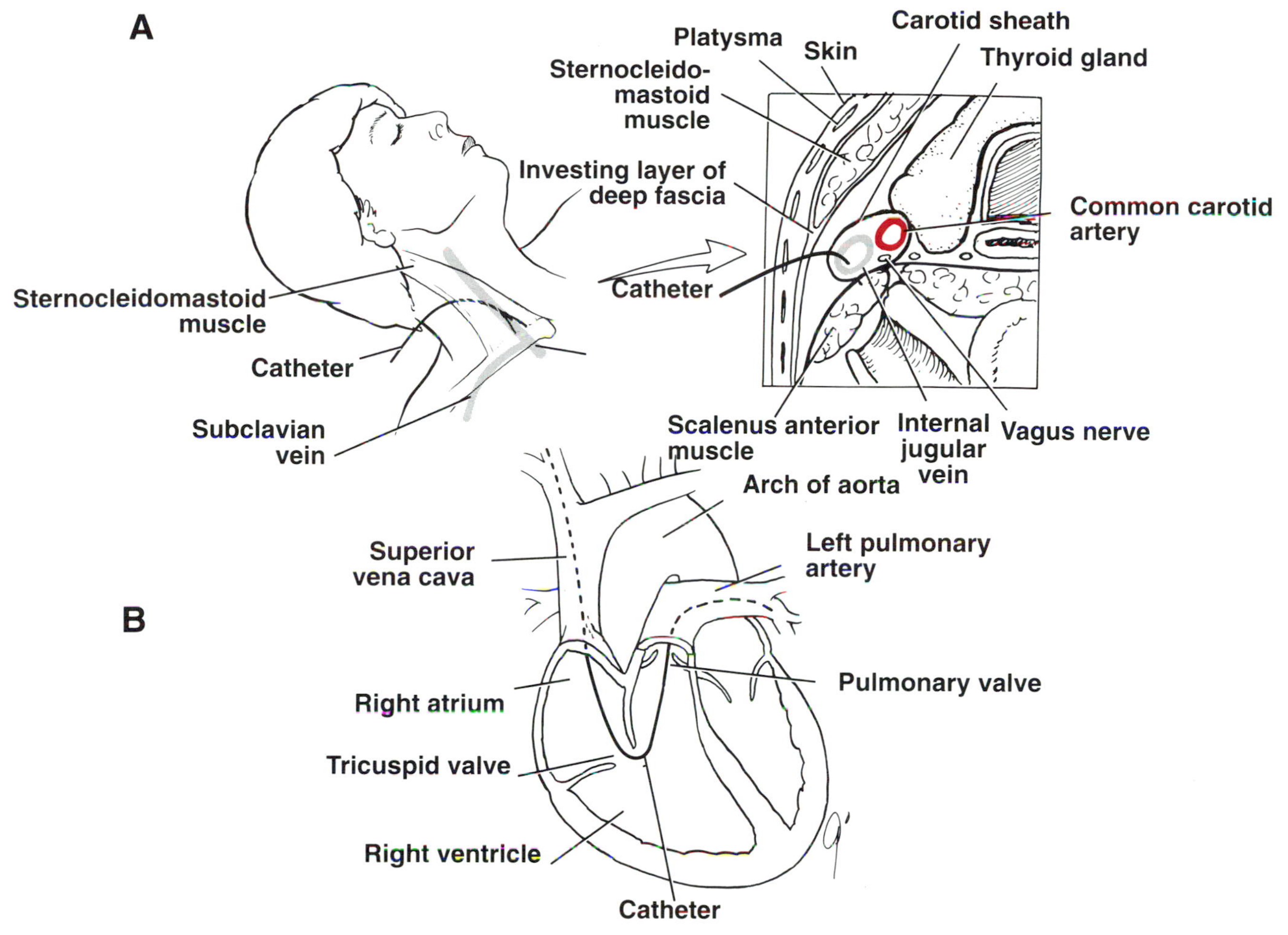

FIG 4–21.
Pulmonary artery (Swan-Ganz) catheterization. **A,** the catheter has been introduced through the right internal jugular vein (posterior approach). **B,** diagram showing the course taken by the catheter through the right side of the heart into the pulmonary trunk and the left pulmonary artery.

Since with partial arterial injury the arterial wall is still intact, blood flow continues into the distal end, and a distal pulse is usually recognizable.

Intimal Damage Only.—In an artery with intimal damage secondary to external blunt trauma, excessive stretching, or internal damage from a catheter, there is a reduction in blood flow and an absence of external hemorrhage. Later, as the result of progressive thrombosis or bleeding into the wall at the site of injury, the blood flow becomes diminished and the distal pulses disappear.

Anatomy of Arterial Injuries in the Upper Extremity.—Many of the injuries are directly related to the anatomy of the upper limb. The extreme mobility of the limb at the shoulder joint permits the forearm and hand to be raised as a shield to ward off an attack. This position of the arm commonly results in laceration of the blood vessels.

Injuries to the axillary artery are often caused by penetrating wounds in the pectoral region, fractures of the surgical neck of the humerus, or by excessive stretching following anterior dislocations of the shoulder joint (see Fig 4–8). Damage to the branches of the brachial plexus may be an added complication.

Injuries to the brachial artery may follow supracondylar fractures (see Chapter 15, p. 612) of the lower end of the humerus (especially in children), a site where the artery is close to the shaft of the humerus as it lies on the brachialis muscle (see Fig 4–11). Damage to the adjacent median nerve may also occur. Severe dislocations of the elbow joint may damage the artery in the cubital fossa. Since the brachial artery is located superficially in the upper part of the arm, it is a common site for arterial catheterization. Frequently the tunica intima is damaged on the wall opposite the penetration site, and arterial thrombosis may follow.

Injuries to the radial artery are common and occur in the lacerations of the front of the forearm. The close relation of the artery to the radial nerve and the forearm tendons means that these structures are also commonly damaged (see Fig 4–13). Catheter injuries are also common just proximal to the wrist joint.

Injuries to the ulnar artery are relatively common and occur in lacerations in the front of the wrist and flexor retinaculum. Here the artery is superficial and easily cut in glass or knife wounds (see Fig 4–13). The close relationship of the artery to the ulnar nerve results in frequent involvement of the nerve.

Intra-arterial Injection by Drug Abusers.—In cases of intra-arterial injection by drug abusers, the patient presents with a swollen and extremely painful hand (because of the fear of being apprehended, usually several hours after the injection). The drug is often injected into the radial or ulnar arteries. The pharmacological agent and its diluent causes damage at the arteriole level as the result of blockage by crystals, chemical necrosis of the tunica intima, or vasospasm of the smooth muscle in the tunica media. Extensive tissue necrosis or gangrene may occur in the area of anatomical distribution of the artery injected.

Arteriovenous Fistulas.—Arteriovenous fistulas are common complications of arterial injuries in the upper limbs. This results from the close relationship that exists between the arteries and veins in the limbs. The axillary artery has the axillary vein on its medial side; the brachial, radial, and ulnar arteries have venae comitantes running alongside.

Arteriovenous fistulas occur when the penetrating arterial injury also perforates the accompanying vein. Bleeding from the artery follows the path of least resistance, and therefore an arteriovenous communication is established with resulting venous hypertension, varicosities, and edema distal to the communication site.

The Importance of the Collateral Circulation When Ligating Arteries in the Upper Limb.—The arteries of the upper limb may be damaged by penetrating wounds or may require ligation in amputation operations. Because of adequate collateral circulation around the shoulder, elbow, and wrist joints, ligation of the main arteries of the upper limb is not followed by tissue necrosis or gangrene, provided, of course, that the arteries forming the collateral circulation are not diseased and that the patient's general circulation is satisfactory. Nevertheless, it may take days or weeks for the collateral vessels to open up sufficiently to provide the distal part of the limb with the same volume of blood as previously provided by the main artery.

At the time of the injury, where there is a complete interruption of the main arterial supply, the collateral flow may be sufficient to prevent the signs of ischemia; it is rare, however, for the physician to be able to palpate a distal pulse at the initial examination.

Anatomy of the Procedure of Arterial Puncture.—The brachial, radial, and ulna arteries are commonly used for arterial puncture.

Brachial Artery.—The brachial artery is usually cannulated as it descends into the cubital fossa on the medial border of the biceps brachii muscle (see Fig 4–13). Unfortunately, the brachial artery has been associated with a higher incidence of postcatheterization thrombosis than the radial artery. This is probably because of the motility of the brachial artery associated with movements at the elbow joint.

Radial Artery.—With the radial artery it is first essential to determine the adequacy of the collateral circulation of the hand by performing the Allen test. This precaution is necessary in case thrombosis occurs during or after cannulation. The forearm is then supinated and the wrist joint is extended to approximately a 50° angle (see Fig 4–22). The radial artery can then be easily palpated as it lies anterior to the distal end of the radius.

Needle Approach.—The needle is passed through the arterial wall at an angle of about 25° to the anterior surface of the wrist (see Fig 4–22). The catheter can then be advanced into the arterial lumen and the needle withdrawn.

Cutdown Approach.—A transverse incision is made over the radial artery just above the proximal transverse skin crease at the wrist. The artery is gently mobilized on the anterior surface of the radius and the cannula is introduced.

Some physicians make a vertical incision for the radial artery cutdown to lessen the risk of cutting a nerve and also to avoid giving the patient a horizontal cut that could potentially be misinterpreted (as a suicide attempt) on the volar aspect of the forearm.

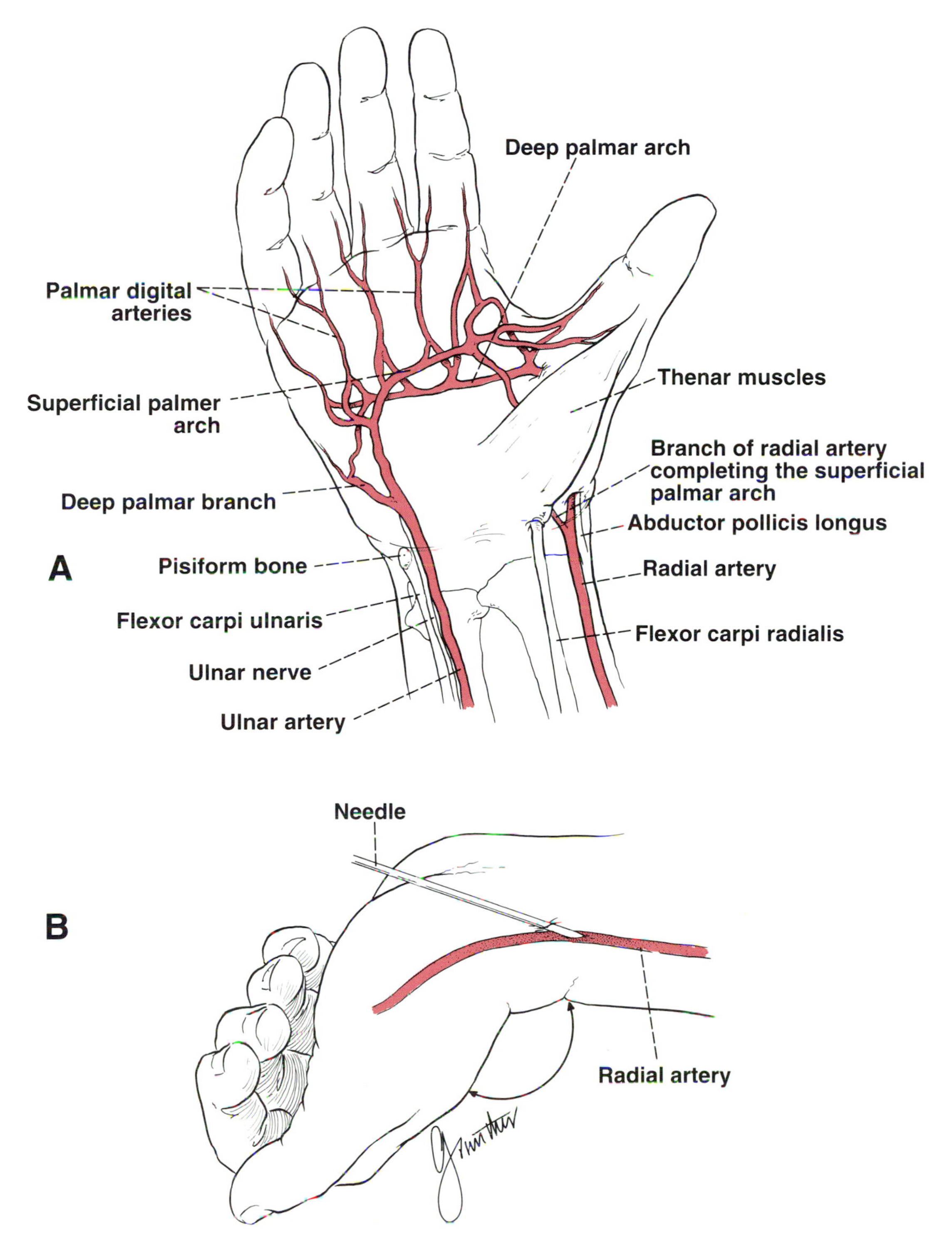

FIG 4–22.
A, the positions of the radial and ulnar arteries in front of the wrist. Note that the superficial palmar arch is formed mainly from the ulnar artery and that the deep palmar arch receives its major contribution from the radial artery. **B,** the wrist joint extended during cannulation of the radial artery.

Clinical Anatomy of the Veins of the Thorax

Penetrating Injuries to the Veins of the Thorax

As has been emphasized before with arterial injuries, the highest mortality can occur in penetrating injuries in the region of the superior mediastinum. Behind the manubrium sterni lie not only the aortic arch and its large branches but also the right and left brachiocephalic veins and the superior vena cava. Moreover, the thoracic cage may hide the extent of the bleeding, which may take place entirely within the thoracic cavity. The cage also renders the veins relatively inaccessible to the operating physician.

Migrating Bullets.—Bullets entering a large artery or vein may migrate with the blood from their site of entrance. Bullets in the aorta can migrate to the distal branches until they become wedged, causing blockage and ischemia. In the same manner a bullet entering a pulmonary vein can migrate to the left atrium, left ventricle, and enter the systemic circulation. A bullet entering the superior vena cava can migrate into the right atrium and right ventricle and enter the pulmonary circulation. Paradoxical movement of bullets through a patent atrial septum has been reported.

Superior Vena Caval or Brachiocephalic Vein Thrombosis

This condition usually results from compression of the veins by tumors in the superior mediastinum; enlarging lymph node metastases secondary to a bronchial carcinoma is a common cause. Venous blockage results in engorgement of the veins of the head and neck.

Important Connection Between the Superior and Inferior Venae Cavae

In obstruction of the superior or inferior venae cavae, the azygos veins provide an alternative pathway for the return of venous blood to the right atrium of the heart. This is possible since these veins and their tributaries connect the superior and inferior venae cavae.

Clinical Anatomy of the Veins of the Upper Extremity

Thrombosis of the Superficial Veins

Prolonged intravenous infusion and, rarely, bacterial cellulitis of the superficial fascia can produce thrombosis of the superficial veins. In both cases injury to the tunica intima is the initiating factor.

Axillary-Subclavian Vein Thrombosis

Spontaneous thrombosis of the axillary and or subclavian veins occasionally occurs following excessive and unaccustomed use of the arm at the shoulder joint. The close relationship of these veins to the first rib and the clavicle and the possibility of repeated minor trauma from these structures is probably a factor in its development.

Secondary thrombosis of axillary and/or subclavian veins is a common complication of an indwelling venous catheter. Rarely the condition may follow a radical mastectomy with a block dissection of the lymph nodes of the axilla. In all cases of thrombosis of these veins the initial lesion is damage to the tunica intima. Persistent pain, heaviness, or edema of the upper limb, especially after exercise, is a complication of this condition.

Finding the Superficial Veins of the Upper Extremity

The cephalic, basilic, median cubital, median cephalic, median basilic, and median veins of the forearm can all be used for venipuncture and blood transfusion. These veins are fairly large and relatively constant in position. Unfortunately, in obese persons the veins cannot always be seen. The surface anatomy of the superficial veins is given on p. 152.

Anatomy of Basilic and Cephalic Vein Catheterization

The median basilic or basilic veins are the veins of choice for central venous catheterization because from the cubital fossa until the basilic vein reaches the axillary vein, the basilic vein increases in diameter and is in direct line with the axillary vein (see Fig 4–7). The valves in the axillary vein may be troublesome, but abduction of the shoulder joint may permit the catheter to move past the obstruction.

The cephalic vein does not increase in size as it ascends the arm, and frequently divides into small branches as it lies within the deltopectoral groove. One or more of these branches may ascend over the clavicle and join the external jugular vein. In its usual method of termination, the cephalic vein pierces the clavipectoral fascia in the deltopectoral groove and joins the axillary vein at a right angle. It may be difficult to maneuver the catheter around this angle.

Anatomy of Subclavian Vein Catheterization

The subclavian vein is located in the lower anterior corner of the posterior triangle of the neck (Fig 4–23), where it lies immediately posterior to the medial third of the clavicle.

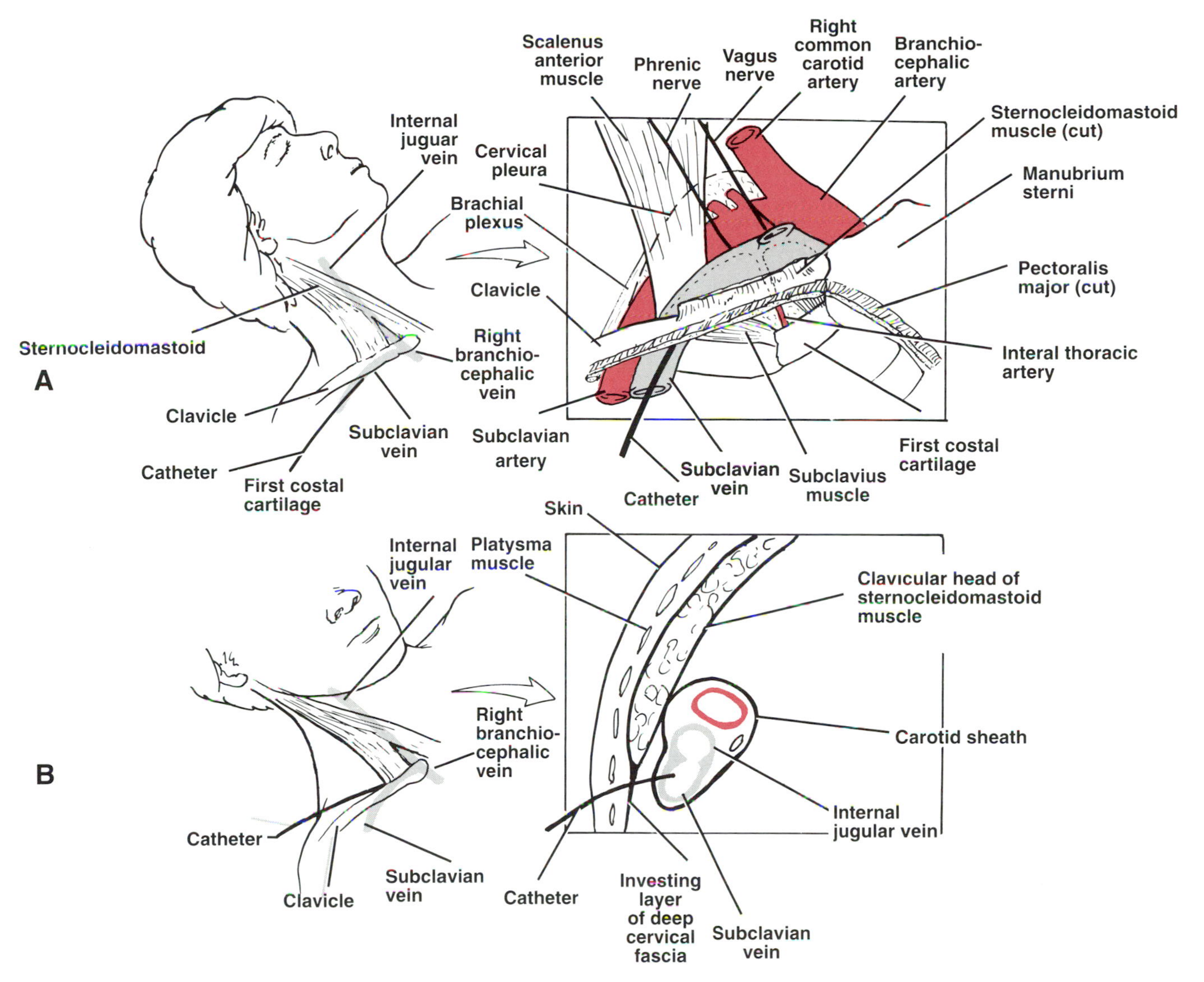

FIG 4–23.
Subclavian vein catheterization. **A,** infraclavicular approach. Note the many important anatomical structures located in this region. **B,** superclavicular approach. The catheter enters the subclavian vein close to its junction with the internal jugular vein to form the brachiocephalic vein.

Infraclavicular Approach.—Since the subclavian vein lies close to the undersurface of the medial third of the clavicle (see Fig 4–23), this is a relatively safe site for catheterization. The vein is slightly more medially placed on the left side than on the right side.

Anatomy of Procedure (see Fig 4–23).—The needle should be inserted through the skin just below the lower border of the clavicle at the junction of the medial third and outer two thirds (coinciding with the posterior border of the origin of the clavicular head of the sternocleidomastoid muscle on the upper border of the clavicle).

The needle pierces the following structures.

1. Skin.
2. Superficial fascia.
3. Pectoralis major muscle (clavicular head).
4. Clavipectoral fascia and subclavius muscle.
5. Wall of subclavian vein.

The needle is pointed upward and posteriorly toward the middle of the suprasternal notch.

Anatomy of Problems.—The following problems may be encountered.

1. Hitting the clavicle. The needle may be "walked" along the lower surface of the clavicle until its posterior edge is reached.

2. Hitting the first rib. The needle may hit the first rib if the needle is pointed downward and not upward.

3. Hitting the subclavian artery. A pulsatile resistance and bright red blood flow indicates that the needle has passed posterior to the scalenus anterior muscle and perforated the subclavian artery.

Anatomy of Complications (see Fig 4–23).—The following complications may be encountered.

1. Pneumothorax. The needle may pierce the cervical dome of the pleura, permitting air to enter the pleural cavity. This complication is more common in children, in whom the pleural reflection is higher than in adults.

2. Hemothorax. The catheter may pierce the posterior wall of the subclavian vein and the pleura.

3. Subclavian artery puncture. The needle pierces the arterial wall of the artery during its insertion.

4. Internal thoracic (internal mammary) artery injury. Hemorrhage may occur into the superior mediastinum.

5. Diaphragmatic paralysis. This occurs when the needle damages the phrenic nerve.

Infraclavicular Approach in Children.—The needle pierces the skin in the deltopectoral groove about 2 cm from the clavicle. The catheter is tunneled beneath the skin to enter the subclavian vein at the point where the clavicle and the first rib cross. The more oblique approach in children minimizes the possibility of entering the subclavian artery.

Supraclavicular Approach.—This approach (see Fig 4–23) is preferred by many for the following anatomical reasons.

1. The site of penetration of the vein wall is larger since it lies at the junction of the internal jugular vein and the subclavian vein, which makes the procedure easier.

2. The needle is pointed downward and medially toward the mediastinum away from the pleura, avoiding the complication of pneumothorax.

3. The catheter is inserted along a more direct course into the brachiocephalic vein and superior vena cava.

Anatomy of the Procedure (see Fig 4–23).—With the patient in the Trendelenburg or simple supine position and the head turned to the opposite side, the posterior border of the clavicular origin of sternocleidomastoid is palpated. The needle is inserted through the skin at the site where the posterior border of the clavicular origin of sternocleidomastoid is attached to the upper border of the clavicle. At this point the needle lies lateral to the lateral border of scalenus anterior muscle and above the first rib.

The needle pierces the following structures (see Fig 4–23).

1. Skin.
2. Superficial fascia and platysma muscle.
3. Investing layer of deep cervical fascia.
4. Wall of the subclavian vein.

The needle is directed downward in the direction of the opposite nipple. The needle enters the junction of the internal jugular vein and the subclavian vein. It is important that the operator understands that the pleura is not being penetrated and that it is possible for the needle to lie in a zone between the chest wall and the cervical dome of the parietal pleura but outside the pleural space (cavity).

Anatomical Complications.—The following complications (see Fig 4–23) may occur as the result of damage to neighboring anatomical structures.

1. Paralysis of the diaphragm. This is caused by injury to the phrenic nerve as it descends posterior to the internal jugular vein on the anterior surface of the scalenus anterior muscle.

2. Pneumothorax or hemothorax. This is caused by damage to the pleura and or internal thoracic artery by the needle passing posteriorly and downwards.

3. Brachial plexus injury. This is caused by the needle passing posteriorly into the roots or trunks of the plexus.

Clinical Anatomy of Arteriovenous Shunts and Fistulas in the Upper Limb for Hemodialysis

Vascular access for hemodialysis can be provided by the construction of an external arteriovenous shunt or an internal arteriovenous fistula. Most methods can be performed under local anesthesia.

External Arteriovenous Shunt

This is commonly used when a short period of treatment is required. The shunt is constructed if possible in the non dominant limb. The radial artery or the ulnar artery, and the cephalic vein or the basilic vein, may be used. The procedure is as follows.

1. The branches of the lateral and medial cutaneous nerves of the forearm are blocked with a local anesthetic.

2. A midline incision is made on the anterior surface of the distal part of the forearm.

3. The cephalic vein or the basilic vein is located in the superficial fascia as it winds round from the dorsum of the hand to ascend the front of the forearm (see Fig 4–17).

4. The deep fascia is incised and the artery is located. The radial artery can be palpated (see Fig 4–12) as it lies anterior to the distal third of the radius and between the tendons of flexor carpi radialis (medially) and the brachioradialis (laterally). The ulnar artery can be felt just lateral to the pisiform bone and can be traced proximally into the forearm (see Fig 4–13).

5. The appropriate artery and vein are then connected to the dialyzer.

In those patients in whom the distal vessels have been previously used, the same vessels can be cannulated at a more proximal site.

In children, if the distal arteries and veins are too small, a shunt can be constructed between the brachial artery and the cephalic vein just proximal to the cubital fossa.

Internal Arteriovenous Fistula

This procedure is most often used when it is necessary to have a prolonged period of hemodialysis. The procedure is as follows.

1. The branches of the lateral and medial cutaneous nerves of the forearm are blocked with a local anesthetic.

2. A midline incision is made on the anterior surface of the distal part of the forearm.

3. The cephalic vein is located in the superficial fascia.

4. The deep fascia is incised and the radial artery is located in front of the distal end of the radius, as described in the previous section.

5. A side-to-side anastomosis is performed between the radial artery and the cephalic vein. Alternatively, an end-to-end or end of vein to side of artery anastomosis can be constructed. The peripheral circulation is maintained by the extensive anastomoses from the ulnar artery around the wrist and through the palmar arches.

6. The vein quickly becomes arterialized and distended and can be easily punctured with a cannula. The cannula from the dialyzer is inserted into the distended vein, and the cannula to the dialyzer is inserted into the fistula to enter the radial artery.

A similar arrangement can be made using the ulnar artery and the basilic vein.

Saphenous Vein Grafts in the Upper Limb

When there are no suitable veins available in the distal forearm, a saphenous vein graft from the lower limb can be placed subcutaneously between the radial or ulnar arteries in the forearm and one of the superficial veins in the anticubital fossa. Another procedure uses the saphenous vein to connect up the distal part of the brachial artery to the cephalic or basilic veins in the upper part of the arm. In this case the saphenous vein is tunneled through the subcutaneous tissue anterior to the biceps muscle.

ANATOMICAL-CLINICAL PROBLEMS

1. Name the common sites on the thoracic aorta where damage occurs in blunt trauma. Between 80% and 90% of such injuries result in immediate death. Explain in anatomical terms the path commonly taken by the escaping blood in cases of traumatic rupture of the thoracic aorta. Name the tissues that can sometimes temporarily control the leak, thus permitting the patients to be taken to the emergency department alive.

2. Why is the radial artery chosen in preference to the ulnar artery or brachial artery for direct blood pressure monitoring? Why are the upper limb arteries used in preference to the dorsalis pedis artery of the foot? What are the important anatomic relations of the radial artery at the site of cannulation? Why is it necessary to extend the wrist joint when the cannula is introduced?

3. During an emergency procedure it was decided to monitor central venous pressure via peripheral access. Why is the basilic vein more often used to establish a central venous pressure line than the cephalic vein?

4. A 15-year-old girl, on examination in the emergency department, was found to have absent pulses in both femoral arteries. In addition, her blood pressure was higher in both upper limbs than in both lower limbs. An anteroposterior radiograph of the chest showed notching of the necks of the up-

per ribs on both sides. What is your diagnosis? Why is there notching of the ribs?

5. A 8-year-old boy, who had been running along a concrete path with a jam jar containing goldfish, slipped and fell. The glass from the broken jar pierced the skin on the front of his right wrist. On examination in the emergency department, a small wound was seen on the front of the flexor retinaculum just to the lateral side of the pisiform bone. Bright red blood spurted from the wound each time the dressing was removed. There appeared to be diminished skin sensation over the palmar surface of the ulnar one and a half fingers and the ulnar side of the palm. There was also sensory loss over the ulnar third of the dorsal surface of the hand and the dorsal surface of the ulnar one and a half fingers. Using your anatomical knowledge, name the damaged structures.

6. A 24-year-old woman was seen in the emergency department complaining of severe pain and discoloration of the fourth and fifth fingers of both hands. She said that she had had similar symptoms before and that they always occurred in very cold weather. Initially, her fingers turned white on exposure to cold and then became deep blue in color. The color change was confined to the distal half of each finger and was accompanied by an aching pain. Placing her hands in hot water was the only treatment that relieved the pain. As the pain disappeared, she said, her fingers became red and swollen. Using your knowledge of anatomy, make the diagnosis. What is the autonomic nerve supply to the blood vessels of the upper limb?

7. A 10-year-old girl fell off a swing and sustained a supracondylar fracture of her right humerus. Following reduction of the fracture in the emergency department, the child was admitted for observation. Three hours later, the child complained of pain in the right forearm, which persisted. Several hours later, the child's right hand looked dusky white, and the pain in the forearm was still present. On examination, there was found to be a complete loss of cutaneous sensation of the hand. The pulse of the radial and ulnar arteries could not be palpated. Every possible effort was made to restore the circulation of the forearm, without avail. Using your knowledge of anatomy, try to explain the cause of the absent radial and ulnar pulses. What deformity would you expect this child to have 1 year later?

8. A 50-year-old man was seen in the emergency department complaining of swelling of both arms. On questioning, he said that he first noticed that his hands were swollen about 2 weeks earlier. He said his wife commented that his face looked puffy, especially around the eyes. He admitted being a heavy smoker and had on several occasions coughed up blood-stained sputum. On examination, pitting edema was present in both the upper limbs, face, and neck. With the patient in the recumbent position, numerous dilated superficial veins were seen over the chest wall and abdomen. Later a chest radiograph revealed a large opacity in the upper lobe of the right lung. A diagnosis of advanced bronchogenic carcinoma of the right upper lobe was made.

Can you explain the presence of edema in both the upper limbs, face, and neck? What is the cause of the dilated superficial veins of the chest and abdominal walls? Is there normally communication between the main veins draining the upper part of the body with those draining the lower half of the body?

9. A 45-year-old man was admitted to the emergency department complaining of a sudden excruciating pain in the front of his chest. He described the pain as knife-like. During the course of the examination, he said that he could also feel the pain in his back between his shoulder blades. On close questioning he said he felt no pain down his arms or in his neck. The blood pressure readings showed 200/110 mm Hg in the right arm and 120/80 mm Hg in the left arm. A diagnosis of aortic dissection of the descending thoracic aorta was made. In anatomical terms explain the difference in the areas of referred pain found with a dissecting aneurysm and a myocardial infarction. Why is there a difference in the blood pressure readings in the right and left arms in this patient?

10. The problem of quickly finding a suitable vein in the upper limb for starting an intravenous transfusion in a life-threatening situation is common in an emergency department. The situation may be critical in an obese individual. Name the sites where superficial veins are consistently found in the upper limb.

11. A 23-year-old medical student decided to assist his father in building a garden shed. Unfortunately, much of the wood had to be cut to length by using a hand saw. He noticed on the third day that his right arm felt heavy and that his right hand was swollen. At the emergency department, a diagnosis of right subclavian vein thrombosis was made. Can you explain the possible anatomical reasons why thrombosis occurred in this vein in a healthy individual?

12. Name the anatomical structures that are pierced by the needle when performing an infraclav-

icular subclavian vein catheterization. Is there any difference in the anatomical position of the vein relative to the clavicle on the two sides? Name the anatomical structures that might be injured by this procedure. Which structure is particularly vulnerable in children?

ANSWERS

1. Blunt traumatic injury to the thoracic aorta involving horizontal deceleration occurs most commonly just distal to the origin of the left subclavian artery. This site is vulnerable since the heart and the aortic arch are mobile and the descending aorta is fixed (see p. 153). Sudden vertical deceleration, as in a fall, may result in an intimal tear at the root of the ascending aorta; the momentum of the heart filled with blood is sufficient to produce the tear.

Rupture of the ascending aorta occurs into the pericardial cavity, producing immediate cardiac tamponade and death. Rupture of the descending thoracic aorta frequently occurs into the left pleural cavity. The tear initially occurs in the tunica intima; the tunica media and adventitia and the surrounding connective tissue and the pleura may delay the complete rupture or temporarily control the leak. If untreated, delayed rupture and death usually occur in these cases within 2 weeks.

2. The radial artery has a lower incidence of arterial thrombosis than the brachial artery possibly because the tunica intima of the brachial artery is more likely to be damaged by the point of the catheter, since the brachial artery is more difficult to immobilize because of the movements at the elbow joint. The dorsalis pedis artery can be easily cannulated. It has a higher incidence of thrombosis, however, and sometimes the circulation of the foot is compromised by the inadequate collateral circulation. The radial artery is usually cannulated 2 or 3 cm proximal to the distal transverse crease of the wrist. Here the artery lies anterior to the distal third of the shaft of the radius, medial to the tendon of the brachioradialis and lateral to the tendon of flexor carpi radialis. It is covered anteriorly by skin and fascia. The forearm is supinated, and the wrist is extended approximately to a 50° angle. Extension stretches and stabilizes the artery during the process of introducing the needle and the catheter.

3. The basilic vein is used more often than the cephalic vein for the following reasons: (1) The basilic vein increases progressively in diameter from the cubital fossa to the axillary and subclavian veins, whereas the diameter of the cephalic vein increases only slightly as it ascends the upper extremity. (2) The basilic vein is in line with the axillary vein, whereas the cephalic vein opens into the axillary vein at a right angle. (3) The basilic vein is directly continuous with the end of the axillary vein, whereas the cephalic vein may bifurcate into several small veins near its termination or may join the external jugular vein.

4. Coarctation of the aorta is a narrowing of the aorta just proximal, opposite, or distal to the site of attachment of the ligamentum arteriosum. It is believed to result from the presence of an unusual quantity of ductus arteriosus muscle tissue incorporated in the wall of the aorta. When the ductus arteriosus contracts after birth, the ductus muscle in the aortic wall also contracts and the aortic lumen becomes narrowed. Later, fibrosis occurs and permanent narrowing takes place. The notching of the lower borders of the ribs is caused by the opening up of the collateral circulation through the subclavian, internal thoracic, and posterior intercostal arteries to carry blood from above the coarctation to the distal part of the aorta; it is the dilated posterior intercostal arteries that notch the ribs.

5. The glass fragment had severed the ulnar nerve and artery on the right side. Remember that the ulnar nerve passes with the ulnar artery anterior to the flexor retinaculum just lateral to the pisiform bone. (The artery lies on the lateral or radial side of the nerve). Both structures are covered in this location only by skin and superficial fascia.

6. This patient had Raynaud's disease. The initial palor of the fingers is due to spasm of the digital arterioles. The cyanosis that follows is due to local capillary dilatation caused by an accumulation of metabolites. Since there is no blood flow through the capillaries, deoxygenated hemoglobin accumulates within them. It is during this period of prolonged cyanosis that the patient experiences severe aching pain. On exposing the fingers to warmth, the vasospasm disappears, and oxygenated blood flows back into the very dilated capillaries. Reactive hyperemia is responsible for the swelling of the affected fingers.

The arteries and veins of the upper extremity are innervated by sympathetic nerves. The preganglionic fibers originate from the second to the eighth thoracic segments of the spinal cord. They ascend in the sympathetic trunk to synapse in the middle cervical, inferior cervical, and first thoracic or stellate ganglia. The postganglionic fibers join the nerves that form the brachial plexus and are distributed to the digital arteries within the ulnar, median, and radial nerves.

7. At the time of the supracondylar fracture of the humerus, or during the reduction of the fracture, the brachial artery may have been injured. Since the brachial artery is normally closely related to the anterior surface of the shaft of the humerus in this region, it is possible that a bone fragment damaged the arterial wall and precipitated the spasm of the smooth muscle in the tunica media. During the following months, the signs and symptoms of Volkmann's ischemic contracture developed.

8. The edema of both upper extremities and the head and neck, and the engorgement of the superficial veins of the chest and abdominal walls clearly indicate the presence of a superior vena caval obstruction. This obstruction was caused by the expanding metastases in the mediastinal lymph nodes secondary to the bronchogenic carcinoma. The dilated superficial veins included the lateral thoracic vein, a tributary of the axillary vein; lumbar veins, tributaries of the inferior vena cava; and the superficial epigastric vein, a tributary of the great saphenous vein that drains into the femoral vein. These venous channels provide an alternative pathway in superior vena caval obstruction, permitting superior vena caval blood to return to the heart via the inferior vena cava. The superior vena cava normally communicates with the inferior vena cava through the azygos veins. However, in this case the tumor was pressing on the superior vena cava proximal to the entrance of the azygos vein.

9. The pain impulses originating in the thoracic aortic wall are believed to ascend to the central nervous system along sympathetic nerves and enter the spinal cord through the posterior roots of the segmental spinal nerves. The pain is then referred along the somatic spinal nerves of the same segments. The number of dermatomes involved will depend on the extent of the dissection. In this patient the fifth, sixth, and seventh thoracic dermatomes were involved throughout their extent i.e., skin of anterior and posterior chest wall.

Pain pathways from the heart are described on p. 125. The upper four thoracic dermatomes are involved (mainly the anterior part of the chest wall). Because the second thoracic dermatome extends down the medial side of the upper part of the arm (intercostobrachial nerve, T2), cardiac pain is often referred down the arm. Furthermore, the cephalid central spread of the pain impulses in the nervous system appears to be more common in cardiac pain (higher thoracic spinal segments), which would explain the referral of cardiac pain to the neck in some patients.

In this patient the aortic dissection had involved the origin of the left subclavian artery causing its partial occlusion. This would explain the lower arterial blood pressure recorded in the left arm.

10. The surface anatomy of the superficial veins of the upper limb are fully described on p. 152. The cephalic vein is constantly found on the posterior surface of the styloid process of the radius just before it winds around the lateral side of the forearm. The median cubital vein or the median basilic vein can usually be found without too much difficulty in the cubital fossa (see p. 152).

11. The subclavian vein is closely related to the upper surface of the first rib and to the posterior surface of the medial third of the clavicle. Repeated minor trauma to the vein wall by these bones during the movements of the right shoulder while sawing resulted in damage to the tunica intima, followed by thrombosis. This problem is particularly likely to occur in an individual who is unused to this type of excessive movement of the shoulder. Usually there is some preexisting compression.

12. In infraclavicular catheterization of the subclavian vein, the needle pierces the following structures: skin, superficial fascia, pectoralis major, clavipectoral fascia, subclavius muscle, and the wall of the subclavian vein. The subclavian vein is slightly more medially placed on the left side than on the right side. The anatomical structures that might be damaged include (1) the pleura, with resulting pneumothorax or hemothorax; (2) the subclavian artery which may be punctured; (3) the internal thoracic artery which may tear and hemorrhage into the superior mediastinum; and (4) the diaphragm, which may be paralyzed from damage to the phrenic nerve.

In children the cervical dome of the pleura is higher than in the adult, and consequently it is more at risk of being pierced.

5

The Neck

A large number of vital structures are present in the neck. The mandible, the clavicle, the vertebral column, and the musculature provide only partial protection to these structures, leaving the anterior aspect very vulnerable to injury. Blunt or penetrating wounds to the neck are potentially life threatening.

Injuries to the larynx or the trachea can compromise the airway. The pharynx or esophagus can be violated, and serious injury can result. Vascular structures, including the jugular veins and their tributaries and the carotid arteries and their branches, may be pierced or torn. Neurological injury following damage to the cervical spine is always possible. Nerves may be injured, including those cranial nerves that descend through the neck, namely, the glossopharyngeal, vagus, accessory, and hypoglossal nerves. The sympathetic trunk could be damaged by a penetrating wound, producing Horner's syndrome; the phrenic nerve could be injured, producing a paralysis of the hemidiaphragm. Many signs and symptoms of emergency conditions of the neck are determined by the anatomical arrangement of the deep fascia, which not only supports the various structures but determines the degree and direction of the spread of blood, pus, and air.

This chapter provides a review of the anatomy of the neck so that emergency neck problems may be more easily assessed and correctly treated.

BASIC ANATOMY

Basic Anatomy of the Triangles of the Neck

For purposes of description, the neck is commonly divided by the sternocleidomastoid muscle into anterior and posterior triangles (Fig 5–1). The *anterior triangle* is bounded by the body of the mandible above, the sternocleidomastoid muscle posteriorly, and the midline anteriorly. The *posterior triangle* is bounded posteriorly by the trapezius muscle, anteriorly by the sternocleidomastoid muscle, and inferiorly by the clavicle.

The anterior triangle is further subdivided into the *carotid triangle,* the *digastric triangle,* and *submental triangle,* and the *muscular triangle.* The position and boundaries of these triangles are shown in Figure 5–1.

The posterior triangle of the neck is subdivided by the inferior belly of the omohyoid muscle into a

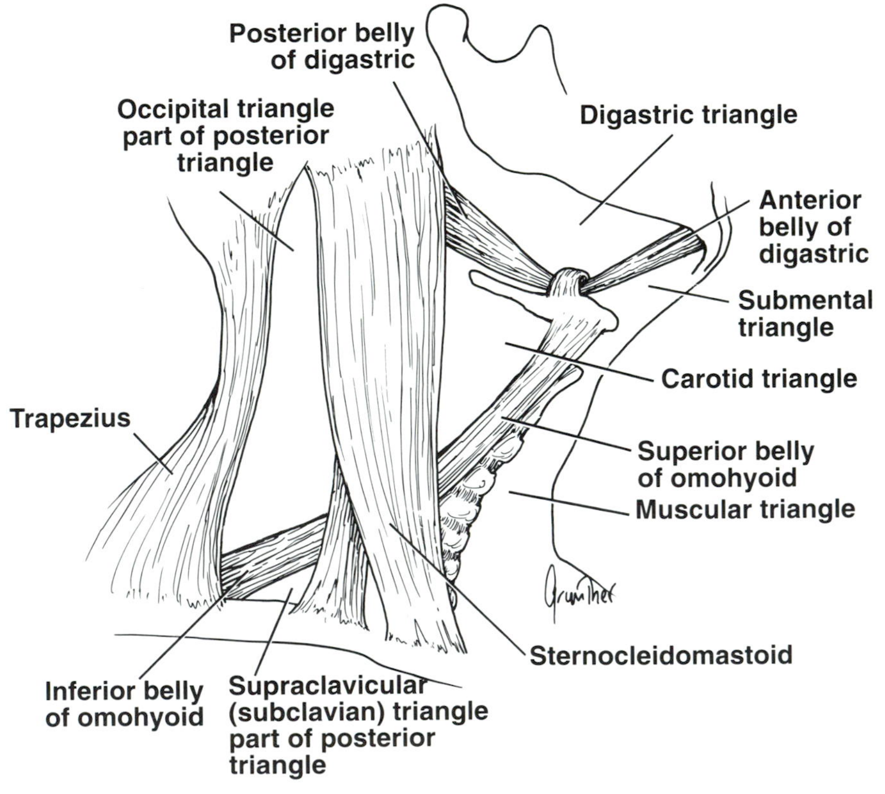

FIG 5–1.
Muscular triangles of the neck.

large *occipital triangle* above and a small *supraclavicular triangle* below (see Fig 5–1).

Basic Anatomy of the Fascia of the Neck

Superficial Fascia

The superficial fascia of the neck is a thin layer of connective tissue uniting the skin to the deep fascia. It contains adipose tissue and encloses the platysma muscle (Fig 5–2). The cutaneous nerves, superficial veins, and superficial lymph nodes are embedded within it.

Deep Fascia

The deep fascia of the neck consists of areolar tissue that supports the muscles, vessels, and viscera of the neck (see Fig 5–2). In certain areas it is condensed to form well-defined fibrous sheets called the investing layer, the pretracheal layer, and the prevertebral layer. It is condensed around the carotid vessels to form the carotid sheath.

Investing Layer.—This layer encircles the neck, splitting to ensheath the trapezius and sternocleido-

Clinical Note

Platysma tone and neck incisions.—The platysma arises from the deep fascia in the pectoral region and extends superiorly over the clavicle to cover the front of the neck. It is inserted into the mandible and blends with the muscles of facial expression. In lacerations or surgical incisions in the neck it is very important that the subcutaneous layer with the platysma be carefully sutured, since the tone of the platysma can pull on the scar tissue, resulting in broad, unsightly scars.

Clinical Note

Innervation of the platysma and neck incisions.—The platysma muscle is innervated by the cervical branch of the facial nerve. This nerve emerges from the lower end of the parotid gland and travels forward to the platysma; it then sometimes crosses the lower border of the mandible to supply the depressor anguli oris muscle. Skin lacerations over the mandible or upper part of the neck may distort the shape of the mouth.

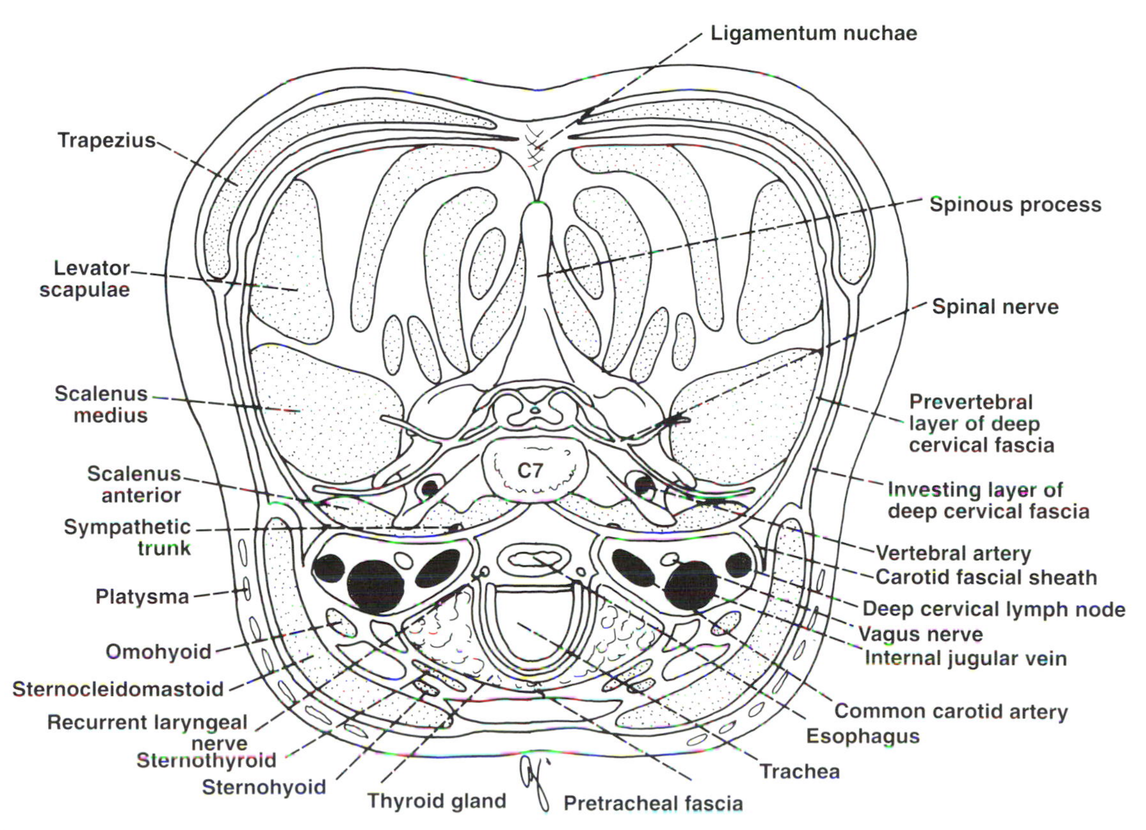

FIG 5–2.
Cross section of the neck at the level of the seventh cervical vertebra showing the arrangement of the layers of deep fascia.

mastoid muscles (see Fig 5–2). It thus roofs over the anterior and posterior triangles. The fascia is attached posteriorly to the ligamentum nuchae and above to the mandible, the zygomatic arch, and the superior nuchal line of the occipital bone. Below it splits into two layers that are attached to the anterior and posterior borders of the manubrium and the clavicle. Above the manubrium, the two layers enclose a slitlike space called the *suprasternal space,* which contains the jugular venous arch and sometimes a lymph node. The investing layer splits to enclose the submandibular and parotid salivary glands.

Pretracheal Layer.—This layer is thin and is attached above to the laryngeal cartilages. Below it extends into the thorax to blend with the fibrous pericardium. Laterally it is attached to the investing layer of deep cervical fascia beneath the sternocleidomastoid muscle (see Fig 5–2). It surrounds the thyroid and parathyroid glands, forming a sheath for them, and invests the infrahyoid (strap) muscles of the neck.

Prevertebral Layer.—This layer passes like a dense septum across the neck between the pharynx, esophagus, and vertebral column (see Fig 5–2). It covers the prevertebral muscles and is attached posteriorly to the ligamentum nuchae. Superiorly, it is attached to the base of the skull; inferiorly, it enters the thorax and blends with the anterior longitudinal ligament of the vertebral column.

Axillary Sheath.—As the subclavian artery and the brachial plexus emerge in the interval between the scalenus anterior and the scalenus medius muscles, they carry a sheath of prevertebral fascia, which extends into the axilla, called the axillary sheath.

Carotid Sheath.—The *carotid sheath* is a condensation of deep fascia in which the common and internal carotid arteries, the internal jugular vein, and the vagus nerve are embedded (see Fig 5–2). The deep cervical lymph nodes form a chain along the internal jugular vein and are also embedded in the carotid sheath. The carotid sheath blends in front with the pretracheal and investing layers of deep cervical fascia, and blends behind with the prevertebral layer of deep fascia.

Fascial Spaces

Between the more dense layers of deep fascia in the neck is loose connective tissue which forms potential spaces that are clinically important.

Visceral Space.—This space lies between the pretracheal and the prevertebral layers of deep fascia (Figs 5–3 and 5–4). It contains the larynx and trachea, the lower part of the pharynx and the esophagus, the

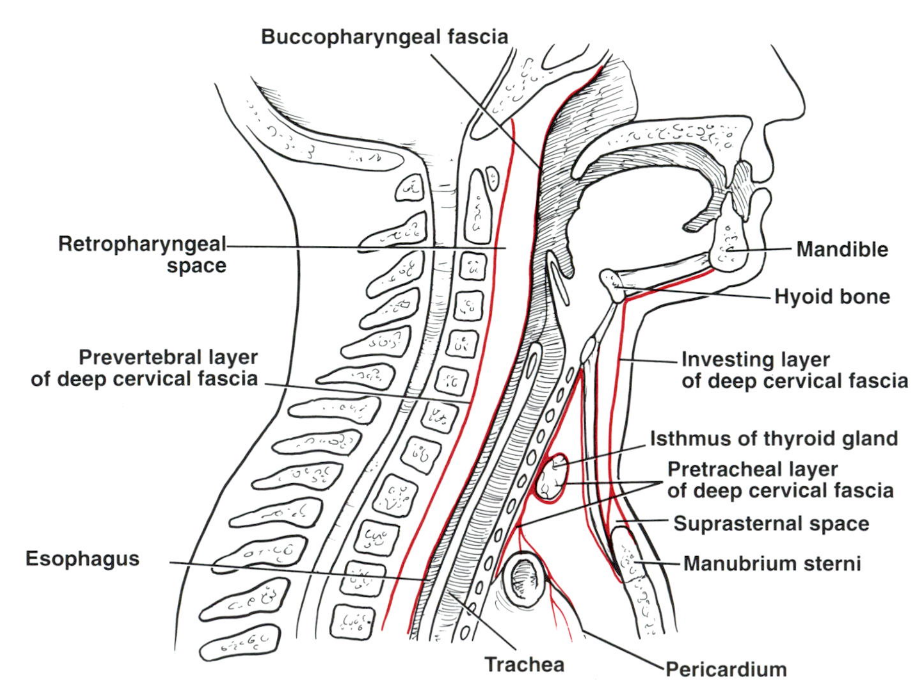

FIG 5–3.
Sagittal section of the head and neck showing the arrangement of the deep cervical fascia. Note the position of the retropharyngeal space, which has been enlarged for clarity; a suprasternal space is also shown.

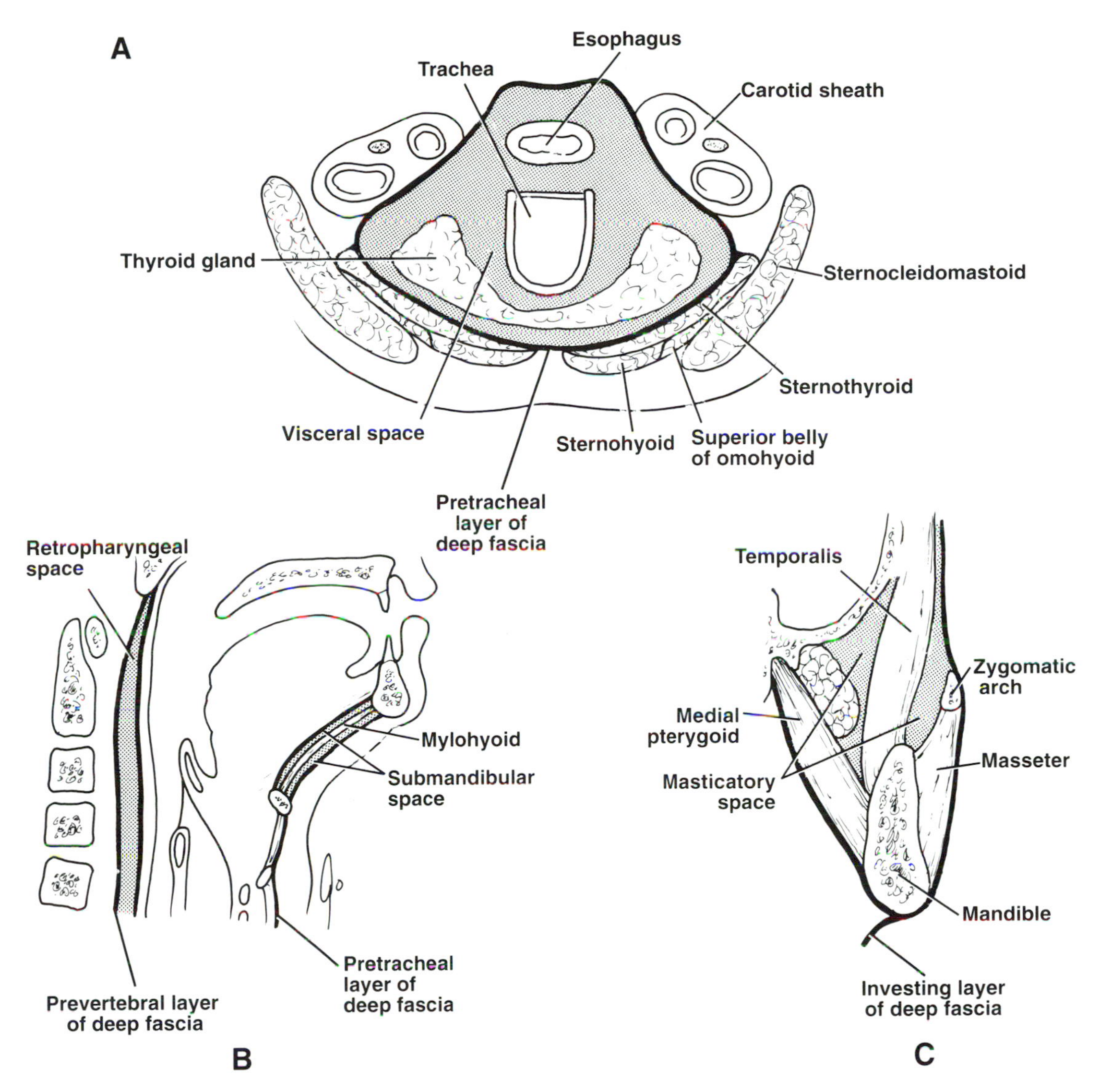

FIG 5–4.
A, cross section of the neck showing the visceral space. **B,** sagittal section of the neck showing the positions of the retropharyngeal space and the submandibular space. **C,** vertical section of the body of the mandible close to the angle showing the masticatory space.

thyroid gland, and the many arteries and veins of the neck. The space is limited above and in front by the attachment of the pretracheal fascia to the thyroid cartilage and hyoid bone; above and behind it is continuous with the retropharyngeal space, and through this space it extends up to the base of the skull. Below the visceral space is continuous with the superior mediastinum. It follows that blood, air, or pus in this space could migrate downward into the mediastinum.

Retropharyngeal Space.—This space lies between the pharynx and the prevertebral layer of deep cervical fascia (see Fig 5–3). It extends upward to the base of the skull, and downward it is continuous with the visceral space; through the visceral space the retropharyngeal space is continuous with the mediastinum. This space is regarded as a possible route for the spread of infection from the pharynx into the mediastinum.

Submandibular Space.—This space lies between the investing layer of fascia and the fascia covering the mylohyoid muscle (see Fig 5–4). It is located between the hyoid bone and the mandible and is limited above by the oral mucous membrane and the tongue.

Parotid Space.—This closed space, which surrounds the parotid salivary gland, is formed by the splitting of the investing layer of deep cervical fascia at the lower end of the gland. The deep layer extends upward to the base of the skull, and the superficial layer passes superiorly to be attached to the zygoma.

Masticatory Space.—This space surrounds the muscles of mastication and is formed by the splitting of the investing layer of deep cervical fascia at the lower margin of the mandible. The deep layer is attached to the base of the skull, and the superficial layer is attached above to the zygoma. The space contains the masseter, the medial pterygoid, and the lower part of the temporalis (see Fig 5–4).

Basic Anatomy of the Sternocleidomastoid Muscle

The sternocleidomastoid muscle forms an oblique band across the side of the neck (Fig 5–5). Not only has it an important function as a muscle, but it serves to protect many of the vital soft structures in the front of the neck.

Origin
This muscle originates from a rounded tendon at the front of the upper part of the manubrium sterni and by a muscular head from the medial third of the upper surface of the clavicle.

Insertion
The two heads join each other, and the muscle is inserted into the mastoid process of the temporal bone and the lateral part of the superior nuchal line of the occipital bone.

Nerve Supply
The nerve supply comes from the spinal part of the accessory nerve; it also receives proprioceptive fibers from the second and third cervical nerves.

Action
Both sternocleidomastoid muscles, acting together, extend the head at the atlanto-occipital joint and flex the cervical part of the vertebral column. The contraction of one muscle pulls the ear down to the tip of the shoulder on the same side and rotates the head so that the face looks upward to the opposite side (i.e., it pulls the mastoid process of the same side down toward the sternum).

If the head is fixed by contracting the prevertebral and postvertebral muscles, the two sternocleidomastoid muscles can act as accessory muscles of inspiration.

Important Relations.—The following relations exist.

Superficial.—Skin, superficial fascia, platysma, external jugular vein, and cutaneous nerves.

Deep.—Strap muscles (sternohyoid, sternothyroid, and omohyoid); common carotid, internal carotid, external carotid, and subclavian arteries; internal jugular vein; deep cervical lymph nodes; cranial nerves—vagus and spinal part of accessory nerves; and nerve plexuses—cervical plexus and phrenic nerve and upper part of the brachial plexus.

Basic Anatomy of the Larynx and the Trachea

These structures are fully considered in Chapter 1, pp. 19 and 27.

Basic Anatomy of the Pharynx

The pharynx is described in Chapter 1, p. 14.

Basic Anatomy of the Esophagus

The esophagus is a muscular tube about 25 cm (10 in.) long that extends from the pharynx to the stomach (Fig 5–6). It begins at the level of the cricoid cartilage in the neck and descends in the midline behind the trachea (see Fig 5–2). In the thorax, it passes downward through the mediastinum and enters the abdominal cavity by piercing the diaphragm at the level of the tenth thoracic vertebra. The esophagus has a short course of about 2.5 cm (1 in.) before it enters the right side of the stomach.

The esophagus, when empty, is flattened anteroposteriorly; when it is distended, it is irregularly cylindrical. The esophagus is constricted at its origin, where it is crossed by the left bronchus in the thorax and where it passes through the diaphragm into the stomach.

Relations of the Esophagus in the Neck.—The following relations exist.

Anteriorly.—Trachea and recurrent laryngeal nerves (see Fig 5–2).

Posteriorly.—Prevertebral muscles and vertebral column.

Laterally.—Thyroid gland, carotid sheath and its contents (carotid arteries, internal jugular vein, vagus nerve, and deep cervical lymph nodes), and on the left side the thoracic duct.

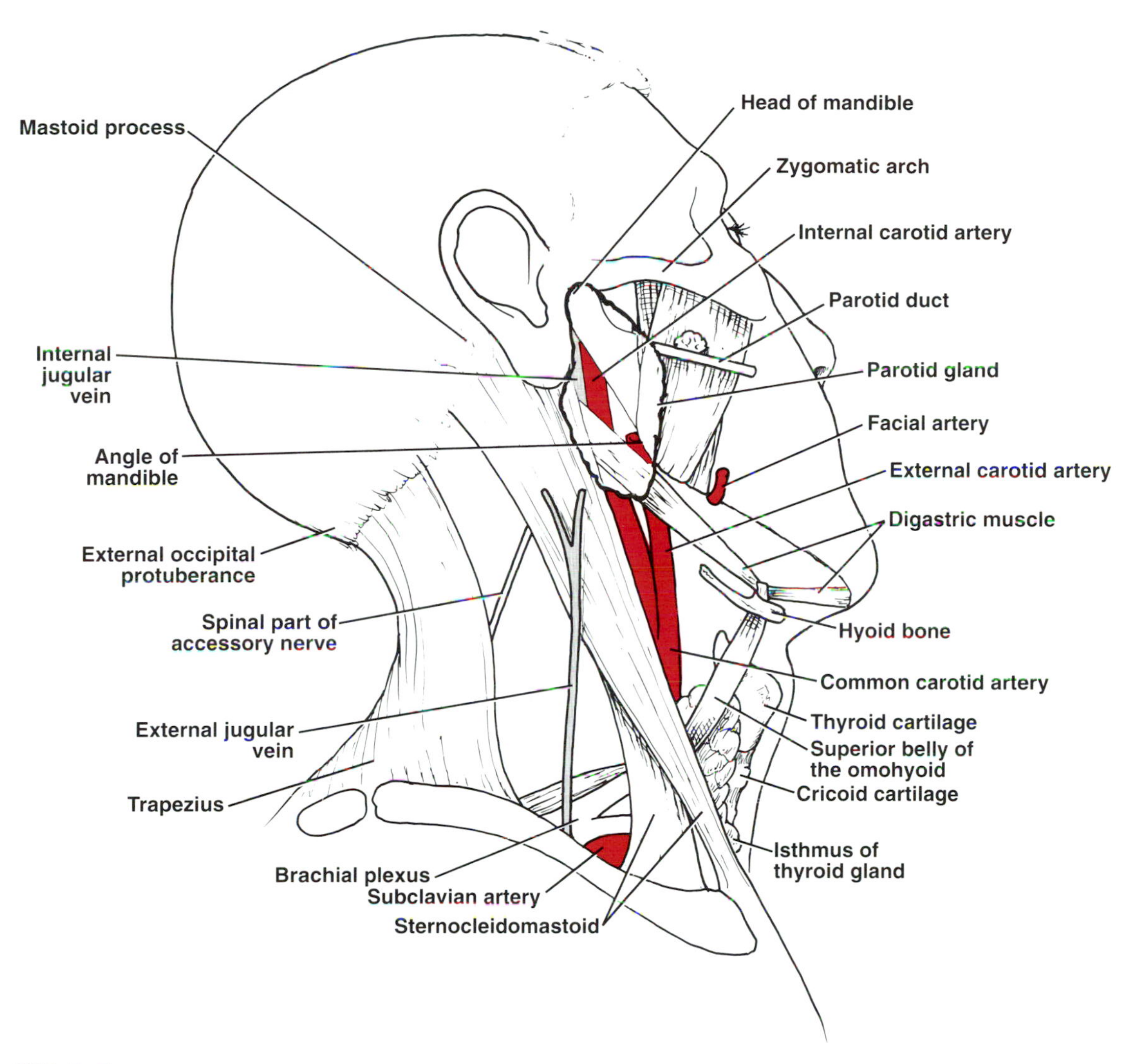

FIG 5–5.
Surface anatomy of the neck from the lateral aspect; the outline of the parotid gland is also shown.

Blood Supply of the Esophagus in the Neck.—Inferior thyroid arteries and the inferior thyroid veins.

Lymphatic Drainage of the Esophagus in the Neck.—Deep cervical lymph nodes.

Nerve Supply of the Esophagus in the Neck.—Recurrent laryngeal nerves and branches from the sympathetic trunks.

Basic Anatomy of the Thyroid Gland

The thyroid gland is situated in the neck and is bound down to the larynx and the trachea by the pretracheal layer of deep cervical fascia. It consists of *right* and *left lobes* connected by a narrow *isthmus* (Fig 5–7). The gland has a fibrous capsule. Each lobe of the gland is pear shaped, with its apex being directed upward along the lateral side of the thyroid cartilage; its base lies below, alongside the trachea. The isthmus extends across the midline in front of the second, third, and fourth tracheal rings (opposite the body of the seventh cervical vertebra). A *pyramidal lobe* is often present, which projects upward from the isthmus. A muscular band, the *levator glandulae thyroideae*, often connects the pyramidal lobe to the hyoid bone.

The thyroid gland is a very vascular organ and is surrounded by a sheath formed of the pretracheal layer of deep cervical fascia (see Fig 5–2).

Relations of the Lobes of the Thyroid Gland.—The following relations exist.

Anterolaterally.—Infrahyoid muscles (sternothyroid, sternohyoid, and superior belly of the omohyoid) and the anterior border of the sternocleidomastoid muscle (see Figs 5–2 and 5–7).

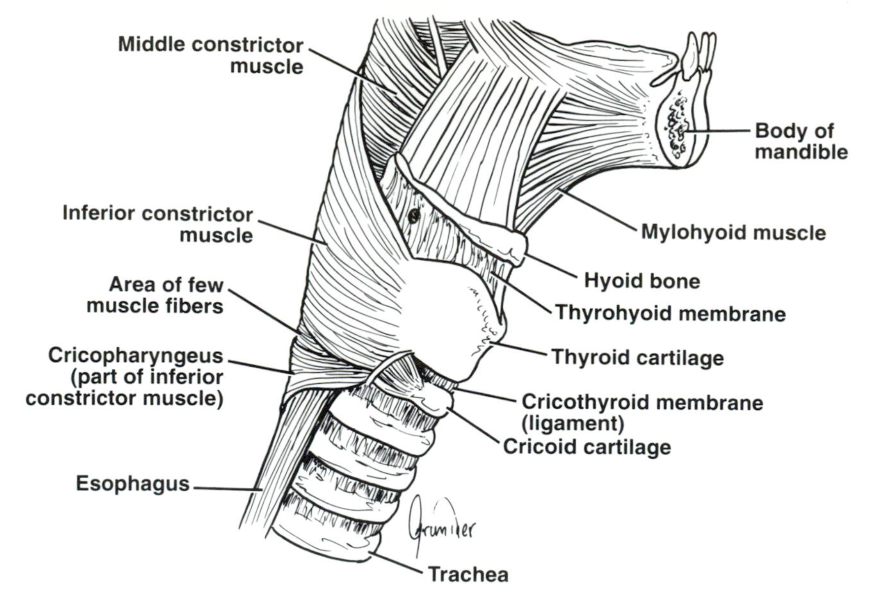

FIG 5–6.
The pharynx, esophagus, larynx, and trachea as seen from the lateral aspect.

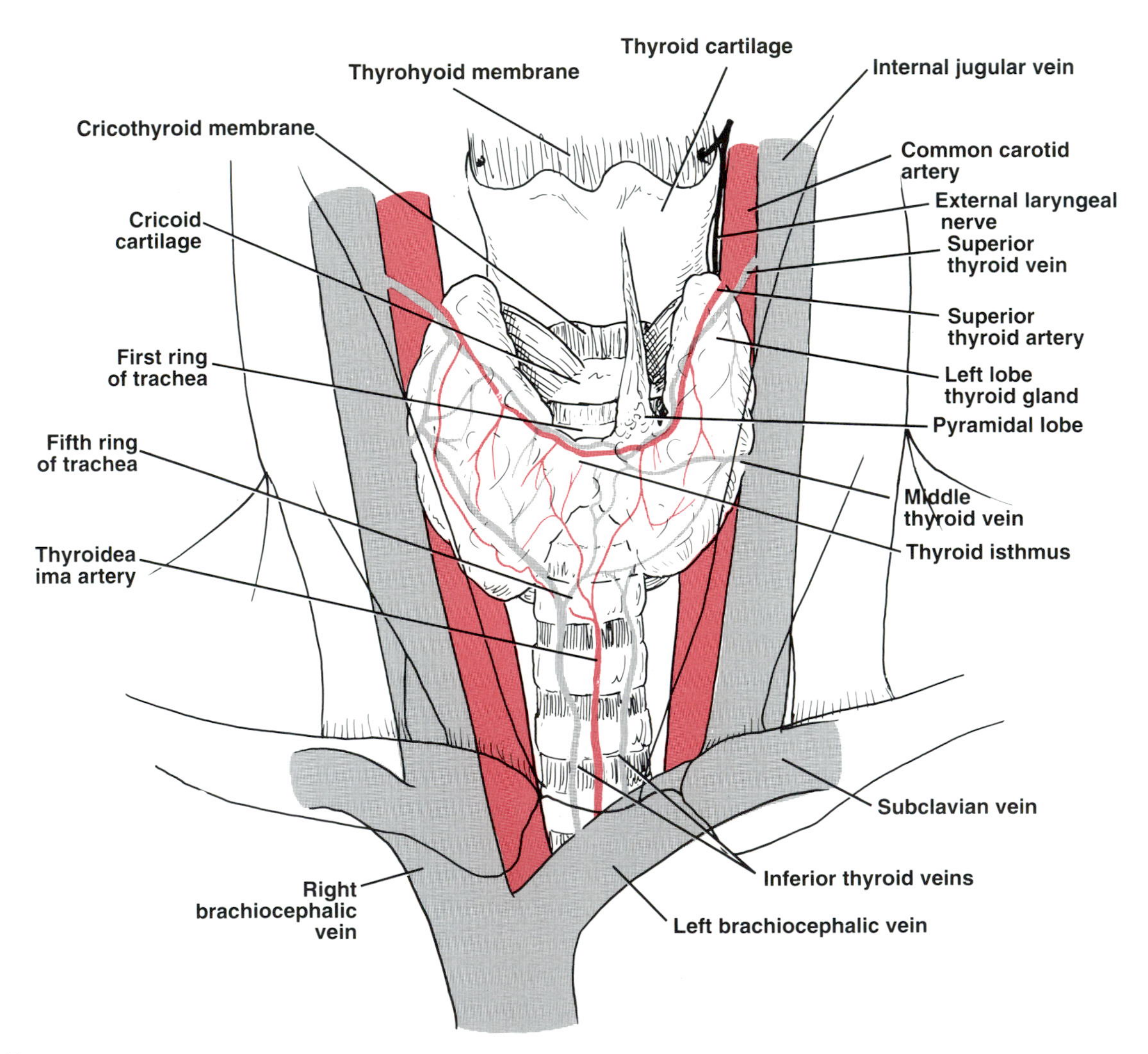

FIG 5–7.
The thyroid gland, its blood supply and venous drainage. Note the structures that lie immediately anterior to the airway.

Clinical Note

Movement of the thyroid on swallowing—The sheath attaches the gland to the larynx and the trachea so that when the patient swallows, the gland moves upward with the larynx. This fact may be helpful in the differential diagnosis of lumps in the neck; thyroid lumps move with the larynx when swallowing takes place.

Posterolaterally.—Common carotid artery, internal jugular vein, vagus nerve, and deep cervical lymph nodes (Fig 5–8).

Medially.—Larynx, trachea, pharynx, esophagus, external laryngeal nerve (supplies cricothyroid muscle), and recurrent laryngeal nerve (supplies all intrinsic muscles of larynx except the cricothyroid).

Posteriorly.—Superior and inferior parathyroid glands (Fig 5–8).

Relations of the Thyroid Isthmus.—The following relations exist.

Anteriorly.—Sternothyroid and sternohyoid muscles, anterior jugular vein, fascia, and skin.

Posteriorly.—Second, third, and fourth tracheal rings (see Figs 5–7 and 5–8).

Blood Supply of the Thyroid Gland

Arteries.—Superior thyroid (related to the external laryngeal nerve) from the external carotid, inferior thyroid (related to the recurrent laryngeal nerve) from the thyrocervical trunk, and thyroidea ima (if present) from the brachiocephalic or aortic arch.

Veins.—The superior thyroid vein drains into the internal jugular vein, the middle thyroid vein drains into the internal jugular vein, and the inferior thyroid vein drains into the left brachiocephalic vein.

Lymphatic Drainage of the Thyroid Gland

The thyroid gland drains via deep cervical and paratracheal lymph nodes.

Nerve Supply of the Thyroid Gland

The nerve supply includes the superior, middle, and inferior cervical sympathetic ganglia.

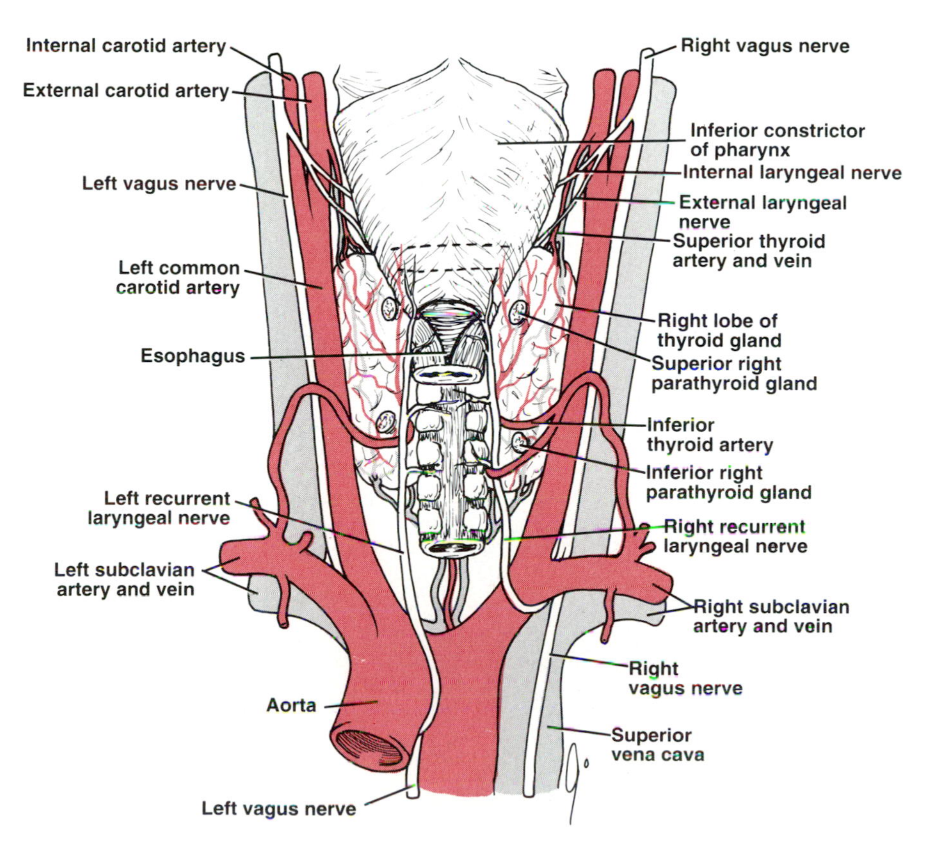

FIG 5–8.
The pharynx, esophagus, and trachea as seen from the posterior aspect. Note the positions of the thyroid and parathyroid glands and the relationships of the large blood vessels and nerves. The dotted lines indicate the level of the cricothyroid membrane.

Basic Anatomy of the Main Arteries of the Neck

Common Carotid Artery

The right common carotid artery arises from the brachiocephalic artery behind the right sternoclavicular joint (Fig 5–9). The left common carotid artery arises from the arch of the aorta in the superior mediastinum. The common carotid artery runs upward and backward through the neck, under cover of the anterior border of the sternocleidomastoid muscle (see Fig 5–7). At the upper border of the thyroid cartilage it divides into the external and internal carotid arteries.

Clinical Note

The carotid pulse.—The bifurcation of the common carotid artery can be easily felt just beneath the anterior border of the sternocleidomastoid muscle at the level of the superior border of the thyroid cartilage. This is a convenient site to take the carotid pulse.

The common carotid artery, at its point of division or the beginning of the internal carotid artery, shows a localized dilatation, called the *carotid sinus* (see Fig 5–9). Here the tunica media is thin and the tunica adventitia is thick and contains nerve endings from the glossopharyngeal nerve (see Fig 5–14). The carotid sinus is a baroreceptor and assists in the regulation of the blood pressure in the cerebral arteries.

Clinical Note

Carotid sinus hypersensitivity.—In this condition minor pressure on one or both carotid sinuses may cause excessive slowing of the heart rate, a fall in blood pressure, and cerebral ischemia with fainting.

The *carotid body* is a small structure that lies posterior to the common carotid artery at its point of bifurcation. It is innervated by the glossopharyngeal nerve and is a chemoreceptor, being sensitive to excessive carbon dioxide and reduced oxygen tensions in the blood circulating through the organ. When the chemoreceptors are stimulated, the heart and respiratory rates are increased.

The common carotid artery ascends the neck in the *carotid facial sheath* and is closely related to the internal jugular vein laterally; the vagus nerve lies between these two structures (see Fig 5–2).

External Carotid Artery

The external carotid artery is one of the terminal branches of the common carotid artery (see Fig 5–9).

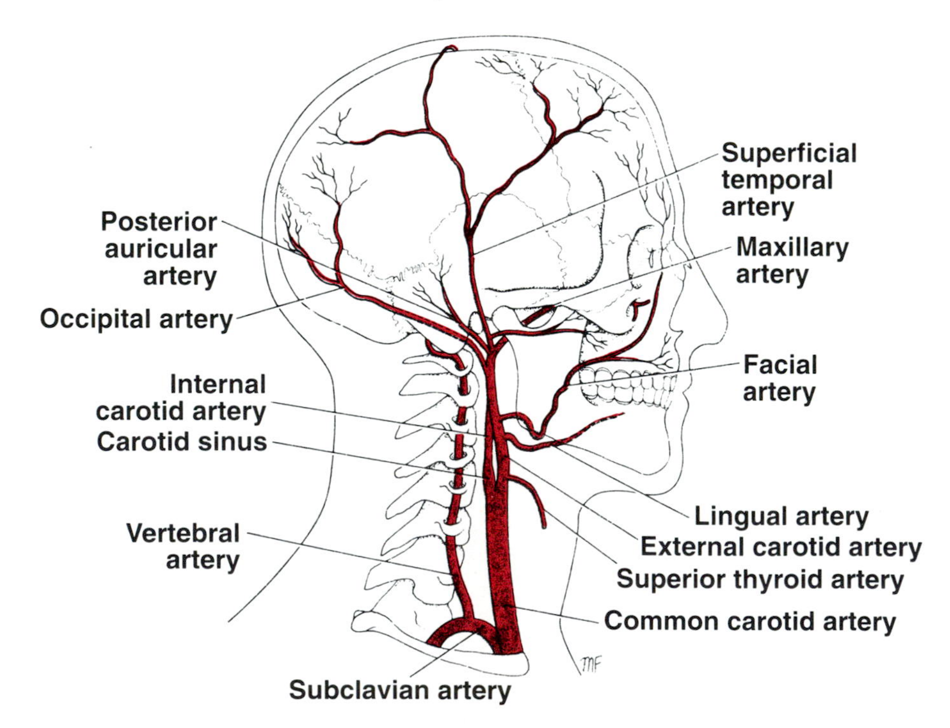

FIG 5–9.
Main arteries of the head and neck. For clarity, the thyrocervical trunk, the costocervical trunk, and the internal thoracic artery branches of the subclavian artery have not been shown.

It begins at the level of the upper border of the thyroid cartilage under cover of the anterior border of the sternocleidomastoid muscle. It terminates in the substance of the parotid salivary gland behind the neck of the mandible by dividing into the superficial temporal and maxillary arteries.

Close to its origin, the artery emerges from under cover of the sternocleidomastoid muscle, where its pulsations can be felt. At first it lies medial to the internal carotid artery, but as it ascends it passes backward and lies lateral to it.

Branches:

1. Superior thyroid artery
2. Ascending pharyngeal artery
3. Lingual artery
4. Facial artery
5. Occipital artery
6. Posterior auricular artery
7. Superficial temporal artery
8. Maxillary artery.

The origin and distribution of these branches is shown in Figure 5–9. The maxillary artery gives rise to an important branch, the *middle meningeal artery*, which enters the skull. Its further course is described in Chapter 8.

Internal Carotid Artery

The internal carotid artery begins at the bifurcation of the common carotid artery at the level of the upper border of the thyroid cartilage (see Fig 5–9). It ascends the neck in the carotid sheath accompanied by the internal jugular vein and the vagus nerve. At first it lies superficially; it then passes deep to the parotid salivary gland, the styloid process of the temporal bone, and the muscles attached to it.

The internal carotid artery leaves the neck by passing through the carotid canal in the petrous part of the temporal bone to enter the cranial cavity. It passes upward and forward in the cavernous sinus, where it is covered by connective tissue and the endothelial lining of the sinus. The artery then leaves the sinus and passes upward again medial to the anterior clinoid process. The internal carotid artery then gives off its first branch, the *ophthalmic artery;* the internal carotid artery terminates by dividing into the anterior and middle cerebral arteries.

Branches.—The cervical portion of the internal carotid artery has *no* branches in the neck. The remainder of the artery within the skull gives off numerous branches that supply the brain, the eye, and the skin of the face above the orbital margin.

Subclavian Artery

The *right subclavian artery* arises from the brachiocephalic artery, behind the right sternoclavicular joint (Fig 5–10). It arches upward and laterally over the pleura and comes to lie on the first rib between the scalenus anterior and scalenus medius muscles. At the outer border of the first rib it becomes continuous with the axillary artery.

The *left subclavian artery* arises from the arch of the aorta behind the origin of the left common carotid artery. It ascends to the root of the neck and then arches laterally in a manner similar to that of the right subclavian artery (see Fig 5–10).

The scalenus anterior muscle passes anterior to the artery on each side and divides it into three parts. The first part is from the origin of the artery to the medial border of the scalenus anterior. The second part includes the portion that lies posterior to the scalenus anterior. The third part is from the lateral border of scalenus anterior to the lateral border of the first rib (see Fig 5–10).

Branches: First part: *Vertebal artery, thyrocervical trunk*, and the *internal thoracic artery.*
Second part: *Costocervical trunk.*
Third part: Usually has no branches; occasionally the *superficial cervical* or *suprascapular arteries* arise from this part of the subclavian artery.

Basic Anatomy of the Main Veins of the Neck

External Jugular Vein

The external jugular vein arises just behind the angle of the mandible by the union of the posterior auricular vein with the posterior division of the retromandibular vein (Fig 5–11). It descends obliquely in the superficial fascia across the sternocleidomastoid muscle. It runs beneath the platysma muscle, and finally pierces the deep fascia to drain into the subclavian vein behind the middle of the clavicle.

Valves.—The external jugular vein has two sets of valves—one pair lies at its entrance into the subclavian vein and the other pair lies about 4 cm superior to the clavicle. The valves are usually incompetent.

Tributaries.—The external jugular vein receives the *posterior external jugular vein* from the back of the scalp, the *superficial cervical vein* from the root of the

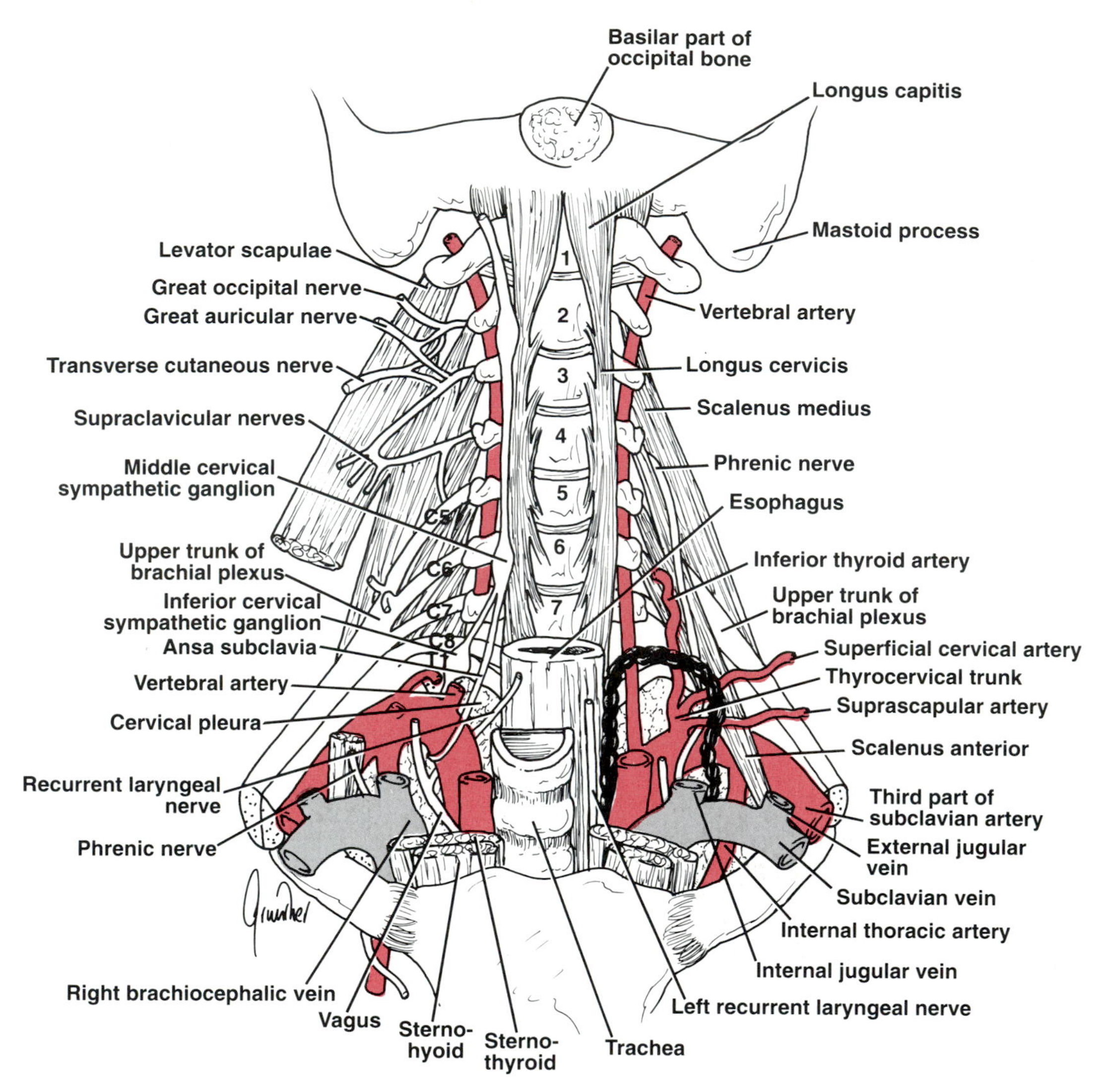

FIG 5–10.
The prevertebral region and the root of the neck. Note the relationships of the various structures at the root of the neck.

neck, and the *suprascapular vein* from the scapular region; it also receives the *anterior jugular vein.*

The anterior jugular vein (see Fig 5–11) runs down the front of the neck close to the midline. Just above the suprasternal notch, the veins of the two sides are united by a transverse trunk called the *jugular arch.* The vein then turns sharply laterally and passes deep to the sternocleidomastoid muscle to drain into the external jugular vein.

Internal Jugular Vein

The internal jugular vein is a large vein that drains blood from the brain, the face, and the neck (see Fig 5–11). It begins at the jugular foramen in the skull as a continuation of the sigmoid sinus. The vein descends through the neck lateral and then anterolateral to the internal carotid and common carotid arteries (see Fig 5–2). It lies within the carotid sheath lateral to the vagus nerve and is overlapped on its lateral surface for the lower part of its course by the sternocleidomastoid muscle (see Fig 5–7). The vein ends by joining the subclavian vein to form the brachiocephalic vein behind the medial end of the clavicle. The internal jugular vein is closely related to the deep cervical lymph nodes throughout its course.

Valves.—The internal jugular vein has a pair of valves just above the junction of the internal jugular vein with the subclavian vein to form the brachiocephalic vein.

Tributaries.—These include the *inferior petrosal sinus, facial vein, pharyngeal veins, lingual veins, superior thyroid veins, and middle thyroid vein.*

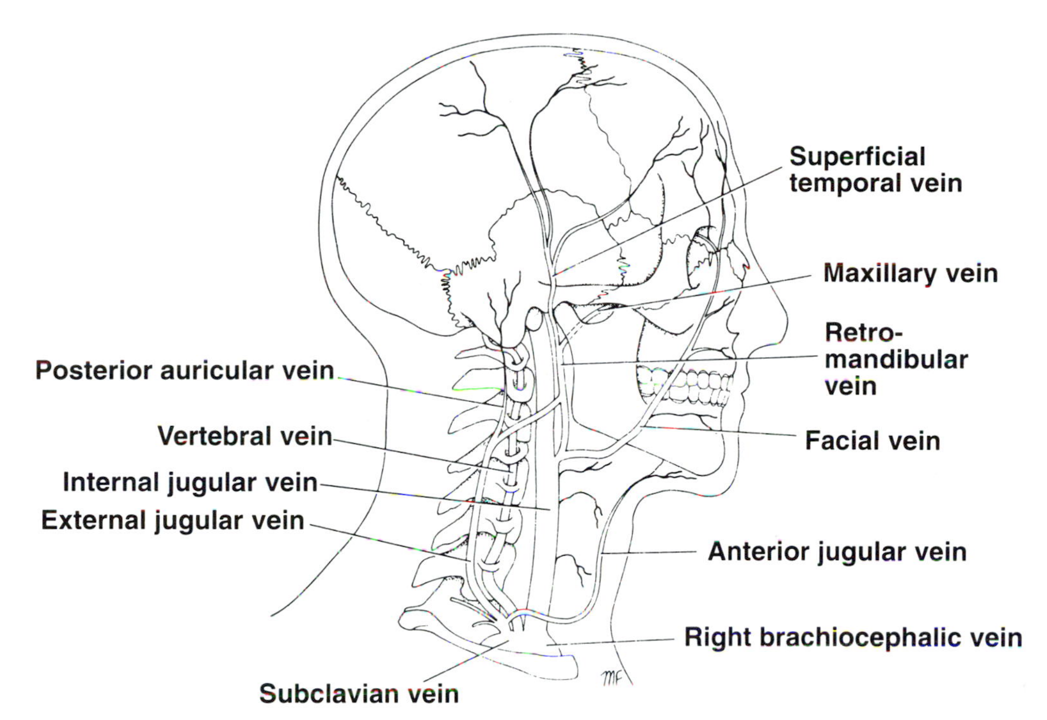

FIG 5–11.
Main veins of the head and neck.

Basic Anatomy of the Lymphatic Drainage of the Head and Neck

Lymph Nodes

The lymph nodes of the head and the neck are arranged as a regional collar that extends from below the chin to the occipital region and as a deep vertical terminal group that is embedded in the carotid sheath (Fig 5–12).

Regional Nodes.—The regional nodes are arranged as follows.

1. *Occipital nodes.*—These are situated over the occipital bone of the skull at the apex of the posterior triangle. When enlarged, these nodes are easily palpable. They receive lymph from the back of the scalp.

2. *Retroauricular (mastoid) nodes.*—These lie behind the ear over the mastoid process. When enlarged, these nodes are easily palpable. They receive lymph from the scalp above the ear, the auricle, and the posterior wall of the external auditory meatus.

3. *Parotid (preauricular) nodes.*—These are situated on or within the parotid salivary gland. When enlarged, these nodes are palpable. They receive lymph from the scalp above the parotid gland, the lateral ends of the eyelids, the parotid gland, the auricle, and the anterior wall of the external auditory meatus.

4. *Buccal (facial) nodes.*—One or two nodes lie on the buccinator muscle in the wall of the cheek. When enlarged, they are palpable. These nodes drain lymph that ultimately passes into the submandibular nodes.

5. *Submandibular nodes.*—These lie superficial to the submandibular salivary gland, just below the lower border of the body of the mandible. When enlarged, these nodes are easily palpable. They receive lymph from the front of the scalp; the nose; the cheek; the upper and lower lips (except the center part of the lower lip); the frontal, maxillary, and ethmoidal sinuses; the upper and lower teeth (except the lower incisors); the anterior two thirds of the tongue (except the tip); the floor of the mouth; and the vestibule and gums.

6. *Submental nodes.*—These lie just below the chin and above the body of the hyoid bone. When enlarged, these nodes are easily palpable. They receive lymph from the tip of the tongue, the floor of the anterior part of the mouth, the incisor teeth, the center part of the lower lip, and the skin over the chin.

7. *Anterior cervical lymph nodes.*—These lie along the course of the anterior jugular veins in the superficial fascia. When enlarged, they are palpable. These nodes receive lymph from the skin over the angle of the jaw, the skin over the lower part of the parotid gland, and the lobe of the ear.

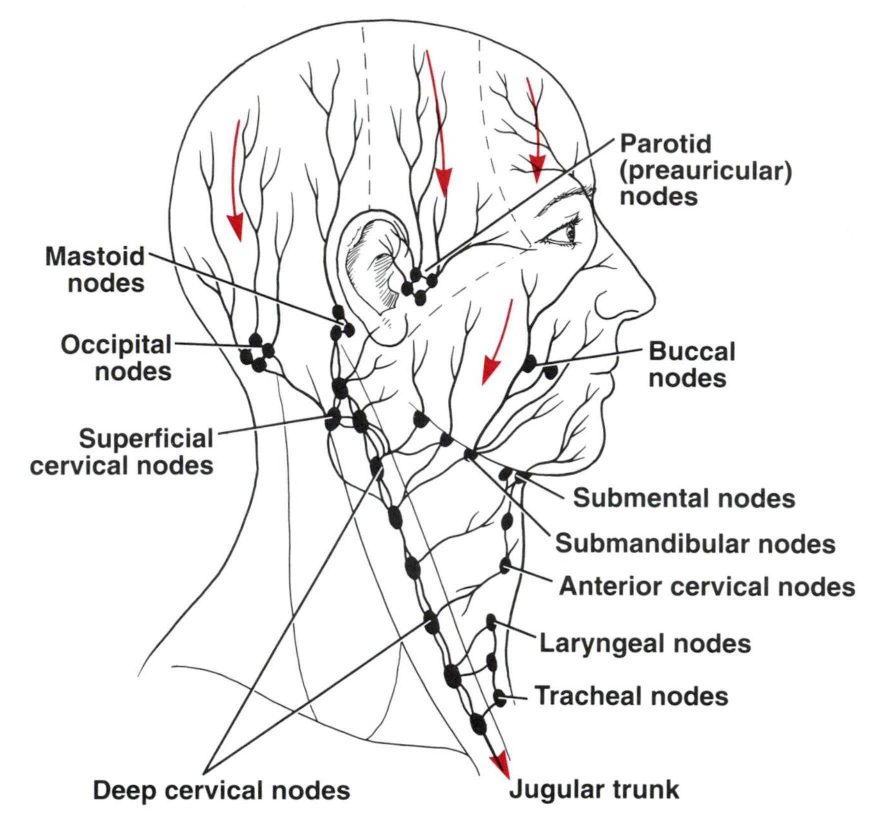

FIG 5–12.
The lymphatic drainage of the head and neck.

8. *Superficial cervical lymph nodes.*—These lie along the course of the external jugular vein in the superficial fascia. When enlarged, they are palpable. These nodes receive lymph from the skin over the angle of the jaw, the skin over the lower part of the parotid gland, and the lobe of the ear.

9. *Retropharyngeal lymph nodes.*—These lie behind the pharynx and in front of the vertebral column. They receive lymph from the pharynx, the auditory tube, and the vertebral column.

10. *Laryngeal lymph nodes.*—These lie in front of the larynx. When enlarged, they may be palpable. The nodes receive lymph from the larynx.

11. *Tracheal (paratracheal) lymph nodes.*—These lie alongside the trachea and when enlarged may be palpable. The nodes receive lymph from neighboring structures, including the thyroid gland.

Deep Cervical Lymph Nodes.—The deep cervical lymph nodes form a vertical chain along the course of the internal jugular vein within the carotid sheath (see Fig 5–12). When these nodes enlarge, they are palpable. They receive lymph from all the above 11 groups of regional nodes. The *jugulodigastric node,* which is located below and behind the angle of the jaw, is mainly concerned with drainage of the tonsil and tongue (so named because of its relationship to the posterior belly of the digastric muscle). The *jugulo-omohyoid node* is mainly associated with drainage of the tongue (so named because of its close relationship to the superior belly of the omohyoid muscle).

The efferent lymph vessels from the deep cervical nodes join to form the *jugular trunk,* which drains into the thoracic duct or the right lymphatic duct (see Fig 5–12).

The lymph drainage of some of the common sites of clinical concern are shown in Table 5–1.

Basic Anatomy of the Main Nerves of the Neck

Glossopharyngeal Nerve

The glossopharyngeal nerve is a motor and a sensory nerve. It leaves the skull by passing through the jugular foramen, where the superior and inferior glossopharyngeal sensory ganglia are situated on the nerve. The nerve then descends through the upper part of the neck (Fig 5–13) in company with the internal jugular vein and the internal carotid artery

TABLE 5–1.
Lymph Drainage of Some Sites in the Head and Neck That Are Clinically Important*

Anatomical Site	Lymph Nodes
Scalp	
Anterior region	Submandibular
Side	Parotid (preauricular), mastoid (postauricular)
Posterior region	Occipital
Face	
Forehead and cheeks, nose, upper lip, lateral lower lip	Submandibular
Central lower lip	Submental
Tongue	
Tip	Submental
Anterior ⅔ lateral edges	Submandibular
Posterior ⅓	Deep cervical
Teeth	
Upper jaw	Submandibular
Lower jaw	
Incisors	Submental
Lateral lower teeth	Submandibular
Palate	
Hard	Submandibular
Soft	Deep cervical
Pharynx	
Oral	
Posterior wall	Retropharyngeal, deep cervical
Tonsil	Deep cervical (jugolodigastric, below and behind angle of jaw)
Piriform fossa	Deep cervical
Larynx	Laryngeal, deep cervical

*All the above regional lymph nodes ultimately drain into the deep cervical nodes.

to reach the posterior border of the stylopharyngeus muscle. It passes forward between the superior and middle constrictor muscles of the pharyngeal wall to be distributed to the pharynx and back of the tongue. A diagram summarizing the branches of the glossopharyngeal nerve is shown in Figure 5–14.

The motor fibers supply the stylopharyngeus muscle (which elevates the pharynx during swallowing); the parasympathetic secretomotor fibers supply the parotid salivary gland. The sensory fibers, which are concerned with general sensation and taste, pass to the posterior third of the tongue and the pharynx; they also innervate the carotid sinus and carotid body.

The glossopharyngeal nerve thus assists swallowing and promotes salivation. It also conducts sensation from the pharynx and the back of the tongue and carries impulses, which influence the arterial blood pressure and respiration, from the carotid sinus and carotid body.

Vagus Nerve

The vagus nerve is composed of both motor and sensory fibers. It leaves the skull through the jugular foramen with the glossopharyngeal and accessory nerves. Just below the skull, the vagus nerve has two sensory ganglia (equivalent to the posterior root ganglia of spinal nerves). The nerve passes vertically down the neck within the carotid sheath, lying at first between the internal jugular vein and the internal carotid artery and then between the vein and the common carotid artery (see Fig 5–13). At the root of the neck the nerve accompanies the common carotid artery and lies anterior to the first part of the subclavian artery (see Fig 5–10). The vagus nerve then enters the thorax. A diagram summarizing the branches of the vagus nerve in the neck is shown in Figure 5–15.

The vagus nerve innervates the heart and great vessels within the thorax, larynx, trachea, bronchi, lungs, and much of the alimentary tract from the pharynx to the distal part of the transverse colon. It also supplies glands associated with the alimentary tract, such as the liver and pancreas.

The vagus nerve has the most extensive distribution of all the cranial nerves and supplies the structures named above with afferent and efferent fibers.

Accessory Nerve

The accessory nerve is composed only of motor fibers. It is formed by the union of cranial and spinal roots. Both cranial and spinal roots come together and leave the skull through the jugular foramen. The *cranial root* separates from the spinal root and joins the vagus (see Fig 5–13) and is distributed in its branches to the muscles of the soft palate and pharynx (via the pharyngeal branch) and to the muscles of the larynx (via the recurrent laryngeal nerve) except the cricothyroid muscle.

The *spinal root* runs downward and laterally in the neck and crosses the internal jugular vein (see Fig 5–13). It reaches the deep surface of the sternocleidomastoid muscle, which it supplies. The nerve emerges above the middle of the posterior border of the sternocleidomastoid muscle and crosses the posterior triangle of the neck to supply the trapezius muscle (see Fig 5–5).

The accessory nerve thus brings about movements of the soft palate, pharynx, and larynx and controls the movements of the sternocleidomastoid and trapezius muscles, two large muscles in the neck.

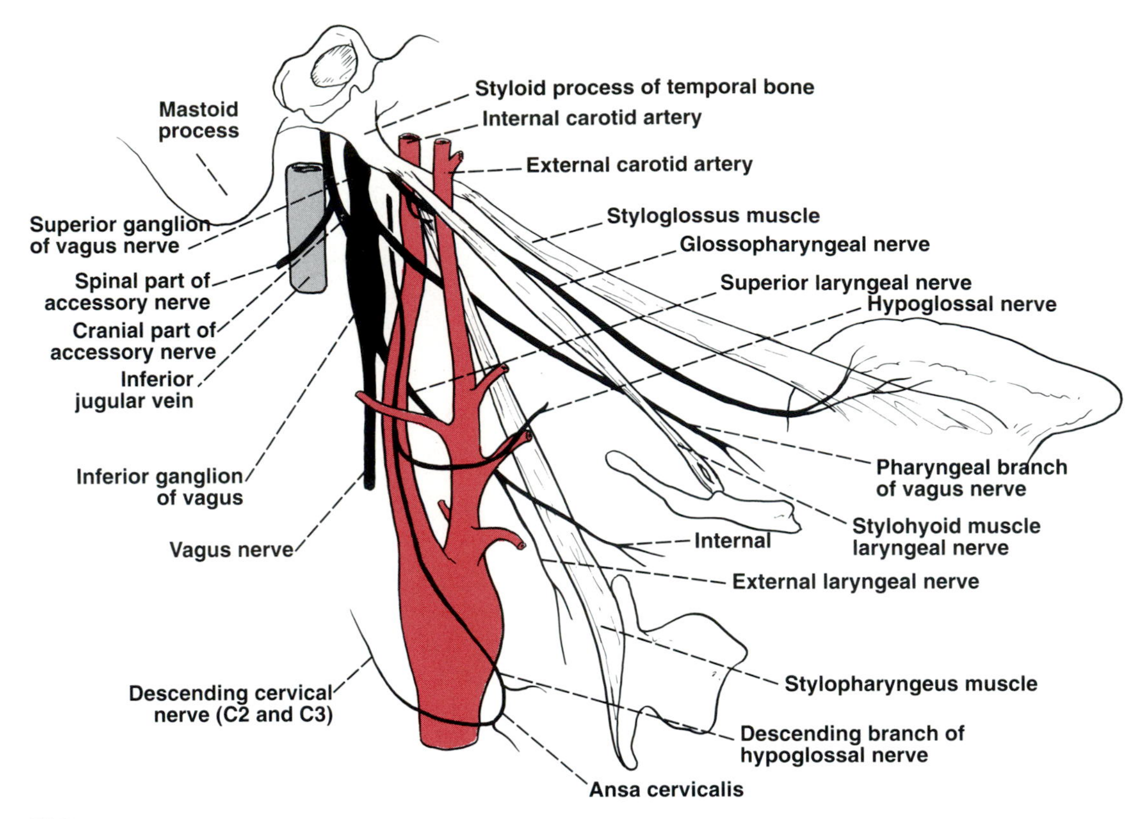

FIG 5–13.
The main blood vessels and cranial nerves in the neck.

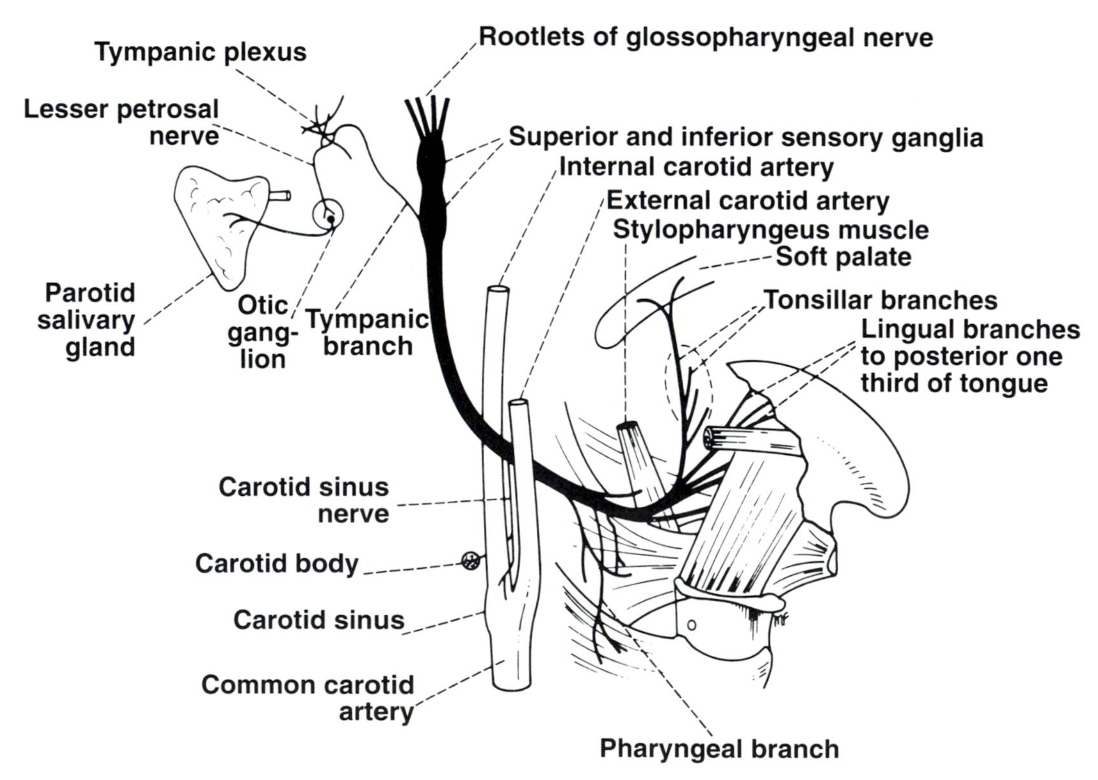

FIG 5–14.
The origin and distribution of the glossopharyngeal nerve.

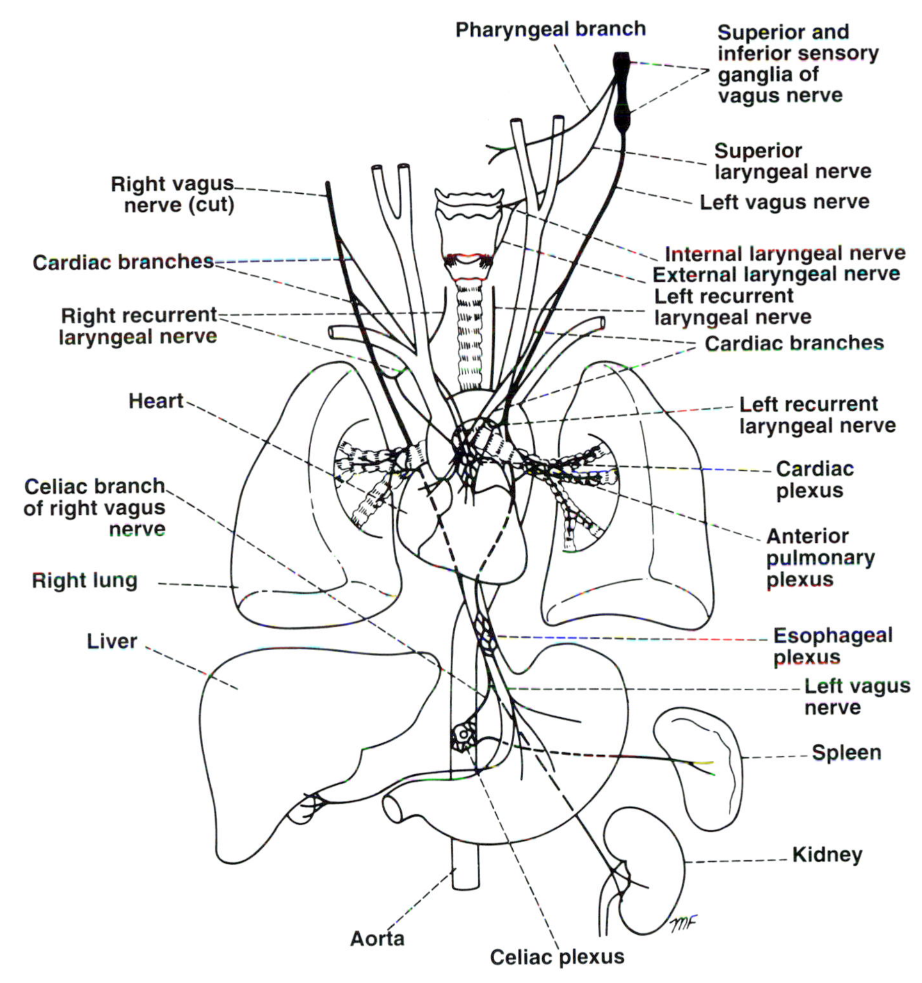

FIG 5–15.
The distribution of the vagus nerve.

Hypoglossal Nerve

The hypoglossal nerve is the motor nerve to the tongue muscles. It leaves the skull through the hypoglossal canal in the occipital bone. It then comes into close relationship with the glossopharyngeal, vagus, and accessory nerves, the internal carotid artery, and the internal jugular vein (see Fig 5–13). It descends between the internal carotid artery and the internal jugular vein until it reaches the lower border of the posterior belly of the digastric muscle (just above and behind the tip of the greater cornu of the hyoid bone). The nerve then turns forward and medially, crossing the internal and external carotid arteries and the loop of the lingual artery. Here, it is crossed by the facial vein. The nerve passes forward and upward, deep to the posterior belly of the digastric and the stylohyoid muscle. It enters the tongue in the submandibular region by passing deep to the posterior border of the mylohyoid muscle (Fig 5–16).

In the upper part of its course, the hypoglossal nerve is joined by a small branch from the cervical plexus (C1 and sometimes C2). This branch later leaves the hypoglossal nerve as its descending branch (with descending cervical nerve [C2 and C3] supplies the infrahyoid muscles), the nerve to the thyrohyoid muscle, and the nerve to the geniohyoid muscle.

The hypoglossal nerve innervates the muscles of the tongue and thus controls the shape and movements of the tongue; it also serves as a conduit for C1 and possibly C2 nerve fibers.

Cervical Part of the Sympathetic Trunk

The cervical part of the sympathetic trunk extends upward to the base of the skull and below to the neck of the first rib, where it becomes continuous with the thoracic part of the sympathetic trunk. The cervical trunk lies directly behind the internal and common carotid arteries and medial to the vagus nerve. The trunk is embedded in deep fascia between the carotid sheath and the prevertebral layer of deep fascia (see Fig 5–2). The sympathetic trunk possesses three ganglia—the superior, middle, and inferior cervical ganglia (see Fig 5–10). The inferior ganglion is most commonly fused with the first thoracic ganglion to form the stellate ganglion. For details concerning branches of the sympathetic ganglia, see Chapter 9.

Cervical Plexus

At the roots of the limbs, the anterior rami of the spinal nerves join in a complicated manner to permit a redistribution of the nerve fibers within the different peripheral nerves. The cervical and brachial plexus are found at the root of the upper limb.

The cervical plexus is deeply placed in the neck and is formed by the anterior rami of the first four cervical spinal nerves (Fig 5–17). The rami are joined by connecting branches, which form loops that lie in front of the origins of the levator scapulae and the scalenus medius muscles (see Fig 5–10). The cervical plexus gives rise to motor nerve fibers to muscles in the neck and to the muscle of the diaphragm. It also gives rise to sensory fibers to the skin of the neck, the shoulder, the upper part of the chest, the pericardium, the pleura, and the peritoneum.

Phrenic Nerve.—The phrenic nerve (C3, C4, and C5) is a large and very important branch of the cervical plexus; it is the only motor nerve supply to the diaphragm. The phrenic nerve runs vertically downward through the neck on the scalenus anterior muscle (see Fig 5–10). Because of the obliquity of the scalenus anterior muscle, the nerve crosses the muscle from its lateral to its medial border. The nerve enters the thorax by passing anterior to the subclavian artery and behind the subclavian vein; it crosses the internal thoracic artery from lateral to medial.

In addition to the motor fibers to the diaphragm, the phrenic nerve contains sensory fibers from the pericardium and the mediastinal parietal pleura and the pleura and the peritoneum covering the upper and lower surfaces of the central part of the diaphragm. Figure 5–17 summarizes the branches of the cervical plexus and their distribution.

Radiological Appearances of the Neck

The radiological appearances of some of the main structures seen in the neck on lateral view are shown in Figure 5–18.

SURFACE ANATOMY

Surface Anatomy of the Neck

Preferably, the neck should be palpated while the patient is in a supine position, when the muscles

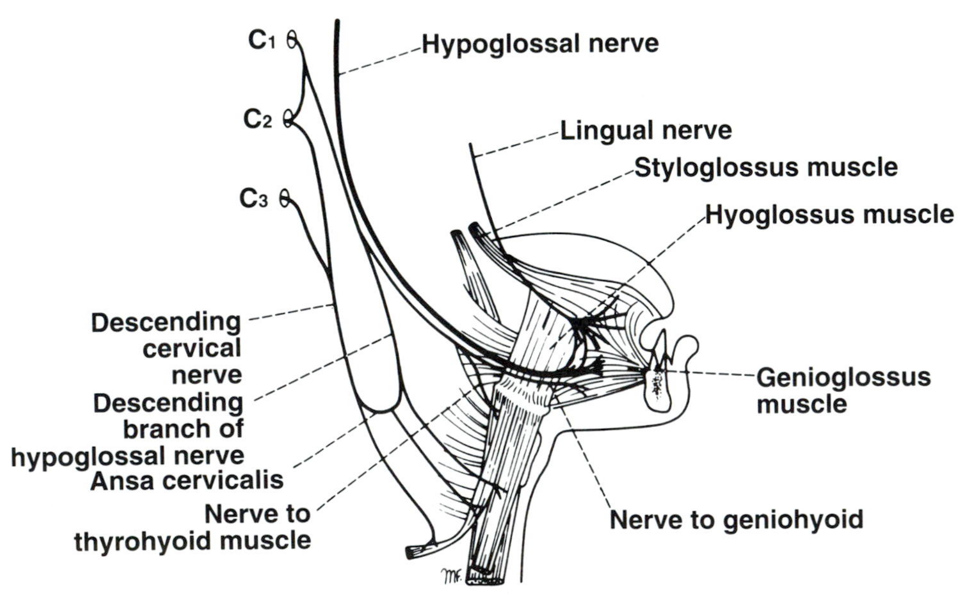

FIG 5–16.
The distribution of the hypoglossal nerve.

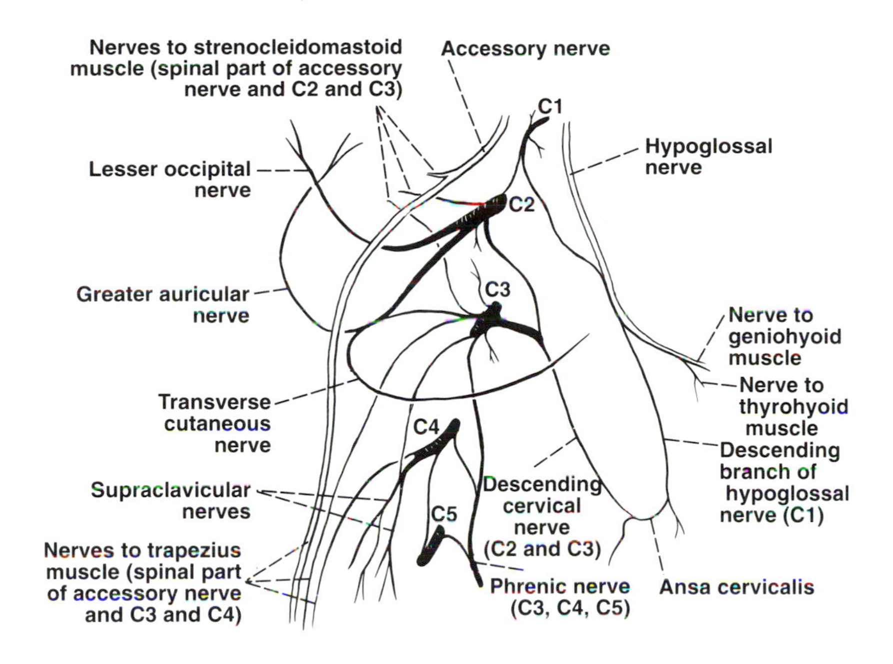

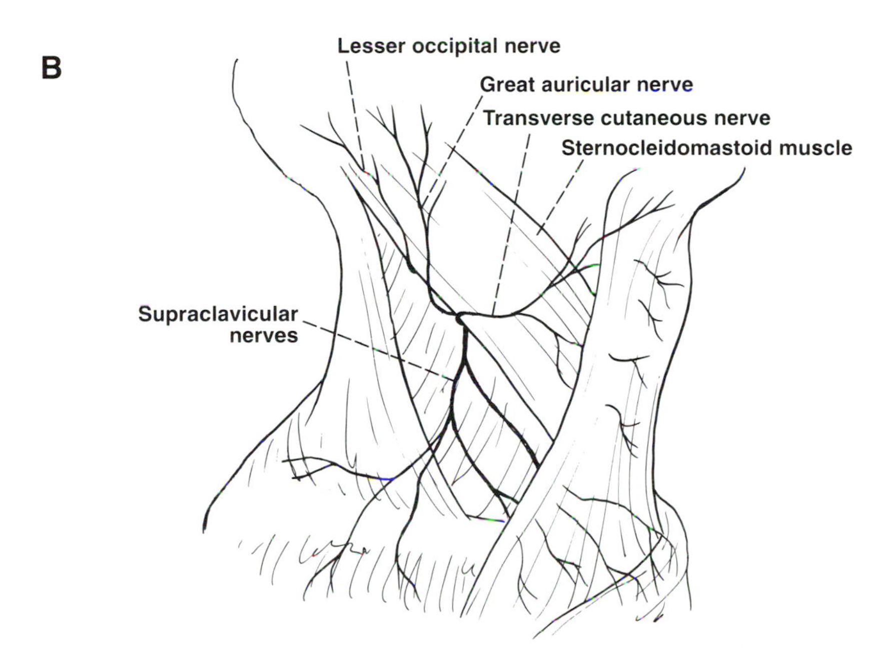

FIG 5–17.
A, the cervical plexus and its main branches. **B,** the superficial branches of the cervical plexus in the neck.

overlying the deeper structures are relaxed and the structures become easier to feel.

Midline Anteriorly

In the midline anteriorly, the following structures may be palpated from above downward.

Symphysis Menti.—The lower margin of the symphysis menti may be felt where the two halves of the body of the mandible unite in the midline (Fig 5–19).

Submental Triangle.—The submental triangle lies between the symphysis menti and the body of the hyoid bone (Fig 5–20). This triangle is bounded in front by the midline of the neck, laterally by the anterior belly of the digastric muscle, and inferiorly by the body of the hyoid bone. The floor is formed by the mylohyoid muscle. The *submental lymph nodes* are located in this triangle. *The body of the hyoid bone* lies opposite the third cervical vertebra (Fig 5–21). The hyoid bone is a horseshoe-shaped structure, and the greater cornua can be felt on each side of the neck between the finger and thumb. The hyoid bone moves superiorly when the patient swallows.

Clinical Note

The hyoid bone provides an important landmark when viewing a lateral radiograph of the neck (see Fig 5–18).

Thyrohyoid Membrane.—This occupies the interval between the hyoid bone and the thyroid cartilage (see Fig 5–20).

Thyroid Cartilage.—The notched upper border of the thyroid cartilage lies opposite the fourth cervical vertebra (Figs 5–19 and 5–20). The anterior border of the thyroid cartilage is more prominent in male adults than in female adults.

Cricothyroid Membrane (Ligament).—This fills in the interval between the cricoid cartilage and the thyroid cartilage (see Fig 5–20).

Clinical Note

The cricothyroid artery, a small branch of the superior thyroid artery, crosses in front of the cricothyroid membrane and may be lacerated during a cricothyroidotomy.

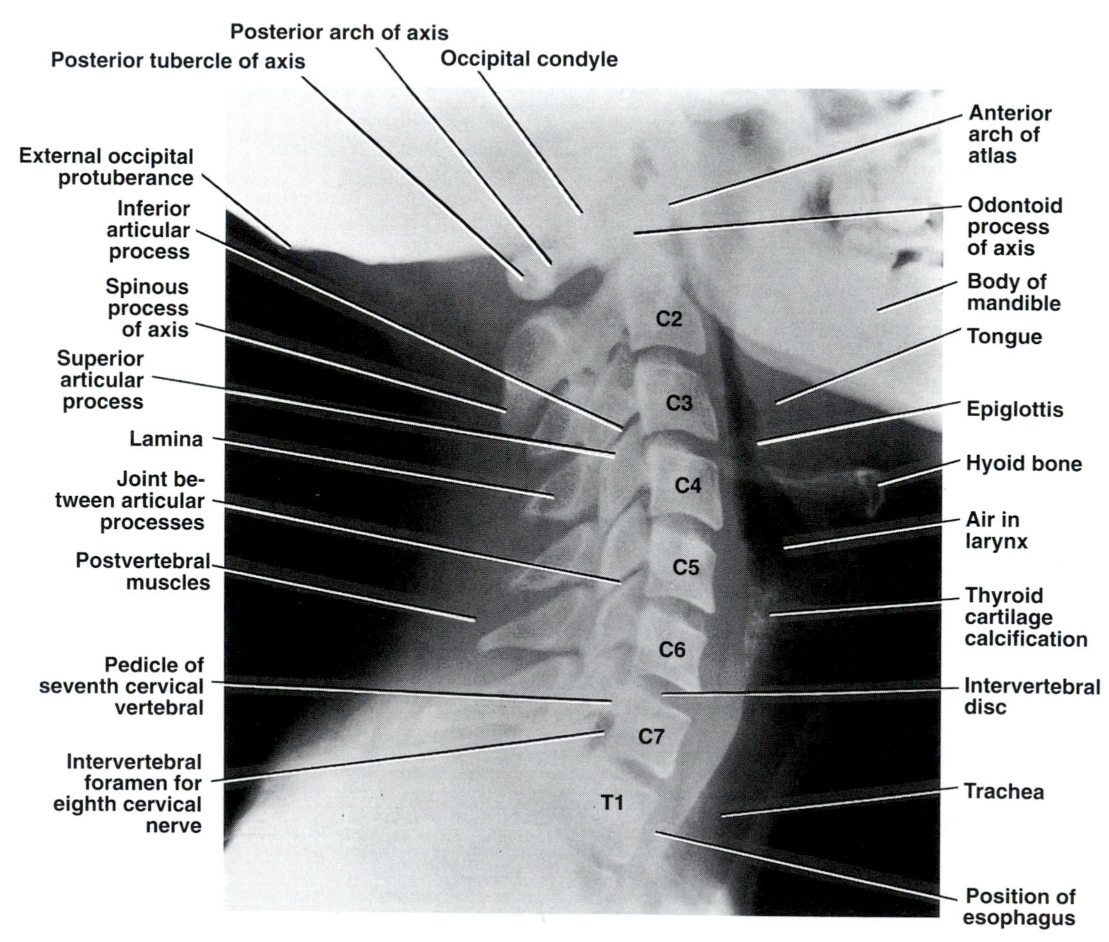

FIG 5–18.
Lateral radiograph of the neck.

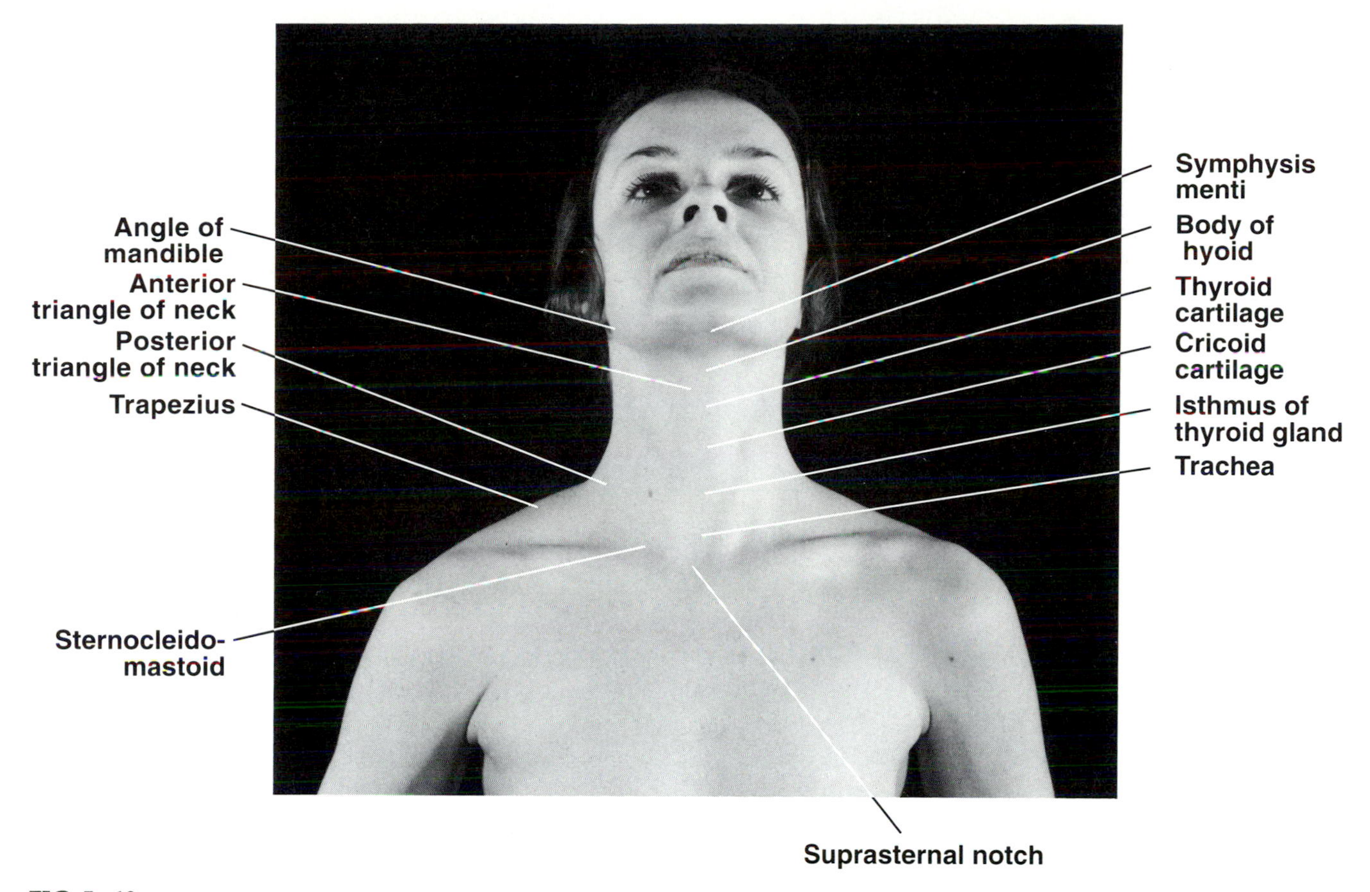

FIG 5–19.
Anterior view of the head and neck. The atlanto-occipital joints and the cervical part of the vertebral column are partially extended for full exposure of the front of the neck.

Cricoid Cartilage.—This lies at the level of the sixth cervical vertebra at the junction of the larynx with the trachea (see Fig 5–20 and 5–21). It is not as easy to feel as the thyroid cartilage, but it can be identified with gentle palpation when the patient is asked to swallow and the cartilage rises in the neck. In the unconscious, noncooperative, or combative patient who will not follow instructions and in whom it is vital to provide an emergency airway, the cricoid cartilage can be located about four fingerbreadths above the suprasternal notch.

Cricotracheal Ligament.—This fills in the interval between the cricoid cartilage and the first ring of the trachea (see Fig 5–20).

First Ring of the Trachea.—With careful palpation, this can usually be identified a few millimeters below the cricoid cartilage.

Isthmus of the Thyroid Gland.—This can be recognized as a soft structure crossing in front of the second, third, and fourth rings of the trachea (see Fig 5–19).

Inferior Thyroid Veins and the Thyroidea Ima Artery.—When these are present (although not palpable), they lie in front of the fifth, sixth, and seventh rings of the trachea.

Trachea.—The trachea is now receding from the surface as it approaches the root of the neck (see Fig 5–19).

Brachiocephalic Artery, the Left Brachiocephalic Vein, the Thymus Gland, and the Upper Margin of the Arch of the Aorta.—These are sometimes present in front of the trachea just above the suprasternal notch in young children.

Jugular Arch.—Although not palpable, this vein connects the two anterior jugular veins just above the suprasternal notch.

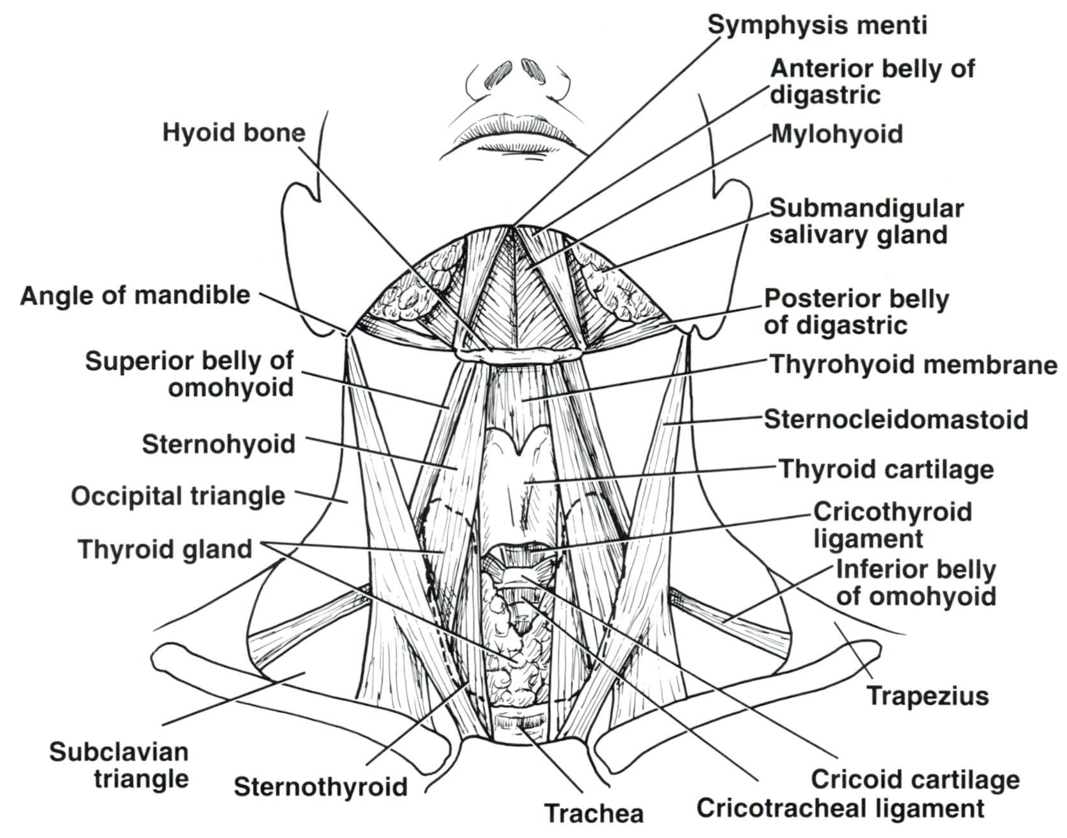

FIG 5–20.
Surface anatomy of the neck from the front.

Suprasternal Notch.—This is the upper margin of the manubrium sterni and can be easily felt between the anterior ends of the clavicles (see Fig 5–19). It lies opposite the lower border of the body of the second thoracic vertebra in the midrespiratory position.

Clinical Note

Tracheal deviation.—The trachea in the neck normally lies in the midline, and it should be possible for the examiner to confirm this by inserting the index and middle finger into the grooves on either side of the trachea between the trachea and the sternocleidomastoid muscles. Deviations of the trachea can usually be easily detected by this examination technique. In adults the trachea may measure as much as 1 in. (2.5 cm) in diameter, whereas in infants it is much narrower (8 mm).

Midline Posteriorly

In the midline posteriorly, the following structures can be palpated with the patient sitting or standing.

External Occipital Protuberance.—The external occipital protuberance lies in the midline at the junction of the head and neck (see Fig 5–5).

Nuchal Groove.—If the index finger is placed on the skin in the midline, it can be drawn downward in the nuchal groove. The first spinous process to be felt is the *seventh cervical vertebra (vertebra prominens).* Cervical spines 1 through 6 are covered by the ligamentum nuchae.

Side of the Neck

On the side of the neck the following structures can be palpated.

Sternocleidomastoid Muscle.—The sternocleidomastoid muscle can be palpated throughout its length as it passes upward from the sternum and clavicle to the mastoid process (see Fig 5–5). The muscle can be made to stand out by asking the patient to approximate his ear to the ipsilateral shoulder and at the same time rotate his head so that his face looks upward toward the contralateral side

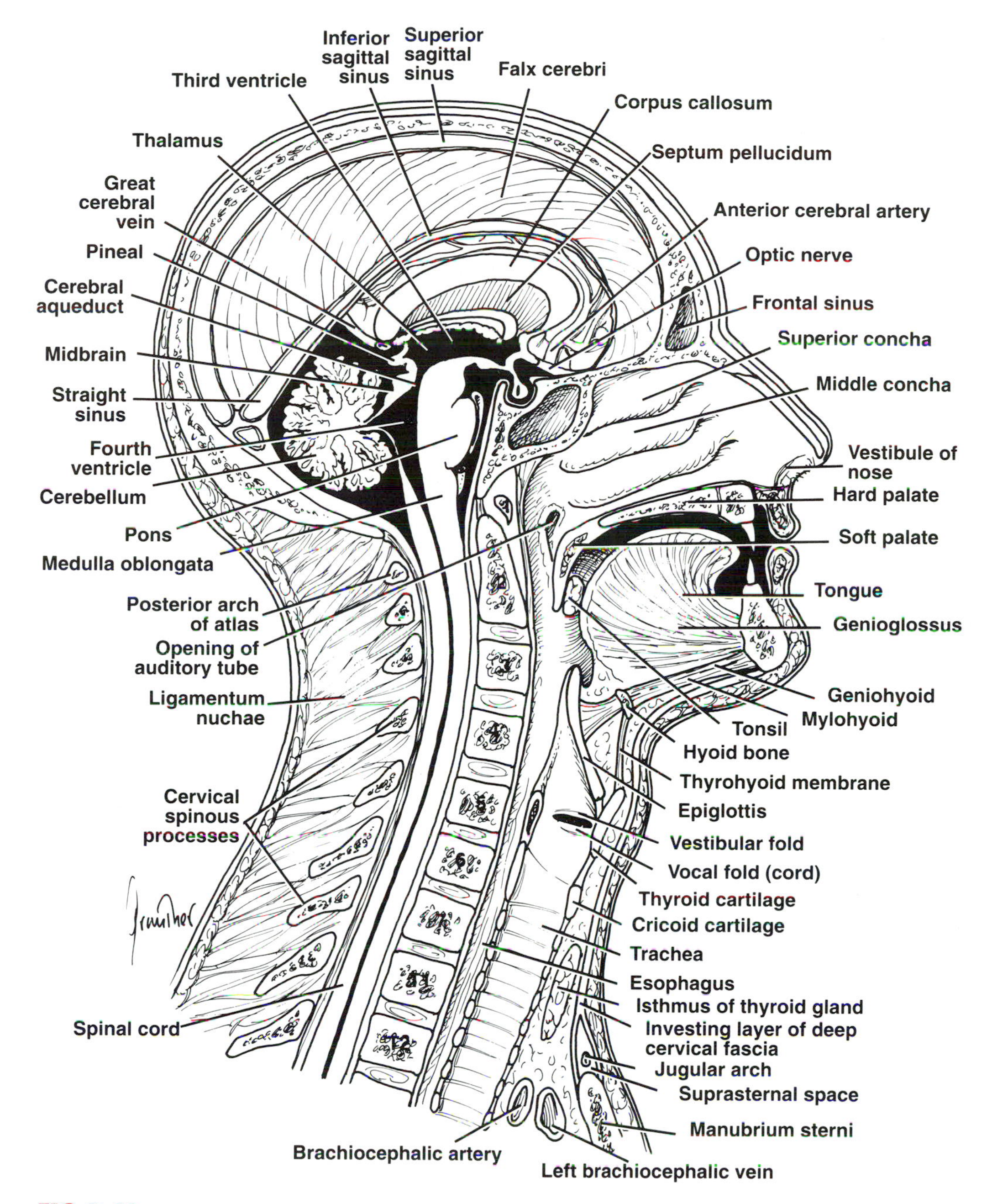

FIG 5–21.
Sagittal section of the head and neck.

(simply, the right sternocleidomastoid muscle turns the face and head to the left side). If the movement is carried out against resistance, the muscle will be felt to contract, and its anterior and posterior borders will be defined.

The sternocleidomastoid muscle, as explained previously, is used for descriptive purposes to divide the neck into anterior and posterior triangles (see Fig 5–1).

Trapezius Muscle.—The anterior border of the trapezius muscle (see Fig 5–5) may be felt by asking the patient to shrug his shoulders. The muscle will be seen to extend from the superior nuchal line of the occipital bone, downward and forward to the posterior border of the lateral third of the clavicle.

Platysma Muscle.—The platysma can be seen as a sheet of muscle just beneath the skin by having the

Clinical Note

Between the sternal and clavicular origins of the sternocleidomastoid muscle and the clavicle at the root of the neck lies a small triangular depression that can be seen and palpated. It is an important landmark, since it is here that central venous lines are often introduced.

---

patient clench his jaws firmly. The muscle extends from the body of the mandible downward over the clavicle onto the anterior thoracic wall.

Anterior Triangle

In the anterior triangle the following structures can be palpated.

Thyroid Gland.—The *isthmus of the thyroid gland,* as described previously, lies in front of the second, third, and fourth rings of the trachea (see Fig 5–19). The *lateral lobes of the thyroid gland* may be palpated deep to the sternocleidomastoid muscles. This is most easily carried out by standing behind the seated patient and asking him to flex his neck forward and so relax the overlying muscles. Then both lobes can be examined simultaneously with the finger tips of both hands. The gland moves superiorly on swallowing.

Carotid Sheath.—The carotid sheath, which contains the *carotid arteries,* the *internal jugular vein,* the *vagus nerve,* and the *deep cervical lymph nodes,* can be marked out by a line joining the sternoclavicular joint to a point midway between the tip of the mastoid process and the angle of the jaw. (*Note,* the sympathetic trunk does not lie within the carotid sheath, but is situated behind it.) At the level of the upper border of the thyroid cartilage, the *common carotid artery* bifurcates into the *internal* and *external carotid arteries* (see Fig 5–5).

Anterior Tubercle (Tubercle of Chassaignac) of the Transverse Process of the Sixth Cervical Vertebra.—This tubercle can be felt by deep palpation medial to the anterior border of the sternocleidomastoid muscle at the level of the cricoid cartilage. The common carotid artery can be compressed against it.

Posterior Triangle

In the posterior triangle the following surface markings are important.

Clinical Note

Arterial Pulse.—The pulsations of the three carotid arteries may be easily palpated at the anterior border of the sternocleidomastoid muscle at the level of the upper border of the thyroid cartilage.

Accessory Nerve.—This nerve is relatively superficial as it emerges from beneath the posterior border of the sternocleidomastoid muscle and runs downward and backward to pass beneath the anterior border of the trapezius (see Fig 5–5). The course of the nerve may be indicated as follows. Draw a line from the angle of the mandible to the tip of the mastoid process. Bisect this line at right angles and extend the second line downward across the posterior triangle; the second line indicates the course of the nerve.

Roots and Trunks of the Brachial Plexus.—These are situated in the lower anterior angle of the posterior triangle between the posterior border of the sternocleidomastoid and the clavicle (see Fig 5–5). The upper limit of the plexus may be indicated by a line drawn downward from the cricoid cartilage to the middle of the clavicle.

Third Part of the Subclavian Artery.—The artery occupies the lower anterior angle of the posterior triangle (see Fig 5–5). Its course may be indicated by a curved line, which passes upward from the sternoclavicular joint for about 0.5 in. (1.3 cm) and then downward to the middle of the clavicle. It is here, where the artery lies on the upper surface of the first rib, that its pulsations can be easily felt. The *subclavian vein* lies behind the clavicle and does not enter the neck.

External Jugular Vein.—This vein lies in the superficial fascia deep to the platysma muscle. It passes downward from the region of the angle of the jaw to the middle of the clavicle (see Fig 5–5). It perforates the deep fascia just above the clavicle and drains into the subclavian vein.

Surface Markings of the Salivary Glands

The *parotid salivary gland* lies below the ear in the interval between the mandible and the anterior border of the sternocleidomastoid muscle (see Fig 5–5). The *parotid duct* runs forward from the gland one fingerbreadth below the zygomatic arch (see Fig 5–5). It can be rolled beneath the examining finger at the anterior border of the masseter muscle as it turns medially and opens into the mouth opposite the upper second molar tooth (see Fig 1–10).

The *submandibular gland* (superficial part) lies beneath the lower margin of the body of the mandible (see Fig 5–20). The deep part of the gland, the submandibular *duct,* and the *sublingual gland* may be palpated through the mucous membrane covering the floor of the mouth in the interval between the tongue and the lower jaw. The submandibular duct opens into the mouth on the side of the frenulum of the tongue (see Fig 1–6).

CLINICAL ANATOMY

Clinical Anatomy of the Deep Cervical Fascia and the Fascial Spaces

The investing layer, the pretracheal layer, and the prevertebral layers of deep cervical fascia have been fully described on p. 169. Between the various layers is loose connective tissue that forms potential fascial spaces. Among the more important spaces clinically are the visceral, retropharyngeal, submandibular, and masticatory spaces (see Figs 5–3 and 5–4).

The deep fascia and the fascial spaces are important because organisms originating in the mouth, teeth, pharynx, and esophagus may spread among the fascial planes and spaces, and the tough fascia may determine the direction of the spread of infection and the path taken by pus. It is possible for blood, pus, or air in the retropharyngeal space or the visceral space to spread downward into the superior mediastinum. For example, mediastinitis may occur secondary to local infection following cricothyroidotomy or tracheostomy. Pneumomediastinum may be the only evidence that the trachea has been violated high up in the neck.

Acute Infections of the Fascial Spaces of the Neck

Infection is commonly caused by hemolytic streptococci or a mixture of aerobic and anaerobic organisms.

Dental Infections.—These most commonly involve the lower molar teeth. The infection spreads medially from the mandible into the submandibular and masticatory spaces. The inflammatory edema extends beneath the floor of the mouth and pushes the tongue forward and upward. Fever, difficulty in swallowing, and trismus are common presenting signs and symptoms.

The infection may extend upward into the fascia around the temporalis muscle, causing an obvious swelling that extends from below the jaw up onto the side of the head. Downward spread may involve the visceral space, which may lead to laryngeal edema and airway obstruction.

Fascial space infection as a complication of a dental infection is serious and requires immediate attention.

Ludwig's Angina.—This condition refers to an acute infection of the submandibular fascial space; it is commonly secondary to a dental infection, as described above.

Chronic Infection of the Fascial Spaces of the Neck

Chronic infections are much less common than acute infections.

Tuberculous Lymphadenitis.—Tuberculous infection of the deep cervical lymph nodes (secondary to infection of the tonsils) may result in liquefaction and destruction of one or more lymph nodes. The pus is first limited by the investing layer of deep fascia. Later, this becomes eroded and the pus passes into the less restricted superficial fascia. A dumbbell or collar-stud abscess is then present, having a superficial and deep component.

Tuberculous Osteomyelitis.—Tuberculous infection of the upper cervical vertebrae may give rise to pus that is trapped behind the prevertebral layer of deep cervical fascia. A midline swelling is formed, which bulges forward in the posterior wall of the pharynx. The pus may track laterally and downward behind the carotid sheath to reach the posterior triangle.

It is important to distinguish this condition from an acute abscess involving the *retropharyngeal lymph nodes.* These nodes lie behind the pharynx and in front of the prevertebral fascia. Such an abscess usually points on the posterior pharyngeal wall and, if untreated, ruptures into the pharyngeal cavity.

Clinical Anatomy of the Sternocleidomastoid Muscle

Protection

This strong, thick muscle crossing the side of the neck protects the underlying soft structures from blunt trauma. Suicide attempts by cutting one's throat often fail because the individual first extends the neck prior to making several horizontal cuts with a knife. Extension of the cervical part of the vertebral column and extension of the head at the atlanto-occipital joint cause the carotid sheath with its contained large blood vessels to slide posteriorly beneath the sternocleidomastoid muscle. To achieve the desired result with the head and neck fully extended, some individuals have to make several attempts and only succeed when the larynx and the greater part of the sternocleidomastoid muscles have been severed. The common sites for the wounds are immediately above and below the hyoid bone.

Congenital Torticollis

Most cases of congenital torticollis are a result of excessive stretching of the sternocleidomastoid muscle during a difficult labor. Hemorrhage occurs into the muscle and may be detected as a small rounded "tumor" during the early weeks after birth. Later, this becomes invaded by fibrous tissue, which contracts and shortens the muscle. The mastoid process is thus pulled down toward the sternoclavicular joint of the same side, the cervical spine is flexed, and the face looks upward to the opposite side. If left untreated, asymmetrical growth changes will occur in the face, and the cervical vertebrae may become wedge shaped.

Clinical Anatomy of the Esophagus

Esophageal Constrictions

The esophagus has three anatomical and physiological constrictions (Fig 5–22). The first constriction is located where the pharynx joins the upper end (site of cricopharyngeus muscle, see p. 14), the second is where the aortic arch and the left bronchus cross its anterior surface, and the third occurs where the esophagus passes through the diaphragm into the stomach. These constrictions are of considerable clinical importance since they are sites where swallowed foreign bodies, such as buttons, coins,* small toys, or dentures, may lodge or through which it may be difficult to pass an esophagoscope. Since a slight delay in the passage of food or fluid occurs at these levels, strictures develop following the drinking of caustic fluids. These constrictions are also the common sites of carcinoma of the esophagus. It is useful to remember that the respective distances of these three constriction sites from the upper incisor teeth are 6 in. (15 cm), 10 in. (25 cm), and 16 in. (41 cm), respectively (see Fig 5–22).

A patient can often accurately indicate the level of esophageal obstruction on the body surface where the food appears to be arrested, i.e., in obstruction at the upper end of esophagus, the patient will often point to the midline of the neck at the level of the cricoid cartilage; when the food is arrested where the aorta and left bronchus cross the esophagus, the patient will point to the anterior chest wall in the midline at the level of the sternal angle; and when the food is blocking the lower end of the esophagus, the patient will point to the lower end of the sternum or the epigastrium.

Esophagus and Cricothyroidotomy

Since the beginning of the esophagus lies behind the cricoid cartilage, it is possible, if special care is not exercised, for the scalpel blade to pass too deeply and perforate the esophagus.

Foreign Body Obstruction of the Esophagus

The great majority of foreign bodies lodge high up in the esophagus at the level of the cricopharyngeus muscle (see p. 14). If a foreign body is impacted in the esophagus at this site of the first constriction behind the cricoid cartilage, an acute respiratory obstruction may occur. The foreign body may be cleared with forceps under direct laryngoscopy when intubation is attempted.

Esophageal Injuries

Blunt and penetrating injuries are rare. Endoscopy and ingested foreign bodies may perforate the relatively thin esophageal wall. Ingestion of caustic substances or the sudden violent increase in abdominal pressure against a closed glottis may perforate the esophagus.

Clinical Anatomy of the Thyroid Gland

Thyroid Gland and the Airway

The close relationship between the trachea and the lobes of the thyroid gland commonly results in pressure on the trachea in patients with pathological enlargement of the thyroid gland.

Retrosternal Goiter

The attachment of the sternothyroid muscles to the thyroid cartilage effectively binds down the thyroid

*Swallowed coins that get stuck lie with their face in an anterior or posterior direction.

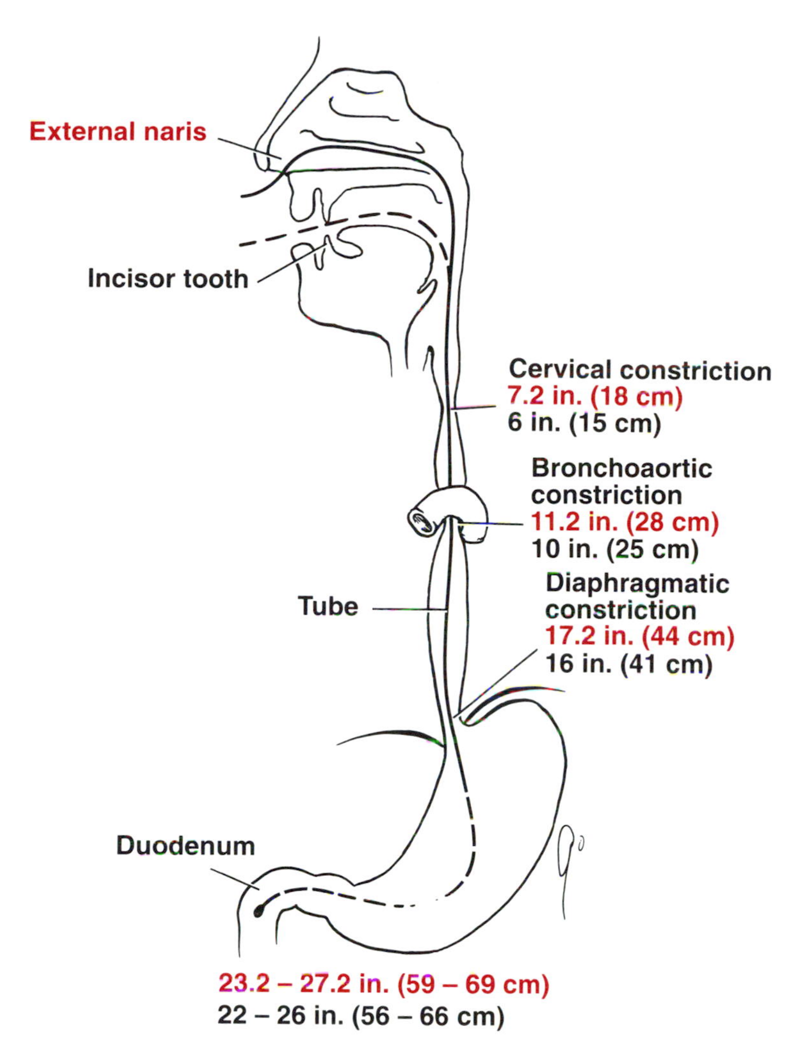

FIG 5–22.
Diagram showing the approximate approximate respective distances from the incisor teeth (black) and the external naris (red) to the normal three constrictions of the esophagus. To assist in the passage of a tube to the duodenum, the distances to the first part of the duodenum have also been included.

gland to the larynx and limits the upward expansion of the gland. Since there is no limitation to downward expansion, it is common for a pathologically enlarged gland to extend downward behind the sternum. A retrosternal goiter may compress the trachea and compromise the airway; it may also cause severe venous compression.

Thyroid Arteries and Important Nerves

The superior thyroid artery on each side is related to the external laryngeal nerve, which supplies the cricothyroid muscle. Damage to this nerve results in hoarseness and an inability to make the vocal cords tense. The inferior thyroid artery on each side is related to the recurrent laryngeal nerve. For the effects of damage to this nerve, see p. 37.

Thyroglossal Cyst

Initially, the thyroid gland develops as an outgrowth from the midline of the floor of the pharynx, and the *thyroglossal duct* is formed. Later, this duct elongates and its distal end becomes bilobed to form the thyroid gland. The remains of the duct connecting the thyroid gland with the tongue normally disappear.

A thyroglossal cyst may appear in childhood or adulthood. It is caused by a persistence of a segment of the thyroglossal duct. Such a cyst occurs in the midline of the neck at any point along the thyroglossal tract.

Clinical Anatomy of the Main Arteries of the Neck

Atherosclerosis of the Carotid Arteries and Neurologic Injury

Approximately 70% of strokes are caused by extracranial arteriosclerosis of the carotid and/or vertebral arteries. In the carotid arteries the atherosclerotic thromboses usually form slowly in the distal part of the common carotid and the first parts of the external and internal carotid arteries. The result is a diminished blood flow to the central nervous system. Less commonly, emboli, formed of fragments of plaques or blood clots, are carried distally to lodge in the ipsilateral central artery of the retina or in the smaller branches of the middle cerebral artery. Such a situation could produce the classic syndrome of ipsilateral blindness and contralateral hemiplegia, although it is unusual to have both at the same time.

Penetrating Injuries to the Carotid Arteries

Hemorrhage may be severe, with consequent hypotension or shock. An enlarging hematoma may press on the larynx or trachea, compromising the airway. Injuries to the internal carotid artery are usually associated with a central neurological deficit. Injuries to the common carotid arteries are less likely to cause neurological problems, provided there is adequate collateral circulation through the external carotid arteries and their branches.

Evaluation of arterial injuries in the neck may require angiography. In penetrating neck wounds that extend into the mediastinum below the level of the suprasternal notch (zone 1), an aortogram should be performed preoperatively. This would show the brachiocephalic, left common carotid, and left subclavian branches of the aortic arch, the vertebral arteries, and branches of the subclavian arteries (see Fig 4–4). Penetrating wounds between the suprasternal notch and the angle of the mandible (zone 2) that involve the carotid arteries do not usually require angiography, since anatomically they are readily exposed surgically. In penetrating wounds above the angle of the mandible (zone 3), a carotid arteriogram may be necessary to determine the extent of an internal carotid artery injury and the feasibility of performing an arterial repair or ligation; the status of the cerebral circulation can also be determined.

Aneurysms of the Carotid Arteries

Aneurysms of the carotid arteries in the neck are rare and are usually caused by arteriosclerosis. They are commonly located at the bifurcation of the common carotid artery. Expansion of the aneurysm may create pressure on the vagus nerve, (causing hoarseness), glossopharyngeal nerve (causing dysphagia), or hypoglossal nerve (causing weakness of the tongue); pressure on the sympathetic trunk as it lies behind the carotid sheath may cause Horner's syndrome. Rupture of the aneurysm resulting in hemorrhage may exert pressure on surrounding structures and compromise the airway.

Clinical Anatomy of the External Jugular Vein

Visibility of External Jugular Vein

This vein is usually less obvious in children and women because their subcutaneous tissue tends to be thicker than the tissue in men. In obese individuals, the vein may be difficult to identify even when they are asked to hold their breath, which impedes the venous return to the right side of the heart and distends the vein.

The superficial veins of the neck tend to be enlarged and often tortuous in professional singers due to prolonged periods of raised intrathoracic pressure.

Venous Manometer

The external jugular vein serves as a useful venous manometer. Normally, when the patient is lying at a horizontal angle of 30° the level of the blood in the external jugular veins reaches about a third of the way up the neck. As the patient sits up, the blood level falls until it is no longer visible behind the clavicle.

Anatomy of External Jugular Vein Catheterization

The external jugular vein may be used for catheterization, but the presence of valves or tortuosity may make the passage of the catheter difficult. Because the right external jugular vein is in the most direct line with the superior vena cava, it is the one most commonly used (Fig 5–23).

The patient is inclined with the head at an angle of 30°; his head is turned to the opposite side. Since the valves are usually incompetent, the external jugular vein becomes distended. The vein may be further distended by obstructing the outflow of the vein by placing a finger on the vein above the clavicle. The vein is catheterized about halfway between the level of the cricoid cartilage and the clavicle. The passage of the catheter should be performed during inspiration when the valves are open. Getting the catheter past the valves and around the angle of union of the vein with the subclavian vein is often facilitated by passing a flexible angiographic wire through the catheter lumen to serve as a guide.

The following anatomical structures are pierced by the needle and catheter:

1. Skin
2. Superficial fascia and platysma
3. Wall of the external jugular vein.

Clinical Anatomy of the Internal Jugular Vein

Penetrating Wounds of the Internal Jugular Vein

The hemorrhage of low-pressure venous blood into the loose connective tissue beneath the investing layer of deep cervical fascia may present as a large, slowly expanding hematoma. *Air embolism* is a serious complication of a lacerated wall of the internal jugular vein. Because the wall of this large vein contains very little smooth muscle, its injury is not followed by contraction and retraction (as occurs with

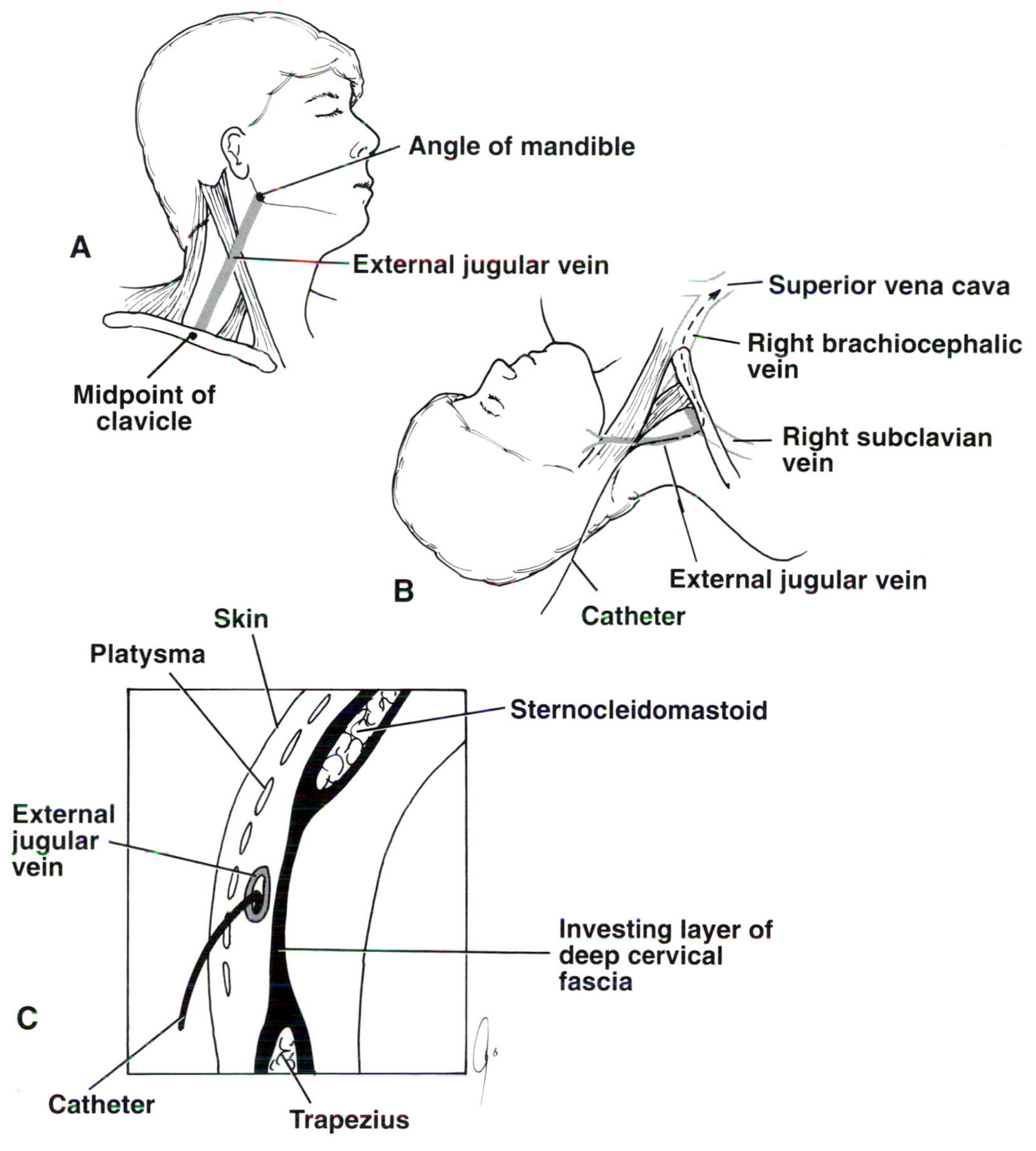

FIG 5–23.
Catheterization of the right external jugular vein. **A,** surface marking of the vein. **B,** site of catheterization. Note how the external jugular vein joins the subclavian vein at a right angle. **C,** cross section of neck showing relationships of external jugular vein as it crosses the posterior triangle of the neck.

arterial injuries). Moreover, the adventitia of the vein wall is attached to the deep fascia of the carotid sheath which hinders the collapse of the vein. Probing neck wounds that penetrate beneath the platysma muscle is contraindicated in the emergency department because of the risk of dislodging a blood clot and provoking hemorrhage. Blind clamping of venous wounds is also prohibited since the vagus and hypoglossal nerves are in the vicinity.

Anatomy of Internal Jugular Vein Catheterization

The internal jugular vein is remarkably constant in position. It descends through the neck from a point halfway between the tip of the mastoid process and the angle of the jaw to the sternoclavicular joint (Fig 5–24). The vein runs in the carotid sheath with the carotid arteries on its medial side. Above it is overlapped by the anterior border of the sternocleidomastoid muscle, and below it is covered laterally by this muscle. Just above the sternoclavicular joint the vein lies beneath a skin depression between the sternal and clavicular heads of the sternocleidomastoid muscle (see Fig 5–24).

Catheterization of the right internal jugular vein is preferred to the left for the following reasons.

1. The right internal jugular vein, the right brachiocephalic vein, the superior vena cava, and the right atrium are in a straight line. (The left internal jugular vein joins the left brachiocephalic vein at an angle, and the left brachiocephalic vein, in turn, joins the superior vena cava at an angle).
2. The right internal jugular vein in adults is larger than the left vein.
3. The cervical dome of the right pleural cavity is lower than that on the left side, and therefore it is less likely to be damaged.
4. The thoracic duct lies on the left side of the root of the neck and could be injured.

Posterior Approach.—The tip of the needle and the catheter are introduced into the vein about two fingerbreadths above the clavicle at the posterior border of the sternocleidomastoid muscle (Fig 5–25). The needle and catheter are directed toward the suprasternal notch. The following anatomical structures are pierced by the needle and catheter:

1. Skin
2. Superficial fascia and platysma muscle
3. Investing layer of deep cervical fascia (two layers—it is split here)
4. Wall of the internal jugular vein.

Anterior Approach.—With the patient's head turned to the left, the triangle formed by the sternal and clavicular heads of the sternocleidomastoid muscle and the medial end of the clavicle are identified (see Fig 5–25). A shallow skin depression usually overlies the triangle. The patient is inclined 30° in the head-down position to distend the internal jugular vein. The needle and catheter are inserted into the vein at the apex of the triangle at a 30° angle in a caudal direction. Identical structures are pierced by the needle as in the posterior approach.

Anatomy of Complications of Catheterization.—Complications are as follows.

1. Hematoma can be caused by damage to the wall of the internal jugular vein or injury to the common carotid artery. Since the common carotid artery lies medial to the vein in the carotid sheath, it can be damaged by a badly positioned needle.
2. Pneumothorax or hemothorax can occur secondary to puncture of the cervical dome of the pleura. It is more common on the left side, since the pleura rises to a higher level there.
3. Thoracic duct injury is rare and occurs if the needle pierces the duct while performing the catheterization on the left side.
4. Pneumomediastinum has been reported as a complication in infants. The needle is directed too far medially, causing damage to the trachea. The very small area and the extremely small structures involved in this age group require very careful handling.
5. On rare occasions it is impossible to pass the wire even when the needle has been correctly inserted. The wire probably has become stuck on one of the valve cusps. A pair of valve cusps is normally situated within the lumen of the internal jugular vein just above its junction with the subclavian vein to form the brachiocephalic vein.

Clinical Anatomy of the Cervical Lymph Nodes

Examination of the Lymph Nodes Relative to the Salivary Glands

The differential diagnosis between an enlarged lymph node and a swelling of the underlying salivary gland is sometimes difficult, even though the nodes have a firmer consistency. The superficial parotid (preauricular) nodes lie immediately in front of the tragus of the ear. When palpably enlarged they are usually mobile and can be distinguished from

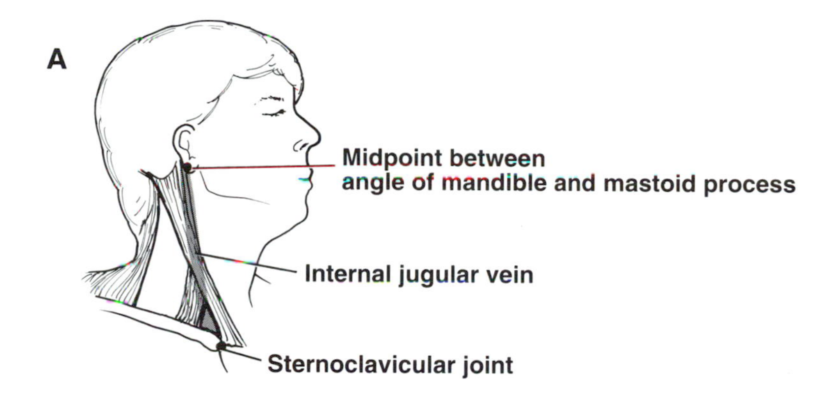

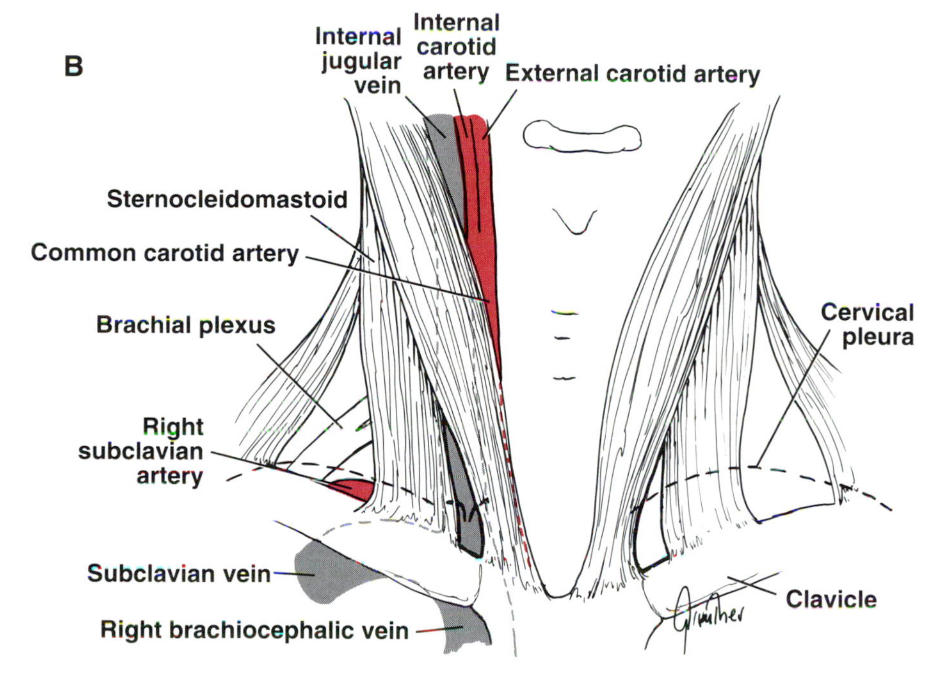

FIG 5–24.
Catheterization of the right internal jugular vein. **A,** surface marking of the vein, side view. Note the position of the skin depression over the triangular interval between the sternal and clavicular heads of the sternocleidomastoid and the clavicle. **B,** surface marking of the vein, anterior view. Note the position of the valve cusps of the internal jugular vein just above its junction with the subclavian vein to form the right brachiocephalic vein. Note also the positions of the common carotid artery and the cervical pleura.

the underlying gland. There are usually three submandibular nodes that lie superficial to the salivary gland but deep to the investing layer of deep cervical fascia. When palpably enlarged, these nodes can usually be recognized as separate entities and can be moved on the underlying salivary gland. In cases where the nodes become invaded by metastases, they are hard, and if the neoplasm has spread into the neighboring salivary gland, the nodes become fixed.

Examination of the Deep Cervical Lymph Nodes
These are best examined from behind the patient. The examination is made easier by asking the patient to flex the neck slightly to slacken the investing layer of deep cervical fascia and reduce the tension in the infrahyoid muscles and the sternocleidomastoid muscles. The nodes are examined progressively from above downward on the two sides.

Since all the lymph from the head and neck ultimately drains into the deep cervical nodes, a patho-

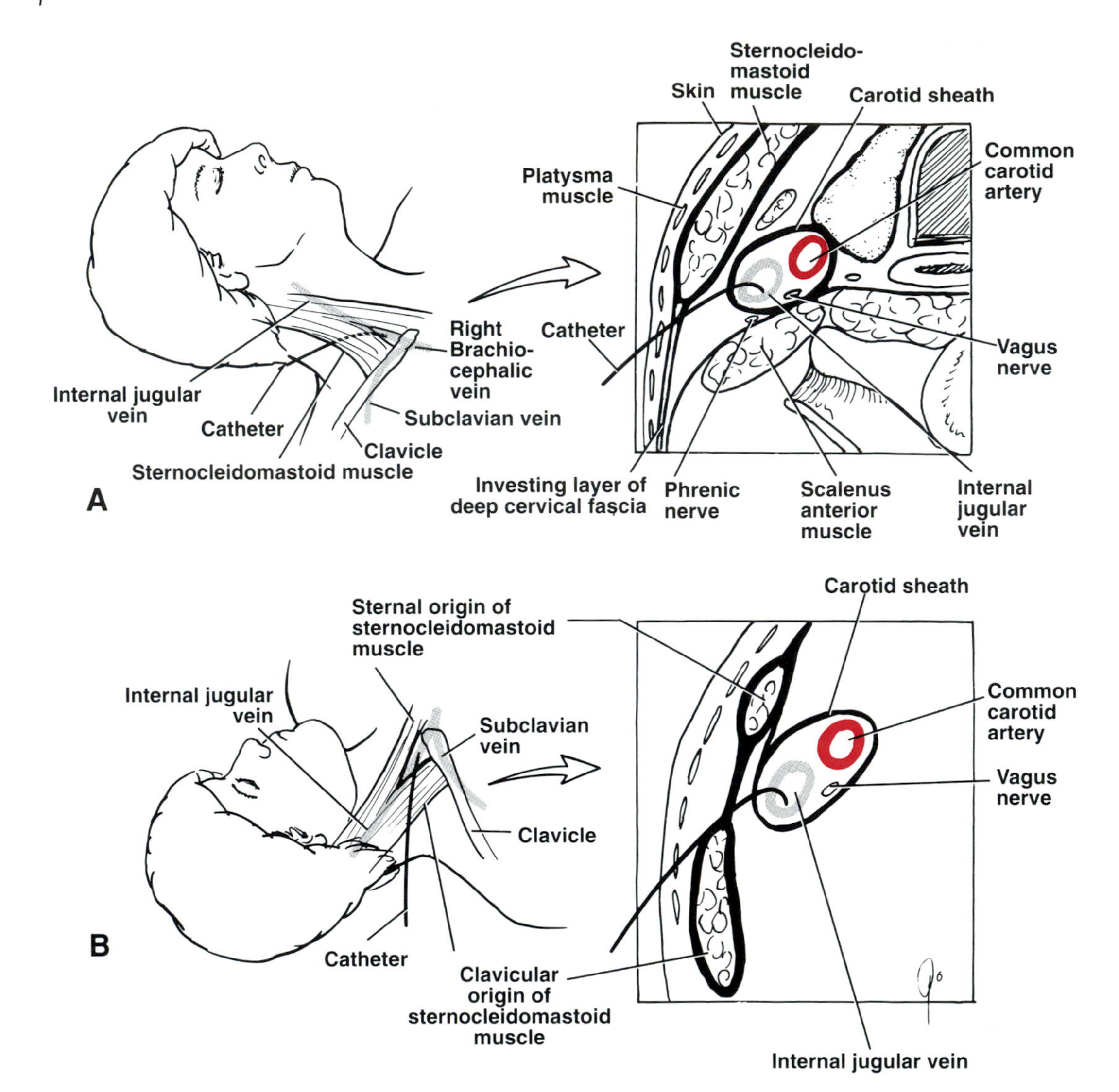

FIG 5–25.
Catheterization of the right internal jugular vein. **A,** posterior approach. Note the position of the catheter relative to the sternocleidomastoid muscle and the common carotid artery. **B,** anterior approach. Note that the catheter is inserted into the vein close to the apex of the triangle formed by the sternal and clavicular heads of sternocleidomastoid muscle and the clavicle.

logical enlargement of a deep cervical node dictates that a wide area of the head and neck has to be meticulously examined to find the cause (see Fig 5–12). In cases with metastases the physician should remember certain sites where the primary growth may be small or overlooked, for example, in the larynx, the pharynx, the cervical part of the esophagus, and the external auditory meatus. The bronchi, breast, and stomach are sometimes the site of the primary tumor. In these cases the secondary growth has spread far beyond the local lymph nodes.

Clinical Anatomy of the Glossopharyngeal Nerve

Testing the Integrity of the Glossopharyngeal Nerve

This nerve may be evaluated by testing the patient's sensation of touch and taste on the posterior third of the tongue. Normally, touching the posterior part of the tongue induces the gag reflex.

Clinical Anatomy of the Vagus Nerve

Testing the Integrity of the Vagus Nerve

Clinical examination of the vagus nerve depends on testing the function of the branches to the pharynx, soft palate, and the larynx. The pharyngeal reflex may be tested by touching the lateral wall of the pharynx with a tongue blade. This should immediately cause the patient to gag, i.e., the pharyngeal muscles will contract.

The innervation of the soft palate may be tested by asking the patient to say "ah." Normally, the soft palate rises and the uvula moves backward in the midline.

Since all the muscles of the larynx are supplied by branches of the vagus nerve, hoarseness or absence of the voice may occur following injury to the nerve. Laryngoscopic examination may reveal abductor paralysis in a lesion involving the recurrent laryngeal branch of the vagus (see p. 37).

Clinical Anatomy of the Accessory Nerve

Testing the Integrity of the Accessory Nerve

The accessory nerve (spinal root) supplies the sternocleidomastoid and trapezius muscles. Clinical examination of this nerve involves asking the patient to rotate his head to one side against resistance, causing the sternocleidomastoid of the opposite side to come into action. Then the patient is asked to shrug his shoulders, causing the trapezius muscles to come into action.

Clinical Anatomy of the Hypoglossal Nerve

Testing the Integrity of the Hypoglossal Nerve

The hypoglossal nerve supplies the muscles of the tongue. To test the integrity of the hypoglossal nerve, the patient is asked to extend his tongue. If there is a lower motor nerve lesion, it will be noted that the tongue deviates toward the paralyzed side. The tongue will be smaller on the side of the lesion owing to muscle atrophy, and fibrillation may accompany or precede the atrophy. If a patient has a lesion of the cerebral cortex or the corticonuclear fibers, there will be no atrophy or fibrillation of the tongue, and on protrusion, the tongue will deviate to the side opposite the lesion. The explanation for these phenomena is as follows. The right and left genioglossus muscles pull the tongue forward in the midline. Should there be paralysis (lower motor neuron), the tongue will be deviated to the side of the lesion. The hypoglossal nerve nucleus receives corticonuclear fibers from both cerebral hemispheres, but the portion of the nucleus that supplies the genioglossus muscle receives corticonuclear fibers only from the opposite cerebral cortex. It therefore follows that if the right corticonuclear fibers are sectioned, the left genioglossus muscle will be paralyzed (upper motor neuron), and the tongue will point to the left side.

Clinical Anatomy of the Cervical Part of the Sympathetic Trunk

Horner's Syndrome

Horner's syndrome includes (1) constriction of the pupil, (2) ptosis, and (3) enophthalmos and is caused by an interruption of the sympathetic nerve supply to the orbit. Pathologic causes include lesions in the brainstem or cervical part of the spinal cord that interrupt the reticulospinal tracts descending from the hypothalamus to the sympathetic outflow in the lateral gray column of the first thoracic segment of the spinal cord. Such lesions include multiple sclerosis and syringomyelia. Traumatic injury to the cervical part of the sympathetic trunk, traction on the stellate ganglion caused by a cervical rib, or involvement of the ganglion in a metastatic lesion may interrupt the peripheral part of the sympathetic pathway to the orbit.

The anatomical explanation for the central and peripheral forms of Horner's syndrome is fully described in Chapter 9, p. 343.

Clinical Anatomy of the Phrenic Nerve in the Neck

Injury

The phrenic nerve can be injured by penetrating wounds in the neck. If that occurs, the paralyzed half of the diaphragm relaxes and is pushed up into the thorax by the positive abdominal pressure. Consequently, the lower lobe of the lung on that side may collapse.

Injury to the phrenic nerve may also occur during attempts at subclavian vein catheterization. The phrenic nerve runs vertically downward through the neck on the scalenus anterior muscle. Because of the obliquity of the scalenus anterior muscle, the nerve crosses the muscle from its lateral to its medial border. The nerve enters the thorax by passing anterior to the subclavian artery and behind the subclavian vein (see Fig 5–10). For nerve damage to occur the needle would have to pierce the posterior wall of the vein or miss the vein altogether. This complication can be avoided by directing the needle as anteriorly as possible when entering the vein (see p. 160).

Clinical Anatomy of the Brachial Plexus in the Neck

Clinical anatomy of the brachial plexus, including a discussion of the Duchenne-Erb palsy, may be seen in Chapter 15, p. 617.

Anatomical Structures That May Be Responsible for Cervical Swellings

Enlarged Cervical Lymph Nodes

Enlargement of the cervical lymph nodes is caused by infection or neoplasm. The possible areas of infection or the site of a primary growth should be carefully examined. The scalp, face, tongue, mouth, tonsil, ear, etc. all drain into the nodes. If no source of the enlargement is found, remember the hidden areas that also drain into the nodes—the nasopharynx, laryngeal recesses, thyroid gland, lungs, and external auditory meatus.

Pathology of the Thyroid Gland

Adenoma or adenocarcinoma of the thyroid gland produces an asymmetrical enlargement of the gland, and if it is visible it will move upward when the patient swallows. This movement results from the pretracheal fascia binding the gland to the larynx.

Conditions Related to the Persistence of the Thyroidglossal Duct

These conditions usually appear in childhood, adolescence, or in young adults.

Thyroglossal Cyst.—Cysts may occur at any point along the course of the thyroglossal tract and have been referred to previously. They occur most commonly in the region below the hyoid bone. Such a cyst occupies the midline and develops as a result of persistence of a small area of epithelium that continues to secrete mucus. As the cyst enlarges it is prone to infection.

Thyroglossal Fistula.—Occasionally a thyroglossal cyst ruptures spontaneously, producing a midline fistula.

Congenital Anomaly of the Lymphatic Vessels

Cystic hygroma.—If, in the formation of the lymphatic vessels, the mesodermal spaces fail to unite and join with the central system, a cystic hygroma results. The cystic hygroma gradually enlarges because the lymph collected in the cystic spaces cannot drain away owing to the atresia of the collecting vessels. This condition is commonly found in the neck, axilla, mediastinum, and tongue.

Congenital Anomalies of the Branchial System

Branchial Cyst.—Normally the second, third, and fourth pharyngeal clefts are buried by the downgrowth of the second pharyngeal arch. The space beneath becomes obliterated. Should the space remain, it forms a branchial cyst. Such cysts are not apparent at birth but gradually enlarge during childhood and become conspicuous in early adult life. They are found most commonly in the upper part of the neck, protruding from beneath the anterior border of the sternocleidomastoid muscle. They gradually increase in size and may become infected.

Branchial Sinus.—Should the second pharyngeal arch fail to bury the pharyngeal clefts completely, they will communicate with the surface, forming a branchial sinus. A branchial sinus may be found anywhere along the anterior border of the sternocleidomastoid muscle, especially at the lower part of the neck.

Branchial Fistula.—A branchial fistula is formed when the sinus extends up the neck between the internal and external carotid arteries and opens into the pharynx just behind the tonsil as well as opening onto the skin surface.

Branchial Cartilage.—An irregular mass of branchial cartilage several millimeters in diameter may be found in the subcutaneous tissues of the neck, usually just anterior to the lower third of the sternocleidomastoid muscle. It is thought that such pieces of cartilage are derived from one of the pharyngeal arches.

Aneurysm of the Common Carotid Artery

This produces a swelling beneath the anterior border of the sternocleidomastoid muscle. On palpation it pulsates, and a systolic bruit may be present.

Carotid Body Tumor

The tumor usually first becomes apparent in middle life. It is situated at the bifurcation of the common carotid artery at the level of the upper border of the thyroid cartilage. The swelling can be moved horizontally with ease but has little vertical mobility.

Pharyngeal Pouch
This condition is a posterior herniation of the mucous membrane of the pharynx between the lower border of the inferior constrictor muscle and the cricopharyngeus muscle. It usually occurs on the left side and gradually enlarges down the neck, producing a visible swelling. The swelling is soft and cystic, and when pressed upon, the contents are emptied into the pharynx. The main symptoms are dysphagia and regurgitation.

Neurofibroma
This may arise as a slowly growing tumor of the glossopharyngeal nerve, vagus nerve, the hypoglossal nerve, the sympathetic trunk, or the roots or trunks of the brachial plexus. The swelling may be moved across the course of the nerve but not in the direction of the course of the nerve.

Sternocleidomastoid 'Tumor' in Infants
This condition is caused by hemorrhage into the sternocleidomastoid muscle during a difficult delivery. Later invasion by fibrous tissue may lead to congenital torticollis.

ANATOMICAL-CLINICAL PROBLEMS

1. A 36-year-old woman was seen in the emergency department complaining of a swelling on the front of her neck and the sudden onset of breathlessness. On examination, a solitary swelling that was firm to touch was found on palpation to the right of the midline of the neck. The swelling moved upward and downward with swallowing. On careful palpation, it appeared to be continuous with the lower pole of the right lobe of the thyroid gland. A diagnosis of adenoma of the thyroid gland was eventually made. From your knowledge of anatomy, explain why the tumor moved upward with swallowing. What is the possible explanation for the sudden onset of breathlessness in this patient? Which part of the airway was being compromised? If this were an adenocarcinoma, which group of lymph nodes would you examine for evidence of metastases?

2. A 5-year-old girl was seen in the emergency department for a greenstick fracture of her right radius. During the course of the examination, it was noticed that she held her head to one side. The neck was held in the position of slight flexion; the left ear was held nearer the shoulder on the left side than normal. She tended to hold her head inclined so that she looked upward to the right. The mother said that she has always held her head slightly to one side. What is the diagnosis? Which muscle or muscles are involved? Are the right or left muscles shortened in this patient?

3. A 10-year-old boy was examined in the emergency department and found to have a painless superficial, fluctuant swelling below and behind the angle of the jaw on the left side. The skin over the swelling was cool to touch and showed no redness. Palpation of the neck revealed three tender, firm swellings matted together beneath the anterior border of the sternocleidomastoid muscle on the left side. The right side of the neck was normal. Examination of the palatine tonsils showed moderate hypertrophy on both sides and yellow exudate draining from the tonsillary crypts on the left side. A diagnosis of tuberculous cervical lymphadenitis was eventually made. Using your knowledge of anatomy, name the group of lymph nodes involved in the disease. What structures present in the neck would tend to limit the spread of the disease in the neck? Since the lymph node showing the most advanced stage of the disease was below and behind the angle of the mandible, which organ in the oral part of the pharynx was most likely to have served as the portal of entry to the tubercle bacilli?

4. A 3-year-old girl was seen in the emergency department with a fever, stridor, and hypertension of the neck. The pharynx looked normal except for what seemed to be fullness of the posterior pharyngeal wall. A lateral radiograph showed some enlargement of the retropharyngeal space. The epiglottis was normal. A diagnosis of retropharyngeal abscess was made. In anatomical terms explain the position of the abscess relative to the constrictor muscles of the pharynx, the fascia, and the vertebral column. Where is the abscess most likely to rupture?

5. During the physical examination of a 30-year-old woman, the physician noticed the presence of a swelling on the front of her neck. The swelling was about 1 in. (2.5 cm) in diameter, painless, and situated just beneath the upper third of the anterior border of the sternocleidomastoid muscle. On palpation, the swelling fluctuated in size and did not appear to be attached to surrounding structures. On questioning the patient, she said that she had noticed the swelling many years previously and that it was gradually increasing in size. What is your diagnosis? Can you explain its presence?

6. A 60-year-old man was seen in the emer-

gency department complaining that he had a piece of meat stuck in his throat. On being asked where exactly he though it was stuck, the patient pointed to his neck just below the thyroid cartilage. The patient appeared to be a little dyspneic and was slightly wheezing. Using your knowledge of anatomy, describe the sites in the esophagus where foreign bodies are commonly trapped.

7. The jugular veins are commonly used to establish a central venous line. Why is it sometimes difficult to pass a catheter from the external jugular vein into the right atrium? What is the surface marking of the internal jugular vein in the neck? What anatomical structures are likely to be damaged when a faulty technique is used in cannulating the internal jugular vein?

8. In deep penetrating injuries of the neck involving the common carotid artery, the status of the collateral circulation determines the feasibility of ligation versus a reconstructive procedure. Describe the collateral circulation of the common carotid artery. How may injury to the artery produce loss of sight on the same side and contralateral hemiplegia? How may a carotid artery injury produce airway obstruction?

9. A 58-year-old man presented with a small swelling below his chin. On examination, a single small, hard swelling could be palpated in the submental triangle. It was mobile on the deep tissues and not tethered to the skin. A diagnosis of a malignant secondary deposit in a submental lymph node was considered. Where would you look for the primary carcinoma? Where do the submental nodes drain?

10. A 17-year-old boy was seen in the emergency department following a stab wound at the front of the neck. The knife entrance wound was located on the left side of the neck just lateral to the tip of the greater cornu of the hyoid bone. During the physical examination the patient was asked to protrude his tongue, which deviated to the left. In anatomical terms can you explain this finding?

11. A 43-year-old woman was seen in the emergency department with a large abscess in the middle of the right posterior triangle of the neck. The abscess was red, hot, and fluctuant. The center of the abscess showed evidence that it was pointing and about to rupture. The physician decided to incise the abscess. He found the interior of the abscess was extensive, and he inserted a drain. Several days later the patient returned to the department for the dressings to be changed. She stated that she felt much better and that her neck was no longer painful. However, there was one thing that she could not understand. She could no longer raise her right hand above her head to brush her hair. In anatomical terms can you explain this patient's disability?

ANSWERS

1. The thyroid gland is surrounded by the pretracheal layer of deep cervical fascia, which binds the gland to the larynx and the trachea. As the larynx moves upward on swallowing, the thyroid and the adenoma also moved upward. Hemorrhage into an adenoma may occur, causing it to expand suddenly and press upon neighboring structures. In this patient the trachea was compressed, compromising the airway. The thyroid gland drains its lymph mainly into the deep cervical lymph nodes, and therefore this group of nodes would be examined for evidence of metastases.

2. The diagnosis is congenital torticollis secondary to hemorrhage into the left sternocleidomastoid muscle during birth. Closer examination of this patient showed that the anterior border of the left sternocleidomastoid muscle was more prominent than the border on the right muscle, and thus the left muscle was diseased. An understanding of the actions of this muscle (see p. 172) will explain the deformity. The deformity is only noticed when the neck begins to elongate as the child grows. Other muscles of the neck show secondary shortening, and wedge-shaped deformities of the cervical vertebrae may occur. Failure to correct the deformity surgically leads to asymmetrical changes in the face because the eyes attempt to work on the same horizontal plane.

3. The deep cervical lymph nodes were involved; the jugulodigastric member of this group was the most extensively infected. Tuberculous infection may remain localized to one node for some time or involve other neighboring nodes, as in this case, which then become matted together. Later the caseating material liquifies and breaks through the capsule of the node. To begin with, the cold abscess is localized beneath the investing layer of deep cervical fascia. Erosion of the fascia eventually occurs and a large abscess is formed beneath the skin. This may become secondarily infected and rupture through the skin, to form a discharging sinus. The left palatine tonsil was the site of entry of the tubercle bacilli in this patient. The organisms had spread to the jugulodigastric lymph node with the lymph.

4. A retropharyngeal abscess is secondary to a

pharyngitis and occurs most commonly in infants and young children. It is caused by a breakdown of the retropharyngeal lymph nodes that drain the pharynx that are situated in the retropharyngeal fascial space. The space is bounded anteriorly by the constrictor muscles of the pharynx covered by pharyngeal fascia. Posteriorly the space is limited by the prevertebral layer of deep cervical fascia. The abscess, if untreated, tends to point anteriorly, rupturing through the posterior pharyngeal wall.

5. This patient has a branchial cyst. Normally the space beneath the downgrowth of the second branchial arch (cervical sinus) becomes obliterated by apposition and fusion of its walls. Should this fail to occur, a branchial cyst forms. Such cysts are lined with stratified squamous epithelium and are filled with a creamy fluid containing cholesterol crystals. Because the cysts tend to increase gradually in size and may become infected, they should be removed surgically.

6. The esophagus has the following three anatomical and physiological constrictions: (1) where the pharynx joins the upper end of the esophagus behind the cricoid cartilage, (2) where the left bronchus and the aortic arch cross the anterior surface of the esophagus, and (3) where the esophagus passes through the diaphragm to join the stomach. Foreign bodies, including the large piece of meat in this patient, frequently become stuck at the level of the cricopharyngeus muscle at the upper end of the esophagus. Here the foreign body may be so large that it presses forward on the cricoid cartilage and the beginning of the trachea, causing respiratory obstruction, as in this patient.

7. The following difficulties may be encountered when a catheter is advanced down the external jugular vein. (1) The catheter tip may catch in the valves. The vein has two pairs of valves, one pair about 4 cm above the clavicle and the lower pair at its entrance into the subclavian vein. (2) The vein may be small, tortuous, duplicated, and difficult to negotiate. By having the patient in a slight head-down position or applying a positive airway pressure, the vein will become distended. (3) The external jugular vein is not in direct line with the superior vena cava.

The surface marking of the internal jugular vein is from a point midway between the tip of the mastoid process and the angle of the jaw down to the sternoclavicular joint.

The following anatomic structures are closely related to the internal jugular vein and may be damaged by a poorly attempted venous catheterization: (1) internal and common carotid arteries, (2) cervical dome of the parietal pleura, (3) thoracic duct on the left side (rare), and (4) pneumomediastinum in children (see p. 196).

8. The common carotid artery has no branches in the neck except for its terminal branches, the internal and external carotid arteries. The internal carotid artery has no branches in the neck, but at the base of the brain it takes part in the arterial circle (of Willis), where it anastomoses with the branches of the vertebral arteries and the internal carotid artery of the opposite side. The external carotid artery, however, gives off numerous important branches in the neck that anastomose with the fellow branches from the opposite side; these branches are the superficial temporal and maxillary arteries, posterior auricular, occipital, facial, lingual, superior thyroid, and ascending pharyngeal arteries.

Since the internal carotid artery gives off the ophthalmic artery and the middle cerebral artery, severe damage or blockage of the common carotid artery could cause ipsilateral blindness and contralateral hemiplegia.

The common carotid artery is contained within the carotid sheath beneath the investing and pretracheal layers of deep cervical fascia. Hemorrhage beneath the deep fascia could spread medially and compress the larynx or the trachea, thus compromising the airway.

9. The submental lymph nodes drain the tip of the tongue, the floor of the mouth in the region of the frenulum of the tongue, the gums and incisor teeth, the middle third of the lower lip, and the skin over the chin. On examination of the inside of the patient's mouth, a small hard-based carcinomatous ulcer was found on the left side of the tip of the tongue. The submental lymph nodes drain into the deep cervical lymph nodes.

10. The protruded tongue pointing to the left side indicated a lesion of the left hypoglossal nerve. The hypoglossal nerve descends in the neck between the internal carotid artery and the internal jugular vein. At about the level of the tip of the greater cornua of the hyoid bone it turns forward and medially and crosses the internal and external carotid arteries and the lingual artery. In this patient the point of the knife blade severed the nerve in this location.

11. The spinal part of the accessory nerve crosses the posterior triangle of the neck in a comparatively superficial position, being only covered by skin, superficial fascia, and the investing layer of deep cervical fascia. The surface marking of the

nerve is as follows: Bisect at right angles a line joining the angle of the jaw to the tip of the mastoid process. Continue the second line downward and backward across the posterior triangle, and it closely follows the course of the nerve. The knife opening the abscess in this patient had also cut the nerve, thus paralyzing the trapezius muscle. In order to raise the hand above the head, it is necessary for the trapezius muscle, assisted by the serratus anterior, to contract and rotate the scapula so that the glenoid cavity faces upward.

6

The Face, Scalp, and Mouth

Facial, scalp, and mouth injuries are commonly encountered in the practice of emergency medicine and vary in seriousness from a small skin laceration to major maxillofacial trauma. One of the tenets of trauma resuscitation is that however grotesque the facial injuries might look, few are immediately life threatening. Attention should properly be concentrated on the integrity of the airway, the cardiovascular and respiratory systems, and neurologic function. Nevertheless, the face is one of the most important areas of the body; it is the window into the individual's personality. Inadequate or untimely delay in the treatment of facial injuries may later lead to irreversible damage and disfigurement. Disfiguring scars may have a devastating effect on the psychosocial relationship of the person with his or her family or peers. This is especially important in the child, since an unsightly injury may result in maladjustment in adulthood. Furthermore, the presence of facial injuries may be a clue to the presence of brain and cervical spine injuries.

Against this background, this chapter discusses the basic anatomy of the face, scalp, and mouth. Emphasis is placed on the clinical relevance of the structures covered and the normal radiology of the region. The chapter specifically excludes consideration of the neurocranium and brain.

BASIC ANATOMY

Basic Anatomy of the Face

Skin

The skin of the face shows considerable variation in thickness, being thinnest over the eyelids. At the lips, the external nares of the nose, and the margins of the eyelids, the skin of the face is continuous with the mucous membranes lining the mouth, the nasal cavities, and the conjunctival sac. The skin of the face contains numerous sweat and sebaceous glands and is very mobile because of a separation from the underlying bones by a layer of superficial fascia.

Wrinkle lines of the face result from the repeated folding of the skin perpendicular to the long axis of the underlying contracting muscles, coupled with the loss of youthful skin elasticity. Thus we find horizontal wrinkles on the brow, crow's feet wrinkles at the lateral angles of the eyes, and vertical wrinkles above and below both lips (see Fig 6–6).

Clinical Note

Scars.—Surgical incisions of the face heal with less scarring if they are made along wrinkle lines. A knowledge of the position of the wrinkle lines or, as sometimes stated, "relaxed lines of tension," is also important when closing facial lacerations. Sutured lacerations that lie parallel to the tension lines spread less than lacerations that run at right angles to the tension lines.

Subcutaneous Tissue

This loose connective tissue connects the skin to the underlying bones. The muscles of facial expression, blood and lymphatic vessels, and nerves are embedded in it; the parotid duct also is located here.

Deep Fascia

Apart from the delicate connective tissue that surrounds the muscles of facial expression, there is *no deep fascia in the face.*

Basic Anatomy of the Sensory Nerve Supply of the Face

The skin of the face is supplied by branches of the three divisions of the trigeminal nerve, except for the small area over the angle of the mandible and the parotid gland (Fig 6–1), which is supplied by the great auricular nerve (C2 and C3). The overlap of the three divisions of the trigeminal nerve is slight in comparison to the considerable overlap of adjacent dermatomes of the trunk and limbs.

Sensory nerves not only supply the skin of the face but also supply proprioceptive fibers to the underlying muscles of facial expression. They are, in addition, the sensory nerve supply to the mouth, teeth, nasal cavities, and paranasal sinuses.

Ophthalmic Nerve

The ophthalmic nerve supplies the skin of the forehead, the upper eyelid, the conjunctiva, and the side of the nose down to and including the tip. The following five branches of the nerve pass to the skin.

1. *Lacrimal nerve.* This supplies the skin and conjunctiva of the lateral part of the upper eyelid (Fig 6–2).

2. *Supraorbital nerve.* This nerve winds around the upper margin of the orbit at the supraorbital notch (see Fig 6–2). It supplies the skin and con-

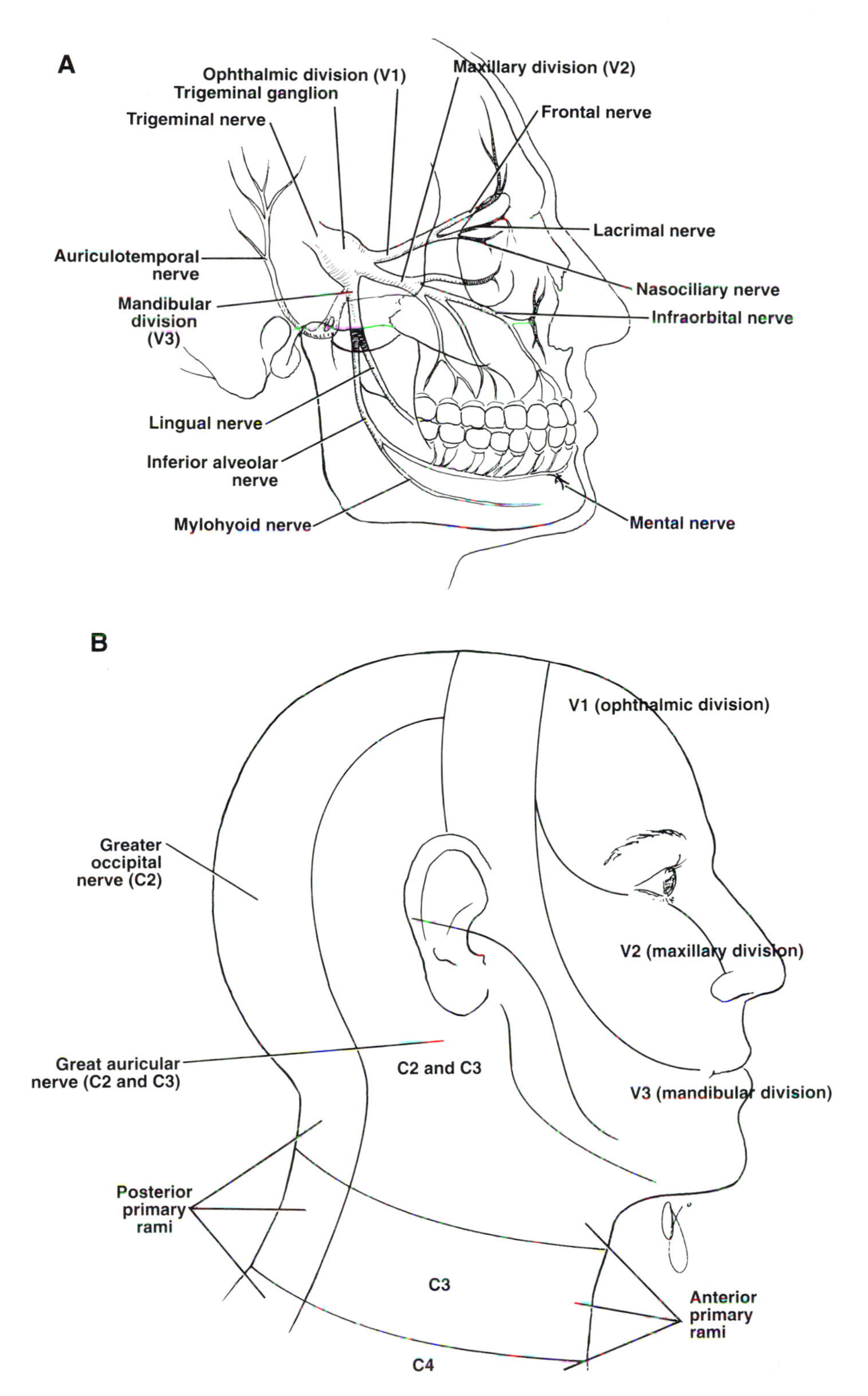

FIG 6–1.
A, the distribution of the trigeminal nerve. **B,** the facial cutaneous distribution of the ophthalmic (V1), the maxillary (V2), and the mandibular (V3) divisions of the trigeminal nerve. The skin over the angle of the jaw is supplied by the great auricular nerve (C2 and C3 segments of the spinal cord).

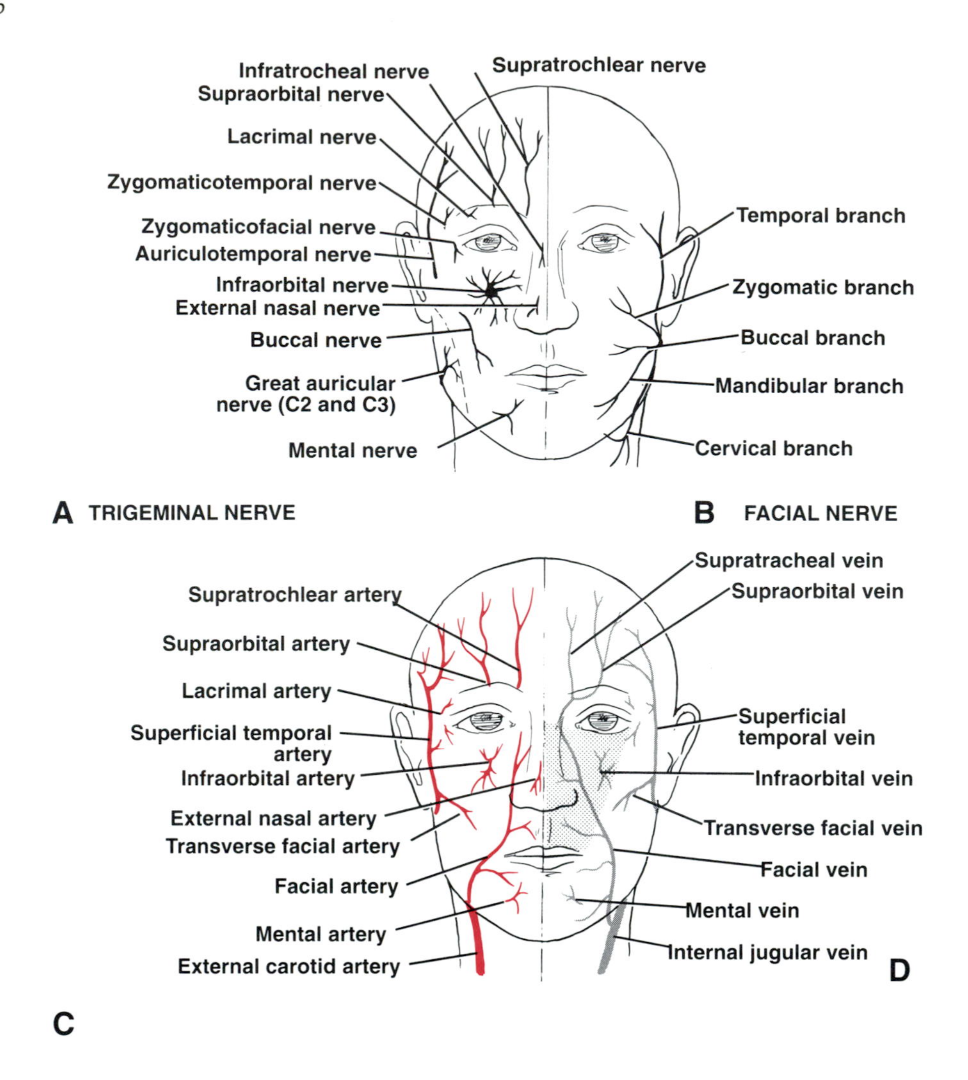

FIG 6–2.
A, sensory nerves supplying the skin of the face. **B,** branches of the facial nerve (seventh cranial nerve) to the muscles of facial expression. **C,** arterial supply of the face. **D,** venous drainage of the face. Note the potentially dangerous triangular zone (dotted area) of skin from which infection may spread via the anterior facial vein to the cavernous sinus inside the skull.

junctiva on the central part of the upper eyelid; it also supplies the skin of the forehead.

3. *Supratrochlear nerve.* This nerve winds around the upper margin of the orbit medial to the supraorbital nerve (see Fig 6–2). It supplies the skin and conjunctiva on the medial part of the upper eyelid and the skin over the lower part of the forehead, close to the midline.

4. *Infratrochlear nerve.* This nerve leaves the orbit below the pulley of the superior oblique muscle. It supplies the skin and conjunctiva on the medial part of the upper eyelid and the adjoining part of the side of the nose (see Fig 6–2).

5. *External nasal nerve.* This nerve leaves the nose by emerging between the nasal bone and the upper nasal cartilage. It supplies the skin on the side of the nose down as far as the tip (see Fig 6–2).

Clinical Note

Sensory innervation of the cornea and iris is also from the ophthalmic nerve through its nasociliary branch. These nerves provide the afferent pathway for the corneal reflex and also transmit pain sensations from corneal ulcers and iritis to the central nervous system.

Maxillary Nerve

The maxillary nerve supplies the skin on the posterior part of the side of the nose, the lower eyelid, the cheek, the upper lip, and the lateral side of the orbital opening. The following three branches of the nerve pass to the skin.

1. *Infraorbital nerve.* This nerve is a direct continuation of the maxillary nerve. It appears on the face through the infraorbital foramen. It immediately divides into numerous branches that supply the skin of the lower eyelid and cheek, the side of the nose, and the upper lip, including the gum (see Fig 6–2).
2. *Zygomaticofacial nerve.* This nerve passes onto the face through the small foramen on the lateral side of the zygomatic bone. It supplies the skin over the prominence of the cheek (see Fig 6–2).
3. *Zygomaticotemporal nerve.* This nerve emerges in the temporal fossa through a small foramen on the posterior surface of the zygomatic bone. It supplies the skin over the temple (see Fig 6–2).

Mandibular Nerve

The mandibular nerve supplies the skin of the lower lip, the lower part of the face, the temporal region, and part of the auricle. It then passes upward to the side of the scalp. The following three branches of the nerve pass to the skin.

1. *Mental nerve.* This nerve emerges from the mental foramen of the mandible and supplies the skin of the lower lip and chin (see Fig 6–2).
2. *Buccal nerve.* This nerve emerges from beneath the anterior border of the masseter muscle and supplies the skin over a small area of the cheek (see Fig 6–2).
3. *Auriculotemporal nerve.* This nerve ascends from the upper border of the parotid gland between the superficial temporal vessels and the auricle. It supplies the skin of the auricle, the external auditory meatus, the outer surface of the tympanic membrane, and the skin of the scalp above the auricle (see Fig 6–2).

Basic Anatomy of the Arterial Supply of the Face

The face receives a rich arterial supply, with the vessels situated in the superficial fascia. The two main arteries, the facial and the superficial temporal arteries, both branches of the external carotid artery, are supplemented by a number of small arteries that accompany the sensory nerves.

Facial Artery

This artery arises from the anterior aspect of the external carotid artery (see Fig 6–2). It ascends over the submandibular salivary gland and crosses the lower margin of the body of the mandible at the anterior border of the masseter muscle. The artery then runs upward in a tortuous course toward the angle of the mouth and is covered by the platysma and the risorius muscles. It then ascends to the medial angle of the eye, where it anastomoses with the terminal branches of the ophthalmic artery, a branch of the internal carotid artery (see Fig 6–2).
Branches:

1. *Submental artery.* This runs along the lower border of the body of the mandible.
2. *Inferior labial artery.* This arises near the angle of the mouth and runs medially in the lower lip.
3. *Superior labial artery.* This arises near the angle of the mouth and runs medially in the upper lip; it gives branches to the nasal septum and the ala of the nose.
4. *Lateral nasal artery.* This arises as the facial artery ascends alongside the nose.

Superficial Temporal Artery

This artery is the smaller terminal branch of the external carotid artery. It leaves the parotid salivary gland and ascends in front of the auricle to supply the scalp (see Fig 6–2).

Branch.—The *transverse facial artery* runs forward across the cheek just above the parotid duct (see Fig 6–2).

Supraorbital and Supratrochlear Arteries

These are branches of the ophthalmic artery and supply the skin of the forehead (see Fig 6–2).

Clinical Note

Arterial supply to facial skin.—The profuse arterial supply may cause severe bleeding from comparatively small injuries. The great vascularity often permits large flaps of skin resulting from severe injury to be sutured back in position without necrosis occurring.

Arterial pulse.—The superficial temporal artery, as it crosses the zygomatic arch in front of the ear, and the facial artery, as it winds around the lower margin of the body of the mandible, can be used to take the patient's pulse.

Basic Anatomy of the Venous Drainage of the Face

Facial Vein

This vein is formed at the medial angle of the eye by the union of the supraorbital and supratrochlear veins (see Fig 6–2). It descends behind the facial artery to the lower margin of the body of the mandible and then crosses superficial to the submandibular salivary gland, where it is joined by the anterior division of the retromandibular vein. The facial vein ends by draining into the internal jugular vein.

Tributaries.—The facial vein receives tributaries that correspond to the branches of the facial artery. It is joined to the pterygoid venous plexus in the infratemporal fossa by the *deep facial vein,* and is joined to the cavernous sinus in the middle cranial fossa by the superior ophthalmic vein.

Clinical Note

Facial vein and cavernous sinus thrombosis.—The facial vein is connected to the cavernous sinus within the skull via the supraorbital vein and the superior ophthalmic vein. This valveless connection is of great clinical importance, since it provides a pathway for the spread of infection from the face to the cavernous sinus. It follows that the *central triangular area* of the skin bounded by the nose, eye, and upper lip is a potentially dangerous zone to have an infection (see Fig 6–2).

Basic Anatomy of the Lymphatic Drainage of the Face

The forehead and the anterior part of the face drain into the submandibular lymph nodes (Fig 6–3). A few buccal lymph nodes may be present along the course of these lymph vessels. The lateral part of the face, including the lateral parts of the eyelids, is drained by lymph vessels that end in the parotid lymph nodes. The central part of the lower lip and the skin of the chin is drained into the submental lymph nodes.

Basic Anatomy of the Parotid Duct

The parotid duct leaves the anterior margin of the gland and passes forward over the *lateral* surface of the masseter muscle below the transverse facial artery (see Fig 6–7). At the anterior border of the muscle it turns sharply medially and pierces the buccal pad of fat and the buccinator muscle. It then passes forward for a short distance between the muscle and the mucous membrane and finally opens into the vestibule of the mouth upon a small papilla, opposite the upper second molar tooth (see Fig 1–10). The oblique passage of the duct forward between the mucous membrane and the buccinator muscle serves as a valve-like mechanism and prevents the inflation of the duct during violent blowing. If an accessory part of the gland is present, it is drained by a small duct that opens into the upper border of the parotid duct.

Basic Anatomy of the Bones of the Face

The bones of the face are shown in Figure 6–3.

Frontal Bone

The frontal bone curves downward to make the upper margins of the orbits. The *superciliary arches* can be seen on either side (Fig 6–4). The *supraorbital notch,* or *foramen,* can be recognized, which is located on a line with the neutral positioned pupil. Medially, the frontal bone articulates with the frontal processes of the maxillae and with the nasal bones. Laterally, the frontal bone articulates with the zygomatic bone.

The *orbital margins* are bounded by the frontal bone superiorly, the zygomatic bone laterally, the maxilla inferiorly, and the processes of the maxilla and frontal bone medially.

The *frontal air sinuses,* which are two hollow spaces within the frontal bone lined with mucous membrane, lie just above the orbital margins.

Nasal Bones

The two nasal bones form the bridge of the nose. They articulate above with the frontal bone and below with the maxilla (see Fig 6–4). The *anterior nasal aperture* is thus formed by the lower borders of the nasal bones and the maxillae. The nasal cavity is divided into two by the bony nasal septum, which is largely formed by the *vomer.* The *superior* and *middle conchae* jut into the nasal cavity from the *ethmoid bone* on each side; the *inferior conchae* are separate bones. The anterior nonbony part of the nose is completed by the *upper* and *lower plates* of cartilage, the septal cartilage, and the small cartilages of the ala of the nose (see Fig 1–1).

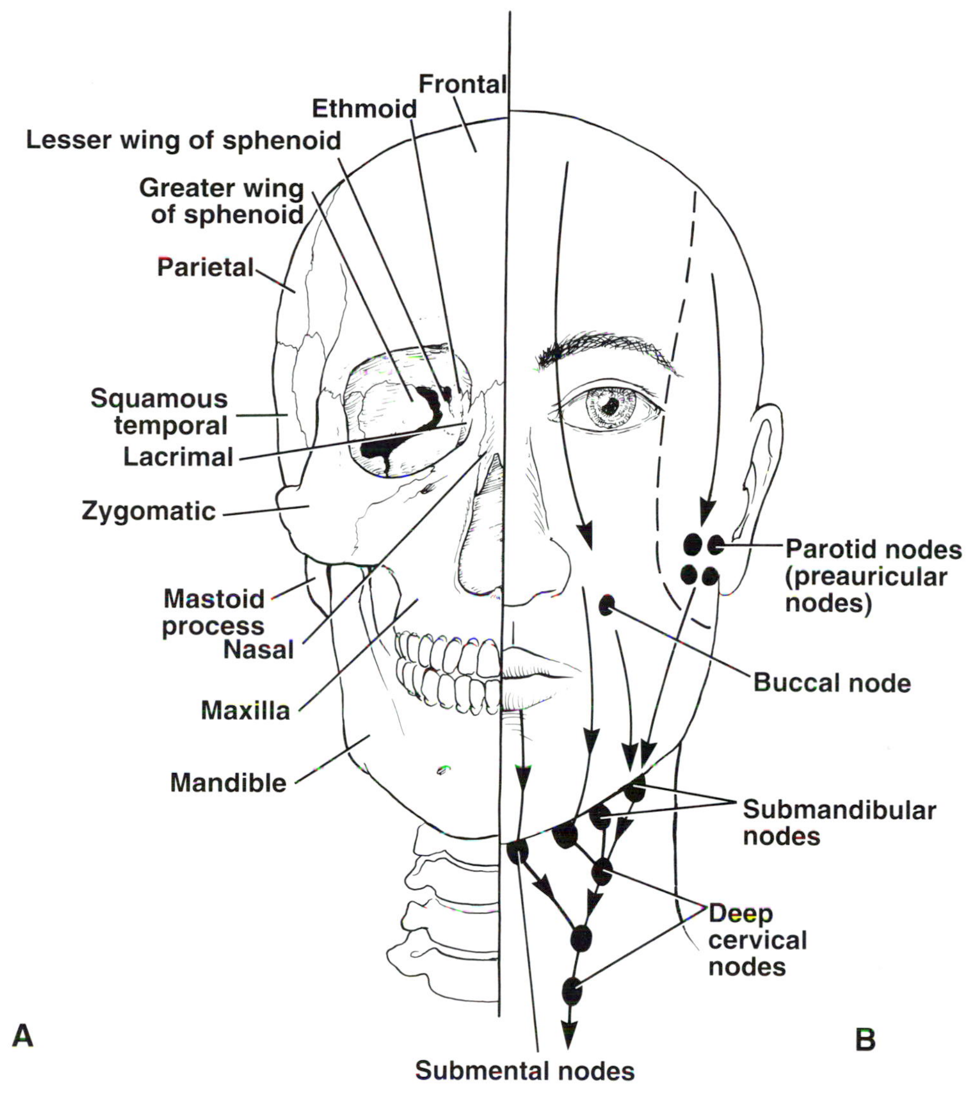

FIG 6–3.
A, bones of the face. **B,** lymph drainage of the face.

Maxilla

The two maxillae form the important middle part of the face (see Fig 6–4). They form the upper jaw, the anterior part of the hard palate, part of the lateral walls of the nasal cavities, and part of the floors of the orbital cavities. The two bones meet in the midline at the *intermaxillary suture* and form the lower margin of the nasal aperture. Here there is a pointed process, which with the fellow of the opposite side forms the *anterior nasal spine* (see Fig 6–4). Below the orbit, the maxilla is perforated by the *infraorbital foramen* for the infraorbital vessels and nerve. The *alveolar process* projects downward and, together with the fellow of the opposite side, forms the *alveolar arch,* which carries the upper teeth. Within each maxilla is the large pyramid-shaped *maxillary sinus* lined with mucous membrane.

Zygomatic Bone

The zygomatic bone (zygoma) forms the prominence of the cheek and part of the lateral wall and floor of the orbital cavity (see Fig 6–4). Medially it articulates with the maxilla, superiorly with the frontal bone, and laterally with the zygomatic process of the temporal bone to form the *zygomatic arch.* The zygomatic bone is perforated by two foramina for the zygomaticofacial and zygomaticotemporal nerves.

Mandible

The mandible, or lower jaw, consists of a horizontal horseshoe-shaped *body* and a pair of vertical *rami.* The body of the mandible meets the ramus at the *angle* (see Fig 6–4).

The body of the mandible, on its external surface in the midline, has a faint ridge, the *symphysis menti,*

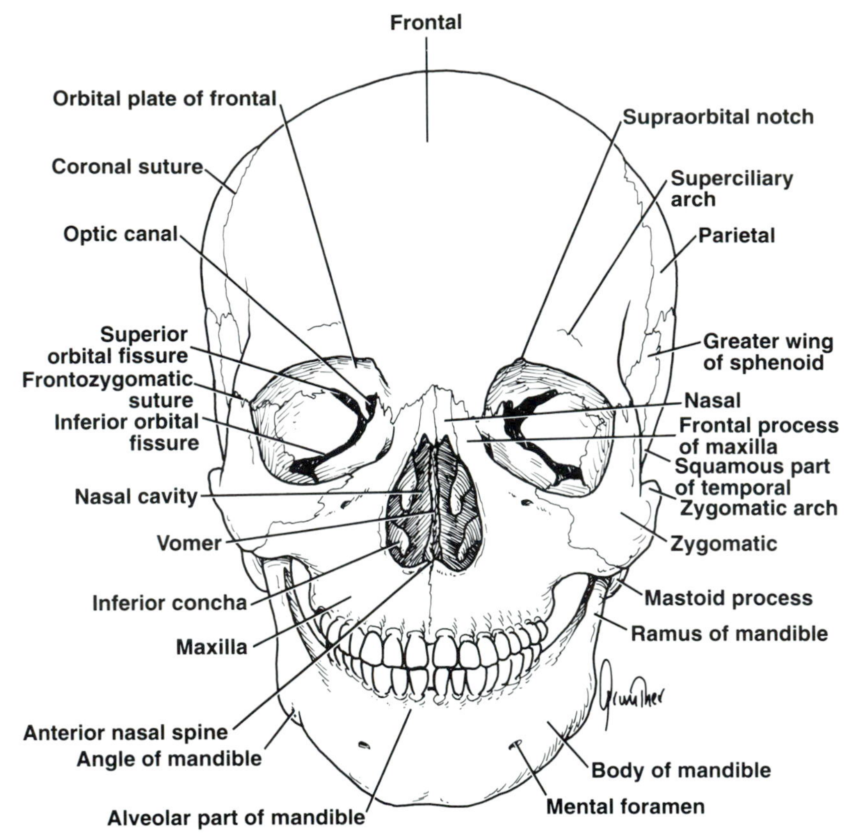

FIG 6–4.
Bones of the anterior aspect of the skull.

indicating the line of fusion of the two halves. The *mental foramen* is below the second premolar tooth and lies on a line that passes through the neutrally positioned pupil and the supraorbital and infraorbital foramina; the mental foramen transmits the mental nerve and vessels. The upper border of the body of the mandible is called the *alveolar* part and contains the roots of the teeth.

The ramus of the mandible is vertically placed and has an anterior *coronoid process* and a posterior *condyloid process,* or *head* (see Fig 6–12); the two processes are separated by the *mandibular notch.* The coronoid process lies medial to the zygomatic arch. The head of the mandible articulates above with the temporal bone, forming the *temporomandibular joint.*

Basic Anatomy of the Temporomandibular Joint

Articulation

The articular tubercle and the mandibular fossa of the temporal bone lie above, and the condyle of the mandible lies below. The surfaces are covered with fibrocartilage (Fig 6–5).

Type

The temporomandibular joint is synovial. The fibrocartilaginous disc divides the joint into upper and lower cavities.

Capsule

The joint is enclosed and lined by *synovial membrane.*

Ligaments

These include the *lateral temporomandibular ligament,* the *sphenomandibular ligament,* and the *stylomandibular ligament.* The *articular disc* is an oval disc of fibrocartilage (see Fig 6–5) that permits gliding movement in the upper part of the joint and hinge movement in the lower part of the joint.

Nerve Supply

The nerve supply includes the auriculotemporal nerve and the masseteric nerve, branches of the mandibular division of the trigeminal nerve.

Movements

Protrusion.—The condyle of the mandible and the articular disc can be protruded forward by the lateral

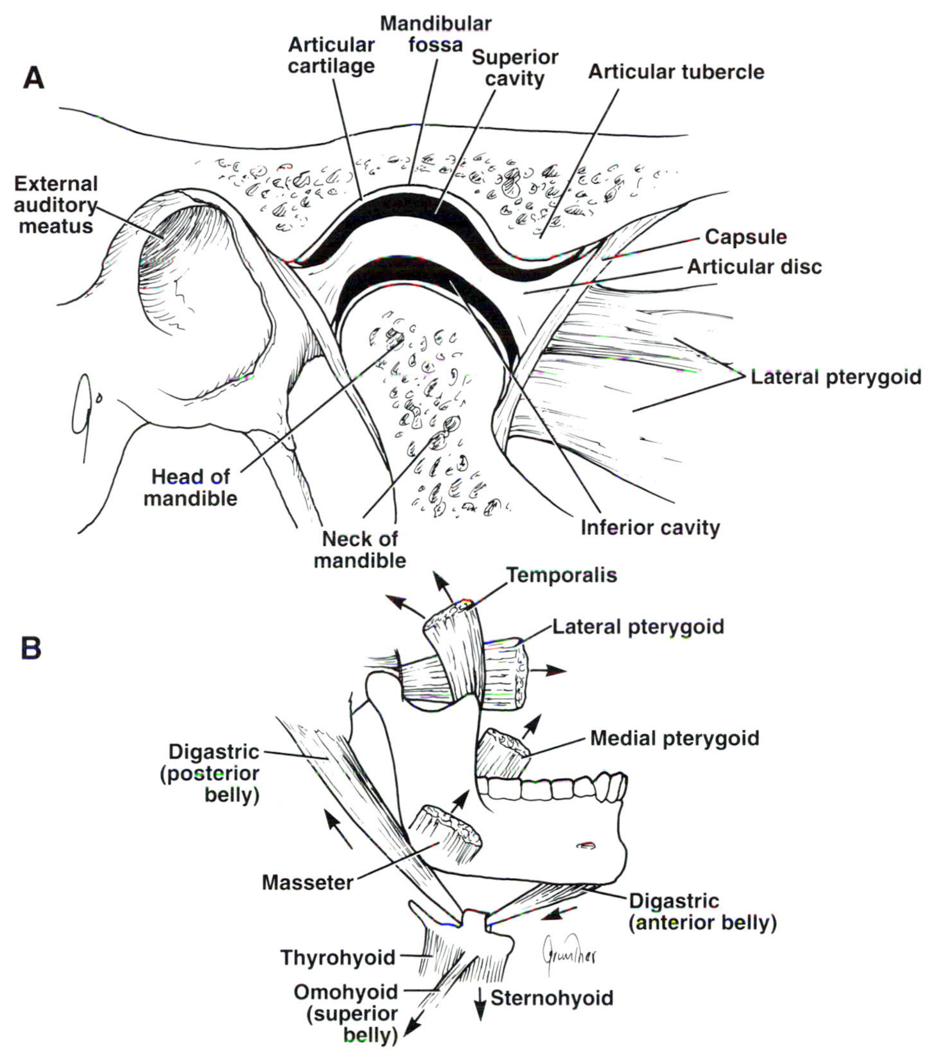

FIG 6–5.
A, the temporomandibular joint (mouth in closed position). **B,** the attachments of the muscles of mastication to the mandible; arrows indicate the direction of their actions.

pterygoid muscle, which originates on the greater wing of the sphenoid bone and the lateral pterygoid plate of the sphenoid and has its insertion onto the neck of the mandible and the articular disc (see Fig 6–5).

Retraction.—The condyle of the mandible and the articular disc can be pulled backward by the posterior fibers of the temporalis muscle, which originates on the outer surface of the parietal bone and the squamous part of the temporal bone and has its insertion onto the coronoid process of the mandible.

Opening the Mouth.—This is accomplished by the condyle of the mandible rotating on the undersurface of the articular disc around a horizontal axis. The digastrics, the geniohyoids, and the mylohyoids depress the mandible. At the same time the lateral pterygoid muscle pulls the neck of the mandible and the articular disc forward (see Fig 6–5).

Elevation of the Mandible.—The temporalis, the masseter, and the medial pterygoid muscles contract to elevate the mandible (see Fig 6–5). The head of the mandible is retracted by the posterior fibers of the temporalis, and the articular disc is pulled backward by the fibroelastic tissue, which connects the disc to the temporal bone posteriorly.

Lateral Chewing.—These movements are brought about by alternately protruding and retracting the mandible on each side.

The action of the various muscles acting on the

temporomandibular joint can be understood by studying Figure 6–5.

Basic Anatomy of the Muscles of the Face (Muscles of Facial Expression)

The muscles of the face are thin and flat and are embedded in the superficial fascia. The majority of these muscles arise from the bones of the skull and are inserted into the skin of the face (Fig 6–6). The facial muscles serve as sphincters or dilators to the orifices of the face, namely, the orbit, nose, and mouth. A secondary function is to modify the expression of the face. All muscles of the face are developed from the second pharyngeal arch, and they are therefore supplied by the nerve of that arch, the facial nerve. The proprioceptive nerve fibers of the facial muscles are branches of the trigeminal nerve.

Basic Anatomy of the Facial Nerve

The facial nerve supplies all the muscles of facial expression, supplies the anterior two thirds of the tongue with taste fibers, and is secromotor to the lacrimal, submandibular, and sublingual glands. *The facial nerve does not supply sensation of the skin of the face* (the trigeminal nerve does).

The facial nerve nuclei lie within the pons in the brain stem; their central connections are summarized in Figure 6–22. The facial nerve emerges from the junction of the pons and the medulla oblongata, crosses the posterior cranial fossa with the vestibulocochlear nerve, and enters the internal acoustic meatus in the temporal bone. At the bottom of the meatus, the nerve enters the facial canal and runs laterally through the inner ear. On reaching the medial wall of the middle ear, the nerve expands to form

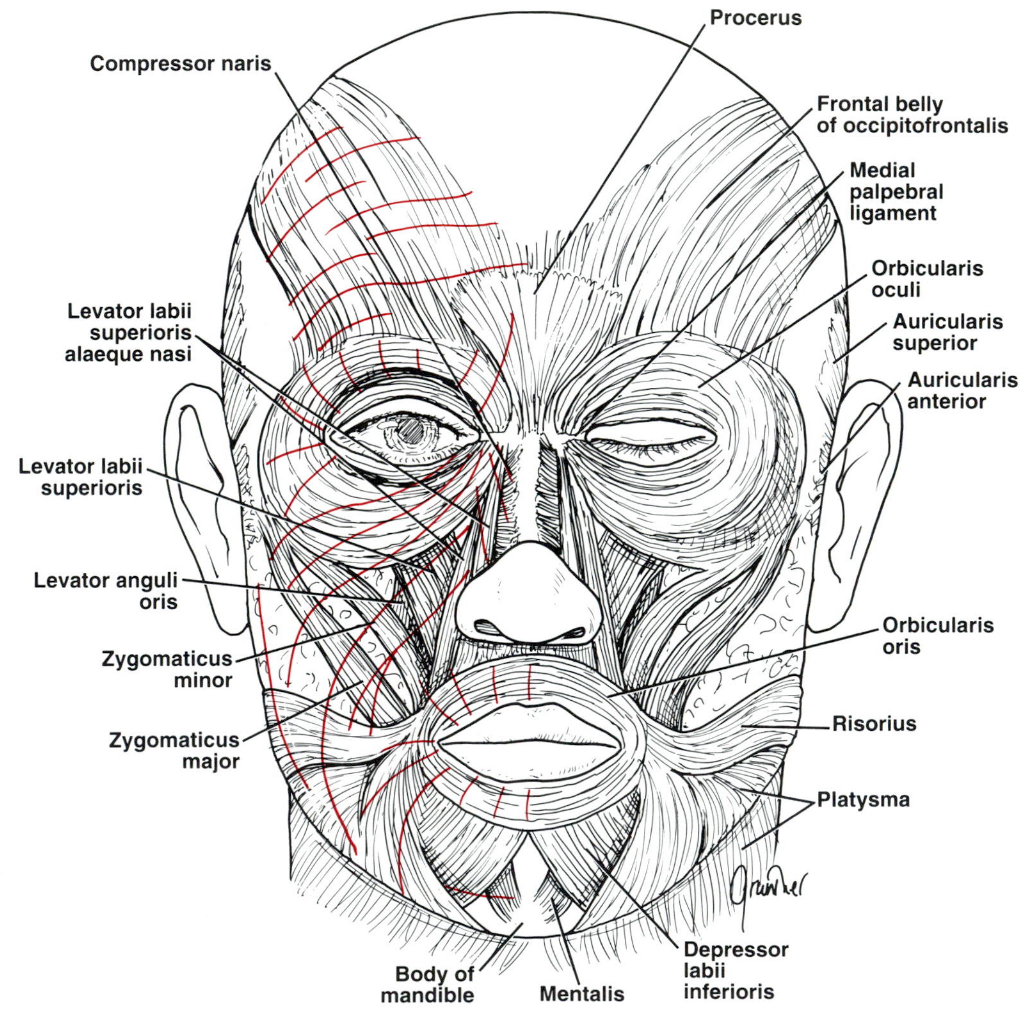

FIG 6–6.
Muscles of facial expression. On the left, the wrinkle lines have been superimposed on the muscles. Note that the wrinkle lines run at right angles to the direction of the muscle fibers.

the sensory *geniculate ganglion.* The facial nerve then turns sharply backward. At the posterior wall of the tympanic cavity, the nerve turns downward (supplies the stapedius muscle and then gives off the chorda tympani nerve) and emerges from the stylomastoid foramen. The facial nerve then passes forward through the substance of the parotid salivary gland, where it divides into five terminal branches that are distributed to the muscles of facial expression (Fig 6–7).

Branches:

1. *Temporal branch.* This emerges from the upper border of the parotid gland and supplies the anterior and superior auricular muscles, the frontal belly of the occipitofrontalis, the orbicularis oculi (which closes the eye), and the corrugator supercilii (which creases the forehead skin).

2. *Zygomatic branch.* This emerges from the anterior border of the gland and supplies the orbicularis oculi muscle.

3. *Buccal branch.* This emerges from the anterior border of the parotid gland. Branches of the buccal nerve run above and below the parotid duct and supply the buccinator muscle and the muscles of the upper lip and nostril.

4. *Mandibular branch.* This emerges from the anterior border of the gland and supplies the muscles of the lower lip.

5. *Cervical branch.* This emerges from the lower border of the gland and passes forward in the neck below the mandible to supply the platysma muscle; it may cross the lower margin of the body of the mandible to supply the depressor anguli oris muscle.

Basic Anatomy of the External Ear

The external ear consists of the auricle and the external auditory meatus (Fig 6–8).

The *auricle* consists of a thin plate of elastic carti-

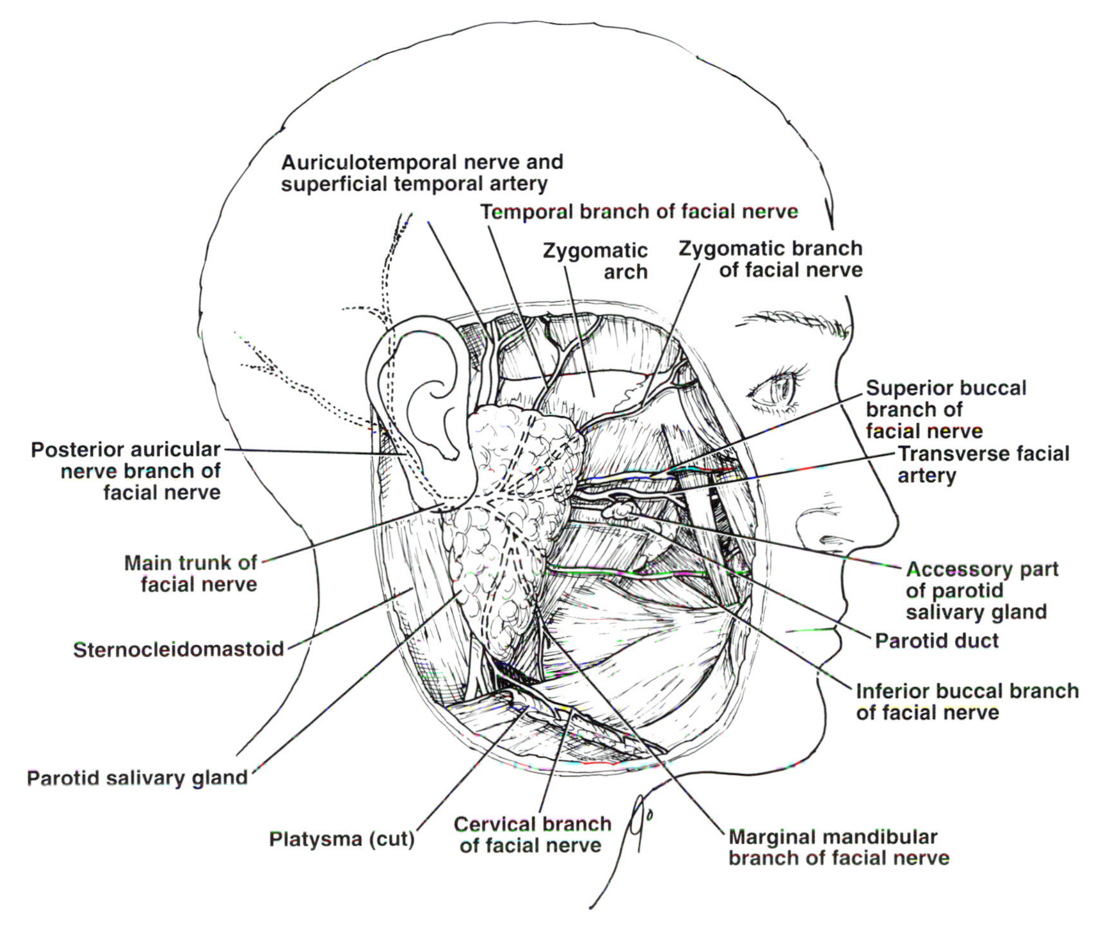

FIG 6–7.
The parotid region and the side of the face showing the position of the parotid duct and the terminal branches of the facial nerve. The facial nerve lies within the parotid gland and as it passes forward it divides into five terminal branches that emerge from the gland.

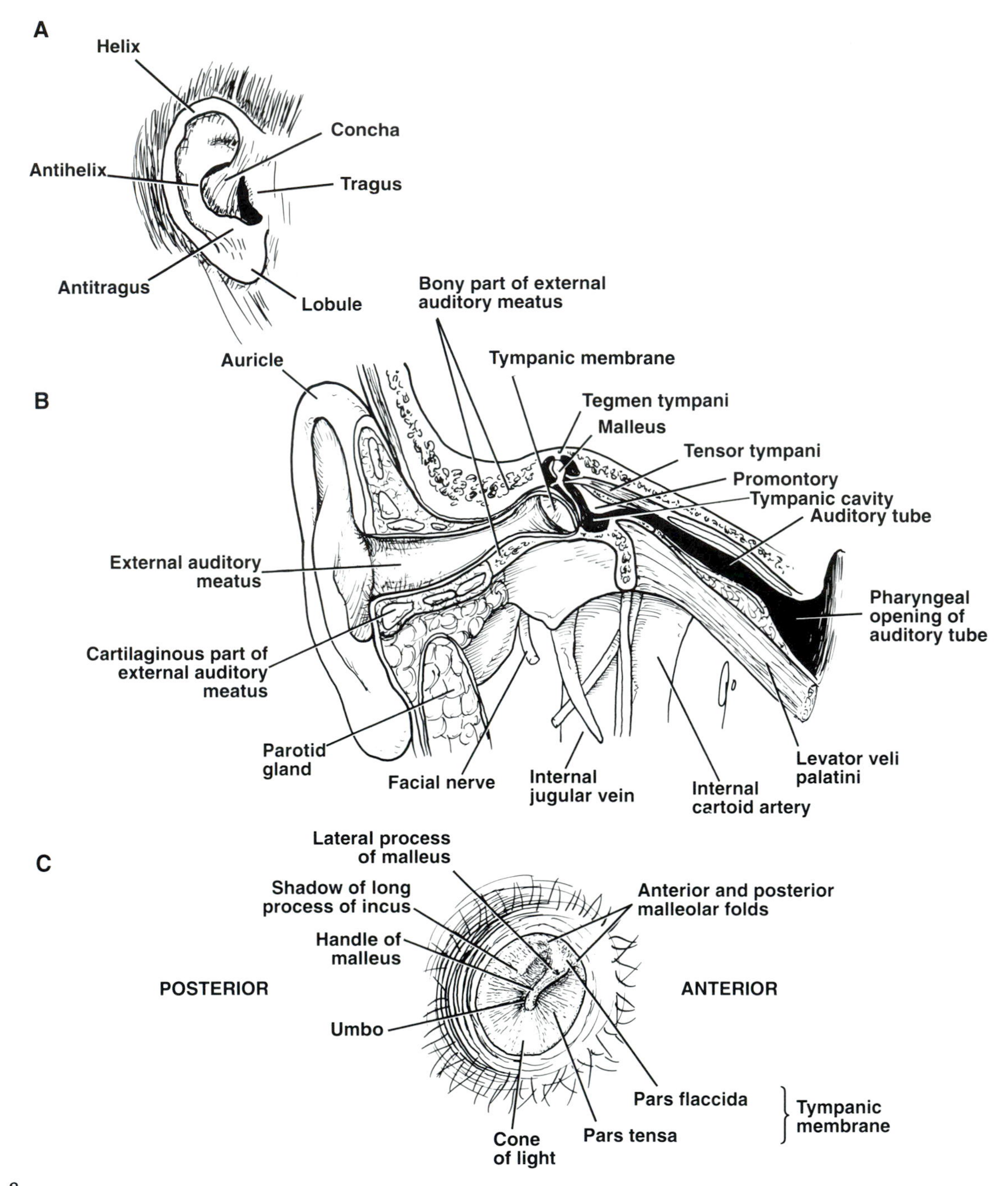

FIG 6–8.
A, different parts of auricle of right ear. **B**, external and middle portions of the right ear, viewed from in front. **C**, right tympanic membrane as seen through an otoscope.

lage covered by thin skin that adheres to the cartilage. The skin possesses fine hairs with sebaceous glands. The auricle has both intrinsic and extrinsic muscles (see Fig 6–6) that are innervated by the facial nerve.

The *external auditory meatus* is a curved tube that leads from the auricle to the tympanic membrane. In the adult the meatus measures about 1 in. (2.5 cm) long and may be straightened for the insertion of an otoscope by pulling the auricle upward and backward. In young children the meatus may be straightened by pulling the auricle directly backward or downward and backward. The meatus is narrowest about 5 mm from the tympanic membrane. The framework of the outer third of the meatus is elastic cartilage, whereas the inner two thirds is bone. The meatus is lined by skin, and its outer third is provided with *hairs, sebaceous glands,* and *ceruminous glands.*

The *tympanic membrane* (see Fig 6–8) is a thin, fibrous membrane that is pearly gray. It is covered on the outer surface with stratified squamous epithelium and on the inner surface with low columnar epithelium. The membrane is obliquely placed—facing downward, forward, and laterally. It is concave laterally, and at the depth of the concavity is a small depression, the *umbo,* produced by the tip of the handle of the malleus. When the membrane is illuminated through an otoscope, the concavity produces a "cone of light," which radiates anteriorly and inferiorly from the umbo.

The tympanic membrane is circular and measures about 1 cm in diameter. The greater part of its circumference is thickened, and it is slotted into a groove in the bone. The groove, or *tympanic sulcus,* is deficient superiorly, which forms a notch. From the sides of the notch, two bands, termed the *anterior* and *posterior malleolar folds,* pass to the lateral process of the malleus. The small triangular area on the tympanic membrane, called the pars flaccida (see Fig 6–8), that is bounded by the folds is slack. The remainder of the membrane, called the *pars tensa,* is tense. The handle of the malleus is bound down to the inner surface of the tympanic membrane by the mucous membrane.

Vibrations of the tympanic membrane are transmitted to the inner ear by the movements of the ossicles, namely, the malleus, incus, and stapes. As the result of the movement of the ossicles, the leverage increases at a rate of 1.3 to 1. Moreover, the area of the tympanic membrane is about 17 times greater than that of the base of the stapes, causing the effective pressure on the perilymph of the inner ear to increase by a total of 22 to 1.

Sensory Nerve Supply to the External Ear

Auricle.—Branches from the great auricular nerve (C2 and C3) supply most of the medial surface and the lower part of the lateral surface. Branches from the lesser occipital nerve (C2) supply the upper part of the medial surface. Branches from the auriculotemporal nerve (from the mandibular division of the trigeminal nerve) supply the lateral surface.

External Auditory Meatus.—Branches from the auriculotemporal nerve supply the anterior and superior walls. The auricular branch of the vagus nerve supplies the posterior and inferior walls.

Tympanic Membrane.—Branches from the auriculotemporal nerve and the auricular branch of the vagus nerve supply the lateral surface; the tympanic branch of the glossopharyngeal nerve supplies its medial surface.

Basic Anatomy of the Scalp

Structure of the Scalp

The scalp consists of the following five layers, the first three of which are intimately bound together and move as a unit (Fig 6–9).

1. The *skin* is thick, hair-bearing, and contains numerous sweat glands.
2. The *connective tissue* of the superficial fascia is fibrofatty, with the fibrous septa uniting the skin to the underlying aponeurosis of the occipitofrontalis muscle (see Fig 6–9). Numerous arteries and veins ramify in this layer. The arteries are branches of the external and internal carotid arteries, and a free anastomosis takes place between them.
3. The *aponeurosis* of the occipitofrontalis muscle is a thin, tough, fibrous sheet (*galea*) that unites the occipital and frontal bellies of the occipitofrontalis muscle (see Fig 6–9). The lateral margins of the aponeurosis are attached to the deep fascia covering the temporalis muscles.
4. *Loose areolar tissue* occupies the subaponeurotic space (see Fig 6–9) and loosely connects the epicranial aponeurosis with the periosteum of the skull (the pericranium). The areolar tissue contains a few small arteries but also contains some important *emissary veins.* The emissary veins are valveless and connect the superficial veins of the scalp with the *diploic veins* of the skull bones and with the intracranial venous sinuses (see Fig 6–9).
5. The *pericranium* is the periosteum covering the outer surface of the skull bones. At the sutures between the individual skull bones, the pericranium becomes continuous with the periosteum on the inner surface of the skull bones (see Fig 6–9).

Muscle of the Scalp: Occipitofrontalis

Origin.—The two occipital bellies arise from the highest nuchal line on the occipital bone; the two frontal bellies arise from the skin and superficial fascia of the eyebrows (see Fig 6–6).

Insertion.—The occipital bellies pass forward and are attached to the frontal bellies by means of the strong aponeurosis.

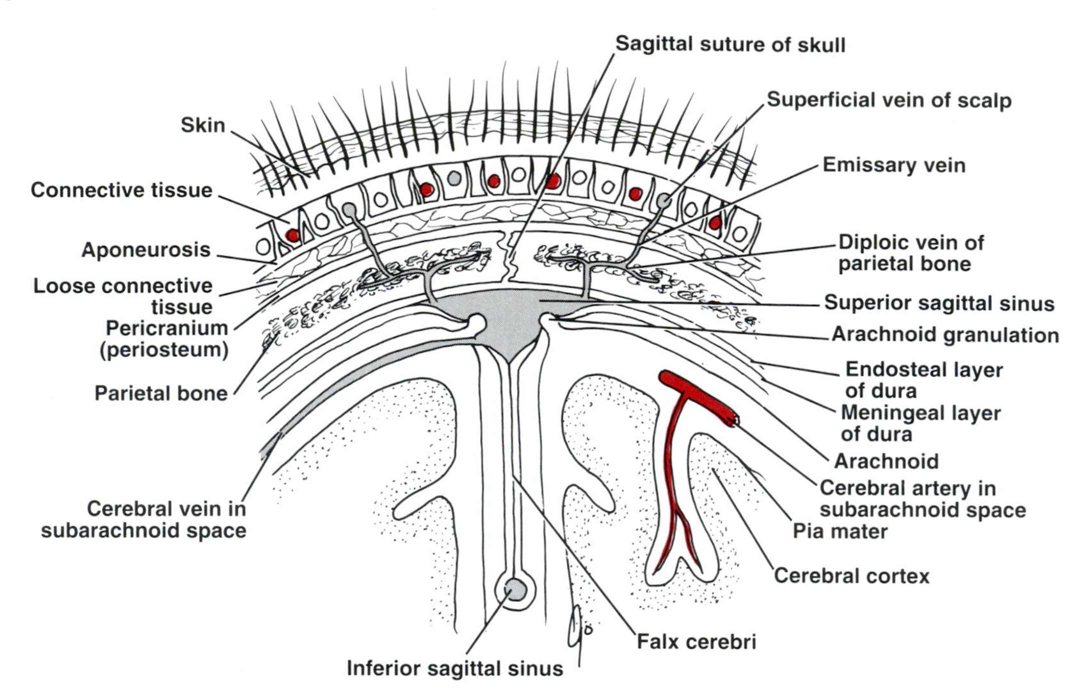

FIG 6–9.
Coronal section of the upper part of the head showing layers of the scalp, sagittal suture of skull, falx cerebri, superior and inferior sagittal venous sinuses, arachnoid granulations, emissary veins, and relation of cerebral blood vessels to subarachnoid space.

Nerve Supply.—The occipital bellies are attached by the posterior auricular branch of the facial nerve; the frontal bellies are attached by the temporal branch of the facial nerve.

Action.—Contraction of the frontal bellies moves the scalp forward; it can also raise the eyebrows in expressions of surprise or horror. The occipital bellies pull the scalp backward. The loose areolar tissue beneath the aponeurosis permits the first three layers of the scalp to move on the pericranium.

Basic Anatomy of the Sensory Nerve Supply of the Scalp

The main trunks of the sensory nerves lie in the superficial fascia of the scalp. Moving laterally from the midline anteriorly, the following nerves are present.

Supratrochlear Nerve

This nerve, which is a branch of the ophthalmic division of the trigeminal nerve, winds around the superior orbital margin and supplies the scalp (Fig 6–10). It extends backward close to the midline nearly as far as the vertex.

Supraorbital Nerve

This nerve, which is a branch of the ophthalmic division of the trigeminal nerve, winds around the superior orbital margin and supplies the scalp as far backward as the vertex (see Fig 6–10).

Zygomaticotemporal and Zygomaticofacial Nerves

These nerves, which are branches of the maxillary division of the trigeminal nerve, supply the scalp over the temple (see Fig 6–10).

Auriculotemporal Nerve

This nerve, which is a branch of the mandibular division of the trigeminal nerve, ascends over the side of the head in front of the ear and supplies the scalp over the temporal region (see Fig 6–10).

Lesser Occipital Nerve (C2)

This nerve, which is a branch of the cervical plexus, supplies the scalp over the lateral part of the occipital region (see Fig 6–10).

Greater Occipital Nerve

This nerve is a branch of the second cervical nerve; it supplies the scalp as far forward as the vertex (see Fig 6–10).

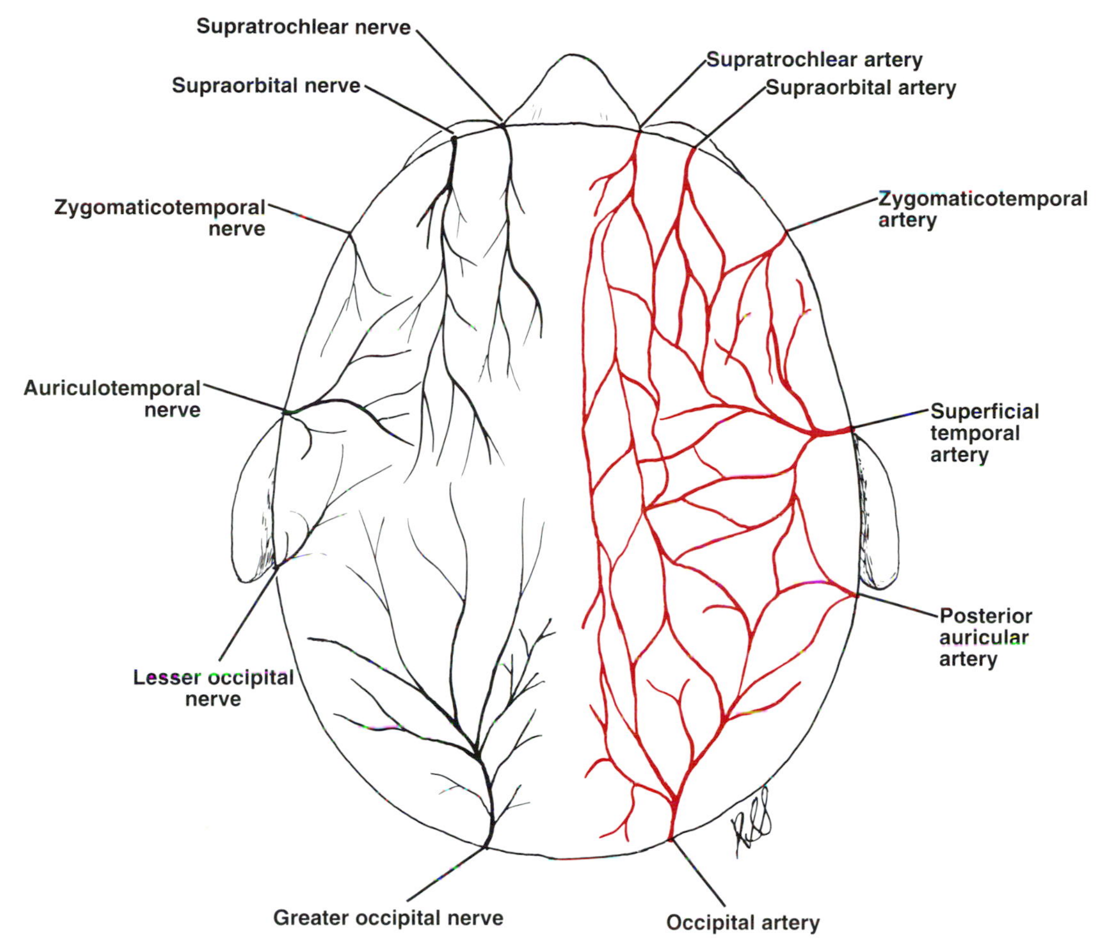

FIG 6–10.
Scalp viewed from above. The left side shows the sensory nerves, and the right side shows the arterial supply.

Basic Anatomy of the Arterial Supply of the Scalp

The scalp has a very rich blood supply that nourishes the hair follicles. The arteries lie in the superficial fascia. Moving laterally from the midline anteriorly, the following arteries are present.

Supratrochlear and Supraorbital Arteries

These arteries are branches of the ophthalmic artery from the internal carotid artery. They ascend over the forehead with the nerves of the same name (see Fig 6–10).

Superficial Temporal Artery

This artery is the smaller terminal branch of the external carotid artery. It ascends in front of the auricle with the auriculotemporal nerve (see Fig 6–10). The artery divides into anterior and posterior branches, which supply the scalp over the frontal and temporal regions.

Posterior Auricular Artery

This artery is a branch of the external carotid artery. It ascends behind the auricle to supply the scalp above and behind the auricle (see Fig 6–10).

Occipital Artery

This artery is a branch of the external carotid artery. It ascends in a tortuous course from the back of the neck with the greater occipital nerve (see Fig 6–10). It supplies the back of the scalp as far up as the vertex.

Basic Anatomy of the Veins of the Scalp

The veins of the scalp freely anastomose with one another and are connected to the diploic veins within the skull bones; they are also connected to the intracranial venous sinuses by the valveless emissary veins.

Clinical Note

Intracranial spread of infections of the scalp.—Unlike scalp arteries, the scalp veins communicate with intracranial veins. Emissary veins can provide a route for the spread of scalp infections to the skull bones, producing osteomyelitis; they can also provide a channel for infected blood to pass to the intracranial venous sinuses (the cavernous, superior sagittal, and lateral venous sinuses) and can cause venous sinus thrombophlebitis.

Basic Anatomy of the Lymphatic Drainage of the Scalp

The anterior part of the scalp and forehead drain into the submandibular lymph nodes (see Fig 6–3).

Clinical Note

Scalp infections and enlarged lymph nodes.—Sometimes the site of infection that is the cause of regional lymphadenitis is difficult to find. An infected hair follicle under a thick mass of hair may be responsible.

The lateral part of the scalp above the ear drains into the superficial parotid preauricular nodes; the part of the scalp above and behind the ear drains into the mastoid nodes. The back of the scalp drains into the occipital nodes (see Fig 5–12).

Basic Anatomy of the Mouth

The basic anatomy of the mouth has been considered in detail in Chapter 1, p. 6. Only a brief overview will be given here. The *mouth proper* has a roof formed by the hard palate in front and the soft palate behind. The floor is formed largely by the anterior two thirds of the tongue and by the reflection of the mucous membrane from the sides of the tongue to the gum of the mandible. A fold of mucous membrane called the *frenulum of the tongue* connects the undersurface of the tongue in the midline to the floor of the mouth (Fig 6–11).

The *orifice of the submandibular duct* opens onto the floor of the mouth on the summit of a small papilla on either side of the frenulum of the tongue (see Fig 6–11). The *sublingual salivary gland* projects up into the mouth, producing a low fold of mucous

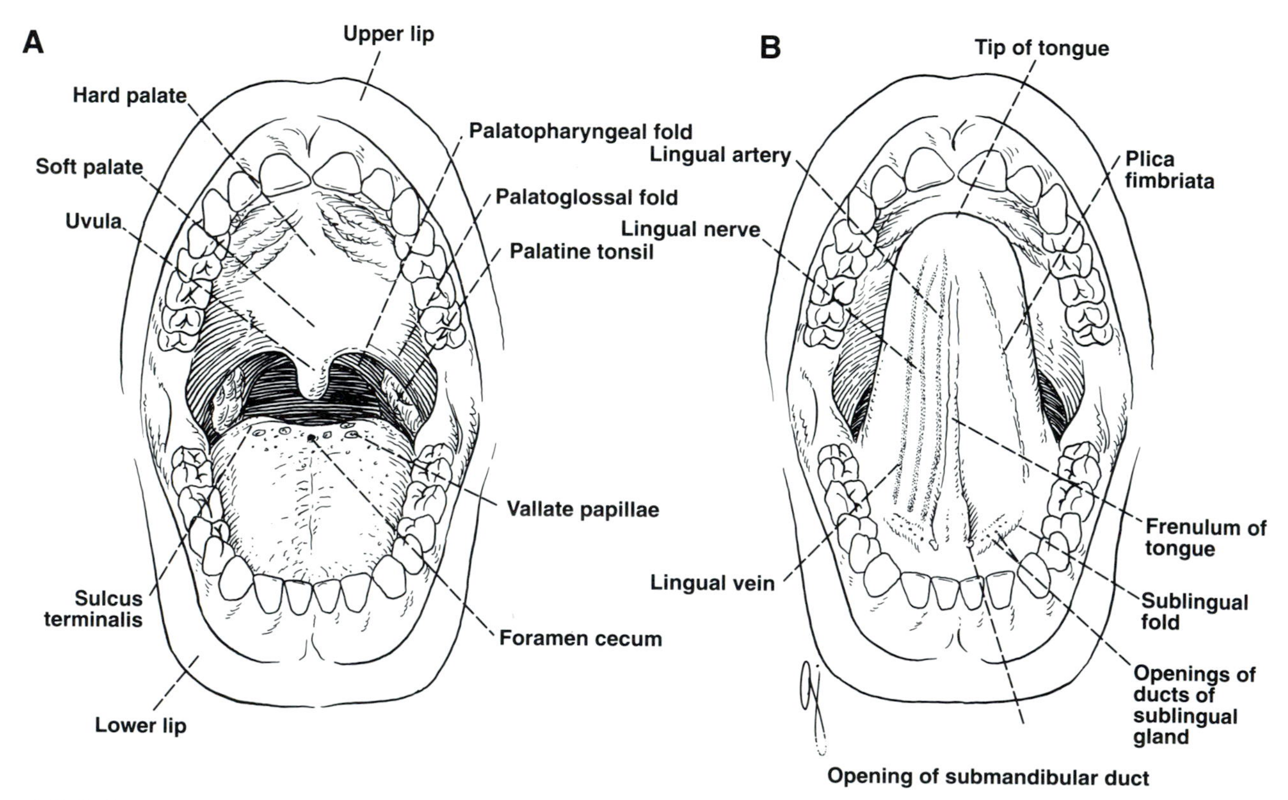

FIG 6–11.
A, mouth open to display cavity of mouth and oral pharynx. **B,** undersurface of tongue.

membrane that extends laterally from the submandibular papilla. There is no single duct; its numerous ducts open on the summit of the fold.

The *vestibule of the mouth* lies between the lips and the cheeks externally and the gums and the teeth internally. This slitlike space communicates with the exterior through the *oral fissure*. The lateral wall of the vestibule is formed by the *cheek*, which is made up of the buccinator muscle and lined with mucous membrane. The tone of the buccinator muscle and that of the muscles of the lips keeps the walls of the vestibule in contact with one another. The *duct of the parotid salivary gland* opens on a small papilla into the vestibule opposite the upper second molar tooth.

There are 20 *deciduous teeth*—four incisors, two canines, and four molars in each jaw. They begin to erupt about 6 months after birth and have all

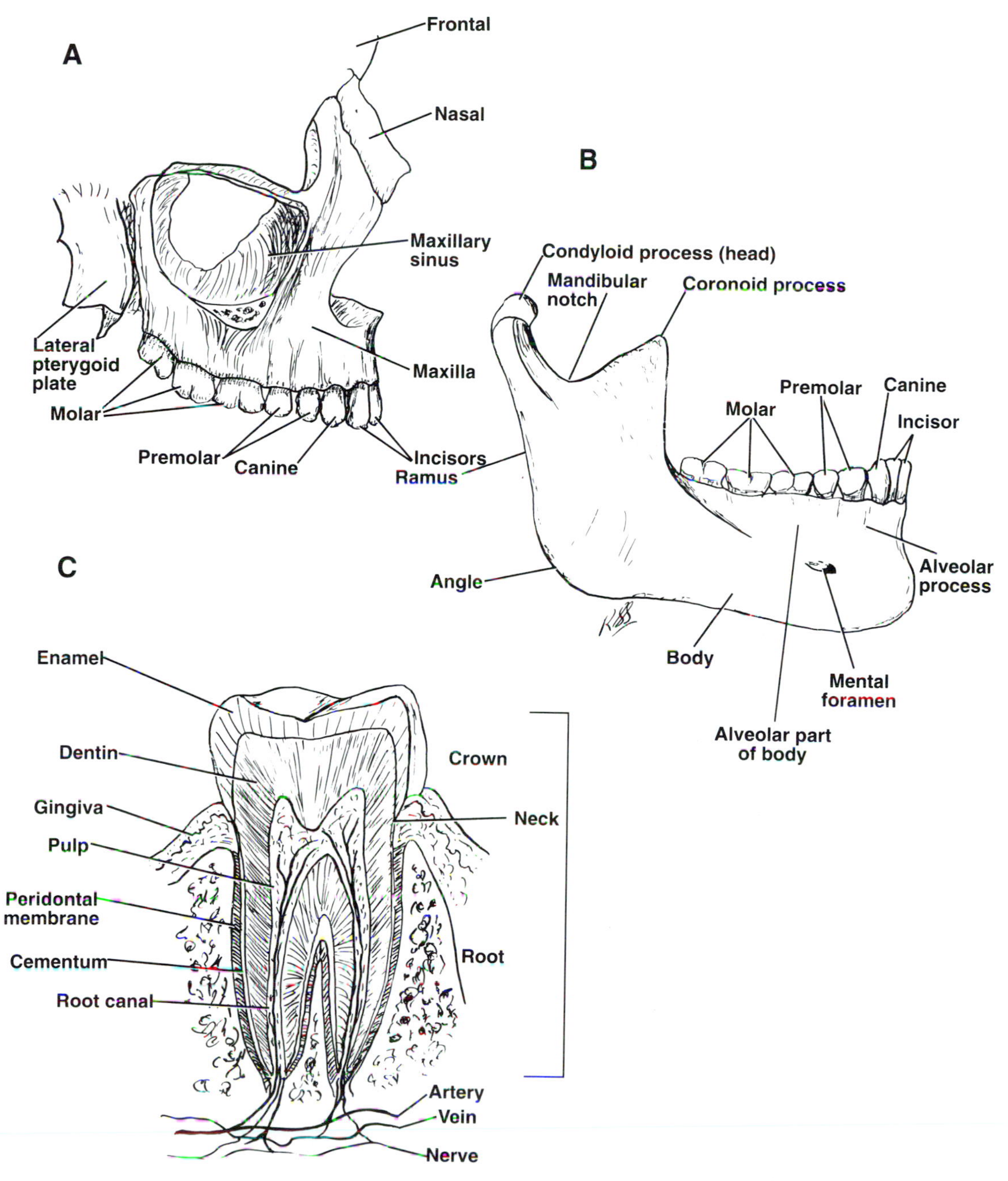

FIG 6–12.
A, permanent teeth in the maxilla. **B,** permanent teeth in mandible. **C,** longitudinal section of a tooth.

erupted by the end of 2 years. The teeth of the lower jaw usually appear before those of the upper jaw.

There are 32 *permanent teeth*—four incisors, two canines, four premolars, and six molars in each jaw (Fig 6–12). They begin to erupt at the age of 6 years. However, the last tooth to erupt—the third molar—may occur between the ages of 17 and 30 years. The teeth of the lower jaw usually appear before those of the upper jaw.

Basic Anatomy of the Tooth

The teeth are set in sockets on the alveolar processes of the mandible and the maxillae. The alveolar processes are covered by a mucoperiosteum, known as the *gums* or *gingivae* (see Fig 6–12). The part of the tooth that projects above the gums is the *crown;* the portion that lies within the socket is the *root* (see Fig 6–12). In the center of each tooth is the *pulp cavity,* which is filled with connective tissue containing nerves, blood, and lymph vessels. Extensions of the pulp cavity pass into the roots of the tooth and are called *root canals.* The wall of the tooth is largely composed of a calcified connective tissue called *dentin.* This is protected in the crown by a covering of tough *enamel,* which is composed of calcium phosphate. The dentin in the root of the tooth is covered by a calcified connective tissue called *cementum* (see Fig 6–12).

Tooth Innervation

The *upper teeth* are supplied by the *anterior, middle, and posterior superior alveolar nerves,* which are branches of the maxillary division of the trigeminal nerve.

The *lower teeth* are supplied by the *inferior alveolar nerve,* a branch of the mandibular division of the trigeminal nerve.

Tooth Arterial Supply

Small arteries enter the pulp cavity through the apical foramen, while others ascend in the peridental ligament (see Fig 6–12).

Clinical Note

Avulsed permanent teeth should be replaced in their sockets as soon as possible without dehydration and with the least disturbance to the peridontal ligament (not scrubbed). The peridontal fibers are responsible for reattachment of the tooth in the socket. Most replaced teeth ultimately undergo avascular necrosis unless replaced in a few minutes.

SURFACE ANATOMY

Surface Anatomy of the Face, Scalp, and Mouth

Nasion

This is the depression in the midline at the root of the nose marking the frontonasal suture (Fig 6–13).

Inion (External Occipital Protuberance)

This is a bony prominence in the middle of the flattened part of the occipital bone. It lies in the midline at the junction of the head and neck and gives attachment to the ligamentum nuchae, which connects the skull to the spinous processes of the cervical vertebrae.

Superior Sagittal Sinus and the Longitudinal Cerebral Fissure

A line joining the nasion to the inion over the superior aspect of the skull would indicate the position of the underlying superior sagittal sinus and the longitudinal cerebral fissure, which separates the right and left cerebral hemispheres.

Vertex

This is the highest point on the skull in the sagittal plane.

Anterior Fontanelle

In infants the anterior fontanelle lies between the two halves of the frontal bone in front and the two parietal bones behind (Fig 6–14). It is usually not palpable after 18 months.

Posterior Fontanelle

In infants the posterior fontanelle lies between the squamous part of the occipital bone and the posterior borders of the two parietal bones (see Fig 6–14). It is usually closed by the end of the first year of life.

Clinical Note

Palpation of Fontanelles.—Palpation of the fontanelles may be useful in obtaining the following information: (1) the progress of growth of the surrounding bones; (2) the degree of hydration of the infant, i.e., a depressed fontanelle indicates that the infant is dehydrated; and (3) the state of the intracranial pressure, i.e., a bulging fontanelle would indicate a raised intracranial pressure.

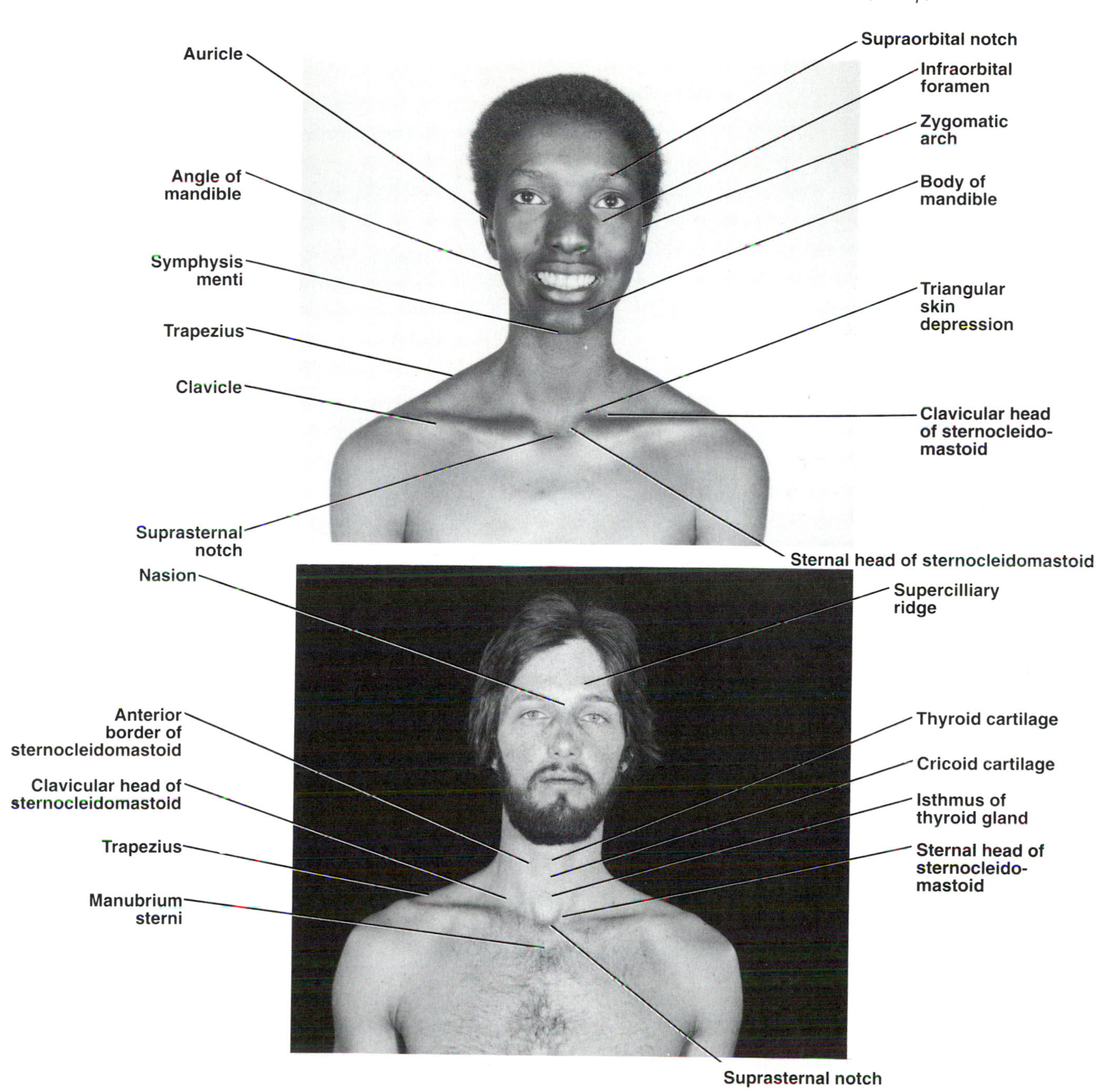

FIG 6–13.
Anterior views of the head and neck.

Superciliary Ridges

These two ridges are prominent on the frontal bones above the upper margin of the orbit (see Fig 6–13). Deep to these ridges lie the *frontal sinuses.*

Mastoid Process of the Temporal Bone

The mastoid process projects downward and forward from behind the ear. It is undeveloped in newborns and grows only as the result of the pull of the sternocleidomastoid muscle as the head is moved. It may be recognized as a bony projection at the end of the second year of life.

Auricle and External Auditory Meatus

The external auditory meatus is about 1 in. (2.5 cm) long and forms an S-shaped curve. As previously mentioned, the outer surface of the tympanic membrane in adults may be visualized through an oto-

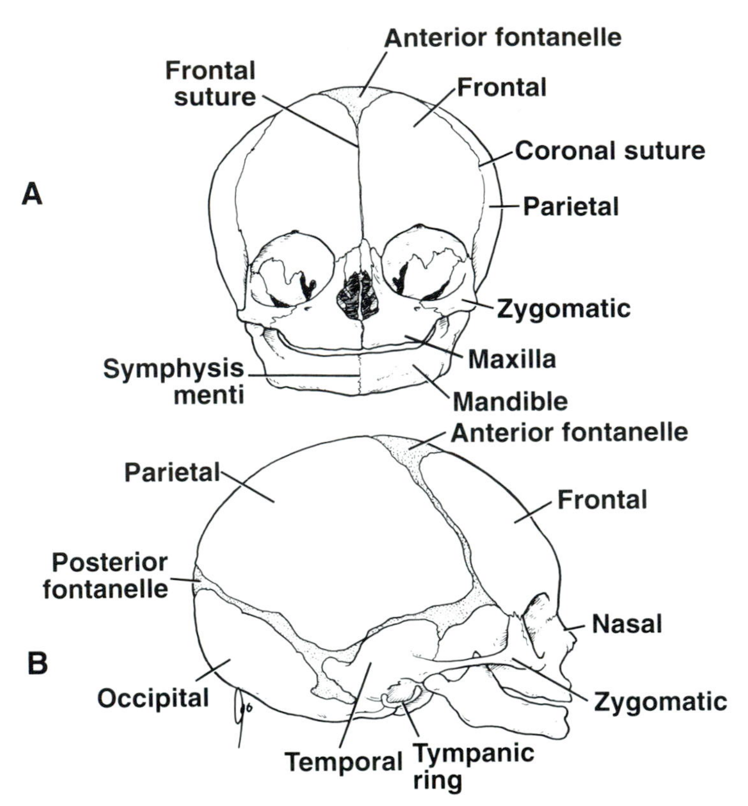

FIG 6–14.
Fetal head. **A,** anterior view. **B,** lateral view.

scope by pulling the auricle upward and backward. In children, the auricle is pulled straight back or downward and backward.

Zygomatic Arch

The zygomatic arch (see Fig 6–13), which consists of the temporal process of the zygomatic bone and the zygomatic process of the temporal bone, can be palpated throughout its length and extends forward in front of the ear and ends in front in the zygomatic bone (*zygoma*). Above the zygomatic arch is the *temporal fossa,* which is filled with the *temporalis muscle.* Attached to the lower margin of the zygomatic arch is the *masseter muscle.* Contraction of both the temporalis and masseter muscles may be felt by clenching the teeth.

Clinical Note

Testing the mandibular division of the trigeminal nerve.— Since both the temporalis and masseter muscles are innervated by the mandibular division of the trigeminal nerve, the integrity of the motor fibers of the nerve can be tested by asking the patient to clench his teeth.

Superficial Temporal Artery

The pulsations of the superficial temporal artery may be felt as it crosses the zygomatic arch, immediately in front of the auricle (see Fig 6–7). Its anterior branch may be seen in older individuals as it makes a wavy course across the temple.

Pterion

This is a surface landmark that is not marked by a depression or an eminence. It lies 1.5 in. (4 cm) above the midpoint of the zygomatic arch and is where the greater wing of the sphenoid bone meets the anterior-inferior angle of the parietal bone. The pterion is important, since beneath it lies the *anterior branch of the middle meningeal artery* (see Fig 8–12).

Suprameatal Triangle

This forms the bony lateral wall of the *mastoid antrum.* It is bounded behind by a line drawn vertically upward from the posterior margin of the external auditory meatus, above by the suprameatal crest of the temporal bone, and below by the external auditory meatus.

Temporomandibular Joint
This joint can be easily palpated in front of the auricle. As the mouth opens the *condyle* (head) of the mandible rotates and moves forward below the tubercle of the zygomatic arch.

Anterior Border of the Ramus of the Mandible
This may be felt deep to the masseter muscle. The *coronoid process* of the mandible may be felt with the finger inside the mouth.

Posterior Border of the Ramus of the Mandible
This is overlapped above by the parotid gland (see Fig 6–7), but below, it is easily felt through the skin. The outer surface of the ramus is covered by the masseter muscle and can be felt on deep palpation when the muscle is relaxed.

Body of the Mandible
This best examined by having one finger inside the mouth and another on the outside. Thus, it is possible to examine the mandible from the symphysis menti as far backward as the *angle* of the mandible.

Anterior Border of the Masseter Muscle
This may be easily felt by clenching the teeth. The masseter produces the fullness of the posterior region of the cheek.

Buccopharyngeal Fat Pad
This is especially large in infants and is responsible for the fullness of the anterior part of the cheek.

Facial Artery
The pulsations of the facial artery may be felt as it crosses the lower margin of the body of the mandible at the anterior border of the contracted masseter muscle (see Fig 6–2).

Parotid Duct
The parotid duct runs forward from the parotid gland one fingerbreadth below the zygomatic arch (see Fig 6–7). It can be rolled beneath the examining finger at the anterior border of the masseter as it turns medially to pierce the buccinator muscle to open into the mouth opposite the upper second molar tooth (see Fig 1–10).

Clinical Note

A clue as to where the duct perforates the cheek: drop a vertical line down from the lateral canthus of the eye. This can be useful in cases of lacerations of the cheek; if the laceration is anterior to that line, the duct is not affected.

The Eye
The surface anatomy of the eye will be described in Chapter 7.

Orbital Margin
This well-defined margin is easily felt. It is formed by the frontal, zygomatic, and maxillary bones (see Fig 6–4).

Supraorbital Notch
If present (sometimes it is a foramen), this can be felt at the junction of the medial and intermediate thirds of the upper margin of the orbit.

Supraorbital Nerve
This nerve, which is a branch of the frontal nerve, a branch of the ophthalmic division of the trigeminal nerve, can usually be rolled against the bone in the supraorbital notch (see Fig 6–2).

Infraorbital Foramen
This lies 5 mm below the lower margin of the orbit (see Fig 6–4) on a line drawn downward from the supraorbital notch to the interval between the two lower premolar teeth. The supraorbital notch and the infraorbital foramen also lie on a line that passes through the middle of the pupil when the eyes are in a neutral position.

Infraorbital Nerve
This nerve, a continuation of the maxillary division of the trigeminal nerve, emerges from the foramen and supplies the skin of the face and upper gum (see Fig 6–2). Deep pressure on the nerve produces discomfort over the area of the cheek.

Clinical Note

The upper gum is where the sensory distribution of the infraorbital nerve should be tested. Traumatic swelling of the skin of the face can lead to diminished sensation of the facial skin and result in the incorrect diagnosis of a lesion of the trunk of the infraorbital nerve.

Maxillary Air Sinus
This is situated within the maxillary bone and lies deep to the infraorbital foramen on each side (see Fig 1–5).

Frontal Air Sinus
This is situated within the frontal bone and lies deep to the superciliary ridge on each side (see Fig 1–5).

The Lips
The line of contact between the lips is opposite the cutting edges of the upper incisor teeth. The angle of the mouth usually lies in front of the first premolar tooth. In the center of the upper lip is a shallow vertical groove, the *philtrum,* which ends inferiorly as a small swelling called the *tubercle.*

CLINICAL ANATOMY

Clinical Anatomy of the Trigeminal Nerve and the Face

Testing for the Integrity of the Trigeminal Nerve
The trigeminal nerve has both sensory and motor functions. The sensory function may be tested by using a cotton wisp over each area of the face supplied by the ophthalmic, maxillary, and mandibular divisions (see Fig 6–1). There is little overlap of the dermatomal areas of each division.

The motor function may be tested by asking the patient to clench his teeth. The masseter and temporalis muscles, which are innervated by the mandibular division of the trigeminal nerve, can be felt to harden as they contract.

Trigeminal Neuralgia
This is a severe, stabbing pain over the face of unknown cause that involves the pain fibers of the trigeminal nerve. Pain is felt most commonly over the skin areas innervated by the mandibular and maxillary divisions of the trigeminal nerve; only rarely is pain felt in the area supplied by the ophthalmic division.

Clinical Anatomy of the Trigeminal Nerve Blocks

The blocking of the supraorbital nerve, the supratrochlear nerve, the infratrochlear nerve, the external nasal nerve, the infraorbital nerve, the auriculotemporal nerve, and the mental nerve are fully described in Chapter 19.

Clinical Anatomy of the Parotid Duct

The parotid duct can be injured in deep lacerations of the face. An injury is suspected when pressure on the parotid gland does *not* result in a globule of saliva being expelled through the previously dried off papilla of the parotid duct in the mouth. Repair of the duct consists of identification of the cut ends, which are then sutured together over a small catheter that opens into the mouth.

Clinical Anatomy of Facial Lacerations

Wound anesthesia is easily achieved by infiltration of a local anesthetic agent around the site of the laceration. Additional anesthesia may be done by blocking the branches of the trigeminal nerve as they exit through the bony foramena, as noted above.

Lip Lacerations
The wound is closed in layers, starting with the deepest layer first. Attention is made to ensure that the orbicularis oris is carefully sutured to avoid a defect when the lip is moved. Special care is taken to suture the vermillion border accurately.

Cheek Lacerations
Through-and-through lacerations, with the mouth cavity opening onto the surface, should be repaired in layers starting with the mucous membrane, then the buccinator muscle, and lastly the skin. Cheek lacerations may involve branches of the facial nerve and the parotid duct; the integrity of these structures must always be assessed.

A lesion of the facial nerve will produce a lower motor neuron type of paralysis of the facial muscles. If possible, the following five branches of the facial nerve should be tested for their functional integrity.

1. The temporal branch innervates the frontalis, which raises the eyebrows and wrinkles the forehead.
2. The zygomatic branch innervates the orbicularis oculi and closes the eyelids.
3. The buccal branch innervates the buccinator muscle and the nasal muscles; it wrinkles the nose.
4. The mandibular branch innervates the circumoral muscles and puckers and depresses the lips.
5. The cervical branch innervates the platysma and wrinkles the skin of the neck.

Lesions of the facial nerve should be repaired to avoid distortion of the face with permanent deformity. It is unnecessary to attempt repair of the fine terminal branches of the facial nerve, since regeneration and cross innervation will take place.

Lesions of the parotid duct are discussed on pp. 13 and 225.

Nose Lacerations

The wound is closed in layers, starting with the deepest layer first. Exposed cartilage must be covered. Unfortunately, the skin of the nose contains many sebaceous glands and is prone to scarring.

Ear Lacerations

These are described on p. 236.

Eyebrow Injuries

The borders of the eyebrows are correctly aligned before suturing. Disfigurement is avoided by removing the minimum of tissue. If debridement is required, incisions are made parallel to the long axis of the hair follicles and not at right angles to the skin surface. In this way the hair follicles are preserved and an unsightly bald area is avoided.

Clinical Anatomy of the Bones of the Face

Pneumatization of the Air Sinuses

The maxillary sinus is present at birth and rapidly increases in size with the development of the teeth. The frontal sinus, which is not present at birth, starts to develop in infancy.

Skeletal Facial Development and Bone Injuries

The developing bones of a child's face are more pliable than an adult's and fractures may be incomplete or greenstick. In adults the presence of well-developed, air-filled sinuses and the mucoperiosteal surfaces of the alveolar parts of the upper and lower jaws means that most facial fractures should be considered to be open fractures, susceptible to infection, and requiring antibiotic therapy.

Anatomy of Common Facial Fractures

Automobile accidents, fisticuffs, and falls are common causes of facial fractures. Fortunately, the upper part of the skull is developed from membrane, and therefore in children this part of the skull is relatively flexible and can absorb considerable force associated with deceleration injuries without resulting in a fracture. The muscles of the face are thin and weak and cause little displacement of the bone fragments. Once a fracture of the maxilla has been reduced, there is no need for prolonged fixation. However, in the case of the mandible, the strong muscles of mastication can create considerable displacement, requiring long periods of fixation.

The most common facial fractures involve the nasal bones, followed by the zygomatic bone and then the mandible. To fracture the maxillary bones and the supraorbital ridges of the frontal bones, an enormous force is required.

Nasal Fractures.—Fractures of the nasal bones, because of the prominence of the nose, are the most common facial fractures. Because the bones are lined with mucoperiosteum, the fracture is considered open; the overlying skin may also be lacerated. The nasal processes of the frontal bone and the frontal process of the maxilla may also be fractured, and the nasal septum may be injured. The bones of the nasal walls and septum are fully described in Chapter 1.

Anatomical Complications.—*Bleeding* from the nose results from tearing of the mucous membrane; the blood supply of the nose is described on p. 2. *Septal hematoma* secondary to injury of the septal arterial supply must be recognized. It is seen as a bulging tense swelling, and, if missed, necrosis of the septal cartilage or bone may occur. *Cerebrospinal rhinorrhea* secondary to fracture of the cribriform plate of the ethmoid bone also occurs.

Anatomy of Local Anesthesia of the Nose.—This is fully described in Chapter 19, p. 814.

Maxillofacial Fractures.—These usually occur as the result of massive facial trauma. There is extensive facial swelling, midface mobility on palpation, malocclusion of the teeth with anterior open bite, and possibly cerebrospinal rhinorrhea secondary to fracture of the cribriform plate of the ethmoid bone. Diplopia may be present due to orbital wall damage, and involvement of the infraorbital nerve with anesthesia or parethesia of the skin of the cheek and upper gum may occur in fractures of the body of the maxilla. Epistaxis usually occurs in nasal injuries, but it may also occur in maxillary fractures. Blood enters into the maxillary air sinus and then leaks into the nasal cavity.

The sites of the fractures were classified by Le Fort as Le Fort I, II, or III fractures (Fig 6–15).

Le Fort I fractures involve the upper alveolar arch (above teeth), the palate, lower parts of the pterygoid plates of the sphenoid, and a portion of the wall of each maxillary sinus.

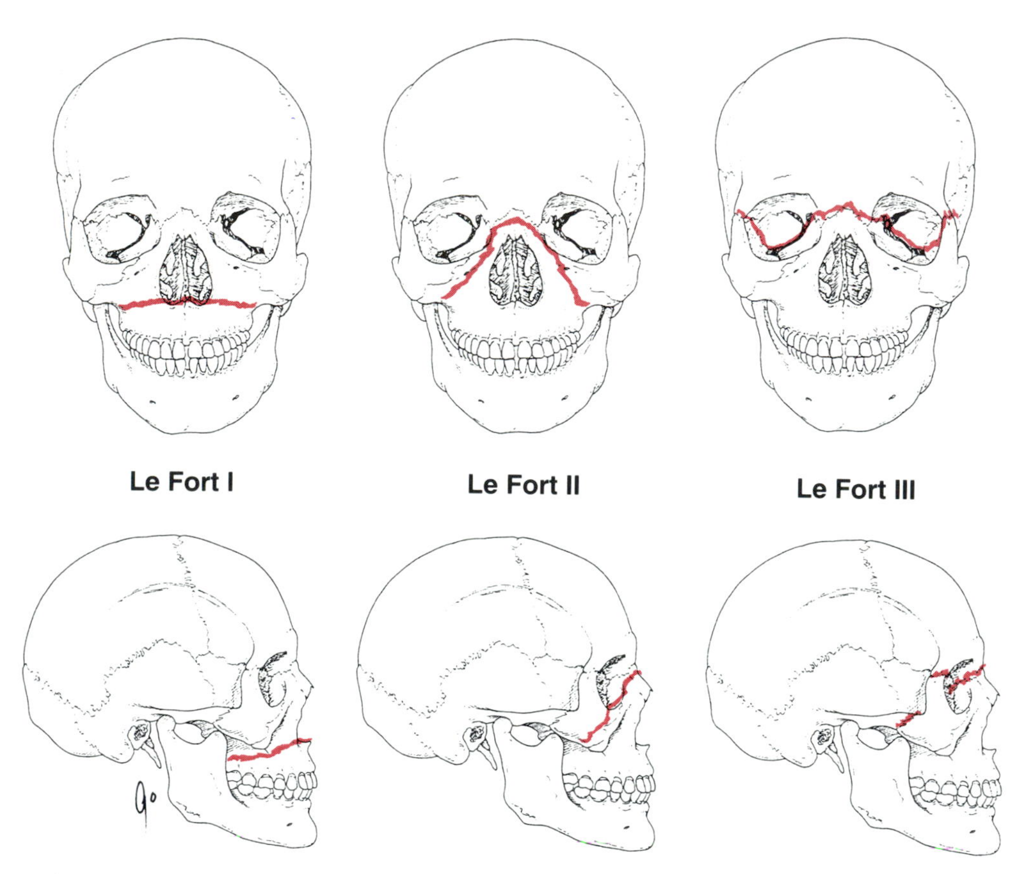

FIG 6–15.
Le Fort classification of maxillofacial fractures. See text for detailed description.

Le Fort II fractures involve the nasal bones and the frontal processes of the maxilla. The zygoma bones are usually not displaced with this type of fracture, but the ethmoid air sinuses are often destroyed. Separation of the eyes and widening of the bridge of the nose usually occur with this type of fracture.

Le Fort III fractures involve the nasal bones, the maxillae, and the zygoma bones, which are separated as a single unit from their attachments.

The force applied to the face penetrates into the depth of the bony structure and often involves the ethmoid and sphenoid bones (Fig 6–16). Many combinations of these fractures may be found, for example, a Le Fort I fracture on one side and a Le Fort II fracture on the opposite side.

Radiographic examination is essential to confirm the sites of the fractures and the extent of the displacement. The Water's view, which is a posterior-anterior view of the face with the head tilted up so that the petrous parts of the temporal bones are projected below the maxillary sinus; the lateral facial views are most helpful. The Water's view should be performed with the patient upright so that the air fluid levels in the maxillary sinus may be appreciated. Normal radiographs of the facial region are shown in Figures 6–17 through 6–22).

Blowout Fractures of the Maxilla.—A severe blow to the orbit may cause the contents of the orbital cavity to explode downward through the floor of the orbit into the maxillary sinus. In a pure blowout fracture of the orbit, the lower rim of the orbit remains intact. In addition, the orbital contents may explode medially through the paper-thin medial wall of the orbit formed by the ethmoid bone.

Damage to the infraorbital nerve crossing the roof of the maxillary sinus may result in altered sensation to the cheek, upper lip, and upper gum. Entrapment of the inferior rectus muscle of the eye between the bony fragments may restrict eye movements, particularly upward gaze; the patient may also complain of diplopia. Orbital fat may herniate into the maxillary sinus, resulting in enophthalmos. Air from the maxillary sinus may pass upward into the upper and lower eyelids and produce subcutaneous emphysema.

Zygoma Fractures.—The zygoma bone (see Fig 6–16) is extremely strong, but when struck by an ob-

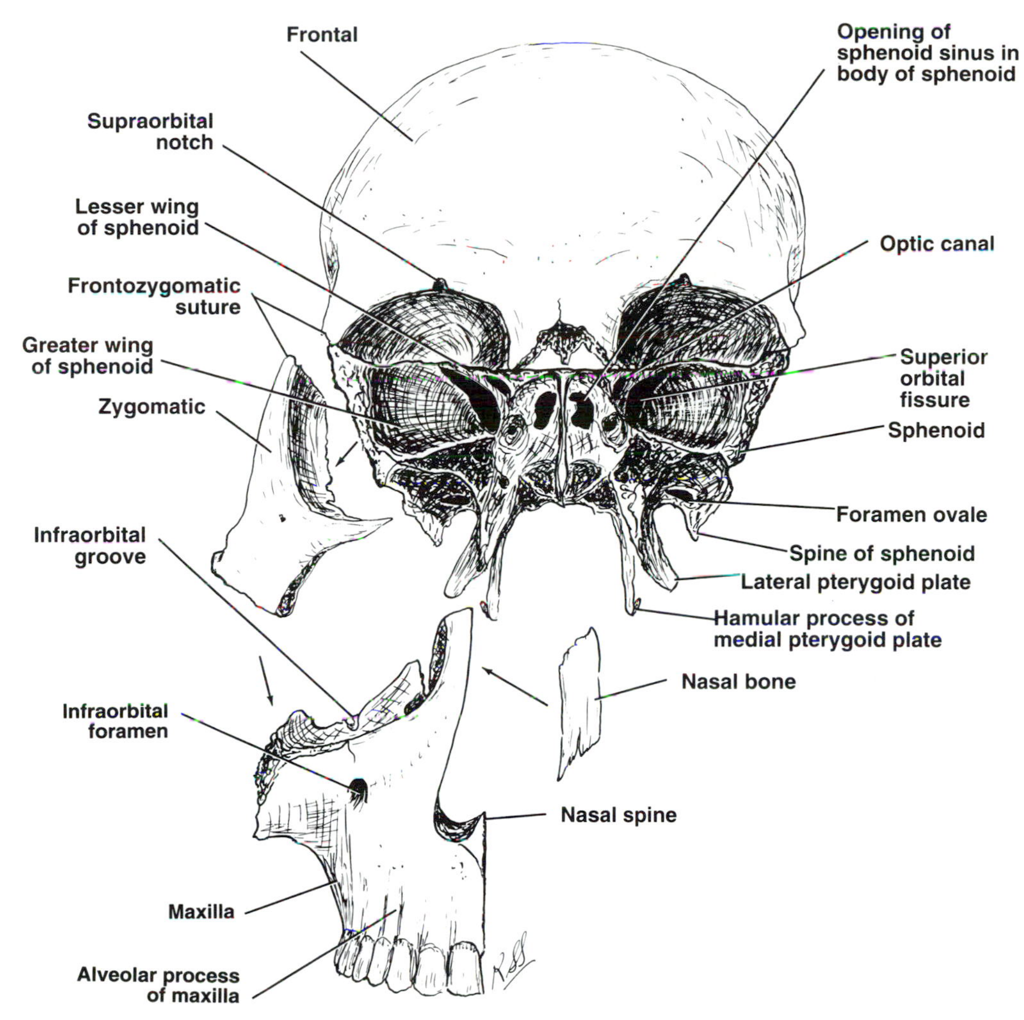

FIG 6–16.
Exploded view of anterior surface of skull shows deep position of the sphenoid bone. For clarity, the ethmoid bone, the lacrimal bone, and the palatine bone have been omitted. Note how the greater and lesser wings of the sphenoid contribute to the orbital cavity.

ject it can be twisted off its tripod articulation with the frontal bone, maxilla, and zygomatic process of the temporal bone. (Such a fracture should more accurately be called a quadrapod fracture of the zygoma, because the zygoma has two articulations with the maxilla—at the infraorbital rim and at the lateral wall of the maxillary sinus.) The normal frontal-zygoma suture line can sometimes be mistaken for a fracture.

The cheek may appear flattened, irregularity of the orbital margin may be felt, and the zygoma is tender on infraorbital palpation. Diplopia may be present. Tearing of the mucous membrane lining the maxillary sinus can cause bleeding from the branches of the maxillary artery, and blood can enter the nose through the maxillary sinus, producing epistaxis. Involvement of the infraorbital nerve may result in altered sensation to the cheek, upper lip, and upper gum.

The zygomatic arch can be fractured from a direct blow to the side of the head. Apart from the area of depression over the side of the head, the movements of the temporomandibular joint are often compromised because of the impingement of the depressed arch against the coronoid process of the mandible. A fractured zygomatic arch impeding motion of the coronoid process of the mandible may cause a persistent open bite. A zygomatic arch fracture is best seen on a "bucket handle," "jug handle," or submental vertex radiograph.

Mandibular Fractures.—The mandible is horseshoe shaped and is formed into a bony ring by th two temporomandibular joints and the base of the skull. Traumatic impact is transmitted around the

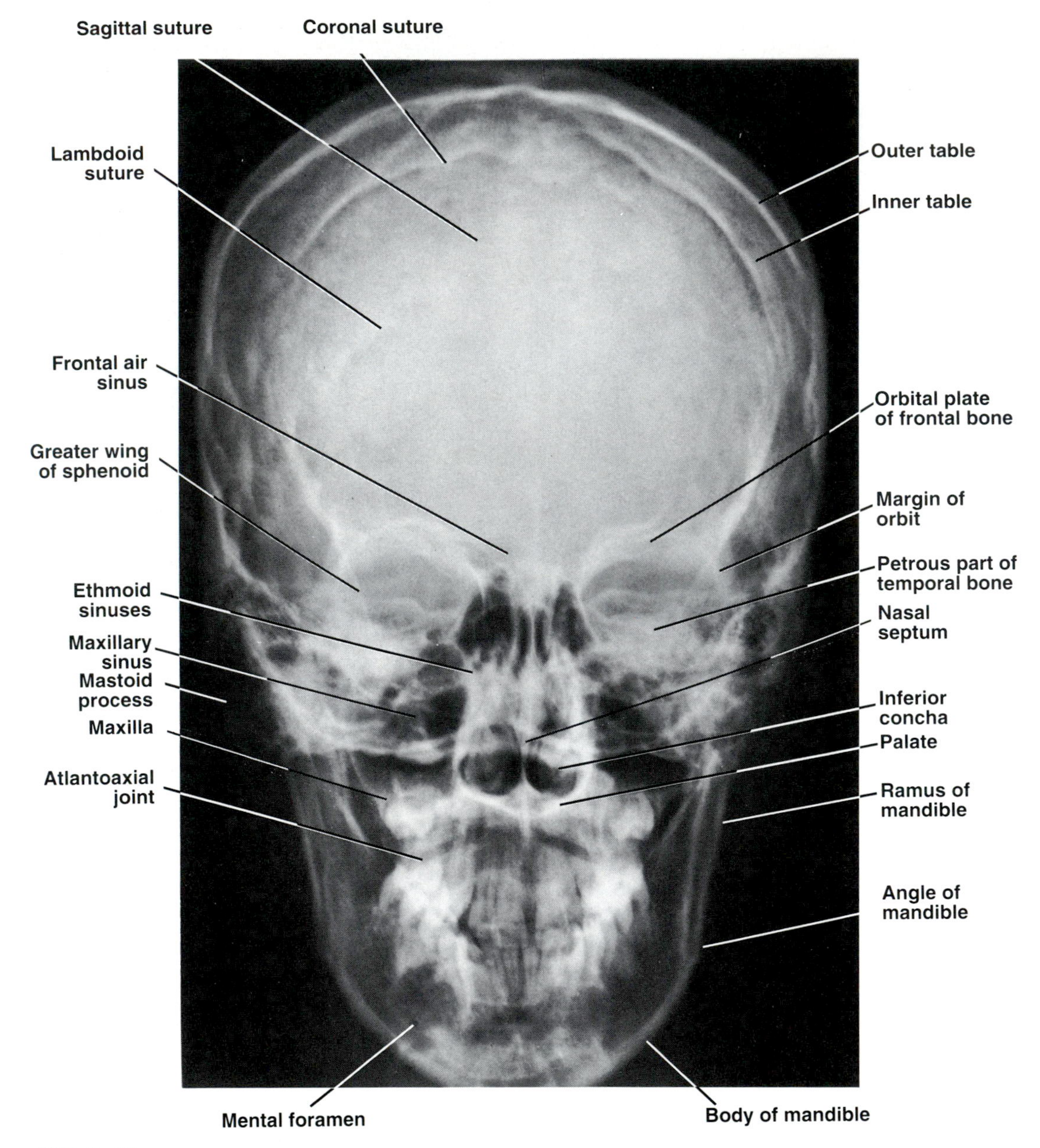

FIG 6–17.
Posteroanterior radiograph of the skull.

ring, causing a single or multiple fractures often far removed from the point of impact. In descending order of frequency, the following mandibular fractures occur singly or in different combinations: (1) condyle; (2) coronoid process, angle, and ramus; (3) body near mental foramen; and (4) body near symphysis. The mandible often fractures in two different places at one time. Disruption of the mandibular anatomy produces the following signs.

1. *Malocclusion.* This is especially severe in bilateral or unilateral condylar fractures.

2. *Bleeding into the mouth.* Minor bleeding usually occurs due to tearing of the mucoperiosteum. Extensive bleeding occurs if the inferior alveolar artery, a branch of the maxillary artery, is lacerated. All mandibular fractures anterior to the angle should be considered to be open.

3. *Displacement of bone fragments.* Anterior fragments are depressed by the pull of the anterior belly of the digastric, geniohyoid, and mylohyoid muscles. Posterior fragments are elevated by the masseter, medial pterygoid, and temporalis muscles. These displacements will not occur if the line of frac-

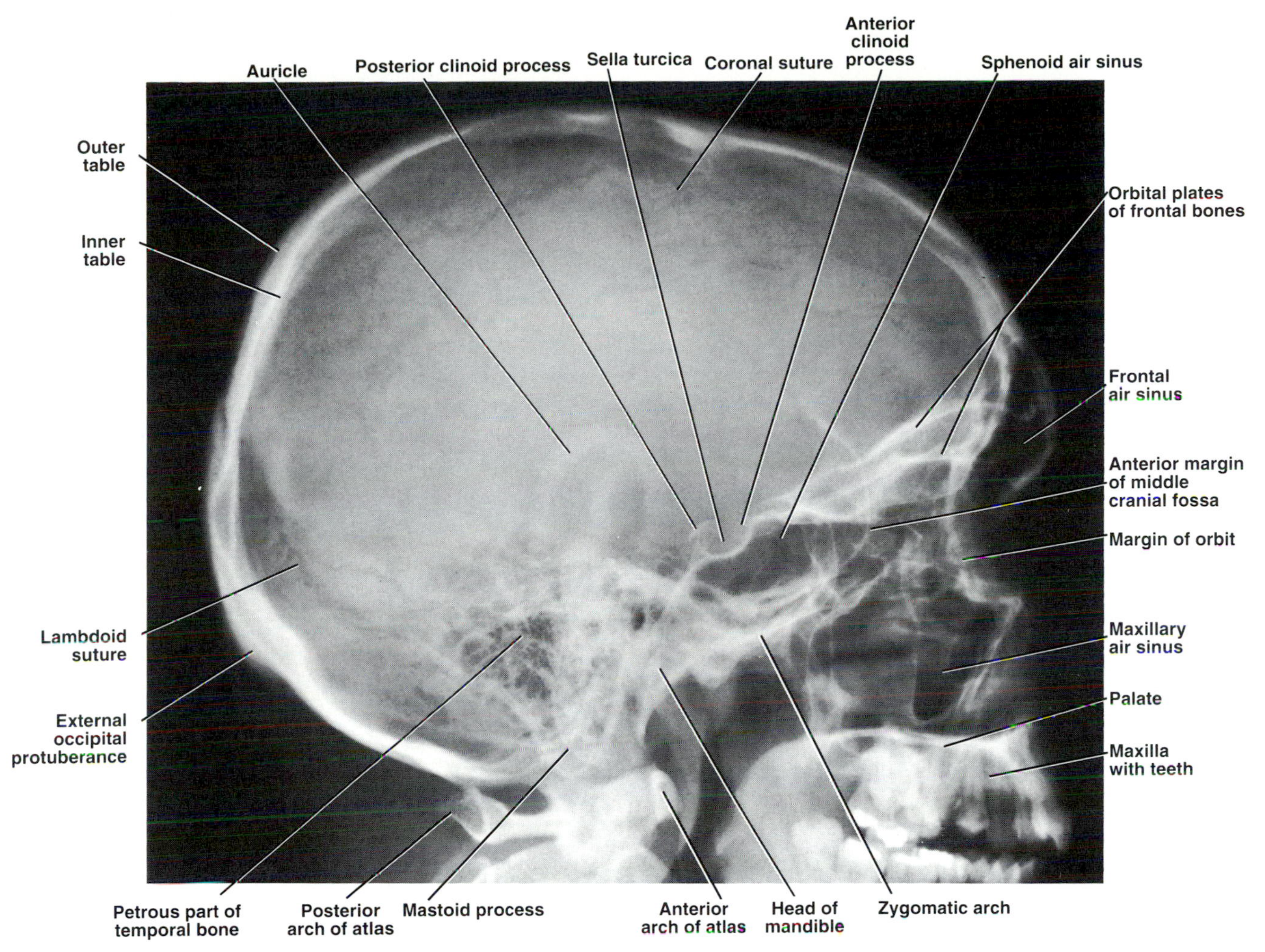

FIG 6–18.
Lateral radiograph of the skull.

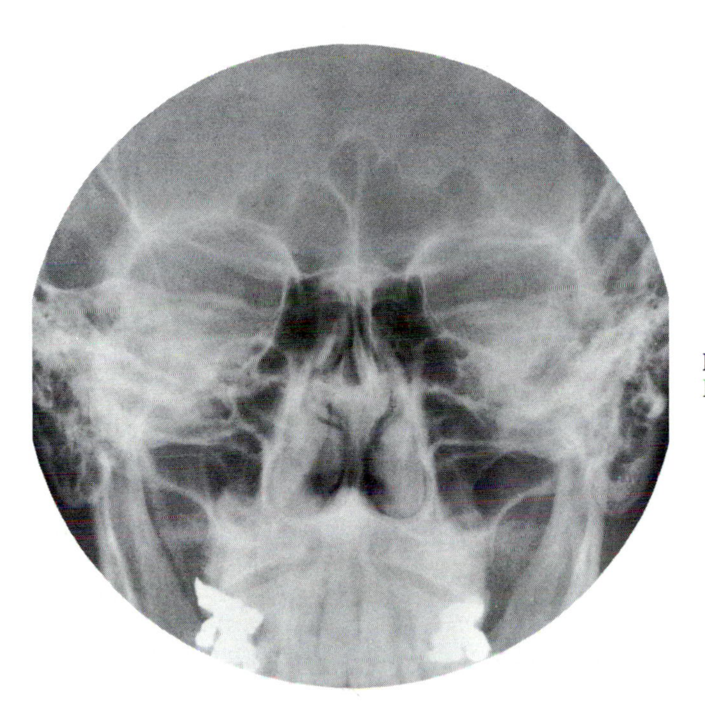

FIG 6–19.
Posteroanterior radiograph of the paranasal sinuses.

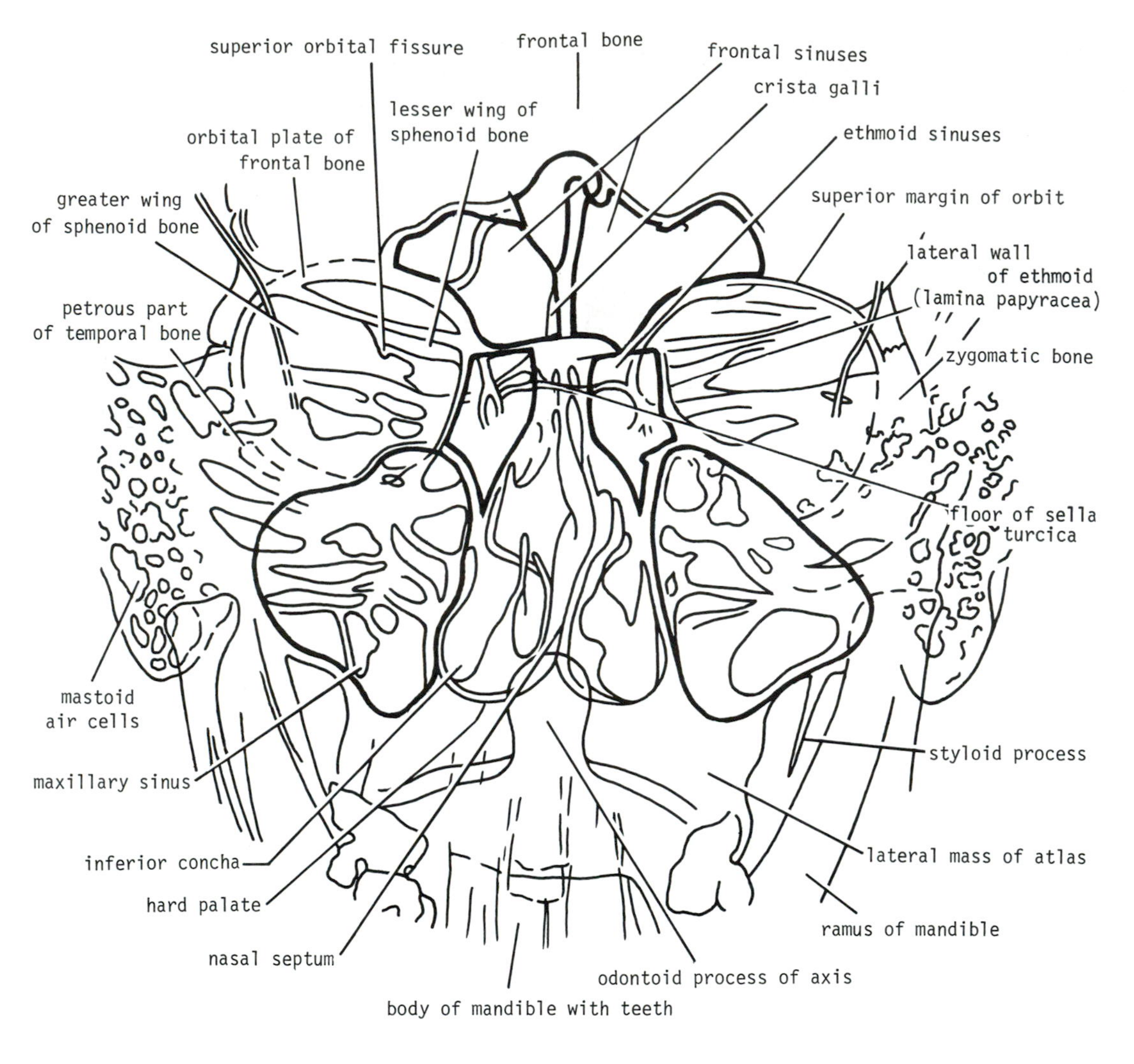

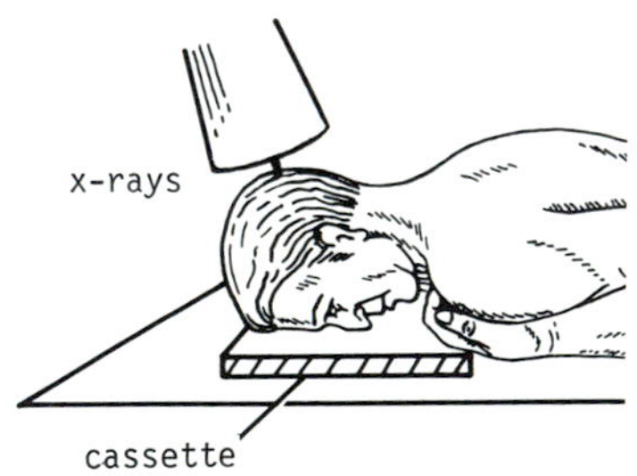

FIG 6–20.
Main features that can be seen in the posteroanterior radiograph of the paranasal sinuses in Figure 6–19.

ture is angled in such a way as to prevent such movement thus the classification of mandibular fractures into "favorable" and "unfavorable."

4. *Loose or missing teeth.* This occurs if the line of fracture involves the alveolar margin. Beware of the possibility of aspiration of a tooth.

5. *Skin anesthesia or paresthesia.* If the fracture line involves the inferior alveolar canal, the inferior alveolar nerve, or the mental nerve, anesthesia or paresthesia of the skin over the lower lip, chin, or lower gums may result.

6. *Airway obstruction.* Bilateral fractures of the body of the mandible may cause instability of the tongue, since the genioglossus muscles will no longer have a stable origin. The tongue may fall back and compromise the airway. Pulling the chin for-

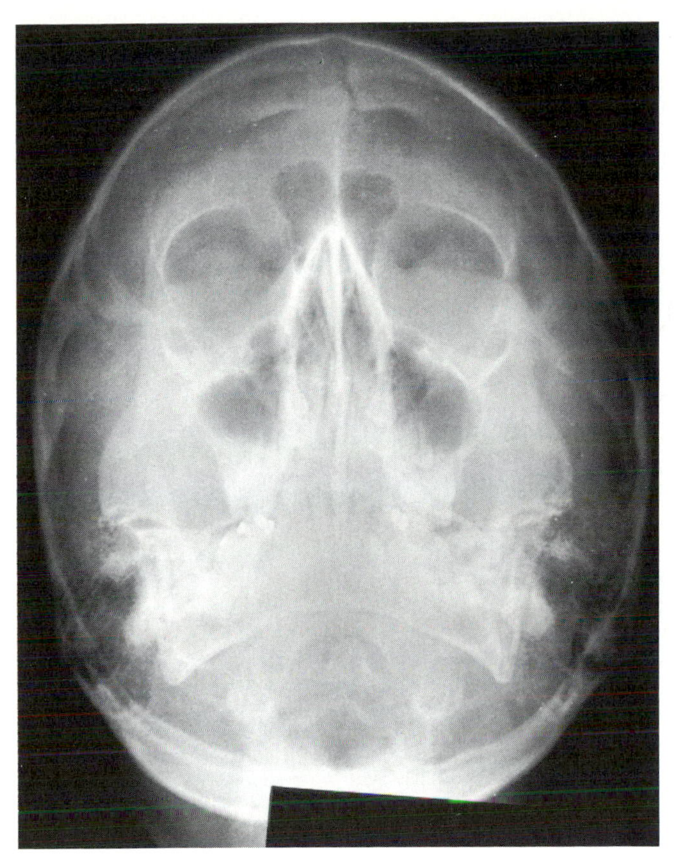

FIG 6–21.
Posteroanterior (Water's) radiograph of the paranasal sinuses of a 9-year-old girl.

ward (the origin of the genioglossus muscles) should keep the tongue anterior and out of the way.

Clinical Anatomy of the Temporomandibular Joint

Dislocations of the temporomandibular joint may occur following trauma or the simple act of yawning. The dislocation may be unilateral or bilateral. The mandible always dislocates anteriorly.

The temporomandibular joint lies immediately in front of the external auditory meatus. Fortunately, the great strength of the lateral temporomandibular ligament prevents the condyle of the mandible from passing backward and fracturing the bony part of the external auditory meatus when a severe blow falls on the chin. Traumatic anterior dislocation must always be differentiated from condylar or neck fractures of the mandible whose clinical features it can mimic.

Excessive opening of the mouth, as may occur in yawning, causes the condyle of the mandible and the articular disc to move forward onto the summit of the articular tubercle. Here the joint is unstable, and a sudden contraction of the lateral pterygoid muscles may pull the disc beyond the summit so that the two condyles of the mandible lie fixed in front of the articular tubercle. The mouth is thus fixed in a persistent open bite, and the accompanying spasm of the muscles may cause the patient severe pain.

Reduction of the dislocation involves exerting downward pressure on the jaw, usually from behind the molar teeth, in conjunction with forward rotation. The downward pressure overcomes the tension of the temporalis, masseter, and medial pterygoid muscles, and the forward rotation unlocks the condyles from the articular tubercles.

Clinical Anatomy of the Muscles of the Face

In repose, the tone of the muscles of the face are in a state of balance with one another—the dilators are in balance with the constrictors, and the right muscles are in balance with the left muscles. Weakness of facial muscles is particularly noticeable around the mouth and eye, where it may be recognized by a sagging corner of the mouth or a drooping lower eyelid. Lesions of the facial nerve and their effect on the facial muscles are discussed below.

Clinical Anatomy of the Facial Nerve

Testing the Integrity of the Facial Nerve

If intact, the facial nerve functions by closing the eyes, opening the lips, and enabling tasting on the anterior two thirds of the tongue. The patient is asked to close his eyes firmly; the observer then attempts to open the eyes by gently raising the patient's upper lids. Asymmetry of the mouth can be tested by clenching the teeth and separating the lips. Taste on each half of the anterior two thirds of the tongue can then be tested with suitable substances that are sweet, salty, sour, and bitter.

Facial Nerve Lesions

The facial nerve may be injured or may become dysfunctional anywhere along its long course from the brain stem to the face. Its anatomical relationship to other structures greatly assists in the localization of the lesion. If the abducent nerve (lateral rectus muscle of the eye) and the facial nerve are not functioning, this suggests a lesion in the pons of the brain stem, where their nuclei are adjacent. If the vestibu-

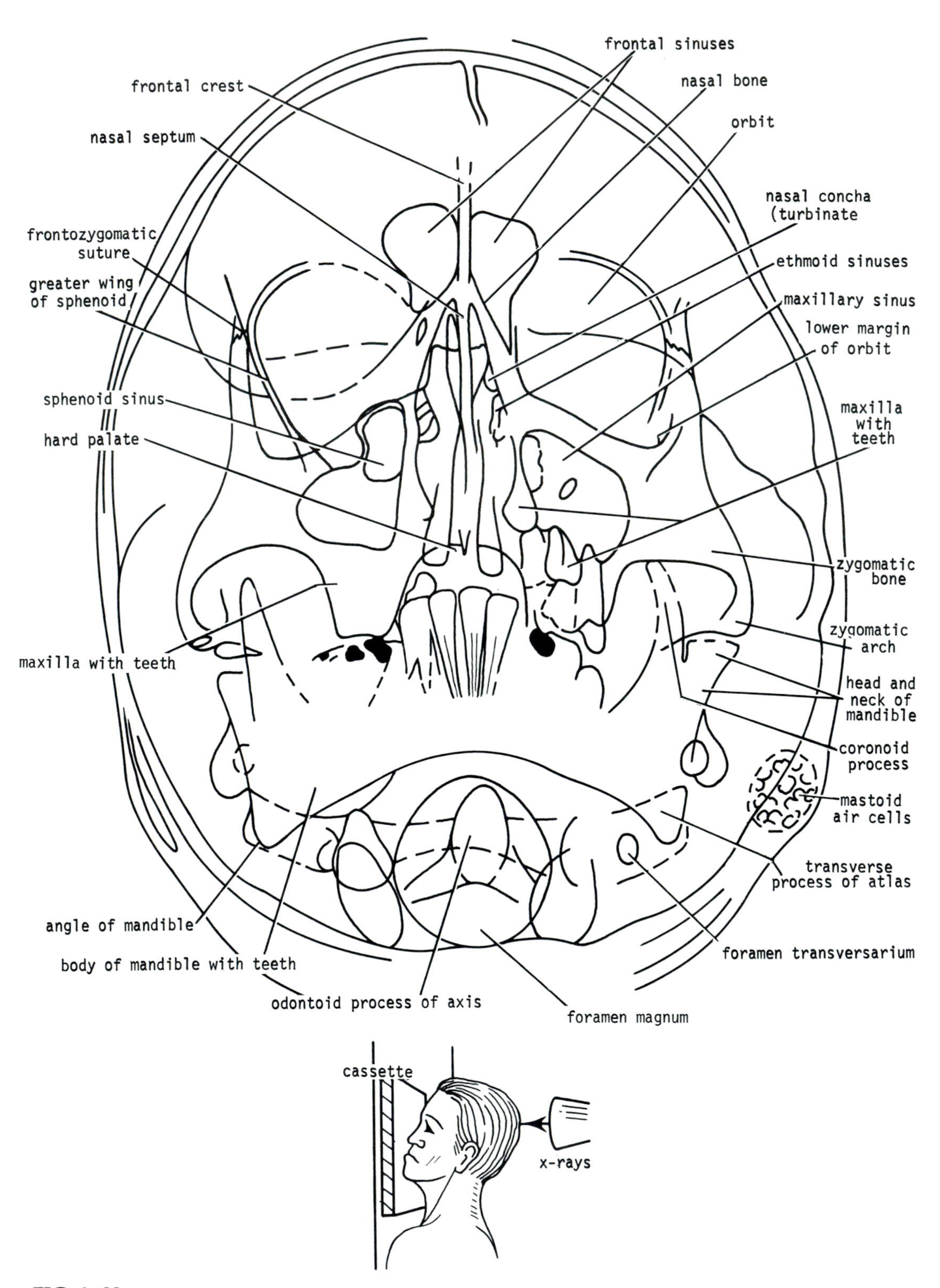

FIG 6–22.
Main features that can be seen in posteroanterior radiograph of paranasal sinuses in Figure 6–21.

locochlear nerve (for balance and hearing) and the facial nerve are not functioning, this suggests a lesion in the internal acoustic meatus. If the patient is excessively sensitive to sound in one ear, the lesion probably involves the nerve to the stapedius muscle, which arises from the facial nerve in the facial canal. Loss of taste over the anterior two thirds of the tongue indicates that the facial nerve is damaged proximal to the point where it gives off the chorda tympani in the facial canal. A firm swelling of the parotid salivary gland associated with impaired function of the facial nerve strongly indicates carcinoma of the parotid gland with involvement of the nerve within the gland.

Deep lacerations of the face may involve branches of the facial nerve.

Lower Motor Nerve Lesions

A lesion of the facial nerve nucleus or the facial nerve, i.e., a lower motor neuron lesion, will cause paralysis of all the muscles on the affected side of the face (Fig 6–23). On the side of the lesion, the lower eyelid will droop, and the angle of the mouth will sag. The cheek puffs out with respiration, and food tends to accumulate between the teeth and cheek on the affected side because the buccinator muscle is paralyzed. Tears will flow over the lower eyelid, and saliva will dribble from the corner of the mouth. The patient will be unable to close his eye or expose his teeth fully on the affected side. The affected side will demonstrate Bell's phenomenon—his eyeball will rotate upward and slightly medially on attempts to close the eyelids, owing to relaxation of the inferior rectus and contraction of the superior rectus (a normal phenomenon that becomes apparent because of the paralysis of the eyelid closure).

Upper Motor Nerve Lesions

In lesions of the upper motor neuron that produce facial muscle paralysis (e.g., cortical stroke), only the muscles of the lower part of the face will be paralyzed (see Fig 6–23). The part of the facial nucleus that controls the muscles of the upper part of the face receives corticonuclear fibers from both cerebral hemispheres, whereas the part of the facial nucleus that controls the muscles of the lower part of the face receives corticonuclear fibers only from the contralateral cerebral hemisphere.

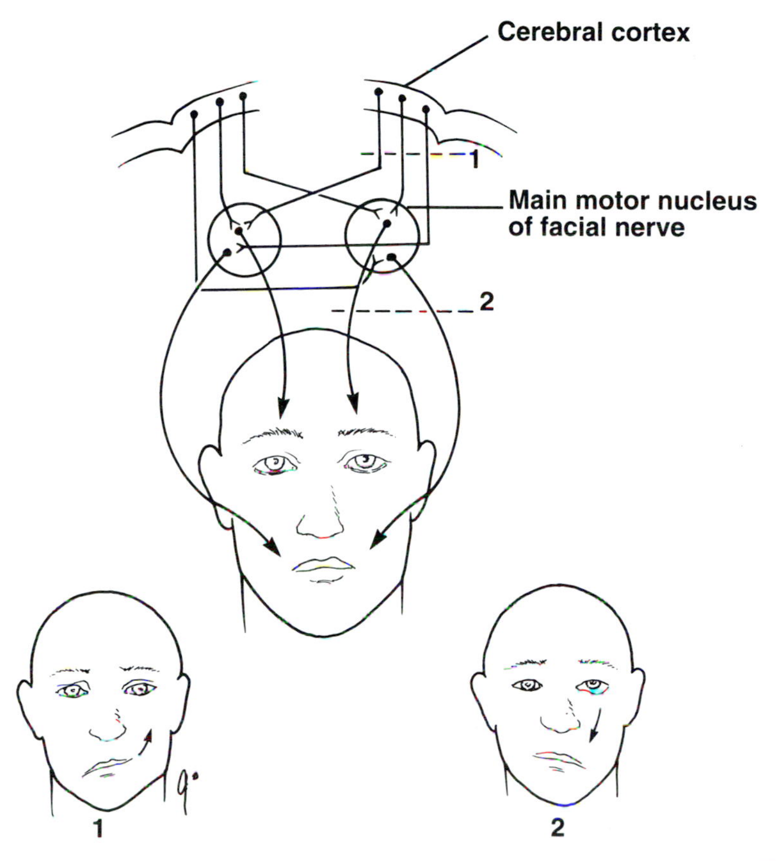

FIG 6–23.
Facial expression defects associated with lesions of the facial nerve. 1 = upper motor neurons; 2 = lower motor neurons.

In patients with an upper motor neuron facial paralysis, the emotional movements of the face are usually preserved. This indicates that the upper motor neurons controlling these *mimetic movements* have a course separate from that of the main corticobulbar fibers. A lesion involving this separate pathway alone results in a loss of emotional movements, but voluntary movements are preserved. A more extensive lesion will produce both mimetic and voluntary facial paralysis.

Bell's Palsy

Bell's palsy is a dysfunction of the facial nerve as it lies within the facial canal; it is usually unilateral. The site of the dysfunction will determine the aspects of facial nerve function that do not work.

The pressure on the nerve temporarily interrupts the function of the nerve, causing a lower motor neuron type of facial paralysis. Weakness of the facial muscles occurs usually in the upper and lower halves of the face; occasionally the lower half of the face may be more affected than the upper half or visa versa.

When the lesion is distal to the geniculate ganglion of the facial nerve, i.e., distal to the point of origin of the greater petrosal nerve (containing parasympathetic fibers to the lacrimal gland, see Fig 9–2), the lacrimal secretion is normal, but there is an excessive collection of tears in the conjunctival sac because the tears have not been expressed into the lacrimal duct by movements of the eyelids. A lesion of the facial nerve proximal to the ganglion will cause a diminished secretion of tears.

The corneal reflex is absent due to paralysis of the upper lid. However, since the corneal sensation is preserved, the contralateral upper lid blinks.

If the lesion is proximal to the origin of the chorda tympani (see Fig 9–2), a loss of submandibular and sublingual salivary secretion will occur on that side (owing to loss of parasympathetic secretomotor supply); a unilateral loss of taste sensation on the anterior two thirds of the tongue will also be observed. Hyperacusis may occur if the facial nerve lesion is proximal to the innervation of the stapedius muscle.

With Bell's palsy, partial or complete recovery occurs. When there is partial recovery, the paralyzed muscles often show contractures, so that a superficial examination may give the impression that the normal nonparalyzed muscles are weak. Twitching of the facial muscles and excessive production of tears on the affected side may be further sequelae.

Clinical Anatomy of the External Ear

Ear Pain

Both traumatic injury to the auricle and otitis externa are extremely painful because of the close attachment of the skin to the underlying perichondrium. This leaves little room in the subcutaneous tissue for the accumulation of edema fluid. Pain is referred along the distribution of the auriculotemporal, great auricular, and lesser occipital nerves. Otitis media can produce severe pain over the ear region. The sensory innervation of the mucous membrane is from the tympanic branch of the glossopharyngeal nerve and the auriculotemporal nerve (tympanic membrane). Presumably the pain is referred along the distribution of the auriculotemporal nerve in these cases.

Referred Ear Pain

A pathological process in the throat (usually near the pyriform fossa) is frequently referred as pain in the ipsilateral ear. The sensory nerve supply to the mucous membrane lining the fossa is the internal laryngeal branch of the vagus. Pain arising from this site is referred to the ear along the auricular branch of the vagus nerve.

Referred ear pain may also occur in tooth infections and diseases of the temporomandibular joint. The teeth receive their sensory innervation from the maxillary division (teeth of the upper jaw) and the mandibular division (teeth of the lower jaw) of the trigeminal nerve. The temporomandibular joint receives its sensory innervation from the mandibular division of the trigeminal nerve. Pain impulses arising at these sites are referred along the auriculotemporal nerve to the region of the external ear and the temporal region of the scalp.

Foreign Bodies

Foreign bodies in the external auditory meatus can cause problems because of the ear canal's delicate, sensitive lining and the fact that the deep end is closed by the tympanic membrane.

Ear Injuries

Because of its exposed position, the auricle is frequently injured by blunt and sharp traumas. Since it has a plentiful blood supply, only minimal debridement should be performed. For the same reason, partial amputations can be managed by primary wound closure. Injury to the cartilage requires careful suturing of the perichondrium, since it contains the arterial supply to the underlying cartilage.

Blunt trauma may cause auricular hematomas. The blood tends to accumulate between the perichondrium and the cartilage. Failure to evacuate a hematoma adequately may lead to necrosis of the cartilage, because the blood supply, which is located in the perichondrium, becomes separated from the underlying cartilage that it nourishes. Necrosis of the cartilage followed by fibrous replacement results in deformity, i.e., cauliflower ear.

Clinical Anatomy of the Sensory Nerve Block of the Scalp

The sensory nerve block of the scalp is fully described in Chapter 19, p. 814.

Clinical Anatomy of the Scalp

Lacerations of the Scalp

The scalp has a profuse blood supply. Even a small laceration of the scalp may cause severe loss of blood. It is often difficult to stop the bleeding of a scalp wound because the arterial walls are attached to fibrous septa in the subcutaneous tissue and are unable to contract or retract to allow blood clotting.

Most scalp lacerations can be closed primarily. There is little risk of infection because of the profuse blood supply. When a deep laceration extends across the aponeurosis of the occipitofrontalis muscles, it is essential to suture this layer separately to overcome the muscle tone that will cause a gaping wound.

Large areas of the scalp may be cut off the skull if a person is projected forward at high speed through a window. Because of the rich blood supply, it is often possible to replace large areas of scalp that are only attached to the skull by a narrow pedicle and suture the scalp in position without necrosis occurring. A completely avulsed scalp can be replaced by reuniting the cut arteries using microsurgical techniques.

Often blunt trauma to the scalp produces a laceration that closely resembles an incised wound because the scalp is split against the unyielding skull, and the pull of the occipitofrontalis muscles causes a gaping wound. This anatomical fact may be of considerable forensic importance.

Blood or Pus in the Scalp

Blood or pus may collect in the potential space beneath the epicranial aponeurosis. It tends to spread over the calvaria, being limited in front by the orbital margin, behind by the nuchal lines on the occipital bone, and laterally by the attachment of the aponeurosis to the temporal fascia. On the other hand, subperiosteal blood or pus is limited to one skull bone because of the attachment of the periosteum to the sutural ligaments.

Scalp Infections

Scalp infections of the sebaceous or sweat glands tend to remain localized and are usually painful because of the presence of abundant fibrous tissue in the subcutaneous layer. For a discussion of the intracranial spread of infection, see p. 220.

Clinical Anatomy of the Mouth

Intraoral nerve blocks are fully described in Chapter 19, p. 812.

Tooth Pain

The nature of tooth pain can be helpful in locating the cause. An acute pulpitis, for example, can produce a sharp, stabbing pain, but a disease in the periodental tissue produces a dull, throbbing ache. Since the teeth of both the upper and lower jaws are innervated by the trigeminal nerve, disease in a lower tooth may give rise to pain in the upper jaw.

Referred Tooth Pain

Dental disease can give rise to referred pain throughout the distribution of the trigeminal nerve, as noted above. Moreover, disease in structures other than teeth, such as in the air sinuses or the temporomandibular joint, i.e., structures that are innervated by the trigeminal nerve, can give rise to referred pain in the teeth. Even trigeminal neuralgia may be the cause of apparent tooth problems.

Dental Trauma

The incisor and canine teeth are the most commonly damaged because of their anterior position and their anatomical structure. Enamel and dentine chips produce a mild discomfort. The exposure of the pulp with its contact with the lip or tongue, however, produces extreme pain. Complete avulsion of teeth is common and is often associated with fracture of the alveolar bone.

Tongue Lacerations

The basic anatomy of the tongue is described on p. 9. The tongue is a muscular organ with a profuse arterial supply from the lingual artery (from the external carotid artery), the tonsillary artery (from the facial artery), and the ascending pharyngeal artery (from the external carotid artery). Lacerations of the tongue are common in children. Bleeding from small lacerations can often be controlled by compressing the tongue. Large lacerations often require suturing of the intrinsic musculature and the mucous membrane under local or general anesthesia.

ANATOMICAL-CLINICAL PROBLEMS

1. A 42-year-old man was seen in the emergency department after being knocked down in a street brawl. He had received a blow on the back of the head with an empty bottle. On examination, the patient was found to be conscious and had a large doughlike swelling over the back of the head. The skin was intact, and the swelling fluctuated on palpation. No other abnormal signs were present. A diagnosis of hematoma of the scalp was made.

Name the layer of the scalp in which the hematoma was situated (1) if the swelling was extensive and extended over more than one skull bone, (2) if the swelling was limited to the area occupied by an underlying skull bone, and (3) if the swelling was small, superficial, and with no fluctuation.

2. A 16-year-old girl visited her dermatologist because of severe acne on her face. On examination, it was found that a small abscess was present on the right side of the nose. The patient was given antibiotics and warned not to press the abscess. Three days later she was seen in the emergency department with a swollen right eye, proptosis, and edema of both eyelids. She was obviously sick and had a fever of 101° F. After careful examination, a diagnosis of thrombophlebitis of the cavernous sinus was made. Why is it dangerous to squeeze, prick, or incise a boil in the area between the eye and the upper lip or between the eye and the side of the nose? What is the connection between a facial boil and cavernous sinus thrombosis?

3. Following a powerful forehand drive from his opponent during a game of tennis, a 34-year-old man received a severe blow to the right orbit from the tennis ball. When he was examined in the emergency department, his right eye and periorbital area were found to be contused and swollen. His right upper lid felt like "rice krispies." Nine days later, the man complained of double vision and was unable to rotate his right eye upward. He also had hyperesthesia of the skin below his right orbit. Using your knowledge of anatomy, make a diagnosis and explain the signs and symptoms.

4. A 14-year-old boy was thrown from his bicycle, and his left cheek was hit by the end of the handle bars. He was rushed to the emergency department bleeding profusely from a deep laceration of his left cheek. On examination, it was found that the cheek had been penetrated, and the vestibule of the mouth was open through the wound to the exterior. Why does a relatively small wound on the face bleed so extensively? Which layers of the cheek have to be sutured separately? Since it is possible that the parotid duct might have been damaged, give the surface marking of the duct.

5. Explain how knowledge of the positions of the wrinkle lines and tension lines is important in making facial incisions and in the suturing of facial lacerations.

6. A 55-year-old man came to the emergency department complaining of an agonizing, stabbing pain over the middle part of the left side of his face. The stabs would last a few seconds and were repeated several times. "The pain is the worst I have ever experienced," he told the physician. A draft of cold air on his face or the touching of a few hairs in the temporal region of his scalp could trigger the pain. Physical examination revealed no sensory or motor loss of the trigeminal nerve. Using your knowledge of anatomy, can you make a diagnosis?

7. A 25-year-old student had been working as a waiter until 1:30 A.M. As he was changing his clothes to go home, he involuntarily opened his mouth wide and yawned. To his great consternation, he found that he could not close his mouth. His jaw was stuck, and he was in great pain over the both sides of his face. What is the diagnosis? Using your knowledge of anatomy, explain how the treatment is carried out. Why did the patient experience bilateral pain over the face?

8. A 40-year-old man was seen in the emergency department with a large boil over the right parietal region of his scalp. Why are such infections potentially dangerous? What is the lymphatic drainage of this area of the scalp?

9. A 64-year-old woman woke up one morning to find the left side of her face paralyzed. When examined in the emergency department, she was found to have complete paralysis of the entire left side of the face. She was also found to have severe hypertension. The patient talked with a slightly slurred speech. The physician told the patient that she had suffered a mild stroke and that she should be treated in bed. Do you think the diagnosis was correct?

10. A 26-year-old man was hit on the nose when he fell from his bicycle. On examination, there was

profuse bleeding from both nostrils and extensive bruising over the bridge of the nose. There was obvious displacement of both nasal bones. A diagnosis of bilateral nasal fracture was made. The fractures were confirmed on radiographic examination. It was decided to reduce the nasal fracture in the emergency department. What is the sensory nerve supply to the skin of the nose and the mucous membrane lining the nose?

11. In a patient with multiple deep lacerations of the face, how would you test for the integrity of the different branches of the facial nerve? Should all severed branches of the facial nerve be routinely sutured irrespective of their size?

12. A 53-year-old man was seen in the emergency department complaining of a severe toothache in the right lower jaw. What is the innervation of the teeth in that situation? Can tooth pain be referred to areas away from the diseased tooth? Name other structures that if diseased may refer their pain to the teeth and confuse the physician.

13. A 13-year-old boy was struck on his right ear by another boy's fist during a fight. By the time the boy was examined, the ear was extremely swollen and bluish and very painful. Explain in anatomical terms where the blood and edema fluid collect in such a case. Can the ear be treated conservatively or should the hematoma be drained? What is a cauliflower ear?

14. What is the sensory nerve supply to the skin of the auricle of the ear?

15. How would you test the integrity of the trigeminal nerve in a patient with severe lacerations of the face? Is there any part of the face that is not supplied by the fifth cranial nerve? Do the dermatomes overlap to any extent in the face?

16. A 26-year-old woman was thrown through a car windshield in an automobile accident. Several deep lacerations of the scalp were present, and each was bleeding profusely. What is the blood supply to the scalp? How is it possible to stop the bleeding in such cases?

17. A 5-year-old child slipped as he was playing on his jungle gym. As he fell he hit his chin on a wooden bar and bit his tongue. On examination in the emergency department, the child and mother were very distressed, and the child was spitting blood-stained saliva. Explain the anatomy of the suturing of a lacerated tongue. What is the blood supply of the tongue?

18. A 38-year-old man was hit over the head with a baseball bat. On examination, a 4-in. long incised wound was found over the vertex of the skull. Can you explain how trauma of the scalp with a blunt object can produce a wound that looks like a knife cut? Name the scalp layers that require suturing in this patient.

19. A 45-year-old man was thrown from his car in a high-speed automobile accident. On admission to the emergency department he was conscious; his face was swollen, bruised, and showed multiple abrasions and small lacerations. On examination, there was bleeding into the mouth, and the upper teeth projected forward with extensive malocclusion. Careful palpation of the left orbital margin showed ridging of the lower margin near the medial corner. Pressure on the left infraorbital area revealed a slight but definite sensation of crepitus.

Following the clinical examination, the patient underwent radiography of the face. The patient was found to have a Le Fort I fracture on the right side and a modified Le Fort II fracture on the left side. In anatomical terms, explain the differences between Le Fort I, II, and III fractures. What are the common projections used when a patient suspected of having a midface fracture undergoes radiography? Can you explain why some patients with Le Fort III fractures have cerebrospinal rhinorrhea?

20. A 27-year-old woman was seen in the emergency department complaining of a severe aching pain in the upper jaw on the right side. She said it was similar to a toothache. On questioning, she said that she was still recovering from a nasty cold that she had had for 3 weeks. It also was found that her teeth, apart from a few old fillings, appeared to be normal. The pain was not accentuated by palpation or tapping of the teeth with forceps. Deep palpation of her right cheek, however, accentuated the pain. A diagnosis of right-sided maxillary sinusitis was made. What is the nerve supply of the maxillary sinus? Can you explain in anatomical terms why the patient experienced an aching pain over the right side of her face that she thought was caused by dental problems?

ANSWERS

1. (1) If the swelling is very extensive, the hematoma is situated beneath the epicranial aponeurosis and is limited only by the attachment of the aponeurosis to the skull. (2) If the swelling is large but restricted to one bone, the hematoma is situated beneath the periosteum of the skull bone and is limited by the attachment of the periosteum to the sutural ligaments. (3) If the swelling is small, superficial, and tense, the hematoma probably lies in the subcutaneous tissue and its spread is limited by the fi-

brous tissue that binds the skin to the epicranial aponeurosis.

2. The "danger area" of the face is drained by the facial vein. Interference with a boil may lead to spread of infection and thrombosis of the facial vein. Since the facial vein is devoid of valves, the pathogenic organisms may then spread via the ophthalmic veins to the cavernous sinus and cause thrombophlebitis there. Thrombosis of the cavernous sinus is a serious condition resulting in cerebral edema and possibly death. In this patient the organism was resistant to the antibiotic initially prescribed. The antibiotic was immediately changed, and the patient recovered rapidly and the proptosis and edema quickly subsided.

3. This patient had a blowout fracture of his right maxillary sinus caused by the injury from the tennis ball. The orbital contents had herniated downward into the sinus, entrapping the inferior rectus muscle and causing diplopia on upward gaze. The fracture of the floor of the orbit also damaged the infraorbital nerve, which explained the right-sided hyperesthesia of the facial skin. The patient's inability to rotate his right eyeball upward was caused by the entrapment of the inferior rectus muscle between the bony fragments in the orbital floor. The subcutaneous emphysema was caused by air percolating up from the maxillary sinus. The radiographs confirmed the fracture and showed multiple spicules of bone projecting upward into the right orbital cavity.

4. Lacerations of the face bleed profusely because of the very rich blood supply of the face. Numerous large arteries are necessary to provide nourishment to the muscles of facial expression and the many hair follicles. Lacerations of the cheek are sutured in the following three layers from deep to superficial: (1) the mucous membrane, (2) the buccinator muscle, and (3) the skin and subcutaneous tissue.

The surface marking of the parotid duct is one finger breadth below the zygomatic arch; it may be rolled on the anterior border of the masseter muscle as it turns medially to pierce the buccinator muscle. It opens into the mouth opposite the upper second molar tooth. Its entry site into the mouth is posterior to a line dropped down from the lateral canthus of the eye.

5. Surgical incisions of the face heal with less scarring if they are made along wrinkle lines. A knowledge of the position of the wrinkle lines and tension lines is also important when closing facial lacerations. The closure of lacerations that lie parallel to the tension lines can be achieved by using absorbable sutures combined with external support, since there is little tension across the wound.

6. In this patient, who was suffering from trigeminal neuralgia, the temporal region of the scalp was the trigger area, which on stimulation initiated the intense stabs of pain in the distribution of the maxillary division of the trigeminal nerve.

7. The student had dislocated his temporomandibular joints on both sides. When he yawned, his lateral pterygoid muscles reflexly contracted forcibly and pulled the head of the mandible and the articular disc forward over the summit of the articular tubercle in each joint. Reduction of the dislocation involves exerting downward pressure on the mandible, usually from behind the molar teeth, in conjunction with a forward rotation. The downward pressure overcomes the tension of the temporalis, masseter, and the medial pterygoid muscles, and the forward rotation unlocks the condyles of the mandible from the articular tubercles of the temporal bones.

The bilateral pain was caused by the spasm of the muscles of mastication; the pain was referred bilaterally along the branches of trigeminal nerve.

8. Infections of the scalp are potentially dangerous because the scalp veins communicate with intracranial veins through valveless emissary veins. The emissary veins join the diploic veins within the skull bones, and spread of infection via these veins may lead to osteomyelitis. The emissary veins also provide a channel for infected blood to pass to the intracranial venous sinuses (the cavernous, superior sagittal, and lateral venous sinuses), where venous sinus thrombophlebitis may take place.

The lymphatic drainage of the parietal area of the scalp is forward and downward into the submandibular nodes and laterally and downward into the superficial parotid nodes.

9. The physician grouped together the facial paralysis, the slurred speech, and the hypertension, and, in the absence of other findings, made the incorrect diagnosis of cerebral hemorrhage. A lesion of the corticonuclear fibers on one side of the brain will cause paralysis only of the muscles of the lower part of the opposite side of the face. This patient had complete paralysis of the entire left side of the face, which could only be caused by a lesion of the lower motor neuron. The correct diagnosis is Bell's palsy, an inflammation of the connective tissue sheath of the facial nerve.

10. Local anesthesia of the nose may be used for the reduction of nasal fractures in the adult. The fol-

lowing sensory nerves need to be blocked. (1) In the skin, the supratrochlear, the external nasal, and the infraorbital nerves are infiltrated with anesthetic (see Chapter 19, p. 805). (2) In the mucous membrane, topical anesthesia is applied to the lateral and septal walls. The lateral wall is innervated by the anterior ethmoidal nerve, branches of the olfactory nerve, lateral posterior nasal nerves, and the pharyngeal nerve (see Fig 1–3, p. 5). The septal wall is innervated by the anterior ethmoidal nerve, branches of the olfactory nerve, the medial posterior superior nasal nerves, and the nasopalatine nerves.

11. The five branches of the facial nerve may be tested for their functional integrity. The details of the facial movements are given on p. 214. Lesions of the facial nerve should be repaired to avoid distortion of the face with permanent deformity. It is unnecessary to attempt repair of the fine terminal branches of the facial nerve, since regeneration and cross innervation will take place.

12. The teeth on the right side of the lower jaw are innervated by the inferior alveolar nerve, a branch of the mandibular division of the right trigeminal nerve. The teeth in the upper and lower jaws, the temporomandibular joint, the skin of the greater part of the face and scalp, and the air sinuses are all innervated by the trigeminal nerve. Dental pain can be referred to any of these areas. Moreover, disease in the nondental sites can give referred pain to the teeth.

13. In auricular hematomas the blood tends to accumulate between the perichondrium and the underlying cartilage. The edema fluid has little room to spread since the skin is tightly bound down to the perichondrium. Failure to aspirate the hematoma may lead to necrosis of the cartilage, since it has been deprived of its blood supply due to separation of the perichondrium from the cartilage. If cartilaginous necrosis takes place, fibrous replacement occurs followed by contraction and deformity, the so-called cauliflower ear.

14. Field nerve block of the ear, which is described in Chapter 19, p. 814, involves the subcutaneous infiltration of anesthetic circumferentially around the ear. This blocks the branches of the auriculotemporal nerve, the great auricular nerve, and the lesser occipital nerve, which innervate the skin of the ear.

15. The testing of the integrity of the trigeminal nerve is fully described on p. 226. The only part of the face that is not sensory innervated by the trigeminal nerve is the area over the angle of the jaw, which is supplied by the great auricular nerve (C2 and C3).

There is very little overlap of the areas of the facial skin supplied by the ophthalmic, maxillary, and mandibular divisions of the trigeminal nerve.

16. The blood supply of the scalp is from the following arteries moving laterally from the midline anteriorly: supratrochlear, supraorbital, superficial temporal, posterior auricular, and occipital. Bleeding of scalp injuries is stopped by applying pressure over the artery that has been severed. In extensive lacerations involving different areas of the scalp, bleeding may be temporarily halted by applying a tourniquet around the circumference of the scalp.

17. The tongue is essentially a muscular organ that is covered by mucous membrane. It has a very rich blood supply. Lacerations of the tongue require a minimum of debridement because of its profuse blood supply. The gap in the muscle is closed with deep sutures followed by closure of the mucous membrane. The tongue receives its arterial supply from the lingual, facial, and ascending pharyngeal arteries.

18. Wounds of the scalp caused by blunt objects often closely resemble an incised wound. This is because the epicranial aponeurosis of the scalp is split against the unyielding skull, and the pull of the occipitofrontalis muscles causes a gaping wound. Thus, if the epicranial aponeurosis has been divided, the wound in the skin, superficial fascia, and the epicranial aponeurosis have to be sutured to close the gap satisfactorily. All of these three layers are bound together by fibrous septa in the superficial fascia.

19. The different types of midfacial fractures, as classified by Le Fort, are described on p. 227. The most useful radiographic projection used in suspected midface fractures is the upright Waters projection (see Fig 6-21), p. 233. Cerebrospinal rhinorrhea in facial injuries indicates that the cribriform plate of the ethmoid has been fractured and cerebrospinal fluid is escaping through the roof of the nose.

20. Maxillary sinusitis commonly gives rise to a deep aching pain over the infraorbital part of the cheek. However, since the sinus mucous membrane is innervated by the infraorbital nerve (a continuation of the maxillary division of the trigeminal nerve), it is not surprising to find that some patients complain of pain that is referred to the teeth of the upper jaw. For a discussion of referred pain to the teeth see p. 237.

7

The Eye and Orbit

The spectrum of emergency eye problems ranges from minor to major—conjunctivitis, corneal abrasions, corneal foreign bodies, acute iritis, and eyelid lacerations. Sudden vision loss, chemical trauma, hyphema, globe perforations, blow-out fractures, and acute glaucoma may be confronted. Moreover, preexisting eye problems and the ocular manifestations of systemic disease also have to be recognized. In isolated communities, the emergency physician often has to manage ocular problems without assistance from an ophthalmologist.

In this chapter the normal anatomy of the eye is reviewed, including the arrangement and action of both the intraocular and extraocular muscles. The anatomy of the eye as viewed with a direct ophthalmoscope is also considered together with the bony and radiological anatomy of the orbital cavity.

BASIC ANATOMY

Basic Anatomy of the Structure of the Eyelids

From superficial to deep layers, each eyelid consists (Fig 7–1) of (1) skin, (2) subcutaneous tissue, (3) striated muscle fibers of the orbicularis oculi, (4) orbital septum and tarsal plates, (5) smooth muscle, and (6) conjunctiva. The upper lid also receives the insertion of the levator palpebrae superioris muscle into the superior tarsal plate.

Skin

The skin is very thin and folds easily. For a full examination of the skin of the eyelids, the eye should be closed to erase the folds. The skin becomes continuous with the conjunctiva just in front of the pos-

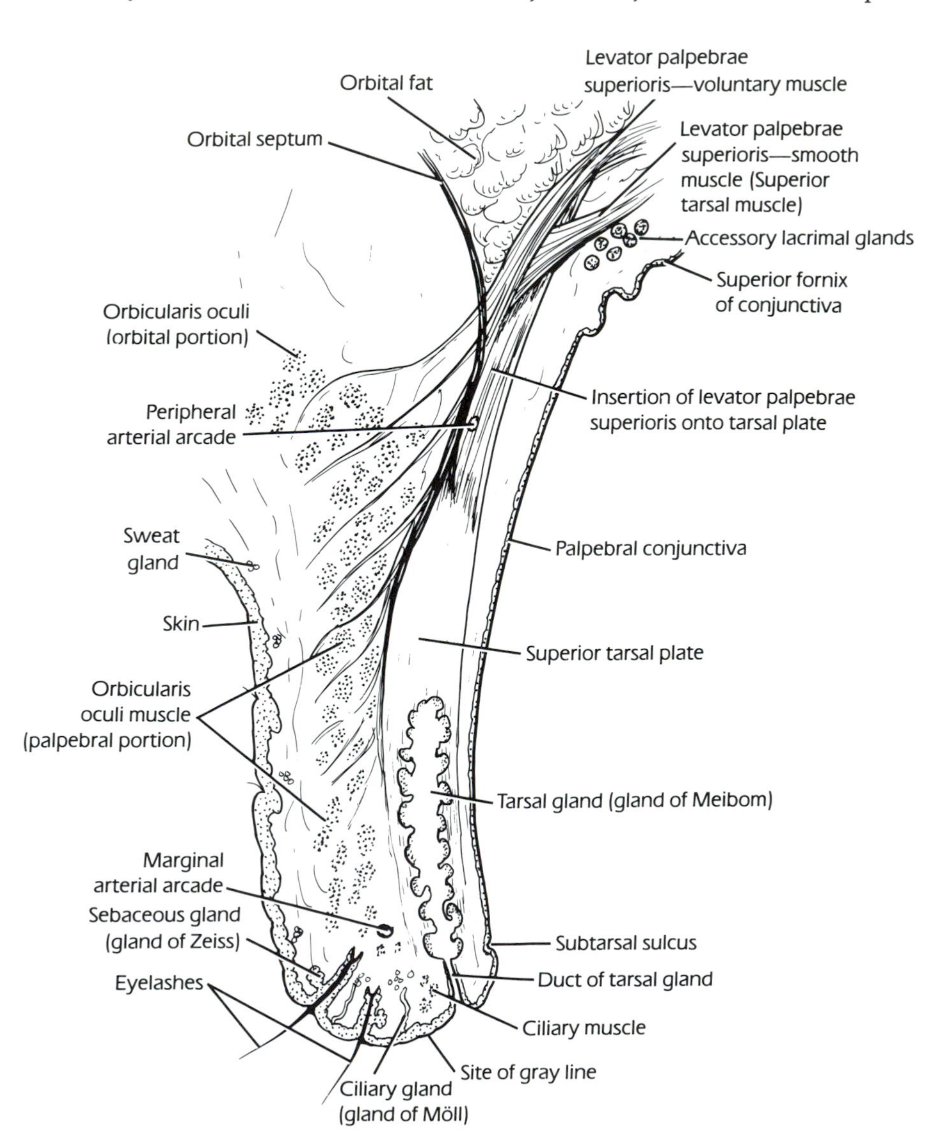

FIG 7–1.
Structure of the upper eyelid as seen in vertical section.

terior edge at the site of the orifices of the tarsal glands (see Fig 7–1).

The *eyelashes* are short, thick, curved and more numerous on the upper lid. The hair follicles are arranged in two or three rows along the anterior edge of the eyelids and do not possess erector pili muscles. The *sebaceous glands of Zeis* open into each follicle.

Behind and between the follicles, modified sweat glands, the *ciliary glands of Moll,* open into the follicles or onto the eyelid margin.

Subcutaneous Tissue

The subcutaneous tissue is very loose and rich in elastic fibers.

Clinical Note

Eyelid swelling.—The looseness of the subcutaneous tissue explains why edema fluid can rapidly accumulate, causing extensive swelling of both lids.

Orbicularis Oculi Muscle

The orbicularis oculi muscle is a flat muscle that surrounds the orbital margin (Fig 7–2). It extends onto the temporal region and cheek (orbital part); it also extends into the eyelids (palpebral part) and farther, behind the lacrimal sac (lacrimal part). It is composed of striated muscle and supplied by the temporal and zygomatic branches of the facial nerve.

Action.—The orbital portion pulls on the skin of the forehead, temple, and cheek like a purse string and draws it toward the medial angle of the orbit. The puckering of the skin over the eyelids protects the underlying eye. This part of the muscle is largely under voluntary control, although it may be made to contract reflexly.

The palpebral portion closes the eyelids. Its action is both voluntary and involuntary. The blinking reflex ensures that a film of tears wipes over the cornea. This reflex is initiated by drying of the cornea. When strongly contracted, both lids are pulled medially.

The lacrimal portion pulls on the lacrimal sac, causing it to dilate. This provides the pumping mechanism for the tears. The tone of the muscle keeps the puncta lacrimalia applied (see Fig 7–6) to the lacus lacrimalis.

The antagonist to the orbital portion of this muscle is the occipitofrontalis muscle (see p. 214); that of the palpebral portion is the levator palpebrae superioris muscle (see p. 246).

Orbital Septum and Tarsal Plates

The fibrous framework of the eyelids is formed by a membranous sheet, the orbital septum (see Fig 7–1). This is attached to the periosteum at the orbital margin. The orbital septum is thickened within the eyelids to form the *tarsal plates.* These are crescent-shaped laminae of dense fibrous tissue, and the superior plate is larger. The lateral ends of the plates

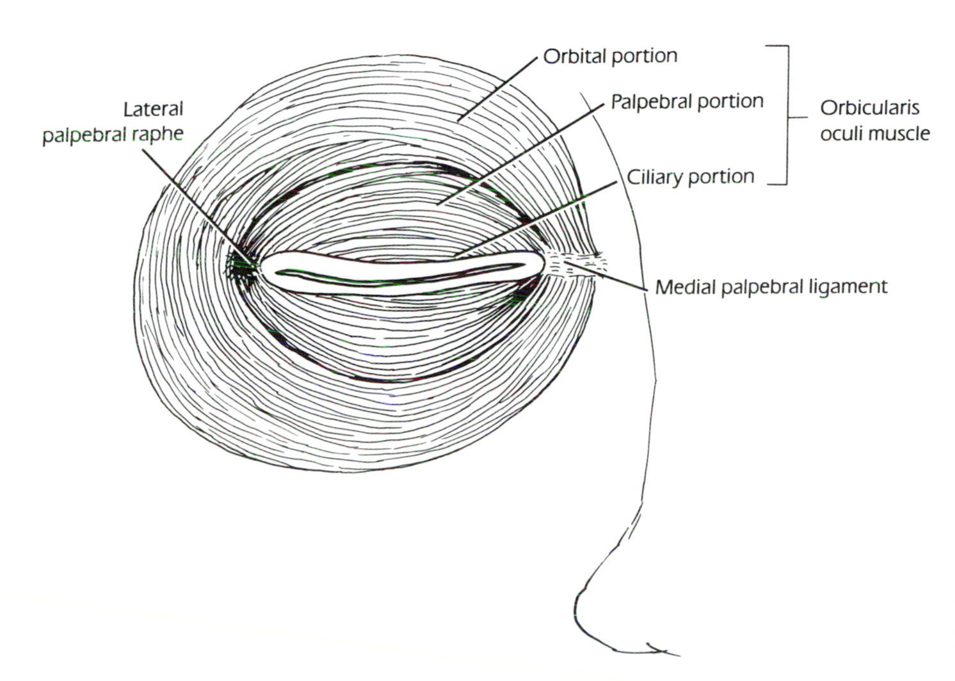

FIG 7–2.
The different parts of the orbicularis oculi muscle. Note the medial palpebral ligament and the lateral palpebral raphe.

are attached by a band, the *lateral palpebral ligament,* to a bony tubercle just within the orbital margin. The medial ends of the plates are attached by a band, the *medial palpebral ligament,* to the crest of the lacrimal bone (Fig 7–3). The medial palpebral ligament lies anterior to the lacrimal sac.

Attached to the upper edge of the superior tarsal plate are the orbital septum and the smooth-muscle fibers of the levator palpebrae superioris muscle (see Fig 7–1).

Tarsal Glands (Meibomian Glands)

The tarsal glands are embedded in the posterior surface of the tarsal plates, and their ducts discharge secretion onto the eyelid margin (see Fig 7–1). When the eyelid is everted, they can be seen as long yellow structures beneath the conjunctiva. The tarsal glands are modified sebaceous glands whose secretion is oily and prevents the overflow of tears, makes the closed eyelids airtight, and contributes to the precorneal tear film; it hinders the evaporation of tears.

Smooth Muscle

The smooth muscle forms the superior and inferior tarsal muscles. The *superior tarsal muscle* (Müller's muscle) forms part of the levator palpebrae superioris muscle; it assists the striated muscle of the levator in raising the upper lid. The *inferior tarsal muscle* is attached to the lower margin of the inferior tarsal plate, and its function is to lower the lower lid. The two tarsal muscles are innervated by sympathetic nerves from the superior cervical sympathetic ganglion.

Levator Palpebrae Superioris Muscle

The upper lid, as distinct from the lower lid, contains the insertion of the powerful striated muscle, the levator palpebrae superioris.

Origin.—This muscle originates from the undersurface of the lesser wing of the sphenoid bone at the back of the orbital cavity above and in front of the optic canal.

Insertion.—The tendon of insertion is a wide aponeurosis that is attached to the superior tarsal plate

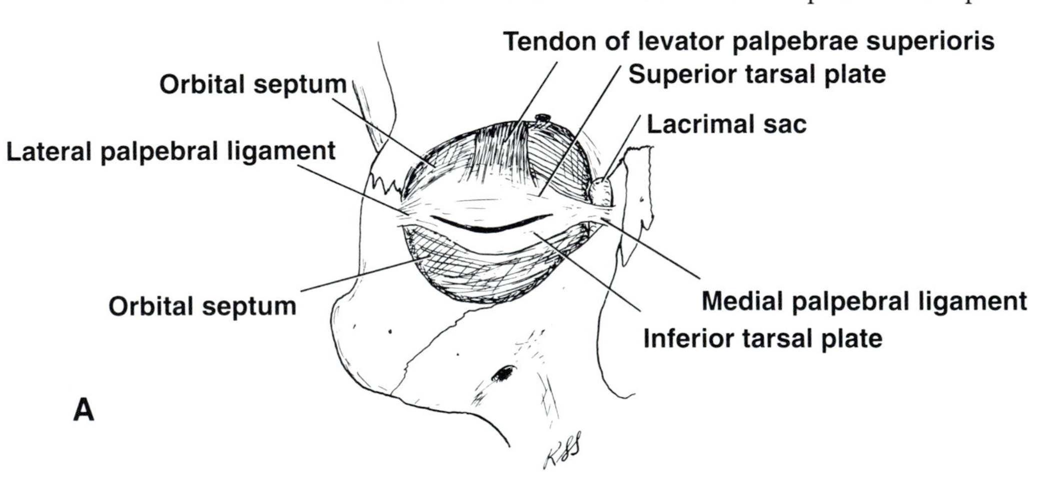

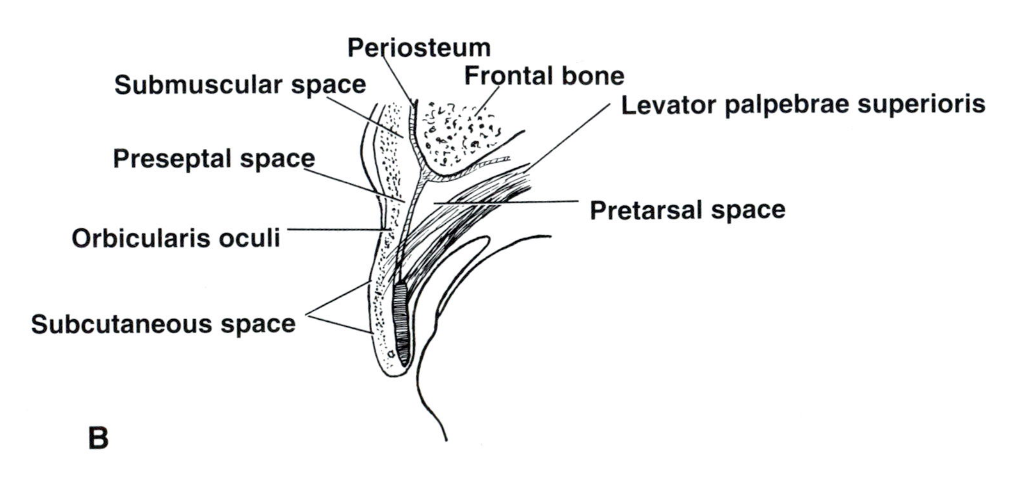

FIG 7–3.
A, the right orbital margin, with the orbital septum and the superior and inferior tarsal plates in position. Note the relationship between the medial palpebral ligament and the lacrimal sac; note also the tendinous fibers of the levator palpebrae superioris piercing the orbital septum in the upper eyelid. **B,** sagittal section of the upper eyelid showing the different fascial spaces.

(see Figs 7–3 and 7–4). The edges of the aponeurosis are attached to the lateral and medial palpebral ligaments. A thin sheet of smooth muscle arises from the inferior surface of the aponeurosis and is inserted into the upper edge of the superior tarsal plate; this muscle is the *superior tarsal muscle.*

Nerve Supply.—The main striated part is supplied by the oculomotor nerve; the smooth-muscle part (superior tarsal muscle) is supplied by sympathetic fibers from the superior cervical sympathetic ganglion.

Action.—The muscle raises the upper lid. Further elevation of the upper lid as occurs with fear or excitement is produced by the contraction of the smooth muscle.

Clinical Note

Ptosis.—In Horner's syndrome, ptosis is caused by loss of sympathetic innervation of the smooth muscle of the superior tarsal muscle (see p. 343). In third cranial nerve dysfunction, ptosis is due to paralysis of the striated muscle of the levator palpebrae superioris (see p. 279).

Conjunctiva

The conjunctiva lines the eyelids and is reflected onto the anterior surface of the eyeball (see Fig 7–4). A shallow groove on the back of the lid, the *subtarsal sulcus,* lies about 2 mm from the posterior edge of the lid margin.

Clinical Note

Subtarsal sulcus.—The sulcus tends to trap small foreign particles introduced into the conjunctival sac.

Fascial Spaces of the Eyelids

A number of potential fascial spaces exist in the eyelids. These are named according to their position—the *subcutaneous, submuscular, pretarsal,* and *preseptal spaces* (see Fig 7–3).

Arterial Supply

The arterial supply comes from the *lateral* and *medial palpebral arteries,* which are from the ophthalmic artery.

Lymphatic Drainage

The lateral two thirds of the eyelids drains into the superficial parotid preauricular lymph nodes. The medial one third drain into the submandibular nodes (Fig 7–5).

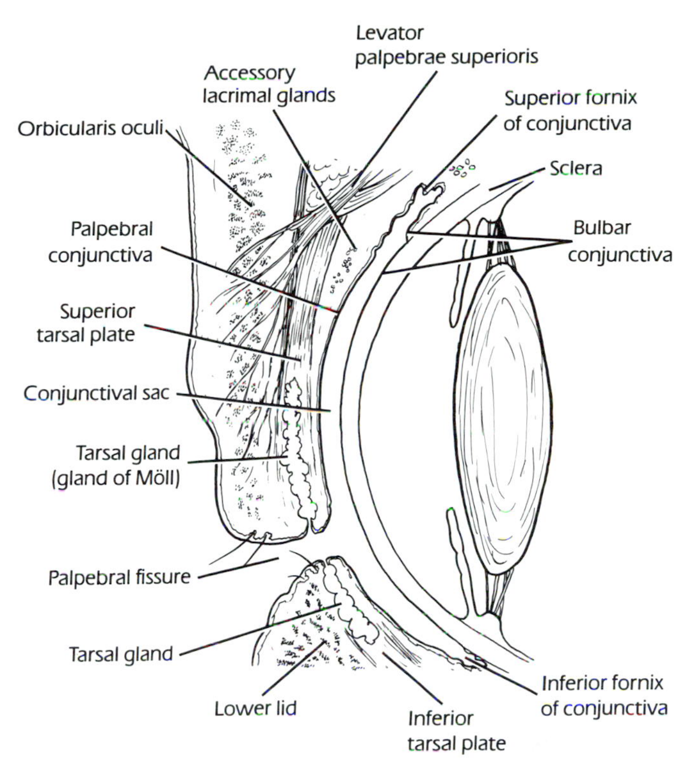

FIG 7–4.
Sagittal section of the eyelids and anterior portion of the eyeball showing the conjunctival sac and the different parts of the conjunctiva.

Sensory Nerve Supply

The upper lid is innervated by the lacrimal, supraorbital, supratrochlear, and infratrochlear nerves from the ophthalmic division of the trigeminal nerve (see Fig 7–5). The lower lid is innervated by the infraorbital nerve and the terminal portion of the maxillary division of the trigeminal nerve, and the medial end is supplied by the infratrochlear nerve.

Movements of the Eyelids

The position of the eyelids depends on the tone of the orbicularis oculi, the levator palpebrae superioris, and the superior tarsal muscles, as well as the position of the eyeball.

Looking Forward With the Eyes Open.—The upper lid covers about half the width of the superior portion of the iris, while the lower lid crosses the lower edge of the cornea.

Fear or Excitement.—The palpebral fissure is widened by the increased tone of the smooth-muscle fi-

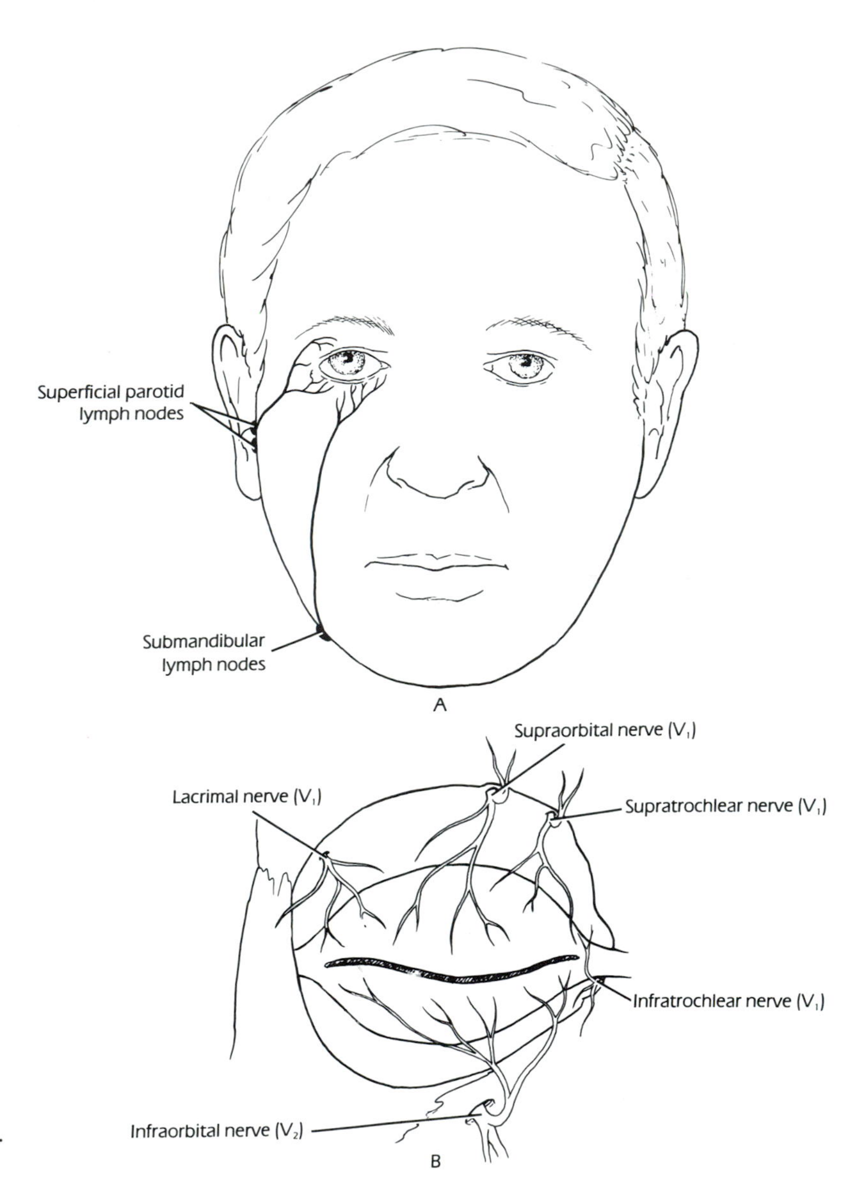

FIG 7–5.
A, the lymphatic drainage of the eyelids. **B,** the sensory nerve supply to the eyelids.

bers of the superior tarsal muscle (part of the levator palpebrae superioris) and the inferior tarsal muscle.

Closing the Eyelids.—The orbicularis oculi is contracted, and the levator palpebrae superioris muscle is relaxed.

Opening the Eye.—The levator palpebrae superioris contracts, raising the upper lid.

Looking Upward.—The levator palpebrae superioris contracts, and the upper lid moves with the eyeball. The lower lid rises slightly but lags behind the eyeball. The raising of the lower lid is believed to occur as the result of the pull of the conjunctiva, which is attached to the sclera and the lower lid.

Looking Downward.—Both lids move, the upper lid continues to cover the upper part of the cornea, and the levator palpebrae superioris relaxes. The lower lid is pulled downward slightly by the conjunctiva, which is attached to the sclera and the lower lid. It is the contraction of the inferior rectus muscle that pulls the conjunctiva downward.

Basic Anatomy of the Conjunctival Sac

The conjunctiva is a thin mucous membrane that lines the eyelids and is reflected at the *superior* and *inferior fornices* onto the anterior surface of the eyeball (see Fig 7–4). The conjunctival epithelium is continuous with the epidermis of the skin at the lid margin and with the corneal epithelium at the limbus. The conjunctiva thus forms a potential space, the *conjunctival sac,* which is open at the palpebral fissure.

The conjunctiva that covers the anterior surface of the eyeball is thin and translucent, and the underlying white sclera is clearly visible.

The specialized areas of the conjunctiva, namely, the *lacus lacrimalis,* the *lacrimal caruncle,* and the fold of conjunctiva, the *plica semilunaris,* are described on p. 268.

Basic Anatomy of the Lacrimal Apparatus

Lacrimal Gland

The lacrimal gland consists of a large *orbital part* and a small *palpebral part,* which are continuous with each other around the lateral edge of the aponeurosis of the levator palpebrae superioris. The gland is above the eyeball in the anterior and upper part of the orbit posterior to the orbital septum (Fig 7–6). About 12 ducts open from the lower surface of the gland into the lateral part of the superior fornix of the conjunctiva.

Histologically, the lacrimal gland is a tubuloacinar structure. The secretion—tears—produced by the acinar cells is poured into the conjunctival sac and contains lysozyme (antibacterial enzyme), IgA (immunoglobulin), and B lysin (bactericidal protein), which serve as a defense against microorganisms.

Nerve Supply.—The parasympathetic secretomotor supply is derived from the *lacrimal nucleus* of the facial nerve. The preganglionic fibers reach the ptery-

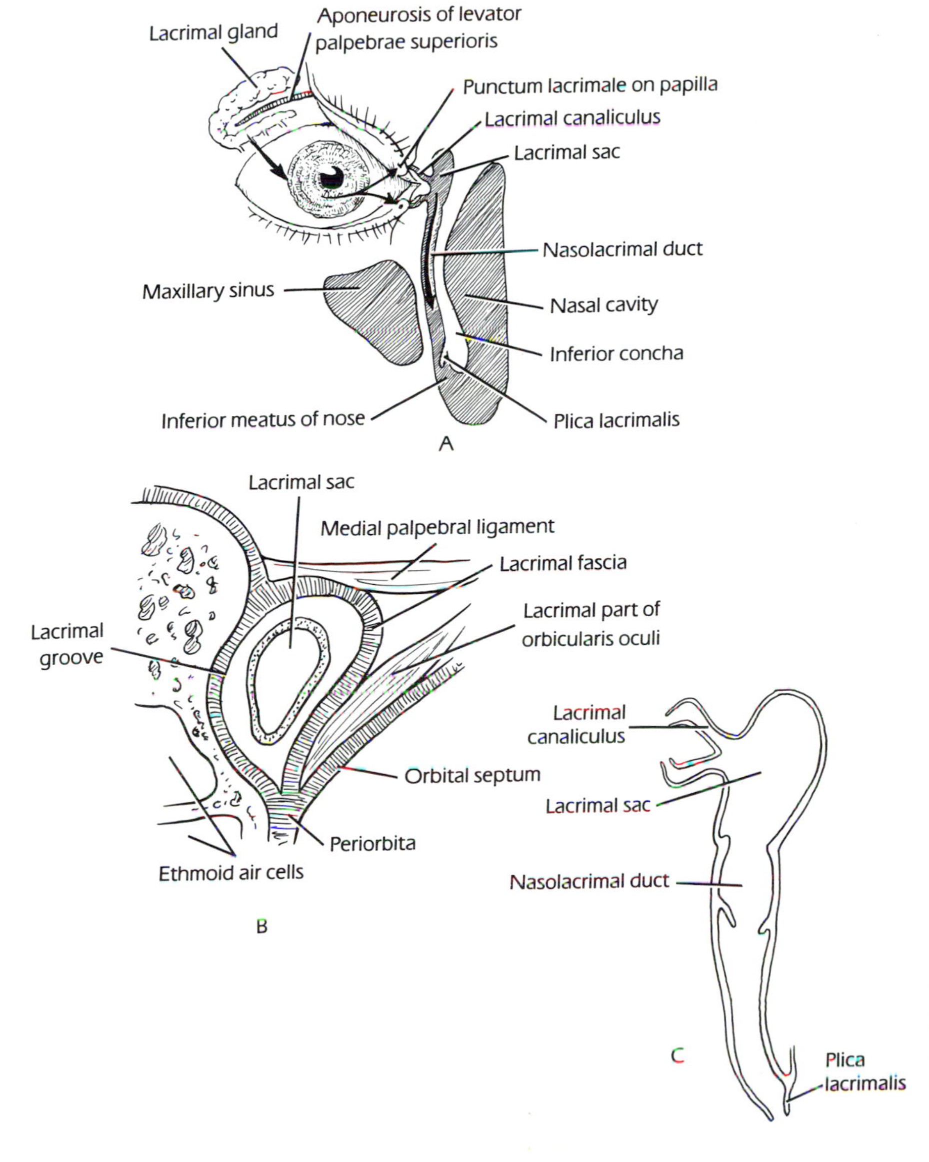

FIG 7–6.
A, the formation of tears from the lacrimal gland and their passage across the front of the eye to drain into the nose. **B,** a cross section of the lacrimal sac showing its relationship to the medial palpebral ligament, the lacrimal part of the orbicularis oculi, the lacrimal fascia, and the periorbita lining the lacrimal groove. **C,** the different drainage ducts connecting the conjunctival sac with the inferior meatus of the nose.

gopalatine ganglion (sphenopalatine ganglion) through the greater petrosal nerve and the nerve of the pterygoid canal. The postganglionic fibers leave the ganglion and pass via the maxillary nerve, the zygomatic branch, the zygomaticotemporal nerve, and the lacrimal nerve to reach the lacrimal gland.

The sympathetic postganglionic fibers arise from the superior cervical sympathetic ganglion and travel to the orbit via the plexus of nerves around the internal carotid artery. They then travel via the deep petrosal nerve, the nerve of the pterygoid canal, the maxillary nerve, the zygomatic nerve, the zygomaticotemporal nerve, and finally the lacrimal nerve. The sympathetic nerves influence the secretion of the lacrimal gland indirectly by causing vasoconstriction of the arterial supply.

Circulation of Tears and Lacrimal Drainage

The tears circulate across the cornea as the result of the physical influence of capillarity and by the blinking movements of the eyelids. The tears accumulate in the *lacus lacrimalis.* From here, the tears enter the *canaliculi lacrimales* through the *puncta lacrimalia.* The canaliculi lacrimales pass medially and open into the *lacrimal sac* (see Fig 7–6). This lies in the lacrimal groove behind the medial palpebral ligament and is the upper blind end of the nasolacrimal duct.

The *nasolacrimal duct* is about ½ in. (1.3 cm) long and emerges from the lower end of the lacrimal sac (see Fig 7–6). The duct descends downward, backward, and laterally in an osseous canal and opens into the inferior meatus of the nose. The opening is guarded by a fold of mucous membrane known as the *lacrimal fold.* This prevents air from being forced up the duct into the lacrimal sac on blowing the nose. The passage of tears down the nasolacrimal duct occurs as the result of gravity and the evaporation of the fluid at the orifice into the nose.

Basic Anatomy of the Eyeball

The eyeball is recessed into a bony socket, the *orbital cavity.* Only about one sixth of the eye is exposed; the remainder is embedded in orbital fat.

The eyeball is spherical, with the segment of a smaller sphere, the cornea, superimposed anteriorly (Fig 7–7). Because the lateral orbital margin is the least prominent, it is consequently the lateral surface of the eyeball that is most exposed.

Layers of the Eyeball

The eyeball consists of three layers (see Fig 7–7), which from without inward are (1) the fibrous layer, (2) the vascular pigmented layer, and (3) the nervous layer.

Fibrous Layer.—The fibrous layer is made up of a posterior, opaque part, the sclera, and an anterior, transparent part, the cornea (see Fig 7–7).

Basic Anatomy of the Sclera

The sclera is white and is composed of dense fibrous tissue. Posteriorly, it is pierced by the optic nerve and is fused with the dural sheath of that nerve. The *lamina cribrosa* is the area of the sclera that is pierced by the nerve fibers of the optic nerve and is relatively weak.

Clinical Note

Lamina cribrosa and cerebrospinal fluid pressure.—The weak lamina cribrosa can be made to bulge out by a rise in pressure inside the eyeball, as in glaucoma, producing a *cupped disc.* A rise of cerebrospinal fluid pressure in the tubular extension of the subarachnoid space that surrounds the optic nerve may cause the optic disc to bulge into the eyeball. See p. 227 for the description of papilledema.

The sclera is pierced by the ciliary nerves and arteries, and their associated veins. The sclera is directly continuous in front with the cornea at the *corneoscleral junction,* or *limbus.*

Function of the Sclera

The tough fibrous structure of the sclera protects the intraocular contents from trauma and mechanical displacement. The firmness and strength of the sclera, together with the intraocular pressure, preserve the shape of the eyeball and maintains the exact position of the different parts of the optic system. The sclera also provides a rigid insertion for the extraocular muscles.

Basic Anatomy of the Cornea

The transparent cornea is largely responsible for the refraction of the light entering the eye. It consists of the following layers (Fig 7–8), from front to back: (1) the corneal epithelium, which is continuous with that of the conjunctiva; (2) the Bowman's layer (membrane); (3) the substantia propria or stroma; (4) Descemet's membrane; and (5) the endothelium, which is in contact with the aqueous humor.

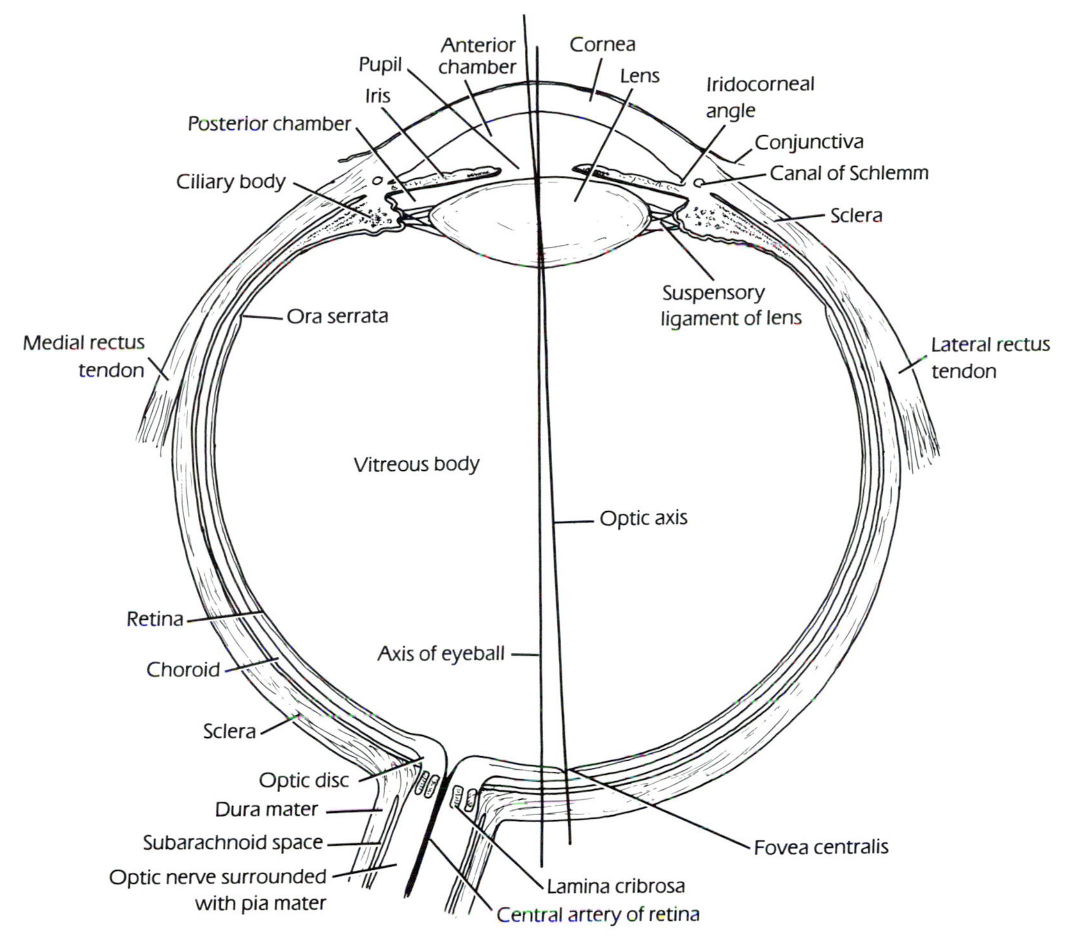

FIG 7–7.
Horizontal section through the eyeball at the level of the optic nerve. The optic axis and the axis of the eyeball are included.

Blood Supply and Lymphatic Drainage

The cornea is avascular and devoid of lymphatic drainage. The capillaries of the conjunctiva and sclera end at the circumference of the cornea. The cornea is nourished by diffusion from the aqueous humor and from the capillaries at its edge.

Nerve Supply

The nerves are branches of the long ciliary nerves from the ophthalmic division of the trigeminal nerve. There are no specialized nerve endings, and the naked axons are sensitive to pain and cold.

Function of the Cornea

The cornea is the most important refractive medium of the eye. This refractive power occurs on the anterior surface of the cornea, where the refractive index of the cornea (1.38) differs greatly from that of the air. The importance of the tear film in maintaining

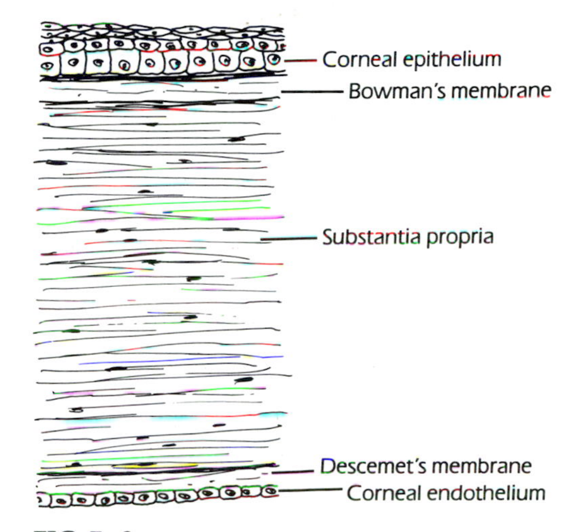

FIG 7–8.
Diagram showing the structure of the cornea.

the normal environment for the corneal epithelial cells should be stressed.

Clinical Note

Cornea and the slit lamp.—When the slit-lamp beam passes through the cornea, it is possible to recognize the anterior and posterior surfaces and the corneal stroma. The anterior surface corresponds to the corneal epithelium; the posterior surface corresponds to the endothelium. The anterior epithelium is smooth and translucent, and on it the examiner can make out mucus in the precorneal tear film. The corneal stroma reflects some light and has a milky, reticular appearance. The posterior endothelial cells are yellowish and hexagonal.

Basic Anatomy of the Vascular Pigmented Layer

The vascular pigmented coat, or uveal tract, consists, from back to front, of the choroid, the ciliary body, and the iris (see Fig 7–7).

Choroid

The choroid is composed of an outer pigmented layer and an inner, highly vascular layer.

Ciliary Body

The ciliary body lies behind the peripheral margin of the iris and is continuous posteriorly with the choroid (see Fig 7–7). It is a complete ring that runs around inside the sclera and is composed of the ciliary processes and the ciliary muscle. The *ciliary processes* are radially arranged folds, or ridges, that are connected to the suspensory ligaments of the lens.

The *ciliary muscle,* which is responsible for changing the shape of the lens, is composed of meridional and circular fibers of smooth muscle. The meridional fibers run backward from the region of the corneoscleral junction to the ciliary processes. The circular fibers run around the eyeball within the ciliary body.

Nerve Supply of the Ciliary Muscle.—Parasympathetic fibers within the oculomotor nerve synapse in the ciliary ganglion. Postganglionic fibers reach the eyeball in the short ciliary nerves (see Fig 7–23).

Action of the Ciliary Muscle.—The ciliary muscle pulls the ciliary body forward. This slackens the suspensory ligaments, and the elastic lens becomes more convex and the refractive power is increased.

Iris

The iris is the colored part of the eye and consists of a thin, contractile, pigmented sheet with a central hole, the *pupil* (Fig 7–9). The iris is suspended in the aqueous humor between the cornea and the lens. The periphery of the iris joins the wall of the eyeball at the *iridocorneal angle* and is attached to the anterior surface of the ciliary body. It divides the space between the cornea and the lens into an *anterior* and a *posterior chamber* (see Fig 7–7).

The smooth muscle of fibers of the iris consist of circular and radiating fibers. The circular fibers form the *sphincter pupillae* and are arranged around the margin of the pupil. The radial fibers form the *dilator pupillae.*

Nerve Supply of the Smooth Muscle of the Iris.—The sphincter pupillae is supplied by parasympathetic fibers from the oculomotor nerve. After synapsing in the ciliary ganglion, the postganglionic fibers reach the eyeball in the short ciliary nerves (see Fig 7–23). The dilator pupillae is supplied by sympathetic nerve fibers. This sympathetic pathway originates in cells in the T1 segment of the spinal cord and synapses in the superior cervical sympathetic ganglion. The postganglionic fibers reach the orbit along the internal carotid and ophthalmic arteries and then pass to the eyeball in the short and long ciliary nerves.

Function of the Iris.—The sphincter pupillae constricts the pupil in the presence of excessively bright light. The dilator pupillae dilates the pupil in the presence of low-intensity light or excessive sympathetic activity. During accommodation for near vision, the pupil also constricts, thus restricting the incoming light to the center part of the lens, so that spherical aberration is diminished.

Clinical Note

Flashlight examination of the pupil.—Normally, the pupils should be of equal or nearly equal diameter (within 1 to 2 mm in diameter is normal). They should be round and react to light and accommodation.

Basic Anatomy of the Nervous Layer—The Retina

The retina is the internal layer of the eyeball and consists of an anterior one fourth that is insensitive

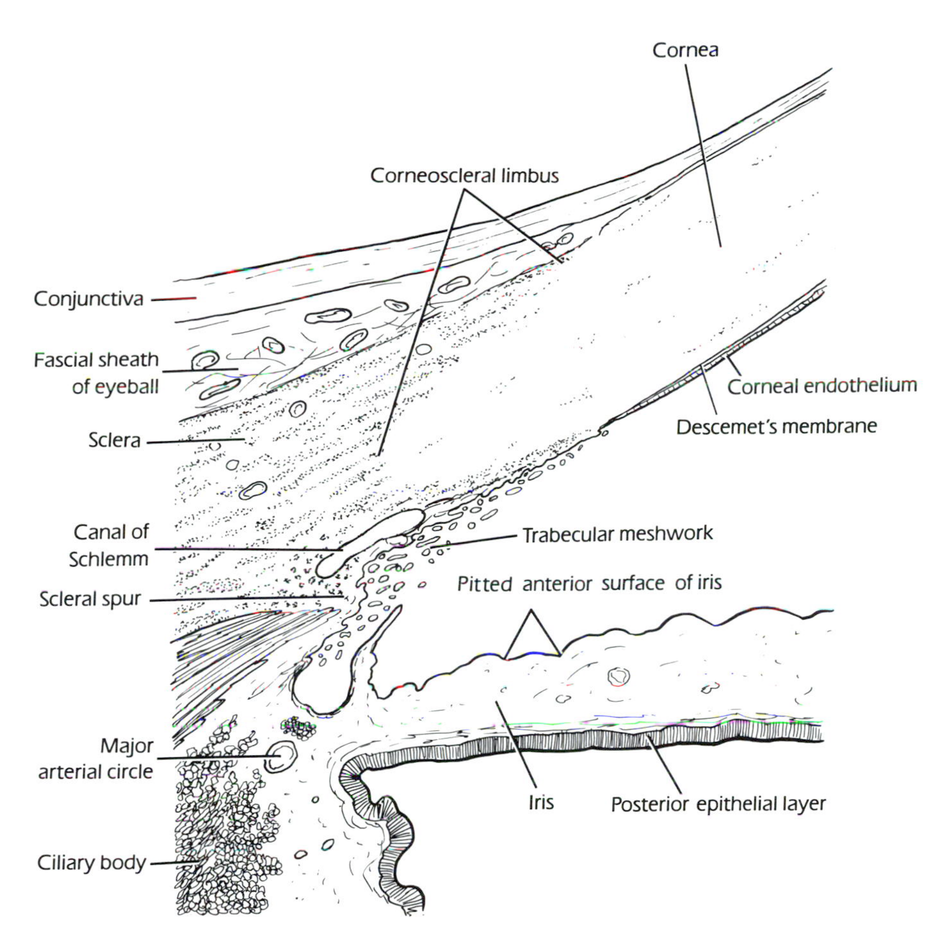

FIG 7–9.
Diagram of structures seen in the anterior portion of the eyeball in the region of the corneoscleral junction.

and a posterior three fourths that is the photoreceptor organ. The retina consists of an outer *pigmented layer* and an inner *nervous layer* (Fig 7–10). Its outer surface is in contact with the choroid, and its inner surface is in contact with the vitreous body (see Fig 7–7).

At the center of the posterior part of the retina is an oval yellowish area, the *macula lutea*. It has a central depression, the *fovea centralis* (see Fig 7–7), which is the area of the retina for the most distinct vision. Here the photoreceptors are present in greater numbers than elsewhere in the retina, and the inner layers of the retina are displaced laterally (thus the depression), allowing an unobstructed passage of the light rays to the layer of photoreceptors.

The optic nerve leaves the retina about 3 mm to the medial side of the macula lutea by the optic disc. The *optic disc* is slightly depressed at its center, where it is pierced by the *central artery of the retina* (Fig 7–11). At the optic disc the *rods* and *cones* are absent, so that the nerve is insensitive to light; this is referred to as the *blind spot*.

Basic Anatomy of the Blood Supply of the Retina

The blood supply of the retina is from two sources. The outer layers of the retina, containing the rods and cones and the outer nuclear layer, are supplied by the choroidal capillaries; the vessels do not enter the retina but tissue fluid exudes between the cells. The inner layers of the retina are supplied by the central artery of the retina. The integrity of the retina depends on both of these circulations, neither of which alone is sufficient.

Central Artery of the Retina

The central artery of the retina is a branch of the ophthalmic artery, which is the first branch of the internal carotid artery. It enters the optic nerve about 12 mm behind the eyeball (see Fig 7–11). To do so, it has to first pierce the dura and arachnoid sheaths of the optic nerve. The central artery then passes forward and pierces the lamina cribrosa to enter the eyeball. It now divides into equal superior and inferior branches (Fig 7–12). These branches then divide dichotomously into superior and inferior nasal and temporal branches. The four arteries now

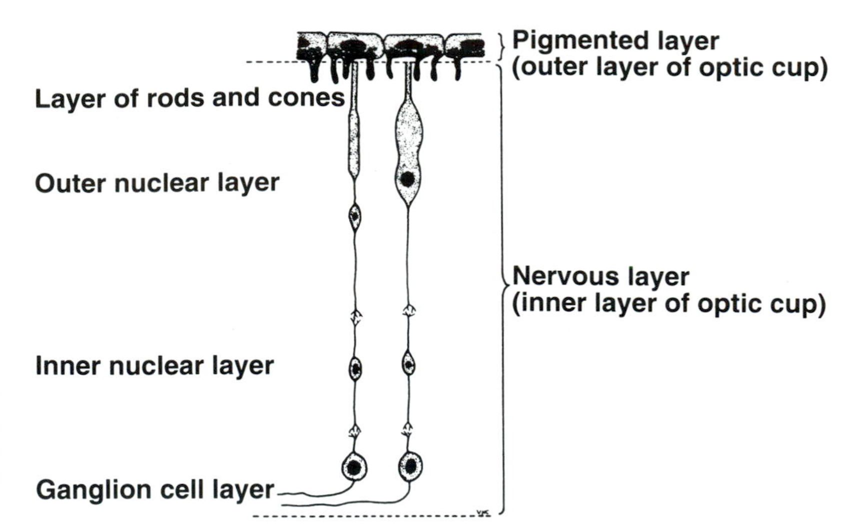

FIG 7–10.
Diagram of the retina showing the various layers. The pigmented layer is derived from the outer layer of the optic cup, and the nervous layer is derived from the inner layer of the optic cup.

supply a quadrant of the retina; there is no overlap, and no anastomosis occurs between branches within a quadrant. The arterioles are distributed throughout the different layers of the neural retina, reaching as far as the internal nuclear layer.

Central Vein of the Retina

The central vein is formed by tributaries that accompany the arteries (see Fig 7–12). The diameter of the vein is about one third larger than that of the corresponding artery. The pattern of the veins, although similar, is not identical to that of the arteries. The arteries tend to lie superficial (i.e., toward the vitreal surface) to the veins and thus cross superficial to the veins. The central vein leaves the eyeball through the lamina cribrosa accompanied by the central artery. The vein crosses the subarachnoid space (see Fig 7–11) and drains either directly into the cavernous venous sinus or into the superior ophthalmic vein.

Basic Anatomy of the Contents of the Eyeball

The contents of the eyeball consist of the refractive media—the aqueous humor, the vitreous body, and the lens.

Aqueous Humor

The aqueous humor is a clear fluid that fills the anterior and posterior chambers of the eyeball (see Fig 7–7) and nourishes the cornea and the lens. The aqueous humor is kept under constant pressure and assists in the maintenance of the normal shape of the eyeball. The aqueous humor is secreted by the epithelium covering the ciliary processes, from which it enters the posterior chamber. It then flows

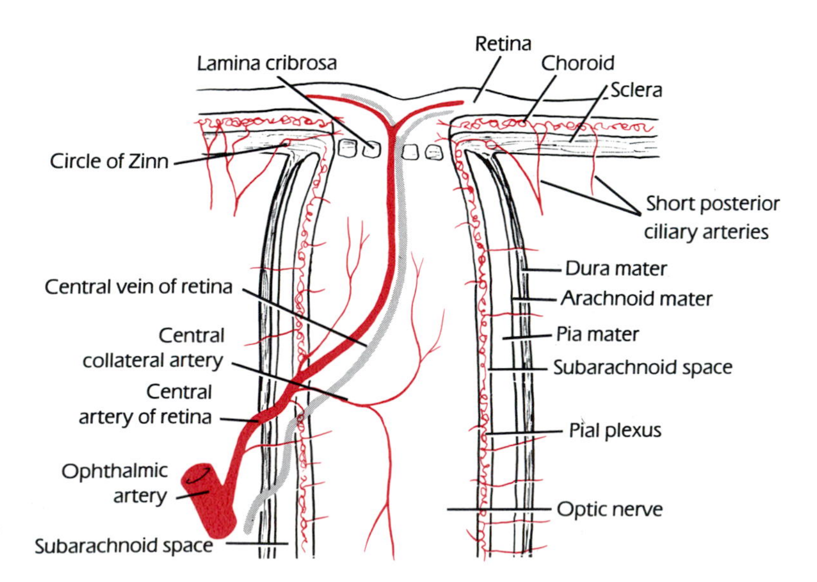

FIG 7–11.
Diagram showing the arterial supply and venous drainage of the optic disc and optic nerve.

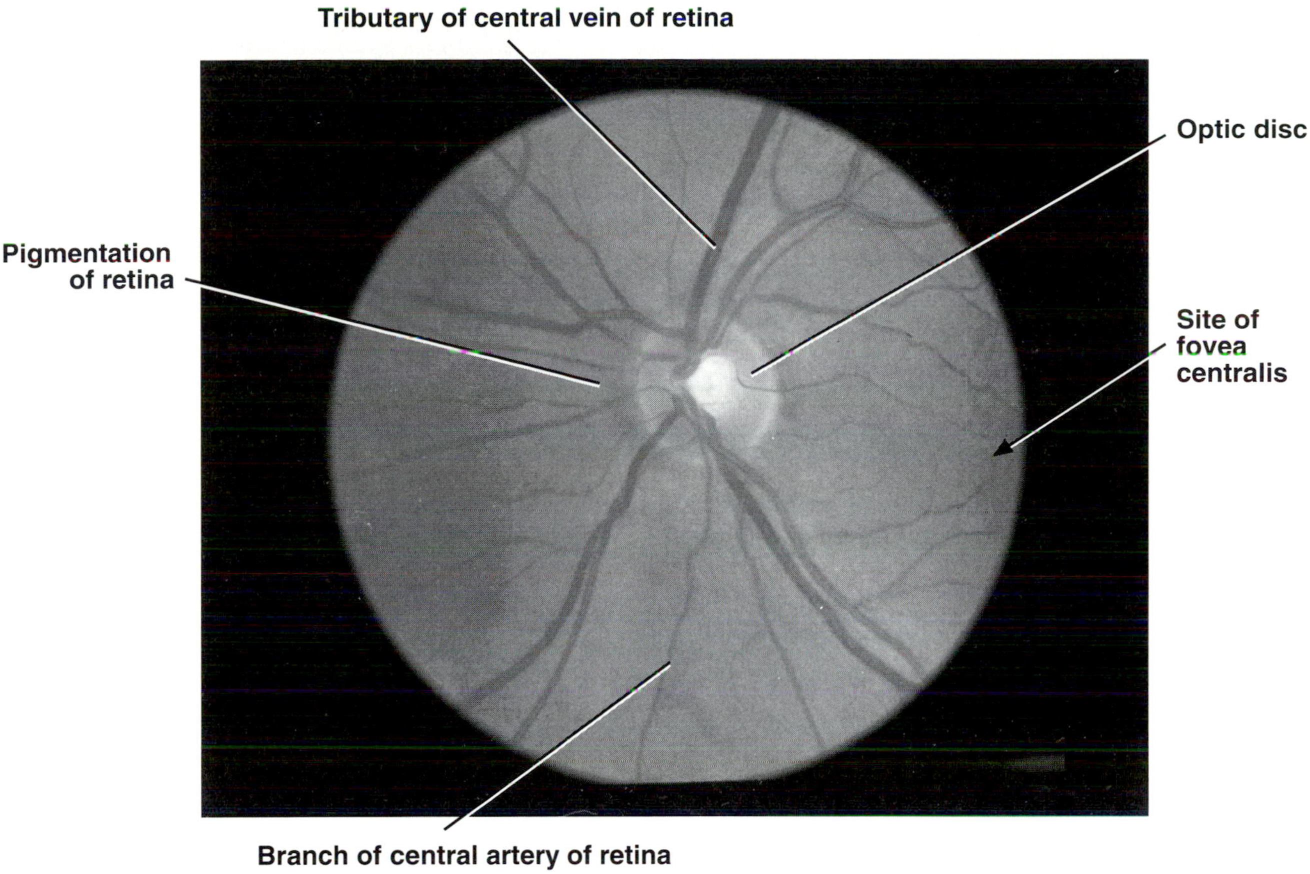

FIG 7–12.
Left ocular fundus, as seen with an ophthalmoscope.

into the anterior chamber through the pupil and is drained away through the *canal of Schlemm* at the iridocorneal angle into a venous sinus (Fig 7–13).

Vitreous Body

The vitreous body fills the eyeball behind the lens (see Fig 7–7). Its function is to maintain the shape of the eyeball and assist in the support of the lens and the retina; in this latter function it keeps the retina against the wall of the eyeball. It differs from the aqueous humor in that it cannot be replaced if lost as the result of injury because it is not continuously being formed. The vitreous body is a transparent gel enclosed by the *vitreous membrane.*

In front, in the region of the margin of the lens, the vitreous membrane is thickened and consists of two layers. The posterior layer covers the vitreous body; the anterior layer consists of a series of delicate, radially arranged fibers. Collectively, the fibers form the *suspensory ligament of the lens;* they are attached laterally to the ciliary processes and centrally to the capsule of the lens in the region of the equator (see Fig 7–13).

Basic Anatomy of the Lens

The lens (see Fig 7–7) is a transparent, biconvex structure enclosed in a transparent capsule. It is situated behind the iris and in front of the vitreous body and is encircled by the ciliary processes. The lens has considerable flexibility and consists of (1) an elastic *capsule,* (2) a *cuboidal epithelium,* which is confined to the anterior surface of the lens, and (3) *lens fibers.*

The elastic capsule envelops the entire lens and is under tension, so that the lens is constantly trying to assume a globular rather than a disc shape. The equatorial region of the lens, or circumference, is attached to the ciliary processes of the ciliary body by the suspensory ligament. The pull of the radiating fibers of the suspensory ligament tends to keep the elastic lens flattened, so that the eye may focus on distant objects.

Accommodation of the Lens

To accommodate the eye for close objects, the ciliary muscle contracts and pulls the ciliary body forward

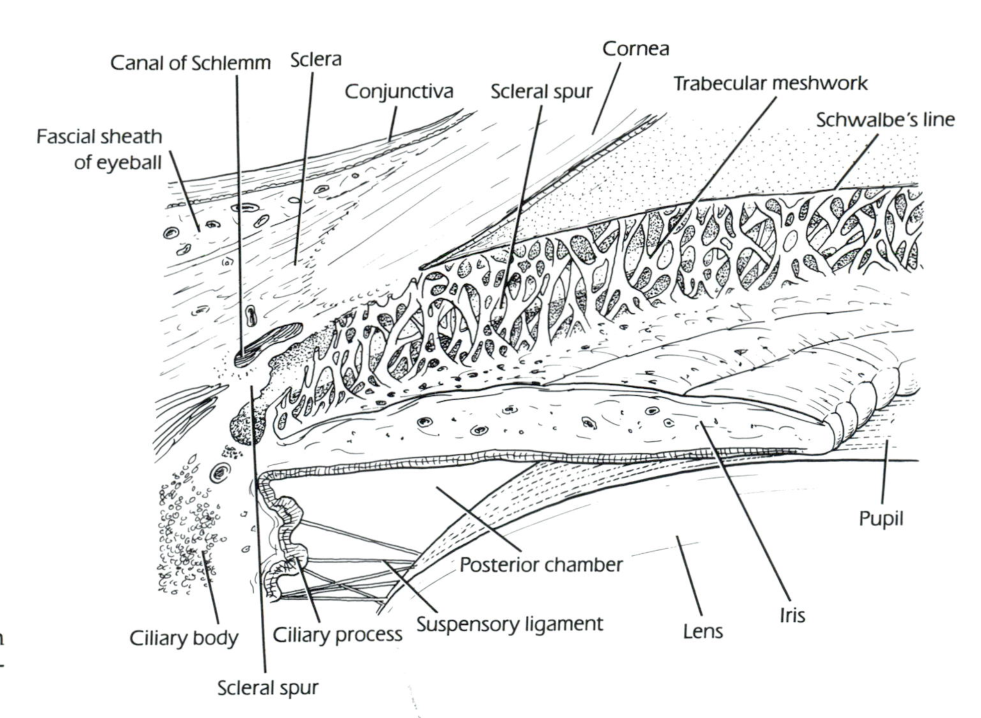

FIG 7–13.
Diagram of structures present in the corneoiridial angle of the anterior chamber.

and inward, so that the radiating fibers of the suspensory ligament are relaxed. This process allows the elastic lens to assume a nearly globular shape. With advancing age, the lens becomes denser and less elastic, and, as a result, the ability to accommodate is lessened (presbyopia).

Constriction of the Pupil During Accommodation of the Lens

To insure that the light rays pass through the center part of the lens so spherical aberration is diminished during accommodation for near objects, the sphincter pupillae muscle contracts so the pupil becomes smaller.

Convergence of the Eyes During Accommodation of the Lens

In humans, the retinae of both eyes focus on only one set of objects (single binocular vision). When an object moves from a distance toward an individual, the eyes converge so that a single object, not two, is seen. Convergence of the eyes results from the coordinated contraction of the medial rectus muscles (see below).

Basic Anatomy of the Fascial Sheath of the Eyeball

The eyeball is enveloped by a thin membrane, the *fascial sheath or Tenon's capsule,* which separates it from the orbital fat. The sheath thus forms a socket for the eyeball and permits its free movement. Anteriorly, the fascial sheath is firmly attached to the sclera just posterior to the corneoscleral junction. Posteriorly, the sheath fuses with the meninges around the optic nerve. The sheath is perforated by the tendons of the six extrinsic muscles of the eye as they pass to their insertion on the eyeball and is reflected onto each of them as a tubular sheath.

Basic Anatomy of Movements of the Eyeball

Terms Used in Describing Eye Movements

The center of the cornea or the center of the pupil is used as the anatomic "anterior pole" of the eye. All movements of the eye are then related to the direction of the movement of the anterior pole as it rotates on any one of the three axes of Fick (horizontal, vertical, and sagittal) (Fig 7–15). The terminology then becomes as follows: *elevation* is the rotation of the eye upward, *depression* is the rotation of the eye downward, *abduction* is the rotation of the eye laterally, and *adduction* is the rotation of the eye medially (Fig 7–14).

Rotatory movements of the eyeball use the upper rim of the cornea (or pupil) as the marker. This point is sometimes referred to as the 12-o'clock position. The terms *inward rotation, medial rotation,* and *intorsion* all mean the rotation of the eye medially. The term *external rotation, lateral rotation,* and *extorsion* mean rotation of the eye laterally (see Fig 7–14).

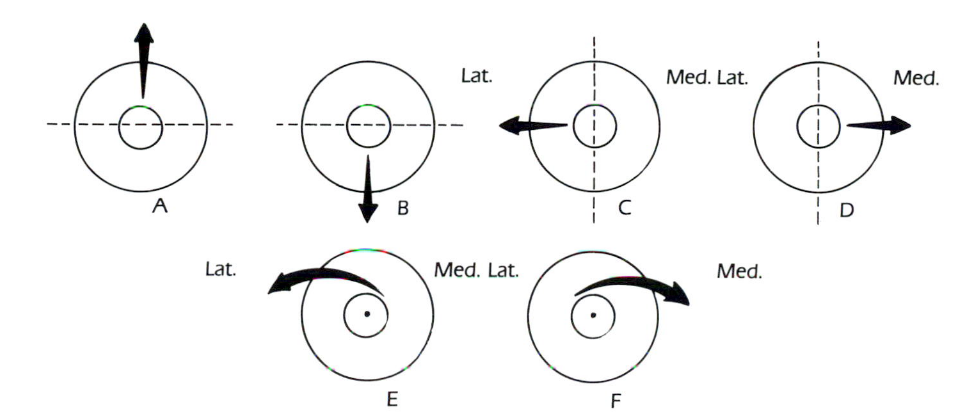

FIG 7–14.
Diagrams illustrating the different eye movements. **A,** elevation (raising the cornea). **B,** depression (lowering the cornea. **C,** abduction (lateral movement of the cornea). **D,** adduction (medial movement of the cornea. **E,** extorsion (lateral rotation or external rotation). **F,** intorsion (medial rotation or inward rotation).

Basic Anatomy of the Muscles Producing Movement of the Eyeball

The six voluntary extraocular muscles of the orbit that produce eye movements are the superior, inferior, medial, and lateral rectus muscles and the superior and inferior oblique muscles.

The Four Rectus Muscles

Origin.—The four recti muscles, (Fig 7–16) arise from a fibrous ring called the *common tendinous ring,* which surrounds the optic canal and bridges the superior orbital fissure. The superior rectus arises from the upper part of the ring, the inferior rectus from the lower part of the ring, and the medial rectus from the medial part of the ring. The lateral rectus arises by two heads from the lateral part of the ring.

Insertion.—As each rectus passes forward, it becomes wider and diverges from its neighbor. Together, they form a muscular cone that encloses the optic nerve and the posterior part of the eyeball. The tendon of each muscle pierces the fascial sheath of the eyeball and is inserted into the sclera about 6 mm behind the margin of the cornea.

Nerve Supply.—The superior, inferior, and medial recti are supplied by the oculomotor nerve; the lateral rectus is supplied by the abducent nerve.

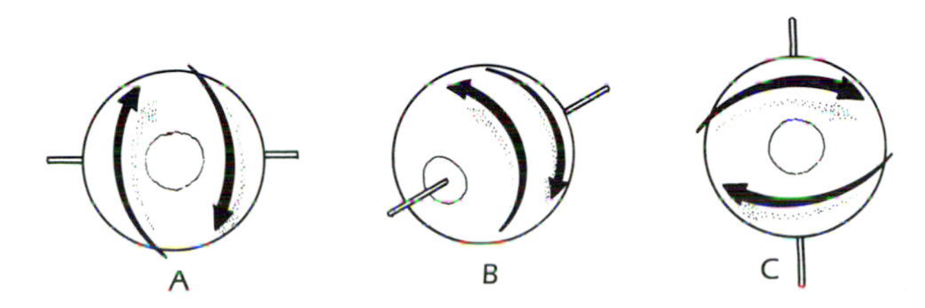

FIG 7–15.
Diagram showing the axes of rotation. **A,** horizontal or transverse axis. **B,** sagittal axis. **C,** vertical axis.

Superior Oblique Muscle

Origin.—From the body of the sphenoid, above and medial to the optic canal (see Figs 7–17 and 7–25).

Insertion.—A slender tendon passes forward and enters a fibrocartilaginous pulley attached to the frontal bone (see Fig 7–17). The tendon then turns backward and laterally and is inserted into the sclera beneath the superior rectus. It is attached to the sclera behind the coronal equator of the eyeball, and the line of pull of the tendon passes medial to the vertical axis.

Nerve Supply.—Trochlear nerve.

The superior oblique muscle is thus attached via a pulley to the eyeball, so despite an origin posterior to the coronal axis of the eyeball, the main direction of motion it causes is depression.

Inferior Oblique Muscle

Origin.—From the anterior part of the floor of the orbit (see Fig 7–17).

Insertion.—The muscle passes backward and laterally below the inferior rectus. It is inserted into the sclera behind the coronal equator, and the line of pull of the tendon passes medial to the vertical axis (see Fig 7–17).

Clinical Note

Cardinal movements of oblique muscles.—"Down and in" movement is the "cardinal movement" caused by the superior oblique muscle. This can be explained by the fact that its insertion is posterior and medial to the vertical axis of the eyeball, when the eye is rotated medially, the muscle overlies the vertical axis and can exert its maximal effect (see above).

"Up and in" movement is the "cardinal movement" caused by the inferior oblique muscle. This

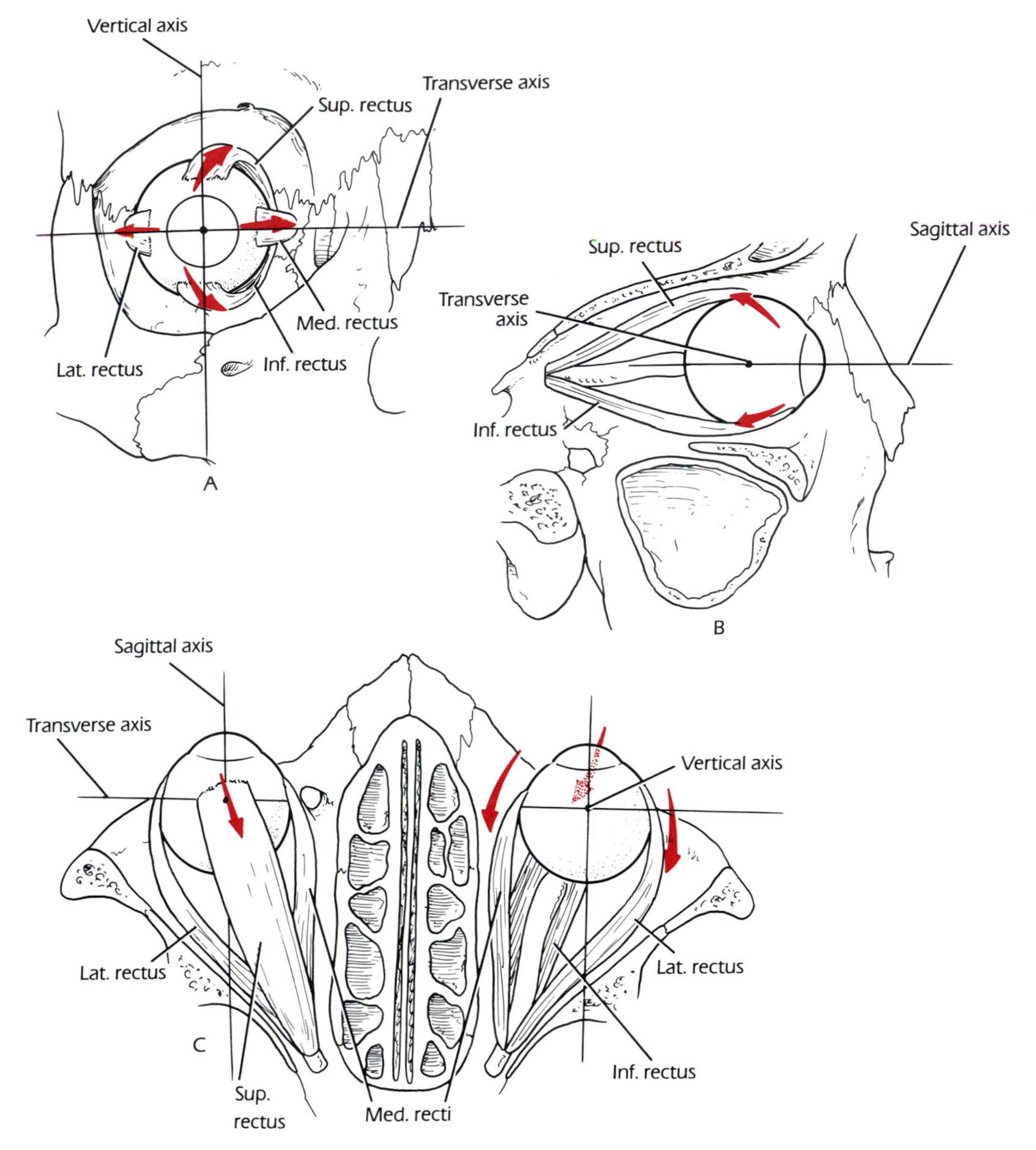

FIG 7–16.
A–C, diagrams showing the actions of the four recti muscles producing movements of the eyeball.

can be explained by the fact that its insertion is posterior and medial to the vertical axis of the eyeball; when the eye is rotated medially, the muscle overlies the vertical axis and can exert its maximal effect (see above).

Nerve Supply.—Oculomotor nerve.

The inferior oblique muscle has no pulley. However, its origin is anterior to the axis of the eyeball so none is needed for it to carry out its function of assisting in elevating the eye.

Basic Anatomy of Eye Movements and the Muscles Used

Abduction

This lateral rotation of the eyeball around an imaginary vertical axis is caused by the contraction of the lateral rectus muscle and the relaxation of the medial rectus muscle (Fig 7–18).

Adduction

This medial rotation of the eyeball around an imaginary vertical axis is caused by the contraction of the medial rectus muscle and the relaxation of the lateral rectus muscle.

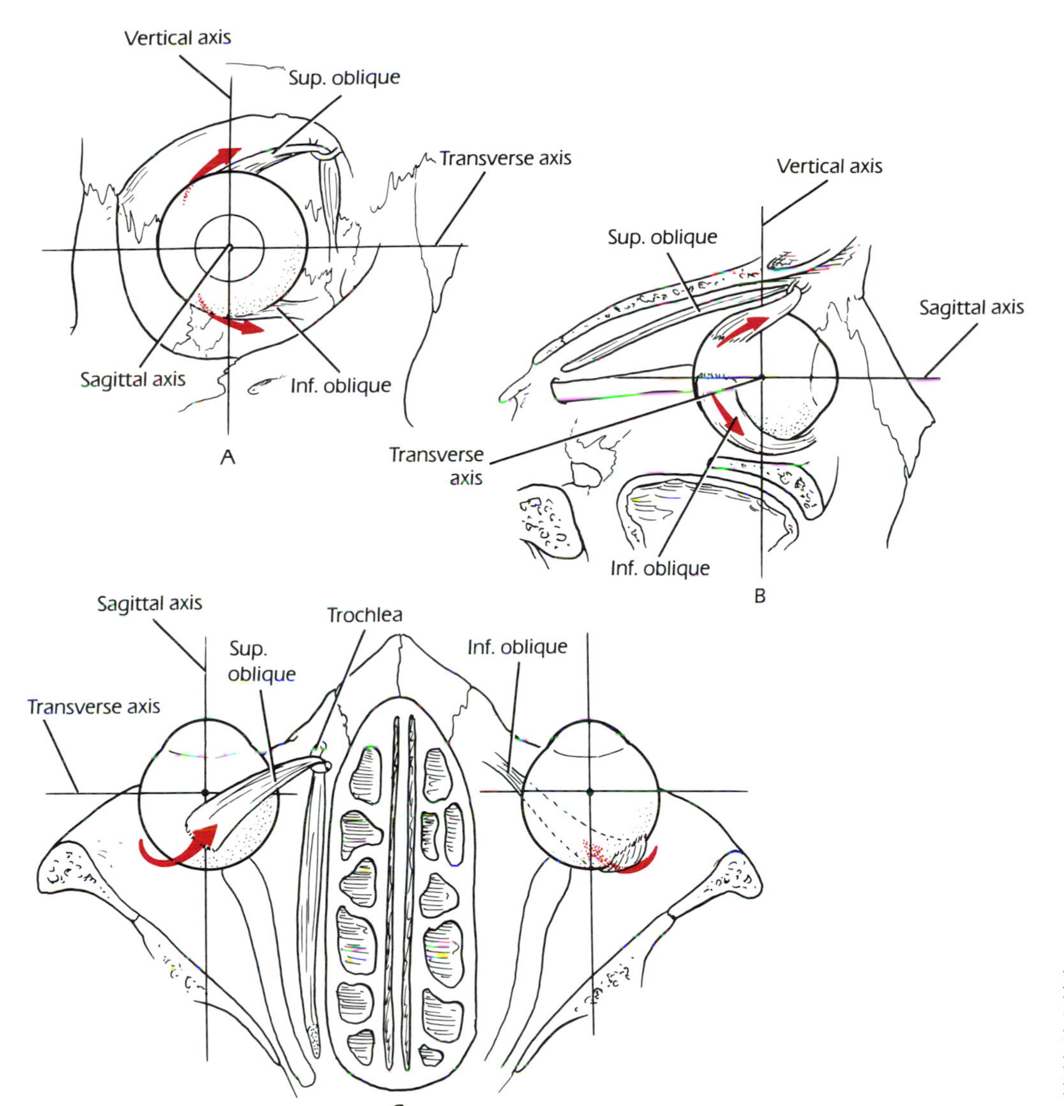

FIG 7–17.
Diagrams showing the actions of the superior and inferior oblique muscles in producing movements of the eyeball.

Elevation

This is the direct vertical movement upward of the eyeball. It is accomplished by the contraction of the superior rectus muscle assisted by the inferior oblique muscle. Because the superior rectus is obliquely placed in the orbit, the direction of pull is not only upward but posteromedial. Thus, the primary action is elevation on the transverse axis, slight medial rotation on the vertical axis, and very slight intorsion on the anteroposterior axis. The tendency to cause medial rotation and intorsion is countered by the simultaneous contraction of the inferior oblique muscle, which also elevates but tends to rotate laterally and cause extorsion. The eye rotates vertically upward from the sum of all these actions (see Fig 7–18).

Depression

This is the direct downward rotation of the eyeball. It is brought about by the contraction of the inferior rectus muscle assisted by the superior oblique muscle. Because the inferior rectus is obliquely placed in the orbit, the direction of pull is not only downward but posteromedial. Thus, the primary action is depression on the transverse axis, slight medial rotation on the vertical axis, and very slight extorsion on the anteroposterior axis. The tendency to cause medial rotation and extorsion is countered by the simultaneous contraction of the superior oblique muscle. The direction of the tendon from the trochlea to its global insertion results in depression of the cornea and also lateral rotation and intorsion. The sum of all these actions produces rotation of the eye vertically downward (see Fig 7–18).

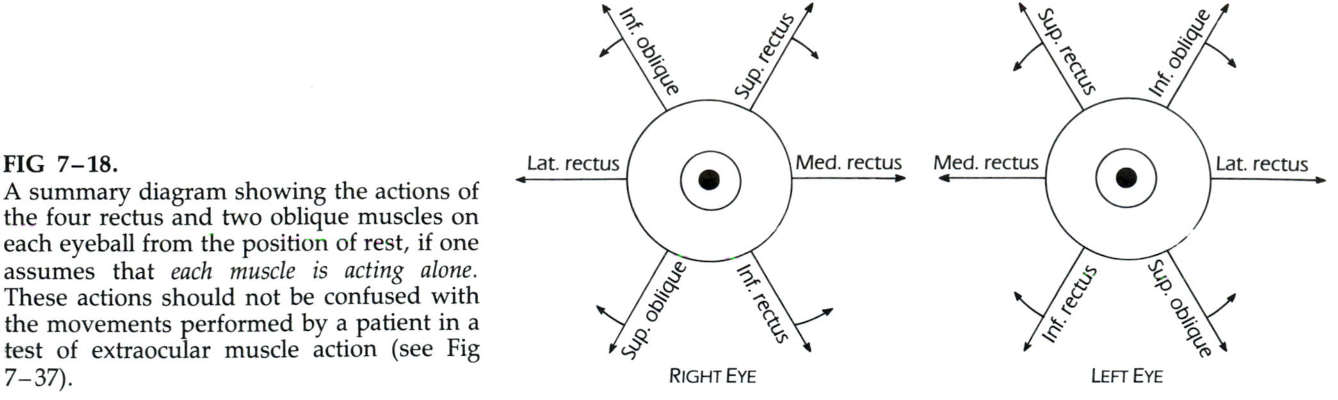

FIG 7–18.
A summary diagram showing the actions of the four rectus and two oblique muscles on each eyeball from the position of rest, if one assumes that *each muscle is acting alone.* These actions should not be confused with the movements performed by a patient in a test of extraocular muscle action (see Fig 7–37).

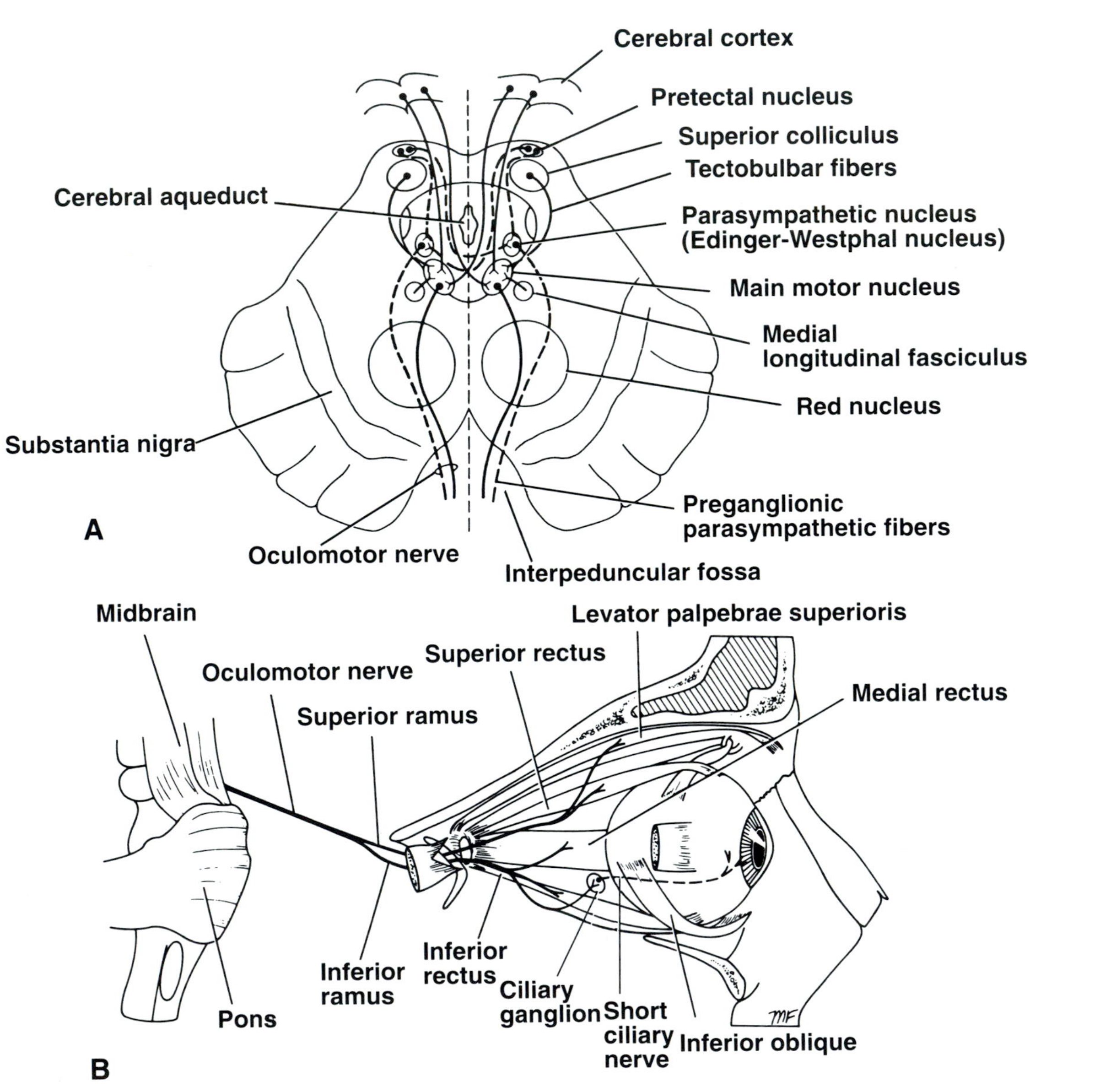

FIG 7–19.
A, oculomotor (III) nerve nuclei and their central connections. **B,** the distribution of the oculomotor nerve.

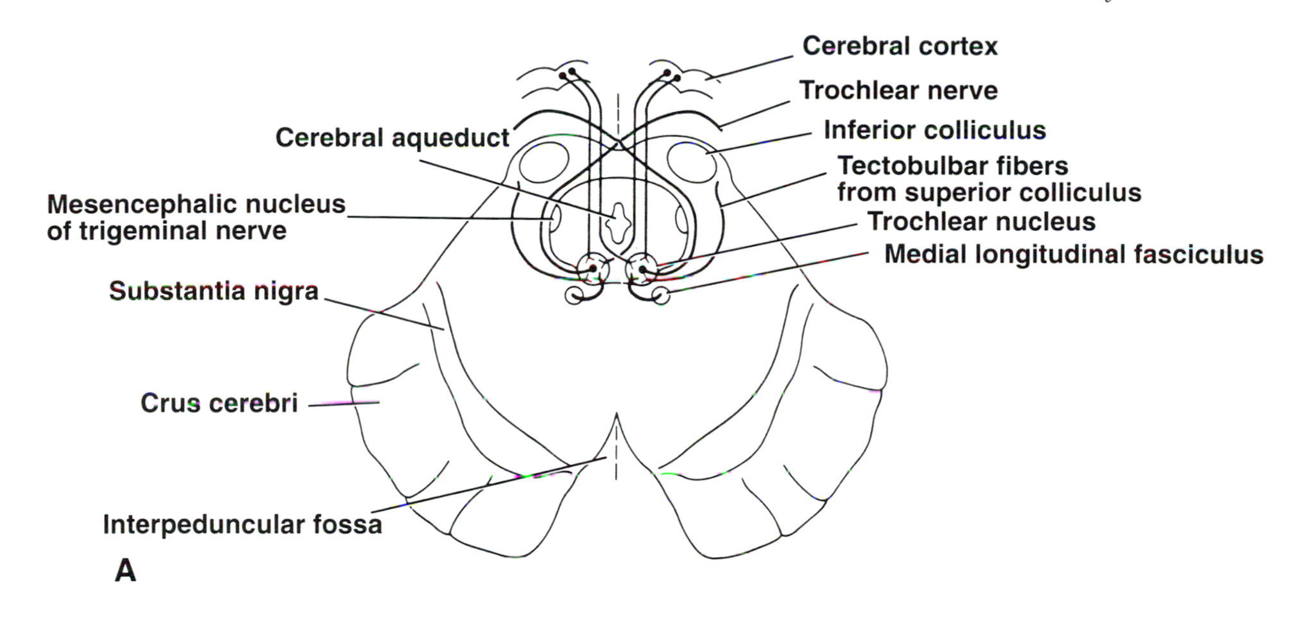

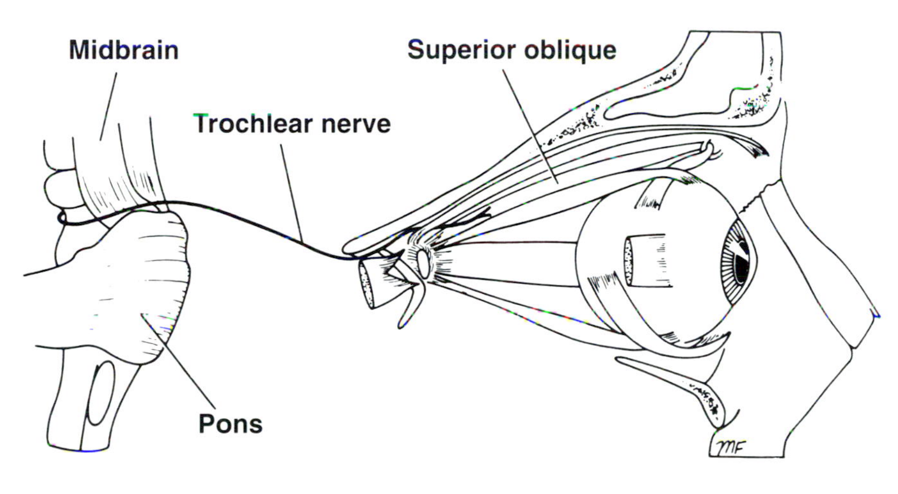

FIG 7–20.
A, trochlear (IV) nerve nucleus and its central connections. **B,** the distribution of the trochlear nerve.

Basic Anatomy of the Nervous Pathways and Cranial Nerve Nuclei Controlling Eye Movements

Figures 7–19 through 7–21 show the nuclei of the oculomotor, trochlear, and abducent nerves and their central connections. The nuclei are connected by the medial longitudinal fasciculus, which ensures yoked eye movements, e.g., when the right medial rectus contracts the left lateral rectus also contracts, and the eyes move to the left together.

The *voluntary fixation movements* are controlled by nervous impulses that originate in the cortical eye field situated in the frontal lobes. The nerve pathways descend to the oculomotor, trochlear, and abducent nuclei in the brain stem.

The *involuntary fixation movements* are controlled by the secondary visual area of the occipital cortex. Here again the pathways descend to the cranial nerve nuclei controlling the eye muscles. The occipital eye field locks the eyes automatically on a given spot and thus prevents the movement of the image across the retina. It is only when the voluntary frontal eye field comes into action that the automatic mechanism is unlocked.

Basic Anatomy of the Visual Pathway

The visual pathway is illustrated in Figure 7–22. Note the following important anatomical points.

1. The fibers of the optic nerve originate from the neurons forming the ganglionic layer of the ret-

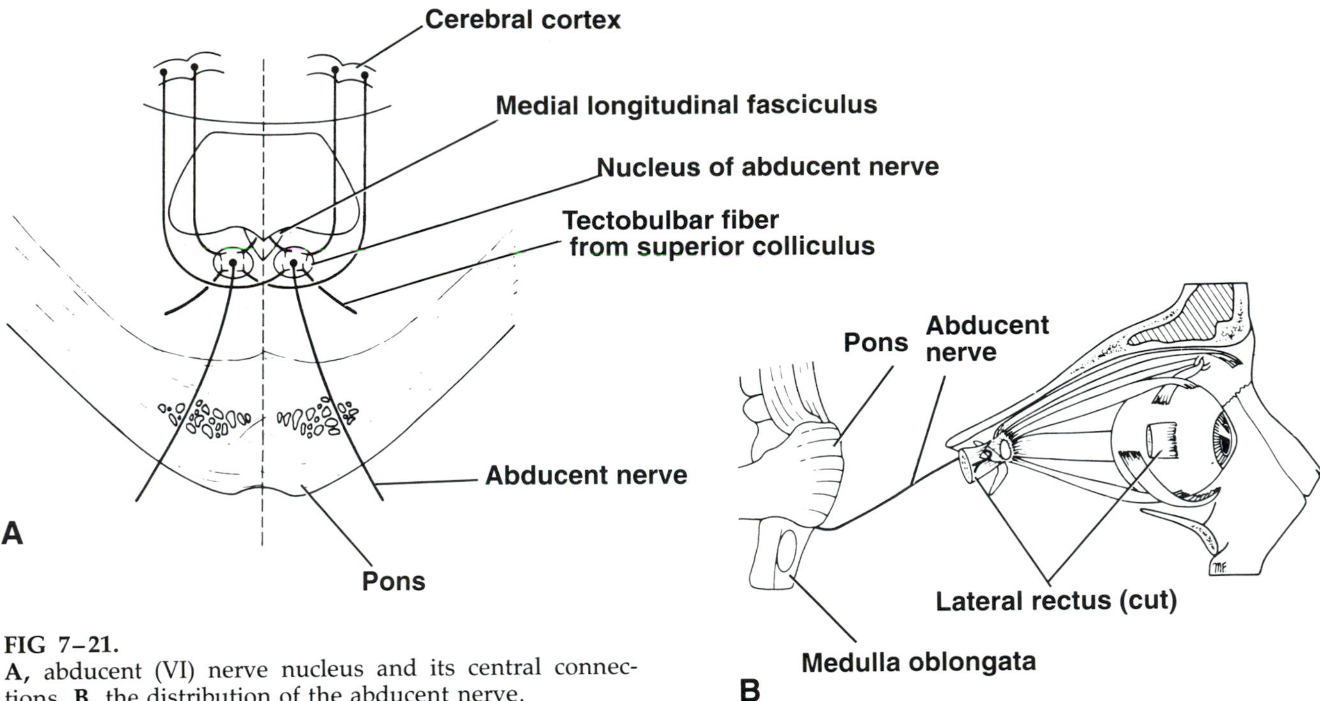

FIG 7–21.
A, abducent (VI) nerve nucleus and its central connections. **B,** the distribution of the abducent nerve.

ina, and they exit from the eyeball at the optic disc. The nerve fibers are myelinated but cannot regenerate if damaged, since the optic nerve is an outgrowth of the central nervous system.

2. In the optic chiasma, the fibers from the nasal half of each retina, including the nasal half of the macula, cross the midline and enter the optic tract of the opposite side, while fibers from the temporal half of each retina, including the temporal half of each macula, pass posteriorly in the optic tract of the same side.

3. Most of the fibers of the optic tract terminate by synapsing with nerve cells in the lateral geniculate body, a projection from the thalamus. A few of the fibers, however, pass to the pretectal nucleus and the superior colliculus in the midbrain and are concerned with mediating the pathway for the pupillary light reflexes (Fig 7–23).

4. The fibers emerge from the lateral geniculate body and form the optic radiation. They pass posteriorly through the internal capsule and terminate in the visual cortex, which occupies the upper and lower lips of the calcarine sulcus on the medial surface of the cerebral hemisphere (see Fig 7–22).

5. In binocular vision, the right and left fields of vision are projected on portions of both retinae. The lower retinal quadrants (upper field of vision) project on the lower wall of the calcarine sulcus, while the upper retinal quadrants (lower field of vision) project on the upper wall of the sulcus.

Clinical Note

Visual Field Defects.—Common visual field defects associated with lesions of the visual pathways are shown in Figure 7–24. Depending on the location of the lesion, the visual field defect may be unilateral, bilateral, homonymous, or bilateral heteronymous.

Basic Anatomy of Visual Reflexes

The various nervous pathways involved in the *direct* and *consensual light reflexes* and the *accomodation reflex* are summarized in Figure 7–23. If a light is shone into one eye, the pupils of both eyes normally constrict. The constriction of the pupil upon which the light is shone is called the *direct light reflex;* the constriction of the opposite pupil, even though no light fell upon that eye, is called the *consensual light reflex* (see Fig 7–23).

The afferent impulses travel through the optic nerve, optic chiasma, and optic tract. Here a small number of fibers leave the optic tract and synapse on nerve cells in the pretectal nucleus, which lies close to the superior colliculus. The impulses are passed by axons of the pretectal nerve cells to the

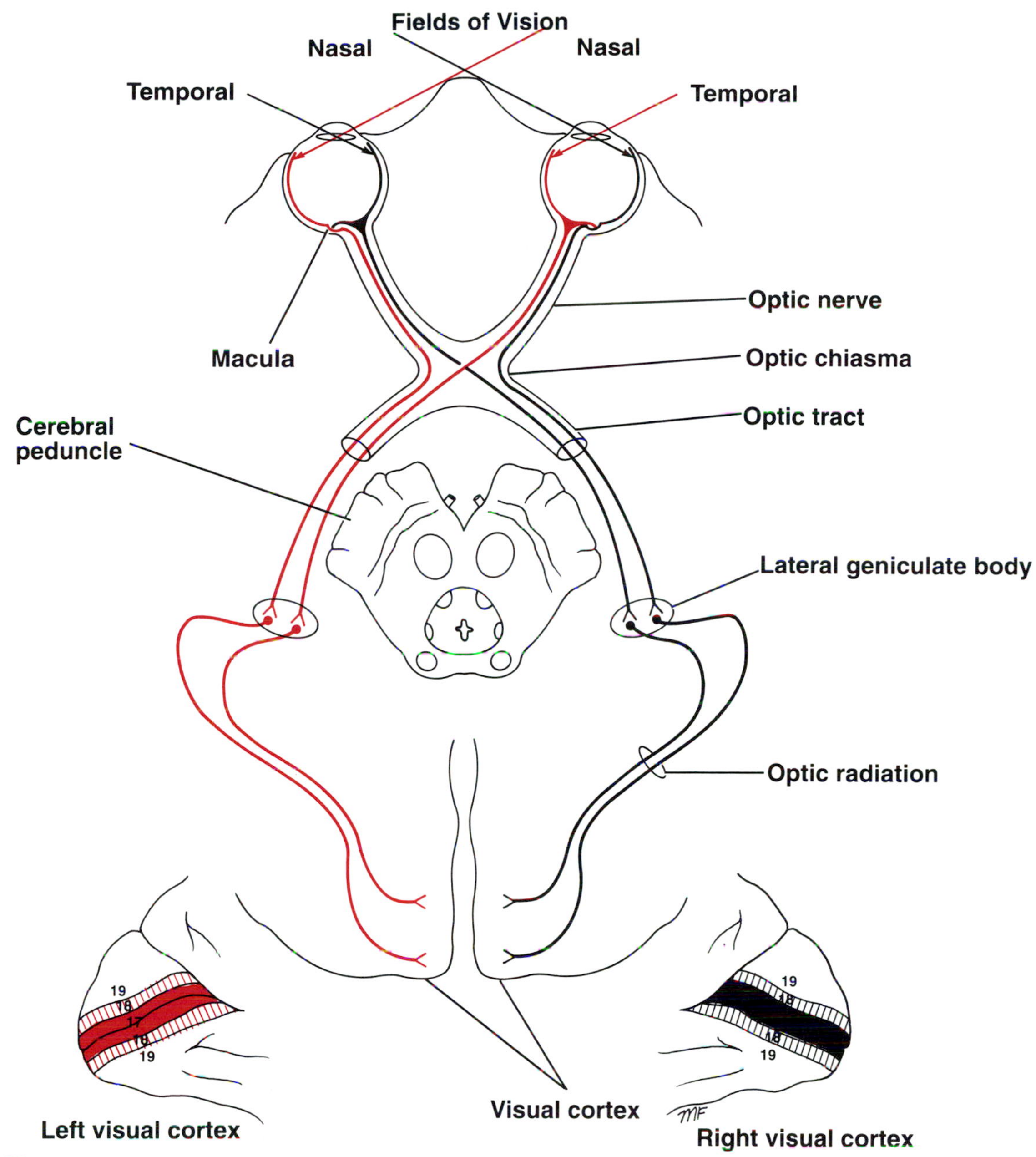

FIG 7–22.
The optic pathway. The inserts show the left and right primary cortices on the medial surface of the cerebral hemispheres.

parasympathetic nuclei (Edinger-Westphal nuclei) of the third cranial nerve on both sides. Here the fibers synapse and the parasympathetic nerves travel through the third cranial nerve to the ciliary ganglion in the orbit (see Fig 7–23). Finally, postganglionic parasympathetic fibers pass through the short ciliary nerves to the eyeball and the constrictor pupillae muscle of the iris. Both pupils constrict in the consensual light reflex because the pretectal nucleus sends fibers to the parasympathetic nuclei on both sides of the midbrain (see Fig 7–23). The fibers that cross the median plane do so close to the cerebral aqueduct of the midbrain.

Corneal Reflex

Afferent impulses from the cornea or conjunctiva travel through the ophthalmic division of the trigeminal nerve to the sensory nucleus of the trigeminal nerve. Connector neurons link this nucleus to the motor nucleus of the facial nerve on both sides. The facial nerve supplies the orbicularis oculi muscle, which causes closure of the eyelids.

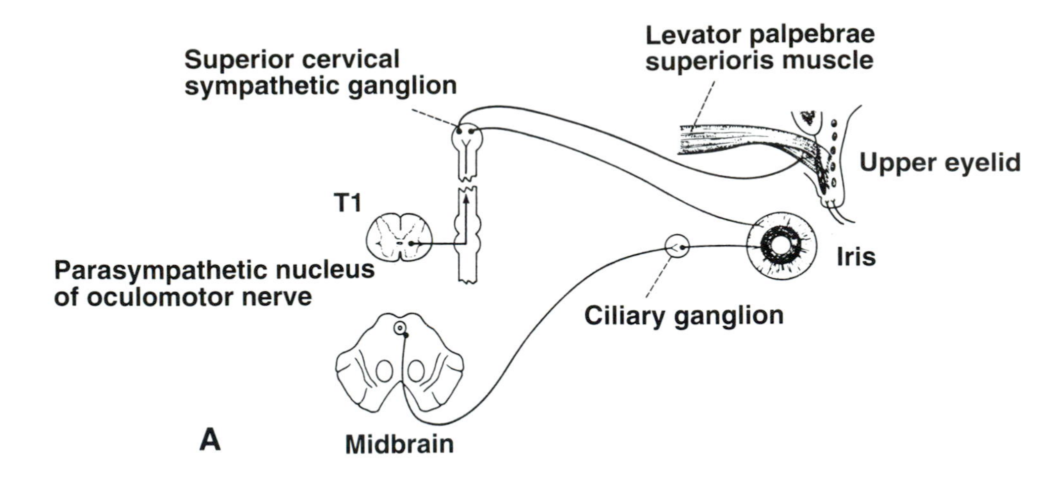

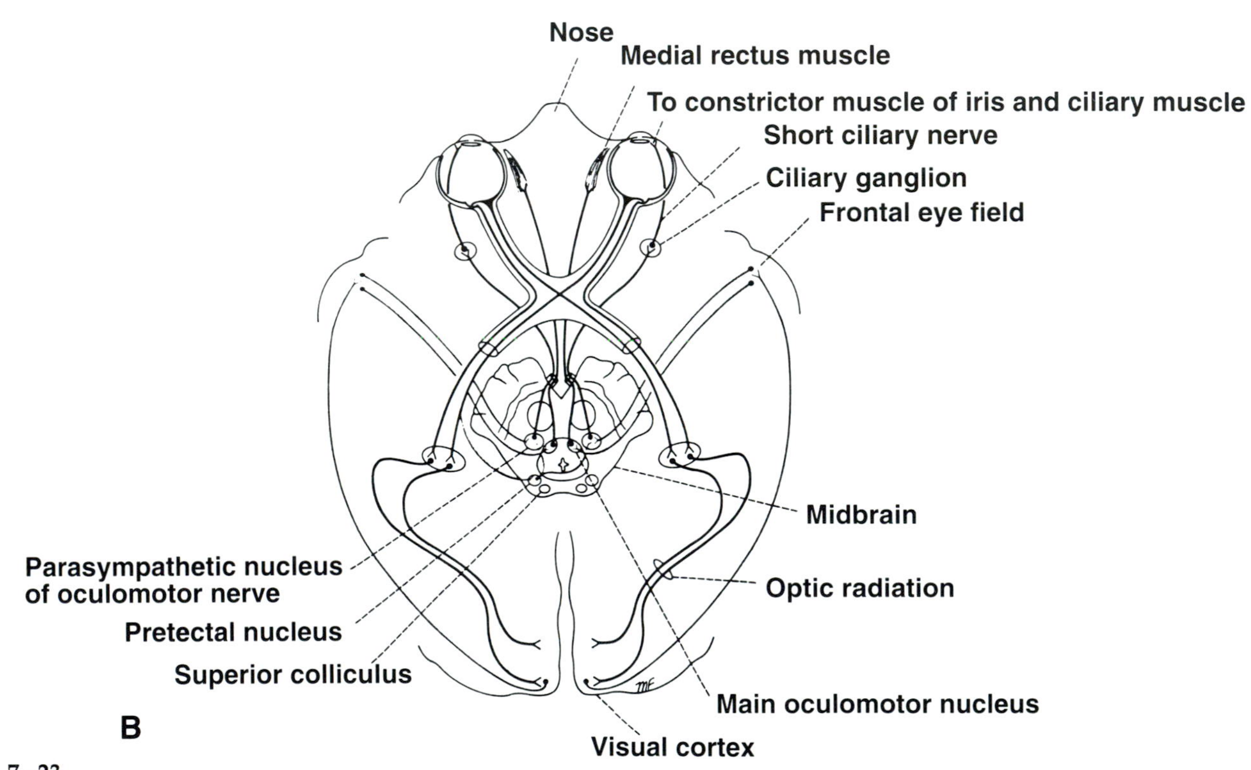

FIG 7–23.
A, the autonomic innervation of the upper lid and the iris. **B,** the optic pathway and the neurons involved in the visual reflexes.

Clinical Note

Argyll Robertson pupil.—This condition is characterized by a small pupil of fixed size that does not react to light but contracts with accommodation. It is usually caused by a neurosyphilitic lesion interrupting fibers that run from the pretectal nucleus to the parasympathetic nuclei (Edinger-Westphal nuclei) of the oculomotor nerve on both sides. The fact that the pupil constricts with accommodation implies that the connections between the parasympathetic nuclei and the constrictor pupillae muscle of the iris are intact.

Basic Anatomy of the Orbital Cavity

The orbital cavities are a pair of large bony sockets that contain the eyeballs, their associated muscles, nerves, vessels, and fat, and most of the lacrimal apparatus. Each cavity is pear shaped and its apex is directed posteriorly. The bones forming the walls of the orbital cavity and the foramina that pierce the walls are shown in Figure 7–25.

Relations of the Bony Orbit

Superior.—Frontal air sinus, frontal lobe of cerebral hemisphere.

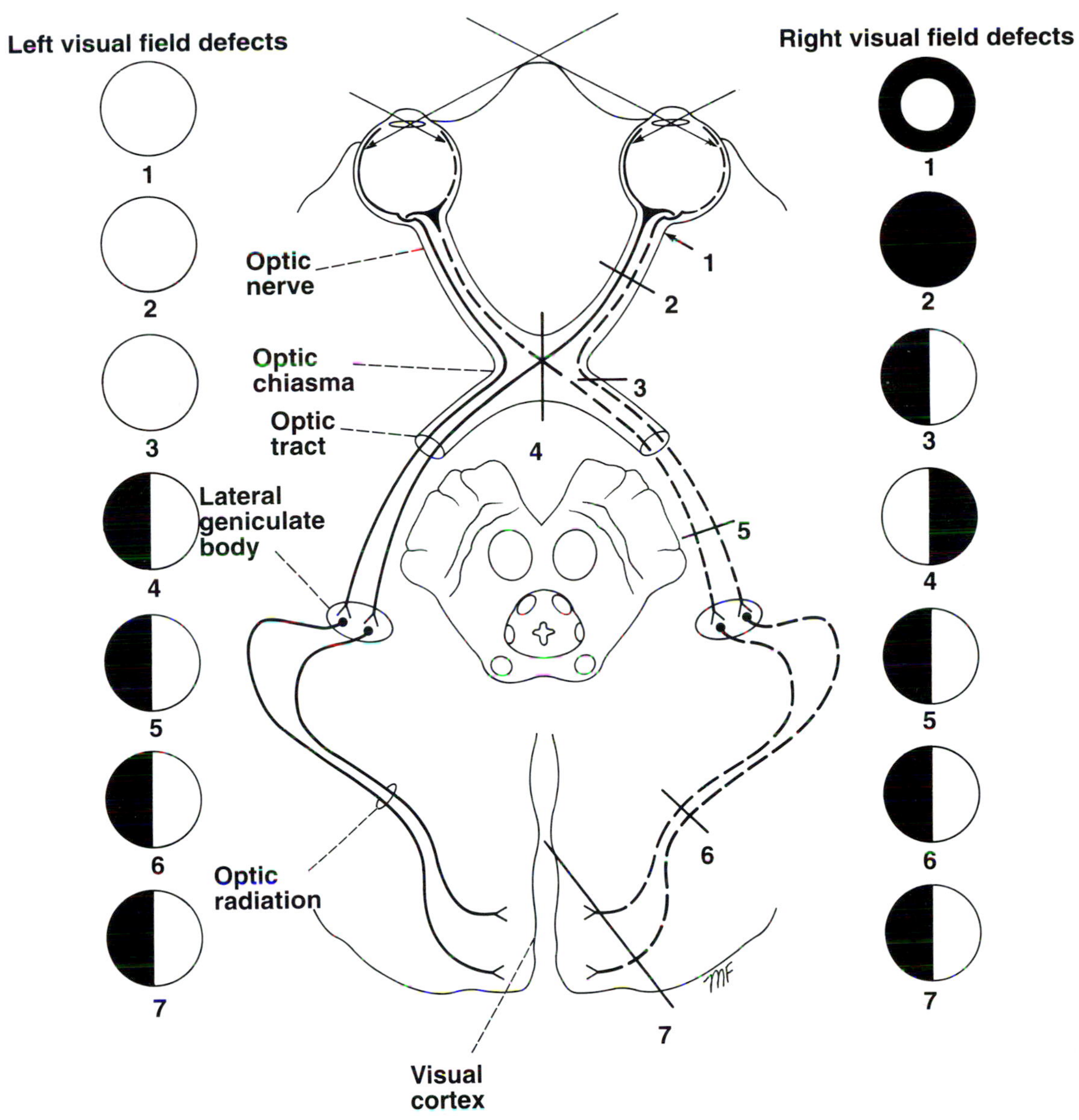

FIG 7–24.
Visual field defects associated with lesions of the optic pathways. *1* = right-sided circumferential blindness due to retrobulbar neuritis; *2* = total blindness of right eye due to division of right optic nerve; *3* = right nasal hemianopia due to a partial lesion of right side of optic chiasma; *4* = bitemporal hemianopia due to a complete lesion of the optic chiasma; *5* = left temporal hemianopia and right nasal hemianopia due to a lesion of the right optic tract; *6* = left temporal and right nasal hemianopia due to a lesion of right optic radiation; *7* = left temporal and right nasal hemianopia due to a lesion of the right visual cortex.

Inferior.—Maxillary air sinus, infraorbital nerve.

Lateral.—Temporal fossa (anteriorly), middle cranial fossa and temporal lobe of cerebral hemisphere (posteriorly).

Medial.—Nasal cavity, ethmoid air sinuses.

Radiographic Appearances of the Orbital Cavity
Appearances of the orbital cavity are shown in Figures 7–26 through 7–30.

SURFACE ANATOMY

The Eye and the Periorbital Structures

Superciliary Ridges
These prominent ridges lie above the upper margin of the orbit (Fig 7–31). Deep to the ridge on either side of the midline lie the frontal air sinuses.

Eyebrow
The medial end of each eyebrow usually lies just inferior to the superior orbital margin, and the lateral

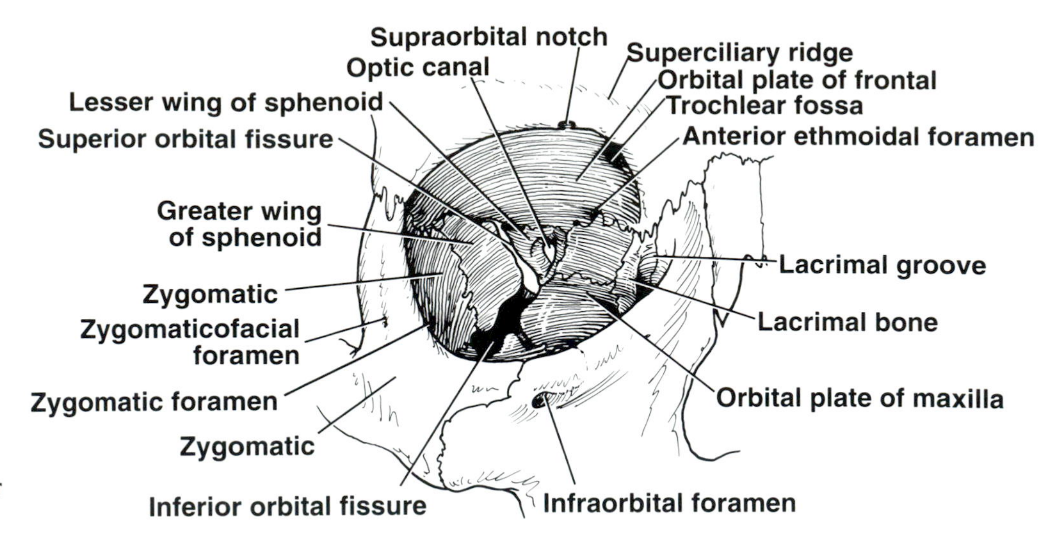

FIG 7–25.
The right orbit, anterior view.

end lies above the orbital margin (see Fig 7–31). The hairs are thick and directed horizontally laterally. Raising the eyebrows can be accomplished by contracting the frontalis muscle, lowering the eyebrows can be done by contracting the orbicularis oculi, and drawing the eyebrows medially can be done by contracting the corrugator supercilii muscle. All these muscles are supplied by the facial nerve.

Clinical Note

Blunt trauma to the eyebrow.— This may result in a gaping wound because of the pull of the underlying corrugator supercilii muscle.

Hypothyroidism.—Hypothyroidism may cause the loss of hair in the outer third of the eyebrow.

Orbital Margins

The frontal, maxillary, and zygomatic bones forming the margins can be easily felt (see Fig 7–31). The *supraorbital margin* is formed by the frontal bone having a sharp lateral two thirds and a rounded medial third. At the junction of the two areas is the *supraorbital notch or foramen* for passage of the supraorbital vessels and nerve (see Fig 7–31). The sharp *infraorbital margin* is formed laterally by the zygomatic bone and medially by the maxilla. The *lateral margin,* the strongest part of the orbital margin, is formed by the frontal process of the zygomatic bone below and the zygomatic process of the frontal bone above (see Fig 7–31). The *frontozygomatic suture* can be clearly felt along the lateral margin. The *medial margin* is formed above by the maxillary process of the frontal bone and below by the lacrimal crest of the frontal process of the maxilla (see Fig 7–31).

Clinical Note

Exposure of the eyeball to trauma.—Although the eyeball is reasonably well protected by the surrounding bony orbit, it is protected anteriorly only from large objects, tennis ball size, which tend to strike the orbital margin but not the globe. The bony margin provides no protection from small objects, such as golf balls, which may cause severe damage to the eye. Careful examination of the eyeball relative to the orbital margins shows that it is least protected from the lateral side. Rupture of the eyeball most commonly occurs from a blow directed from below and laterally.

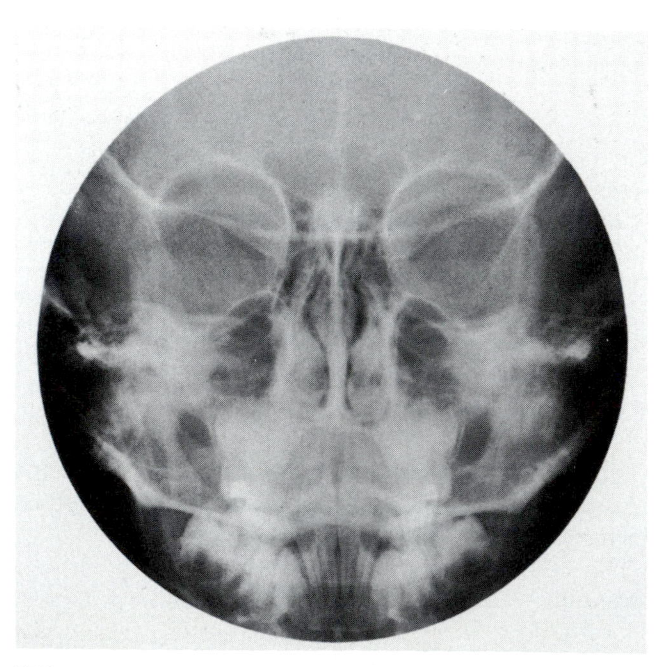

FIG 7–26.
Posteroanterior radiograph of orbit and paranasal sinuses.

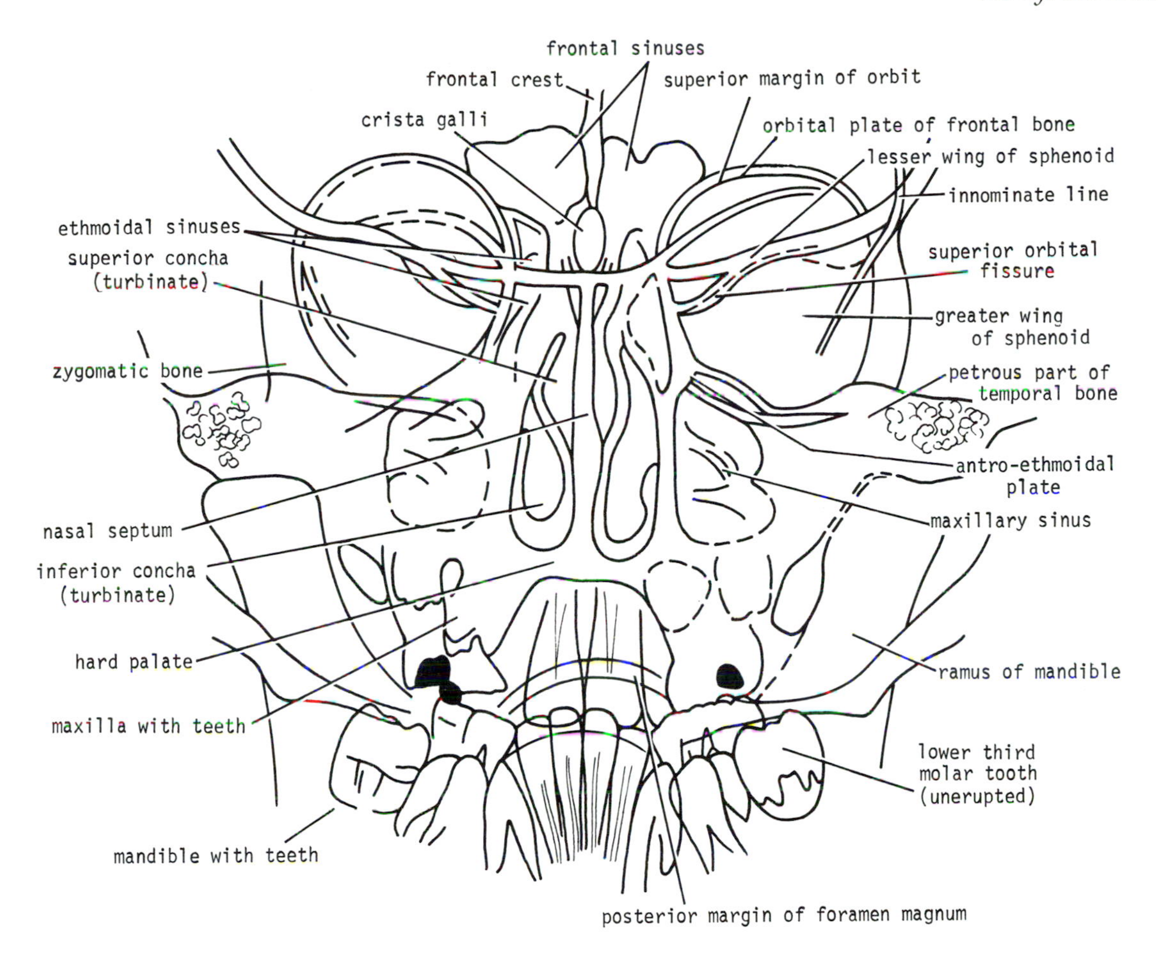

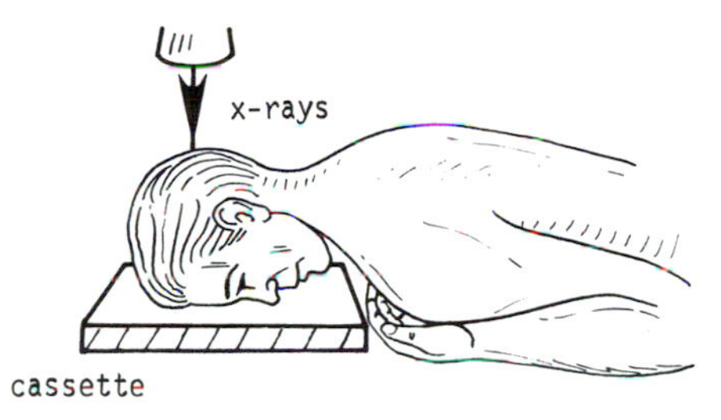

FIG 7–27.
Main features that can be seen in a posteroanterior radiograph of the orbit and paranasal sinuses in Figure 7–26.

Trochlea
The trochlea, or pulley, of the superior oblique tendon can be felt with the fingertip within the superomedial part of the orbital margin (see Fig 7–31).

Lateral Palpebral Ligament
The lateral palpebral ligament can be felt on deep pressure when the eyes are closed; it connects the lateral bony margin of the orbit to the lateral end of the eyelids (see Fig 7–31).

Medial Palpebral Ligament
The medial palpebral ligament can be felt connecting the medial bony margin of the orbit to the medial ends of the eyelids (see Fig 7–31). The lower free border of the ligament can be more easily felt if the eyelids are gently pulled laterally.

Lacrimal Groove
The lacrimal groove for the lacrimal sac can be felt along the medial part of the orbital margin.

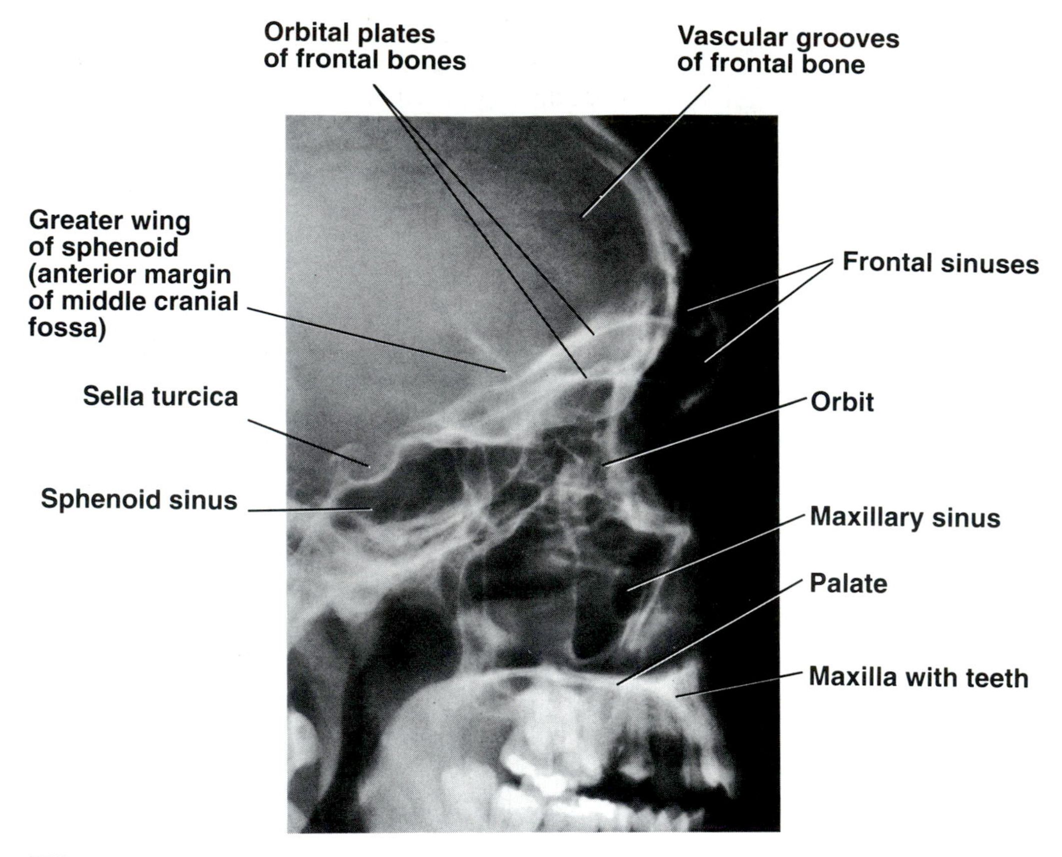

FIG 7–28.
Lateral radiograph of the orbit and paranasal sinuses.

Infraorbital Foramen
The infraorbital foramen lies about 5 mm below the lower margin of the orbit (see Fig 7–31). Its sharp upper margin can be recognized on deep pressure with the fingertip.

Eyelids
Each eyelid is divided by a horizontal furrow, the *palpebral sulcus,* into an *orbital* and a *tarsal part* (Fig 7–32).

The upper eyelid is larger and more mobile than the lower eyelid. The eyelids meet at the medial and lateral *angles or canthi.* The *palpebral fissure* is the elliptical opening between the eyelids into the *conjunctival sac.* In whites and blacks, the palpebral fissure is widest at the junction of the medial third with the lateral two thirds; in Asians, the fissure is widest halfway along its length. In Asians the medial angle is overlapped by a vertical skin fold, the *epicanthus.*

When the eye is closed, the upper eyelid completely covers the cornea of the eye. When the eye is open and looking straight ahead, the upper lid just covers the upper margin of the cornea (see Fig 7–31). The lower lid lies just below the cornea when the eye is open and rises only slightly when the eye is closed.

The lateral angle of the eye is directly in contact with the eyeball, whereas the medial rounded angle lies about 6 mm medially from the eyeball (see Fig 7–31).

Lacus Lacrimalis
The lacus lacrimalis is a small triangular space situated where the medial ends of the two eyelids are separated (see Figs 7–32 and 7–34). In the center of the space is a small pinkish elevation, the *caruncula lacrimalis.* A semilunar fold, called the *plica semilunaris,* lies on the lateral side of the caruncle.

Eyelashes
The eyelashes are short, curved hairs, found on the margins of the eyelids from the lateral angle of the eye to the lacrimal papilla.

Lacrimal Papilla
The lacrimal papilla is a small elevation on the margin of the eyelid about 5 mm from the medial angle (see Fig 7–32). On the summit of the papilla is the

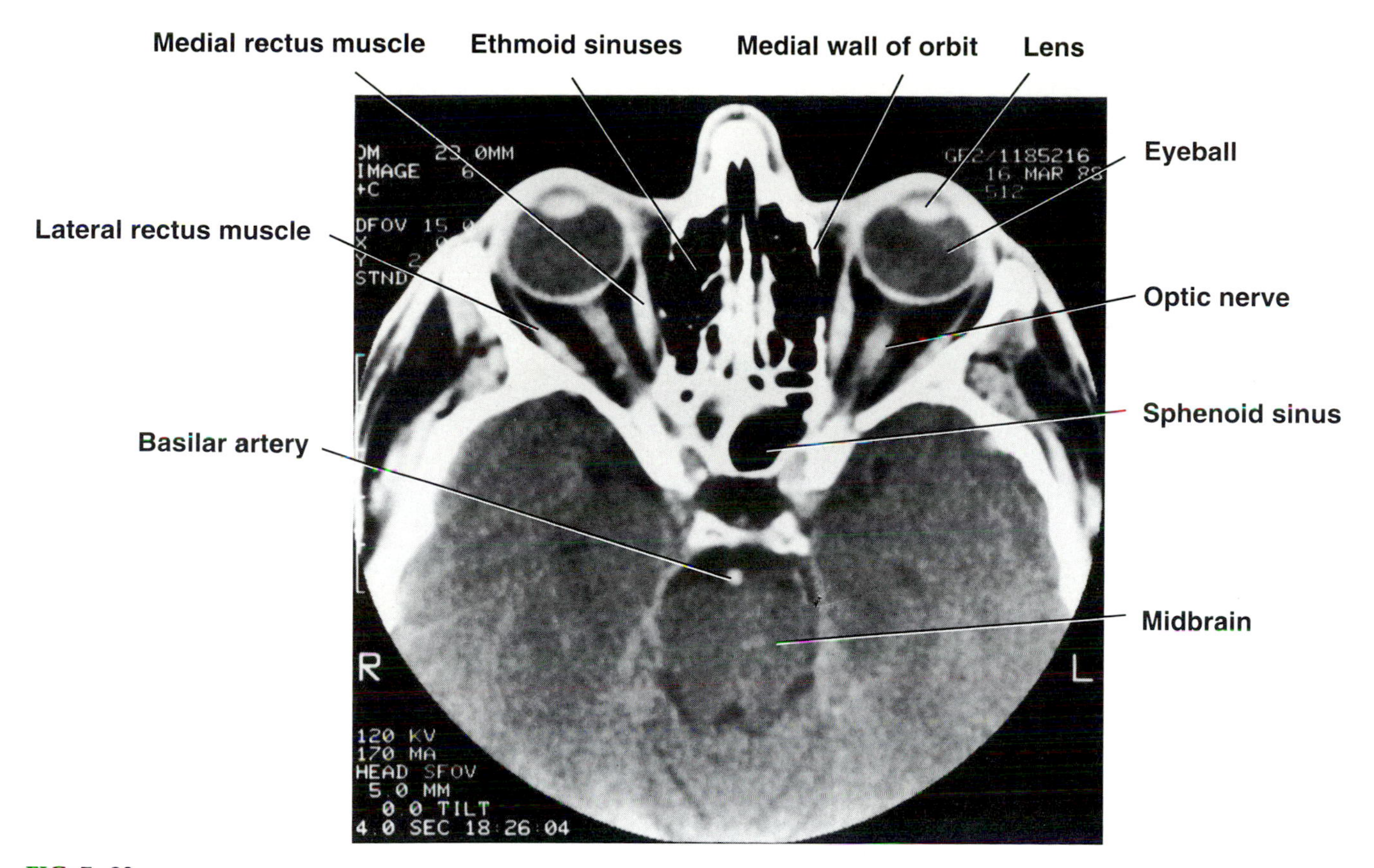

FIG 7–29.
Computed tomographic scan of the skull at the level of the orbital cavities and the tympanic cavities. The walls of the orbit and the eyeball can be seen.

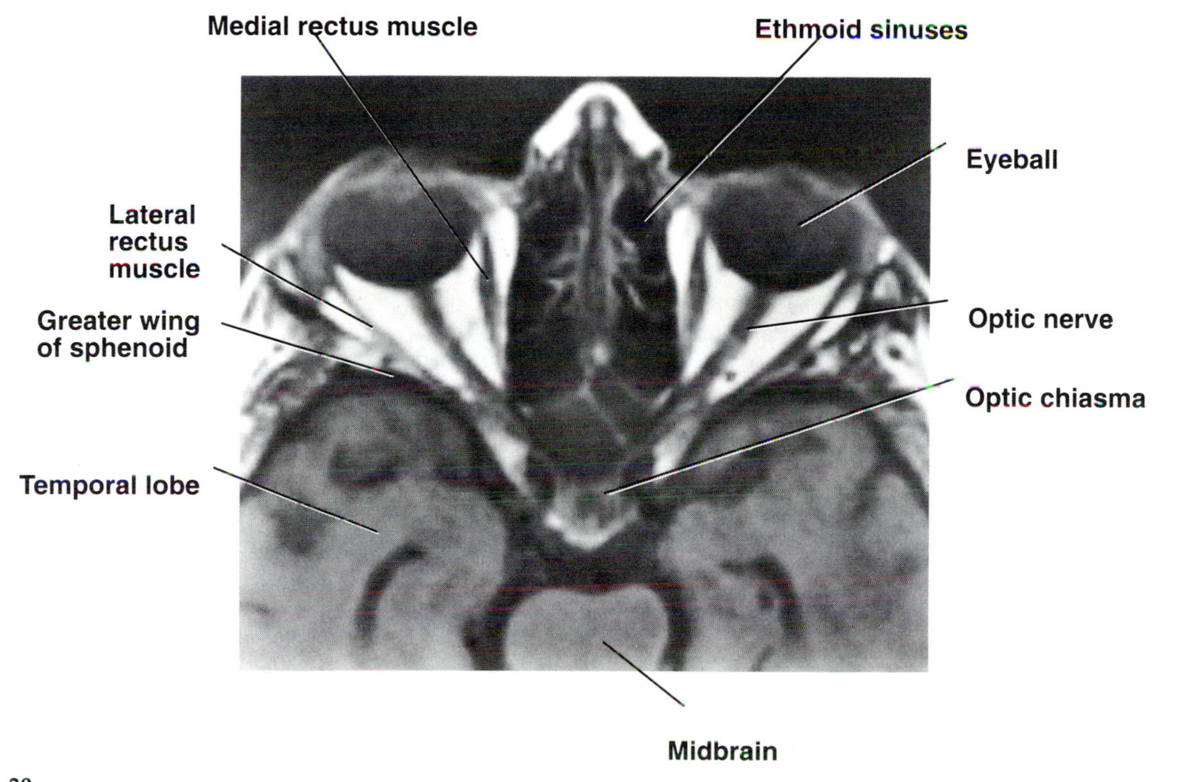

FIG 7–30.
Magnetic resonance image showing the contents of the orbital and cranial cavities. The eyeballs, optic nerves, optic chiasma, and extraocular muscles can be identified.

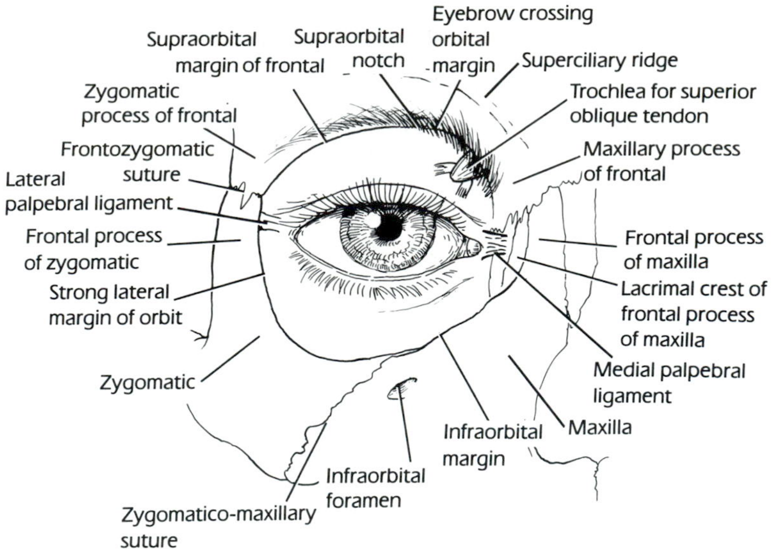

FIG 7–31.
Surface anatomy of the right orbital margin showing positions of eyebrow, eyelids, and trochlea of the superior oblique tendon.

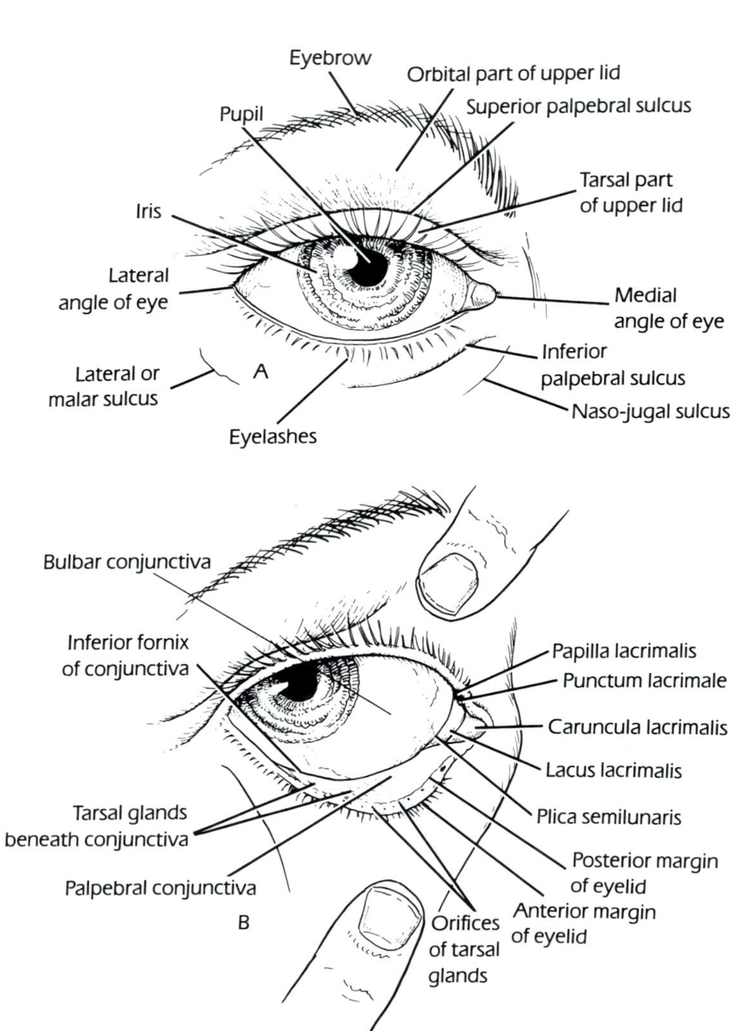

FIG 7–32.
A, right eye showing parts of eyelids, furrows, and sulci. With the eye open the upper lid just covers the upper margin of the cornea. **B,** right eye with its lids everted to show the conjunctival sac, the lacus lacrimalis, and the caruncula lacrimalis. Note the position of the tarsal glands beneath the conjunctiva and the puncta lacrimalia.

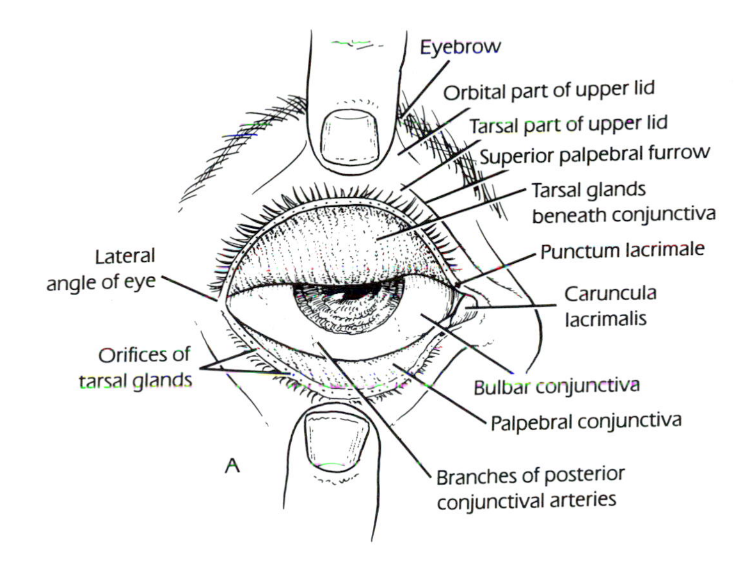

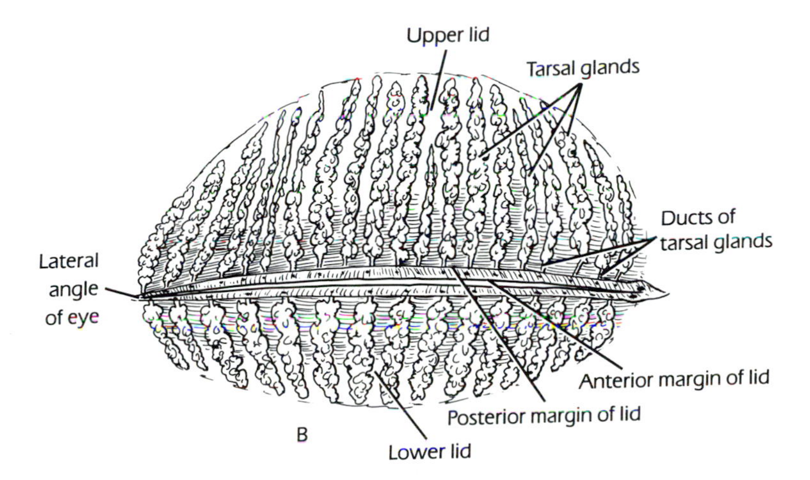

FIG 7–33.
A, complete eversion of the upper eyelid of the right eye made possible by stiffness of the superior tarsal plate; the lower eyelid is pulled downward. Note the orifices of the tarsal glands and the puncta lacrimalia. **B,** posterior view of the eyelids with the upper and lower lids nearly closed. Note the tarsal glands with their short ducts and orifices. In this diagram the conjunctiva has been removed from the back of the eyelids to reveal the tarsal glands in situ.

punctum lacrimale, which leads into the *canaliculus lacrimalis.* The papilla lacrimalis projects into the medial triangular lacus, and the punctum and the canaliculus serve to carry the tears down into the nose.

Tarsal Glands (Meibomian Glands)

The orifices of the tarsal glands can be seen just in front of the posterior edge of the margin of the lids (see Fig 7–33). The tarsal glands can be seen as yellowish lines on the inner surface of the everted lid.

Gray Line

A grayish line or slight sulcus can sometimes be seen running along the eyelid margin between the eyelashes and the openings of the tarsal glands. This represents the line of demarcation between the anterior portion of the eyelid formed by the skin and muscle and the posterior portion formed by the tarsal plate and the conjunctiva.

Eyeball Examination

The eyeball can be gently ballotted through the eyelids. It is possible to feel for the normal tension of the globe by this means.

The detection of unilateral hypotonia by gentle ballottement suggests the presence of a perforated or ruptured eyeball. The detection of hypertonia suggests glaucoma.

Anatomy of the Eye as Seen With the Direct Ophthalmoscope

Red Reflex

On looking through the ophthalmoscope, holding it about 1 ft away from the patient, the examiner notes that the fundus appears red (see Fig 7–12). The fundus shows red because the light is being reflected back from the blood in the choroidal blood vessels, with the intervening retina being transparent.

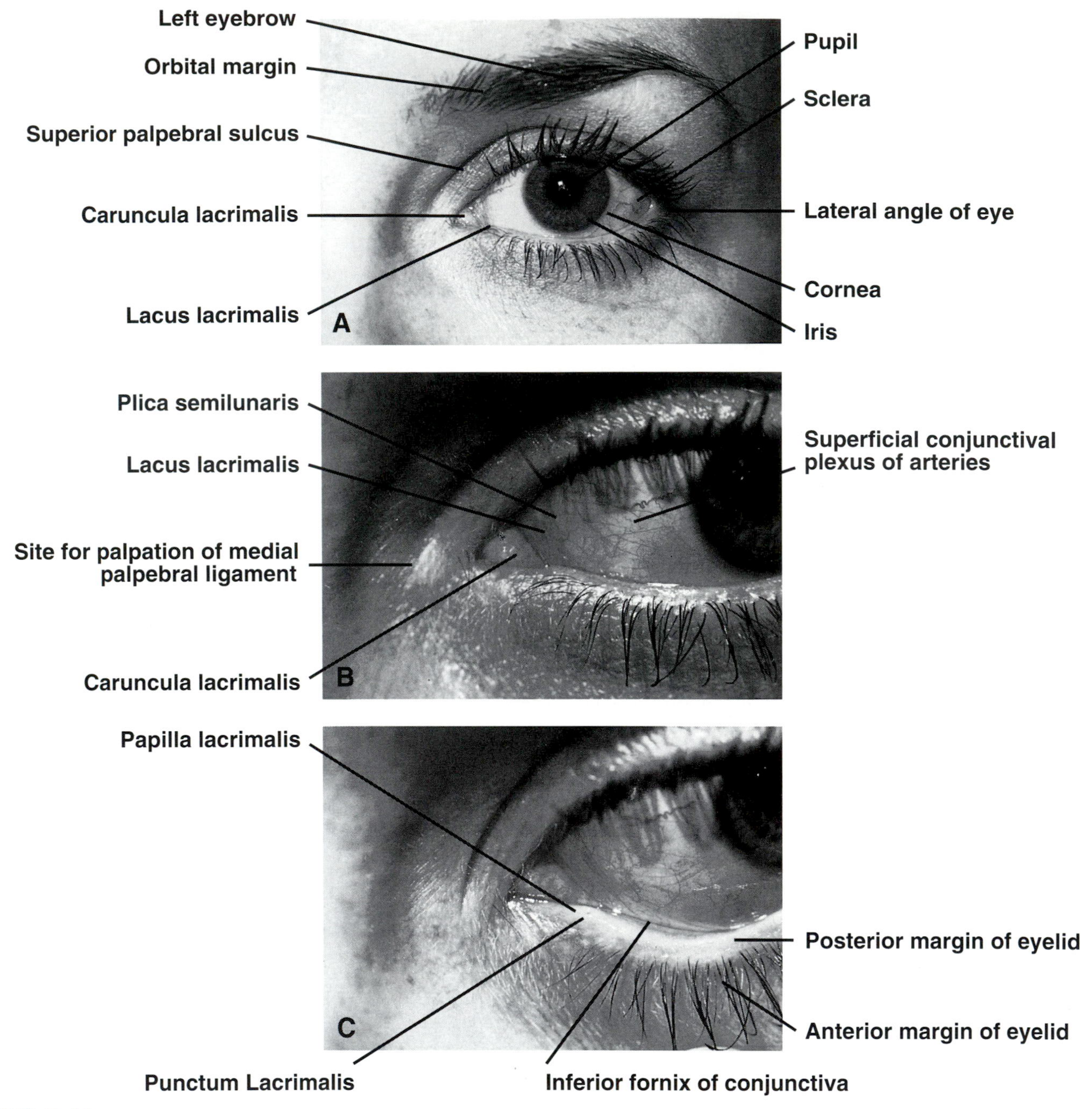

FIG 7-34.
Left eye of a 29-year-old woman. **A**, the names of structures seen in examining the eye. **B**, an enlarged view of the medial angle between the eyelids. **C**, the lower eyelid, pulled downward and slightly everted to reveal the punctum lacrimale.

Clinical Note

Absent red reflex.—This means that either there is an opacity in the refractive media or the retina is not against the choroid. The possible opacities include a cataract, a vitreous hemorrhage, and a detached retina.

Fundus Examination

Without pupillary dilatation only about 15% of the fundus can be seen. With full pupillary dilation, about 50% of the fundus can be viewed, but the area between the equator of the eyeball and the orra serrata cannot be seen.

Optic Disc

This structure is circular or vertically oval with a vertical orientation (see Fig 7–12). It is pink, with the temporal side slightly lighter than the nasal side. The disc measures about 1.5 mm in diameter and can be used as a unit of measurement. The center of the disc has a pale, almost white, depression called the *physiologic cup.* The edge of the disc is usually flat and sharply defined. In some individuals in whom the retina does not quite reach the margin of the disc, an arc of choroid pigment may be visible.

The bright red central artery of the retina becomes visible on the disc surface emerging from the optic cup, where it divides into its superior and inferior branches. The arteries do not normally pulsate. The darker red main tributaries of the central vein of the retina pass into the cup and unite in the cup or deeper out of site within the optic nerve.

Clinical Note

Cupped disc in glaucoma.—In glaucoma the increase in intraocular pressure leads to atrophy of the optic nerve and defects in the visual field. Since the lamina cribrosa of the sclera at the optic disc is a weak area, a rise in intraocular pressure can cause it to bulge outward, producing a cupped disc that can be seen with an ophthalmoscope.

Clinical Note

Pulsations of central vein.—Pulsations of the central vein can often be seen on fundoscopy. A pulsating central vein ensures that the intracranial pressure is not elevated.

Retinal Arteries and Veins

The arteries are bright red; the veins are darker red (see Fig 7–12). The arteries are smaller than the veins (about a 3:4 ratio). The arteries have thicker walls, which reflect the light as a shiny central reflex stripe. The walls of the arteries and the veins are transparent, so that the examiner observes a moving column of blood. The arteries usually cross the veins on their superficial or vitreal surface, and normally the arteries do not compress or nick the veins at the site of crossing. The branching of the vessels is variable.

Macula

The macular area lies about 2 disc diameters on the temporal side of the optic disc (see Fig 7–12). It is darker than the surrounding retina. The superior and inferior temporal blood vessels arch above and below the macular area, and no blood vessels are visible in the center of the macula. The center of the macula shows a small, dark red area called the *fovea centralis.* A small white-yellow light reflex can be detected at the center of the fovea, caused by the reflection of the ophthalmoscope light from the concavity of the fovea.

CLINICAL ANATOMY

Clinical Anatomy of the Eyebrow

Eyebrow Lacerations

The position of the hairs guides the approximation of the cut edges. Since the eyebrow is usually burst open as the result of blunt trauma striking the skin against the underlying frontal bone, the eyebrow should be closed in layers starting with the muscle (frontalis and or orbicularis), followed by the subcutaneous tissue, and finally the skin.

Clinical Anatomy of the Eyelids

Clinical Examination of the Eyelids

Retraction of the Eyelids.—Both the upper and lower lids can be retracted by applying the thumbs to the patient's forehead and the cheek just above and below the orbital margins. The lids are then gently retracted away from the eyeball, and the lower lid may be easily everted.

Eversion of the Upper Lid.—Eversion of the upper lid is more difficult than that of the lower lid because of its size and muscular attachments. It is performed for purposes of inspection of the eye and removing a superficial conjunctival foreign body. With the patient looking downward, the upper lid is grasped by the central eyelashes and pulled downward while a cotton-tipped applicator is applied centrally to the skin on the upper surface of the upper lid. With the superior tarsal plate serving to stiffen the upper lid, the upper lid is gently flipped upward over the applicator tip so that the conjunctival surface is fully exposed (Fig 7–33). In the procedure of double eversion of the upper lid for the purpose of examining the superior fornix of the conjunctiva for foreign bodies, the applicator is pushed anteriorly. This separates the upper lid from the eyeball and opens up the superior fornix for inspection.

The Cornea and the Palpebral Fissure
In an emergency, when the eyes reflexly close as a protective mechanism against trauma, chemicals, or fire, the last part of the cornea to be covered by the lids is just below the center. The damaged area may be even lower than expected since the cornea tends to move upward with approaching danger. In the elderly, corneal damage may be more extensive because of poor tone of the orbicularis oculi.

Lid Glands
Chalazion.— This is a localized, progressive, painless swelling of the lid resulting from chronic inflammation of a tarsal gland. Since the gland lies on the conjunctival surface of the tarsal plate, the swelling should be incised through the conjunctival surface of the lid.

Hordeolum (Stye).— An external hordeolum is an acute infection of a lash follicle or a sebaceous gland (of Zeis) or a ciliary sweat gland (of Moll); all drain externally to the skin surface of the lid. An internal hordeolum is an acute infection of a tarsal gland (of Meibom). The tarsal glands usually drain through the conjunctival surface of the lid.

Lid Muscle
Orbicularis Oculi Muscle Paralysis and Facial Nerve Lesion.— Paralysis of the orbicularis oculi muscle from a lesion of the facial nerve prevents closure of the eye and permits the lower lid to sag away from the eyeball (ectropion). The puncta of the lacrimal canaliculi are no longer kept in the lacus lacrimalis, so that tears escape over the lower lid (epiphora). Senile weakness of the orbicularis may also cause the above conditions.

Paralysis of the Levator Palpebrae Superioris Muscle.— Paralysis of the levator palpebrae superioris muscle from a lesion of the oculomotor nerve, myasthenia gravis, or senile dehiscence of the aponeurosis produces severe drooping of the upper eyelid (ptosis) and loss of the superior palpebral fold and horizontal furrow.

Paralysis of the smooth muscle of the levator (superior tarsal muscle) can occur following development of a lesion in the cervical part of the sympathetic nervous system. A less severe form of ptosis occurs. When associated with constriction of the pupil, the condition is known as *Horner's syndrome.*

Orbital Septum, Herniation of Orbital Fat, and Infection
The orbital septum separates the lid connective tissue spaces from the orbital contents. Not only does it serve to hold the orbital fat in position, but it forms a barrier to prevent infection from passing from the eyelids into the orbital cavity (or vice versa). Lacerations of the eyelid should be assessed for the presence of orbital fat, which would indicate violation/involvement of the orbital septum. Herniation of the orbital fat into the eyelid is most commonly seen in the elderly. The septum is weakest on the medial side of the lower lid.

Tarsal Plates
The tarsal plates, which consist of dense connective tissue, give the lids shape and support. Lacerations of the tarsal plates should be repaired by a specialist.

Fascial Spaces of the Lids
The subcutaneous, pretarsal, and preseptal spaces of the lids are potential areas for accumulation of blood or inflammatory exudate (see Fig 7–3).

Eyelid Lacerations
Correct alignment of the eyelid margin is essential to avoid deformity. The insertion of a traction suture at the start of the repair preserves this alignment. Through-and-through lacerations are then closed in layers, with the deepest layer consisting of the conjunctiva and the tarsal plate being sutured first. The muscle and skin are then sutured as a single layer.

Lacerations involving the lacrimal canaliculus require that a stent be placed within the lumen. If the medial or lateral palpebral ligaments are lacerated, they have to be carefully repaired.

Lacerations parallel with the lid margin can be closed with simple interrupted sutures. Since the pull of the orbicularis muscle is in line with the laceration, an unsightly scar is unlikely. Excessive debridement is unnecessary since the lids have a rich blood supply; moreover, debridement should be avoided to prevent ectropion.

Anesthesia of the Eyelids
The branches of the lacrimal, supraorbital, supratrochlea, infratrochlea, and the infraorbital nerves to the eyelids run in the connective tissue beneath the orbicularis oculi muscle. The nerves can be successfully blocked by injecting the anesthetic deep to the muscle (see Chapter 19, p. 805).

Clinical Anatomy of the Conjunctiva

Palpebral Conjunctiva

This is thin, translucent, and firmly attached to the deep surface of the tarsal plates of the eyelid. The thinness and transparency allow the physician, by everting the lids, to visualize the yellowish tarsal glands arranged in rows embedded in the posterior surface of the tarsal plates. Many blood vessels can also be seen, and the color of the palpebral conjunctiva can be used clinically in the diagnosis of anemia.

Pinguecula

This yellowish spot of proliferation seen on the nasal side of the conjunctiva covers the bulb near the corneoscleral junction (limbus); it is commonly found in the elderly.

Foreign Bodies in the Conjunctival Sac

Foreign bodies in the conjunctival sac produce severe pain and reflex tearing. The superior and inferior fornices can be examined for the foreign bodies by everting the eyelids as described previously. Small particles often migrate and become lodged in the subtarsal sulcus. *Corneal abrasions* may occur as the result of the foreign body being carried across the cornea with the movement of the eyelids.

Arterial Supply of the Bulbar Conjunctiva and Conjunctivitis

The arterial supply of the conjunctiva arises from the palpebral arteries in each eyelid and from the anterior ciliary arteries (both from the ophthalmic artery).

The peripheral bulbar conjunctiva is supplied by small arteries that can only just be seen. In acute conjunctivitis these vessels dilate, and, because they are superficial, they appear bright red. Since they are situated in the connective tissue of the conjunctiva, they move with the conjunctiva and can be constricted with a topical sympathomimetic.

The central bulbar conjunctiva, i.e., the area adjacent to the corneoscleral limbus, is supplied by arteries that form a *superficial pericorneal plexus.* In inflammatory disease of the cornea, this superficial plexus becomes dilated and the vessels appear bright red. Because they are superficial, they move with the conjunctiva and can be constricted with a topical sympathomimetic.

In inflammatory disease of the iris or ciliary body or in closed-angle glaucoma, a *deep episcleral pericorneal plexus* becomes dilated. The vessels appear dull red because of their deep location; they do not move with the conjunctiva, but the redness disappears on pressure; they cannot be constricted with a topical sympathomimetic.

Clinical Anatomy of the Lacrimal Apparatus

Traumatic Injury to the Lacrimal Canaliculi

Lacerations of the medial ends of the eyelids can involve the lacrimal papilla and the lacrimal canaliculi. Lacerations of the lower papilla and canaliculus must be repaired or epiphora results.

Dry Eyes

Drying of the cornea can result from a deficiency of the watery component of tears, secondary to disease of the lacrimal gland. It can also occur when there is a deficiency of the mucin component of tears, as in disease of the goblet cells of the conjunctiva in cases of hypovitaminosis A, Stevens-Johnson disease, and chemical burns of the eye.

Inflammation of the Lacrimal Gland

Enlargement of the lacrimal gland, secondary to inflammation, may be seen as a tender swelling above the eyeball in the anterior and upper part of the orbit posterior to the upper eyelid.

Inflammation of the Lacrimal Sac

Enlargement of the lacrimal sac secondary to inflammation may be seen as a tender swelling above the upper margin of the medial palpebral ligament. Gentle pressure on the sac may result in a yellowish discharge emerging from the puncta lacrimalia.

Clinical Anatomy of the Eyeball

Trauma

Although the eyeball is reasonably well protected by the surrounding bony orbit, it is protected anteriorly only from large objects, such as tennis balls, which tend to strike the orbital margin but not the globe. The bony orbit provides no protection from small objects, such as golf balls, which may cause severe damage to the eye. Careful examination of the eyeball relative to the orbital margins shows that it is least protected from the lateral side. Rupture of the eyeball most commonly occurs from a blow directed from below and laterally and usually takes place at the sites of muscle insertion, where the sclera is weakest. Rupture of the sclera is almost always associated with damage to the underlying choroid and retina.

Clinical Anatomy of the Cornea

Astigmatism
Often the cornea is not the section of a perfect sphere so that the refractive power is not the same in all directions, a condition known as astigmatism.

Aging Changes in the Cornea
With advancing years, the cornea becomes less translucent, and dustlike opacities may occur in the deeper parts of the substantia propria. *Arcus senilis* appears as white arcs near the edge of the cornea and is caused by an extracellular infiltration of lipid; it is present in almost every person over 60 years old.

Trauma and the Cornea
Because a portion of the cornea is exposed between the eyelids, injuries from *foreign bodies* or abrasions are very common. Damage to the corneal epithelium causes considerable pain, reflex tearing, and vasodilatation of the conjunctival capillaries. Later, edema of the lids will be apparent.

The stratified squamous epithelium covering the anterior surface of the cornea is capable of rapid regeneration after an abrasion. When the Bowman's layer is damaged, it is not reformed but is replaced by fibrous tissue. Damage to Descemet's membrane results in its replacement by the endothelial cells. It is a strong membrane and resistant to trauma, and when torn it curls up in the anterior chamber. The cells of the endothelium tend to be stripped off as a sheet.

Foreign bodies driven with great force may penetrate the cornea and enter the anterior chamber or even the deepest parts of the eyeball.

Corneal Ulcers
Corneal ulcers are caused by a bacterial invasion of the cornea with the formation of a stromal abscess. The ability of the cornea to resist bacterial invasion depends on the cleansing action of the tears and their normal circulation and the integrity of the corneal epithelium. A breakdown in this mechanism can occur as the result of mild trauma, such as that which occurs when soft corneal lenses are worn for an excessive period of time, or in the presence of chronic disease. Corneal ulcers often appear opaque on direct visualization, distinguishing them from corneal abrasions, which are transparent.

Vascularization of the Cornea
The cornea is an avascular structure. Inflammation and infection can result in invasion of the cornea at its circumference by new blood vessels. Unfortunately, if the condition persists, there is a loss of transparency, and the cornea may become opaque.

Clinical Anatomy of the Iris

Iritis
Acute inflammation of the iris or ciliary body may occur spontaneously or after blunt trauma. There is a deep boring pain, photophobia, and redness of the eye. The conjunctival redness involves the deep plexus of vessels around the cornea and overlying the ciliary body (see p. 275).

Horner's Syndrome
This is fully described on p. 343.

Malignant Melanoma and the Uveal Tract
The uveal tract is the most common site of malignant intraocular tumors.

Clinical Anatomy of the Retina

Detachment of the Retina
The neural retina is firmly attached to the underlying pigment epithelium (Fig 7–35) at the optic disc and the ora serrata.

Pathological separation of the two layers of the retina may follow trauma to the eyeball or degeneration of the neural retina. Vitreous traction on the retina, or the presence of a hole or tear, allows accumulation of fluid between the pigment epithelium and the neural retina, causing the layers to separate or become detached.

Clinical Anatomy of the Blood Supply of the Retina

Central Retinal Artery Occlusion
At the point where the central artery pierces the lamina cribrosa, it is subject to atherosclerosis and can undergo complete or partial occlusion. Disease changes in the arteriolar wall can be seen with the ophthalmoscope where the arterioles cross the veins as a nicking or narrowing of the venous blood column.

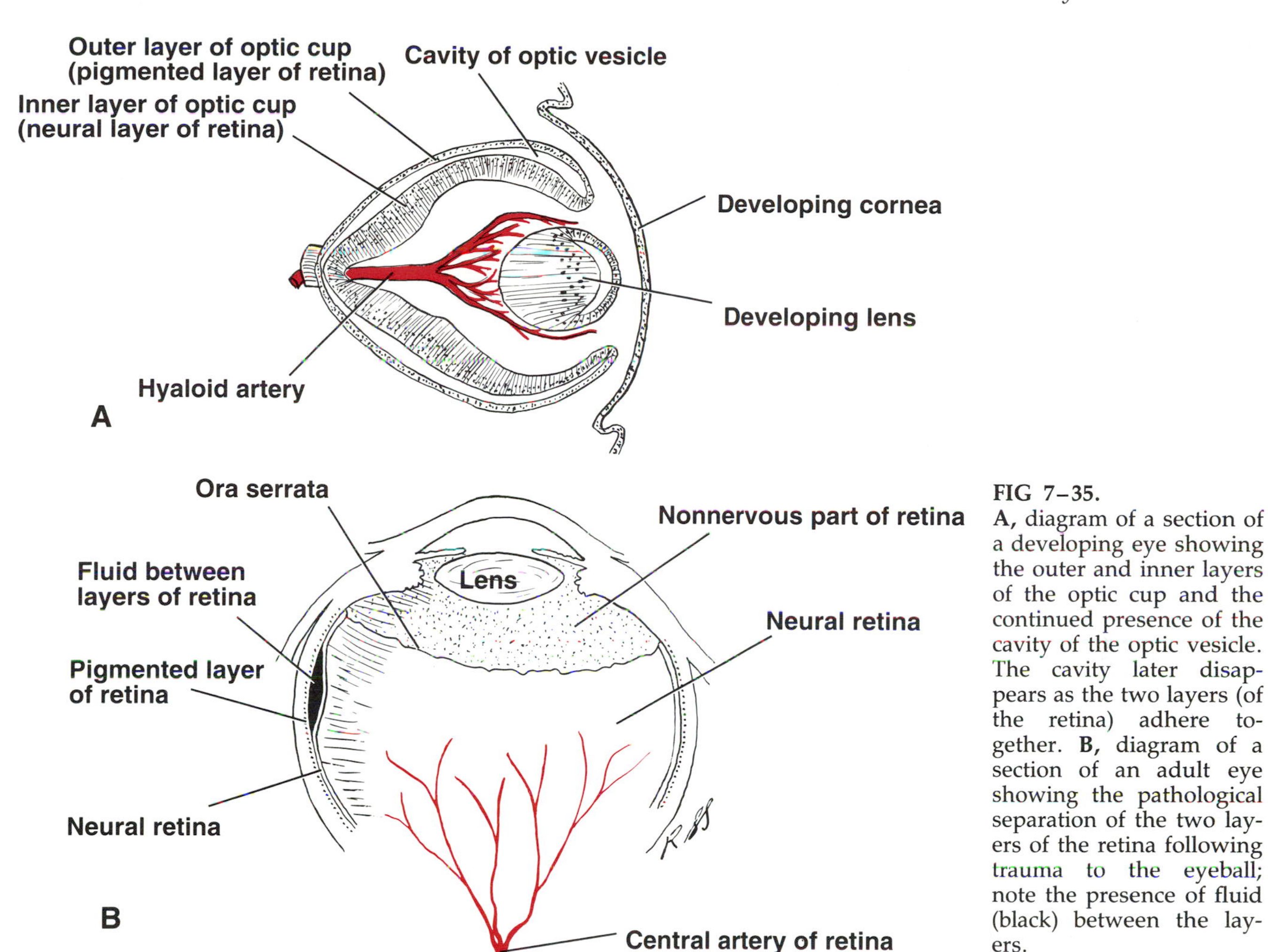

FIG 7–35.
A, diagram of a section of a developing eye showing the outer and inner layers of the optic cup and the continued presence of the cavity of the optic vesicle. The cavity later disappears as the two layers (of the retina) adhere together. **B,** diagram of a section of an adult eye showing the pathological separation of the two layers of the retina following trauma to the eyeball; note the presence of fluid (black) between the layers.

In complete central artery occlusion there is a sudden onset of unilateral blindness. In branch arteriole occlusion there is a partial loss of sight corresponding to the sector supplied by the arteriole. Total arterial occlusion lasting longer than 1½ hours can produce irreversible retinal degeneration.

Papilledema, the Central Vein, and Increased Cerebrospinal Fluid Pressure
Since the optic nerve is surrounded by the dura and arachnoid sheaths, an increase in the intracranial pressure is transmitted through the cerebrospinal fluid along the extension of the subarachnoid space to the lamina cribrosa of the eyeball. Because the central artery and vein of the retina cross the subarachnoid space to enter or leave the optic nerve, they will be subject to a rise in cerebrospinal fluid pressure. The thick-walled artery is unaffected, but the thin-walled vein may be compressed, causing congestion of the retinal veins and edema of the retina; bulging of the optic disc may also occur.

Clinical Anatomy of the Aqueous Humor

Circulation of Aqueous Humor and Glaucoma
Two areas of resistance to the circulation of aqueous humor normally exist (Fig 7–36): (1) where the anterior surface of the lens is in contact with the iris, and (2) where the aqueous leaves the anterior chamber to enter the veins. The iris-lens resistance may increase from such factors as age, diabetes mellitus, miotics, and increased viscosity caused by hemorrhage or inflammation. The drainage of aqueous into the venous system can be inhibited by such factors as age and melanin debris that blocks the meshwork at the drainage site.

In *glaucoma,* the intraocular pressure becomes pathologically high and may lead to blindness. In *open-angled glaucoma* the relationship between the iris root, the trabecular meshwork, and the corneal is normal. However, the drainage resistance is increased within the meshwork or the venous channels. Hemorrhage into the anterior chamber fol-

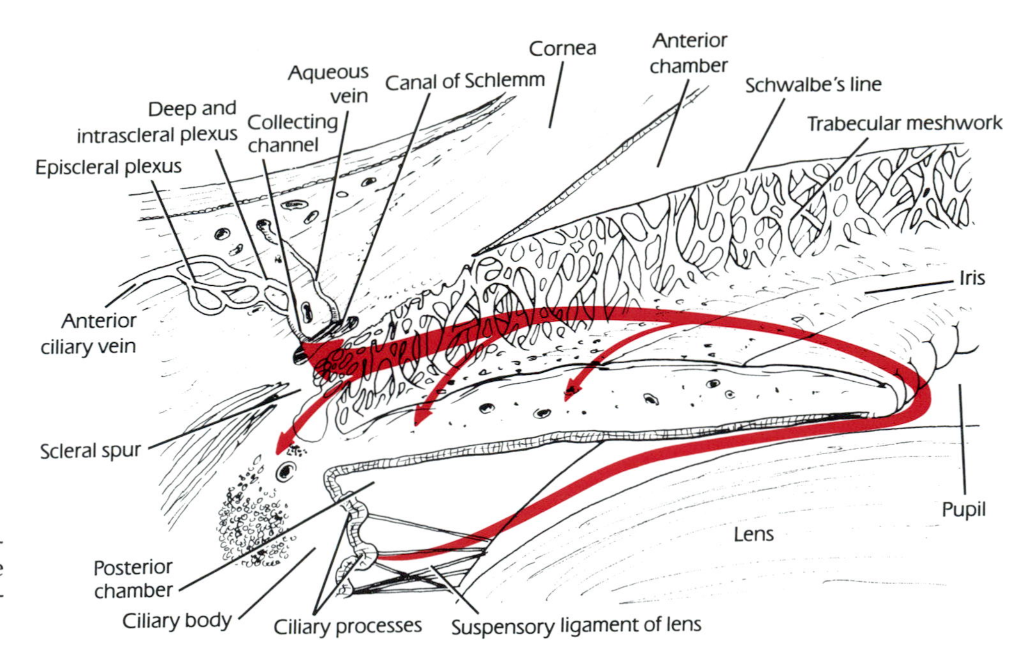

FIG 7–36.
Diagram of the anterior portion of the eye showing the origin, circulation, and drainage of the aqueous humor.

lowed by blockage of the meshwork or venous channels is an example. Open-angled glaucoma is a chronic and insidious process and causes painless loss of vision.

In *closed-angle glaucoma,* the iris root is displaced forward, blocking the meshwork and obstructing the outflow. This occurs acutely, and the patient may present with mild discomfort and a feeling of fullness of the eye to the most intense pain referred along the distribution of the ophthalmic division of the trigeminal nerve. Edema of the cornea produces blurring of vision. The sudden rise in intraocular pressure may reflexly stimulate the autonomic nervous system, producing bradycardia, excessive sweating, nausea, and vomiting.

Since in closed-angle glaucoma the iris root is displaced forward, it follows that an attack may be precipitated if a patient enters a darkened room when the pupil dilates and causes further crowding and impairment of drainage. Constriction of the pupil, as occurs in a brightly illuminated room or following the administration of miotics, tends to open up the drainage angle and lessen the intraocular pressure.

Clinical Anatomy of the Lens

Cataract

In this condition the lens becomes opaque. Metabolic products accumulate within the lens fibers. Senile cataract is the most common form; its cause is unknown.

Clinical Anatomy of the Fascial Sheath

Prosthetic Eyeball

A well-fitted prosthesis that fits snugly into the socket formed by the fascial sheath will move with the contractions of the extraocular muscles because the muscle tendons are attached to the fascial sheath.

Clinical Anatomy of the Muscles Producing Eye Movements

Clinical Testing for the Actions of the Recti and the Oblique Muscles

Since the actions of the superior and inferior recti and the superior and inferior oblique muscles are complicated when a patient is asked to look vertically upward or vertically downward, the emergency physician tests the eye movements in which the single action of each muscle predominates.

The origins of the superior and inferior recti are situated about 23° medial to their insertion and, therefore, when the patient is asked to turn the cornea laterally, these muscles are placed in the optimum position to raise the cornea (superior rectus) or lower it (inferior rectus). To test the superior rectus, have the patient look up and laterally; to test the inferior rectus, have the patient look down and out.

Using the same rationale, the examiner tests the superior and the inferior oblique muscles. The pulley of the superior oblique and the origin of the inferior oblique muscles lie medial and anterior to their insertions. The opthalmologist tests the action of these muscles by asking the patient first to look medially, thus placing these muscles in the optimum position to lower the cornea (superior oblique) or to raise it (inferior oblique). In other words, when you ask a patient to look medially and downward at the tip of his nose, you are testing the superior oblique at its best position. Conversely, by asking the patient to look medially and upward, you are testing the inferior oblique at its best position.

Because the lateral and medial recti are neutrally placed relative to the axes of the eyeball, asking the patient to turn his cornea directly laterally tests the lateral rectus, and turning the cornea directly medially tests the medial rectus.

The cardinal positions of the eyes and the actions of the recti and oblique muscles are shown in Figure 7–37.

Lesions of the Cranial Nerves That Control Eye Movements

Oculomotor Nerve Lesions.—In a complete lesion of the oculomotor nerve, the eye cannot be moved upward, downward, or inward. At rest, the eye looks laterally (external strabismus) owing to the activity of the lateral rectus and downward owing to the activity of the superior oblique. The patient has diplopia. There is ptosis of the upper lid due to paralysis of the levator palpebrae superioris. The pupil is widely dilated and nonreactive to light owing to paralysis of the sphincter pupillae and the unopposed action of the dilator (supplied by the sympathetic). Accommodation of the eye is paralyzed.

Incomplete lesions of the oculomotor nerve are common and may spare the extraocular muscles or the intraocular muscles. The condition in which the innervation of the extraocular muscles is spared with

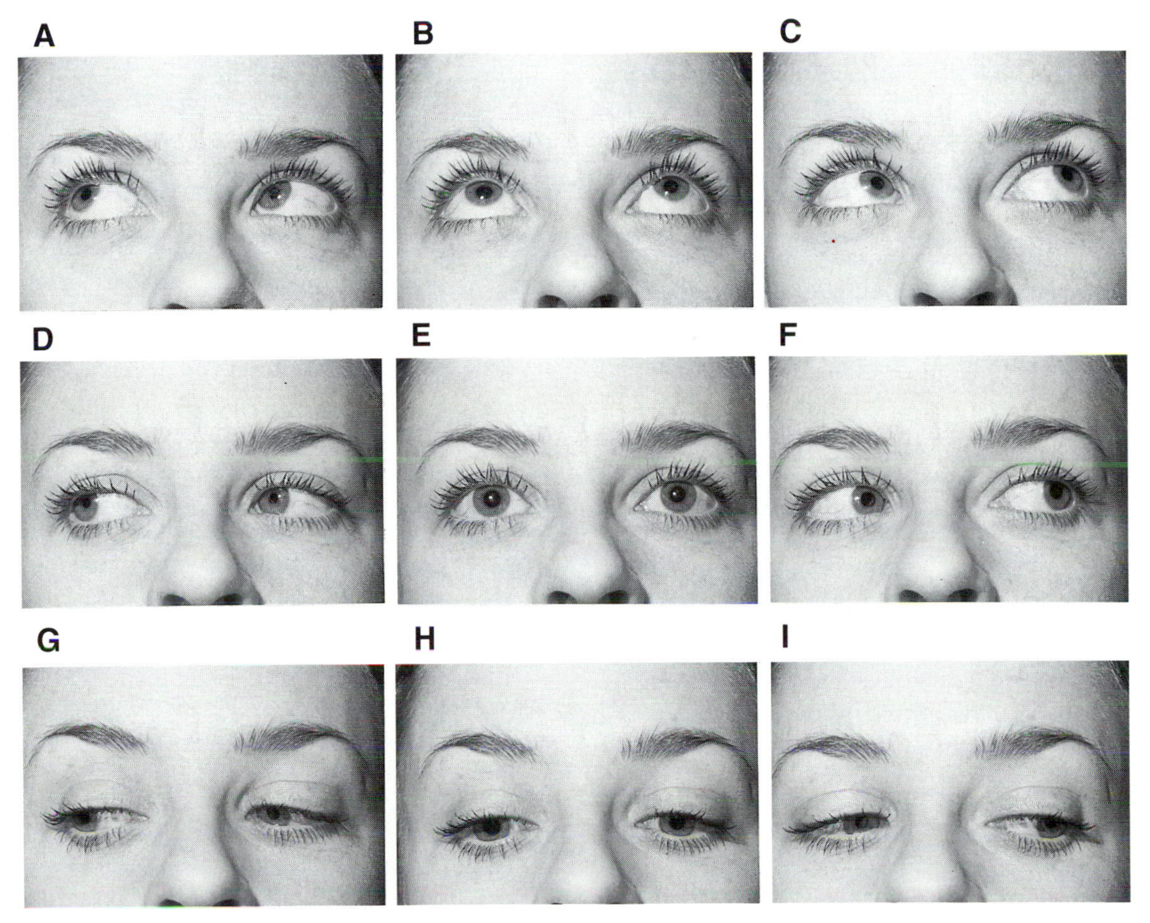

FIG 7–37.
The cardinal positions of the right and left eyes and the actions of the rectus and oblique muscles *principally* responsible for the movements of the eyes. **A,** right eye, superior rectus muscle; left eye, inferior oblique muscle **B.** Both eyes, superior recti and inferior oblique muscles. **C,** right eye, inferior oblique muscle; left eye, superior rectus muscle. **D,** right eye, lateral rectus muscle; left eye, medial rectus muscle. **E,** primary position, with the eyes fixed on a distant fixation point. **F,** right eye, medial rectus muscle; left eye, lateral rectus muscle. **G,** right eye, inferior rectus muscle; left eye, superior oblique muscle. **H,** both eyes, inferior recti and superior oblique muscles. **I,** right eye, superior oblique muscle; left eye, inferior rectus muscle.

selective loss of the automomic innervation of the sphincter pupillae and ciliary muscle is called *internal ophthalmoplegia.* The condition in which the sphincter pupillae and ciliary muscle are spared with paralysis of the extraocular muscles is called *external ophthalmoplegia.* It has been suggested that in the precavernous sinus course of the oculomotor nerve, the parasympathetic autonomic fibers are superficially placed within the nerve and are likely to be first affected by compression, whereas distally they are deeply placed. The nature of the disease also plays a role. For example, in diabetic neuropathy the pupil and ciliary muscle invariably remain unaffected, and in lesions of the oculomotor nerve nucleus the autonomic fibers usually are not affected.

The conditions most often affecting the oculomotor nerve are diabetes, aneurysm, tumor, trauma, inflammation, and vascular disease.

Trochlear Nerve Lesions.—In lesions of the trochlear nerve the patient complains of double vision because the images of the two eyes are tilted relative to each other. Unopposed action of the antagonistic inferior oblique muscle will cause the paretic eye to be turned upward in the primary position. The patient characteristically carries his head tilted toward the noninvolved side with the chin depressed. It is an unconscious attempt on the patient's part to compensate for the diplopia, particularly the torsional effects of muscle paresis.

Lesions of the trochlear nerve include stretching or bruising as a complication of head injuries (the nerve is long and slender), cavernous sinus thrombosis, aneurysm of the internal carotid artery, and vascular lesions of the dorsal part of the midbrain.

Abducent Nerve Lesions.—In a lesion of the abducent nerve the patient cannot turn the eye laterally. When the patient is looking straight ahead, the lateral rectus is paralyzed and the unopposed medial rectus pulls the eyeball medially, causing esotropia (internal strabismus).

Lesions of the abducent nerve include damage in head injuries (the nerve is long and slender), cavernous sinus thrombosis or aneurysm of the internal carotid artery, and vascular lesions of the pons.

Internuclear Ophthalmoplegia.—Lesions of the medial longitudinal fasciculus will disconnect the oculomotor nucleus that innervates the medial rectus muscle from the abducent nucleus that innervates the lateral rectus muscle. When the patient is asked to look laterally to the right or the left, the ipsilateral lateral rectus contracts, turning the eye laterally, but the contralateral medial rectus fails to contract and the eye remains in the primary position.

Bilateral internuclear ophthalmoplegia may occur in multiple sclerosis, occlusive vascular disease, trauma, or brain stem tumors. Unilateral internuclear ophthalmoplegia may follow an infarct of a small branch of the basilar artery.

Clinical Anatomy of the Orbit

Blunt Trauma to the Orbit

The orbital margin is very strong and not easily fractured. However, severe injury, as in an automobile accident, may involve the medial margin and the nose. Fractures of the superior margin, although rare, may damage or displace the trochlea, producing symptoms of superior oblique paralysis. Fractures of the lateral margin involving the zygoma result in depression of the prominence of the cheek. Comminuted fractures of the lower margin also occur.

Blow-out fractures of the orbital floor involving the maxillary sinus commonly occur as the result of blunt force to the face. If the force is applied to the eye, the orbital fat explodes inferiorly into the maxillary sinus, fracturing the orbital floor. A severe blow to the maxilla may explode superiorly into the orbital cavity. Not only can blow-out fractures cause displacement of the eyeball, with resulting symptoms of diplopia, but the fracture may injure the infraorbital nerve, producing hyperesthesia of the skin of the cheek and ipsilateral gum. Entrapment of the inferior rectus muscle in the fracture may limit upward gaze. Air from the maxillary sinus may percolate into the eyelids and cause subcutaneous emphysema.

Penetrating Wounds

Sticks, pointed metal objects, and umbrellas may pierce the thin roof of the orbit and enter the cranial cavity and the frontal lobe of the brain. The thin medial wall can also be pierced by pointed objects.

Sinusitis and the Orbit

Infection of the air sinuses is a common cause of orbital cellulitis. The extremely thin medial orbital wall formed by the ethmoid bone is a common route by which infection enters the orbital cavity as the result of infection of the ethmoid sinus.

ANATOMICAL-CLINICAL PROBLEMS

1. A 30-year-old man was knocked off his bicycle by a passing car, and he landed in the gutter on the right side of his face. On examination by the emergency physician, it was found that the right side of his face was severely bruised; there were numerous abrasions. He was bleeding profusely from a vertical laceration located in the medial part of his right eyebrow, measuring about ½ in. long. Using your knowledge of anatomy, explain why the lacerated eyebrow bled excessively. Which nerves carry sensation from this area? Considering the direction of the laceration, would you have to take special precautions in placing the sutures to prevent gaping of the wound? What anatomical landmarks would assist you in obtaining a good cosmetic result? Name the layers of the eyebrow that require individual suturing.

2. An 18-year-old student went to the emergency department complaining of an acute tender area on the middle of his right upper eyelid. Examination revealed a localized red, indurated area on the eyelid margin. Close inspection showed a yellowish spot in the center of the swelling. Gentle eversion of the lid showed no evidence of swelling on its posterior surface. What is the diagnosis? Which anatomic structure(s) is (are) involved in the inflammatory process? On which part of the eyelid does the abscess tend to point?

3. An 83-year-old man was being examined in the emergency department for a Colles' fracture of his right wrist. The physician noticed a small ulcerated nodule on the outer part of his left upper eyelid. On questioning the patient, he said he had noticed the swelling about 6 months ago but because it did not hurt him he had not sought medical advice. On examination, the nodule was about 4 mm in diameter, and the edges of the ulcer were hard. The base of the ulcer was tethered to deeper structures. What is your diagnosis? If you suspect that the ulcer is malignant, which lymph nodes would you examine for evidence of metastases?

4. Two 16-year-old boys were exploring the attic in their grandmother's house when they came upon two Civil War swords. It was not long after the discovery that they started to engage in a fencing game. One of the boys lunged forward with his sword and nicked the lower eyelid of his opponent's right eye. The parents rushed the boy to the nearest hospital, and he was examined in the emergency department. A clean horizontal laceration about 1 in. long was found just below the margin of the medial half of the right lower lid. Which nerves carry sensation from this area of the eyelid? Name the layers of the lid that would require separate sutures. Name the muscles of the lower eyelid. Would the pull of these muscles tend to widen the scar in this patient?

5. A 15-year-old schoolboy was hit in the left eye by another boy during recess. During the next hour, both the eyelids of the victim swelled up until he could barely see. Examination in the emergency department revealed a bluish-red discoloration of both eyelids of his left eye with narrowing of the palpebral fissure. The discoloration extended to the forehead and the left cheek. Careful separation of the eyelids showed a localized hemorrhage of the inferolateral part of the bulbar conjunctiva. When the conjunctiva was gently moved with the tip of the examiner's little finger, the hemorrhage moved also. When the patient was asked to look medially, the physician could clearly see the posterior limit of the conjunctival hemorrhage. Does this patient have a simple "black eye," or is this a fracture of his anterior cranial fossa? What role does the orbital septum play in enabling one to distinguish between these lesions? Is the appearance of the conjunctival hemorrhage important in making the diagnosis?

6. A 13-year-old girl was waiting for the school bus when a sudden gust of wind carried some dust into her right eye and she experienced intense pain. Although she wiped her eye and repeatedly blew her nose, the discomfort persisted. On examination, she was found to have a small foreign body in her right conjunctival sac. Using your knowledge of anatomy, describe the different regions of the sac. What is the sensory nerve supply to the conjunctiva? Where do foreign bodies frequently lodge beneath the upper eyelid? In order to remove the foreign body from beneath the upper lid, the eyelid must be carefully everted. What structure is present within the upper lid that preserves its shape and form and assists the physician in keeping the lid everted?

7. A 46-year-old man was seen in the emergency department because he had developed double vision. He was watching his favorite television program when he first noticed double vision, which persisted and worsened throughout the day. He had no other symptoms. After a complete physical examination, it was found that his right eye, when at rest, was turned medially, and he was unable to turn it laterally. The pupillary reaction to light was normal.

What nerve is involved? What often causes this type of injury?

8. A 6-month-old girl was seen in the emergency department because her mother had noticed a yellowish, sticky discharge from the baby's left eye. On questioning, the mother said that she had first noticed the condition that morning when her daughter woke up. She also said that she had noticed that her daughter's left eye watered excessively when she cried ever since birth. The physician confirmed the epiphora of the left eye and noted the emergence of yellow pus into the lacus lacrimalis from the puncta when gentle pressure was exerted over the medial palpebral ligament. What is the diagnosis? What is the most likely cause in a child of this age? What are the posterior relations of the medial palpebral ligament? Describe the anatomy of the drainage passages of the conjunctival sac and give the direction and length of each of the tubes.

9. When a patient is asked to gaze vertically upward and downward during an ophthalmologic examination, it is noted that the eyelids move also to some extent with the movements of the eyeball. Can you explain this anatomically?

10. A group of children were playing with some garden sticks. Suddenly, one child screamed and ran to his mother with a stick projecting from his left eye. On examination in the emergency department, the physician found a sharp pointed stick had penetrated the left upper eyelid and was directed upward, backward, and medially. Using your knowledge of anatomy, name the structures that the stick might have penetrated, having entered the orbital cavity.

11. An inflammation of the ciliary body and iris may give rise to a so-called ciliary injection. Describe the arteries that give rise to the ciliary injection.

12. A 12-year-old girl was seen in the emergency department with a grossly swollen right eye. The eye was extremely painful, and the pain was made worse when she moved her eye. On examination, her left eye appeared normal. The upper and lower lids of her right eye were red and swollen with edema of the skin; the conjunctiva was markedly injected. Both eyelids were tender to touch, and there was evidence of slight proptosis. She had a slight pyrexia (99.8° F). Her parents stated that her general health had been good, but for the past few weeks she had a severe cold. Using your knowledge of anatomy, explain the proptosis. Is there any connection between her cold and her eye problem?

13. Describe the formation, circulation, and drainage of the aqueous humor. Where do areas of resistance to the drainage normally occur? What is the difference between open-angled and closed-angled glaucoma?

14. The optic nerve is surrounded by a sheath formed by the meninges of the brain. What part does the sheath play in the production of papilledema in patients with raised intracranial pressure?

15. A 47-year-old man was walking along a country lane when a sudden gust of wind blew some dirt into the air from a freshly plowed field. Suddenly the man experienced an acute stabbing pain in his right eye. Although he blinked his eye several times, the pain persisted. Because of the severity of the pain and the excessive tearing, the man immediately went to the emergency department of a local hospital. On examination of the cornea with a slit lamp, the physician identified a small abraded area where the corneal epithelium was missing. Does the cornea heal quickly? Describe the structure and function of the cornea. How does the cornea receive its nourishment?

16. Describe the structure of the iris. What is responsible for its color? What is the innervation of the sphincter and dilator pupillae muscles?

17. An 82-year-old woman suddenly experiences complete blindness of her left eye. The diagnosis is central retinal artery occlusion. Describe the blood supply of the neural retina.

18. A 54-year-old blind woman was admitted to the emergency department because she had sustained a severe fall in the street. During her physical examination, it was noted that both her pupils reacted normally to light even though she was totally blind in both eyes. Is it possible to have a normal direct and consensual light reflex in a totally blind person?

ANSWERS

1. All parts of the face, including the forehead, have a rich blood supply to nourish the muscles of facial expression. The eyebrow receives its arterial supply from the supratrochlear and supraorbital arteries, branches of the ophthalmic artery. The supratrochlear and supraorbital branches of the ophthalmic division of the trigeminal nerve are sensory to this area of the eyebrow. A vertical laceration of the eyebrow is in the line of pull of the frontalis and orbicularis oculi muscles; however, the pull of the corrugator supercilii muscle is at right angles to the

line of the laceration. It is the pull of the latter muscle that should concern the physician when suturing the wound. The hairs of the eyebrow serve as an important landmark for cosmetic excellence during wound closure. The hairs should therefore not be shaved off, and the margins of the hair-bearing area should be used to line up the edges of the laceration correctly and avoid discontinuity. The subcutaneous tissue, the muscle, and the skin should be sutured separately.

2. The student had a hordeolum, or stye, in his right eye. The usual cause is a staphylococcal infection of the eyelash follicle, the sebaceous gland of Zeis, or the ciliary gland of Moll. The suppurative infection tends to point on the anterior part of the lid margin. Repeated multiple styes tend to occur as the result of spread of infection along the eyelid margin.

3. The small chronic ulcer on the outer part of the left upper eyelid with an indurated edge and a fixed base was found on pathological examination to be a low-grade squamous cell carcinoma. The lymphatic drainage of the lateral part of both the upper and lower eyelids is into the superficial parotid lymph nodes. Clinically, there was no evidence of lymphatic metastases in this patient.

4. The infraorbital nerve is the sensory nerve supply to this area of the lower eyelid. Any wound involving the medial canthal region of the eyelids requires that the integrity of the lacrimal canaliculi be inspected. Should the canaliculus be severed it should be carefully repaired around a stent, usually by an ophthalmologist. Furthermore, the integrity of the medial palpebral ligament, if cut, must be repaired. The conjunctiva and the tarsal plate are sutured in one layer, the orbicularis oculi muscle in one layer, and the subcutaneous tissue and skin are sutured together as one layer.

The muscles of the lower lid include the orbicularis oculi (voluntary muscle) and the inferior tarsal (smooth muscle). The pull of the strong orbicularis muscle is in line with the laceration and therefore would not cause gaping of the wound. The inferior tarsal muscle is a weak muscle and would not affect the wound.

5. This schoolboy has a severe "black eye." In this patient the contusion involved not only the eyelids but the skin of the cheek and forehead. In anterior cranial fossa fractures, the hemorrhage occurs into the orbital cavity and is limited anteriorly by the attachment of the orbital septum to the orbital margin. In such cases the discoloration tends to be circular. In fractures of the anterior cranial fossa, because the bleeding is deeply placed, it tends to be purplish from the start, whereas with a black eye the color is initially red.

If the conjunctiva is traumatized in a so-called black eye, the hemorrhage tends to be localized and is into the conjunctiva and moves with the conjunctiva. In fractures of the anterior cranial fossa the hemorrhage extends forward under the conjunctiva; there is no posterior edge, and the blood does not move with the conjunctiva.

6. The conjunctival sac is arbitrarily divided into the palpebral conjunctiva (lining the eyelids), the superior and inferior conjunctival fornices (reflection of conjunctiva from eyelids onto the eyeball), and the bulbar conjunctiva (attached to eyeball).

The sensory nerve supply for the conjunctiva is derived superiorly from the supraorbital and supratrochlear nerves from the ophthalmic division of the trigeminal nerve, and is derived inferiorly from the infraorbital nerve a continuation of the maxillary division of the trigeminal nerve. The circumcorneal region of the bulba conjunctiva is supplied by the long ciliary nerves from the ophthalmic division of the trigeminal nerve.

Foreign bodies frequently move around the conjunctival sac, and if they do not reach the lacus lacrimalis, they often lodge in the subtarsal sulcus, which runs close to and parallel with the posterior edge of the upper lid margin.

The superior tarsal plate, formed of dense fibrous tissue and semilunar in shape, serves as a skeleton to the upper lid to maintain its shape. It is easily recognized through the conjunctiva of the everted upper lid.

7. The convergent strabismus of this patient's right eye, the diplopia, and the inability to turn the right eye laterally were due to paralysis of the right lateral rectus muscle, caused by a lesion of the abducent nerve. This lesion of the right abducent nerve palsy is commonly caused by diabetes mellitus.

8. This girl was suffering from chronic dacrocystitis secondary to congenital obstruction of the nasolacrimal duct. The obstruction results from failure of the nasolacrimal duct to open up and drain into the inferior meatus of the nose (see p. 250).

The posterior relation of the medial palpebral ligament is the lacrimal sac.

The drainage passages start at the puncta on the tip of the lacrimal papilla. The canaliculi first pass vertically in the eyelids for about 2 mm and then turn sharply at right angles and run medially for about 8 mm to enter the lacrimal sac. The lower end of the lacrimal sac is connected to the inferior meatus of the nose by the nasolacrimal duct, which is about 15 mm long. The duct passes downward, backward, and laterally.

9. The tendon of the superior rectus muscle is connected to the aponeurosis of the levator palpebrae superioris and the superior fornix of the conjunctiva by a fascial band. As a result, when the two muscles contract, they act together and raise the upper lid and the superior fornix of the conjunctiva when a person gazes upward. Similar connections are present in the lower lid so that the contraction of the inferior rectus causes the lower lid to move downward with the eyeball and the conjunctiva.

10. Posteroanterior and lateral radiographs of the child's head showed that the point of the stick had traversed the thin roof of the orbit formed by the orbital plate of the frontal bone. In its passage the point had pierced the frontal air sinus and the meninges, including the subarachnoid space, and was situated in the frontal lobe of the left cerebral hemisphere. (Usually a wooden stick is not radiopaque, but in this case a faint shadow could be visualized.)

11. Ciliary injection is a clinical condition in which the blood vessels of the deep pericorneal plexus become dilated. The plexus is fed by branches of the anterior ciliary arteries. In the normal eye the plexus can only just be seen. Inflammation of the ciliary body and iris can cause ciliary injection.

12. This patient had orbital cellulitis secondary to an acute infection of her ethmoidal air sinuses. The severe cold was complicated by the spread of the infection into the ethmoidal sinuses. The infection had then invaded the soft tissues of the orbital cavity. The edema and swelling of the extraocular muscles worsened the pain when the eye was moved. The edema of the retrobulbar tissues in the confined bony space had resulted in proptosis. The inflammatory reaction had extended forward through the orbital septum to produce redness and swelling of the eyelids.

13. A detailed description of the formation, circulation, and drainage of the aqueous humor is given on p. 254. The areas offering resistance to the circulation are (1) where the lens is in contact with the posterior surface of the iris, and (2) where the aqueous leaves the anterior chamber to enter the canal of Schlemm.

In open-angle glaucoma, the relationship between the iris root, the trabecular meshwork, and the cornea is normal. The block in the circulation is occurring in the drainage system (see p. 277).

In closed-angle glaucoma, the iris root is displaced forward and blocks the trabecular meshwork, thus obstructing the outflow (see p. 278).

14. The optic nerve is surrounded by a tubular sheath of the meninges of the brain that fuse with the sclera at the back of the eyeball. There is thus an extension of the subarachnoid space to the lamina cribrosa. The thin-walled central vein of the retina and the thick-walled central artery of the retina cross the space to enter the optic nerve. A rise in cerebrospinal fluid pressure will compress the vein, causing edema of the retina and engorgement of the veins. In addition, the axon flow of the optic nerve will be impeded by the external pressure on the optic nerve and will cause the optic disc to bulge into the eye. By this mechanism the signs of papilledema are established.

15. Abrasions of the cornea heal quickly unless complicated by infection. The structure of the cornea is fully described on p. 250. The cornea is the most refractive medium of the eye. The refractive power occurs on the anterior surface of the cornea. The cornea receives its nourishment by diffusion from the aqueous humor and from the capillaries at its edge. The center part of the cornea receives its oxygen directly from the air.

16. The structure of the iris is described on p. 252. The melanin pigment granules present in the melanocytes are responsible for its color.

The sphincter pupillae is innervated by parasympathetic postganglionic fibers from the ciliary ganglion. This pathway originates in the Edinger-Westphal nucleus of the oculomotor nerve. The dilator pupillae is innervated by postganglionic sympathetic fibers from the superior cervical sympathetic ganglion.

17. The blood supply of the neural retina is from the following two sources: (1) the outer laminae, including the rods and cones and the outer nuclear layer, supplied from the choroidal capillaries, and (2) the inner laminae, supplied by the central artery of the retina.

The retinal arteries are anatomic end arteries; there are no arteriovenous anastomoses. The integrity of the retina depends on both of the above circulations, neither of which alone is sufficient.

18. The nerve fibers that form the afferent pathway of the pupillary light reflex leave the optic tract

just before they reach the lateral geniculate body. They pass to the pretectal nucleus of the midbrain where they synapse. From there new nerve fibers travel to the parasympathetic nuclei of both oculomotor nerves that supply the constrictor pupillae muscles of both eyes. Consequently, it is possible for a patient who is totally blind, secondary to lesions of the lateral geniculate bodies, optic radiations, or calcarine cortices, to have normal direct and consensual light reflexes.

8

The Skull, the Meninges, and the Blood Supply of the Brain Relative to Trauma and Intracranial Hemorrhage

Head injuries from blunt trauma and penetrating missiles are associated with a high mortality and disabling morbidity. The cerebrovascular accident (stroke) still remains as the third leading cause of morbidity and death in the United States. Headaches are usually caused by nonserious conditions, but they can represent the earliest manifestations of a life-threatening process.

The purpose of this chapter is to review briefly the anatomy of the skull and its contents and to highlight those areas that are important to understanding the pathophysiology of head injuries, intracranial hemorrhage, and headaches.

BASIC ANATOMY

Basic Anatomy of the Skull

The bones forming the anterior wall, lateral wall, and base of the skull are shown in Figures 8–1 through 8–3.

Tables of the Skull

The skull bones are made up of *external* and *internal tables* of compact bone, separated by a layer of spongy bone called the *diploë*. The bones are covered on the outer and inner surfaces with periosteum.

Sutures of the Skull

The bones of the skull are united at immobile joints called *sutures*. The *coronal suture* lies between the frontal bones and the parietal bones, the *lambdoid suture* lies between the parietal bones and the occipital bone, and the *sagittal suture* lies between the two parietal bones (see Figs 8–1, and 8–2).

Fontanelles

At birth areas of membrane still remain between the bones; these soft areas are known as fontanelles. The *anterior fontanelle* is diamond shaped and lies between the two halves of the frontal bone and the two parietal bones (see Fig 6–14). The anterior fontanelle is closed by 18 months of age. The *posterior fontanelle* is triangular and lies between the two parietal bones and the occipital bone. The posterior fontanelle is usually closed by the end of the first year of life.

Base of the Skull

The interior of the base of the skull is conveniently divided up into three cranial fossae—anterior, mid-

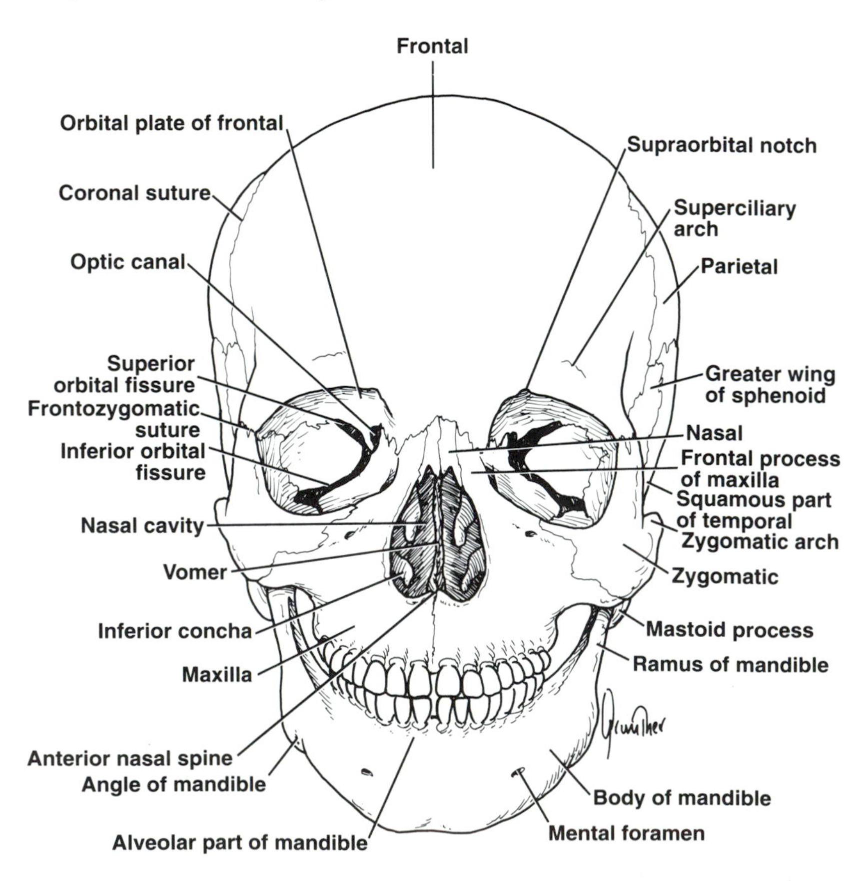

FIG 8–1.
Bones of the anterior aspect of the skull.

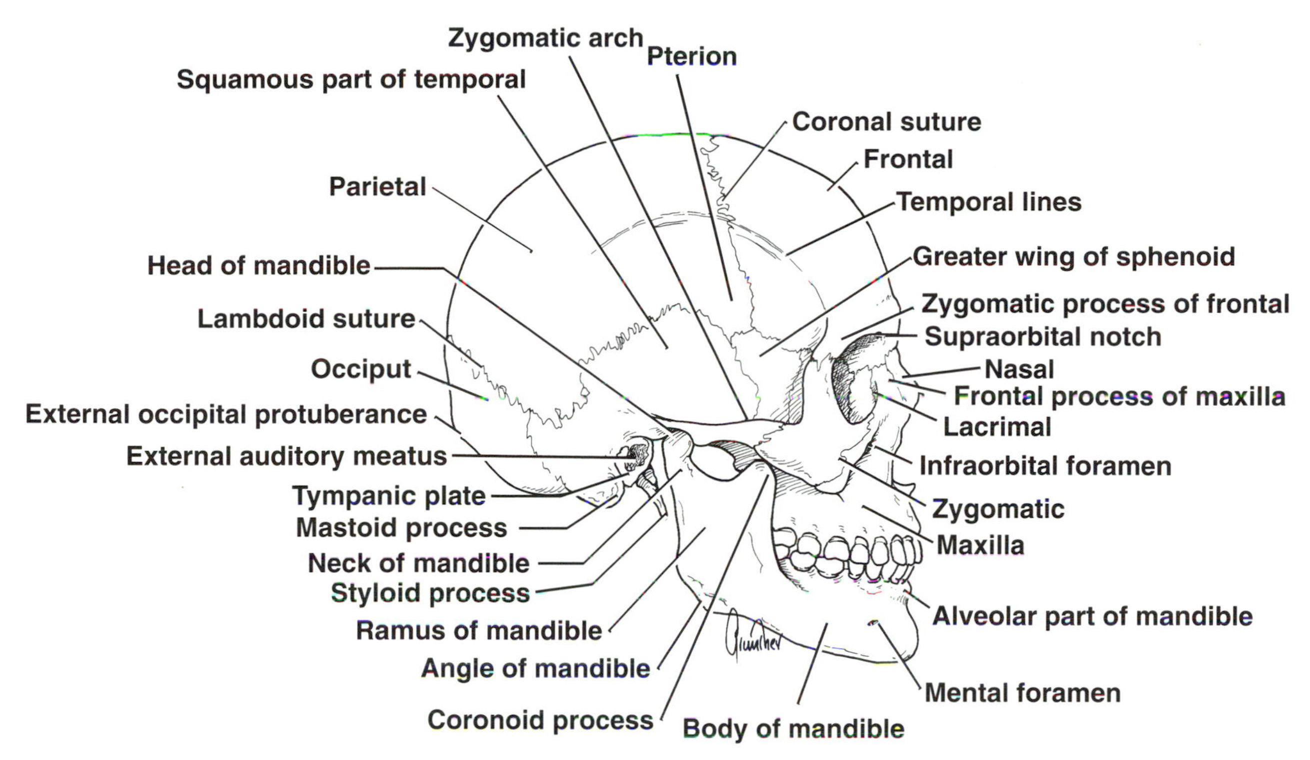

FIG 8–2.
Bones of the lateral aspect of the skull.

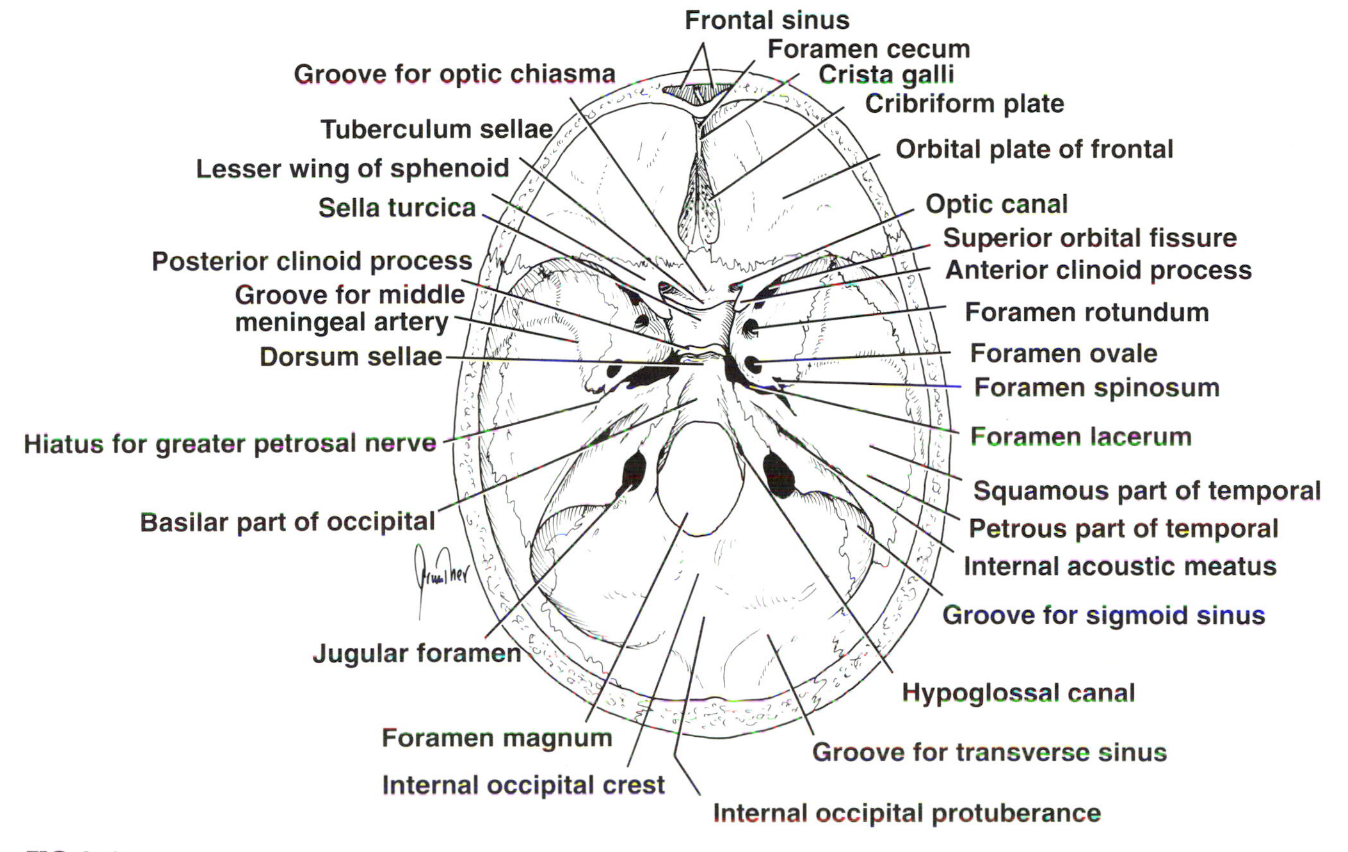

FIG 8–3.
The internal surface of the base of the skull.

Clinical Note

The thinnest part of the lateral wall of the skull is where the anteroinferior corner of the parietal bone articulates with the greater wing of the sphenoid; this point is known as the *pterion.* This is a very important area since it overlies the anterior division of the *middle meningeal artery and vein* (see epidural hemorrhage, p. 310).

---

dle, and posterior (see Fig 8–3). The anterior cranial fossa is separated from the middle cranial fossa by the lesser wing of the sphenoid, and the middle cranial fossa is separated from the posterior cranial fossa by the petrous portion of the temporal bone.

The *anterior cranial fossa* lodges the frontal lobes of the cerebral hemispheres, the lateral parts of the *middle cranial fossa* lodge the temporal lobes of the cerebral hemispheres, and the very deep *posterior cranial fossa* lodges parts of the cerebellum, the pons, and the medulla oblongata.

The *sphenoid bone* occupies the central position in the cranial floor. It has a centrally placed *body* with *greater* and *lesser wings* that are outstretched on each side. The sphenoid bone stabilizes the center of the skull by being attached by sutures to the frontal, parietal, occipital, and ethmoid bones. The body of the sphenoid contains the *sphenoid air sinuses.*

The following important foramina can be identified in Figure 8–3.

In the anterior cranial fossa, the *perforations of the cribriform plate of the ethmoid* can be seen; these transmit the olfactory nerves. In the middle cranial fossa the *optic canal* is present in the lesser wing of the sphenoid; this transmits the optic nerve and the ophthalmic artery. The slitlike *superior orbital fissure* that exists between the lesser and greater wings of the sphenoid transmits the oculomotor, trochlear, branches of the ophthalmic division of the trigeminal, and the abducent nerves. The *foramen rotundum* in the greater wing of the sphenoid transmits the maxillary division of the trigeminal nerve. The *foramen ovale* perforates the greater wing and transmits the mandibular division of the trigeminal nerve. The small *foramen spinosum,* which is also in the greater wing, transmits the middle meningeal artery. The large irregular *foramen lacerum* lies between the greater wing of the sphenoid and the petrous part of the temporal bone and allows the passage of the internal carotid artery from the carotid canal into the cranial cavity.

In the posterior cranial fossa the large *foramen magnum* in the occipital bone transmits the medulla oblongata. Here the medulla becomes continuous with the spinal cord. The foramen also allows the passage of the spinal roots of the accessory nerves and the two vertebral arteries.

The *hypoglossal canal* transmits the hypoglossal nerve, and the *jugular foramen* transmits the glossopharyngeal, vagus, and accessory nerves. Here the sigmoid venous sinus leaves the skull to become the internal jugular vein.

The *internal acoustic meatus* pierces the posterior surface of the petrous part of the temporal bone and transmits the vestibulocochlear nerve and the facial nerve.

Radiographic Appearance of the Skull

The *straight posteroanterior view* of the skull (Fig 8–4) is taken with the forehead and nose against the film cassette and the X-ray tube positioned behind the head, perpendicular to the film and in line with the external auditory meatus and the palpebral fissure. Unfortunately, in this position the petrous parts of the temporal bones are superimposed on the lower halves of the orbits. The different parts of the skull that are visible are shown in Figure 8–4. Note that the sagittal, coronal, and lambdoid sutures may be seen. The frontal sinuses, the upper and lower margins of the orbit, the nasal septum and the conchae, the maxillary sinuses, and the maxillary teeth can be identified. The rami and body of the mandible are easily recognized. The sphenoidal and ethmoidal air sinuses produce a composite shadow.

The *lateral view of the skull* (Figs 8–5 and 8–6) is taken with the sagittal plane of the skull parallel with the film cassette. The X-ray tube is centered over the region of the sella turcica. The different parts of the vault and base of the skull are well shown in Figure 8–5. The zygomatic and maxillary bones are superimposed on each other and are not clear. The coronal, squamosal (between the squamous part of the temporal bone and the parietal bone), and lambdoid sutures can be recognized. The inner and outer tables of the skull bones and the intervening diploë can be seen. Depressions on the inner table are commonly seen in children and are produced by the underlying cerebral convolutions.

The grooves produced by the anterior and posterior branches of the middle meningeal vessels may be seen running posteriorly across the parietal bones. A wide groove for the transverse sinus may also be identified as it crosses the occipital bone. Diploic vessels may be recognized as branching dark lines.

The pineal gland, if calcified, can be seen as a

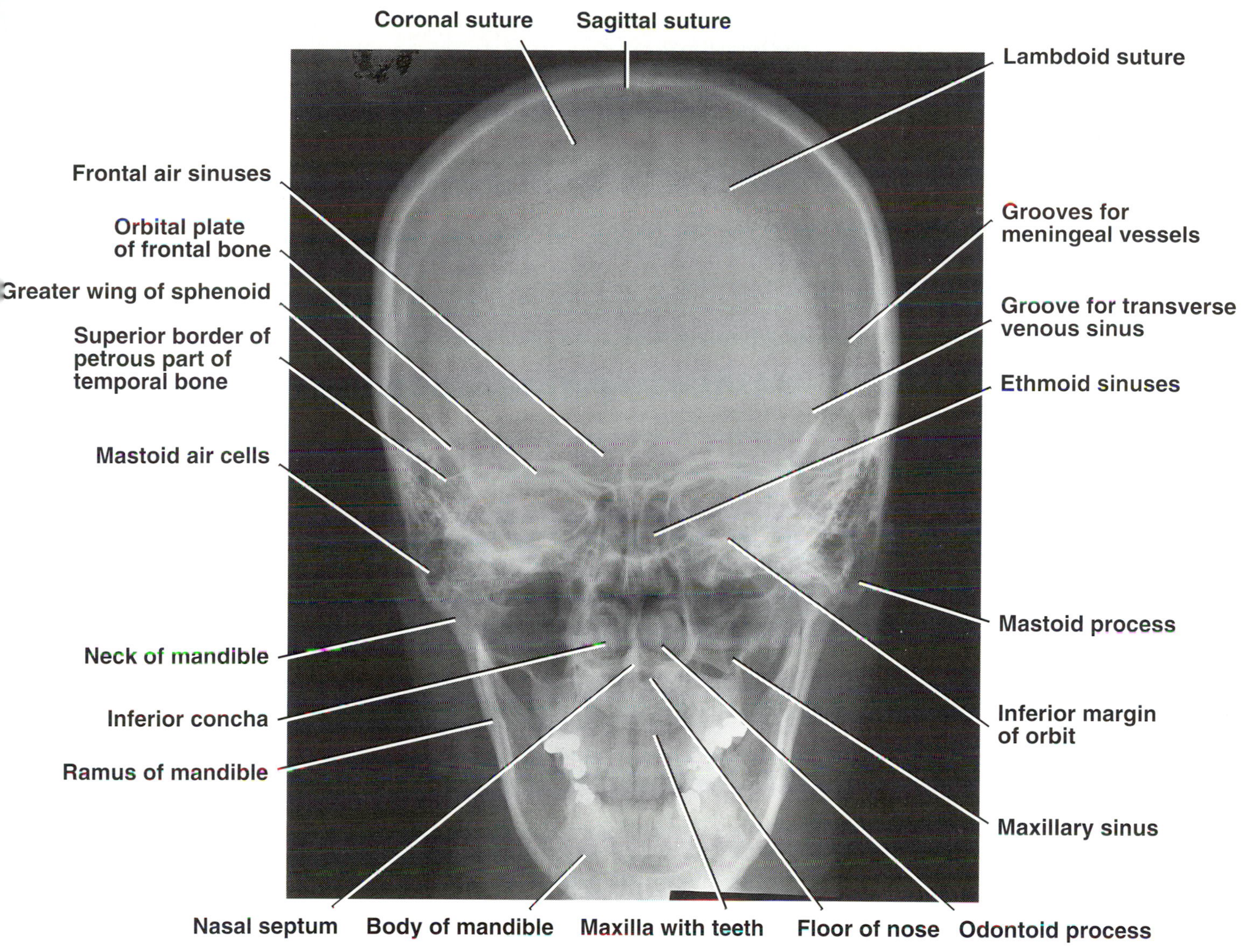

FIG 8–4.
Posteroanterior radiograph of the skull.

small shadow above and behind the external auditory meatus.

Anteriorly, the frontal sinuses are clearly shown superimposed on one another. Behind them are the two orbital plates of the frontal bones that form the roofs of the orbits. Behind these are the lesser wings of the sphenoid, the anterior clinoid processes, and the sella turcica. The curved lines of the greater wings of the sphenoid and the sphenoid air sinuses should also be recognized.

Behind the sella turcica, the dorsum sellae and the posterior clinoid processes are clearly seen (see Fig 8–5). The two petrous parts of the temporal bones are superimposed and form a dense shadow between the middle and posterior cranial fossae. Translucent areas formed by the external auditoy meatus and, behind them, the mastoid air cells can be identified. The auricle of the external ear frequently produces a curved shadow above the petrous parts of the temporal bones. The temporomandibular joint can be recognized in front of the external auditory meatus.

The nasal bones, the cribriform plate, the hard palate, the maxillary air sinus, and the teeth of the upper and lower jaws can all be seen. The ramus and body of the mandible, the hyoid bone, and the upper part of the cervical vertebral column can be identified.

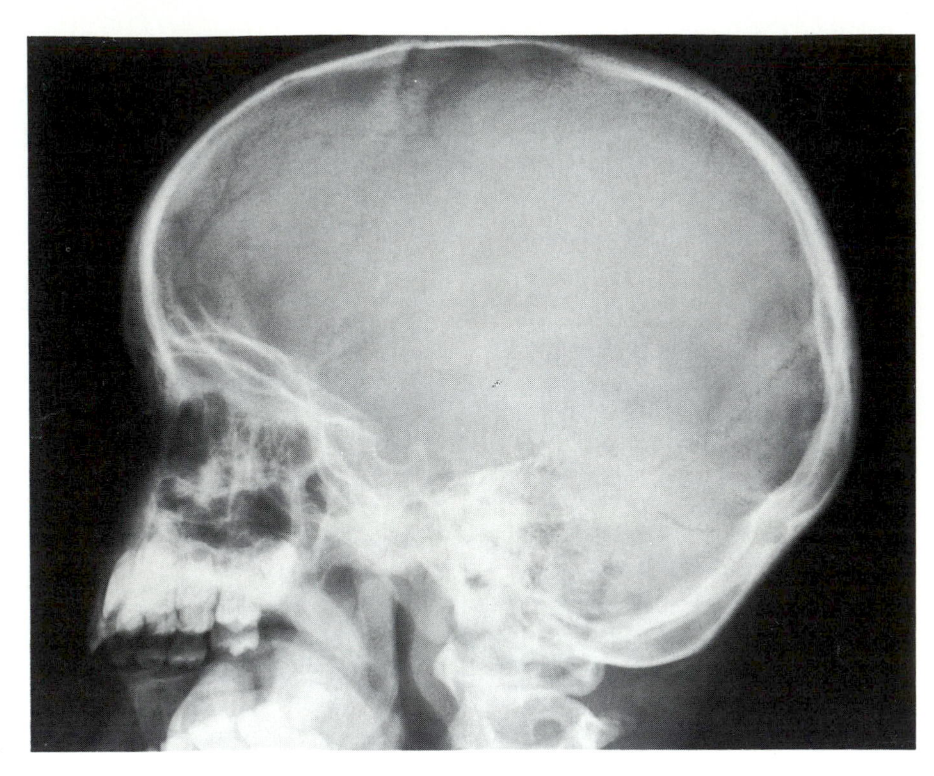

FIG 8–5.
Lateral radiograph of the skull.

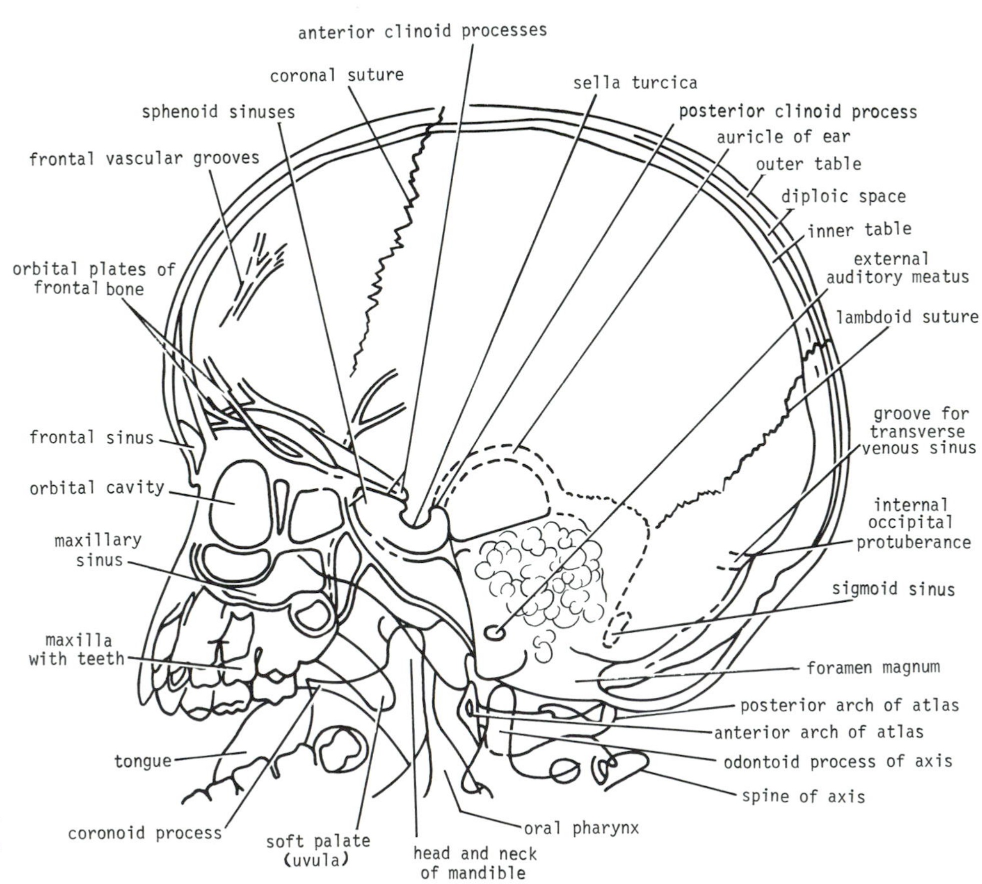

FIG 8–6.
Diagram showing the main features seen in the radiograph in Figure 8–5.

Computed Tomography and Magnetic Resonance Imaging of the Skull

Examples of computed tomographic scans and magnetic resonance images of the head are shown in Figures 8–7 through 8–9.

Basic Anatomy of the Meninges of the Brain

The brain, like the spinal cord, is surrounded by three membranes—the dura mater, the arachnoid mater, and the pia mater.

Dura Mater

The dura mater of the brain is formed of two layers, the endosteal layer and the meningeal layer. These are closely united except along certain lines, where they separate to form *venous sinuses.*

Endosteal Layer.—The endosteal layer is nothing more than the ordinary periosteum covering the inner surface of the skull bones. It *does not extend* through the foramen magnum to become continuous with the dura mater of the spinal cord. Around the margins of all the foramina in the skull it becomes continuous with the periosteum on the outside of the skull bones. At the sutures it is continuous with the sutural ligaments. It is most strongly adherent to the bones over the base of the skull.

Meningeal Layer.—The meningeal layer is the dura mater proper. A dense, strong, fibrous membrane covering the brain, it is continuous through the foramen magnum with the dura mater of the spinal cord. The meningeal layer sends four septa inward—the falx cerebri, the tentorium cerebelli, the falx cerebelli, and the diaphragma sellae—which divide the cranial cavity into the freely communicating spaces that lodge the subdivisions of the brain. The function of the septa is to restrict rotatory displacement of the brain.

Falx Cerebri.—The falx cerebri is a sickle-shaped fold of dura mater that lies in the midline between the two cerebral hemispheres (Fig 8–10). Its narrow end in front is attached to the internal frontal crest and the crista galli projecting upward from the ethmoid bone. Its broad posterior part blends in the midline with the upper surface of the tentorium cerebelli. The superior sagittal sinus passes forward in its upper fixed margin, the inferior sagittal sinus is located in its lower concave free margin, and the straight sinus runs forward along its attachment to the tentorium cerebelli.

Tentorium Cerebelli.—The tentorium cerebelli is a crescent-shaped fold of dura mater that roofs over the posterior cranial fossa (see Fig 8–10). It covers the upper surface of the cerebellum and supports the occipital lobes of the cerebral hemispheres. In front there is a gap, the *tentorial notch,* for the passage of the midbrain (Fig 8–11). The outer border is attached to the posterior clinoid processes, the superior borders of the petrous bones, and the margins of the grooves for the transverse sinuses on the occipital bone. The free inner border runs forward at its two ends, crosses the attached outer border, and is affixed to the anterior clinoid process on each side. At the point where the two borders cross, the oculomotor and trochlear nerves pass forward to enter the lateral wall of the cavernous sinus (see Fig 8–11).

Close to the apex of the petrous part of the temporal bone, the lower layer of the tentorium is pouched forward beneath the superior petrosal sinus to form a recess for the trigeminal nerve and the trigeminal ganglion (see Fig 8–11).

The falx cerebri and the falx cerebelli are attached to the upper and lower surfaces of the tentorium, respectively. The straight sinus runs along its attachment to the falx cerebri, the superior petrosal sinus along its attachment to the petrous bone, and the transverse sinus along its attachment to the occipital bone (see Fig 8–10).

Falx Cerebelli.—The falx cerebelli is a small, sickle-shaped fold of dura mater that is attached to the internal occipital crest and projects forward between the two cerebellar hemispheres. Its posterior fixed margin contains the occipital sinus.

Diaphragma Sellae.—The diaphragma sellae is a small circular fold of dura mater that forms the roof for the sella turcica (see Fig 8–10). A small opening in its center allows passage of the stalk of the hypophysis cerebri.

Dural Nerve Supply.—Branches of the trigeminal, vagus, and first three cervical nerves and branches from the sympathetic system pass to the dura.

The dura possesses numerous sensory endings that are sensitive to stretching, which produces the sensation of headache. Stimulation of the sensory endings of the trigeminal nerve above the level of the tentorium cerebelli produces referred pain to an area of skin on the same side of the head. Stimulation of the dural endings below the level of the tentorium produces pain referred to the back of the neck and the back of the scalp along the distribution of the greater occipital nerve.

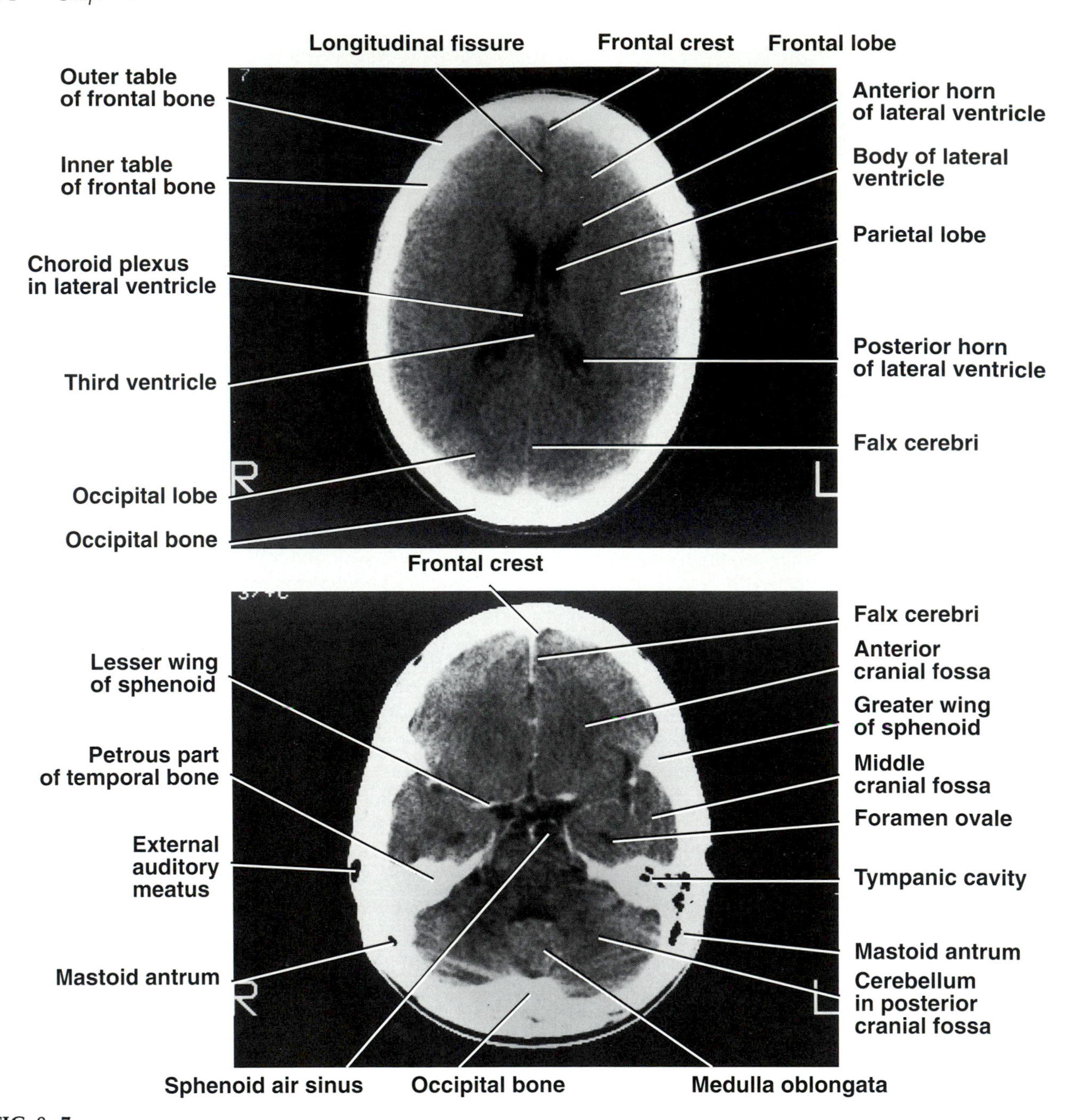

FIG 8–7.
Axial (horizontal) computer tomographic scans of the skull. **A,** the skull bones and the brain and the different parts of the lateral ventricles. **B,** a cut made at a lower level showing the three cranial fossae.

Dural Arterial Supply.—Numerous arteries supply the dura mater from the internal carotid, maxillary, ascending pharyngeal, occipital, and vertebral arteries. From the clinical standpoint, the most important of these arteries is the middle meningeal artery, which can be damaged in head injuries.

Middle Meningeal Artery.—The middle meningeal artery arises from the maxillary artery (which is a terminal branch of the external carotid artery) in the infratemporal fossa and enters the skull through the foramen spinosum. It runs forward and laterally in a groove on the upper surface of the squamous part of the temporal bone. It *lies between the meningeal and endosteal layers* of dura (Fig 8–12). The anterior branch deeply grooves or tunnels the anteroinferior angle of the parietal bone, and its course corresponds roughly to the line of the underlying precentral gyrus of the brain. The poste-

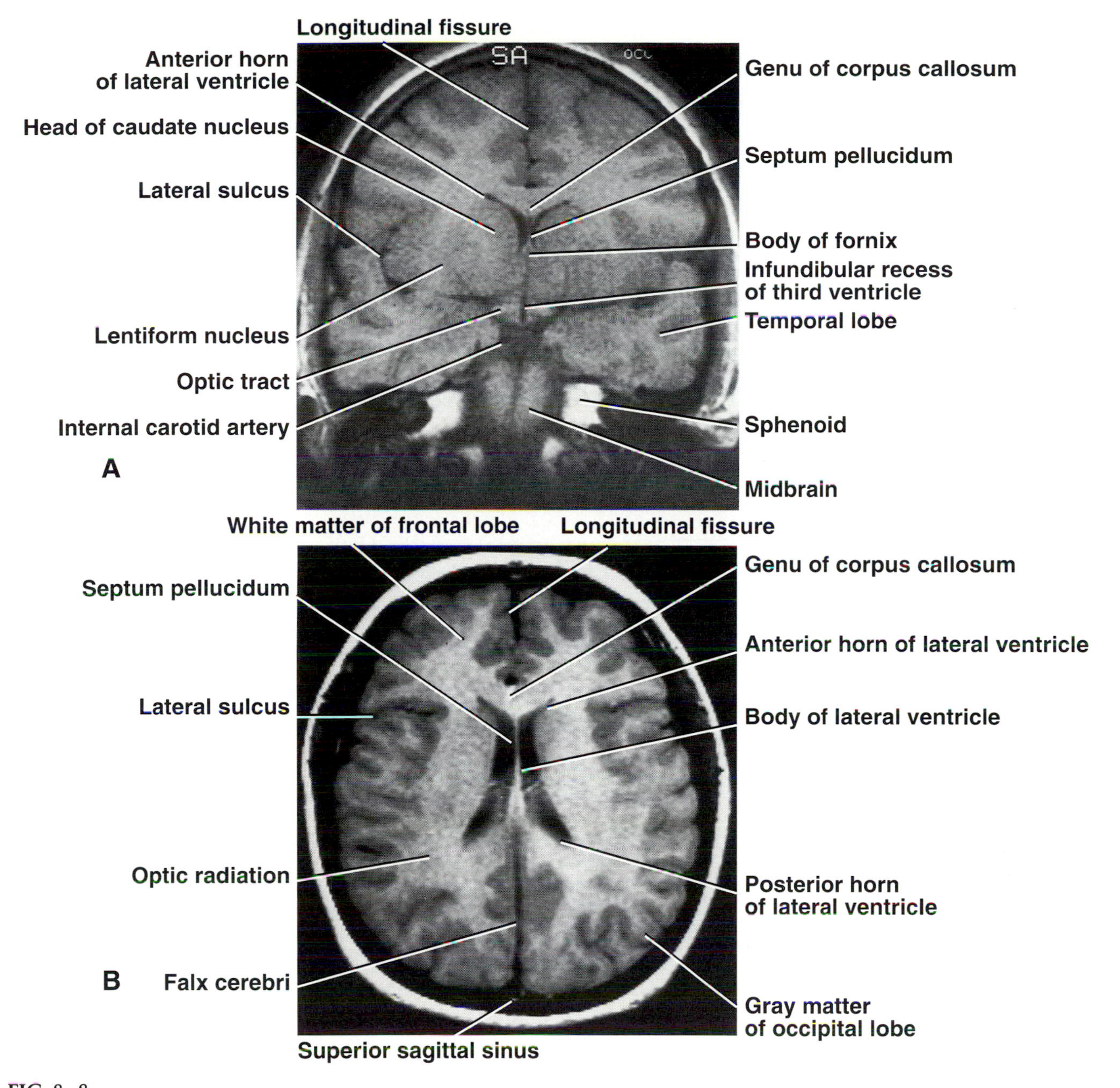

FIG 8–8.
Magnetic resonance images of the skull. **A,** coronal image through the frontal lobe of the brain showing the anterior horn of the lateral ventricle. **B,** axial image of the brain showing the different parts of the lateral ventricle and the lateral sulcus. Note the improved contrast between the gray and white matter as compared with the computed tomographic scans seen in Figure 8–7.

rior branch curves backward and supplies the posterior part of the dura.

Meningeal Veins.—These lie in the endosteal layer of dura. The middle meningeal vein follows the branches of the middle meningeal artery and drains into the pterygoid venous plexus or the sphenoparietal sinus. The veins lie lateral to the arteries.

Arachnoid Mater

The arachnoid mater is a delicate, impermeable membrane covering the brain and lying between the pia mater internally and the dura mater externally (Fig 8–13). It is separated from the dura by a potential space, the *subdural space,* and from the pia by the *subarachnoid space,* which is filled with *cerebrospinal fluid.*

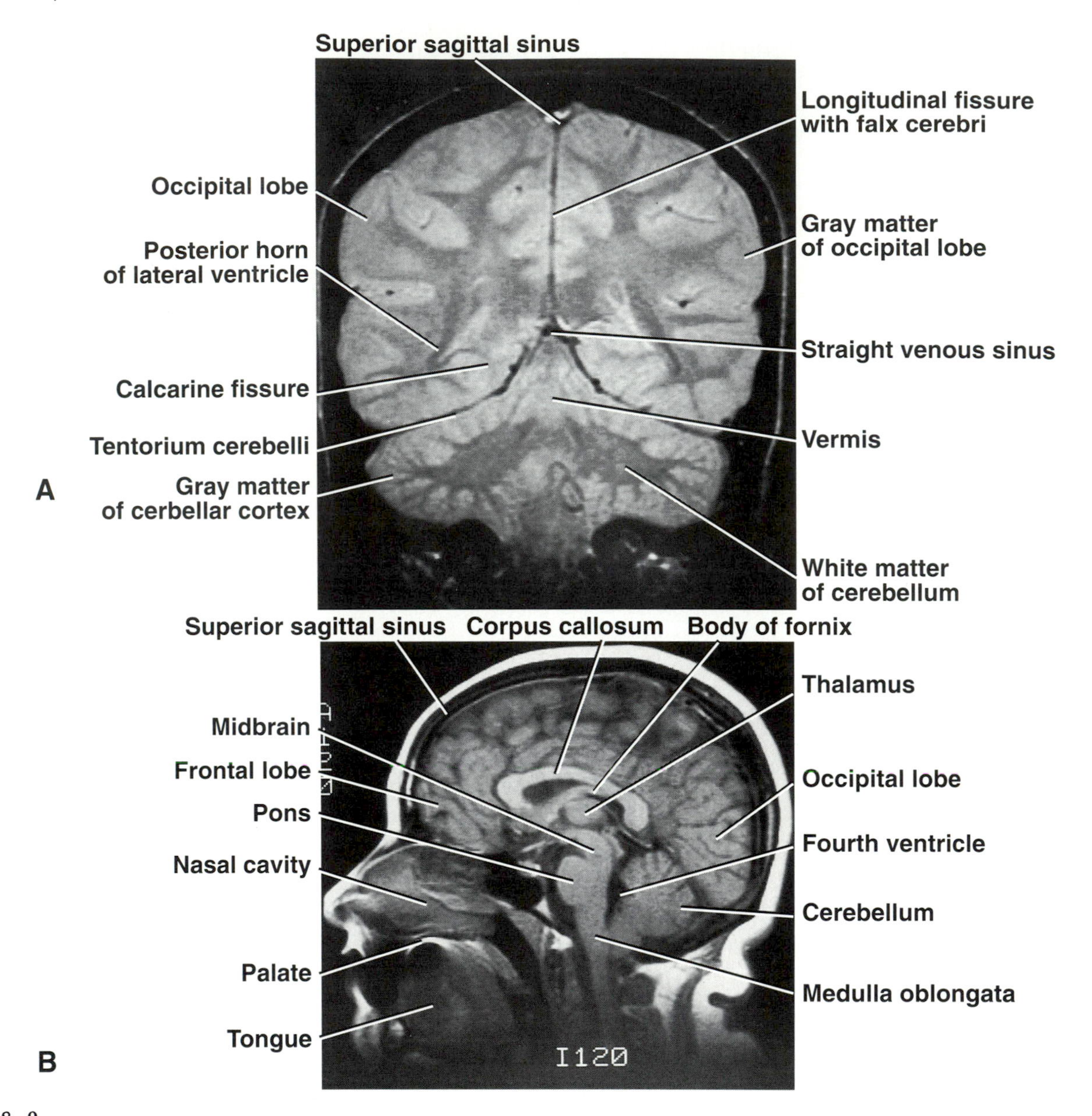

FIG 8–9.
Magnetic resonance images of the skull. **A,** coronal image through the occipital lobes of the brain showing the posterior horn of the lateral ventricle and the cerebellum. **B,** sagittal image showing the different parts of the brain and the nasal and mouth cavities.

The arachnoid bridges over the sulci on the surface of the brain, and in certain situations the arachnoid and pia are widely separated to form spaces called *subarachnoid cisternae.* In certain areas the arachnoid projects into the venous sinuses to form *arachnoid villi.* The arachnoid villi are most numerous along the superior sagittal sinus. Aggregations of arachnoid villi are referred to as *arachnoid granulations* (see Fig 8–13). Arachnoid villi serve as sites where the cerebrospinal fluid diffuses into the bloodstream.

Subarachnoid Space.—This space lies between the arachnoid and pia and is filled with cerebrospinal fluid. Structures passing to and from the brain to the skull or its foramina must pass through the subarachnoid space. All the cerebral arteries and veins lie in this space, as do the cranial nerves. For the optic nerve, the arachnoid forms a sheath that extends into the orbital cavity through the optic canal and fuses with the sclera of the eyeball (see Fig 7–7).

Inferiorly, the subarachnoid space around the brain is continuous through the foramen magnum

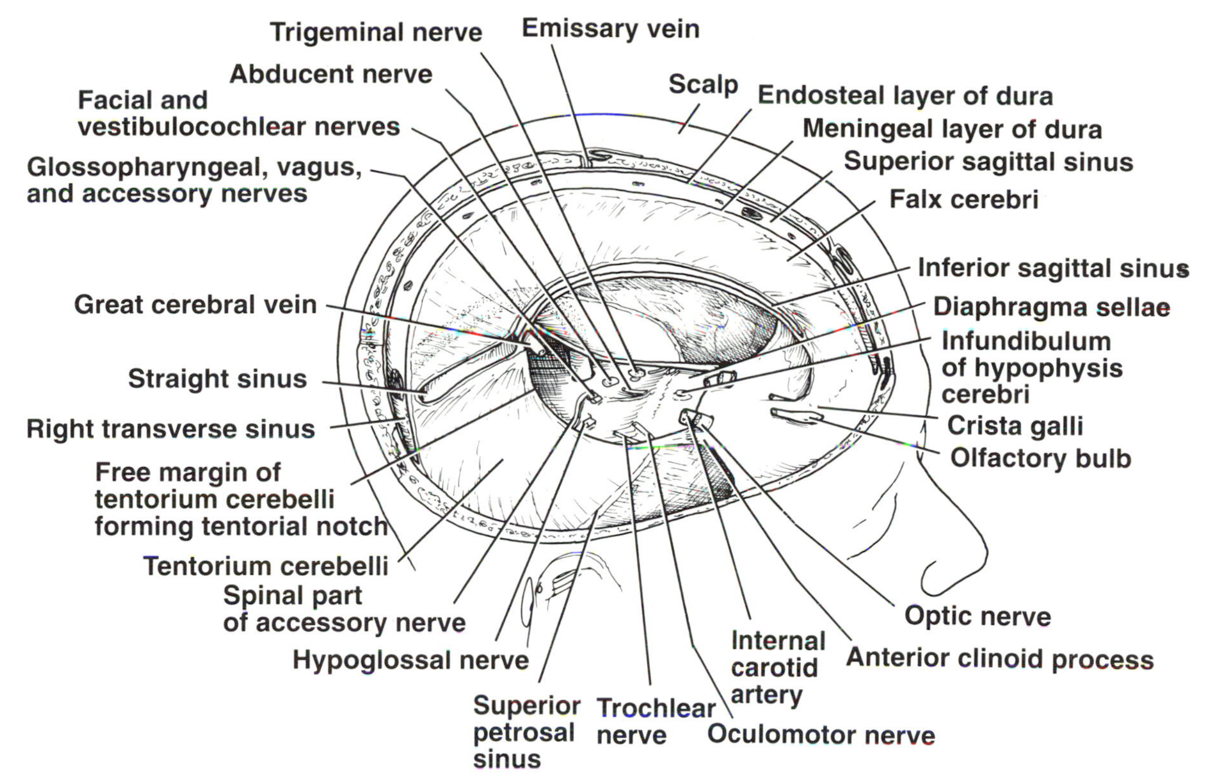

FIG 8–10.
Interior of skull showing the dura mater and its contained venous sinuses.

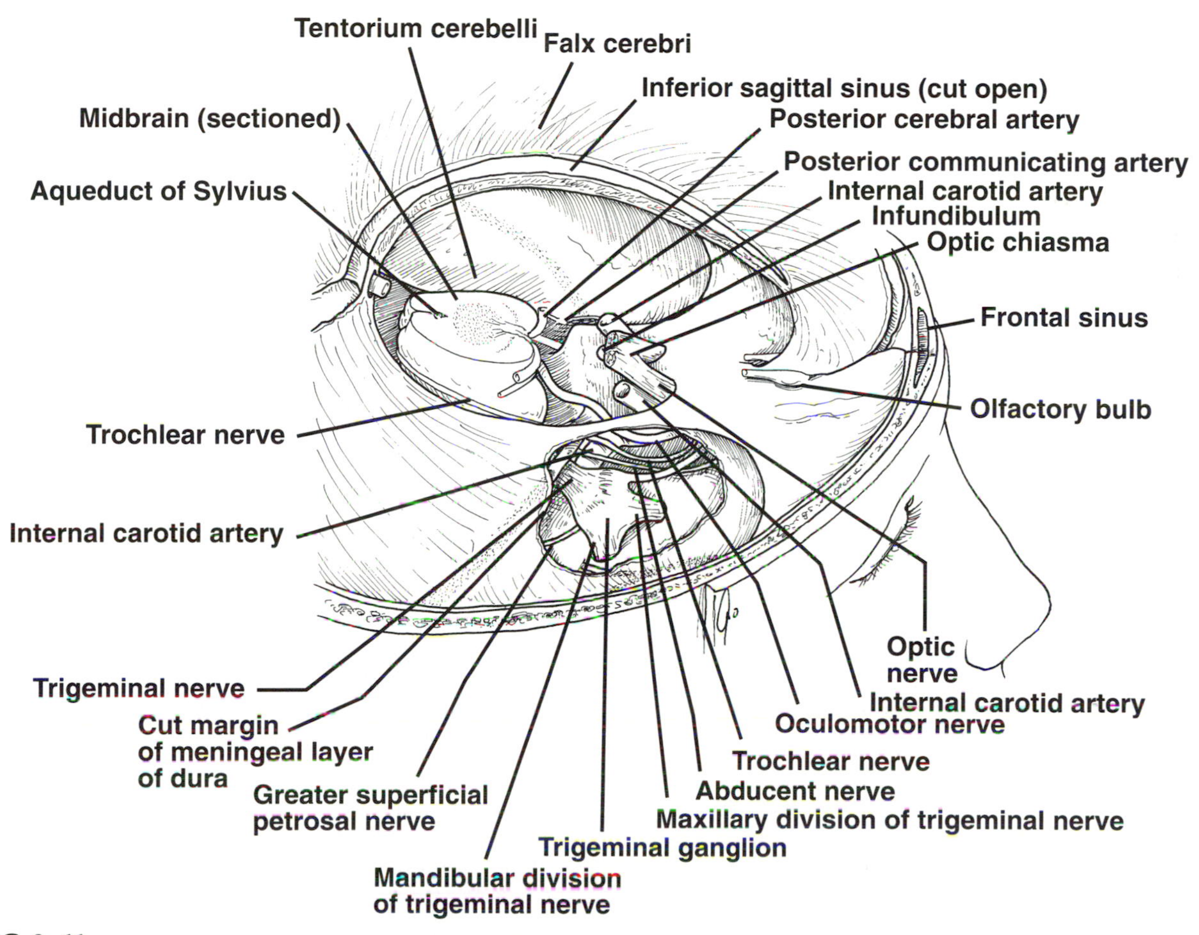

FIG 8–11.
Lateral view of interior of skull showing the falx cerebri, tentorium cerebelli, brain stem, and trigeminal ganglion.

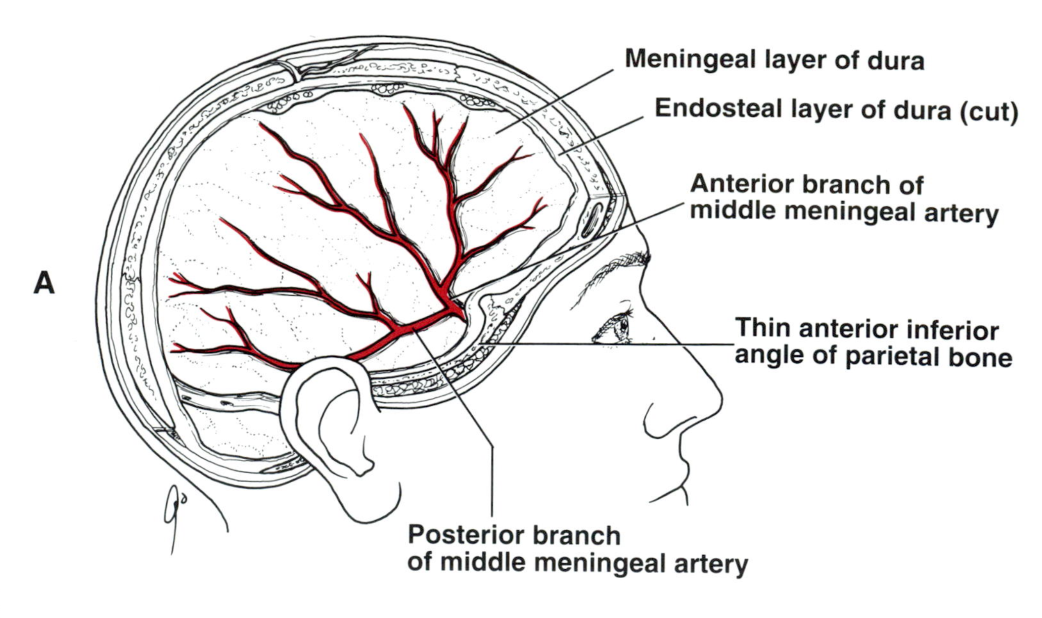

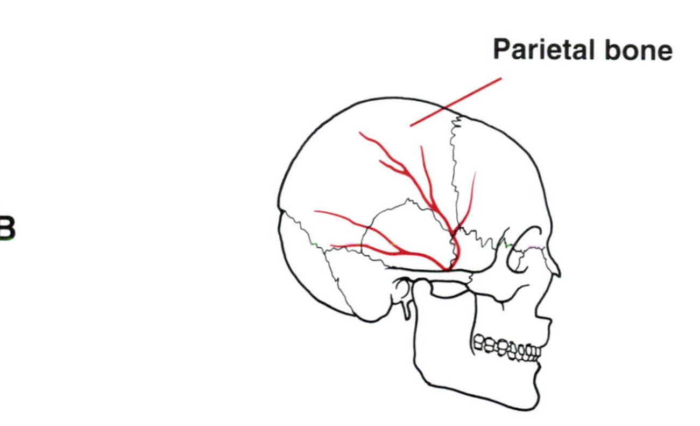

FIG 8–12.
Right side of the head. **A,** the skull bones and periosteal (endosteal) layer of dura have been removed to reveal the middle meningeal artery, which lies between the endosteal and meningeal layers of dura. **B,** the course of the middle meningeal artery in relation to the skull bones. The anterior branch is related to the anterior inferior angle of the parietal bone and its course corresponds to the underlying precentral gyrus (motor area) of the brain.

with the subarachnoid space around the spinal cord (see p. 305). Below, it extends beyond the lower end of the spinal cord and invests the cauda equina (see Fig 8–13). The subarachnoid space ends inferiorly at the level of the interval between the second and third sacral vertebrae.

Pia Mater

The pia mater is a vascular membrane that closely invests the brain, covering the gyri and descending into the deepest sulci (see Fig 8–13). It extends over the cranial nerves and fuses with their epineurium. The cerebral arteries entering the substance of the brain carry a sheath of pia with them.

Venous Sinuses

The venous sinuses of the cranial cavity are blood-filled spaces located between the layers of the dura mater; they are lined with endothelium. Their walls are thick and composed of fibrous tissue; there is no muscular tissue. The sinuses have no valves. They receive the tributaries from the brain, from the diploë of the skull, from the orbit, and from the internal ear. The position and arrangement of the sinuses are shown in Figure 8–14.

Cavernous Sinuses.—Each sinus is situated on the side of the body of the sphenoid bone. The internal carotid artery, surrounded by its sympathetic nerve plexus, runs forward through the cavernous sinus (Fig 8–15). The abducent nerve also passes through the sinus. The artery and the nerves are separated from the blood by an epithelial covering.

The oculomotor, trochlear, and the ophthalmic and maxillary divisions of the trigeminal nerve run forward in the lateral wall of the cavernous sinus embedded in the meningeal layer of dura (see Fig 8–15).

The superior and inferior ophthalmic veins, the cerebral veins, the sphenoparietal sinus, and the central vein of the retina drain into the cavernous sinus. The sinus, in turn, drains into the petrosal sinuses and the pterygoid venous plexus.

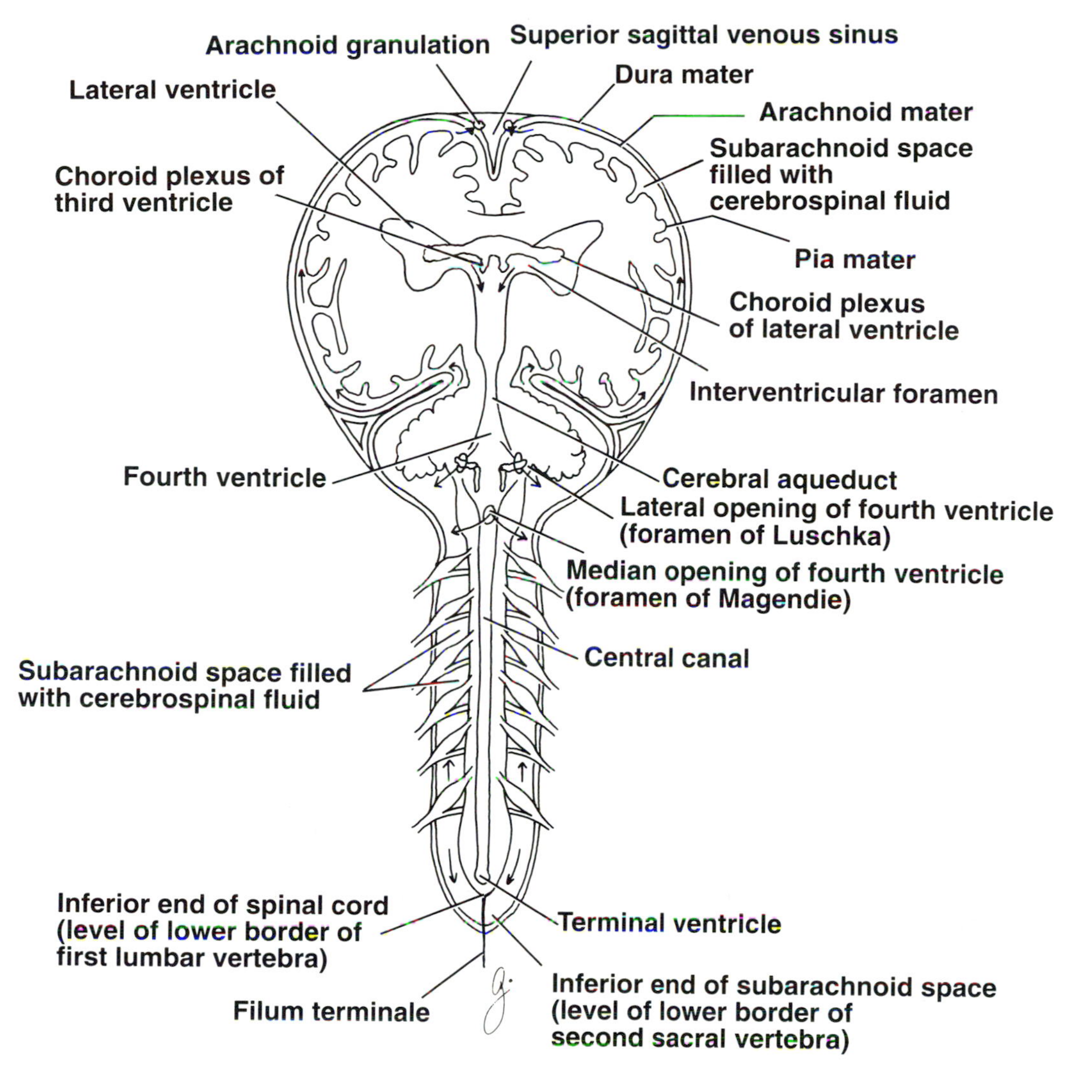

FIG 8–13.
Diagram showing the origin and circulation of the cerebrospinal fluid.

Clinical Note

Infections of the face and cavernous sinus thrombosis.—This has been discussed in detail on p. 210.

Basic Anatomy of the Blood Supply of the Brain

Arteries of the Brain

The brain is supplied by the two internal carotid and the two vertebral arteries (Fig 8–16). The four arteries lie within the subarachnoid space, and their branches anastomose on the inferior surface of the brain to form the *circulus arteriosus* (circle of Willis).

Internal Carotid Artery.—The internal carotid artery perforates the base of the skull by passing through the carotid canal in the temporal bone. The artery then runs horizontally forward through the cavernous sinus and emerges on the medial side of the anterior clinoid process. It now enters the subarachnoid space and turns posteriorly at the medial end of the lateral cerebral sulcus, where it divides into the anterior and middle cerebral arteries (see Fig 8–16).

Branches.—The branches of the internal carotid artery include the following.

1. The *ophthalmic artery* supplies the eye and other orbital structures.
2. The *posterior communicating artery,* a small branch, originates from the internal carotid artery close to its terminal bifurcation (see Fig 8–16). It runs posteriorly above the oculomotor nerve to join

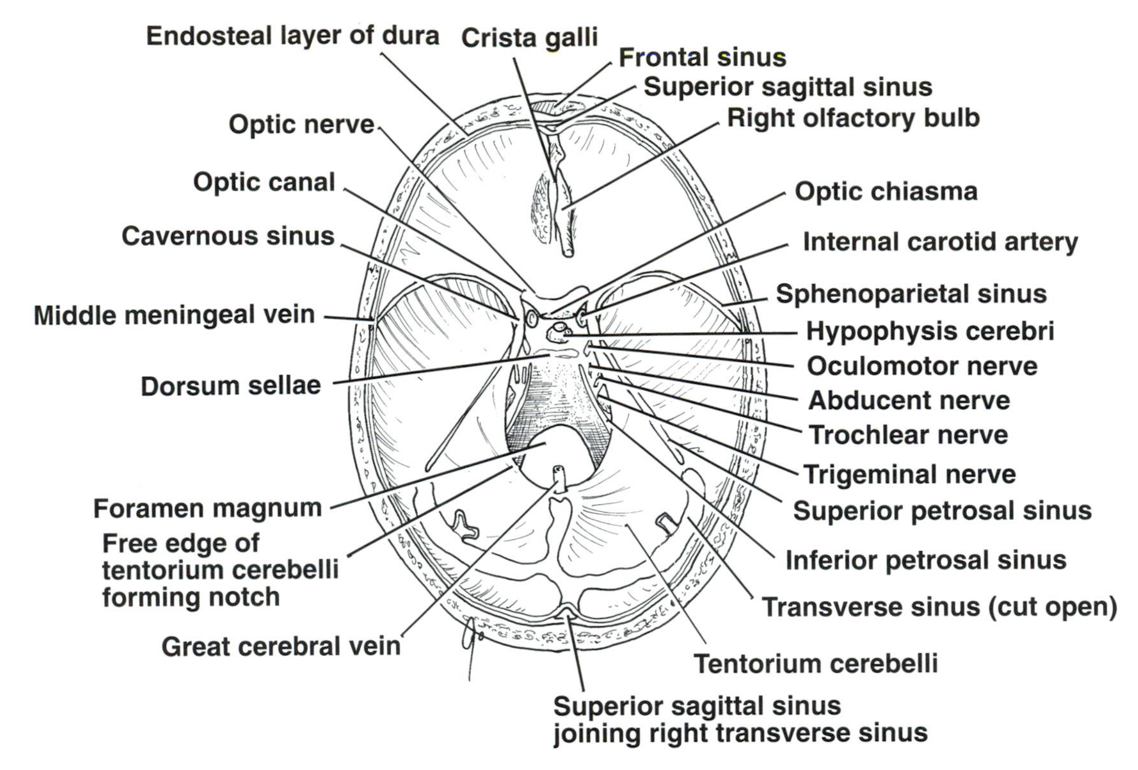

FIG 8–14.
The diaphragma sellae and the tentorium cerebelli. Note the position of the venous sinuses.

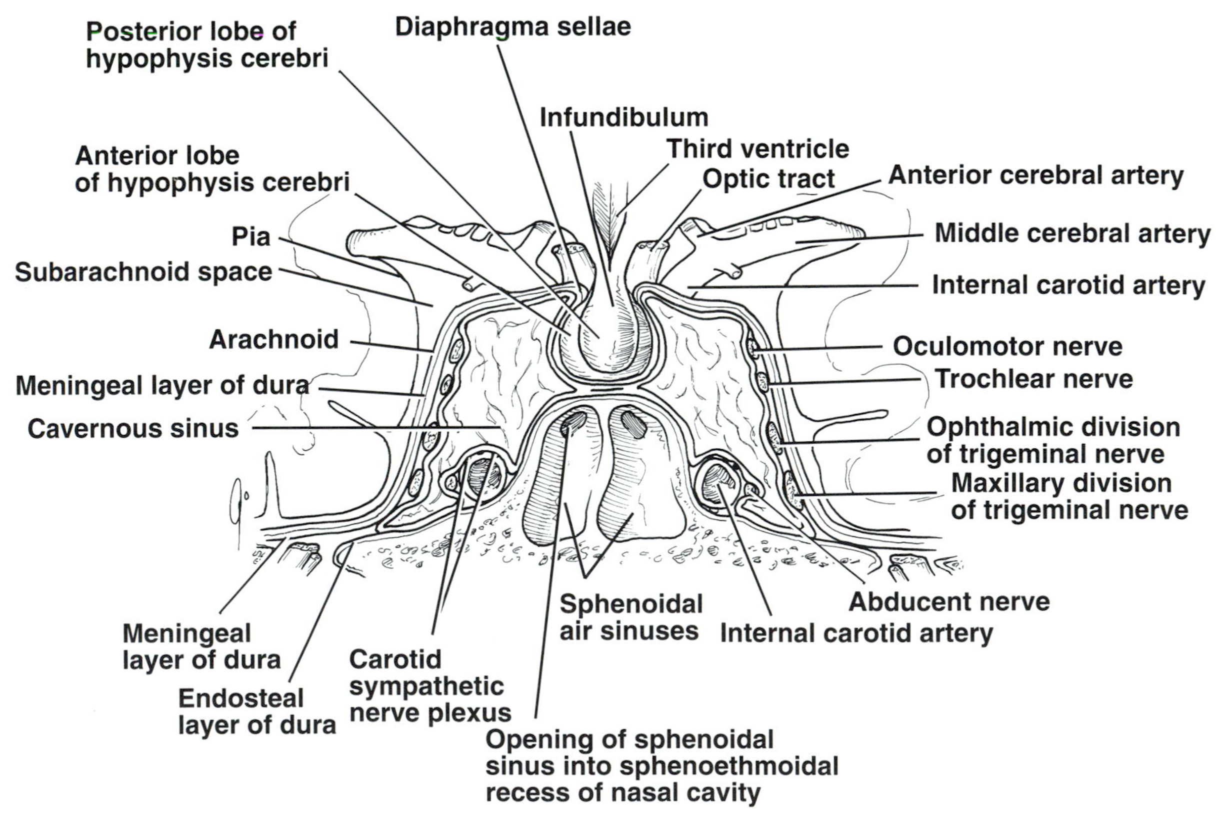

FIG 8–15.
Coronal section through the body of the sphenoid showing the hypophysis cerebri and cavernous sinuses. Note the position of the internal carotid artery and cranial nerves.

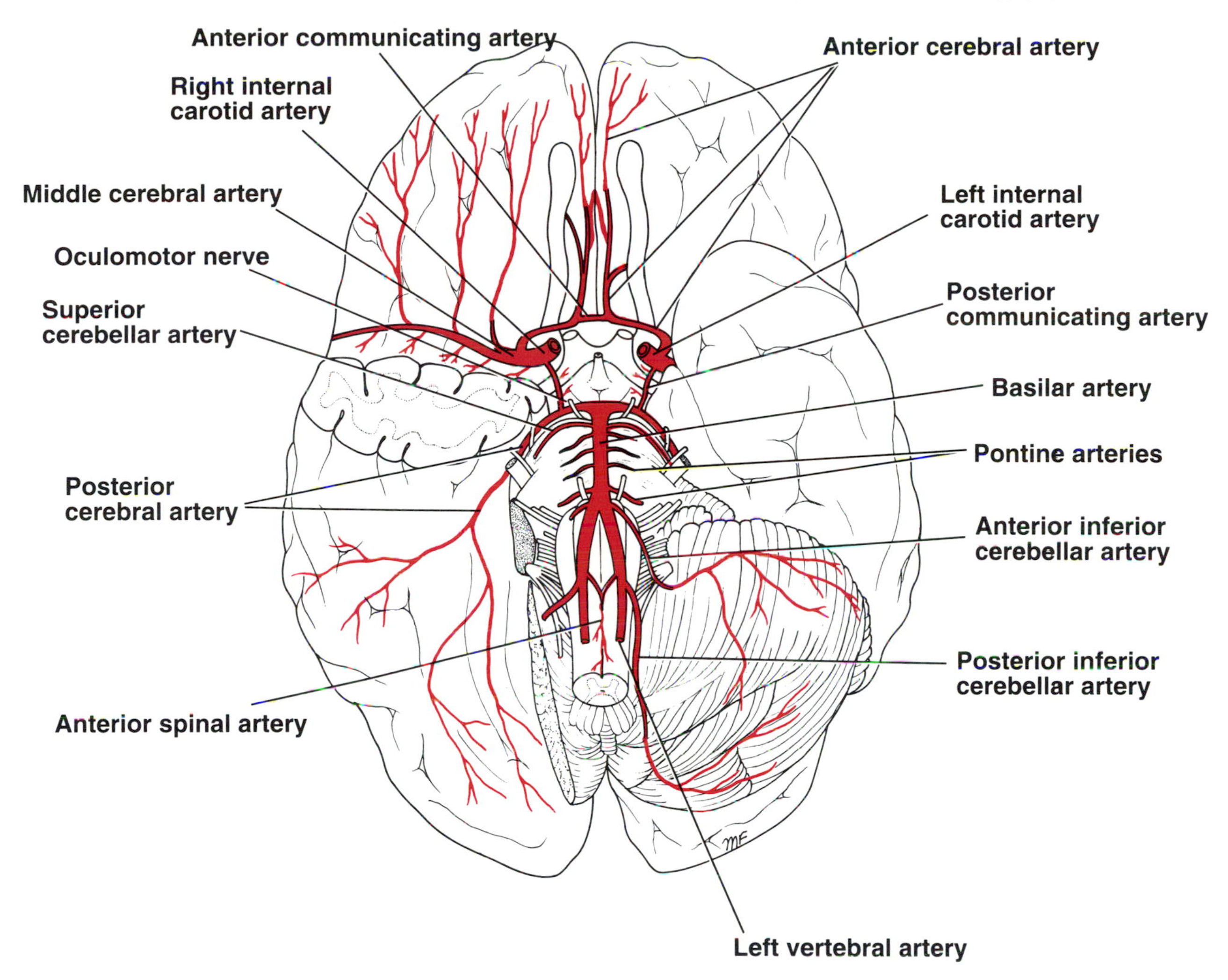

FIG 8–16.
The arteries of the inferior surface of the brain. Note the formation of the circulus arteriosus (circle of Willis). The anterior pole of the right temporal lobe has been removed to show the course of the right middle cerebral artery.

the posterior cerebral artery, thus forming part of the circulus arteriosus.

3. The *choroidal artery,* which is a small branch, enters the lateral ventricle and ends in the choroid plexus; it also supplies the optic tract and the internal capsule.

4. The *anterior cerebral artery* is the smaller terminal branch of the internal carotid artery. It runs forward and enters the longitudinal fissure of the cerebrum. Here it is joined to the anterior cerebral artery of the opposite side by the *anterior communicating artery* (see Fig 8–16). It curves backward over the corpus callosum, and, finally, anastomoses with the posterior cerebral artery.

The *cortical branches* supply all the medial surface of the cerebral cortex as far back as the parieto-occipital sulcus (Fig 8–17). They also supply a strip of cortex about 1 in. (2.5 cm) wide on the adjoining lateral surface. The anterior cerebral artery thus supplies the "leg area" of the precentral gyrus.

The *central branches* help supply parts of the lentiform and caudate nuclei and the internal capsule.

5. The *middle cerebral artery,* the largest branch of the internal carotid artery (see Fig 8–16), runs laterally in the lateral cerebral sulcus.

The *cortical branches* supply the entire lateral surface of the hemisphere, except for the narrow strip supplied by the anterior cerebral artery, and the occipital pole and inferolateral surface of the hemisphere, which are supplied by the posterior cerebral artery (see Fig 8–17). The middle cerebral artery thus supplies all the motor area except the "leg area."

The *central branches* supply the lentiform and caudate nuclei and the internal capsule.

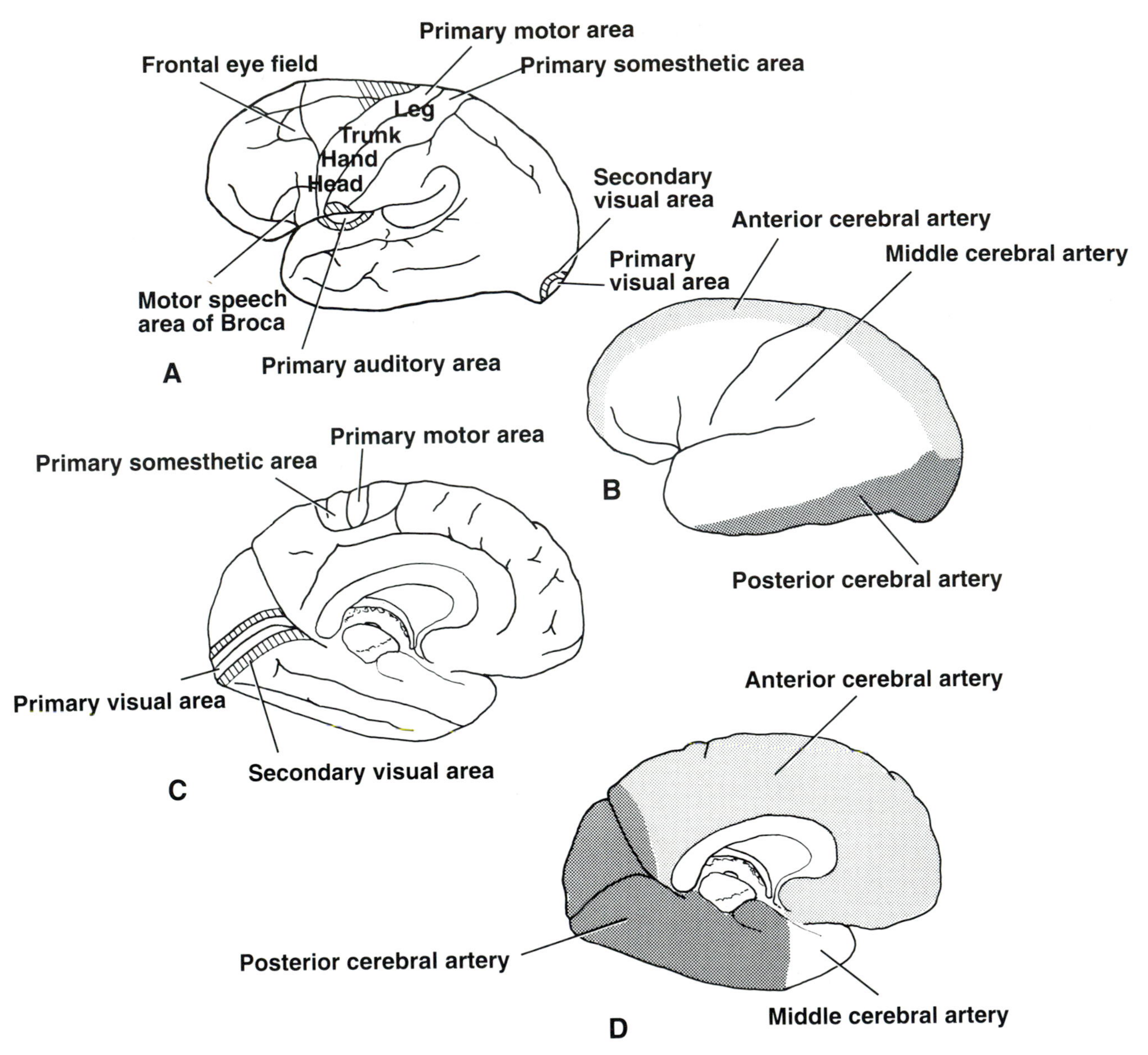

FIG 8–17.
Areas of the cortex supplied by the cerebral arteries. **A,** functional localization of the lateral surface of the left cerebral cortex. **B,** arterial supply of the lateral surface of the left cerebral cortex. **C,** functional localization of the medial surface of the left cerebral cortex. **D,** arterial supply of the medial surface of the left cerebral cortex.

Vertebral Artery.—The vertebral artery enters the skull through the foramen magnum and enters the subarachnoid space. It then ascends on the anterior surface of the medulla oblongata, and at the lower border of the pons it joins the vessel of the opposite side to form the *basilar artery* (see Fig 8–16).

Branches.—The branches of the vertebral artery include the *meningeal branches, posterior spinal artery, anterior spinal artery, posterior inferior cerebellar artery,* and *medullary arteries.*

Basilar Artery.—The basilar artery is formed by the union of the two vertebral arteries. It is located in a groove on the anterior surface of the pons.

Branches.—The branches of the basilar artery include the *pontine arteries, labyrinthine artery, anterior inferior cerebellar artery, superior cerebellar artery,* and *posterior cerebral artery.*

The posterior cerebral artery passes backward around the midbrain and is joined by the posterior communicating branch of the internal carotid artery (see Fig 8–16).

The *cortical branches* supply the inferolateral surface and medial surfaces of the temporal lobe and the lateral and medial surfaces of the occipital lobe (see Fig 8–17). Thus the posterior cerebral artery supplies the visual cortex.

The *central branches* supply parts of the thalamus and the lentiform nucleus and the midbrain. A choroidal branch supplies the choroid plexuses in the lateral and third ventricles.

Circulus Arteriosus (Circle of Willis).—The circulus arteriosus lies at the base of the brain in the interpeduncular fossa. It is formed by the anastomosis between the two internal carotid arteries and the two vertebral arteries (Figs 8–16 and 8–18). The circulus arteriosus allows blood that enters by either internal carotid or vertebral arteries to be distributed to any part of both cerebral hemispheres. Cortical and central branches arise from the circle and supply the brain substance.

Veins of the Brain

The veins of the brain have no muscular tissue in their very thin walls, and they have no valves. They emerge from the brain and lie in the subarachnoid space. They drain into the cranial venous sinuses (see Fig 8–14).

Clinical Note

The third cranial nerve passes forward between the posterior cerebral artery and the superior cerebellar artery, and at that point lies close to the posterior communicating artery (see Fig 8–16). An aneurysm in any one of these arteries can press on the third cranial nerve and cause an extraocular movement problem or a defect in pupilloconstriction.

Basic Anatomy of the Ventricular System of the Brain

The ventricles of the brain are the lateral ventricles, the third ventricle, and the fourth ventricle (Fig 8–19). The two *lateral ventricles* communicate through the *interventricular foramina (of Monro)* with the *third ventricle.* The third ventricle is connected to

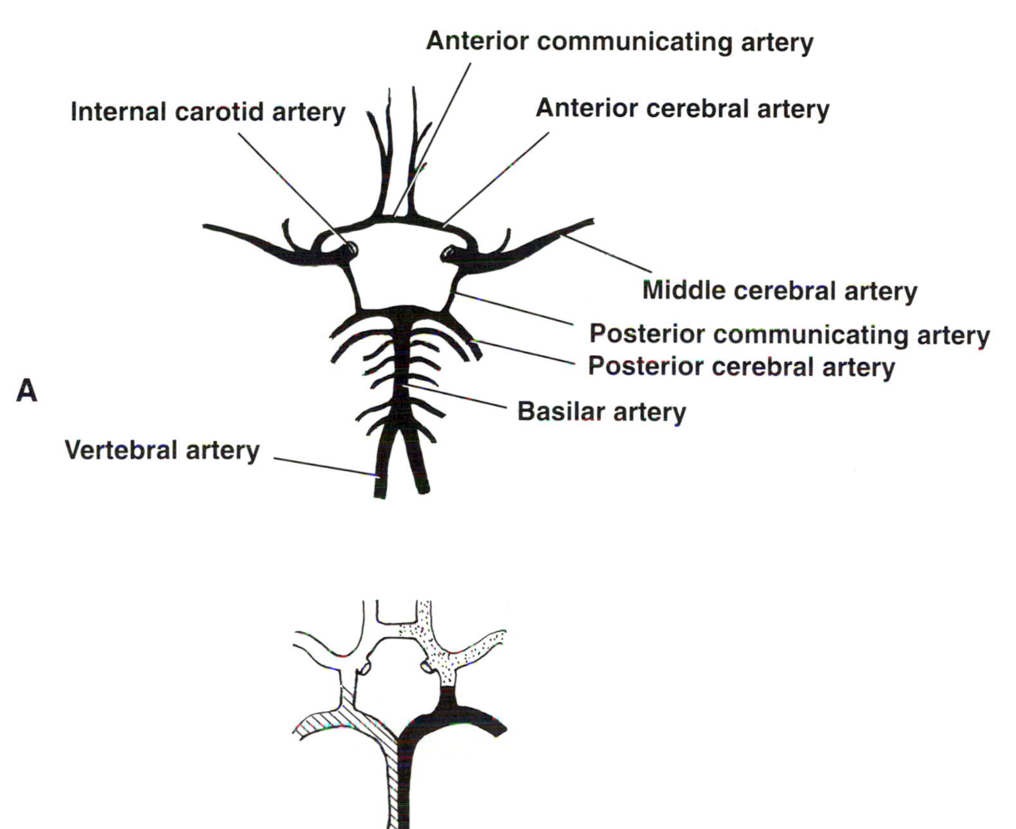

FIG 8–18.
A, the formation of the circus arteriosus (circle of Willis) from the two internal carotid and two vertebral arteries. **B,** the distribution of blood from the four main arteries.

the *fourth ventricle* by the *cerebral aqueduct (aqueduct of Sylvius)*. The fourth ventricle, in turn, is continuous with the narrow *central canal* of the spinal cord and, through the three foramina in its roof, with the subarachnoid space (see Fig 8–19).

The ventricles are lined throughout with *ependyma* and are filled with *cerebrospinal fluid*.

The size and shape of the cerebral ventricles may be visualized clinically using computed tomography (see Fig 8–7).

Basic Anatomy of the Cerebrospinal Fluid

Cerebrospinal fluid is a clear, colorless fluid. It possesses, in solution, inorganic salts similar to those in the blood plasma. The glucose content is about half that of blood, and there is only a trace of protein. In the lateral recumbent position, cerebrospinal fluid pressure, as measured by lumbar puncture, is about 60 to 150 mm of water. This pressure may be easily raised by straining or coughing or compressing the internal jugular veins in the neck.

Functions of the Cerebrospinal Fluid

Cerebrospinal fluid, which bathes the external and internal surfaces of the brain and spinal cord, serves as a cushion between the central nervous system and the surrounding bones, thus protecting the brain against mechanical trauma. The close relationship of the fluid to the nervous tissue and the blood enables the fluid to serve as a reservoir and assist in the regulation of the contents of the skull. If the brain volume or the blood volume increases, the

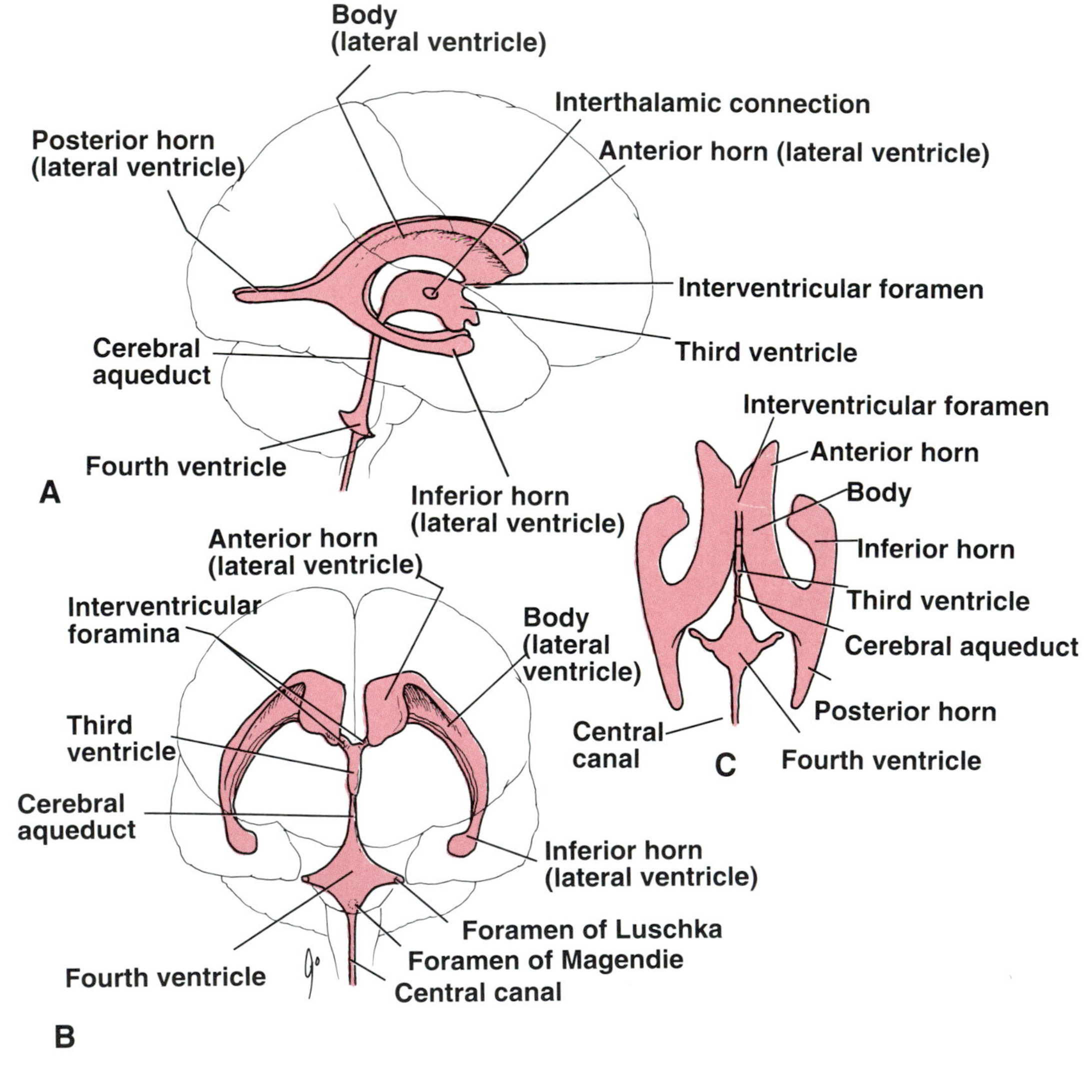

FIG 8–19.
The shape and position of the ventricular cavities of the brain. **A,** right lateral view. **B,** anterior view. **C,** superior view.

cerebrospinal fluid volume decreases. Since the cerebrospinal fluid is an ideal physiological substrate, it probably plays an active part in the nourishment of the nervous tissue; it almost certainly assists in the removal of products of neuronal metabolism. It is possible that the secretions of the pineal gland influence the activities of the pituitary gland by circulating through the cerebrospinal fluid in the third ventricle.

Formation of Cerebrospinal Fluid

Cerebrospinal fluid is formed mainly in the *choroid plexuses* (see Fig 8–13) of the lateral, third, and fourth ventricles; some may originate as tissue fluid formed in the brain substance. The cuboidal epithelium covering the surface of the choroidal plexuses actively secretes cerebrospinal fluid.

Circulation of Cerebrospinal Fluid

The circulation begins with its secretion from the choroid plexuses in the ventricles and its production from the brain surface. The fluid passes from the lateral ventricles into the third ventricle through the interventricular foramina (see Figs 8–13 and 8–19). It then passes into the fourth ventricle through the cerebral aqueduct. The circulation is aided by the arterial pulsations of the choroid plexuses.

From the fourth ventricle, the fluid passes through the median aperture and the lateral foramina in the roof of the fourth ventricle and enters the subarachnoid space. The fluid flows superiorly through the notch in the tentorium cerebelli to reach the inferior surface of the cerebrum (see Fig 8–13). It now moves superiorly over the lateral aspect of each cerebral hemisphere. Some of the cerebrospinal fluid moves inferiorly in the subarachnoid space around the spinal cord and cauda equina. It is believed that the pulsations of the cerebral and spinal arteries and the movements of the vertebral column facilitate this gradual flow of fluid.

Absorption of Cerebrospinal Fluid

The main sites for the absorption of the cerebrospinal fluid are the *arachnoid villi* that project into the dural venous sinuses, especially the superior sagittal sinus (Fig 8–20). The arachnoid villi tend to be grouped together to form *arachnoid granulations.* The arachnoid granulations increase in number and size with age and tend to become calcified with advanced age.

Absorption of cerebrospinal fluid into the venous sinuses occurs when the pressure of the fluid exceeds that of the blood in the sinus. Some of the cerebrospinal fluid probably is absorbed directly into the veins in the subarachnoid space and some possibly escapes through the perineural lymphatic vessels of the cranial and spinal nerves.

Clinical Note

Turnover of cerebrospinal fluid.—Cerebrospinal fluid is produced continuously at a rate of about 0.5 mL/min and with a total volume of about 130 mL; this corresponds to a turnover time of about 5 hours.

Basic Anatomy of Consciousness

A conscious person is awake and aware. For normal consciousness, the active functioning of two main parts of the nervous system is necessary—the reticular formation (in the brain stem) and the cerebral cortex. The reticular formation is responsible for the state of wakefulness; the cerebral cortex is necessary for the state of awareness, i.e., the state in which the individual can respond to stimuli and interact with the environment. Eye opening is a brain stem function; speech is a cerebral cortex function.

The reticular formation consists of a deeply situated continuous network of nerve cells and fibers extending from the spinal cord through the medulla, pons, midbrain, subthalamus, hypothalamus, and thalamus. Inferiorly, the reticular formation is continuous with the interneurons of the gray matter of the spinal cord, whereas superiorly, impulses are relayed to the whole cerebral cortex and to the cerebellum.

Multiple ascending pathways carrying sensory information to higher centers are channeled through the reticular formation, which, in turn, projects this information to different parts of the cerebral cortex, causing a sleeping person to awaken. It is now believed that the state of wakefulness is dependent on the continuous projection of sensory information to the cortex from the reticular formation *(reticular activating system).* Normally, different degrees of wakefulness seem to depend on the degree of activity of the reticular formation.

A person can have an intact reticular formation but a nonfunctioning cerebral cortex. That person is awake but not aware, i.e., the eyes are open and move around, and the person has sleep-awake cycles; however, the person has no awareness and therefore cannot respond to stimuli such as a verbal command or pain. It is not possible, however, to have awareness without wakefulness. The cerebral

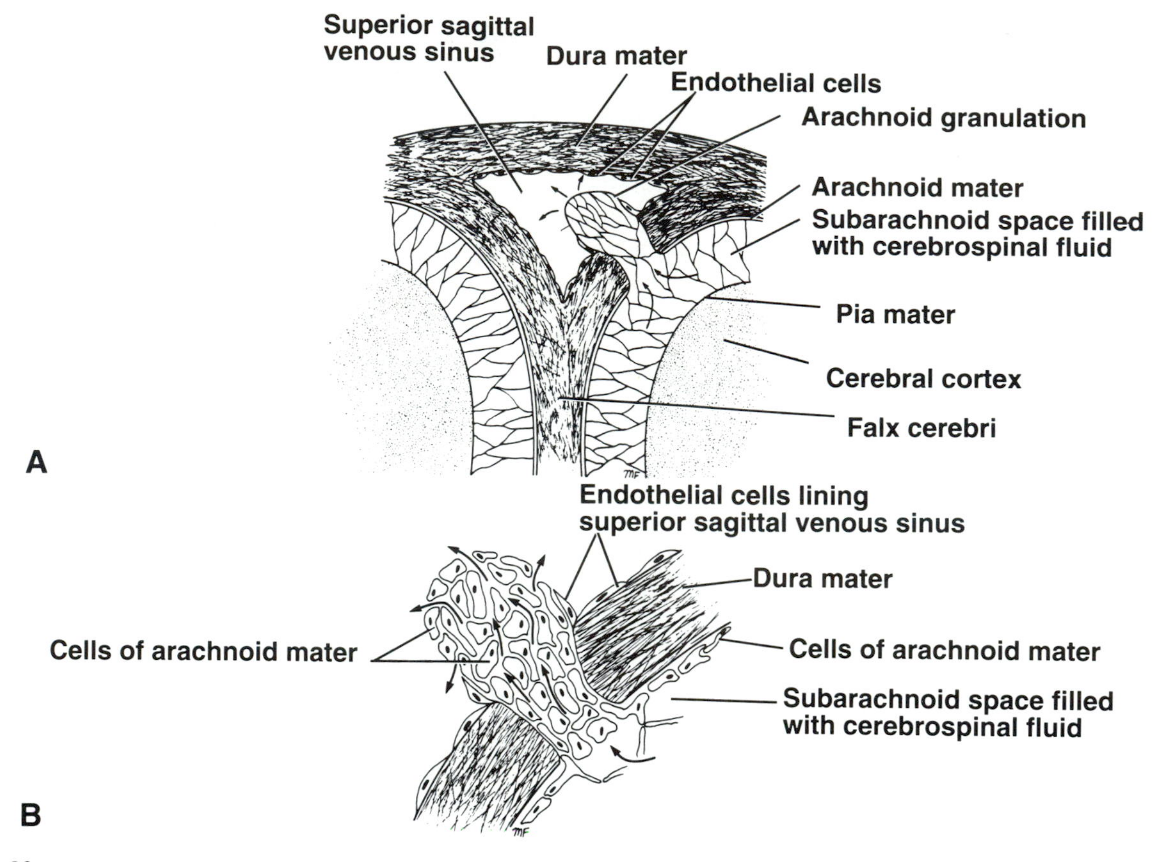

FIG 8–20.
Arachnoid granulation. **A,** coronal section of superior sagittal sinus showing an arachnoid granulation. **B,** magnified view of arachnoid granulation showing path taken by cerebrospinal fluid between the cells to enter the venous blood of the superior sagittal sinus.

cortex requires the substrate of input from the reticular formation in order to function.

SURFACE ANATOMY

Surface Anatomy of the Skull

The surface landmarks of the skull have already been extensively discussed relative to the face (Chapter 6, p. 222) and the orbit (Chapter 7, p. 265). However, certain important landmarks will be repeated here.

Nasion
This is the depression in the midline of the root of the nose (Fig 8–21).

External Occipital Protuberance
This is a bony prominence in the middle of the squamous part of the occipital bone (see Fig 8–21). It lies in the midline at the junction of the head and neck and gives attachment to the ligamentum nuchae.

Falx Cerebri, Superior Sagittal Sinus, and the Longitudinal Cerebral Fissure Between the Cerebral Hemispheres
The position of these structures can be indicated by passing a line over the vertex of the skull in the sagittal plane that joins the nasion to the external occipital protuberance.

Parietal Eminence
This is a raised area on the lateral surface of the parietal bone and can be felt about 2 in. (5 cm) above the auricle. It lies close to the lower end of the *central cerebral sulcus of the brain* (see Fig 8–21).

Pterion
This is the point where the greater wing of the sphenoid meets the anteroinferior angle of the parietal bone. Lying 1½ in. (4 cm) above the midpoint of the zygomatic arch (see Fig 8–21), it is not marked by an eminence or a depression, but it is important since the *anterior branches of the middle meningeal artery and vein lie beneath it.*

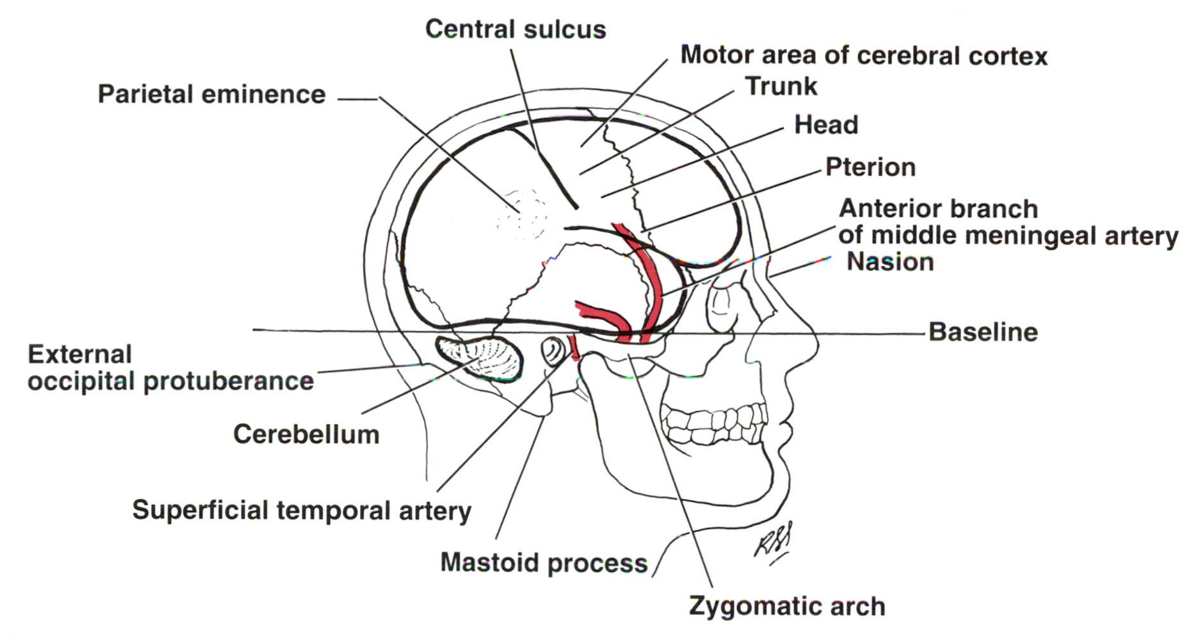

FIG 8–21.
Surface landmarks on the right side of the head. The relations of the middle meningeal artery and brain to the surface of the skull are shown.

Mastoid Process of the Temporal Bone

The mastoid process projects downward and forward from behind the ear (see Fig 8–21). It is undeveloped in the newborn and grows only as the result of the pull of the sternocleidomastoid muscle, as the child moves his head. It may be recognized as a bony projection at the end of the second year of life.

Zygomatic Arch

The zygomatic arch extends forward in front of the ear and ends in front in the zygomatic bone (see Fig 8–21). Above the zygomatic arch is the *temporal fossa,* which is filled with the *temporalis muscle.* Attached to the lower margin of the zygomatic arch is the masseter muscle. Contraction of both the temporalis and masseter muscles (testing for the integrity of the motor part of the mandibular division of the trigeminal nerve) may be felt by clenching the teeth.

Anatomical Baseline of the Skull

This baseline extends from the lower margin of the orbit backward through the upper margin of the external auditory meatus. The *cerebrum* lies entirely above the line, and the *cerebellum* lies in the posterior cranial fossa below the posterior third of the line (see Fig 8–21).

CLINICAL ANATOMY

Clinical Anatomy of the Skull

Fractures of the Skull

The type of fracture that occurs in the skull will depend on the age of the patient, the severity of the blow, and the area of the skull receiving the trauma. Fractures of the skull are more common in adults than in young children.

The *adult skull* possesses a certain limited resilience beyond which it splinters. The inner table of the skull is particularly brittle. Moreover, the sutural ligaments begin to ossify during middle age. A severe, localized blow will produce a local indentation, often accompanied by splintering of the bone. Blows to the vault often result in a series of linear fractures that radiate out through the thin areas of bone. The petrous parts of the temporal bones and the occipital crests strongly reinforce the base of the skull and tend to deflect the linear fractures.

In the *young child,* the skull is more resilient than that of the adult, and the bones are separated by fibrous sutural ligaments. A localized blow produces a depression without splintering, which is commonly referred to as a "pond" fracture.

Complications of Skull Fractures.—Fractures of the skull are of minor importance compared with the possible associated damage to the underlying brain. It is possible to have multiple skull fractures without damage to the brain or cranial nerves. However, when skull damage due to compression injury is combined with motion, the intracranial soft tissues are liable to be injured (see discussion of injuries of the brain, below).

Complications of Linear Skull Fractures.—Linear skull fractures are usually diagnosed by radiographic examination and seldom give rise to complications. However, they may extend into air spaces, such as the middle ear or the nose, and effectively become open fractures or they may cross the path of the middle meningeal vessels and cause an epidural hemorrhage (epidural hematoma).

Cerebrospinal Fluid Rhinorrhea.—This occurs when the fracture line involves the cribriform plate of the ethmoid or the wall of the frontal air sinus. Note that the cribriform plate is a very thin perforated plate of bone that forms the floor of the anterior cranial fossa and the roof of the nasal cavity. Unfortunately, fractures of the cribriform plate or the wall of the frontal sinus provide an ascending pathway for infection into the subarachnoid space.

Cerebrospinal Fluid Otorrhea.—This occurs when the fracture line extends into the petrous part of the temporal bone and cerebrospinal fluid escapes into the tympanic cavity and/or the external auditory meatus. Here, again, there is a risk of infection entering the subarachnoid space.

Epidural Hemorrhage.—Often a relatively minor blow to the side of the head may result in the fracture line in the region of the anterior inferior angle of the parietal bone, severing or tearing the middle meningeal vessels. Epidural hemorrhage is discussed on p. 310.

Complications of Basal Skull Fractures.—In anatomical terms, the base of the skull is the lowest part of the cranium, and, when viewed from the interior, is divided up into three cranial fossae—anterior, middle, and posterior (see p. 288).

Fractures of the Anterior Cranial Fossa.—*Epistaxis and Cerebrospinal Rhinorrhea.*—The cribriform plate of the ethmoid may be damaged. The tearing of the overlying meninges and the underlying mucoperiosteum will produce epistaxis and cerebrospinal rhinorrhea.

Subconjunctival Hemorrhage and Periorbital Ecchymosis.—Fractures involving the orbital plate of the frontal bone will result in hemorrhage into the orbital cavity, which may be extensive and cause proptosis ("racoon eyes"); it may also extend forward around the globe of the eye under the conjunctiva, producing a type of subconjunctival hemorrhage. In these cases there may also be extensive bruising around the eye, which has a sharply defined margin produced by the attachments of the orbital septum (see p. 245). The frontal sinus may also be involved, with resulting epistaxis.

Anosmia.—Unilateral or bilateral anosmia may follow fracture of the cribriform plate of the ethmoid. This loss is difficult to test in the traumatized patient.

Fractures of the Middle Cranial Fossa.—Fractures of the middle cranial fossa are common, since this is the weakest part of the skull. Anatomically, this weakness is caused by the presence of numerous foramina and canals in this region; the cavities of the middle ear and the sphenoidal sinuses are particularly vulnerable.

The Leakage of Cerebrospinal Fluid and Blood From the Ear.—The cerebrospinal fluid and blood can leak from the external auditory meatus after a fracture. *Battle's sign* is an ecchymosis over the mastoid process secondary to the leakage of blood to the surface and usually takes at least 24 to 36 hours to develop. It is *not* a result of a direct blow to the mastoid process.

Cranial Nerve Damage.—The facial and vestibulocochlear nerves may be damaged as they pass through the petrous part of the temporal bone. Injury to these nerves produces peripheral facial palsy, hearing loss, and signs of vestibular dysfunction (vertigo or nystagmus). If the lateral wall of the cavernous sinus is torn, the oculomotor, trochlear, and abducent nerves may be damaged, resulting in ophthalmoplegia.

Fractures of the Posterior Cranial Fossa.—*Cranial Nerve Damage.*—In fractures involving the jugular foramen (see Fig 8–3), the glossopharyngeal, vagus, and accessory nerves may be damaged. The strong bony wall of the hypoglossal canal usually protect the hypoglossal nerve from injury.

Penetrating Wounds of the Skull.—Stab wounds that penetrate the skull may damage only local tissue and are often not fatal. Bullet wounds, on the other hand, cause extensive local damage to the brain, and the shock wave causes destruction of nervous tissue at some distance from the point of entry.

Clinical Anatomy of the Meninges and Venous Sinuses

Movements of the Brain Relative to the Skull and Meninges in Head Injuries

Brain injuries are produced by displacement and distortion of the neuronal tissues at the moment of impact (Fig 8–22). The brain, which is incompressible,

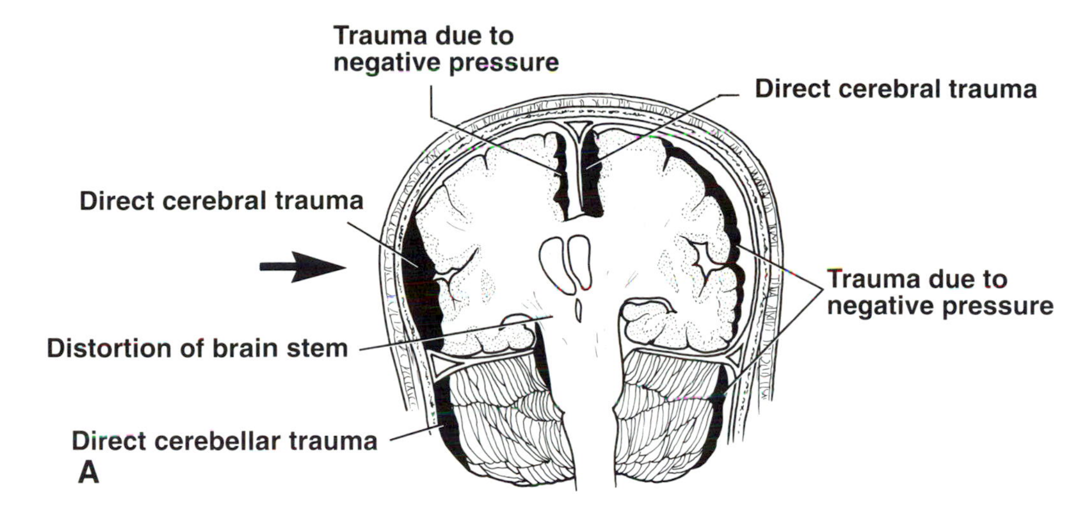

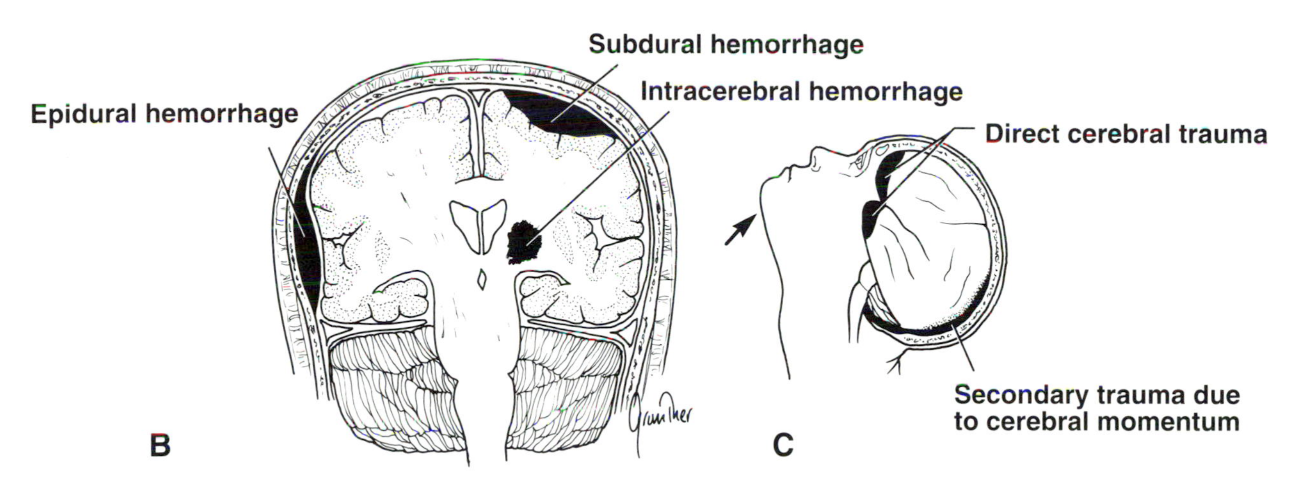

FIG 8–22.
A, mechanisms of acute cerebral injury when blow is applied to the lateral side of the head. **B,** varieties of intracranial hemorrhage. **C,** mechanism of cerebral trauma following a blow on the chin. The movement of the brain within the skull can also tear the cerebral veins.

may be likened to a log soaked with water floating submerged in water. The brain is floating in the cerebrospinal fluid in the subarachnoid space and is capable of a certain amount of anteroposterior and lateral gliding movement. The anteroposterior movement is limited by the attachment of the superior cerebral veins to the superior sagittal sinus. In lateral movements, the lateral surface of one hemisphere hits the side of the skull, and the medial surface of the opposite hemisphere hits the side of the falx cerebri (see Fig 8–22). In superior movements, the superior surfaces of the cerebral hemispheres hit the vault of the skull, and the superior surface of the corpus callosum may hit the sharp free edge of the falx cerebri; the superior surface of the cerebellum presses against the inferior surface of the tentorium cerebelli.

It follows from these anatomical facts that blows on the front or back of the head lead to displacement of the brain, which may produce severe cerebral damage, stretching and distortion of the brain stem, and stretching and even tearing of the commissures of the brain. Blows to the side of the head produce less cerebral displacement, and the injuries to the brain consequently tend to be less severe. The falx cerebri, however, is a tough structure and may cause considerable damage to the softer brain tissue in cases where there has been a severe blow to the side of the head (see Fig 8–22). Furthermore, glancing blows to the head may result in considerable rotation of the brain, causing shearing strains and distortion, particularly in areas where further rotation is prevented by bony prominences in the anterior and middle cranial fossae. Brain lacerations are likely to occur when the brain is forcibly thrown against the sharp edges of bone within the skull—the lesser wing of the sphenoid, for example.

When the brain is suddenly given momentum within the skull, the part of the brain that moves away from the skull wall is subjected to diminished pressure, because the cerebrospinal fluid has not had time to accommodate to the brain movement (see Fig 8–22). This results in a suction effect on the brain surface, with rupture of surface blood vessels.

A sudden severe blow on the head, as in an automobile accident, may result in damage to the brain at the following two sites: (1) at the point of impact, and (2) at the pole of the brain opposite the point of impact, where the brain is thrown against the skull wall. This is referred to as *contrecoup injury.*

Movements of the brain relative to the skull and dural septa may seriously injure the cranial nerves that are tethered as they pass through the various foramina. This particularly applies to the long, slender nerves, such as the trochlear, abducent, and oculomotor nerves. Furthermore, the fragile cortical veins that tether the brain and drain into the dural sinuses may be torn, resulting in severe subdural or subarachnoid hemorrhage. The large arteries found at the base of the brain are tortuous, and this, coupled with their strong walls, explains why they are seldom damaged.

Intracranial Hemorrhage and the Meninges

Although the brain is cushioned by the surrounding cerebrospinal fluid in the subarachnoid space, any severe hemorrhage within the rigid skull will ultimately exert pressure on the brain.

Intracranial hemorrhage may result from trauma or cerebral vascular lesions (see Fig 8–22).

Epidural Hemorrhage.—Epidural hemorrhage results from injuries to the meningeal arteries or veins. The most common artery damaged is the *anterior division of the middle meningeal artery.* A blow to the side of the head, resulting in fracture of the skull in the region of the anterior inferior portion of the parietal bone, may sever the artery. Arterial or venous injury is especially liable to occur if the vessels enter a bony canal in this region. Bleeding occurs and strips the meningeal layer of dura from the internal surface of the skull (Fig 8–23). The intracranial pressure rises, and the enlarging blood clot exerts local pressure on the underlying motor area in the precentral gyrus. Blood may also pass laterally through the fracture line to form a soft swelling under the temporalis muscle. To stop the hemorrhage, the torn artery must be ligated through a temporal craniectomy (see p. 318).

In patients with epidural hematomas from sagittal sinus or other venous sinus tears, the bleeding has to be stopped through an appropriately placed craniotomy.

Subdural Hemorrhage.—Subdural hemorrhage results from tearing of the superior cerebral veins at their point of entrance into the superior sagittal sinus. The cause is usually a blow on the front or the back of the head, causing excessive anteroposterior displacement of the brain within the skull.

This condition, which is more common than middle meningeal hemorrhage, can be produced by a sudden minor blow. Once the vein is torn, the blood under low pressure begins to accumulate in the potential space between the meningeal layer of dura and the arachnoid (see Fig 8–23). In a few cases the condition is bilateral.

Acute and chronic forms of the clinical condition occur, depending on the speed of accumulation of fluid in the subdural space. For example, if the patient starts to vomit, the venous pressure will rise as the result of a rise in the intrathoracic pressure. Under these circumstances, the subdural blood clot will rapidly increase in size and produce acute symptoms. In the chronic form, over a course of several months, the small blood clot will attract fluid by osmosis, so that a hemorrhagic cyst is formed, which gradually expands and produces pressure symptoms.

Computed Tomographic Scans of Epidural and Subdural Hematomas.—The different appearances of the blood clots in these two conditions as seen on computed tomographic scans is related to the anatomy of the area (see Fig 8–23). In an epidural hemorrhage the blood strips up the meningeal layer of the dura from the endosteal layer of dura (periosteum of the skull), producing a *lens-shaped* hyperdense collection of blood that compresses the brain and displaces the midline structures to the opposite side. The shape of the blood clot is determined by the adherence of the meningeal layer of dura to the periosteal layer of dura.

In patients with subdural hematoma the blood accumulates in the extensive potential space between the meningeal layer of dura and the arachnoid, producing a long *crescent-shaped,* hyperdense rim of blood that extends from anterior to posterior along the inner surface of the skull. With a large hematoma, the brain sulci are obliterated, and the midline structures are displaced to the opposite side.

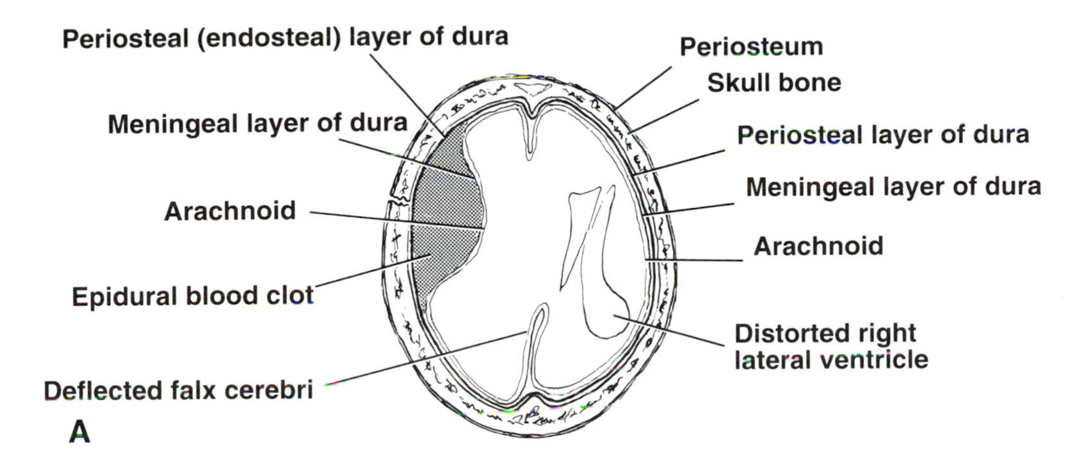

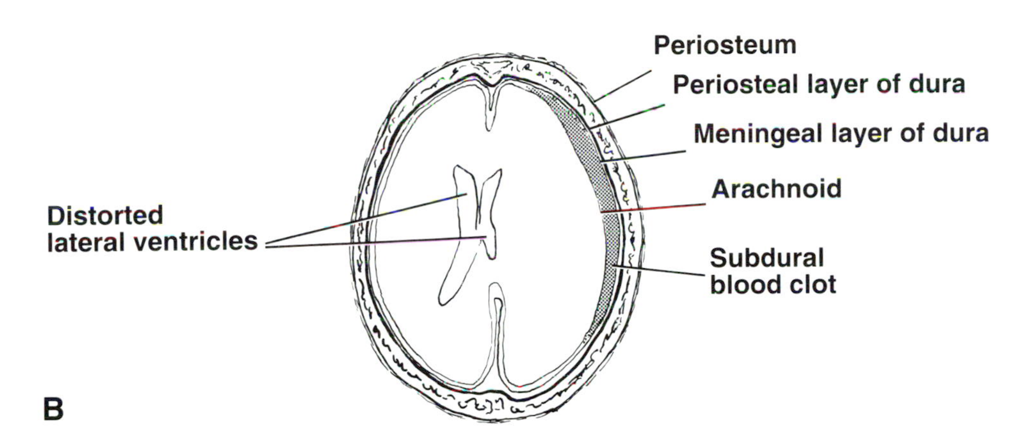

FIG 8–23.
Diagrammatic representation of an epidural hemorrhage and a subdural hemorrhage. **A,** epidural hemorrhage from the middle meningeal artery or vein on the left side. The hematoma is lens shaped and occupies the space between the endosteal layer of dura (periosteum of the skull) and the meningeal layer of dura (true dura—hence the term epidural). **B,** subdural hemorrhage from the cerebral veins at the site of entrance into the venous sinus on right side. The hematoma is crescent shaped and occupies the space between the meningeal layer of dura and the arachnoid, i.e., beneath the dura.

Subarachnoid and cerebral hemorrhages are described below.

Clinical Anatomy of the Arterial Supply to the Brain

Congenital Aneurysms

Congenital aneurysms occur most commonly at the site where two arteries join in the formation of the circulus arteriosus. At this point, there is a deficiency in the tunica media that so weakens the arterial wall that an aneurysm develops. The enlarging aneurysm may press on neighboring structures, such as the optic, oculomotor, trochlear, and abducent nerves, and produce signs and symptoms, or may suddenly rupture into the subarachnoid space.

Subarachnoid Hemorrhage

Subarachnoid hemorrhage usually results from nontraumatic leakage or rupture of a congenital aneurysm on the cerebral arterial circle or less commonly from an arteriovenous malformation.

Cerebral Hemorrhage

Spontaneous intracerebral hemorrhage is most common in patients with hypertension. It usually occurs in middle-aged individuals and often involves a rupture of the thin-walled *lenticulostriate artery,* a branch of the middle cerebral artery. The important corticonuclear and corticospinal fibers in the internal capsule are damaged, producing hemiplegia on the opposite side of the body. In some cases the hemorrhage bursts into the lateral ventricle, resulting in deeper unconsciousness and corticospinal lesions on both sides of the body. Hemorrhage may also occur into the pons and cerebellum.

Cerebral Ischemia

It has been pointed out that there are two distinct, yet interconnected vascular systems supplying the brain. The carotid arteries are the major suppliers of the cerebral hemispheres, and the basilar and vertebral arteries are the major suppliers of the brain stem and cerebellum. The neurological deficit following blockage of one of the intracranial vessels will depend on the location of the blockage and the status of the collateral circulation. The blood supply to the functional areas of the cerebral cortex are shown in Figure 8–17.

Anterior Cerebral Artery Occlusion.—If the occlusion of the artery is proximal to the anterior communicating artery, the collateral circulation is usually adequate to preserve the circulation. Occlusion distal to the communicating artery may produce the following signs and symptoms.

1. Contralateral hemiparesis and hemisensory loss involving mainly the leg and foot.
2. Inability to identify objects correctly, apathy, and personality changes.

Middle Cerebral Artery Occlusion.—Occlusion of the artery may produce the following signs and symptoms.

1. Contralateral hemiparesis and hemisensory loss.
2. Aphasia if the left hemisphere is affected (rarely if the right hemisphere is affected).
3. Homonymous hemianopia.
4. Anosognosia if the right hemisphere is affected (rarely if the left hemisphere is affected).

Internal Carotid Artery Occlusion.—This may produce all the symptoms and signs of anterior and middle cerebral artery occlusion, depending on the degree of collateral circulation at the circulus arteriosus (circle of Willis); in addition, the following may be seen.

1. Loss of vision on the same side as the internal carotid artery occlusion due to blockage of the ophthalmic artery.
2. Decreased level of consciousness.

Vertebral Artery Occlusion.—This produces a variable clinical picture and may include the following signs and symptoms.

1. Ipsilateral pain and temperature sensory loss of the face and contralateral pain and temperature sensory loss of the body.
2. Ipsilateral loss of the gag reflex, dysphagia, and hoarseness as the result of lesions of the nuclei of the glossopharyngeal and vagus nerves.
3. Vertigo, nystagmus, nausea, and vomiting.
4. Ipsilateral Horner's syndrome.
5. Ipsilateral ataxia.

If the lesion is more extensive, the corticospinal tracts may be involved, producing contralateral hemiparesis of the body. Contralateral loss of position and vibration sense may also be lost due to damage to the medial lemniscus.

Basilar Artery Occlusion.—Since this artery gives off numerous branches to the pons, cerebellar peduncles, and cerebellum, total blockage of this artery can produce lesions of the trigeminal, abducent, and facial nerve nuclei, quadriplegia, and coma (reticular formation). If occlusion is restricted to branches of the basilar artery, there may be contralateral hemiparesis, contralateral sensory loss, or evidence of cerebellar dysfunction.

Central Branch Artery Occlusion.—Small artery occlusion will cause discrete areas of brain necrosis. The signs and symptoms produced will obviously depend on the area involved. For example, a lesion of the internal capsule may result in contralateral hemiplegia.

Transient Ischemic Attacks.—These are brief, self-limited focal neurologic deficits caused by embolic or thrombotic occlusion of arteries supplying the brain. The signs and symptoms will depend on the area of brain involved.

Clinical Anatomy of the Ventricles and Cerebrospinal Fluid

Hydrocephalus

Hydrocephalus is an abnormal increase in the volume of the cerebrospinal fluid within the skull. If hydrocephalus is accompanied by raised cerebrospinal fluid pressure, it is caused by either (1) an abnormal increase in fluid formation, (2) blockage of the fluid circulation, or (3) diminished absorption of the fluid. Rarely, hydrocephalus occurs with normal cerebrospinal fluid pressure, and in these patients compensatory hypoplasia or atrophy of the brain substance exists.

When the block of the movement of cerebrospinal fluid lies within the brain, the hydrocephalus is the *noncommunicating* type (i.e., the cerebrospinal fluid inside the brain does not communicate with that on the outside). If the fluid is able to pass out through the roof of the fourth ventricle into the subarachnoid space and cannot be absorbed by the arachnoid villi, the hydrocephalus is the *communicating* type (i.e., the cerebrospinal fluid inside the brain communicates with that on the outside).

Hydrocephalus Resulting From Excessive Formation of Cerebrospinal Fluid.—This condition is rare and may occur when there is a tumor of the choroid plexuses.

Hydrocephalus Resulting From Blockage of Cerebrospinal Fluid Circulation.—An obstruction of the interventricular foramen by a tumor will block the drainage of the lateral ventricle on that side. The continued production of cerebrospinal fluid by the choroid plexus of that ventricle will cause distention of that ventricle and atrophy of the surrounding neural tissue.

An obstruction in the cerebral aqueduct may be congenital or result from inflammation or pressure from a tumor. This causes a symmetrical distension of both lateral ventricles and distension of the third ventricle.

Obstruction of the foramina in the roof of the fourth ventricle by inflammatory exudate, or by tumor growth, will produce symmetrical dilatation of both lateral ventricles and the third and fourth ventricle.

Sometimes inflammatory exudate secondary to meningitis will block the subarachnoid space and obstruct the flow of cerebrospinal fluid over the outer surface of the cerebral hemispheres. Here, again, the entire ventricular system of the brain will become distended.

Hydrocephalus Resulting From Diminished Absorption of Cerebrospinal Fluid.—Interference with the absorption of cerebrospinal fluid at the arachnoid granulations can be caused by inflammatory exudate, venous thrombosis or pressure on the venous sinuses, or obstruction of the internal jugular vein.

Clinical Anatomy of Lumbar Puncture (Spinal Tap)

See p. 396.

Clinical Anatomy of Consciousness

Persistent Vegetative State

This condition is usually seen following severe head injury or from an anoxic cerebral insult. The reticular formation in the brain stem has survived but the cerebral cortex is dead. The patients are awake but not aware. They spontaneously open their eyes, move their eyes around, have sleep-awake cycles, but they do not respond to commands, they do not speak, and they do not respond appropriately to pain. Unfortunately, the lay observer thinks the patient is "conscious."

"Locked In" Syndrome

This condition occurs following severe head injury or from an anoxic cerebral insult. The cortex is functioning, the reticular formation is partially functioning, but the corticobulbar and corticospinal tracts have been destroyed (i.e., low midbrain?). The patient is awake and aware but cannot respond; the patient is "locked in." The patient lies mute, still, facially motionless, and quadriplegic. The patient's only possible means of communicating with the external environment is through eye movements and blinking. One must assume in these cases that the corticobulbar fibers to the oculomotor, trochlear, and abducent nerve nuclei are partially intact and that the blinking eye reflex involving the trigeminal and facial nerves is functioning.

Coma

Coma is a state of pathologic and nonwillful unresponsiveness in which the patient lies quietly with the eyes closed. The patient must either have reticular formation damage or bilateral cortical damage; unilateral cortical damage does not cause coma.

Lesions of the Reticular Formation

The reticular formation in the brain stem can be damaged directly by trauma, hemorrhage, or ischemia or indirectly by pressure or traction or metabolic problems.

Because the cranial cavity is enclosed by nondistensible bone, any rise in intracranial pressure by a space-occupying lesion will result in displacement of brain tissue.

Rises in Supratentorial Pressure.—The common causes of a rise in supratentorial pressure are intracerebral hemorrhage, subarachnoid hemorrhage, subdural hemorrhage, epidural hemorrhage, and cerebral edema. Two forms of caudal herniation of the brain through the tentorial notch of the tentorium cerebelli can occur as the result of the raised supratentorial pressure.

Central Herniation Syndrome.—In this syndrome the thalamus and midbrain are pushed caudally through the tentorial notch.

Uncal Herniation Syndrome.—In this syndrome the uncus of the temporal lobe is displaced medially and pushes the midbrain against the opposite sharp edge of the tentorial notch. At the same time the displaced uncus presses on the ipsilateral oculomotor nerve at the notch, resulting in a sluggishly reactive and dilated pupil (Fig 8–24).

Rises in Subtentorial Pressure.— Posterior cranial fossa lesions such as those in cerebellar hemorrhage cause a rise in pressure that can directly compress the brain stem or its blood supply. Indirect compression can follow upward herniation of the cerebellum through the tentorial notch or downward herniation of the cerebellar tonsils through the foramen magnum. In the latter instance the medulla will also be displaced and pressed upon. The problem can be compounded by pressure on the cerebral aqueduct or the roof of the fourth ventricle, producing an acute obstructive hydrocephalus.

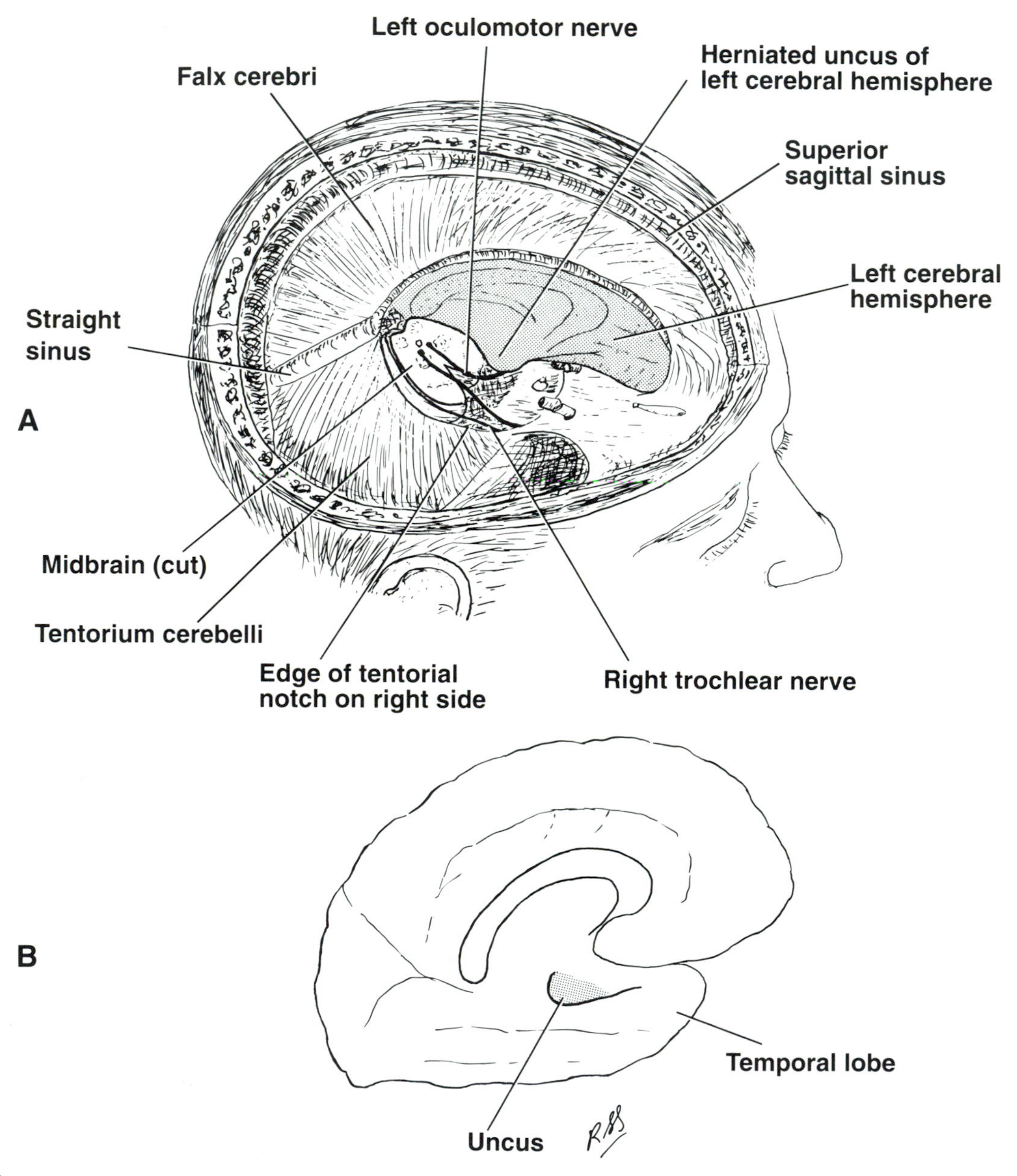

FIG 8–24.
A, lateral view of interior of skull showing the falx cerebri, tentorium cerebelli, and the brain stem. As the result of an abnormal supratentorial pressure, the uncus of the left cerebral hemisphere has herniated down through the tentorial notch of the tentorium cerebelli and is pressing on the left oculomotor nerve. **B,** the position of the uncus on the temporal lobe of the left cerebral hemisphere in a normal brain.

Lesions of Both Cerebral Cortices

Bilateral cerebral cortex dysfunction can be caused by seizure activity or by diffuse metabolic insult, although in the latter case there is probably brain stem dysfunction as well.

Clinical Anatomy of Scales That Assess Unresponsiveness

The Glasgow Coma Scale is used to assess level of responsiveness in a head trauma patient. The scale has three component parts—verbal behavior, eye opening, and motor response.

Verbal activity measures cerebral cortical activity. *Eye opening activity* measures brain stem activity. *Motor responses* measure the activity of the cerebral cortex and the brain stem and specifically the integrity of the corticospinal tract.

Eye Signs

Eyelid Opening.—The nervous pathways involved in this movement are summarized in Figure 7–19. If the patient opens his eyes spontaneously or in response to stimuli, it means that areas contiguous to the reticular formation are functioning and that because the neuroanatomic loci are close together in this part of the brain stem, it would be virtually impossible for the eyelids to open unless the reticular formation were functioning. Note that severe phencyclidine intoxication can result in a mute patient with a blank stare.

Fundus Examination.—The extension of the subarachnoid space to the back of the eyeball is shown in Figure 7–11. Intracranial hypertension may be revealed by the presence of bilateral *papilledema.* The raised pressure extends along the outside of the optic nerve in the tubular extension of the subarachnoid space. This causes the lamina cribrosa (see p. 277) to bulge forward. In addition, the central vein of the retina, which crosses the subarachnoid space to enter the optic nerve, is compressed, causing congestion of the retinal veins with retinal hemorrhages and edema of the optic nerve head.

Pupil Examination.—The innervation of the pupil is summarized in Figure 7–23. The size of the resting pupil depends on the balance between the tone of the parasympathetic innervated pupilloconstrictor muscle and the sympathetic innervated pupillodilator muscle.

Normal Parasympathetic Innervation of the Pupilloconstrictor Muscle.—The parasympathetic control of the pupilloconstrictor muscle begins in the hypothalamus. Nerve fibers descend to the Edinger-Westphal nucleus located in the midbrain. The preganglionic fibers emerge in the oculomotor nerve and pass between the posterior cerebral and superior cerebellar arteries. Here they are vulnerable to compression by a posterior communicating artery aneurysm. The oculomotor nerve then passes forward along the tentorium cerebelli, close to the tentorial notch, where it is susceptible to pressure from a herniating uncus. The oculomotor nerve then enters the lateral wall of the cavernous sinus and passes forward into the orbit to synapse in the ciliary ganglion.

Dysfunction of Pupilloconstriction.—Lesions of the oculomotor nerve result in impaired pupilloconstriction with a dilated, sluggishly reactive, and then fixed pupil. The oculomotor nerve also supplies four of the six extraocular muscles as well as the levator palpebrae superioris muscle, which opens the eye. A complete oculomotor nerve lesion therefore results in a dilated pupil, an abducted eyeball (lateral rectus is supplied by the abducent nerve and is unapposed by the medial rectus, which is innervated by the oculomotor nerve), and a ptotic eyelid.

Lesions Outside the Brain Stem.—Compression of the oculomotor nerve outside the brain stem by a herniating uncus or an expanding aneurysm results first in dysfunction of pupilloconstriction prior to dysfunction of the extraocular movements or eyelid opening. The pupilloconstrictor fibers within the nerve are located near the surface and are thus more susceptible to pressure. Conversely, diabetic oculomotor palsy affects the extraocular muscles and spares the pupil. An uncal herniation produces a dilated pupil on the same side as the herniating uncus, although the hemiparesis that results may be on the same side as the pupil if the contralateral corticospinal tract is compressed against the contralateral tentorial notch (see p. 316) in Kernohan's notch syndrome.

Normal Sympathetic Innervation of the Dilator Pupillae Muscle.—The sympathetic control of the dilator pupillae muscle is a three-neuron pathway that begins in the hypothalamus. The nerve fibers descend through the brain stem, the cervical part of the spinal cord, until they reach the first thoracic segment of the cord. Here they synapse with the connector neurons in the lateral gray column. The preganglionic fibers then travel to the sympathetic trunk and finally synapse in the superior cervical sympathetic ganglion. The postganglionic fibers travel with the internal carotid artery and enter the orbit with the ophthalmic division of the trigeminal nerve; they reach the eyeball in the nasociliary nerve.

Dysfunction Caused by Brain Stem Lesions.—A lesion in the midbrain may destroy both the pupilloconstrictor and pupillodilator pathways. An extensive lesion at this level produces the midposition pupils that are nonreactive (the pupils of death). A lesion in the pons destroys the descending sympathetic fibers from the hypothalamus but leaves the parasympathetic fibers intact because they have already left the midbrain at a higher level; the pupil shows unopposed constriction (bilateral pinpoint pupils, "pontine pupils").

Lesions Outside the Cerebrospinal Axis.—A lesion in the neck involving the sympathetic trunk will produce classic ipsilateral Horner's syndrome (described elsewhere). The pupil is dilated, and there is ptosis of the upper lid.

Motor Signs That Are Associated With Pupillary Lesions.—Weakness of the contralateral arm and leg may accompany the pupillary sign either because of direct brain stem damage to the corticospinal tract or because of damage to the ipsilateral motor area in the precentral gyrus. Sometimes the patient's weakness appears on the same side as the dilated pupil. This represents Kernohan's notch syndrome, in which the contralateral cerebral peduncle is compressed against the edge of the notch of the tentorium cerebelli.

Extraocular Movements.—Patients who can conjugately move their eyes in a horizontal direction in a yoked manner have effectively demonstrated the intactness of the oculomotor nucleus in the midbrain (supplying the medial rectus and adducting the eye), the abducent nerve nucleus in the pons (supplying the lateral rectus and abducting the eye), and the medial longitudinal bundle that connects the nuclei. It would be virtually impossible to have this extensive an area of the brain stem intact in the presence of substantial damage to the brain stem because these cranial nerve nuclei and their connections are located close to the reticular formation.

Oculocephalic Reflex (Doll's Eye Reflex).—This reflex is elicited by rapidly rotating the patient's head to the right or to the left and observing the position of the eyes. If the eyes stay fixed with respect to the orbit, i.e., move with the head as it is turned, then doll's eyes are said to be "absent," and the intactness of the conjugate gaze has not been demonstrated. If the eyes move with respect to the orbit but stay fixed with respect to the room, then the intactness of the conjugate gaze has been demonstrated, and doll's eyes are said to be present. A patient whose brain stem is not functioning will not have doll's eyes. A patient whose cerebral cortex is not functioning but whose brain stem is functioning will have doll's eyes. The patient who has cortical function (either because the unresponsiveness is psychogenic in origin or because the degree of brain injury is minimal), however, may be able to effect a cortical override of the doll's eye reflex. Unless the examiner is astute and follows an absent doll's eye result with oculovestibular testing, the degree of patient unresponsiveness may be overestimated. Elicitation of the oculocephalic reflex should not be attempted if there is any possibility the patient has a cervical vertebral injury.

The Oculovestibular Reflex (Caloric Stimulation).—This reflex is elicited by injecting ice water into the external auditory meatus to stimulate the horizontal semicircular canal. If the patient is placed 30° up from the supine position, the horizontal semicircular canal will be placed vertically, and gravity will help reinforce the response.

The following three responses are possible.

1. Patients with both an intact brain stem and an intact cerebral cortex will have nystagmus, with the fast component directed away from the side of the cold water stimulus and the slow component directed toward it.
2. Patients with an intact brain stem and a damaged cerebral cortex will have only the slow component of nystagmus present, i.e., there will be tonic deviation of the eyes toward the side of the cold stimulus.
3. Patients with a damaged brain stem will have eyes that remain looking straight ahead.

The validity of this as a test of brain stem function depends on the intactness of the afferent part of the reflex, namely, the vestibular component of the vestibulocochlear nerve, and on an external auditory meatus free of cerumen and able to transmit the temperature stimulus across an intact tympanic membrane.

The Isolated Oculomotor or Abducent Nuclear Palsy.—The patient will have medial deviation at rest with abducent nerve palsy and lateral deviation at rest with oculomotor palsy.

A patient with an isolated lesion of the medial longitudinal fasciculus (as in multiple sclerosis) will have eyes that are pointed straight ahead at rest, but there will be paralysis of adduction (moving medially) of the ipsilateral (to the medial longitudinal fasciculus) eye and nystagmus of the abducting eye on attempts at conjugate lateral gaze.

Clinical Anatomy of Headache

Pain-sensitive structures that can be the site of headaches can be divided into extracranial and intracranial sites.

Extracranial Sites for Pain

These sites include the following.

1. Periosteum of the skull.
2. Scalp—skin, sensory nerves, blood vessels, and occipitofrontalis muscles.
3. Postvertebral muscles of the neck.
4. Ears.
5. Eyes.
6. Teeth.
7. Nose.
8. Paranasal sinuses.
9. Temporomandibular joint.

Intracranial Sites for Pain

These sites include the following.

1. Meninges, including dural arteries, and venous sinuses.
2. Intracerebral arteries.

Clinical Note

Brain parenchyma and pain.—The substance of the brain and most of the covering meninges are *not* pain sensitive. Pain from many of the sites listed above is referred along the distribution of the branches of the trigeminal nerve (head area of pain) or the cutaneous branches of the second and third cervical nerves to the neck.

Muscle Contraction Headache

The pain is usually bandlike and its onset is gradual. Bilateral tenderness over the involved muscles is common. The involved muscles are usually bilateral and may include the occipitofrontalis, temporalis, and postvertebral muscles of the neck. If the neck muscles are the source of the pain, movements of the cervical part of the vertebral column accentuate the discomfort. Electromyographic evidence of sustained muscle contraction is often absent in these patients so that the cause is difficult to explain.

Vascular Dilatation Headaches

These may be bilateral or unilateral and may involve the extracranial or intracranial blood vessels.

Bilateral Vascular Headache.—The vasculature dilatation can be produced by toxins (endogenous or exogenous) and by hypertension.

Unilateral Vascular Headache.—The intracranial and extracranial vascular dilation can be produced in *migraine.* Initially the arteries are believed to undergo vasoconstriction, which is followed by vasodilation. In the so-called *cluster headaches* the vasodilatation involves the external or internal carotid arteries or their branches.

Headache of Subarachnoid Hemorrhage

The sudden displacement of the intracranial blood vessels is believed to be responsible for the initial headache. Later, the presence of blood in the subarachnoid space produces a chemical meningitis and vasodilation of dural blood vessels.

Headache of Epidural and Subdural Hematomas

The presence of an expanding hematoma exerts traction on the pain-sensitive dura or blood vessels.

Headaches of Cerebrovascular Accidents or Tumors

The expanding blood mass or tumor stretches the pain-sensitive dura and intracranial blood vessels. Supratentorial space-occupying lesions stimulate pain receptors in the dura and send their impulses via the trigeminal nerve, and the headache is referred to the forehead and face. Pain impulses from below the tentorium in the posterior cranial fossa travel via the upper three cervical nerves, and the headache is referred to the back of the head and neck.

Headache of Meningitis

The headache of meningitis is felt over the entire head and may be associated with retrobulbar pain of the eye. There is stiffness of the neck and a definite fever. It is believed that the headache originates from the pain-sensitive endings in the dura, since the pia and arachnoid are not sensitive to pain.

Clinical Anatomy of Techniques for Treating Intracranial Hematomas

Burr Holes

Indications for Burr Holes.—Cranial decompression is performed in a patient with a history of progressive neurologic deterioration and signs of brain herniation, in spite of adequate medical treatment. The presence of a hematoma should, if possible, be confirmed by a computed tomographic scan.

Anatomy of the Technique for a Temporal Burr Hole.—The technique involves the following steps.

1. The patient is placed in a supine position with the head rotated so that the side for the burr hole is uppermost. For example, in a patient with a right-sided fixed dilated pupil, indicating herniation of the right uncus with pressure on the right oculomotor nerve, a hematoma on the right side must be presumed, and a burr hole is placed on the right side.

2. The temporal skin is shaved and prepared for surgery in the usual way.

3. A 3-cm vertical skin incision is made two fingerbreadths anterior to the tragus of the ear and three fingerbreadths above this level (Fig 8–25).

4. The following structures are then incised:

a. Skin

b. Superficial fascia containing small branches of the superficial temporal artery

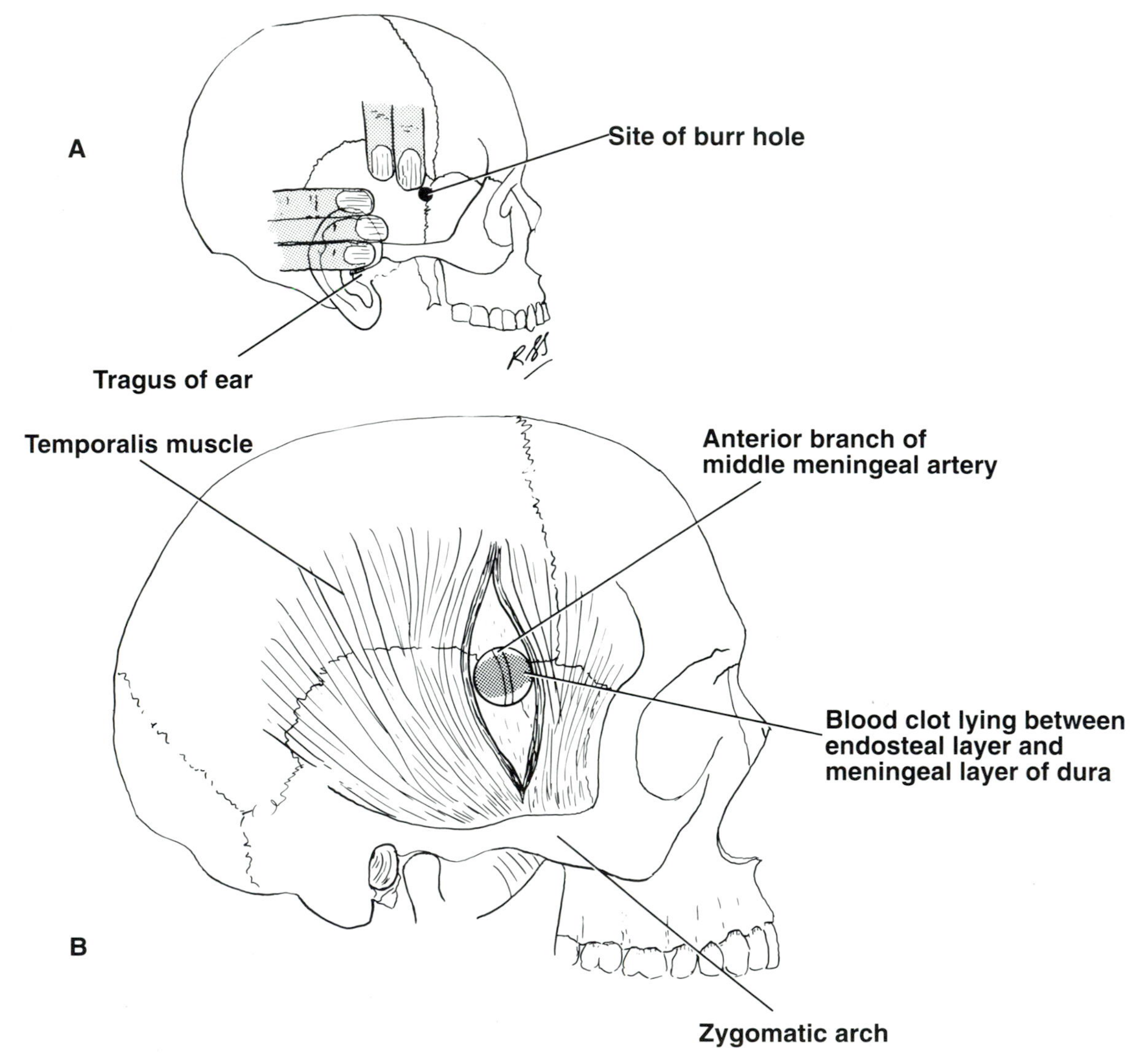

FIG 8–25.
A, surface landmarks for a temporal burr hole. **B,** the vertical incision passes through the temporalis muscle down to bone. The middle meningeal artery lies between the endosteal and meningeal layers of dura and is embedded in the endosteal layer of dura or lies in a bony tunnel.

c. Deep fascia covering the outer surface of the temporalis muscle

d. The temporalis muscle is then incised vertically down to the periosteim of the squamous part of the temporal bone (see Fig 8–25)

e. The temporalis muscle is elevated from its attachment to the skull and a retractor is positioned (some muscular bleeding will be encountered)

f. A small hole is then drilled through the outer and inner tables of the skull at right angles to the skull surface, and the hole is enlarged with a burr (unless a blood clot is present between the inner table and the endosteal layer of dura)

g. The white meningeal layer of dura is flexible and gives slightly on gentle pressure

h. The hole may be enlarged with a curette, and bleeding from the diploë may be controlled with bone wax.

The surgical wound is closed in layers with interrupted sutures placed in the temporalis muscle, the deep fascia covering the temporalis muscle, and the scalp.

Burr Hole for Epidural Hematoma.—Once the inner table of the squamous part of the temporal bone (or the anterior inferior angle of the parietal bone) is pierced with a small bit and enlarged with a burr, the dark red clotted blood beneath the endosteal layer of dura is usually easily recognized. However, bright red liquid blood means that the middle meningeal artery or one of its branches is bleeding. The meningeal artery is located deep to the clot and between the endosteal layer of dura and the meningeal layer of dura or in the substance of the endosteal layer of dura; or it may lie in a tunnel of bone.

Burr Hole for Subdural Hematoma.—When the squamous part of the temporal bone is penetrated, as described above, the endosteal layer of dura will be exposed. In this case there is no blood clot between the endosteal layer of dura and the meningeal layer of dura, but both fused layers of dura will be dark bluish. The dura (endosteal and meningeal layers) is gently incised to enter the space between the meningeal layer of dura and the arachnoid mater. The subdural blood usually gushes out, leaving the unprotected brain covered only by arachnoid and pia mater in the depths of the hole.

Anatomy of the Complications of a Temporal Burr Hole.—The complications include the following:

1. *Facial nerve injury.* The skin incision may extend too far inferiorly and cross the zygomatic arch, where it may cut the zygomatic branch of the facial nerve.

2. *Superficial temporal vessel injury.* The skin incision should be made two fingerbreadths anterior to the tragus to avoid the superficial temporal artery and vein and the auriculotemporal nerve. These latter structures ascend from the infratemporal fossa just behind the temporomandibular joint; the pulsations of the artery can easily be felt.

3. *Brain injury.* Unless great care is taken when penetrating the skull bone with a drill bit, the end of the bit can plunge through the meninges into the brain substance. In the case of subdural hematoma, the brain may be damaged when the meningeal layer of dura is incised to open up the space between the dura and the arachnoid mater. Excessive suction during the removal of the blood clot may also damage the underlying brain.

4. *Uncontrollable bleeding from the middle meningeal artery.* The branches of the artery are located between the endosteal layer of dura and the meningeal layer of dura. Should the artery be embedded in the endosteal layer of dura, it may be difficult to secure without cautery; the placement of the artery in a bony tunnel may also make ligation of the vessel difficult.

5. *Bleeding from meningeal veins.* Uncontrollable bleeding from large meningeal veins may be a problem in cases of subdural hematoma when excessive suction is used.

Anatomy of the Technique for Frontal and Parietal Burr Holes.—The anterior burr hole is made three fingerbreadths from the midline and three fingerbreadths back from the hair line. The posterior burr hole is made four fingerbreadths posterior to the anterior hole. The incision is made through the layers of the scalp (see Fig 6–9) to the periosteum of the skull. The temporalis muscle is not present in these areas. Penetration of the skull bones and the meninges is identical to that described for the temporal burr hole. One of the possible complications is injury to the superior sagittal sinus and the arachnoid granulations, should the burr hole be made too close to the midline.

Clinical Anatomy of the Technique of Ventriculostomy

Indications for Ventriculostomy

Ventriculostomy is indicated in acute hydrocephalus, in which there is a sudden obstruction to the flow of cerebrospinal fluid.

Anatomy of the Technique of Ventriculostomy

The needle is inserted into the lateral ventricle through either a frontal or parietal burr hole. The anatomy of these burr holes has been described previously. The needle is inserted through the burr hole using the following anatomical landmarks.

1. *Frontal approach.* The needle is inserted through the frontal burr hole and is directed downward and forward in the direction of the inner canthus of the ipsilateral eye (Fig 8–26).

2. *Parietal approach.* The needle is inserted through the parietal burr hole and is directed downward and forward in the direction of the pupil of the ipsilateral eye (see Fig 8–26).

The needle is inserted to a depth of about 5.5 cm from the skull opening; in cases of chronic hydrocephalus with gross dilatation of the ventricles, the depth of penetration to the ventricular cavity may be much less.

Anatomy of Complications.—Complications include the following.

1. *Infection.* Infection through the wound.

2. *Hemorrhage.* Bleeding from the surface meningeal vessels and hemorrhage into the ventricles from the branches of the cerebral arteries.

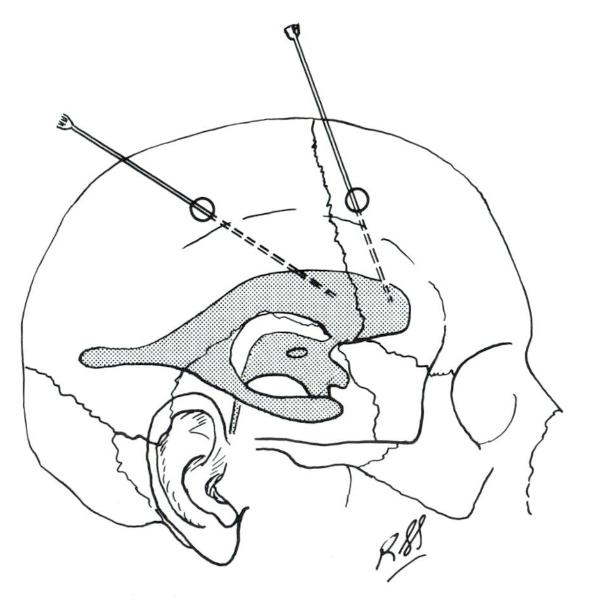

FIG 8–26.
Ventriculostomy. Needles passing through frontal or parietal burr holes to enter the lateral ventricle are shown. The needle is inserted to a depth of about 5.5 cm from the skull opening in order to enter the lateral ventricle.

Clinical Anatomy of the Technique for Using Gardner-Wells Tongs

Indications

Skull tongs are used in patients who have unstable fractures of the cervical part of the vertebral column that require traction in the long axis of the column (see Chapter 10, p. 386).

Anatomy of the Procedure

The scalp is shaved and prepared above both ears and a local anesthetic is infiltrated to block the sensory nerves (branches of the auriculotemporal nerve from the trigeminal nerve, and branches of the lesser occipital nerve from the cervical plexus).

The points of the tongs are applied above and just behind the auricle and below the equator of the skull, i.e., below the widest point on the circumference of the skull (Fig 8–27). This corresponds to a level just below the temporal lines and is about 5 cm above the tip of the mastoid process. The points of the tongs pierce the layers of the scalp and come to rest on the outer table of the skull with its covering of periosteum. The periosteum is richly supplied with sensory nerves that need to be adequately anesthetized.

Anatomy of the Complications

These are described in Chapter 10, p. 386.

ANATOMICAL-CLINICAL PROBLEMS

1. A 27-year-old man was admitted to the emergency department unconscious. He had been hit on the side of the head by a car while crossing the road. Within an hour, his state of unconsciousness deepened. On examination, he was found to have a large, doughlike swelling over the left temporalis muscle. He also had the signs of left-sided hemiplegia. Later, a left-sided, fixed, dilated pupil developed. A lateral radiograph of the skull showed a linear fracture running downward and forward across the pterion. His coma deepened, and he died 4 hours after the accident. Using your knowledge of anatomy, make a diagnosis. Explain the clinical findings. How would you explain the homolateral hemiplegia?

2. There is no doubt that the disabilities experienced by a patient following a head injury are likely to be greater following damage to the dominant hemisphere than to the opposite hemisphere, and that the effect is much greater in adults than in chil-

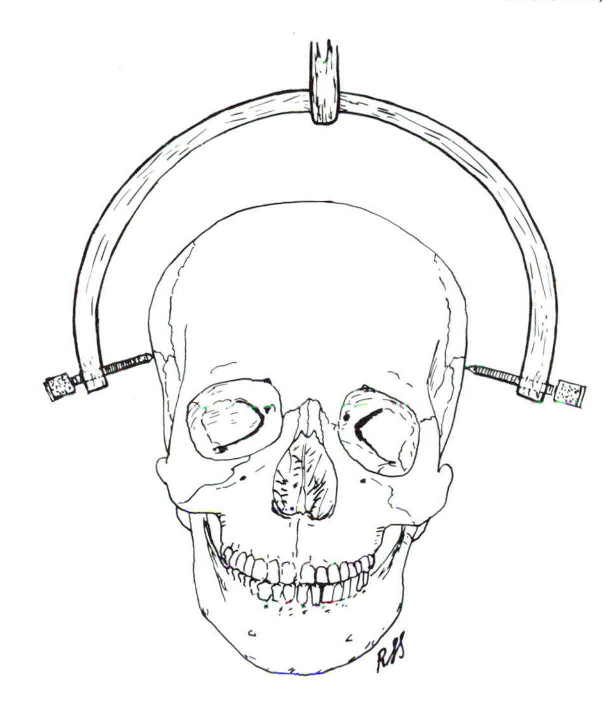

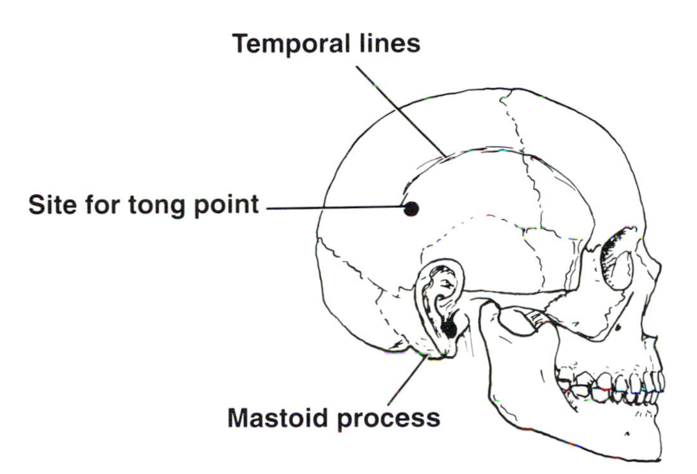

FIG 8–27.
Surface landmarks for the application of Gardner-Wells tongs. The points of the tongs are applied just above and behind the auricle and below the equator of the skull, i.e., just below the temporal lines and about 5 cm above the tip of the mastoid process.

dren. What structures exist within the skull to limit damage to the cerebral hemispheres and other parts of the brain? Which blood vessels are damaged more commonly, the cerebral arteries or the cerebral veins? Which cranial nerves are likely to be damaged in head injuries? What is the reason for their susceptibility?

3. A 27-year-old woman was found, on examination of her right eye, to have paralysis of the lateral rectus muscle; the right pupil was dilated but reacted slowly to light, and there was some anesthesia of the skin over the right forehead. A carotid arteriogram revealed the presence of an aneurysm of the right internal carotid artery situated in the cavernous sinus. Using your knowledge of anatomy, can you explain the clinical findings on physical examination?

4. A 56-year-old woman was examined in the emergency department complaining of a severe headache. She said that the headache had started about 1 hour after she had hit her head on the mantlepiece of a fireplace after bending down to poke the fire. Four hours later it was noticed that she was becoming mentally confused and was developing a left-sided hemiplegia on the side opposite the head injury. Her deep reflexes were exaggerated, and she had a positive Babinski response on the left side. Computed tomographic scan demonstrated a right subdural hematoma. Explain in anatomical terms the development of a subdural hematoma.

5. List the common extracranial sites that can cause headaches and state their sensory innervation. To which areas of the head is the pain referred to in each case?

6. In anatomical terms explain the signs and complications of basal fractures of the skull involving the anterior cranial fossa.

7. Which part of the base of the skull is most prone to fracture? Can you give an anatomical reason for this answer?

8. For a person to be conscious they have to be awake and aware. Which part of the brain is responsible for each of these characteristics? Describe in simple terms the afferent pathways in the brain that are associated with wakefulness.

9. Explain the doll's eye reflex and the cold caloric test. Explain in anatomical terms the significance of these tests.

10. It is not uncommon to read in newspapers of the survival of a baby that has fallen from a great height, such as a third-floor window, and yet it is known that if an adult falls from a similar height, it would be fatal. Have you an anatomical explanation, based on age, for this difference in survival?

11. Name the structures within the skull that are sensitive to pain. What is their nerve supply? Is the brain substance sensitive to pain?

12. Using your knowledge of the anatomical pathways along which the cerebrospinal fluid flows, name the sites at which pathological blockage may occur.

13. There are no anastomoses of clinical importance between the terminal end arteries within the brain substance, but there are many important anastomoses between the large arteries, both within and outside the skull, and these may play a major role in determining the extent of brain damage in cerebral vascular disease. Name the sites at which important arterial anastomoses take place.

14. A 59-year-old woman experienced paralysis on the left side of her body, mainly involving the lower limb. She also had some sensory loss on the left side of the body. She was able to swallow normally and did not appear to have difficulty with her speech. What is the anatomic diagnosis?

15. A 38-year-old man was seen in the emergency department with a history of a sudden excruciating, generalized headache while gardening. Ten minutes later the patient collapsed to the ground in a state of unconsciousness. After being carried indoors and placed on a settee, he regained consciousness but appeared confused. He complained of a severe headache and a stiff neck. Physical examination revealed some rigidity of the neck but nothing further. A careful neurological examination 3 days later revealed some loss of tone of the muscles of the left leg. Using your knowledge of anatomy, make a diagnosis. What caused the neck rigidity?

16. Can you explain why patients with a thrombosis of the middle cerebral artery often present with homonymous hemianopia as well as hemiplegia and hemianesthesia?

17. A 60-year-old man was walking to work when he collapsed in the street. He complained of a sudden severe headache. While he was being examined in the emergency department, it was noted that his face began to sag on the right side and his speech became slurred. His right arm and leg were found to be weaker than his left arm and leg, and the muscles were hypotonic. His eyes were deviated to the left. Later, his right arm and leg showed complete paralysis and were insensitive to pinprick. A Babinski sign was present on the right side. Three hours later, the patient relapsed into a deep coma with bilateral dilated fixed pupils. Later, his respirations became deep and irregular, and he died 7 hours later. Using your knowledge of anatomy, make a diagnosis. Explain each of the clinical findings in anatomical terms.

18. Explain the possible cause of headache in patients with meningitis.

19. In anatomical terms explain Battle's sign in head injuries.

ANSWERS

1. The initial loss of consciousness was due to cerebral trauma. The swelling over the left temporalis muscle and the radiographic finding of a linear fracture over the anterior inferior angle of the left parietal bone (pterion) would suggest that the left middle meningeal artery had been damaged and an extradural hemorrhage had occurred. Blood had extravasated through the fracture line into the overlying temporalis muscle and soft tissue. The left homolateral hemiplegia was due to the compression of the right cerebral peduncle against the edge of the tentorial notch of the tentorium cerebelli. This is unusual. A right hemiplegia due to pressure on the left precentral gyrus is more common.

The left-sided, fixed, dilated pupil was due to the pressure on the left oculomotor nerve by the hippocampal gyrus, which had herniated through the tentorial notch.

2. Loss of function of the dominant hemisphere

results in not only the loss of acquired skills of the dominant hand but also the loss of certain mental functions. The degree of disability depends on the degree of dominance of one hemisphere over the other. A child's brain possesses a degree of flexibility that is not present in the adult, so that good recoveries can be expected in young children.

The meninges and the cerebrospinal fluid afford a remarkable degree of protection to the brain tissue. The dural partitions limit the extent of brain movement within the skull.

The thin-walled cerebral veins are liable to be damaged during excessive movements of the brain relative to the skull, especially at the point where the veins join the dural venous sinuses. The thick-walled cerebral arteries are rarely damaged.

The small-diameter cranial nerves of long length are particularly prone to damage during head injuries. The trochlear, abducent, and oculomotor nerves are commonly injured.

3. The internal carotid artery passes forward on the lateral surface of the body of the sphenoid within the cavernous sinus. An aneurysm of the artery may press on the abducent nerve (Fig 8–15) and cause paralysis of the lateral rectus muscle. Further expansion of the aneurysm may cause compression of the oculomotor nerve and the ophthalmic division of the trigeminal nerve as they lie in the lateral wall of the cavernous sinus. This patient had right lateral rectus paralysis and paralysis of the right pupillary constrictor muscle owing to the involvement of the abducent and oculomotor nerves, respectively. The slight anesthesia of the skin over the right forehead was due to pressure on the ophthalmic division of the right trigeminal nerve.

4. A subdural hematoma is an accumulation of blood clot in the interval between the meningeal layer of dura and the arachnoid mater. It results from tearing of the superior cerebral veins at their point of entrance into the superior sagittal sinus. The cause is usually a blow on the front or the back of the head, resulting in excessive anteroposterior displacement of the brain within the skull.

5. (1) Periosteum of the skull (trigeminal and C2 and C3 spinal nerves).
(2) Scalp.—sensory nerves and blood vessels (ophthalmic and mandibular divisions of trigeminal nerve, greater occipital nerve (C2), and sympathetic nerves); occipitofrontalis muscle (motor innervation-facial nerve, sensory innervation-trigeminal nerve).
(3) Postvertebral muscles of neck (cervical spinal nerves).
(4) Ears.—middle ear (glossopharyngeal nerve); tympanic membrane (auriculotemporal nerve and auricular branch of the vagus).
(5) Eyes.—globe and extraocular muscles (ophthalmic division of the trigeminal nerve).
(6) Teeth.—upper jaw (maxillary division of trigeminal nerve); lower jaw (mandibular division of trigeminal nerve).
(7) Nose.—smell (Olfactory nerve); general sensations (ophthalmic and maxillary divisions of the trigeminal nerve).
(8) Paranasal sinuses (ophthalmic and maxillary divisions of the trigeminal nerve).
(9) Temporomandibular joint (mandibular division of the trigeminal nerve).

Pain originating in structures innervated by the trigeminal nerve will be referred to the trigeminal dermatomes (see Fig 6–1). Similarly, pain originating in structures innervated by cervical nerves will be referred to cervical dermatomes.

6. The signs and complications of fractures of the anterior cranial fossa include epistaxis and cerebrospinal rhinorrhea, subconjunctival hemorrhage and ecchymosis of eyelids, and anosmia. These conditions are described on p. 308.

7. The middle cranial fossa is the part of the skull most prone to fracture, since it possesses numerous foramina, canals, and has air spaces, namely, the sphenoid air sinus and the tympanic cavity. The foramen magnum in the posterior cranial fossa is very large but its boundaries are extremely thick.

8. Awakeness requires the normal functioning of the reticular formation in the brain stem. Awareness requires the normal functioning of both cerebral cortices. The reticular formation, which is the important part of the awakening system of the brain, receives branches from all the ascending sensory pathways as they pass up to the thalamus to be relayed to the cerebral cortex.

9. The doll's eye reflex and the caloric test and their significance are fully described on p. 316.

10. In infants the skull bones are more resilient than in adults, and they are separated by fibrous sutural ligaments. In adults the inner table of the skull is particularly brittle and the sutural ligaments begin to ossify during middle age (see also p. 307).

11. (1) Meninges.—The dura mater above the tentorium cerebelli receives its innervation from the trigeminal nerve, and a headache is referred to the skin of the forehead and face. Below the level of the tentorium the dura is innervated by the vagus nerve and the upper three cervical nerves. Pain from this

area is referred to the back of the head and neck via the cervical nerves.

(2) Intracerebral arteries.—These arteries are innervated by sympathetic nerves, and it is not known whether these nerves carry pain impulses. The brain substance is insensitive to pain.

12. The common sites for blockage of the flow of cerebrospinal fluid are where the passages are narrowest, namely, the interventricular foramina (foramina of Monro), the cerebral aqueduct, the median aperture, and the lateral apertures in the roof of the fourth ventricle. It is possible for inflammatory exudate secondary to meningitis or a cerebral tumor to narrow down or even obliterate the opening in the tentorial notch so that the passage of the cerebrospinal fluid to the outer surface of the cerebral hemisphere is impeded or stopped. Inflammatory exudate may also block the drainage of the fluid into the superior sagittal sinus at the arachnoid villi.

13. Once the terminal branches of the cerebral arteries enter the brain substance, no further anastomoses occur. Blockage of such end arteries by disease is quickly followed by neuronal death and necrosis. The following important anastomoses exist between the cerebral arteries: (1) the circle of Willis, (2) anastomoses between the branches of the cerebral arteries on the surface of the cerebral hemispheres and the cerebellar hemispheres, and (3) anastomoses between the branches of the internal and external carotid arteries at their origin at the common carotid artery, at the anastomosis between the branches of the ophthalmic artery within the orbit and the facial and maxillary arteries, and between the meningeal branches of the internal carotid artery and the middle meningeal artery.

14. Left-sided hemiplegia mainly involving the leg strongly suggests cerebrovascular disease with a lesion involving the right cerebral hemisphere. The paralysis mainly involving the leg would indicate that the right anterior cerebral artery or one of its branches was blocked by a thrombus or embolus.

15. This patient had a congenital aneurysm of the anterior communicating artery. The sudden onset of a severe headache, which is often so dramatic that the patient feels as though he has been hit on the head, is characteristic of rupture of a congenital aneurysm into the subarachnoid space. The stiff or rigid neck is due to meningeal irritation caused by the presence of blood in the subarachnoid space. This patient had no evidence of previous pressure on the optic nerve leading to unilateral visual defect, which sometimes occurs when the aneurysm is situated on the anterior part of the circle of Willis. The loss of tone in the left leg muscles is difficult to explain, although it may be due to penetration of the hemorrhage into the right cerebral hemisphere.

16. The middle cerebral artery, in addition to giving off cortical branches, gives off central branches that supply part of the posterior limb of the internal capsule and the optic radiation. Occlusion of these branches will cause contralateral homonymous hemianopia.

17. The sudden onset of severe headache, slurring of speech, right lower facial weakness, right-sided hemiplegia, a right positive Babinski sign, right-sided hemianesthesia, and deviation of the eyes to the left side are all diagnostic of a cerebrovascular accident involving the left cerebral hemisphere. The perforating central branches of the left middle cerebral artery were found at autopsy to be extensively affected by atherosclerosis. One of these arteries had ruptured, resulting in a large hemorrhage into the lentiform nucleus and left internal capsule. The combination of hypertension and atherosclerotic degeneration of the artery was responsible for the fatal hemorrhage. The dilated fixed pupils, the irregularity in breathing, and, finally, death were due to the raised pressure within the hemisphere causing downward pressure effects within the brain stem.

18. The headache of meningitis originates in the pain-sensitive endings in the dura mater.

19. Battle's sign is an ecchymosis seen over the mastoid process 24 to 48 hours after a head injury and is secondary to a fracture of the petrous part of the temporal bone (see p. 308).

9

Overview of the Autonomic Nervous System

The autonomic nervous system exerts control over the functions of many organs and tissues in the body. The emergency physician needs a sound grounding in the structure and function of this system to apply physiology and pharmacology in daily clinical practice. This chapter briefly reviews the autonomic system and gives some insight into the different types of receptors and the action of neurotransmitters.

GENERAL ARRANGEMENT OF THE AUTONOMIC NERVOUS SYSTEM

The autonomic nervous system, along with the endocrine system, controls and brings about the fine internal adjustments necessary for the optimal internal environment of the body.

The autonomic nervous system, like the somatic nervous system, has afferent, connector, and efferent neurons. The afferent impulses originate in visceral receptors and travel via afferent pathways to the central nervous system, where they are integrated through connector neurons at different levels and then leave via efferent pathways to visceral effectors.

The efferent pathways of the autonomic system are made up of preganglionic and postganglionic neurons. The cell bodies of the preganglionic neurons are situated in the lateral gray column of the spinal cord and in the motor nuclei of the third, seventh, ninth, and tenth cranial nerves. The axons of these cell bodies synapse on the cell bodies of the postganglionic neurons that are collected together to form *ganglia* outside the central nervous system.

The control of the autonomic system is widespread; one preganglionic axon may synapse with several postganglionic neurons. Large collections of afferent and efferent nerve fibers and their associated ganglia form *autonomic plexuses* in the thorax, abdomen, and pelvis.

The visceral receptors include chemoreceptors, baroreceptors, and osmoreceptors. Pain receptors are present in viscera, and certain types of stimuli, such as a lack of oxygen, stretch, may cause extreme pain.

BASIC ANATOMY

Basic Anatomy of the Autonomic Nervous System

The autonomic nervous system innervates involuntary structures such as the heart, smooth muscles, and glands. The system is distributed throughout the central and peripheral nervous systems, is divided into two parts—the *sympathetic* and the *parasympathetic* —and, as emphasized above, consists of both afferent and efferent nerve fibers. This division between sympathetic and parasympathetic is made on the basis of differences in anatomy, differences in neurotransmitters, and differences in physiologic effects (Table 9–1).

Sympathetic Part of the Autonomic System

The sympathetic system is the larger of the two parts of the autonomic system and is widely distributed throughout the body, innervating the heart and lungs, the muscle in the walls of many blood vessels, the hair follicles and the sweat glands, and many abdominopelvic viscera.

The function of the sympathetic system is to prepare the body for an emergency. The heart rate is increased, arterioles of the skin and intestine are constricted, those of the skeletal muscles are dilated, and blood pressure is raised. The blood is redistributed so that it leaves the skin and gastrointestinal tract and passes to the brain, heart, and skeletal muscle. In addition, the sympathetic nerves dilate the pupils; inhibit the smooth muscle of the bronchi, intestine, and bladder wall; and close the sphincters. The hair is made to stand on end, and sweating occurs.

The sympathetic system includes the efferent outflow from the spinal cord, two ganglionated sympathetic trunks, important branches, plexuses, and regional ganglia.

Efferent Nerve Fibers (Sympathetic Outflow).—The lateral gray columns (horns) of the spinal cord from the first thoracic segment to the second lumbar segment (and sometimes the third lumbar segment) possess the cell bodies of the sympathetic connector neurons (Fig 9–1). The myelinated axons of these cells leave the cord with the anterior nerve roots and pass into the spinal nerve trunks. They then leave the nerve via the *white rami communicantes* (the white rami are white because the nerve fibers are covered with white myelin) and join the *paravertebral ganglia* of the *sympathetic trunk.* Once these fibers (preganglionic) reach the ganglia in the sympathetic trunk, the distribution is as follows.

1. They synapse with an excitor neuron in the ganglion (Figs 9–1 and 9–2). The gap between the two neurons is bridged by the neurotransmitter *acetylcholine.* The postganglionic nonmyelinated axons leave the ganglion and pass to the thoracic spinal

TABLE 9–1.
Comparison of Anatomical, Physiological, and Pharmacological Characteristics of the Sympathetic and Parasympathetic Parts of the Autonomic Nervous System

Characteristics	Sympathetic	Parasympathetic
Action	Prepares body for emergency	Conserves and restores energy
Outflow	T1–L2 (3)	Cranial nerves 3, 7, 9, and 10; spinal nerves S2 through S4
Preganglionic fibers	Myelinated	Myelinated
Ganglia	Paravertebral (sympathetic trunks); prevertebral (e.g., celiac, superior mesenteric, inferior mesenteric)	Small ganglia close to viscera (e.g., otic, ciliary) or ganglion cells in plexuses (e.g., cardiac, pulmonary)
Neurotransmitter within ganglia	Acetylcholine	Acetylcholine
Ganglion-blocking agents	Hexamethonium and tetraethylammonium by competing with acetylcholine	Hexamethonium and tetraethylammonium by competing with acetylcholine
Postganglionic fibers	Long, nonmyelinated	Short, nonmyelinated
Characteristic activity	Widespread due to many postganglionic fibers and liberation of epinephrine and norepinephrine from suprarenal medulla	Discrete action with few postganglionic fibers
Neurotransmitter at postganglionic endings	Norepinephrine at most endings and acetylcholine at a few endings (sweat glands)	Acetylcholine at all endings
Blocking agents on receptors of effector cells	Alpha-adrenergic receptors, phenoxybenzamine, beta-adrenergic receptors, propranolol	Atropine, scopolamine
Agents inhibiting synthesis and storage of neurotransmitter at postganglionic endings	Reserpine	
Agents inhibiting hydrolysis of neurotransmitter at site of effector cells		Acetylcholinesterase blockers (e.g., neostigmine)
Drugs mimicking autonomic activity	Sympathomimetic drugs—phenylephrine: alpha receptors; isoproterenol: beta receptors	Parasympathomimetic drugs—pilocarpine, methylcholine
Higher control	Hypothalamus	Hypothalamus

nerves as *gray rami communicantes* (the gray rami are gray because the nerve fibers are devoid of myelin). They are distributed in branches of the spinal nerves to smooth muscle in the blood vessel walls, sweat glands, and arrector pili muscles of the skin.

2. Axons may travel cephalad in the sympathetic trunk (see p. 329) to synapse in ganglia in the cervical region (see Fig 9–2). The postganglionic nerve fibers pass via gray rami communicantes to join the cervical spinal nerves. Many of the preganglionic fibers entering the lower part of the sympathetic trunk from the lower thoracic and upper two lumbar segments of the spinal cord travel caudad to synapse in ganglia in the lower lumbar and sacral regions. Here, again, the postganglionic nerve fibers pass via gray rami communicantes to join the lumbar, sacral, and coccygeal spinal nerves (see Fig 9–2).

3. Preganglionic nerve fibers may pass through the ganglia of the sympathetic trunk without synapsing. These myelinated fibers leave the sympathetic trunk as the *greater splanchnic, lesser splanchnic, and lowest or least splanchnic nerves.* The greater splanchnic nerve is formed from branches of the 5th

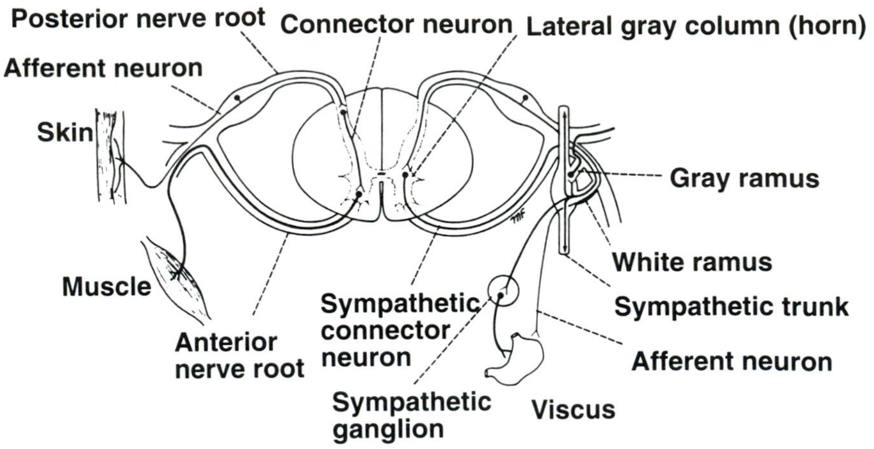

Posterior nerve root
Connector neuron
Lateral gray column (horn)
Afferent neuron
Skin
Gray ramus
White ramus
Muscle
Sympathetic trunk
Sympathetic connector neuron
Anterior nerve root
Afferent neuron
Sympathetic ganglion
Viscus

FIG 9–1.
General arrangement of the somatic part of the nervous system (at left) compared with autonomic part of nervous system (at right).

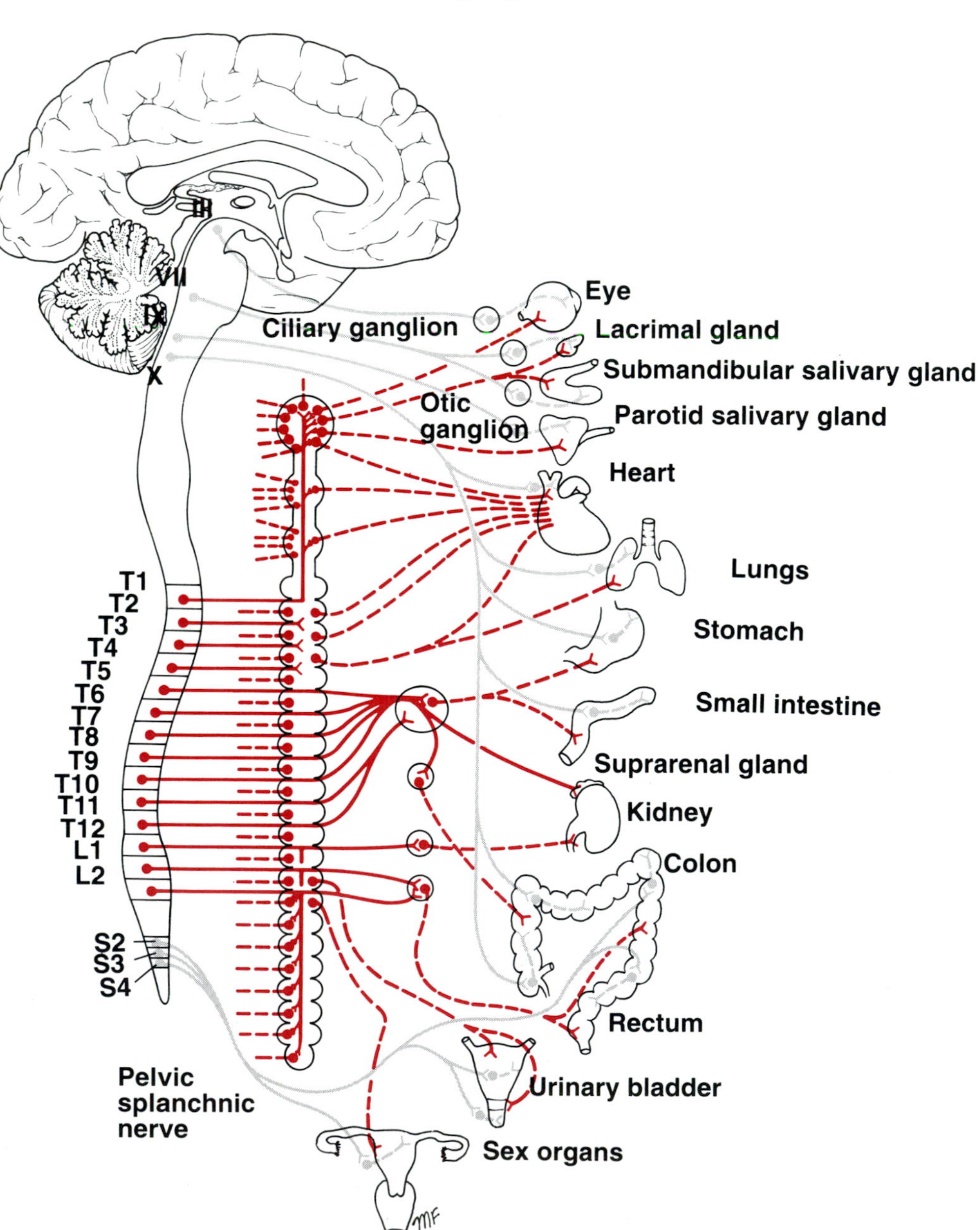

FIG 9–2.
Efferent part of the autonomic nervous system. Preganglionic sympathetic fibers (solid red) and postganglionic sympathetic fibers (interrupted red) are shown. Preganglionic parasympathetic fibers (solid gray) and postganglionic parasympathetic fibers (interrupted gray) are shown.

to 9th thoracic ganglia. It descends obliquely on the side of the bodies of the thoracic vertebrae and pierces the crus of the diaphragm to synapse with excitor cells in the ganglia of the *celiac plexus,* the *renal plexus,* and the suprarenal medulla. The lesser splanchnic nerve is formed from branches of the 10th and 11th thoracic ganglia. It descends with the greater splanchnic nerve and pierces the diaphragm to join excitor cells in ganglia in the lower part of the *celiac plexus.* The lowest splanchnic nerve (when present) arises from the 12th thoracic ganglion, pierces the diaphragm, and synapses with excitor neurons in the ganglia of the *renal plexus.*

The splanchnic nerves, therefore, are formed of preganglionic fibers. The postganglionic fibers arise from the excitor cells in the peripheral plexuses and are distributed to the smooth muscle and glands of the viscera. A few preganglionic fibers, traveling in the greater splanchnic nerve, end directly on the cells of the *suprarenal medulla* (see Fig 9–2). These medullary cells, which may be regarded as modified sympathetic excitor neurons, are responsible for epinephrine and norepinephrine secretion.

Afferent Nerve Fibers.—The afferent myelinated nerve fibers travel from the viscera through the sympathetic ganglia without synapsing. They pass to the spinal nerve via white rami communicantes and reach their cell bodies in the posterior root ganglion of the corresponding spinal nerve (see Fig 9–1). The central axons then enter the spinal cord and may form the afferent component of a local reflex arc or ascend to higher centers, such as the hypothalamus.

Sympathetic Trunks.—The sympathetic trunks are two ganglionated nerve trunks that extend the whole length of the vertebral column (see Fig 9–2). In the neck, each trunk has 3 ganglia; in the thorax, 11 or 12; in the lumbar region, 4 or 5; and in the pelvis, 4 or 5. In the neck, the trunks lie anterior to the transverse processses of the cervical vertebrae; in the thorax, they are anterior to the heads of the ribs or lie on the sides of the vertebral bodies; in the abdomen, they are anterolateral to the sides of the bodies of the lumbar vertebrae; and in the pelvis, they are anterior to the sacrum. Below, the two trunks end by joining together to form a single ganglion, the *ganglion impar.*

Cervical Part of the Sympathetic Trunk.—This extends upward to the base of the skull and below to the first rib, where it becomes continuous with the thoracic part of the sympathetic trunk. The trunk lies behind the carotid sheath. The cervical part of the trunk possesses three ganglia—the superior, middle, and inferior cervical ganglia (Figs 9–2 and 9–3). This part of the trunk sends gray rami communicantes to the spinal nerves, but there are no white rami communicantes.

Superior Cervical Ganglion.—This lies immediately below the skull and is joined to the middle ganglion by the connecting trunk.

Branches.—These include the *internal carotid nerve, gray rami communicantes, arterial branches, cranial nerve branches, pharyngeal branches,* and the *superior cardiac branch.*

Middle Cervical Ganglion.—This lies at the level of the cricoid cartilage. It is the smallest of the three cervical ganglia and is connected to the inferior ganglion by two or more nerve bundles. The anterior bundle crosses anterior to the first part of the subclavian artery, lying medial to the scalenus anterior muscle, and is known as the *ansa subclavia.*

Branches.—These include the *gray rami communicantes, thyroid branches, middle cardiac branch,* and *tracheal and esophageal branches.*

Inferior Cervical Ganglion.—In most subjects (80%), this is fused with the first thoracic ganglion to form the large *stellate ganglion.* It is located between the transverse process of the seventh cervical vertebra and the neck of the first rib. Inferior to the ganglion lies the cervical dome of the pleura.

Branches.—These include the *gray rami communicantes, arterial branches,* and the *inferior cardiac branch.*

The preganglionic sympathetic fibers that supply the head and neck leave the spinal cord through the upper four thoracic nerves (mainly the upper two or three). They ascend the sympathetic trunk to synapse with cells in the cervical ganglia. The postganglionic fibers are distributed as branches of the ganglia as described above.

Thoracic Part of the Sympathetic Trunk.—This at first runs downward in front of the heads of the ribs and then passes medially to lie on the side of the vertebral bodies. The thoracic part of the sympathetic trunk lies behind the costal pleura. It leaves the thorax by passing behind the medial arcuate ligament of the diaphragm to become continuous with the lumbar part of the sympathetic trunk. The trunk has 11 or 12 segmentally arranged ganglia (Figs 9–4 and 9–5).

Branches.—These include the *white rami communicantes* (white ramus contains preganglionic nerve fibers and afferent sensory nerve fibers), *gray rami communicantes* (gray ramus contains postganglionic

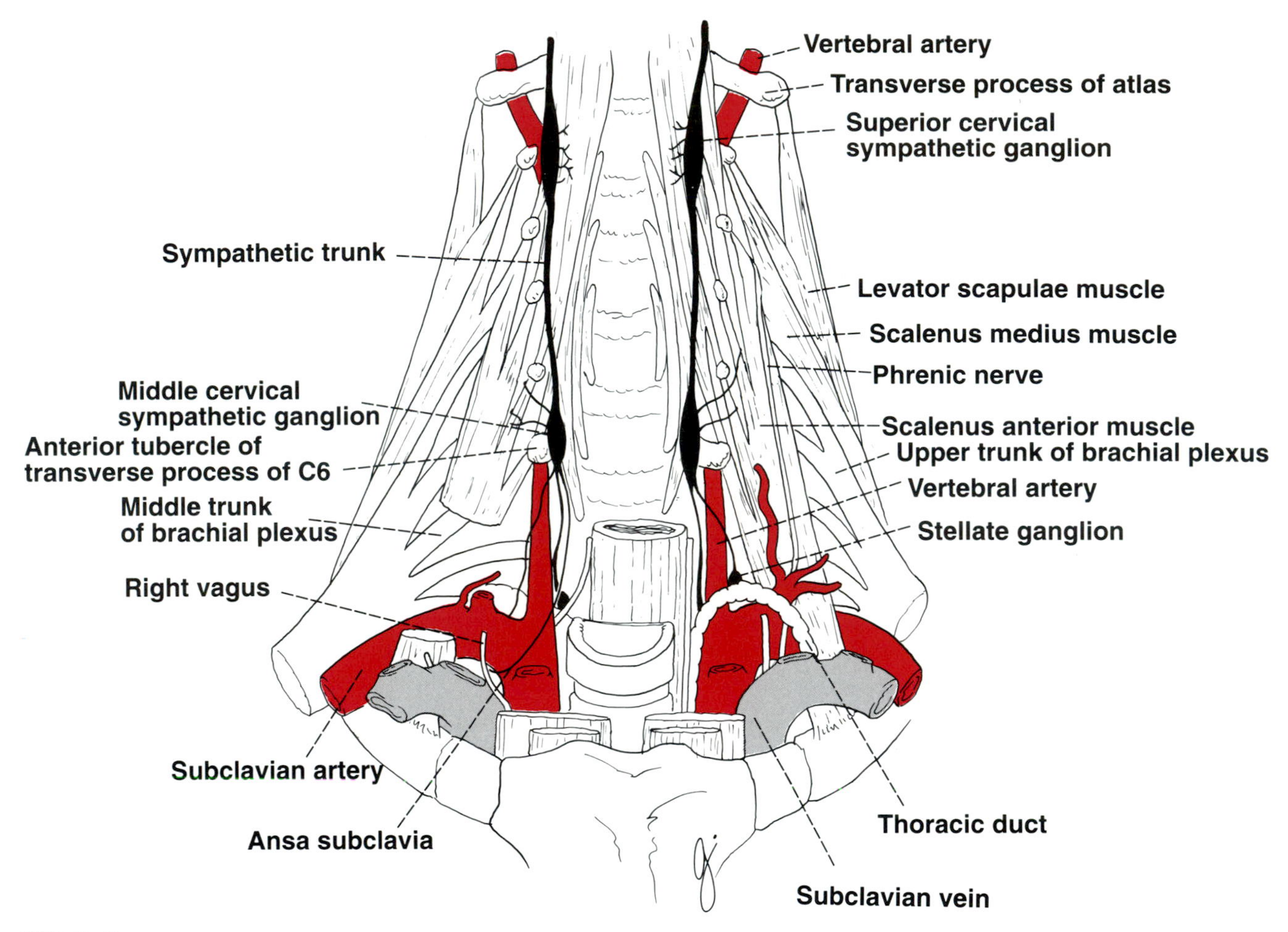

FIG 9–3.
The root of the neck showing the cervical parts of the sympathetic trunks.

nerve fibers), *visceral branches,* and *splanchnic branches.*

The splanchnic nerves supply the abdominal viscera and enter the abdomen by piercing the crura of the diaphragm. The *greater splanchnic nerve* is formed from branches of ganglia 5 through 9, the *lesser splanchnic nerve* is formed from branches of ganglia 10 and 11, and the *lowest splanchnic nerve* is a branch of the last thoracic ganglion (see Figs 9–4 and 9–5).

Lumbar Part of the Sympathetic Trunk.—This runs downward on the bodies of the lumbar vertebrae in the retroperitoneal connective tissue along the medial border of the psoas muscle (Fig 9–6). It is continuous above with the thoracic part of the trunk behind the medial arcuate ligament and below with the pelvic part of the trunk behind the common iliac vessels. The right trunk lies posterior to the right margin of the inferior vena cava, and the left trunk lies close to the left margin of the aorta and the lateral aortic lymph nodes, anterior to the lumbar vessels. Each trunk has four or five segmentally arranged ganglia (see Fig 9–6).

Branches.—These include the *white rami communicantes, gray rami communicantes, aortic branches,* and *pelvic branches.*

Pelvic Part of the Sympathetic Trunk.—This is continuous above, behind the common iliac vessels, with the lumbar part of the trunk (Fig 9–7). Below, the two trunks come together in front of the coccyx to form the *ganglion impar.* Each trunk descends in the retroperitoneal connective tissue behind the rectum and in front of the sacrum and medial to the anterior sacral foramina. The sympathetic trunk has four or five segmentally arranged ganglia (see Fig 9–7).

Branches.—These include *gray rami communicantes* and *hypogastric branches.*

Medulla of the Suprarenal Gland.—Preganglionic sympathetic fibers descend to the gland in the

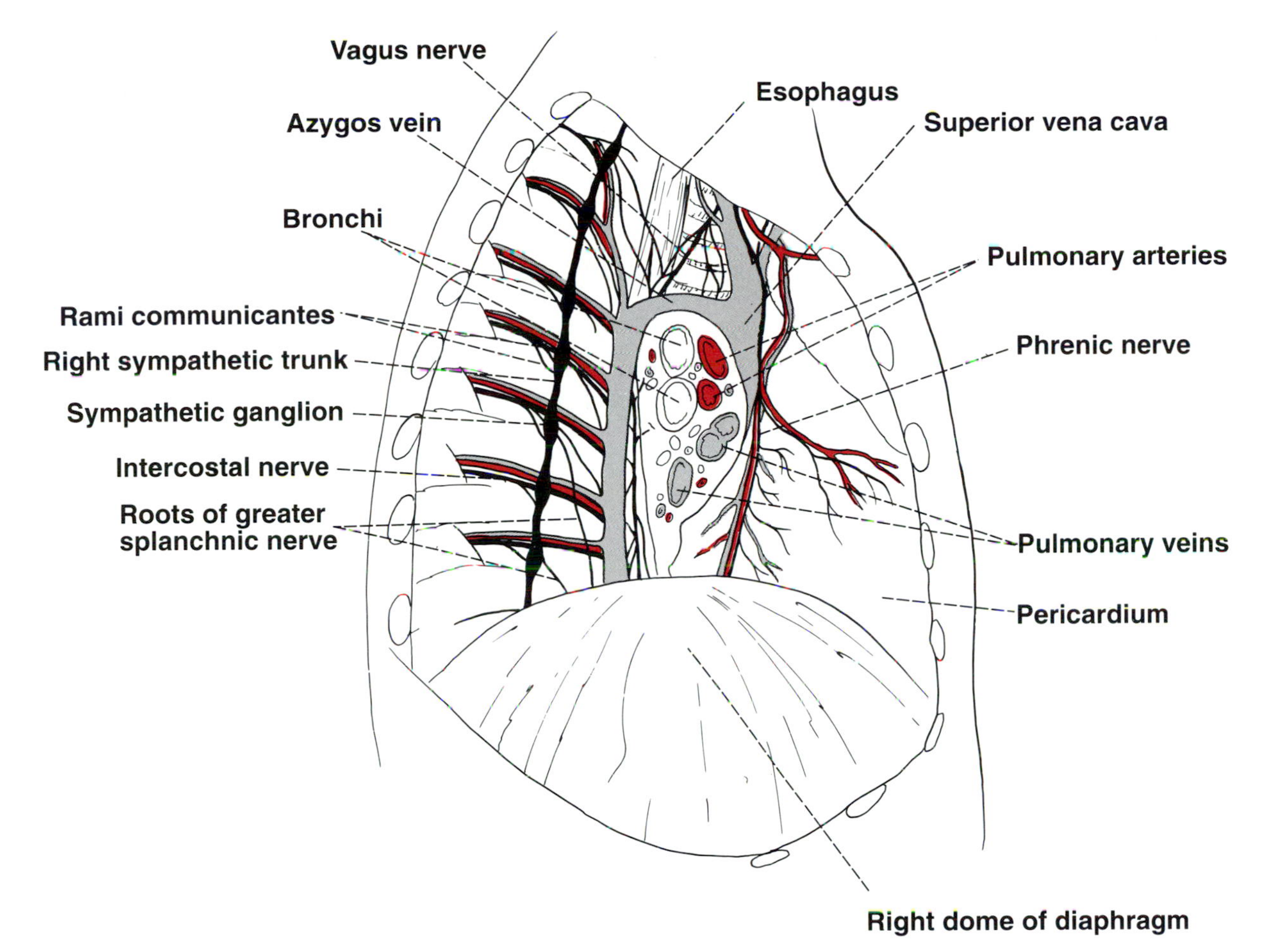

FIG 9–4.
Right side of the mediastinum showing the thoracic part of the sympathetic trunk and its main branches.

greater splanchnic nerve, a branch of the thoracic part of the sympathetic trunk (see Fig 9–2). The nerve fibers terminate on the secretory cells of the medulla, which are comparable to postganglionic neurons. Acetylcholine is the transmitter substance between the nerve endings and the secretory cells, as at all other preganglionic endings. The sympathetic nerves stimulate the secretory cells of the medulla to increase the output of epinephrine and norepinephrine. There is no parasympathetic innervation.

The spinal segmental levels of the sympathetic connector neurons associated with different organs in the body are shown in Figure 9–8.

Parasympathetic Part of the Autonomic System

The activities of the parasympathetic part of the autonomic system are directed toward conserving and restoring energy. In this regard, the peristaltic and glandular activity of the gut is increased. In addition, the pupils constrict, the heart rate is slowed, the sphincters are opened, and the bladder wall contracts.

Efferent Nerve Fibers (Craniosacral Outflow).—The parasympathetic part of the autonomic nervous system is located in the nuclei of the oculomotor (Edinger-Westphal nucleus), facial (superior salivatory and lacrimatory nuclei), glossopharyngeal (inferior salivatory nucleus), and vagus (dorsal nucleus) cranial nerves and the second, third, and fourth sacral segments of the spinal cord (see Fig 9–2). The axons of the cranial connector nerve cells are myelinated and emerge from the brain within the cranial nerves. The sacral connector nerve cells give rise to myelinated axons that leave the spinal cord in the anterior nerve roots of the corresponding spinal nerves. They then leave the sacral nerves and form the *pelvic splanchnic nerves* (see Fig 9–2).

The myelinated efferent fibers of the craniosacral outflow are preganglionic and synapse in peripheral

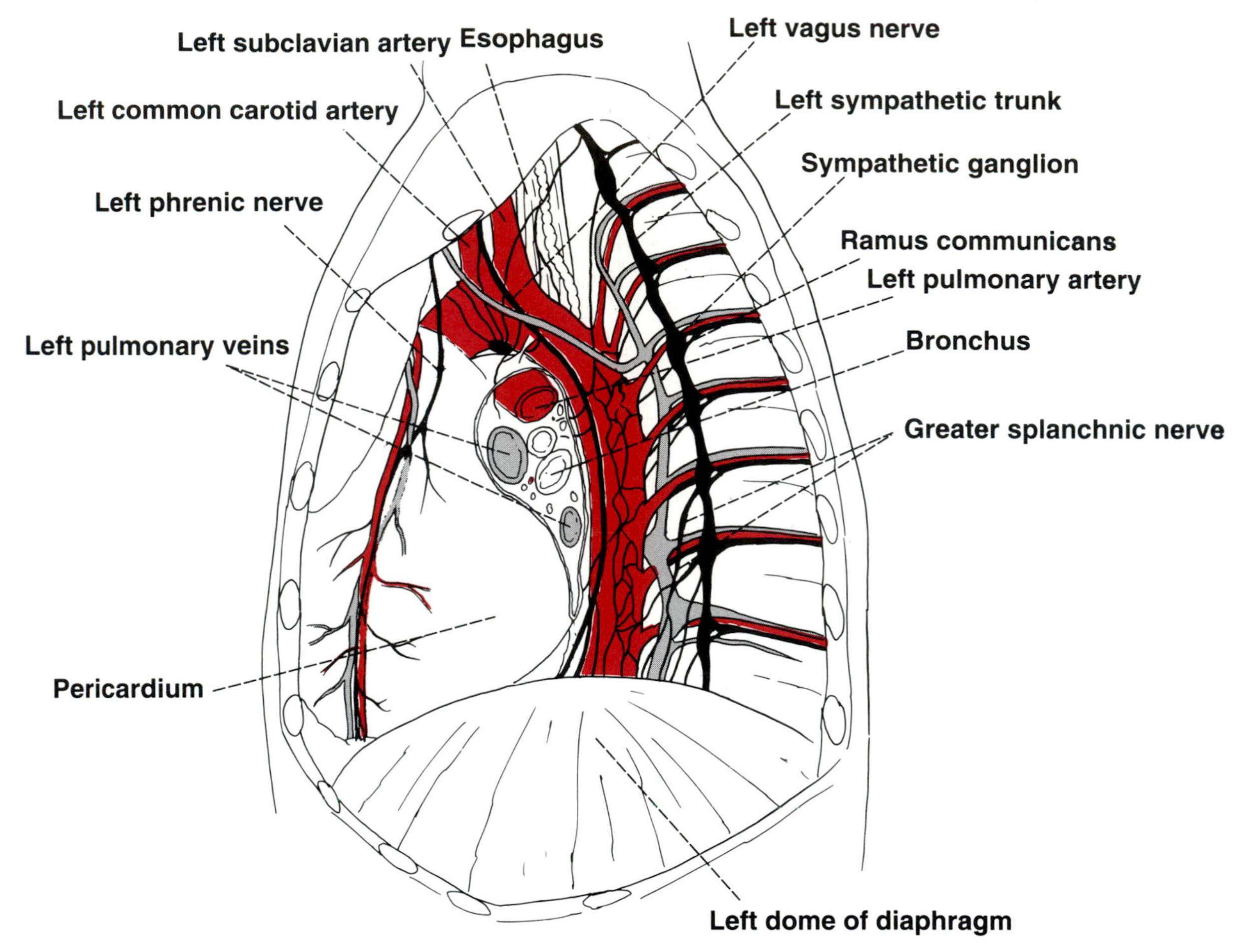

FIG 9–5.
Left side of the mediastinum showing the thoracic part of the sympathetic trunk and its main branches.

ganglia located close to the viscera they innervate. Here, again, acetylcholine is the neurotransmitter. The cranial parasympathetic ganglia are the *ciliary, pterygopalatine, submandibular, and otic.* In certain locations the ganglion cells are placed in nerve plexuses, such as the *cardiac plexus, pulmonary plexus, myenteric plexus (Auerbach's plexus),* and *mucosal plexus (Meissner's plexus)* ; the last two plexuses are associated with the gastrointestinal tract. The pelvic splanchnic nerves synapse in ganglia in the hypogastric plexuses. The postganglionic parasympathetic fibers are nonmyelinated and short in length.

Parasympathetic Ganglia.—These include the ciliary, pterygopalatine, submandibular, and otic ganglia.

Ciliary Ganglion.—This ganglion, about the size of a pin's head, is situated at the back of the orbital cavity between the optic nerve and the lateral rectus muscle.

Branches.—These include the following.

1. *Parasympathetic preganglionic fibers* arise from the Edinger-Westphal nucleus of the oculomotor nerve. The fibers reach the ganglion via the oculomotor nerve and its branch to the inferior oblique muscle.

2. *Parasympathetic postganglionic fibers* arise from the ganglion nerve cells and pass to the eyeball via the short ciliary nerves. The fibers innervate the ciliary muscle (which controls the curvature of the lens in accommodation) and the sphincter pupillae muscle, which constricts the pupil.

3. *Sensory fibers from the eyeball* reach the ganglion via the short ciliary nerves and pass through it without interruption to join the nasociliary branch of the ophthalmic division of the trigeminal nerve. They ascend in the trigeminal nerve to reach the central nervous system.

4. *Sympathetic postganglionic fibers* arise from the superior cervical sympathetic ganglion and reach the ciliary ganglion via the internal carotid plexus. The

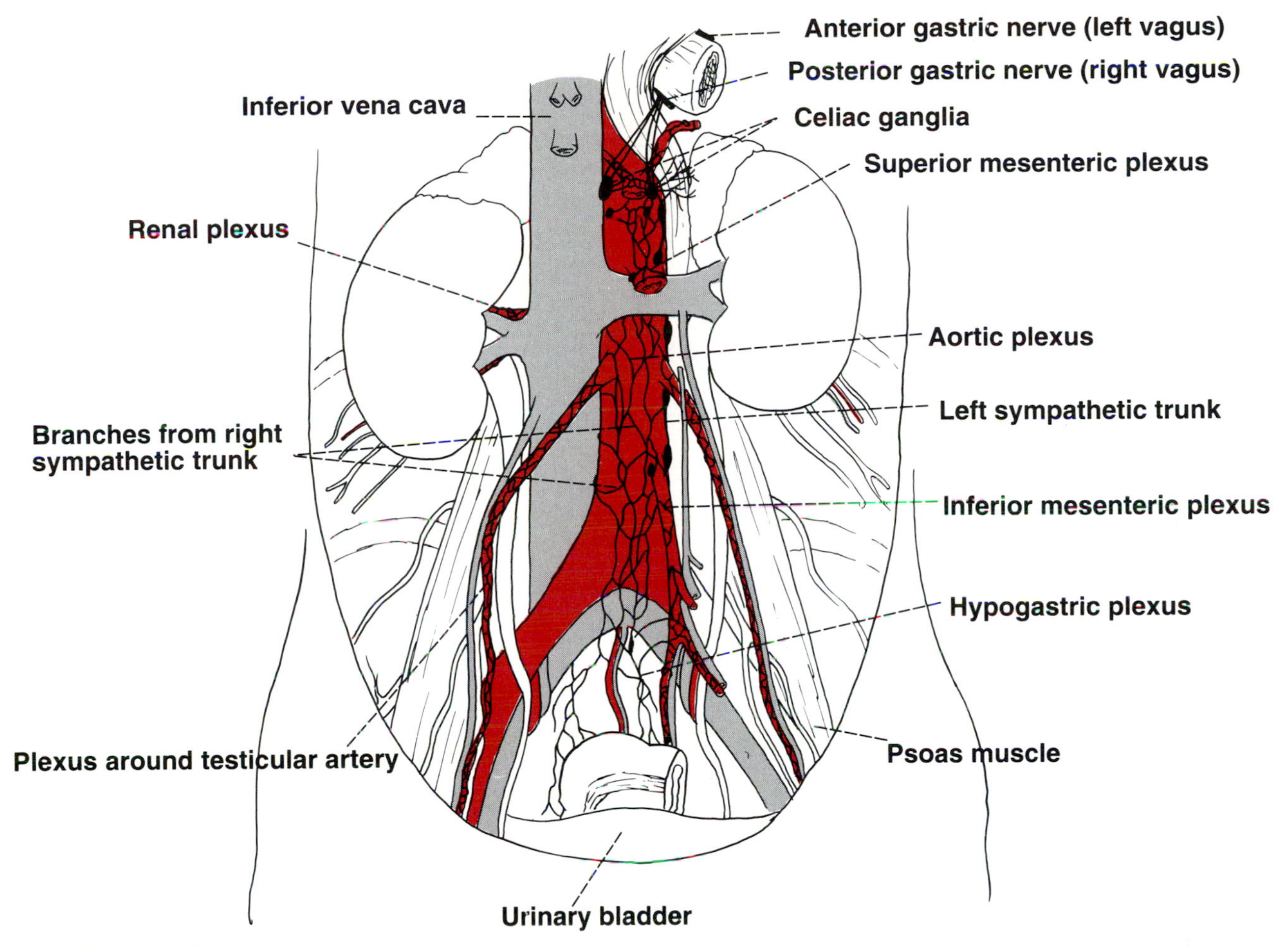

FIG 9–6.
Posterior abdominal wall showing the aorta and related sympathetic plexuses.

fibers pass through the ganglion without interruption to the eyeball via the short ciliary nerves and supply the dilator pupillae muscle.

Pterygopalatine (Sphenopalatine) Ganglion.—This is located in the pterygopalatine fossa (behind the maxilla) and is suspended from the lower border of the maxillary nerve.

Branches.—These include the following.

1. *Parasympathetic preganglionic fibers* arise from the lacrimatory nucleus of the facial nerve. The fibers run in the facial nerve and then travel in the *greater petrosal branch* of the nerve and the *nerve of the pterygoid canal.* The latter nerve then joins the ganglion.
2. *Parasympathetic postganglionic fibers,* which arise from the ganglion nerve cells, pass to the maxillary nerve, its zygomatic branch, the lacrimal nerve, and supply the lacrimal gland. Other postganglionic fibers supply the glands of the pharynx, palate, and nose.

Submandibular Ganglion.—This lies on the side of the tongue suspended from the lingual nerve.

Branches.—These include the following.

1. *Parasympathetic preganglionic fibers* arise from the superior salivatory nucleus of the facial nerve, pass with the nerve and its chorda tympani branch, and join the lingual nerve. The fibers leave the lingual nerve to enter the ganglion.
2. *Parasympathetic postganglionic fibers* pass from the ganglion nerve cells to the submandibular and sublingual salivary glands.

Otic Ganglion.—This lies just below the foramen ovale on the medial side of the mandibular division of the trigeminal nerve.

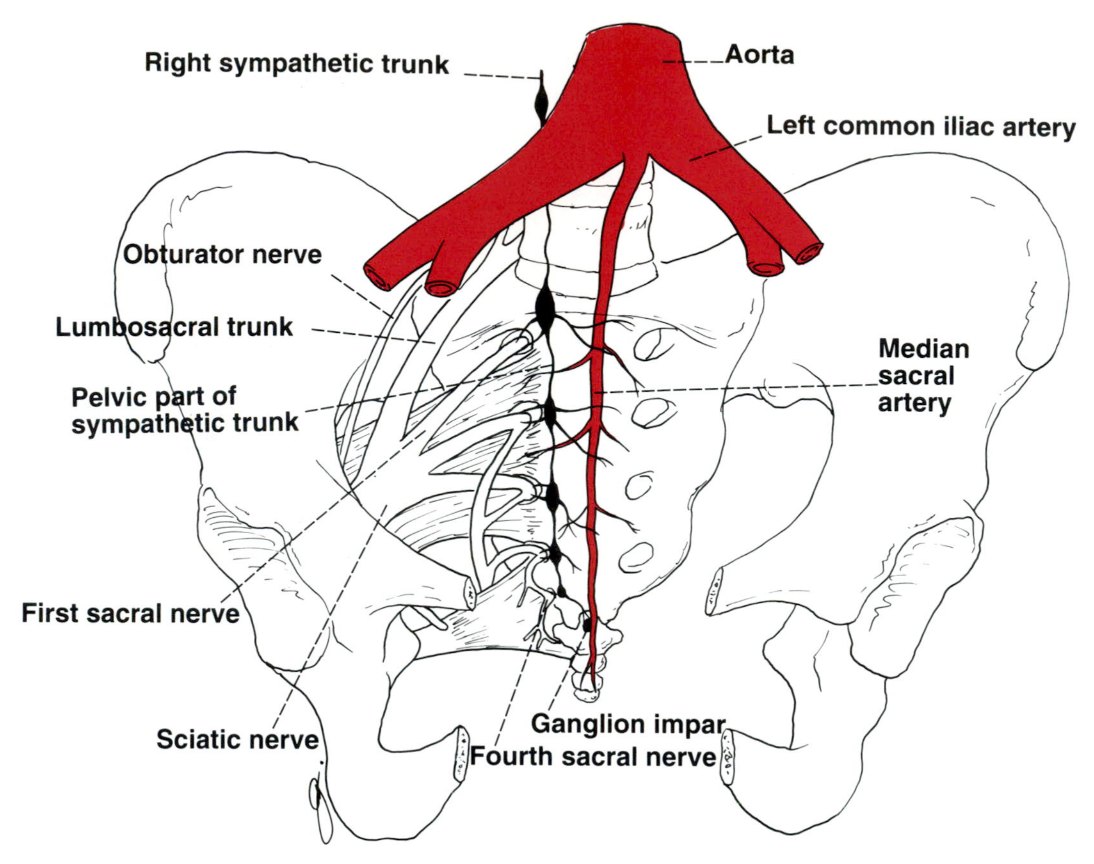

FIG 9–7.
Posterior pelvic wall showing the right sacral plexus and the pelvic part of the right sympathetic trunk.

Branches.—These include the following.

1. *Parasympathetic preganglionic nerve fibers* arise from the inferior salivatory nucleus of the glossopharyngeal nerve and pass via the tympanic branch of the glossopharyngeal nerve to the tympanic plexus in the middle ear. They leave the plexus via the *lesser petrosal nerve* to reach the ganglion.
2. *Parasympathetic postganglionic fibers* arise from the ganglion nerve cells and travel via the auriculotemporal nerve to the parotid salivary gland.

Afferent Nerve Fibers.—The afferent myelinated fibers leave the viscera and reach their cell bodies in the sensory ganglia of cranial nerves or in posterior root ganglia of the sacral spinal nerves. The central axons then enter the central nervous system and form regional reflex arcs or ascend to higher centers, such as the hypothalamus. Once the afferent fibers gain entrance to the spinal cord or brain, they are thought to travel alongside, or mix with, the somatic afferent fibers.

*Large Autonomic Plexuses**

Large collections of sympathetic and parasympathetic efferent nerve fibers and their associated ganglia, together with visceral afferent fibers, form autonomic nerve plexuses in the thorax, abdomen, and pelvis. Branches from these plexuses innervate the viscera. In the abdomen the plexuses are associated with the aorta and its branches; thus, the subdivisions of the autonomic plexuses are named according to the branch of the aorta along which they lie. The following are those plexuses, which are not described here: the cardiac, pulmonary, celiac, superior mesenteric, inferior mesenteric, aortic, and superior and inferior hypogastric plexuses.

Autonomic Ganglia

The autonomic ganglion is located where preganglionic fibers synapse on postganglionic neurons. Gan-

*A *nerve plexus* is a collection of nerve fibers that form a network; nerve cells may be present within such a network. A *ganglion* is a knotlike mass of nerve cells found outside the central nervous system. The term must be distinguished from the term ganglion within the central nervous system consisting of nuclear groups, e.g., basal ganglia.

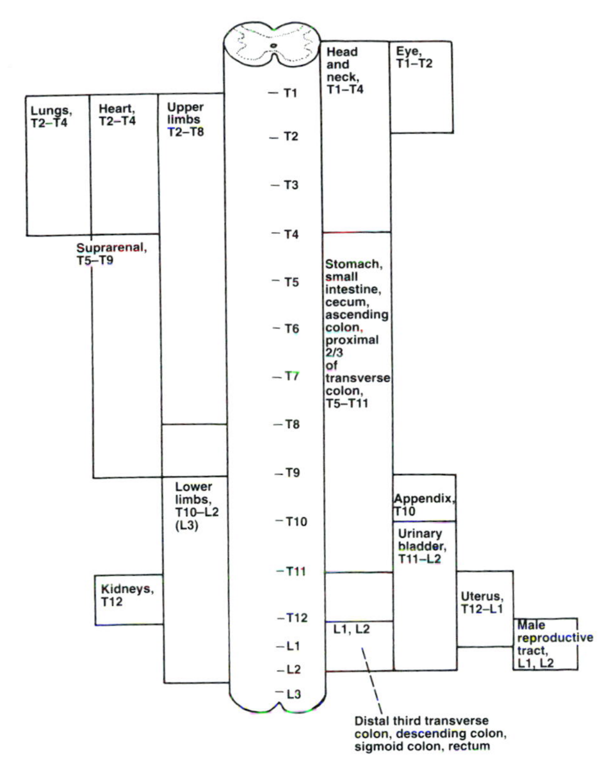

FIG 9–8.
Spinal segmental levels of sympathetic connector neurons associated with different organs in the body.

glia are situated along the course of efferent nerve fibers of the autonomic nervous system. Sympathetic ganglia form part of the sympathetic trunk or are prevertebral in position (e.g., celiac ganglia). Parasympathetic ganglia, on the other hand, are situated close to or within the walls of the viscera.

Preganglionic fibers are myelinated, small, and relatively slow-conducting B fibers. The postganglionic fibers are unmyelinated, smaller, and slower-conducting C fibers.

Preganglionic Transmitters.—The synaptic transmitter that excites the postganglionic neurons in both sympathetic and parasympathetic ganglia is *acetylcholine.* The action of acetylcholine in autonomic ganglia is terminated by acetylcholinesterase. The small ganglionic interneurons contain *dopamine,* which is thought to act as a transmitter.

Ganglion-Blocking Agents.—There are two types of ganglion-blocking agents—depolarizing and nonpolarizing. *Nicotine* acts as a blocking agent in high concentrations by first stimulating the postganglionic neuron and causing depolarization, and then by maintaining depolarization of the excitable membrane. *Hexamethonium* and *tetraethylammonium* block ganglia by competing with acetylcholine at the receptor sites.

Postganglionic Nerve Endings.—The postganglionic fibers terminate on the effector cells without special discrete endings. The axons run between the gland cells and the smooth and cardiac muscle fibers and lose their covering of Schwann cells. At sites where transmission occurs, clusters of vesicles are present within the axoplasm. The transmission site on the axon may lie at some distance from the effector cell so that the transmission time may be slow at these endings. The diffusion of the transmitter through the large extracellular distance also permits a given nerve to have an action on a large number of effector cells.

Postganglionic Transmitters.—The parasympathetic postganglionic nerve endings liberate *acetylcholine* as their transmitter substance. The acetylcholine traverses the synaptic cleft and binds reversibly with the cholinergic receptor on the postsynaptic membrane. Within 2 to 3 msec it is hydrolyzed into acetic acid and choline by the enzyme acetylcholinesterase, which is located on the surface of the nerve and receptor membranes. The choline is reabsorbed into the nerve ending and used again for synthesis of acetylcholine.

Most sympathetic postganglionic nerve endings liberate *norepinephrine* as their transmitter substance. Some sympathetic postganglionic nerve endings, particularly those that end on cells of sweat glands, release *acetylcholine.*

Sympathetic endings that use norepinephrine are called *adrenergic endings.* There are two major kinds of receptors in the effector organs, called *alpha* and *beta receptors.* Two subgroups of alpha receptors (alpha-1 and alpha-2 receptors) and two subgroups of beta-receptors (beta-1 and beta-2 receptors) have been described. Norepinephrine has a greater effect on alpha receptors than on beta receptors. Phenylephrine is a pure alpha stimulator. The bronchodilating drugs such as metaproterenol and albuterol mainly act on beta-2 receptors. As a general rule, alpha-receptor sites are associated with most of the excitatory functions of the sympathetic system (for example, smooth-muscle contraction, vasoconstriction, diaphoresis), whereas the beta-receptor sites are associated with most of the inhibitory functions (for example, smooth-muscle relaxation). Beta-2 receptors are mainly in the lung, and stimulation results in bronchodilatation. Beta-1 receptors are in the myocardium, where they are associated with excitation.

The action of norepinephrine on the receptor site of the effector cell is terminated by re-uptake into the nerve terminal, where it is stored in presynaptic vesicles for reuse. While it is stored in the presynaptic vesicles, it is protected from inactivation by the enzyme monoamine oxidase.

Blocking of the Cholinergic Receptors.—In the case of the parasympathetic and the sympathetic postganglionic nerve endings that liberate acetylcholine as the transmitter substance, the receptors on the effector cells are *muscarinic.* This means that the action can be blocked by *atropine.* Atropine competitively antagonizes the muscarinic action by occupying the cholinergic receptor sites on the effector cells.

Blocking of the Adrenergic Receptors.—Phenoxybenzamine, phentolamine, and possibly chlorpromazine are examples of drugs that block peripheral norepinephrine receptors.

Higher Control of the Autonomic Nervous System

The sympathetic outflow in the spinal cord (T1 to L2 or L3) and the parasympathetic craniosacral outflow (cranial nerves 3, 7, 9, and 10 and spinal nerves S2 through S4) are controlled by the hypothalamus. The hypothalamus appears to integrate the autonomic and neuroendocrine systems, thus preserving body homeostasis. It receives signals from all parts of the nervous system, afferent input from the viscera, and information concerning blood hormone levels. Within the hypothalamus, this input is integrated and transmitted to the lower autonomic centers in the brain stem and spinal cord by descending tracts of the reticular formation. In a similar manner, *releasing factors* or *release-inhibiting factors* are liberated into the circulation from the hypothalamus to affect hormone levels and endocrine secretions, thus influencing organ activity.

Functions of the Autonomic Nervous System

The autonomic nervous system, along with the endocrine system, maintains body homeostasis. The endocrine control is slower and exerts its influence by means of bloodborne hormones.

The sympathetic and parasympathetic components of the autonomic system cooperate in maintaining the stability of the internal environment. The sympathetic part prepares and mobilizes the body in an emergency, when there is sudden severe exercise, fear, or rage. The parasympathetic part aims at conserving and storing energy, for example, in the promotion of digestion and the absorption of food by increasing secretions of the glands of the gastrointestinal tract and stimulating peristalsis.

The sympathetic and parasympathetic parts of the autonomic system usually have antagonistic control over a viscus. For example, sympathetic activity will increase the heart rate, whereas parasympathetic activity will slow the heart rate. Sympathetic activity will make the bronchial smooth muscle relax, but the muscle is contracted by parasympathetic activity.

Many viscera, however, do not possess this fine dual control from the autonomic system. For example, arrector pili have sympathetic innervation only.

The activities of some viscera are kept under a constant state of inhibition by one or other components of the autonomic system. The heart in a trained athlete is maintained at a slow rate by the activities of the parasympathetic system, thus permitting adequate diastolic filling of the ventricles.

For important anatomical, physiological, and pharmacological differences between the sympathetic and the parasympathetic parts of the autonomic system, see Table 9–1. The sympathetic part of the system has a widespread action on the body as the result of the preganglionic fibers synapsing on many postganglionic neurons and the suprarenal medulla releasing norepinephrine and epinephrine into the bloodstream. The parasympathetic has a more discrete control, since the preganglionic fibers synapse on only a few postganglionic neurons and there is no comparable organ to the suprarenal medulla.

The effects of the autonomic system on body organs is summarized in Table 9–2.

Important Autonomic Innervations

Eye.—The autonomic innervations of parts of the eye are as follows.

Upper Lid.—The upper lid is raised by the levator palpebrae superioris. The major part of this muscle consists of skeletal fibers innervated by the oculomotor nerve. A small part consists of smooth-muscle fibers innervated by sympathetic postganglionic fibers from the superior cervical sympathetic ganglion (Fig 9–9).

Iris.—The sphincter pupillae is supplied by parasympathetic fibers from the parasympathetic nucleus (Edinger-Westphal nucleus) of the oculomotor nerve (see Fig 9–9). After synapsing in the *ciliary ganglion,* the postganglionic fibers pass forward to

TABLE 9–2.
Effects of Autonomic Nervous System on Organs of the Body

Organ	Sympathetic Action	Parasympathetic Action
Eye		
Pupil	Dilates	Constricts
Ciliary muscle	Relaxes	Contracts
Glands		
Lacrimal, parotid, submandibular, sublingual, nasal	Reduce secretion by causing vasoconstriction of blood vessels	Increases secretion
Heart		
Cardiac muscle	Increases force of contraction	Decreases force of contraction
Lung		
Bronchial muscle	Relaxes (dilates bronchi)	Contracts (constricts bronchi)
Bronchial secretion		Increases secretion
Bronchial arteries	Constricts	Dilates
Gastrointestinal tract		
Muscle in walls	Decreases peristalsis	Increases peristalsis
Muscle in sphincters	Contracts	Relaxes
Glands	Reduces secretion by vasoconstriction of blood vessels	Increases secretion
Liver	Breaks down glycogen into glucose	
Gallbladder	Relaxes	Contracts
Kidney	Decreases output due to constriction of arteries	
Urinary bladder		
Detrusor	Relaxes	Contracts
Sphincter	Contracts	Relaxes
Erectile tissue of penis and clitoris		Relaxes, causes erection
Ejaculation	Contracts smooth muscle of vas deferens, seminal vesicles, and prostate	
Systemic arteries		
Skin	Constricts	
Abdominal	Constricts	
Muscle	Constricts (alpha receptors), dilates (beta receptors), dilates (cholinergic)	
Erector pili muscles	Contracts	
Suprarenal		
Cortex	Stimulates	
Medulla	Liberates epinephrine and norepinephrine	

the eyeball in the short ciliary nerves (the ciliary muscle of the eye is also supplied by the short ciliary nerves).

The dilator pupillae is supplied by postganglionic fibers from the superior cervical sympathetic ganglion (see Fig 9–9), which reach the orbit along the internal carotid and ophthalmic arteries. They pass uninterrupted through the ciliary ganglion and reach the eyeball in the short and long ciliary nerves.

Heart.—Sympathetic postganglionic fibers arise from cervical and upper thoracic portions of the sympathetic trunks (Fig 9–10). Postganglionic fibers reach the heart from the *superior, middle, and inferior cardiac branches* of the cervical portion of the sympathetic trunk and from a number of *cardiac branches* from the thoracic portion of the sympathetic trunk. The fibers pass via the *cardiac plexuses* to terminate on the *sinoatrial* and *atrioventricular nodes*, cardiac muscle fibers, and coronary arteries. Activation of

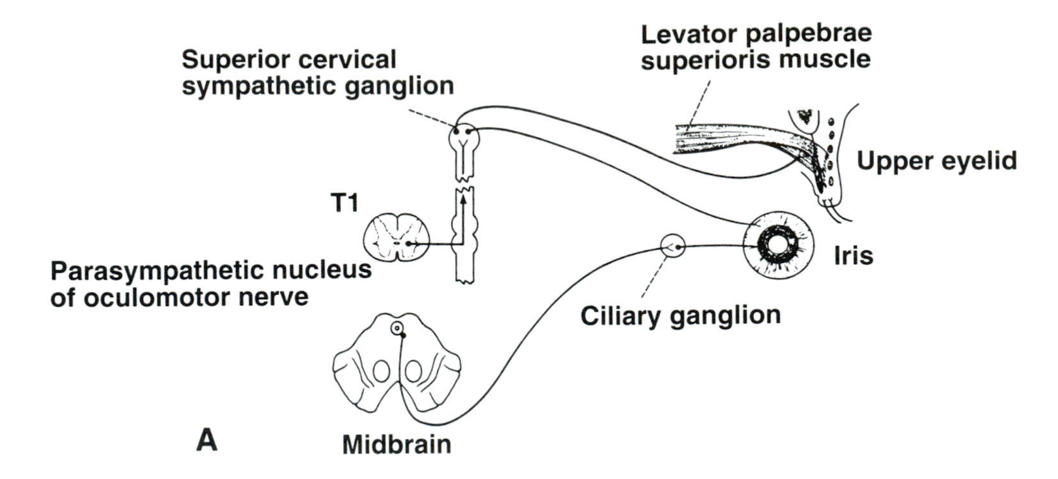

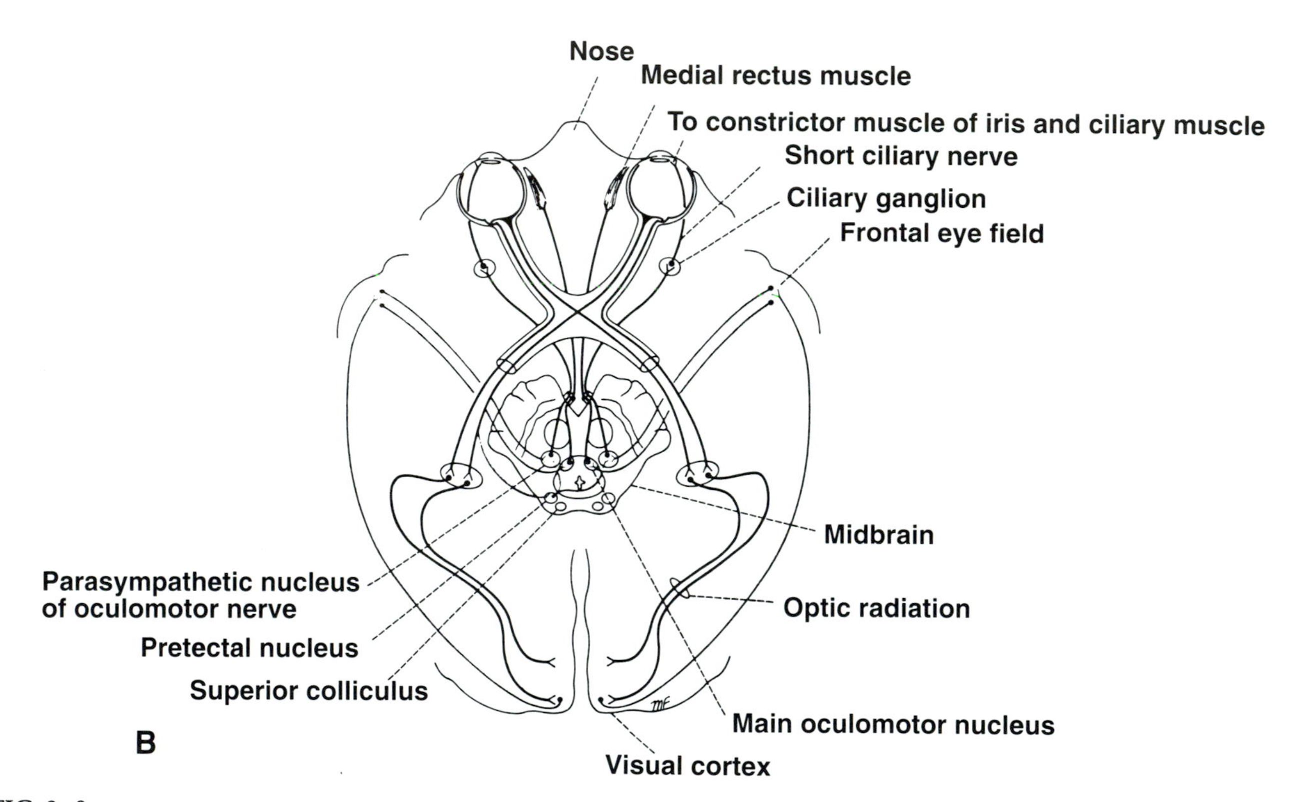

FIG 9–9.
A, the autonomic innervation of the upper eyelid and the iris. **B,** the optic pathway and the visual reflexes.

these nerves results in cardiac acceleration, increased contractile force of the cardiac muscle, and dilatation of the coronary arteries. Coronary dilatation, however, results mainly in response to local metabolic needs, rather than by direct nerve stimulation.

The parasympathetic preganglionic fibers originate in the *dorsal nucleus of the vagus nerve* and descend into the thorax in the vagus nerves. The fibers terminate by synapsing with postganglionic neurons in the *cardiac plexuses.* Postganglionic fibers terminate on the *sinoatrial* and *atrioventricular nodes* and on the coronary arteries. Activation of these nerves results in a reduction in the rate and force of contraction of the myocardium and in constriction of the coronary arteries. Here, again, coronary constriction is produced mainly by reduction in local metabolic needs rather than by neural effects.

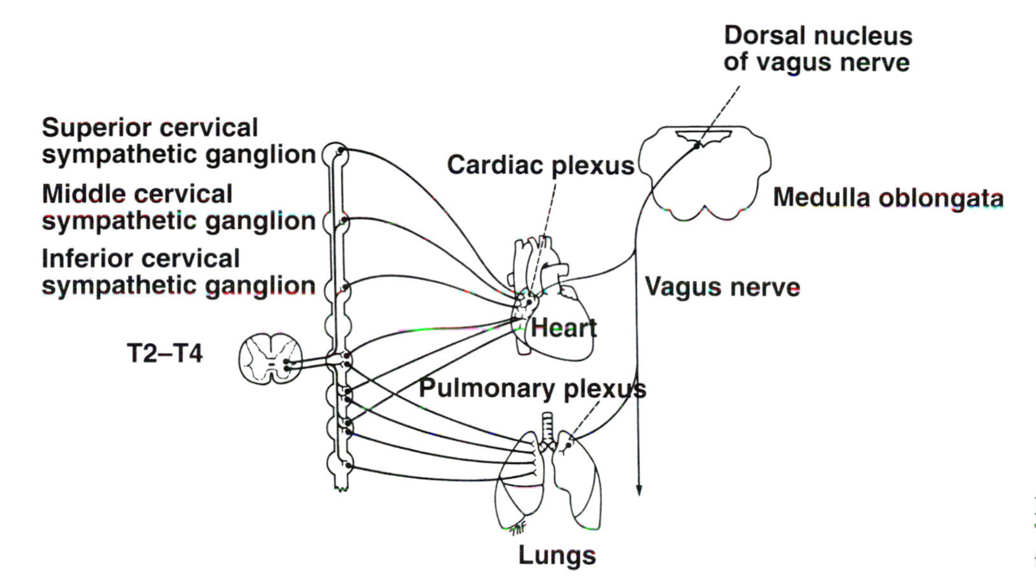

FIG 9–10.
The autonomic innervation of the heart and lungs.

Lungs.—Sympathetic postganglionic fibers arise from the second to fifth thoracic ganglia of the sympathetic trunk (see Fig 9–10). They pass through the pulmonary plexuses and enter the lung to form networks around the bronchi and blood vessels. The sympathetic fibers produce bronchodilatation and slight vasoconstriction.

Parasympathetic preganglionic fibers arise from the *dorsal nucleus of the vagus nerve* and descend to the thorax within the vagus nerves. The fibers terminate by synapsing with postganglionic neurons in the pulmonary plexuses that enter the lung, where they form networks around the bronchi and blood vessels. The parasympathetic fibers produce bronchoconstriction, slight vasodilation, and increased glandular secretion.

Gastrointestinal Tract.—The autonomic innervation of the gastrointestinal tract is shown in Figure 9–11.

Gallbladder and Biliary Ducts.—The gallbladder and biliary ducts receive postganglionic parasympathetic and sympathetic fibers from the hepatic plexus. Parasympathetic fibers derived from the vagus are thought to be motor to the smooth muscle of the gallbladder and bile ducts and inhibitory to the sphincter of Oddi. Autonomic afferent fibers are also present. Some of these fibers are believed to leave the hepatic plexus and join the right phrenic nerve, thus explaining the phenomenon of referred shoulder pain in gallbladder diseases.

Urinary Bladder.—Sympathetic postganglionic fibers originate in the first and second lumbar ganglia of the sympathetic trunk and travel to the *hypogastric plexuses* in the pelvis. From there the fibers pass to the bladder wall (Fig 9–12). The sympathetic nerves inhibit contraction of the detrusor muscle of the bladder wall and stimulate closure of the *sphincter vesicae.* The sympathetic nerves to the bladder are believed to play no role in the process of micturition, but by causing the contraction of the sphincter vesicae, they do prevent the reflux of semen into the bladder during ejaculation.

The parasympathetic preganglionic fibers arise as the *pelvic splanchnic nerves* from the second, third, and fourth sacral nerves; they pass through the hypogastric plexuses to reach the bladder wall, where they synapse with postganglionic neurons (see Fig 9–12). The parasympathetic nerves stimulate the contraction of the detrusor muscle of the bladder wall and inhibit the action of the sphincter vesicae.

Afferent nerve fibers from the bladder run with both the parasympathetic and sympathetic nerves.

Micturition.—The process of micturition is essentially a spinal reflex mechanism that is facilitated and inhibited by higher centers in the brain. Bladder dysfunction can occur as the result of interruption of the afferent nerves from the bladder, the interruption of parasympathetic and sympathetic efferent nerves to the bladder, and the interruption of voluntary descending fibers in the brain and spinal cord.

Involuntary Internal Anal Sphincter.—The smooth-muscle sphincter is innervated by postganglionic sympathetic fibers from the *hypogastric plexuses* (see Fig 9–12). Each hypogastric plexus receives sympathetic fibers from the *aortic plexus* and from the lum-

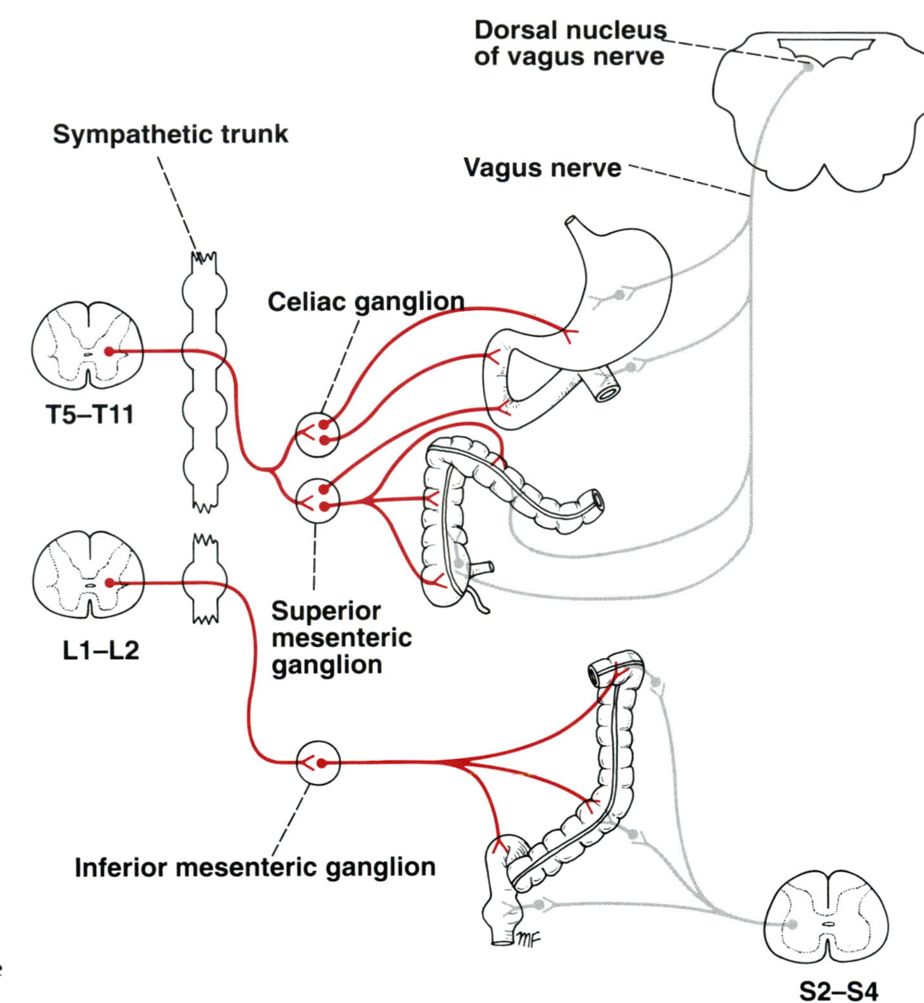

FIG 9–11.
The autonomic innervation of the gastrointestinal tract.

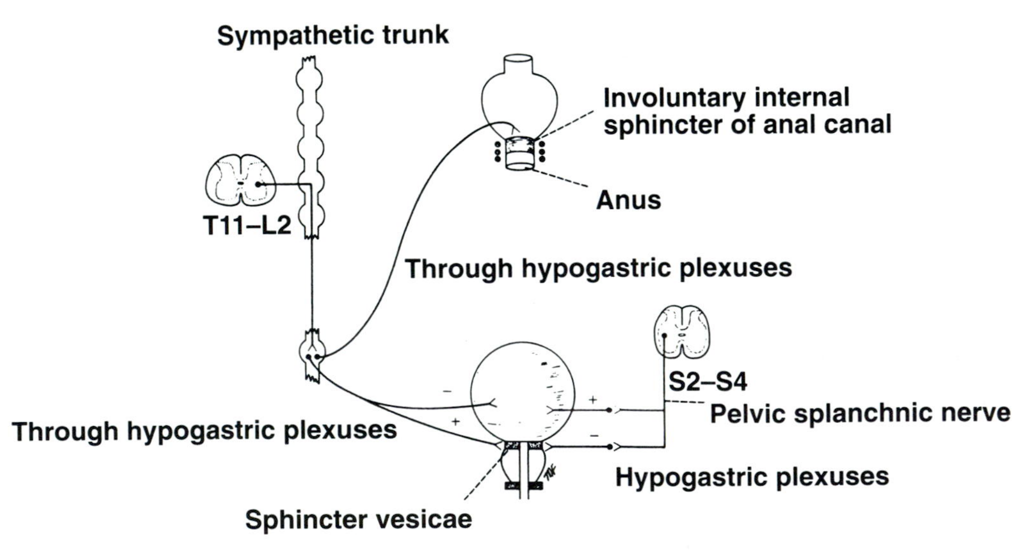

FIG 9–12.
The autonomic innervation of the sphincters of the anal canal and the urinary bladder.

bar and pelvic parts of the sympathetic trunks. Sympathetic nerves cause the internal anal sphincter to contract.

Defecation.—Defecation involves a coordinated reflex that results in the emptying of the descending colon, pelvic colon, rectum, and anal canal. It is assisted by a rise in the intra-abdominal pressure brought about by contraction of the muscles of the anterior abdominal wall. The desire to defecate is initiated by stimulation of the stretch receptors in the wall of the rectum.

Erection of the Penis and Clitoris.—Vascular engorgement is controlled by parasympathetic fibers that originate in the second, third, and fourth sacral segments of the spinal cord (Fig 9–13). The fibers enter the pelvic plexuses in the extraperitoneal connective tissue of the pelvis and synapse on the postganglionic neurons. Postganglionic fibers join the internal pudendal arteries and are distributed along their branches, which enter the erectile tissue. Parasympathetic fibers cause vasodilatation of the arteries and greatly increase the blood flow to the erectile tissue.

Ejaculation.—Preganglionic sympathetic fibers leave the spinal cord at the first and second lumbar segments (see Fig 9–13). Many of these fibers synapse in the first and second lumbar ganglia with postganglionic neurons. Other fibers may synapse in ganglia in the lower lumbar or pelvic parts of the sympathetic trunks. The postganglionic fibers are then distributed to the vas deferens, seminal vesicles, and prostate through the hypogastric plexuses. The sympathetic nerves stimulate the contraction of the smooth muscle in the walls of these structures and cause the spermatozoa, together with the secretions of the seminal vesicles and prostate, to be discharged into the urethra.

Uterus.—The uterus is innervated by branches of the inferior hypogastric plexus situated in the base of the broad ligament (Fig 9–14). This part of the plexus is often referred to as the *uterovaginal plexus.* The nerves are sympathetic, parasympathetic, and afferent sensory. Sympathetic preganglionic fibers leave the spinal cord at segmental levels T12 and L1 and are believed to synapse with ganglion cells in the uterovaginal plexus. Parasympathetic preganglionic fibers leave the spinal cord at levels S2, S3, and S4 and relay in the uterovaginal plexus. Although it is recognized that the uterus is largely under hormonal control, sympathetic innervation may cause uterine contraction and vasoconstriction, whereas parasympathetic fibers have the opposite effect.

Afferent pain fibers from the fundus and the body of the uterus ascend to the spinal cord through the hypogastric plexuses and enter the spinal cord through the posterior roots of the 10th, 11th, and 12th thoracic spinal nerves (see Fig 9–14). Fibers from the cervix run in the pelvic splanchnic nerves and enter the spinal cord through the posterior roots of the 2nd, 3rd, and 4th sacral nerves.

The *sympathetic innervations of the arteries of the upper and lower limbs* are summarized in Figure 9–15.

Important Autonomic Reflexes

Visual Reflexes.—These include the direct and consensual light reflexes and the accommodation reflex.

Direct and Consensual Light Reflexes.—Afferent nervous impulses travel from the retina through the optic nerve, optic chiasma, and optic tract (see Fig

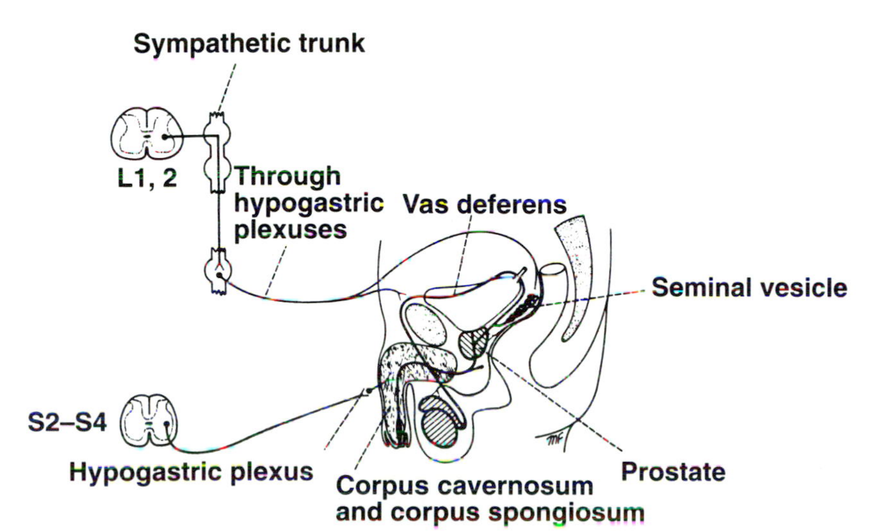

FIG 9–13.
The autonomic innervation of the male reproductive tract.

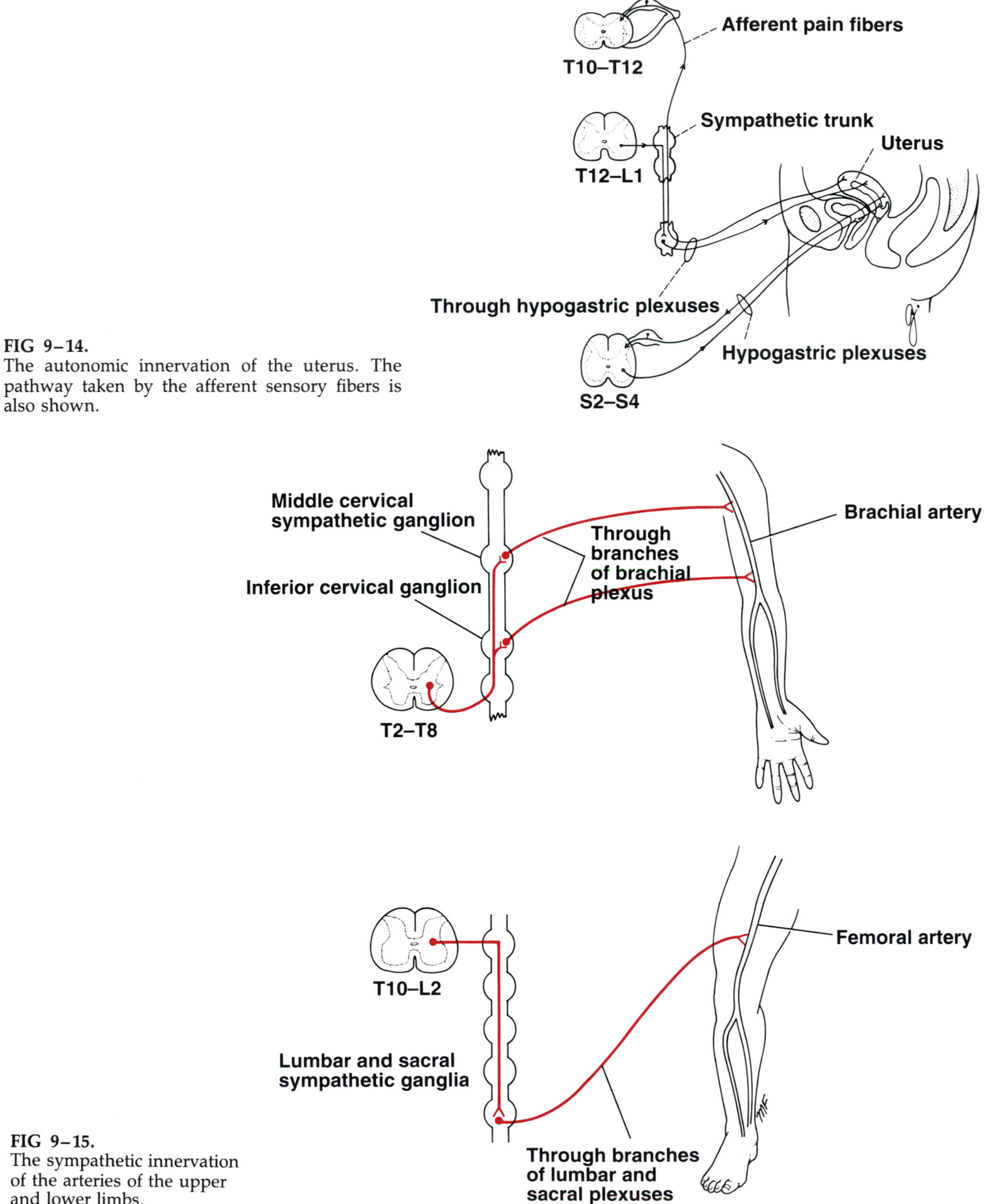

FIG 9–14.
The autonomic innervation of the uterus. The pathway taken by the afferent sensory fibers is also shown.

FIG 9–15.
The sympathetic innervation of the arteries of the upper and lower limbs.

9–9). A small number of fibers leave the optic tract and synapse on nerve cells in the *pretectal nucleus,* which lies close to the superior colliculus. The impulses are passed by axons of the pretectal nerve cells to the parasympathetic nuclei (Edinger-Westphal nuclei) of the oculomotor nerve on both sides. Here the fibers synapse and travel through the oculomotor nerve to the *ciliary ganglion* in the orbit. Finally, postganglionic parasympathetic fibers pass through the short ciliary nerves to the eyeball to the constrictor pupillae muscle of the iris. Both pupils constrict in the consensual light reflex because the pretectal nucleus sends fibers to the parasympathetic nuclei on both sides of the midbrain.

Accommodation Reflex.—When the eyes are redirected from a distant to a near object, contraction of the medial recti brings about convergence of the ocular axes, the lens thickens to increase its refractive power by contraction of the ciliary muscle, and the pupils constrict to limit the light waves to the thickest central part of the lens. The afferent visual impulses travel through the optic nerve, optic chiasma, optic tract, lateral geniculate body, and optic radiation to the visual cortex (see Fig 9–9). The visual cortex is connected to the eye field of the frontal cortex. From here, cortical fibers descend via the internal capsule to the oculomotor nuclei in the midbrain. The oculomotor nerve travels to the medial recti muscles. Some of the descending cortical fibers synapse with the parasympathetic nuclei (Edinger-Westphal nuclei) of the oculomotor nerve on both sides. The parasympathetic preganglionic fibers then travel through the oculomotor nerve to the ciliary ganglion in the orbit where they synapse. Finally, postganglionic parasympathetic fibers pass through the short ciliary nerves to the ciliary muscle and the constrictor pupillae muscle of the iris.

Cardiovascular Reflexes.—These include the carotid sinus and aortic arch reflexes and the Bainbridge right atrial reflex.

Carotid Sinus and Aortic Arch Reflexes.—The carotid sinus, located at the bifurcation of the common carotid artery, and the aortic arch serve as baroreceptors. As the blood pressure rises, nerve endings in the walls of these vessels are stimulated. Afferent fibers from the carotid sinus ascend in the glossopharyngeal nerve and terminate in the *nucleus solitarius* in the brain stem. Afferent fibers from the aortic arch ascend in the vagus nerve. Connector neurons in the medulla oblongata activate the parasympathetic nucleus (dorsal nucleus) of the vagus, which slows the heart rate. At the same time, reticulospinal fibers descending to the spinal cord inhibit the preganglionic sympathetic outflow to the heart and cutaneous arterioles. The combined effect of stimulation of the parasympathetic action on the heart and peripheral blood vessels reduces the rate and force of contraction of the heart and peripheral resistance. Consequently, the blood pressure falls. A person's blood pressure is thus modified by afferent information received from baroreceptors. The modulator of the autonomic nervous system, namely, the hypothalamus, in turn, can be influenced by other, higher centers in the central nervous system.

Bainbridge Right Atrial Reflex.—This reflex is initiated when the nerve endings in the wall of the right atrium and venae cavae are stimulated by a rise in venous pressure. The afferent fibers ascend in the vagus to the medulla oblongata and terminate on the *nucleus of the tractus solitarius.* Connector neurons inhibit the parasympathetic (dorsal) nucleus of the vagus, and reticulospinal fibers stimulate the thoracic sympathetic outflow to the heart, resulting in cardiac acceleration.

CLINICAL ANATOMY

Clinical Anatomy of the Autonomic Nervous System

Horner's Syndrome

This syndrome consists of (1) constriction of the pupil (miosis), (2) mild ptosis, (3) enophthalmos,* (4) skin vasodilation, and (5) anhydrosis—all resulting from an interruption of the sympathetic nerve supply to the head and neck. Pathologic causes include lesions in the brain stem or cervical part of the spinal cord that interrupt the reticulospinal tracts descending from the hypothalamus to the sympathetic outflow in the lateral gray column of the first thoracic segment of the spinal cord. Such lesions include *multiple sclerosis* and *syringomyelia.* Traction on the stellate ganglion due to a *cervical rib,* or involvement of the ganglion in a metastatic lesion, may interrupt the peripheral part of the sympathetic pathway.

All cases of Horner's syndrome have miosis and ptosis. However, a distinction should be made between lesions occurring at the first neuron (within the central nervous system), the second neuron (preganglionic), and the third neuron (postganglionic). For example, the clinical signs suggestive of a

*The enophthalmos of Horner's syndrome is often apparent but not real and caused by the ptosis. However, the smooth muscle, the orbitalis, situated at the back of the orbit, is paralyzed and may be responsible.

first neuronal defect (central Horner's syndrome) include contralateral hyperesthesia of the body, loss of sweating of the entire half of the body, or vertigo. Signs suggesting a second neuronal involvement (preganglionic Horner's syndrome) include anhydrosis limited to the face and neck, and the presence of flushing or blanching of the face and neck. Signs suggesting a third neuronal involvement (postganglionic Horner's syndrome) include facial pain or ear, nose, or throat disease.

The presence or absence of other signs and symptoms may assist in differentiating the three types of Horner's syndrome. For example, in the preganglionic syndrome a thyroidectomy scar (due to previous thyroid surgery), hoarseness, cervical osteoarthritis, thoracic surgery, Pancoast syndrome, and brachial plexus lesions may be evident.

Argyll Robertson Pupil

This condition is characterized by a small pupil, which is of fixed size and does not react to light but does contract with accommodation. It is usually caused by a neurosyphilitic lesion interrupting fibers that run from the pretectal nucleus to the parasympathetic nuclei (Edinger-Westphal nuclei) of the oculomotor nerve on both sides. The fact that the pupil constricts with accommodation implies that the connections between the parasympathetic nuclei and the constrictor pupillae muscle of the iris are intact.

Adie's Tonic Pupil Syndrome

In this condition the pupil has a decreased or absent light reflex, a slow or delayed contraction to near vision, and a slow or delayed dilatation in the dark. This benign syndrome, which probably results from a disorder of the parasympathetic innervation of the constrictor pupillae muscle, must be distinguished from the Argyll Robertson pupil (above), which is caused by neurosyphilis. Adie's syndrome can be confirmed by looking for hypersensitivity to cholinergic agents. Drops commonly used for this test are 2.5% methacholine (Mecholyl) or 1.25% pilocarpine. The Adie's tonic pupil should constrict when these drops are put in the hypersensitive eye, but the small doses will not cause constriction in the normal eye. Moreover, these cholinergic agents in small doses do not cause pupillary constriction in mydriasis caused by oculomotor lesions or in drug-related mydriasis because of the absence of hypersensitivity present in Adie's syndrome.

Frey's Syndrome

This is an interesting complication that sometimes follows penetrating wounds of the parotid gland. During the process of healing, the postganglionic parasympathetic secretomotor fibers traveling in the auriculotemporal nerve grow out and join the distal end of the great auricular nerve, which supplies the sweat glands of the overlying facial skin. By this means, a stimulus intended to stimulate saliva flow instead produces sweat.

A similar syndrome may follow injury to the facial nerve. During the process of regeneration, parasympathetic fibers normally destined for the submandibular and sublingual salivary glands are diverted to the lacrimal gland. This produces watering of the eyes associated with salivation, so-called *crocodile tears.*

Hirschsprung's Disease (Megacolon)

This is a congenital condition in which the myenteric plexus (Auerbach's plexus) fails to develop in the distal part of the colon. The involved part of the colon possesses no parasympathetic ganglion cells, and peristalsis is absent. This effectively blocks the passage of feces, and the proximal part of the colon becomes enormously distended.

Clinical Anatomy of Bladder Dysfunction Following Spinal Cord Injuries

Injuries to the spinal cord (or cauda equina) are followed by disruption of the nervous control of micturition.

The atonic bladder occurs during the phase of spinal shock immediately following the injury and may last from a few days to several weeks. The bladder wall muscle is relaxed, the sphincter vesicae are tightly contracted (loss of inhibition from higher levels), and the sphincter urethrae are relaxed. The bladder becomes greatly distended and finally overflows. Depending on the level of the cord injury, the patient may or may not be aware that the bladder is full; there is no voluntary control.

The *automatic reflex bladder* occurs after the patient has recovered from spinal shock, provided that the cord lesion lies above the level of the parasympathetic outflow (S2 through S4). This is the type of bladder normally found in infancy. Since the descending fibers in the spinal cord are sectioned, there is no voluntary control. The bladder fills and empties reflexly. Stretch receptors in the bladder wall are stimulated as the bladder fills, and the afferent impulses pass to the spinal cord (S2 through S4). Efferent impulses pass down to the bladder muscle, which contracts; the sphincter vesicae and urethral sphincter both relax. This simple reflex occurs every 1 to 4 hours.

The autonomous bladder occurs if the sacral segment of the spinal cord is destroyed or if the cauda equina is severed. The bladder has no reflex control or voluntary control. The bladder wall is flaccid, and the capacity of the bladder is greatly increased. It fills to capacity and overflows, which results in continual dribbling. The bladder may be emptied by manual compression of the lower part of the anterior abdominal wall, but infection of the urine and back pressure effects on the ureters and kidneys are inevitable.

Clinical Anatomy of Defecation Following Spinal Cord Injuries

Following severe spinal cord injuries (or cauda equina injuries), the patient is not aware of rectal distention. Moreover, the parasympathetic influence on the peristaltic activity of the descending colon, pelvic colon, and rectum is lost. In addition, control over the abdominal musculature and sphincters of the anal canal may be severely impaired. The rectum, now an isolated structure, responds by contracting when the pressure within its lumen rises. This local reflex response is much more efficient if the sacral segments of the spinal cord and the cauda equina are intact. At best, however, the force of the contractions of the rectal wall is small, resulting in constipation and impaction.

Botulinum Toxin

A very small amount of this toxin binds irreversibly to the nerve plasma membranes and prevents the release of acetylcholine at cholinergic synapses and neuromuscular junctions, producing an atropine-like syndrome with skeletal muscle weakness.

Black Widow Spider Venom

The venom causes a brief release of acetylcholine at the nerve endings followed by a permanent blockade.

Anticholinesterase Agents

Acetylcholinesterase, which is responsible for hydrolyzing and limiting the action of acetylcholine at nerve endings, can be blocked by certain drugs. Physostigmine, neostigmine, pyridostigmine, and carbamate and organophosphate insecticides are effective acetylcholinesterase inhibitors. Their use results in an excessive stimulation of the cholinergic receptors, producing the *"SLUD syndrome"* —salivation, lacrimation, urination, and defecation.

Injuries to the Autonomic Nervous System

Sympathetic Injuries.—The sympathetic trunk in the neck may be injured by stab and bullet wounds. Traction injuries to the first thoracic root of the brachial plexus can damage sympathetic nerves destined for the stellate ganglion. All these conditions may produce a preganglionic type of Horner's syndrome (see p. 343). Injuries to the spinal cord or cauda equina can disrupt the sympathetic control of the bladder, as discussed on p. 344.

Parasympathetic Injuries.—The oculomotor nerve is vulnerable in head injuries (herniated uncus) and may be damaged by compression by aneurysms in the junction between the posterior cerebral artery and posterior communicating artery. The preganglionic parasympathetic fibers traveling in this nerve are situated in the periphery of the nerve and may be damaged. Surface aneurysmal compression characteristically causes dilatation of the pupil and loss of the visual light reflexes.

The autonomic fibers in the facial nerve may be damaged in fractures of the skull involving the temporal bone. The vestibulocochlear nerve is closely related to the facial nerve in the internal acoustic meatus so that clinical findings involving both nerves are common. Involvement of the parasympathetic fibers in the facial nerve may produce impaired lacrimation in addition to the facial paralysis.

The glossopharyngeal and vagus nerves are at risk in stab and bullet wounds of the neck. The parasympathetic secretomotor fibers to the parotid salivary gland leave the glossopharyngeal nerve just below the skull so that they are rarely damaged.

The parasympathetic outflow in the sacral region of the spinal cord (S2 through S4) may be damaged in spinal cord and cauda equina injuries leading to disruption of bladder, rectal, and sexual functions (see pp. 339 and 341).

ANATOMICAL-CLINICAL PROBLEMS

1. A 41-year-old bricklayer was working on the outside wall on the sixth story of a building when he lost his footing and slipped on the scaffolding. As he began to fall, he grabbed at the scaffolding on the floor below with his right hand and managed to hang on until he was rescued. He was rushed by ambulance to the emergency department of a neighboring hospital. Examination of his right upper limb showed paralysis of the flexor carpi ulnaris and flexor digitorum profundus, and weakness of the palmar and dorsal interossei and the thenar and hy-

pothenar muscles. There was also some loss of sensation on the ulnar side of his forearm and hand. The pupil of the right eye was constricted, and there was ptosis of the right upper lid. A slight degree of right-sided enophthalmos was also present. The skin on his right cheek felt warmer and drier than the skin on his left cheek. How can you account for these widespread physical signs?

2. A 15-year-old boy was peddling drugs on a street corner when a rival dealer suddenly pulled a gun and shot him in the abdomen at point-blank range. The boy was rushed to a hospital and examined in the emergency department. The boy had an entrance wound in the lower part of the right side of the abdomen, but there was no exit wound. Radiographs revealed the presence of a bullet lodged in the vertebral canal at the level of the second lumbar vertebra. After the patient recovered from surgery, in which numerous small-bowel perforations were repaired, a careful neurological examination revealed that he had also suffered a complete lesion of the cauda equina. Is this patient going to have normal bladder function? What is the autonomic innervation to the urinary bladder?

3. Explain the following anatomic facts concerning the autonomic nervous system: (1) the parasympathetic nerve supply to the large intestine down as far as the splenic flexure is from the vagus nerves, although the vagal nerve trunks apparently come to an end very soon after piercing the diaphragm with the esophagus; (2) there are sympathetic nerve fibers present in parasympathetic ganglia; and (3) how does the sacral parasympathetic outflow reach the splenic flexure of the colon?

4. During a physical examination of a 20-year-old woman, the physician noted that her left pupil failed to react to the direct and consensual light reflexes. Moreover, the same pupil contracted very slowly when the patient was asked to focus on a near object. The pupillary reflexes were normal in her right eye. Can you explain the possible underlying anatomic defect in this condition? Can the condition become bilateral? How does this condition differ from an Argyll Robertson pupil?

5. What transmitter substances are liberated at the following nerve endings: (1) preganglionic sympathetic, (2) preganglionic parasympathetic, (3) postganglionic parasympathetic, (4) postganglionic sympathetic fibers to the heart muscle, and (5) postganglionic sympathetic fibers to the sweat glands of the hand.

6. What is meant by the terms alpha and beta receptors at sympathetic postganglionic endings?

7. A 2-year-old boy was seen in the emergency department with complete intestinal obstruction. The child had not passed flatus or feces for 3 days. The child's mother said that he had had bowel problems since birth, with vomiting, abdominal distension, and failure to pass stools. On several occasions she had to take him to a pediatrician for an enema to evacuate his bowel. On physical examination, the child's abdomen was distended, and a dough-like mass could be palpated in the left iliac region. Examination of the rectum showed it to be empty and not dilated. Following an enema and a subsequent radiological examination, it was found that the descending colon was grossly distended, and there was an abrupt change in diameter where the descending colon joined the sigmoid colon. What is the possible diagnosis?

8. Give some examples of drugs that act as acetylcholinesterase inhibitors. Explain the mechanism by which botulinum toxin produces irreversible blockade of cholinergic junctions.

9. An obese 47-year-old woman with five children was seen in the emergency department complaining of severe colicky pain beneath the right costal margin. She said that she had had similar attacks of pain previously, and on two occasions the pain had radiated through to her back and up to the right shoulder. What is the diagnosis? In anatomical terms explain the referred pain to the back and shoulder.

10. A 55-year-old man was seen in the emergency department complaining of a severe aching pain down the inside of his left arm. He said that the pain starts when he walks upstairs. What is the autonomic innervation of the heart? Can you explain why a patient with angina pectoris often experiences pain down the left arm?

11. A 36-year-old man was seen in the emergency department following an accident in an insecticide factory. The patient's arms and face had been splashed with a solution containing an organophosphate compound. On examination the patient exhibited the signs and symptoms of bradycardia, diaphoresis, and nausea and vomiting. His pupils were of normal size. Can you explain why this patient, who clearly was demonstrating cholinergic overdrive, had a normal-sized pupil?

12. A 25-year-old woman with a history of a fracture of the shaft of the left humerus was seen in the emergency department complaining of pain and stiffness of the left shoulder joint. On examination, the left shoulder joint showed some discomfort on movement, and there was limitation in the range of

movement. Her left hand appeared cold on palpation and was wet to touch. Can you explain the findings on her left hand?

ANSWERS

1. This man had sustained a severe traction injury of the eighth cervical and first thoracic roots of the brachial plexus. The various paralyzed muscles and the sensory loss are characteristic of Klumpke's paralysis. In addition, the white ramus communicans passing from the first thoracic nerve to the stellate ganglion was torn, cutting off the preganglionic sympathetic fibers to the head and the neck, producing Horner's syndrome.

2. The urinary bladder receives its sympathetic innervation from the first and second lumbar segments of the spinal cord and its parasympathetic fibers from the second, third, and fourth sacral segments of the cord. In this patient, with a complete section of the cauda equina at the level of the second lumbar vertebra, the preganglionic sympathetic fibers that descend in the anterior root of the first lumbar nerve were left intact, as they leave the vertebral canal to form the first lumbar nerve above the level of the bullet. The preganglionic parasympathetic fibers were, however, sectioned as they descended in the vertebral canal within the anterior roots of the second, third, and fourth sacral nerves. The patient would, therefore, have an autonomous bladder and would be without any external reflex control. The bladder would fill to capacity and then overflow. The patient could activate micturition by powerful contraction of the abdominal muscles, assisted by manual pressure on the lower anterior abdominal wall.

A full discussion of the innervation of the bladder and bladder dysfunction following spinal cord or cauda equina injuries is given on p. 344.

3. (1) The vagal nerve trunks, on reaching the abdominal cavity, split up into their terminal branches after a short course with the esophagus. The posterior vagal trunk (right vagus) gives off an important branch that passes to the celiac and superior mesenteric plexuses. The terminal fibers are distributed with the branches of the celiac and superior mesenteric arteries to the small and large intestine as far as the splenic flexure. (2) It is not uncommon to find sympathetic preganglionic and postganglionic fibers passing through a parasympathetic ganglion without interruption. The nerve fibers are merely using the ganglion as a conduit en route to their destination. Visceral sensory nerve fibers travel in a similar manner. (3) The sacral parasympathetic outflow (S2 through S4) leaves the anterior rami of the sacral nerves as the pelvic splanchnic nerve. These preganglionic fibers pass through the hypogastric and aortic plexuses to reach the inferior mesenteric plexus. The fibers are then distributed to the splenic flexure and descending colon along with the branches of the inferior mesenteric artery.

4. This young woman had a condition known as Adie's tonic pupil syndrome. The syndrome can be confirmed by looking for hypersensitivity to cholinergic agents such as 2.5% Mecholyl or 1.5% pilocarpine. The Adie's pupil should constrict when the drops are put in the eye. These cholinergic agents do not cause pupillary constriction in mydriasis caused by oculomotor lesions or in drug-related mydriasis. Adie's syndrome is a benign disorder, which is probably caused by a lesion of the parasympathetic innervation of the constrictor pupillae muscle. The condition can become bilateral, although initially it is uniocular. The Argyll Robertson pupil is caused by neurosyphilis, with the lesion interrupting the nerve fibers that run from the pretectal nucleus to the parasympathetic nuclei of the oculomotor nerve on both sides (see p. 262). It is characterized by a small, fixed pupil; the pupil does not react to light but does contract with accommodation.

5. (1) acetylcholine, (2) acetylcholine, (3) acetylcholine, (4) norepinephrine, and (5) acetylcholine.

6. Sympathetic endings that use norepinephrine as a transmitter substance are called adrenergic endings. There are two major kinds of receptors—alpha and beta receptors—for adrenergic endings. For details see p. 335.

7. This child has Hirschsprung's disease, a congenital condition in which there is a failure of development of the myenteric plexus (Auerbach's plexus) in the distal part of the colon. The proximal part of the colon is normal but becomes greatly distended because of accumulated feces. In this patient, the lower sigmoid colon, later during surgery, was shown to have no parasympathetic ganglion cells. Thus, this segment of the bowel had no peristalsis and effectively blocked the passage of the feces. Once the diagnosis had been confirmed by taking a biopsy specimen of the distal segment of the bowel, the treatment was to remove the aganglionic segment of the bowel by surgical resection.

8. Physostigmine, neostigmine, pyridostigmine, and carbamate and organophosphate insecticides are examples of effective acetylcholinesterase inhibitors.

Botulinum toxin binds irreversibly to the nerve

plasma membranes and prevents the release of acetylcholine at cholinergic synapses and neuromuscular junctions.

9. The patient was suffering from gallstone colic. The visceral pain originated from the cystic duct or common bile duct and was due to stretching or spasm of the smooth muscle in its wall. The afferent pain fibers pass through the celiac ganglia and ascend in the greater splanchnic nerve to enter the fifth to ninth thoracic segments of the spinal cord. The pain was referred to the fifth through ninth thoracic dermatomes on the right side. Some of the ascending fibers are believed to join the right phrenic nerve (C3, C4, and C5); this would explain the referral of the pain to the shoulder dermatomes via the supraclavicular nerves (C3 and C4). For a full discussion of referred pain, see p. 459.

10. The autonomic innervation of the heart is fully described on p. 337. Cardiac pain is usually experienced over the middle of the sternum and may radiate down the inside of either arm or up into the neck. The nervous pathways taken by the afferent pain fibers ascend to the central nervous system through the cardiac branches of the sympathetic trunk and enter the spinal cord through the posterior roots of the upper four thoracic nerves. The pain is referred to the skin areas supplied by the corresponding spinal nerves. The skin areas supplied by the upper four intercostal nerves and by the intercostobrachial nerve (T2) are therefore affected. The intercostobrachial nerve is distributed to skin on the medial side of the upper part of the arm. A certain amount of spread of nervous information must occur within the central nervous system, for the pain is sometimes felt in the neck and jaw.

11. Organophosphates, which are readily absorbed through intact skin, cause a cholinergic overdrive due to the blockade of acetylcholinesterase, resulting in the accumulation of excess acetylcholine at cholinergic nerve endings. This patient demonstrated such an overdrive. However, the normal-sized pupils, which are sometimes seen in these cases, can be explained by the effect of the chemical blocking agent on the acetylcholinesterase in the sympathetic ganglia, resulting in the stimulation of the postganglionic neurons supplying the dilator pupillae muscles. In other words, both the constrictor pupillae and the dilator pupillae muscles were stimulated in this patient and the pupils remained neutrally placed, without being constricted or dilated. Some patients with organophosphate poisoning also fail to develop bradycardia due to this sympathetic stimulation as well as parasympathetic stimulation.

12. This patient has sympathetic reflex dystrophy. Following an injury to an extremity, which is often minor, several changes take place in the limb that show exaggerated sympathetic activity. The onset of symptoms is often associated with disuse of a limb, and should the patient regain the use of the limb early, the condition may rapidly subside. In addition to cold skin (caused by vasoconstriction) and hyperhidrosis (caused by excessive sympathetic stimulation of the sweat glands), pain in neighboring joints may occur, which may be accompanied by joint effusions. The basic mechanism for the disorder is not understood, although the establishment of abnormal connections between efferent autonomic sympathetic nerves in patients with partial nerve injuries has been suggested. As in the condition of causalgia, sympathectomy may relieve the symptoms.

10

The Vertebral Column, the Spinal Cord, and the Spinal Nerves

Injuries to the vertebral column commonly occur in automobile and motorcycle accidents, falls, sports injuries, and gunshot wounds. Spinal cord and spinal nerve damage may be associated with vertebral fractures and herniated intervertebral discs. Back injuries range from a simple acute back strain to a catastrophic injury of the spinal cord or cauda equina.

Because unprotected movement of a damaged vertebral column during initial medical care can result in injury to the delicate spinal cord, emergency personnel must be familiar with the overall anatomy of this region. The assessment of neurological damage requires an understanding of the main nervous pathways and their position in the spinal cord as well as an ability to correlate radiological evidence of bone injury with segmental levels of the spinal cord and subtle neurologic deficits.

The purpose of this chapter is to review the basic anatomy of the vertebral column and related soft nervous tissue structures.

BASIC ANATOMY

Basic Anatomy of the Vertebral Column

The vertebral column is the central bony pillar of the body. It supports the skull, pectoral girdle, upper limbs, thoracic cage, and, by way of the pelvic girdle, transmits body weight to the lower limbs. Within its cavity lie the spinal cord, the roots of the spinal nerves, and the covering of the meninges, to which the vertebral column gives great protection.

Composition of the Vertebral Column

The vertebral column (Fig 10–1) is composed of 33 vertebrae—7 cervical, 12 thoracic, 5 lumbar, 5 sacral (fused), and 4 coccygeal (fused). Because it is segmented and made up of vertebrae, joints, and pads of fibrocartilage called intervertebral discs, it is a flexible structure. The intervertebral discs form about one fourth of the length of the column.

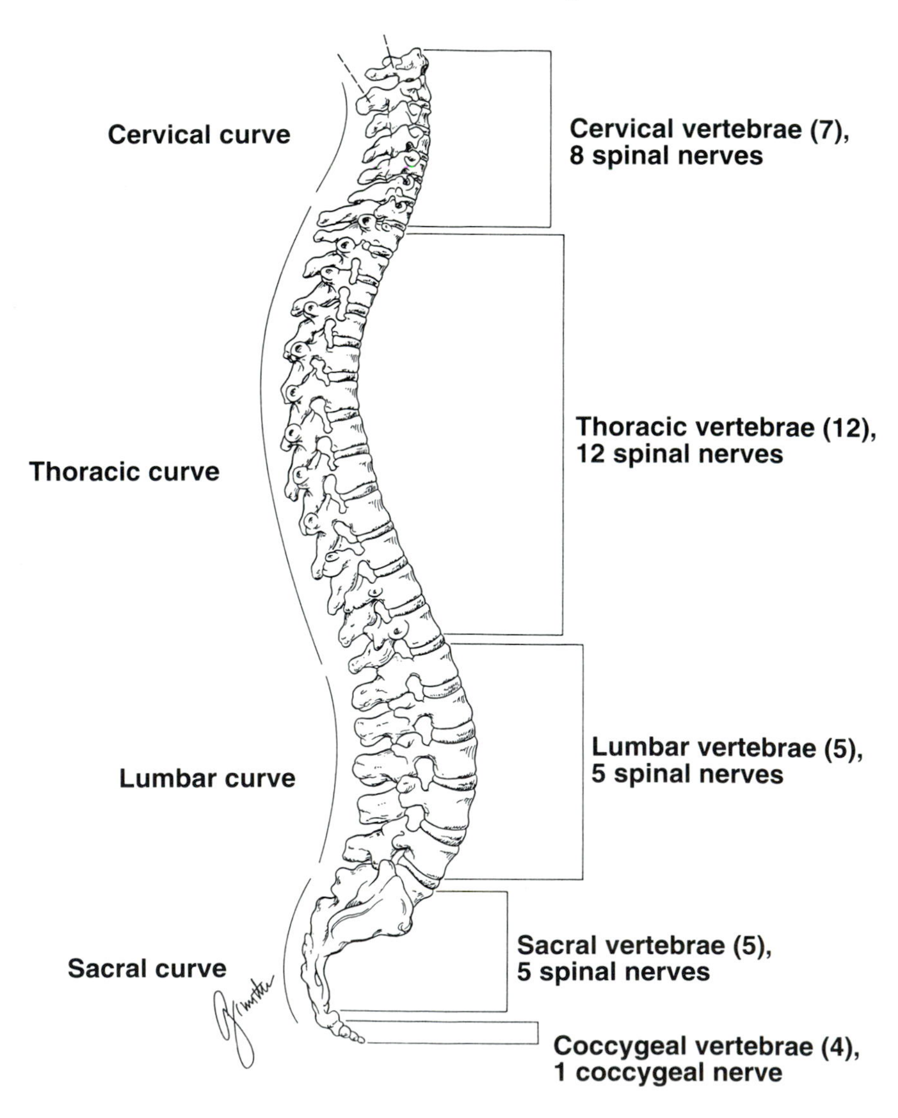

FIG 10–1.
Lateral view of the vertebral column.

Curves of the Vertebral Column

In the standing adult (Fig 10–1), the vertebral column exhibits, in the sagittal plane and from a posterior perspective, the following regional curves: cervical concavity, thoracic convexity, lumbar concavity, and sacral convexity. The degree of curvature varies greatly from one individual to another. During the later months of pregnancy, with the increase in size and weight of the fetus, the posterior lumbar concavity in women increases in an attempt to preserve their center of gravity. In the elderly, the intervertebral discs atrophy and the vertebral bodies become wedge shaped due to osteoporosis, resulting in a loss of height and a gradual return of the vertebral column to a continuous posterior convexity seen in the fetus.

General Characteristics of a Vertebra

Although vertebrae show regional differences, they all possess a common pattern (Fig 10–2).

A *typical vertebra* consists of a rounded *body* anteriorly and a *vertebral arch* posteriorly. These enclose a space, called the *vertebral foramen,* through which the spinal cord and its coverings run. The vertebral arch consists of a pair of cylindrical *pedicles,* which form the sides of the arch, and a pair of flattened *laminae,* which complete the arch posteriorly. The vertebral arch gives rise to seven processes—one spinous, two transverse, and four articular (see Fig 10–2). The so-called *posterior elements* of a vertebra include all parts of the vertebra that are situated posterior to the body of the vertebra.

The *spinous process,* or *spine,* is directed posteriorly from the junction of the two laminae. The *transverse processes* are directed laterally from the junction of the laminae and the pedicles. Both the spinous and transverse processes serve as levers and receive attachments of muscles and ligaments. The *articular processes* are vertically arranged and consist of two superior and two inferior processes. They arise from the junction of the laminae and the pedicles, and their articular surfaces are covered with hyaline cartilage. The two superior articular processes of one vertebral arch articulate with the two inferior articular processes of the arch above, forming two synovial joints (see Fig 10–9).

The pedicles are notched on their upper and lower borders, forming the *superior and inferior vertebral notches.* On each side, the superior notch of one vertebra and the inferior notch of an adjacent vertebra together form an *intervertebral foramen.* These foramina, in an articulated skeleton, serve to transmit the spinal nerves and blood vessels. The anterior

Clinical Note

It is important to delineate the boundaries of an intervertebral foramen from the perspective of the anterior and posterior nerve roots and the segmental spinal nerve.

Anterior. Intervertebral disc and lower part of vertebral body.

Posterior. Articular processes and joints of processes.

Superior. Inferior notch of pedicle of vertebra above.

Inferior. Superior notch of pedicle of vertebra below.

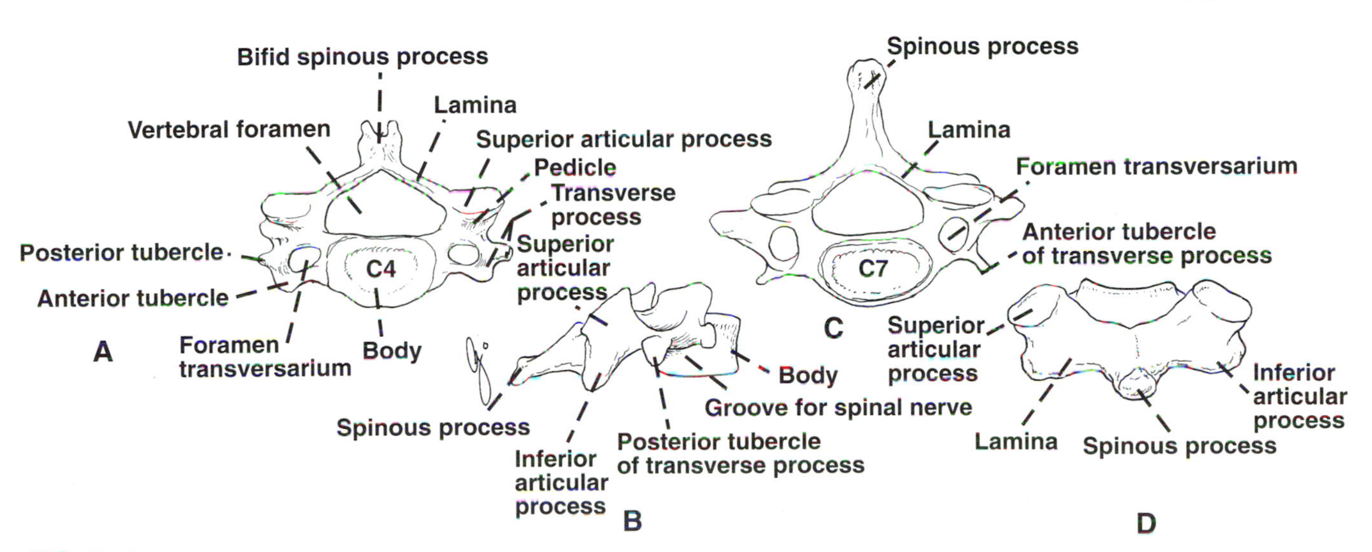

FIG 10–2.
Fourth cervical vertebra: **A,** superior view. **B,** lateral view. Seventh cervical vertebra: **C,** superior view. **D,** posterior view.

and posterior nerve roots of a spinal nerve unite within these foramina with their coverings of dura to form the segmental spinal nerves.

Characteristics of a Typical Cervical Vertebra
A typical cervical vertebra has the following characteristics (see Fig 10–2).

1. The transverse processes possess a *foramen transversarium* for the passage of the vertebral artery and veins.
2. The spines are small and bifid.
3. The body is small and broad from side to side.
4. The vertebral foramen is large and triangular.
5. The superior articular processes face backward and upward; the inferior articular processes have facets that face downward and forward.

Characteristics of Atypical Cervical Vertebrae
The first, second, and seventh cervical vertebrae are atypical.

The *first cervical vertebra or atlas* (Fig 10–3):

1. Does not possess a body.
2. Does not have a spinous process.
3. Has an anterior and posterior arch, and
4. Has a lateral mass on each side with articular surfaces on its upper surface for articulation with the occipital condyles and articular surfaces on its lower surface for articulation with the axis.

The *second cervical vertebra or axis* (Fig 10–4) has a peglike *odontoid process* that projects from the superior surface of the body (representing the body of the atlas that has fused with the body of the axis).

The *seventh cervical vertebra,* or *vertebra prominens* (see Fig 10–2), has the longest spinous process, and the process is not bifid.

Clinical Note

It is a useful landmark for counting the lower vertebrae.

The transverse process is large but the foramen transversarium is small (it does not transmit the vertebral artery).

Characteristics of a Typical Thoracic Vertebra
A typical thoracic vertebra has the following characteristics (Fig 10–5):

1. The body is medium sized and heart shaped.
2. The vertebral foramen is small and circular.
3. The spines are long and inclined downward.
4. Costal facets are present on the sides of the bodies for articulation with the heads of the ribs.
5. Costal facets are present on the transverse processes for articulation with the tubercles of the ribs.
6. The superior articular processes bear facets that face backward and laterally, whereas the facets on the inferior articular processes face forward and medially.

Characteristics of a Typical Lumbar Vertebra
A typical lumbar vertebra has the following characteristics (Fig 10–6):

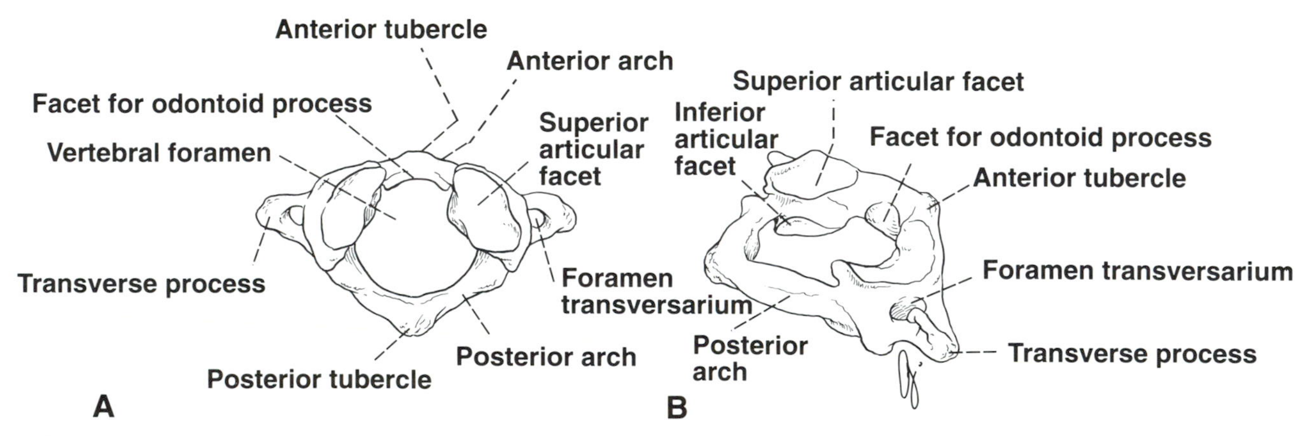

FIG 10–3.
Atlas. **A,** superior view. **B,** oblique view.

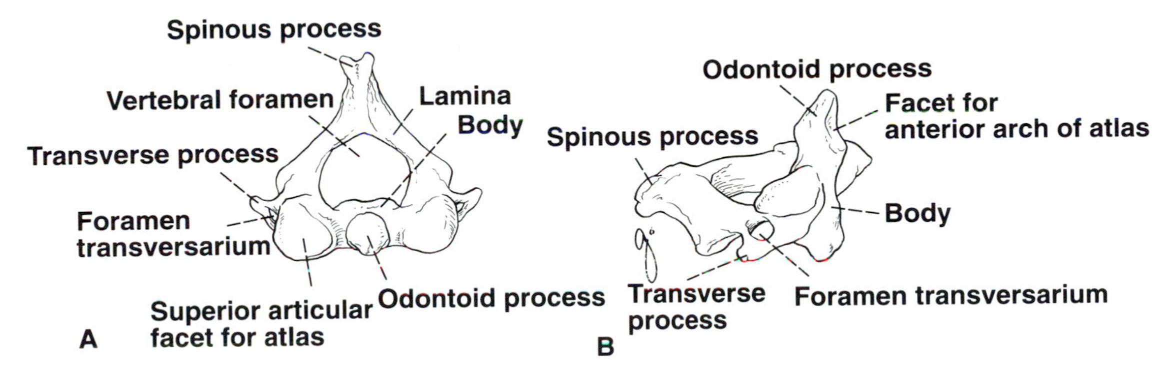

FIG 10–4.
Axis. **A,** superior view. **B,** oblique lateral view.

1. The body is large and kidney shaped.
2. The pedicles are strong and directed backward.
3. The laminae are thick.
4. The vertebral foramina are triangular.
5. The transverse processes are long and slender.
6. The spinous processes are short, flat, and quadrangular and project directly backward.
7. The articular surfaces of the superior articular processes face medially, and those of the inferior articular processes face laterally.

Sacrum

The sacral part of the vertebral column shows a forward concavity (see Fig 10–1) and is made up of five rudimentary vertebrae that are fused together to form a single wedge-shaped bone, the *sacrum.* The upper border, or base, articulates with the fifth lumbar vertebra and the narrow inferior border with the coccyx. Laterally, the sacrum articulates with the two iliac bones to form the sacroiliac joints. The anterior and upper margin of the first sacral vertebra bulges forward as the posterior margin of the pelvic inlet and is known as the *sacral promontory* (Fig

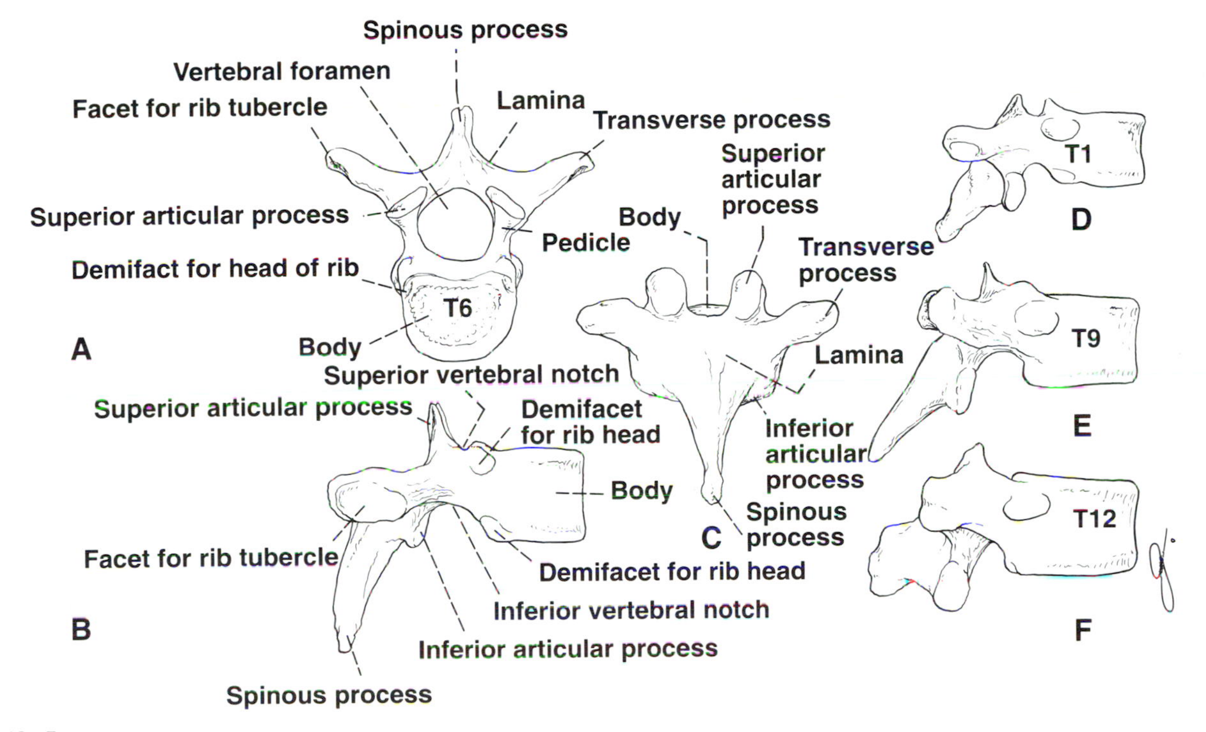

FIG 10–5.
Sixth thoracic vertebra: **A,** superior view. **B,** lateral view. **C,** posterior view. First thoracic vertebra: **D,** lateral view. Ninth thoracic vertebra: **E,** lateral view. Twelfth thoracic vertebra: **F,** lateral view.

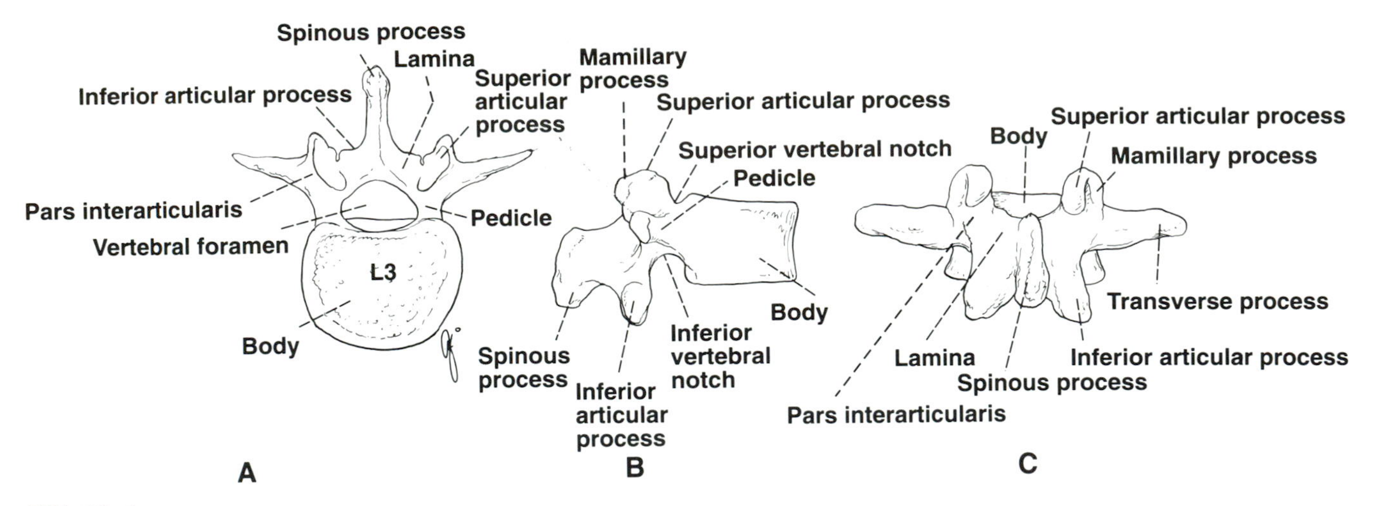

FIG 10–6.
Third lumbar vertebra. **A,** superior view. **B,** lateral view. **C,** posterior view. Note the position of the pars interarticularis.

10–7). This is an important obstetric landmark used when measuring the size of the pelvis.

The vertebral foramina are present and together form the *sacral canal.* The laminae of the fifth sacral vertebra, and sometimes those of the fourth, fail to meet in the midline, forming the *sacral hiatus* (see Fig 10–7). The *sacral canal* contains the anterior and posterior roots of the lumbar, sacral, and coccygeal spinal nerves, the filum terminale, and fibrofatty material. It also contains the lower part of the subarachnoid space down as far as the lower border of the second sacral vertebra.

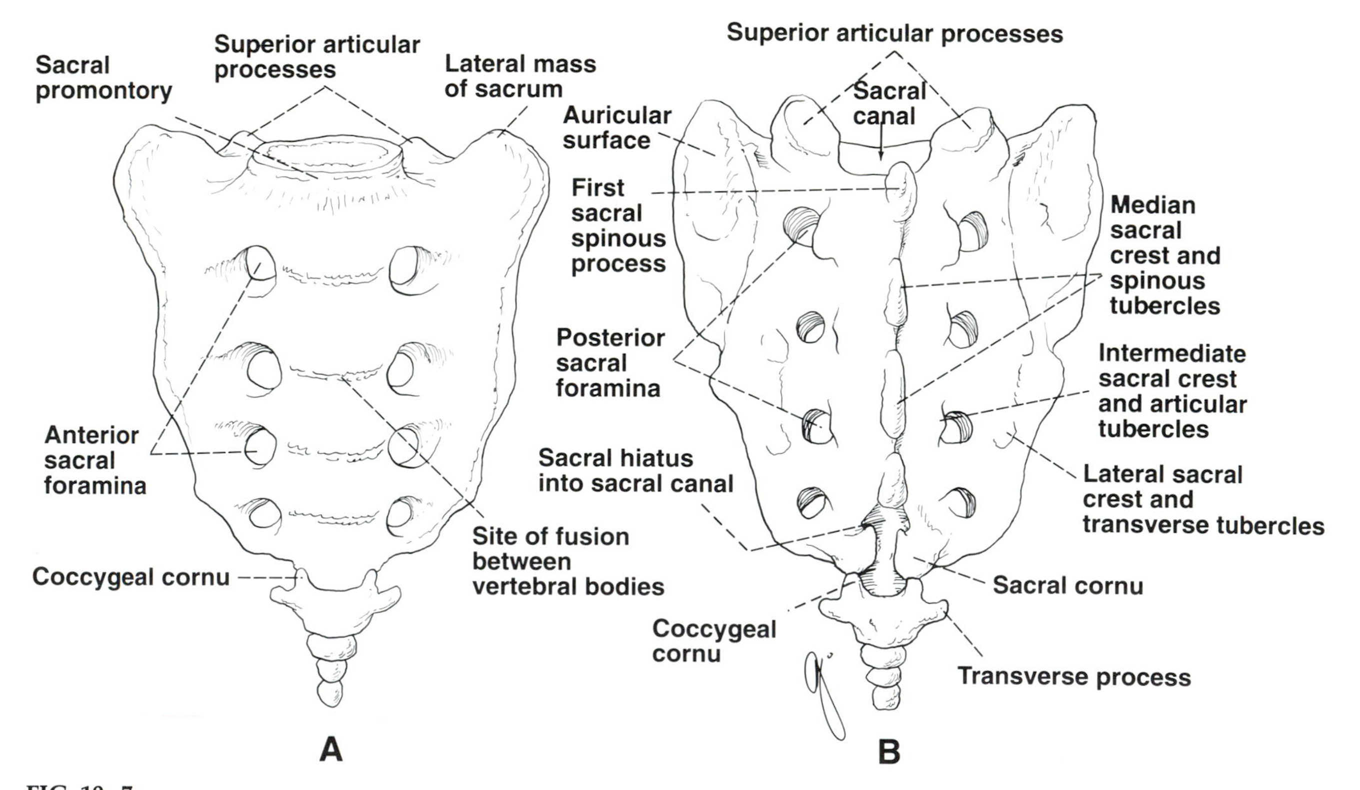

FIG 10–7.
Sacrum and coccyx in the adult male. **A,** anterior view. **B,** posterior view. The median sacral crest is formed by the fusion of the spinous processes; the tips of the spinous processes form the spinous tubercles. The intermediate sacral crest is formed by the fused articular processes; small projections remain as the articular tubercles. The lateral sacral crest is formed by the fused transverse processes, and their remains are known as the transverse tubercles.

The anterior and posterior surfaces of the sacrum each have four foramina bilaterally for the passage of the anterior and posterior rami of the upper four sacral spinal nerves.

Coccyx

The coccyx consists of four vertebrae fused together to form a single, small triangular bone that articulates at its base with the lower end of the sacrum (see Fig 10–7). The first coccygeal vertebra is usually not fused, or is incompletely fused with the second vertebra.

A knowledge of the preceding basic anatomy of the vertebral column is important when interpreting radiographs and when noting the precise sites of bony pathologic features relative to soft-tissue injury.

Important Variations in the Vertebrae

The number of cervical vertebrae is constant, but the seventh cervical vertebrae may possess a *cervical rib* (see p. 87). The thoracic vertebrae may be increased in number by the addition of the first lumbar vertebra, which may have a rib. The fifth lumbar vertebra may be incorporated into the sacrum; this is usually incomplete and may be limited to one side. The first sacral vertebra may remain partially or completely separate from the sacrum and resemble a sixth lumbar vertebra. A large extent of the posterior wall of the sacral canal may be absent because the laminae and spines fail to develop.

The coccyx, which usually consists of four fused vertebrae, may have three or five vertebrae. The first coccygeal vertebra may be separate. In this condition, the free vertebra usually projects downward and anteriorly from the apex of the sacrum.

Clinical Note

The laminae may fail to develop adequately and not meet in the midline, producing *spina bifida*. This is most commonly seen in the lumbosacral region. Suppression of the development of half a vertebra or the presence of an additional half vertebra will produce a *hemivertebra*, causing severe scoliosis.

Basic Anatomy of the Joints of the Vertebral Column

Atlanto-occipital Joints

The structure of these joints is shown in Figure 10–8.

Movements.— Flexion, extension, and lateral flexion; there is no rotation.

Atlantoaxial Joints

The structure of these joints is shown in Figure 10–8. The odontoid process of the axis is kept closely in contact with the anterior arch of the atlas by the transverse ligament.

Movements.— Extensive rotation of the atlas and thus of the head on the axis.

Joints of the Vertebral Column Below the Axis

With the exception of the first two cervical vertebrae, the remainder of the mobile vertebrae articulate with each other by means of cartilaginous joints between their bodies and by synovial joints between their articular processes (Fig 10–9).

Joints Between Two Vertebral Bodies.— The upper and lower surfaces of the bodies of adjacent vertebrae are covered by thin plates of hyaline cartilage. An intervertebral disc of fibrocartilage is sandwiched between the plates of hyaline cartilage (see Fig 10–9). The collagen fibers of the disc strongly unite the bodies of the two vertebrae.

In the lower cervical region, small synovial joints are present at the sides of the intervertebral bodies of the vertebrae.

Ligaments.—The *anterior and posterior longitudinal ligaments* run as continuous bands down the anterior and posterior surfaces of the vertebral column from the skull to the sacrum (see Fig 10–9). These ligaments hold the vertebrae firmly together but at the same time permit a small amount of movement to take place between them.

Intervertebral Discs.—The intervertebral discs are responsible for one fourth of the length of the vertebral column (see Fig 10–1). They are thickest in the cervical and lumbar regions, where the movements of the vertebral column are greatest. Each disc consists of a peripheral part, the anulus fibrosus, and a central part, the nucleus pulposus (see Fig 10–9).

The *anulus fibrosus* is composed of fibrocartilage in which the collagen fibers are arranged in concentric layers or sheets. The collagen bundles pass obliquely between adjacent vertebral bodies, and their inclination is reversed in alternate sheets. The more peripheral fibers are strongly attached to the anterior and posterior longitudinal ligaments of the vertebral column.

The *nucleus pulposus* in children is an ovoid mass of gelatinous material containing a large amount of

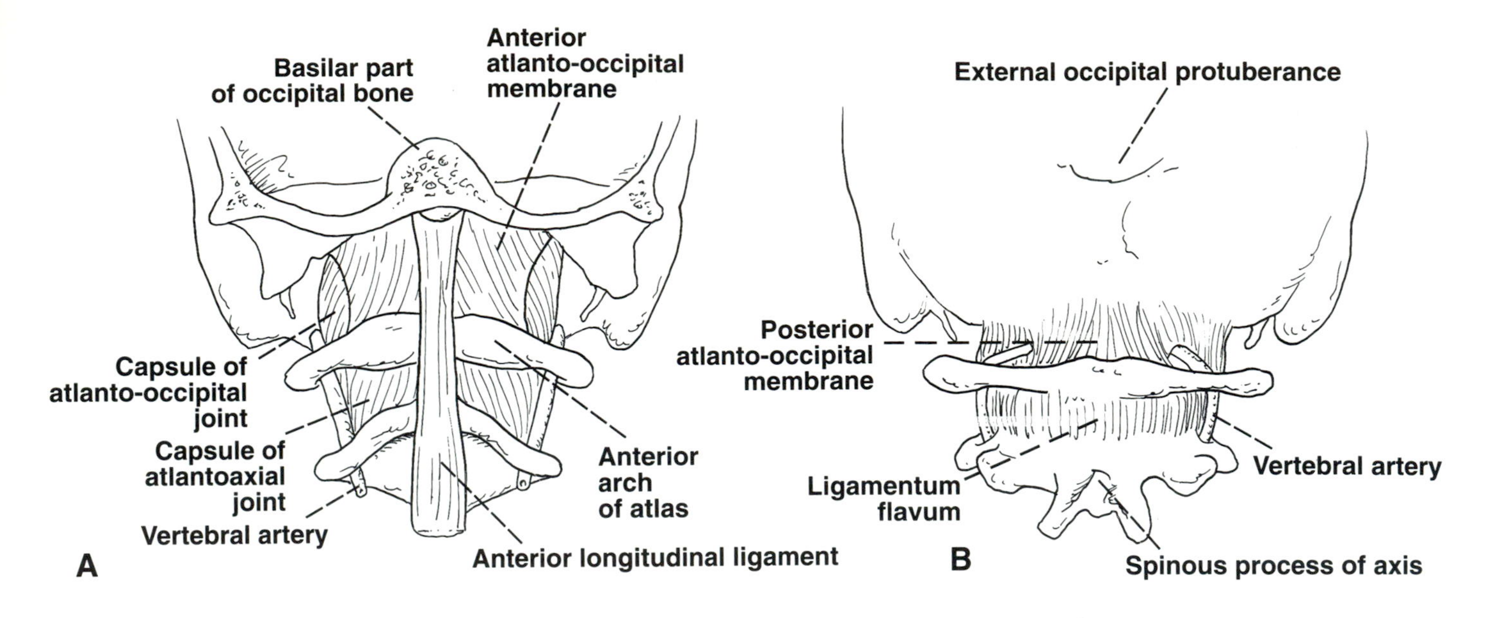

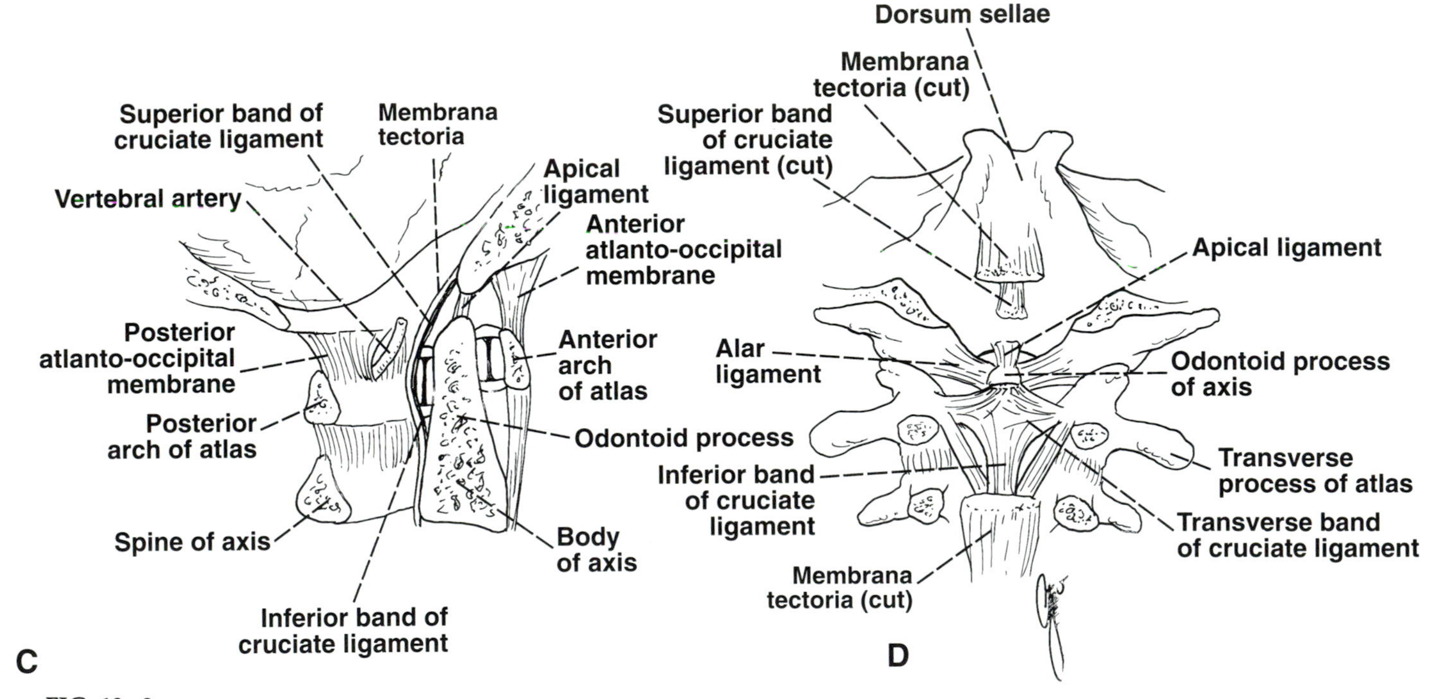

FIG 10–8.
Atlanto-occipital joints: **A,** anterior view. **B,** posterior view. Atlantoaxial joints: **C,** sagittal section. **D,** posterior view. The posterior arch of the atlas and the laminae and the spine of the axis have been removed.

water, a small number of collagen fibers, and a few cartilage cells. It is normally under pressure and situated slightly nearer to the posterior margin of the disc than to the anterior margin.

The upper and lower surfaces of the bodies of adjacent vertebrae that abut onto the disc are covered with thin plates of hyaline cartilage. The semifluid nature of the nucleus pulposus allows it to change shape and permits one vertebra to rock forward or backward on another, as in flexion and extension of the vertebral column. With advancing age the water content of the nucleus pulposus diminishes and is replaced by fibrocartilage. In the elderly, the discs are thin and less elastic, and it is no longer possible to distinguish the nucleus from the anulus.

No discs are found between the first two cervical vertebrae or in the sacrum or coccyx.

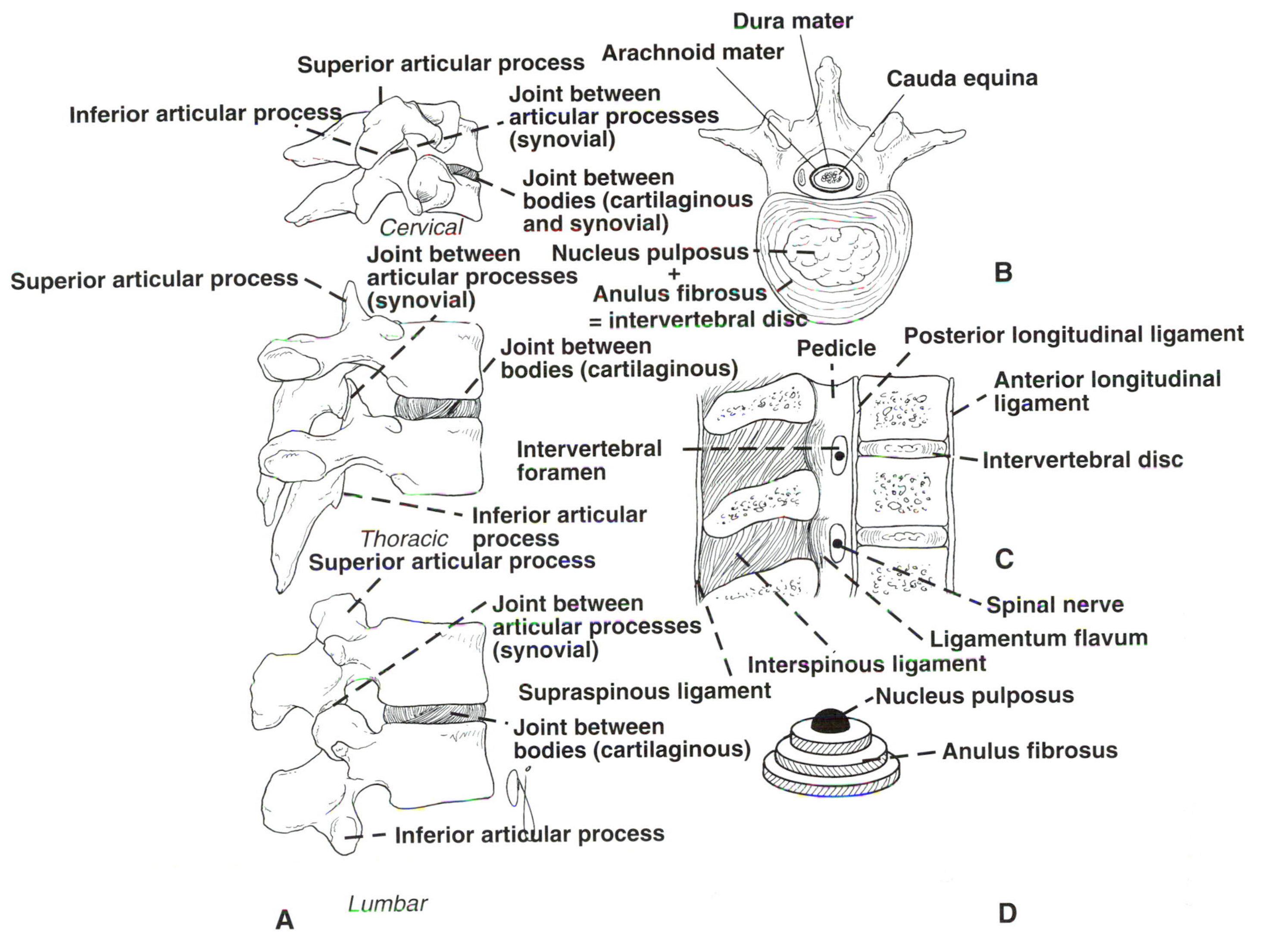

FIG 10–9.
A, joints in cervical, thoracic, and lumbar regions of the vertebral column. **B,** third lumbar vertebra viewed from above showing the relationship between the intervertebral disc and the cauda equina. **C,** sagittal section through three lumbar vertebrae showing ligaments and intervertebral discs. **D,** structure of an intervertebral disc.

Clinical Note

Rupture of Anulus Fibrosus. —A sudden increase in the compression load on the vertebral column, as when carrying a heavy load or when suddenly flexing the vertebral column, may cause the anulus fibrosus to rupture. This allows the nucleus pulposus to herniate and protrude into the vertebral canal, where it may press on the spinal nerve roots, the spinal nerve, or even the spinal cord (see p. 382). It is possible for a disc in which the anulus fibrosus has degenerated to rupture spontaneously without an increased compression load.

Joints Between Two Vertebral Arches.—The joints between two vertebral arches consist of synovial joints between the superior and inferior articular processes of adjacent vertebrae. The articular facets are covered with hyaline cartilage, and the joints are surrounded by a capsular ligament.

Ligaments.—These include the following.

1. *Supraspinous ligament* (see Fig 10–9). This runs between adjacent spines.
2. *Interspinous ligament* (see Fig 10–9). This connects adjacent spines.
3. *Intertransverse ligaments.* These run between adjacent transverse processes.

4. *Ligamentum flavum* (see Fig 10–9). This connects the laminae of adjacent vertebrae.

In the cervical region, the supraspinous and interspinous ligaments are greatly thickened to form the *ligamentum nuchae.* The latter extends from the spine of the seventh cervical vertebra to the external occipital protuberance of the skull, with its anterior border being strongly attached to the cervical spines in between.

Nerve Supply of Vertebral Joints

The joints between the vertebral bodies are innervated by the meningeal branches of each spinal nerve (Fig 10–10). The joints between the articular processes are innervated by branches of the posterior rami of the spinal nerves (see Fig 10–10).

Clinical Note

Back pain and the innervation of vertebral structures. — The vertebrae, ligaments, joint capsules, discs, and postvertebral muscles have sensory innervation that is derived from spinal nerves of the same and adjoining segments. Since one is normally not conscious of each anatomical structure, pain due to disease of these structures is usually referred to the appropriate dermatome. Also, internal organs, such as the pancreas, uterus, ovaries, uterine tubes, prostate, kidney, or aorta (dissection and aneurysm), may give rise to pain referred to the back.

It is believed that the nerve supply of the vertebral joints at any particular level is derived from two segmental spinal nerves.

Muscles Acting on the Vertebral Column

Postvertebral Muscles.—In the standing position, the line of gravity (Fig 10–11) passes through the odontoid process of the axis, behind the centers of the hip joints, and in front of the knee and ankle joints. It follows that when the body is in this position, the greater part of its weight falls in front of the vertebral column. It is, therefore, not surprising to find that the postvertebral muscles of the back are well developed in humans. The postural tone of these muscles is the major factor responsible for the maintenance of the normal curves of the vertebral column.

The deep muscles of the back form a broad, thick column of muscle tissue that occupies the hollow on each side of the spinous processes (see Fig 10–11). These muscles extend from the sacrum to the skull and lie beneath the thoracolumbar fascia. This complicated muscle mass is composed of many separate muscles of varying length. Each individual muscle may be regarded as a string, which, when pulled on, causes one or several vertebrae to be extended or rotated on the vertebra below. Because the origins and insertions of the different groups of muscles overlap, entire regions of the vertebral column can be made to move smoothly.

The spines and the transverse processes of the

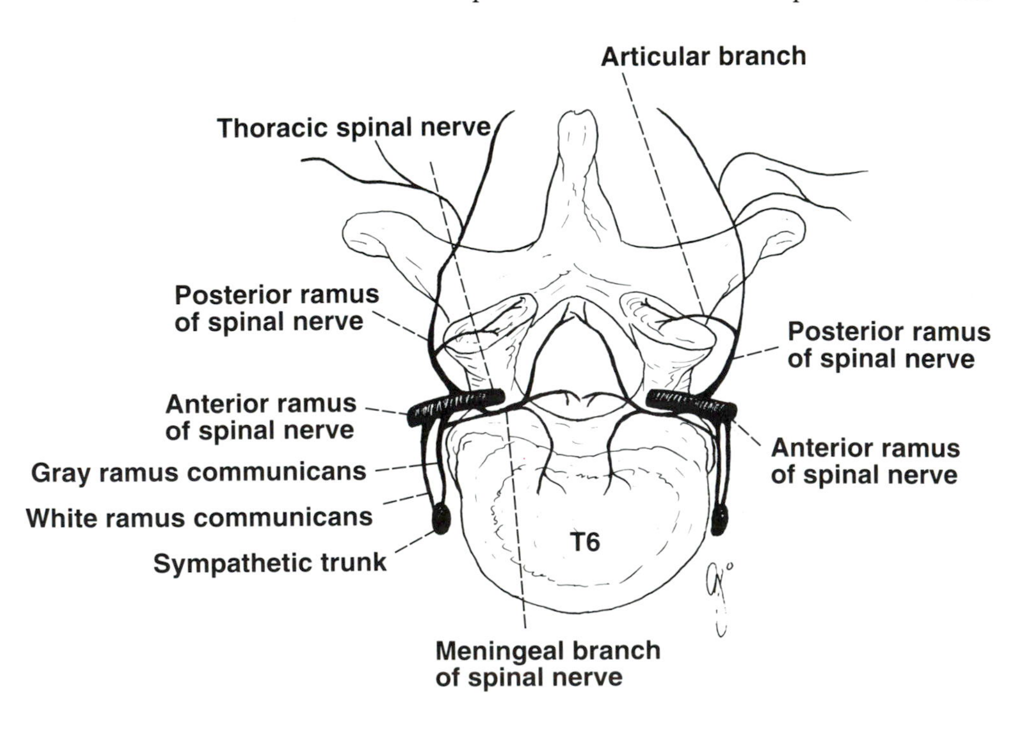

FIG 10–10.
Diagram showing the innervation of vertebral joints. At any particular vertebral level the joints receive nerve fibers from two adjacent spinal nerves.

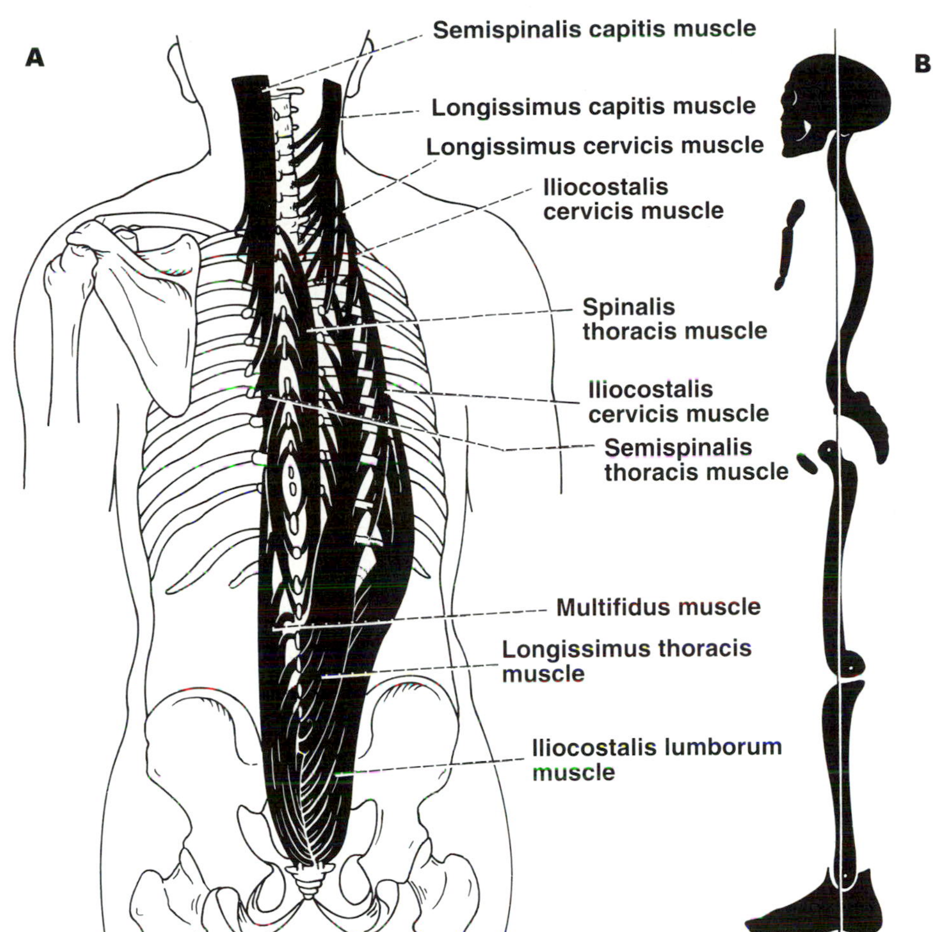

FIG 10–11.
A, arrangement of the deep muscles of the back. **B,** lateral view of the skeleton showing the line of gravity.

vertebrae serve as levers that facilitate muscular actions. The muscles of longest length lie superficially and run vertically from the back of the sacrum to the rib angles, the transverse processes, and the upper vertebral spines (see Fig 10–11). The muscles of intermediate length run obliquely from the transverse processes to the spines. The shortest and deepest muscle fibers run vertically between the spines and between the transverse processes of adjacent vertebrae.

The muscles of the back are classified as follows: The *superficial vertically running muscles* include the erector spinae (the iliocostalis, longissimus, and spinalis muscles); the *intermediate oblique running muscles* include the transversospinalis (the semispinalis, multifidus, and rotatores muscles); and the *deepest muscles* include the interspinales and intertransversales muscles.

Figure 10–11 shows the position of the more important postvertebral muscles.

Radiographic Appearances of the Vertebral Column

Cervical Region

In the cervical region, radiographs are usually taken in the anteroposterior view, the lateral view, and the oblique view.

In the *anteroposterior view,* the atlantoaxial articulation (C1 and C2) may be demonstrated by asking the patient to keep the mandible in motion while the film is being exposed or by directing the X-ray tube through an open mouth (Fig 10–12). If the latter method is used, the entire length of the odontoid process may be visualized lying between the lateral masses of the atlas.

Below the level of the third cervical vertebra, the bodies of the vertebrae are seen and the spines are clearly visualized (Fig 10–13). The laminae can be identified. The transverse processes overlap one another and are difficult to distinguish separately.

In the *lateral view,* it is important to demonstrate the alignment of all seven vertebrae, including the

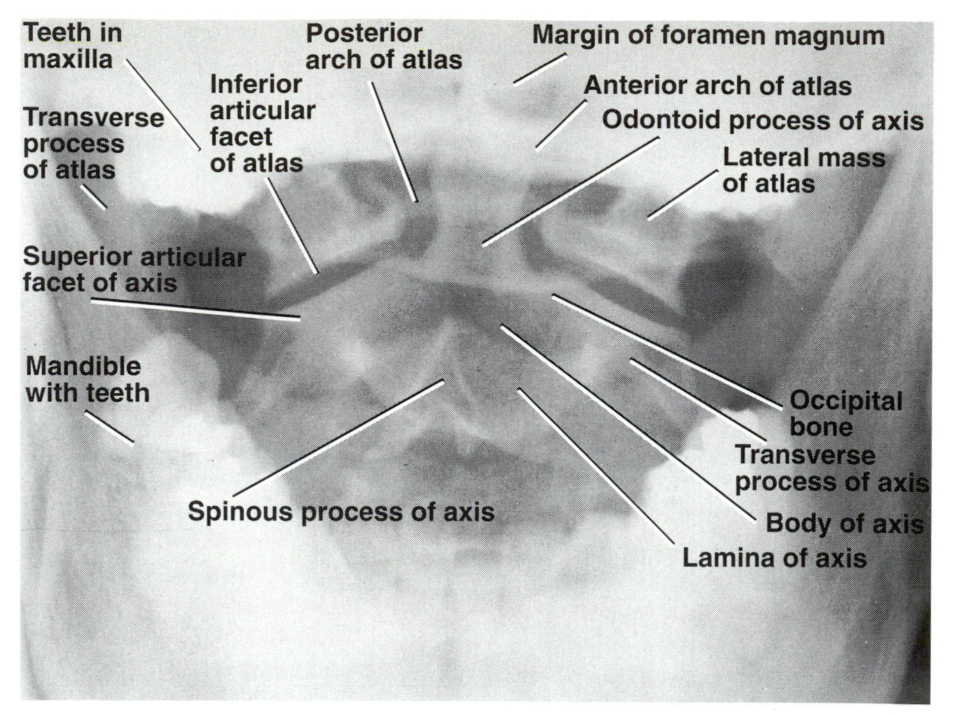

FIG 10–12.
Anteroposterior radiograph of the upper cervical region of the vertebral column with the patient's mouth open to show the odontoid process of the axis.

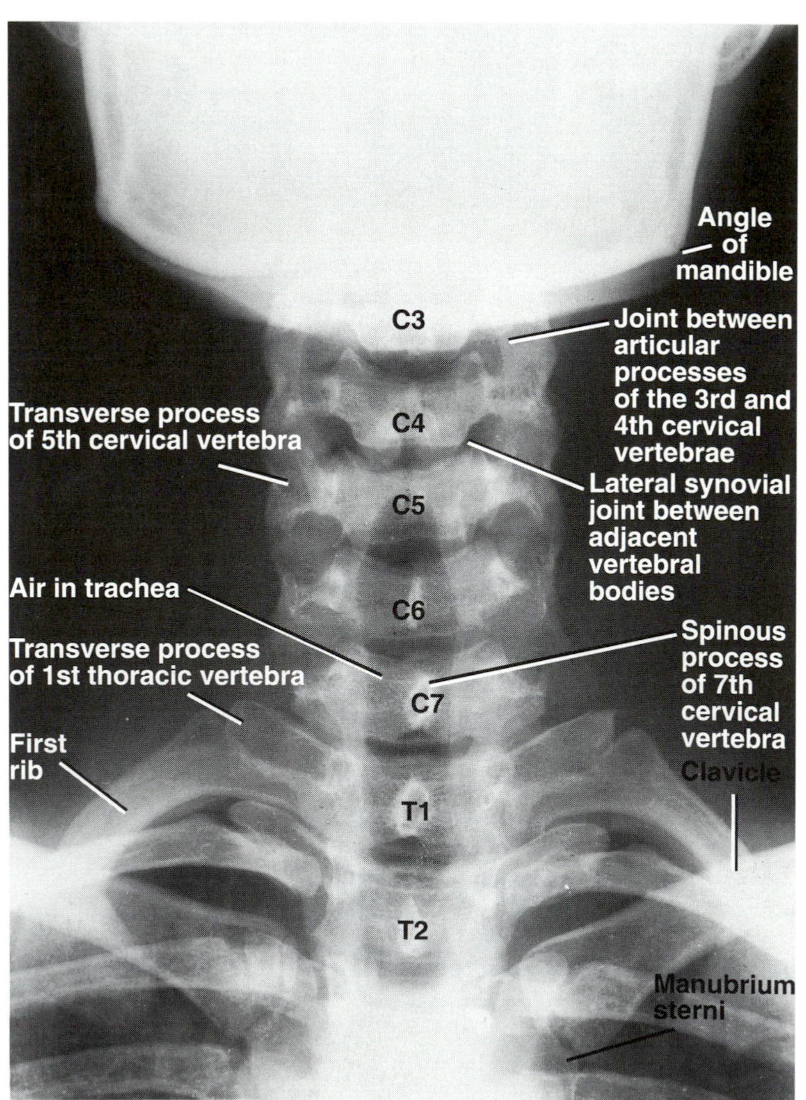

FIG 10–13.
Anteroposterior radiograph of the cervical region of the vertebral column.

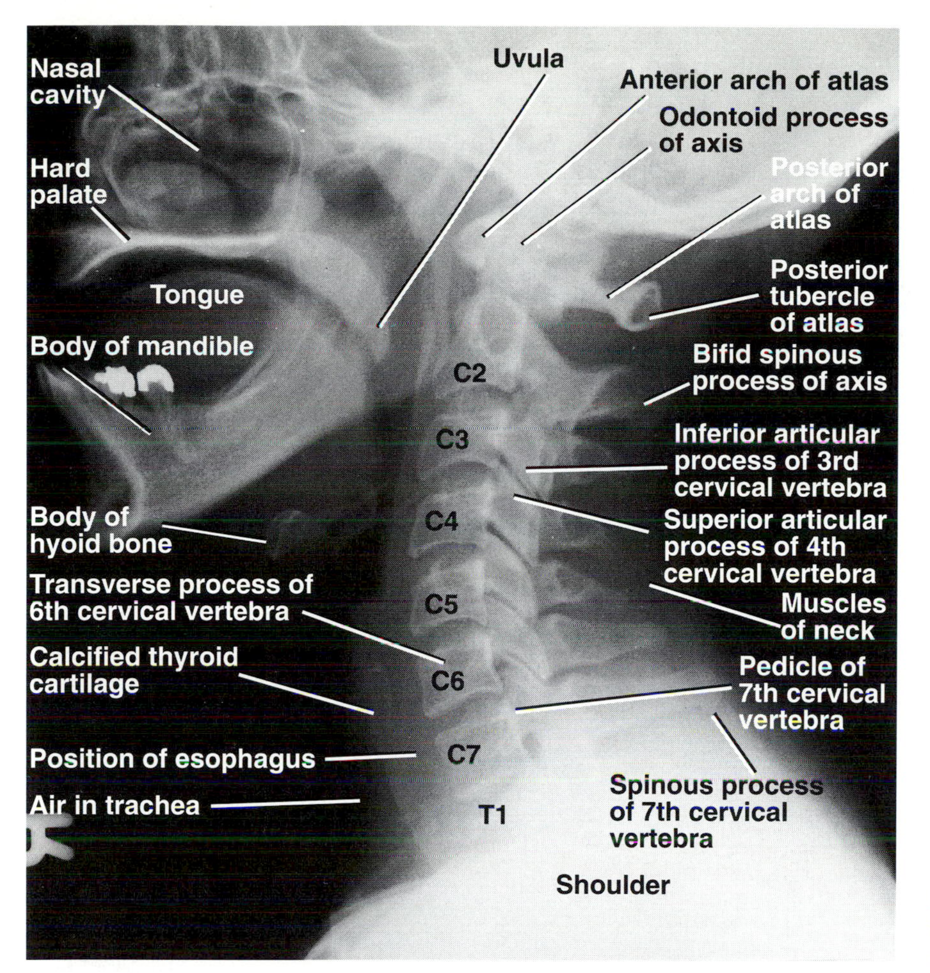

FIG 10–14.
Lateral radiograph of the cervical region of the vertebral column.

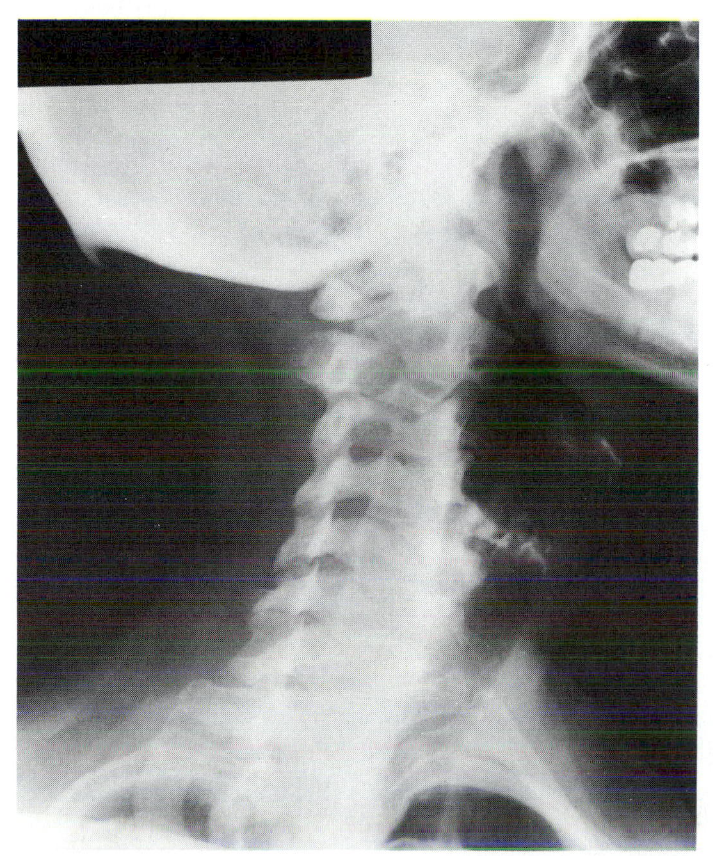

FIG 10–15.
Oblique radiograph of the cervical region of the vertebral column.

articulation between the seventh cervical and the first thoracic vertebrae (Fig 10–14). This can be accomplished by exerting downward traction on the patient's internally rotated arms so that all seven cervical vertebrae are exposed, or by shooting a "swimmer's" view, in which the arm closest to the X-ray beam is abducted at the shoulder and the beam is aimed through the axilla.

The *oblique view* of the cervical spine is used to show the intervertebral foramina (Figs 10–15 and 10–16) and to demonstrate the possible presence of osteophytes that might be encroaching on the opening.

The atlanto-occipital joint is difficult to delineate. The anterior and posterior arches of the atlas are well shown (see Figs 10–14 and 10–15), and the

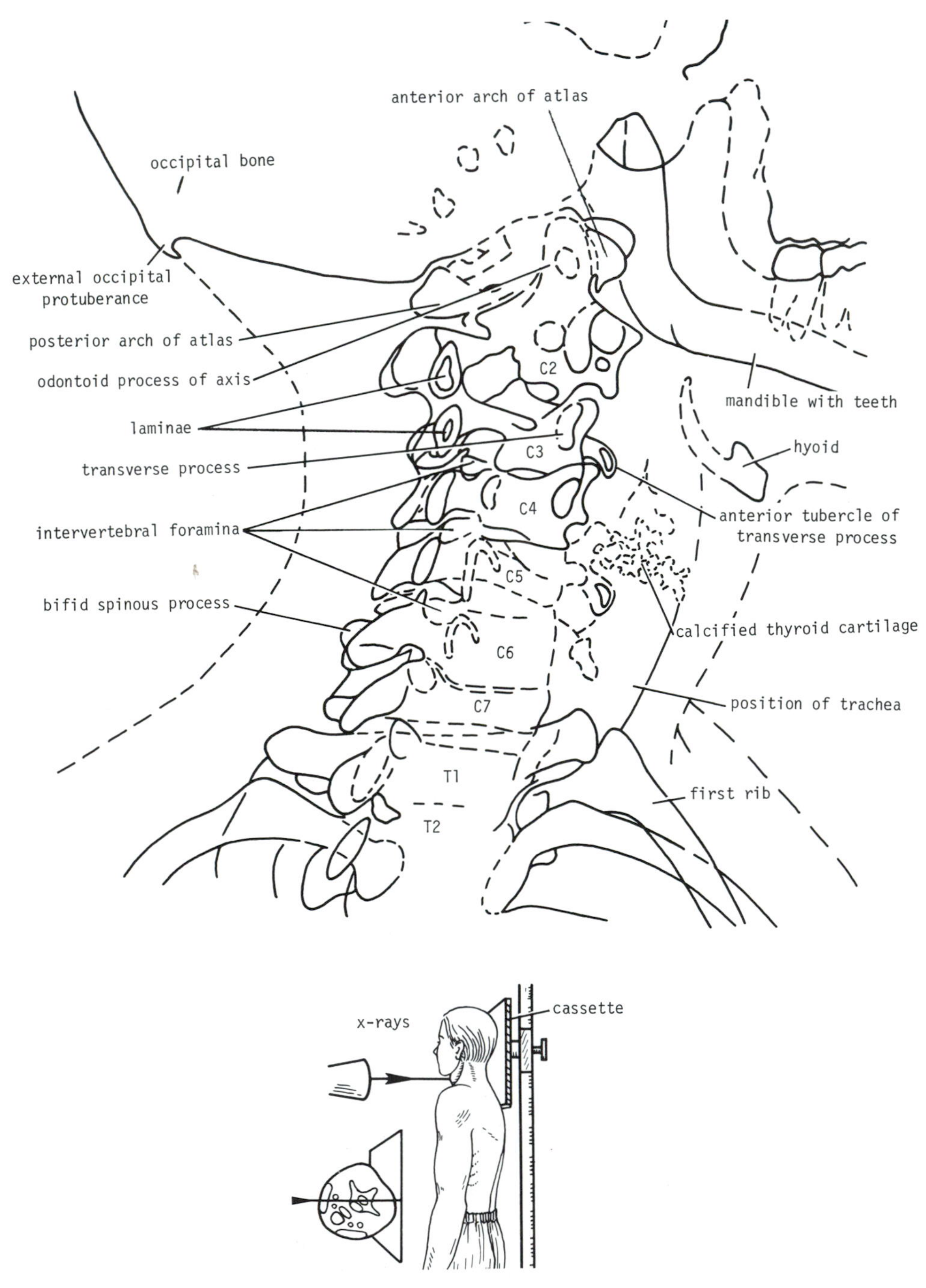

FIG 10–16.
Diagram of the main features seen in an oblique radiograph of the cervical region of the vertebral column shown in Figure 10–15.

body of the axis is easily identified. The odontoid process of the axis extends upward, close to the posterior margin of the anterior arch of the atlas; it is held in place by the transverse ligament.

Clinical Note

Ossification centers and fractures of the odontoid process.—Two ossification centers are present in the base of the odontoid process and fuse at birth; a third center appears at the apex and fuses with those of the base at 12 years of age. The centers for the odontoid process remain separated from the center of the body of the axis by a cartilaginous plate until adulthood. The gaps that normally exist between these various centers during development may be mistaken for a fracture.

The articular processes of the cervical vertebrae are well shown, and the spinous processes can be clearly seen. The transverse processes are difficult to see in the lateral view because they are superimposed on the vertebral bodies. The pedicles are seen as short connections between the vertebral bodies and the articular processes. The intervertebral disc spaces between the bodies of adjacent vertebrae are easily defined and are of equal height.

The anterior and posterior surfaces of the vertebral bodies and the posterior wall of the vertebral canal form smooth curved lines that are roughly parallel (see Fig 10–14).

Thoracic Region

In the thoracic region, radiographs are usually taken in the anteroposterior view and the lateral view.

In the *anteroposterior view,* because of the curvature of the thoracic part of the vertebral column, the upper and lower margins of the bodies of adjacent vertebrae overlap (Fig 10–17). The spinous processes and laminae are superimposed on the bodies.

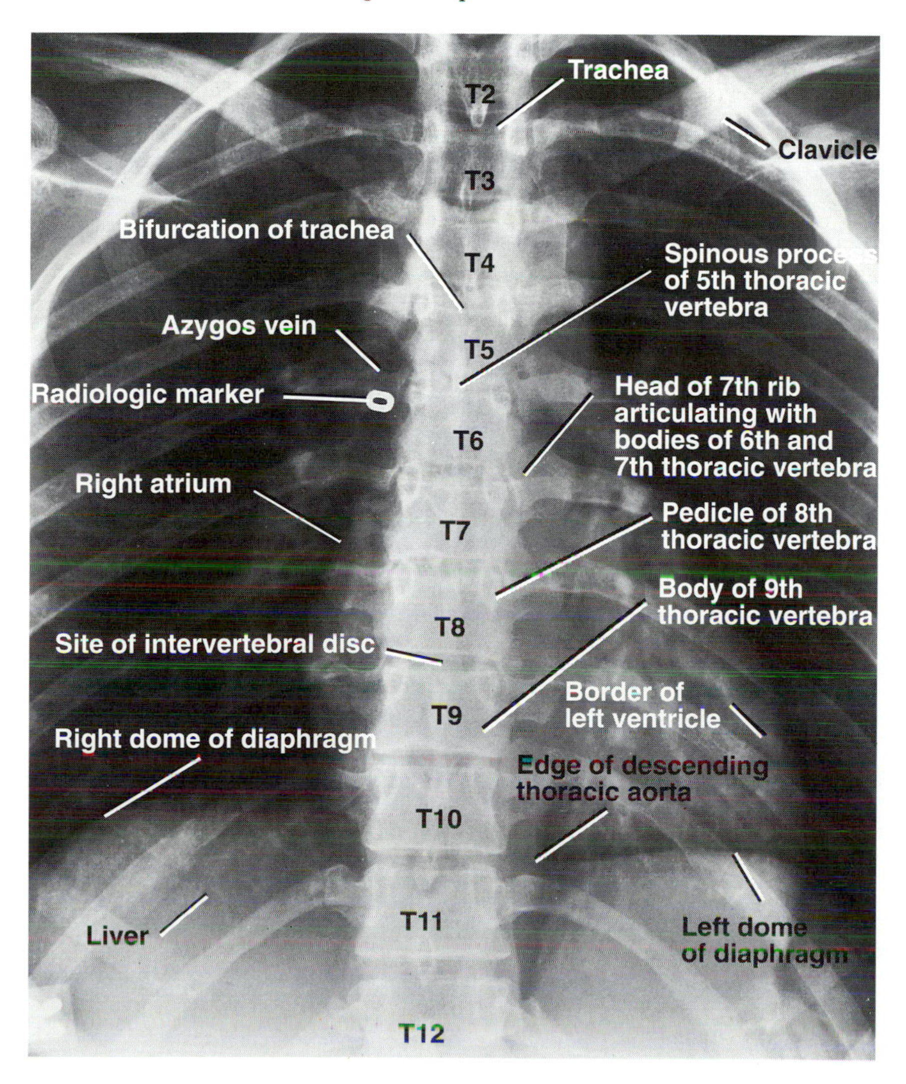

FIG 10–17.
Anteroposterior radiograph of the thoracic region of the vertebral column.

The transverse processes can be identified, but they are obscured by the heads and necks of the ribs. The 1st rib and the 10th, 11th, and 12th ribs on each side articulate only with the bodies of the 1st, 10th, 11th, and 12th thoracic vertebrae, respectively; all the other ribs articulate with two vertebrae. The pedicles are clearly seen as ovoid structures that are superimposed on the lateral parts of the bodies.

The translucent trachea and the heart shadow are superimposed on the thoracic vertebrae.

In the *lateral view,* the rectangular vertebral bodies and the intervertebral disc spaces are clearly seen, even though the ribs and lungs are superimposed on them. The upper four vertebrae are obscured by the shadows of the shoulder girdle.

The pedicles and intervertebral foramina are well demonstrated. The spinous processes, laminae, transverse processes, and ribs are superimposed on one another, however, and their detail is obscured. The vertebral canal is well shown.

Lumbosacral Region

In the lumbosacral region, radiographs are usually taken in the anteroposterior view and the lateral view. Oblique views are used to visualize the pedicles and the pars interarticularis.

In the *anteroposterior view,* the bodies, transverse processes, spinous processes, laminae, and the intervertebral disc spaces are clearly seen (Fig 10–18). The pedicles produce ovoid shadows, and the artic-

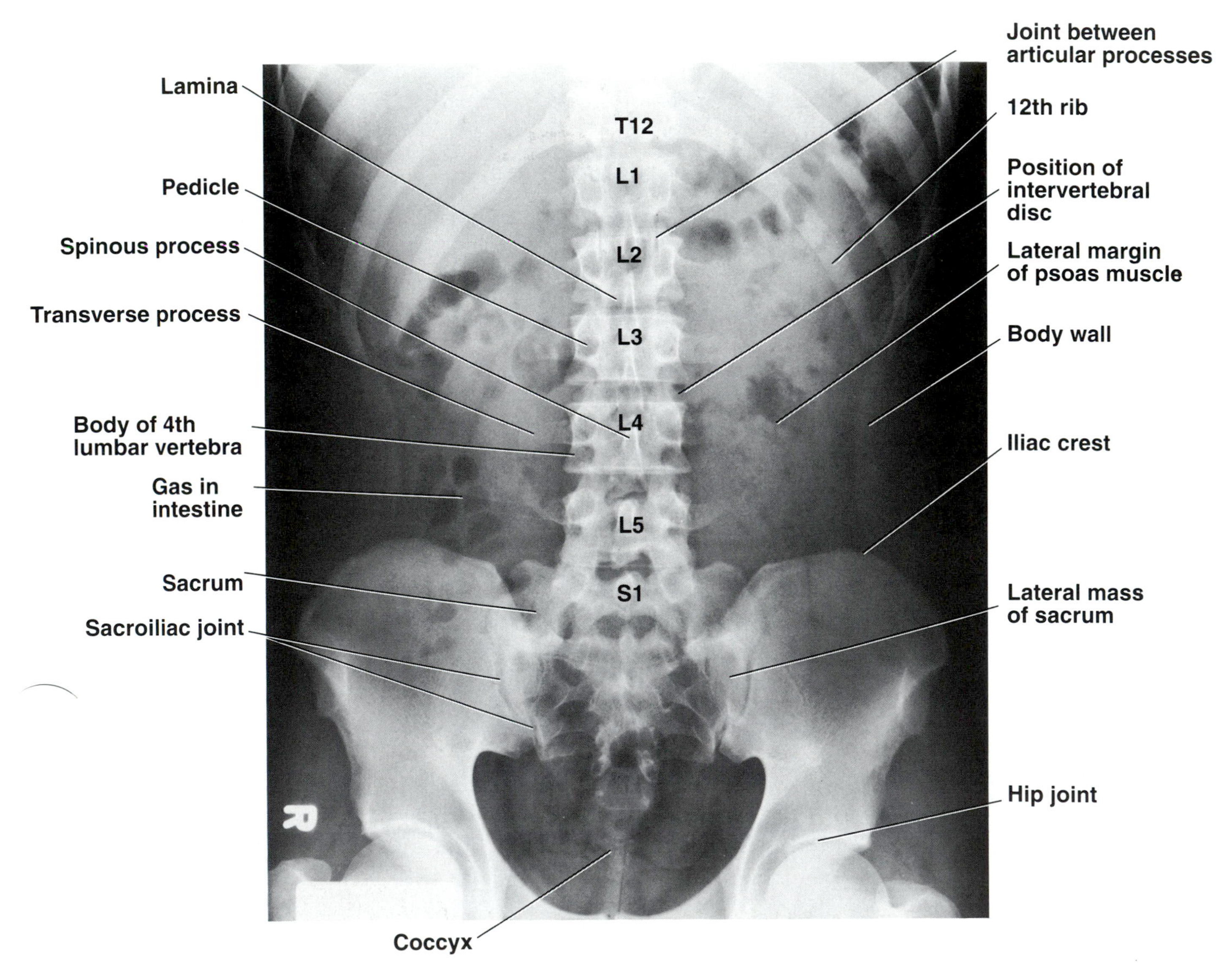

FIG 10–18.
Anteroposterior radiograph of the lumbar region of the vertebral column; parts of the thoracic, sacral, and coccygeal regions are also shown.

ular processes and posterior intervertebral joints can be delineated.

Because of its obliquity, the sacroiliac joint is visualized as two lines—the lateral one corresponding to the anterior margin and the medial one corresponding to the posterior margin (see Fig 10–18). The lower segments of the sacrum and the coccyx are tilted posteriorly and are usually overlapped by the symphysis pubis. In addition, the presence of gas and fecal material in the rectum and sigmoid colon commonly obscures the sacrum. To demonstrate the sacrum in a more direct anteroposterior view, the X-ray tube may be tilted.

In the *lateral view,* the large vertebral bodies, the intervertebral disc spaces, and the intervertebral foramina are clearly seen (Fig 10–19). The pedicles, the articular processes, and the spinous processes are easily visualized. The transverse processes can be identified, but they are superimposed on the sides of the preceding structures. The anterior and posterior surfaces of the vertebral bodies and the posterior wall of the vertebral canal form smooth curved lines that are roughly parallel.

The sacrum on lateral view shows the promontory, the sacral canal, and the fused sacral bodies and the spinous processes. Note the localized anterior angulation between the body of the fifth lumbar vertebra and the first sacral vertebra.

In the oblique view of the lumbar spine, the transverse process, the superior articular process,

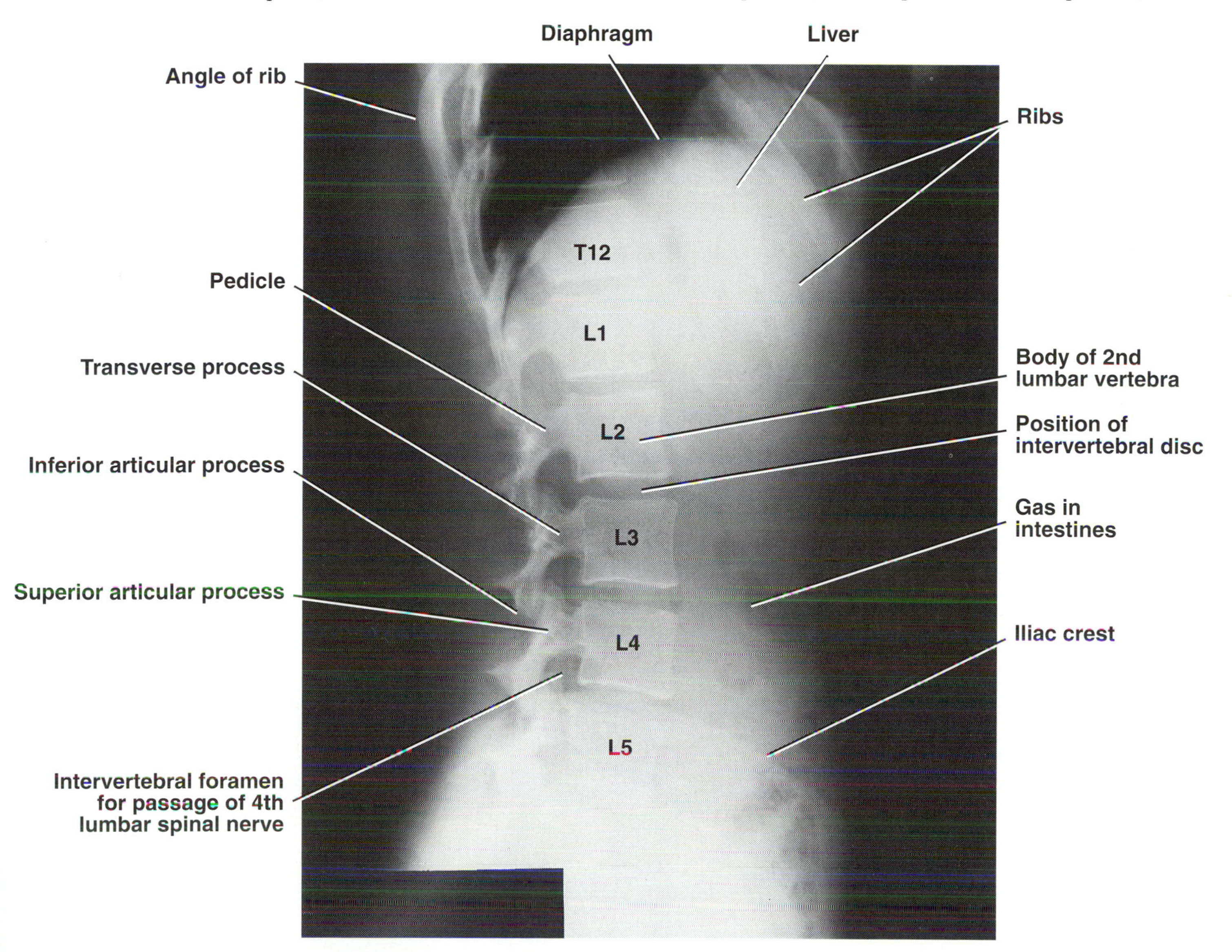

FIG 10–19.
Lateral radiograph of the lumbar region of the vertebral column.

and the joint between the articular process give the appearance of a "Scotty dog" (Figs 10–20 and 10–21). The ear of the dog is the superior articular process. The nose of the dog is the ipsilateral transverse process. The eye of the dog is the ipsilateral pedicle (the X-ray beam "shoots" down the pedicle). The front leg of the dog is the inferior articular process. The body of the dog is the contralateral pedicle. The neck of the dog is the pars interarticularis, i.e., the area where the pedicle, lamina, and articular processes all come together. The presence of a "collar" on the dog's neck indicates a defective area of ossification or a traumatic break in the pars interarticularis, called *spondylolysis.* If the higher vertebra has slipped forward on the lower vertebra along this defect, then *spondylolisthesis* exists.

Coccyx

The coccyx is not well shown on routine anteroposterior and lateral radiographs because of its oblique position relative to the film and to the presence of gas and feces in the rectum and sigmoid colon. These difficulties may be partially overcome by tilting the X-ray tube.

Computed tomographic scans and magnetic resonance images of the vertebral column are shown in Figures 10–22 through 10–25.

Basic Anatomy of the Spinal Cord

The spinal cord is cylindrical (Fig 10–26) and begins superiorly at the foramen magnum, where it is continuous with the medulla oblongata of the brain. At birth, the lower end of the spinal cord ends at the level of the third lumbar vertebra. In adults, it ends at a higher level, the lower border of the first lumbar vertebra (see Fig 10–27).

The spinal cord thus occupies the upper two thirds of the vertebral canal and is surrounded by three meninges—the *dura mater,* the *arachnoid mater,* and the *pia mater* (see Fig 10–26). Further protection is provided by the *cerebrospinal fluid,* which surrounds the spinal cord in the *subarachnoid space.*

Enlargements of the Spinal Cord

In the cervical region, where the spinal cord gives origin to the brachial plexus, and in the lower tho-

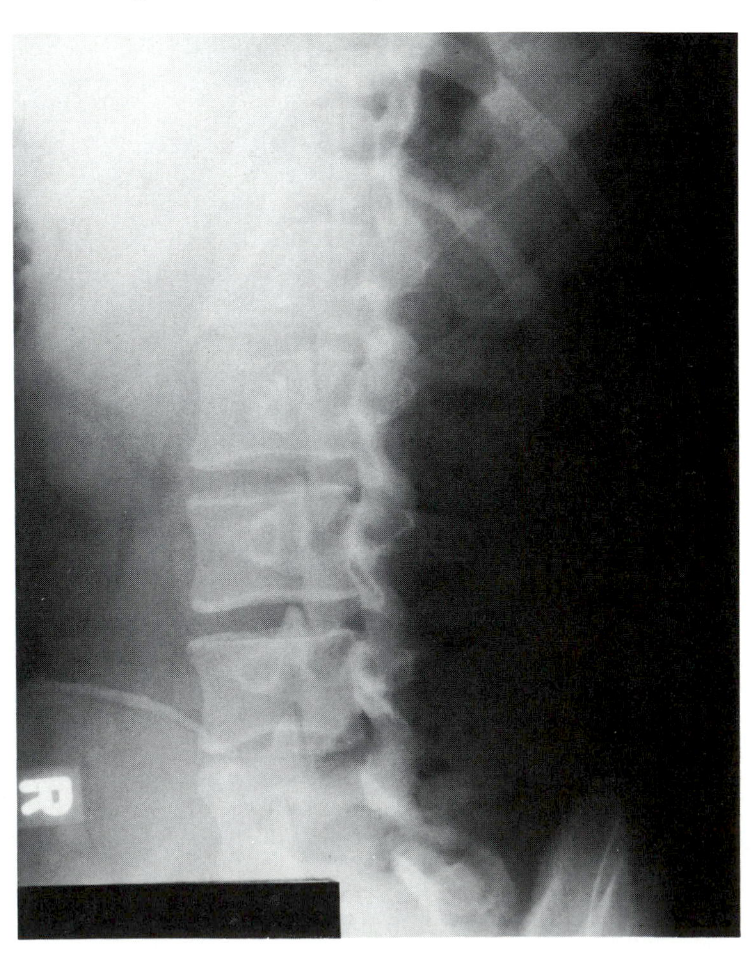

FIG 10–20.
Oblique radiograph of the lumbar region of the vertebral column.

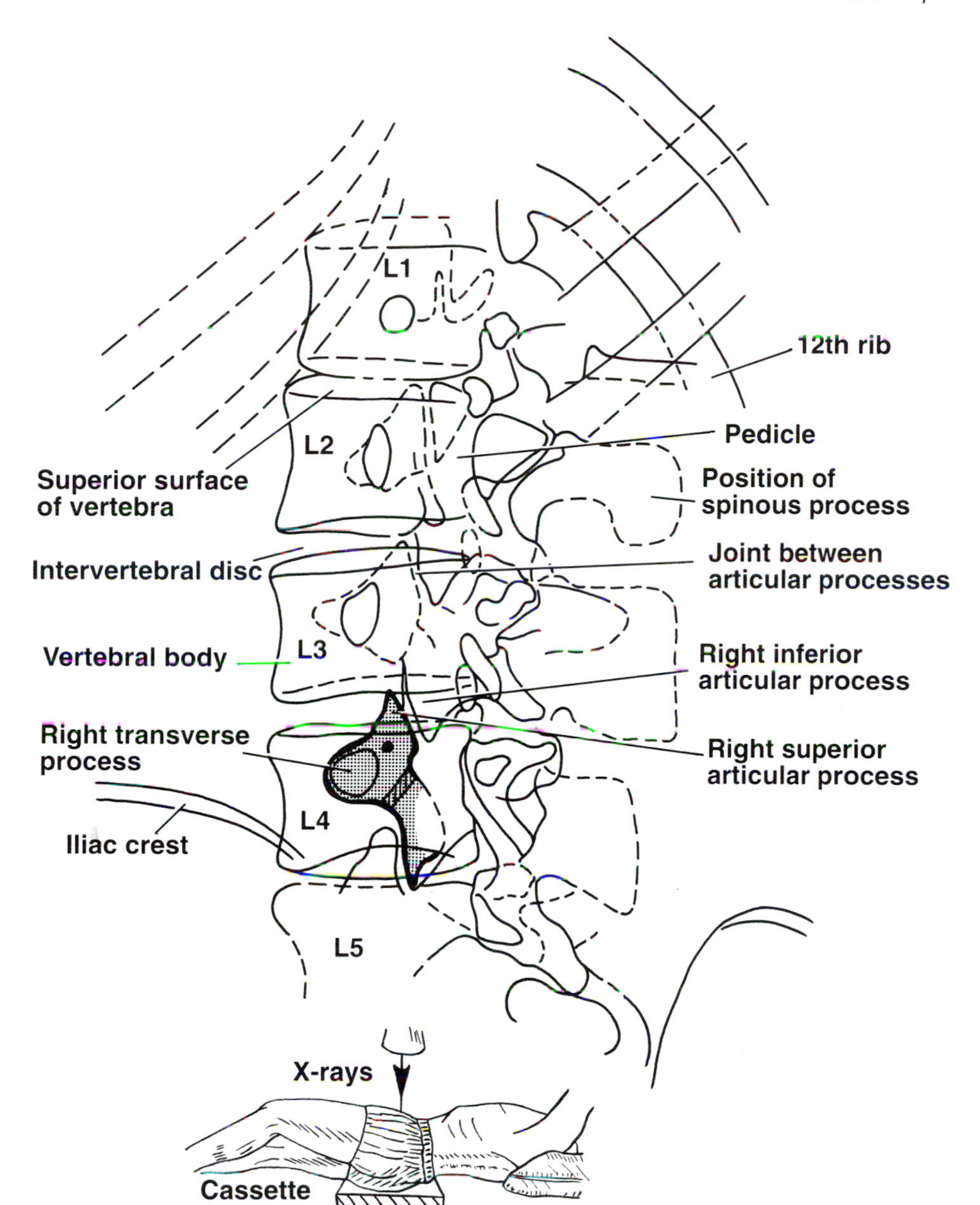

FIG 10–21.
Diagram of the main features seen in an oblique radiograph of the lumbar region of the vertebral column in Figure 10–20. The stippled "Scotty dog" appearance is seen on the fourth lumbar vertebra, as described in the text.

racic and lumbar regions, where the spinal cord gives origin to the lumbosacral plexus, there are fusiform enlargements called the *cervical* and *lumbar enlargements* (see Fig 10–26). Inferiorly, the lumbar enlargement tapers off into the *conus medullaris;* from its apex, a prolongation of the pia mater called the *filum terminale* descends to be attached to the back of the coccyx.

Roots of the Spinal Nerves

Along the entire length of the spinal cord is attached 31 pairs of spinal nerves by the *anterior or motor roots* and the *posterior or sensory roots* (see Fig 10–26). Each root is attached to the cord by a series of *rootlets,* which extend the whole length of the corresponding segment of the cord. Each posterior nerve root possesses a *posterior root ganglion.*

The spinal nerve roots pass laterally from each spinal cord segment to the level of their respective intervertebral foramina, where they unite to form a *spinal nerve* (see Figs 10–26 and 10–27). Here the motor and sensory fibers become mixed, so that a spinal nerve is made up of a mixture of motor and sensory fibers.

Because of the disproportionate growth in length of the vertebral column during development compared with that of the spinal cord, the length of the roots increases progressively from above downward (see Fig 10–42). In the upper cervical region the spinal nerve roots are short and run almost hor-

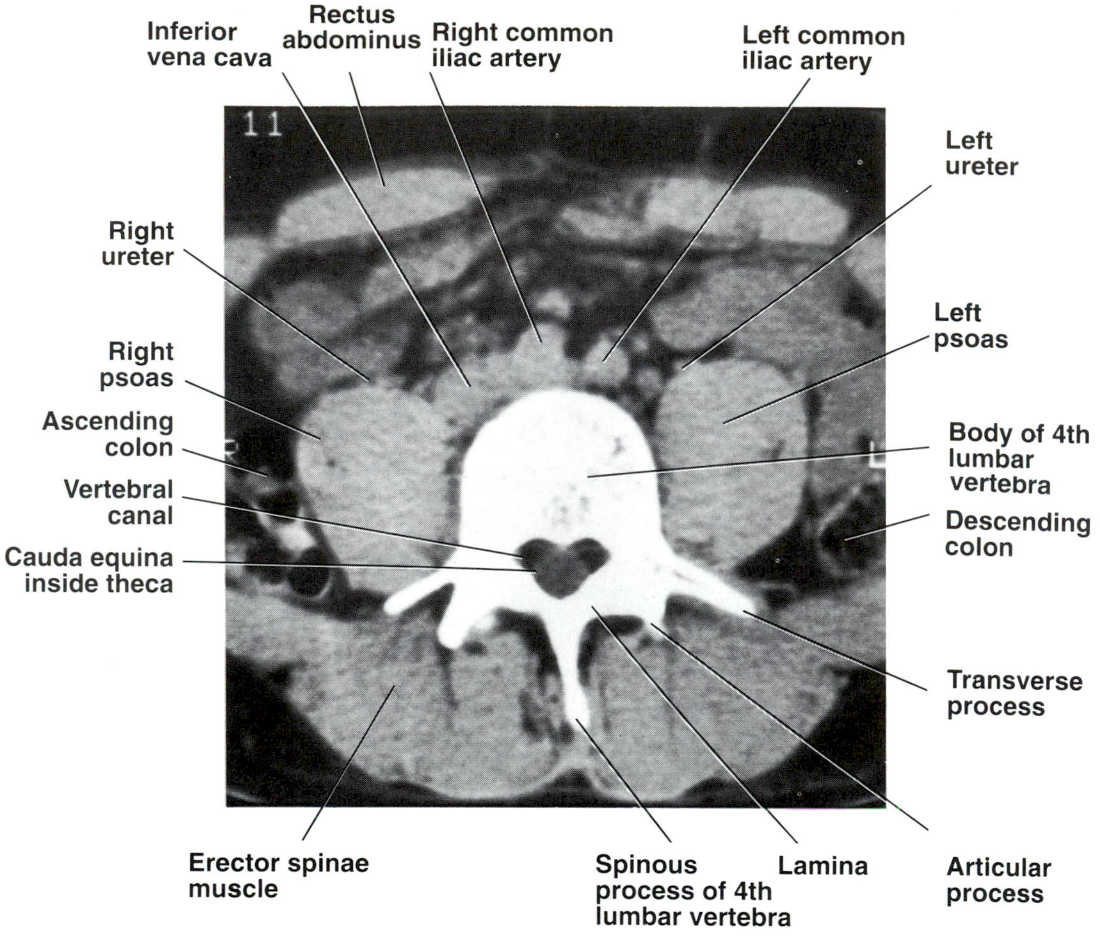

FIG 10–22.
An axial computed tomographic scan of the fourth lumbar vertebra. The different parts of the vertebra and the presence of the cauda equina are seen in the thecal sac (subarachnoid space) in the vertebral canal.

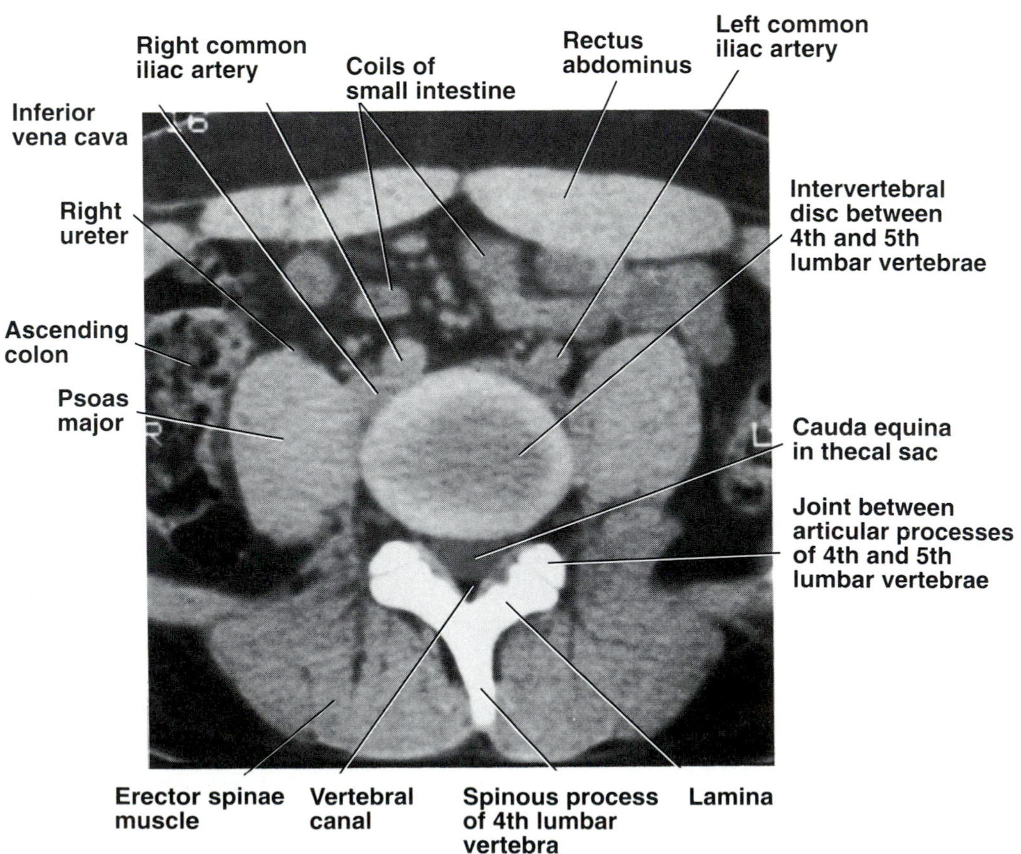

FIG 10–23.
An axial computed tomographic scan of the lumbar part of the vertebral column passing through the intervertebral disc between the fourth and fifth lumbar vertebrae.

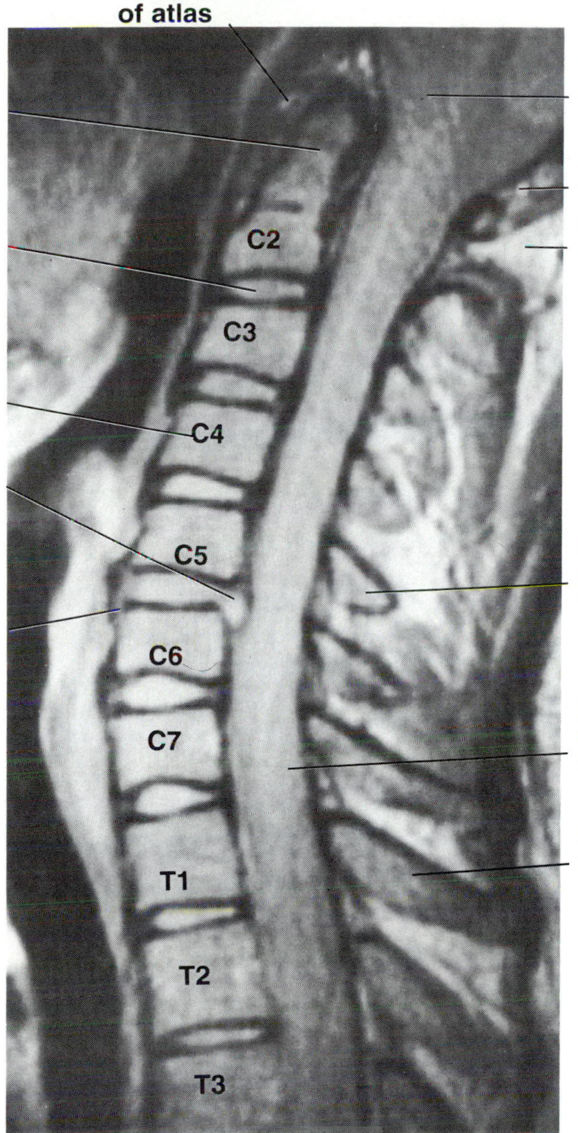

FIG 10–24.
A sagittal magnetic resonance image of the cervical part of the vertebral column. Note the demonstration of hard- and soft-tissue structures, including the spinal cord, vertebral ligaments, and the intervertebral discs. The herniated disc is seen between C5 and C6 vertebrae impinging on the spinal cord and meninges. (Courtesy of Dr Pait.)

izontally, but the roots of the lumbar and sacral nerves below the level of the termination of the spinal cord (lower border of the first lumbar vertebra in adults) form a vertical leash of nerves around the *filum terminale* (see Fig 10–27). Together these lower nerve roots are called the *cauda equina.*

After emerging from the intervertebral foramen, each spinal nerve immediately divides into a large *anterior ramus* and a smaller *posterior ramus,* which contain both motor and sensory fibers.

Spinal Part of the Accessory Nerve (11th Cranial Nerve).—The cervical roots of the spinal part of the accessory nerve emerge from the upper five cervical segments of the spinal cord and run upward through the foramen magnum into the skull, where they join the cranial part of the nerve (see p. 181).

Structure of the Spinal Cord

The spinal cord is composed of an inner core of *gray matter* that is surrounded by an outer covering of *white matter* (see Fig 10–26). The gray matter is gray to the naked eye because of the presence of large numbers of nerve cells; the white matter appears white because of the white myelin of the nerve fibers. The gray matter is seen on cross section as an H-shaped pillar with *anterior* and *posterior gray columns or horns* united by a thin *gray commissure* containing the *central canal.*

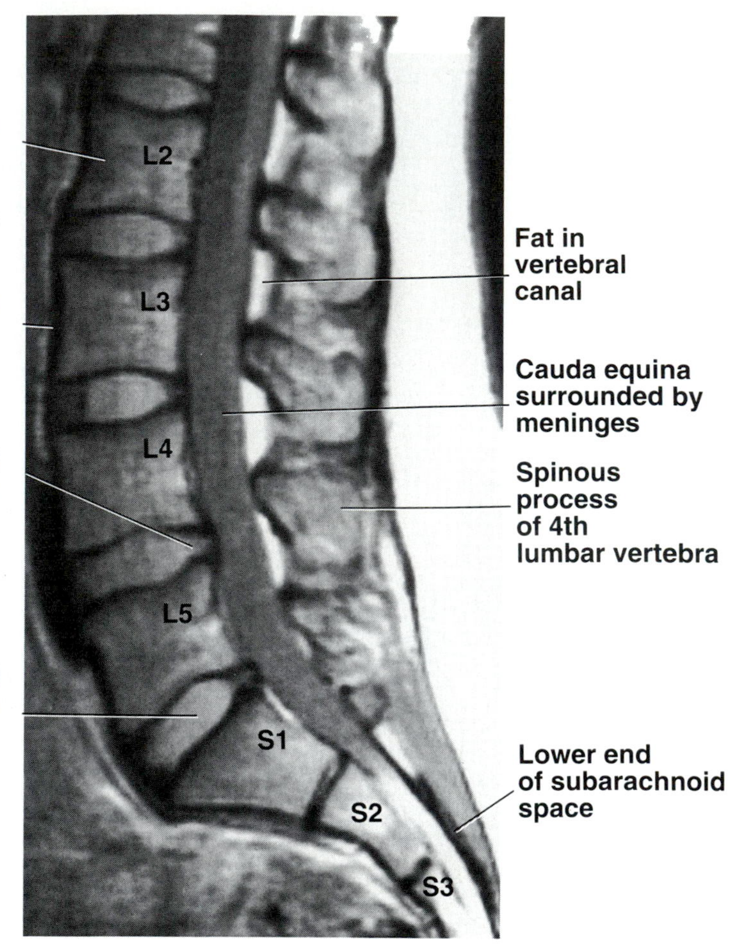

FIG 10–25.
A sagittal magnetic resonance image of the lumbosacral region of the vertebral column demonstrating the presence of herniated discs between L4 and L5 and between L5 and S1 vertebrae. (Courtesy of Dr Pait.)

Organization of the Gray Matter.—The gray matter consists of nerve cells embedded in neuroglia. For purposes of description, it is divided into anterior, posterior, and lateral gray columns (see Fig 10–26).

Anterior Gray Columns.—The majority of the nerve cells are large and their axons pass out in the anterior roots of the spinal nerves to innervate skeletal muscles.

Posterior Gray Columns.—The nerve cells are sensory and receive somatic and visceral afferent information.

Lateral Gray Column.—This column extends from the first thoracic to the second or third lumbar segments of the spinal cord (see Fig 10–26). The cells give rise to the preganglionic sympathetic fibers. A similar group of cells found in the second, third, and fourth sacral segments of the spinal cord give rise to preganglionic parasympathetic fibers. The cranial parasympathetic outflow occurs in cranial nerves 3, 7, 9, and 10.

Central Canal.—The central canal is present throughout the spinal cord (see Fig 10–26). Superiorly, it is continuous with the central canal of the medulla oblongata, and above this it opens into the fourth ventricle. It is filled with cerebrospinal fluid and is lined with *ependyma,* a type of neuroglial cell. Below, the central canal terminates in the conus medullaris. The cerebrospinal fluid is made to circulate by the cilia of the ependymal cells aided by the gross movements of the spinal cord within the vertebral column. The cerebrospinal fluid finally escapes from the fourth ventricle into the subarachnoid space through the holes in the roof of the fourth ventricle.

Organization of the White Matter.—The white matter consists of nerve fibers embedded in neuroglia. For purposes of description, it is divided into anterior, lateral, and posterior white columns (see Fig 10–26).

Nerve Fiber Tracts of the Spinal Cord.—A simplified diagram showing the arrangement of the nerve fiber tracts as seen on cross section is shown in Figure 10–28.

Ascending Tracts.—The *lateral spinothalamic tract*

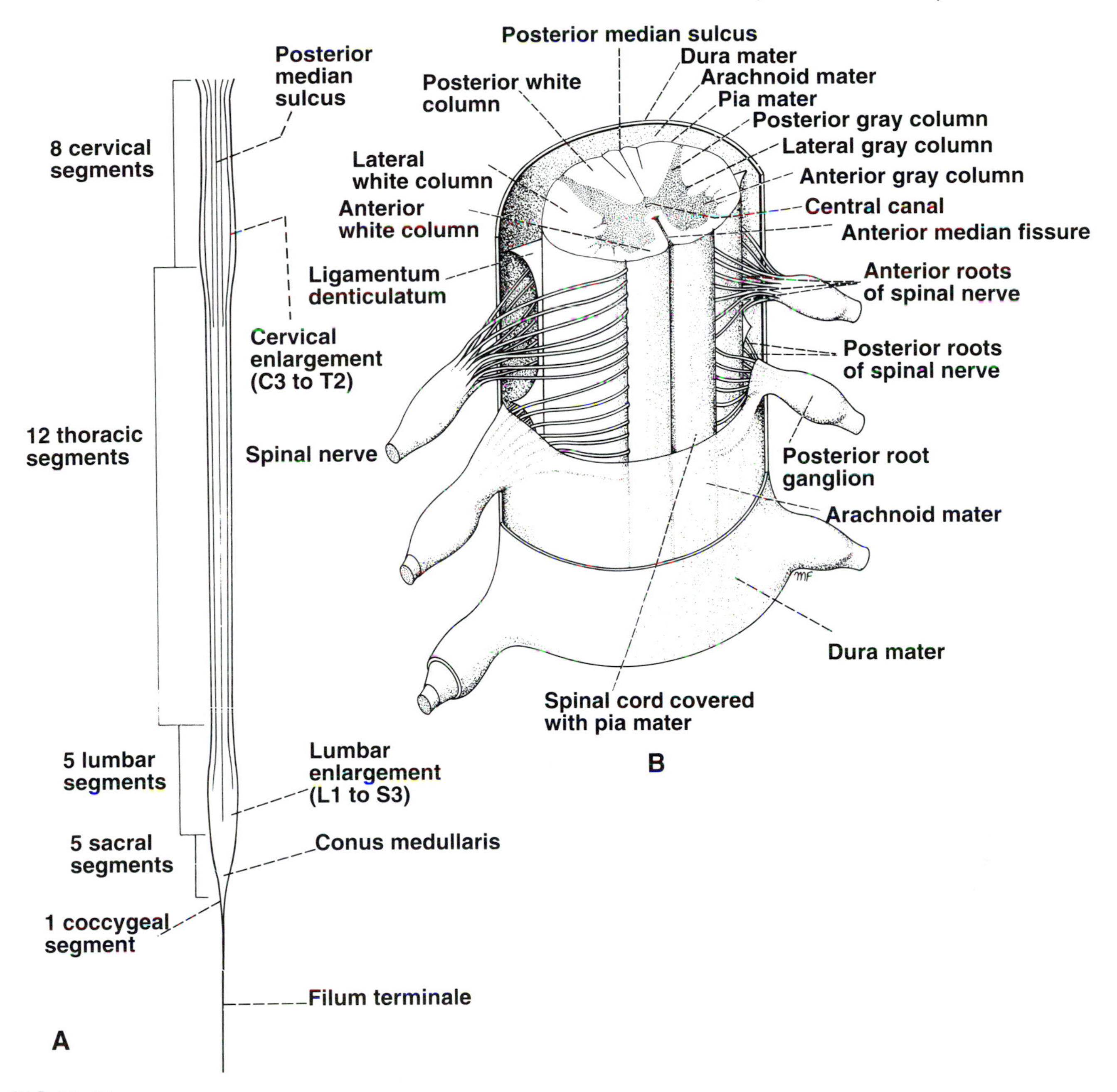

FIG 10–26.
Spinal cord. **A,** posterior view showing cervical and lumbar enlargements. **B,** three segments of the spinal cord showing the coverings of dura mater, arachnoid mater, and pia mater.

(pain and temperature pathways), the *anterior spinothalamic tract* (light touch and pressure pathways), the *fasciculus gracilis and fasciculus cuneatus* (discriminative touch, vibratory sense, and conscious muscle joint sense pathways), and the *posterior and anterior spinocerebellar tracts* (muscle joint sense pathways to the cerebellum) are shown in Figures 10–29 through 10–32.

Descending Tracts.—The motor neurons in the anterior gray columns of the spinal cord send axons to innervate skeletal muscle through the anterior roots of the spinal nerves. These neurons are sometimes referred to as the *lower motor neurons* and constitute the final common pathway to the muscles.

The lower motor neurons are constantly bombarded by nervous impulses that descend from the medulla, pons, midbrain, and cerebral cortex, as well as those that enter along sensory fibers from

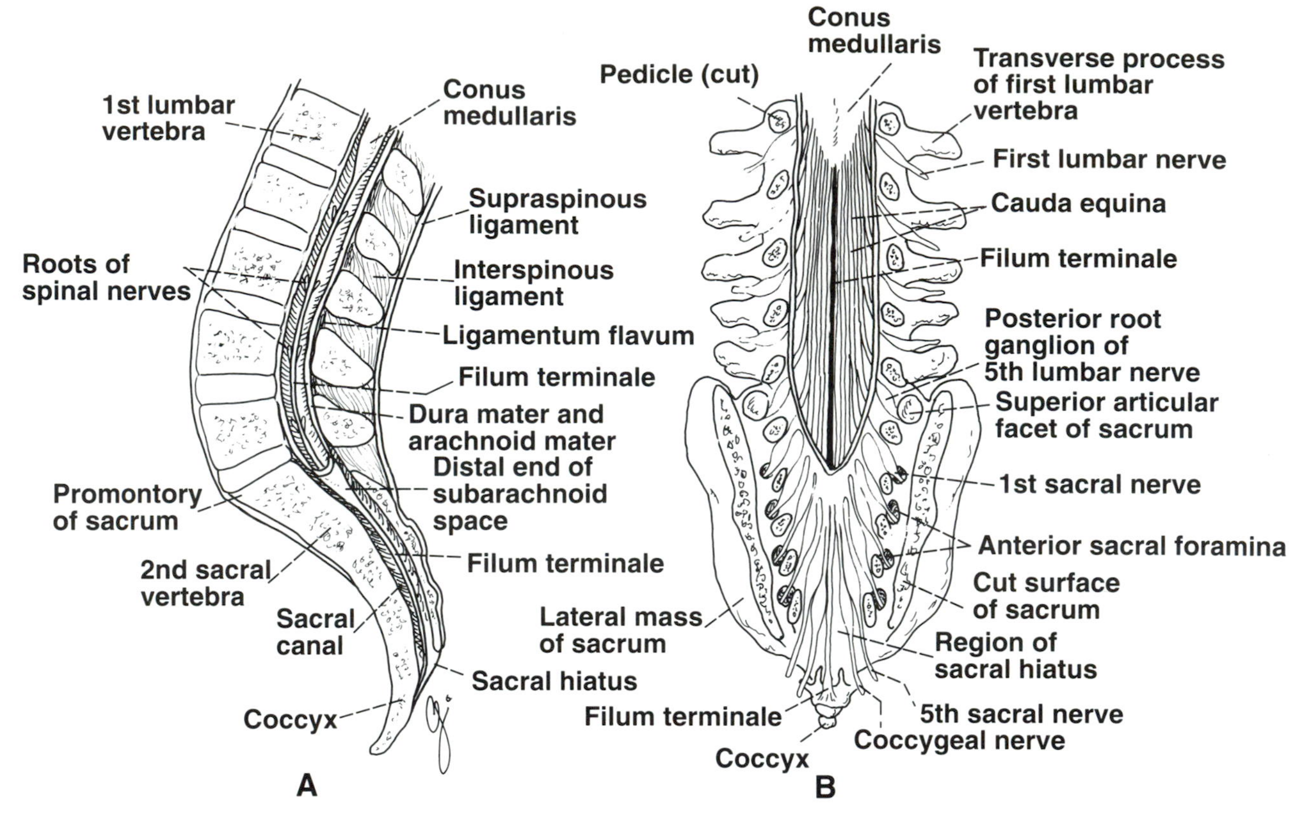

FIG 10–27.
A, median sagittal section through the lumbosacral region showing the conus medullaris, the filum terminale, and the lower limit of the subarachnoid space in the adult. **B,** lumbar vertebrae and sacrum, posterior view. The laminae and spines have been removed to show the lower end of the spinal cord and the cauda equina lying within the subarachnoid space. The meninges (thecal sac) have been cut away. Note the filum terminale tethering the lower end of the conus medullaris to the coccyx.

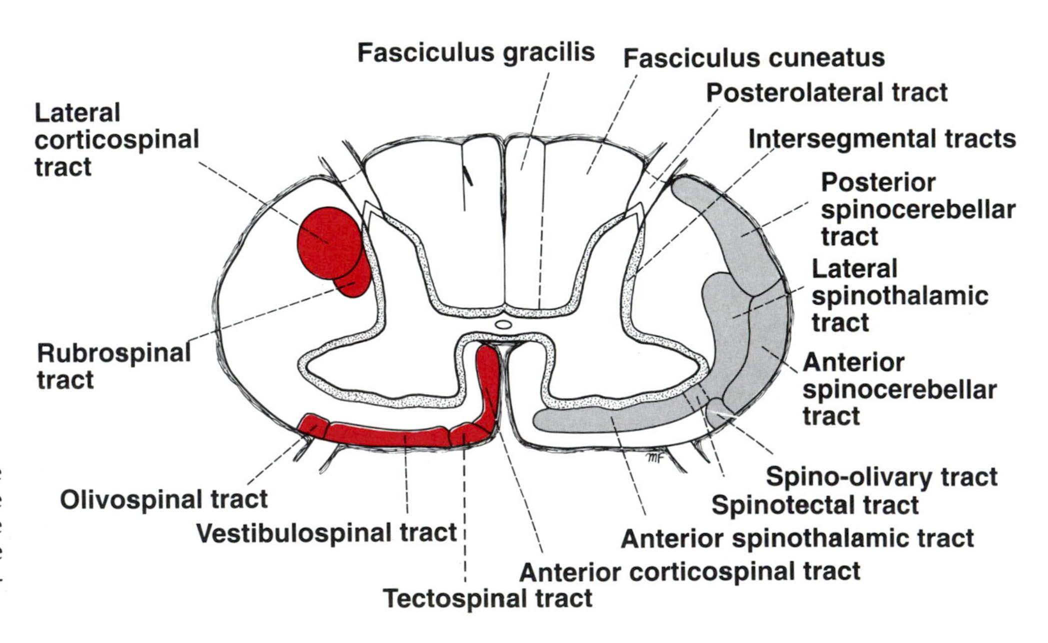

FIG 10–28.
Transverse section of the spinal cord showing the general arrangement of the ascending tracts on the right side and the descending tracts on the left side.

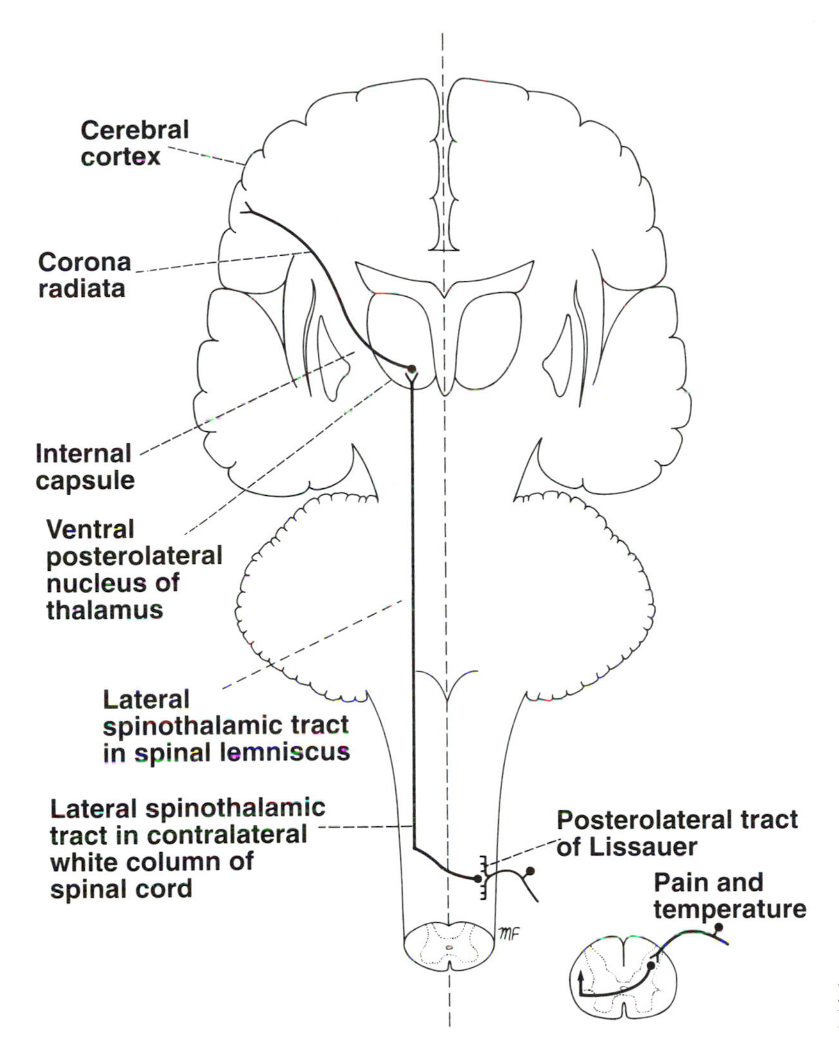

FIG 10–29.
Pain and temperature pathways.

the posterior roots. The nerve fibers that descend in the white matter from different supraspinal nerve centers are segregated into nerve bundles called the *descending tracts.* These supraspinal neurons and their tracts sometimes are referred to as the *upper motor neurons,* and they provide numerous separate pathways that can influence motor activity.

The descending tracts include the *corticospinal tracts* (Fig 10–33), (pathway that confers speed and agility to voluntary movements; used in performing rapid skilled movements), the *reticulospinal tracts* (influence voluntary movements and reflex activity), the *tectospinal tract* (concerned with reflex postural movements in response to visual stimuli), the *rubrospinal tract* (facilitates the activity of the flexor muscles and inhibits the activity of the extensor or antigravity muscles), the *vestibulospinal tract* (the inner ear and the cerebellum through this tract facilitate the activity of the extensor muscles and inhibit the activity of the flexor muscles to maintain balance).

The *descending autonomic fibers* form the pathway through which higher centers in the central nervous system control autonomic activity. These tracts probably descend in the lateral white column of the spinal cord and terminate by synapsing on the autonomic motor cells in the lateral gray columns in the thoracic, upper lumbar (sympathetic outflow), and midsacral (parasympathetic outflow) levels of the spinal cord.

Spinal Reflex Arcs

In its simplest form, a spinal reflex arc consists of the following anatomical structures: (1) a receptor organ, (2) an afferent neuron, (3) an effector neuron, and (4) an effector organ, such as a skeletal muscle. Such a reflex arc involving only one synapse is referred to as a *monosynaptic reflex arc.* Interruption of the reflex arc at any point along its course would abolish the response of the muscle.

Most skeletal muscles are innervated by two, three, or four spinal nerves and therefore are inner-

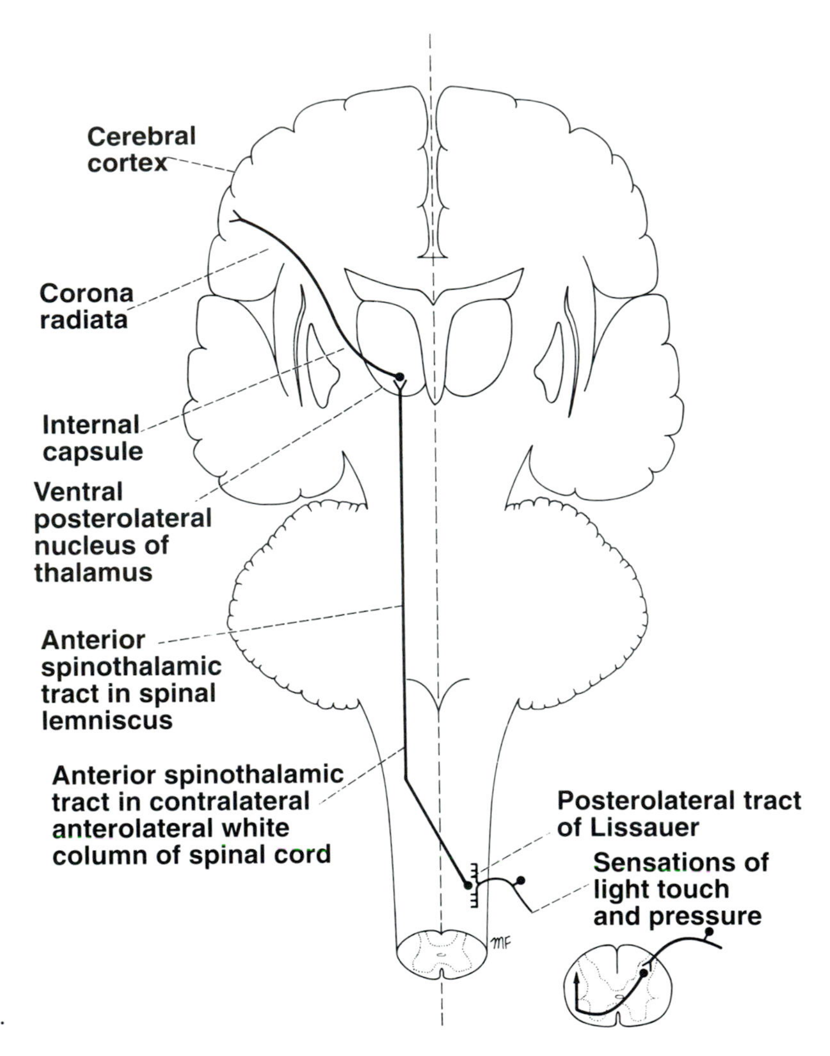

FIG 10–30.
Light touch and pressure pathways.

vated by the same number of segments of the spinal cord. It is possible to test the segmental innervation of many muscles by eliciting simple muscle reflexes in the patient as follows.

1. *Biceps brachii tendon reflex.* C5 and **C6** (flexion of the elbow joint by tapping the biceps tendon).

2. *Triceps tendon reflex.* C6, **C7**, and **C8** (extension of the elbow joint by tapping the triceps tendon).

3. *Brachioradialis tendon reflex.* C5, **C6**, and C7 (supination of the radioulnar joints by tapping the insertion of the brachioradialis tendon).

4. *Abdominal superficial reflexes* (contraction of underlying abdominal muscles by stroking the skin). Upper abdominal skin, T6 and T7; middle abdominal skin, T8 and T9; and lower abdominal skin, T10 through T12.

5. *Patellar tendon reflex* (knee jerk). L2, **L3**, and **L4** (extension of knee joint on tapping the patellar tendon).

6. *Achilles tendon reflex* (ankle jerk). **S1** and **S2** (plantar flexion of ankle joint on tapping the Achilles tendon).

Blood Supply of the Spinal Cord

The spinal cord receives its arterial supply from three small longitudinally running arteries—the two *posterior spinal arteries* and one *anterior spinal artery.* These arteries are reinforced by small segmentally arranged arteries that arise from arteries outside the vertebral column and enter the vertebral canal through the intervertebral foramina. These vessels anastomose on the surface of the cord and send branches into the substance of the white and gray matter. Considerable variation exists as to the size and segmental levels at which reinforcing arteries occur.

Posterior Spinal Arteries.—The posterior spinal arteries arise either directly from the vertebral arteries inside the skull or indirectly from the posterior infe-

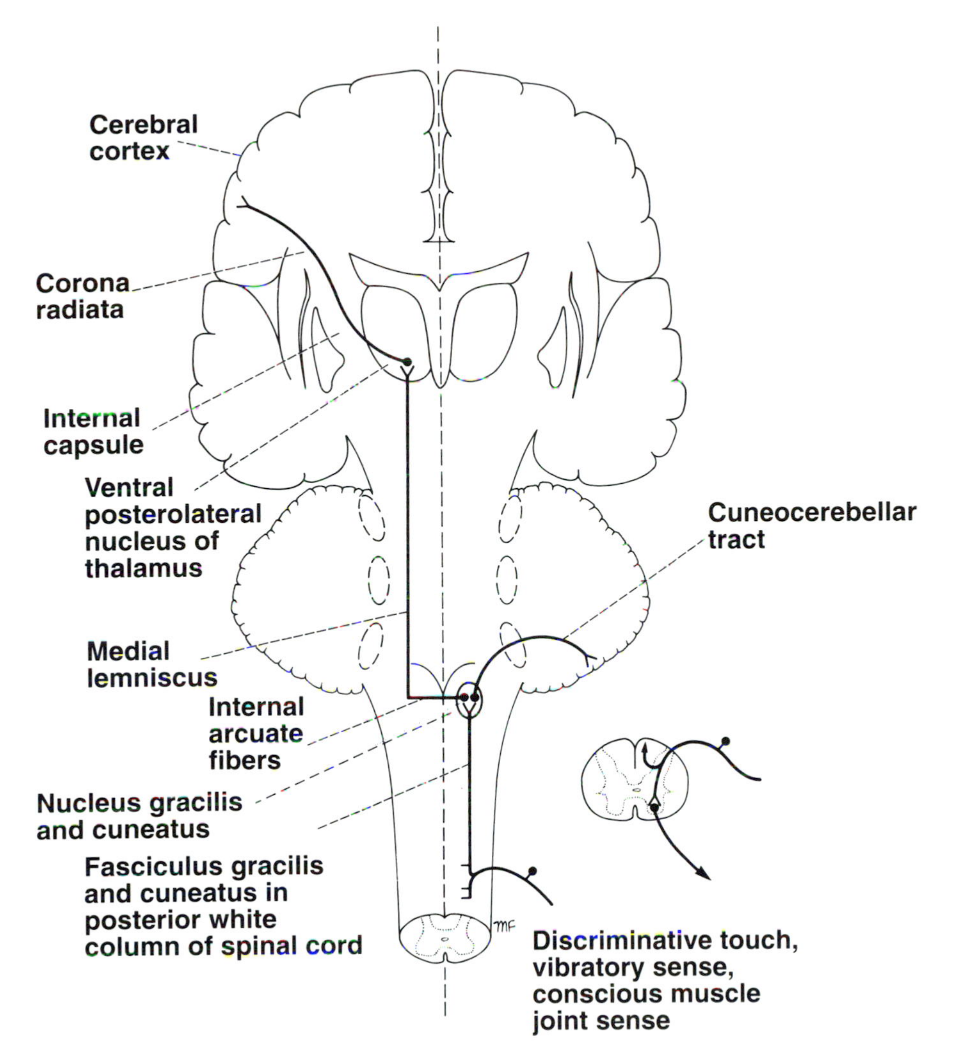

FIG 10–31.
Pathways for discriminative touch, vibratory sense, and conscious muscle joint sense.

rior cerebellar arteries. Each artery descends on the posterior surface of the spinal cord close to the posterior nerve roots and gives off branches that enter the substance of the cord (Fig 10–34). The posterior spinal arteries supply the posterior third of the spinal cord.

The posterior spinal arteries are small in the upper thoracic region, and the first three thoracic segments of the spinal cord are particularly vulnerable to ischemia should the segmental or radicular arteries in this region be occluded.

Anterior Spinal Arteries.—The anterior spinal arteries arise from the two vertebral arteries within the skull and unite to form a single artery. The anterior spinal artery then descends on the anterior surface of the spinal cord (see Fig 10–34). The artery, although larger than the posterior spinal arteries, varies in diameter as it descends, being smallest in the midthoracic region. Branches from the anterior spinal artery enter the substance of the cord and supply the anterior two thirds of the spinal cord.

In the upper and lower thoracic segments of the spinal cord the anterior spinal artery may be extremely small. Should the segmental or radicular arteries be occluded in these regions, the fourth thoracic and the first lumbar segments of the spinal cord would be particularly liable to ischemic necrosis.

Segmental Spinal Arteries.—At each intervertebral foramen the longitudinally running posterior and anterior spinal arteries are reinforced by small segmental arteries on both sides (see Fig 10–34). The arteries are branches of arteries outside the vertebral column. Having entered the vertebral canal, each segmental spinal artery gives rise to *anterior and posterior radicular arteries* that accompany the anterior and posterior nerve roots to the spinal cord.

Additional feeder arteries enter the vertebral ca-

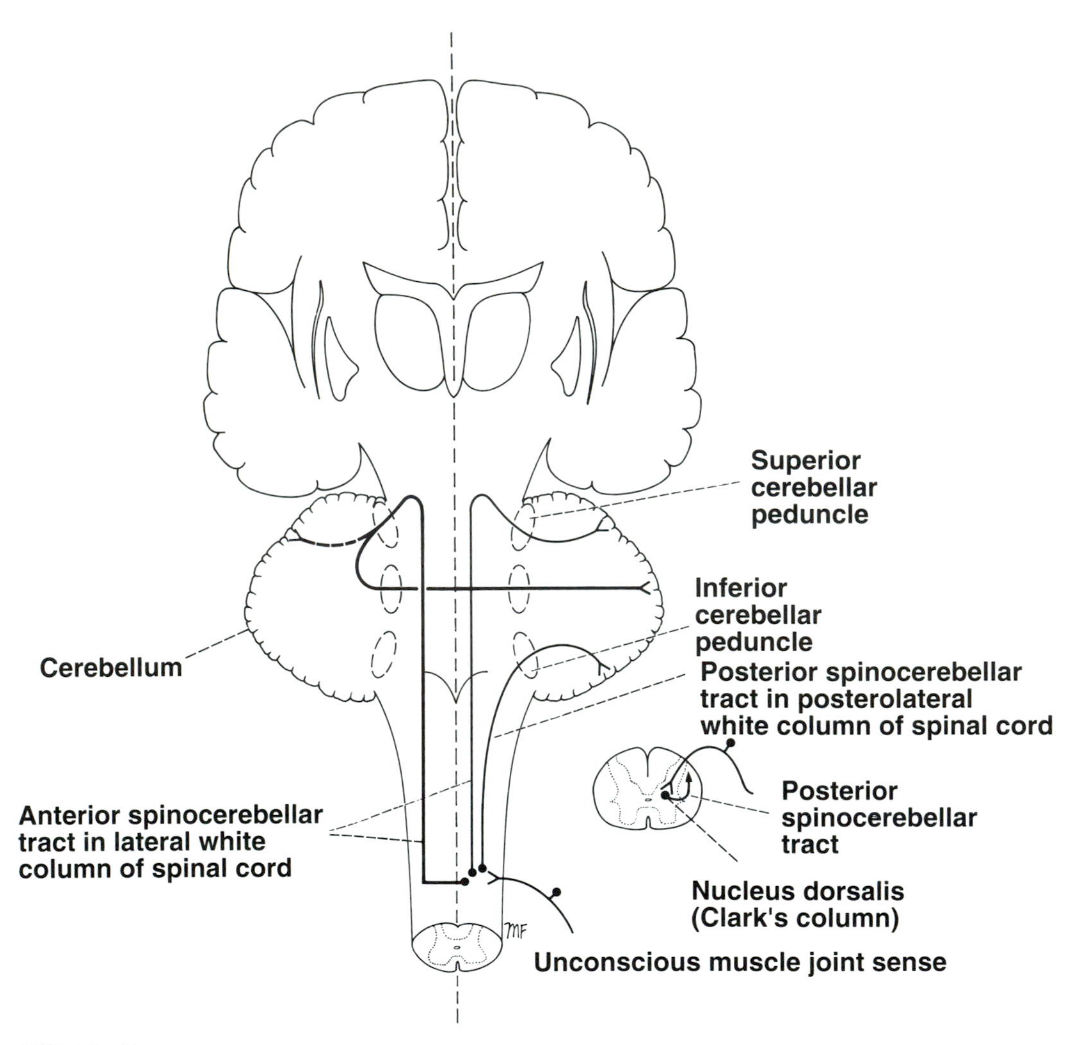

FIG 10–32.
Muscle joint sense pathways to cerebellum.

nal and anastomose with the anterior and posterior spinal arteries. The number and size of the feeder arteries vary considerably from individual to individual. One large, important feeder artery, the *great anterior medullary artery of Adamkiewicz,* enters the vertebral canal in the lower thoracic or upper lumbar regions. It arises from the aorta, is unilateral, and in the majority of persons enters the lumbar segments of the spinal cord from the left side. This artery is important because it may be the major source of blood to the lower two thirds of the spinal cord.

Clinical Note

Spinal cord ischemia and thoracic aortic dissection. —In the thoracic region the spinal cord receives its segmental arteries from the posterior intercostal arteries, which arise directly from the thoracic aorta. The origins of the posterior intercostal arteries may become blocked as the process of aortic dissection progresses.

Clinical Note

Spinal cord ischemia as a complication of a leaking abdominal aortic aneurysm. —In the lumbar region the spinal cord receives its segmental arteries from the lumbar arteries, which are branches of the abdominal aorta. The effect of direct pressure on the lumbar arteries by the leaking aneurysm or the formation of a mural thrombus may interfere with the blood supply to the cord.

Meninges of the Spinal Cord

The spinal cord, like the brain, is covered by three meninges—the dura mater, arachnoid mater, and pia mater (see Fig 10–26).

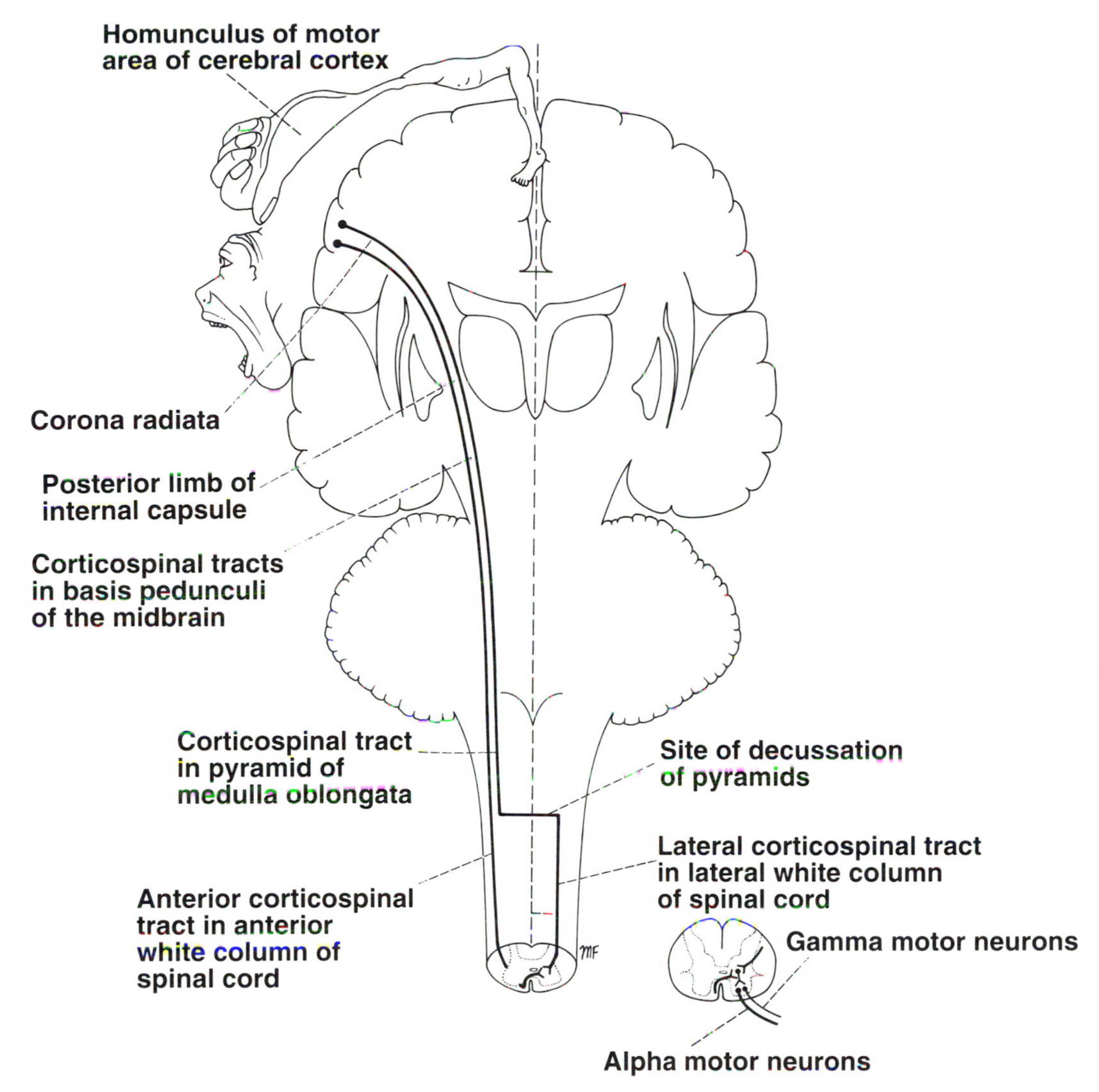

FIG 10–33.
Corticospinal tracts.

Dura Mater.—This is the most external membrane and is a dense fibrous sheet that encloses the spinal cord and cauda equina. It is continuous above with the meningeal layer of dura covering the brain. Below, the spinal dura ends on the filum terminale at about the level of the lower border of the second sacral vertebra (see Fig 10–27).

The dura mater extends along each nerve root and becomes continuous with connective tissue surrounding each spinal nerve at the intervertebral foramen. The inner surface of the dura mater is separated from the arachnoid mater by the potential *subdural space.* The dural sheath is separated from the walls of the vertebral canal by the extradural space (epidural space). This contains loose fatty areolar tissue and the internal vertebral venous plexus.

Arachnoid Mater.—This is a delicate nonvascular, impermeable membrane that lies within the dura and outside the pia (see Fig 10–26). It is separated from the dura by the subdural space that contains a thin film of tissue fluid. The arachnoid is separated from the pia mater by a wide space, the *subarachnoid space,* which is filled with *cerebrospinal fluid.* The arachnoid is continuous above through the foramen magnum, with the arachnoid covering the brain. Inferiorly, it ends on the filum terminale at about the level of the lower border of the second sacral vertebra. Between the levels of the conus medullaris and the lower end of the subarachnoid space lie the nerve roots of the cauda equina bathed in cerebrospinal fluid. Throughout the length of the subarachnoid space the arachnoid sends out small di-

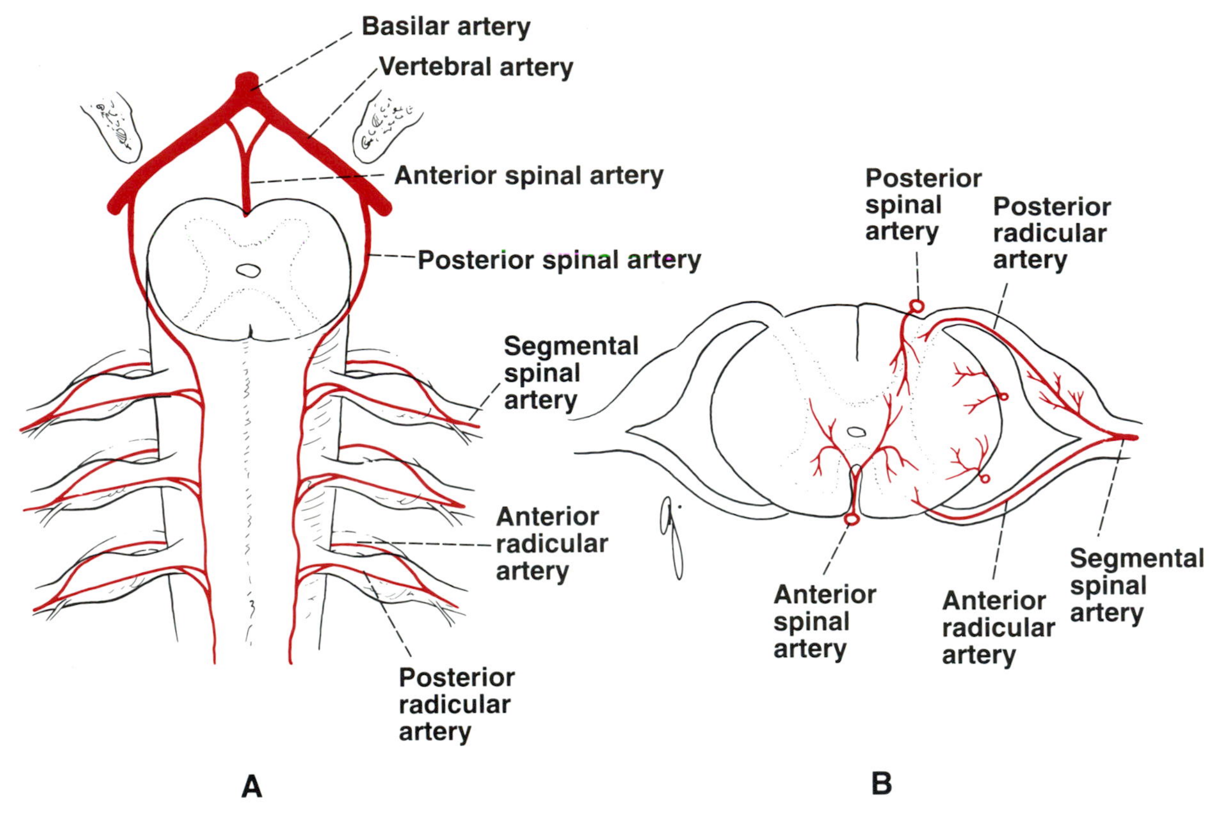

FIG 10–34.
A, arterial supply of spinal cord showing formation of two posterior arteries and one anterior spinal artery. **B,** transverse section of the spinal cord showing the segmental spinal arteries and the radicular arteries.

verticula along each spinal nerve root, forming small lateral extensions of the subarachnoid space (see Fig 10–36).

Pia Mater.—This is a vascular membrane that closely covers the spinal cord (see Fig 10–26). It has lateral extensions on either side of the spinal cord that lie between the nerve roots and form the *ligamentum denticulatum.* This ligament is attached laterally to the dura mater and suspends the spinal cord within the dural sheath. The pia mater extends along each nerve root as far as the spinal nerve. Inferiorly, it becomes the *filum terminale.*

Cerebrospinal Fluid

The cerebrospinal fluid is a clear, colorless fluid formed mainly by the *choroid plexuses* of the lateral, third, and fourth ventricles. The fluid circulates through the ventricular system and enters the subarachnoid space through the three foramina in the roof of the fourth ventricle (Fig 10–35). It ascends in the subarachnoid space over the outer surface of the cerebrum. Some of the fluid moves inferiorly in the subarachnoid space around the spinal cord and cauda equina. The cerebrospinal fluid is mainly absorbed through the *arachnoid villi* that project into the dural venous sinuses, in particular the *superior sagittal venous sinus.* For further details on the composition, formation, circulation, absorption, and function of the cerebrospinal fluid, see pp. 304 and 305.

Radiographic Examination of the Spinal Cord

Computed Tomographic Scans and Magnetic Resonance Images

Computed tomography and magnetic resonance imaging are now commonly used to identify lesions involving the spinal cord that are not visible on plain radiographs. Figures 10–23 and 10–24 show a computed tomographic scan and a magnetic resonance image.

Myelography

Myelography is the radiographic study of the subarachnoid space by the injection of a contrast medium through a lumbar puncture. A normal myelo-

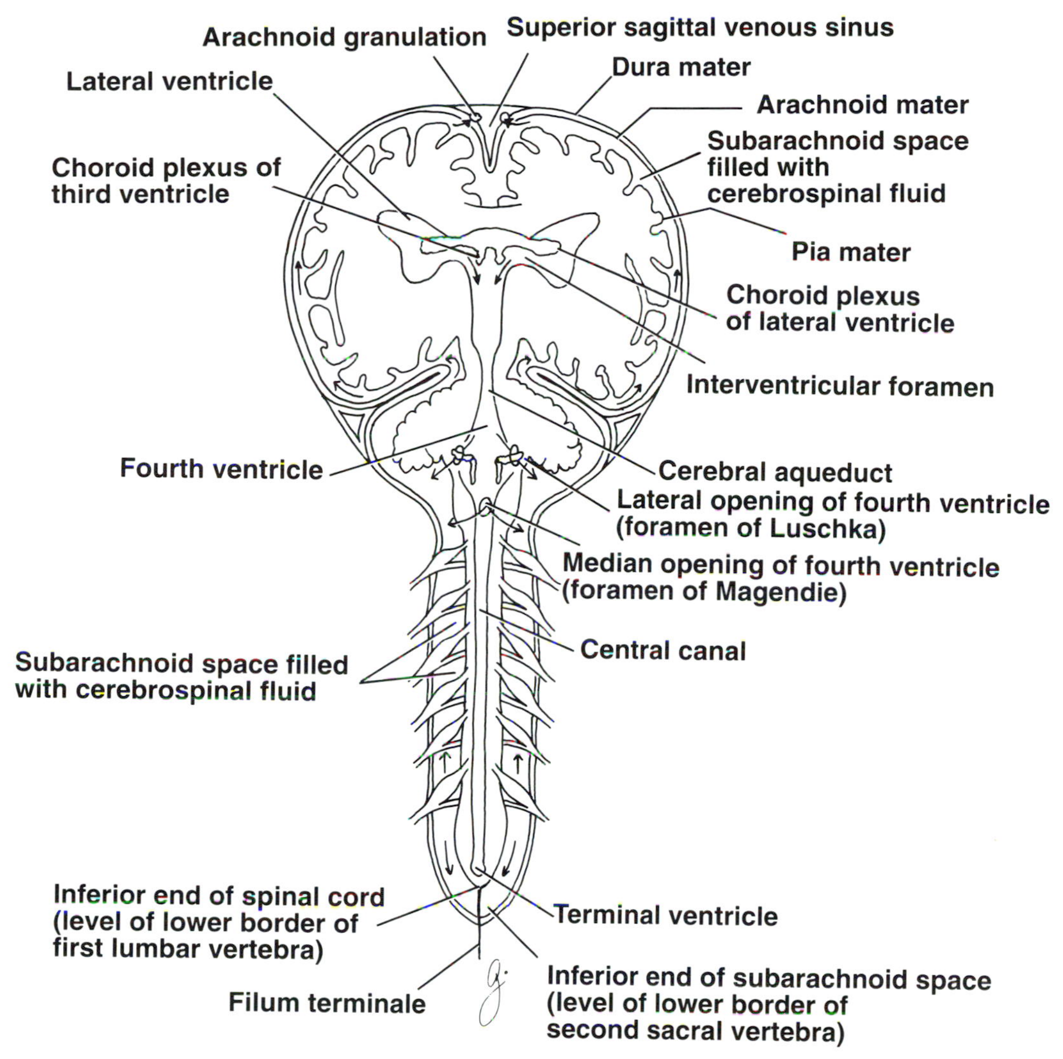

FIG 10–35.
Diagram showing the origin and circulation of the cerebrospinal fluid.

gram will show pointed lateral projections at regular intervals at the intervertebral space levels because the opaque medium fills the lateral extensions of the subarachnoid space around each spinal nerve. The presence of a tumor or a prolapsed intervertebral disc may obstruct the movement of the medium from one spinal region to another when the patient is tilted. An example of a normal myelogram is shown in Figure 10–36.

SURFACE ANATOMY

Surface Anatomy of the Vertebral Column and Back

External Occipital Protuberance

The external occipital protuberance lies at the junction of the head and the neck. If the index finger is placed on the skin in the midline, the finger can be drawn downward from the protuberance in the *nuchal groove.*

Cervical Vertebrae

The most prominent spinous process that can be felt in the neck (Fig 10–37) is that of the *seventh cervical vertebra (vertebra prominens).* Cervical spines 1 through 6 are covered by the *ligamentum nuchae,* a large ligament that runs down the back of the neck, connecting the skull to the spinous processes of the cervical vertebrae.

The *transverse processes* are short but easily palpable from the lateral side in a thin neck. The *anterior tubercle of the sixth cervical transverse process (tubercle of Chassaignac)* can be palpated medial to the sternocleidomastoid and against it the common carotid artery can be compressed.

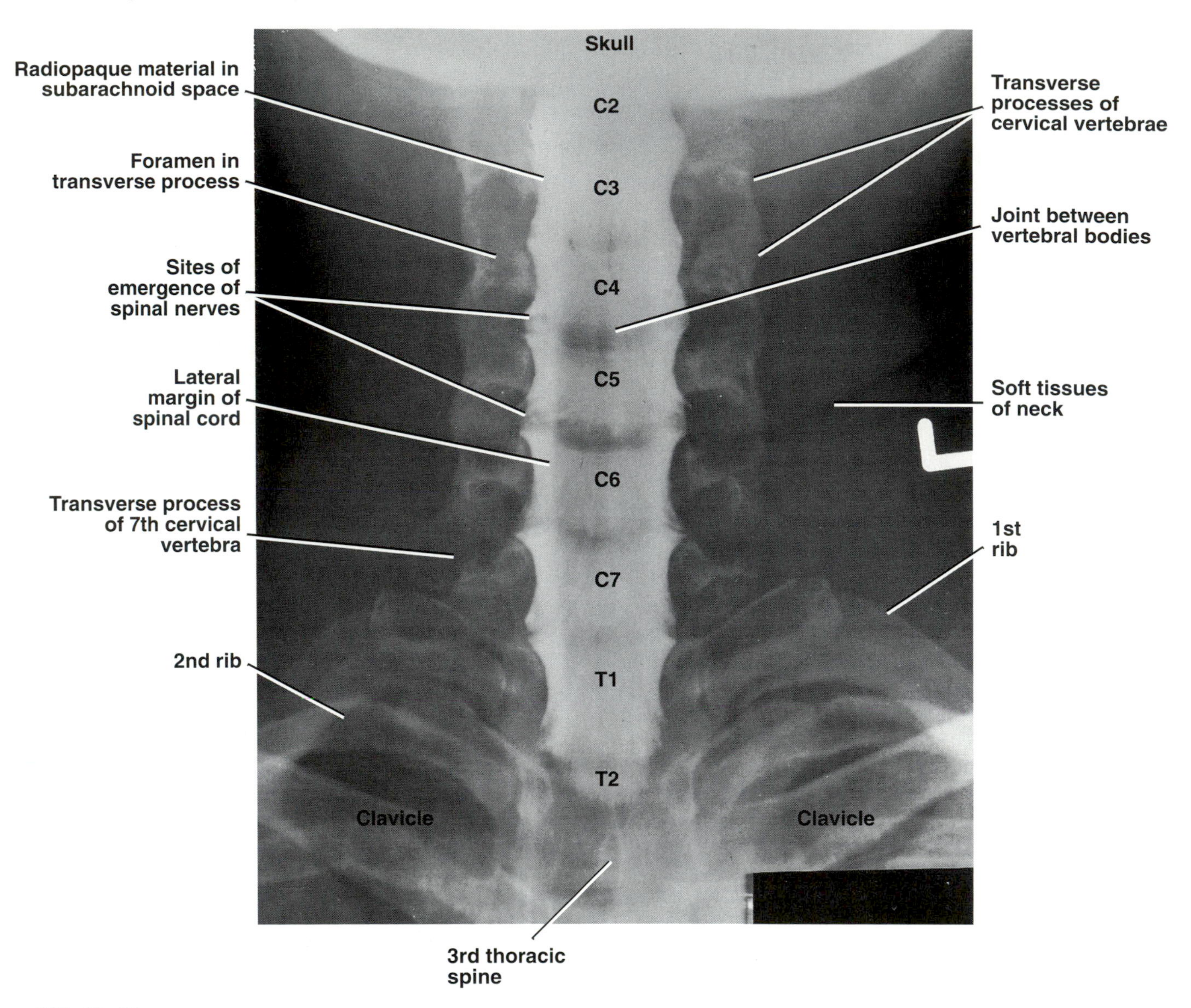

FIG 10–36.
Posteroanterior myelogram of the cervical region. Note the lateral margin of the spinal cord and the sites where the spinal nerves emerge.

Thoracic and Lumbar Vertebrae

The nuchal groove is continuous below with a furrow that runs down the middle of the back over the tips of the spines of all the thoracic vertebrae and the upper four lumbar vertebrae. The most prominent spinous process that can be felt in the thoracic region is that of the first thoracic vertebra; the others may be easily recognized when the trunk is bent forward.

Sacrum

The *spines of the sacrum* are fused with each other in the midline to form the *median sacral crest*. The crest can be felt beneath the skin in the uppermost part of the groove between the buttocks.

The *sacral hiatus* is situated on the posterior aspect of the lower end of the sacrum, and it is here that the *extradural space* (epidural space) terminates. The hiatus lies about 2 in. (5 cm) above the tip of the coccyx and beneath the skin in the groove between the buttocks.

Coccyx

The inferior surface and tip of the coccyx can be palpated in the natal cleft about 1 in. (2.5 cm) behind the anus. The anterior surface of the coccyx

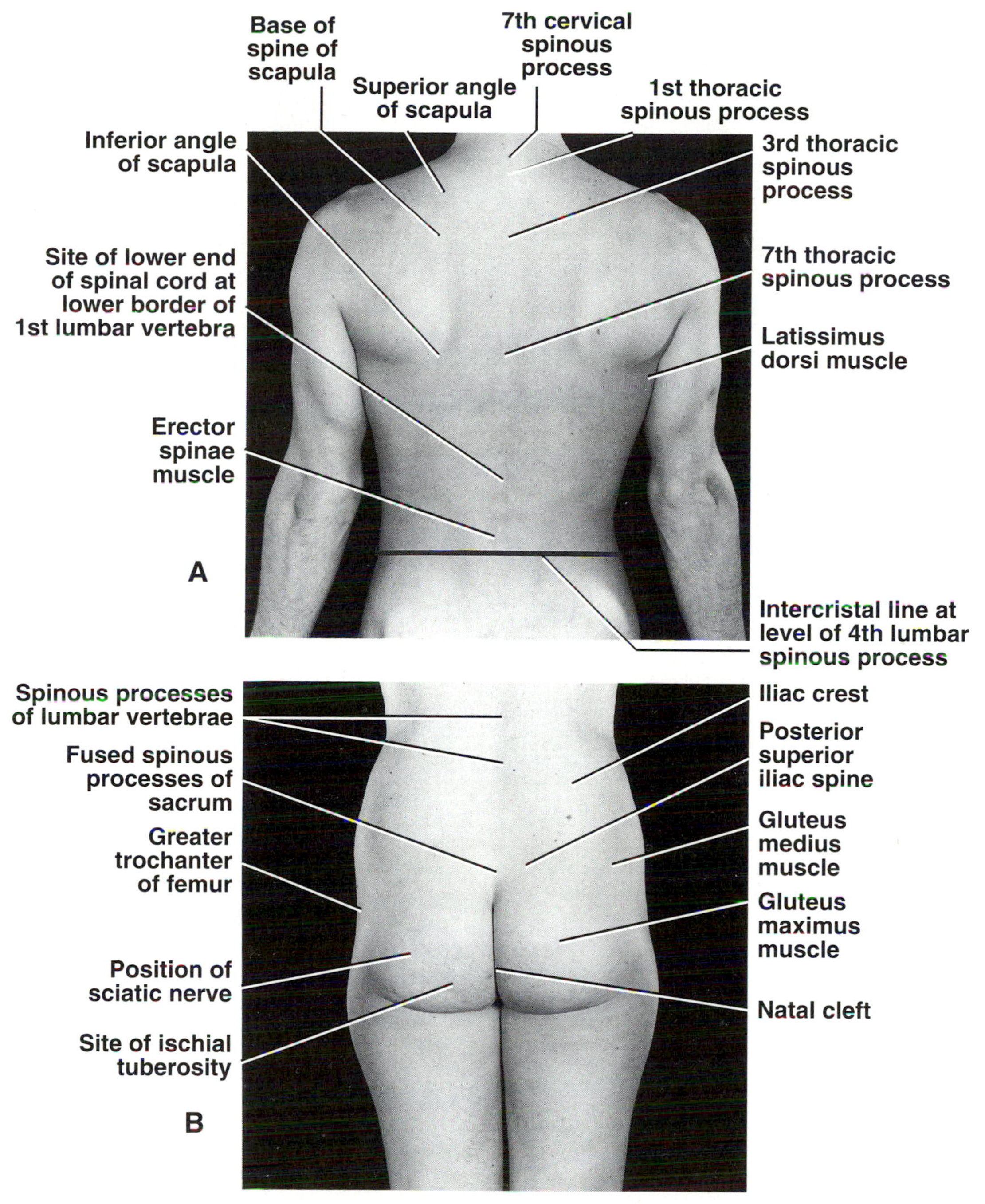

FIG 10–37.
A, the back of a 27-year-old male. **B,** the back and gluteal region of a 25-year-old female.

may be palpated with a gloved finger in the anal canal.

Upper Lateral Part of the Thorax
The upper lateral part of the thorax is covered by the scapula and its associated muscles. The scapula lies posterior to the first to the seventh ribs.

Scapula
The *medial border* of the scapula forms a prominent ridge, which ends above at the superior angle and below at the inferior angle (see Fig 10–37).

The *superior angle* can be palpated opposite the first thoracic spine, and the inferior angle can be palpated opposite the seventh thoracic spine.

The *crest of the spine of the scapula* can be palpated and traced medially to the medial border of the scapula, which it joins at the level of the third thoracic spine (see Fig 10–37).

The *acromion process of the scapula* forms the lateral extremity of the spine of the scapula. It is subcutaneous and easily located.

Lower Lateral Part of the Back

The lower lateral part of the back is formed by the posterior aspect of the upper part of the bony pelvis (false pelvis) and its associated gluteal muscles.

Iliac Crests

The iliac crests are easily palpated along their entire length (see Fig 10–37). They lie at the level of the fourth lumbar spine and are used as a landmark when performing a lumbar puncture. Each crest ends in front at the *anterior superior iliac spine* and behind at the *posterior superior iliac spine;* the latter lies beneath a skin dimple at the level of the second sacral vertebra and the middle of the sacroiliac joint.

Spinal Cord and Subarachnoid Space

The *spinal cord* in adults extends down to the level of the lower border of the spine of the first lumbar vertebra (see Fig 10–37). In young children it may extend to the third lumbar spine.

The *subarachnoid space,* with its *cerebrospinal fluid,* extends down to the lower border of the second sacral vertebra, which lies at the level of the posterior superior iliac spine (see Fig 10–37).

CLINICAL ANATOMY

Clinical Anatomy of the Vertebral Column

Herniated Intervertebral Discs

The physical properties of the intervertebral disc allow one vertebra to rock on another while serving as an efficient shock-absorbing mechanism. The resistance of these discs to compression forces is substantial. Nevertheless, the discs are vulnerable to sudden shocks, particularly if the vertebral column is flexed and the disc is undergoing degenerative changes, that result in herniation of the nucleus pulposus.

The discs most commonly affected by herniation are those in areas where a mobile part of the column joins a relatively immobile part, i.e., the cervicothoracic junction and the lumbosacral junction. In these areas the posterior part of the anulus fibrosus ruptures, and the nucleus pulposus is forced posteriorly, like toothpaste out of a tube. This herniation may result either in a central protrusion in the midline under the posterior longitudinal ligament of the vertebrae or in a lateral protrusion at the side of the posterior ligament close to the intervertebral foramen (Fig 10–38).

Cervical disc herniations are less common than those in the lumbar region (see Fig 10–24). The discs most susceptible to this condition are those between the fifth and sixth and the sixth and seventh vertebrae. Lateral protrusions cause pressure on a spinal nerve or its roots. Each spinal nerve emerges above the corresponding vertebra; thus, the protrusion of the disc between the fifth and sixth cervical vertebrae may compress the C6 spinal nerve or its roots (see Fig 10–38). Pain is felt near the lower part of the back of the neck and shoulder and along the area in the distribution of the spinal nerve involved. Central protrusions may press on the spinal cord and the anterior spinal artery and involve the various spinal tracts.

Lumbar disc herniations are more common than cervical disc herniations (see Fig 10–38). The discs usually affected are those between the fourth and fifth lumbar vertebrae and between the fifth lumbar vertebra and the sacrum (see Fig 10–25). In the lumbar region, the roots of the cauda equina run posteriorly over a number of intervertebral discs (see Fig 10–38). A lateral herniation may press on one or two roots and often involves the nerve root going to the intervertebral foramen just below. However, because C8 nerve roots exist and there is no C8 vertebral body, the thoracic and lumbar roots exit *below* the vertebra of the corresponding number. Thus, the L5 nerve root exits between the L5 and S1 vertebrae. Moreover, because the nerve roots move laterally as they pass toward their exit, the root corresponding to that disc space (L4 in the case of the L4-L5 disc) is already too lateral to be impinged upon by the herniated disc. Herniation of the L4-L5 disc usually gives rise to symptoms referable to the L5 nerve roots, even though the L5 root exits between L5 and S1 vertebrae. The nucleus pulposus occasionally herniates directly backward and, if it is a large herniation, the whole cauda equina may be compressed, producing paraplegia.

In lumbar disc herniations, pain is referred down the leg and foot in the distribution of the affected nerve. Since the sensory posterior roots most commonly pressed upon are the fifth lumbar and first sacral, pain is usually felt down the back and lateral side of the leg, radiating to the sole of the foot

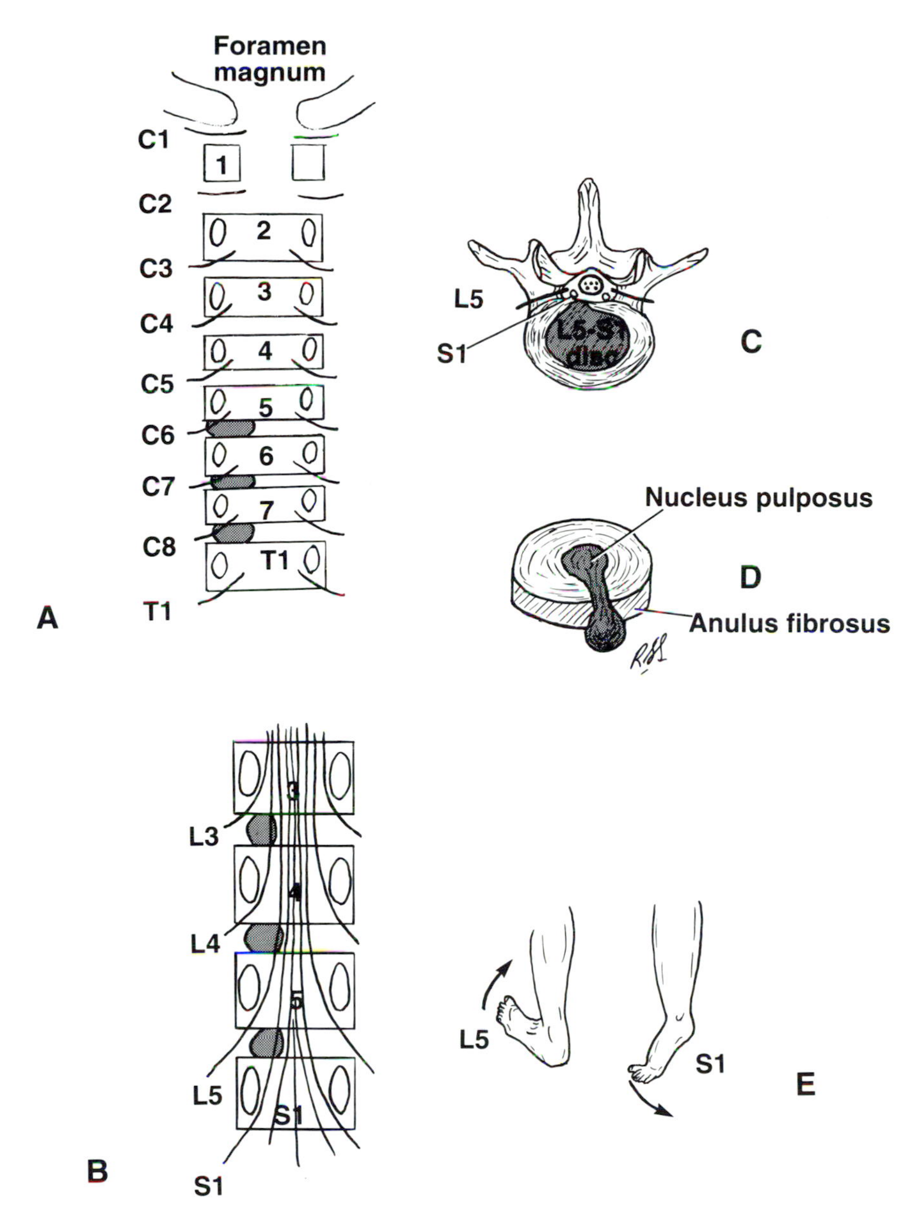

FIG 10–38.
A and **B**, posterior views of vertebral bodies in the cervical and lumbar regions showing the relationship that might exist between herniated nucleus pulposus and spinal nerve roots. Note there are eight cervical nerves and only seven cervical vertebrae. In the lumbar region, for example, the emerging L4 nerve roots pass out laterally close to the pedicle of the fourth lumbar vertebra and are not related to the intervertebral disc between the fourth and fifth lumbar vertebrae. **C**, posterolateral herniation of nucleus pulposus of intervertebral disc between the fifth lumbar vertebra and the first sacral vertebra showing pressure on the S1 nerve root. **D**, an intervertebral disc that has herniated its nucleus pulposus posteriorly. **E**, pressure on the L5 motor nerve root produces weakness of dorsiflexion of ankle; pressure on the S1 motor nerve root produces weakness of plantar flexion of the ankle joint.

(*sciatica*). In severe cases there may be paresthesia or actual sensory loss.

Pressure on the anterior motor roots causes muscle weakness. Involvement of the fifth lumbar motor root produces weakness of dorsiflexion of the ankle, while pressure on the first sacral motor root causes weakness of plantar flexion and a diminished or absent ankle jerk reflex (see Fig 10–38).

A correlation between the disc lesion, the nerve roots involved, the pain dermatome, the muscle weakness, and the missing or diminished reflex is shown in Table 10–1.

Cauda Equina Compression.—This condition is usually caused by a massive centrally placed herniation of a lumbar intervertebral disc. The symptoms include low back pain, bilateral sciatica, bilateral motor weakness of the lower limbs, saddle anesthesia, and bowel and bladder incontinence. In extreme cases the patient may exhibit paraplegia.

Compression of the cauda equina can be caused by spinal stenosis. The compression produces a syndrome called *spinal claudication* because pain occurs down the legs with exertion (see p. 566). Cauda equina compression from a herniated disc must be distinguished from spinal stenosis and pseudoclaudication.

Disease and the Intervertebral Foramina

The intervertebral foramina (see Fig 10–9) transmit the spinal nerves and the small segmental arteries and veins, all of which are embedded in areolar tis-

TABLE 10–1.
Summary of Important Features Found in Cervical and Lumbosacral Root Syndromes

Root Injury	Dermatome Pain	Muscle Supplied	Movement Weakness	Reflex Involved
C5	Lower lateral upper part of arm	Deltoid and biceps	Shoulder abduction, elbow flexion	Biceps
C6	Lateral forearm	Extensor carpi radialis longus and brevis	Wrist extensors	Brachioradialis
C7	Middle finger	Triceps and flexor carpi radialis	Extension of elbow and flexion of wrist	Triceps
C8	Medial forearm	Flexor digitorum superficialis and profundus	Finger flexion	None
L1	Groin	Iliopsoas	Hip flexion	Cremaster
L2	Anterior part of thigh	Iliopsoas, sartorius, hip adductors	Hip flexion, hip adduction	Cremaster
L3	Medial knee	Iliopsoas, sartorius, quadriceps, hip adductors	Hip flexion, knee extension, hip adduction	Patellar
L4	Medial calf	Tibialis anterior, quadriceps	Foot inversion, knee extension	Patellar
L5	Lateral lower leg and dorsum of foot	Extensor hallucis longus, extensor digitorum longus	Toe extension, ankle dorsiflexion	None
S1	Lateral edge of foot	Gastrocnemius, soleus	Ankle plantar flexion	Ankle jerk
S2	Posterior part of thigh	Flexor digitorum longus, flexor hallucis longus	Ankle plantar flexion, toe flexion	None

sue. Each foramen is bounded above and below by the pedicles of adjacent vertebrae—in front by the lower part of the vertebral body and by the intervertebral disc, and behind by the articular processes and the joint between them. In this situation the spinal nerve is very vulnerable and may be pressed upon or irritated by disease of the surrounding structures. Herniation of the intervertebral disc, fractures of the vertebral bodies, and osteoarthritis involving the joints of the articular processes or the joints between the vertebral bodies may all result in pressure, stretching, or edema of the emerging spinal nerve. Such pressure would give rise to dermatomal pain, muscle weakness, and diminished or absent reflexes.

Dislocations of the Vertebral Column

Pure dislocations occur only in the cervical region, because the inclination of the articular processes of the cervical vertebrae permits dislocation to take place without fracture of the processes. In the thoracic and lumbar regions, dislocations can occur only if the vertically positioned articular processes are fractured.

Dislocations commonly occur between the fourth and fifth or fifth and sixth cervical vertebrae, where mobility is greatest (Fig 10–39). In unilateral dislocations the inferior articular process of one vertebra is forced forward over the anterior margin of the superior articular process of the vertebra below. Since the articular processes normally overlap, they become locked in the dislocated position. The spinal nerve on the same side is usually nipped in the intervertebral foramen, producing severe pain. Fortunately, the large size of the vertebral canal allows the spinal cord to escape damage in most cases.

Bilateral cervical dislocations result in the movement of 50% of the vertebral body width of one body on the other and are almost always associated with severe injury to the spinal cord (see Fig 10–39). Death occurs immediately if the upper cervical vertebrae are involved, since the respiratory muscles, including the diaphragm (phrenic nerves C3 through C5), are paralyzed.

Fractures of the Vertebral Column

Fractures of the Spinous Processes, Transverse Processes, or Laminae.—These fractures are caused by direct injury or, in rare cases, by severe muscular activity.

Anterior and Lateral Compression Fractures.—Anterior compression fractures of the vertebral bodies are usually caused by an excessive flexion-compression type of injury and take place at the sites of maximum mobility or at the junction of the mobile and

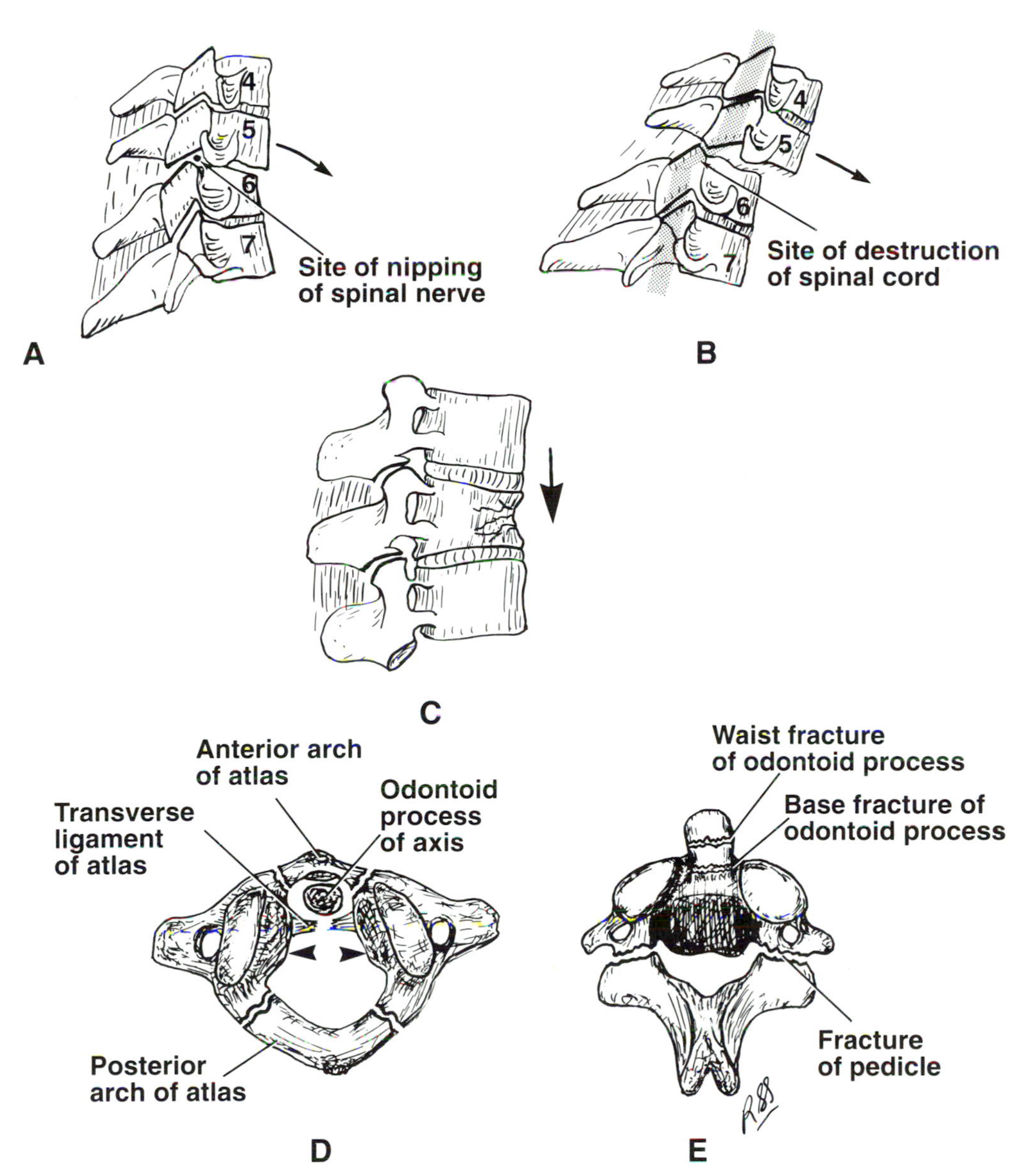

FIG 10–39.
Dislocations and fractures of the vertebral column. **A,** unilateral dislocation of the fifth on the sixth cervical vertebra; note the forward displacement of the inferior articular process over the superior articular process of the vertebra below. **B,** bilateral dislocation of the fifth on the sixth cervical vertebra; note that 50% of the vertebral body width has moved forward on the vertebra below. **C,** flexion compression-type fracture of the vertebral body in the lumbar region. **D,** Jefferson's-type fracture of the atlas. **E,** fractures of the odontoid process and the pedicles (Hangman's fracture) of the axis.

fixed regions of the column. In the cervical region, this frequently occurs where the lower mobile vertebrae (C5 through C7) articulate with the relatively fixed thoracic vertebrae. In the lumbar region (see Fig 10–39), fractures are common in the upper lumbar vertebrae that are related to the relatively fixed thoracic spine and the lower lumbar vertebrae that are related to the fixed sacrum. In the pure flexion type of injury, the anterior part of the body is crushed while the strong posterior longitudinal ligament remains intact. The vertebral arches remain unbroken and the intervertebral ligaments remain intact, so that vertebral displacement and spinal cord injury do not occur. When the injury causes excessive lateral flexion in addition to excessive flexion, the lateral part of the body is also crushed.

Fracture Dislocations.—Fracture dislocations are usually caused by a combination of flexion and rotation type of injury; the upper vertebra is excessively flexed and twisted on the lower vertebra. Here, again, the site is usually where there is maximum mobility, as in the lumbar region, or at the junction of the mobile and fixed region of the column, as in the lower lumbar vertebrae. Since the articular processes are fractured and the ligaments are torn, the vertebrae involved are unstable, and the spinal cord is usually severely damaged or severed, with accompanying paraplegia.

Vertical Compression Fractures.—These fractures occur in the cervical and lumbar regions, where it is possible to fully straighten the vertebral column. In the cervical region, with the neck straight, an excessive vertical force applied from above will cause the ring of the atlas to be disrupted (see Fig 10–39) and the lateral masses to be displaced laterally (Jefferson's fracture). If the neck is slightly flexed the lower cervical vertebrae remain in a straight line and the compression load is transmitted to the lower vertebrae, commonly the fifth or sixth, and the intervertebral disc is disrupted, causing the body of one of the vertebrae to break up. Although the posterior longitudinal ligament remains intact, pieces of the shattered body are frequently forced back into the spinal cord.

It is possible for nontraumatic compression fractures to occur in severe cases of osteoporosis and for pathological fractures to take place.

In the straightened lumbar region an excessive force from below may cause the vertebral body to break up, with protrusion of fragments posteriorly into the spinal canal.

Fractures of the Odontoid Process of the Axis

Fractures of the odontoid process are relatively common and result from falls or blows on the head. The site of the fracture may occur just beneath the tip (waist) or at the base where the process becomes continuous with the body of the axis (see Fig 10–39). Normally, the odontoid process is kept in position against the posterior surface of the anterior arch of the atlas by the transverse ligament and by the two alar ligaments. Excessive mobility of the odontoid caused by fracture or rupture of the transverse ligament may result in compression injury to the spinal cord. When a lateral radiograph of the region is viewed, it is useful to remember the "rule of thirds"; the odontoid process, spinal cord, and empty space each occupy one third of the spinal canal at the level of the arch of the atlas.

Fracture of the Pedicles of the Axis (Hangman's Fracture)

Severe extension injury of the neck, such as might occur in an automobile accident or a fall, is the usual cause. Sudden overextension of the neck, as produced by the knot of a hangman's rope beneath the chin, is the reason for the alternative name. Since the vertebral canal is enlarged by the forward displacement of the vertebral body of the axis, the spinal cord is rarely compressed (see Fig 10–39).

Clinical Anatomy of the Technique for Gardner-Wells Tongs

Indications

Unstable fractures of the cervical vertebrae must be stabilized; dislocations of the vertebrae must be reduced. This may be accomplished with traction and a tong device, such as Gardner-Wells tongs or by surgical intervention. Gardner-Wells tongs consist of a horseshoe-shaped bar with two sharp traction pins at the open end of the horseshoe (Fig 10–40).

Anatomy of the Procedure

The scalp is shaved and prepared above both ears, and a local anesthetic is infiltrated to block the sensory nerves (branches of the auriculotemporal nerve, from the trigeminal nerve and the lesser occipital nerve from the cervical plexus).

The points of the tongs are applied above and just behind the auricle and below the equator of the skull, i.e., below the widest point of the circumference of the skull (see Fig 10–40). This corresponds to a level just below the temporal lines and is about 5 cm above the tip of the mastoid process. The points of the tongs pierce the following structures as they are advanced to the skull.

1. Skin.
2. Superficial fascia.
3. Deep fascia covering the temporalis muscle.
4. Upper margin of the origin of the temporalis muscle.
5. Periosteum.
6. Outer table of parietal bone of skull.

The periosteum is richly supplied with sensory nerves that need to be adequately blocked with the anesthetic agent. To obtain correct anatomical reduction in cases of fractures of the odontoid process of the axis, it is important to prevent rotation of the cervical vertebrae at the atlantoaxial joint.

Anatomy of Complications

1. Wrongful placement of the points of the tongs. For example, if the tongs are placed immediately above the external auditory meatus, the line of traction will be too far anterior.

2. Application of too much weight will cause excessive distraction and tearing of intervertebral ligaments and possibly injure the spinal cord. Only the minimum amount of traction should be applied to reduce the displaced bones and keep them in their

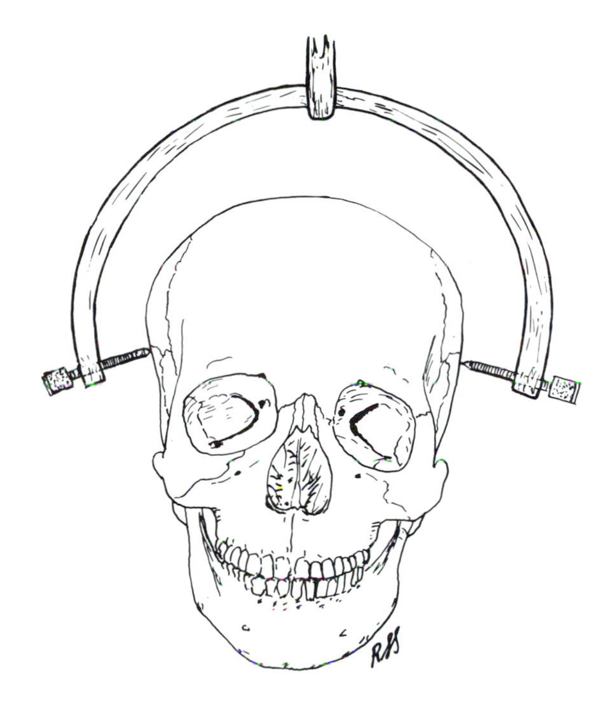

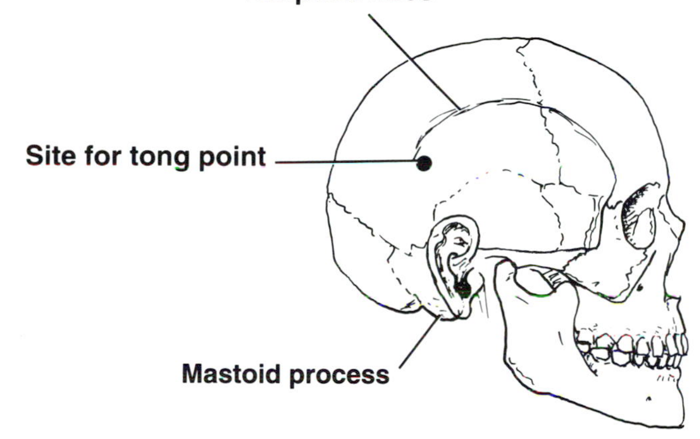

FIG 10–40.
Surface landmarks for the application of Gardner-Wells tongs. The points of the tongs are applied just above and behind the auricle and below the equator of the skull, i.e., just below the temporal lines and about 5 cm above the tip of the mastoid process.

normal anatomical position. Weights of 10 to 35 lb should be used, depending on the vertebral level of the injury; the head of the patient's bed should be raised by about 20°. Follow-up radiographs are taken.

3. Tong points may pull out of the skull if they have not been correctly placed and applied with the correct amount of pressure.

4. Tong points may penetrate the inner table of the skull because of excessive tightening.

5. Infection of the scalp, osteomyelitis of the skull, and even brain abscess have been reported.

Spondylolisthesis

Spondylolisthesis must be distinguished from *spondylosis,* which is a general term used for degenerative changes in the vertebral column caused by osteoarthritis and from *spondylolysis,* which may be defined as a break in a pedicle caused by a congenital defect and associated with stress. In spondylolysis there is no separation of one vertebra on another. In spondylolisthesis the body of a lower lumbar vertebra, usually the fifth, moves forward on the body of the vertebra below and carries with it the whole of the upper portion of the vertebral column (Fig 10–41). The essential defect is in the pedicles of the migrating vertebra. It is now generally believed that in this condition the pedicles are abnormally formed, and accessory centers of ossification are present that fail to unite. A stress fracture develops in the inherently weakened area. The spine, laminae, and inferior articular processes remain in posi-

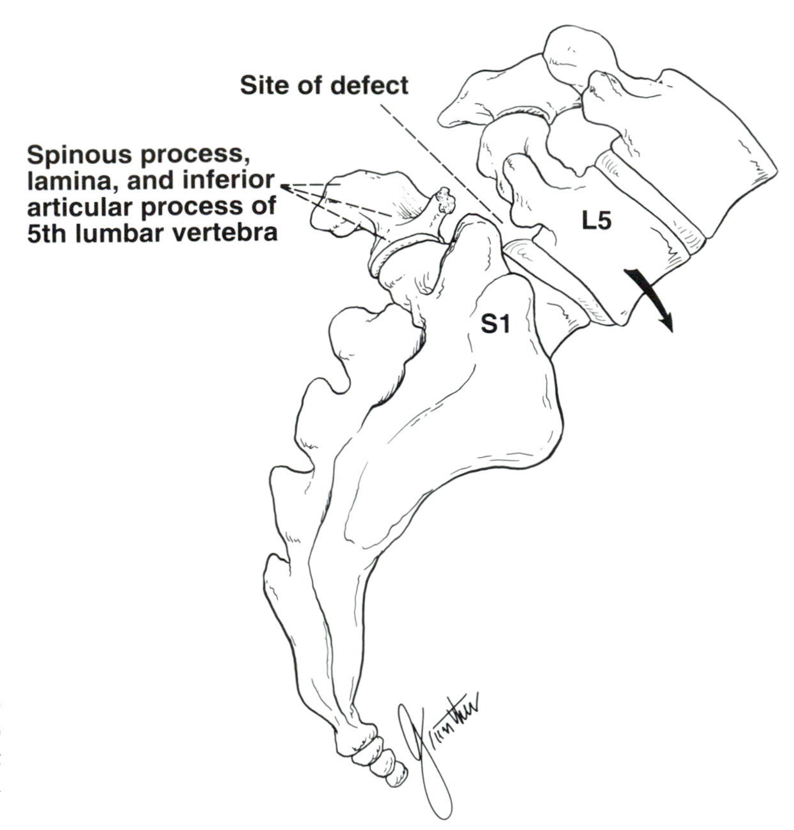

FIG 10–41.
Spondylolisthesis. Defect in the pedicles of the fifth lumbar vertebra allow the vertebral body to move anteriorly and inferiorly, carrying with it the whole of the upper portion of the vertebral column.

tion, while the remainder of the vertebra, having lost the restraining influence of the inferior articular processes, slips forward. Since the laminae are left behind, the vertebral canal is not narrowed, but the nerve roots may be pressed upon, causing low backache and pain down the distribution of the sciatic nerve. In severe cases the trunk becomes shortened, and the lower ribs contact the iliac crest.

Osteophytes and Nerve Root Pain

Spinal nerve roots exit from the vertebral canal through the intervertebral foramina. Each foramen is bounded superiorly and inferiorly by the pedicles, anteriorly by the intervertebral disc and the vertebral body, and posteriorly by the articular processes and joints (see Fig 10–9). In the lumbar region, the largest foramen is between the first and second lumbar vertebrae and the smallest is between the fifth lumbar and first sacral vertebra.

One of the complications of osteoarthritis of the vertebral column is the growth of osteophytes, which commonly encroach on the intervertebral foramina, causing pain along the distribution of the segmental nerve. The fifth lumbar spinal nerve is the largest of the lumbar spinal nerves, and it exits from the vertebral column through the smallest intervertebral foramen. For this reason it is the most vulnerable.

Osteoarthritis as a cause of root pain is suggested by the patient's age, its insidious onset, and a history of back pain of long duration; this diagnosis is made only when all other causes have been excluded. For example, a prolapsed disc usually occurs in a younger age group and often has an acute onset.

Sacroiliac Joint Disease

The basic anatomy of this joint is described on p. 681 The clinical aspects of this joint are considered here since disease of this joint can cause low back pain and may be confused with disease of the lumbosacral joints. The sacroiliac joint is innervated by the lower lumbar and sacral nerves, so that disease in the joint may produce low back pain and sciatica.

The sacroiliac joint is inaccessible to clinical examination. However, a small area located just medial to and below the posterior superior iliac spine is where the joint comes closest to the surface. In disease of the lumbosacral region, movements of the vertebral column in any direction cause pain in the lumbosacral part of the column. In sacroiliac disease, pain is extreme on rotation of the vertebral column and is worst at the end of forward flexion. The latter movement causes pain, since the hamstrings hold the hip bones in position while the sacrum is rotating forward as the vertebral column is flexed.

Narrowing of the Spinal Canal

After about the fourth decade of life the spinal canal becomes narrowed by aging. Osteoarthritic changes in the joints of the articular processes with the formation of osteophytes, together with degenerative changes in the intervertebral discs and the formation of large osteophytes between the vertebral bodies, can lead to narrowing of the spinal canal and intervertebral foramina. In persons in whom the spinal canal was originally small, significant stenosis in the cauda equina area can lead to neurological compression. Symptoms vary from mild discomfort in the lower back to severe pain radiating down the leg with the inability to walk. Since the neural elements tend to be stretched in extension, it is found that these patients get some relief when the vertebral column is flexed.

Pseudoclaudication may develop in patients with advanced spinal stenosis. Pain is experienced in the buttock, thigh, or leg when standing or walking and is relieved by rest in the presence of a normal and adequate arterial supply to the lower limb. Here, again, flexion of the lumbar vertebral column relieves the symptoms.

Clinical Anatomy of the Spinal Cord

Relationships of Spinal Cord Segments to Vertebral Numbers

Since the spinal cord is shorter than the vertebral column, the spinal cord segments do not correspond numerically with the vertebrae that lie at the same level (see Fig 10–42). The following table will help determine which spinal segment is contiguous with a given vertebral body.

Vertebrae	Spinal Segment
Cervical	Add 1
Upper thoracic	Add 2
Lower thoracic (T7-T9)	Add 3
10th thoracic	L1 and L2 cord segments
11th thoracic	L3 and L4 cord segments
12th thoracic	L5 cord segment
1st lumbar	Sacral and coccygeal cord segments

Spinal Cord and Muscle Tone

Muscle tone is a state of continuous partial contraction of a muscle and is dependent on the integrity of a monosynaptic reflex arc. The activity of this reflex arc is influenced by impulses received through the descending tracts from supraspinal levels. Muscle tone is abolished if any part of that simple reflex arc is destroyed. An atonic muscle feels soft and flabby and atrophies rapidly.

Lesions of the Anterior and Posterior Nerve Roots

A lesion of a posterior spinal nerve root will produce pain in the area of skin innervated by that root and also in the muscles that receive their sensory nerve supply from that root. Movements of the vertebral column in the region of the lesion will heighten the pain, and coughing and sneezing will also make it worse by raising the pressure within the vertebral canal. Before there is actual loss of sensation in the dermatome, there may be evidence of hyperalgesia and hyperesthesia.

A lesion of an anterior root will result in paralysis of any muscle that is supplied exclusively by that root and a partial paralysis of any muscle that is supplied partially by that root. In both cases, fasciculation and muscle atrophy occur.

Table 10–1 summarizes the relationship between the nerve roots, the dermatomes, the muscles, and the reflexes of the upper and lower extremities.

Summary of the Principal Sensory Pathways

Conscious Sensory Information.— Pain and temperature sensations cross obliquely to the contralateral side of the cord within one spinal segment of their entrance. They ascend in the lateral spinal thalamic tract and eventually reach the postcentral gyrus of the cerebral cortex on the contralateral side of the body (see Fig 10–29).

Discriminative touch, vibratory sense, and muscle joint sense ascend in the fasciculus gracilis and fasciculus cuneatus on the ipsilateral side of the spinal cord and eventually reach the postcentral gyrus on the opposite side of the body; this pathway crosses the midline in the medulla oblongata (see Fig 10–31).

Light touch and pressure sensations cross very obliquely to the contralateral side of the cord within several spinal segments of their entrance. They ascend in the anterior spinothalamic tract and eventually reach the postcentral gyrus on the contralateral side of the body (see Fig 10–30).

Unconscious Sensory Information.— Sensations from the muscles, tendons, and joints ascend in the posterior spinocerebellar tracts on the ipsilateral side of the spinal cord and in the anterior spinocerebellar tracts on both sides of the cord (Fig 10–32). Al-

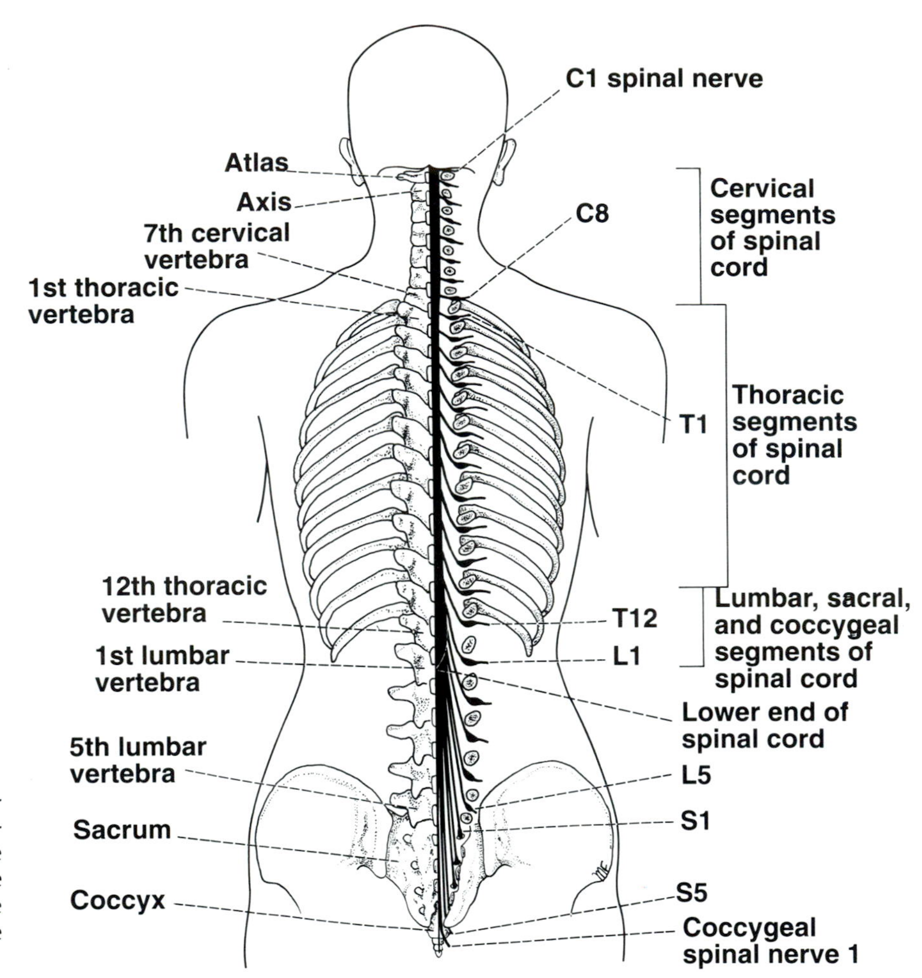

FIG 10–42.
Posterior view of the spinal cord showing the origins of the roots of the spinal nerves and their relationship to the different vertebrae. On the right, the laminae have been removed to expose the right half of the spinal cord and the nerve roots.

though most of the fibers in the anterior spinocerebellar tracts have crossed from the opposite side, there is clinical evidence to suggest that the fibers cross back to their original side within the cerebellum. Thus, the above proprioceptive information is conveyed ultimately to the cerebellar hemisphere of the same side of the body.

Injury to the Ascending Tracts Within the Spinal Cord

Lateral Spinothalamic Tract.—Destruction of this tract produces contralateral loss of pain and thermal sensibilities below the level of the lesion.

Anterior Spinothalamic Tract.—Destruction of this tract produces contralateral loss of light touch and pressure sensibilities below the level of the lesion. Remember that discriminative touch (two-point discrimination) will still be present, because this information is conducted through the fasciculus gracilis and fasciculus cuneatus. The patient will not feel the light touch of a piece of cotton placed against the skin or feel pressure from a blunt object placed against the skin.

Fasciculus Gracilis and Fasciculus Cuneatus.—Destruction of these tracts cuts off the supply of information from the muscles and joints to consciousness; thus, the individual does not know about the position and movements of his ipsilateral limbs below the level of the lesion. With the eyes closed, the patient is unable to tell the location of a limb or extremity part. For example, if the patient's big toe is passively dorsiflexed by the examiner, the patient is unable to tell whether the toe is pointing upward or downward. The patient has impaired muscular control, and his movements are jerky or ataxic.

The patient also loses vibration sense below the level of the lesion on the same side. This is easily tested by applying a vibrating tuning fork to a bony prominence such as the lateral malleolus of the fibula or the styloid process of the radius.

Two-point tactile discrimination on the side of the lesion will also be lost. In a healthy individual, the points have to be separated by about 3 to 4 mm before they are recognized as separate points on the tips of the fingers.

The sense of general light touch remains unaffected, because these impulses ascend in the anterior spinothalamic tracts.

It is extremely rare for a lesion of the spinal cord to be so localized as to affect one sensory tract only. It is more common to have several ascending and descending tracts involved.

Summary of the Principal Motor Pathways

The descending pathways from the cerebral cortex and the brain stem, i.e., the upper motor neurons, influence the activity of the lower motor neurons either directly or through internuncial neurons.

Corticospinal Tracts.—The corticospinal tracts (see Fig 10–33) arise from pyramidal cells in the precentral gyrus (two thirds) and from the postcentral gyrus (one third; these do not control motor activity but influence sensory input). The descending fibers become grouped together on the anterior surface of the medulla oblongata to form the *pyramids.* At the junction of the medulla and the spinal cord, most of the fibers cross the midline at the *decussation of the pyramids* and enter the lateral white column of the spinal cord as the *lateral corticospinal tracts.* The remaining fibers do not cross but descend in the anterior white column as the *anterior corticospinal tracts.* The lateral corticospinal fibers end by synapsing with motor neurons in the anterior gray horn throughout the entire length of the spinal cord. The anterior corticospinal fibers have a similar termination but extend down only as far as the cervical and upper thoracic levels of the cord.

The corticospinal tracts are believed to control the prime mover muscles, especially those responsible for the highly skilled movements of the distal parts of the limbs.

Reticulospinal Tracts.—The reticulospinal fibers from the reticular formation cross the midline or remain uncrossed and descend through the anterior and lateral white columns of the spinal cord. They end by synapsing with the motor neurons in the anterior gray horns of the spinal cord, where they inhibit or facilitate the activity of these cells.

Tectospinal Tracts.—The tectospinal fibers arise from neurons in the superior colliculi of the midbrain. Most of the fibers cross the midline and descend in the anterior white column of the spinal cord. They end by synapsing with motor neurons in the anterior gray horns of the spinal cord, where they are believed to be concerned with reflex postural movements in response to visual stimuli.

Rubrospinal Tract.—The rubrospinal fibers arise from the red nucleus, cross the midline, and descend in the lateral white column of the spinal cord. The fibers end by synapsing with motor neurons in the anterior gray horns of the spinal cord. This is believed to be an important pathway by which the cerebral cortex and the cerebellum can influence the activity of the motor neurons of the spinal cord. They facilitate the activity of the flexor muscles and inhibit the activity of the extensor or antigravity muscles.

Vestibulospinal Tracts.—The vestibulospinal fibers arise from the lateral vestibular nucleus in the medulla and descend on the ipsilateral side of the spinal cord in the lateral white column. The fibers end by synapsing with motor neurons in the anterior gray horns of the spinal cord. This tract permits the inner ear and the cerebellum to facilitate the activity of the extensor muscles and inhibit the activity of the flexor muscles in association with the maintenance of balance.

These supraspinal descending tracts (other than the corticospinal tracts) play a major role in simple basic voluntary movements and, in addition, bring about an adjustment of the muscle tone so that easy, rapid movements of the joints can take place.

Pyramidal and Extrapyramidal Tracts.—The term *pyramidal tract* is commonly used and refers specifically to the corticospinal tracts. The term came into common usage when it was learned that the corticospinal fibers become concentrated on the anterior part of the medulla oblongata in an area referred to as the *pyramids.*

The term *extrapyramidal tracts* refers to all the descending tracts other than the corticospinal tracts and includes the reticulospinal, tectospinal, rubrospinal, vestibulospinal, and olivospinal tracts.

Lesions of the Upper Motor Neurons

Lesions of the Corticospinal Tracts (Pyramidal Tracts).—Lesions of the corticospinal tracts may result in the following.

1. Present *Babinski sign.* The great toe becomes dorsally flexed in response to scratching the skin along the lateral aspect of the sole of the foot. The

normal response is plantar flexion of the toes. The Babinski sign normally is present during the first year of life because the corticospinal tract is not myelinated until the end of that year.

2. Absent *superficial abdominal reflexes.* The abdominal muscles fail to contract when the skin of the abdomen is scratched. This reflex depends on the integrity of the corticospinal tracts, which exert a tonic excitatory influence on the internuncial neurons.

3. Absent *cremasteric reflex.* The cremaster muscle fails to contract when the skin on the medial side of the thigh is stroked. This reflex arc passes through the first and second lumbar segments of the spinal cord. This reflex is dependent on the integrity of the corticospinal tracts, which exert a tonic excitatory influence on the internuncial neurons.

4. *Loss of performance of fine-skilled voluntary movements.* This occurs especially at the distal end of the limbs.

Lesions of the Descending Tracts Other Than the Corticospinal Tracts (Extrapyramidal Tracts)

Lesions of the descending extrapyramidal tracts result in the following.

1. *Severe paralysis* with little or no muscle atrophy (except secondary to disuse).

2. *Spasticity or hypertonicity* of the muscles. Initially, there is flaccidity, but this followed by hypertonicity. The lower limb is maintained in extension, and the upper limb is maintained in flexion.

3. *Exaggerated deep muscle reflexes* and *clonus* may be present in the flexors of the fingers, the quadriceps femoris, and the calf muscles (initially, the reflexes may be decreased).

4. *Clasp-knife reaction.* When passive movement of a joint is attempted, there is resistance owing to spasticity of the muscles. The muscles, on stretching, suddenly give way; hence the name.

It is very rare for a lesion to be restricted only to the pyramidal tracts or the extrapyramidal tracts. Usually, both sets of tracts are affected to a variable extent, producing both groups of clinical signs. As the pyramidal tracts normally tend to increase muscle tone and the extrapyramidal tracts inhibit muscle tone, the balance between these opposing effects is altered with the upper motor neuron injury, producing different degrees of muscle tone.

Spinal Cord Lesions Involving the Lower Motor Neuron

Trauma, infection (poliomyelitis), vascular disorders, degenerative diseases, and neoplasms may all produce a lesion of the lower motor neuron by destroying the nerve cell body in the anterior gray column or its axon in the anterior root or spinal nerve.

The following clinical signs are present in the lower motor neuron lesions.

1. *Flaccid paralysis* of muscles supplied.
2. *Atrophy* of muscles supplied.
3. *Loss of reflexes* of muscles supplied.
4. *Muscular fasciculation.* This twitching of muscles is seen when there is slow destruction of the lower motor neuron.

Spinal Shock Syndrome

This is a clinical condition that follows severe damage to the spinal cord. A cord lesion will eventually result in upper motor neuron effects on the segments below—spasticity, exaggerated reflexes, and clasp-knife rigidity. Initially, however, a flaccid paralysis occurs. The segmental spinal reflexes are depressed due to the removal of influences from the higher centers that are mediated through the corticospinal, reticulospinal, tectospinal, rubrospinal, and vestibulospinal tracts. In most patients the shock persists for less than 24 hours, whereas in others it may persist for as long as a week.*

The presence of spinal shock may be determined by testing for the activity of the anal sphincter reflex. The reflex may be elicited by placing a gloved finger in the anal canal and stimulating the anal sphincter to contract by squeezing the glans penis or clitoris, or gently tugging on an inserted Foley catheter. A cord lesion involving the sacral segments of the cord would of course nullify this test, since the neurons giving rise to the inferior hemorrhoidal nerve to the anal sphincter (S2 through S4) would be nonfunctioning.

Destructive Lesions of the Spinal Cord

The degree of spinal cord injury at different vertebral levels is governed largely by anatomical factors. In the cervical region, dislocation or fracture dislocation is common, but the large size of the vertebral canal often prevents severe injury to the spinal cord. However, when there is considerable displacement, the cord is sectioned. Respiration ceases if the lesion occurs above the segmental origin of the phrenic nerves (C3 through C5).

In fracture dislocations of the thoracic region, displacement is often considerable, and the small

*Note that "spinal shock" has another meaning also—the hypotension that occurs from loss of vasomotor tone when high cord injury occurs.

size of the vertebral canal results in severe injury to the spinal cord.

In fracture dislocations of the lumbar region, two anatomical facts aid the patient. First, the spinal cord in the adult extends down only as far as the level of the lower border of the first lumbar vertebra (see Fig 10–27). Second, the large size of the vertebral foramen in this region gives the roots of the cauda equina ample room. Nerve injury may therefore be minimal in this region.

Injury to the spinal cord may produce partial or complete loss of function at the level of the lesion and partial or complete loss of function of afferent or efferent nerve tracts below the level of the lesion.

Destructive Spinal Cord Syndromes.—When neurologic impairment is identified following the disappearance of spinal shock, it can often be categorized into one of the following syndromes: (1) complete cord transection syndrome, (2) anterior cord syndrome, (3) central cord syndrome, and (4) Brown-Séquard syndrome or hemisection of the cord. The clinical findings often indicate a combination of lower motor neuron injury (at the level of the cord destruction) and an upper motor neuron injury for those cord segments below the level of destruction.

Complete Cord Transection Syndrome.—This transection results in complete loss of all sensibility and voluntary movement below the level of the lesion. It may be caused by fracture dislocation of the vertebral column, by a bullet or stab wound, or by an expanding tumor. The following characteristic clinical features will be seen *after* the period of spinal shock has ended.

1. Bilateral lower motor neuron paralysis and muscular atrophy in the segment of the lesion. This results from damage to the neurons in the anterior gray columns (i.e., lower motor neuron) and possibly from damage to the nerve roots of the same segment.
2. Bilateral spastic paralysis below the level of the lesion. A bilateral Babinski sign is present and, depending on the level of the segment of the spinal cord damaged, bilateral loss of the superficial abdominal and cremaster reflexes occurs. All these signs are caused by an interruption of the corticospinal tracts on both sides of the cord. The bilateral spastic paralysis is produced by the cutting of the descending tracts other than the corticospinal tracts.
3. Bilateral loss of all sensation below the level of the lesion. The loss of tactile discrimination and vibratory and proprioceptive sense is due to bilateral destruction of the ascending tracts in the posterior white columns. The loss of pain, temperature, and light touch is caused by the section of the lateral and anterior spinothalamic tracts on both sides. Because these tracts cross obliquely, the loss of thermal sensations and light touch occurs two or three segments below the lesion distally.
4. Bladder and bowel functions are no longer under voluntary control, since all the descending autonomic fibers have been destroyed.

If there is a complete fracture dislocation at the L2-L3 vertebral level, i.e., a level below the lower end of the cord in the adult, no cord injury occurs and neural damage is confined to the cauda equina, and lower motor neuron, autonomic, and sensory fibers are involved.

Anterior Cord Syndrome.—The anterior cord syndrome (Fig 10–43) may be caused by cord contusion during vertebral fractures or dislocations, from injury to the anterior spinal artery or its feeder arteries with resultant ischemia of the cord, or by a herniated intervertebral disc. The following characteristic clinical features will be seen *after* the period of spinal shock has ended.

1. Bilateral lower motor neuron paralysis in the segment of the lesion and muscular atrophy. This is caused by damage to the neurons in the anterior gray columns (i.e., lower motor neuron) and possibly by damage to the anterior nerve roots of the same segment.
2. Bilateral spastic paralysis below the level of the lesion; the extent of the paralysis depends on the size of the injured area of the cord. These signs are caused by the interruption of the anterior corticospinal tracts on both sides of the cord. The bilateral spastic paralysis is produced by the interruption of the tracts other than the corticospinal tracts.
3. Bilateral loss of pain, temperature, and light touch. These signs are caused by the interruption of the anterior and lateral spinothalamic tracts on both sides.
4. Tactile discrimination and vibratory and proprioceptive sense are preserved because the posterior white columns on both sides are undamaged.

Central Cord Syndrome.—This syndrome is most often caused by hyperextension of the cervical spine (see Fig 10–43). The cord is pressed upon anteriorly by the vertebral bodies and posteriorly by the bulging of the ligamentum flavum, causing damage to the central region of the spinal cord. The radiographs of these injuries are often normal because no fracture or dislocation has occurred.

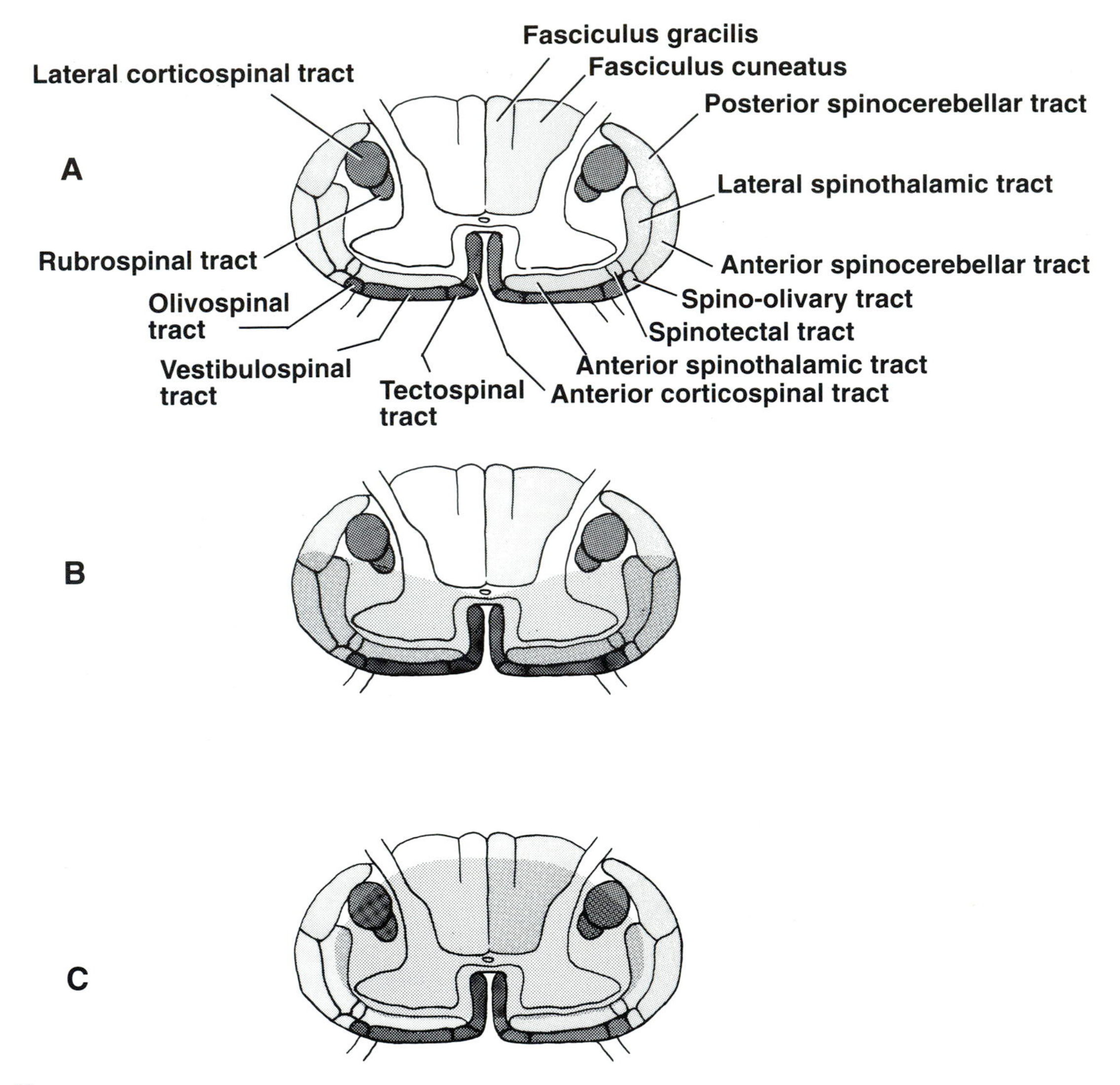

FIG 10–43.
A, transverse section of the spinal cord at the midcervical level showing the ascending tracts on the right and the descending tracts on the left. **B,** shaded area shows tracts that might be damaged in an anterior spinal cord syndrome; note the sparing of the posterior white column. **C,** shaded area shows tracts that might be damaged in a central cord syndrome.

The following characteristic clinical features will be seen *after* the period of spinal shock has ended.

1. Bilateral lower motor neuron paralysis in the segment of the lesion and muscular atrophy. This is caused by damage to the neurons in the anterior gray columns (i.e., lower motor neuron) and possibly by damage to the nerve roots of the same segment.

2. Bilateral spastic paralysis below the level of the lesion with characteristic sacral "sparing." The lower limb fibers are affected less than the upper limb fibers because the descending fibers in the lateral corticospinal tracts are laminated, with the upper limb fibers located medially and the lower limb fibers located laterally.

3. Bilateral loss of all sensation below the level of the lesion with characteristic sacral "sparing." Because the ascending fibers in the ascending tracts are also laminated with the upper limb fibers located laterally and the lower limb fibers located medially, the upper limb fibers are more susceptible to damage than those of the lower limb.

It follows from this discussion that the clinical picture of a patient with a history of a hyperextension injury of the neck, presenting with motor and sensory tract injuries involving principally the upper limb, would strongly suggest central cord syndrome. The sparing of the lower part of the body may be evidenced by (1) the presence of perianal sensation, (2) good anal sphincter tone, and (3) the ability to move the toes slightly. In those patients whose damage is caused by edema alone, prognosis is often very good. A mild central cord syndrome may occur, which consists only of paresthesias of the upper part of the arm and some mild arm and hand weakness.

Brown-Séquard Syndrome or Hemisection of the Cord.—Hemisection of the spinal cord may be caused by fracture dislocation of the vertebral column, by a bullet or stab wound, or by an expanding tumor. Incomplete hemisection is common; complete hemisection is rare. The following characteristic clinical features will be seen in patients with a complete hemisection of the cord (Fig 10–44) *after* the period of spinal shock has ended.

1. Ipsilateral lower motor neuron paralysis in the segment of the lesion and muscular atrophy. These signs are caused by damage to the neurons on the anterior gray column and possibly by damage to the nerve roots of the same segment.

2. Ipsilateral spastic paralysis below the level of the lesion. An ipsilateral Babinski sign is present, and, depending on the segment of the cord damaged, an ipsilateral loss of the superficial abdominal reflexes and cremasteric reflex occurs. All these signs are due to loss of the corticospinal tracts on the side of the lesion. The spastic paralysis is produced by the interruption of the descending tracts other than the corticospinal tracts.

3. Ipsilateral band of cutaneous anesthesia in the segment of the lesion. This results from the destruction of the posterior root and its entrance into the spinal cord at the level of the lesion.

4. Ipsilateral loss of tactile discrimination and of vibratory and proprioceptive sense below the level of the lesion. These signs are caused by the destruction of the ascending tracts in the posterior white column on the same side of the lesion.

5. Contralateral loss of pain and temperature sense below the lesion. This is due to destruction of the crossed lateral spinothalamic tracts on the same

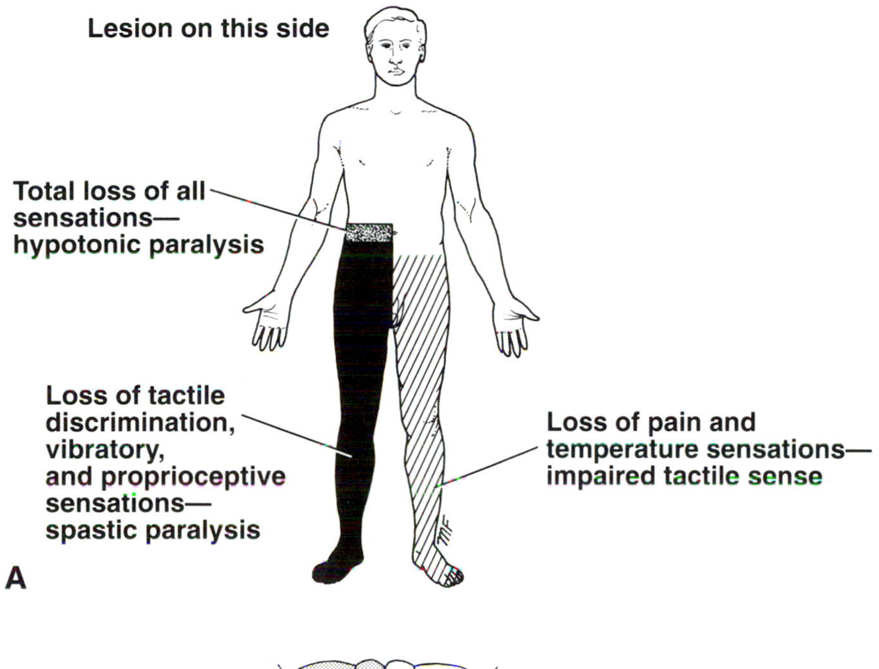

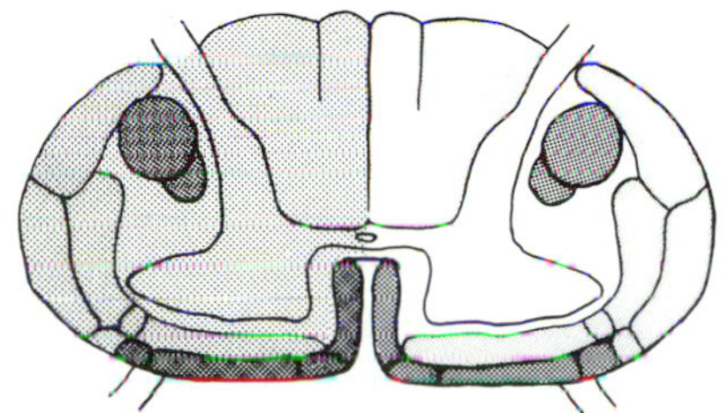

FIG 10–44.
A, Brown-Séquard syndrome with spinal cord lesion at the right tenth thoracic level showing motor and sensory losses. **B,** Brown-Séquard syndrome with spinal cord lesion at the right midcervical level; shaded area shows the various tracts that are involved. It is extremely rare to have a lesion that damages exactly half of the spinal cord, as depicted in the diagram.

side of the lesion. Because the tracts cross obliquely, the sensory loss occurs two or three segments below the lesion distally.

6. Contralateral but not complete loss of tactile sense below the lesions. This condition is brought about by the destruction of the crossed anterior spinothalamic tracts on the side of the lesion. Here, again, because the tracts cross obliquely, the sensory impairment occurs two or three segments below the lesion distally. The contralateral loss of tactile sense was incomplete because discriminative touch traveling in the ascending tracts in the contralateral posterior white column remained intact.

Syringomyelia

This disease is caused by a developmental abnormality in the formation of the central canal of the spinal cord. It is accompanied by gliosis and cavitation of the cervical segments of the spinal cord and brain stem. It occasionally occurs at other levels of the spinal cord. The condition interrupts the lateral and anterior spinothalamic tracts as they cross the spinal cord. The patient has segmental losses of pain and thermal sensibility, which are often bilateral, and some impairment of touch sensation. As the cavitation expands, other tracts and nerve cells become involved.

When the condition involves the cervical region, there is often a capelike distribution of the sensory deficits. This can be explained by the destruction of the spinothalamic tracts carrying information derived from the supraclavicular nerves (C3 and C4) and upper lateral cutaneous nerve of the arm (C5 and C6), which supply the skin over the shoulder region.

Ischemia of the Spinal Cord

The blood supply to the spinal cord is surprisingly meager, considering the importance of this nervous tissue. The longitudinally running arteries are small and variable in diameter, and the reinforcing segmental arteries vary in number and size. Ischemia of the spinal cord can easily follow minor damage to the arterial supply, as the result of nerve block procedures, aortic surgery, or any operation in which severe hypotension occurs. Watershed areas of the spinal cord are where the blood supply is very poor; such areas occur at the fourth thoracic and first lumbar segments of the cord.

Spinal Subarachnoid Hemorrhage

Spinal subarachnoid hemorrhage may be caused by spinal neoplasms, vascular malformations, and blood dyscrasia. The symptoms of sudden low back pain at the level of the lesion may be due to blood in the subarachnoid space or the entrance of blood into the substance of the spinal cord. Spinal cord dysfunction often accompanies the pain.

Spinal Epidural Abscess

This abscess is located outside the dura but within the vertebral canal. Commonly occurring in the midthoracic region, purulent necrosis of the epidural fat occurs that extends over several or many segments of the spinal cord. The pus is usually on the posterior surface of the cord. Necrosis of the spinal cord occurs secondary to pressure from the abscess or as the result of interference with its blood supply. Apart from the fever and malaise that accompanies the infection, symptoms and signs will appear that result from the necrosis of the ascending and descending tracts of the affected segments of the cord. Root pain and lower motor neuron muscular weakness due to local pressure will also be present.

Anatomy of Lumbar Puncture

Indications.—The diagnostic procedure may be used in the following.

1. CNS infection.
2. Subarachnoid hemorrhage.
3. Myelography.

The therapeutic procedure may be used in the following.

1. Instillation of chemotherapy.
2. Spinal anesthesia.

Lumbar Puncture Technique.—Lumbar puncture may be performed to withdraw a sample of cerebrospinal fluid for examination. Fortunately, the spinal cord terminates (see Fig 10–27) at the level of the lower border of the first lumbar vertebra in adults. (In infants, it may reach as low as the third lumbar vertebra; see also p. 366). The subarachnoid space extends down as far as the lower border of the second sacral vertebra. The lower lumbar part of the vertebral canal is thus occupied by the subarachnoid space, which contains the cauda equina, i.e., the lumbar and sacral nerve roots and the filum terminale. A needle introduced into the subarachnoid space in this region usually pushes the nerve roots to one side without causing damage.

With the patient in the lateral prone position or in the upright sitting position, with the vertebral column well flexed, the space between adjoining laminae in the lumbar region is opened to a maximum (Fig 10–45). An imaginary line joining the highest points on the iliac crests (see Fig 10–37) passes over the fourth lumbar spine (in the female, the line crosses the fourth lumbar spine at a slightly lower level). With a careful aseptic technique and under local anesthesia, the lumbar puncture needle, fitted with a stylet, is passed into the vertebral canal above or below the fourth lumbar spine (see Fig 10–45). The needle will pass through the following anatomical structures before it enters the subarachnoid space: (1) skin, (2) superficial fascia, (3) supraspinous ligament, (4) interspinous ligament, (5) ligamentum flavum, (6) areolar tissue containing the internal vertebral venous plexus in the epidural space, (7) dura mater, (8) arachnoid mater, and (9) cerebrospinal fluid. The depth to which the needle will have to pass will vary from an inch or less in children to as much as 4 in. (10 cm) in obese adults.

If the entering needle should stimulate one of the nerve roots of the cauda equina, the patient will experience a fleeting discomfort in one of the dermatomes, or a muscle will twitch, depending on whether a sensory or a motor root was impaled. If the needle should be pushed too far anteriorly, it may hit the body of the fourth lumbar vertebra (see Fig 10–45).

The cerebrospinal fluid pressure may be measured by attaching a manometer to the needle. In the recumbent position, the normal pressure is about 120 mm H_2O.

During the procedure, it has been suggested that the bevel of the needle should lie at right angles to the long axis of the patient's vertebral column. This permits the needle to pass between the transversely oriented fibers of the dura mater rather than cut through them. This theoretically lessens the size of the hole in the dura and prevents excess cerebrospinal fluid from escaping when the needle is withdrawn.

Anatomy of "Not Getting In".—If bone is encountered, the needle should be withdrawn as far as the subcutaneous tissue and the angle of insertion changed. The commonest bone to be encountered is the spinous process of the vertebra above or below the path of insertion. If the needle is directed laterally rather than in the midline, it may hit the lamina or an articular process.

Anatomy of Complications of Lumbar Puncture.—Complications include the following.

1. *Post–lumbar puncture headache.* This headache starts up to 48 hours after the procedure and lasts for 24 to 48 hours, but rarely longer. The cause is a leak of cerebrospinal fluid through the dural puncture and usually follows the use of a wide-bore needle. The leak reduces the volume of cerebrospinal fluid, which, in turn, causes a downward displacement of the brain and stretches the nerve-sensitive meninges—a headache follows. The headache is relieved by assuming the recumbent position. Using small-gauge styletted needles and avoiding multiple dural holes reduce the incidence of headache.

2. *Intraspinal epidermoid tumors.* The absence of a

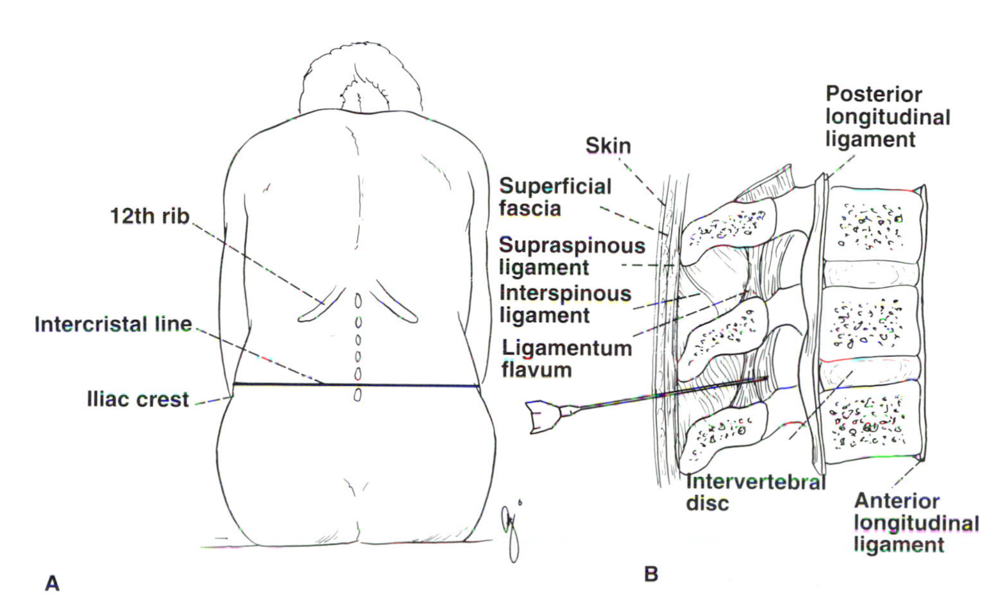

FIG 10–45.
A, important anatomic landmarks when performing a lumbar puncture. **B,** structures penetrated by the lumbar puncture needle before it reaches the dura mater. A lumbar puncture is usually performed with the patient in a lateral recumbent position.

stylet may result in the implantation of epidermal skin fragments into the vertebral canal.

3. *Infection.* Local infection of the skin or an epidural abscess is absolute contraindication to performing a lumbar puncture. The needle would carry the infection to the meninges, causing meningitis.

4. *Brain herniation.* Lumbar puncture is contraindicated in cases in which there is a markedly raised intracranial pressure. Large supratentorial mass lesions with a high intracranial pressure may result in a caudal displacement of the uncus through the tentorial notch or a dangerous displacement of the medulla through the foramen magnum, when the lumbar cerebrospinal fluid pressure is reduced. Computed tomographic scans may assist in the recognition of patients who are at risk.

5. *Injury to the intervertebral disc.* Excessively deep penetration by the needle has resulted in damage to the anulus fibrosus of the intervertebral disc.

6. *Injury to the nerve roots of the cauda equina.* Transient sensory symptoms in one of the dermatomes may follow stimulation by the needle of one of the posterior spinal nerve roots. Similar stimulation of an anterior root may cause a muscle to twitch.

7. *Injuries to the vertebral body or the aorta.* These are extremely rare complications and are caused by the needle passing anteriorly well beyond the subarachnoid space.

Anatomy of Caudal Anesthesia

Anesthetic solutions injected into the sacral canal through the sacral hiatus pass upward in the loose connective tissue and bathe the spinal nerves as they emerge from the dural sheath (Fig 10–46).

Indications.—The following are indications for caudal anesthesia.

1. Operations in the sacral region, including anorectal surgery and culdoscopy.
2. Childbirth, during the first and second stages of labor.

Technique of Caudal Anesthesia.—The sacral hiatus is palpated as a distinct depression in the midline about 4 cm above the tip of the coccyx in the upper part of the cleft between the buttocks. The hiatus is triangular or U shaped and is bounded laterally by the sacral cornua (see Fig 10–46).

The size and shape of the hiatus depend on the number of laminae that fail to fuse in the midline posteriorly. The common arrangement is for the hiatus to be formed by the nonfusion of the fifth and sometimes the fourth sacral vertebrae.

With a careful aseptic technique and under local anesthesia, the needle, fitted with a stylet, is passed into the vertebral (sacral) canal through the sacral hiatus.

The needle pierces the following: (1) skin and fascia and the (2) sacrococcygeal membrane that fills in the sacral hiatus (see Fig 10–46). The membrane is formed of dense fibrous tissue and represents the fused supraspinous and interspinous ligaments as well as the ligamentum flavum. A distinct feeling of "give" is felt when the ligament is penetrated.

Note that the sacral canal is curved and follows the general curve of the sacrum (see Fig 10–46). The anterior wall, formed by the fusion of the bodies of the sacral vertebrae, is rough and ridged. The posterior wall, formed by the fusion of the laminae, is smooth. The average distance between the sacral hiatus and the lower end of the subarachnoid space at the second sacral vertebra is about 47 mm in adults.

Note also that the sacral canal contains (1) the dural sac (containing the cauda equina), which is tethered to the coccyx by the filum terminale, (2) the sacral and coccygeal nerves as they emerge from the dural sac surrounded by their dural sheath, and (3) the thin-walled veins of the vertebral venous plexus.

ANATOMICAL-CLINICAL PROBLEMS

1. A 12-year-old boy was showing off in front of friends by diving into the shallow end of a swimming pool. After one particularly daring dive, he surfaced quickly and climbed out of the pool, holding his head between his hands. He said he had hit the bottom of the pool with his head and now had severe pain in the root of the neck, which was made worse when he tried to move his neck. An ambulance was called, and the boy was immobilized on a long spine backboard and taken to the emergency department. It was observed that the boy held his head rotated to the left and complained of severe pain in the region of the back of the neck and right shoulder. The boy reported that any movement of the neck in any direction greatly accentuated the pain. The deep muscles on the right side of the back of the neck were tender and in spasm. No other neurological signs or symptoms were present. What is your diagnosis?

2. A 37-year-old coal miner was crouching at the coal face when a large rock suddenly became dislodged from the roof of the mine shaft and struck

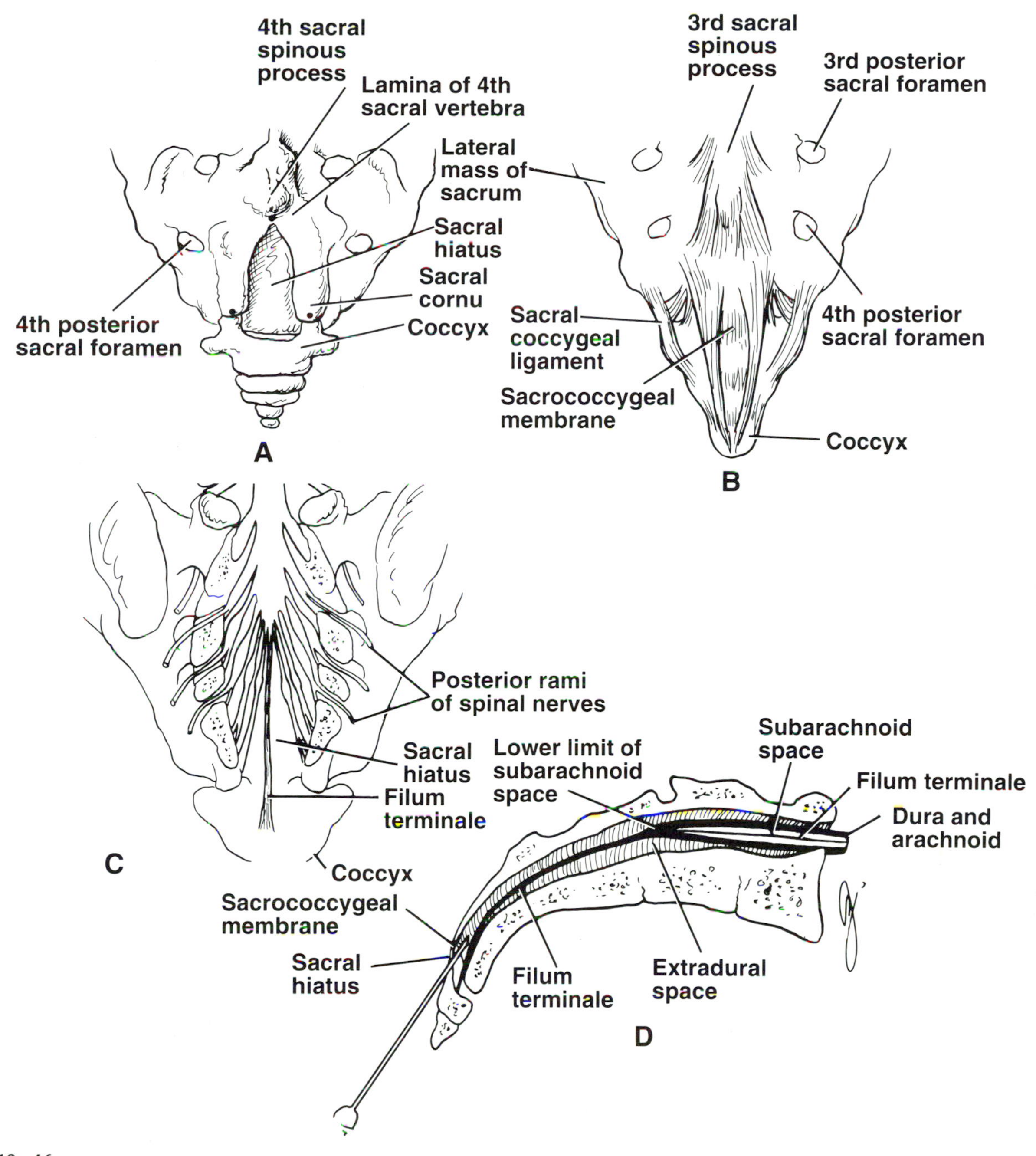

FIG 10–46.
A, the sacral hiatus. Black dots indicate the position of important bony landmarks. **B,** posterior surface of the lower end of the sacrum and the coccyx showing the sacrococcygeal membrane covering the sacral hiatus. **C,** the dural sheath (thecal sac) around the lower end of the spinal cord and spinal nerves in the sacral canal; the laminae have been removed. **D,** longitudinal section through the sacrum showing the anatomy of caudal anesthesia.

him on the upper part of his back. The emergency department physician suspected displacement of the upper thoracic spines on the eighth thoracic spine. What anatomical factors in the thoracic region determine the degree of injury that may occur to the spinal cord?

3. A 64-year-old man was seen in the emergency department complaining of a burning pain over his right shoulder region and the upper part of his right arm. The pain had started 2 days previously and had progressively worsened. The pain was accentuated by moving his neck and by coughing. Two years previously he had been treated for osteoarthritis of his vertebral column. Physical ex-

amination revealed weakness, wasting, and fasciculation of the right deltoid and biceps brachii muscles. The right biceps tendon reflex was absent. Radiological examination showed extensive spur formation on the bodies of the fourth, fifth, and sixth cervical vertebrae. The patient also had hyperesthesia in the skin over the lower part of the right deltoid and down the radial side of the arm. Using your knowledge of neuroanatomy, make the diagnosis. How is the pain produced? Why is the pain made worse by coughing?

4. You are performing a lumbar puncture. Name the anatomical structures that your needle will pierce before you enter the subarachnoid space. Why is it necessary to flex the vertebral column well? What surface markings would you use to determine the correct vertebral level to carry out this procedure? At what level does the spinal cord end interiorly in the average adult? Where does the subarachnoid space end inferiorly? What is the normal cerebrospinal fluid pressure in a patient in the lateral prone position?

5. A 19-year-old student joined a group of other students who were attempting to lift a car out of a snowdrift. Suddenly, he experienced an acute pain in the back. On examination in the emergency department, the deep muscles of the back in the left lumbar region felt firmer than normal and were tender to touch. He said his pain was accentuated by coughing. Evaluation of the muscles of the lower limbs showed some slight weakness in the left quadriceps femoris and a diminished left knee jerk. A diagnosis of herniated intervertebral disc between the third and fourth lumbar vertebrae was made. Which spinal nerve exits between these two vertebrae? Which spinal nerve roots are most likely to be damaged by an L3-L4 herniated disc?

6. What is the blood supply of the spinal cord? Which areas of the spinal cord are supplied by the anterior spinal artery? Which regions of the spinal cord are most susceptible to ischemia?

7. A 48-year-old emergency medicine physician woke up one morning with a severe pain at the root of the back of the neck and the left shoulder. The previous evening, he had participated in a hard-fought tennis tournament at his club. On questioning, it was found that the pain was also referred along the lateral side of the left arm. Movement of the neck caused an increase in the intensity of the pain, which was accentuated by coughing. A lateral radiograph of the neck showed slight narrowing of the space between the fifth and sixth cervical vertebral bodies. What is the likely cause of the condition? Which nerve root was involved?

8. A 60-year-old man was reaching for a box on a high shelf as he balanced on a chair. He lost his balance and fell to the floor, catching his left lumbar region on the edge of the chair. On examination in the emergency department 3 hours later, a large swollen, bruised area was found in the left lumbar region, which was extremely tender to touch. What structures might be injured?

9. A 29-year-old woman was involved in an automobile accident. She was the driver and was thrown forward on impact. Although her seat belt restrained her trunk, her head and neck were excessively flexed. On initial evaluation in the emergency department, she was found to have signs and symptoms of a neurological deficit in the upper and lower extremities. A lateral radiograph of the cervical spine showed fragmentation of the body of the fourth cervical vertebra. In order to stabilize the fracture and prevent further neurologic damage, it was decided to apply skeletal traction with Gardner-Wells tongs. State where you would place the points of the tongs on the head and what anatomical structures would be pierced by the points of the tongs. Indicate the factors that prevent the tong points from pulling out when traction is applied.

10. A 26-year-old woman was seen in the emergency department complaining of low-back pain off-and-on for the past 6 years, which she said had suddenly gotten worse. On examination, the patient was found to have a severe lumbar lordosis, with excessive prominence of the first sacral spinous process. A prominent fold of skin was seen on either side above the iliac crests, and the last ribs appeared to rest on the iliac crests. Using your knowledge of anatomy, make the diagnosis. What are the underlying embryological reasons for the condition?

11. A 21-year-old student was driving home from a party and crashed his car head-on into a brick wall. On examination in the emergency department, he was found to have a fracture dislocation of the seventh thoracic vertebra, with signs and symptoms of severe damage to the spinal cord. Thirty-six hours later, when the spinal shock had passed, he had an upper motor neuron paralysis of the left leg and loss of muscle joint sense of the left leg. When his cutaneous sensibility was tested, he had a band of cutaneous hyperesthesia extending around the abdominal wall on the left side at the level of the umbilicus. Just below this, he had a narrow band of anesthesia and analgesia. On the right side, there was total an-

algesia, thermoanesthesia, and partial loss of tactile sense of the skin of the abdominal wall below the level of the umbilicus and involving the whole of the right leg. Using your knowledge of neuroanatomy, state the level at which the spinal cord was damaged. Was the spinal cord completely sectioned? If not, on which side did the hemisection occur? Explain the sensory losses found on examination in this patient.

12. A 39-year-old woman was seen in the emergency department complaining that she had severely burnt the index finger of her left hand on a hot stove. She stated that she was unaware that the burn had taken place until she smelled the burning skin. On examination she was found to have considerably reduced pain and temperature sense involving the sixth and seventh cervical dermatomes of the left hand. However, her sense of tactile discrimination was normal in these areas. Examination of the right upper extremity showed a similar but much less severe dissociated sensory loss involving the same areas. No further abnormal signs were discovered. What is the diagnosis? Can you name the tract or tracts that were involved in this disease?

ANSWERS

1. On striking the bottom of the pool, the boy's head forced his neck into excessive flexion. The right inferior articular process of the fifth cervical vertebra was forced over the anterior margin of the right superior articular process of the sixth cervical vertebra, producing a unilateral dislocation and thus the rotation of the head to the left. The tearing of the capsular ligaments and the nipping of the sixth right cervical nerve caused spasm of the neck muscles and extreme pain in the sixth right cervical dermatome. A lateral radiograph of the cervical vertebral column would reveal the dislocation. The large size of the vertebral canal in the cervical region permitted the spinal cord to escape injury.

2. This patient had a severe fracture dislocation between the seventh and eighth thoracic vertebrae. The vertical arrangement of the articular processes and the low mobility of this region because of the thoracic cage mean that a dislocation can occur in this region only if the articular processes are fractured by a great force. The small circular vertebral canal leaves little space around the spinal cord, so that a severe injury to the cord is certain.

3. This patient was suffering from cervical spondylosis with pressure on the anterior and posterior roots of the fifth and sixth spinal nerves. As the result of repeated trauma and of aging, degenerative changes occurred at the articulating surfaces of the fourth, fifth, and sixth cervical vertebrae. Extensive spur formation resulted in narrowing of the intervertebral foramina with pressure on the nerve roots. The burning pain and hyperesthesia were due to pressure on the posterior roots, and the weakness, wasting, and fasciculation of the deltoid and biceps brachii muscles were due to pressure on the anterior roots. Movements of the neck presumably intensified the symptoms by exerting further pressure or traction on the nerve roots. Coughing or sneezing raised the pressure within the vertebral canal and resulted in further pressure on the roots.

4. The lumbar puncture needle will pass through the following structures before it enters the subarachnoid space: (1) skin, (2) superficial fascia, (3) supraspinous ligament, (4) interspinous ligament, (5) ligamentum flavum, (6) areolar tissue containing the internal vertebral venous plexus, (7) dura mater, and (8) arachnoid mater.

With the patient in the lateral prone or upright sitting position, the vertebral column should be well flexed in order to separate the laminae of adjacent vertebrae.

An imaginary line joining the highest points of the iliac crests passes over the fourth lumbar spine; the needle should be inserted above or below this spine in the adult.

The spinal cord ends below in the adult at the level of the lower border of the first lumbar vertebra. The subarachnoid space ends below at the level of the lower border of the second sacral vertebra.

With the patient in the lateral prone position, the normal cerebrospinal fluid pressure is about 120 mm of water.

5. A herniated intervertebral disc between the third and fourth lumbar vertebrae may press on the third lumbar spinal nerve, which exits from between these vertebrae. However, the roots of the fourth lumbar spinal nerve are more likely to be compressed as they cross the posterior surface of the disc to exit between the fourth and fifth lumbar vertebrae.

6. The blood supply to the spinal cord is described on p. 374. The anterior spinal artery supplies the anterior two thirds of the spinal cord.

In the upper and lower thoracic segments of the spinal cord the anterior spinal artery may be extremely small. Should the segmental or radicular arteries be occluded in these watershed areas, the fourth thoracic and the first lumbar segments of the spinal cord would be particularly liable to ischemic necrosis.

7. This patient had symptoms suggestive of irritation of the left sixth cervical nerve root. The radiographic evidence of narrowing of the space between the fifth and sixth cervical vertebral bodies would suggest a herniation of the nucleus pulposus of the intervertebral disc at this level. Disc space narrowing, however, is a very poor predictor of disc herniation.

8. Anteroposterior and lateral radiographs of the lumbar vertebral column confirmed the presence of a fracture of the left transverse process of the third lumbar vertebra. The transverse processes of the lumbar vertebrae are long and tapered and are often fractured by direct trauma. The process of the third lumbar vertebra is the longest and least protected and therefore the most vulnerable. It is often necessary to "hot light" the radiograph to see the fracture.

Torn muscles in this region, such as the erector spinae or quadratus lumborum, are often responsible for protracted pain in the elderly. Although the bulk of the left kidney lies superior and anterior to the transverse process of the third lumbar vertebra, it may be injured. Contusion of the bowel may also occur and can result in ileus.

9. The points of the Gardner-Wells tongs are applied 5 cm above the tip of the mastoid process, i.e., below the widest part of the calvarium. The points of the tongs pierce the skin, the superficial fascia, the deep fascia covering the origin of the temporalis muscle, the upper edge of the temporalis muscle, the periosteum, and they are finally embedded in the outer table of the parietal bone.

If it is assumed that the tongs have been applied correctly with the right amount of pressure, the upward slant of the points of the tongs and the attachment of the points to the skull below the widest part of the calvarium (see Fig 10–40) ensure that the tongs do not pull out when traction is applied.

10. This patient had spondylolisthesis involving the fifth lumbar vertebra. During the 6 years that she had experienced back pain, the body of the fifth lumbar vertebra had slowly dislocated in front of the sacrum. The details of this condition are given on p. 387.

11. A fracture dislocation of the seventh thoracic vertebra would result in severe damage to the tenth thoracic segment of the spinal cord. The unequal sensory and motor losses on the two sides indicate a left hemisection of the cord. The narrow band of hyperesthesia on the left side was caused by the irritation of the cord immediately above the site of the lesion. The band of anesthesia and analgesia was caused by the destruction of the cord on the left side at the level of the tenth thoracic segment; all afferent fibers entering the cord at that point were interrupted. The loss of pain and thermal sensibilities and the loss of light touch below the level of the umbilicus on the right side were caused by the interruption of the lateral and anterior spinothalamic tracts on the left side of the cord.

12. This patient has the early signs and symptoms of syringomyelia. The gliosis and cavitation had resulted in interruption of the lateral and anterior spinothalamic tracts as they decussated in the spinal cord at the level of the sixth and seventh cervical segments. Because of the uneven growth of the cavitation, the condition was worse on the left side than on the right. Since tactile discrimination was normal in both upper extremities, the fasciculus cuneatus in both posterior white columns was unaffected. This dissociated sensory loss is characteristic of syringomyelia.

11

The Abdominal Wall, the Peritoneal Cavity, the Gastrointestinal Tract and Associated Viscera, and the Retroperitoneum

Acute abdominal pain, blunt and penetrating trauma to the abdominal wall, and gastrointestinal bleeding are common problems facing the emergency physician. The problems are complicated by the fact that the abdomen contains multiple organ systems, and in many patients more than one system is involved.

Knowledge of spatial relationships of the different abdominal organs is essential to making an accurate and complete diagnosis. Children with abdominal pain present a special diagnostic challenge; many diseases of childhood produce the symptoms of abdominal pain.

This chapter provides an overview of the anatomy of the abdomen with special reference to the abdominal wall, gastrointestinal tract, liver and biliary ducts, pancreas, spleen, and peritoneal cavity.

BASIC ANATOMY

Basic Anatomy of the Structure of the Anterior Abdominal Wall

The anterior abdominal wall is made up of skin, superficial fascia, deep fascia, muscles, extraperitoneal fascia, and parietal peritoneum.

Skin

Nerve Supply.—The cutaneous nerve supply to the anterior abdominal wall is derived from the anterior rami of the lower six thoracic and the first lumbar nerves (Fig 11–1). The thoracic nerves are the lower five intercostal and the subcostal nerves; the first lumbar nerve is represented by the iliohypogastric and inguinal nerves, branches of the lumbar plexus. The dermatome of T7 is located in the epigastrium over the xiphoid process, that of T10 includes the umbilicus, and that of L1 lies just above the inguinal ligament and the symphysis pubis.

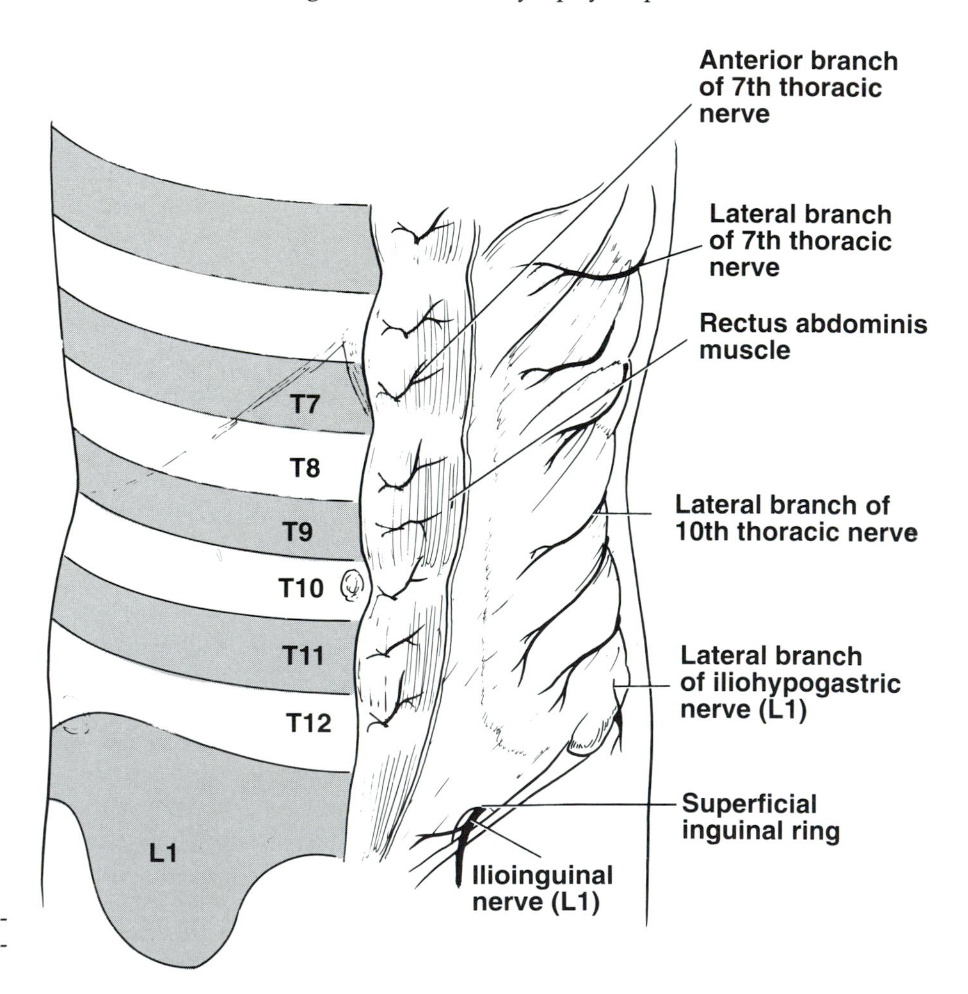

FIG 11–1.
Dermatomes and distribution of cutaneous nerves on the anterior abdominal wall.

Blood Supply.— The skin near the midline is supplied by branches of the superior epigastric artery— a branch of the internal thoracic artery and the inferior epigastric artery— a branch of the external iliac artery. The skin of the flanks is supplied by branches from the intercostal, lumbar, and deep circumflex iliac arteries (Fig 11–2).

The venous drainage passes above to the axillary vein via the lateral thoracic vein and below into the femoral vein via the superficial epigastric and great saphenous veins.

Clinical Note

Caval-caval anastomoses and paraumbilical veins.— Note the important indirect connection between the superior and inferior venae cavae. This may permit the reversal of blood flow in patients with an obstructed vena cava caused by a large mediastinal or abdominal tumor. Note also the presence of small *paraumbilical veins* that connect the systemic skin veins in the region of the umbilicus along the ligamentum teres to the portal vein. This may provide an important portal-systemic anastomosis in patients with obstruction of the portal vein, as in cirrhosis of the liver.

Lymph Drainage.— The cutaneous lymph vessels above the level of the umbilicus drain upward into the anterior axillary lymph nodes. The vessels below this level drain downward into the superficial inguinal nodes.

Superficial Fascia

The superficial fascia may be divided into the superficial *fatty layer (fascia of Camper)* and a deep *membranous layer (Scarpa's fascia)* (Fig 11–3).

The *fatty layer of superficial fascia* is continuous

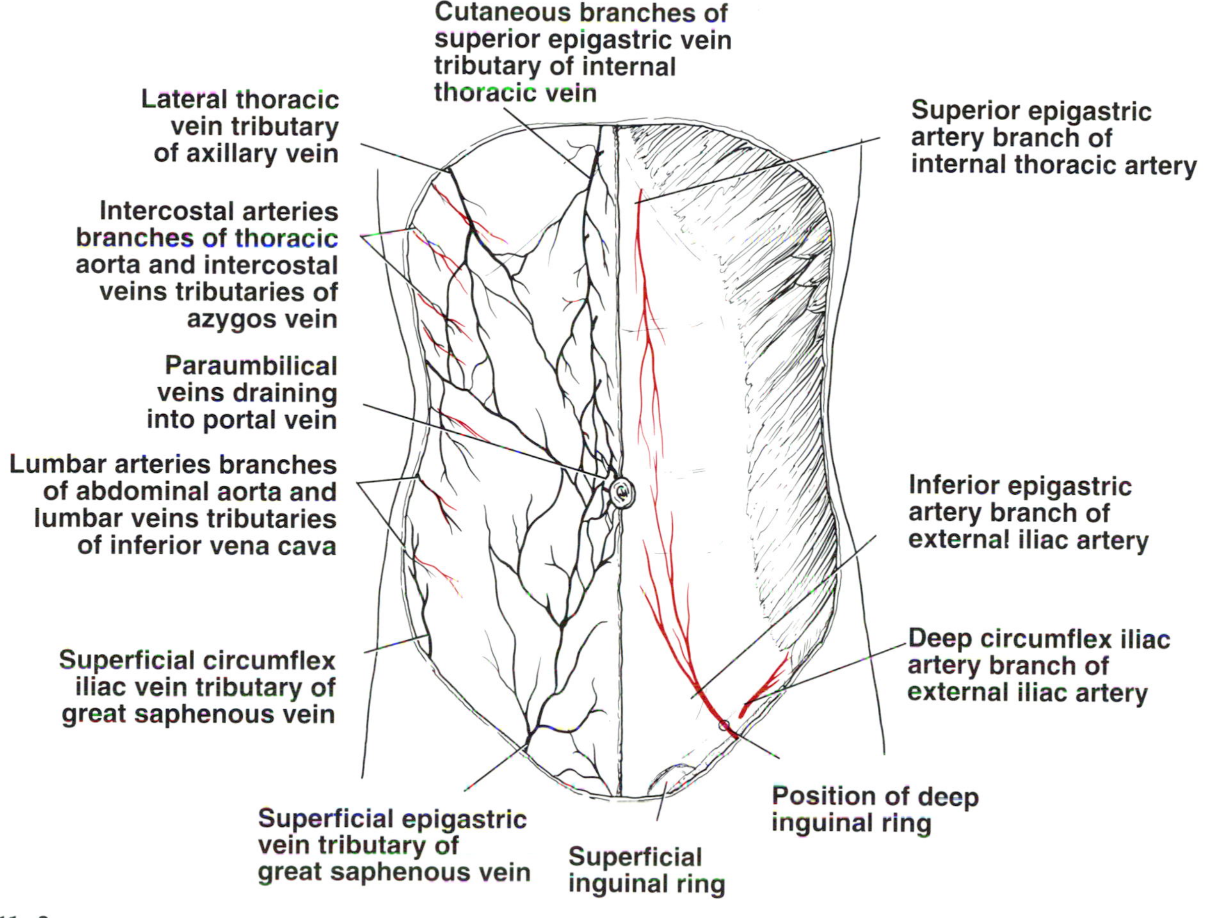

FIG 11–2.
On the left, venous drainage of the anterior abdominal wall. On the right, arterial supply to the anterior abdominal wall.

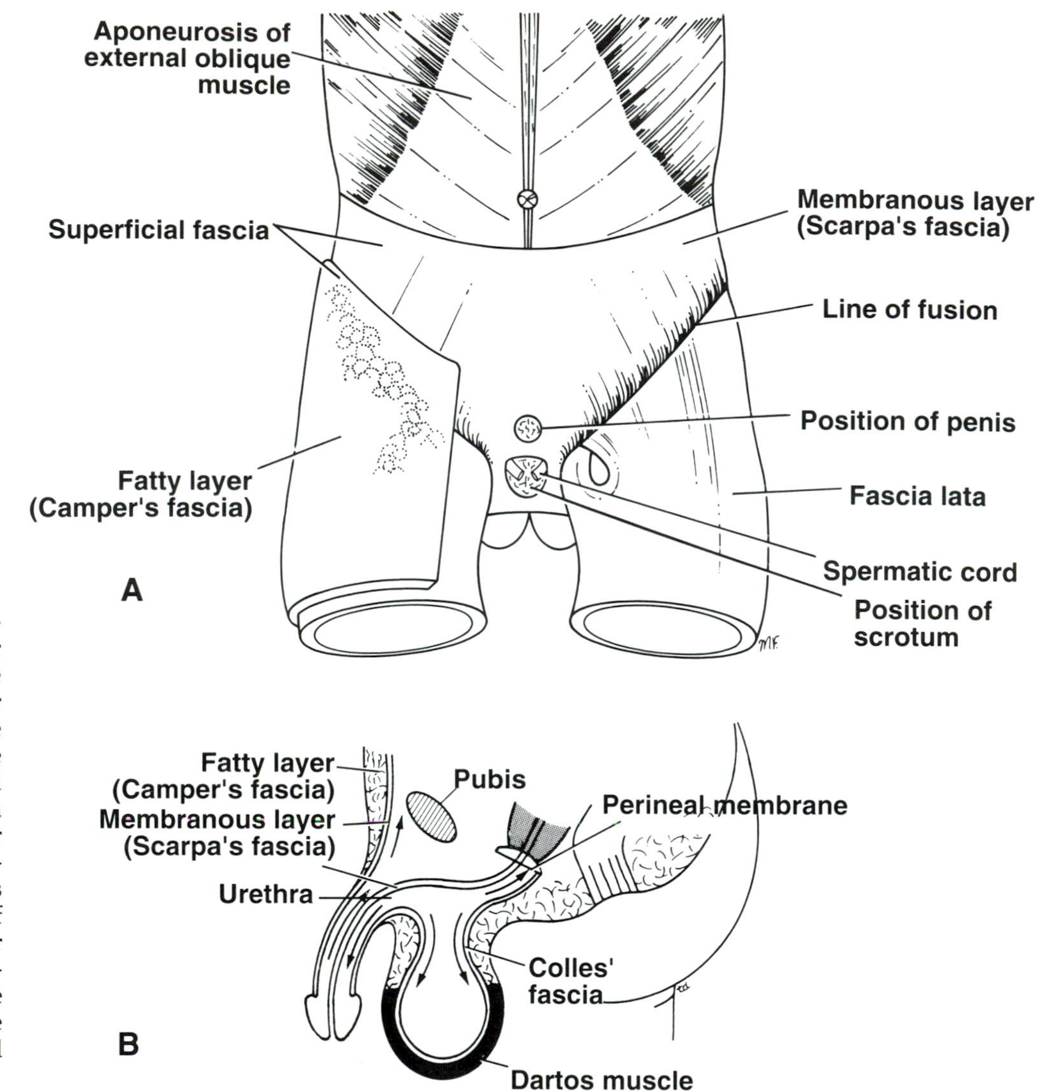

FIG 11–3.
Arrangement of fatty layer and membranous layer of superficial fascia in the lower part of the anterior abdominal wall. Note the line of fusion between the membranous layer and deep fascia of the thigh (fascia lata). In the lower diagram, note the attachment of the membranous layer (here called fascia of Colles') to the posterior margin of perineal membrane. Arrows indicate the path taken by the urine in cases of ruptured urethra.

with the superficial fascia over the rest of the body and in obese individuals may be extremely thick.

The *membranous layer of superficial fascia* is thin and fades out laterally and above, where it becomes continuous with the superficial fascia of the back and the thorax, respectively. Inferiorly, the membranous layer passes over the inguinal ligament to fuse with the thick deep fascia of the thigh (fascia lata) about one fingerbreadth below the inguinal ligament (see Fig 11–3). In the midline inferiorly, the membranous layer of fascia is not attached to the pubis, but forms a tubular sheath for the penis (clitoris). Below in the perineum, it enters the scrotal wall and from there passes to be attached on each side to the margins of the pubic arch; it is here referred to as the *Colles' fascia.* Posteriorly in the perineum, it fuses with the perineal body and the posterior margin of the perineal membrane (see Fig 11–3).

In the scrotum the fatty layer of the superficial fascia is represented as a thin layer of smooth muscle, the *dartos muscle.* The membranous layer of the superficial fascia persists as a separate layer.

Deep Fascia

The deep fascia in the anterior abdominal wall is merely a thin layer of connective tissue covering the muscles; it lies immediately deep to the membranous layer of superficial fascia.

Muscles of the Anterior Abdominal Wall

The muscles of the anterior abdominal wall consist of three broad, thin sheets that are aponeurotic in front; from exterior to interior they are the *external oblique, internal oblique, and transversus* (Fig 11–4). On either side of the midline anteriorly there is in addition a wide vertical muscle, the *rectus abdominis* (Fig

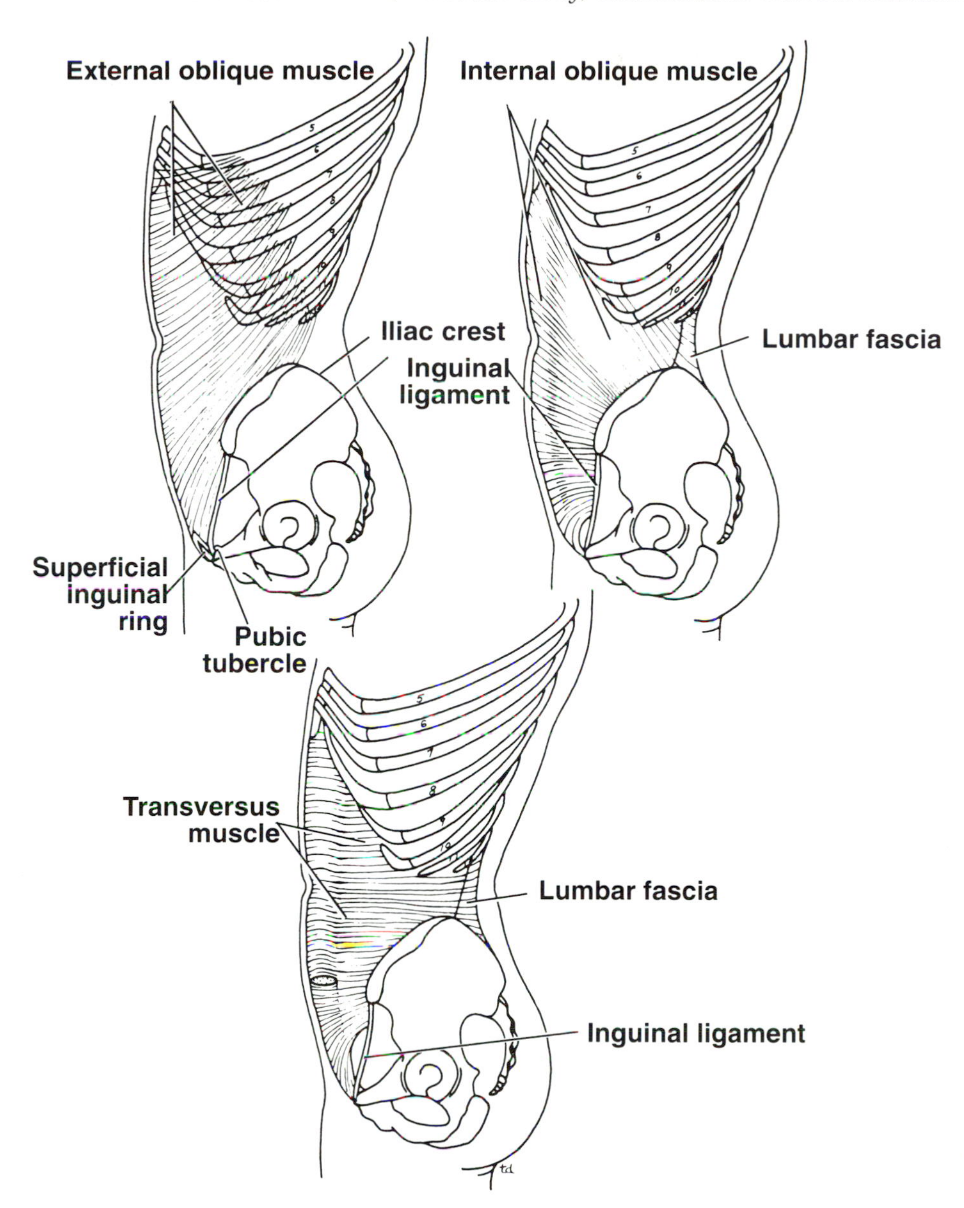

FIG 11–4.
External oblique, internal oblique, and transversus muscles of the anterior abdominal wall.

11–5). As the aponeuroses of the three sheets pass forward, they enclose the rectus abdominis to form the *rectus sheath.*

In the lower part of the rectus sheath, a small muscle called the *pyramidalis* may be present.

The *cremaster muscle,* which is derived from the lower fibers of the internal oblique, passes inferiorly as a covering of the spermatic cord and enters the scrotum.

The origins, insertions, nerve supply, and actions of the muscles of the anterior abdominal wall are shown in Table 11–1. The muscles are innervated by the lower six thoracic nerves and the first lumbar nerve (Fig 11–6), which pass forward in the interval between the internal oblique and transversus muscles.

Rectus Sheath

The rectus sheath (Figs 11–5 and 11–7) is a long fibrous sheath that encloses the rectus abdominis muscle and pyramidalis muscle (if present) and contains the anterior rami of the lower six thoracic nerves and the superior and inferior epigastric vessels and lymph vessels. It is formed mainly by the aponeuroses of the three anterior abdominal muscles. The internal oblique aponeurosis splits at the lateral edge of the rectus abdominis to form two laminae—one passes anterior to the rectus and one passes posterior. The aponeurosis of the external oblique fuses with the anterior lamina, and the transversus aponeurosis fuses with the posterior lamina. Below the umbilicus and at the level of the anterior superior iliac spines (S1 vertebral level), all three

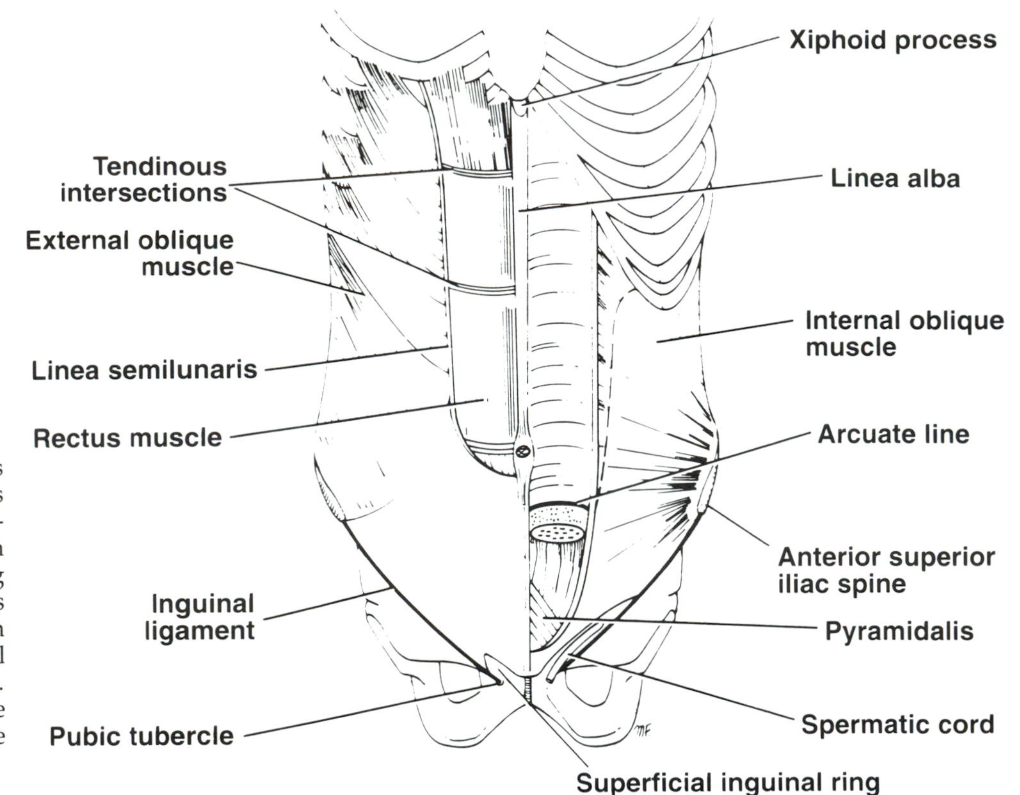

FIG 11–5.
Anterior view of the rectus abdominis muscle and rectus sheath. On the left, the anterior wall of sheath has been partly removed, revealing the rectus muscle with its tendinous intersections. On the right, the posterior wall of rectus sheath is shown. The edge of the arcuate line is shown at the level of the anterior superior iliac spine.

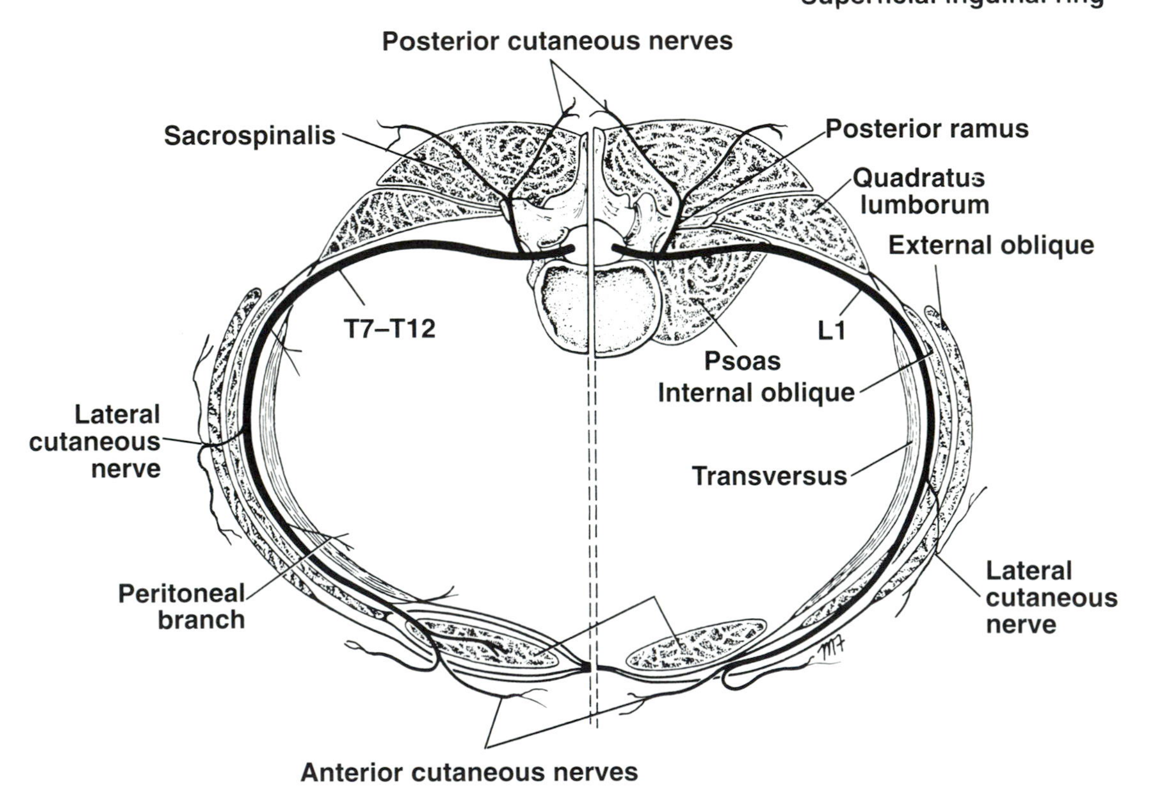

FIG 11–6.
Cross section of the abdomen showing courses of lower thoracic and first lumbar nerves.

TABLE 11–1.
Muscles of the Anterior Abdominal Wall

Muscle	Origin	Insertion	Nerve Supply	Action
External oblique	Lower eight ribs	Xiphoid process, linea alba, pubic crest, pubic tubercle, iliac crest	Lower six thoracic nerves and iliohypogastric and ilioinguinal nerves (L1)	Supports abdominal contents; compresses abdominal contents; assists in flexing and rotation of trunk; assists in forced expiration, micturition, defecation, parturition, and vomiting
Internal oblique	Lumbar fascia, iliac crest, lateral ⅔ of inguinal ligament	Lower three ribs and costal cartilages, xiphoid process, linea alba, symphysis pubis	Lower six thoracic nerves, iliohypogastric and ilioinguinal nerves (L1)	As above
Transversus	Lower six costal cartilages, lumbar fascia iliac crest, lateral third of inguinal ligament	Xiphoid process, linea alba, symphysis pubis	Lower six thoracic nerves, iliohypogastric and ilioinguinal nerves (L1)	Compresses abdominal contents
Rectus abdominis	Symphysis pubis, pubic crest	Fifth, sixth, and seventh costal cartilages and xiphoid process	Lower six thoracic nerves	Compresses abdominal contents and flexes vertebral column; accessory muscle of expiration
Pyramidalis (if present)	Anterior surface of pubis	Linea alba	Twelfth thoracic nerve	Tenses the linea alba

aponeuroses pass anterior to the rectus muscle, leaving the sheath deficient posteriorly below this level. The lower crescent-shaped edge of the posterior wall of the sheath is called the *arcuate line.* All three aponeuroses fuse with each other and with their fellows of the opposite side in the midline between the right and left recti muscles to form a fibrous band called the *linea alba.* The linea alba extends from the xiphoid process above to the pubic symphysis below (Figs 11–7 and 11–8).

Clinical Note

Tendinous intersections.—The transverse tendinous intersections divide up the rectus abdominis into distinct segments. There are usually three—one at the level of the xiphoid process, one at the level of the umbilicus, and one between these two. The tendinous intersections are firmly attached to the anterior wall of the rectus sheath, whereas the posterior wall of the sheath is not attached.

Linea Semilunaris

The linea semilunaris is the lateral edge of the rectus abdominis muscle and crosses the costal margin at the tip of the ninth costal cartilage.

Conjoint Tendon

The internal oblique has a lower free border that arches over the spermatic cord (or round ligament of the uterus) and then descends behind it to be attached to the pubic crest and the pectineal line on the pubic bone. Near their insertion, the lowest tendinous fibers are joined by similar fibers from the transversus abdominis to form the *conjoint tendon.* The conjoint tendon strengthens the medial half of the posterior wall of the inguinal canal (Fig 11–9).

Inguinal Ligament

The inguinal ligament (see Fig 11–9) connects the anterior superior iliac spine to the pubic tubercle. It is formed by the lower border of the aponeurosis of the external oblique muscle, which is folded back upon itself. From the medial end of the ligament, the *lacunar ligament* (see Fig 11–8) extends backward and upward to the pectineal line on the superior ramus of the pubis, where it becomes continuous with the *pectineal ligament* (thickening of the periosteum). The lower border of the inguinal ligament is attached to the deep fascia of the thigh, the *fascia lata.*

Fascia Transversalis

This is a thin layer of fascia that lines the transversus muscle and is continuous with a similar layer lining the diaphragm and the iliacus muscle. The fascia

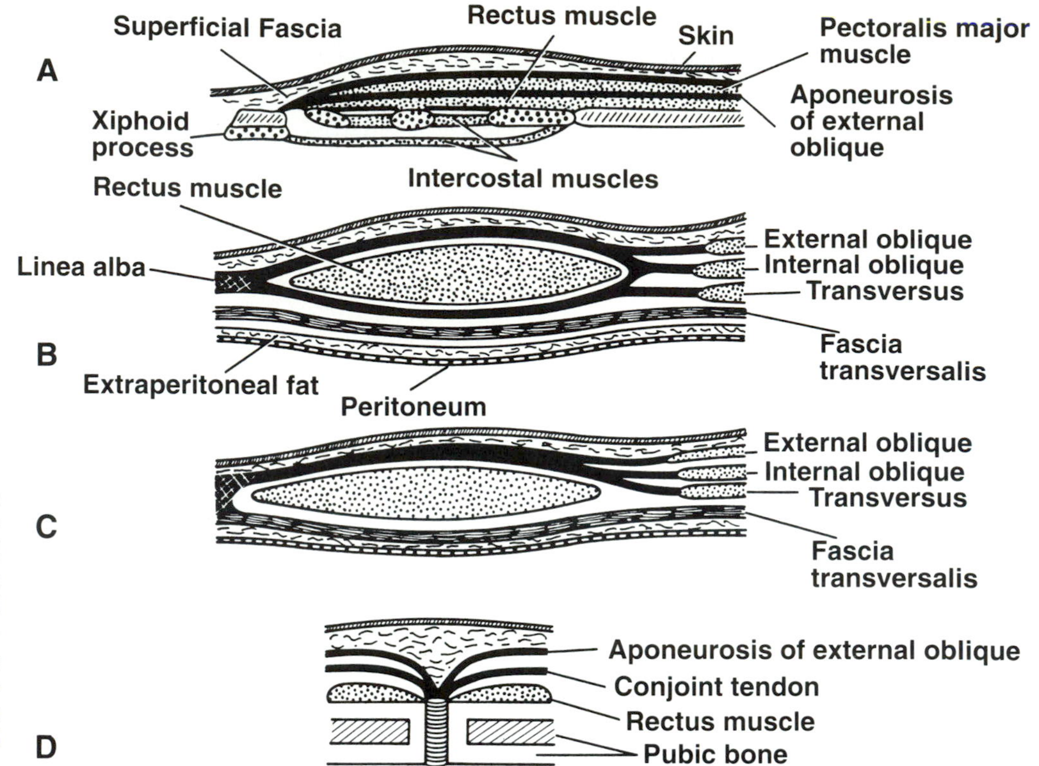

FIG 11–7.
Transverse sections of the rectus sheath seen at four levels. **A**, above the costal margin. **B**, between the costal margin and the level of the anterior superior iliac spine. **C**, below the level of the anterior superior iliac spine and above the pubis. **D**, at the level of the pubis.

transversalis, the diaphragmatic fascia, the iliacus fascia, and the pelvic fascia form one continuous lining to the abdominal and pelvic cavities.

The *femoral sheath* for the femoral vessels is formed from the fascia transversalis and the fascia iliaca that covers the iliacus muscle (see p. 448).

Extraperitoneal Fat

The fascia transversalis is separated from the peritoneum by a thin layer of connective tissue that contains a variable amount of fat (see Fig 11–7).

Parietal Peritoneum

The walls of the abdomen are lined with parietal peritoneum. This is a thin serous membrane and is continuous below with the parietal peritoneum lining the pelvis. For further details see p. 451.

Nerve Supply.—The parietal peritoneum lining the anterior abdominal wall is supplied segmentally by intercostal and lumbar nerves, which also supply the overlying muscles and skin (see Fig 11–6).

Basic Anatomy of the Inguinal Canal

The inguinal canal (Figs 11–10 and 11–11) is an oblique passage through the lower part of the anterior abdominal wall and is present in both sexes. It allows structures to pass to and from the testis to the abdomen in males. In females it permits the passage of the round ligament of the uterus from the uterus to the labium majus.

The canal is about 4 cm (1½ in.) long in adults and extends from the deep inguinal ring, a hole in the fascia transversalis, downward and medially to the superficial inguinal ring, a hole in the aponeurosis of the external oblique muscle. It lies parallel to and immediately above the inguinal ligament.

Deep Inguinal Ring

This is an oval opening in the fascia transversalis and lies about ½ in. (1.3 cm) above the inguinal ligament midway between the anterior superior iliac spine and the symphysis pubis (see Fig 11–11). Related to it medially are the inferior epigastric vessels, which pass upward from the external iliac vessels. The margins of the ring give attachment to the *internal spermatic fascia*.

Superficial Inguinal Ring

This is a triangular defect in the aponeurosis of the external oblique muscle and lies immediately above and medial to the pubic tubercle (see Fig 11–10). The margins of the ring, sometimes called the *crura*, give attachment to the *external spermatic fascia*.

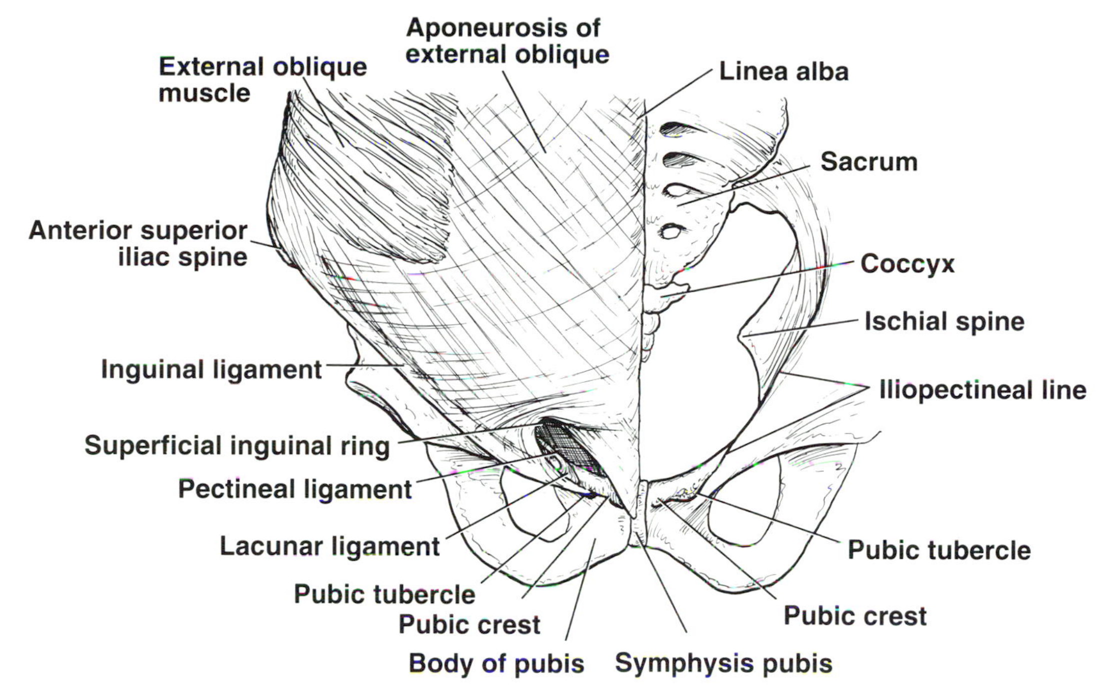

FIG 11–8.
Aponeurosis of the external oblique muscle of the abdomen; the inguinal, lacunar, and pectineal ligaments; and the superficial inguinal ring.

Walls of the Inguinal Canal

Anterior Wall.—External oblique aponeurosis, reinforced laterally by the origin of the internal oblique from the inguinal ligament (see Figs 11–10 and 11–11).

Posterior Wall.—Conjoint tendon medially, fascia transversalis laterally (see Figs 11–10 and 11–11).

Roof or Superior Wall.—Arching fibers of internal oblique and transversus (see Fig 11–10).

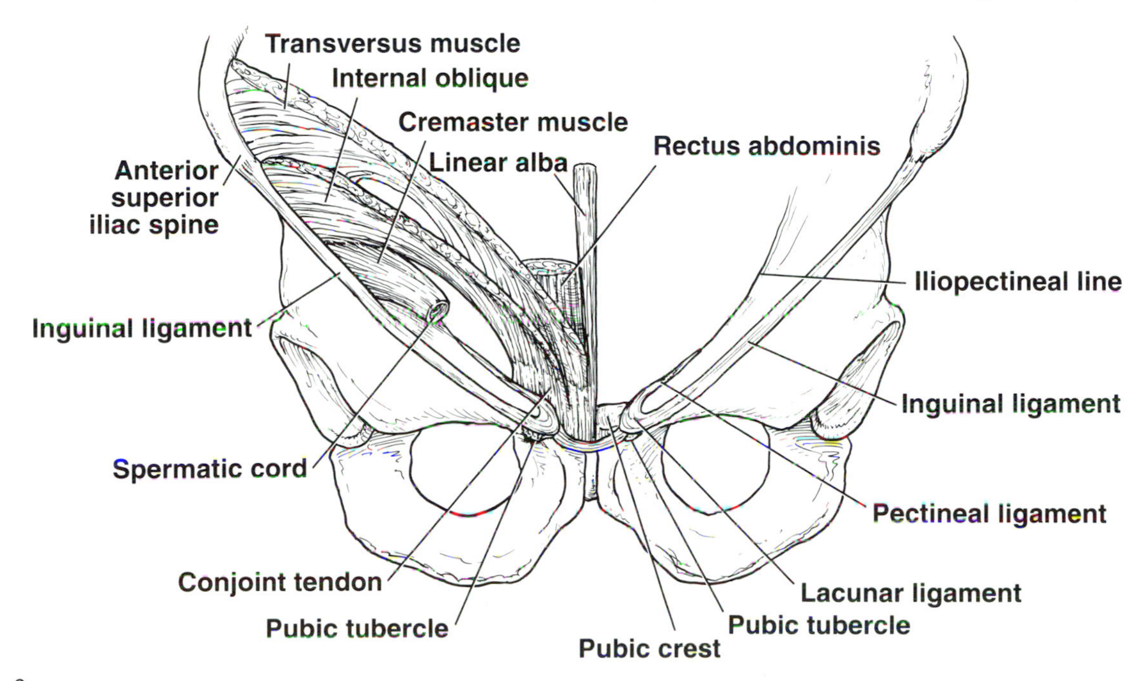

FIG 11–9.
Anterior view of the pelvis showing attachment of conjoint tendon to the pubic crest and adjoining part of the pectineal line.

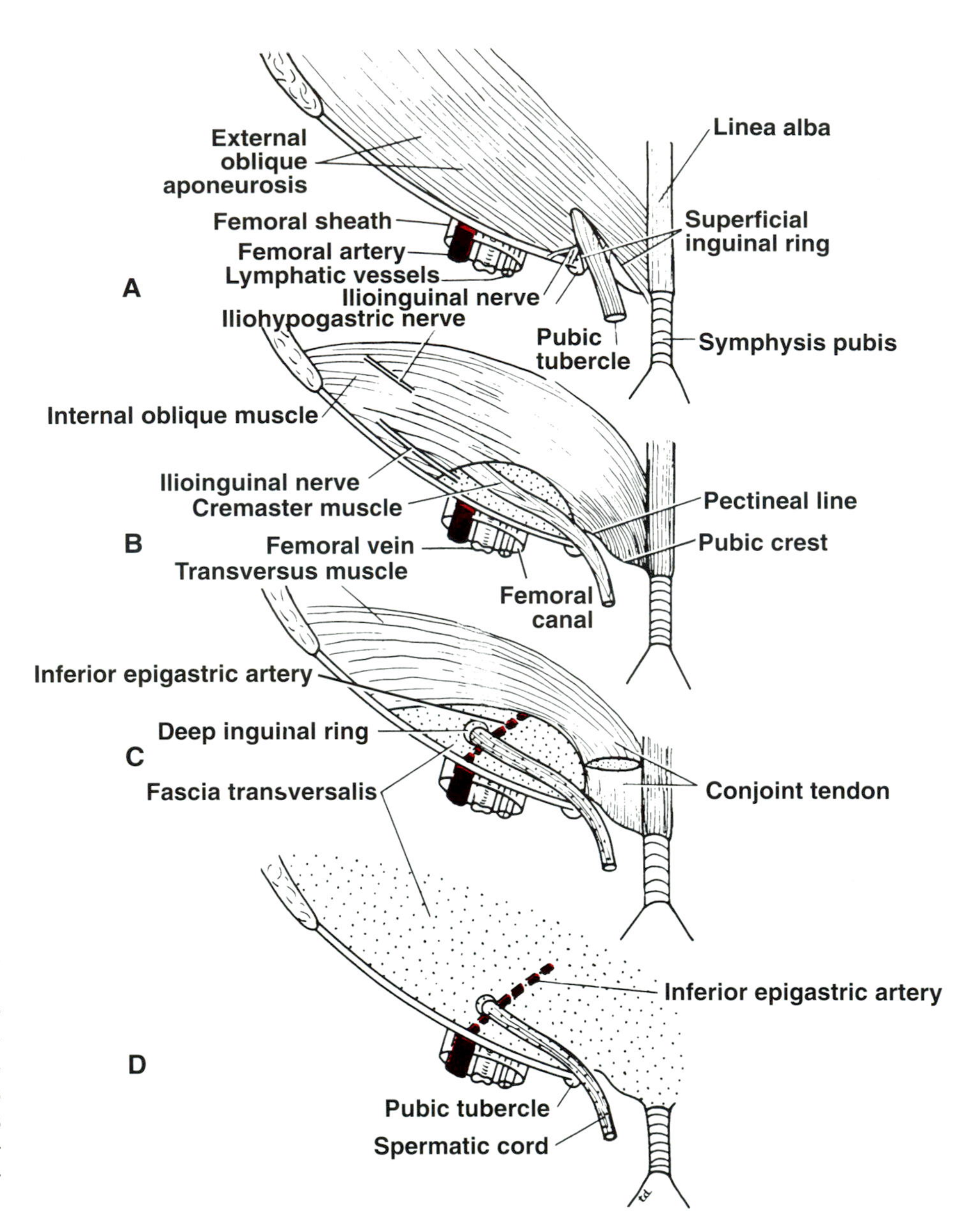

FIG 11–10.
Inguinal canal showing the following arrangement. **A,** external oblique muscle. **B,** internal oblique muscle. **C,** transversus muscle. **D,** fascia transversalis. The anterior wall of the canal is formed by the external oblique and internal oblique, and the posterior wall is formed by the fascia transversalis and conjoint tendon. The deep inguinal ring lies lateral to the inferior epigastric artery.

Floor or Inferior Wall.—Rolled-under inferior edge of the inguinal ligament and the lacunar ligament (see Fig 11–9).

Function of the Inguinal Canal
The inguinal canal allows the structures of the spermatic cord to pass to and from the testis to the abdomen in the male. (Normal spermatogenesis takes place only if the testis leaves the abdominal cavity to enter a cooler environment in the scrotum.) In females the smaller canal permits the passage of the round ligament of the uterus from the uterus to the labium majus. In both sexes the canal also transmits the ilioinguinal nerve.

Mechanics of the Inguinal Canal
The inguinal canal is a site of potential weakness in both sexes. On coughing and straining, as in micturition, defecation, and parturition, the arching lowest fibers of the internal oblique and transversus abdominis muscles contract, flattening out the arch so that the roof of the canal is lowered toward the floor and the canal is virtually closed.

Basic Anatomy of the Spermatic Cord

The spermatic cord is a collection of the following structures that pass through the inguinal canal to and from the testis.

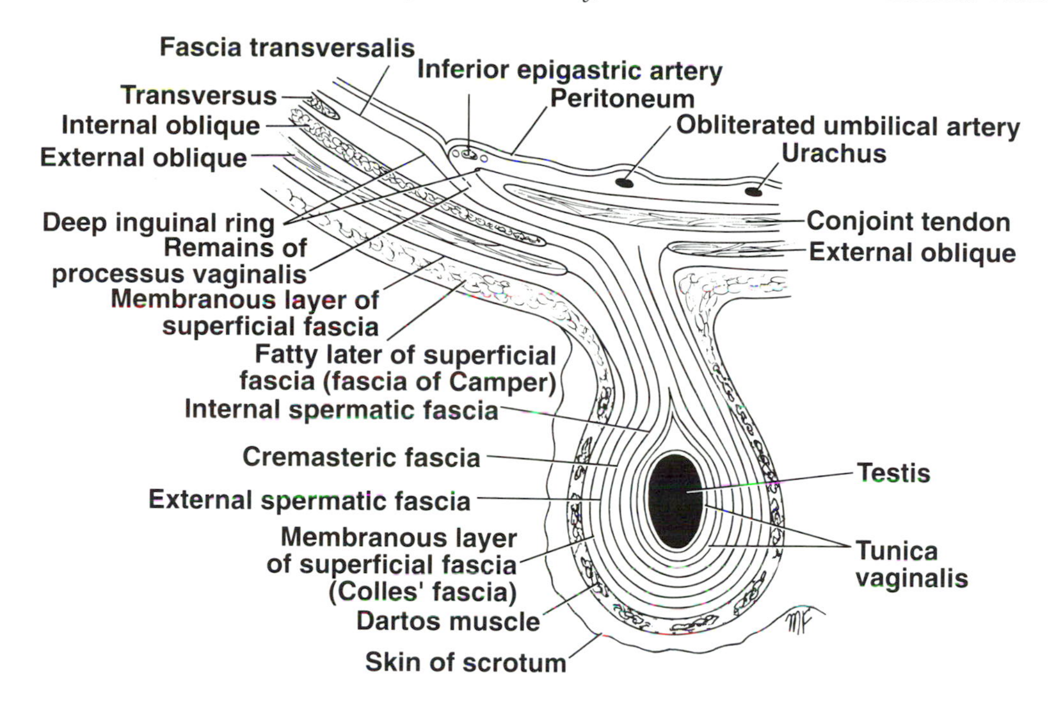

FIG 11–11.
Continuity of the different layers of the anterior abdominal wall with the coverings of spermatic cord and the wall of the scrotum; the skin and superficial fascia of the scrotum have been included, and the tunica vaginalis is also shown.

The structures include the (1) vas deferens, (2) testicular artery, (3) testicular veins (pampiniform plexus), (4) testicular lymph vessels, (5) autonomic nerves, (6) processus vaginalis (remains of), (7) cremasteric artery, (8) artery of the vas deferens, and (9) genital branch of the genitofemoral nerve that supplies the cremaster muscle.

Coverings of the Spermatic Cord

These are three concentric layers of fascia derived from the layers of the anterior abdominal wall (see Fig 11–11). During development, each covering is acquired as the fingerlike process of peritoneum, the processus vaginalis, descends into the scrotum through the layers of the abdominal wall.

The coverings are as follows.

1. *External spermatic fascia* derived from the external oblique muscle and attached to the margins of the superficial inguinal ring.
2. *Cremasteric fascia* derived from the internal oblique muscle.
3. *Internal spermatic fascia* derived from the fascia transversalis and attached to the margins of the deep inguinal ring.

Basic Anatomy of the Peritoneal Cavity

Peritoneum

The peritoneum is the serous membrane lining the abdominal and pelvic cavities and clothing the viscera (Fig 11–12). It may be regarded as a "balloon" into which organs are pressed into from the outside. The *parietal layer* lines the walls of the abdominal and pelvic cavities, and the *visceral layer* covers the organs. The potential space between the parietal and visceral layers, which is in effect the inside space of the balloon, is called the *peritoneal cavity*. In males this is a closed cavity, but in the females there is a communication with the exterior through the uterine tubes, the uterus, and the vagina.

The peritoneal cavity may be divided into two parts, the greater sac and the lesser sac (see Fig 11–12). The *greater sac* is the main compartment of the peritoneal cavity and extends from the diaphragm down into the pelvis. The *lesser sac* is smaller and lies behind the stomach. The greater and lesser sacs are in free communication with one another through the *epiploic foramen*. The peritoneum secretes a small amount of serous fluid, which lubricates the surfaces of the peritoneum and facilitates free movement between the viscera.

Intraperitoneal and Retroperitoneal Relationships

The terms intraperitoneal and retroperitoneal are used to describe the relationship of various organs to their peritoneal covering. An organ is said to be intraperitoneal when it is almost totally covered with visceral peritoneum. The stomach, jejunum, ileum, and spleen are good examples of intraperitoneal organs. Retroperitoneal organs are those that lie behind the peritoneum and are only partially covered with visceral peritoneum. The pancreas and the ascending and descending parts of the colon are examples of retroperitoneal organs. No organ, however, is actually within the peritoneal cavity. An "intraperitoneal" organ, such as the stomach, appears to be surrounded by the peritoneal cavity, but it is covered with visceral peritoneum and is attached to other organs by omenta.

Peritoneal Ligaments, Omenta, and Mesenteries

Peritoneal ligaments are two-layered folds of peritoneum that connect solid viscera to the abdominal walls. The liver, for example, is connected to the diaphragm by the *falciform ligament*, the *coronary ligament*, and the *right and left triangular ligaments* (see Fig 11–12).

Omenta are two-layered folds of peritoneum that connect the stomach to another viscus. The *greater omentum* connects the greater curvature of the stomach to the transverse colon (see Fig 11–12). It hangs down like an apron in front of the coils of the small intestine and is folded back on itself. The *lesser omentum* suspends the lesser curvature of the stomach from the fissure for the ligamentum venosum and the porta hepatis of the liver (see Fig 11–12). The *gastrosplenic omentum* (ligament) connects the stomach to the hilus of the spleen.

Mesenteries are two-layered folds of peritoneum connecting parts of the intestines to the posterior abdominal wall, for example, the *mesentery of the small intestine*, the *transverse mesocolon*, and the *sigmoid mesocolon* (see Fig 11–12).

The peritoneal ligaments, omenta, and mesenteries permit blood, lymphatic vessels, and nerves to reach the viscera.

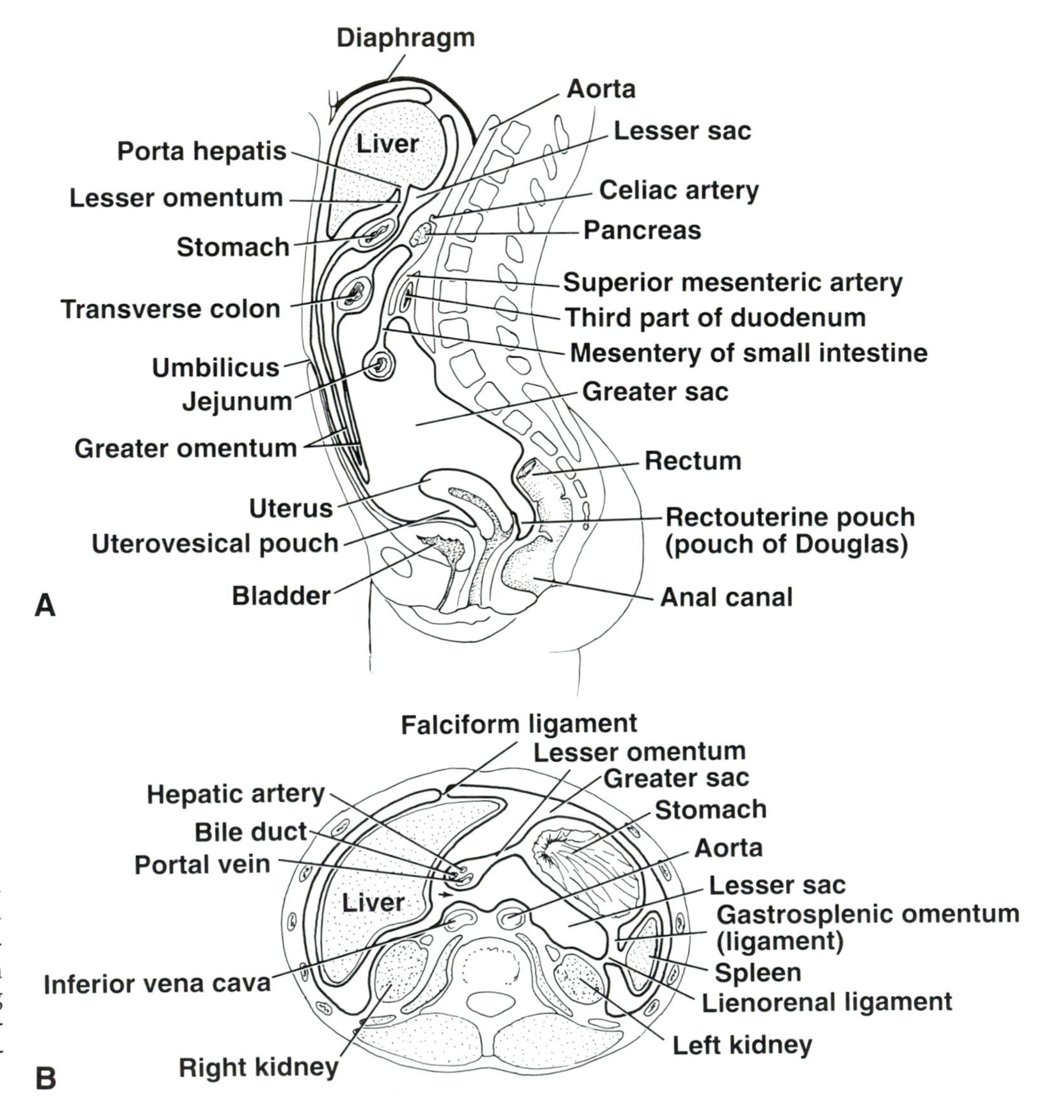

FIG 11–12.
A, sagittal section of the female abdomen showing arrangement of the peritoneum. **B**, transverse section of the abdomen showing arrangement of the peritoneum. This anatomic section is viewed from below.

Peritoneal Fossae, Spaces, and Gutters

Duodenal Fossae.— Close to the duodenojejunal junction there may be four small pouches of peritoneum called the *superior duodenal fossa, inferior duodenal fossa,* and *paraduodenal fossae.*

Cecal Fossae.— Folds of peritoneum close to the cecum produce three peritoneal fossae called the *superior ileocecal, inferior ileocecal,* and *retrocecal fossae.* These fossae are depressions in the peritoneal lining, and occasionally they form deep intraperitoneal pouches.

Clinical Note

Internal herniae.— The fossae are important clinically since a loop of small intestine may become incarcerated within one of them. This may result in intestinal obstruction and/or compromising the blood supply to the bowel.

Subphrenic Spaces.— These spaces lie between the diaphragm and the liver and are called the *right and left anterior* and *posterior subphrenic spaces.*

Clinical Note

Subphrenic Abscess— These spaces are important clinically since they may provide sites for the accumulation of pus beneath the diaphragm.

Paracolic Gutters.— These gutters lie on the lateral and medial sides of the ascending and descending colons, respectively.

Clinical Note

Spread of infection.— The gutters provide channels for the movement of infected fluid in the peritoneal cavity.

Nerve Supply of the Peritoneum

Parietal Peritoneum (For pain, temperature, touch, and pressure).— This is supplied by the lower six thoracic and first lumbar nerves. The parietal peritoneum in the pelvis is mainly supplied by the obturator nerve, a branch of the lumbar plexus.

Visceral Peritoneum.— This is sensitive only to stretch and tearing and is not sensitive to touch, pressure, or temperature. It is supplied by autonomic afferent nerves that supply the viscera or are traveling in the mesenteries.

Basic Anatomy of the Gastrointestinal Viscera

Esophagus (Abdominal Portion)

The esophagus enters the abdomen through an opening in the right crus of the diaphragm (Fig 11–13). After a short course of about 1 in. (2.5 cm), it enters the stomach on its right side. It is covered on its anterior and lateral surfaces by peritoneum.

Important Relations.— The following relations are important.

Anteriorly.—Left lobe of the liver.

Posteriorly.—Left crus of the diaphragm.

The left and right vagi lie on the anterior and posterior surfaces of the esophagus, respectively, as the anterior and posterior gastric nerves.

Gastroesophageal Sphincter

No anatomical sphincter exists at the lower end of the esophagus. However, there is no doubt that the circular layer of smooth muscle in this region serves as a physiological sphincter. Tonic contraction of this muscle prevents the stomach contents from regurgitating into the esophagus. The closure of the sphincter is under vagal control, and this can be augmented by the hormone gastrin and reduced in response to secretin, cholecystokinin, and glucagon.

Blood Supply.— This includes the following.

Arteries.—Branches from the left gastric artery (see Fig 11–16).

Veins.—These drain into the left gastric vein, a tributary of the portal vein.

Basic Anatomy of the Stomach

Location and Description

The stomach is the dilated portion of the alimentary canal and is situated in the upper part of the abdomen (Figs 11–13 and 11–14). It is roughly J shaped and has two openings, the *cardiac and pyloric orifices;* two curvatures, the *greater and lesser curvatures;* and two surfaces, an *anterior* and a *posterior surface* (Fig 11–15).

The stomach may be divided into the following parts.

Fundus.— This is dome shaped and projects upward and to the left of the cardiac orifice. It is usually full of gas.

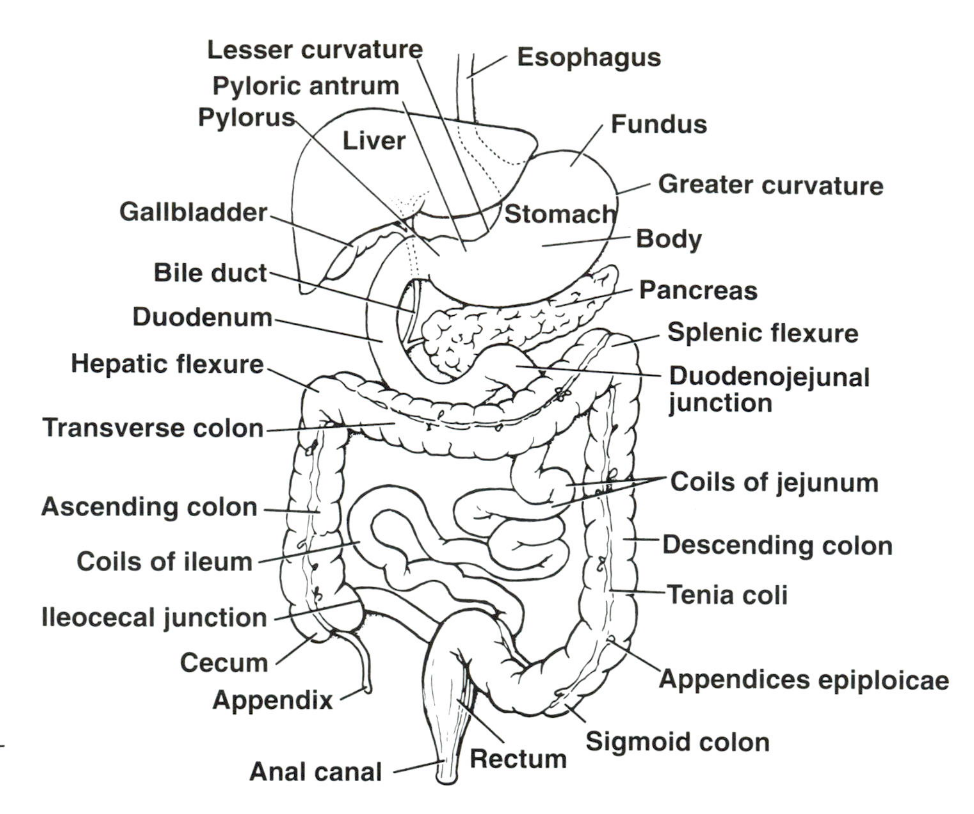

FIG 11–13.
General arrangement of the abdominal viscera.

Body.—This extends from the cardiac orifice to the *incisura angularis,* a constant notch in the lower part of the lesser curvature (see Fig 11–15).

Pyloric Antrum.—This extends from the incisura angularis to the pylorus.

Pylorus.—This is the tubular part of the stomach. It has a thick muscular wall called the *pyloric sphincter.* The cavity of the pylorus is the *pyloric canal.*

Lesser Curvature.—This forms the right border of the stomach and is connected to the liver by the lesser omentum.

Greater Curvature.—This is much longer than the lesser curvature and extends from the left of the cardiac orifice, over the dome of the fundus, and along the left border of the stomach. The gastrosplenic omentum (ligament) extends from the upper part of the greater curvature to the spleen, and the greater omentum extends from the lower part of the greater curvature to the transverse colon.

Cardiac Orifice.—This is where the esophagus enters the stomach. Although no anatomical sphincter can be demonstrated here, a physiologic mechanism exists that prevents regurgitation of the stomach contents into the esophagus.

Pyloric Orifice.—This is formed by the pyloric canal. The circular muscle coat of the stomach is much thicker here and forms the anatomic and physiologic *pyloric sphincter.*

Peritoneal Covering.—For practical purposes the stomach is completely covered with visceral peritoneum. The visceral peritoneum leaves the lesser curvature as the lesser omentum and the greater curvature as the gastrosplenic omentum and the greater omentum.

Important Relations.—These include the following (vary with filling).

Anteriorly.—Left costal margin, anterior abdominal wall, diaphragm, left pleura, base of left lung, pericardium, quadrate and left lobes of liver

Posteriorly.—Lesser sac, pancreas (body and tail), splenic artery, diaphragm, left suprarenal gland, and upper part of the left kidney, spleen, transverse mesocolon (Fig 11–16).

Blood Supply of the Gastrointestinal Tract.—Before the blood supply to the stomach is considered, it might be helpful to recognize the blood supply of the different parts of the gastrointestinal tract relative to their development. The *celiac artery,* a branch of the abdominal part of the aorta, is the artery of the foregut and supplies the gastrointestinal tract

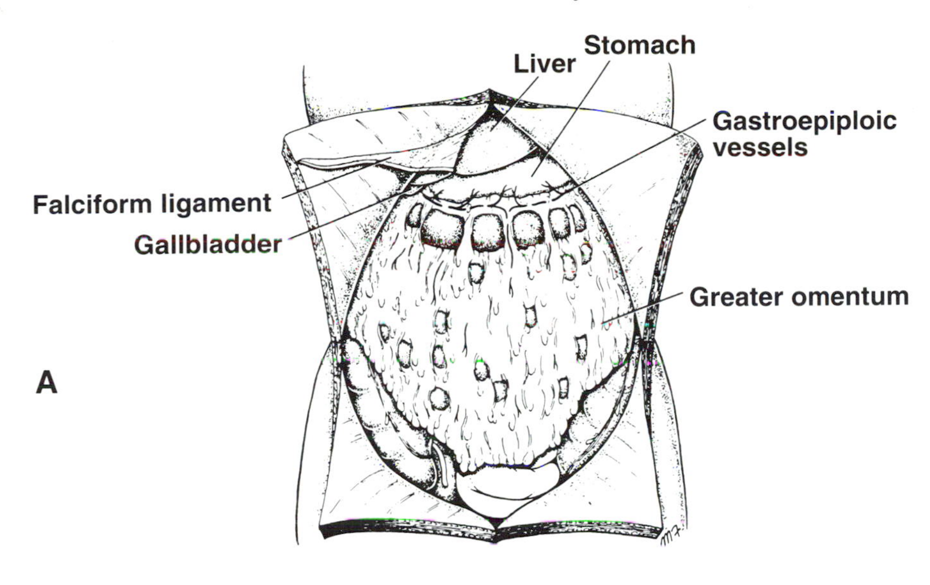

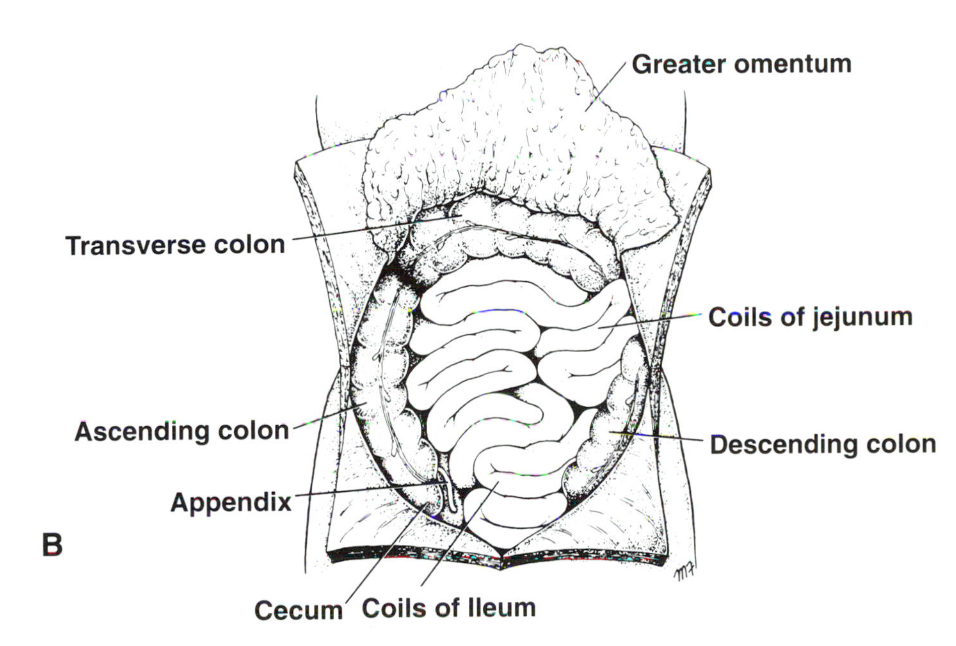

FIG 11–14.
A, abdominal organs in situ. The greater omentum hangs down in front of the small and large intestines. **B,** abdominal contents after the greater omentum has been turned upward. Coils of small intestine occupy the central part of the abdominal cavity, while ascending, transverse, and descending parts of the colon are located at the periphery.

from the lower third of the esophagus down as far as the middle of the second part of the duodenum. The *superior mesenteric artery,* a branch of the abdominal aorta, is the artery of the midgut and supplies the gastrointestinal tract from the middle of the second part of the duodenum as far as the distal third of the transverse colon. The *inferior mesenteric artery,* also a branch of the abdominal aorta, is the artery of the hindgut and supplies the large intestine from the distal third of the transverse colon to halfway down the anal canal.

Blood Supply of the Stomach.—This includes the following.

Arteries.—The right and left gastric arteries supply the lesser curvature. The right and left gastroepiploic arteries supply the greater curvature. The short gastric arteries, from the splenic, supply the fundus (Fig 11–17).

Veins.—These drain into the portal circulation. The right and left gastric veins drain into the portal vein, the short gastric and left gastroepiploic veins drain into the splenic vein, and the right gastroepiploic vein drains into the superior mesenteric vein.

Lymphatic Drainage.—The lymph vessels follow the arteries into the left and right gastric nodes, the left and right gastroepiploic nodes, and the short

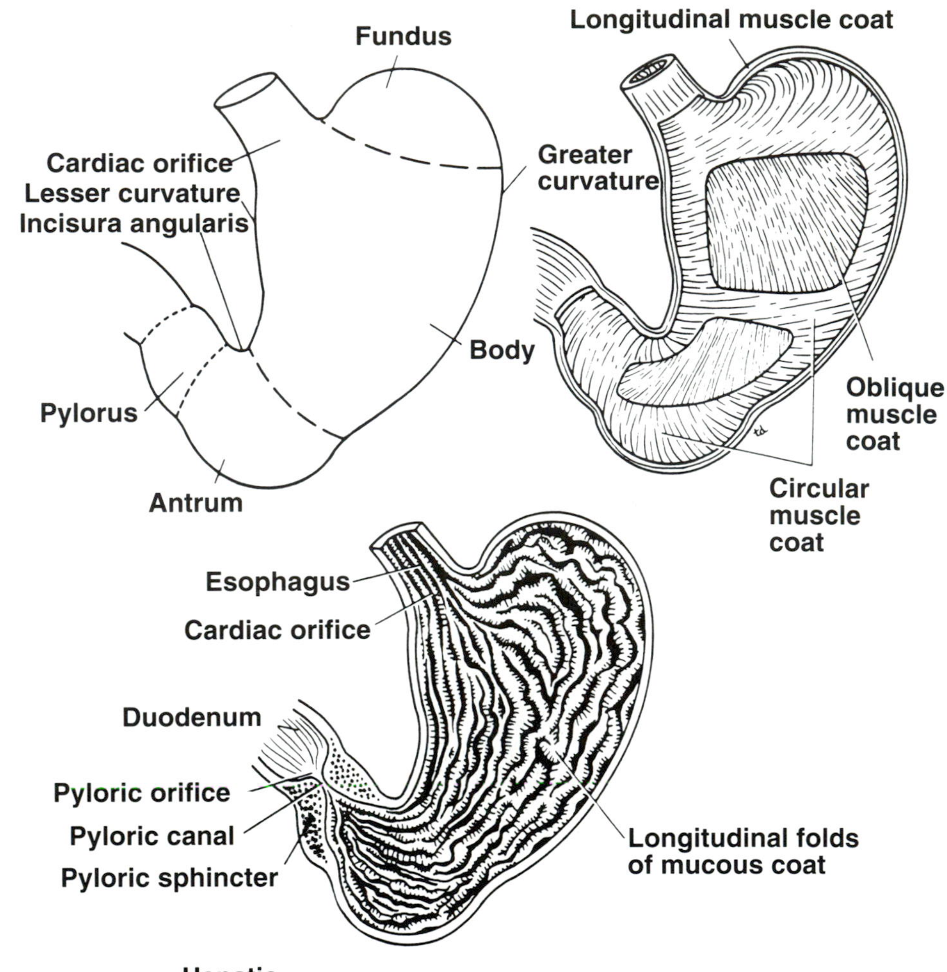

FIG 11–15.
Stomach showing different parts, muscular coats, and mucosal lining. Note the increased thickness of the circular muscle forming the pyloric sphincter.

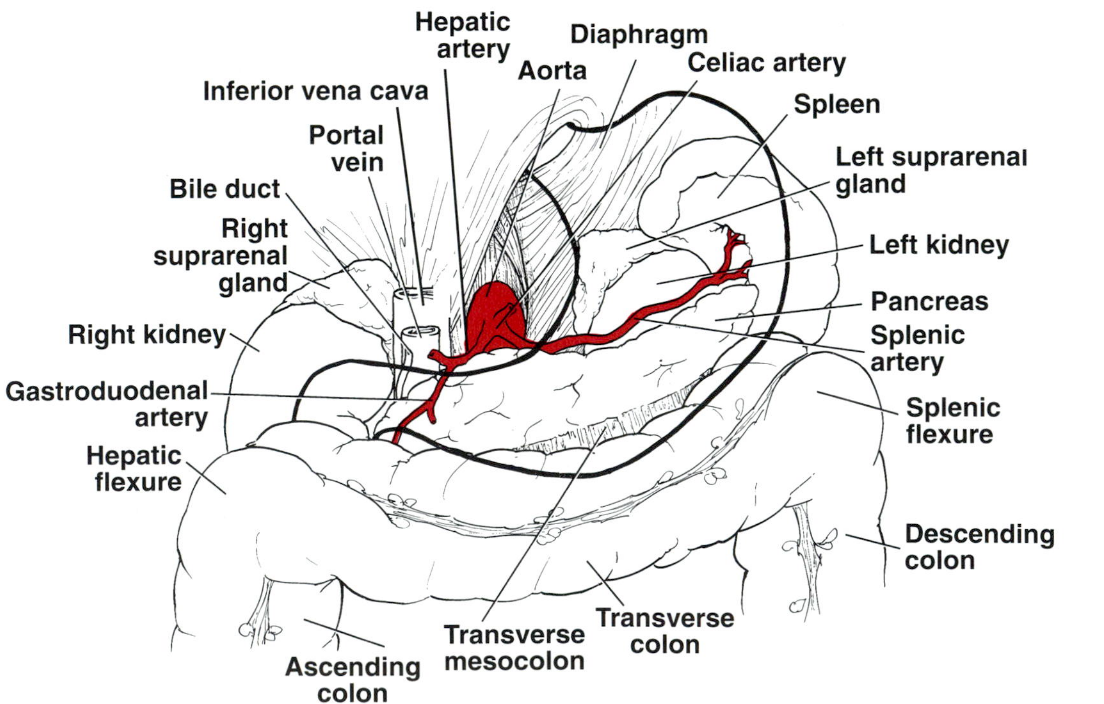

FIG 11–16.
Structures situated on the posterior abdominal wall behind the stomach.

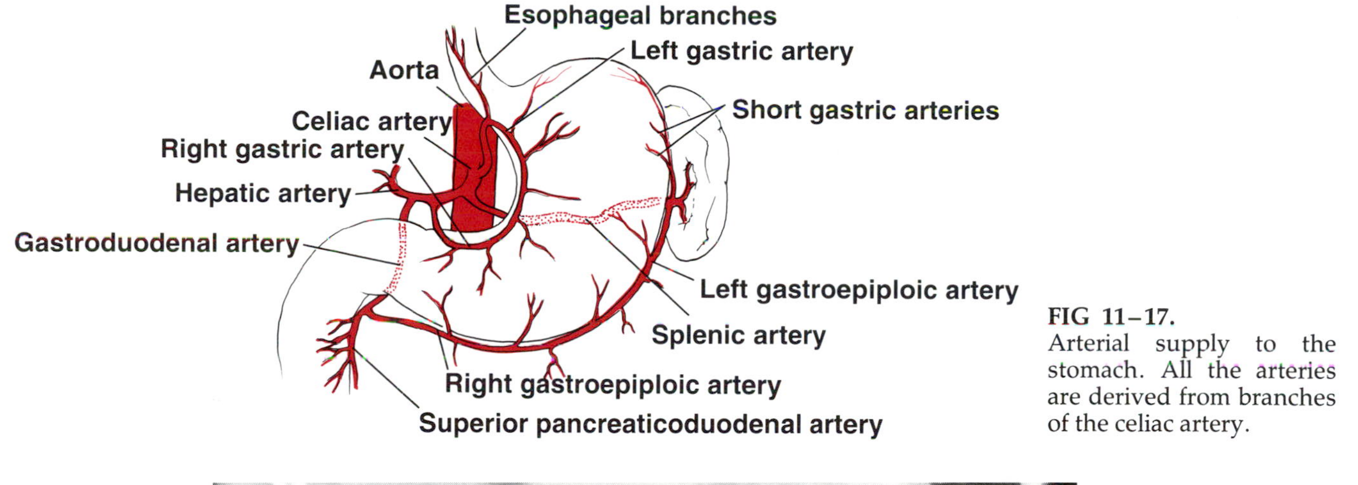

FIG 11–17.
Arterial supply to the stomach. All the arteries are derived from branches of the celiac artery.

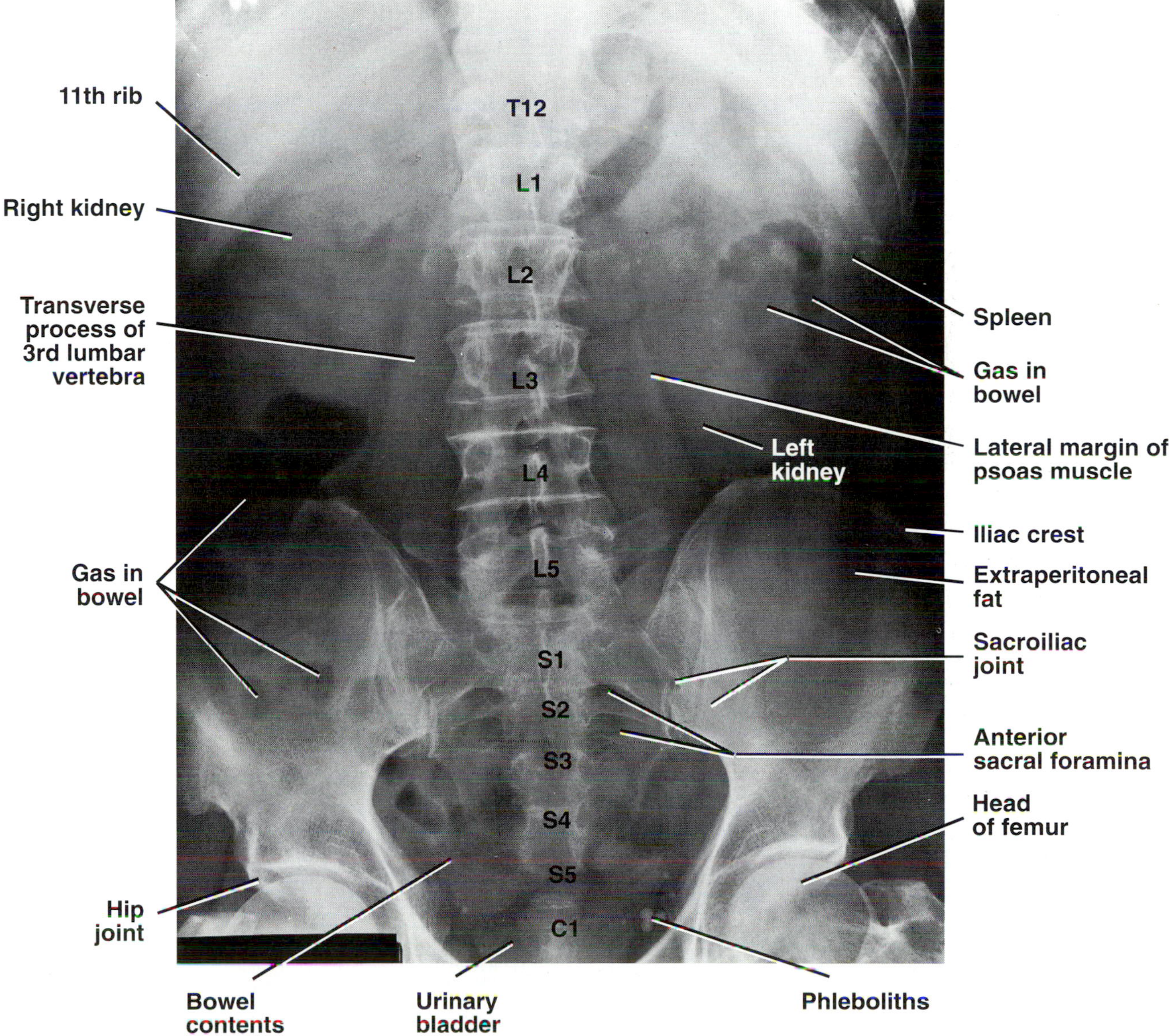

FIG 11–18.
Anteroposterior radiograph of the abdomen.

gastric nodes. All lymph from the stomach eventually passes to the celiac nodes located around the root of the celiac artery on the posterior abdominal wall.

Nerve Supply.—This includes sympathetic fibers from the celiac plexus and parasympathetic fibers from the vagus nerves.

Radiological Appearances

The normal radiological appearances of the stomach are shown in Figures 11–18 and 11–19.

Basic Anatomy of the Duodenum

The duodenum is a C-shaped tube about 10 in. (25 cm) long that curves around the head of the pancreas (Fig 11–20). It begins opposite the right side of the first lumbar vertebra at the pyloric sphincter of the stomach; it ends at the left of the second lumbar vertebra by becoming continuous with the jejunum at the duodenojejunal flexure. Although the first inch of the duodenum is covered on its anterior and posterior surfaces with peritoneum, and it has the lesser omentum attached to its upper border and the greater omentum attached to its lower border, the

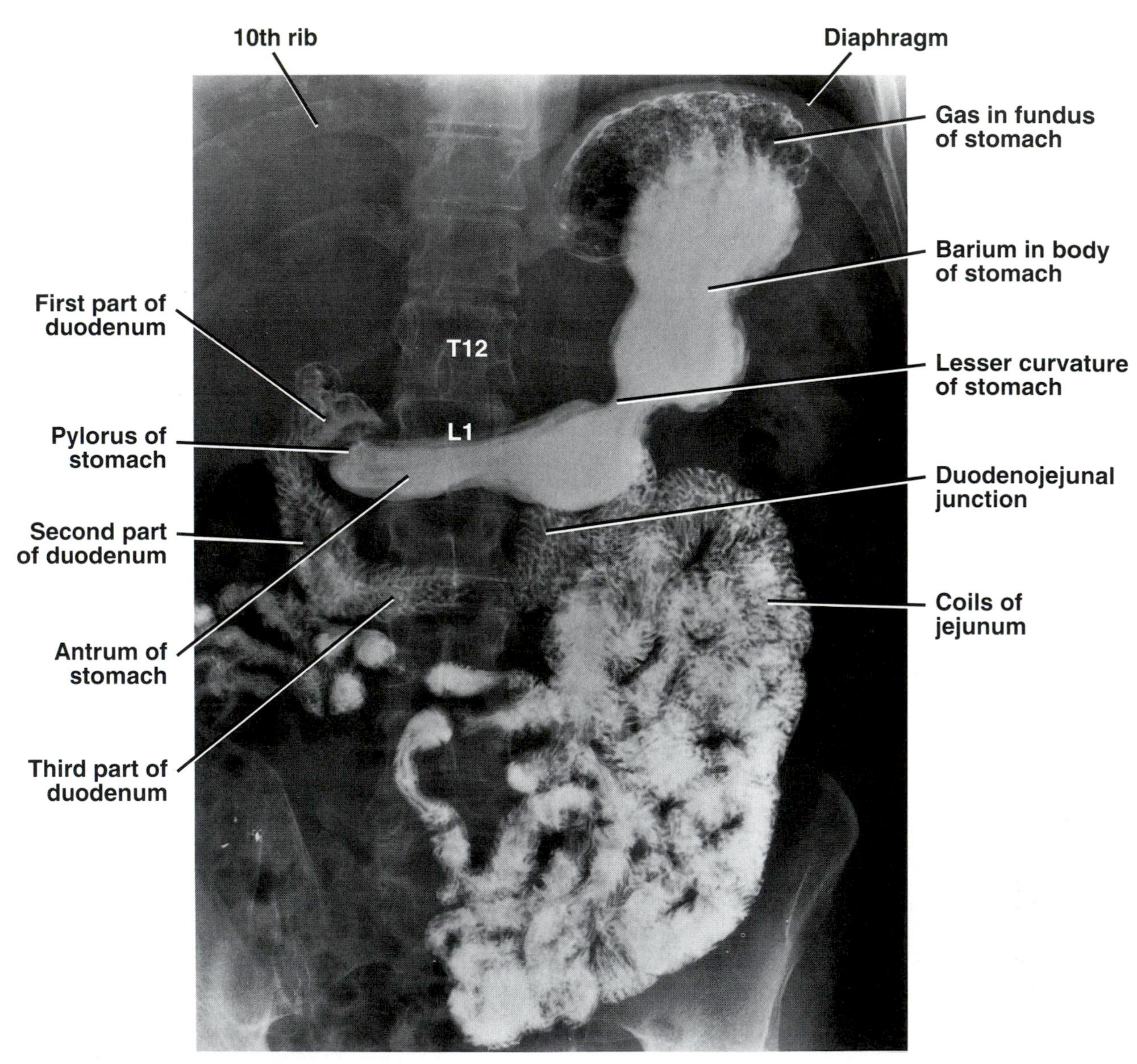

FIG 11–19.
Anteroposterior radiograph of the stomach and small intestine following ingestion of a barium meal.

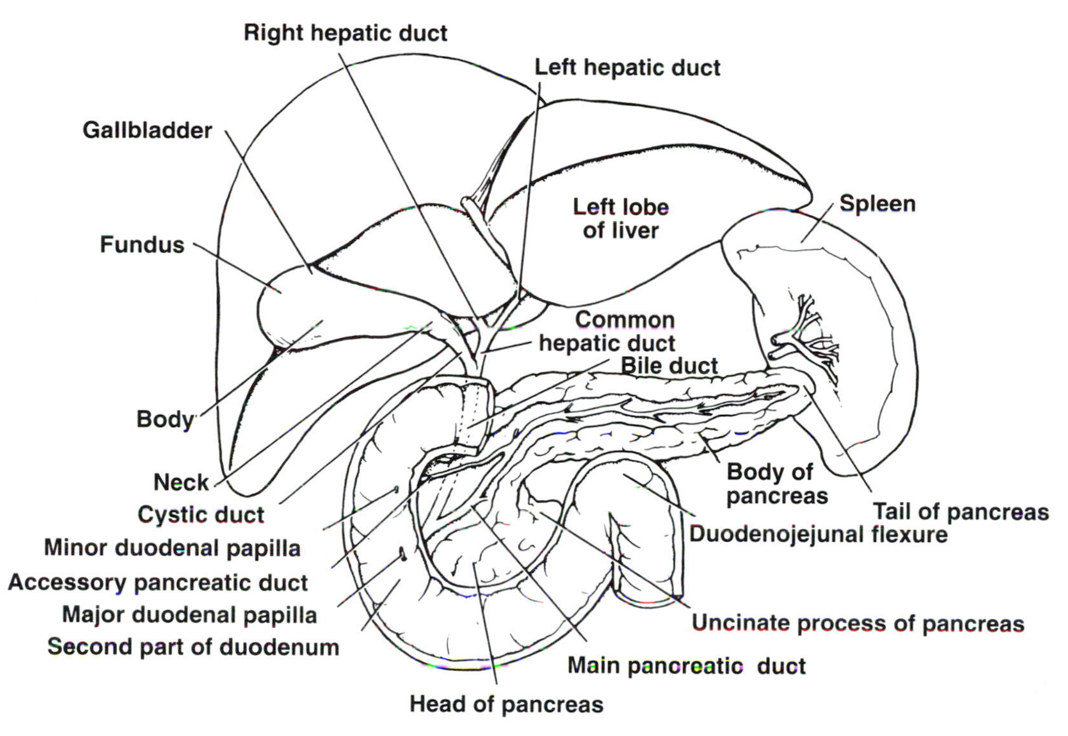

FIG 11–20.
Liver, biliary ducts, pancreas, and spleen; the anterior wall of the duodenum has been removed to reveal the openings of the duodenal papillae, and the pancreas has been dissected to show the pancreatic ducts.

remainder of the duodenum is retroperitoneal.

The duodenum is divided into the following four parts.

First Part

This runs upward and backward on the transpyloric plane at the level of the first lumbar vertebra.

Second Part

This runs vertically downward, and the bile and main pancreatic ducts pierce the medial wall about halfway down. They unite to form the ampulla that opens on the summit of a *major duodenal papilla* (see Fig 11–20). The accessory pancreatic duct, if present, opens into the duodenum on a *minor duodenal papilla* about ¾ in. (1.9 cm) above the major duodenal papilla.

Third Part

This passes horizontally in front of the vertebral column. It is crossed anteriorly by the root of the mesentery of the small intestine and the superior mesenteric vessels.

Fourth Part

This runs upward and to the left to the *duodenojejunal junction.* The junction is held in position by the *ligament of Treitz,* which is attached to the right crus of the diaphragm.

Important Relations.—These relations are as follows.

First Part.—Anteriorly: quadrate lobe of the liver, gallbladder. Posteriorly: lesser sac (first inch only), gastroduodenal artery, bile duct, portal vein, inferior vena cava.

Second Part.—Anteriorly: fundus of the gallbladder, right lobe of the liver, transverse colon, coils of the small intestine. Posteriorly: hilus of the right kidney. Medially: head of the pancreas, bile duct, and pancreatic ducts.

Third Part.—Anteriorly: root of the mesentery of the small intestine, superior mesenteric vessels, coils of the jejunum. Posteriorly: right ureter, inferior vena cava, aorta. Superiorly: head of the pancreas.

Fourth Part.—Anteriorly: beginning of the root of the mesentery and coils of the jejunum. Posteriorly: left margin of the aorta.

Blood Supply.—This includes the following.

Arteries.—The upper half is supplied by the superior pancreaticoduodenal artery, a branch of the gastroduodenal artery. The lower half is supplied by the inferior pancreaticoduodenal artery, a branch of the superior mesenteric artery.

Veins.—The superior pancreaticoduodenal vein drains into the portal vein; the inferior vein joins the superior mesenteric vein.

Lymphatic Drainage.—The lymph vessels drain upward via pancreaticoduodenal nodes to gastroduodenal nodes and celiac nodes, and drain downward via pancreaticoduodenal nodes to the superior mesenteric nodes.

Nerve Supply.—This includes the sympathetic and vagus nerves via the celiac and superior mesenteric plexuses.

Radiological Appearances

The normal radiological appearances of the duodenum are shown in Figure 11–19. A barium meal passes into the first part of the duodenum and forms a triangular homogeneous shadow, the *duodenal cap,* which has its base toward the pylorus (see Fig 11–19). The barium quickly leaves the duodenal cap and passes rapidly through the remaining portions of the duodenum. The outline of the barium shadow in the first part of the duodenum is smooth because of the absence of mucosal folds. In the remainder of the duodenum, the presence of plicae circulares breaks up the barium emulsion, giving it a floccular appearance.

Basic Anatomy of the Jejunum and Ileum

Location and Description

The small bowel extends from the duodenojejunal flexure to the cecum and is freely mobile (see Fig 11–13). The jejunum measures about 8 ft (2.5 m) long, and the ileum measures about 12 ft (3.6 m) long. The jejunum begins in the upper part of the abdominal cavity to the left of the midline. It is wider in diameter, thicker walled, and redder than the ileum. The coils of ileum occupy the lower right part of the abdominal cavity and tend to hang down into the pelvis. The ileum ends at the ileocecal junction. The coils of the jejunum and the ileum are suspended from the posterior abdominal wall by a fan-shaped fold of peritoneum called the *mesentery of the small intestine.*

Important Relations.—These are as follows.

Anteriorly.—Anterior abdominal wall and greater omentum, which usually covers over the coils.

Posteriorly.—Posterior abdominal wall and retroperitoneal structures.

Blood Supply.—This includes the following.

Arteries.—Branches of the superior mesenteric artery (Fig 11–21) that anastomose with one another to form arcades.

Veins.—Drain into the superior mesenteric vein.

Lymphatic Drainage.—The lymph passes to the superior mesenteric nodes via a large number of intermediate mesenteric nodes.

Nerve Supply.—Sympathetic and vagus nerve fibers from the superior mesenteric plexus.

Radiological Appearances

Radiological appearances of the jejunum and ileum are shown in Figure 11–19. On entering the jejunum and ileum, the barium shadow is scattered by the mucosal folds and the peristaltic activity. In the last part of the ileum, the barium meal tends to form a continuous mass of barium.

Basic Anatomy of the Large Intestine

The large intestine extends from the ileum to the anus. It is divided into the cecum, the vermiform appendix, the ascending colon, the transverse colon, the descending colon, the sigmoid colon, the rectum, and the anal canal. The rectum and anal canal will be considered in Chapter 13.

External Differences Between the Small and Large Intestines

This information may be helpful in patients with prolapse of the gut through an abdominal wound.

1. The small intestine is more mobile (exception—the duodenum), whereas the ascending and descending parts of the colon are fixed.
2. The small intestine has a mesentery (except the duodenum), whereas the large intestine is retroperitoneal (except the transverse colon and sigmoid colon).
3. The diameter of a full small intestine is smaller than that of a full large intestine.
4. The longitudinal muscle of the small intestine forms a continuous layer around the gut, whereas in the large intestine (with the exception of the appendix, rectum, and anal canal) the longitudinal muscle forms three visible bands, the *teniae coli.*
5. The small intestine has no fatty tags attached

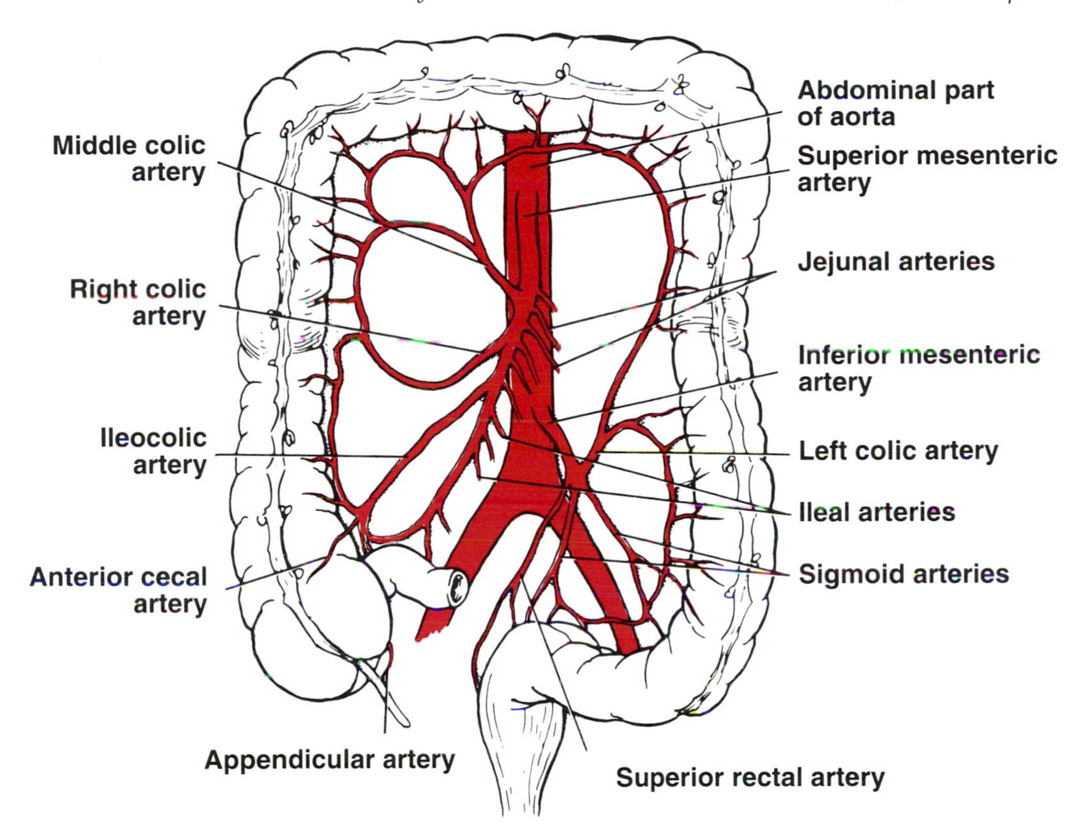

FIG 11–21.
The arterial supply to the small and large intestines. Note the branches of the superior and inferior mesenteric arteries.

to its wall, whereas the large intestine has the *appendices epiploicae.*

6. The wall of the small intestine is smooth, whereas that of the large intestine is sacculated.

Basic Anatomy of the Cecum and Appendix

Cecum

The cecum is a blind-ended pouch that lies within the right iliac fossa and is completely covered with peritoneum (see Fig 11–13). At the junction of the cecum with the ascending colon, it is joined on the left side by the terminal part of the ileum. Attached to its posteromedial surface is the appendix.

Important Relations.— These relations are as follows.

Anteriorly.—Anterior abdominal wall in the right iliac region, coils of small intestine.

Posteriorly.—Iliopsoas muscle.

Blood Supply.— This involves the following.

Arteries.—Anterior and posterior cecal arteries from the ileocolic artery (see Fig 11–21), a branch of the superior mesenteric artery.

Veins.—Drain into the superior mesenteric vein.

Lymphatic Drainage.— Mesenteric nodes and superior mesenteric nodes.

Nerve Plexuses.— Sympathetic and vagus nerves via the superior mesenteric plexus.

Ileocecal Valve

A rudimentary structure, this valve consists of two horizontal folds of mucous membrane that project around the orifice of the ileum. The valve plays little or no part in the prevention of reflux of cecal contents into the ileum. The circular muscle of the lower end of the ileum (called the *ileocecal sphincter* by physiologists) serves as a sphincter and controls the flow of contents from the ileum into the colon. The smooth-muscle tone is reflexly increased when the cecum is distended; the hormone gastrin, which is produced by the stomach, causes relaxation of the muscle tone.

Radiological Appearances

Radiological appearances of the cecum are shown in Figures 11–23 and 11–24.

Appendix

The appendix (see Fig 11–13) is a narrow, muscular tube containing a large amount of lymphoid tissue in its wall. It is attached to the posteromedial surface of the cecum about 1 in. (2.5 cm) below the ileocecal junction. It has a complete peritoneal covering, which is attached to the mesentery of the small intestine by a short mesentery of its own, the *mesoappendix.* The mesoappendix contains the appendicular vessels and nerves.

The appendix lies in the right iliac region, and in relation to the anterior abdominal wall its base is situated one third of the way up the line joining the right anterior superior iliac spine to the umbilicus (McBurney's point). Inside the abdomen the base of the appendix can be easily located by tracing the teniae coli of the cecum and following them to the base of the appendix, where they converge to form a continuous longitudinal muscle coat.

Common Positions of Tip of the Appendix.—The tip of the appendix is subject to a considerible range of movement and may be found in the following positions: (1) hanging down into the pelvis against the right pelvic wall, (2) coiled up behind the cecum, (3) projecting upward along the lateral side of the cecum, or (4) in front of or behind the terminal part of the ileum. The first and second positions are the commonest sites.

Blood Supply.—This involves the following.

Arteries.—Appendicular artery, a branch of the posterior cecal artery (see Fig 11–21).

Veins.—Appendicular vein drains into the posterior cecal vein.

Lymphatic Drainage.—One or two nodes in the mesoappendix and then eventually into the superior mesenteric lymph nodes.

Nerve Supply.—Sympathetic and vagus nerves from the superior mesenteric plexus.

Basic Anatomy of the Colon

Ascending Colon

The ascending colon is about 5 in. (13 cm) long and extends upward from the cecum to the inferior surface of the right lobe of the liver (see Fig 11–13). Here it turns to the left, forming the *right colic flexure,* and becomes continuous with the transverse colon. The peritoneum covers the front and sides of the ascending colon, binding it to the posterior abdominal wall.

Important Relations.—These relations are as follows. **Anteriorly:** coils of the small intestine, greater omentum, anterior abdominal wall. **Posteriorly:** iliacus muscle, iliac crest, quadratus lumborum, lower pole of right kidney (Fig 11–22).

Blood Supply.—This involves the following.

Arteries.—Ileocolic and right colic branches of the superior mesenteric artery (Fig 11–21).

Veins.—Drain into the superior mesenteric vein.

Lymphatic Drainage.—Colic lymph nodes and superior mesenteric nodes.

Nerve Supply.—Sympathetic and vagus nerves from the superior mesenteric plexus.

Transverse Colon

The transverse colon is about 15 in. (38 cm) long and passes across the abdomen, occupying the umbilical and hypogastric regions (see Fig 11–13). It begins at the right colic flexure below the right lobe of the liver and hangs downward, suspended by the transverse mesocolon from the pancreas. It then ascends to the *left colic flexure* below the spleen. The left colic flexure is higher than the right colic flexure and is held up by the *phrenicocolic ligament.* The *transverse mesocolon,* or mesentery of the transverse colon, is attached to the superior border of the transverse colon, and the posterior layers of the greater omentum are attached to the inferior border.

Important Relations.—These relations are as follows. **Anteriorly:** greater omentum, anterior abdominal wall (umbilical and hypogastric regions). **Posteriorly:** duodenum (second part), head of the pancreas, coils of the small intestine.

Blood Supply.—This involves the following.

Arteries.—The proximal two thirds is supplied by the middle colic artery (see Fig 11–21), a branch of the superior mesenteric artery. The distal third is supplied by the left colic artery, a branch of the inferior mesenteric artery.

Veins.—Drain into the superior and inferior mesenteric veins.

Lymphatic Drainage.—The proximal two thirds drain into the colic nodes and then into the superior

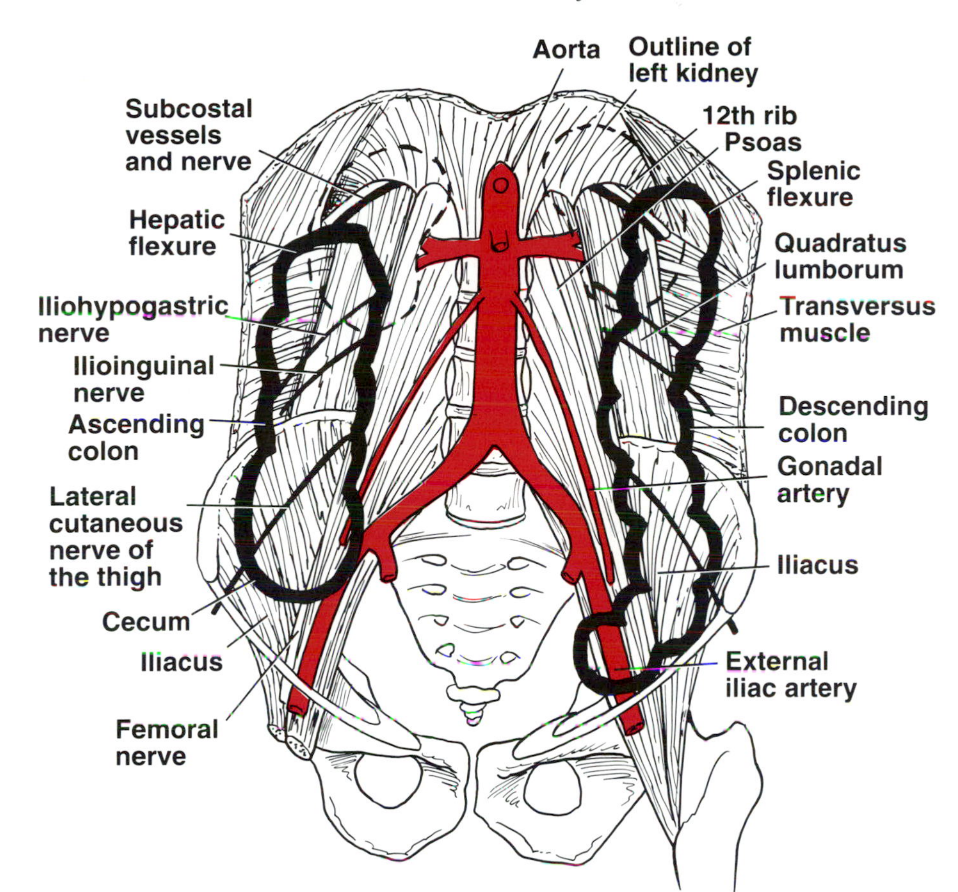

FIG 11–22.
Posterior abdominal wall showing posterior relations of the kidneys and colon.

mesenteric nodes; the distal third drains into the colic nodes and then the inferior mesenteric nodes.

Nerve Supply.—The proximal two thirds is innervated by sympathetic and vagal nerves through the superior mesenteric plexus; the distal third is innervated by sympathetic and parasympathetic pelvic splanchnic nerves through the inferior mesenteric plexus.

Descending Colon

The descending colon is about 10 in. (25 cm) long and extends downward from the left colic flexure to the pelvic brim, where it becomes continuous with the sigmoid colon (see Fig 11–13). The peritoneum covers the front and sides and binds it to the posterior abdominal wall.

Important Relations.—These relations are as follows. **Anteriorly:** coils of the small intestine, greater omentum, anterior abdominal wall. **Posteriorly:** left kidney, quadratus lumborum muscle, iliac crest, iliacus muscle (see Fig 11–22).

Blood Supply.—This involves the following.

Arteries.—Left colic branch and sigmoid branches of the inferior mesenteric artery (see Fig 11–21).

Veins.—Inferior mesenteric vein.

Lymphatic Drainage.—Colic nodes and inferior mesenteric nodes.

Nerve Supply.—Sympathetic and parasympathetic pelvic splanchnic nerves through the inferior mesenteric plexus.

Sigmoid Colon

The sigmoid colon is about 10 to 15 in. (25 to 38 cm) long and begins as a continuation of the descending colon in front of the pelvic brim (see Fig 11–13). Below, it becomes continuous with the rectum in front of the third sacral vertebra. The sigmoid colon is attached to the posterior pelvic wall by the fan-shaped *sigmoid mesocolon.* Since the colon is suspended by the sigmoid mesocolon, it is mobile and hangs down into the pelvic cavity in the form of a loop; its curves vary, but it usually curves to the right of the midline before joining the rectum.

Important Relations.— These relations are as follows. **Anteriorly:** in males, urinary bladder; in females, uterus and upper part of the vagina. **Posteriorly:** rectum and sacrum.

Blood Supply.— This involves the following.

Arteries.—Sigmoid branches of the inferior mesenteric artery (see Fig 11–21).

Veins.—Inferior mesenteric vein.

Lymphatic Drainage.— Colic nodes and inferior mesenteric nodes.

Nerve Supply.— Sympathetic and parasympathetic pelvic splanchnic nerves through the inferior hypogastric plexuses.

Radiological Appearances

Radiological appearances of the large intestine are shown in Figures 11–23 and 11–24. Following a bar-

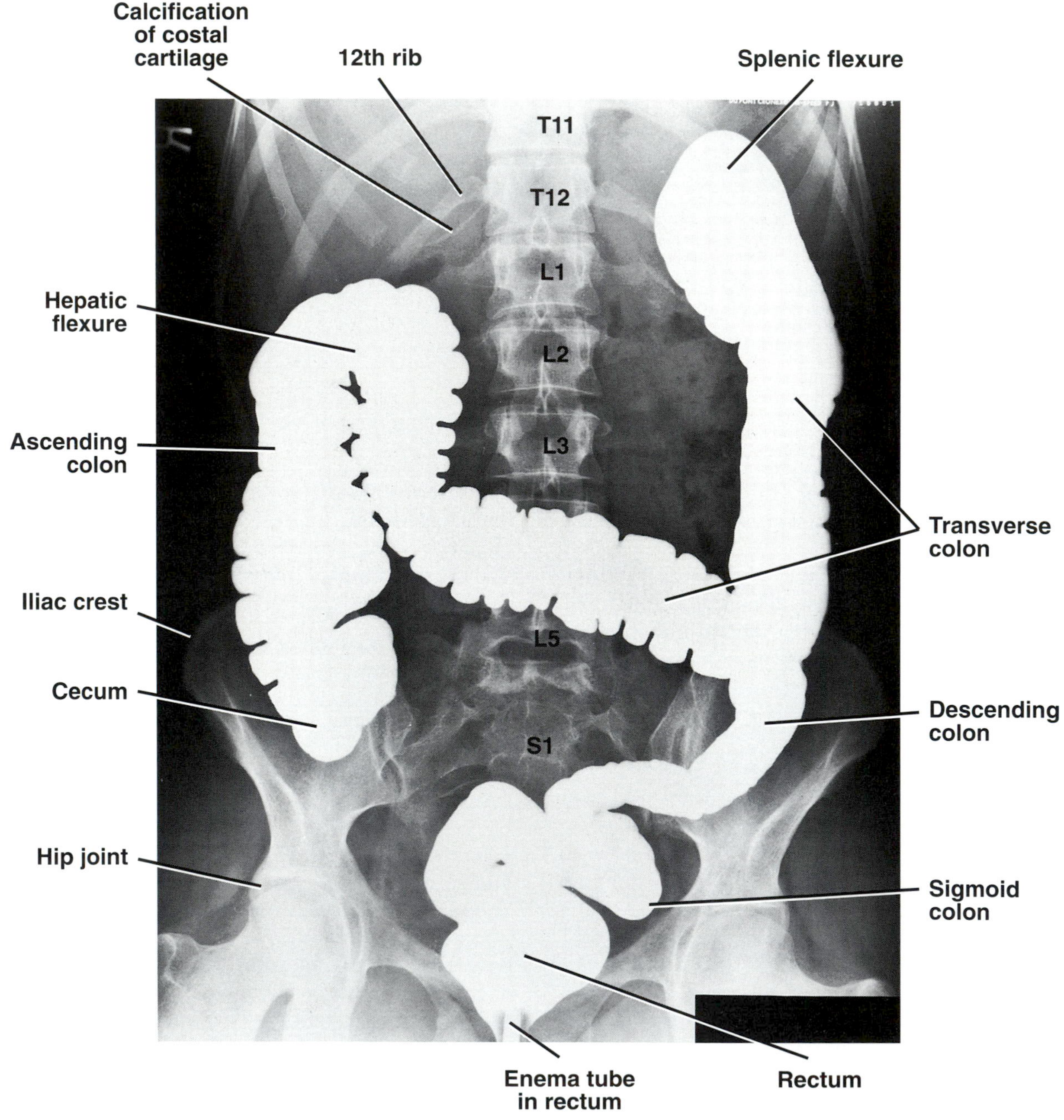

FIG 11–23.
Radiograph of the large intestine following a barium enema examination.

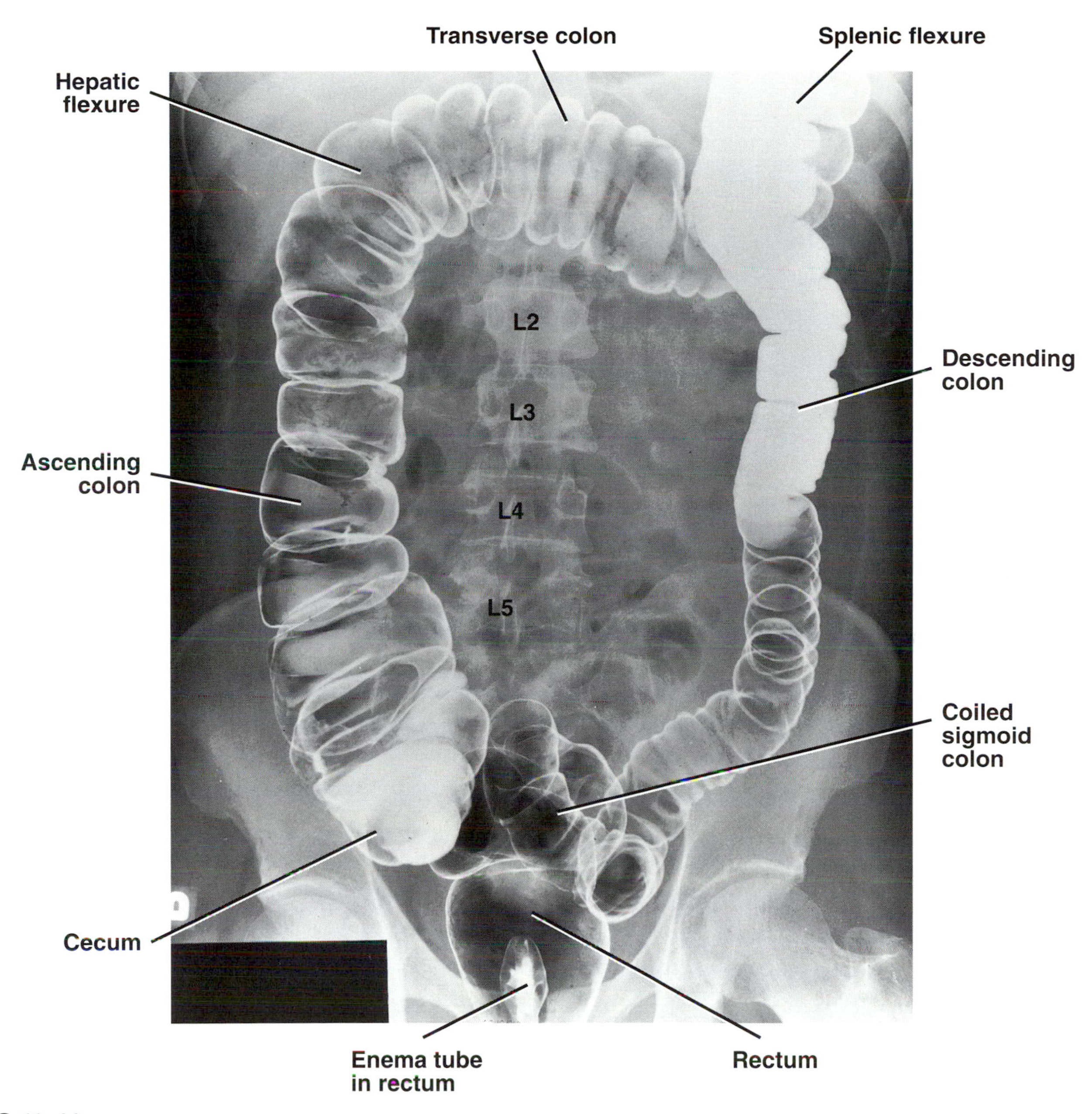

FIG 11–24.
Radiograph of the large intestine following a barium enema examination. Air has been introduced into the intestine through the enema tube after evacuation of most of the barium (contrast enema).

ium enema, the entire outline may be seen in an anteroposterior projection. Oblique and lateral views of the colic flexures may be necessary. The characteristic sacculations are well seen when the bowel is filled, and, after the enema has been evacuated, the mucosal pattern is clearly demonstrated (see Fig 11–24).

The appendix frequently fills with barium after an enema. The appearances of the sigmoid colon are similar to those seen in the proximal parts of the colon, but a distended sigmoid colon usually shows no sacculations. The rectum is seen to have a wider caliber than the colon.

A contrast enema is sometimes very useful for examining the mucous membrane of the colon. The barium enema is partly evacuated and air is injected into the colon. By this means the walls of the colon become clearly outlined (see Fig 11–24).

Computed tomographic scans of the abdomen showing the colon are seen in Figure 11–25.

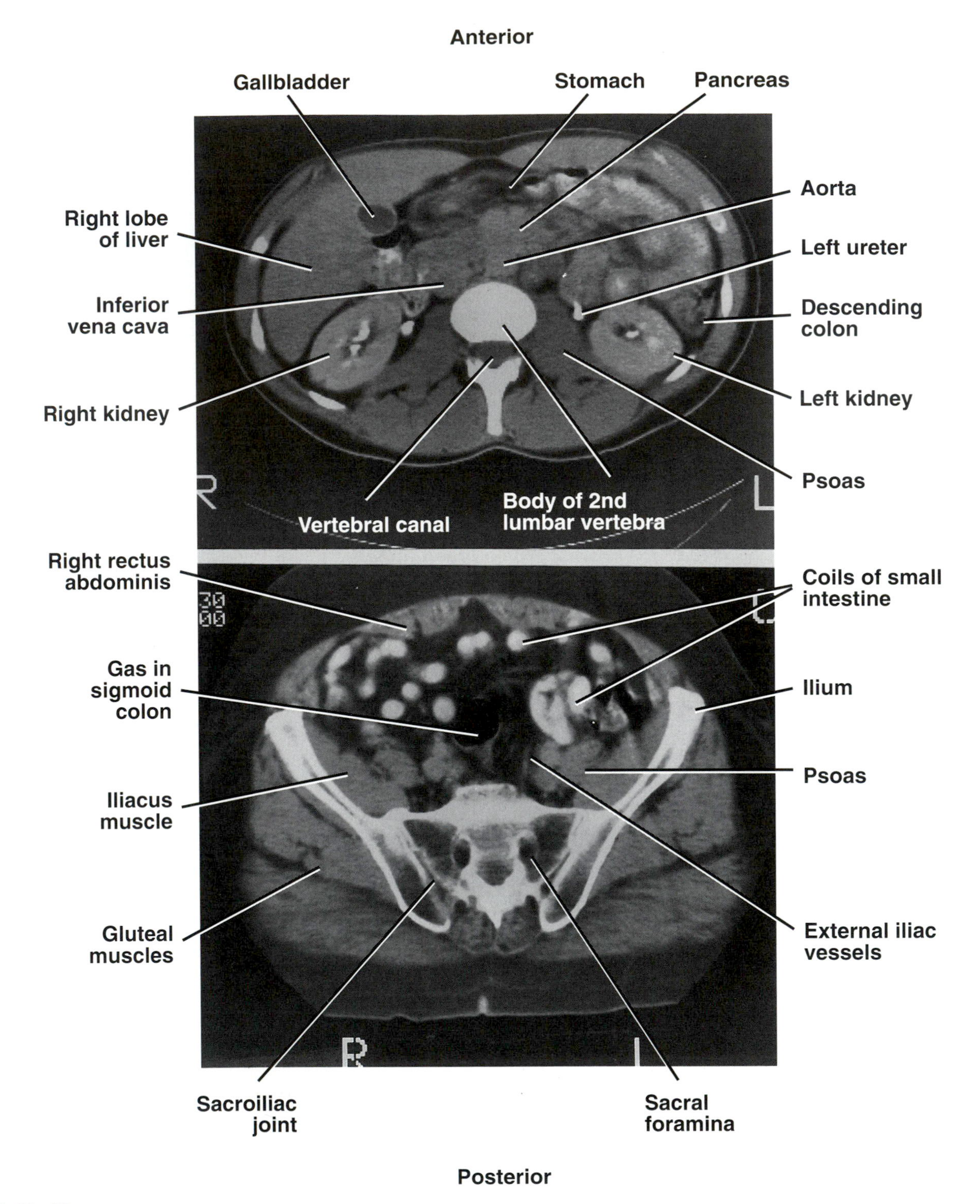

FIG 11–25.
Top, computed tomographic (CT) scan of the abdomen at the level of the second lumbar vertebra following an intravenous pyelogram. The radiopaque material can be seen in the renal pelvis and the ureters. The section is viewed from below. **Bottom,** CT scan of the lower abdomen at the level of the sacrum following a barium meal. Note the presence of barium in the small intestine.

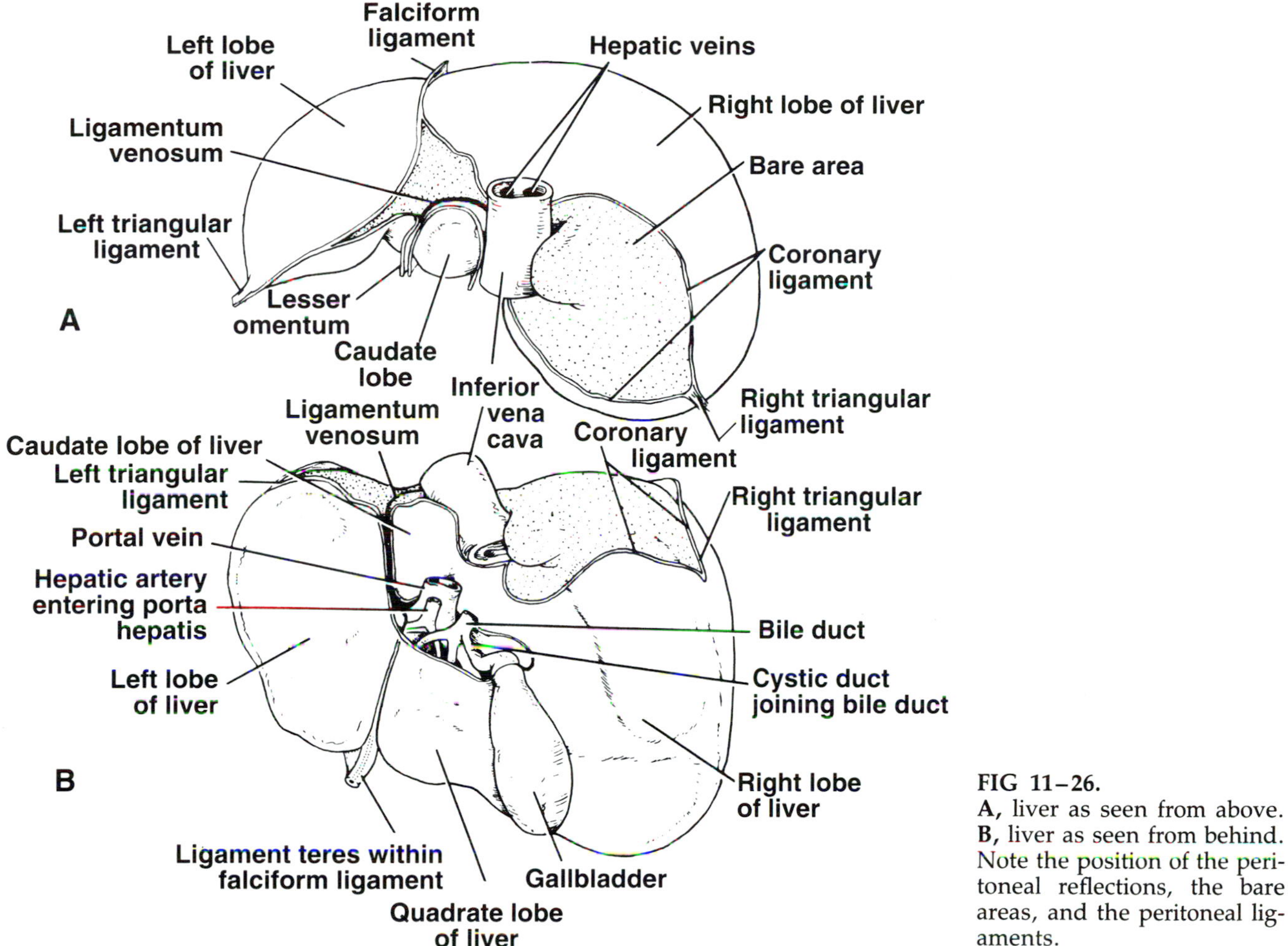

FIG 11–26.
A, liver as seen from above. **B,** liver as seen from behind. Note the position of the peritoneal reflections, the bare areas, and the peritoneal ligaments.

Basic Anatomy of the Liver

Location and Description

The liver is the largest organ in the body, and it occupies the upper part of the abdominal cavity just beneath the diaphragm (see Fig 11–13). The greater part of the liver is situated between the right hemidiaphragm and the right costal margin. It extends to the left across the epigastric region to reach the left hemidiaphragm.

The liver may be divided into a large *right lobe* and a small *left lobe* by the attachment of the two-layered fold of peritoneum, the falciform ligament (Fig 11–26). The right lobe is further subdivided into a *quadrate lobe* and a *caudate lobe* by the presence of the gallbladder, the fissure for the ligamentum teres, the inferior vena cava, and the fissure for the ligamentum venosum (see Fig 11–26).

The liver is completely surrounded by a fibrous capsule but only partially covered by peritoneum. The *porta hepatis* or hilus of the liver is found on the posteroinferior surface (see Fig 11–26). Within the porta hepatis lie the right and left hepatic ducts, the right and left branches of the hepatic artery and portal vein, nerves, and lymph vessels. Attached to the liver are various peritoneal ligaments including the *falciform ligament*, the *coronary ligament*, the *right and left triangular ligaments*, and the *lesser omentum* (see Fig 11–26). The lesser omentum connects the liver to the lesser curvature of the stomach.

Important Relations.—These relations are as follows.

Anteriorly.—Diaphragm, right and left costal margins, right and left pleura and lower margins of both lungs, xiphoid process, and anterior abdominal wall in the subcostal angle.

Posteriorly.—Diaphragm, right kidney, hepatic flexure of the colon, duodenum, gallbladder, inferior vena cava, and esophagus and fundus of the stomach.

Blood Supply.—This includes the following.

Arteries.—The hepatic artery, a branch of the celiac artery, divides into right and left terminal branches that enter the porta hepatis.

Veins.—The portal vein divides into right and left terminal branches that enter the porta hepatis behind the arteries. The hepatic veins (three or more) emerge from the posterior surface of the liver and drain into the inferior vena cava.

Lymphatic Drainage.—The lymph enters lymph nodes in the porta hepatis and then drains to the celiac nodes. Some lymph passes through the diaphragm to enter the posterior mediastinal nodes.

Nerve Supply.—Sympathetic and parasympathetic (vagal) fibers from the celiac plexus. The left vagus gives rise to a large hepatic branch, which travels directly to the liver.

Basic Anatomy of the Portal Vein

This important vein is about 2 in. (5 cm) long and is formed behind the neck of the pancreas by the union of the superior mesenteric vein and the splenic vein (Fig 11–27). It ascends behind the first part of the duodenum and enters the lesser omentum. At the porta hepatis it divides into right and left terminal branches.

The portal vein drains blood from the gastrointestinal tract from the lower end of the esophagus to halfway down the anal canal; from the pancreas, gallbladder, and the bile ducts; and from the spleen (see Fig 11–27). For further details concern-

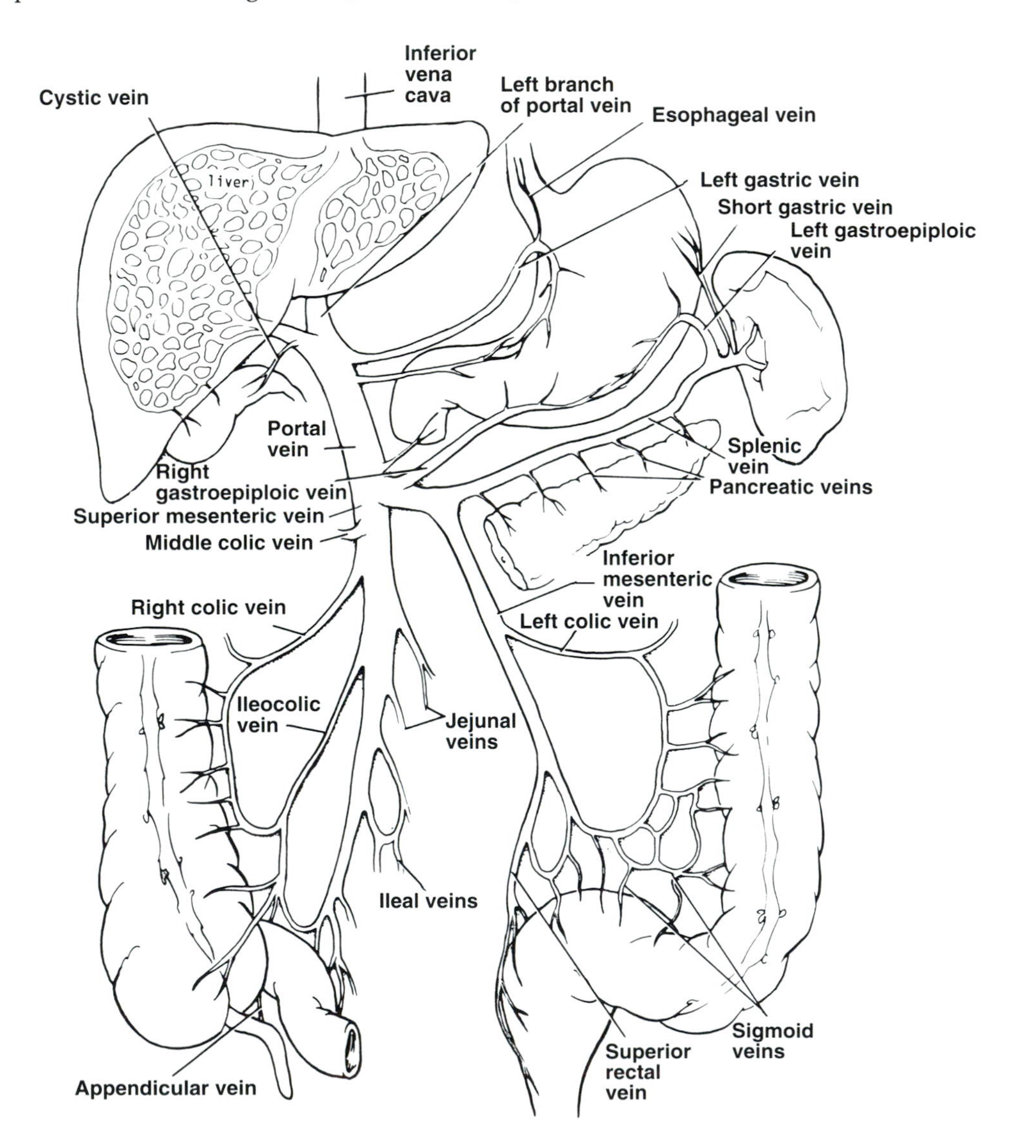

FIG 11–27.
Tributaries of the portal vein.

ing the basic anatomy of the portal vein, see p. 545, Chapter 14.

Basic Anatomy of the Gallbladder and Bile Ducts

Gallbladder

The gallbladder is a pear-shaped sac lying on the undersurface of the liver (see Fig 11–26). It is divided into a *fundus, body,* and *neck.* It has a capacity of about 30 mL and stores bile, which it concentrates by absorbing water. The neck is continuous with the cystic duct (Fig 11–28).

Blood Supply.—This involves the following.

Arteries.—Cystic artery, a branch of the right hepatic artery.

Veins.—Cystic vein drains into the portal vein.

Lymphatic Drainage.—The lymph drains into the cystic lymph node near the neck of the gallbladder, then the hepatic nodes in the porta hepatis, and finally the celiac nodes.

Nerve Supply.—Sympathetic and parasympathetic vagal fibers from the celiac plexus. The gallbladder contracts in response to the hormone *cholecystokinin,* which is produced by the mucous membrane of the duodenum on the arrival of food from the stomach.

Bile Ducts

Hepatic Ducts.—The *right* and *left hepatic ducts* emerge from the right and left lobes of the liver in the porta hepatis. Each hepatic duct has been formed by the union of small bile ducts (bile canaliculi) within the liver. The *common hepatic duct* is formed by the union of the right and left hepatic ducts. It is joined on the right side by the cystic duct from the gallbladder to form the bile duct (see Fig 11–28).

Cystic Duct.—The cystic duct is an S-shaped duct that connects the neck of the gallbladder to the common hepatic duct to form the bile duct (see Fig 11–28). The mucous membrane is raised to form a spiral fold (spiral valve), whose function is to keep the lumen constantly open.

Bile Duct (Common Bile Duct).—The bile duct is formed by the union of the cystic duct with the common hepatic duct (see Fig 11–28). It runs in the right free margin of the lesser omentum, having the portal vein behind and the hepatic artery on the left. It descends in front of the opening into the lesser sac and passes behind the first part of the duodenum and then the head of the pancreas. The bile duct ends below by piercing the medial wall of the duodenum about halfway down its length (see Fig 11–20). It is usually joined by the main pancreatic duct, and together they open into a small ampulla in

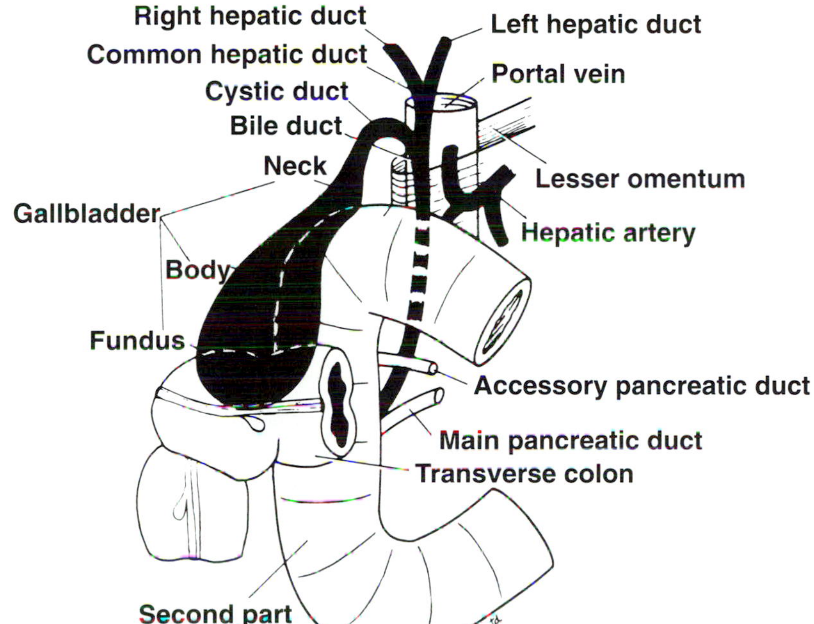

FIG 11–28.
The gallbladder and the different biliary ducts. Note the relations of the gallbladder to the duodenum and transverse colon.

the duodenal wall, the *ampulla of Vater*. The ampulla opens into the lumen of the duodenum by means of a small papilla, the *major duodenal papilla* (see Fig 11–20). The terminal parts of both ducts and the ampulla are surrounded by circular smooth muscle called the *sphincter of Oddi*. Occasionally, the bile and pancreatic ducts open separately into the duodenum.

Radiological Appearances

Radiological appearances of the gallbladder and bile ducts are shown in Figure 11–29. A *sonogram of the gallbladder* is shown in Figure 11–30.

Basic Anatomy of the Pancreas

The pancreas is both an exocrine and an endocrine gland. It is an elongated structure that lies in the epigastrium and left hypochondrium. It is situated on the posterior abdominal wall behind the stomach and behind the peritoneum. The pancreas may be divided into a *head, neck, body,* and *tail* (see Fig 11–20).

Pancreatic Ducts

The *main pancreatic duct* runs the length of the gland and opens into the second part of the duodenum with the bile duct (common bile duct) on the major duodenal papilla (see Fig 11–20). Sometimes the main duct drains separately into the duodenum. The *accessory duct,* when present, drains the upper part of the head and opens into the duodenum on the minor duodenal papilla.

Basic Anatomy of the Spleen

Location and Description

The spleen is the largest single mass of lymphoid tissue in the body (Fig 11–31). It lies in the left hypo-

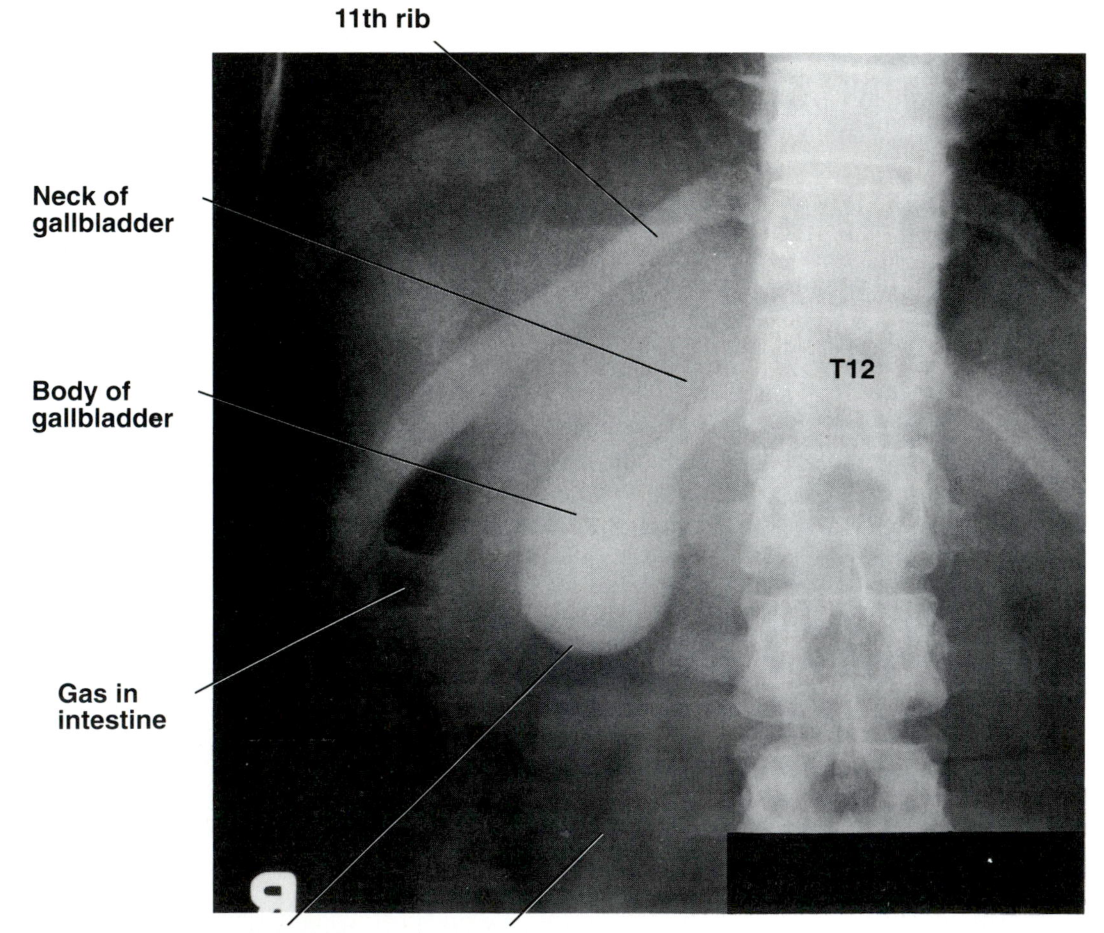

FIG 11–29.
A cholecystogram.

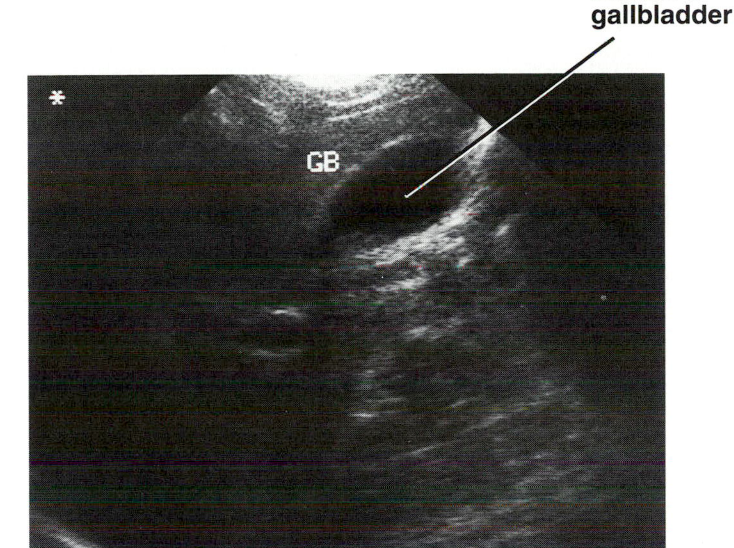

FIG 11–30.
Sonogram of the upper part of the abdomen showing the lumen of the gallbladder. Courtesy of Dr. M. C. Hill.

chondrium beneath the left half of the diaphragm close to the 9th, 10th, and 11th ribs.

The spleen is ovoid, with a notched anterior border. It is surrounded by peritoneum that passes from the hilus to the stomach as the *gastrosplenic omentum* (ligament) and to the left kidney as the *lienorenal ligament* (see Fig 11–12). The gastrosplenic omentum contains the short gastric and left gastroepiploic vessels, and the lienorenal ligament contains the splenic vessels and the tail of the pancreas.

Important Relations.—These relations are as follows.

Anteriorly.—Stomach, tail of pancreas, left colic flexure; the left kidney along its medial border.

Posteriorly.—Diaphragm, left lung and pleura; 9th, 10th, and 11th ribs.

Blood Supply.—This includes the following.

Artery.—Large splenic artery, a branch of the celiac artery (see Fig 11–31).

Vein.—The splenic vein joins the superior mesenteric vein to form the portal vein.

Basic Anatomy of the Retroperitoneal Space

The retroperitoneal space lies on the posterior abdominal wall behind the parietal peritoneum. It extends from the 12th thoracic vertebra and the 12th rib above to the sacrum and iliac crests below (Fig 11–32).

Clinical Note

Clinically the retroperitoneal space is located in the lumbar and iliac regions, and when viewed from the back, it lies lateral to the postvertebral muscles in the costovertebral angle.

The floor or posterior wall of the space is formed from medial to lateral by the psoas and quadratus lumborum muscles and the origin of the transversus abdominis muscle (see Fig 11–32). Each of these muscles is covered on the anterior surface by a definite layer of fascia. In front of the fascial layers is a variable amount of fatty areolar tissue that forms a bed for the suprarenal glands, the kidneys, the ascending and descending parts of the colon, and the duodenum. The retroperitoneal space also contains the ureters and the renal and gonadal blood vessels.

SURFACE ANATOMY

Surface Anatomy of the Abdominal Wall

General Appearances

The normal abdominal wall is soft and pliable and undergoes inward and outward excursion with respiration. The contour is subject to considerable vari-

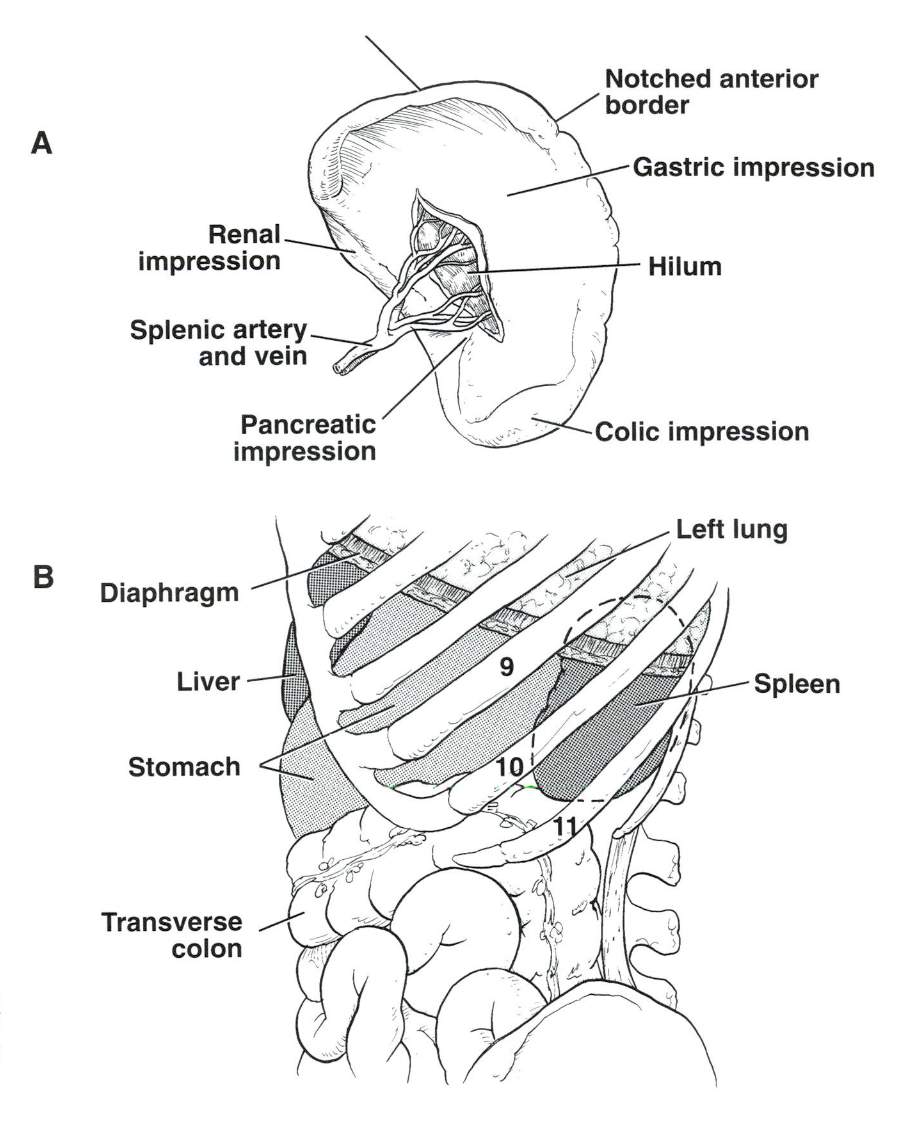

FIG 11–31.
A, spleen as seen from in front; note the notched anterior border. **B,** left side of the lower thorax and abdomen showing the relationships of the spleen.

ation and depends on the tone of its muscles and the amount of fat in the subcutaneous tissue. Well-developed muscles or an abundance of fat can prove to be a severe obstacle to the palpation of the abdominal contents.

Surface Landmarks

Xiphoid Process.—This thin cartilaginous lower part of the sternum is easily palpated in the depression where the costal margins meet in the upper part of the anterior abdominal wall (Fig 11–33). The *xiphisternal junction* lies opposite the body of the ninth thoracic vertebra and is identified by feeling the lower edge of the body of the sternum.

Costal Margin.—This is the curved lower margin of the thoracic wall and is formed in front by the cartilages of the 7th, 8th, 9th, and 10th ribs (see Fig 11–33) and formed behind by the cartilages of the 11th and 12th ribs. The costal margin reaches its lowest level at the 10th costal cartilage, which lies opposite the body of the third lumbar vertebra. The 12th rib may be short and difficult to palpate.

Iliac Crest.—This may be felt along its entire length and ends in front at the *anterior superior iliac spine* (see Fig 11–33) and behind at the *posterior superior iliac spine* (see Fig 11–33). Its highest point lies opposite the body of the fourth lumbar vertebra.

About 2 in. (5 cm) posterior to the anterior superior iliac spine, the outer margin projects to form the *tubercle of the crest* (Fig 11–34). The tubercle lies at the level of the body of the fifth lumbar vertebra.

Symphysis Pubis.—This cartilaginous joint lies in the midline between the bodies of the pubic bones

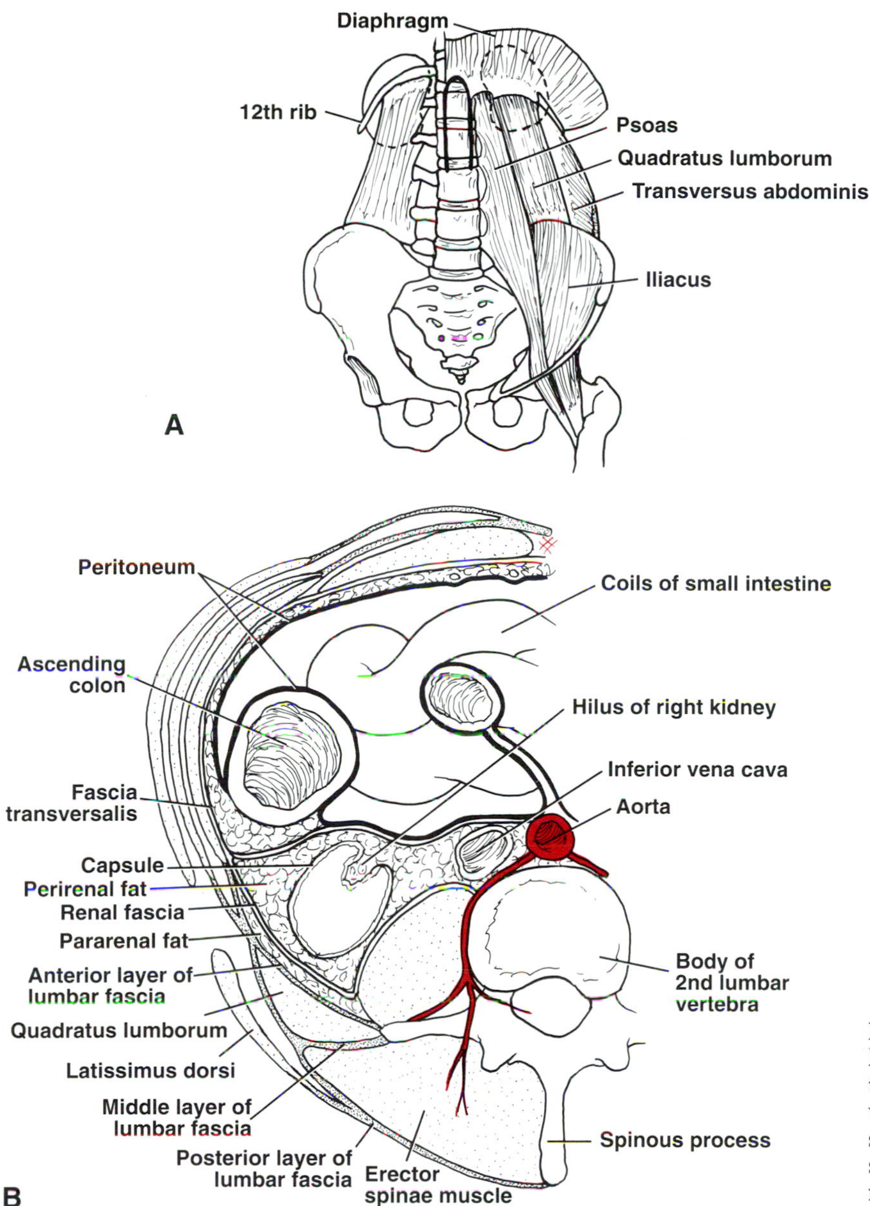

FIG 11–32.
Retroperitoneal space. **A,** shows structures present on the posterior abdominal wall behind the peritoneum. **B,** transverse section of the posterior abdominal wall showing the structures in the retroperitoneal space as seen from below.

(see Fig 11–8). It is felt as a solid structure beneath the skin at the lower extremity of the anterior abdominal wall. The *pubic crest* is the ridge on the superior surface of the pubic bones medial to the pubic tubercle (see Fig 11–8). The *pubic tubercle* may be identified as protuberance along the superior surface of the pubis (see Fig 11–8).

Inguinal Ligament.—This ligament lies beneath a skin crease in the groin. It is the rolled-under inferior margin of the aponeurosis of the external oblique muscle (see Fig 11–8). The inguinal ligament is attached laterally to the anterior superior iliac spine and curves downward and medially, to be attached to the pubic tubercle.

Superficial Inguinal Ring.—This triangular aperture in the aponeurosis of the external oblique muscle is situated above and medial to the pubic tubercle (see Fig 11–34). In adult males, the margins of the ring can be felt by invaginating the skin of the upper part of the scrotum with the tip of a finger. The soft tubular *spermatic cord* can be felt emerging from the ring and descending over or medial to the pubic tubercle into the scrotum (see Fig 11–10). The *vas deferens,* which is part of the spermatic cord, can be felt as a firm cord as it leaves the epididymis in the upper part of the scrotum.

In females the superficial inguinal ring is smaller and difficult to palpate; it transmits the *round ligament of the uterus.*

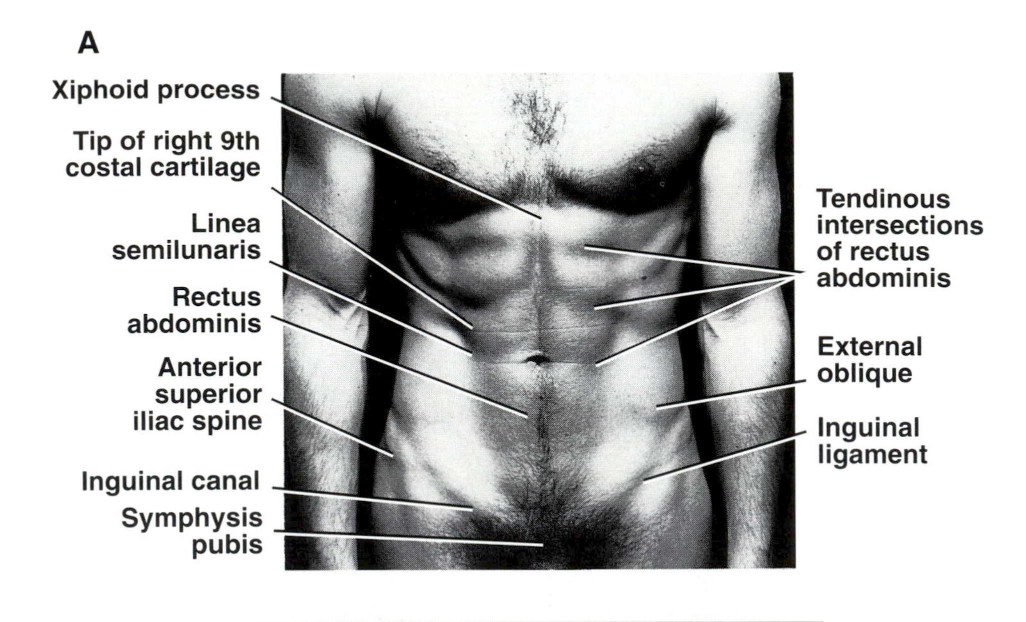

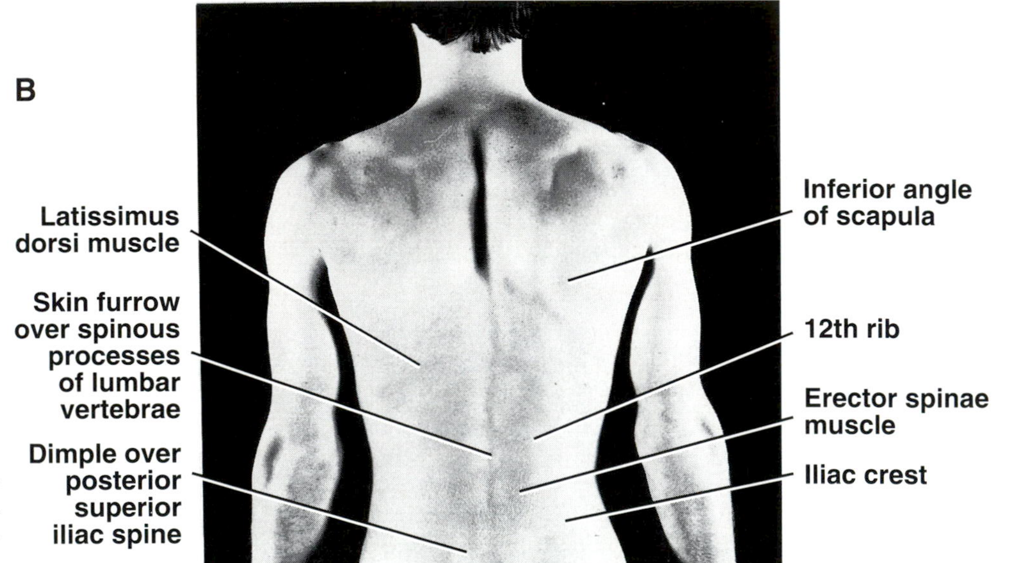

FIG 11–33.
Surface landmarks of the abdomen of a male adult. **A**, anterior abdominal wall. **B**, posterior abdominal wall.

Umbilicus.—This puckered scar, which is the site of attachment of the umbilical cord in the fetus, lies in the linea alba and is inconstant in position.

Rectus Abdominis.—The rectus abdominis muscles lie on either side of the linea alba and run vertically in the abdominal wall; they can be made prominent by asking the patient to raise his shoulders while in a supine position without using his arms.

Linea Semilunaris.—This is the lateral edge of the rectus abdominis muscle and crosses the costal margin at the tip of the ninth costal cartilage (see Fig 11–34). To accentuate the semilunar lines, the patient is asked to contract the rectus abdominis muscles so that their lateral edges stand out.

Tendinous Intersections of the Rectus Abdominis.—These three intersections run across the rectus abdominis muscle; in muscular individuals they can be palpated as transverse depressions at the level of the tip of the xiphoid process, at the umbilicus, and halfway between the two (see Fig 11–33).

Abdominal Lines and Planes

Vertical lines and horizontal planes (see Fig 11–34) are commonly used to facilitate the description of

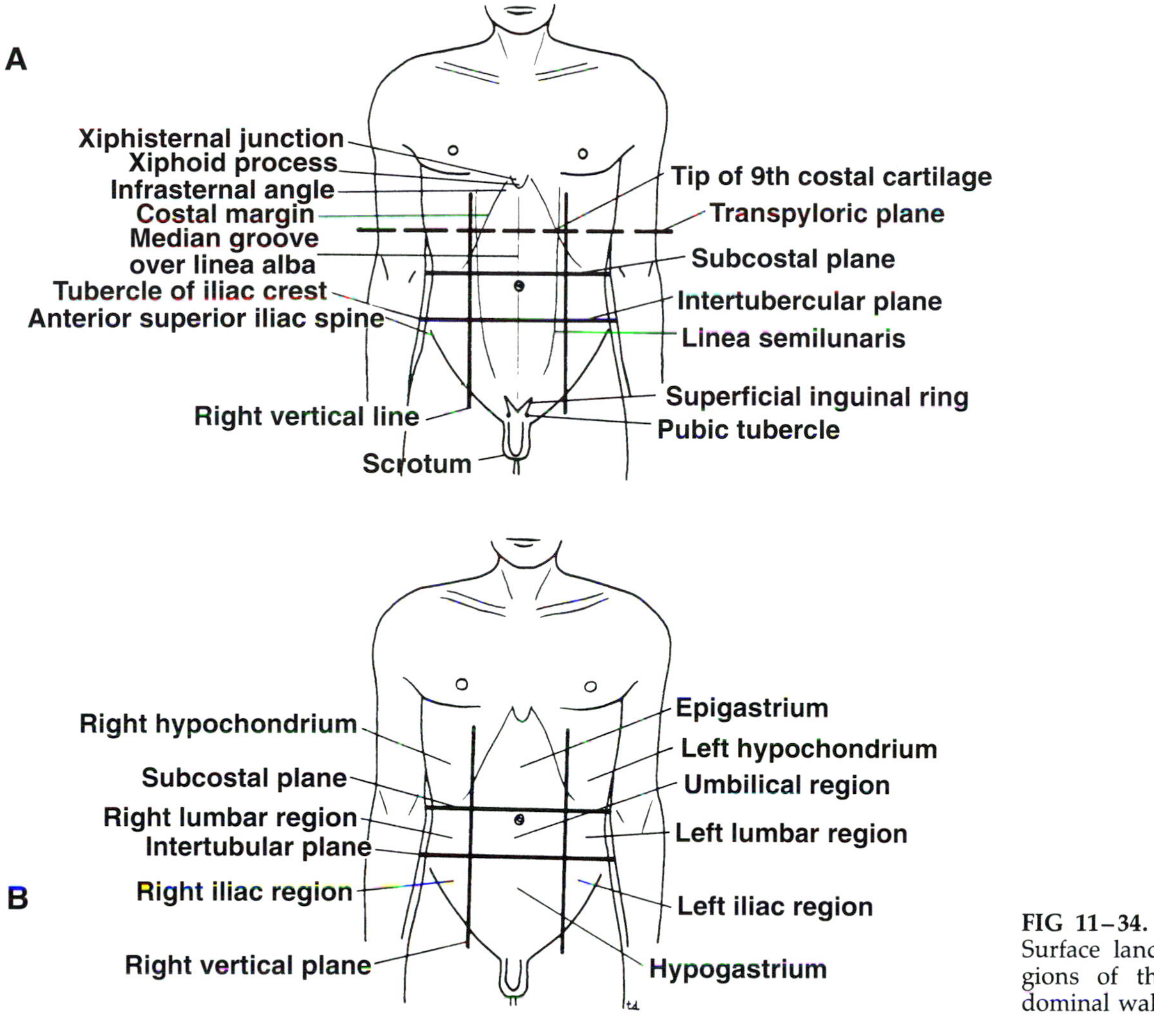

FIG 11–34.
Surface landmarks and regions of the anterior abdominal wall.

the location of diseased structures or the performing of abdominal procedures.

Vertical Lines.—These two lines pass through the midpoint between the anterior superior iliac spine and the symphysis pubis.

Transpyloric Plane.—This horizontal plane passes through the tips of the ninth costal cartilages on the two sides, i.e., the point where the lateral margin of the rectus abdominis (linea semilunaris) crosses the costal margin. It lies at the level of the body of the first lumbar vertebra. This plane passes through the pylorus, the duodenojejunal junction, the neck of the pancreas, and the hili of the kidneys.

Subcostal Plane.—This horizontal line joins the lowest point of the costal margin on each side, i.e., the tenth costal cartilage. The subcostal plane lies at the level of the third lumbar vertebra.

Intercristal Plane.—This plane passes across the highest points on the iliac crests and lies on the level of the body of the fourth lumbar vertebra.

Clinical Note

Intercristal plane.—This is commonly used as a surface landmark when performing a lumbar spinal tap (see p. 396).

Intertubercular Plane.—This horizontal line joins the tubercles on the iliac crests. This plane lies at the level of the fifth lumbar vertebra.

Abdominal Quadrants

It is common practice to divide up the abdomen into quadrants by using a vertical and horizontal line that intersect at the umbilicus. The quadrants are the upper right, the upper left, the lower right, and the lower left. The term *periumbilical* is loosely used to indicate the area around the umbilicus.

Abdominal Regions
Using the abdominal lines and planes, it is also possible to divide up the abdomen into nine regions (see Fig 11–34).

In the upper abdomen, the right hypochondrium, epigastrium, and left hypochondrium are located.

In the middle abdomen, the right lumbar, umbilical, and left lumbar are located.

In the lower abdomen, the right iliac region, hypogastrium, and left iliac region are located.

Abdominal Viscera
The positions of the majority of the abdominal viscera show individual variations as well as variations in the same person at different times. Posture and respiration have a profound influence on the position of viscera.

The following organs are more-or-less fixed, and their surface markings are of clinical value.

Liver.—The liver lies under cover of the lower ribs, and most of its bulk lies in the right upper quadrant and epigastrium. In infants, until about the end of the third year of life, the lower margin of the liver extends one or two fingerbreadths below the costal margin. In the adult who is obese or has a well-developed right rectus abdominis muscle, the liver is not palpable. In a thin adult the lower edge of the liver may be felt a fingerbreadth below the costal margin. It is most easily felt when the patient inspires deeply and the diaphragm contracts and pushes down the liver.

Gallbladder.—The fundus of the gallbladder lies opposite the tip of the right ninth costal cartilage, i.e., where the lateral edge of the right rectus abdominis muscle crosses the costal margin (see Fig 11–34).

Spleen.—The spleen is situated in the left hypochondrium and lies under cover of the 9th, 10th, and 11th ribs (see Fig 11–31). Its long axis corresponds to that of the 10th rib, and in adults the spleen does not normally project forward in front of the midaxillary line. In infants the lower pole of the spleen may just be felt.

Pancreas.—The pancreas lies across the transpyloric plane. The head lies below and to the right, the neck lies on the plane, and the body and tail lie above and to the left.

Kidneys.—The right kidney lies at a slightly lower level than the left kidney (due to the bulk of the right lobe of the liver), and the lower pole may be palpated in the right lumbar region at the end of deep inspiration in a person with poorly developed abdominal muscles. Each kidney moves about 1 in. (2.5 cm) in a vertical direction during full respiratory movement of the diaphragm. The normal left kidney, which is higher than the right kidney, is impalpable.

On the anterior abdominal wall the hilus of each kidney lies on the transpyloric plane, about three fingerbreadths from the midline. On the back, the kidneys extend from the 12th thoracic spine to the 3rd lumbar spine, and the hili are opposite the 1st lumbar vertebra.

Stomach.—The *cardioesophageal junction* lies about three fingerbreadths below and to the left of the xiphisternal junction (esophagus pierces diaphragm at level of the tenth thoracic vertebra). The *pylorus* lies on the transpyloric plane just to the right of the midline. The *lesser curvature* lies on a curved line joining the cardioesophageal junction and the pylorus. The *greater curvature* has an extremely variable position in the umbilical and hypogastric regions.

Duodenum (First Part).—This lies on the transpyloric plane about four fingerbreadths to the right of the midline.

Cecum.—The cecum is situated in the right iliac region (see Fig 11–34). It is often distended with gas and gives a resonant sound on percussion. It can be palpated through the anterior abdominal wall.

Appendix.—The appendix lies in the right iliac region. The base of the appendix is situated one third of the way up the line, joining the anterior superior iliac spine to the umbilicus (McBurney's point). The position of the free end of the appendix is very variable.

Ascending Colon.—The ascending colon extends upward from the cecum on the lateral side of the right vertical line and disappears under the right costal margin. It can be palpated through the anterior abdominal wall.

Transverse Colon.—The transverse colon extends across the abdomen, occupying the umbilical and hypogastric regions. It hangs downward and, because it has a mesentery, its position is variable.

Descending Colon.—The descending colon extends downward from the left costal margin on the lateral side of the left vertical line (see Fig 11–34). In the left iliac region it curves medially and downward to become continuous with the sigmoid colon. The descending colon has a smaller diameter than the ascending colon and can be palpated through the anterior abdominal wall.

Aorta

The aorta lies in the midline of the abdomen and bifurcates below into the right and left common iliac arteries opposite the fourth lumbar vertebra, i.e., on the intercristal plane. The pulsations of the aorta may be easily palpated through the upper part of the anterior abdominal wall.

External Iliac Artery

The pulsations of this artery may be felt as it passes under the inguinal ligament to become continuous with the femoral artery. It may be located at a point halfway between the anterior superior iliac spine and the symphysis pubis.

Urinary Bladder and Pregnant Uterus

The full bladder and pregnant uterus may be palpated through the lower part of the anterior abdominal wall in the hypogastrium (see pp. 480 and 512).

CLINICAL ANATOMY

Clinical Anatomy of the Anterior Abdominal Wall

Skin

The skin is loosely attached to the underlying structures except at the umbilicus, where it is tethered to the scar tissue.

Lines of Cleavage.—The lines of cleavage on the anterior abdominal wall run downwards and forwards. If possible, all surgical incisions should be made in the lines of cleavage where the bundles of collagen fibers in the dermis run in parallel rows; this will give the best cosmetic result. Incisions that cross the cleavage lines tend to disrupt the rows of collagen and result in the excess production of fresh collagen and the formation of a broad, unsightly scar.

Lymphatic Drainage of the Skin.—Above the level of the umbilicus the lymph drains upward and laterally into the anterior axillary (pectoral) nodes, which can be palpated just beneath the lower border of the pectoralis major muscle. Below the umbilicus the lymph drains downward and laterally into the superficial inguinal nodes situated a fingerbreadth below the inguinal ligament. The skin of the back above the level of the iliac crests drains upward into the posterior axillary group of nodes palpated on the posterior wall of the axilla; below the level of the iliac crests the lymph from the skin drains downward into the superficial inguinal lymph nodes (Fig 11–35).

Superficial Veins

The superficial veins around the umbilicus and the paraumbilical veins connecting them to the portal vein may become grossly distended in cases of portal vein obstruction (see Fig 11–35). The distended subcutaneous veins radiate out from the umbilicus, producing in severe cases the clinical picture referred to as the *caput medusae.* If there is obstruction to the superior vena cava or inferior vena cava, the venous blood causes distension of the veins running from the anterior chest wall to the thigh. The lateral thoracic vein, a tributary of the axillary vein, anastomoses with the superficial epigastric vein, a tributary of the great saphenous vein of the leg. In these circumstances a tortuous vein may extend from the lower abdomen to the axilla.

Nerves of the Anterior Abdominal Wall

The nerves of the anterior abdominal wall supply the skin, the muscles, and the parietal peritoneum. They are derived from the anterior rami of the lower six thoracic nerves and the first lumbar nerves. (The skin of the back is supplied by the posterior rami of the same spinal nerves.) The 7th to the 11th thoracic anterior rami are intercostal nerves and also supply the skin, intercostal muscles, and parietal pleura of the thoracic wall. For further information on the nerves of the anterior abdominal wall, see abdominal pain, p. 458.

Anatomy of an Anterior Abdominal Wall Nerve Block

The anatomy of the anterior abdominal wall nerve block is fully discussed in Chapter 19, p. 825.

Umbilicus

The umbilicus is a consolidated scar representing the site of attachment of the umbilical cord in the fetus;

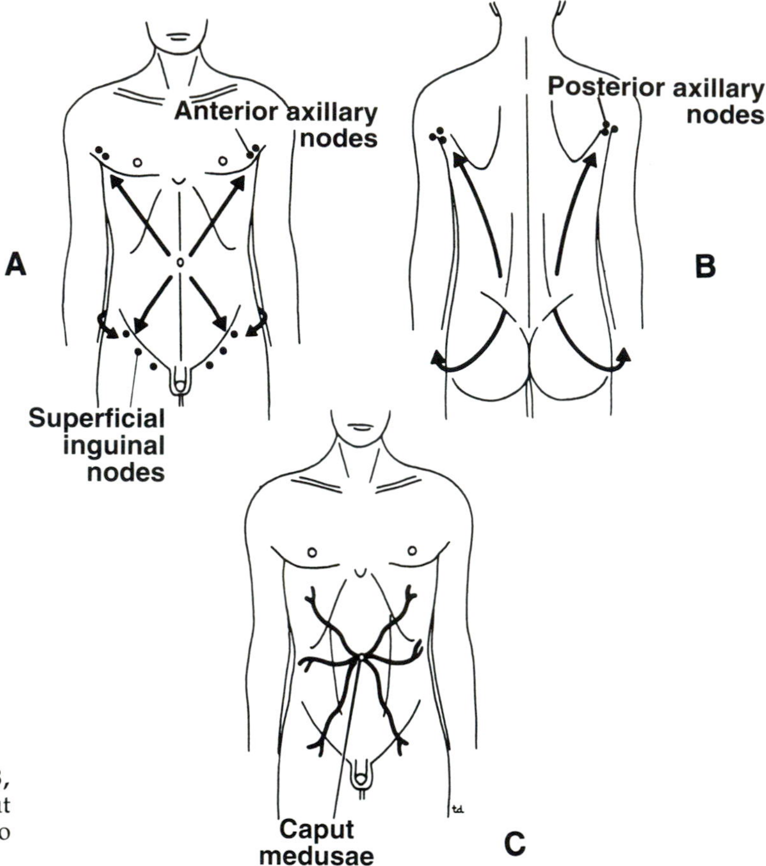

FIG 11–35.
Lymphatic drainage. **A,** anterior abdominal wall. **B,** posterior abdominal wall. **C,** an example of caput medusae in a patient with portal obstruction due to cirrhosis of the liver.

it is situated in the linea alba. It possesses a number of embryological remains that may give rise to clinical problems.

Patent Urachus.—The urachus is the remains of the allantois of the fetus and normally persists as a fibrous cord that runs from the apex of the bladder to the umbilicus. Occasionally, the cavity of the allantois persists and urine passes from the bladder through the umbilicus. In newborns it usually reveals itself when there is congenital urethral obstruction. More often, it remains undiscovered until old age, when enlargement of the prostate may obstruct the urethra.

Vitellointestinal Duct.—The vitelline duct in the early embryo connects the developing gut to the yolk sac. Normally, as development proceeds the duct is obliterated, severs its connection with the small intestine, and disappears. Persistence of the vitellointestinal duct may result in an umbilical fecal fistula. If the duct remains as a fibrous band, a loop of bowel may become wrapped around it, causing intestinal obstruction. For a discussion of *Meckel's diverticulum,* see p. 454.

Clinical Anatomy of Umbilical Vessel Catheterization.—The umbilical cord at birth is a twisted, tortuous structure that measures about ¾ in. (2 cm) in diameter and about 20 in. (50 cm) long. The cord is surrounded by amnion and contains a connective tissue core, called *Wharton's jelly.* Embedded in this jelly are the remains of the vitellointestinal duct and the allantois, and the single umbilical vein and the two umbilical arteries (Fig 11–36). The vein is a larger thin-walled vessel and is located at the 12-o'clock position when facing the umbilicus; the two arteries, which lie adjacent to one another and are located at the 4-o'clock and 8-o'clock positions when facing the umbilicus, are smaller and thick walled.

At birth, the umbilical cord is ligatured, and the umbilical vessels constrict and thrombose.

Indications for Umbilical Artery Catheterization.—Indications are as follows.

1. Administration of fluids or blood for resuscitation purposes.
2. Arterial blood gas and blood pressure monitoring. The umbilical arteries may be cannulated most easily during the first few hours after birth, but they may be cannulated up to 6 days after delivery.

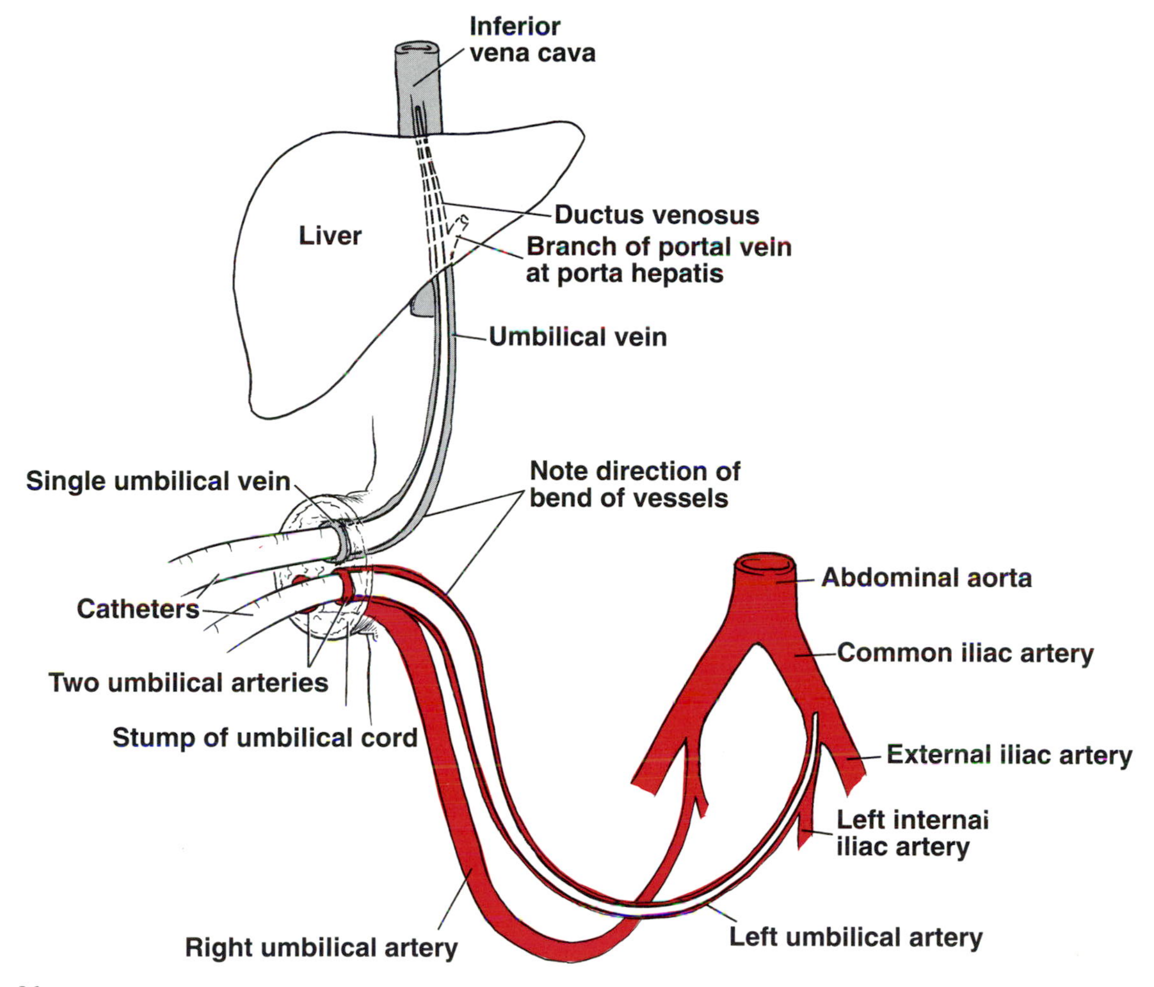

FIG 11–36.
Catheterization of the umbilical blood vessels. Arrangement of the single umbilical vein and the two umbilical arteries in the umbilical cord and the paths taken by the catheter in the umbilical vein and the umbilical artery.

Anatomy of the Procedure.—The following procedure should be used.

1. The umbilical stump is suitably treated with bacteriocidal solution and anchored by the insertion of a pursestring suture or hemostat.
2. One of the small, thick-walled arteries is identified in the Wharton's jelly and dilated with forceps.
3. Since the umbilical arteries are branches of the internal iliac arteries in the pelvis, the catheter is introduced and advanced slowly in the direction of the feet.
4. Sometimes slight resistance to forward advance is felt about 1 cm into the artery because the artery changes direction at this point and turns caudally toward the pelvis. The resistance may be overcome by directing the catheter more toward the feet and exerting steady gentle pressure.

The catheter can be inserted for about 7 cm in a premature infant and 12 cm in a full-term infant. The course of the catheter can be confirmed on a radiograph and is as follows: (1) umbilical artery (directed downward into the pelvis), (2) internal iliac artery (acute turn into this artery), and (3) common iliac artery and aorta.

Anatomy of the Complications.—The following complications may occur.

1. Catheter perforates arterial wall at a point where the artery turns downward toward the pelvis at the anterior abdominal wall.
2. Catheter enters the thin-walled wider vein instead of the thick-walled smaller artery.
3. Catheter enters the thin-walled persistent urachus (urine returned).

4. Vasospasm of the umbilical and iliac arteries cause blanching of an extremity.

5. Perforation of arteries distal to the umbilical artery, e.g., the iliac arteries or even the aorta.

Other complications include thrombosis, emboli, and infection of the umbilical stump.

Indications for Umbilical Vein Catheterization.—These include the following.

1. Administration of fluids or blood for resuscitation purposes.
2. Exchange transfusions—the umbilical vein may be cannulated up to 7 days after birth.

Anatomy of the Procedure for Umbilical Vein Catheterization.—The procedure is as follows.

1. The procedure is similar to that described for arterial catheterization. The umbilical vein is located in the cord stump at the 12-o'clock position, as described previously, and is easily recognized because of its thin wall and large lumen.
2. The catheter is advanced gently and is directed toward the head, since the vein runs in the free margin of the falciform ligament to join the ductus venosus at the porta hepatis.

The catheter may be advanced about 5 cm in a full-term infant. The course of the catheter may be confirmed by radiography and is as follows: (1) the umbilical vein, (2) the ductus venosus, and (3) the inferior vena cava (10 to 12 cm).

Anatomy of the Complications of Umbilical Vein Catheterization.—The following complications may occur.

1. Catheter perforates the venous wall. This is most likely to occur where the vein turns cranially at the abdominal wall.

Other complications include liver necrosis, hemorrhage, and infection.

Superficial Fascia

Membranous Layer of Superficial Fascia (Scarpa's Fascia).—The membranous layer is important clinically, since beneath it lies a potential closed space that does not open into the thigh, but is continuous with the superficial perineal pouch via the penis and scrotum. Rupture of the penile urethra may be followed by *extravasation of urine* into the scrotum, perineum, and penis and then up into the lower part of the anterior abdominal wall deep to the membranous layer of fascia. The urine is excluded from the thigh because of the attachment of the fascia to the fascia lata (see Fig 11–3).

Muscles

Hematoma of the Rectus Sheath.—This condition is uncommon but important since it is often overlooked. It occurs most often on the right side below the level of the umbilicus. The source of the bleeding is the inferior epigastric vein or, more rarely, the inferior epigastric artery. These vessels may be stretched during a severe bout of coughing or in the later months of pregnancy, which may predispose to the condition. The cause is usually blunt trauma to the anterior abdominal wall, such as a fall or a kick. The symptoms that follow the trauma include midline abdominal pain. An acutely tender mass confined to one rectus sheath is diagnostic.

Abdominal Wall Trauma

Thoracoabdominal Trauma.—Since the diaphragm may rise on full expiration to as high as the level of the fourth intercostal space, it follows that many stab or gunshot wounds to the lower part of the thorax involve not only the diaphragm but the upper abdominal viscera. These viscera include the liver, the stomach, and the spleen. Peritoneal lavage may be necessary to make the diagnosis of peritoneal violation (see p. 443).

Anatomy of Abdominal Stab Wounds.—Abdominal stab wounds may or may not penetrate the parietal peritoneum and violate the peritoneal cavity, and consequently may or may not significantly damage the abdominal viscera. The structures in the various layers through which an abdominal stab wound penetrates will depend on the anatomical location.

Lateral to the rectus sheath (see Fig 11–38) are the following: (1) skin, (2) fatty layer of superficial fascia, (3) membranous layer of superficial fascia, (4) thin layer of deep fascia, (5) external oblique muscle or aponeurosis, (6) internal oblique muscle or aponeurosis, (7) transversus abdominis muscle or aponeurosis, (8) fascia transversalis, (9) extraperitoneal connective tissue (often fatty), and (10) parietal peritoneum.

Anterior to the rectus sheath (see Fig 11–38) are the following: (1) skin, (2) fatty layer of superficial fascia, (3) membranous layer of superficial fascia, (4) thin layer of deep fascia, (5) anterior wall of rectus sheath, (6) rectus abdominis muscle, with segmental nerves and epigastric vessels lying behind the muscle, (7) posterior wall of rectus sheath, (8) fascia transversalis, (9) extraperitoneal connective tissue (often fatty), and (10) parietal peritoneum.

In the midline (Fig 11–37) are the following: (1) skin, (2) fatty layer of superficial fascia, (3) membranous layer of superficial fascia, (4) thin layer of deep fascia, (5) linea alba, (6) fascia transversalis, (7) extraperitoneal connective tissue (often fatty), and (8) parietal peritoneum.

In a stab wound, peritoneal lavage may be used to determine whether any damage to viscera or vasculature has occurred.

Anatomy of Abdominal Gunshot Wounds.—Gunshot wounds are much more serious than stab wounds; in most patients, the peritoneal cavity has been entered, and significant visceral damage has ensued.

Anatomy of Peritoneal Lavage

Peritoneal lavage is used to sample the intraperitoneal space for evidence of damage to viscera and blood vessels. It is generally employed as a diagnostic technique in certain cases of blunt abdominal trauma. In nontrauma situations, peritoneal lavage has been used to confirm the diagnosis of acute pancreatitis and primary peritonitis, to correct hypothermia, and to conduct peritoneal dialysis.

Midline Incision Technique.—The following technique should be used.

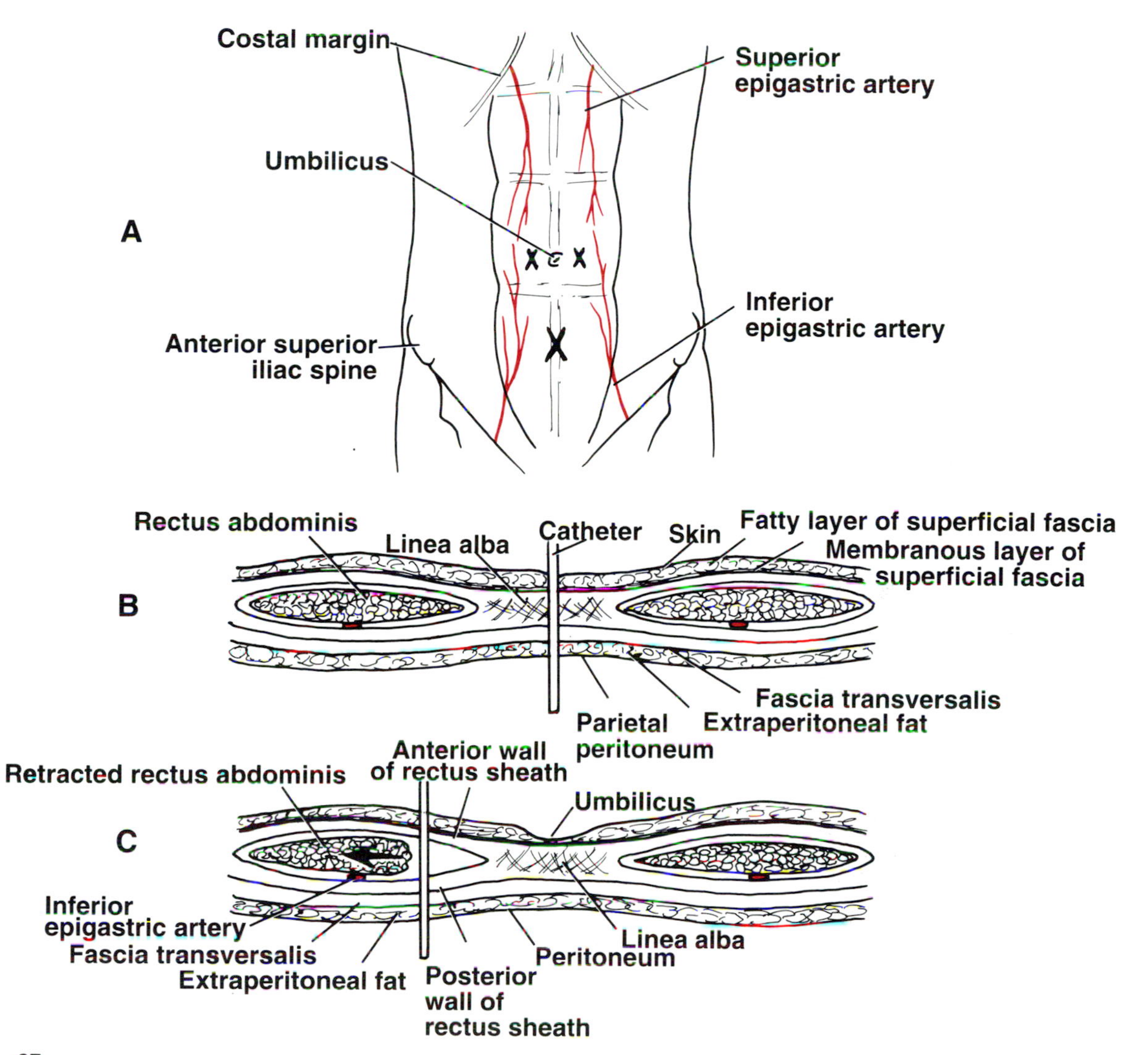

FIG 11–37.
Peritoneal lavage. **A,** the two common sites used in this procedure. Note the positions of the superior and inferior epigastric arteries in the rectus sheath. **B,** cross section of the anterior abdominal wall in the midline; note the structures pierced by the catheter. **C,** cross section of the anterior abdominal wall just lateral to the umbilicus; note the structures pierced by the catheter. The rectus muscle has been retracted laterally.

1. The patient is in a supine position.
2. The urinary bladder is emptied by catheterization. In a small child the bladder is an abdominal organ; in adults the full bladder may rise out of the pelvis and reach as high as the umbilicus.
3. The stomach is emptied by a nasogastric tube because a distended stomach may extend to the anterior abdominal wall.
4. The infraumbilical skin is anesthetized by local infiltration with an anesthetic agent.
5. A 3-cm vertical incision is made in the midline below the level of the umbilicus.

The following anatomical structures are penetrated in order to reach the parietal peritoneum (see Fig 11–37): (1) skin, (2) fatty layer of superficial fascia, (3) membranous layer of superficial fascia, (4) thin layer of deep fascia, (5) linea alba, (6) fascia transversalis, (7) extraperitoneal fat, and (8) parietal peritoneum.

Paraumbilical Technique.—The following technique should be used.

1. The patient is in a supine position and the urinary bladder and stomach are emptied as indicated previously.
2. A 3-cm vertical incision is made lateral to the umbilicus.

The following anatomical structures are penetrated in order to reach the parietal peritoneum (see Fig 11–37): (1) skin, (2) fatty layer of superficial fascia, (3) membranous layer of superficial fascia, (4) thin layer of deep fascia, (5) anterior wall of rectus sheath, (6) the rectus abdominis muscle is retracted laterally, (7) the posterior wall of the rectus sheath, (8) fascia transversalis, (9) extraperitoneal fat, and (10) parietal peritoneum.

It is important that all the small blood vessels in the superficial fascia be secured, since bleeding into the peritoneal cavity might produce a *false-positive result.* These vessels are the terminal branches of the superficial and deep epigastric arteries and veins.

Anatomy of the Complications of Peritoneal Lavage.—The following complications may occur.

1. Failure to enter the peritoneal cavity exactly in the midline (midline technique). If this occurs, the incision or trochar may traverse the rectus sheath, the vascular rectus abdominis muscle, and encounter branches of the inferior epigastric vessels. Bleeding from this source could produce a false-positive result.
2. Perforation of the gut by the scalpel or trocar.
3. Perforation of the mesenteric blood vessels or vessels on the posterior abdominal wall or pelvic walls.
4. Perforation of a full bladder.
5. Wound infection.

Anatomy of Abdominal Paracentesis

Paracentesis of the abdomen may be necessary to withdraw excessive collections of peritoneal fluid, as in ascites secondary to cirrhosis of the liver or malignant ascites secondary to an advanced ovarian carcinoma.

A catheter or a needle is inserted through the anterior abdominal wall. The underlying coils of intestine are not damaged, since they are mobile and are pushed away by the cannula.

If the cannula is inserted in the midline (Fig 11–38), it will pass through the following anatomical structures: (1) skin, (2) superficial fascia, (3) deep fascia (very thin), (4) linea alba (virtually bloodless), (5) fascia transversalis, (6) extraperitoneal connective tissue (fatty), and (7) parietal peritoneum.

If the cannula is inserted in the flank (Fig 11–38) lateral to the inferior epigastric artery and above the deep circumflex iliac artery, it will pass through the following structures: (1) skin, (2) superficial fascia, (3) deep fascia (very thin), (4) aponeurosis or external oblique muscle, (5) internal oblique muscle, (6) transversus abdominis muscle, (7) fascia transversalis, (8) extraperitoneal connective tissue (fatty), and (9) parietal peritoneum.

Anatomy of Abdominal Herniae

A hernia is the protrusion of part of the abdominal contents beyond the normal confines of the abdominal wall (Fig 11–39). It consists of three parts: the sac, the contents of the sac, and the coverings of the sac. The *hernial sac* is a diverticulum of peritoneum and has a neck and a body. The *hernial contents* may consist of any structure found within the abdominal cavity and may vary from a small piece of omentum to a large viscus such as the kidney. The *hernial coverings* are formed from the layers of the abdominal wall through which the hernial sac passes.

Abdominal herniae are of the following types.

1. Inguinal, which may be either indirect or direct (see p. 447).
2. Femoral (see p. 448).

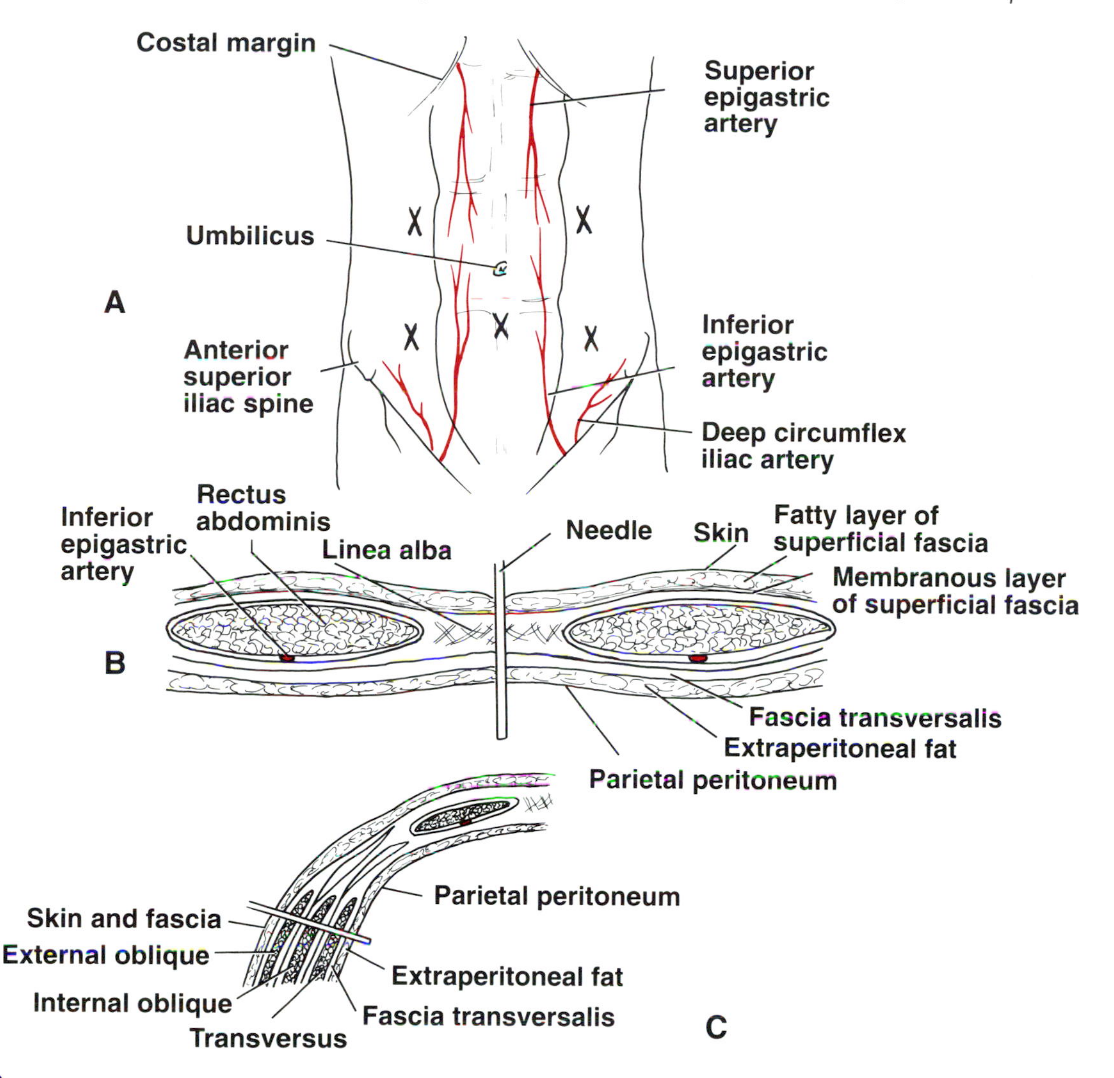

FIG 11–38.
Abdominal paracentesis. **A,** common sites used in this procedure. Note the position of the superior and inferior epigastric arteries in the rectus sheath and the deep circumflex iliac arteries. **B,** cross section of the anterior abdominal wall showing structures pierced by the needle in the midline approach. **C,** cross section of the anterior abdominal wall showing structures pierced by the needle in the lateral approach.

3. Umbilical, which may be either congenital or acquired.
4. Epigastric.
5. Separation of the recti abdominis.
6. Incisional hernia.
7. Hernia of the linea semilunaris (Spigel's hernia).
8. Lumbar hernia (Petit's triangle hernia).
9. Internal hernia.

Anatomy of Umbilical Herniae.— The following types exist.

Congenital Umbilical Hernia or exomphalos (omphalocele).—This is caused by a failure of part of the midgut to return to the abdominal cavity from the extraembryonic coelom during fetal life. For a diagram of the hernial sac and its relationship to the umbilical cord, see Figure 11–40.

Acquired Infantile Umbilical Hernia.—This is a small hernia that sometimes occurs in children and is caused by a weakness in the scar of the umbilicus in the linea alba (see Fig 11–40). The majority become smaller and disappear without treatment as the abdominal cavity enlarges.

Acquired Umbilical (Paraumbilical) Hernia.—The hernial sac does not protrude through the umbilical scar, but through the linea alba in the region of the umbilicus (see Fig 11–40). Paraumbilical herniae gradually increase in size and hang downward. The neck

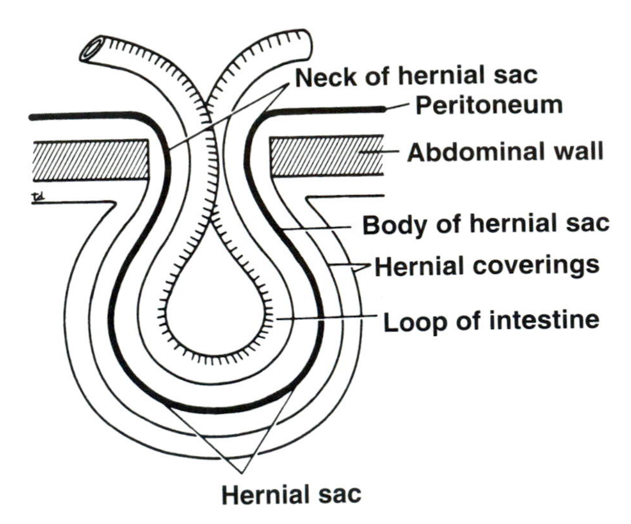

FIG 11–39.
Different parts of a hernia.

of the sac may be narrow, but the body of the sac often contains coils of the small and large intestine and omentum. Paraumbilical herniae are much more common in women than in men.

Anatomy of Epigastric Hernia.—Epigastric hernia occurs through the widest part of the linea alba, anywhere between the xiphoid process and the umbilicus (see Fig 11–40). The hernia is usually small and starts off as a small protrusion of extraperitoneal fat between the fibers of the linea alba. During the following months or years, the fat is forced farther through the linea alba and eventually drags behind it a small peritoneal sac. The body of the sac often contains a small piece of greater omentum. This type of hernia is common in middle-aged manual workers.

Anatomy of the Separation of the Recti Abdominis.—Separation of the recti abdominis occurs in elderly multiparous women with weak abdominal muscles (see Fig 11–40). In this condition, the aponeuroses forming the rectus sheath become excessively stretched. When the patient coughs or strains, the recti separate widely, and a large hernial sac, containing abdominal viscera, bulges forward between the medial margins of the recti.

Incisional Hernia.—A postoperative incisional hernia is most likely to occur in patients in whom it was necessary to cut one of the segmental nerves supplying the muscles of the anterior abdominal wall; postoperative wound infection with necrosis of the abdominal musculature is also a common cause. The

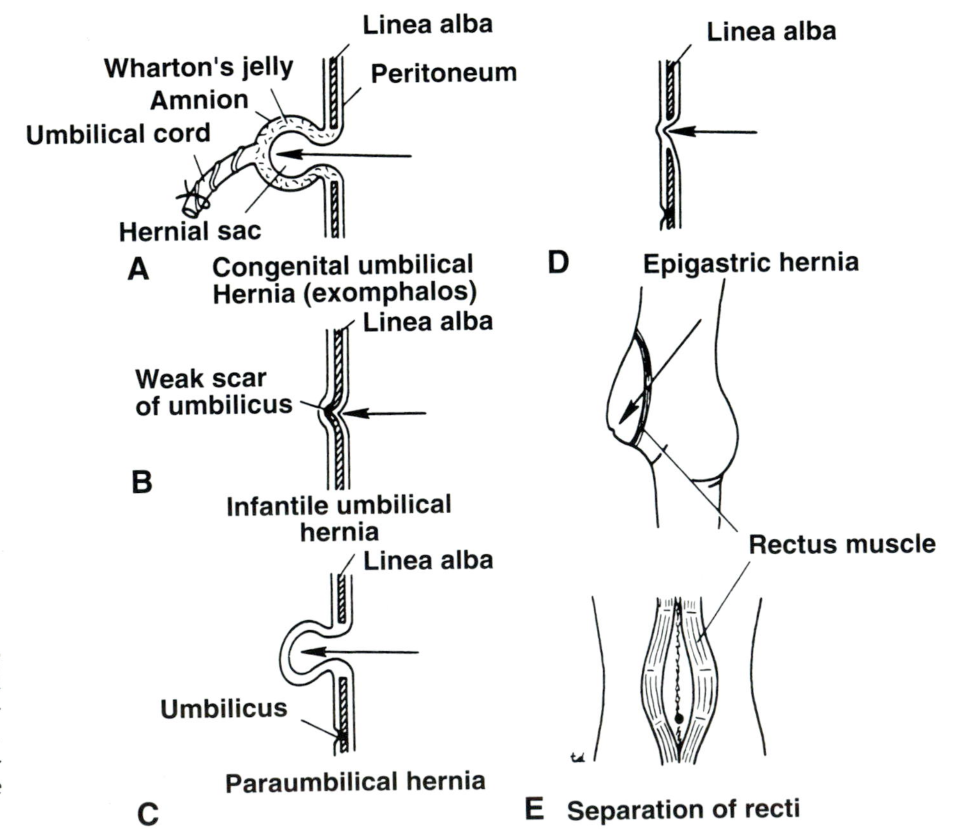

FIG 11–40.
Examples of midline herniae through the anterior abdominal wall. **A,** congenital umbilical hernia. **B,** infantile umbilical hernia. **C,** paraumbilical hernia. **D,** epigastric hernia. **E,** separation of the recti abdominis muscles.

neck of the sac is usually large, and incarceration and strangulation are rare complications. In very obese individuals the extent of the abdominal wall weakness is often difficult to access.

Anatomy of the Hernia of the Linea Semilunaris (Spigel's Hernia).—This uncommon hernia occurs through the aponeurosis of the transversus abdominis just lateral to the lateral edge of the rectus sheath. They usually occur just below the level of the umbilicus. The neck of the sac is usually narrow so that incarceration and strangulation are common complications.

Anatomy of the Lumbar Hernia.—This rare hernia occurs through the lumbar triangle. The lumbar triangle (Petit's triangle) is a weak area in the posterior part of the abdominal wall. It is bounded anteriorly by the posterior margin of the external oblique muscle, posteriorly by the anterior border of the latissimus dorsi muscle, and inferiorly by the iliac crest. The floor of the triangle is formed by the internal oblique and transversus abdominis muscles. The neck of the hernia is usually large, and the incidence of strangulation is low.

Anatomy of the Internal Hernia.—Occasionally, a loop of intestine enters a peritoneal fossa (for example, the lesser sac or the duodenal fossae) and becomes strangulated at the edges of the fossa (see p. 415).

Clinical Anatomy of the Inguinal Canal

Anatomy of the Indirect Inguinal Hernia

Indirect inguinal hernia is the most common form of hernia and is believed to be congenital in origin (Fig 11–41). The hernial sac is the remains of the processus vaginalis. (This is an outpouching of peritoneum that in the fetus is responsible for the formation of the inguinal canal.) It follows that the sac enters the inguinal canal through the deep inguinal ring *lateral to the inferior epigastric vessels* (Fig 11–42). It may extend part of the way along the canal or the full length, as far as the superficial inguinal ring. If the processus vaginalis has undergone no obliteration, then the hernia will be complete and will extend through the superficial inguinal ring into the scrotum or labium majus. Under these circumstances the neck of the hernial sac will lie at the deep inguinal ring lateral to the inferior epigastric vessels, and the body of the sac will reside in the inguinal canal and scrotum (base of the labium majus).

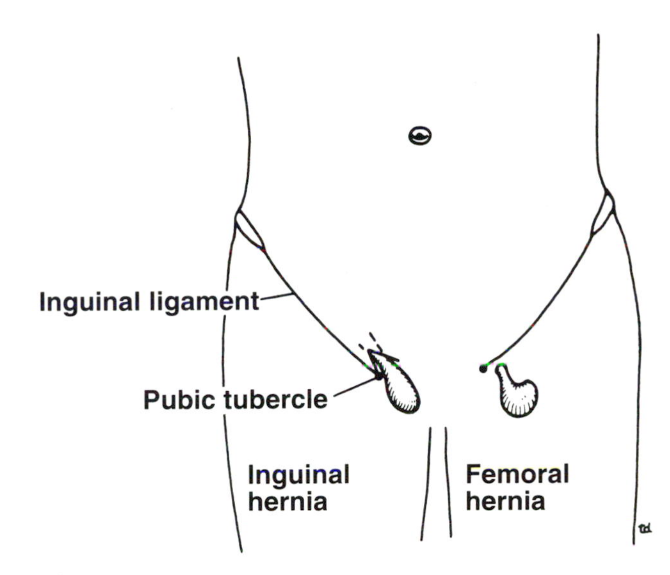

FIG 11–41.
Relation of inguinal and femoral hernial sacs to pubic tubercle.

An indirect inguinal hernia is about 20 times more common in males than in females, and nearly one third are bilateral. It is more common on the right side (normally, the right processus vaginalis becomes obliterated after the left; the right testis descends later than the left). It is most common in children and young adults.

The indirect inguinal hernia may be summarized as follows.

1. It is the remains of the processus vaginalis and therefore is congenital in origin.
2. It is more common than a direct inguinal hernia.
3. It is much more common in males compared with females.
4. It is more common on the right side.
5. It is most common in children and young adults.
6. The hernial sac enters the inguinal canal through the deep inguinal ring and *lateral* to the inferior epigastric vessels. The neck of the sac is narrow.
7. The hernial sac may extend through the superficial inguinal ring above and medial to the pubic tubercle. (Femoral hernia is located below and lateral to the pubic tubercle.)
8. The hernial sac may extend down into the scrotum or labium majus.

Anatomy of the Direct Inguinal Hernia

The direct inguinal hernia comprises about 15% of all inguinal hernias. The sac of a direct hernia bulges

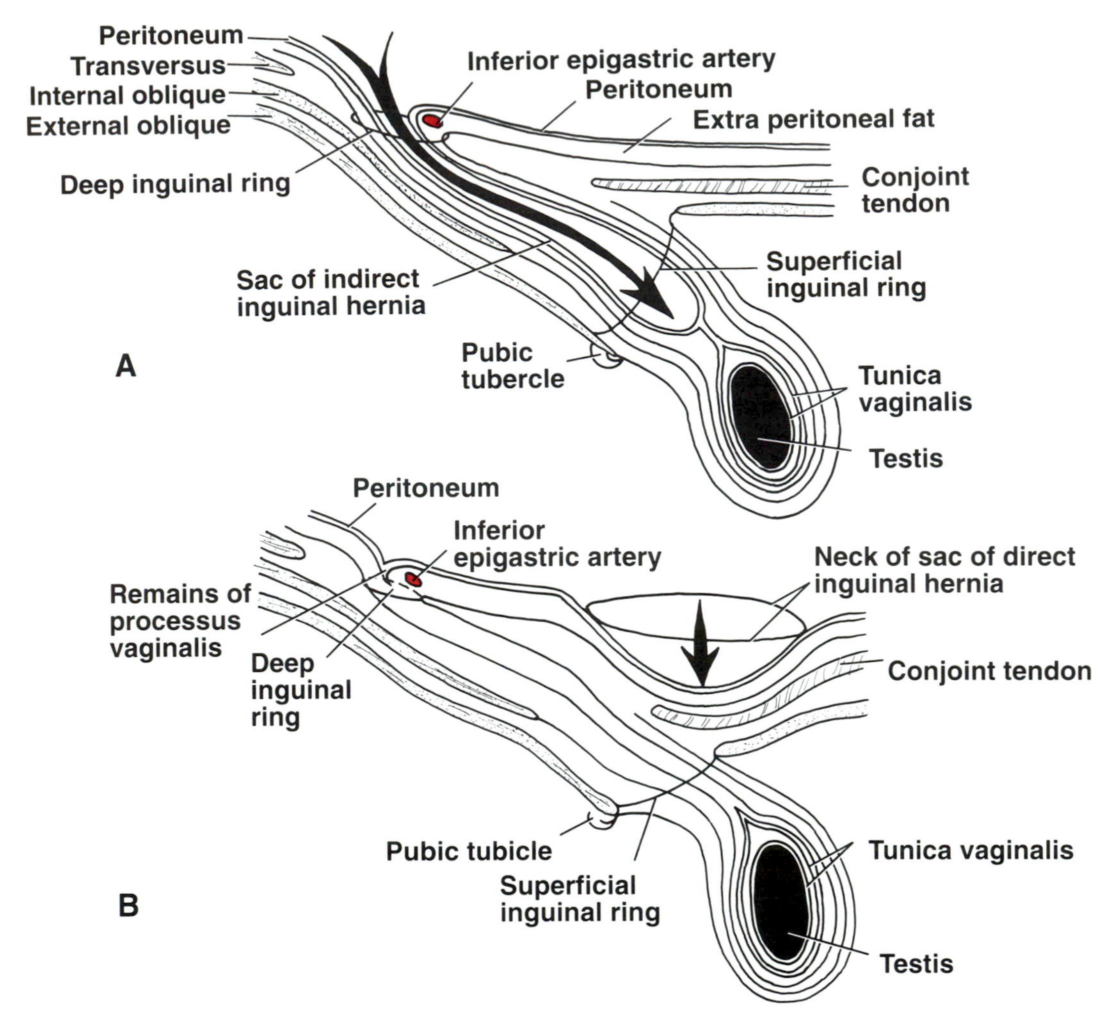

FIG 11–42.
A, indirect inguinal hernia. **B,** direct inguinal hernia. The neck of the indirect inguinal hernia lies lateral to the inferior epigastric artery, and the neck of the direct inguinal hernia lies medial to the inferior epigastric artery.

directly anteriorly through the posterior wall of the inguinal canal *medial to the inferior epigastric vessels* (see Fig 11–42). Because of the presence of the strong conjoint tendon (combined tendons of insertion of the internal oblique and the transversus muscles), this hernia is usually nothing more than a generalized bulge, and therefore the neck of the hernial sac is wide.

Direct inguinal hernias are rare in women, and the majority are bilateral. It is a disease of old men with weak abdominal muscles.

A direct inguinal hernia may be summarized as follows.

1. It is common in old men with weak abdominal muscles and rare in women.
2. The hernial sac bulges forward through the posterior wall of the inguinal canal *medial* to the inferior epigastric vessels.
3. The neck of the hernial sac is wide.

Anatomy of a Femoral Hernia

The hernial sac descends through the femoral canal within the femoral sheath.

The femoral sheath is a prolongation downward into the thigh of the fascial lining of the abdomen. It surrounds the femoral vessels and lymphatic vessels for about 1 in. (2.5 cm) below the inguinal ligament (see Fig 11–10). The *femoral artery,* as it enters the thigh below the inguinal ligament, occupies the lateral compartment of the sheath. The *femoral vein,* which lies on its medial side and is separated from it by a fibrous septum, occupies the intermediate compartment. The *lymphatics,* which are separated from the vein by a fibrous septum, occupy the most medial compartment.

The *femoral canal,* the compartment for the lymphatic vessels, occupies the medial part of the sheath. It is about ½ in. (1.3 cm) long, and its upper opening is referred to as the *femoral ring.* The *femoral septum,* which is a condensation of extraperitoneal tissue, plugs the opening of the femoral ring.

A femoral hernia is more common in women than in men (possibly due to a wider pelvis and femoral canal). The hernial sac passes down the femoral canal, pushing the femoral septum before it. On escaping through the lower end of the femoral canal, it expands to form a swelling in the upper part of the thigh deep to the deep fascia. With further expansion the hernial sac may turn upward to cross the anterior surface of the inguinal ligament.

The neck of the sac always lies below and lateral to the pubic tubercle (see Fig 11–41), which serves to distinguish it from an inguinal hernia. The neck of the sac is narrow and lies at the femoral ring and is therefore related to the following structures: *anteriorly,* the inguinal ligament; *posteriorly,* the pectineal ligament and pubis; *medially,* the sharp edge of the lacunar ligament; and *laterally,* the femoral vein.

Because of the presence of these structures, the neck of the sac is unable to expand. Once an abdominal organ has passed through the neck into the body of the sac, it may be difficult to push the organ up and return it to the abdominal cavity *(irreducible hernia).* Furthermore, after straining or coughing, a piece of bowel may be forced through the neck, and its blood vessels may be compressed by the femoral ring, seriously impairing the blood supply *(strangulated hernia).*

A femoral hernia may be summarized as follows.

1. A protrusion of abdominal parietal peritoneum down through the femoral canal to form the hernial sac.
2. It is more common in women than in men.
3. The neck of the hernial sac lies below and lateral to the pubic tubercle.
4. The neck of the hernial sac lies at the femoral ring and at that point is related anteriorly to the inguinal ligament, posteriorly to the pectineal ligament and pubis, laterally to the femoral vein, and medially to the sharp free edge of the lacunar ligament.

In the differential diagnosis of femoral hernia, remember the possibility of confusing the swelling of the hernial sac with an enlarged superficial inguinal lymph node. There are five or six superficial inguinal nodes in the horizontal group that lie in the superficial fascia about a fingerbreadth below the inguinal ligament at the crease in the groin.

Clinical Anatomy of the Spermatic Cord

Congenital Anomalies of the Processus Vaginalis

The processus vaginalis is a peritoneal diverticulum formed in the fetus that passes through the layers of the lower part of the anterior abdominal wall to form the inguinal canal. The *tunica vaginalis* is the lower expanded part of the processus vaginalis. Normally just before birth the cavity of the tunica vaginalis becomes shut off from the upper part of the processus and the peritoneal cavity. The tunica vaginalis is thus a closed sac, invaginated from behind by the testis. The following anomalies may occur.

1. *Preformed sac of indirect inguinal hernia.* The processus may persist partially or in its entirety as a hernial sac (Fig 11–43).
2. *Congenital hydrocele.* The processus vaginalis becomes narrowed but not obliterated and remains in communication with the abdominal cavity. Peritoneal fluid accumulates in it, forming a hydrocele (see Fig 11–43).
3. *Encysted hydrocele of the cord.* The upper and lower ends of the processus become obliterated, leaving a small intermediate encysted area. An encysted hydrocele presents as a small fluctuant swelling in the inguinal region (often within the inguinal canal) that moves medially on gentle pulling of the testis and the covering of tunica vaginalis downwards (see Fig 11–43).

Clinical Anatomy of the Peritoneal Cavity

Peritoneal Fluid

The peritoneal fluid not only lubricates the surfaces of the viscera but also contains leukocytes and antibodies that have remarkable powers of resisting infection.

The peritoneal fluid circulates around the peritoneal cavity and quickly finds its way into the lymphatics of the diaphragm. While it is probable that peritoneal fluid can be absorbed at other sites in the peritoneal cavity, it is generally accepted that absorption from under the diaphragm is the most rapid route.

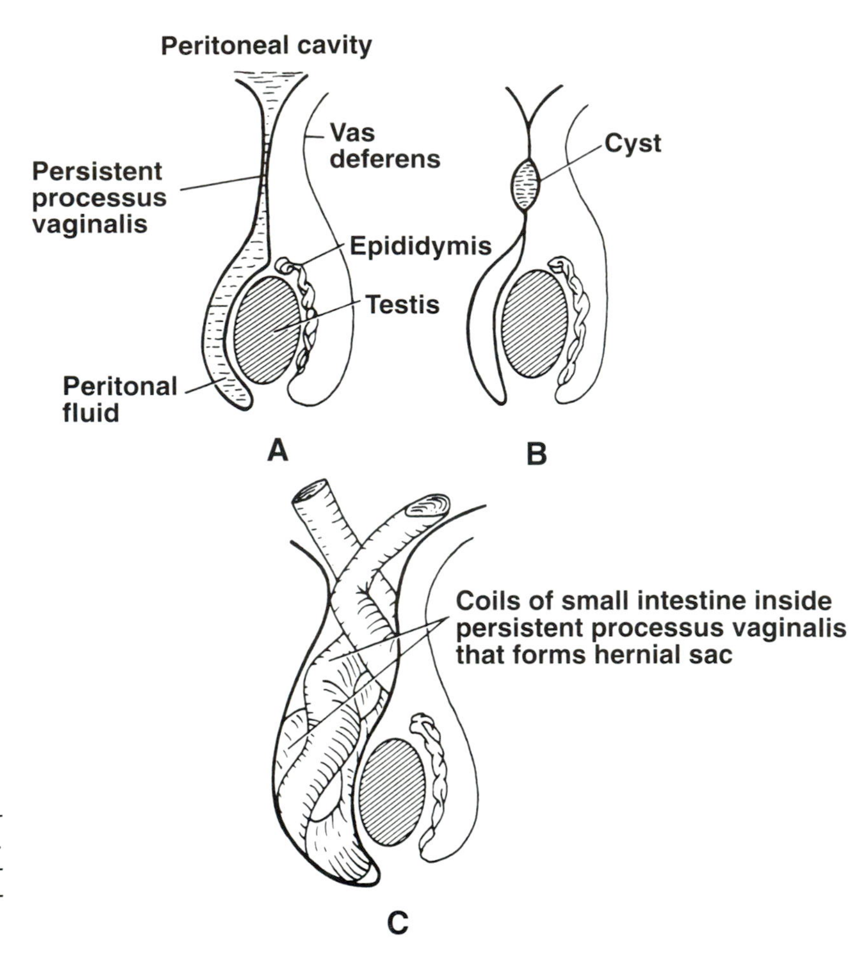

FIG 11–43.
Common congenital anomalies of the processus vaginalis. **A,** congenital hydrocele. **B,** encysted hydrocele of cord. **C,** preformed hernial sac for indirect inguinal hernia.

Ascites.—Ascites can occur secondary to hepatic cirrhosis, malignant disease, or congestive heart failure. It is essentially an excessive accumulation of peritoneal fluid within the peritoneal cavity. In a thin patient, as much as 1,500 mL has to accumulate before ascites can be recognized clinically. In obese subjects a far greater amount has to collect before it may be detected. The withdrawal of peritoneal fluid from the peritoneal cavity is described on p. 444.

Peritoneal Cavity

The peritoneal cavity is divided into an upper part within the abdomen and a lower part in the pelvis. The abdominal part is further subdivided by the many peritoneal reflections into important fossae and spaces, which, in turn, are continued into the paracolic gutters. The attachment of the transverse mesocolon and the mesentery of the small intestine to the posterior abdominal wall provides natural peritoneal barriers that may hinder the movement of infected peritoneal fluid from the upper to the lower part of the peritoneal cavity.

With the patient in a supine position, the right subphrenic peritoneal space and the pelvic cavity are the lowest areas of the peritoneal cavity, and the region of the pelvic brim is the highest area.

Peritoneal Infection.—Infection may gain entrance to the peritoneal cavity through a number of routes—(1) from the interior of the gastrointestinal tract and gallbladder, (2) through the anterior abdominal wall, (3) via the uterine tubes in females (gonococcal peritonitis in adults and pneumococcal peritonitis in children occur through this route), and (4) from the blood.

Collection of infected peritoneal fluid in one of the subphrenic spaces is often accompanied by infection of the pleural cavity. It is common to find a localized empyema in a patient with a subphrenic abscess. It is believed that the infection spreads from the peritoneum to the pleura via the diaphragmatic lymphatics. A patient with a subphrenic abscess may complain of pain over the shoulder.* The skin of the shoulder is supplied by the supraclavicular

*This also holds true for collections of blood under the diaphragm, which irritate the parietal diaphragmatic peritoneum.

nerves (C3 and 4), which have the same segmental origin as the phrenic nerve, which supplies the peritoneum in the center of the undersurface of the diaphragm.

Peritoneal Pain

Parietal Peritoneum.—The parietal peritoneum is supplied by the lower six thoracic and first lumbar nerves. Abdominal pain originating from the parietal peritoneum is therefore of the somatic type and can be *precisely localized;* it is usually severe (see section on abdominal pain, p. 459).

The parietal peritoneum in the pelvis is innervated by the obturator nerve and can be palpated by means of a rectal or vaginal examination. An inflamed appendix may hang down into the pelvis and irritate the parietal peritoneum. A pelvic examination may detect extreme tenderness of the parietal peritoneum.

Visceral Peritoneum.—The visceral peritoneum, including the mesenteries, is innervated by autonomic afferent nerves. Stretch caused by overdistension of a viscus or pulling on a mesentery, will give rise to the sensation of pain. Since the gastrointestinal tract arises embryologically as a midline structure and receives a bilateral nerve supply, pain is referred to the midline. Pain arising from an abdominal viscus is dull and *poorly localized* (see section on abdominal pain, p. 459).

Peritoneal Dialysis

Because the peritoneum is a semipermeable membrane, it allows rapid bidirectional transfer of substances across itself. Because the surface area of the peritoneum is enormous, this transfer property has been made use of in patients with acute renal insufficiency. The efficiency of this method is only a fraction of that achieved by hemodialysis.

A watery solution, the dialysate, is introduced through a catheter through a small midline incision below the umbilicus. The technique is the same as for peritoneal lavage.

Clinical Anatomy of the Abdominal Part of the Esophagus

Achalasia

The cause of this condition is unknown, but it is associated with a degeneration of the parasympathetic plexus (Auerbach's plexus) in its wall. The primary site of the disorder may be in the innervation of the cardioesophageal sphincter by the vagus nerves. Dysphagia and regurgitation are common symptoms that are later accompanied by proximal dilatation and distal narrowing of the esophagus.

Perforation of the Abdominal Esophagus

Rupture of the esophagus from nonpenetrating external trauma is rare. Hiccuping, seizures, or vomiting (Boerhaave's syndrome) may cause perforation of the wall of the lower end of the esophagus, which may extend superiorly into the left thoracic cavity. Explosive vomiting against a closed esophageal sphincter may produce linear tears of the lower esophagus (Mallory-Weiss syndrome).

Penetrating injuries to the esophagus associated with gunshot wounds to the lower thorax or upper abdomen are becoming increasingly common. Foreign-body ingestion also is responsible in a few cases. Once lodged in the narrow lower end of the esophagus, the foreign body may ulcerate through the esophageal wall.

Bleeding Esophageal Varices

At the lower third of the esophagus there is an important *portal-systemic venous anastomosis* (for other portal-systemic anastomoses, see p. 456). Here the esophageal tributaries of the azygos veins (systemic veins) anastomose with the esophageal tributaries of the left gastric vein (which drains into the portal vein). Should the portal vein become obstructed, as, for example, in *cirrhosis of the liver, portal hypertension develops,* resulting in dilatation and varicosity of the portal systemic anastomoses. Varicosed esophageal veins may rupture, causing severe hematemesis.

Anatomy of the Insertion of the Sengstaken-Blakemore Balloon for Esophageal Hemorrhage

The use of this triple-lumen tube is reserved for cases of massive esophageal hemorrhage from esophageal varices in which conservative treatment has failed. A gastric balloon anchors the tube against the esophageal-gastric junction. An esophageal balloon occludes the esophageal varices by counterpressure. The tube can be inserted through the nose or by using the oral route.

1. The patient is placed in the left lateral position or can be sitting up.
2. The lubricated tube is inserted through the external naris of the nose or through the mouth.
3. The tube is passed into the stomach, and the gastric balloon is inflated. In the average adult the distance between the external nares and the stomach is 44 cm (17.2 in.), and the distance between the in-

cisor teeth and the stomach is 41 cm (16 in.). Auscultation over the epigastrium confirms that the air is entering the gastric balloon and that the balloon is in the stomach and not in the esophagus.

Anatomy of the Complications.— Complications are as follows.

1. Difficulty in passing the tube through the nose (see passage of nasogastric tube below).
2. Damage to the esophagus, including rupture, can result from overinflation of the esophageal tube.
3. Aspiration of gastric contents during the passage of the tube. This can be partially avoided by the prior insertion of a nasogastric tube and aspiration of the gastric contents.
4. Pressure on neighboring mediastinal structures as the esophagus is expanded by balloon within its lumen. Cardiac arrythmias and pulmonary edema have been reported.
5. Persistent hiccups caused by irritation of the diaphragm by the distended esophagus and irritation of the stomach by the blood.

Clinical Anatomy of the Stomach

Congenital Hypertrophic Pyloric Stenosis
This relatively common emergency presents itself between the ages of 3 and 6 weeks. The child ejects his stomach contents with considerable force (projectile vomiting). The exact cause of the stenosis is unknown, although there is evidence that the number of autonomic ganglion cells in the stomach wall are fewer than normal. This possibly leads to prenatal neuromuscular incoordination and localized muscular hypertrophy and hyperplasia of the pyloric sphincter. It is much more common in male children.

Trauma
Apart from its attachment to the esophagus at the cardia and its continuity with the duodenum at the pylorus, the stomach is relatively mobile. It is protected on the left by the lower part of the rib cage. These factors greatly protect the stomach from blunt trauma to the abdomen. However, its extensive area makes it vulnerable to gunshot wounds.

Gastric Ulcer
Gastric ulcers occur in the alkaline-producing mucosa of the stomach, usually on or close to the lesser curvature. A chronic ulcer invades the muscular coats and will, in time, involve the peritoneum so that the stomach will adhere to neighboring structures.

Penetrating and Perforating Gastric Ulcers
An ulcer situated on the posterior wall of the stomach may penetrate into the lesser sac or become adherent to the pancreas (see Fig 11–16). Erosion into the pancreas will produce pain referred to the back. A perforating ulcer of the anterior stomach wall may result in the escape of stomach contents into the greater sac, producing diffuse peritonitis. The anterior stomach wall may adhere to the liver, and the chronic ulcer may penetrate the liver substance.

Hemorrhage From the Splenic Artery
Erosion of the left gastric artery is a cause of bleeding from a gastric ulcer. A penetrating ulcer on the posterior stomach wall may erode the splenic artery. This artery runs along the upper border of the pancreas (see Fig 11–16), and its erosion may produce a massive and even fatal hemorrhage. Even when bleeding occurs from a large artery, it is usually periodic and stops when the blood pressure falls and a clot forms. Bleeding can be so rapid that red blood passes per rectum caused by bleeding in the upper gastrointestinal tract.

Gastric Pain
Stomach pain is caused by the stretching or spasmodic contraction of the smooth muscle in its walls and is referred to the epigastrium. It is believed that the pain-transmitting fibers leave the stomach in company with the sympathetic nerves. They pass through the celiac ganglia and reach the spinal cord via the greater splanchnic nerves.

Anatomy of Nasogastric Intubation

1. The patient is placed in the semiupright position or left decubitus position to avoid aspiration.
2. The well-lubricated tube is inserted through the wider external naris and is directed backward (not upward as it may become caught on the nasal chonchae) along the nasal floor.
3. Once the tube has passed the soft palate and entered the oral pharynx, there is a feeling of decreased resistance, and the conscious patient will feel like "gagging."
4. Some important distances in the adult may be useful— external nares to cardiac orifice of the stomach is about 17.2 in. (44 cm), and the cardiac orifice to pylorus of the stomach is about 12 to 14 cm (4.8 to 5.6 in.). The curved course taken by the tube

from the cardiac orifice to the pylorus is usually longer–15 to 25 cm (6.0 to 10.0 in.).

Anatomical Structures That May Impede the Passage of the Nasogastric Tube.—The structures are as follows.

1. Deviated septum, making passage of the tube difficult on the narrower side.
2. Choanal deviation.
3. Three sites of esophageal narrowing may offer resistance to the nasogastric tube—at the beginning of the esophagus behind the cricoid cartilage (18 cm, or 7.2 in.), where the left bronchus and the arch of the aorta cross the front of the esophagus (28 cm, or 11.2 in.), and where the esophagus enters the stomach (44 cm, or 17.2 in.). The upper esophageal narrowing may be overcome by grasping the ala or wings of the thyroid cartilage between the thumb and the index finger and gently pulling the larynx forward. This manuever opens up the normally collapsed esophagus and permits the tube to pass down without further delay.

Anatomy of Complications.—The following complications may occur.

1. Entrance of the nasogastric tube went into the larynx instead of the esophagus; the patient will immediately start to cough. The tube is withdrawn into the oral pharynx, and the procedure is attempted again.
2. Bleeding from the choanae. This may occur following the rough insertion of the tube superiorly rather than inferiorly.
3. Ulceration of the mucous membrane of the nose, pharynx, and esophagus.
4. Penetration of the wall of the esophagus or stomach.
5. Intracranial perforation of the cribriform plate of the ethmoid (extremely rare). To avoid this complication, gastric tubes should be inserted orally in patients with facial fractures.

Clinical Anatomy of the Duodenum

Trauma

Apart from the first inch, the duodenum is rigidly fixed to the posterior abdominal wall by peritoneum and therefore cannot move away from crush injuries. In severe crush injuries to the anterior abdominal wall, the third part of the duodenum may be severely contused or torn against the third lumbar vertebra. Hemorrhage into the duodenal wall can be so large as to obstruct its lumen.

Because the duodenum is largely retroperitoneal, injury from blunt trauma commonly results in the escape of the contents into the retroperitoneal tissues, which may produce minimal signs and symptoms. Pain referred to the back is a common finding.

The escape of alkaline duodenal contents into the general peritoneal cavity causes immediate chemical irritation and produces the symptoms and signs of chemical peritonitis.

Duodenal Ulcer

This chronic disease occurs more commonly in men than women. The pain occurs in the epigastrium or to the right of the midline. The ulcer occurs most commonly on the anterior or posterior wall of the first part of the duodenum. It is in this region that the acid chyme is squirted into the stomach as the stomach empties.

Perforated Duodenal Ulcer

An ulcer of the anterior wall of the first inch of the duodenum may perforate into the upper part of the greater sac, above the transverse colon. The escaping fluid may localize in the right subhepatic space, or the transverse colon may direct it into the right lateral paracolic gutter and thus down to the right iliac fossa. If the latter occurs, the differential diagnosis between a perforated duodenal ulcer and a perforated appendix can sometimes be very difficult.

Hemorrhage From the Gastroduodenal Artery

An ulcer of the posterior wall of the first part of the duodenum may penetrate the wall and erode the relatively large gastroduodenal artery, causing a very severe gastrointestinal hemorrhage. The gastroduodenal artery is a branch of the hepatic artery, a branch of the celiac trunk (see Fig 11–17).

Clinical Anatomy of the Jejunum and Ileum

Trauma

Because of its extent and position, the small intestine is commonly damaged by trauma. The extreme mobility and elasticity permit the coils to move freely over one another in instances of blunt trauma. Small penetrating injuries may self-seal as the result of mucosal prolapse and contraction of the muscle wall. Large wounds leak freely into the peritoneal cavity. The presence of the vertebral column and the sacral promontary may provide a firm background for intestinal crushing in cases of midline crush injuries.

Small-bowel contents have a nearly neutral pH and produce only slight chemical irritation to the peritoneum.

Mesentery of the Small Intestine
The line of attachment of the small intestine to the posterior abdominal wall should be remembered. It extends from a point just to the left of the midline about 2 in. below the transpyloric plane (L1) downwards to the right iliac fossa. A tumor or cyst of the mesentery, when palpated through the anterior abdominal wall, will be more mobile in a direction at right angles to the line of attachment than along the line of attachment.

Pain Fibers From the Jejunum and Ileum
These nerve fibers traverse the superior mesenteric sympathetic plexus and pass to the spinal cord via the splanchnic nerves. Pain referred from this segment of the gastrointestinal tract is felt in the dermatomes supplied by the 9th, 10th, and 11th thoracic nerves. For example, strangulation of a coil of small intestine in an inguinal hernia first gives rise to pain in the region of the umbilicus. Only later, when the parietal peritoneum of the hernial sac becomes inflamed, does the pain become intense and localized to the inguinal region (see abdominal pain section, p. 458.

Mesenteric Arterial Occlusion
The superior mesenteric artery, a branch of the abdominal aorta, supplies an extensive territory of the gut, from halfway down the second part of the duodenum to the left colic flexure. Occlusion of the artery or one of its branches will result in death of all or part of this segment of the gut. The occlusion may occur as the result of an embolus, a thrombus, an aortic dissection, or an abdominal aneurysm. The embolus may originate from a mural thrombus on an atherosclerotic artery (aorta), from endocarditis, or from a prosthetic heart valve. The main stem of the artery may become involved in an aneurysm of the abdominal aorta (superior mesenteric artery arises from aorta at level of L2), or from an aortic dissection of the thoracic aorta. Severe impairment of the circulation of the superior mesenteric artery can accompany hypovolemic shock, congestive heart failure, or myocardial infarction and may cause intestinal infarction.

Mesenteric Vein Thrombosis
The superior mesenteric vein, which drains the same area of the gut as supplied by the superior mesenteric artery, may undergo thrombosis following stasis of the venous bed. Cirrhosis of the liver with portal hypertension, peritonitis, or polycythemia vera may predispose to the condition.

Meckel's Diverticulum
This congenital anomaly represents a persistent portion of the vitellointestinal duct, which connects the midgut to the yolk sac in the developing embryo. The diverticulum is located on the antimesenteric border of the ileum about 2 ft from the ileocecal junction. It is about 2 in. long and occurs in about 2% of individuals. The diverticulum is important clinically, since it may possess a small area of gastric mucosa, and bleeding may occur from a "gastric" ulcer in its mucous membrane. Should a fibrous band connect the diverticulum to the umbilicus, a loop of small bowel may become wrapped around it, causing intestinal obstruction.

Clinical Anatomy of the Appendix

Variability of Position
The inconstancy of the position of the appendix should be kept in mind when attempting to diagnose appendicitis. A retrocecal appendix, for example, may lie behind a cecum distended with gas, and thus it may be difficult to elicit tenderness on palpation in the right iliac region. Irritation of the psoas muscle, on the other hand, may cause the patient to keep the right hip joint flexed.

An appendix hanging down in the pelvis may result in absent abdominal tenderness in the right iliac region but deep tenderness in the hypogastric region. Rectal or vaginal examination may elicit tenderness of the peritoneum in the pelvis on the right side.

A pelvic appendicitis may give rise to symptoms related to the rectum (painful defecation), bladder (frequency of micturition and painful micturition), and uterine tubes or ovaries (dysmenorrhea).

Predisposition of the Appendix to Infection
The following factors contribute to why the appendix is prone to infection: (1) it is a long, narrow, blind-ended tube, (2) it has a large amount of lymphoid tissue in its wall, and (3) the lumen has a tendency to become obstructed by enteroliths.

Predisposition of the Appendix to Perforation

The appendix is supplied by a long artery that does not anastomose with other arteries. The blind end of the appendix is supplied by the terminal branches of the appendicular artery. Inflammatory edema of the appendicular wall compresses the blood supply to the appendix and often leads to thrombosis of the appendicular artery. These conditions commonly result in necrosis or gangrene of the appendicular wall, with perforation.

Appendicitis and the Greater Omentum

As the appendix becomes acutely inflamed, a sticky inflammatory exudate is formed on the peritoneal surface of the appendix. Often the greater omentum becomes adherent and wraps itself around the infected organ. By this means the infection is often localized to a small area of the peritoneal cavity, thus saving the patient from a serious diffuse peritonitis.

Pain of Appendicitis

Visceral pain in the appendix is produced by distension of its lumen or spasm of its muscle. The afferent pain fibers enter the spinal cord at the level of the tenth thoracic segment, and a *vague referred pain* is felt in the region of the umbilicus. Later, the pain shifts to where the inflamed appendix irritates the parietal peritoneum. Here, the pain is *precise, severe,* and *localized* (see abdominal pain section, p. 458).

Clinical Anatomy of the Colon

Trauma

Blunt or penetrating injuries to the colon occur. Blunt injuries are not common and usually occur in automobile accidents when seat belts have been improperly placed around the waist. The common sites of injury are where the mobile parts of the colon (transverse and sigmoid) join the fixed parts (ascending and descending).

Penetrating injuries following stab wounds or gunshot wounds to the colon are very common. In fact, the colon is second only to the liver in its frequency of injury. Because of the large bacterial content of the colon, a severe form of peritonitis will follow perforation unless rapid treatment, in the form of antibiotics and surgery, takes place. Once the leakage of bacteria and feces are eliminated, the peritoneal cavity possesses remarkable powers of recovery.

The multiple anatomical relationships of the different parts of the colon explain why isolated colonic trauma is unusual. Coils of small intestine, liver, stomach, kidneys, spleen, pancreas, diaphragm, and even the urinary bladder may also be damaged. The morbidity and mortality of patients with colonic injuries rises rapidly as the number of other organs injured increases.

Diverticulosis and Diverticulitis

Hypertrophy of the colonic muscle associated with dietary fiber deficiency is probably responsible for the high intraluminal pressure found in patients with diverticular disease. A herniation of the mucous membrane lining through the circular muscle occurs between the teniae coli (bands of longitudinal muscle). The hernial site is at points of anatomical weakness, i.e., where the blood vessels pierce the circular muscle (Fig 11–44). The accumulation of fecal material or a fecolith in a mucosal diverticulum, which has no muscle in its wall, results in stagnation of its contents and predisposes to diverticulitis.

The emergency physician may be faced with the complications of diverticulitis, which include (1) a tender abdominal or pelvic mass along the course of the colon (walling off of the inflammatory process by the adherence of surrounding structures, including the omentum, coils of small intestine, mesocolon, bladder, and abdominal wall), (2) fistula (rupture of a pericolic abscess into adjacent anatomical structures, such as the small intestine or bladder), (3) peritonitis (rupture of a pericolic abscess into the peritoneal cavity).

Volvulus

Because of its extreme mobility, the sigmoid colon sometimes rotates around its mesentery. This may correct itself spontaneously, or the rotation may continue until the blood supply of the gut is cut off completely. The rotation commonly occurs in a counterclockwise direction. Cecal volvulus sometimes occurs, especially if the peritoneal attachment of the cecum to the posterior abdominal wall is excessively slack. In these conditions the cecum becomes twisted on its long axis.

Intussusception

This is the telescoping of a proximal segment of the bowel into the lumen of an adjoining distal segment, with the risk of ischemia and gangrene. It is a common form of intestinal obstruction in children. Ileocolic, colocolic, and ileoileal forms of intussusception do occur, but ileocolic is the most common.

The high incidence of this condition in young

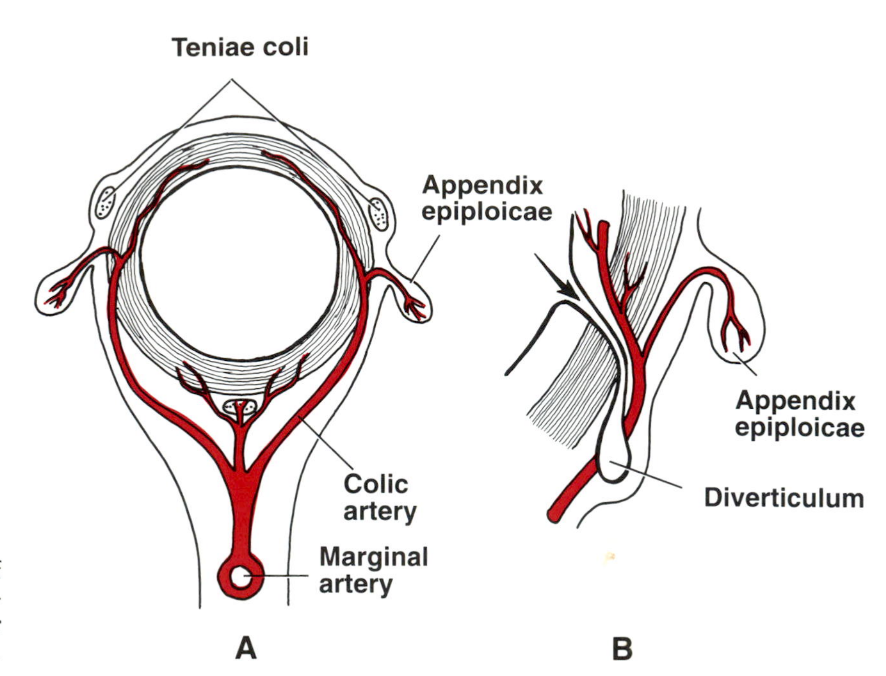

FIG 11–44.
A, blood supply to colon. **B,** formation of diverticulum. Note passage of the mucosal diverticulum through the circular muscle coat along the course of the artery.

children may be because the large bowel is relatively larger when compared with the small intestine at this time of life. Swelling of the Peyer's patches in the distal part of the ileum secondary to infection may also play a role. Any lesion on the bowel wall, such as a polyp, a carcinoma, or even a Meckel's diverticulum, may be sufficient to stimulate peristalsis and start the invaginating process.

Clinical Anatomy of the Liver

Trauma

The liver is the largest organ in the abdomen and is frequently the site of major, exsanguinating injury. Blunt traumatic injury, secondary to automobile accidents, is very common; penetrating injuries caused by gunshot or stab wounds are also common. Since the liver lies within the lower part of the rib cage, it must not be forgotten when rib fractures or penetrating wounds of the chest are assessed.

The liver is a soft, friable structure enclosed in a fibrous capsule. Liver injuries may be (1) transcapsular, (2) subcapsular, and (3) central. In transcapsular tears, the visceral layer of peritoneum is also damaged, and blood and bile escapes into the peritoneal cavity, giving rise to some distension and tenderness in the right upper quadrant. Because of its size, the right lobe of the liver is involved much more often than the left lobe.

Because the bile ducts, hepatic arteries, and the portal vein are distributed in a segmental manner, appropriate ligation of these structures allows the surgeon to remove large portions of the liver in patients with severe traumatic lacerations.

Peritoneal Spaces Around the Liver

The important *subphrenic spaces* have already been referred to on p. 415. Under normal conditions these are potential spaces only, and the peritoneal surfaces are in contact. An abnormal accumulation of gas or fluid is necessary for the separation of the peritoneal surfaces.

Clinical Anatomy of the Portal Vein

Blood Flow in the Portal Vein

The portal vein conveys about 70% of blood to the liver. The remaining 30% is oxygenated blood, which passes to the liver via the hepatic artery. The wide angle of union of the splenic vein with the superior mesenteric vein to form the portal vein leads to streaming of the blood flow in the portal vein. The right lobe of the liver receives blood mainly from the intestine, whereas the left lobe plus the quadrate lobe and the caudate lobe receive blood from the stomach and spleen. This distribution of blood may explain the distribution of secondary malignant deposits in the liver.

Portal-Systemic Anastomoses

Under normal conditions, the portal venous blood traverses the liver and drains into the inferior vena cava of the systemic venous circulation by way of the hepatic veins. This is the direct route. However, other, smaller communications exist between the portal and systemic systems, which become important should the direct route become blocked.

These communications are as follows.

1. At the lower end of the esophagus, the esophageal branches of the left gastric vein (portal tributary) anastomose with the esophageal veins, draining the middle third of the esophagus into the azygos veins (systemic tributary).

2. Halfway down the anal canal, the superior rectal veins (portal tributary) draining the upper half of the anal canal anastomose with the middle and inferior rectal veins (systemic tributaries), which are tributaries of the internal iliac and internal pudendal veins, respectively.

3. The paraumbilical veins connect the left branch of the portal vein with the superficial veins of the anterior abdominal wall (systemic tributaries). The paraumbilical veins travel in the falciform ligament.

4. The veins of the ascending colon, descending colon, duodenum, pancreas, and liver (portal tributaries) anastomose with the renal, lumbar, and phrenic veins (systemic tributaries).

Portal hypertension secondary to cirrhosis of the liver is a common clinical condition and causes enlargement of the portal-systemic anastomoses. See p. 451 for a discussion of esophageal varices.

Clinical Anatomy of the Gallbladder and Bile Ducts

Trauma

Penetrating injuries of the gallbladder are rare and usually follow gunshot or stab wounds. Because of the location of the greater part of the gallbladder (body and neck) beneath the visceral surface of the liver, in the majority of cases liver damage also is present. The escape of sterile bile into the peritoneal cavity produces a chemical peritonitis with pain and guarding in the upper quadrants.

Cholelithiasis

Gallstones are usually asymptomatic; however, they may give rise to gallstone colic or produce acute cholecystitis.

Biliary colic is caused by spasm of the smooth muscle of the wall of the gallbladder in an attempt to expel a gallstone. Afferent nerve fibers ascend through the celiac plexus and the greater splanchnic nerves to the thoracic segments of the spinal cord. Referred pain is felt in the right upper quadrant or the epigastrium (T7 through T9 dermatomes).

Acute cholecystitis is usually a febrile condition with discomfort in the right upper quadrant or epigastrium (see referred pain, above). Inflammation of the gallbladder may cause irritation of the subdiaphragmatic parietal peritoneum, which is supplied in part by the phrenic nerve (C3 through C5). This may give rise to referred pain over the shoulder, since the skin in this area is supplied by the supraclavicular nerves (C3 and 4).

Clinical Anatomy of the Pancreas

The deep location of the pancreas sometimes gives rise to problems of diagnosis for the following reasons. (1) Pain from the pancreas is commonly referred to the back. (2) Since the pancreas lies behind the stomach and transverse colon, disease of the gland may be confused with that of the stomach or transverse colon. (3) Inflammation of the gland may cause inflammation of the peritoneum of the lesser sac, with the development of a pseudocyst.

Trauma

The pancreas is deeply placed within the abdomen and is well protected by the costal margin and the anterior abdominal wall. It is most commonly damaged by gunshot or stab wounds. Such injuries are usually accompanied by damage to adjacent organs such as the liver, stomach, duodenum, bowel, kidney, and spleen.

The fixation of the pancreas behind the peritoneum makes it vulnerable to blunt trauma. A sports injury in which there is a sudden blow to the abdomen may compress the pancreas against the vertebral column. The neck may be torn as it lies in front of the rigid first and second lumbar vertebrae.

Damaged pancreatic tissue releases activated pancreatic enzymes that produce the signs and symptoms of an acute peritonitis. Because of its position behind the lesser sac, these are not immediately evident and tend to develop slowly over a 12-hour period.

Acute Pancreatitis

The cause of this disease appears to be related to the anatomy of the duct system and its relationship with the biliary ducts. The following facts have been suggested. (1) A reflux of bile into the main pancreatic duct, thus activating the pancreatic enzymes, may occur secondary to alcoholic relaxation of the sphincter of Oddi. Blockage of the ampulla of Vater with a gallstone or spasm of the sphincter of Oddi could also cause bile to enter the pancreatic duct in cases where the bile and pancreatic ducts unite at the ampulla. (2) Spread of infection from the biliary system into the pancreatic ducts activating the pancreatic enzymes can occur.

Because of the close relationship of the pancreas to the lesser sac, necrosis of the gland results in a serohemorrhagic exudate into the peritoneal sac, producing severe epigastric pain radiating straight through to the back.

Clinical Anatomy of the Spleen

Trauma

Penetrating injury to the spleen can occur from gunshot or stab wounds. The entrance wound may be in the lower left thorax or upper left abdomen. Because the spleen is closely related to other abdominal organs, these may be damaged also, including the stomach, the left kidney, the pancreas, and the splenic flexure of the colon. A thoracic wound also involves the left lung and pleura and the left dome of the diaphragm.

Blunt trauma to the spleen can occur from automobile accidents, falls, and during contact sports. Although the spleen is often an isolated injury, other organs may also be damaged, including the liver, left lung, small intestine, pancreas, and stomach. Fractures of the left lower ribs are frequently associated with a ruptured spleen. Any spleen that is pathologically enlarged (as in infectious mononucleosis or malaria, for example) is more prone to injury than a normal spleen.

Splenic injuries include *linear* and *stellate tears of the capsule* and *subcapsular hematomas.* Because of the extremely friable texture of the splenic pulp and the large size of the splenic artery, severe hemorrhage into the peritoneal cavity can occur. The hemorrhage is immediate when the capsule is torn or delayed in cases of subcapsular hematoma.

Pain and tenderness in the upper quadrants (especially the left) are common clinical findings caused by hemorrhagic irritation of the parietal peritoneum. If the blood migrates up beneath the diaphragm, it may irritate the central parietal peritoneum innervated by the phrenic nerve (C3 through C5), giving referred pain to the skin over the left shoulder (innervated by the supraclavicular nerves, C3 and C4). Depending on the severity of the tear, the patient will exhibit different degrees of hypovolemia.

Clinical Anatomy of the Retroperitoneal Space

Trauma

Retroperitoneal trauma may present very difficult problems in diagnosis. Palpation of the anterior abdominal wall in the lumbar and iliac regions may give rise to signs indicative of peritoneal irritation, including tenderness and rigidity. Palpation of the back may reveal tenderness in the costovertebral angle, suggestive of kidney disease. Abdominal radiographs may demonstrate intraperitoneal air under the diaphragm or air in the extraperitoneal tissues, indicating perforation of a viscus. Computed tomographic scans, especially when combined with contrast media, can often accurately define the extent of injury to extraperitoneal organs.

Abscess Formation

Infection originating in retroperitoneal organs, such as the kidney (perinephric abscess), lymph nodes (para aortic nodes), and appendix (retrocecal appendicitis), may extend widely into the fatty areolar tissue, producing diffuse abscesses.

Leaking Aortic Aneurysms

The blood may first be confined to the retroperitoneal space or may escape into the peritoneal cavity. Aortic aneurysms are fully discussed in Chapter 15.

Clinical Anatomy of Abdominal Pain

One of the most important problems facing the emergency physician is the accurate assessment of the patient with abdominal pain. This section is intended to provide a detailed anatomical basis for such an assessment.

Pain Receptors

Free nerve endings are the pain receptors. They are found between the epithelial cells of the skin, in the alimentary tract, and in connective tissues, including the dermis, fascia, and muscle. The terminal endings are devoid of myelin sheaths, and there are no Schwann cells covering their tips.

The pain receptors are stimulated by mechanical damage, by excessive heat or cold, and by chemicals. They are unique among sensory receptors in having very little ability to adapt, and in fact they become increasingly sensitive if the stimulus is continued. The process of depolarization of the terminal nerve fibers by the injury is probably brought about by the release of substances, such as bradykinin, arachidonic acid, or prostaglandins.

Abdominal Pain

Three distinct forms of abdominal pain exist—somatic pain, visceral pain, and referred pain.

Somatic abdominal pain in the abdominal wall may arise from the skin, fascia, muscles, and parietal peritoneum. It may be severe and can be precisely localized. When the origin is on one side of the midline, the pain is also lateralized. The somatic pain impulses from the abdomen reach the central nervous system in the following segmental spinal nerves: *central part of diaphragm,* phrenic nerve (C3 through C5); *peripheral part of diaphragm,* intercostal nerves (T7 through T11); *anterior abdominal wall,* thoracic nerves (T7 through T12), and first lumbar nerve; and *pelvic wall,* obturator nerve (L2 through L4).

The inflamed parietal peritoneum is extremely sensitive, and, because the full thickness of the abdominal wall is innervated by the same nerves, it is not surprising to find cutaneous hyperesthesia and tenderness. Local reflexes involving the same nerves bring about a protective phenomenon in which there is an increased tone of the abdominal muscles. This increased tone or *rigidity,* sometimes called *guarding,* is an attempt to rest and localize the inflammatory process.

Rebound tenderness occurs when the parietal peritoneum is inflamed. Any movement of that inflamed peritoneum, even when that movement is elicited by removing the examining hand from a site distant from the inflamed peritoneum, brings about tenderness.

Examples of acute, severe, localized pain originating in the parietal peritoneum are seen in the later stages of appendicitis. Cutaneous hyperesthesia, tenderness, and muscular rigidity occur in the right lower quadrant of the anterior abdominal wall. A perforated peptic ulcer, in which there is chemical irritation of the parietal peritoneum, will produce the same symptoms and signs but will involve the right upper and right lower quadrants.

Visceral abdominal pain arises in abdominal organs, visceral peritoneum, and the mesenteries. The causes of visceral pain include stretching, distension, ischemia, and chemical damage. Pain arising from an abdominal viscus is dull and poorly localized. The sensory endings have a high threshold and a very slow rate of adaptation. Visceral pain is referred to the midline, probably because the viscera develop embryologically as midline structures and receive a bilateral nerve supply.

Colic is a form of visceral pain produced by the violent contraction of smooth muscle; it is commonly caused by luminal obstruction in the intestine, biliary ducts, or ureters.

Many of the visceral afferent fibers that enter the spinal cord participate in reflex activity. Reflex sweating, salivation, nausea, vomiting, and tachycardia may accompany visceral pain.

The sensations that arise in viscera reach the central nervous system in afferent nerves that accompany the sympathetic nerves and enter the spinal cord through the posterior roots. The significance of this pathway is better understood when referred visceral pain is discussed below.

Referred abdominal pain is the feeling of pain at a location other than that of the site of origin of the stimulus, but in an area supplied by the same or adjacent segments of the spinal cord. Both somatic and visceral structures can produce referred pain.

In the case of referred somatic pain, the possible explanation is that the nerve fibers from the diseased structure and the area where the pain is felt ascend in the central nervous system along a common pathway, and the cerebral cortex is incapable of distinguishing between the sites. Examples of referred somatic pain are as follows. Pleurisy involving the lower part of the costal parietal pleura may give rise to referred pain in the abdomen because the lower parietal pleura receives its sensory innervation from the lower five intercostal nerves, which also innervate the skin and muscles of the anterior abdominal wall.

Visceral pain from the stomach is commonly referred to the epigastrium (Fig 11–45). The afferent pain fibers from the stomach ascend in company with the sympathetic nerves and pass through the celiac plexus and the greater splanchnic nerves. The sensory fibers enter the spinal cord at segments T5 through T9 and give rise to referred pain in the dermatomes T5 through T9 on the lower chest and abdominal walls.

Visceral pain from the appendix (see Fig 11–45), which is produced by distension of its lumen or spasm of its muscle, travels in nerve fibers that accompany sympathetic nerves through the superior mesenteric plexus and the lesser splanchnic nerve to the spinal cord (T10 segment). The vague referred pain is felt in the region of the umbilicus (T10 dermatome). Later, when the inflammatory process involves the parietal peritoneum, the severe somatic pain dominates the clinical picture and is localized precisely to the right lower quadrant (see above).

Visceral pain impulses from the gallbladder (acute cholecystitis, gallstone colic) travel in nerve fibers that accompany sympathetic fibers through the celiac plexus and the greater splanchnic nerves to the spinal cord (segments T5 through T9). The vague referred pain is felt in the dermatomes (T5 through T9) on the lower chest and upper abdominal walls (see Fig 11–45). Should the inflammatory process spread to involve the parietal peritoneum of the anterior abdominal wall or peripheral diaphragm, the severe somatic pain will be felt in the right upper quadrant and through to the back below the inferior angle of the scapular. Involvement of the central diaphragmatic parietal peritoneum, which is innervated by the phrenic nerve (C3 through C5), may give rise to referred pain over the shoulder, since the skin in this area is innervated by the supraclavicular nerves (C3 and C4).

ANATOMICAL-CLINICAL PROBLEMS

1. An obese 40-year-old woman was seen in the emergency department complaining of a severe pain over the right shoulder and in her right side and back below the right shoulder blade. She said that she had experienced the pain on several occasions before and that when she ate fatty foods, it seemed to make the pain worse. Ultrasound demonstrated the presence of gallstones. Her condition was diagnosed as cholelithiasis, and the pain was attributed to gallstone colic. Where would you attempt to pal-

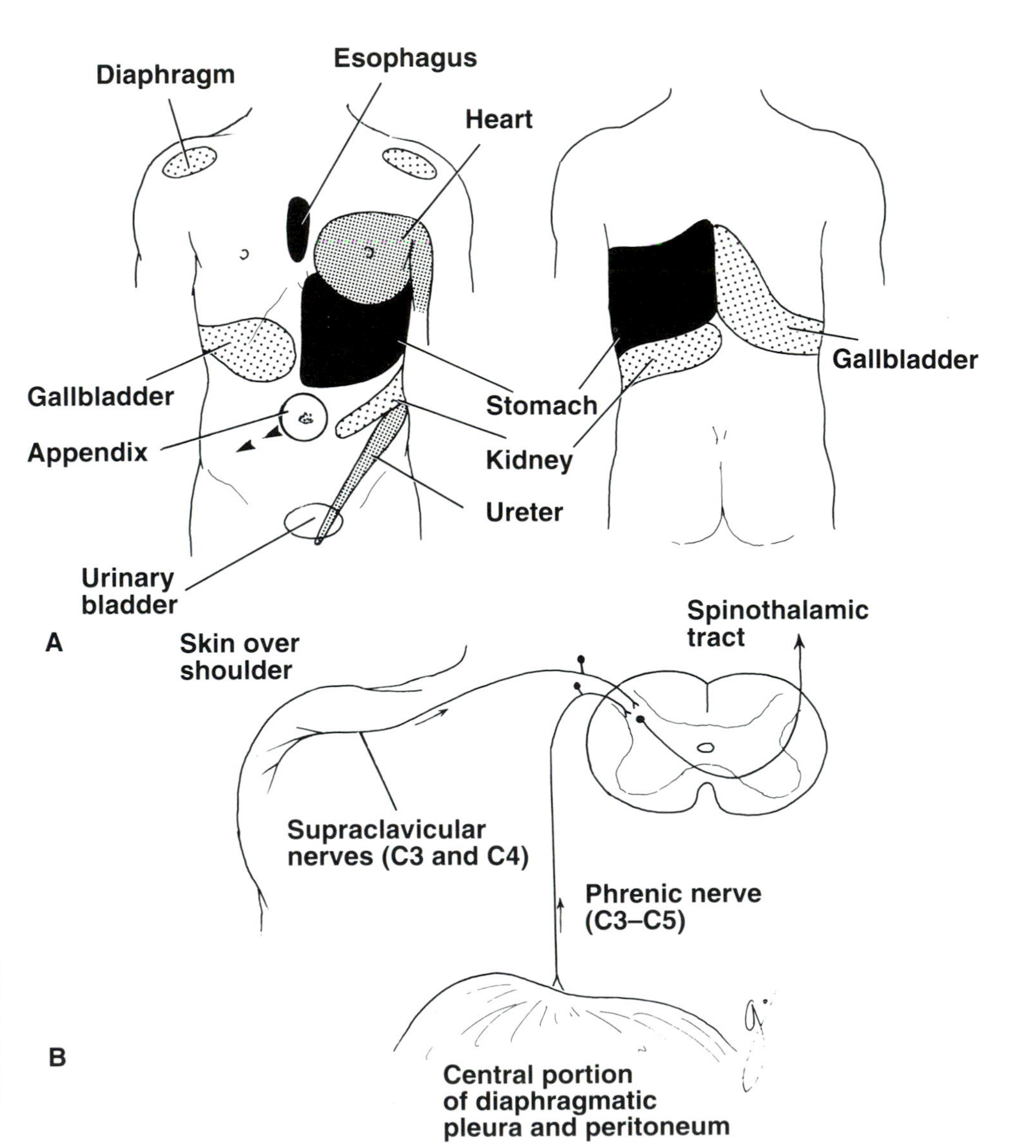

FIG 11–45.
A, some important skin areas involved in referred visceral pain. **B,** convergence of somatic and visceral afferent pain fibers in the posterior gray column of the spinal cord.

pate the gallbladder? In anatomical terms, explain the distribution of the pain to the right shoulder, right side, and back.

2. A 10-year-old boy was seen in the emergency department with a temperature of 101° F, a furred tongue, and pain in the right iliac region. On examination, the skin in the right lower quadrant was hyperesthetic and tender to touch; the abdominal muscles showed rigidity and guarding. A diagnosis of acute appendicitis was made. In anatomical terms, explain the presence of pain, tenderness, hyperesthesia, and muscle rigidity in the right iliac region in this patient.

3. A workman engaged in demolishing a building lost his balance and fell astride a girder on the floor below. On examination he was found to have extensive swelling of his perineum, scrotum, and penis. He was unable to urinate normally, passing only a few drops of blood-stained urine. The lower part of the anterior abdominal wall was also swollen, but his thighs were normal. Can you explain the distribution of the swelling in anatomical terms?

4. A 25-year-old woman was brought by ambulance to the emergency department from a local riding stable. She had apparently received a severe blow on the anterior abdominal wall from the hind leg of a horse. On examination, there was extensive bruising and swelling of the skin over the lower part of the left rectus muscle. On gentle palpation, a deep swelling confined to the left rectus sheath was felt. Using your knowledge of anatomy, make a diagnosis. Which blood vessel was likely to have been ruptured?

5. A 37-year-old man was involved in a shootout in a liquor store 1 year previously, when a bullet had ricocheted off the wall and entered his left flank. Fortunately, the bullet had not entered the peritoneal cavity. Recently, in addition to diminished skin sensation over the left lumbar region and umbilicus, he noticed a bulging forward of the left side of his anterior abdominal wall. Can you explain these defects in anatomical terms?

6. A 28-year-old man was seen in the emergency department with a stab wound in the right inguinal region. Name the nerves that innervate the skin and deep structures in the inguinal region. From which segment of the spinal cord do they originate? Which important surface landmarks would you use when blocking these nerves?

7. A 41-year-old woman noticed a painful swelling in her right groin after helping her husband move some heavy furniture. On examination, a small tender swelling was noted in the right groin, situated below and lateral to the pubic tubercle. What is the likely diagnosis? Give your reasons in anatomical terms.

8. A 25-year-old man involved in purchasing drugs was knifed in the abdomen in the left upper quadrant. On examination in the emergency department, it was difficult to determine whether the knife had penetrated into the peritoneal cavity. It was decided to do a midline peritoneal lavage below the umbilicus to see if there was any free blood in the peritoneal cavity. Name the layers of tissue that the trocar and cannula would have to penetrate in order to enter the peritoneal cavity. Name the possible anatomical complications.

9. While a 70-year-old man was being examined in the emergency department for abdominal pain, it was noticed that he had a bulge in his left groin. On closer examination, an elongated swelling was seen above the medial end of the left inguinal ligament. When the patient coughed, the swelling enlarged, but did not descend into the scrotum. The patient had weak abdominal muscles. What is the cause of the swelling? Why did the swelling enlarge when the patient coughed?

10. A 60-year-old man with a long history of duodenal ulcer was seen in the emergency department after vomiting blood-stained fluid and exhibiting all the signs and symptoms of severe hypovolemia. Which blood vessel was likely to have been eroded? Where in the duodenum was the ulcer likely to have been situated?

11. The differential diagnosis between a perforated duodenal ulcer and a perforated appendix may prove difficult. Can you explain this in anatomical terms?

12. The pain of acute appendicitis is commonly first felt in the umbilical region and later in the right iliac region. Can you explain this in anatomical terms?

13. A 52-year-old man was seated in a restuarant when he suddenly started to vomit blood. He was taken to the emergency department of a local hospital. On examination, he had all the signs of severe hypovolemic shock. His wife, who had ridden with him in the ambulance, said that he was a chronic alcoholic and that he had vomited blood before, 6 months previously, and had nearly died. On palpation of the anterior abdominal wall the right lobe of the liver was felt to extend four fingerbreadths below the costal margin. A number of superficial veins around the umbilicus were seen to be enlarged. How might the previous findings give a clue to the cause of his bleeding?

14. When a nasogastric tube is passed, what are the anatomical areas of narrowing that impair the forward movement of the tube down into the stomach? How far is the entrance into the stomach from the external nares? Approximately, how far is the entrance into the duodenum from the external nares?

15. A 45-year-old woman with a history of flatulent dyspepsia suddenly experienced an excruciating colicky pain across the upper part of the abdomen. On examination in the emergency department, she was found to have some rigidity and tenderness in the right hypochondrium. A diagnosis of biliary colic was made. What is the mechanism responsible for pain in gallstone colic?

16. Following a terrorist bomb attack, a young child was seen in the emergency department with a gaping hole in the anterior abdominal wall, and coils of gut were exposed on the surface. How would you determine by inspection whether the gut was small or large intestine?

17. A 20-year-old football player was accidently kicked on the left side of his chest. On returning to the locker room he said he felt faint and collapsed to the floor. On examination in the emergency department, he was found to be in hypovolemic shock. He had tenderness and guarding in the left upper quadrant of his abdomen. He also had extreme local tenderness over his left tenth rib in the midaxillary line. A diagnosis of a ruptured spleen and the possibility of a fractured tenth rib was made. Explain the tenderness and guarding in the abdomen in this patient.

18. A 21-year-old man was seen in the emergency department complaining of a painful swelling in the left groin; he had vomited three times in the last 2 hours. On examination, he was seen to be dehydrated, and his abdomen was moderately distended. Auscultation of the anterior abdominal wall revealed moderate borborygmi. A large, tense swelling, which was very tender on palpation, was seen in the left groin. The swelling extended from above and medial to the pubic tubercle down into the scrotum. An attempt to gently push the contents of the swelling back into the abdomen was found to be impossible. A diagnosis of a left complete indirect inguinal hernia was made, with incarceration of some small intestinal loops. Explain the origin of the hernial sac. Where is the neck of the hernial sac in relation to the inguinal canal? What is the relationship between the neck of the sac and the inferior epigastric artery? Which is the most common—a strangulated indirect inguinal hernia or a strangulated femoral hernia? Does a direct inguinal hernia ever strangulate?

ANSWERS

1. The fundus of the gallbladder lies against the anterior abdominal wall next to the tip of the right ninth costal cartilage, i.e., where the linea semilunaris crosses the costal margin on the right side. The parietal peritoneum in this area is innervated by the eighth and ninth intercostal nerves, which give rise to referred pain in the eighth and ninth dermatomes, thus explaining the pain in the side and back. The parietal peritoneum on the undersurface of the diaphragm (central part) is supplied by the phrenic nerve (C3 through C5). It must be assumed that the patient had cholecystitis, in addition to the gallstone colic, and that the inflamed gallbladder was irritating this area of the peritoneum also. The pain was referred to the shoulder along the supraclavicular nerves (C3 and C4), which supply the skin of this region.

2. An acutely inflamed appendix produces a peritonitis, which is localized at first, but may become generalized later if the appendix ruptures. Inflammation of the parietal peritoneum causes pain over the area and reflex spasm of the anterior abdominal muscles. The parietal peritoneum, the abdominal muscles, and the overlying skin are supplied by the same segmental nerves (T11 and T12 and L1). This is essentially a protective mechanism and an attempt to keep that area of the abdomen at rest so that the inflammatory lesion will remain localized.

3. The patient's fall ruptured the urethra in the perineum. When he attempted to micturate, the urine extravasated beneath the Colles' fascia (within the superficial perineal pouch see p. 406). The urine then passed over the scrotum and penis under the membranous layer of the superficial fascia and up onto the anterior abdominal wall. It did not pass backward because of the attachment of the fascia to the posterior edge of the perineal membrane (see Fig 11–3) and did not extend into the thighs because of the attachment of the fascia to the fascia lata, just below the inguinal ligament.

4. The sudden unexpected blow to the anterior abdominal wall caused the left inferior epigastric artery to rupture, and bleeding occurred into the left rectus sheath.

5. The bullet cut the ninth and tenth intercostal nerves just inferior to the costal margin on the left side. The diminished skin sensation was due to the loss of the sensory nerve supply to the ninth and tenth thoracic dermatomes, i.e., a band of skin extending forward to the region of the umbilicus. Portions of the olique, transversus, and rectus abdominis muscles on the left side were paralyzed. Atrophy of these muscles resulted in loss of support to the abdominal viscera, which consequently sagged forward (visceroptosis).

6. The iliohypogastric and ilioinguinal nerves innervate the inguinal region. They originate from the L1 segment of the spinal cord. The anterior superior iliac spine and the umbilicus are the landmarks used in blocking these nerves. The needle is introduced about 1 in. (2.5 cm) medial along a line joining the spine and the umbilicus. Local spread of the anesthetic may also block the superficial terminal branches of T12 and possibly T11 and the genital branch of the genital femoral nerve.

7. Following excessive exertion and an increase in intra-abdominal pressure, a hernial sac (formed of parietal peritoneum) was forced down through the right femoral canal. The patient had a right-sided femoral hernia. The neck of a femoral hernial sac is always situated below and lateral to the pubic tubercle.

8. In a midline subumbilical peritoneal lavage, the trocar and cannula would pierce the following structures: skin, fatty layer of superficial fascia, membranous layer of superficial fascia, deep fascia, linea alba, fascia transversalis, extraperitoneal tissue, and the parietal peritoneum.

Possible anatomical complications include bleeding from small blood vessels in the superficial fascia, giving a false-positive result; failure to enter the peritoneal cavity in the midline so that the trocar pierces the vascular rectus muscle and sheath instead of the avascular linea alba; perforation of the gut; perforation of the mesenteric vessels or the blood vessels on the posterior abdominal wall; and perforation of the full bladder.

9. The inguinal swelling was a direct inguinal hernia caused by the weak abdominal muscles. Since the hernial sac of a direct inguinal hernia is wide and in direct communication with the peritoneal cavity, a rise in intra-abdominal pressure on coughing will cause the hernial swelling to expand.

10. Hemorrhage from a duodenal ulcer most commonly involves the gastroduodenal artery, a branch of the hepatic artery. Since the artery passes down behind the first part of the duodenum, the ulcer is most likely situated on the posterior wall of the first part of the duodenum.

11. A small perforation of the first part of the duodenum may result in the contents of the duodenum running down the right paracolic gutter to the right iliac fossa. The signs then closely resemble those of a perforated appendix.

12. The initial pain of acute appendicitis is a vague discomfort and is referred to the umbilical region. The afferent nerve fibers accompany the sympathetic nerves to the superior mesenteric plexus. They enter the thorax via the splanchnic nerves and enter the spinal cord at the level of the tenth thoracic segment. The tenth thoracic intercostal nerve supplies the skin of the umbilicus, and for this reason pain is commonly referred to the umbilicus. Once the inflammatory process has extended beyond the confines of the appendix and has involved the parietal peritoneum, a severe localized pain is felt in the right iliac region. The parietal peritoneum, the overlying muscles of the anterior abdominal wall, and the covering skin are all supplied by the first lumbar nerve.

13. The diagnosis is cirrhosis of the liver secondary to chronic alcoholism. The normal flow of portal venous blood through the liver is impaired in liver cirrhosis. Portal hypertension develops, and the venous blood returns to the general circulation via the enlarged and congested portal systemic anastomoses. One such anastomosis is located at the lower end of the esophagus. Here the esophageal tributaries of the azygos vein (systemic vein) are joined to the esophageal tributaries of the left gastric vein (portal vein). Rupture of a varicosity in the congested anastomosis was responsible for the severe hematemesis in this case.

The paraumbilical veins link the superficial abdominal wall veins (systemic) to the portal vein. With portal hypertension these veins become congested and visible around the umbilicus.

14. The passage of a nasogastric tube is fully described on p. 452. The distance between the stomach and the external nares is about 44 cm (17.2 in.), and the distance between the duodenum and the external nares is about 59–69 cm (23.2–27.2 in.).

15. Pain in gallstone colic is caused by spasm of the smooth muscle in the wall of the gallbladder and distension of the bile ducts by the stones.

16. A list of the naked-eye differences between the small and large intestine are given on p. 422.

17. Initially in this patient, the spleen underwent a subcapsular hemorrhage, and later, in the locker room, the capsule gave way, allowing the

blood to escape into the peritoneal cavity. The presence of blood in the peritoneal cavity irritated the parietal peritoneum, causing tenderness in the left upper quadrant and reflex guarding of the muscles in the same area.

18. An indirect inguinal hernia is now recognized as a congenital anomaly in which the hernial sac is the remains of the processus vaginalis. The processus fails to close, thus leaving the tunica vaginalis in free communication with the peritoneal cavity. In an indirect inguinal hernia the neck of the sac lies at the deep inguinal ring, an opening in the fascia transversalis. The neck of the sac lies lateral to the inferior epigastric artery.

An indirect inguinal hernia is much more common than a femoral hernia. However, the inflexible boundaries of the neck of the hernial sac in the femoral hernia (lacunar ligament, femoral vein, inguinal ligament, and the pubic bone) make the femoral hernia much more likely to strangulate. A direct inguinal hernia is a weakness of the abdominal wall, and the neck of the sac is very wide. For this reason strangulation of a direct inguinal hernia does not occur.

12

Urinary System and Male Organs of Reproduction, Including the Penis and Scrotum

Emergency problems involving the urinary system are common and may involve the kidney, ureter, urinary bladder, and urethra. The patient may present with diverse symptoms, ranging from excruciating pain, painless hematuria, to failure to void urine. Traumatic injury to the renal system occurs in about 10% of patients with abdominal injury.

The reproductive system in males may present the emergency physician with a variety of conditions, from urethral obstruction, traumatic rupture of the urethra, to infections of the epididymis, testis, or prostate.

The purpose of this chapter is to review the significant anatomy of the two systems relative to emergency problems. Emphasis will be placed on age and sexual differences; for example, in young children the urinary bladder is an abdominal organ rather than a pelvic organ and is consequently more prone to abdominal injuries than in adults; in females, cystitis is much more common than in males

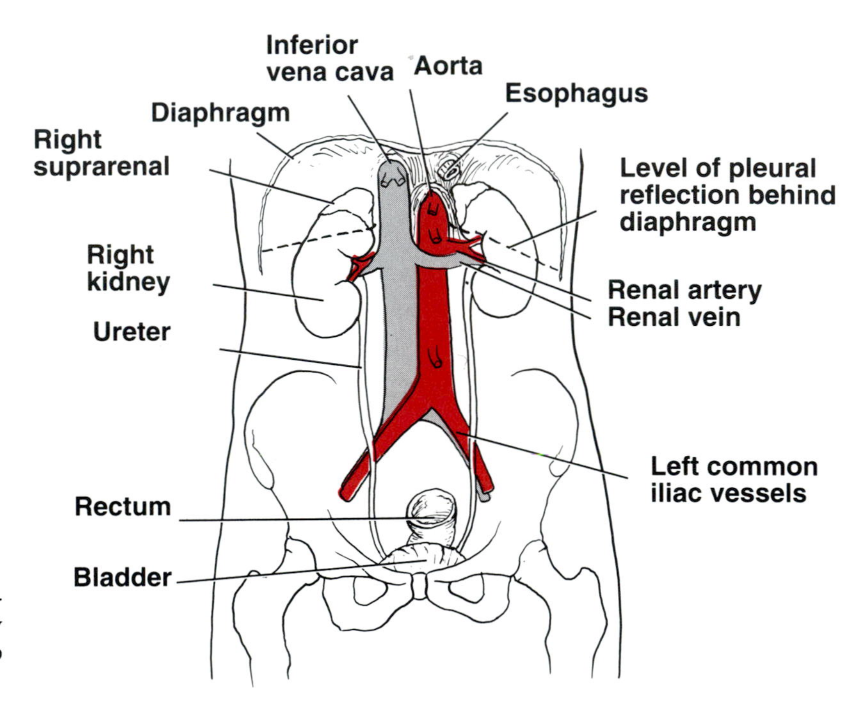

FIG 12–1.
Posterior abdominal wall showing the position of the kidneys and ureters; the urinary bladder and the pelvic bones are also shown.

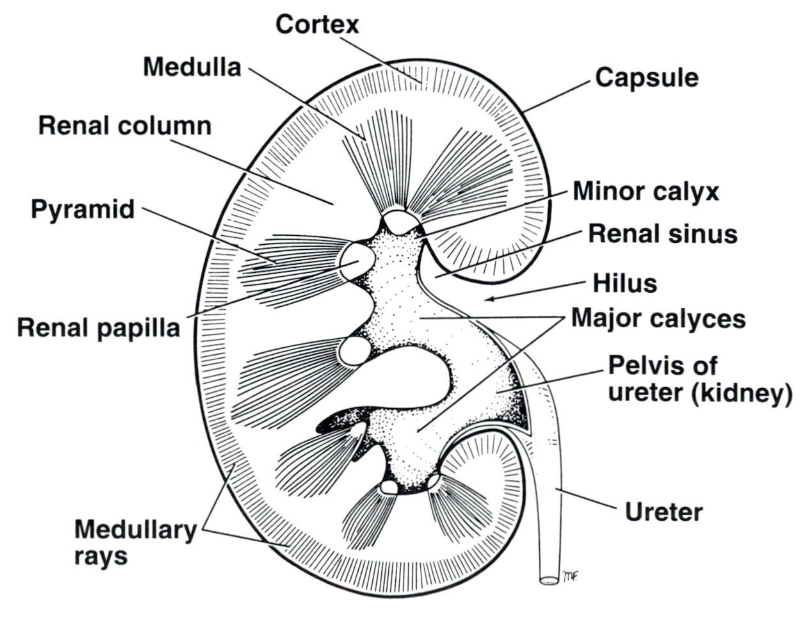

FIG 12–2.
Longitudinal section through the kidney showing the cortex, medulla, pyramids, renal papillae, and calyces. Note the hilus and renal sinus.

because the urethra is much shorter and ascending infection is more likely.

BASIC ANATOMY

Basic Anatomy of the Kidneys

Position

The kidneys are paired organs that lie behind the peritoneum high up on the posterior abdominal wall on either side of the vertebral column (Fig 12–1). The right kidney lies slightly lower than the left kidney because of the large size of the right lobe of the liver. With contraction of the diaphragm during respiration, both kidneys move downward in a vertical direction by as much as (2.5 cm) 1 in. On the medial concave border of each kidney is the *hilus,* which extends into a large cavity called the renal sinus (Fig 12–2). The hilus transmits the renal pelvis, the renal artery, the renal vein, and sympathetic nerve fibers.

Coverings

The kidneys have the following coverings (Fig 12–3).

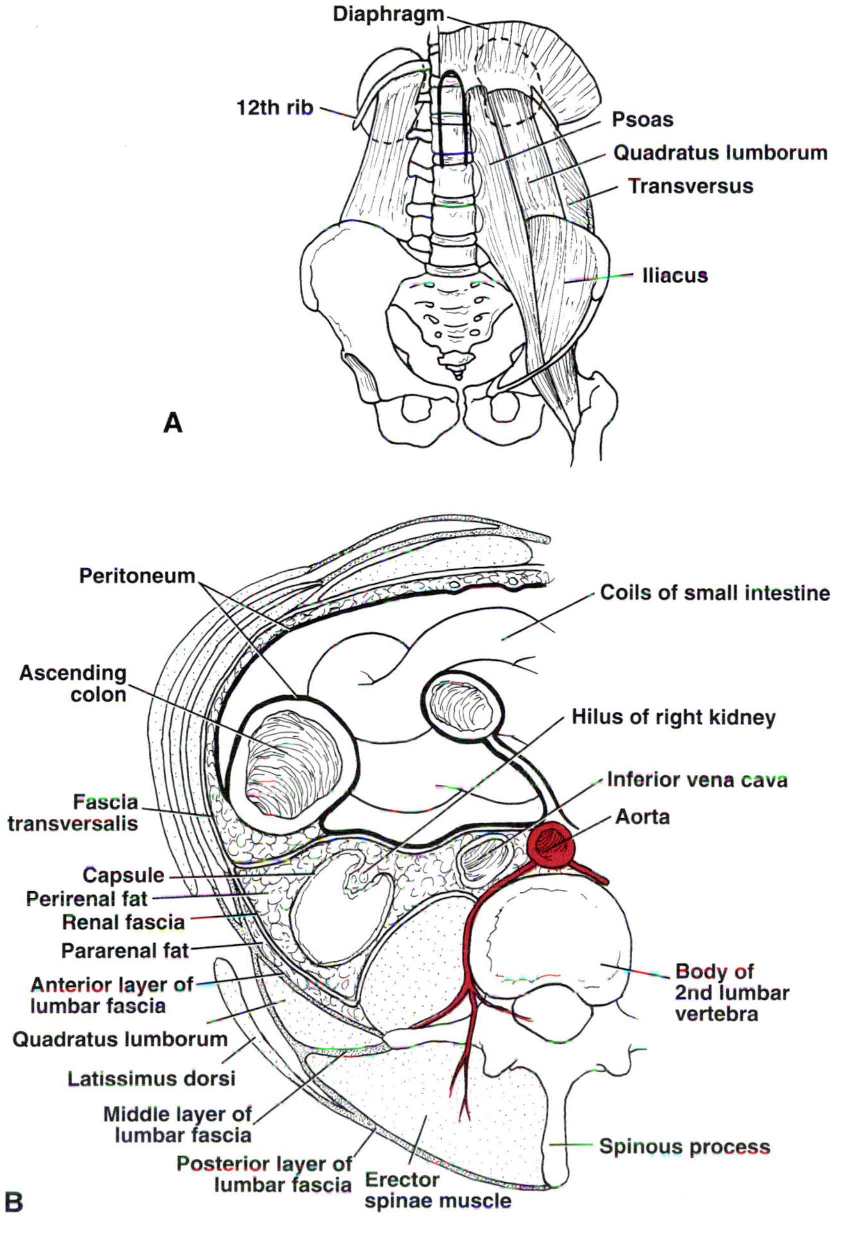

FIG 12–3.
A, structures on the posterior abdominal wall behind the kidneys. **B,** transverse section through the abdomen at the level of the second lumbar vertebra showing the peritoneum and fascia associated with the right kidney.

1. A *fibrous capsule* that is closely applied to its outer surface.
2. *Perirenal fat* that covers the fibrous capsule.
3. *Renal fascia*, a condensation of areolar tissue that lies outside the perirenal fat and encloses the kidneys and suprarenal glands.
4. *Pararenal fat*, which lies external to the renal fascia and is often in large quantity. It forms part of the retroperitoneal fat (see p. 433).

The perirenal fat, renal fascia, and pararenal fat support the kidneys and hold them in position on the posterior abdominal wall.

Important Relations (Figs 12–3 and 12–4).—These include the following.

Right Kidney.*—Anteriorly:* the suprarenal gland, the liver, the second part of the duodenum, and the right colic flexure. *Posteriorly:* the diaphragm, the costodiaphragmatic recess of the pleura, the 12th rib, the psoas, quadratus lumborum, the transversus abdominis muscle. The subcostal (T12), iliohypogastric, and ilioinguinal nerves (L1) run downward and laterally.

Left Kidney.*—Anteriorly:* the suprarenal gland, the spleen, the stomach, the pancreas, the left colic flexure, and coils of jejunum. *Posteriorly:* the diaphragm, the costodiaphragmatic recess of the pleura, the 11th (the left kidney is higher) and 12th ribs, and the psoas, quadratus lumborum, and transversus abdominis muscles. The subcostal (T12), iliohypogastric, and ilioinguinal nerves (L1) run downward and laterally.

Renal Structure

There is a dark brown outer *cortex* and a light brown inner *medulla* in each kidney. The medulla is composed of about a dozen *renal pyramids,* each having its base oriented toward the cortex and its apex, the *renal papilla,* projecting medially (see Fig 12–2). The cortex extends into the medulla between adjacent pyramids as the *renal columns.* Extending from the bases of the renal pyramids into the cortex are striations known as *medullary rays.*

On the medial border of each kidney is a vertical slit, which is bounded by thick lips of renal substance and is called the *hilus* (see Fig 12–2). The hilus transmits, from the front backward, the renal

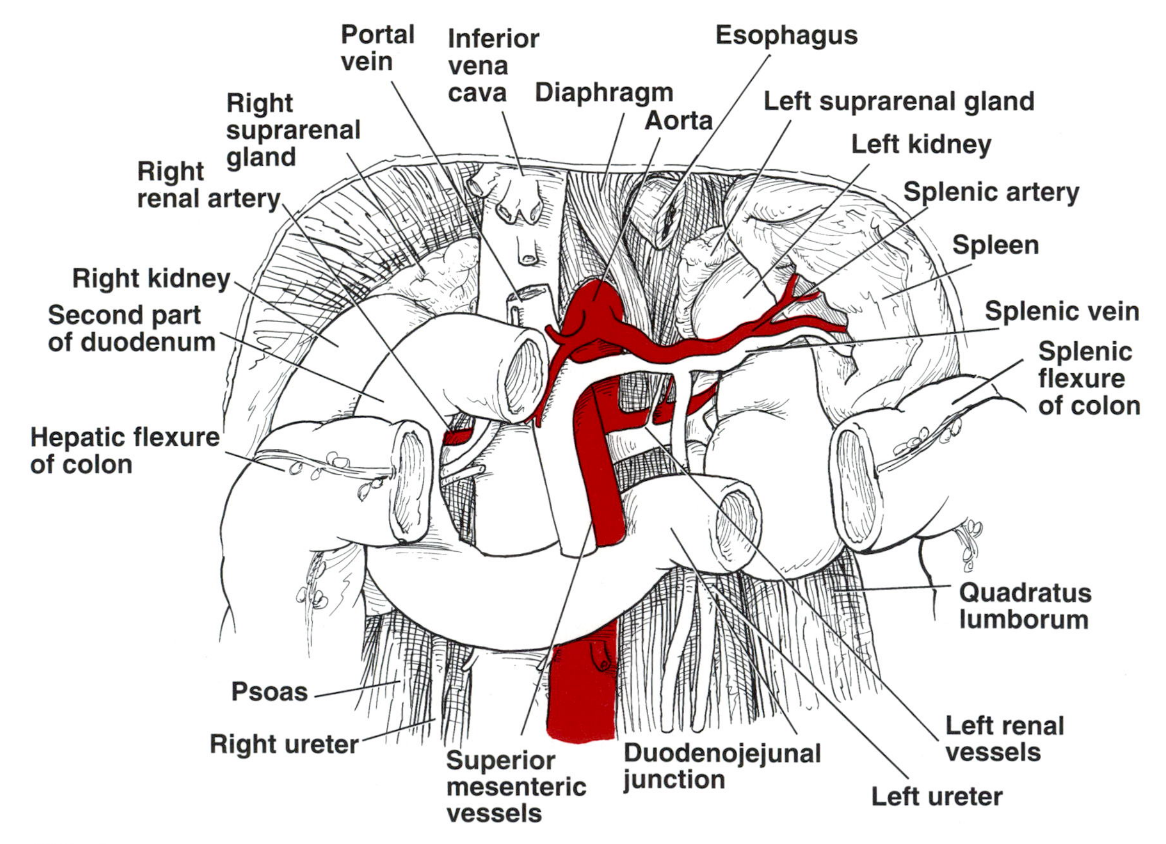

FIG 12–4.
Posterior abdominal wall showing the anterior relations of the kidneys.

vein, two branches of the renal artery, the ureter, and a further branch of the renal artery (VAUA). Lymph vessels and sympathetic nerves also pass through the hilus. The space within the hilus is referred to as the *renal sinus.* Within the renal sinus, the upper expanded end of the ureter, the *renal pelvis,* divides into two or three *major calyces,* each of which divides into two or three *minor calyces* (see Fig 12–2). Each minor calyx is indented by the apex of the renal pyramid, the *renal papilla.*

Blood Supply.—This includes the following.

Arteries.—Renal artery, a branch of the aorta.

Veins.—Renal vein into the inferior vena cava.

Lymphatic Drainage.—Lateral aortic lymph nodes around the origin of the renal artery.

Nerve Supply.—Renal sympathetic plexus. The sympathetic postganglionic fibers are distributed mainly to the afferent and efferent arterioles of the glomeruli.

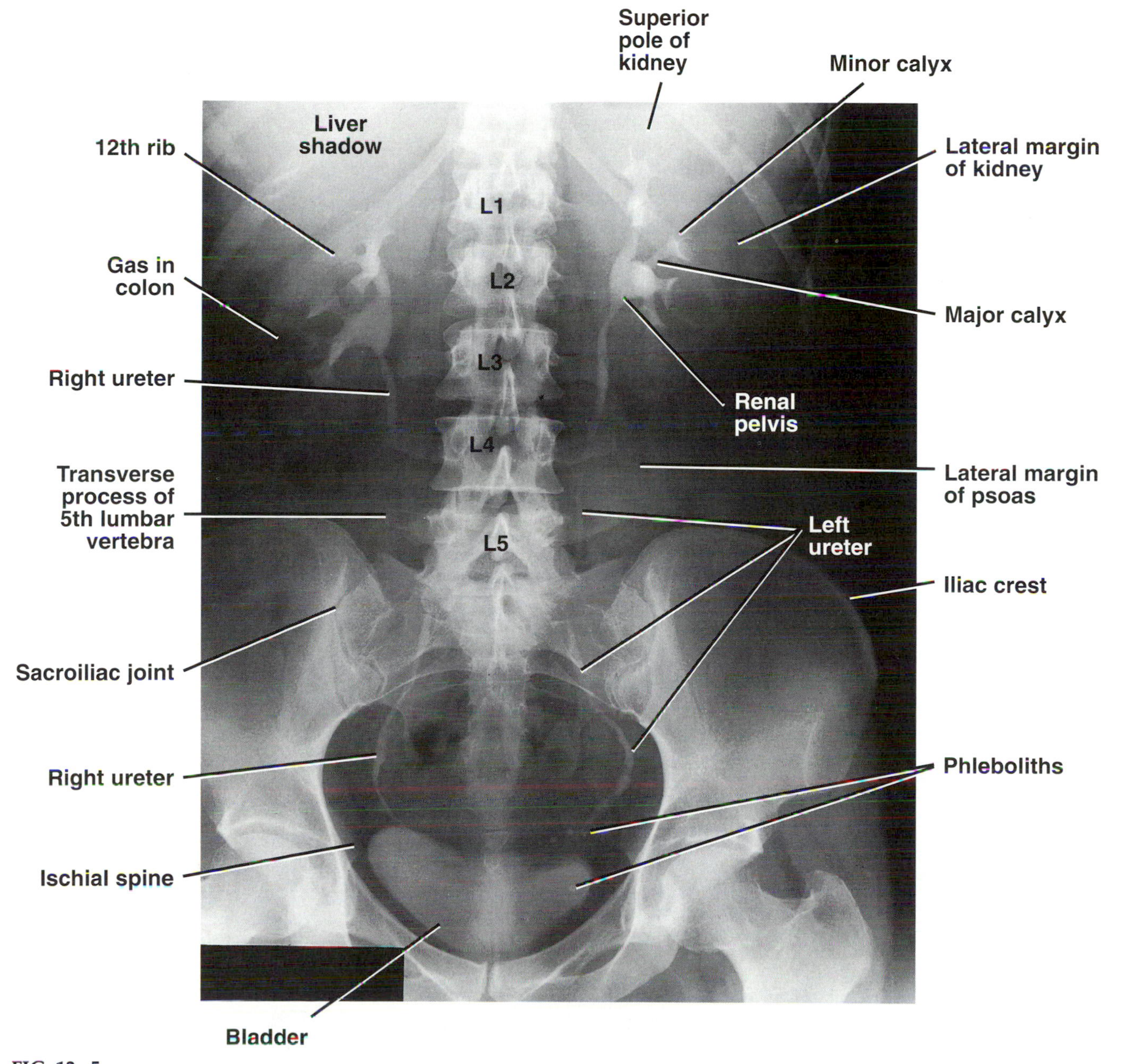

FIG 12–5.
An intravenous pyelogram obtained 30 minutes after the injection of a suitable contrast medium (female aged 39 years).

Radiographic Appearances

Radiographic appearances of the kidney are shown in Figures 12–5 and 12–6. A *computed tomographic scan of the kidneys* is shown in Figure 12–7.

Basic Anatomy of the Ureters

Position

The two ureters are muscular tubes that extend from the kidneys to the posterior surface of the urinary bladder (see Fig 12–1). Each ureter measures about (25 cm) 10 in. long and has an upper expanded end called the *renal pelvis.* The renal pelvis lies within the hilus of the kidney, where it receives the major calyces.

Ureteric Constrictions

These constrictions are (1) where the renal pelvis joins the ureter, (2) where it is kinked as it crosses the pelvic brim to enter the pelvis, and (3) where it pierces the bladder wall.

The ureter emerges from the hilus of the kidney and runs vertically downward behind the parietal peritoneum (adherent to it) on the psoas muscle, which separates it from the tips of the transverse processes of the lumbar vertebrae. It enters the pelvis by crossing the bifurcation of the common iliac artery in front of the sacroiliac joint (see Fig 12–1). The ureter then runs down the lateral wall of the pelvis to the region of the ischial spine and turns forward to enter the lateral angle of the bladder. The ureter passes obliquely through the wall of the bladder for about ¾ in. (1.9 cm) before opening into the bladder cavity.

In males the ureter is crossed near its termination by the vas deferens (see Fig 12–17). In females the ureter descends behind the ovary and on reaching the ischial spine turns forward beneath the base of the broad ligament, where it is crossed by the uterine artery (see Fig 13–10). The ureter then runs forward, lateral to the lateral fornix of the vagina, to enter the bladder.

Blood Supply.—This includes the following.

Arteries.—(1) Upper end, the renal artery, (2) middle portion, the testicular or ovarian artery, and (3) inferior end, superior vesical artery.

Veins.—Veins that correspond to the arteries.

Nerve Supply.—Renal, testicular (or ovarian), and hypogastric plexuses.

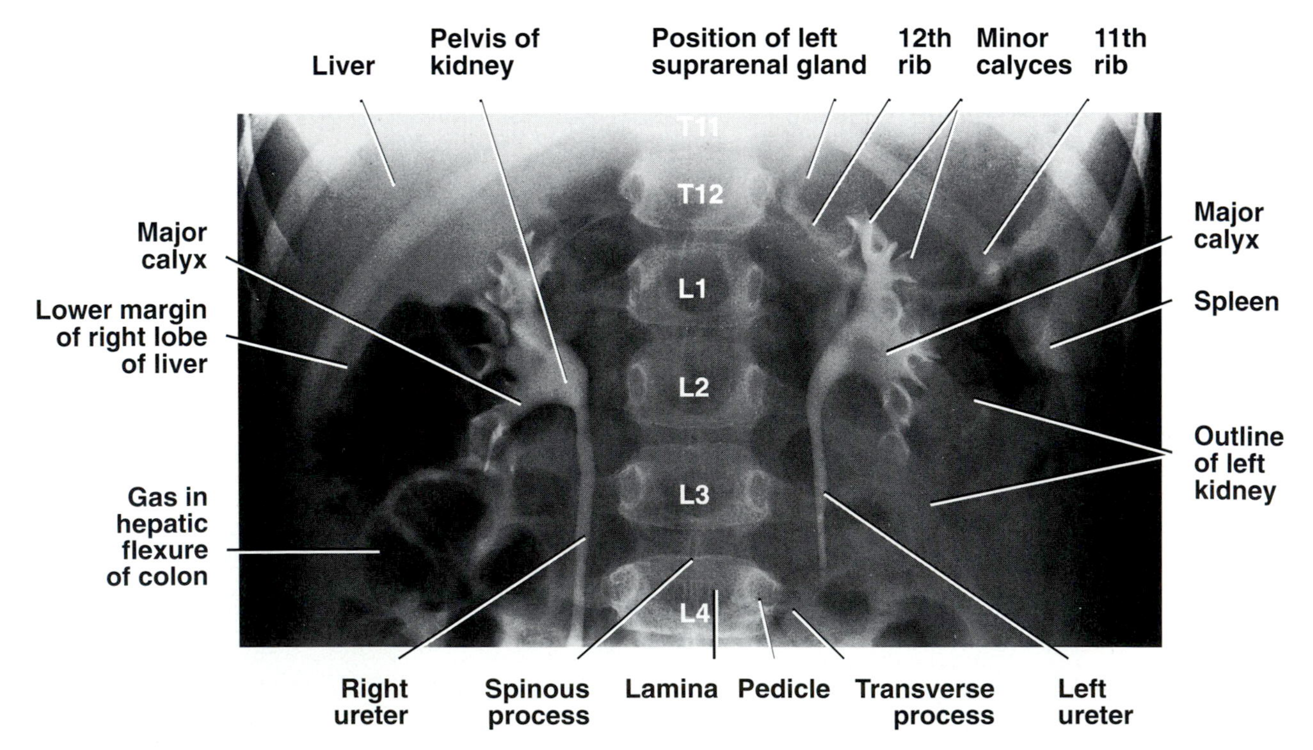

FIG 12–6.
An intravenous pyelogram obtained 15 minutes after the injection of a suitable contrast medium (female aged 5 years).

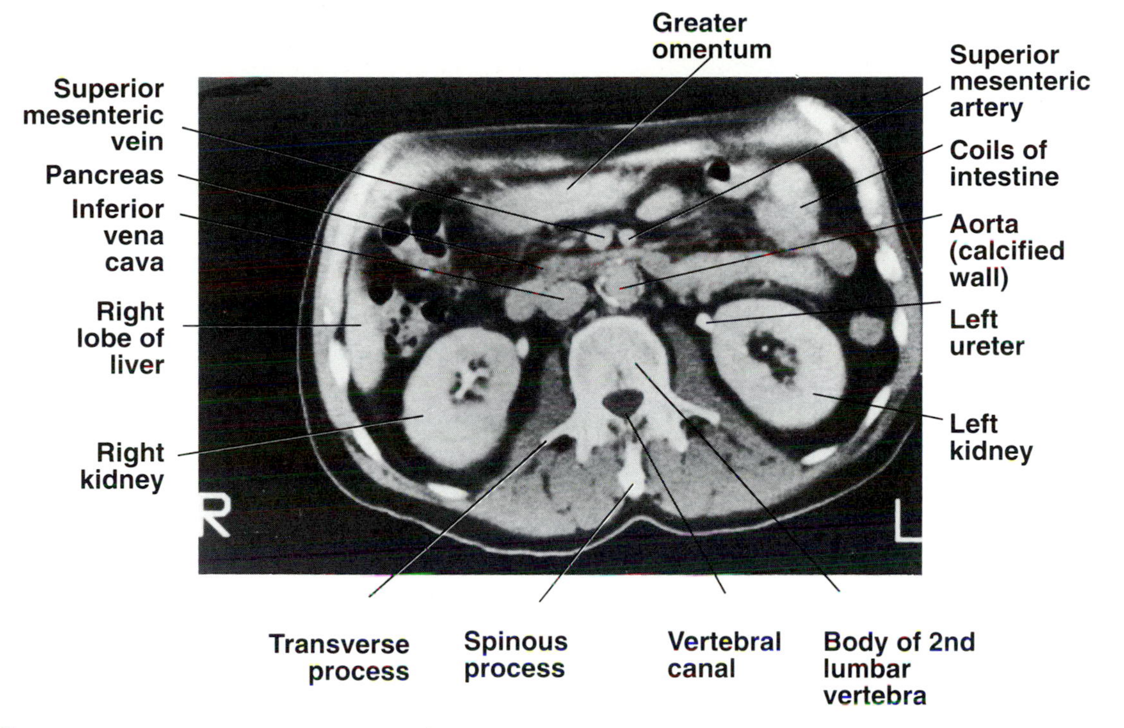

FIG 12–7.
Computed tomographic scan of the upper abdomen at the level of the second lumbar vertebra following an intravenous pyelogram. Note the position of the kidneys and the ureters and the presence of calcification in the aortic wall.

Radiological Appearances

Radiological appearances of the ureter are shown in Figures 12–5 and 12–6.

Basic Anatomy of the Urinary Bladder

Position and Shape

The urinary bladder is located immediately behind the pubic bones within the pelvis (see Fig 12–1). The normal bladder has a capacity of about 500 mL.

Clinical Note

Bladder distension.—In the presence of urinary obstruction in males, the bladder may become greatly distended without permanent damage to the bladder wall; in such cases, it is routinely possible to drain 1,000 to 1,200 mL of urine through a catheter.

The empty bladder is pyramidal, having an apex, a base, and a superior and two inferolateral surfaces (Fig 12–8); it also has a neck. When the bladder fills it becomes ovoid, and the superior surface rises into the abdomen. In young children the empty bladder projects upward into the abdomen; later when the pelvis enlarges, the bladder sinks to become a pelvic organ.

The *apex* of the bladder points anteriorly and is connected to the umbilicus by the *median umbilical ligament* (remains of urachus). The *base* of the bladder faces posteriorly and is triangular. The ureters enter the supralateral angles, and the urethra leaves the inferior angle. The *superior surface* of the bladder is covered with peritoneum, which is reflected laterally onto the lateral pelvic walls. As the bladder fills, the superior surface bulges upward into the abdominal cavity, peeling the peritoneum off the lower part of the anterior abdominal wall. The *neck* of the bladder points inferiorly.

Interior of the Bladder

The internal surface of the base of the bladder is called the *trigone* (see Fig 12–8). Here the mucous membrane is firmly adherent to the underlying muscle and is always smooth. The trigone has at its lateral angles the small, slitlike openings of the ureters and below the crescentic opening of the urethra. The *interureteric ridge* runs from one ureteric orifice to the other. It is caused by the underlying muscle and forms the upper limit of the trigone. In males the median lobe of the prostate bulges upward into the bladder slightly, behind the urethral orifice, producing a swelling, the *uvula vesicae*.

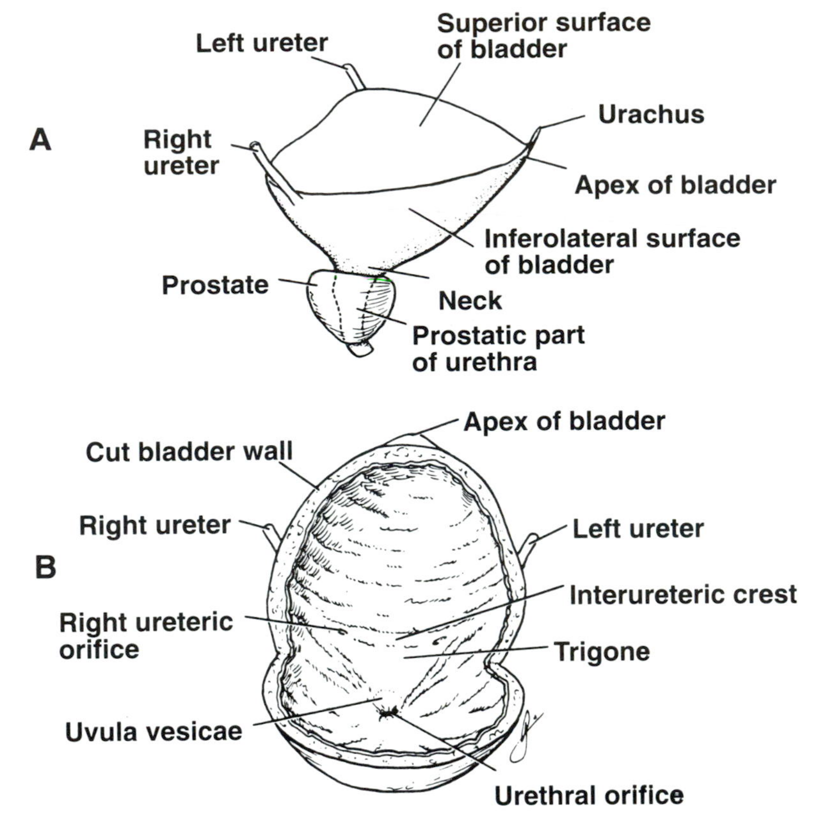

FIG 12–8.
A, urinary bladder (partly filled) showing the general shape and surfaces. Note the positions of the apex anteriorly and the neck inferiorly. **B,** interior of the urinary bladder in the male, as seen from in front. The middle lobe of the prostate causes the uvula vesicae.

Muscle Coat of the Bladder Wall
The muscle coat, the *detrusor muscle,* consists of three layers of smooth-muscle fibers, an outer longitudinal layer, a middle circular layer, and an inner longitudinal layer. At the neck of the bladder, the middle layer of circular muscle forms the *sphincter vesicae.*

Important Relations.— These are as follows.

***In* Males** (Fig 12–9).—*Anteriorly:* symphysis pubis, retropubic pad of fat, anterior abdominal wall. *Posteriorly:* rectovesical pouch, vasa deferentia, the seminal vesicles, rectovesical fascia, and the rectum. *Laterally:* obturator internus muscle above and the levator ani muscle below. *Superiorly:* peritoneal cavity, coils of ileum, and sigmoid colon. *Inferiorly:* prostate.

***In* Females.**—Because of the absence of the prostate, the bladder lies at a lower level in the female pelvis than in the male pelvis, and the neck rests directly on the urogenital diaphragm. The close relation of the bladder to the uterus and the vagina is of considerable clinical importance (Fig 12–10). *Anteriorly:* symphysis pubis, retropubic pad of fat, anterior abdominal wall. *Posteriorly:* separated from the rectum by the vagina. *Laterally:* obturator internus muscle above and the levator ani muscle below. *Superiorly:* uterovesical pouch of peritoneum and the body of the uterus. *Inferiorly:* urogenital diaphragm.

Blood Supply.— This includes the following.

Arteries.—Superior and inferior vesical arteries, branches of the internal iliac artery.

Veins.—Vesical veins drain into the internal iliac veins.

Lymphatic Drainage.— Internal and external iliac nodes.

Nerve Supply.— Sympathetic fibers* travel from the hypogastric plexuses in the pelvis to the bladder wall. The sympathetic nerves inhibit contraction of the detrusor muscle of the bladder wall and stimulate closure of the sphincter vesicae.

The parasympathetic fibers travel from the hy-

*The sympathetic nerves to the detrusor muscle are now thought to have little or no action on the smooth muscle of the bladder wall and are distributed mainly to the blood vessels. The sympathetic nerves to the sphincter vesicae are thought to play only a minor role in causing contraction of the sphincter in maintaining urinary continence. However, in males, the sympathetic innervation of the sphincter causes active contraction of the bladder neck during ejaculation (brought about by sympathetic action), thus preventing seminal fluid from entering the bladder.

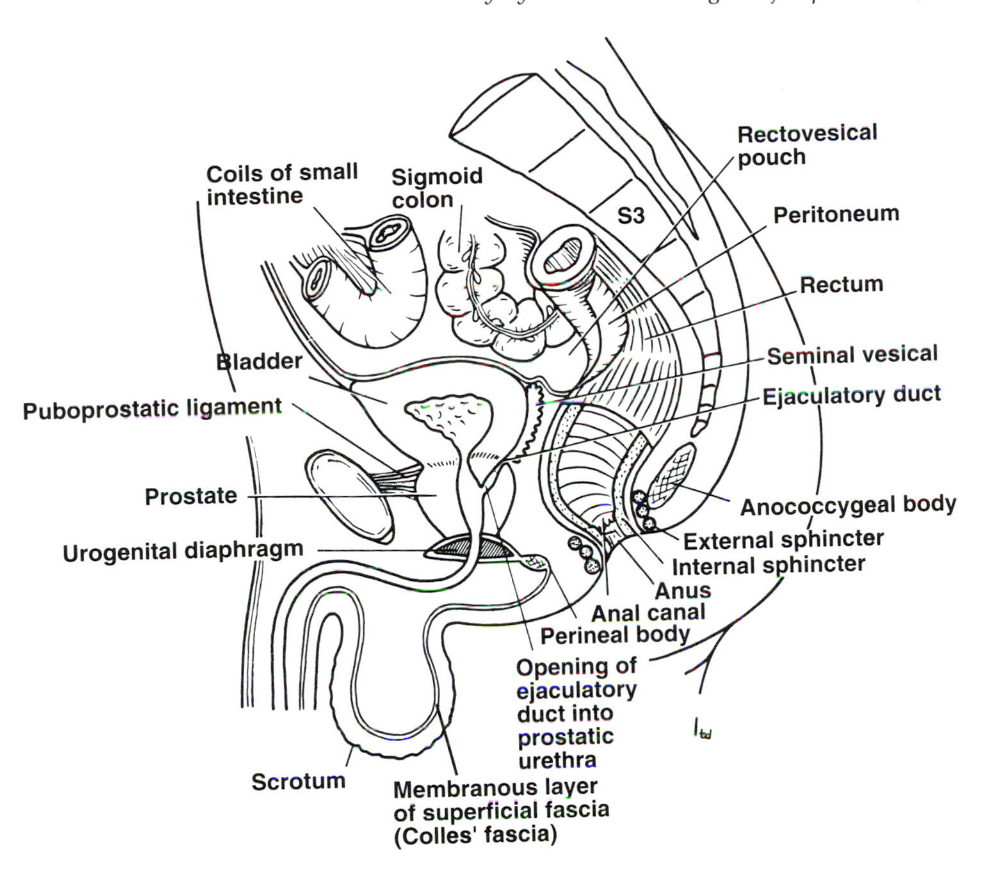

FIG 12–9.
Sagittal section of the male pelvis showing the relations of the bladder.

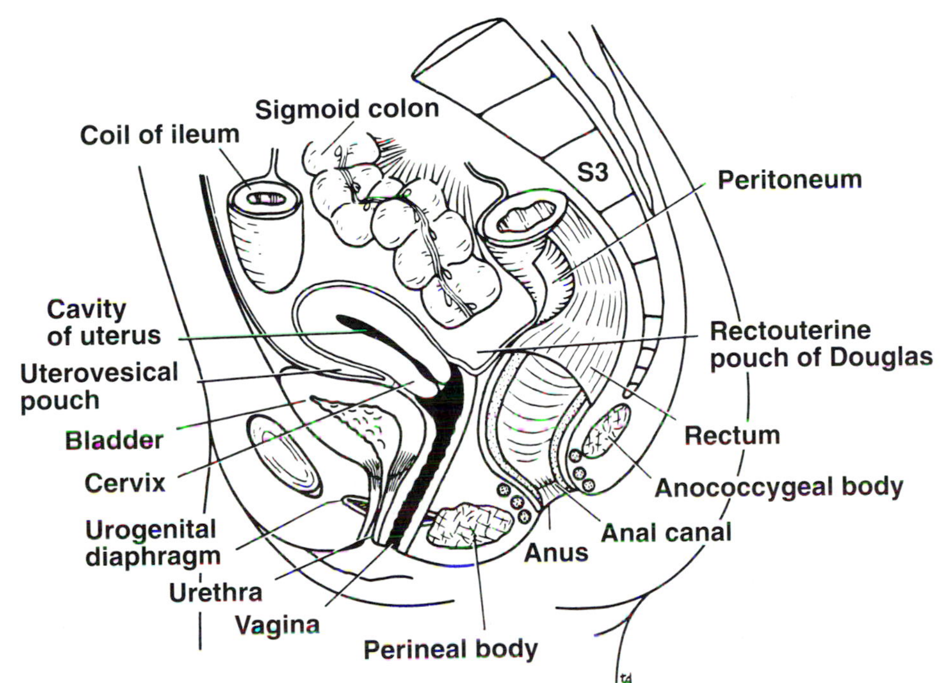

FIG 12–10.
Sagittal section of the female pelvis showing the relations of the bladder.

pogastric plexuses to the bladder wall. The parasympathetic nerves stimulate the contraction of the detrusor muscle of the bladder wall and inhibit the action of the sphincter vesicae (Fig 12–11).

Micturition

In adults, when the volume of urine reaches about 300 mL, stretch receptors in the wall of the bladder transmit impulses to the central nervous system, and the individual has a conscious desire to micturate.

The majority of afferent sensory fibers arising in the bladder reach the central nervous system along with the parasympathetic nerves (pelvic splanchnic nerves); they enter the second, third, and fourth sacral segments of the spinal cord (see Fig 12–11). Some afferent fibers travel with the sympathetic nerves via the hypogastric plexuses and enter the first and second lumbar segments of the spinal cord.

Efferent parasympathetic impulses leave the cord from the second, third, and fourth sacral segments and pass via the hypogastric plexuses to the bladder wall, where they synapse with postganglionic neurons (see Fig 12–11). The detrusor muscle then contracts and the sphincter vesicae relaxes. Efferent impulses also pass to the urethral sphincter via the pudendal nerve, which then relaxes. Micturition can be assisted by contracting the abdominal muscles so as to raise the intra-abdominal and pelvic pressures and exert external pressure on the bladder.

Voluntary control of micturition is accomplished by contracting the sphincter urethrae; this is assisted by the sphincter vesicae.

Basic Anatomy of the Urethra

Male Urethra

The male urethra is about 8 in. (20 cm) long and extends from the neck of the bladder to the external meatus on the glans penis (see Fig 12–9). It is divided into three parts—prostatic, membranous, and penile.

Prostatic Urethra.—The prostatic urethra is about 1¼ in. (3 cm) long and passes through the prostate from the base to the apex (Fig 12–12). It is the widest and most dilatable portion of the entire urethra. On the posterior wall there is a longitudinal ridge called the *urethral crest* (see Fig 12–15). On each side of this ridge is a groove called the *prostatic sinus;* the prostatic glands open into these grooves. On the summit of the urethral crest is a depression, the *prostatic utricle,* which is an analog of the uterus and vagina in females. On the edge of the mouth of the utricle are the openings of the two ejaculatory ducts.

Membranous Urethra.—The membranous urethra is about ½ in. (1.25 cm) long and lies within the urogenital diaphragm surrounded by the sphincter urethrae muscle (see Fig 12–12). It is the least dilatable portion of the urethra.

Sphincter Urethrae Muscle.—The sphincter urethrae muscle surrounds the urethra in the deep

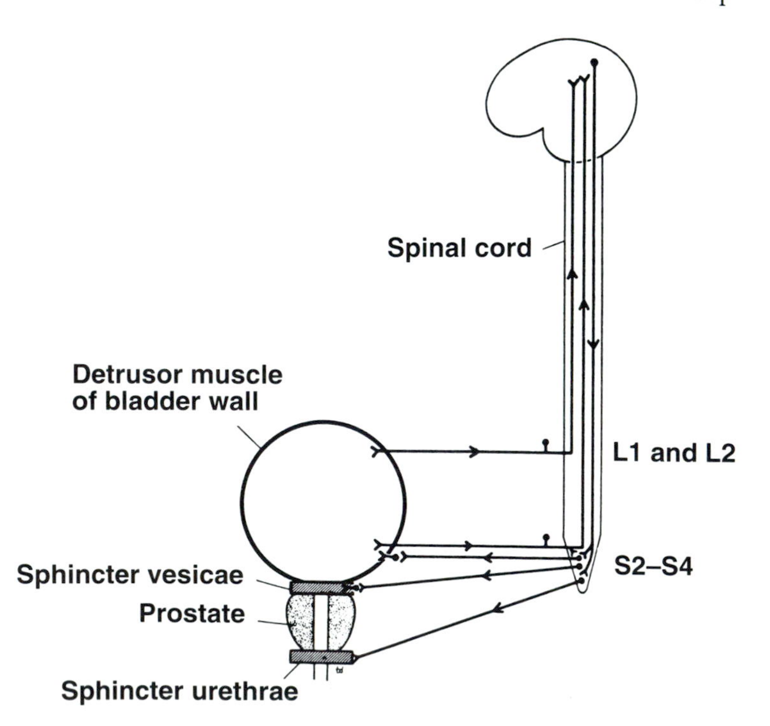

FIG 12–11.
Nervous control of the bladder. Note the afferent sensory nerve fibers leaving the bladder and entering the central nervous system and the parasympathetic innervation of the bladder. For the sake of simplification, the sympathetic fibers have been omitted.

perineal pouch (see Fig 12–12). It arises from the pubic arch on the two sides and the fibers pass medially to encircle the urethra; some fibers arise from neighboring fascia and pass to the perineal body. *Nerve supply:* This is from the perineal branch of the pudendal nerve (S2 through S4). *Action:* The muscle compresses the membranous part of the urethra and relaxes during micturition. It is the means by which voluntary micturition can be voluntarily stopped.

Penile Urethra.—The penile urethra is about 6 in. (15.75 cm) long and passes through the bulb (expanded posterior end) and shaft of the corpus spongiosum of the penis (Fig 12–13). The *external meatus* is the narrowest part of the entire urethra. The part of the urethra that lies within the glans penis is dilated to form the *fossa terminalis (navicular fossa).* The *bulbourethral glands* open into the penile urethra below the urogenital diaphragm.

Female Urethra

The female urethra is only 1½ in. (3.8 cm) long and extends from the bladder neck to the *external meatus* (see Fig 12–10). Throughout its course it lies in close relationship to the anterior wall of the vagina. The urethra passes through the urogenital diaphragm, where it traverses the sphincter urethrae and opens onto the surface below the clitoris and in front of the vagina. At the sides of the external urethral meatus are the small openings of the ducts of the *paraurethral glands.* The urethra can be dilated relatively easily.

Basic Anatomy of the Male External Organs of Reproduction

Anatomy of the Penis

The penis has a fixed root and a body, which hangs free (see Fig 12–13).

Root of the Penis.—This is made up of three masses of erectile tissue, which are the *bulb of the penis* and the right and left *crura of the penis* (Figs 12–13 and 12–14). The bulb is situated in the midline and is attached to the undersurface of the urogenital diaphragm. It is traversed by the urethra and is covered on its outer surface by the *bulbospongiosus muscles.* Each crus is attached to the side of the pubic arch

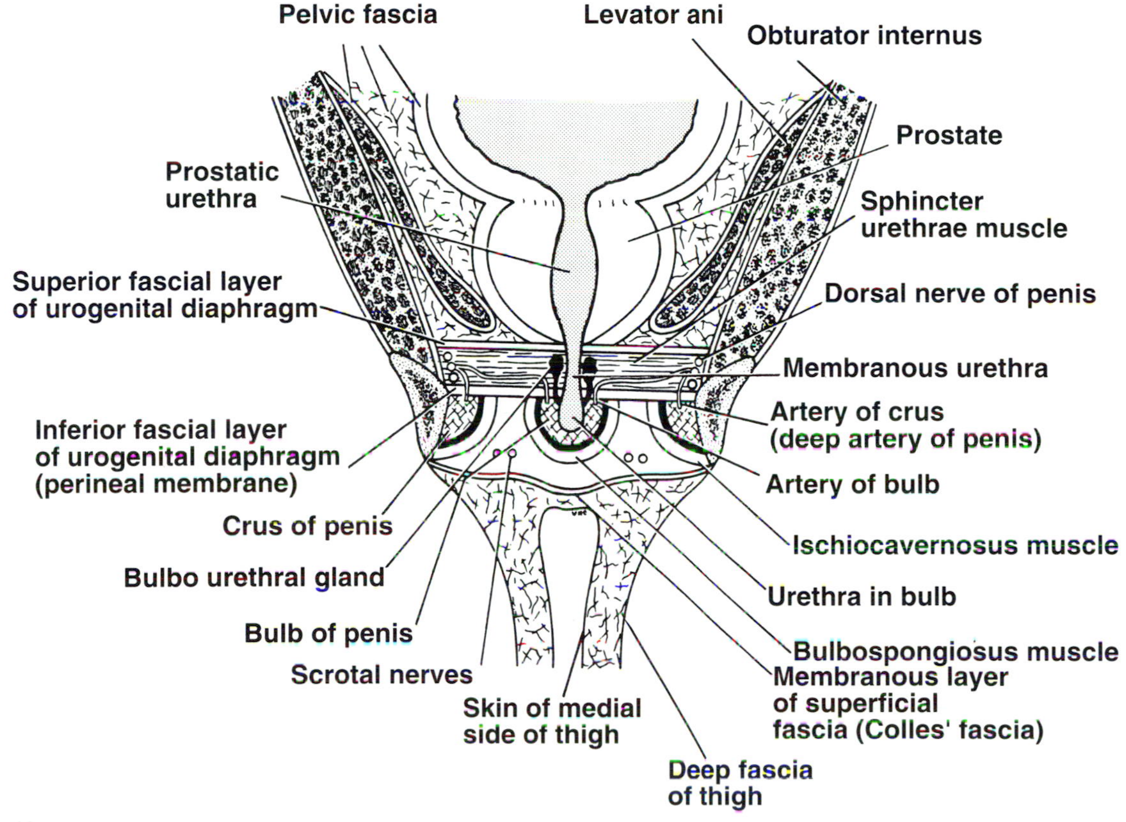

FIG 12–12.
Coronal section of the male pelvis showing the neck of the bladder, the prostatic urethra, the membranous part of the urethra, and the urethra in the bulb. Note that the urethra in the bulb lies below the urogenital diaphragm in the superficial perineal pouch and that the pouch is bounded by the membranous layer of superficial fascia.

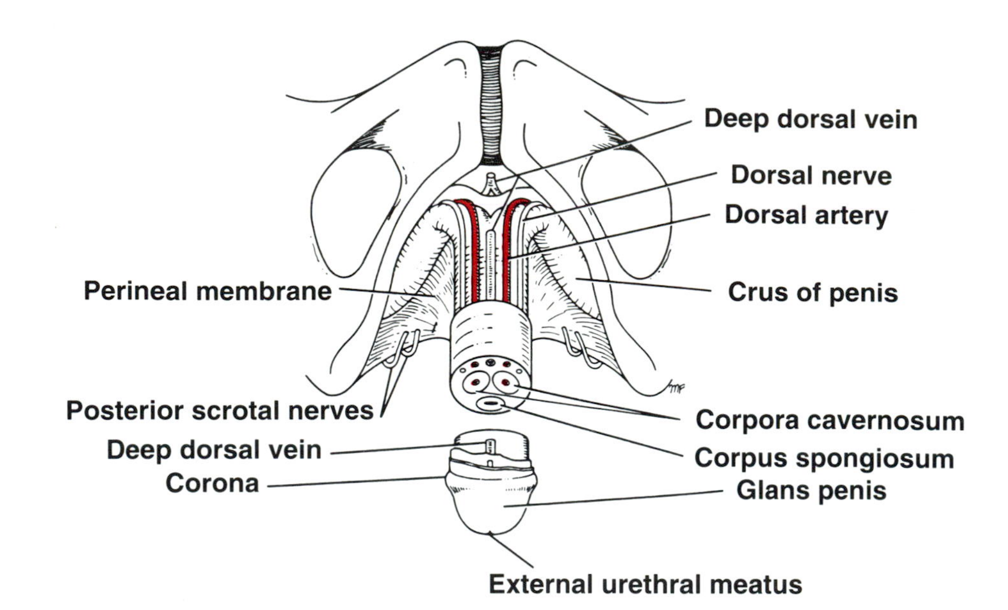

FIG 12–13.
The root and body of the penis.

and is covered on its outer surface by the *ischiocavernosus muscle.* The bulb is continued forward into the body of the penis and forms the *corpus spongiosum* (see Figs 12–13 and 12–14). The two crura converge anteriorly and come to lie side-by-side in the dorsal part of the body of the penis, forming the *corpora cavernosa* (see Fig 12–14).

Body of the Penis.—This is essentially composed of three cylinders of erectile tissue enclosed in a tubular sheath of fascia (Buck's fascia). The erectile tissue is made up of two dorsally placed corpora cavernosa (which communicate with each other) and a single corpus spongiosum applied to their ventral surface (see Fig 12–14). The outer coat of each cylinder of

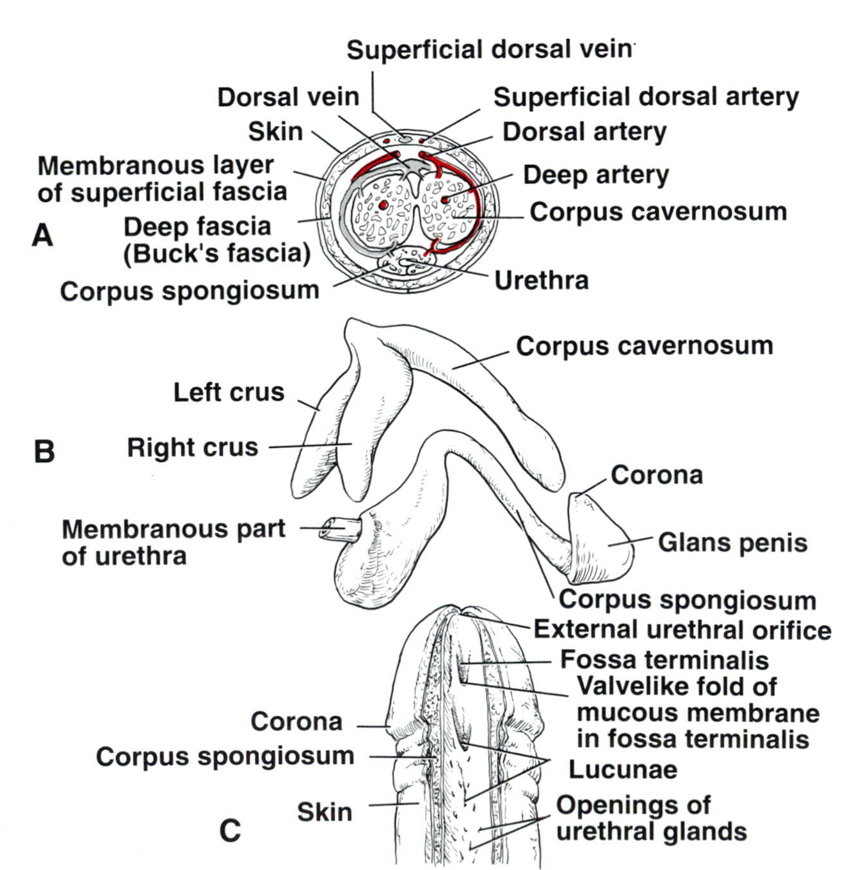

FIG 12–14.
The penis. **A** and **B,** the three bodies of erectile tissue, the two corpora cavernosa and the corpus spongiosum with the glans. **C,** the penile urethra slit open to show the folds of mucous membrane and glandular orifices in the roof of the urethra.

erectile tissue is composed of dense fibrous tissue called the *tunica albuginea.* At its distal extremity, the corpus spongiosum expands to form the *glans penis,* which covers the distal ends of the corpora cavernosa. On the tip of the glans penis is the slitlike orifice of the urethra, called the *external urethral meatus.*

Prepuce.—This is a hoodlike fold of skin that covers the glans. It is connected to the glans just below the urethral orifice by a fold called the *frenulum.*

The body of the penis is supported by two bands of deep fascia that extend downward from the linea alba and symphysis pubis to be attached to the fascia of the penis.

Blood Supply of the Penis.—This includes the following.

Arteries.—The corpora cavernosa are supplied by the deep arteries of the penis; the corpus spongiosum is supplied by the artery of the bulb. In addition, there is the dorsal artery of the penis and the external pudendal artery on each side (see Fig 12–14).

Lymphatic Drainage of the Penis.—The skin of the penis is drained into the medial group of superficial inguinal nodes. The deep structures of the penis are drained into the internal iliac nodes.

Anatomy of the Scrotum and Its Contents

The scrotum is an outpouching of the lower part of the anterior abdominal wall and contains the testes, epididymes, and lower ends of the spermatic cords (see Fig 12–15).

The wall of the scrotum has the following layers: (1) skin, (2) superficial fascia (dartos muscle, which is smooth muscle, replaces the fatty layer of the anterior abdominal wall, and Scarpa's fascia [membranous layer] is now called Colles' fascia), (3) external spermatic fascia derived from the external oblique, (4) cremasteric fascia derived from the internal oblique, (5) internal spermatic fascia derived from the fascia transversalis, and (6) tunica vaginalis; this is a closed sac that covers the anterior, medial, and lateral surfaces of each testis.

Lymphatic Drainage of the Scrotal Wall.—This involves a medial group of superficial inguinal nodes. Note that the testis is drained by lymph vessels into the paraortic nodes on the posterior abdominal wall at the level of the first lumbar vertebra.

Testes.—Each testis is a firm, mobile organ lying within the scrotum. The left testis usually lies at a lower level than the right. The *tunica albuginea* is the outer fibrous capsule of the testis (see Fig 12–15). Normal spermatogenesis will only take place at a temperature lower than that of the abdominal cavity, which explains the descent of the testes into the scrotum.

Epididymis.—This is a firm structure lying posterolaterally to the testis with the vas deferens lying on its medial side (see Fig 12–15). The epididymis has a *head,* a *body,* and a pointed *tail* inferiorly and consists of a long coiled tube embedded in fascia. The long length of the duct of the epididymis provides storage space for the spermatozoa and allows them to mature. One of the main functions of the epididymis is the absorption of fluid. Another function may be the addition of substances to the seminal fluid to nourish the maturing sperm.

Blood Supply.—The testicular artery is a branch of the abdominal aorta. The testicular veins emerge from the testis and the epididymis as a venous network, the *pampiniform plexus.* This becomes reduced to a single vein as it ascends through the inguinal canal. The right testicular vein drains into the inferior vena cava, and the left vein joins the left renal vein.

Lymphatic Drainage.—Paraortic lymph nodes on the side of the aorta at the level of the first lumbar vertebra.

Basic Anatomy of the Male Internal Organs of Reproduction

Anatomy of the Prostate

The prostate is a glandular structure that surrounds the prostatic urethra. It lies below the neck of the bladder and above the urogenital diaphragm (Fig 12–16). The prostate has a fibrous capsule and is surrounded by a fibrous sheath that is part of the visceral layer of pelvic fascia. The prostate has a *base,* which lies superiorly against the bladder neck, and an *apex,* which lies inferiorly against the urogenital diaphragm.

The two ejaculatory ducts pierce the upper part of the posterior surface of the prostate to open into the prostatic urethra at the lateral margins of the opening of the *prostatic utricle* (see Fig 12–16). The ejaculatory ducts are formed on each side by the union of the vas deferens and the duct of the semi-

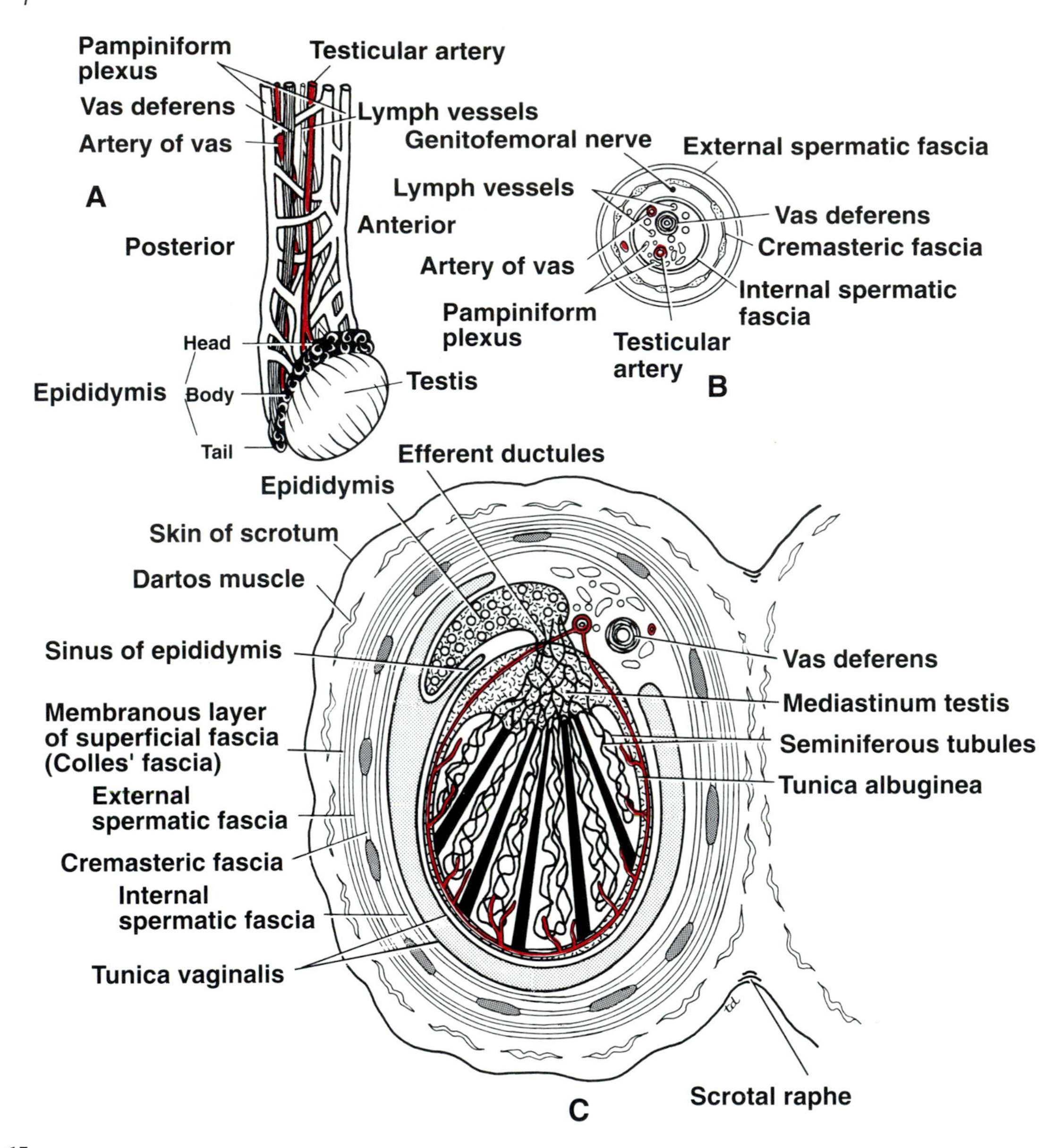

FIG 12–15.
A, the testis and epididymis at the lower end of the spermatic cord. Note that the vas deferens lies posteriorly in close relationship with the testicular artery and pampiniform plexus of veins. **B,** cross section of the spermatic cord showing its content and coverings. **C,** horizontal section of the scrotum showing the testis and epididymis. Note the various layers that form the wall of the scrotum and the position of the tunical vaginalis.

nal vesicle (Fig 12–17); their function is to drain the seminal fluid into the prostatic urethra.

The numerous glands of the prostate are embedded in a mixture of smooth muscle and connective tissue, and their ducts open into the prostatic urethra. The prostate may be divided into a number of lobes (see Fig 12–16). The *anterior lobe* lies in front of the urethra; the *middle or median lobe* lies behind the urethra and above the ejaculatory ducts. The *right and left lateral lobes* lie on either side of the urethra.

The function of the prostate is the production of a thin, milky fluid containing citric acid and acid phosphatase. It is added to the seminal fluid at the time of ejaculation. The smooth muscle in the capsule and stroma contract, and the secretion from the many glands is squeezed into the prostatic urethra. The prostatic secretion is alkaline and helps neutralize the acidity in the vagina.

Blood Supply.—This includes the following.

Arteries.—Branches of the inferior vesical and middle rectal arteries.

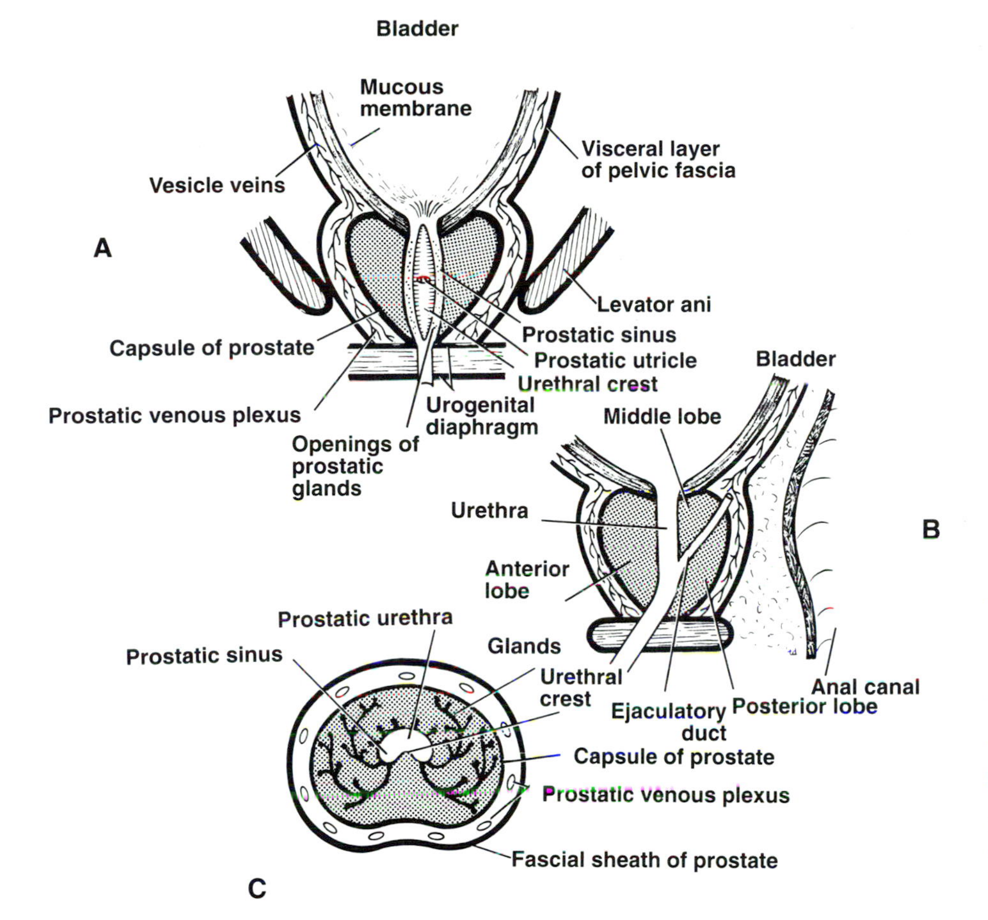

FIG 12–16.
Prostate. **A,** coronal section taken below the neck of the bladder. The apex of the prostate rests on the upper surface of the urogenital diaphragm. **B,** sagittal section showing the urethra and the ejaculatory duct. Note how these structures serve to divide up the prostate into lobes. **C,** horizontal section showing the various prostatic glands opening into the prostatic sinus of the prostatic urethra.

Veins.—Prostatic venous plexus that drains into the internal iliac veins.

Lymphatic Drainage.—Internal iliac nodes.

Nerve Supply.—Inferior hypogastric plexus. The sympathetic nerves stimulate the smooth muscle of the prostate during ejaculation. For details concerning the *prostatic sinus,* the *prostatic utricle,* and the *urethral crest,* see the section on the prostatic urethra, p. 474.

Anatomy of the Seminal Vesicles

The seminal vesicles are two lobulated sacs about 2 in. (5 cm) long, lying on the posterior surface of the bladder (Fig 12–17). Their upper ends are widely separated, and their lower ends are close together. On the medial side of each vesicle lies the terminal part of the vas deferens. Posteriorly, the seminal vesicles are related to the rectum (see Fig 12–9). Inferiorly, each seminal vesicle narrows and joins the vas deferens of the same side to form the *ejaculatory duct.* The two ejaculatory ducts pierce the prostate and open into the prostatic urethra, close to the margins of the prostatic utricle.

Each seminal vesicle consists of a much coiled tube embedded in connective tissue. The function of the seminal vesicles is to produce a secretion that is added to the seminal fluid. The secretions contain substances that are essential for the nourishment of the spermatozoa. The walls of the seminal vesicles contract during ejaculation and expel their contents into the ejaculatory ducts, thus washing the spermatozoa out of the urethra.

SURFACE ANATOMY

Surface Anatomy of the Urinary System

Kidneys

The right kidney lies at a slightly lower level than the left kidney (due to the bulk of the right lobe of the liver), and its lower pole may be palpated in the right lumbar region at the end of deep inspiration in a person with poorly developed abdominal muscles. Each kidney moves about 1 in. (2.5 cm) downward during the descent of the diaphragm on inspiration. Because the normal left kidney, which also moves downward on inspiration, is situated higher up on

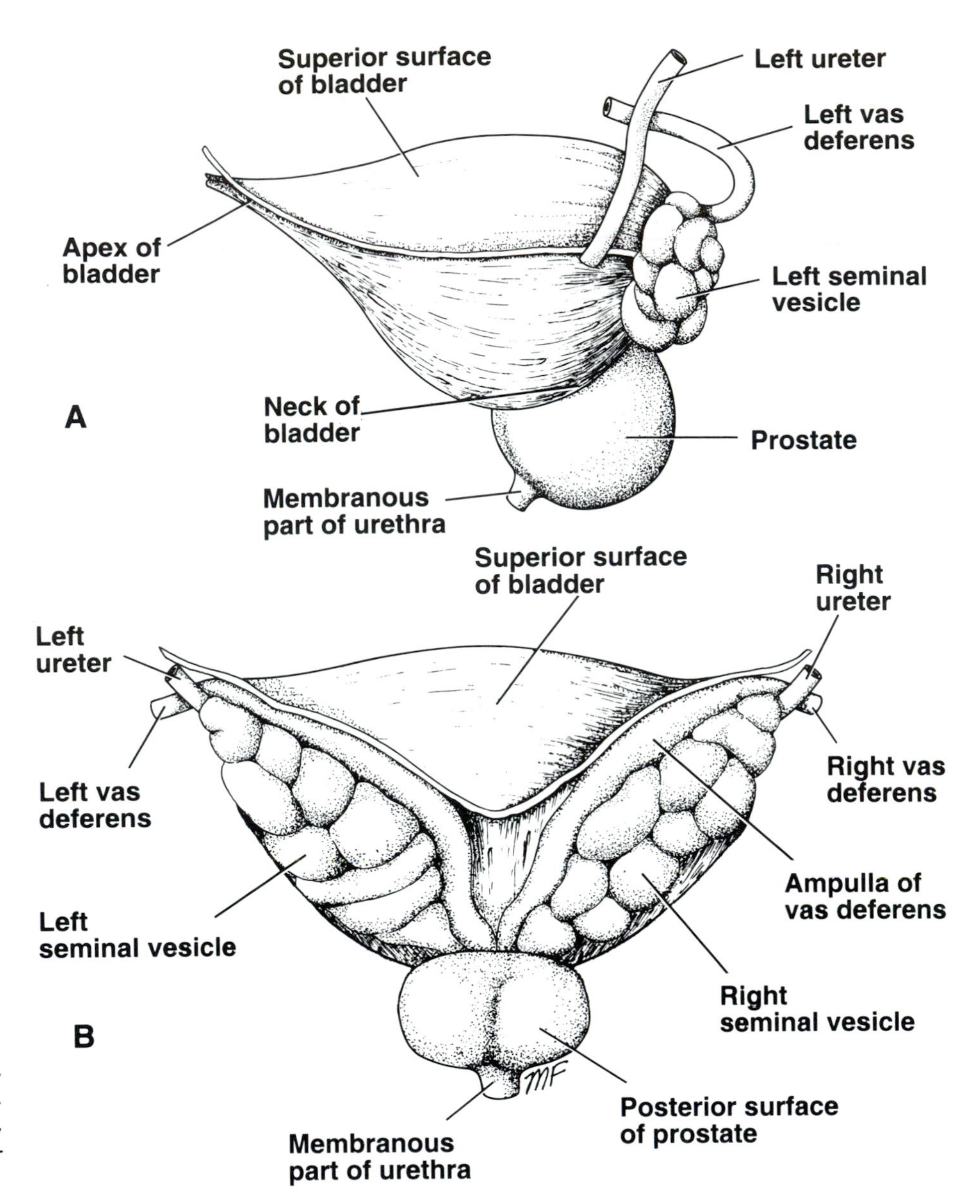

FIG 12–17.
A, lateral view of the bladder, prostate, and left seminal vesicle. **B,** posterior view of the bladder, prostate, vasa deferentia, and seminal vesicles.

the posterior abdominal wall than the right kidney, it is impalpable.

On the anterior abdominal wall the hilus of each kidney lies on the transpyloric plane (i.e., at the level of the tips of the ninth costal cartilages), about three fingerbreadths from the midline (Fig 12–18). On the back, the kidneys extend from the 12th thoracic spine to the third lumbar spine, and the hili are opposite the first lumbar spine (see Fig 12–18).

Ureter

On the anterior abdominal wall the ureter may be indicated by a line drawn downwards from the transpyloric plane at a distance of about (5 cm) 2.5 in. from the midline (see Fig 12–18). At the level of the anterior superior iliac spine the pelvic portion of the ureter may be indicated by curving the line downward and medially to the pubic tubercle.

On the posterior abdominal wall the abdominal portion of the ureter may be indicated by a line drawn downward from the level of the first lumbar spine to the posterior inferior iliac spine at a distance of (5 cm) 2 in. from the midline.

Urinary Bladder

In adults, the empty bladder is a pelvic organ and lies posterior to the pubic symphysis. As the bladder fills it rises up out of the pelvis and comes to lie in the abdomen. The peritoneum covering the distended bladder becomes peeled off from the anterior

A

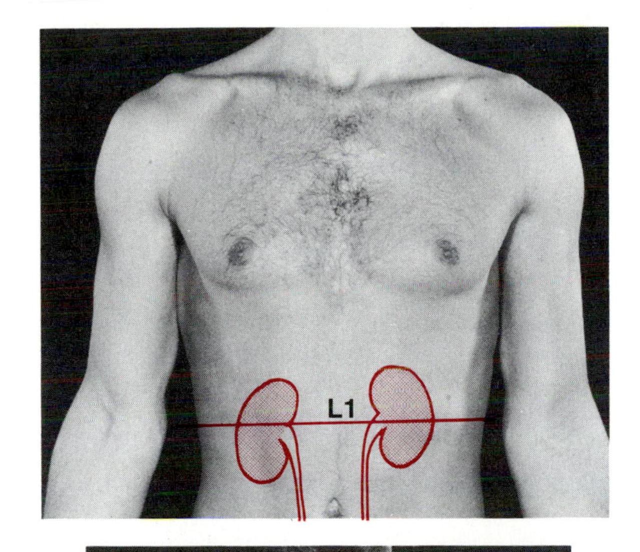

B

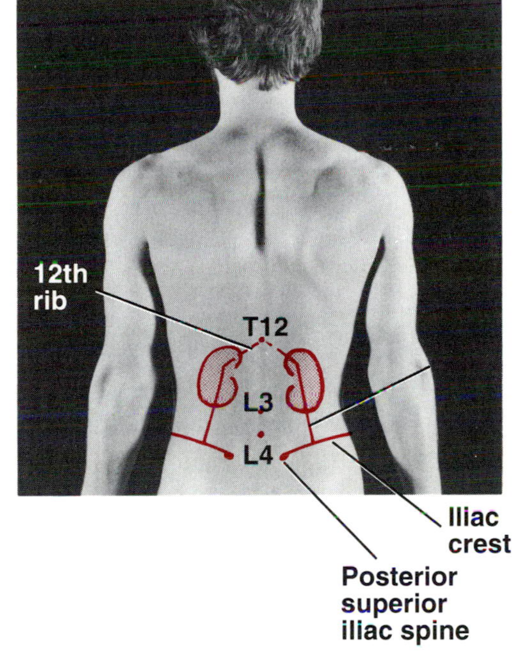

FIG 12–18.
A, surface anatomy of the kidneys and ureters on the anterior abdominal wall. Note the relationship of the hilum of each kidney to the transpyloric plane. **B,** surface anatomy of the kidneys on the posterior abdominal wall.

abdominal wall so that the front of the bladder comes to lie in direct contact with the abdominal wall, as previously discussed.

Clinical Note

Suprapubic aspiration.—The distended bladder in an emergency can be aspirated suprapubically because of this altered position of the peritoneum.

In children until the age of 6 years, the bladder is an abdominal organ even when empty, since the capacity of the pelvic cavity is not great enough to contain it. The neck of the bladder lies just below the level of the upper border of the symphysis pubis. In infants, suprapubic aspiration is a common procedure to obtain a urine sample in situations of a febrile infant who needs a urine analysis as part of a septic workup.

Urethra

The *male* urethra is about (20 cm) 8 in. long and extends from the neck of the bladder to the external meatus on the glans penis. It is divided into three parts—prostatic, membranous, and penile (see Fig 12–9).

The prostatic and membranous parts are deeply placed and cannot be palpated directly. The penile part lies within the bulb and shaft of the corpus spongiosum and can be felt throughout its course. The external meatus is the narrowest part of the entire urethra.

The *female* urethra is about (3.8 cm) 1½ in. long. It extends from the neck of the bladder to the vestibule of the vulva, where it opens about (2.5 cm) 1 in. below the clitoris.

Surface Anatomy of the Penis and Scrotum

Penis

The *root of the penis* consists of three masses of erectile tissue, called the *bulb of the penis* and the *right* and *left crura of the penis.* The bulb may be felt on deep palpation in the midline of the perineum, posterior to the scrotum.

The *body of the penis* is the free portion of the penis that is suspended from the symphysis pubis. Note that the dorsal surface (anterior surface of the flaccid organ) usually possesses an easily recognized *superficial dorsal vein* in the midline (see Fig 12–14).

The *glans penis* forms the extremity of the body of the penis. At the summit of the glans is the *external urethral meatus.* Extending from the lower margin of the external meatus is a fold called the *frenulum.* The edge of the base of the glans is called the *corona* (see Fig 12–14). The *prepuce,* or *foreskin,* is formed by a fold of skin attached to the neck of the penis. The prepuce covers the glans for a variable extent, and it should be possible to retract it over the glans.

Scrotum

The sac of skin and fascia contains the testes and epididymes. The skin of the scrotum is rugose and is covered with sparse hairs. The bilateral origin of the scrotum is indicated by the presence of a dark line in the midline, called the *scrotal raphe,* along the line of fusion.

The *testis* on each side is a firm ovoid body and lies free within the tunical vaginalis (see Fig 12–15); it is not tethered to the subcutaneous tissue or skin. The *epididymis* is an elongated structure that lies posterolateral to the testis (see Fig 12–15). The enlarged upper end forms the *head,* and below it is the *body;* the body narrows at its lower end to form the *tail.* The cordlike *vas deferens* emerges from the tail and ascends medial to the epididymis to enter the spermatic cord at the upper end of the scrotum.

CLINICAL ANATOMY

Clinical Anatomy of the Kidneys

Congenital Anomalies

Each kidney develops as a pelvic organ and only later ascends into the abdomen to take up its final position. Rarely is the ascent arrested.

Renal Cysts

Renal cysts occur in two main forms—polycystic kidney and solitary cyst of the kidney.

Polycystic Kidney.—This hereditary disease can be transmitted by either parent. It may be associated with congenital cysts of the liver, pancreas, and lung. Both kidneys are enormously enlarged and riddled with cysts of varying size. Polycystic kidney is thought to be caused by a failure of union between the developing convoluted tubules and the collecting tubules. The accumulation of urine in the proximal tubules results in the formation of retention cysts.

Solitary Cyst of the Kidney.—In this condition more than one cyst may be found, so the term is misleading. However, the cysts usually are larger and much fewer in number than those found in polycystic disease. The cause of the condition is thought to be the same as that of polycystic kidney. For a representation of *horseshoe kidney* and other congenital anomalies, see Figure 12–19.

Supernumerary Renal Arteries.—These are relatively common. They represent persistent fetal renal arteries, which grow in sequence from the aorta to supply the kidney as it ascends from the pelvis. Their occurrence is clinically important, since a supernumerary artery may cross the pelviureteral junction and obstruct the outflow of urine, producing *hydronephrosis* (see Fig 12–19).

Kidney Trauma

Although the kidneys are well protected by the lower costal margin, the lumbar muscles, and the vertebral column, they are commonly injured in major blunt trauma. Laceration of the kidneys from fractures of the 11th or 12th ribs is a rare occurrence since fractures of these ribs are uncommon. Because 25% of the cardiac outflow passes through the kidneys, renal injury can result in a very rapid blood loss.

Blunt injuries to the kidney occur because the

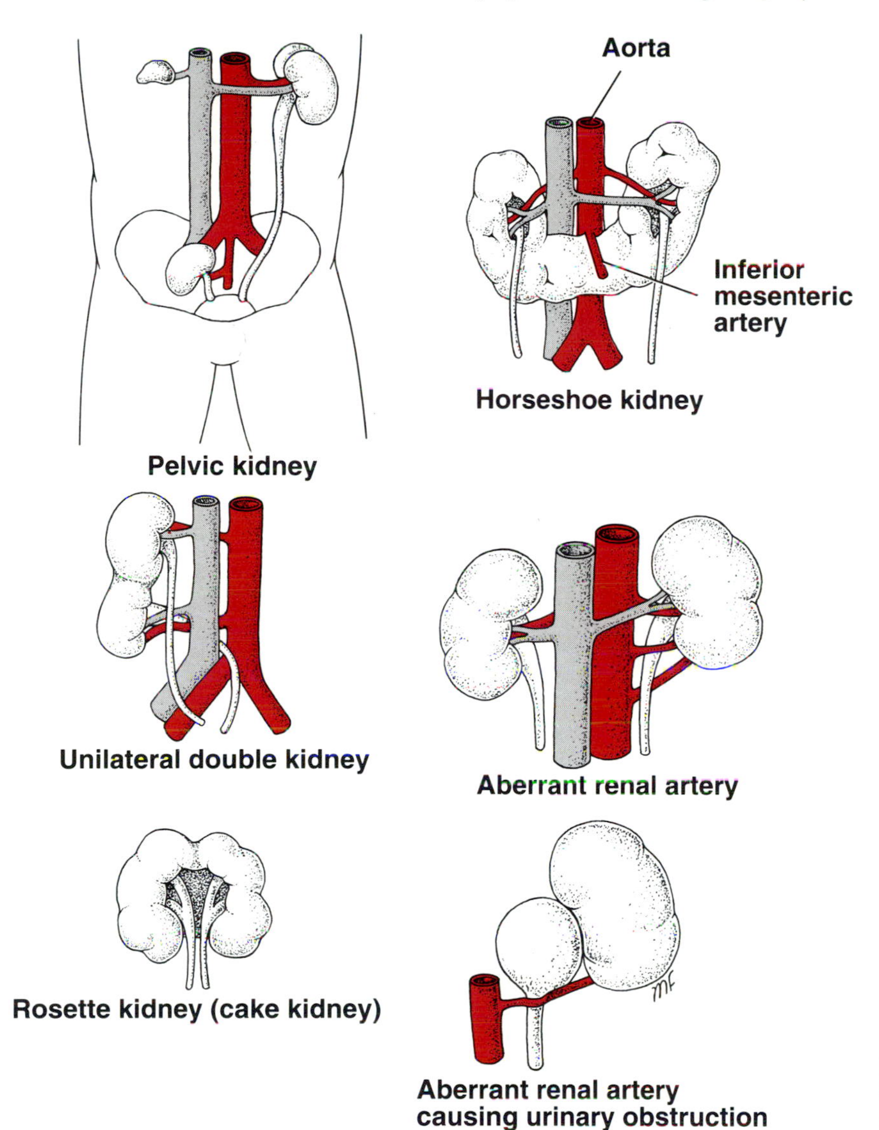

FIG 12–19.
Some common congenital anomalies of the kidney.

kidney is crushed against the last rib and the vertebral column. Injury to the kidney from blunt trauma is classified as follows (Fig 12–20):

1. *Contusions. The renal capsule remains intact.* The kidney is bruised, and in severe cases there may be an accumulation of blood beneath the renal capsule.
2. *Tearing of the renal capsule.* Hemorrhage extends into the surrounding perirenal and pararenal fat. If the laceration of the kidney extends into the calyces, urine as well as blood may extravasate into the pararenal fat and into the peritoneal cavity.
3. *Shattered kidney.* The entire kidney is lacerated with extensive hemorrhage and extravasation of urine.
4. *Pedicle injury.* The renal artery or vein may be injured or actually torn in blunt trauma.

Penetrating injuries to the kidney are usually caused by stab wounds or gunshot wounds with points of entrance through the abdomen in front and lower thorax and flank from the back. In nearly half of the cases, other viscera are injured also, most commonly the spleen, liver, lung, duodenum, and pancreas.

Renal Pain

Renal pain varies from a dull ache to a severe sharp pain in the flank that may radiate downward into the lower abdomen. Renal pain may result from stretching the capsule or spasm of the smooth muscle in the renal pelvis. The afferent fibers pass through the *renal plexus* around the renal artery and ascend to the spinal cord through the lowest tho-

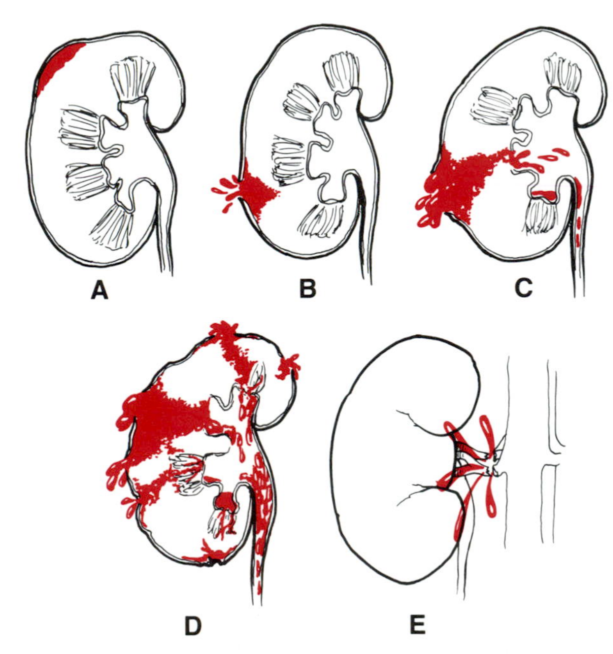

FIG 12–20.
Injuries to the kidney. **A,** contusion, with hemorrhage confined to the cortex beneath the intact fibrous capsule. **B,** tearing of the capsule and cortex with bleeding occurring into the perirenal fat. **C,** tearing of the capsule, the cortex, and the medulla. Note the escape of blood into the calyces and therefore the urine. Urine as well as blood may extravasate into the perirenal and pararenal fat and into the peritoneal cavity. **D,** shattered kidney with extensive hemorrhage and extravasation of blood and urine into the perirenal and pararenal fat; blood also enters the calyces and appears in the urine. **E,** injury to the renal pedicle involving the renal vessels and possibly the renal pelvis.

racic splanchnic nerve and the lower part of the sympathetic trunk. They enter the spinal cord at the level of T12. Pain is referred along the distribution of the subcostal nerve to the flank and the anterior abdominal wall.

Location of Transplanted Kidneys

The iliac fossa (which is the region of the internal concavity of the ilium on the posterior abdominal wall) is the usual site chosen for transplantation of the kidney.

The fossa is exposed through an incision in the anterior abdominal wall just above the inguinal ligament. The iliac fossa is approached retroperitoneally. The kidney is positioned and the vascular anastomosis constructed. The renal artery is anastomosed end-to-end to the internal iliac artery and the renal vein end-to-side to the external iliac vein (Fig 12–21). There is a sufficient anastomosis of the branches of the internal iliac arteries on the two sides so that the pelvic viscera on the side of the renal arterial anastomosis are not at risk. Ureteronephrostomy is then performed by opening the bladder and providing a wide entrance for the ureter through the bladder wall.

Clinical Note

Children usually receive adult kidneys, and the large adult blood vessels can be anastomosed end-to-side into their small vessels. Alternatively, adult vessels are anastomosed end-to-side into the aorta and inferior vena cava.

Accessory renal arteries in the transplanted kidney are not to be ligated since they are end arteries and a section of the kidney may become infarcted. In these circumstances, the donor renal arteries are anastomosed to one another and then a single vessel is implanted into the recipient vessel.

Clinical Anatomy of the Ureters

Traumatic Ureteral Injuries

Because of its protected position and small size, injuries to the ureter are rare. The great majority of injuries are caused by gunshot wounds and, in a few individuals, penetrating stab wounds. In children, where the kidneys are more mobile than in adults, the ureter can be torn from the ureteropelvic junction; this rare injury occurs with severe blunt trauma. Since the ureters are retroperitoneal in position, urine may escape into the retroperitoneal tissues on the posterior abdominal wall.

Ureteric Stones

Ureteric stones are a common problem mainly affecting males between the ages of 30 and 50 years. Passage of the stone through the ureter may be arrested at one of three anatomical sites, namely, the pelviureteral junction, the pelvic brim, and where the ureter enters the bladder (Fig 12–22).

The majority of calculi, although radiopaque, are small enough to be impossible to see definitely along the course of the ureter on plain radiographic examination. An intravenous pyelogram is usually necessary. The ureter runs down in front of the tips of the transverse processes of the lumbar vertebrae, crosses the region of the sacroiliac joint, swings out to the ischial spine, and then turns medially to the bladder.

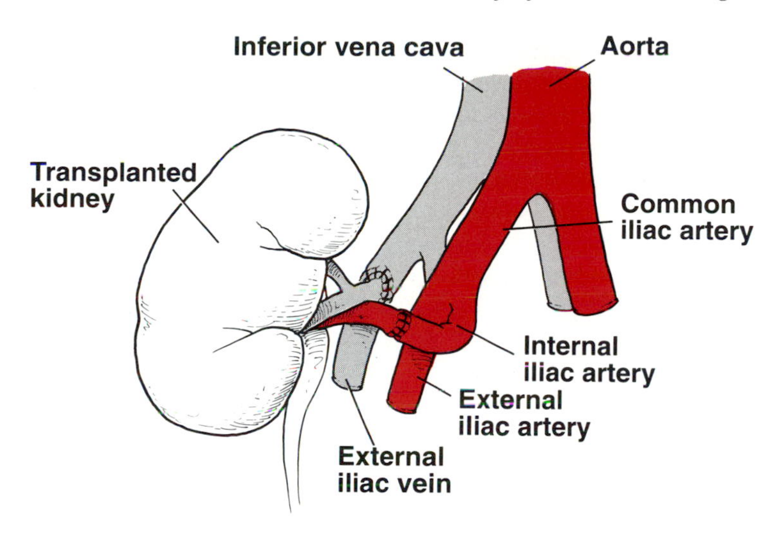

FIG 12–21.
The transplanted kidney.

Renal Colic.—All renal stones originate in the kidney. Here they may grow and reach a large size without causing pain. However, once a small stone enters the lumen of the ureter, the smooth muscle goes into spasm in an attempt to move the stone onward, which causes an intense, agonizing, colicky pain. The pain begins abruptly in the flank and courses laterally around the abdomen to the groin.

The *pain fibers* from the renal pelvis and the ureter enter the spinal cord at segments T11, T12, L1, and L2. The pain is then referred to the skin areas that are supplied by these segments of the spinal cord, namely, the flank, loin, and groin. When a calculus enters the lower part of the ureter, the pain is felt at a lower level and is often referred to the testis or the tip of the penis in the male and the labium majus in the female. Sometimes ureteral pain is referred along the femoral branch of the genitofemoral nerve (L1 and L2), so that pain is experienced in the front of the thigh. The pain is often so severe that the afferent pain impulses spread within the central nervous system, giving rise to nausea.

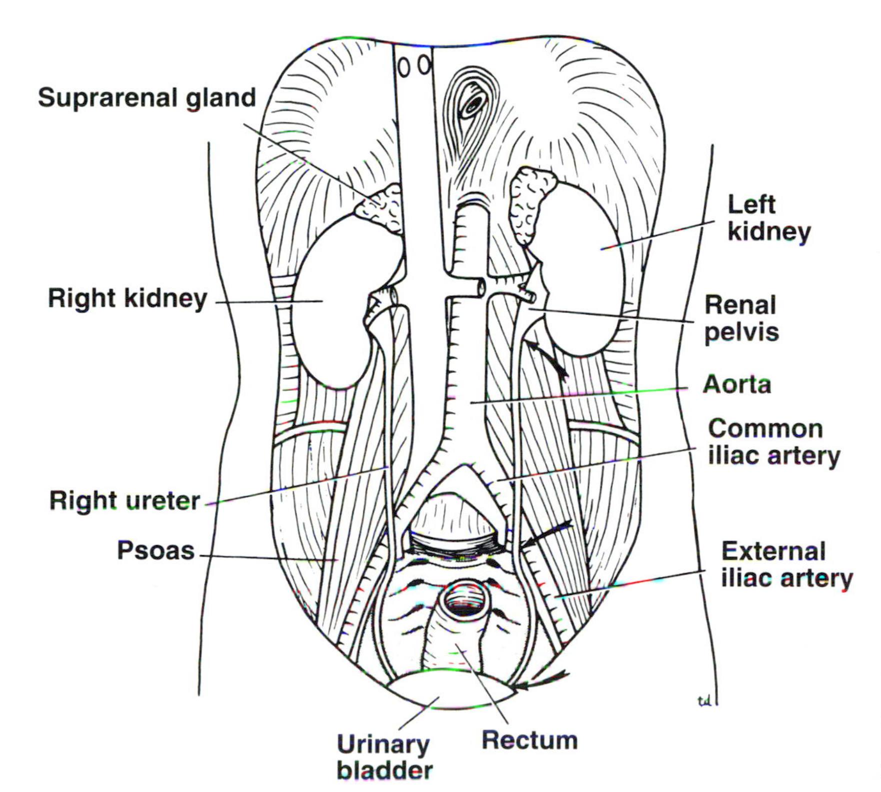

FIG 12–22.
Posterior abdominal wall showing the kidneys and ureters in situ. The arrows indicate the three sites where the ureter is narrowed.

Radiological Examination.—On a plain film, ureteric calculi may be confused with a (1) a calcified costal cartilage, (2) a phlebolith in a pelvic vein, (3) a calcified lymph node, or (4) a tip of a transverse process of a lumbar vertebra. Opaque stones may be difficult to identify if they are situated in front of the bony pelvis. An intravenous pyelogram is necessary in most cases to decide whether a given radiopacity is indeed a ureteric stone.

Clinical Anatomy of the Urinary Bladder

Bladder Injuries

The bladder may rupture intraperitonally or extraperitonally. Intraperitoneal rupture usually involves the superior wall of the bladder and occurs most commonly when the bladder is full and has extended up into the abdomen. Urine and blood escape freely into the peritoneal cavity. Extraperitoneal rupture involves the anterior part of the bladder below the peritoneal reflection; it most commonly occurs in fractures of the pelvis when bony fragments pierce the bladder wall. Lower abdominal pain and hematuria are found in the great majority of patients.

In young children, the bladder is an abdominal organ, so abdominal trauma can injure the empty bladder.

Urinary Retention

In adult males, urinary retention is commonly caused by obstruction to the urethra by a benign or malignant enlargement of the prostate. An acute urethritis or prostatitis can also be responsible. Acute retention occurs much less frequently in females. The only anatomic cause of urinary retention in females is acute inflammation around the urethra (e.g., from herpes).

Spinal Cord Injuries

Following injuries to the spinal cord, the nervous control of micturition is disrupted. The innervation of the normal bladder has been fully described in Chapter 9, p. 339. The nerve supply may be summarized as follows.

Sympathetic outflow is from the first and second lumbar segments of the spinal cord. The sympathetic nerves* inhibit contraction of the detrusor muscle of the bladder wall and stimulate closure of the sphincter vesicae.

Parasympathetic outflow is from the second, third, and fourth sacral segments of the spinal cord. The parasympathetic nerves stimulate the contraction of the detrusa muscle of the bladder wall and inhibit the action of the sphincter vesicae.

Sensory nerve fibers enter the spinal cord at the above segments.

The normal process of micturition has been described on p. 474.

Disruption of the process of micturition by spinal cord injuries may produce the following types of bladder.

Atonic Bladder.—The atonic bladder occurs during the phase of spinal shock immediately following the injury and may last for a few days to several weeks. The bladder wall muscle is relaxed, the sphincter vesicae tightly contracted, and the sphincter urethrae relaxed. The bladder becomes greatly distended and finally overflows. Depending on the level of the cord injury, the patient may or may not be aware that the bladder is full.

Autonomic Reflex Bladder.—The autonomic reflex bladder (Fig 12–23) occurs after the patient has recovered from spinal shock, provided that the cord lesion lies above the level of the parasympathetic outflow (S2 through S4). Efferent impulses pass down to the bladder muscle, which contracts; the sphincter vesicae and the urethral sphincter both relax. This simple reflex occurs every 1 to 4 hours.

Autonomous Bladder.—The autonomous bladder (see Fig 12–23) is the condition that occurs if the sacral segments of the spinal cord are destroyed. The sacral segments of the spinal cord are situated in the upper part of the lumbar region of the vertebral column. The bladder is without any external reflex control. The bladder wall is flaccid, and its capacity is greatly increased. It merely fills to capacity and overflows, which results in continual dribbling. The bladder may be partially emptied by manual compression of the lower part of the anterior abdominal wall, but infection of the urine and back pressure effects on the ureters and kidneys are inevitable.

Clinical Anatomy of the Male Urethra

Infection

The many glands that open into the urethra, including those of the prostate, the bulbourethral glands, and the many small penile urethral glands, are com-

*See footnote, p. 472.

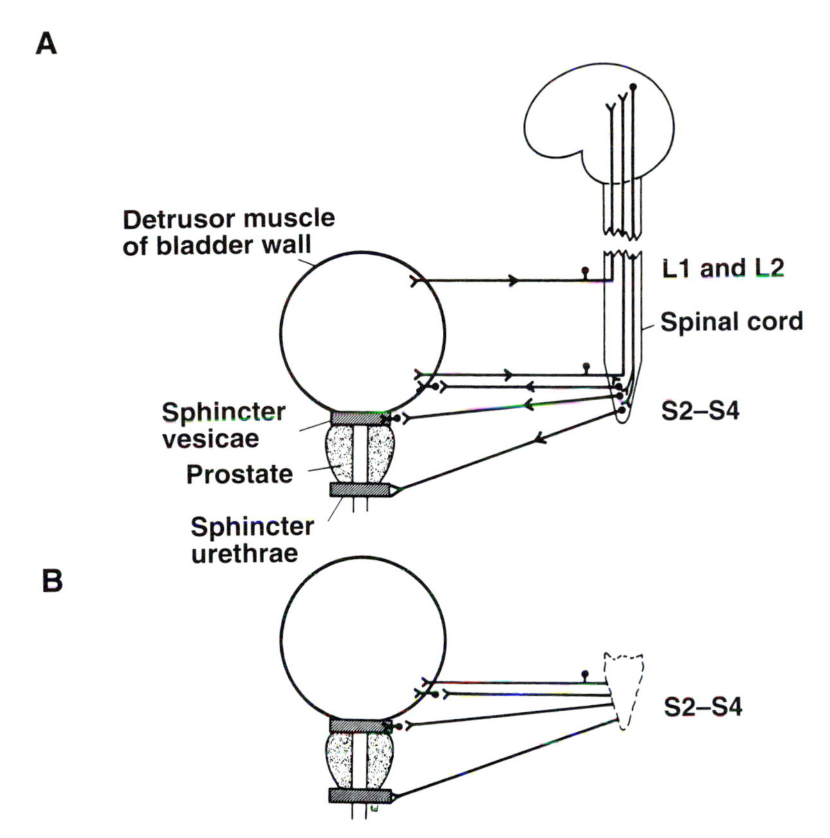

FIG 12–23.
Nervous control of the bladder following **A**, section of the spinal cord in the high thoracic region. **B**, destruction of the sacral segments of the spinal cord. The diagram shows the afferent sensory fibers from the bladder entering the central nervous system and the parasympathetic efferent fibers passing to the bladder; the sympathetic fibers have been omitted for clarity.

mon sites for chronic *Chlamydia* and gonococcal infection (see Figs 12–14 and 12–16).

The most dependent part of the urethra is that which lies within the bulb of the penis. Here it is subject to chronic inflammation and stricture formation.

Injuries to the Urethra

Injuries to the male urethra are relatively rare. The clinical implications vary with the part injured.

Prostatomembranous Injuries.—Fractures of the pelvis are the usual cause. A shearing force on the urethra just above the urogenital diaphragm causes complete or incomplete laceration of the urethra. When the urethra is completely severed, the bladder and prostate are displaced upward and the empty cavity quickly fills with blood clot and urine. In addition, bony fragments may be thrust backward, penetrating the bladder wall.

Bulbar Injuries (Fixed Part of the Penile Urethra).

This usually results from severe perineal contusion by falling astride an object. The urethra is crushed against the inferior surface of the urogenital diaphragm, and the laceration may or may not be complete (the bladder and prostate remain in their normal relation to the urogenital diaphragm). If the deep fascia (Buck's fascia) around the bulb is disrupted, blood and urine will extravasate into the superficial perineal pouch and will be limited only by the attachments of Colles' fascia in the perineum and Scarpa's fascia on the anterior abdominal wall (see Fig 11–3). If treatment is delayed, the extravasated fluid can spread over the penis and scrotum onto the anterior abdominal wall.

Penile Injuries (Pendulous Part of the Urethra).

The penile urethra is rarely injured because of the extreme mobility of the penis. Foreign bodies inserted into the external meatus may damage the urethra. Periurethral hemorrhage and extravasation of urine may occur.

Catheterization of the Male Urethra

The following anatomical facts should be remembered before passing a catheter or other instrument along the male urethra.

1. The external meatus at the glans penis is the narrowest part of the entire urethra.
2. Within the glans the urethra dilates to form the fossa terminalis (navicular fossa).
3. Near the posterior end of the fossa, a fold of

mucous membrane projects into the lumen from the roof (see Fig 12–14).

4. The membranous part of the urethra is narrow and fixed.

5. The prostatic part of the urethra is the widest and most dilatable part of the urethra.

6. By holding the penis upward, the S-shaped curve to the urethra is converted into a J-shaped curve.

Anatomy of the Procedure of Catheterization.—The procedure is as follows.

1. The patient lies in a supine position.
2. The penis is held erect with gentle traction at right angles to the anterior abdominal wall. The lubricated catheter is passed through the narrow external urethral meatus. The catheter should pass easily along the penile urethra. On reaching the membranous part of the urethra a slight resistance is felt because of the tone of the urethral sphincter and the surrounding rigid perineal membrane.
3. The penis is then lowered toward the thighs, and the catheter is gently pushed through the sphincter.
4. The passage of the catheter through the prostatic urethra and bladder neck should not present any difficulty.

Anatomy of False Passage.—The passage of an instrument in patients with an urethral stricture may result in the creation of an artificial channel outside the urethral lumen. Since the urethra is surrounded by the delicate erectile tissue of the corpus spongiosum, little resistance is encountered, and a false passage may be created. The new passage may then become epithelialized and persist. The use of a filiform to identify the correct passage and then the passage of followers will enable the stricture to be dilated.

Clinical Anatomy of the Female Urethra

Risk of Ascending Infection

The short length of the female urethra predisposes to ascending infection; consequently, *cystitis* is more common in females than males.

Injuries to the Urethra

Because of the short length of the urethra, injuries are very rare. In fractures of the pelvis the urethra may be damaged by shearing forces as it emerges from the fixed urogenital diaphragm.

Catheterization of the Female Urethra

Because the female urethra is shorter, wider, and more dilatable, catheterization is much easier than in males. Moreover, the urethra is straight, and only minor resistance is felt as the catheter passes through the urethral sphincter.

Clinical Anatomy of the Male External Genitalia

Penis

Injuries to the penis may occur as the result of blunt trauma, penetrating trauma, or strangulation. As a result, the skin, the erectile tissue, and the urethra may be damaged. Fortunately, the anatomical position of the penis and its extreme mobility in the flaccid state provide some natural protection.

Blunt Trauma.—This may cause penile fracture when it is in the erect state. A transverse laceration of the fibrous envelopes of the erectile tissue of the penis occurs followed by the formation of a large hematoma; the base of the penis is the common site of the injury. Rupture of the penile urethra may also occur and is accompanied by a bloody urethral discharge.

Penetrating Trauma.—This may injure the skin, fascia, the erectile tissue, and the urethra. Amputation of the entire penis should be repaired by anastomosis using microsurgical techniques to restore continuity of the main blood vessels.

Strangulation.—Strangulation of the penis by means of a ring or ligature may cause ischemia of the entire penis. It is imperative that the constriction be removed without delay to avoid compromising the blood supply.

Priapism.—This is a prolonged, painful erection that is not necessarily related to sexual stimulation. The condition involves the corpora cavernosa, but the corpus spongiosum and the glans are soft. In the majority of patients the cause is unknown, but priapism occurs with increased frequency in patients with sickle cell disease, leukemia, polycythemia, and malignancy. Abnormalities of the autonomic control of erection may occur in patients with paraplegia.

Phimosis.—In this condition the opening in the prepuce is narrowed so that it is impossible to retract the prepuce over the glans penis. Occasionally the narrowing is so extreme that the urinary flow is obstructed. The usual cause of the condition is infection under the prepuce, causing fibrosis and subsequent contraction of the prepuce.

Paraphimosis.—In this condition the prepuce has become stuck in the retracted position proximal to the glans. Edema of the prepuce causes it to swell, which constricts the penis, causing pain and congestion of the glans.

Erection and Ejaculation Following Spinal Cord Injuries.—Erection of the penis is controlled by the parasympathetic nerves that originate from the second, third, and fourth sacral segments of the spinal cord. Bilateral damage to the reticulospinal tracts (see p. 373) in the spinal cord will result in loss of erection. Later, when the effects of spinal shock have disappeared, spontaneous or reflex erection may occur if the sacral segments of the spinal cord are intact.

Ejaculation is controlled by sympathetic nerves that originate in the first and second lumbar segments of the spinal cord. Ejaculation brings about a flow of seminal fluid into the prostatic urethra. The final ejection of the fluid from the penis is the result of the rhythmic contractions of the bulbospongiosus muscles, which compress the urethra. The bulbospongiosus muscles are innervated by the pudendal nerve (S2 through S4). Discharge of seminal fluid into the bladder is prevented by the contraction of the sphincter vesicae at the bladder neck, which is innervated by the sympathetic nerves (L1 and L2). As in the case of erection, severe bilateral damage to the spinal cord results in loss of ejaculation. Later, reflex ejaculation may be possible in patients with spinal cord transections in the thoracic or cervical regions. Some individuals have a normal ejaculation without external emission, and the seminal fluid passes into the bladder owing to paralysis of the sphincter vesicae.

Clinical Anatomy of the Scrotum

Injuries to the Scrotum and Testicle.—Because of the mobility and the protected position of the scrotum, these are uncommon. In penetrating injuries to the skin, superficial fascia (dartos muscle), external spermatic fascia, cremasteric fascia, and the internal spermatic fascia may be damaged, producing a large hematoma in the loose tissue. If the penetration pierces the outer layer of the tunica vaginalis, blood will collect within the tunica, forming a *hematocele.* Further penetration through the tunica albuginea of the testis will cause the seminiferous tubules to extrude through the wound.

Fournier's Gangrene of the Scrotum.—This is an acute ascending infection of the scrotal wall or perineal area in which the patient is very ill with fever, anorexia, and pain. Because of its loose fascial layers, the scrotal wall rapidly becomes edematous, swollen, and discolored; this is later followed by extensive necrosis. Infection may spread up onto the anterior abdominal wall; the condition carries a high mortality rate.

The patient is usually in poor health and may have diabetes mellitus or chronic alcoholism. The source of infection may be derived from an indwelling Foley catheter with chronic urethritis or infection in the ischiorectal region. The various fascial layers of the scrotal wall provide pathways for the rapid spread of infection.

Hydrocele.—This is an accumulation of fluid within the tunica vaginalis. Although the great majority of hydroceles are idiopathic, an inflammation of the testis or epididymis can produce the condition.

Varicocele.—In this condition there is an elongation and dilation of the veins of the pampiniform plexus. It is a common disorder found in adolescents and young adults. The great majority occur on the left side because the right testicular vein drains into the low-pressure inferior vena cava, whereas the left vein drains into the left renal vein, in which the venous pressure is higher. Very rarely, a malignant tumor of the left kidney with invasion of the left renal vein may block the exit of the testicular vein.

Congenital Anomalies of the Testis and Epididymis

Imperfect Descent (Cryptorchidism).—This includes the following.

1. *Incomplete descent.* The testis, although traveling down its normal path, fails to reach the floor of the scrotum. It may be found within the abdomen, the inguinal canal, at the superficial inguinal ring, or high up in the scrotum.
2. *Maldescent.* The testis travels down an abnormal path and fails to reach the scrotum. It may be found in the superficial fascia of the anterior abdominal wall, in front of the pubis, in the perineum, or in the thigh. A maldescended testis is very susceptible to traumatic injury, and it is believed that there

is a greater than normal incidence of tumors in such testes. For these reasons, maldescended testes should be surgically placed in the scrotum.

Appendix of the Testis and the Epididymis.—Cysts of these embryological remnants are occasionally found at the upper poles of these organs.

Torsion of the Testis.—This condition is a rotation of the testis around the spermatic cord within the scrotum. It is often associated with an excessively large tunica vaginalis. The testis may rotate or spin several times around on its axis; the rotation is usually in a medial direction (i.e., counterclockwise on the right, clockwise on the left, when looking from above).

Torsion commonly occurs in active young men and children. The pain, which is severe, occurs in the testis and ascends along the afferent autonomic nerves to the upper lumbar spinal segments of the cord. Here the impulses enter the spinal cord, and the pain is referred to the first lumbar dermatome, i.e., the flank. Pain from a kidney or ureteral stone may also be referred to the flank or to the testicle; this is because some of the afferent sensory fibers from the kidney and ureter also enter the spinal cord at the first lumbar segment.

Anatomy of Scrotal Swellings

Indirect Inguinal Hernia.—Persistence of the processus vaginalis gives rise to the preformed sac of the hernia (see p. 447). The swelling lies in front of the testis and continues up into the inguinal canal through the superficial inguinal ring (Fig 12–24).

Epididymitis.—Infection gains entrance to the epididymis via the vas deferens or the lymphatic vessel around this structure. The site of the primary infection is usually the prostate, urethra, or bladder. The swelling lies behind the testis (see Fig 12–24). In the acute form, the epididymis is very tender and the scrotal wall is red and edematous. In the chronic form (tuberculous), the epididymis is not tender and is hard and irregular; the swelling is often continuous with irregular beading of the vas deferens due to involvement of the lymphatic vessels around the vas deferens.

Hydrocele.—The accumulation of excess fluid in the tunica vaginalis; the cause may be unknown or secondary to infection of the epididymis, trauma, or orchitis secondary to mumps. The swelling lies in front and on the sides of the testis (see Fig 12–24). If the swelling is very large, it may be difficult to palpate the testis.

Varicocele.—This involves dilatation and varicosity of the pampiniform plexus, usually on the left side. The dilated veins give the impression that the scrotum is full of "worms" (see Fig 12–24).

Spermatocele.—This is a cyst of the efferent duct of the rete testis, which lies within the mediastinum testis (see Fig 12–15). The swelling lies above and behind the testis in association with the epididymis (see Fig 12–24).

Orchitis.—This may occur secondary to mumps and usually follows subsidence of the acute parotitis. The testis is swollen but is not accompanied by scrotal edema.

Testicular Tumor.—This is usually a hard, irregular, nontender mass involving the testis (see Fig 12–24).

Torsion of the Testis.—The testis is swollen and extremely tender. The testis may appear elevated due to the twisting of the spermatic cord, and the epididymis may be anterior in position, depending on the degree of rotation.

Skin Swellings.—These include sebaceous cysts and benign and malignant tumors.

Clinical Anatomy of the Prostate and Seminal Vesicles

Examination of the Prostate

The prostate can be examined clinically by palpation through the anterior rectal wall. The normal prostate feels firm and elastic. Each of its lateral lobes is separated by a median furrow. At the lower end of the furrow, a soft area is formed by the membranous part of the urethra. On each side of the midline the bulbourethral glands are situated and can be felt if enlarged. At the upper end of the median furrow lie the seminal vesicles on each side of the midline. These cannot be felt unless distended with fluid. Between the seminal vesicle and the lateral lobe of the prostate is a groove in which run the lymphatic drainage of the prostate.

Benign Enlargement of the Prostate

Benign enlargement of the prostate is common in men over 50 years. The median or middle lobe of the prostate, which contains the most glandular tissue,

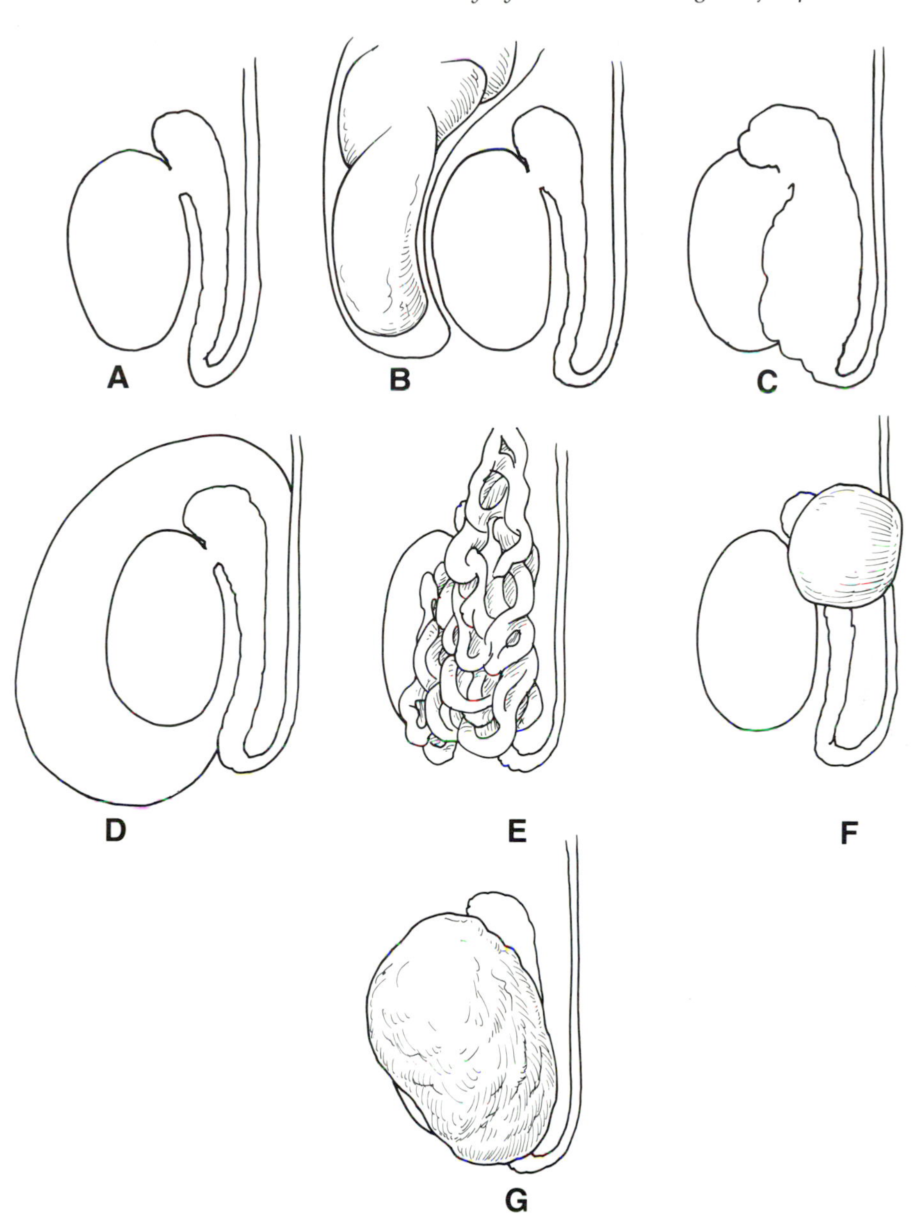

FIG 12–24.
Some common causes of scrotal swellings. **A,** normal testis and epididymis. **B,** indirect inguinal hernia. **C,** epididymitis, **D,** hydrocele. **E,** varicocele. **F,** spermatocele. **G,** tumor of the testis.

enlarges and interferes with the normal function of the urinary tract. The median lobe enlarges upward and encroaches within the sphincter vesicae in the neck of the bladder. The resulting leakage of urine into the prostatic urethra causes an intense reflex desire to micturate.

The prostatic urethra becomes elongated, distorted, and compressed by the enlarging median and lateral lobes of the gland so that the patient experiences difficulty in passing urine, and the stream is weak. A further slight enlargement of the prostate, due to congestion caused by alcohol or by an acute prostatitis, may result in total obstruction to urinary flow and force the patient to seek treatment in the emergency department.

The projection upward of the median lobe into the trigone of the bladder results in the formation of a pouch of stagnant urine behind the urethral orifice within the bladder (Fig 12–25). The stagnant urine frequently becomes infected, and the inflamed bladder (cystitis) adds to the patient's symptoms.

Carcinoma of the Prostate

Carcinoma of the prostate is very common in men over 50 years of age and usually arises in the posterior part of the periphery of the gland. The lesion is an adenocarcinoma and is stony hard and irregular on palpation. Because of its peripheral origin, it only causes urethral obstruction late in the disease and,

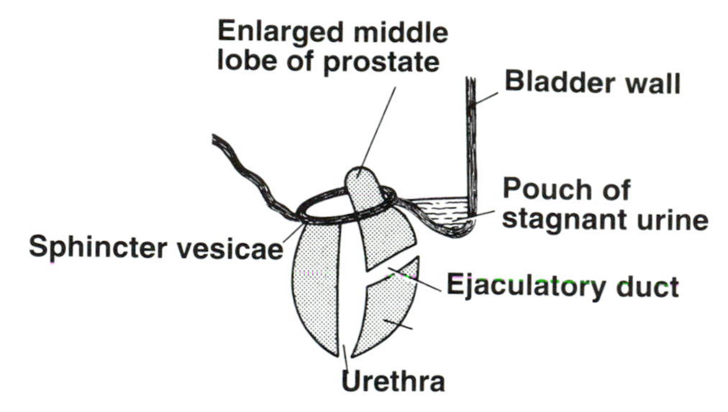

FIG 12–25.
Sagittal section of the prostate that had undergone benign enlargement of the middle lobe. Note how the function of the sphincter vesicae can be interfered with by the enlarging gland; a bladder pouch is filled with stagnant urine behind the enlarged prostate.

for this reason, is unlikely to be seen in an emergency department.

There is a direct communication between the prostatic venous plexus and the vertebral veins. This anatomical fact is of significance in carcinoma since it explains the route taken by the malignant cells in establishing metastases in the vertebral column and pelvic bones.

ANATOMICAL-CLINICAL PROBLEMS

1. A 32-year-old man was seen in the emergency department with hematuria and left flank pain following an automobile accident. During patient workup, an intravenous pyelogram was performed that revealed the left kidney was in its normal position and appeared to be functioning satisfactorily; the right kidney was located in front of the right sacroiliac joint and had a poor output. Can you explain the low position of the right kidney on embryological grounds?

2. A 17-year-old boy received a severe kick in the left flank while playing football at school. On examination in the emergency department, his left flank was severely bruised, and his left costovertebral angle was extremely tender on palpation. A specimen of urine showed microscopic hematuria. Explain in anatomical terms how blunt trauma is likely to damage the kidney. Does abdominal examination following kidney trauma ever reveal a palpable mass? Describe the type of kidney damage that is likely to produce hematuria.

3. A 16-year-old boy was involved in a gang fight. It started as an argument but quickly degenerated into a street brawl with the use of knives. He was examined in the emergency department with a bleeding stab wound in his left flank. A urine specimen revealed frank blood. Since stab wounds involving the kidneys involve other abdominal organs in a high percentage of cases, name the various abdominal viscera that are closely related to the anterior surface of the left kidney.

4. Renal pain is a common symptom faced by emergency personnel. Describe the path taken by pain fibers from the kidneys. In which regions of the body is pain commonly referred to?

5. A 50-year-old woman was rolling on the kitchen floor, crying out from agonizing pain in her abdomen, when she was found by her husband. The pain came in waves and extended down from the right loin to the groin and to the right thigh. A right ureteral calculus was suspected. Can you explain the following? (1) What causes the pain in patients with ureteric calculus? (2) Why is the pain felt in such an extensive area? (3) Where does one look for the course of the ureter in a radiograph? (4) Where along the course of the ureter is a calculus likely to be held up?

6. In relation to abdominal trauma, can you explain the differences between the bladder position in a child compared with that of an adult? Does the degree of filling of the adult bladder affect the signs and symptoms presented by a patient with a ruptured bladder? Can you explain how it is possible to pass an aspirating needle through the anterior abdominal wall into the full bladder in an adult without entering the peritoneal cavity?

7. A 41-year-old man was admitted to the emergency department following a gunshot wound to the lower part of his back. Radiographic examination revealed that the bullet was lodged in the vertebral canal at the level of the third lumbar vertebra. A comprehensive neurological examination indicated that a complete lesion of the cauda equina had occurred. What is the nerve supply to the bladder? Is this patient going to have any interference with bladder function?

8. When catheterizing a male patient, it is important to know (1) which is the narrowest part of the entire urethra, (2) which is the narrowest and least dilatable portion of the urethra, (3) which is the widest and most dilatable portion of the urethra, and (4) are there any mucosal folds present in the urethra that are likely to impede the passage of the catheter?

9. A 21-year-old male cyclist was taking part in an international race in Europe. He was on the last

lap of the race and extremely fatigued when, on reaching a steep incline, he stood up on his pedals to increase his speed. Unfortunately, his right foot slipped off the pedal and he fell violently, with his perineum hitting the bar of the bicycle. Several hours later he was admitted to an emergency department unable to micturate. On examination, his penis and scrotum were found to have extensive swelling. The diagnosis of ruptured urethra was made. Which part of the urethra was likely to have been damaged? What is the cause of the extensive swelling of the penis and scrotum?

10. Why is acute cystitis more common in females than in males? Can the female urethra be easily dilated? In anatomical terms, explain stress incontinence.

11. Describe in anatomical terms the structures that one can feel through the anterior rectal wall when performing a rectal examination in the male.

12. A 64-year-old man was seen in the emergency department with acute retention of urine. A hard swelling was found on the posterior surface of his prostate. A radiograph of the pelvis revealed multiple bony metastases. Explain in anatomical terms the route by which an adenocarcinoma of the prostate spreads to involve the skeletal system.

13. A 55-year-old man was examined for a large, tense, painful swelling in the scrotum. The swelling was located in front of his left testis and fluctuated on palpation. The patient said that he had first noticed the swelling 6 months ago and that it had gradually increased in size. A diagnosis of hydrocele was made. Name the layers of the scrotal wall that cover the hydrocele. Name the anatomical structure in which the fluid was situated.

14. A 30-year-old man presented in the emergency department with a large scrotal swelling. Name the anatomical structures that lie within the scrotum that could give rise to a large swelling.

15. What is the lymphatic drainage of the testis? Which group of lymph nodes would you palpate if (1) a malignant disease involving the testis was confined to the testis and (2) if the disease had ulcerated through the scrotal wall?

16. A 26-year-old man was seen in the emergency department with a fractured penis. During sexual intercourse he had heard a sudden cracking sound that was followed by a severe pain and detumescence. On examination, the entire shaft of the penis was swollen and discolored due to extensive hematoma formation. Name the structures that are likely to have been damaged in this patient. What is Buck's fascia? What is the tunica albuginea?

ANSWERS

1. Both kidneys originate in the pelvis and with development rise up on the posterior abdominal wall until the hili lie opposite the first or second lumbar vertebra. Occasionally, one of the kidneys fails to reach its normal position.

2. In blunt trauma the kidney tends to be crushed between the last rib and the vertebral column; rarely the kidney is injured by fractures of the 11th or 12th ribs. In most cases the kidney contusions are mild and result in nothing more than microscopic hematuria; the intravenous pyelogram shows nothing abnormal. In severe kidney lacerations, where there is extensive hemorrhage and extravasation of blood and urine into the pararenal fat behind the peritoneum, a palpable mass may be present.

The types of renal injury that may be associated with hematuria include (1) severe contusion, (2) tearing of the renal capsule, (3) shattered kidney, and (4) tearing of the renal pedicle (see also p. 483).

3. The following structures are closely related to the anterior surface of the left kidney: stomach, spleen, body of pancreas, left colic flexure, and coils of jejunum. (The left suprarenal gland overlaps the upper part of the medial border of the left kidney.) See also Figure 12–4.

4. Renal pain may result from stretching the renal capsule or spasm of the smooth muscle in the renal pelvis. The afferent nerve fibers pass through the renal plexus around the renal artery and ascend to the spinal cord through the lowest thoracic splanchnic nerve and the sympathetic trunk. They enter the 12th thoracic spinal nerve through the white rami communicantes and the 12th thoracic segment of the spinal cord through the posterior root of the spinal nerve. The pain fibers are then believed to ascend to the brain in the lateral spinal thalamic tracts. The pain is referred along the distribution of the subcostal nerve (T12) to the flank and the anterior abdominal wall.

5. (1) The cause of pain in patients with ureteric calculus is spasm of the smooth muscle in the wall of the renal pelvis and ureter as it attempts to move the calculus down the urinary tract. (2) Afferent pain fibers from the ureter enter the spinal cord in the first and second lumbar segments. The anterior rami of the first lumbar nerves are distributed to the skin in the lumbar region and groin as the iliohypogastric and ilioinguinal nerves. The pain experienced in the front of the thigh was referred along the femoral

branch of the genitofemoral nerve (L1 and L2). (3) The ureter descends in front of the tips of the transverse processes of the lumbar vertebrae, passes in front of the sacroiliac joint, and then passes forward close to the ischial spine to enter the bladder. (4) A calculus is likely to be held up at sites where the ureter is narrowed, namely, where the renal pelvis joins the upper end of the ureter, where the ureter is kinked as it crosses the sacroiliac joint to descend into the pelvis, and where the ureter passes through the bladder wall.

6. In young children, the empty bladder lies in the abdomen; later, as the pelvis enlarges, the bladder sinks to become a pelvic organ.

In adults, the full bladder lies behind the lower part of the anterior abdominal wall. Severe trauma to the lower abdomen in patients with a full bladder may cause the superior wall of the bladder to rupture into the peritoneal cavity. The blood and urine irritates the peritoneum, causing lower abdominal tenderness and later muscle rigidity.

When the bladder is empty, the anterior wall of the bladder lies behind the symphysis pubis. Trauma to the lower abdomen in these circumstances may not damage the bladder. However, if the pubic bones are fractured, the anterior wall of the bladder may be damaged by bone fragments, and urine escapes below the reflection of the peritoneum and does not enter the peritoneal cavity. Lower abdominal pain and hematuria are present in the majority of such patients.

It is possible to pass an aspirating needle through the lower part of the anterior abdominal wall into the full bladder in an adult without entering the peritoneal cavity. This can be explained because as the bladder fills, the superior wall rises and strips the peritoneum off the lower part of the anterior abdominal wall. Thus, the aspirating needle will enter the bladder below the peritoneal reflection and avoid the peritoneal cavity.

7. The vesical sphincter is innervated by sympathetic fibers from the first and second lumbar segments of the spinal cord; the detrusor muscle of the bladder wall is innervated by parasympathetic fibers from the second, third, and fourth sacral segments of the spinal cord. Sensory nerve fibers from the bladder enter the spinal cord at the same segmental levels.

The cauda equina consists of anterior and posterior spinal nerve roots below the level of the first lumbar segment of the spinal cord. In this patient, the cauda equina was sectioned at the level of the third lumbar vertebra. This meant that the preganglionic sympathetic fibers to the vesical sphincter that descend in the anterior roots of the first and second lumbar nerves were left intact, since they leave the vertebral canal to form the appropriate spinal nerves above the level of the bullet.

The preganglionic parasympathetic fibers to the detrusor muscle were, however, sectioned as they descended in the vertebral canal within the anterior roots of the second, third, and fourth sacral nerves. The patient would, therefore, have an autonomous bladder and would be without any external reflex control. The bladder would fill to capacity and then overflow. Micturition could be activated by powerful contraction of the abdominal muscles by the patient assisted by manual pressure on his anterior abdominal wall in the suprapubic region.

8. (1) The narrowest part of the entire male urethra is the external orifice at the glans penis. (2) The membranous part of the urethra is the narrowest and least dilatable part of the urethra, once the catheter has passed through the external meatus. (3) The prostatic part of the urethra is the widest and most dilatable part of the urethra. (4) Near the posterior end of the fossa terminalis is a fold of mucous membrane that projects into the urethral lumen from the roof.

9. The bulbous part of the urethra or the membranous part may be damaged in accidents of this nature. There is usually extensive extravasation of urine into the superficial perineal pouch. In this patient the urine passed forward over the scrotum and penis deep to the membranous layer of superficial fascia.

10. Acute cystitis is much more common in females than males because the urethra is much shorter. In females the urethra measures 1 ½ in. long, whereas in males the urethra measures 8 in. long, and thus bacteria have a shorter distance to travel in the female. The female urethra is easily dilatable.

Stress incontinence usually follows a difficult childbirth, where there has been injury to the pelvic floor. This results in an alteration in the position of the bladder neck relative to the urethra.

11. The following structures can be felt through the anterior rectal wall in the male. (1) Opposite the terminal phalanx are the contents of the rectovesical pouch, the posterior surface of the bladder, the seminal vesicles, and the vasa deferentia. (2) Opposite the middle phalanx is the prostate (see also p. 490). (3) Opposite the proximal phalanx are the perineal

body, the urogenital diaphragm, and the bulb of the penis.

12. The prostatic venous plexus is drained into the internal iliac veins. Large valveless veins also connect the plexus to the valveless vertebral veins. On coughing or sneezing, the blood may be forced from the prostatic plexus in the pelvis into the vertebral veins. Dislodged prostatic carcinoma cells may be carried along this route to the vertebral column. In a similar manner malignant cells may be carried into the marrow of the pelvic bones.

13. The following layers of the scrotal wall cover the hydrocele: skin, dartos muscle, membranous layer of superficial fascia (Colles' fascia), external spermatic fascia, cremasteric fascia, internal spermatic fascia, and the parietal layer of the tunical vaginalis that encloses the hydrocele fluid.

14. Persistent processus vaginalis may give rise to a preformed sac of an indirect inguinal hernia; epididymis swells with acute and chronic infections; tunica vaginalis containing excessive fluid and forming a hydrocele; testis may undergo torsion, infection, or tumor formation; pampiniform plexus may become dilated and varicosed, forming a varicocele. Rete testis tubules in the mediastinum testis may form a spermatocele. The scrotal skin rarely gives rise to large swellings; inflammatory edema of the skin secondary to acute infection of the epididymis, however, may contribute to the size of a scrotal swelling.

15. The lymphatic drainage of the testis is into the paraortic nodes at the level of the first lumbar vertebra, i.e., at the level of the transpyloric plane. If the malignant tumor was locally confined to the testis, then secondaries would spread to the paraortic nodes. Once the tumor had extended locally to involve the scrotal wall, then metastasis may occur in the medial group of superficial inguinal lymph nodes.

16. Fracture of the penis occurs during the erect stage and commonly takes place as a transverse laceration across the cylinders of erectile tissue at the root of the penile shaft.

Buck's fascia is the deep fascia that completely surrounds the three cylinders of erectile tissue. The tunica albuginea is the outer fibrous coat of the corpora cavernosa and spongiosa; it is this layer that is torn when the penis is fractured.

13

The Perineum, the Female Organs of Reproduction, and Childbirth

Infections, injuries, and prolapses involving the anorectum and the external genitalia are frequently encountered in the emergency department. The female reproductive system may present the emergency physician with a variety of conditions, such as unsuspected pregnancy, ectopic pregnancy, spontaneous abortion, abnormal uterine bleeding, or acute pelvic inflammatory disease. The purpose of this chapter is to provide an overview of the anatomy relevant to common clinical conditions.

BASIC ANATOMY

Basic Anatomy of the Perineum

The cavity of the pelvis is divided by the pelvic diaphragm into the main pelvic cavity above and the perineum below (Fig 13–1). When seen from below with the thighs abducted, the perineum is diamond shaped and is bounded anteriorly by the symphysis pubis, posteriorly by the tip of the coccyx, and laterally by the ischial tuberosities (Fig 13–2).

Pelvic Diaphragm

The pelvic diaphragm is formed by the important levatores ani muscles and the small coccygeus muscles and their covering fasciae (Figs 13–1 and 13–3). It is incomplete anteriorly to allow for the passage of the urethra in males and females and, in the female, the vagina also.

Levator Ani Muscle.—The levator ani muscle is a wide, thin sheet that has a linear origin from the back of the body of the pubis, a tendinous arch formed by a thickening of the pelvic fascia covering the obturator internus muscle, and the spine of the ischium (see Fig 13–1). From this extensive origin, groups of fibers sweep downward and medially to their insertion (see Fig 13–3), as follows.

1. *Anterior fibers.* The *levator prostatae* or *sphincter vaginae* forms a sling around the prostate or vagina and is inserted into a mass of fibrous tissue, called the *perineal body,* in front of the anal canal.
2. *Intermediate fibers.* The *puborectalis* forms a sling around the junction of the rectum and anal canal. The *pubococcygeus* passes posteriorly to be inserted into a small fibrous mass, called the *anococcygeal body,* between the tip of the coccyx and the anal canal.
3. *Posterior fibers.* The *iliococcygeus* is inserted into the anococcygeal body and the coccyx.

Action.—The levatores ani muscles of the two sides form an efficient muscular sling that supports and maintains the pelvic viscera in position. They

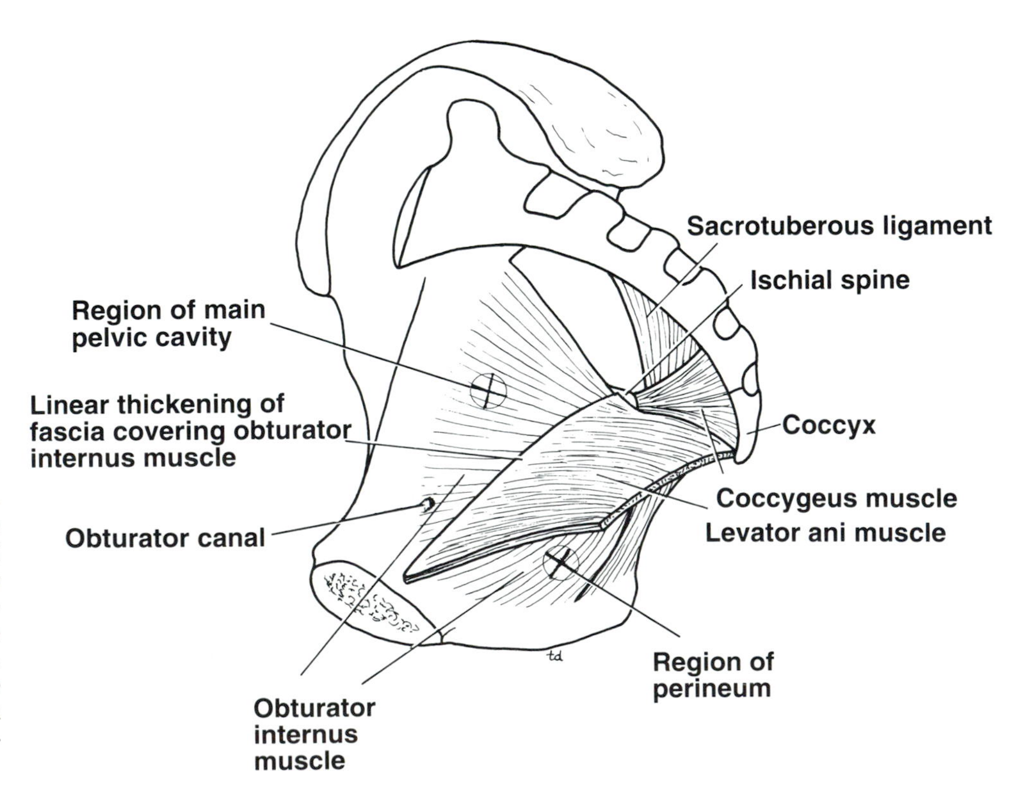

FIG 13–1.
Right half of the pelvis showing the muscles forming the pelvic floor. The levator ani muscle arises from the body of the pubis, a linear thickening of the fascia covering the obturator internus muscle, and the ischial spine. The coccygeus muscle arises from the ischial spine and is inserted into the coccyx.

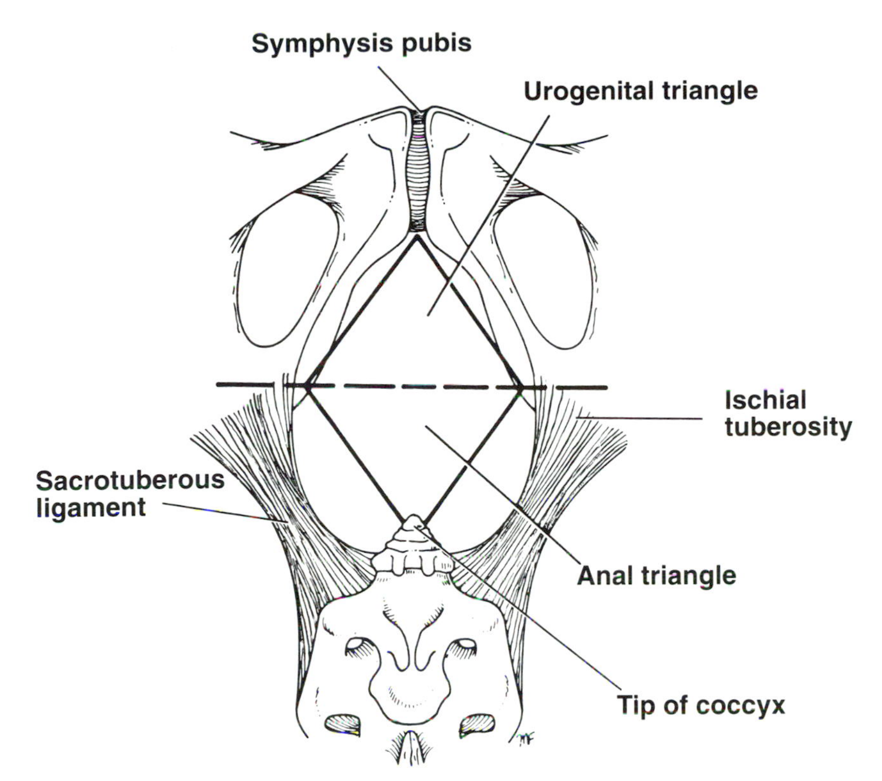

FIG 13–2.
The perineum is shaped like a diamond and may be divided by a broken line into the urogenital triangle anteriorly and the anal triangle posteriorly.

resist the rise in intrapelvic pressure during straining and expulsive efforts of the abdominal muscles, as occurs in coughing. They also have an important sphincter action on the anorectal junction, and in the female they serve also as a sphincter of the vagina.

Nerve Supply.—Perineal branch of the fourth sacral nerve and from the perineal branch of the pudendal nerve.

Coccygeus Muscle.—This small triangular muscle arises from the spine of the ischium and is inserted into the lower end of the sacrum and into the coccyx.

Action.—The two muscles assist the levatores ani in supporting the pelvic viscera.

Nerve Supply.—Branches of the fourth and fifth sacral nerves.

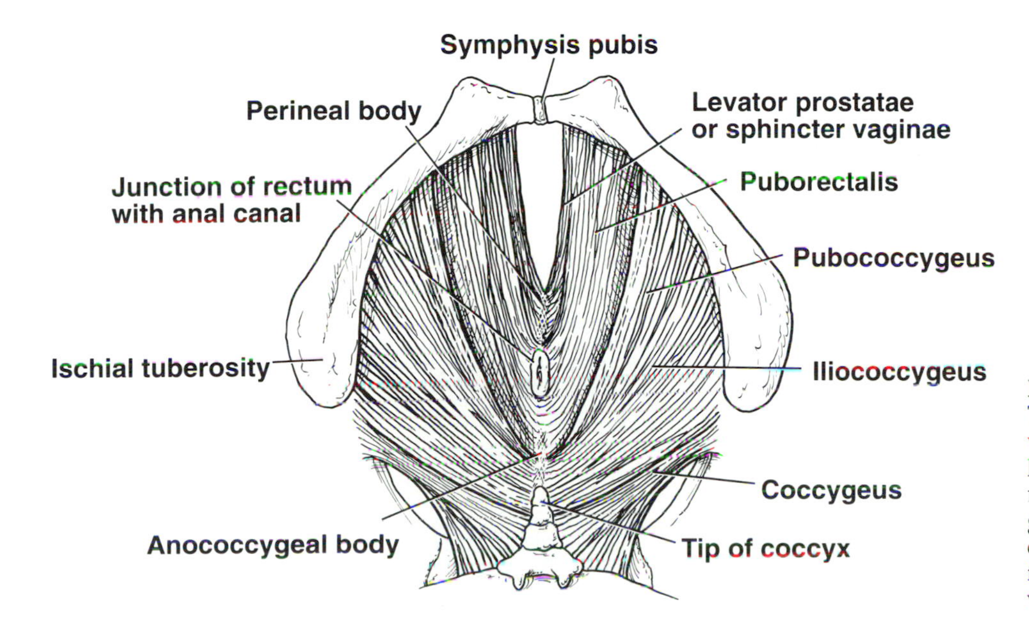

FIG 13–3.
The muscular floor of the pelvis as seen from below. The levator ani is made up of a number of different muscle groups. The levator ani and coccygeus muscles with their fascial coverings form the pelvic diaphragm.

Basic Anatomy of the Rectum

Position and Description

Although the rectum is a pelvic organ and no part of it lies below the pelvic diaphragm, it will be considered here with the anal canal.

The rectum is about 5 in. (13 cm) long and begins in front of the third sacral vertebra as a continuation of the sigmoid colon (Fig 13–4). It passes downward, following the curve of the sacrum and coccyx, and ends in front of the tip of the coccyx by piercing the pelvic floor and becoming continuous with the anal canal. The lower part of the rectum is dilated to form the *rectal ampulla* (Fig 13–5).

The peritoneum covers the anterior and lateral surfaces of the first third of the rectum and only the anterior surface of the middle third, leaving the lower third devoid of peritoneum (see Fig 13–5). The mucous membrane of the rectum, together with the circular muscle layer, form three semicircular folds—two (the upper and the lower) are placed on the left rectal wall and the middle one on the right rectal wall. They are called the *transverse folds of the rectum;* they vary in number and position.

Important Relations.—These include the following.

In the Male.—*Anteriorly:* upper two thirds includes sigmoid colon and coils of ileum that occupy the rectovesical pouch of peritoneum. Lower third includes posterior surface of the bladder, termination of the vas deferens and the seminal vesicles on

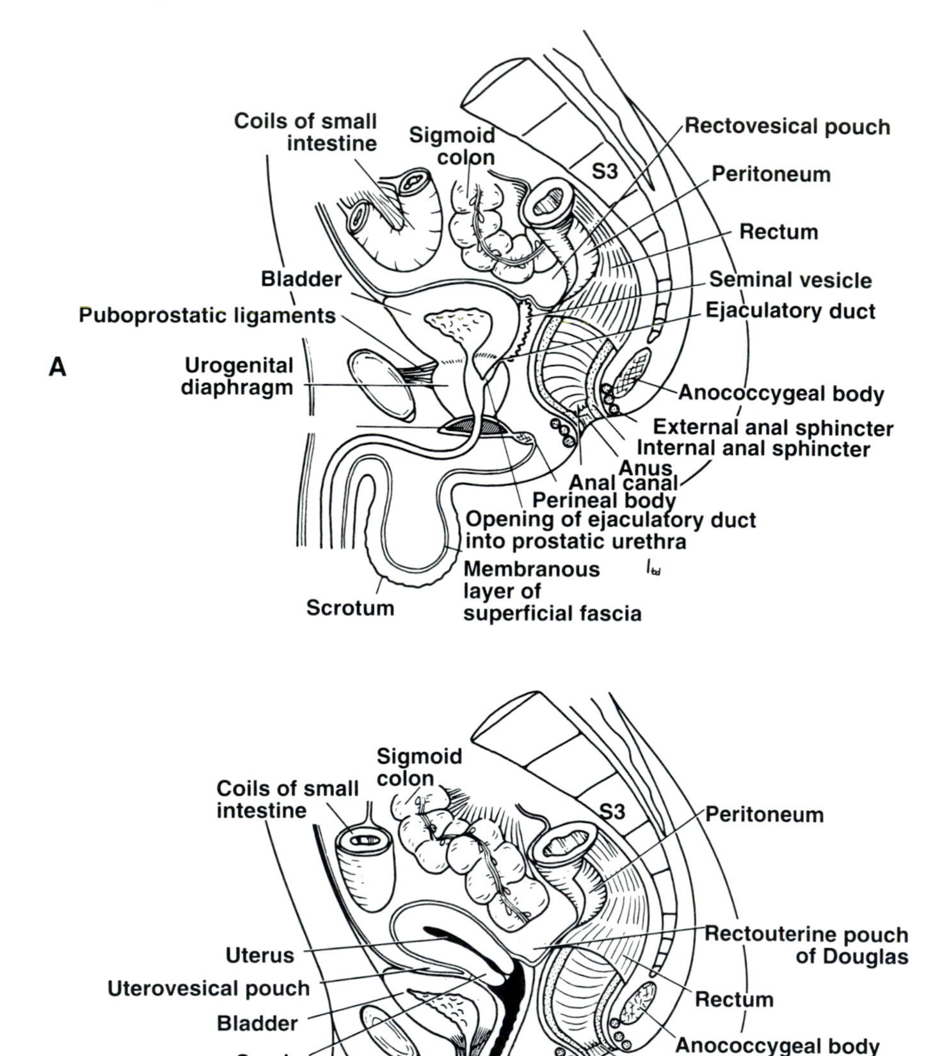

FIG 13–4.
A, sagittal section of the male pelvis. **B,** sagittal section of the female pelvis.

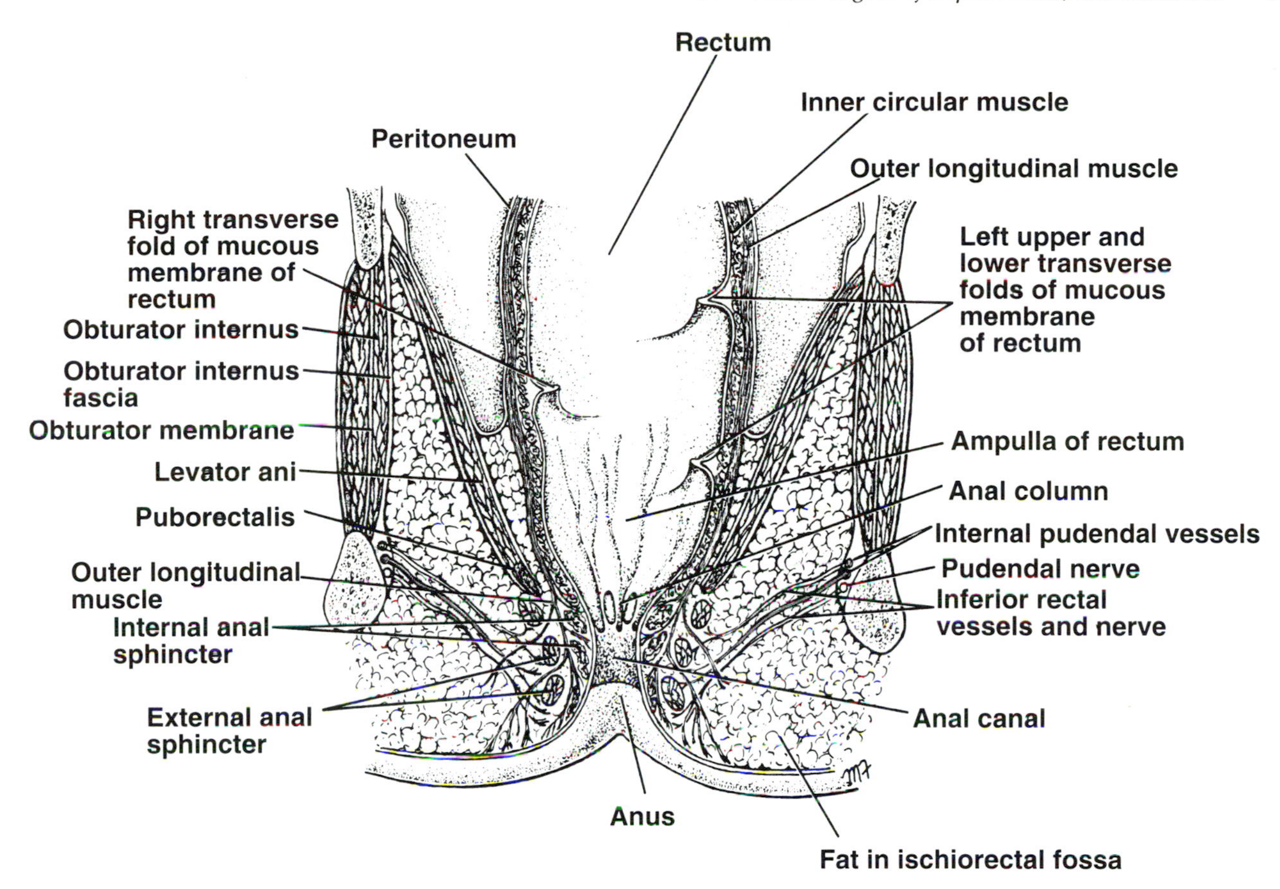

FIG 13–5.
Coronal section through the pelvis showing the rectum, anal canal, and pelvic floor. Note the presence of the inferior rectal vessels and nerve, and the fat in each ischiorectal fossa.

each side, and the prostate. These structures are embedded in visceral pelvic fascia (see Fig 13–4).

In the Female.—Anteriorly: upper two thirds includes sigmoid colon and coils of ileum that occupy the rectouterine pouch of Douglas. Lower third includes posterior surface of the vagina (see Fig 13–4).

In Both Sexes.—Posteriorly: the rectum is in contact with the sacrum and coccyx, the piriformis, coccygeus, and the levator ani muscles, the sacral plexus, and the sympathetic trunks (see Fig 13–4).

Blood Supply of the Rectum.— This includes the following.

Arteries.—Superior rectal artery, direct continuation of the inferior mesenteric artery; middle rectal artery, a branch of the internal iliac artery; inferior rectal artery, a branch of the internal pudendal artery.

Veins.—The venous blood drains into the portal vein by the superior rectal vein and into the systemic system by the middle and inferior rectal veins. The anastomosis between the rectal veins is an important portal-systemic anastomosis.

Lymphatic Drainage of the Rectum.— The lymph passes to *pararectal nodes* then upward to the *inferior mesenteric nodes.* Some lymph vessels pass to the *internal iliac nodes.*

Nerve Supply of the Rectum.— Sympathetic and parasympathetic nerves through the hypogastric plexuses. The rectum is only sensitive to stretch.

Basic Anatomy of the Anal Canal

Position and Description

The anal canal is about 1½ inches (4 cm) long and lies below the pelvic floor (Figs 13–5 and 13–6). It passes downward and backward from the rectal ampulla to open onto the surface at the *anus.* Except during defecation, its lateral walls are kept in apposition by the levatores ani muscles and the anal sphincters.

The *mucous membrane of the upper half of the anal canal* is derived from hindgut entoderm. It has the following important anatomical features.

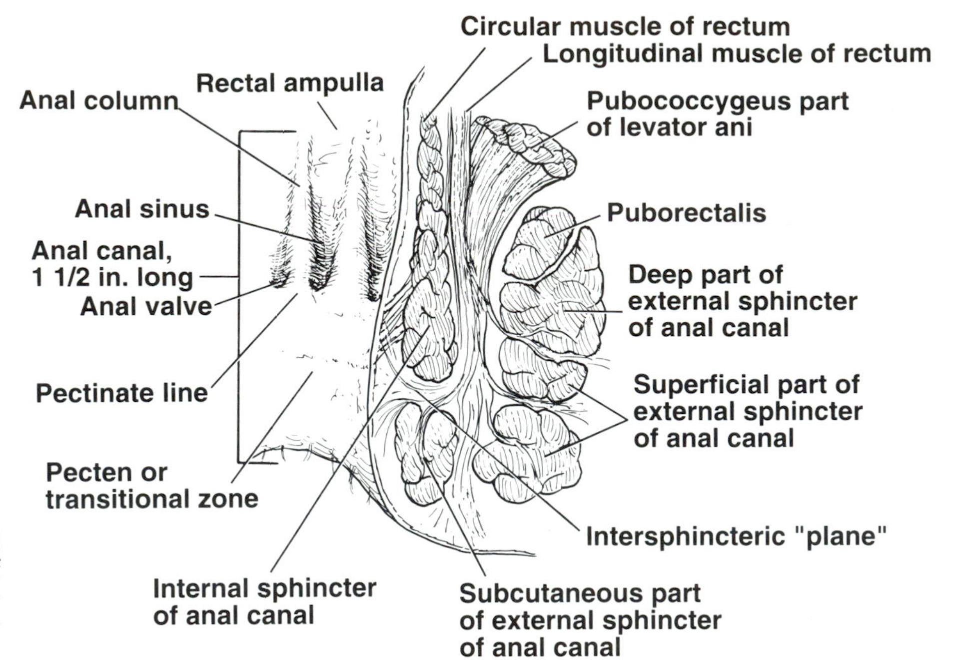

FIG 13–6.
Coronal section of the anal canal showing the detailed anatomy of the mucous membrane and the arrangement of the internal and external anal sphincters.

1. It is lined by columnar epithelium.
2. It is thrown into vertical folds called *anal columns,* which are joined together at their lower ends by small semilunar folds, called *anal valves* (see Fig 13–6).
3. The nerve supply is the same as that for the rectal mucosa and is derived from the autonomic hypogastric plexuses. It is sensitive only to stretch (Fig 13–7).
4. The arterial supply is that of the hindgut, namely, the superior rectal artery, a branch of the inferior mesenteric artery (see Fig 13–7). The venous drainage is mainly by the superior rectal vein, a tributary of the inferior mesenteric vein (see Fig 13–7).
5. The lymphatic drainage is mainly upward along the superior rectal artery to the pararectal nodes and then eventually to the inferior mesenteric nodes (see Fig 13–7).

The *mucous membrane of the lower half of the anal canal* is derived from ectoderm of the proctodeum. It has the following important features.

1. It is lined by stratified squamous epithelium that gradually merges at the anus with the perianal skin (see Fig 13–7).
2. There are *no* anal columns.
3. The nerve supply is from the somatic inferior rectal nerve; it is thus sensitive to pain, temperature, touch, and pressure (see Fig 13–7).
4. The arterial supply is the inferior rectal artery, a branch of the internal pudendal artery (see Fig 13–7). The venous drainage is by the inferior rectal vein, a tributary of the internal pudendal vein, which drains into the internal iliac vein (see Fig 13–7).
5. The lymphatic drainage is downward to the medial group of superficial inguinal nodes (see Fig 13–7).

The *muscle coat* is divided into an outer longitudinal and inner circular layer of smooth muscle. The circular layer is thickened at the upper end of the anal canal to form the *involuntary internal sphincter* (see Fig 13–5). Surrounding the internal sphincter of smooth muscle is a collar of striped muscle called the *voluntary external sphincter* (see Figs 13–5). The external sphincter is divided into the following three parts.

1. A *subcutaneous part,* which encircles the lower end of the anal canal and has no bony attachments (see Fig 13–7).
2. A *superficial part,* which is attached to the coccyx behind and the perineal body in front (see Fig 13–7).
3. A *deep part,* which encircles the upper end of the anal canal and has no bony attachments (see Fig 13–7).

The *puborectalis* fibers of the two levator ani muscles blend with the deep part of the external sphincter (see Fig 13–6). The puborectalis fibers of the two sides form a sling, which is attached in front to the

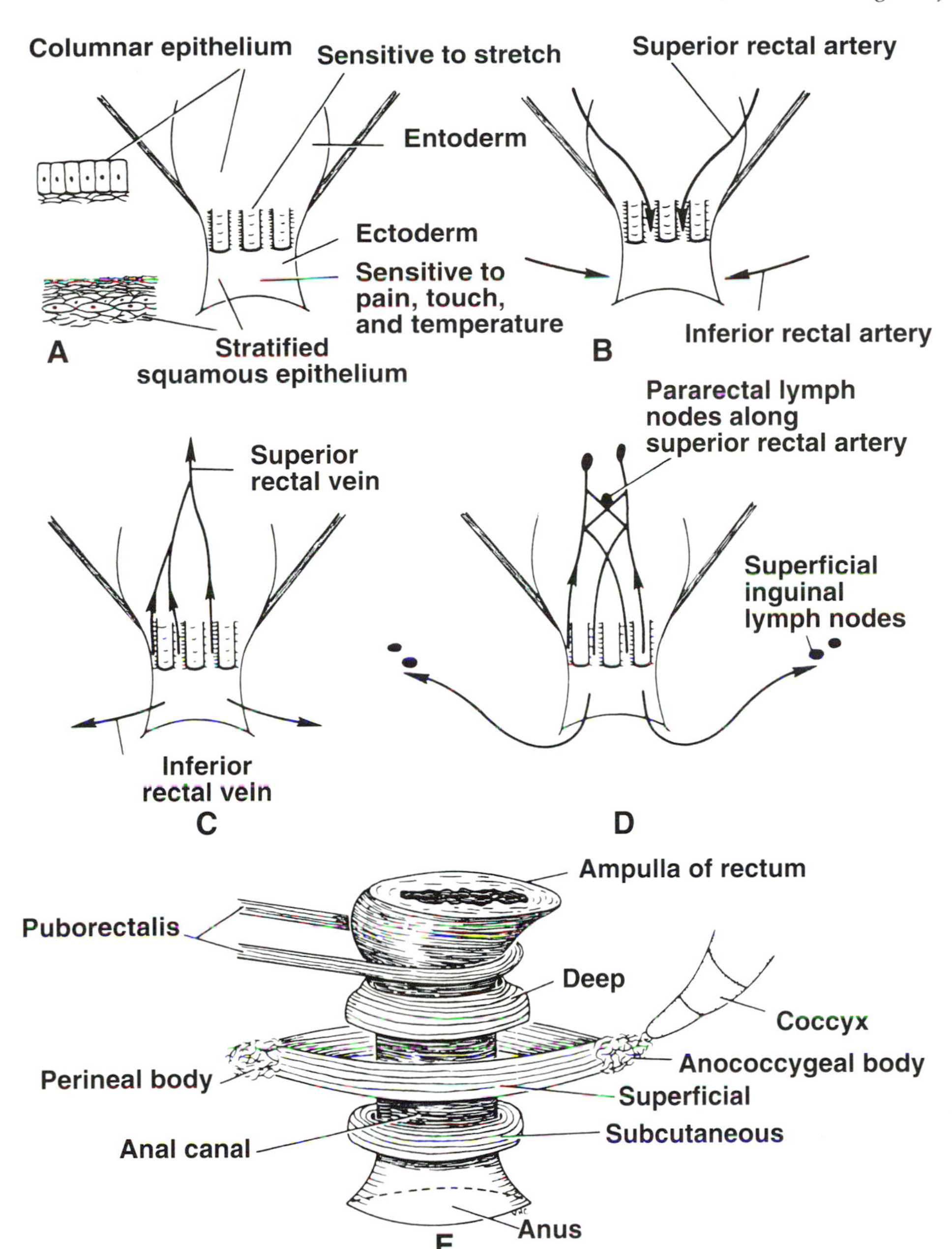

FIG 13–7.
Upper and lower halves of the anal canal showing the following. **A,** embryological origin and lining epithelium. **B,** arterial supply. **C,** venous drainage. **D,** lymphatic drainage. **E,** arrangement of the muscle fibers of the puborectalis muscle and the different parts of the external anal sphincter.

pubic bones and passes around the junction of the rectum and the anal canal, pulling the two forward at an acute angle (see Fig 13–3).

The longitudinal smooth muscle of the anal canal is continuous above with that of the rectum. It forms a continuous coat around the anal canal and descends in the interval between the internal and external sphincters. Some of the longitudinal fibers are attached to the mucous membrane of the anal canal, while others pass laterally into the ischiorectal fossa, or are attached to the perianal skin (see Fig 13–6).

At the anorectal junction, the internal sphincter, the deep part of the external sphincter, and the puborectalis form a distinct ring, called the *anorectal ring,* which can be felt on rectal examination (see Fig 13–7).

Important Relations.—These are as follows.

In the Male.—Anteriorly: the perineal body, the urogenital diaphragm, the membranous part of the urethra, and the bulb of the penis (see Fig 13–4).

In the Female.—Anteriorly: the perineal body, the urogenital diaphragm, and the lower part of the vagina (see Fig 13–4).

In Both Sexes.—Posteriorly: the anal canal is related posteriorly to the *anococcygeal body,* which is a mass of fibrous tissue lying between the anal canal and the coccyx (see Fig 13–4).

Blood Supply of the Anal Canal.—This includes the following.

Arteries.—The superior rectal artery supplies the upper half, and the inferior rectal artery supplies the lower half (see Fig 13–7).

Veins.—The upper half is drained by the superior rectal vein into the inferior mesenteric vein, the lower half is drained by the inferior rectal vein into the internal pudendal vein. The anastomosis between the rectal veins forms an important portal-systemic anastomosis.

Lymphatic Drainage of the Anal Canal.—The upper half drains into the pararectal nodes and then the inferior mesenteric nodes. The lower half drains into the medial group of superficial inguinal nodes (see Fig 13–7).

Nerve Supply of the Anal Canal.—The mucous membrane of the upper half is sensitive to stretch and is innervated by sensory fibers that ascend through the hypogastric plexuses. The lower half is sensitive to pain, temperature, touch, and pressure and is innervated by the inferior rectal nerves (see Fig 13–7).

The internal anal sphincter is supplied by sympathetic nerves from the hypogastric plexus. The voluntary external anal sphincter is supplied by the inferior rectal nerves.

Defecation

The desire to defecate is initiated by stimulation of the stretch receptors in the wall of the rectum by the presence of feces in the lumen. The act of defecation involves a coordinated reflex that results in the emptying of the descending colon, sigmoid colon, rectum, and anal canal. It is assisted by a rise in intra-abdominal pressure brought about by contraction of the muscles of the anterior abdominal wall.

The involuntary internal sphincter of the anal canal is innervated by postganglionic sympathetic fibers from the hypogastric plexuses, and the voluntary external sphincter of the anal canal is innervated by the inferior rectal nerve. The tonic contraction of the internal and external anal sphincters, including the puborectalis muscles, is voluntarily inhibited.

The feces are evacuated through the anal canal. Depending on the laxity of the submucous coat, the mucous membrane of the lower part of the anal canal is extruded through the anus ahead of the fecal mass. At the end of the act, the mucosa is returned to the anal canal by the tone of the longitudinal muscle of the anal walls and the contraction and upward pull of the puborectalis muscle. The empty anal canal is then closed by the tonic contraction of the anal sphincters.

Basic Anatomy of the Ischiorectal Fossa

This wedge-shaped space is located on each side of the anal canal (see Fig 13–5). The base of the wedge is superficial and formed by the skin. The edge of the wedge is formed by the junction of the medial and lateral walls. The medial wall is formed by the sloping levator ani muscle and the anal canal. The lateral wall is formed by the lower part of the obturator internus muscle covered with pelvic fascia.

The fossa is filled with fat that supports the anal canal. Its function is to allow the anal canal to distend during the process of defecation. The pudendal nerve and the internal pudendal vessels lie in a fascial canal, the *pudendal canal,* on the medial side of the ischial tuberosity.

Basic Anatomy of the Female External Genitalia (Vulva)

The vulva includes the mons pubis (hair-bearing skin in front of the pubis), the labia majora, the labia minora, the clitoris, and the greater vestibular glands (Bartholin's glands).

Labia Majora

The labia majora are prominent folds of skin extending from the mons pubis to unite in the midline posteriorly (Fig 13–8). They contain fat and are covered with hair on their outer surfaces. (They are equivalent to the scrotum in the male.)

Labia Minora

The labia minora are two smaller folds of skin devoid of hair that lie between the labia majora (see Fig 13–8). Their posterior ends are united to form a sharp fold, the *fourchette.* Anteriorly, they split to enclose the clitoris, forming an anterior *prepuce* and a posterior *frenulum.*

Vestibule of the Vagina

The vestibule of the vagina is the space between the labia minora. It has the clitoris at its apex and has

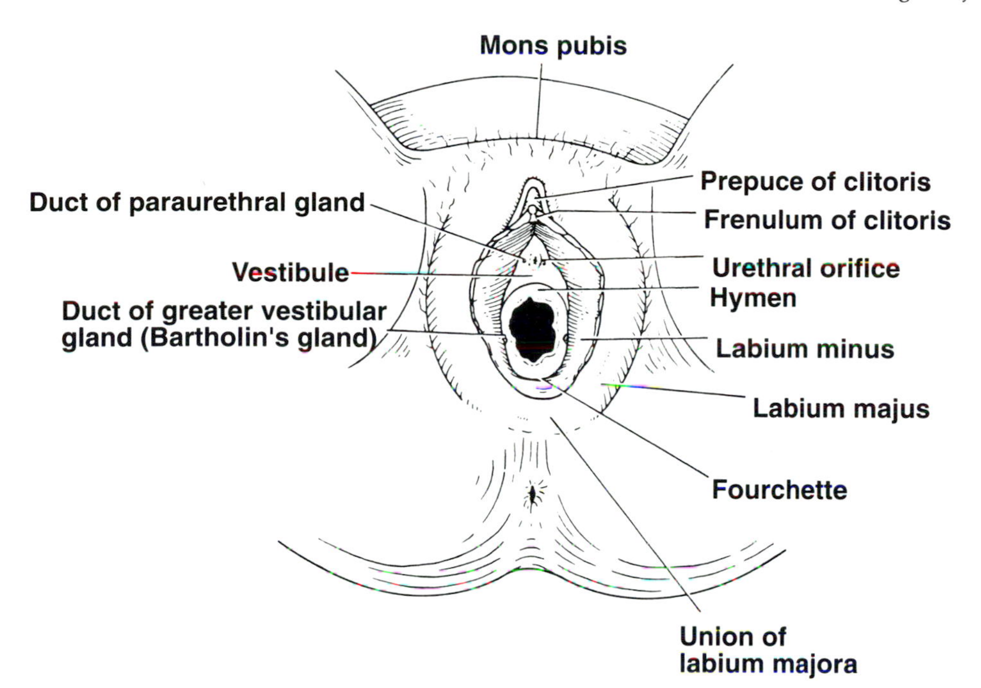

FIG 13–8.
The vulva. Note the openings of the ducts of the paraurethral glands and the greater vestibular glands.

the openings of the urethra, the vagina, and the ducts of the greater vestibular glands in its floor (see Fig 13–8).

Clitoris

The clitoris corresponds to the penis in the male. The *glans* of the clitoris is partly hidden by the *prepuce.* The *root of the clitoris* is made up of three masses of erectile tissue, which are called the bulb of the vestibule and the right and left crura of the clitoris (Fig 13–9). The *bulb of the vestibule* corresponds to the bulb of the penis, but because of the presence of the vagina, it is divided into two halves. It is attached to the undersurface of the urogenital diaphragm and is covered by the *bulbospongiosus muscles.* Anteriorly, the two halves unite to form the glans clitoris. The *crura of the clitoris* correspond to the crura of the penis. They are covered by the *ischiocavernosus muscles.*

Greater Vestibular Glands (Bartholin's Glands)

The greater vestibular glands are a pair of mucus-secreting glands that lie under cover of the posterior parts of the bulb of the vestibule and the labia major (see Fig 13–9). The duct of each gland opens into the groove between the hymen and the posterior part of the labium minus. The glands secrete a lubricating mucus during sexual intercourse.

Urethra

The female urethra is only 1½ in. (3.8 cm) long and extends from the bladder neck to the external meatus. It passes through the urogenital diaphragm, where it traverses the sphincter urethrae and is closely related to the anterior surface of the vagina (see Fig 13–4). The urethra opens onto the surface below the clitoris and in front of the vagina.

Paraurethral Glands

The paraurethral glands, which correspond to the prostate in the male, open into the vestibule by small ducts on either side of the urethral orifice (see Fig 13–8).

Vaginal Orifice

The vaginal orifice possesses a thin mucosal fold called the *hymen,* which is perforated at its center. The blood supply to the hymen enters its circumference at the 4- and 8-o'clock positions (this information may be helpful when patients present with vaginal bleeding following tears during sexual intercourse).

While the lower half of the vagina lies within the perineum, the upper half lies above the pelvic floor, and the posterior fornix is related to the rectouterine pouch (pouch of Douglas) in the peritoneal cavity (see Fig 13–4).

Lymphatic Drainage of the Vulva.—The lymphatic drainage of the vulva, including the lower third of the vagina, is into the medial group of superficial inguinal nodes.

The relationship of the various structures of the

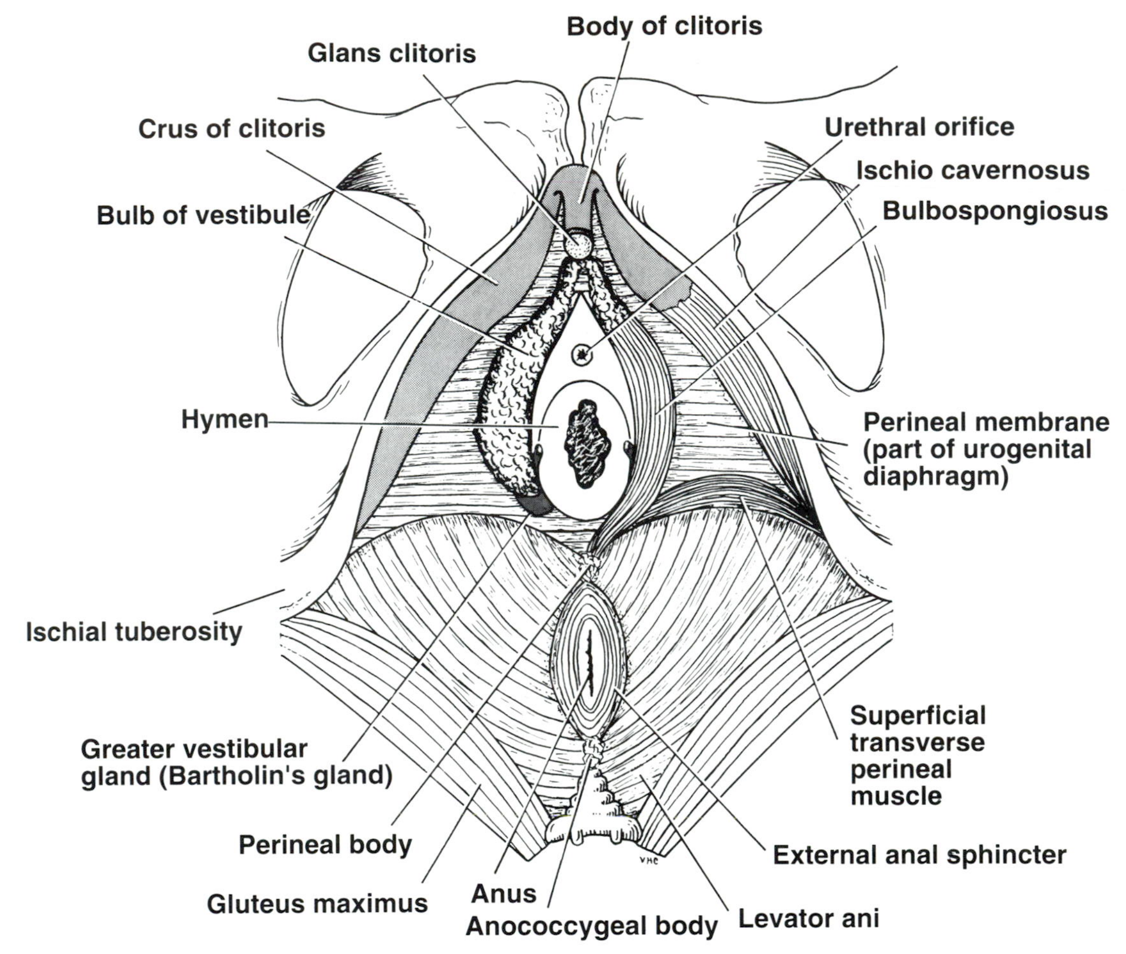

FIG 13–9.
The root and body of the clitoris and the perineal muscles.

vulva to one another and to the urogenital diaphragm is shown in Figure 13–9.

Basic Anatomy of the Female Internal Organs of Reproduction

Basic Anatomy of the Ovaries

Each ovary* is attached to the back of the broad ligament by the *mesovarium* (Fig 13–10). The ovary usually lies near the lateral wall of the pelvis in a depression called the *ovarian fossa,* bounded by the external and internal iliac arteries.

Blood Supply.—The *ovarian artery* is a branch of the abdominal aorta. The *ovarian vein* drains into the inferior vena cava on the right side and into the left renal vein on the left side. The blood vessels reach the ovary by passing through the lateral part of the broad ligament (called the *suspensory ligament).*

*During the early stages of development, the ovary is covered completely by peritoneum. Later, the covering cells become cuboid and are then referred to as germinal epithelium; this layer of cells separates the mature ovary from the peritoneal cavity.

Basic Anatomy of the Uterine Tubes

Position and Description

There are two uterine tubes (see Fig 13–10); each one lies within the broad ligament in its upper border. The uterine tube connects the peritoneal cavity in the region of the ovary with the cavity of the uterus. The tube may be divided into the following four parts. (1) The *infundibulum* is the funnel-shaped lateral end that has fingerlike processes, known as *fimbrae,* that are draped over the ovary. (2) The *ampulla* is the widest part of the tube. (3) The *isthmus* is the narrowest part of the tube and lies just lateral to

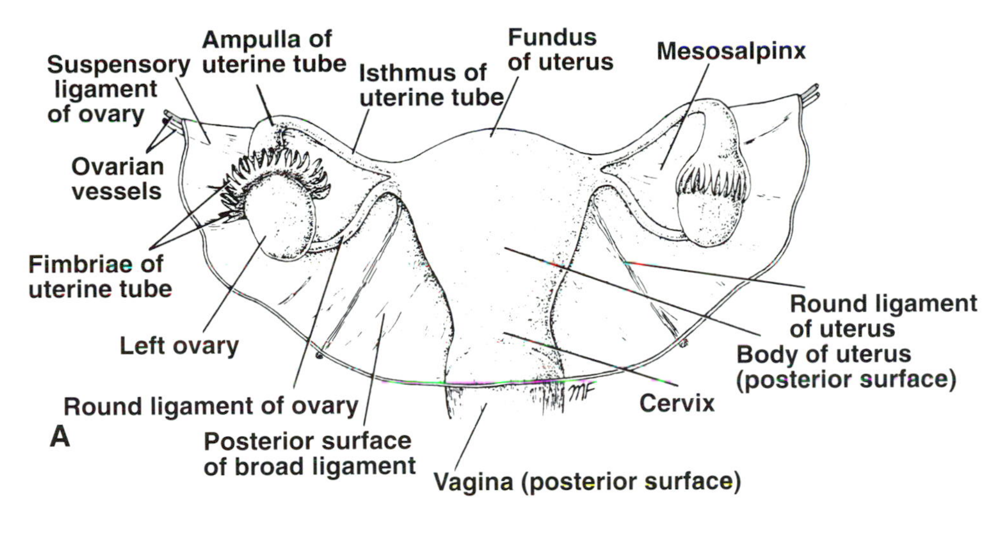

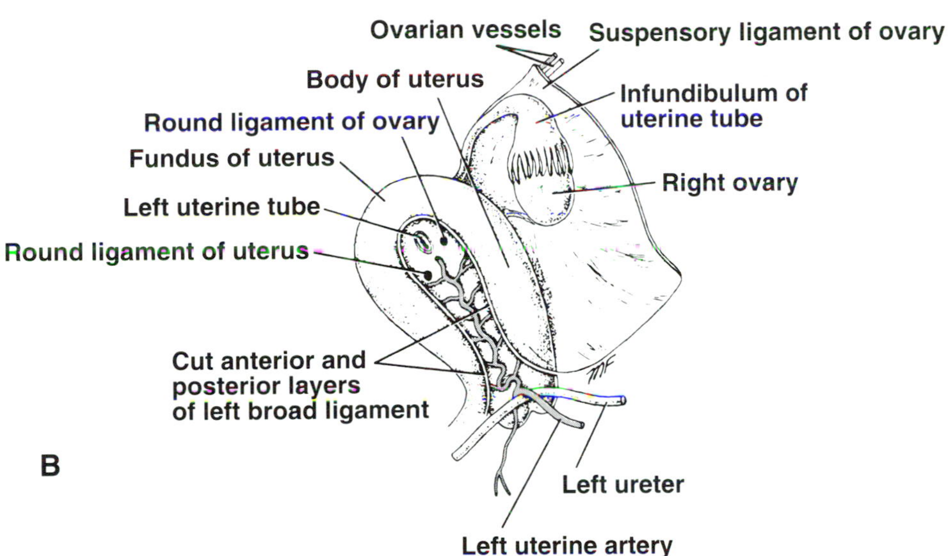

FIG 13–10.
A, posterior surface of the uterus and the broad ligaments; note the position of the ovaries and the different parts of the uterine tubes. **B,** lateral view of the uterus; note the structures that lie within the broad ligament and also the close relationship between the ureter and the uterine artery and the cervix.

the uterus, and (4) the *intramural part* is the segment that pierces the uterine wall (Fig 13–11).

The uterine tube provides a site where fertilization of the ovum can take place (usually in the ampulla). It provides nourishment for the fertilized ovum, and transports the ovum to the cavity of the uterus. The tube serves as a conduit along which the spermatozoa travel to reach the ovum.

Blood Supply.—This includes the following.

Arteries.—Uterine artery from the internal iliac artery and ovarian artery from the abdominal aorta.

Veins.—Uterine and ovarian veins.

Nerve Supply.—Sympathetic and parasympathetic nerves from the superior and inferior hypogastric plexuses.

Basic Anatomy of the Uterus and Vagina

Uterus

The uterus is divided into the fundus, body, and cervix (see Fig 13–11). The *fundus* is the part of the uterus that lies above the entrance of the uterine tubes. The *body* is the part of the uterus that lies below the entrance of the uterine tubes. It narrows below, where it becomes continuous with the *cervix.* The cervix pierces the anterior wall of the vagina. The cavity of the cervix, the *cervical canal,* communicates with the cavity of the body through the *internal os,* and with that of the vagina, through the *external os.*

Position of the Uterus.—The uterus may be anteverted, midposition, or retroverted (see Fig 13–11). These terms describe the bending of the

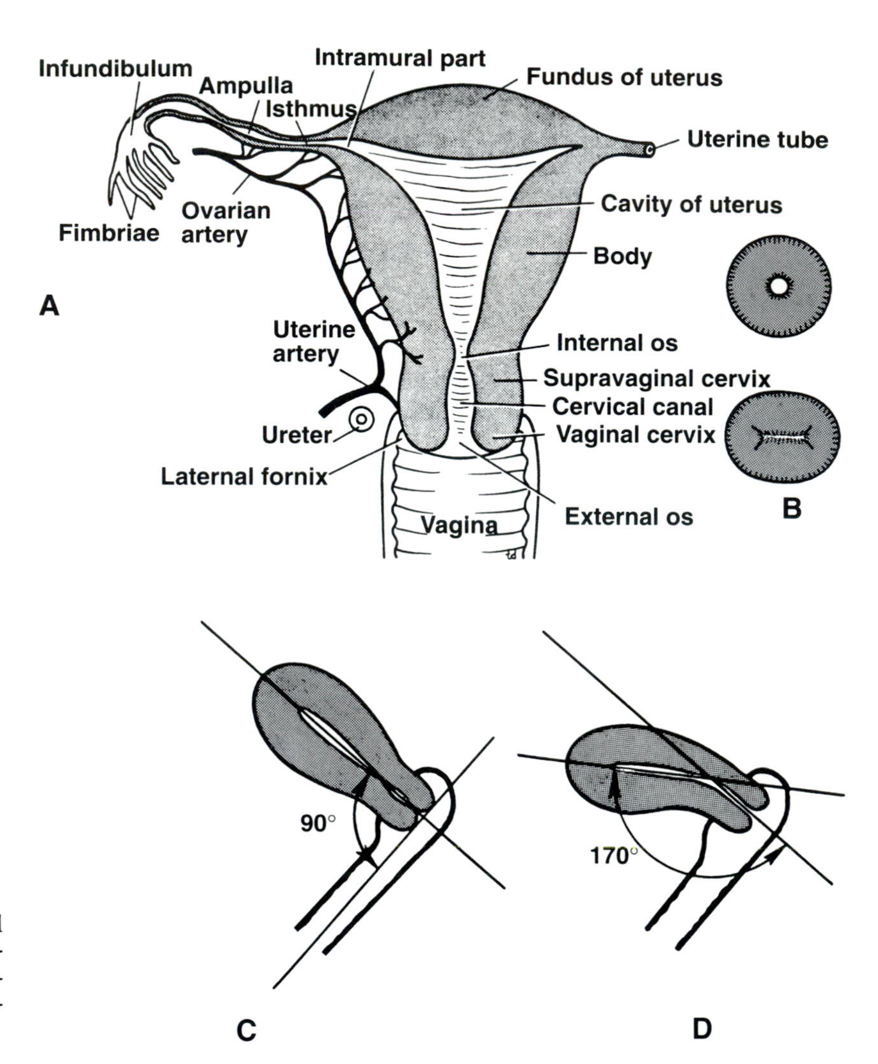

FIG 13–11.
A, different parts of the uterine tube and the uterus. **B,** external os of cervix: nulliparous (above), parous (below). **C,** anteverted position of the uterus. **D,** anteverted and anteflexed positions of uterus.

uterus on the long axis of the vagina. Anteflexion and retroflexion are terms used to describe the bending of the body of the uterus on the cervix.

Supports of the Uterus.—The main supports include the following (Fig 13–12).

1. The *pelvic diaphragm* consists of the levatores ani and the coccygeus muscles and their fascia, which form the pelvic floor.
2. The *perineal body* consists of fibromuscular tissue that is located in the perineum between the vagina and the anal canal; it is supported by the levatores ani muscles.
3. The *transverse cervical (cardinal) ligaments* that attach the cervix and upper part of the vagina to the lateral pelvic walls.
4. The *pubocervical ligaments* that attach the cervix to the pubic bones.
5. The *sacrocervical ligaments* that attach the cervix and the upper part of the vagina to the lower end of the sacrum.

Broad Ligaments.—These two layered folds of peritoneum extend across the pelvic cavity from the lateral margins of the uterus to the lateral pelvic walls (see Fig 13–10). Each broad ligament contains the uterine tube, the round ligaments of the ovary and uterus, the uterine and ovarian blood vessels, lymphatics, and nerves. The broad ligaments give little support to the uterus.

Round Ligament of the Ovary.—This extends from the medial margin of the ovary to the lateral wall of the uterus (see Fig 13–10). It has no function.

Round Ligament of the Uterus.—This extends from the superolateral angle of the uterus, through the inguinal canal, in the anterior abdominal wall, to the subcutaneous tissue of the labium majus (see Fig

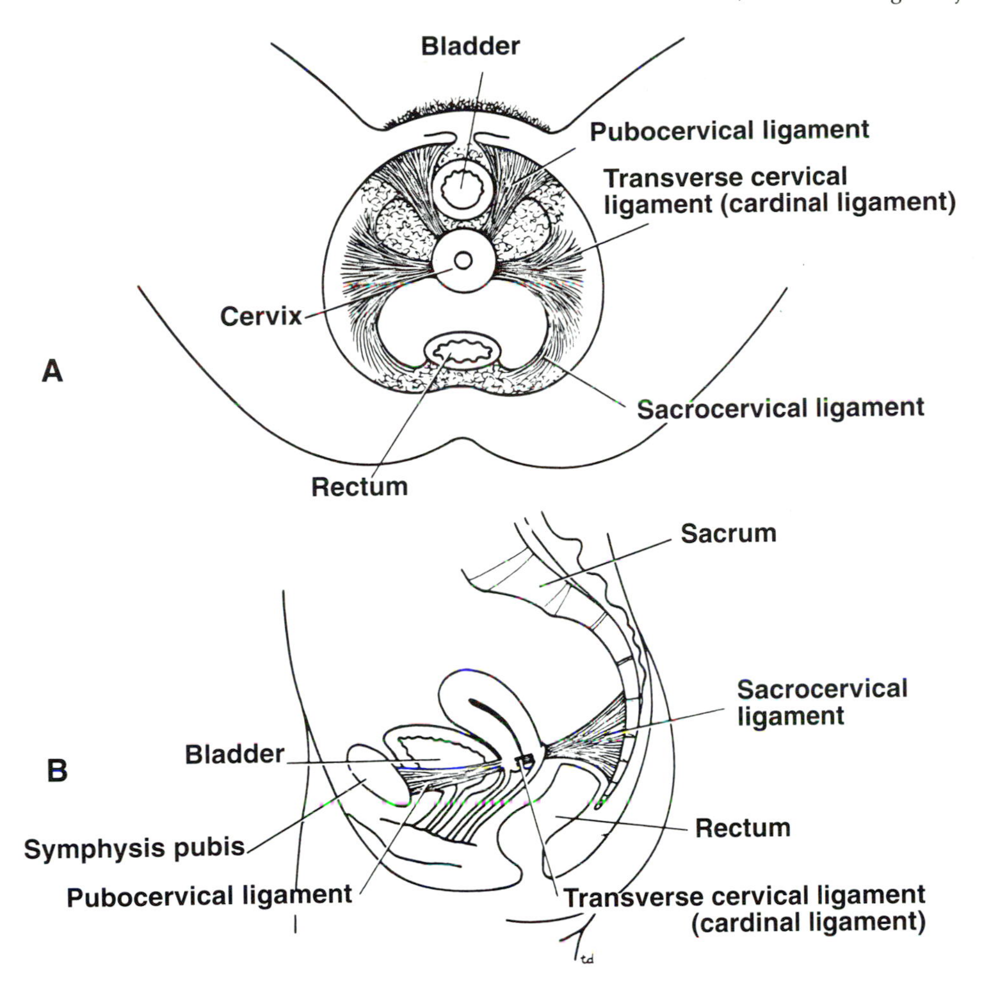

FIG 13–12.
The ligaments supporting the uterus. **A,** as seen from below. **B,** lateral view.

13–10). It helps keep the uterus anteverted (tilted forward) and anteflexed (bent forward), but is considerably stretched during pregnancy.

Blood Supply.—*Uterine artery* from the internal iliac artery and the *ovarian artery*. The *veins* correspond to the arteries.

Lymphatic Drainage.—From the fundus the lymph vessels follow the ovarian artery to the para-aortic nodes at the level of the first lumbar vertebra. From the body and cervix they drain into the internal and external iliac nodes. A few lymph vessels pass through the inguinal canal to the superficial inguinal nodes.

Nerve Supply.—Sympathetic and parasympathetic nerves from the inferior hypogastric plexus.

Structure of the Uterine Wall

The wall of the uterus has three coats as follows: (1) the serosa, (2) the muscular coat, or myometrium, and (3) the mucous membrane, or endometrium.

Serous Coat.—The serous coat consists of the peritoneum. It covers the uterus and is reflected onto the bladder anteriorly. Posteriorly it continues onto the vagina and from there is reflected onto the rectum, forming the *pouch of Douglas* (see Fig 13–4). Laterally the peritoneum extends to the lateral wall of the pelvis, forming the broad ligament (see Fig 13–10).

Muscular Coat.—The muscular coat or myometrium, is very thick and is composed of smooth-muscle fibers and connective tissue (Fig 13–13).

Mucous Membrane, or Endometrium.—The endometrium is continuous above with the mucous membrane of the uterine tubes and below with the mucous membrane of the vagina. The endometrium is applied directly to the muscle since there is no submucosa.

The endometrium in the body of the uterus and the upper third of the cervix, from puberty until the menopause, undergoes extensive changes in structure during the menstrual cycle in response to the ovarian hormones. Basically, the endometrium is lined with a simple columnar epithelium resting on a layer of connective tissue (see Fig 13–13). There are numerous simple tubular glands whose mouths open into the cavity of the uterus. The glands extend almost to the myometrium and are lined with columnar mucus-secreting cells. The endometrium can be divided into a thick, superficial part that undergoes functional and structural changes during the menstrual cycle and a thin, basal part that changes little during the menstrual cycle and is responsible for re-forming the superficial layer after the menstrual flow has ceased.

In the cervix of the uterus, the endometrium is lined with ciliated columnar epithelium. In the lower part, the cilia are absent, and at the external os, the columnar cells give way to stratified squamous epithelium that lines the vagina. The glands of the endometrium of the cervix are large and secrete a thick, alkaline mucus.

Cyclical Changes in the Structure of the Endometrium.—The duration of the menstrual cycle varies in women, and perfectly healthy women may regularly have cycles of 26 to 32 days. In this account, we will use a cycle of 28 days as an example.

A menstrual (endometrial) cycle may be divided into the following three phases: menstrual, proliferative, and secretory.

Menstruation.—Menstruation is the shedding of the superficial zone of the endometrium into the uterine cavity, leaving only the basal zone (Fig 13–13). The *menstrual flow* thus consists of blood, partially disintegrated epithelium and connective tissue, and secretions of the endometrial glands. Men-

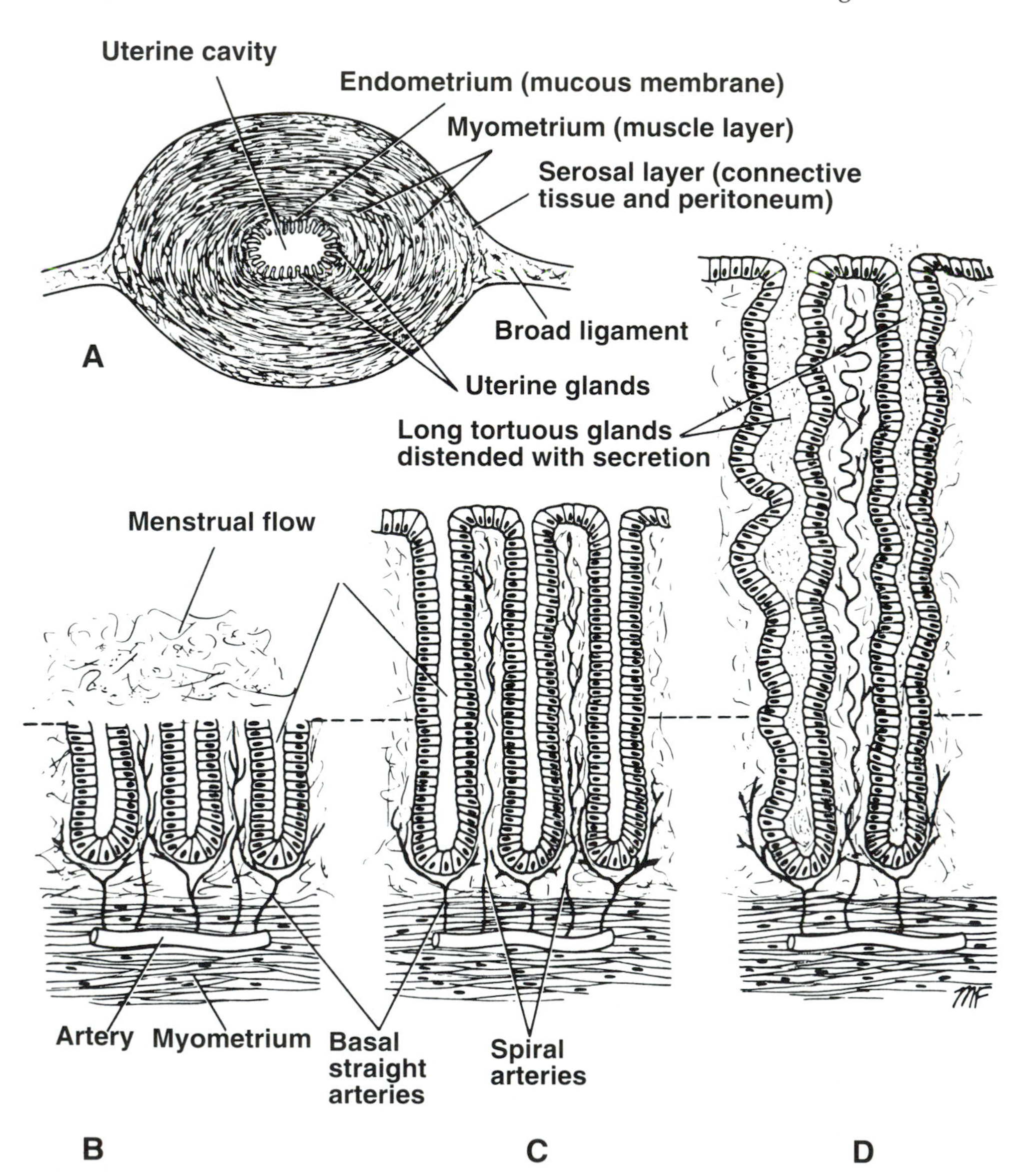

FIG 13–13.
A, general structure of the body of the uterus as seen in cross section. **B**, structure of the endometrium during menstruation. **C**, structure of the endometrium during the proliferative phase of the menstrual cycle. **D**, structure of the endometrium during the secretory phase of the cycle. In **B**, **C**, and **D**, note the basal straight arteries and the spiral arteries.

struation is produced by the sudden reduction in both estrogens and progesterone at the end of the ovarian cycle. The endometrium, suddenly deprived of the supporting action of these hormones, and the spiral arteries that supply the superficial zone of the endometrium constrict. After a variable time, the constricted arteries open up, the walls of the damaged vessels in the superficial part of the endometrium rupture, and blood pours into the connective tissue stroma, disrupting the tissue. The process leads to hemorrhage into the uterine cavity and the shedding of all but the deepest part of the endometrium. Menstruation normally lasts about 4 to 6 days.

Proliferative Phase.—At the end of menstruation, only the deep, or basal, parts of the endometrium remain. These are supplied by their own arteries, the basal arteries, which do not undergo constriction. The epithelial cells that form the blind ends of the glands in the basal endometrium then multiply and repair the denuded surface of the endometrium. This process is controlled by the secretions of estrogens by the ovaries. On about the 6th day of the menstrual cycle, when the damage resulting from the menstrual period has been repaired, proliferative changes begin. The endometrium initially is thin and consists of columnar epithelium that dips down into loose connective tissue to form simple tubular glands (see Fig 13–13). From the 6th to the 14th day, the endometrium thickens. This thickening is caused by increasing amounts of connective tissue, elongation of the tubular glands, and growth of the blood vessels; all these proliferative changes are controlled by the ovarian estrogens. At the time of ovulation, which in a 28-day cycle occurs on day 14, plus or minus 1 day, the endometrium is 2 to 3 mm thick.

Secretory Phase.—Once ovulation has occurred, the corpus luteum is formed and starts to secrete progesterone and estrogens. The combined action of these two hormones produces the secretory changes in the endometrium. The endometrium thickens further, and the glands become long and tortuous and distended with secretion (see Fig 13–13). The connective tissue between the glands proliferates, and the arteries, known as the spiral arteries, become highly coiled and congested. At the end of this phase of the cycle, the endometrium is 5 to 8 mm thick.

Thus, the various changes that take place in the endometrium during the second half of the menstrual cycle may be regarded as preparing the uterine lining for the nourishment and reception of the fertilized ovum (blastocyst). On entering the uterine cavity, the fertilized ovum lies free within the secretion produced by the highly developed endometrial glands. This secretion provides an excellent medium for the rapidly dividing cells in the blastocyst.

Should fertilization not occur, the corpus luteum of the ovary degenerates, and the endometrium is once again deprived of the supporting action of estrogen and progesterone. Menstruation then begins again.

Radiographic Appearances

Radiographic appearances of the uterus and uterine tubes are shown in Figure 13–14.

Vagina

The vagina is not only the female genital canal but serves as the excretory duct for the menstrual flow and forms part of the birth canal. This muscular tube extends between the vulva and uterus. The cervix of the uterus pierces its anterior wall. The vaginal orifice in a virgin possesses a thin mucosal fold called the *hymen* that is perforated at its center.

The area of the vaginal lumen that surrounds the cervix of the uterus is divided into four regions —the *anterior fornix,* the *posterior fornix,* the *right lateral fornix,* and the *left lateral fornix* (see Fig 13–11).

The upper half of the vagina lies within the pelvis between the bladder anteriorly and the rectum posteriorly; the lower half lies within the perineum between the urethra anteriorly and the anal canal posteriorly (see Fig 13–4).

Supports of the Vagina.— Upper third: levatores ani muscles, transverse cervical, pubocervical, and sacrocervical ligaments. Middle third: urogenital diaphragm. Lower third: perineal body.

Blood Supply.—*Vaginal artery,* a branch of the internal iliac artery, *vaginal branch of the uterine artery. Vaginal veins* drain into the internal iliac veins.

Lymphatic Drainage.— Upper third: internal and external iliac nodes. Middle third: internal iliac nodes. Lower third: superficial inguinal nodes.

Nerve Supply.— Inferior hypogastric plexuses.

Basic Anatomy of a Normal Pregnancy

As the fetus enlarges, the uterus increases in size, and the fundus gradually rises out of the pelvic cav-

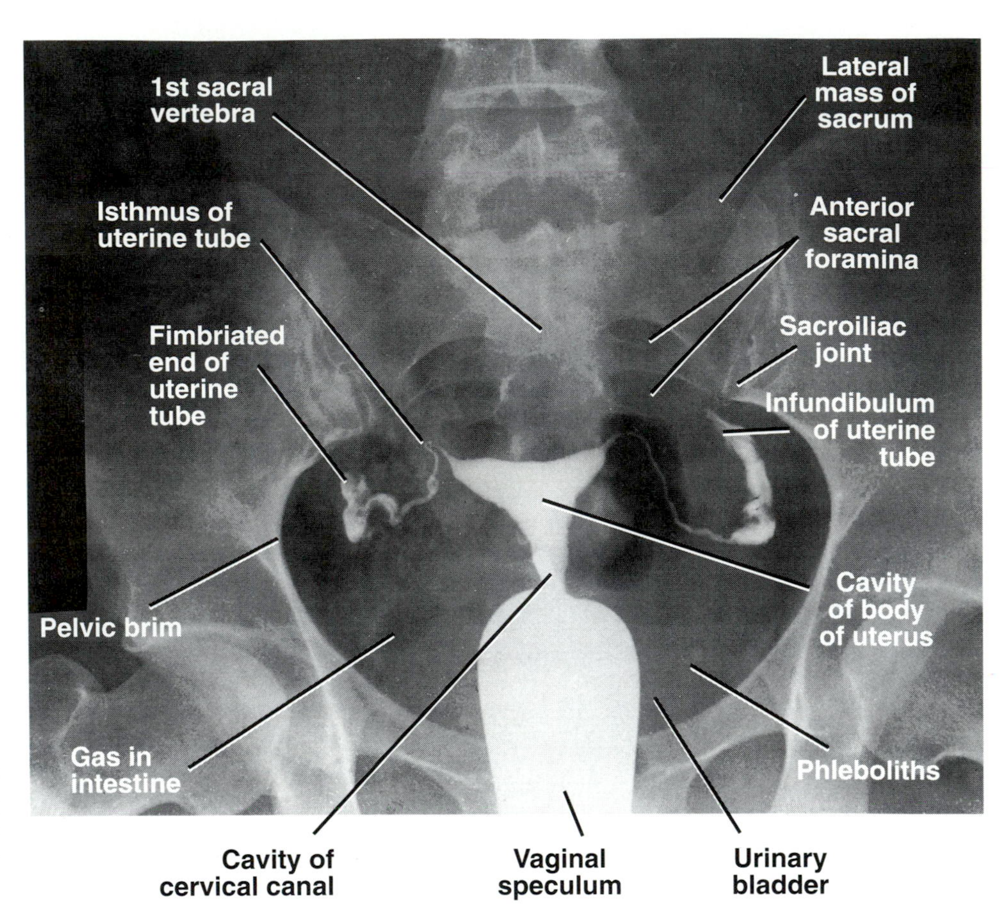

FIG 13–14.
Anteroposterior radiograph of the female pelvis following the injection of radiopaque compound into uterine cavity (hysterosalpingogram).

ity (Fig 13–15). The ascent of the fundus is fairly uniform. At 12 weeks, it lies just above the symphysis pubis; at 20 weeks, it lies about halfway between the symphysis pubis and the umbilicus; at 24 weeks, it is level with the umbilicus; at 32 weeks, it lies halfway between the umbilicus and the xiphoid process; at 36 weeks, it has reached the xiphoid process; and at 40 weeks, in the first pregnancy, the fetus sinks downward as the presenting part enters the pelvic cavity and the fundus again takes up the position halfway between the xiphoid and the umbilicus. In multipara, the presenting part of the fetus descends at a later date.

Further evidence of fetal growth can be determined by the mother first recognizing movements of the fetus (quickening), which usually occurs between the 16th and 20th weeks of pregnancy. The fetal heart can usually be heard through the abdominal wall with a stethoscope during the 20th week of pregnancy. Fetal heart sounds may be heard as early as the 12th week, when a Doppler stethoscope is used.

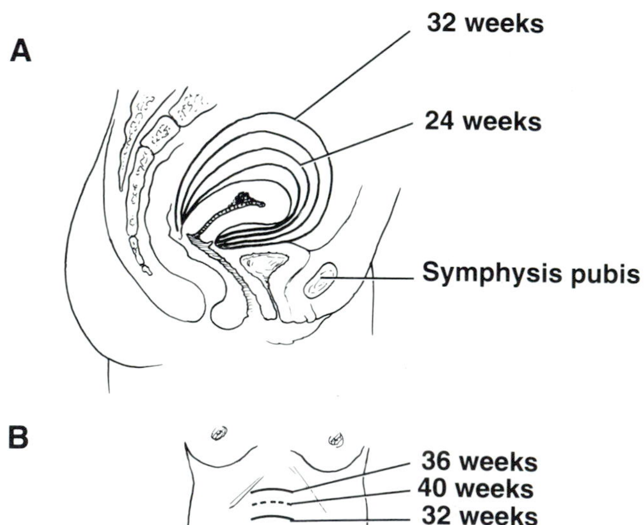

FIG 13–15.
The height of the fundus of the uterus at various weeks of pregnancy. **A,** as seen in sagittal section. **B,** as felt through the anterior abdominal wall.

Ultrasound Anatomy of the Developing Fetus

With sequential longitudinal and transverse scans of the uterus, the developing fetus, amniotic fluid, and placenta can be studied. The gestational sac can be recognized at about 5 weeks after the first day of the last menstrual period. By 6 to 7 weeks the fetal pole (extreme end or ends of fetus) can be visualized, and the beating heart can be seen. Later the fetal head, trunk, and limbs can be identified (Figs 13–16 and 13–17).

Brief Summary of the Implantation of Fertilized Ovum

The blastocyst enters the uterine cavity between the 4th and the 9th days after ovulation. Normal implantation takes place in the endometrium of the body of the uterus, most frequently on the upper part of the posterior wall near the midline (Fig 13–18). As the result of the enzymatic digestion of the uterine epithelium by the trophoblast of the embryo, the blastocyst sinks beneath the surface epithelium and becomes embedded in the stroma by the 11th or 12th day.

Summary of the Formation of the Placenta

The placenta is the organ that carries out respiration, excretion, and nutrition for the embryo, and it is fully formed during the 4th month. The formation of the placenta is complicated and is essentially the development of an organ by mother and child in symbiosis and consists of fetal and maternal parts.

The fetal part develops as follows. The trophoblast becomes a highly developed structure with villi that continue to erode and penetrate deeper into the endometrium. Large irregular spaces, known as lacunae, appear, which become filled with maternal blood. At the center of each villus is connective tissue containing fetal blood vessels that will eventually anastomose with one another and converge to form the umbilical cord (Fig 13–19).

The maternal part develops as follows. Under the influence of progesterone secreted first by the corpus luteum and later by the placenta itself, the

A

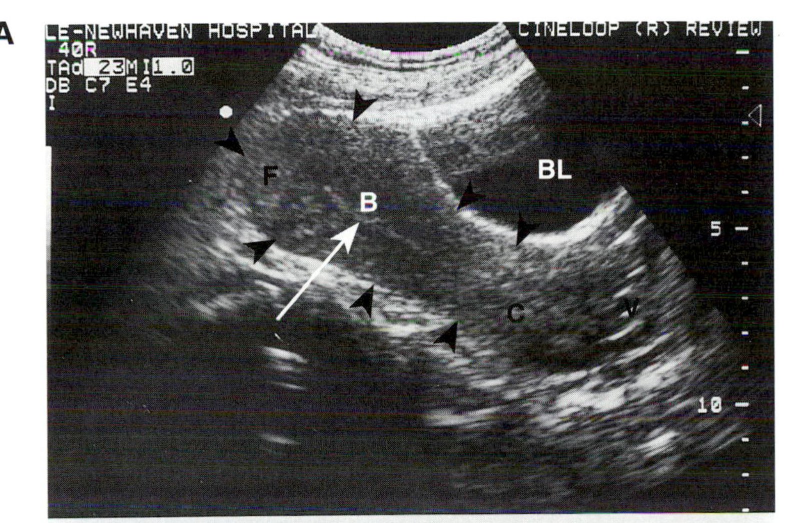

B

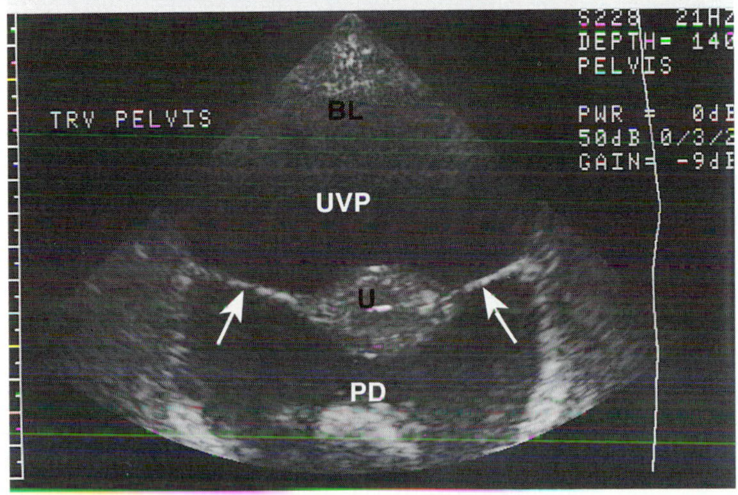

FIG 13–16.
A, longitudinal sonogram of a normal uterus (black arrowheads) showing the fundus *(F)*, body *(B)*, and cervix *(C)*; the vagina *(V)* and the bladder *(BL)* are also shown. Note the linear echogenic line representing the endometrium (white arrow). **B,** transverse sonogram of the pelvis in a woman following an automobile accident in which the liver was lacerated and blood escaped into the peritoneal cavity. Identify the bladder *(BL)*, the body of the uterus *(U)*, and the broad ligaments (white arrows). Note the presence of blood (dark areas) in the uterovesical pouch *(UVP)* and the pouch of Douglas *(PD)*. (Courtesy of Dr Leslie Scoutt.)

A

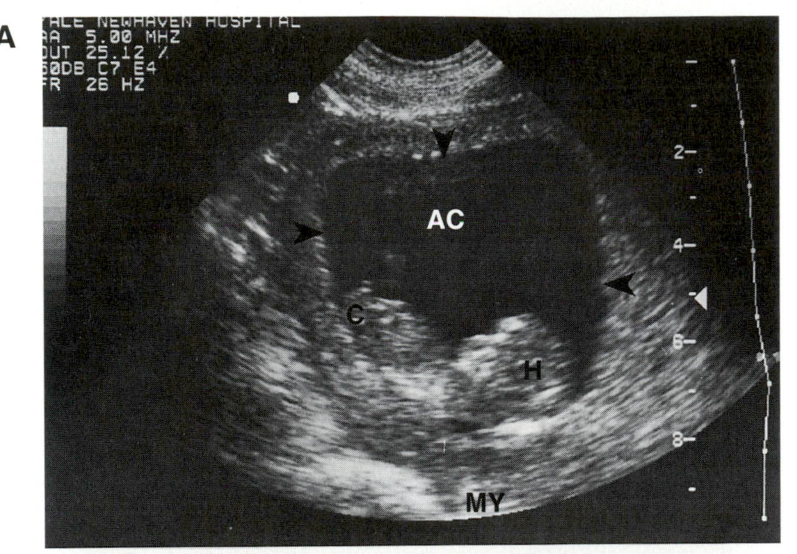

B

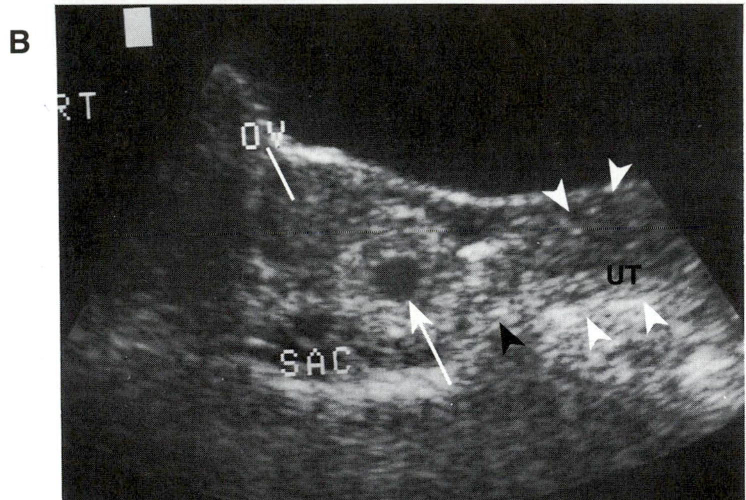

FIG 13–17.
A, longitudinal sonogram of a normal 11-week intrauterine gestational sac (black arrowheads) showing the amniotic cavity *(AC)* filled with amniotic fluid; the fetus is seen in longitudinal section with the head *(H)* and coccyx *(C)* well displayed. The myometrium *(MY)* of the uterus can be identified. **B,** transverse sonogram of a tubal ectopic pregnancy (white arrow); the ovary *(OV)*, the isthmus of the uterine tube (black arrowhead), and the lateral angle of the body of the uterus (white arrowheads) are clearly seen. (Courtesy of Dr Leslie Scoutt.)

endometrium becomes greatly thickened and is known as the decidua. Large areas of the decidua become excavated by the invading trophoblastic villi to form the intervillous spaces. The maternal blood vessels open into the spaces so that the outer surfaces of the villi of the fetal part of the placenta become bathed in oxygenated blood (see Fig 13–19).

By the 4th month of pregnancy, the placenta is a well-developed organ. As the pregnancy continues, the placenta increases in area and thickness. The placental attachment occupies one third of the internal surface of the uterus.

At birth, a few minutes after the delivery of the child, the placenta separates from the uterine wall and is expelled from the uterine cavity as the result of the contractions of the uterine musculature. The line of separation occurs through the spongy layer of the decidua (see Fig 13–19).

Gross Appearance of the Placenta at Birth

At full term, the placenta has a spongelike consistency; is flattened; circular, with a diameter of about 8 in. (20 cm) and a thickness of about 1 in. (2.5 cm); and weighs about 1 lb (500 g). It thins out at the edges, where it is continuous with the fetal membranes (Fig 13–20).

The outer, or maternal, surface of a freshly shed placenta is rough on palpation, dark red, and oozes blood from the torn maternal blood vessels.

The inner, or fetal, surface is smooth and shining and is raised in ridges by the umbilical blood vessels, which radiate from the attachment of the umbilical cord near its center.

The fetal membranes (Fig 13–21), which surround and enclose the amniotic fluid, are continuous with the edge of the placenta and are the amnion and chorion and a small amount of the adherent maternal decidua.

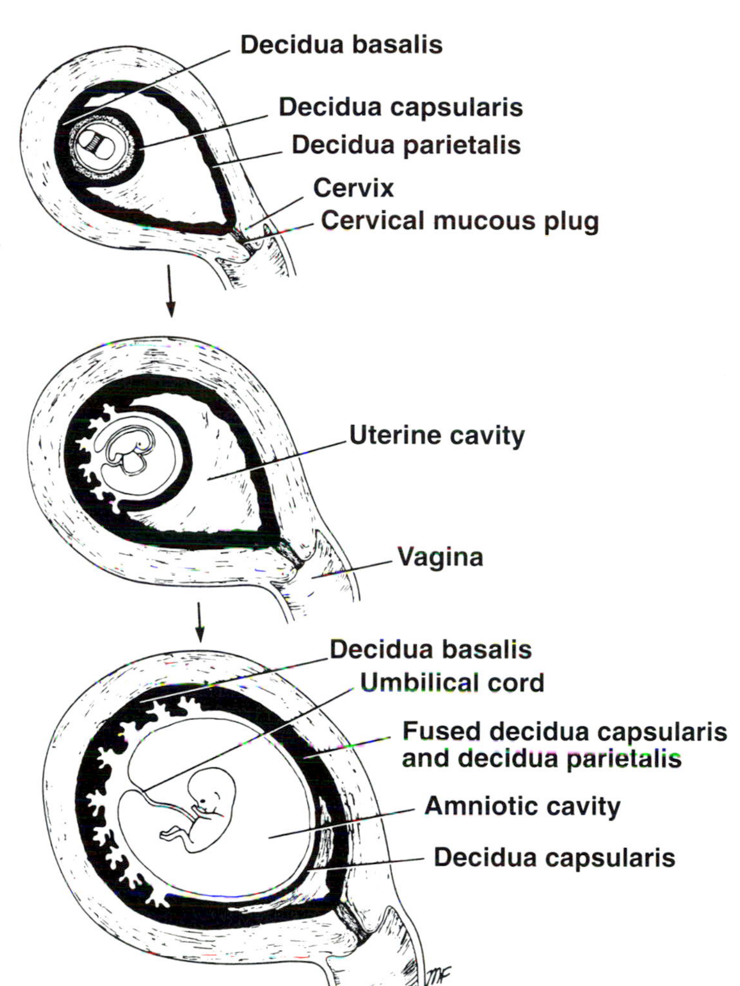

FIG 13–18.
Sagittal section of the uterus showing the developing conceptus expanding into the uterine cavity. The three different regions of the decidua can be recognized. By the 16th week, the uterine cavity is obliterated by the fusion of the decidua capsularis with the decidua parietalis.

Anatomy of Childbirth

Labor Pains.—The underlying cause of labor pains is believed to be the anoxia of the uterine muscle brought about by uterine contractions. Afferent pain fibers from the fundus and body of the uterus ascend to the spinal cord through the hypogastric plexuses, entering the cord through the posterior roots of the 10th, 11th, and 12th thoracic spinal nerves (see Fig 9–14). Sensory fibers from the cervix run in the pelvic splanchnic nerves and enter the spinal cord through the posterior roots of the second, third, and fourth sacral nerves. Contraction pains from the fundus and body are referred to the lower part of the anterior abdominal wall and the lower part of the back (dermatomes T10 through T12).

Labor.—Forceful uterine contractions occurring at regular intervals indicate the onset of labor. The *first stage* is dilatation of the cervix. The *second stage* is expulsion of the fetus. The *third stage* begins immediately following the delivery of the baby and ends with expulsion of the placenta and fetal membranes.

Duration of Labor.—This depends on (1) the strength and frequency of the uterine contractions and (2) the resistance offered to the passage of the baby by the bony pelvis and the soft tissues of the lower part of the genital tract, namely, the cervix, the vagina, the pelvic floor, and the perineum. The approximate length of each stage is as follows: first stage, 9½ hours; second stage, 50 minutes; and third stage, 10 minutes. In the multigravida the duration of the first and second stages is shorter because of the reduced resistance offered by the maternal soft tissues.

Processes of Labor.—During the first stage of labor, the liquor amnii and the fetal membranes are forced down into the cervical canal as a hydrostatic wedge, and the cervix slowly dilates (Fig 13–22). The pulling away of the membranes from the uterine wall in the region of the internal os of the cervix causes a little bleeding, and this, together with the

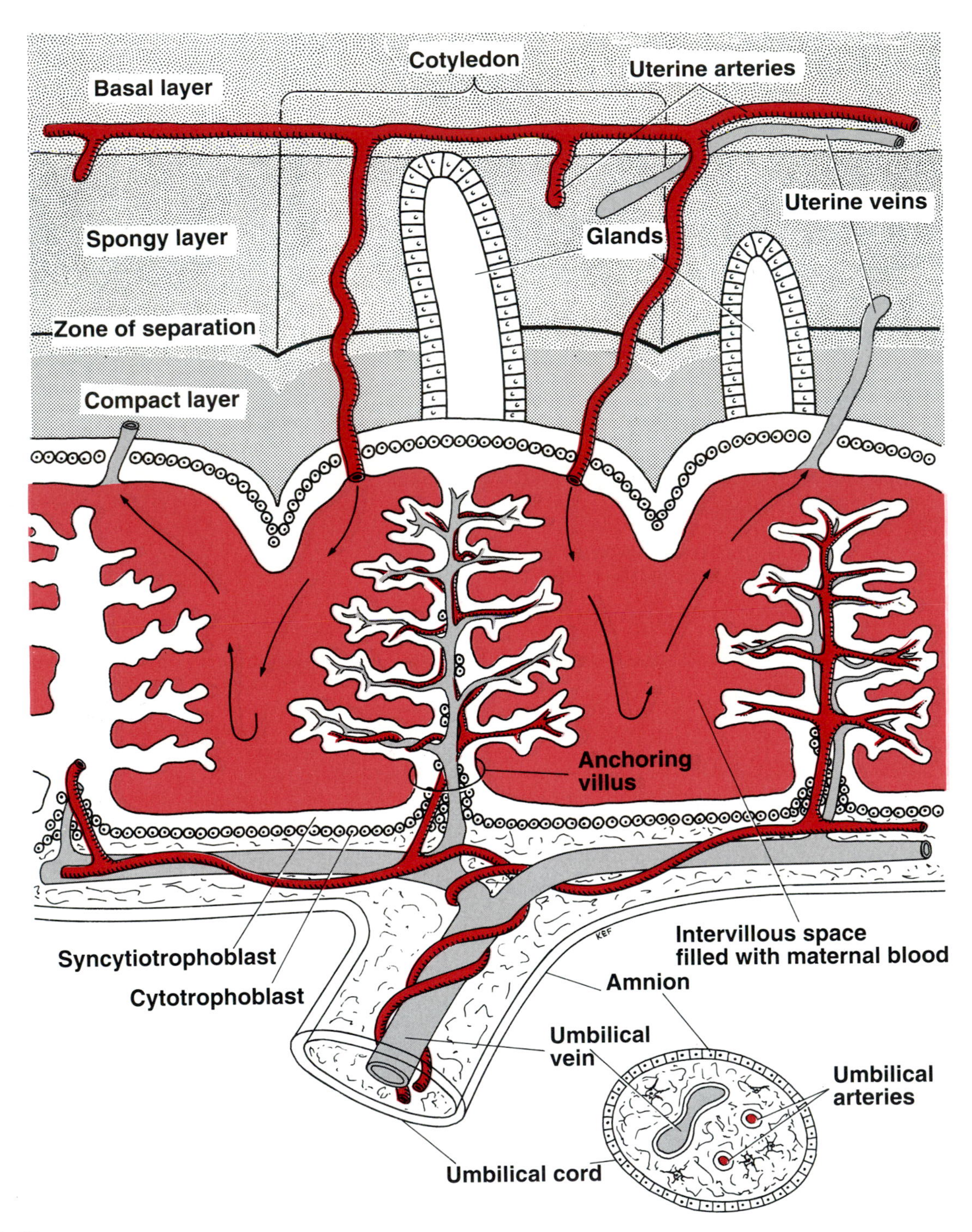

FIG 13–19.
Section through the placenta showing the maternal part (above) and the fetal part (below). Note that the maternal part is divided into the basal layer, the spongy layer, and the compact layer. The heavy solid line in the spongy layer indicates where separation occurs between the maternal and fetal parts of the placenta during the third stage of labor.

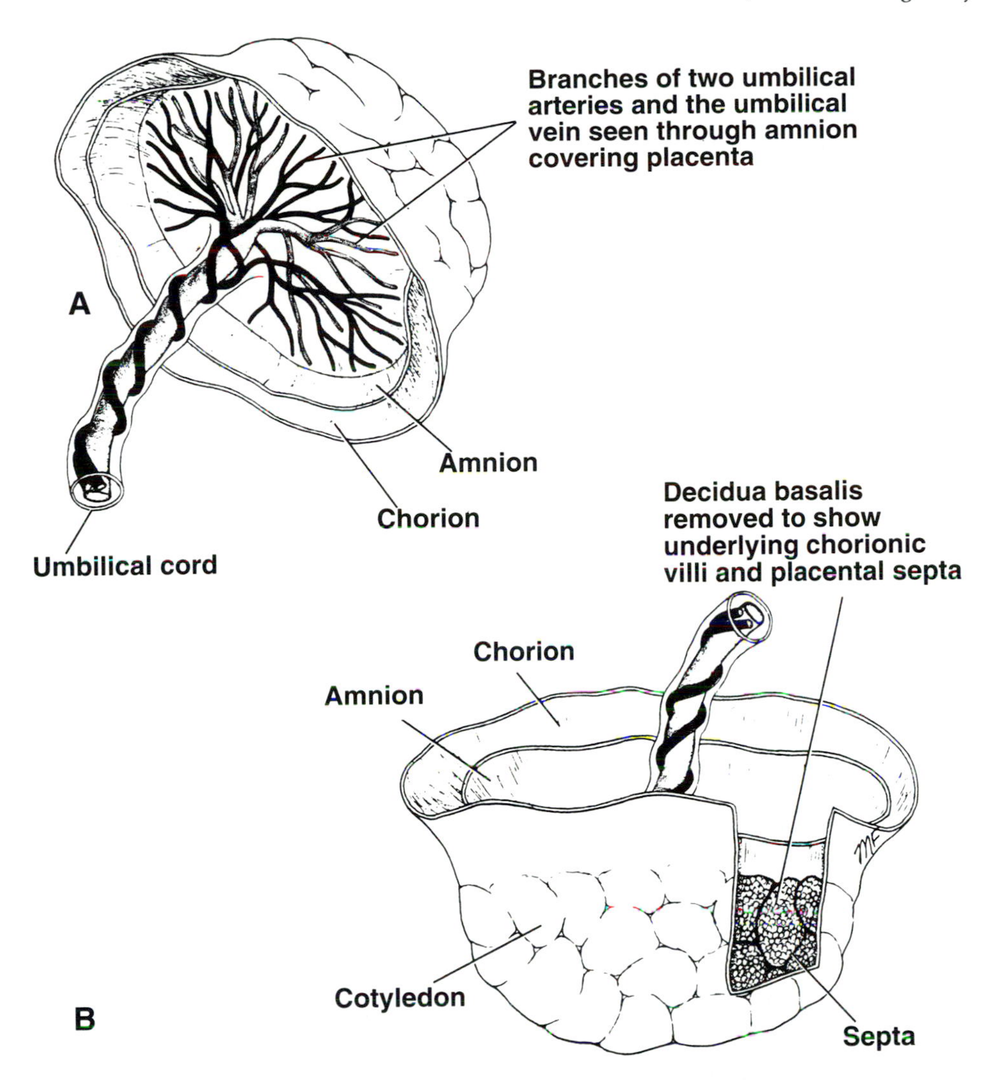

FIG 13–20.
Mature placenta. **A,** as seen from the fetal surface. **B,** as seen from the maternal surface.

cervical plug of mucus that fills the canal of the cervix, forms the so-called *bloody show.* The rupture of the membranes and the escape of the liquor amnii usually occur when the first stage of labor is well advanced. The escape of the liquor amnii allows the uterus to exert pressure directly on the baby and force it down.

During the second stage of labor, the combined actions of the very strong uterine contractions and the reflex and voluntary contractions of the abdominal muscles force the baby through the maternal passages (Fig 13–23).

The pelvic floor, formed by the levatores ani and coccygeus muscles, serves an important function during the second stage of labor. From their origin on the two sides of the pelvis, the muscle fibers slope downward and backward in the midline, producing a gutter that slopes downward and forward. To begin with, the long axis of the baby's head is transversely positioned at the pelvic inlet. When the head reaches the pelvic floor, the gutter shape of the floor tends to cause the baby's head to rotate, so that its long axis comes to lie in the anteroposterior position, with the occiput anteriorly in the majority. The occipital part of the head now moves downward and forward along the gutter until it lies under the pubic arch. As the baby's head passes through the lower part of the birth canal, the small gap that exists in the anterior part of the pelvic floor becomes enormously enlarged, so that the head may slip through into the perineum. Once the baby has passed through the perineum, the levatores ani muscles recoil and take up their previous position.

The third stage of labor then commences, and the uterus contracts downward so that the fundus lies at the level of the umbilicus. Rhythmic uterine contractions continue, as in the second stage, but they are painless. Separation of the placenta from the uterine wall takes place. As the uterus diminishes in size, there is a decrease in the surface area

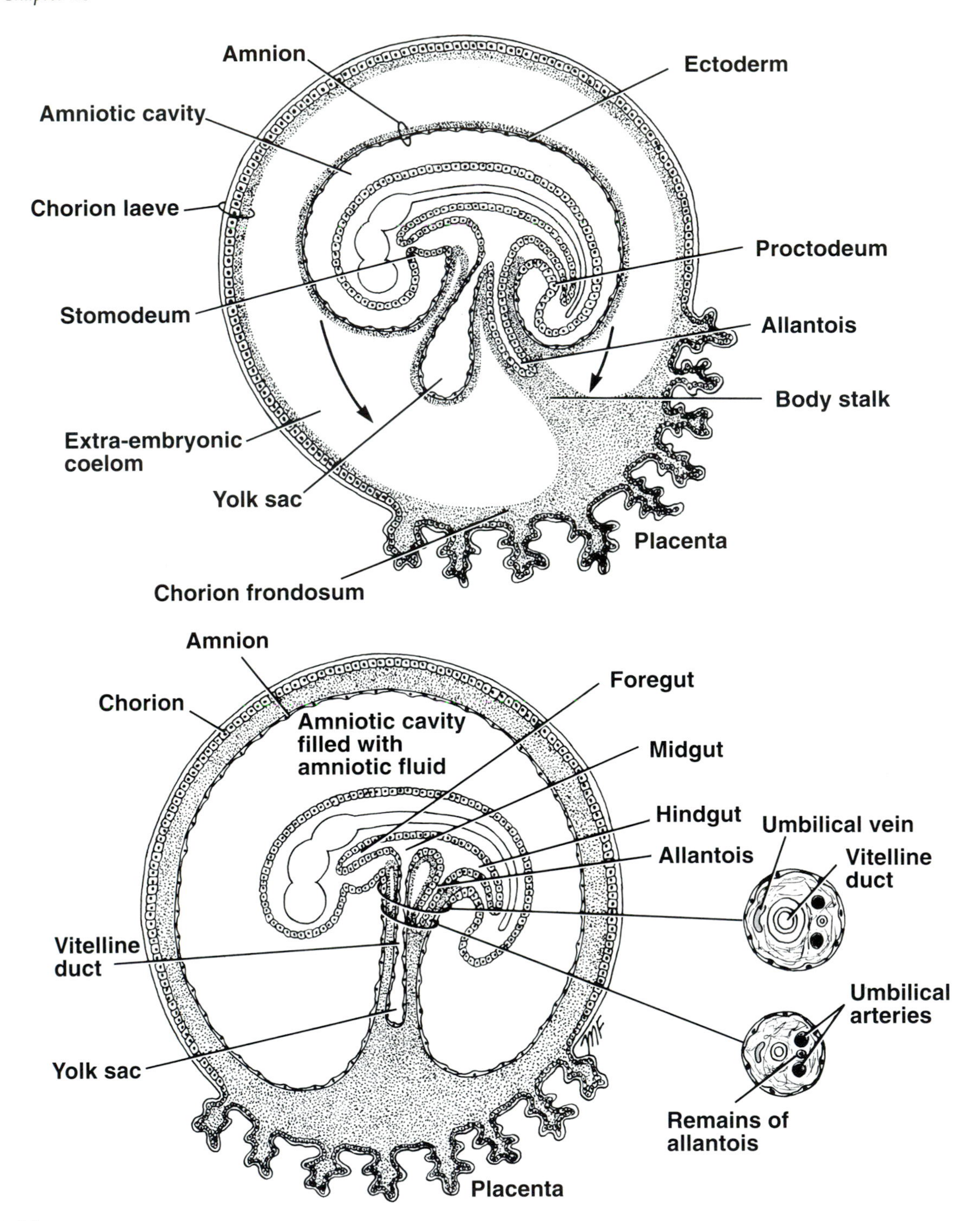

FIG 13–21.
Sagittal sections of the developing embryo showing the fusion of the amnion with the chorion, the expanding amniotic cavity filled with amniotic fluid, and the coming together of the yolk sac with the body stalk to form the umbilical cord.

at the site of placental attachment. The placenta, being unable to accommodate itself to the decreased area, begins to fold up and is squeezed off the uterine wall (Fig 13–24). Separation takes place at the spongy layer of the decidua, and some bleeding occurs. In a similar manner, the fetal membranes and the remains of the decidua are squeezed from the remainder of the uterine wall. As the placenta is finally expelled from the uterus by the uterine contractions, the membranes that are attached to it are

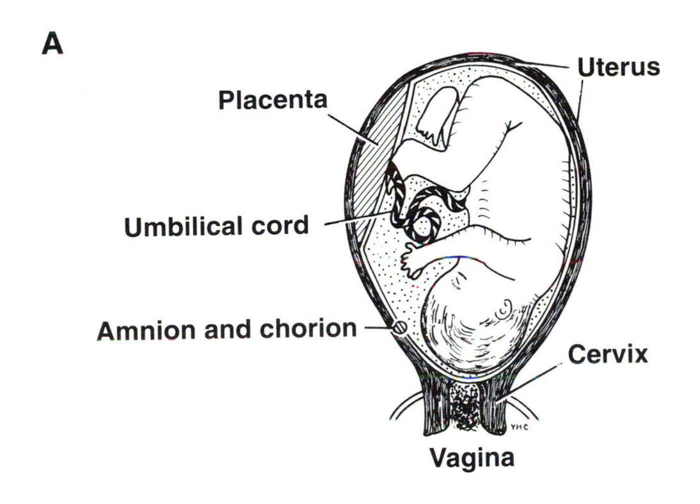

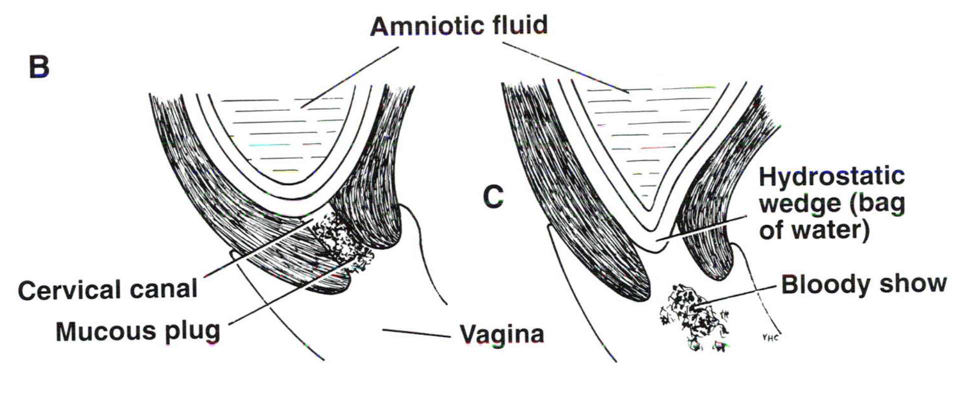

FIG 13–22.
A, the relation of the fetus to the uterus, the placenta, and the fetal membranes at the 40th week of pregnancy. **B** and **C,** the cervix dilating during the first stage of labor as the result of the fetal membranes being forced down into the cervical canal as a hydrostatic wedge. The bloody show consists of the mucous plug from the cervical canal mixed with the small amount of bleeding that occurs following separation of the fetal membranes from the uterine wall in the region of the internal os.

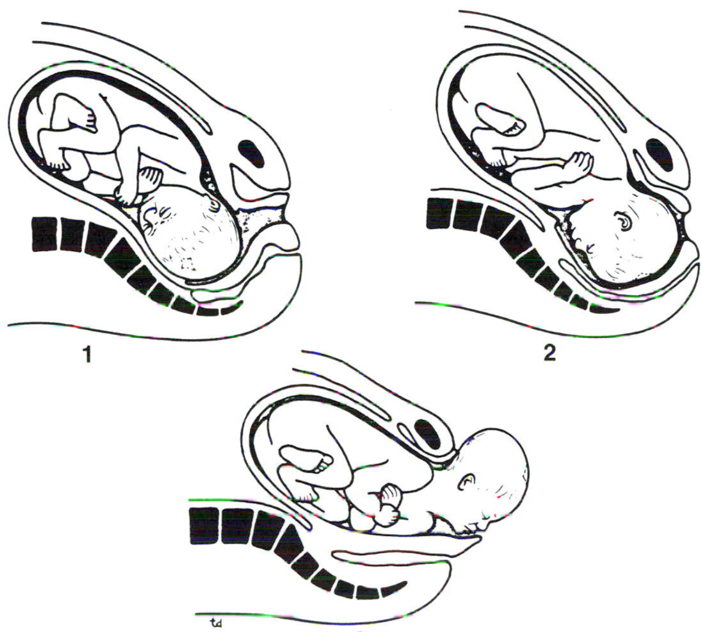

FIG 13–23.
The stages in the rotation of the baby's head during the second stage of labor. The gutter shape of the pelvic floor plays an important part in this process.

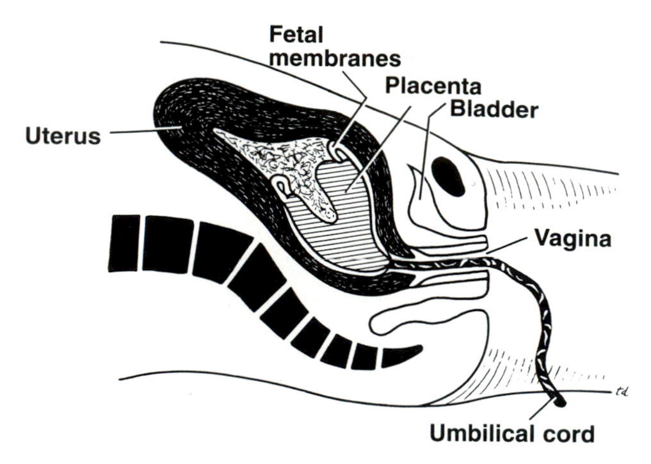

FIG 13–24.
The third stage of labor showing the separated placenta with the attached fetal membranes about to be expelled through the vagina.

dragged after it and peeled off from the inner surface of the uterus. The expelled placenta and fetal membranes are often referred to as the *afterbirth*. The uterine muscle becomes firmly contracted and little bleeding occurs.

SURFACE ANATOMY

Surface Anatomy of the Perineum

Bony Landmarks

The perineum, when seen from below with the thighs abducted (see Fig 13–2), is diamond shaped and is bounded anteriorly by the symphysis pubis, posteriorly by the tip of the coccyx, and laterally by the ischial tuberosities.

Symphysis Pubis.—This is the cartilaginous joint that lies in the midline between the bodies of the pubic bones (see Fig 13–2). It is felt as a solid structure beneath the skin in the midline at the lower extremity of the anterior abdominal wall.

Coccyx.—The inferior surface and tip of the coccyx can be palpated in the cleft between the buttocks about 1 in. (2.5 cm) behind the anus (see Fig 13–2).

Ischial Tuberosity.—This can be palpated in the lower part of the buttock (see Fig 13–2). In the standing position, the tuberosity is covered by the gluteus maximus muscle. In the sitting position, the ischial tuberosity emerges from beneath the lower border of the gluteus maximus and supports the weight of the body.

It is customary to divide the perineum into two triangles by joining the ischial tuberosities by an imaginary line (see Fig 13–2). The posterior triangle, which contains the anus, is called the *anal triangle;* the anterior triangle, which contains the urogenital orifices, is called the *urogenital triangle.*

Anal Triangle

Anus.—The anus lies in the midline, with a reddish-brown margin, and is puckered by the contraction of the external anal sphincter. Around the anal margin are a number of coarse hairs.

Female Urogenital Triangle

Vulva.—This is the name applied to the female external genitalia.

Mons Pubis.—This is the rounded, hair-bearing elevation of skin found anterior to the pubis (see Fig 13–8). The pubic hair in the female has an abrupt horizontal superior margin, whereas in the male it extends upward to the umbilicus.

Labia Majora.—These are prominent, hair-bearing folds of skin extending posteriorly from the mons pubis to unite posteriorly in the midline (see Fig 13–8).

Labia Minora.—These two smaller, hairless folds of soft skin lie between the labia majora (see Fig 13–8). Their posterior ends are united to form a sharp fold, the *fourchette.* Anteriorly they split to enclose the clitoris, forming an anterior *prepuce* and a posterior *frenulum* (see Fig 13–8).

Vestibule.—This is a smooth triangular area bounded laterally by the labia minora, with the clitoris at its apex and the fourchette at its base (see Fig 13–8).

Vaginal Orifice.—This is protected in virgins by a thin mucosal fold called the *hymen,* which is perforated at its center. At the first coitus the hymen tears, usually posteriorly or posterolaterally, and after childbirth only a few tags of the hymen remain (see Fig 13–8).

Orifices of the Ducts of the Greater Vestibular Glands (Bartholin's Glands).—These are small orifices, one on each side, in the groove between the hymen and the posterior part of the labium minus (see Fig 13–8).

Clitoris.—This is situated at the apex of the vestibule anteriorly (see Fig 13–8). The *glans of the clitoris* is partly hidden by the *prepuce*.

CLINICAL ANATOMY

Clinical Anatomy of the Rectum and Anal Canal

Anus

At the anal margin, the hairs, hair follicles, and sweat glands come to an abrupt end; this represents the junction of the perianal skin with the stratified squamous epithelium lining the lower half of the anal canal (see Fig 13–6). Above this level is a transition zone between the stratified squamous epithelium of the lower half of the canal and the columnar epithelium of the upper half of the canal. At the level of the anal valves the stratified squamous epithelium ceases and the simple columnar epithelium begins, and the lining of the upper half of the anal canal has the appearance of bowel mucous membrane. Because of the confusion that occurs when terms such as anocutaneous line, mucocutaneous line, pectinate line, and the white line of Hilton are used, these terms have been purposely avoided.

Anal Columns, Anal Valves, and Anal Ducts or Glands

The *anal columns* are permanent longitudinal folds of mucous membrane present in the upper half of the anal canal (see Fig 13–6). They are bluish because they contain the tributaries of the superior rectal vein. There are usually five to ten columns that are connected together at their lower ends by thin folds of mucous membrane called *anal valves*. In the majority of individuals the *ducts of anal glands* open into the canal behind the anal valves. There are few glands, and the majority are located in the posterior part of the anal canal; they are situated in the submucosa, but some extend outward into the internal anal sphincter. Infection of the anal glands may predispose to the formation of abscesses or fistulae in the perianal region.

Rectal Examination

The following structures can be palpated by the gloved index finger inserted into the anal canal and rectum in the healthy patient (Fig 13–25).

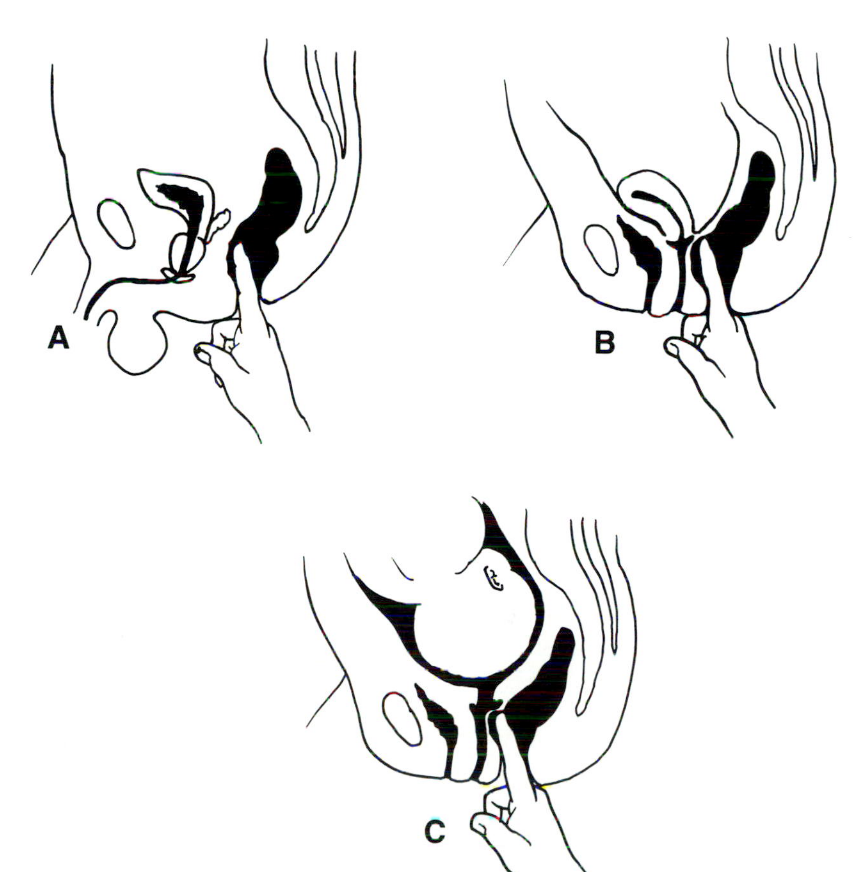

FIG 13–25.
A, rectal examination in the male showing the palpation of the prostate and seminal vesicles through the anterior rectal wall. **B,** rectal examination in the nonpregnant female showing the palpation of the lower part of the pouch of Douglas and the cervix through the anterior rectal wall. **C,** rectal examination in the pregnant female showing the palpation of the external os of the cervix through the anterior rectal wall.

In the Male.—*Anteriorly:* the anterior structures include the following.

1. Opposite the terminal phalanx are the contents of the rectovesical pouch, the posterior surface of the bladder, the seminal vesicles, and the vasa deferentia.
2. Opposite the middle phalanx are the rectoprostatic fascia and the prostate.
3. Opposite the proximal phalanx are the perineal body, the urogenital diaphragm, and the bulb of the penis.

Posteriorly: the posterior structures include the following.

1. The sacrum, coccyx, and anococcygeal body.

Laterally: the lateral structures include the following.

1. The ischiorectal fossae and ischial spines.

In the Female.—*Anteriorly:* the anterior structures include the following.

1. Opposite the terminal phalanx are the rectouterine pouch of Douglas, the vagina, and the cervix (see Fig 13–25).
2. Opposite the middle phalanx are the urogenital diaphragm and the vagina.
3. Opposite the proximal phalanx are the perineal body and the lower part of the vagina.

The posterior and lateral relations are identical to those found in the male.

Rectal Injuries

The management of penetrating rectal injuries will be determined by the site of penetration relative to the peritoneal covering. The upper third of the rectum is covered on the anterior and lateral surfaces by peritoneum, the middle third is covered only on its anterior surface, while the lower third is devoid of a peritoneal covering (see Figs 13–4 and 13–5). The treatment of penetration of the intraperitoneal portion of the rectum is identical to that of the colon, since the peritoneal cavity has been violated. In the case of penetration of the extraperitoneal portion, the rectum is treated by a diverting colostomy, antibiotics, and repair and drainage of the perirectal space in front of the sacrum.

Portal-Systemic Anastomosis in the Rectum

In the submucosa of the anal canal is a plexus of veins that is principally drained upward by the superior rectal vein (Fig 13–26). The small tributaries of the middle and inferior rectal veins communicate with each other and with the superior rectal vein through this plexus. The rectal venous system therefore forms an important portal systemic anastomosis, since the superior rectal vein drains into the portal vein, and the middle and inferior rectal veins ultimately drain into the internal iliac veins.

Hemorrhoids

Internal Hemorrhoids.— These are varicosities of the tributaries of the superior rectal (hemorrhoidal) vein and are covered by mucous membrane (see Fig 13–26). The tributaries of the vein, which lie in the anal columns at the 3-, 7-, and 11-o'clock positions when the patient is viewed in the lithotomy position, are particularly liable to become varicosed. Anatomically, a hemorrhoid is therefore a fold of mucous membrane and submucosa containing a varicosed tributary of the superior rectal vein and a terminal branch of the superior rectal artery. Internal hemorrhoids are initially contained within the anal canal (first degree). As they enlarge (see Fig 13–26), they are extruded from the canal on defecation, but return at the end of the act (second degree). With further elongation, they prolapse on defecation and remain outside the anus (third degree).

Since internal hemorrhoids occur in the upper half of the anal canal, where the mucous membrane is innervated by autonomic afferent nerves, they are painless and are only sensitive to stretch. This may explain why large internal hemorrhoids give rise to an aching sensation rather than acute pain. Internal hemorrhoids occur at the most dependent part of the portal venous system; it is not surprising, therefore, that large internal hemorrhoids may undergo severe bleeding, if ulceration occurs. The mucous membrane covering an internal hemorrhoid contains columnar cells, many of which are goblet mucus-secreting cells. This would explain why in patients with prolapsed internal hemorrhoids there may be extensive mucous soiling of the perineum.

Etiology of Internal Hemorrhoids.—Congenital weakness of the vein wall is a common cause. The superior rectal vein is the most dependent part of the portal circulation and is valveless. The weight of the venous blood is thus greatest in the veins in the upper half of the anal canal. Here the loose connective tissue of the submucosa gives little support to

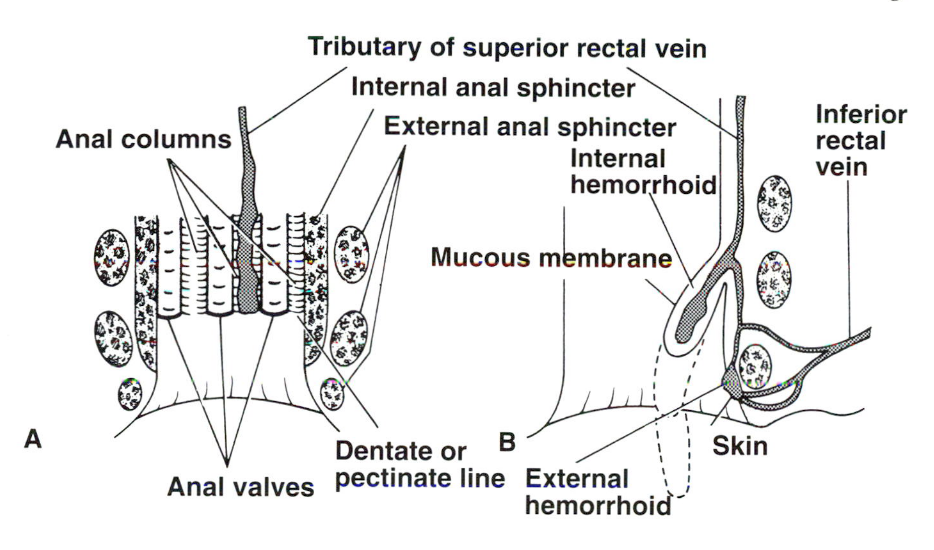

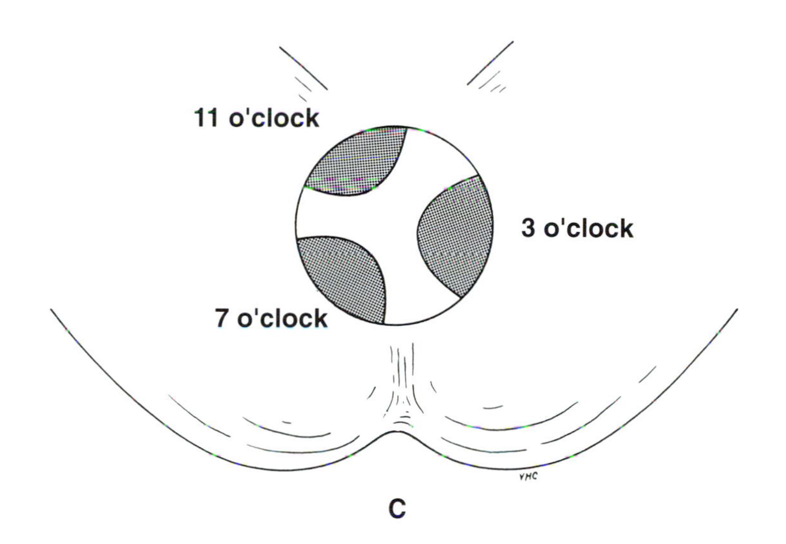

FIG 13–26.
A, normal tributary of superior rectal vein within the anal column. **B,** varicosed tributary of superior rectal vein forming an internal hemorrhoid (dotted lines indicate increasing degrees of severity of condition). **C,** sites of three internal hemorrhoids as seen through proctoscope. (For ease of orientation, the patient in this diagram is shown in the lithotomy position, although a rectal examination is usually made either in the left lateral or knee-chest position).

the walls of the veins. Moreover, the venous return is interrupted by the contraction of the muscular coat of the rectal wall during defecation. Chronic constipation, associated with prolonged straining at stool, is a common predisposing factor. *Pregnancy hemorrhoids* are common due to the pressure on the superior rectal veins by the gravid uterus. Portal hypertension due to cirrhosis of the liver may also cause hemorrhoids. Carcinoma of the rectum blocking the superior rectal vein can also cause hemorrhoids.

External Hemorrhoids.—These are varicosities of the tributaries of the inferior rectal (hemorrhoidal) vein as they run laterally from the anal margin. They are covered by the squamous epithelium of the mucous membrane of the lower half of the anal canal (see Fig 13–26) and by skin; they are commonly associated with well-established internal hemorrhoids.

External hemorrhoids are covered by the mucous membrane of the lower half of the anal canal or the skin, and they are innervated by the inferior rectal nerves. They are sensitive to pain, temperature, touch, and pressure, which explains why external hemorrhoids tend to be painful. Thrombosis of an external hemorrhoid is common. Its cause is unknown, although coughing or straining may produce distension of the hemorrhoid followed by stasis. The presence of a small, acutely tender swelling at the anal margin is immediately recognized by the patient.

Anal Fissure

The lower ends of the anal columns are connected by small folds called *anal valves* (Fig 13–27). In persons suffering from chronic constipation, the anal

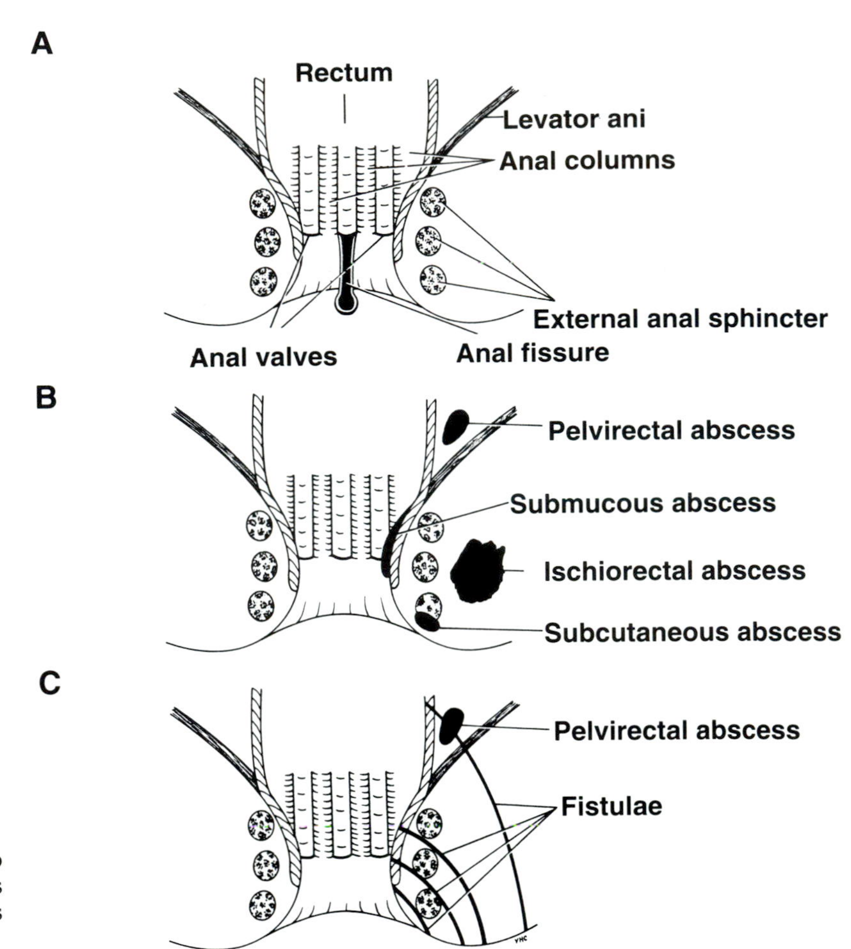

FIG 13–27.
A, downward tearing of the anal valve to form an anal fissure. **B,** common locations of perianal abscesses. **C,** common positions of perianal fistulae.

valves may be torn down to the anus as the result of the edge of the fecal mass catching on the fold of mucous membrane. The elongated ulcer so formed is called an anal fissure (see Fig 13–27) and is extremely painful. The fissure occurs most commonly in the midline posteriorly, or, less commonly, anteriorly, and this may be from the lack of support provided by the superficial part of the external sphincter in these areas. The superficial part of the external sphincter does not encircle the anal canal, but sweeps past its lateral sides (see Fig 13–7).

The site of the anal fissure in the sensitive lower half of the anal canal, which is innervated by the inferior rectal nerve, results in reflex spasm of the external anal sphincter, aggravating the condition. Because of the intense pain, an anal fissure may have to be examined under local anesthesia.

Perianal Abscesses

Perianal abscesses are produced by fecal trauma to the anal mucosa (see Fig 13–27). Infection may gain entrance to the submucosa through a small mucosal tear; the abscess may complicate an anal fissure or the infection of an anal gland. The abscess may be localized to the submucosa *(submucous abscess),* may occur beneath the perianal skin *(subcutaneous abscess),* or may occupy the ischiorectal fossa *(ischiorectal abscess).* Large ischiorectal abscesses sometimes extend circumferentially posteriorly to invade the ischiorectal fossa of the opposite side *(horseshoe abscess).* An abscess may be found in the space between the ampulla of the rectum and the upper surface of the levator ani *(pelvirectal abscess).* Anatomically, these abscesses are closely related to the different parts of the external sphincter and levator ani muscles, as seen in Figure 13–27.

Drainage of Anorectal Abscesses.—A subcutaneous or perianal abscess can be drained under a local anesthetic through an elliptical or cruciate skin incision over the abscess. The wound is then adequately drained. A submucous abscess is drained through a mucosal incision made through a proctoscope. Ischiorectal abscesses, including horseshoe abscesses or pelvirectal abscesses, are larger and more serious conditions. Great care must be taken to preserve the integrity of the anal sphincters, especially the puborectalis muscle, to avoid rectal incontinence.

Anal Fistula

Anal fistulae develop as the result of spread or inadequate treatment of perianal abscesses. The fistula opens at one end at the lumen of the anal canal or lower rectum and at the other end on the skin surface close to the anus (see Fig 13–27). If the abscess opens onto only one surface, it is known as an *anal sinus,* not a fistula. The high-level fistulae are rare and run from the rectum to the perianal skin. They are located above the anorectal ring (see below), and as a result, fecal material constantly soils the clothes. The low-level fistulae occur below the level of the anorectal ring, as shown in Figure 13–27.

The *anorectal ring* is the most important part of the sphincteric mechanism of the anal canal. It consists of the deep part of the external sphincter, the internal sphincter, and the puborectalis part of the levator ani. Surgical operations on the anal canal that result in damage to the anorectal ring will produce fecal incontinence.

Partial and Complete Prolapses of the Rectum

Patients with partial and complete prolapse of the rectum through the anus may present in the emergency department (Fig 13–28).

In partial prolapse, the rectal mucous membrane and submucous coat protrude for a short distance outside the anus. In complete prolapse, the whole thickness of the rectal wall protrudes through the anus. In both conditions, many causative factors may be involved. The most important contributing factors are damage to the levator muscles of the pelvic floor as a result of childbirth and poor muscle tone in the aged. A complete rectal prolapse may be regarded as a sliding hernia through the pelvic floor.

Rectal Incontinence Associated With Prolapse

Rectal incontinence may accompany severe prolapse of long duration. It is thought that the prolonged and excessive stretching of the anal sphincters is the cause of the condition. The condition may be treated by restoring the anorectal angle by tightening the puborectalis part of the levator ani muscles and the external anal sphincters behind the anorectal junction.

Rectal Incontinence Following Trauma

Trauma such as childbirth or damage to the sphincters during surgery of perianal abscesses or fistulae can be responsible.

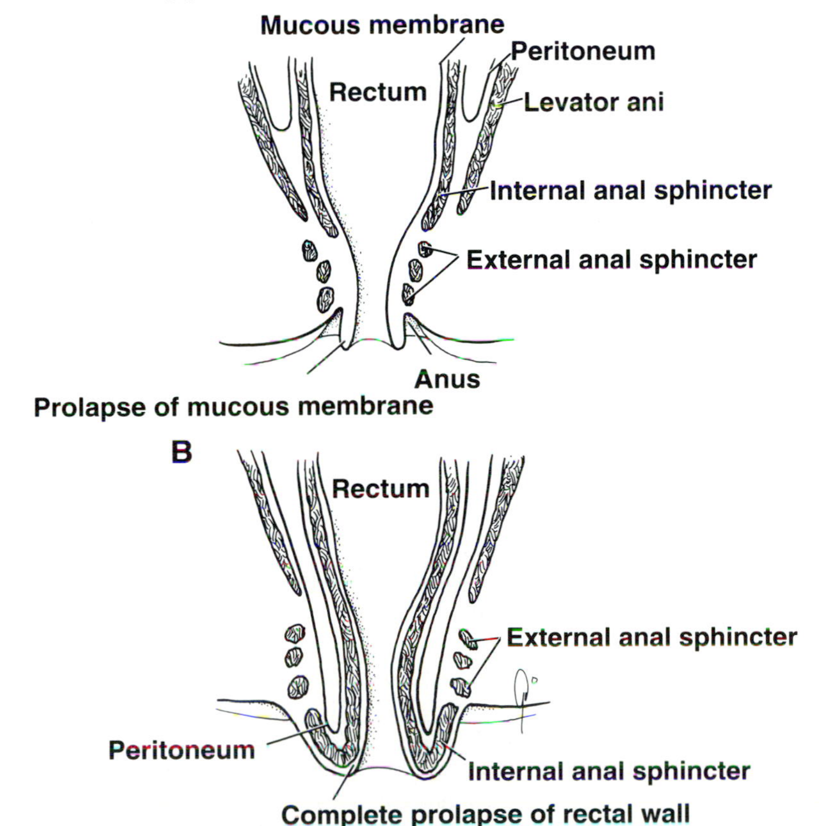

FIG 13–28.
Coronal section of the rectum and anal canal. **A,** incomplete rectal (mucosal prolapse). **B,** complete rectal prolapse.

Rectal Incontinence Following Spinal Cord Injury
Following severe spinal cord injuries, the patient is not aware of rectal distension. Moreover, the parasympathetic influence on the peristaltic activity of the descending colon, sigmoid colon, and rectum is lost. In addition, control over the abdominal musculature and sphincters of the anal canal may be severely impaired. The rectum, now an isolated structure, responds by contracting when the pressure within its lumen rises. This local reflex response is much more efficient if the sacral segments of the spinal cord are spared. At best, however, the force of the contractions of the rectal wall is small, and constipation and impaction are the usual outcome.

Pilonidal Sinus Disease
A pilonidal sinus occurs in the natal cleft overlying the lower sacrum and coccyx. The sinus may have one or more openings onto the skin and commonly contains loose hairs and epithelial debris. Occasionally the sinus does not open onto the surface and becomes a cyst that becomes infected. Initially the condition may commence as a congenital skin pit. However, it is thought that the majority occur as the result of hairs drilling their way into the skin during movements of the buttocks.

The treatment consists of excising the area and allowing free drainage of the wound. All hair and epithelial debris are removed. It is essential that all the pits and sinuses and their openings be removed.

Anatomical Facts Relevant to Sigmoidoscopy

1. The patient is placed in the left lateral position with the left knee flexed and the right knee extended (Fig 13–29). Alternatively, the patient is placed in the Trendelenburg knee-chest position.
2. The sigmoidoscope is gently inserted into the anus and anal canal in the direction of the umbilicus to ensure that the instrument passes along the long axis of the canal. Gentle but firm pressure is applied to overcome the resistance of the anal sphincters.
3. After a distance of about 1½ in. the instrument enters the ampulla of the rectum. At this point the tip of the sigmoidoscope should be directed posteriorly in the midline to follow the sacral curve of the rectum.
4. Slow advancement is made under direct vision. Some slight side-to-side movement may be necessary to bypass the *transverse rectal folds.*
5. At approximately 6½ in. (16.25 cm) from the anal margin, the rectosigmoid junction will be reached. The sigmoid colon here bends forward and to the left, and the lumen appears to end in a blind

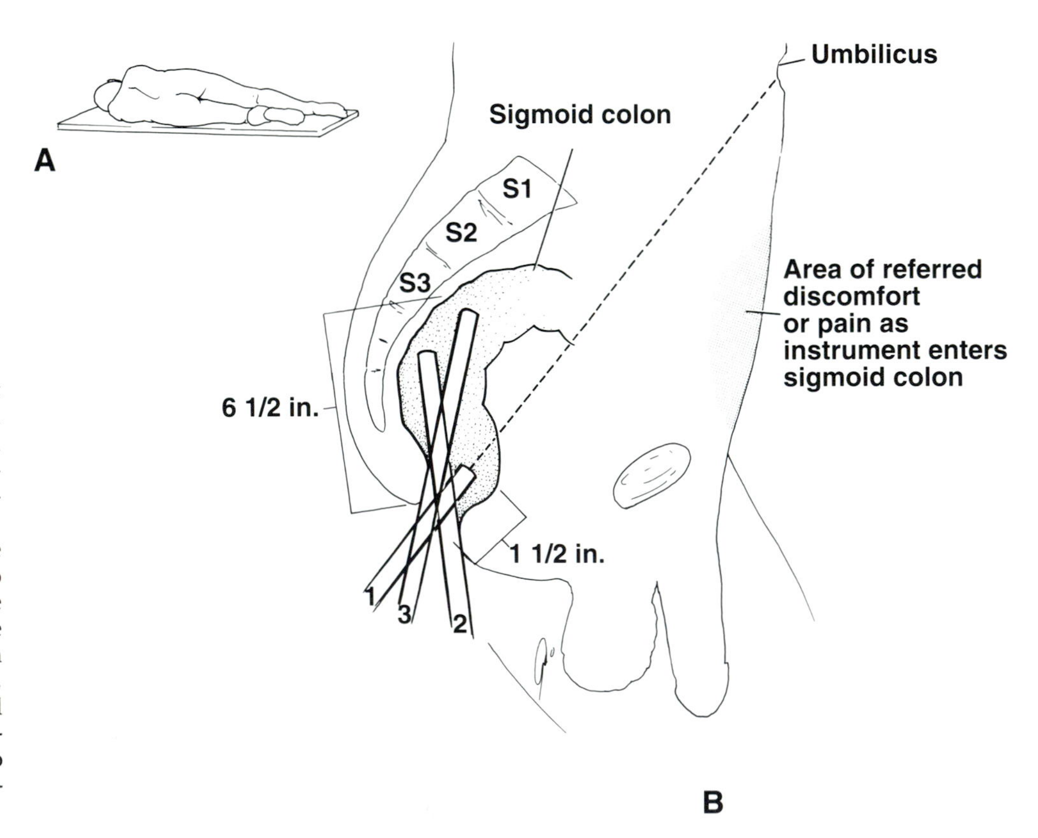

FIG 13–29.
Sigmoidoscopy. **A,** patient in the left lateral position with the left knee flexed and right knee extended. **B,** sagittal section of the male pelvis showing the positions *(1, 2,* and *3)* of the tube of the sigmoidoscope relative to the patient as it ascends the anal canal and rectum. The area of discomfort or pain experienced by the patient, as the tube is negotiated round the bend into the sigmoid colon, is referred to the skin of the hypogastrium as shown.

cul-de-sac. To negotiate this angulation, the tip of the sigmoidoscope must be directed anteriorly and to the patient's left side. This maneuver may cause some discomfort in the anal canal from distortion of the anal sphincters by the shaft of the sigmoidoscope. Another possibility is that the point of the instrument may stretch the wall of the colon, giving rise to colicky pain in the lower abdomen.

6. Once the instrument has entered the sigmoid colon, it should be possible to pass it smoothly along its full extent using the full length of the sigmoidoscope.

7. The sigmoidoscope may now be slowly withdrawn, carefully inspecting the mucous membrane. The normal rectal and colonic mucous membrane is smooth and glistening, it is pale pink with an orange tinge, and the blood vessels in the submucosa can be clearly seen. The mucous membrane is supple and moves easily over the end of the sigmoidoscope.

Anatomy of the Complications of Sigmoidoscopy.— The following complications may occur.

Perforation of the Sigmoid Colon.—Perforation of the bowel at the rectosigmoid junction may occur. This is almost invariably caused by the operator failing to carefully negotiate the curve between the rectum and the sigmoid colon. In some patients the curve is in the form of an acute angulation, which may frustrate the overzealous advancement of the sigmoidoscope. Perforation of the sigmoid colon results in violation of the peritoneal cavity, which is immediately contaminated with bowel contents.

Removal of Anorectal Foreign Bodies

Normally the anal canal is kept closed by the tone of the internal and external anal sphincters and the tone of the puborectalis part of the levator ani muscle. The rectal contents are supported by the levator ani muscles, possibly assisted by the transverse rectal folds. For these reasons the removal of a large foreign body from the rectum may be a formidable problem. The following procedure is usually successful.

1. The foreign body must first be fixed so that the sphinteric tone, together with external attempts to grab the object, do not displace the object further up the rectum.

2. Large, irregular foreign bodies may not be removed so easily, and it may be necessary to paralyze the anal sphincter by giving the patient a general anesthetic or performing an anal sphincter nerve block.

Anatomy of the Anal Sphincter Nerve Block and Anesthetizing the Perianal Skin

The voluntary external anal sphincter and the perianal skin are innervated by the inferior hemorroidal nerve, a branch of the pudendal nerve (S2 through S4), and the perineal branch of the fourth sacral nerve; the coccygeal nerve may also provide additional sensory innervation.

The procedure is as follows.

1. The patient lies on the right side with the knees drawn up.

2. An intradermal wheal is raised 1 in. (2.5 cm) behind the anus in the midline (Fig 13–30).

3. The gloved index finger of the left hand is inserted into the anal canal to serve as a future guide.

4. A longer needle attached to the syringe is then inserted through the cutaneous wheal and passed deeply into the tissues in a forward and upward direction, passing around the left side of the anal canal; the exact position of the needle relative to the anal canal being judged by the left index finger.

5. About 20 mL of anesthetic solution is injected into the sphincter muscles along the left lateral and posterior sides of the anal canal (Fig 13–30). The course of the needle is deep so that no visible swelling is seen on the surface.

6. The needle is then withdrawn, and the procedure is repeated on the right side of the anal canal.

The purpose of the finger in the anal canal is to prevent penetration of the anal mucous membrane with the long anesthetic needle. By blocking the branches of the inferior hemorrhoidal nerve and the perineal branch of the fourth sacral nerve (also C1), the anal sphincters will be relaxed and the perianal skin anesthetized. Should the perianal skin not be completely anesthetized, additional infiltration with local anesthetic in the subcutaneous tissues around the anal margin can be performed.

Clinical Anatomy of the Female External Genitalia

Vulvar Swellings

The presence of the urethral and vaginal orifices and the openings of the ducts of numerous glands (see Fig 13–8) make differential diagnosis of vulval swellings a clinical challenge.

In children, a mucosal swelling protruding from the urethra could be a prolapsing *ureterocele,* where

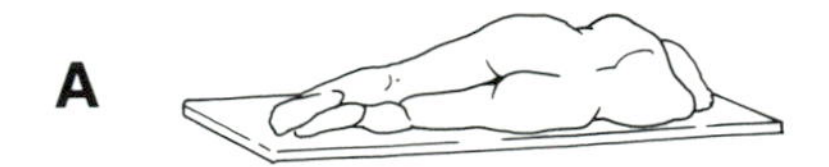

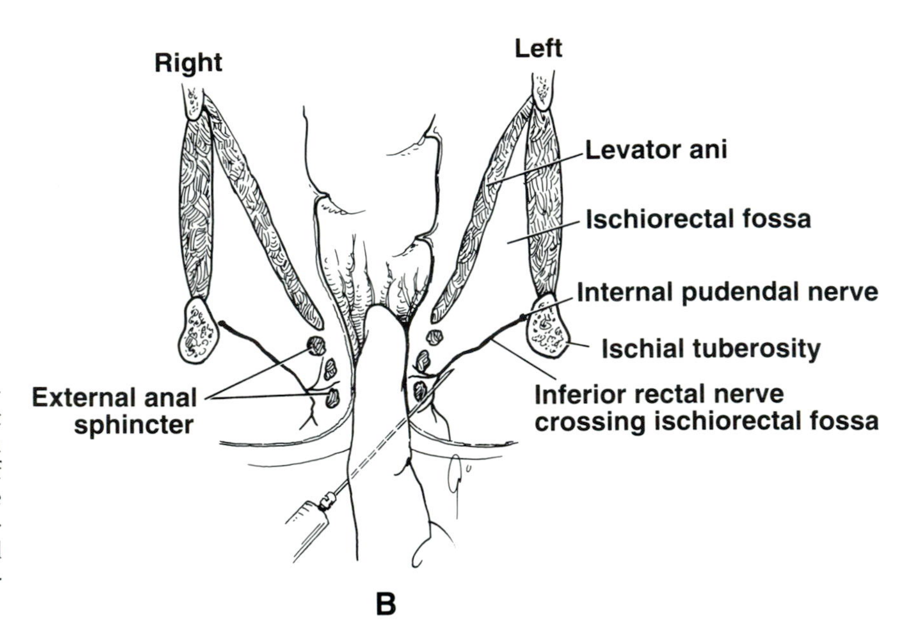

FIG 13-30.
Anal sphincter nerve block and anesthetizing of the perianal skin. **A,** the patient is placed in the right lateral position with both knees flexed. **B,** coronal section of pelvis showing the needle path on the left side of the anal canal. The exact position of the needle relative to the anal canal can be judged by the left index finger in the anal canal.

the distal end of an abnormal ureter has extended down the urethra to the external meatus. Another possibility seen in older girls is a *prolapse of the urethral mucous membrane.* A *hydrocolpos* may present as a bluish swelling in the midline accompanied by a lower abdominal swelling. In infants, the maternal hormones have stimulated the uterine secretions, which have become trapped proximal to an *imperforate hymen.* Following puberty a similar situation may arise due to retention of the menstrual flow.

In older children or adults, the presence of a cyst or abscess of the *greater vestibular glands (Bartholin's cyst)* may present as a swelling deep to the labia majora. Swellings at the vaginal orifice in later life may be caused by an advanced *uterine and vaginal prolapse.* The existence of an *indirect inguinal hernia* where the sac has extended down through the superficial inguinal ring to enter the labium majus must always be considered.

Vulvar Trauma

In any severe perineal trauma, it must be assumed that there is also a possibility of vaginal injury. Moreover, once vaginal injury has been established, the possibility that the peritoneal cavity has been violated must be excluded.

Because of the structure of the labia, which consist of skin filled with fatty areolar tissue, extensive contusions, ecchymoses, and large *hematomas* may sometimes be seen. For the same reasons widespread edema may develop in the vulvar area following severe trauma. One of the major problems may be the development of urinary retention caused by swelling of the urethral orifice.

In skin and mucous membrane lacerations, the bleeding vessels should be ligated and the tissue edges carefully sutured.

Vaginal Trauma

Coital injury, picket fence–type of impalement injury, and vaginal perforation caused by water under pressure, as occurs in water skiing, are common injuries. Lacerations of the vaginal wall involving the posterior fornix may violate the pouch of Douglas of the peritoneal cavity and cause prolapse of the small intestine into the vagina.

Air insinuated into the vagina in cunnilingus may enter the pouch of Douglas through a small rupture of the posterior fornix. Later the patient may complain of abdominal or shoulder pain due to the presence of air in the peritoneal cavity. Rarely, air may enter a vaginal blood vessel and then embolize.

Vaginal Examination

The anatomy of a vaginal examination is described on p. 530.

Bartholin's Cyst and Abscess

Infection of the greater vestibular gland (Bartholin's gland) is usually secondary to a blockage of the gland duct. A retention cyst first develops, which later becomes infected. A painful, tender swelling occurs beneath the labium majus; the labium itself becomes edematous and swollen and the skin reddened. The labium minus, which lies medially, is less involved and tends to cross the swelling longitudinally.

Anatomy of Pudendal Nerve Block

This procedure is fully discussed in Chapter 19, p. 832.

Clinical Anatomy of the Ovary

Position of the Ovary

Most often the ovary lies in the ovarian fossa on the lateral pelvic wall. Following pregnancy the broad ligament is lax, and the ovaries may hang down in the rectouterine pouch (pouch of Douglas). In this position the venous return may be impeded and cause congestion and tenderness of the ovary. An ovary situated in this position may be palpated through the posterior fornix of the vagina and may be responsible for dyspareunia.

Cysts of the Ovary

Follicular cysts are very common and originate in unruptured graafian follicles; they rarely exceed 1.5 cm in diameter. They are located in the cortex of the ovary just beneath the peritoneum. *Luteal cysts* are formed in the corpus luteum during the early stages of its existence. Fluid is retained, and the corpus luteum cannot become fibrosed. Luteal cysts rarely exceed 3 cm in diameter. Cysts associated with ovarian tumors also occur, such as *cystadenomas* or *cystadenocarcinomas*.

Torsion of an Ovarian Cyst.—Torsion of an ovarian cyst gives rise to severe pain in the hypogastrium, which, when occurring on the right side, may be confused with acute appendicitis.

Clinical Anatomy of the Uterine Tubes

Pelvic Inflammatory Disease (Endometritis, Salpingitis, Parametritis)

This is a major health problem in the United States. The pathogenic organism(s) enter the body through sexual contact and come to reside in (1) the glands of the uterine cervix, (2) the periurethral glands, (3) the greater vestibular glands (Bartholin's glands), and (4) the anal mucosal crypts. The pathogenic organisms ascend through the uterus and enter the uterine tubes. Salpingitis may follow, with leakage of pus into the peritoneal cavity, causing pelvic peritonitis. A pelvic abscess usually follows or the infection spreads further, causing a general peritonitis.

Following successful treatment of the acute infection, the patient often enters a chronic phase in which the pelvic viscera are stuck together by adhesions. Permanent damage to the uterine tubes commonly results in sterility or ectopic pregnancy.

Ectopic Pregnancy

Ectopic pregnancy is the implantation and growth of the fertilized ovum outside the uterine cavity (Fig 13–31). It is estimated that 1 in 100 pregnancies is an ectopic pregnancy and is responsible for about 5% to 10% of maternal deaths in the United States each year.

The uterine tube is the most common site of ectopic implantation, which can also take place, although rarely, on the ovary or in the peritoneal cavity.

The causes of tubal ectopic pregnancy include any condition that damages the lining of the uterine

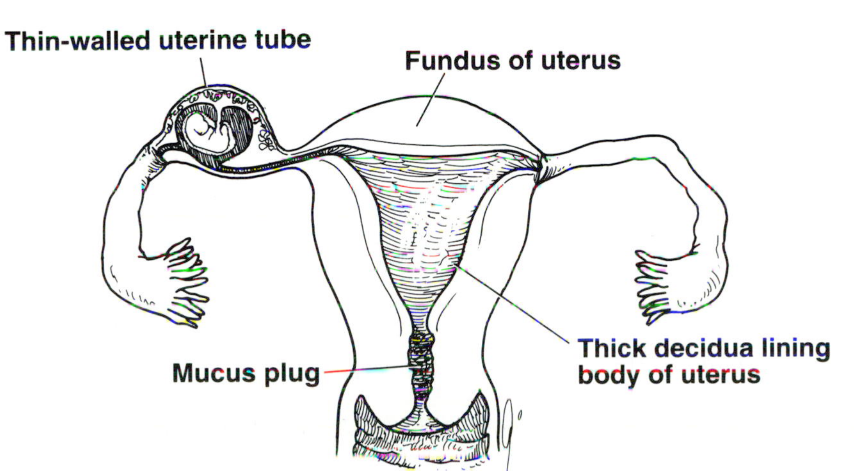

FIG 13–31.
An ectopic pregnancy located where the infundibulum of the uterine tube narrows down to join the isthmus. Note the thin tubal wall as compared with the thick decidua that lines the body of the uterus.

tube or results in the kinking of its lumen, leading to its occlusion, including chronic salpingitis, peritubular adhesions, tubal sterilization, and the use of an intrauterine contraceptive device.

The most common site of tubal pregnancy is where the infundibulum narrows down to join the isthmus (see Fig 13–17). The blastocyst buries itself in the thin tubal wall as the result of the destructive action of the trophoblast. Early abortion then takes place with or without rupture of the vascular uterine tube and can be accompanied by considerable hemorrhage. This usually occurs between the 6th and 10th week of pregnancy.

The congested uterine tube, which receives its arterial supply from enlarged uterine and ovarian arteries, may bleed catastrophically on rupture. The blood pours down into the rectouterine pouch (pouch of Douglas) and/or into the uterovesical pouch. The blood may quickly ascend into the general peritoneal cavity, giving rise to severe abdominal pain, tenderness, and guarding. Irritation of the subdiaphragmatic peritoneum (supplied by the phrenic nerve C3 through C5) may give rise to referred pain to the shoulder skin (supraclavicular nerves C3 and C4).

Clinical Anatomy of the Uterus and Vagina

Vaginal Examination

The following structures may be palpated through the vaginal walls from above downward.

Anteriorly.— Bladder and the urethra.

Posteriorly.— Loops of ileum and sigmoid colon in the rectouterine pouch (pouch of Douglas), the rectal ampulla, and the perineal body.

Laterally.— Ureters, the pelvic fascia and the anterior fibers of the levatores ani muscles, and the urogenital diaphragm (see Fig 13–25).

Prolapse of the Uterus

The tone of the levatores ani muscles is very important in the support of the uterus. The transverse cervical, pubocervical, and the sacrocervical ligaments are important in positioning the cervix within the pelvic cavity. Damage to these structures during childbirth, or general poor body muscular tone, may result in downward displacement of the uterus, or *uterine prolapse.* It most commonly reveals itself after menopause, when the visceral pelvic fascia tends to atrophy along with the pelvic organs. In advanced cases, the cervix descends the length of the vagina and may protrude through the orifice.

Because of the attachment of the cervix to the vaginal vault, it follows that prolapse of the uterus is always accompanied by some prolapse of the vagina.

Prolapse of the Vagina

As noted above, prolapse of the uterus is necessarily associated with some degree of sagging of the vaginal walls. However, if the supports of the bladder, urethra, or anterior rectal wall are damaged in childbirth, prolapse of the vaginal walls can occur with the uterus remaining in its correct position. A *cystocele* is a condition in which the bladder sags downward and results in a bulging of the anterior wall of the vagina. A *rectocele* occurs when the ampulla of the rectum sags against the posterior wall of the vagina.

Endometriosis

Endometriosis is the presence of endometrial glands outside the uterus. The exact pathogenesis of the condition is not known, but the following explanations have been suggested: (1) abnormal differentiation of peritoneal epithelium, (2) regurgitation of endometrial fragments through the uterine tubes during menstruation followed by implantation of the fragments onto the peritoneal walls of the pelvis, or possibly (3) hematogenous or lymphatic spread of endometrial cells. The clinical signs and symptoms are the result of intrapelvic bleeding from the ectopic nests of endometrial cells with each menstrual cycle. The common sites include the ovary, the broad ligament, the pouch of Douglas, and old laparotomy scars.

The bleeding in endometriosis causes extreme dysmenorrhea, and the presence of blood in the pelvis causes adhesions. Eventually the pelvic viscera, including the uterine tubes and ovaries, become embedded in a mass of fibrous tissue, leading to sterility.

Dysfunctional Uterine Bleeding

The normal buildup of the endometrium is carefully controlled by the plasma levels of estrogens and progesterone, and the concentrations of these hormones must rise and fall regularly with each menstrual cycle. A failure of this hormonal control can lead to excessive buildup of the endometrium, atrophy of the endometrium, or premature shedding of the endometrium. Moreover, the cyclical changes may become very irregular and may be accompanied by bleeding between menstrual periods.

In anovulatory menstrual cycles, which are common at the menarche and just before the menopause, the graafian follicle fails to liberate the ovum, and the corpus luteum does not form. Under these circumstances, the endometrium proliferates excessively under the influence of the estrogens alone; finally, when menstruation occurs, the hyperplastic endometrium is cast off, usually accompanied by excessive bleeding. Other endocrine disorders, such as pituitary, thyroid, and adrenal disease, may lead to ovarian dysfunction.

Low production of progesterone by the corpus luteum may lead to inadequate development of the endometrium during the secretory phase. Infertility and excessive menstrual bleeding may be the presenting symptoms and signs in these patients.

Failure of the corpus luteum to degenerate may lead to a prolonged secretory phase of the cycle; when menstruation finally occurs, it is accompanied by excessive bleeding.

Tumors of the Uterus

The two most common uterine tumors are the leiomyomas (fibroids) and the adenocarcinomas.

The *leimyomas* are very common benign tumors arising in the smooth muscle of the myometrium. They are thought to be caused by excessive stimulation of the smooth muscle by estrogens. They may grow rapidly during pregnancy and may complicate labor. Such tumors frequently are responsible for excessive blood loss during the menstrual period.

Adenocarcinoma of the endometrium is a common form of cancer in older women. The cause is thought to be excessive stimulation of the endometrium by estrogens, which are produced outside the ovary. The tumor arises from the columnar cells of the endometrium. The common symptom is irregular vaginal bleeding associated with excessive mucus production.

Inflammation of the Cervix of the Uterus

The cervical canal is about 1 in. (2.5 cm) long; opening into it are numerous long, branching glands. The canal opens at the external os into the vagina, which normally contains many different kinds of bacteria. The trauma of childbirth, the passing of instruments through the cervix in gynecological procedures, and intercourse are some of the factors predisposing to the entrance of pathogenic organisms into the cervix and the development of an acute or chronic cervicitis. Excessive mucoid discharge from the vagina (leukorrhea) is a common sign. Epithelial thickening, inflammation and edema of the underlying connective tissue, and the blockage of the ducts of the mucous glands are common findings. Chronic cervicitis with leukorrhea is a cause of sterility; if it is left untreated, epithelial metaplasia may occur, which may progress to cervical carcinoma.

Carcinoma of the Cervix

Carcinoma of the cervix is a very common condition. It is rare, however, in women who have never had sexual intercourse and in women who have not had children. Early coitus and multiple sexual partners increase the likelihood of this disease, as does herpes virus infection. Chronic cervicitis is believed to be an important predisposing factor.

The great majority of carcinomas of the cervix arise from the squamous epithelium near the squamocolumnar junction at the external os. The remainder are adenocarcinomas that arise from the columnar epithelial cells. It is now generally accepted that carcinoma of the cervix is preceded by dysplasia and an anaplasia of the epithelium, which is a precursor of carcinoma in situ; this later becomes an invasive carcinoma.

Anatomy of Culdocentesis

Because the peritoneal cavity (pouch of Douglas) is very close to the posterior fornix of the vagina (Fig 13–32), it is possible to identify blood or pus in the peritoneal cavity by the passage of a needle through the posterior fornix. This procedure is most commonly used in the diagnosis of a ruptured ectopic pregnancy or a ruptured ovarian cyst.

The procedure is as follows (see Fig 13–32).

1. The patient is in the lithotomy position.
2. By wiggling the cervix in and out with the tenaculum, it is possible to identify the reflection of the vaginal mucous membrane onto the cervix in the posterior fornix.
3. An 18-gauge needle attached to a syringe is inserted in the midline in the posterior fornix. The direction of the needle should be upward parallel with the posterior wall of the uterus. The needle should not be inserted further than 2 cm while suction is applied with the syringe.
4. The patient should sit up slightly to obtain maximum pooling in the pouch of Douglas.

Anatomical Structures Through Which the Needle Passes.—The needle passes through the (1) mucous membrane of the vagina, (2) muscular coat of the va-

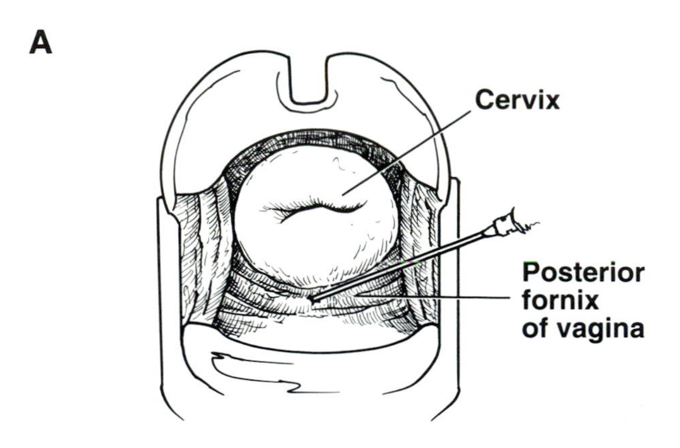

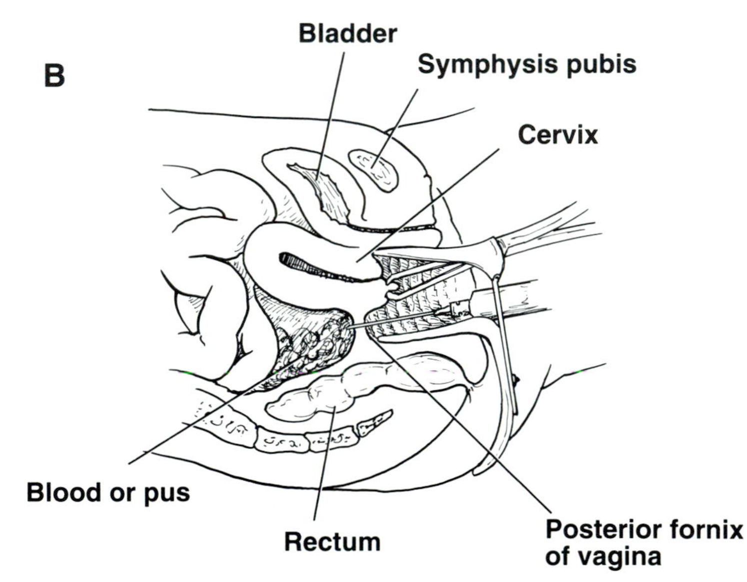

FIG 13–32.
Culdocentesis. **A,** the needle being inserted into the mucous membrane of the posterior fornix of the vagina in the midline. **B,** the needle entering the pouch of Douglas. The direction of the needle is upward, parallel with the posterior wall of the uterus.

gina, (3) connective tissue coat, (4) visceral layer of pelvic fascia, and (5) visceral layer of peritoneum.

Anatomy of the Complications of Culdocentesis.—These complications are as follows.

1. Loops of ileum and the sigmoid colon, structures that are normally present within the pouch of Douglas, could be impaled by the needle. However, the presence of blood or pus within the pouch tends to deflect the viscera superiorly.
2. Occasionally, when the uterus is somewhat retroflexed, the needle may enter the posterior wall of the body of the uterus.

Clinical Anatomy of Pregnancy

Anatomy of Paracervical Nerve Block

This procedure is fully discussed in Chapter 19, p. 832.

Bleeding in Late Pregnancy

The common causes of substantial vaginal bleeding in the third trimester are placenta previa and placental abruption.

Placenta Previa.—Placenta previa occurs once in about every 200 pregnancies. It is more common in multiparous women and in those who have had surgery on their lower uterine segments. Normally the placenta is situated in the upper half of the uterus. Should implantation occur in the lower half of the body of the uterus, the condition is called placenta previa (Fig 13–33).

Three types of placenta previa may be recognized—a central placenta previa, in which the entire internal os is covered by placental tissue; a marginal placenta previa, when the edge of the placenta is encroaching on the internal os; and a low-lying placenta previa, when the placenta lies low down in the uterus lateral to the internal os. Severe, painless

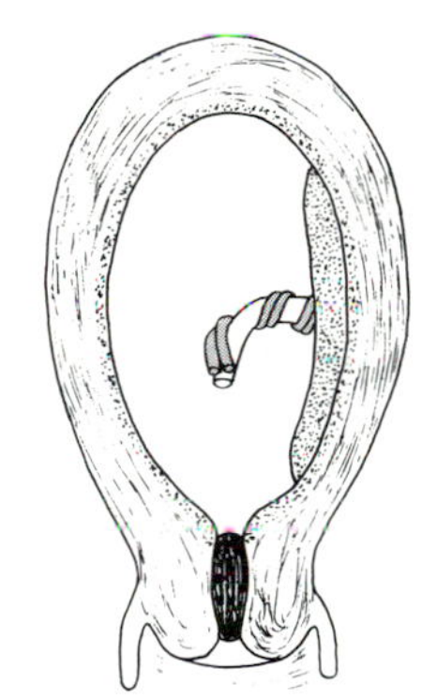

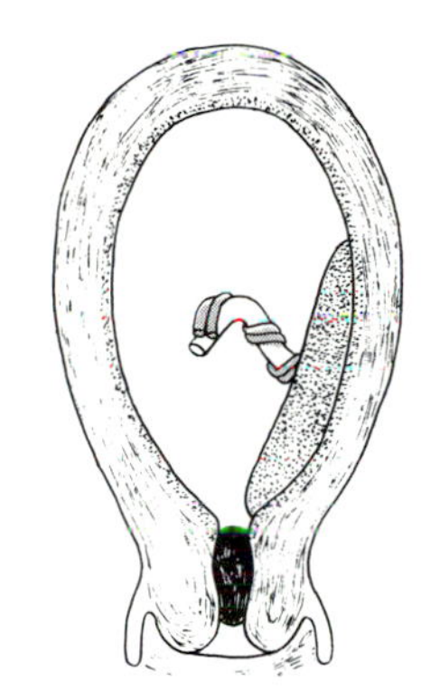

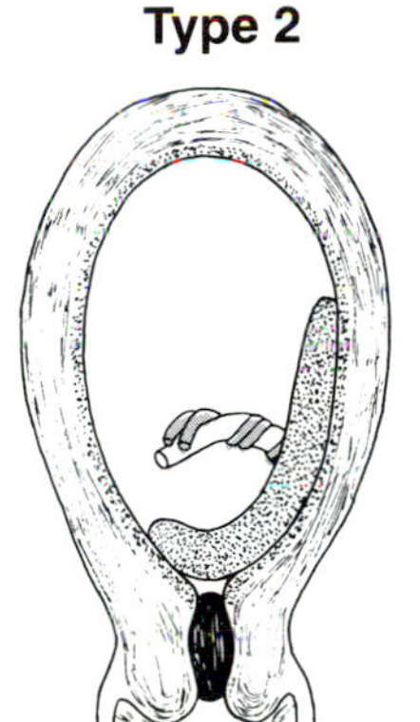

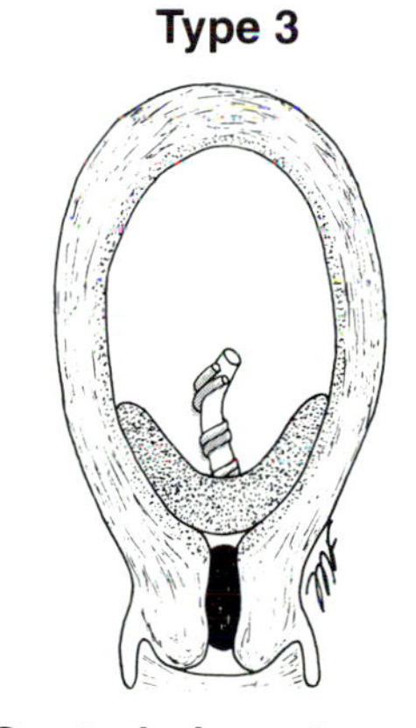

FIG 13–33.
The different types of placenta previa.

hemorrhage occurring from the 28th week onward is the clinical sign of placenta previa and is caused by expansion of the lower half of the uterine wall at this time and by its tearing away from the placenta.

Placental Abruption.—This condition is a premature separation of the placenta where normal implantation has occurred. It occurs in about 1% of pregnancies. It is more common in multiparous women and in women with hypertension in pregnancy. As the placenta separates, hemorrhage occurs and the blood clot dissects the fetal membranes away from the uterine wall. The blood usually escapes through the cervix or ruptures into the amniotic cavity. The blood irritates the myometrium and uterine muscle tone is increased, which results in contractions. The placental circulation is compromised by the placental separation and the increased pressure on the placenta by the increased uterine tone.

Varicosed Veins and Hemorrhoids in Pregnancy

These are common conditions in pregnancy. The following factors probably contribute to their cause: (1) increased progesterone levels in the blood, leading to relaxation of the smooth muscle in the walls of the veins and venous dilatation, and (2) pressure of the gravid uterus on the inferior vena cava and the inferior mesenteric vein, impairing venous return.

Emergency Cesarean Section

An emergency cesarean section is rarely performed in the emergency department. However, the emergency physician may need to perform this surgery in cases where the mother may die following a severe traumatic incident. Following maternal death, placental circulation ceases, and the child must be delivered within 10 minutes; after a delay of more than 20 minutes, neonatal survival is rare.

Technique.—The following steps are taken.

1. The bladder is emptied and an indwelling catheter is left in position. This allows the empty bladder to sink down away from the operating field.

2. A midline skin incision is made that extends from just below the umbilicus to just above the symphysis pubis. The following structures are then incised: (1) superficial fascia, fatty layer and the membranous layer, (2) deep fascia (thin layer), (3) linea alba, (4) fascia transversalis, (5) extraperitoneal fatty layer, and (6) parietal peritoneum.

In order to avoid the possibility of damaging loops of small intestine or the greater omentum that might be lying beneath the parietal peritoneum, a fold of peritoneum is raised between two hemostats; an incision is then made between the hemostats.

3. The bladder is identified, and a cut is made in the floor of the uterovesical pouch. The bladder is then separated from the lower uterine segment and depressed downward into the pelvis.

4. The uterus is palpated to identify the presenting part of the fetus.

5. A transverse incision about 1 in. (2.5 cm) long is made into the exposed lower segment of the uterus. Care has to be exercised so that the uterine wall is not immediately penetrated and the fetus injured.

6. When the uterine cavity is entered, the amniotic cavity is opened and amniotic fluid spurts. The uterine incision is then enlarged sufficiently to deliver the head and trunk of the fetus. Where possible, the large tributaries and branches of the uterine vessels in the myometrial wall are avoided. Great care has to be taken to avoid the large uterine arteries that course along the lateral margin of the uterus.

7. Once the fetus is delivered, the umbilical cord is clamped and divided.

If the mother is dead, the next steps in the procedure are irrelevent.

8. The contracting uterus will cause the placenta to bulge through the uterine incision. The placenta and fetal membranes are then delivered.

9. The uterine incision is closed with a full-thickness continuous suture. The peritoneum over the bladder and lower uterine segment is then repaired to restore the integrity of the uterovesical pouch. Finally, the abdominal wall incision is closed in layers.

Suction Curretage

Suction curretage is a common emergency department procedure. It is used to evacuate the uterus during the first 3 months of pregnancy. The procedure is as follows.

1. The vagina is prepared in the usual way and a speculum is introduced.

2. The anterior lip of the cervix is secured with a tenaculum, and the tip of the suction catheter is introduced through the cervical canal into the body of the uterus. The tip of the catheter is gently advanced following the direction of the uterine cavity until the fundus is reached. The suction begins.

3. The catheter is slowly withdrawn, and the fetal tissue is removed. The procedure is repeated until no further tissue is removed.

ANATOMICAL-CLINICAL PROBLEMS

1. A 23-year-old man involved in a barroom brawl was seen in the emergency department. He was found to have a blood-stained tear on the seat of his trousers and lacerations of the anal margin. During the fight he was knocked down and fell in the sitting position on the leg of an upturned bar stool. Anatomically, explain how he could develop the signs and symptoms of peritonitis. What parts of the rectum are related to the peritoneal cavity?

2. A 47-year-old woman was seen in the emergency department complaining of agonizing pain in the "rectum," which occurred on defecation. She had first noticed the pain 2 days previously when she tried to defecate. The pain lasted for about an hour, then passed off, only to return with the next bowel movement. On questioning the patient, the physician learned that the woman suffered from chronic constipation. Her stools were sometimes streaked with blood. On examination, the anus was found to be tightly closed. An attempt to examine the anal canal digitally failed because of the severe pain it caused the patient. By gently everting the anal margin, however, the lower edge of a linear tear in the posterior wall of the anal canal could be seen; a small tag of skin projected from the lower end of the tear. A diagnosis of anal fissure was made. In anatomical terms, explain the possible causes of anal fissure. What is responsible for the small tag of skin commonly found at the lower end of the fissure? Why is this condition so painful, when, for example, a carcinoma confined to the upper half of the anal canal is painless? Why was the anus tightly closed?

3. A 62-year-old man was seen in the emergency department because he had noticed that his "bowel" was protruding from his anus after defecation, which was causing him considerable discomfort. On questioning, the patient stated that for the past 3 years he had frequently passed blood-stained stools. The blood was bright red but small in amount, and he did not think it was serious enough to consult his physician. Recently, he had noticed that he had intense perianal irritation, especially at night. On physical examination, the perianal skin was noted to be red, moist, and excoriated. Digital examination of the anal canal revealed nothing abnormal. Proctoscopic examination showed that the mucous membrane at the level of the anal valves tended to bulge downward in three areas when the patient strained. The swollen mucous membrane contained large congested veins beneath the surface. What is the diagnosis? With the patient in the lithotomy position, where would you expect the bulges of the mucous membrane to occur?

4. A 21-year-old man was seen in the emergency room complaining that he had something in his "rectum" that he could not remove. Digital examination revealed the presence of a hard, smooth object about 3 in. (8 cm) above the anus that filled the distended lumen of the rectum. Careful examination with a proctoscope showed the presence of a light bulb. What anatomical structures are present in the rectum that might resist the patient's efforts to pass the bulb?

5. An over-enthusiastic physician decided to explore and open up a perianal fistula. Ten days after the operation, he was surprised to learn that the patient was suffering from fecal incontinence. What anatomical structures must be preserved in this region in order to prevent fecal incontinence?

6. A 32-year-old woman was seen in the emergency department complaining of a painful swelling in the region of the anus. On examination, a hot, red, tender swelling was found on the left side of the anal margin. A diagnosis of ischiorectal abscess was made. Can ischiorectal abscess be a complication of anal fissure? Are there any important structures in the ischiorectal fossa that might be damaged by a surgical incision of an abscess in this region? How would you classify the different types of perianal abscess?

7. A patient was seen in the emergency department with a subcutaneous perianal abscess. Prior to incising the abscess, the physician had to anesthetize the skin. What is the nerve supply to the perianal skin?

8. In anatomical terms describe the process of sigmoidoscopy. How long is the anal canal? What is the approximate distance of the rectosigmoid junction from the anus? At what point during the procedure is the lower bowel most likely to be ruptured?

9. A 25-year-old woman was seen in the emergency department complaining of a swelling in the genital region. On examination, a tense cystic swelling was found beneath the posterior two thirds of the right labium majus and minus. What is the likely diagnosis? What is the possible cause? Is infection of the cyst a common complication? What is the lymphatic drainage of the area?

10. A 25-year-old woman was seen in the emergency department complaining of severe pain in the right iliac region. Just prior to admission she had fainted. On physical examination, her abdominal wall was extremely tender on palpation in the right iliac region, and some rigidity and guarding of the lower abdominal muscles were noticed. A vaginal examination revealed a fairly firm cervix with a closed external os. A tender "doughlike mass" could be felt through the posterior fornix. The patient had missed her last period. A diagnosis of a ruptured ectopic pregnancy was made. Using your knowledge of anatomy, (1) name the sites where tubal pregnancies commonly occur; (2) what is the relationship of the uterine tube to the peritoneal cavity? (3) why does an ectopic pregnancy rupture? and (4) why may a woman with an ectopic pregnancy often have a history of vaginal bleeding?

11. Describe the structures that can normally be felt on a vaginal examination.

12. A 56-year-old woman was seen in the emergency department complaining of a bearing-down feeling in the pelvis and of a low backache. On vaginal examination, the external os of the cervix was found to be located just within the vaginal orifice. A diagnosis of uterine prolapse was made. What are the main supports of the uterus?

13. In anatomical terms explain the procedure of culdocentesis.

14. Briefly describe the formation of the placenta. What is a placenta previa?

15. What part does the pelvic floor play in the rotation of the fetal head during a normal delivery?

16. What is believed to be the cause of pain in labor? Describe the pathways taken by the afferent pain fibers in labor and explain the distribution of the areas of referred pain.

ANSWERS

1. This patient had impaled his rectum on the leg of the upturned bar stool. At operation, a laceration of the anterior wall of the upper end of the rectum was found. The pelvic peritoneum was contaminated with rectal contents. The upper third of the rectum is covered on the anterior and lateral surfaces by peritoneum; the middle third is covered on the anterior surface only by peritoneum; and the lower third is devoid of a peritoneal covering.

2. The edge of a hard fecal mass may have caught one of the anal valves and torn it downward as it descended; the skin tag represented the remains of the valve. The mucous membrane lining the anterior and posterior walls of the anal canal at this level is poorly supported by the superficial part of the external anal sphincter. This explains why fissures are commonly found on the anterior and posterior anal walls. The mucous membrane of the lower half of the anal canal is innervated by the inferior rectal nerve, a branch of the pudendal nerve, and is very sensitive to pain. The mucous membrane of the upper half of the canal is supplied by autonomic afferent fibers and is sensitive only to stretch. The external anal sphincter is supplied by the inferior rectal nerve and is reflexly in a state of spasm due to the painful afferent impulses arising from the fissure.

3. This patient had internal hemorrhoids, i.e., varicosities of the tributaries of the superior rectal vein. With the patient in the lithotomy position, the distended tributaries are usually situated beneath the mucous membrane at the 3-, 7-, and 11-o'clock positions. In this patient the varicosities had become pedunculated and remained prolapsed outside the anus after defecation (third-degree internal hemorrhoids). The closure of the anal sphincter on the pedicles of the hemorrhoids caused congestion of the mucous membrane and the production of excessive mucus, which was responsible for the pruritus ani. Abrasion of the hemorrhoids by the feces during defecation was responsible for the bleeding.

4. The transverse mucosal folds of the rectal wall, the posterior pouch of the rectal ampulla, and spasm of the anal sphincters often prevent patients from ridding themselves of the objects.

In some cases it may be necessary to block the inferior rectal nerves to produce relaxation of the external anal sphincter. When it is impossible to deliver the foreign body from below, the abdomen has to be opened under a general anesthetic, so that the object can be pushed down from above.

5. The anorectal ring is composed of the puborectalis part of the levator ani muscle, the internal anal sphincter, and the deep part of the external anal sphincter; the muscle fibers blend together at the anorectal junction. Destruction of the anorectal ring results in fecal incontinence.

6. The infection in the base of an anal fissure can track laterally through the external anal sphincter to enter the ischiorectal fossa. The fat in the fossa is poorly supplied with blood and very prone to abscess formation. The only structures of importance that cross the ischiorectal fossa are the inferior rectal vessels and nerve. The different types of perianal abscess are fully discussed on p. 524.

7. The perianal skin is innervated from branches of the inferior rectal nerve, a branch of the pudendal nerve (S2 through S4) and from branches of the fifth sacral and first coccygeal nerves. The anatomy of the procedure for anesthetizing the perianal skin is described on p. 527.

8. The anatomy of sigmoidoscopy is fully described on p. 526. The anal canal is approximately 1½ in. (4 cm) long. The rectosigmoid junction is approximately 6½ in. (16.25 cm) from the anus. Rupture of the lower bowel with a sigmoidoscope is most likely to take place at the rectosigmoid junction where the bowel bends forward and to the left.

9. This patient had a cyst of the right greater vestibular gland (Bartholin's cyst). Blockage of the duct secondary to infection is a common cause. Infection of the cyst often occurs, forming a painful abscess. The lymphatic drainage of this area is into the medial group of superficial inguinal lymph nodes.

10. (1) Tubal pregnancies commonly occur where the infundibulum narrows to join the isthmus. (2) Each uterine tube is situated in the upper free margin of the broad ligament of peritoneum. (3) An ectopic tubal pregnancy almost invariably results in rupture of the tube with severe intraperitoneal hemorrhage. This happens because there is no decidua formation in the uterine tube, and the eroding action of the trophoblast quickly destroys the wall of the tube. (4) Once the tubal pregnancy dies due to insufficient nourishment, the decidual lining of the uterus begins to be shed because of the lack of hormonal support; this causes vaginal bleeding.

11. The following structures can normally be felt on a vaginal examination from above downward: anteriorly—the bladder and the urethra; posteriorly—loops of ileum and sigmoid colon in the rectouterine

pouch (pouch of Douglas), the rectal ampulla, and the perineal body; and laterally—the ureters, pelvic fascia and anterior fibers of the levator ani muscles, and the urogenital diaphragm.

12. The uterus is mainly supported by the tone of the levator ani muscles and the ligaments of the visceral pelvic fascia, namely, the transverse cervical, sacrocervical, and pubocervical ligaments.

13. The anatomy of the procedure of culdocentesis is fully described on p. 531.

14. The formation of the placenta is briefly described on p. 513. Normally the placenta is formed in the upper half of the uterus. Should implantation occur in the lower half of the uterus, the condition is called placenta previa. For details, see p. 532.

15. The pelvic floor is a gutter-shaped sheet of muscle formed by the levatores ani and coccygeus muscles and their covering fasciae. During the second stage of labor, when the baby's head reaches the pelvic floor, the gutter shape of the floor tends to cause the head to rotate, so that its long axis comes to lie in the anteroposterior position. The occipital part of the head then moves downward and forward along the gutter until it lies under the pubic arch. Finally, the head passes through the lower part of the birth canal to the exterior.

16. The cause of labor pains is fully discussed on p. 515. The path taken by the afferent pain fibers from the uterus is described on p. 515. The areas of the body to which labor pains are referred are also described on p. 515.

14

The Major Blood Vessels of the Abdomen, the Pelvis, and the Lower Extremity

Within the abdomen and pelvis lie the aorta and its branches, the inferior vena cava and its tributaries, and the important portal vein. Because most major blood vessels of the abdomen are retroperitoneal, blunt injuries may not result in an immediate fatal intraperitoneal hemorrhage. Because bleeding may initially be confined to the retroperitoneal space, making the diagnosis may be difficult or delayed. Penetrating injuries, usually, give the blood access to the peritoneal cavity, and the diagnosis of intraperitoneal hemorrhage is readily made.

In the lower extremity, acute limb-threatening ischemia requires a timely diagnosis so that treatment may be quickly implemented and the limb saved. A knowledge of the collateral circulation in these circumstances is imperative. The venous system of the lower limb is the source of many emergency problems. The emergency physician is aided by a knowledge of venous anatomy in diagnosing phlebitis and obtaining lower extremity venous access.

The purpose of this chapter is to review the major arterial and venous vessels of the abdomen and the pelvis and the lower limb relative to emergency situations.

BASIC ANATOMY

Basic Anatomy of the Abdominal Aorta

Location

The aorta enters the abdomen through the aortic opening of the diaphragm in front of the 12th thoracic vertebra (Fig 14–1). It descends behind the peritoneum on the anterior surfaces of the bodies of the lumbar vertebrae. At the level of the fourth lumbar vertebra it divides into the two common iliac arteries.

Relations.— On the right side of the aorta lie the inferior vena cava, the cisterna chyli, and the beginning of the azygos vein. On its left side lies the left

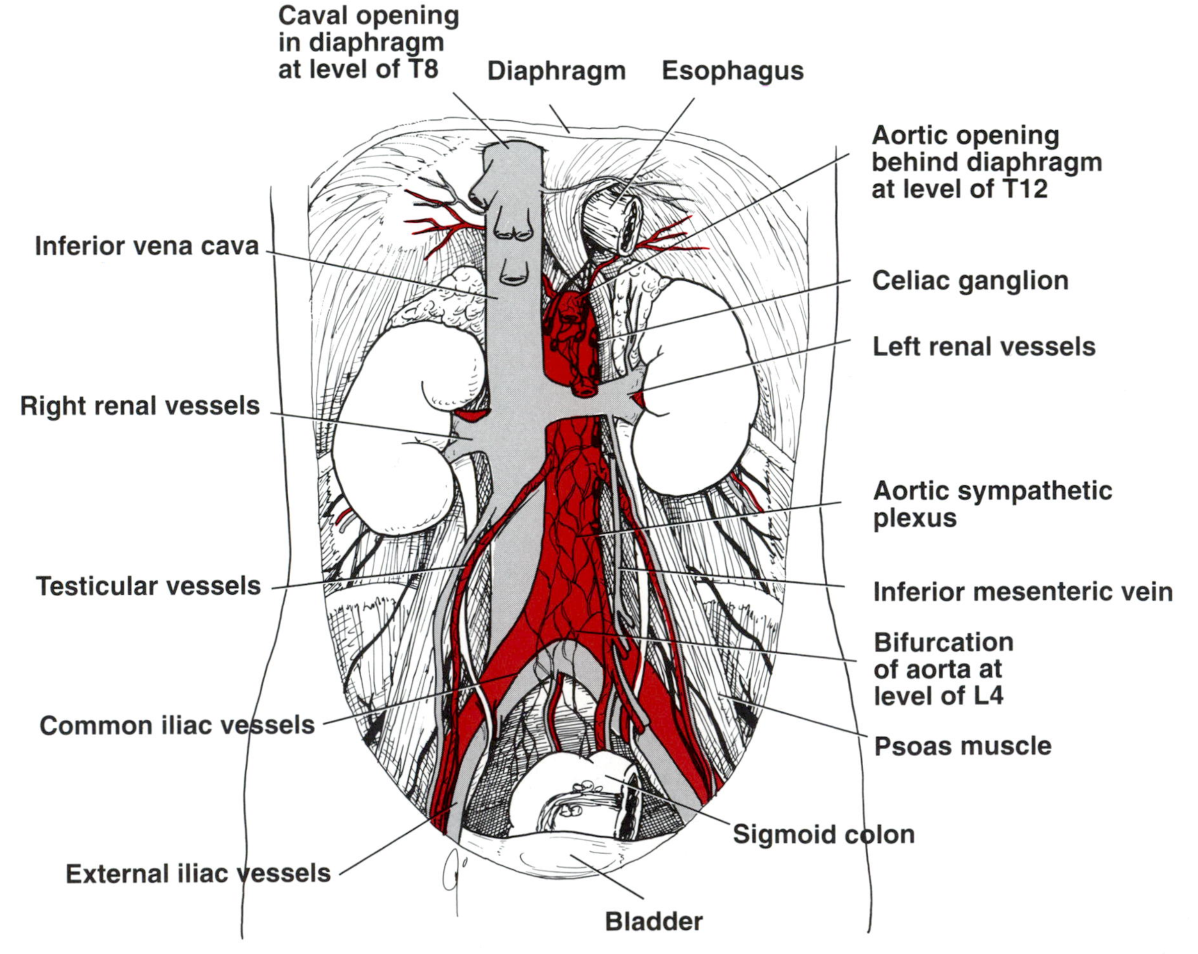

FIG 14–1.
Posterior abdominal wall showing the aorta and inferior vena cava.

sympathetic trunk. On the anterior surface, the aorta is related to the stomach, celiac plexus, pancreas, splenic vein, left renal vein, third part of the duodenum, coils of small intestine, and peritoneum.

Branches (Fig 14–2).—These include the following.

1. Three anterior visceral branches—the celiac artery, superior mesenteric artery, and inferior mesenteric artery.
2. Three lateral branches—the suprarenal artery, renal artery, and testicular or ovarian artery.
3. Five lateral abdominal wall branches—the inferior phrenic artery and four lumbar arteries.
4. Three terminal branches—the two common iliac arteries and the median sacral artery.

Celiac Artery.—The celiac artery or trunk is very short and arises from the commencement of the abdominal aorta at the level of the 12th thoracic vertebra (see Fig 14–2). It supplies the gut from the lower third of the esophagus down as far as halfway down the second part of the duodenum (the foregut). The celiac artery is surrounded by the celiac plexus of nerves and lies behind the lesser sac of peritoneum. It has three terminal branches—the *left gastric, splenic,* and *hepatic arteries* (Fig 14–3).

Superior Mesenteric Artery.—The superior mesenteric artery arises from the front of the abdominal aorta just below the celiac artery at the level of the first lumbar vertebra (see Fig 14–2). It runs downward and to the right behind the neck of the pancreas and in front of the third part of the duodenum. It continues downward to the right between the layers of the mesentery of the small intestine and ends by anastomosing with the ileal branch of its own ileocolic branch (see Fig 14–4).

The superior mesenteric artery supplies the gut from halfway along the duodenum to just proximal to the left colic flexure. It gives rise to the *inferior pancreaticoduodenal artery, middle colic artery, right colic artery, ileocolic artery,* and *jejunal* and *ileal arteries.*

Inferior Mesenteric Artery.—The inferior mesenteric artery arises from the abdominal aorta about 1½ in. (3.8 cm) above its birfurcation at the level of the third lumbar vertebra (see Fig 14–2). The artery runs downward and to the left and crosses the left com-

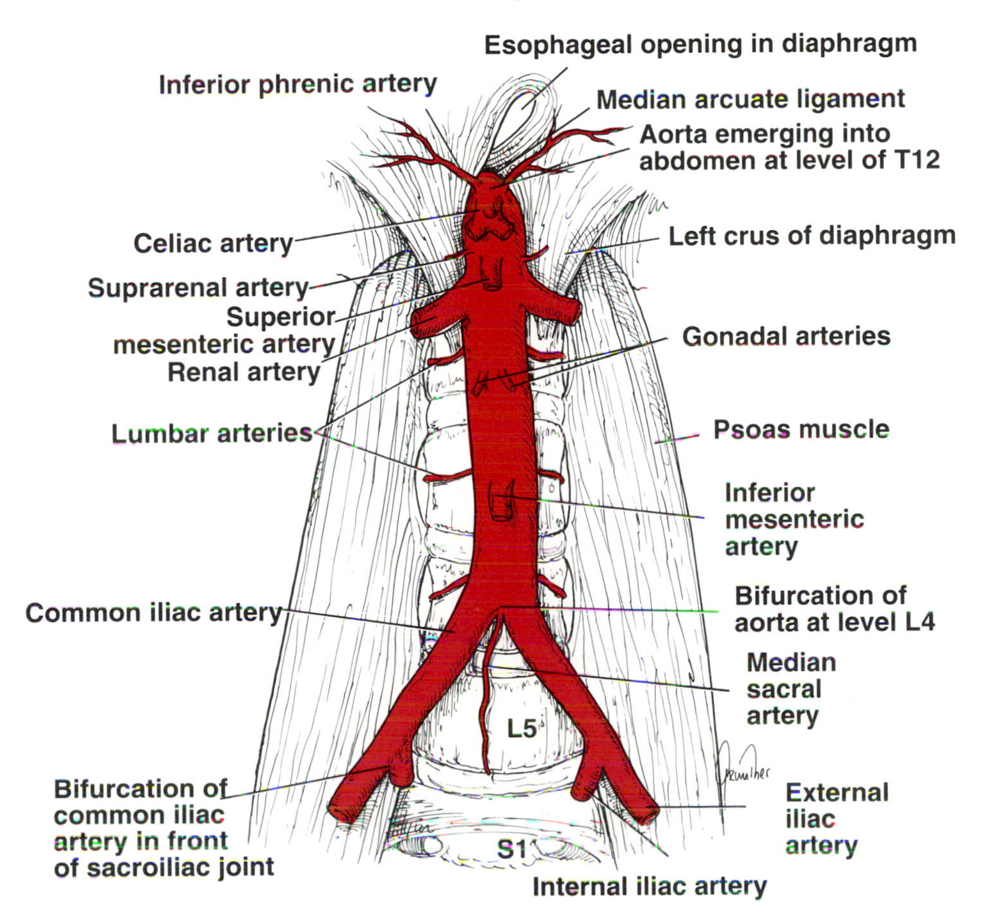

FIG 14–2.
The abdominal part of the aorta and its branches.

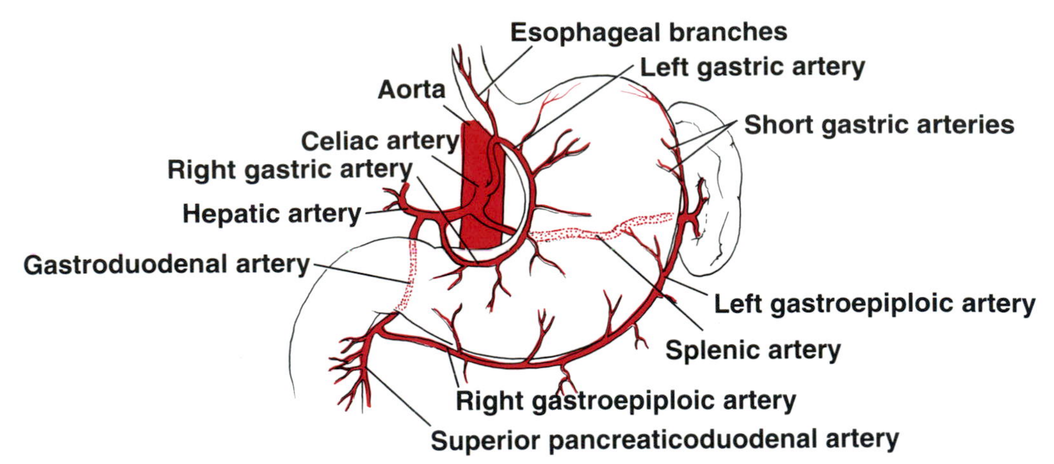

FIG 14–3.
The celiac artery and its branches.

mon iliac artery, where it changes its name and becomes the superior rectal artery (Fig 14–4).

The inferior mesenteric artery supplies the large intestine from the distal third of the transverse colon to halfway down the anal canal (hindgut). It gives rise to the *superior left colic artery* (left colic artery), the two or three *inferior left colic arteries* (sigmoid arteries), and the *superior rectal artery*.

The surface anatomy location of the celiac, superior mesenteric, and inferior mesenteric arteries will be discussed on p. 556.

Common Iliac Arteries.—The right and left common iliac arteries are the terminal branches of the abdominal aorta (see Fig 14–2). They arise at the level of the fourth lumbar vertebra and run downward and laterally to end opposite the sacroiliac joint by dividing into external and internal iliac ar-

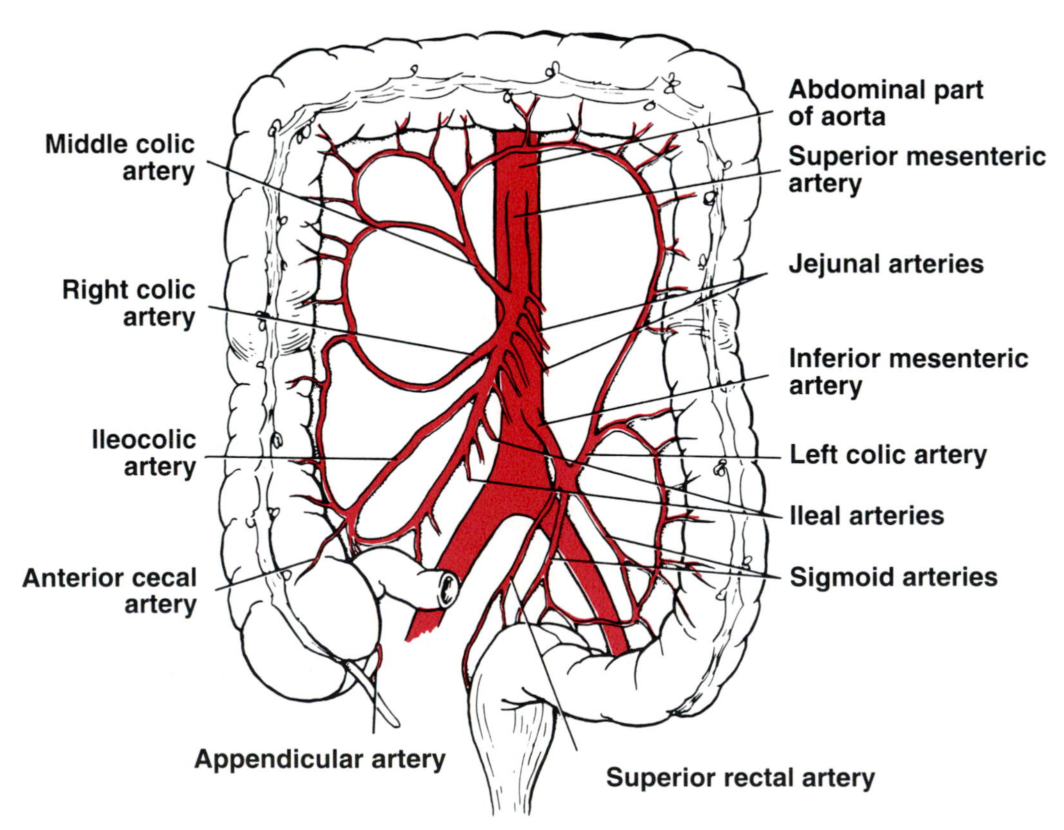

FIG 14–4.
The superior and inferior mesenteric arteries and their branches.

teries. At the bifurcation, the common iliac artery on each side is crossed anteriorly by the ureter (see Fig 14–1).

External Iliac Artery.—The external iliac artery runs along the medial border of the psoas muscle, following the pelvic brim, and gives off the inferior epigastric and deep circumflex iliac branches (Figs 14–1 and 14–5). The artery enters the thigh by passing under the inguinal ligament to become the femoral artery. The *inferior epigastric artery* arises just above the inguinal ligament. It passes upward and medially along the medial margin of the deep inguinal ring (see Fig 11–10) and enters the rectus sheath behind the rectus abdominis muscle. The *deep circumflex iliac artery* arises close to the inferior epigastric artery (see Fig 14–5). It ascends laterally to the anterior superior iliac spine and the iliac crest, supplying the muscles of the anterior abdominal wall.

Internal Iliac Artery.—The internal iliac artery (see Fig 14–5) passes down into the pelvis to the upper margin of the greater sciatic foramen, where it divides into anterior and posterior divisions. The branches of these divisions supply the pelvic viscera, perineum, buttock, and sacral canal.

Branches of the Anterior Division of the Internal Iliac Artery (see Fig 14–5).—These are as follows.

1. *Superior vesical artery* supplies the upper portion of the bladder.
2. *Inferior vesical artery* supplies the base of the bladder, and the prostate and the seminal vesicle in males.
3. *Middle rectal artery* supplies the muscle of the lower rectum and anastomoses with the superior and inferior rectal arteries.
4. *Uterine artery* runs across the floor of the pelvis and crosses the ureter. It passes above the lateral fornix of the vagina to reach the uterus. Here it ascends between the layers of the broad ligament along the lateral margin of the uterus (see Fig 13–10). It also supplies the uterine tube and vagina.
5. *Vaginal artery* usually takes the place of the inferior vesical artery present in males. It supplies the vagina and base of the bladder.
6. *Obturator artery* runs forward along the lateral wall of the pelvis with the obturator nerve. It leaves the pelvis through the obturator canal in the upper part of the obturator membrane to enter the thigh. It gives off muscular branches and an important branch that supplies the head of the femur.
7. *Internal pudendal artery* leaves the pelvis through the greater sciatic foramen to enter the gluteal region. It then enters the perineum through the lesser sciatic foramen. The internal pudendal artery gives off the *inferior rectal artery* that supplies the lower half of the anal canal. It also gives off branches to the scrotum and penis in males and the labia and clitoris in females.
8. *Inferior gluteal artery* leaves the pelvis through

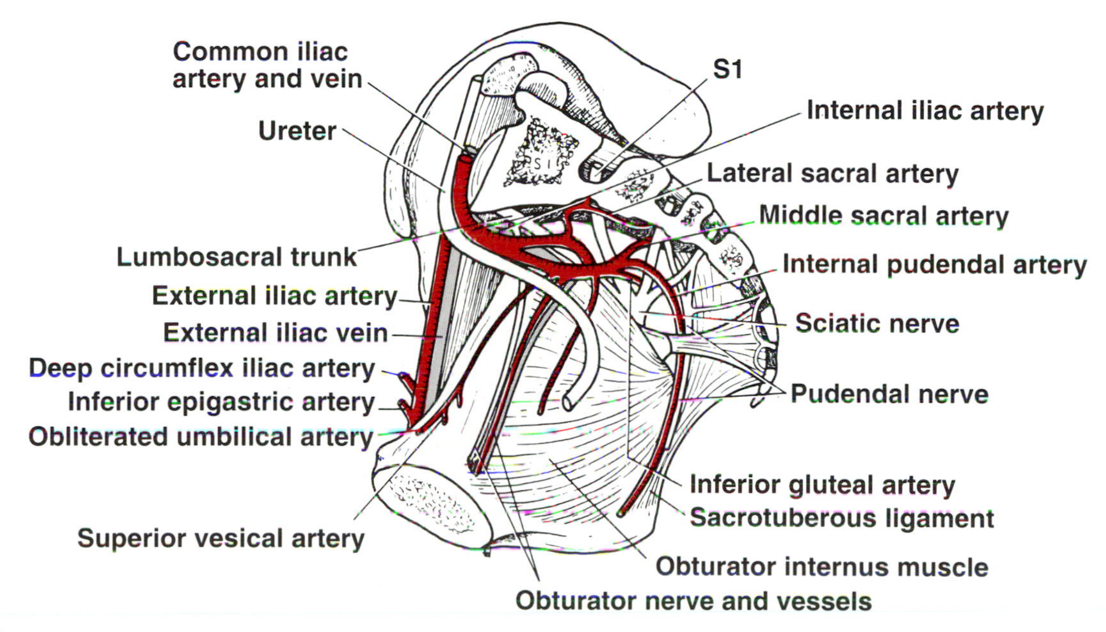

FIG 14–5.
The right half of the pelvis showing the common iliac, external iliac, and internal iliac vessels and their branches. The ureter crosses the bifurcation of the common iliac artery in front of the sacroiliac joint.

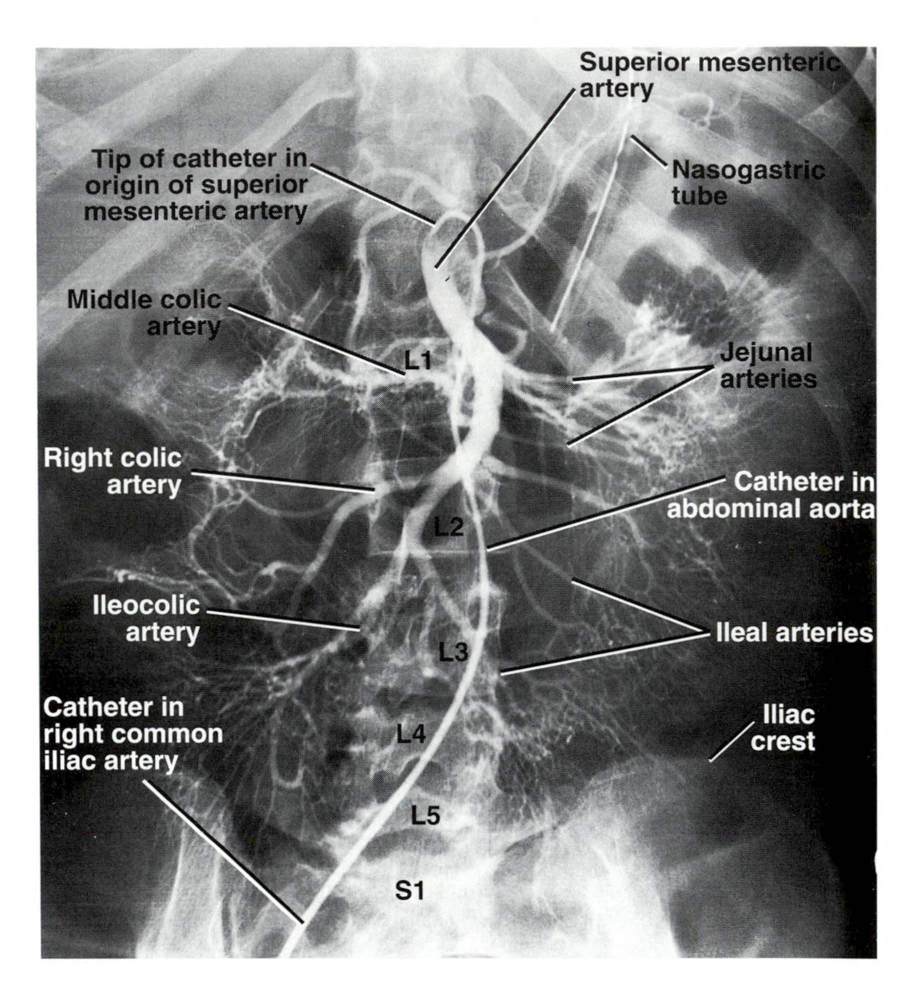

FIG 14–6.
An arteriogram of the superior mesenteric artery. The catheter has been inserted into the right femoral artery and has passed up the external and common iliac arteries to ascend the aorta to the origin of the superior mesenteric artery. A nasogastric tube is also in position.

the greater sciatic foramen and supplies the gluteal region.

Branches of the Posterior Division of the Internal Iliac Artery (see Fig 14–5).—These include the following.

1. *Superior gluteal artery* leaves the pelvis through the greater sciatic foramen and supplies the gluteal region.
2. *Iliolumbar artery* ascends posteriorly to the external iliac vessels, psoas, and iliacus muscles to supply structures in the iliac fossa.
3. *Lateral sacral arteries* descend in front of the sacrum and the sacral plexus, giving off branches to neighboring structures.

Radiographic Appearances of the Abdominal Aorta and Some of its Main Branches

Arteriography of the superior mesenteric artery, inferior mesenteric artery, and iliac arteries is shown in Figures 14–5 through 14–8. Computed tomographic scans of the abdomen showing the aorta and the common iliac vessels are seen in Figures 14–9 and 14–10.

Basic Anatomy of the Inferior Vena Cava

Location

The inferior vena cava conveys blood to the right atrium from all structures situated below the diaphragm. It is formed by the union of the common iliac veins behind the right common iliac artery at the level of the fifth lumbar vertebra (see Fig 14–1). It ascends on the right side of the aorta, pierces the central tendon of the diaphragm at the level of the eighth thoracic vertebra, and drains into the right atrium of the heart.

Relations.—As the inferior vena cava passes up the posterior abdominal wall, it has the following important relations.

Anteriorly.—Coils of small intestine, third part of the duodenum, head of the pancreas, first part of duodenum, entrance into the lesser sac of perito-

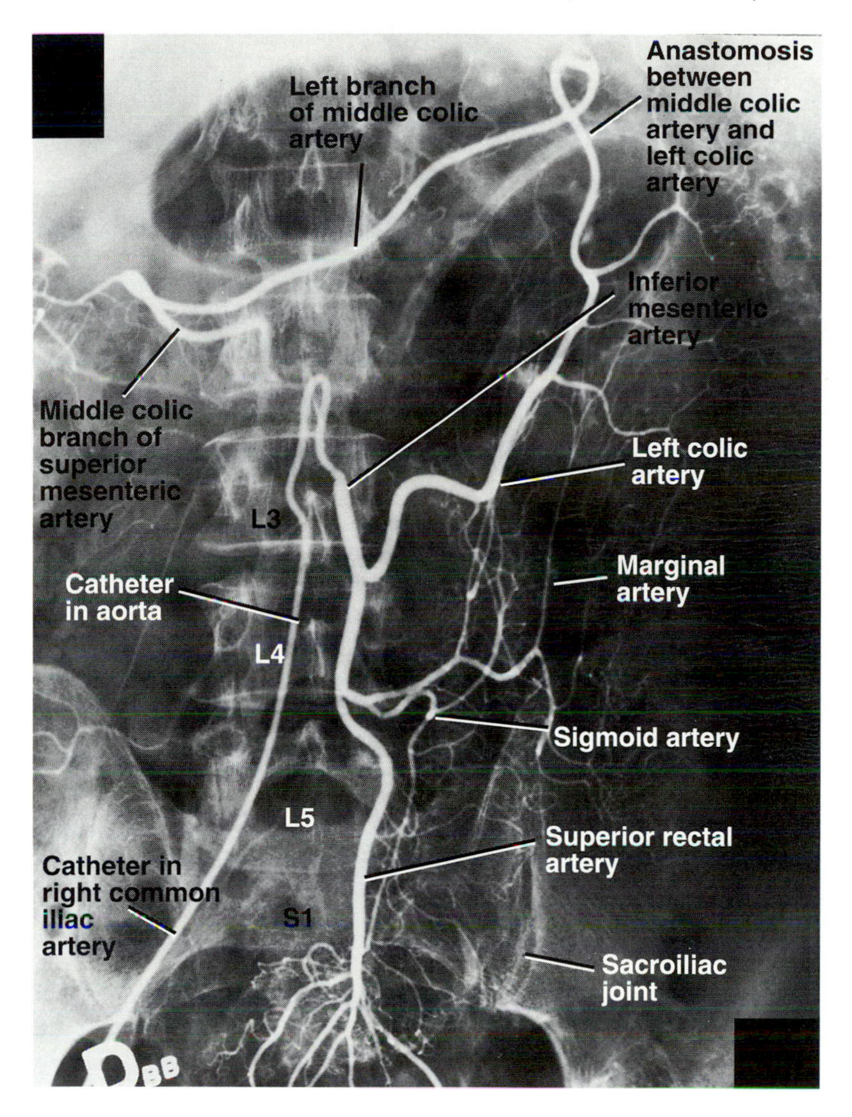

FIG 14–7.
An arteriogram of the inferior mesenteric artery. The catheter has been inserted into the right femoral artery and has passed up the external and common iliac arteries to ascend the aorta to the origin of the inferior mesenteric artery. The radiopaque dye has spread to enter the middle colic branch of the superior mesenteric artery.

neum (which separates the inferior vena cava from the portal vein, common bile duct, and hepatic artery), and the liver.

Laterally.—The right sympathetic trunk and the right ureter.

Medially.—Abdominal aorta.

Tributaries.—The inferior vena cava has the following tributaries (Fig 14–11).

1. Two anterior visceral tributaries—the hepatic veins.
2. Three lateral visceral tributaries—the right suprarenal (left vein drains into the left renal vein), renal veins, and right testicular or ovarian vein (the left vein drains into the left renal vein).
3. Five lateral abdominal wall tributaries—the inferior phrenic vein and four lumbar veins.
4. Three veins of origin—two common iliac veins and the median sacral vein.

Radiographic Appearances of the Inferior Vena Cava

The appearances of the inferior vena cava on a computed tomographic scan is shown in Figure 14–9.

Basic Anatomy of the Portal Vein

Location

This important vein drains blood from the abdominal part of the gastrointestinal tract from the lower third of the esophagus to halfway down the anal canal; it also drains blood from the spleen, pancreas, and gallbladder. The portal vein enters the liver and breaks up into sinusoids, from which blood passes into the hepatic veins that join the inferior vena cava. The portal vein is about 2 in. (5 cm) long and is

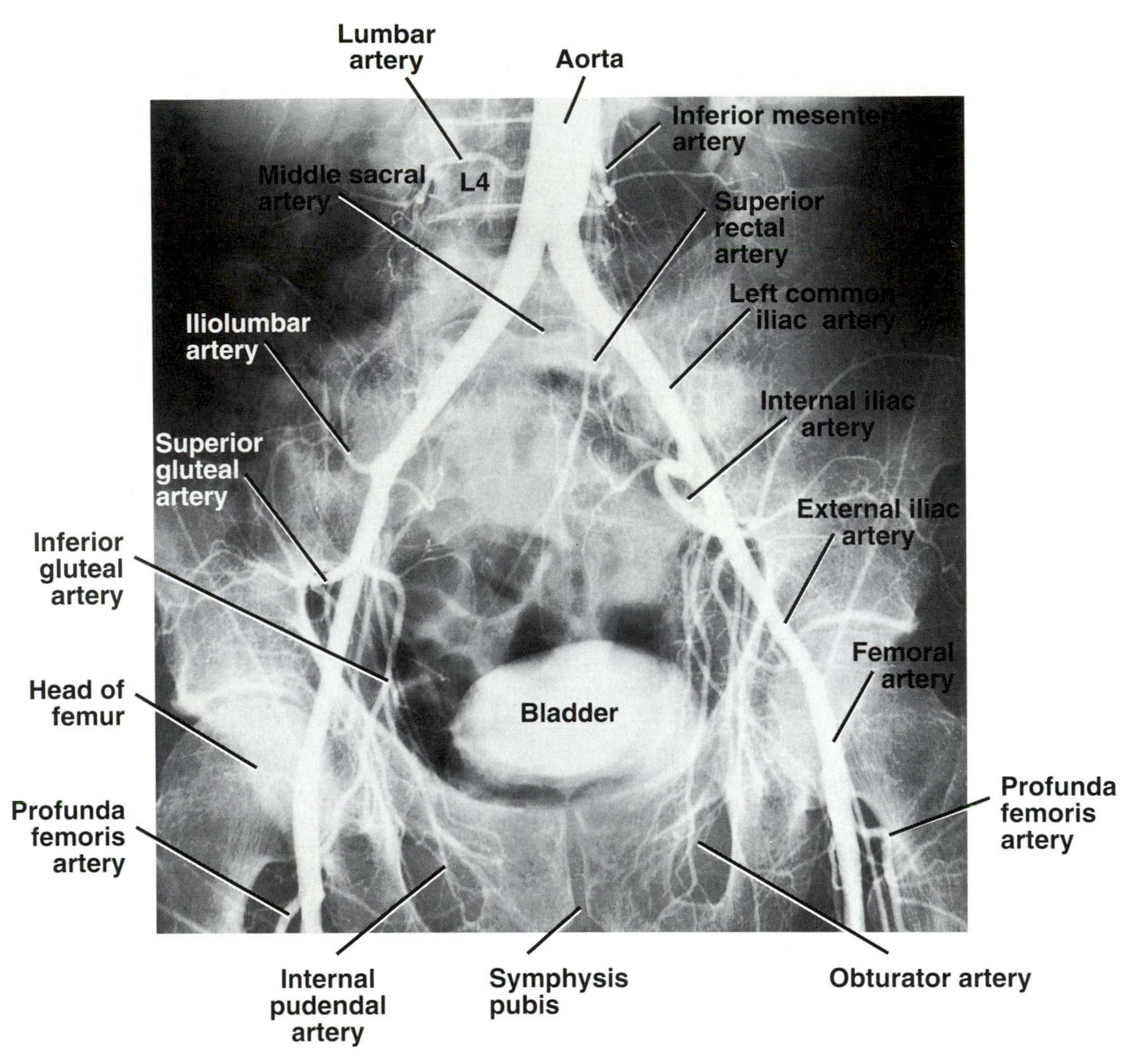

FIG 14–8.
An arteriogram of the lower part of the abdominal aorta showing the iliac arteries and their branches. The catheter (not visible) has been inserted into the left femoral artery. Some of the radiopaque material has already been excreted in the urine and shows in the bladder.

formed behind the neck of the pancreas by the union of the superior mesenteric vein and the splenic vein (Fig 14–12). It ascends behind the first part of the duodenum and enters the lesser omentum. At the porta hepatis it divides into right and left terminal branches.

The tributaries of the portal vein are the splenic vein, superior mesenteric vein, left gastric vein, right gastric vein, and cystic veins.

Splenic Vein.—This vein arises in the hilum of the spleen and passes to the right in the lienorenal ligament. It crosses in front of the left kidney and behind the body of the pancreas. It lies below the splenic artery and ends behind the neck of the pancreas by uniting with the superior mesenteric vein to form the portal vein (see Fig 14–12). It receives the short gastric, left gastroepiploic, inferior mesenteric, and pancreatic veins.

Inferior Mesenteric Vein.—This vein drains blood from the upper half of the anal canal, rectum, sigmoid colon, descending colon, splenic flexure, and distal third of the transverse colon. It ascends on the posterior abdominal wall on the left side of its artery. It passes close to the duodenojejunal flexure and joins the splenic vein behind the body of the pancreas (see Fig 14–12). It receives the superior rectal veins, sigmoid veins, and the left colic vein.

Superior Mesenteric Vein.—The superior mesenteric vein drains the jejunum, ileum, cecum and ap-

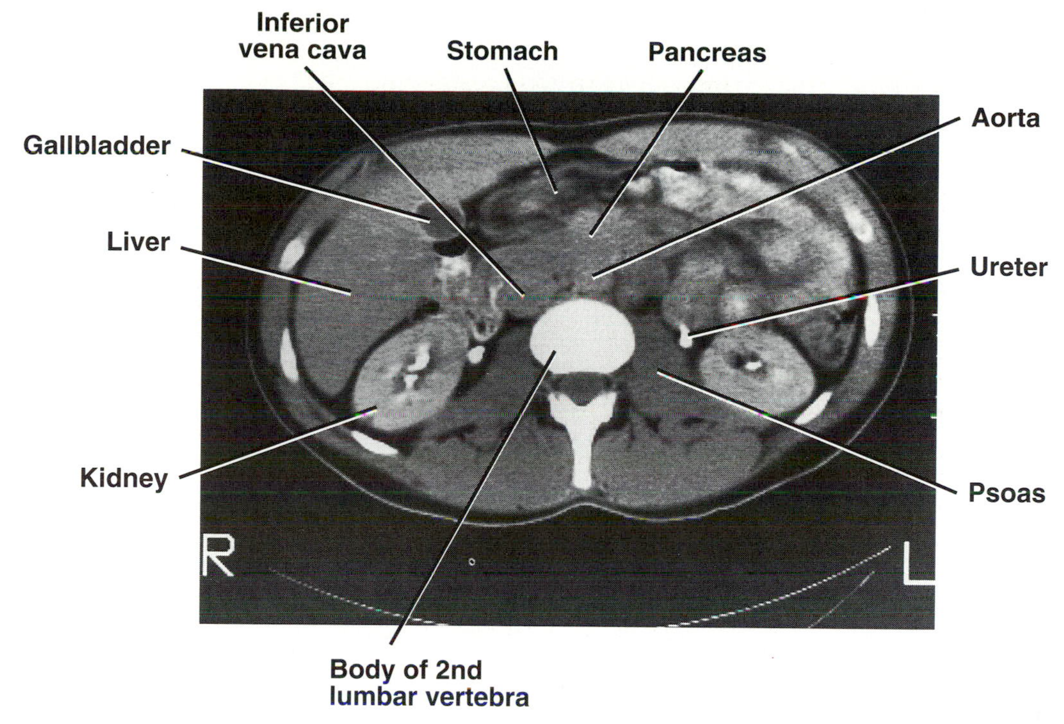

FIG 14–9.
A computed tomographic scan of the abdomen at the level of the second lumbar vertebra following an intravenous pyelogram. The aorta and the inferior vena cava can be seen on the posterior abdominal wall.

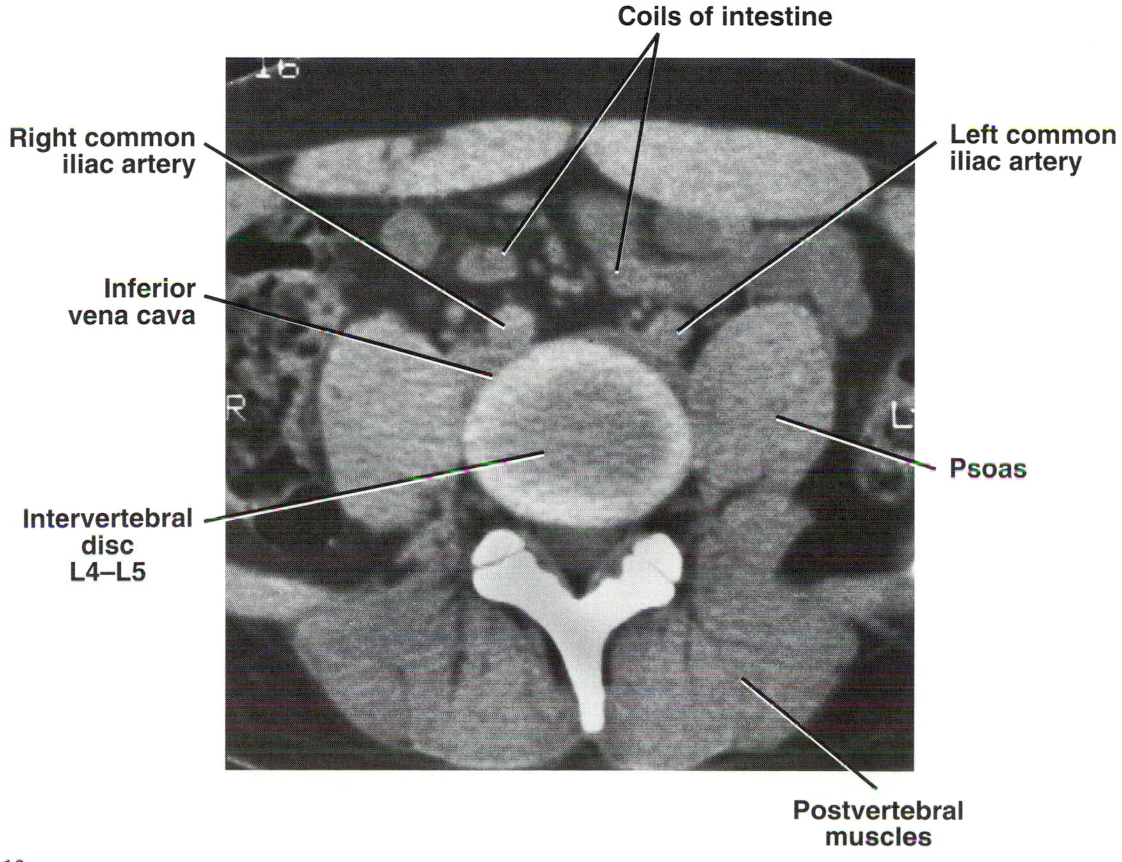

FIG 14–10.
A computed tomographic scan of the abdomen at the level of the intervertebral disc between the fourth and fifth lumbar vertebrae. Shows the right and left common iliac arteries close to their origin and the beginning of the inferior vena cava.

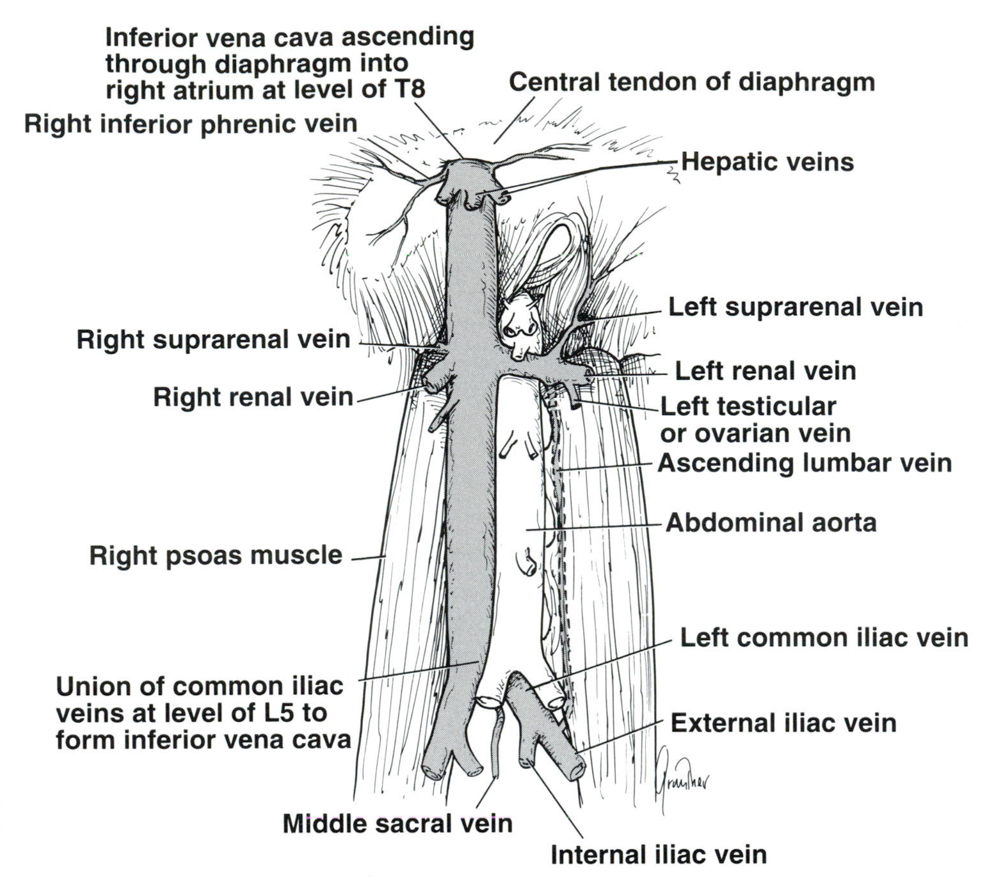

FIG 14–11.
The inferior vena cava and its tributaries; the abdominal part of the aorta is also shown.

pendix, and ascending colon, and the proximal two thirds of the transverse colon. It ascends in the root of the mesentery of the small intestine on the right of the superior mesenteric artery. It then passes in front of the third part of the duodenum and joins the splenic vein behind the neck of the pancreas to form the portal vein (see Fig 14–12). It receives jejunal, ileal, ileocolic, right colic, middle colic, inferior pancreaticoduodenal, and right gastroepiploic veins.

Left Gastric Vein.—This vein drains the left portion of the lesser curvature of the stomach and the distal part of the esophagus.

Right Gastric Vein.—This vein drains the right portion of the lesser curvature of the stomach.

Cystic Veins.—These veins drain the gallbladder either directly into the liver or join the portal vein.

Basic Anatomy of the Major Arteries of the Lower Extremity

Femoral Artery (Superficial Femoral Artery)

The femoral artery is a continuation of the external iliac artery (Fig 14–13). It begins behind the inguinal ligament, where it lies midway between the anterior superior iliac spine and the symphysis pubis. Here its pulsations may be easily palpated, since it may be pressed backward against the pectineus muscle and the superior ramus of the pubis. The artery descends almost vertically toward the adductor tubercle of the femur and ends at the opening of the adductor magnus muscle by entering the popliteal space as the popliteal artery (Fig 14–14).

Relations.—These are as follows.

Anteriorly.—In the upper part of its course in the femoral triangle, it is superficial and is covered by skin and fascia. In the lower part of its course, it lies in the adductor canal (subsartorial canal) and passes behind the sartorius muscle.

Posteriorly.—The psoas muscle, which separates the artery from the hip joint.

Medially.—The artery is related to the femoral vein and the femoral canal in the upper part of its course.

Laterally.—The femoral nerve and its branches.

The femoral artery is accompanied by the femoral vein, which lies on the medial side at the inguinal ligament and posterior to it at the apex of the femoral triangle. At the opening in the adductor magnus, the vein lies on the lateral side of the artery, *i.e., the vein changes its mediolateral relationship to*

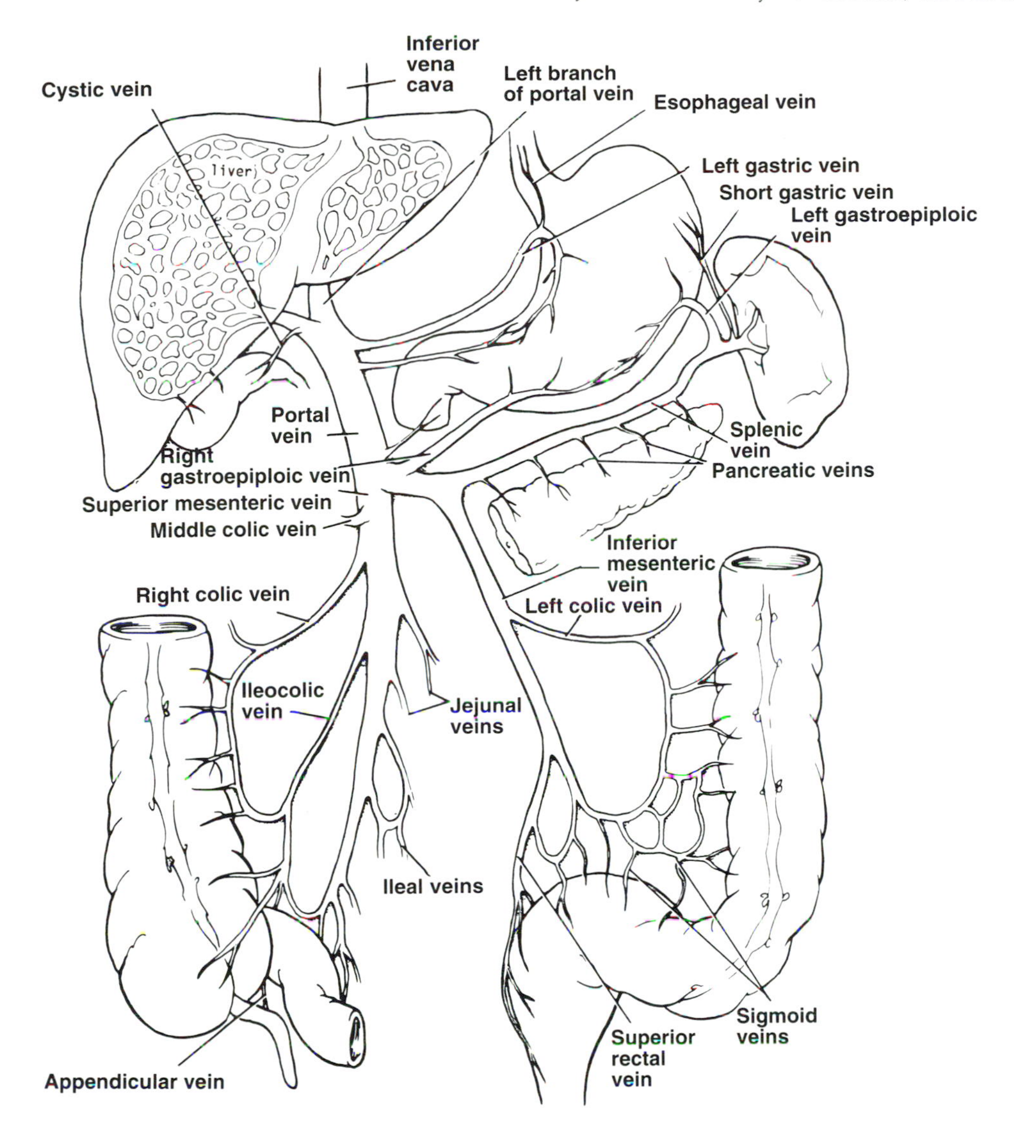

FIG 14–12.
Tributaries of the portal vein.

the artery, moving from being medial at the groin to being lateral at the lower part of the femur.

Branches.—These include the following arteries.

1. *Superficial circumflex iliac artery.* This artery arises just below the inguinal ligament and runs laterally to the region of the anterior superior iliac spine.
2. *Superficial epigastric artery.* This artery arises just below the inguinal ligament and ascends onto the abdominal wall as high as the umbilicus.
3. *Superficial external pudendal artery.* This artery arises just below the inguinal ligament and runs medially to supply the skin of the scrotum or labium majus.
4. *Deep external pudendal artery.* This artery also arises just below the inguinal ligament and runs medially to supply the skin of the scrotum or labium majus.
5. *Profunda femoris artery (deep femoral artery).* This is a large, important branch that arises from the lateral and posterior surface of the femoral artery about 1½ in. (4 cm) below the inguinal ligament (see Fig 14–13). It supplies structures in the anterior, medial, and posterior fascial compartments of the thigh by means of the following branches. At its origin it gives off the *medial and lateral femoral circumflex arteries,* and during its course it gives off *three perforating arteries.* The artery ends by becoming the *fourth perforating artery.*
6. *Descending genicular artery.* This is a small branch that arises from the femoral artery in the subsartorial canal (see Fig 14–14).

Arterial Anastomosis Around the Hip Joint.—To compensate for the narrowing of the femoral artery, which occurs during flexion of the thigh, a profuse anastomosis exists between the internal iliac artery and the femoral artery. The vessels involved are the

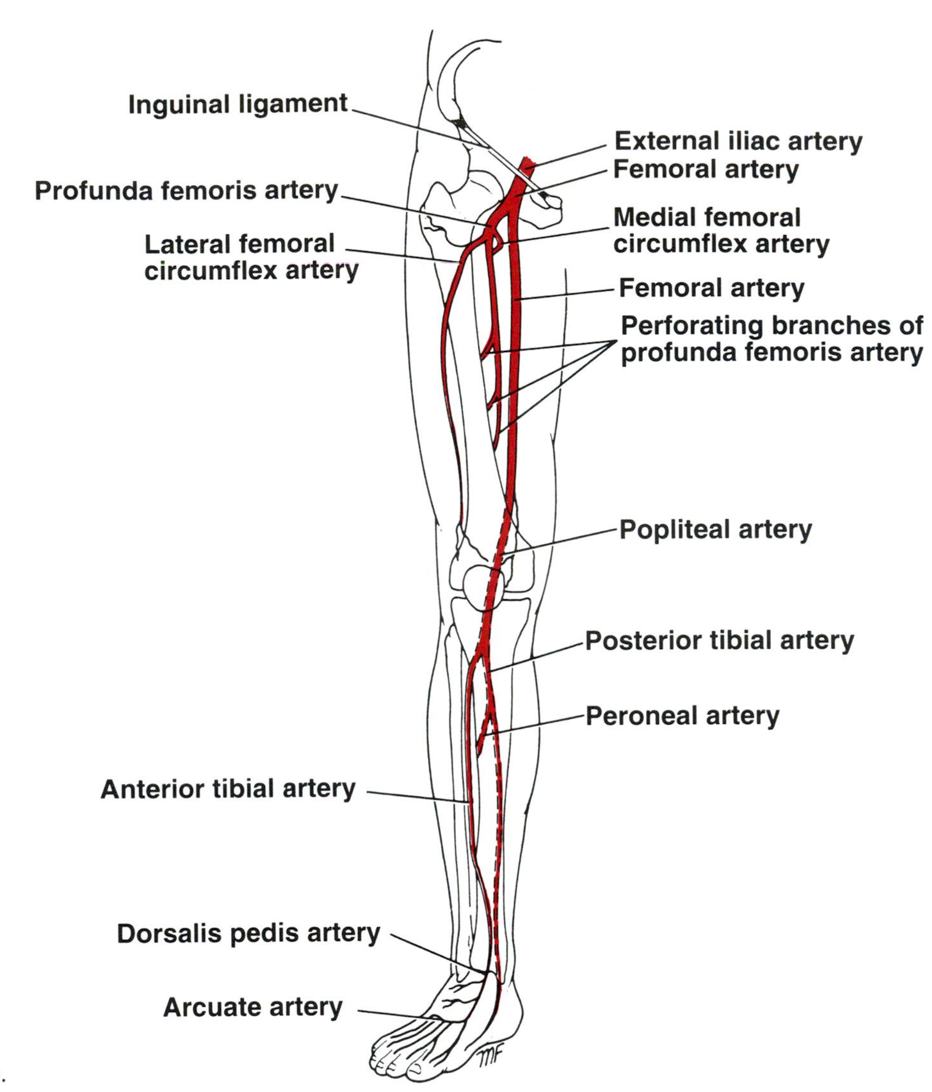

FIG 14–13.
The major arteries of the lower limb.

superior and inferior gluteal arteries, the medial and lateral femoral circumflex arteries, and the first perforating branch of the profunda artery.

Clinical Note

Collateral circulation.—The above arterial anastomosis may provide an adequate alternative circulation in cases of occlusion of the external iliac and femoral arteries.

Popliteal Artery

The popliteal artery is a continuation of the femoral artery and extends from the opening in the adductor magnus to the lower border of the popliteus muscle, where it divides into anterior and posterior tibial arteries (see Fig 14–13). It is deeply placed in the popliteal fossa (Fig 14–15).

Relations.—These include the following.

Anteriorly.—The popliteal surface of the femur, the knee joint, and the popliteus muscle.

Posteriorly.—The popliteal vein, the tibial nerve, fascia, and skin.

Branches.—These are as follows.

1. *Muscular branches.*
2. *Articular branches.*
3. *Terminal branches—anterior and posterior tibial arteries.*

Arterial Anastomosis Around the Knee Joint.—To compensate for narrowing of the popliteal artery, which occurs during flexion of the knee, a profuse arterial anastomosis exists. The vessels involved are the descending genicular of the femoral, the lateral femoral circumflex of the profunda femoris, the ar-

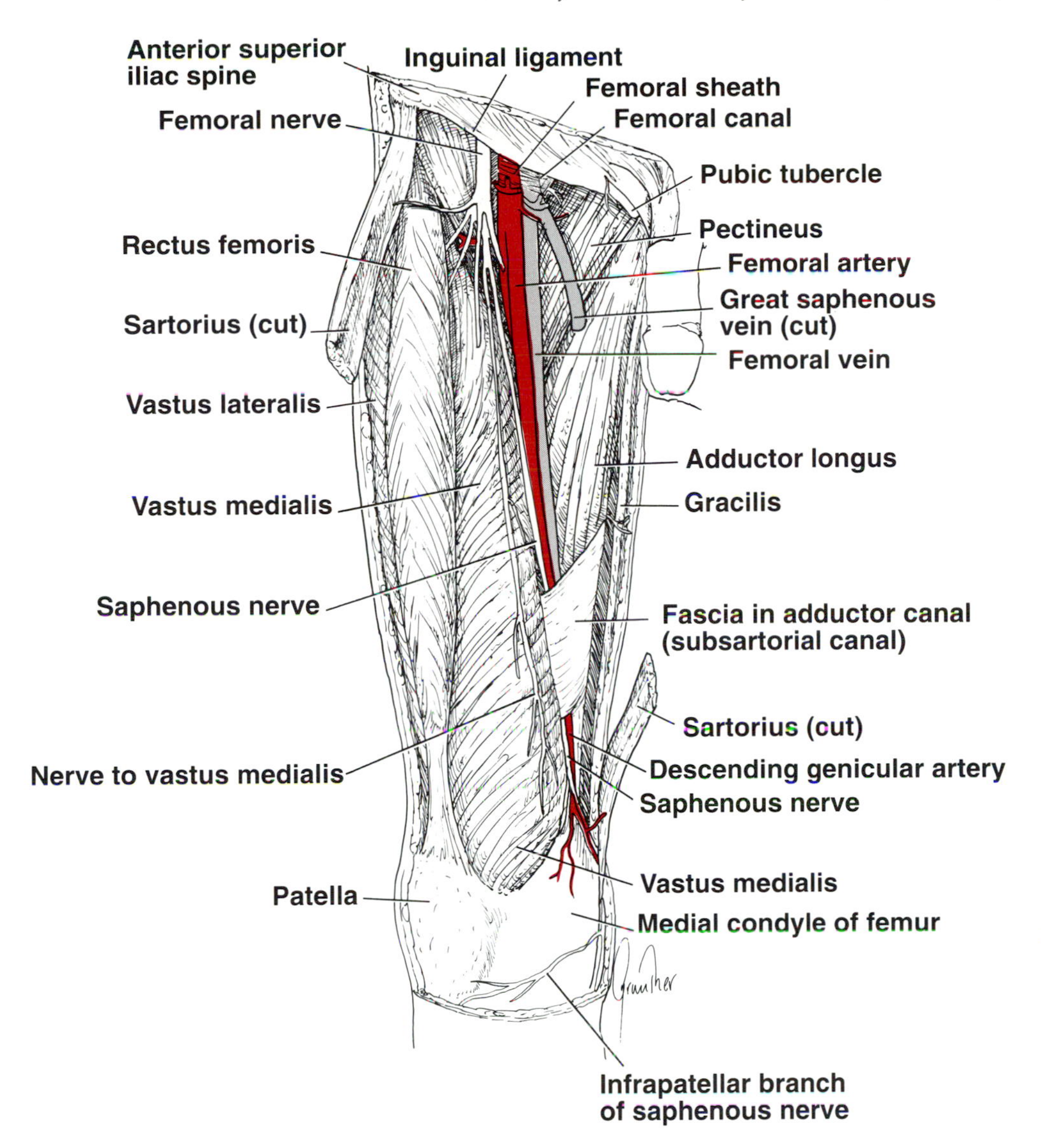

FIG 14–14.
Front of thigh in the right lower limb showing the femoral vessels and nerve. The sartorius muscle has been cut and reflected to show the underlying adductor (subsartorial) canal and its contents.

ticular branches of the popliteal, and branches from the anterior and posterior tibial arteries.

Anterior Tibial Artery

The anterior tibial artery arises at the bifurcation of the popliteal artery at the level of the lower border of the popliteus muscle (see Fig 14–13). It passes forward between the tibia and the fibula through the upper part of the interosseous membrane to enter the anterior compartment of the leg. It descends with the deep peroneal nerve to the front of the ankle joint, where it becomes the dorsalis pedis artery (Fig 14–16).

In the upper part of its course it lies deep beneath the muscles of the anterior compartment. In the lower part of the course it lies superficial in front of the lower end of the tibia. At the ankle it has the tendon of the extensor hallucis longus on the medial side and the tendons of extensor digitorum longus on its lateral side; here the pulsations can easily be felt (see Fig 14–16).

Branches.—These are as follows.

1. *Muscular branches.*
2. *Anastomotic branches,* which anastomose with branches of other arteries around the knee and ankle joints.

Dorsalis Pedis Artery

The dorsalis pedis artery begins in front of the ankle joint as a continuation of the anterior tibial artery (see Fig 14–16). It ends by passing downward into the sole through the proximal part of the space between the first and second metatarsal bones. Having passed through the first dorsal interosseous muscle,

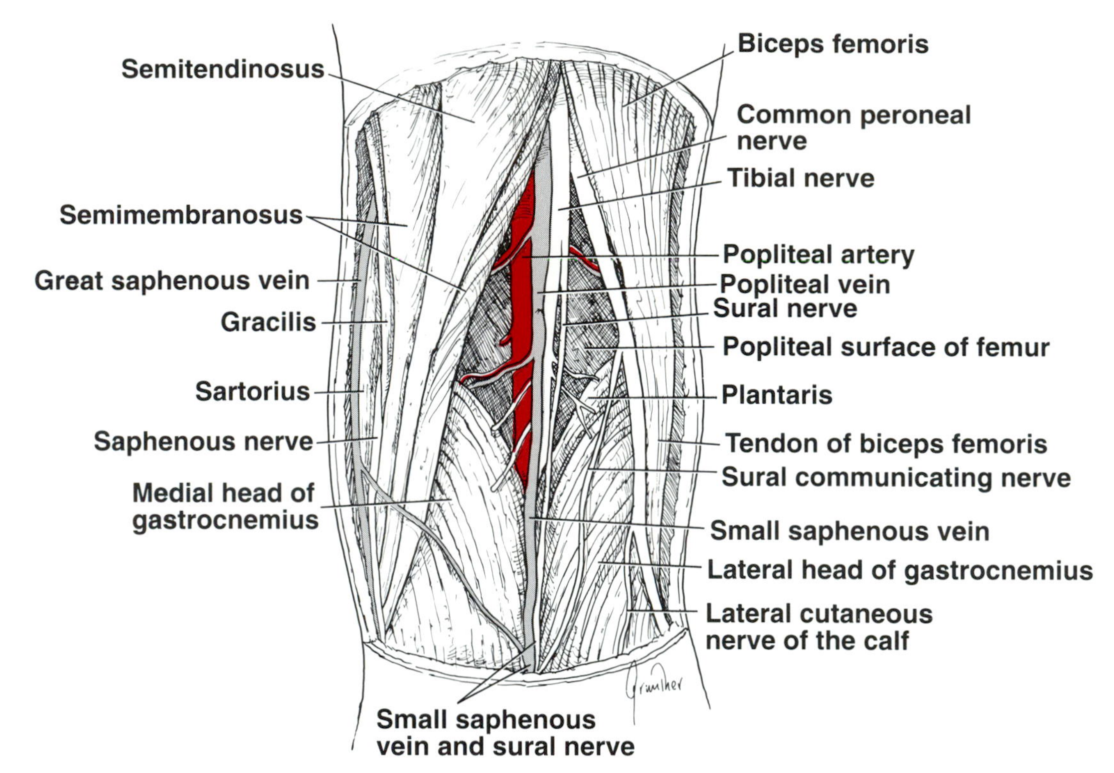

FIG 14–15.
Right popliteal fossa showing the boundaries and contents. The popliteal vessels are displayed. Note the small saphenous vein draining into the popliteal vein.

it joins the lateral plantar artery and completes the plantar arch.

The artery is in a superficial position, covered only by skin and fascia. On its lateral side lie the tendons of the extensor digitorum longus and on the medial side the tendon of extensor hallucis longus (see Fig 14–16). Its pulsations can be easily felt.

Branches.—These are as follows.

1. *Lateral tarsal artery.* Supplies the dorsum of the foot.
2. *Arcuate artery.* Runs laterally across the dorsum of the foot and gives off branches to the toes.
3. *First dorsal metatarsal artery.* Supplies both sides of the big toe.

Posterior Tibial Artery

The posterior tibial artery arises at the bifurcation of the popliteal artery at the level of the lower border of the popliteus muscle (Fig 14–17). It descends in the deep posterior compartment of the leg accompanied by the tibial nerve. The artery terminates behind the medial malleolus by dividing into medial and lateral plantar arteries.

In the upper two thirds of its course it lies deep to the gastrocnemius and soleus muscles. In the lower third of the leg it becomes superficial and lies on the posterior surface of the tibia; here it is covered only by fascia and skin. The pulsations of the artery may be palpated at a spot midway between the medial malleolus and the heel (Fig 14–18).

Branches.—These are as follows.

1. *Peroneal artery.* This is a large artery that arises close to the origin of the posterior tibial artery (see Fig 14–18). It descends behind the fibula, giving off a nutrient branch and numerous muscular branches. It takes part in anastomoses around the ankle joint.
2. *Muscular branches.*
3. *Nutrient artery to the tibia.*
4. *Anastomotic branches around the ankle joint.*
5. *Medial and lateral plantar arteries.*

Plantar Arch

The plantar arch is formed by the terminal part of the lateral plantar artery and is completed medially by the dorsalis pedis artery. The plantar arch gives off numerous branches that include perforating and metatarsal arteries. The metatarsal arteries give rise to digital arteries that supply the lateral four toes.

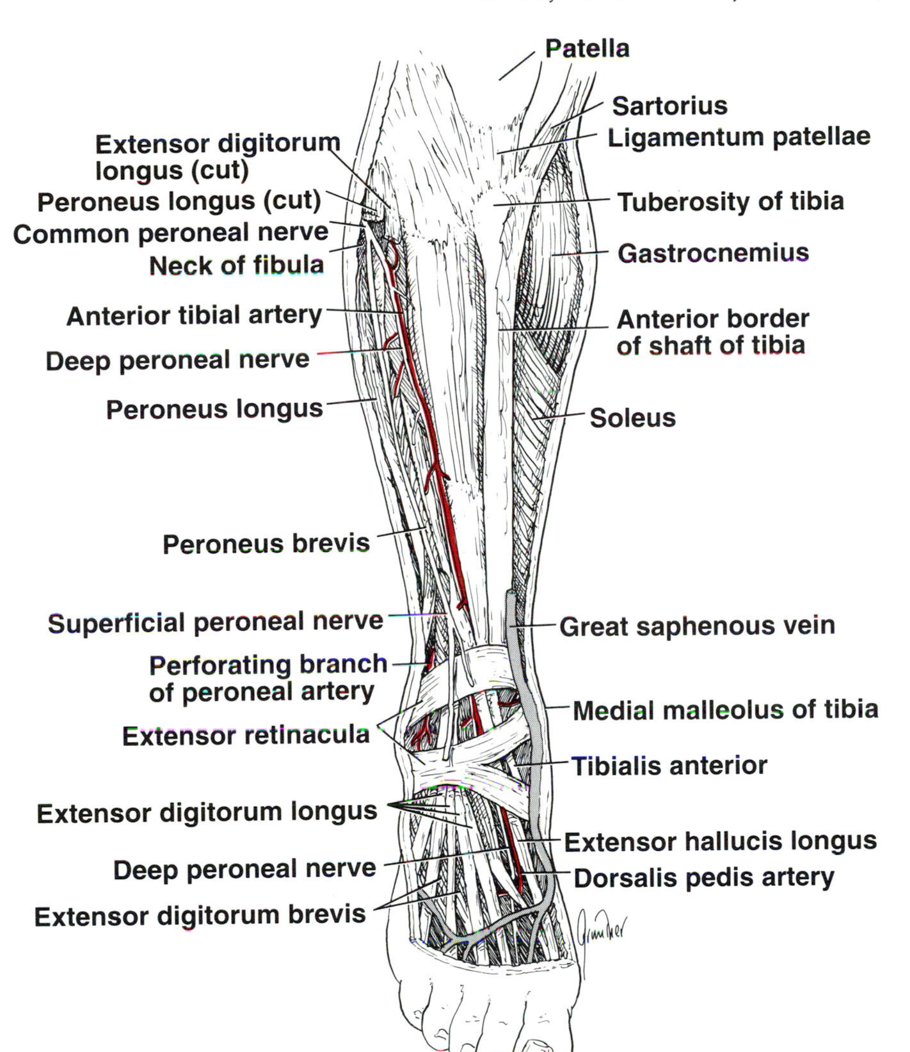

FIG 14–16.
Structures present on the anterior and lateral aspects of right leg and on the dorsum of the foot. Portions of the peroneus longus and the extensor digitorum longus muscles have been removed to display the common peroneal, deep peroneal, and superficial peroneal nerves.

Radiographic Appearances of the Arteries of the Lower Extremity

The radiographic appearances of the main arteries of the lower extremity are shown in Figures 14–8 and 14–19.

Basic Anatomy of the Major Veins of the Lower Extremity

The veins of the lower extremity may be divided into superficial and deep groups (Fig 14–20). The superficial veins lie in the superficial fascia and have relatively thick muscle walls, and the deep veins accompany the main arteries and have thin walls.

Superficial Veins

Dorsal Venous Arch.—This venous arch lies on the dorsum of the foot (see Fig 14–20). The greater part of the blood from the whole foot drains into the arch via digital veins and communicating veins that pass through the interosseous spaces. The dorsal venous arch is drained on the medial side by the great saphenous vein and on the lateral side by the small saphenous vein.

Great Saphenous Vein.—The great saphenous vein arises from the medial end of the dorsal venous arch of the foot and ascends directly *in front of* the medial malleolus (see Fig 14–20). It is accompanied by the saphenous nerve and lies in the superficial fascia.

Clinical Note

Great saphenous vein variation.—Occasionally the great saphenous vein at the medial malleolus is replaced by several small veins instead of a single large one. This possibility is of great clinical importance when performing a venous cut down.

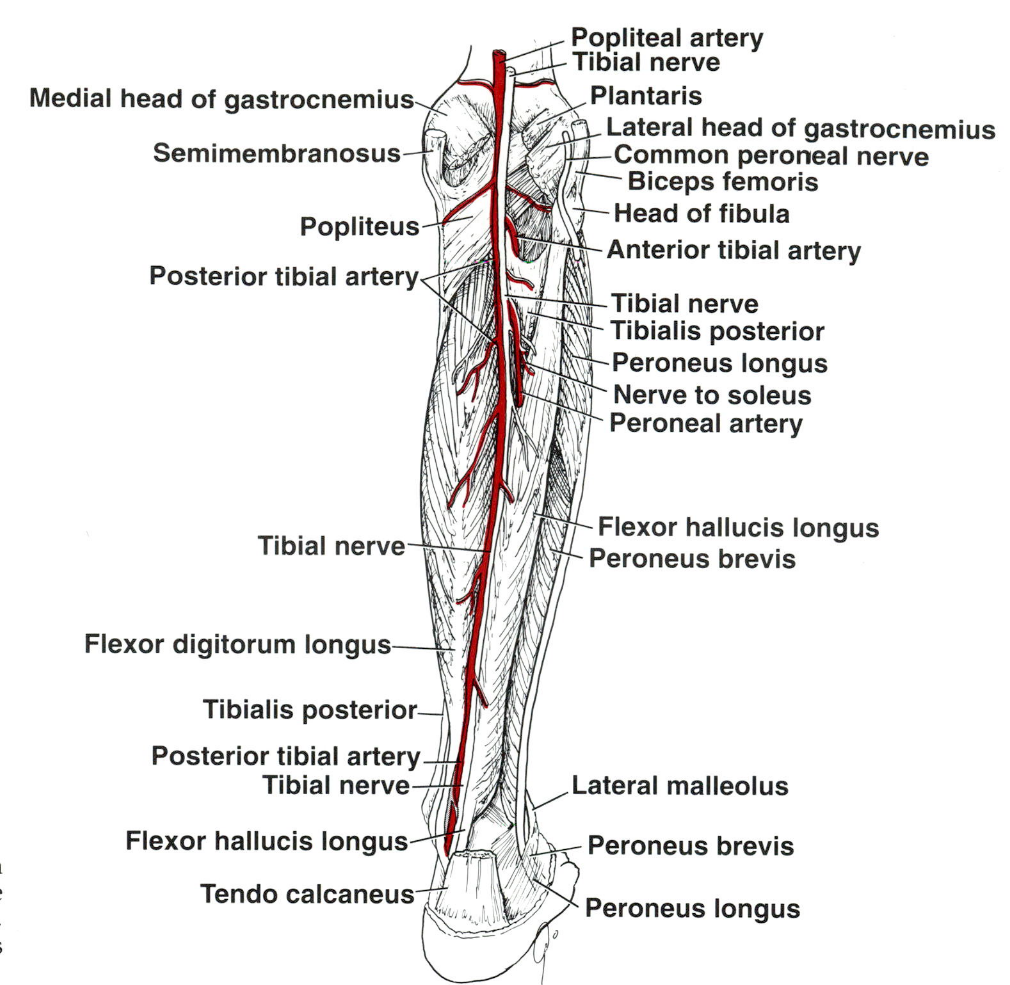

FIG 14–17.
Deep structures present on the posterior surface of the right leg. The gastrocnemius, soleus, and plantaris muscles have been removed.

The great saphenous vein passes up the medial side of the leg, behind the knee, and curves forward around the medial side of the thigh. It then runs through the saphenous opening in the deep fascia and joins the femoral vein about 1½ in. (4 cm) below and lateral to the pubic tubercle. The great saphenous vein possesses numerous valves and is connected to the small saphenous vein by branches that pass behind the knee. A number of *perforating veins* connect the great saphenous vein with the deep veins along the medial side of the calf. The perforating veins possess valves that are arranged to prevent the flow of blood from the deep to the superficial veins (see Fig 14–20).

The great saphenous vein receives a variable number of subcutaneous tributaries and near its termination the following small named veins: (1) *superficial circumflex iliac vein,* (2) *superficial epigastric vein,* and (3) *superficial external pudendal vein.*

Small Saphenous Vein.—The small saphenous vein arises from the lateral side of the dorsal venous arch of the foot (see Fig 14–20). It ascends *behind* the lateral malleolus in company with the sural nerve. It then passes up the back of the calf and pierces the deep fascia at about the middle of the leg. It enters the popliteal fossa by passing between the two heads of the gastrocnemius muscle and drains into the popliteal vein. The small saphenous vein communicates with the deep veins by *perforating veins* and with the great saphenous vein.

Deep Veins

Venae Comitantes.—The deep veins accompany the respective arteries as venae comitantes. The venae comitantes of the anterior and posterior tibial arteries unite in the popliteal fossa to form the popliteal vein.

Popliteal Vein.—The popliteal vein (see Fig 14–15), formed by the union of the venae comitantes of the anterior and posterior tibial arteries, ascends through the popliteal space behind the artery, and at the opening in the adductor magnus becomes the

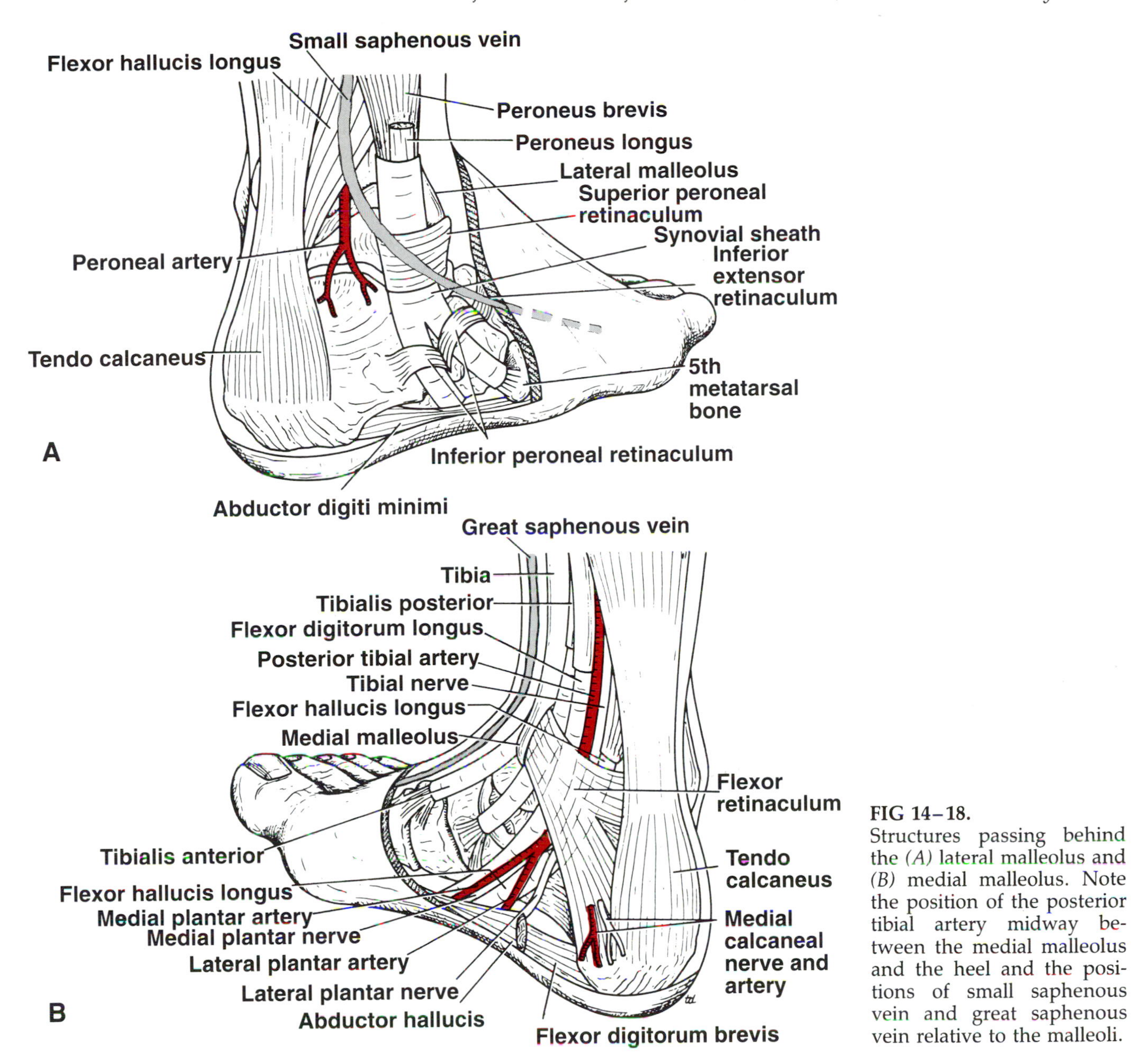

FIG 14–18.
Structures passing behind the *(A)* lateral malleolus and *(B)* medial malleolus. Note the position of the posterior tibial artery midway between the medial malleolus and the heel and the positions of small saphenous vein and great saphenous vein relative to the malleoli.

femoral vein. The popliteal vein receives numerous tributaries, with the small saphenous vein draining into it at the lower end of the popliteal space.

Femoral Vein.—The femoral vein is a continuation of the popliteal vein at the opening in the adductor magnus. It ascends through the thigh (see Fig 14–14) and comes to lie on the medial side of the femoral artery in the intermediate compartment of the femoral sheath. It ends by becoming continuous with the external iliac vein behind the inguinal ligament. The femoral vein receives the great saphenous vein and veins that correspond to branches of the femoral artery. The great saphenous vein enters the femoral vein 1½ in. below and lateral to the pubic tubercle (Fig 14–21).

SURFACE ANATOMY

Surface Anatomy of the Blood Vessels of the Abdomen

Abdominal Aorta

The pulsations of the abdominal aorta can be felt on palpation of the anterior abdominal wall (Fig 14–22). The aorta is a midline structure that enters the abdomen through the aortic hiatus in the diaphragm at the level of the 12th thoracic vertebra. This level may

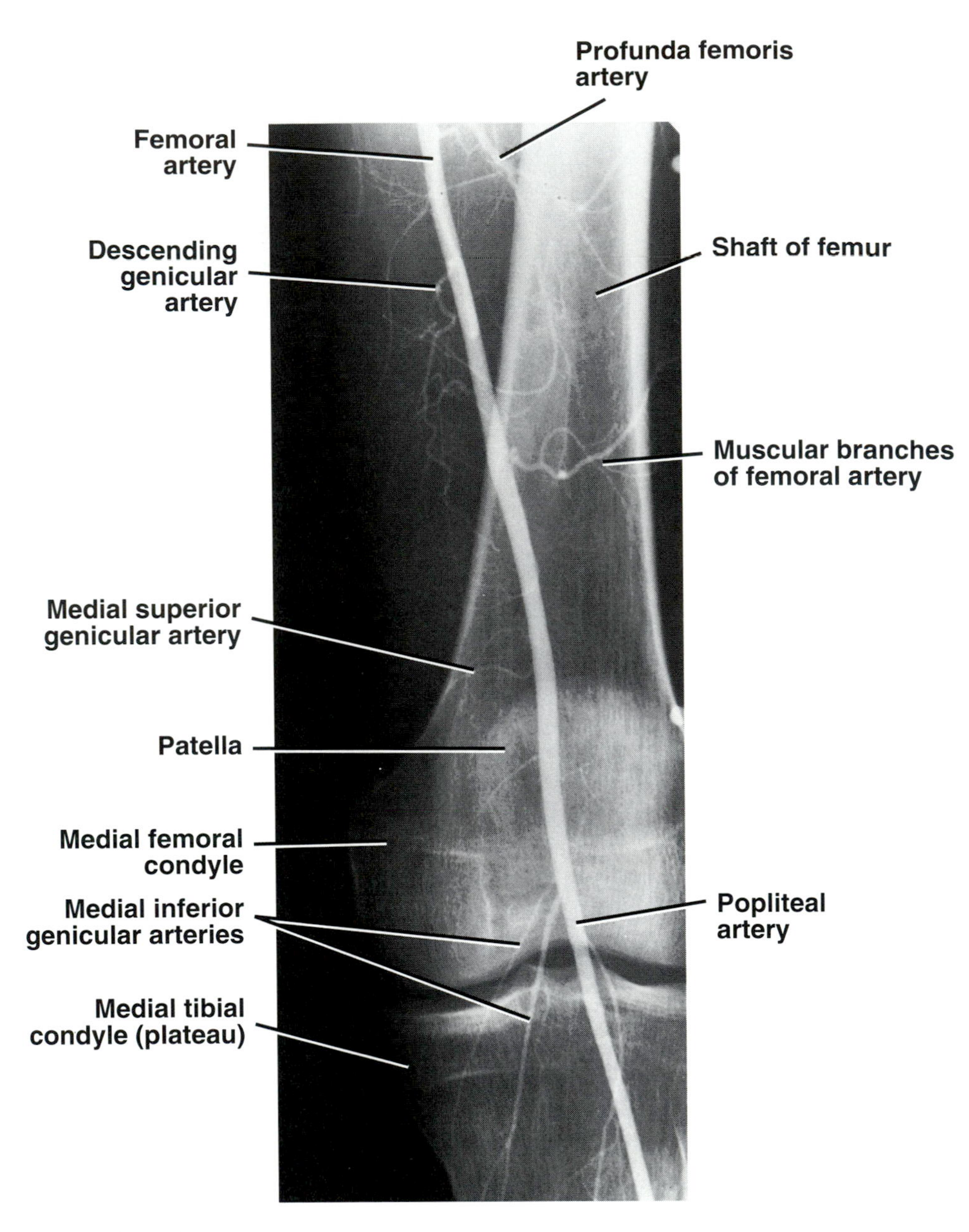

FIG 14–19.
Arteriogram of the femoral and popliteal arteries and their branches; the catheter (not visible) was inserted into the femoral artery.

be projected on the anterior abdominal wall as lying just above the transpyloric plane (L1), which lies at the level of the ninth costal cartilage, i.e., the point where the lateral margin of the rectus abdominis crosses the costal margin. Below, the aorta bifurcates into the right and left common iliac arteries opposite the fourth lumbar vertebra, i.e., on a line joining the summits of the iliac crests (see Fig 14–22).

The approximate vertebral level of the origin of the main branches of the aorta is as follows: *celiac artery* (T12), *superior mesenteric artery* (L1), *inferior mesenteric artery* (L3), and the *renal arteries* (L1).

Common Iliac Artery
This artery runs along the upper part of a line that extends from the middle of the anterior abdominal wall at the level of the summits of the iliac crests, to a point midway between the anterior superior iliac spine and the symphysis pubis (see Fig 14–22).

External Iliac Artery
This artery is a continuation of the common iliac artery. Its pulsations may be felt as it passes under the inguinal ligament to become continuous with the femoral artery. The lower end of the external iliac artery may be located at a point midway between the anterior superior iliac spine and the symphysis pubis (Fig 14–23).

Inferior Vena Cava
The inferior vena cava lies just to the right of the midline on the right side of the abdominal aorta (see

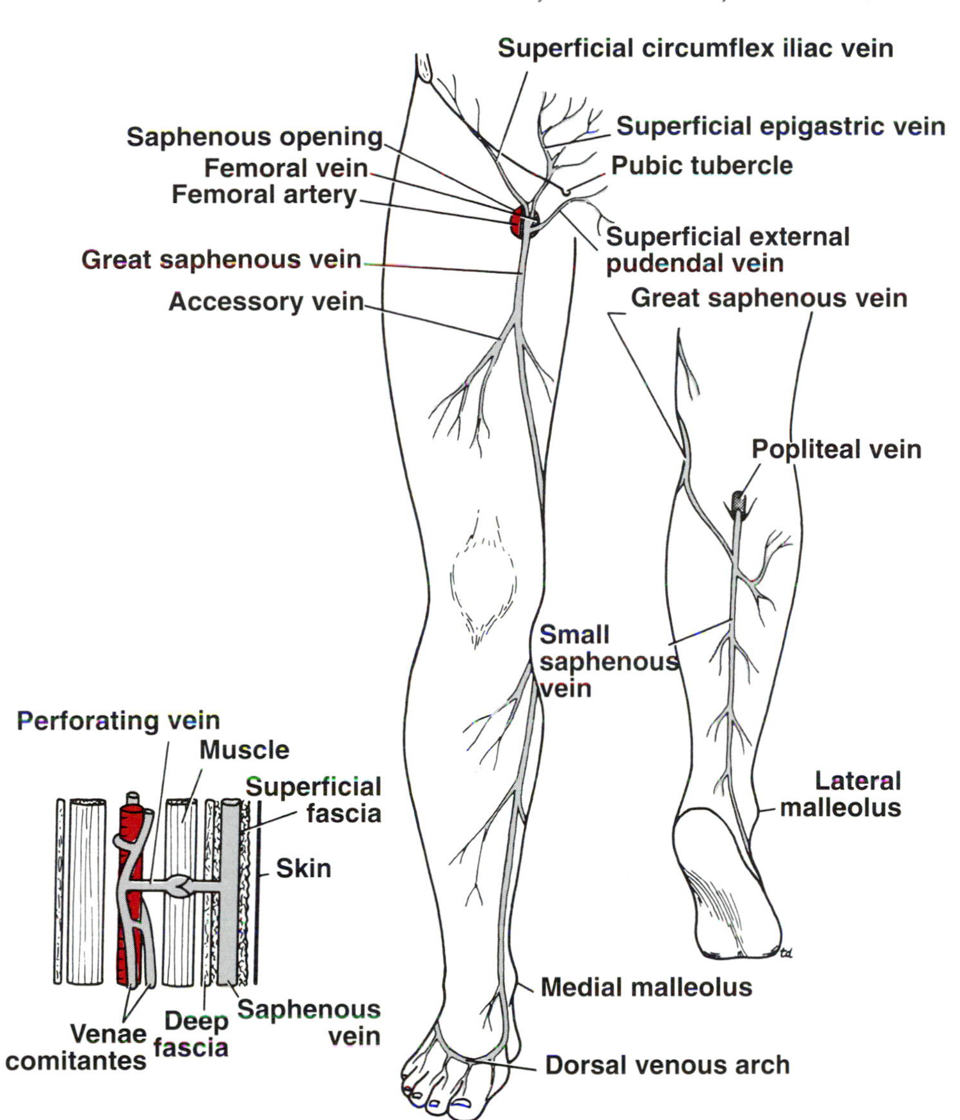

FIG 14–20.
Superficial veins of the right lower limb. The valves in the perforating veins normally allow blood to pass from the superficial veins into the deep veins but prevent the blood in the deep veins, when they are compressed by muscles, from entering the superficial veins.

Fig 14–22). It is formed by the union of the right and left common iliac veins in front of the fifth lumbar vertebra, i.e., on a line joining the tubercles of the iliac crests (about 1 in. below the level of the iliac crests). It leaves the abdomen by passing through the caval opening in the diaphragm at the level of the eighth thoracic vertebra, i.e., just above and to the right of the xiphisternal junction (T9).

Portal Vein
The portal vein is about 2 in. (5 cm) long. It is formed behind the neck of the pancreas by the union of the superior mesenteric and splenic veins at the level of the transpyloric plane (L1). It runs upward and to the right and divides into right and left terminal branches in the porta hepatis of the liver.

Surface Anatomy of the Blood Vessels of the Lower Extremity

Arteries
Femoral Artery.—The femoral artery enters the thigh behind the inguinal ligament (see Fig 14–23) at the midpoint of a line joining the symphysis pubis to the anterior superior iliac spine; its pulsations are easily felt. The artery descends almost vertically toward the adductor tubercle of the femur. The adductor tubercle is a small elevation of bone just proximal to the medial femoral condyle, and the tendon of adductor magnus muscle is attached to it.

Popliteal Artery.—The popliteal artery can be felt pulsating in the depths of the popliteal fossa, provided that the deep fascia covering the fossa is fully relaxed by passive flexion of the knee joint. Gentle

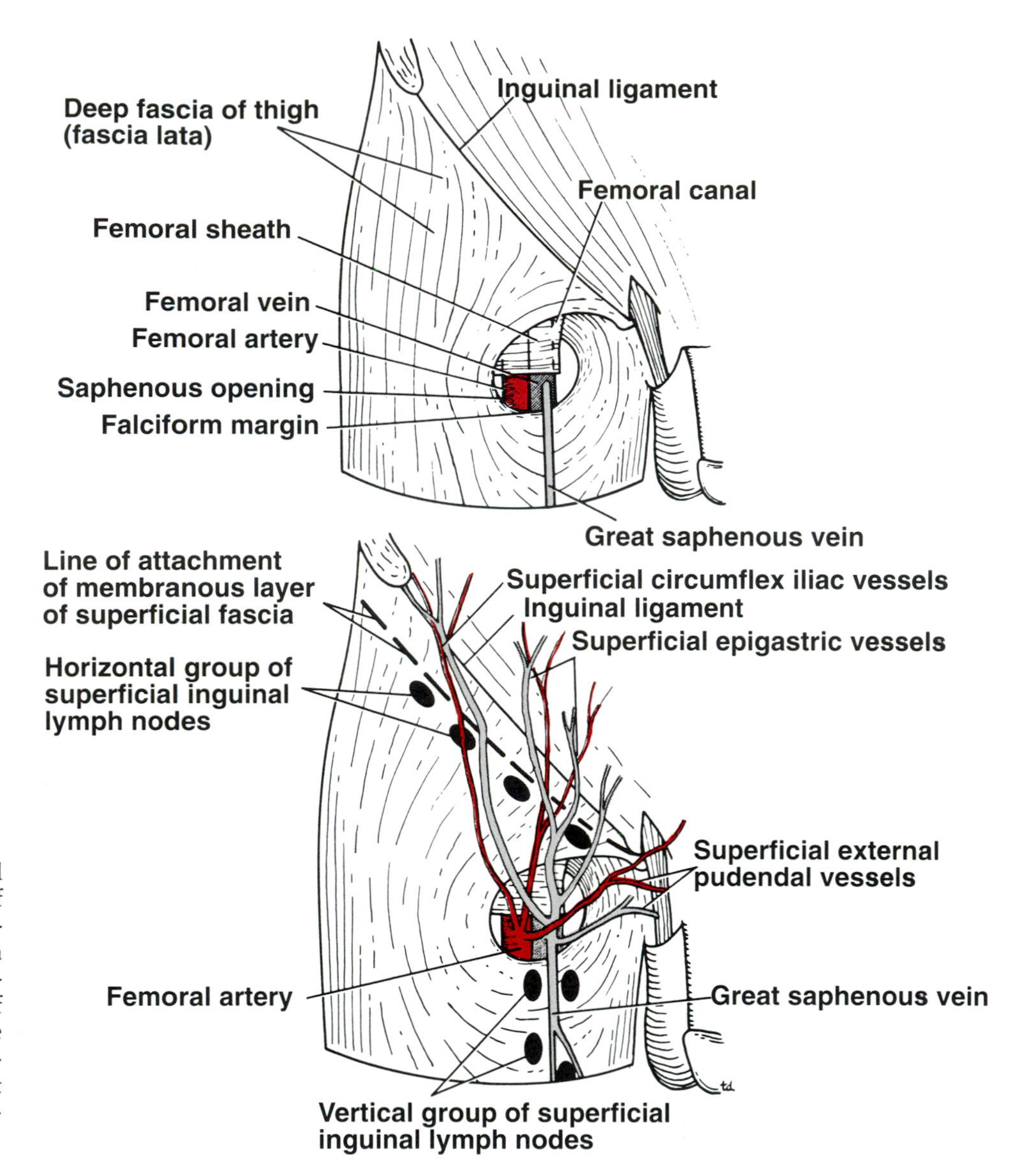

FIG 14–21.
Superficial veins, arteries, and lymph nodes on the front of the right thigh. Note saphenous opening in deep fascia and its relationship to the femoral sheath and the position of the line of attachment of the membranous layer of superficial fascia to deep fascia, about a fingerbreadth below the inguinal ligament.

but firm pressure on the center of the fossa by the palpating finger is required to detect the pulse through the overlying structures.

Posterior Tibial Artery.—The posterior tibial artery can be felt pulsating behind the medial malleolus halfway between the malleolus and the heel (see Fig 14–18).

Dorsalis Pedis Artery.—The dorsalis pedis artery can be felt pulsating on the front of the ankle joint midway between the two malleoli (see Fig 14–16). The artery is situated between the tendons of the extensor hallucis longus and the extensor digitorum longus.

Veins

Femoral Vein.—The femoral vein leaves the thigh by passing behind the inguinal ligament medial to the pulsating femoral artery (see Fig 14–23). If the femoral pulse is difficult to feel, the position of the vein can be determined to be deep to the inguinal ligament, about one fingerbreadth medial to the midpoint of a line joining the anterior superior iliac spine to the symphysis pubis.

Dorsal Venous Arch or Plexus.—The dorsal venous arch can usually be seen in the subcutaneous tissue on the dorsum of the foot proximal to the toes (see Fig 14–20).

Great Saphenous Vein.—The great saphenous vein leaves the medial part of the dorsal venous arch and

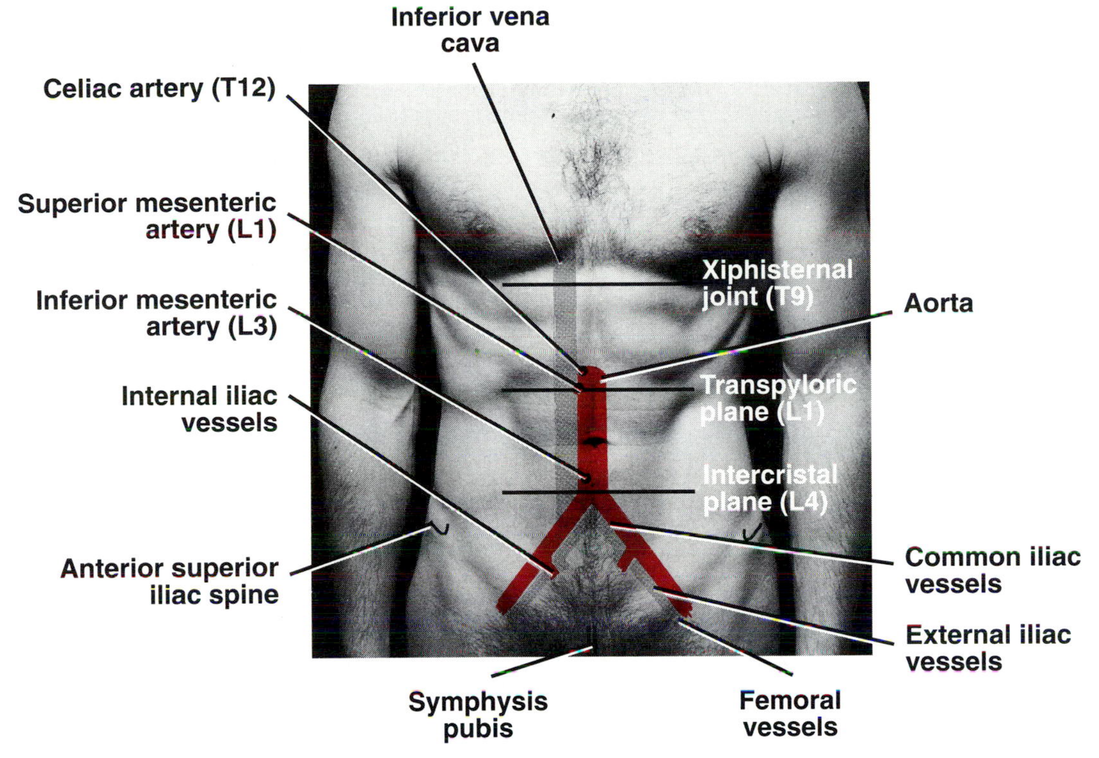

FIG 14–22.
Surface markings of large blood vessels on the anterior abdominal wall.

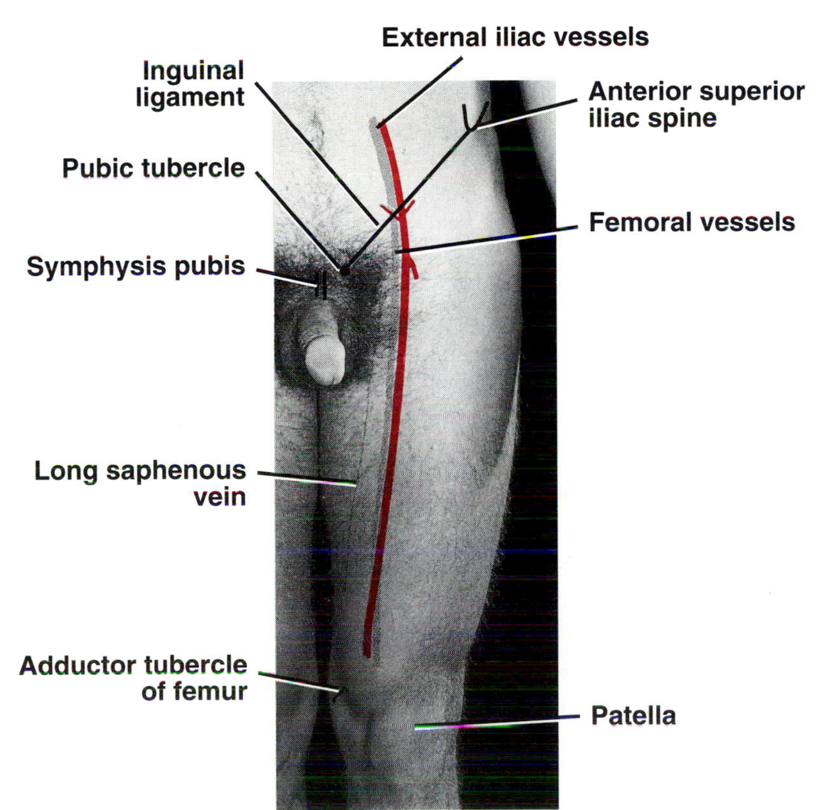

FIG 14–23.
Surface markings of major blood vessels on the anterior surface of the right thigh.

passes upward *in front of* the medial malleolus of the tibia (see Fig 14–20). Its further course up the leg can usually not be seen. It drains into the femoral vein about 1½ in. (two fingerbreadths) below and lateral to the pubic tubercle.

Small Saphenous Vein.—The small saphenous vein leaves the lateral part of the dorsal venous arch and passes upward *behind* the lateral malleolus of the fibula (see Fig 14–20). Its further course up the leg can usually not be seen. It drains into the popliteal vein in the popliteal fossa.

CLINICAL ANATOMY

Clinical Anatomy of the Abdominal Aorta

Traumatic Injury to the Abdominal Aorta

Blunt Injury.—Because of the deep position of the aorta on the posterior abdominal wall behind the peritoneum (Figs 14–24 and 14–25), blunt injuries to the aorta are relatively rare. In children the elasticity of the aortic wall and the usual absence of atherosclerosis make the condition even more rare.

In blunt trauma from an automobile accident, the abdominal aorta can be injured by the crossing band of a seat belt. The tunica intima is commonly damaged just distal to the origin of the inferior mesenteric artery at the level of the third lumbar vertebra. The diagnosis is difficult since occlusion due to thrombosis may occur at the time of the trauma or be delayed for several months. In the presence of complete occlusion, the femoral pulses are absent, and motor and sensory deficits may be present in the lower limbs due to ischemia of the peripheral nerves.

Deceleration injuries to the renal vessels may occur as the body stops its forward motion. The abdominal aorta, like the descending thoracic aorta, is tightly attached to the vertebral column by connective tissue. The kidney, however, is relatively mobile and continues to move forward after body impact, being finally restrained by the attachment of the renal artery to the aorta. Excessive stretching of the renal artery may cause intimal damage with clot formation. The renal vessels may be avulsed from the hilum of the kidney, or the renal artery may be torn free from the aorta. The hemorrhage may be contained in the retroperitoneal space, and hypovolemic shock may not immediately occur.

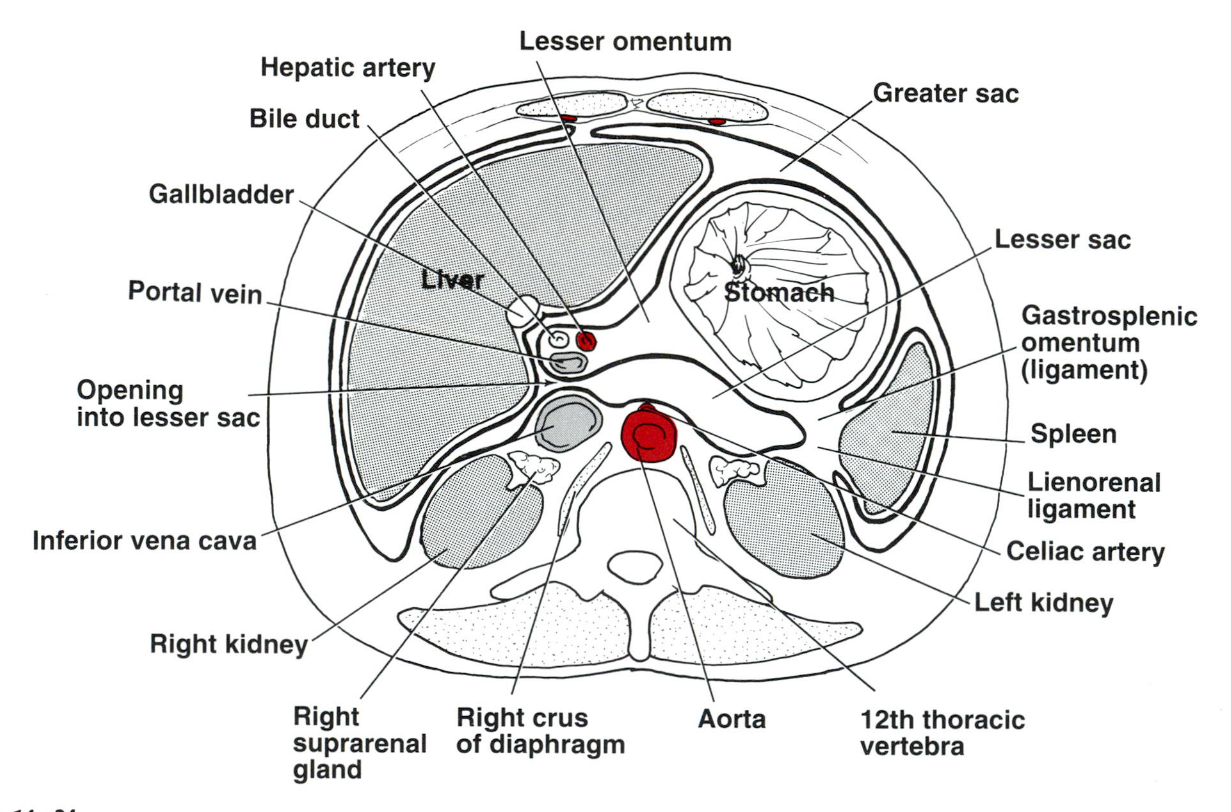

FIG 14–24.
Cross section of the abdomen at the level of the 12th thoracic vertebra as seen from below. Note the positions of the aorta and inferior vena cava relative to the other structures.

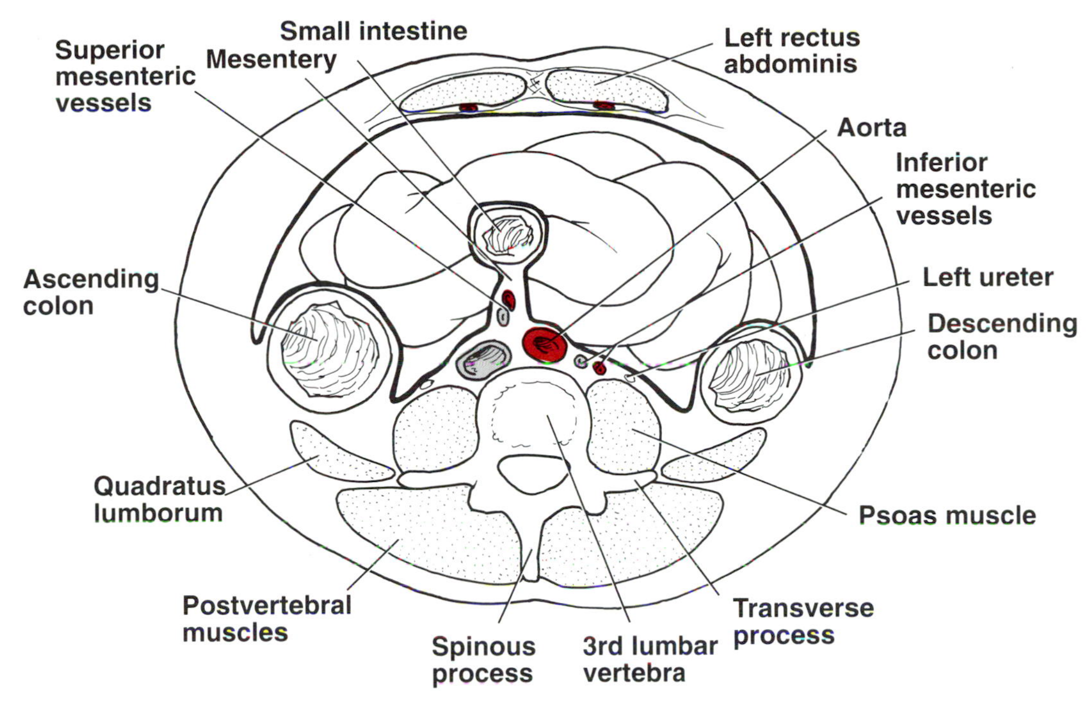

FIG 14–25.
Cross section of the abdomen at the level of the third lumbar vertebra as seen from below. Note the positions of the aorta and inferior vena cava relative to the other structures.

Penetrating Injury.—Penetrating trauma to the abdominal aorta is common and is usually associated with multiple intra-abdominal injuries. The peritoneum has been violated, and hemorrhage occurs directly into the peritoneal cavity. The signs of a distended abdomen associated with those of hypovolemic shock make the diagnosis relatively simple. However, when the arterial leak is small, peritoneal lavage may be necessary to confirm the diagnosis. Penetrating injuries through the back or flank (especially when directed from the left) may cause a retroperitoneal injury to the aorta; the blood may be contained within the retroperitoneal space and delay the onset of hypovolemic shock.

Leaking Aortic Aneurysms

The abdominal aorta is the most common site of aneurysmal disease, and the majority occur below the origin of the renal arteries (Fig 14–26). The cause in most cases is atherosclerotic disease and subsequent weakening of the arterial wall. Occasionally, an aortic aneurysm may extend distally to involve the common iliac arteries. The average age of occurrence is 65 to 75 years. The majority of abdominal aortic aneurysms that are larger than 8 cm will rupture within 5 years, unless resected.

Many aortic aneurysms are asymptomatic and are found during a physical examination of the abdomen or detected by chance by an abdominal radiograph, a computed tomographic scan, or a magnetic resonance image.

Symptomatic nonruptured aneurysms may be encountered in the emergency department. The symptoms occur as the result of the expansion of the aneurysm and the pressure on neighboring structures. Vague gastrointestinal symptoms may be caused by the displacement of the duodenum; the involvement of the inferior mesenteric artery by the aneurysm may give rise to altered bowel habits. Back pain caused by pressure on the lumbar vertebrae is also a common symptom. Abdominal examination may reveal a pulsatile swelling along the course of the aorta. The swelling is located in the midline and may be situated in the epigastrium, the umbilical region, or lower down at the level of the iliac crests, where the aorta bifurcates.

A leaking aneurysm usually causes severe abdominal pain that often extends through to the back. Sometimes the pain is referred along the branches of the lumbar plexus, especially along the femoral nerve, to the groin or thigh, most often the left. Upon rupture, the blood may first be confined to the retroperitoneal space and may track around the abdominal wall, giving rise to ecchymosis in the flanks (Gray-Turner sign) or in the periumbilical region (Cullen's sign). The blood may escape through the

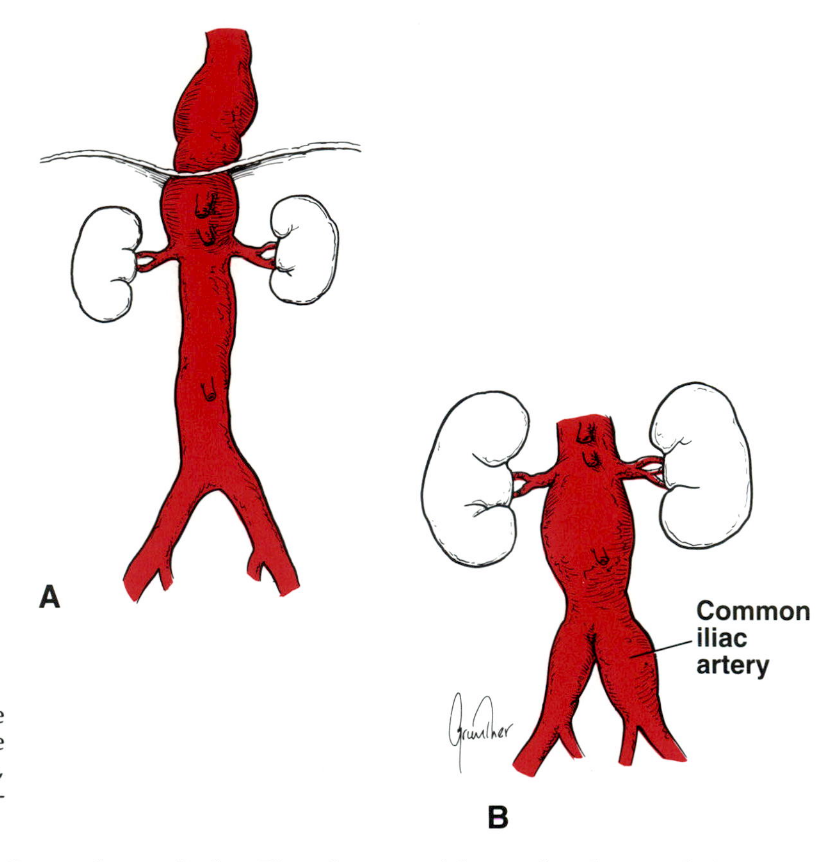

FIG 14–26.
Aneurysms of the abdominal aorta. **A,** above the origin of the renal arteries. **B,** below the level of the renal arteries. The latter arteries, which are the most common, may extend inferiorly to involve the common iliac arteries.

peritoneum on the posterior abdominal wall into the peritoneal cavity. Once the latter occurs, peritoneal irritation causes reflex rigidity of the abdominal muscles. Another possibility is that the aneurysm may rupture into the inferior vena cava, which lies along its right side, or it may rupture into the third part of the duodenum, which crosses its anterior surface. In the latter situation, the patient may present with gastrointestinal bleeding. Usually an aorta that ruptures or erodes into the gastrointestinal tract does so at the site of a previously placed aortic graft.

Embolic Blockage of the Abdominal Aorta

The aorta narrows at its bifurcation into the common iliac arteries, and this is the usual site for lodgment of arterial emboli. If the blockage is complete (aortic saddle embolus) and the diagnosis is delayed, the patient may lose both lower extremities.

Aortic emboli may originate from an auricular thrombosis associated with atrial fibrillation, from thrombi on diseased mitral or aortic valves, from a mural thrombus following myocardial infarction, or from atherosclerotic plaques on the wall of the aorta. Unlike the patient who has a gradual occlusion of the aorta due to atherosclerotic disease and is able to develop an adequate collateral circulation (Fig 14–27), the sudden blockage of the aorta by an embolus produces immediate ischemia of the lower limbs. The absence of femoral pulses and the presence of cold, pale lower extremities should make the diagnosis relatively straightforward. Anesthesia and muscle paralysis of the lower limbs may be present (secondary to peripheral nerve ischemia). Most often, the blockage is incomplete and only one leg is affected. The absence of a history of intermittent claudication is evidence against a gradual occlusion of the aortic bifurcation by a growing thrombus. Immediate thromboembolectomy is the treatment of choice.

Thrombotic Obliteration of the Abdominal Aorta and Iliac Arteries (Leriche Syndrome)

This chronic process is unlikely to present itself in the emergency department. The basic cause is an atherosclerotic lesion in the wall of the terminal part of the abdominal aorta with superimposed thrombosis. The result is a gradual diminished blood flow to both lower limbs and the pelvic viscera, including the penis. The symptoms include fatigue and weakness of both lower limbs due to ischemic atrophy of the muscles, pallor of both lower limbs due to reduced cutaneous blood flow, and an inability to maintain a stable erection due to a lack of blood flow. On examination, there is an absence of pulses distal to the point of obstruction. Since the progress of the disease is slow, some collateral circulation is established but is physiologically inadequate. How-

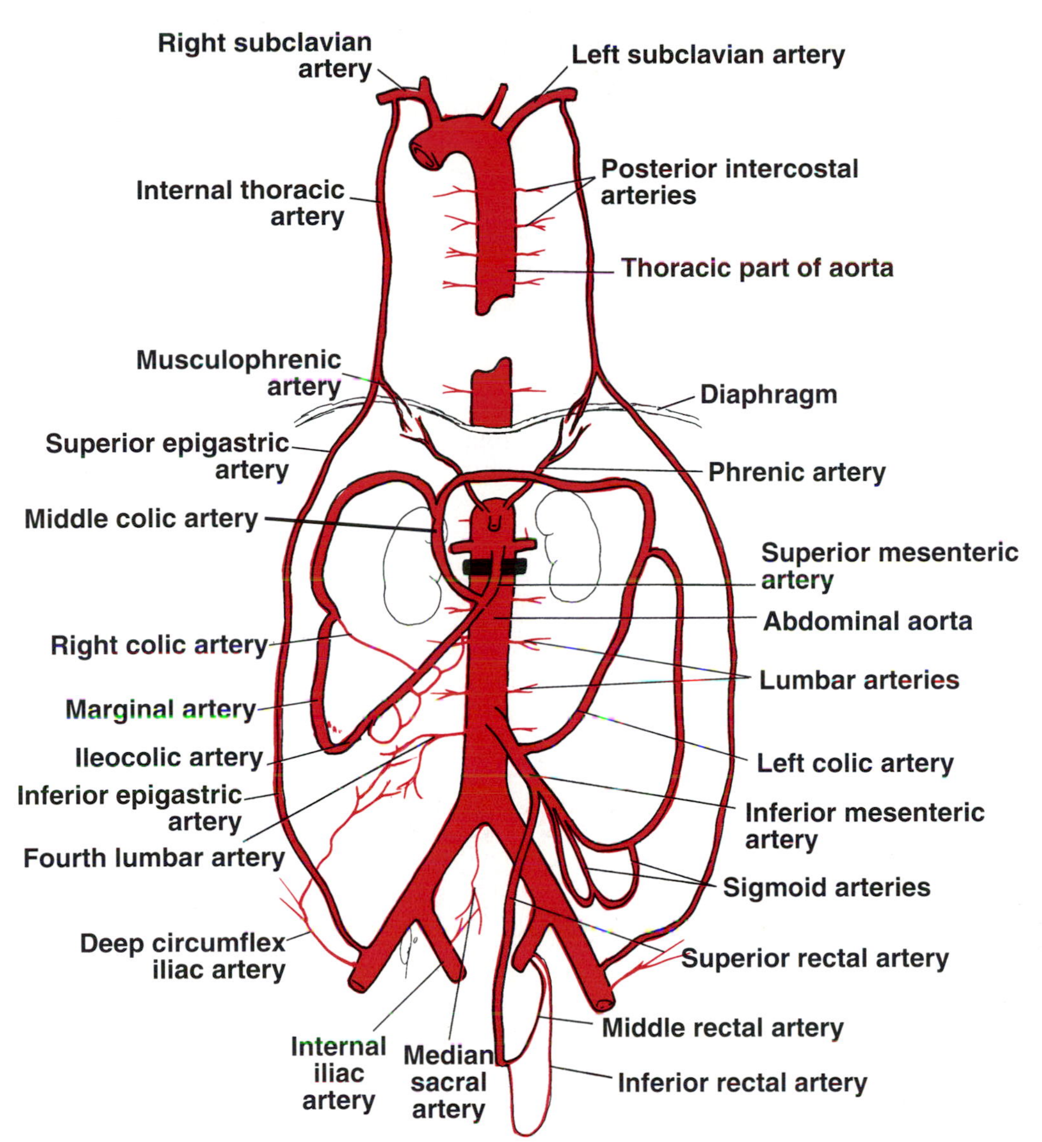

FIG 14–27.
This diagram shows the possible collateral circulations of the abdominal aorta. Note the great dilatation of the mesenteric arteries and their branches that occurs should the aorta be slowly blocked just below the level of the renal arteries (black bar).

ever, the collateral blood flow does prevent ischemic necrosis of both lower limbs, although ischemic ulcers may occur.

The *collateral circulation* of the abdominal aorta (see Fig 14–27) includes the following arteries.

1. The fourth lumbar artery, a branch of the aorta, anastomoses with the deep circumflex iliac branch of the external iliac artery and via the iliolumbar artery anastomoses with the internal iliac artery.
2. The median sacral artery, a branch of the aorta, anastomoses with the lateral sacral artery, a branch of the internal iliac artery.

Mesenteric Artery Occlusion

Occlusion of the superior or inferior mesenteric arteries with intestinal ischemia has been discussed in Chapter 11. The occlusive process commonly occurs at the origin of the artery or in the proximal 1 to 2 cm of the artery and may be caused by an embolus, a thrombus, or by trauma. Occlusive disease of the superior mesenteric artery is much more common than that of the inferior mesenteric artery, which may be explained by the angle of takeoff of the superior artery from the aorta. In cases of embolus of the superior mesenteric artery, the embolus usually lodges in the region of the middle colic artery so that the jejunum may be spared.

Clinical Anatomy of the Inferior Vena Cava

Trauma

Blunt and Penetrating Injuries.—Although the inferior vena cava contains blood under low pressure, injuries to this vessel are commonly lethal because hemorrhage from the upper part of the inferior vena cava is difficult to control. The anatomical inaccessibility of the vessel behind the liver, duodenum, and mesentery, and the blocking presence of the right costal margin make the surgical approach a difficult one. In addition, the thin wall of the vena cava makes it prone to tears that may be extensive. Moreover, traumatic injuries to the vena cava are commonly associated with extensive hepatic injuries. One of the most common sites for injury is where the hepatic veins drain into the vena cava just below the diaphragm.

Because of the extensive anastomoses of the tributaries of the inferior vena cava, it is possible to ligate the inferior cava below the level of the renal veins without serious adverse results.* The anastomoses of the vena cava are as follows (Fig 14–28).

*The majority of patients will suffer venous congestion of both lower limbs.

1. The lumbar veins, which are tributaries of the inferior vena cava, anastomose behind the diaphragm with the azygos and hemiazygos veins, which are tributaries of the superior vena cava.

2. The lumbar veins also anastomose with the superficial veins of the trunk, which eventually drain into the superior vena cava via the lateral thoracic veins, tributaries of the axillary veins.

Thrombosis of the Inferior Vena Cava

Thrombosis of the inferior vena cava or external compression by a retroperitoneal tumor results in the opening up of an alternative venous pathway (see the venous anastomoses listed above). This is commonly referred to as a *caval caval shunt.* The same pathway comes into effect in patients with a superior mediastinal tumor compressing the superior vena cava. Clinically, the enlarged subcutaneous anastomosis between the lateral thoracic vein, a

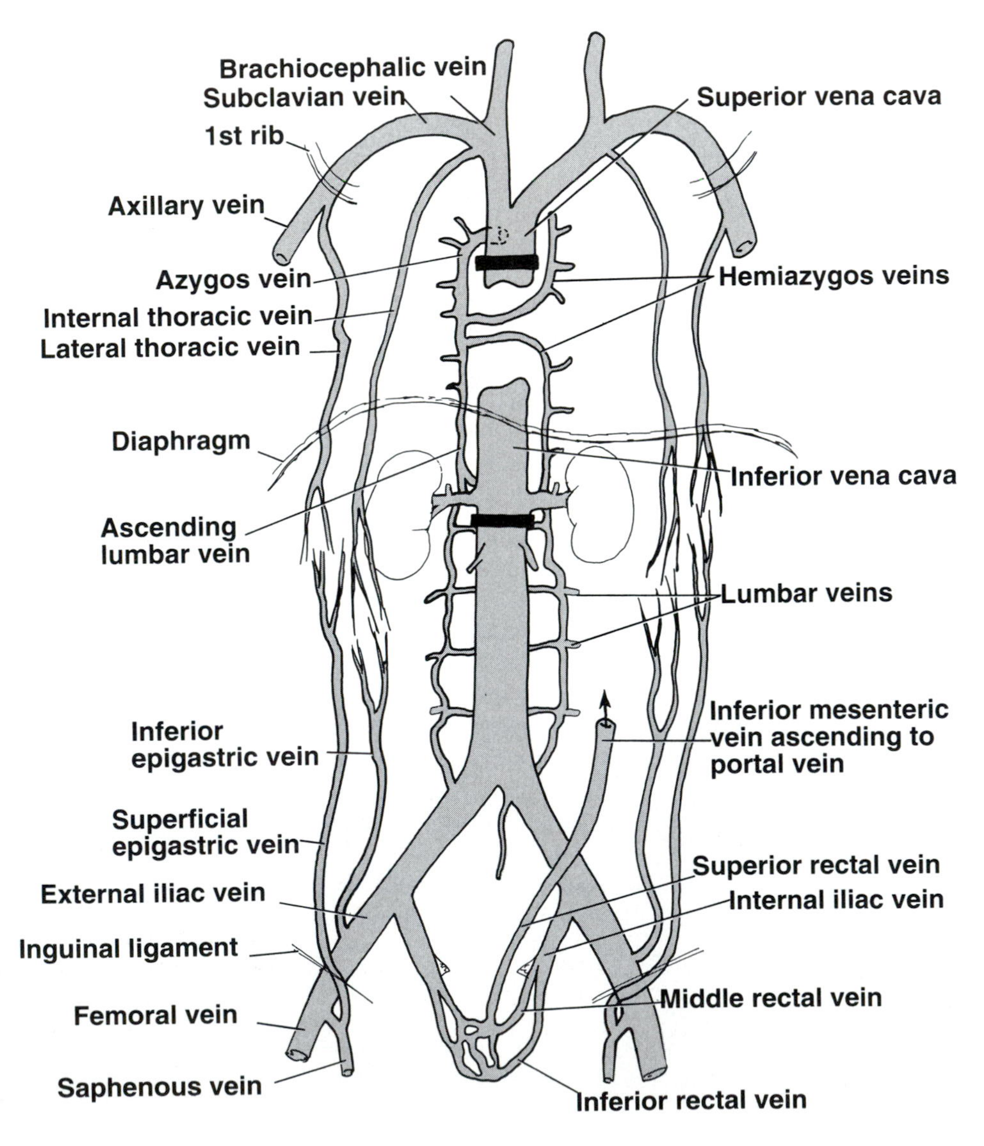

FIG 14–28.
Diagram showing the possible collateral circulations of the superior and inferior vena cavae. Note the alternative pathways that exist for blood to return to the right atrium should the superior vena cava become blocked below the entrance of the azygos vein (black bar). Similar pathways exist should the inferior vena cava become blocked. Note also the connections that exist between the portal circulation and the systemic veins in the anal canal.

tributary of the axillary vein, and the superficial epigastric vein, a tributary of the femoral vein, may be seen on the thoracoabdominal wall (see Fig 14–28).

Aortoinferior Vena Caval and Other Abdominal Vessel Fistulae

Aortoinferior vena caval fistulae usually follow penetrating wounds to the abdomen; rarely they follow erosion of the venous wall by an expanding aortic aneurysm. The close anatomical relationship of the aorta to the left side of the inferior vena cava, the renal artery to the renal vein, and the iliac arteries to the iliac veins facilitates the formation of arteriovenous fistulae. The clinical findings include a raised pulse rate, increased cardiac size, lowered diastolic blood pressure with increased pulse pressure, the presence of a machinelike murmur, a thrill, and the formation of an aneurysm at the site of the arteriovenous communication.

Clinical Anatomy of the Portal Vein

Blood flow in the portal vein, portal systemic anastomoses, and *portal hypertension* are fully discussed in Chapter 11.

Trauma

Penetrating Injuries to the Portal Vein.—Penetrating injuries to the portal vein are life threatening and are usually associated with multiple abdominal injuries. A deep penetrating abdominal wound on the transpyloric plane, about two fingerbreadths to the right of the midline, could easily penetrate the liver and perforate the first part of the duodenum, the portal vein, and the inferior vena cava.

Clinical Anatomy of the Arteries of the Lower Extremity

Collateral Circulation

If the arterial supply to the lower extremity is occluded, necrosis or gangrene will follow unless there is an adequate collateral circulation bypassing the obstruction. Sudden occlusion of the femoral artery by ligature or embolism, for example, is usually followed by gangrene. However, gradual occlusion such as occurs in atherosclerosis is less likely to be followed by necrosis, since the collateral blood vessels have time to dilate fully. The collateral circulations around the hip and knee joints are described on p. 549.

Traumatic Injury

Injury to the large femoral artery can cause rapid exsanguination of the patient. Unlike in the upper extremity, arterial injuries of the lower limb do not have a good prognosis. The collateral circulations around the hip and knee joints, although present, are not as adequate as those around the shoulder and elbow. The presence of existing hemodynamic impairment from atherosclerosis worsens the chances of saving the distal part of the limb.

Traumatic injuries to the arteries may be direct from the trauma or indirect from neighboring bone. In those situations where the artery lies close to a bone or joint, for example the popliteal artery and the knee joint, a fracture or dislocation may cause considerable damage to the vessel. In such cases, especially in the presence of weak or absent distal pulses, arteriography should be performed.

In patients with penetrating wounds, penetration, transection, intimal flap raising, and laceration are the usual forms of arterial injury. Blunt trauma may damage the tunica intima, resulting in thrombosis and occlusion. Shock waves from high-velocity bullets when passing close to an artery produce an area of arterial contusion with damage to the tunica intima.

The concomitant damage to a neighboring large vein further complicates the situation and causes further impairment of the circulation to the distal part of the limb. Inadequate venous return produces edema and fascial compartmental problems, leading to muscle necrosis.

The restoration of an adequate circulation should be undertaken by an experienced surgeon and depends on the following factors. (1) Rapid repair of the damaged artery should be done by simple suture or resection and anastomosis; if this is impossible the continuity of the artery may be restored using an autogenous vein graft. It is uneccessary to repair arteries distal to the popliteal artery unless more than one of the arteries is damaged. (2) Repair of the accompaning vein should be done to improve the circulation and reduce the incidence of compartmental syndromes and venous emboli. (3) Extensive fasciotomy may be necessary in the presence of marked edema; all four osteofascial compartments of the distal leg should be opened.

Anatomy of the Complications of Arterial Injury

Arteriovenous Fistulae.—A single perforating injury to an artery in the lower limb may also perforate an accompanying vein and establish an acute arteriovenous fistula. This complication is common in regions where large vessels run close together, such as the femoral artery and vein in the femoral triangle and the subsartorial canal, and the popliteal artery and vein behind the knee. Such a shunt produces a continuous machinerylike murmur over the fistula with the later development of varicosities and edema of the distal part of the limb.

False Aneurysms.— Anatomically, a true aneurysm is one whose wall contains all three layers of the arterial wall, namely, the intima, media, and adventitia. A false aneurysm, sometimes callled a *pulsating hematoma,* is one whose wall contains only the tunica adventitia. An arterial perforation may lead to a false aneurysm that becomes walled off by the adventitia and surrounding tissues. Pressure on neighboring nerves may give rise to neurological symptoms.

Compartment Syndromes.— Delay in the diagnosis and repair of injured arteries, especially when accompanied by vein damage, may lead to muscle necrosis and compartmental edema. The compartment syndromes are discussed in Chapter 17.

Intermittent Claudication and Arterial Occlusive Disease of the Lower Extremity

Atherosclerosis of the lower limb arteries is common in men. Ischemia of the muscles produces a cramplike pain with exercise. *Intermittent claudication* is a condition characterized by calf muscle cramping pain on exertion that is relieved by rest. Thrombosis in a diseased segment of the artery may lead to a sudden worsening of the symptoms and even cause nocturnal ischemic pain. The common sites for occlusion are the femoral, popliteal, and tibioperoneal arteries. If the obstruction occurs more proximally in the aorta or iliac arteries, impotence is common.

Popliteal, Anterior, and Posterior Tibial Artery Occlusions

Popliteal artery occlusion occurs just below the beginning of the artery (just below the opening in the adductor magnus muscle). In some cases the occlusion extends distally to involve the origins of the anterior and posterior tibial arteries and even the peroneal artery. Symptoms include intermittent claudication, night cramps, and rest pain caused by ischemic neuritis. Signs include impaired or absent arterial pulses, lowered skin temperature, color changes, muscular weakness, and trophic changes.

Sympathetic Innervation of the Arteries of the Lower Extremity

The arteries of the lower extremity are innervated from the lower three thoracic and upper two or three lumbar segments of the sympathetic outflow of the spinal cord. The preganglionic fibers pass to the lower thoracic and upper lumbar ganglia via white rami. The fibers synapse in the lumbar and sacral ganglia, and the postganglionic fibers reach the blood vessels via branches of the lumbar and sacral plexuses. For example, the femoral artery receives its sympathetic fibers from the femoral and obturator nerves. The more distal arteries receive their postganglionic fibers via the common peroneal and tibial nerves.

Lumbar Sympathectomy and Occlusive Arterial Disease

Lumbar sympathectomy may be advocated as a form of treatment in occlusive arterial disease of the lower extremity in order to increase the blood flow through the collateral circulation. Preganglionic sympathectomy is performed by removing the upper three lumbar ganglia and the intervening parts of the sympathetic trunk.

Femoropopliteal and Femorotibial Bypass for Lower Extremity Vascular Insufficiency

Bypasses have been used with success in patients with severe distal extremity ischemia. In the presence of gangrene or severe rest pain, the distal limb has been successfully salvaged. A reversed ipsilateral autogenous saphenous vein, a contralateral saphenous vein, or a cephalic vein graft has been used to connect up the proximal femoral artery or popliteal artery to the distal popliteal or tibial or peroneal arteries.

Aneurysms of the Lower Extremity

These occur much less frequently than abdominal aortic aneurysms and are usually caused by atherosclerosis. Most patients are over 50 years of age, and the common sites are the femoral and popliteal arteries. The diagnosis is usually made by finding an expansile swelling along the course of the artery. Patients may present in the emergency department with complications, which include sudden embolic obstruction to arteries distal to the aneurysm or sudden thrombotic occlusion of the aneurysm. Pressure on neighboring nerves may give rise to symptoms; for example, an enlarging popliteal aneurysm may press on the tibial nerve, causing pain in the foot. Rupture of femoral or popliteal aneurysms is rare.

Femoral Artery Catheterization

Anatomy of the Technique.—The femoral artery is first located just below the inguinal ligament midway between the symphysis pubis and the anterior superior iliac spine (see Fig 14–23). The needle or catheter is then inserted into the artery. The following structures are pierced: (1) skin, (2) superficial fascia, (3) deep fascia, and (4) anterior layer of the femoral sheath.

Anatomy of the Complications.—The femoral vein lies immediately medial to the artery and may be entered in error. The nonpulsatile nature of the vein on palpation should exclude this possibility. Since the hip joint lies posterior to the femoral artery, the erroneous passage of the needle through the posterior arterial wall may cause it to pierce the psoas muscle and enter the joint cavity. Some difficulty may be experienced in passing the catheter up the femoral artery if the artery is tortuous or if there is extensive atherosclerosis of the arterial wall.

Clinical Anatomy of the Veins of the Lower Extremity

Venous Function

While it is generally understood that the veins of the lower extremity serve as blood conduits, thermoregulators, and storage vessels, the importance of the venous pump is not always appreciated.

Venous Pump of the Lower Limb

Within the closed fascial compartments of the lower limb, the thin-walled, valved venae comitantes are subjected to intermittent pressure both at rest but especially during exercise. The pusations of the adjacent arteries help move blood up the limb. The contractions of the large muscles within the compartments during exercise compress these deeply placed veins and force blood up the limb.

The superficial saphenous veins, except near their termination, lie within the superficial fascia and are not subject to these compression forces. The valves in the perforating veins prevent the high-pressure venous blood from being forced outward into the low-pressure superficial veins. Moreover, as the muscles within the closed fascial compartment relax, venous blood is sucked from the superficial to the deep veins.

Venous Valve Incompetence

The competency of the valves in the great saphenous vein may be determined by using the *Trendelenburg test*. With the patient in a supine position, the lower extremity is raised to allow the blood to drain out of the superficial vein into the femoral vein. The fingers are then placed firmly over the great saphenous vein about 1½ in. (two fingerbreadths) below and lateral to the pubic tubercle, i.e., at a point where the saphenous vein joins the femoral vein. With firm pressure still in place, the patient then assumes the upright position and the finger pressure is suddenly removed. If the great saphenous vein fills immediately from above downward, it is clear that the valves in the vein are incompetent.

Varicosed Veins

A varicosed vein is one that has a larger diameter than normal and is elongated and often tortuous. The condition commonly occurs in the superficial veins of the lower limb, and although not a life-threatening disease, it is responsible for considerable discomfort and pain.

Varicosed veins have many causes, including hereditary weakness of the vein walls and incompetent valves, elevated intra-abdominal pressure as the result of multiple pregnancies or abdominal tumors, and thrombophlebitis of the deep veins, which results in the superficial veins becoming the main pathway for the lower limb.

Varicose Leg Ulcers.—These occur in the region of the medial malleolus, are caused by venous skin stasis, and may be a complication of varicosed veins; many are caused by postthrombotic incompetent perforating veins in the region.

A venous ulcer must be distinguished from an *arterial ulcer* caused by atherosclerosis of the skin arteries. An arterial ulcer tends to occur on the lateral side of the distal leg, and the leg is often pulseless and cool. A venous ulcer occurs on the medial side of the distal leg because skin venous stasis tends to be more severe on the medial side in the presence of varicosed veins. The explanation for the laterally placed arterial ulcer is that the skin over the lateral malleolus receives a poorer arterial supply than that over the medial malleolus.

Traumatic Bleeding From a Varicosed Vein.—Profuse bleeding from a pierced varicosed vein may cause a patient to seek treatment in the emergency department.

Superficial Thrombophlebitis

Thrombosis of the superficial veins of the lower limb (see Fig 14–20) is often associated with varicosed veins. The condition is painful, and the thrombosed vein is tender to touch; the overlying skin is reddened and edematous. The thrombus is usually strongly adherent to the wall of the vein so that emboli are rarely formed. However, should the thrombosis extend to the deep veins through a perforating vein, embolic formation in the deep veins can be a serious, although rare, complication.

Superficial thrombophlebitis must be distinguished from *superficial lymphangitis* involving the superficial lymphatic vessels of the lower extremity. These vessels ascend from the foot in the superficial fascia accompanying the great and small saphenous veins. In acute lymphangitis a red skin line or lines due to the presence of the underlying inflammed lymphatic vessels can often be seen coursing from the site of infection in the foot to the regional lymph nodes in the inguinal region.

Deep Thrombophlebitis

Thrombosis of the deep veins can occur at any time, but significant predisposition is immobility of the lower limbs in bed or in a cast. The common site where the process starts is the veins draining the soleus muscles in the calf. It must be assumed that the pressure of the bed on the calf veins damages the tunica intima, and this together with certain predisposing factors, such as surgical trauma, malignant disease, pregnancy, or estrogen therapy, initiates thrombus formation. Once formed, the thrombus may extend proximally into the popliteal and femoral veins and even higher into the iliac veins. The symptoms include discomfort and tightness in the calf, especially when the patient is using the calf muscles, as in standing and walking. Tenderness of the calf muscles may be apparent, and edema of the ankles, pretibial area, or thigh may be present. The superficial veins may be dilated and more obvious than normal. The great danger of deep vein thrombosis is the high incidence of pulmonary embolism. A secondary problem is residual chronic venous insufficiency of the lower extremities.

Great Saphenous Vein Cut Down

This procedure is usually performed at the ankle (Fig 14–29). Unfortunately, this site has the disadvantage that phlebitis is a potential complication. The great saphenous vein also may be entered at the groin where phlebitis is relatively rare; the larger diameter of the vein at this site permits the use of larger catheters and the rapid infusion of large volumes of fluids.

Anatomy of Ankle Vein Cut Down.—The procedure is as follows.

1. The sensory nerve supply to the skin immediately in front of the medial malleolus of the tibia is from branches of the saphenous nerve, a branch of the femoral nerve. The saphenous nerve branches are blocked with local anesthetic.
2. A transverse incision is made through the skin and subcutaneous tissue across the long axis of the vein just anterior and superior to the medial malleolus (see Fig 14–29). Although the vein may not be visible through the skin, it is *constantly* found at this site.
3. The vein is easily identified, and the *saphenous nerve* should be recognized; the nerve usually lies just anterior to the vein (see Fig 14–29).

Anatomy of Groin Vein Cut Down.—The procedure is as follows.

1. The area of thigh skin below and lateral to the scrotum or labium majus is supplied by branches of the ilioinguinal nerve and the intermediate cutaneous nerve of the thigh. The branches of these nerves are blocked with local anesthetic.
2. A transverse incision is made through the skin and subcutaneous tissue centered on a point about 1½ in. (4 cm) below and lateral to the pubic tubercle (see Fig 14–29). If the femoral pulse can be felt, the incision is carried medially just medial to the pulse.
3. The great saphenous vein lies in the subcutaneous fat and passes posteriorly through the saphenous opening in the deep fascia to join the femoral vein about 1½ in. (two fingerbreadths) below and lateral to the pubic tubercle.

Two potential problems may occur with saphenous vein cut down at the groin. The operator may look for the vein too far laterally and may cut too deeply and penetrate the delicate cribriform fascia (see Saphenous Opening below) or the deep fascia. The vein lies superficially in the subcutaneous fat (which may be thick in obese individuals) and is usually easy to recognize. The deep fascia has a shiny appearance and is tough in consistency.

Anatomy of Groin Vein Cut Down in Relation to the Saphenous Opening.—This relationship is as follows.

Saphenous Opening.—The saphenous opening (fossa ovalis) is an oval gap in the deep fascia of

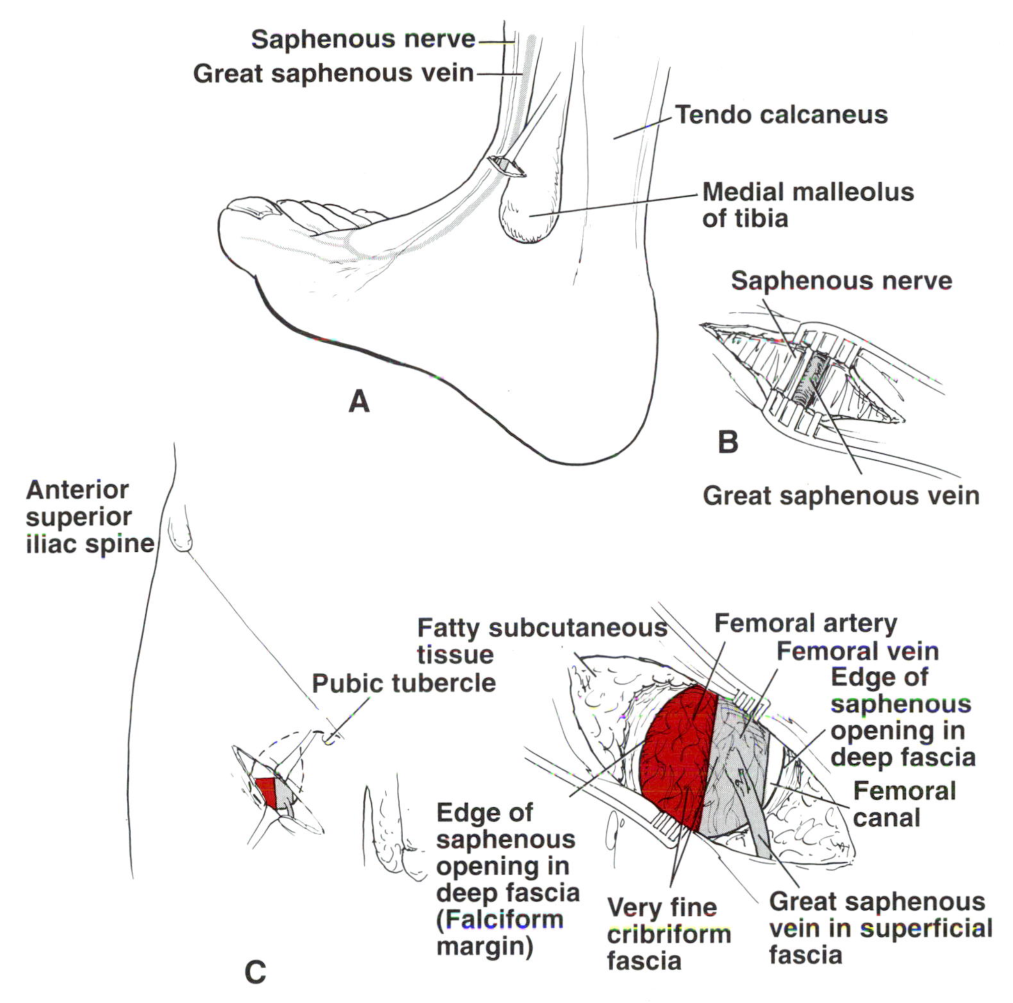

FIG 14–29.
Great saphenous vein cut down. **A** and **B,** at the ankle. The great saphenous vein is constantly found in front of the medial malleolus of the tibia. **C** and **D,** at the groin. The great saphenous vein drains into the femoral vein two fingerbreadths below and lateral to the pubic tubercle.

the thigh, which transmits the great saphenous vein, some small branches of the femoral artery, and lymph vessels (see Fig 14–29). It is situated about 1½ in. (two fingerbreadths) below and lateral to the pubic tubercle. The *falciform margin* is the lower lateral border of the opening, which lies anterior to the femoral vessels (see Fig 14–29). The medial border of the opening curves around behind the femoral vessels. It is thus seen that the deep fascia is attached to the whole length of the inguinal ligament above, and the falciform margin of the saphenous opening sweeps downward and laterally from the pubic tubercle. The border of the opening then curves upward and medially and then laterally behind the femoral vessels, to be attached to the pectineal line of the superior ramus of the pubis.

The saphenous opening is filled with loose connective tissue called the *cribriform fascia.* It is important to understand that the great saphenous vein passes through the saphenous opening to gain entrance to the femoral vein. However, the size and shape of the opening are subject to variation.

Femoral Vein Catheterization

This procedure is used when there is a need for rapid access to a large vein. The femoral vein has a constant relationship to the medial side of the femoral artery just below the inguinal ligament and is easily cannulated. However, because of the high incidence of thrombosis with the possibility of fatal pulmonary embolism, the catheter should be removed once the patient is stabilized.

Anatomy of the Procedure.—The procedure is as follows.

1. The skin of the thigh below the inguinal ligament is supplied by the genitofemoral nerve; this nerve is blocked with a local anesthetic.
2. The femoral pulse is palpated, and the femoral vein lies immediately medial to it. Should the pa-

tient be pulseless, the position of the femoral artery can be determined by taking the midpoint of a line joining the anterior superior iliac spine to the symphysis pubis; the femoral vein lies a fingerbreadth medial to this point.

3. At a site about two fingerbreadths below the inguinal ligament, the needle is inserted into the femoral vein.

Anatomy of the Complications.— These include the following.

1. Thrombophlebitis of the femoral vein. This is a common complication and occurs following prolonged femoral catheterization. Pulmonary embolism is a further complication.

2. Hematoma in the femoral triangle. Poor technique and excessive damage to the wall of the femoral vein are the causes.

3. Infection of the hip joint. This occurs if the needle is pushed too deeply and pierces the femoral vein or misses the vein altogether. The needle traverses the psoas muscle and pierces the capsule of the hip joint, which lies posterior to the vein.

4. Rarely, the femoral nerve is damaged because the needle is inserted too far laterally. The surface marking of the femoral nerve as it emerges from beneath the inguinal ligament is the midpoint of a line joining the anterior superior iliac spine to the pubic tubercle.

5. Rarely, if the catheter is advanced too rapidly, the tip may penetrate the wall of the external iliac vein and enter the peritoneal cavity.

Intraosseous Infusion in the Infant

The technique may be used for the infusion of fluids and blood when it has been found impossible to obtain an intravenous line. The procedure is easy and rapid to perform as follows.

1. With the distal leg adequately supported, the anterior subcutaneous surface of the tibia is palpated.

2. The skin is anesthetized about 1 in. (2.5 cm) distal to the tibial tuberosity, thus blocking the infrapatellar branch of the saphenous nerve (see Fig 14–14).

3. The bone marrow needle is directed at right angles through the skin, superficial fascia, deep fascia, and tibial periosteum and the cortex of the tibia. Once the needle tip reaches the medulla and bone marrow, the operator senses a feeling of "give." The position of the needle in the marrow can be confirmed by aspiration. The needle should be directed slightly caudad to avoid injury to the epiphyseal plate of the proximal end of the tibia. The transfusion may then commence.

Doppler Ultrasound Examination of Venous Flow in the Lower Extremity

Posterior Tibial Veins.— The probe is applied to the skin just posterior to the medial malleolus of the tibia. Here the posterior tibial veins accompany the posterior tibial artery between the tendon of the flexor digitorum longus and the posterior tibial nerve (Fig 14–30).

Popliteal Vein.— The probe is applied to the skin over the popliteal space with the knee partly flexed to relax the deep fascia. The flow signal is best heard over the vein just lateral to the popliteal artery (see Fig 14–30).

Femoral Vein (Superficial Femoral Vein).— The probe is placed over the vein at midthigh as it lies in the subsartorial canal with the femoral artery (see Fig 14–30).

Femoral Vein (Common Femoral Vein).— The probe is applied to the skin in the femoral triangle just below the inguinal ligament. (At this site the deep femoral vein [profunda vein] has joined the superficial femoral vein [femoral vein] to form the common femoral vein, using the old terminology.) The femoral vein lies medial to the femoral artery (see Fig 14–30). The pulse of the artery can easily be felt at the midpoint between the anterior superior iliac spine and the symphysis pubis.

ANATOMICAL-CLINICAL PROBLEMS

1. A 55-year-old man was involved in a head-on automobile accident. When seen in the emergency department, he was in hypovolemic shock and showed signs of extensive bruising on the lower part of the anterior abdominal wall. He was wearing a seat belt at the time of the accident. On examination, his abdomen was distended and tense; he had hypotension and tachycardia. A diagnosis of ruptured abdominal aorta was made during an emergency laparotomy. In cases of blunt traumatic injury to the abdominal aorta, do all patients become hypotensive immediately? Explain the possible role that the kidneys may play in causing damage to the aorta in deceleration injuries.

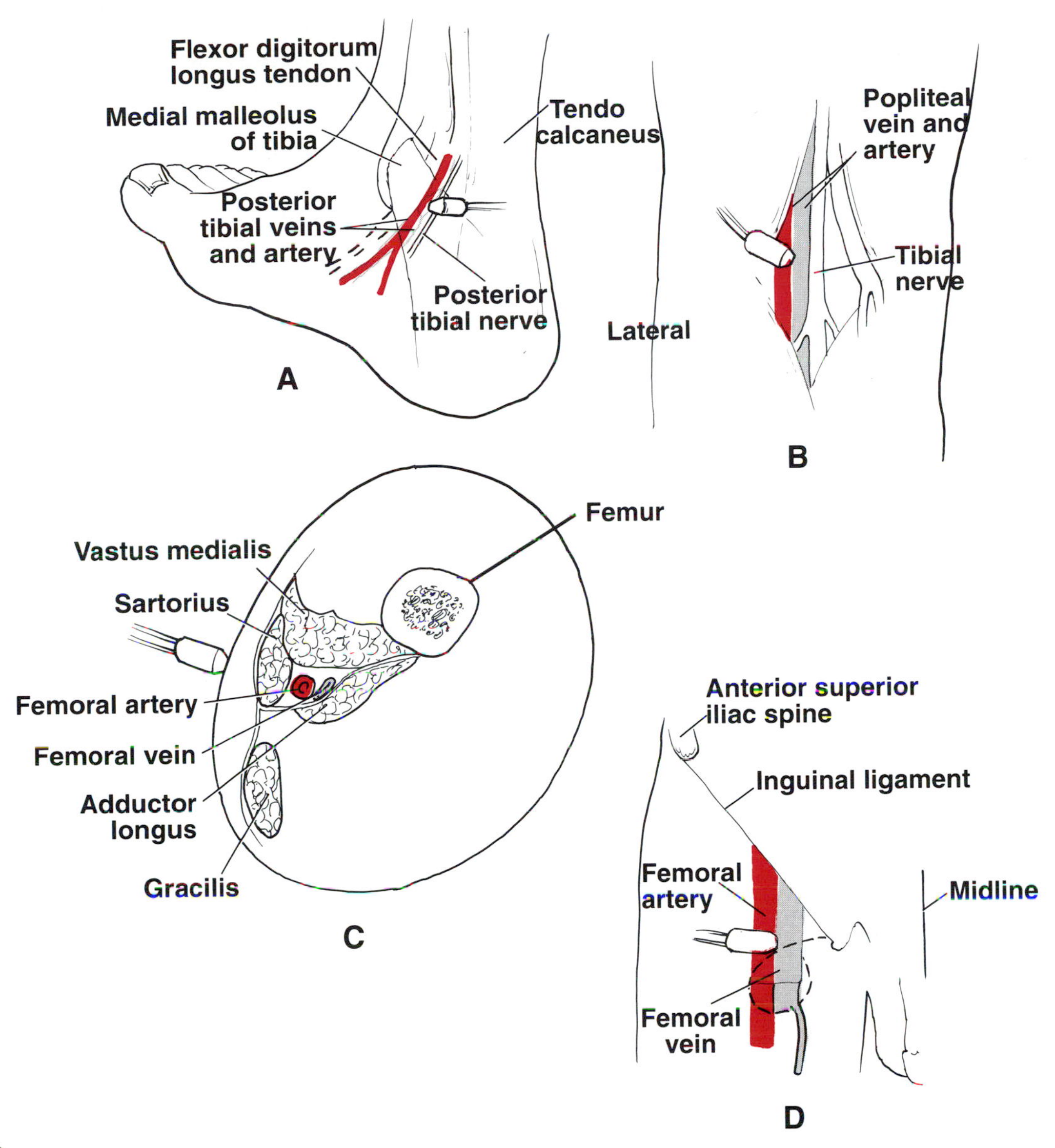

FIG 14–30.
Doppler ultrasound of venous flow in the lower limb. **A,** posterior tibial veins at the ankle. **B,** popliteal vein behind the knee. **C,** femoral vein at the subsartorial (adductor) canal in the midthigh. **D,** femoral vein just below the inguinal ligament.

2. A 70-year-old man was seen in the emergency department complaining of the sudden onset of severe lumbar back pain. Two years previously he had had a myocardial infarction. On questioning, the patient admitted that he often experienced mild back and hip pains on getting up in the morning, but never had he experienced such a severe back pain. On examination, a somewhat tender pulsatile swelling could be felt in the abdomen at the level of the umbilicus. Both femoral pulses were present. A diagnosis of abdominal aortic aneurysm was made. What is the surface marking of the abdominal aorta? Explain why the back pain had started so suddenly and its significance. When an abdominal aneurysm ruptures does an immediate fatal outcome always occur? Into which hollow viscera or blood vessels may an aortic aneurysm rarely rupture into?

3. A 59-year-old man was seen in the emergency department with a fracture of the right ankle. He said he had had difficulty in walking for the past year and that day he had stumbled on an uneven sidewalk. On questioning, he gave a history of progressive weakness of both lower limbs and stated that sometimes during a walk he experienced pain

in his left calf that disappeared with rest. On close questioning, he admitted that he had been impotent for 2 years. On examination, both legs had a thin, pale appearance. Both femoral pulses were barely palpable, and the popliteal and dorsalis pedis pulses were absent in both legs. If it is assumed that the patient had a chronic arterial obstruction to both legs, explain (1) the muscular weakness in both lower limbs and (2) the impotence. If it is assumed that this patient had thromboembolic disease at the aortic bifurcation, explain how the collateral circulation prevented the onset of gangrene of both lower extremities.

4. Explain in anatomical terms why penetrating injuries to the inferior vena cava are commonly fatal. Explain how it is possible to ligate the inferior vena cava below the level of the renal veins without adverse effects.

5. It is generally known that penetrating wounds to the arteries of the lower limb have a worse prognosis than those of the upper limb. In anatomical terms explain this statement. Name the situations in the lower limb where large arteries travel close to bones and may be damaged by a bone fracture. What is the collateral circulation around the knee joint? Where in the lower limb do large arteries and veins travel close to one another and are possible sites of arteriovenous fistulae?

6. A 65-year-old man was seen in the emergency department complaining of the onset of a sudden pain in his right foot. He said that for the past 6 months he had experienced some aching pain in the lower part of the right leg, but the foot pain had occurred quite suddenly and was different. On examination, a tender pulsatile swelling could be palpated in the right popliteal space. In anatomical terms explain the chronic aching pain in the right lower leg. What is your explanation for the sudden onset of pain in the right foot?

7. A 45-year-old woman was seen in the emergency department with a painfully swollen left leg. On examination she was found to have extensive varicosed veins in both lower extremities. Her left leg was swollen, and her leg skin was discolored brown, especially in the region of the medial malleolus. Several of the varicosed veins below the knee were tender to touch, and the overlying skin was reddened and warm. The tenderness showed a definite linear course along the superficial veins. What is the diagnosis? Is this condition likely to spread to her deep veins? Is pulmonary embolism a common complication in this condition?

8. Explain the significance of the valved perforating veins of the lower limbs. What is the surface marking of the small saphenous vein at the ankle?

9. What is the surface anatomy of the femoral artery? Name the structures that are pierced by the insertion of a catheter into the femoral artery. What is the relationship of the femoral artery to the femoral vein and the hip joint?

10. Compare in anatomical and practical terms the advantages and disadvantages of cutting down on the great saphenous vein at the groin and ankle. What is the surface marking of the great saphenous vein in the groin and at the ankle? What is the relationship of the saphenous nerve to the great saphenous vein at the ankle?

11. What are the common complications of femoral vein catheterization? What is the surface marking of the femoral vein?

12. Any treatment that involves prolonged immobilization of the lower limb runs the risk of thrombophlebitis of the deep veins of the leg. Where is the common site for the thrombosis to start?

ANSWERS

1. Frequently, patients with blunt rupture to the abdominal aorta may not immediately show signs of hypovolemic shock because the aorta is situated behind the peritoneum in the retroperitoneal space and the blood may not escape immediately into the peritoneal cavity.

On impact, the patient may be held stationary by the seat belt, but the kidneys may continue forward until restrained by the vascular pedicles. Avulsion of the renal artery from the side of the aorta may take place under these circumstances.

2. The abdominal aorta is a midline structure that enters the abdomen at the level of the 12th thoracic vertebra, and its entrance may be projected onto the anterior abdominal wall just above the transpyloric plane (see p. 555). The vessel extends downward to its bifurcation into the common iliac arteries at the level of the summit of the iliac crests.

The sudden onset of severe back pain can be explained by the aneurysm suddenly expanding or rupturing and pressing on the vertebral column, which lies immediately posterior to the aorta.

Death does not always immediately follow an abdominal aneurysm rupture. This can be explained by the fact that the hemorrhage may be initially confined to the retroperitoneal space, and a tamponade effect may temporarily prevent further bleeding.

The abdominal aorta is crossed by the third part

of the duodenum, and cases have been reported of an aneurysm rupturing into the duodenal lumen. The inferior vena cava lies along the right side of the aorta, and an aneurysm has been known to rupture into it producing a massive arteriovenous fistula.

3. The chronic arterial insufficiency of both lower limbs over a period of months or years had resulted in extensive muscular atrophy with accompanying weakness. The gradual blockage of the aorta at the bifurcation had closed the entrance into both common iliac arteries and substantially reduced the blood flow through the internal iliac arteries and its branches to the penis—thus the loss of erection.

The collateral circulation of the abdominal aorta is fully described on p. 563. The collateral circulation allowed sufficient blood to enter the femoral arteries in both lower limbs, distal to the obstruction, to prevent the onset of gangrene of the limbs.

4. Penetrating injuries of the upper part of the inferior vena cava are commonly fatal because (1) the site of the injury is inaccessible behind the liver, duodenum, and the mesentery of the small intestine; (2) the presence of the right costal margin makes surgical access difficult; (3) the thin walls of the vena cava are likely to tear extensively and make repair difficult; and (4) the almost certain possibility that the liver is also damaged.

The extensive anastomosis of the lumbar veins with other retroperitoneal veins ensures that the blood is able to bypass the obstruction should the inferior vena cava be ligated below the level of the renal veins.

5. Although the femoral and popliteal arteries have a collateral circulation, they are not so extensive as those of the main arteries of the upper limb around the shoulder and elbow joints. Moreover, the arteries of the lower limb, including those of the collateral circulation, are more subject to atherosclerotic narrowing than those of the upper limb.

The popliteal artery lies in contact with the popliteal surface of the femur and may easily be damaged in fractures involving the distal end of the femur.

The collateral circulation around the knee joint is fully described on p. 550.

In the femoral triangle and the subsartorial canal, the femoral artery and vein lie in close contact with one another; in the popliteal space, the popliteal artery and vein lie close together.

6. The chronic aching pain in the right lower leg could be explained by the pressure of the expanding popliteal aneurysm on the tibeal nerve in the popliteal space. The sudden onset of severe pain in the foot could be explained by the lodging of an embolus in one of the arteries in the foot. The embolus could have originated as a thrombus in the wall of the popliteal aneurysm.

7. The diagnosis is a thrombophlebitis of the superficial veins of the left lower limb. Superficial thrombosis only rarely extends to the deep veins via perforating veins with incompetent valves. Pulmonary embolism is extremely rare in patients with thrombophlebitis of the superficial veins.

8. Normally, the valved perforating veins drain the superficial veins through the deep fascia into the deep veins. Incompetence of these important veins permits reflux of deep venous blood into the superficial veins and commonly results in the formation of local superficial varices.

The small saphenous vein drains the lateral end of the dorsal venous arch of the foot and ascends in the superficial fascia *posterior* to the lateral malleolus of the fibula. Here the position is constant and it can be readily seen.

9. The femoral artery enters the thigh beneath the inguinal ligament at a point midway between the anterior superior iliac spine and the symphysis pubis.

The following structures are pierced by a catheter entering the femoral artery in the thigh just below the inguinal ligament: (1) skin, (2) superficial fascia, (3) deep fascia, and (4) anterior layer of the femoral sheath.

The femoral vein lies along the medial side of the femoral artery within the femoral sheath. The cavity of the hip joint lies posterior to the femoral artery, separated by the psoas muscle and the fibrous joint capsule.

10. The advantages of great saphenous vein cut down at the ankle are: (1) the position of the vein in front of the medial malleolus is constant; (2) apart from the presence of the saphenous nerve, there are no other anatomical structures to damage—the cut down is made over bone. The disadvantages are as follows: (1) phlebitis is a common complication; (2) the small diameter precludes the rapid instillation of large volumes of fluid; in young children the small diameter of the vein sometimes makes it difficult to identify.

The advantages of great saphenous cut down in the groin are: (1) the larger diameter of the vein at this site permits the rapid instillation of large volumes of fluid, and (2) there is easier recognition of the vein at this site. The disadvantages of the groin site are: (1) the great saphenous vein lies in thick subcutaneous fat about 1½ in. below and lateral to the pubic tubercle; its identification may prove difficult in obese patients; and (2) other important structures may be damaged, including the femoral artery and vein, if the procedure is carried out by an inexperienced individual.

The saphenous nerve usually lies just anterior to the great saphenous vein as it ascends anterior to the medial malleolus of the tibia.

11. The common complications of femoral vein catheterization are: (1) thrombophlebitis of the femoral vein, especially if the catheterization is prolonged (since the catheter entering the saphenous vein at the groin also goes into the femoral vein, there is risk of phlebitis with saphenous catheterization as well); (2) hematoma formation if the procedure is poorly carried out and the vein wall is torn; (3) infection of the hip joint if an infected catheter pierces the femoral vein completely or misses the vein and traverses the psoas muscle and the anterior part of the capsule of the hip joint; and (4) damage to the femoral nerve, which normally lies some distance laterally to the femoral artery (midpoint between the anterior superior iliac spine and the pubic tubercle.

The surface marking of the femoral vein is just medial to the pulsating femoral artery below the inguinal ligament. If the artery is pulseless, the position of the artery may be determined as being midway between the anterior superior iliac spine and the symphysis pubis; the vein lies just medial to it.

12. The majority of cases of thrombosis of the deep veins of the lower limb start in the large veins draining the soleus muscle. From here the thrombus may extend proximally into the popliteal and femoral veins. In a smaller number of cases the thrombus originates in the iliac or femoral veins and extends downward.

15

The Shoulder and Upper Extremity

Upper limb pain, fractures, dislocations, and nerve injuries are all commonly seen in the emergency department. This chapter is primarily concerned with the presentation of the basic anatomy of the upper limb to assist the physician in making a diagnosis and initiating prompt treatment. The wrist and hand are covered in depth in Chapter 16.

BASIC ANATOMY

Basic Anatomy of the Arrangement of the Upper Extremity

The upper limb may be regarded as a multijointed lever that is freely movable on the trunk at the shoulder joint. At the distal end of the upper limb is the important prehensile organ, the hand.

Lines of Force

A force applied to the hand is transmitted to the distal end of the radius. From there the force passes to the ulna across the interosseous membrane. At the elbow joint the ulna articulates snugly with the distal end of the humerus. Once the force reaches the shoulder joint it is transmitted to the scapula. The force is now carried through the very strong coracoclavicular ligament to the clavicle. The clavicle serves as a strut and holds the scapula and the upper extremity away from the chest wall. At the medial end of the clavicle the strong articular disc transmits the force to the sternum at the sternoclavicular joint.

Organization of the Upper Extremity

The pectoral girdle consisting of the clavicle and scapula is concerned with suspending the upper limb from the trunk. Since there is only one small joint, the sternoclavicular joint, attaching the shoulder girdle to the trunk, the upper limb has a tremendous degree of freedom of movement. The arm and forearm are compartmentalized by fibrous sheets of deep fascia. Each compartment has muscles, blood vessels, and nerves, whose function is to enable movements of the elbow and wrist joints. In the arm the anterior and posterior compartments contain the

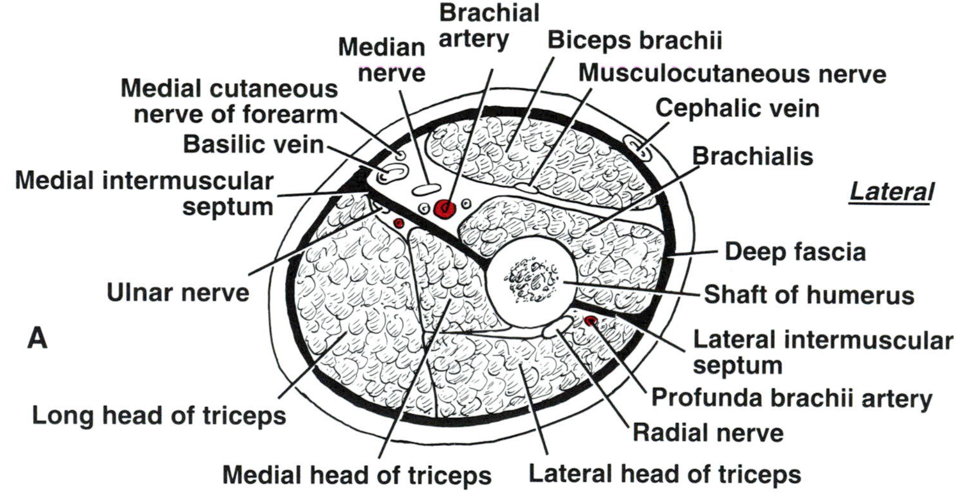

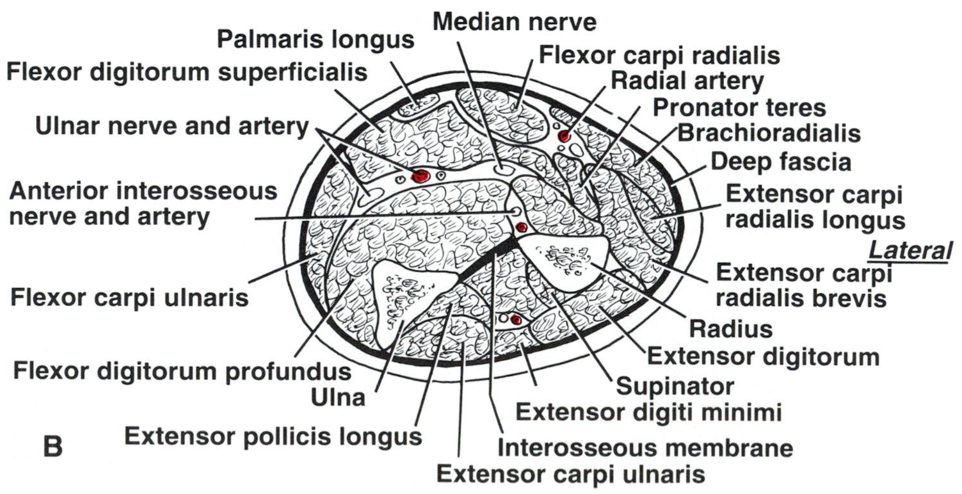

FIG 15–1.
A, cross section of the upper part of the arm just below the level of insertion of the deltoid muscle showing division of the arm by the humerus and medial and lateral intermuscular septa into anterior and posterior compartments. **B,** cross section of the forearm at the level of insertion of the pronator teres showing division of the forearm by the radius and ulna and interosseous membrane into anterior, lateral, and posterior compartments.

flexor and extensor muscles of the elbow joint, respectively (Fig 15–1). In the forearm, the radius and ulna are connected together by a strong interosseous membrane, and there are three fascial compartments (see Fig 15–1). The anterior compartment contains the muscles that flex the wrist and fingers, a posterior compartment whose muscles extend the wrist and fingers, and a lateral compartment whose muscles flex the elbow and extend and abduct the hand.

Much of the importance of the hand is dependent on the pincer action of the thumb, which enables one to grasp objects between the thumb and index finger. The extreme mobility of the metacarpal of the thumb makes the thumb functionally as important as all the remaining fingers combined.

Basic Anatomy of the Bones of the Upper Extremity

Clavicle

The clavicle is a long, slender, S-shaped bone that lies horizontally just beneath the skin (Fig 15–2). It articulates with the sternum and first costal cartilage medially and with the acromion process of the scapula laterally. The clavicle acts as a strut that holds the upper limb away from the trunk. It also serves to transmit forces from the upper limb to the axial skeleton and provides attachment for muscles.

Scapula

The scapula is a flat triangular bone that lies on the posterior thoracic wall between the second and seventh ribs. The different parts of the scapula are shown in Figure 15–2. The *glenoid cavity,* or *fossa,* is pear shaped and lies on the superolateral angle of the scapula; it articulates with the head of the humerus at the shoulder joint.

Humerus

The humerus articulates with the scapula at the shoulder joint and the radius and the ulna at the elbow joint. The different parts of the humerus are illustrated in Figure 15–3.

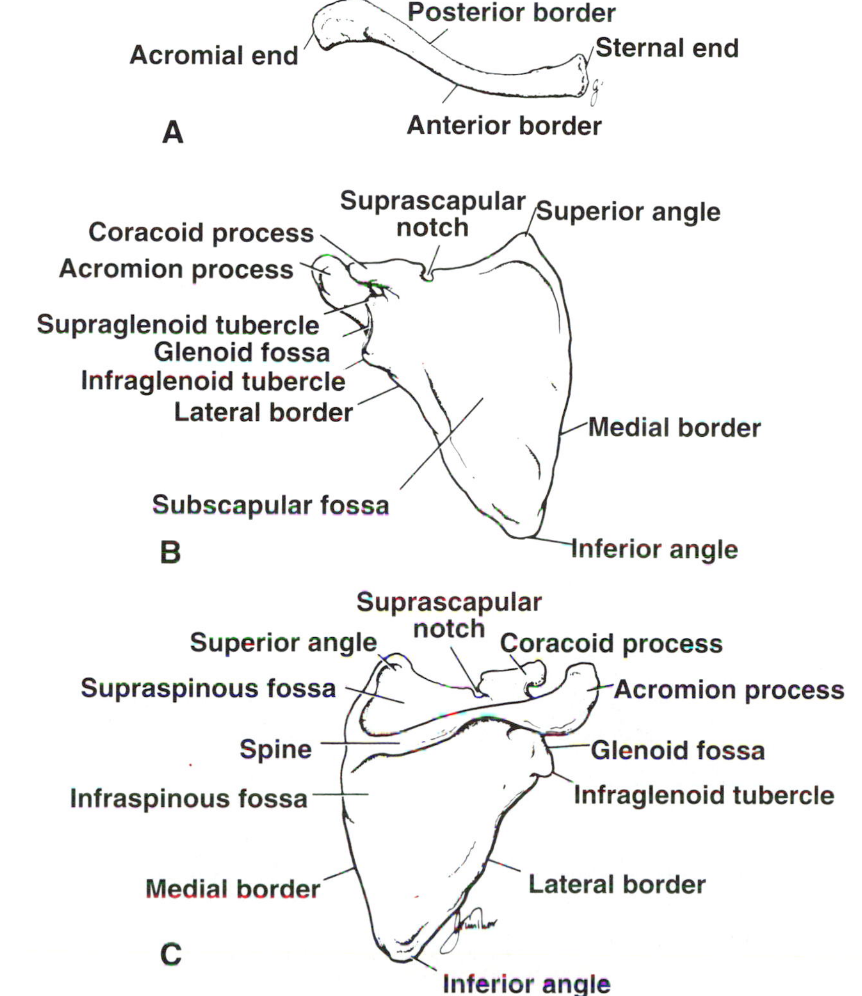

FIG 15–2.
A, superior surface of the right clavicle. **B,** anterior surface of the right scapula. **C,** posterior surface of the right scapula.

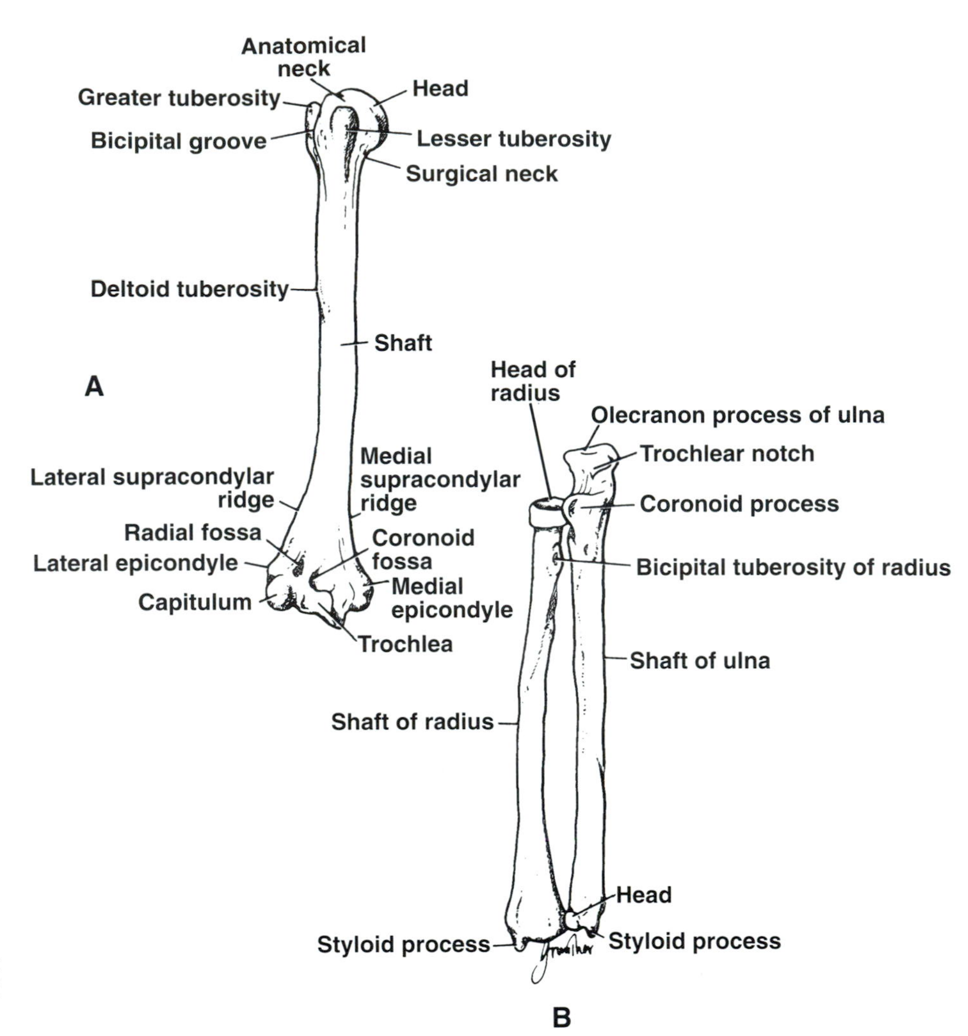

FIG 15–3.
A, anterior surface of the right humerus. **B,** anterior surface of the right radius and ulna.

Radius and Ulna Bones

The different parts of the radius and ulna bones are shown in Figure 15–3.

Basic Anatomy of the Muscles of the Upper Extremity

Shoulder Region

Axilla.—The axilla, or armpit, is a pyramid-shaped space between the upper part of the arm and the side of the chest (see Fig 15–21). The upper end, or *apex,* is directed into the root of the neck and is bounded in front by the clavicle, behind by the upper border of the scapula, and medially by the outer border of the first rib. The lower end, or *base,* is bounded in front by the anterior axillary fold (formed by the lower border of the pectoralis major muscle), behind by the posterior axillary fold (formed by the tendon of latissimus dorsi and the teres major muscle), and medially by the chest wall.

The axilla contains the principal vessels and nerves to the upper limb (see Fig 15–21) and many lymph nodes.

The muscles connecting the upper limb and the vertebral column, the muscles connecting the upper limb and the thoracic wall, and the scapular muscles are shown in Tables 15–1 through 15–3.

Rotator Cuff.—The four muscles—the supraspinatus, infraspinatus, teres minor, and subscapularis—form what is termed the *rotator cuff.* The tone of these muscles assists in holding the head of the humerus in the glenoid cavity of the scapula during movements of the shoulder joint. Therefore, they as-

Clinical Note

Rotator cuff.—The clinical importance of this structure cannot be overemphasized. See rotator cuff tendinitis, p. 609.

TABLE 15–1.
Muscles Connecting the Upper Limb and the Vertebral Column*

Name of Muscle	Origin	Insertion	Nerve Supply	Nerve Root*	Action
Trapezius	Occipital bone, ligamentum nuchae, spines of all thoracic vertebrae	Upper fibers into lateral third of clavicle; middle and lower fibers into spine of scapula	Spinal part of accessory nerve and C3 and C4 (sensory)	XI cranial nerve (spinal part)	Upper fibers elevate the scapula; middle fibers pull scapula medially; lower fibers pull medial border of scapula downward
Latissimus dorsi	Iliac crest, lumbar fascia, spines of lower six thoracic vertebrae, lower three or four ribs, inferior angle of scapula	Floor of bicipital groove of humerus	Thoracodorsal nerve	**C6,** C7, C8	Extends, adducts, and medially rotates the arm
Levator scapulae	Transverse process of first four cervical vertebrae	Medial border of scapula	C3 and C4 and dorsal scapular nerve	C3, C4, C5	Raises medial border of scapula
Rhomboid minor	Ligamentum nuchae and spines of seventh cervical and first thoracic vertebrae	Medial border of scapula	Dorsal scapular nerve	C4, C5	Raises medial border of scapula
Rhomboid major	Second to fifth thoracic spines	Medial border of scapula	Dorsal scapular nerve	C4, C5	Raises medial border of scapula

*The predominant nerve root supply is indicated by boldface type.

sist in stabilizing the shoulder joint. The cuff lies on the anterior, superior, and posterior aspects of the joint (Fig 15–4). The cuff is deficient inferiorly, which is a site of potential weakness.

Quadrilateral Space.—The quadrilateral space is a potential intermuscular space found in the shoulder region (see Fig 15–7). The space is bounded above by the subscapularis in front and teres minor behind, and between these two muscles is the capsule of the shoulder joint. The space is bounded below by the teres major muscle. It is bounded medially by the long head of the triceps and laterally by the surgical neck of the humerus.

TABLE 15–2.
Muscles Connecting the Upper Limb and the Thoracic Wall*

Name of Muscle	Origin	Insertion	Nerve Supply	Nerve Root*	Action
Pectoralis major	Clavicle, sternum, and upper six costal cartilages	Lateral lip of bicipital groove of humerus	Medial and lateral pectoral nerves from brachial plexus	C5, **C6, C7, C8,** T1	Adducts arm and rotates it medially; clavicular fibers also flex arm
Pectoralis minor	Third, fourth, and fifth ribs	Coracoid process of scapula	Medial pectoral nerve from brachial plexus	C6, **C7,** C8	Depresses point of shoulder; if the scapula is fixed, it elevates ribs of origin
Subclavius	First costal cartilage	Clavicle	Nerve to subclavius from upper trunk of brachial plexus	**C5,** C6	Depresses the clavicle and steadies this bone during movements of the shoulder girdle
Serratus anterior	Upper eight ribs	Medial border and inferior angle of scapula	Long thoracic nerve	C5, **C6, C7**	Draws the scapula forward around the thoracic wall; rotates scapula

*The predominant nerve root supply is indicated by boldface type.

TABLE 15–3.
Muscles of the Scapula*

Name of Muscle	Origin	Insertion	Nerve Supply	Nerve Root*	Action
Deltoid	Lateral third of clavicle, acromion process, spine of scapula	Middle of lateral surface of shaft of humerus	Axillary nerve	**C5,** C6	Abducts arm; anterior fibers flex arm; posterior fibers extend arm
Supraspinatus	Supraspinous fossa of scapula	Greater tuberosity of humerus	Suprascapular nerve	C4, **C5,** C6	Abducts arm and stabilizes head of humerus in glenoid cavity of scapula
Infraspinatus	Infraspinous fossa of scapula	Greater tuberosity of humerus	Suprascapular nerve	(C4), **C5,** C6	Laterally rotates arm
Teres major	Lower third lateral border of scapula	Medial lip of bicipital groove of humerus	Lower subscapular nerve	**C6,** C7	Medially rotates and adducts arm
Teres minor	Upper two thirds, lateral border of scapula	Greater tuberosity of humerus	Axillary nerve	(C4), **C5,** C6	Laterally rotates arm
Subscapularis	Subscapular fossa	Lesser tuberosity of humerus	Upper and lower subscapular nerves	C5, **C6,** C7	Medially rotates arm

*The predominant nerve root supply is indicated by boldface type.

The axillary nerve and the posterior circumflex humeral vessels pass backward through this space (see Fig 15–7).

Upper Part of the Arm

Fascial Compartments of the Upper Part of the Arm.—The upper part of the arm is enclosed in a sheath of deep fascia (see Fig 15–1). Two fascial septa, one on the medial and one on the lateral side, extend from this sheath and are attached to the medial and lateral supracondylar ridges of the humerus, respectively. By this means, the upper part of the arm, as noted previously, is divided into an anterior and a posterior fascial compartment, each having its muscles, nerves, and arteries (see Fig 15–1).

The muscles of the upper part of the arm are shown in Table 15–4. See also Figures 15–22 and 15–24.

TABLE 15–4.
Muscles of the Upper Part of the Arm*

Name of Muscle	Origin	Insertion	Nerve Supply	Nerve Root*	Action
		Muscles of the Anterior Fascial Compartment			
Biceps brachii					
Long head	Supraglenoid tubercle of scapula	Tuberosity of radius and bicipital aponeurosis into deep fascia of forearm	Musculocutaneous nerve	C5, **C6**	Supinator of forearm and flexor of elbow joint; also weak flexor of shoulder joint
Short head	Coracoid process of scapula				
Coracobrachialis	Coracoid process of scapula	Medial aspect of shaft of humerus	Musculocutaneous nerve	C5, **C6,** C7	Flexes arm and also weak adductor
Brachialis	Front of lower half of humerus	Coronoid process of ulna	Musculocutaneous nerve and radial nerves	C5, **C6**	Flexor of elbow joint
		Muscles of the Posterior Fascial Compartment			
Triceps					
Long head	Infraglenoid tubercle of scapula				
Lateral head	Upper half of posterior surface of shaft of humerus	Olecranon process of ulna	Radial nerve	C6, C7, **C8**	Extensor of the elbow joint
Medial head	Lower half of posterior surface of shaft of humerus				

*The predominant nerve root supply is indicated by boldface type.

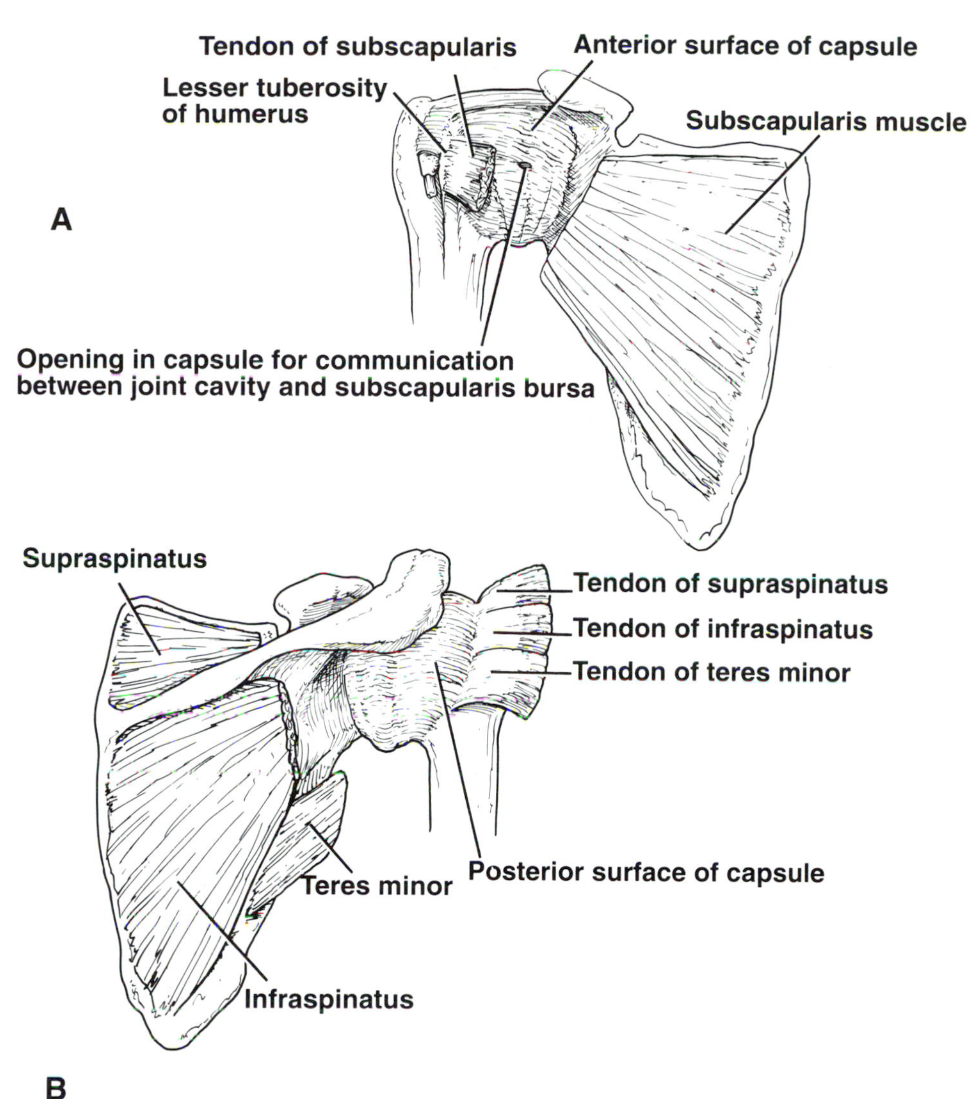

FIG 15–4.
Rotator cuff. **A,** anterior view of the shoulder joint showing the tendon of insertion of the subscapularis muscle. **B,** posterior view of the shoulder joint showing the tendons of insertion of the supraspinatus, infraspinatus, and teres minor muscles. The tendons of the above muscles strengthen the capsule of the joint and collectively form the rotator cuff.

Cubital Fossa.—The cubital fossa is a triangular skin depression that lies in front of the elbow (see Fig 15–25). It has the following boundaries—laterally, the brachioradialis muscle and medially, the pronator teres muscle. The base of the triangle is formed by an imaginary line drawn between the two epicondyles of the humerus. The floor of the fossa is formed by the supinator muscle laterally and the brachialis muscle medially. The roof is formed by the skin and fascia and is reinforced by the bicipital aponeurosis.

The cubital fossa contains the following structures, enumerated from the medial to the lateral side—the median nerve, the bifurcation of the brachial artery into the ulnar and radial arteries, the tendon of the biceps muscle, and the radial nerve and its deep branch (see Fig 15–25).

Lying in the superficial fascia covering the cubital fossa are the important superficial veins—the cephalic and the basilic veins and their tributaries.

Forearm

Fascial Compartments of the Forearm.—The forearm is enclosed in a sheath of deep fascia, which is attached to the periosteum of the posterior subcutaneous border of the ulna (see Fig 15–1). This fascial sheath, together with the interosseous membrane and fibrous intermuscular septa, divides the forearm into a number of compartments, each having its own muscles, nerves, and blood supply.

Interosseous Membrane.—The interosseous membrane is a thin but strong membrane uniting the radius and the ulna; it is attached to their interosseous borders. Its fibers are taut, and therefore the forearm is most stable when it is in the midprone position, i.e., the position of function. The in-

TABLE 15–5.
Muscles of the Anterior Fascial Compartment of the Forearm*

Name of Muscle	Origin	Insertion	Nerve Supply	Nerve Root*	Action
Pronator teres					
Humeral head	Medial epicondyle of humerus	Lateral aspect of shaft of radius	Median nerve	C6, **C7**	Pronation and flexion of forearm
Ulnar head	Coronoid process of ulna				
Flexor carpi radialis	Medial epicondyle of humerus	Bases of second and third metacarpal bones	Median nerve	C6, **C7**	Flexes and abducts hand at wrist joint
Palmaris longus (often absent)	Medial epicondyle of humerus	Flexor retinaculum and palmar aponeurosis	Median nerve	C7, C8	Flexes hand
Flexor carpi ulnaris					
Humeral head	Medial epicondyle of humerus	Pisiform bone, hook of the hamate, base of fifth metacarpal bone	Ulnar nerve	C8, T1	Flexes and adducts the hand at the wrist joint
Ulnar head	Olecranon process and posterior border of ulna				
Flexor digitorum superficialis					
Humeroulnar head	Medial epicondyle of humerus, coronoid process of ulna	Middle phalanx of middle four fingers	Median nerve	C7, **C8,** T1	Flexes middle phalanx of fingers and assists in flexing proximal phalanx and hand
Radial head	Oblique line on anterior surface of shaft of radius				
Flexor pollicis longus	Anterior surface of shaft of radius	Distal phalanx of thumb	Anterior interosseous branch of median nerve	**C8,** T1	Flexes distal phalanx of thumb
Flexor digitorum profundus	Anterior surface of shaft of ulna; interosseous membrane	Distal phalanges of medial four fingers	Ulnar (medial half) and median (lateral half) nerves	**C8,** T1	Flexes distal phalanx of the fingers; then assists in flexion of middle and proximal phalanges and the wrist
Pronator quadratus	Anterior surface of shaft of ulna	Anterior surface of shaft of radius	Anterior interosseous branch of median nerve	**C8,** T1	Pronates forearm

*The predominant nerve root supply is indicated by boldface type.

terosseous membrane provides attachment for neighboring muscles.

The muscles of the anterior fascial compartment of the forearm are shown in Table 15–5. The muscles are also seen in Figures 15–25 and 15–26. The muscles of the lateral fascial compartment are shown in Table 15–6. The muscles are also seen in Figures 15–1 and 15–25. The muscles of the posterior fascial compartment are shown in Table 15–7. The muscles are also seen in Figures 15–1 and 15–27.

TABLE 15–6.
Muscles of the Lateral Fascial Compartment of the Forearm*

Name of Muscle	Origin	Insertion	Nerve Supply	Nerve Root*	Action
Brachioradialis	Lateral supracondylar ridge of humerus	Styloid process of radius	Radial nerve	C5, **C6,** C7	Flexes forearm at elbow joint; rotates forearm to midprone position
Extensor carpi radialis longus	Lateral supracondylar ridge of humerus	Base of second metacarpal bone	Radial nerve	C6, C7	Extends and abducts hand at wrist joint

*The predominant nerve root supply is indicated by boldface type.

TABLE 15–7.
Muscles of the Posterior Fascial Compartment of the Forearm*

Name of Muscle	Origin	Insertion	Nerve Supply	Nerve Root*	Action
Extensor carpi radialis brevis	Lateral epicondyle of humerus	Base of third metacarpal bone	Deep branch of radial nerve	**C7,** C8	Extends and abducts the hand at wrist joint
Extensor digitorum	Lateral epicondyle of humerus	Middle and distal phalanges of the medial four fingers	Deep branch of radial nerve	**C7,** C8	Extends fingers and hand
Extensor digiti minimi	Lateral epicondyle of humerus	Extensor expansion of little finger	Deep branch of radial nerve	**C7,** C8	Extends metacarpophalangeal joint of little finger
Extensor carpi ulnaris	Lateral epicondyle of humerus	Base of fifth metacarpal bone	Deep branch of radial nerve	C7, **C8**	Extends and adducts hand at the wrist joint
Anconeus	Lateral epicondyle of humerus	Olecranon process of ulna	Radial nerve	C7, C8, (T1)	Extends elbow joint
Supinator	Lateral epicondyle of humerus, anular ligament of superior radioulnar joint and ulna	Neck and shaft of ulna	Deep branch of radial nerve	C5, C6	Supination of forearm
Abductor pollicis longus	Shafts of radius and ulna	Base of first metacarpal bone	Deep branch of radial nerve	C7, **C8**	Abducts and extends thumb
Extensor pollicis brevis	Shaft of radius and interosseous membrane	Base of proximal phalanx of thumb	Deep branch of radial nerve	C7, **C8**	Extends metacarpophalangeal joints of thumb
Extensor pollicis longus	Shaft of ulna and interosseous membrane	Base of distal phalanx of thumb	Deep branch of radial nerve	C7, **C8**	Extends distal phalanx of thumb
Extensor indicis	Shaft of ulna and interosseous membrane	Extensor expansion of index finger	Deep branch of radial nerve	C7, **C8**	Extends metacarpophalangeal joint of index finger

*The predominant nerve root supply is indicated by boldface type.

The *flexor and extensor retinacula* at the wrist, the *carpal tunnel*, the *fibrous flexor sheaths* and the *synovial flexor sheaths* of the fingers, the *insertion of the long flexor tendons* into the fingers, and the *insertion of the long extensor tendons* into the fingers are described in Chapter 16.

Basic Anatomy of the Clavicular Joints

Sternoclavicular Joint

Articulation.—This occurs between the medial end of the clavicle, the manubrium sterni, and the first costal cartilage (Fig 15–5).

Type.—Synovial gliding joint.

Capsule.—Encloses the joint.

Ligaments.—The capsule is reinforced above, in front of, and behind the joint by strong ligaments, the *interclavicular* and the *sternoclavicular ligaments* (see Fig 15–5). The joint is divided by the *articular disc* into two compartments—lateral and medial. The disc is attached by its circumference to the capsule, to the superior margin of the articular surface of the clavicle, and to the first costal cartilage below. The disc serves as a strong intra-articular ligament. The *costoclavicular ligament* is strong and runs from the junction of the first rib with the first costal cartilage to the inferior surface of the clavicle (see Fig 15–5).

Synovial Membrane.—Lines the capsule.

Nerve Supply.—Supraclavicular nerve and the nerve to the subclavius muscle.

Movements.—Anterior and posterior movements of the clavicle take place in the medial compartment of the sternoclavicular joint (Fig 15–6). Elevation and depression of the clavicle take place in the lateral compartment; they are more extensive than the anterior and posterior movements and may be as much as 60°. All these movements are associated with movements of the scapula.

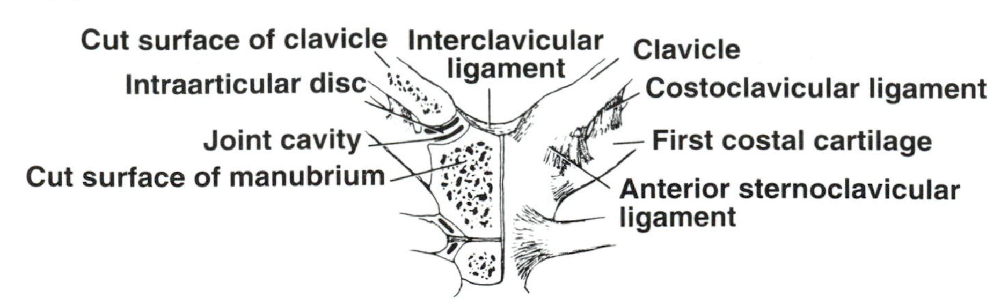

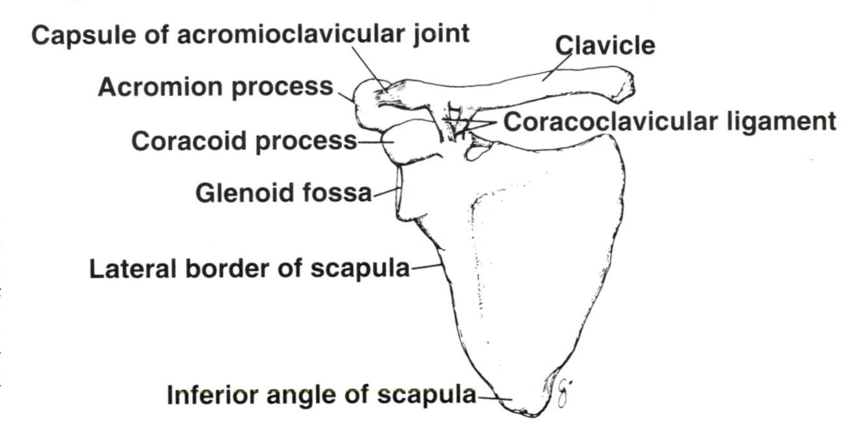

FIG 15–5.
A, anterior view of the sternoclavicular joints; the right joint has been opened to reveal the joint cavity and the disc. **B,** anterior view of the right acromioclavicular joint. The important coracoclavicular ligament consists of the medial conoid and lateral trapezoid parts.

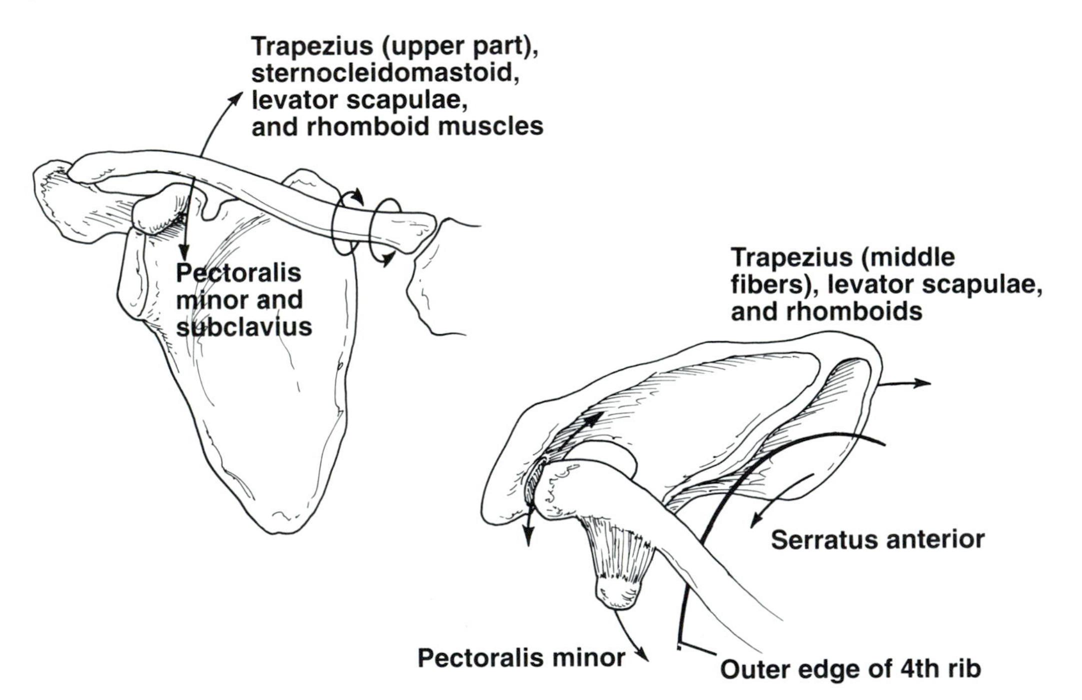

FIG 15–6.
Wide range of movements possible at the sternoclavicular and acromioclavicular joints, giving great mobility to the clavicle and the upper limb.

Muscles Producing Movements.—These include the following.

Anterior Movement.—Serratus anterior muscle.

Posterior Movement.—Trapezius and rhomboid muscles.

Elevation.—Trapezius, sternocleidomastoid, levator scapulae, and rhomboid muscles.

Depression.—Pectoralis minor and subclavius muscles.

Acromioclavicular Joint

Articulation.—This occurs between the acromion process of the scapula and the lateral end of the clavicle (see Fig 15–5).

Type.—Synovial gliding joint.

Capsule.—Encloses the joint.

Ligaments.—The *acromioclavicular ligament* strengthens the capsule superiorly. The *disc* is incomplete, wedge shaped, and composed of fibrocartilage; it projects into the joint cavity from above and is attached to the capsule superiorly. The *coracoclavicular ligament* connects the coracoid process of the scapula to the inferior surface of the clavicle (see Fig 15–5); it has *conoid and trapezoid parts* and is largely responsible for suspending the weight of the scapula and the upper limb from the clavicle.

Synovial Membrane.—Lines the capsule.

Nerve Supply.—Suprascapular nerve.

Movements.—A gliding movement when the scapula rotates or when the clavicle is elevated or depressed.

Basic Anatomy of the Shoulder Joint

Description

Articulation.—The rounded head of the humerus and the shallow, pear-shaped glenoid cavity of the

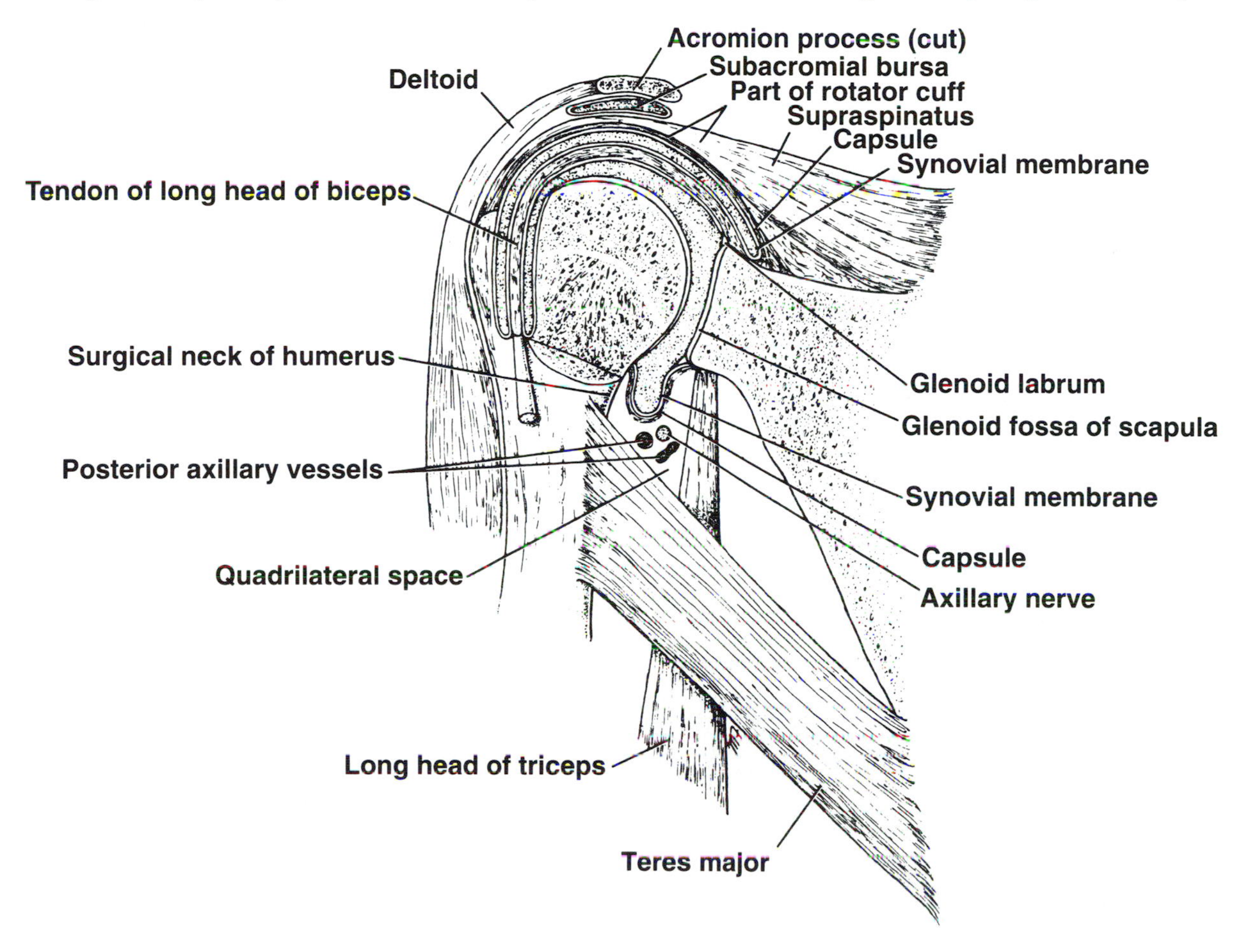

FIG 15–7.
Interior of the right shoulder joint. The head of the humerus and the scapula have been cut to reveal the interior of the joint.

scapula. The glenoid cavity is deepened by the presence of a fibrocartilaginous rim called the *glenoid labrum* (Fig 15–7).

Type.—Synovial ball and socket joint.

Capsule.—This surrounds the joint and is thin and lax, allowing a wide range of movement. It is strengthened by the tendons of the short muscles that surround the joint and strengthen the capsule (Figs 15–4 and 15–8).

Ligaments.—*Glenohumeral ligaments* include three weak bands that strengthen the interior and anterior parts of the capsule (see Fig 15–8). The *transverse humeral ligament* strengthens the capsule and bridges the gap between the greater and lesser tuberosities of the humerus. It holds the tendon of the long head of the biceps muscle in place. The *coracohumeral ligament* strengthens the capsule above and extends from the root of the coracoid process to the greater tuberosity of the humerus (see Fig 15–8). The *coracoacromial ligament* lies outside the joint and extends between the coracoid process and the acromion pro-

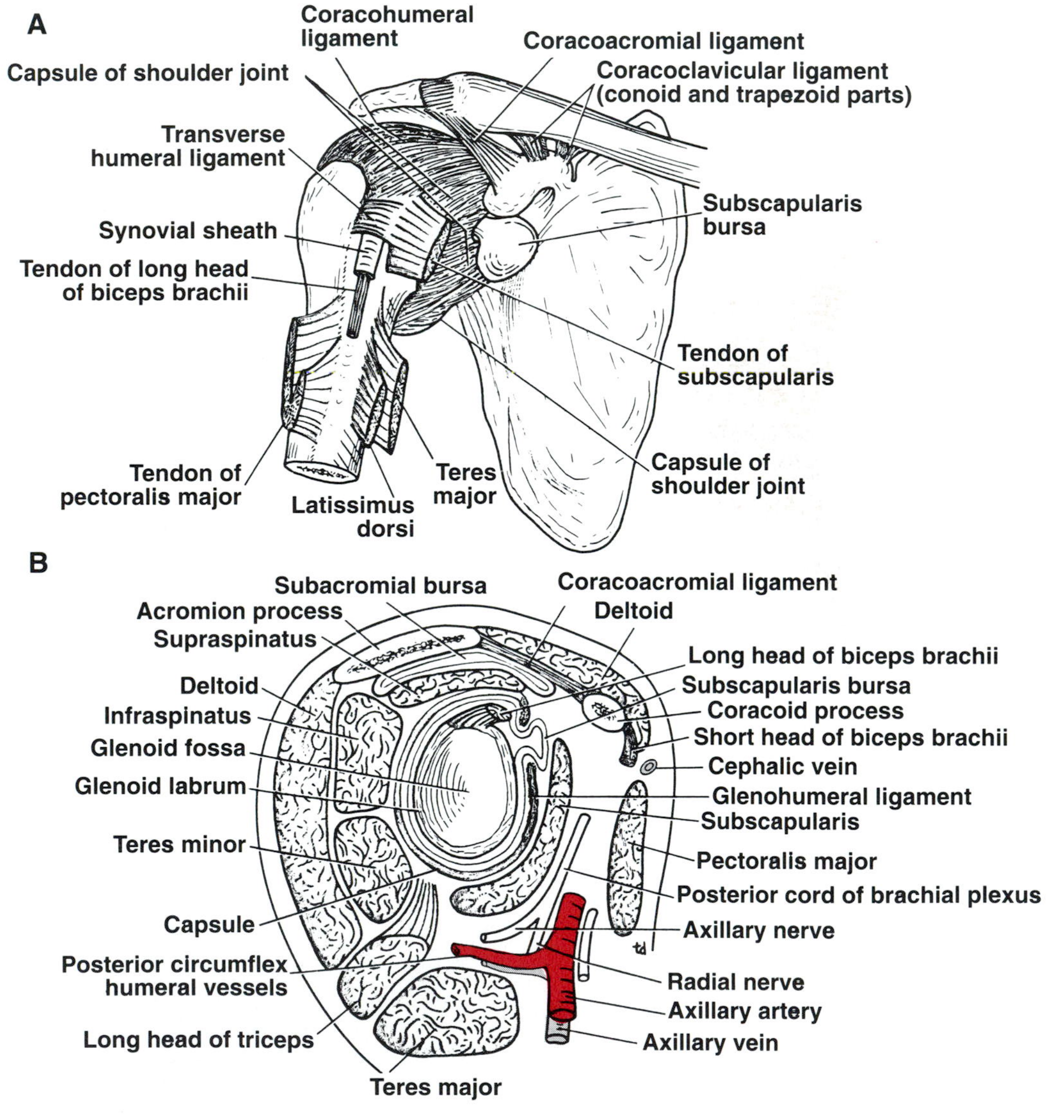

FIG 15–8.
A, anterior view of the right shoulder joint showing the capsule strengthened by the coracohumeral and transverse humeral ligaments; the coracoacromial ligament is also shown. Note the subscapularis bursa and the synovial sheath around the tendon of the long head of the biceps brachii muscle. **B,** sagittal section of the right shoulder joint showing its relations and the axillary nerve passing backward through the quadrilateral space below the joint.

cess (see Fig 15–8). It protects the superior aspect of the joint.

Synovial Membrane.—This lines the capsule. It surrounds the tendon of the biceps and also protrudes forward through the anterior wall of the capsule to form a bursa, which lies beneath the subscapularis muscle (see Fig 15–8).

Clinical Note

The subacromial bursa.—The subacromial bursa is completely separate from the cavity of the shoulder joint. It lies completely outside the shoulder joint between the acromion process and the supraspinatus muscle. Both the subacromial bursa and the shoulder joint can be aspirated and injected. The approaches are different for each, and are described in Chapter 18, pp. 769 and 779.

Nerve Supply.—Axillary and suprascapular nerves.

Movements.—The shoulder joint has a wide range of movement, and the stability of the joint has been sacrificed to permit this (Fig 15–9). The strength of the joint depends on the tone of the short muscles that cross in front, above, and behind the joint, namely, the subscapularis, supraspinatus, infraspinatus, and teres minor. The tendons of insertion of these muscles are fused to the capsule of the joint and together form the important *rotator cuff* (see Fig 15–4).

When the joint is abducted, the lower surface of the head of the humerus is supported by the long head of the triceps, which bows downward because of its length and gives little actual support to the humerus. In addition, the inferior part of the capsule is the weakest area.

Flexion.—Normal flexion is about 90°. Involves

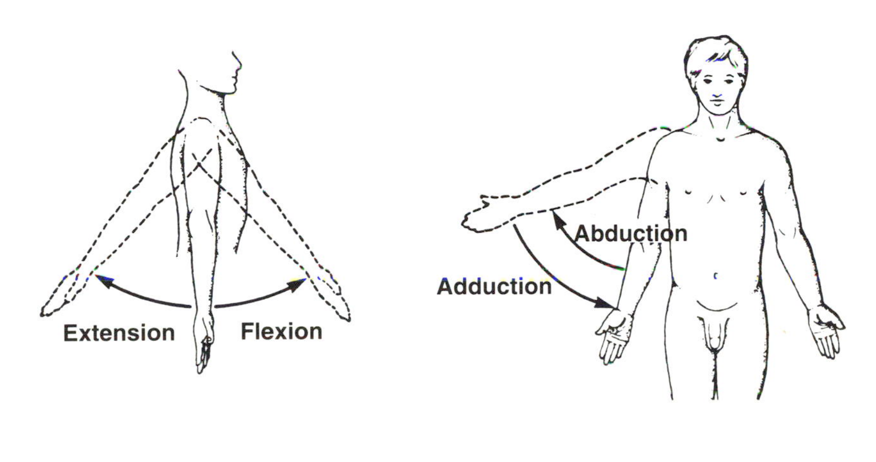

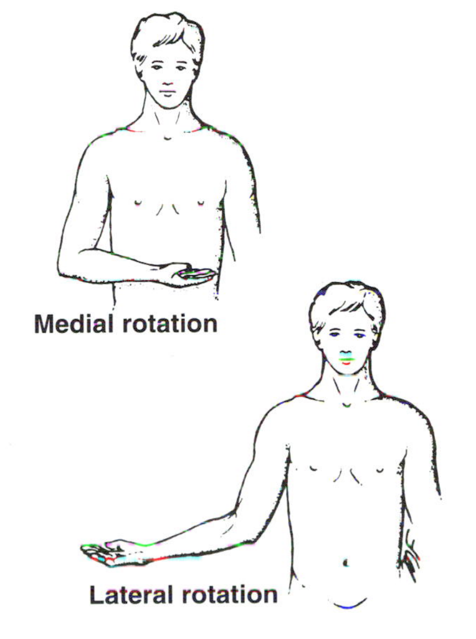

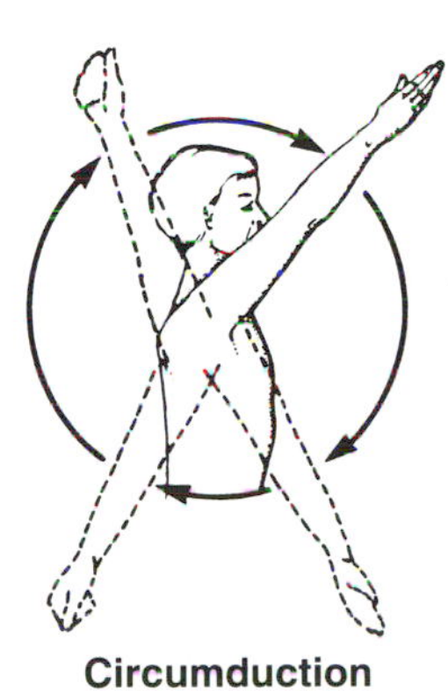

FIG 15–9.
The movements possible at the shoulder (glenohumeral) joint. Pure glenohumeral abduction is only possible as far as about 120°; further movement of the upper limb above the level of the shoulder requires the rotation of the scapula (see text).

the anterior fibers of deltoid, pectoralis major, biceps, and coracobrachialis.

Extension.—Normal extension is about 45°. Involves the posterior fibers of deltoid, latissimus dorsi, and teres major.

Abduction.—Abduction of the upper limb occurs both at the shoulder joint and between the scapula and the thoracic wall (see scapular-humeral mechanism, p. 590). Middle fibers of the deltoid, assisted by the supraspinatus, are involved. The supraspinatus muscle initiates the movement of abduction and holds the head of the humerus against the glenoid fossa of the scapula; this latter function allows the deltoid muscle to contract and abduct the humerus at the shoulder joint.

Adduction.—Normally the upper limb can be swung 45° across the front of the chest. Involves the pectoralis major, latissimus dorsi, teres major, and teres minor.

Lateral Rotation.—Normal lateral rotation is about 40° to 45°. Involves the infraspinatus, teres minor, posterior fibers of deltoid.

Medial Rotation.—Normal medial rotation is about 55°. Involves the subscapularis, latissimus dorsi, teres major, anterior fibers of deltoid.

Circumduction.—A combination of the above movements.

Important Relations (see Fig 15–8).—These include the following.

Anteriorly.—Axillary vessels and the brachial plexus.

Inferiorly.—Axillary nerve and the posterior circumflex humeral vessels as they lie in the quadrilateral space.

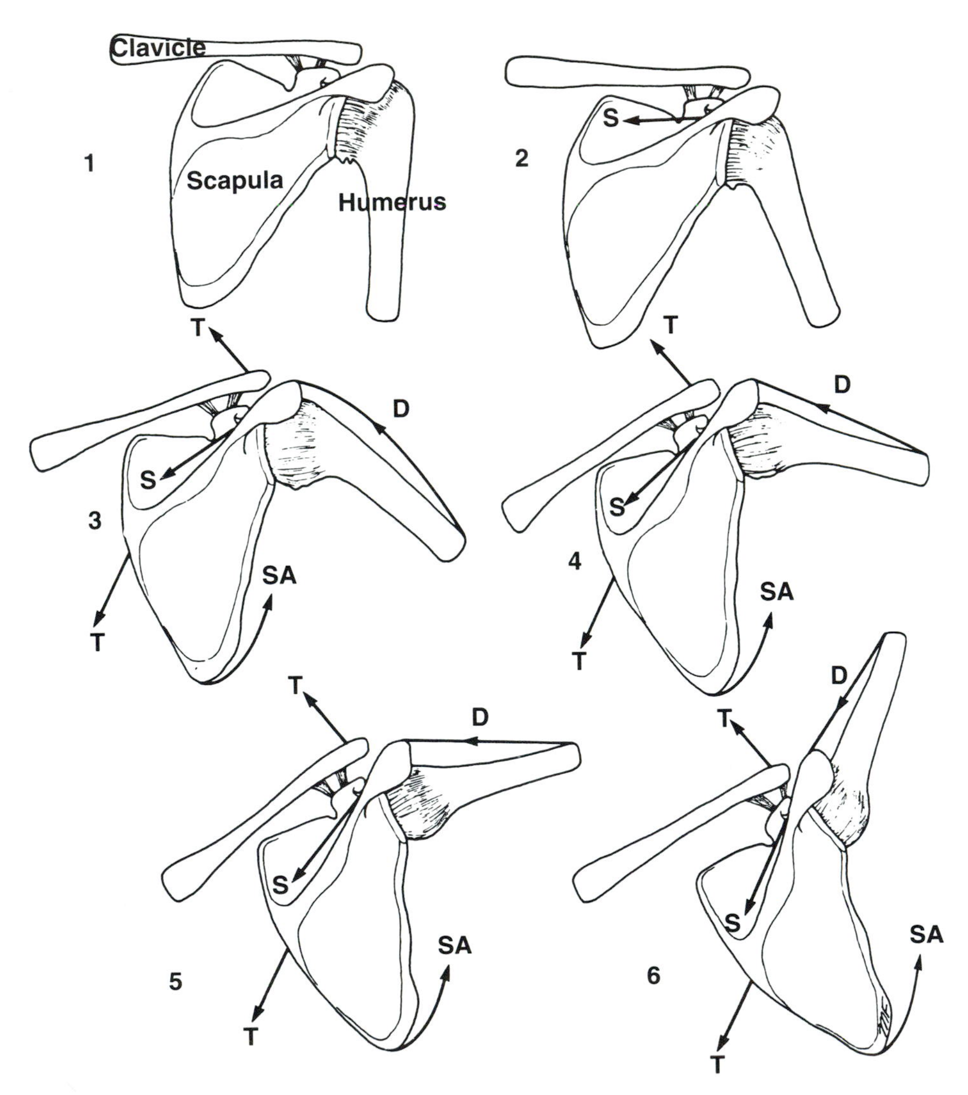

FIG 15–10.
Movements of abduction of the upper limb at the shoulder and the muscles producing these movements (1 through 6). For every 3° of abduction of the arm, a 2° abduction occurs in the shoulder (glenohumeral) joint, and at the same time a 1° abduction occurs by rotation of the scapula. At about 120° of abduction at the glenohumeral joint, the greater tuberosity of the humerus comes in contact with the lateral edge of the acromion process of the scapula. Further elevation of the arm above the head is accomplished solely by rotating the scapula. *S* = supraspinatus; *D* = deltoid; *T* = trapezius; *SA* = serratus anterior.

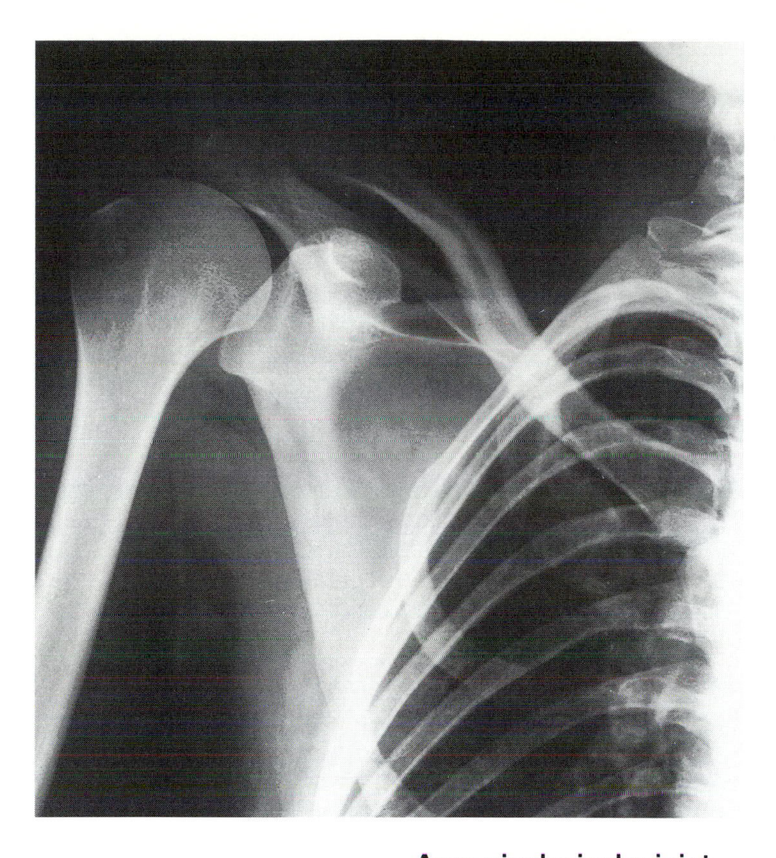

FIG 15–11.
Anteroposterior radiograph of the shoulder region.

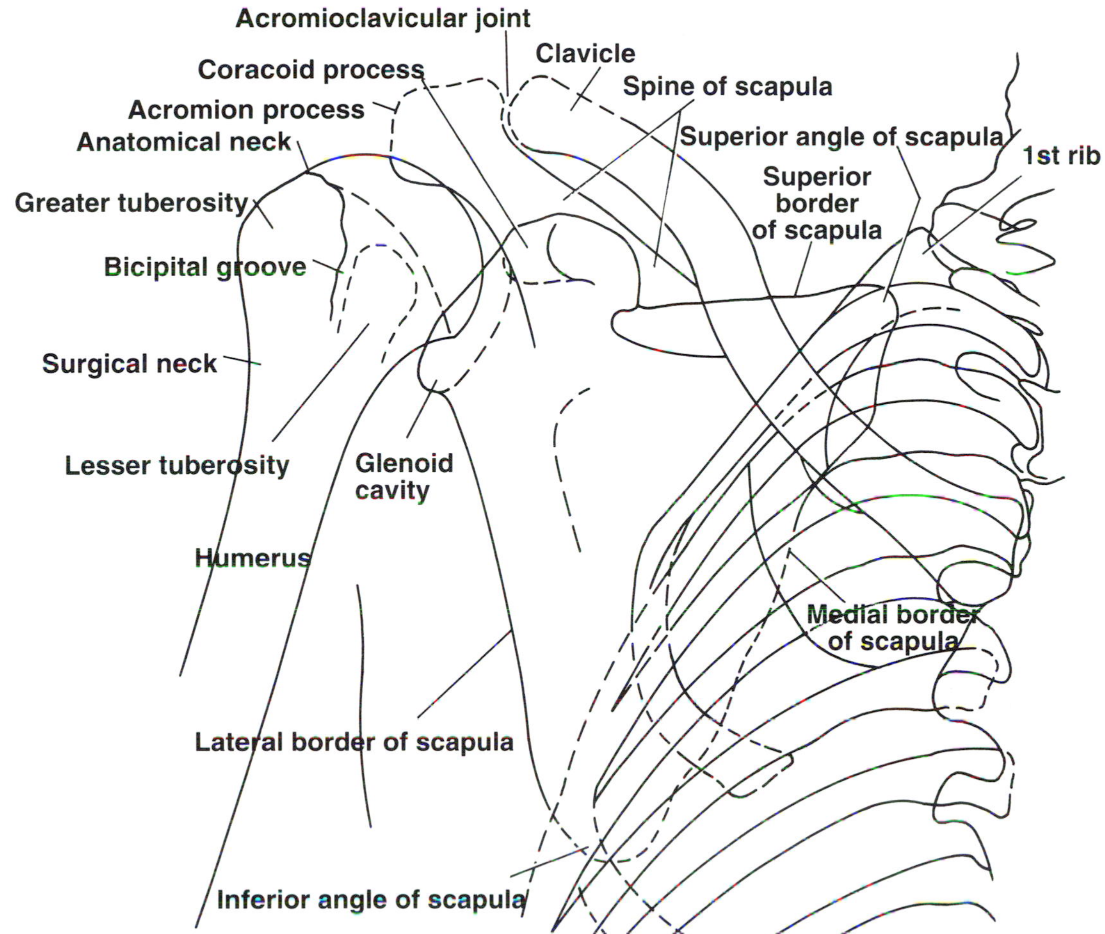

FIG 15–12.
Main features that can be seen in the anteroposterior radiograph of the shoulder region in Figure 15–11.

Basic Anatomy of the Scapular-Humeral Mechanism

The weight of the scapular and upper limb is suspended from the clavicle by the very strong *coracoclavicular ligament* assisted by the tone of muscles. When the scapula rotates on the chest wall so that the position of the glenoid cavity may be altered, the axis of rotation may be considered to pass through the coracoclavicular ligament.

The movement of abduction of the arm involves the rotation of the scapula as well as movement at the shoulder joint. For every 3° of abduction of the arm, a 2° abduction occurs in the shoulder joint and 1° abduction occurs by rotation of the scapula. At about 120° of abduction of the arm, the greater tuberosity of the humerus comes into contact with the lateral edge of the acromion process. Further elevation of the arm above the head is accomplished by rotating the scapula. Figure 15–10 summarizes the movements of abduction of the arm and shows the direction of pull of the muscles responsible for these movements.

Radiographic Appearances of the Shoulder Region
The views of the shoulder region commonly used are the anteroposterior and the inferosuperior (axil-

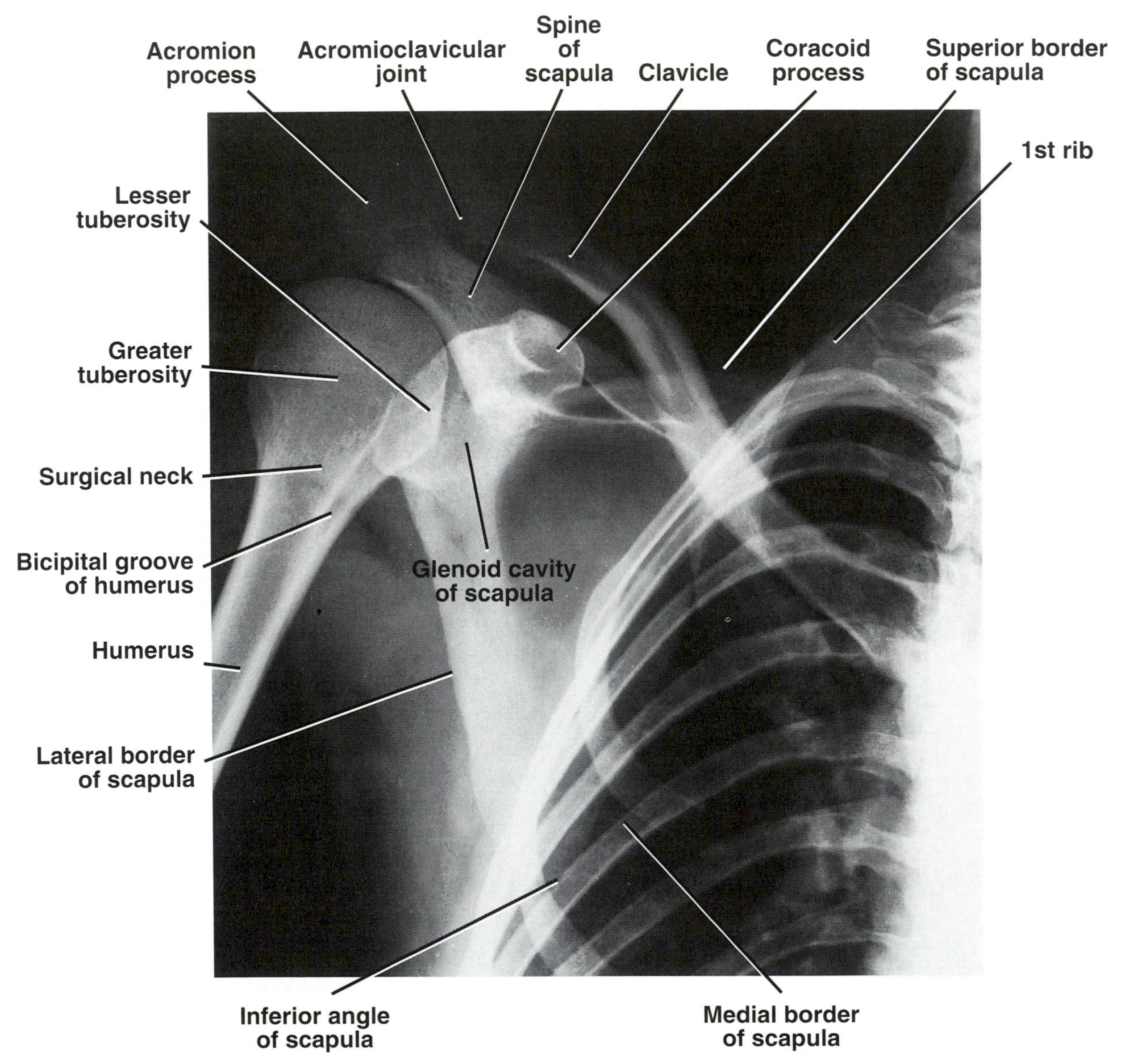

FIG 15–13.
Anteroposterior radiograph of the shoulder region with medial rotation of the humerus.

lary) positions. Examples of such views are shown in Figures 15–11 through 15–13.

Basic Anatomy of the Elbow Joint

Description

Articulation.—This occurs between the trochlea and the capitulum of the humerus and the trochlear notch of the ulna and the head of the radius (Fig 15–14).

Type.—Synovial hinge joint.

Capsule.—Encloses the joint.

Ligaments.—The *lateral collateral ligament* is triangular and is attached by its apex to the lateral epicondyle of the humerus and is attached by its base to the superior margin of the anular ligament and to the ulna (see Fig 15–14). The *medial collateral ligament* is triangular and consists of three bands—(1) the anterior band passes from the medial epicondyle of the humerus to the medial margin of the coronoid process of the ulna, (2) the posterior band that connects the medial epicondyle of the humerus to the olecranon, and (3) the transverse or oblique band that passes between the ulnar attachments of the two preceding bands (see Fig 15–14).

Synovial Membrane.—This lines the capsule and is continuous below with the synovial membrane of the superior radioulnar joint.

Nerve Supply.—Median, ulnar, musculocutaneous, and radial nerves.

Movements and Muscles Producing Movements.—These include the following.

Flexion.—Brachialis, biceps, brachioradialis, and pronator teres.

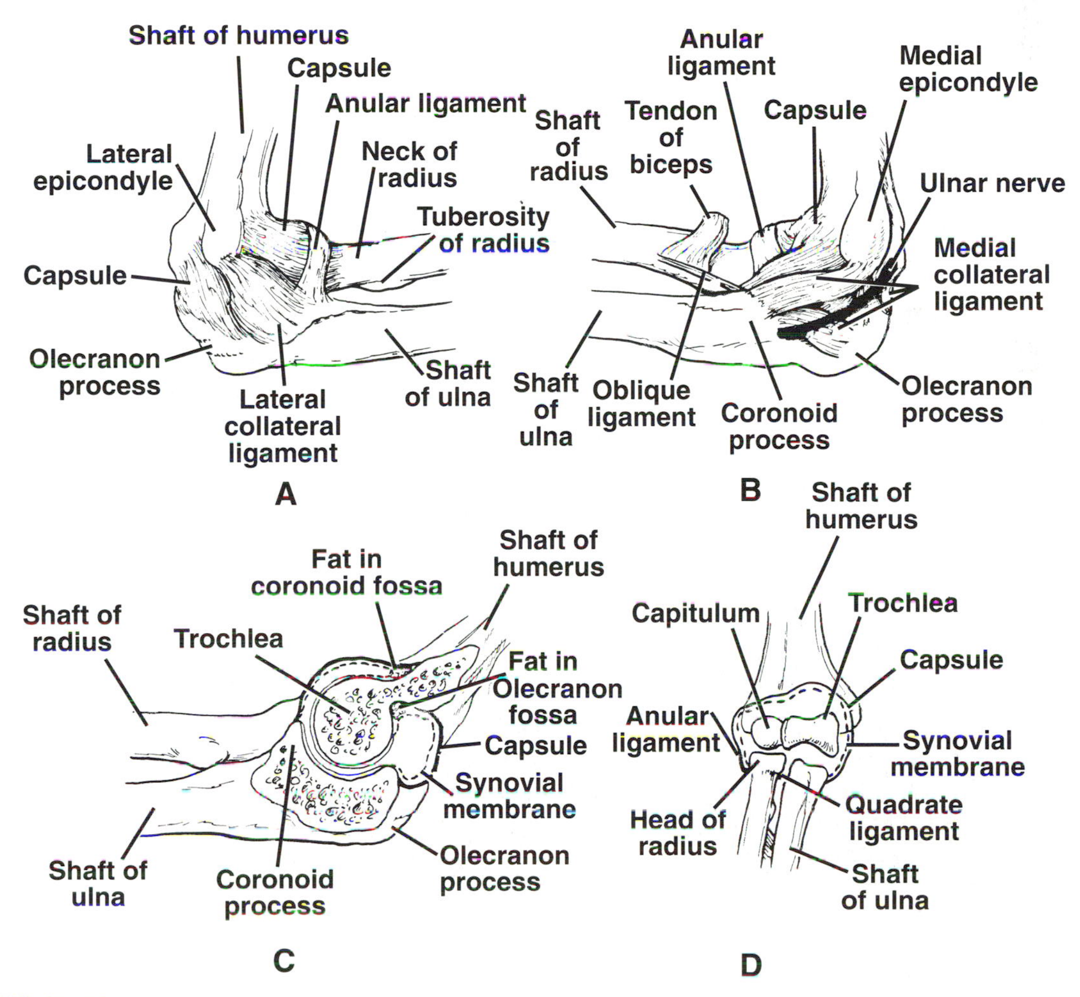

FIG 15–14.
Right elbow joint. **A**, lateral view. **B**, medial view. **C**, sagittal section. **D**, anterior view of the interior of the joint.

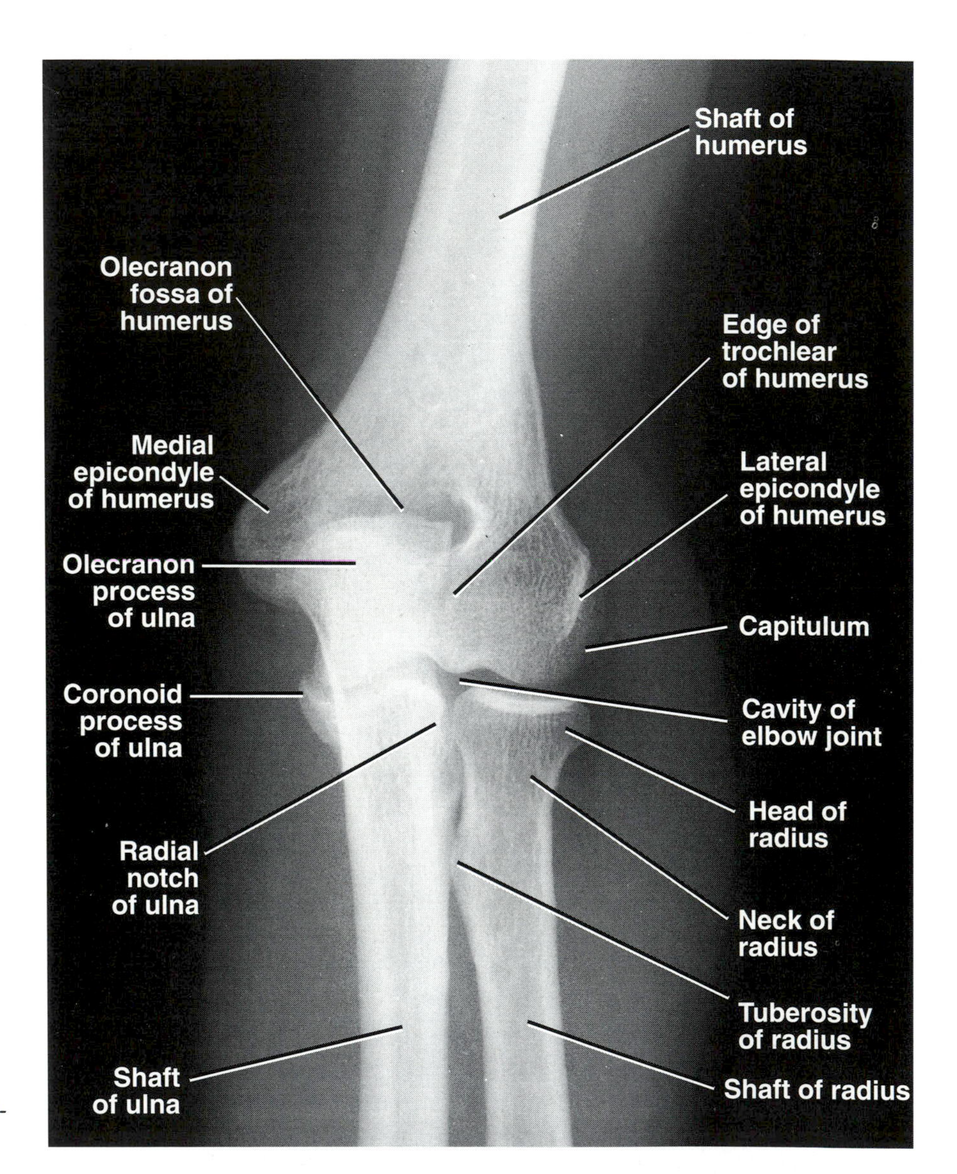

FIG 15–15.
Anteroposterior radiograph of the elbow region.

Extension.—Triceps and anconeus.

A summary of the origins, insertions, nerve supply, and actions of the muscles acting on the elbow joint is given in Tables 15–4 through 15–7.

Clinical Note

Carrying angle of forearm.—The extended forearm lies at an angle to the upper part of the arm. This angle, which opens laterally, is called the *carrying angle* and is about 170° in males and 167° in females. The angle disappears when the elbow joint is fully flexed.

Important Relations.—These include the following.

Anteriorly.—Median nerve and brachial artery (see Fig 15–25).

Medially.—Ulnar nerve that lies behind the medial epicondyle of the humerus (see Fig 15–14).

Radiographic Appearances of the Elbow Region

The views of the elbow region commonly used are the anteroposterior, lateral, and oblique positions and are shown in Figures 15–15 through 15–19.

Basic Anatomy of the Brachial Plexus

The brachial plexus is formed in the posterior triangle of the neck by the union of the anterior rami of the fifth, sixth, seventh, and eighth cervical and first thoracic nerves (Fig 15–20). The plexus may be divided up into *roots, trunks, divisions, and cords.* The roots of C5 and C6 unite to form the *upper trunk,* the root of C7 continues as the *middle trunk,* the roots of C8 and T1 unite to form the *lower trunk.* Each trunk then divides into *anterior and posterior divisions.* The anterior divisions of the upper and middle trunks unite to form the *lateral cord,* the anterior division of the lower trunk continues as the *medial cord,* and the posterior divisions of all three trunks join to form the *posterior cord.*

The roots, trunks, and divisions of the brachial plexus reside in the lower part of the posterior triangle of the neck, whereas the cords and most of the branches of the plexus lie in the axilla (Fig 15–21). The brachial plexus is surrounded by a sheath of fascia, the *axillary sheath,* which is derived from the prevertebral layer of deep cervical fascia (see p. 170). A

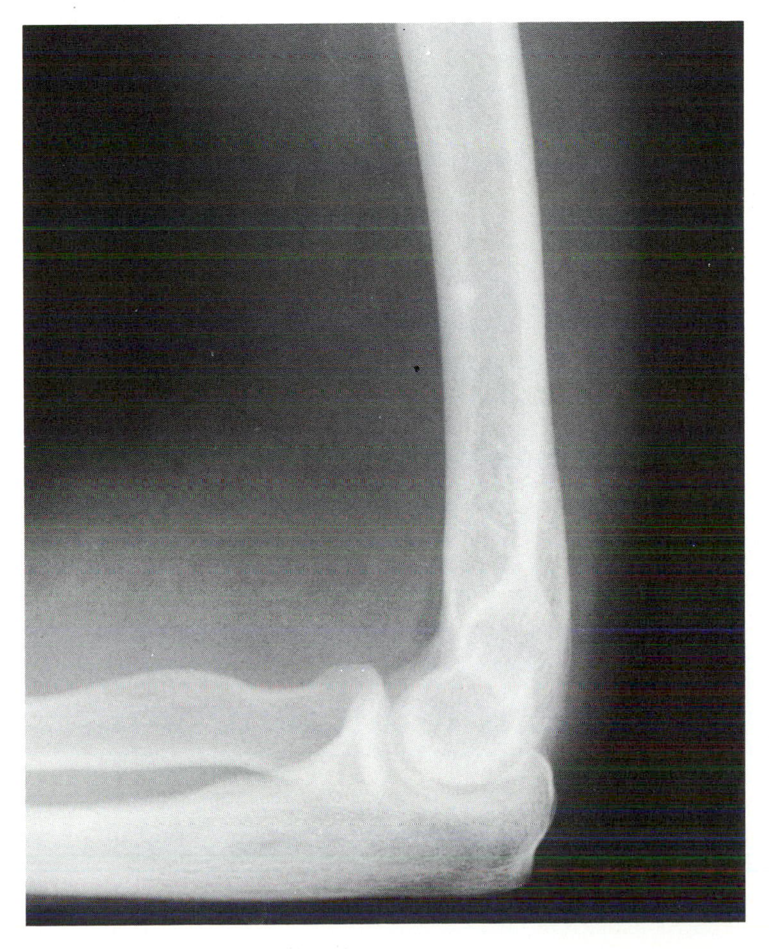

FIG 15–16.
Lateral radiograph of the elbow region.

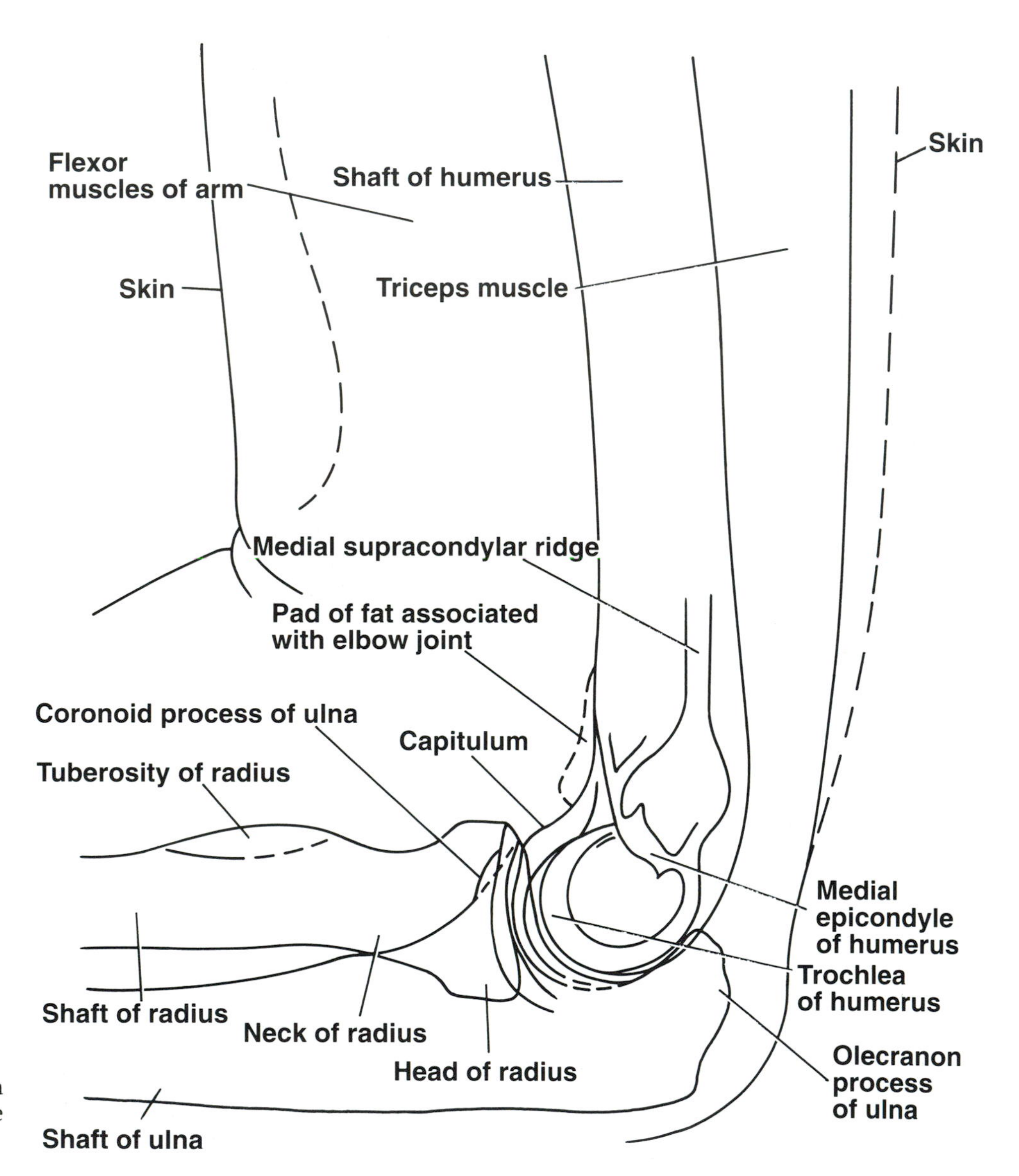

FIG 15–17.
Main features that can be seen in a lateral radiograph of the elbow region in Figure 15–16.

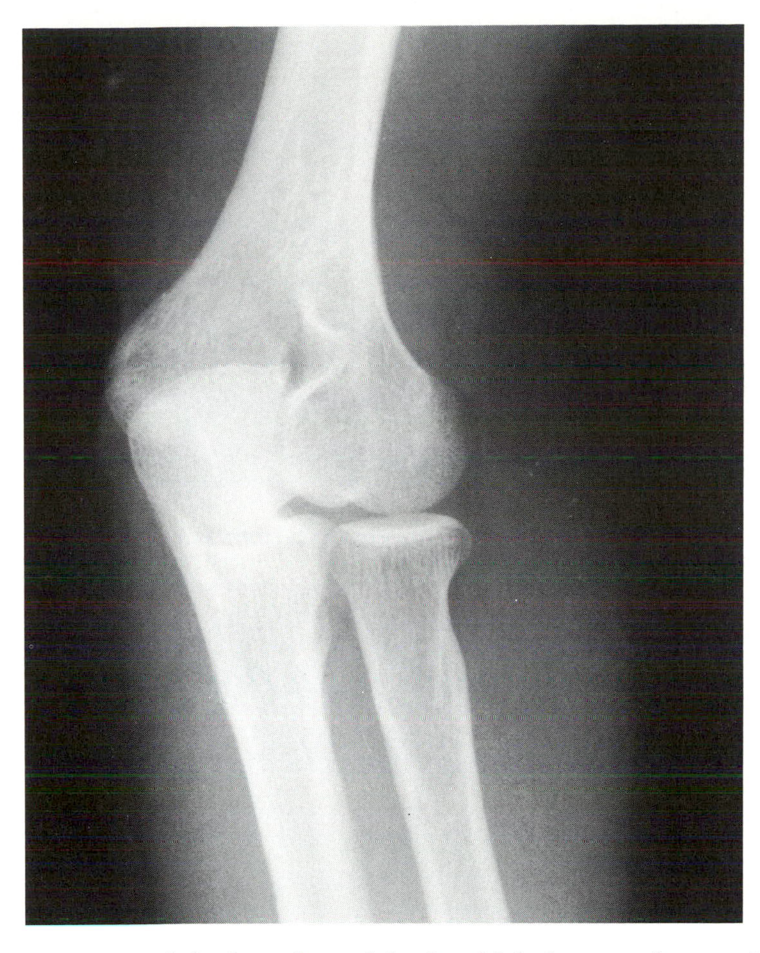

FIG 15–18.
Oblique radiograph of the elbow region.

summary of the branches of the brachial plexus and their distribution is shown in Table 15–8.

Basic Anatomy of the Musculocutaneous Nerve

The musculocutaneous nerve (see Figs 15–20 and 15–22) arises from the lateral cord of the brachial plexus (C5 through C7). It pierces the coracobrachialis muscle and descends between the biceps and the brachialis muscles. It pierces the deep fascia in the region of the elbow and is distributed along the lateral side of the forearm as the *lateral cutaneous nerve of the forearm* (Fig 15–23). The musculocutaneous nerve supplies the coracobrachialis, both heads of the biceps, and the greater part of the brachialis muscles.

Basic Anatomy of the Median Nerve

The median nerve (see Fig 15–20) arises from the medial and lateral cords of the brachial plexus (C5 through C8 and T1). The nerve descends on the lateral side of the axillary artery and at first on the lateral side of the brachial artery (see Fig 15–21). At the middle of the arm, it crosses the brachial artery to reach its medial side (see Fig 15–22). The median nerve gives off no cutaneous or motor branches in the axilla or in the arm.

The median nerve enters the forearm between the two heads of pronator teres muscle and descends on the deep surface of the flexor digitorum superficialis muscle lying on the flexor digitorum profundus muscle (Fig 15–25). In the upper third of the front of the forearm, by unnamed branches or by its anterior interosseous branch (Fig 15–26), it supplies all the muscles of the front of the forearm except the flexor carpi ulnaris and the medial half of the flexor digitorum profundus, which are supplied by the ulnar nerve.

At the wrist, it lies posterior to the tendon of the palmaris longus and between the tendons of the flexor carpi radialis and flexor digitorum superficialis muscles (see Fig 15–25). Here it gives rise to a *palmar cutaneous branch*, which crosses in front of the flexor retinaculum and supplies the skin on the radial half of the palm (see Fig 15–23).

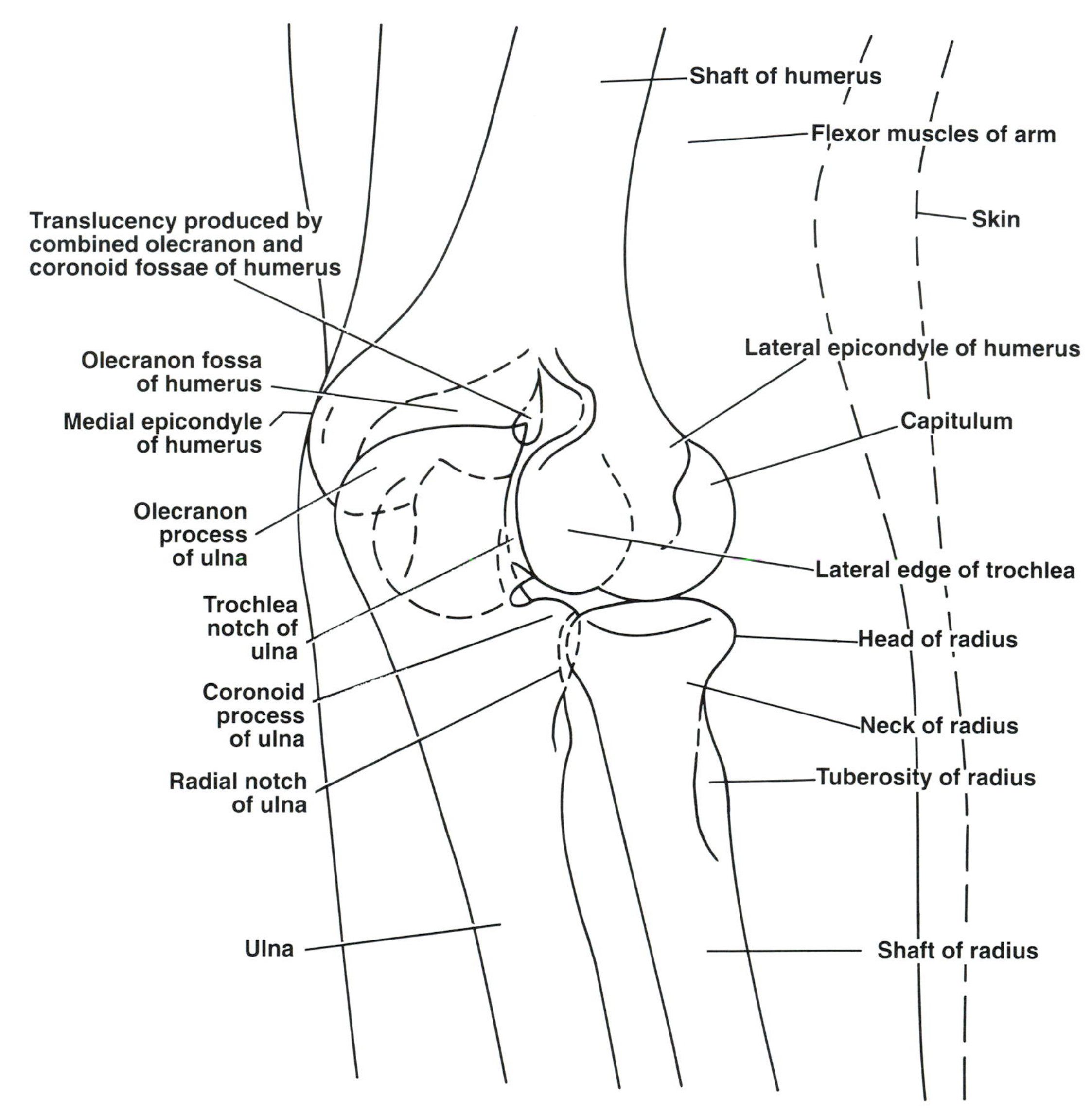

FIG 15–19.
Main features that can be seen in an oblique radiograph of the elbow region in Figure 15–18.

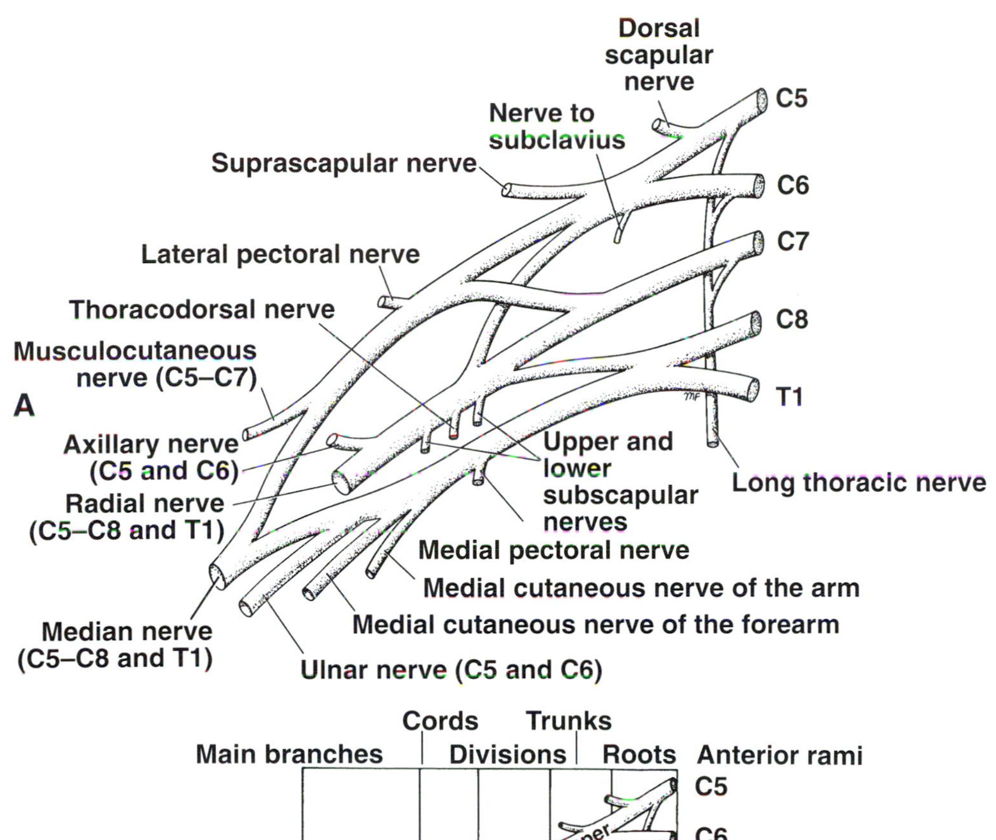

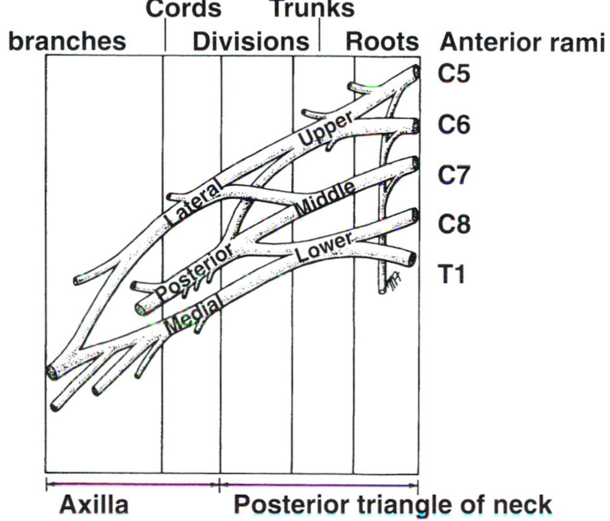

FIG 15–20.
Brachial plexus. **A,** the roots, trunks, divisions, cords, and terminal branches of brachial plexus. **B,** the location of the different parts of the brachial plexus.

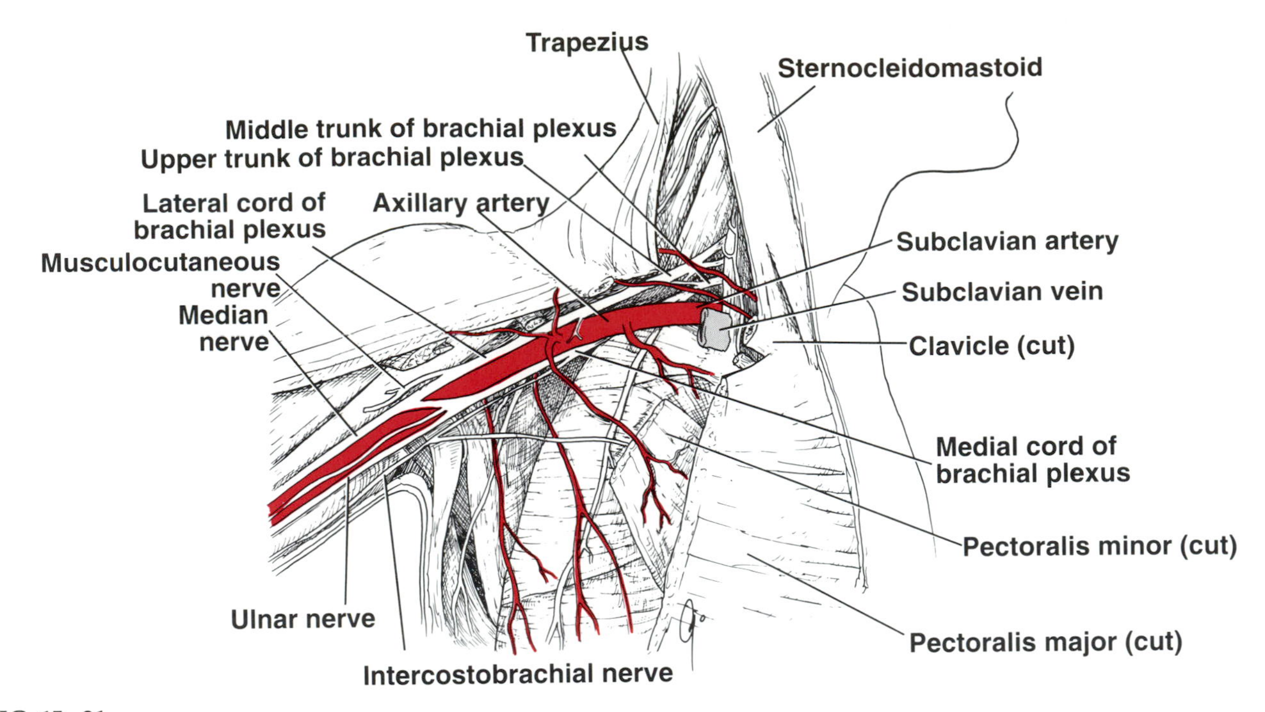

FIG 15–21.
The nerves and blood vessels in the right axilla. Portions of the clavicle, the pectoralis major and minor muscles, and the clavipectoral fascia have been removed to display underlying structures.

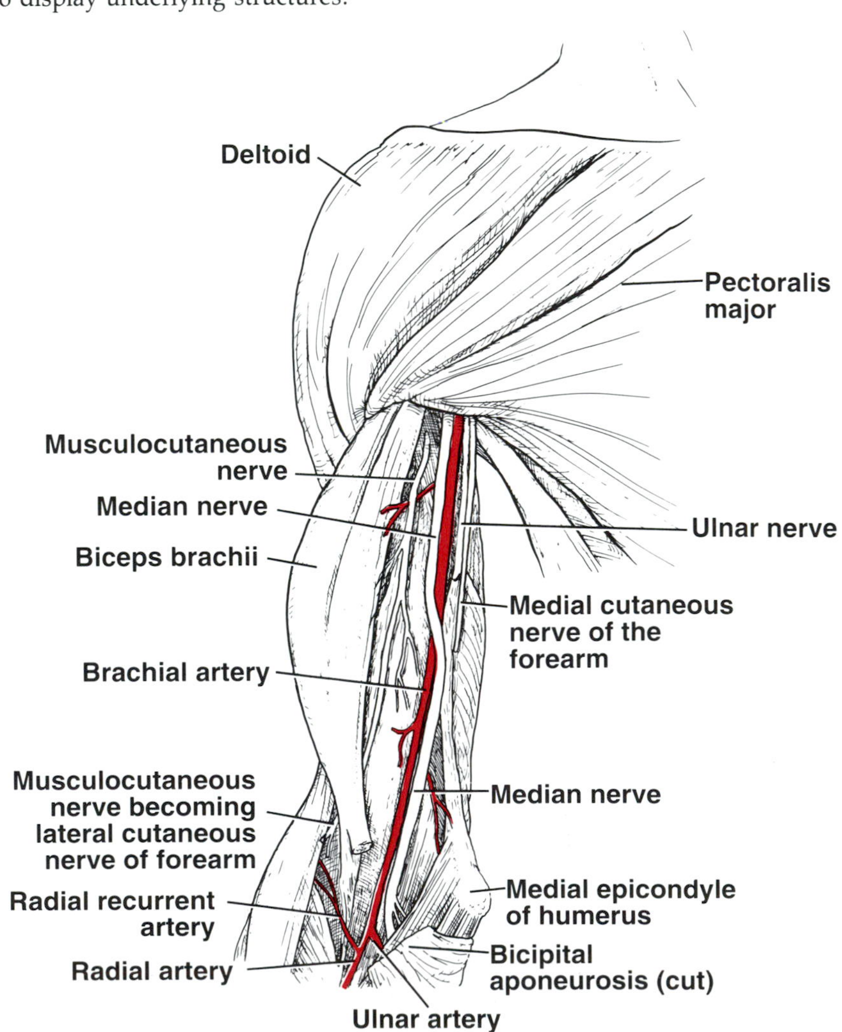

FIG 15–22.
Anterior view of the upper part of the arm. The biceps brachii has been pulled laterally to show the musculocutaneous nerve and the brachial artery.

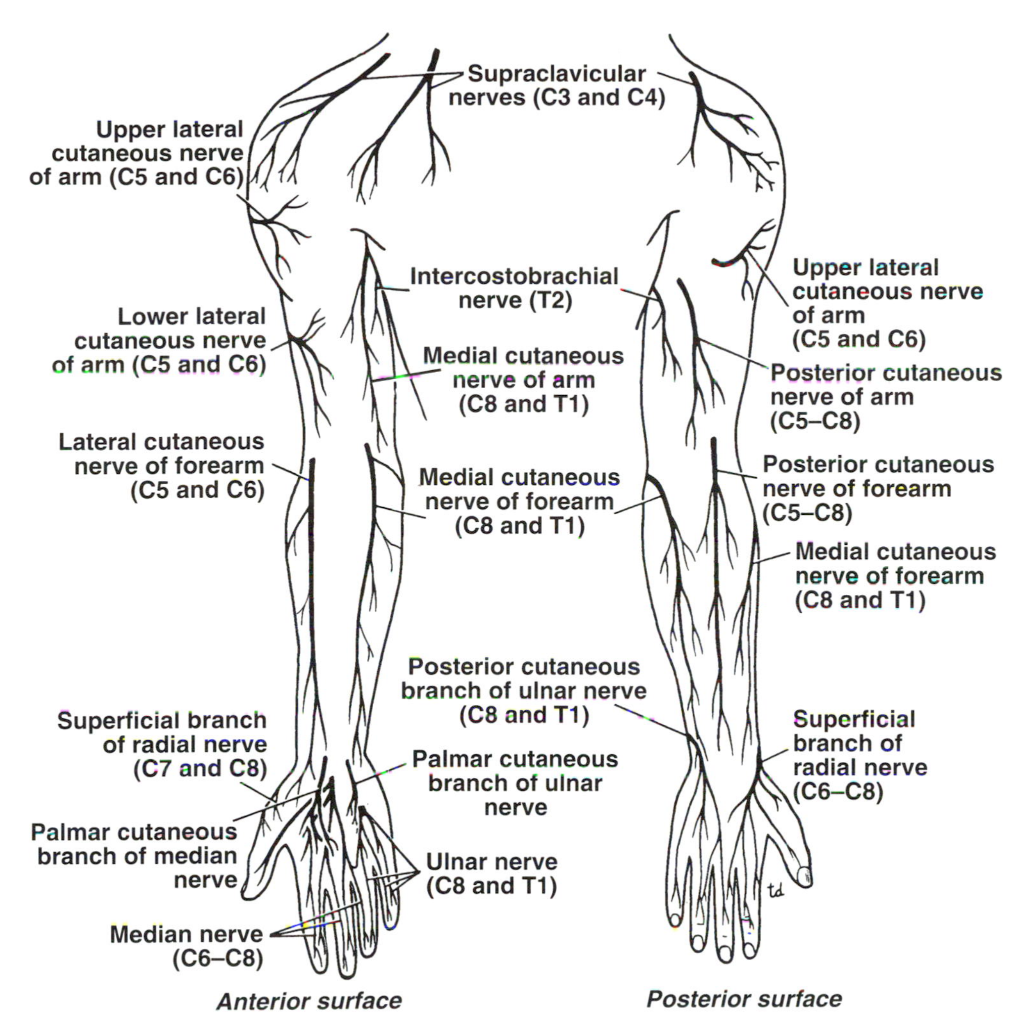

FIG 15–23.
Cutaneous innervation of the upper limb.

The median nerve enters the palm by passing *underneath* the flexor retinaculum and through the carpal tunnel (see Fig 15–25). In the palm the median nerve supplies the three muscles of the thenar eminence (abductor pollicis brevis, flexor pollicis brevis, and the opponens pollicis) and the first two lumbricals and gives sensory innervation to the skin of the palmar aspect of the radial 3½ fingers, including the nail beds on the dorsum.

The distribution of the median nerve is summarized in Table 15–8.

Basic Anatomy of the Ulnar Nerve

The ulnar nerve (see Fig 15–20) arises from the medial cord of the brachial plexus (C8 and T1). It descends along the medial side of the axillary and brachial arteries in the anterior compartment of the arm (see Fig 15–22). At the middle of the arm it pierces the medial intermuscular septum to enter the back of the arm and passes down *behind* the medial epicondyle of the humerus (see Fig 15–14). It gives off no cutaneous or motor branches in the axilla or arm.

The ulnar nerve then enters the anterior compartment of the forearm (see Fig 15–26) by passing between the two heads of flexor carpi ulnaris muscle, where it supplies the flexor carpi ulnaris and the medial part of the flexor digitorum profundus. The nerve descends behind the flexor carpi ulnaris medial to the ulnar artery (see Fig 15–25). In the distal third of the forearm, it gives off its palmar and posterior cutaneous branches (see Fig 15–23).

The *palmar cutaneous branch* supplies the skin over the hypothenar eminence; the *posterior cutaneous branch* supplies the skin over the ulnar third of the dorsum of the hand and the ulnar 1½ fingers. The posterior branch commonly supplies 2½ instead of 1½ fingers. It does not supply the skin over the distal part of the dorsum of these fingers.

At the wrist, the ulnar nerve passes *anterior* to the flexor retinaculum, lateral to the pisiform bone and medial to the hook of the hamate (see Fig 15–25). Here, a superficial part of the flexor retinac-

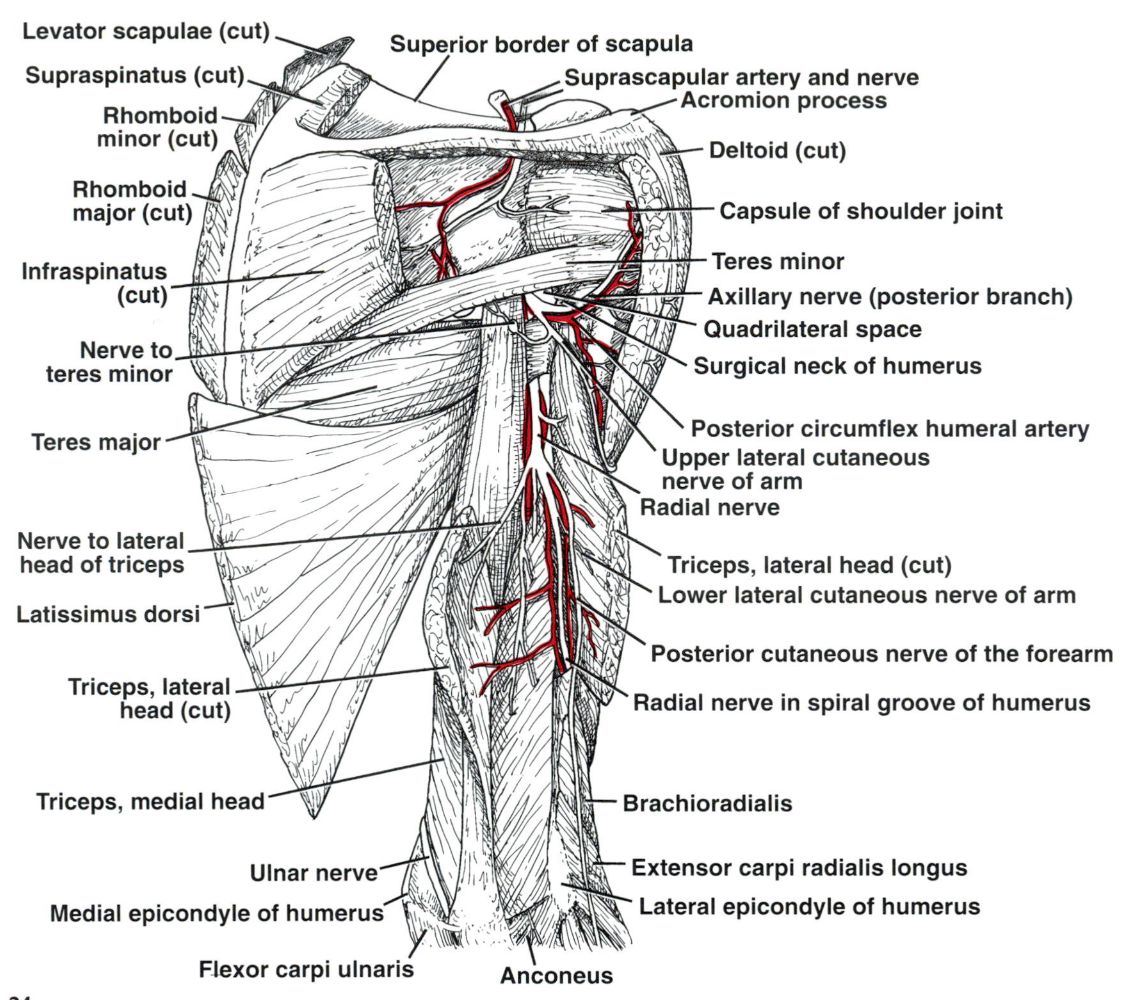

FIG 15–24.
Posterior view of the scapular region and the posterior compartment of the upper part of the arm showing the muscles, nerves, and blood vessels.

ulum passes between the two bones in front of the nerve, creating a fibrous-osseous canal called the *tunnel of Guyon* (a common site for nerve compression). The nerve then divides into superficial and deep terminal branches. The *superficial branch* supplies the skin of the palmar surface of the ulnar 1½ fingers, including the nail beds; it also supplies the palmaris brevis muscle. The *deep branch* supplies all the small muscles of the hand except the three muscles of the thenar eminence and the first two lumbricals, which are supplied by the median nerve.

The distribution of the ulnar nerve is summarized in Table 15–8.

Basic Anatomy of the Radial Nerve

The radial nerve (see Fig 15–20) arises from the posterior cord of the brachial plexus (C5 through C8 and T1). It descends behind the axillary and brachial arteries and passes posteriorly between the long and medial heads of the triceps muscle to enter the posterior compartment of the arm (see Fig 15–24). The radial nerve then winds round in the spiral groove on the back of the humerus with the profunda brachii vessels. Piercing the lateral intermuscular septum just above the elbow, it continues downward into the cubital fossa between the brachialis and the brachioradialis muscles (see Fig 15–26). At the level of the lateral epicondyle it divides into superficial and deep terminal branches.

The radial nerve characteristically gives off its branches some distance proximal to the part to be innervated.

In the axilla it gives off three branches—the posterior cutaneous nerve of the arm, which supplies the skin on the back of the arm down to the elbow; the nerve to the long head of the triceps; and the nerve to the medial head of the triceps.

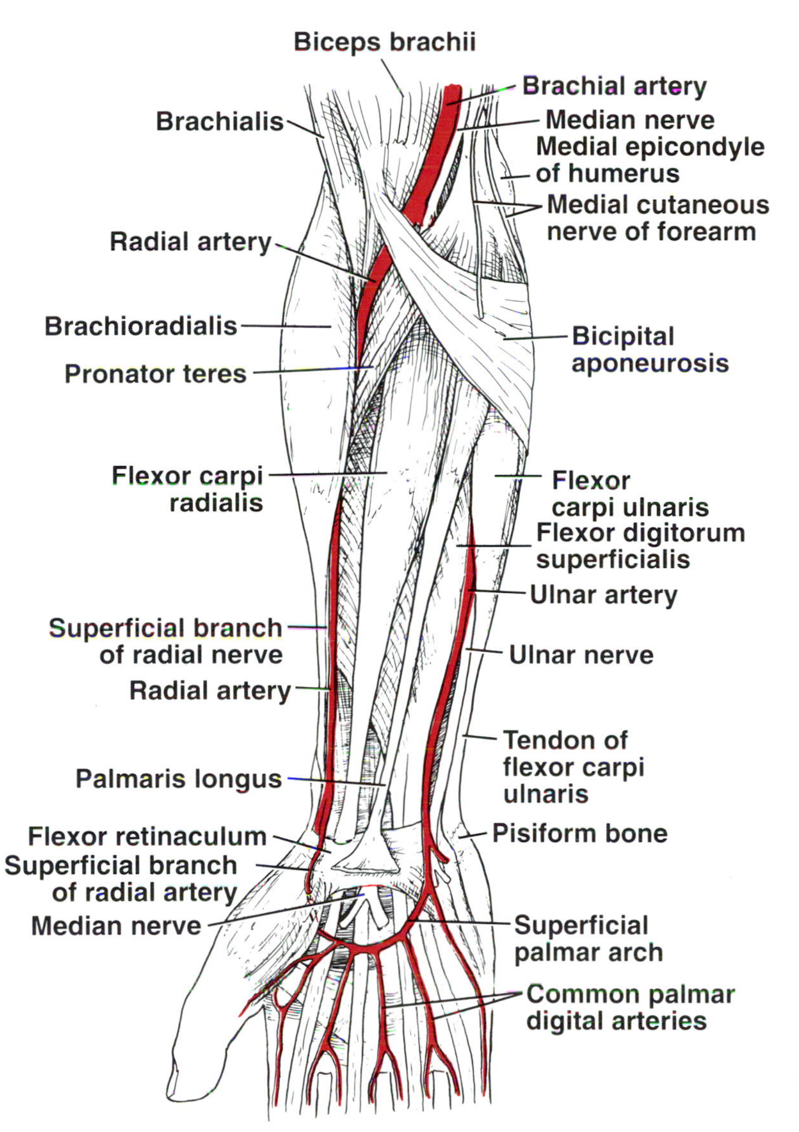

FIG 15–25.
Anterior view of the forearm and hand.

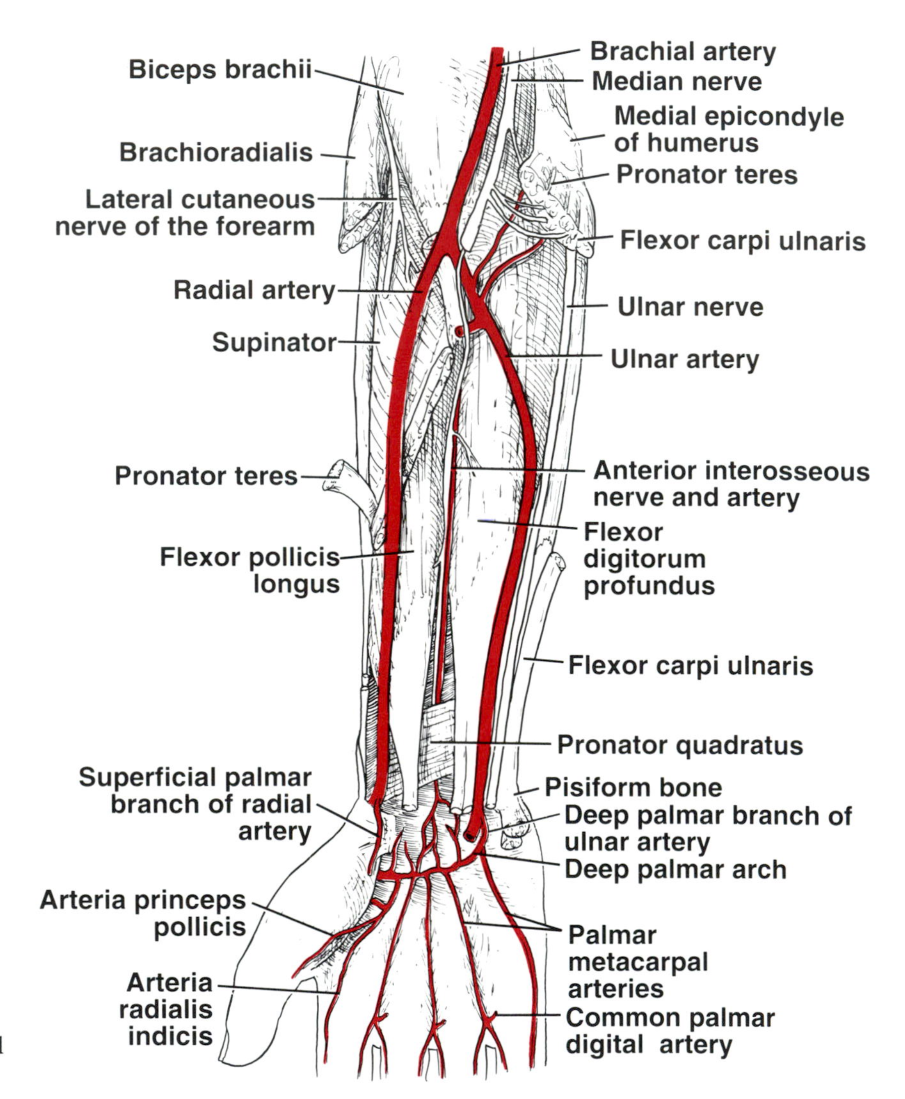

FIG 15–26.
Anterior view of the forearm and hand showing deep structures.

TABLE 15–8.
Summary of the Branches of the Brachial Plexus and Their Distribution

Branches	Distribution
Roots	
Dorsal scapular nerve (C5)	Rhomboid minor, rhomboid major, levator scapulae muscles
Long thoracic nerve (C5, C6, C7)	Serratus anterior muscle
Upper trunk	
Suprascapular nerve (C5, C6)	Supraspinatus and infraspinatus muscles
Nerve to subclavius (C5, C6)	Subclavius
Lateral cord	
Lateral pectoral nerve (C5, C6, C7)	Pectoralis major muscle
Musculocutaneous nerve (C5, C6, C7)	Coracobrachialis, biceps brachii, brachialis muscles; supplies skin along lateral border of forearm when it becomes the lateral cutaneous nerve of forearm
Lateral root of median nerve (C5, C6, C7)	See Medial root of median nerve
Posterior cord	
Upper subscapular nerve (C5, C6)	Subscapularis muscle
Thoracodorsal nerve (C6, C7, C8)	Latissimus dorsi muscle
Lower subscapular nerve (C5, C6)	Subscapularis and teres major muscles
Axillary nerve (C5, C6)	Deltoid and teres minor muscles; upper lateral cutaneous nerve of arm supplies skin over lower half of deltoid muscle
Radial nerve (C5, C6, C7, C8, T1)	Triceps, anconeus, part of brachialis, extensor carpi radialis longus; via deep radial nerve branch supplies extensor muscles of forearm: supinator, extensor carpi radialis brevis, extensor carpi ulnaris, extensor digitorum, extensor digiti minimi, extensor indicis, abductor pollicis longus, extensor pollicis longus, extensor pollicis brevis; skin, lower lateral cutaneous nerve of arm, posterior cutaneous nerve of arm, and posterior cutaneous nerve of forearm; skin on radial side of dorsum of hand and dorsal surface of radial 3½ fingers; articular branches to elbow, wrist, and hand
Medial cord	
Medial pectoral nerve (C8, T1)	Pectoralis major and minor muscles
Medial cutaneous nerve of arm joined by intercostobrachial nerve from second intercostal nerve (C8, T1, T2)	Skin of medial side of arm
Medial cutaneous nerve of forearm (C8, T1)	Skin of medial side of forearm
Ulnar nerve (C8, T1)	Flexor carpi ulnaris and medial half of flexor digitorum profundus, flexor digiti minimi, opponens digiti minimi, abductor digiti minimi, adductor pollicis, third and fourth lumbricals, interossei, palmaris brevis, skin of ulnar half of dorsum of hand and palm, skin of palmar and dorsal surfaces of ulnar 1½ fingers
Medial root of median nerve (with lateral root) forms median nerve (C5, C6, C7, C8, T1)	Pronator teres, flexor carpi radialis, palmaris longus, flexor digitorum superficialis, abductor pollicis brevis, flexor pollicis brevis, opponents pollicis, first two lumbricals (by way of anterior interosseous branch), flexor pollicis longus, flexor digitorum profundus (lateral half), pronator quadratus; palmar cutaneous branch to radial half of palm and digital branches to palmar surface of radial 3½ fingers; articular branches to elbow, wrist, and carpal joints

In the spiral groove of the humerus it gives off four branches—the lower lateral cutaneous nerve of the arm, which supplies the lateral surface of the arm down to the elbow; the posterior cutaneous nerve of the forearm, which supplies the skin down the middle of the back of the forearm as far as the wrist; the nerve to the lateral head of the triceps; and the nerve to the medial head of the triceps and the anconeus.

In the anterior compartment of the arm above the lateral epicondyle it gives off the three branches—the nerve to a small part of the brachialis; the nerve to the brachioradialis; and the nerve to the extensor carpi radialis longus.

In the cubital fossa it gives off the deep branch of the radial nerve and continues as the superficial radial nerve. The deep branch supplies the extensor carpi radialis brevis, the supinator in the cubital fossa, and all the extensor muscles in the posterior compartment of the forearm (see Fig 15–27). The superficial radial nerve is sensory and supplies the skin over the radial part of the dorsum of the hand

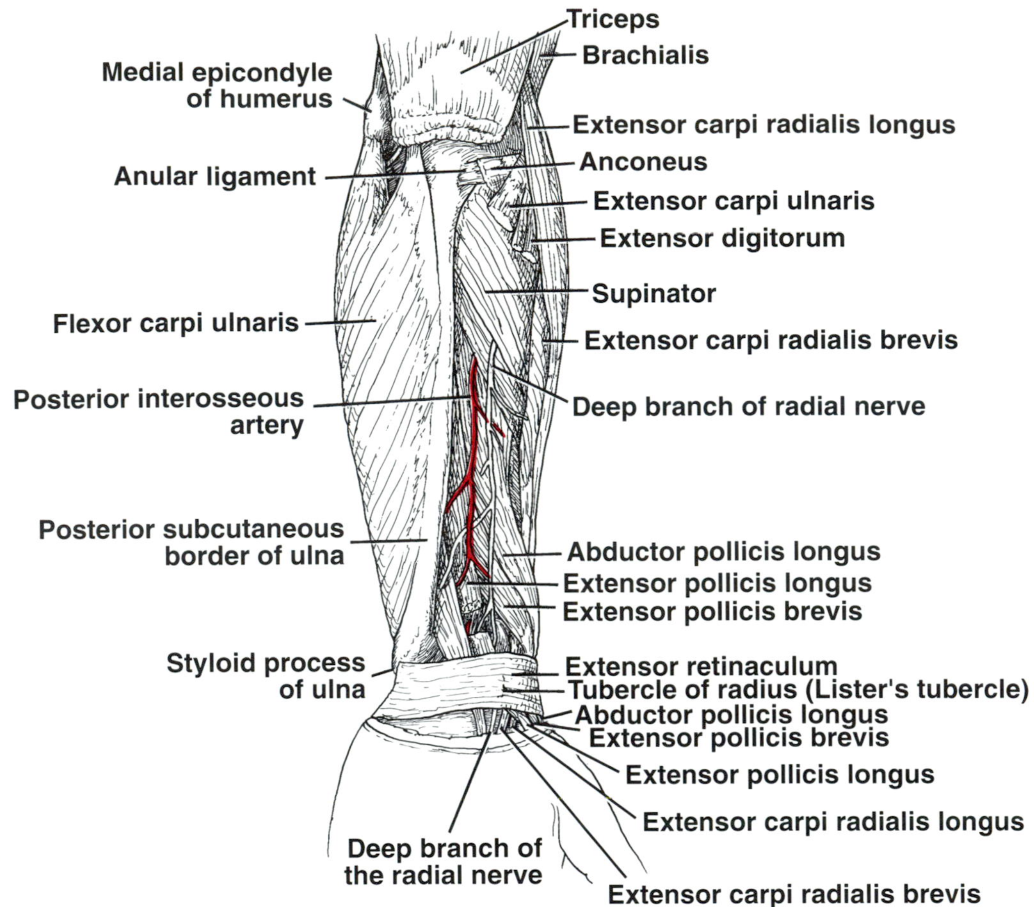

FIG 15–27.
Posterior view of the forearm. The superficial muscles have been removed to display the deep branch of the radial nerve.

(see Fig 15–23) and the dorsal surface of the radial 3½ fingers proximal to the nail beds. (The ulnar nerve supplies the ulnar part of the dorsum of the hand and the dorsal surface of the ulnar 1½ fingers; the exact cutaneous areas innervated by the radial and ulnar nerves on the hand are subject to variation.)

A summary of the distribution of the radial nerve is shown in Table 15–8.

Basic Anatomy of the Axillary Nerve

The axillary nerve (see Fig 15–20) arises from the posterior cord of the brachial plexus (C5 and C6). It passes backward through the quadrilateral space below the shoulder joint in company with the posterior circumflex humeral vessels (see Fig 15–24). The main trunk supplies the shoulder joint, and its *anterior terminal branch* winds around the surgical neck of the humerus beneath the deltoid muscle; the anterior branch supplies the deltoid muscle and skin that covers its lower half. A *posterior terminal branch* supplies the teres minor muscle, the deltoid muscle, and then becomes the *upper lateral cutaneous nerve of the arm,* which supplies the skin over the lower part of the deltoid muscle.

A summary of the distribution of the axillary nerve is shown in Table 15–8.

SURFACE ANATOMY

Surface Anatomy of the Shoulder Region

Clavicles
The clavicle is situated at the root of the neck and can be palpated throughout its length (Fig 15–28). The positions of the *sternoclavicular* and *acromioclavicular joints* can be easily identified. Note that the medial end of the clavicle projects above the margin of the manubrium sterni.

Deltopectoral Triangle
This small, triangular depression is situated below the outer third of the clavicle; it is bounded by the pectoralis major and the deltoid muscles (see Fig 15–28).

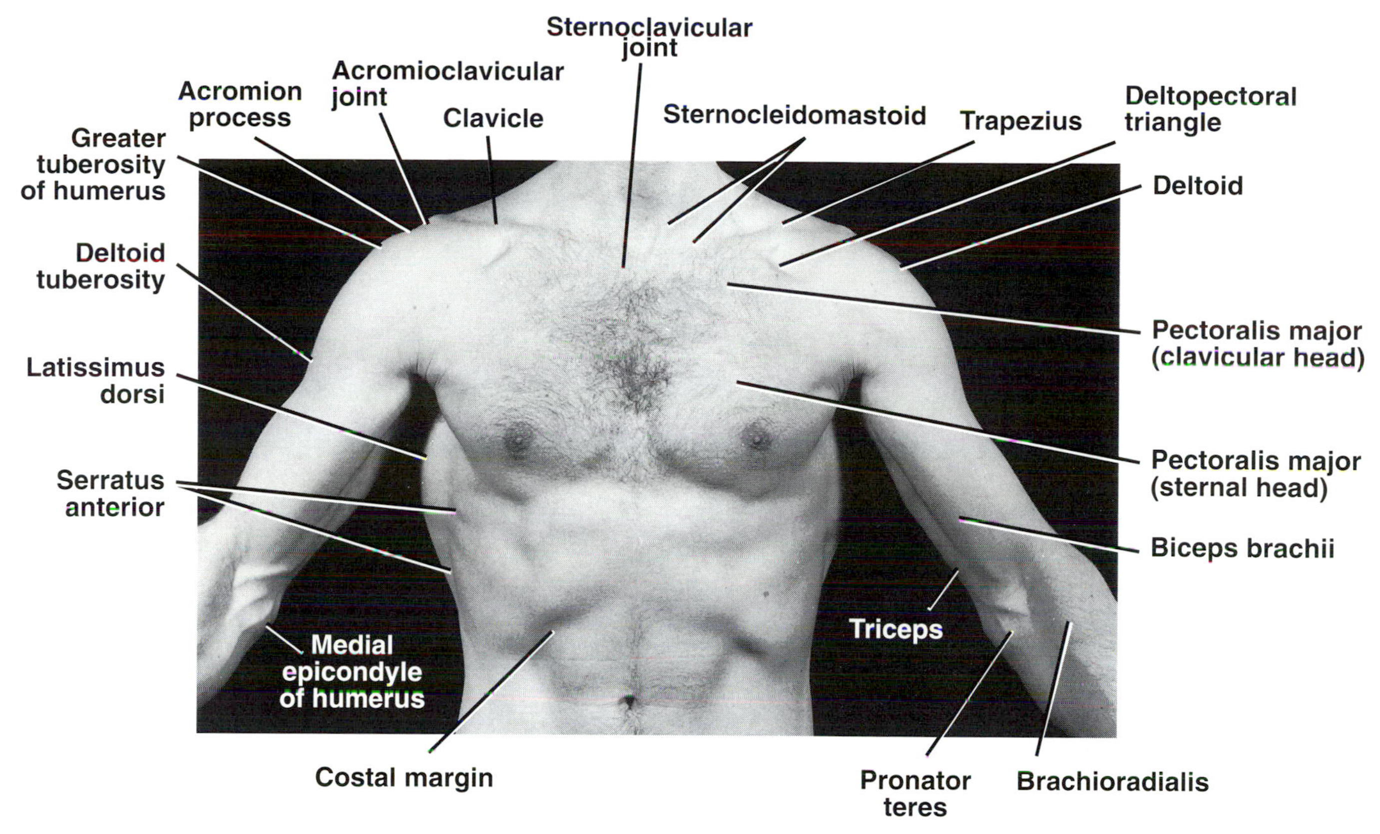

FIG 15–28.
The front chest of a 27-year-old man.

Scapulae

The tip of the *coracoid process* of the scapula can be felt on deep palpation in the lateral part of the deltopectoral triangle; it is covered by the anterior fibers of the deltoid.

The *acromion process* of the scapula forms the lateral extremity of the spine of the scapula. It is subcutaneous and easily located (see Fig 15–28).

Immediately below the lateral edge of the acromion process, the smooth, rounded curve of the shoulder is produced by the *deltoid muscle,* which covers the *greater tuberosity of the humerus* (see Fig 15–28).

The *crest of the spine of the scapula* can be palpated and traced medially to the medial border of the scapula, which it joins at the level of the third thoracic spine.

The *inferior angle of the scapula* can be palpated opposite the seventh thoracic spine (Fig 15–29).

Anterior Axillary Fold

The anterior axillary fold, which is formed by the lower margin of the pectoralis major muscle, can be palpated between the finger and thumb (Fig 15–30).

Posterior Axillary Fold

The posterior axillary fold, which is formed by the tendon of latissimus dorsi winding around the lower border of the teres major muscle, can be similarly palpated between the finger and thumb (see Fig 15–30).

Axilla

The axilla should be examined with the forearm supported and the pectoral muscles relaxed. With the arm by the side, the inferior part of the *head of the humerus* can be easily palpated through the floor of the axilla. The pulsations of the *axillary artery* may be felt high up in the axilla, and around the artery the *cords of the brachial plexus* may be palpated. The medial wall of the axilla is formed by the *upper ribs* covered by the *serratus anterior muscle,* the serrations of which can be seen and felt in a muscular patient (see Fig 15–28). The lateral wall is formed by the *coracobrachialis* and *biceps brachii muscles* and the *bicipital groove of the humerus.*

Surface Anatomy of the Elbow Region

The *medial and lateral epicondyles of the humerus* and the *olecranon process of the ulna* can be palpated (Figs

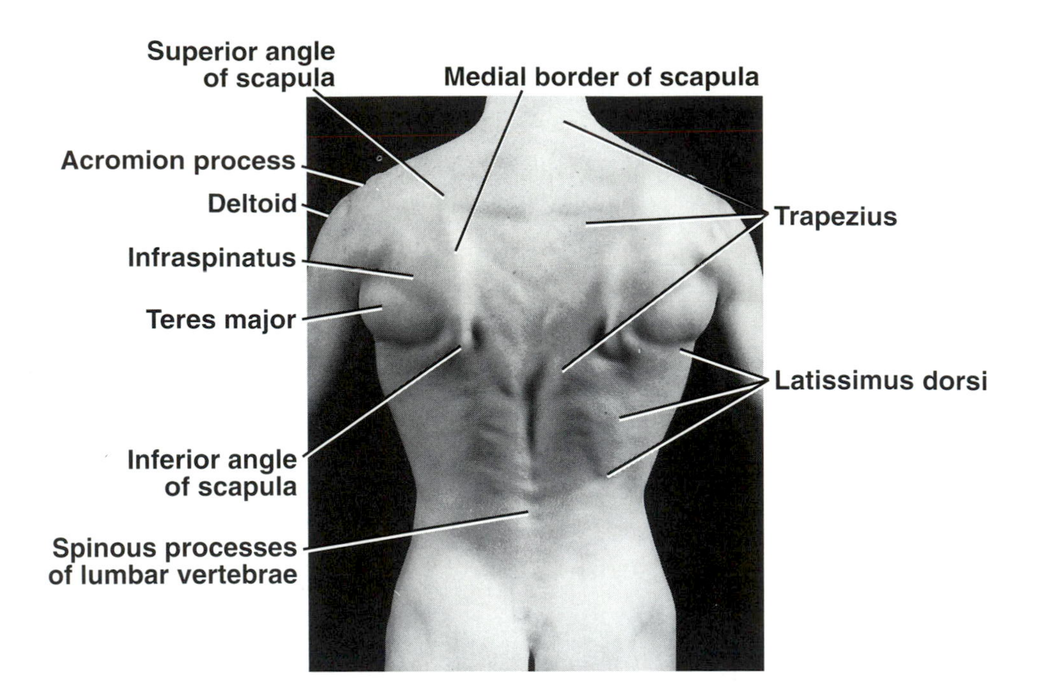

FIG 15–29.
The back of a 27-year-old man.

15–30 and 15–31). When the elbow joint is extended, these bony points lie on the same straight line; when the elbow is flexed, these three points form the boundaries of an equilateral triangle.

The *head of the radius* can be palpated in a depression on the posterolateral aspect of the extended elbow (see Fig 15–30), distal to the lateral epicondyle. The head of the radius can be felt to rotate during pronation and supination of the forearm.

The posterior border of the *ulna bone* is subcutaneous and can be palpated along its entire length.

The boundaries of the *cubital fossa* in front of the elbow joint (see Fig 15–31), can be seen and felt; the brachioradialis muscle forms the radial (lateral) boundary, and the pronator teres forms the ulnar (medial) boundary. The *tendon of the biceps muscle* can be palpated as it passes downward into the fossa, and the *bicipital aponeurosis* can be felt as it leaves the tendon to join the deep fascia of the forearm. The tendon and aponeurosis are most easily felt if the elbow joint is flexed against resistance.

The *ulnar nerve* can be palpated as a rounded cord where it lies behind the medial epicondyle of the humerus.

The *brachial artery* can be palpated as it descends into the cubital fossa overlapped by the medial border of the biceps muscle (see Fig 15–31).

CLINICAL ANATOMY

Clinical Anatomy of the Clavicle and the Clavicular Joints

Fractures of the Clavicle

The clavicle is one of the most commonly fractured bones in the body. Because of its position, it is exposed to trauma and transmits forces from the upper limb to the trunk. The fracture usually occurs as the result of a fall on the shoulder or outstretched hand. The force is transmitted along the clavicle, which breaks at its weakest point, the junction of the middle and outer thirds. Fractures of the medial or lateral end are usually the result of direct trauma.

Following fractures of the middle portion of the clavicle, the lateral fragment is depressed by the weight of the arm and it is pulled medially and forward by the strong adductor muscles of the shoulder joint, especially the pectoralis major. The medial end is tilted upward by the sternocleidomastoid muscle. The figure-of-eight bandage, commonly used in the treatment of clavicular fractures, tends to counteract further displacement of the clavicular fragments.

Anatomy of the Complications.—The position of the clavicle relative to the first rib and the scalenus anterior muscle (see Fig 15–21) explains the occasional compression of the subclavian vein, subclavian artery, and the brachial plexus that may occur in fractures of the clavicle. Severe depression of the

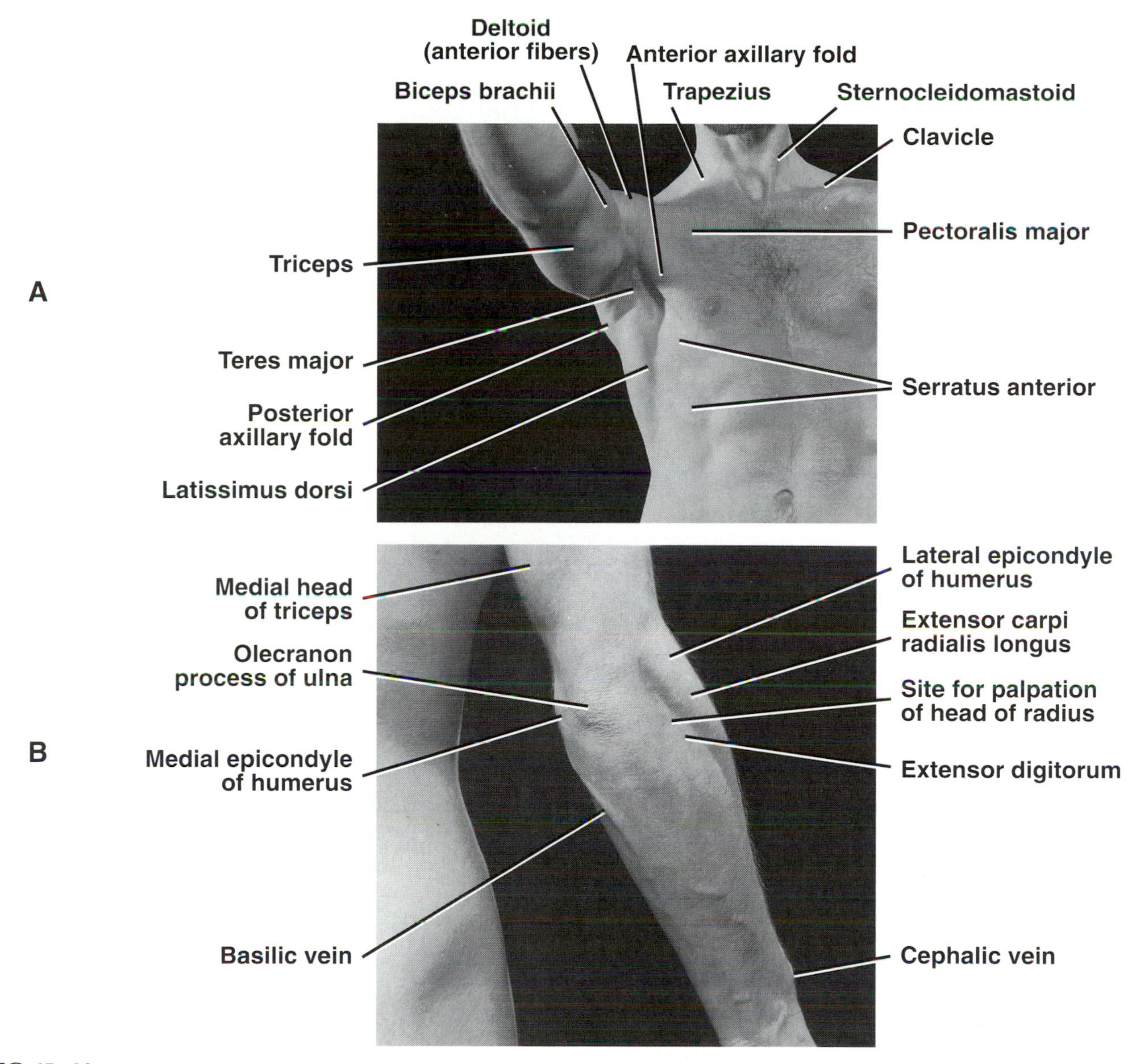

FIG 15–30.
The upper extremity in a 27-year-old man. **A,** the axilla and anterolateral view of the trunk. **B,** the elbow region, posterior view.

outer clavicular fragment is responsible. The close relationship of the supraclavicular nerves as they descend superficial to the clavicle to supply the skin of the upper part of the chest may rarely result in their involvement in the callus; this may be the cause of persistent pain over the side of the neck.

Sternoclavicular Joint Injuries

Injuries to the sternoclavicular joint are rare and occur as the result of violent forces directed along the long axis of the clavicle; usually the clavicle fractures, but occasionally only the joint is injured. The strength of the joint depends on the integrity of the very strong costoclavicular ligament, the intra-articular disc, and the sternoclavicular ligaments. Minor tearing of these ligaments leads to *sprains* or *subluxation* of the joint; major tears lead to a dislocation.

Anterior Dislocation.—The medial end of the clavicle projects forward beneath the skin and may be pulled upward by the sternocleidomastoid muscle.

Posterior Dislocation.—This form of dislocation usually follows direct trauma applied to the front of the joint that drives the clavicle backward. This type is the more serious since the displaced clavicle may

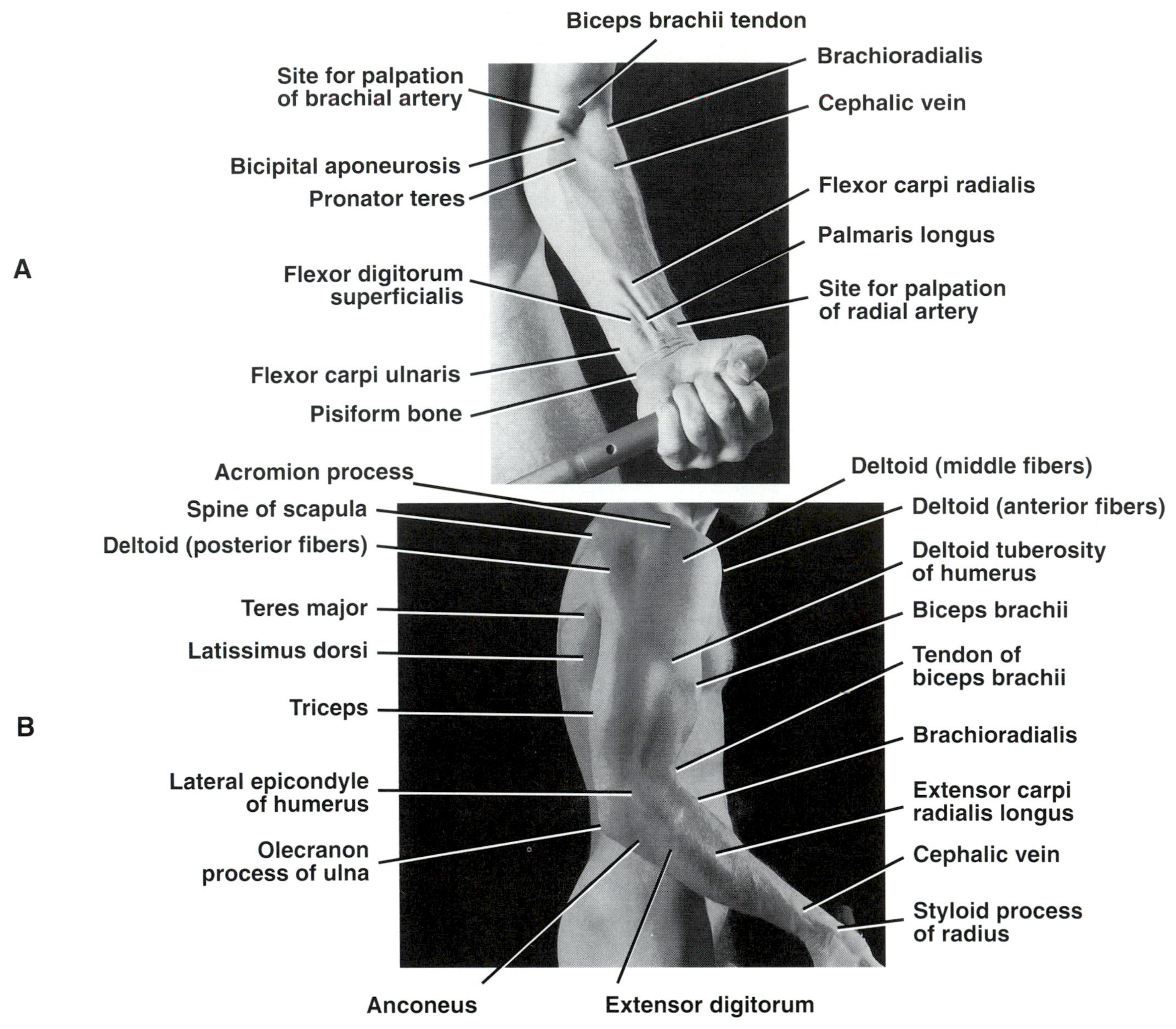

FIG 15–31.
A, left upper limb, anterior view. **B,** right upper limb, lateral view.

press upon the trachea, esophagus, and the brachiocephalic veins, causing ipsilateral venous congestion of the upper limb.

Should the costoclavicular ligament rupture completely, it is difficult to maintain the normal position of the clavicle once reduction has been accomplished.

Acromioclavicular Joint Injuries

The plane of the articular surfaces of the acromioclavicular joint passes downward and medially, so that there is a tendency for the lateral end of the clavicle to ride up over the upper surface of the acromion. The strength of the joint is dependent on the very strong coracoclavicular ligament, which binds the coracoid process to the undersurface of the lateral part of the clavicle. Injuries to the joint range from a *minor sprain,* a *subluxation,* to a complete dislocation.

Acromioclavicular Dislocation.—A severe blow on the point of the shoulder, as is incurred during blocking or tackling in football or any severe fall, may result in the acromion being thrust beneath the lateral end of the clavicle, tearing the coracoclavicular ligament. This condition is known as *shoulder separation*. Three degrees of separation are usually described as follows: (1) partial tear of the acromioclavicular ligaments with minimal separation of the bones, usually diagnosed clinically by tenderness over the acromioclavicular joint in the absence of a distal clavicular fracture; (2) an additional partial tear of the coracoclavicular ligament that causes the clavicle to displace upward when the upper limbs are weight bearing; and (3) the acromioclavicular ligaments and both the conoid and trapezoid components of the coracoclavicular ligament are torn; the outer end of the clavicle is high, riding above the acromion. As in the case of the sternoclavicular joint, the dislocation is easily reduced, but withdrawal of support results in immediate redislocation.

Arthrocentesis of the Clavicular Joints

This is fully described in Chapter 18, pp. 768 and 769.

Clinical Anatomy of the Shoulder Joint

Rotator Cuff Tendinitis

The rotator cuff, consisting of the tendons of the subscapularis, supraspinatus, infraspinatus, and the teres minor muscles, which are fused to the underlying capsule of the shoulder joint, plays a very important role in stabilizing the shoulder joint. Lesions of the cuff are a common cause of pain in the shoulder region.

Excessive overhead activity of the upper limb may be implicated in tendinitis, although many cases appear spontaneously. During the movement of abduction of the shoulder joint, the supraspinatus tendon is exposed to friction against the acromion process (Fig 15–32). Under normal conditions the amount of friction is reduced to a minimum by the large subacromial bursa, which extends laterally beneath the deltoid muscle. Degenerative changes in the bursa are followed by degenerative changes in the underlying supraspinatus tendon, and these may extend into the other tendons of the rotator cuff. The condition is known as *subacromial bursitis, supraspinatus tendinitis*, or *pericapsulitis*.

The condition is characterized by the presence of a spasm of pain in the middle range of abduction, when the diseased area impinges on the acromion process (see Fig 15–32). Often the pain becomes less when the arm passes beyond the painful arc. There is tenderness over the greater tuberosity of the humerus.

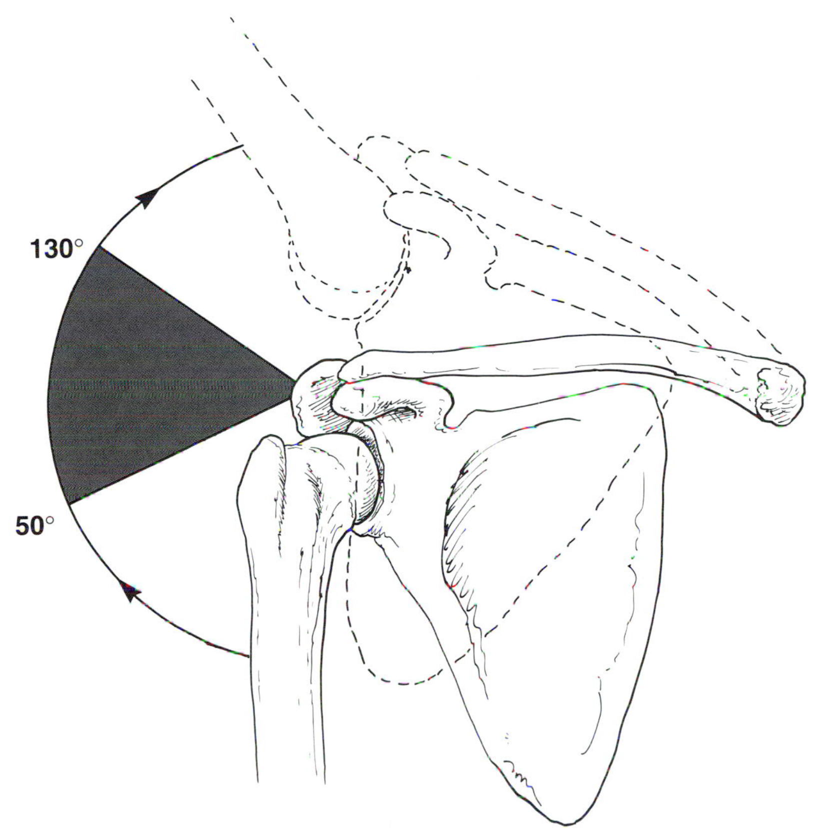

FIG 15–32.
Subacromial bursitis, supraspinatus tendinitis, or pericapsulitis. Diagram showing the painful arc in the middle range of abduction, when the diseased area impinges on the acromion process.

Rotator Cuff Tears

In advanced cases of supraspinatus tendinitis, the tendon may become calcified or may rupture. Rupture of the tendon seriously interferes with the normal abduction movement of the shoulder joint. The normal function of the supraspinatus muscle is to hold the head of the humerus in the glenoid fossa at the commencement of abduction. The patient with a ruptured supraspinatus tendon is unable to initiate abduction of the arm. However, if the arm is passively assisted for the first 15° of abduction, the deltoid can then take over and complete the movement to a right angle.

Biceps Tendinitis

The site of this condition is where the tendon leaves the shoulder joint and lies within the bicipital groove of the humerus between the greater and lesser tuberosities. The tendinitis occurs after strenuous exercise and produces pain located over the anterior part of the joint. Extreme tenderness is found on palpation of the tendon in the bicipital groove (see p. 616 for rupture of biceps tendon).

Dislocations of the Shoulder Joint

The shoulder joint is the most commonly dislocated large joint. While dislocations may occur in any direction, anterior-inferior dislocations are by far the most common.

Anterior-Inferior Dislocation.— Sudden violence applied to the humerus with the joint fully abducted tilts the humeral head downward onto the inferior weak part of the capsule, which tears, and the humeral head comes to lie inferior to the glenoid cavity. During this movement the acromion has acted as a fulcrum. The strong flexors and adductors of the shoulder joint now usually pull the humeral head forward and upward into the subcoracoid position. The patient holds the injured arm with the opposite hand, and the dislocated arm is slightly medially rotated. Pain and spasm of the muscles surrounding the joint are present.

On inspection of the patient with an anterior shoulder dislocation, the normal rounded appearance of the shoulder is seen to be lost, since the greater tuberosity of the humerus is no longer bulging laterally beneath the deltoid muscle. The humeral head can often be palpated in the deltopectoral triangle.

Posterior Dislocation.— These are rare and are usually due to direct violence to the front of the joint or can occur during a seizure. Clinical examination usually finds the joint to be internally rotated, and passive external rotation is impossible. The diagnosis is often missed because the anteroposterior view of the shoulder using radiography on first inspection appears fairly normal. Note the ice cream cone shape of the dislocated humeral head and the overlap between the humeral head and the glenoid fossa that looks abnormal; moreover, it is impossible to take the anteroposterior view with the humerus externally rotated. An axillary view will reveal the posterior dislocation.

Anatomy of the Complications.— Subglenoid displacement of the head of the humerus into the quadrilateral space (see Fig 15–7) may cause damage to the axillary nerve, as indicated by paralysis of the deltoid muscle and loss of skin sensation over the *lower half* of the deltoid. Downward displacement of the humerus may also damage the radial nerve. The radial nerve, which is wrapped around the back of the shaft of the humerus (see Fig 15–24), may be pulled downward, stretching the nerve excessively. The clinical findings of radial nerve injury include weakness or loss of active extension of the elbow joint, wrist joint, and the fingers (see p. 621).

Tears of the rotator cuff may also occur, and fractures of the humeral head and edge of the glenoid fossa are relatively common complications.

Anatomy of Reducing Shoulder Dislocations

Reduction of Anterior Dislocations.— The status of the joint bones after dislocation is as follows. The head of the humerus lies outside the glenoid cavity and is either anterior to it below the coracoid process (subcoracoid dislocation), inferior to the glenoid (subglenoid dislocation), or inferior to the clavicle (subclavicular dislocation).

Hennepin Technique.—The elbow is flexed to 90° and kept in the adducted position against the trunk. The technique is explained to the patient, who is encouraged to relax. Using the forearm as an indicator and a lever, the humerus is then gently, steadily, externally rotated to stretch the subscapularis muscle. Should pain be experienced, the physician should pause and then continue again until the arm is fully rotated. By this time the head of the humerus is out from under the glenoid rim and may have already slipped back into the glenoid cavity. If this technique does not immediately succeed, slight abduction of the fully externally rotated arm may be all that is necessary to complete the reduction. This latter maneuver further stretches the subscapularis muscle and the anterior capsule and facilitates the reduction. The arm is then fully medially rotated against the chest wall, ensuring that the humeral head resides within the glenoid cavity.

Stimson Technique.—The patient lies prone on the examining table with the affected upper limb hanging vertically downward. Weights are attached to the forearm to gently elongate the shoulder muscles that are in spasm. Within 20 to 30 minutes the muscles should have sufficiently relaxed to allow the humeral head to be displaced away from and then into the glenoid cavity.

Traction and Countertraction Technique.—This method is advocated for dislocations that are seen late. Anatomically the technique relies on the opposing forces to stretch the surrounding shoulder muscles so that the humeral head will slip back into the glenoid cavity. While the physician applies traction to the long axis of the humerus by pulling on the forearm, the assistant applies countertraction on the trunk and scapula by pulling on a folded sheet wrapped around the upper part of the chest. When the muscles have elongated, slight lateral traction on the proximal end of the humerus generally allows the humeral head to slip back into the glenoid cavity.

Scapular Rotation Technique.—With the patient prone (preferably with the arm hanging over the side of the stretcher), the inferior angle of the scapular is rotated medially. At the same time the superior angle of the scapula is rotated laterally so that the glenoid fossa is moved down towards the humeral head. Often an audible clunk is heard as the reduction is completed.

Posterior Dislocations.—In posterior dislocations the head of the humerus lies behind the glenoid cavity. With the patient in a supine position, traction is applied along the longitudinal axis of the humerus by pulling on the forearm. Lateral traction to the upper part of the humerus is provided by an assistant using a folded towel. This maneuver elongates the shortened muscles and the humeral head is pulled forward into the glenoid cavity. Here, again, sedation may be required.

Arthrocentesis of the Shoulder Joint

The anatomy of this procedure is fully described in Chapter 18, p. 769.

Clinical Anatomy of Shoulder Pain

When the differential diagnosis of shoulder pain is considered, a number of conditions must be included—supraspinatus tendinitis, subdeltoid bursitis, and rotator cuff disease, which are all part of same degenerative process, and acromioclavicular joint injuries, biceps tendinitis, dislocation of the shoulder joint, and humeral head fracture. All these conditions involve structures that are in or around the shoulder joint and cause painful reflex spasm of the surrounding muscles; the acromioclavicular joint is innervated by the supraclavicular nerve, which is also sensory to the skin over the point of the shoulder.

In addition, visceral pain referred from myocardial ischemia occasionally gives rise to shoulder pain, although it is usually referred along the medial side of the upper part of the arm via the intercostobrachial nerve (see p. 125). Referred pain from irritation of the diaphragmatic pleura (pleurisy) or peritoneum (peritonitis) must not be overlooked; the pain is transmitted via the phrenic nerve and the supraclavicular nerves (see p. 125). Prolapsed cervical discs may compress or irritate cervical nerve roots and give rise to referred pain to the shoulder and down the arm, involving the appropriate dermatomes.

Clinical Anatomy of the Humerus

Fractures of the Humerus

Fractures of the Proximal End of the Humerus.—These include the following.

Humeral Head Fractures (Fig 15–33).—These fractures may occur during the process of anterior and posterior dislocations of the shoulder joint. The glenoid rim of the scapula produces the fracture, and the rim may become jammed in the defect, making reduction of the shoulder joint difficult. Rarely the fracture is more complete and occurs at the anatomical neck of the humerus; the head then lies free and may require reduction.

Greater Tuberosity Fractures.—The greater tuberosity of the humerus may be fractured by direct trauma, displaced by the glenoid rim during dislocation of the shoulder joint, or avulsed by violent contractions of the supraspinatus muscle. The bone fragment will have the attachments of the supraspinatus, teres minor, and infraspinatus muscles, whose tendons form part of the rotator cuff. When associated with a shoulder dislocation, severe tearing of the cuff with the fracture may result in the greater tuberosity remaining displaced posteriorly after the shoulder joint has been reduced. In this situation, open reduction of the fracture is necessary to attach the rotator cuff back into place.

Lesser Tuberosity Fractures.—Occasionally a lesser tuberosity fracture accompanies posterior dislocations of the shoulder joint. The bone fragment

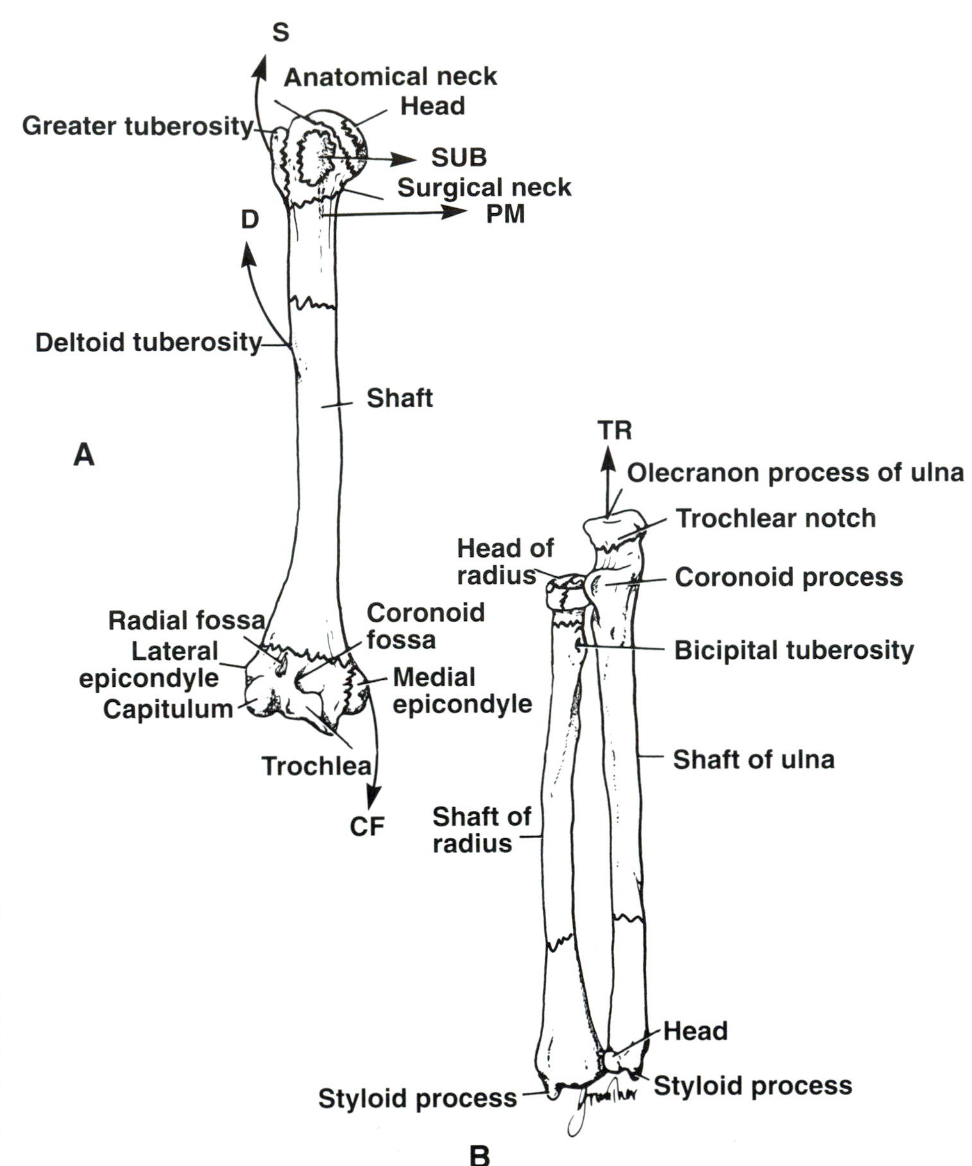

FIG 15–33.
A, common fractures of the humerus. **B,** common fractures of the radius and ulna. The displacement of the bony fragments depends on the site of the fracture line and the pull of the muscles. *S* = supraspinatus; *D* = deltoid; *PM* = pectoralis major; *CF* = pull of common flexure muscles; *TR* = triceps; SUB =subscapularis.

receives the insertion of the subscapularis tendon (see Fig 15–33), a part of the rotator cuff.

Surgical Neck Fractures.—The surgical neck of the humerus (see Fig 15–33), which lies immediately distal to the lesser tuberosity, may be fractured by a direct blow on the lateral aspect of the shoulder or in an indirect manner by the person falling on the outstretched hand. In children, injury to the proximal epiphyseal cartilage may be followed by retardation of growth in the length of the humerus.

Fractures of the Shaft of the Humerus.—These fractures are common, with the displacement of the fragments dependent on the relation of the site of fracture to the insertion of the deltoid muscle (see Fig 15–33). When the fracture line is proximal to the deltoid insertion, the proximal fragment is adducted by the pectoralis major, latissimus dorsi, and the teres major muscles; the distal fragment is pulled proximally by the deltoid, biceps, and triceps. When the fracture is distal to the deltoid insertion, the proximal fragment is abducted by the deltoid and the distal fragment is pulled proximally by the biceps and triceps. The radial nerve may be damaged where it lies in the spiral groove on the posterior surface of the humerus under cover of the triceps muscle.

Fractures of the Distal End of the Humerus.—Supracondylar fractures (see Fig 15–33) are common in children and occur when the child falls on the outstretched hand with the elbow joint partially flexed.

Injuries to the median, radial, and ulnar nerves are not uncommon, although function usually quickly returns after reduction. Damage to or pressure on the brachial artery may occur at the time of the fracture or from swelling of the surrounding tissues; the circulation to the forearm may be compromized, leading to Volkmann's ischemic contracture.

The medial epicondyle (see Fig 15–33) can be avulsed by the medial collateral ligament of the elbow joint if the forearm is forcibly abducted. The ulnar nerve may be injured at the time of the fracture, may become involved in the callus formation, or may undergo irritation on the irregular bony surface after the bone fragments are reunited.

Clinical Anatomy of the Elbow Region

The elbow joint is stable because of the wrench-shaped articular surface of the olecranon and the pulley-shaped trochlea of the humerus; it also has very strong medial and lateral collateral ligaments. When the elbow joint is examined, the normal relations of the bony points should be noted. In extension, the medial and lateral epicondyles and the top of the olecranon process are in a straight line; in flexion, the bony points form the boundaries of an equilateral triangle. The close relationship of the median and ulnar nerves and the brachial artery to the joint often results in their damage in dislocations of the joint.

The presence of ossification centers at the distal end of the humerus and the proximal ends of the radius and ulna may cause confusion when reading radiographs in this region. The order of ossification can be committed to memory using the word CRITOE: *C* = capitulum, year 1; *R* = radial head, years 5 through 7; *I* = internal (medial) epicondyle of humerus, years 5 through 9; *T* = trochlea of humerus, year 10; *O* = olecranon process of ulna, year 11; and *E* = extermal (lateral) epicondyle of humerus, year 14. These centers of ossification fuse around 14 to 15 years of age in females and between 18 and 21 years in males.

Dislocations of the Elbow Joint

Elbow dislocations are common, and the majority are posterior.

Posterior Dislocation.—This most commonly follows falling on an outstretched hand. The distal end of the humerus is forced through the anterior part of the capsule and tears the overlying brachialis muscle; the triceps insertion into the olecranon process may be stripped off the bone. The medial and lateral collateral ligaments are often torn. On examination, the elbow joint is held in partial flexion, and the forearm appears foreshortened. The olecranon process is prominent posteriorly, and the relationship between the olecranon and the epicondyles of the humerus is altered. Injuries to the median and ulnar nerves and the brachial artery may occur. Fractures of the medial epicondyle, the head of the radius, or coronoid process may complicate the dislocation.

Anterior Dislocation.—This is a rare injury. It is caused by a violent blow to the olecranon process with the elbow in flexion. Severe injury to the median and ulnar nerves and brachial artery is common.

Anatomy of the Reduction of Posterior Dislocations of the Elbow.—The status of the bones is as follows—the olecranon process of the ulna projects posteriorly, the trochlea of the humerus is no longer in the trochlea notch of the ulna but resides anterior to it on the coronoid process of the ulna, and the head of the radius lies below and behind the capitulum of the humerus.

The associated trauma to soft tissues usually causes considerable local swelling. For this reason, immediate reduction is desirable with analgesia. Traction is applied to the supinated forearm while an assistant applies countertraction to the humerus. Pressure is applied to the proximal forearm to displace the ulna downwards away from the humerus so that the coronoid process of the ulna disengages from the trochlea of the humerus and slips forward into its normal position. As this takes place the trochlea of the humerus slips into the trochlea notch of the ulna, and the olecranon moves into its normal position. Alternatively, the olecranon process can be pushed inferiorly and anteriorly to effect the reduction. Medial and lateral displacements are corrected. The integrity of the lateral and medial collateral ligaments are assessed after reduction by applying appropriate medial and lateral forces to the forearm. Neurovascular function is always checked following reduction of elbow dislocations.

Subluxation of the Radial Head in Children

The strength of the superior radioulnar joint depends on the integrity of the strong anular ligament that binds the head of the radius to the radial notch of the ulna (Fig 15–34). In young children between 1 and 6 years old, the size of the radial head is relatively small, and a sudden jerk on the forearm may

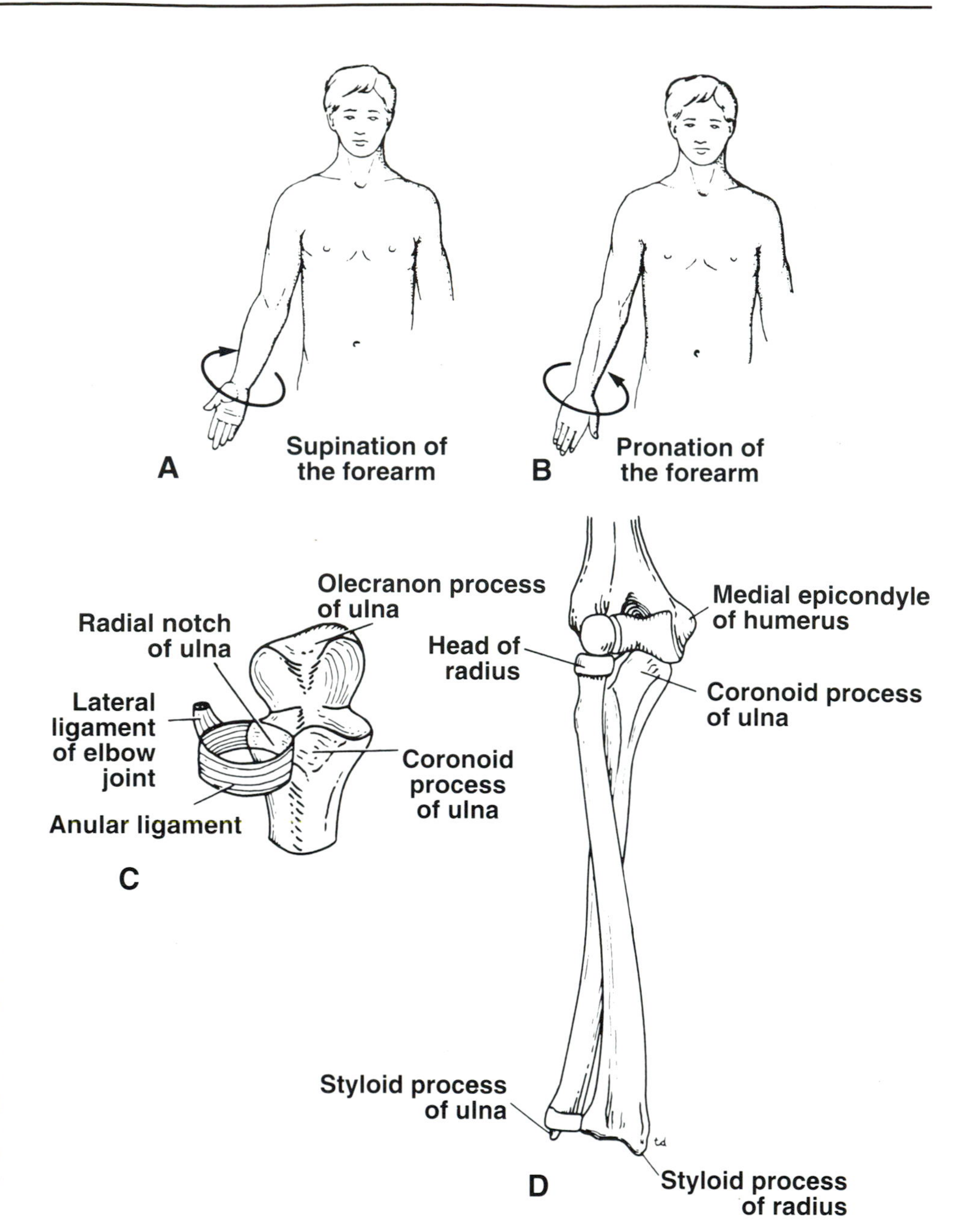

FIG 15–34.
A and **B,** movements of supination and pronation of the forearm that take place at the superior and inferior radioulnar joints. **C,** ligaments and articular surface of ulna in superior radioulnar joint. The anular ligament binds the head of the radius to the radial notch of the ulna. **D,** relative positions of the radius and ulna when the forearm is fully pronated.

pull the radial head down through the anular ligament. The upper edge of the anular ligament folds over the upper surface of the radial head and becomes trapped within the elbow joint between the radial head and the capitulum. The forearm is held pronated with the elbow joint partially flexed. Slight flexion of the elbow joint combined with a full supination of the forearm returns the ligament and the radial head to their correct positions.

Arthrocentesis of the Elbow Joint
The anatomy of arthrocentesis of the elbow joint is fully described in Chapter 18.

Olecranon Bursitis
A small subcutaneous bursa is present over the olecranon process of the ulna, and repeated trauma often produces chronic bursitis. The anatomy of olecranon bursitis and the aspiration of the bursa are described in Chapter 18, p. 780.

Clinical Anatomy of the Radius and Ulna

Fractures of the Radius and Ulna
Fractures of the Head of the Radius.—These can occur from a fall on an outstretched hand. As the force is transmitted along the radius, the head of the ra-

dius is driven sharply against the capitulum, splitting or splintering the head (see Fig 15–33).

Fat Pads of the Elbow Joint.—Between the capsule and the synovial membrane of the elbow joint are three fat pads located in the fossae of the distal end of the humerus. The largest pad is in the olecranon fossa and is pressed into the fossa by the contracting triceps muscle; the other two are situated in the radial and coronoid fossae and are pressed into the fossae by the brachialis muscle during elbow extension. The position of these fat pads can be of great importance when viewing radiographs for possible fractures in the elbow region. The posterior fat pad is not normally visible on radiographs. If it is visible, then a fracture should be suspected. The anterior fat pad is normally seen as a thin layer of lucency against the humerus. In the presence of a fracture, the anterior fat pad "sails out."

Fractures of the Neck of the Radius.—These commonly occur in young children from falls on an outstretched hand (see Fig 15–33).

Fractures of the Shafts of the Radius and Ulna.—These may or may not occur together (see Fig 15–33). Displacement of the fragments is usually considerable and depends on the pull of the attached muscles. The proximal fragment of the radius is supinated by the supinator and the biceps brachii muscles (Fig 15–35). The distal fragment of the radius is pronated and pulled medially by the pronator quadratus muscle. The strength of the brachioradialis and extensor carpi radialis longus and brevis shortens and angulates the forearm.

In fractures of the ulna, the ulna angulates posteriorly. In order to restore the normal movements of pronation and supination, the normal anatomical relationship of the radius, ulna, and interosseous membrane must be regained.

A fracture of one forearm bone may be associated with a dislocation of the other bone. In *Monteggia's fracture,* for example, the shaft of the ulna is fractured by a force applied from behind. There is a bowing forward of the ulna shaft and an anterior dislocation of the radial head. In *Galeazzi's fracture* the proximal third of the radius is fractured and is associated with a dislocation of the distal end of the ulna at the inferior radioulnar joint.

Fractures of the *olecranon process* may result from a fall on the flexed elbow or from a direct blow. Depending on the location of the fracture line, the bony fragment may be displaced by the pull of the triceps muscle, which is inserted on the olecranon process (see Fig 15–33). Avulsion fractures of part of the olecranon process can be produced by the pull of the triceps muscle. Good functional return in all these fractures depends on the accurate anatomical reduction of the fragment.

Colles' fracture, Smith's fracture, and radioulnar dislocation at the wrist are described in Chapter 16, p. 659.

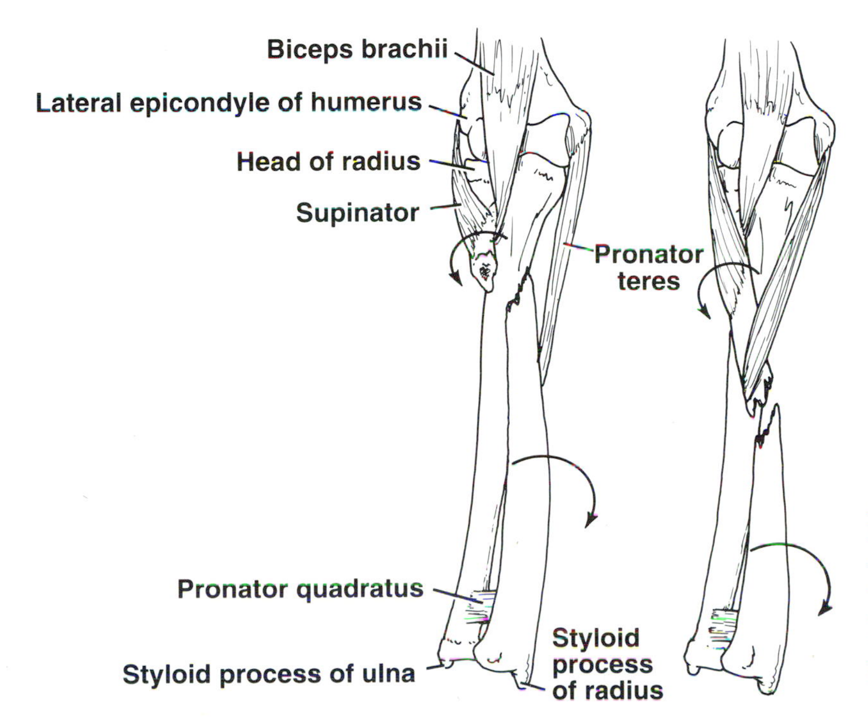

FIG 15–35.
Fracture of the upper part of the shaft of the radius. The proximal bony fragment is supinated by the supinator and the powerful biceps brachii, and the distal fragment is pronated by both the pronator teres and the pronator quadratus or the pronator quadratus alone.

Clinical Anatomy of the Compartment Syndrome of the Forearm

The forearm is enclosed in a sheath of deep fascia, which is attached to the periosteum of the posterior subcutaneous border of the ulna (see Fig 15–1). This fascial sheath, together with the interosseous membrane and fibrous intermuscular septa, divides the forearm into a number of compartments, each having its own muscles, nerves, and blood supply.

Contents of Anterior Fascial Compartment

Muscles.—These include the following groups.

Superficial Group.—Pronator teres, flexor carpi radialis, palmaris longus, and flexor carpi ulnaris.

Intermediate Group.—Flexor digitorum superficialis.

Deep Group.—Flexor pollicis longus, flexor digitorum profundus, and pronator quadratus.

Arteries.—Ulnar and radial arteries.

Nerves.—All the muscles of the anterior compartment are supplied by the median nerve and its branches, except the flexor carpi ulnaris and the medial part of flexor digitorum profundus, which are supplied by the ulnar nerve.

Contents of the Lateral Fascial Compartment

Muscles.—Brachioradialis and extensor carpi radialis longus.

Arteries.—Radial and brachial arteries.

Nerves.—Radial nerve.

Contents of the Posterior Fascial Compartment

Muscles.—These include the following groups.

Superficial Group.—Extensor carpi radialis brevis, extensor digitorum, extensor digiti minimi, extensor carpi ulnaris, and anconeus. These muscles have a common tendon of origin that is attached to the lateral epicondyle of the humerus.

Deep Group.—Supinator, abductor pollicis longus, extensor pollicis brevis, extensor pollicis longus, extensor indicis.

Arteries.—Posterior and anterior interosseous arteries.

Nerves.—Deep branch of the radial nerve.

The fascial sheath of the forearm encloses both the flexor and extensor muscles in a common sheath (see Fig 15–1). Unfortunately, there is very little room within each fascial compartment, so that any edema can cause secondary vascular compression of the traversing blood vessels; the veins are first affected and later, the arteries.

Soft-tissue injury is a common cause, and early diagnosis is critical. Early signs include paresthesia (due to ischemia of sensory nerves traversing the compartment), pain disproportionate to the injury (due to pressure on nerves within the compartment), pain on passive stretching of muscles that pass through the compartment or pain on attempted active resisted contraction of the muscles that pass through the compartment (due to ischemia), induration and tenderness of skin over the compartment (a late sign caused by edema), absence of capillary refill in the nail beds (due to pressure on arteries within the compartment), and finally, as a late sign, absent distal arterial pulses (due to occlusion of arteries within the compartment). Once the diagnosis is made, fasciotomy is the treatment of choice to decompress the affected compartment. A delay of as little as 4 hours may cause irreversible damage to the muscles.

Clinical Anatomy of Volkmann's Ischemic Contracture

Although not often seen in the emergency department, Volkmann's ischemic contracture may result from poor initial emergency treatment. This condition is essentially a contracture of the muscles of the forearm that commonly follows fractures of the lower end of the humerus or fractures of the radius and ulna. In this syndrome a localized segment of the brachial artery goes into spasm, reducing the arterial flow to the flexor and extensor muscles so that they undergo ischemic necrosis. The flexor muscles are larger than the extensor muscles, and therefore they are the ones mainly affected. The muscles are replaced by fibrous tissue, which contracts and produces the deformity. The arterial spasm is usually caused by an overtight cast, but in some cases the fracture itself may be responsible.

The condition can be avoided in most cases by the careful reduction of the fracture, the avoidance of overtight casts, and the careful follow-up of the neurovascular integrity of the limb distal to the fracture site.

Clinical Anatomy of Rupture of the Biceps Tendon

The tendon of the long head of the biceps arises from the supraglenoid tubercle of the scapula within the shoulder joint; it is surrounded by a synovial sheath. Advanced osteoarthritic changes in the joint may lead to erosion and fraying of the tendon by osteophytic outgrowths, and rupture of the tendon occurs. The condition often takes place suddenly with an audible snap, and the muscle mass expands in the lower part of the arm. Since the biceps has a short head that is attached to the scapula at the coracoid process, there is only a minimal loss of function.

Clinical Anatomy of Tennis Elbow and Golfer's Elbow

Both these conditions are caused by repeated overuse of the forearm muscles, resulting in microtraumatic injury to their origins. Tennis elbow is caused by overloading the forearm extensor muscles, producing acute localized pain at their point of origin on the back of the lateral epicondyle of the humerus. The pain is exentuated by dorsiflexing the wrist against resistance.

Golfer's elbow is a similar condition involving the origin of the flexor muscles on the front of the medial epicondyle of the humerus. The pain is exentuated by palmar flexing the wrist against resistance. Of course any activity that involves these muscles, such as excessive painting, can produce the same syndromes.

Clinical Anatomy of the Rupture of the Extensor Pollicis Longus Tendon

Rupture of this tendon can occur following fracture of the distal third of the radius. Roughening of the dorsal tubercle of the radius by the fracture line can cause excessive friction on the tendon, which may then rupture. Rheumatoid arthritis can also cause rupture of this tendon.

Clinical Anatomy of the Brachial Plexus

Complete lesions involving all the roots of the brachial plexus are rare. Incomplete injuries are common and are usually caused by traction or pressure; individual nerves may be divided by stab wounds.

Upper Lesions of the Brachial Plexus (Erb-Duchenne Palsy)—Roots C5 and C6

Upper lesions of the brachial plexus result from excessive displacement of the head to the opposite side and depression of the shoulder on the same side. This causes excessible traction or even tearing of the C5 and C6 roots of the plexus. It occurs in infants during a difficult delivery or in adults following a blow or fall on the shoulder. The suprascapular nerve, the nerve to subclavius, and the musculocutaneous and axillary nerves all possess nerve fibers derived from C5 and C6 roots and will therefore be functionless. The following muscles will consequently be paralyzed: (1) the supraspinatus (abductor of shoulder) and infraspinatus (lateral or external rotator of shoulder), (2) the subclavius (depresses the clavicle), (3) the biceps brachii (supinator of forearm, flexor of elbow, weak flexor of shoulder) and the greater part of the brachialis (flexor of elbow) and coracobrachialis (flexes shoulder), and (4) the deltoid (abductor of shoulder) and the teres minor (lateral rotator of shoulder). Thus, the limb will hang limply by the side, medially rotated by the unopposed sternocostal part of the pectoralis major; the elbow will be extended and the forearm will be pronated due to loss of the action of the biceps. The position of the upper limb in this condition has been likened to that of a porter or waiter hinting for a tip (Fig 15–36). In addition, there will be a loss of sensation down the lateral side of the arm.

Lower Lesions of the Brachial Plexus (Klumpke Palsy)—Roots T1 and (C8)

Lower lesions of the brachial plexus are usually traction injuries caused by excessive abduction of the arm, as occurs when a person falling from a height clutches at an object to stop the fall. The first thoracic nerve is usually torn. The nerve fibers from this segment run in the ulnar and median nerves to supply *all the small muscles of the hand.* The hand has a clawed appearance due to hyperextension of the metacarpophalangeal joints and flexion of the interphalangeal joints (see Fig 15–36). The extensor digitorum is unopposed by the lumbricals and interossei and extends the metacarpophalangeal joints; the flexor digitorum superficialis and profundus are unopposed by the lumbricals and interossei and flex the middle and terminal phalanges, respectively. There also will be a loss of sensation along the medial side of the arm. If the eighth cervical nerve is also damaged, the extent of anesthesia will be greater and will involve the medial side of the forearm, hand, and medial two fingers.

Lower lesions of the brachial plexus may also be produced by the presence of a cervical rib (see p. 87) or metastases from the lungs in the lower deep cervical lymph nodes.

Horner's syndrome may be associated with this condition (see p. 343).

The distinguishing of a brachial plexopathy from

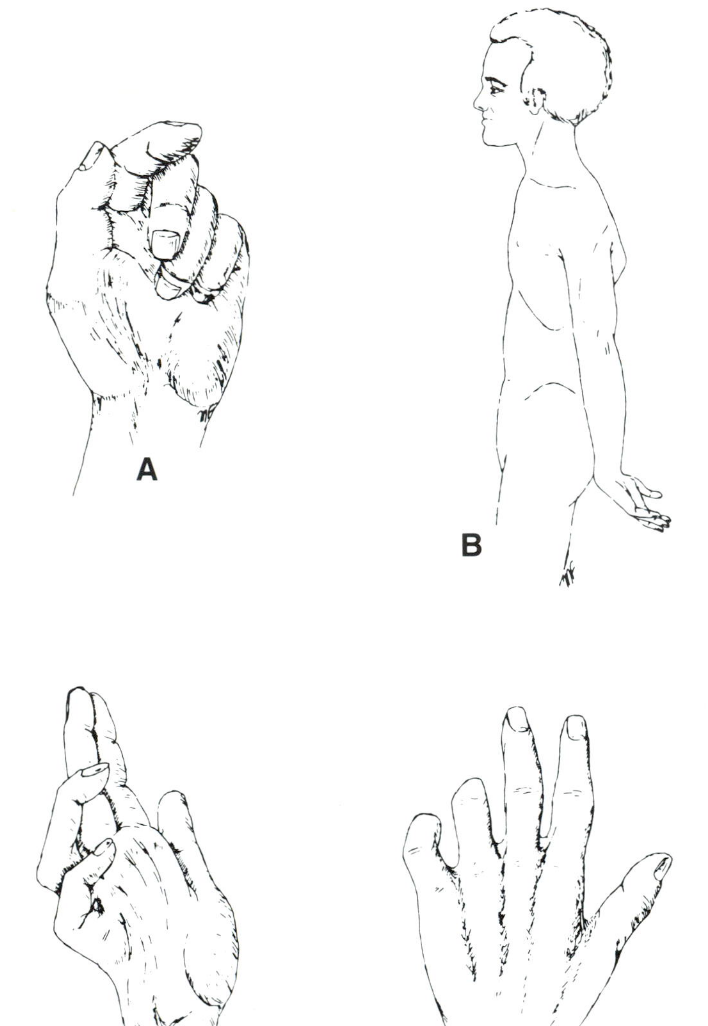

FIG 15–36.
A, median nerve palsy. **B,** Duchenne-Erb palsy (waiter's tip). **C,** ulnar nerve palsy.

a root lesion (caused by a cervical disc) or from a peripheral neuropathy will be discussed in Chapter 16.

Brachial Plexus Nerve Block

The anatomy of interscalene block, supraclavicular block, infraclavicular block, and axillary approach block are fully described in Chapter 19, p. 814.

Clinical Anatomy of the Musculocutaneous Nerve

The musculocutaneous nerve (C5 through C7) is a branch of the lateral cord of the brachial plexus (Fig 15–37). It is rarely injured because of its protected position beneath the biceps brachii muscle. If the nerve is injured high up in the arm, the biceps and coracobrachialis are paralyzed and the brachialis muscle is much weakened (the latter muscle is also supplied by the radial nerve). Flexion of the elbow joint is then produced by the remainder of the brachialis muscle and the flexors of the forearm. When the forearm is in the prone position, the extensor carpi radialis longus and the brachioradialis muscles assist in flexion of the forearm. Also, sensory loss occurs along the lateral side of the forearm.

Wounds or cuts of the forearm can sever the lateral cutaneous nerve of the forearm (continuation of the musculocutaneous nerve beyond the cubital fossa), resulting in sensory loss along the lateral side of the forearm.

Musculocutaneous Nerve Block

The anatomy of musculocutaneous nerve block is fully described in Chapter 19, p. 819.

Clinical Anatomy of the Median Nerve

The median nerve (C5 through C8 and T1) is formed from the lateral and medial cords of the brachial plexus (see Fig 15–37). The median nerve is occasionally injured in the elbow region in supracondylar fractures of the humerus. It is more commonly injured by stab wounds or broken glass just proximal to the flexor retinaculum, where it lies in the interval between the tendons of the flexor carpi radialis and flexor digitorum superficialis, overlapped by the palmaris longus (if present).

The clinical findings in injuries to the median nerve are as follows.

Injuries to the Median Nerve at the Elbow

Motor.—The pronator muscles of the forearm and the long flexor muscles of the wrist and fingers, with the exception of the flexor carpi ulnaris and the medial half of the flexor digitorum profundus, will be paralyzed. As a result, the forearm is kept in a supinated position; wrist flexion is weak and is accompanied by adduction. The latter deviation is due to paralysis of the flexor carpi radialis and the strength of the flexor carpi ulnaris and the medial half of the flexor digitorum profundus. No flexion is possible at the interphalangeal joints of the index and middle fingers, although weak flexion of the metacarpophalangeal joints of these fingers is attempted by the interossei. When the patient tries to make a fist, the index and, to a lesser extent, the middle finger tend to remain straight, while the ring and little fingers flex

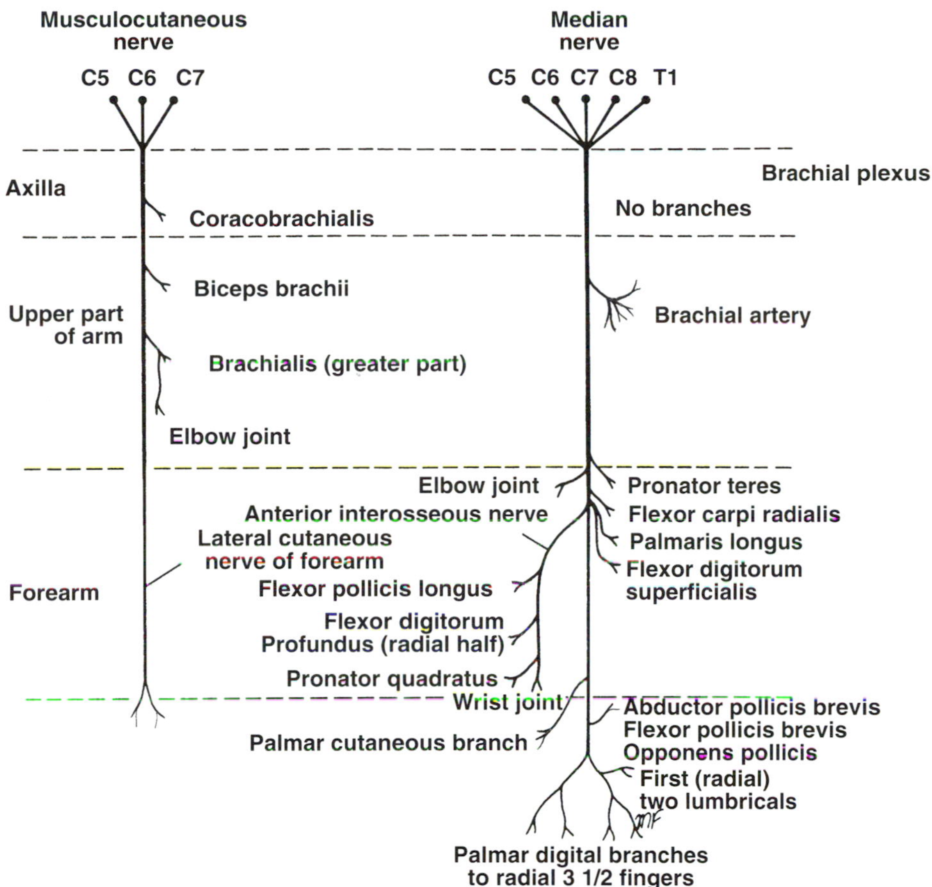

FIG 15–37.
Summary diagrams of the main branches of the musculocutaneous and median nerves.

because the flexor digitorum profundus of these two fingers is innervated by the ulnar nerve (Fig 15–36). However, flexion of these two fingers is weakened because of the loss of the flexor digitorum superficialis, which is innervated by the median nerve.

Flexion of the terminal phalanx of the thumb is lost due to paralysis of the flexor pollicis longus. The muscles of the thenar eminence are paralyzed and later become wasted, so that the eminence is flattened. The thumb is laterally rotated and adducted, and the hand becomes flattened and "apelike."

Sensory.—There is loss of skin sensation of the lateral half or less of the palm of the hand and the palmar aspect of the radial 3½ fingers. There is also sensory loss of the skin of the distal parts of the dorsal surfaces of the radial 3½ fingers. The area of total anesthesia is considerably less due to the overlap of adjacent nerves.

Vasomotor Changes.—The skin areas involved in sensory loss are warmer and drier than normal because of the arteriolar dilatation and absence of sweating resulting from loss of sympathetic control.

Trophic Changes.—In long-standing cases, the trophic changes are found in the hand and fingers. The skin is dry and scaly, the nails crack easily, and the pulp of the fingers is atrophied.

Injuries of the Median Nerve at the Wrist

These are described in detail in Chapter 16, p. 660.

Perhaps the most serious disability of all in median nerve injuries is the loss of ability to oppose the thumb to the other fingers and the loss of sensation over the radial fingers. The delicate pincerlike action of the hand is no longer possible.

Carpal Tunnel Syndrome

This is described in detail in Chapter 16, p. 655.

Median Nerve Block

The anatomy of median nerve block is fully described in Chapter 19, p. 819.

Clinical Anatomy of the Ulnar Nerve

The ulnar nerve (C8 and T1) arises from the medial cord of the brachial plexus (Fig 15–38). The nerve is

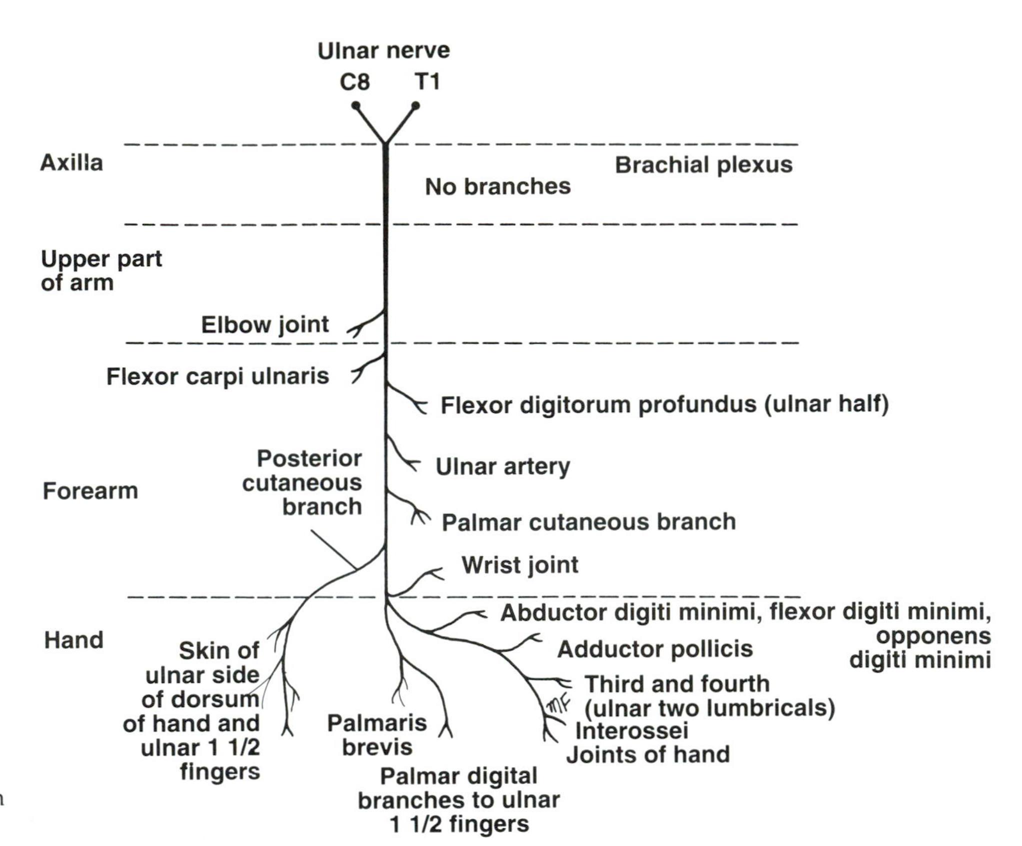

FIG 15–38.
Summary diagram of the main branches of the ulnar nerve.

most commonly injured at the elbow, where it lies behind the medial epicondyle, and at the wrist, where it lies with the ulnar artery in front of the flexor retinaculum. The injuries at the elbow are usually associated with fractures of the medial epicondyle. The superficial position at the wrist makes it very vulnerable to damage from cuts and stab wounds.

The clinical findings in injury to the ulnar nerve are as follows.

Injuries to the Ulnar Nerve at the Elbow

Motor.—The flexor carpi ulnaris and the ulnar half of the flexor digitorum profundus muscles are paralyzed. The paralysis of the flexor carpi ulnaris can be observed by asking the patient to make a tightly clenched fist. Normally, the synergistic action of the flexor carpi ulnaris tendon can be observed as it passes to the pisiform bone; the tightening of the tendon will be absent if the muscle is paralyzed. The profundus tendons to the ring and little fingers will be functionless, and the terminal phalanges of these fingers are therefore not capable of being markedly flexed. Flexion of the wrist joint will result in radial deviation owing to paralysis of the flexor carpi ulnaris. The medial border of the front of the forearm will later show flattening due to wasting of the underlying ulnaris and profundus muscles.

The small muscles of the hand will be paralyzed, except the muscles of the thenar eminence and the first two lumbricals, which are supplied by the median nerve. The patient is unable to adduct and abduct the fingers and consequently is unable to grip a piece of paper placed between the fingers. Remember that the extensor digitorum can abduct the fingers to a small extent, but only when the metacarpophalangeal joints are hyperextended.

It is impossible to adduct the thumb, because the adductor pollicis muscle, which is ulnar innervated, is paralyzed. A patient asked to grip a piece of paper between the thumb and the index finger does so by strongly contracting the flexor pollicis longus and flexing the terminal phalanx (Froment's sign).

The metacarpophalangeal joints become hyperextended due to the paralysis of the lumbrical and interosseous muscles, which normally flex these joints. Since the first and second lumbricals are not paralyzed (they are supplied by the median nerve), the hyperextension of the metacarpophalangeal joints is most prominent in the fourth and fifth fingers. The interphalangeal joints are flexed, again due to the paralysis of the lumbrical and interosseous muscles, which normally extend these joints through the extensor expansion. The flexion deformity at the interphalangeal joints of the fourth and fifth fingers is more obvious than the deformity at the index and middle fingers because the lumbrical muscles of the index and middle fingers are innervated by the median nerve and are not paralyzed. In long-standing cases the hand assumes the characteristic "claw" deformity (main en griffe). Wasting of the paralyzed muscles results in flattening of the hypothenar eminence and loss of the convex curve to the medial border of the hand. Examination of the dorsum of the hand in these cases will show hollowing between the metacarpal bones due to wasting of the dorsal interosseous muscles (see Fig 15–36).

Sensory.—Loss of skin sensation will be observed over the anterior and posterior surfaces of the ulnar third of the hand and the ulnar 1½ fingers.

Vasomotor Changes.—The skin areas involved in sensory loss are warmer and drier than normal because of the arteriolar dilatation and absence of sweating resulting from loss of sympathetic control.

Injuries to the Ulnar Nerve at the Wrist

These are described in detail in Chapter 16, p. 661.

Unlike median nerve injuries, lesions of the ulnar nerve leave a relatively efficient hand. The sensation over the lateral part of the hand is intact, and the pincerlike action of the thumb and index finger is reasonably good, although some weakness exists due to loss of the adductor pollicis.

How to Distinguish an Ulnar Nerve Problem From Klumpke Paralysis

Lesions involving the lower roots of the brachial plexus (C8 and T1), i.e., a Klumpke type, are most often restricted to the T1 root with sparing of the C8 dermatome along the medial side of the forearm, hand, and medial two fingers. A complete Klumpke lesion involving both roots will present signs and symptoms that are identical to a complete lesion of the ulnar nerve at the elbow.

Ulnar Nerve Block

The anatomy of ulnar nerve block is fully described in Chapter 19, p. 820.

Clinical Anatomy of the Radial Nerve

The radial nerve (C5 through C8 and T1) arises from the posterior cord of the brachial plexus (Fig 15–39). The radial nerve is commonly stretched or contused,

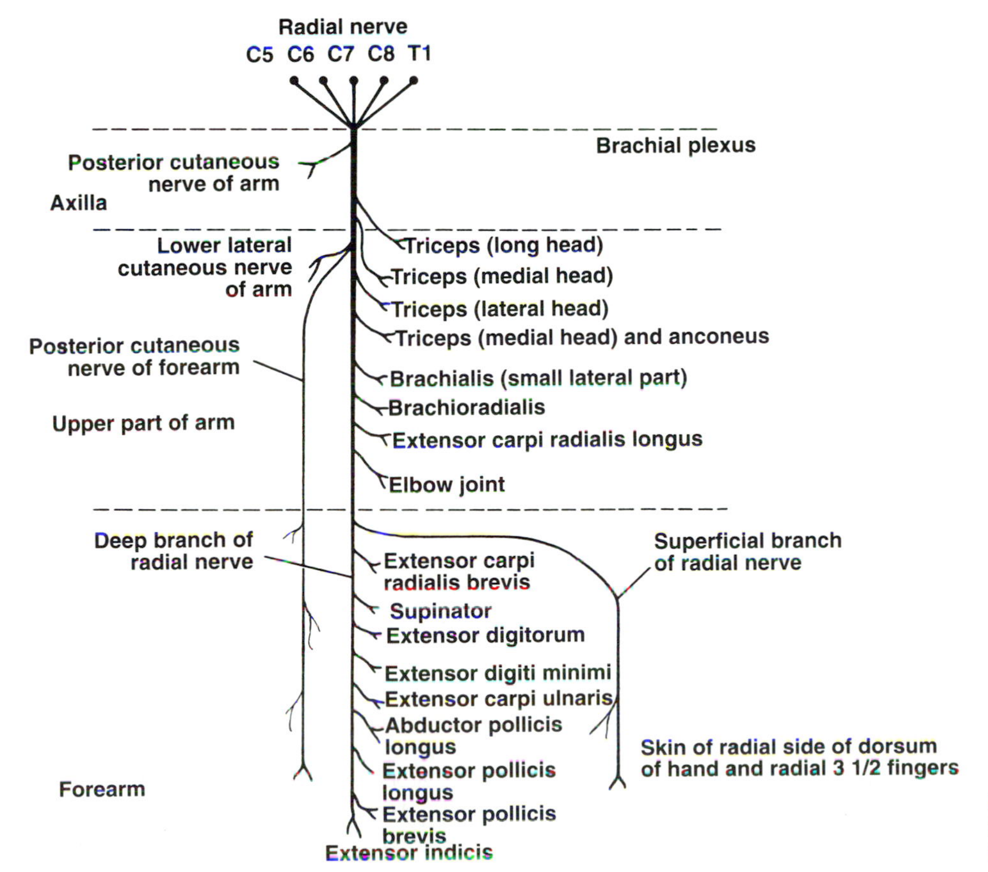

FIG 15–39.
Summary diagram of the main branches of the radial nerve.

but because of its protected position, rarely lacerated. For this reason, sensory and motor function can be expected to return in 3 or 4 months after injury. The radial nerve is often damaged in the axilla and as it courses in the spiral groove around the humerus.

Injuries to the Radial Nerve in the Axilla
The nerve may be injured by the pressure of the upper end of a badly fitting crutch pressing up into the armpit, or by falling asleep with the arm over the back of a chair. It may also be badly damaged in the axilla by fractures and dislocations of the upper end of the humerus. When the humerus is displaced downward in dislocations of the shoulder, the radial nerve, which is wrapped around the back of the shaft of the bone, is pulled downward, stretching the nerve in the axilla excessively.

The clinical findings in injury to the radial nerve in the axilla are as follows.

Motor.—The triceps, the anconeus, and the long extensors of the wrist are paralyzed; the patient is unable to extend the elbow joint, the wrist joint, and fingers. *Wristdrop,* or flexion of the wrist, occurs as the result of the action of the unopposed flexor muscles of the wrist. Wristdrop is very disabling because wrist extension is needed to obtain the strong finger flexion necessary for firmly gripping an object. If the wrist and proximal phalanges are passively extended by holding them in position with the opposite hand, the middle and distal phalanges of the fingers can be extended by the action of the lumbricals and interossei, which are inserted into the extensor expansions.

The brachioradialis and supinator muscles are also paralyzed, but supination is still performed well by the biceps brachii innervated by the musculocutaneous nerve.

Sensory.—A small loss of skin sensation occurs down the posterior surface of the lower part of the arm and down a narrow strip on the back of the forearm. There is also a variable area of sensory loss on the radial side of the dorsum of the hand and the dorsal surface of the proximal parts of the radial 3½ fingers. The area of total anesthesia is relatively small due to the overlap of sensory innervation by adjacent nerves.

Trophic Changes.—These are slight.

Injuries to the Radial Nerve in the Spiral Groove
This may occur at the time of fracture of the shaft of the humerus, or subsequently involved during the formation of the callus. The pressure of the back of the arm on the edge of the examining table in an unconscious patient has been known to injure the nerve at this site. The prolonged application of a tourniquet to the arm in a person with a slender triceps muscle may be followed by temporary radial palsy.

The injury to the radial nerve occurs most commonly in the distal part of the groove, beyond the origin of the nerves to the triceps and the anconeus and beyond the origin of the cutaneous nerves.

Motor.—The patient is unable to extend the wrist and the fingers, and there is wrist drop (see above).

Sensory.—A variable small area of anesthesia is present over the root of the dorsal surface of the hand and the dorsal surface of the proximal parts of the radial 3½ fingers.

Trophic Changes.—These are very slight or absent.

Injuries to the Deep Branch of the Radial Nerve
The deep branch of the radial nerve is a motor nerve to the extensor muscles in the posterior compartment of the forearm. It may be damaged in fractures of the proximal radius, during dislocation of the radial head at the time of fracture of the ulnar shaft, or by a penetrating injury to the dorsum of the forearm.

The nerve supply to the supinator and the extensor carpi radialis longus will be undamaged (they are innervated in the cubital fossa), and because the latter muscle is powerful, it will keep the wrist joint extended, and wrist drop will not occur.

Injuries to the Superficial Radian Nerve
This sensory nerve may be damaged by a stab wound and will result in a variable small area of anesthesia over the dorsum of the hand and the dorsal surface of the proximal parts of the radial 3½ fingers.

Radial Nerve Block
The anatomy of the radial nerve block is fully described in Chapter 19, p. 821.

Clinical Anatomy of the Axillary Nerve

The axillary nerve (C5 and C6) arises from the posterior cord of the brachia plexus. The axillary nerve is particularly vulnerable as it passes backward through the quadrilateral space (see Fig 15–7) (formed by the teres minor and major muscles, the long head of triceps muscle, and the surgical neck of the humerus). Downward displacement of the humeral head during shoulder dislocations or fracture of the surgical neck of the humerus may injure the nerve. Axillary nerve injury results in paralysis of the deltoid muscle with loss of function of the upper lateral cutaneous nerve of the arm. The paralyzed muscle wastes rapidly, and the underlying greater tuberosity of the humerus can be readily palpated. Since the supraspinatus is the only other abductor of the shoulder, this movement is much impaired. There is loss of skin sensation over the *lower half of the deltoid muscle.*

Clinical Anatomy of the Dermatomes of the Upper Extremity

The integrity of the spinal cord segments C3 through T1 may be tested by examining the dermatomes of the upper extremity. The diagram in Figure 15–40 shows the arrangement of the dermatomes of the upper limb. The dermatomes for the upper cervical segments C3 to C6 are located along the radial margin of the upper limb; the C7 dermatome is situated on the middle finger, and the dermatomes for C8, T1, and T2 are along the ulnar margin of the limb. The nerve fibers from a particular segment of the spinal cord, although they exit from the cord in a spinal nerve of the same segment, pass to the skin in two or more different cutaneous nerves.

The skin over the point of the shoulder and halfway down the lateral surface of the deltoid muscle is supplied by the supraclavicular nerves (C3 and C4). Pain may be referred to this region from the peripheral diaphragmatic pleura and peritoneum. The afferent stimuli reach the spinal cord via the phrenic nerves (C3 through C5). Pleurisy, peritonitis, subphrenic abscess, or gallbladder disease may therefore be responsible for shoulder pain (see pp. 611 and 125 for a full discussion of referred pain).

Clinical Anatomy of the Tendon Reflexes of the Upper Extremity

Skeletal muscle receives segmental innervation. Most muscles are innervated by several spinal nerves and therefore by several segments of the spinal cord. The segmental innervation of the following muscles should be known, since it is possible to test them by eliciting simple muscle reflexes in a clinical setting.

1. *Biceps brachii tendon reflex* **C5** and C6 (flexion of the elbow joint by tapping the biceps tendon).
2. *Triceps tendon reflex* C6, **C7,** and C8 (extension of the elbow joint by tapping the triceps tendon).
3. *Brachioradialis tendon reflex* C5, **C6,** and C7 (supination of the radioulnar joints by tapping the insertion of the brachioradialis tendon).

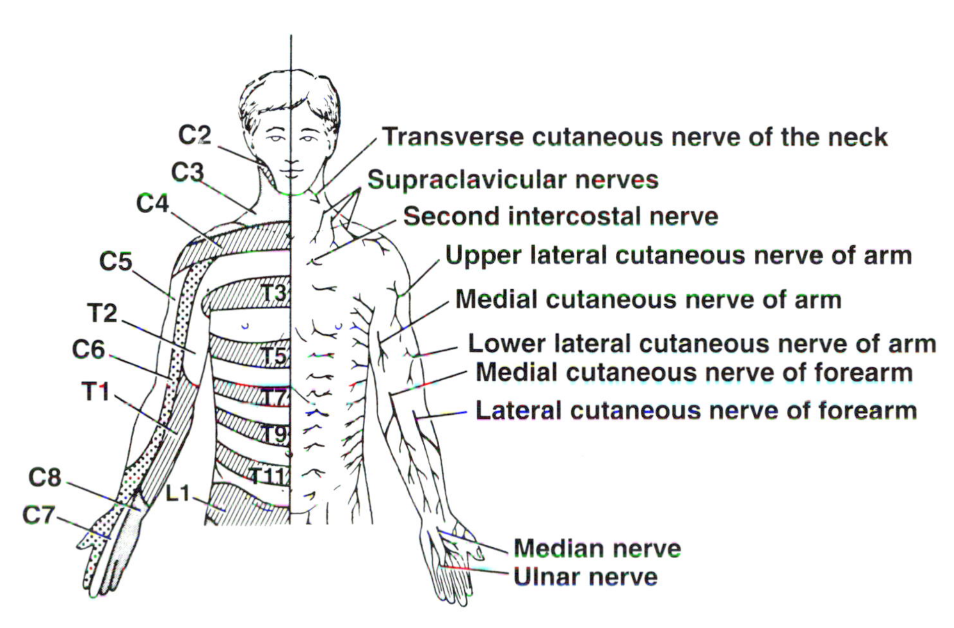

FIG 15–40.
Dermatomes and distribution of cutaneous nerves on the anterior aspect of the upper limb.

ANATOMICAL-CLINICAL PROBLEMS

1. A 15-year-old girl, while demonstrating to her friends her proficiency at standing on her hands, suddenly went off balance and put all her body weight on her left outstretched hand. A distinctive cracking noise was heard, and she felt a sudden pain in her left shoulder region. On examination in the emergency department, the smooth contour of her left shoulder was absent. The shoulder and the lateral end of the clavicle were depressed, and the medial part of the clavicle was elevated. The clavicle was obviously fractured, and the edges of the bony fragments could be palpated. Using your knowledge of anatomy, state the following: (1) Which part of the clavicle is most commonly fractured? (2) Why are the bony fragments displaced? (3) Why do the mechanical forces usually fracture the clavicle instead of dislocating the sternoclavicular or acromioclavicular joints? (4) What structures are likely to be damaged in a simple fracture of the clavicle?

2. Discuss acromioclavicular joint separations. Explain why such separations are unstable after reduction.

3. A 32-year-old man was seen in the emergency department complaining of severe pain over his right shoulder. The pain started after he had been sawing a large amount of wood with a hand saw. On examination, the shoulder was found to be extremely tender just in front and lateral to the acromion process. Passive movements of the joint caused extreme discomfort, especially abduction. When the shoulder was actively abducted, it was found that the pain was most intense during the middle range of the movement (50° to 120°). A diagnosis of supraspinatus tendinitis was made. Explain the structure and function of the rotator cuff. What is the difference between supraspinatus tendinitis and pericapsulitis? Could this patient have subdeltoid bursitis?

4. A 60-year-old woman fell down the stairs and was admitted to the emergency department with severe right shoulder pain. On examination, she was sitting up with her right arm by her side and her right elbow flexed and supported by her left hand. Inspection of the right shoulder showed loss of the normal rounded curvature and evidence of a slight swelling below the right clavicle. Any attempt at active or passive movement of the shoulder joint was stopped by severe pain in the shoulder. The examiner then tested the cutaneous sensibility over the lower part of the right deltoid muscle and found it to be normal. What is the diagnosis? Can you explain the infraclavicular swelling? What was the purpose of testing the skin sensitivity over the lower deltoid? Could the same test have been performed over the upper part of the deltoid?

5. A 60-year-old man slipped on the ice and dislocated his left shoulder joint. It was relocated successfully. Several days after the dislocation had been reduced, it was noted that the patient had signs and symptoms indicative of radial nerve damage in the axilla. Name the muscles that would be paralyzed and the area of the skin that would show sensory loss. Describe the deformity of the upper limb in a person with a long-standing radial nerve palsy following damage to the nerve in the axilla.

6. A father, seeing his 3-year-old daughter playing in the garden, ran up and picked her up by both hands and swung her around in a circle. The child's enjoyment suddenly turned to tears, and she said her left elbow hurt. On examination, the child held her left elbow joint semiflexed and her forearm pronated. What is the diagnosis?

7. An 8-year-old boy was climbing a tree when he slipped and fell. The emergency physician diagnosed a supracondylar fracture of the right humerus. A few hours later, the child complained of pain in the forearm, which persisted. Four hours later the child's right hand looked dusky white, and the pain in the forearm was still present. On examination, there was found to be a complete loss of cutaneous sensation of the hand. The color of the skin was bluish-white, and the pulse of the radial and ulnar arteries could not be palpated. Every possible effort was made to restore the circulation of the forearm, without avail. Explain the possible cause for the lack of circulation in this patient. What deformity might you expect the child to have 1 year later? Explain the deformity in anatomical terms.

8. A 19-year-old woman was involved in an automobile accident. A radiological examination of her right upper extremity showed an oblique fracture of the humerus just proximal to the deltoid tuberosity. Using your knowledge of anatomy, explain the possible displacement of the proximal and distal fragments that might have occurred. Which nerve might have been damaged in a fracture of this type?

9. A 35-year-old man received a stab wound in the lower part of the neck on the right side. On examination, a small wound was found just above the right clavicle in the posterior triangle of the neck. Miraculously, the subclavian vessels appeared to have been missed by the knife blade, and there was little bleeding from the wound. The right hemithorax showed normal breath sounds, and respiration

appeared to be normal. Further examination of the right upper extremity showed evidence of damage to the first thoracic spinal nerve. Describe exactly the position of the fingers and thumb that you might find in such a patient. Would you expect to find any sensory deficiency?

10. During a physical examination of a 67-year-old man, the emergency physician tested the biceps and triceps reflexes on both upper extremities. The reflexes were normal on the left side but absent on the right side. In anatomical terms explain the possible significance of this finding.

11. Following severe injury to the forearm involving extensive damage to the soft tissues, the physician was concerned with the possibility of the development of a compartmental syndrome. What compartments exist in the forearm? What is their clinical significance?

12. An 18-year-old man, riding pillion on a snowmobile, was involved in an accident. The machine was traveling at high speed when it hit a tree stump buried in snow. The man was thrown 12 ft and landed on his right shoulder and the right side of his head. After 3 weeks of hospitalization, it was noticed that he kept his right arm internally rotated by his side with the forearm pronated. An area of anesthesia was present along the radial side of the upper part of the arm. A diagnosis of damage to the upper part of the brachial plexus was made. In anatomical terms, explain the position adopted by the right arm in this patient. The precise nature of the nerve lesions and the muscles paralyzed should be stated.

13. A 24-year-old woman was thrown from her bicycle and landed on her right outstretched hand. On examination in the emergency department she was found to have a severe posterior fracture dislocation of her right elbow joint. Not only was her forearm foreshortened and her olecranon process excessively prominent posteriorly, but the medial epicondyle was fractured. A complete neurovascular examination showed evidence of severe damage to the ulnar nerve at the elbow. Using your knowledge of anatomy, explain the signs and symptoms that you would expect to find in a patient with this type of nerve injury.

14. A 12-year-old boy was injured in an automobile accident and sustained a fracture of the lower end of his right humerus. He visited his physician 30 years later, complaining of a "pins and needles" feeling and numbness over the front and back surfaces of the little and ring fingers of his right hand. He had also noticed a weakness of his fingers of the right hand, especially when performing the upstroke in writing. On examination, his carrying angle on the right side was noted to be greater than on the left side (cubitus valgus), and there was some flattening along the medial border of the right forearm. There was also flattening of the hypothenar eminence and a loss of curvature along the ulnar border of the hand. On the dorsum of the right hand there was evidence of "hollowing out" between the metacarpal bones. The patient could draw his thumb across the palm, but if asked to pinch a piece of paper between his thumb and fingers, the terminal phalanx of the thumb had to be flexed to grip the paper; the movement of adduction of the thumb was extremely weak. He was unable to grip a piece of paper by adducting the index and middle fingers. He also had loss of touch over the ulnar 1½ fingers, both anteriorly and posteriorly.

Using your knowledge of anatomy, explain this patient's disability. Why was the medial border of the forearm flattened? What muscles were wasted? Why was the hypothenar eminence flattened? Why was there a hollowing out on the dorsum of the hand? Why could the patient not adduct his thumb to his index and middle fingers?

ANSWERS

1. The clavicle is one of the most common bones in the body to be fractured. The violent force applied to the clavicle is usually indirect, namely, from the arm through the scapula. (1) Anatomically, the weakest part of the clavicle is the junction of the middle and lateral thirds, and this is where a fracture usually occurs. Fractures of the medial and lateral ends are usually caused by direct trauma. (2) The lateral bony fragment is displaced downward by the weight of the arm and pulled forward and medially by the pectoral muscles. The medial fragment is elevated by the sternocleidomastoid muscle. (3) Dislocation of the sternoclavicular joint is prevented by the ligaments of the joint, especially the very strong costoclavicular ligament. The very strong conoid and trapezoid ligaments (coracoclavicular ligament) assist in preventing the dislocation of the acromioclavicular joint. If the mechanical force was great enough, and the clavicle strong enough, dislocation of one or other of these joints would occur. (4) The supraclavicular nerves, which descend over the clavicle from the cervical plexus, or a communicating vein between the cephalic and internal jugular vein, may be damaged by the bone fragments. Posterior displacement of the clavicle could compromise the

subclavian vessels and the brachial plexus between the clavicle and the first rib.

2. Subluxation of a joint implies that the ligaments and capsule are stretched or torn but the damage is not so severe that the articulating surfaces lose contact with one another. In the acromioclavicular joint the lateral end of the clavicle elevates and becomes more prominent than normal; there is a definite step down onto the acromion. A dislocation occurs when the damage to the restraining structures is more severe and the articulating surfaces lose contact with one another. In the case of the acromioclavicular joint, the clavicle rises above the acromion and the joint is very unstable.

The main strength of the acromioclavicular joint depends on the integrity of the strong coracoclavicular ligament. Should this ligament be disrupted, the acromioclavicular joint dislocates; the lateral end of the clavicle rides over the acromion process and the upper limb is depressed.

3. This patient had supraspinatus tendinitis. During the middle range of abduction, the tendon of the supraspinatus impinges against the outer border of the acromion. Normally, the large subacromial bursa intervenes and ensures that the movement is relatively free of friction and is painless. In this condition, the bursa has degenerated and the supraspinatus tendon exhibits a localized area of collagen degeneration.

The rotator cuff consists of the tendons of the supraspinatus, the infraspinatus, the teres minor, and the subscapularis, which are fused to the capsule of the shoulder joint. The tone of these muscles assists in holding the head of the humerus in the glenoid cavity of the scapula during movements of the shoulder joint. Note that the cuff lies on the anterior, superior, and posterior aspects of the joint. The cuff is deficient inferiorly, which is a site of potential weakness.

Supraspinatus tendinitis and pericapsulitis are the same condition; the latter term implies that the degenerative condition is more extensive.

The subdeltoid bursa is a lateral extension of the subacromial bursa.

4. This patient had a subcoracoid dislocation of the right shoulder joint. The head of the humerus was dislocated downward through the weakest part of the capsule of the joint. It was then displaced medially in front of the scapula by the pull of the subscapularis and the pectoralis major muscles. The greater tuberosity of the humerus no longer displaced the deltoid muscle laterally, and the normal curve of the shoulder was therefore lost. The head of the humerus had come to rest behind the subscapularis muscle and below the coracoid process of the scapula, and was responsible for the fullness felt below the lateral end of the clavicle.

The purpose of testing the skin sensation over the lower half of the deltoid muscle was to determine whether the displaced humeral head had damaged the axillary nerve as it passes backward through the quadrilateral space below the shoulder joint. The integrity of the axillary nerve cannot be examined by performing the skin sensitivity test over the upper half of the deltoid, since the cutaneous nerve supply here is provided by the supraclavicular nerves. Note that although the deltoid muscle receives its motor nerve supply from the axillary nerve, the patient is in too much pain with a dislocated shoulder to actively contract the deltoid.

5. The motor and sensory defects and the deformity that follow damage to the radial nerve in the axilla are described on p. 622.

6. The sudden traction on the wrist resulted in the small head of the radius being partially pulled out of the anular ligament. This accident occurs only when the head of the radius is relatively small, as compared with the size of the anular ligament, and is almost entirely confined to children younger than 6 years.

7. At the time of the supracondylar fracture of the humerus, the distal fragment may tilt forward and impinge on the brachial artery, causing it to go into spasm; sometimes during the reduction of such a fracture trauma to the brachial artery may also cause it to go into spasm. The forearm muscles then underwent ischemic necrosis followed by fibrous replacement; a typical Volkmann's contracture followed. The anatomical changes in this condition are described on p. 616.

8. With fractures of the shaft of the humerus, the displacement of the bone fragments depends on the relation of the site of the fracture to the insertion of the deltoid. In this patient the fracture line was proximal to the insertion. The proximal fragment was adducted by the pectoralis major, latissimus dorsi, and the teres major muscles, and the distal fragment was pulled proximally by the deltoid, biceps, and triceps muscles. The radial nerve, as it lies in the spiral groove of the humerus, may be damaged in such fractures. It was unharmed in this patient.

9. This is a case of Klumpke palsy, in which the first thoracic nerve has been damaged. This nerve

supplies all the small muscles of the hand via the median and ulnar nerves. The condition is fully described on p. 617.

10. Absence of the tendon reflexes of the biceps brachii and the triceps muscles of the right arm would indicate the presence of disease in the C5 through C8 segments of the spinal cord or in the motor or sensory nerve fibers passing to or from these muscles.

11. The deep fascia of the forearm is attached to the posterior subcutaneous border of the ulna and encircles the forearm. This fascia, together with the interosseous membrane and the intermuscular septa, divides the forearm into a number of fascial compartments that contain muscles, nerves, and blood vessels. There are three fascial compartments in the forearm—anterior, lateral, and posterior (for details, see p. 616). The clinical significance of such an arrangement is the lack of room for expansion, should edema fluid accumulate in one of the compartments. Should a compartment of the forearm be severely injured, there is a strong possibility that the blood supply would be compromised, resulting in ischemic necrosis. The compartment syndrome is described on p. 616.

12. This patient has a right-sided Erb-Duchenne palsy, i.e., a lesion of the fifth and the sixth roots of the brachial plexus. The suprascapular nerve, the nerve to the subclavius, and the musculocutaneous and axillary nerves all possess nerve fibers derived from C5 and C6 roots and will therefore be functionless. The following muscles will consequently be paralyzed: (1) supraspinatus and infraspinatus, (2) the subclavius, (3) the biceps brachii and the greater part of the brachialis and coracobrachialis, and (4) the deltoid and the teres minor. Thus, the right limb will hang limply by the side, medially rotated by the unopposed sternocostal part of the pectoralis major; the forearm will be pronated due to loss of the action of the biceps. The position of the upper limb in this condition has been likened to that of a porter or waiter hinting for a tip. In addition, there will be a loss of sensation down the radial side of the right arm.

13. The motor and sensory defects following ulnar nerve injury at the elbow are fully described on p. 620.

14. This patient's old supracondylar fracture of the right humerus had increased the carrying angle on the right side to such an extent that the ulnar nerve was running around the medial epicondyle like a string around a pulley when the elbow joint was flexed and extended. Repeated friction caused interstitial neuritis of the ulnar nerve and consequent interference with the motor and sensory functions of the nerve. The effects of ulnar nerve palsy are fully described on p. 620.

The upstroke of writing is produced by flexion of the metacarpophalangeal joint and extension of the interphalangeal joints; both movements are normally carried out by the lumbricals and the interossei, which are supplied by the ulnar nerve. For a full discussion of this complicated process, see Chapter 16, p. 647.

16

The Wrist and Hand

Wrist and hand injuries occur frequently. The goal of management is preservation of as much function as possible. Particular attention should be paid to the thumb; the pincer action between the thumb and index finger is central to fine motor movement and depends on the unique ability of the thumb to be drawn across the palm and opposed to the other fingers.

The emergency physician must be knowledgeable about all the hand structures that may be injured, including the skin, nerves, bones, joints, tendons, and blood and lymphatic vessels and their anatomical relationships. The purpose of this chapter is to provide an overview of the anatomy of the wrist and hand with emphasis on their function.

BASIC ANATOMY

Basic Anatomy of the Region of the Wrist

Flexor and Extensor Retinacula

The flexor and extensor retinacula are bands of deep fascia that function to hold underlying tendons in position and serve as pulleys around which the tendons may move.

Flexor Retinaculum.—This stretches across the front of the wrist and converts the concave anterior surface into an osteofascial tunnel, called the *carpal tunnel,* for the passage of the median nerve and the flexor tendons of the thumb and fingers (Fig 16–1). The flexor retinaculum is attached ulnarly to the pisiform bone and the hook of the hamate and is attached radially to the tubercle of the scaphoid and the trapezium bones.

The proximal border of the retinaculum corresponds to the distal transverse crease in front of the wrist and is continuous with the deep fascia of the forearm. The distal border is attached to the palmar aponeurosis (see Fig 16–4).

Extensor Retinaculum.—This stretches across the back of the wrist (see Fig 16–1). It converts the grooves on the back of the distal ends of the radius and ulna into six separate tunnels for the passage of the long extensor tendons. The retinaculum is attached ulnarly to the pisiform bone and the hook of the hamate and is attached radially to the distal end of the radius.

The proximal and distal borders of the retinaculum are continuous with the deep fascia of the forearm and hand, respectively.

The arrangement of the tendons, arteries, and nerves in the region of the wrist joint is shown in Figures 16–5 and 16–13.

Structures on the Anterior Aspect of the Wrist

The following structures from ulnar to radial pass superficial to the flexor retinaculum (see Fig 16–1).

1. *Flexor carpi ulnaris tendon,* ending on the pisiform bone. (This tendon does not actually cross the flexor retinaculum but is included for the sake of completeness.)
2. *Ulnar nerve* lies lateral to the pisiform bone.
3. *Ulnar artery* lies lateral to the ulnar nerve.
4. *Palmar cutaneous branch of ulnar nerve.*
5. *Palmaris longus tendon,* passing to its insertion into the flexor retinaculum and the palmar aponeurosis.
6. *Palmar cutaneous branch of the median nerve.*

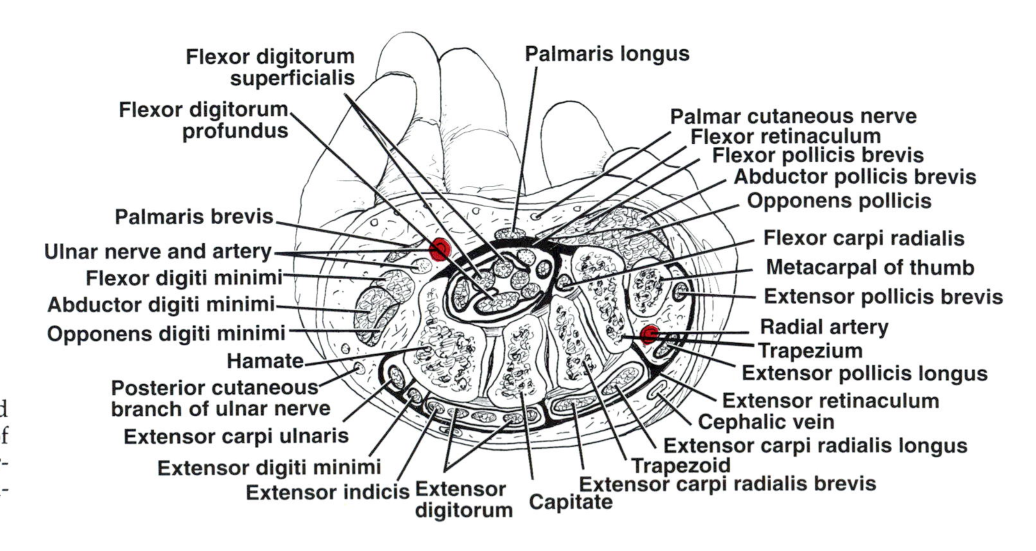

FIG 16–1.
Cross section of the hand showing the relationship of the tendons, nerves, and arteries to the flexor and extensor retinacula.

The following structures pass beneath the flexor retinaculum from ulnar to radial (see Fig 16–1).

1. *Flexor digitorum superficialis tendons,* and posterior to these, the *tendons of flexor digitorum profundus;* both groups of tendons share a common synovial sheath.
2. *Median nerve.*
3. *Flexor pollicis longus tendon* surrounded by a synovial sheath.
4. *Flexor carpi radialis tendon* passes between a superficial and a deep layer of the flexor retinaculum. The tendon is surrounded by a synovial sheath.

Structures on the Posterior Aspect of the Wrist

The following structures pass superficial to the extensor retinaculum, from ulnar to radial (see Fig 16–1).

1. *Dorsal (posterior) cutaneous branch of the ulnar nerve.*
2. *Basilic vein.*
3. *Cephalic vein.*
4. *Superficial branch of the radial nerve.*

Beneath the extensor retinaculum, fibrous septa pass to the underlying radius and ulna and form six compartments that contain the tendons of the extensor muscles. Each compartment is provided with a synovial sheath, which extends proximal and distal to the retinaculum.

From ulnar to radial, these compartments contain the following.

1. *Extensor carpi ulnaris tendon.*
2. *Extensor digiti minimi tendon.*
3. *Extensor digitorum and extensor indicis tendons,* which share a common synovial sheath.
4. *Extensor pollicis tendon,* which winds around the medial side of the dorsal tubercle (Lister's tubercle) found on the dorsal surface of the distal end of the radius (see Fig 16–13).
5. *Extensor carpi radialis longus* and *brevis tendons,* which share a common synovial sheath.
6. *Abductor pollicis longus* and *extensor pollicis brevis tendons,* which have separate synovial sheaths but share a common compartment.

The *radial artery* reaches the dorsum of the hand by passing between the wrist joint and the tendons of the abductor pollicis longus and extensor pollicis brevis.

Wrist Joint (Radiocarpal Joint)

Articulation.—The proximal articular surface is formed by the lower end of the radius and the triangular cartilaginous disc connecting together the lower ends of the radius and ulna (Fig 16–2). The distal articular surface is formed by the scaphoid, lunate, and triquetral bones. The proximal articular surface forms an ovoid concave surface, which is adapted to the distal ovoid convex surface.

Type.—Synovial condyloid joint.

Capsule.—The capsule encloses the joint.

Ligaments.—*Anterior and posterior ligaments* strengthen the capsule. *Medial ligament* connects the styloid process of the ulna to the triquetral bone. *Lateral ligament* connects the styloid process of the radius to the scaphoid bone.

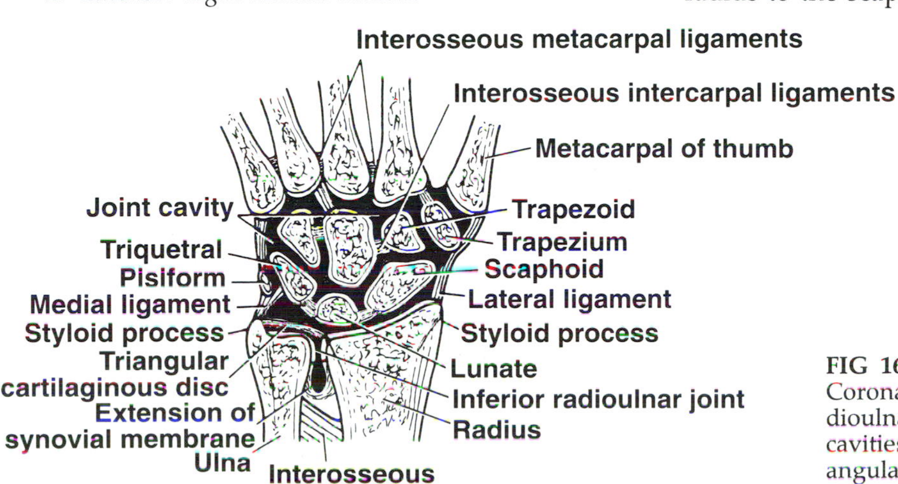

FIG 16–2.
Coronal section of the right wrist, inferior radioulnar and carpal joints showing the joint cavities (black area). The position of the triangular cartilaginous disc separates the head of the ulna from the wrist joint.

TABLE 16–1.
Muscles of the Anterior Fascial Compartment of the Forearm That Act on the Hand

Name of Muscle	Origin	Insertion	Nerve Supply	Nerve Roots*	Action
Flexor carpi radialis	Medial epicondyle of humerus	Bases of second and third metacarpal bones	Median nerve	C6, **C7**	Flexes and abducts hand at wrist joint
Palmaris longus (often absent)	Medial epicondyle of humerus	Flexor retinaculum and palmar aponeurosis	Median nerve	C7, C8	Flexes hand
Flexor carpi ulnaris					
Humeral head	Medial epicondyle of humerus	Pisiform bone, hook of the hamate, base of fifth metacarpal bone	Ulnar nerve	C8, T1	Flexes and adducts the hand at the wrist joint
Ulnar head	Olecranon process and posterior border of ulna				
Flexor digitorum superficialis					
Humeroulnar head	Medial epicondyle of humerus, coronoid process of ulna		Median nerve	C7, **C8,** T1	Flexes middle phalanx of fingers and assists in flexing proximal phalanx and hand
Radial head	Oblique line on anterior surface of shaft of radius	Middle phalanx of middle four fingers			
Flexor pollicis longus	Anterior surface of shaft of radius	Distal phalanx of thumb	Anterior interosseous branch of median nerve	**C8,** T1	Flexes distal phalanx of thumb
Flexor digitorum profundus	Anterior surface of shaft of ulna; interosseous membrane	Distal phalanges of medial four fingers	Ulnar (medial half) and median (lateral half) nerves	**C8,** T1	Flexes distal phalanx of the fingers; then assists in flexion of middle and proximal phalanges and the wrist

*The predominant nerve root supply is indicated by boldface type.

Synovial Membrane.—Lines the capsule.

Nerve Supply.—Anterior interosseous nerve from the median and deep branches of the radial and ulnar nerves.

Movements.—Flexion, extension, abduction (radial deviation), adduction (ulnar deviation), and circumduction. Rotation is not possible because the articular surfaces are ovoid. The lack of rotation is compensated for by the movements of pronation and supination of the forearm. The muscles producing the movements at the wrist joint are shown in Tables 16–1 through 16–3.

Basic Anatomy of the Hand

Palm of the Hand

Skin.—The skin of the palm is thick compared with the dorsal skin. It is bound down to the underlying deep fascia by numerous fibrous bands.

TABLE 16–2.
Muscle of the Lateral Fascial Compartment of the Forearm That Acts on the Hand

Name of Muscle	Origin	Insertion	Nerve Supply	Nerve Roots	Action
Extensor carpi radialis longus	Lateral supracondylar ridge of humerus	Base of second metacarpal bone	Radial nerve	C6, C7	Extends and abducts hand at wrist joint

TABLE 16–3.
Muscles of the Posterior Fascial Compartment of the Forearm That Act on the Hand

Name of Muscle	Origin	Insertion	Nerve Supply	Nerve Roots*	Action
Extensor carpi radialis brevis	Lateral epicondyle of humerus	Base of third metacarpal bone	Deep branch of radial nerve	**C7,** C8	Extends and abducts the hand at wrist joint
Extensor digitorum	Lateral epicondyle of humerus	Middle and distal phalanges of the medial four fingers	Deep branch of radial nerve	**C7,** C8	Extends fingers and hand
Extensor digiti minimi	Lateral epicondyle of humerus	Extensor expansion of little finger	Deep branch of radial nerve	**C7,** C8	Extends metacarpophalangeal joint of little finger
Extensor carpi ulnaris	Lateral epicondyle of humerus	Base of fifth metacarpal bone	Deep branch of radial nerve	C7, **C8**	Extends and adducts hand at the wrist joint
Abductor pollicis longus	Shafts of radius and ulna	Base of first metacarpal bone	Deep branch of radial nerve	C7, **C8**	Abducts and extends thumb
Extensor pollicis brevis	Shaft of radius and interosseous membrane	Base of proximal phalanx of thumb	Deep branch of radial nerve	C7, **C8**	Extends metacarpophalangeal joints of thumb
Extensor pollicis longus	Shaft of ulna and interosseous membrane	Base of distal phalanx of thumb	Deep branch of radial nerve	C7, **C8**	Extends distal phalanx of thumb
Extensor indicis	Shaft of ulna and interosseous membrane	Extensor expansion of index finger	Deep branch of radial nerve	C7, **C8**	Extends metacarpophalangeal joint of index finger

*The predominant nerve root supply is indicated by boldface type.

Clinical Note

Palmar infection.—The presence of the fibrous bands limits the development of inflammatory edema of the palm, and, consequently, palmar inflammation often reveals itself as a swelling on the dorsal surface of the hand.

Sweat glands are present in large numbers. A diminished amount of subcutaneous tissue covers the tendons on the fingers at the sites of the joints. Clinically, this is important since a laceration of the skin in this area could damage the tendons or their sheaths.

Sensory Innervation.—The sensory nerves are derived from the *palmar cutaneous branch of the median nerve,* which crosses in front of the flexor retinaculum and supplies the radial part of the palm, and the *palmar cutaneous branch of the ulnar nerve;* the latter nerve also crosses in front of the flexor retinaculum and supplies the ulnar part of the palm (Fig 16–3). The skin over the base of the thenar eminence is supplied by the *superficial branch of the radial nerve.*

Deep Fascia

The deep fascia of the palm is thickened to form the *palmar aponeurosis,* which is triangular and occupies the central area of the palm (see Fig 16–5). The apex of the aponeurosis is attached to the distal border of the flexor retinaculum and receives the insertion of the palmaris longus tendon. The base of the aponeurosis divides at the bases of the fingers into four slips.

The medial and lateral borders of the palmar aponeurosis are continuous with the thinner deep fascia covering the hypothenar and thenar muscles. From each of these borders, fibrous septa pass posteriorly into the palm and take part in the formation of the palmar fascial spaces (see p. 643).

Bones of the Hand

The bones of the hand include the carpal bones, the metacarpal bones, and the phalanges (Fig 16–4).

Carpal Bones.—There are eight small carpal bones in the region of the wrist that are strongly united to one another by ligaments. The bones are made up of two rows of four (see Fig 16–4). The proximal row

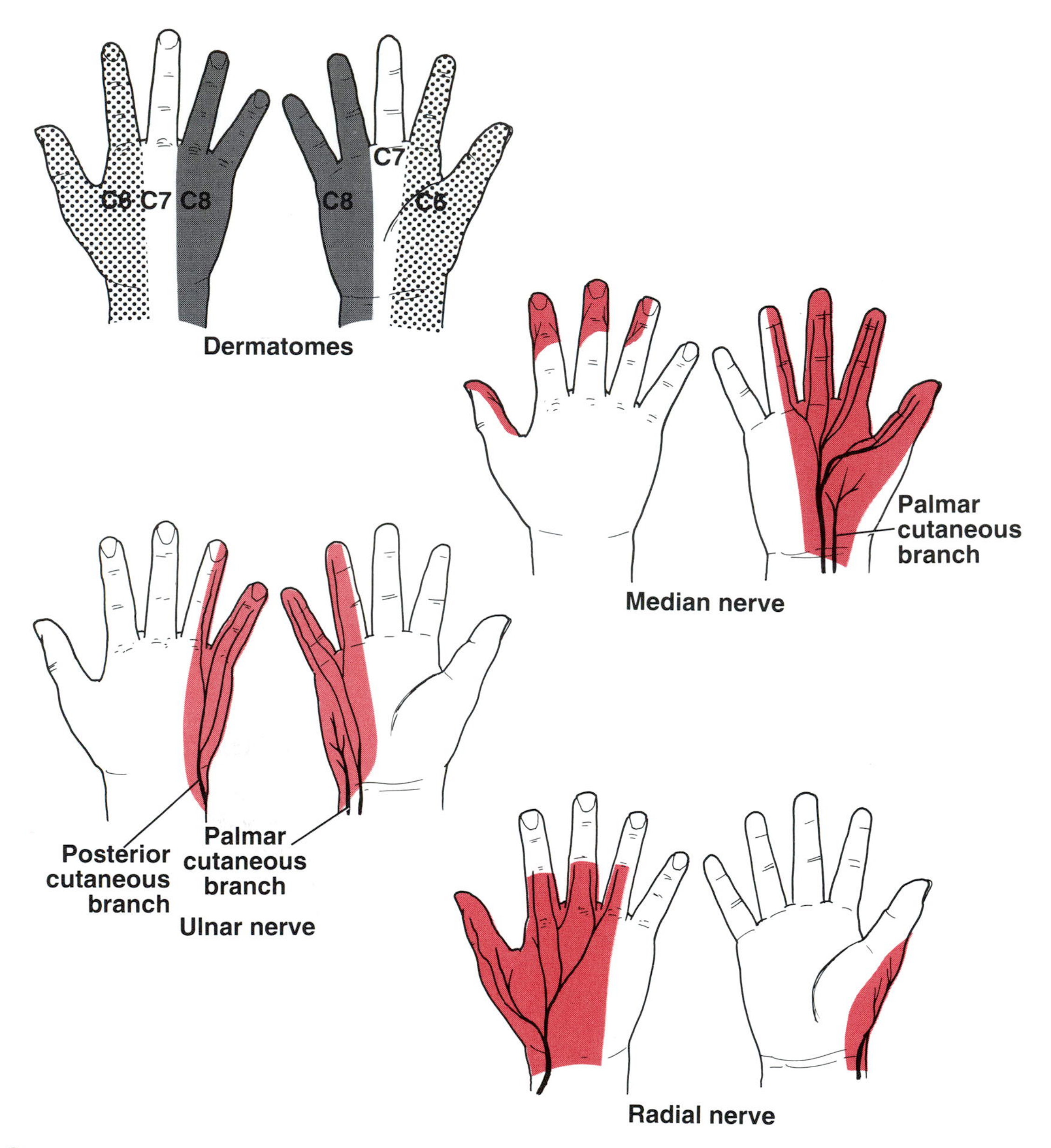

FIG 16–3.
Sensory innervation of the skin of the volar and dorsal aspects of the hand; the arrangement of the dermatomes is also shown.

consists of (from radial to ulnar) the *scaphoid (navicular), lunate, triquetral,* and *pisiform* bones. The distal row consists of (from radial to ulnar) the *trapezium, trapezoid, capitate,* and *hamate* bones. Together the bones of the carpus present a concavity on their anterior surface. The lateral and medial edges of this concavity are attached to a strong membranous band, the *flexor retinaculum,* which forms a bridge. The bridge and the bones thus form a tunnel, known as the *carpal tunnel,* for the passage of the median nerve and the long flexor tendons of the fingers (see Fig 16–1).

The carpal bones ossify in an orderly spiral sequence as follows: capitate and hamate, 1st year; triquetral, 3rd year; lunate, 4th year; scaphoid, 5th year; trapezoid and trapezium, 6th year; and the pisiform, 12th year.

Metacarpals and Phalanges.—There are five metacarpal bones in the hand, each of which has a proximal *base,* a *shaft,* and a distal *head* (see Fig 16–4). The metacarpal bones are numbered from one to five, starting with the metacarpal bone of the thumb. The first metacarpal bone of the thumb is

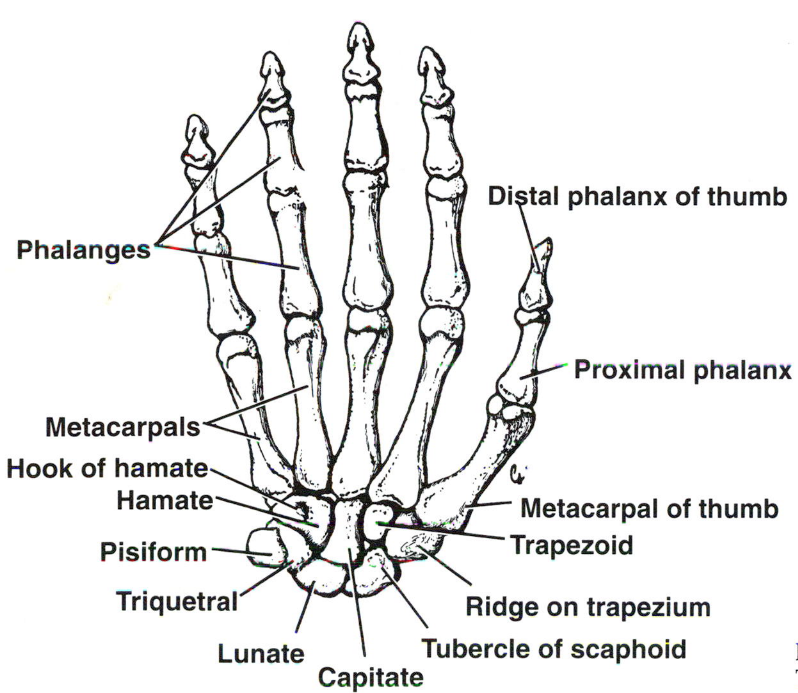

FIG 16–4.
The bones of the hand, anterior aspect.

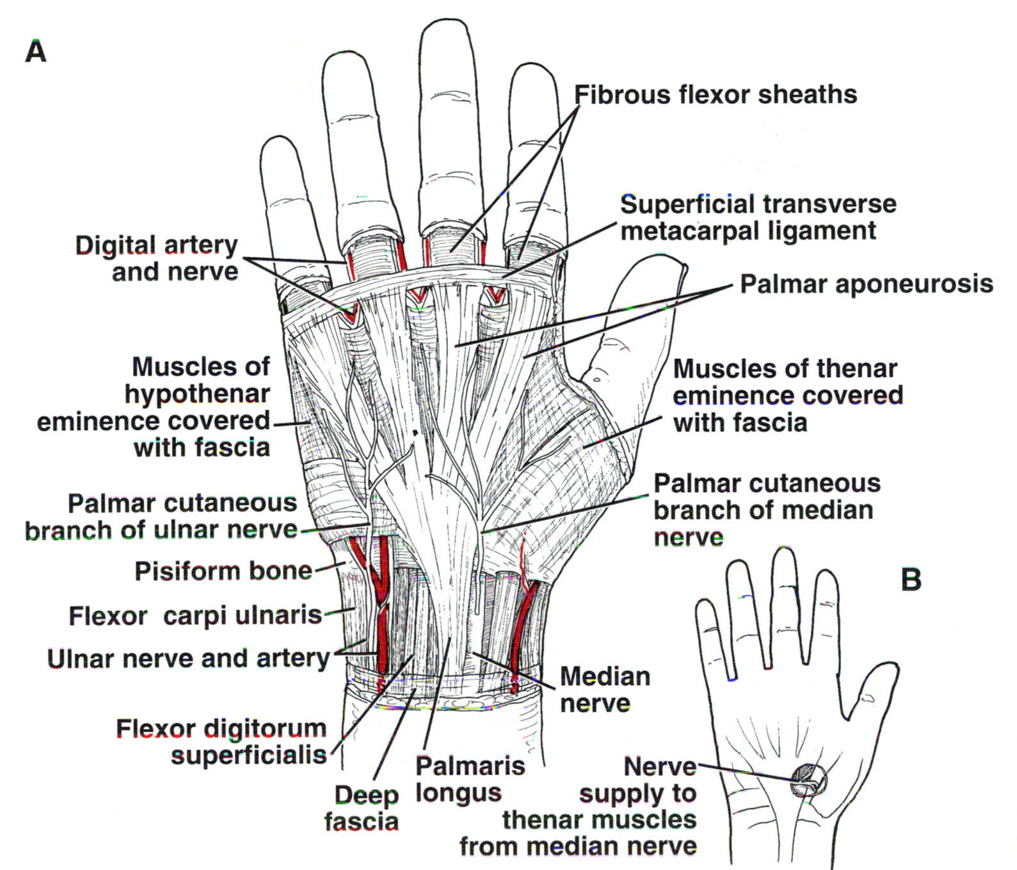

FIG 16–5.
Anterior view of the palm of the hand. **A,** palmar aponeurosis. **B,** nerve supply to muscles of the thenar eminence.

the shortest and most mobile. The bases of the metacarpal bones articulate with the distal row of the carpal bones; the heads, which form the knuckles, articulate with the proximal phalanges (see Fig 16–4).

There are three phalanges for each of the fingers, but only two for the thumb (see Fig 16–4). Each phalanx has a proximal *base,* a *shaft,* and a distal *head.*

Note that the epiphyseal plates are situated at the distal end of each metacarpal bone and at the proximal end of each phalanx. However, the metacarpal bone of the thumb resembles a phalanx in that the epiphyseal plate is located at the proximal end (see Fig 16–19).

Fibrous Flexor Sheaths

The anterior surface of each finger, from the head of the metacarpal to the base of the distal phalanx, is provided with a strong fibrous sheath that is attached to the sides of the phalanges. The sheath, together with the anterior surfaces of the phalanges and the interphalangeal joints, forms a blind tunnel in which the flexor tendons of the fingers lie (Figs 16–5 through 16–7).

Synovial Flexor Sheaths

The crowded long flexor tendons emerge from the carpal tunnel and diverge as they pass down into the hand.

The flexor pollicis longus tendon enters the osseofibrous tunnel of the thumb and is inserted into the base of the distal phalanx (see Fig 16–7). The tendon is surrounded by a synovial sheath, which extends into the forearm for a distance of about one fingerbreadth proximal to the flexor retinaculum, and distally it extends to the insertion of the flexor pollicis longus tendon.

The eight tendons of the flexor digitorum superficialis and profundus invaginate a common synovial sheath from the radial side (see Fig 16–7). This com-

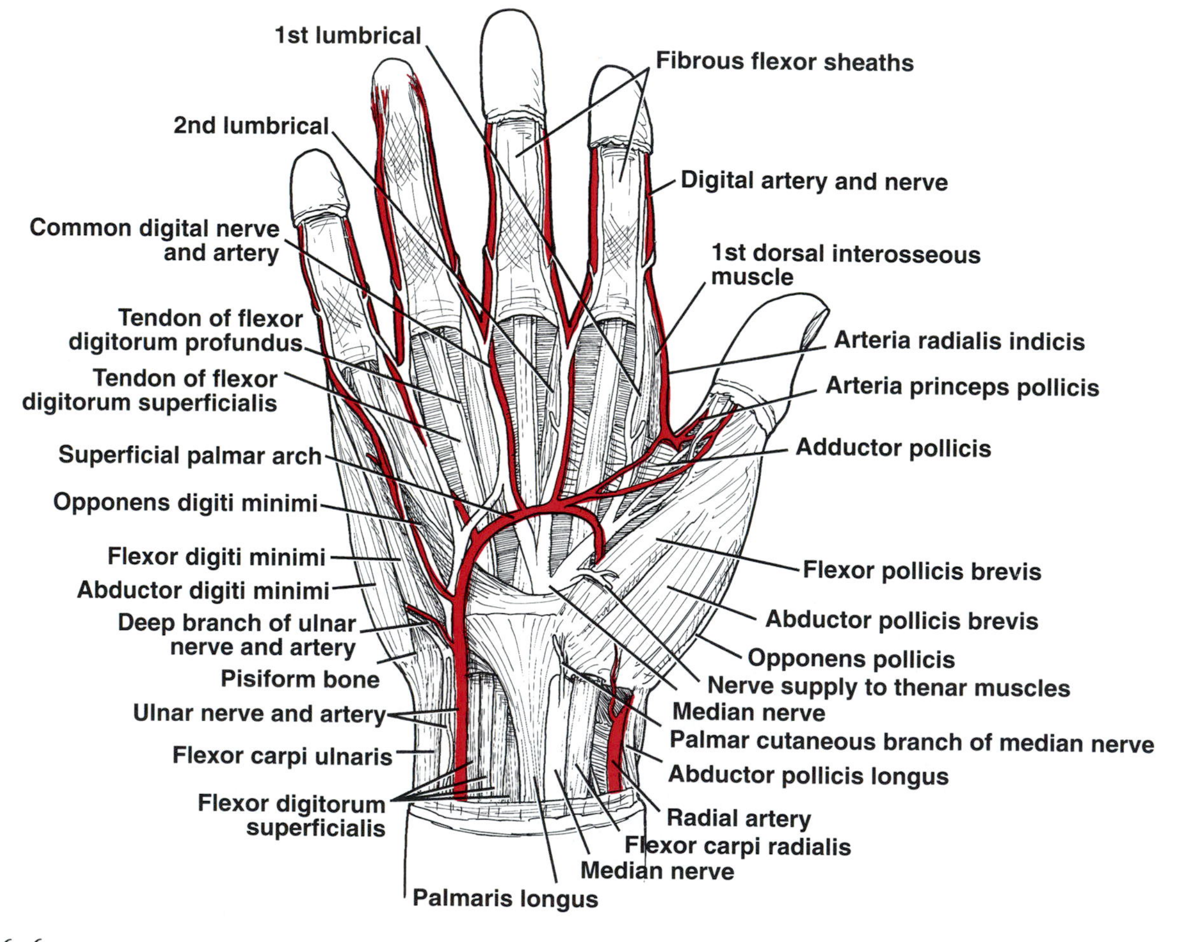

FIG 16–6.
Anterior view of the palm of the hand; the palmar aponeurosis has been removed to display the superficial palmar arch, median nerve, and the long flexor tendons.

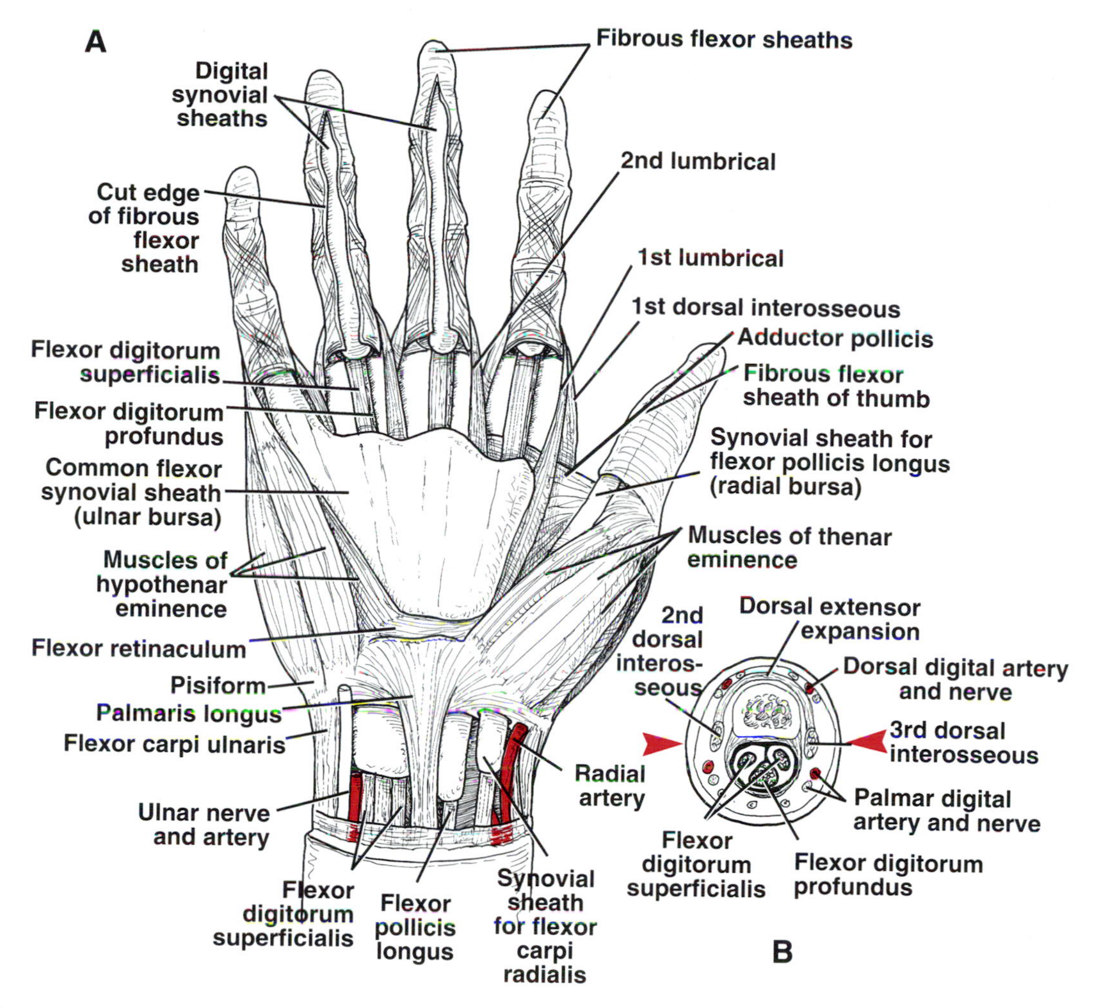

FIG 16–7.
A, anterior view of the palm of the hand showing the synovial sheaths. Note the positions of the ulna and radial bursae. **B,** cross section of middle finger at the level of the proximal phalanx showing the digital synovial sheath and the digital nerves and arteries. The black pointer heads indicate the position of the so-called "neutral" lines, where lateral incisions may be made and avoid damage to the palmar and dorsal digital nerves and arteries.

mon sheath extends proximally into the forearm for about a fingerbreadth proximal to the flexor retinaculum. Distally, the ulnar part of the sheath continues downward without interruption on the tendons of the little finger as far as the base of the distal phalanx. The remainder of the sheath ends blindly at approximately the level of the proximal transverse crease of the palm.

The distal ends of the flexor tendons of the index, middle, and ring fingers have *digital synovial sheaths,* which commence just about the level of the distal transverse crease of the palm and end at the bases of the distal phalanges (see Fig 16–7). Thus, for a short length between the distal transverse crease of the palm and the commencement of the digital synovial sheaths, the tendons for these three fingers are devoid of a synovial covering.

The synovial sheath of the flexor pollicis longus (sometimes known as the *radial bursa*) communicates with the common synovial sheath of the superficialis and profundus tendons (sometimes known as the *ulnar bursa*) at the level of the wrist in about 50% of patients.

Vincula longa and *brevia* are small vascular folds of synovial membrane that connect the tendons to the anterior surface of the phalanges (see Fig 16–10). They resemble mesenteries and convey blood vessels to the tendons.

Clinical Note

Tendon necrosis.—In cases of severe tenosynovitis the vincula may be destroyed, resulting in avascular necrosis of the tendons.

Insertion of the Long Flexor Tendons
The flexor pollicis longus tendon is inserted onto the base of the distal phalanx of the thumb (Fig 16–8).

Each tendon of the flexor digitorum superficialis enters the fibrous flexor sheath; opposite the proximal phalanx it divides into two halves, which pass around the profundus tendon and meet on its deep or posterior surface, where partial decussation of the fibers takes place (see Fig 16–8). The superficialis tendon, having united again, divides almost at once into two further slips, which are attached to the borders of the middle phalanx.

Each tendon of the flexor digitorum profundus, having passed through the division of the superficialis tendon, continues distally, to be inserted into the base of the distal phalanx (see Fig 16–10).

Small Muscles of the Hand
Lumbrical Muscles.—The four lumbrical muscles (see Fig 16–8) arise from the tendons of the flexor digitorum profundus in the palm and pass distally into the fingers to insert into the radial side of the corresponding extensor expansion (see Fig 16–10).

The radial two lumbricals are innervated by the median nerve, and the ulnar two are innervated by the ulnar nerve.

Clinical Note

No man's land of long flexor tendon injuries.—This is located between the distal transverse crease of the palm and the midportion of the middle phalanx. It is the site where the single tendon of flexor digitorum superficialis and the single tendon of flexor digitorum profundus, surrounded by their digital synovial sheath, lie close together within the fibro-osseous tunnel in each finger. This is the site where even in the best circumstances, the restoration of good digital function following tendon injury is extremely difficult.

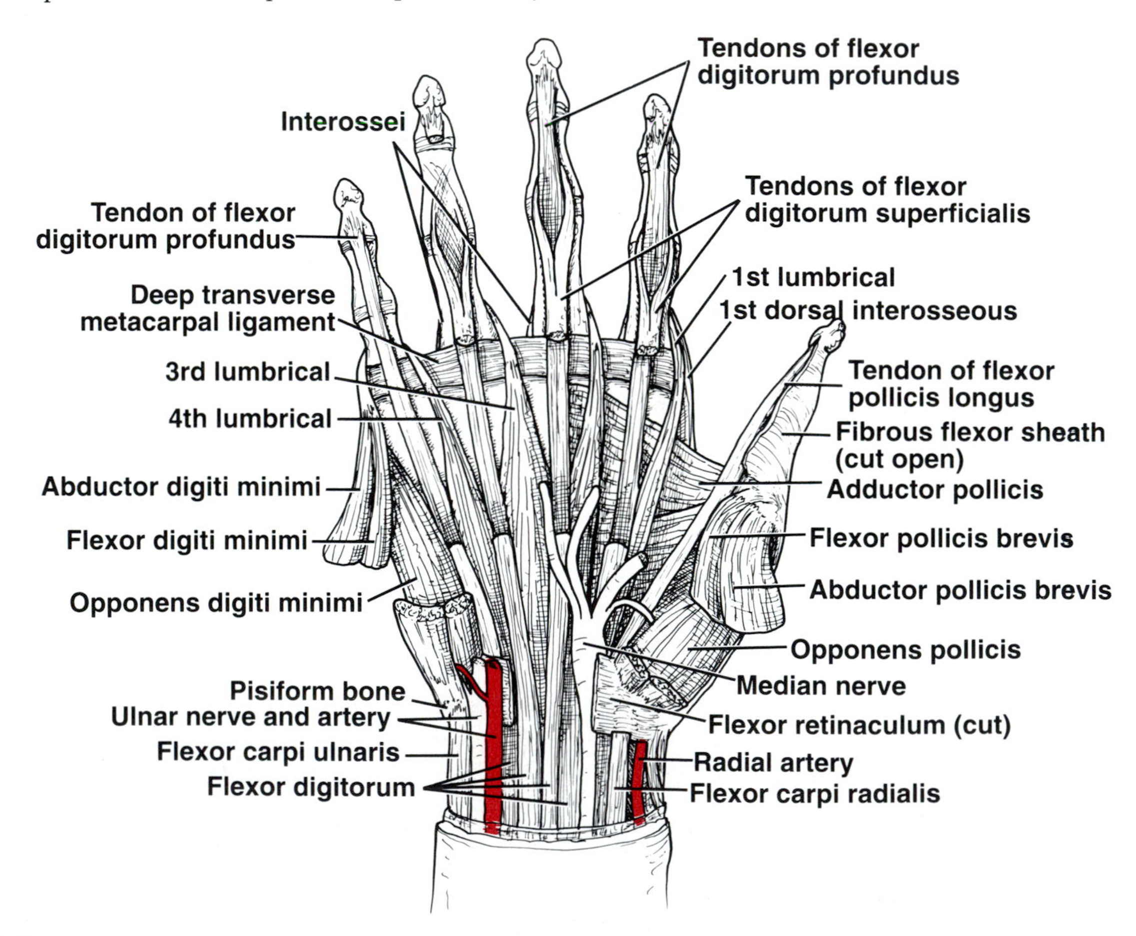

FIG 16–8.
Anterior view of the palm of the hand. The palmar aponeurosis and the greater part of the flexor retinaculum have been removed to display the median nerve, the long flexor tendons, and the lumbrical muscles. Segments of tendons of the flexor digitorum superficialis muscle have been removed to show underlying tendons of the flexor digitorum profundus muscle.

Interossei Muscles.—The eight interossei,* consisting of four volar and four dorsal muscles, arise from the metacarpal bones and are inserted into the fingers.

The volar (palmar) interossei have the following origins and insertions (Figs 16–9 and 16–10). The first arises from the medial side of the base of the first metacarpal bone. The second, third, and fourth arise from the anterior surfaces of the second, fourth, and fifth metacarpal bones, respectively. The first is inserted into the ulnar side of the base of the proximal phalanx of the thumb. The second is inserted into the ulnar side of the base of the proximal phalanx of the index finger. The third and fourth are inserted into the radial side of the corresponding bones of the ring finger and little finger, respectively. In addition, all the interossei are inserted into the extensor expansion of the digit on which they act (see Fig 16–9).

The four dorsal interossei have the following origins and insertions (see Figs 16–9 and 16–10). The four dorsal interossei arise from the contiguous sides of the first and second, second and third, third and fourth, and fourth and fifth metacarpal bones, respectively. The first dorsal interosseous (see Fig 16–10) is inserted into the radial side of the base of the proximal phalanx of the index finger; the second, into the radial side of the base of the proximal phalanx of the middle finger (see Fig 16–10); the third, on the ulnar side of the same bone; and the fourth, on the ulnar side of the base of the proximal phalanx of the ring finger. In addition, all the interossei are inserted into the extensor expansion of the digit on which they act (see Fig 16–9).

*Some authors describe only three palmar interossei and state that the first volar interosseous is a second head to the flexor pollicis brevis; others believe that it is part of the adductor pollicis muscle.

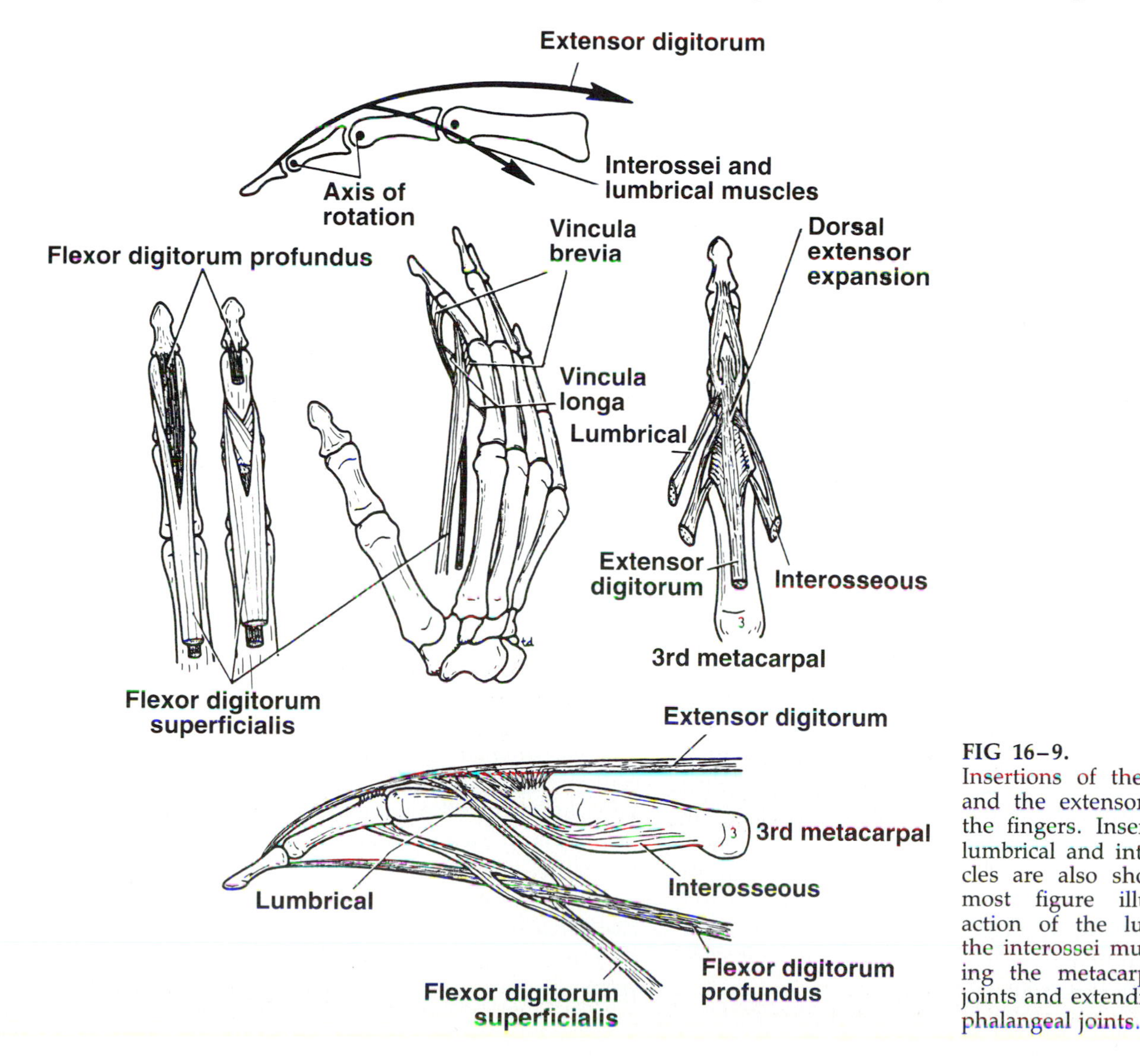

FIG 16–9.
Insertions of the long flexor and the extensor tendons in the fingers. Insertions of the lumbrical and interossei muscles are also shown. Uppermost figure illustrates the action of the lumbrical and the interossei muscles in flexing the metacarpophalangeal joints and extending the interphalangeal joints.

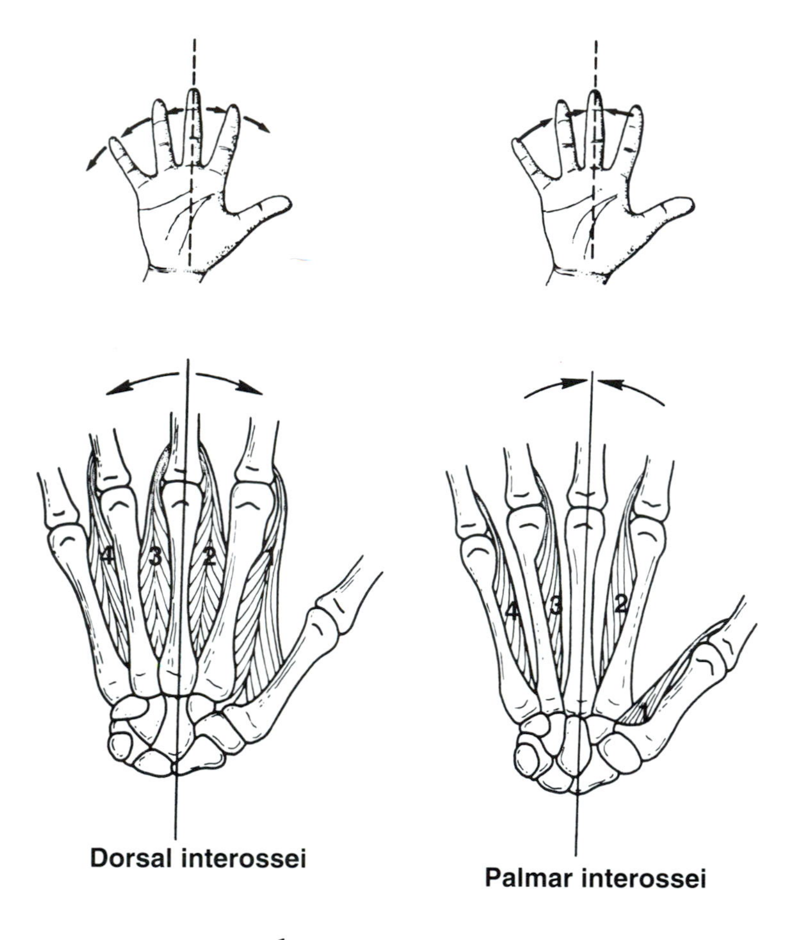

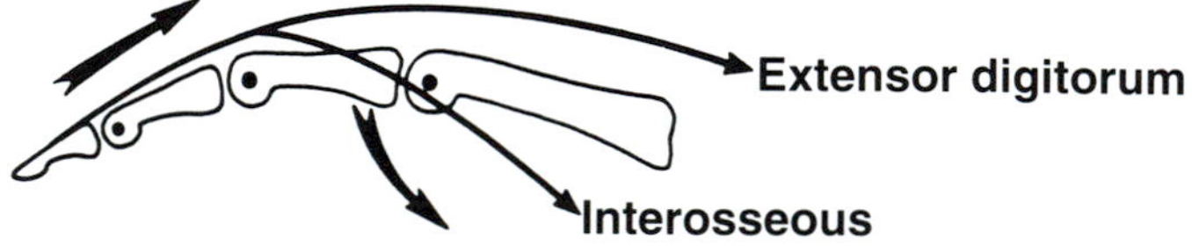

FIG 16–10.
Origins and insertions of the palmar and dorsal interossei muscles. The actions of these muscles are also shown.

All the interossei are innervated by the ulnar nerve.

Short (Intrinsic) Muscles of the Thumb.—The short muscles of the thumb are the abductor pollicis brevis, the flexor pollicis brevis, the opponens pollicis, and the adductor pollicis (Figs 16–7, 16–8, and 16–11). The first three of these muscles form the *thenar eminence.* All the short muscles of the thumb are innervated by the median nerve except the adductor pollicis, which is innervated by the ulnar nerve.

Short (Intrinsic) Muscles of the Little Finger.—The short muscles of the little finger are the abductor digiti minimi, the flexor digiti minimi, and the opponens digiti minimi, all of which together form the *hypothenar eminence* (see Figs 16–7, 16–8, and 16–11). All the short muscles of the little finger are innervated by the ulnar nerve.

The detailed origin, insertion, nerve supply, and action of these muscles are summarized in Table 16–4.

Arteries of the Palm of the Hand

Ulnar Artery.—The ulnar artery enters the hand *anterior* to the flexor retinaculum on the radial side of the ulnar nerve and the pisiform bone. The artery gives off a deep branch and then continues into the palm as the superficial palmar arch.

Superficial Palmar Arch.—This is a direct continuation of the ulnar artery (see Fig 16–6). It curves laterally behind the palmar aponeurosis and in front of the long flexor tendons. The arch is completed on the radial side by a branch of the radial artery. The curve of the arch lies across the palm, level with the distal border of the fully extended thumb. *Four digital arteries* arise from the arch and pass to the fingers.

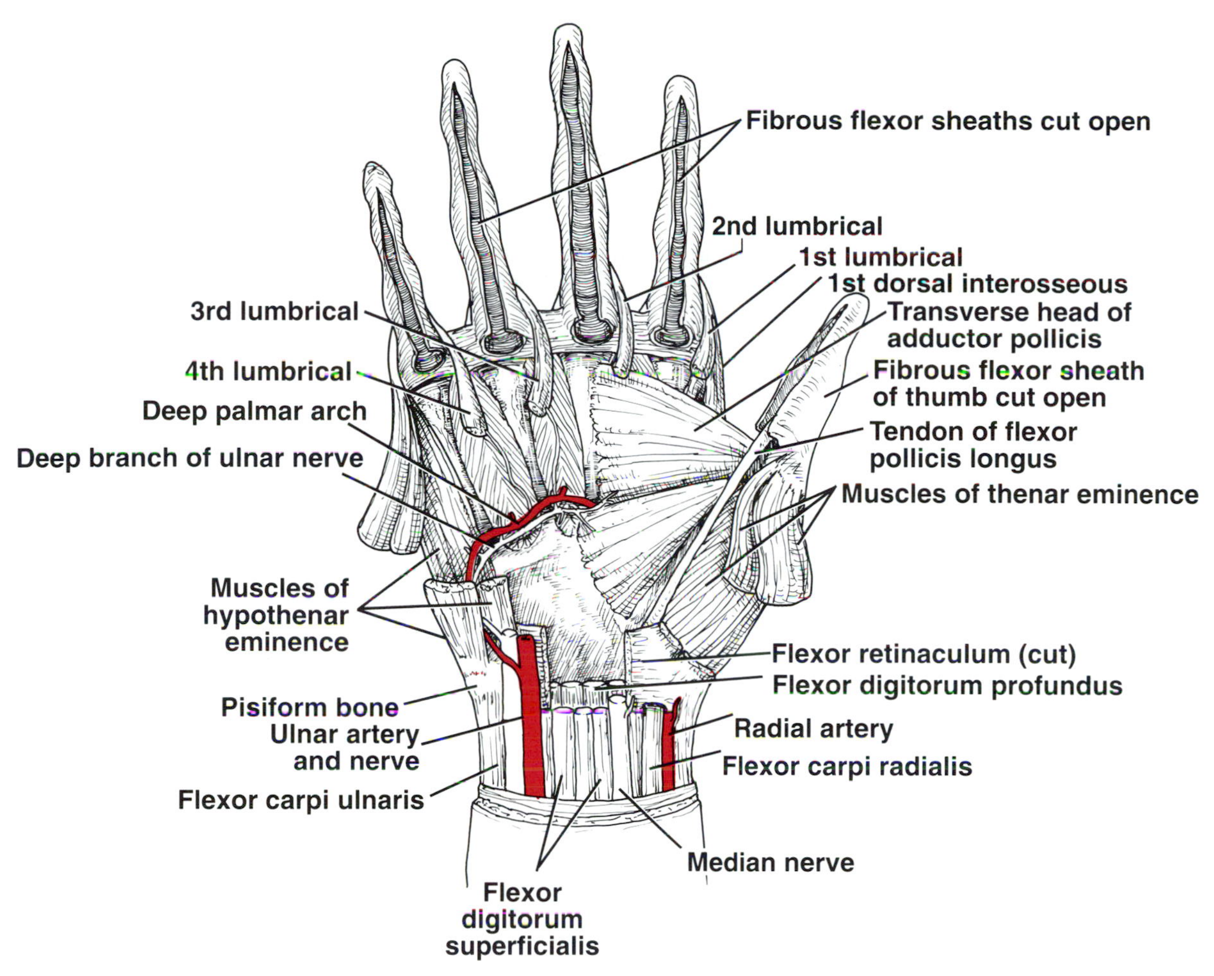

FIG 16–11.
Anterior view of the palm of the hand. The long flexor tendons have been removed. Note the position of the deep palmar arch and the adductor pollicis muscle.

Deep Branch of the Ulnar Artery.—This arises in front of the flexor retinaculum and passes between the muscles of the hypothenar eminence. It ends by joining the radial artery to complete the deep palmar arch (see Fig 16–11).

Radial Artery.—The radial artery lies on the anterior aspect of the radius in the distal part of the forearm (see Fig 16–7). It winds round the lateral side of the wrist onto the dorsum of the hand by passing deep to the tendons of abductor pollicis longus and the extensor pollicis brevis. The radial artery then leaves the dorsum of the hand by turning forward between the proximal ends of the first and second metacarpal bones. On entering the palm, it curves medially and continues as the deep palmar arch (see Fig 16–11).

Deep Palmar Arch.—This is a continuation of the radial artery. It curves medially posterior to the long flexor tendons and in front of the metacarpal bones (Fig 16–11). The arch is completed on the medial side by the deep branch of the ulnar artery. The deep branch of the ulnar nerve accompanies the arch. The curve of the arch lies across the palm at a level with the proximal border of the fully extended thumb.

The superficial and deep palmar arterial arches are so named because they are superficial and deep to the long flexor tendons and their synovial sheaths.

Lymphatic Drainage of the Hand

The lymphatic vessels of the fingers pass along their borders to reach the webs. From there the vessels ascend onto the dorsum of the hand. Lymphatic vessels on the palm form a plexus that is drained by vessels that ascend in front of the forearm or pass around the medial and lateral borders to join vessels on the dorsum of the hand.

TABLE 16–4.
Small Muscles of the Hand

Name of Muscle	Origin	Insertion	Nerve Supply	Nerve Roots*	Action
Lumbricals	Tendons of flexor digitorum profundus	Extensor expansion of medial four fingers	First and second, i.e., lateral two, median nerve; third and fourth ulnar nerve	C8, **T1**	Flex metacarpophalangeal joints and extend interphalangeal joints of fingers except thumb
Interossei					
Palmar (4)	First, second, fourth, and fifth metacarpal bones	Base of proximal phalanges of fingers; extensor expansion	Deep branch of ulnar nerve	C8, **T1**	Palmar interossei adduct fingers toward center of third finger
Dorsal (4)	Contiguous sides of 5 metacarpal bones	Base of proximal phalanges of fingers; extensor expansion	Deep branch of ulnar nerve	C8, **T1**	Dorsal interossei abduct fingers from center of third finger; both palmar and dorsal flex the metacarpophalangeal joints and extend the interphalangeal joints
Palmaris brevis	Flexor retinaculum and palmar aponeurosis	Skin of palm	Superficial branch of ulnar nerve	C8, **T1**	Corrugates the skin and improves grip of palm
Short Muscles of Thumb					
Abductor pollicis brevis	Scaphoid, trapezium, flexor retinaculum	Base of proximal phalanx of thumb	Median nerve	**C8,** T1	Abduction of thumb
Flexor pollicis brevis	Flexor retinaculum	Base of proximal phalanx of thumb	Median nerve	**C8,** T1	Flexes thumb
Opponens pollicis	Flexor retinaculum	Shaft of metacarpal bone of thumb	Median nerve	**C8,** T1	Pulls thumb medially and forward across palm
Adductor pollicis	Oblique head: second and third metacarpal bones; Transverse head: third metacarpal bone	Base of proximal phalanx of thumb	Deep branch of ulnar nerve	C8, **T1**	Adduction of thumb
Short Muscles of Little Finger					
Abductor digiti minimi	Pisiform bone	Base of proximal phalanx of little finger	Deep branch of ulnar nerve	C8, **T1**	Flexes little finger
Flexor digiti minimi	Flexor retinaculum	Base of proximal phalanx of little finger	Deep branch of ulnar nerve	C8, **T1**	Flexes little finger
Opponens digiti minimi	Flexor retinaculum	Shaft of metacarpal bone of little finger	Deep branch of ulnar nerve	C8, **T1**	Flexes little finger and pulls fifth metacarpal bone forward as in cupping the hand

*The predominant nerve root supply is indicated by boldface type.

The lymph from the ulnar side of the hand ascends in vessels that accompany the basilic vein; they drain into the *supratrochlear nodes* and then ascend to drain into the *lateral axillary nodes.* The lymph from the radial side of the hand ascends in vessels that accompany the cephalic vein; they drain into the *infraclavicular nodes,* and some drain into the *lateral axillary nodes.*

Nerves of the Palm.—These include the following.

Median Nerve.—The median nerve enters the palm by passing *behind* the flexor retinaculum and through the carpal tunnel (see Fig 16–1). It gives off the following branches.

1. *Muscular branches,* including the abductor pollicis brevis, flexor pollicis brevis, opponens pollicis, and the first and second lumbrical muscles. The branch to the muscles of the thenar eminence takes a recurrent course around the lower border of the flexor retinaculum and lies about one fingerbreadth distal to the tubercle of the scaphoid (see Fig 16–5).
2. *Cutaneous branches* to the palmar aspect of the radial 3½ fingers and the distal half of the dorsal aspect of each of these 3½ fingers (see Fig 16–3).

Note that the *palmar cutaneous branch* of the median nerve given off in front of the forearm crosses *anterior* to the flexor retinaculum and supplies the skin over the radial part of the palm (see Fig 16–3).

Ulnar Nerve.—The ulnar nerve enters the palm *anterior (volar)* to the flexor retinaculum, ulnar to the ulnar artery, and radial to the pisiform bone (see Fig 16–5). The ulnar nerve then divides into superficial and deep terminal branches.

Clinical Note

Tunnel of Guyon.—In the depression between the pisiform bone and the hook of the hamate, the ulnar nerve and artery may lie in a fibro-osseous tunnel, the *tunnel of Guyon,* created by fibrous tissue derived from the superficial part of the flexor retinaculum. Here the nerve may be subject to entrapment.

Superficial Terminal Branch.—The following branches are involved.

1. *Muscular branch,* including the palmaris brevis (a small, unimportant muscle that arises from the flexor retinaculum (see Fig 16–1) and is inserted into the skin of the palm)
2. *Cutaneous branches* to the skin over the palmar aspect of the ulnar 1½ fingers including the nail beds (see Fig 16–3). (The proximal half of the dorsal surface of each of these fingers receives its cutaneous innervation from the dorsal digital nerves, which are terminal branches of the posterior cutaneous branch of the ulnar nerve that arises from the parent trunk above the level of the wrist joint.)

Deep Terminal Branch.—The following branches are involved.

1. *Muscular branches,* including the abductor digiti minimi, flexor digiti minimi, opponens digiti minimi, all the palmar and dorsal interossei, the third and fourth lumbricals, and adductor pollicis.
2. *Articular branches,* including the carpal joints.

Note that the *palmar cutaneous branch of the ulnar nerve* given off in front of the forearm crosses *anterior* to the flexor retinaculum and supplies the skin over the ulnar part of the palm (see Fig 16–3).

Fascial Spaces of the Palm.—The fascial spaces of the palm are potential spaces filled with loose connective tissue. Their boundaries are important clinically, since they may limit the spread of infection.

From the ulnar border of the palmar aponeurosis, a fibrous septum passes backward and is attached to the fifth metacarpal bone (Fig 16–12). From the lateral border of the aponeurosis, a second fibrous septum passes obliquely backward to the third metacarpal bone. A further septum passes from the radial border of the aponeurosis to the first metacarpal bone. In this manner these oblique septa divide the palm into a *thenar space,* which lies radially, and a *midpalmar space,* which lies ulnarly (see Fig 16–12). The posterior boundary of these spaces is formed by the volar surfaces of the shafts of the metacarpal bones and the fascia covering the interosseous muscles. These spaces are distinct and different from the spaces containing the thenar and hypothenar muscles.

Proximally, the thenar and midpalmar spaces are closed off by fascia in the carpal tunnel. Distally, the two spaces are continuous with the appropriate fascial spaces around the lumbrical muscles.

Pulp Spaces of the Fingers.—The deep fascia of the pulp of each finger fuses with the periosteum of the terminal phalanx distal to the insertion of the long flexor tendons and closes off a fascial compartment, known as the *pulp space* (see Fig 16–12). Each pulp space is subdivided by numerous fibrous septa into compartments filled with fat. The diaphysis of the

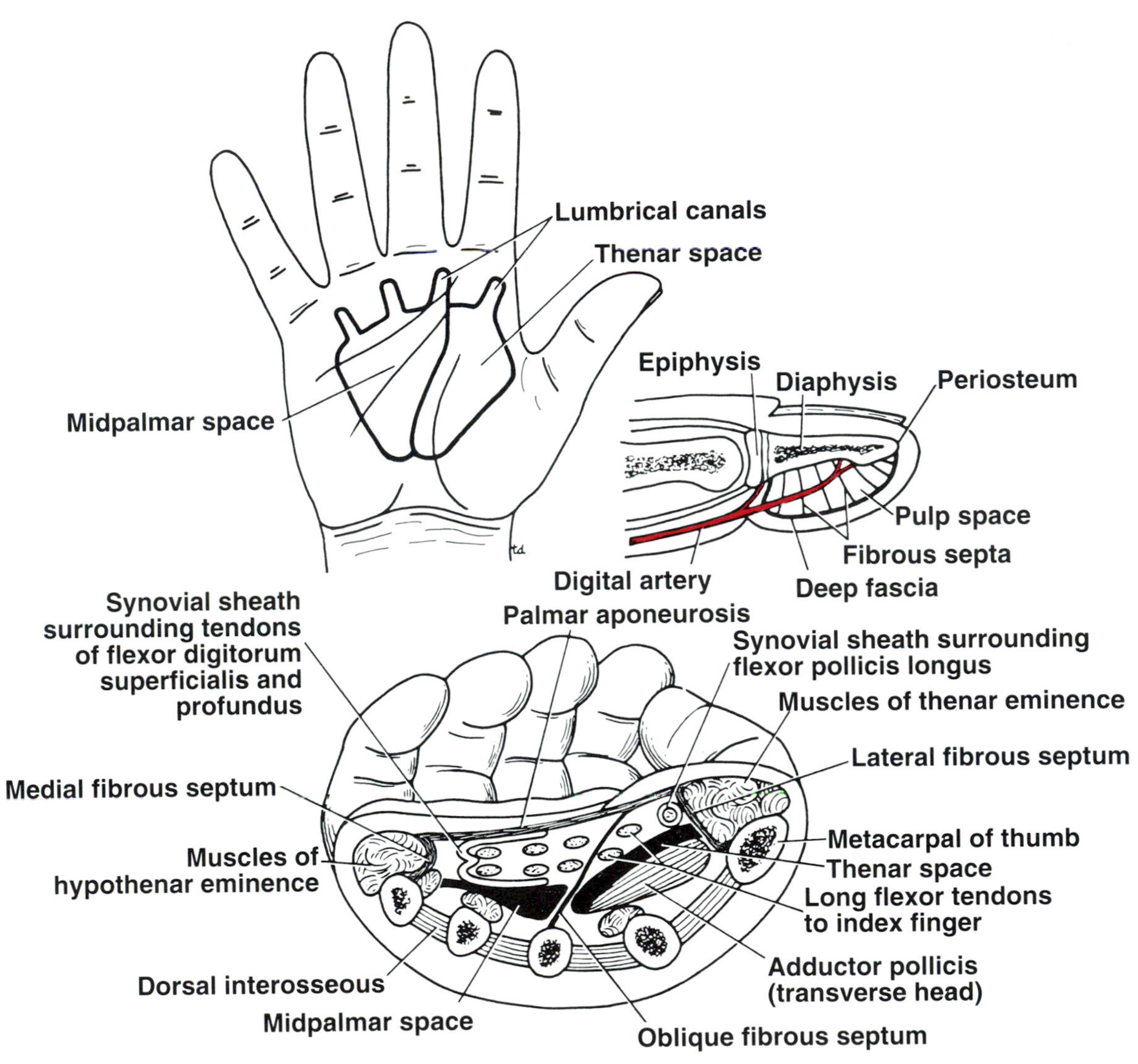

FIG 16–12.
The palmar and pulp fascial spaces.

Clinical Note

Thenar and midpalmar spaces and the flexor synovial sheaths.—These important fascial spaces are dorsally placed and must not be confused with the volarly placed flexor synovial sheaths. Although the fascial spaces do not communicate with one another, the flexor synovial sheath for the flexor pollicis longus tendon (radial bursa) communicates with the synovial sheath for the tendons of the flexor digitorum superficialis and the flexor digitorum profundus (ulnar bursa) in 50% of subjects.

distal phalanx receives its arterial supply from a terminal branch of a digital artery that passes through the pulp space. The epiphysis of the distal phalanx (which is located proximally) receives its blood supply proximal to the pulp space.

Clinical Note

Avascular necrosis of diaphysis.—Thrombosis of the arterial supply to the diaphysis of the distal phalanx, secondary to infection of the pulp space, may lead to avascular necrosis of the diaphysis in children.

Dorsum of the Hand

Skin.—The skin on the dorsum of the hand is thin, hairy, and freely mobile on the underlying tendons and bones. The subcutaneous tissue contains the *dorsal venous arch or network* that lies just proximal to the metacarpophalangeal joints. The network is drained on the radial side by the cephalic vein and on the ulnar side by the basilic vein (see p. 655).

Sensory Innervation.—The skin is supplied on the dorsum of the hand by the superficial branch of the radial nerve and the posterior cutaneous branch of the ulnar nerve (see Fig 16–3).

The superficial branch of the radial nerve descends superficial to the extensor retinaculum and supplies the lateral two thirds of the dorsum of the hand and the proximal parts of the dorsal surface of the lateral 3½ fingers (see Fig 16–3).

The posterior cutaneous branch of the ulnar nerve descends superficial to the extensor retinaculum and supplies the medial third of the dorsum of the hand and the proximal parts of the dorsal surfaces of the medial 1½ fingers (see Fig 16–3).

The distribution of the superficial radial nerve is variable in extent, and this is compensated for by the posterior cutaneous branch of the ulnar nerve. Also, the distal part of the dorsum of each finger receives its nerve supply from the palmar digital nerves (see Fig 16–3), which are supplied by the median and ulnar nerves.

Insertion of the Long Extensor Tendons.—The four tendons of the extensor digitorum emerge from under the extensor retinaculum and fan out over the dorsum of the hand (Fig 16–13). Strong oblique fibrous bands connect the tendons to the little, ring, and middle fingers. The tendon to the index finger is joined on its *radial* side by the tendon of the extensor indicis, and the tendon to the little finger is joined on its *ulnar* side by the two tendons of the extensor digiti minimi (see Fig 16–13).

On the posterior surface of each finger, the extensor tendon joins the fascial expansion called the *extensor expansion* (see Fig 16–13). Near the proximal interphalangeal joint, the extensor expansion splits into three parts—a central part, which is inserted into the base of the middle phalanx, and two lateral parts, which converge to be inserted into the base of the distal phalanx (see Fig 16–13).

The dorsal extensor expansion receives the tendon of insertion of the corresponding interosseous

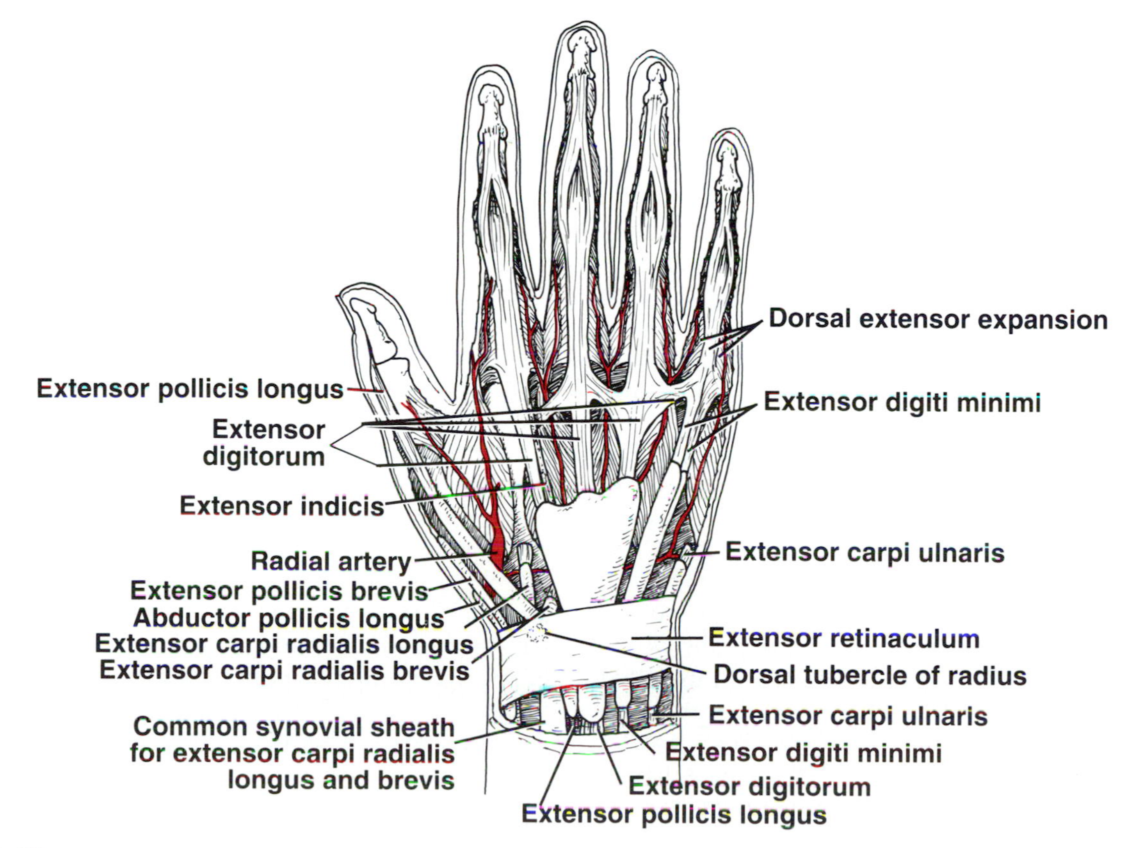

FIG 16–13.
Dorsal surface of the hand showing the long extensor tendons and their insertions. Note the position of the synovial sheaths.

muscle on each side, and farther distally receives the tendon of the lumbrical muscle on the radial side (see Fig 16–9).

Clinical Note

Connections between extensor tendons making precise diagnosis of lacerated tendons difficult.—The strong fibrous bands connecting together the tendons to the little, ring, and middle fingers sometimes mask the diagnosis and make it difficult to identify the fully lacerated extensor tendon in patients with a dorsal laceration of the proximal hand.

The Hand as a Functional Unit

Much of the importance of the hand depends on the pincers action of the thumb, which enables objects to be grasped between the thumb and index finger. The extreme mobility of the first metacarpal bone makes the thumb functionally as important as all the remaining fingers combined.

Position of the Hand.—The hand is able to perform its most delicate movements with the forearm in the semipronated position and the wrist joint partially extended. The reasons for this are as follows. In the semiprone or semisupinated position, the interosseous membrane binding the radius and ulna together is taut and therefore in this position the forearm is most stable. With the wrist joint partially extended, the long flexor and extensor tendons are working to their best mechanical advantage; at the same time, the flexors and extensors of the carpus can exert a balanced fixator action on the wrist joint, ensuring a stable base for the movements of the fingers.

Position of Rest.—This is the position adopted by the hand when the fingers are at rest and the hand is relaxed (Fig 16–14). The forearm is in the semiprone position; the wrist joint is slightly extended; the second, third, fourth, and fifth fingers are partially flexed, although the index finger is not flexed as much as the others; and the plane of the thumbnail lies at right angle to the plane of the other fingernails.

Position of Function.—This is the posture adopted by the hand when it is about to grasp an object between the thumb and index finger (see Fig 16–14). The forearm is in the semiprone position, the wrist joint is partially extended (more so than in the position of rest), and the fingers are partially flexed, with the index finger being flexed as much as the others. The metacarpal bone of the thumb is rotated in such a manner that the plane of the thumbnail lies parallel with that of the index finger, and the pulp of the thumb and index finger are in contact.

Clinical Note

Hand position and immobilization.—The position of function is most important clinically. Should the hand require immobilization for the treatment of disease of any part of the upper limb, it should be immobilized (if possible) in the position of function. This means that if there is loss of movement at the wrist joint, or at the joints of the hand or fingers, the patient will at least have a hand that is in a position of mechanical advantage and one that can serve a useful purpose. As a general principle, in order to retain adequate movements of the hand, the hand should only be splinted when really necessary.

Movements of the Fingers.—This section should assist the physician when evaluating nerve or tendon injuries involving loss of finger function. Only the primary action of the individual muscles is considered.

Movements of the Thumb.—These include the following.

Flexion.—This is the movement of the thumb across the palm that maintains the plane of the thumbnail at right angles to the plane of the other fingernails (see Fig 16–14). The movement takes place between the trapezium and the first metacarpal bone at the metacarpophalangeal and interphalangeal joints. The flexor pollicis longus and brevis and the opponens pollicis muscles produce the movement.

Extension.—This is the movement of the thumb in a lateral or coronal plane away from the palm that maintains the plane of the thumbnail at right angles to the plane of the other fingernails (see Fig 16–14). The movement takes place between the trapezium and the first metacarpal bone at the metacarpophalangeal and interphalangeal joints. The extensor pollicis longus and brevis muscles produce the movement.

Abduction.—This is the movement of the thumb in an anteroposterior plane away from the palm, with the plane of the thumbnail being kept at right angles to the plane of the other nails (see Fig 16–14). The movement takes place between the trapezium and the first metacarpal bone at the metacarpophalangeal and interphalangeal joints. The extensor pol-

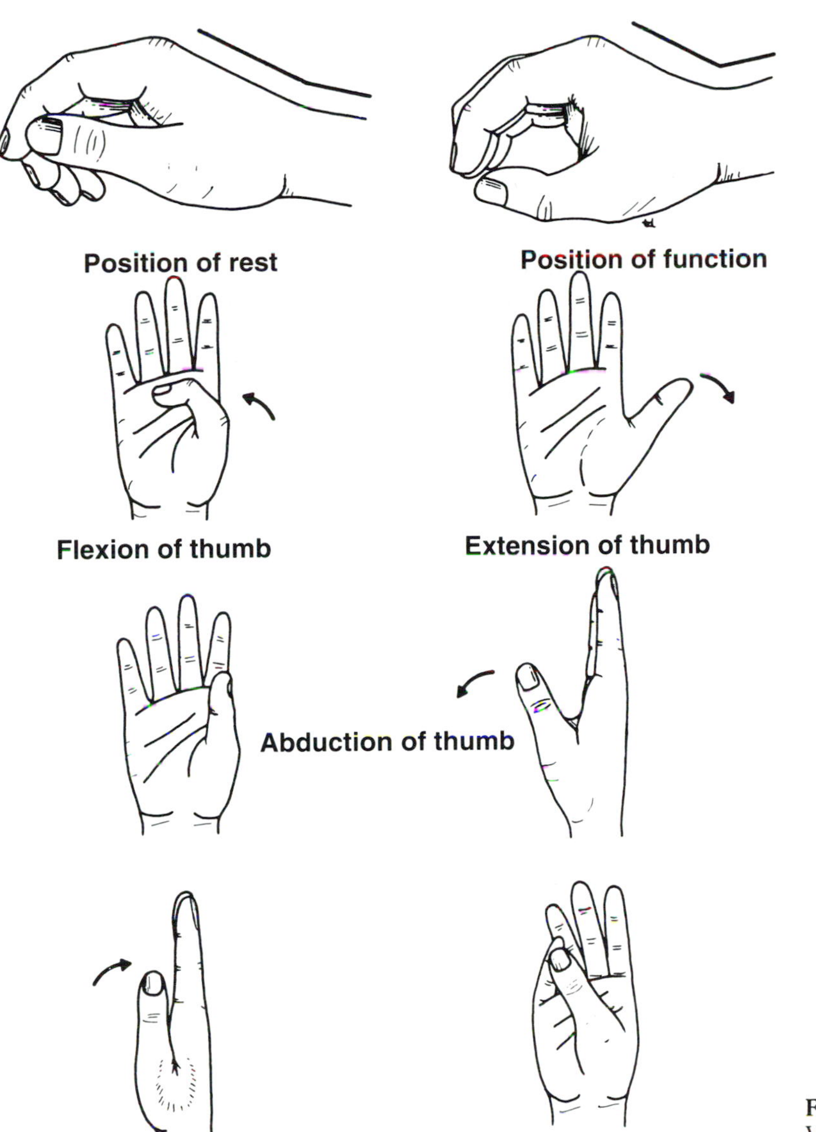

FIG 16–14.
Various positions of the hand and movements of the thumb.

licis longus and brevis and the abductor pollicis muscles produce the movement.

Adduction.—This is the movement of the thumb in an anteroposterior plane toward the palm, with the plane of the thumbnail being kept at right angles to the plane of the other fingernails (see Fig 16–14). The movement takes place between the trapezium and the first metacarpal bone. The muscle producing the movement is the adductor pollicis.

Opposition.—This is the movement of the thumb across the palm so that the anterior surface of the tip comes into contact with the anterior surface of the tip of any of the other fingers (see Fig 16–14). The movement is accomplished by the medial rotation of the first metacarpal bone and the attached phalanges on the trapezium. The plane of the thumbnail comes to lie parallel with the plane of the nail of the opposed finger. The opponens pollicis muscle produces the movement.

Movements of the Index, Middle, Ring, and Little Fingers.—These include the following.

Flexion.—This is the forward movement of the finger in an anteroposterior plane. The movement takes place at the interphalangeal and metacarpophalangeal joints. The distal phalanx is flexed by the flexor digitorum profundus, the middle phalanx by the flexor digitorum superficialis, and the proximal phalanx by the lumbricals and the interossei.

Extension.—This is the backward movement of the finger in an anteroposterior plane. The movements take place at the interphalangeal and metacarpophalangeal joints. The distal phalanx is extended by the lumbricals and interossei, the middle phalanx by the lumbricals and interossei, and the proximal phalanx by the extensor digitorum (in addition, by the extensor indicis for the index finger and the extensor digiti minimi for the little finger). However, when the metacarpophalangeal joint is flexed, which is a function of the lumbricals and interossei, extension of the proximal and distal interphalangeal joints is carried out by the extensor digitorum.

Abduction.—This is the movement of the fingers (including the middle finger) away from the imaginary midline of the middle finger (see Fig 16–10). The movement takes place at the metacarpophalangeal joint. The dorsal interossei muscles produce the movement; the abductor digiti minimi abducts the little finger.

Adduction.—This is the movement of the fingers toward the midline of the middle finger (see Fig 16–10). The movement takes place at the metacarpophalangeal joint. The palmar interossei muscles produce the movement.

Abduction and adduction of the fingers are possible only in the extended position. In the flexed position of the finger, the articular surface of the base of the proximal phalanx lies in contact with the flattened anterior surface of the head of the metacarpal bone. The two bones are held in close contact by the collateral ligaments, which are taut in this position. In the extended position of the metacarpophalangeal joint, the base of the phalanx is in contact with the rounded part of the metacarpal head, and the collateral ligaments are slack.

Basic Anatomy of the Metacarpophalangeal Joints

Description

Articulation.—Convex heads of the metacarpal bones and the concave bases of the proximal phalanges.

Type.—Synovial condyloid joints.

Capsule.—Encloses the joint.

Ligaments.—*Palmar ligaments (volar plate)* are strong and contain some fibrocartilage. *Collateral ligaments* are cordlike bands that join the head of the metacar-

Clinical Note

Splinting of fingers and shortening of metacarpophalangeal ligaments.—When possible, the metacarpophalangeal joints should be splinted in flexion to avoid shortening of the metacarpophalangeal ligaments. Once contracted, these ligaments may never be stretched back to their original length.

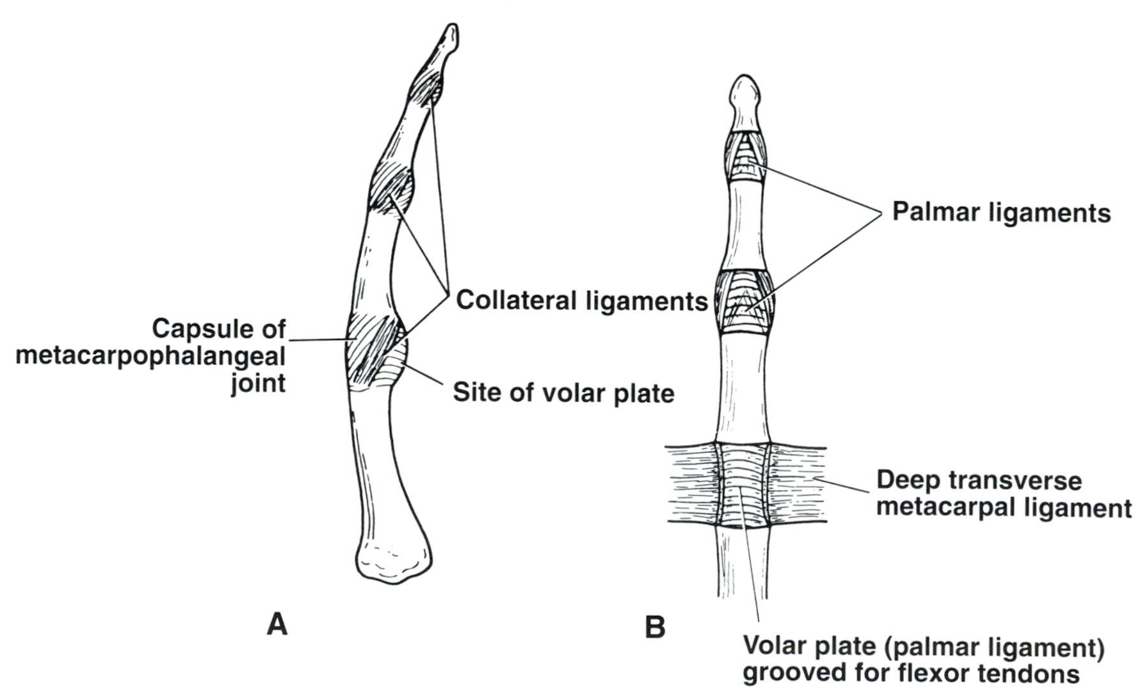

FIG 16–15.
Index finger showing the capsules and ligaments of the metacarpophalangeal and interphalangeal joints. **A,** side view. **B,** anterior view.

pal bone to the base of the phalanx (Fig 16–15). The collateral ligaments are taut when the joint is flexed and lax when the joint is extended. The fingers are thus locked in flexion but can be abducted and adducted in extension.

Synovial Membrane.—Lines the capsule.

Movements.—The following muscles produce the movements.

Flexion.—Lumbricals and interossei assisted by the flexor digitorum superficialis and profundus.

Extension.—Extensor digitorum, extensor indicis, and extensor digiti minimi.

Abduction (Movement away from the midline of the third finger).—Dorsal interossei.

Adduction (Movement toward the midline of the third finger).—Palmar interossei.

In the metacarpophalangeal joint of the thumb, *flexion* is performed by the flexor pollicis longus and brevis and *extension* is performed by the extensor pollicis longus and brevis. The movements of abduction and adduction are performed at the carpometacarpal joint.

Clinical Note

Gamekeeper's thumb.—This relatively common condition, a chronic instability of the metacarpophalangeal joint of the thumb, is caused by a tear of the ulnar collateral ligament.

Acute tear of the ulnar collateral ligament (which is not technically a gamekeeper's thumb) is an important finding in the emergency department. It is diagnosed by firmly holding and stabilizing the metacarpal bone of the thumb while the examiner abducts the proximal and distal phalanges. With a complete tear, the thumb moves 90°, and the pain is not severe. If the tear is incomplete, the remaining intact fibers of the ligament prevent more than 10° to 20° of painful movement.

Basic Anatomy of the Interphalangeal Joints

The interphalangeal joints are synovial hinge joints that have a structure similar to that of the metacarpophalangeal joints (see Fig 16–15). The interphalangeal joints have a volar plate and collateral ligaments like the metacarpophalangeal joints, and these are often torn in interphalangeal dislocations.

Basic Anatomy of the Nail

The nails are flattened horny plates situated on the dorsal surface of the distal parts of the fingers (Fig 16–16). The proximal part of the nail is called the *root;* it emerges from a groove in the skin to form the *body* of the nail, which is exposed. The root of the nail is covered by a fold in the skin called the *proxi-*

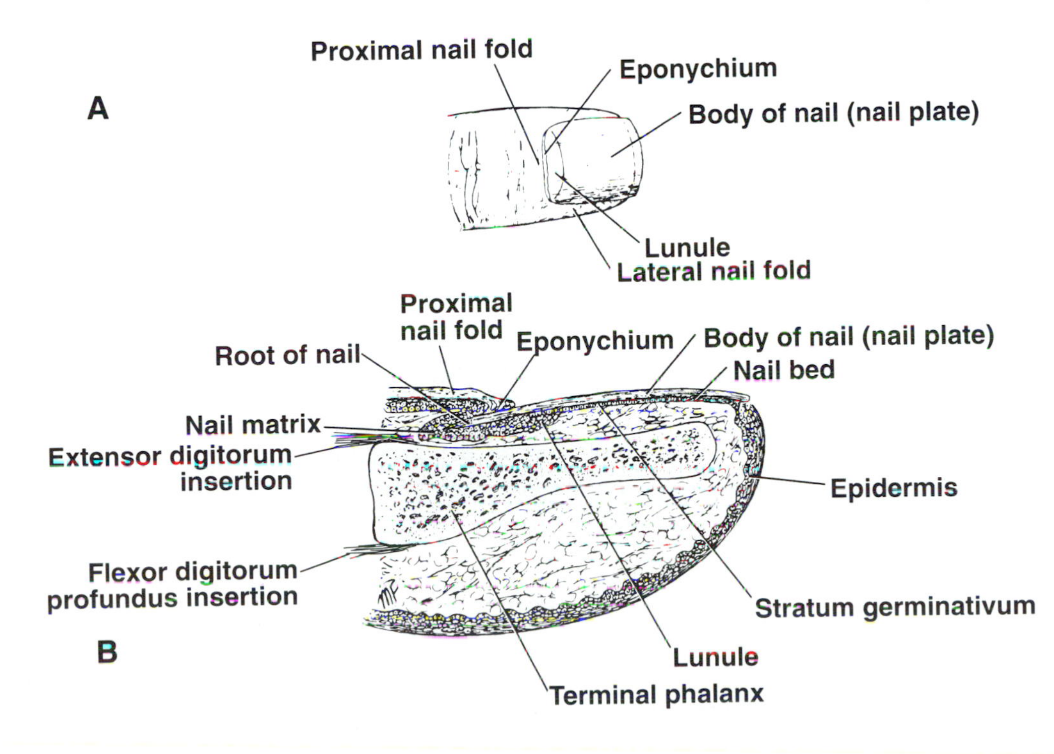

FIG 16–16.
A, nail on the dorsal surface of the finger. **B,** structure of the different parts of the nail, as seen in longitudinal section of the distal part of the finger.

mal nail fold. The stratum corneum of the epidermis of the nail fold extends out over the body of the nail for a short distance to form the *cuticle, or eponychium.* The lateral borders of the nail body are overlapped by a fold of skin called the *lateral nail fold.* The *nail bed* is the skin beneath the body of the nail, where the epidermis consists only of the stratum germinativum (basal layer of the epidermis). Under the proximal part of the body of the nail, the stratum germinativum is thickened and opaque and forms the *lunule,* which is most easily seen in the thumb. The stratum germinativum in the lunule is actively proliferative and is responsible for the growth of the nail. This region of activity is known as the *nail matrix.* As the epidermal cells are formed, they become tightly packed keratinized cells that do not desquamate. With continued growth of the nail, the body slides distally over the nail bed. The nail bed does *not* contribute to the nail but provides a smooth surface for the nail to glide over. Fingernails grow at the rate of 0.5 to 1.2 mm per week (toenails grow much more slowly).

In the nail bed, the dermis is bound down to the underlying periosteum of the phalanx. The dermal papillae run in the long axis of the nail and are highly vascular and produce the pink color seen through the translucent nail. The lunule is white because the dermal papillae in this region are less vascular and the stratum germinativum is thick and opaque.

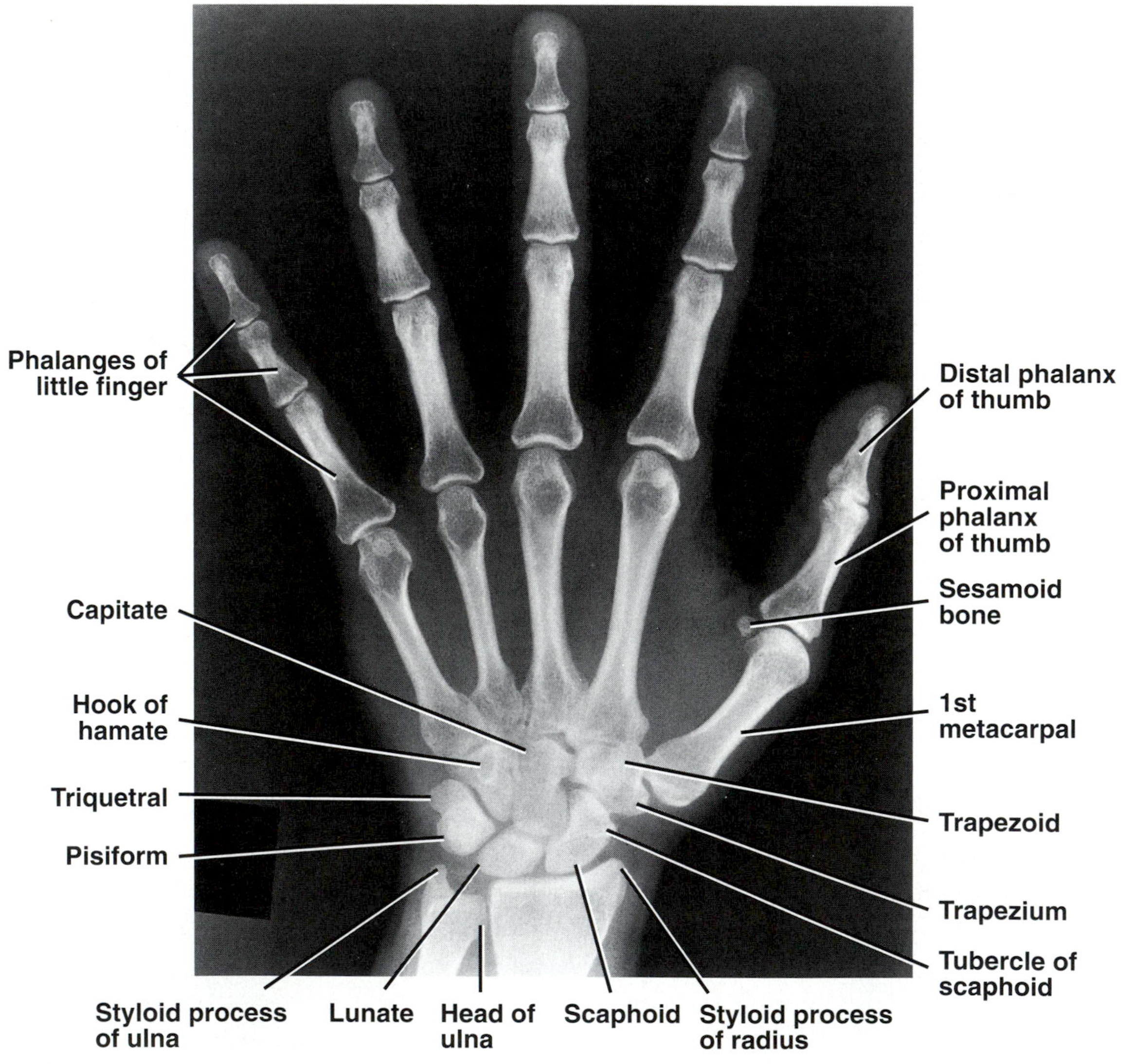

FIG 16–17.
Posteroanterior radiograph of the wrist and hand with the forearm pronated.

Radiographic Appearances of the Wrist and Hand
The views commonly used are posteroanterior and lateral; oblique views are often used for the scaphoid bone.

Posteroanterior View.—This is taken with the forearm pronated and the fingers partially flexed. The film cassette is placed against the palm of the hand, and the X-ray tube is directed onto the dorsal surface of the hand. The lower ends of the radius and ulna, with their styloid processes, can be seen, and the radial styloid process is seen to extend farther distally than that of the ulna (Figs 16–17 through 16–19). The proximal row of carpal bones is seen with the pisiform bone superimposed on the triquetral bone. The distal row of carpal bones is also seen, and the hook of the hamate can be visualized as a small oval area of increased density. The joint spaces of the carpal, wrist, and inferior radioulnar joints can be seen. The space between the scaphoid and the lunate should not exceed 3 mm. Any space larger than 3 mm between the scaphoid and the lunate is indicative of a scapholunate dislocation.

The different parts of the metacarpal bones and phalanges may also be seen. The sesamoid bones of the abductor pollicis brevis and flexor pollicis brevis tendons and the tendons of the adductor pollicis and the first palmar interosseous muscle can usually be recognized. The sesamoid bones overlap the first metacarpophalangeal joint.

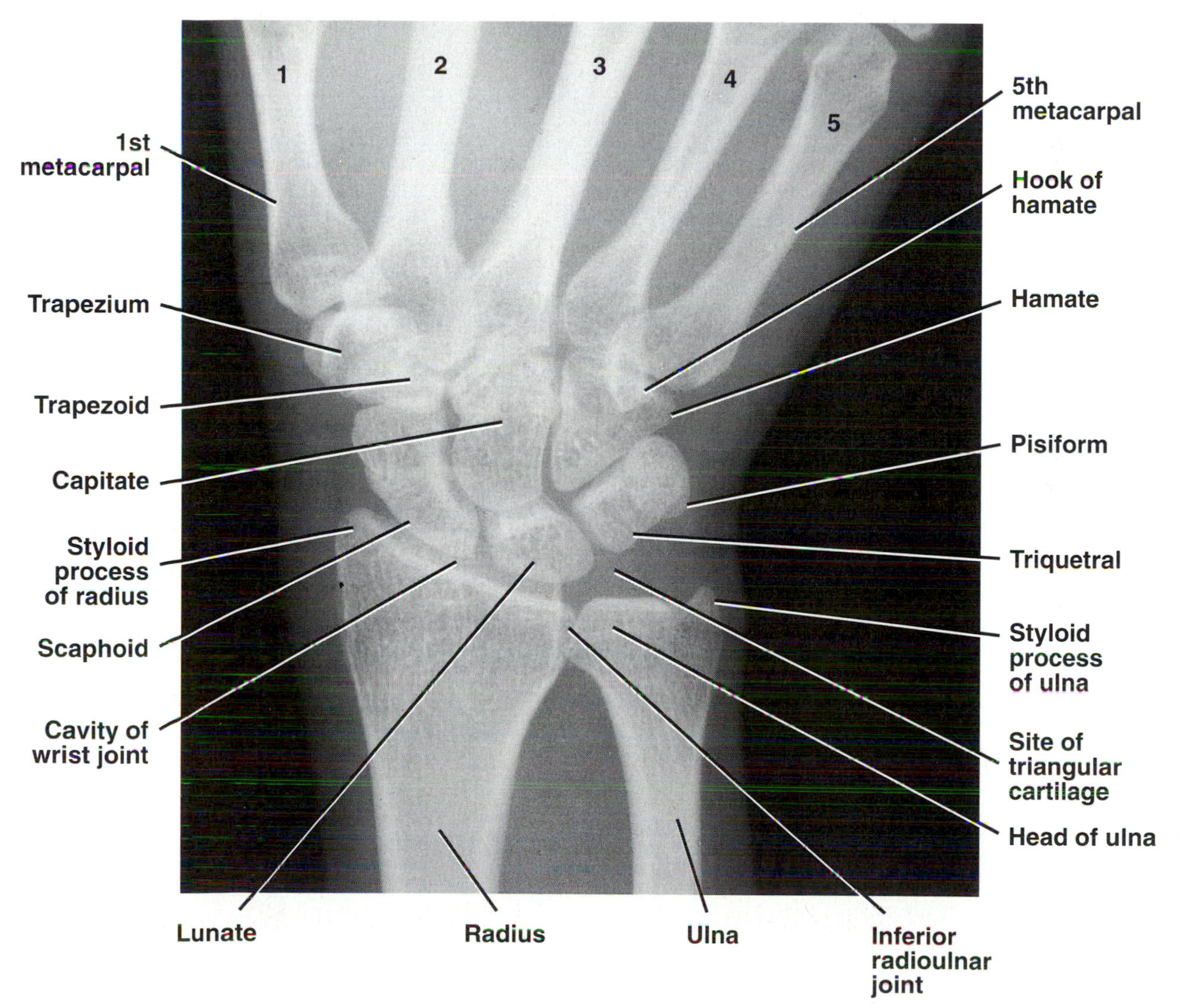

FIG 16–18.
Posteroanterior radiograph of the wrist with the forearm pronated.

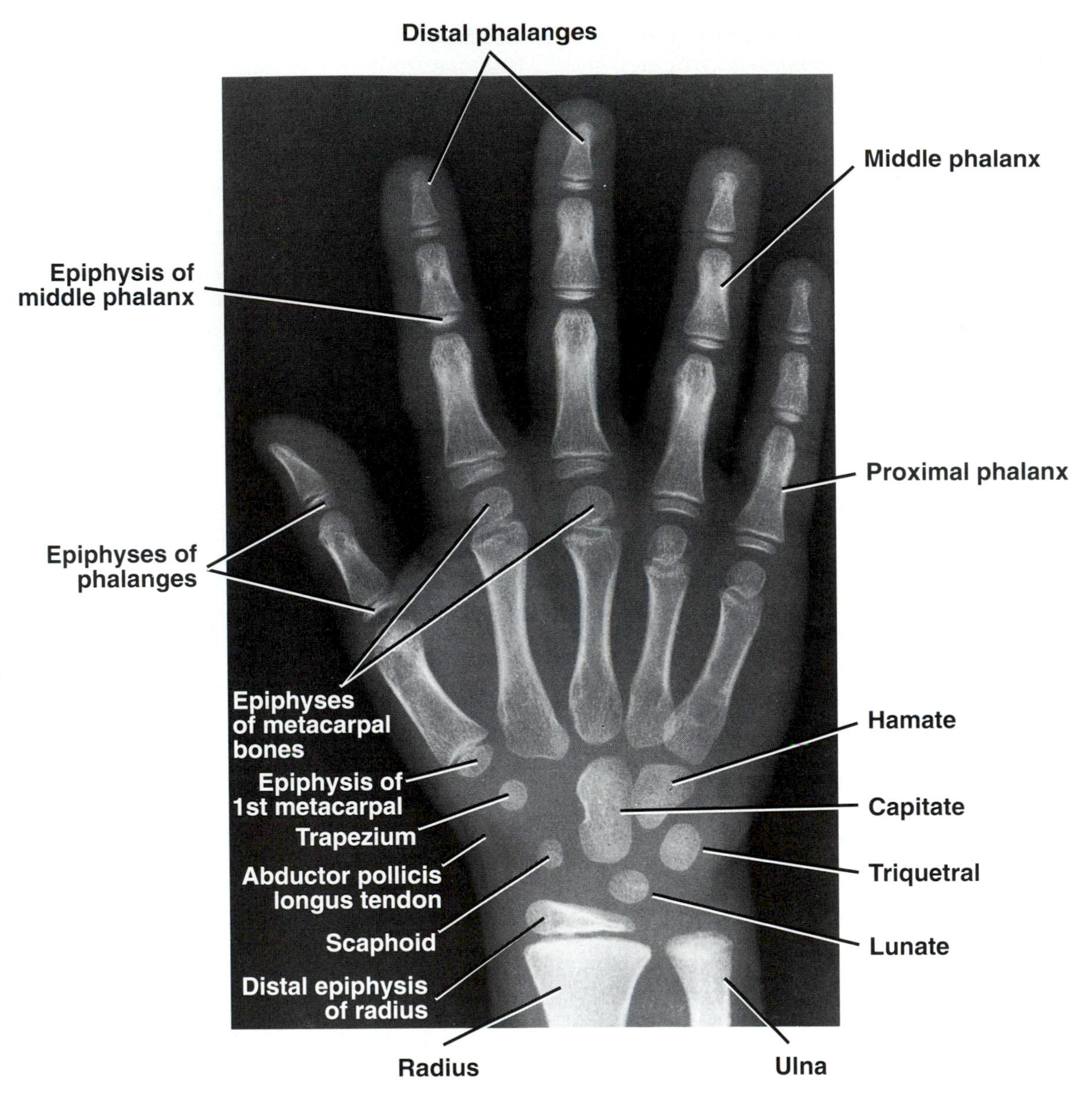

FIG 16–19.
Posteroanterior radiograph of the wrist and hand of an 8-year-old boy.

Lateral View.—This is taken with the forearm in the semiprone position. The film cassette is placed against the ulnar border of the hand, and the X-ray tube is directed through the carpus (Figs 16–20 and 16–21). The articulation of the radius with the lunate is well shown on this view. The midplane of the carpus lies at an angle of 15° volar to the long axis of the forearm. The concave distal surface of the lunate articulating with the capitate is also seen. The pisiform bone is visualized anteriorly and may overlap the scaphoid bone.

The scaphoid is the most frequently fractured bone of the wrist. Because of the difficulty experienced in visualizing its midsection radiographically, oblique views of the carpus are sometimes necessary (Figs 16–22 and 16–23). Triquetral fractures are best seen on a lateral radiograph.

SURFACE ANATOMY

Styloid Processes of the Radius and Ulna

These can be palpated at the wrist (Fig 16–24). The styloid process of the radius lies about ¾ in. (1.9 cm) distal to that of the ulna.

Dorsal Tubercle of the Radius (Lister's Tubercle)

This is palpable on the posterior surface of the lower end of the radius (see Fig 16–24). It is in line with

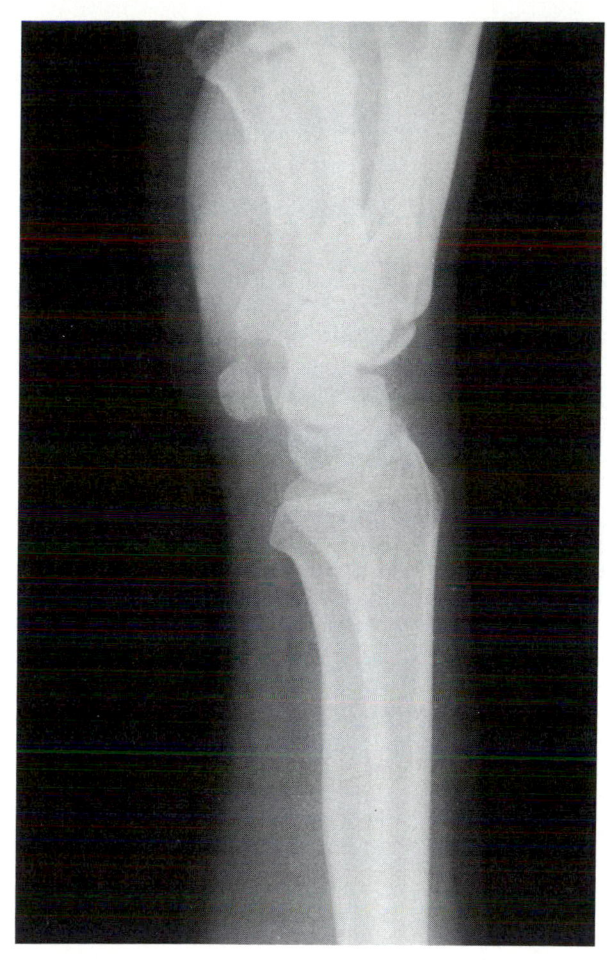

FIG 16–20.
Lateral radiograph of the wrist. Note the normal anterior angulation of the distal end of the radius on the shaft.

the cleft between the index and middle fingers. Just distal and ulnar to Lister's tubercle is the skin depression over the lunate bone and the site for wrist arthrocentesis. Just radial to Lister's tubercle pass the extensor carpi radialis brevis and longus tendons.

Head of the Ulna

This is most easily felt with the forearm pronated; the head then stands out prominently on the ulnar side of the wrist (see Fig 16–24). The rounded head can be distinguished from the more distal pointed styloid process.

Pisiform Bone

This can be felt on the ulnar side of the anterior aspect of the wrist between the two transverse creases (see Fig 16–24).

Hook of the Hamate

This can be felt on deep palpation of the hypothenar eminence, a fingerbreadth distal and radial to the pisiform bone.

Transverse Creases at the Wrist

These creases seen in front of the wrist are important landmarks (Figs 16–25 and 16–26). The proximal transverse crease lies at the level of the wrist joint. The distal transverse crease corresponds to the proximal border of the flexor retinaculum.

Important Structures Lying in Front of the Wrist

Radial Artery.—The pulsations of this artery can easily be felt anterior to the distal third of the radius, where it lies just beneath the skin and fascia between the tendons of brachioradialis and flexor carpi radialis muscles (see Fig 16–24).

Tendon of Flexor Carpi Radialis.—This lies medial to the pulsating radial artery (see Fig 16–24).

Tendon of the Palmaris Longus (If Present).—This tendon lies medial to the tendon of flexor carpi radialis and overlies the *median nerve* (see Figs 16–24 and 16–25).

Tendons of Flexor Digitorum Superficialis.—This group of four tendons lies medial to the tendon of palmaris longus and can be seen moving beneath the skin when the fingers are flexed and extended (see Fig 16–24).

Tendon of Flexor Carpi Ulnaris.—This is the most medially placed tendon on the front of the wrist and can be followed distally to its insertion on the pisiform bone (see Fig 16–24). The tendon can be made prominent by asking the patient to clench his fist or spread his fingers (the muscle contracts to assist in fixing and stabilizing the wrist joint).

Ulnar Artery.—The pulsations of this artery can be felt just radial to the tendon of flexor carpi ulnaris (Fig 16–24).

Ulnar Nerve.—The ulnar nerve lies immediately medial to the ulnar artery (see Figs 16–24 and 16–25).

Important Structures Lying on the Lateral Side of the Wrist

"Anatomical Snuffbox."—This important area is a skin depression that lies distal to the styloid process

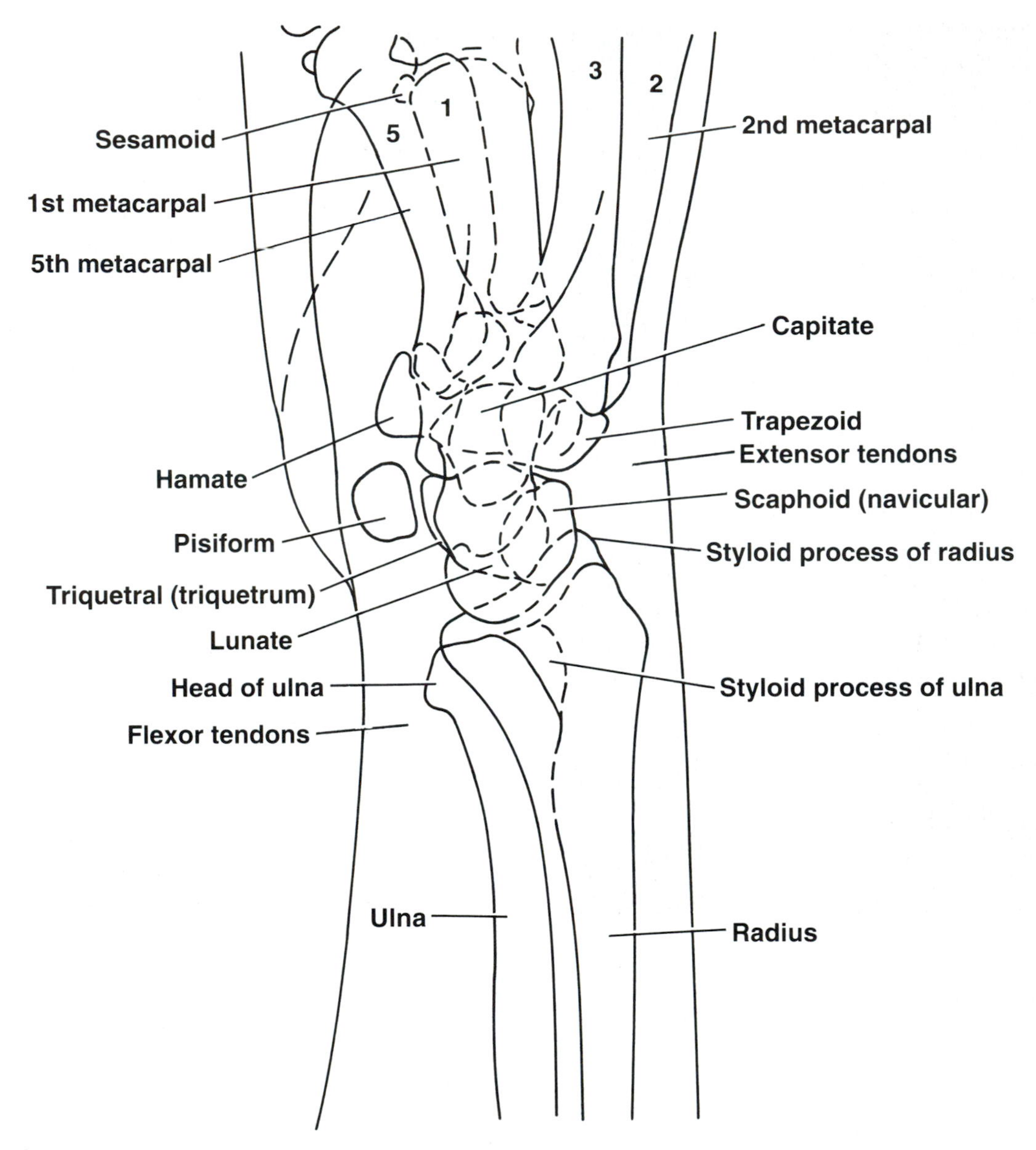

FIG 16–21.
Diagram showing the main features seen in the lateral radiograph of the wrist seen in Figure 16–20.

of the radius (see Fig 16–24). It is bounded ulnarly by the *tendon of extensor pollicis longus* and radially by the *tendons of abductor pollicis longus* and *extensor pollicis brevis.* In its floor may be palpated the *styloid process of the radius* (proximally) and the *base of the first metacarpal bone* of the thumb (distally); between these bones beneath the floor lie the *scaphoid* and the *trapezium* (felt but not identifiable). The *radial artery* may be palpated within the snuffbox as the artery winds around the lateral margin of the wrist to reach the dorsum of the hand (see Fig 16–13). The *cephalic vein* can also sometimes be recognized crossing the snuffbox as it ascends the forearm.

Important Structures Lying on the Back of the Wrist

Tendons of Extensor Carpi Radialis Longus and Brevis.—These can be felt if the fist is tightly clenched; they can be traced distally to their insertions into the bases of the second and third metacarpal bones, respectively.

Lunate.—This lies in the proximal row of carpal bones. It can be palpated just distal to the dorsal tubercle of the radius when the wrist joint is flexed.

Important Structures Lying on the Dorsum of the Hand

Tendons of Extensor Digitorum, the Extensor Indicis, and the Extensor Digiti Minimi.—These can be

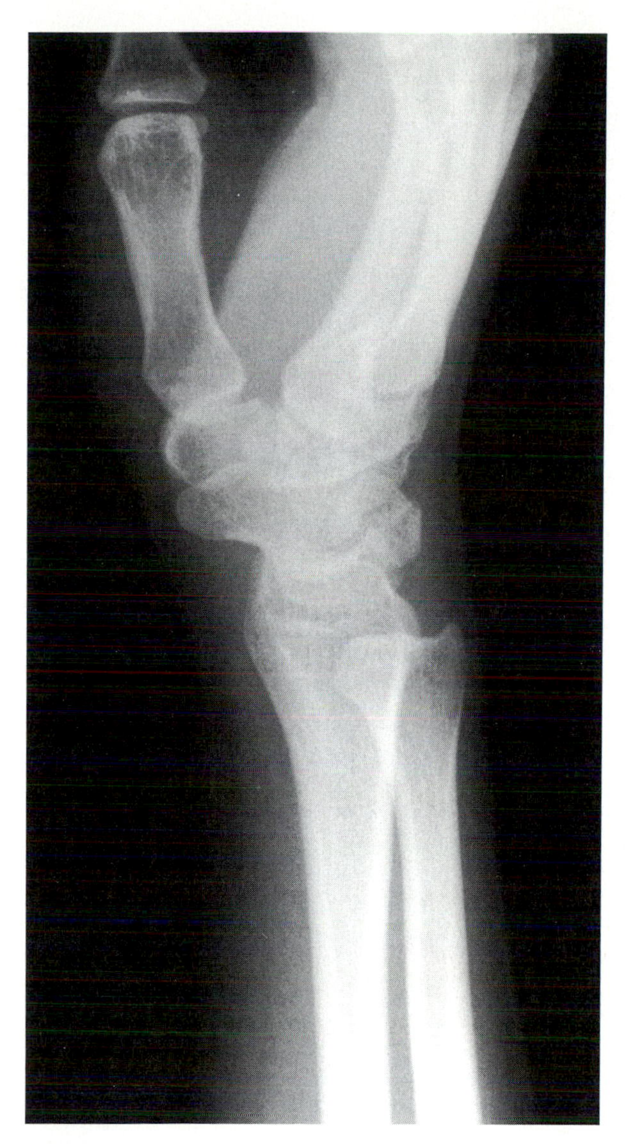

FIG 16–22.
Oblique radiograph of the wrist.

seen and felt as they pass distally to the bases of the fingers (see Fig 16–24).

Dorsal Venous Network of Superficial Veins.—These can be seen on the dorsum of the hand. The network drains upward into a radial cephalic vein and a ulnar basilic vein. The *cephalic vein* crosses the anatomical snuffbox and winds around onto the anterior aspect of the forearm. The *basilic vein* can be traced from the dorsum of the hand around the medial side of the forearm and reaches the anterior aspect just below the elbow.

Skin Creases.—The relationship of skin creases and the underlying joints are shown in Figure 16–26.

CLINICAL ANATOMY

Clinical Anatomy of the Region of the Wrist

Stenosing Tenosynovitis (de Quervain's Disease)

The abductor pollicis longus and the extensor pollicis brevis tendons share a common synovial sheath as they pass beneath the extensor retinaculum on the posterior surface of the distal end of the radius. Excessive use of the wrist or thumb can cause inflammation of the tendons and their synovial sheath. This produces localized pain over the lateral aspect of the wrist, which is made worse by grasping objects or clenching the fist. On examination, pain can be elicited in the region of the swollen synovial sheath by forced ulnar deviation of the wrist with the thumb adducted and flexed (Finkelstein's test).

Carpal Tunnel Syndrome

The carpal tunnel, formed by the concave anterior surface of the carpal bones and closed by the flexor retinaculum, is tightly packed with the long flexor tendons of the fingers, with their surrounding synovial sheaths, and the median nerve. Clinically, the syndrome consists of a burning pain or a "pins and needles" feeling along the distribution of the median nerve to the radial 3½ fingers and weakness of the thenar muscles. Occasionally, sensory symptoms are referred up the forearm.

Carpal tunnel syndrome is produced by compression of the median nerve within the tunnel. As expected, there is no paresthesia over the thenar eminence, since this area of skin is supplied by the palmar cutaneous branch of the median nerve, which passes superficially to the flexor retinaculum. The symptoms of carpal tunnel syndrome are often exacerbated by percussion over the median nerve as it passes underneath the flexor retinaculum just radial to the palmaris longus tendon (Tinel's sign). Phalen's test, in which the wrist joint is hyperextended, also reproduces the symptoms.

Carpal tunnel syndrome is dramatically relieved by decompressing the tunnel through a longitudinal incision through the flexor retinaculum.

Guyon's Canal Syndrome

As the ulnar nerve and artery enter the palm anterior to the flexor retinaculum, they pass between the pisiform bone and the hook of the hamate. Here a superficial part of the flexor retinaculum passes between the two bones anterior to the nerve and the artery, creating a fibro-osseous canal called the *tun-*

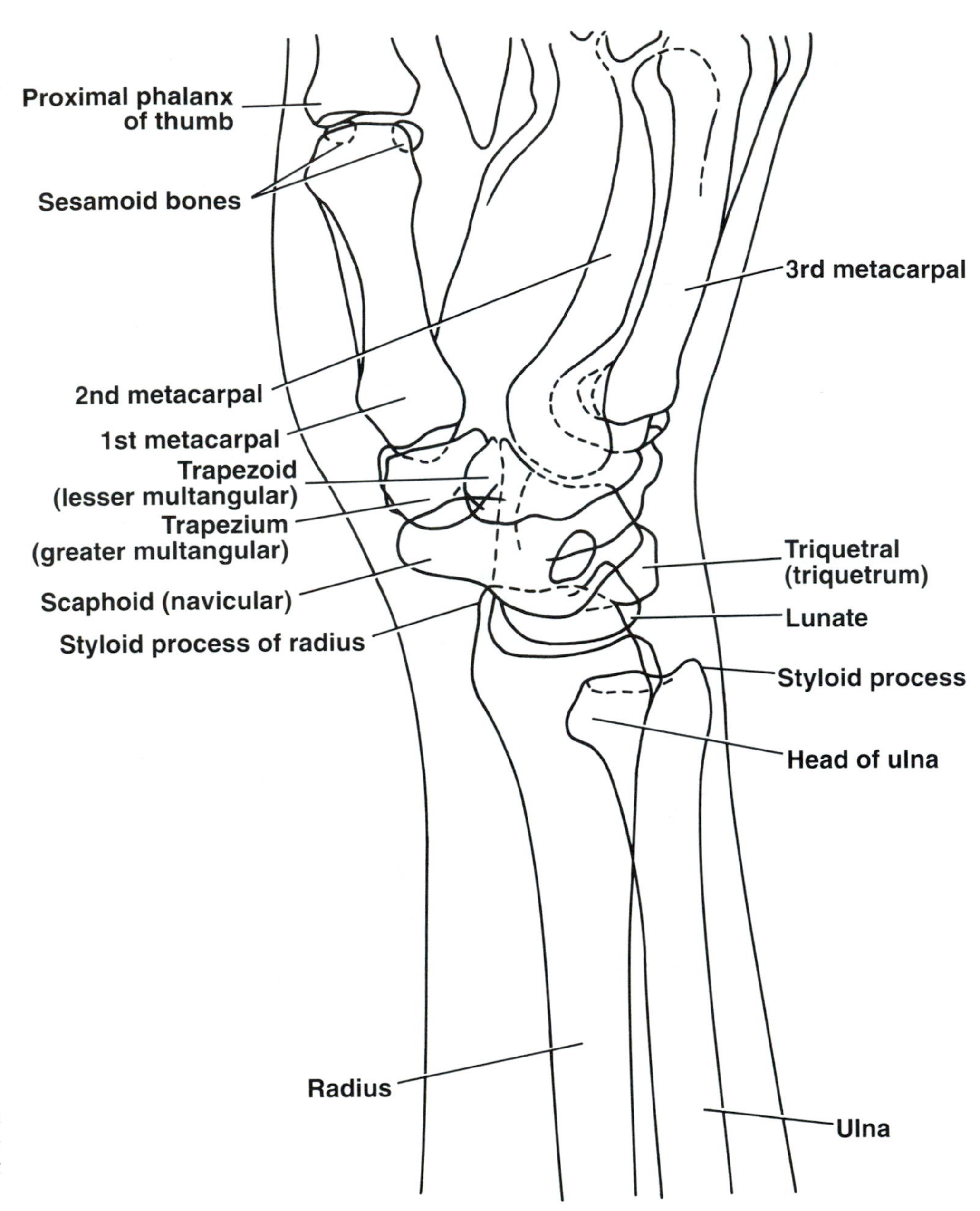

FIG 16–23.
Diagram showing the main features seen in the oblique radiograph of the wrist seen in Figure 16–22.

nel of Guyon. This is the common site for compression of the ulnar nerve brought about by repetitive external trauma. The patient complains of impairment of sensation to the palmar surface of the ulnar 1½ fingers and to the dorsal aspects of the middle and distal phalanges of the same fingers. There is usually no impairment of motor activity. On examination, the ulnar nerve is often tender on deep pressure at the point where it lies within the tunnel.

Injuries to the Wrist

Open injuries to the wrist may involve tendons, nerves, and blood vessels.

Tendon Injury.—The flexor and extensor tendons are protected to some extent by the presence of the flexor and extensor retinacula, but severe incisive trauma can produce partial or complete tendon lacerations. In order to make a correct diagnosis as to the extent of the tendon injury, the position of the tendons at the time of injury must be known. A lacerated tendon is diagnosed in two ways—by inspection of the fingers to see if they are all in their correct "at rest" position and by systematically testing the strength of flexion and extension against resistance at each of the three joints of each finger.

On the anterior surface of the wrist there are 12

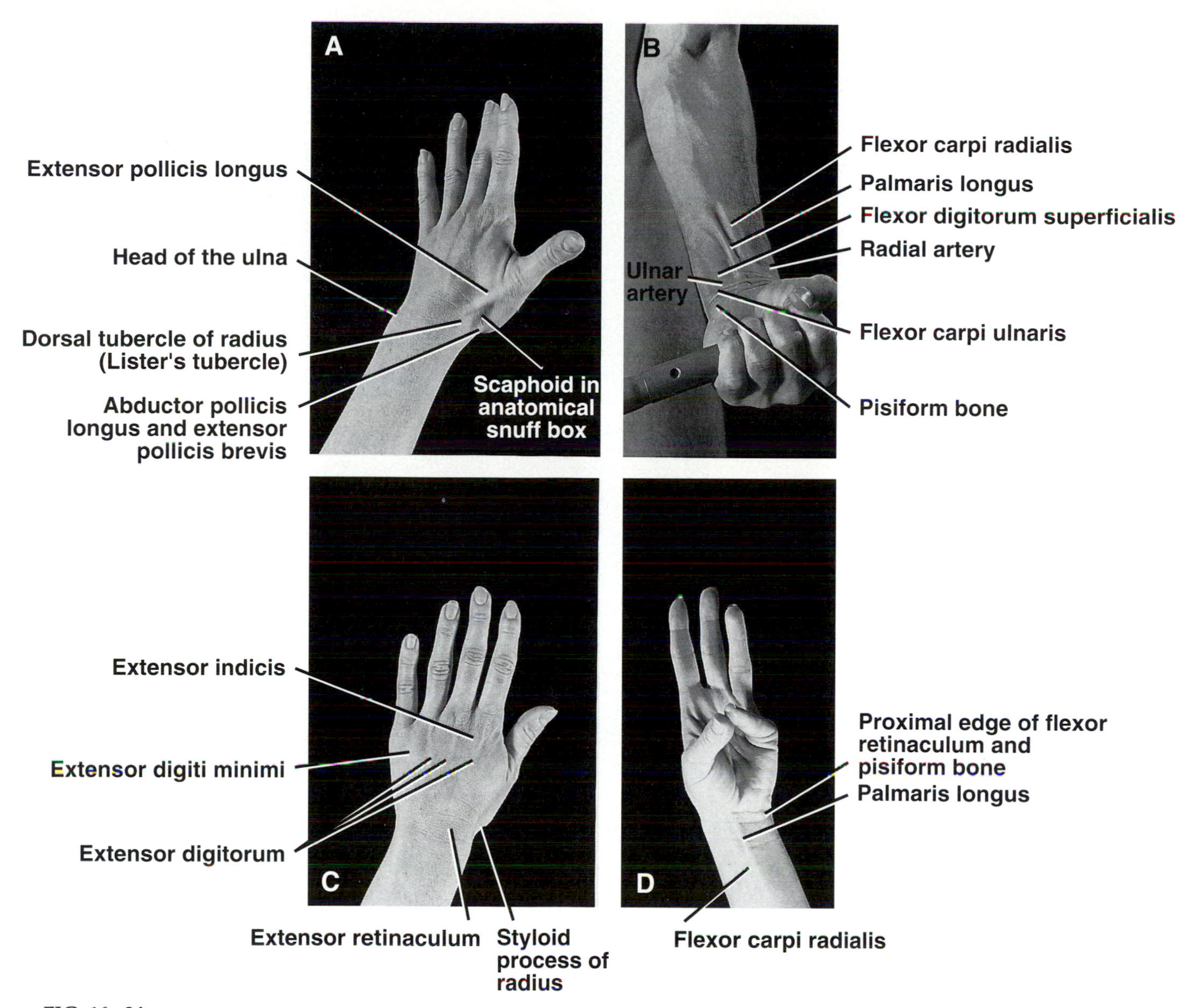

FIG 16–24.
Some important surface landmarks of the wrist and hand. Note in **D** the movement of opposition of the thumb.

flexor tendons—3 to the wrist, 2 to each finger, and 1 to the thumb. The position of the tendons are shown in Figures 16–5 and 16–6, and their functions are summarized in Table 16–1. Remember that the primary action of the superficialis tendon is flexion of the middle phalanx of the finger and that of the profundus tendon is flexion of the terminal phalanx. If this is not remembered, some confusion may occur when the wound involves only one of these tendons.

On the posterior surface of the wrist there are 12 extensor tendons—3 to the wrist, 1 to each finger with an additional extensor indicis passing to the index finger and the extensor digiti minimi to the little finger, and 3 to the thumb. The position of the tendons are shown in Figure 16–13, and their functions are summarized in Tables 16–2 and 16–3. Remember that the primary action of the extensor digitorum muscle is extension of the metacarpal phalangeal joint and that primary extension of the interphalangeal joints is produced by the interossei and lumbrical muscles. Also, the bands of fibrous tissue connect the extensor tendons to each other on the dorsum of the hand; therefore, some weak active extension of a metacarpophalangeal joint may occur in the presence of its severed extensor digitorum tendon, so long as its companion tendons remain intact.

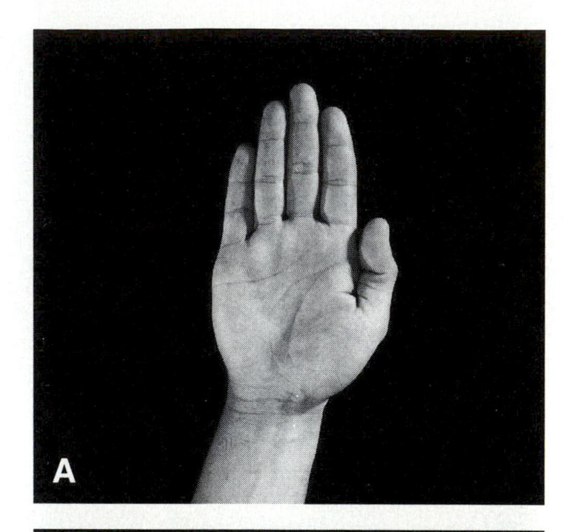

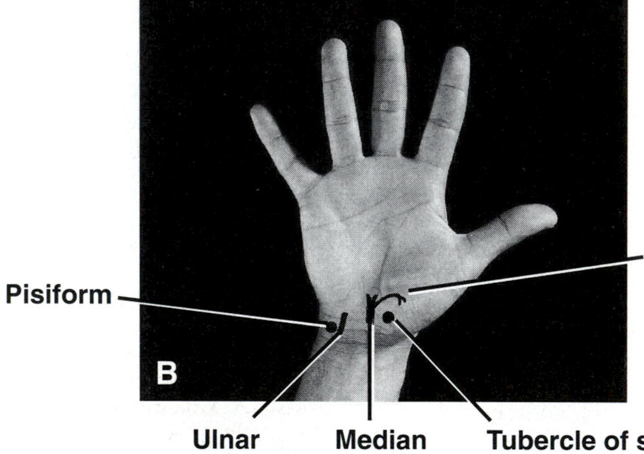

FIG 16–25.
A, adduction of the fingers produced by the palmar interossei. **B,** abduction of the fingers produced by the dorsal interossei. The surface landmark for the ulnar nerve is on the radial side of the pisiform bone; also, the surface landmark for the motor branch of the median nerve to the muscles of the thenar eminence is about a fingerbreadth distal to the tubercle of the scaphoid bone.

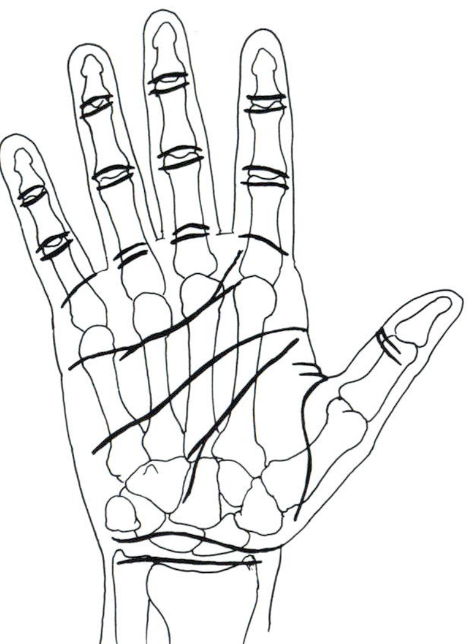
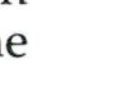

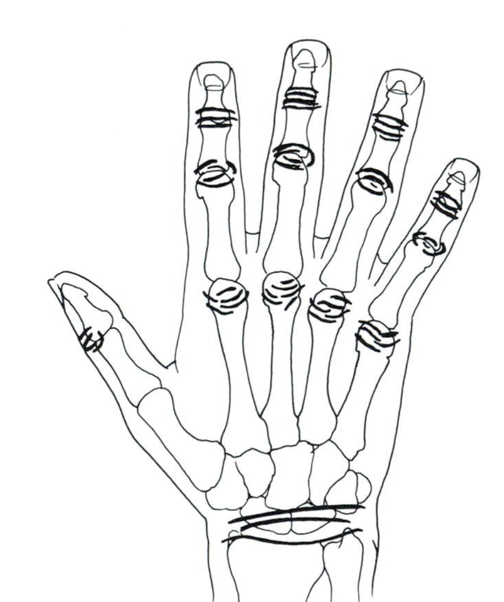

FIG 16–26.
The relationships of the main skin creases of the hand and fingers to the underlying bones.

Nerve Injuries.— These are described on p. 660.

Arterial Injuries.— Severed radial and ulnar arteries rarely spout blood in the emergency department because the arteries usually rapidly contract and retract after injury, and major bleeding usually ceases. The circulation of the hand can be maintained with only one of the two arteries intact.

Intra-arterial drug injection may result in complete arterial occlusion and destruction of the hand. A white, painful, swollen hand in the absence of a surface wound is strongly suggestive of an intra-arterial drug injection.

Fractures of the Distal End of the Radius.— These include the following.

Colles' Fracture.—This results from falling on an outstretched hand and commonly occurs in patients over the age of 50 years. The force drives the distal fragment of the radius posteriorly and superiorly, and the distal articular surface is inclined posteriorly (Fig 16–27). This posterior displacement produces a posterior bump, sometimes referred to as the "dinner-fork deformity," since the forearm and wrist resemble the shape of a dinner fork.

There are three independent components in a Colles' fracture— dorsal angulation of distal radial fragment with angulation of distal articular surface, dorsal displacement of bone fragment, and shortening of wrist region as the result of dorsal angulation and displacement. The ulnar styloid process also may sometimes be fractured.

On examination it is important to exclude the possibility of damage to the median nerve. When the fracture is being reduced, the styloid process of the radius has to lie ¾ in. (1.9 cm) distal to the ulnar styloid. Failure to restore the distal articular surface

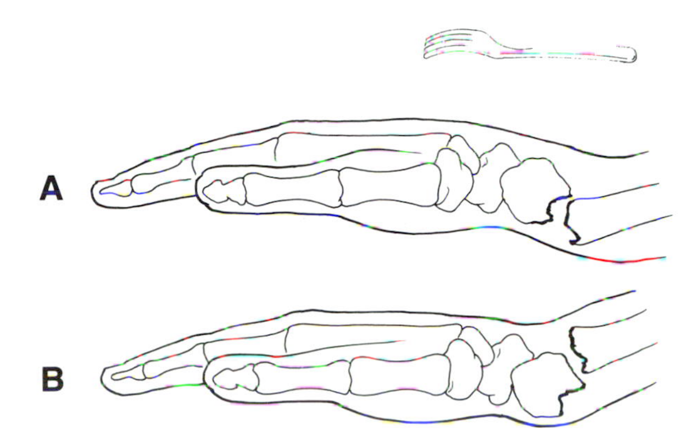

FIG 16–27.
Fractures of the distal end of the radius. **A,** Colles' fracture. **B,** Smith's fracture.

to its normal position will severely limit the range of flexion of the wrist joint. The normal articular surface of the radius is angled volarly 15°.

Smith's Fracture.—This is a fracture of the distal end of the radius, occurring from a fall on the back of the hand. It is a "reversed Colles' fracture" because the distal fragment is displaced anteriorly (see Fig 16–27).

Injuries to the Wrist Joint

A fall on an outstretched hand may strain the anterior ligament of the wrist joint, producing synovial effusion, joint pain, and limitation of movement. These symptoms and signs must not be confused with those produced by a fractured scaphoid or dislocation of the lunate bone, which are very similar.

Clinical Anatomy of the Hand

General Principles

The hand is one of the most important parts of the body. From the purely mechanical point of view, the hand may be regarded as a pincerlike mechanism between the thumb and fingers, situated at the end of a multijointed lever. The most important part of the hand is the thumb (Fig 16–28), and its function should be preserved as much as possible so that the pincerlike mechanism can be maintained. The pincerlike action of the thumb largely depends on its unique ability to be drawn across the palm and opposed to the other fingers (see Fig 16–24). Unfortunately, this movement alone, although important, is insufficient for the mechanism to work effectively. The opposing skin surfaces must have tactile sensation, which explains why median nerve palsy is so much more disabling than ulnar nerve palsy.

The importance of immobilizing the hand in the position of function has already been mentioned. Should the patient suffer loss of the movement of the joints of the hand or fingers as a result of immobilization, at least the patient will have a hand that can still serve a useful purpose.

Also, when the fingers (excluding the thumb) are flexed individually into the palm, they all point toward the tubercle of the scaphoid. This fact helps diagnose rotational or angular deformities after an injury. A rotational deformity is one in which the plane of the affected nail is not parallel to the plane of the other nails; the deformity may exacerbate with flexion. In an angular deformity, the distal part of the finger is canted ulnarly or radially; angular deformities may interfere with the making of a fist. An

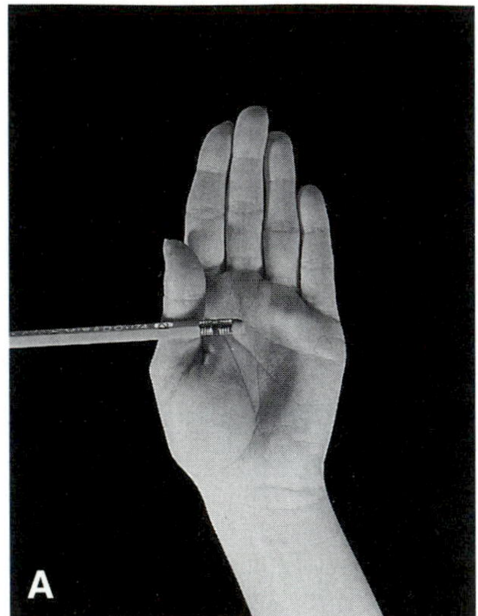

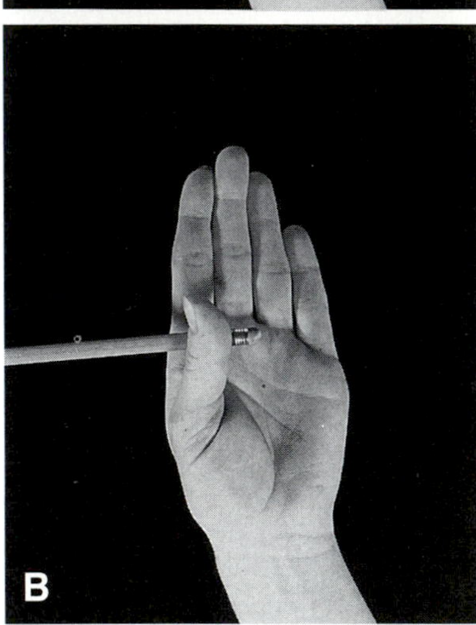

FIG 16–28.
Testing for abduction **(A)** and adduction **(B)** movements of the thumb.

angular deformity of the pinky finger may interfere with placing one's hands in one's pocket.

Skin

Palmar Skin.—Since the palmar skin is tightly bound down to the underlying palmar aponeurosis, a massive force is required to avulse the skin. For the same reason swelling of the palmar skin and subcutaneous tissue is limited. Lacerations across flexion creases almost invariably result in contracture, which is due to the fixation of the skin to the fibrous aponeurosis and its subsequent contracture with wound healing. Because the skin and underlying aponeurosis are relatively immobile, adherence of the long flexor tendons to the deep surface will result in severe loss of tendon movement.

Dorsal Skin.—The dorsal skin is loosely attached by areolar tissue to the extensor tendons and metacarpal bones. Severe avulsion injuries of the skin from relatively minor trauma are common. Because of this looseness of the skin and fascia, inflammatory disease in the palm is first seen as a swelling on the dorsum of the hand. Consequently, dorsal swelling does not necessarily mean a dorsal infection. Lacerations of the skin are not followed by contracture because the skin is mobile and can adapt.

The adherence of the skin to the underlying extensor tendons is not followed by impairment of function, since the tendons have short excursions and the skin is mobile.

Trauma

Lacerations.—The treatment of lacerations of the hand will not be discussed here. However, there are certain anatomical principles that should be noted.

1. The hand has little expendable tissue and therefore anything that might be viable for future function should be saved.
2. The pliable skin on the back of the hand can be readily mobilized to close a wound, whereas the palmar skin is tethered to the palmar aponeurosis and difficult to mobilize.
3. Bone fragments, even if dirty and completely detached from soft tissue, can be cleaned and returned to their correct anatomical position.
4. Nerves and tendons, even though they are badly traumatized and dirty, are often salvageable and should not be removed.

Nerve Injuries.—These include the median, ulnar, and superficial radial nerves.

Median Nerve.—Injuries include the following.

Injuries to the Median Nerve at the Wrist.—The median nerve is most commonly injured by stab wounds or broken glass just proximal to the flexor retinaculum, where it lies in the interval between the tendons of the flexor carpi radialis and flexor digitorum superficialis, overlapped by the palmaris longus (see Fig 16–8).

The clinical findings in injury to the median nerve are as follows.

1. *Motor changes.*—The muscles of the thenar eminence are paralyzed and later become wasted so that the eminence becomes flattened or scooped out (Fig 16–29). The thumb becomes laterally rotated so that its volar pad is more in the same plane as the

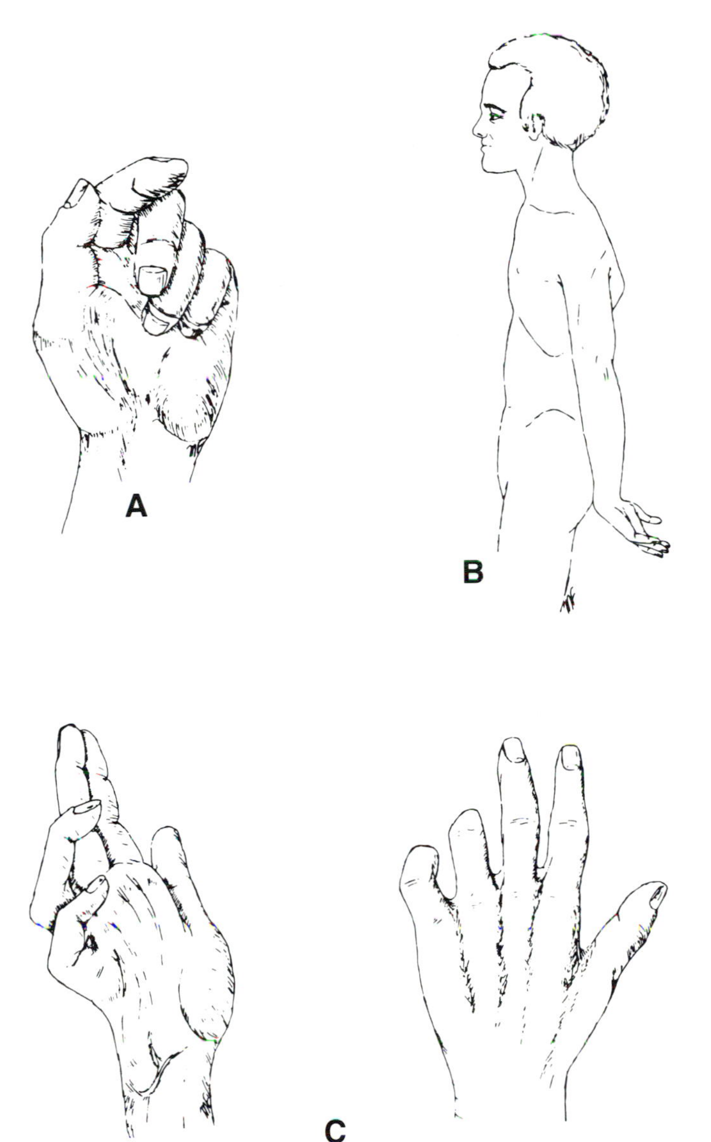

FIG 16–29.
A, median nerve palsy. **B,** Erb-Duchenne palsy (Waiter's tip). **C,** ulnar nerve palsy.

volar pads of the other fingers (loss of opponens pollicis); the thumb is also adducted (loss of abductor pollicis brevis). With the passage of time the hand looks flattened and "apelike." Opposition movement of the thumb is impossible (loss of opponens pollicis). The first two lumbricals are paralyzed, which can be recognized clinically when the patient is asked to make a fist slowly, and the index and middle fingers tend to lag behind the ring and little fingers. The index and middle fingers will eventually flex at the metacarpophalangeal joints even in the absence of lumbrical action because of the intact function of the flexor digitorum superficialis and the flexor digitorum profundus.

2. *Sensory.*—There is loss of skin sensation of the radial half or less of the palm of the hand (palmar cutaneous branch of median nerve) and the palmar aspect of the radial 3½ fingers. There is also sensory loss of the skin of the distal parts of the dorsal surfaces of the radial 3½ fingers (see Fig 16–3). The area of total anesthesia is considerably less because of the overlap of adjacent nerves.

3. *Vasomotor changes.*—The skin areas involved in sensory loss are warmer and drier than normal because of the arteriolar dilatation and absence of sweating resulting from loss of sympathetic control.

4. *Trophic changes.*—In long-standing cases, changes are found in the hand and fingers. The skin is dry and scaly, the nails crack easily, and there is atrophy of the pulp of the fingers.

Stab Wound Injury to the Motor Branch of the Median Nerve to the Thenar Eminence.—Stab wounds to the proximal part of the palm at about a fingerbreadth directly distal to the tubercle of the scaphoid bone commonly cut the motor nerve supply to the muscles of the thenar eminence, leaving the sensory innervation intact.

The Median Nerve and Carpal Tunnel Syndrome.—The clinical syndrome is fully described on p. 655.

Ulnar Nerve.—Injuries include the following.

Injuries to the Ulnar Nerve at the Wrist.—The superficial position of the ulnar nerve at the wrist (anterior to the flexor retinaculum) makes it very vulnerable to damage from lacerations (see Fig 16–8). The clinical findings in injury to the ulnar nerve are as follows.

1. *Motor.*—The small muscles of the hand will be paralyzed, except the muscles of the thenar eminence and the first two lumbricals, which are supplied by the median nerve. The patient is unable to adduct and abduct the fingers (loss of interossei) and consequently is unable to grip a piece of paper placed between the fingers (palmar interossei). Remember that the extensor digitorum can abduct the fingers to a small extent, but only when the metacarpophalangeal joints are hyperextended. It is impossible to adduct the thumb because the adductor pollicis muscle is paralyzed. If the patient is asked to grip a piece of paper between the thumb and index finger, he does so by strongly contracting his flexor pollicis longus and flexing the terminal phalanx *(Froment's sign).* The metacarpophalangeal joints become hyperextended due to paralysis of the lumbrical and interosseous muscles, which normally flex these

joints. Since the first and second lumbricals are not paralyzed (they are supplied by the median nerve), the hyperextension of the metacarpophalangeal joints is most prominent in the fourth and fifth fingers. The interphalangeal joints are flexed, again due to the paralysis of the lumbrical and interosseous muscles, which normally extend these joints through the extensor expansion. The flexion deformity at the interphalangeal joints of the fourth and fifth fingers is more obvious because the first and second lumbricals of the index and middle fingers are not paralyzed. In long-standing cases the hand assumes the characteristic "claw" deformity ("main en griffe"). Wasting of the paralyzed muscles results in flattening of the hypothenar eminence and loss of the convex curve to the medial border of the hand. Examination of the dorsum of the hand will show hollowing between the metacarpal bones due to wasting of the dorsal interosseous muscles (see Fig 16–29).

2. *Sensory loss.*—The main ulnar nerve and its palmar cutaneous branch are usually severed; the posterior cutaneous branch, which arises from the ulnar nerve trunk about 2½ in. (6 cm) above the pisiform bone, is usually unaffected. The sensory loss will therefore be confined to the palmar surface of the medial third of the hand and the medial 1½ fingers and to the dorsal aspects of the middle and distal phalanges of the same fingers (see Fig 16–3).

3. *Vasomotor changes.*—The skin areas involved in sensory loss are warmer and drier than normal. This is caused by arteriolar dilatation and the absence of sweating resulting from loss of sympathetic control.

In patients with old wrist lesions of the ulnar nerve, the clawing deformity of the hand is more severe than with old elbow lesions (or more proximal injuries) of the nerve; this is because with the lesion at the wrist the nerve supply to the flexor digitorum profundus, which arises from the ulnar nerve in the region of the elbow, is intact.

Ulnar nerve lesions, unlike median nerve injuries, leave a relatively efficient hand. The sensation over the radial part of the hand is intact, and the pincerlike action of the thumb and index finger is reasonably good, although there is some weakness due to loss of the adductor pollicis.

Superficial Radial Nerve.—Injuries include the following. Division of the superficial radial nerve, which is entirely sensory, results in a *variable* area of sensory loss on the radial part of the dorsum of the hand and the dorsal surface of the lateral 3½ fingers over the proximal phalanges and the base of the thumb (see Fig 16–3). The area of total anesthesia is relatively small because of the overlap of sensory innervation by adjacent nerves.

Is It a Radicular Injury, Brachial Plexus Injury, or Peripheral Nerve Injury at the Wrist?

This is a common problem that challenges the emergency physician. The following is a brief explanation of the various dysfunctional changes that may be found with the different nerve injuries. Only by carefully examining the upper extremity, and in particular the hand, can an accurate diagnosis be made.

Segmental or Radicular Lesions Causing Dysfunction of the Hand.—These involve the spinal roots and nerves.

Anatomy of the Spinal Roots and Spinal Nerves.—Each segment of the spinal cord is connected to a spinal nerve by an anterior or motor root and a posterior or sensory root. The mixed spinal nerve then leaves the spinal canal of the vertebral column by passing through an intervertebral foramen. After emerging from the foramen, the spinal nerve divides into a smaller posterior ramus, which supplies the muscles and skin of the back, and a larger anterior ramus, which supplies the muscles and skin of the trunk and the upper extremity. The first thoracic spinal nerve is also connected to the stellate ganglion by white and gray rami.

Injuries to Spinal Roots and Spinal Nerves.—These include the following.

1. *Compression* in and about the intervertebral foramen caused by prolapsed intervertebral discs, osteoarthritis of intervertebral joints, hypertrophy of the ligamentum flavum, and primary or secondary tumors.
2. *Traction* of the spinal nerves caused by excessive movements of vertebral column or upper extremity.
3. *Penetrating injuries* caused by missiles or sharp objects.
4. *General neurological diseases,* including Guillain-Barré syndrome, Lyme disease, herpes zoster, and diabetes may involve the spinal nerve roots.

Clinical Evaluation.—Spinal root lesions and spinal nerve lesions require a precise knowledge of the dermatome or the myotome supplied by each root or spinal nerve. The following symptoms and signs may be present.

1. *Motor.*—Anterior root lesions produce weakness and atrophy in the muscles containing the myotome supplied by the affected root. Muscle fasciculations may also be present.

2. *Sensory.*—Sharp or burning pain referred to a specific dermatome that is accentuated by any movement that stretches the posterior nerve root. Coughing or sneezing that causes an increased intraspinal pressure may also precipitate the onset of pain. Dermatomal paresthesia or dyesthesias occur when there is a progressive destructive lesion of the nerve root or spinal nerve. Because of dermatomal overlap, total sensory loss is only evident when at least three segmental nerve roots are destroyed.

3. *Reflex loss.*—Any lesion that breaks the integrity of the reflex arc controlling a myotome will result in impairment or loss of a specific stretch reflex. It follows, therefore, that a lesion of an anterior root, a posterior root, or the spinal nerve will affect the reflex activity of the muscle served by that segmental nerve.

Table 16–5 summarizes the important features found in cervical root lesions involving the upper extremity; only the major muscles involved are shown.

The segmental nerve fibers reach the hand via the median nerve (C6 through C8 and T1), the ulnar nerve (C8 and T1), and the radial nerve (C6 through C8 and T1). The segmental sensory nerves that travel to the skin of the hand do so in the median, ulnar, and radial nerves; the segmental motor nerves travel in the median and ulnar nerves. In fact, the motor innervation of the hand can be summarized by stating that all the small muscles of the hand are supplied by the branches of the ulnar nerve except the muscles of the thenar eminence and the first two lumbricals, which are supplied by the median nerve. The radial nerve supplies no motor innervations to the intrinsic muscles of the hand.

Brachial Plexus Lesions Causing Dysfunction of the Hand.—These lesions include the following.

Lower Lesions of the Brachial Plexus (Klumpke Palsy).—Lower lesions of the brachial plexus (C8 and T1) are usually traction injuries caused by excessive abduction of the arm, as occurs when a person falling from a height clutches at an object to stop the fall. The first thoracic nerve is usually torn.

1. *Motor.*—The nerve fibers from this segment run in the median and ulnar nerves to supply *all the small muscles of the hand.* The hand has a clawed appearance due to hyperextension of the metacarpophalangeal joints and flexion of the interphalangeal joints. The extensor digitorum is unopposed by the lumbricals and interossei and extends the metacarpophalangeal joints; the flexor digitorum superficialis and profundus are unopposed by the lumbricals and interossei and flex the middle and terminal phalanges, respectively.

2. *Sensory.*—There will be a loss of sensation along the ulnar side of the arm. If the eighth cervical nerve is also damaged, the extent of anesthesia will be greater and will involve the medial side of the forearm, hand, and medial two fingers.

Median Nerve Injuries at the Elbow Causing Dysfunction of the Hand.—These injuries include the following.

1. *Motor.*—The pronator muscles of the forearm and the long flexor muscles of the wrist and fingers,

TABLE 16–5.
Summary of Important Features Found in Cervical Root Syndromes

Root Injury	Disc Level	Dermatome Pain	Muscle Supplied	Movement Weakness	Reflex Involved
C5	C4, C5	Radial side of lower arm	Deltoid, biceps	Shoulder abduction, elbow flexion	Biceps
C6	C5, C6	Radial side of forearm, wrist, thumb, and index finger	Extensor carpi radialis longus and brevis	Wrist extensors	Brachioradialis
C7	C6, C7	Middle finger	Triceps and flexor carpi radialis	Extension of elbow and flexion of wrist	Triceps
C8	C7, T1	Ulnar side of forearm, ring, and little fingers	Flexor digitorum superficialis and profundus	Finger flexion	. . .
T1*	T1, T2	Ulnar side of elbow	Small muscles of hand (interossei)	Abduction, adduction of fingers	. . .

*Note that sympathetic white rami passing to the stellate ganglion from the first thoracic spinal nerve may be interrupted, resulting in ipsilateral Horner's syndrome.

with the exception of the flexor carpi ulnaris and the ulnar half of the flexor digitorum profundus, will be paralyzed. As a result, the forearm is kept in a supine position; wrist flexion is weak and is accompanied by adduction (ulnar deviation). The latter deviation is caused by paralysis of the flexor carpi radialis and the strength of the flexor carpi ulnaris and the medial half of the flexor digitorum profundus. No flexion is possible at the interphalangeal joints of the index and middle fingers, although weak flexion of the metacarpophalangeal joints of these fingers is attempted by the interossei. When the patient tries to make a fist, the index and, to a lesser extent, the middle finger tend to remain straight, whereas the ring and little fingers flex (see Fig 16–29). The latter two fingers are, however, weakened by the loss of the flexor digitorum superficialis. Flexion of the terminal phalanx of the thumb is lost due to paralysis of the flexor pollicis longus. The muscles of the thenar eminence are paralyzed and later become wasted, so that the eminence is flattened. The thumb is laterally rotated and adducted. The hand in long-standing cases looks flattened and "apelike."

2. *Sensory.*—There is loss of skin sensation of the lateral half or less of the palm of the hand and the palmar aspect of the lateral 3½ fingers. There is also sensory loss of the skin of the distal parts of the dorsal surfaces of the radial 3½ fingers. The area of total anesthesia is considerably less due to overlap of adjacent nerves.

3. *Vasomotor changes.*—The skin areas involved in sensory loss are warmer and drier than normal. This is due to the arteriolar dilatation and absence of sweating resulting from loss of sympathetic control.

4. *Trophic changes.*—In long-standing cases, changes are found in the hand and fingers. The skin is dry and scaly, the nails crack easily, and there is atrophy of the pulp of the fingers.

Ulnar Nerve Injuries at the Elbow Causing Dysfunction of the Hand.—These injuries involve the following.

1. *Motor.*—The flexor carpi ulnaris and the medial half of the flexor digitorum profundus muscles are paralyzed. The paralysis of the flexor carpi ulnaris can be observed by asking the patient to make a tightly clenched fist. Normally, the synergistic action of the flexor carpi ulnaris tendon can be observed as it passes to the pisiform bone; the tightening of the tendon will be absent if the muscle is paralyzed. The profundus tendons to the ring and little fingers will be functionless, and the terminal phalanges of these fingers are therefore not capable of being markedly flexed. Flexion of the wrist joint will result in abduction, owing to paralysis of the flexor carpi ulnaris. In long-standing cases, the ulnar border of the front of the forearm will eventually show flattening because of the wasting of the underlying ulnaris and profundus muscles. The small muscles of the hand will be paralyzed, except the muscles of the thenar eminence and the first two lumbricals, which are supplied by the median nerve. The patient is unable to adduct and abduct the fingers and consequently is unable to grip a piece of paper placed between the fingers. The extensor digitorum can abduct the fingers, but only when the metacarpophalangeal joints are hyperextended. It is impossible to adduct the thumb because the adductor pollicis muscle is paralyzed. If the patient is asked to grip a piece of paper between the thumb and the index finger, he strongly contracts his flexor pollicis longus and flexes the terminal phalanx (Froment's sign). The metacarpophalangeal joints become hyperextended due to paralysis of the lumbrical and interosseous muscles, which normally flex these joints. Since the first and second lumbricals are not paralyzed (they are supplied by the median nerve), the hyperextension of the metacarpophalangeal joints is most prominent in the fourth and fifth fingers. The interphalangeal joints are flexed, again due to paralysis of the lumbrical and interosseous muscles, which normally extend these joints through the extensor expansion. The flexion deformity at the interphalangeal joints of the fourth and fifth fingers is more obvious, because the first and second lumbrical muscles of the index and middle fingers are not paralyzed. In long-standing cases the hand assumes the characteristic "claw" deformity ("main en griffe"). Wasting of the paralyzed muscles results in flattening of the hypothenar eminence and loss of the convex curve to the medial border of the hand. Examination of the dorsum of the hand will show hollowing between the metacarpal bones due to wasting of the dorsal interosseous muscles (see Fig 16–29).

2. *Sensory.*—Loss of skin sensation will be observed over the anterior and posterior surfaces of the medial third of the hand and the medial 1½ fingers.

3. *Vasomotor changes.*—The skin areas involved in sensory loss are warmer and drier than normal because of the arteriolar dilatation and absence of sweating resulting from loss of sympathetic control.

Radial Nerve Injuries in the Axilla Causing Dysfunction of the Hand.—The injuries involve the following.

1. *Motor.*—The triceps, the anconeus, and the long extensors of the wrist are paralyzed. The patient is unable to extend the elbow joint, the wrist joint, and the fingers. Wristdrop, or flexion of the wrist, occurs as the result of the unopposed flexor muscles of the wrist. Wristdrop is very disabling, since the fingers cannot be flexed strongly with the wrist fully flexed. If the wrist and proximal phalanges are passively extended by holding them in position with the opposite hand, the middle and distal phalanges of the fingers can be extended by the action of the lumbricals and interossei, which are inserted into the extensor expansions. The brachioradialis and supinator muscles are also paralyzed, but supination is still performed well by the biceps brachii.

2. *Sensory.*—There is a small loss of skin sensation down the posterior surface of the lower part of the arm and down a narrow strip on the back of the forearm. Also, a variable area of sensory loss exists on the lateral part of the dorsum of the hand and base of the thumb (see Fig 16–3). The area of total anesthesia is relatively small due to overlap of sensory innervation by adjacent nerves.

3. *Trophic changes.*—These are slight.

Tendon Injuries

Tendons are commonly injured by stab or laceration wounds. Although some tendons may be completely divided, others may be only partially severed. In the latter case, active movements are weak and painful.

Flexor Tendon Injuries.—In the normal hand in the resting position, the fingers are partially flexed, with the index finger being flexed the least and the remaining fingers showing increasing degrees of flexion, with the little finger being flexed the most—"a cascade of curvature." If one of the fingers does not show this pattern, i.e., it "hangs out," a tendon injury should be suspected. Movements of the wrist also normally affect finger position. When the wrist joint is passively flexed, the fingers should extend, and when the wrist is passively extended, the fingers should flex.

Flexor Digitorum Profundus Tendons.—Intactness of tendon function can be tested by asking the patient to flex the terminal phalanx in each finger in turn against resistance.

Flexor Digitorum Superficialis Tendons.—Intact tendons can be tested by asking the patient to flex the middle phalanx while the examiner holds the remaining three fingers in full extension. Maintaining extension of these other fingers is important because it prevents the patient from flexing the terminal phalanx of the tested finger (since all flexor digitorum profundus tendons, to some extent, contract together) and effectively "curling down" the middle phalanx in the absence of flexor digitorum superficialis function. This is a subtle but important distinction because the flexor digitorum superficialis tendons are anatomically anterior to the flexor digitorum profundus and so may be lacerated without any injury to the flexor digitorum profundus. This isolation of the flexor digitorum superficialis works best in the middle and index fingers.

Flexor Pollicis Longus Tendon.—The tendon can be tested by asking the patient to flex the terminal phalanx of the thumb while immobilizing the metacarpophalangeal joint.

Flexor Carpi Radialis, Palmaris Longus, and Flexor Carpi Ulnaris Tendons.—These tendons can be felt to move when the patient is asked to flex the wrist joint.

Extensor Tendon Injuries.—The extensor mechanism of the fingers is much more complicated than the flexor mechanism, and it is important to get the patient to actively extend each of the phalanges in turn for a tendon injury to be detected.

Extension of the Metacarpophalangeal Joints of the Fingers.—This is accomplished by the tendons of extensor digitorum. With the index finger, the digitorum tendon is assisted from the ulnar side by the tendon of extensor indicis. Note that both tendons lie alongside one another and join the dorsal extensor expansion at the base of the proximal phalanx (see Fig 16–13). With the little finger, the digitorum tendon is assisted from the ulnar side by the *two* tendons of extensor digiti minimi, and all three join the dorsal extensor expansion at the base of the proximal phalanx (see Fig 16–13).

Extension of the Interphalangeal Joints of the Fingers.—This is brought about by the lumbricals and the interossei tendons that are inserted into the dorsal extensor expansion derived from the extensor digitorum tendon (see Fig 16–9). The lumbricals and interossei flex the metacarpophalangeal joints and extend the interphalangeal joints (when the metacarpophalangeal joint is in extension).

Extension of the Metacarpophalangeal Joint of the Thumb.—This is accomplished by the tendon of extensor pollicis brevis (see Fig 16–13) assisted by the tendon of extensor pollicis longus (after the interphalangeal joint has been extended).

Extension of the Interphalangeal Joint of the Thumb.—This is brought about by the tendon of extensor pollicis longus, which can be seen and felt as it contracts (see Fig 16–13).

Extension of the Wrist Joint.—With the fingers flexed, to exclude the action of the extensor digitorum tendons, the wrist joint is actively extended. This is accomplished by the tendons of extensor carpi ulnaris and extensor carpi radialis longus and brevis (see Fig 16–13). These tendons can be palpated on the posterior surface of the wrist as they become taut.

Mallet Finger.—Avulsion of the insertion of one of the extensor tendons into the distal phalanges can occur if the distal phalanx is forcibly flexed when the extensor tendon is taut. Sometimes a fragment of bone is pulled off the distal phalanx by the tendon. The last 20° of active extension is lost, though full passive extension is possible; the resulting condition is known as mallet finger (Fig 16–30).

Boutonnière Deformity.—Avulsion of the central slip of the extensor tendon proximal to its insertion into the base of the middle phalanx results in the lateral slips subluxing volar to the axis of motion of the proximal interphalangeal joint (see Fig 16–30). The deformity results from the paradoxical flexing of the proximal interphalangeal joint and hyperextension of the distal interphalangeal joint. This injury can result from direct end-on trauma to the finger, direct trauma over the back of the proximal interpha-

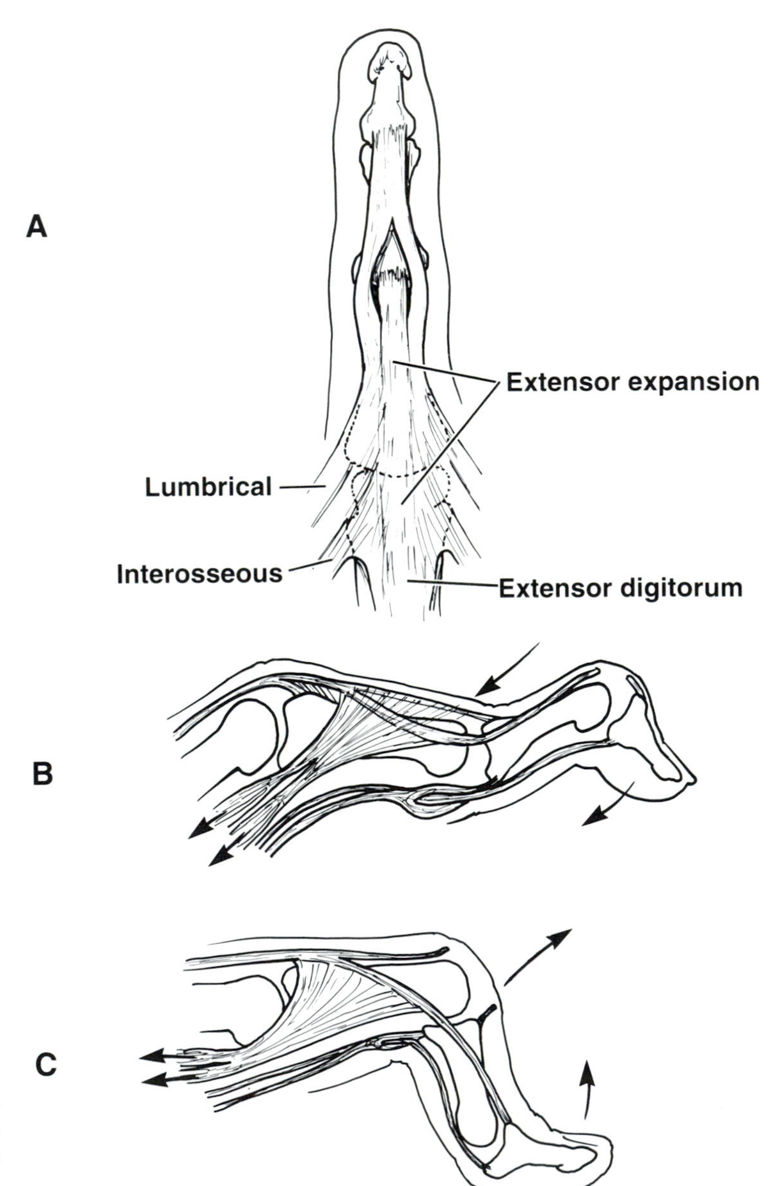

FIG 16–30.
A, posterior view of normal dorsal extensor expansion. The extensor expansion near the proximal interphalangeal joint splits into three parts—a central part, which is inserted into the base of the middle phalanx, and two lateral parts, which converge to be inserted into the base of the distal phalanx. **B,** mallet, or baseball, finger. The insertion of the extensor expansion into the base of the distal phalanx is ruptured; sometimes a flake of bone of the base of the phalanx is pulled off. **C,** boutonnière deformity. The insertion of the extensor expansion into the base of the middle phalanx is ruptured. The arrows indicate the direction of the pull of the muscles and the deformity.

langeal joint, or from laceration of the dorsum of the finger.

Trigger Finger.—This condition is occasionally seen in the emergency department. A palpable and even audible snapping occurs when a patient is asked to flex and extend the fingers. This is caused by the presence of a localized swelling of one of the long flexor tendons that catches on a narrowing of the fibrous flexor sheath anterior to the metacarpophalangeal joint. The snapping and limitation of motion may take place either in flexion or in extension.

Clinical Anatomy of Fractures and Dislocations of the Carpal Bones

Fractured Scaphoid (Navicular)

Fracture of the scaphoid bone is common in young adults and, unless treated effectively, the fragments will not unite and permanent weakness, pain, and disability of the wrist will result, with the subsequent development of osteoarthritis and possible aseptic necrosis of the proximal fragment. The fracture line usually goes through the narrowest part of the bone, which, because of its location in a carpal joint, becomes bathed in synovial fluid. This fact alone may interfere with bony union. The blood vessels to the scaphoid enter its proximal and distal ends, although the blood supply is occasionally confined to its distal end. If the latter occurs, a fracture deprives the proximal fragment of its arterial supply, and this fragment may undergo avascular necrosis.

The scaphoid may be palpated in the anatomical snuffbox. Tenderness in the anatomical snuffbox may indicate a scaphoid fracture.

Fracture of the Triquetral Bone.—This fracture is usually caused by a direct blow to the hand or it occurs in association with a transscaphoid perilunar dislocation (see below). Tenderness occurs on the dorsal surface of the wrist just distal and radial to the styloid process of the ulna. This fracture is best seen on a lateral radiograph of the wrist.

Dislocation of the Lunate.—Dislocation of the lunate is uncommon. It usually occurs following a fall on an outstretched hand. Injury to the median nerve is a common complication.

Dislocations of the lunate may be restricted to ligamentous disruption or combined with fractures. In a lunate dislocation, the lunate is rotated forward into the palm, leaving the radius and capitate in their normal positions; the lunate is typically rotated 180° (Fig 16–31). In perilunate dislocations of the carpus, the lunate remains in its normal anatomic position with respect to the radius but the capitate, other bones of the distal carpal row, and the fingers are dislocated dorsally to the lunate (see Fig 16–31).

Clinical Anatomy of Carpal Arthrocentesis

This is fully described in Chapter 18, p. 773.

Clinical Anatomy of Finger Fractures

Epiphyseal Plate Injuries

In children the epiphyseal plate, which is responsible for the longitudinal growth of the bone, is weaker than bone and ligaments. Consequently, when a child's finger is subject to tensile or shearing strains, an epiphyseal injury may occur. The interference to the local blood supply of the epiphyseal cartilage may lead to diminished or, sometimes, increased or irregular growth, with later development of finger deformity.

In children the epiphyseal plates for the metacarpal bones are located distally, while those for the phalanges are located proximally. The metacarpal of the thumb is similar to a phalanx, and the epiphyseal plate is located at the proximal end.

Fractures of the Distal Phalanx.—These are common and usually result from direct crush injuries with comminution of the distal end; nail injuries frequently accompany such fractures.

Fractures of the Middle Phalanx.—These usually result from a direct crush injury and may be stable or

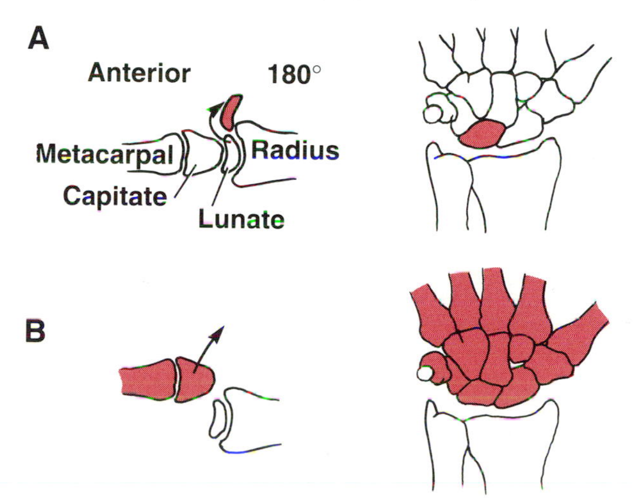

FIG 16–31.
A, lunate dislocation. **B,** perilunate dislocation.

unstable. When the fragments are angulated, the forces acting on the phalanx consist of the flexor digitorum superficialis, which is situated on the palmar surface and is inserted onto the borders of the phalanx, and the central slip of the dorsal extensor expansion, which inserts onto the dorsal surface of the base of the middle phalanx. The direction of the deformity will depend on the site of the fracture line relative to the tendon attachments.

Fractures of the Proximal Phalanx.— Direct injury is usually the cause, and volar angulation is common. The angulation is produced by the pull of the palmar and dorsal interossei, which are attached to the base of the proximal phalanx and flex the proximal fragment; the insertion of the dorsal extensor expansion onto the dorsal aspect of the middle and distal phalanges causes dorsal displacement of the distal fragment (Fig 16–32).

Fractures of the Metacarpal Bones.— These usually result from direct violence, such as the clenched fist striking a hard object. The "boxer's fracture"* frequently involves the neck of the fifth and sometimes the fourth metacarpal bones. The fracture always angulates dorsally (see Fig 16–32). The distal fragment is pulled forward anteriorly by the lumbricals and interossei assisted by the long flexor tendons.

The bases of the fourth and fifth metacarpal bones have hinge joints with the hamate bone of the carpus; this means that normally both these bones can move freely forward and backward on the carpus, the fourth less than the fifth (the fifth metacarpal normally has a 30° range of mobility). The base of the third metacarpal has a flat articular surface for the capitate of the carpus and therefore the movements are restricted. The base of the second metacarpal bone is mortised between the capitate and the

*The "boxer's fracture" should really be called the "street fighter's fracture" because a professional boxer knows that force is best delivered through the second and third metacarpals, which connect on a direct line through the capitate, lunate, and distal end of the radius. The fourth and fifth metacarpal bones, because of their relative mobility, are prone to fracture at the metacarpal neck.

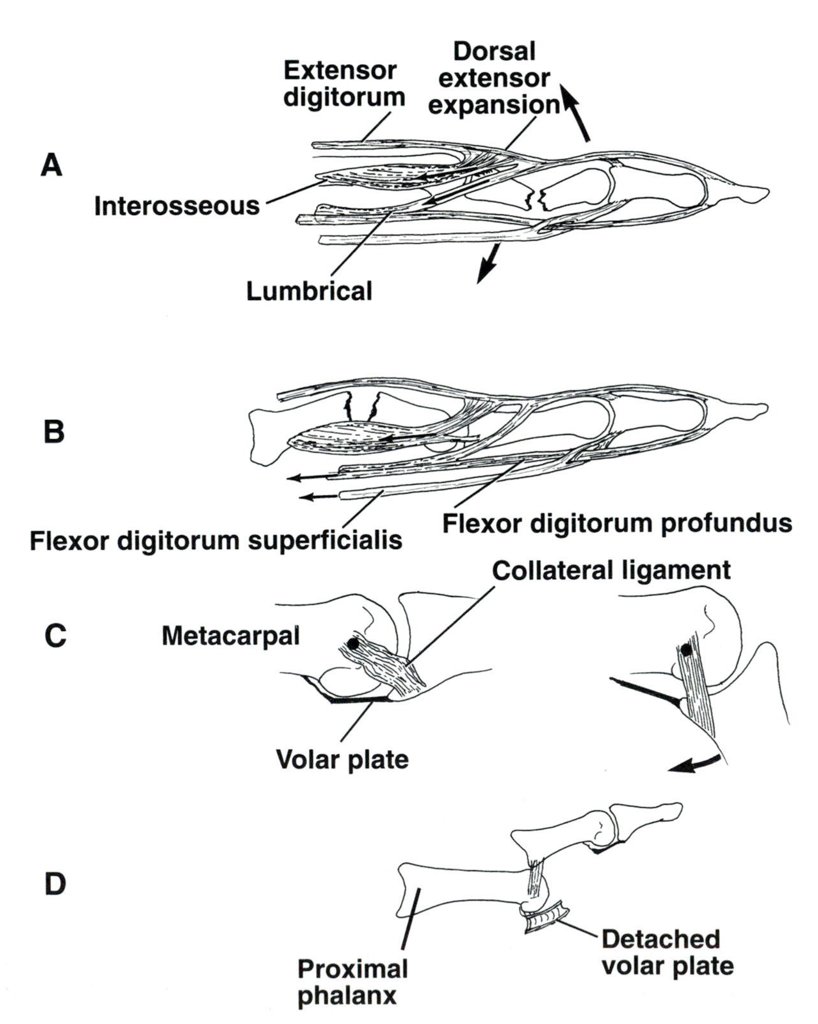

FIG 16–32.
A, fractures of the shaft of the proximal phalanx angulate volarly. **B,** fractures of the shaft of the metacarpal angulate dorsally. **C,** normally the collateral ligaments of the metacarpophalangeal joint are slack in extension and tight in flexion due to the shape of the metacarpal head. **D,** dorsal dislocation of the middle phalanx at the proximal interphalangeal joint with detachment of the volar plate.

trapezium and also articulates with the trapezoid; consequently, this metacarpal is almost immobile.

Fracture of the Thumb Metacarpal (Bennett's Fracture).—This is a fracture of the base of the metacarpal and is caused when violence is applied along the long axis of the thumb or the thumb is forcibly abducted. The fracture is oblique and enters the carpometacarpal joint. The fracture is unstable. The small fragment is attached by a strong ligament to the tubercle of the trapezium but the tendon of the abductor pollicis longus muscle pulls the base of the metacarpal dorsally and laterally. Operative reduction is required.

In all fractures involving the bones of the fingers and thumb, it is imperative that rotational displacement be corrected. Failure to do so will result in deformity of the fingers. Rotational malalignment can be checked by comparing the fingernail position with that of the opposite hand held in the same position. If the finger has no rotational or angular deformity, the tip of the distal phalanx should point to the tubercle of the scaphoid when it is flexed.

Angulation is corrected wherever possible so that functional deficits are minimal. In this connection, it is important to remember the degree of mobility that normally exists between the metacarpal bones and the carpus (the fourth and fifth are freely mobile, but the second and third are almost immobile; see p. 668 for discussion). This explains why in the metacarpal of the index finger an angulation of greater than 10° is unacceptable, and in the fourth and fifth metacarpals an angulation greater than 30° is unacceptable. Failure to follow this general rule will result in impaired function at the metacarpophalangeal joints and a poor cosmetic outcome.

Clinical Anatomy of Dislocations of the Fingers

Dorsal dislocations of the metacarpophalangeal and interphalangeal joints are similar. The palmar (volar) ligament is avulsed from its attachment distally and the capsule is torn, leaving the cordlike collateral ligaments intact (see Fig 16–32). In these circumstances the joint is stable on reduction of the dislocation. However, if tearing of one or both collateral ligaments occurs, joint instability is present. Dorsal dislocations of the distal interphalangeal joints are unusual; dorsal dislocations of the proximal interphalangeal joints are common.

Occasionally, the dorsal dislocation of the metacarpophalangeal joint may have complications that will necessitate an open reduction. The head of the metacarpal bone may become trapped between the flexor tendons or the lumbricals, or the palmar ligament may tear away from the phalanx and turn into the joint, so that it becomes interposed between the metacarpal head and the base of the phalanx.

Palmar dislocation of the proximal interphalangeal joint is a serious condition due to damage to the dorsal extensor expansion and tearing of the collateral ligaments. Open reduction and ligament fixation are required.

Clinical Anatomy of Nail Injuries

General Principles

1. The nail matrix must be preserved since this is the site of nail growth. If the nail matrix is destroyed, the nail will never grow back. If the nail matrix is torn, it must be carefully repaired with fine sutures, otherwise, a *split nail* may result.

2. The grooves that exist around the three sides of a nail beneath the proximal and lateral nail folds must be preserved to allow the nail to grow out normally over the nail bed. Longitudinal and transverse lacerations involving both the nail bed and the nail folds must be sutured separately to preserve these grooves, otherwise *adhesions* between the nail bed and the nail folds and distorted nails will form. Failure to suture lacerations of the nail bed may result in ridging of the nails.

3. Severe bleeding beneath the body of the nail or avulsion of the nail plate from under the eponychium invariably means that the underlying nail bed is lacerated. To repair this laceration, the nail must first be removed.

4. When avulsion of the proximal portion of the nail bed occurs out from underneath the eponychium, it is necessary to first reattach, by suturing, the proximal edge of the bed back underneath the eponychium. A half-buried horizontal mattress suture works perfectly in this situation. The remainder of the bed can then be sutured, if necessary, into its correct anatomic position.

5. Severe pain is experienced with *subungual hematoma* because the dermis of the nail bed is extremely sensitive. The pain is easily relieved by making a small hole in the body of the nail plate to reduce the pressure.

Clinical Anatomy of Hand Infections

Paronychia

A paronychia is an infection between the nail and the nail fold; it is one of the most common infections of the hand. To begin with, swelling, redness, and pain are found at the corner of the eponychium, between the base and side of the nail. Pus then starts to accumulate and spreads between the eponychium and the base and side of the nail (Fig 16–33). The pus may eventually track between the nail plate and the nail bed (starting at the base of the nail plate) and forms a proximally spreading lake that is essentially a subungual abscess.

Anatomy of Treatment.—In the early stages the pus may be released by inserting a scalpel blade along the surface of the nail and uplifting the lateral part of the proximal nail folds (see Fig 16–33). In the later stages it is necessary to remove the base of the nail plate to provide adequate drainage.

Pulp Space Infection (Felon)

Infection of the pulp space is usually caused by a penetrating wound, such as an infected needle. It is an extremely painful condition due to the high concentration of sensory nerve endings in the overlying skin. The fibrous septa that tether the skin to the terminal phalanx and limit skin mobility and slippage unfortunately restrict the expansion of the inflammatory exudate, and the pressure in the pulp space quickly rises. If the infection is left without decompression, osteomyelitis of the terminal phalanx can occur. In children, the blood supply to the diaphysis of the phalanx passes through the pulp space (Fig 16–12), and, consequently, avascular necrosis of the diaphysis may occur. Proximal extension of the infection can involve the flexor synovial sheath of the long flexor tendon, which is inserted on the base of the terminal phalanx. This complication may, in turn, spread to involve the fascial spaces of the palm. This extensive spread can occur without the pulp space abscess pointing and spontaneously draining onto the surface.

Anatomy of Treatment.—The abscess must be drained to avoid the complications listed above. If

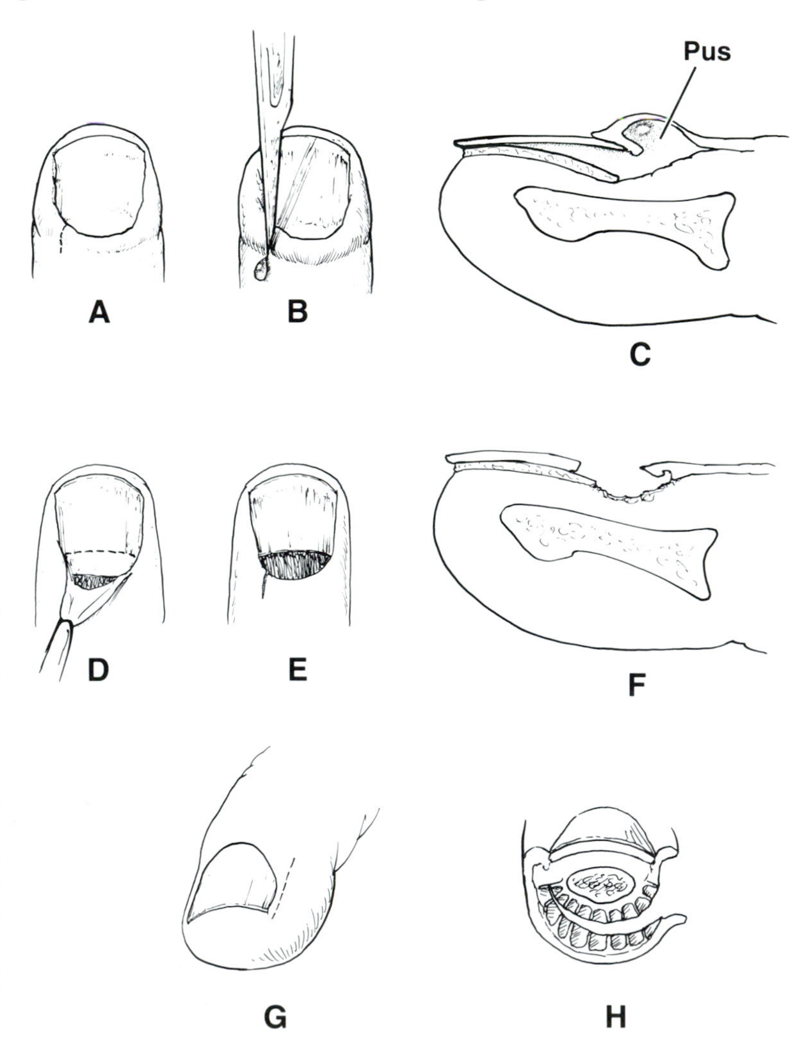

FIG 16–33.
A, classical signs of a paronychia; the dotted line indicates the site for incision. **B,** a bead of pus emerging through the scalpel wound. **C,** sagittal section through the nail and terminal phalanx showing the extent of the abscess in a severe paronychia. **D–F,** in cases of severe paronychia, the removal of the proximal portion of the nail may be necessary to provide free drainage of the abscess. **G,** felon, or pulp space, infection. The dotted line indicates the site of the incision for drainage. The incision is *not* carried around the distal end of the nail so as not to interfere with the sensations of touch at the finger end. **H,** the scalpel blade is directed across the pulp to break down the fascial septa and open up the fascial spaces for drainage.

the abscess is pointing, the incision should be made at the site of the point. Otherwise, it is important to avoid scar formation on the sensitive working surface of the fingertip. A longitudinal incision is made along the side of the finger just volar to the distal phalanx (see Fig 16–33). The incision is used to access the pulp space, cut through the fibrous septa, and provide free drainage. In the more advanced cases with tissue necrosis, the dead tissue is removed.

The side of the finger on which the incision is made is important and is related to the position of the ultimate scar and the occupation of the individual. The index, middle, and ring fingers are usually incised on the ulnar side and the thumb and little finger on the radial side.

Tenosynovitis of the Flexor Synovial Sheaths

The detailed anatomy of the flexor synovial sheaths is fully described on p. 636 and is summarized in Figure 16–7.

The sheaths become infected through penetrating wounds, such as a needle or thorn, or may become involved from spread of a pulp space infection or a felon. A flexor tenosynovitis may be distinguished from a subcutaneous abscess by the presence of the following four cardinal signs of Kanavel: (1) The painful finger is swollen and held rigidly in the semiflexed position; in this position the sensitive synovial sheath is subject to the least amount of tension. (2) Gentle passive extension of the finger causes increased pain, which stretches the tendon and sheath. (3) There is a fusiform swelling along the course of the sheath that (4) is acutely tender to touch.

If the digital sheaths of the little finger and thumb are infected, the ulnar and radial bursae are quickly involved. In 50% of individuals, the radial and ulnar bursae communicate with each other at their proximal ends so that infection may rapidly spread from one to the other.

As the inflammatory process continues, the pressure within the digital sheath rises and may compromise the blood supply to the tendons that travel in the vincula longa and brevia (see Fig 16–9); sloughing or later severe scarring of the tendons may follow.

A further increase in pressure may cause the sheath to rupture at its proximal end. Unfortunately, such a rupture is often accompanied by a decrease in pain, which may fool the patient and the physician into thinking that the condition is improving. Anatomically, the digital sheath of the index finger is related proximally to the thenar space (see Fig 16–12), whereas that of the ring finger is related to the midpalmar space. The sheath for the middle finger is related to both the thenar and midpalmar spaces. These relationships explain how infection can extend from the digital synovial sheaths and involve the palmar fascial spaces.

Should infection in the ulnar and radial bursae be neglected, pus may burst through the proximal ends of these bursae and enter the fascial space of the forearm *(space of Parona)*. This space lies between the tendons of flexor digitorum profundus anteriorly and the pronator quadratus and the interosseous membrane posteriorly.

Anatomy of Treatment.—In order to prevent tendon necrosis and reduce the possibility of fascial space involvement, the tendon sheaths are decompressed (see Fig 16–34 for drainage sites).

Palmar Fascial Space Infections

The anatomy of the midpalmar and thenar spaces has been fully described on p. 643 and is summarized in Figure 16–12. These spaces may become infected and distended with pus as the result of the spread of infection in acute suppurative tenosynovitis; rarely they may become infected following penetrating wounds such as falling on a dirty nail.

Anatomy of Treatment.—The midpalmar space may be drained through a curved incision 1 to 2 mm proximal to the distal transverse crease of the palm, extending from the third to the fifth metacarpal bone. The palmar aponeurosis is incised, and care is taken to avoid the digital nerves and arteries and the flexor tendons and lumbrical muscles (see Fig 16–34).

The thenar space may be drained through a curved transverse incision on the dorsal surface of the thumb web; this permits direct access to the thenar space.

Clinical Anatomy of Nerve Blocks to the Hand and Fingers

These are fully described in Chapter 19, pp. 819 to 821.

Clinical Anatomy of Arthrocentesis of the Joints of the Wrist and Fingers

This is fully described in Chapter 18, pp. 772 to 774.

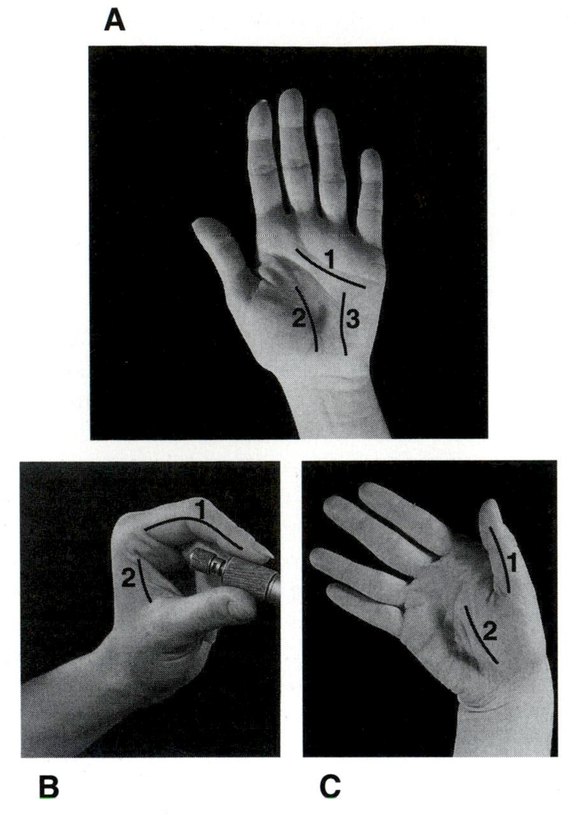

FIG 16–34.
Incisions in the hand and fingers for drainage of infection. **A,** *(1)* midpalmar space infection. The incision is made 1 to 2 mm proximal to the distal transverse crease of the palm and extends from the third to the fifth metacarpal. *(2)* thenar muscle space infection (This is rare and must not to be confused with the midpalmar space infection.). The incision is made along the ulnar border of the thenar eminence. *(3)* tenosynovitis of the flexor digitorum superficialis and profundus tendons (ulnar bursa). The incision is made on the radial side of the hypothenar eminence. **B,** *(1)* tenosynovitis of the flexor digitorum superficialis and profundus tendons. The incision runs just dorsal to the most dorsal part of the flexion creases of the flexed interphalangeal joints, thus avoiding the digital nerve and artery (so-called neutral line). *(2)* thenar space infection. A curved incision is made through the dorsal part of the thumb web. **C,** *(1)* tenosynovitis of the flexor pollicis longus tendon. The incision runs just dorsal to the most dorsal part of the flexion creases of the flexed interphalangeal joint, thus avoiding the digital nerve and artery (so-called neutral line). *(2)* in severe cases the incision is made along the ulnar side of the thenar eminence and the synovial sheath is approached from the ulnar side.

ANATOMICAL-CLINICAL PROBLEMS

1. A 26-year-old man was drying a glass after it had been washed. The glass suddenly broke into many pieces, and a long sliver pierced his skin on the volar aspect of his left wrist. On examination in the emergency department, a small, clean wound was seen on the front of his left wrist at the level of the proximal transverse crease. On gentle palpation over the wound, a gap was felt in the palmaris longus tendon, which enlarged when the patient flexed his wrist. His thumb was found to be laterally rotated and adducted, and he was unable to oppose his thumb to the other fingers. When he was asked to make a fist slowly, the flexion of the index and middle fingers lagged behind the ring and little fingers. There was diminished skin sensation over the radial half of the palm of the hand and the palmar aspect of the radial 3½ fingers. There was also sensory loss of the skin of the distal parts of the dorsal surface of the radial 3½ fingers. Using your anatomical knowledge, name the structure or structures that had been injured.

2. A 20-year-old woman was closing a window when a gust of wind swung it back toward her face. In order to protect herself, she held out her right hand, which smashed through the glass. Her friend found her sitting on the floor, bleeding profusely from a superficial laceration in front of her right wrist. On examination in the emergency department, she was found to have sensory loss over the palmar aspect of the medial 1½ fingers, but normal sensation on the back of these fingers. She was unable to grip a piece of paper between her right index and middle fingers. All her long flexor tendons were intact. Which artery and nerve were cut in the accident?

3. A 45-year-old man was out in his yard chopping firewood when suddenly a large splinter of wood pierced his skin on the palm of his right hand at the root of the thumb. On examination in the emergency department, a small skin wound was found over the thenar eminence about one fingerbreadth distal to the tubercle of the scaphoid. On testing for skin sensation, it was found to be normal on the palmar and dorsal surfaces of the hand and fingers. The thumb was laterally rotated and adducted, and the patient could not oppose the thumb to the other fingers. The other fingers moved normally, both actively and passively. Using your knowledge of anatomy, make the diagnosis.

4. A 60-year-old woman was seen in the emergency department complaining of a "pins and needles" feeling in her right hand at the index and middle fingers. The patient admitted that she had had the condition for several months and it was becoming progressively worse, especially at night. She said that she had experienced difficulty in buttoning up her clothes when dressing. On examination the pa-

tient showed no sensory deficits over the areas where she experienced the discomfort. The muscles of the thenar eminence appeared to be functioning normally, and there was no loss of power or wasting of the thenar eminence. What anatomical structure is diseased in this patient?

5. A 23-year-old medical student fell off his bicycle onto his outstretched hand. He thought he had sprained his wrist joint and treated himself with aspirin and bound his wrist with an elastic bandage. Three weeks later, he still experienced pain on moving his wrist and decided to visit the emergency department. On examination of the dorsal surfaces of both hands, with the fingers and thumbs fully extended, a localized tenderness could be felt over the scaphoid bone in the anatomical snuffbox of his right hand. A diagnosis of a fractured right scaphoid bone was made. For what anatomical reasons do fractures of this bone sometimes fail to unite?

6. A 54-year-old woman, while pruning some rose bushes, stuck a thorn into the anterior surface of the middle finger of her right hand. Four days later the entire finger became swollen, red, and painful. On examination in the emergency department, the whole finger was found to be swollen and was held in a semiflexed position; it was tender, especially along the line of the flexor tendons. Flexion of the finger was difficult, and passive extension caused extreme pain. The site of entry of the thorn was seen as a small black dot over the anterior surface of the middle phalangeal region. What is the diagnosis? If the diseased finger is left untreated, where is the infection likely to spread and what complications to the finger can occur? Name the anatomical structures that should be avoided when making the incision.

7. A 19-year-old man was involved in a fight. His opponent suddenly drew a knife and slashed him on the back of his right wrist. Name the structures that descend to the hand on the dorsal surface of the wrist. How would you test for the integrity of the tendons in this region? Are there any important arteries and nerves in this area?

8. A 69-year-old man fell down a flight of stairs and sustained a fracture of the lower end of the right radius. On examination, the distal end of the radius was displaced posteriorly. What is the name of this fracture? What structures may be damaged with such a fracture? When the fracture is reduced, what is the correct relationship of the radial styloid to the ulnar styloid and what is the correct angle of the radiolunate line? The wrist joint and hand are splinted in the position of function. What is the position of function? What is the clinical importance of this position?

9. A 30-year-old machinist was working at his lathe when a sliver of metal entered the pulp of his right index finger. After thoroughly washing his hands, he thought that he had removed the metal fragment. Thirty-six hours later, he woke up in the night complaining of severe pain in the affected finger. On examination in the emergency department, the skin over the anterior surface of the terminal phalanx of the right index finger was swollen, red, and extremely tender to touch. Over the center of the pulp was a small area of yellow devitalized skin. What is the diagnosis? From your knowledge of anatomy, explain the dangers of infections in the pulp space in children. Where is the infection likely to spread if treatment is delayed? Explain the anatomical rationale of the surgical treatment required.

10. A 46-year-old woman slipped on a shiny floor and sustained a fracture of the fifth metacarpal bone on her left hand. What type of angulation of the fragments is commonly found in fractures at this site? When a splint is applied with the little finger flexed, in which direction should the little finger be pointing?

11. A 15-year-old girl was seen in the emergency department complaining of severe localized pain at the root and side of the nail of her right index finger. Examination showed redness and swelling of the skin of the lateral and proximal nail folds. The pulp of the finger was normal and free from tenderness, and the movements of the fingers were normal. The line of the flexor tendons on the affected finger also showed no evidence of tenderness. What is the likely diagnosis? Explain the anatomy of the area where the infection had entered the finger. Where would you examine the patient for evidence of lymph node involvement?

12. Fascial space infections of the palm are noted for causing permanent disability of the hand if not promptly and adequately treated. Where are the fascial spaces situated in the palm? Name the structures that lie within each space. On the basis of your knowledge of anatomy, explain where incisions would be made to drain the spaces.

13. A 25-year-old maid was making a bed in a motel when she caught the end of her middle finger of her left hand in the fold of a sheet. She experienced a sudden severe pain over the base of the terminal phalanx. The next day when the pain had diminished, she noted that the end of the finger was swollen and she could not extend the terminal phalangeal joint completely. What is your diagnosis?

ANSWERS

1. The glass fragment had severed the palmaris longus tendon and the underlying median nerve as it lay between the tendons of flexor digitorum superficialis and the flexor carpi radialis muscles. The palmar cutaneous branch of the median nerve had also been severed. The effects of median nerve palsy are fully described on p. 660.

2. The ulnar artery and nerve of the right hand were transected in front of the flexor retinaculum.

3. The wood splinter had entered the palm at the point where the median nerve gives off its motor branch to the muscles of the thenar eminence. The point of the splinter had pierced the palmar aponeurosis and severed the motor nerve. The thumb was laterally rotated because the medial rotator, the opponens pollicis, was paralyzed. The thumb was adducted because the abductor pollicis brevis was paralyzed.

4. This patient was suffering from carpal tunnel syndrome, in which there is pressure exerted on the main trunk of the median nerve as it passes beneath the flexor retinaculum. Altered sensation was felt in the skin areas supplied by the digital branches of the median nerve. Although the thenar muscles, which are supplied by the median nerve, did not appear to be weakened, it is clear from her statement about the difficulty she experienced in buttoning up her clothes that the muscles were in fact not functioning normally.

5. (1) The fracture line on the scaphoid bone may deprive the proximal fragment of its arterial supply and result in ischemic necrosis of this fragment. (2) Because of the articulation of the scaphoid with other bones, the fracture line may enter a joint and be bathed in synovial fluid. The presence of synovial fluid may inhibit union between the bone fragments. (3) The scaphoid bone is a difficult bone to immobilize because of its position and small size. Very often the correct diagnosis is made late and the start of immobilization is consequently delayed. All these factors contribute to the high incidence of nonunion in fractures of the scaphoid.

6. This patient had acute suppurative tenosynovitis of the digital sheath of the middle finger of her right hand. In this condition, if the tension within the sheath is not relieved, it is likely to rupture at its proximal end, with discharge of pus into the midpalmar fascial space. If the hand remains untreated, the infection of the midpalmar space may spread to the thenar fascial space (50% communicate with each other) and could spread up into the forearm to enter the space of Parona. Very high pressure within the synovial sheath may occlude the small arteries to the tendons of flexor digitorum superficialis and profundus, which travel in the vincula longa and brevia; avascular necrosis of the tendons may follow, resulting in great loss of function.

7. The structures that descend to the hand on the dorsum of the wrist are fully described on p. 631. The method of testing the action of the extensor tendons are described on pp. 646 and 657. The posterior cutaneous nerve, a branch of the ulnar nerve, and the terminal part of the superficial radial nerve descend onto the dorsum of the hand. The important radial artery crosses the anatomical snuffbox in this region.

8. This patient had sustained a Colles' fracture of the distal part of the radius. Occasionally the styloid process of the ulna is also fractured. The median nerve may be injured at the time of the fall. When the fracture is reduced, the styloid process of the radius should come to lie about ¾ in. (1.9 cm) distal to that of the ulna. The radiolunate line normally angulates 15° volarly (anteriorly) (see Fig 16–20). The fracture produces posterior or dorsal angulation. Neutral angulation is acceptable in a reduction. The position of function of the hand is described on p. 646. It is clinically important because when the upper limb is immobilized for whatever cause, the hand should be splinted, if possible, in the position of function. This ensures that if there is a loss of movement at the wrist joint, or at the joints of the hand or fingers, the patient will have a hand that is in a position of mechanical advantage, one that can serve a useful purpose.

9. This man had a pulp-space infection of the right index finger. If treatment is delayed, the infection may spread proximally and involve the synovial sheath of the flexor digitorum profundus tendon; further spread may involve the thenar fascial space of the palm. In children the danger involves the rise in tension within the pulp space and occlusion of the blood supply to the diaphysis of the terminal phalanx. Osteomyelitis of the terminal phalanx may also occur.

The purpose of incising an established pulp space infection is to reduce tension within the space and provide adequate drainage of the numerous compartments between the vertical septa (see p. 670).

10. Fractured metacarpal bones show dorsal angulation caused by forward pull of the long flexor tendons and the lumbricals and interossei on the distal fragment. When flexed individually, all fingers

(excluding the thumb) point toward the tubercle of the scaphoid. When a finger is unstable following a fracture, it should be aligned so that its tip points to the scaphoid tubercle; failure to achieve this will result in malfunction.

11. This girl had a paronychia. The common site for entry of pathogenic organisms is the junction between the lateral edge of the nail, the nail bed, and the lateral nail fold. Once established, the infection spreads beneath the proximal nail fold and may sometimes extend underneath the proximal edge of the nail.

The lymphatic drainage of the index finger follows the cephalic vein and ultimately drains into the deltopectoral nodes; a few may enter the lateral axillary nodes.

12. The midpalmar and thenar spaces of the palm are situated deep to the palmar aponeurosis and in front of the metacarpal bones (for details see p. 643). The structures lying within the spaces are described on p. 644). The midpalmar space may be drained through an incision 1 or 2 mm proximal to the distal transverse crease of the palm and should extend from the third to the fifth metacarpal bones. The thenar space is opened through a curved incision on the dorsal surface of the thumb web.

13. This patient avulsed the insertion of the dorsal extensor expansion into the distal phalanx of the left middle finger. The fold of the sheet suddenly caused flexion of the phalanx while the extensor tendons going to the extensor expansion were taut. The last 20° of active extension was lost in the terminal interphalangeal joint, producing mallet finger. The finger is splinted in extension for 6 weeks. Too much hyperextension, however, may produce ischemia and necrosis of the thin eponychial skin.

17

The Bony Pelvis and the Lower Extremity

The main functions of the bony pelvis are to transmit the weight of the body from the vertebral column to the femurs to contain, support, and protect the pelvic viscera and to provide attachment for trunk and lower limb muscles. Structurally the pelvis is very strong, and it takes a tremendous amount of force to fracture the pelvis. For this reason, when fractures of the pelvis do occur, the likelihood of associated extensive injuries of both bone and soft tissues is high. The common complications include hemorrhage, bladder and urethral injuries, rectal injuries, and other skeletal injuries. Men between 20 and 40 years old are at highest risk, whereas young children with deformable pelves are at least risk.

Lower extremity problems are some of the most common dealt with by an emergency physician. Fractures, dislocations, sprains, lacerations, knee effusions, leg pain, ankle injuries, and peripheral nerve injuries are just a few of the conditions seen in the emergency department.

In this chapter the basic anatomy of the bony pelvis and the lower extremity is reviewed in relation to common clinical situations. The vascular anatomy of the area has been dealt with in Chapter 14.

BASIC ANATOMY

Basic Anatomy of the Bony Pelvis

The bony pelvis provides a strong, stable connection between the trunk and the lower extremities. It is composed of four bones—the two *innominate bones*, or *hip bones*, which form the lateral and anterior walls, and the *sacrum* and the *coccyx*, which are part of the vertebral column, form the back wall (Fig 17–1).

The two innominate bones articulate with each other anteriorly at the *symphysis pubis* and posteri-

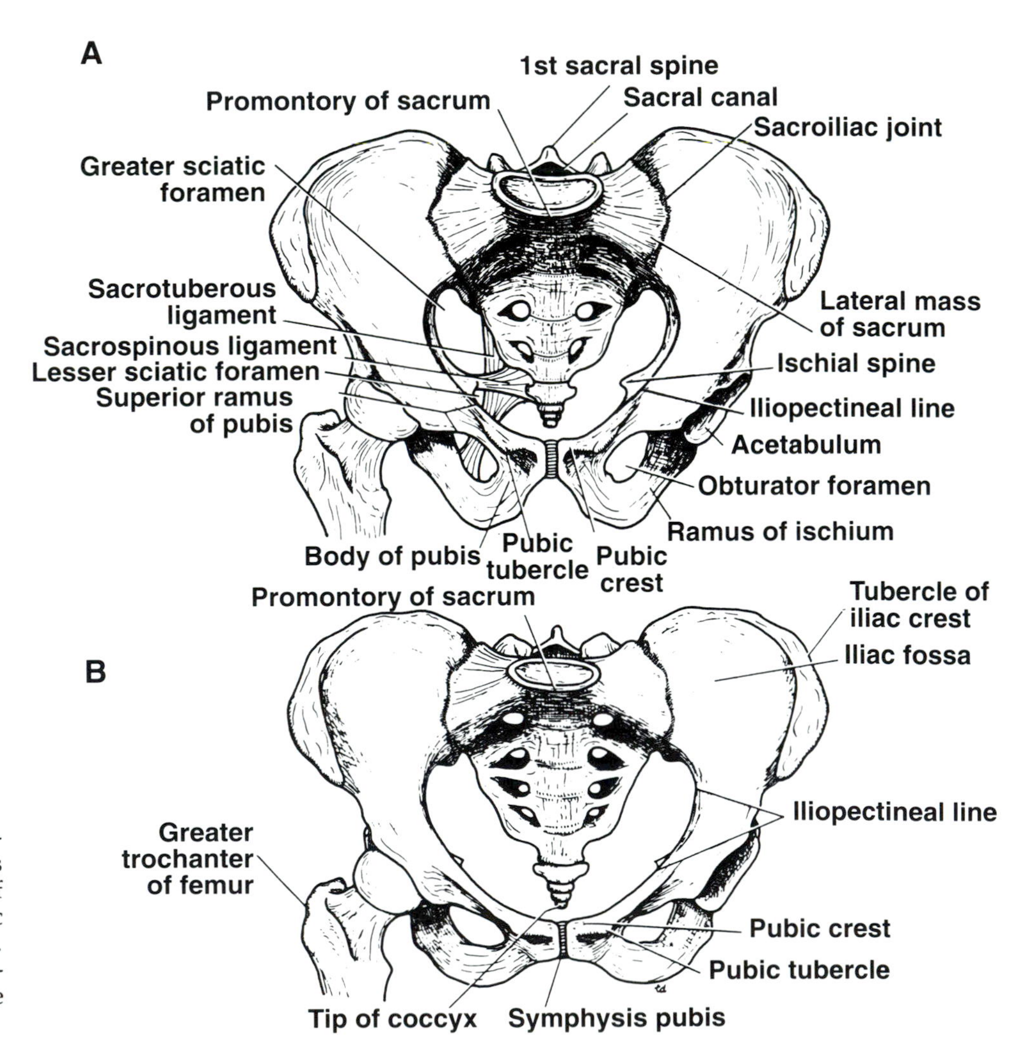

FIG 17–1.
A, male pelvis showing sacrotuberous and sacrospinous ligaments. Note position of the greater and lesser sciatic foramina. **B,** female pelvis. Note position of the iliopectineal line and the wide subpubic angle.

orly with the sacrum at the *sacroiliac joints*. The bony pelvis with its joints form a strong basin-shaped structure that contains and protects the lower parts of the intestinal and urinary tracts and the internal organs of reproduction.

The pelvis is divided into two parts by the *pelvic brim*, which is formed by the *sacral promontory* (anterior and upper margin of first sacral vertebra) behind, the *iliopectineal lines* (a line that runs downward and forward around the inner surface of the ileum) laterally, and the *symphysis pubis* (joint between bodies of pubic bones) anteriorly. Above the brim is the *false pelvis*, which forms part of the abdominal cavity. Below the brim is the *true pelvis*.

True Pelvis

This is a bowel-shaped structure that has an inlet, an outlet, and a cavity (Fig 17–2). The *pelvic inlet* is bounded posteriorly by the sacral promontory, laterally by the iliopectineal lines, and anteriorly by the symphysis pubis. The *pelvic outlet* is bounded posteriorly by the coccyx, laterally by the ischial tuberosities, and anteriorly by the pubic arch. The pelvic outlet does not present a smooth outline but has three wide notches. Anteriorly the *pubic arch* lies between the ischiopubic rami, and laterally there are the sciatic notches. The sciatic notches are divided by the *sacrotuberous* and *sacrospinous ligaments* (see Fig 17–2) into *greater* and *lesser sciatic foramina*.

Clinical Note

Pelvic outlet.—From an obstetrical standpoint, since the sacrotuberous and sacrospinous ligaments are strong and relatively inflexible, they should be considered to form part of the perimeter of the pelvic outlet. Thus, the outlet is diamond shaped, with the ischiopubic rami and the symphysis pubis forming the boundaries in front, and the sacrotuberous ligaments and the coccyx forming the boundaries behind.

The *pelvic cavity* lies between the inlet and the outlet. It is a short, curved canal with a shallow anterior wall and a much deeper posterior wall.

Innominate or Hip Bone

In children, each innominate bone consists of the ilium, which lies superiorly, the ischium, which lies posteriorly and inferiorly, and the pubis, which lies anteriorly and inferiorly (Fig 17–3). At puberty these three bones fuse together to form one large, irregular bone. The hip bones articulate with the sacrum at the sacroiliac joints and form the anterolateral walls of the pelvis; they also articulate with one another anteriorly at the symphysis pubis.

On the outer surface of the innominate bone is a deep depression, the *acetabulum*, which articulates with the hemispherical head of the femur (see Fig

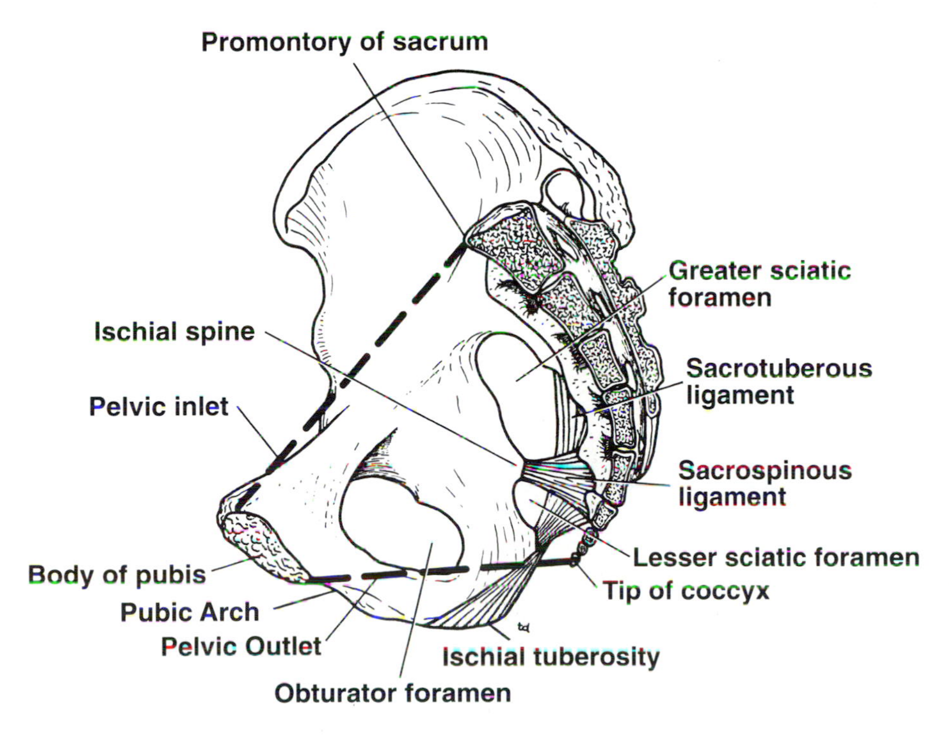

FIG 17–2.
Right half of pelvis showing the pelvic inlet, the pelvic outlet, and the sacrotuberous and sacrospinous ligaments.

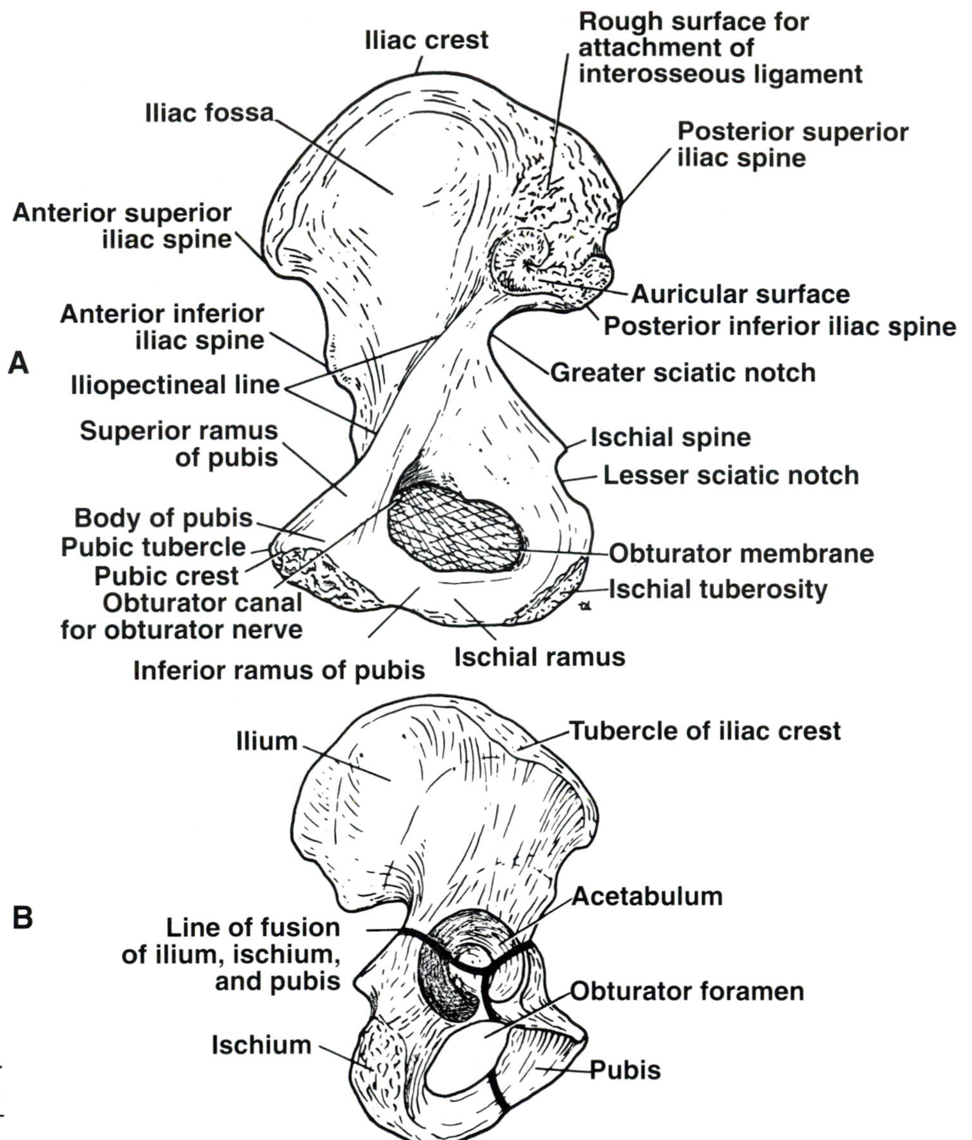

FIG 17–3.
Right innominate (hip) bone. **A,** medial surface. **B,** lateral surface. Note the lines of fusion between the three bones—the ilium, ischium, and the pubis.

17–3). Behind the acetabulum is a large notch, the *greater sciatic notch,* which is separated from the *lesser sciatic notch* by the *spine of the ischium.* As mentioned previously, the sciatic notches are converted into the *greater* and *lesser sciatic foramina* by the presence of the *sacrotuberous* and *sacrospinous ligaments* (see Fig 17–2). The inferior margin of the acetabulum is deficient and is marked by the *acetabular notch.* The articular surface of the acetabulum is limited to a horseshoe-shaped area and is covered with hyaline cartilage. The floor of the acetabulum is nonarticular and is called the *acetabular fossa.*

The *ilium,* which is the upper flattened part of the hip bone, possesses the *iliac crest* (see Fig 17–3). The *iliac crest* runs between the *anterior* and *posterior superior iliac spines.* Below these spines are the corresponding *anterior* and *posterior inferior iliac spines.* On the inner surface of the ilium is the large *auricular surface* for articulation with the sacrum. The *iliopectineal line* runs downward and forward around the inner surface of the ilium and serves to divide the false from the true pelvis.

The *ischium* is the inferior and posterior part of the innominate bone and possesses an *ischial spine* and an *ischial tuberosity* (see Fig 17–3).

The *pubis* is the anterior part of the innominate bone and has a *body* and *superior* and *inferior pubic rami.* The body of the pubis bears the *pubic crest* and the *pubic tubercle* and articulates with the pubic bone of the opposite side at the *symphysis pubis* (see Fig 17–1).

In the lower part of the innominate bone is a

large opening, the *obturator foramen,* which is bounded by the parts of the ischium and pubis (see Fig 17–1). The obturator foramen is filled in by the *obturator membrane* (see Fig 17–3).

Sacrum and Coccyx

These are considered in detail with the vertebral column in Chapter 10.

The Pelvis as a Basin With Holes in Its Walls

The walls of the pelvis are formed by bones and ligaments; these are partly lined with muscles (obturator internus and piriformis) covered with fascia and parietal peritoneum. On the outside of the pelvis are the attachments of the gluteal muscles and the obturator externus muscle. The greater part of the bony pelvis is thus sandwiched between inner and outer muscles.

The pelvis has anterior, posterior, and lateral walls, and it has an inferior wall or floor formed by the important levator ani and coccygeus muscles (see Chapter 13).

The posterior wall has holes on the anterior surface of the sacrum, the *anterior sacral foramina,* for the passage of the anterior rami of the sacral spinal nerves. The two ligaments, the *sacrotuberous* and *sacrospinous ligaments,* convert the greater and lesser sciatic notches into the *greater and lesser sciatic foramina.* The greater sciatic foramen provides an exit from the pelvis into the gluteal region for the sciatic nerve, the pudendal nerve, and the gluteal nerves and vessels; the lesser sciatic foramen provides an entrance into the perineum from the gluteal region for the pudendal nerve and internal pudendal vessels.

The lateral pelvic wall has a large hole, the *obturator foramen,* which is closed by the *obturator membrane* except for a small opening that permits the obturator nerve to leave the pelvis and enter the thigh.

Attached to the iliac spines and the ischial tuberosity are strong muscles that can, with extreme traction, pull off the bony projections. The sartorius muscle is attached to the anterior superior iliac spine, the straight head of the rectus femoris is attached to the anterior inferior iliac spine, and the hamstring muscles are attached to the ischial tuberosity.

Joints of the Bony Pelvis

Sacroiliac Joint.—This includes the following.

Articulation.—The articular surfaces of the sacrum and the ilium (Fig 17–4). The articular surfaces are irregular and interlock with one another.

Type.—Synovial gliding joint.

Capsule.—Encloses the joint.

Ligaments.—The *anterior sacroiliac ligament* is a thickening of the capsule. The *interosseous sacroiliac ligament* is very strong. It is situated above and behind the joint cavity and binds the bones together. The *posterior sacroiliac ligament* overlies the interosseous ligament (see Fig 17–4).

Accessory Ligaments.—The *sacrotuberous ligament* connects the back of the sacrum to the ischial tuberosity (see Fig 17–4). The *sacrospinous ligament* lies anterior to the sacrotuberous ligament and connects the sacrum to the ischial spine. The *iliolumbar ligament* connects the tip of the transverse process of the fifth lumbar vertebra to the iliac crest.

Synovial Membrane.—Lines the capsule.

Nerve Supply.—Branches of the sacral spinal nerves.

Movements.—A small amount of movement is possible. During pregnancy the ligaments undergo softening in response to hormones, thus increasing the mobility and increasing the potential size of the pelvis during childbirth. Obliteration of the joint cavity occurs in both sexes after middle age.

Symphysis Pubis.—This includes the following.

Articulation.—Apposed surfaces of the bodies of the pubic bones (see Fig 17–4).

Type.—Secondary cartilaginous joint.

Ligaments.—*Superior* and *inferior ligaments* extend from one pubic bone to the other. The *interpubic disc* is composed of fibrocartilage and connects the two pubic bones together.

Movements.—Almost no movement is possible. During pregnancy the ligaments undergo softening in response to hormones, thus increasing the mobility and the potential size of the pelvis during childbirth.

Radiographic Appearances of the Pelvis

The radiographic appearances of the pelvis are shown in Figures 17–19 and 17–20.

Basic Anatomy of the Arrangement of the Lower Extremity

The primary function of the lower limbs is to support the weight of the body and to provide a stable foundation in standing, walking, and running; they have become specialized for locomotion.

Because the two hip bones articulate posteriorly with the trunk at the strong sacroiliac joints and an-

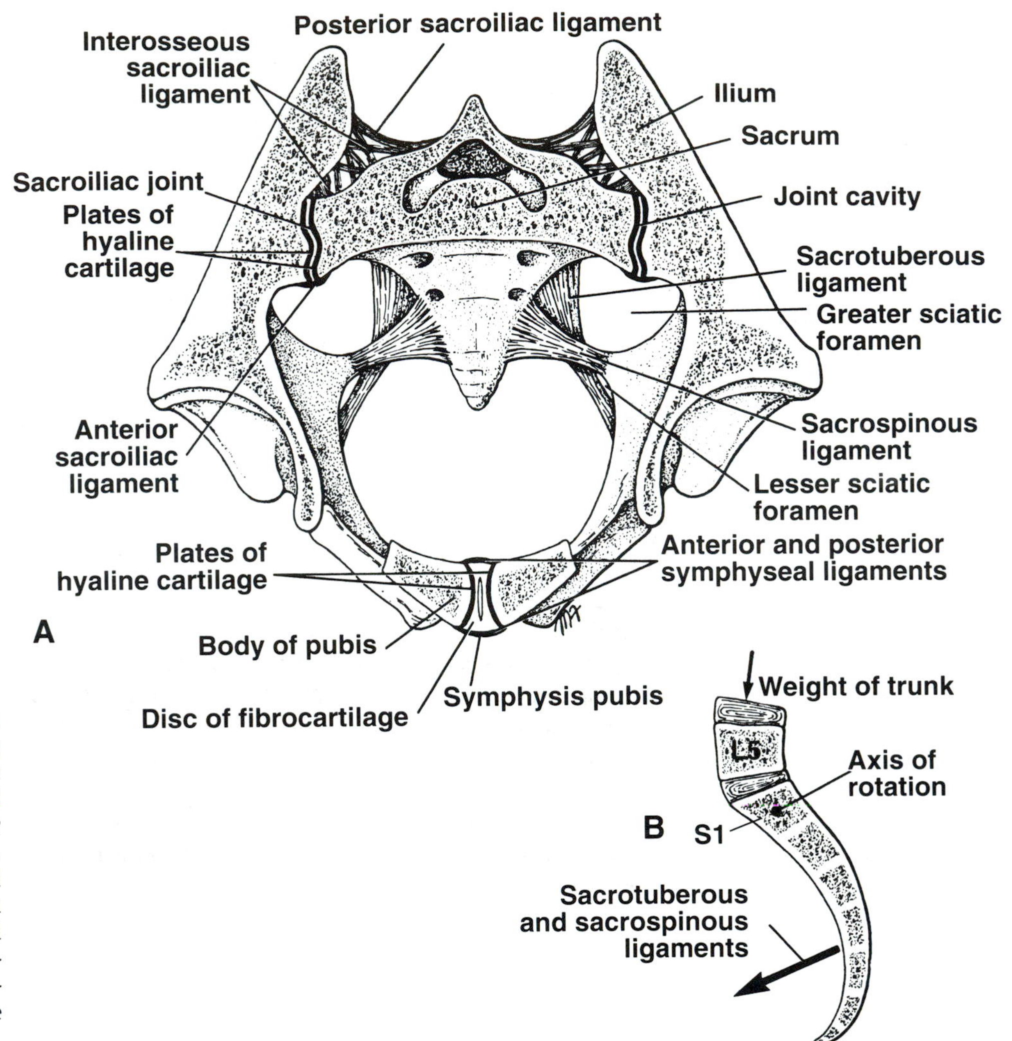

FIG 17–4.
A, horizontal section through the pelvis showing the sacroiliac joints and the symphysis pubis. **B,** vertical section through the lower part of the vertebral column, including the sacrum and coccyx, showing the function of the sacrotuberous and sacrospinous ligaments in resisting the rotation force exerted on the sacrum by the weight of the trunk.

teriorly with each other at the symphysis pubis, the lower extremities are very stable and can bear the weight of the body.

Organization of the Lower Extremity

The lower extremities are divided into different regions and compartments. Each compartment has muscles that perform group functions, and each has a distinct nerve and blood supply.

Gluteal Region.—This is bounded above by the iliac crests and inferiorly by the fold of the buttock. The region is largely made up of the gluteal muscles, supplied by the superior and inferior gluteal nerves, and a thick layer of superficial fascia.

Thigh.—The deep fascia of the thigh encloses it like a sheath. Three fascial septa pass from the inner aspect of the sheath to be attached to the posterior surface of the femur (Fig 17–5). By this means the thigh is divided into three compartments, each having muscles, nerves, and arteries. The bulk of the femur thus lies in the anterior compartment of the thigh. The *anterior compartment* contains the flexor muscles of the hip joint that are innervated by the femoral nerve. The *medial compartment* contains the adductor muscles of the hip joint that are innervated by the obturator nerve. The *posterior compartment* contains the hamstring muscles that flex the knee joint and extend the hip joint; they are supplied by the sciatic nerve. There is no lateral compartment for the muscles that abduct the hip joint, namely, the gluteus medius and minimus muscles, the sartorius, the tensor fasciae latae, and the piriformis muscles.

Leg.—The deep fascia of the leg encloses it as a sheath that is attached to the anterior and medial borders of the shaft of the tibia. Two intermuscular

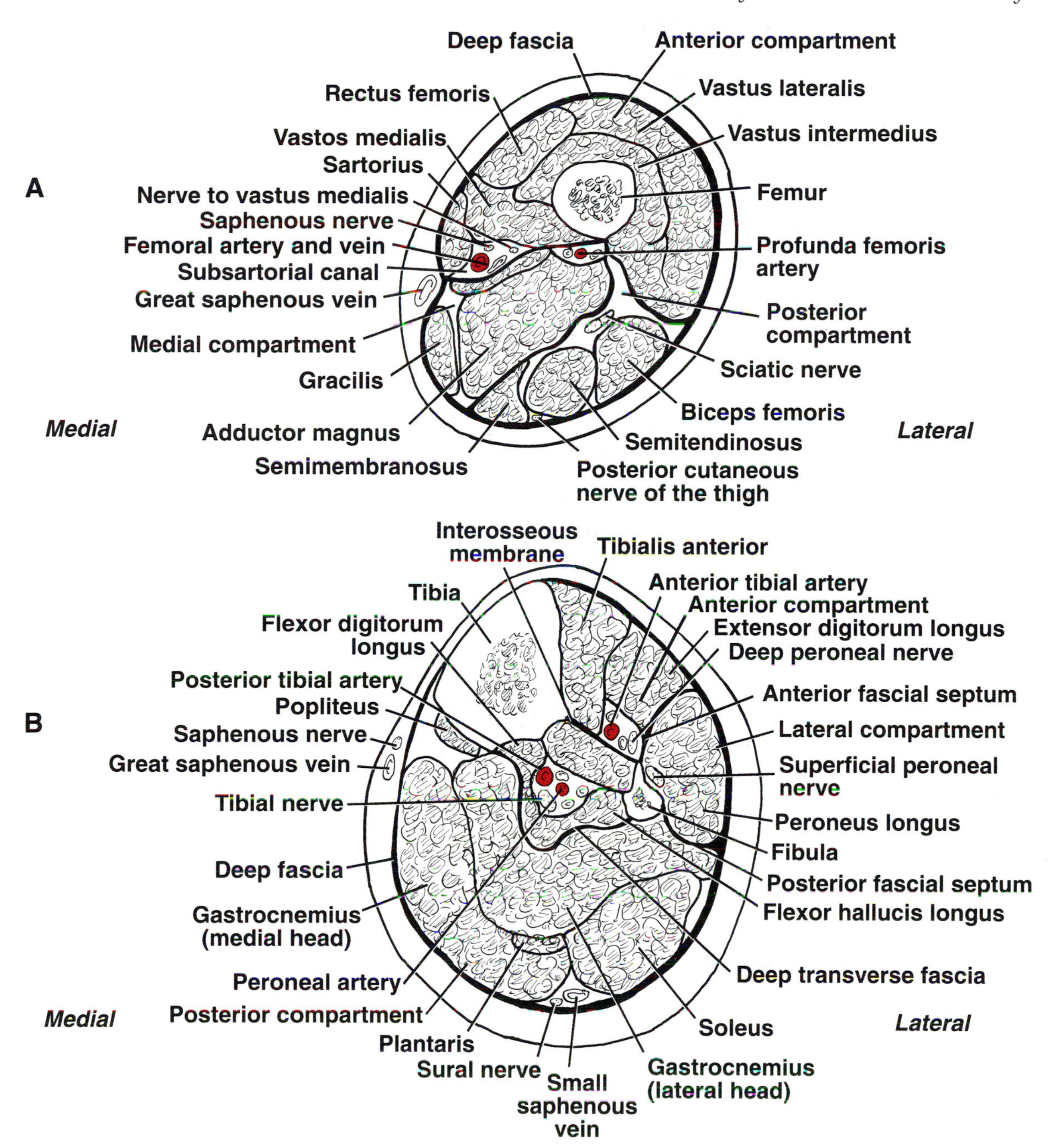

FIG 17–5.
A, transverse section through the middle of the right thigh as seen from above. Note the positions of the three compartments and their contents. **B,** transverse section through the middle of the right leg as seen from above. Note the position of the three compartments and their contents.

septa pass from its deep aspect to be attached to the fibula (see Fig 17–5). These components, together with the interosseous membrane, divide the leg into three compartments. The *anterior compartment* contains the dorsiflexors of the ankle joint and some muscles that extend the toes; they are innervated by the deep peroneal nerve. The *lateral compartment* contains the plantar flexors of the ankle joint and the evertors of the foot; they are innervated by the superficial peroneal nerve. The *posterior compartment* contains the plantar flexors of the ankle joint; some of the deeper muscles flex the toes and invert the foot. The muscles of the posterior compartment are innervated by the tibial nerve.

Basic Anatomy of the Bones of the Lower Extremity

Femur

The femur articulates above with the acetabulum to form the hip joint, and below it articulates with the tibia and the patella to form the knee joint. The upper end of the femur has a head, neck, and greater and lesser trochanters (Fig 17–6). The *head* forms about two thirds of a sphere and articulates with the acetabulum of the hip bone to form the hip joint. In the center of the head there is a depression, called the *fovea capitis,* for the attachment of the ligament of the head.

The *neck,* which connects the head to the shaft, passes downward, backward, and laterally, makes an angle of about 160° in young children and about 125° in adults (slightly less in the female) with the long axis of the shaft. The neck of the femur receives a profuse blood supply from the branches of the medial femoral circumflex artery. These branches pierce the capsule of the hip joint and ascend the neck deep to the synovial membrane. As long as the epiphyseal cartilage remains, there is no communica-

---

Clinical Note

Blood supply of the femoral head.—Part of the blood supply from the obturator artery to the femoral head is conveyed along this ligament and enters the bone at the fovea (see Fig 17–16). This fact is important in the understanding of avascular necrosis of the femoral head.

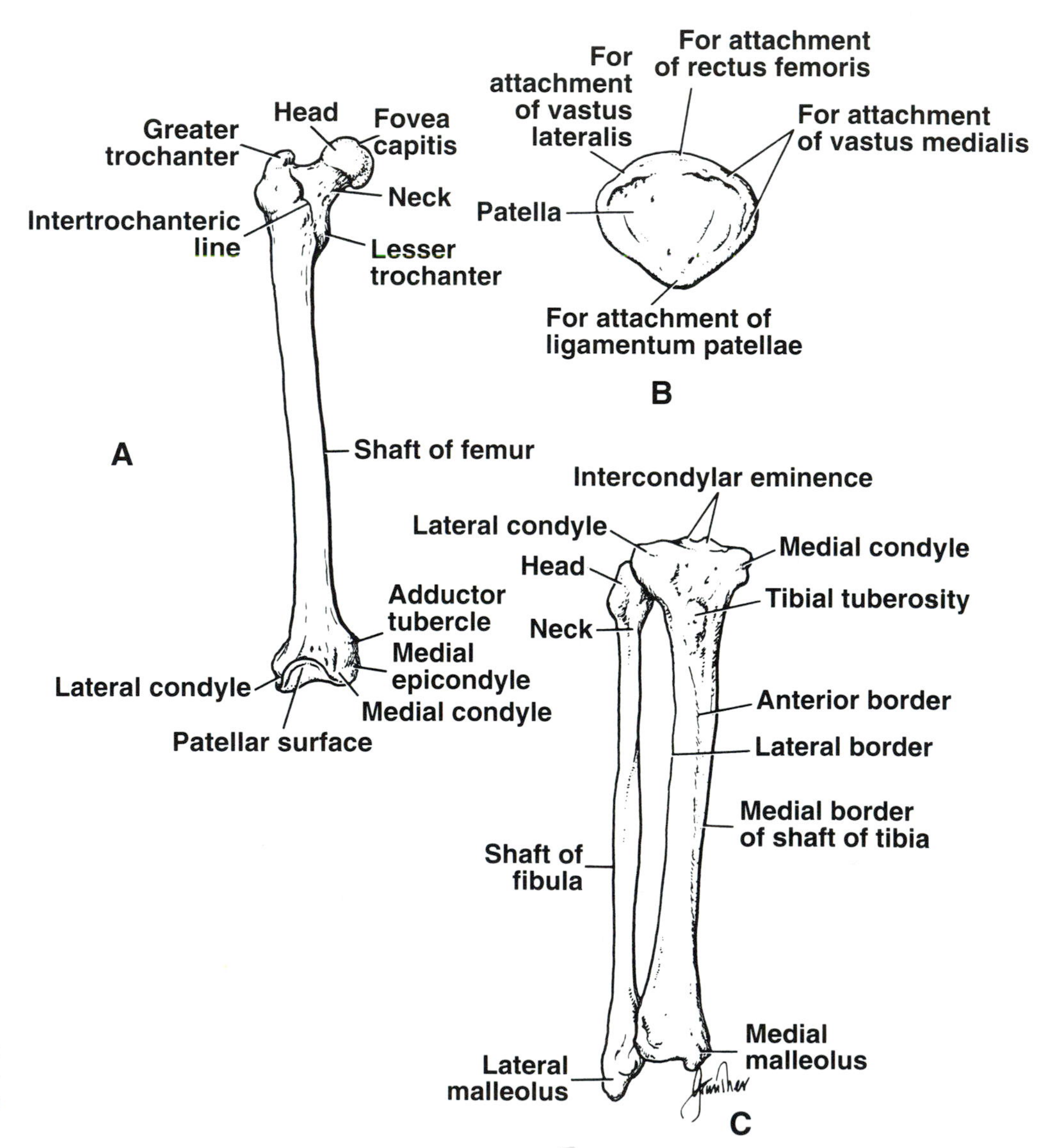

FIG 17–6.
A, right femur, anterior view. **B,** right patella, anterior view. **C,** right tibia and fibula, anterior view.

tion between the blood supply to the femoral head from the obturator artery and the blood supply to the neck (see p. 730).

Clinical Note

Angle of the femoral neck.—The size of this angle may change with conditions such as a fractured femoral neck, slipped femoral epiphysis, or congenital dislocation of the hip (see p. 728).

The *greater* and *lesser trochanters* are large eminences situated at the junction of the neck and the shaft (see Fig 17–6). Connecting the two trochanters are the *intertrochanteric line,* anteriorly, and a prominent *intertrochanteric crest,* posteriorly, on which lies the *quadrate tubercle* for the attachment of the quadratus femoris muscle.

The *shaft* of the femur shows a general forward convexity. It is smooth on its anterior surface but posteriorly has a ridge, the *linea aspera.* The margins of the linea aspera diverge above and below. The medial margin continues below as the *medial supracondylar ridge* to the *adductor tubercle* on the medial condyle (Fig 17–6). The lateral margin becomes continuous below with the *lateral supracondylar ridge.* On the posterior surface of the shaft below the greater trochanter is the *gluteal tuberosity* for the insertion of the gluteus maximus muscle. The shaft becomes broader toward its distal end and forms a flat triangular area on its posterior surface, called the *popliteal surface.*

The lower end of the femur has *lateral* and *medial condyles* separated posteriorly by the *intercondylar notch.* The anterior surfaces of the condyles are joined by an articular surface for the patella. Above the condyles are the *medial* and *lateral epicondyles* (see Fig 17–6). The *adductor tubercle* is continuous with the medial epicondyle.

Patella

The patella (see Fig 17–6) is a sesamoid bone lying within the quadriceps tendon that is situated in an exposed position in front of the knee. It is triangular and its apex lies inferiorly; the apex is connected to the tuberosity of the tibia by the ligamentum patellae.

The posterior surface articulates with the condyles of the femur. The upper, lateral, and me-

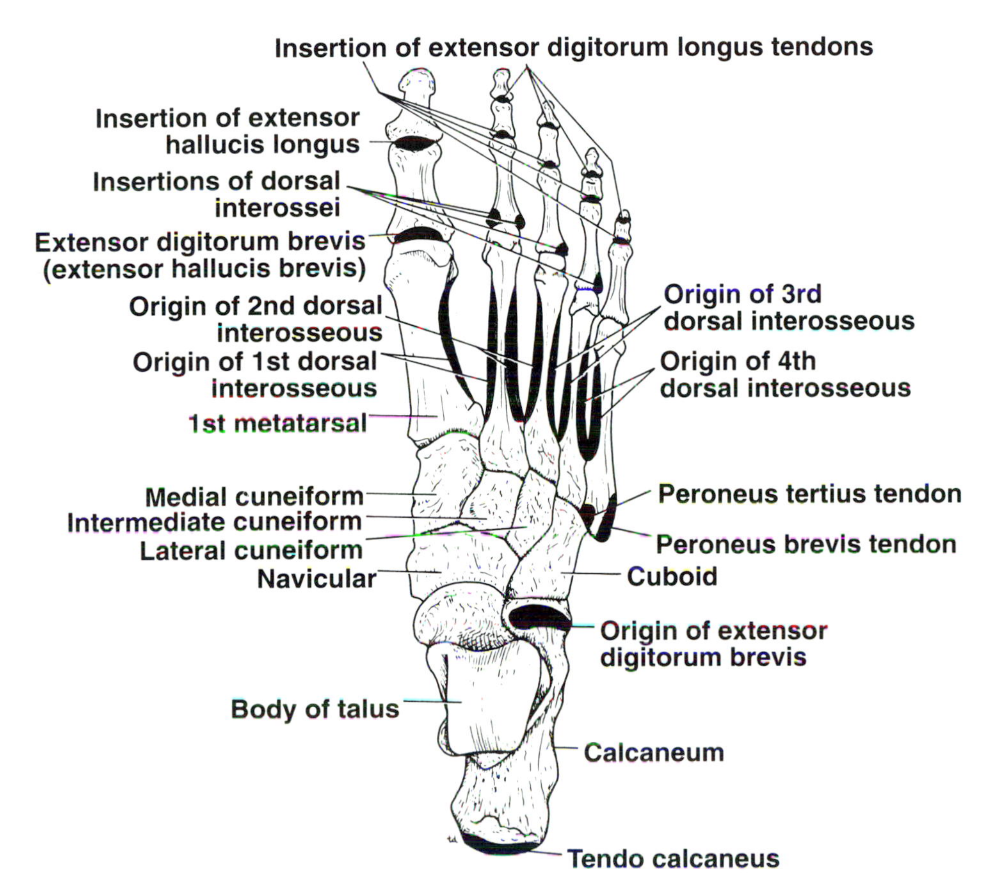

FIG 17–7.
Dorsal aspect of bones of the right foot showing the muscle attachments.

dial margins of the patella give attachment to the different parts of the quadriceps femoris muscle.

Clinical Note

Displacement of the patella.—Normally the patella is prevented from displacement laterally during the action of the quadriceps muscle. This is due to the action of the lower horizontal fibers of the vastus medialis, which pull the patella medially, and by the large size of the lateral condyle of the femur, which prevents lateral displacement of the patella (see Fig 17–52).

Tibia

The tibia is the large medial bone of the leg (see Fig 17–6). It supports the greater part of the weight on the leg. The tibia articulates with the condyles of the femur and the head of the fibula above, and with the talus and the distal end of the fibula below. It has an expanded upper end, a smaller lower end, and a shaft.

At the upper end the *lateral and medial condyles* articulate with the lateral and medial condyles of the femur, with the *lateral and medial meniscal cartilages* intervening. The intercondylar eminence separates the upper articular surfaces of the tibial condyles. The lateral condyle possesses on its lateral aspect an *oval articular facet for the head of the fibula.*

The *shaft of the tibia* is triangular in cross section, presenting three borders and three surfaces. Its anterior and medial borders, with the medial surface between them, are subcutaneous. The *tuberosity* is at the junction of the anterior border with the upper end of the tibia; it receives the attachment of the ligamentum patellae. The anterior border becomes rounded below, where it becomes continuous with the medial malleolus (see Fig 17–6). The lateral or interosseous border gives attachment to the *interosseous membrane,* which binds the tibia and the fibula together. The posterior surface of the shaft shows an oblique line, the *soleal line,* for the attachment of the soleus muscle.

The lower end of the tibia is slightly expanded. On its inferior aspect it shows a saddle-shaped articular surface for the talus. The lower end is prolonged downward and medially to form the *medial malleolus.* The lateral surface of the medial malleolus articulates with the talus. The lower end of the tibia shows a wide, rough depression on its lateral surface for articulation with the fibula.

Fibula

The fibula is the slender, needle-shaped bone on the lateral side of the leg (see Fig 17–6). It takes no part in the articulation at the knee joint, but below it forms the lateral malleolus of the ankle joint. It has an expanded upper end, a shaft, and a lower end.

The *upper end, or head,* has a *styloid process* and possesses an *articular surface* for articulation with the lateral condyle of the tibia.

The *shaft of the fibula* is long and slender. The medial or *interosseous border* gives attachment to the interosseous membrane.

The *lower end of the fibula* forms the *lateral malleolus,* which is subcutaneous. On the medial surface of the lateral malleolus is an *articular facet* for articulation with the lateral aspect of the talus.

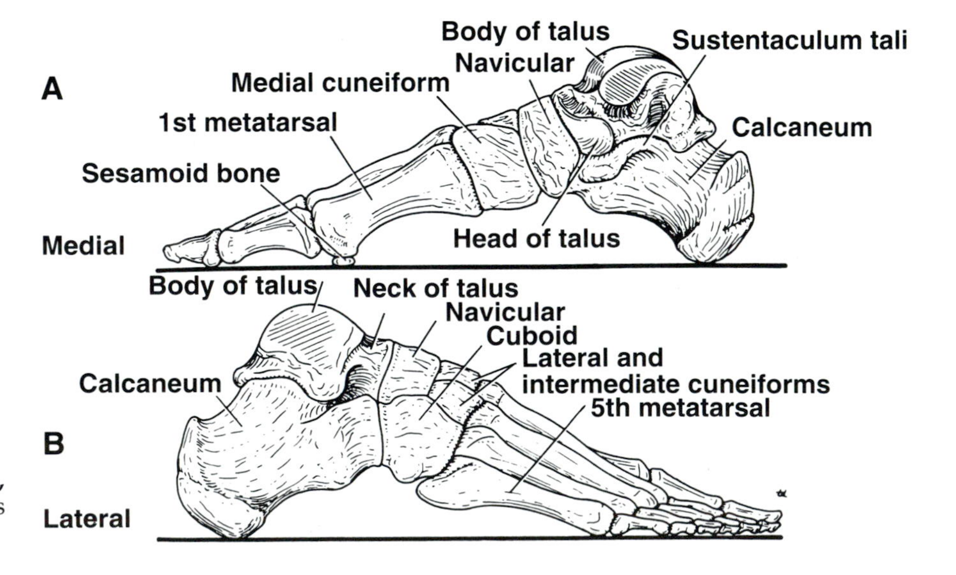

FIG 17–8.
Articulated bones of the right foot. **A,** as seen from the medial aspect. **B,** as seen from the lateral aspect.

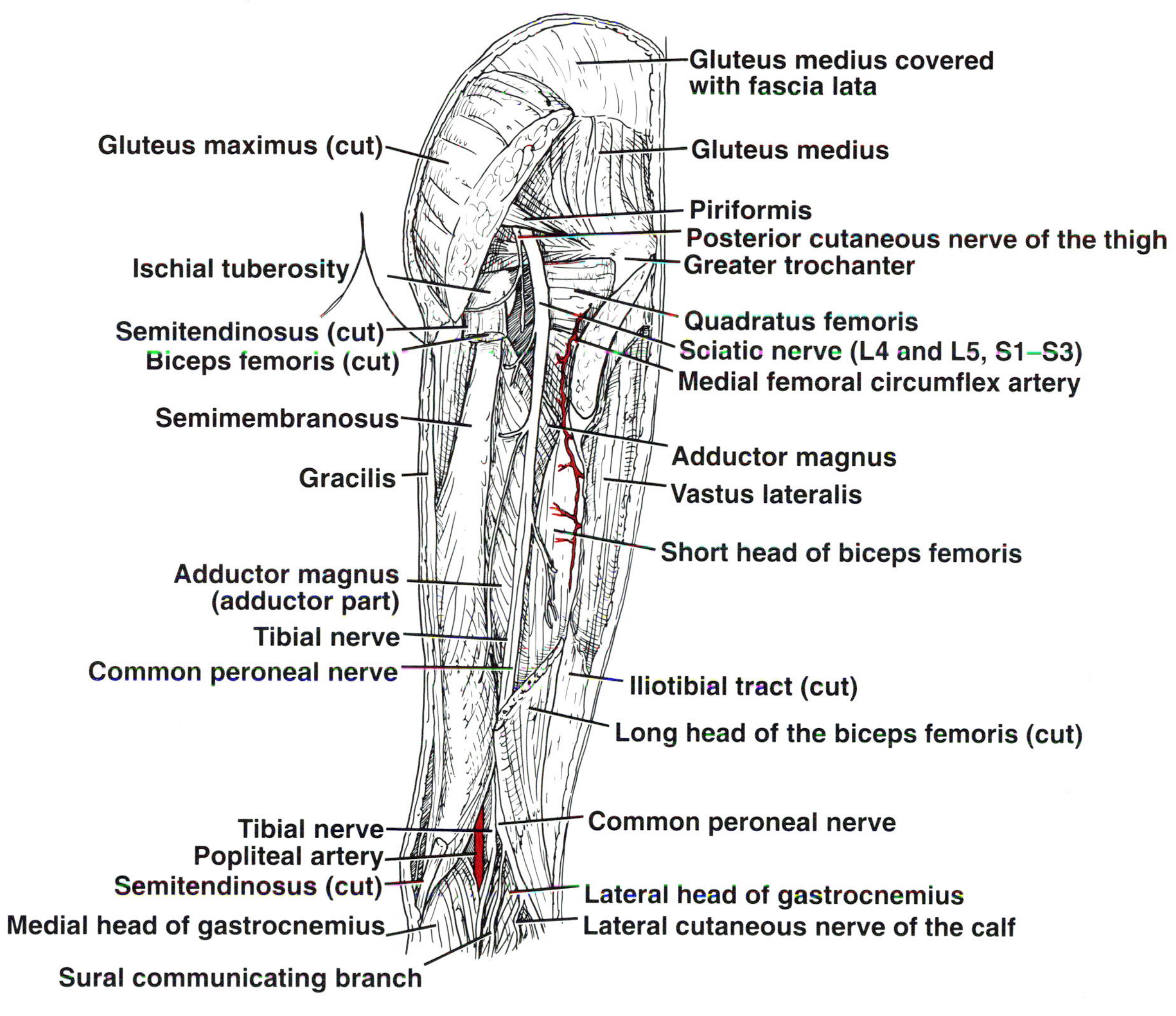

FIG 17–9.
Right gluteal region and the posterior view of the thigh. The greater part of the gluteus maximus and the long head of the biceps femoris have been removed to display the sciatic nerve; the semitendinosus muscle has also been removed.

The Bones of the Foot

The anatomy of the different bones of the foot is summarized in Figures 17–7 and 17–8.

Basic Anatomy of the Muscles of the Lower Extremity

Gluteal Region

The muscles of the gluteal region occupy the buttock and are shown in Table 17–1 (see also Fig 17–9).

Thigh Region

Deep Fascia of the Thigh (Fascia Lata).—The deep fascia encloses the thigh like a trouser leg, and at its upper end it is attached to the pelvis and its associated ligaments. On its lateral aspect it is thickened to form the *iliotibial tract,* which is attached above to the tubercle of the iliac crest and below to the lateral condyle of the tibia. The iliotibial tract receives the insertion of the tensor fasciae latae and the greater part of the gluteus maximus muscle. The *saphenous opening* (see p. 568) is a gap in the deep fascia in the front of the thigh that transmits the great saphenous vein, some small branches of the femoral artery, and lymph vessels.

Fascial Compartments of the Thigh.—Three fascial septa pass from the inner aspect of the deep fascial sheath of the thigh to the linea aspera of the femur (see Fig 17–5). By this means, the thigh is divided into three compartments, already referred to, each having muscles, nerves, and arteries. The compartments are anterior, medial, and posterior.

TABLE 17–1.
Muscles of the Buttock or Gluteal Region

Name of Muscle	Origin	Insertion	Nerve Supply	Nerve Root*	Action
Gluteus maximus	Outer surface of ilium, sacrum, coccyx, sacrotuberous ligament	Iliotibial tract and gluteal tuberosity of femur	Inferior gluteal nerve	L5, **S1, S2**	Extends and laterally rotates thigh at hip joint; through iliotibial tract it extends knee joint
Gluteus medius	Outer surface of ilium	Greater trochanter of femur	Superior gluteal nerve	**L5,** S1	Abducts thigh at hip joint; tilts pelvis when walking
Gluteus minimus	Outer surface of ilium	Greater trochanter of femur	Superior gluteal nerve	**L5,** S1	Abducts thigh at hip joint; anterior fibers medially rotate thigh
Tensor fasciae latae	Iliac crest	Iliotibial tract	Superior gluteal nerve	L4, L5	Assists gluteus maximus in extending the knee joint
Piriformis	Anterior surface of sacrum	Greater trochanter of femur	First and second sacral nerves	L5, **S1,** S2	Lateral rotator of thigh at hip joint
Obturator internus	Inner surface of obturator membrane	Greater trochanter of femur	Sacral plexus	L5, **S1**	Lateral rotator of thigh at hip joint
Gemellus superior	Spine of ischium	Greater trochanter of femur	Sacral plexus	L5, S1	Lateral rotator of thigh at hip joint
Gemellus inferior	Ischial tuberosity	Greater trochanter of femur	Sacral plexus	L5, S1	Lateral rotator of thigh at hip joint
Quadratus femoris	Ischial tuberosity	Quadrate tubercle on upper end of femur	Sacral plexus	L5, S1	Lateral rotator of thigh at hip joint
Obturator externus	Outer surface of obturator membrane	Greater trochanter of femur	Obturator nerve	L3, **L4**	Lateral rotator of thigh at hip joint

*The predominant nerve root supply is indicated by boldface type.

TABLE 17–2.
Muscles of the Anterior Fascial Compartment of the Thigh

Name of Muscle	Origin	Insertion	Nerve Supply	Nerve Root*	Action
Sartorius	Anterior superior iliac spine	Upper medial surface shaft of tibia	Femoral nerve	L2, L3	Flexes, abducts, laterally rotates thigh at hip joint; flexes and medially rotates leg at knee joint
Iliacus	Iliac fossa	With psoas into lesser trochanter of femur	Femoral nerve	**L2,** L3	Flexes thigh on trunk; if thigh is fixed, it flexes the trunk on the thigh as in sitting up from lying down
Psoas	Twelfth thoracic vertebral body; transverse processes, bodies and intervertebral discs of the five lumbar vertebrae	With iliacus into lesser trochanter of femur	Lumbar plexus	**L1, L2,** L3	Flexes thigh on trunk; if thigh is fixed, it flexes the trunk on the thigh as in sitting up from lying down
Pectineus	Superior ramus of pubis	Upper end shaft of femur	Femoral nerve	**L2,** L3	Flexes and adducts thigh at hip joint
Quadriceps femoris					
Rectus femoris	Straight head: anterior inferior iliac spine; reflected head: ilium above acetabulum	Quadriceps tendon into patella	Femoral nerve	L2, **L3, L4**	Extension of leg at knee joint
Vastus lateralis	Upper end and shaft of femur	Quadriceps tendon into patella	Femoral nerve	L2, **L3, L4**	Extension of leg at knee joint
Vastus medialis	Upper end and shaft of femur	Quadriceps tendon into patella	Femoral nerve	L2, **L3, L4**	Extension of leg at knee joint
Vastus intermedius	Shaft of femur	Quadriceps tendon into patella	Femoral nerve	L2, **L3, L4**	Extension of leg at knee joint

*The predominant nerve root supply is indicated by boldface type.

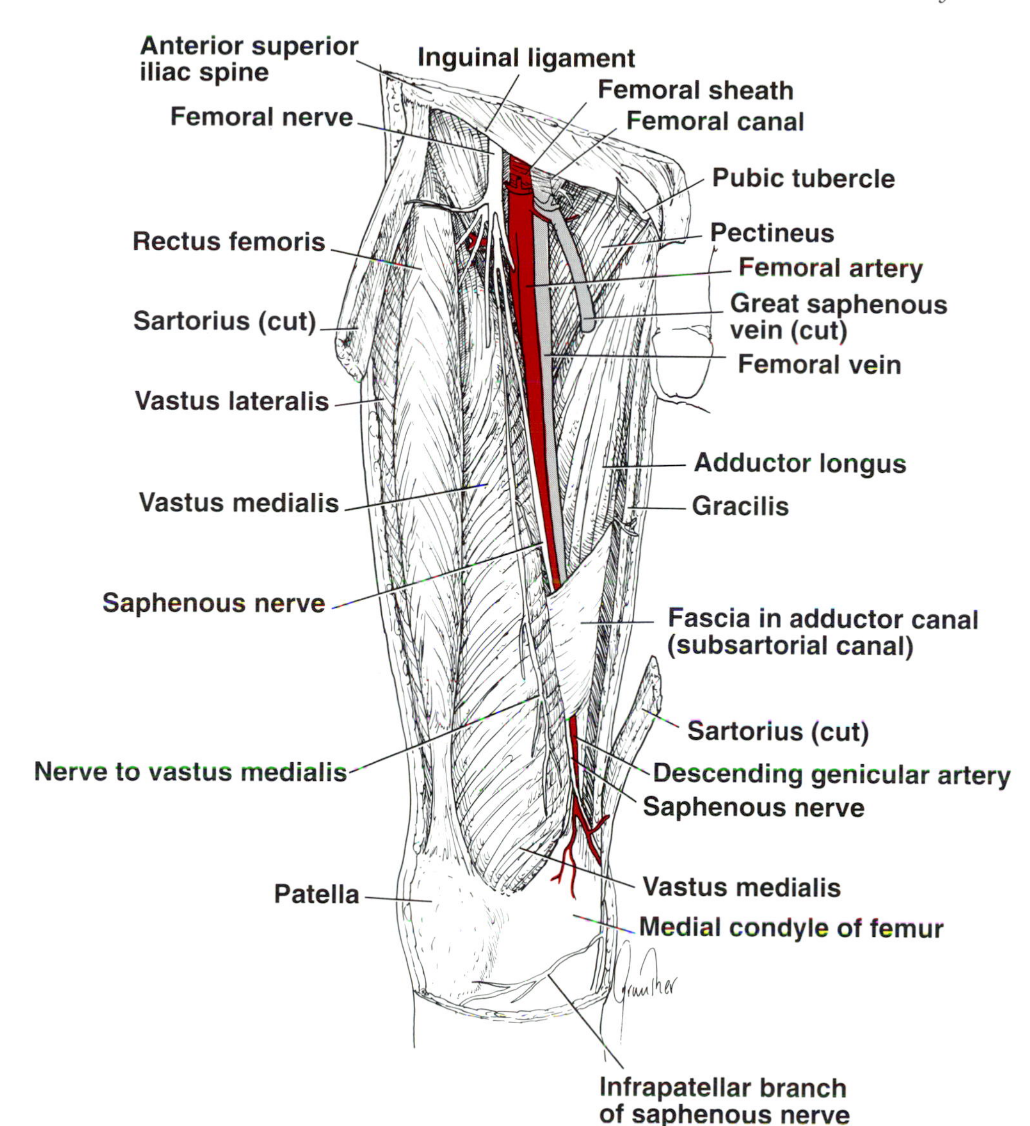

FIG 17–10.
Front of the thigh in the right lower limb showing the femoral vessels and nerve. The sartorius muscle has been cut and reflected to show the underlying adductor (subsartorial) canal and its contents.

Muscles of the Thigh.—The muscles of the anterior compartment of the thigh are shown in Table 17–2 (see also Fig 17–10). The muscles of the medial or adductor compartment of the thigh are shown in Table 17–3 (see also Fig 17–11). The muscles of the posterior fascial compartment of the thigh are shown in Table 17–4 (see also Fig 17–9).

Femoral Triangle.—The femoral triangle is a triangular area situated in the upper part of the medial aspect of the thigh. Its boundaries are as follows: base, inguinal ligament; lateral border, sartorius muscle; and medial border, adductor longus muscle. The femoral triangle contains the terminal part of the femoral nerve and its branches, the femoral sheath, the femoral artery and its branches, the femoral vein and its tributaries, and the inguinal lymph nodes (see Fig 17–10).

Adductor Canal (or Subsartorial Canal).—The adductor canal is an intermuscular cleft situated on the medial aspect of the middle third of the thigh beneath the sartorius muscle (see Fig 17–5). It contains the femoral artery and vein, the deep lymph vessels, and the saphenous nerve. The femoral artery leaves the adductor canal by passing through an opening in the adductor magnus muscle.

Clinical Note

Popliteal aneurysm.—The pulsations of the wall of the femoral artery against the tendon of adductor magnus at the opening of the adductor magnus is thought to contribute to the cause of popliteal aneurysms.

TABLE 17–3.
Muscles of the Medial Fascial Compartment of the Thigh

Name of Muscle	Origin	Insertion	Nerve Supply	Nerve Root*	Action
Gracilis	Inferior ramus of pubis; ramus of ischium	Upper part of shaft of tibia on medial surface	Obturator nerve	**L2**, L3	Adducts thigh at hip joint; flexes leg at knee joint
Adductor longus	Body of pubis	Posterior surface of shaft of femur	Obturator nerve	L2, **L3, L4**	Adducts thigh at hip joint and assists in lateral rotation
Adductor brevis	Inferior ramus of pubis	Posterior surface of shaft of femur	Obturator nerve	L2, **L3, L4**	Adducts thigh at hip joint and assists in lateral rotation
Adductor magnus	Inferior ramus of pubis; ramus of ischium, ischial tuberosity	Posterior surface of shaft of femur; adductor tubercle of femur	Obturator nerve and sciatic nerve (hamstring part)	L2, **L3, L4**	Adducts thigh at hip joint and assists in lateral rotation; hamstring part extends thigh at hip joint

*The predominant nerve root supply is indicated by boldface type.

Knee Region

Popliteal Fossa.—The popliteal fossa is a diamond-shaped intermuscular space situated at the back of the knee (Fig 17–12). It contains the popliteal vessels, the small saphenous vein, the common peroneal and tibial nerves, the posterior cutaneous nerve of the thigh, connective tissue, and lymph nodes. Its boundaries are as follows: laterally, the biceps femoris above and the lateral head of the gastrocnemius and plantaris below; and medially, the semimembranosus and semitendinosus above and the medial head of the gastrocnemius below (see Fig 17–12).

Leg Region

Deep Fascia of the Leg and Foot.—The deep fascia surrounds the leg and is continuous above with the deep fascia of the thigh. Below the tibial condyles it is attached to the anterior and medial borders of the tibia (see Fig 17–5).

Fascial Compartments of the Leg.—Two intermuscular septa pass from the deep aspect of the deep fascia to be attached to the fibula. These, together with the interosseous membrane, divide the leg into three compartments—anterior, lateral, and posterior, each having its own muscles, blood supply, and nerve supply. The posterior compartment is usually divided by a transverse septum into a superficial compartment and a deep compartment (see Fig 17–5).

Interosseous Membrane.—The interosseous membrane is a strong membrane connecting the interosseous borders of the tibia and the fibula. The interosseous membrane binds the tibia and the fibula together and provides attachment for neighboring muscles (see Fig 17–5).

Ankle Region

Retinacula of the Ankle.—In the region of the ankle joint, the deep fascia is thickened to form a series of

TABLE 17–4.
Muscles of the Posterior Fascial Compartment of the Thigh

Name of Muscle	Origin	Insertion	Nerve Supply	Nerve Root*	Action
Biceps femoris	Long head: ischial tuberosity; short head: shaft of femur	Head of fibula	Sciatic nerve (long head: tibial nerve; short head: common peroneal nerve)	L5, **S1**, S2	Flexes and laterally rotates leg at knee joint; long head also extends thigh at hip joint
Semitendinosus	Ischial tuberosity	Upper part medial surface of shaft of tibia	Sciatic nerve (tibial portion)	**L5, S1**, S2	Flexes and medially rotates leg at knee joint; extends thigh at hip joint
Semimembranosus	Ischial tuberosity	Medial condyle of tibia; forms oblique popliteal ligament	Sciatic nerve (tibial portion)	**L5, S1**, S2	Flexes and medially rotates leg at knee joint; extends thigh at hip joint
Adductor magnus (hamstring portion)	Ischial tuberosity	Adductor tubercle of femur	Sciatic nerve (tibial portion)	L2, **L3, L4**	Extends thigh at hip joint

*The predominant nerve root supply is indicated by boldface type.

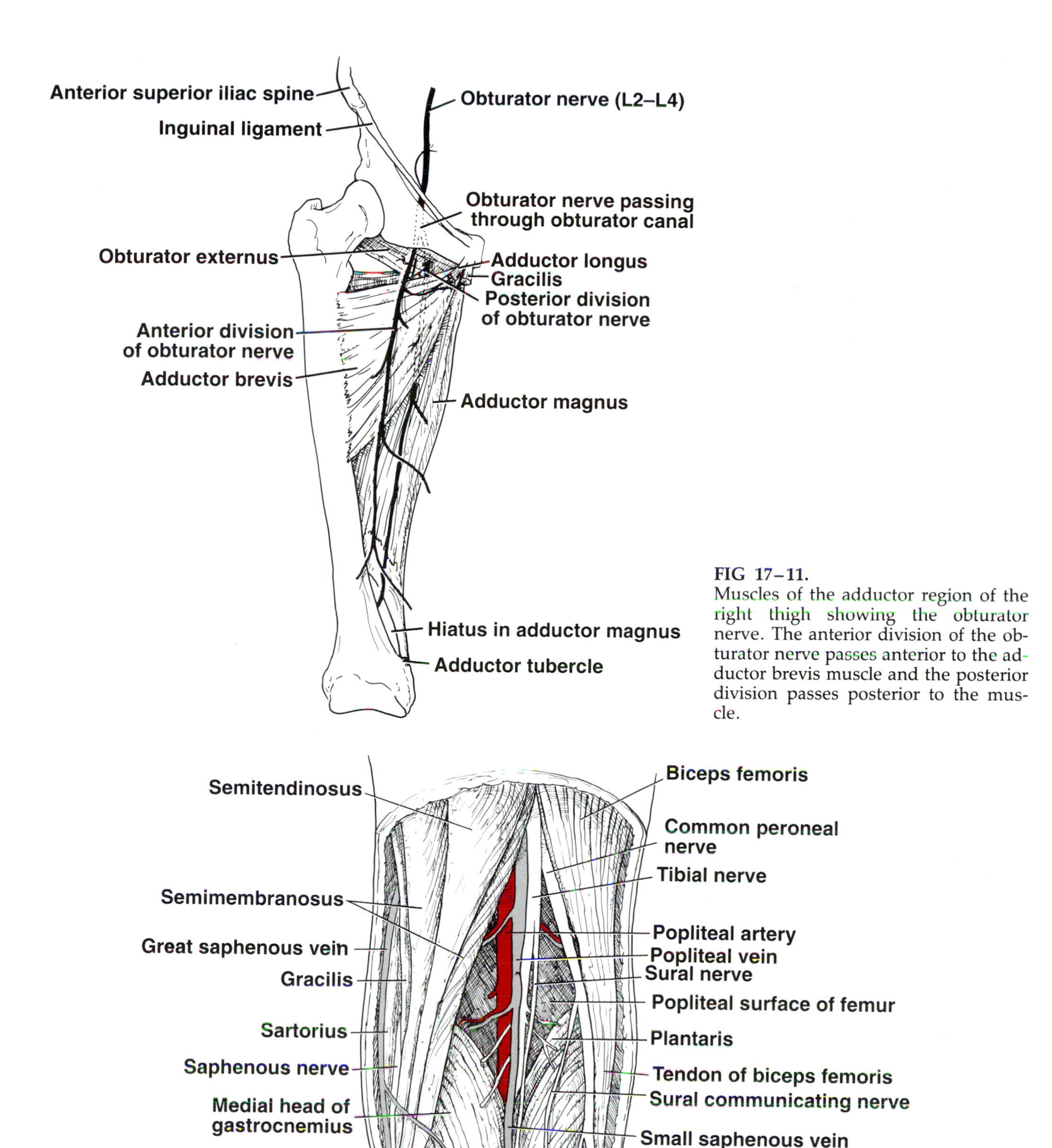

FIG 17–11.
Muscles of the adductor region of the right thigh showing the obturator nerve. The anterior division of the obturator nerve passes anterior to the adductor brevis muscle and the posterior division passes posterior to the muscle.

FIG 17–12.
Right popliteal fossa showing the boundaries and contents. The popliteal vessels, the common peroneal nerve, and the tibial nerve are displayed.

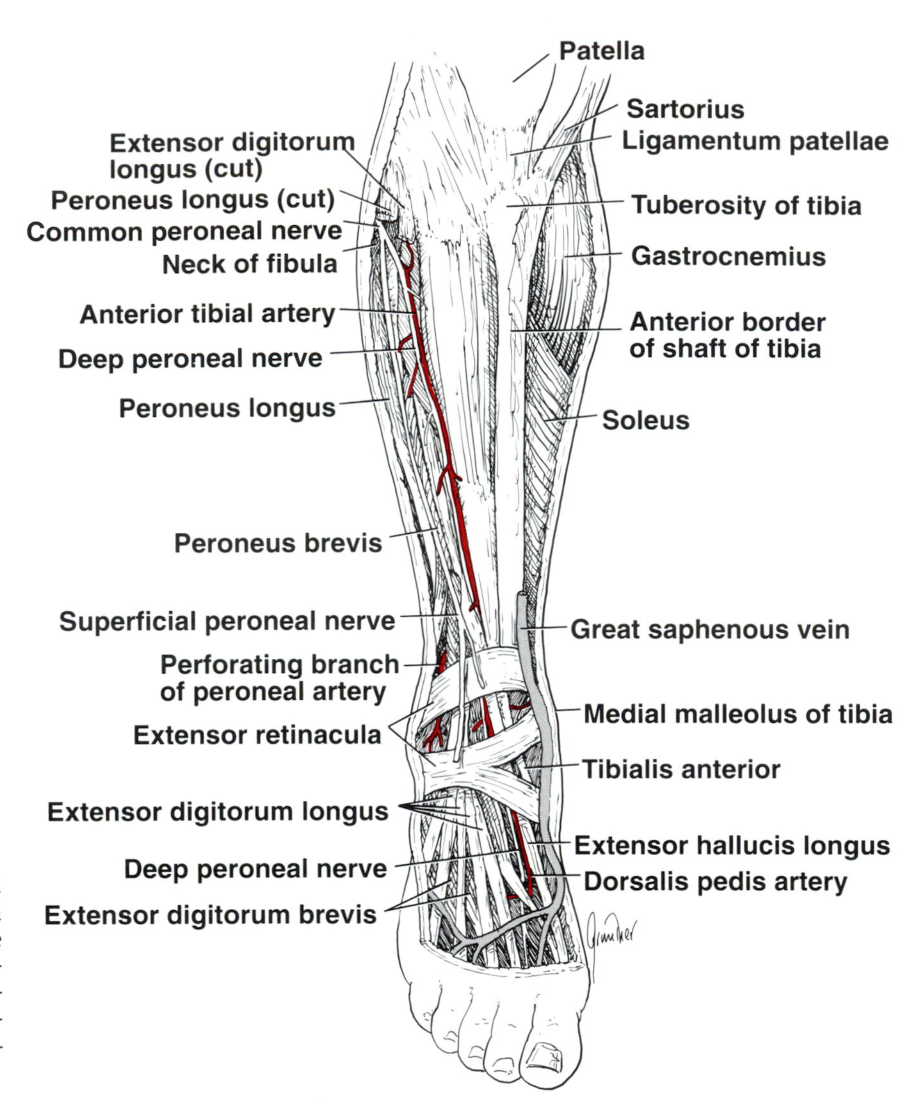

FIG 17–13.
Structures present on the anterior and lateral aspects of the right leg and on the dorsum of the foot. Portions of the peroneus longus and the extensor digitorum longus muscles have been removed to display the common peroneal, the deep peroneal, and the superficial peroneal nerves.

bands or retinacula, which serve to keep the long tendons in position and act as modified pulleys. The *superior extensor retinaculum* is attached to the distal ends of the anterior borders of the fibula and the tibia (Fig 17–13). The *inferior extensor retinaculum* is a Y-shaped band located in front of the ankle joint (see Fig 17–13). The *flexor retinaculum* extends from the medial malleolus downward and backward to be attached to the medial surface of the calcaneum. It binds the tendons of the deep muscles to the medial side of the ankle as they pass forward from behind the medial malleolus to enter the sole of the foot (see Fig 17–14). The *superior peroneal retinaculum* extends from the lateral malleolus downward and backward to be attached to the lateral surface of the calcaneum (Fig 17–14). It binds the tendons of the peroneus longus and brevis to the lateral side of the ankle. The *inferior peroneal retinaculum* is attached to the calcaneum above and below the peroneal tendons.

The muscles of the anterior, lateral, and posterior fascial compartments of the leg are shown in Tables 17–5 through 17–7 (see also Figs 17–13 and 17–15). The muscle on the dorsum of the foot is shown in Table 17–8 (see also Fig 17–13).

Sole of the Foot

The *deep fascia* is thickened to form the *plantar aponeurosis,* which is triangular and attached by its apex to the medial and lateral tubercles of the calcaneum. The base of the aponeurosis divides into five slips that pass into the toes.

The muscles of the sole are conveniently described in four layers from the inferior layer superiorly. The muscles of the sole are shown in Table 17–9.

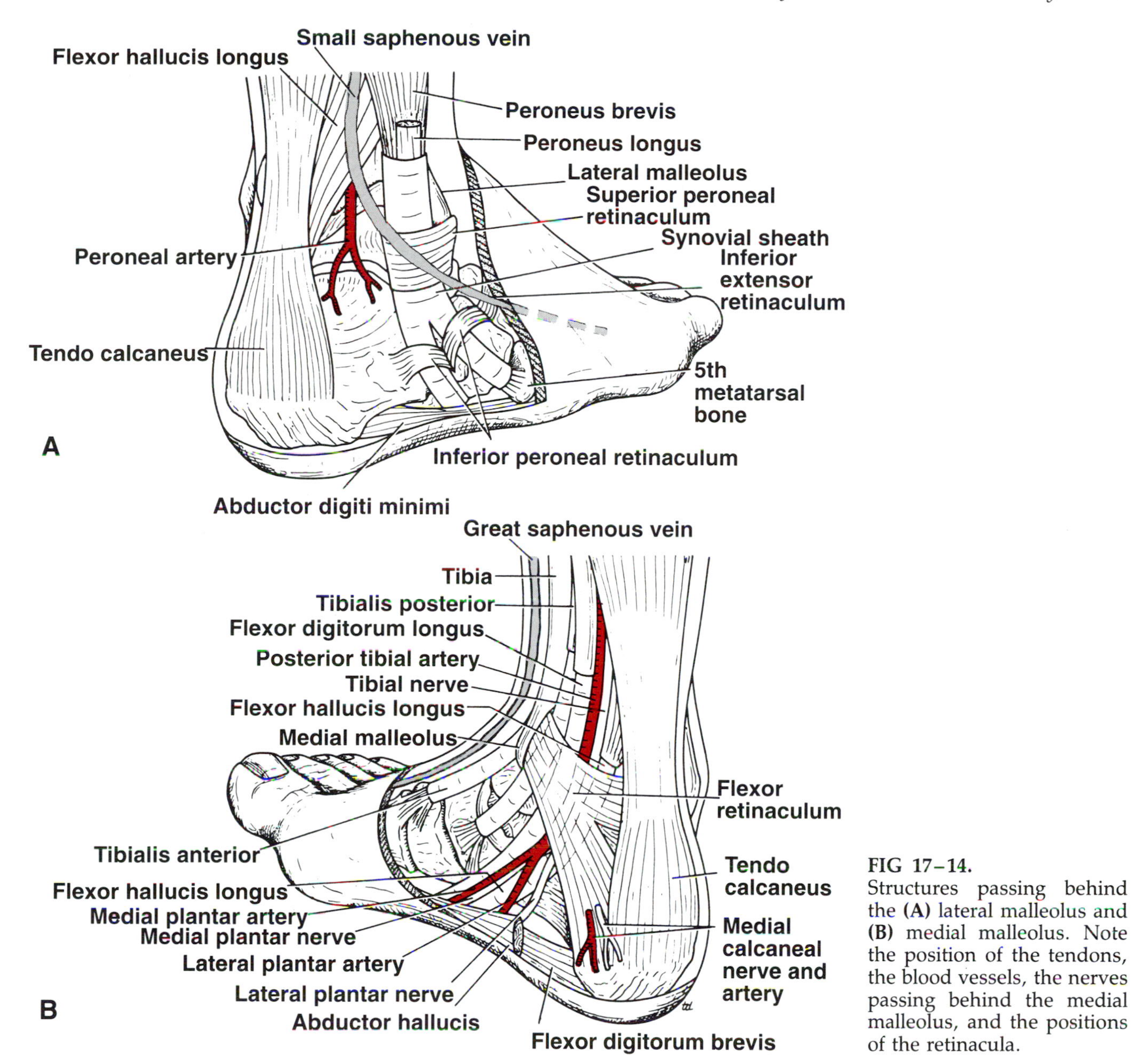

FIG 17–14.
Structures passing behind the **(A)** lateral malleolus and **(B)** medial malleolus. Note the position of the tendons, the blood vessels, the nerves passing behind the medial malleolus, and the positions of the retinacula.

Basic Anatomy of the Hip Joint

Articulation.—The hemispherical head of the femur articulates with the cup-shaped acetabulum of the hip bone (Fig 17–16). The articular surface of the acetabulum is horseshoe shaped and is deficient inferiorly at the *acetabular notch*. The cavity of the acetabulum is deepened by the presence of a fibrocartilaginous rim called the *acetabular labrum*. The labrum bridges across the acetabular notch and is here called the *transverse acetabular ligament*. The articular surfaces are covered with hyaline cartilage.

Type.—Synovial ball-and-socket joint.

Capsule.—This encloses the joint and is attached medially to the acetabular labrum (see Fig 17–16). Laterally it is attached to the intertrochanteric line of the femur in front and halfway along the posterior aspect of the neck of the bone behind.

Ligaments.—These include the following.

Iliofemoral Ligament.—This is very strong and shaped like an inverted Y (Fig 17–17). Its base is attached to the anterior inferior iliac spine above, and

TABLE 17–5.
Muscles of the Anterior Fascial Compartment of the Leg

Name of Muscle	Origin	Insertion	Nerve Supply	Nerve Root*	Action
Tibialis anterior	Shaft of tibia and interosseous membrane	Medial cuneiform and base of first metatarsal bone	Deep peroneal nerve	**L4,** L5	Extends† the foot at ankle joint; inverts foot at subtalar and transverse tarsal joints; holds up medial longitudinal arch of foot
Extensor digitorum longus	Shaft of fibula and interosseous membrane	Extensor expansion of lateral four toes	Deep peroneal nerve	L5, S1	Extends toes; dorsiflexes foot at ankle joint
Peroneus tertius	Shaft of fibula and interosseous membrane	Base of fifth metatarsal bone	Deep peroneal nerve	L5, S1	Dorsiflexes foot at ankle joint; everts foot at subtalar and transverse tarsal joints
Extensor hallucis longus	Shaft of fibula and interosseous membrane	Base of distal phalanx of great toe	Deep peroneal nerve	L5, S1	Extends big toe; dorsiflexes foot at ankle joint; inverts foot at subtalar and transverse tarsal joints

*The predominant nerve root supply is indicated by boldface type.
†Extension, or dorsiflexion, of the ankle is the movement of the foot away from the ground.

the two limbs of the Y are attached to the upper and lower parts of the intertrochanteric line of the femur.

Pubofemoral Ligament.—This is triangular (see Fig 17–17). The base is attached to the superior ramus of the pubis, and the apex is attached below to the lower part of the intertrochanteric line.

Ischiofemoral Ligament.—This spiral-shaped ligament is attached to the body of the ischium and attached laterally to the greater trochanter (see Fig 17–17).

Ligament of the Head of the Femur.—This is flat and triangular (see Fig 17–16). It is attached by its apex to the fovea capitis of the femur and by its base to the transverse acetabular ligament, which lies within the joint and is ensheathed by synovial membrane (see Fig 17–16).

Synovial Membrane.—This lines the capsule and is attached to the margins of the articular surfaces (see Fig 17–16). It covers the portion of the neck of the femur that lies within the joint capsule. It ensheathes the ligament of the head of the femur and covers the fat in the acetabular fossa. It frequently communicates with the *psoas bursa* beneath the psoas tendon.

Nerve Supply.—Femoral, obturator, and sciatic nerves and the nerve to the quadratus femoris.

TABLE 17–6.
Muscles of the Lateral Fascial Compartment of the Leg

Name of Muscle	Origin	Insertion	Nerve Supply	Nerve Root*	Action
Peroneus longus	Shaft of fibula	Base of first metatarsal and the medial cuneiform	Superficial peroneal nerve	**L5, S1,** S2	Plantar flexes foot at ankle joint; everts foot at subtalar and transverse tarsal joints; supports lateral longitudinal and transverse arches of foot
Peroneus brevis	Shaft of fibula	Base of fifth metatarsal bone	Superficial peroneal nerve	**L5, S1,** S2	Plantar flexes foot at ankle joint; everts foot at subtalar and transverse tarsal joint; holds up lateral longitudinal arch of foot

*The predominant nerve root supply is indicated by boldface type.

TABLE 17–7.
Muscles of the Posterior Fascial Compartment of the Leg

Name of Muscle	Origin	Insertion	Nerve Supply	Nerve Root*	Action
Superficial Group					
Gastrocnemius	Medial and lateral condyles of femur	Via Achilles tendon into calcaneum	Tibial nerve	**S1,** S2	Plantar flexes foot at ankle joint; flexes knee joint
Plantaris	Lateral supracondylar ridge of femur	Calcaneum	Tibial nerve	**S1,** S2	Plantar flexes foot at ankle joint; flexes knee joint
Soleus	Shafts of tibia and fibula	Via Achilles tendon into calcaneum	Tibial nerve	S1, **S2**	Together with gastrocnemius and plantaris is powerful plantar flexor of ankle joint; provides main propulsive force in walking and running
Deep Group					
Popliteus	Lateral condyle of femur	Shaft of tibia	Tibial nerve	L4, L5, S1	Flexes leg at knee joint; unlocks knee joint by lateral rotation of femur on tibia and slackens ligaments of joint
Flexor digitorum longus	Shaft of tibia	Distal phalanges of lateral four toes	Tibial nerve	**S2,** S3	Flexes distal phalanges of lateral four toes; plantar flexes foot; supports medial and lateral longitudinal arches of foot
Flexor hallucis longus	Shaft of fibula	Base of distal phalanx of big toe	Tibial nerve	**S2,** S3	Flexes distal phalanx of big toe; plantar flexes foot at ankle joint; supports medial longitudinal arch of foot
Tibialis posterior	Shafts of tibia and fibula and interosseous membrane	Tuberosity of navicular bone	Tibial nerve	L4, L5	Plantar flexes foot at ankle joint; inverts foot at subtalar and transverse tarsal joints; supports medial longitudinal arch of foot

*The predominant nerve root supply is indicated by boldface type.

Clinical Note

Referred pain.—The femoral nerve not only supplies the hip joint but, via the intermediate and medial cutaneous nerves of the thigh, also supplies the skin of the front and medial side of the thigh. It is not surprising, therefore, for pain originating in the hip joint to be referred to the front and medial side of the thigh. The posterior division of the obturator nerve supplies both the hip and knee joints. This would explain why hip joint disease sometimes gives rise to pain in the knee joint.

Movements.—The hip joint has a wide range of movement, but less so than the shoulder joint. The strength of the joint depends largely on the shape of the bones taking part in the articulation and on the very strong ligaments. When the knee is flexed, flexion is limited by the anterior surface of the thigh coming into contact with the anterior abdominal wall. When the knee joint is extended, flexion of the hip joint is limited by the tension of the hamstring group of muscles on the back of the thigh. Extension, which is the movement of the flexed thigh backward to the anatomical position, is limited by the tension of the iliofemoral ligament. Abduction is limited by the tension of the pubofemoral ligament, and adduction is limited by contact with the opposite limb and by the tension in the ligament of the head of the femur. Lateral (external) rotation is limited by the tension in the iliofemoral and pubofemoral ligaments, and medial (internal) rotation is limited by the ischiofemoral ligament. The following movements take place.

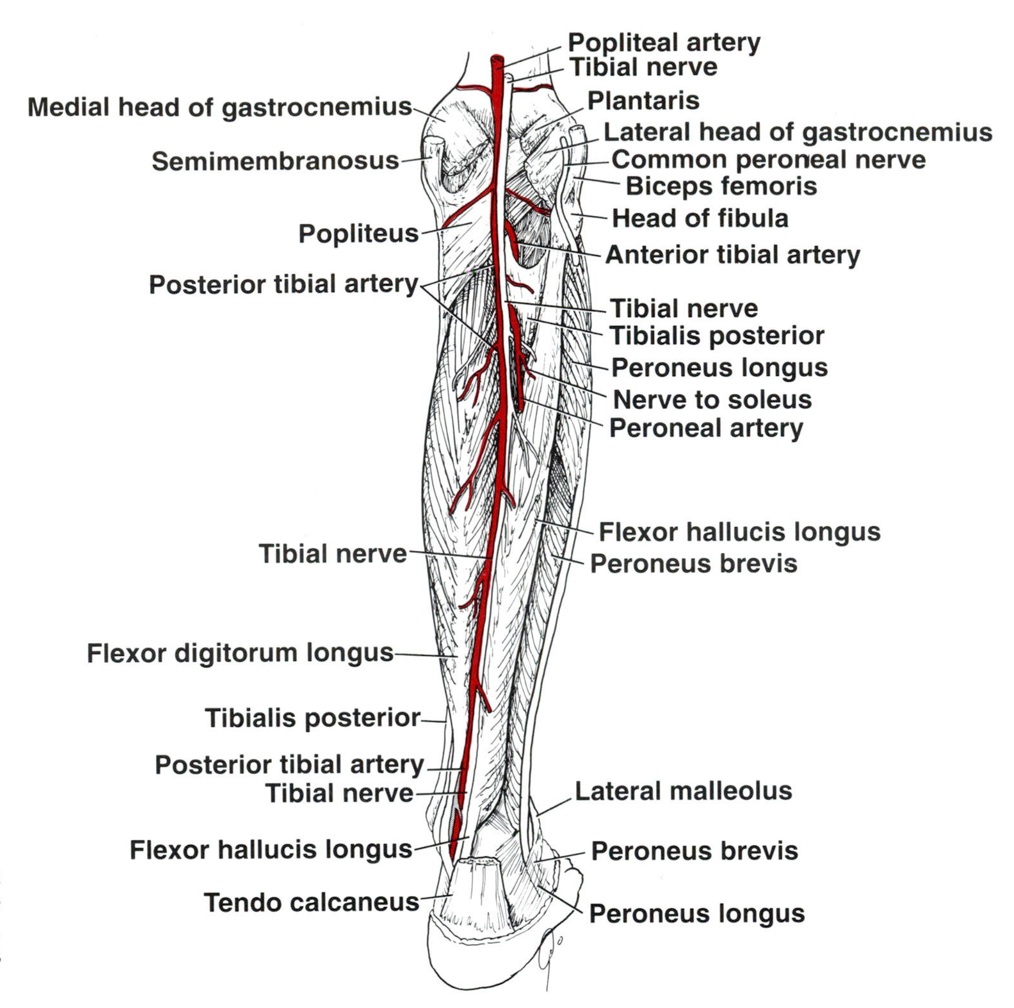

FIG 17–15.
Deep structures present on the posterior surface of the right leg. The gastrocnemius, soleus, and plantaris muscles have been removed.

Flexion.—Iliopsoas, rectus femoris, sartorius, and adductor muscles.

Extension.—Gluteus maximus and hamstring muscles.

Abduction.—Gluteus medius and minimus, sartorius, tensor fasciae latae, and piriformis.

Adduction.—Adductor longus and brevis, adductor fibers of adductor magnus, pectineus, and gracilis.

Lateral Rotation.—Piriformis, obturator internus and externus, superior and inferior gemelli, quadratus femoris, and gluteus maximus.

Medial Rotation.—Anterior fibers of gluteus medius and minimus, and tensor fasciae latae.

Circumduction.—A combination of the above movements.

Important Relations.—These include the following.

Anteriorly.—Femoral vessels and nerve.

Posteriorly.—Sciatic nerve (Fig 17–18).

Radiographic Appearances of the Hip Region

The views commonly used are the anteroposterior and the lateral views.

TABLE 17–8.
Muscle on the Dorsum of the Foot

Name of Muscle	Origin	Insertion	Nerve Supply	Nerve Root	Action
Extensor digitorum brevis	Calcaneum	By four tendons into the proximal phalanx of big toe and long extensor tendons to second, third, and fourth toes	Deep peroneal nerve	S1, S2	Extends toes

TABLE 17–9.
Muscles of the Sole

Name of Muscle	Origin	Insertion	Nerve Supply	Nerve Root*	Action
First Layer					
Abductor hallucis	Medial tubercle of calcaneum, flexor retinaculum	Medial side, base proximal phalanx big toe	Medial plantar nerve	S2, **S3**	Flexes, abducts big toe; supports medial longitudinal arch
Flexor digitorum brevis	Medial tubercle of calcaneum	Middle phalanx of four lateral toes	Medial plantar nerve	S2, **S3**	Flexes lateral four toes; supports medial and lateral longitudinal arches
Abductor digiti minimi	Medial and lateral tubercles of calcaneum	Lateral side base proximal phalanx fifth toe	Lateral plantar nerve	S2, **S3**	Flexes, abducts fifth toe; supports lateral longitudinal arch
Second Layer					
Flexor accessorius (quadratus plantae)	Medial and lateral sides calcaneum	Tendon flexor digitorum longus	Lateral plantar nerve	S2, **S3**	Aids long flexor tendon to flex lateral four toes
Flexor digitorum longus	(See Table 17–7)	Base of distal phalanx of lateral four toes	Tibial nerve	**S2,** S3	Flexes distal phalanges of lateral four toes; plantar flexes foot; supports longitudinal arches
Lumbricals	Tendons of flexor digitorum longus	Dorsal extensor expansion of lateral four toes	First lumbrical medial plantar; remainder deep branch lateral plantar nerve	S2, **S3**	Extends toes at interphalangeal joints
Flexor hallucis longus	(See Table 17–7)	Base of distal phalanx of big toe	Tibial nerve	**S2,** S3	Flexes distal phalanx of big toe; plantar flexes foot; supports medial longitudinal arch
Third Layer					
Flexor hallucis brevis	Cuboid, lateral cuneiform bones; tibialis posterior insertion	Medial and lateral sides of base of proximal phalanx of big toe	Medial plantar nerve	S2, **S3**	Flexes metatarsophalangeal joint of big toe; supports medial longitudinal arch
Adductor hallucis					
Oblique head	Bases second, third, and fourth metatarsal bones	Lateral side base proximal phalanx big toe	Deep branch lateral plantar	S2, **S3**	Flexes big toe, supports transverse arch
Transverse head	Plantar ligaments				
Flexor digiti minimi brevis	Base of fifth metatarsal bone	Lateral side base of proximal phalanx big toe	Superior branch lateral plantar nerve	S2, **S3**	Flexes little toe
Fourth Layer					
Interossei					
Dorsal (4)	Adjacent sides of metatarsal bones	Bases of phalanges and dorsal expansion of corresponding toes	Lateral plantar nerve	S2, **S3**	Abduct toes from second toe; flex metatarsophalangeal joints; extend interphalangeal joints
Plantar (3)	Third, fourth, and fifth metatarsal bones	Bases of phalanges and dorsal expansion of corresponding toes	Lateral plantar nerve	S2, **S3**	Adduct toes to second toe; flex metatarsophalangeal joints; extend interphalangeal joints
Peroneus longus	(See Table 17–6)				
Tibialis posterior	(See Table 17–7)				

*The predominant nerve root supply is indicated by boldface type.

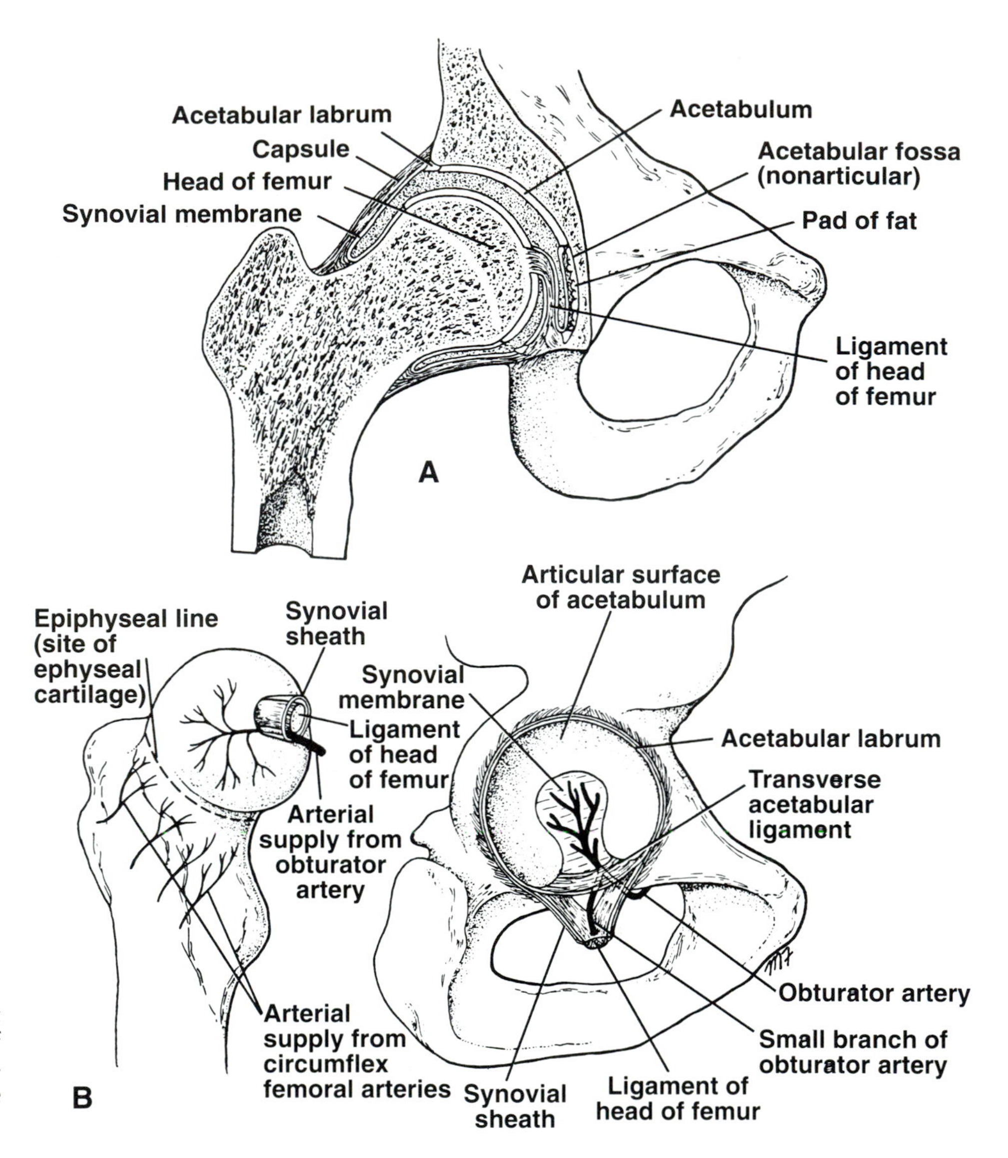

FIG 17–16.
A, coronal section of the right hip joint. **B,** articular surfaces of the right hip joint. Note the arterial supply of the head of the femur.

Anteroposterior View.—This is taken with the patient in a supine position. The film cassette is placed behind the hip, and the x-ray tube is positioned in front of the hip, centered over a point 1 in. (2.5 cm) below the midpoint of the inguinal ligament. The thigh is medially rotated slightly so that the toes touch; this is important so that the full length of the neck of the femur is visualized and it is not foreshortened. It may be desirable to view the whole pelvis so that the two hips may be compared. In this case, the entire pelvis must be symmetrical, and the x-ray tube must be centered over a point about 1 in. (2.5 cm) above the symphysis pubis.

The relevant features seen in the pelvis are first examined (Figs 17–19 and 17–20). The sacrum and sacroiliac joints should be recognized. The iliopectineal line and the symphysis pubis are well shown. The boundaries of the obturator foramen and the ischial tuberosity can be identified (Fig 17–21). The superior shelving margin of the acetabulum can be seen. The articulating surfaces of the hip joint are seen to be parallel and separated by a narrow space occupied by radiotranslucent articular cartilage. The head, neck, greater and lesser trochanters, and intertrochanteric crest of the femur can all be visualized. The spacial relationships of the hip joint should be studied (see Fig 17–21). The inferior margin of the neck of the femur should form a smooth, continuous curve with the superior margin of the obturator foramen *(Shenton's line).* The angle formed by the long axis of the neck of the femur with the long axis of the shaft of the femur is well shown.

Lateral View.—This is taken with the patient in a supine position with the x-ray tube directed either

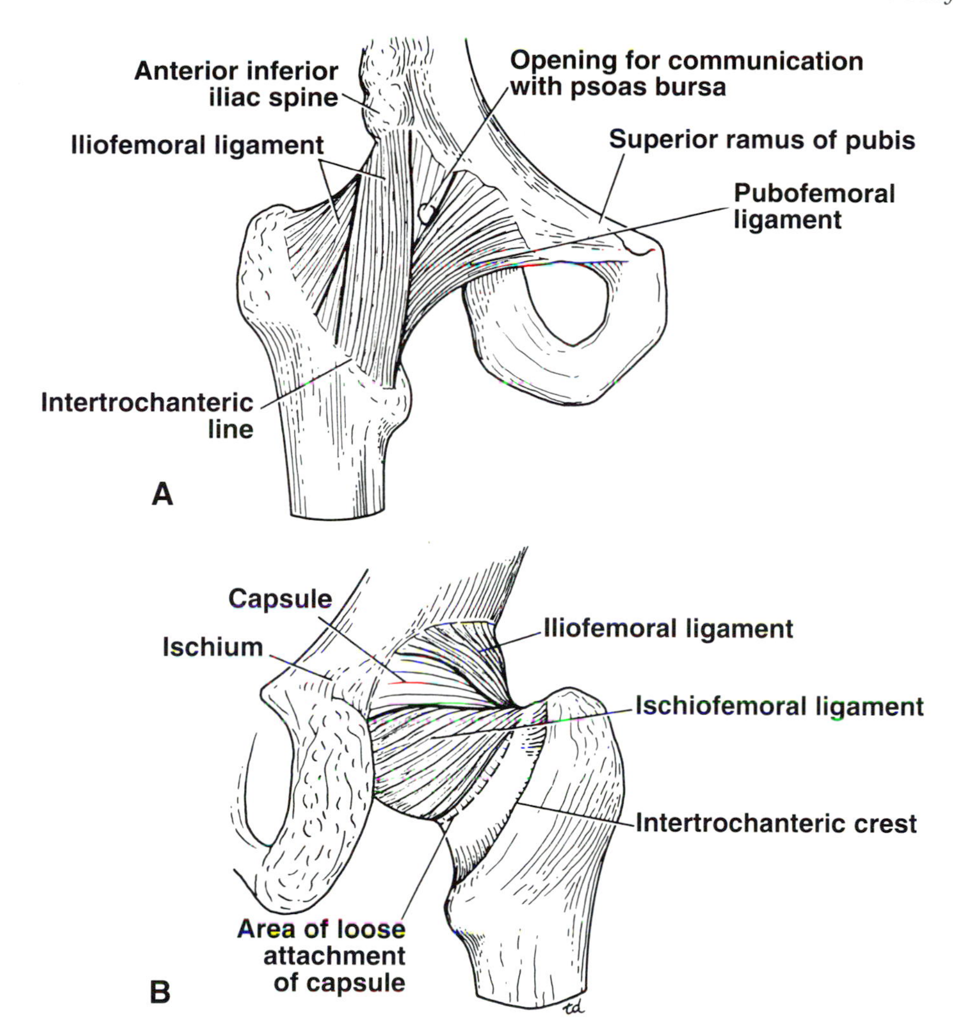

FIG 17–17.
Right hip joint. **A,** anterior aspect. **B,** posterior aspect.

from the medial or lateral aspect of the thigh. A horizontal x-ray beam is employed. The film cassette is placed perpendicular to the tabletop. Alternatively, the "frog leg" position may be used (Fig 17–22). The patient externally rotates both thighs and allows the soles of both feet to touch one another.

As many of the relevant parts of the pelvis as possible are first recognized. The obturator foramen, the ischial spine and tuberosity, the pubic ramus, and the body of the pubis may all be recognized (Figs 17–22 and 17–23). The acetabular rims and the head and the whole neck of the femur are demonstrated. The greater and lesser trochanters and the proximal part of the shaft are visualized (see Fig 17–22).

Basic Anatomy of the Knee Joint

Articulation.—Above are the rounded condyles of the femur; below are the condyles of the tibia and their menisci (semilunar cartilages) (Fig 17–24); in front is the articulation between the lower end of the femur and the patella. The articular surfaces are covered with hyaline cartilage.

Type.—Between the femur and tibia is a synovial joint of the hinge variety, but some degree of rotatory movement is possible. A synovial gliding joint is between the patella and femur.

Capsule.—This encloses the joint. On the front of the joint, however, the capsule is absent, permitting the synovial membrane to pouch upward beneath the quadriceps tendon, forming the *suprapatellar bursa* (see Fig 17–24).

Ligaments.—These may be divided into ligaments that lie outside the capsule and those that lie within the capsule.

Extracapsular Ligaments.—These involve the following.

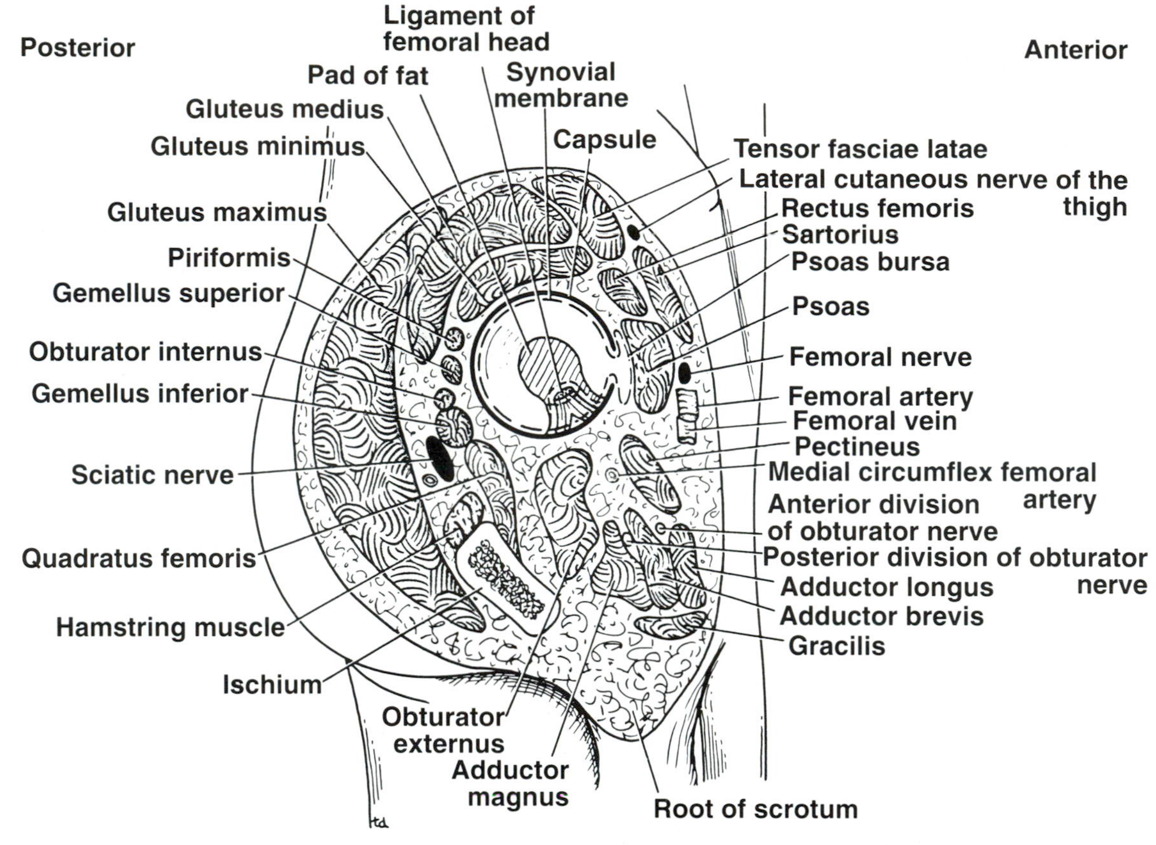

FIG 17–18.
Anteroposterior section through the right hip joint showing the relations. Note the position of the sciatic nerve on the posterior aspect of the joint and the femoral vessels and nerve on the anterior aspect of the joint.

Ligamentum Patellae.—This is attached above to the lower border of the patella and below to the tubercle of the tibia. It is a continuation of the tendon of the quadriceps femoris muscle.

Lateral Collateral Ligament.—This is cordlike and is attached above to the lateral condyle of the femur and below to the head of the fibula. The tendon of the popliteus muscle separates the ligament from the lateral meniscus (Figs 17–24 and 17–25).

Medial Collateral Ligament.—This is a flat band that is attached above to the medial condyle of the femur and below to the medial surface of the shaft of the tibia. It is strongly attached to the medial meniscus (see Fig 17–24).

Oblique Popliteal Ligament.—This is a tendinous expansion derived from the semimembranosus muscle. It strengthens the posterior part of the capsule.

Intracapsular Ligaments.—These two very strong *cruciate ligaments* cross each other within the joint cavity (see Fig 17–24). They are named anterior and posterior, according to their tibial attachments.

Anterior Cruciate Ligament.—This is attached to the anterior intercondylar area of the tibia and passes upward, backward, and laterally to be attached to the posterior part of the medial surface of the lateral femoral condyle (see Fig 17–24). This ligament prevents posterior displacement of the femur on the tibia.

Posterior Cruciate Ligament.—This is attached to the posterior intercondylar area of the tibia and passes upward, forward, and medially to be attached to the anterior part of the lateral surface of the medial femoral condyle (see Fig 17–24). This ligament prevents anterior displacement of the femur on the tibia.

Menisci (Semilunar Cartilages).—The menisci are C-shaped sheets of fibrocartilage. The peripheral border of each is thick and attached to the capsule, and the inner border is thin and concave and forms a free edge (see Fig 17–25). The upper surfaces are in contact with the femoral condyles. Their function

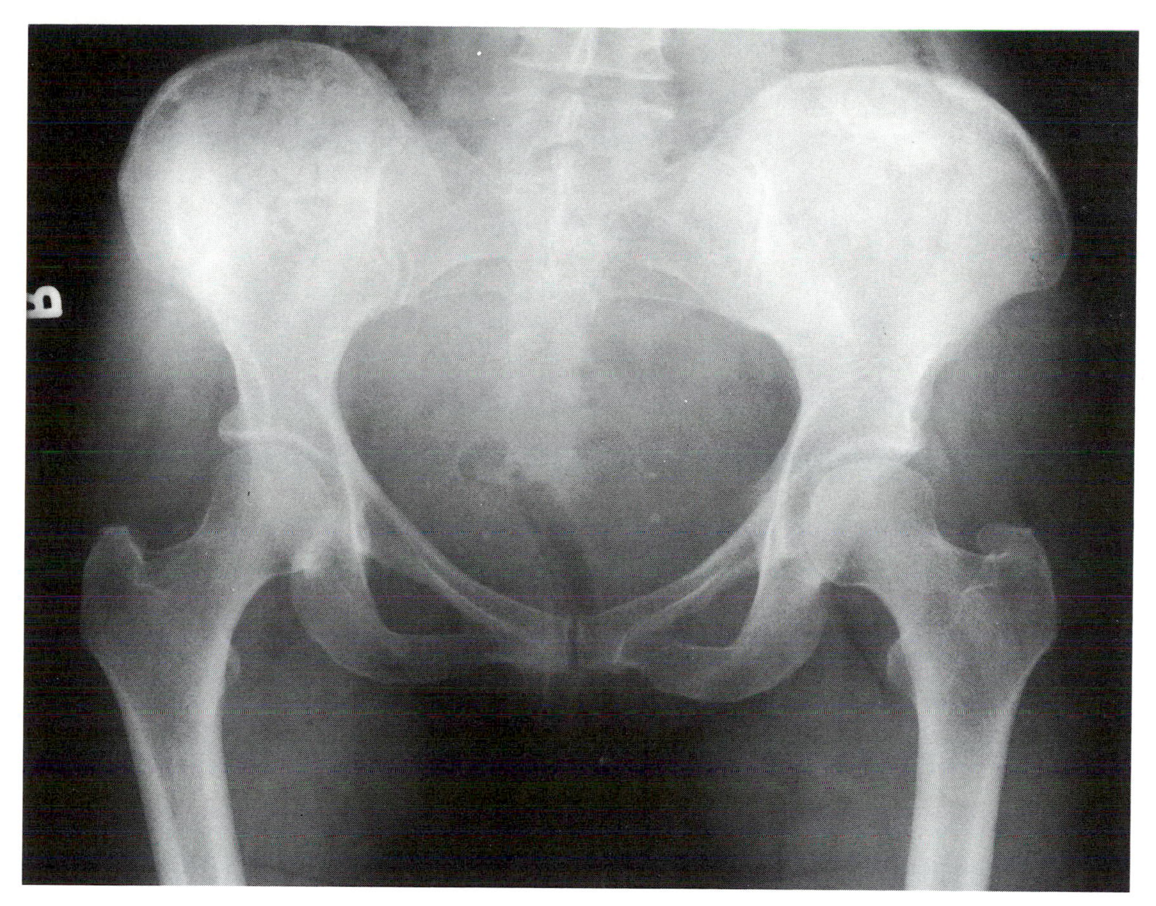

FIG 17–19.
Anteroposterior radiograph of the adult female pelvis.

is to deepen the articular surfaces of the tibial condyles to receive the convex femoral condyles. Each meniscus is attached to the upper surface of the tibia by anterior and posterior horns. Because the medial meniscus is also attached to the medial collateral ligament, it is relatively immobile.

Synovial Membrane.—The synovial membrane lines the capsule and is attached to the margins of the articular surfaces. On the front and above the joint it forms a pouch that extends up beneath the quadriceps femoris muscle for three fingerbreadths above the patella, forming the *suprapatellar bursa* (see Fig 17–24).

At the back of the joint the synovial membrane is prolonged downward on the deep surface of the tendon of the popliteus, forming the *popliteal bursa*. The synovial membrane is reflected forward from the posterior part of the capsule around the front of the cruciate ligaments. As a result the cruciate ligaments lie behind the synovial cavity and are not bathed in synovial fluid.

In the anterior part of the joint the synovial membrane is reflected backward from the ligamentum patellae to form the *infrapatellar fold;* the free borders of the fold are called the *alar folds*.

Bursae Related to the Knee Joint.—Numerous bursae are related to the knee joint. They are found wherever skin, muscle, or tendon rubs against bone. The following bursae always communicate with the joint cavity—the *suprapatellar bursa* that lies beneath the quadriceps muscle and the *popliteal bursa* that surrounds the tendon of the popliteus muscle as it leaves the joint.

The *semimembranosus bursa,* which lies between the tendon of this muscle and the medial condyle of the tibia, may communicate with the joint cavity.

Nerve Supply.—Femoral, obturator, common peroneal, and tibial nerves.

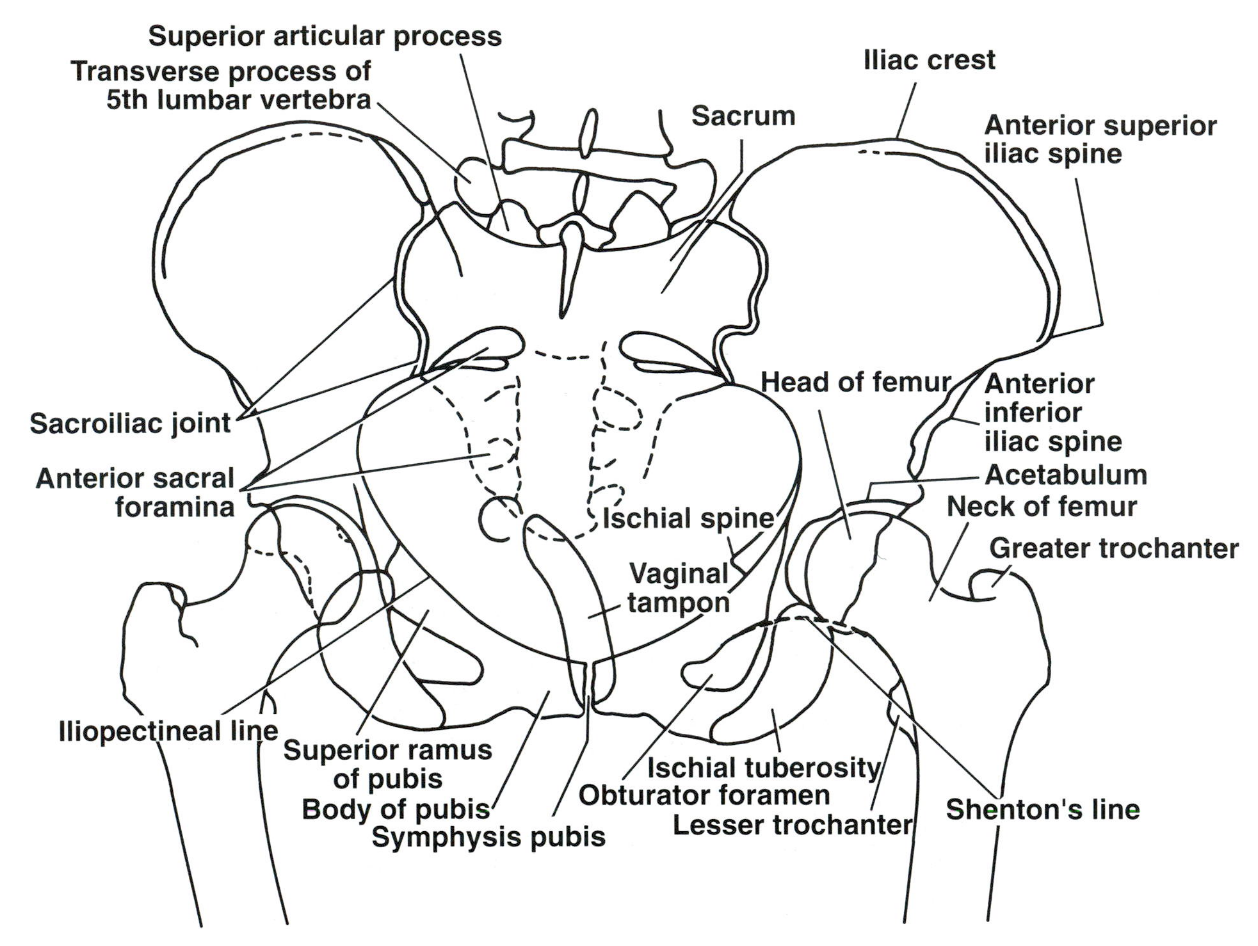

FIG 17–20.
Diagram showing the main features of the pelvis seen in the radiograph in Figure 17–19.

Clinical Note

Semimembranosus bursa swelling.—This is the most common swelling found in the popliteal space. It is made tense by extending the knee joint and becomes flaccid when the joint is flexed. It should be distinguished from a *Baker's cyst,* which is centrally located and arises as a pathological (osteoarthritis) diverticulum of the synovial membrane through a hole in the back of the capsule of the knee joint.

Movements.—The knee joint can flex, extend, and rotate. As the knee joint assumes the position of full extension,* medial (internal) rotation of the femur results in a twisting and tightening of all the major ligaments of the joint, and the knee becomes a mechanically rigid structure; the cartilaginous menisci are compressed like rubber cushions between the femoral and tibial condyles. The extended knee is said to be in the locked position.

Before flexion of the knee joint can occur, it is essential that the major ligaments be untwisted and slackened to permit movements between the joint surfaces. This unlocking or untwisting process is accomplished by the popliteus muscle, which laterally (externally) rotates the femur on the tibia. Once again the menisci have to adapt their shape to the changing contour of the femoral condyles.

When the knee joint is flexed to a right angle, a considerable range of rotation is possible. In the flexed position, the tibia can also be moved passively forward and backward on the femur. This is possible because the major ligaments, especially the cruciate ligaments, are slack in this position. The fol-

*Note that when the foot is firmly planted on the ground when a person is standing, the femur is internally rotated on the tibia to lock and stabilize the knee joint. However, if the foot is raised off the ground, the tibia may be externally rotated on the femur to lock the knee joint.

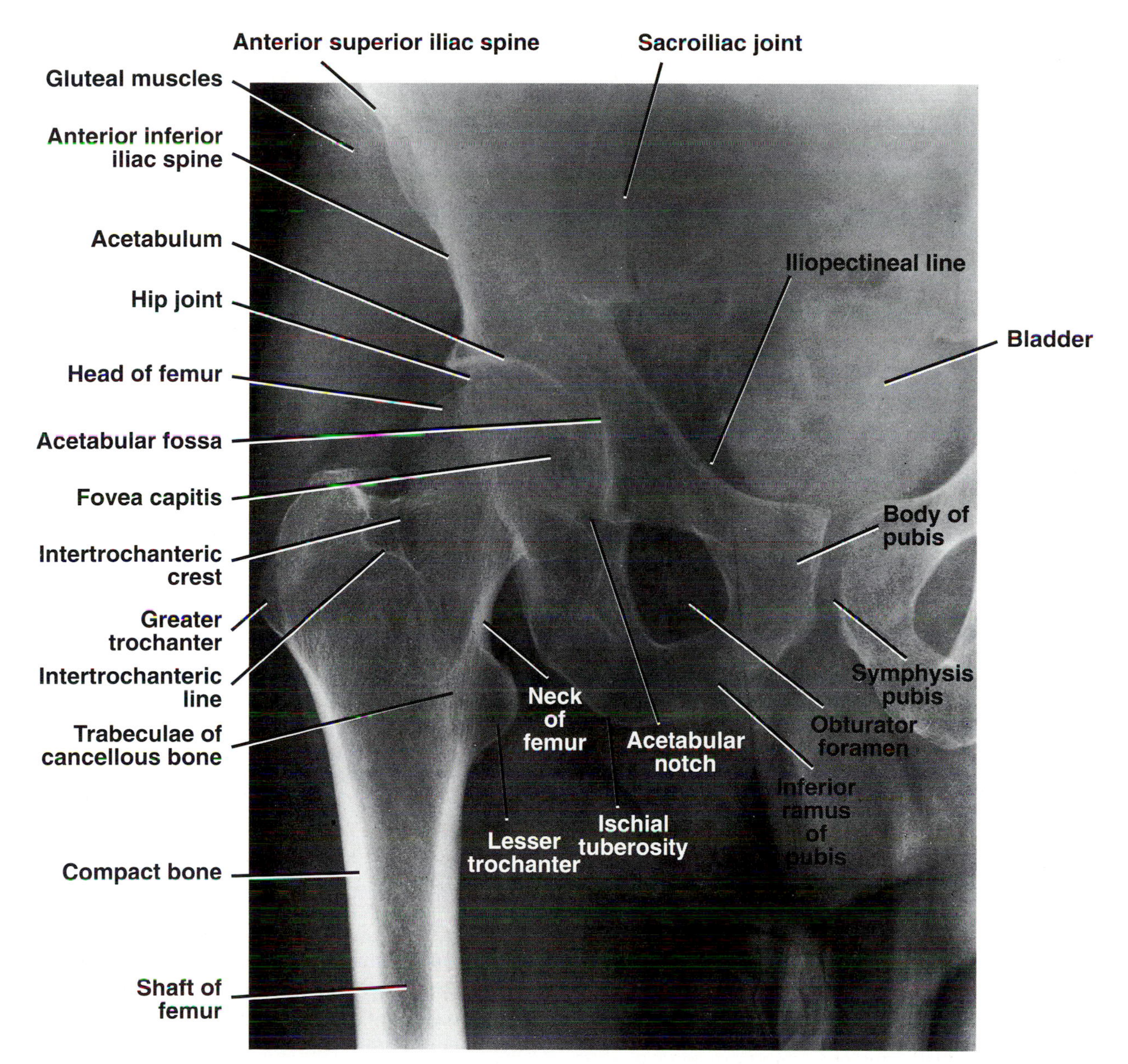

FIG 17–21.
Anteroposterior radiograph of the hip joint.

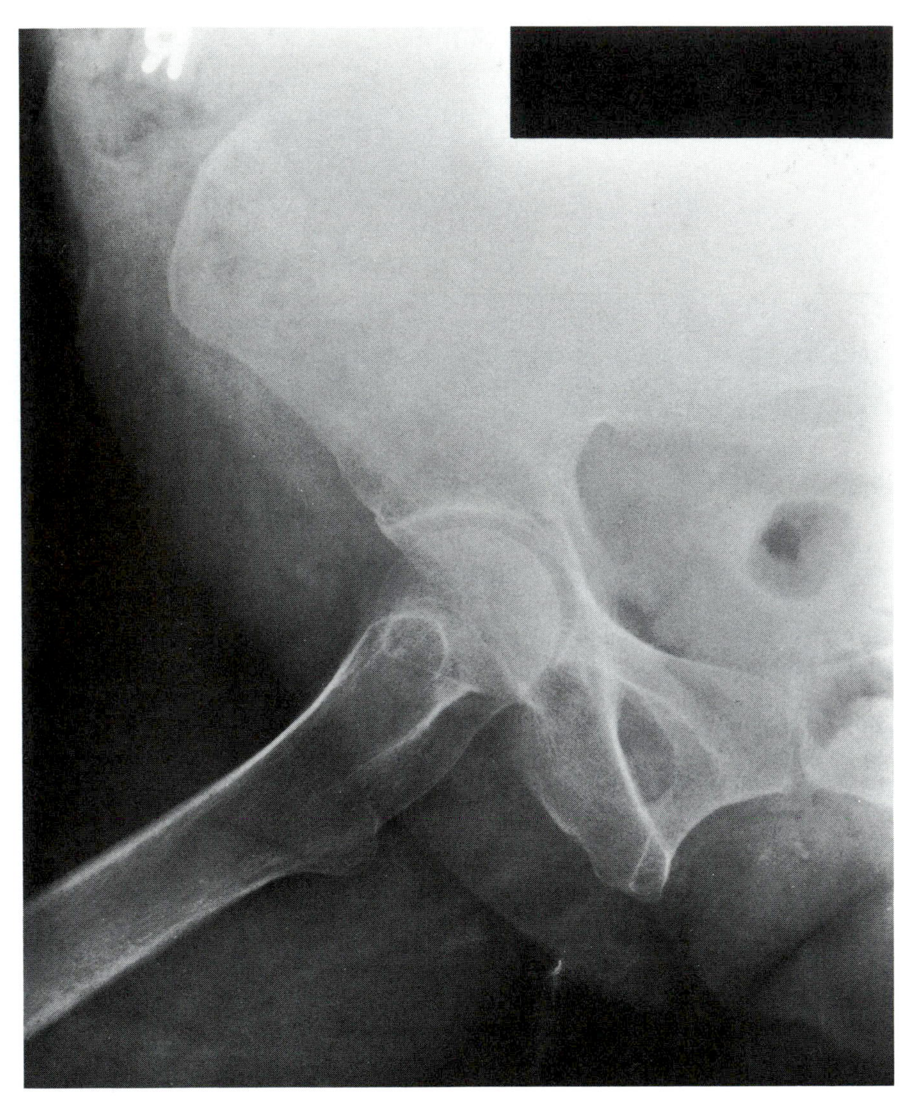

FIG 17–22.
Lateral radiograph of the hip joint.

lowing muscles produce the movements of the knee joint.

Flexion.—Biceps femoris, semitendinosus, and semimembranosus.

Extension.—Quadriceps femoris.

Medial Rotation.—Sartorius, gracilis, and semitendinosus.

Lateral Rotation.—Biceps femoris.

Radiographic Appearances of the Knee Region

The views commonly used are anteroposterior and lateral.

Anteroposterior View.— This view is taken with the patient in a supine position with the film cassette placed behind the knee. The x-ray tube is placed in front of the knee and centered over a point about ½ in. (1.3 cm) below the apex of the patella.

The lower part of the shaft of the femur, the lateral and medial epicondyles, and the adductor tubercle are easily visualized (Fig 17–26). The patella is seen superimposed in front of the lateral and medial femoral condyles. The *fabella,* a sesamoid bone in the lateral head of the gastrocnemius muscle, is sometimes seen superimposed on the lateral femoral condyle. The parallel joint surfaces, separated by a wide space occupied by the articular cartilage and the menisci, which cast no shadow, are easily recognized. The intercondylar notch of the femur and the intercondylar eminence of the tibia are well shown. The medial and lateral condyles of the tibia are seen. The head of the fibula partly overlaps the lateral

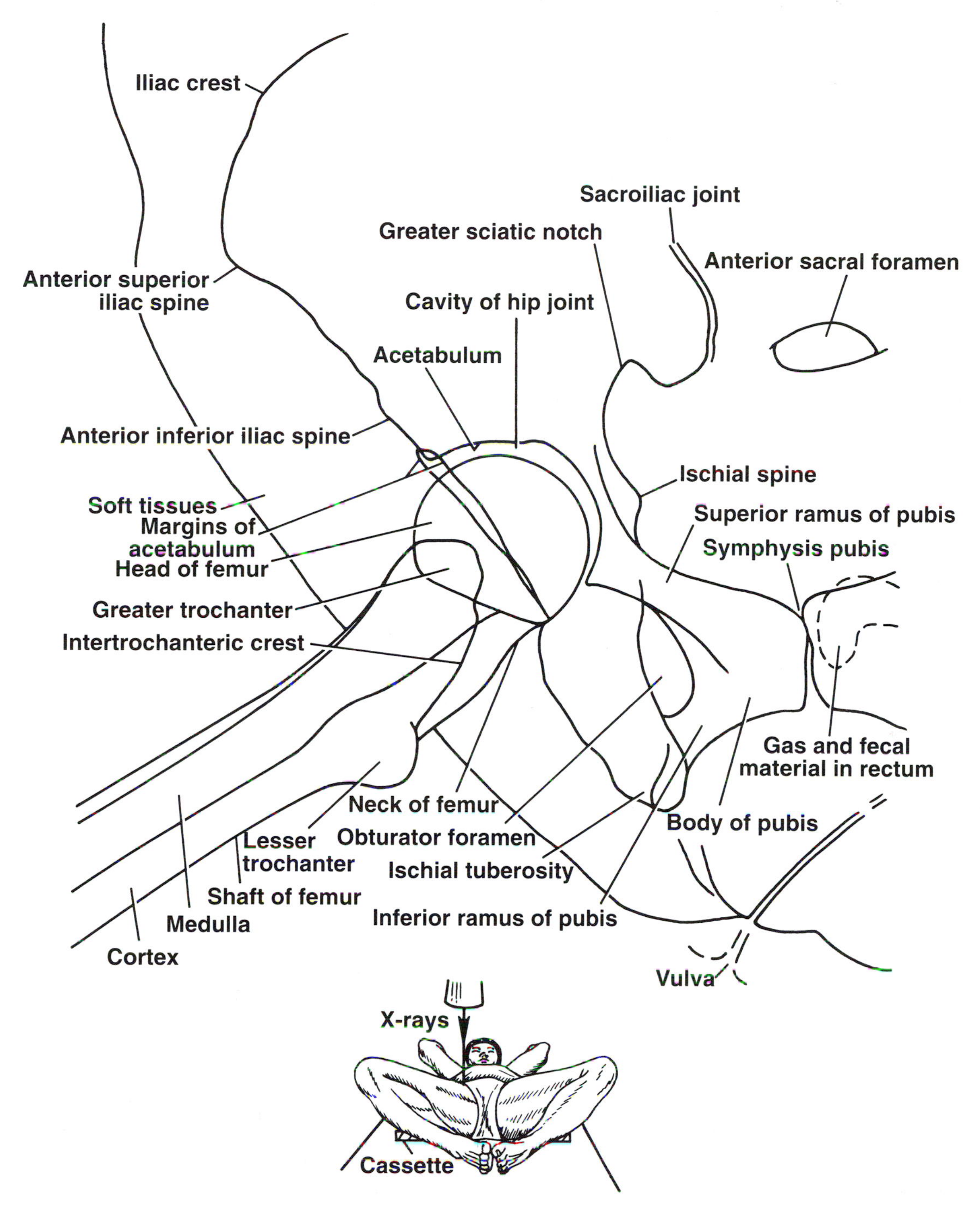

FIG 17–23.
The main features of the hip joint seen in the radiograph in Figure 17–22.

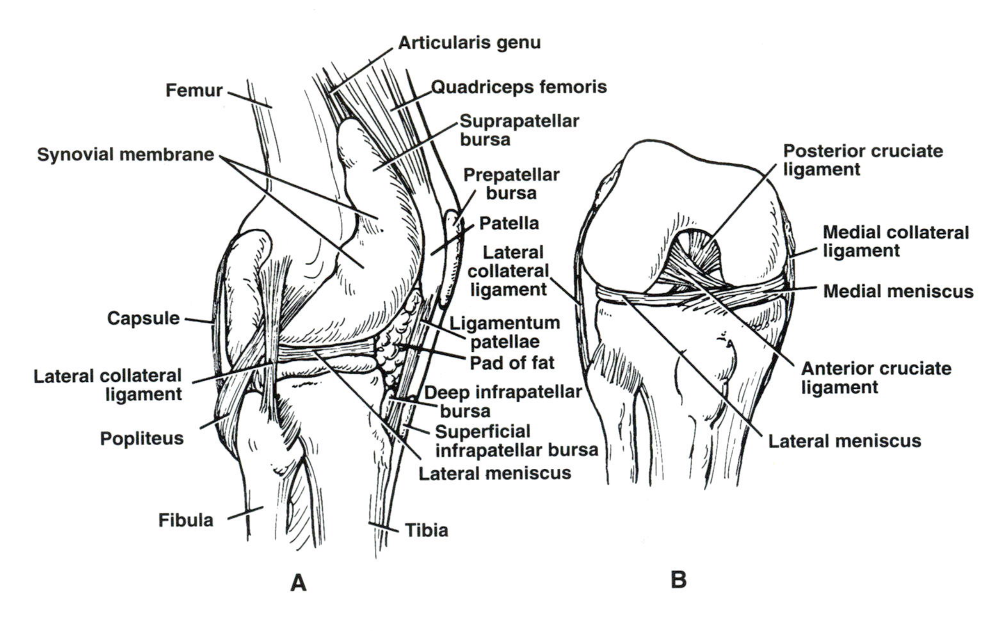

FIG 17–24.
Right knee joint. **A,** lateral aspect showing the extensive synovial membrane communicating with the suprapatellar bursa. **B,** Anterior aspect with the joint flexed and showing the cruciate ligaments and the medial and lateral menisci. **C,** posterior aspect. **D,** posterior aspect showing the cruciate ligaments and the menisci.

condyle of the tibia. The neck of the fibula and the upper parts of the shafts of the fibula and tibia are usually clearly seen.

Lateral View.—This is taken with the knee joint partially flexed. The film cassette is placed against the lateral aspect of the joint, and the x-ray tube is centered on the medial side of the joint line. The patient reclines on his side on the table.

The lower part of the shaft of the femur is seen, and the lateral and medial femoral condyles are partly superimposed on each other (Fig 17–27). The patella is clearly visualized in front of the femoral condyles. Below the patella the infrapatella fat pad can sometimes be recognized. The intercondylar eminence of the tibia projects upward into the intercondylar notch of the femur, and its summit is overlapped by the femoral condyles. The lateral and medial tibial condyles are superimposed, and the tibial turberosity is seen on the anterior surface of the bone.

The head, neck, and upper part of the shaft of the fibula are seen, with the fibula overlapping the tibia to some extent. A tangential view of the patella is shown in Fig 17–28.

Basic Anatomy of the Ankle Joint

Articulation.—Between the lower end of the tibia and the two malleoli above, and the body of the talus below (Fig 17–29), the *inferior transverse tibiofibular ligament* deepens the socket into which the body of the talus fits snugly.

Type.—Synovial hinge joint.

Capsule.—Encloses the joint.

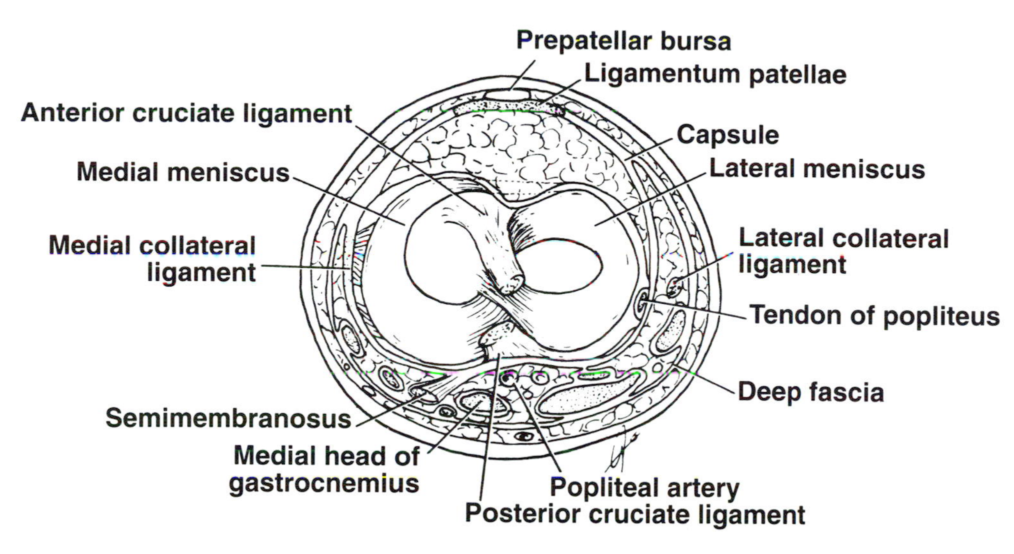

FIG 17–25.
Cross section of the right knee joint showing the position of the ligaments and the menisci.

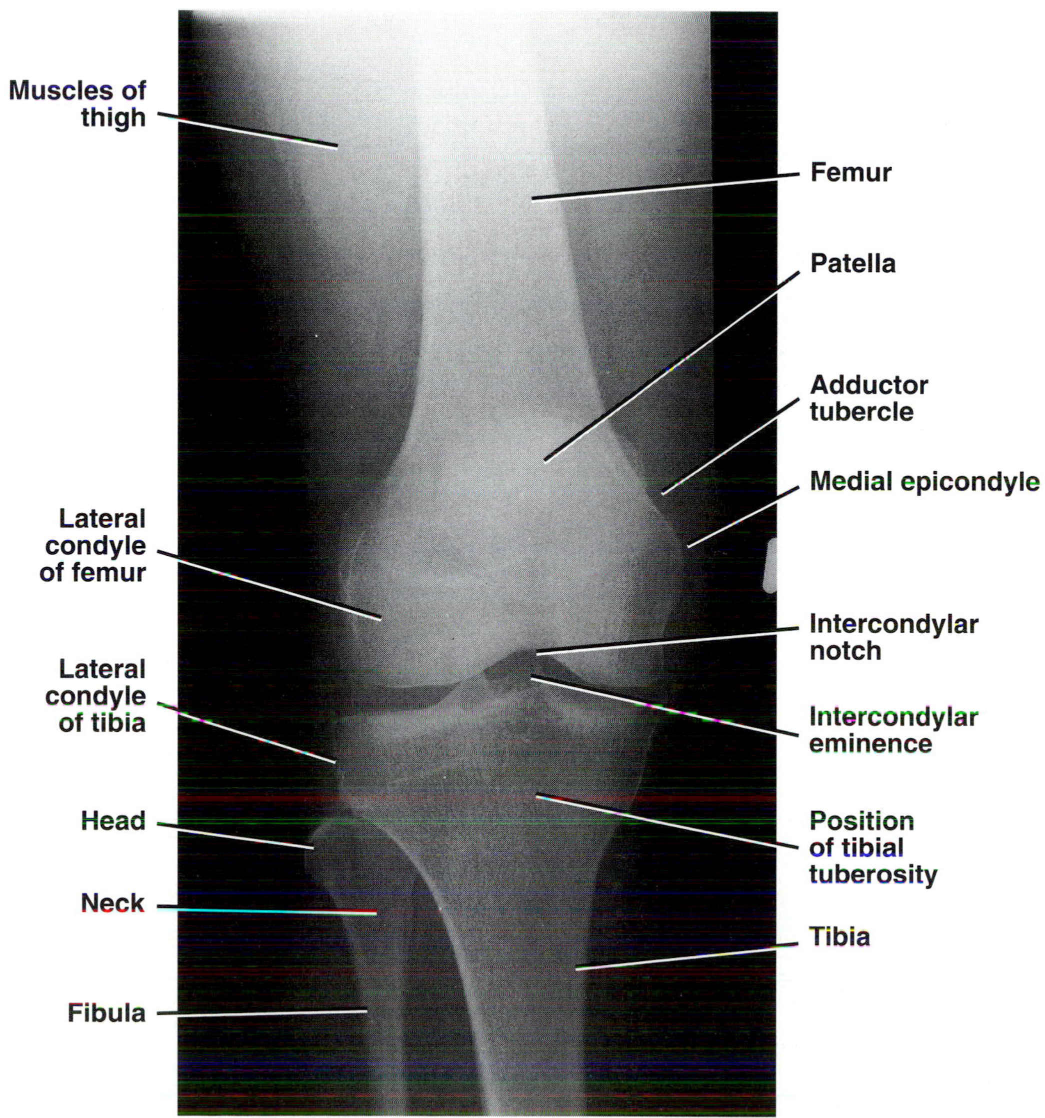

FIG 17–26.
Anteroposterior radiograph of the knee joint.

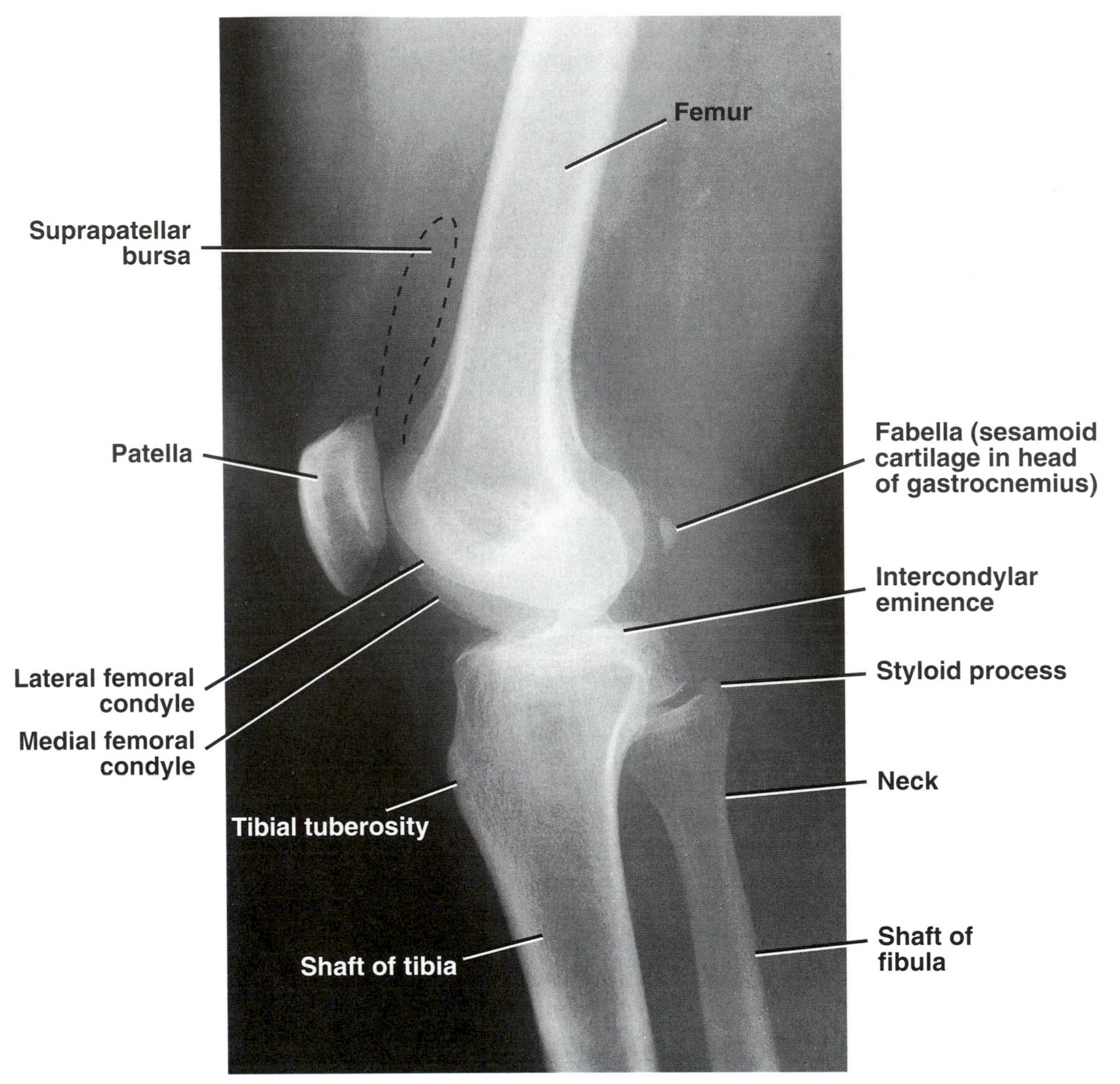

FIG 17–27.
Lateral radiograph of the knee joint.

Ligaments.—These include the following.

Medial (Deltoid) Ligament.—This very strong ligament is attached by its apex to the tip of the medial malleolus (see Fig 17–29). The deep fibers are attached below to the medial surface of the body of the talus, the superficial fibers are attached to the medial side of the talus, the sustentaculum tali, the plantar calcaneonavicular ligament, and the tuberosity of the navicular bone.

Lateral Ligament.—This is weaker than the medial ligament and consists of three bands. The *anterior talofibular ligament* runs from the lateral malleolus to the lateral surface of the talus. The *calcaneofibular ligament* runs from the lateral malleolus to the lateral surface of the calcaneum. The *posterior talofibular ligament* runs from the lateral malleolus to the posterior tubercle of the talus (see Fig 17–29).

Synovial Membrane.—This lines the capsule.

Nerve Supply.—Deep peroneal and tibial nerves.

Movements.—These include dorsiflexion (toes pointing upward) and plantar flexion. The movements of inversion and eversion take place at the tarsal joints and *not at the ankle joint.* The following muscles perform the movements.

Dorsiflexion.—Tibialis anterior, extensor hallucis

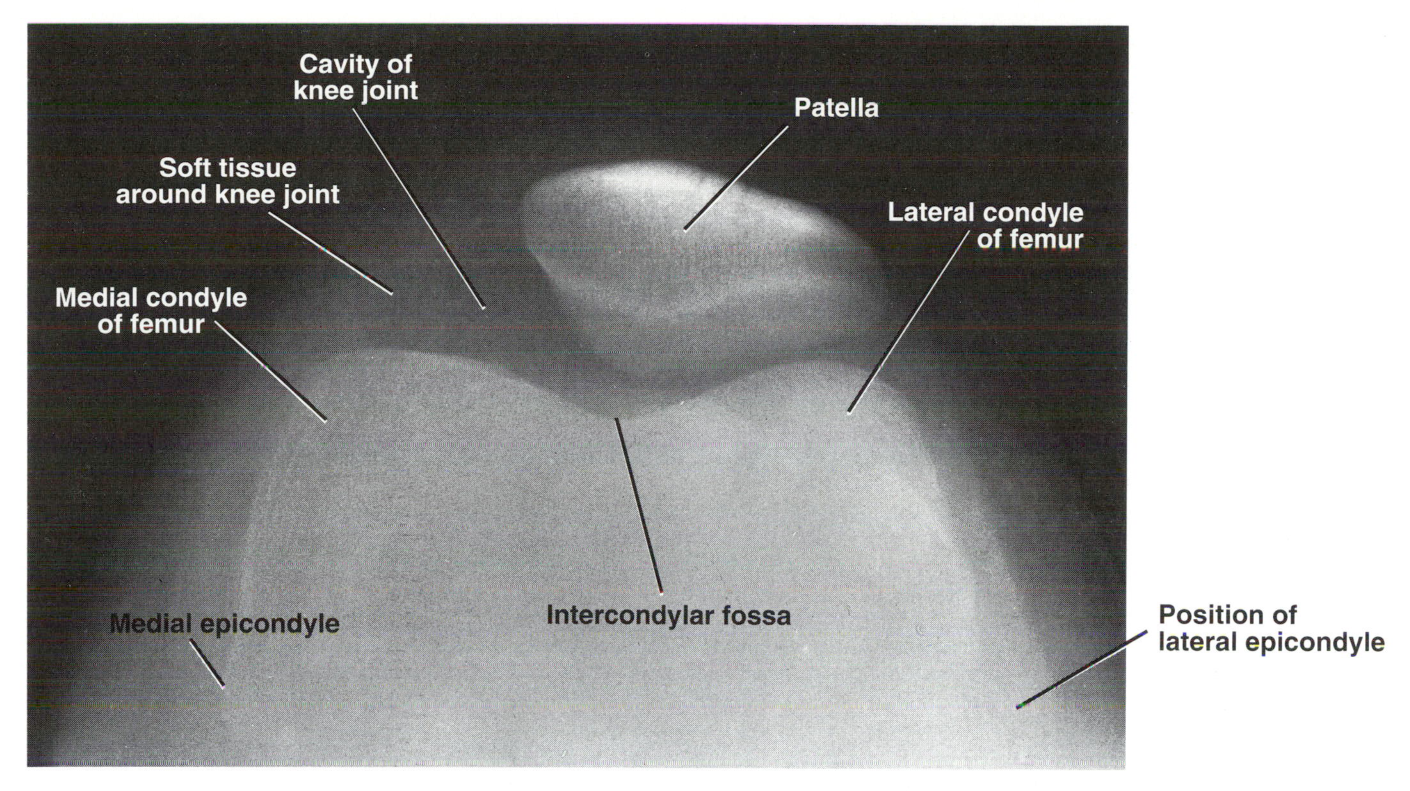

FIG 17–28.
Tangential (rising sun) radiograph of the patella.

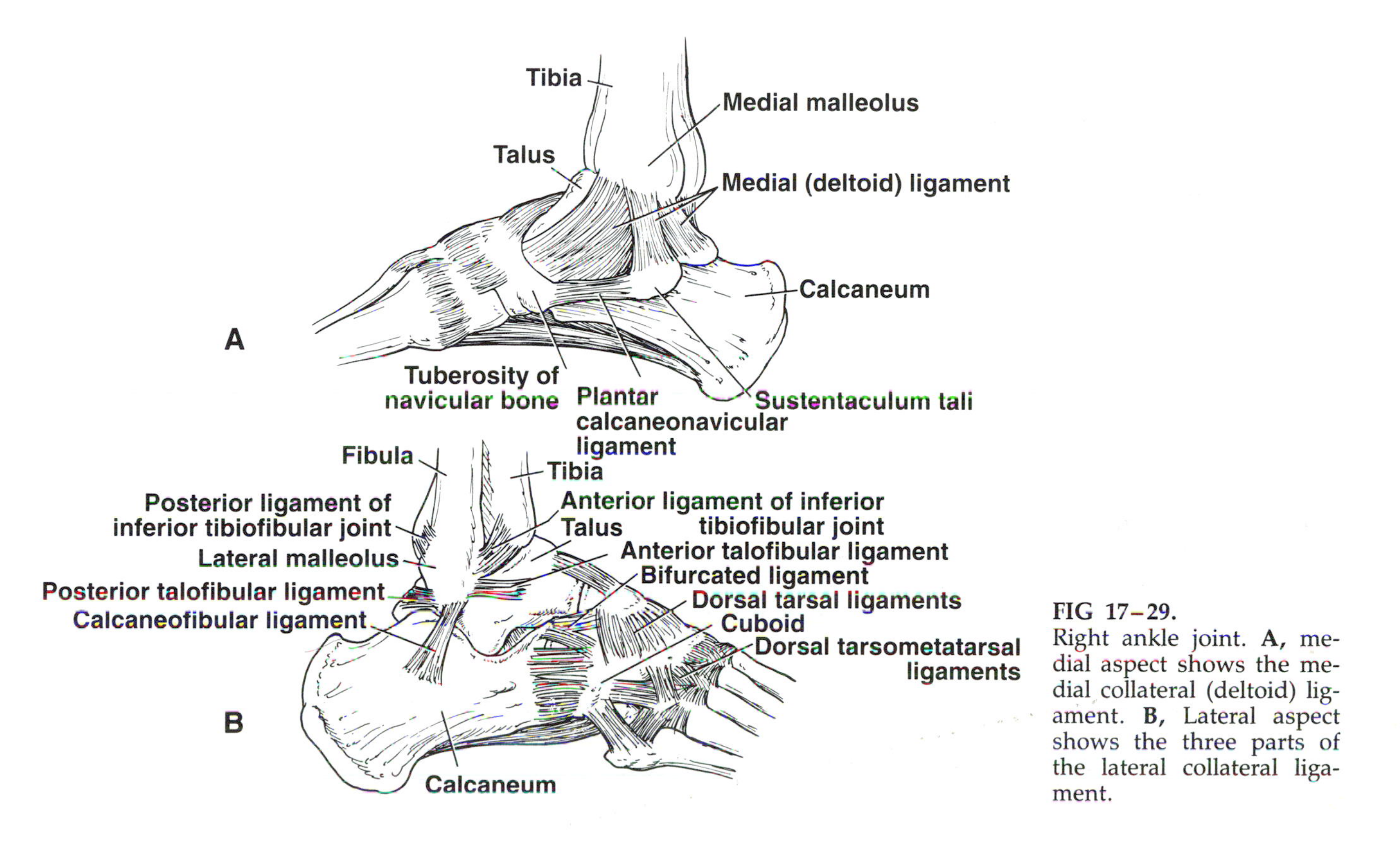

FIG 17–29.
Right ankle joint. **A,** medial aspect shows the medial collateral (deltoid) ligament. **B,** Lateral aspect shows the three parts of the lateral collateral ligament.

longus, extensor digitorum longus, and peroneus tertius.

Plantar Flexion.—Gastrocnemius, soleus, plantaris, peroneus longus, peroneus brevis, tibialis posterior, flexor digitorum longus, and flexor hallucis longus.

Important Relations.—These involve the following.

Anteriorly.—Anterior tibial vessels and the deep peroneal nerve.

Posteriorly.—Achilles tendon.

Posterolaterally (Behind the Lateral Malleolus).—Tendons of peroneus longus and brevis muscles (Fig 17–30).

Posteromedially (Behind the Medial Malleolus).—Posterior tibial vessels and the tibial nerve and the long flexor tendons of the foot.

Radiographic Appearances of the Ankle Region

The views commonly used are anteroposterior, lateral, and oblique (mortise).

Anteroposterior View.—This is taken with the patient in a supine position. The ankle joint is dorsiflexed to a right angle, and the big toe is pointed slightly medially. The film cassette is placed behind the ankle joint, and the x-ray tube is centered over the front of the ankle joint.

The lower ends of the tibia and fibula and the inferior tibiofibular joint are well shown (Fig 17–31). The medial and lateral malleoli and the articular surfaces of the tibia and the body of the talus are easily seen. The lateral malleolus usually partly overlaps the lateral aspect of the talus. The articular surfaces of the lower end of the tibia and the superior surface of the talus are seen to be parallel and separated by a narrow space occupied by the articular cartilage, which is radiotranslucent. Other than the talus, the tarsal bones are not clearly visualized.

Lateral View.—This is taken with the lateral malleolus against the film cassette. It is important that the sagittal plane of the leg be parallel with the plane of

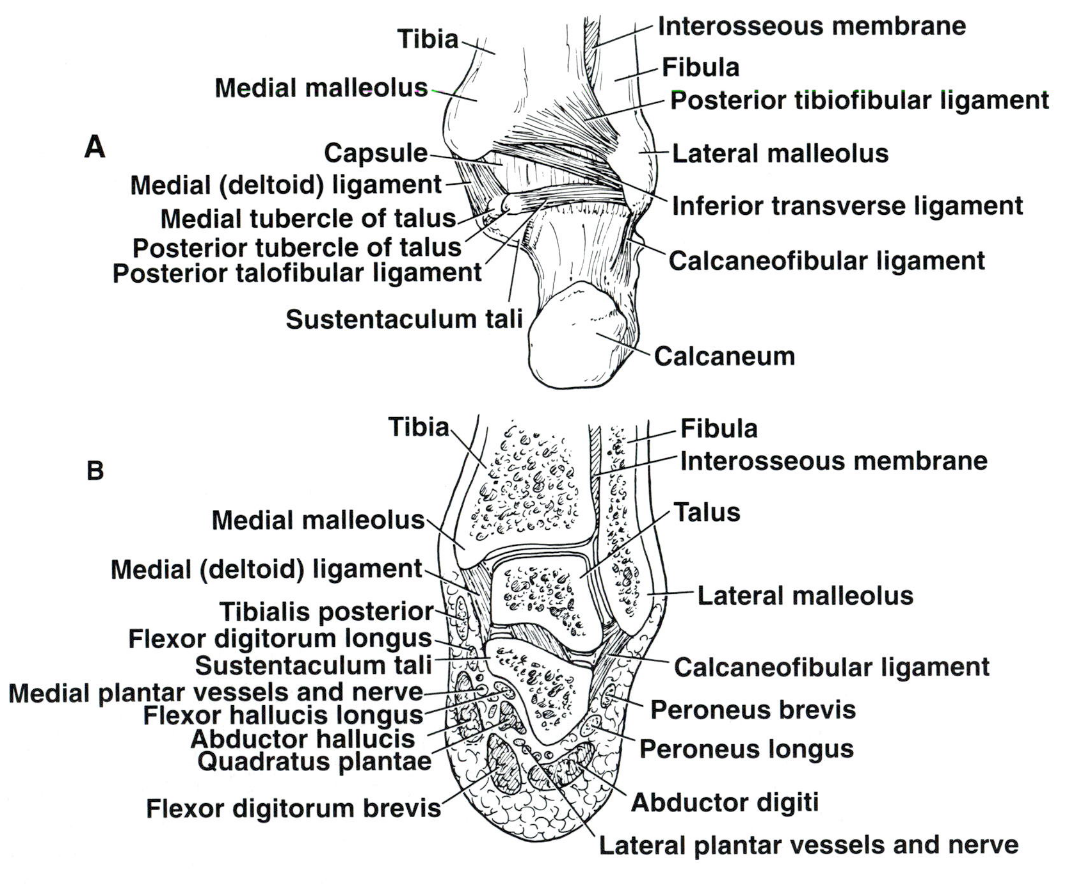

FIG 17–30.
Right ankle joint. **A,** posterior aspect. **B,** coronal section. Note the relationship between the talus and the underlying calcaneum.

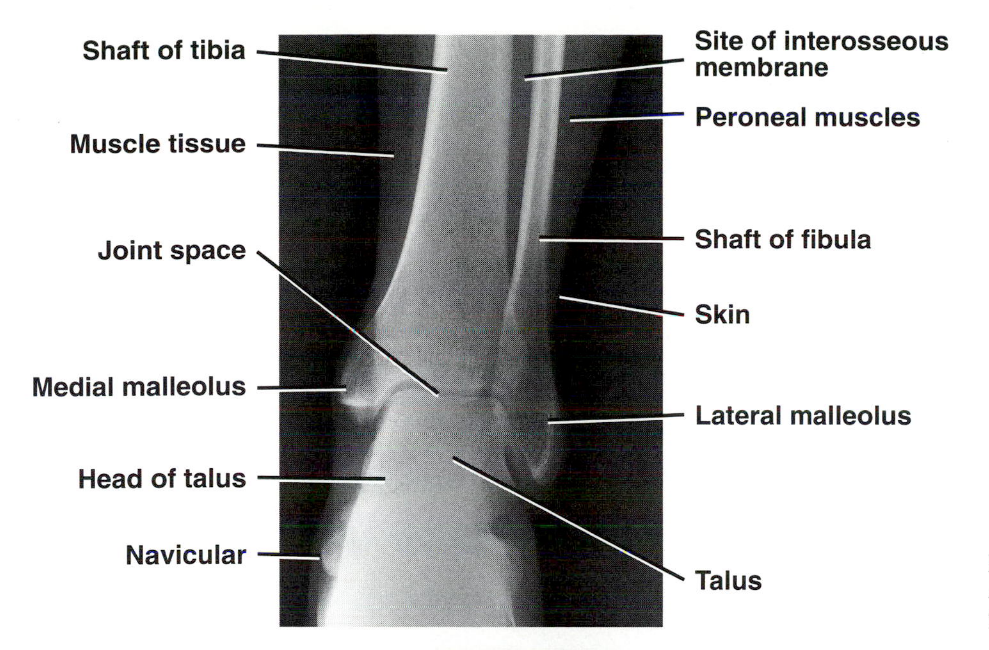

FIG 17–31.
Anteroposterior radiograph of the ankle joint.

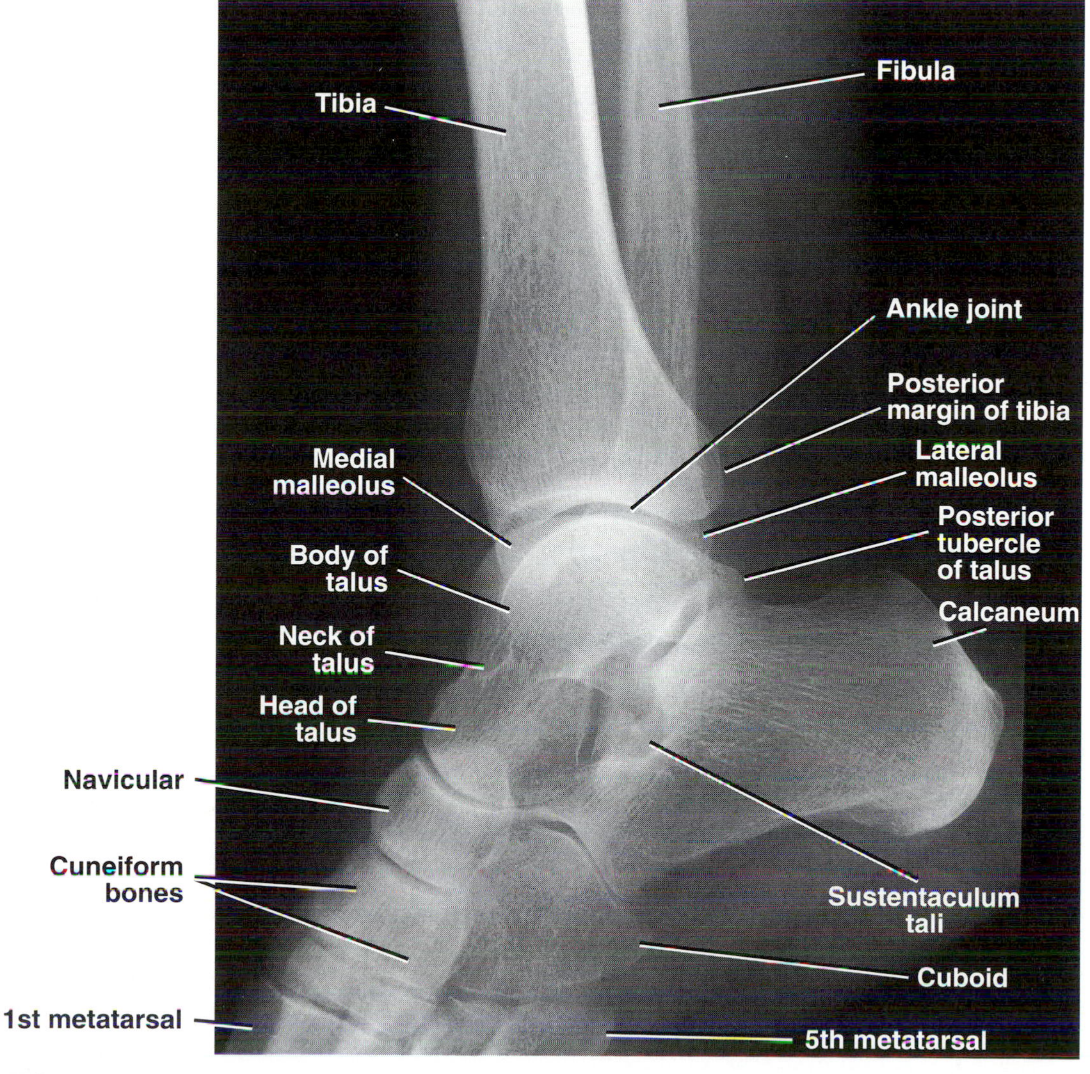

FIG 17–32.
Lateral radiograph of the ankle joint.

the film. The x-ray tube is centered over a point about ¾ in. (1.9 cm) proximal to the tip of the lateral malleolus.

This view shows the lower ends of the tibia and fibula; the lateral and medial malleoli are superimposed (Fig 17–32). It should, however, be possible to make out the anterior and posterior margins of both malleoli. The articular surfaces of the ankle joint are clearly visualized. The talus and calcaneum are seen in profile, and the subtalar and transverse tarsal joints can be identified. The cuneiform bones and cuboid are overlapped and not clearly seen.

Oblique View.—This is taken with 15° of internal rotation of the foot on the leg (Figs 17–33 and 17–34). The view shows the articular surface of the lateral malleolus clearly so that the whole ankle joint space is symmetrically displayed.

Clinical Note

Fifth metatarsal bone.—A good ankle radiograph series should include the base of the fifth metatarsal bone so that a *Jones* fracture is not inadvertently missed.

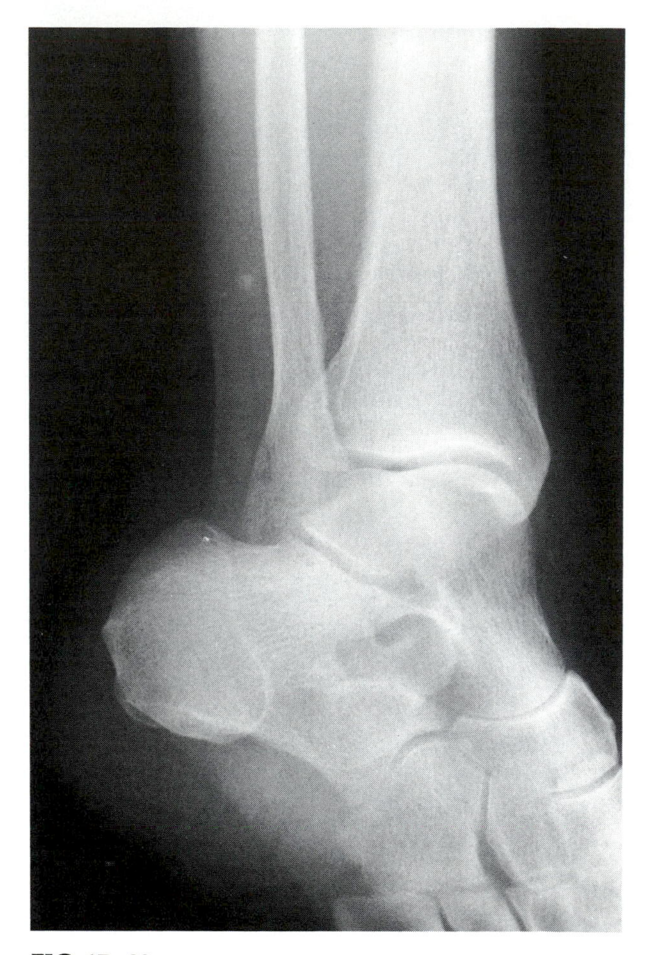

FIG 17–33.
Internal oblique radiograph of the ankle joint.

Radiographic Appearances of the Tarsus, Metatarsus, and Phalanges

The views commonly used are the anteroposterior, lateral, and oblique.

The particular view used will depend on which bone needs to be visualized the best (Figs 17–35 and 17–36). The oblique view of the metatarsal bones is often of greater value than the lateral view since, in the latter, the bones are superimposed. In the anteroposterior view, the film cassette is placed in contact with the sole. The tarsal bones, the metatarsals, and the phalanges are seen (see Fig 17–35). The two sesamoid bones of the big toe overlap the head of the first metatarsal bone.

Basic Anatomy of the Femoral Sheath and Femoral Canal

Femoral Sheath

The femoral sheath is a downward protrusion from the abdomen into the thigh of the fascia transversalis and fascia iliaca (Fig 17–37). The sheath surrounds the femoral blood vessels and lymph vessels for about 1 in. (2.5 cm) below the inguinal ligament. The *femoral artery,* as it enters the thigh beneath the inguinal ligament, occupies the *lateral compartment* of the sheath. The *femoral vein* occupies the *intermediate compartment,* and the lymphatic vessels (and usually one lymph node) occupy the most *medial compartment.*

Femoral Canal

The femoral canal is the small medial compartment of the femoral sheath for the lymphatics (see Fig 17–37). It is about ½ in. (1.3 cm) long. The canal is a potentially weak area in the wall of the abdomen. A protrusion of peritoneum could be forced down the femoral canal to form a femoral hernia. The upper opening of the femoral canal is called the *femoral ring.* It is filled by a plug of extraperitoneal fat called the *femoral septum.* The lower end of the canal is normally closed by the adherence of its medial wall to the tunica adventitia of the femoral vein. It lies close to the saphenous opening in the deep fascia of the thigh for the great saphenous vein (see Fig 17–48).

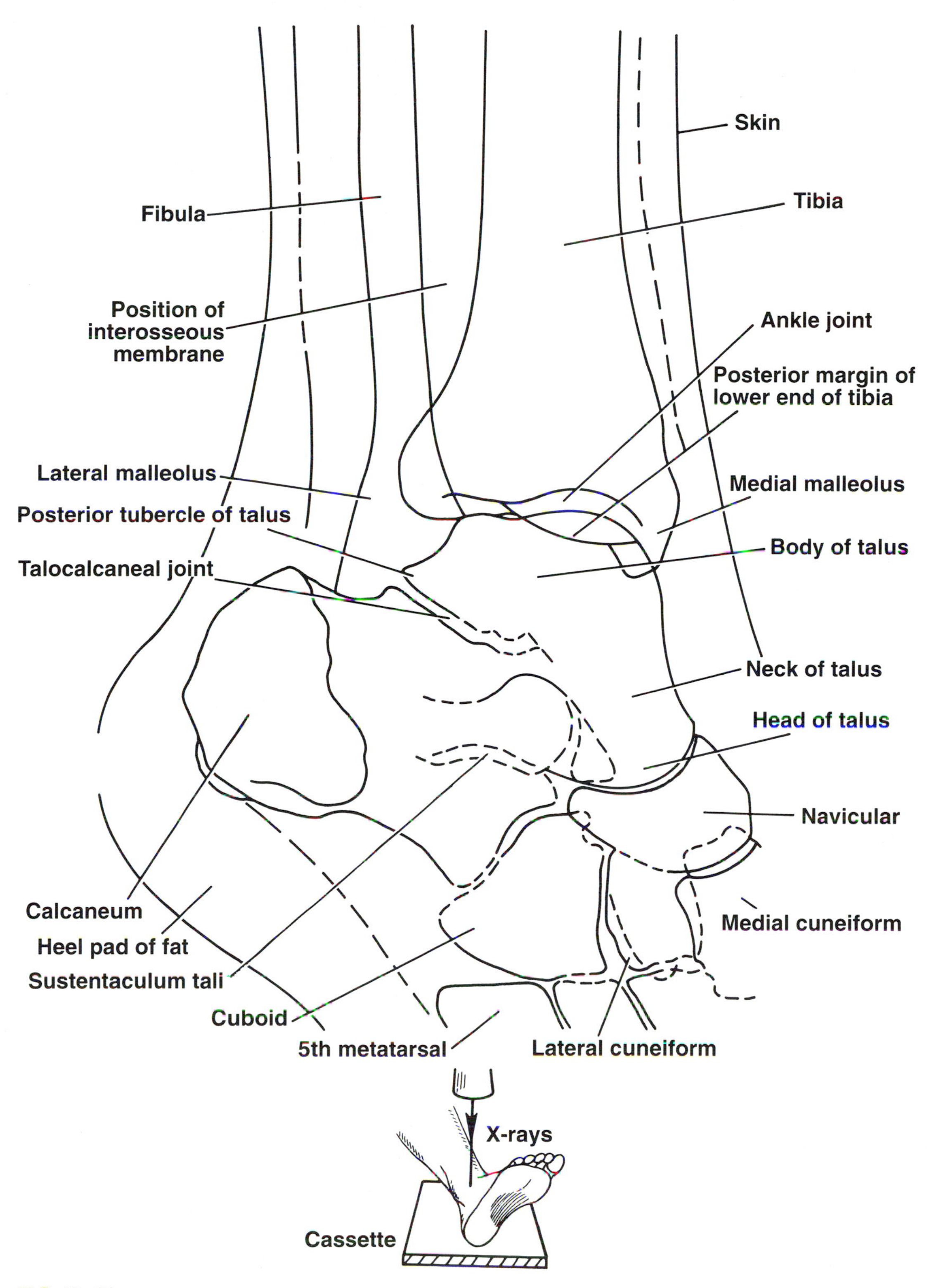

FIG 17–34.
The main features seen on the radiograph in Figure 17–33.

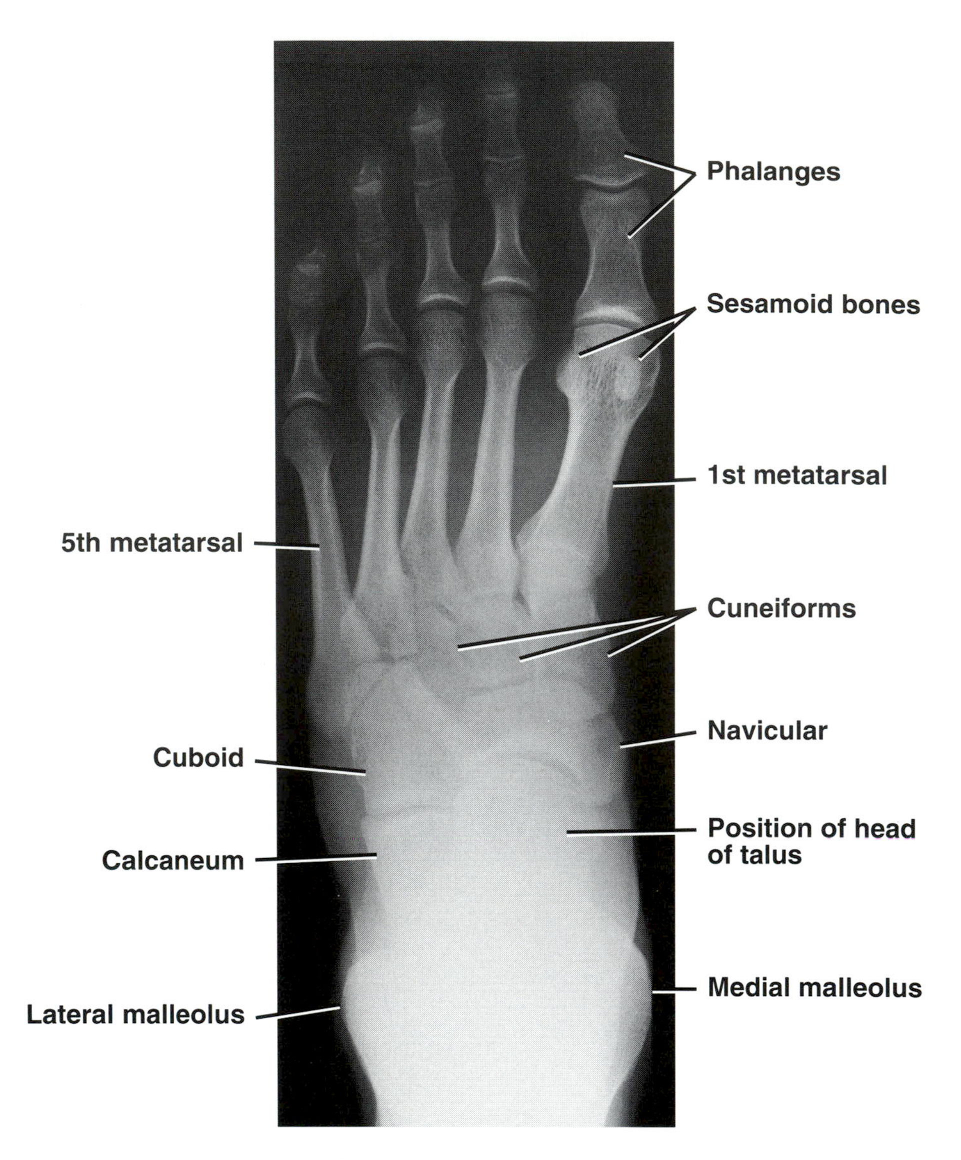

FIG 17–35.
Anteroposterior radiograph of the foot.

Clinical Note

Pubic tubercle and femoral hernia.—The lower end of the femoral canal lies below and lateral to the pubic tubercle. This fact is of great clinical significance, since a femoral hernia will emerge from the canal *below and lateral to the pubic tubercle,* whereas an indirect inguinal hernia will emerge above and medial to the pubic tubercle.

Contents of the Femoral Canal.—This includes the following.

1. Fatty connective tissue.
2. All the efferent lymph vessels from the deep inguinal lymph nodes.
3. One of the deep inguinal lymph nodes.

Important Relations of the Femoral Ring (see Fig 17–37).—These include the following.

Anteriorly.—Inguinal ligament.

Posteriorly.—Superior ramus of pubis and the perineal ligament (thickening of periosteum and lateral continuation of the lacunar ligament, see p. 409).

Laterally.—Femoral vein.

Medially.—Lacunar ligament (an extension of the medial end of the inguinal ligament, see p. 409).

Basic Anatomy of the Lymphatic Drainage of the Lower Extremity

The lymph vessels of the lower limb are arranged as superficial and deep sets. The superficial vessels ascend the limb with the superficial veins (Fig 17–38). The deep lymph vessels lie deep to the deep fascia

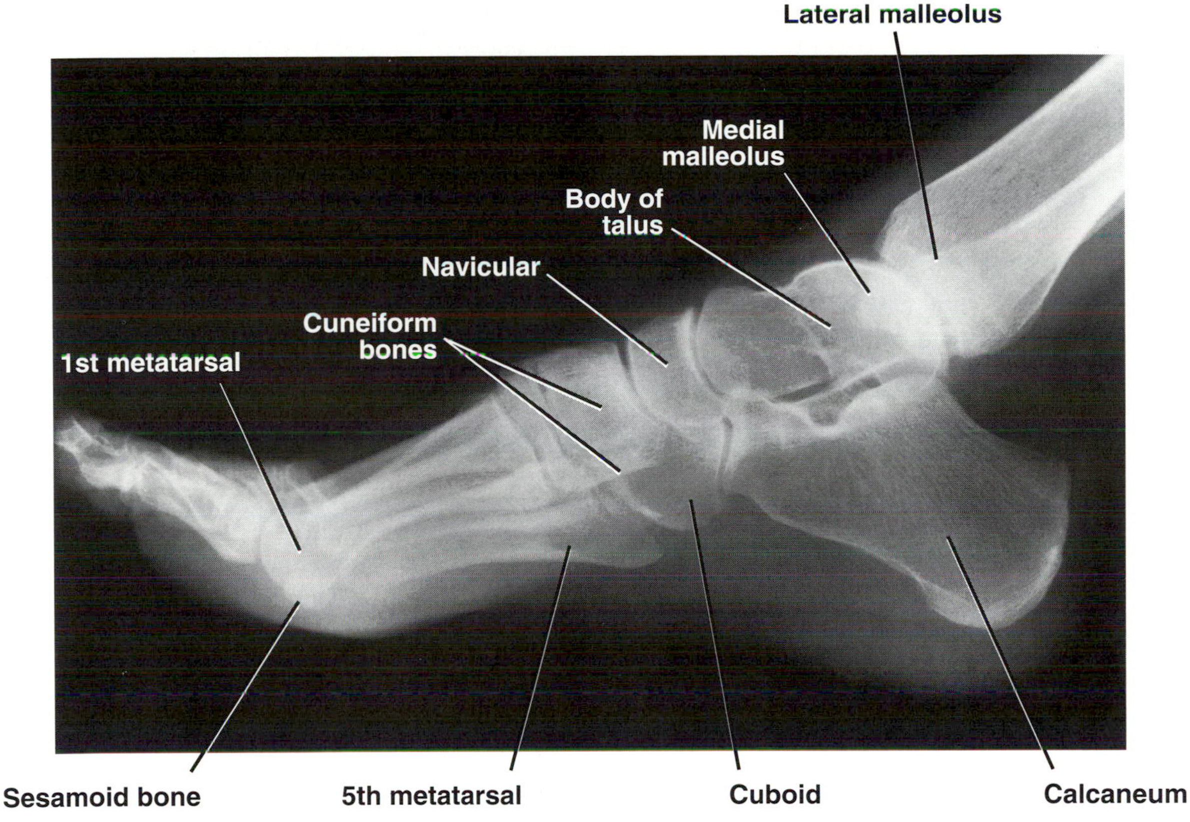

FIG 17–36.
Lateral radiograph of the foot.

and follow the deep arteries and veins. All the lymph vessels from the lower extremity drain into the deep inguinal group of nodes that are situated in the groin.

Superficial Inguinal Lymph Nodes

These nodes lie in the superficial fascia just below the inguinal ligament. They may be divided into a horizontal and vertical group. The *horizontal group* receives lymph from the superficial lymph vessels of the anterior abdominal wall below the umbilicus, from the perineum, the external genitalia of both sexes (but not the testes), and the lower half of the anal canal (see Fig 17–38). They also receive lymph from the superficial lymph vessels of the buttocks.

The *vertical group* lies along the terminal part of the great saphenous vein and receives the majority of the superficial lymph vessels of the lower limb except from the back and lateral side of the calf and lateral side of the foot (see Fig 17–38). The superficial nodes all drain into the deep inguinal nodes.

Deep Inguinal Lymph Nodes

There are usually three nodes situated along the medial side of the femoral vein and in the femoral canal (see Fig 17–38). They receive all the lymph from the superficial inguinal nodes and from all the deep structures of the lower extremity. The efferent lymph vessels ascend through the femoral canal into the abdominal cavity and drain into the external iliac nodes.

Popliteal Lymph Nodes

These nodes lie in the popliteal fossa (see Fig 17–38). They receive superficial lymph vessels that accompany the small saphenous vein from the lateral side of the foot and the back and lateral side of the calf. They also receive lymph from the deep structures of the leg below the knee. The efferent vessels drain upward to the deep inguinal nodes.

Basic Anatomy of the Peripheral Nerves of the Lower Extremity

Anatomy of the Sciatic Nerve

The sciatic nerve (Fig 17–39) arises from L4 and L5 and S1 through S3. It passes out of the pelvis into the gluteal region through the greater sciatic fora-

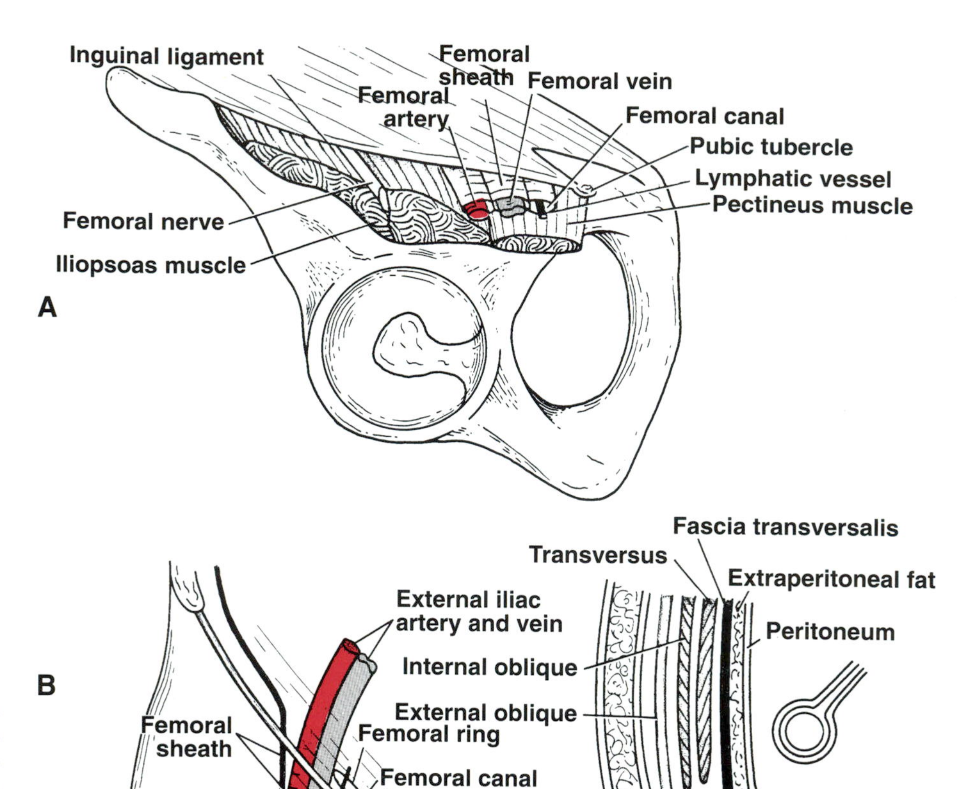

FIG 17–37.
Right femoral sheath. **A,** inferior view of the femoral sheath as it emerges into the thigh from beneath the inguinal ligament. The femoral nerve does not lie within the femoral sheath. **B,** anterior view of the femoral sheath showing the femoral canal and femoral ring. **C,** vertical section of the femoral sheath showing the femoral canal and its contents. The femoral sheath is formed from the fascia transversalis and the fascia iliaca.

men. The nerve curves laterally and downward beneath the gluteus maximus muscle and behind the hip joint (see Fig 17–9). It is situated at first midway between the posterior superior iliac spine and the ischial tuberosity, and lower down it is situated midway between the tip of the greater trochanter and the ischial tuberosity. The nerve then descends in the midline of the back of the thigh beneath the long head of the biceps femoris muscle (see Fig 17–9). At a variable site above the popliteal fossa, it divides into the tibial and common peroneal nerves.

Branches in the Gluteal Region: Articular Branches.—Hip joint.

Branches in the Back of the Thigh: Muscular Branches.—These arise just below the level of the ischial tuberosity and supply the biceps femoris (long head), semitendinosus, semimembranosus, and the hamstring part of adductor magnus.

Anatomy of the Tibial Nerve

The tibial nerve (L4 and L5 and S1 through S3), one of the two main branches of the sciatic nerve (the other being the common peroneal), runs downward through the popliteal fossa (see Fig 17–39) crossing the popliteal artery from the lateral to the medial side; it then passes deep to the gastrocnemius and soleus muscles and descends to the interval between the medial malleolus and the heel (see Fig 17–12). It is covered here by the flexor retinaculum and divides into the medial and lateral plantar nerves (Fig 17–14).

Branches of the Tibial Nerve (see Fig 17–39).—These include the following.

1. *Cutaneous branches.* The *sural nerve* is usually joined by the sural communicating branch of the common peroneal nerve. It supplies the skin of the calf and the back of the leg and accompanies the

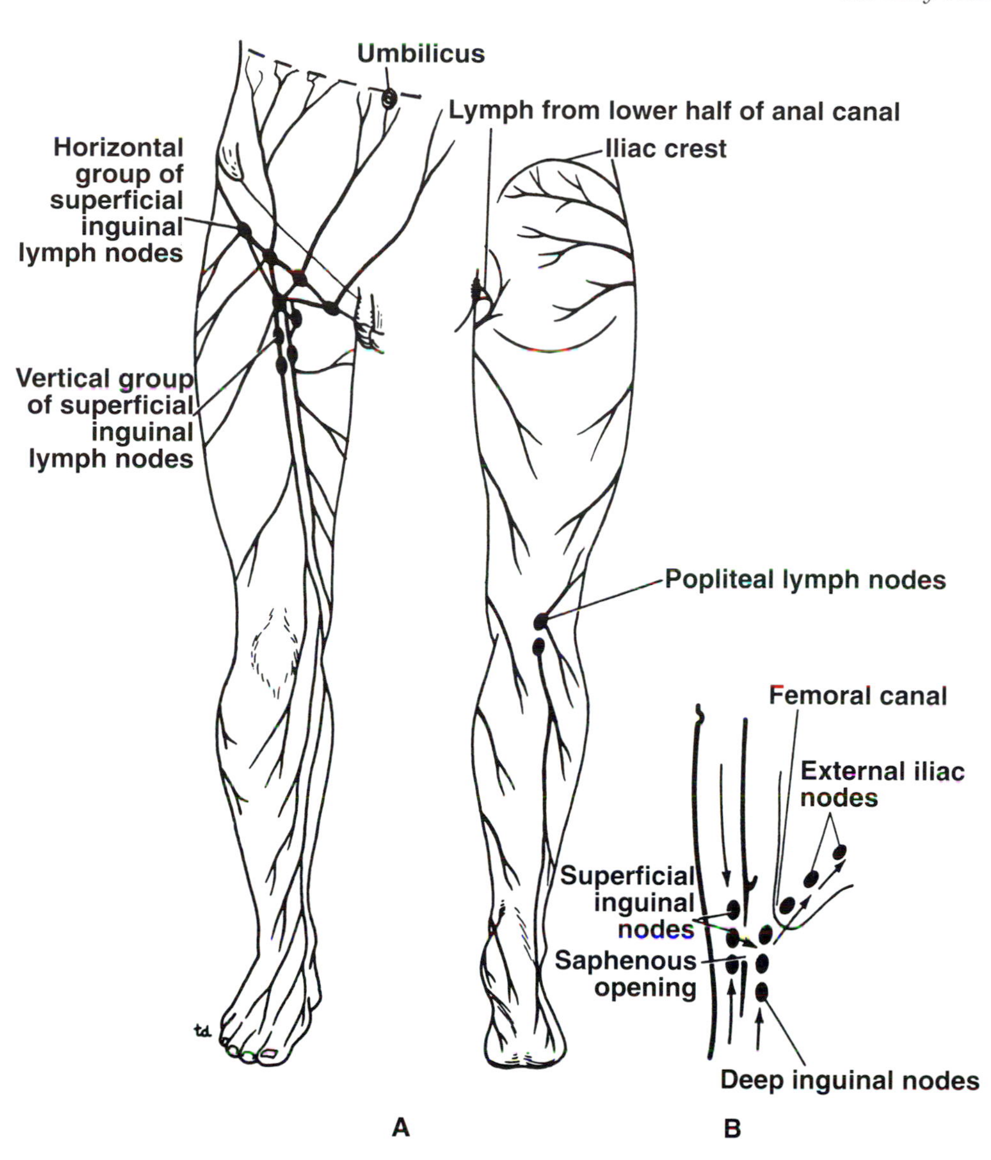

FIG 17–38.
Lymph drainage of the superficial tissues of the right lower limb and abdominal walls below the level of the umbilicus and the iliac crest. Note the arrangement of the superficial **(A)** and the deep **(B)** inguinal lymph nodes and their relationship to the saphenous opening in the deep fascia. Also, all the lymph from these nodes ultimately drains into the external iliac nodes via the lymphatic vessels in the femoral canal.

small saphenous vein behind the lateral malleolus to supply the skin of the lateral border of the foot and the lateral side of the little toe. The *medial calcaneal branch* supplies the skin over the medial surface of the heel.

2. *Muscular branches.* The gastrocnemius, plantaris, soleus, popliteus, flexor digitorum longus, flexor hallucis longus, and tibialis posterior (all muscles of the posterior compartment of the leg).

3. *Articular branches.* Knee and ankle joints.

4. *Medial plantar nerve.* Runs forward deep to the abductor hallucis with the medial plantar artery. *Branches of the medial plantar nerve* include the following: (1) *cutaneous* (the medial part of sole and plantar digital nerves supply the medial 3½ toes and the nail beds); and (2) *muscular* (the abductor hallucis, flexor digitorum brevis, flexor hallucis brevis, and first lumbrical).

5. *Lateral plantar nerve.* Runs forward deep to the abductor hallucis and flexor digitorum brevis in company with the lateral plantar artery. *Branches of the lateral plantar nerve* include the following: (1) *cutaneous* (skin of lateral part of the sole and plantar digital branches to the lateral 1½ toes and the nail beds); and (2) *muscular* (the flexor digitorum accessorius, abductor digitis minimi, flexor digiti minimi brevis, adductor hallucis, interosseous muscles, and second, third, and fourth lumbricals).

Anatomy of the Common Peroneal Nerve

The common peroneal nerve (L4, L5, S1, and S2), a branch of the sciatic nerve (Fig 17–40), runs down-

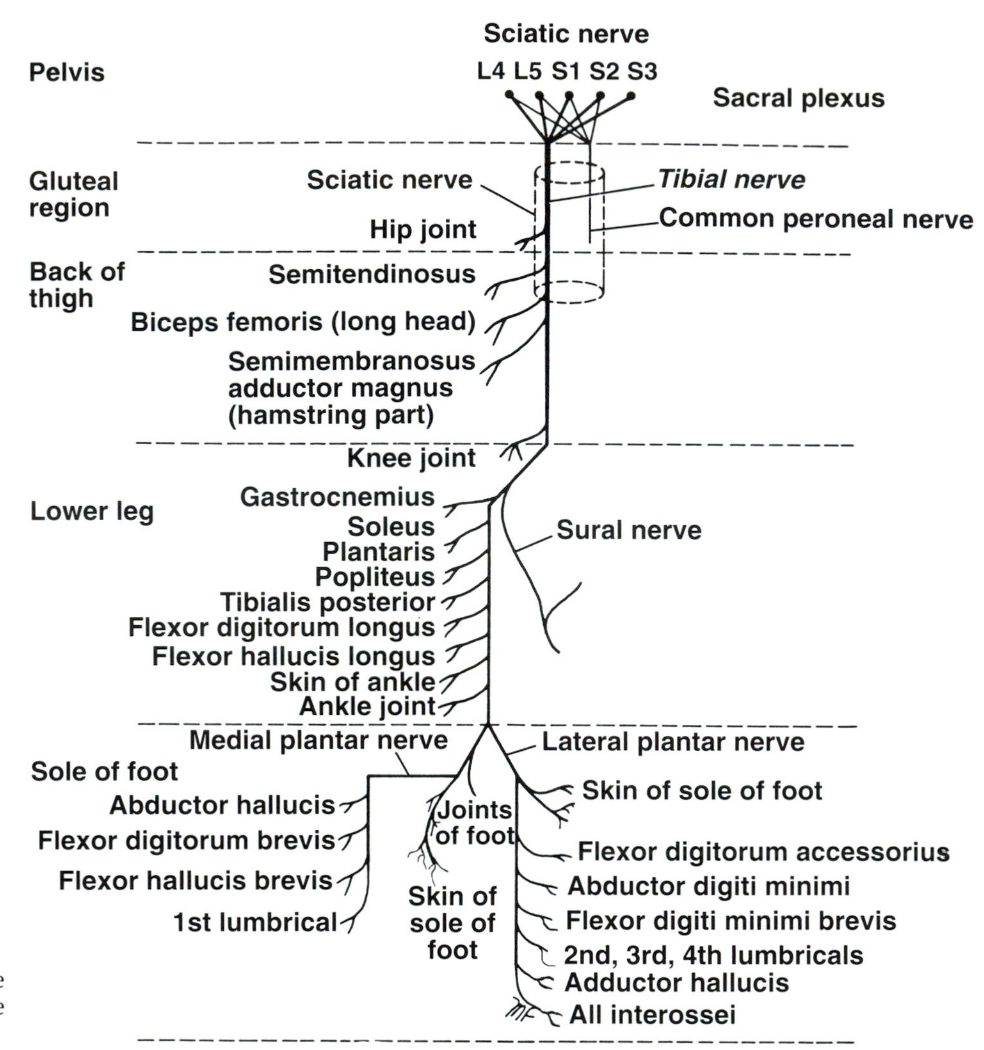

FIG 17–39.
The origin of the sciatic nerve and the main branches of the tibial nerve.

ward through the popliteal fossa following the tendon of the biceps femoris muscle (see Fig 17–12). It then passes laterally around the neck of the fibula, pierces the peroneus longus muscle, and divides into the superficial peroneal nerve and the deep peroneal nerve.

Branches of the Common Peroneal Nerve (see Fig 17–40).—These include the following.

1. *Cutaneous branches.* The *sural communicating branch* joins the sural nerve. The *lateral cutaneous nerve of the calf* supplies the skin on the lateral side of the back of the leg.
2. *Muscular branch.* The short head of the biceps femoris.
3. *Articular branches.* Knee joint.
4. *Superficial peroneal nerve.* This descends between the peroneus longus and brevis and then between the peroneus brevis and the extensor digitorum longus (see Fig 17–13). *Branches of the superficial peroneal nerve* (see Fig 17–40) include the following: (1) *cutaneous* (medial and lateral branches are distributed to skin on the front of leg and dorsum of the foot except the cleft between the big and second toes); and (2) *muscular* (peroneus longus and brevis, muscles of the lateral compartment of the leg).
5. *Deep peroneal nerve.* It leaves the peroneus longus and enters the anterior compartment of the leg (see Fig 17–13). It descends deep to the extensor digitorum longus muscle and is accompanied by the anterior tibial vessels. The nerve enters the dorsum of the foot on the lateral side of the dorsalis pedis artery and divides into medial and lateral terminal branches. *Branches of the deep peroneal nerve* (see Fig 17–40). These include the following: (1) *cutaneous* (adjacent sides of the big and second toes); (2) *muscular* the (tibialis anterior, extensor digitorum longus, peroneus tertius, extensor hallucis longus, and extensor digitorum brevis, all muscles of the anterior

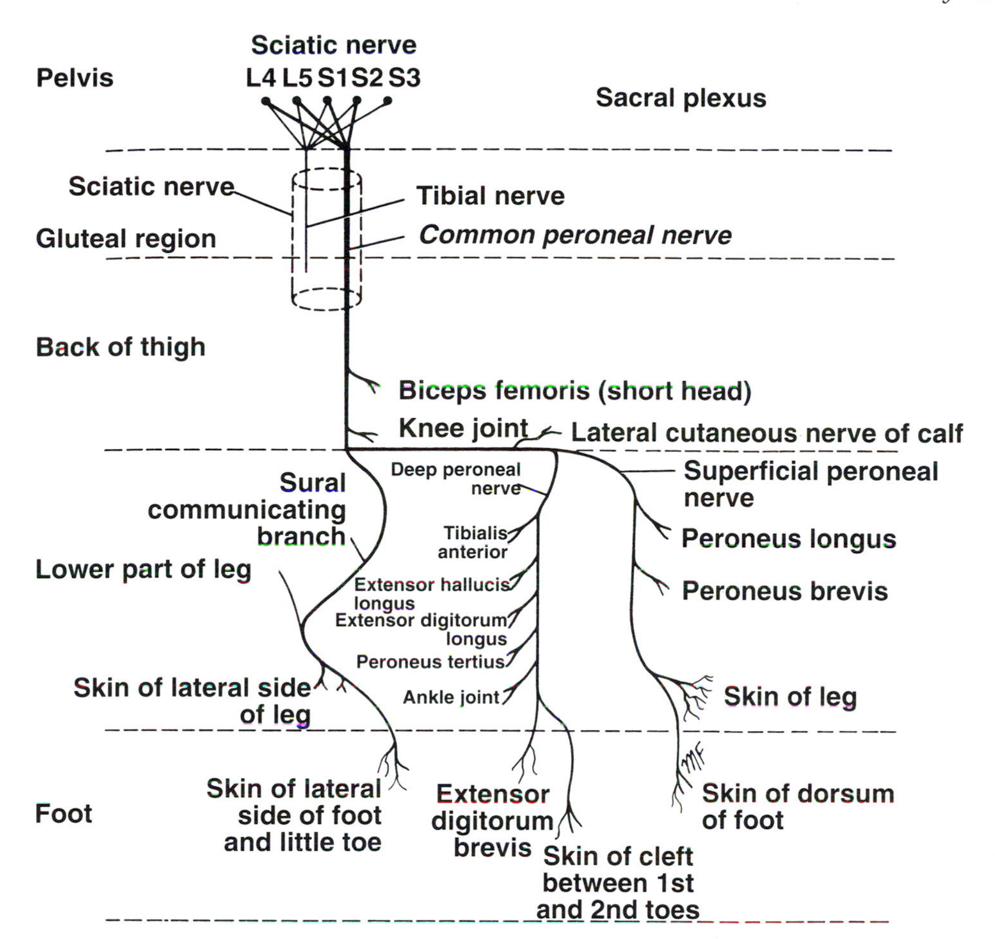

FIG 17–40.
The origin of the sciatic nerve and the main branches of the common peroneal nerve.

compartment of the leg); and (3) *articular* (ankle joint and tarsal joints).

Anatomy of the Femoral Nerve

The femoral nerve is the largest branch of the lumbar plexus and arises from the L2 through L4 lumbar nerves (Fig 17–41). It emerges from the lateral border of the psoas muscle in the abdomen and descends between the psoas and the iliacus muscles. It enters the thigh behind the midpoint of the inguinal ligament and lies lateral to the femoral vessels and the femoral sheath (see Fig 17–10). About 1.5 in. (4 cm) below the inguinal ligament, it terminates by dividing into anterior and posterior divisions.

Branches of the Femoral Nerve in the Abdomen (see Fig 17–41).—These include the following.

1. *Muscular branches* to the iliacus.

Branches of the Femoral Nerve in the Thigh (see Fig 17–41).—These include the following.

1. *Cutaneous.* The *medial cutaneous nerve of the thigh* supplies the skin on the medial side of the thigh and knee. The *intermediate cutaneous nerve of the thigh* supplies the skin on the anterior surface of the thigh. (The lateral cutaneous nerve of the thigh is a direct branch from the lumbar plexus, L2 and L3.) The *saphenous nerve* descends into the adductor canal (subsartorial canal), crossing the femoral artery on its anterior surface (see Fig 17–10). The nerve emerges on the medial side of the knee joint between the tendons of the sartorius and gracilis muscles. It descends along the medial side of the leg in company with the great saphenous vein. It passes *anterior* to the medial malleolus and along the medial border of the foot as far as the ball of the big toe.

2. *Muscular branches.* The sartorius, pectineus, and quadriceps femoris. The muscular branch to the rectus femoris also supplies the hip joint; the branches to the three vasti muscles also supply the knee joint.

Anatomy of the Obturator Nerve

The obturator nerve (Fig 17–42) arises from the lumbar plexus (L2 through L4). It emerges on the medial border of the psoas muscle within the abdomen. The nerve descends and crosses the pelvic brim to run

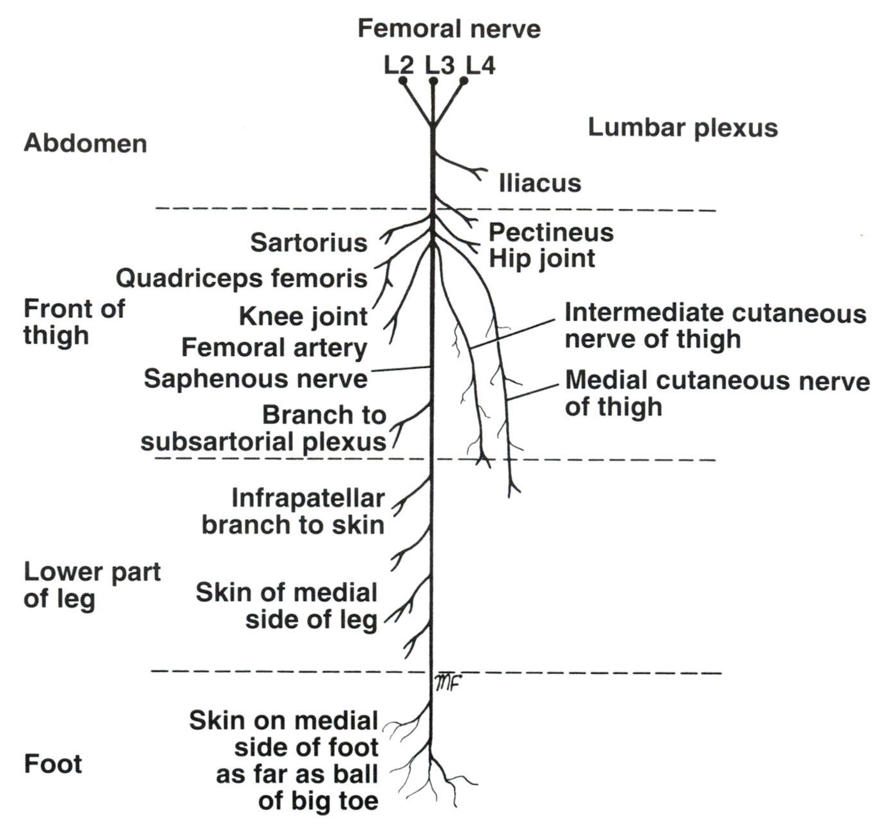

FIG 17–41.
The origin and the main branches of the femoral nerve.

downward and forward on the lateral pelvic wall. On reaching the obturator canal (upper part of the obturator foramen), it divides into anterior and posterior divisions (see Fig 17–11).

Branches of the Obturator Nerve (see Fig 17–42).—These include the following.

1. *Parietal peritoneum.* Sensory fibers to the parietal peritoneum on the lateral wall of pelvis.

2. *Anterior division.* Descends into the thigh anterior to the obturator externus and adductor brevis.

a. *Muscular branches.* The gracilis, adductor

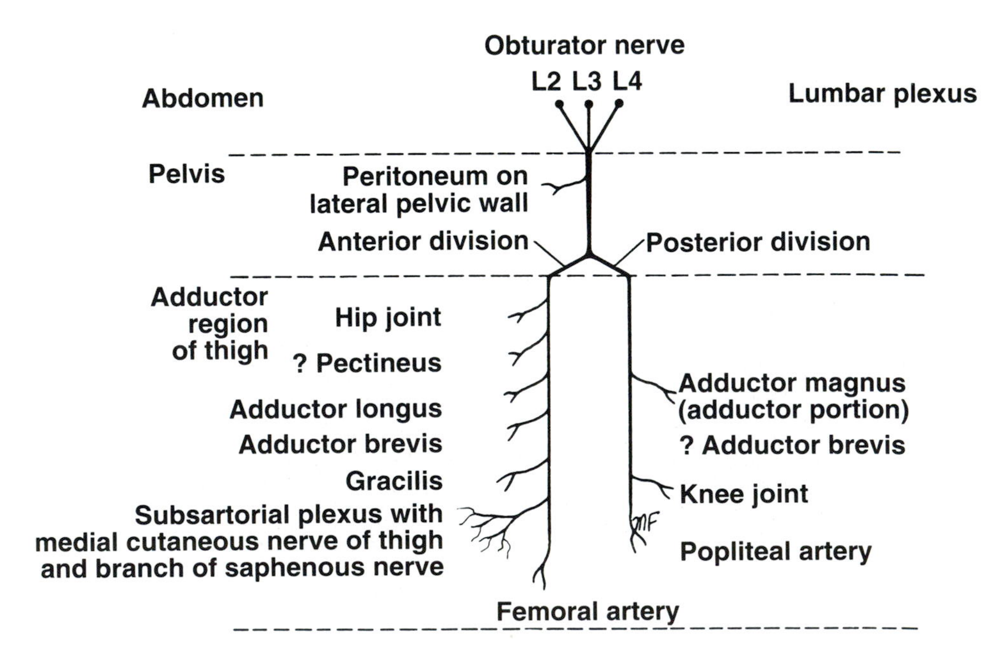

FIG 17–42.
The origin and main branches of the obturator nerve.

brevis, adductor longus, and sometimes the pectineus.

b. *Cutaneous branch.* Skin on the medial side of the thigh.

c. *Articular branch.* Hip joint.

3. *Posterior division.* Descends through the obturator externus and passes behind the adductor brevis and in front of the adductor magnus.

a. *Muscular branches.* The obturator externus, adductor magnus (adductor part), and sometimes the adductor brevis.

b. *Articular branch.* Knee joint.

SURFACE ANATOMY

Surface Anatomy of the Hip Region

Iliac Crest

This can be felt through the skin along its entire length (see Fig 17–44).

Anterior Superior Iliac Spine

This is situated at the anterior end of the iliac crest and lies at the upper lateral end of the fold of the groin (Fig 17–43). The lateral end of the inguinal ligament is attached to it, and below it originates the sartorius muscle.

Posterior Superior Iliac Spine

This is situated at the posterior end of the iliac crest (Fig 17–44). It lies at the bottom of a small skin dimple and on a level with the second sacral spine, which coincides with the lower limit of the subarachnoid space; it also coincides with the level of the middle of the sacroiliac joint. The sacrotuberous ligament and the gluteus maximus muscle are attached to the spine.

Pubic Tubercle

This can be felt on the upper border of the pubis (see Fig 17–43). Attached to it is the medial end of the inguinal ligament. The tubercle can be palpated easily in males by invaginating the scrotum from below with the examining finger. In females the pubic tubercle can be palpated through the lateral margin of the labium majus.

Pubic Crest

This is the ridge of bone on the superior surface of the pubic bone, medial to the pubic tubercle (see Fig 17–3).

Inguinal Ligament

This lies beneath the fold of the groin and may be felt along its entire length. It is attached laterally to

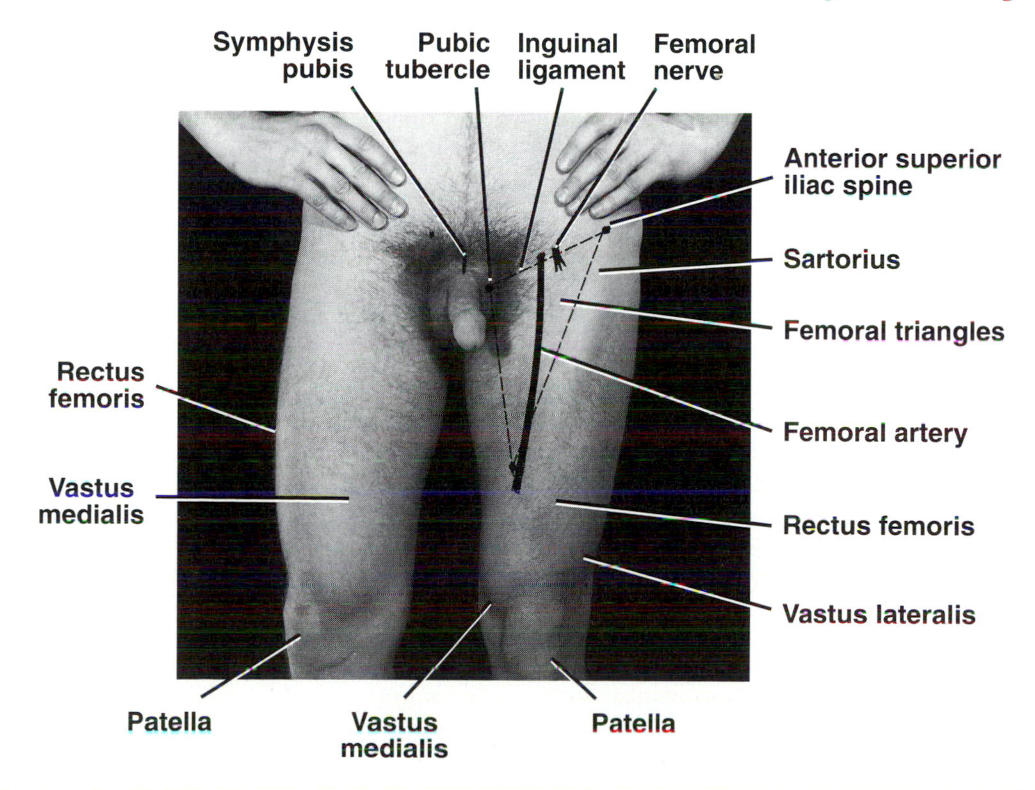

FIG 17–43.
Anterior aspect of the thighs. The broken lines indicate the boundaries of the left femoral triangle; note the positions of the femoral artery and femoral nerve. The right leg is laterally rotated at the hip joint.

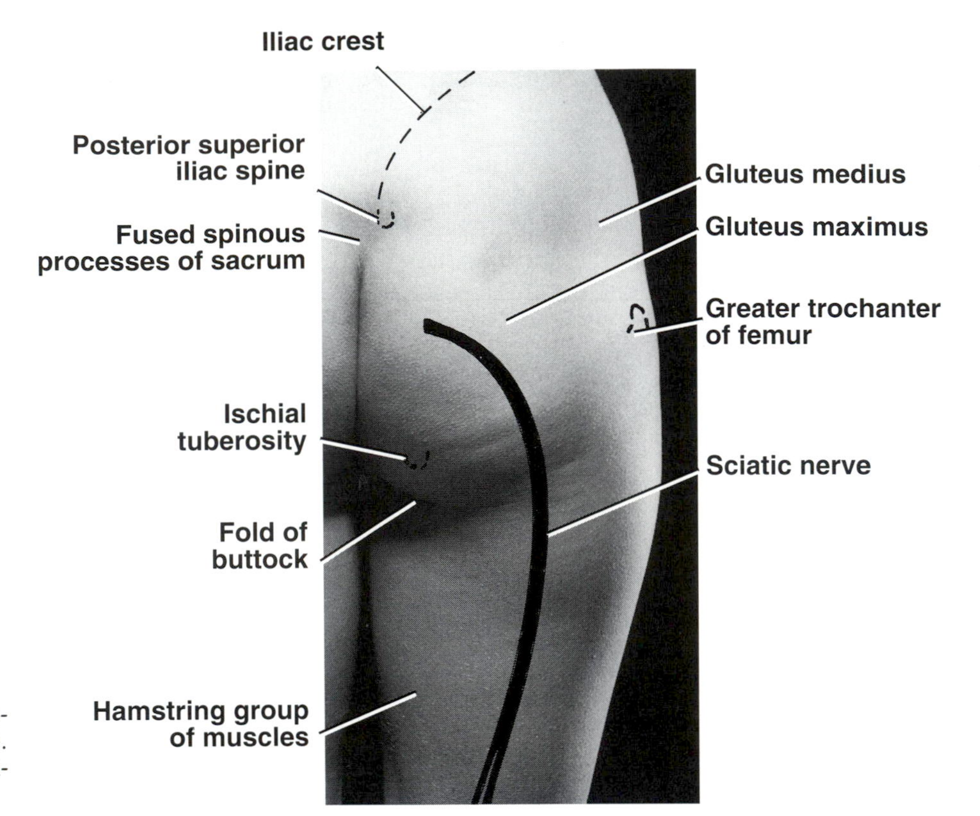

FIG 17–44.
The gluteal region and posterior aspect of the right thigh in a female. Note the surface marking of the sciatic nerve.

the anterior superior iliac spine and medially to the pubic tubercle (see Fig 17–43).

Symphysis Pubis

This is the cartilaginous joint that lies in the midline between the bodies of the pubic bones (see Fig 17–43). It can be palpated as a solid structure through the fat that is present in this region.

Femoral Triangle

This muscular triangle can be seen as a depression below the fold of the groin in the upper part of the thigh (see Fig 17–43). In a thin, muscular person, the boundaries of the triangle can be identified when the thigh is flexed, abducted, and laterally rotated. The base of the triangle is formed by the inguinal ligament, the lateral border by the sartorius muscle, and the medial border by the adductor longus muscle.

Femoral Artery

The femoral artery enters the thigh behind the inguinal ligament (see Fig 17–43) at the midpoint of a line joining the symphysis pubis to the anterior superior iliac spine; its pulsations are easily felt.

Femoral Vein

The femoral vein leaves the thigh by passing behind the inguinal ligament medial to the pulsating femoral artery (see Fig 17–43). If the arterial pulse cannot be felt, use the surface marking for the artery and allow one fingerbreadth for the width of the artery.

Femoral Canal

The lower opening of the femoral canal lies just below and lateral to the pubic tubercle (the bony projection on the upper border of the body of the pubis). Normally the function of the canal is to conduct all the lymphatic vessels of the lower limb up into the pelvis.

Clinical Note

Femoral hernia.—The lower opening of the femoral canal is the site where a femoral hernia sac enters the thigh.

Femoral Nerve

The femoral nerve enters the thigh (see Fig 17–43) behind the midpoint of the inguinal ligament, i.e., midway between the anterior superior iliac spine and the pubic tubercle (the nerve is lateral to the artery).

Great Saphenous Vein
This vein drains into the femoral vein 1½ in. (4 cm) below and lateral to the pubic tubercle (see Fig 17–48).

Superficial Inguinal Lymph Nodes
The horizontal group of superficial inguinal lymph nodes may be palpated in the superficial fascia just below and parallel to the inguinal ligament (see Fig 17–38).

Spinous Processes of the Sacrum
These processes (see Fig 17–44) are fused with each other in the midline to form the *median sacral crest.* The crest can be felt beneath the skin in the uppermost part of the natal cleft between the buttocks.

Sacral Hiatus
This is situated on the posterior aspect of the lower end of the sacrum, where the extradural space terminates. The hiatus lies about 2 in. (5 cm) above the tip of the coccyx and beneath the skin of the natal cleft.

Coccyx
The inferior surface and tip of the coccyx can be palpated in the cleft between the buttocks about 1 in. (2.5 cm) behind the anus. The anterior surface of the coccyx may be palpated with the gloved finger in the anal canal.

Ischial Tuberosity
This can be palpated in the lower part of the buttock (see Fig 17–44). In the standing position, the tuberosity is covered by the gluteus maximus. In the sitting position, the ischial tuberosity emerges from beneath the lower border of the gluteus maximus and supports the weight of the body; in this position, the tuberosity is separated from the skin by only a bursa and a pad of fat.

Greater Trochanter of the Femur
This can be felt on the lateral surface of the thigh (see Fig 17–44) and moves beneath the examining finger as the hip joint is flexed and extended. It is important to verify that in the normal hip joint, the upper border of the greater trochanter lies on a line connecting the anterior superior iliac spine to the ischial tuberosity.

Clinical Note

Level of greater trochanter relative to the pelvis.—If the greater trochanter does not lie on a line connecting the anterior superior iliac spine and the ischial tuberosity, the relationship between the femur and the pelvis is abnormal. For example, the upper border of the greater trochanter lies above the line in patients with congenital dislocation of the hip (p. 729), coxa vara, or slipped femoral epiphysis (p. 728).

The Fold of the Buttocks
This is most prominent in the standing position; its lower border does not correspond to the lower border of the gluteus maximus muscle.

Sciatic Nerve
The sciatic nerve lies under cover of the gluteus maximus muscle. As it curves laterally and downward, it is situated at first midway between the posterior superior iliac spine and the ischial tuberosity and lower down, midway between the tip of the greater trochanter and the ischial tuberosity (see Fig 17–44).

Surface Anatomy of the Knee Region

Patella and the Ligamentum Patellae
These structures can readily be palpated in front of the knee joint; the ligamentum patellae can be traced downward to its attachment to the *tuberosity of the tibia* (Fig 17–45).

Condyles of the Femur and Tibia
These can be recognized on the sides of the knee, and the *joint line* can be identified between them (see Fig 17–45).

Collateral Ligaments
The bandlike medial collateral ligament and the rounded lateral collateral ligament can be palpated on the sides of the joint line (see Fig 17–25); they can be followed above and below to their bony attachments. Because the ligaments cover the joint line, the joint line cannot be palpated at the sites of the collateral ligaments.

Menisci (Semilunar Cartilages)
These are located in the interval between the femoral and tibial condyles (see Fig 17–24). Although not recognizable, the outer edges of the medial and lat-

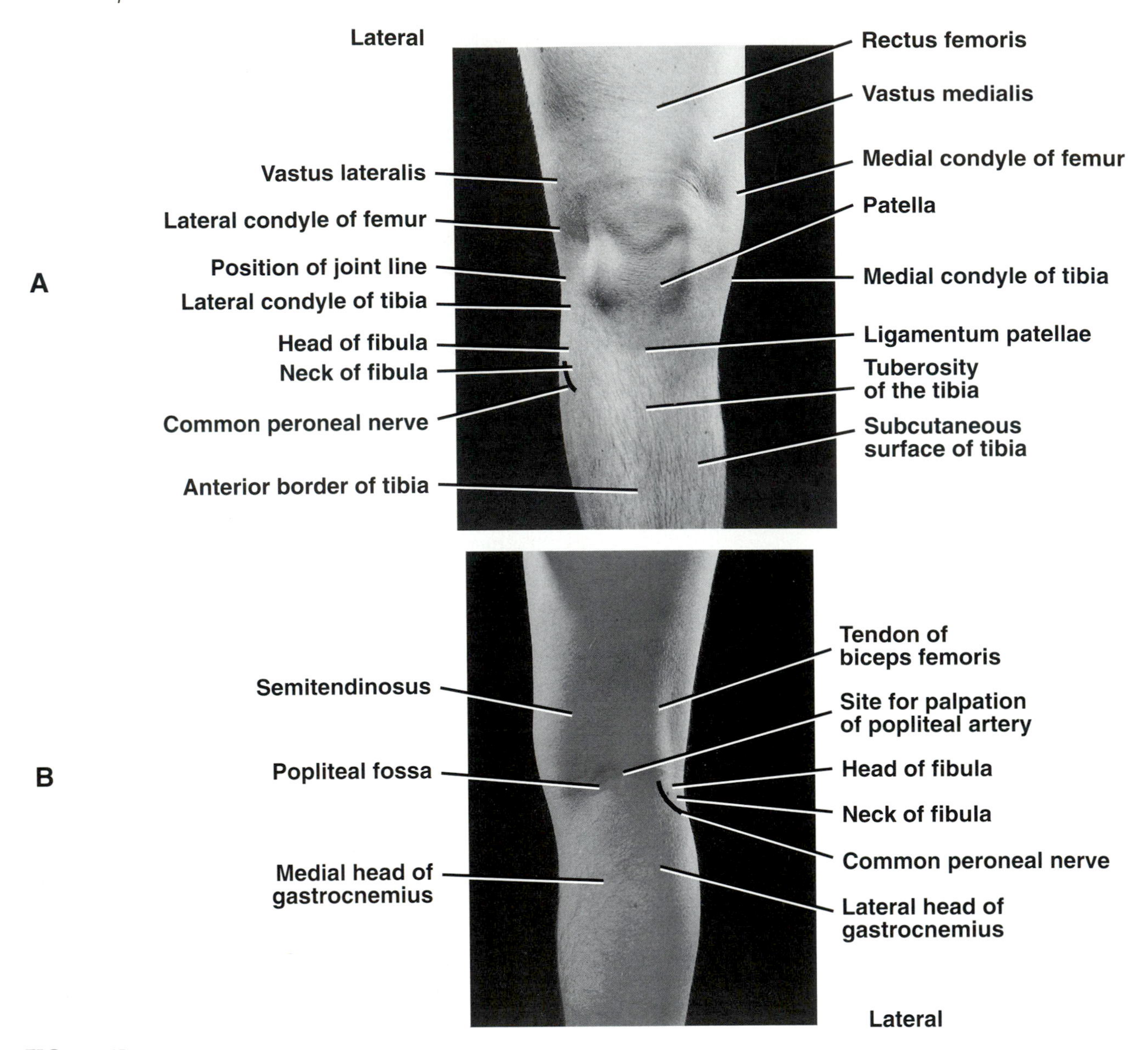

FIG 17–45.
Right knee. **A**, anterior aspect. **B**, posterior aspect. Note surface markings of the popliteal artery and the common peroneal nerve.

eral menisci may be palpated on the joint line between the ligamentum patellae and the medial and lateral collateral ligaments, respectively.

Tendon of Biceps Femoris
This can be felt as a rounded structure on the lateral aspect of the knee and can be traced down to the *head of the fibula* (see Fig 17–45).

Common Peroneal Nerve
This can be rolled beneath the examining finger just below the head of the fibula (see Fig 17–45); here it passes forward round the lateral side of the bone.

Clinical Note

Common peroneal nerve.—The common peroneal nerve is extremely vulnerable to injury at this site and is exposed to direct trauma or involvement in fractures of the upper part of the fibula. Injury to the common peroneal nerve causes *foot drop.*

Adductor Tubercle
This can be palpated on the medial aspect of the femur just above the medial condyle; the *hamstring part* of *adductor magnus* can be felt as it passes to its insertion onto it.

Popliteal Fossa
This can be identified as a diamond-shaped depression between the muscles behind the knee (see Fig 17–45). When the knee is flexed, the deep fascia, which roofs over the fossa, is relaxed, and the boundaries are easily defined. Its upper part is bounded laterally by the tendon of the biceps femoris muscle and medially by the tendons of the semimembranosus and semitendinosus muscles. Its lower part is bounded on each side by one of the heads of the gastrocnemius muscle.

Common Peroneal Nerve
This can be palpated on the medial side of the tendon of the biceps femoris (see Fig 17–45), as the latter passes to its insertion on the head of the fibula. With the knee joint partially flexed, the nerve can be rolled beneath the finger.

Popliteal Artery
The popliteal artery can be felt by gentle palpation in the depths of the popliteal fossa, provided that the deep fascia is fully relaxed by passively flexing the knee joint.

Tibia
The medial surface and anterior border of the tibia are subcutaneous and can be felt throughout their length (see Fig 17–45).

Surface Anatomy of the Ankle Region and Foot

Lateral Malleolus
The lateral malleolus is formed by the lower subcutaneous portion of the fibula (Fig 17–46).

Medial Malleolus
The medial malleolus of the tibia is subcutaneous, and its tip lies about ½ in. (1.3 cm) proximal to the level of the tip of the lateral malleolus (see Fig 17–46).

Posterior Tibial Artery
The pulsations of the posterior tibial artery may be felt midway between the medial malleolus and the heel (see Fig 17–46).

Medial Malleolar-Calcaneal Interval
The following structures lie in this interval in this order from anterior to posterior: the tibialis anterior tendon, the flexor digitorum longus tendon, the posterior tibial vessels, the posterior tibial nerve, and the flexor hallucis tendon.

Lateral Malleolar-Calcaneal Interval
The tendons of peroneus longus and brevis lie just posterior to the lateral malleolus (see Fig 17–46).

Tendon of Tibialis Anterior
On the anterior surface of the ankle joint the tibialis anterior tendon can be seen when the foot is dorsiflexed and inverted (see Fig 17–46). The *tendon of extensor hallucis longus* lies lateral to it and can be made to stand out by extending the big toe (see Fig 17–46). Lateral to the extensor hallucis longus lie the *tendons of extensor digitorum longus* and *peroneus tertius.*

Dorsalis Pedis Artery
The pulsations may be felt between the tendons of the extensor hallucis and extensor digitorum longus, midway between the two malleoli on the front of the ankle (see Fig 17–46). The two tendons can be made prominent by actively dorsiflexing the toes.

Calcaneum
This bone forms the prominence of the heel on the posterior surface of the ankle joint. Above the heel is the *Achilles tendon (tendo calcaneus)* (see Fig 17–46).

Head of the Talus
This can be felt on the dorsum of the foot just in front of the malleoli.

Peroneal Tubercle
This can be palpated on the lateral side of the foot about 1 in. (2.5 mm) below and in front of the tip of the lateral malleolus. Above the peroneal tubercle, the *tendon of the peroneus brevis,* which is a plantar flexor and strong evertor of the foot, passes forward to its insertion on the base of the *fifth metatarsal bone* (see Fig 17–46). Below the tubercle, the *tendon of the peroneus longus,* which is a plantar flexor and strong evertor of the foot, passes forward to enter the groove on the under aspect of the cuboid bone.

Sustentaculum Tali
This bony shelf, which assists in holding up the talus, can be felt on the medial side of the ankle about

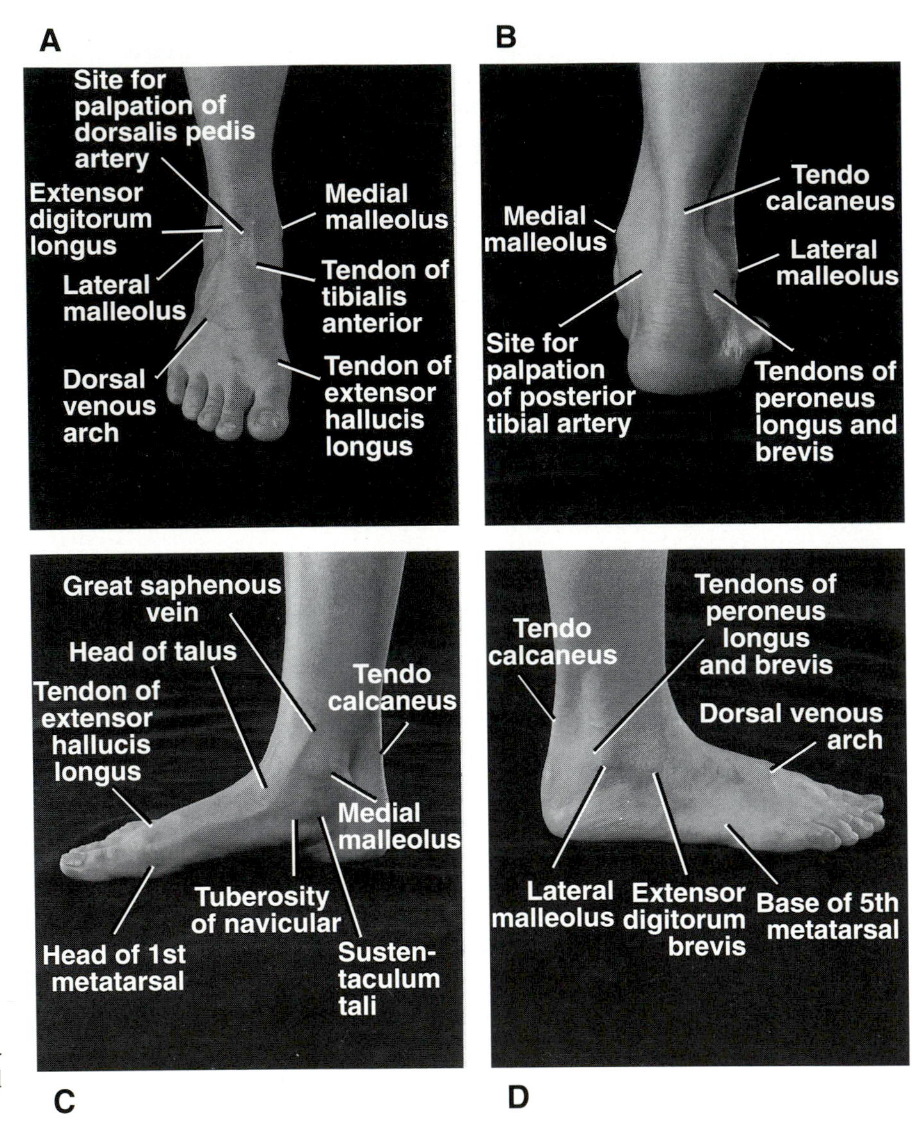

FIG 17–46.
Right foot and ankle. **A,** anterior aspect. **B,** posterior aspect. **C,** medial aspect. **D,** lateral aspect.

1 in. (2.5 cm) below the tip of the medial malleolus (see Fig 17–46).

Tuberosity of the Navicular Bone
This can be seen and palpated in front of the sustentaculum tali (see Fig 17–46).

Dorsal Venous Arch or Plexus
This can be seen on the dorsal surface of the foot just proximal to the toes (see Fig 17–46). The *great saphenous vein* leaves the medial part of the plexus and passes upward in the subcutaneous tissue *in front of the medial malleolus* (see Fig 17–46). The *small saphenous vein* drains the lateral part of the plexus and passes up *behind* the lateral malleolus.

CLINICAL ANATOMY

Clinical Anatomy of the Bony Pelvis

Fractures of the False Pelvis
These fractures occasionally occur and are caused by severe direct trauma. The upper part of the ilium is seldom displaced because of the attachments of the iliacus muscle on the inside and the gluteal muscles on the outside.

Fractures of the True Pelvis
These are usually caused by severe crushing injuries. The mechanism of these fractures can be better understood if the pelvis is regarded not only as a basin but as a rigid ring (Fig 17–47). The ring is made up of the pubic rami, the ischium, the acetabulum,

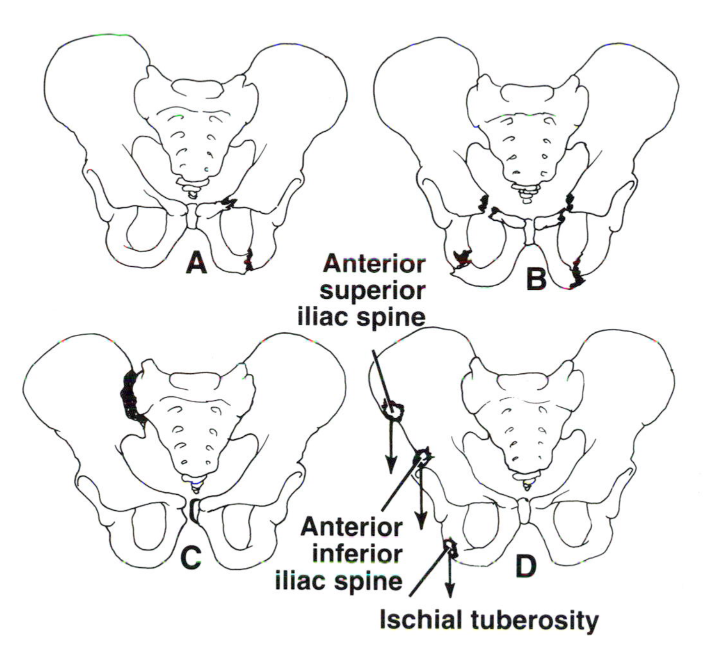

FIG 17–47.
A, B, and **C** show different types of fractures of the pelvic basin. **D** shows avulsion fractures of the pelvis. The sartorius muscle is responsible for the avulsion of the anterior superior iliac spine, the straight head of the rectus femoris muscle for the avulsion of the anterior inferior iliac spine, and the hamstring muscles for avulsion of the ischial tuberosity.

the ilium, and the sacrum, joined by strong fibrous ligaments at the sacroiliac and symphyseal joints. If there is a break in the ring at any one point, the fracture will be stable and no displacement will occur. However, should two breaks occur in the ring, the fracture will be unstable and displacement will occur, since the postvertebral and abdominal muscles will shorten and elevate the lateral part of the pelvis (see Fig 17–47). The break in the ring may occur not as the result of a fracture but as the result of disruption of the sacroiliac or symphyseal joints (see Fig 17–47). Fracture of bone on either side of the joint is more common than disruption of the joint.

The forces responsible for the disruption of the bony ring may be anteroposterior compression, lateral compression, or shearing.

Fractures of the Sacrum and Coccyx

Fractures of the lateral mass of the sacrum may occur as part of a pelvic fracture. Fractures of the coccyx are rare. However, *coccydynia* is common and is usually caused by direct trauma to the coccyx, as in falling down a flight of concrete steps. The anterior surface of the coccyx may be palpated with a rectal examination.

Minor Fractures of the Pelvis

The anterior superior iliac spine may be avulsed by the forcible contraction of the sartorius muscle in athletes (see Fig 17–47). In a similar manner the anterior inferior iliac spine may be avulsed by the contraction of the rectus femoris muscle (origin of the straight head). The ischial tuberosity can be avulsed by the contraction of the hamstring muscles. Healing may occur by fibrous union, possibly resulting in elongation of the muscle unit and some reduction in muscular efficiency.

Acetabular Fractures.—These are described with the hip joint on p. 728.

Anatomy of the Complications of Pelvic Fractures

Fractures of the true pelvis are commonly associated with injuries to the soft pelvic tissues.

The thin pelvic veins, namely, the internal iliac veins and their tributaries, that lie in the fascia beneath the parietal peritoneum, if damaged, can be the source of a massive hemorrhage, which may be life-threatening. Thrombophlebitis and pulmonary embolism are additional complications of pelvic vein damage.

The male urethra is often damaged, especially in vertical shear fractures that may disrupt the urogenital diaphragm (see p. 487).

The bladder that lies immediately behind the pubis in both sexes is occasionally damaged by spicules of bone; a full bladder is more likely to be injured than an empty bladder (see p. 486).

The rectum lies within the concavity of the sacrum and is protected and rarely damaged. Fragments of the sacrum or ischial spine may be thrust into the pelvic cavity, tearing the rectum.

Nerve injuries may follow sacral fractures; fibrosis around the anterior or posterior nerve roots or the branches of the sacral spinal nerves may result in persistent pain.

Damage to the sciatic nerve may occur in fractures involving the boundaries of the greater sciatic notch. The peroneal part of the sciatic nerve is most often involved, resulting in the inability to dorsiflex the ankle joint in a conscious patient or failure to reflexly plantar flex (ankle jerk) the foot in an unconscious patient.

Clinical Anatomy of the Hip Joint

Hip Joint Stability

The stability of the hip joint depends on the ball-and-socket arrangement of the articular surfaces and the very strong ligaments.

The stability of the hip joint when a person stands on one leg with the foot of the opposite leg raised above the ground depends on the following three factors.

1. The gluteus medius and minimus must be functioning normally.
2. The head of the femur must be located normally within the acetabulum.
3. The neck of the femur must be intact and must have a normal angle with the shaft of the femur.

If any one of these factors is defective, then the pelvis will sink downward on the opposite, unsupported side. The patient is then said to exhibit a *positive Trendelenberg's sign.*

Congenital Dislocation of the Hip

This condition has an incidence of 1 per 1,000 live births in the general population. It occurs more commonly in female than male infants, and bilateral involvement occurs in 25% of cases. Although an abnormal capsule and ligament laxity may play a role in this condition, the failure of adequate development of the upper lip of the acetabulum occurs in many cases. The femoral head, having no stable platform under which it can lodge, rides up out of the acetabulum onto the outer surface of the ilium.

A patient with a congenital dislocated hip will exhibit a positive Trendelenberg's sign (see above), and the unsupported, i.e., contralateral, side of the pelvis will sink below the horizontal line. If the patient is asked to walk, he will show the characteristic "dipping gait" on the contralateral side. In patients with bilateral congenital dislocation of the hip, the gait is typically "waddling."

Traumatic Dislocations of the Hip Joint

These are usually caused by motor vehicle accidents. Since the joint is so strong, large forces are necessary, and dislocation is frequently associated with fractures of the acetabulum. The dislocations may be classified into posterior, anterior, and central categories.

Posterior Dislocation of the Hip.—This is the most common dislocation and occurs when the joint is flexed and adducted. The head of the femur is displaced posteriorly out of the acetabulum, and it comes to rest under the gluteal muscles on the outer surface of the ilium. The thigh is shortened, medially rotated, and adducted. If the hip joint is in the partially abducted position at the time of injury, the femoral head may fracture the posterosuperior border of the acetabulum as it is driven posteriorly.

Anatomy of the Complications.—(1) The close relation of the sciatic nerve to the posterior surface of the hip joint makes it prone to injury in posterior dislocations (see Fig 17–18). (2) Fracture of the posterosuperior border of the acetabulum, as noted above. (3) Avascular necrosis of the femoral head is always a serious possible complication, and its incidence is a function of how long the head stays out of the acetabulum before being reduced (see p. 730).

Anatomy of the Reduction of the Posterior Dislocation.—With the patient in a supine position, the pelvis is stabilized by one or more assistants. The examiner applies longitudinal traction in line with the adducted leg, and the knee joint is gently flexed to 90°. Firm traction is applied to the flexed knee to pull the femoral head anteriorly. On slight external rotation, the head usually slips into the acetabulum.

In another method of hip reduction the patient is placed over the side of the bed with his leg dangling down. An assistant stabilizes the torso, and the operator pulls down on the leg.

Anterior Dislocation of the Hip.—This may occur from a blow on the back of the joint while the individual is squatting. The femoral head is forced forward out of the acetabulum and through the anterior wall of the capsule. The patient is seen with the hip abducted, externally rotated, and flexed. The complications include injury to the femoral vessels and nerve (see Fig 17–18).

Central Dislocation of the Hip.—This follows severe lateral to medial trauma to the greater trochanter of the femur; the femoral head is forced through the floor of the acetabulum, which is fractured. Complications include injury to the pelvic viscera from spicules of bone, sciatic nerve injury, and severe intrapelvic hemorrhage.

Septic Arthritis of the Hip Joint

This occurs most commonly in young children. The infection reaches the joint via the blood stream, or by extension from an osteomyelitis of the metaphysis of the femur or by contamination from femoral artery or femoral vein cannulation. The child holds the hip in the position of partial flexion, externally rotated and abducted. In this position the capacity of the joint cavity is maximum to accommodate the inflammatory exudate.

Legg-Calvé-Perthes Disease

This condition is a form of avascular necrosis of the femoral head. It is uncommon and usually occurs in children between the ages of 4 and 10 years, more often in boys. The cause of Legg-Calvé-Perthes disease is not known. The patient presents with a limp and later pain in the joint, or referred pain along the medial side of the thigh or knee (medial cutaneous nerve of the thigh from the femoral nerve). Physical examination reveals limitation of medial rotation and abduction of the hip joint. The diagnosis is made by radiography.

Slipped Femoral Epiphysis

This condition occurs most commonly in male adolescents. The cause is unknown. Most patients are overweight, which may be a factor in the etiology. The perichondrium around the femoral neck gives the epiphysis its main support. The perichondrium is thinner during adolescence, and thus the bone-cartilage junction is weakened at this time.

The patient may present with an acute slippage secondary to trauma or the patient may have experienced symptoms for months. The symptom is pain in the groin, which may be referred to the medial side of the thigh and knee. The patient limps. Because of the altered position of the slipped femoral head relative to the femoral neck, the joint is held in lateral rotation and the degree of medial rotation is limited. In fact, as the hip joint is flexed, it also externally rotates. The diagnosis is made by radiography.

Arthrocentesis of the Hip

This is fully described in Chapter 18, p. 774.

Clinical Anatomy of the Femoral Hernia

A femoral hernia is more common in women than in men (possibly because in women the pelvis and femoral canal are wider). The hernial sac passes down the femoral canal, pushing the femoral septum before it. On escaping through the lower end, it lies beneath the saphenous opening in the deep fascia of the thigh (Fig 17–48). With further expansion, the hernial sac may turn forward and upward to form a swelling in the upper part of the thigh (see Fig 17–48).

The neck of the sac always lies below and lateral to the pubic tubercle (see Fig 17–48). This serves to distinguish it from an inguinal hernia, which lies above and medial to the pubic tubercle. The neck of the hernial sac is narrow and lies at the femoral ring. The ring is related anteriorly to the inguinal ligament; posteriorly, to the pectineal ligament and the superior ramus of the pubis; medially, to the sharp free edge of the lacunar ligament; and laterally, to the femoral vein. *Because of these anatomical structures,* the neck of the hernial sac is unable to expand. Once an abdominal viscus has passed through the neck into the body of the sac, it may be difficult to push it up and return it to the abdominal cavity *(irreducible hernia).* Furthermore, should the patient strain or cough, a piece of bowel may be forced through the neck, and its blood supply compromised *(strangulated hernia).* A femoral hernia is a dangerous condition and should be treated surgically.

Anatomy of the Differential Diagnosis of a Femoral Hernia

When the differential diagnosis of a femoral hernia is considered, it is important to include diseases that may involve other anatomical structures close to the inguinal ligament as follows.

1. *Inguinal canal.* The swelling of an inguinal hernia lies above the medial end of the inguinal ligament (see p. 447). Should the hernial sac emerge through the superficial inguinal ring to start its descent into the scrotum, the swelling will lie above and medial to the pubic tubercle. The sac of a femoral hernia lies below and lateral to the pubic tubercle.

2. *Superficial inguinal nodes.* Usually more than one lymph node is enlarged. In patients with lymphadenitis, the entire field of the body that drains into the nodes should be carefully inspected. A small, unnoticed skin abrasion may be found. Never forget to inspect the umbilicus and always examine the mucous membrane of the lower half of the anal canal—it may have an undiscovered carcinoma; both these sites drain into the superficial inguinal nodes.

3. *Great saphenous vein.* A localized dilatation of the terminal part of the great saphenous vein *(saphenous varix),* although rare, may cause confusion. A hernia and varix increase in size when the patient is asked to cough (elevated intra-abdominal pressure drives the blood downward). The presence of vari-

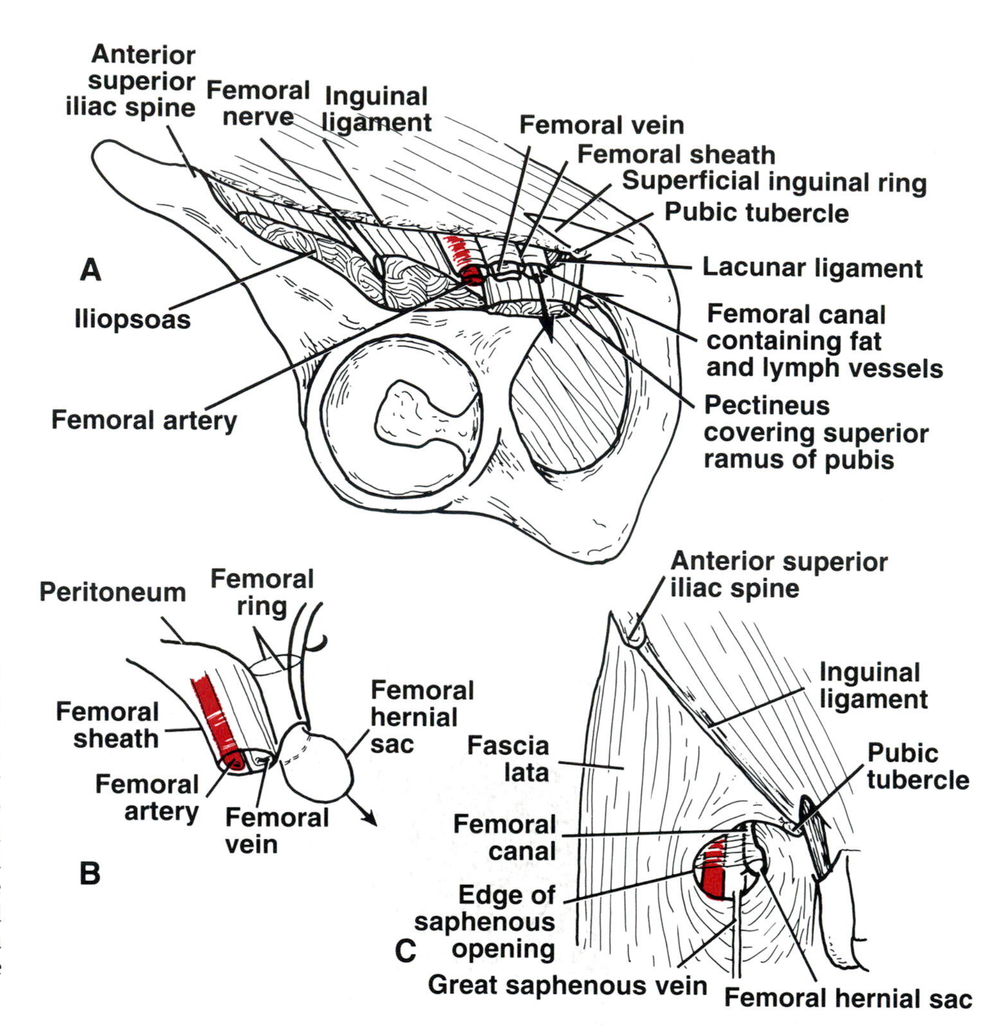

FIG 17–48.
Femoral hernia. **A,** normal right femoral sheath and its contents as seen from below. Arrow emerging from the femoral canal indicates the path taken by the femoral hernial sac. **B,** the hernial sac descending through the femoral canal. The femoral ring is the upper opening of the femoral canal. **C,** the femoral hernial sac emerging through the saphenous opening in the deep fascia in the thigh.

cose veins elsewhere in the leg should assist in the diagnosis.

4. *Psoas sheath.* Tuberculous osteomyelitis of a lumbar vertebra may rarely result in the extravasation of pus down the psoas sheath into the thigh, a swelling occurs above and below the inguinal ligament.

5. *Femoral artery.* An expansile swelling lying along the course of the femoral artery that fluctuates with the pulse rate should suggest the diagnosis of *aneurysm of the femoral artery.*

Clinical Anatomy of the Femur

Head of the Femur

The head of the femur, i.e., the part that is not intra-acetabular, can be palpated on the anterior aspect of the thigh just inferior to the inguinal ligament and just lateral to the pulsating femoral artery.

Blood Supply to the Femoral Head

The anatomy of the blood supply to the femoral head explains why avascular necrosis of the head of the femur may occur following fractures of the neck of the femur.

In young persons, the epiphysis of the femoral head is supplied by a small branch of the obturator artery that passes to the head along the ligament of the femoral head. The upper part of the neck of the femur receives a profuse blood supply from the branches of the medial femoral circumflex artery. These branches pierce the capsule and ascend the neck deep to the synovial membrane. As long as the epiphyseal cartilage remains, there is no communication between the two sources of blood. In adults, after the epiphyseal cartilage disappears, an anastomosis between the two sources of blood supply is established. It is not difficult to understand that fractures of the femoral neck will interfere with or completely interrupt the blood supply from the root of the femoral neck to the femoral head. The scant blood flow along the small artery that accompanies the round ligament may be insufficient to sustain the viability of the femoral head, and ischemic necrosis gradually takes place.

Coxa Valga and Coxa Vara

The neck of the femur is inclined at an angle with the shaft; the angle is about 160° in young children and about 125° in adults (slightly less in females). An increase in this angle is referred to as *coxa valga,* and it occurs, for example, in cases of congenital dislocation of the hip. In this condition, adduction of the hip joint is limited. A decrease in this angle is referred to as *coxa vara,* which occurs in fractures of the neck of the femur and in slipping of the femoral epiphysis. In this condition, abduction of the hip joint is limited. Shenton's line is a useful means of assessing the angle of the femoral neck on a radiograph of the hip region (see p. 698).

Fractures of the Femur

Fractures of the Neck of the Femur.—These common fractures are of two types—subcapital and trochanteric.

The subcapital fracture occurs in the elderly and is usually produced by a minor trip or stumble. Subcapital femoral fractures are particularly common in women following menopause. Avascular necrosis of the head is a common complication, with nonunion and late collapse frequently occurring. If the fragments are not impacted, there is considerable displacement. The strong muscles of the thigh (Fig 17–49), including the rectus femoris, the adductor muscles, and the hamstring muscles, pull the distal fragment upward so that the leg is shortened (as measured from the anterior superior iliac spine to the adductor tubercle or medial malleolus). The gluteus maximus, the piriformis, the obturator internus, the gemelli, and the quadratus femoris rotate the distal fragment externally (as seen by the toes pointing laterally).

Trochanteric fractures commonly occur in young and middle-aged persons as the result of direct trauma. The fracture line is extracapsular, and the fragments have a profuse blood supply. The anatomical classification depends on the number of free bony fragments (Fig 17–50). Type 1 involves the head and neck as one fragment and the trochanters and the femoral shaft as the other fragment; type 2 involves the head, neck, and lesser trochanter as one fragment, the greater trochanter as the second fragment, and the shaft as the third fragment; and type 3 involves the head and neck as one fragment, the lesser trochanter as the second fragment, the greater trochanter as the third fragment, and the shaft as the fourth fragment.

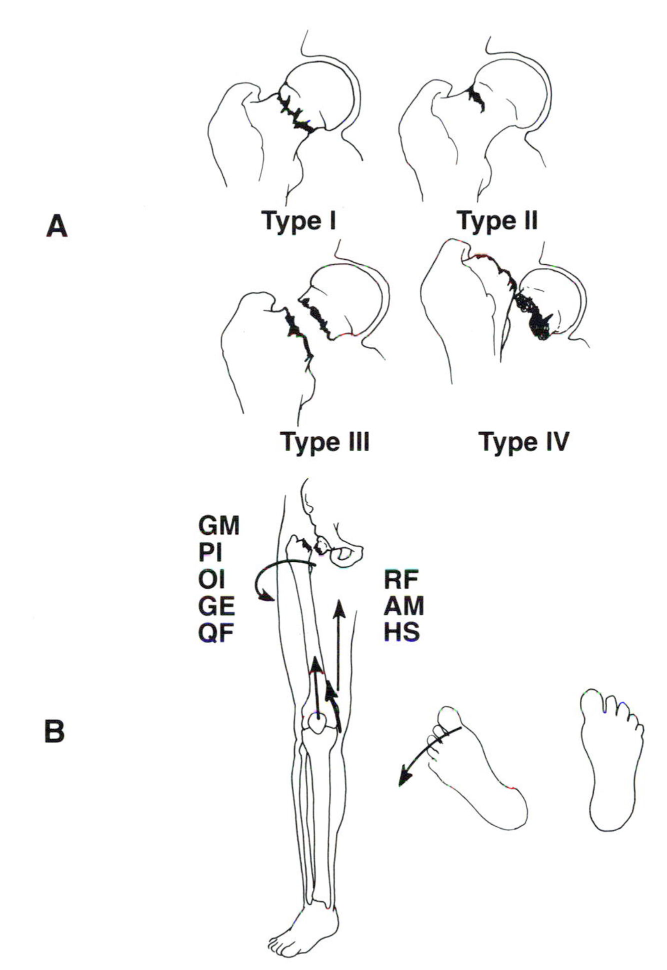

FIG 17–49.
A, fractures of the neck of the femur. **B,** displacement of the lower bone fragment caused by the pull of the powerful muscles. Note in particular the outward rotation of the leg so that the foot characteristically points laterally. *GM*=gluteus maximus; *PI*=piriformis; *OI*=obturator internus; *GE*=gemelli; *QF*=quadratus femoris; *RF*=rectus femoris; *AM*=adductor muscles; *HS*=hamstring muscles.

If the bone fragments are not impacted, the pull of the strong muscles will produce shortening and lateral rotation of the leg, as previously explained.

Fractures of the Upper Third of the Shaft of the Femur.—These fractures usually occur in young and healthy persons. The proximal fragment is flexed by the iliopsoas, abducted by the gluteus medius and minimus, and laterally rotated by the gluteus maximus, the piriformis, the obturator internus, the ge-

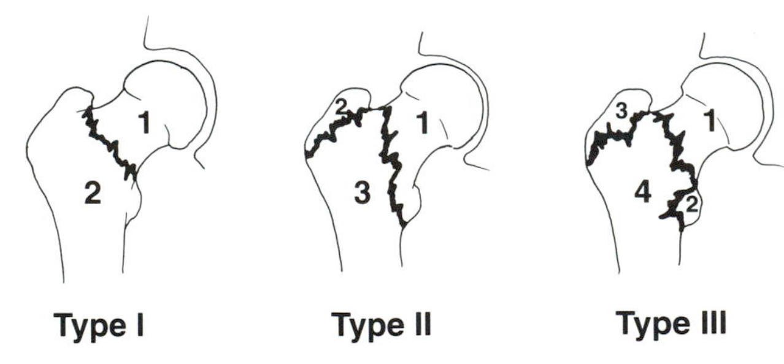

FIG 17–50.
Intertrochanteric fractures of the femur.

melli, and the quadratus femoris (Fig 17–51). The lower fragment is adducted by the adductor muscles, pulled upward by the hamstrings and quadriceps, and externally (laterally) rotated by the adductors and the weight of the foot (see Fig 17–51).

Fractures of the Middle Third of the Shaft of the Femur.—These fractures usually occur in young and healthy persons. The distal fragment is pulled upward by the hamstrings and the quadriceps (see Fig 17–51) so that there is considerable shortening. The distal fragment is also rotated backward by the pull of the two heads of the gastrocnemius (see Fig 17–51).

Fractures of the Distal Third of the Shaft of the Femur.—Here, again, these fractures usually occur in young and healthy persons. The same displacement of the distal fragment occurs as seen in fractures of the middle third of the shaft. However, the distal fragment is smaller and is rotated backward by the gastrocnemius muscle (see Fig 17–51) to a greater degree and may exert pressure on the popliteal artery and compromise blood flow through the leg and foot.

From the above accounts of the femoral fractures, it is clear that the displacements of the bony fragments depend on the action of many powerful muscles. Considerable traction is usually required to overcome the pull of these muscles to restore the

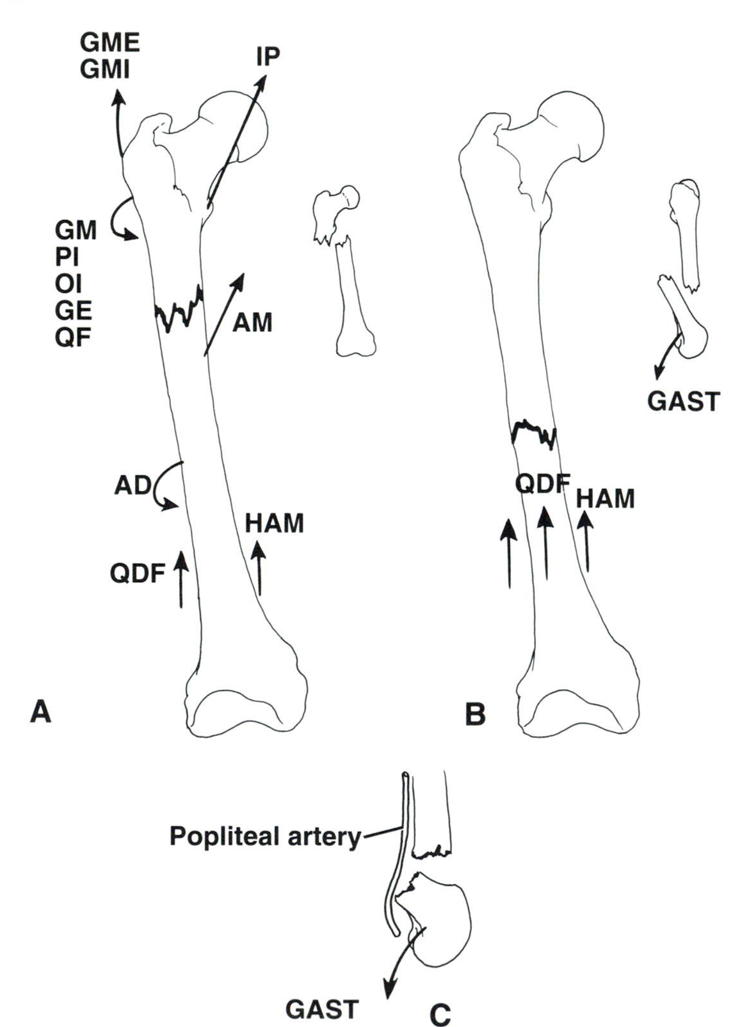

FIG 17–51.
Fractures of the shaft of the femur. **A,** upper third of the femoral shaft; note the displacement caused by the pull of the powerful muscles. **B,** middle third of the femoral shaft; note the posterior displacement of the lower fragment caused by the gastrocnemius muscle. **C,** lower third of the femoral shaft; note the excessive displacement of the lower fragment caused by the pull of the gastrocnemius muscle threatening the integrity of the popliteal artery. *IP* = iliopsoas; *GME* = gluteus medius; *GMI* = gluteus minimus; *GM* = gluteus maximus; *PI* = piriformis; *OI* = obturator internus; *GE* = gemelli; *QF* = quadratus femoris; *AM* = adductor muscles; *QDF* = quadriceps femoris; *HAM* = hamstrings; *GAST* = gastrocnemius.

limb to its correct length prior to manipulation and operative therapy.

Anatomy of the Complications of Femoral Fractures

1. *Avascular necrosis of the femoral head* in fractures of the neck of the femur. This condition has already been considered (see p. 730).
2. *Neurovascular damage* of the femoral nerve and femoral vessels in fractures of the upper third of the femoral shaft. This occurs during flexion displacement of the proximal fragment. Injury to the popliteal artery may occur when the distal fragment is displaced posteriorly in fractures of the distal third of the femoral shaft.

Clinical Anatomy of the Patella

Patellar Fractures

Direct violence to the patella, as occurs when a person strikes the dashboard in an automobile accident, shatters it into a number of small pieces. Since the patella lies within the quadriceps femoris tendon, little separation of the fragments takes place. The close relationship of the patella to the overlying skin may result in the fracture being open.

Indirect violence to the patella caused by the sudden contraction of the quadriceps muscle causes snapping of the patella across the femoral condyles. The knee is in the semiflexed position, and the fracture line is transverse. Separation of the fragments usually occurs.

A simple transverse fracture may be treated by wiring the two pieces together. Only by doing such a procedure is the quadriceps extensor mechanism restored.

A complication of patellar fractures is the disruption of the tendinous (retinacula) and muscular attachments of the quadriceps muscle to the lateral, superior, and medial borders of the patella. Failure to reattach the quadriceps to the patella will result in weakness of the extensor mechanism of the knee joint and a likelihood of recurrent dislocations of the patella.

Patellar Ossification—Accessory Centers Sometimes Misinterpreted as Fractures

Between 3 and 6 years of age, several centers of ossification appear that later fuse. Accessory marginal centers may appear that remain separate throughout life and may be confused with a fracture. The upper outer quadrant is a common site for such a marginal center.

Patellar Dislocation

The lower horizontal fibers of the vastus medialis muscle and the large size of the lateral condyle of the femur help prevent lateral displacement of the patella.

Congenital Recurrent Dislocations of the Patella.—These are found in young patients and are usually due to underdevelopment of the lateral condyle of the femur, which may be associated with some valgus deformity of the femur on the tibia.

Traumatic Dislocation of the Patella.—This is caused by direct trauma to the quadriceps attachments to the patella (especially the vastus medialis), with or without fracture to the patella. The patella always dislocates laterally because of its asymmetric shape and because of the normal upward and lateral pull of the quadriceps on the patella. The patient usually presents with the knee slightly flexed. If the knee can be fully extended or hyperextended, the patella can usually be nudged back into position.

Clinical Anatomy of the Knee Joint

The strength of the knee joint depends on the strength of the ligaments that bind the femur to the tibia and on the tone of the muscles acting on the joint. The most important muscle group is the quadriceps femoris, and, provided this is well developed, the knee can be stabilized even in the presence of torn ligaments.

Quadriceps Mechanism

The quadriceps femoris muscle, consisting of the rectus femoris, the vastus intermedius, the vastus lateralis, and the vastus medialis, is inserted into the patella and, via the ligamentum patellae, is attached to the tibial tuberosity. Together they provide a powerful extensor of the knee joint. Some of the tendinous fibers of the vastus lateralis and medialis form bands, or *retinacula,* that join the capsule of the knee joint and strengthen it. The lowest muscle fibers of the vastus medialis are almost horizontal and prevent the patella from being pulled laterally during contraction of the quadriceps muscle (see Fig 17–52).

As stated previously, the tone of the quadriceps muscle greatly strengthens the knee joint. Remember that in knee joint disease, the vastus medialis is the first muscle to atrophy and the last to recover.

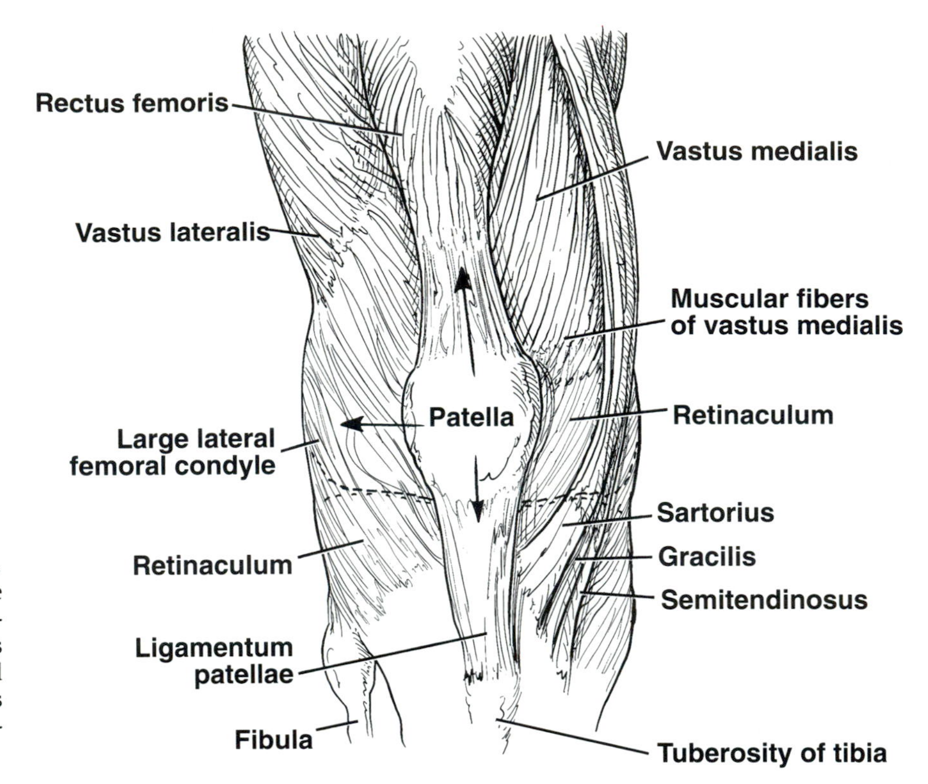

FIG 17–52.
The quadriceps femoris mechanism. The lateral and upward pull of the powerful rectus femoris and the vastus lateralis muscles on the patella is counteracted by the lowest horizontal muscular fibers of the vastus medialis and the large lateral condyle of the femur, which projects forward.

Rupture of the Rectus Femoris
The rectus femoris may rupture in sudden violent extension movements of the knee joint. The muscle belly retracts proximally, leaving a gap that may be palpable on the front of the thigh.

Rupture of the Ligamentum Patellae
This may occur when a sudden flexing force is applied to the knee joint when the quadriceps mechanism is actively contracting. The ligament may be avulsed from its attachment to the patella. This condition can be diagnosed by the inability on the part of the patient to actively extend the leg against gravity.

Synovial Membrane
The synovial membrane of the knee joint is very extensive and, in injuries to the joint, produces excessive amounts of synovial fluid that distends the large synovial cavity. The wide communication between the suprapatellar bursa, which lies beneath the quadriceps muscle, and the joint cavity results in this structure also becoming distended. The swelling of the knee extends some three or four fingerbreadths, above the patella and laterally beneath the aponeuroses (retinacula) of the insertion of the vastus lateralis and medialis. The knee effusion thus causes loss of definition and surface architecture of the kneecap.

Semimembranosus Bursitis.—This is a distension of the semimembranosus bursa that lies between the semimembranosus muscle and the medial head of the gastrocnemius muscle. The bursa usually communicates with the cavity of the knee joint. The bursitis presents as a discrete, painless, mobile swelling in the popliteal fossa behind the joint. The cyst can be made more prominent by extending the knee joint and thus expressing the synovial fluid from the joint cavity into the bursa. Unless very large, the cyst disappears when the knee is flexed.

Ligamentous Injury of the Knee Joint
There are four ligaments that are commonly injured in the knee—the medial collateral ligament, lateral collateral ligament, anterior cruciate ligament, and posterior cruciate ligament.

Medial Collateral Ligament.—Injuries to this ligament are common and are caused by valgus stress, i.e., forced abduction of the leg at the knee, and are usually associated with some degree of twisting.

Sprains of the ligament are caused by minor trauma and present with pain and localized tenderness over the ligament on the medial side of the joint. The pain is accentuated by the examiner abducting the leg at the knee.

Tearing of the medial ligament is a more severe condition. The tear may be incomplete or complete, and the pain and tenderness is felt over one or other of the bony attachments or at the joint line. If the tear is complete (Fig 17–53), the patient experiences joint instability when he attempts to bear weight on the knee; this can be confirmed by abducting the leg at the knee.

Lateral Collateral Ligament.—Injuries to this ligament are less common than the medial ligament and are caused by varus stress, i.e., forced adduction of the leg at the knee (see Fig 17–53).

Sprains and tears occur depending on the degree of force applied. Avulsion of the fibular attachment often takes place, and damage to the common peroneal nerve may occur as a complication (foot drop and loss of cutaneous sensation on the front of the ankle and dorsum of the foot, including the cleft between the first and second toes). Complete ligamentous tears can be confirmed by adducting the leg at the knee.

Note on Testing the Collateral Ligaments.—Whether one is testing for abduction or adduction instability, the knee joint should be held in 20° of flexion to relax the cruciate ligaments and the gastrocnemius muscles; spasm of the latter muscles can exert a stabilizing influence on the joint. The detection of a torn collateral ligament can be impaired by the camouflaging action of intact cruciate ligaments. If the joint opens up for a distance of greater than 1 cm, it is likely that the anterior cruciate ligament is also torn.

Tears of the Cruciate Ligaments.—These occur when severe force is applied to the knee joint and may be partial or complete depending on the size of the force. The injury is always accompanied by damage to other knee structures, including the synovial membrane, the collateral ligaments, and the capsule. The joint cavity usually quickly fills with blood (hemarthrosis), and the joint is swollen. However, occasionally the examiner can be fooled by an absence of joint swelling due to damage to the capsule and synovial membrane, allowing the joint effusion to escape. In acute injuries the joint is extremely painful, making examination difficult. The more complete the tear of the ligaments of the knee joint, the less pain experienced by the patient. This apparent contradiction can be explained by the fact that intact lig-

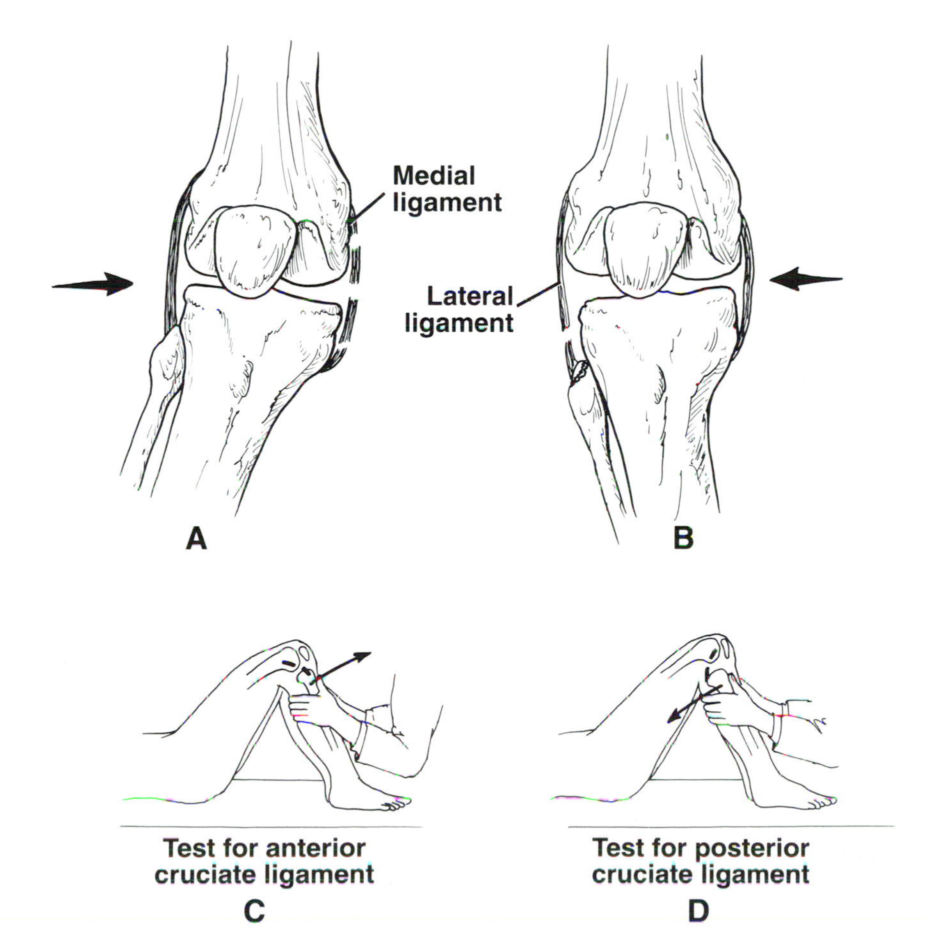

FIG 17–53.
Ruptured ligaments of the knee joint. **A**, three sites where the medial ligament is commonly ruptured; note the direction of the force that causes the rupture. **B**, the site where the lateral ligament may be ruptured; note the direction of the force that causes the rupture. **C**, ruptured anterior cruciate ligament; note that the tibia can be pulled excessively forward on the femur. **D**, ruptured posterior cruciate ligament; note that the tibia can be made to move excessively backward on the femur.

amentous fibers still have intact pain stretch receptors present, while a totally disrupted joint may have few functioning pain receptors.

Tears of the Anterior Cruciate Ligament.—These occur when excessive force is applied to the knee, usually when it is flexed. Examination of patients with a ruptured anterior cruciate ligament shows that the tibia can be pulled excessively forward on the femur (see Fig 17–53).

Tears of the Posterior Cruciate Ligament.—These are extremely rare. Examination shows that the tibia can be made to move excessively backward on the femur (see Fig 17–53).

Meniscal or Semilunar Cartilage Injury of the Knee Joint

Injuries of the menisci are very common.

Injury to the Medial Meniscus.—The medial cartilage is damaged much more frequently than the lateral cartilage, probably because of its strong attachment to the medial collateral ligament of the knee joint, which restricts its mobility. The injury occurs when the femur is rotated on the tibia, or the tibia is rotated on the femur, with the knee joint partially flexed and taking the weight of the body. The tibia is usually abducted on the femur, and the medial cartilage is pulled into an abnormal position between the femoral and tibial condyles (Fig 17–54). A sudden movement between the condyles results in the cartilage being subjected to a severe grinding force, and it splits along its length. When the torn part of the cartilage becomes wedged between the articular surfaces, further movement is impossible, and the joint is said to "lock."

Injury to the Lateral Meniscus.—Injury to the lateral cartilage is less common than injury to the medial meniscus. This is probably because the lateral cartilage is not attached to the lateral collateral ligament and, consequently, is more mobile. The popliteus muscle sends a few of its fibers into the lateral cartilage, and these may pull the cartilage into a more favorable position during sudden movements of the knee joint.

Dislocations of the Knee Joint

Anterior Dislocation.—This follows extreme hyperextension caused by a large force. The posterior part of the capsule tears first, followed by disruption of the anterior cruciate ligament. The tibia moves forward on the femur and comes to rest in front of the femoral condyles.

Posterior Dislocation.—This occurs when the force pushes the proximal end of the tibia posteriorly, rupturing the posterior cruciate ligament. Occasionally, the femoral condyles pierce the retinacula of

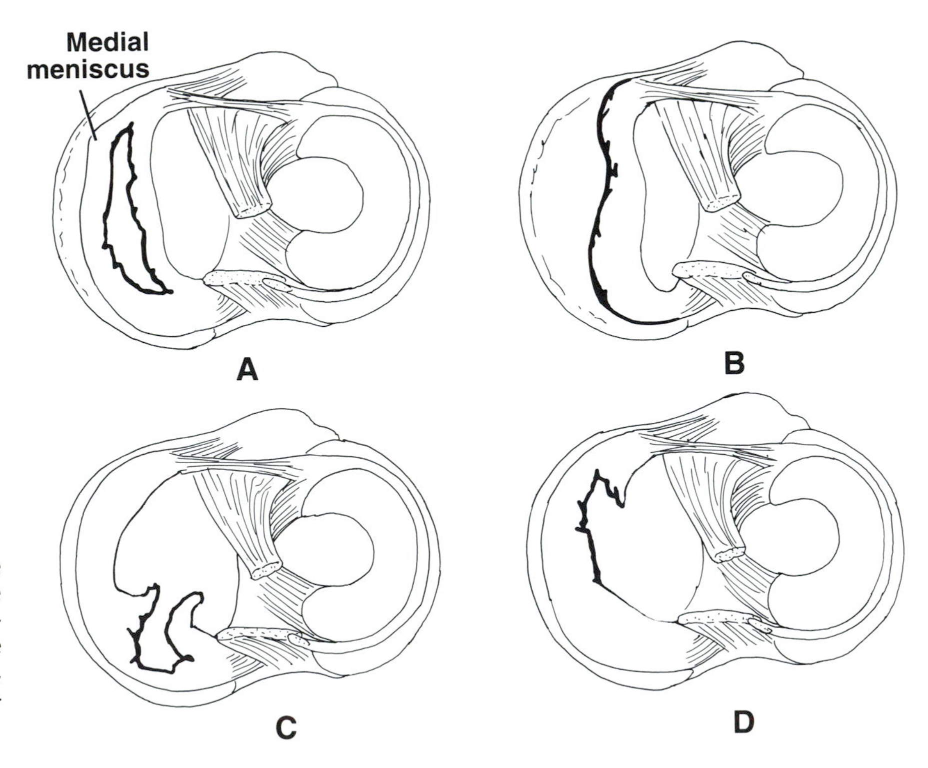

FIG 17–54.
Tears of the medial meniscus of the knee joint. **A,** complete bucket handle tear. **B,** meniscus is torn from its peripheral attachment. **C,** tear of the posterior portion of the meniscus. **D,** tear of the anterior portion of the meniscus.

the quadriceps muscle and disrupt the extensor mechanism of the knee. The tibial condyles come to rest behind the femoral condyles.

Anatomy of the Complications.—In dislocations of the knee the excessive movements of the tibia on the femur may severely compromise the popliteal artery as it lies close behind the joint in the popliteal fossa.

Arthrocentesis of the Knee

The anatomy of this procedure is fully described in Chapter 18, p. 775.

Clinical Anatomy of the Tibia and Fibula

Fractures

Fractures of the Tibia and Fibula.—Fortunately, if only one bone is fractured, the other acts as a splint, and there is often minimal displacement. However, a one-bone fracture may result in a compartmental syndrome, which occurs less frequently when both bones are fractured.

Fractures of the shaft of the tibia are often open, since the entire length of the medial surface is covered by skin and superficial fascia. Fractures of the distal third of the shaft of the tibia are sometimes oblique, and in these cases the pull of the strong calf muscles results in considerable overriding and anterior angulation. In addition, fractures of this part of the tibia are prone to delayed or nonunion. This may be due to tearing of the nutrient artery at the line of fracture, with consequent reduction in blood flow to the distal fragment; it is also possible that the splintlike action of the fibula prevents the proximal and distal fragments from coming into apposition.

Fractures of the Proximal End of the Tibia (Fractures of the Tibial Plateau).—Fractures of the tibial condyles are common in middle-aged and elderly persons; they usually result from direct violence to the lateral side of the knee joint, as when a person is hit by the bumper of an automobile. The tibial condyle may show a split fracture or be broken up, or the fracture line may pass between both condyles in the region of the intercondylar eminence. As the result of forced abduction of the knee joint, the medial collateral ligament may also be torn or ruptured.

Fractures of the Distal End of the Tibia.—These are considered with the ankle joint (see p. 741).

Osgood-Schlatter's Disease.—In this condition pain and tenderness occurs over the attachment of the patellar ligament into the tibial tuberosity. Osgood-Schlatter's disease commonly occurs in adolescence and often follows excessive traction from the quadriceps muscle on the tibial tuberosity. The tibial tuberosity usually arises as a downward protrusion from the upper tibial epiphysis. However, it may have an additional center of ossification that extends up from the shaft. The condition may be likened to a slipped femoral epiphysis, and the traction of the ligamentum patellae pulls up and detaches the epiphyseal plate at the site of the tibial tuberosity.

Anatomy of the Intraosseous Infusion Technique in the Tibia

Intraosseous infusion is useful for immediate vascular access during cardiopulmonary resuscitation and is indicated when venous access is not readily available. It is most often used during cardiac arrest in an infant or child. The most common site is the proximal end of the tibia; the distal end may also be used (Fig 17–55).

The bone marrow of the tibia receives a large *nutrient artery* from the posterior tibial artery that enters the bone high up on its posterior surface. On reaching the medullary cavity, the artery divides into ascending and descending branches. These branches give rise to arterioles, capillaries, and medullary sinusoids and eventually join a large central venous sinus that is drained by nutrient veins into the general venous circulation (see Fig 17–55). The medullary sinusoids will accept fluids and drugs during intraosseous infusion.

Anatomy of the Procedure for the Proximal Tibia.—The tibial tuberosity is identified by palpation, and the medial border of the shaft of the tibia is located. Halfway between these two points and one fingerbreadth distally is the ideal site for inserting a sturdy needle fitted with a stylet (Fig 17–55). The site is distal to the epiphyseal growth plate, and the bone is soft enough to permit easy access of the needle. Once the needle reaches the periosteum, a twisting motion ensures that it pierces the periosteum and hard bony cortex. The trabeculae of the spongy bone provide less resistance.

Anatomy of the Procedure for the Distal Tibia.—The medial malleolus is palpated and the needle is inserted at the site where the medial malleolus becomes continuous with the shaft of the tibia (see Fig 17–55).

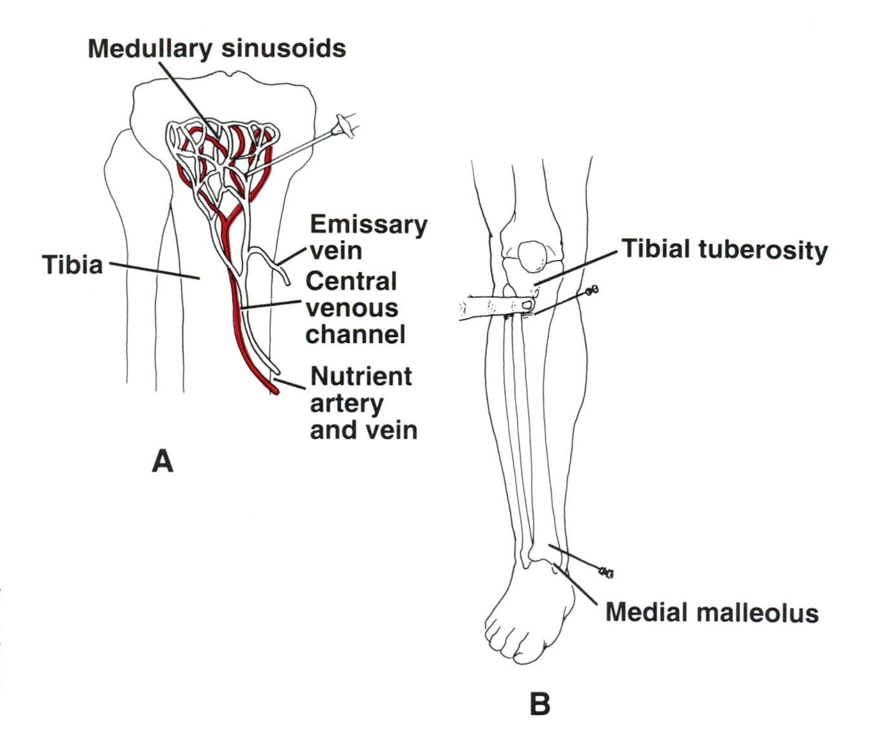

FIG 17–55.
Intraosseous infusion. **A,** the arrangement of the blood supply to the tibia. **B,** the two common sites used for intraosseous infusion of the tibia. In children the needle is directed away from the epiphyseal cartilage.

Anatomy of the Complications.— These include the following.

1. Penetration of the bone due to excessive force applied to the needle.
2. Blockage of the needle by bony trabeculae.
3. Infection of the soft tissues and bone.
4. Damage to the epiphyseal growth plate.
5. Fat embolism. This complication is rare and only occurs in adults who have yellow bone marrow. The bone marrow of the tibia remains hemopoetic until the age of 7 years, when yellow marrow starts to replace the red marrow at the distal end and gradually moves proximally.

Clinical Anatomy of Compartment Syndromes of the Leg

The leg is enclosed in a sheath of deep fascia, which is attached to the anterior and medial borders of the tibia, where it is fused with the periosteum (Fig 17–56). Two intermuscular septa pass from its deep aspect to be attached to the fibula. These septa, together with the interosseous membrane, divide the leg below the knee into three compartments— anterior, lateral, and posterior, with each compartment having its own muscles, nerves, and blood supply.

Contents of the Anterior Fascial Compartment of the Leg (see Fig 17–56)
Muscles.— Tibialis anterior, extensor digitorum longus, peroneus tertius, and extensor hallucis longus.

Blood Supply.— Anterior tibial artery.

Nerve Supply.— Deep peroneal nerve (cutaneous distribution— adjacent sides of the big and second toes).

Contents of the Lateral Fascial Compartment of the Leg (see Fig 17–56)
Muscles.— Peroneus longus and peroneus brevis.

Blood Supply.— Branches from the peroneal artery.

Nerve Supply.— Superficial peroneal nerve (cutaneous distribution— front of leg and dorsum of foot).

Contents of the Posterior Fascial Compartment of the Leg (see Fig 17–56)
A *deep transverse septum* of fascia divides the muscles of the posterior compartment into superficial and deep groups.

Muscles.— These include the following.
Superficial Group of Muscles.—Gastrocnemius, plantaris, and soleus.

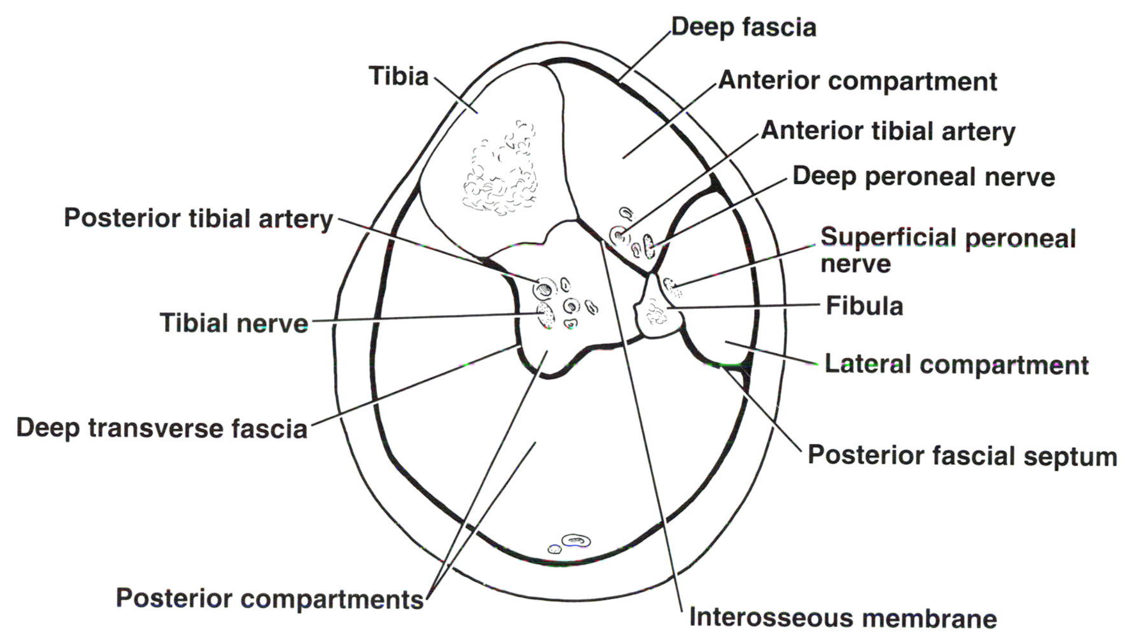

FIG 17–56.
Compartments of the leg.

Deep Group of Muscles.—Popliteus, flexor digitorum longus, flexor hallucis longus, and tibialis posterior.

Blood Supply.—Posterior tibial artery.

Nerve Supply.—Tibial nerve (cutaneous distribution—via sural branch, calf and back of the leg and lateral border of the foot and lateral side of the little toe; via medial plantar nerve, medial part of the sole and medial 3½ toes and the nail beds; and via lateral plantar nerve, lateral part of the sole and lateral 1½ toes and the nail beds).

The anterior tibiofibular compartment is the one most often afflicted by this syndrome. Unfortunately, as with the forearm, there is very little room within the compartments besides the structures mentioned. The development of edema can cause secondary vascular compression, with the veins being first affected and, later, the arteries.

Soft-tissue injury in association with bone fractures is a common cause, and early diagnosis is critical. The early signs include paresthesia (due to ischemia of the sensory nerves traversing the compartment), pain disproportionate to the injury (due to pressure on the nerves within the compartment), pain on passive stretching or an active contraction of the muscles that pass through the compartment (due to muscle ischemia), induration and tenderness of the skin over the compartment (a late sign caused by edema), absence of capillary refill in the nail beds (due to pressure on the arteries within the compartment), and, finally, as a late sign, absent distal arterial pulses (due to occlusion of arteries within the compartment). Once the diagnosis is made, fasciotomy is the treatment of choice to decompress the affected compartment. In the case of the anterior tibiofibular compartment, a delay of as little as 4 hours may cause irreversible damage to the muscles, resulting in foot drop.

Anatomy of Compartmental Pressure Measurement Techniques

Basically the techniques involve the introduction of a needle or probe into one of the anatomical compartments of the leg while avoiding damage to neurovascular structures. With readily identifiable surface landmarks, such as the anterior and posterior borders of the shaft of the tibia, the head and neck of the fibula, and the medial and lateral malleoli, the desired compartment is entered. The operator should keep in mind a clear understanding of the arrangement of each compartment and its contents as viewed on cross section (see Fig 17–56). All the leg compartments are superficial except for the deep posterior compartment, which requires the use of a long spinal needle.

Clinical Anatomy of Pain in the Back of the Calf

A consideration of the various anatomical structures located in the calf will assist in the differential diagnosis of calf pain.

Nerves

Dermatomes over the calf (see Figs 2–32 and 2–33) include L4, L5, and S2. The sensory innervation is provided by the lateral cutaneous nerve of the calf (common peroneal nerve), the sural nerve (tibial nerve), and the saphenous nerve (femoral nerve). Pain from disease of the knee joint and, rarely, the hip joint can be referred to the calf. The knee joint is innervated by the femoral, obturator, common peroneal, and tibial nerves, whereas the hip joint is innervated by the femoral, obturator, and sciatic nerves. At a higher level, lesions of the sciatic nerve (L4 and L5 and S1 through S3) in the pelvis can give rise to referred pain down the back of the calf and onto the dorsum of the foot. Nerve root involvement, as in lumbar disc lesions or lumbar spinal stenosis, could also produce calf pain. Patients with vascular neuropathy, particularly diabetic neuropathy, may present with a sciatic or posterior tibial syndrome. Finally, intraspinal tumors pressing on nerve roots may first reveal their presence by giving sciatic nerve symptoms.

Arteries

Atherosclerotic disease of the aortoiliac, femoral, and popliteal-tibial arterial systems can produce claudication in the calf. Occlusive disease involving the more proximal arteries may also give rise to pain in the thigh and, occasionally, the buttock.

Veins

Superficial thrombophlebitis of the short saphenous vein may produce pain in the calf. The swelling and inflammatory signs along the course of the vein and its main tributaries are evident. Moreover, the cord-like thrombosed veins can often be felt. Deep vein thrombophlebitis in the veins of the soleus muscle give rise to mild pain or tightness in the calf and calf muscle tenderness. However, deep vein thrombosis also can occur with no signs or symptoms.

Lymphatics

Lymphangitis associated with cellulitis can usually be recognized by the presence of red skin streaks. These are caused by the inflamed lymphatic vessels passing up the back of the calf to the popliteal space to enter the popliteal nodes. A high fever and leukocytosis assists in the distinction between this condition and superficial thrombophlebitis.

Muscles

Tearing of the gastrocnemius or soleus muscles or rupture of the plantaris will produce severe localized pain and tenderness over the damaged muscle. Swelling may be present, and the finding of ecchymosis at the sides of the Achilles tendon at the ankle is diagnostic. The differential diagnosis of a tearing of the gastrocnemius, soleus, or plantaris is a rupture of the Achilles tendon, which is discussed on p. 743.

Simple spontaneous night cramps may leave the patient with an extremely sore calf. The pain usually disappears quickly with rest. The strange condition of "crazy legs," in which the patient complains of jumps or restlessness of the legs following a period of fatigue or stress, may present with tender calf muscles. The tenderness disappears with rest.

Clinical Anatomy of the Ankle Joint

The ankle joint is a hinge joint possessing great stability. The deep mortise formed by the lower end of the tibia and the medial and lateral malleoli securely holds the talus in position.

Acute Sprains of the Lateral Ankle

These injuries are more common than those on the medial side of the ankle. Excessive inversion of the foot with plantar flexion of the ankle or attempted medial rotation of the ankle result in damage to the lateral ligament (usually the anterior talofibular ligament). Pain and later swelling of the lateral side of the ankle occur. Points of extreme tenderness can be detected over the particular part of the lateral ligament that is damaged.

Rupture of the Lateral Ligament of the Ankle

Following excessive force, parts or all of the lateral ligament may be ruptured, giving instability to the joint on examination (Fig 17–57). If the anterior talofibular component of the lateral ligament is ruptured, the instability will show as an anterior displacement of the foot on the ankle.

Acute Sprains of the Medial Ankle

Excessive eversion of the foot with some lateral rotation and dorsiflexion may damage the deltoid ligament. The ligament, which is very strong, may be

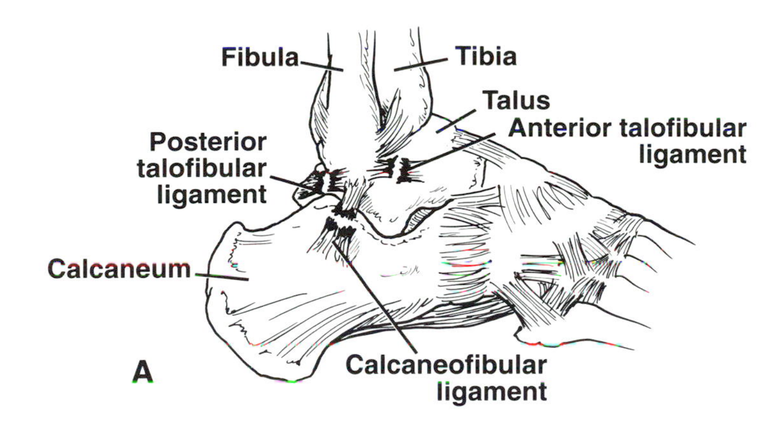

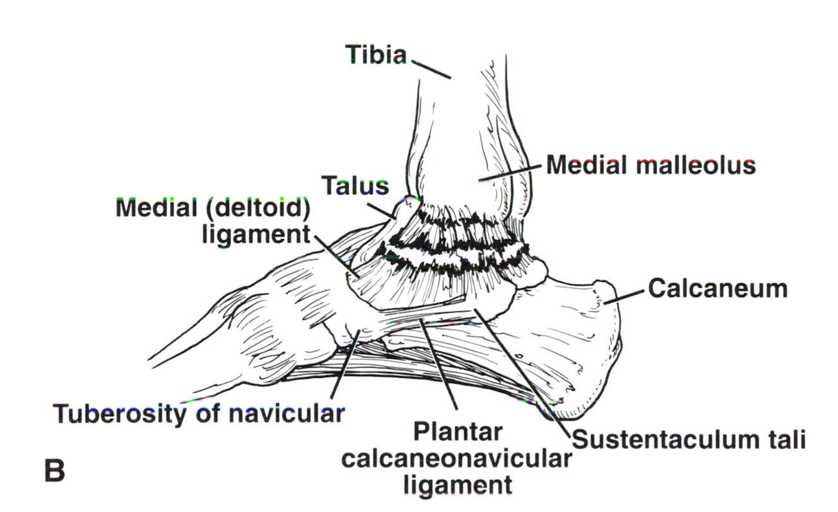

FIG 17–57.
Sprains and ruptures of the ankle ligaments. **A,** ankle bones, lateral view, showing rupture of the lateral ligaments. **B,** ankle bones, medial view, showing rupture of the medial ligament; sometimes the tip of the medial malleolus is avulsed.

stretched, torn, or *ruptured;* sometimes the tip of the medial malleolus is pulled off (see Fig 17–57). Pain and later swelling of the medial side of the ankle takes place, and there is tenderness over the damaged ligament. Rupture of the medial ligament gives rise to joint instability with displacement of the foot laterally at the ankle.

Fracture Dislocations of the Ankle Joint

These are common and are caused by a forced external rotation and overeversion of the foot. The talus is externally rotated forcibly against the lateral malleolus of the fibula. The torsion effect on the lateral malleolus causes it to fracture spirally. If the force continues, the anterior talofibular ligament is torn and the mortise of the ankle joint tends to open up anteriorly. The talus can then move laterally, and the medial ligament of the ankle joint becomes taut and pulls off the tip of the medial malleolus. If the talus is forced to move still farther, its rotary movement results in its violent contact with the posterior inferior margin of the tibia (sometimes referred to as the posterior malleolus), which shears off.

Other less common types of fracture dislocation are due to forced overeversion (without rotation), in which the talus presses the lateral malleolus laterally and causes it to fracture transversely. Overinversion (without rotation), in which the talus presses against the medial malleolus, will produce a vertical fracture through the base of the medial malleolus.

Anatomy of the Complications of Ankle Fracture Dislocations

The isolated lateral malleolar fracture is the common fracture and in most cases is easily reduced without problems. However, in some patients the failure to diagnose additional damage to the medial ligament may result in the patient being left with an unstable joint.

The failure to restore joint congruency will result in poor function and the later development of osteoarthritis. The inability to restore the bony fragments to their correct anatomical position by closed manipulation would indicate the necessity to have open reduction and internal fixation.

Anatomy of the Closed Reduction of the Simple Lateral Malleolar Fractures
The patient lies in a supine position on the table with the involved leg hanging free over the edge in front of the operator. The knee joint is flexed to a right angle. The foot is inverted to stretch the lateral ligament of the ankle joint. Further inversion causes the lateral ligament to pull the lateral malleolus distally. The operator's fingertips then pull the lateral malleolus forward and medially rotate it into its correct anatomical position. The forefoot is supported to prevent plantar flexion and further inversion. A below-knee cast with a walking heel is then applied.

Dislocations of the Ankle Joint
These are classified according to the abnormal position of the talus relative to the distal ends of the tibia and fibula (Fig 17–58).

Anterior Dislocation.—In anterior dislocations the talus is displaced anteriorly, which usually occurs as the result of forced dorsiflexion of the ankle joint or a blow to the heel with the foot dorsiflexed; they may be associated with malleolar fractures.

Posterior Dislocation.—In posterior dislocations the talus is displaced posteriorly, which usually occurs as the result of forced plantar flexion of the ankle joint; they are often accompanied by fractures of the malleoli or fracture of the posterior inferior margin of the tibia.

Superior Dislocation.—In superior dislocations the talus is displaced superiorly, disrupting the inferior talofibular joint. It commonly results from persons landing on their feet from a height, when the upward directed force drives the talus into the inferior tibiofibular joint.

Lateral Dislocations.—These types of dislocations usually follow excessive inversion or eversion of the foot and are commonly associated with fractures of one or both malleoli (see above).

Anatomical Complications of Dislocations of the Ankle Joint.—Excessive displacement of the talus relative to the distal ends of the tibia and fibula can compromise the integrity of the anterior tibial artery (or dorsalis pedis artery), which lies anterior to the joint, or the posterior tibial artery, which lies behind the medial malleolus (Fig 17–59). Disruption or excessive stretching of the ligaments attached to the talus, which carry the arterial supply to the talus, can lead to avascular necrosis of the talus.

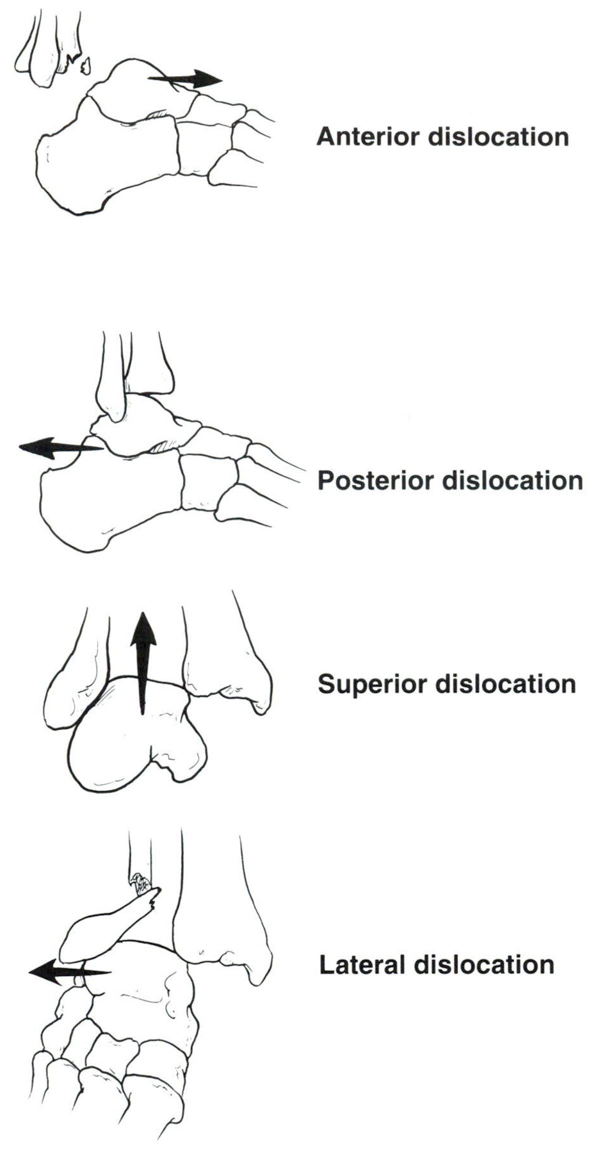

FIG 17–58.
Dislocations of the ankle joint.

Arthrocentesis of the Ankle
The anatomy of ankle arthrocentesis is fully described in Chapter 18, p. 777.

Clinical Anatomy of the Achilles Tendon

Achilles Tendinitis
This is usually caused by overuse of the tendon and is common in runners. There is pain in the area of the tendon, which is tender on palpation. The problem is relieved by rest and elevation of the heel.

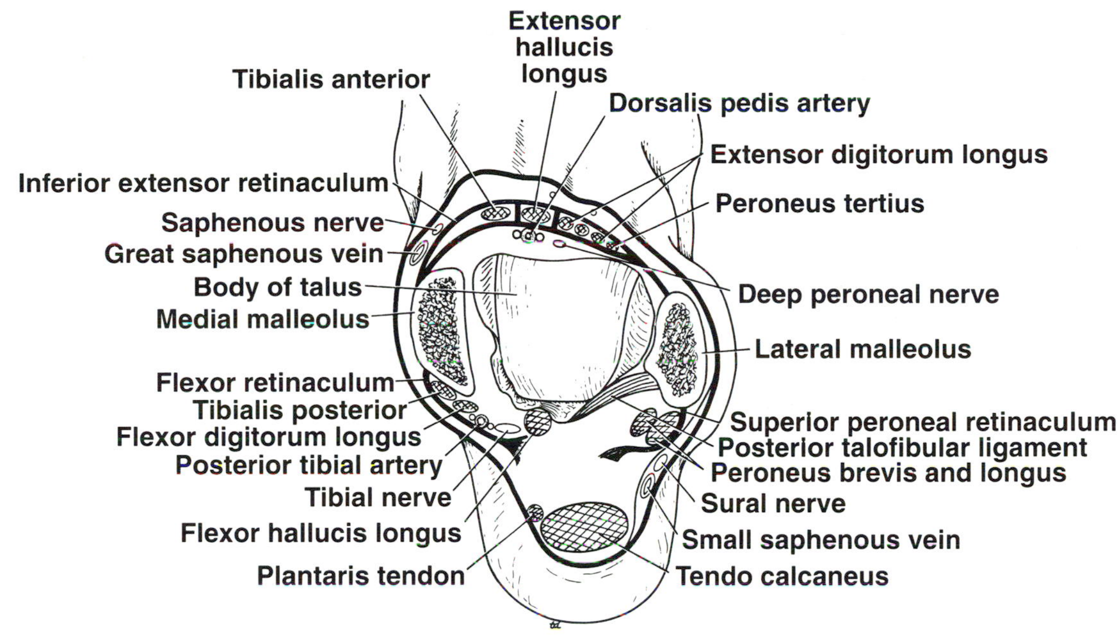

FIG 17–59.
Cross section of the right ankle joint showing the relations.

Rupture of the Achilles Tendon
This is common in middle-aged men. The rupture occurs at its narrowest part, about 2 in. (5 cm) above its insertion onto the back of the calcaneum. A sudden sharp pain is accompanied by immediate disability. The gastrocnemius and soleus muscles retract proximally, leaving a palpable gap in the tendon. It is impossible for the patient to actively plantar flex the foot. The *Thompson test* is positive. The patient is asked to kneel on a stretcher with the foot and ankle hanging over the side. The operator squeezes and compresses the calf muscles. Ordinarily, this would result in foot plantar flexing. With a ruptured Achilles tendon, the foot does not move.

Rupture of the Plantaris Tendon
Rupture of the plantaris tendon is not uncommon, although tearing of the fibers of the soleus or partial tearing of the Achilles tendon is frequently diagnosed as such a rupture. The patient complains of a sharp pain in the calf, as if he were kicked.

Clinical Anatomy of the Peroneus Longus and Brevis Tendons

Tenosynovitis can affect the tendon sheaths of the peroneus longus and brevis tendons as they pass posterior to the lateral malleolus.

Dislocation of the peroneus longus and brevis tendons can occur following traumatic tearing of the superior peroneal retinaculum; the tendons are displaced forward from behind the lateral malleolus. The patient experiences difficulty in everting the foot at the subtalar and transverse tarsal joints and has some weakness of plantar flexing the ankle joint.

Clinical Anatomy of the Bones of the Foot

Fractures of the Foot Bones
Fractures of the Talus.—These occur at the neck or body of the talus (Fig 17–60). Neck fractures occur during violent dorsiflexion of the ankle joint when the neck is driven against the anterior edge of the distal end of the tibia. The body of the talus can be fractured by jumping from a height, although the two malleoli prevent diplacement of the fragments. The posterior process of the body of the talus may have its own center of ossification that remains separate as an *os trigonum*. This may lead to radiological misinterpretation as a fracture.

Fractures of the Calcaneum.—Compression fractures of the calcaneum result from falls from a height (see Fig 17–60). The weight of the body drives the talus downward into the calcaneum, crushing it in such a way that it loses vertical height and becomes wider laterally. This fracture reduces the normal angle *(Boehler's angle)* between a line drawn along the posterior superior margin of the calcaneum and a line along the superior margin of

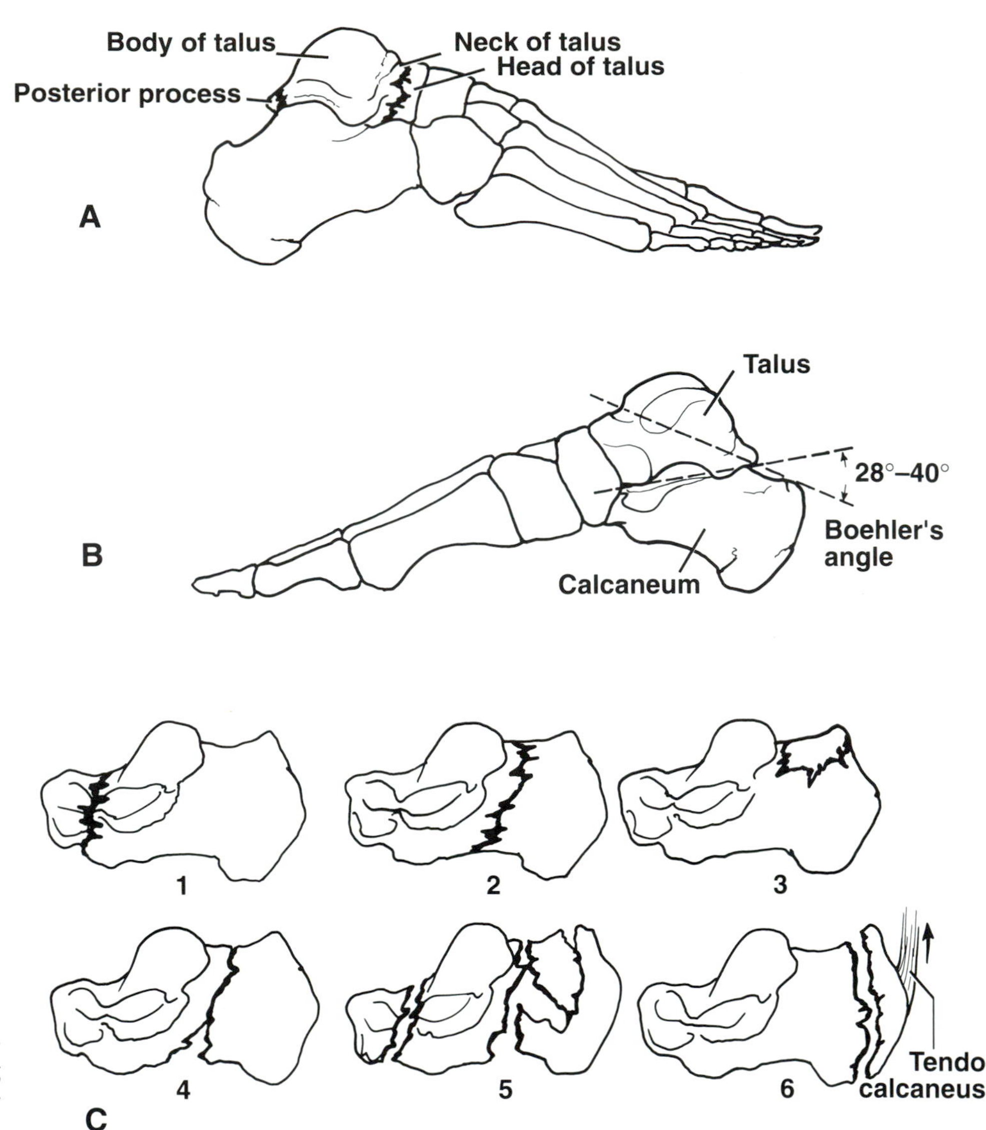

FIG 17–60.
A, fractures of the talus; the bones of the foot are viewed from the lateral side. **B,** Boehler's angle; the bones of the foot are viewed from the medial side. **C,** fractures of the calcaneum. In type 6 the tendo calcaneus is displacing the bony fragment of the calcaneum superiorly.

the calcaneum, as seen on a lateral radiograph of the ankle (see Fig 17–60). An angle of less than 28° is abnormal. The posterior portion of the calcaneum above the insertion of the Achilles tendon can be fractured by the posterior displacement of the talus. The sustentaculum tali, which projects from the medial surface of the calcaneum, can be fractured by forced inversion of the foot.

Fractures of the Metatarsal Bones.—The base of the fifth metatarsal can be fractured during forced inversion of the foot, at which time the tendon of insertion of the peroneus brevis muscle pulls off the base of the metatarsal. Note that the base of the fifth metatarsal has a long triangular facet for articulation with the cuboid, and this facet may during development separate off from the rest of the bone as a separate ossicle. Also, in young persons the proximal epiphyseal line may be mistaken for a fracture line (the epiphysis unites with the diaphysis between 17 and 20 years of age). The epiphyseal line is oblique, whereas the fracture line is usually transverse.

Stress fractures of a metatarsal bone are common in soldiers after long marches; it can also occur in joggers, nurses, and hikers. It occurs most frequently in the distal third of the second, third, or fourth metatarsal bone. Minimal displacement of the fragments occurs because of the attachment of the interosseous muscles.

Clinical Anatomy of Plantar Aponeurosis

Plantar fasciitis can occur in individuals who do a great deal of standing or walking. Repeated minor trauma to the plantar aponeurosis produces pain and tenderness of the sole of the foot. Multiple attacks of this condition induce ossification in the posterior attachment of the aponeurosis to the calcaneum, leading to the formation of a *calcaneal spur.*

Clinical Anatomy of Heel Pain

The causes of heel pain may be directly related to structures situated in this region.

Adventitial bursa (see p. 768) sometimes develops between the skin and the insertion of the Achilles tendon and the calcaneum. With badly fitting shoes, the bursa may become inflamed.

Achilles bursitis is an inflammation of the bursa that lies between the Achilles tendon and the upper part of the posterior surface of the calcaneum.

Achilles tendinitis is described on p. 742.

Sever's disease is an aseptic necrosis of the calcaneum at the site of insertion of the Achilles tendon in adolescents.

Calcaneal spur is described above.

Plantar fasciitis is described on p. 744.

Clinical Anatomy of the Lymphatic Drainage of the Lower Extremity

The superficial lymph nodes not only drain lymph from the lower limb, but also drain lymph from the skin of the anterior and posterior abdominal walls below the level of the umbilicus; lymph from the external genitalia and the mucous membrane of the lower half of the anal canal also drains into these nodes.

Lymphangitis in the Ankle Region

The superficial lymphatic vessels leaving the foot ascend with the superficial veins in front of the medial malleolus and behind the lateral malleolus. The red lines of acute lymphangitis may therefore be found at these sites.

Clinical Anatomy of the Sciatic Nerve

The nerve may be injured by penetrating wounds, fractures of the pelvis, or dislocations of the hip joint. It is most frequently injured by badly placed intramuscular injections in the gluteal region. (To avoid this, the intramuscular injections should be made into the gluteus maximus or medius well forward on the upper outer quadrant of the buttock.)

The majority of the sciatic nerve lesions are incomplete, and in 90% of injuries, the common peroneal part of the nerve is the most affected. This can probably be explained by the fact that the common peroneal nerve fibers lie most superficial in the sciatic nerve trunk. The following clinical features are present (for summary diagrams of the sciatic nerve, see Figs 17–39 and 17–40).

Motor.—The hamstring muscles are paralyzed, but weak flexion of the knee is possible because of the action of the sartorius (femoral nerve) and gracilis (obturator nerve). All the muscles below the knee are paralyzed, and the weight of the foot causes it to assume the plantar-flexed position, or *footdrop.*

Sensory.—There is loss of sensation below the knee, except for a narrow area down the medial side of the lower part of the leg and along the medial border of the foot as far as the ball of the big toe, which is supplied by the saphenous nerve (branch of the femoral nerve).

Sciatica

This term is used to describe the condition in which pain is felt along the sensory distribution of the sciatic nerve. Thus the pain is experienced on the posterior aspect of the thigh, the posterior and lateral sides of the leg, and the lateral part of the foot. Sciatica pain can be caused by prolapse of an intervertebral disc (see p. 382), with pressure on one or more roots of the lower lumbar and sacral spinal nerves, pressure on the sacral plexus or sciatic nerve by an intrapelvic tumor, or in inflammation of the sciatic nerve or its terminal branches.

Clinical Anatomy of the Tibial Nerve

The tibial nerve (see Fig 17–39) leaves the popliteal fossa by passing deep to the gastrocnemius and soleus muscles. Because of its deep and protected position, it is rarely injured. Complete division results in the following clinical features.

Motor.—All the muscles in the back of the leg and the sole of the foot are paralyzed. The opposing muscles dorsiflex the foot at the ankle joint and evert the foot at the subtalar and transverse tarsal joints, referred to as *calcaneovalgus.*

Sensory.—There is loss of skin sensation on the sole of the foot; later trophic ulcers will develop.

Clinical Anatomy of the Common Peroneal Nerve

The common peroneal nerve (see Fig 17–40) is in a very exposed position as it leaves the popliteal fossa and winds around the neck of the fibula to enter the peroneus longus muscle.

The nerve is commonly injured in fractures of the neck of the fibula and by pressure from casts or splints. The following clinical features are present.

Motor.—The muscles of the anterior and lateral compartments of the leg are paralyzed, namely, the tibialis anterior, the extensor digitorum longus and brevis, the peroneus tertius, the extensor hallucis longus (supplied by the deep peroneal nerve), and the peroneus longus and brevis (supplied by the superficial peroneal nerve). As a result, the opposing muscles, the plantar flexors of the ankle joint, and the invertors of the subtalar and transverse tarsal joints cause the foot to be plantar flexed *(footdrop)* and inverted, which is referred to as *equinovarus.*

Sensory.—There is loss of sensation down the anterior and lateral sides of the leg and dorsum of the foot and toes, including the medial side of the big toe. The lateral border of the foot and the lateral side of the little toe are virtually unaffected (sural nerve, mainly from tibial nerve). The medial border of the foot as far as the ball of the big toe is completely unaffected (saphenous nerve, a branch of the femoral nerve).

When the injury to the common peroneal nerve occurs distal to the site of origin of the lateral cutaneous nerve of the calf, the loss of sensibility is confined to the area of the foot and toes.

Clinical Anatomy of the Femoral Nerve

The femoral nerve enters the thigh from behind the inguinal ligament at a point midway between the anterior superior iliac spine and the pubic tubercle; it lies about a fingerbreadth lateral to the femoral pulse. About 1.5 in. (4 cm) below the inguinal ligament, the nerve splits into its terminal branches (see Fig 17–41).

The femoral nerve may be injured by stab and gunshot wounds, but complete division of the nerve is rare. The following clinical features are present when the nerve is completely divided.

Motor.—The quadriceps femoris muscle is paralyzed, and the knee joint cannot be extended. In walking, this is compensated for to some extent by the use of the adductor muscles.

Sensory.—There is loss of skin sensation over the anterior and medial sides of the thigh, the lower part of the leg, and along the medial border of the foot as far as the ball of the big toe; this latter area is normally supplied by the saphenous nerve.

Clinical Anatomy of the Obturator Nerve

The obturator nerve is deeply placed and enters the thigh through the upper part of the obturator foramen as anterior and posterior divisions. They descend beneath the floor of the femoral triangle and in front of the adductor magnus muscle (see Fig 17–42).

It is very rarely injured in penetrating wounds to the thigh, in anterior dislocations of the hip joint, or in obturator herniae. It may be pressed upon by the fetal head during parturition. The following clinical features occur.

Motor.—All the adductor muscles are paralyzed except the hamstring part of adductor magnus, which is supplied by the sciatic nerve.

Cutaneous.—The sensory loss is minimal on the medial aspect of the thigh.

Clinical Anatomy of the Dermatomes of the Lower Extremity

The integrity of the spinal cord segments L1 through S3 may be tested by examining the dermatomes of the lower extremity. The diagrams in Figures 2–32 and 2–33 show the arrangement of the dermatomes of the lower extremity. Unfortunately, the dermatome pattern appears to be more complicated than that of the upper limb because during fetal development, the lower limb underwent medial rotation as it grew out from the trunk, so that the big toe came to lie on the medial side of the foot; this accounts for the spiraling pattern of the dermatomes.

Tendon Reflexes of the Lower Extremity

Skeletal muscle receives a segmental innervation, and it is possible to test the nerves in the lower limb by illiciting simple reflexes.

Patellar Tendon Reflex (Knee Jerk).—L2, **L3**, and **L4** (extension of the knee joint on tapping the patellar tendon).

Achilles Tendon Reflex (Ankle Jerk).—S1 and S2 (plantar flexion of ankle joint on tapping the Achilles tendon). There is no L5 reflex in the lower extremity.

ANATOMICAL-CLINICAL PROBLEMS

1. A 50-year-old man was running down a flight of stone steps when he misjudged the position of one of the steps and fell down hard onto his buttocks. Following the fall he complained of severe pain in the lower part of his spine. On returning home the pain persisted, and he had great difficulty sitting down. He was examined in the emergency department about 2 hours after the fall and was found to be acutely tender on palpation about 2 in. posterior to the anus. A rectal examination revealed an acutely tender area posteriorly. What is the most likely diagnosis? Will the pain from this condition disappear quickly?

2. A 37-year-old woman was involved in a light plane accident. She and her husband were flying home from a business trip when they had to make a forced landing in a field due to fog. On landing, the plane hit a tree and came to rest on its nose. Her husband was killed on impact and she was thrown from the cockpit. She was evaluated in the emergency department with multiple injuries. Radiographic examination of her pelvis showed a fracture of her left ilium and iliac crest. From your knowledge of anatomy, would you expect much displacement of the bone fragments?

3. A 46-year-old man, who had been drinking heavily, staggered away from the bar and attempted to cross a busy street. On stepping off the sidewalk, he was hit and run over by a passing bus. On admission to the emergency department, he was unconscious and showed signs of severe hypovolemic shock. He had extensive bruising of the lower part of the anterior abdominal wall, and the body of the pubis was prominent on the left side. On examination of the penis, it was possible to express a drop of blood-stained fluid from the external meatus. What is the most likely diagnosis in this case? Should one look for more than one fracture of the pelvis in cases such as this? How would you explain the blood-stained discharge from the penis? Even though there was no evidence of an external hemorrhage in this patient, how would you explain the hypovolemic shock?

4. A 24-year-old man was seen in the emergency department with the signs and symptoms consistent with a right ureteric calculus. This was confirmed on a radiograph of the abdomen, and, since the calculus was small and likely to pass spontaneously, the nurse was instructed to give the patient an intramuscular injection of an analgesic into the left buttock. Ten days later, the patient returned to the emergency department stating that he had successfully passed the calculus. However, he was concerned that he was experiencing numbness and tingling sensation down the anterior and lateral sides of the left leg and dorsum of the foot. He also stated that his left leg felt heavy and tended to catch on the edges of carpets. On examination, there was evidence of impaired skin sensation on the anterior and lateral sides of the left leg and the dorsum of the left foot. The patient held his foot slightly plantar flexed and slightly inverted. When the strengths of the plantar and dorsiflexors of the ankle joint on both legs were compared, it was found that the dorsiflexors of the left ankle were weaker than normal. The evertor muscles were also weaker on the left side. Using your knowledge of anatomy, explain this patient's new signs and symptoms. What steps would you take to prevent this serious problem from occurring again in future patients?

5. A 48-year-old man was seen in the emergency department complaining of a lump in the right groin. He first noticed the lump 4 months earlier and that it was gradually getting larger. He said that the lump caused him no pain or discomfort. On examination, a small discrete lump was found about 1 in. (2.5 cm) below and lateral to the right pubic tubercle on the front of the thigh. The lump was hard and freely movable. Careful palpation of the area revealed three other small lumps with similar characteristics. The skin on the anterior and posterior surfaces of the abdomen below the level of the umbilicus was inspected and nothing abnormal found. The external genitalia were carefully examined, and a rectal examination was performed. Nothing abnormal was discovered. A close examination of the skin of the right lower extremity showed a small dark mole over the medial malleolus; it was raised above the surface and showed evidence of recent bleeding. Using your knowledge of anatomy, make a diagnosis. What is the connection between the swellings in the right groin and the pigmented lesion over the medial malleolus?

6. A 20-year-old woman was involved in an automobile accident. Her car skidded into a utility pole, and she was thrown forward, striking her right knee on the dashboard. On examination in the emergency department, she was found to have a posterior fracture dislocation of her right hip joint. What bone or bones are likely to be fractured? What

anatomical structures are likely to be compromised in a dislocation of this type?

7. A 70-year-old woman was seen in the emergency department, having tripped over a rug in her apartment. Her right leg was found to be shortened, and her right foot was turned out laterally. On careful examination of the right hip region, the greater trochanter was found to be situated above a line joining the ischial tuberosity to the anterior superior iliac spine. An anteroposterior radiograph of the right hip joint showed that Shenton's line was not intact on the right side. What is the diagnosis? What is Shenton's line? Why is avascular necrosis of the femoral head common in patients with this problem?

8. A 37-year-old woman was seen in the emergency department complaining of severe colicky pains in the abdomen and repeated vomiting. The pain was most severe in the region of the umbilicus. On examination the patient was found to be dehydrated. On listening to the abdomen with a stethoscope, loud bowel sounds could be heard. A small, tense swelling was found in the front of the left thigh. When the patient was asked to cough, there was no enlargement of the swelling. The swelling was situated below and lateral to the left pubic tubercle. When the swelling was pointed out to the patient, she said she had noticed a small swelling there for about 2 years, but since it did not cause pain, she had ignored it. However, she said that 2 days ago she had had a bout of coughing and following this the swelling had increased in size and become tender. Using your knowledge of anatomy, make a diagnosis. Can you explain the sudden change in the patient's condition following the coughing episode?

9. A 33-year-old man was seen in the emergency department following an automobile accident. He was found to have many superficial abrasions and a fracture of the lower third of his right femur. On examination his right leg was 3 in. (7.5 cm) shorter than his left. A lateral radiograph of the region of the knee showed overlap of the fragments, with the distal fragment rotated backward. Examination of the dorsalis pedis artery showed a very poor pulse. The posterior tibial artery was not palpable. Can you explain the shortening of the leg? What was responsible for the backward rotation of the distal fragment? What complications are likely to occur in such cases?

10. Two football players collided with one another. As one of the players fell to the ground, the left knee, which was taking the weight of his body, was partially flexed, the femur was rotated medially, and the lower leg abducted on the thigh. A sudden severe pain was felt in the left knee joint, and he was unable to extend it. He was carried off the field and taken to the emergency department of the nearest hospital. On examination, the patient was unable to extend his left knee, which was greatly swollen. Severe local tenderness was found along the joint line on the medial side of the joint between the medial collateral ligament and the ligamentum patellae. A stress test to the lateral collateral ligament was negative; a similar test carried out on the medial collateral ligament showed some pain, but there was no evidence of ligamentous rupture. What is the diagnosis? Why was the knee locked? Why was the swelling so extensive over the knee and above the patella?

11. A 25-year-old medical student was seen in the emergency department following a motorcycle accident. Apparently, while weaving in and out of cars in a traffic jam, he struck the outer side of his right knee a severe glancing blow on a car headlamp. Name the important nerve that lies on the lateral side of the knee region. Assuming that the nerve was severely damaged, describe the disability likely to be experienced by this student.

12. A 49-year-old man was advised by his physician at his annual check-up examination to reduce his weight, go on a diet, and exercise more. One morning while jogging, he heard a sharp snap and felt a sudden pain in his right lower calf. On examination in the emergency department, it was noted that the upper part of the right calf was swollen, and a gap was apparent between the swelling and the right heel. With the patient lying on his back, the physician gently squeezed the upper part of the right calf and noted no evidence of plantar flexion of the foot. What is the diagnosis?

13. A patient is suspected of having torn the anterior cruciate ligament of the right knee joint. In anatomical terms explain how you would test for the integrity of this ligament. What are the bony attachments of this ligament? If a knee joint is severely disrupted, is it possible to have little or no swelling of the knee? If so, explain this phenomenon.

14. A 22-year-old woman was running across a field when she caught her foot in a hole in the ground. As she fell, the right foot was violently rotated laterally and excessively everted. On attempting to stand, she found she could not place any weight on her right foot. Later, she was examined in the emergency department, and the right ankle was found to be swollen, especially on the lateral side.

The right heel was excessively prominent. Anteroposterior and lateral radiographs of the ankle showed a spiral fracture of the distal third of the fibula and a fracture of the tip of the medial malleolus of the tibia. What is the diagnosis? If the rotational force had been greater in this case, what additional fracture might have occurred? What type of fracture usually occurs involving the medial malleolus? Is it transverse, spiral, or oblique?

15. Name the anatomical structures in the region of the heel that can become diseased and painful and may present in the emergency department.

16. A 17-year-old patient received a knife wound to the front of the thigh. What signs and symptoms would be present if the femoral nerve was severed? What is the surface marking of the femoral nerve in the thigh?

ANSWERS

1. The coccyx is commonly bruised or fractured in injuries of this type. The bone can be palpated beneath the skin in the natal cleft behind the anus. A gloved finger in the anal canal can also palpate the anterior surface of this bone. Unfortunately, traumatic injuries to the coccyx are nearly always followed by pain and discomfort extending over a period of weeks or months. The posttraumatic pain is sometimes referred to as coccydynia.

2. Most fractures of the upper part of the ilium have little displacement of the bone fragments. This is because the iliacus muscle is attached to the inner surface and the gluteal muscles are attached to the outer surface. Splinting the bones is unnecessary because of the attachment of these muscles.

3. This man had a dislocation of the symphysis pubis and a linear fracture through the lateral part of the sacrum on the left side. The true pelvis may be likened to a bony ring, and if a fracture breaks the ring in one place with displacement, there is always a second fracture or joint dislocation in a second place (see discussion, p. 726).

In this patient the urethra was also damaged by the shearing forces, which explains the blood-stained fluid from the external urethral meatus.

The bony walls of the true pelvis are lined with muscles, fascia, and peritoneum. Embedded in the pelvic fascia are the large, thin-walled pelvic veins that are commonly damaged in pelvic fractures. A large hemorrhage can easily occur into the extraperitoneal fascia, causing severe hypovolemic shock, as in this patient.

4. This patient sustained damage to the left sciatic nerve as a direct result of the incorrect administration of the intramuscular injection into the left buttock. The common peroneal branch of the sciatic nerve was damaged, resulting in the loss of skin sensation in the areas normally supplied by the lateral cutaneous nerve of the calf and the superficial peroneal nerve. The muscles of the anterior and lateral compartments of the leg were partially paralyzed. The unopposed plantar flexors and invertors of the foot caused the patient to hold his left foot in equinovarus.

Injections into the buttock are often incorrectly given because the operator confines the injection area to the summit of the buttock, which actually overlies the sciatic nerve. Intramuscular injections should be restricted to the *upper outer quadrant of the buttock.* Alternate buttocks, or other sites for injection, should be used when there are multiple injections extending over many weeks.

5. This patient had a malignant melanoma of the skin over the medial malleolus of the right leg, which had spread by way of the lymphatics into the vertical group of superficial inguinal nodes.

6. In posterior dislocations of the hip joint with the joint flexed and partially abducted, the common fracture is through the posterior rim of the acetabulum; the trauma may also cause a fracture to the femoral head.

The sudden posterior displacement of the femoral head may damage the sciatic nerve, which lies directly behind the joint. Pressure on the sciatic nerve would give rise to pain in the anterior and lateral aspects of the leg, the dorsum and lateral border of the foot, and the toes, including the medial side of the big toe.

7. The patient had a fracture of the neck of the right femur. Shenton's line is seen on an anteroposterior radiograph of the hip region in a normal individual and consists of a continuous curved line made up of the inferior margin of the superior ramus of the pubis and the lower margin of the neck of the femur. In cases of fracture of the neck of the femur, the continuity of the line is broken.

The femoral head receives its blood supply from two sources—a small artery that runs with the round ligament of the femoral head and a profuse blood supply from the medial femoral circumflex femoral artery, branches of which ascend the femoral neck beneath the synovial membrane. Fracture of the femoral neck may deprive the femoral head of part or all the blood from the medial femoral circumflex femoral artery, and avascular necrosis will occur.

8. This patient had acute intestinal obstruction secondary to a strangulated left femoral hernia. The pain was felt in the region of the umbilicus, which is the area of the tenth thoracic dermatome. Pain originating in the small intestine is referred to the umbilical region (for details, see p. 459). When the patient coughed, a loop of ileum was forced down into a preexisting femoral hernial sac. The unyielding nature of the femoral ring resulted in venous congestion of the gut and, later, arterial occlusion, at which point peristalsis ceased (paralytic ileus) and intestinal obstruction occurred.

9. The shortening of the leg was caused by the upward pull of the hamstring and the quadriceps femoris muscles on the distal fragment of the femur. The backward rotation of the distal fragment was caused by the pull of the two heads of the gastrocnemius muscle. The backward rotation of the distal fragment may easily impinge on the popliteal artery in the popliteal space and compromise the blood flow to the leg and foot. The poor pulse of the dorsalis pedis artery and the absent posterior tibial pulse would indicate that this complication had occurred.

10. As the football player fell, the medial meniscus was drawn laterally within the knee joint. The sudden movement of the knee joint, which occurred on striking the ground, resulted in the grinding of the relatively immobile medial meniscus between the medial femoral and tibial condyles. The cartilage split along part of its length, and the detached portion became jammed, like a wedge, between the articular surfaces, limiting further extension, i.e., "locking" the joint. The tenderness was experienced when the examiner palpated the medial meniscus on the joint line.

The trauma to the medial meniscus stimulated the production of excessive amounts of synovial fluid and blood, which filled the joint cavity, the semimembranosus bursa and the suprapatellar bursa. It was the filling of the latter bursa that was responsible for the swelling above the patella.

11. The right common peroneal nerve was severely injured in this patient. The signs and symptoms of common peroneal damage are fully described on p. 746.

12. This patient had ruptured the Achilles tendon of his right leg. The bellies of the gastrocnemius and soleus muscles retracted upward, leaving a gap between the divided ends of the tendon.

13. The cruciate ligaments form one of the main bonds of union between the femur and the tibia at the knee joint. To test for the integrity of the anterior cruciate ligament, the knee joint is gently flexed and the tibia is pulled forward by the examiner. Excessive forward movement (as compared with the normal knee) is indicative of a torn anterior cruciate ligament.

The anterior cruciate ligament is attached below to the anterior part of the intercondylar area of the tibia; it passes upward, backward, and laterally to be attached to the posterior part of the medial surface of the lateral condyle of the femur.

If the capsule and synovial membrane are ruptured by severe trauma to the knee joint, the synovial fluid will escape into the surrounding tissues so that the usual swelling of the knee would not be as apparent.

14. This patient had a fracture dislocation of the right ankle joint. In a fracture dislocation that involves rotation and overeversion of the foot, there may be, in addition to the spiral fracture of the lateral malleolus of the fibula, a transverse fracture of the medial malleolus (the medial collateral ligament pulls off the tip of the medial malleolus), and in very severe cases where the rotational force continues, a fracture of the posterior margin of the lower end of the tibia.

15. The common structures that can become diseased and cause heel pain are described on p. 745 and include adventitial bursa formation at the heel, Achilles bursa, Achilles tendon, plantar fascia, and calcaneum with spur formation.

16. The signs and symptoms found in a patient with a complete femoral nerve palsy are described on p. 746. It is rare for all the branches of the femoral nerve to be cut. The surface marking of the femoral nerve trunk in the upper part of the thigh is midway between the anterior superior iliac spine and the pubic tubercle; it divides into its terminal branches about 1½ in. (4 cm) below the inguinal ligament.

18

Joint Spaces and Bursae Relative to Arthrocentesis and Inflammation

In the emergency department the injection of medications into synovial joints is now rarely performed. However, the aspiration of joint cavities for the relief of pain or for the assessment of joint fluid for bacteria or crystals is a common procedure.

Bursae are similar to synovial joints and tendon sheaths in that they are prone to inflammation resulting from trauma due to overuse or pressure; they may also be subject to infections and metabolic disorders.

In this chapter, the basic anatomy of each joint is briefly reviewed and the sites where each joint can be most easily penetrated by a needle are discussed. Emphasis is placed on the bony landmarks that can be used when choosing each aspiration site.

The location and extent of the numerous bursae that are commonly subject to bursitis are also reviewed. Here, again, emphasis is placed on their surface anatomy so that aspiration or injection sites can be chosen with confidence.

BASIC ANATOMY

Basic Anatomy of Joints

Classification of Joints

Joints may be classified into three types—fibrous, cartilaginous, and synovial.

Fibrous Joints.—A fibrous joint has no joint cavity and the bones are held together by dense fibrous tissue. The sutures between the skull bones and the inferior tibiofibular joints are good examples. Little or no movement is possible.

Cartilaginous Joints.—A cartilaginous joint has no joint cavity and the bones are held together by plates of cartilage. The joints between the bodies of adjacent vertebra and between the pubic bones at the symphysis pubis are good examples. A small amount of movement is possible.

Synovial Joints.—A synovial joint has a joint cavity and the articular surfaces are covered by a thin layer of hyaline cartilage. The general arrangement of a synovial joint is shown in Figure 18–3. A synovial joint can be freely mobile. The joint cavity is lined by *synovial membrane,* which extends from the margin of one articular surface to those of the other. The synovial membrane is protected on the outside by a tough fibrous membrane referred to as the *capsule* of the joint. The articular surfaces are covered and are lubricated by a thin layer of viscous fluid called *synovial fluid.* In certain synovial joints, for example in the knee joint, discs or wedges of fibrocartilage are interposed between the articular surfaces of the bones. These are referred to as *articular discs.*

The degree of movement in a synovial joint is limited by the shape of the bones participating in the joint, the coming together of adjacent anatomical structures (for example, the thigh against the anterior abdominal wall on flexing the hip joint), and the presence of fibrous *ligaments* uniting the bones. Most ligaments lie outside the joint capsule and are known as *extracapsular ligaments.* Examples of extracapsular ligaments are the strong iliofemoral, pubofemoral, and ischiofemoral ligaments of the hip joint. *Intracapsular ligaments* lie within the joint capsule but they lie external to the joint cavity; they are excluded from the joint cavity by coverings of synovial membrane. Examples of intracapsular ligaments are the cruciate ligaments found in the knee joint (see Fig 18–8).

Types of Synovial Joints.—Synovial joints may be classified according to the arrangement of the articular surfaces and the types of movement that are possible.

Hinge Joints.—A hinge joint permits movement in only one plane so that flexion and extension movements are possible. The elbow and knee joints are good examples.

Gliding Joints.—A gliding joint is one in which the flat articular surfaces slide one upon the other. The acromioclavicular and sternoclavicular joints are good examples.

Pivot Joints.—A pivot joint is one in which one bone is permitted to rotate on another. The head of the radius rotates on the ulna in the superior radioulnar joint, permitting pronation and supination; the atlas rotates on the axis in the upper part of the vertebral column.

Ball and Socket Joints.—A ball and socket joint is one in which a hemispherical-shaped end of a bone fits into a cup-shaped cavity of another bone. The hip joint and the shoulder joint are two good examples. A considerable amount of movement is permitted at this type of joint.

Condyloid Joints.—These joints have two distinct convex surfaces that articulate with two concave surfaces. The movements of flexion, extension, abduction, and adduction are possible together with a small amount of rotation. The metacarpophalangeal joints are good examples.

Ellipsoid Joints.—In these joints, there is an elliptical convex articular surface that fits into an elliptical concave articular surface. The movements of flexion, extension, abduction, and adduction can take place, but rotation is impossible. The wrist joint is a good example.

Saddle Joints.—In saddle joints the articular surfaces are reciprocally concavoconvex and resemble a saddle on a horse's back. The movements of flexion, extension, abduction, adduction, and rotation can take place. The carpometacarpal joint of the thumb is the best example.

Anatomy of the Temporomandibular Joint

Articulation.—This is formed above by the articular tubercle and the mandibular fossa of the temporal bone and below by the head of the mandible; the surfaces are covered with fibrocartilage (Fig 18–1).

Type of Joint.—Synovial. A fibrocartilaginous articular disc divides the joint into upper and lower cavities.

Capsule.—Encloses the joint.

Ligaments.—The *lateral temporomandibular ligament* is located on the lateral surface of the joint and is attached above to the articular tubercle at the root of the zygomatic arch and below to the neck of the mandible. The fibers extend downward and backward. The *sphenomandibular ligament* is located on the medial surface of the joint and is attached above to the spine of the sphenoid and below to the entrance of the mandibular foramen. The *articular disc* is located within the joint and divides the cavity into upper and lower parts (see Fig 18–1). This oval disc of fibrocartilage is attached at its circumference to the capsule. It is pulled forward by the lateral pterygoid muscle when the mouth is opened. The disc permits gliding forward and backward movement in the upper part of the joint and hinge movements in the lower part of the joint.

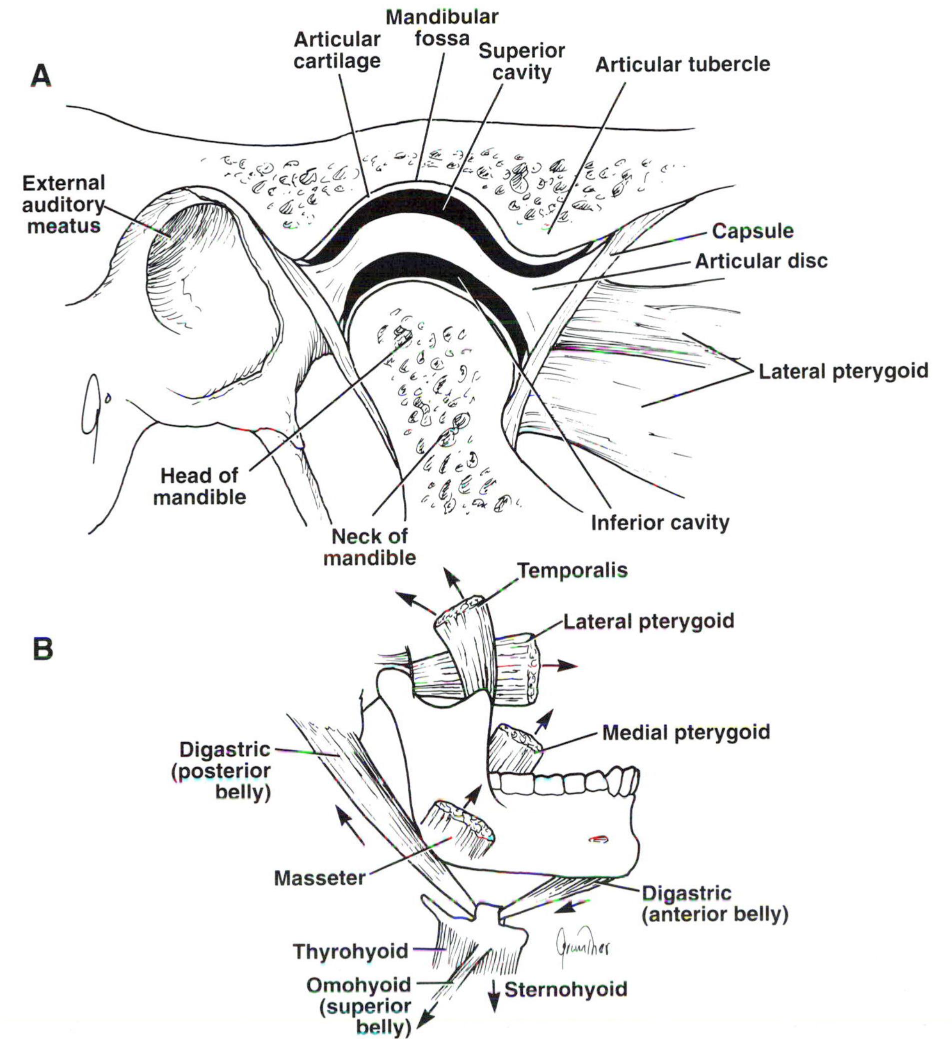

FIG 18–1.
A, the temporomandibular joint; the capsule has been removed to reveal the joint cavities and the cartilaginous articular disc (mouth in closed position). **B**, the attachments of the muscles of mastication to the mandible; the arrows indicate the direction of their actions.

Synovial Membrane.—This lines the capsule.

Nerve Supply.—Auriculotemporal and masseteric branches of the mandibular division of the trigeminal nerve.

Movements.—These include the following:

Protrusion.—The head of the mandible and the articular disc move forward.

Retraction.—The head of the mandible and the articular disc move backward.

The *mouth is opened* by the head of the mandible rotating on the undersurface of the articular disc around a horizontal axis. At the same time the mandible is protruded forward with the disc.

The *mouth is closed* by the head of the mandible rotating on the undersurface of the articular disc and at the same time the mandible is retracted backward with the disc.

Table 18–1 gives a summary of the movements possible at the temporomandibular joint and the muscles producing those movements.

Anatomy of the Sternoclavicular Joint

Articulation.—This occurs between the medial end of the clavicle, the manubrium sterni, and the first costal cartilage (Fig 18–2).

Type.—Synovial gliding joint.

Capsule.—Encloses the joint.

Ligaments.—The *interclavicular* and the *sternoclavicular ligaments* (see Fig 18–2) reinforce the joint capsule above, in front of, and behind the joint. The *articular disc* divides the joint into a medial and a lateral compartment. The disc is attached by its circumference to the capsule, to the superior margin of the articular surface of the clavicle, and to the first costal cartilage below. The disc serves as a strong intraarticular ligament. The strong *costoclavicular ligament* lies outside the joint (see Fig 18–2). It runs from the junction of the first rib with the first costal cartilage to the inferior surface of the clavicle.

Synovial Membrane.—This lines the capsule in each compartment of the joint.

Nerve Supply.—The supraclavicular nerve (C3 and C4) from the cervical plexus and the nerve to subclavius (C5 and C6) from the brachial plexus.

Movements.—Anterior and posterior movements of the clavicle take place in the medial compartment. Elevation and depression of the clavicle take place in the lateral compartment. All these movements are associated with movements of the scapula (see p. 590).

Posterior Relations.—On the right the brachiocephalic artery and the right brachiocephalic vein, and on the left the left common carotid artery and left brachiocephalic vein.

Anatomy of the Acromioclavicular Joint

Articulation.—This occurs between the acromion process of the scapula and the lateral end of the clavicle (see Fig 18–2).

TABLE 18–1.
Summary of the Movements of the Temporomandibular Joint and the Muscles Producing Those Movements

Movements	Muscles	Origin	Insertion	Nerve Supply
Elevation	Masseter	Zygomatic arch	Lateral surface ramus of mandible	Mandibular division of cranial nerve
	Temporalis	Floor of temporal fossa and covering fascia	Coronoid process of mandible	Mandibular division of cranial nerve V
	Medial pterygoid	Tuberosity of maxilla and medial pterygoid plate	Medial surface angle of mandible	Mandibular division of cranial nerve V
Depression (assisted by digastric, geniohyoid, and mylohyoid muscles)	Lateral pterygoid	Greater wing of sphenoid and lateral pterygoid plate	Neck of mandible and articular disc	Mandibular division of cranial nerve V
Protrusion	Lateral and medial pterygoids of both sides	See above		
Retraction	Temporalis (posterior fibers)	See above		
Lateral movements	Lateral and medial pterygoids of both sides acting alternatively	See above		

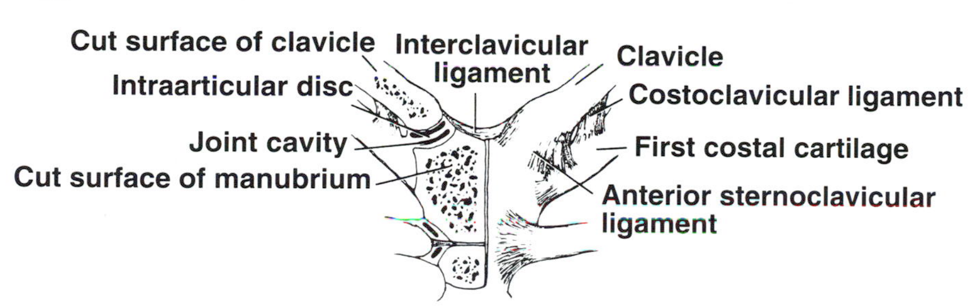

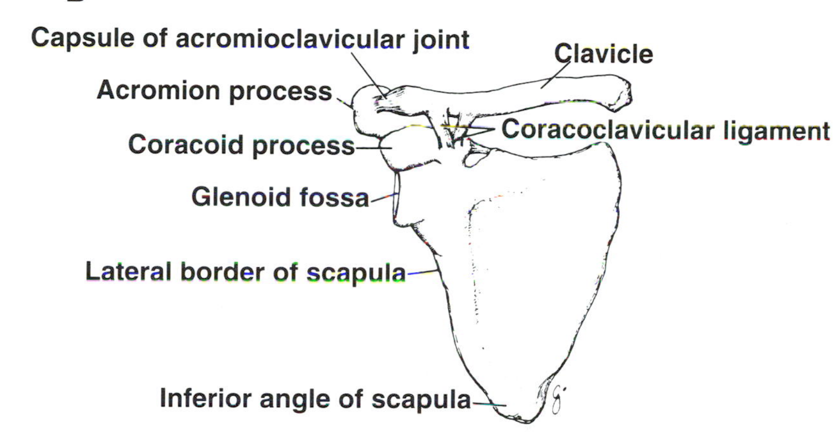

FIG 18–2.
A, anterior view of sternoclavicular joints; the right joint has been opened to reveal the joint cavity and the disc. **B,** anterior view of the right acromioclavicular joint. Note the important coracoclavicular ligament consisting of the medial conoid and lateral trapezoid parts.

Type.—A synovial gliding joint.

Capsule.—Surrounds the joint.

Ligaments.—The *acromioclavicular ligament* strengthens the capsule superiorly. The *articular disc* is an incomplete wedge of fibrocartilage that projects into the joint cavity from above. It is attached to the capsule superiorly. The strong *coracoclavicular ligament* connects the coracoid process to the inferior surface of the clavicle (see Fig 18–2); it is largely responsible for suspending the weight of the scapula and the upper limb from the clavicle.

Synovial Membrane.—This lines the capsule.

Nerve Supply.—Supraclavicular nerve (C3 and C4) from the cervical plexus.

Movements.—A gliding movement when the scapula rotates, or when the clavicle is elevated or depressed (see p. 590).

Anatomy of the Shoulder Joint

Articulation.—This occurs between the rounded head of the humerus and the shallow, pear-shaped glenoid cavity of the scapula. The glenoid cavity is deepened by the presence of a fibrocartilaginous rim called the *glenoid labrum* (see Fig 18–3).

Type.—Synovial ball and socket

Capsule.—Surrounds the joint; it is thin and lax, allowing a wide range of movement. It is strengthened by the tendons of the short muscles around the joint, which together form the *rotator cuff.*

Ligaments.—*Glenohumeral ligaments* are three bands of fibrous tissue that lie inside the anterior part of the capsule and strengthen it. The *transverse humeral ligament* strengthens the capsule and holds the tendon of the long head of biceps muscle in place between the greater and lesser tuberosities. The *coracohumeral ligament* strengthens the capsule above. The *coracoacromial ligament* lies outside the joint and extends between the coracoid process and the acromion; it protects the superior aspect of the joint.

Synovial Membrane.—This lines the capsule. The synovial membrane surrounds the tendon of the

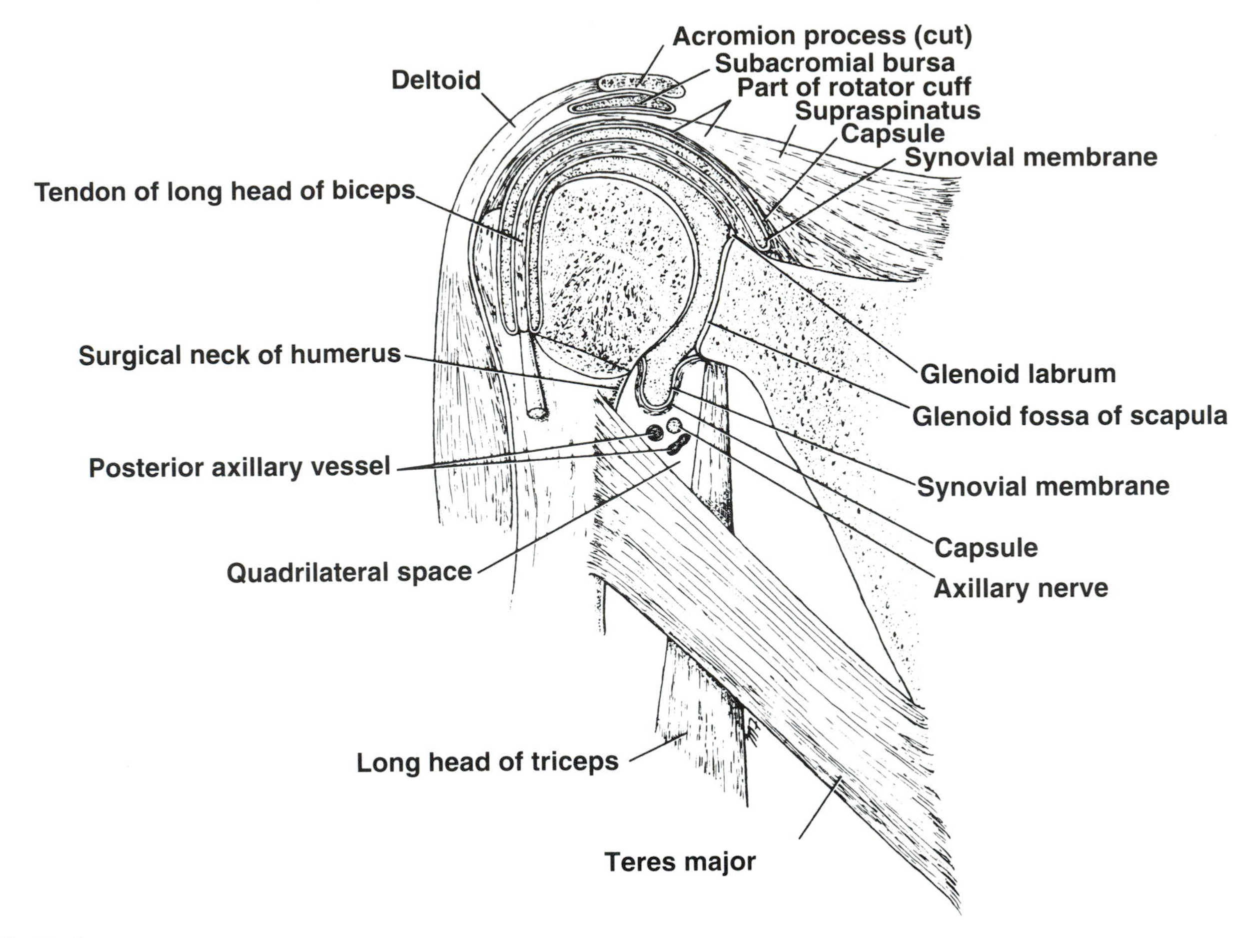

FIG 18–3.
Interior of the right shoulder joint. The head of the humerus and the scapula have been cut to reveal the interior of the joint.

long head of the biceps muscle and provides it with lubrication as the tendon emerges from the capsule between the greater and lesser tuberosities of the humerus. The synovial membrane also protrudes forward through a hole in the anterior wall of the capsule to form a bursa beneath the subscapularis muscle.

Nerve Supply.—Axillary nerve (C5 and C6) and the suprascapular nerve (C5 and C6) from the brachial plexus.

Movements.—Flexion, extension, abduction, adduction, lateral rotation, medial rotation, and circumduction.

Table 18–2 gives a summary of the movements possible at the shoulder joint and the muscles producing those movements.

Important Relations.—These include the following.

Anteriorly.—Brachial plexus and axillary vessels.

Inferiorly.—Axillary nerve and posterior circumflex humeral vessels.

Anatomy of the Elbow Joint

Articulation.—The trochlea of the humerus articulates with the trochlea notch of the ulna, and the capitulum of the humerus articulates with the head of the radius (Fig 18–4).

Type.—Synovial hinge joint.

Capsule.—Encloses the joint.

Ligaments.—The *lateral collateral ligament* is triangular and is attached by its apex to the lateral epicondyle of the humerus and by its base to the superior margin of the anular ligament and to the ulna (see Fig 18–4). The *medial collateral ligament* is also triangular and consists of three bands—the anterior band that connects the medial epicondyle of the humerus to the coronoid process of the ulna, the pos-

TABLE 18–2.

Summary of the Movements of the Shoulder Joint and the Muscles Producing Those Movements*

Movements	Muscles	Origin	Insertion	Nerve Supply	Segmental Nerve†
Flexion	Deltoid (anterior fibers)	Clavicle	Middle of lateral surface of shaft of humerus	Axillary nerve	**C5,** C6
	Pectoralis major (clavicular part)	Clavicle	Lateral lip bicipital groove of humerus	Medial and lateral pectoral nerves from brachial plexus	C5, **C6**
	Biceps brachii				
	Long head	Supraglenoid tubercle of scapula	Tuberosity of radius, deep fascia of forearm	Musculocutaneous nerve	C5, **C6**
	Short head	Coracoid process of scapula			
	Coracobrachialis	Coracoid process of scapula	Medial aspect of shaft of humerus	Musculocutaneous nerve	C5, **C6,** C7
Extension	Deltoid (posterior fibers)	Spine of scapula	Middle of lateral surface of shaft of humerus	Axillary nerve	**C5,** C6
	Latissimus dorsi	Iliac crest, lumbar fascia, spines of lower 6 thoracic vertebrae, lower 3 or 4 ribs, and inferior angle of scapula	Floor of bicipital groove of humerus	Thoracodorsal nerve	**C6, C7,** C8
	Teres major	Lower third lateral border of scapula	Medial lip of bicipital groove of humerus	Lower subscapular nerve	**C6,** C7
Abduction	Middle fibers of deltoid	Acromion process of scapula	Middle of lateral surface of shaft of humerus	Axillary nerve	C5, **C6**
	Supraspinatus	Supraspinous fossa of scapula	Greater tuberosity of humerus	Suprascapular nerve	C4, **C5,** C6
Adduction	Pectoralis major (sternal part)	Sternum and upper 6 costal cartilages	Lateral lip of bicipital groove of humerus	Medial and lateral pectoral nerves	**C7, C8,** T1
	Latissimus dorsi	Iliac crest, lumbar fascia, spines of lower 6 thoracic vertebrae, lower 3 or 4 ribs, inferior angle of scapula	Floor of bicipital groove of humerus	Thoracodorsal nerve	**C6, C7,** C8
	Teres major	Lower third lateral border of scapula	Medial lip of bicipital groove of humerus	Lower subscapular nerve	**C6,**C7
	Teres minor	Upper ⅔ lateral border of scapula	Greater tuberosity of humerus	Axillary nerve	**C5,** C6
Lateral rotation	Infraspinatus	Infraspinous fossa of scapula	Greater tuberosity of humerus	Suprascapular nerve	**C5,** C6
	Teres minor	Upper ⅔ lateral border of scapula	Greater tuberosity of humerus	Axillary nerve	**C5,** C6
	Deltoid (posterior fibers)	Spine of scapula	Middle of lateral surface of shaft of humerus	Axillary nerve	**C5,** C6
Medial rotation	Subscapularis	Subscapular fossa	Lesser tuberosity of humerus	Upper and lower subscapular nerves	C5, **C6**
	Latissimus dorsi	Iliac crest, lumbar fascia, spines of lower 3 or 4 ribs, inferior angle of scapula	Floor of bicipital groove of humerus	Thoracodorsal nerve	**C6, C7,** C8
	Teres major	Lower third lateral border of scapula	Medial lip bicipital groove of humerus	Lower subscapular nerve	**C6,** C7
	Deltoid (anterior fibers)	Clavicle	Middle of lateral surface of shaft of humerus	Axillary nerve	**C5,** C6

*Circumduction is a combination of all the movements described.
†The predominant segmental nerve supply is indicated by boldface type.

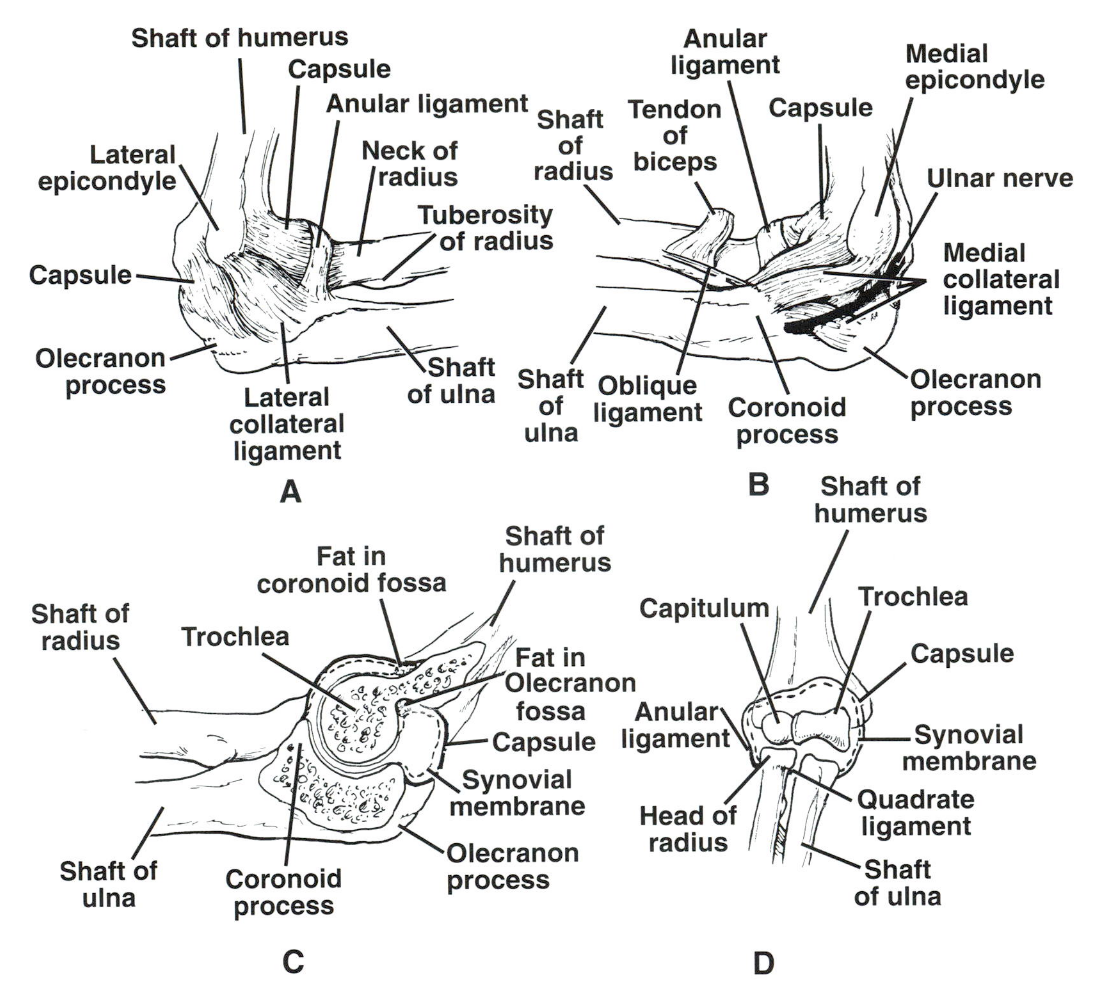

FIG 18–4.
Right elbow joint. **A,** lateral view. **B,** medial view. **C,** sagittal section. **D,** anterior view of interior of joint.

terior band that connects the medial epicondyle of the humerus to the olecranon process of the ulna, and the oblique band that connects the ulnar attachments of the two previous bands (see Fig 18–4).

Synovial Membrane.—This lines the capsule and is continuous below with the synovial membrane lining the superior radioulnar joint.

Nerve Supply.—Median nerve (C5 through C8 and T1), ulnar nerve (C8 and T1), musculocutaneous nerve (C5 through C7), and radial nerve (C5 through C8 and T1); all branches of the brachial plexus.

Movements Possible.—Flexion and extension. The extended forearm lies at an angle to the upper part of the arm. This angle, which opens laterally, is called the *carrying angle* and is about 170° in males and 167° in females; the angle disappears when the elbow joint is fully flexed.

Table 18–3 gives a summary of the movements possible at the elbow joint and the muscles producing those movements.

Anatomy of the Wrist Joint

Articulation.—This occurs between the lower end of the radius and the triangular articular disc above and the scaphoid, lunate, and triquetral bones below (Fig 18–5). The proximal articular surface forms an ovoid concave surface, which is adapted to the distal ovoid convex surface.

Type.—Synovial condyloid joint.

Capsule.—The capsule encloses the joint.

Ligaments.—*Anterior and posterior radiocarpal ligaments* strengthen the capsule. The *medial ligament* connects the styloid process of the ulna to the tri-

TABLE 18–3.
Summary of the Movements of the Elbow Joint and the Muscles Producing Those Movements

Movements	Muscles	Origin	Insertion	Nerve Supply	Segmental Nerve*
Flexion	Brachialis	Front of lower half of humerus	Coronoid process of ulna	Musculocutaneous nerve	C5, **C6**
	Biceps brachii				
	Long head	Supraglenoid tubercle of scapula	Tuberosity of radius, deep fascia of forearm	Musculocutaneous nerve	C5, **C6**
	Short head	Coracoid process of scapula			
	Brachioradialis	Lateral supracondylar ridge of humerus	Styloid process of radius	Radial nerve	C5, **C6,** C7
	Pronator teres				
	Humeral head	Medial epicondyle of humerus	Lateral aspect of shaft of radius	Median nerve	C6, **C7**
	Ulnar head	Coronoid process of ulna			
Extension	Triceps				
	Long head	Infraglenoid tubercle of scapula	Olecranon process of ulna	Radial nerve	C6, **C7,** C8
	Lateral head	Posterior surface of shaft of humerus			
	Medial head	Lower half posterior surface of shaft of humerus			
	Anconeus	Lateral epicondyle of humerus	Olecranon process of ulna	Radial nerve	C7, C8, T1

*The predominant segmental nerve supply is indicated by boldface type.

quetral bone. The *lateral ligament* connects the styloid process of the radius to the scaphoid bone.

Synovial Membrane.—This lines the capsule.

Nerve Supply.—Anterior interosseous nerve, a branch of the median nerve (C5 through C8 and T1), and deep branches of the radial nerve (C5 through C8 and T1) and ulnar nerve (C8 and T1).

Movements.—Flexion, extension, abduction, adduction, and circumduction. Rotation is not possible because the articular surfaces are ovoid. The lack of rotation is compensated for by the movements of pronation and supination of the forearm.

Table 18–4 gives a summary of the movements possible at the wrist joint and the muscles producing those movements.

Anatomy of the Carpometacarpal Joint of the Thumb

Articulation.—This occurs between the trapezium and the saddle-shaped base of the first metacarpal bone.

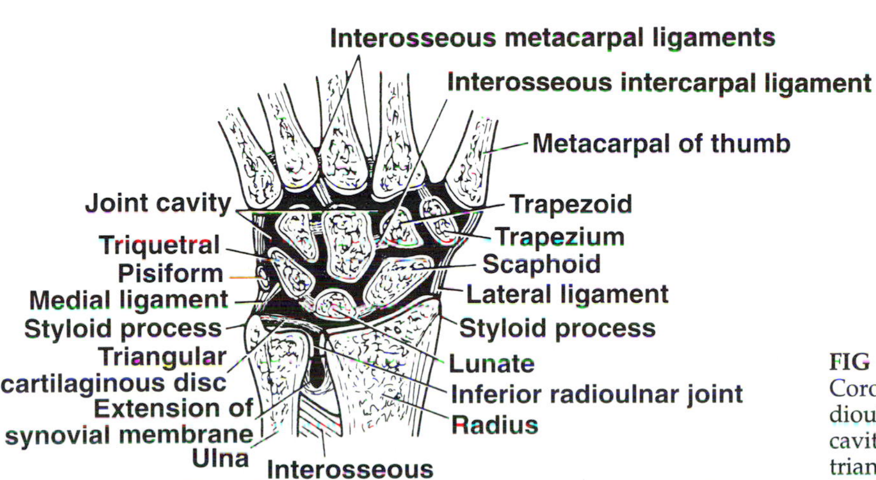

FIG 18–5.
Coronal section of the right wrist, inferior radioulnar and carpal joints, showing the joint cavities *(black area).* Note the position of the triangular cartilaginous disc that separates the head of the ulna from the wrist joint.

TABLE 18–4.
Summary of the Movements of the Wrist Joint and the Muscles Producing Those Movements

Movements	Muscles	Origin	Insertion	Nerve Supply	Segmental Nerve*
Flexion	Flexor carpi radialis	Medial epicondyle of humerus	Bases of second and third metacarpal bones	Median nerve	C6, **C7**
	Flexor carpi ulnaris				
	Humeral head	Medial epicondyle of humerus	Pisiform bone, hook of hamate, base of fifth metacarpal bone	Ulnar nerve	C7, **C8**
	Ulnar head	Olecranon process, posterior border of ulna			
	Palmaris longus	Medial epicondyle of humerus	Flexor retinaculum, palmar aponeurosis	Median nerve	C7, C8
	Flexor digitorum superficialis				
	Humeroulnar head	Medial epicondyle of humerus, coronoid process of ulna	Middle phalanx of medial 4 fingers	Median nerve	C7, **C8,** T1
	Radial head	Oblique line anterior surface shaft of radius			
	Flexor digitorum profundus	Anterior surface shaft of ulna, interosseous membrane	Distal phalanx of medial 4 fingers	Ulnar half—ulnar nerve, radial half—median nerve	**C8,** T1
	Flexor pollicis longus	Anterior surface shaft of radius	Distal phalanx of thumb	Anterior interosseous branch of median	**C8,** T1
Extension	Extensor carpi radialis longus	Lateral supracondylar ridge of humerus	Base of second metacarpal bone	Radial nerve	C6, C7
	Extensor carpi radialis brevis	Lateral epicondyle of humerus	Base of third metacarpal bone	Deep branch of radial nerve	**C7,** C8
	Extensor carpi ulnaris	Lateral epicondyle of humerus	Base of fifth metacarpal bone	Deep branch of radial nerve	C7, **C8**
	Extensor digitorum	Lateral epicondyle of humerus	Middle and distal phalanges of medial 4 fingers	Deep branch of radial nerve	**C7,** C8
	Extensor indicis	Shaft of ulna and interosseous membrane	Extensor expansion of index finger	Deep branch of radial nerve	C7, **C8**
	Extensor digiti minimi	Lateral epicondyle of humerus	Extensor expansion of little finger	Deep branch of radial nerve	**C7,**C8
	Extensor pollicis longus	Shaft of ulna and interosseous membrane	Base of distal phalanx of thumb	Deep branch of radial nerve	C7, **C8**
Abduction	Flexor carpi radialis	Medial epicondyle of humerus	Bases of second and third metacarpal bones	Median nerve	C6, **C7**
	Extensor carpi radialis longus	Lateral supracondylar ridge of humerus	Base of second metacarpal bone	Radial nerve	C6, C7
	Extensor carpi radialis brevis	Lateral epicondyle of humerus	Base of third metacarpal bone	Deep branch of radial nerve	**C7,** C8
	Abductor pollicis longus	Shafts of radius and ulna	Base of first metacarpal bone	Deep branch of radial nerve	C7, **C8**
	Extensor pollicis longus	Shaft of ulna and interosseous membrane	Base of distal phalanx of thumb	Deep branch of radial nerve	C7, **C8**
	Extensor pollicis brevis	Shaft of radius and interosseous membrane	Base of proximal phalanx of thumb	Deep branch of radial nerve	C7, **C8**
Adduction	Flexor carpi ulnaris				
	Humeral head	Medial epicondyle of humerus	Pisiform bone, hook of hamate, base of fifth metacarpal bone	Ulnar nerve	C7, **C8**
	Ulnar head	Olecranon process of ulna			
	Extensor carpi ulnaris	Lateral epicondyle of humerus	Base of fifth metacarpal bone	Deep branch of radial nerve	C7, **C8**

*The predominant segmental nerve supply is shown in boldface type.

Type.—Synovial saddle joint (biaxial joint).

Capsule.—Encloses the joint.

Synovial Membrane.—Lines the capsule and forms separate joint cavity.

Nerve Supply.—Radial (C5 through C8 and T1) and median nerves (C5 through C8 and T1).

Movements.—Flexion, extension, abduction, adduction, and rotation (opposition) (see p. 646).

Anatomy of the Metacarpophalangeal Joints

Articulation.—This occurs with the convex heads of metacarpal bones and the concave bases on the proximal phalanges (Fig 18–6).

Type.—Synovial condyloid joints.

Capsule.—Encloses the joint.

Ligaments.—*Palmar ligaments* are strong and contain some fibrocartilage. *Collateral ligaments* are cordlike bands that join the head of the metacarpal bone to the base of the phalanx. The collateral ligaments are taut when the joint is flexed and lax when the joint is extended. When in flexion, the fingers are locked, but they can be abducted and adducted when in extension.

Synovial Membrane.—Lines the capsule.

Movements.—Flexion, extension, abduction, and adduction (see p. 649).

Basic Anatomy of the Interphalangeal Joints

The interphalangeal joints are synovial hinge joints that have a similar structure to that of the metacarpophalangeal joints.

Anatomy of the Hip Joint

Articulation.—This occurs with the hemispherical head of the femur and the cup-shaped acetabulum of the hip bone (Fig 18–7). The articular surface of the acetabulum is horseshoe shaped and is deficient inferiorly at the *acetabular notch.* The cavity of the acetabulum is deepened by the fibrocartilaginous rim, the *acetabular labrum.* The labrum bridges across the acetabular notch, where it is called the *transverse acetabular ligament.*

Type.—Synovial ball and socket.

Capsule.—This encloses the joint and is attached medially to the acetabular labrum. Laterally it is attached to the intertrochanteric line of the femur in front and halfway along the posterior surface of the neck of the femur behind.

Ligaments.—The *iliofemoral ligament* is strong and shaped like an inverted Y. Its base is attached to the

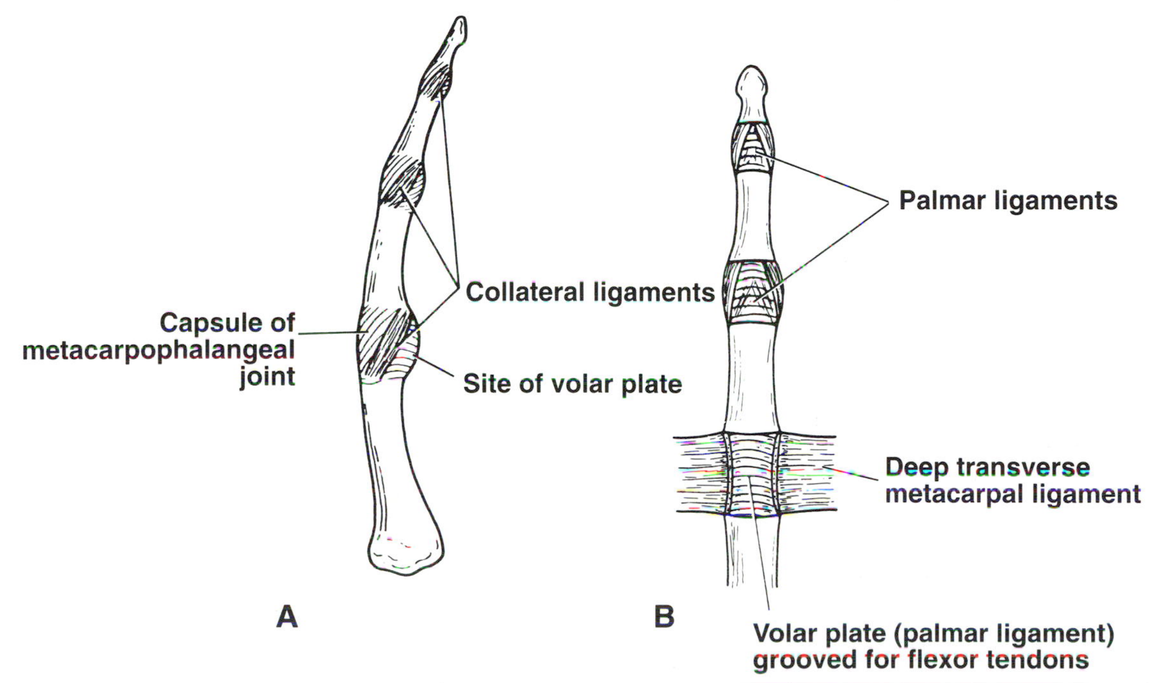

FIG 18–6.
Index finger showing the capsules and ligaments of the metacarpophalangeal and interphalangeal joints. **A,** side view. **B,** anterior view.

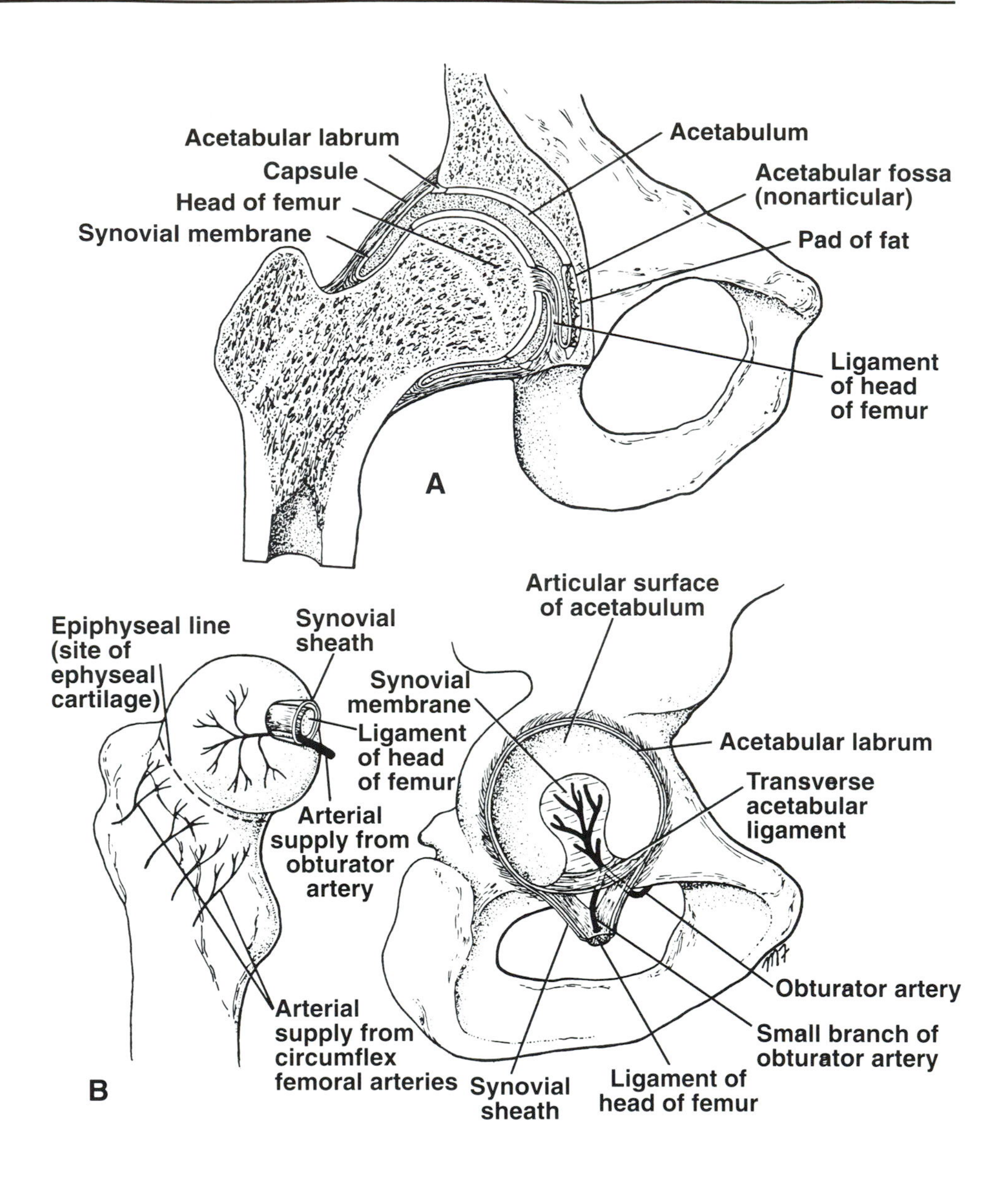

FIG 18–7.
A, coronal section of the right hip joint. **B,** articular surfaces of the right hip joint. Note the arterial supply of the head of the femur.

anterior inferior iliac spine above, and the two limbs of the Y are attached to the upper and lower parts of the intertrochanteric line of the femur. The *pubofemoral ligament* is triangular. The base is attached to the superior ramus of the pubis, and the apex is attached below to the lower part of the intertrochanteric line. The spiral-shaped *ischiofemoral ligament* is attached to the body of the ischium and laterally to the greater trochanter. The *ligament of the head of the femur* lies within the joint and is covered with synovial membrane; it is attached to the fovea capitis of the femur and to the transverse acetabular ligament (Fig 18–7).

Synovial Membrane.—Lines the capsule and may communicate with the *psoas bursa* (15%) beneath the psoas tendon.

Nerve Supply.—Femoral nerve (L2 through L4), obturator nerve (L2 through L4), and sciatic nerve (L4 and L5, S1 through S3).

Movements.—Flexion, extension, abduction, adduction, lateral rotation, medial rotation, and circumduction.

Table 18–5 gives a summary of the movements possible at the hip joint and the muscles producing those movements.

Anatomy of the Knee Joint

Articulation.—Rounded condyles of the femur and the condyles of the tibia and their semilunar cartilages; in front is the articulation between the patella and the lower end of the femur (Fig 18–8).

TABLE 18–5.

Summary of the Movements of the Hip Joint and the Muscles Producing Those Movements*

Movements	Muscles	Origin	Insertion	Nerve Supply	Segmental Nerve†
Flexion	Iliacus	Iliac fossa	Lesser trochanter of femur	Femoral nerve	**L2**, L3
	Psoas	Body of 12th thoracic vertebra, transverse processes, bodies and intervertebral discs of the 5 lumbar vertebrae	Lesser trochanter of femur	Lumbar plexus	**L1, L2,** L3
	Rectus femoris				
	Straight head	Anterior inferior iliac spine	Patella	Femoral nerve	L2, **L3, L4**
	Reflected head	Ilium above acetabulum			
	Sartorius	Anterior superior iliac spine	Upper medial surface of shaft of tibia	Femoral nerve	L2, L3
Extension (a posterior movement of the flexed thigh)	Gluteus maximus	Outer surface of ilium, sacrum, coccyx, sacrotuberous ligament	Iliotibial tract, gluteal tuberosity of femur	Inferior gluteal nerve	L5, **S1, S2**
	Biceps femoris	Long head: ischial tuberosity	Head of fibula	Tibial nerve (sciatic nerve)	L5, **S1,** S2
	Semitendinosus	Ischial tuberosity	Upper part of medial surface of shaft of tibia	Tibial nerve (sciatic nerve)	**L5,** S1, S2
	Semimembranosus	Ischial tuberosity	Medial condyle of tibia	Tibial nerve (sciatic nerve)	**L5, S1,** S2
	Adductor magnus	Ischial tuberosity	Adductor tubercle of femur	Tibial nerve (sciatic nerve)	L2, **L3, L4**
Abduction	Gluteus medius	Outer surface of ilium	Greater trochanter of femur	Superior gluteal nerve	**L5,** S1
	Gluteus minimus	Outer surface of ilium	Greater trochanter of femur	Superior gluteal nerve	**L5,** S1
	Sartorius	Anterior superior iliac spine	Upper medial surface of shaft of tibia	Femoral nerve	L2, L3
	Tensor fasciae latae	Iliac crest	Iliotibial tract	Superior gluteal nerve	L4, L5
	Piriformis	Anterior surface of sacrum	Greater trochanter of femur		L5, **S1,** S2
Adduction	Adductor longus	Body of pubis	Posterior surface of shaft of femur	Obturator nerve	L2, **L3,** L4
	Adductor brevis	Inferior ramus of pubis	Posterior surface of shaft of femur	Obturator nerve	L2, L3, L4
	Adductor magnus (adductor fibers)	Inferior ramus of pubis, ramus of ischium, ischial tuberosity	Posterior surface of shaft of femur, adductor tubercle of femur	Obturator nerve	L2, **L3,** L4
	Pectineus	Superior ramus of pubis	Upper end of shaft of femur	Femoral nerve	L2, L3
	Gracilis	Inferior ramus of pubis, ramus of ischium	Upper part of shaft of tibia on medial surface	Obturator nerve	**L2,** L3
Lateral rotation	Piriformis	Anterior surface of sacrum	Greater trochanter of femur	. . .	L5, **S1,** S2
	Obturator internus	Inner surface of obturator membrane	Greater trochanter of femur	Sacral plexus	L5, **S1**
	Obturator externus	Outer surface of obturator membrane	Greater trochanter of femur	Obturator nerve	L3, **L4**
	Superior gemellus	Spine of ischium	Greater trochanter of femur	Sacral plexus	L5, S1
	Inferior gemellus	Ischial tuberosity	Greater trochanter of femur	Sacral plexus	L5, S1
	Quadratus femoris	Ischial tuberosity	Quadrate tubercle on upper end of posterior surface of femur	Sacral plexus	L5, S1
	Gluteus maximus	Outer surface of ilium, sacrum, coccyx, sacrotuberous ligament	Iliotibial tract, gluteal tuberosity of femur	Inferior gluteal nerve	L5, **S1, S2**
Medial rotation	Gluteus medius	Outer surface of ilium	Greater trochanter of femur	Superior gluteal nerve	**L5,** S1
	Gluteus minimus	Outer surface of ilium	Greater trochanter of femur	Superior gluteal nerve	**L5,** S1
	Tensor fasciae latae	Iliac crest	Iliotibial tract	Superior gluteal nerve	L4, L5

*Circumduction is a combination of all the movements described.

†The predominant segmental nerve supply is indicated by boldface type.

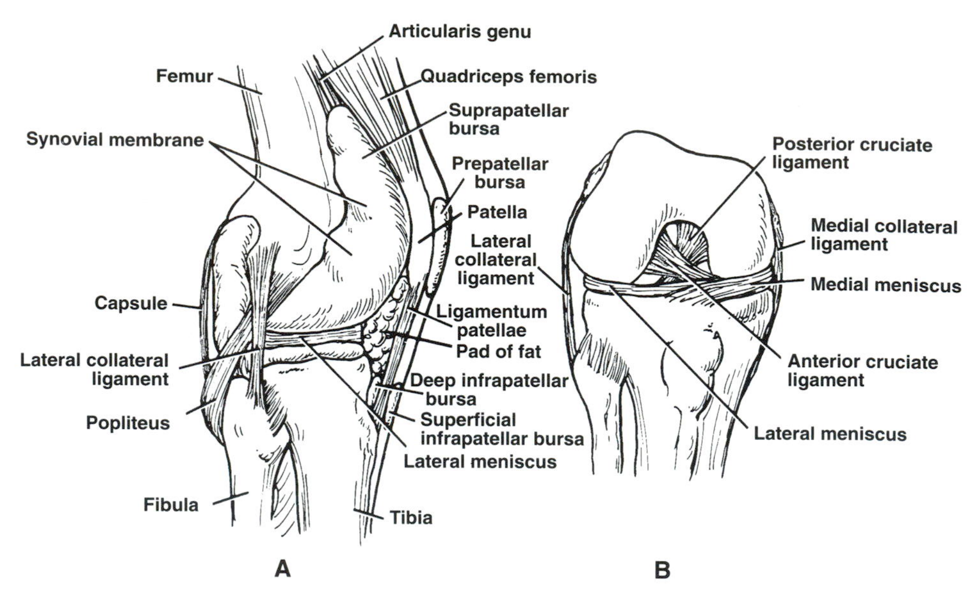

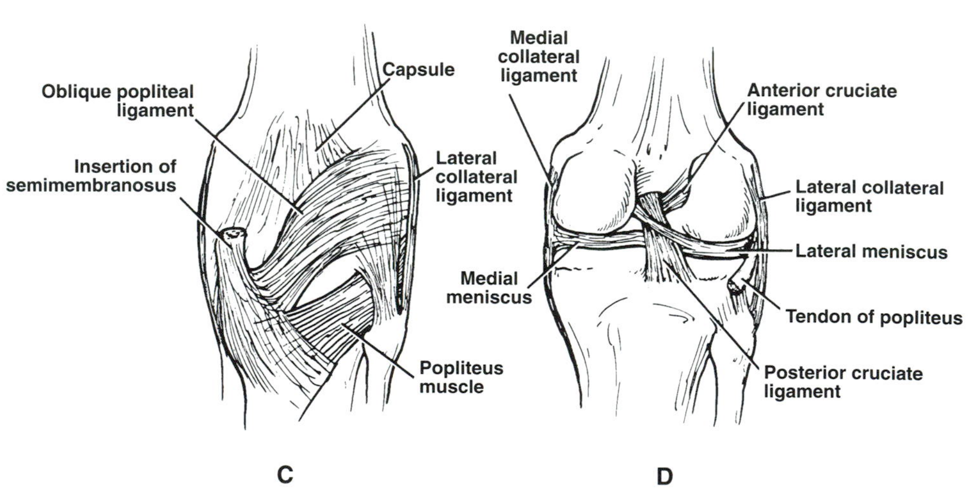

FIG 18–8.
Right knee joint. **A,** lateral aspect showing the extensive synovial membrane communicating with the suprapatellar bursa. **B,** anterior aspect with the joint flexed and showing the cruciate ligaments and the medial and lateral menisci. **C,** posterior aspect. **D,** posterior aspect showing the cruciate ligaments and the menisci.

Type.—Main joint is a synovial hinge; subsidiary joint between patella and the femur is a synovial gliding joint.

Capsule.—Encloses the joint but is absent in front above the patella, permitting the synovial membrane to pouch upward beneath the quadriceps tendon to form the *suprapatellar bursa* (see Fig 18–8).

Ligaments.—*Ligamentum patellae* connects the patella to the tuberosity of the tibia; it is a continuation of the insertion of the quadriceps tendon. The *lateral collateral ligament* is cordlike and connects the lateral femoral condyle to the head of the fibula; it prevents the movement of adduction of the knee joint. The *medial collateral ligament* is a flat band that connects the medial femoral condyle to the medial surface of the shaft of the tibia; it is strongly attached to the medial semilunar cartilage and prevents the movement of abduction of the knee joint. *Anterior and posterior cruciate ligaments* are two very strong ligaments that are located within the joint capsule but outside the synovial cavity (see Fig 18–8). They connect the intercondylar areas of the femur and tibia. The anterior cruciate ligament limits the posterior movement of the femur on the tibia. The posterior cruciate ligament limits the anterior movement of the femur on the tibia.

Medial and Lateral Semilunar Cartilages.—These are C-shaped sheets of fibrocartilage that lie between the femoral and tibial condyles. They are attached to the upper surface of the tibia by anterior and posterior horns. Because the medial cartilage is also attached to the medial collateral ligament, it is relatively immobile and is commonly damaged.

Synovial Membrane.—This lines the capsule and extends up beneath the quadriceps femoris muscle for three fingerbreadths above the patella, forming the *suprapatellar bursa* (see Fig 18–8). At the back of the joint it is continuous with the *popliteal bursa,* and it may communicate with the *semimembranosus bursa.* Below the patellar it covers the infrapatellar pad of fat.

Nerve Supply.—Femoral nerve (L2 through L4), obturator nerve (L2 through L4), common peroneal nerve (L4 and L5, S1 and S2), and tibial nerve (S2 and S3).

Movements.—Flexion, extension, medial rotation, and lateral rotation.

Table 18–6 gives a summary of the movements possible at the knee joint and the muscles producing those movements.

Anatomy of the Ankle Joint

Articulation.—This occurs between the lower end of the tibia, the two malleoli, and the body of the talus (Fig 18–9). The *inferior transverse tibiofibular ligament* deepens the socket into which the body of the talus fits snugly.

Type.—Synovial hinge joint.

Capsule.—Encloses the joint.

Ligaments.—The *medial, or deltoid, ligament* is very strong and connects the apex of the medial malleolus to the medial side of the talus and the sustentaculum tali and the tuberosity of the navicular bone (see Fig 18–9). The *lateral ligament* consists of three parts that connect the lateral malleolus to the lateral surface of the talus, the lateral surface of the calcaneum, and the back of the talus (see Fig 18–9).

Synovial Membrane.—Lines the capsule.

Nerve Supply.—Deep peroneal and tibial nerves.

Movements.—Dorsiflexion (toes pointing upward) and plantarflexion (toes pointing downward). The movements of inversion and eversion take place at the tarsal joints and *not at the ankle joint.*

Table 18–7 gives a summary of the movements possible at the ankle joint and the muscles producing those movements.

Important Relations.—These include the following.

Anteriorly.—Dorsalis pedis vessels and the deep peroneal nerve.

Posteriorly.—Tendo calcaneus.

TABLE 18–6.
Summary of the Movements of the Knee Joint and the Muscles Producing Those Movements

Movements	Muscles	Origin	Insertion	Nerve Supply	Segmental Nerve*
Flexion	Biceps femoris				
	Long head	Ischial tuberosity	Head of fibula	Tibial nerve	L5, **S1**, S2
	Short head	Shaft of femur		Common peroneal nerve	
	Semitendinosus	Ischial tuberosity	Upper part of medial surface of shaft of tibia	Tibial nerve	**L5, S1**, S2
	Semimembranosus	Ischial tuberosity	Medial condyle of tibia	Tibial nerve	**L5, S1**, S2
	Gastrocnemius	Medial, lateral condyles of femur	Via Achilles tendon into calcaneum	Tibial nerve	**S1**, S2
Extension	Quadriceps femoris: rectus femoris				
	Straight head	Anterior inferior iliac spine;	Patella	Femoral nerve	L2, **L3, L4**
	Reflected head	Ilium above acetabulum			
	Vastus lateralis	Upper end and shaft of femur	Patella	Femoral nerve	L2, **L3, L4**
	Vastus medialis	Upper end and shaft of femur	Patella	Femoral nerve	L2, **L3, L4**
	Vastus intermedius	Shaft of femur	Patella	Femoral nerve	L2, **L3, L4**
Medial rotation	Sartorius	Anterior superior iliac spine	Upper medial surface of shaft of tibia	Femoral nerve	L2, L3
	Gracilis	Inferior ramus of pubis, ramus of ischium	Upper part of shaft of tibia on medial surface	Obturator nerve	**L2**, L3
Lateral rotation	Biceps femoris				
	Long head	Ischial tuberosity	Head of fibula	Tibial nerve	L5, **S1**, S2
	Short head	Shaft of femur	. . .	Common peroneal nerve	L5, **S1**, S2

*The predominant segmental nerve supply is indicated by boldface type.

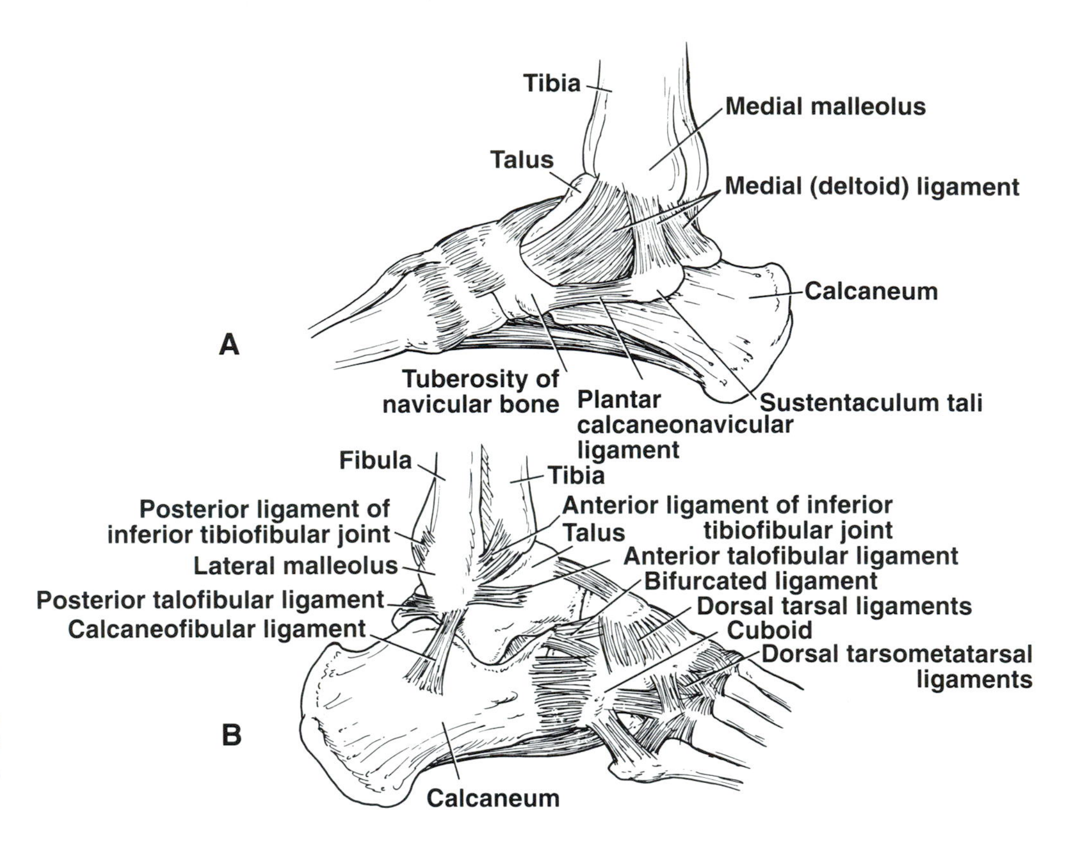

FIG 18–9.
Right ankle joint. **A,** medial aspect shows the medial collateral (deltoid) ligament. **B,** lateral aspect shows the three parts of the lateral collateral ligament.

Behind the Lateral Malleolus.—Tendons of peroneus longus and brevis muscles.

Behind the Medial Malleolus.—Posterior tibial vessels and the tibial nerve and the long flexor tendons to the foot.

Anatomy of the Subtalar Joint

Articulation.—This occurs between the body of the talus above and the calcaneum below.

Type.—Synovial.

Capsule.—Encloses the joint.

Ligaments.—*Medial and lateral talocalcaneal ligaments* strengthen the capsule. *Interosseous talocalcaneal ligament* is very strong and binds the bones together.

Synovial Membrane.—Lines the capsule.

Movements.—Inversion and eversion.

Anatomy of the Metatarsophalangeal and Interphalangeal Joints

These joints are similar in structure to those of the hand (see pp. 648 and 649).

Basic Anatomy of Bursae

A bursa is a lubricating device consisting of a closed fibrous sac lined with a delicate smooth membrane formed of synovial-like cells. Its walls are separated by a film of viscous fluid. Bursae are found wherever tendons rub against bones, ligaments, or other tendons. They are commonly found close to joints where the skin rubs against underlying bony structures, e.g., the prepatellar bursa (see Fig 18–21). Occasionally, the cavity of a bursa communicates with the cavity of a synovial joint. For example, the suprapatellar bursa communicates with the knee joint (see Fig 18–8), and the subscapularis bursa communicates with the shoulder joint.

Anatomy of the Subacromial Bursa

The subacromial bursa lies beneath the deltoid muscle, the acromion process, and the coracoacromial ligament above and the shoulder joint below (see Fig 18–3). The bursa does not communicate with the shoulder joint. Its function is to serve as a lubricating device between the greater tuberosity of the humerus and the acromion process when the shoulder joint is fully abducted. It is closely related to the supraspinatus tendon, which is here fused with the

TABLE 18–7.
Summary of the Movements of the Ankle Joint and the Muscles Producing Those Movements*

Movements	Muscles	Origin	Insertion	Nerve Supply	Segmental Nerve
Dorsiflexion	Tibialis anterior	Shaft of tibia, interosseous membrane	Medial cuneiform, base of first metatarsal bone	Deep peroneal nerve	**L4,** L5
	Extensor hallucis longus	Shaft of fibula, interosseous membrane	Base of distal phalanx of great toe	Deep peroneal nerve	L5, S1
	Extensor digitorum longus	Shaft of fibula, interosseous membrane	Dorsal extensor expansion of lateral four toes	Deep peroneal nerve	L5, S1
	Peroneus tertius	Shaft of fibula, interosseous membrane	Base of fifth metatarsal bone	Deep peroneal nerve	L5, S1
Plantar flexion	Gastrocnemius	Medial, lateral condyles of femur	Via Achilles tendon into calcaneum	Tibial nerve	**S1,** S2
	Soleus	Shaft of tibia and fibula	Via Achilles tendon into calcaneum	Tibial nerve	S1, **S2**
	Plantaris	Lateral supracondylar ridge of femur	Calcaneum	Tibial nerve	**S1,** S2
	Peroneus longus	Shaft of fibula	Base of first metatarsal and medial cuneiform	Superficial peroneal nerve	**L5, S1,** S2
	Peroneus brevis	Shaft of fibula	Base of fifth metatarsal bone	Superficial peroneal nerve	**L5,** S1, S2
	Tibialis posterior	Shafts of tibia, fibula, interosseous membrane	Tuberosity of navicular	Tibial nerve	L4, L5
	Flexor digitorum longus	Shaft of tibia	Distal phalanges of lateral 4 toes	Tibial nerve	**S2,** S3
	Flexor hallucis longus	Shaft of fibula	Base of distal phalanx of big toe	Tibial nerve	**S2,** S3

*The predominant segmental nerve supply is indicated by boldface type.

upper part of the capsule of the shoulder joint. Supraspinatus tendinitis or tears of the supraspinatus tendon invariably involve the bursa.

Subacromial bursitis causes severe pain during the process of abduction of the shoulder joint. The pain is most intense at the point when the greater tuberosity impinges on the inflamed bursa beneath the acromion process. The point of maximum tenderness is usually just lateral to the anterolateral corner of the acromion.

Anatomy of the Olecranon Bursa

The olecranon bursa lies in the superficial fascia beneath the skin on the posterior surface of the olecranon process of the ulna. It does not communicate with the elbow joint (see Fig 18–19). Because of its exposed position, it is subject to trauma and the spread of infection from the skin.

Anatomy of the Greater Trochanter Bursa

The greater trochanter bursa is usually large and multilocular. It lies on the lateral surface of the greater trochanter of the femur beneath the overlying gluteus maximus muscle. The bursa does not communicate with the hip joint (see Fig 18–20).

Anatomy of the Prepatellar Bursa

The prepatellar bursa is large and lies in the superficial fascia between the skin and the front of the patellar and the patellar ligament (see Fig 18–21). The bursa does not communicate with the knee joint. The function of the bursa is to allow the skin to glide over the patella and withstand pressure when kneeling. The superficial position of the bursa predisposes it to traumatic inflammation and infection from the skin.

Anatomy of the Pes Anserine Bursa

This bursa lies between the tendons of insertion of the sartorius, gracilis, and semitendinosus muscles at their point of attachment to the upper part of the medial surface of the shaft of the tibia (see Fig 18–21). The bursa does not communicate with the knee joint. Its function is to reduce friction between these tendons, as their bony attachments are very close together. Bursitis is common in cyclists and runners.

Anatomy of the Tendo Calcaneus Bursa

The tendo calcaneus (tendo Achillis) is inserted into a roughened area on the middle third of the posterior surface of the calcaneum. The bursa is located between the tendon and the upper third of the posterior surface of the calcaneum (see Fig 18–21). It does not communicate with the ankle joint. Long-distance runners frequently experience traumatic inflammation of this bursa.

The tendo calcaneus bursa must not be confused with an adventitious "bursa," or blister, which may develop in the superficial fascia over the tendo calcaneus. Such a subcutaneous bursa may form in response to the trauma associated with badly fitting shoes.

CLINICAL ANATOMY

Clinical Anatomy of Arthrocentesis

Basic Principles

When performing an arthrocentesis, the following basic principles may be helpful.

1. *Use of extensor surface of joint.* Since the synovial cavity is often closest to the skin surface on the extensor aspect of the joint, and the large blood vessels and nerves usually travel on the flexor surface of the joint, the extensor surface is most often chosen as the optimal site for arthrocentesis. When, possible, the needle should go around rather than through the tendon to enter a joint.
2. *Joint position.* It is possible in most joints, by careful positioning, to enlarge the joint space to maximum advantage. By flexing a joint slightly the extensor aspect of the joint space is often enlarged and easier to enter.
3. *Joint traction.* Traction to the distal component of a joint along its long axis separates the joint surfaces and opens up the joint space. This technique is particularly helpful in the small joints of the fingers and toes.
4. *Joint compression.* The application of external pressure to a joint either manually or by means of an elastic bandage will often squeeze the joint fluid from the peripheral parts of the joint space into the main joint cavity. This may greatly facilitate the aspiration of fluid in a joint where the volume of available fluid is small.
5. *Needle direction.* It is important that the physician has a good understanding of the spatial relationships of the articular surfaces of the joint so that the needle point may be directed away from the articular cartilage and avoid damage to the articular surface.
6. *Prevention of infection.* Adequate preparation of the overlying skin greatly reduces the possibility of joint infection. Also, aspiration should be avoided, if possible, through potentially cellulitic areas.

Anatomy of Temporomandibular Arthrocentesis

This joint can easily be palpated on its lateral surface in front of the auricle by asking the patient to open and close the mouth. As the mouth opens, the head (condyle) of the mandible rotates and moves forward below the tubercle of the zygomatic arch. The pulsations of the superficial temporal artery should be palpated just anterior to the auricle (tragus) and behind the temporomandibular joint (Fig 18–10). The superficial temporal artery and the auriculotemporal nerve that lies immediately posterior to the artery have to be avoided in this procedure.

The needle may be introduced ½ in. (1.3 cm) anterior to the auricle (see Fig 18–10) and above the head of the mandible; the needle should be directed medially for about ½ in.

The following structures are pierced by the needle before it enters the joint cavity (see Fig 18–10).

1. Skin.
2. Superficial and deep layers of fascia.
3. Lateral temporomandibular ligament.
4. Joint capsule and synovial lining.

Theoretically, it is possible to impale the disc with an arthrocentesis needle. In practical terms the needle penetration of the disc would not permanently damage the joint.

Anatomy of Sternoclavicular Arthrocentesis

The sternoclavicular joint can easily be palpated at the medial end of the clavicle. In order to open up the joint space to its maximum, the patient should be placed in a supine position with the shoulder joint abducted to 90° and extended (allowing the weight of the limb to extend the arm backward over the table edge). This maneuver brings the medial end of the clavicle forward. The needle is inserted into the joint lateral to the suprasternal notch and is directed posteriorly (Fig 18–11). Remember that the brachiocephalic artery and the right brachiocephalic vein lie immediately behind the joint on the right side and the left common carotid artery and the left brachiocephalic vein lie behind the joint on the left side.

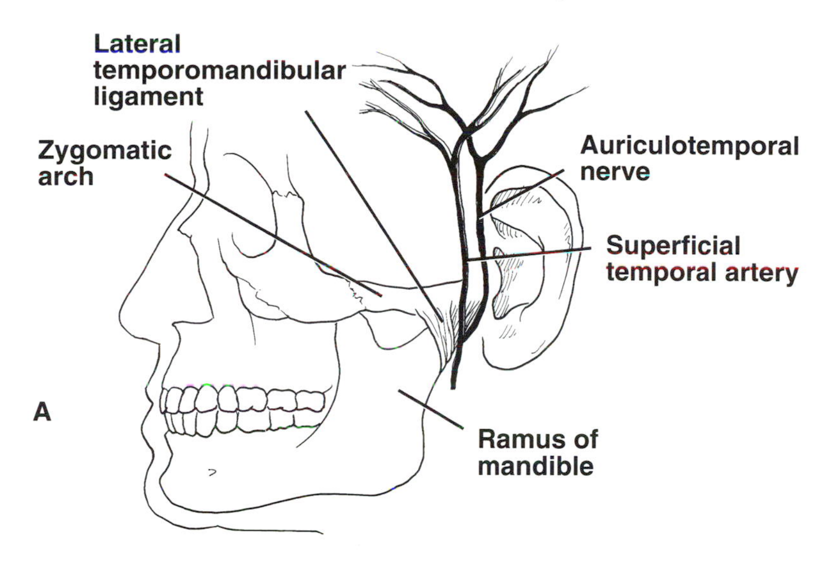

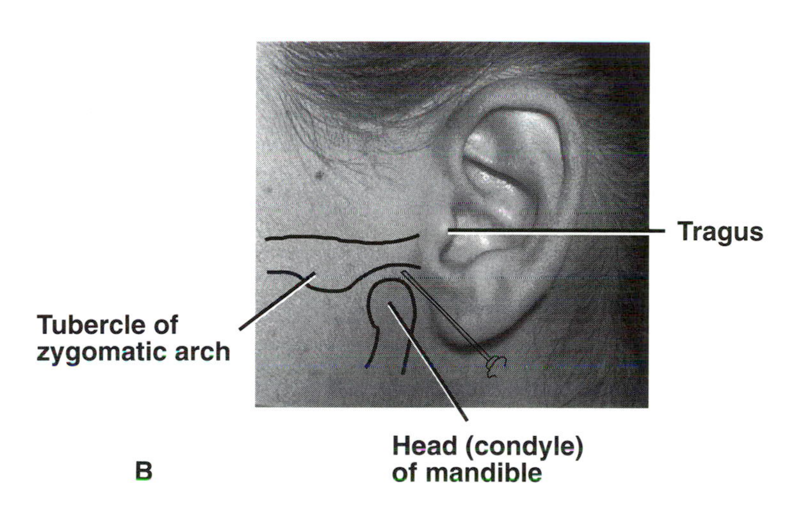

FIG 18–10.
Temporomandibular arthrocentesis. **A,** the left temporomandibular joint. Note the lateral temporomandibular ligament and the positions of the superficial temporal artery and the auriculotemporal nerve. **B,** the needle is inserted about 1 in. anterior to the tragus of the auricle and above the head of the mandible and is directed medially into the joint cavity.

The following structures are pierced by the needle before it enters the joint cavity.

1. Skin.
2. Superficial and deep layers of fascia.
3. Anterior sternoclavicular ligament.
4. Joint capsule and lining of synovial membrane.

Anatomy of Acromioclavicular Arthrocentesis

The acromioclavicular joint is easily palpated at the lateral end of the clavicle. The joint space may be opened to its maximum by placing the patient in a supine position with the shoulder joint extended over the table edge with no abduction. The needle is then inserted on the upper surface of the joint and directed downward and medially following the inclination of the joint surfaces (see Fig 18–11).

The following structures are pierced by the needle before it enters the joint cavity.

1. Skin.
2. Superficial and deep layers of fascia.
3. Acromioclavicular ligament.
4. Joint capsule and lining of synovial membrane.

Anatomy of Shoulder Arthrocentesis

This may be performed through the anterior or posterior aspects of the joint. (Note the difference in approach when aspirating the subacromial bursa, see p. 779).

Anterior Approach.—The patient is positioned sitting up with the arm hanging down by the side. The weight of the limb tends to open up the joint cavity (further opening of the joint cavity may be obtained

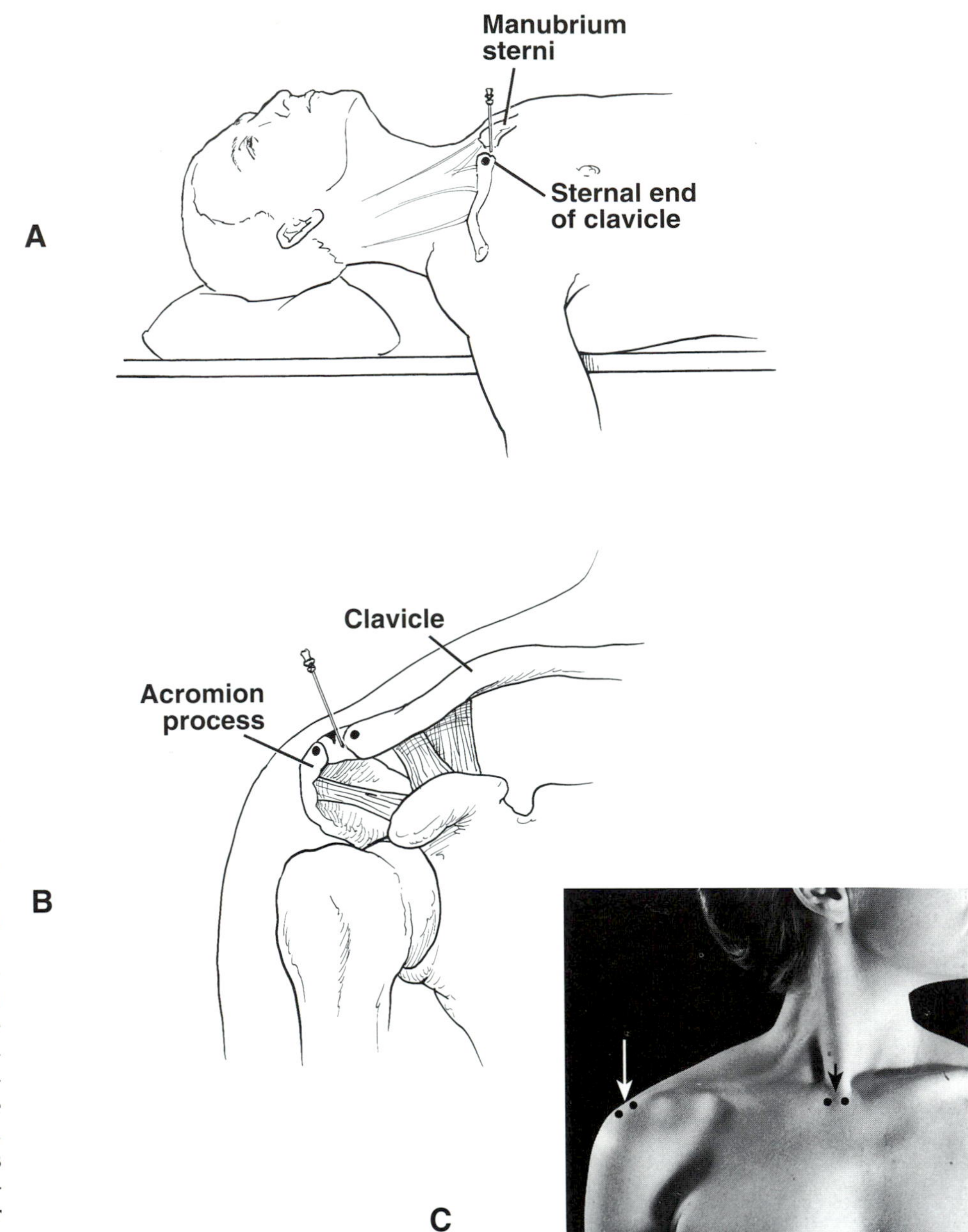

FIG 18–11.
A, sternoclavicular arthrocentesis. The needle is inserted medial to the sternal end of the clavicle and lateral to the suprasternal notch; it is directed posteriorly into the joint cavity. **B,** acromioclavicular arthrocentesis. The needle is inserted over the joint line between the acromion process and the lateral end of the clavicle; it is directed downward through the upper surface of the joint. **C,** the important bony landmarks for arthrocentesis of the sternoclavicular and acromioclavicular joints.

by exerting gentle downward traction on the arm). The tip of the coracoid process is then palpated in the lateral part of the deltopectoral triangle (triangular depression situated below the outer third of the clavicle and bounded by the pectoralis major and the deltoid muscles) just beneath the anterior border of the deltoid muscle (Fig 18–12). The head of the humerus is palpated. The needle is then inserted midway between the tip of the coracoid process and the head of the humerus, and the tip of the needle is directed backward and laterally (see Fig 18–12). Note that the brachial plexus and axillary vessels are located some distance medially to the point of penetration (see Fig 18–12).

The following structures are pierced by the needle before it enters the joint cavity.

1. Skin.
2. Superficial and deep layers of fascia.
3. Pectoralis major muscle (clavicular head).
4. ? Coracobrachialis and the short head of biceps.
5. Subscapularis muscle.
6. ? Subscapularis bursa.

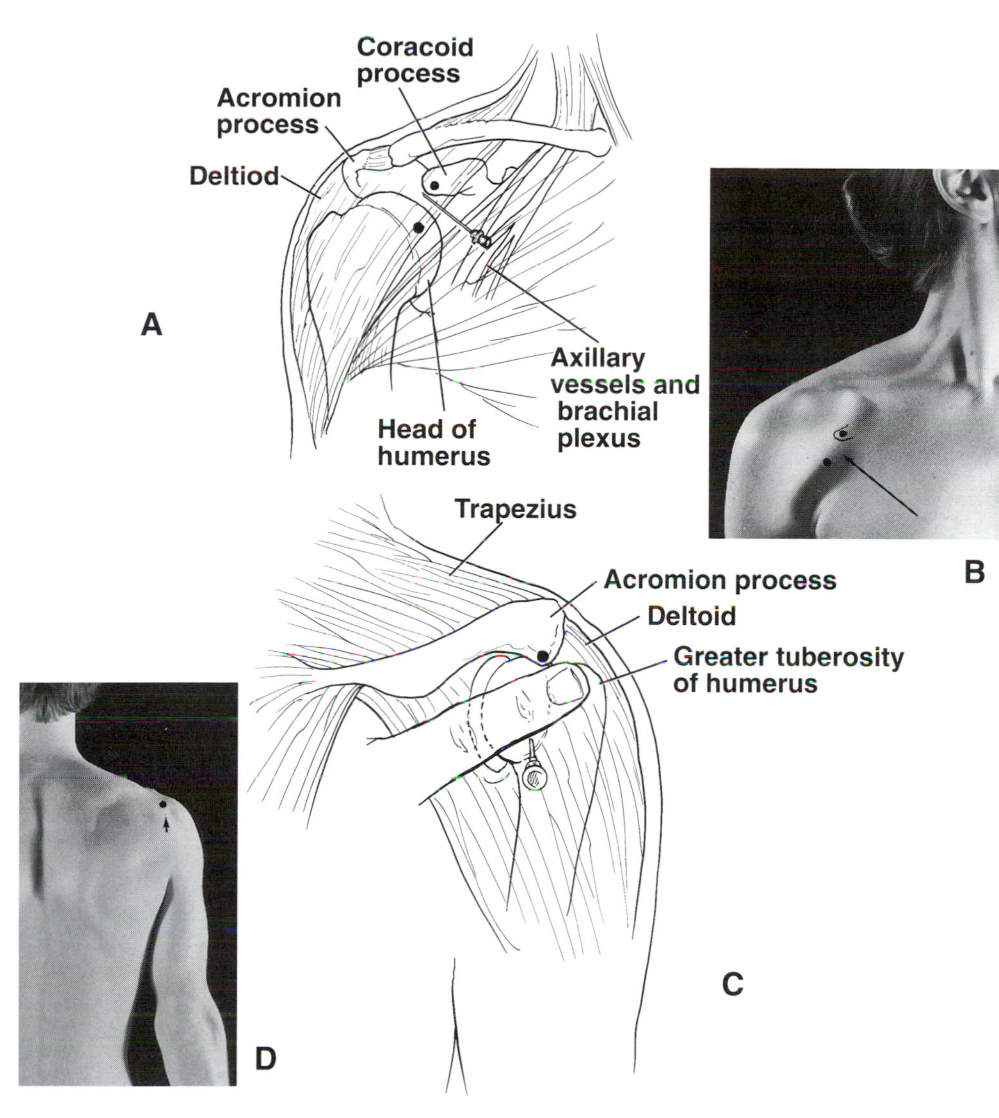

FIG 18–12.
Shoulder arthrocentesis. **A** and **B,** anterior approach. The needle is inserted midway between the tip of the coracoid process and the head of the humerus and is directed backward and laterally. *Arrow* indicates direction of needle. **C** and **D,** posterior approach. The needle is inserted at a point one fingerbreadth below the posterior corner of the acromion process and is directed horizontally forward. *Arrow* indicates site of insertion of needle.

7. Joint capsule thickened by glenohumeral ligaments.
8. Synovial membrane.

Posterior Approach.—The patient is positioned as before. The posterior fibers of the joint capsule are tightened by medially rotating the shoulder joint. The posterior corner of the lateral edge of the acromion process is palpated. The needle is inserted at a point one fingerbreadth below the posterior corner of the acromion process and is directed horizontally forward (see Fig 18–12). It passes between the head of the humerus and the edge of the glenoid cavity.

The needle pierces the following structures before entering the joint cavity.

1. Skin.
2. Superficial and deep layers of fascia.
3. Posterior fibers of the deltoid muscle.
4. Tendon of infraspinatus (part of rotator cuff).
5. Joint capsule and lining of synovial membrane.

The posterior approach has the following advantages over the anterior approach. (1) The only muscles that have to be pierced are the deltoid and the infraspinatus, whereas the pectoralis major, the coracobrachialis and biceps, and the subscapularis muscles have to be pierced with the anterior approach. (2) The posterior wall of the capsule is thin, whereas the anterior wall of the capsule is thickened by the glenohumeral ligaments. (3) Penetration of the brachial plexus and the axillary vessels is of no concern when performing the posterior approach.

Anatomy of Elbow Arthrocentesis

Three bony landmarks are first palpated (Fig 18–13)—the lateral epicondyle of the humerus, the head of the radius, and the tip of the olecranon process of the ulna. With the elbow joint partially extended to 135° and the forearm held in the midprone position, i.e., halfway between full supination and full pronation, the tip of the needle is inserted in the middle of the isosceles triangle formed by connecting the three bony points. The needle should be directed medially to enter the joint cavity between the humerus and the olecranon process (see Fig 18–13).

The following structures are pierced by the needle before entering the joint cavity.

1. Skin.
2. Superficial and deep layers of fascia.
3. Anconeus muscle.
4. Lateral collateral ligament.
5. Joint capsule and lining of synovial membrane.

Arthrocentesis of the elbow joint on the medial side is not recommended because of the location of the ulnar nerve behind the medial epicondyle of the humerus.

Anatomy of Wrist Arthrocentesis

Since the flexor (anterior) aspect of the wrist joint is covered with numerous tendons, blood vessels, and nerves, this surface should not be used for arthrocentesis. On the extensor (posterior) aspect of the wrist, there are many tendons but few important blood vessels and nerves. Palpate the dorsal tubercle (Lister's tubercle) on the radius (see Fig 18–13) and note that the tendon of the extensor pollicis longus

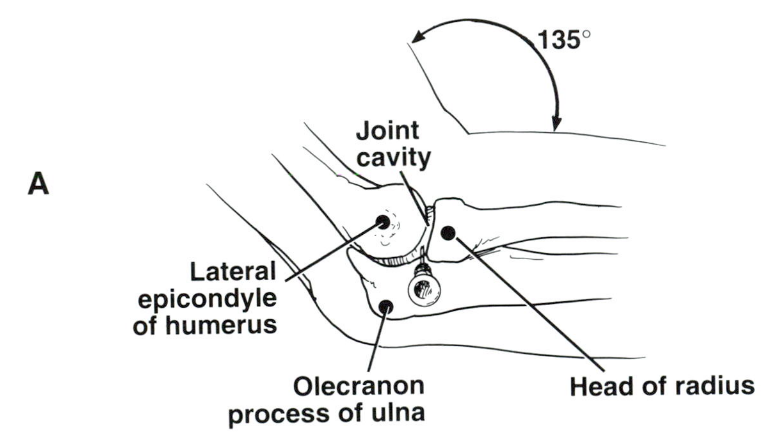

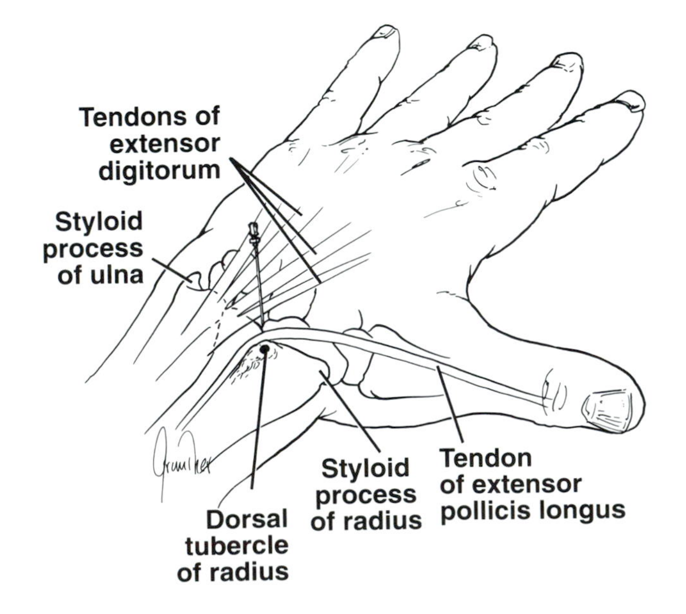

FIG 18–13.
A, elbow arthrocentesis. The needle is inserted in the middle of the isosceles triangle formed by connecting the three bony points—the lateral epicondyle of the humerus, the head of the radius, and the tip of the olecranon process of the ulna. The needle is directed medially to enter the joint cavity. **B,** wrist arthrocentesis. The needle is inserted distal to the dorsal tubercle of the radius on the joint line and just ulnar to the tendon of extensor pollicis longus. The needle should be directed anteriorly.

passes distally around the ulnar border of the tubercle. Extending the thumb makes the tendon stand out. The joint line may be identified as running between the radial and ulnar styloid processes, which are easily palpated. The wrist joint should be flexed and deviated toward the ulnar side to open up the joint cavity. The needle is inserted distal to the dorsal tubercle of the radius and just ulnar to the tendon of the extensor pollicis longus on the joint line. The needle should be directed anteriorly between the tendons of extensor pollicis longus and the extensor digitorum tendons into a depression that is easily palpated (see Fig 18–13). The tip of the needle will then pass forward between the extensor digitorum tendons and the tendon of extensor carpi radialis brevis. The needle will enter the joint cavity between the distal end of the radius and the lunate bone.

The needle pierces the following structures before it enters the joint cavity.

1. Skin.
2. Superficial fascia.
3. Extensor retinaculum.
4. Posterior radiocarpal ligament.
5. Posterior surface of the joint capsule and the synovial membrane.

Anatomy of Carpometacarpal Arthrocentesis of the Thumb

This small joint cavity has to be opened to the maximum extent by applying traction to the thumb such that the wrist joint is flexed and adducted (i.e., moved in an ulnar direction). The proximal end of the dorsal aspect of the first metacarpal bone is palpated in the anatomical snuff box (Fig 18–14). The pulsations of the radial artery are carefully identified in order to locate the artery as it crosses the snuff box. The needle is inserted into the joint line on the dorsal and lateral aspect, and great care is taken to avoid the radial artery (see Fig 18–14).

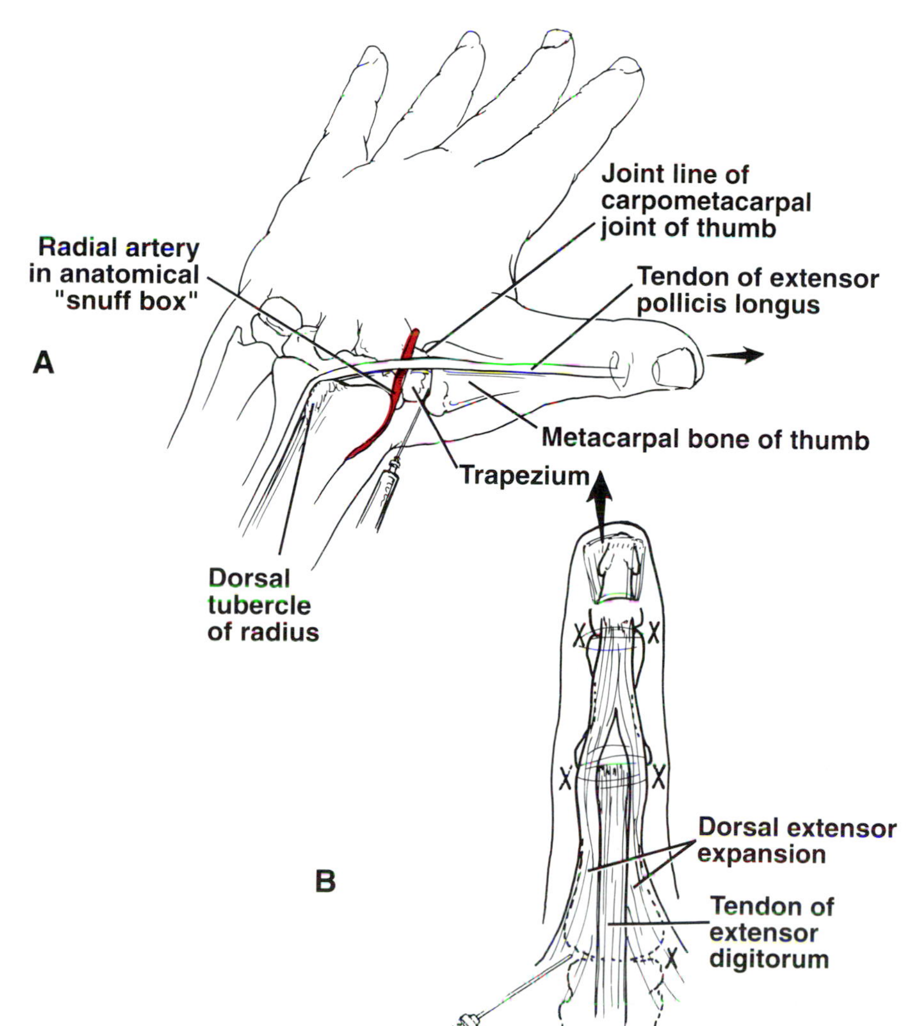

FIG 18–14.
Carpometacarpal, metacarpophalangeal, and interphalangeal arthrocenteses. **A,** carpometacarpal arthrocentesis of the thumb. The proximal end of the dorsal aspect of the first metacarpal bone is identified in the anatomical snuffbox. The needle is inserted into the joint line on the dorsal and lateral aspect. Note the position of the radial artery. **B,** metacarpophalangeal and interphalangeal arthrocenteses. The needle is inserted on the dorsal surface of the joint line on the radial or ulnar side of the extensor tendon *(X)*. With the joint cavity opened up as much as possible, the needle is directed anteriorly. The *arrow* indicates the direction of traction.

The following structures are pierced by the needle before it enters the joint cavity.

1. Skin.
2. Superficial and deep layers of fascia.
3. Tendons of abductor pollicis longus and or extensor pollicis brevis.
4. Joint capsule and lining of synovial membrane.

Anatomy of Metarcarpophalangeal and Interphalangeal Arthrocentesis

The palmar surface of these joints has tendons, synovial sheaths, digital vessels, digital nerves, and strong fibrocartilaginous palmar ligaments. Therefore the dorsal aspect of the joint is used. The dorsal surface of the joint line should be palpated by recognizing the head of the metacarpal bone (knuckle) or base of phalanx on each side of the extensor tendon. Traction should be applied to the finger, and the joint should be slightly flexed to open up the small cavity as much as possible. The needle is directed anteriorly into the joint line just radial or ulnar to the extensor tendon (see Fig 18–14).

The needle pierces the following structures before entering the joint cavity.

1. Skin.
2. Superficial and deep layers of fascia.
3. Fibrous dorsal extensor expansion associated with the extensor tendon.
4. Joint capsule and lining of synovial membrane.

Anatomy of Hip Arthrocentesis

This is a difficult procedure because the joint is deeply placed and surrounded by thick muscles. The procedure is usually performed under fluoroscopy. The presence of the femoral vessels and nerve in front of the joint and the sciatic nerve on the back of the joint increase the risk of complications. Two approaches may be used—the anterior approach and the lateral approach.

Anterior Approach.—With the patient in a supine position, the following important landmarks must first be identified. The anterior superior iliac spine can be easily palpated at the anterior end of the iliac crest (Fig 18–15). The pubic tubercle can be felt as a small protuberance at the lateral end of the upper border of the body of the pubis. (In males, the tubercle may be most easily felt by invaginating the scrotum with the examining finger; in females, the tubercle is felt by pressing on the labium majus from the lateral side.) The femoral pulse should be identified just below the inguinal ligament at a point midway between the anterior superior iliac spine and the symphysis pubis (see Fig 18–15).

With the hip joint extended and medially rotated, the needle is inserted 1 to 1½ in (2.5 to 3.8 cm) distal to (below) the midpoint on a line drawn between the anterior superior iliac spine and the pubic tubercle (see Fig 18–15). The needle is directed posteromedially at a 60° angle with the skin. The needle is advanced until it pierces the capsule and the bone is reached.

The needle pierces the following structures before entering the joint cavity.

1. Skin.
2. Superficial fascia.
3. Deep fascia (fascia lata).
4. Iliopsoas muscle.
5. Psoas bursa.
6. Joint capsule and lining of synovial membrane.

The femoral artery must be avoided, and its position should be verified by detecting the femoral pulse.

Lateral Approach.—With the patient in a supine position and the hip joint extended, the superior margin of the greater trochanter of the femur is palpated. The identification of the greater trochanter is made easier by bringing it forward from under cover of the gluteus maximus muscle. This is done by medially rotating the hip joint. The needle is inserted superior to the middle of the upper margin of the greater trochanter. The needle is directed at right angles to the long axis of the shaft of the femur and parallel to the table (see Fig 18–15).

The needle pierces the following structures before entering the joint cavity.

1. Skin.
2. Superficial and deep layers of fascia.
3. Gluteus medius muscle.
4. Iliofemoral or ischiofemoral ligaments.
5. Joint capsule and lining of synovial membrane.

The advantage of this approach is there are no large vessels or nerves in the area. Small terminal branches of the superior gluteal vessels are the only vessels in the vicinity.

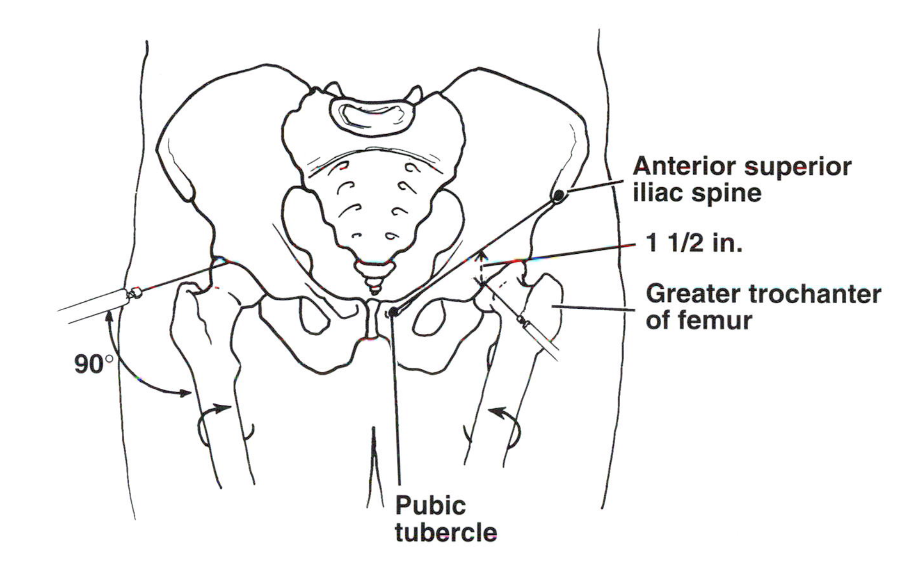

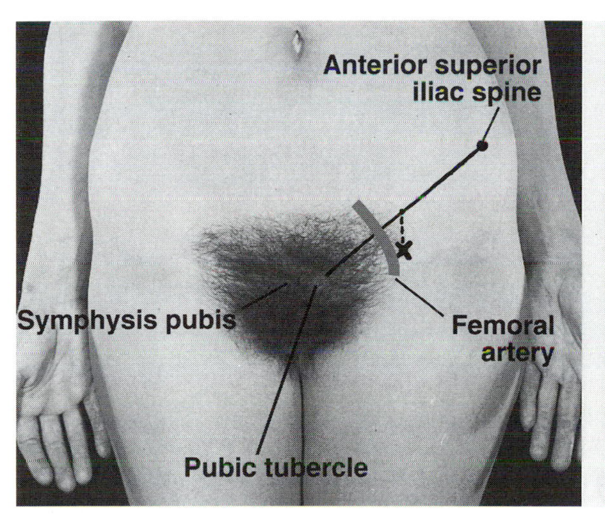

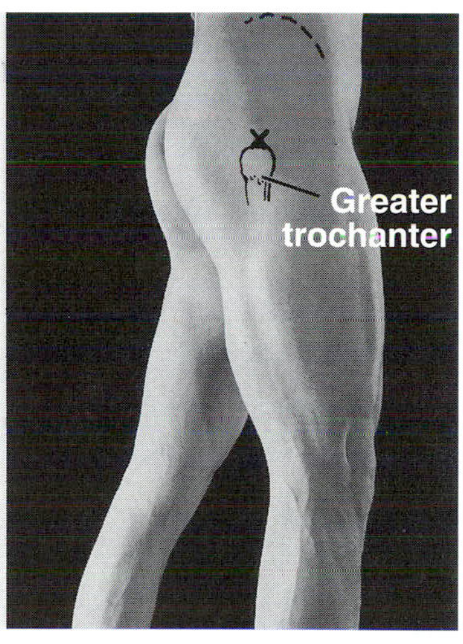

FIG 18–15.
Hip arthrocentesis. **A** and **B**, anterior approach. The needle is inserted 1½ in. below the midpoint between the anterior superior iliac spine and the pubic tubercle. The needle is directed posteromedially at a 60° angle with the skin. Note the position of the femoral artery. **A** and **C**, lateral approach. The needle is inserted above the middle of the upper margin of the greater trochanter of the femur *(X)*. The needle is directed at right angles to the long axis of the shaft of the femur. (The patient is placed in a supine position.)

Anatomy of Knee Arthrocentesis

The knee joint is a frequent site of inflammation and is the most commonly aspirated joint in the emergency department. Because the cavity is located close to the surface it is the easiest joint to aspirate. Three approaches are commonly used—(1) suprapatellar, (2) medial and lateral parapatellar, and (3) infrapatellar.

Suprapatellar Approach.—With the patient in a supine position, the knee joint is fully extended. The needle is inserted at the midpoint of the upper border of the patella and is directed between the posterior surface of the patella and the intercondylar notch of the femur. The angle of approach should avoid scraping the articular cartilage. The aspirator must be aware of the wedge shape of the patella (Fig 18–16), so that the angle of the needle is appropriate as the needle is advanced into the joint.

The needle pierces the following structures before entering the joint cavity.

1. Skin.
2. Superficial and deep layers of fascia.
3. Quadriceps tendon
4. Synovial membrane of the suprapatellar bursa.

The knee can accommodate as much as 50 to 70 mL of fluid. In addition, the suprapatellar bursa is an upward extension of the synovial lining of the knee joint beneath the quadriceps muscle. It rises to a height of approximately three fingerbreadths above the upper margin of the patella. If there is a large

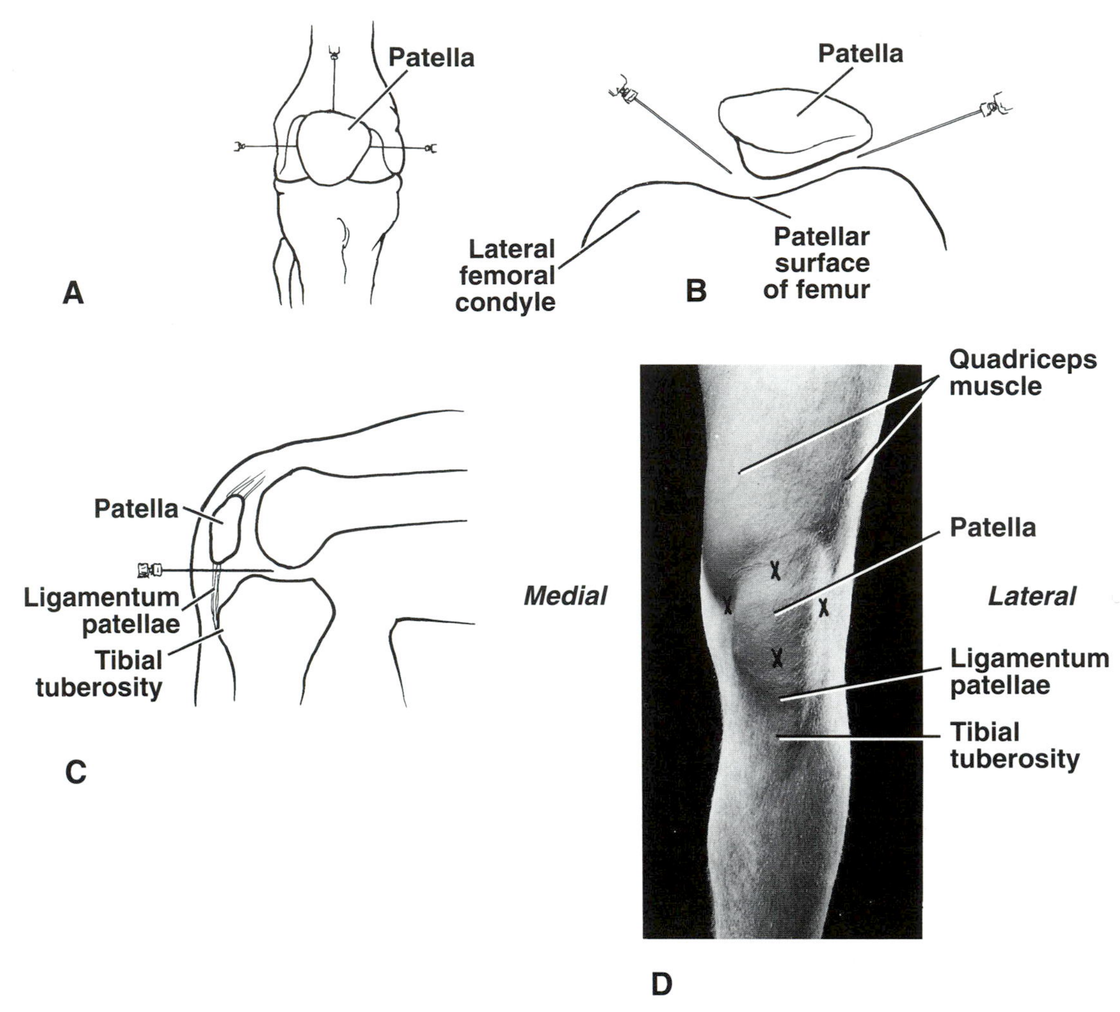

FIG 18–16.
Knee arthrocentesis. **A** and **D,** suprapatellar approach. The needle is inserted at the midpoint of the upper border of the patella and is directed between the posterior surface of the patella and the intercondylar notch of the femur. With the medial and lateral parapatellar approach, the needle is inserted at the medial or lateral borders of the patella and is directed between the patella and the anterior surface of the patellar surface of the femur. **B,** the wedge shape of the patella as seen on section; this determines the angle of the needle as it is advanced into the joint. **C** and **D,** infrapatellar approach. The needle is inserted into the skin over the ligamentum patellae just below the patella at the joint line and is directed posteriorly at right angles to the shaft of the tibia. *X* indicates the needle insertion site.

volume of fluid present in the suprapatellar bursa, it is possible to aspirate fluid from the bursa by inserting the needle directly posteriorly through the quadriceps tendon.

Medial and Lateral Parapatellar Approach.—With the patient in a supine position, the knee joint is fully extended. The needle is inserted at the medial or lateral border of the patella and is directed laterally or medially, respectively, between the patella and the patella surface of the lower end of the femur (see Fig 18–16). In patients in whom there is little fluid present in the joint, the suprapatellar bursa may be compressed by an assistant or by the use of an elastic bandage applied around the thigh just proximal to the patella. This will empty the bursa of fluid into the main joint cavity.

The needle pierces the following structures before entering the joint cavity.

1. Skin.
2. Superficial and deep layers of fascia.
3. Fibrous extensions (patellar retinacula) from the vastus medialis or the vastus lateralis fused with the joint capsule.
4. Synovial membrane lining the capsule.

There are no large vessels or nerves in this area of the joint.

Infrapatellar Approach.—The knee joint is flexed to 90° by having the patient hang his legs over the end of the table. The patella can easily be palpated, and the ligamentum patellae can be felt as it passes from the inferior margin of the patella downward to be attached to the tuberosity of the tibia (see Fig 18–16). On the sides of the knee, the condyles of the femur and tibia can be felt and the joint line can be identified. The needle is inserted into the skin over the middle of the ligamentum patellae, just below the inferior margin of the patella at the level of the joint line. The needle is directed posteriorly at right angles to the shaft of the tibia (see Fig 18–16). The needle enters the joint cavity between the condyles of the femur. The weight of the lower leg separates the tibia from the femur and opens up the joint cavity. Here, again, simultaneous compression of the suprapatellar bursa empties the joint fluid into the main joint cavity.

The needle pierces the following structures before entering the joint cavity.

1. Skin.
2. Superficial and deep layers of fascia.
3. Tough fibrous ligamentum patellae.
4. Infrapatellar pad of fat.
5. Synovial membrane.

There are no large blood vessels or nerves in this area of the joint. Moreover, since the tip of the needle enters the intercondylar area of the femur, there is less chance of damaging the articular cartilage covering the femoral condyles.

Anatomy of Ankle Arthrocentesis

Two approaches are commonly used—anteromedial and anterolateral.

Anteromedial Approach.—The patient is placed in a supine position. The medial malleolus of the tibia is palpated (Fig 18–17). The ankle joint is then plantar flexed (which opens up the joint space), and the subtalar joint is inverted. This latter maneuver makes the tibialis anterior tendon stand out on the dorsum of the foot. The needle is then inserted into the groove between the tendon of tibialis anterior and the medial malleolus. Alternatively, the tendon of the extensor hallucis longus can be used as a landmark in place of the tibialis anterior. The extensor tendon is made to stand out by asking the patient to extend the big toe. This time the needle is inserted medial to the extensor tendon and in line with the medial malleolus (see Fig 18–17).

The needle is directed posteriorly at right angles to the table top, and the tip of the needle should enter the joint cavity just distal to the lower edge of the tibia.

The dorsalis pedis vessels (see Fig 18–17) and the deep peroneal nerve are located on the lateral side of the tendon of extensor hallucis longus.

Anterolateral Approach.—The patient is placed in a supine position and the ankle joint is plantar flexed. The lateral malleolus of the fibula is then palpated (see Fig 18–17). The patient then extends the toes and makes the tendons of extensor digitorum longus stand out on the dorsum of the foot. The needle is inserted midway between the lateral malleolus and the lateral margin of the tendons of the extensor digitorum longus (see Fig 18–17). Here, again, the needle point enters the joint cavity just distal to the lower edge of the tibia.

In both the anteromedial approach and the anterolateral approach, the needle pierces the following structures before entering the joint cavity.

1. Skin.
2. Superficial fascia.
3. Deep fascia (superior band of extensor retinaculum).
4. Joint capsule lined by synovial membrane.

The dorsalis pedis vessels (see Fig 18–17) and the deep peroneal nerve are located on the lateral side of the extensor hallucis longus tendon and on the medial side of the tendons of the extensor digitorum longus.

Anatomy of Subtalar Arthrocentesis

This is a difficult procedure, since the joint space is small. The patient is placed in a supine position with the foot held at right angles to the leg. The lateral malleolus is palpated, and the needle is inserted just below the tip. The needle direction is perpendicular to the skin and enters the joint space between the talus and the calcaneum (see Fig 18–17).

The needle pierces the following structures before entering the joint cavity.

1. Skin.
2. Superficial and deep layers of fascia.
3. Lateral talocalcaneal ligament.
4. Joint capsule and lining of synovial membrane.

There are no large vessels or nerves present in the area.

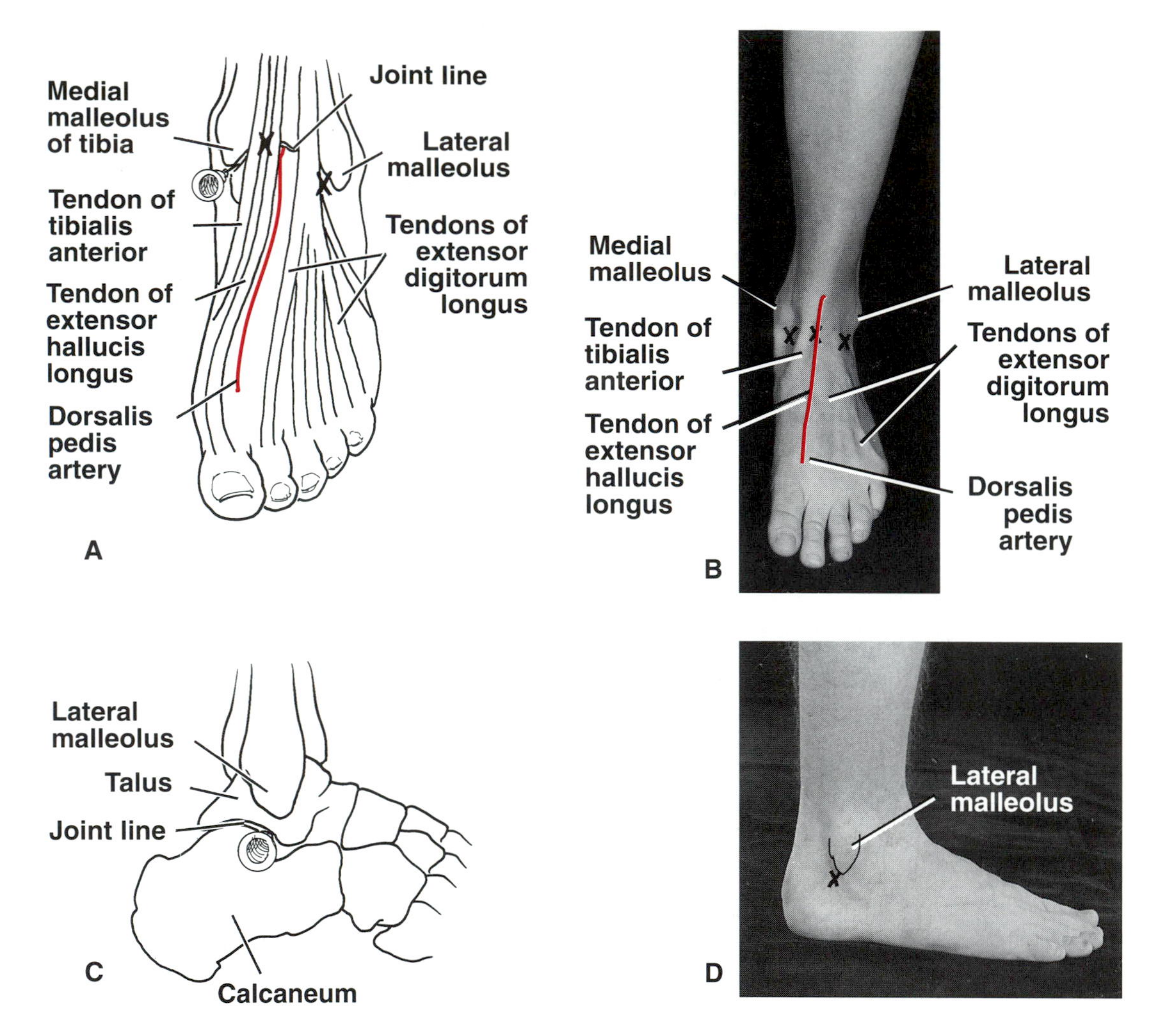

FIG 18–17.
Ankle arthrocentesis. **A** and **B,** anteromedial approach. The needle is inserted into the groove between the tendon of tibialis anterior and the medial malleolus or medial to the tendon of extensor hallucis longus. The needle is directed posteriorly just distal to the lower edge of the tibia. **A** and **B,** anterolateral approach. The needle is inserted midway between the lateral malleolus and the tendons of the extensor digitorum longus. The needle is directed posteriorly just distal to the lower edge of the tibia. **C** and **D,** subtalar arthrocentesis. The needle is inserted just below the tip of the lateral malleolus and is directed perpendicular to the skin to enter the space between the talus and the calcaneum.

Anatomy of Metatarsophalangeal Arthrocentesis

The joint spaces are small but may be enlarged to some extent by applying distal traction to the toe (Fig 18–18). The metatarsal head and the base of the proximal phalanx are easily palpated to ascertain the position of the joint line. The needle is inserted perpendicular to the toe on its dorsal aspect at the joint line. In the case of the big toe, the point of the needle is inserted just medial or lateral to the tendon of extensor hallucis longus (see Fig 18–18). Arthrocentesis of the other toes is carried out in a similar manner, and the needle is inserted on the medial or lateral side of the tendon of extensor digitorum longus.

Anatomy of Interphalangeal Arthrocentesis

Here, again, the joint spaces are very small and may be enlarged by applying distal traction to the toe. The joint line is located by identifying the head and base of the phalanges forming the joint. The needle is inserted perpendicular to the toe on its dorsal aspect at the joint line. The needle is made to penetrate the dorsal extensor expansion from the medial or lateral side to enter the joint cavity (see Fig 18–18).

The needle pierces the following structures before entering the joint cavity when performing a metatarsophalangeal or interphalangeal arthrocentesis.

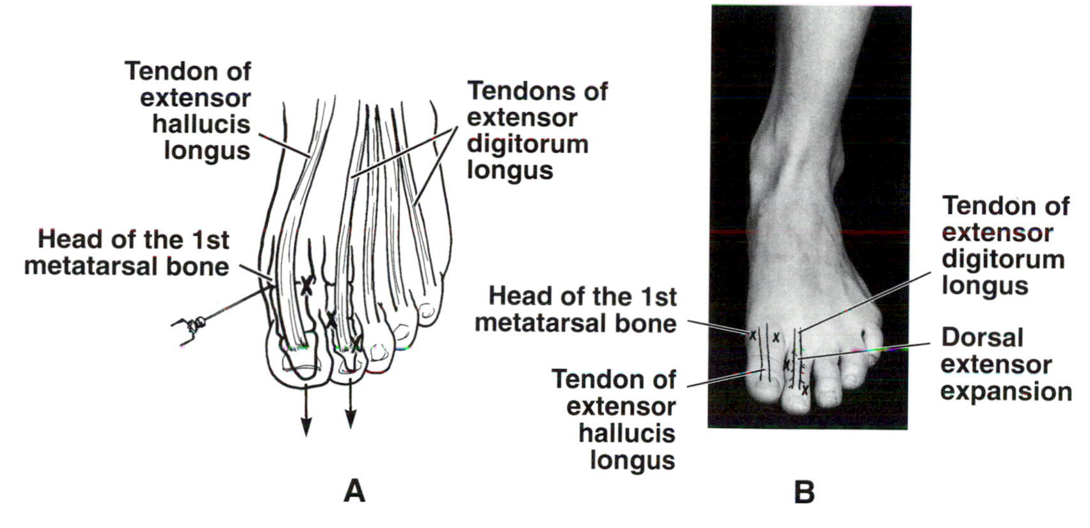

FIG 18–18.
Metarsophalangeal arthrocentesis. **A** and **B,** in the big toe the needle is inserted on the dorsal aspect on the medial or lateral side of the tendon of extensor hallucis longus at the joint line; it is directed perpendicular to the skin to enter the joint cavity. In the other toes the needle is inserted on the dorsal aspect on the medial or lateral aspect of the tendon of the extensor digitorum longus at the joint line. In interphalangeal arthrocentesis, the needle is inserted on the dorsal aspect of the toe and penetrates the dorsal extensor expansion on the medial or lateral side; the needle is directed perpendicular to the skin to enter the joint cavity *(X).*

1. Skin.
2. Superficial and deep layers of fascia.
3. Fibrous dorsal extensor expansion associated with the tendons of extensor digitorum longus (extensor hallucis longus in the case of the big toe).
4. Joint capsule and lining of synovial membrane.

The small dorsal digital vessels and nerve may be damaged with these procedures. However, at the interphalangeal joints these structures may be avoided if the needle is inserted from the medial or lateral side of the toe.

Clinical Anatomy of Bursae

Bursitis

Bursitis, or inflammation of a bursa, can be caused by acute or chronic trauma, crystal disease, infection, or disease of a neighboring joint that communicates with the bursa. Bursitis most commonly occurs in the subacromial bursa, the olecranon bursa, the greater trochanter bursa, the prepatellar bursa, and the anserine bursa on the medial aspect of the proximal tibia.

Clinical Anatomy for the Aspiration or Injection of Bursae

Anatomy for Aspiration of the Subacromial (Subdeltoid) Bursa

The patient is placed in a sitting position with the arm at the side and the hand across the lap. The spine of the scapula is identified and followed upward and laterally to the acromion process. Downward traction is applied to the flexed elbow to separate as much as possible the greater tuberosity of the humerus from the acromion process and the coracoacromial ligament. The lateral edge of the acromion process is palpated. The needle is inserted horizontally (Fig 18–19) from the lateral side just below the lateral edge of the acromion process (this point usually corresponds to the point of maximum tenderness). The needle then enters the distended bursa.

The needle pierces the following structures before entering the cavity of the bursa.

1. Skin.
2. Superficial and deep layers of fascia.
3. Deltoid muscle.
4. Fibrous wall of bursa.

There are no important blood vessels or nerves in this area.

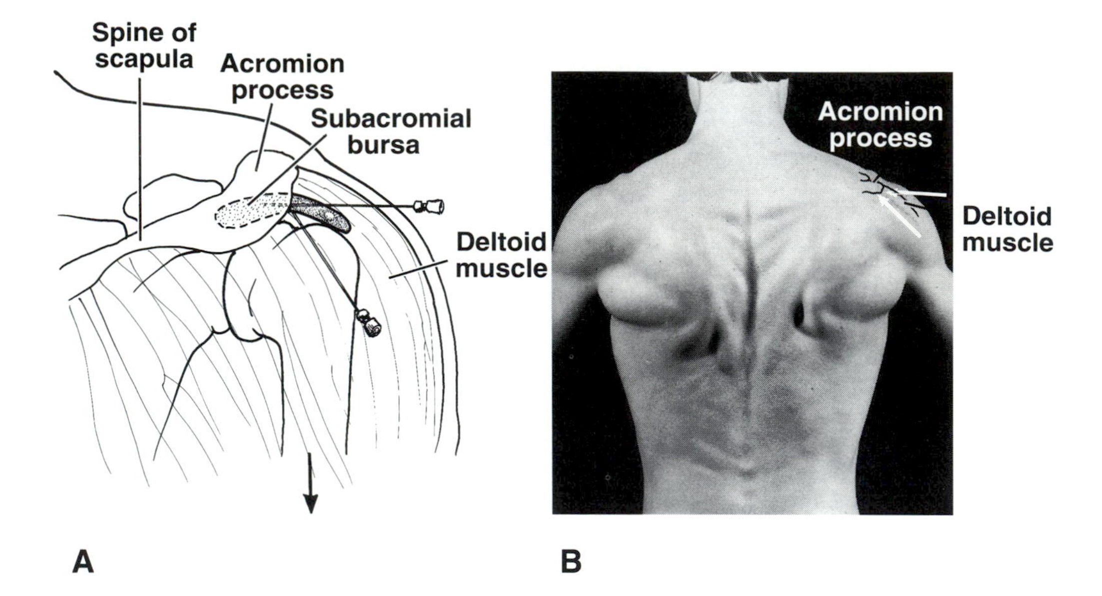

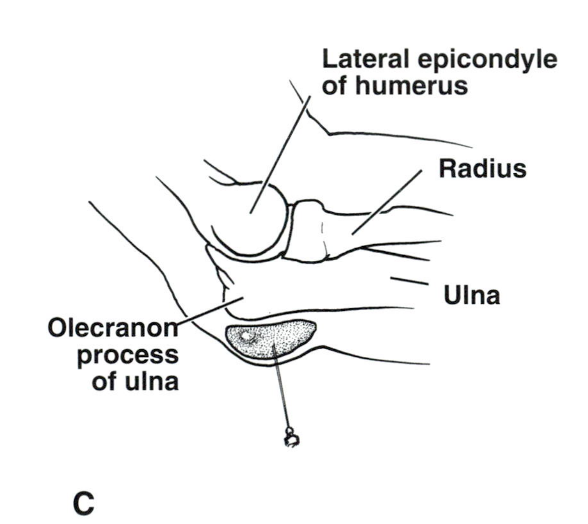

FIG 18–19.
A and **B**, aspiration of subacromial (subdeltoid) bursa. With the lateral approach, the needle is inserted horizontally from the lateral side and is directed just below the lateral edge of the acromion process. With the posterior approach, the needle may be inserted from behind just below the posterior border of the acromion process. **C**, aspiration of olecranon bursa. The needle is inserted at right angles to the skin over the olecranon process of the ulna and directed into the cavity of the bursa.

Anatomy for Aspiration of the Olecranon Bursa

Because of its superficial position over the olecranon process of the ulna, the full extent of the bursa can be seen when it is tensely swollen with an effusion. The patient is placed in a sitting position with his arm at the side, the elbow joint flexed, and the hand placed in the lap. The needle is inserted at right angles to the skin and directed through the subcutaneous tissue to enter the bursa (see Fig 18–19).

Anatomy for Aspiration of the Greater Trochanter Bursa

The greater trochanter of the femur can be palpated on the lateral surface of the thigh and can be felt to move beneath the examining fingers as the hip joint is flexed and extended. The patient is placed in a supine position with the hip joint extended and medially rotated. The needle is inserted at right angles to the skin over the greater trochanter at the point of maximum tenderness (Fig 18–20).

The needle pierces the following structures before entering the cavity of the bursa.

1. Skin.
2. Superficial fascia.
3. Deep fascia (fascia lata).
4. Gluteus maximus muscle.
5. Outer wall of bursa.

Anatomy for Aspiration of the Prepatellar Bursa

The patient is placed in a supine position with the hip and knee joints extended. The patella and the

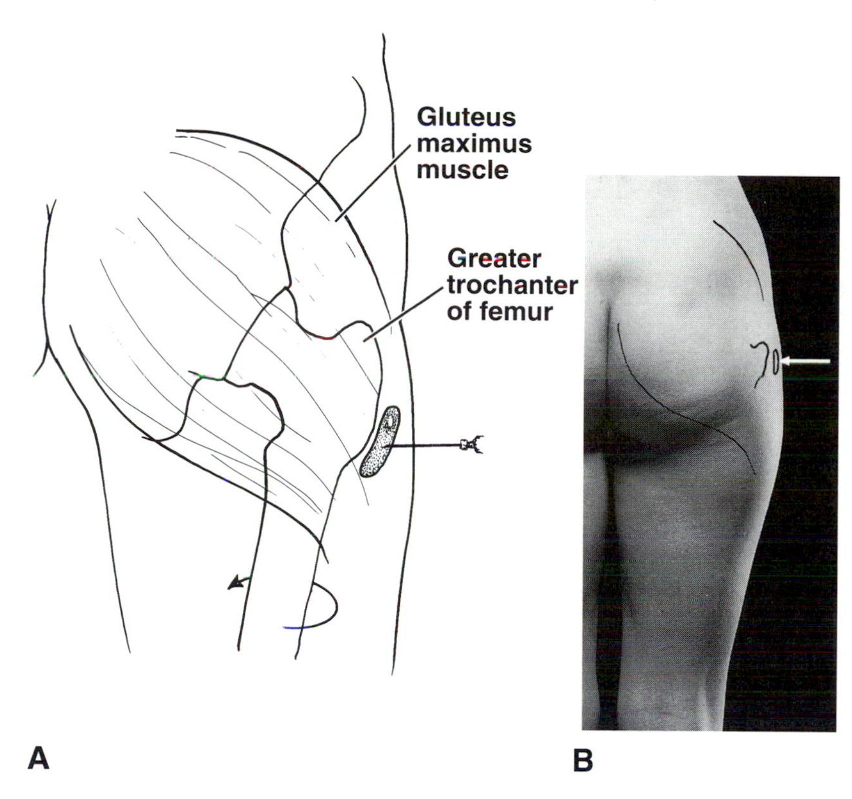

FIG 18–20.
Aspiration of the greater trochanter bursa. The needle is inserted at right angles to the skin over the greater trochanter.

patellar ligament are easily palpated. The needle is inserted perpendicular to the skin over the area of maximum tenderness (see Fig 18–21).

Aspiration of the Pes Anserine Bursa

The patient is placed in a supine position with the hip and knee joints extended. The tuberosity of the tibia and the medial tibial condyle can be easily palpated; the medial surface of the shaft of the tibia is subcutaneous throughout its length and can be identified. The bursa is located at the point of insertion of the sartorius, gracilis, and semitendinosus muscles on the upper part of the medial surface of the tibial shaft. The needle is inserted perpendicular to the skin over the point of maximum tenderness (see Fig 18–21).

The needle pierces the following structures before entering the cavity of the bursa.

1. Skin.
2. Superficial and deep layers of fascia.
3. Wall of bursa.

Anatomy for Aspiration of the Tendo Calcaneus Bursa

The patient is placed in a prone position, and the tendo calcaneus and the calcaneum are palpated. The bursa is situated just anterior to the tendon just above its insertion into the calcaneum. The needle is inserted into the skin on the lateral side of the tendon and directed toward the point of maximum tenderness, avoiding the tendon (see Fig 18–21).

ANATOMICAL-CLINICAL PROBLEMS

1. When arthrocentesis of the temporomandibular joint is performed, what are the important structures that lie in front of the auricle of the ear that must be avoided?

2. Either an anterior or a posterior approach may be used when performing an arthrocentesis of the shoulder joint. Using your knowledge of anatomy, explain why the posterior approach is easier and safer.

3. What is the difference between the subacromial and the subdeltoid bursae? Name the structures that are pierced by the needle when injecting steroids into the subacromial bursa.

4. Name the bony landmarks that are used when aspirating the elbow joint. What is the danger of entering the elbow joint on the medial side?

5. Describe the anatomy of the olecranon bursa. Does this bursa communicate with the elbow joint?

6. Describe the two approaches that one might use in performing an arthrocentesis of the hip joint.

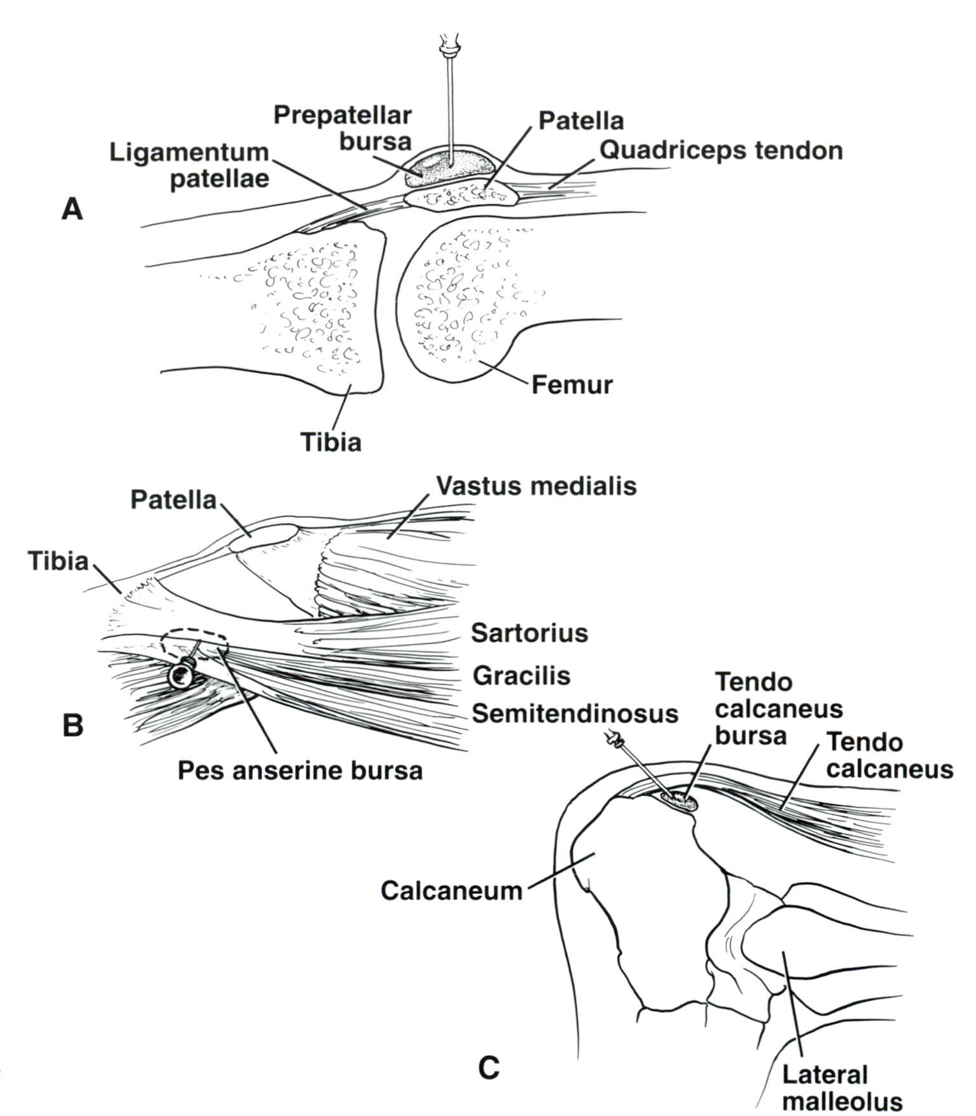

FIG 18–21.
A, aspiration of the prepatellar bursa. The needle is inserted perpendicular to the skin over the area of maximum tenderness. **B,** aspiration of the pes aserine bursa. The needle is inserted perpendicular to the skin over the point of maximum tenderness at the place of insertion of the sartorius, gracilis, and semitendinosus muscles. **C,** aspiration of the tendo calcaneus bursa. The needle is inserted into the skin on the lateral side of the tendon and is directed toward the point of maximum tenderness.

In each case name the structures that the needle would pierce before entering the joint cavity. Does the cavity of the hip joint communicate with any bursae?

7. Describe the site of entrance of the needle when performing an arthrocentesis on the metacarpophalangeal joint of the thumb.

8. The wrist joint is related to many important tendons, blood vessels, and nerves. Name the site where a needle may be passed into the joint cavity without causing damage to these structures. Name the landmarks used in the procedure.

9. A patient is seen with a swollen knee. What maneuver would you use to make sure that as much synovial fluid as possible is present in the main joint cavity when performing an arthrocentesis?

10. Name the bony landmarks that are useful when aspirating the ankle joint.

11. What is the pes anserine bursa? What is its function? Describe the type of activity that is commonly the cause for inflammation of this bursa.

12. When arthrocentesis of the metatarsophalangeal joint of the big toe is performed, is the needle directed into the joint from the dorsal or plantar aspect? Which tendon is used as a surface landmark for the site of the needle entrance?

ANSWERS

1. The superficial temporal artery and the auriculotemporal nerve lie between the temporomandibular joint and the tragus of the auricle. The pulsations of the artery can easily be felt and the nerve lies immediately posterior to the artery.

2. The posterior approach for arthrocentesis of

the shoulder joint has the following advantages over the anterior approach: (1) The only muscles that have to be pierced by the needle are the deltoid and the infraspinatus, whereas the pectoralis major, the coracobrachialis and the biceps, and the subscapularis muscles have to be pierced with the anterior approach. (2) The posterior wall of the capsule is thin, whereas the anterior wall is thickened by the glenohumeral ligaments. (3) The brachial plexus and the axillary vessels are of no concern with the posterior approach since they lie in front of the joint.

3. The subacromial and subdeltoid bursae are all part of the same bursa. The subacromial part of the bursa lies beneath the acromion process and the coracoacromial ligament, whereas the subdeltoid bursa is a lateral extension of the subacromial bursa and lies beneath the deltoid muscle.

The following structures are pierced by the needle when it is passed into the subacromial bursa: skin, superficial and deep layers of fascia, and the deltoid muscle.

4. The bony landmarks used when entering the elbow joint from the lateral side are the lateral epicondyle of the humerus, the head of the radius, and the tip of the olecranon process of the ulna. The needle is directed medially through the center of the triangle formed by the union of these bony points. The medial side of the elbow joint should be avoided because of the presence of the ulnar nerve.

5. The olecranon bursa lies just beneath the skin in the superficial fascia on the posterior surface of the olecranon process of the ulna; it does not communicate with the elbow joint.

6. The two approaches that might be used in performing an arthrocentesis of the hip joint—anterior and lateral—are fully described on p. 774. With the anterior approach, the needle pierces the following structures before entering the joint: skin, superficial fascia, deep fascia (fascia lata), iliopsoas, ? psoas bursa, joint capsule, and lining of synovial membrane. With the lateral approach, the needle pierces the following structures before entering the joint: skin, superficial and deep layers of fascia, gluteus medius, iliofemoral or ischiofemoral ligaments, joint capsule, and lining of synovial membrane. The cavity of the hip joint may communicate with the bursa beneath the psoas muscle.

7. The dorsal approach to the metacarpophalangeal joint of the thumb is used because the palmar surface is crowded with structures that include the tendon of the flexor pollicis longus muscle with its synovial sheath, the palmar digital vessels and nerves, and the strong fibrocartilaginous palmar ligament. The site for the insertion of the needle is the joint line, which can be identified by palpating the head of the first metacarpal and the base of the proximal phalanx. The needle is directed into the joint on the medial or lateral side of the tendons of the extensor pollicis longus and extensor pollicis brevis muscles.

8. The posterior approach to the wrist joint is used since there are few large blood vessels and nerves on this aspect of the joint. The needle is inserted distal to the dorsal tubercle of the radius and ulnar to the tendon of the extensor pollicis longus on the joint line. The needle is directed anteriorly between the tendons of extensor pollicis longus and the extensor digitorum tendons. If the tip of the needle is kept close to the extensor digitorum tendons, it will then pass forward between the extensor digitorum tendons and the deeper placed tendon of extensor carpi radialis brevis. The needle will enter the joint cavity between the distal end of the radius and the lunate bone.

The following landmarks are used: (1) the dorsal tubercle of the radius, (2) the styloid processes of the radius and ulna to determine the level of the joint line, and (3) the tendons of the extensor pollicis longus and extensor digitorum longus.

9. The suprapatellar bursa, which communicates with the synovial cavity of the knee joint, is compressed. This may be accomplished by pressing upon the quadriceps muscle above the patellar by hand or using an elastic bandage. The suprapatellar bursa extends about three fingerbreadths above the superior margin of the patella.

10. The medial malleolus of the tibia for the anteromedial approach and the lateral malleolus of the fibula for the anterolateral approach.

11. The pes anserine bursa is located on the upper part of the medial surface of the shaft of the tibia. The bursa lies between the tendons of insertion of the sartorius, gracilis, and the semitendinosus muscles. The name is derived from the fact that the arrangement of the tendons at this site resembles a goose's foot. The function of the bursa is to reduce friction between the tendons when the muscles are active. Inflammation of this bursa commonly occurs in runners and cyclists.

12. The dorsal aspect of the metatarsophalangeal joint of the big toe is the surface used to introduce a needle into the joint cavity. The plantar surface is related to the flexor hallucis longus tendon and its synovial sheath and the strong plantar ligament. The needle is inserted on the medial side of the tendon of extensor hallucis longus at the joint line.

19

Peripheral Nerves and Peripheral Nerve Blocks

Local anesthesia is extensively used in the emergency department for the repair of lacerations, the drainage of abscesses, and the reductions of fractures and fracture-dislocations. Three procedures are commonly utilized—local infiltration, field block anesthesia, and regional nerve blocks.

Local infiltration anesthesia is the production of analgesia by the infiltration of the wound with the anesthetic agent. This method is very commonly used but it may distort the wound margins; it is also less effective in producing analgesia than nerve blocks.

Field block anesthesia is the production of regional analgesia by the injection of the anesthetic agent around the wounded area. This method blocks the terminal branches of the local sensory nerves. Unfortunately the procedure often involves the use of multiple injection sites and the administration of large volumes of anesthetic.

Regional nerve blocks involve the production of regional analgesia by the injection of the anesthetic agent around the main sensory nerve or nerves supplying the region. This form of anesthetic procedure has the following advantages. (1) The anesthetic agent is usually injected at some distance from the wound site and therefore does not cause distortion of the wound edges and does not cause local tissue anoxia by producing vasoconstriction. (2) Since analgesia can be produced over a large area, this method is preferred to giving a general anesthetic, especially in patients with cardiac or respiratory problems. (3) It is particularly helpful in the elderly. (4) It avoids the use of multiple injection sites and the need for using large volumes of the anesthetic agent.

The purpose of this chapter is to provide an overview of the anatomy of the peripheral nerves and their branches that are commonly blocked in an emergency center. Special attention is paid to the important anatomical landmarks used when performing a nerve block. The advantages and disadvantages of the different techniques used and the many anesthetic agents available are adequately covered in standard textbooks of emergency medicine and have been purposely omitted here.

BASIC ANATOMY

Basic Anatomy of the Trigeminal Nerve

The trigeminal nerve is the largest cranial nerve. It has three major branches—*ophthalmic (V1), maxillary (V2),* and *mandibular (V3).* It is a mixed cranial nerve and contains both sensory and motor fibers; the sensory fibers are distributed mainly to the skin of the face (Fig 19–1), and the motor fibers innervate the muscles of mastication.

The trigeminal nerve leaves the pons of the brain as a small motor root and a large sensory root. The nerve passes forward out of the posterior cranial fossa. It then rests on the apex of the petrous part of the temporal bone in the middle cranial fossa (Fig 19–2). The large sensory root now expands to form the crescent-shaped *trigeminal ganglion* that lies within a pouch of the dura mater, called the *trigeminal or Meckel's cave.* The ophthalmic, maxillary, and mandibular nerves arise from the convex anterior border of the ganglion.

The trigeminal ganglion has the following relations. Medially it is related to the posterior part of the cavernous sinus and the internal carotid artery. Inferiorly lies the motor root of the trigeminal nerve and the greater petrosal nerve, the apex of the petrous part of the temporal bone, and the cartilage filling in the foramen lacerum. Superiorly lies the temporal lobe of the brain.

Ophthalmic Nerve (V1)

The ophthalmic nerve is the smallest division of the trigeminal nerve and is entirely sensory. It runs forward in the lateral wall of the cavernous sinus in the middle cranial fossa and divides into three branches—the lacrimal, the frontal, and the nasociliary nerves, which enter the orbital cavity through the superior orbital fissure (see Figs 19–1 and 8–15).

Lacrimal Nerve.—The lacrimal nerve runs forward along the lateral wall of the orbit and enters the lacrimal gland. It is distributed to the conjunctiva and the skin of the upper eyelid.

Frontal Nerve.—The frontal nerve runs forward beneath the roof of the orbit and divides into the *supraorbital* and *supratrochlear nerves* (see Fig 19–1). The larger supraorbital nerve emerges from the orbital cavity by passing through the supraorbital notch (or foramen) and supplies the skin of the forehead and the scalp as far as the vertex (Fig 19–3); it also supplies the frontal air sinus. The smaller supratrochlear nerve passes above the pulley for the superior oblique muscle and winds around the upper margin of the orbital cavity to supply the skin of the forehead medial to the area supplied by the supraorbital nerve (see Fig 19–3).

Nasociliary Nerve.—The nasociliary nerve crosses the optic nerve and runs forward on the medial wall

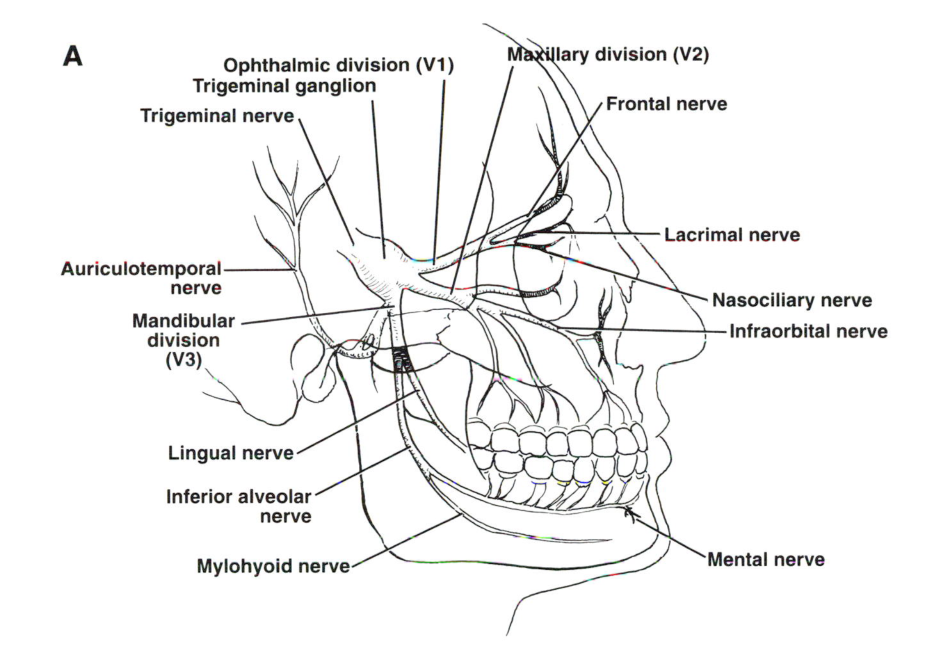

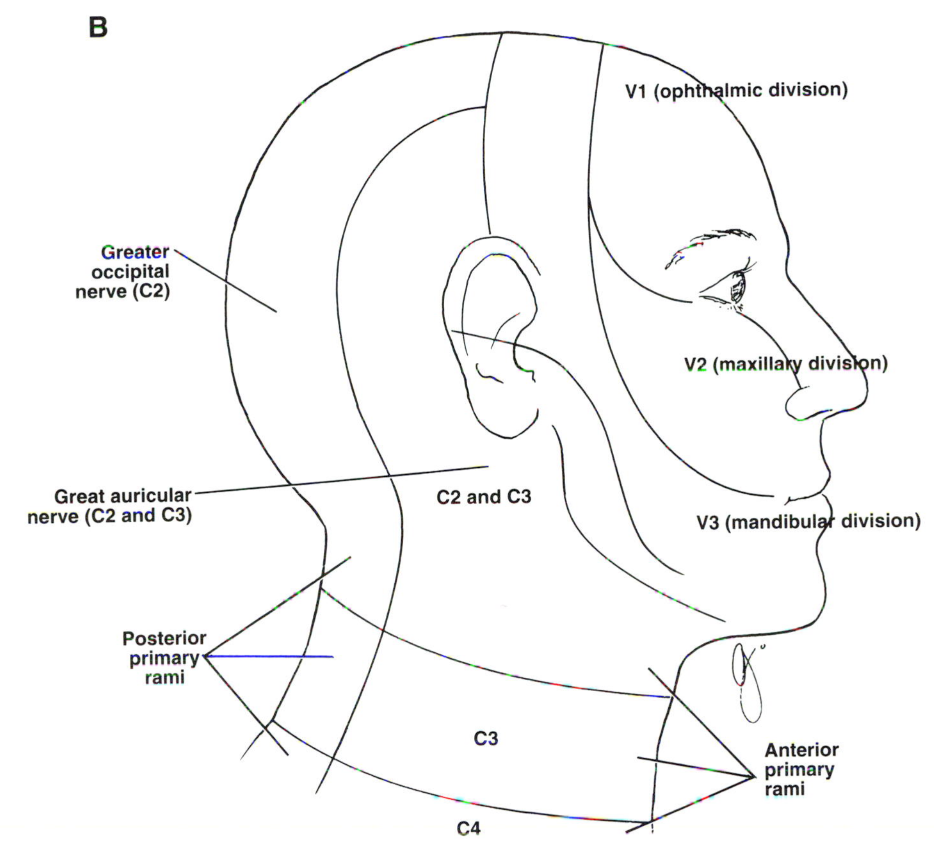

FIG 19–1.
A, the distribution of the trigeminal nerve. **B,** the facial cutaneous distribution of the ophthalmic (V1), the maxillary (V2), and the mandibular (V3) divisions of the trigeminal nerve. The skin over the angle of the jaw is supplied by the great auricular nerve (C2 and C3 segments of the spinal cord).

of the orbit (see Fig 19–1). It continues as the *anterior ethmoidal nerve* through the anterior ethmoidal foramen to enter the cranial cavity. It then descends through a slit in the crista galli to enter the nasal cavity. It gives off two *internal nasal branches* and then continues as the *external nasal nerve.* The external nasal nerve appears on the face between the nasal bone and the upper lateral nasal cartilage (see Fig 19–3). It supplies the skin down as far as the tip of the nose.

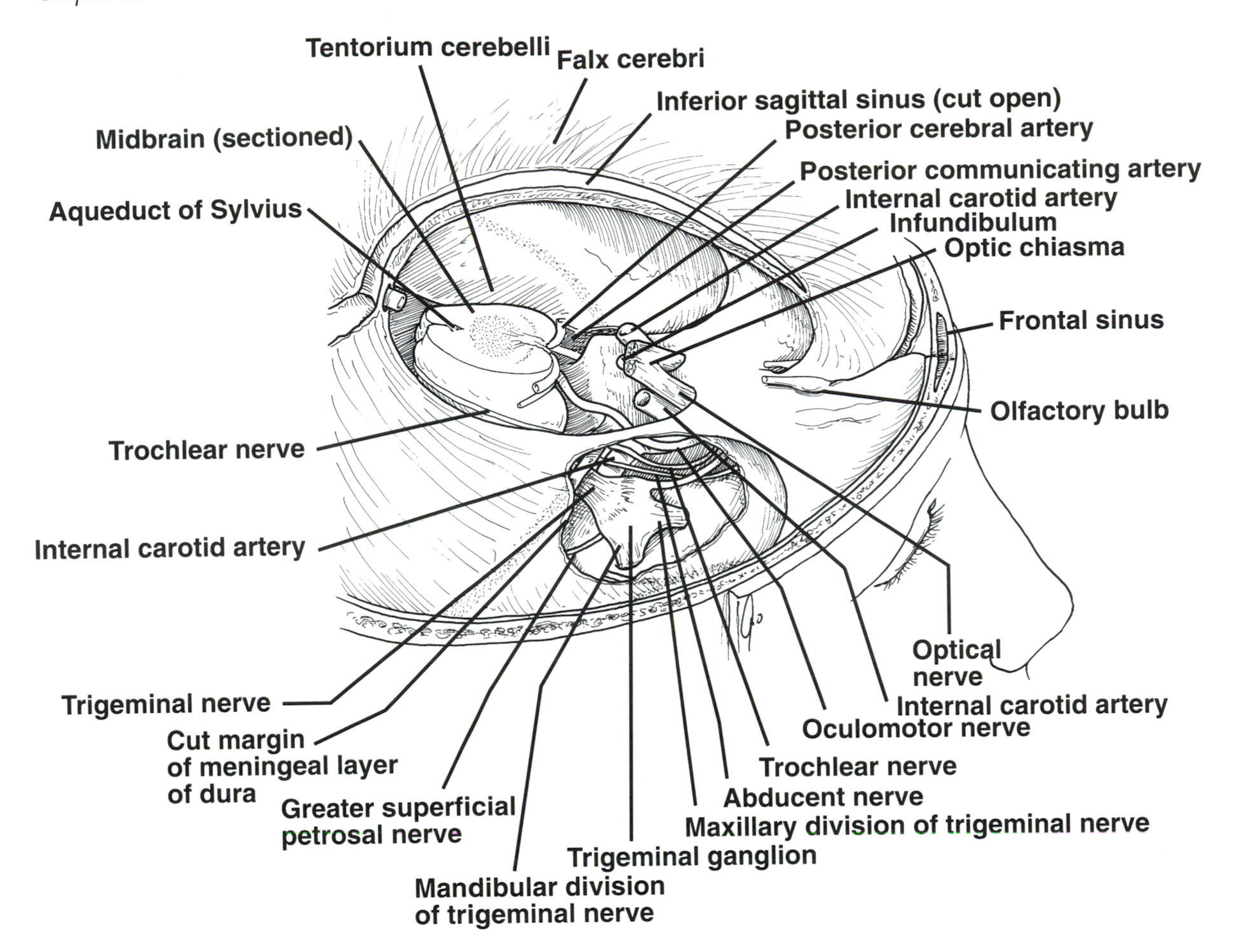

FIG 19–2.
Lateral view of the interior of the skull showing the trigeminal ganglion and the ophthalmic, maxillary, and mandibular divisions of the trigeminal nerve. Note that the dura on the floor of the middle cranial fossa has been cut away to reveal these structures; the lateral wall of the cavernous sinus has also been dissected to reveal the third, the fourth, the opthalmic division of the fifth, and the sixth cranial nerves and the internal carotid artery.

Branches of the Nasociliary Nerve.—These include the following.

1. *Sensory fibers* to the ciliary ganglion.
2. *Long ciliary nerves* that contain sympathetic nerves to the dilator pupillae muscle and sensory fibers to the cornea.
3. *Infratrochlear nerve* supplies the skin of the medial part of the upper eyelid and the adjacent part of the nose.
4. *Posterior ethmoidal nerve* to the ethmoidal and sphenoid sinuses.

Maxillary Nerve (V2)

The maxillary nerve is purely sensory. It passes forward in the lateral wall of the cavernous sinus and leaves the middle cranial fossa through the foramen rotundum. It then crosses the pterygopalatine fossa

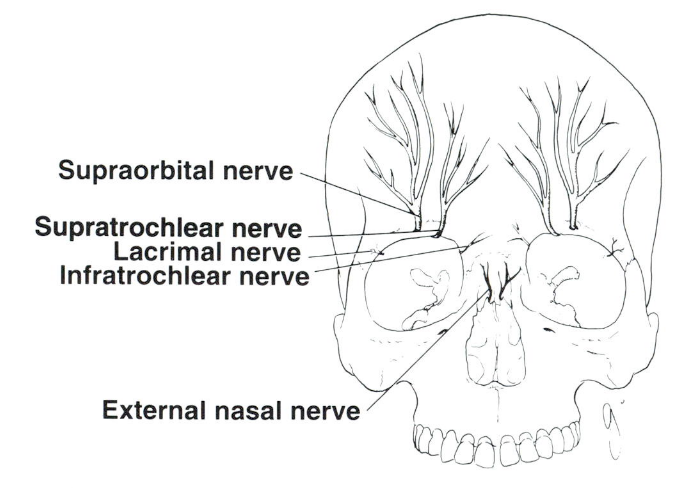

FIG 19–3.
Anterior view of the skull showing the branches of the ophthalmic (V1) division of the trigeminal nerve emerging onto the face.

to enter the orbit through the inferior orbital fissure (see Fig 19–1). It continues as the *infraorbital nerve* in the infraorbital groove and emerges onto the face through the infraorbital foramen (Fig 19–4). It gives sensory fibers to the skin of the face, side of the nose, and the upper gum (see Fig 19–22).

Branches of the Maxillary Nerve.—These include the following.

1. *Meningeal branches.*
2. *Zygomatic branch.* This divides into the *zygomaticofacial* and *zygomaticotemporal branches.* The zygomaticofacial nerve passes onto the face through a small foramen on the lateral side of the zygomatic bone. It supplies the skin over the prominence of the cheek. The zygomaticotemporal nerve emerges in the temporal fossa through a small foramen on the posterior surface of the zygomatic bone and supplies the skin over the temple. The zygomaticotemporal branch gives parasympathetic secretomotor fibers to the lacrimal nerve for the lacrimal gland.
3. *Ganglionic branches.* Two short nerves suspend the pterygopalatine ganglion in the pterygopalatine fossa (see Fig 19–4). They contain sensory fibers that have passed through the ganglion from the nose, the palate, and the pharynx. They also contain postganglionic parasympathetic fibers that are going to the lacrimal gland.
4. *Posterior superior alveolar nerve* innervates the maxillary sinus and the upper molar teeth and adjoining parts of the gum and the cheek (see Fig 19–4).
5. *Middle superior alveolar nerve* goes to the maxillary sinus and the upper premolar teeth and the gums and the cheek (see Fig 19–4).
6. *Anterior superior alveolar nerve* provides sensation to the maxillary sinus and upper canine and incisor teeth (see Fig 19–4).

Pterygopalatine Ganglion.—The pterygopalatine ganglion is a parasympathetic ganglion that is suspended from the maxillary nerve in the pterygopalatine fossa (see Fig 19–4). It is secretomotor to the lacrimal and nasal glands (see p. 331).

Branches of the Pterygopalatine Ganglion.—These include the following.

1. *Orbital branches.* These enter the orbit through the inferior orbital fissure and innervate the periosteum of the orbital cavity.
2. *Greater and lesser palatine nerves.* These go to the mucous membrane of the palate, the tonsil, and the nasal cavity (see Fig 19–23).
3. *Nasal branches.* These enter the nose through the sphenopalatine foramen. They make up the *short medial* and *lateral posterior superior nasal nerves* and the *long nasopalatine nerve.* The latter nerve runs down

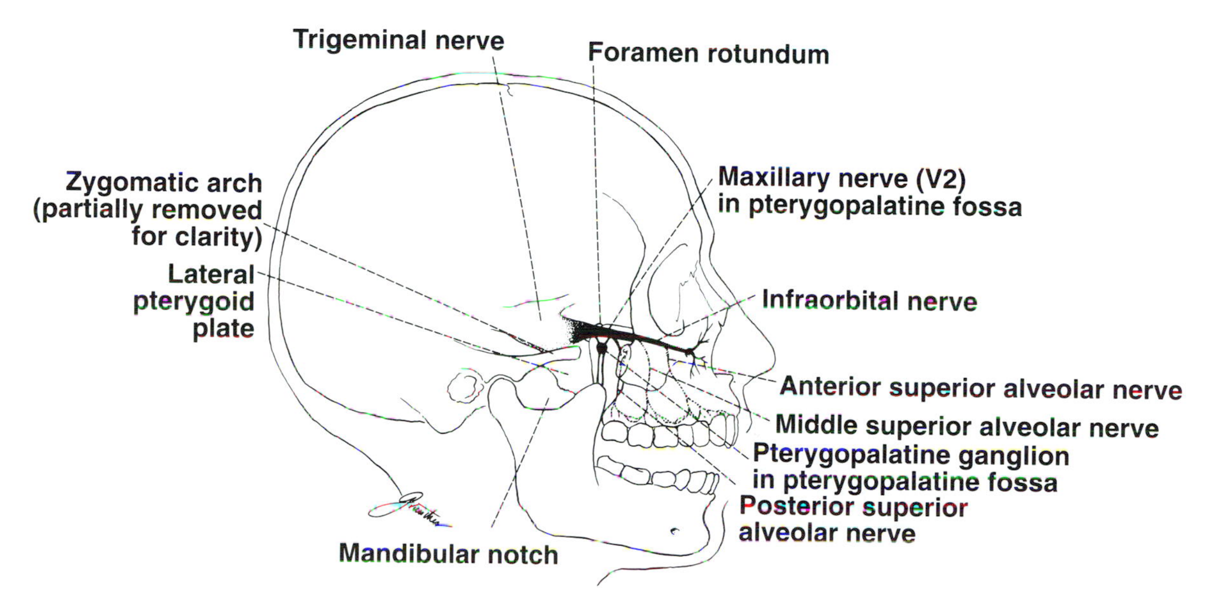

FIG 19–4.
Lateral view of the skull showing the maxillary (V2) division of the trigeminal nerve leaving the trigeminal ganglion and passing forward to become the infraorbital nerve, which emerges on the face. Note the location of the pterygopalatine (sphenopalatine) ganglion (parasympathetic) in the pterygopalatine fossa. Note also the sensory innervation of the teeth of the upper jaw.

on the nasal septum and supplies the hard palate by passing through the incisive foramen.

4. *Pharyngeal branch.* This innervates the roof of the nasopharynx.

Mandibular Nerve (V3)

The mandibular nerve is the largest division of the trigeminal nerve and is motor and sensory. The sensory root leaves the trigeminal ganglion and passes out of the skull through the foramen ovale (see Fig 19–1 and 19–2). The motor root of the trigeminal nerve also leaves the skull through the foramen ovale and almost immediately joins the sensory root to form the trunk of the mandibular nerve. The nerve trunk then divides into a small anterior and a large posterior division.

Branches from the Main Trunk of the Mandibular Nerve.—These include the following.

1. *Meningeal branch.*
2. *Nerve to the medial pterygoid muscle.* This also innervates the tensor tympani and the tensor veli palatini muscles.

Branches from the Anterior Division of the Mandibular Nerve.—These include the following.

1. *Motor nerves.* These go to the muscles of mastication, including the masseter, the temporalis, and the lateral pterygoid muscles.
2. *Sensory nerve.* The *buccal nerve* passes to the skin of the cheek and the mucous membrane lining the cheek.

Branches from the Posterior Division of the Mandibular Nerve.—These include the following.

1. *Auriculotemporal nerve.* This ascends behind the tempomandibular joint on the medial surface of the parotid gland (see Fig 19–1). It then crosses the root of the zygomatic arch posterior to the superficial temporal vessels (see Fig 19–24). It supplies the skin of the auricle, the external auditory meatus, the temporomandibular joint, and the scalp. The nerve conveys postganglionic parasympathetic secretomotor fibers from the otic ganglion to the parotid salivary gland.
2. *Lingual nerve.* This descends (see Fig 19–1) and enters the mouth to pass along the side of the tongue to reach its tip. It is joined by the chorda tympani nerve. The lingual nerve supplies the mucous membrane of the anterior two thirds of the tongue and the floor of the mouth. The fibers from the chorda tympani provide important preganglionic parasympathetic secretomotor fibers to the submandibular ganglion for the submandibular and sublingual salivary glands. Other fibers of the chorda tympani conduct taste impulses from the anterior two thirds of the tongue (except for the vallate papillae that are supplied by the glossopharyngeal nerve).
3. *Inferior alveolar nerve.* This descends posterior to the lingual nerve (see Fig 19–1) and enters the mandibular canal to supply the teeth of the lower jaw. It gives off branches to the molar and premolar teeth and the adjacent gingiva. It ends by dividing into an *incisive* and *mental branch.* The incisive branch innervates the canine and incisive teeth, and the mental branch emerges through the mental foramen as the *mental nerve* (see Fig 19–1) to supply the skin of the face over the chin (see Fig 19–26). The inferior alveolar nerve gives off the *mylohyoid nerve* that innervates the mylohyoid muscle and the anterior belly of the digastric muscle (see Fig 19–1).

Otic Ganglion.—This small parasympathetic ganglion is located medial to the mandibular nerve just below the foramen ovale. The preganglionic fibers originate in the glossopharyngeal nerve and reach the ganglion via the lesser petrosal nerve. The postganglionic fibers reach the parotid salivary gland via the auriculotemporal nerve.

Submandibular Ganglion.—This small parasympathetic ganglion lies between the tongue and the submandibular salivary gland. Preganglionic parasympathetic fibers reach the ganglion via the chorda tympani and the lingual nerves. Postganglionic secretomotor fibers pass to the submandibular and sublingual salivary glands.

Basic Anatomy of the Cervical Plexus

The cervical plexus is formed by the anterior rami of the first four cervical nerves (Fig 19–5). The rami are joined by connecting branches which form loops that lie in front of the origins of the levator scapulae and the scalenus medius muscles. The cervical plexus supplies the skin and the muscles of the head, neck, and shoulders.

Phrenic Nerve

The phrenic nerve (C3 through C5) is a large and very important branch of the cervical plexus; it is the only motor nerve supply to the diaphragm. The phrenic nerve runs vertically downward through the

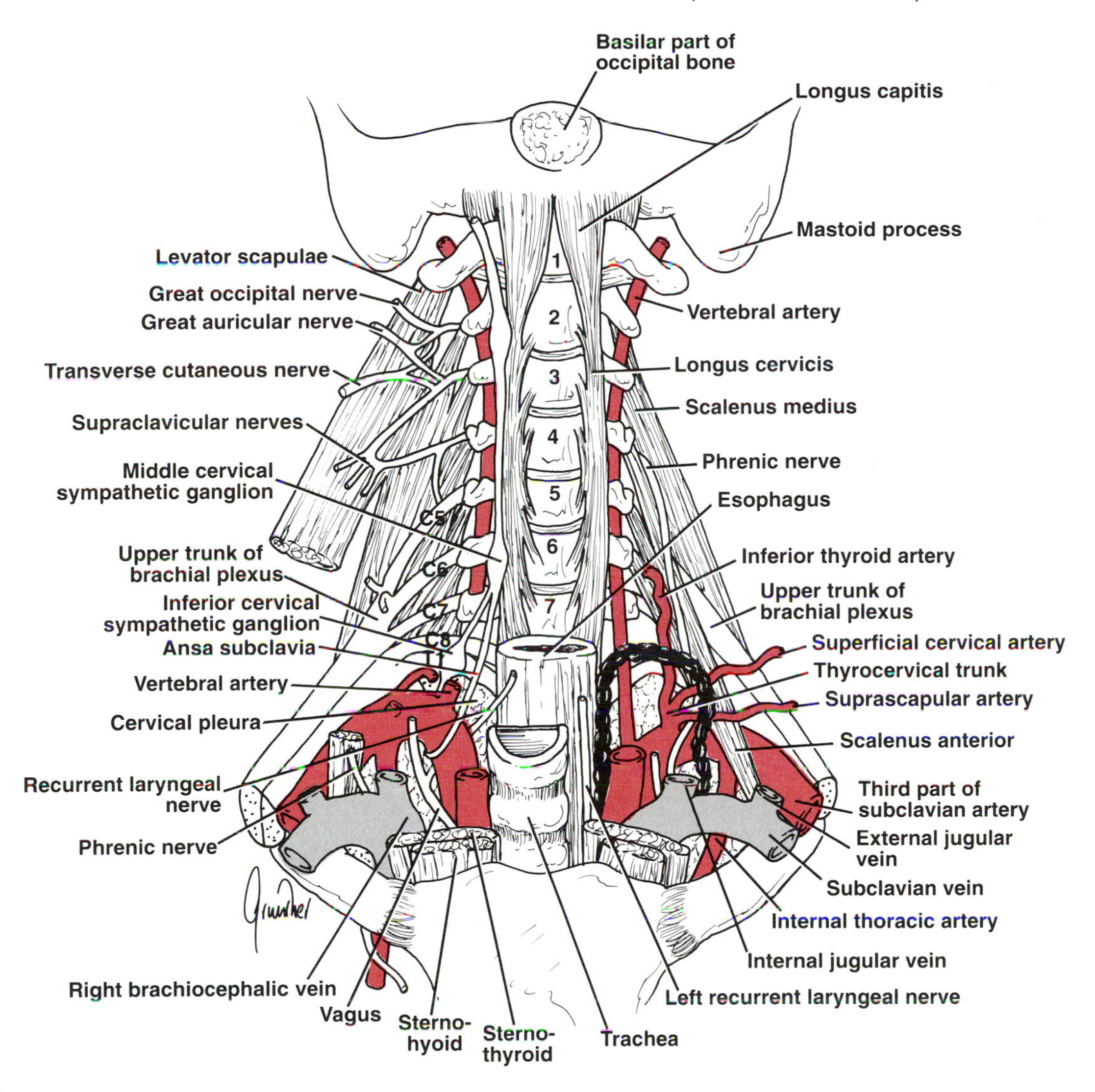

FIG 19–5.
The prevertebral region and the root of the neck. Note the position of the cervical plexus in front of the scalenus medius muscle.

neck on the scalenus anterior muscle (see Fig 19–5). Because of the obliquity of the scalenus anterior muscle, the nerve crosses the muscle from its lateral to its medial border. The nerve enters the thorax by passing anterior to the subclavian artery, and it crosses the internal thoracic artery from lateral to medial.

The phrenic nerves then descend through the mediastinum in front of the roots of the lungs to the diaphragm. In addition to the motor fibers to the diaphragm, the phrenic nerve contains sensory fibers from the pericardium and the mediastinal parietal pleura and the pleura and the peritoneum covering the upper and lower surfaces of the central part of the diaphragm.

A summary of the branches of the cervical plexus and their distribution can be seen in Figure 5–17.

Basic Anatomy of the Brachial Plexus

The brachial plexus is formed in the posterior triangle of the neck by the union of the anterior rami of the fifth, sixth, seventh, and eighth cervical and the first thoracic spinal nerves (Fig 19–6). The plexus may be divided up into *roots, trunks, divisions, and cords.* The roots of C5 and C6 unite to form the upper trunk, the root of C7 continues as the *middle trunk,* the roots of C8 and T1 unite to form the *lower*

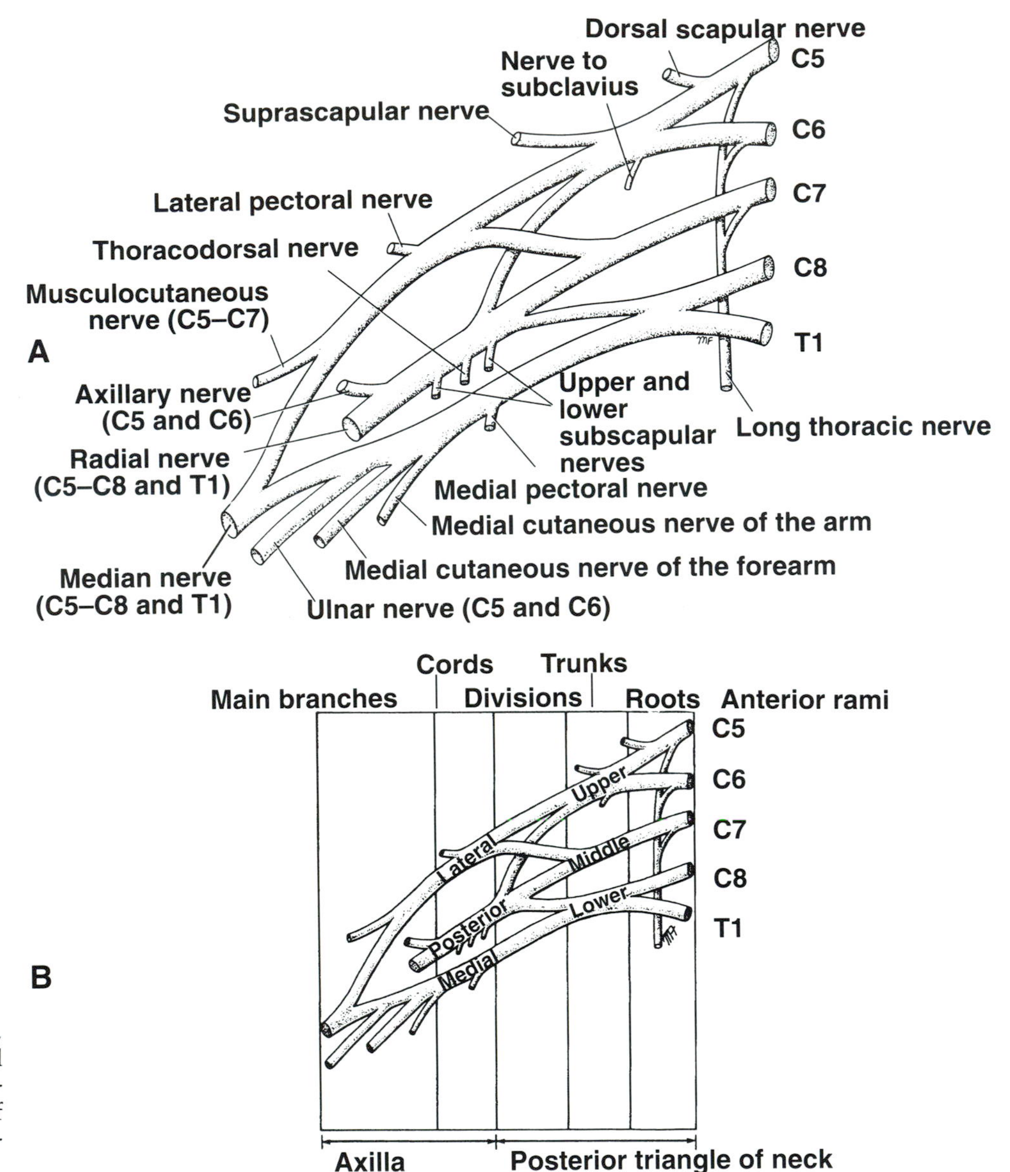

FIG 19–6.
Brachial plexus. **A,** the roots, trunks, divisions, cords, and terminal branches of the brachial plexus. **B,** the location of the different parts of the brachial plexus.

trunk. Each trunk now divides into *anterior and posterior divisions.* The anterior divisions of the upper and middle trunks unite to form the *lateral cord,* the anterior division of the lower trunk continues as the *middle cord,* and the posterior divisions of all three trunks join to form the *posterior cord.*

The roots of the brachial plexus enter the posterior triangle (see Fig 15–21) at the base of the neck between the scalenus anterior and scalenus medius muscles and behind the sternocleidomastoid muscle. Here they are surrounded by a sheath of fascia, the *axillary sheath,* which is derived from the prevertebral layer of the deep cervical fascia. The trunks and divisions cross the posterior triangle of the neck and the cords become arranged around the axillary artery within the axillary sheath in the axilla (Fig 19–7).

Table 19–1 summarizes the branches of the brachial plexus and their distribution.

Basic Anatomy of the Musculocutaneous Nerve

The musculocutaneous nerve (Fig 19–8) arises from the lateral cord of the brachial plexus (C5 through C7). It pierces the coracobrachialis muscle and descends between the biceps and the brachialis muscles. It pierces the deep fascia in the region of the elbow and is distributed along the lateral side of the

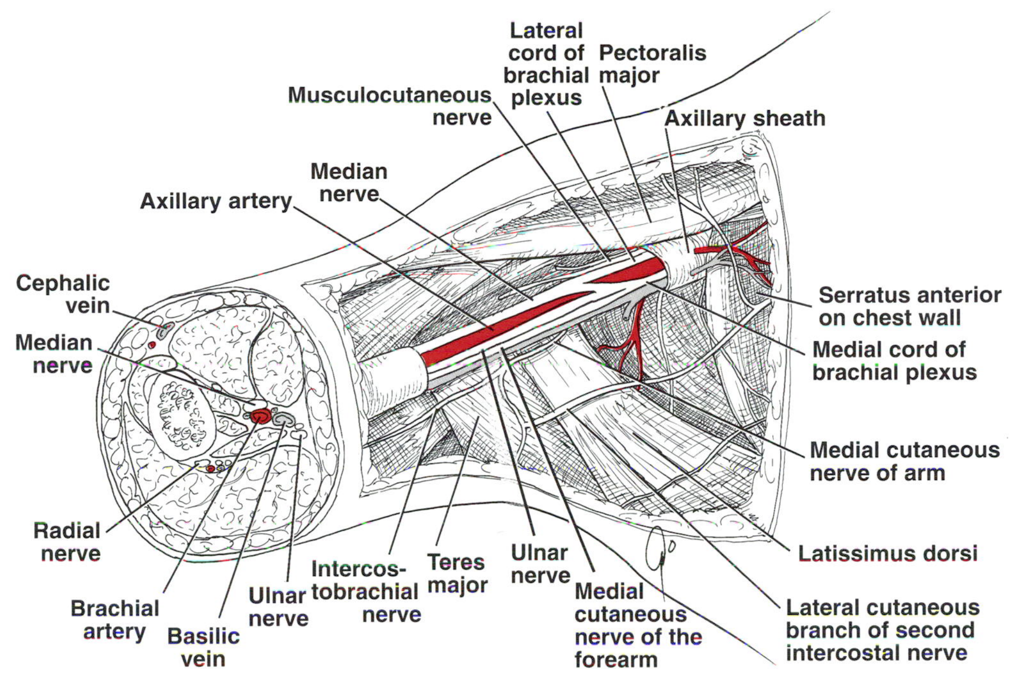

FIG 19–7.
The right armpit (axilla), inferior view, shows the brachial plexus and its branches in relation to the axillary vessels. The main nerves and axillary blood vessels are surrounded by the fascial axillary sheath that extends down from the root of the neck into the upper part of the arm.

forearm as the *lateral cutaneous nerve of the forearm.* The musculocutaneous nerve supplies the coraco-brachialis, both heads of the biceps, and the greater part of the brachialis muscles (see p. 595).

Basic Anatomy of the Median Nerve

The median nerve (see Fig 19–8) arises from the medial and lateral cords of the brachial plexus (C5 through C8 and T1). The nerve descends on the lateral side of the axillary artery and at first on the lateral side of the brachial artery. At the middle of the arm, it crosses the brachial artery to reach its medial side. The nerve enters the forearm between the two heads of pronator teres muscle and descends on the posterior surface of the flexor digitorum superficialis muscle lying on the flexor digitorum profundus muscle. At the wrist, it lies posterior to the tendon of palmaris longus and between the tendons of the flexor carpi radialis and flexor digitorum superficialis muscles. The median nerve enters the palm by passing *behind* the flexor retinaculum and through the carpal tunnel.

Branches of the Median Nerve in the Forearm.—These include the following.

1. *Muscular branches.* These go to the pronator teres, flexor carpi radialis, palmaris longus, and flexor digitorum superficialis (see p. 595).
2. *Articular branches.* These go to the elbow joint.
3. *Anterior interosseous nerve.* Includes *muscular branches* to the flexor pollicis longus, pronator quadratus, and lateral half of flexor digitorum profundus and *articular branches* to the wrist and carpal joints.
4. *Palmar branch.* This is distributed to the skin over the lateral part of the palm.

Branches of the Median Nerve in the Palm.—These include the following.

1. *Muscular branches.* These lead to the abductor pollicis brevis, flexor pollicis brevis, opponens pollicis, and the first and second lumbrical muscles. The branch to the muscles of the thenar eminence takes a recurrent course around the lower border of the flexor retinaculum and lies about one fingerbreadth distal to the tubercle of the scaphoid (see Fig 16–5).
2. *Cutaneous branches.* These lead to the palmar aspect of the lateral 3½ fingers and the distal half of the dorsal aspect of each of those fingers (see Fig 19–36).

TABLE 19–1.
Summary of the Branches of the Brachial Plexus and Their Distribution

Branches	Distribution
Roots	
Dorsal scapular nerve (C5)	Rhomboid minor, rhomboid major, levator scapulae muscles
Long thoracic nerve (C5, C6, C7)	Serratus anterior muscle
Upper trunk	
Suprascapular nerve (C5, C6)	Supraspinatus and infraspinatus muscles
Nerve to subclavius (C5, C6)	Subclavius
Lateral cord	
Lateral pectoral nerve (C5, C6, C7)	Pectoralis major muscle
Musculocutaneous nerve (C5, C6, C7)	Coracobrachialis, biceps brachii, brachialis muscles; supplies skin along lateral border of forearm when it becomes the lateral cutaneous nerve of forearm
Lateral root of median nerve C(5), C6, C7	See Medial root of median nerve
Posterior cord	
Upper subscapular nerve (C5, C6)	Subscapularis muscle
Thoracodorsal nerve (C6, C7, C8)	Latissimus dorsi muscle
Lower subscapular nerve (C5, C6)	Subscapularis and teres major muscles
Axillary nerve (C5, C6)	Deltoid and teres minor muscles; upper lateral cutaneous nerve of arm supplies skin over lower half of deltoid muscle
Radial nerve (C5, C6, C7, C8, T1)	Triceps, anconeus, part of brachialis, extensor carpi radialis longus; via deep radial nerve branch supplies extensor muscles of forearm: supinator, extensor carpi radialis brevis, extensor carpi ulnaris, extensor digitorum, extensor digiti minimi, extensor indicis, abductor pollicis longus, extensor pollicis longus, extensor pollicis brevis; skin, lower lateral cutaneous nerve of arm, posterior cutaneous nerve of arm, and posterior cutaneous nerve of forearm; skin on lateral side of dorsum of hand and dorsal surface of lateral 3½ fingers; articular branches to elbow, wrist, and hand
Medial cord	
Medial pectoral nerve (C8, T1)	Pectoralis major and minor muscles
Medial cutaneous nerve of arm joined by intercostal brachial nerve from second intercostal nerve (C8, T1, T2)	Skin of medial side of arm
Medial cutaneous nerve of forearm (C8, T1)	Skin of medial side of forearm
Ulnar nerve (C8, T1)	Flexor carpi ulnaris and medial half of flexor digitorum profundus, flexor digiti minimi, opponens digiti minimi, abductor digiti minimi, adductor pollicis, third and fourth lumbricals, interossei, palmaris brevis, skin of medial half of dorsum of hand and palm, skin of palmar and dorsal surfaces of medial 1½ fingers
Medial root of median nerve (with lateral root) forms median nerve (C5, C6, C7, C8, T1)	Pronator teres, flexor carpi radialis, palmaris longus, flexor digitorum superficialis, abductor pollicis brevis, flexor pollicis brevis, opponens pollicis, first two lumbricals (by way of anterior interosseous branch), flexor pollicis longus, flexor digitorum profundus (lateral half), pronator quadratus; palmar cutaneous branch to lateral half of palm and digital branches to palmar surface of lateral 3½ fingers; articular branches to elbow, wrist, and carpal joints

Basic Anatomy of the Ulnar Nerve

The ulnar nerve (see Fig 19–6) arises from the medial cord of the brachial plexus (C8 and T1). It descends along the medial side of the axillary and brachial arteries in the anterior compartment of the arm (see Fig 19–7). At the middle of the arm it pierces the medial intermuscular septum and passes down *behind* the medial epicondyle of the humerus (see Fig 19–34). It then enters the anterior compartment of the forearm by passing between the two heads of flexor carpi ulnaris muscle. It descends behind the flexor carpi ulnaris ulnar to the ulnar artery (Fig 19–9). At the wrist, it passes *anterior* to the flexor retinaculum, lateral to the pisiform bone. It then divides into superficial and deep terminal branches.

Branches of the Ulnar Nerve in the Forearm.—These include the following.

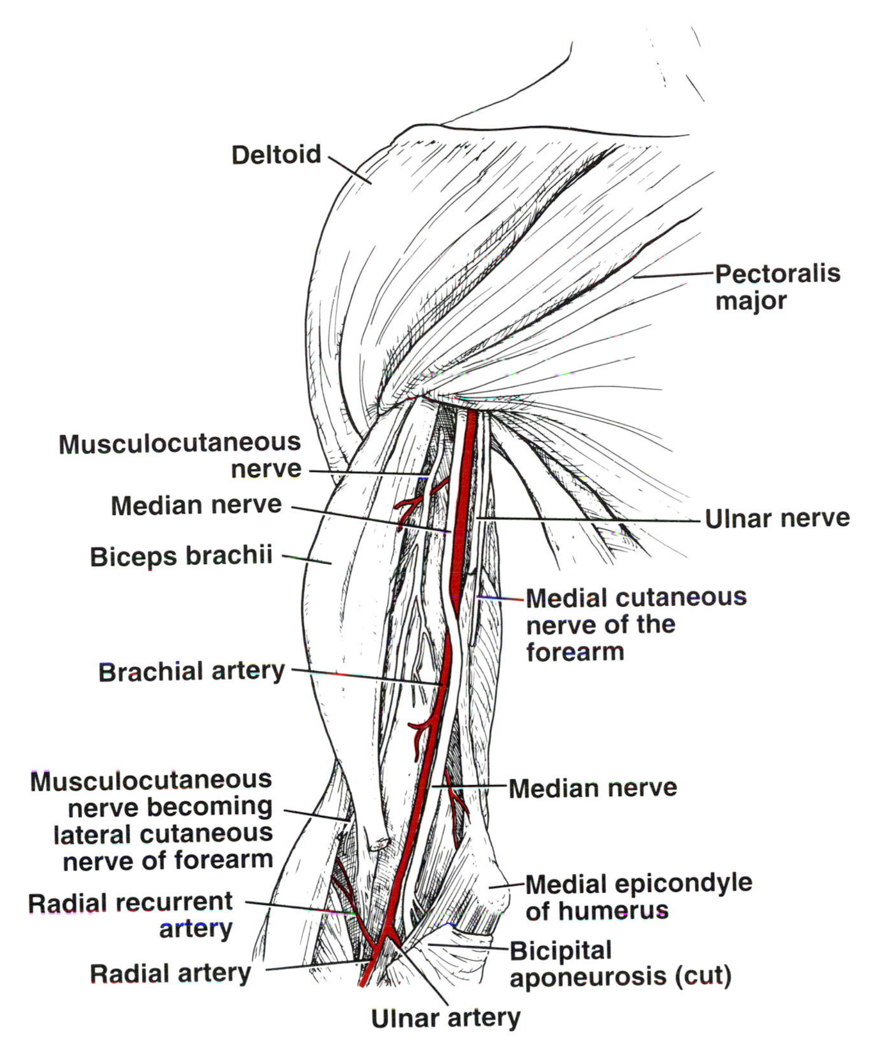

FIG 19–8.
Anterior view of the right upper part of the arm. The biceps brachii has been pulled laterally to show the musculocutaneous nerve.

1. *Muscular branches.* These go the flexor carpi ulnaris and the medial half of flexor digitorum profundus.
2. *Articular branches.* These go to the elbow joint.
3. *Palmar branch.* Supplies the skin over the hypothenar eminence.
4. *Dorsal cutaneous branch.* Supplies the skin over the medial side of the back of the hand and the back of the medial 1½ fingers over the proximal phalanges.

Branches of the Ulnar Nerve in the Hand.—The superficial terminal branch descends into the palm and gives off the following branches (see Fig 16–5).

1. *Muscular branch.* Goes to the palmaris brevis.
2. *Cutaneous branches.* Supply the skin over the palmar aspect of the medial 1½ fingers, including their nail beds.

The *deep terminal branch* runs backward between the abductor digiti and the flexor digiti minimi muscles and pierces the opponens digiti minimi. It gives off the following branches.

1. *Muscular branches.* These lead to the abductor digiti minimi, the flexor digiti minimi, the opponens digiti minimi, all the palmar and all the dorsal interossei, the third and fourth lumbricals, and the adductor pollicis.
2. *Articular branches.* These go to the carpal joints.

Basic Anatomy of the Radial Nerve

The radial nerve (see Fig 19–6) arises from the posterior cord of the brachial plexus (C5 through C8 and T1). It descends behind the axillary and brachial arteries and passes posteriorly between the long and medial heads of the triceps muscle to enter the pos-

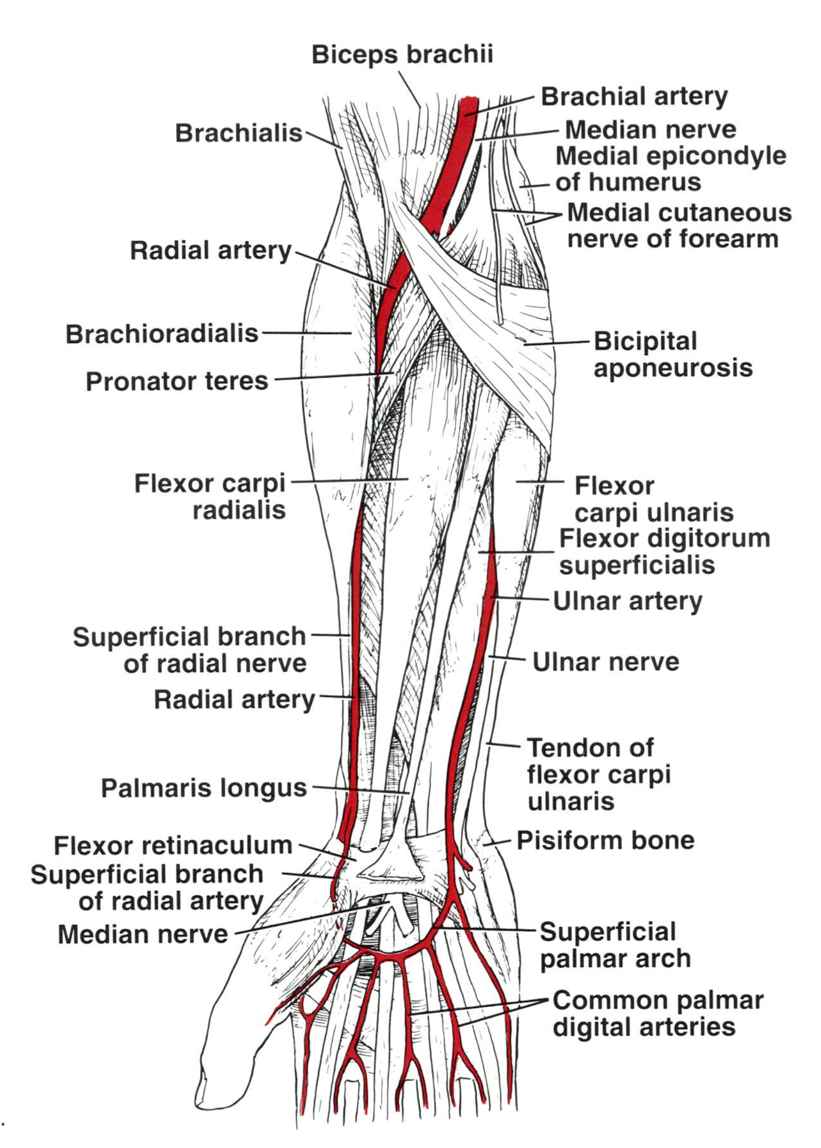

FIG 19–9.
Anterior view of the forearm and hand.

terior compartment of the arm. The radial nerve then winds round in the spiral groove on the back of the humerus with the profunda brachii vessels (Fig 19–10). Piercing the lateral intermuscular septum just above the elbow, it continues downward into the cubital fossa between the brachialis and the brachioradialis muscles. At the level of the lateral epicondyle it divides into superficial and deep branches.

Branches of the Radial Nerve in the Axilla.—These include the following.

1. *Muscular branches.* Supply the long and medial heads of the triceps.
2. *Cutaneous branch.* The posterior cutaneous nerve of the arm.

Branches of the Radial Nerve in the Spiral Groove.—These include the following.

1. *Muscular branches.* Supply the lateral and medial heads of the triceps and the anconeus.
2. *Cutaneous branches.* The lower lateral cutaneous nerve of the arm and the posterior cutaneous nerve of the forearm.

Branches of the Radial Nerve in the Anterior Compartment of the Arm.—These include the following.

1. *Muscular branches.* Supply the brachialis, the brachioradialis, and the extensor carpi radialis longus.
2. *Articular branches.* Pass to the elbow joint.

Superficial Branch of the Radial Nerve.—This is the direct continuation of the radial nerve at the elbow. It descends under cover of the brachioradialis muscle on the lateral side of the radial artery. In the lower part of the forearm it passes backward under

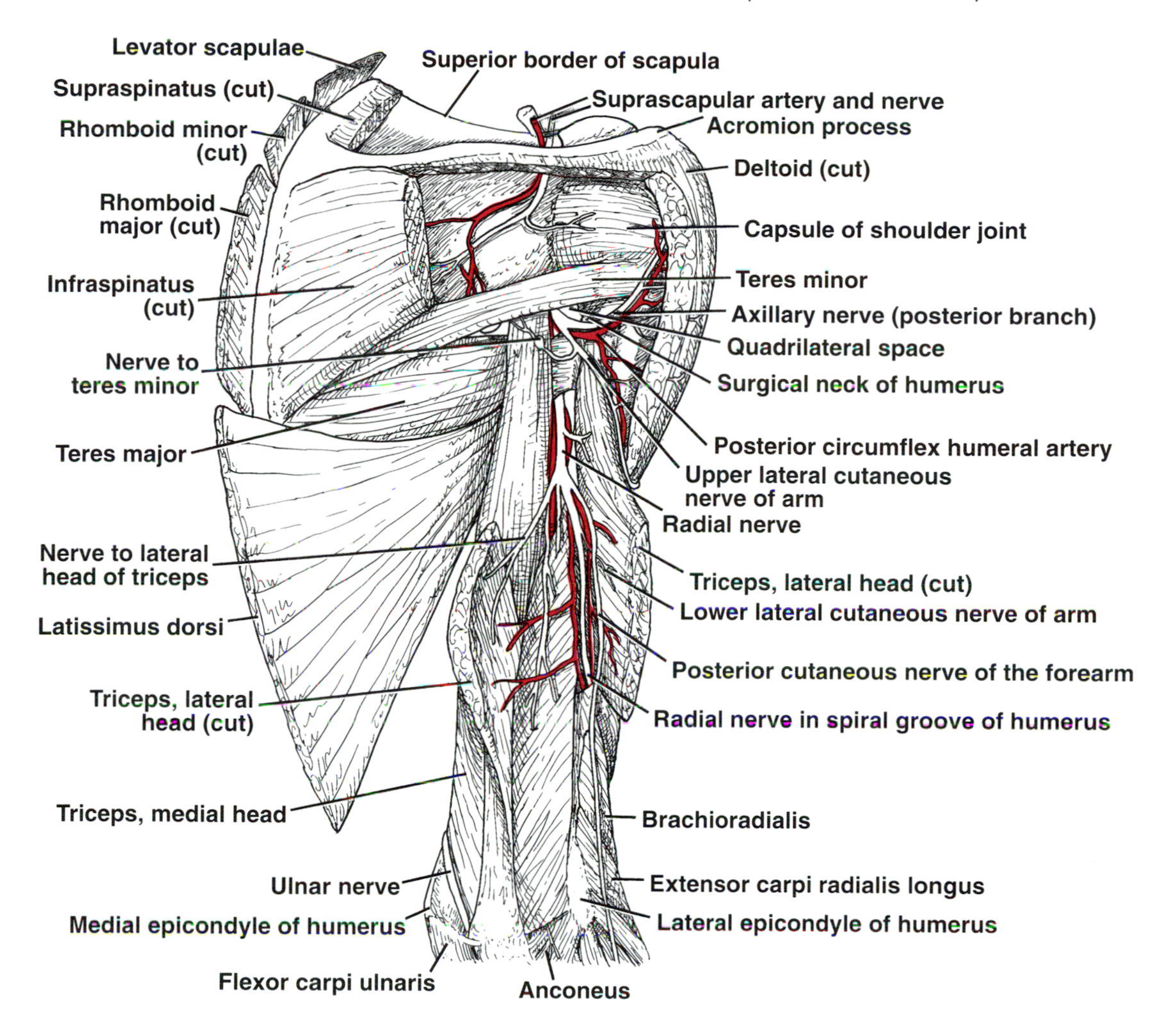

FIG 19–10.
Posterior view of the right scapular region and the posterior compartment of the upper part of the arm showing the muscles, nerves, and blood vessels.

the tendon of the brachioradialis muscle and descends on to the back of the hand (see Fig 19–35).

Branches of the superficial branch of the radial nerve.—These include the following.

1. *Cutaneous branches.* These lead to the lateral two thirds of the posterior surface of the hand (variable area) and the posterior surface of the lateral 2½ fingers over the proximal phalanges.

Deep Branch of the Radial Nerve.—The deep branch of the radial nerve winds around the lateral side of the neck of the radius in the supinator muscle to enter the posterior compartment of the forearm. It descends between the superficial and deep layers of muscle and reaches the back of the interosseous membrane.

Branches of the deep branch of the radial nerve.—These include the following.

1. *Muscular branches.* These supply the extensor carpi radialis brevis, the supinator, the extensor digitorum, the extensor digiti minimi, the extensor carpi ulnaris, the abductor pollicis longus, the extensor pollicis brevis, the extensor pollicis longus, and the extensor indicis.
2. *Articular branches.* These pass to the wrist and carpal joints.

Basic Anatomy of the Thoracic Spinal Nerves

There are 12 thoracic spinal nerves. The posterior rami supply the deep back muscles and the skin over the posterior surface of the back (Fig 19–11). The anterior rami of the first 11 thoracic spinal

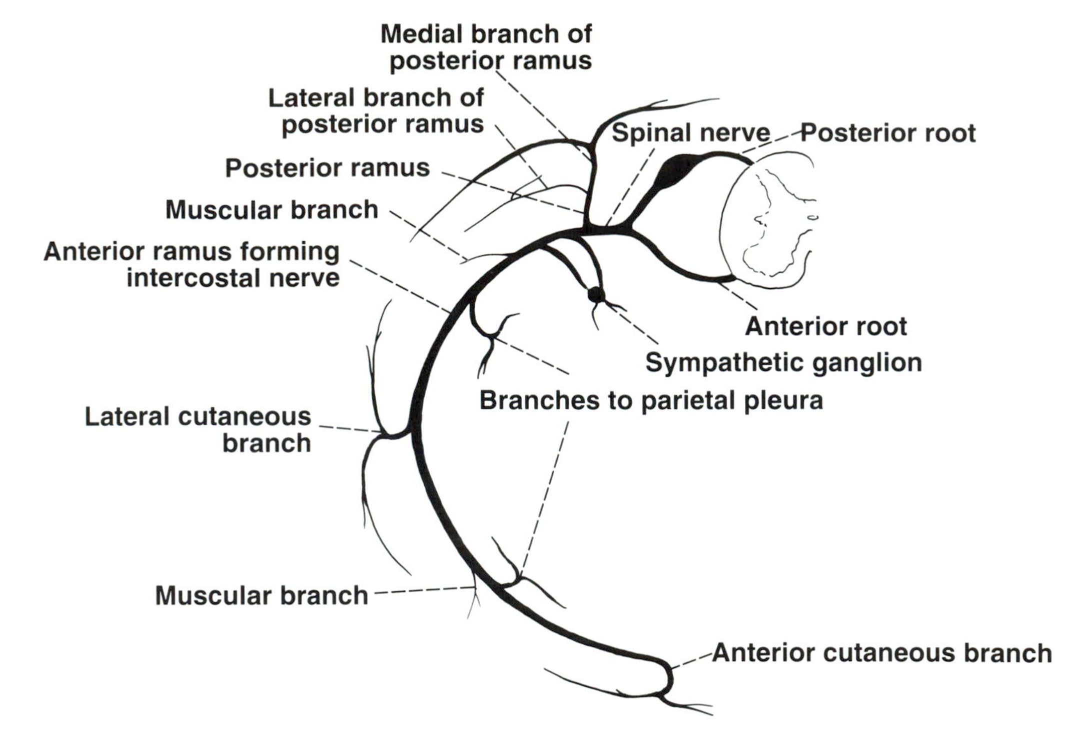

FIG 19–11.
The origin and distribution of a typical thoracic spinal nerve.

nerves run forward between the ribs in intercostal spaces and are known as *intercostal nerves.* The anterior ramus of the 12th thoracic nerve lies below the 12th rib in the abdomen and runs forward in the abdominal wall.

Intercostal Nerves

The intercostal nerves are the anterior rami of the first 11 thoracic spinal nerves. Each intercostal nerve enters an intercostal space (see Fig 19–38) and then runs forward inferiorly to the intercostal vessels in the subcostal groove of the corresponding rib, between the transverse thoracic and the internal intercostal muscles. The first six nerves are distributed within their intercostal spaces. The 7th to 9th intercostal nerves leave the anterior ends of their intercostal spaces by passing deep to the costal cartilages, to enter the abdominal wall (Fig 19–12). In the case of the 10th and 11th nerves, as the corresponding ribs are floating, these nerves pass directly into the abdominal wall.

Branches of the Intercostal Nerves.—These include the following.

1. *Collateral branch.* Runs forward below the main nerve in the lower part of the intercostal space.
2. *Lateral cutaneous branch.* Arises in the midaxillary line and divides into anterior and posterior branches that supply the skin (see Fig 19–11).
3. *Anterior cutaneous branch.* The terminal part of the main trunk. It reaches the skin near the midline and divides into a medial and a lateral branch (see Fig 19–11).
4. *Muscular branches.* These are given off by the main nerve and its collateral branch; they innervate the intercostal muscles and the levator costarum and the serratus posterior muscles.
5. *Pleural branches.* Lead to the parietal pleura and *peritoneal branches* (7th through 11th intercostal nerves only) to the parietal peritoneum. These are sensory nerves (see Fig 19–11).

The first six intercostal nerves give off numerous branches that supply (1) the skin and the parietal pleura covering the outer and inner surfaces of each intercostal space, respectively, and (2) the intercostal muscles of each intercostal space. In addition, the 7th to 11th intercostal nerves supply (1) the skin and the parietal peritoneum covering the outer and inner surfaces of the abdominal wall, respectively, and (2) the anterior abdominal muscles, which include the external oblique, internal oblique, transverse abdominis, and rectus abdominis muscles (see Fig 19–12). The first and second intercostal nerves, however, are an exception.

First Intercostal Nerve.—This nerve is joined to the brachial plexus by a large branch that is equivalent

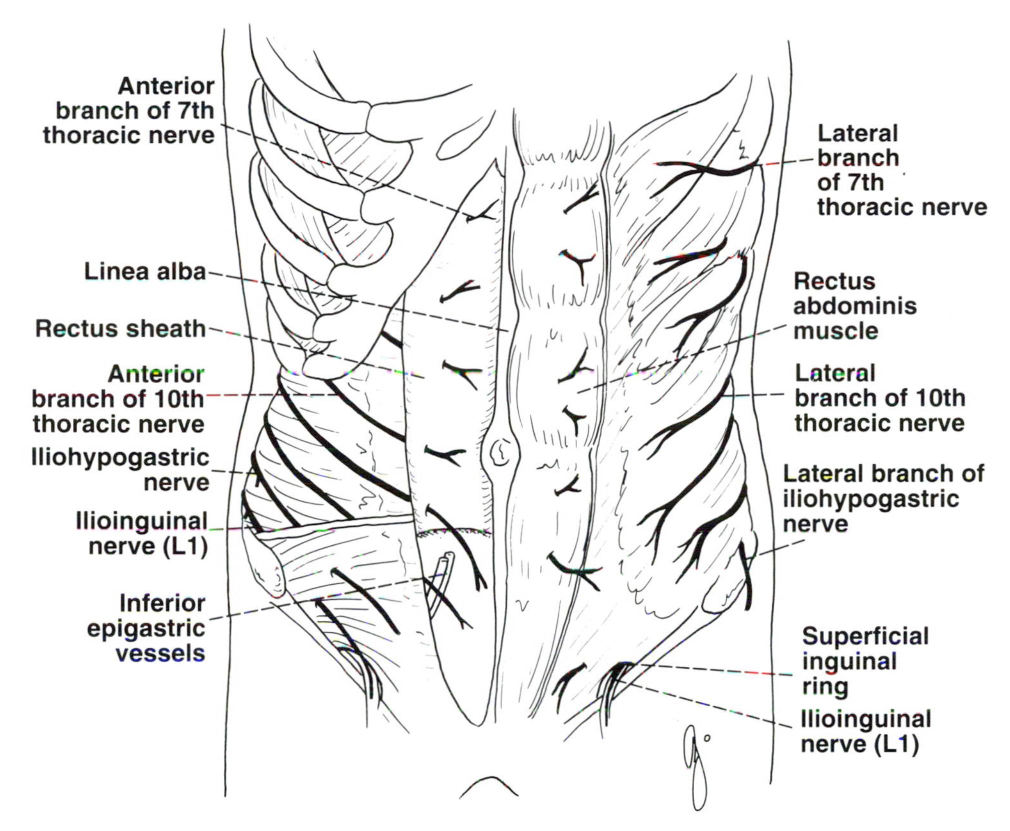

FIG 19–12.
The distribution of the thoracic and lumbar nerves in the anterior abdominal wall. On the right side the rectus sheath has been opened and the rectus abdominis muscle has been removed to show the nerves passing through the sheath.

to the lateral cutaneous branch of typical intercostal nerves. The remainder of the first intercostal nerve is small.

Second Intercostal Nerve.—A branch of this nerve, called the *intercostobrachial nerve,* runs across the axilla to join the medial cutaneous nerve of the arm (see Fig 19–7). The second intercostal nerve, therefore, supplies the skin of the armpit and the upper medial side of the arm.

Clinical Note

Intercostobrachial nerve and heart disease.—In coronary heart disease, pain is referred along this nerve to the medial side of the arm (see p. 125).

Basic Anatomy of the Lumbar Plexus

The lumbar plexus, which is one of the main nervous pathways supplying the lower limb, is formed inside the psoas muscle from the anterior rami of the upper four lumbar nerves (Fig 19–13). The branches of the plexus emerge from the lateral and medial borders of the muscle and from its anterior surface. The courses of the more important nerves and their branches are as follows.

Femoral Nerve

This is the largest branch of the lumbar plexus and arises from L2 through L4 lumbar nerves. It emerges from the lateral border of the psoas muscle in the abdomen and descends between the psoas and the iliacus muscles. It enters the thigh behind the inguinal ligament and lies lateral to the femoral vessels and the femoral sheath (Fig 19–14). About 1.5 in. (4 cm) below the inguinal ligament, it terminates by dividing into anterior and posterior divisions (see Fig 19–40).

Branches of the Femoral Nerve in the Abdomen.—These include the following.

1. *Muscular branches.* These supply the iliacus.

Branches of the Femoral Nerve in the Thigh.—These include the following.

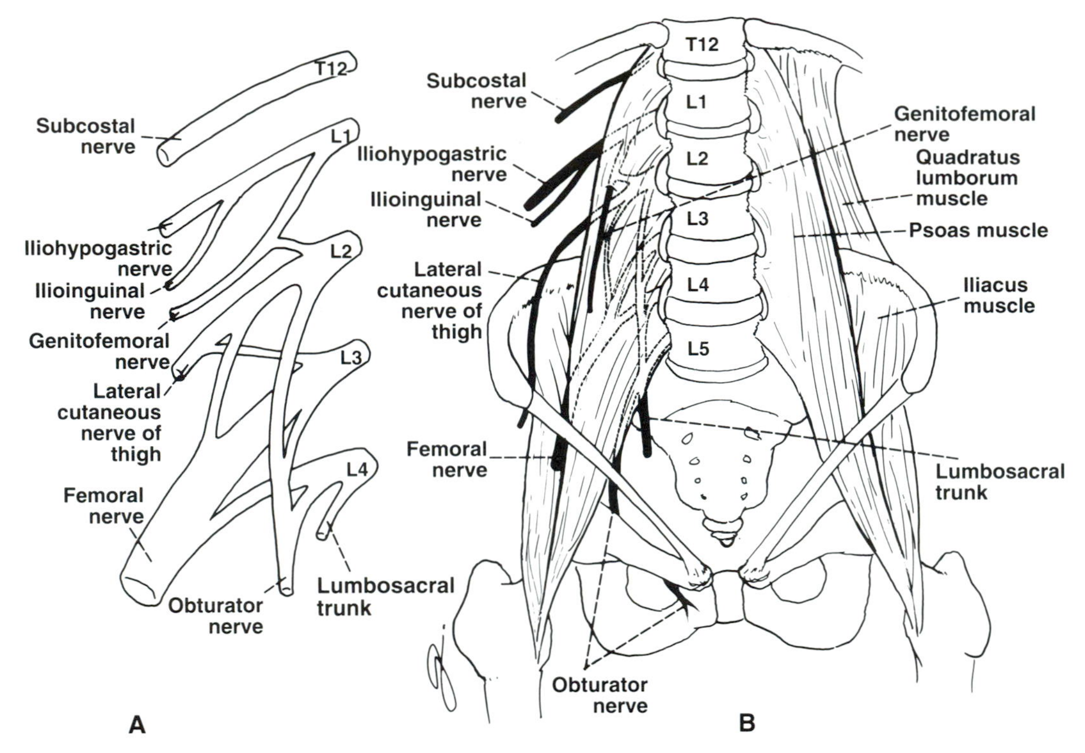

FIG 19–13.
A, the lumbar plexus and its main branches. **B,** the lumbar plexus and its branches on the posterior abdominal wall.

1. *Cutaneous branches.* The *medial cutaneous nerve of the thigh* supplies the skin on the medial side of the thigh. The *intermediate cutaneous nerve of the thigh* supplies the skin on the anterior surface of the thigh. The *saphenous nerve* descends into the adductor canal; crossing the femoral artery on its anterior surface (see Fig 19–14). The nerve emerges on the medial side of the knee joint between the tendons of the sartorius and gracilis muscles. It descends along the medial side of the leg in company with the great saphenous vein. It passes *anterior* to the medial malleous and along the medial border of the foot as far as the ball of the big toe.

2. *Muscular branches.* These supply the sartorius, pectineus, and quadriceps femoris. The muscular branch to the rectus femoris also supplies the hip joint; the branches to the three vasti muscles also supply the knee joint.

Lateral Cutaneous Nerve of the Thigh

The lateral cutaneous nerve of the thigh arises from the lumbar plexus (L2 and L3). It runs downward on the iliacus muscle on the posterior abdominal wall and enters the thigh behind the lateral end of the inguinal ligament. Having divided into anterior and posterior branches, it supplies the skin of the lateral aspects of the thigh and knee (see Fig 19–40). It also supplies the skin of the lower lateral quadrant of the buttock.

Basic Anatomy of the Sacral Plexus

The sacral plexus is formed from the anterior rami of the fourth and fifth lumbar nerves and the anterior rami of the first, second, third, and fourth sacral nerves (Fig 19–15). The contribution from the fourth lumbar nerve joins the fifth lumbar nerve to form the *lumbosacral trunk.* The lumbosacral trunk passes down into the pelvis and joins the sacral nerves. Parasympathetic branches arise from the second, third, and fourth sacral nerves to form the *pelvic splanchnic nerves.*

The courses of the most important nerves and their branches are as follows.

Sciatic Nerve

The sciatic nerve (Fig 19–16) is the largest nerve in the body and arises from L4 and L5 and S1 through S3. It passes out of the pelvis into the gluteal region through the greater sciatic foramen. The nerve emerges below the piriformis muscle and curves downward and laterally beneath the gluteus maximus muscle to reach the back of the adductor mag-

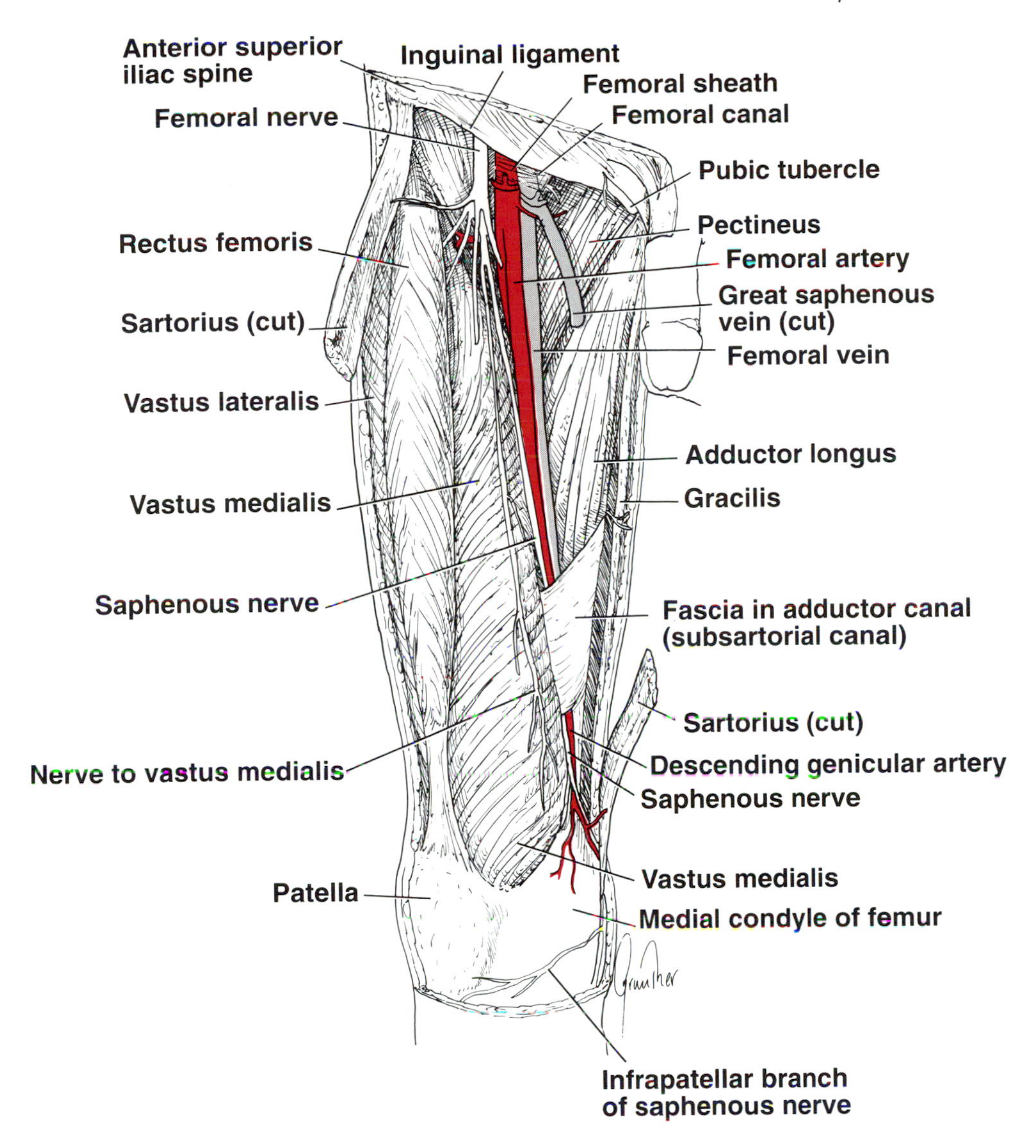

FIG 19–14.
Front of the thigh in the right lower limb shows the femoral vessels and femoral nerve. The sartorius muscle has been cut and reflected to show the underlying adductor (subsartorial) canal and its contents.

nus muscle. Here it is covered by the long head of the biceps femoris muscle. In the lower third of the thigh (occasionally at a higher level), it ends by dividing into the tibial and common peroneal nerves.

Clinical Note

Nerves below knee.—All the nerves below the knee come from the sciatic nerve except the saphenous nerve, which comes from the femoral nerve.

Branches of the Sciatic Nerve.——These include the following.

1. *Muscular branches.* These supply the biceps femoris (long head), semitendinosus, semimembranosus, and the hamstring part of adductor magnus.
2. *Articular branches.* These pass to the hip joint.

Tibial Nerve

The tibial nerve (L4 and L5 and S1 through S3) runs downward through the popliteal fossa, crossing the popliteal artery from the lateral to the medial side (Fig 19–17). It then passes deep to the gastrocnemius and soleus muscles and descends to the interval between the medial malleolus and the heel (see Fig 19–42). It is covered here by the flexor retinaculum and divides into the medial and lateral plantar nerves.

Branches of the Tibial Nerve.—These include the following.

1. *Cutaneous branches.* The *sural nerve* is usually joined by the sural communicating branch of the common peroneal nerve. It supplies the skin of the calf and the back of the leg and accompanies the small saphenous vein behind the lateral malleolus to

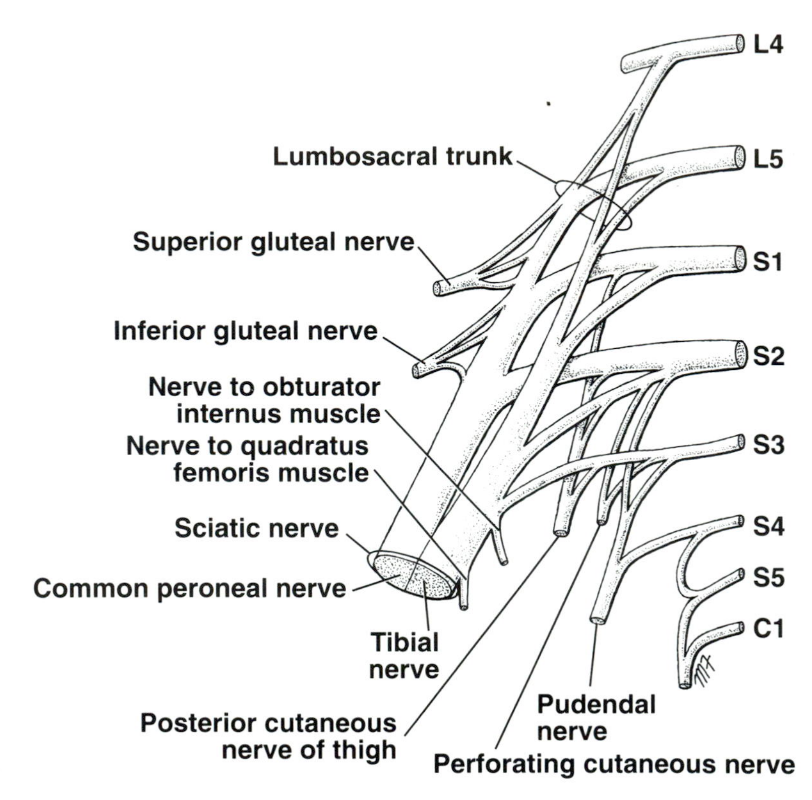

FIG 19–15.
The sacral plexus and its main branches.

supply the skin of the lateral border of the foot and the lateral side of the little toe. The *medial calcaneal branch* supplies the skin over the medial surface of the heel.

2. *Muscular branches.* These supply the gastrocnemius, plantaris, soleus, popliteus, flexor digitorum longus, flexor hallucis longus, and tibialis posterior.

3. *Articular branches* These pass to the knee and ankle joints.

Medial Plantar Nerve

The medial plantar nerve runs forward deep to the abductor hallucis muscle with the medial plantar artery.

Branches of the Medial Plantar Nerve.—These include the following.

1. *Cutaneous branches.* These lead to the medial part of the sole (see Fig 19–44) and digital nerves to the medial 3½ toes and the nail beds.

2. *Muscular branches.* These supply the abductor hallucis, flexor digitorum brevis, flexor hallucis brevis, and the first lumbrical.

Lateral Plantar Nerve

The lateral plantar nerve runs forward deep to the abductor hallucis and flexor digitorum brevis in company with the lateral plantar artery.

Branches of the Lateral Plantar Nerve.—These include the following.

1. *Cutaneous branches.* These lead to the lateral part of the sole (see Fig 19–44) and plantar digital nerves to the lateral 1½ toes and the nail beds.

2. *Muscular branches.* These supply the flexor digitorum accessorius, abductor digiti minimi, flexor digiti minimi brevis, adductor hallucis, interrosseous muscles, and the second, third, and fourth lumbricals.

Common Peroneal Nerve

The common peroneal nerve (L4 and L5 and S1 and S2) runs downward through the popliteal fossa following the tendon of the biceps femoris muscle (see Fig 19–17). It then passes laterally around the neck of the fibula, pierces the peroneus longus muscle, and divides into the superficial peroneal nerve and the deep peroneal nerve (Fig 19–18).

Branches of the Common Peroneal Nerve.—These include the following.

1. *Cutaneous branches. Sural communicating branch* joins the sural nerve. *Lateral cutaneous nerve of the calf* supplies the skin on the lateral side of the back of the leg.

2. *Muscular branch.* Supplies the short head of the biceps femoris.

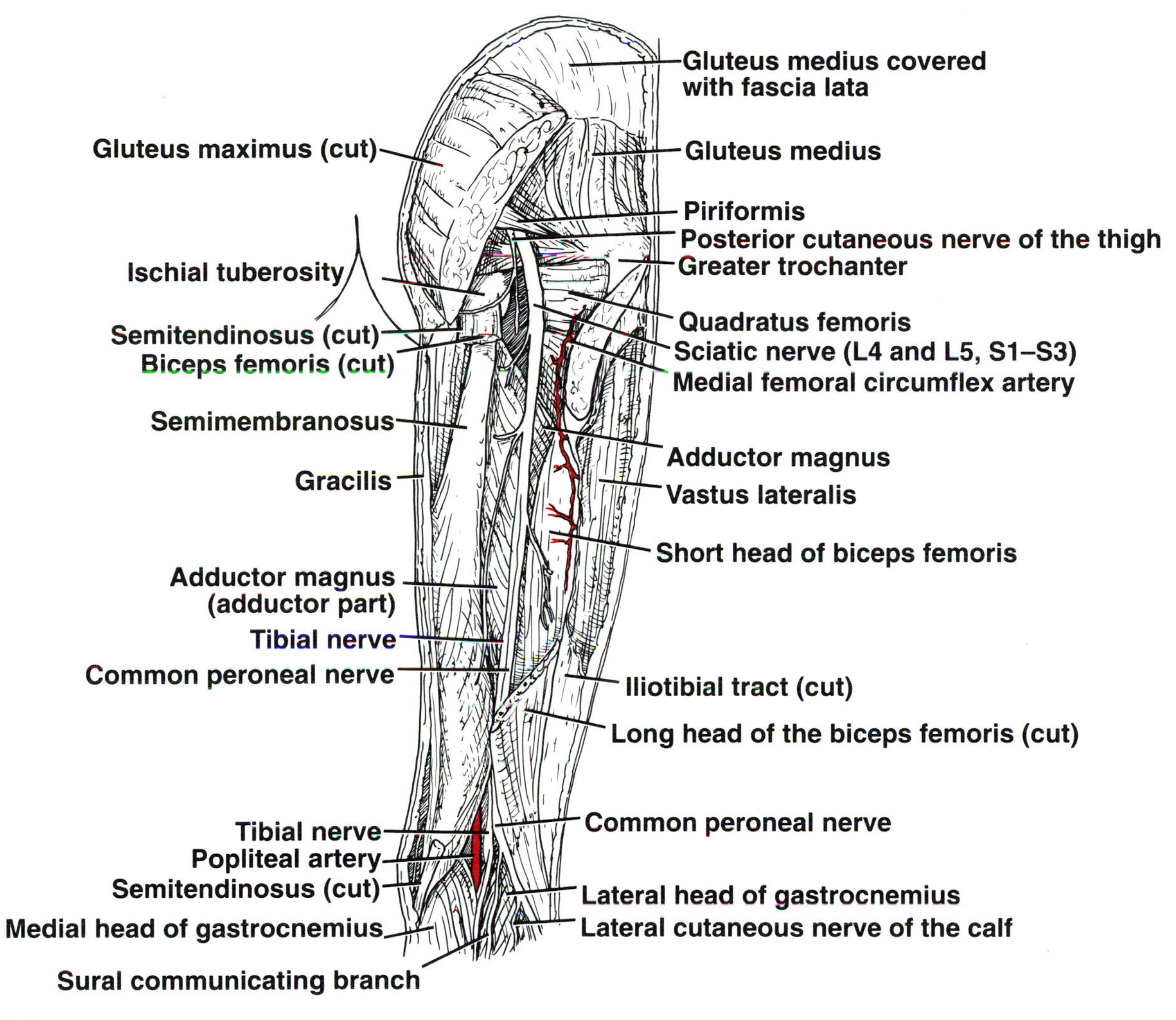

FIG 19–16.
Right gluteal region and the posterior view of the thigh. The greater part of the gluteus maximus muscle and the long head of the biceps femoris muscle have been removed to display the sciatic nerve; the semitendinosus muscle has also been removed.

3. *Articular branches.* Lead to the knee joint.
4. *Terminal branches.*

Superficial Peroneal Nerve

The superficial peroneal nerve descends at first deeply in the leg between the peroneus longus and the peroneus brevis muscles and then between the peroneus brevis and the extensor digitorum longus muscles (see Fig 19–18). In the lower third of the leg it becomes superficial and supplies the skin.

Branches of the Superficial Peroneal Nerve.—These include the following.

1. *Cutaneous branches.* These pass to the lower third of the leg and the dorsum of the foot, except the cleft between the big and second toes (see Fig 19–44).
2. *Muscular branches.* These lead to the peroneus longus and brevis.

Deep Peroneal Nerve

The deep peroneal nerve (S2 through S4) leaves the peroneus longus muscle and descends deep to the extensor digitorum longus muscle on the interosseous membrane (see Fig 19–18). It is accompanied by the anterior tibial vessels. The nerve enters the dorsum of the foot on the lateral side of the dorsalis pedis artery and divides into medial and lateral terminal branches.

Branches of the Deep Peroneal Nerve.—These include the following.

1. *Cutaneous branch.* Passes to the adjacent sides of the big and second toes.

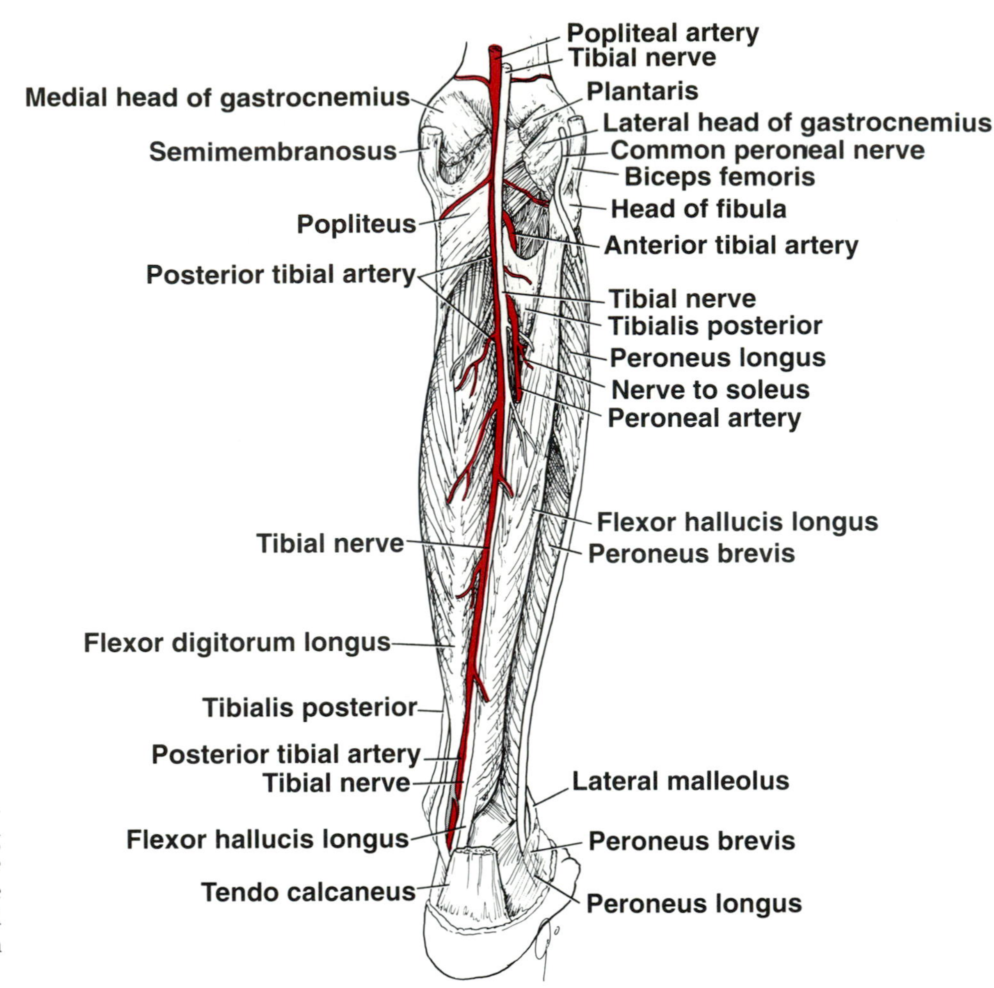

FIG 19–17.
Deep structures on the posterior surface of the right leg show the tibial nerve and the posterior tibial artery. The gastrocnemius, soleus, and plantaris muscles have been removed.

2. *Muscular branches.* These supply the tibialis anterior, extensor digitorum longus, peroneus tertius, extensor hallucis longus, and the extensor digitorum brevis.

3. *Articular branches.* These pass to the ankle and tarsal joints.

Pudendal Nerve

The pudendal nerve arises from the S2 through S4 spinal nerves. It leaves the pelvis through the greater sciatic foramen and, after a brief course in the gluteal region, enters the perineum through the lesser sciatic foramen. The nerve passes forward in the *pudendal canal* with the internal pudendal vessels on the lateral wall of the ischiorectal fossa. It ends by dividing into the perineal nerve and the dorsal nerve of the penis (clitoris).

Branches of the Pudendal Nerve.—These include the following.

1. *Inferior rectal nerve.* Crosses the ischiorectal fossa to the anal canal. The *sensory branch* goes to the mucous membrane of the lower half of the anal canal and the perianal skin, and the *muscular branch* supplies the external anal sphincter.

2. *Perineal nerve.* The *cutaneous branches—the posterior scrotal (labial) nerves*—lead to the posterior surface of the scrotum or labia majora. The *muscular branches* go to the superficial and deep transverse perineal muscles, bulbospongiosus, ischiocavernosus, sphincter urethrae, and the levator ani.

3. *Dorsal nerve of the penis (clitoris).* Supplies the skin and deeper structures of the penis (clitoris).

Anatomy of Dermatomes

The skin covering the whole body is supplied segmentally by spinal nerves. The area of skin supplied by a single spinal nerve, and therefore a single segment of the spinal cord, is called a *dermatome* (Figs 19–19 and 19–20). On the trunk, adjacent dermatomes overlap considerably, therefore, to produce a region of complete anesthesia. A total of at least three spinal nerves have to be blocked. In the upper

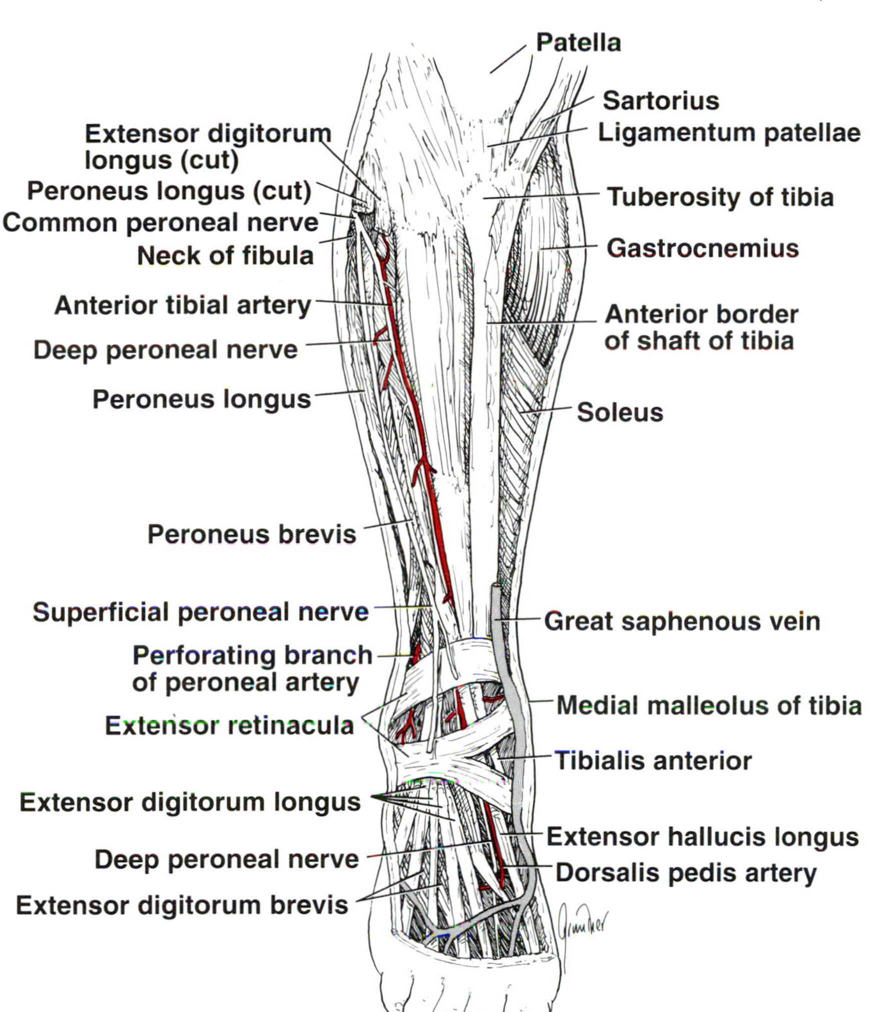

FIG 19–18.
Structures present on the anterior and lateral aspects of the right leg and on the dorsum of the foot. Portions of the peroneus longus and the extensor digitorum longus muscles have been removed to display the common peroneal, the deep peroneal, and the superficial peroneal nerves.

and lower limbs, the arrangement of the dermatomes is more complicated, which is caused by the embryological changes that take place as the limbs grow out from the body wall. The skin of the face is largely supplied by the three divisions of the trigeminal cranial nerve, where there is little or no overlap of the areas of skin supplied by each of the divisions.

CLINICAL ANATOMY

Clinical Anatomy of Trigeminal Nerve Blocks

Ophthalmic Nerve Block

The terminal branches of the ophthalmic division of the trigeminal nerve that emerge onto the face and the nose include the supraorbital, supratrochlear, infratrochlear, external nasal, and lacrimal nerves (see Fig 19–3). The latter nerve gives off only a few branches to the skin and is seldom blocked.

Supraorbital Nerve Block.—This involves the following.

Area of Anesthesia.—Skin of the upper eyelid, the forehead, and the scalp as far back as the vertex (Fig 19–21).

Indications.—Repair of lacerations of the upper eyelid, forehead, and scalp.

Procedure.—The supraorbital nerve emerges from the orbital cavity in the same vertical plane as the pupil when the patient is looking straight ahead (see Fig 19–21). If the nerve passes through the supraorbital notch, this can easily be palpated on the supraorbital margin, which is the site for injection. If the notch is a foramen, however, it is small

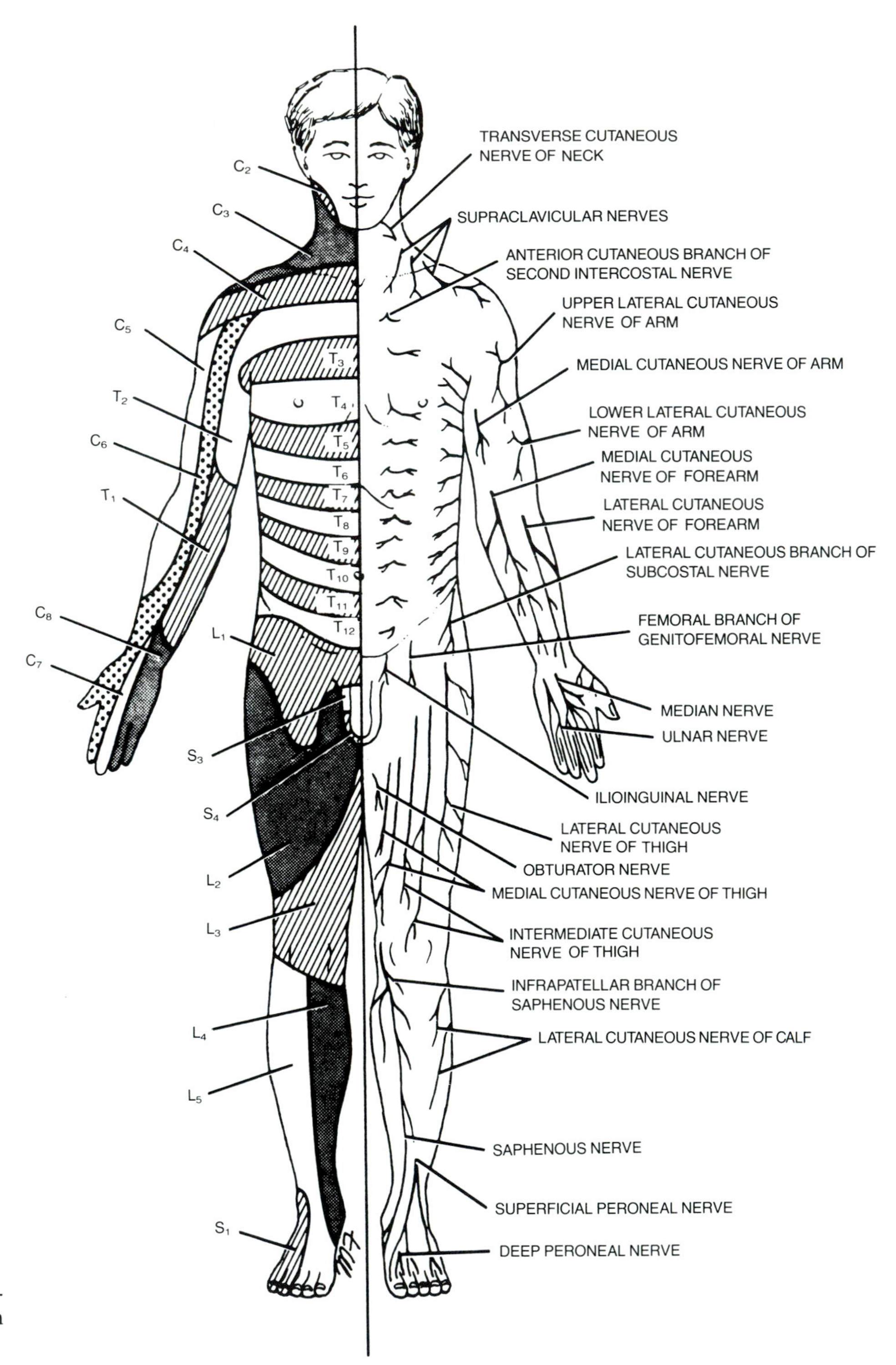

FIG 19–19.
The dermatomes and the distribution of cutaneous nerves on the anterior aspect of the body.

and difficult to feel, and the needle is inserted into skin over the supraorbital margin in line with the pupil.

Supratrochlear Nerve Block.—This involves the following.

Area of Anesthesia.—Skin of the upper eyelid and the lower forehead close to the midline (see Fig 19–21).

Indications.—Repair of lacerations of the medial end of the upper eyelid and the forehead close to the midline.

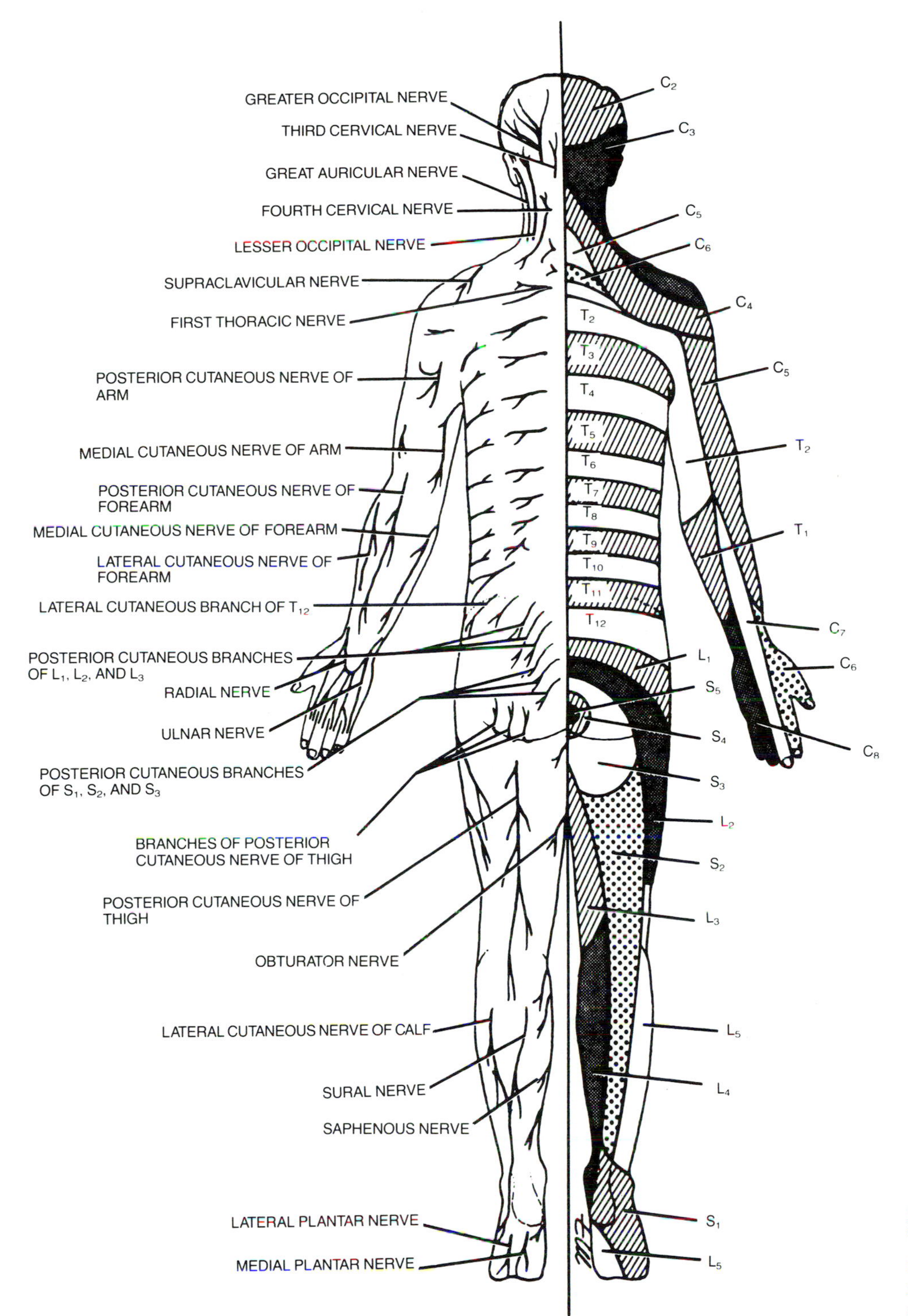

FIG 19–20.
The dermatomes and the distribution of the cutaneous nerves on the posterior aspect of the body.

Procedure.—The supratrochlear nerve winds around the supraorbital margin about a fingerbreadth medial to the supraorbital nerve. The needle is inserted at the point where the bridge of the nose meets the supraorbital margin (see Fig 19–21).

Infratrochlear Nerve Block.—This involves the following.

Area of Anesthesia.—Skin of the medial ends of the eyelids and the side of the root of the nose (see Fig 19–21).

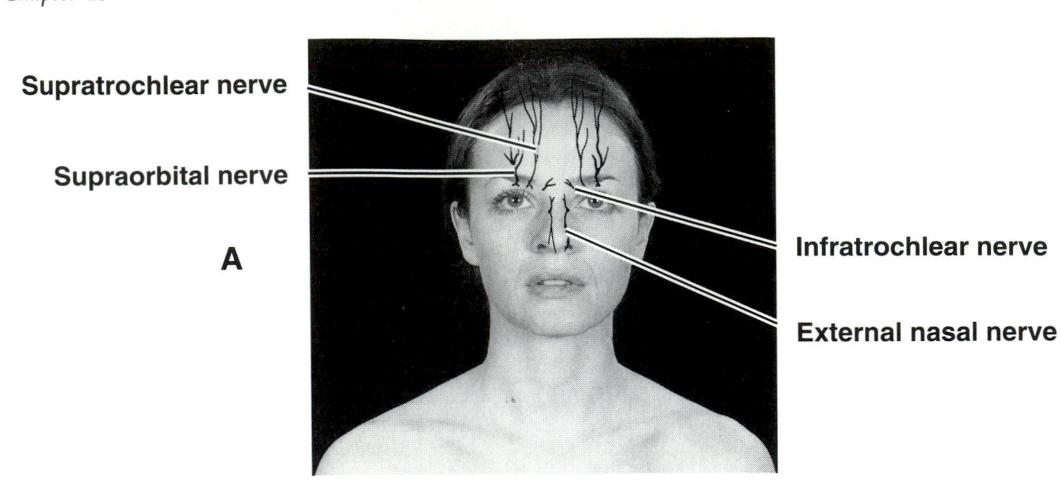

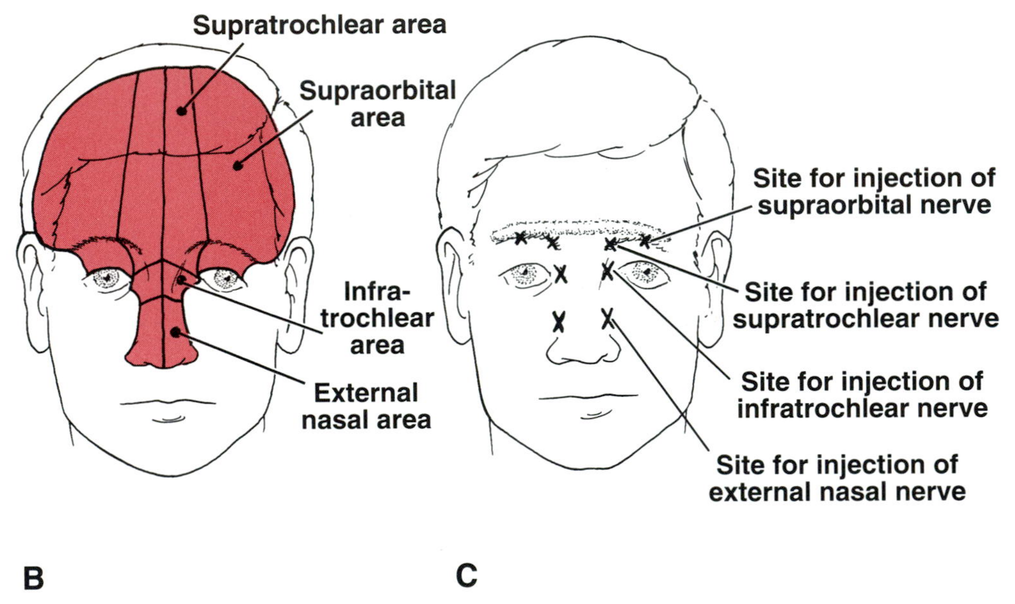

FIG 19–21.
Supraorbital, supratrochlear, infratrochlear, and external nasal nerve blocks. **A,** the positions of the nerves on the face; the supraorbital nerve emerges from the orbital cavity in the same vertical plane as the pupil. **B,** the areas of skin anesthetized by blocking these nerves. **C,** sites where the nerves may be blocked. The supraorbital and supratrochlear nerves may be blocked by raising a horizontal wheal of anesthetic solution above the orbital margin.

Indications.—Repair of lacerations of the medial eyelids and the root of the nose.

Procedure.—The infratrochlear nerve emerges from the orbital cavity at the junction of the superior and medial walls, which is the site for injection (see Fig 19–21).

External Nasal Nerve Block.—This involves the following.

Area of Anesthesia.—Skin of the side of the nose down as far as the tip (see Fig 19–21).

Indications.—Repair of lacerations of the skin of the nose.

Procedure.—The external nasal nerve is blocked at the point where it emerges between the nasal bone and the upper lateral nasal cartilage (see Fig 19–21).

Maxillary Nerve Block

This block is not used in emergency medicine. The infraorbital nerve, which is a continuation of the maxillary nerve onto the face, is commonly blocked. Occasionally, the pterygopalatine ganglion is blocked.

Infraorbital Nerve Block.—This involves the following.

Area of Anesthesia.—Skin of the lower eyelid, the lateral nose, the cheek, and the skin and mucous membrane of the upper lip and the upper gingiva. Since the anesthetic agent also blocks the anterior and middle superior alveolar nerves, the upper incisor, canine, and premolar teeth are also affected (see Fig 19–4).

Indications.—Lacerations of the cheek, side of the nose, and the upper lip.

Procedure.—The infraorbital nerve emerges from the infraobital foramen as a direct continuation of the maxillary nerve (Fig 19–22). The opening of the foramen is situated about 1 cm below the midpoint of the lower border of the orbit and faces downward and medially.

Intraoral Method.—With the index finger of the left hand palpating the infraorbital foramen through the skin of the cheek and serving as a guide, the needle is inserted into the reflection of the mucous membrane from the upper lip onto the gingiva (see Fig 19–22). The site for the needle insertion is just posterior to the canine tooth and is directed upward to the infraorbital foramen.

Extraoral Method.—The infraorbital foramen is palpated below the lower margin of the orbit, and the needle is inserted through the skin and is directed upward and outward toward the foramen (see Fig 19–22).

Pterygopalatine (Sphenopalatine) Ganglion Block.—The pterygopalatine ganglion is a small parasympa-

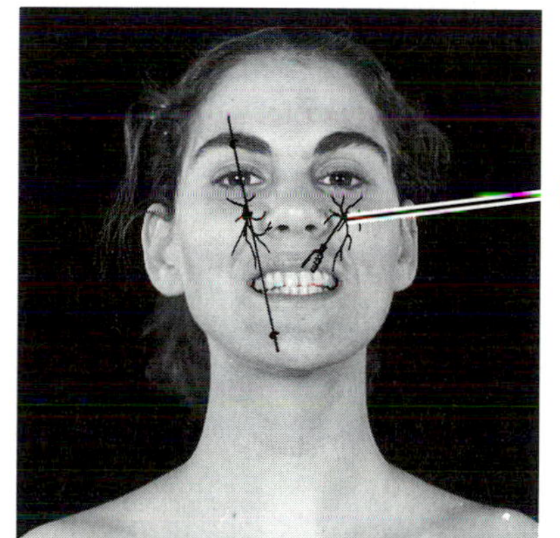

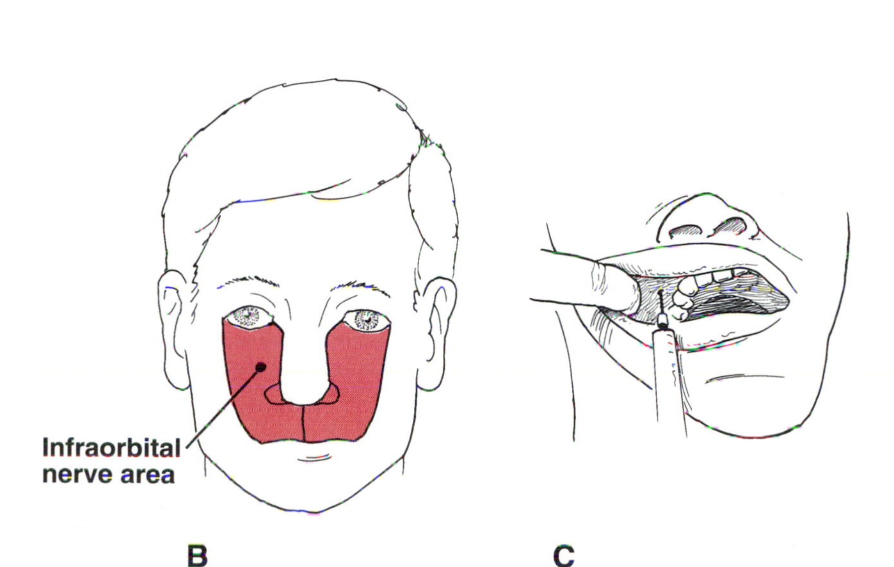

FIG 19–22.
Infraorbital nerve block. **A,** extraoral method shows the infraorbital nerve emerging from the infraorbital foramen. The infraorbital foramen lies on the same vertical line that passes through the supraorbital notch, the mental foramen, and the first premolar tooth. The blocking needle is inserted in the direction of the infraorbital foramen just below the lower margin of the orbit. **B,** the area of skin anesthetized by blocking the infraorbital nerve. **C,** intraoral method. The needle is inserted into the reflection of the mucous membrane from the upper lip onto the gingiva just posterior to the canine tooth and is directed toward the infraorbital foramen.

thetic ganglion and is suspended from the lower border of the maxillary nerve in the pterygopalatine fossa (Fig 19–23). Passing through the ganglion without interruption are the sensory fibers from the orbit, the nose, the hard and soft palate, the gums, and the tonsillar region of the pharynx.

Area of Anesthesia.—The lower nasal cavity, hard and soft palates, the upper gum, the teeth of the upper jaw, and the tonsillar region of the pharynx.

Indications.—Repair of lacerations involving the palate.

Procedure.—The ganglion and, therefore, the sensory fibers may be blocked by inserting a long-angled needle into the greater palatine foramen with the mouth wide open (see Fig 19–23). The foramen is located at the posterior portion of the hard palate just medial to the gumline of the third molar tooth. The greater palatine foramen leads superiorly into the pterygopalatine fossa. Injection of the anesthetic blocks the greater and lesser palatine nerves, the orbital nerves, the nasal nerves, and the pharyngeal nerves.

Mandibular Nerve Block

This block is not used in emergency medicine. The auriculotemporal nerve, the lingual nerve, and the inferior alveolar nerve, which are branches of the mandibular nerve, are commonly blocked.

Auriculotemporal Nerve Block.—This involves the following.

Area of Anesthesia.—The external auditory meatus, the tympanic membrane, the upper part of the

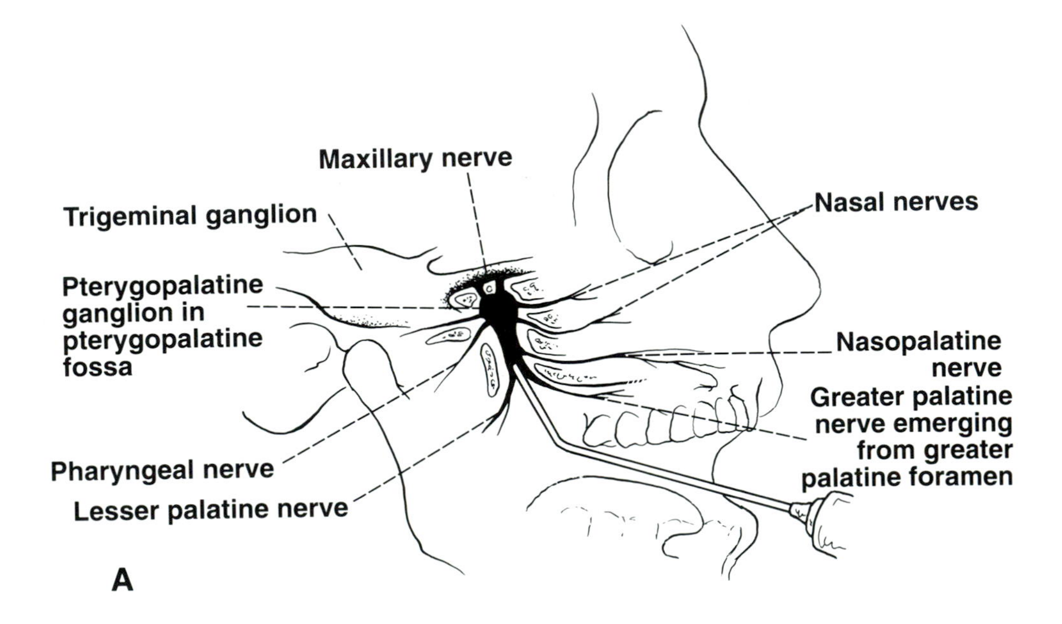

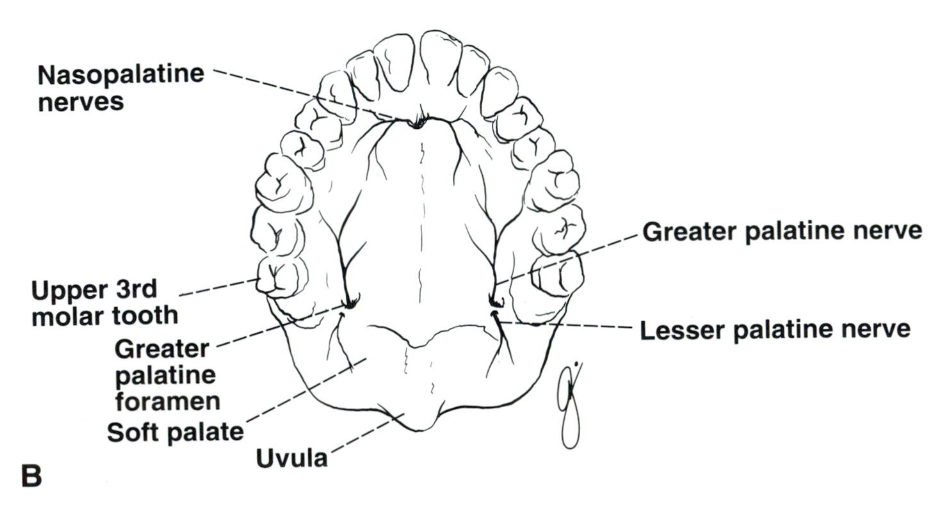

FIG 19–23.
Blocking the pterygopalatine (sphenopalatine) ganglion. **A**, lateral view of the skull showing the insertion of the needle into the greater palatine foramen in order to block the ganglion. **B**, the undersurface of the palate showing the position of the greater palatine foramen in relation to the upper third molar tooth and the soft palate. Note the distribution of the branches of the maxillary nerve through the ganglion to the walls of the nose and the palate.

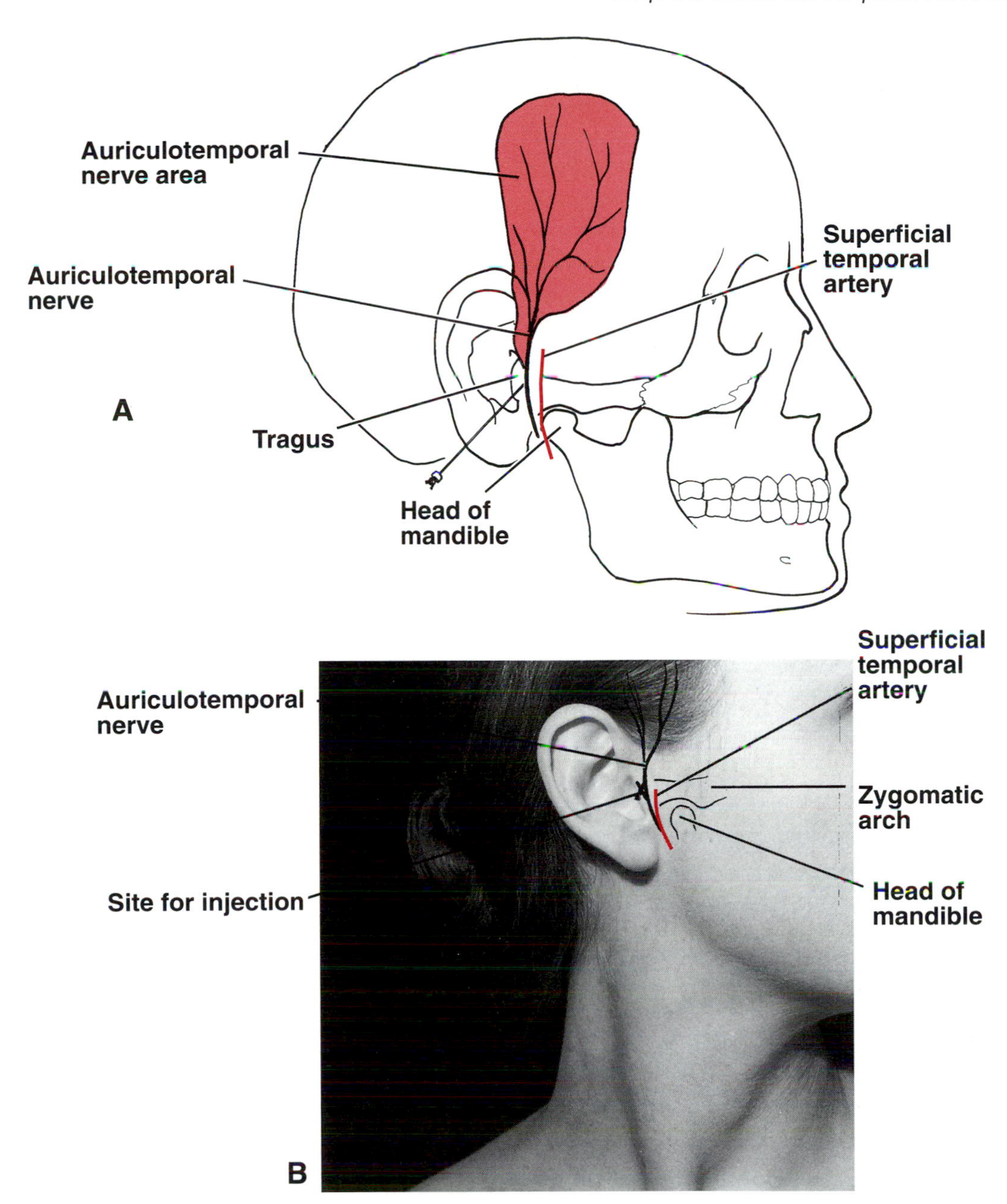

FIG 19–24.
Blocking the auriculotemporal nerve. **A,** the relationship of the auriculotemporal nerve to the superficial temporal artery and the tragus of the ear. Note the area of skin supplied by this sensory nerve. The needle is inserted just behind the pulsating superficial temporal artery and in front of the tragus; the needle is directed horizontally medially. **B,** the surface marking of the auriculotemporal nerve and its relationship to the superficial temporal artery and the temporomandibular joint. X marks the site for injection.

auricle, and the scalp in the temporal region (Fig 19–24).

Indications.—Repair of lacerations of the auricle and scalp.

Procedure.—The auriculotemporal nerve is easily blocked as it ascends in front of the auricle over the posterior root of the zygoma, behind the superficial temporal artery (see Fig 19–24).

Anatomy of Complications.—The superficial temporal artery may be pierced if the needle is inserted too far anteriorly.

Lingual Nerve and Inferior Alveolar Nerve Blocks.— These involve the following.

Area of Anesthesia.—The lingual nerve supplies the mucous membrane of the anterior two thirds of the tongue and the floor of the mouth (taste is supplied by the chorda tympani branch of the facial nerve), and the lower gums. The inferior alveolar nerve supplies the lower teeth and gums and the skin of the lower lip and chin.

Indications.—Repair of lacerations of the tongue, floor of the mouth, and the lower lip and chin.

Procedure.—Both the lingual and the inferior alveolar nerves may be blocked as they pass downward and forward in the infratemporal fossa on the lateral surface of the medial pterygoid muscle and on the medial surface of the ramus of the mandible (Fig 19–25). With the patient's mouth wide open, the anterior border of the ramus of the mandible is palpated just above the third molar tooth. The blocking needle is inserted above the palpating finger and between the mucosa and the inner surface of the ramus of the mandible, and the barrel of the syringe lies in line with the interval between the bicuspids on the opposite side of the mandible (see Fig 19–25). The needle is advanced posteriorly and slightly superiorly until the tip lies in close proximity to the mandibular foramen. The anesthetic solution will then infiltrate around the nerves.

Mental Nerve Block.— This involves the following.

Area of Anesthesia.—The lower lip and gums.

Indications.—Repair of lacerations of the lower lip.

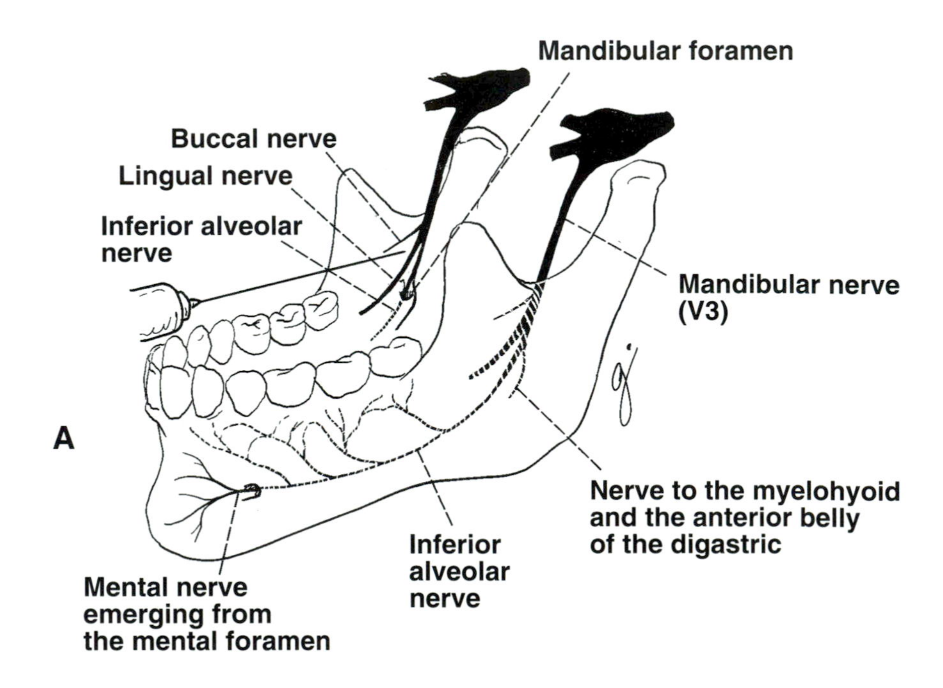

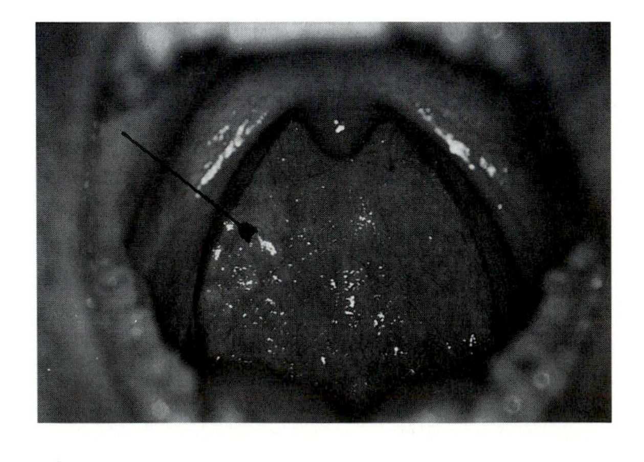

FIG 19–25.
Lingual and inferior alveolar nerve blocks. **A,** the location of the buccal, lingual, and inferior alveolar nerves in relation to the mandible. **B,** the needle is inserted just above the lower third molar tooth and directed between the mucosa and the inner surface of the ramus of the mandible; the barrel of the syringe lies in line with the interval between the bicuspids on the opposite side of the mandible. The needle is advanced posteriorly and slightly superiorly until the tip lies in close proximity to the mandibular foramen.

Procedure.—The mental nerve may be blocked as it emerges from the mental foramen on the body of the mandible (Fig 19–26). The foramen lies on the same vertical line that passes through the supraorbital notch, the infraorbital foramen, and the first premolar tooth.

Intraoral Method.—The left index finger palpates the position of the mental foramen. The needle is inserted through the reflection of the mucous membrane from the lower lip onto the gum between the apices of the premolar teeth (see Fig 19–26). The point of the needle is directed toward the mental foramen.

Extraoral Method.—The mental foramen is palpated, and the needle is inserted through the skin. When the mandible is contacted with the needle, the point is directed toward the mental foramen (see Fig 19–26).

Special Areas for Nerve Blocks

Tooth Nerve Blocks.—Two techniques are commonly used—supraperiosteal infiltration and dental nerve blocks.

Supraperiosteal Infiltration.—This technique is commonly used in the emergency department for the relief of toothache. The anesthetic solution is applied directly to the outer surface of the periosteum opposite the apices of the roots of the teeth (see Fig 19–26). The anesthetic diffuses through the periosteum and the alveolar bone to reach the dental nerve fibers entering the apices of the dental roots; it also reaches the nerves supplying the mucoperiosteum of the gums and the peridontal membrane.

The labiogingival or buccalgingival folds, where the mucous membrane lining the lips or cheek are reflected onto the gums, are identified. This may be

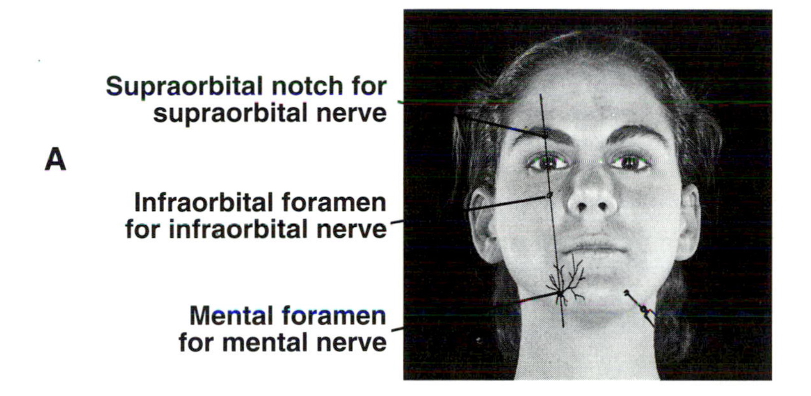

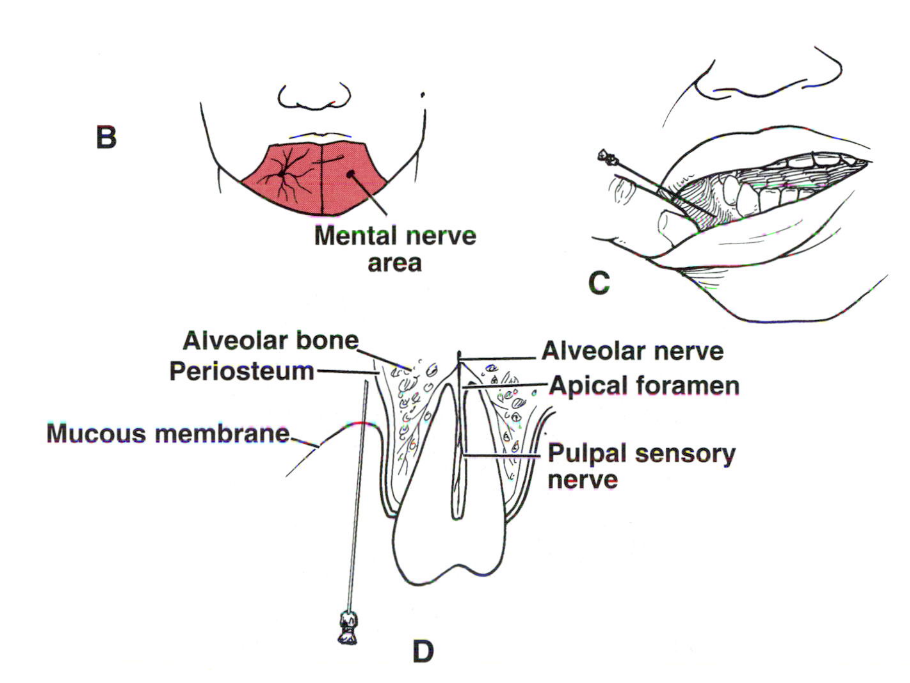

FIG 19–26.
Mental nerve and tooth nerve blocks. **A,** extraoral mental nerve block. The surface marking of the mental nerve as it emerges from the mental foramen. The mental foramen lies on the same vertical plane as the supraorbital notch, the pupil (when the patient is looking straight ahead), the infraorbital foramen, and the first premolar tooth. The mental foramen is palpated, and the needle is inserted through the skin and is directed toward the foramen. **B,** area of skin anesthesia produced by blocking the mental nerve. **C,** intraoral mental nerve block. The needle is inserted through the reflection of mucous membrane between the apices of the premolar teeth and is directed toward the mental foramen. **D,** supraperiosteal infiltration. Needle is inserted through the mucous membrane with the bevel against the periosteum and is advanced until it reaches the level of the apex of the tooth.

accomplished in the maxilla by pulling the upper lip downward, and in the case of the mandible by pulling the lower lip upward. At the point where the mucous membrane becomes fused with the periosteum to form the mucoperiosteum of the gum, the needle is inserted with the bevel against the periosteum (see Fig 19–26). The needle is advanced until it reaches the level of the apex of the root of the tooth, and the anesthetic solution is injected. If anesthesia is not produced, it may be necessary to repeat the injection on the palatal surface of the gums. Failure to produce an adequate nerve block may be due to the needle tip being inserted too far away from the apex of the tooth, i.e., too far away from the periosteum or too far above or below the tooth apex. The technique is less successful for mandibular teeth because of the density of the bone structure of the mandible.

The upper teeth are innervated by the anterior, middle, and superior alveolar branches of the maxillary division of the trigeminal nerve. The buccal nerve, a branch of the mandibular division of the trigeminal nerve, supplies the lateral surface of the gum and the greater palatine, and the nasopalatine nerves, from the maxillary nerve, supply the medial surface of the gum.

The lower teeth are innervated by the inferior alveolar nerve, a branch of the mandibular division of the trigeminal nerve. The buccal nerve supplies the lateral surface of the gums, and the lingual nerve, a branch of the mandibular nerve, supplies the medial surface.

Dental Nerve Blocks.—For the maxillary teeth the anterior and middle superior alveolar nerves are blocked along with the infraorbital nerve as described on p. 809. For the mandibular teeth, the inferior alveolar nerve is blocked as described on p. 812.

Anesthesia of the Nose.—This involves the following.

Nasal Interior.—The lateral wall of the nose is innervated by the anterior ethmoidal branch of the nasociliary branch of the ophthalmic division of the trigeminal nerve, from branches of the maxillary division of the trigeminal nerve, and from the olfactory nerve (see p. 5, Chapter 1). The nasal septum is innervated by branches of the anterior ethmoidal nerve, branches of the maxillary nerve, and from the olfactory nerve.

Analgesia of the mucous membrane can easily be obtained by placing pledgets soaked in local anesthetic in the nose between the conchae and the septum for 5 to 10 minutes.

Nasal Exterior.—The skin of the nose is innervated by the supratrochlear and infratrochlear branches of the ophthalmic division of the trigeminal nerve and the infraorbital nerve, a continuation of the maxillary division of the trigeminal nerve (Fig 19–27). Skin analgesia is obtained by infiltrating first the base of the nose and then the nasofacial groove, thus blocking the terminal branches of the ophthalmic and maxillary nerves (see Fig 19–27).

Anesthesia of the Ear.—The auricle is innervated by the greater auricular nerve (C2 and C3), a branch of the cervical plexus (Fig 19–28). This nerve mainly supplies the skin on the medial and lateral surfaces of the inferior part of the auricle. The auriculotemporal branch of the mandibular division of the trigeminal nerve supplies the lateral and upper part of the auricle. The lesser occipital nerve (C2) may also supply a small area on the medial surface. The external auditory meatus is also innervated by the auricular branch of the vagus nerve.

Anesthesia of the skin is obtained by multiple subcutaneous injections along a line that is continued circumferentially around the auricle (see Fig 19–28). The external auditory meatus may be anesthetized by using a four-quadrant block of the canal; in addition, several drops of anesthetic solution may be instilled into the canal to anesthetize the tympanic membrane.

Anesthesia of the Scalp.—The anterior part of the scalp extending back as far as the vertex is innervated by the supraorbital and supratrochlear branches of the ophthalmic division of the trigeminal nerve (Fig 19–29). The posterior part of the scalp is innervated by the greater occipital nerve (C2) and the lesser occipital nerve (C2). The lateral part of the scalp is supplied by the auriculotemporal branch of the mandibular division of the trigeminal nerve. A small area over the temple is supplied by the zygomaticotemporal nerve from the maxillary division of the trigeminal nerve.

A subcutaneous infiltration with anesthetic solution is made around the circumference of the head from just above the nose and eyebrows to the ear and back to the external occipital protuberance (see Fig 19–29). A large volume of anesthetic is required to completely anesthetize the scalp.

Clinical Anatomy of Brachial Plexus Block

The brachial plexus in the neck occupies the lower part of the posterior triangle (Fig 19–30). It lies be-

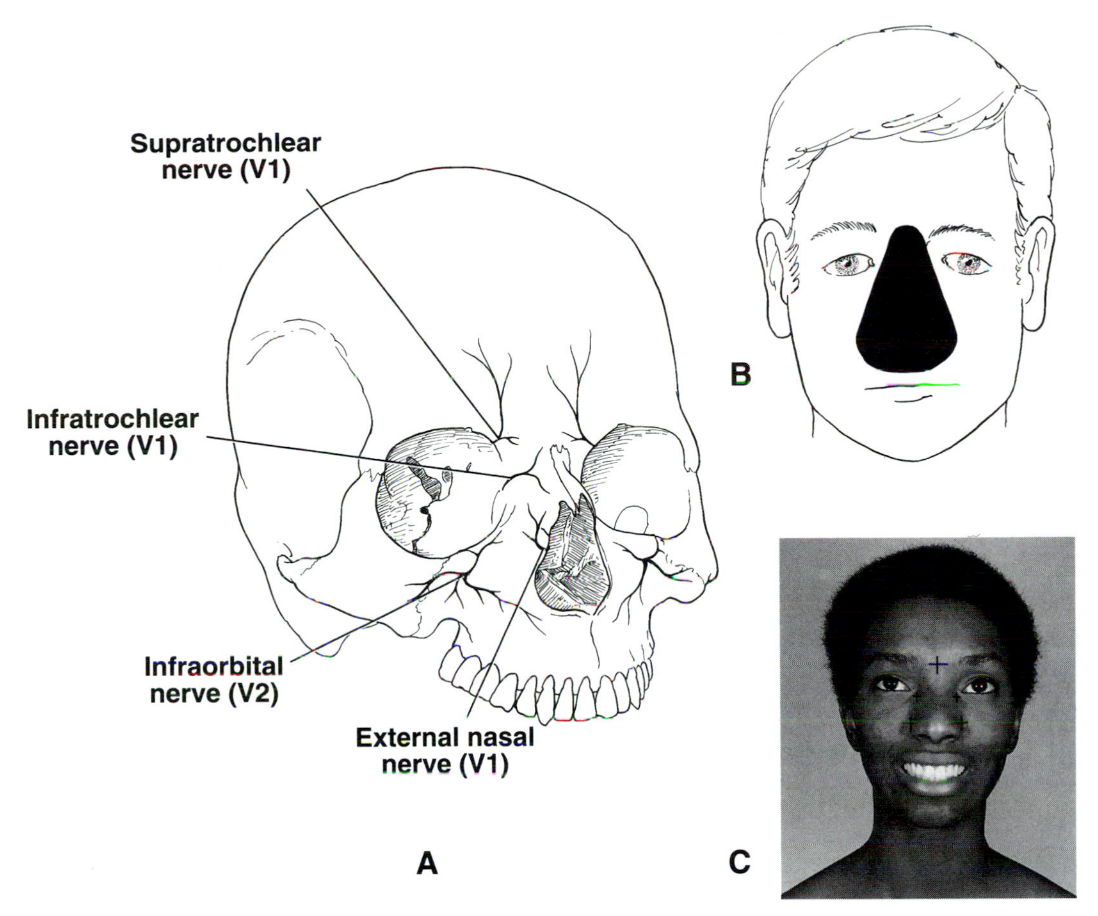

FIG 19–27.
Anesthesia of the external nose. **A,** the sensory nerves that supply the skin of the nose emerging from the skull. **B,** the extent of the skin supplied by these nerves. **C,** the sites *(X)* where the needle is introduced to produce anesthesia over the area shown in **B.**

low and anterior to a line connecting the cricoid cartilage to the midpoint of the clavicle. In the axilla, the brachial plexus and its branches are arranged within the axillary sheath around the axillary artery, which can be palpated.

Although brachial plexus nerve blocks are not generally performed in the emergency department, four techniques can be used—interscalene block, supraclavicular block, infraclavicular block, and axillary block.

Interscalene Block

Procedure.—At the level of the cricoid cartilage (C6), the posterior border of the sternocleidomastoid muscle can be palpated. With the head turned laterally and upward from the side of the block, the palpating finger can feel the groove between the scalenus anterior and the scalenus medius muscles just lateral to the sternocleidomastoid muscle (see Fig 19–30). The blocking needle is inserted into the interval between the scalene muscles, and the roots of the upper part of the brachial plexus can be blocked.

Supraclavicular Block

Procedure.—The trunks of the brachial plexus can be blocked as they cross the first rib and enter the axilla (see Fig 19–30). The third part of the subclavian artery is first palpated in the lower anterior corner of the posterior triangle. The posterior border of the sternocleidomastoid muscle is then felt. A blocking needle is inserted at the level of the cricoid cartilage between the scalenus anterior and the scalenus medius muscles and directed caudally behind the subclavian artery toward the upper surface of the first rib (see Fig 19–30). It is here that the brachial plexus is very compact, consisting of the upper, middle, and lower trunks.

Anatomy of Complications.—The subclavian artery lies between the scalenous anterior and scalenus me-

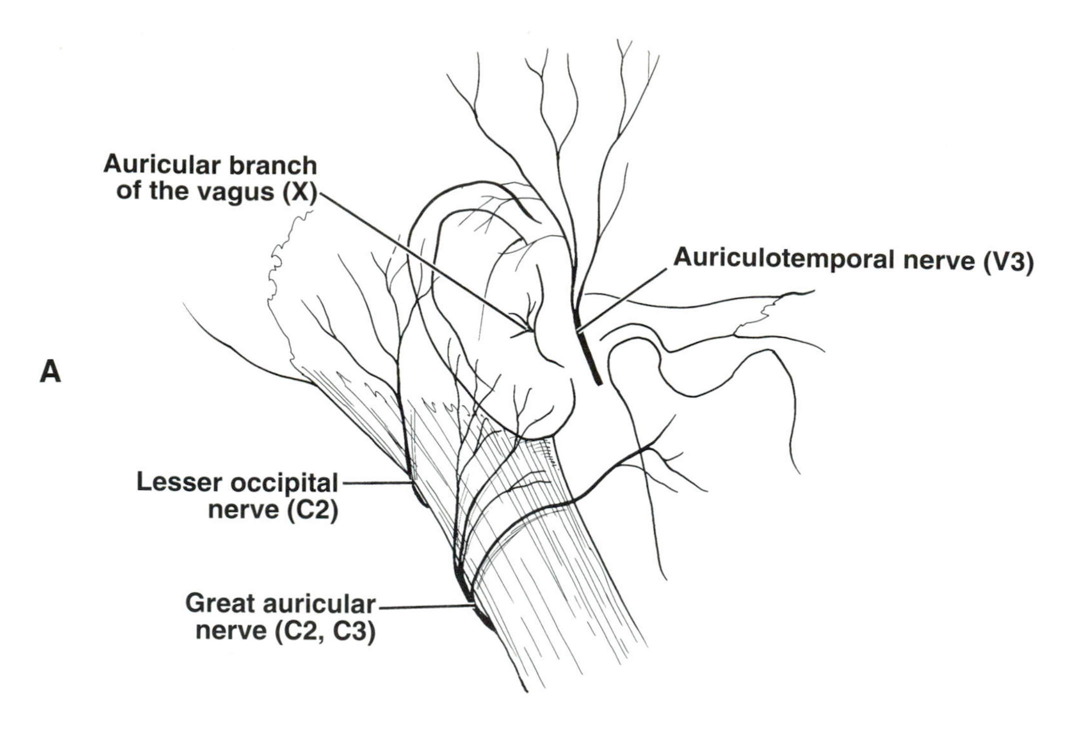

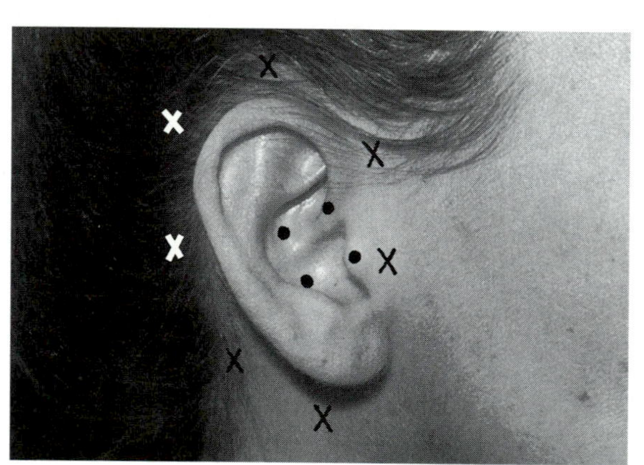

FIG 19–28.
Ear nerve blocks. **A,** the sensory innervation of the auricle; note the auricular branch of the vagus nerve that supplies part of the external auditory meatus. **B,** the sites *(X)* at which multiple subcutaneous injections may be made circumferentially around the auricle to block the sensory nerves. The external auditory meatus may be anesthetized by using a four-quadrant block (dots).

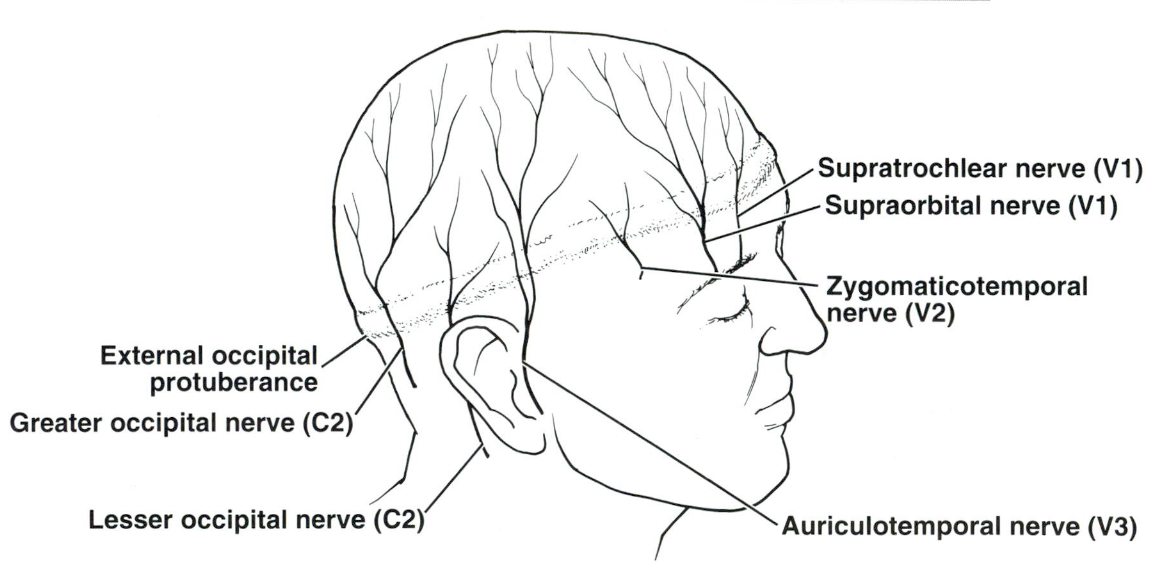

FIG 19–29.
Scalp nerve block. A subcutaneous infiltration with anesthetic solution is made around the circumference of the head from just above the eyebrows to the region of the external occipital protuberance.

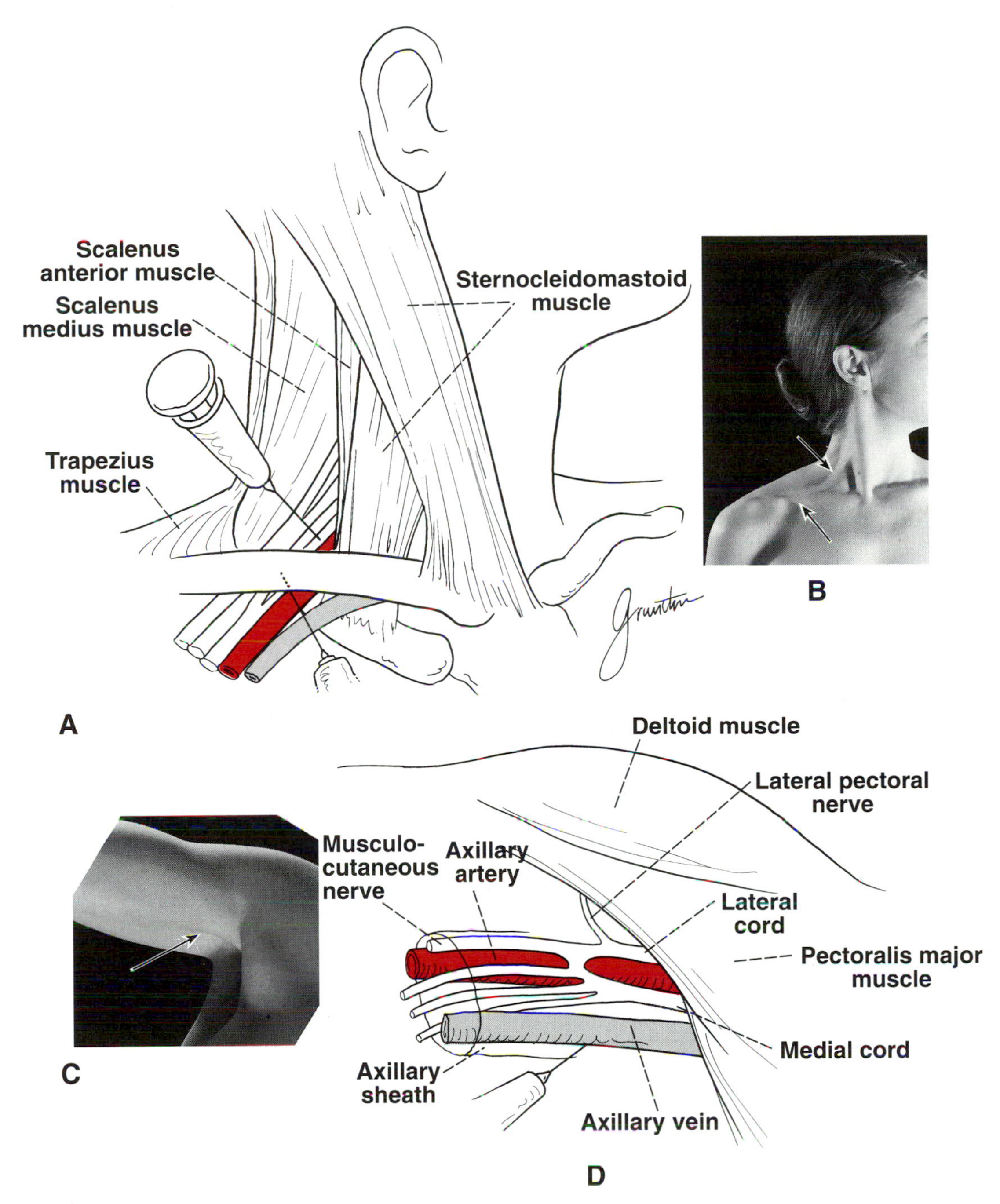

FIG 19–30.
A, brachial plexus block. Supraclavicular block. The needle is inserted just posterior to the posterior border of the sternocleidomastoid muscle at the level of the cricoid cartilage at the lower anterior corner of the posterior triangle. The needle is directed caudally behind the palpable subclavian artery. In an infraclavicular block, the needle is inserted inferior to the midpoint of the clavicle and is directed laterally in the direction of the head of the humerus and toward the subclavian artery. **B,** sites of supraclavicular and infraclavicular brachial plexus blocks. **C,** axillary nerve block. The axillary artery is palpated and the needle is inserted into the axillary sheath (see text). **D,** site of the axillary brachial plexus block.

dius muscles just behind the clavicle and may be pierced by the needle. The cervical dome of parietal pleura is situated close to the brachial plexus, and a pneumothorax can be caused if the needle enters the pleural cavity.

Infraclavicular Block

Procedure.—The middle of the clavicle is identified. A blocking needle is inserted 1 in. (2.5 cm) inferior to it and directed laterally in the direction of the head of the humerus and toward the subclavian artery. The anesthetic solution is infiltrated around the trunks of the brachial plexus (see Fig 19–30).

Anatomy of Complications.—The close relationship of the axillary vessels to the brachial plexus within the axillary sheath means that vessel puncture and hematoma formation may occur. Frequent aspirations are necessary before the anesthetic agent is injected.

Axillary Block

Procedure.—With the arm abducted to an angle greater than 90°, the axillary artery within the axillary sheath may be palpated high up in the axilla (see Fig 19–30). The artery is compressed, and a blocking needle is inserted just proximal to the point of compression into the axillary sheath. The cords and branches of the plexus lie within the sheath along with the artery. The disadvantage of this approach is the difficulty in blocking the musculocutaneous nerve. The object of compressing the artery distal to the point of injection is to close off the axillary sheath distally so that the anesthetic agent may rise in the sheath to the musculocutaneous nerve.

Anatomy of Complications.—The close relationship of the axillary vessels to the brachial plexus within the axillary sheath means that vessel puncture and hematoma formation may occur. Frequent aspira-

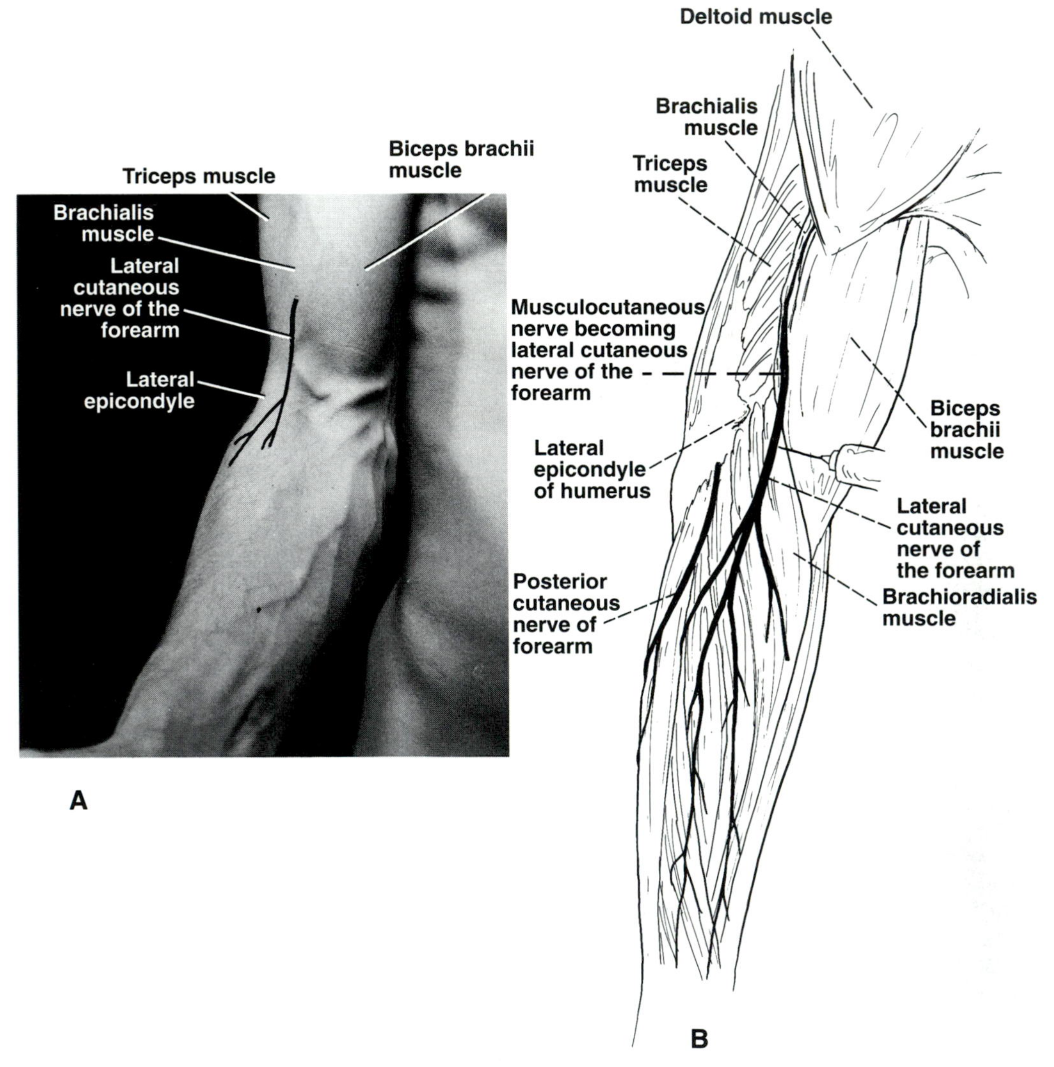

FIG 19–31.
Lateral cutaneous nerve of the forearm block. **A** and **B** show the musculocutaneous nerve becoming the lateral cutaneous nerve of the forearm. The nerve is blocked just lateral to the tendon of the biceps brachii muscle on a line between the two epicondyles of the humerus.

tions are necessary before the anesthetic agent is injected.

Clinical Anatomy of Musculocutaneous Nerve Block

Area of Anesthesia.—The anterior and posterior surfaces of the lateral border of the forearm down as far as the thenar eminence (see Figs 19–19 and 19–20).

Indications.—Repair of lacerations on the lateral border of the forearm.

Procedures.—These include the following.

Brachial Plexus Block.—The musculocutaneous nerve trunk may be blocked with the rest of the brachial plexus as it arises from the lateral cord of the brachial plexus (see Fig 19–30). The infraclavicular or axillary approach is used; in the axillary approach great care has to be taken to ensure that the anesthetic agent rises sufficiently high in the axillary sheath to block the musculocutaneous nerve.

Lateral Cutaneous Nerve of the Forearm Approach.—The musculocutaneous nerve may also be blocked as it emerges between the biceps and the brachialis muscles just above the lateral epicondyle of the humerus, where it becomes the lateral cutaneous nerve of the forearm (Fig 19–31). The needle is inserted just lateral to the tendon of the biceps muscle on a line between the two epicondyles of the humerus.

Clinical Anatomy of Median Nerve Block

Area of Anesthesia.—The skin on the radial half of the palm, the palmar aspect of the radial 3½ fingers, including the nail beds on the dorsum (see Fig 19–36).

Indications.—Repair of lacerations of the palm and fingers.

Procedures.—These include the following.

Block at the Elbow.—With the elbow joint extended, the brachial artery can easily be palpated in the cubital fossa on the medial side of the tendon of the biceps muscle. The needle is inserted on the ulnar side of the brachial artery (Fig 19–32).

Block at the Wrist.—Here the median nerve lies on the ulnar side of the tendons of the flexor carpi radialis and to the radial side of the flexor digitorum superficialis (Fig 19–33); it usually lies posterior and radial to the tendon of the palmaris longus muscle (sometimes absent). The nerve may be infiltrated

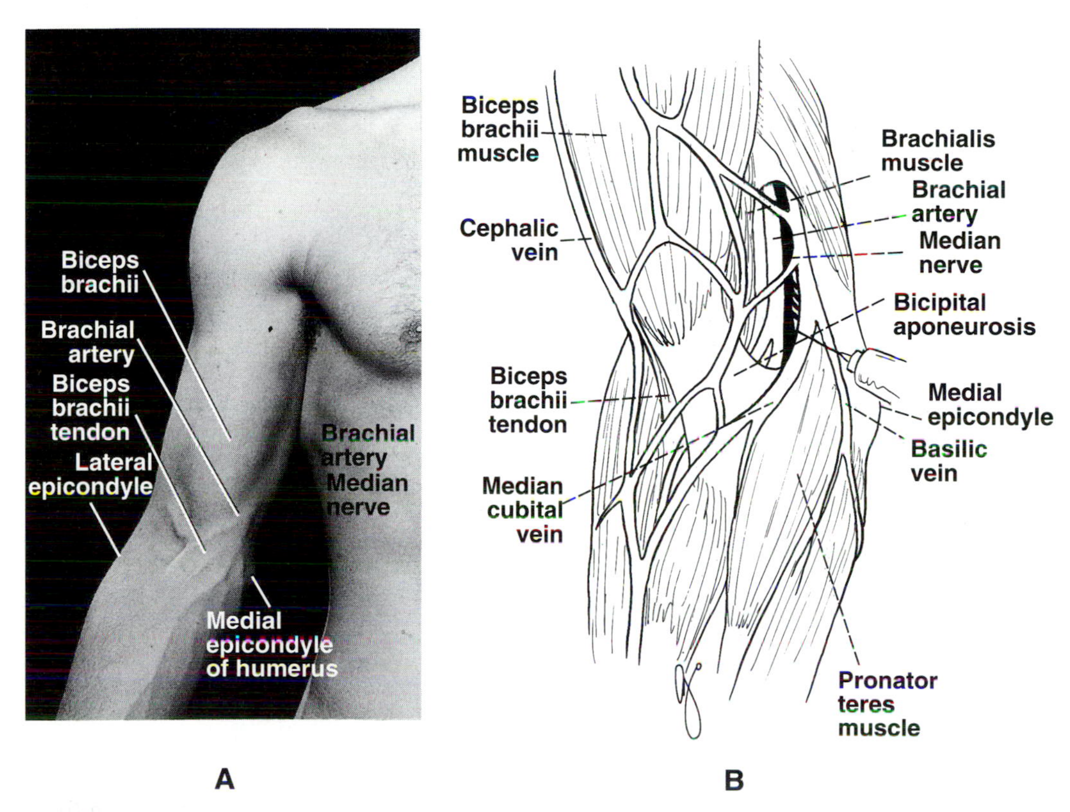

FIG 19–32.
A and **B**, median nerve block at the elbow. The brachial artery is identified in the cubital fossa, and the needle is inserted on the ulnar side of the artery.

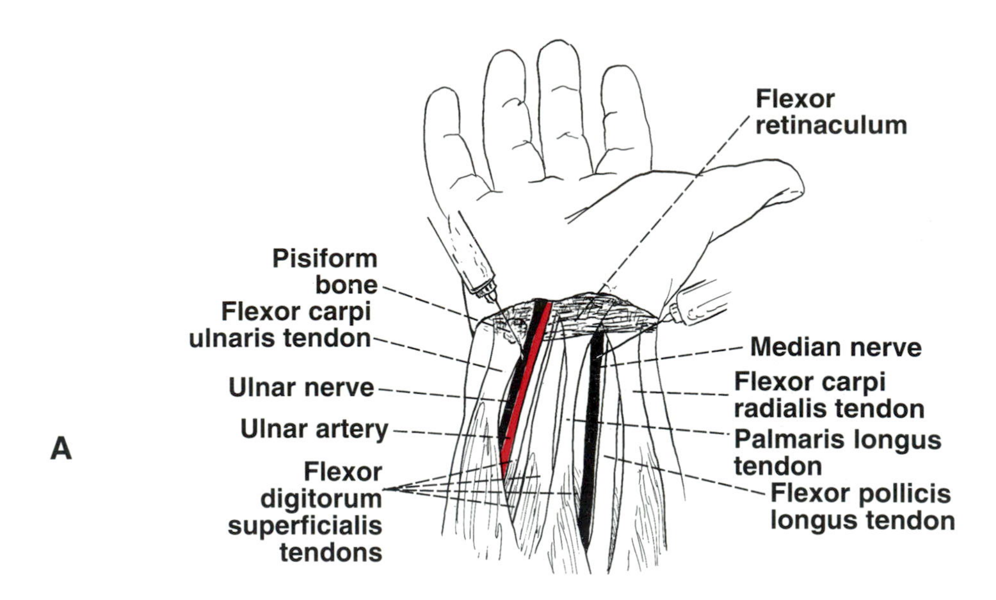

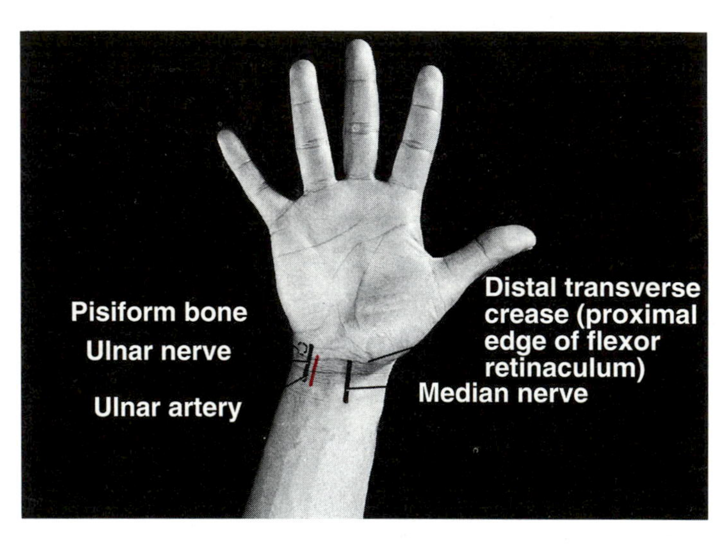

FIG 19–33.
A and **B**, median and ulnar nerve blocks at the wrist. In a median nerve block, the needle is inserted just proximal to the distal transverse crease of the wrist between the tendons of flexor carpi radialis and the palmaris longus. In an ulnar nerve block, the needle is inserted just radial to the flexor carpi ulnaris tendon at the level of the distal transverse crease of the wrist.

here with local anesthetic just proximal to the flexor retinaculum, i.e., just proximal to the distal transverse crease in front of the wrist. The needle is inserted for ⅜ in. between the tendons of flexor carpi radialis and the palmaris longus.

The palmar cutaneous branch of the median nerve leaves the main trunk just proximal to the distal transverse crease of the palm, and, unless this nerve is also blocked, the sensation to the skin on the lateral part of the palm will remain intact.

Clinical Anatomy of Ulnar Nerve Block

Area of Anesthesia.—The skin of the ulnar one third of the palmar and dorsal surfaces of the hand and the palmar and dorsal surfaces of the ulnar 1½ fingers (see Fig 19–36).

Indications.—Repair of lacerations of the hand and fingers.

Procedures.—These involve the following.

Block at the Elbow.—At the elbow, the ulnar nerve enters the forearm between the olecranon process of the ulna and the medial epicondyle of the humerus (see Fig 19–34). Here the nerve may be palpated and infiltrated with an anesthetic agent.

Block at the Wrist.—At the wrist, the ulnar nerve enters the hand *anterior* to the flexor retinaculum and radial to the tendons of the flexor carpi ulnaris muscle and the pisiform bone (see Fig 19–33). The ulnar artery lies on the radial side of the ulnar nerve. The needle is inserted just radial to the flexor carpi ulnaris tendon at the level of the distal transverse crease of the wrist.

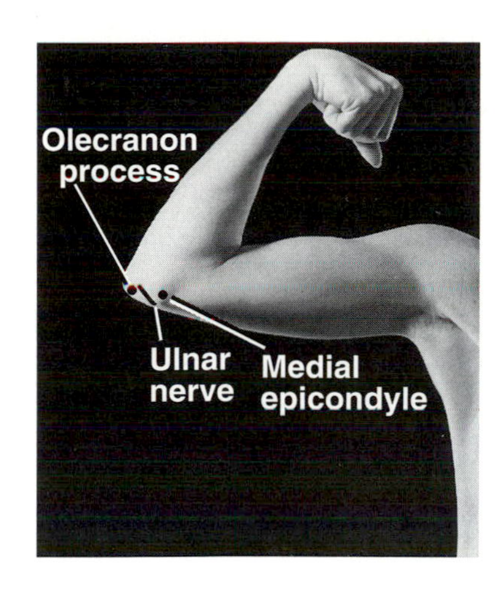

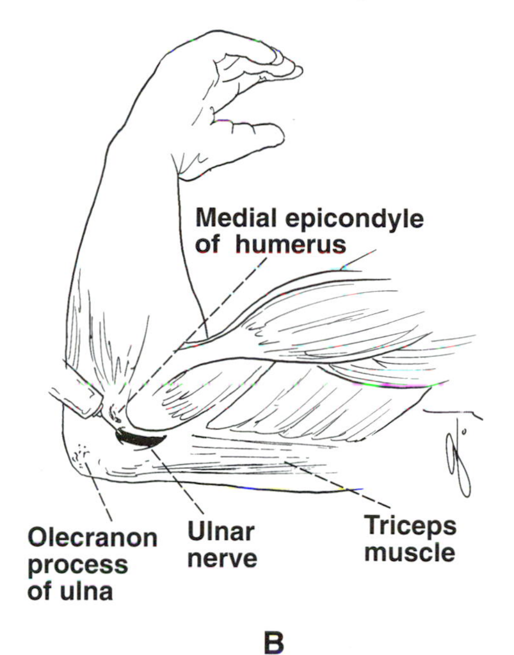

FIG 19–34.
A and **B**, ulnar nerve block at the elbow. The ulnar nerve may be palpated between the olecranon process of the ulna and the medial epicondyle of the humerus, where it may be infiltrated with anesthetic.

Note that the dorsal cutaneous branch of the ulnar nerve (see Fig 19–36) leaves the main trunk about 2 in. (5 cm) proximal to the wrist; an ulnar nerve block at the wrist will therefore leave the skin sensation on the dorsum of the hand and fingers intact.

Also, the superficially placed palmar cutaneous branch of the ulnar nerve leaves the nerve trunk a variable distance proximal to the transverse crease of the wrist; unless this nerve is also infiltrated with anesthetic, the ulnar block at the wrist will leave the sensory nerve supply to the ulnar part of the palm intact.

Clinical Anatomy of Radial Nerve Block

Area of Anesthesia.—The skin of the back of the arm down as far as the elbow, the skin of the lower lateral surface of the arm down to the elbow, the skin down the middle of the posterior surface of the forearm as far as the wrist, the skin of the radial half of the dorsal surface of the hand, and the dorsal surface of the radial 3½ fingers proximal to the nail beds (see Figs 19–19 and 19–20).

Indications.—Repair of lacerations of the hand.

Procedures.—These involve the following.

Block at the Elbow.—At the elbow, the radial nerve descends anterior to the lateral epicondyle of the humerus in the interval between the brachialis and the brachioradialis muscles (see Fig 19–35). With the elbow joint extended, the lateral edge of the biceps tendon is easily palpated. The needle is inserted halfway between the tendon and the tip of the lateral epicondyle, and the local anesthetic is injected at this point.

Block at the Wrist.—Just proximal to the wrist, the superficial branch of the radial nerve lies lateral to the radial artery. The nerve leaves the artery and passes laterally and backward under the tendon of the brachioradialis to reach the posterior surface of the wrist (see Fig 19–35). At the level of the proximal transverse flexor crease on the radial side of the radial artery, the nerve may be infiltrated with an anesthetic solution. Since other terminal branches of the radial nerve run in the subcutaneous tissue on the dorsum of the wrist, a subcutaneous wheal of local anesthetic is necessary; this should run across the lateral half of the dorsum of the wrist.

Clinical Anatomy of Digital Nerve Blocks

Area of Anesthesia.—Skin of the fingers. Each finger is supplied by four digital nerves at the 2-o'clock, 5-o'clock, 7-o'clock, and 10-o'clock positions. The palmar digital nerves are derived from the ulnar and median nerves; the dorsal digital nerves are derived from the ulnar and radial nerves (Fig 19–36).

The palmar digital nerves, which arise from the superficial terminal branch of the ulnar nerve in the hand, supply the palmar surface of the ulnar 1½ fingers, including their nail beds. The dorsal digital nerves, which arise from the dorsal cutaneous branch of the ulnar nerve in the forearm, supply the

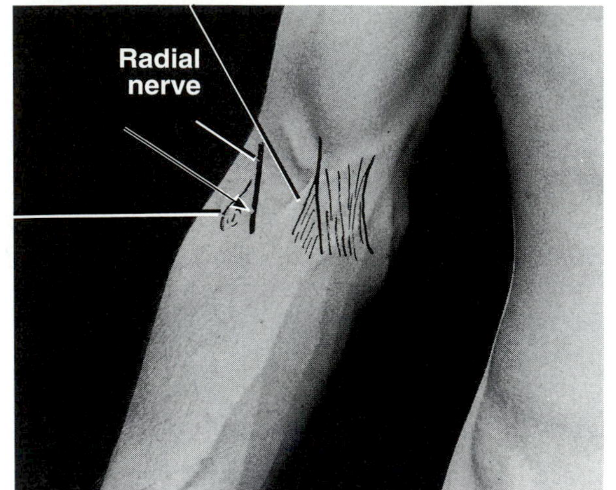

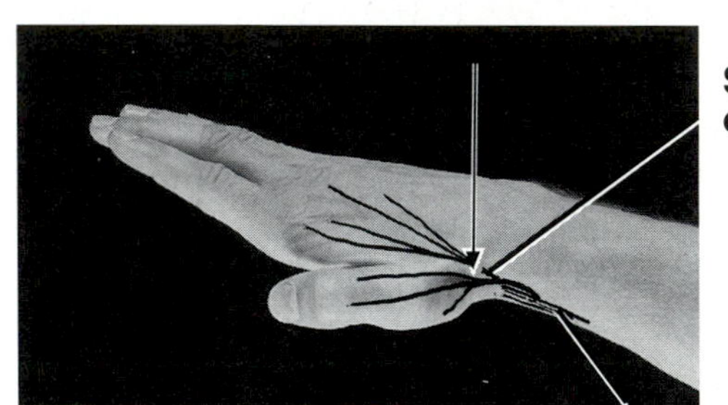

FIG 19–35.
A, radial nerve block at the elbow. The needle is inserted halfway between the tendon of the biceps brachii and the tip of the lateral epicondyle of the humerus. **B,** radial nerve block at the wrist. The needle is inserted into the subcutaneous tissue on the dorsum of the hand, and a wheal of anesthetic is placed across the lateral half of the dorsum of the wrist.

dorsal surface of the proximal parts of the ulnar 1½ fingers (see Fig 19–36).

The palmar digital nerves, which arise from the median nerve in the palm, supply the palmar surface of the radial 3½ fingers, including their nail beds. The dorsal digital nerves, which arise from the superficial branch of the radial nerve, supply the dorsal surface of the proximal parts of the radial 3½ fingers (see Fig 19–36).

Note that the origins of the dorsal digital nerves from the ulnar and radial nerves are subject to variation.

Indications.—Repair of lacerations involving individual fingers; removal of nails.

Procedures.—These involve the following.

Web Space Method.—At the web space, the digital nerves are about to enter the fingers (Fig 19–37). The needle is inserted about 0.5 cm, and the nerves are infiltrated with the anesthetic agent. A block on both sides of the finger adequately deals with the four digital nerves supplying the finger. When blocking the digital nerves of the index and little fingers, a subcutaneous wheal of anesthetic solution must be deposited on the radial side of the index finger and the ulnar side of the little finger.

Dorsal Metacarpal Method.—A skin wheal of anesthetic is raised between the metacarpal bones on the dorsum of the hand (see Fig 19–37). The needle is inserted through the wheal and advanced slowly forward between the metacarpal bones, stopping just short of the palmar skin. The anesthetic solution will block the common palmar and dorsal digital nerves.

Metacarpal Head Block.—The needle is inserted on the palm at the metacarpal head and directed slightly distally and ulnarly and then slightly distally and radially in order to block the volar digital nerves (see Fig 19–37). The dorsal digital nerves, if they remain unblocked, are handled with an injection across the base of the dorsum of the proximal phalanx.

Ring Block.—Since the digital nerves lie in pairs on either side of the proximal phalanges, they are easily blocked at the base of the digit (see Fig 19–36). The needle is inserted on both sides of the base of the proximal phalanx, and the anesthetic agent is infiltrated around the nerves between the bone and the skin; this produces a half-ring block on either side of the finger (Fig 19–37).

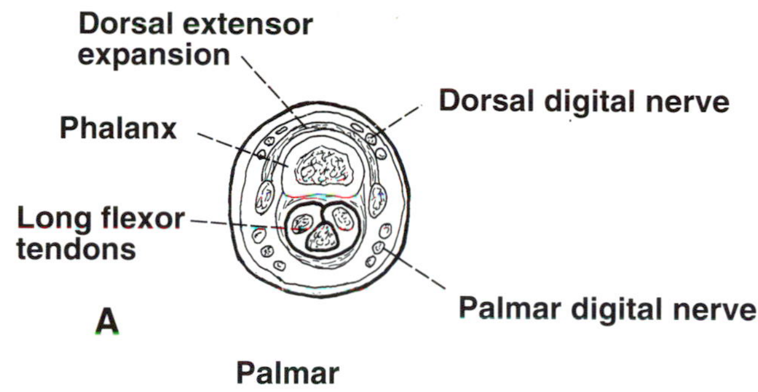

B

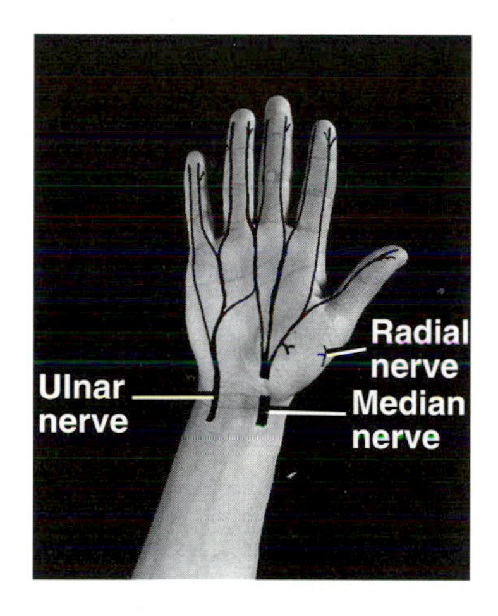

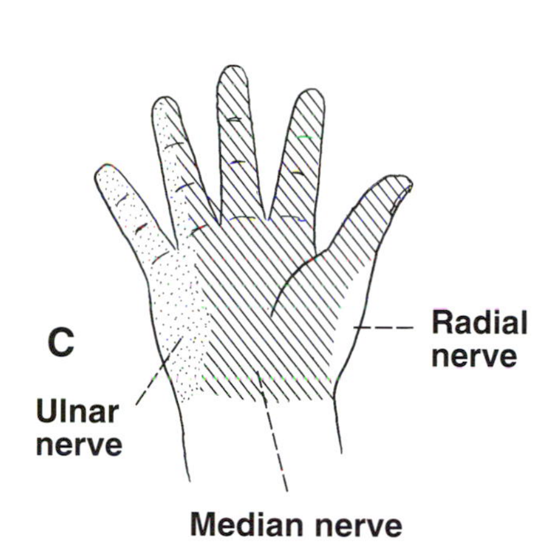

D

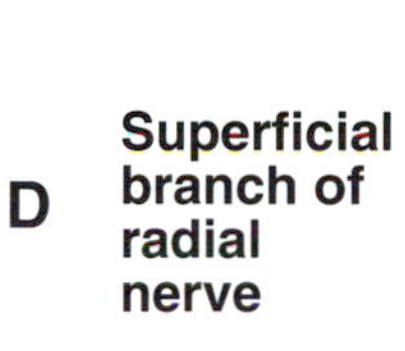

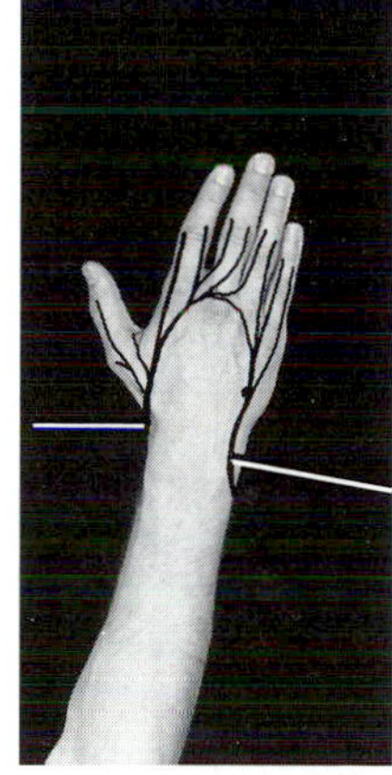

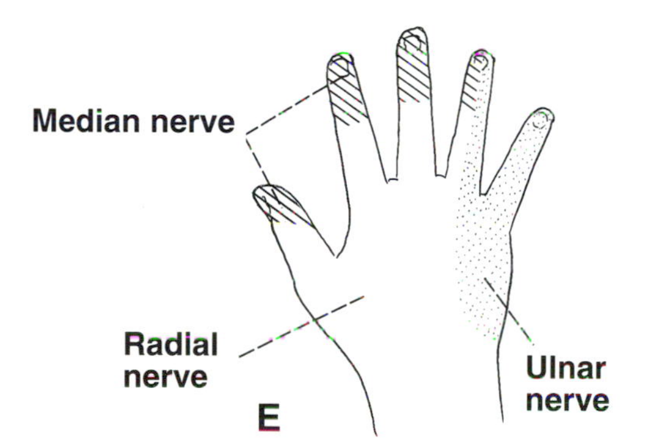

FIG 19–36.
A, transverse section of the finger showing the location of the dorsal and palmar digital nerves. **B,** palmar surface of hand and fingers showing the origin of the palmar digital nerves. **C,** palm of the surface of the hand and fingers showing the areas of skin supplied by the ulnar, median, and radial nerves. **D,** dorsal surface of hand and fingers showing the origin of the dorsal digital nerves. **E,** dorsal surface of the hand and fingers showing the areas of skin supplied by the radial, ulnar, and median nerves.

Clinical Anatomy of Thoracic Spinal Nerve Blocks

Intercostal Nerve Block

Area of Anesthesia.—The skin and the parietal pleura covering the outer and inner surfaces of each intercostal space respectively; the 7th to 11th intercostal nerves supply the skin and the parietal peritoneum covering the outer and inner surfaces of the abdominal wall, respectively; therefore these areas will also be anesthetized. In addition, the periosteum of the adjacent ribs is anesthetized.

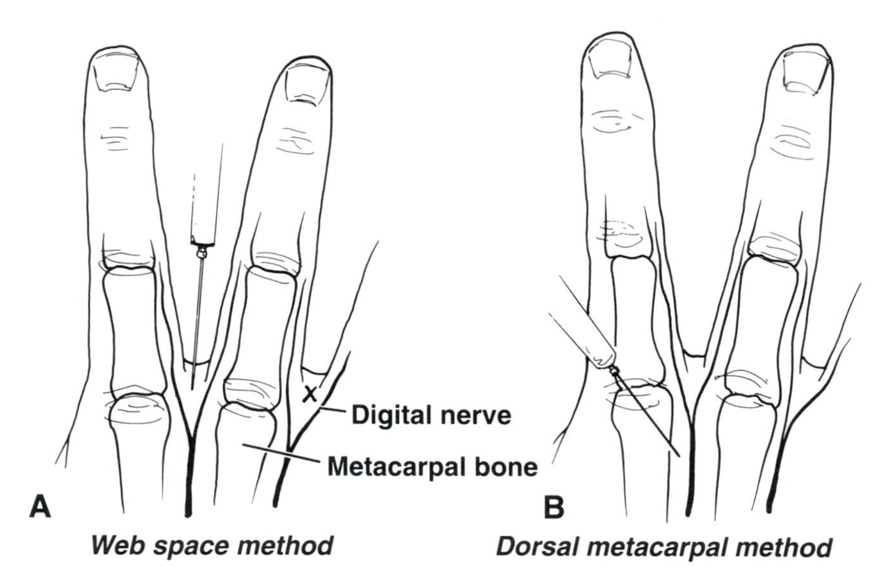

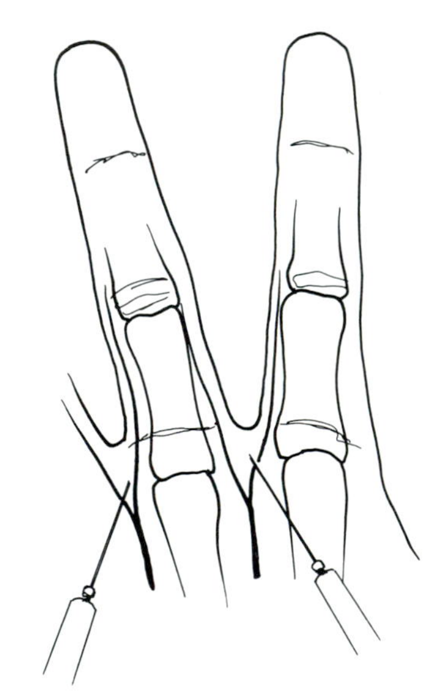

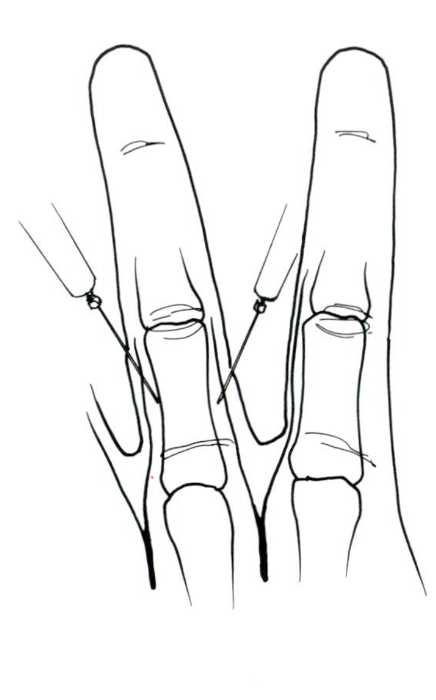

FIG 19–37.
Digital nerve blocks. **A,** web space method. The needle is inserted about 0.5 cm into the web space. A block on both sides of the finger anesthetizes the four digital nerves to the finger. **B,** dorsal metacarpal method. The needle is directed forward between the metacarpal bones, stopping just short of the palmar skin. The four digital nerves to each finger are blocked. **C,** metacarpal head block. The needle is inserted on the palm at the level of the metacarpal head and directed slightly distally and ulnarly and then slightly distally and radially. **D,** ring block. The needle is inserted on both sides of the base of the proximal phalanx to produce a half-ring block on either side of the finger.

Indications.—Repair of lacerations of the thoracic and abdominal walls; relief of pain in rib fractures and to allow pain-free respiratory movements.

Procedure.—To produce analgesia of the anterior and lateral thoracic and abdominal walls, the intercostal nerve should be blocked before the lateral cutaneous branch arises at the midaxillary line. The ribs may be identified by counting up from the 12th or down from the 2nd (opposite sternal angle). The needle is directed toward the rib near the lower border (Fig 19–38), and the tip comes to rest near the subcostal groove, where the local anesthetic is infiltrated around the nerve. Remember that the order of structures lying in the neurovascular bundle from above downward is intercostal vein, artery, and nerve, and that these structures are situated between the posterior intercostal membrane of the internal intercostal muscle and the parietal pleura. Furthermore, laterally the nerve lies between the internal intercostal muscle and the subcostals and the innermost intercostals (i.e, transverse thoracic muscle).

Anatomy of Complications.—These include the following.

Pneumothorax.—This complication can occur if the needle point misses the subcostal groove and penetrates too deeply through the parietal pleura.

Hemorrhage.—Puncture of the intercostal blood vessels. This is a common complication so that aspiration should always be performed before injecting the anesthetic. A small hematoma may result.

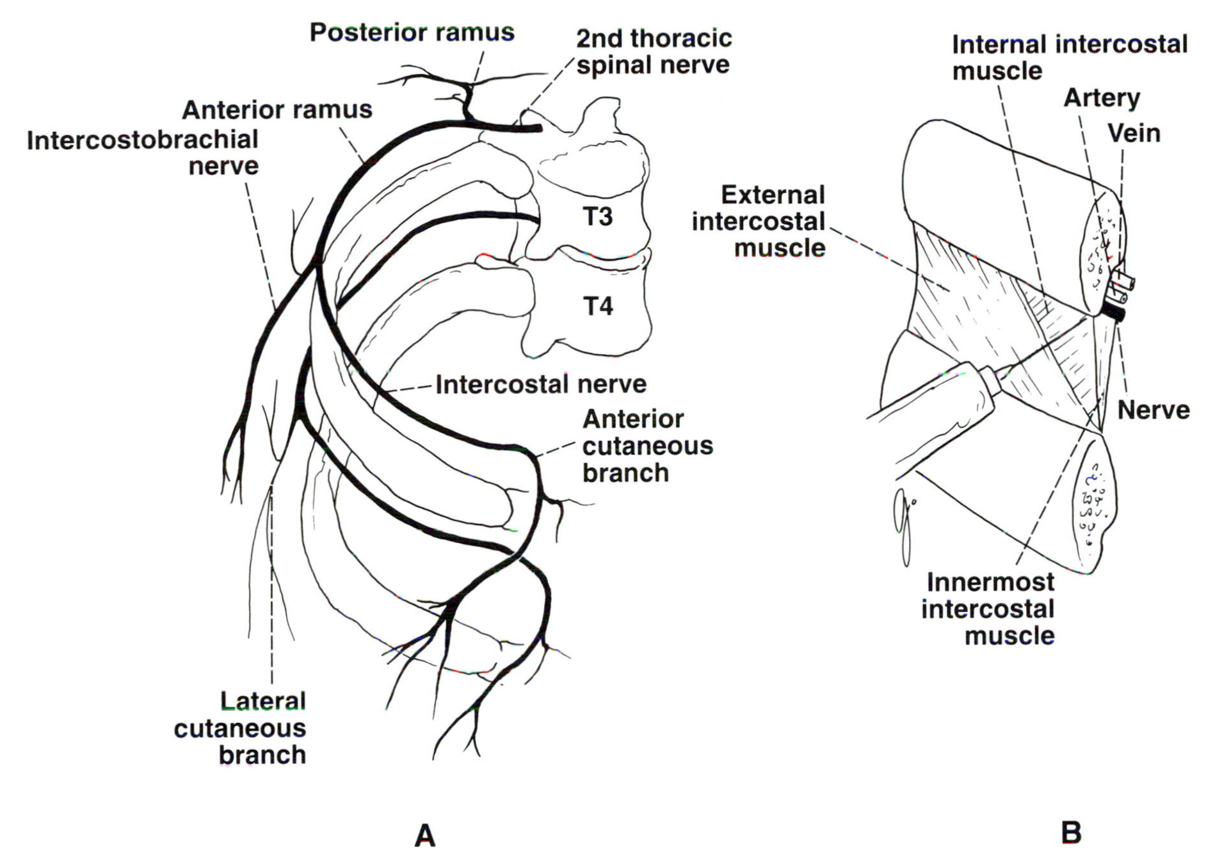

FIG 19–38.
Intercostal nerve block. **A,** the distribution of two intercostal nerves relative to the rib cage. **B,** section through an intercostal space showing the positions of the intercostal nerve, artery, and vein relative to the intercostal muscles. The needle is directed toward the rib near the lower border, and the tip comes to rest near the subcostal groove.

Anterior Abdominal Nerve Block

Area of Anesthesia.—Skin of the anterior abdominal wall. The nerves of the anterior and lateral abdominal walls are the anterior rami of the 7th through the 12th thoracic nerves and the 1st lumbar nerve.

Indications.—Repair of lacerations of the anterior abdominal wall.

Procedure.—The anterior ends of the intercostal nerves T7 through T11 enter the abdominal wall by leaving the intercostal spaces and passing posterior to the costal cartilages (Fig 19–39). An abdominal field block is most easily carried out along the lower border of the costal margin and then infiltrating the nerves as they emerge between the xiphoid process and the 10th or 11th rib along the costal margin.

Clinical Anatomy of Lumbar Spinal Nerve Blocks

Ilioinguinal and Iliohypogastric Nerve Blocks

Area of Anesthesia.—Skin of the lower part of the anterior abdominal wall.

Indications.—Repair of lacerations on the anterior abdominal wall.

Procedure.—The ilioinguinal nerve (L1) passes forward around the anterior abdominal wall, through the inguinal canal, and emerges through the superficial inguinal ring to supply the skin of the groin and part of the scrotum or labium majus. The iliohypogastric nerve (L1) passes around the abdominal wall and pierces the external oblique aponeurosis above the superficial inguinal ring to supply the skin (see Fig 19–39).

The two nerves are easily blocked by inserting the anesthetic needle 1 in. (2.5 cm) above the anterior superior iliac spine on the spinoumbilical line (see Fig 19–39).

Genitofemoral Nerve Block

Area of Anesthesia.—Small area of skin of the thigh below the inguinal ligament and adjacent part of the scrotum or labium majus.

Indications.—Repair of lacerations in the thigh and genital area.

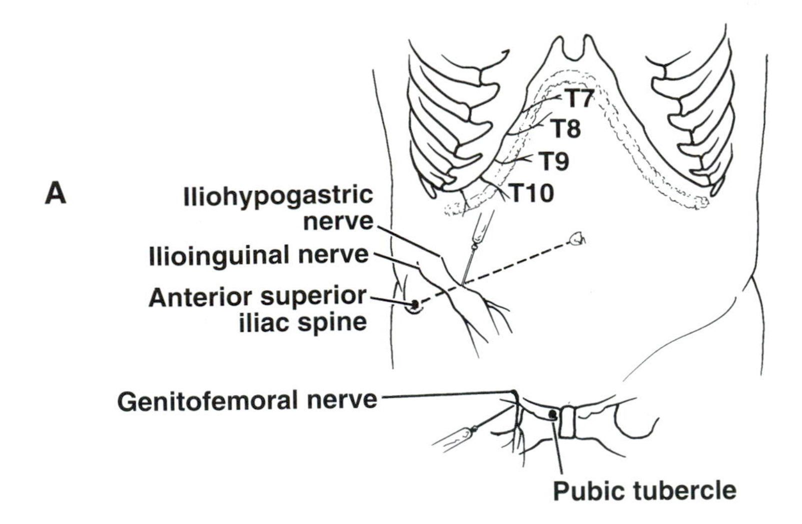

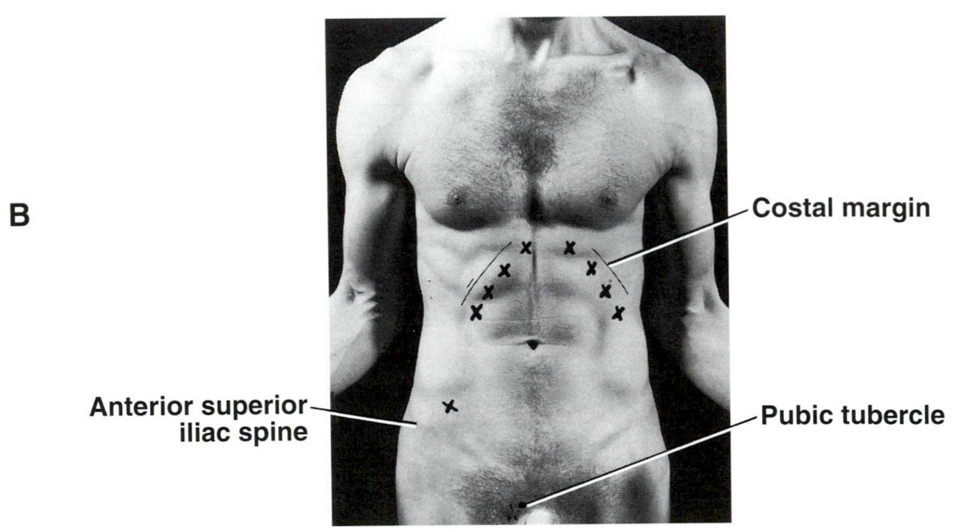

FIG 19–39.
Anterior abdominal wall nerve blocks. **A** and **B**, T7 through T11 intercostal nerves are blocked *(X)* as they emerge from beneath the costal margin. The iliohypogastric and ilioinguinal nerves are blocked by inserting the needle about 1 in. above the anterior superior iliac spine on the spinoumbilical line *(X)*. The terminal branches of the genitofemoral nerve are blocked by inserting the needle through the skin just lateral to the pubic tubercle and infiltrating the subcutaneous tissue with anesthetic solution *(X)*.

Procedure.—The genitofemoral nerve (L1 and L2) runs downward under the inguinal ligament at a point halfway between the anterior superior iliac spine and the symphysis pubis. The terminal branches are most easily blocked by infiltrating the subcutaneous tissue through a needle inserted through the skin lateral to the pubic tubercle (see Fig 19–39).

Femoral Nerve Block

(Rarely performed in the emergency department.)

Area of Anesthesia.—Skin of the front and medial side of the thigh, extending down the medial side of the knee and leg, and along the medial border of the foot as far as the ball of the big toe.

Indications.—Repair of lacerations of the thigh, medial side of the leg, and medial side of the foot.

Procedure.—The femoral nerve (L2 through L4) enters the thigh beneath the inguinal ligament at a point midway between the anterior superior iliac spine and the pubic tubercle (Fig 19–40). Here it lies lateral to the femoral artery.

The nerve may be blocked by introducing the anesthetic needle just below the midpoint of the inguinal ligament and lateral to the femoral artery (see Fig 19–40).

Saphenous Nerve Block

Area of Anesthesia.—Skin of the medial side of leg and the medial border of the foot down as far as the ball of the big toe.

Indications.—Repair of lacerations on the medial side of the leg and the medial side of the foot.

Procedure.—The saphenous nerve is a continuation of the femoral nerve and becomes superficial on the medial side of the knee after emerging between the tendons of sartorius and gracilis muscles (Fig 19–41).

The nerve may be blocked by inserting the anesthetic needle on the medial side of the knee joint ei-

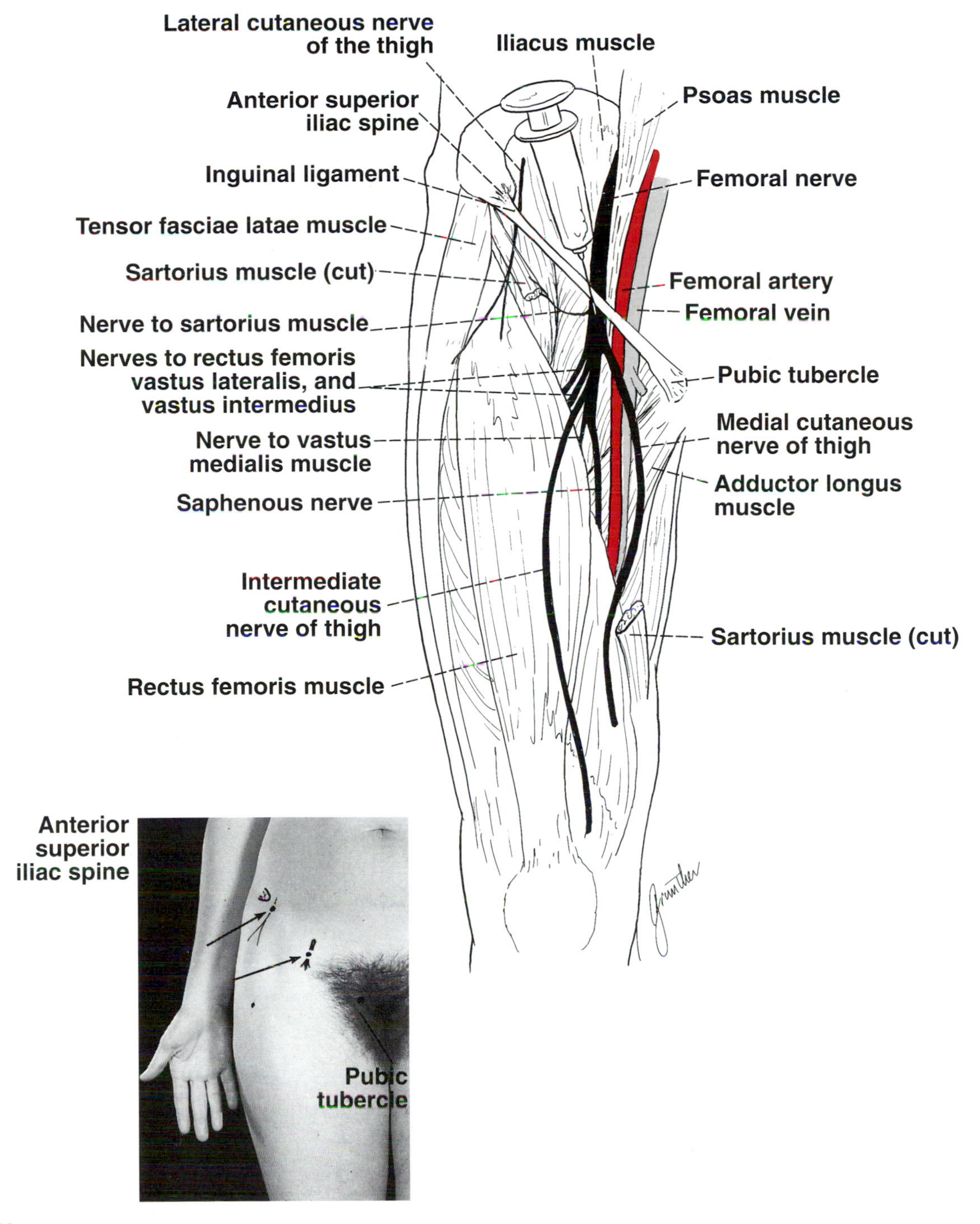

FIG 19–40.
Femoral nerve block. The needle is directed posteriorly just below the midpoint of the inguinal ligament and lateral to the femoral artery. In a lateral cutaneous nerve of the thigh block, the needle is directed posteriorly just inferior to the inguinal ligament about 0.5 in. (1.3 cm) medial to the anterior superior iliac spine.

ther over the medial femoral condyle or lower down over the condyle of the tibia (see Fig 19–41). Care should be taken to avoid the great saphenous vein. The nerve may also be blocked at the ankle where it passes anterior to the medial malleolus (see Fig 19–41).

Lateral Cutaneous Nerve of the Thigh Block

Area of Anesthesia.—Skin of the anterolateral surface of the thigh down to the lateral side of the knee.

Indications.—Repair of lacerations on the anterolateral surface of the thigh.

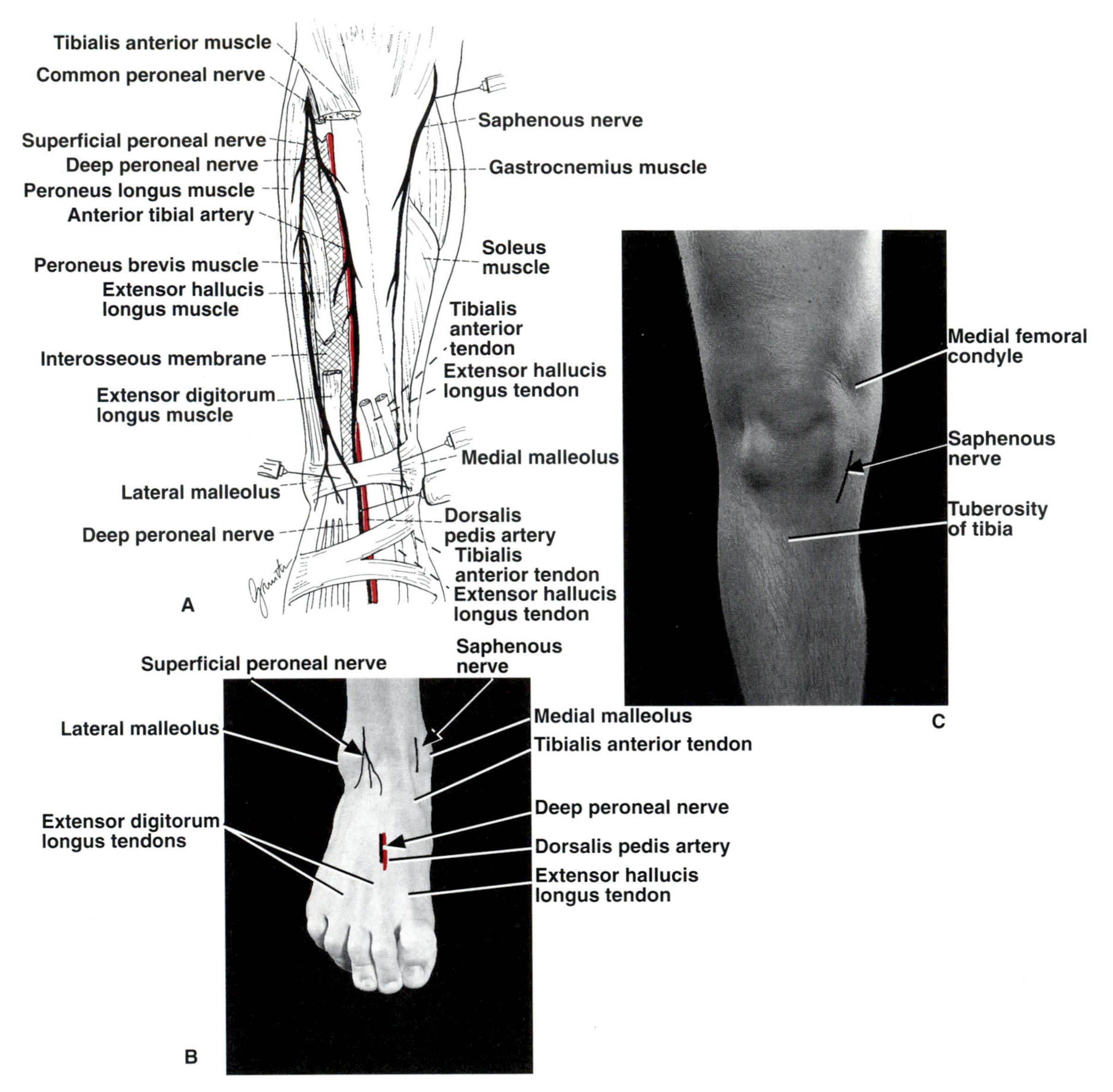

FIG 19–41.
Nerve blocks on the front of the lower part of the leg and dorsum of foot. **A,** the saphenous nerve, the superficial peroneal nerve, and the deep peroneal nerve in relation to other important anatomical structures. **B,** the surface landmarks in the ankle region and on the dorsum of the foot necessary for performing superficial peroneal nerve block, saphenous nerve block, and deep peroneal nerve block. **C,** the surface landmarks used for performing saphenous nerve block on the medial side of the knee.

Procedure.—The lateral cutaneous nerve of the thigh (L2 and L3) enters the thigh behind (or through) the lateral end of the inguinal ligament just medial to the anterior superior iliac spine (see Fig 19–40). It then descends anterior or through the sartorius muscle and divides into terminal anterior and posterior branches.

The nerve may be blocked by inserting the anesthetic needle just inferior to the inguinal ligament about 0.5 in. (1.3 cm) medial to the anterior superior iliac spine.

Sacral Spinal Nerve Blocks

Tibial Nerve Block

Area of Anesthesia.—Skin of the sole of the foot (medial and lateral plantar nerves).

Indications.—Repair of lacerations on the sole of foot.

Procedure.—The tibial nerve (L4 and L5 and S1 through S3) is the largest terminal branch of the sciatic nerve. At the ankle, the nerve, accompanied by the posterior tibial artery, becomes superficial. It lies behind the medial malleolus, between the tendons of the flexor digitorum longus and the flexor hallucis longus muscles, and is covered by the flexor retinaculum (Fig 19–42).

The tibial nerve may be blocked as it lies behind the medial malleolus. By careful palpation, the pulsations of the posterior tibial artery can be felt midway between the medial malleolus and the heel. The nerve lies immediately posterior to the artery, and the anesthetic needle can be inserted at this location (see Fig 19–42).

Sural Nerve Block

Area of Anesthesia.—Skin of the lateral border of foot and lateral side of the little toe.

Indications.—Repair of lacerations on the lateral side of the foot and little toe.

Procedure.—The sural nerve is a branch of the tibial nerve in the popliteal space. It descends superficially in the calf accompanied by the small saphenous vein. It courses behind the lateral malleolus and passes to its distribution along the lateral border of the foot and little toe (Fig 19–43).

The sural nerve may be blocked by inserting the anesthetic needle midway between the lateral malleolus and the tendo calcaneus (Achillis) and infiltrating the subcutaneous tissue with anesthetic solution (see Fig 19–43).

Common Peroneal Nerve Block

Area of Anesthesia.—Skin on the anterior and lateral sides of the leg and the dorsum of the foot and toes, including the medial side of the big toe.

Indications.—Repair of lacerations on the anterior and lateral sides of the leg and the dorsum of the foot and toes.

Procedure.—The common peroneal nerve (L4 and L5 and S1 and S2) is the smaller of the terminal branches of the sciatic nerve. It winds laterally around the neck of the fibula to pierce the peroneus longus muscle (see Fig 19–43). As the nerve lies on the lateral side of the neck of the fibula, it is subcutaneous and can easily be rolled against the bone.

The common peroneal nerve can be blocked by first palpating the nerve below the head of the fibula and infiltrating the tissue around the nerve with a local anesthetic solution.

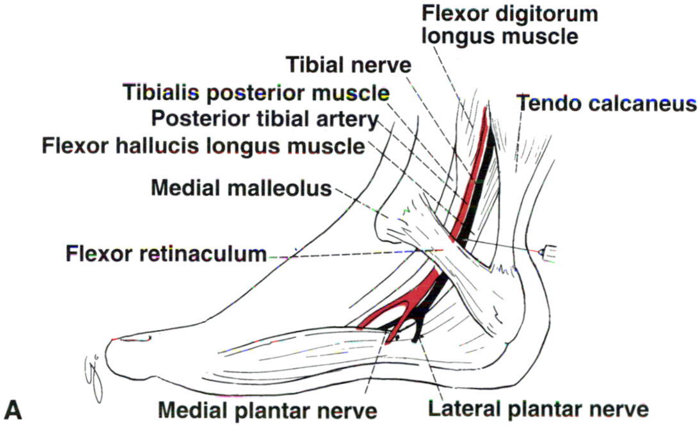

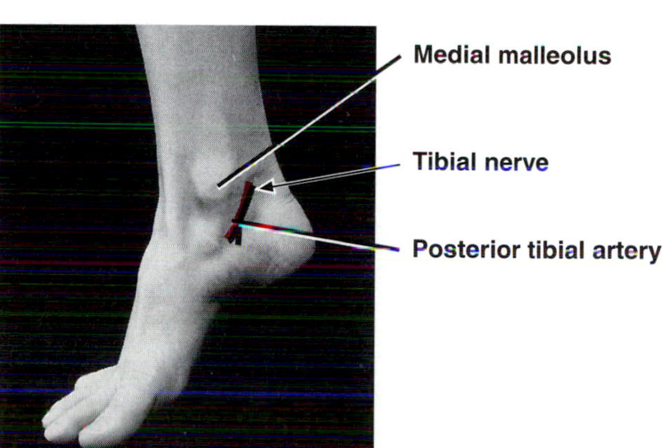

FIG 19–42.
Tibial nerve block. **A,** the tibial nerve and its relationships as it passes behind the medial malleolus of the tibia. **B,** the important surface landmarks used when blocking the tibial nerve at the ankle.

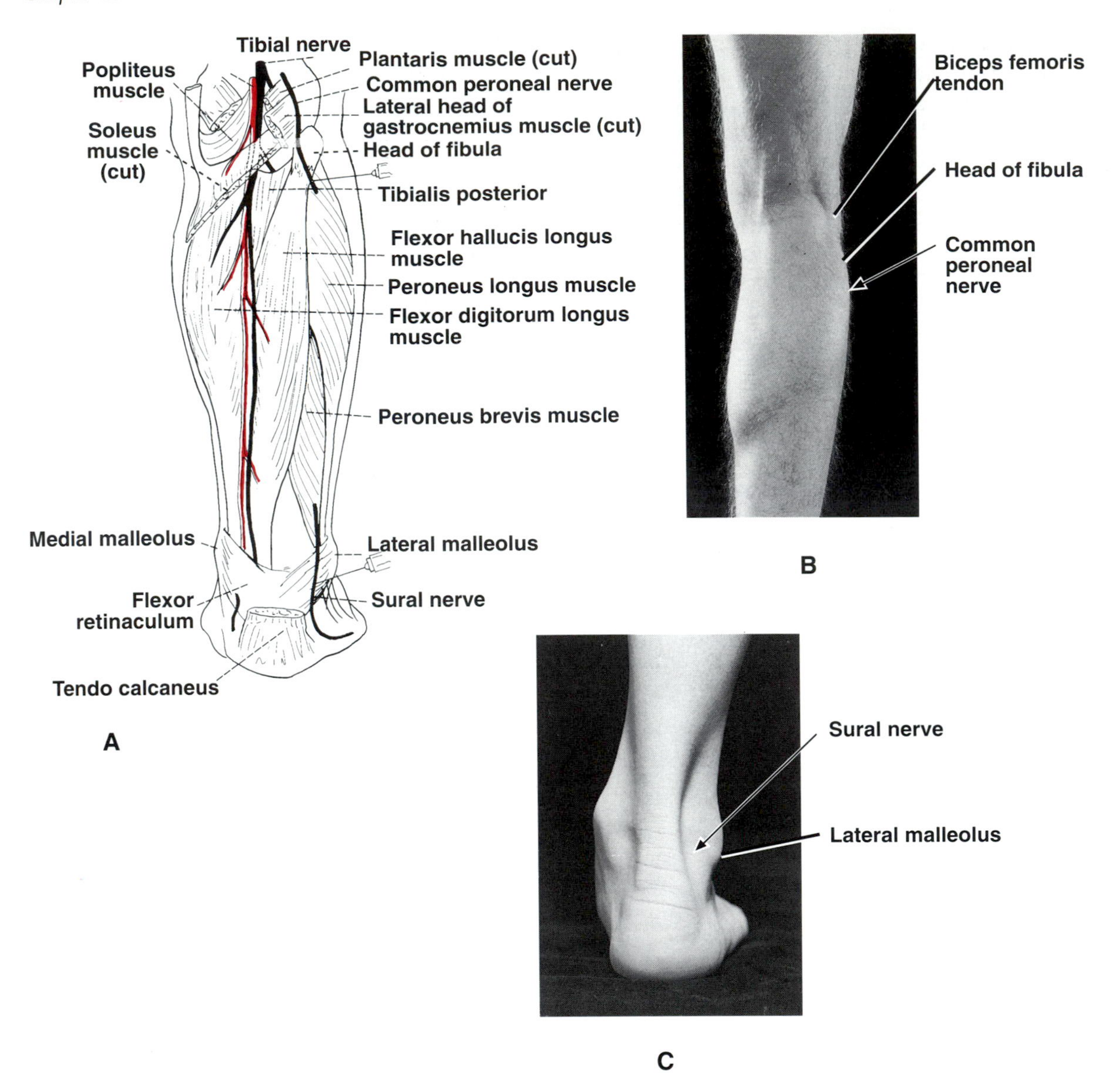

FIG 19–43.
A, the important anatomical relationships of the common peroneal nerve at the back of the knee joint and the sural nerve behind the lateral malleolus of the fibula. **B,** the surface landmarks used for performing a common peroneal nerve block at the knee. **C,** the surface landmarks used for performing a sural nerve block at the ankle.

Superficial Peroneal Nerve Block

Area of Anesthesia.—Skin on the lower anterior and lateral sides of the leg and the dorsum of the foot and toes (except for the cleft between the first and second toes, which is innervated by the deep peroneal nerve and the lateral side of the little toe, which is supplied by the sural nerve).

Indications.—Repair of lacerations in the area of its cutaneous distribution.

Procedure.—The superficial peroneal nerve is a branch of the common peroneal nerve. In the lower third of the leg it becomes superficial and its terminal branches pass to their distribution on the dorsum of the foot and toes (see Fig 19–41).

The superficial peroneal nerve is easily blocked in the lower part of the leg by infiltrating the anesthetic in the subcutaneous tissue along a transverse line connecting the medial and lateral malleoli (see Fig 19–41).

Deep Peroneal Nerve Block

Area of Anesthesia.—Skin in the cleft between the big and second toes.

Indications.—Repair of lacerations in the cleft between the big and second toes.

Procedure.—The deep peroneal nerve is a terminal branch of the common peroneal nerve. It descends in the anterior compartment of the leg and at the ankle it passes onto the dorsum of the foot. Here the nerve lies on the lateral side of the dorsalis pedis artery and is superficially placed between the tendons of extensor digitorum longus and the extensor hallucis longus muscles (see Fig 19–41).

First, the dorsalis pedis artery is palpated midway between the medial and lateral malleoli. With the foot actively dorsiflexed, the tendons of extensor digitorum longus and extensor hallucis longus muscles can be seen. The nerve lies on the lateral side of the artery between these tendons (see Fig 19–41). The needle is then inserted over the nerve, and the surrounding tissues are infiltrated with anesthetic.

Toe Nerve Blocks

Area of Anesthesia.—Skin of the toes. Each toe is supplied by four digital nerves at the 2-o'clock, 5-o'clock, 7-o'clock, and 10-o'clock positions. The plantar digital nerves are derived from the medial and lateral plantar nerves (Fig 19–44), the dorsal digital nerves are from the superficial peroneal nerve (except the cleft between the big and second toes, which is supplied by the deep peroneal nerve, and the lateral side of the little toe, which is supplied by the sural nerve).

Indications.—Repair of lacerations of the toes, removal of foriegn bodies, and removal of nails.

Procedure.—The nerves are easily blocked with small volumes of anesthetic solution injected subcu-

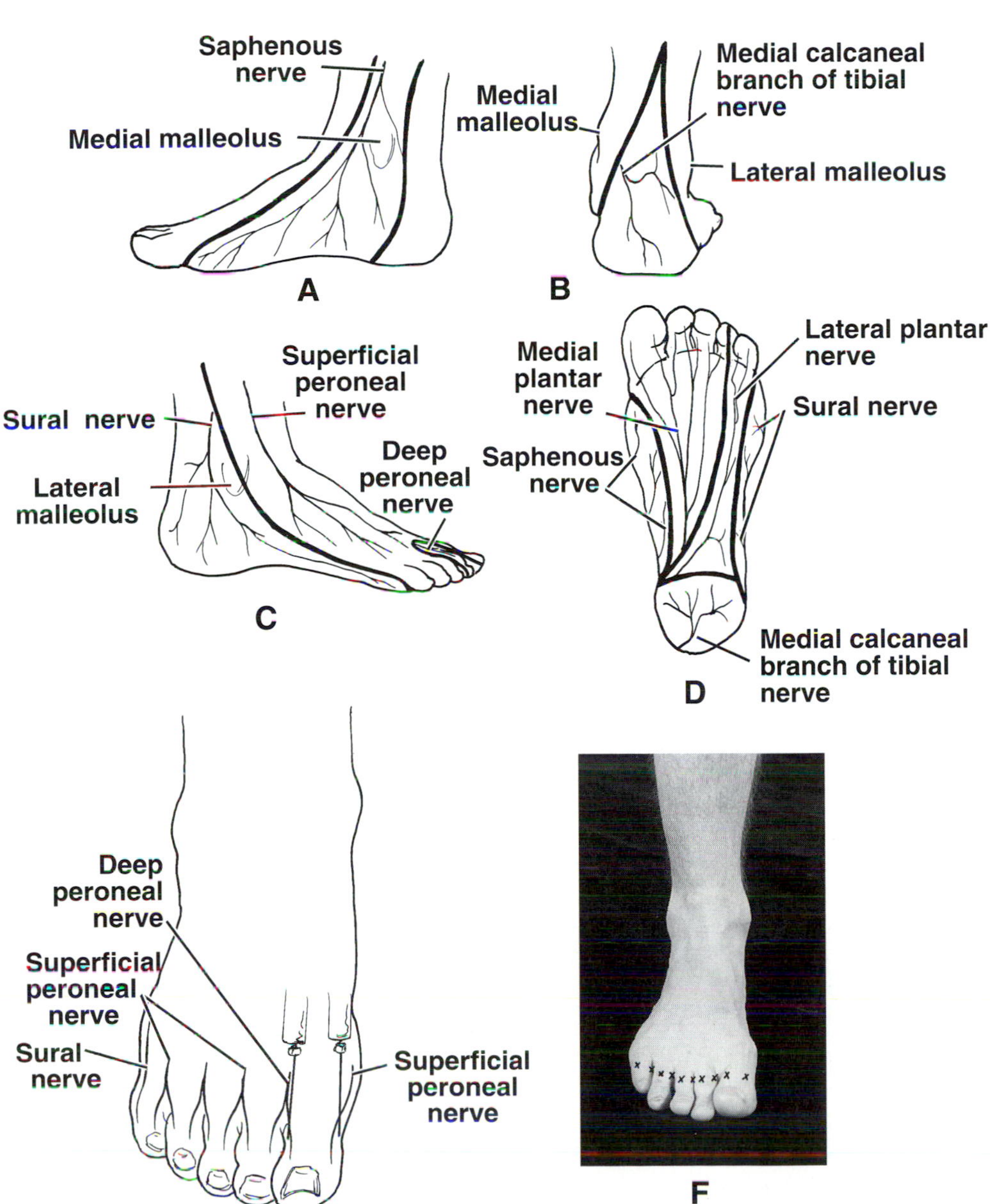

FIG 19–44.
Toe nerve blocks. **A, B, C,** and **D,** the sensory nerve supply to the foot and toes; the heavy lines indicate the boundaries between the different areas of innervation. The plantar digital nerves are derived from the medial and lateral plantar nerves, the dorsal digital nerves are from the superficial peroneal nerve (except the cleft between the big and second toes, which are supplied by the deep peroneal nerve, and the lateral side of the little toe, which is supplied by the sural nerve. **E** and **F,** the sites at the base of each toe *(X)* where the anesthetic solution may be injected subcutaneously.

taneously and circumferentially around the base of each toe (see Fig 19–44).

Pudendal Nerve Block

Area of Anesthesia.—Skin of the perineum; it does not, however, abolish sensation from the anterior part of the perineum, which is innervated by terminal branches of the ilioinguinal nerve and the genitofemoral nerve. Needless to say, it does not abolish pain from uterine contractions that ascend to the spinal cord via the sympathetic afferent nerves.

Indications.—Forceps delivery and episiotomy repair.

Procedures.—These involve the following.

Transvaginal Method.—The bony landmark used is the ischial spine (Fig 19–45). The index finger is inserted through the vagina to palpate the ischial spine. The needle of the syringe is then passed through the vaginal mucous membrane toward the ischial spine. On passing through the sacrospinous ligament there is a feeling of "give." The anesthetic solution is then infiltrated into the tissues around the pudendal nerve (see Fig 19–45).

Perineal Method.—The bony landmark is the ischial tuberosity (see Fig 19–45). The tuberosity is palpated subcutaneously through the buttock, and the needle is introduced into the pudendal canal along the medial side of the tuberosity. The canal lies about 1 in. (2.5 cm) deep to the free surface of the ischial tuberosity. The local anesthetic is then infiltrated around the pudendal nerve.

Clinical Anatomy of the Paracervical Nerve Block

The afferent nerve fibers from the uterus (see p. 341) pass through the inferior hypogastric plexus and are concentrated in the loose connective tissue around the supravaginal part of the cervix. Paracervical nerve block can be used to relieve the pain experienced during the first stage of labor. Fetal blood oxygenation is, however, reduced in many cases, and this can be explained by the decrease in placental blood flow that occurs following vasoconstriction of the uterine arteries that lie close by (Fig 19–46).

Technique

The local anesthetic is introduced as follows.

1. A 5-in. (12.5-cm) needle is used and introduced into the vagina along the index and middle fingers, using them as a guide. The needle is advanced to the lateral fornix on the side of the cervix.
2. The tip of the needle then pierces the mucous membrane of the lateral vaginal fornix and is advanced for about ½ in. (1.3 cm). The needle is now in the connective tissue containing the inferior hypogastric plexus (see Fig 19–46).
3. After careful aspiration to determine that the needle is not in the uterine vessels, about 10 mL of local anesthetic is infiltrated into the tissue. The procedure is then repeated on the other side.

Anatomy of Complications.—Because of the possibility of extensively lowering the fetal blood oxygen,

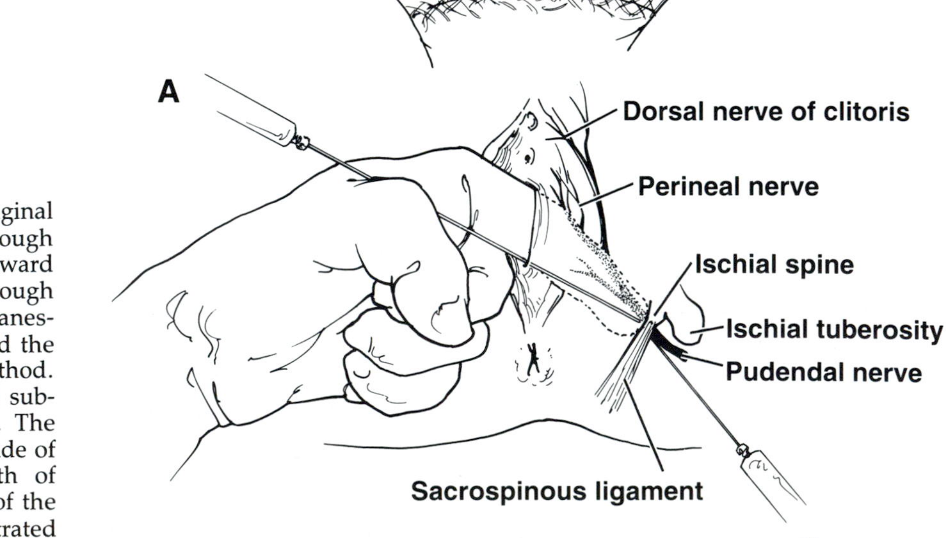

FIG 19–45.
Pudendal nerve blocks. **A,** transvaginal method. The needle is passed through the vaginal mucous membrane toward the ischial spine. On passing through the sacrospinous ligament the anesthetic solution is infiltrated around the pudendal nerve. **B,** perineal method. The ischial tuberosity is palpated subcutaneously through the buttock. The needle is inserted on the medial side of the ischial tuberosity to a depth of about 1 in. from the free surface of the tuberosity. The anesthetic is infiltrated around the pudendal nerve.

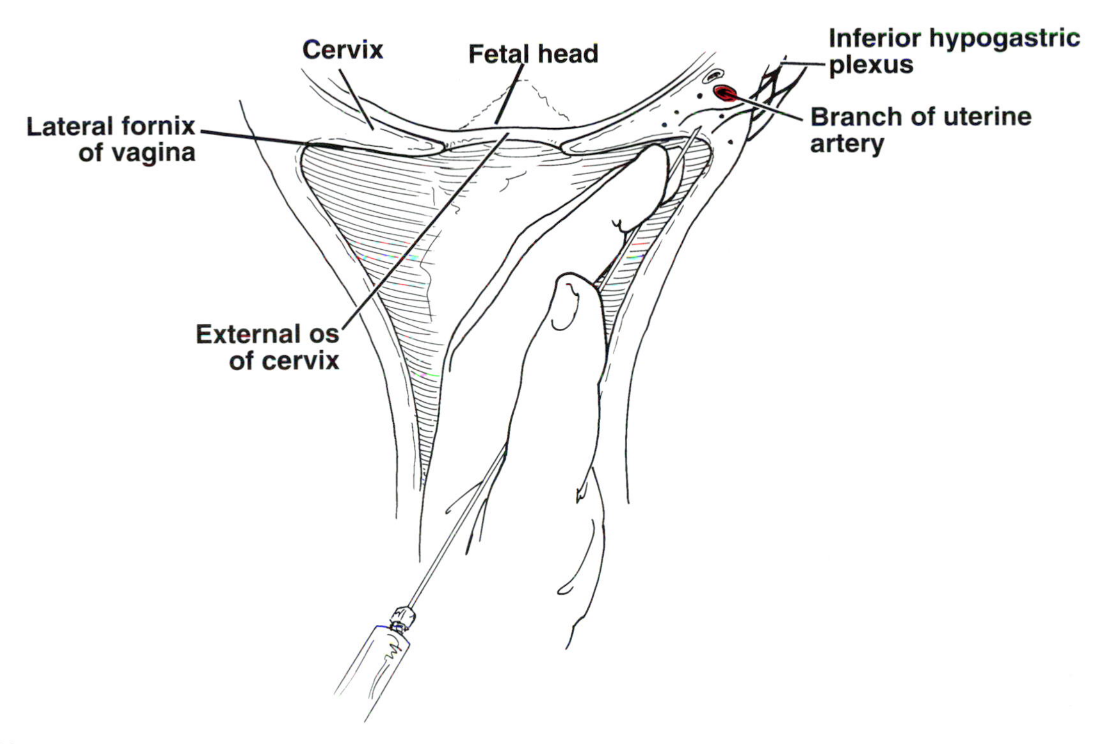

FIG 19–46.
Paracervical nerve block. The needle is introduced into the lateral fornix of the vagina and is advanced through the mucous membrane to reach the connective tissue containing the inferior hypogastric plexus.

the procedure should not be used where fetal distress is already in evidence.

Introduction of the local anesthetic directly into the uterine artery could compromise the fetus by substantially reducing the placental blood flow as indicated previously.

ANATOMICAL-CLINICAL PROBLEMS

1. A 36-year-old woman fell off her bicycle and lacerated the skin of her forehead just above the right eyebrow. What is the sensory innervation of the skin of the forehead? Where may these nerves be blocked?

2. A 45-year-old man was about to have his lower jaw wired with intermaxillary fixation as treatment for his fractured jaw. It was decided to perform the procedure with a mandibular nerve block. Where does the mandibular nerve exit from the skull? What is the sensory distribution of the mandibular division of the trigeminal nerve? If the block is successful, which muscles would be paralyzed? As the needle is advanced medially across the infratemporal fossa, it encounters a bone on the medial wall. What is the name of that bone? What is the possible complication if the needle is advanced too deeply?

3. A patient sustained a tear of the ulnar collateral ligament of the thumb metacarpophalangeal joint. It was decided to repair the laceration under a brachial plexus block. Which spinal nerves contribute to the brachial plexus? Are the roots of the plexus formed by anterior or posterior rami of the spinal nerves? What forms the axillary sheath? When the axillary approach is used, which branch of the plexus is most likely to remain unblocked? Name the anatomical structures that are likely to be inadvertently invaded by the blocking needle and cause complications during an interscalene block when the needle is used in a horizontal direction. When the supraclavicular approach is made to block the brachial plexus, the entire upper limb is usually anesthetized except for a small area of skin on the medial side of the arm. What is the name of the nerve that supplies this area of skin? A supraclavicular brachial plexus block is sometimes complicated by a pneumothorax. Explain in anatomical terms how this can occur.

4. A patient was seen in the emergency department with a laceration of the skin over the middle phalanx of the right index finger. What is the sen-

sory innervation of that finger? Where is it possible to perform a nerve block so that the wound could be sutured?

5. A 34-year-old man caught his left hand in a machine while at work and suffered multiple skin lacerations on the hand. It was decided to suture the lacerations by performing a complete nerve block of the hand above the wrist. What is the innervation of the skin of the hand? Where may these nerves be blocked in the region of the wrist?

6. A 41-year-old woman was involved in an automobile accident and sustained a fracture of her sixth left rib. It was decided to relieve her acute pain caused by the broken rib by performing an intercostal nerve block. Which intercostal nerve innervates the sixth rib? What is the course taken by an intercostal nerve in an intercostal space? From what part of a spinal nerve is an intercostal nerve formed? Name the order of the structures forming the neurovascular bundle in an intercostal space from above downward.

7. A 16-year-old boy received a knife wound on the lateral surface of his right thigh. What is the sensory nerve supply to this region of the thigh? Which important surface landmarks would you use when blocking the nerves?

8. What is the sensory nerve supply to the skin on the dorsum of the foot. Where would you block these nerves? When you perform a sensory nerve block, are any autonomic nerve fibers also blocked?

9. A 25-year-old man was involved in a motorcycle accident and was seen in the emergency department with extensive lacerations of the left ear. It was decided to suture the skin lesions under local anesthesia. How would you anesthetize the ear?

10. What is your definition of a dermatome? What is the difference between a dermatome of the face and the trunk?

11. Describe the anatomical principles involved in performing a paracervical nerve block. Are there any dangers to the fetus when using this procedure?

ANSWERS

1. The skin of the forehead is innervated by the supraorbital and supratrochlear branches of the frontal nerve, a branch of the ophthalmic division of the trigeminal nerve. The nerves may be blocked as they emerge onto the forehead around the superior margin of the orbital cavity (see p. 805 and 806).

2. The mandibular division of the trigeminal nerve exits from the middle cranial fossa through the foramen ovale in the greater wing of the sphenoid bone. The sensory distribution of this nerve is as follows. A small meningeal branch returns to the middle cranial fossa. From the anterior division, the buccal nerve supplies the skin and mucous membrane of the cheek. From the posterior division, the lingual nerve innervates the mucous membrane of the floor of the mouth and the anterior two thirds of the tongue with general sensation. The inferior alveolar nerve supplies the teeth and gums of the lower jaw and the skin over the chin. The auriculotemporal nerve innervates the skin of the auricle, the external auditory meatus, the temporomandibular joint, and the scalp.

A successful mandibular nerve block would result in the paralysis of the muscles of mastication (masseter, temporalis, and medial and lateral pterygoids), the tensor veli palatini, and the tensor tympani.

The otic ganglion is situated close to the main trunk of the mandibular nerve just inferior to the foramen ovale and is likely to be blocked along with the mandibular nerve. Paralysis of this ganglion would result in the blocking of the parasympathetic secretomotor fibers to the parotid salivary gland.

The name of the bone on the medial wall of the infratemporal fossa is the lateral pterygoid plate.

If the anesthetic needle is advanced too far behind the posterior border of the lateral pterygoid plate, it will pierce the superior constrictor muscle and enter the pharynx.

3. The brachial plexus is formed from the anterior rami of C5 through C8 and T1 spinal nerves. The axillary sheath is formed as a prolongation of the prevertebral fascia from the lateral margins of the scalenus anterior and scalenus medius muscles into the axilla; it encloses the axillary artery and the brachial plexus. The musculocutaneous nerve is the most likely nerve to escape blockage during a brachial plexus block approached from the axilla.

If the needle is inserted horizontally when an interscalene block of the brachial plexus is performed, it could pass between the bony transverse processes and pierce the vertebral vessels or enter the dural sheath around a cervical spinal nerve.

The intercostobrachial nerve, a branch of the second intercostal nerve, supplies with the medial cutaneous nerve of the arm a small area of skin on the medial side of the arm. This nerve will not be blocked when the brachial plexus is blocked.

The brachial plexus in the root of the neck

passes downward and laterally over the cervical dome of the pleura. If the vertical approach of introducing the needle is used, it is possible for the needle to travel beyond the trunks of the plexus and pierce the parietal pleura.

4. The index finger receives the following sensory innervation: anterior aspect, two digital branches of the median nerve; posterior aspect, two digital branches of the superficial radial nerve. The nerves are easily blocked by injecting small volumes of anesthetic solution around the base of the finger (see Fig 19–37).

5. The following nerves supply the skin of the hand. The lateral two thirds of the palm and the anterior surface of the lateral 3½ fingers, including the nail beds, are innervated by the median nerve. The medial third of the palm and the anterior surface of the medial 1½ fingers, including the nail beds, are innervated by the ulnar nerve. On the posterior surface of the hand, the lateral two thirds and the lateral 3½ fingers are innervated by the superficial branch of the radial nerve (the areas stated are variable). The medial third of the dorsal surface of the hand and the medial 1½ fingers are supplied by the ulnar nerve.

The median nerve may be blocked just proximal to the flexor retinaculum, as it lies medial to the tendon of the flexor carpi radialis tendon and beneath the palmaris longus tendon (if present). The small palmar cutaneous branch of the median nerve that passes anterior to the flexor retinaculum may be blocked by subcutaneous infiltration in a plane anterior to the median nerve.

The ulnar nerve may be blocked as it lies on the lateral side of the tendon of the flexor carpi ulnaris at its point of insertion onto the pisiform bone. The dorsal cutaneous branch of the ulnar nerve that passes to the back of the hand and fingers is a large branch that arises from the ulnar nerve in the distal third of the forearm. It passes medially between the tendon of the flexor carpi ulnaris muscle and the ulna and may be blocked by subcutaneous infiltration about 2 in. (5 cm) above the wrist.

The superficial branch of the radial nerve may be blocked on the lateral side of the radial artery just proximal to the proximal wrist crease. To block all the small branches given off by the radial nerve proximal to this site, a cuff of anesthetic is usually injected on the lateral and posterior aspects of the wrist at this level.

6. The periosteum of the sixth rib is innervated by the sixth intercostal nerve. (It is conceivable that in the individual who has a collateral branch of the fifth intercostal nerve, the periosteum of the sixth rib is also innervated by this nerve.) Intercostal nerves are formed from the anterior rami of T1 through T11 spinal nerves. An intercostal nerve passes forward around the thoracic wall first, between the posterior intercostal membrane and the parietal pleura, then between the internal intercostal muscle and the intercostalis intimus muscle (part of the transversus thoracis). The nerve lies in the subcostal groove of the rib of its own number. The intercostal vein, the intercostal artery, and the intercostal nerve lie within the subcostal groove in that order from above downward.

7. The sensory innervation of the skin on the lateral surface of the thigh is the lateral cutaneous nerve of the thigh, a branch of the lumbar plexus (L2 and L3). The nerve may be blocked by inserting an anesthetic needle just inferior to the inguinal ligament about 0.5 in. (1.3 cm) medial to the anterior superior iliac spine.

8. The sensory nerve supply to the skin on the dorsum of the foot is from the superficial peroneal nerve. The cleft between the first and second toes is from the deep peroneal nerve.

The superficial peroneal nerve can be blocked by infiltrating the anesthetic in the subcutaneous tissue along a transverse line connecting the medial and lateral malleoli (see p. 830). The deep peroneal nerve is blocked by first palpating the dorsalis pedis artery as it lies between the tendons of extensor hallucis longus and the extensor digitorum longus. The nerve lies on the lateral side of the artery between these tendons. The needle is then inserted over the nerve, and the surrounding tissue is infiltrated with anesthetic.

9. The auricle receives its sensory innervations from the greater auricular nerve (C2 and C3), the auriculotemporal nerve (mandibular division of the trigeminal nerve) and from the auricular branch of the vagus nerve.

Analgesia of the skin can be obtained by multiple subcutaneous injections that are continued circumferentially around the auricle (see p. 814).

10. A dermatome is an area of skin that is supplied by a single spinal nerve, and, therefore, a single segment of the spinal cord.

The skin of the face is mainly supplied by the three divisions of the trigeminal nerve, and there is little or no overlap of the skin areas supplied by each of the divisions. On the trunk, adjacent dermatomes overlap considerably, therefore, to produce a region of complete anesthesia. At least three adjacent spinal nerves have to be blocked.

11. The anatomical principles involved and the technique of paracervical nerve block are described on p. 832. The vasoconstriction of the uterine arteries or their branches that may occur following the introduction of the local anesthetic may seriously lower the level of fetal oxygen. The procedure of paracervical nerve block should therefore not be used where fetal distress is already in evidence.

Index

C

D

I

M

N

Q

R

S